Encyclopedia of Electronics

Stan Gibilisco
Editor-in-Chief

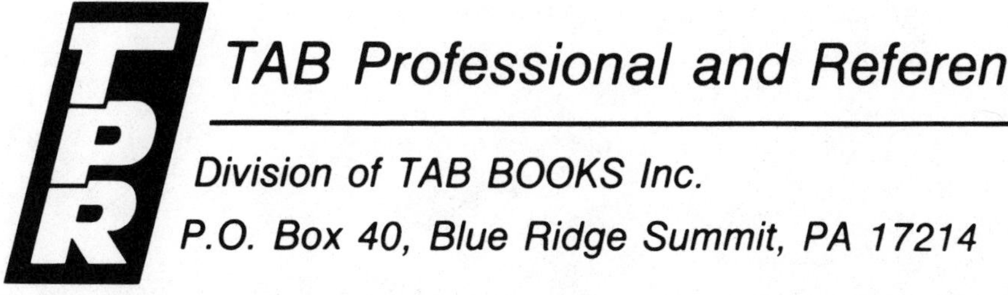
TAB Professional and Reference Books
Division of TAB BOOKS Inc.
P.O. Box 40, Blue Ridge Summit, PA 17214

To Beverly

FIRST EDITION

SECOND PRINTING

Printed in the United States of America

Library of Congress Cataloging in Publication Data

Gibilisco, Stan.
Encyclopedia of electronics.

Includes index.
1. Electronics—Dictionaries. I. Title.
TK7804.G53 1984 621.381'0321 84-16437
ISBN 0-8306-2000-1

Editorial Board

How to Use This Book

This encyclopedia is a complete source of information on all aspects of electricity, electronics, and communications technology. It is intended as a permanent reference source for students, hobbyists, and professionals.

There are more than 3,000 individual articles. An effort is made to present the material in a clear and understandable way without "talking down" or sacrificing thoroughness.

Articles are listed in alphabetical order. Related articles are listed in the text and at the end of each article.

This encyclopedia has three basic parts. Preceding the alphabetically arranged articles, there are lists of articles by category, along with a table of standard schematic symbols. By far the largest part of the book is made of up the articles themselves. A detailed index follows the articles.

If you already know what subject you want to look up, you can simply turn to that term in the main body of articles, just as you would when using a dictionary. If that term is not an article title, check the index. The index contains many more electronic terms and will direct you to pages containing information on a specific term.

If you are interested in a general subject rather than a specific term, turn to the classification list at the front of this book. There you will find various fields in electronics, with specific articles listed underneath each heading. The categories are as follows:

Antennas and Feed Lines. All aspects of radio receiving and transmitting antennas and transmission lines for all frequencies.

Audio Electronics. High fidelity, stereo, recording, reproduction, amplification, and sound processing.

Broadcasting and Communications. Amateur radio, Citizens' Band, commercial broadcasting, shortwave listening, wave propagation, receivers, and transmitters.

Components and Practical Circuits. Characteristics and uses of individual components; design and operation of complete electronic devices.

Computers and Digital Electronics. All types of calculators and computers; digital concepts, theory, and practice.

Electricity and Magnetism. Basic principles of electric and magnetic effects.

Electronic Tubes. Theory, characteristics, and uses of vacuum tubes in electronic circuits.

Mathematical Data. Basic formulas and mathematical techniques as applied to electronics.

Physical Data. Properties and uses of materials; characteristics of matter.

Power Supplies. Design, construction, and operation of all types of power supplies.

Radiolocation, Radionavigation, and Radar. Types of systems; theory and practice.

Solid-State Electronics. Characteristics and applications of semiconductor devices of all kinds, including diodes, transistors, and integrated circuits.

Switching and Control. Manual and electronic switching; mechanical control devices.

Tests and Measurements. Calibration, debugging, repair, testing, and troubleshooting of electronic devices.

Units, Standards, and Constants. Quantity, time, frequency, and other parameters; performance standards.

Wiring and Construction. Power transmission, circuit-building practice.

Miscellaneous Terminology. General properties, characteristics, and terms that do not fit neatly into any other classification.

In some cases, specific articles appear under more than one category listing. Suppose, for example, that you want to know something about various codes used in telecommunications. You would look under the classification Broadcasting and Communications. But these codes also appear under Units, Standards, and Constants. Terms are often listed under two or even three different categories. This redundancy of listings will help you find what you want with comparative ease, minimizing the need for having to look through several different categories.

Consider another example: You want to know how an airline pilot finds his position in the sky. You would look under Radiolocation, Radionavigation, and Radar. If you want to know what auroral propagation is, you look under Communications.

Some terms do not fit well in any specific category. This book takes care of that problem by means of a Miscellaneous Terminology classification.

Suggestions for future revisions are welcome.

Stan Gibilisco
Editor-in-Chief

Categorized List of Articles

Antennas and Feed Lines

UMBRELLA ANTENNA
UNBALANCED LINE
UNBALANCED LOAD
UNBALANCED OUTPUT
UNBALANCED SYSTEM
UNBALANCED TRANSMISSION LINE
UNIDIRECTIONAL ANTENNA
UNIDIRECTIONAL PATTERN
VANE ATTENUATOR
V BEAM
VELOCITY FACTOR
VERTICAL ANTENNA
VERTICAL DIPOLE ANTENNA
VERTICAL POLARIZATION
VOLTAGE FEED
VOLTAGE LOOP
VOLTAGE NODE
WAVEGUIDE
WAVELENGTH
WHIP ANTENNA
WIND LOADING
WINDOM ANTENNA
YAGI ANTENNA
ZEPPELIN ANTENNA

Audio Electronics

ACOUSTIC FEEDBACK
ACOUSTICS
ACOUSTIC TRANSDUCER
ACOUSTIC TRANSMISSION
ACQUISITION TONE
ACTIVE FILTER
AMPLIFICATION
AMPLIFICATION FACTOR
AMPLIFICATION NOISE
AMPLIFIER
ARTIFICIAL EAR
ARTIFICIAL VOICE
ATTENUATION-VS.-FREQUENCY CHARACTERISTIC
AUDIBILITY
AUDIO AMPLIFIER
AUDIO FREQUENCY
AUDIO-FREQUENCY TRANSFORMER
AUDIO IMAGE
AUDIO LIMITER
AUDIO MIXER
AUDIO POWER
AUDIO RESPONSE
AUDIO SIGNAL GENERATOR
AUDIO TAPER
AUDIO-VISUAL COMMUNICATION
BAFFLE
BALANCE
BALANCE CONTROL
BASEBAND
BASS
BASS-REFLEX ENCLOSURE
BASS RESPONSE
BEAT
BIDIRECTIONAL PATTERN
CARDIOID PATTERN
CASSETTE
CERAMIC MICROPHONE
CERAMIC PICKUP
CHANNEL SEPARATION
CONSONANCE
CONSTANT-VELOCITY RECORDING
CONTACT MICROPHONE
CROSSOVER NETWORK
CRYSTAL MICROPHONE
DAMPING FACTOR
DEAD ROOM
DECCA
DECIBEL
DECIBEL METER
DIRECTIVITY INDEX
DISK RECORDING
DISSONANCE
DISTORTION
DOLBY
DROP-IN AND DROP-OUT
DYNAMIC LOUDSPEAKER
DYNAMIC MICROPHONE
DYNAMIC PICKUP
DYNAMIC RANGE
EARPHONE
ECCENTRICITY
EIGHT-TRACK RECORDING
ELASTIC WAVE
ELECTRONIC MUSIC
ELECTROSTATIC MICROPHONE
ELECTROSTATIC SPEAKER
EQUALIZER
FIDELITY
FLAME MICROPHONE
FLUTTER
FOUR-TRACK RECORDING
HALF STEP
HARMONY
HEAD
HEADPHONE
HIGH FIDELITY
IMPEDANCE MATCHING
LABYRINTH SPEAKER
LAVALIER MICROPHONE
LEADER TAPE
LOUDNESS
MAGNETIC RECORDING
MAGNETIC TAPE
MICROPHONE
MICROPHONICS
MIDRANGE
MOOG SYNTHESIZER
MOTORBOATING
MUSIC
MUSIC POWER
OCTAVE
OMNIDIRECTIONAL MICROPHONE
OPTICAL SOUND RECORDING AND REPRODUCTION
PICKUP
PITCH
PSEUDOSTEREOPHONICS
PSYCHOACOUSTICS
QUADROPHONICS
RECORDING HEAD
RECORDING INSTRUMENT

RECORDING DISK
RECORDING TAPE
REEL-TO-REEL TAPE RECORDER
REVERBERATION
ROLLOFF
RUMBLE
SIBILANT
SIMPLE TONE
SONE
SOUND
SPEAKER
STEREO DISK RECORDING
STEREO MULTIPLEX
STEREOPHONICS
STEREO TAPE RECORDING
STYLUS
TAPE RECORDER
TAPE SPLICING
THERMOPLASTIC RECORDING
THRESHOLD OF HEARING
THROAT MICROPHONE
TIMBRE
TREBLE
TREBLE RESPONSE
TRIAXIAL SPEAKER
TUNING FORK
TURNTABLE
TWEETER
TWO-TRACK RECORDING
ULTRASONIC DEVICE
ULTRASONIC HOLOGRAPHY
ULTRASONIC MOTION DETECTOR
ULTRASONIC SCANNER
ULTRASONIC TRANSDUCER
ULTRASOUND
UNIDIRECTIONAL PATTERN
VELOCITY MICROPHONE
VOCODER
VODER
VOICE FREQUENCY CHARACTERISTICS
VOICE PRINT
VOLUME
VOLUME COMPRESSION
VOLUME CONTROL
VOLUME UNIT
VOLUME-UNIT METER
WIRELESS MICROPHONE
WOOFER
WOW

Broadcasting and Communications

ABSORPTION MODULATION
ACTIVE COMMUNICATIONS SATELLITE
ADJACENT-CHANNEL INTERFERENCE
AERONAUTICAL BROADCASTING AND RADIONAVIGATION
ALTERNATE ROUTING
AMATEUR RADIO
AMERICAN MORSE CODE
AMERICAN RADIO RELAY LEAGUE
AMPLIFICATION
AMPLIFICATION FACTOR
AMPLIFICATION NOISE
AMPLITUDE MODULATION
ARTICULATION
ASCII
ASPECT RATIO
ASYMMETRICAL DISTORTION
ASYNCHRONOUS DATA
ATTACK
ATTENDANT EQUIPMENT
ATTENUATION
ATTENUATION DISTORTION
ATTENUATION-VS.-FREQUENCY CHARACTERISTIC
AUDIO-VISUAL COMMUNICATION
AURORA
AURORAL PROPAGATION
AUTOMATIC BIAS
AUTOMATIC BRIGHTNESS CONTROL
AUTOMATIC CONTRAST CONTROL
AUTOMATIC DIALING UNIT
AUTOMATIC DIRECTION FINDER
AUTOMATIC FREQUENCY CONTROL
AUTOMATIC GAIN CONTROL
AUTOMATIC INTERCEPT
AUTOMATIC LEVEL CONTROL
AUTOMATIC MESSAGE HANDLING
AUTOMATIC MODULATION CONTROL
AUTOMATIC NOISE LIMITER
AUTOMATIC SCANNING RECEIVER
AUTOPATCH
AUXILIARY LINK
AVERAGE NOISE FIGURE
AVERAGE POWER
BABBLE SIGNAL
BACKGROUND NOISE
BACKLASH
BACKSCATTER
BALANCE
BALANCED DETECTOR
BALANCED MODULATOR
BAND
BANDPASS FILTER
BANDPASS RESPONSE
BAND-REJECTION FILTER
BAND-REJECTION RESPONSE
BANDWIDTH
BASEBAND
BAUDOT
BAUD RATE
BEACON
BEAT
BEAT-FREQUENCY OSCILLATOR
BELL, ALEXANDER GRAHAM
BIAS DISTORTION
BIT
BIT RATE
BLACK TRANSMISSION
BLANKETING
BLANKING
BREAK
BREAKER
BRIGHTNESS
BROADCAST BAND
BROADCASTING
CABLE COMMUNICATIONS

Components and Practical Circuits

CARCINOTRON
CARD
CARTRIDGE
CAVITY RESONATOR
CELL
CENTRAL PROCESSING UNIT
CERAMIC CAPACITOR
CERAMIC FILTER
CERAMIC MICROPHONE
CERAMIC PICKUP
CERAMIC RESISTOR
CHARACTER GENERATOR
CHEBYSHEV FILTER
CHOKE
CHOKE-INPUT FILTER
CHOPPER
CIRCUIT PROTECTION
CLAPP OSCILLATOR
CLASS-A AMPLIFIER
CLASS-AB AMPLIFIER
CLASS-B AMPLIFIER
CLASS-C AMPLIFIER
CLICK FILTER
CLIPPER
CLOCK
COAXIAL TANK CIRCUIT
CODE TRANSMITTER
COIL
COIL WINDING
COINCIDENCE CIRCUIT
COLPITTS OSCILLATOR
COMMUTATOR
COMPACTRON
COMPANDOR
COMPARATOR
COMPONENT
COMPRESSION CIRCUIT
CONSTANT-K FILTER
CONVERTER
CORE
CORRELATION DETECTION
COUPLER
CROSBY CIRCUIT
CRYSTAL
CRYSTAL DETECTOR
CRYSTAL-LATTICE FILTER
CRYSTAL OSCILLATOR
CRYSTAL OVEN
CRYSTAL TRANSDUCER
CUTOFF ATTENUATOR
DAISY-WHEEL PRINTER
DANIELL CELL
DARLINGTON AMPLIFIER
DC AMPLIFIER
DC ERASING HEAD
DC GENERATOR
DC-TO-AC CONVERTER
DC-TO-DC CONVERTER
DC TRANSDUCER
DECADE BOX
DEFLECTOR
DELAY CIRCUIT
DELAY LINE
DELAY TIMER
DIAPHRAGM
DIELECTRIC AMPLIFIER
DIELECTRIC LENS
DIETZHOLD NETWORK
DIFFERENTIAL AMPLIFIER
DIFFERENTIAL CAPACITOR
DIFFERENTIAL TRANSDUCER
DIFFERENTIAL TRANSFORMER
DIODE DETECTOR
DIODE MIXER
DIODE OSCILLATOR
DIRECTIONAL FILTER
DISCRETE COMPONENT
DISK CAPACITOR
DISPLACEMENT TRANSDUCER
DISPLAY
DISPOSABLE COMPONENT
DISTRIBUTED AMPLIFIER
DISTRIBUTION FRAME
DIVIDER
DOHERTY AMPLIFIER
DOT-MATRIX PRINTER
DOUBLE BALANCED MIXER
DOUBLE-CONVERSION RECEIVER
DOUBLER
DRIVER
DROPPING RESISTOR
DUAL COMPONENT
DUPLEXER
DUST PRECIPITATOR
DYNAMOTOR
ELECTRICAL INTERLOCK
ELECTRIC EYE
ELECTROCHEMICAL TRANSDUCER
ELECTRODE
ELECTROLUMINESCENT CELL
ELECTROLYTIC CAPACITOR
ELECTROLYTIC RECTIFIER
ELECTROLYTIC RESISTOR
ELECTROCHEMICAL AMPLIFIER
ELECTROCHEMICAL RECTIFIER
ELECTROCHEMICAL TRANSDUCER
ELECTRON-BEAM GENERATOR
ELECTRON-COUPLED OSCILLATOR
ELECTRONIC BUZZER
ELECTROSTATIC SHIELDING
ENCODER
ENVELOPE DETECTOR
EXCITER
EXCITRON
EXPANDER
EXTRACTOR
FAN
FARADAY CAGE
FEEDBACK AMPLIFIER
FEEDTHROUGH CAPACITOR
FERRITE BEAD
FERRITE CORE
FERROELECTRIC CAPACITOR
FIBERSCOPE
FINAL AMPLIFIER
FINNED SURFACE
FLIP-FLOP
FLOATING PARAPHASE INVERTER
FLOATING POINT
FLY'S-EYE LENS
FLYWHEEL TUNING
FOLLOWER

FORK OSCILLATOR
FOSTER-SEELEY DISCRIMINATOR
FREQUENCY COMPARATOR
FREQUENCY CONVERTER
FREQUENCY DIVIDER
FREQUENCY MULTIPLIER
FREQUENCY SYNTHESIZER
GANG CAPACITOR
GLOW LAMP
GRADED FILTER
GRATING
GRID-LEAK DETECTOR
HALF-ADDER
HALF-BRIDGE
HALF-WAVE TRANSMISSION LINE
HALL GENERATOR
HARDWARE
HARTLEY OSCILLATOR
HEAT SINK
HELICAL FILTER
HELMHOLTZ COIL
HERTZ OSCILLATOR
HETERODYNE DETECTOR
HETERODYNE REPEATER
HIGHPASS FILTER
HYBRID DEVICES
IGNITRON
IMAGE CONVERTER
INCANDESCENT LAMP
INDUCTION COIL
INDUCTIVE CAPACITOR
INDUCTOR
INDUCTOR TUNING
INFRARED DEVICE
INPUT DEVICE
INPUT TUNING
INSULATOR
INTERFERENCE FILTER
INTERMEDIATE-FREQUENCY AMPLIFIER
INTERPRETER
INTERRUPTER
INVERTER
INVERTING AMPLIFIER
JACK PANEL
KEYER
KINESCOPE RECORDER
KOZANOWSKI OSCILLATOR
LADDER ATTENUATOR
LASER
LATTICE FILTER
LC CIRCUIT
LEYDEN JAR
LIMITER
LINEAR AMPLIFIER
LINEAR TRANSFORMER
LINE PRINTER
LINE TRAP
LINE TUNING
LINK FEEDBACK
LIQUID COOLING
LIQUID-CRYSTAL DISPLAY
LOCAL OSCILLATOR
LOFTIN-WHITE AMPLIFIER
LOWPASS FILTER
LUMISTOR
MAGNET COIL
MAGNETIC CIRCUIT
MAGNETOSTRICTION OSCILLATOR
MARX GENERATOR
MASER
MATCHTONE
M-DERIVED FILTER
MECHANICAL FILTER
MEGGER
MERCURY-VAPOR LAMP
MERCURY-VAPOR RECTIFIER
MESH
MESNY CIRCUIT
MICA CAPACITOR
MICROWAVE OVEN
MICROWAVE REPEATER
MILLER OSCILLATOR
MIXER
MODULATOR
MONOSTABLE MULTIVIBRATOR
MOTOR
MULTIVIBRATOR
MUTING CIRCUIT
MYLAR CAPACITOR
NEON LAMP
NEON-LAMP OSCILLATOR
NETWORK
NEUTRALIZATION
NEUTRALIZING CAPACITOR
NOISE BLANKER
NOISE FILTER
NOISE LIMITER
NONLINEAR CIRCUIT
NONPOLARIZED COMPONENT
NONPOLARIZED ELECTROLYTIC CAPACITOR
NONREACTIVE CIRCUIT
NONRESONANT CIRCUIT
OPERATIONAL AMPLIFIER
OSCILLATOR
OUTPUT DEVICE
PAD
PADDER CAPACITOR
PAPER CAPACITOR
PARAMETRIC AMPLIFIER
PARAPHASE INVERTER
PARASITIC SUPPRESSOR
PASSIVE COMPONENT
PASSIVE FILTER
PERMANENT MAGNET
PHANTASTRON
PHASE COMPARATOR
PHASE INVERTER
PHASE-LOCKED LOOP
PHOTOCELL
PHOTODIODE
PHOTOFET
PHOTORESISTOR
PIERCE OSCILLATOR
PI NETWORK
PLUG-IN COMPONENT
POLARIZED COMPONENT
POLYGRAPH
POLYPHASE RECTIFIER
POLYSTYRENE CAPACITOR
POT CORE
POTENTIAL TRANSFORMER
POTENTIOMETER

POWDERED-IRON CORE
POWER AMPLIFIER
PREAMPLIFIER
PRECISION POTENTIOMETER
PRESELECTOR
PRINTER
PROBE
PRODUCT DETECTOR
PROTECTOR
PULSE AMPLIFIER
PULSE GENERATOR
PULSE TRANSFORMER
PUSH-PULL AMPLIFIER
PUSH-PULL, GROUNDED GRID/BASE/GATE AMPLIFIER
Q MULTIPLIER
QUADRIFILAR WINDING
QUADRUPLER
QUARTER-WAVE TRANSMISSION LINE
RADIO-FREQUENCY PROBE
RADIO-FREQUENCY TRANSFORMER
RATIO DETECTOR
REACTANCE MODULATOR
REACTIVE CIRCUIT
RECEIVER
RECEIVER INCREMENTAL TUNING
REGENERATIVE DETECTOR
REGULATOR
REINARTZ CRYSTAL OSCILLATOR
REPEATER
RESISTANCE-CAPACITANCE CIRCUIT
RESISTANCE-INDUCTANCE CIRCUIT
RESONANT CIRCUIT
RF AMPLIFIER
RF CHOKE
RING MODULATOR
ROTARY DIALER
SATURABLE REACTOR
SCHMITT TRIGGER
SENSE AMPLIFIER
SEPARATOR
SEVEN-SEGMENT DISPLAY
SHELF LIFE
SIDETONE OSCILLATOR
SILVER-MICA CAPACITOR
SINGLE BALANCED MIXER
SINGLE-CONVERSION RECEIVER
SLIDE POTENTIOMETER
SMOKE DETECTOR
SNOOPERSCOPE
SODIUM-VAPOR LAMP
SOLENOID
SPARK TRANSMITTER
SPECTROSCOPE
SQUARE-LAW DETECTOR
SQUELCH STATOR
STENODE CIRCUIT
STEP-DOWN TRANSFORMER
STEP-UP TRANSFORMER
STEP-VOLTAGE REGULATOR
STRIPLINE
SUPERHETERODYNE RECEIVER
SUPERREGENERATIVE RECEIVER
SYNCHRONOUS CONVERTER
SYNCHRONOUS DETECTOR
SYNCHROTRON
SYNTHESIZER
TANTALUM CAPACITOR
TERMINAL UNIT
TERTIARY COIL
TESLA COIL
THERMISTOR
THERMOCOUPLE
THERMOPILE
THRESHOLD DETECTOR
TOROID
TRANSCEIVER
TRANSDUCER
TRANSITRON OSCILLATOR
TRANSMITTER
TRANSPONDER
TRANSVERTER
TRIMMER CAPACITOR
TRIPLER
TRI-TET OSCILLATOR
TUBULAR CAPACITOR
TUNABLE CAVITY RESONATOR
TUNED-INPUT/TUNED-OUTPUT OSCILLATOR
ULTRASONIC MOTION DETECTOR
ULTRASONIC TRANSDUCER
ULTRAVIOLET DEVICE
VACUUM VARIABLE CAPACITOR
VAN DE GRAAFF GENERATOR
VARIABLE CAPACITOR
VARIABLE CRYSTAL OSCILLATOR
VARIABLE-FREQUENCY OSCILLATOR
VARIABLE INDUCTOR
VARIAC
VARIOMETER
VOCODER
VODER
VOICE TRANSMITTER
VOLTAGE-CONTROLLED OSCILLATOR
VOLTAGE-DEPENDENT RESISTOR
VOLTAGE DIVIDER
VOLTAGE DOUBLER
VOLTAGE TRANSFORMER
WIEN-BRIDGE OSCILLATOR
XENON FLASHTUBE
ZENER DIODE

Computer and Digital Electronics

ACCESS TIME
ACQUISITION TONE
ADDRESS
ALGORITHM
ALPHABETIC-NUMERIC
ANALOG COMPUTER
ANALOG-TO-DIGITAL CONVERTER
AND GATE
APPLICATION
ARITHMETIC OPERATION
ARTIFICIAL INTELLIGENCE
ASCII
ASSEMBLER AND ASSEMBLY LANGUAGE
ASSOCIATIVE STORAGE
ASTABLE MULTIVIBRATOR
ATTENDANT EQUIPMENT

NONSATURATED LOGIC
NOR GATE
OCTAL NUMBER SYSTEM
OR GATE
OVERFLOW
PARALLEL DATA TRANSFER
PARITY
POSITIVE LOGIC
PROBLEM-ORIENTED LANGUAGE
PROGRAM
PROGRAMMABLE CALCULATOR
PROGRAMMABLE READ-ONLY MEMORY
PULSE
PUSHDOWN STACK
QUEUING THEORY
RANDOM-ACCESS MEMORY
READ-MAINLY MEMORY
READ-ONLY MEMORY
READ-WRITE MEMORY
REAL-TIME OPERATION
REGISTER
REINITIALIZATION
RELAY LOGIC
RESET
RESISTOR-CAPACITOR-TRANSISTOR LOGIC
RESISTOR-TRANSISTOR LOGIC
RING COUNTER
SATURATED LOGIC
SCANNING
SCHMITT TRIGGER
SCHOTTKY LOGIC
SCRATCH-PAD MEMORY
SEQUENTIAL ACCESS MEMORY
SERIAL DATA TRANSFER
SHIFT REGISTER
SIMULATION
SOFTWARE
SORTING
SPEECH RECOGNITION
SPEECH SYNTHESIS
STACK
STATIC MEMORY
STOCHASTIC PROCESS
STORAGE
STORAGE TIME
STORE
STORING
SUBROUTINE
SYNTHESIZER
TABULATOR
THROUGHPUT
TIME SHARING
VERIFIER
VIDEO DISPLAY TERMINAL
VIDEO GAME
WORD
WORD PROCESSING
WORDS PER MINUTE

Electricity and Magnetism

AC GENERATOR
AC NETWORK
AC POWER TRANSMISSION
AC RIPPLE
AC VOLTAGE
ADMITTANCE
ALKALINE CELL
ALTERNATING CURRENT
AMPERE
AMPERE'S LAW
AMPERE TURN
AMPLIDYNE
ANGLE OF LAG
ANGLE OF LEAD
ANGULAR FREQUENCY
APPARENT POWER
ASSYMETRICAL CONDUCTIVITY
AVERAGE CURRENT
AVERAGE POWER
AVERAGE VOLTAGE
BARKHAUSEN EFFECT
BARNETT EFFECT
BATTERY
B-H CURVE
BUBBLE MEMORY
CAPACITANCE
CAPACITOR
CHARGE
CLOSED CIRCUIT
COERCIVE FORCE
COERCIVITY
CONDUCTANCE
CONDUCTIVE MATERIAL
CONDUCTIVITY
COPPER
CORBINO EFFECT
CORE LOSS
CORE MEMORY
CORE SATURATION
COULOMB
COULOMB, CHARLES AUGUSTIN DE
COULOMB'S LAW
CURRENT
CYCLE
CYCLES PER SECOND
DANIELL CELL
DC COMPONENT
DC GENERATOR
DC POWER
DC SHORT CIRCUIT
DC TRANSMISSION
DEGAUSSING
DEMAGNETIZING FORCE
DIAMAGNETIC MATERIAL
DIELECTRIC
DIELECTRIC ABSORPTION
DIELECTRIC AMPLIFIER
DIELECTRIC BREAKDOWN
DIELECTRIC CONSTANT
DIELECTRIC CURRENT
DIELECTRIC HEATING
DIELECTRIC LENS
DIELECTRIC LOSS
DIELECTRIC POLARIZATION
DIELECTRIC RATING
DIELECTRIC TESTING

Electronic Tubes

A BATTERY
ACORN TUBE
AIR COOLING
ALIGNED-GRID TUBE
ANGLE OF DEFLECTION
ANODE
B BATTERY
BEAM-POWER TUBE
CAMERA TUBE
CATHODE
CATHODE BIAS
CATHODE MODULATION
CATHODE-RAY TUBE
C BATTERY
CHARACTERISTIC CURVE
CHARACTRON
CHILD'S LAW
COLD CATHODE
CONTROL GRID
CONVERGENCE
COOLING
CUTOFF
CUTOFF VOLTAGE
DIODE
DIODE ACTION
DIODE TUBE
DISINTEGRATION VOLTAGE
DISSECTOR TUBE
DOORKNOB TUBE
DYNATRON EFFECT
DYNODE
ELECTRODE
ELECTRODE CAPACITANCE
ELECTRODE CONDUCTANCE
ELECTRODE DARK CURRENT
ELECTRODE IMPEDANCE
ELECTRON-BEAM GENERATOR
ELECTRON-BEAM TUBE
ELECTRON EMISSION
ELECTRON-MULTIPLIER TUBE
END-PLATE MAGNETRON
EXTENDED-CUTOFF TUBE
FILAMENT
FILAMENT CHOKE
FILAMENT SATURATION
FILAMENT TRANSFORMER
FILTER TUBE
FLASH TUBE
FLOOD GUN
FLUORESCENT TUBE
FORCED-AIR COOLING
FORWARD BIAS
FORWARD CURRENT
FORWARD RESISTANCE
FORWARD VOLTAGE DROP
GAS TUBE
GATED-BEAM TUBE
GETTER
GRID
GRID BIAS
GRID MODULATION
HEPTODE
HEXODE
ICONOSCOPE
IMAGE ORTHICON
INDIRECTLY HEATED CATHODE
INTERELECTRODE CAPACITANCE
KLYSTRON
MAGNETRON
NIXIE TUBE
NUVISTOR TUBE
OPERATING BIAS
OPERATING POINT
PARTITION NOISE
PENTAGRID CONVERTER
PENTODE TUBE
PHOTOCATHODE
PHOTOCELL
PHOTOMOSAIC
PHOTOMULTIPLIER
PHOTOTUBE
PICTURE TUBE
PLATE
PLATE CURRENT
PLATE DISSIPATION
PLATE MODULATION
PLATE POWER INPUT
PLATE RESISTANCE
PLATE VOLTAGE
POWER TUBE
PRIMARY EMISSION
PROTECTOR TUBE
RECEIVING TUBE
RECTIFIER TUBE
REVERSE BIAS
SATURATION
SATURATION CURRENT
SATURATION CURVE
SATURATION VOLTAGE
SCREEN GRID
SECONDARY EMISSION
SHOT EFFECT
STIMULATED EMISSION
SUPPRESSOR GRID
SWITCHING TUBE
TETRODE TUBE
THYRATRON
TRANSCONDUCTANCE
TRANSMITTING TUBE
TRAVELING-WAVE TUBE
TRIODE TUBE
TUBE
VIDICON
YOKE
ZERO BIAS

Mathematical Data

ALGEBRA
ALGORITHM
ALPHABETIC-NUMERIC
ARITHMETIC MEAN
ARITHMETIC OPERATION
ARITHMETIC SYMMETRY

AUTOCORRELATION FUNCTION
AXIS
BESSEL FUNCTION
BOOLEAN ALGEBRA
CALCULATION
CALCULUS
CARRY
CARTESIAN COORDINATES
COMPLEMENT
COMPLEX NUMBER
CONE
CONIC SECTION
CONSTANT
CONTINUOUS FUNCTION
COORDINATE SYSTEM
CORRELATION
COSECANT
COSINE
COSINE LAW
COTANGENT
CROSS PRODUCT
CURVE
CURVILINEAR COORDINATES
DE BROGLIE'S EQUATION
DECADE
DECIMAL
DEFINITE INTEGRAL
DE MOIVRE'S EQUATION
DE MORGAN'S THEOREM
DERIVATIVE
DETERMINANT
DIFFERENTIAL EQUATION
DIFFERENTIATION
DIMENSIONLESS QUANTITY
DIRECTED NUMBER
DIRECTION ANGLE AND COSINE
DISCONTINUOUS FUNCTION
DISCRIMINANT
DISTRIBUTIVE LAW
DOMAIN
DOT PRODUCT
DOUBLE INTEGRAL
EQUATION
EXPONENTIAL DISTRIBUTION
EXPONENTIAL FUNCTION
EXTRAPOLATION
FAMILY
FORMULA
FOURIER SERIES
FOURIER TRANSFORM
FUNCTION
FUNCTION TABLE
GAUSSIAN FUNCTION
GEOMETRIC MEAN
GRAPH
GRAPHICAL ANALYSIS
HISTOGRAM
HYPERBOLA
HYPERBOLIC TRIGONOMETRIC FUNCTIONS
INCREMENT
INDEFINITE INTEGRAL
INDEPENDENT VARIABLE
INFORMATION THEORY
INTEGRATION
INTERPOLATION
INVERSE FUNCTION
IRRATIONAL NUMBER
J OPERATOR
LAW OF AVERAGES
LAW OF COSINES
LAW OF SINES
LAW OF TANGENTS
LIMIT
LINE
LINEAR ALGEBRA
LINEAR EQUATION
LINEAR FUNCTION
LOGARITHM
LOGARITHMIC FUNCTION
LOGARITHMIC SCALE
LOGIC
LOGICAL EQUIVALENCE
LOGIC EQUATION
LOGIC FUNCTION
MATHEMATICAL LOGIC
MATRIX
MATRIX ALGEBRA
MEAN
MEDIAN
MESH ANALYSIS
MODULUS
MONTE CARLO METHOD
MULTIPLICATION
NATURAL NUMBER
NOMOGRAPH
NONLINEAR FUNCTION
NORMAL DISTRIBUTION
NUMBER LINE
NUMBER SYSTEMS
ORDER OF MAGNITUDE
PARABOLA
PARTIAL DERIVATIVE
PERIODIC FUNCTION
PERMUTATION
PLANE
POINT
POISSON DISTRIBUTION
POLAR COORDINATES
POLISH NOTATION
POWER SERIES
PROBABILITY
PSEUDORANDOM NUMBERS
PYTHAGOREAN THEOREM
QUADRATIC FORMULA
RADIAN
RANDOM NUMBERS
RANGE
RATIONAL NUMBER
RAYLEIGH DISTRIBUTION
REAL NUMBER
RECIPROCAL
RELATION
REPEATING DECIMAL
RESULTANT VECTOR
ROUNDING OF NUMBERS
SAMPLING THEOREM
SCALAR
SCIENTIFIC NOTATION
SECANT
SEQUENCE
SERIES
SIGNIFICANT FIGURES

SIMILARITY
SINE
SLIDE RULE
SLOPE
SPHERICAL COORDINATES
STANDARD DEVIATION
STATISTICAL ANALYSIS
STEP FUNCTION
STERADIAN
STOCHASTIC PROCESS
SUBSTITUTION METHOD FOR SOLVING EQUATIONS
TANGENT
TANGENT LINE
TAYLOR SERIES
TRANSFORMATION FUNCTIONS
TRIGONOMETRIC FUNCTION
TRIGONOMETRIC IDENTITIES
TRIGONOMETRY
TRUTH TABLE
VARIABLE
VECTOR
VECTOR ADDITION
VECTOR DIAGRAM
VENN DIAGRAM
X AXIS
Y AXIS

Miscellaneous Terminology

ABSORPTANCE
ACTIVATION
ACTIVITY RATIO
ALTERNATE ENERGY
AMPLITUDE
APPLICATION
ARC
ARC RESISTANCE
ARM-LIFT, BACK/CHEST PRESSURE RESUSCITATION
ARTIFICIAL RESPIRATION
BIOELECTRONICS
BIONICS
BODY CAPACITANCE
C
CAPACITIVE REACTANCE
CAPACITY
CARDIOPULMONARY RESUSCITATION
CARRYING CAPACITY
CHARACTERISTIC
CIRCUIT CAPACITY
CIRCUIT EFFICIENCY
CIRCULATING TANK CURRENT
COEFFICIENT OF COUPLING
COMPENSATION THEOREM
COMPLEX STEADY-STATE VIBRATION
COMPLEX WAVEFORM
COMPLIANCE
COMPTON EFFECT
CONJUGATE IMPEDANCE
CONSERVATION OF ENERGY
CONVERSION EFFICIENCY
CONVERSION GAIN AND LOSS
COORDINATION
CORNER EFFECT
CORNER REFLECTOR
CORONA
COUNTER VOLTAGE
COUPLED IMPEDANCE
CREST FACTOR
CRITICAL COUPLING
CROSS COUPLING
CURRENT AMPLIFICATION
CURRENT DRAIN
CYBERNETICS
DAMPED WAVE
DAMPING RESISTANCE
DEBYE LENGTH
DECAY
DECAY TIME
DEFIBRILLATION
DEFLECTION
DEHUMIDIFICATION
DELAY
DELAY DISTORTION
DELAY TIME
DELTA WAVE
DERATING CURVE
DIATHERMY
DIFFERENTIAL VOLTAGE GAIN
DIFFRACTION
DIFFUSION
DIODE ACTION
DIP
DISCONTINUITY
DISPLACEMENT CURRENT
DISSIPATION
DISSIPATION FACTOR
DISSIPATION RATING
DISTORTION
DISTRIBUTION
DISTURBANCE
DRIFT
DROOP
DURATION TIME
DUTY CYCLE
DUTY FACTOR
DYNAMIC CHARACTERISTIC
DYNAMIC RESISTANCE
DYNAMIC SPATIAL RECONSTRUCTOR
DYNATRON EFFECT
EARTH CURRENT
EDGE EFFECT
EDISON EFFECT
EFFICIENCY
EINSTEIN, ALBERT
ELASTANCE
ELECTRICAL ANGLE
ELECTRICAL DISTANCE
ELECTRIC CONSTANT
ELECTRIC SHOCK
ELECTROCARDIOGRAM
ELECTROCAUTERY
ELECTROENCEPHALOGRAPH
ELECTROMAGNETIC DEFLECTION
EMISSIVITY
ENDORADIOSONDE
ENERGY
ENERGY CONVERSION

ENERGY CRISIS
ENERGY DENSITY
ENERGY EFFICIENCY
ENERGY LEVEL
ENERGY STORAGE
ENTLADUNGSSTRAHLEN
ERASE
ESNAULT-PELTERIE FORMULA
EXCITATION
EXTERNAL FEEDBACK
FATIGUE
FEEDBACK
FEEDBACK RATIO
FIGURE OF MERIT
FIRE PROTECTION
FIRST AID
FIRST HARMONIC
FLASH ARC
FLASHBACK
FLASHOVER
FLASHOVER VOLTAGE
FLICKER EFFECT
FLICKER FREQUENCY
FLUCTUATING LOAD
FLUOROSCOPE
FLYBACK
FLYWHEEL EFFECT
FORM FACTOR
FREQUENCY
FREQUENCY TOLERANCE
FUNDAMENTAL FREQUENCY
FUNDAMENTAL SUPPRESSION
GAIN
GALVANISM
GEOTHERMAL ENERGY
GLOW DISCHARGE
GROUND EFFECT
HALF CYCLE
HALL MOBILITY
HARMONIC
HEARTH FIBRILLATION
HOT SPOT
IDEAL
IMAGE IMPEDANCE
IMPULSE
INDICATOR
INDUCED EFFECTS
INDUCTION HEATING
INDUCTION LOSS
INDUSTRIAL ELECTRONICS
INITIAL SURGE
INPUT
INPUT CAPACITANCE
INPUT IMPEDANCE
INPUT RESISTANCE
INSTANTANEOUS EFFECT
INTENSITY
INTERACTION
INTERNAL IMPEDANCE
INVERSE VOLTAGE
IONIZATION VOLTAGE
ISOLATION
JOSEPHSON EFFECT
JUNCTION CAPACITANCE
KICKBACK
KINETIC ENERGY
LEAD CAPACITANCE
LEAKAGE CURRENT
LEAKAGE FLUX
LEAKAGE INDUCTANCE
LEAKAGE REACTANCE
LEAKAGE RESISTANCE
LINEARITY
LINEAR RESPONSE
LINE BALANCE
LOAD CURRENT
LOAD IMPEDANCE
LOAD LOSS
LOAD POWER
LOAD VOLTAGE
LOCAL ACTION
LONGITUDINAL WAVE
LOSS
LOSS ANGLE
LOSS TANGENT
MAXIMUM POWER TRANSFER
MEDICAL ELECTRONICS
MICROELECTRONICS
MICROWAVE
MILLER EFFECT
MOUTH-TO-MOUTH RESUSCITATION
MOUTH-TO-NOSE RESUSCITATION
MU
NATURAL FREQUENCY
NEGATIVE FEEDBACK
NEGATIVE RESISTANCE
NEGATIVE TRANSCONDUCTANCE
NODE
NO-LOAD CURRENT
NO-LOAD VOLTAGE
NOMINAL VALUE
NONLINEARITY
NONSINUSOIDAL WAVEFORM
NUCLEAR ENERGY
NUCLEAR FISSION
NUCLEAR FUSION
NUCLEAR MAGNETIC RESONANCE
NULL
NYQUIST DIAGRAM
OHMIC LOSS
OPERATING ANGLE
OSCILLATION
OUT OF PHASE
OUTPUT
OUTPUT IMPEDANCE
PACEMAKER
PARALLEL RESONANCE
PARAMETER
PATENT
PEAK
PERFORMANCE CURVE
PERIOD
PHASE BALANCE
PHASE DISTORTION
PHASE SHIFT
PHASING
PHOTOCONDUCTIVITY
PIEZOELECTRIC EFFECT
POINT SOURCE
POLE
POTENTIAL ENERGY
POWER DENSITY

POWER FACTOR
POWER GAIN
POWER LOSS
POWER SENSITIVITY
PROXIMITY EFFECT
PULSATION
PULSE
PULSE RISE AND DECAY
QUARTER WAVELENGTH
RADIANT ENERGY
RADIOLOGY
RADIO WAVE
RAY
REACTIVE POWER
REDUNDANCY
REFLECTANCE
REGENERATION
RELAXATION OSCILLATION
RESISTANCE-CAPACITANCE TIME CONSTANT
RESISTANCE-INDUCTANCE TIME CONSTANT
RESISTANCE LOSS
RESOLUTION
RESOLVING POWER
RESONANCE
RESONANCE CURVE
RESONANT FREQUENCY
RESPONSE TIME
RF
RF CURRENT
RF POWER
RF VOLTAGE
RISE
RISE TIME
SECONDARY RADIATION
SELF RESONANCE
SERIES RESONANCE
SHOCK HAZARD
SKIN EFFECT
SLEW RATE
SNELL'S LAW
SOLAR ENERGY
SOLAR POWER
SPARK
SPECIFIC RESISTANCE
SPECTRAL DENSITY
SPECTRAL ENERGY DISTRIBUTION
SPECTRAL RESPONSE
SPECULAR REFLECTION
SPONTANEOUS EMISSION AND ABSORPTION
STABILITY
STATIC CHARACTERISTIC
STIMULATED EMISSION
SUBHARMONIC
SUBLIMINAL COMMUNICATIONS
SURGE
TAIL
TANGENTIAL SENSITIVITY
TEMPERATURE COEFFICIENT
THERMAL AGITATION
THERMAL BREAKDOWN
THERMAL CONDUCTIVITY
THERMAL ENERGY
THERMAL SHOCK
THERMAL STABILITY
THERMAL TIME CONSTANT
THERMIONIC EMISSION
THERMOELECTRIC CONVERSION
THOMSON EFFECT
THRESHOLD
THRESHOLD OF VISIBILITY
TOTAL HARMONIC DISTORTION
TRANSFER IMPEDANCE
TRANSMITTANCE
TRANSVERSE WAVE
TRUE POWER
TRUE POWER GAIN
ULTRASONIC SCANNER
UNITY
VARIABLE FREQUENCY
VOLTA EFFECT
VOLTAGE AMPLIFICATION
WAVELENGTH
WEIGHT
WHITE LIGHT
WIND POWER

Physical Data

ACCELERATION
ACETATE
ALBEDO
ALPHA PARTICLE
ANGLE OF REFLECTION
ANGLE OF REFRACTION
ANGULAR MOTION
ARSENIC
ATMOSPHERE
ATOM
ATOMIC CHARGE
ATOMIC CLOCK
ATOMIC MASS UNIT
ATOMIC NUMBER
ATOMIC WEIGHT
AVOGADRO CONSTANT
BARIUM
BETA PARTICLE
BETATRON
BLACK BODY
BOHR ATOM
BOMBARDMENT
BONDING
BOUND ELECTRON
CADMIUM
CARBON
CENTRIFUGAL FORCE
CERAMIC
CERENKOV RADIATION
CESIUM
CHAIN REACTION
CHROMA
COHERENT LIGHT
COHERENT RADIATION
COLLIMATION
COMPLEMENTARY COLORS
COMPOUND
CONTAMINATION
COPPER
CORPUSCLE

QUARTZ
QUASAR
RADIANCE
RADIATION
RADIOACTIVITY
RADIO ASTRONOMY
RADIO FREQUENCY
RADIO GALAXY
RADIOISOTOPE
RADIOMETRY
RADIOPAQUE SUBSTANCE
RELATIVISTIC SPEED
RELATIVITY THEORY
RIGHT ASCENSION
SAINT ELMO'S FIRE
SCATTERING
SCIENTIFIC NOTATION
SCINTILLATION
SECONDARY COLOR
SECOND LAW OF THERMODYNAMICS
SELENIUM
SEMICONDUCTOR
SIGNIFICANT FIGURES
SILICON
SILICON DIOXIDE
SILICONE
SILICON MONOXIDE
SILICON NITRIDE
SILICON STEEL
SILVER
SODIUM
SOLAR SYSTEM
SOUND
SPACE CHARGE
SPACE TRAVEL
SPECIFICATIONS
SPECIFIC GRAVITY
SPECIFIC HEAT
SPECTRUM
STEATITE
STEFAN-BOLTZMANN LAW
STRAIN
STRESS
SUNSPOT
SUNSPOT CYCLE
SUPERCONDUCTIVITY
SYNCHRONOUS ORBIT
TEMPERATURE
THERMODYNAMICS
THORIUM
TORQUE
TOTAL INTERNAL REFLECTION
TUNGSTEN
ULTRASONIC HOLOGRAPHY
ULTRASOUND
ULTRAVIOLET
VAN ALLEN RADIATION BELTS
VECTOR
VISIBLE SPECTRUM
WIEN'S DISPLACEMENT LAW
WORK
WORK FUNCTION
XENON
X RAY

Power Supplies

A BATTERY
AC RIPPLE
AC-TO-DC CONVERTER
ALKALINE CELL
AMPLIDYNE
ARC
ARC RESISTANCE
AUXILIARY POWER
BATTERY
BATTERY CHARGING
B BATTERY
BOOST CHARGING
B PLUS
BRIDGE CIRCUIT
BRIDGE RECTIFIER
CASCADE VOLTAGE DOUBLER
C BATTERY
CELL
CHARGING
CHOKE-INPUT FILTER
CHOPPER
CHOPPER POWER SUPPLY
CONSTANT-CURRENT SOURCE
CONSTANT-VOLTAGE POWER SUPPLY
CONTROL RECTIFIER
CURRENT LIMITING
CURRENT REGULATION
CUT-IN/CUT-OUT ANGLE
CUTOUT
DC COMPONENT
DC GENERATOR
DC POWER
DC POWER SUPPLY
DC-TO-AC CONVERTER
DC-TO-DC CONVERTER
DC TRANSMISSION
DEPOLARIZATION
DIAC
DIODE
DIODE ACTION
DIODE FEEDBACK RECTIFIER
DIODE RATING
DIODE TUBE
DIODE TYPES
DROPPING RESISTOR
DRY CELL
DYNAMIC REGULATION
DYNAMOTOR
ELECTROLYTIC CAPACITOR
ELECTROLYTIC CELL
ELECTROLYTIC RECTIFIER
ELECTROMECHANICAL RECTIFIER
EMERGENCY POWER SUPPLY
FILTER CAPACITOR
FLYBACK POWER SUPPLY
FULL-WAVE RECTIFIER
FULL-WAVE VOLTAGE DOUBLER
FUSE
GENERATOR
GOLDSCHMIDT ALTERNATOR
GROVE CELL
HALF-WAVE RECTIFIER

HUM
HYDROELECTRIC POWER PLANT
INVERTER
LEAD-ACID BATTERY
LITHIUM CELL
MARX GENERATOR
MECHANICAL RECTIFIER
MEMORY BACKUP BATTERY
MERCURY CELL
NICKEL-CADMIUM BATTERY
OVERCURRENT PROTECTION
OVERVOLTAGE PROTECTION
POLYPHASE RECTIFIER
POWER
POWER PLANT
POWER SUPPLY
POWER TRANSFORMER
QUICK-BREAK FUSE
QUICK CHARGE
RECHARGEABLE CELL
RECTIFICATION
RECTIFIER
RECTIFIER CIRCUITS
RECTIFIER DIODE
RECTIFIER TUBE
REGULATED POWER SUPPLY
REGULATING TRANSFORMER
REGULATION
RETURN CIRCUIT
RIPPLE
SERIES REGULATOR
SHORT-CIRCUIT PROTECTION
SHUNT REGULATOR
SILICON-CONTROLLED RECTIFIER
SILVER-OXIDE CELL
SILVER-ZINC CELL
SLOW-BLOW FUSE
SMOOTHING
SURGE
SURGE SUPPRESSOR
SWINGING CHOKE
SWITCHING POWER SUPPLY
SYNCHRONIZED VIBRATOR POWER SUPPLY
THREE-PHASE ALTERNATING CURRENT
THREE-PHASE RECTIFIER
THREE-QUARTER BRIDGE RECTIFIER
THREE-WIRE SYSTEM
THYRECTOR
THYRISTOR
TOTAL REGULATION
TRANSFORMER
TRANSFORMER EFFICIENCY
TRANSFORMER PRIMARY
TRANSFORMER SECONDARY
TRANSIENT
TRANSIENT PROTECTION
TRANSISTOR BATTERY
TRIAC
TRICKLE CHARGE
TWO-WIRE SYSTEM
VARIAC
VIBRATOR
VIBRATOR POWER SUPPLY
VOLTAGE DIVIDER
VOLTAGE DOUBLER
VOLTAGE DROP
VOLTAGE LIMITING
VOLTAGE MULTIPLIER
VOLTAGE REGULATION
VOLTAGE TRANSFORMER
ZINC CELL

Radiolocation, Radionavigation, and Radar

AERONAUTICAL BROADCASTING AND RADIONAVIGATION
AIR TRAFFIC CONTROL
ANTIHUNT DEVICE
APPROACH CONTROL
A-SCAN
AUTOMATIC DIRECTION FINDER
AVIONICS
AZ-EL
AZIMUTH
AZIMUTH ALIGNMENT
AZIMUTH BLANKING
AZIMUTH RESOLUTION
AZUSA
B DISPLAY
BEACON
BEAMWIDTH
BEARING
BLANKING
BLIP
CHAIN RADAR
CIRCULAR SWEEP
CONDOR
CONE MARKER
CONE OF SILENCE
CONICAL SCANNING
CONTRAN
CROSS BEARING
DEPTH FINDER
DIRECTION FINDER
DISH ANTENNA
DISTANCE RESOLUTION
ECHO
ECHO BOX
ELECTRONIC WARFARE
FIRING ANGLE
F SCAN
GROUND RETURN
GUIDED MISSILE
HOMING DEVICE
HUNTING
INTERCONTINENTAL BALLISTIC MISSILE
I SCAN
JAMMING
J DISPLAY
LORAC
LORAN
MAGNETIC BEARING
MAGNETIC POLE
PIP
RADAR

RADAR ANTENNA
RADAR TELESCOPE
RADIOLOCATION
RADIONAVIGATION
RANGE
SCAN FUNCTIONS
SODAR
SOFAR
SONAR
SONIC ALTIMETER
TACTICAL AIR NAVIGATION

Solid-State Electronics

ALLOY-DIFFUSED SEMICONDUCTOR
ALPHA
ALPHA-CUTOFF FREQUENCY
AVALANCHE
AVALANCHE BREAKDOWN
AVALANCHE IMPEDANCE
AVALANCHE TRANSISTOR
AVALANCHE VOLTAGE
BACK DIODE
BARDEEN, JOHN AND BRATTAIN, WALTER
BARRIER CAPACITANCE
BARRIER VOLTAGE
BASE
BETA
BETA-ALPHA RELATION
BETA-CUTOFF FREQUENCY
BIPOLAR TRANSISTOR
CARRIER
CARRIER MOBILITY
CAT'S WHISKER
CHANNEL
CHANNEL EFFECT
CHARACTERISTIC CURVE
CHIP
COLLECTOR
COLLECTOR CURRENT
COLLECTOR RESISTANCE
COMPLEMENTARY METAL-OXIDE SEMICONDUCTOR
COPPER-OXIDE DIODE
CURRENT HOGGING
CURRENT SATURATION
CUT-IN/CUT-OUT ANGLE
CUTOFF
CUTOFF VOLTAGE
CUTOUT
CZOCHRALSKI METHOD
DARLINGTON AMPLIFIER
DEPLETION MODE
DEPLETION LAYER
DIGITAL INTEGRATED CIRCUIT
DIODE
DIODE ACTION
DIODE CAPACITANCE
DIODE IMPEDANCE
DIODE RATING
DIODE TYPES
DOPING
DRAIN
DRIFT-FIELD TRANSISTOR
DUAL IN-LINE PACKAGE
ELECTROCHEMICAL DIFFUSED-COLLECTOR TRANSISTOR
ELECTRON-HOLE PAIR
EMITTER
EMITTER-BASE JUNCTION
EMITTER CURRENT
EMITTER DEGENERATION
EMITTER RESISTANCE
EMITTER STABILIZATION
EMITTER VOLTAGE
ENHANCEMENT MODE
EPITAXIAL TRANSISTOR
ESAKI DIODE
EXTRINSIC SEMICONDUCTOR
FLAT PACK
FORWARD BIAS
FORWARD BREAKOVER VOLTAGE
FORWARD CURRENT
FORWARD RESISTANCE
FORWARD VOLTAGE FROP
FOUR-LAYER SEMICONDUCTOR
FOUR-LAYER TRANSISTOR
GALENA
GATE
GATE-PROTECTED MOSFET
GERMANIUM
GRADED JUNCTION
GUNN DIODE
HOLE
HOOK TRANSISTOR
HOT-CARRIER DIODE
IMPATT DIODE
IMPURITY
INTEGRATED CIRCUIT
INTRINSIC SEMICONDUCTOR
JUNCTION
JUNCTION CAPACITANCE
JUNCTION DIODE
JUNCTION FIELD-EFFECT TRANSISTOR
JUNCTION TRANSISTOR
KNEE
LARGE-SCALE INTEGRATION
LIGHT-ACTIVATED SILICON-CONTROLLED RECTIFIER
LIGHT-EMITTING DIODE
LINEAR INTEGRATED CIRCUIT
LIQUID-CRYSTAL DISPLAY
MAJORITY CARRIER
MEDIUM-SCALE INTEGRATION
METAL-OXIDE SEMICONDUCTOR
METAL-OXIDE-SEMICONDUCTOR FIELD-EFFECT TRANSISTOR
METAL-OXIDE-SEMICONDUCTOR LOGIC FAMILIES
MICROALLOY TRANSISTOR
MICROCIRCUIT
MICROMINIATURIZATION
MINIATURIZATION
MINORITY CARRIER
NPN TRANSISTOR
N-TYPE SEMICONDUCTOR
OPERATING BIAS
OPERATING POINT
PEAK INVERSE VOLTAGE
PHOTOCELL
PHOTODIODE
PHOTOFET

PHOTOTRANSISTOR
PHOTOVOLTAIC CELL
PINCHOFF
PINCHOFF VOLTAGE
PIN DIODE
PLANAR TRANSISTOR
P-N JUNCTION
PNP TRANSISTOR
POINT-CONTACT JUNCTION
POINT-CONTACT TRANSISTOR
POWER TRANSISTOR
PROGRAMMABLE UNIJUNCTION TRANSISTOR
PROTECTIVE BIAS
P-TYPE SEMICONDUCTOR
RECOMBINATION
RECTIFIER DIODE
REVERSE BIAS
SATURATION
SATURATION CURRENT
SATURATION CURVE
SATURATION VOLTAGE
SCHOTTKY BARRIER DIODE
SECOND BREAKDOWN
SELENIUM
SELENIUM RECTIFIER
SEMICONDUCTOR
SHOCKLEY DIODE
SHOT EFFECT
SIGNAL DIODE
SILICON
SILICON-CONTROLLED RECTIFIER
SILICON DIODE
SILICON DIOXIDE
SILICON NITRIDE
SILICON RECTIFIER
SILICON TRANSISTOR
SILICON UNILATERAL SWITCH
SINGLE IN-LINE PACKAGE
SOLAR CELL
SOLID-STATE ELECTRONICS
SOURCE
SOURCE RESISTANCE
SOURCE VOLTAGE
STABISTOR
STACK
STEP-RECOVERY DIODE
SWITCHING DIODE
SWITCHING TRANSISTOR
TETRODE TRANSISTOR
THYRECTOR
THYRISTOR
TRANSCONDUCTANCE
TRANSISTOR
TRANSIT TIME
TRIAC
TUNNEL DIODE
ULTRA-LARGE-SCALE INTEGRATION
UNIJUNCTION TRANSISTOR
VARACTOR DIODE
VERY-LARGE-SCALE INTEGRATION
VOLTAGE HOGGING
ZENER DIODE
ZERO BIAS

Switching and Control

ACCELERATION SWITCH
AC RELAY
ANALOG CONTROL
ANTENNA TUNING
APPROACH CONTROL
AUDIO TAPER
AUTOMATIC BRIGHTNESS CONTROL
AUTOMATIC CONTRAST CONTROL
AUTOMATIC FREQUENCY CONTROL
AUTOMATIC GAIN CONTROL
AUTOMATIC LEVEL CONTROL
AUTOMATIC MODULATION CONTROL
AUTOMATION
AUTOMOTIVE ELECTRONICS
AUTOPILOT
CENTRIFUGAL SWITCH
CHATTER
CHOPPER
CHOPPER POWER SUPPLY
COMPENSATION
CONTROL
CONTROLLER
CONTROL PANEL
CONTROL RECTIFIER
CONTROL REGISTER
CONTROL UNIT
CROSSBAR SWITCH
DC SHORT CIRCUIT
DECOUPLING STUB
DIAC
DIGITAL CONTROL
DIMMER
DISCONNECT
DISTRIBUTOR
DOUBLE POLE
DOUBLE THROW
DRY-REED SWITCH
DYNAMIC CONTACT RESISTANCE
ELECTRIC EYE
ELECTROMAGNETIC SWITCH
ELECTRONIC CONTROL
ENABLE
FAILSAFE
FLOATING CONTROL
FOCUS CONTROL
FREQUENCY CONTROL
GAIN CONTROL
GANGED CONTROLS
HOLDING CURRENT
INHIBIT
INITIALIZATION
INTERRUPT
JOYSTICK
LIGHT-ACTIVATED SWITCH
LIMIT SWITCH
LINEAR TAPER
LINE TUNING
LOADING
LOGARITHMIC TAPER
MAKE
MASTER
MOTOR-SPEED CONTROL

MULTIPLE POLE
MULTIPLE THROW
POTENTIOMETER
RACK AND PINION
REED RELAY
RELAY
RHEOSTAT
ROBOTICS
ROTARY SWITCH
ROTOR
SELSYN
SERVOMECHANISM
SERVO SYSTEM
SETTLING TIME
SILICON-CONTROLLED RECTIFIER
SILICON UNILATERAL SWITCH
SINGLE THROW
SLIDE SWITCH
SLAVE
SWITCH
SWITCHING
SWITCHING DIODE
SWITCHING POWER SUPPLY
SWITCHING TRANSISTOR
SWITCHING TUBE
SYNCHRO
SYNCHRONIZATION
SYNCHRONOUS MOTOR
THERMOSTAT
THYRATRON
THYRECTOR
THYRISTOR
TIME-DELAY RELAY
TIME SWITCH
TONE CONTROL
TRANSMIT-RECEIVE SWITCH
TRIAC
TUNING
TUNING DIAL
VACUUM VARIABLE CAPACITOR
VARACTOR DIODE
VARIABLE CAPACITOR
VARIABLE INDUCTOR
VARIAC
VARIOMETER
VERNIER
VIBRATOR
VIBRATOR POWER SUPPLY
VOLTAGE-CONTROLLED OSCILLATOR
VOLUME CONTROL
WAFER SWITCH
WEIGHT CONTROL

Tests and Measurements

ABSORPTION WAVEMETER
ACCELEROMETER
ACCURACY
AMMETER
AMPERE-HOUR METER
ANALOG
ANALOG METERING
ANDERSON BRIDGE
APPARENT POWER
AREA REDISTRIBUTION
AUDIO SIGNAL GENERATOR
AVAILABLE POWER
AVAILABLE POWER GAIN
AVERAGE ABSOLUTE PULSE AMPLITUDE
AVERAGE CURRENT
AVERAGE NOISE FIGURE
AVERAGE POWER
AVERAGE VALUE
AVERAGE VOLTAGE
BACK BIAS
BACKLASH
BAR GENERATOR
BENCHMARK
BIAS
BIAS COMPENSATION
BIAS CURRENT
BIAS DISTORTION
BIAS STABILIZATION
BLACK BOX
BLOCKING
BLOCKING CAPACITOR
BREAKDOWN
BREAKDOWN VOLTAGE
BURN-IN
BYPASS CAPACITOR
CALIBRATION
CALORIMETER
CANNIBALIZATION
CAPACITIVE COUPLING
CAPACITIVE FEEDBACK
CAPACITOR TUNING
CARRIER CURRENT
CARRIER POWER
CARRIER VOLTAGE
CASCADE
CASCODE
CATHODE BIAS
CATHODE COUPLING
CATHODE FOLLOWER
CAVITY FREQUENCY METER
CELSIUS TEMPERATURE SCALE
CHARACTERISTIC DISTORTION
CHIRP
CHRONOGRAPH
CIRCUIT ANALYZER
CIRCUIT NOISE
CIRCUIT PROTECTION
COLOR-BAR GENERATOR
COMMON CATHODE/EMITTER/SOURCE
COMMON GRID/BASE/GATE
COMMON PLATE/COLLECTOR/DRAIN
COMPONENT LAYOUT
CONDUCTION COOLING
CONVECTION COOLING
CONVERSION
COOLING
CRITICAL COUPLING
CRITICAL DAMPING
CROSS-CONNECTED NEUTRALIZATION
CROSSED-NEEDLE METER
CRYSTAL CONTROL
CRYSTAL TEST CIRCUIT

CURVE TRACER
DAMPING
D'ARSONVAL
DEBUGGING
DECIBEL METER
DESIGN
DIELECTRIC TESTING
DIFFERENTIAL INSTRUMENT
DIGITAL METERING
DIODE CHECKER
DIODE FIELD-STRENGTH METER
DIRECT COUPLING
DIRECTION FINDER
DISTORTION ANALYZER
DOT GENERATOR
DOUBLE CIRCUIT TUNING
DOWNTIME
DRIVE
DYNAMIC RANGE
ELECTRICAL BANDSPREAD
ELECTRICAL ENGINEERING
ELECTROCARDIOGRAM
ELECTRODYNAMOMETER
ELECTROENCEPHALOGRAPH
ELECTROMECHANICAL FREQUENCY METER
ELECTROMECHANICAL OSCILLOSCOPE
ELECTROMETER
ELECTRON MICROSCOPE
ELECTRON TELESCOPE
ELECTROSCOPE
ELECTROSTATIC INSTRUMENT
EMISSION-TYPE TUBE TESTER
EMITTER COUPLING
EMITTER FOLLOWER
EMITTER STABILIZATION
EMPIRICAL DESIGN
ERROR
ERROR ACCUMULATION
EXPLORING COIL
EXPLORING ELECTRODE
FAHRENHEIT TEMPERATURE SCALE
FAILURE RATE
FEEDBACK CONTROL
FELICI MUTUAL-INDUCTANCE BALANCE
FET VOLTMETER
FIELD-STRENGTH METER
FIXED BIAS
FLUXMETER
FREQUENCY CALIBRATOR
FREQUENCY COMPENSATION
FREQUENCY COUNTER
FREQUENCY MEASUREMENT
FREQUENCY METER
FULL-SCALE ERROR
FUNCTION GENERATOR
GALVANOMETER
GATE-DIP METER
GEIGER COUNTER
GONIOMETER
GRACEFUL DEGRADATION
GREMLIN
GRID-DIP METER
HAZARD RATE
HETERODYNE FREQUENCY METER
HOT-WIRE METER
HUMAN ENGINEERING
ILLUMINATION METER
IMPEDANCE BRIDGE
INDUCTANCE MEASUREMENT
INDUCTIVE FEEDBACK
INSERTION GAIN
INSERTION LOSS
INSTRUMENT
INSTRUMENT AMPLIFIER
INSTRUMENT ERROR
INSTRUMENT TRANSFORMER
INTENSITY MEASUREMENT
INTERCHANGEABILITY
INTERFEROMETER
INTERMITTENT-DUTY RATING
INTERMITTENT FAILURE
INTERSTAGE COUPLING
IRON-VANE METER
KELVIN ABSOLUTE ELECTROMETER
KELVIN BALANCE
KELVIN DOUBLE BRIDGE
KELVIN TEMPERATURE SCALE
LECHER WIRES
LIGHT METER
LINK COUPLING
LISSAJOUS FIGURE
LOOSE COUPLING
MAXWELL BRIDGE
MCPROUD TEST
MECHANICAL BANDSPREAD
METER
METRIC SYSTEM
MICROPHONICS
MODULAR CONSTRUCTION
MURRAY LOOP TEST
NEUTRALIZATION
NOISE GENERATOR
OHMMETER
OHMS PER VOLT
OPTICAL COUPLING
OPTIMIZATION
OVERDRIVE
OVERLOADING
OVERMODULATION
OVERSHOOT
OWEN BRIDGE
PADDING
PANORAMIC RECEIVER
PARASITIC OSCILLATION
PATCH CORD
PEAK CURRENT
PEAK ENVELOPE POWER
PEAK POWER
PEAK-READING METER
PEAK-TO-PEAK VALUE
PEAK VALUE
PEAK VOLTAGE
PEN RECORDER
PERSISTENCE TRACE
PHOTOMETRY
PINCUSHION DISTORTION
PLUNGER-TYPE METER
POWER MEASUREMENT
PROBABLE ERROR
PROTOTYPE
PULLING
QUALITATIVE TESTING

QUANTITATIVE TESTING
RADIO ASTRONOMY
RADIOSONDE
RADIO TELESCOPE
RANKINE TEMPERATURE SCALE
RATIO-ARM BRIDGE
RAWINSONDE
RECORDING INSTRUMENT
RECTANGULAR RESPONSE
REFERENCE FREQUENCY
REFERENCE LEVEL
REFLECTOMETER
REGISTER CONSTANT
RELATIVE MEASUREMENT
RELATIVE-POWER METER
RELIABILITY
RESISTANCE-CAPACITANCE COUPLING
RESISTANCE MEASUREMENT
RESISTANCE THERMOMETER
RETRACE
RF AMMETER
RF VOLTMETER
ROOT MEAN SQUARE
SAFETY FACTOR
SAMPLING OSCILLOSCOPE
SCHERING BRIDGE
SCINTILLATION COUNTER
SECONDARY FAILURE
SELF OSCILLATION
SERVICE EQUIPMENT
SERVICE MONITOR/GENERATOR
SETTLING TIME
SHIELDING EFFECTIVENESS
SHUNT RESISTOR
SIGNAL GENERATOR
SIGNAL MONITOR
SIGNAL-PLUS-NOISE TO-NOISE RATIO
SIGNAL-TO-INTERFERENCE RATIO
SIGNAL-TO-NOISE RATIO
SIGNAL VOLTAGE
SINAD
SINAD METER
S METER
SOURCE COUPLING
SOURCE FOLLOWER
SOURCE STABILIZATION
SPECTRUM ANALYSIS
SPECTRUM ANALYZER
SPECTRUM MONITOR
SPONGE LEAD
SQUARE-LAW METER
STAGGER TUNING
STANDARD CELL
STANDARD CAPACITOR
STANDARD INDUCTOR
STANDARD INTERNATIONAL SYSTEM OF UNITS
STANDARD RESISTOR
STANDARD TEST MODULATION
STORAGE OSCILLOSCOPE
STRING ELECTROMETER
SWEEP
SWEEP-FREQUENCY FILTER OSCILLATOR
SWEEP GENERATOR
SYNCHROSCOPE
SYSTEMS ENGINEERING
TACHOMETER
TANGENTIAL GALVANOMETER
TAUT-BAND SUSPENSION
TEMPERATURE COMPENSATION
TEMPERATURE DERATING
TEMPERATURE REGULATION
TEST
TEST INSTRUMENT
TEST LABORATORY
TEST MESSAGE
TEST PATTERN
TEST PROBE
TEST RANGE
THERMAL PROTECTION
THERMAL RUNAWAY
THERMOMETER
TOLERANCE
TRANSFORMER COUPLING
TRANSISTOR TESTER
TRAPEZOIDAL PATTERN
TRIGGER
TRIGGERING
TROUBLESHOOTING
TWO-TONE TEST
VACUUM-TUBE VOLTMETER
VAR METER
VERNIER
VOLT-AMPERE-HOUR METER
VOLT BOX
VOLTMETER
VOLT-OHM-MILLIAMMETER
VOLUME-UNIT METER
WATT-HOUR METER
WATTMETER
WAVEMETER
WAVE-TRAP FREQUENCY METER
WHEATSTONE BRIDGE
WIEN BRIDGE

Units, Standards, and Constants

AB UNITS
AMERICAN MORSE CODE
AMERICAN NATIONAL STANDARDS INSTITUTE, INC.
AMERICAN WIRE GAUGE
AMPERE
AMPERE TURN
ASCII
ATOMIC MASS UNIT
AVOGADRO CONSTANT
BAND
BAUDOT
BIRMINGHAM WIRE GAUGE
BIT
BIT SLICE
BOLTZMANN CONSTANT
BRITISH STANDARD WIRE GAUGE
BRITISH THERMAL UNIT
BROADCAST BAND
BYTE
CANDELA

CANDLEPOWER
CGS SYSTEM
CHU
CODE
COLOR CODE
CONSTANT
COORDINATED UNIVERSAL TIME
COULOMB
CYCLE
DARAF
dBa
dBd
dBi
dBm
DECIBEL
DEGREE
DIELECTRIC CONSTANT
ELECTROMAGNETIC CONSTANT
ELECTRONIC INDUSTRIES ASSOCIATION
ELECTRON VOLT
ENERGY UNITS
ERG
FARAD
FARADAY
FOOT CANDLE
FOOT LAMBERT
FOOT POUND
FRANKLINE
GAUGE
GAUSS
GIGA
GILBERT
GREENWICH MEAN TIME
HALL CONSTANT
HECTO
HENRY
HERTZ
HOLLERITH CODE
HORSEPOWER
INSTITUTE OF ELECTRICAL AND ELECTRONIC ENGINEERS
INTERNATIONAL MORSE CODE
INTERNATIONAL SYSTEM OF UNITS
JOULE
KILO
KILOGRAM
KILOHERTZ
KILOWATT
KILOWATT HOUR
LAMBERT
LITER
LUMEN
LUMINAIRE
LUX
MAXWELL
MEGA
MEGAHERTZ
MEGAWATT
MEGOHM
MHO
MICRO
MIL
MILLI
MKS SYSTEM
MOLE
MORSE CODE
NANO
NATIONAL BUREAU OF STANDARDS
NATIONAL ELECTRIC CODE
NAUTICAL MILE
NEPER
NEWTON
NYQUIST STABILITY CRITERION
OERSTED
OHM
OUNCE
PHOT
PI
PICO
PLANCK'S CONSTANT
POUND
PREFIX MULTIPLIERS
PRIVATE LINE
Q SIGNAL
RESISTOR COLOR CODE
ROENTGEN
SAFETY FACTOR
SECOND
SECONDARY STANDARD
SPECIFICATIONS
STANDARD INTERNATIONAL SYSTEM OF UNITS
STANDARD TIME THROUGHOUT THE WORLD
STAT UNITS
STATUTE MILE
TEN CODE
TERA
TESLA
THERMAL TIME CONSTANT
TIME CONSTANT
TIME STANDARD
TIME ZONE
TOUCHTONE
TRANSFORMATION FUNCTIONS
TUNING FORK
UNDERWRITERS' LABORATORIES, INC.
VAR
VOLT
VOLT-AMPERE
VOLUME UNIT
WATT
WATT HOUR
WEBER
WORD
WWV/WWVH

Wiring and Construction

ADAPTOR
ALLIGATOR CLIP
ALUMINUM
ALUMINUM-CLAD WIRE
AMERICAN WIRE GAUGE
ARTWORK
BANANA JACK AND PLUG
BELL WIRE
BEZEL
BIFILAR WINDING

BINDING POST
BIRMINGHAM WIRE GAUGE
BLACK BOX
BLOCK DIAGRAM
BNC CONNECTOR
BRAID
BRANCH CIRCUIT
BREADBOARD TECHNIQUE
BREAK
BREAKER
BRIDGET-T NETWORK
BRIDGING CONNECTION
BRITISH STANDARD WIRE GAUGE
BUS
BUSHING
CABLE
CAPACITIVE COUPLING
CAPACITIVE FEEDBACK
CARD
CARTRIDGE FUSE
CASCADE
CASCODE
CATHODE BIAS
CATHODE COUPLING
CATHODE FOLLOWER
CATHODE KEYING
CATHODE MODULATION
CENTER TAP
CHASSIS
CIRCUIT
CIRCUIT BOARD
CIRCUIT BREAKER
CIRCUIT PROTECTION
CLAMPING
CLIP LEAD
CLOSED CIRCUIT
COMMON CATHODE/EMITTER/SOURCE
COMMON GRID/BASE/GATE
COMMON PLATE/COLLECTOR/DRAIN
COMPONENT LAYOUT
CONDUCTION COOLING
CONNECTION
CONNECTOR
CONTACT
CONTACT ARC
CONTACT BOUNCE
CONTACT RATING
CONTINUITY
CONTINUOUS-DUTY RATING
CONVECTION COOLING
COOLING
COPPER
COPPER-CLAD WIRE
CORD
COUPLING
CROSS-CONNECTED NEUTRALIZATION
CROSS COUPLING
DC GROUND
DC SHORT CIRCUIT
DECOUPLING
DESOLDERING TECHNIQUE
DIAGRAM
DIAL SYSTEM
DIP SOLDERING TECHNIQUE
DIRECT COUPLING
DISTRIBUTED ELEMENT
DOWN CONDUCTOR
EDGE CONNECTOR
ELECTRIC COUPLING
ELECTRIC WIRING
ELECTROPLATING
EMITTER FOLLOWER
EMITTER KEYING
ENAMELED WIRE
ENCAPSULATION
ENCLOSURE
ETCHING
FEEDBACK CONTROL
FEEDER CABLE
FEMALE
FLEXIBLE COUPLING
FORCED-AIR COOLING
FRAME
FUSE
GROUND BUS
GROUND FAULT
GROUND LOOP
GROUND ROD
HERMAPHRODITIC CONNECTOR
HERMETIC SEAL
HIGH-TENSION POWER LINE
H NETWORK
HOOKUP WIRE
ICE LOADING
INDUCTIVE FEEDBACK
INSULATED CONDUCTOR
INSULATOR
INTERCONNECTION
INTERSTAGE COUPLING
JACK
JACKET
JACK PANEL
JONES PLUG
JUNCTION
LADDER NETWORK
LEAD INDUCTANCE
LINE
LINE FAULT
LINE FILTER
LINK COUPLING
L NETWORK
LOOP
LOOSE COUPLING
LUG
LUMPED ELEMENT
MAGNETIC COUPLING
MAGNET WIRE
MALE
METAL
MODULAR CONSTRUCTION
N-TYPE CONNECTOR
OPEN CIRCUIT
OPEN LOOP
PARALLEL CONNECTION
PARALLEL LEADS
PARALLEL TRANSISTORS
PARALLEL TUBES
PATCH CORD
PHONE JACK
PHONE PLUG
PHONO JACK
PHONO PLUG

PLUG
PLUG-IN COMPONENT
POINT-TO-POINT WIRING
POWER LINE
POWER PLANT
POWER TRANSFORMER
PRINTED CIRCUIT
PROTECTIVE GROUNDING
PROTOTYPE
PUSH-PULL CONFIGURATION
PUSH-PUSH CONFIGURATION
QUADRIFILAR WINDING
QUICK-BREAK FUSE
REACTANCE GROUNDING
RESISTANCE-CAPACITANCE COUPLING
RETURN CIRCUIT
RIBBON CABLE
SAFETY GROUNDING
SAG
SCHEMATIC DIAGRAM
SCHEMATIC SYMBOLS
SERIES CONNECTION
SERIES-PARALLEL CONNECTION
SHORT CIRCUIT
SILICONE
SILVER
SILVER SOLDER
SLIDE-RULE TUNING
SLOW-BLOW FUSE
SOCKET
SOLDER
SOLDERING GUN
SOLDERING IRON
SOLDERING TECHNIQUE
SOLDER WICK
SOURCE COUPLING
SOURCE FOLLOWER
SOURCE KEYING
SPACER
SPADE LUG
SPAGHETTI
SPHERE GAP
SPONGE LEAD
STAGGER TUNING
STANDOFF INSULATOR
STEEL
STEEL WIRE
STRAIN
STRAIN GAUGE
STRAIN INSULATOR
STRANDED WIRE
STRAPPING
STRESS
SUBMARINE CABLE
SURGE SUPPRESSOR
SWITCH
SWITCHING
TANDEM CONNECTION
TANK CIRCUIT
TEE CONNECTION
TENSILE STRENGTH
THREE-WIRE SYSTEM
TIE STRIP
TIN
TINNING
T NETWORK
TRANSFORMER COUPLING
TRANSIENT PROTECTION
TRANSMISSION LINE
TRIFILAR WINDING
TUNED CIRCUIT
TUNED FEEDERS
TWIN-LEAD
TWO-WIRE SYSTEM
UHF CONNECTOR
UNBALANCED LINE
UNBALANCED LOAD
UNBALANCED OUTPUT
UNBALANCED SYSTEM
UNDERGROUND CABLE
UNDERWRITERS' LABORATORIES, INC.
VARIAC
VENIER
VOLTAGE DIVIDER
WIND LOADING
WIRE
WIRE ANTENNA
WIRE STRETCH
WIRE WRAPPING
WIRING DIAGRAM
WRITING GUN
WYE CONNECTION

Standard Symbols for Use in Schematic Diagrams

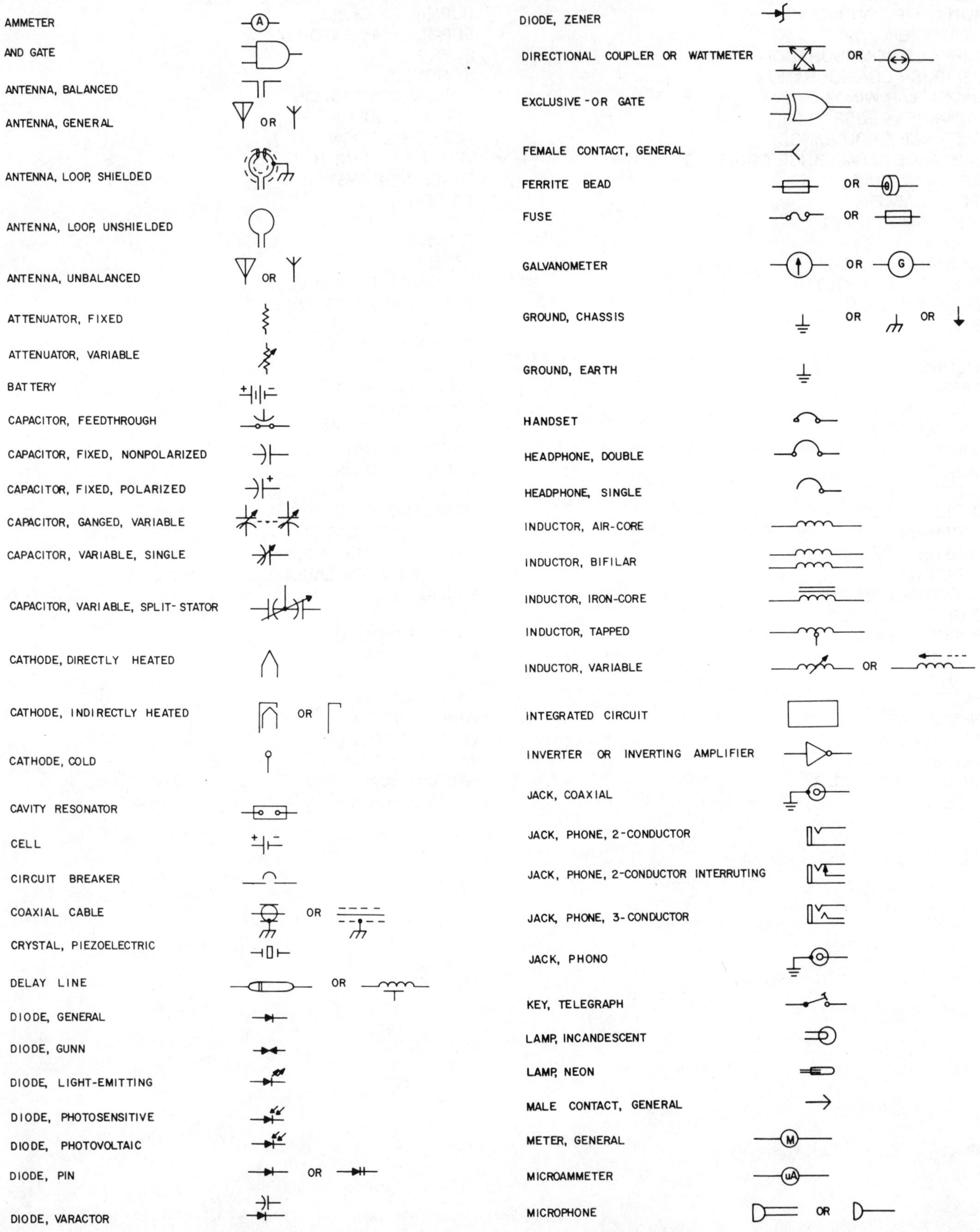

AMMETER

AND GATE

ANTENNA, BALANCED

ANTENNA, GENERAL — OR

ANTENNA, LOOP, SHIELDED

ANTENNA, LOOP, UNSHIELDED

ANTENNA, UNBALANCED — OR

ATTENUATOR, FIXED

ATTENUATOR, VARIABLE

BATTERY

CAPACITOR, FEEDTHROUGH

CAPACITOR, FIXED, NONPOLARIZED

CAPACITOR, FIXED, POLARIZED

CAPACITOR, GANGED, VARIABLE

CAPACITOR, VARIABLE, SINGLE

CAPACITOR, VARIABLE, SPLIT-STATOR

CATHODE, DIRECTLY HEATED

CATHODE, INDIRECTLY HEATED — OR

CATHODE, COLD

CAVITY RESONATOR

CELL

CIRCUIT BREAKER

COAXIAL CABLE — OR

CRYSTAL, PIEZOELECTRIC

DELAY LINE — OR

DIODE, GENERAL

DIODE, GUNN

DIODE, LIGHT-EMITTING

DIODE, PHOTOSENSITIVE

DIODE, PHOTOVOLTAIC

DIODE, PIN — OR

DIODE, VARACTOR

DIODE, ZENER

DIRECTIONAL COUPLER OR WATTMETER — OR

EXCLUSIVE-OR GATE

FEMALE CONTACT, GENERAL

FERRITE BEAD — OR

FUSE — OR

GALVANOMETER — OR

GROUND, CHASSIS — OR — OR

GROUND, EARTH

HANDSET

HEADPHONE, DOUBLE

HEADPHONE, SINGLE

INDUCTOR, AIR-CORE

INDUCTOR, BIFILAR

INDUCTOR, IRON-CORE

INDUCTOR, TAPPED

INDUCTOR, VARIABLE — OR

INTEGRATED CIRCUIT

INVERTER OR INVERTING AMPLIFIER

JACK, COAXIAL

JACK, PHONE, 2-CONDUCTOR

JACK, PHONE, 2-CONDUCTOR INTERRUTING

JACK, PHONE, 3-CONDUCTOR

JACK, PHONO

KEY, TELEGRAPH

LAMP, INCANDESCENT

LAMP, NEON

MALE CONTACT, GENERAL

METER, GENERAL

MICROAMMETER

MICROPHONE — OR

MILLIAMMETER

NAND GATE

NEGATIVE VOLTAGE CONNECTION

NOR GATE

OPERATIONAL AMPLIFIER OR

OR GATE

OUTLET, UTILITY, 117-V

OUTLET, UTILITY, 234-V

PHOTOCELL, TUBE

PLUG, PHONE, 2-CONDUCTOR

PLUG, PHONE, 3-CONDUCTOR

PLUG, PHONO

PLUG, UTILITY, 117-V

PLUG, UTILITY, 234-V

POSITIVE-VOLTAGE CONNECTION

POTENTIOMETER OR

PROBE, RADIO-FREQUENCY OR

RECTIFIER, SEMICONDUCTOR

RECTIFIER, SILICON-CONTROLLED

RECTIFIER, TUBE-TYPE

RELAY, DPDT

RELAY, DPST

RELAY, SPDT

RELAY, SPST

RESISTOR

RESONATOR

RHEOSTAT OR

SATURABLE REACTOR

SHIELDING

SIGNAL GENERATOR

SPEAKER OR

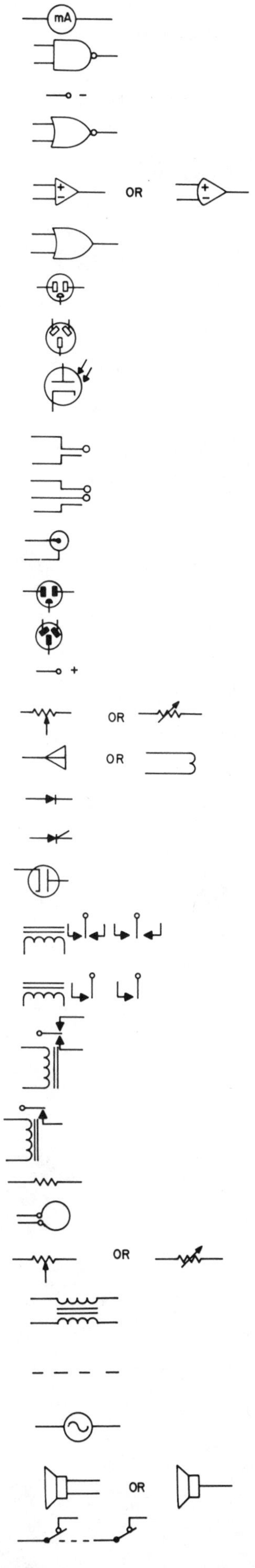

SWITCH, DPDT

SWITCH, DPST

SWITCH, MOMENTARY-CONTACT

SWITCH, ROTARY

SWITCH, SPDT

SWITCH, SPST

TERMINALS, GENERAL, BALANCED

TERMINALS, GENERAL, UNBALANCED

TEST POINT

THERMOCOUPLE OR

THYRISTOR

TRANSFORMER, AIR-CORE

TRANSFORMER, IRON-CORE

TRANSFORMER, TAPPED PRIMARY

TRANSFORMER, TAPPED SECONDARY

TRANSISTOR, BIPOLAR, npn

TRANSISTOR, BIPOLAR, pnp

TRANSISTOR, FIELD-EFFECT. N-CHANNEL

TRANSISTOR, FIELD-EFFECT, P-CHANNEL

TRANSISTOR, METAL-OXIDE, DUAL-GATE OR

TRANSISTOR, METAL-OXIDE, SINGLE-GATE OR

TRANSISTOR, PHOTOSENSITIVE OR

TRANSISTOR, UNIJUNCTION OR

TUBE, DIODE

TUBE, PENTODE

TUBE, PHOTOMULTIPLIER

TUBE, TETRODE

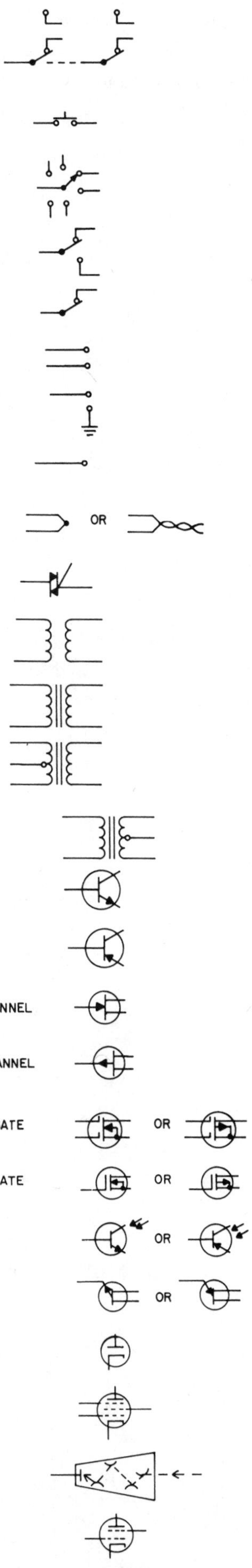

TUBE, TRIODE

UNSPECIFIED UNIT OR COMPONENT

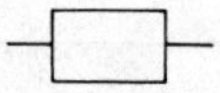

VOLTMETER

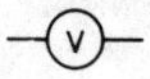

WATTMETER 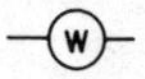 — OR —

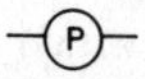

WIRES, CROSSING, CONNECTED  — OR —

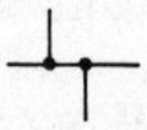

WIRES, CROSSING, NOT CONNECTED  — OR —

WIRES, CONNECTED (3-WAY)

"A" BATTERY

In vacuum-tube equipment, a source of power is required to heat the tube filaments. Normally, this power is supplied by a transformer from the ac mains; however, in tube-type portable equipment, where the power source is derived from batteries, the filament is powered by a battery called the "A" battery.

Virtually all portable equipment today is solid-state. Tubes are hardly ever used. Therefore, the "A" battery is seldom seen.

ABSOLUTE TEMPERATURE SCALE

See KELVIN TEMPERATURE SCALE.

ABSORPTANCE

When visible light strikes a substance, some of that light may pass through the substance or be reflected. The remainder of the energy is absorbed and converted to some other form of energy, usually heat. For transparent or semitransparent materials, absorptance is expressed in decibels-per-unit-length (*see* DECIBEL). For reflective materials, absorptance is simply expressed in decibels. If P_1 is the impinging power and P_2 is the power following absorption so that P_2 is less than P_1, then (see the illustration)

$$\text{absorptance (dB)} = 10 \log_{10} (P_1/P_2)$$

Absorptance may be defined for any electromagnetic wavelength and any material substance. The resulting quantity is dependent on the wavelength of the electromagnetic energy as well as on the substance. Certain ultrapure glass may have absorptance levels as low as 2 dB per kilometer. A perfect vacuum is considered to have an absorptance of 0 dB per kilometer at all wavelengths. *See also* TRANSMITTANCE.

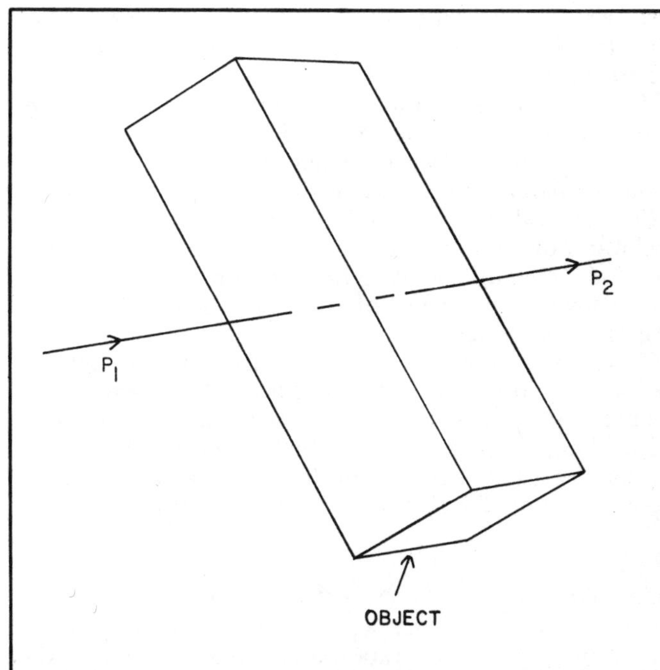

ABSORPTANCE: When visible light, or any other form of electromagnetic wave, strikes a substance, the impinging power (P_1) is always greater than the emerging power (P_2).

ABSORPTION

When any kind of energy or current is converted to some other form of energy, the transformation is referred to as absorption if it is dissipated and not used. Light may be converted to heat, for example, when it strikes a dark surface. Generally, absorption refers to an undesired conversion of energy.

Radio signals encountering the ionosphere undergo absorption as well as refraction. The amount of absorption depends on the wavelength, the layer of the ionosphere, the time of day, the time of year, and the level of sunspot activity. Sometimes most of the radio signal is converted to heat in the ionosphere; sometimes very little is absorbed and most of it is refracted or allowed to continue on into space. See A, B, and C.

The amount of absorption in a given situation is called absorptance, and is expressed in decibels or decibels per unit length. *See also* ABSORPTANCE.

ABSORPTION MODULATION

Absorption modulation is a means of impressing a modulating signal on an RF carrier by absorbing some of the RF carrier and dissipating it in a load other than the antenna or a following stage. It was used by early radio enthusiasts as an inexpensive means of modulating low-power transmitters that had been developed for Morse code (CW) communications. It had the advantage that audio amplifiers and audio transformers were not needed.

The method used was to wind a few turns of wire around the output tuned circuit of the amplifier. These turns were coupled to a carbon microphone as shown. The carbon granules in the microphone acted as a resistive

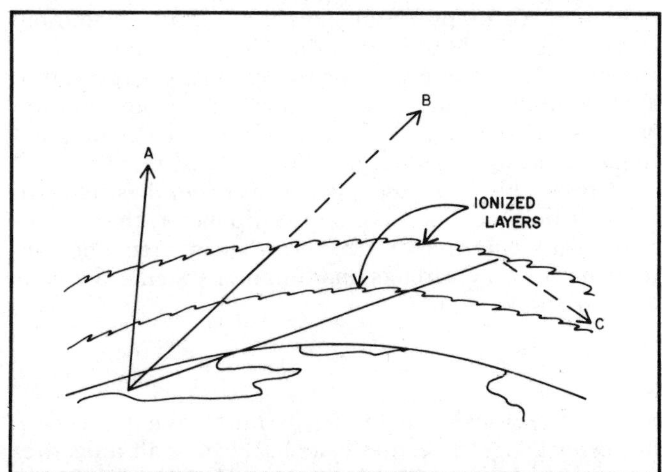

ABSORPTION: Radio waves encountering the ionosphere may be unaffected, passing on into space (A), or they may be absorbed partially (B). Sometimes they are returned to earth (C), but are partly absorbed.

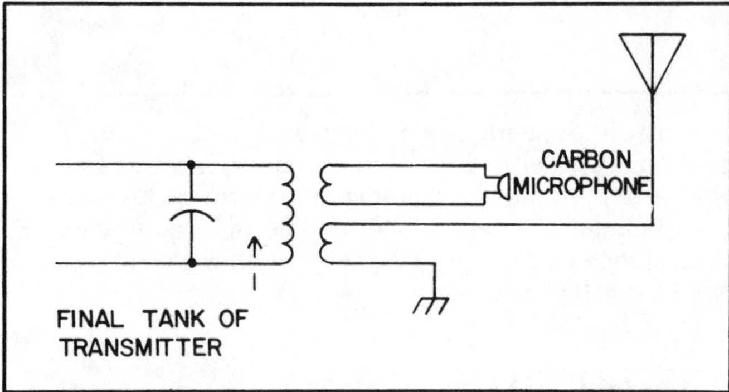

ABSORPTION MODULATION: The carbon microphone absorbs varying amounts of RF energy from the circuit as voice or sounds vibrate the carbon granules.

load, absorbing some of the RF energy from the tuned circuit. When a voice vibrated the diaphragm and disturbed the carbon granules, the resistance through the carbon changed in correlation with the sound waves.

This system had many problems; for example, power from the transmitter had to be very low to prevent burning (or fusing) the granules and ruining the microphone. Modulation percentage was very low because it was not possible to completely cut off the carrier wave (*see* AMPLITUDE MODULATION). Also, the microphone tended to be "hot" with RF energy, which often burned the fingers, lips, or nose of the operator unless great care was taken. It was, however, an interesting step in the quest for a means of transmitting voice-modulated signals in the early days of radio development.

ABSORPTION WAVEMETER

An absorption wavemeter is a device for measuring radio frequencies. It consists of a tuned inductance-capacitance (LC) circuit, which is loosely coupled to the source to be measured. Knowing the resonant frequency of the tuned circuit for various capacitor or inductor settings (by means of a calibrated dial), the LC circuit is adjusted until maximum energy transfer occurs. This condition is indicated by a peaking of RF voltage as shown by a meter. See the illustration.

When using an absorption wavemeter, it is extremely important that it not be too tightly coupled to the circuit under test. An RF probe, or short wire pickup, should be connected to the LC circuit of the absorption wavemeter, and the probe brought near the source of RF energy. If too much coupling occurs, reactance may be introduced into the circuit under test, and this might in turn change its frequency, resulting in an inaccurate measurement.

A special kind of absorption wavemeter has its own oscillator built in. Known as a grid-dip meter, this device enables easy determination of the resonant frequency or frequencies of LC circuits and antenna systems. *See also* GRID-DIP METER.

AB UNITS

Physical and electrical quantities are normally expressed in the meter-kilogram-second system. That is, all units are reducible to these three basic measures. In some instances, however, the centimeter-gram-second system is used. This results in units of different magnitude; a centimeter is

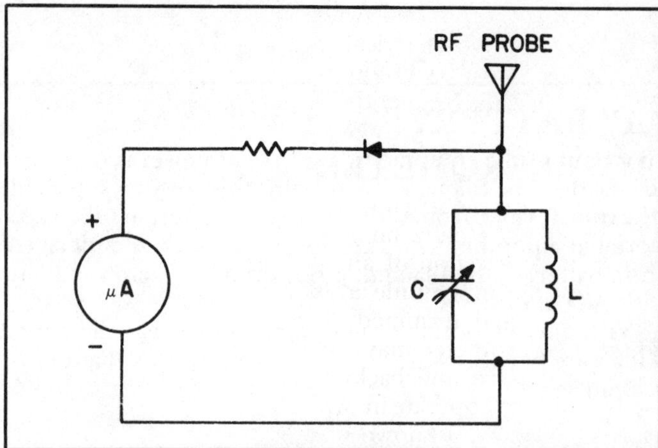

ABSORPTION WAVEMETER: The absorption wavemeter indicates that energy is coupled to a tuned circuit from a tested circuit. When the frequency of the LC circuit is properly adjusted, coupling is greatest, and the meter reading peaks.

1/100 of a meter, but a gram is 1/1000 of a kilogram.

Units in the centimeter-gram-second system are called ab units: the abampere, abvolt, and so on. One abampere is equal to 10 amperes, one abvolt is 10^{-8} volt, one abcoulomb is 10 coulombs, and one abolm is 10^{-9} ohm. Ab units are not generally used in electronics. The meter-kilogram-second (MKS) system is the basis for electrical units.

AC

See ALTERNATING CURRENT.

ACCELERATION

Acceleration is the rate of change in velocity of an object. Velocity, in physics, is determined by a given direction and a given speed. When an object continues in a particular direction at a constant rate of speed, the acceleration is zero. If either the speed or direction (or both) are changing, then the object is accelerating.

Velocity and acceleration are vector quantities, since both have a definite direction and intensity. When a car moves forward at increasing speed, its direction of acceleration is forward, in the same direction as the velocity. When the brakes are applied, the acceleration is directed backward, opposite to the velocity. At a constant cruising speed along a curved road, the direction of the acceleration is toward the center of curvature, at roughly right angles to the velocity.

Electrons and other charged particles may be accelerated by electromagnetic fields. This is how radio-wave propagation occurs. The acceleration may be back and forth, along a straight wire, in which case it is called linear acceleration. The acceleration may also take the form of circular motion (*see* PARTICLE ACCELERATOR).

ACCELERATION SWITCH

A switch that opens or closes in the presence of acceleration is known as an acceleration switch. Acceleration always produces a force opposite to the direction of acceleration. If the force reaches a certain intensity, the acceleration switch is activated.

A simple acceleration switch might consist of a weighted spring with an electrical contact which the weight will touch when the acceleration force is sufficient. Acceleration switches are frequently seen in aerospace apparatus.

ACCELEROMETER

A device for measuring acceleration is called an accelerometer. Actually, an accelerometer measures the intensity and direction of an acceleration-produced force. From this, the magnitude and direction of the acceleration vector may be determined.

An accelerometer may operate in just two directions, such as forward and backward, side to side, or up and down; or it may operate in two dimensions or three dimensions. Generally, a known mass is suspended by springs, and its position is noted under conditions of zero acceleration. Then the acceleration is found by electronically measuring the spring tension and direction of displacement.

Accelerometers are useful in such devices as guided missiles. A computer can determine course errors from the readings and produce instructions for corrective maneuvers.

Two linear (two-direction) accelerometers may be combined at right angles to form a two-dimensional accelerometer; three such linear accelerometers may be combined orthogonally to form a three-dimensional accelerometer. The vector components are determined and then added. (*See* VECTOR, VECTOR ADDITION.)

ACCESS TIME

When accessing an electronic memory, such as in a microcomputer calculator, the data is not received instantaneously. Although the process may appear instantaneous, there is always a slight delay, called the access time. When storing information in an electronic memory, a delay occurs as well between the storage command and actual storage; this, too, is known as access time, or it may be called storage time. If the computer itself activates a memory circuit, the access time is the delay between arrival of the command pulse at the memory circuit and the arrival of the required information.

In small computers and calculators, access time is so brief that it can hardly be noticed. If much information must be stored or retrieved in a large computer, the access time may be much longer. This is especially likely if the computer time is simultaneously shared among many users. *See also* MEMORY.

ACCURACY

Physical quantities such as time, distance, temperature, and voltage can never be determined precisely; there is always some error. Standard instruments or objects are used as the basis for all measurements. The accuracy of a given instrument is expressed as a percentage difference between the reading of that instrument and the reading of a standard instrument. Mathematically, if a standard instrument reads x_s units and the device under test reads x_t units, then the accuracy A in percent is:

$$A = 100\,|x_s - x_t|/x_s$$

For meters, accuracy is usually measured at several points on the meter scale, and the accuracy taken as the greatest percentage error. Sometimes only the full scale reading is used, then the accuracy is stated as a percentage of full scale. For example, a meter may be specified as being accurate to within ±10% (plus or minus 10 percent); this guarantees that the meter reading is within 10 percent of the actual value. If full scale is specified, then the only tested point is at full scale. These figures assume that the meter reading is sufficient to cause significant deflection of the needle.

Electronic components, such as resistors and capacitors, are given tolerance factors in percent, which is an expression of the accuracy of their values. Typical tolerances are 5, 10, and 20 percent. However, certain components are available with much lower tolerances (and therefore greater accuracy).

All units in the United States, and in the engineering world in general, are based on the meter-kilogram-second (MKS) system. The National Bureau of Standards is the ultimate basis in the U.S. for all determinations of accuracy. *See also* CALIBRATION, NATIONAL BUREAU OF STANDARDS, TOLERANCE.

ACETATE

Acetate is the name for a synthetically formulated plastic used in the manufacture of recording disks and tapes. Acetate is extremely durable and resistant to mechanical stress. This acts to preserve the recorded information for a longer time than would be possible without it. This is especially true of phonograph records.

AC GENERATOR

Mechanical energy is converted to electric current, with the aid of a magnetic field, in a device called an electric

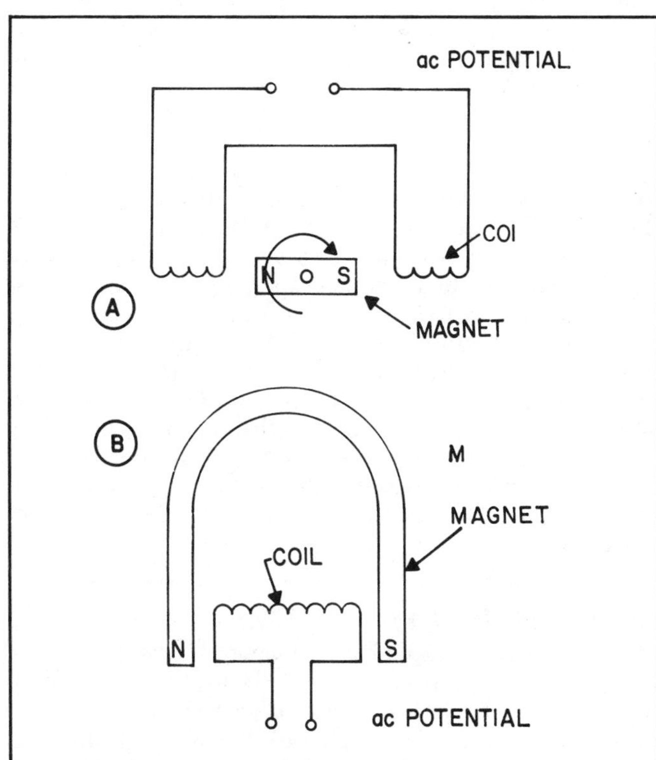

AC GENERATOR: An ac generator may consist of a magnet rotated inside a coil of wire as shown at A. Or, it may have a rotating coil inside a magnet (B). In either case, the magnetic field produces alternating currents in the wire when the wire passes through the field.

generator. Alternating current is produced by an ac generator.

The ac generator is essentially the same device as an ac motor, except that the conversion of energy takes place in reverse. Basic designs of an ac generator are shown in A and B. By rotating the magnet inside the coil of wire, an alternating magnetic field is produced. This change in magnetic flux causes the electrons in the wire to be accelerated, first in one direction and then in the other. The frequency of alternation depends on the speed of rotation of the magnet.

Some ac generators employ a rotating coil inside a magnet. The magnet may be operated from electricity itself. Ac generators may provide only a few milliwatts of power, or they may supply many thousands of watts, as is the case with commercial ac power generators.

Any radio-frequency transmitter or audio oscillator is, theoretically, an ac generator, since it puts out alternating current. *See also* ALTERNATING CURRENT, AUDIO FREQUENCY, RADIO FREQUENCY.

AC NETWORK

An alternating-current network is a circuit containing resistance and reactance. It differs from a direct-current network, where there is only simple resistance. Resistances are provided by all electrical conductors, resistors, coils, lamps, and the like; when a current passes through a resistance, heat is generated. Pure reactances, though, do not convert electrical current into heat. Instead, they store the energy and release it later. In storing energy, a reactance offers opposition to the flow of alternating current. Inductors and capacitors are the most elementary examples of reactances. Reactance is always either positive (inductive) or negative (capacitive). Certain specialized semiconductor circuits can be made to act like coils or capacitors by showing reactance at a particular ac frequency. Shorted or open lengths of transmission line also behave like reactances at some frequencies.

While simple resistance is a one-dimensional quantity, and reactance is also one-dimensional (though it may be positive or negative), their combination in an ac network is two-dimensional. Resistance ranges from zero to unlimited values; reactance ranges from zero to unlimited positive or negative values. Their combination in an alternating-current circuit is called impedance. Any ac network has a net impedance at a given frequency. The impedance generally changes as the frequency changes, unless it is a pure resistance. The impedance is expressed as a single point on a resistance-reactance half-plane as shown. Reactance is multiplied by the imaginary number j, called the j factor, for mathematical convenience. This quantity is defined as the square root of −1, also sometimes denoted by i. *See also* IMPEDANCE, J OPERATOR, REACTANCE.

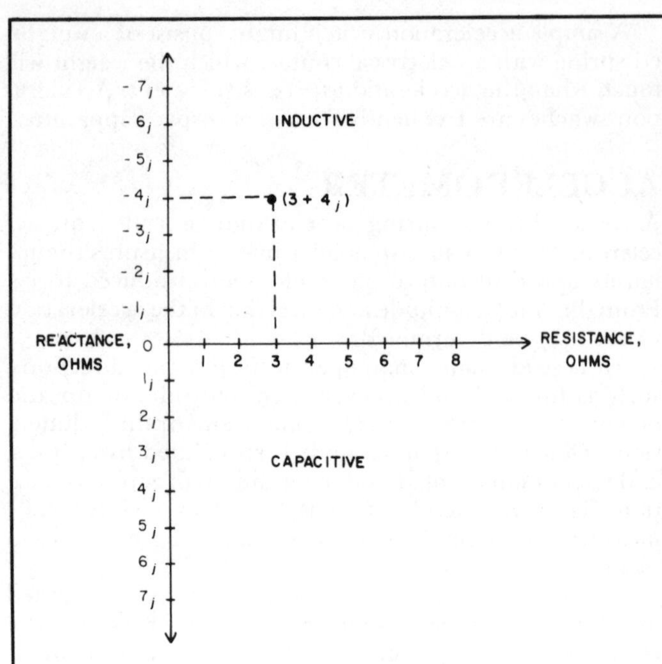

AC NETWORK: Impedance is shown as a combination, geometrically, of resistance and reactance. Any ac network displays an impedance that may be represented as a point on the resistance and reactance plane.

ACORN TUBE

A vacuum tube with no base, designed for use at uhf and microwave frequencies, with leads protruding radially outward, is called an acorn tube. It derives this name from its similarity to an acorn in physical shape. See the illustration.

Most vacuum tubes have leads at the bottom only, for use in sockets. At ultra-high or microwave frequencies, however, where the wavelength may be as small as a few millimeters in free space, the lead length must be reduced to the greatest extent possible. Otherwise, degradation of performance may result from excessive interelectrode capacitance and inductance. In the acorn tube, all esthetic considerations are abandoned. Leads may protrude from the sides, top, and bottom. *See also* TUBE.

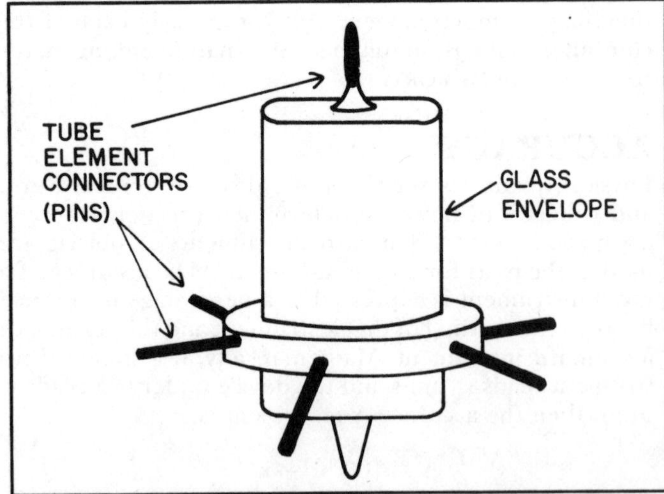

ACORN TUBE: An acorn tube, showing leads protruding from the sides and the top.

ACOUSTIC FEEDBACK

Positive feedback in a circuit, if it reaches sufficient proportions, will cause oscillation of an amplifier. In a public-address system, feedback may occur not only electrically, between the input and output component wiring, but between the speaker(s) and microphone. The result is a loud audible rumble, tone, or squeal. It may take almost any frequency, and totally disables the public-address system for its intended use. This kind of feedback may also occur between a radio receiver and transmitter, if both are voice modulated and operated in close proximity. This is called acoustic feedback.

Another form of acoustic feedback occurs in voice-operated communications systems (*see* VOX). While receiving signals through a speaker, the sound may reach sufficient amplitude to actuate the transmitter switching circuits. This results in intermittent unintended transmissions, and makes reception impossible. Compensating circuits in some radio transceivers equipped with VOX reduce the tendency toward this kind of acoustic feedback.

To prevent acoustic feedback in a public-address system, a directional microphone should be used, and all speakers should be located well outside the microphone response field. The gain (volume) should be kept as low as possible consistent with the intended operation. The room or environment in which the system is located should have sound-absorbing qualities, if at all possible, to minimize acoustic reflection.

ACOUSTICS

The science of sound engineering is known as acoustics. The excellent sound-distributing qualities of the symphony concert hall are no accident. See the illustration. If the room is badly designed acoustically, there will be areas where certain sound frequencies are amplified or attenuated out of proportion. You may then hear the violins but not the cello! Sound from certain parts of the stage may be amplified too much or too little in certain areas of the audience.

Acoustics are important in high-fidelity sound recording. The higher frequencies must be carefully balanced with the low and midrange frequencies. The channel balance in stereo recordings must be accurate. Also, relative loudness must be faithfully retrieved. The science of acoustics is extremely sophisticated and complex, but technology has reached the point today where anyone, with reasonably priced equipment and very little or no knowledge of acoustics, can have realistic concert sound in his own living room. *See also* HIGH FIDELITY.

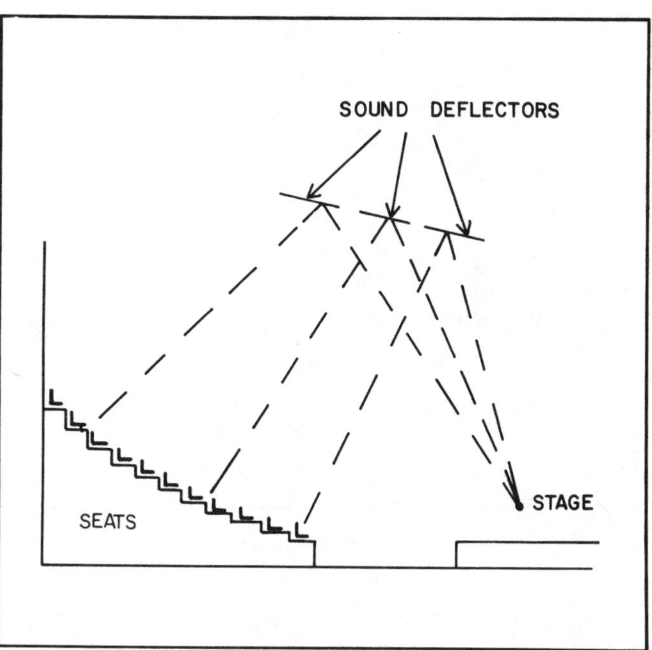

ACOUSTICS: A room designed for excellent sound transmission to specified places is a feat of acoustic engineering.

ACOUSTIC TRANSDUCER

Any device that converts sound energy to another form of energy, or some form of energy to sound energy, is an acoustic transducer. The most common acoustic transducers are the familiar speaker and microphone. Buzzers, bells, beepers, and sound or vibration detectors, however, are also acoustic transducers. *See also* MICROPHONE, SPEAKER, ULTRASONIC DEVICE.

ACOUSTIC TRANSMISSION

Acoustic transmission is the transfer of energy in the form of regular mechanical vibration through a solid, liquid, or gaseous medium. The most familiar form of acoustic transmission is, of course, the sound wave. In air, sound waves travel at approximately 1100 feet per second. In water, sound waves travel considerably faster; in a metallic solid such as steel, sound travels faster still. In general, the speed of acoustic transmission is roughly proportional to the density of the transmitting medium.

Acoustic transmission involves the displacement of individual molecules. As a burst of sound emanates from a given source in air, the area of disturbance expands outward in an ever-growing spherical front. The intensity of the disturbance at any given point diminishes with the distance from the source, although the total amount of energy in the whole wavefront remains essentially constant with time.

Any molecule of air, or of any other substance, is always in random motion. When a burst of acoustic energy passes, a molecule is caused to oscillate back and forth in accordance with the frequency pattern of the disturbance. The oscillation continues for the length of time the disturbance occurs. See A, B, and C. Each molecule vibrates back and forth along a line passing through the source of the disturbance. We hear noise when oscillating molecules cause our eardrums to vibrate in unison with them. A transducer (such as a microphone) converts acoustic disturbances into electrical impulses.

Acoustic transmission cannot take place in a vacuum, since there are no molecules to provide a medium for sound waves.

AC POWER TRANSMISSION

Whenever alternating current is delivered from one place to another, ac power transmission takes place. This may occur in an antenna feed line or waveguide, but the most common example of ac power transmission is the transfer of 60-Hz alternating current in utility wires.

If long-distance power transfer is necessary, the voltage leaving the power generating plant should be as high as possible—hundreds of thousands of volts are not uncommon levels, and occasionally voltages over a million will be found. High voltages are better for long-distance transmission because, for a given amount of power, the current is smaller if the voltage is larger, and the resistance loss in a power line is proportional to the current in the line. By maximizing the voltage, the current is minimized and thus the losses are reduced.

Extremely high voltages are somewhat inconvenient to work with; precautions must be taken to prevent flashover

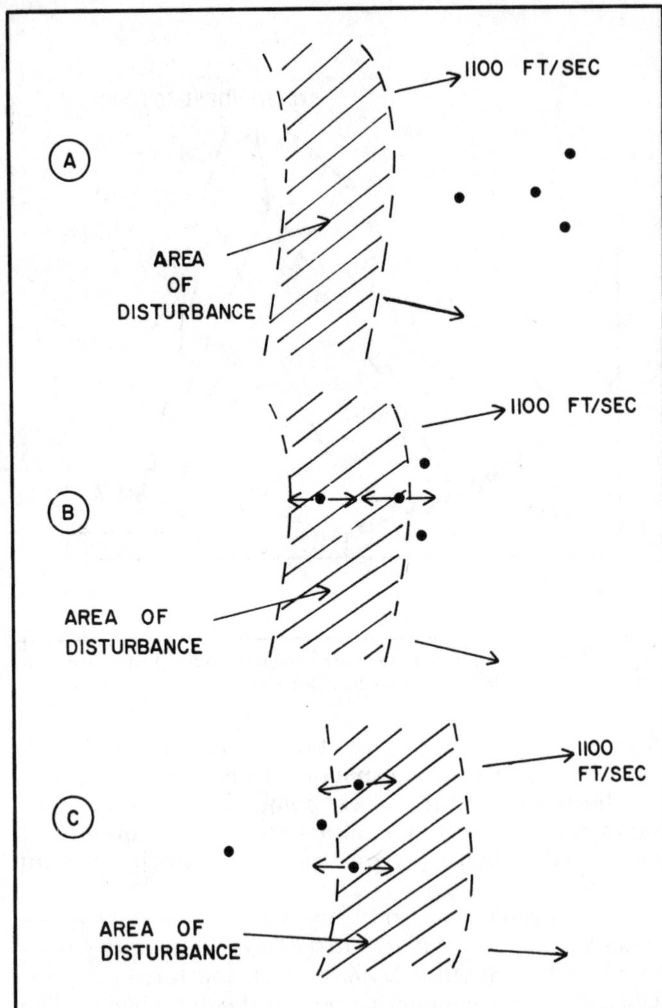

ACOUSTIC TRANSMISSION: Acoustic transmission in air involves the longitudinal vibration of molecules. At A, the disturbance has not yet reached the given molecules. At B and C, the sound disturbancepasses through the vicinity of the molecules.

between the transmission line wires and supporting structures or other wires. This makes large insulators mandatory. High-voltage ac transmission lines must be positioned well above the ground and obstructions such as trees and bushes to keep ground losses at a minimum. This necessitates the use of large supporting poles or towers as shown.

In recent years, technology has been perfected to the point where efficient power transmission is possible over distances of hundreds of miles. *See also* POWER PLANT.

ACQUISITION TONE

An audible tone that verifies entry into a microcomputer or minicomputer is called an acquisition tone. The tone is generated by means of a simple audio oscillator connected to the keyboard input of the device.

Acquisition-tone generators are found in small calculators, electronic games, and some microcomputer-controlled radio devices. Appealing to the electronically-oriented consumer, acquisition-tone generators have evolved quickly from the status of a novelty to a virtual necessity, since a certain amount of psychological reinforcement results from audio as well as video displays of operating functions.

AC POWER TRANSMISSION: A high-voltage power line requires the use of large supporting towers and massive insulators.

AC RELAY

An alternating-current (ac) relay is a device designed for the purpose of power switching from remote points (*see* RELAY). The ac relay differs from the dc relay in that it utilizes alternating current rather than direct current in its electromagnet. This offers convenience, since no special power supply is required if the electromagnet is designed to operate from ordinary 117-volt ac house outlets.

Ac relays are less likely than dc units to get permanently magnetized. This occurs when the core of the electromagnet becomes a magnet itself. In an ac relay, the magnetic field is reversed every time the direction of current flow changes, so that one polarity is not favored over the other. Ac relays must be damped to a certain extent for 60-Hz operation, or the armature will attempt to follow the current alternations and the relay will buzz. The armature and magnetic pole-pieces in ac relays are specially designed to reduce this tendency to buzz.

AC RIPPLE

Alternating-current ripple, usually referred to simply as ripple, is the presence of undesired modulation on a signal or power source. The most common form of ripple is 60-Hz or 120-Hz ripple originating from ac-operated power supplies.

In theory, the output of any power supply will contain some ripple when the supply delivers current. This ripple can, and should, be minimized in practice, because it will cause undesirable performance of equipment if not kept under control. Sufficient filtering, if used, will ensure that the ripple will not appear in the output of the circuit.

Ac ripple is virtually eliminated by using large inductors in series with the output of a power supply, and by connecting large capacitors in parallel with the supply output. The more current the supply is required to deliver, the more inductance and capacitance will be required in the filter stage. *See also* POWER SUPPLY.

ACTIVATION

In electronics and physics, the term activation may refer to several different things. Most obvious, we say that a circuit or device has been activated once it has been rendered operational by applying power or by programming.

Activation also refers to the process by which a substance is made radioactive by artificial means. When bombarded with neutrons, alpha particles, or other atomic nuclei in a particle accelerator, many elements can be made radioactive. (*See* PARTICLE ACCELERATOR.)

The cathode, or electron radiating element, of a tube may be treated to enhance its emission properties. This process is also referred to as activation. (*See* CATHODE.)

Activation also refers to the process of adding electrolyte material to a storage battery, making the battery operational. One example of this is the common lead-acid automobile battery, which is activated by filling it with a solution of sulfuric acid. (*See* LEAD-ACID BATTERY, STORAGE BATTERY.)

ACTIVE COMMUNICATIONS SATELLITE

An active communications satellite is an orbiting, self-sustaining repeater (*see* REPEATER). Active communications satellites receive signals from the earth at one frequency or band of frequencies, and simultaneously retransmit them on another frequency or band of frequencies. (Feedback problems prevent retransmission on the same frequency.) See the photograph of a typical active communications satellite.

ACTIVE COMMUNICATIONS SATELLITE: (photo courtesy National Aeronautics and Space Administration.)

Some communications satellites orbit the earth in synchronous orbits, also sometimes called geostationary orbits (*see* SYNCHRONOUS COMMUNICATIONS SATELLITE). The advantage of the synchronous satellite is the fact that it always remains at the same azimuth and elevation (*see* AZ-EL) for any location on the surface of the earth.

Overseas television transmission is now common via active communications satellite. The picture quality, marginal in the first satellites, now rivals that of a local television station.

Active communications satellites are unaffected by the ionosphere and offer reliable, predictable circuits. Eventually, ionospheric communications via the short waves may become secondary to active communications satellites for amateur radio as well as commercial radio and television use.

ACTIVE COMPONENT

In any electronic circuit, certain components are directly responsible for producing gain, oscillation, switching action, or rectification. If such components draw power from an external source such as a battery or power supply, then they are called active components. Active components include integrated circuits, transistors, vacuum tubes, and some diodes. An active component always requires a source of power to perform its function.

A passive component, by contrast, does its job with no outside source of power. Passive components include resistors, capacitors, inductors, and some diodes.

Some devices may act as either passive or active components, depending on the way they are used. One example is the varactor diode (*see* VARACTOR DIODE). When an audio-frequency voltage is applied across this type of diode, its junction capacitance fluctuates without any source of external power, and thus frequency modulation may be produced in an oscillator circuit. A dc voltage from a power supply, however, may be applied across the same diode, in the same circuit, for the purpose of adjusting the carrier frequency of the oscillator. In the first case, the varactor may be considered a passive component since it requires no battery power to produce FM from an audio signal. But in the second case, dc from an external source is necessary, and the varactor becomes an active component.

ACTIVE FILTER

An active filter is a filter that uses active components to provide selectivity. Generally, active filters are used in the audio range.

Active filters may be designed to have predetermined selective characteristics, and are lightweight and small. Control is easily accomplished by switches and potentiometers. Such filters, being active devices, require a source of dc power, but since the filters consume very little current, a miniature 9-volt transistor-radio type battery is usually sufficient to maintain operation for several weeks. Most audio filters use operational amplifiers (*see* OPERATIONAL AMPLIFIER), or "op amps." Active filters are not often seen at radio frequencies.

ACTIVITY RATIO

In a computer, not all the records in its file are needed or used at any given time. The term activity ratio refers to the proportion of the number of records actually in use at a particular time to the total number of records in the computer file. Activity ratio may be expressed as a percentage:

$$A = 100 (x_u/x_t)$$

where A is the activity ratio in percent, x_u is the number of records in use, and x_t is the total number of records in the file.

AC-TO-DC CONVERTER

More commonly called a dc power supply, an ac-to-dc converter consists of a rectifier circuit and a filtering circuit. The rectifier is made up of diodes. The filter circuit is made up of capacitors, and may include series dropping resistors or chokes. Such a circuit may be as simple as that shown in the schematic diagram, or it may be considerably more sophisticated. *See also* POWER SUPPLY, RECTIFICATION, RECTIFIER.

AC VOLTAGE

There are several different ways of defining voltage in an alternating-current circuit. They are called the peak, peak-to-peak, and root-mean-square (RMS) methods.

An ac waveform does not necessarily look like a simple sine wave. It may be square, sawtoothed, or irregular in shape. But whatever the shape of an ac waveform, the peak voltage is definable as the largest instantaneous value the waveform reaches. See A. The peak-to-peak voltage is the difference between the largest instantaneous values the waveform reaches to either side of zero. See B. Usually, the peak-to-peak voltage is exactly twice the peak voltage. However, if the waveform is not symmetrical, the peak value may be different in the negative direction than in the positive direction, and the peak-to-peak voltage may not be twice the positive peak or negative peak voltage.

The RMS voltage is the most commonly mentioned property of ac voltage. RMS voltage is defined as the dc voltage needed to cause the same amount of heat dissipation, in a simple, nonreactive resistor, as a given alternating-current voltage. For symmetrical, sinusoidal waveforms, the RMS voltage is 0.707 times the peak voltage and 0.354 times the peak-to-peak voltage. *See also* ROOT MEAN SQUARE.

ADAPTOR

Any device that makes two incompatible things work together is an adaptor. In electronics, adaptors are most

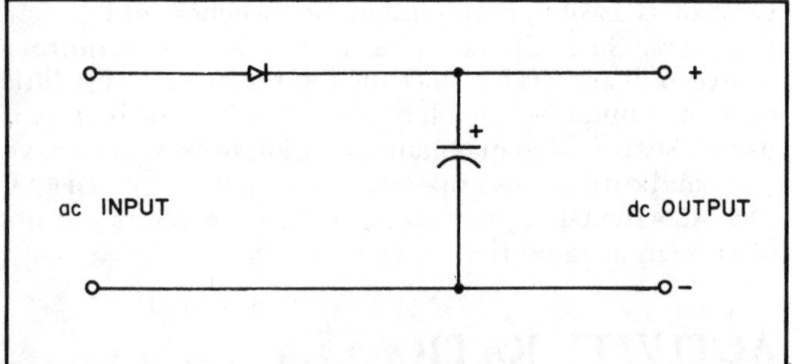

AC-TO-DC CONVERTER: This is a simple half-wave power supply with a single capacitor for filtering.

frequently seen in cable connectors, since there are so many different kinds of connectors. Such cable adaptors are sometimes called "tweenies" in the popular jargon. Two types of coaxial-cable adaptors are shown in the illustration.

Adaptors should be used as sparingly as possible, especially at radio frequencies, because they sometimes produce impedance irregularities along a section of feed line or cable. However, adaptors are invaluable in engineering and test situations as a convenience. Every test or service shop should have a good supply. *See also* CONNECTOR.

ADCOCK ANTENNA

An Adcock antenna is a special type of phased array (*see* PHASED ARRAY) for achieving a bidirectional, or figure-eight, radiation pattern in the horizontal plane.

A pair of vertical antennas separated by ½ wavelength (or occasionally less) is connected in such a way that the signal reaching one antenna is 180 degrees out of phase with the signal reaching the other antenna. This is usually accomplished by making one transmission-line section ½ wavelength longer, electrically, than the other. When this is done, the antenna combination radiates well in horizontal directions that line up with the two antennas, but poorly in directions perpendicular to the line connecting the antenas. See A and B. This is because the signals from the antennas cancel in the latter directions. In inter-

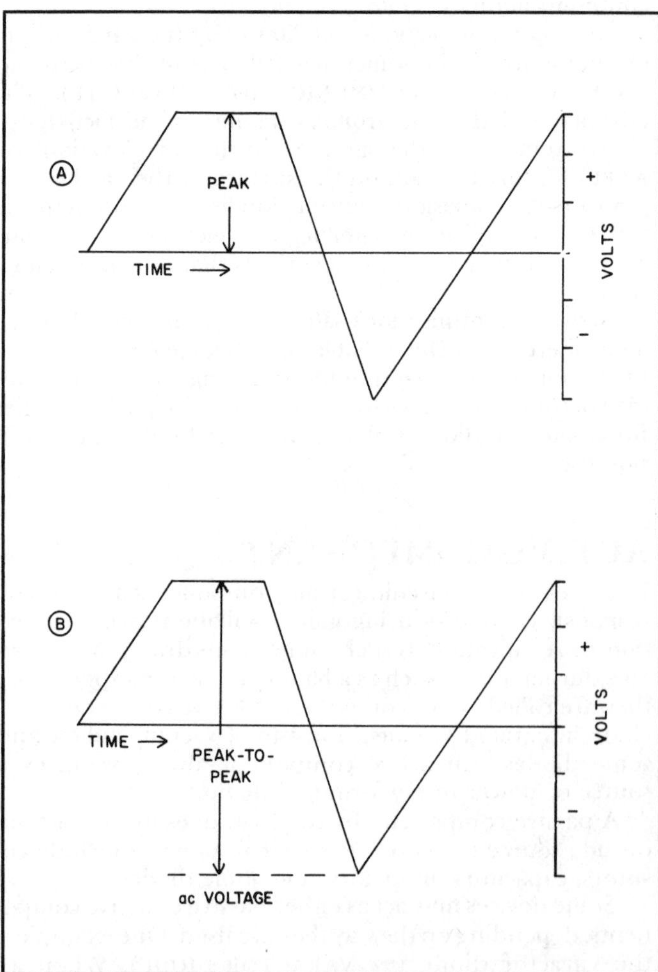

AC VOLTAGE: At A, the positive peak voltage of an irregular ac waveform. At B, the peak-to-peak voltage.

mediate directions, the antenna system radiates, but not as well as in the most favored directions.

Adcock antennas may be used for receiving as well as transmitting. An interfering signal may be attenuated by orienting the Adcock antenna so that the undesired signal falls into the "notch" in the figure-eight pattern. Some AM broadcast stations must use directional transmitting antennas, such as the Adcock antenna, to minimize interference with each other. A nearby station on the same channel, or an adjacent channel, as a given station might otherwise cause interference. Arrays of Adcock antennas are often used as direction-finding systems.

The Adcock antenna displays a different impedance than a simple vertical antenna, and matching systems are therefore required for proper transmitter and feed line operation. *See also* ANTENNA MATCHING.

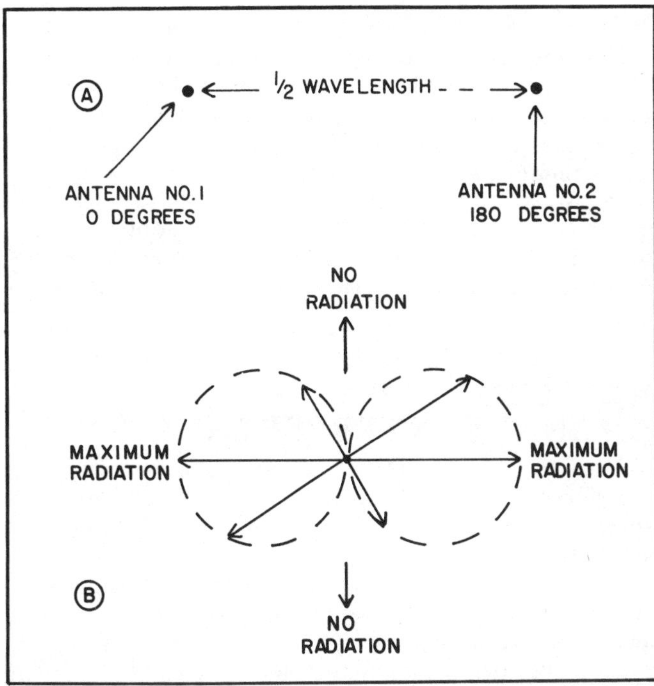

ADCOCK ANTENNA: The Adcock antenna (A) consists of two vertical radiators spaced ½ wavelength (sometimes closer) from each other, but driven 180 degrees out of phase. The best radiation occurs in line with the radiators, and the worst is perpendicular to the line connecting them (B).

ADDER

In digital electronics, an adder is a circuit that forms the sum of two numbers. An adder is also a circuit that combines two binary digits and produces a carry output. Such a combination is simple in binary arithmetic (*see* BINARY-CODED NUMBER). The table shows the carry output as a function of the two added digits.

In color television receivers, the circuit that combines the red, green, and blue signals from the receiver is called an adder. *See also* COLOR TELEVISION.

ADDRESS

Computer memory is stored in discrete packages for easy access. Each memory location bears a designator, usually a number, called the address. By selecting a particular address by number, the corresponding set of memory data is made available for use.

A calculator or small computer may have several different memory channels for storing numbers. Each channel is itself designated by a number, for example, 1 through 8. By actuating a memory-address function control, followed by the memory address number, the contents of the memory channel are called for use. Some radio receivers and transceivers make use of a memory-address function for convenience in calling, or switching among, frequently used frequencies. The memory-address status is shown by a panel indicator such as an LCD display. See the photo. *See also* MEMORY.

ADDER: CARRY OUTPUT IN BINARY ARITHMETIC AS A FUNCTION OF ADDED DIGITS.

Addends	Carry Output
0,0	0
0,1	0
1,0	0
1,1	1

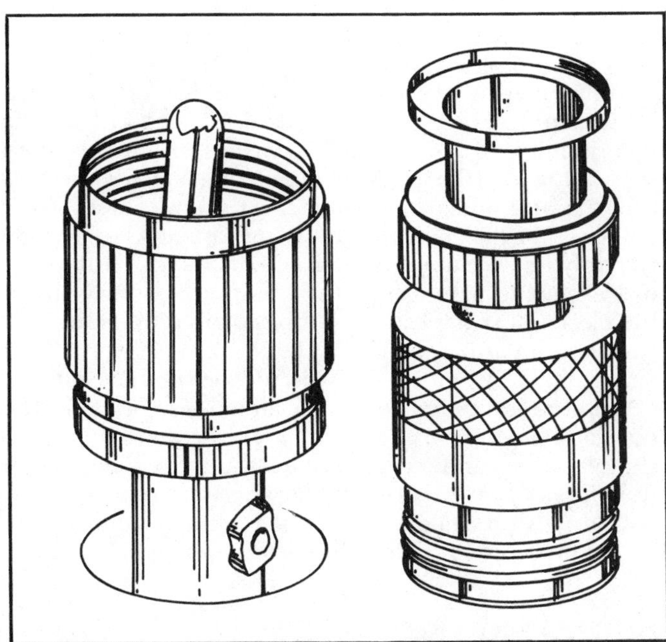

ADAPTOR: Cable adaptors are a handy thing to have around the test bench. At left, an adaptor for BNC female to UHF male; at right an adaptor for UHF female to BNC male.

ADDRESS: Memory address indicators on the front panel of an Amateur-Radio transceiver (courtesy of Amateur-Wholesale Electronics, Inc.).

ADMITTANCE

Admittance is the ease with which current flows in an alternating-current circuit. Admittance is the reciprocal of impedance (*see* IMPEDANCE), and is expressed in units called mhos. The word mho is ohm spelled backwards. A more modern term is Siemens (*see* SIEMENS).

Resistors, inductors, and capacitors all contribute to the admittance in an ac circuit. In a dc circuit, only the resistance determines the admittance. In this specialized case, admittance is identical to conductance (*see* CONDUCTANCE).

ADJACENT-CHANNEL INTERFERENCE

When a receiver is tuned to a particular frequency and interference is received from a signal on a nearby frequency, the effect is referred to as adjacent-channel interference.

To a certain extent, adjacent-channel interference is unavoidable. When receiving an extremely weak signal near an extremely strong one, interference is likely, especially if the stronger signal is voice modulated. No transmitter has absolutely clean modulation, and a small amount of off-frequency emission occurs with voice modulation, especially AM and SSB types. (*See* AMPLITUDE MODULATION, SINGLE SIDEBAND, SPLATTER.)

Adjacent-channel interference may be reduced by using proper engineering techniques in transmitters and receivers. Transmitter audito amplifiers, modulators, and RF amplifiers should produce as little distortion as the state of the art will permit. Receivers should employ selective filters of the proper bandwidth for the signals to be received and the adjacent-channel response should be as low as possible. A flat response in the passband (*see* PASSBAND), and a steep drop-off in sensitivity outside the passband, are characteristics of good receiver design. See A and B. Modern technology has made great advancements in the area of receiver-passband selectivity.

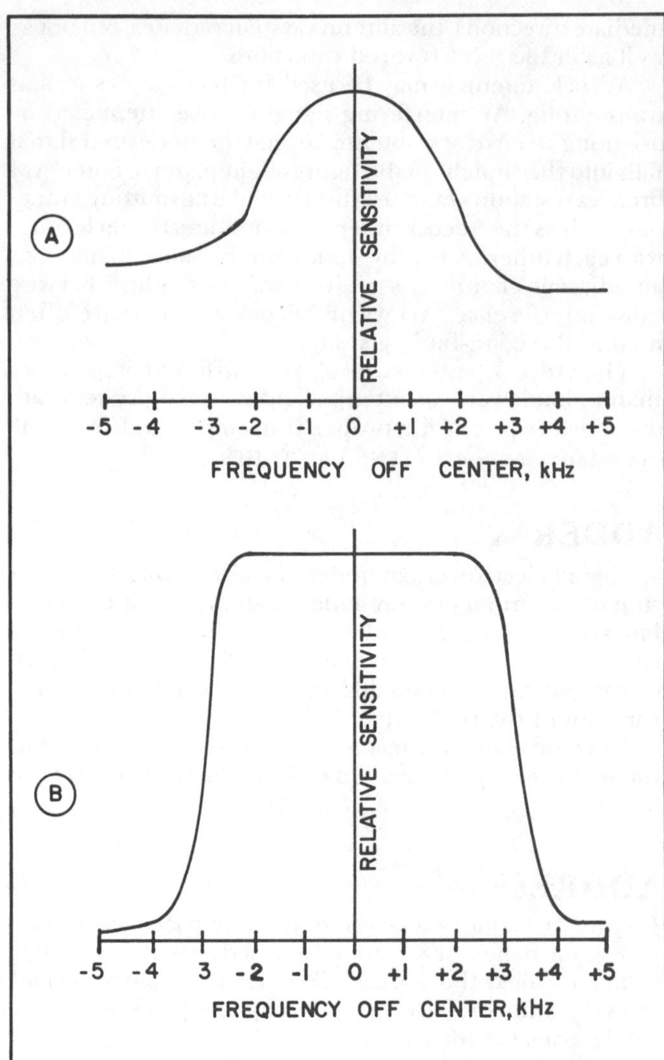

ADJACENT-CHANNEL INTERFERENCE: At A, a poor receiver bandpass response; at B, a good receiver bandpass response. These bandwidths are for AM reception.

AERONAUTICAL BROADCASTING AND RADIONAVIGATION

The Aeronautical Broadcasting Service is committed to the dissemination of information for purposes of air navigation. Air navigation and communication are carried on by an extensive network of transmitting stations, radio beacons, radiolocation stations, two-way voice and digital communications, and radar. *See also* BEACON, RADIOLOCATION, RADAR.

AFC

See AUTOMATIC FREQUENCY CONTROL.

AGC

See AUTOMATIC GAIN CONTROL.

AGING

All electronic components deteriorate with time. This process is called aging. Some components age more rapidly than others. Some, such as wire and carbon-composition resistors, are almost immune to aging.

Aging may be regarded as either mechanical or electrical in nature. Examples of mechanical aging include the deterioration of control shaft bearings, frequently used screw threads, the physical scratching and denting of cabinets, and similar wear and tear. Examples of electrical aging are the dielectric breakdown of capacitors (especially electrolytics), contamination of coaxial cables, "softening" or leakage in vacuum and other things that directly affect the electrical specification of a piece of equipment.

Electronic apparatus should be checked periodically for visible signs of aging. Tests should be conducted to be certain the equipment meets its claimed specifications. If it does not, alignment or repair should be done promptly. Circuit boards and terminal strips should be kept free of dust, grease, and corrosion. Loss tests should be made periodically for coaxial cables, since contamination or corrosion inside such cables is not visibly apparent.

AIR COOLING

Components that generate great amounts of heat, such as vacuum tubes, transistor power amplifiers, and some re-

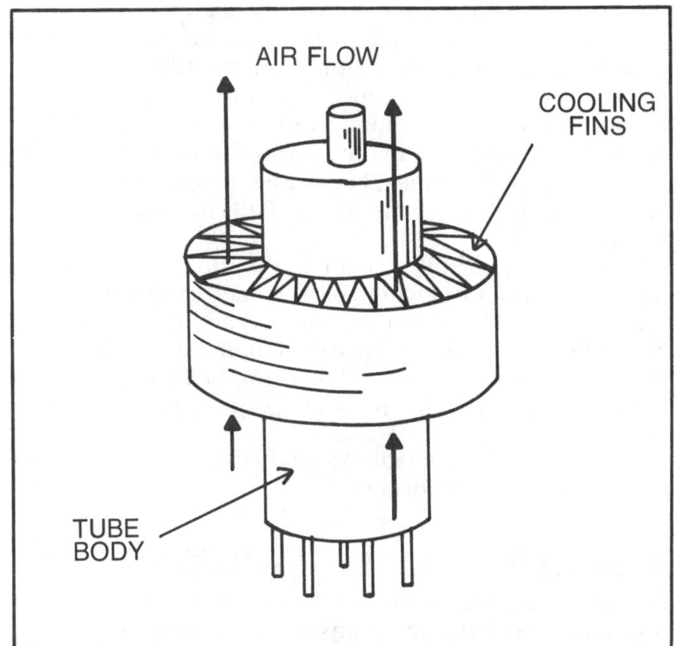

AIR COOLING: Some electronic tubes have fins for the purpose of air cooling. A fan forces air through the fins, keeping the body of the tube within rated temperatures.

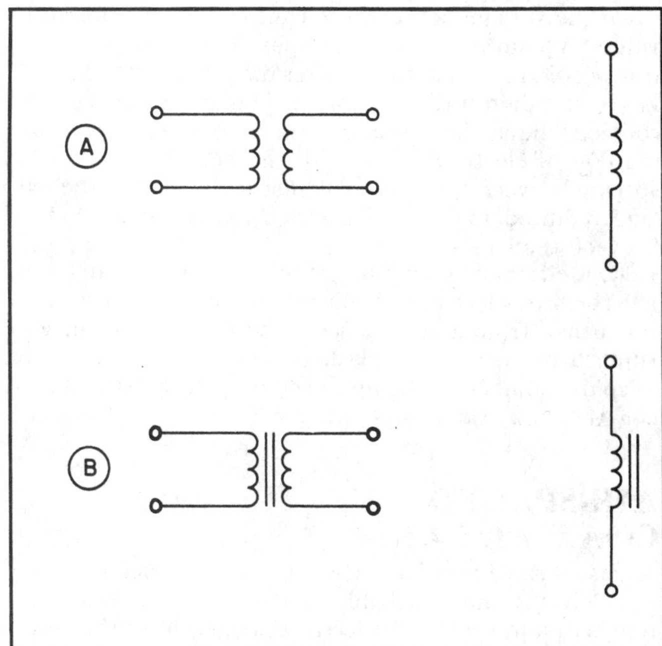

AIR CORE: At A, the schematic symbols for air-core transformers and inductors. At B, the schematic symbols for powdered-iron or ferrite-core transformers and inductors.

sistors, must be provided with some means for cooling or damage may result. Such components may be air cooled or conduction cooled (*see* CONDUCTION COOLING). Air cooling may take place as heat radiation, or as convection.

In high-powered vacuum-tube transmitters, a fan is usually provided to force air over the tubes or through special cooling fins. See illustration. By using such fans, greater heat dissipation is possible than would be the case without them, and this allows higher input and output power levels.

Low-powered transistor amplifiers use small heatsinks to conduct heat away from the body of the transistor (*see* HEATSINK). The heatsink may then radiate the heat into the atmosphere as infrared energy, or the heat may be dissipated into a large, massive object such as a block of metal. Ultimately, however, some of the heat from conduction-cooled equipment is dissipated in the air as radiant heat.

With the increasing use of solid-state amplifiers in radio and electronic equipment, conduction cooling is becoming more common, replacing the air blowers so often seen in tube-type amplifiers. Conduction cooling is quieter and requires no external source of power. *See also* FAN.

AIR CORE

The term air core is usually applied in reference to inductors or transformers. At higher radio frequencies, air-core coils are used because the required inductances is small. Powdered-iron and ferrite cores greatly increase the inductance of a coil as compared to an air core (*see* FERRITE CORE, POWERED-IRON CORE). This occurs because such materials cause a concentration of magnetic flux within the coil. The magnitude of this concentration is referred to as permeability (*see* PERMEABILITY); air is given, by convention, a permeability of 1 at sea level.

Air-core inductors and transformers may be identified in schematic diagrams by the absence of lines near the turns. In a ferrite or powdered-iron core, two parallel straight lines indicate the presence of a permeability-increasing substance in the core (See A and B). Coils are sometimes wound on forms made of dielectric material such as glass or bakelite. Since these substances have essentially the same permeability as air (with minor differences), they are considered air-core inductors in schematic representations.

AIR DIELECTRIC

Air is a dielectric material of variable quality (*see* DIELECTRIC), depending on the amount of dust, water vapor, and other impurities present. Even with considerable dust and humidity, however, the loss caused by air as a dielectric material is small (*see* DIELECTRIC LOSS).

Small-value variable capacitors use air as their dielectric material (*see* AIR-VARIABLE CAPACITOR). For low-power and receiving applications, the plate spacing is small, but at high power levels the plate spacing may be quite large.

Coaxial and parallel-wire transmission lines sometimes have an air dielectric. Some prefabricated lines are sealed and filled with a dry, inert gas such as helium or nitrogen. Such air-dielectric lines have relatively low loss per-unit-length. Dry, pure air is given, by convention, a dielectric constant of 1 at sea level (*see* DIELECTRIC CONSTANT). Air dielectric has a high tolerance for potential differences although at sufficient voltages it will ionize and conduct. A typical and frequent example of the dielectric breakdown of air is a lightning stroke. (*See*DIELECTRIC BREAKDOWN.) The ionization of air is more likely to occur when the relative humidity or dust particle concentration is high. Dielectric breakdown is not as likely when the air is pure and dry.

AIR GAP

An air gap is formed by two pointed contacts, spaced a given distance apart, facilitating discharge through the air when the voltage between the contacts reaches a certain value. A common application of the air gap is in a transmission-line lightning arrestor (*see* LIGHTNING ARRESTOR). When sufficient potential builds up between the contacts, the air ionizes and forms a conductive path for the flow of electrons to allow discharge. The greater the spacing between the pointed contacts, the greater the voltage required to cause dielectric breakdown of the air between the contacts.

Pointed contacts are generally used (see the figure) in air gaps because electrons discharge more easily from a sharp point than from a blunt object. The metal in the air-gap contacts must have a fairly high melting temperature to keep the points from being dulled by repeated arcing. *See also* ARCBACK, ARC RESISTANCE.

AIR-SPACED COAXIAL CABLE

A coaxial cable may have several kinds of dielectric material. Polyethylene is probably the most common. But if loss is to be minimized, air-dielectric coaxial cable is the best.

The principal difficulty in designing air-spaced coaxial cable is the maintenance of proper spacing between the inner conductor and the shield. Usually, disk-shaped pieces of polyethylene or other solid dielectric material are positioned at intervals inside the cable (see illustration). While sharp bends in an air-dielectric coaxial cable cannot be made without upsetting the spacing and possibly causing a short circuit, the disks keep the center conductor properly positioned while affecting very little the low-loss characteristics of the air dielectric. Each disk is kept in place by adhesive material or bumps or notches in the center conductor.

It is important that moisture be kept from the interior of an air-spaced coaxial cable. If water should get into such a cable, the low-loss properties and characteristic impedance are upset. (*See* CHARACTERISTIC IMPEDANCE.)

A solid, rigid, coaxial cable, which may have either air or solid dielectric, is called a hard line (*See* HARD LINE.)

AIR TRAFFIC CONTROL

Hundreds or even thousands of commercial, private, and military aircraft are in the air over the United States at any given time. Most of this concentration is near large cities. While the statistical probability of a midair collision between two planes is small, order must be maintained to assure the safety of the people on the planes. An immensely intricate communications, radionavigation, and radar network keeps things under control. Pilots are warned when they are in the vicinity of other planes, turbulent weather, or unexpected phenomena. Computers keep constant track of the status of all commercial airline flights.

Air-traffic controllers man the equipment at airport towers (see photo), giving specific instructions for landings and takeoffs. Sometimes split-second decisions must be made. Pilots radio in information about weather conditions which may not be visible from the ground, so that advisories may be sent to other pilots approaching the area.

The air-traffic control system makes air travel eminently safe and comfortable.

AIR-VARIABLE CAPACITOR

An air-variable capacitor is a device whose capacitance is adjustable, usually by means of a rotating shaft. One rotating and one fixed set of metal plates are positioned in meshed fashion with rigidly controlled spacing. Air forms the dielectric material for such capacitors. The capacitance is set to the desired value by rotating one set of plates, called the rotor plates, to achieve a certain amount of overlap with a fixed set of plates, called the stator plates. The rotor plates are usually connected electrically to the metal shaft and frame of the unit. The photo shows a common type of air-variable capacitor.

Air-variable capacitors come in many physical sizes and shapes. For receiving, and low-power RF transmitting applications, the plate spacing may be as small as a fraction of a millimeter. At high levels of RF power, the plates may be spaced an inch or more apart. The capacitance range of an air variable has a minimum of a few picofarads (abbreviated pF and equal to 10^{-12} farad or 10^{-6} microfarad) and a maximum that may range from about 10 pF to 1000 pF. The maximum capacitance depends on the size and number of plates used, and on their spacing.

Since the dielectric material in an air-variable capacitor is air, which has a small amount of loss, air variables are efficient capacitors as long as they are not subjected to excessive voltages resulting in flashover. A special kind of variable capacitor is the vacuum-variable, a variable capacitor placed in an evacuated enclosure. Such

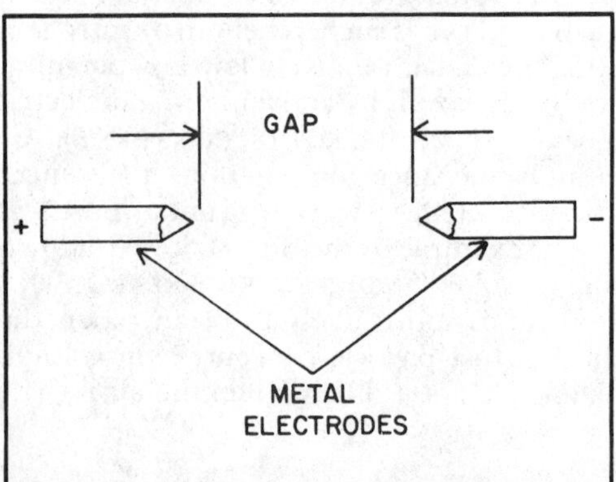

AIR GAP: An air gap consists of two pointed metal electrodes, separated by a predetermined distance to allow spark discharge.

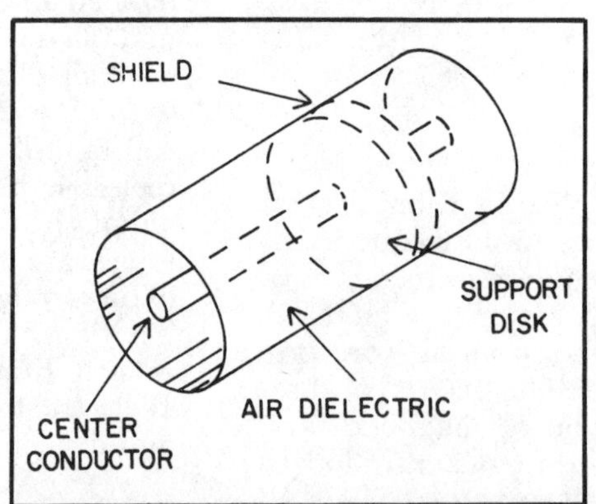

AIR-SPACED COAXIAL CABLE: Supporting disks or beads keep the center conductor spaced away from the shield.

AIR TRAFFIC CONTROL: Air-traffic controllers work in an elvated tower to allow ideal visibility.

capacitors are even more efficient than air variables.

Air-variable capacitors are frequently found in tuned circuits, RF power-amplifier output networks, and antenna matching systems. *See also* ANTENNA MATCHING, TRANSMATCH.

ALARM SYSTEM

For protection against burglary and fire, electronic alarm systems are becoming commonplace in homes and businesses.

Burglar alarm systems, interconnected with the telephone lines, inform a central station when a break-in or suspected break-in occurs (see figure). Ultrasonic devices can be used to detect motion; when suspicious changes in the ultrasonic interference pattern take place, a silent or audible alarm is set off. A silent alarm activates the telephone line to the central station. An audible alarm, in addition, sets off a loud bell or siren.

Another means of detecting unauthorized personnel is the infrared (heat) sensor. This device reacts to the body heat of an intruder but it is immune to mechanical vibrations. Unless an alarm is properly disabled when someone enters the building, the police are summoned by the central station.

Electronic devices such as conducting metallic strips are often placed around windows. When the window glass is broken, so is the metal strip. This triggers the alarm. Some alarms, not connected to the telephone lines, simply set off a loud siren or other audible device, and perhaps turn on strategically positioned lights in an effort to scare a potential intruder away.

Fire alarms operate by detecting smoke or heat. The common battery-operated smoke detector emits a loud buzz or squeal when the concentration of smoke in the air reaches a certain level. Fire alarms generally have power supplies such as batteries, independent of the ac mains, because commercial power is often cut off before the smoke or heat level rises sufficiently to activate the device. *See also* INFRARED DEVICE, SMOKE DETECTOR, ULTRASONIC DEVICE.

ALBEDO

The reflectivity of an object or substance is its albedo. Usually, albedo refers to the visible-light reflectivity, but it can also be specified for infrared, ultraviolet, X rays, radio waves, or even sound waves.

Albedo may be expressed as a ratio or as a percentage. As a percentage, if A represents the albedo, e_i the impinging energy, and e_r the reflected energy, then

$$A = 100\ (e_r/e_i)$$

Thus, albedo may range from zero to 100 percent.

An object with a zero albedo at all electromagnetic wavelengths is called a black body (*see* BLACK BODY). It

AIR-VARIABLE CAPACITOR: An air-variable capacitor forms an important part of this transmitter antenna tuning network.

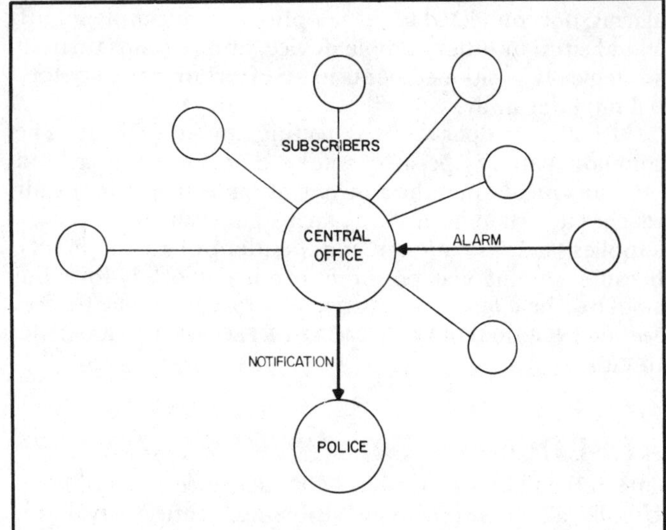

ALARM SYSTEM: An alarm system may have many subscribers, all linked to a central office or station by telephone lines. The central station informs the police when an alarm is registered from a subscriber.

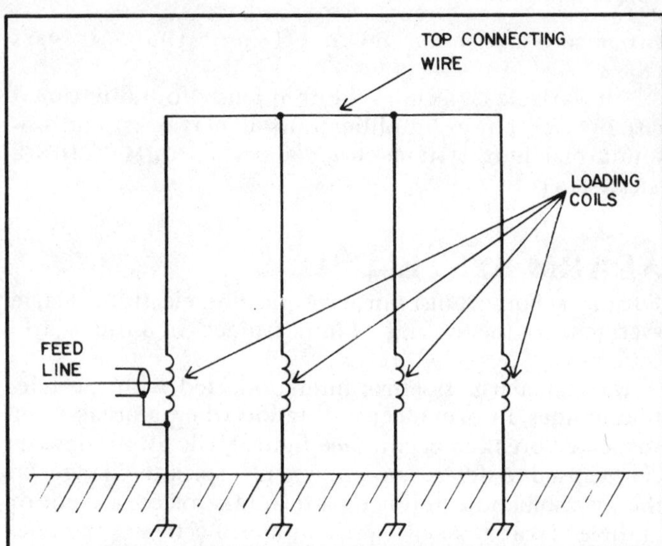

ALEXANDERSON ANTENNA: The Alexanderson antenna consists of several loaded vertical radiators in parallel.

absorbs all the energy it receives. With respect to visible light, freshly fallen snow has an albedo of nearly 100 percent. At radar wavelengths, objects with high albedo show up better than objects with low albedo (*see* RADAR); the "stealth" aircraft, designed to have an extremely low albedo at radar frequencies, is difficult to follow with tracking devices.

The energy not reflected by an object is either absorbed or transmitted through the object. *See also* ABSORPTANCE, TRANSMITTANCE.

ALC

See AUTOMATIC LEVEL CONTROL.

ALEXANDERSON ANTENNA

An antenna for use at low or very low frequencies, the Alexanderson antenna consists of several base-loaded vertical radiators connected together at the top and fed at the bottom of one radiator. The figure illustrates the concept of the antenna.

At low frequencies, the principal problem with transmitting antenna design is the fact that any radiator of practical height has an exceedingly low radiation resistance (*see* RADIATION RESISTANCE), since the wavelength is so large. This results in severe loss, especially in the earth near the antenna system. By arranging several short, inductively loaded antennas in parallel, and coupling the feed line to only one of the radiators, the effective radiation resistance is greatly increased. This improves the efficiency of the antenna, because more of the energy from the transmitter appears across the larger radiation resistance.

The Alexanderson antenna has not been extensively used at frequencies above the standard AM broadcast band. But where available ground space limits the practical height of an antenna and prohibits the installation of a large system of ground radials, the Alexanderson antenna could be a good choice at frequencies as high as perhaps 5 MHz. The Alexanderson antenna requires a far less elaborate system of ground radials than a single-radiator vertical antenna worked against ground. The radiation resistance of an Alexanderson array, as compared to a single radiator of a given height, increases according to the square of the number of elements. *See also* INDUCTIVE LOADING.

ALGEBRA

Algebra is the branch of mathematics dealing with the determination of unknown quantities. One or more variables, usually denoted by lower-case letters p through z, are determined from one or more equations.

Algebraic equations are given an order of magnitude, corresponding to the highest exponential value attached to the variables. An equation of order 2, such as $x^2 + 2x + 1 = 0$, is sometimes referred to as a quadratic equation, and an equation of order 3, such as $x^3 + 3 = 0$, is sometimes called a cubic equation. In general, the higher the order of a single equation, the more difficult it is to solve. Some such as $x^2 = -1$, have no real-number solutions, although they may have solutions in the set of complex numbers (*see* COMPLEX NUMBER, J OPERATOR).

Linear algebra deals with the solutions to sets of first-order, or linear, equations. There may be hundreds of equations and hundreds of variables in such a system. Computers must then be used to arrange the variables and coefficients into a pattern called a matrix. The set of equations is then solved by an algorithm (*see* ALGORITHM).

ALGORITHM

An algorithm is any problem-solving procedure. The term is usually applied to the solution-finding process for a mathematical problem. However, any procedure that can be broken down into discrete steps is an algorithm. The number of steps in an algorithm must be countable, or finite. One example of a rather complicated algorithm for a simple process is the tying of a shoelace. (Try writing down the procedure step by step, without illustrations!)

A mathematical example of an algorithm is procedure of extracting the square root of a number. By repeating, or iterating, this algorithm many times, any desired number of significant digits can be determined.

Algorithms are sometimes written out in the form of computer programs. Sometimes an algorithm for a troubleshooting process is written in the form of a flowchart (*see* FLOWCHART). This makes it easy to localize the cause of a problem in a piece of electronic equipment.

ALIGNED-GRID TUBE

In a vacuum tube with more than one grid (*see* PENTODE TUBE, TETRODE TUBE), it is desirable, in certain applications, to align the grids. The grid structure (*see* GRID) is similar to a screen, with crossed wires in a mesh formation. Electrons pass easily through the openings in this mesh.

When there are two or more grids in a tube, alignment of the mesh wires minimizes the probability that any given electron will strike the second or succeeding grids. This reduces the amount of thermal and electrical noise, and improves the efficiency of the aligned-grid tube compared with an ordinary multigrid vacuum tube. Electrons are directed in more defined beams, and pass with greater ease to the plate.

ALKALINE CELL

The alkaline cell is a power-generating cell similar to the zinc-carbon cell (*see* ZINC CELL). Instead of a paste of sal ammoniac, the alkaline cell uses potassium hydroxide as the electrolyte material. The voltage is approximately the same as that of the zinc-carbon cell, or 1.5 volts. But, alkaline cells have greater energy-storage—as much as twice the amount of the zinc-carbon cell.

Alkaline cells come in a variety of sizes, identical to the sizes of standard cells. Alkaline cells may be used in any application in place of zinc-carbon cells, including small flashlights, radios, and other household devices requiring a few volts of direct-current potential.

ALLIGATOR CLIP

For electronic testing and experimentation, where temporary connections are needed, alligator clips (also known as clip leads, although there are other kinds of clip leads) are often used. They are easy to use and require no modification to the circuit under test.

Alligator clips come in a variety of sizes, ranging from less than ½ inch long to several inches long. They are clamped to a terminal or a piece of bare wire. While such clips are convenient for temporary use, they are not good for long-term installations because of their limited current-carrying capacity and the tendency toward corrosion, especially outdoors.

See the drawing showing common alligator clips. They derive their somewhat humorous name from their visible resemblance to the mouth of an alligator! *See also* CLIP LEAD.

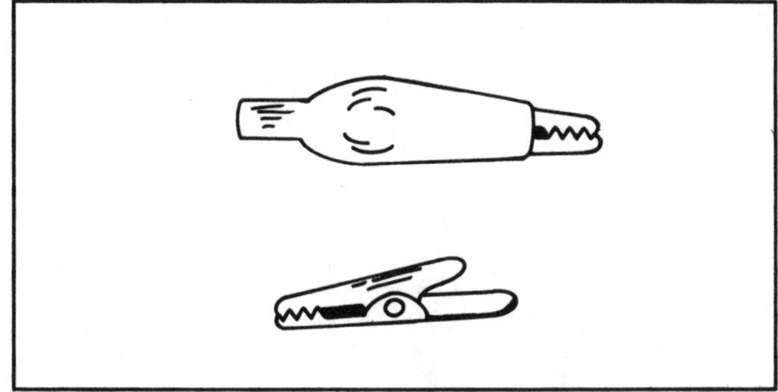

ALLIGATOR CLIP: Alligator clips are often used for temporary wiring or test situations.

ALLOY-DIFFUSED SEMICONDUCTOR

Some semiconductor junctions are formed by a process called alloy diffusion. A semiconductor wafer of P-TYPE or N-type material forms the heart of the device. An impurity metal is heated to its melting point and placed onto the semiconductor wafer. As the impurity metal cools, it combines with the semiconductor material to form a region of the opposite type from the semiconductor wafer.

A transistor may be formed in this manner by starting with a wafer of N-type semiconductor. A small amount of metal such as indium is melted on each side of this wafer as shown, and the melting process is continued so that the indium diffuses into the N-type wafer. This gives the effect of doping, or creating an alloy with, the N-type material next to the indium. (*See* DOPING.) Indium is an acceptor impurity (*see* IMPURITY), and thus two P-type regions are formed on either side of the N-type material. The result is a PNP transistor.

Alloy-diffused semiconductor transistors can be made to have extremely thin base regions. This makes it possible to use the transistor at very high frequencies. *See also* ALPHA-CUTOFF FREQUENCY.

ALL-PASS FILTER

An all-pass filter is a device or network designed to have constant attenuation at all frequencies of alternating current. However, a phase shift may be introduced, and this phase shift is also constant for all alternating-current frequencies. In practice, the attenuation is usually as small as possible (*see* ATTENUATION).

All-pass filters are generally constructed using noninverting operational amplifiers. An example of this is shown in the figure. The amount of phase delay is determined by the values of resistor R and capacitors C. The amount of attenuation is regulated by the values of the other resistors. *See also* OPERATIONAL AMPLIFIER.

ALPHA

In a transistor, the ratio between a change in collector current and a change in emitter current is known as the alpha for that particular transistor. Alpha is represented by the first letter of the Greek alphabet, lower case (α). Alpha is determined in the grounded-base arrangement.

The collector current in a transistor is always smaller than the emitter current. This is because the base draws some current from the emitter-collector path when the transistor is forward biased. Generally, the alpha of a transistor is given as a percentage:

$$\alpha = 100\ (I_C/I_E),$$

where I_C is the collector current and I_E is the emitter current.

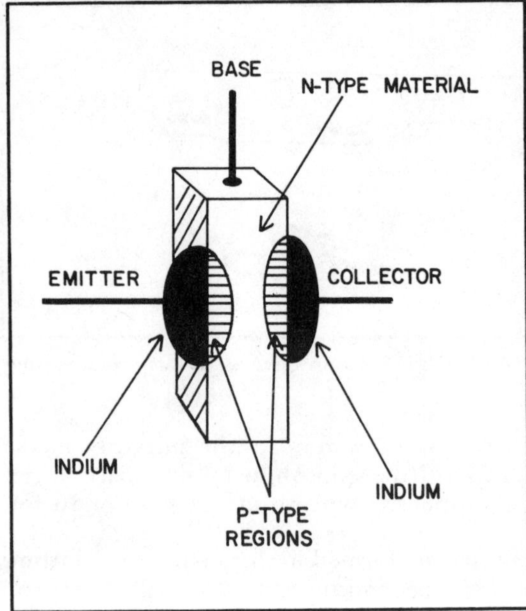

ALLOY-DIFFUSED TRANSISTOR: An alloy-diffused transistor consists of a wafer of N-type material on which two pieces of indium are melted. The resulting doping creates a PNP transistor with a thin base.

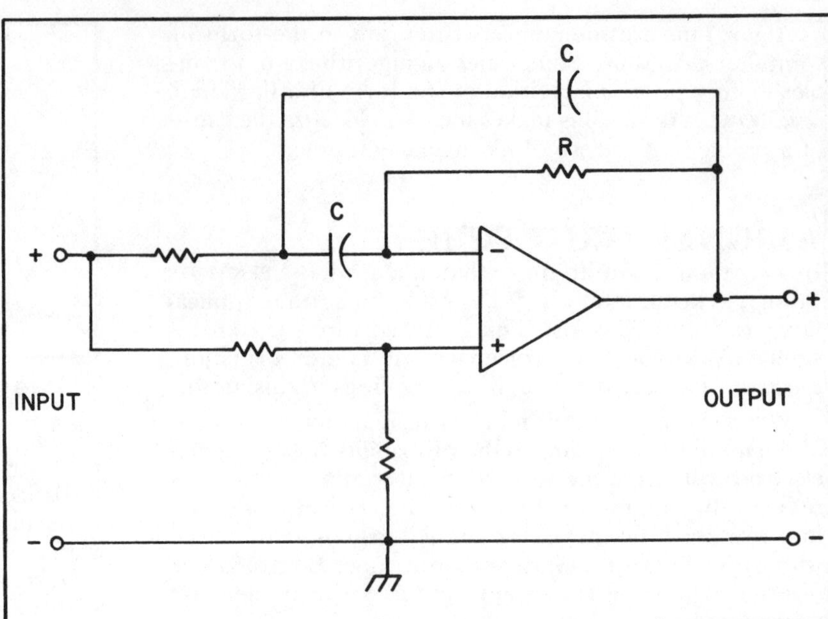

ALL-PASS FILTER: An all-pass filter uses an operational amplifier. The phase delay is determined by the values of R and C.

Transistors typically have alpha values from 95 to 99 percent. Alpha must be measured, of course, with the transistor biased for normal operation.

ALPHA-CUTOFF FREQUENCY

As the frequency through a transistor amplifier is made larger, the amplification factor of the transistor decreases. The current gain, or beta (*see* BETA) of a transistor is measured at a frequency of 1 kHz with a pure sine-wave input for reference when determining the alpha-cutoff frequency. Then, a test generator must be used (*see* SIGNAL GENERATOR) which has a constant output amplitude over a wide range of frequencies. The frequency to the amplifier input is increased until the current gain in the common-base arrangement decreases by 3 dB with respect to its value at 1 kHz. A decrease in current gain of 3 dB represents a drop to 0.707 of its previous magnitude. The frequency at which the beta is 3 dB below the beta at 1 kHz is called the alpha-cutoff frequency for the transistor.

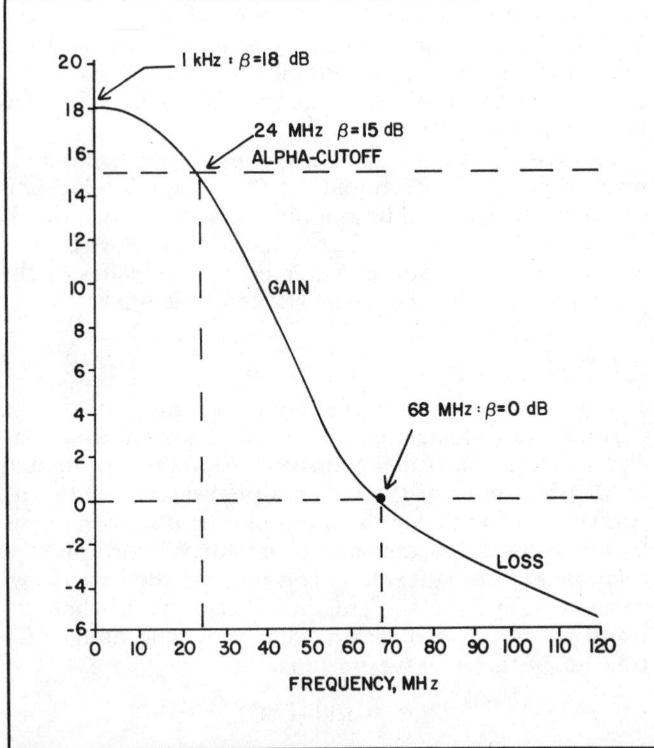

ALPHA CUTOFF FREQUENCY: For this hypothetical transistor, the alpha-cutoff frequency is 24 MHz, where its beta (current gain) is 15 dB.

Depending on the type of transistor involved, the alpha-cutoff frequency might be only a few MHz, or perhaps hundreds of MHz. The alpha-cutoff frequency is an important specification in the design of an amplifier. An alpha-cutoff frequency that is too low for a given amplifier requirement will result in poor gain characteristics. If the alpha-cutoff frequency is unnecessarily high, expense becomes a factor; and under such conditions there is a greater tendency toward unwanted vhf parasitic oscillation (*see* PARASITIC OSCILLATION).

As the input frequency is increased past the alpha-cutoff frequency, the gain of the transistor continues to decrease until it reaches unity, or zero dB. At still higher frequencies, the gain becomes smaller than unity. The drawing illustrates a sample gain-*vs*-frequency curve for a hypothetical transistor, showing the alpha-cutoff and unity-gain frequencies. *See also* DECIBEL, GAIN.

ALPHA PARTICLE

An alpha particle is a nucleus of the helium atom, consisting of two protons and two neutrons. It has a positive charge. Many radioactive substances emit alpha particles as they decay.

An alpha particle has twice the positive charge of a single proton, and four times the atomic mass. When alpha particles are accelerated to high speeds, they are capable of breaking up or modifying the nuclei of heavier atoms. (*See*

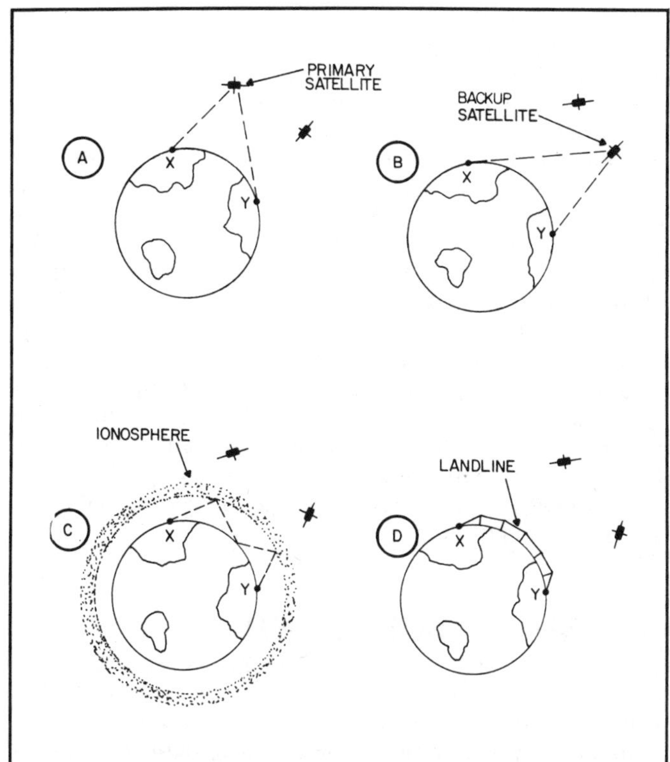

ALTERNATE ROUTING: Alternate routing is needed when the primary means of COMMUNICATION (A) is out of order. Various types of backup systems for a satellite link might include another satellite (B), a shortwave link (C), or a telephone link (D).

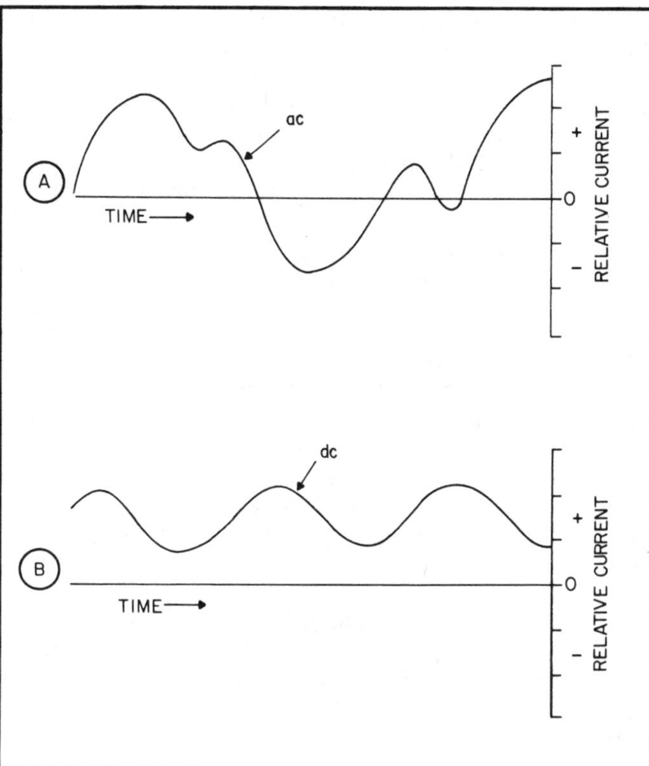

ALTERNATING CURRENT: At A, the current is alternating since the electrons periodically reverse their direction. At B, however, the electrons always flow in the same direction, resulting in a direct current of varying intensity.

PARTICLE ACCELERATOR.) Alpha particles, when numerous and traveling at great speed, are sometimes called alpha rays. They are the least energetic of radioactive emissions. Some alpha particles strike the atmosphere of the earth as they arrive from distant stars and galaxies. Many of these alpha rays are absorbed by the atmosphere before they reach the ground. *See also* COSMIC RADIATION.

ALPHABETIC-NUMERIC

An expression containing letters of the alphabet and numerals 0 through 9 is called an alphabetic-numeric expression. Sometimes this term is shortened to alphanumeric.

Alphabetic-numeric characters are arranged in the order ABCDEFGHIJKLMNOPQRSTUVWXYZ0123 456789. When placing an expression in alphabetic-numeric sequence, the first character is evaluated first, then the second, and so on until a disagreement occurs between that expression and its neighbors. The digits 0 through 9 are usually treated as letters of the alphabet following Z. Alphabetic-numeric expressions are commonly used as designators for variables in computer programs.

ALTERNATE ENERGY

Any source of energy other than fossil fuels, used for constructive purposes, is an alternate energy source. Alternate energy sources include the sun, the internal heat of the earth, nuclear power, hydrogen fuel, wind, water, and many others.

Research is constantly going on in the field of alternate energy, because many scientists believe that such developments are the only real long-term solution to the world's energy problems. *See also* HYDROELECTRIC POWER PLANT, NUCLEAR ENERGY, SOLAR ENERGY.

ALTERNATE ROUTING

When the primary system for communications between two points breaks down, a backup system must be used to maintain the circuit. Such a system, and its deployment, is called alternate routing. Alternate routing may also be used in power transmission, in case of interruption of a major power line, to prevent prolonged and widespread blackouts. Power from other plants is routed to cities affected by the failure of one particular generating plant or transmission line.

As an example, the primary communications link for a particular system might be via a geostationary satellite, as shown in (A). If the satellite fails, another satellite can be used in its place if one is available (B). This is alternate routing. If the second satellite ceases to function or is not available, further backup systems might be used, such as an HF shortwave link (C) or telephone connection (D). Alternate routing systems should be set up and planned in advance, before the primary system goes down, so that communications may be maintained with a minimum of delay.

ALTERNATING CURRENT

Whenever electrons in a conductor flow in both directions, alternating current exists. If the electrons always move in the same direction, even if it is at variable speed, then the current is considered direct (*see* DIRECT CURRENT).

The most familiar example of alternating current is the 60-Hz house current used to operate common appliances. Radio-frequency energy in an antenna or transmission line is also alternating current, as is the audio-frequency energy in a stereo speaker system. There are many different kinds of alternating current (ac).

An ac waveform might be quite simple, such as a sine wave, square wave or sawtooth wave. Or, it might be irregular in shape and perhaps very complicated, such as a voice pattern. The prime criterion for a given current to be regarded as ac is that the electrons repeatedly reverse their direction (*see* A and B). Any alternating current may be converted to direct current if a sufficient steady, direct current is impressed on it. Alternating current can also be converted to direct current by means of a rectifier and filter. This is how a common dc power supply functions from standard house current.

Alternating current has advantages over direct current for power transmission (*see* AC POWER TRANSMISSION). The most significant of these advantages is the ease of voltage transformation (*see* TRANSFORMER). Direct current generates less electromagnetic noise than alternating current and is more efficiently transferred over long distances, but the transformation of voltage is considerably more difficult. *See also* SAWTOOTH WAVE, SINE WAVE, SQUARE WAVE.

ALTERNATOR

See AC GENERATOR.

ALUMINUM

Aluminum is a dull, light, somewhat brittle metal, atomic number 13, atomic weight 27, commonly used as a conductor for electricity. Although it does break rather easily, aluminum is very strong in proportion to its weight, and has replaced much heavier metals such as steel and copper in many applications.

ALUMINUM: An antenna made from aluminum tubing.

Aluminum, like other metallic elements, is found in the earth's crust. It occurs in a rock called bauxite. Recent advances in mining and refining of bauxite have made aluminum one of the most widely used, and inexpensive, industrial metals.

Aluminum is fairly resistant to corrosion, and is an excellent choice in the construction of communications antenna systems. Hard aluminum tubing is available in many sizes and thicknesses. The do-it-yourself electronics hobbyist can build quite sophisticated antennas from aluminum tubing purchased at a hardware store. Most commercially manufactured antennas use aluminum tubing (*see* the photograph).

Soft aluminum wire is used for grounding systems in communications and utility service. Large-size aluminum wire often proves the best economic choice for such applications. Some municipalities, however, require copper wire for grounding. *See also* GROUND CONNECTION, LIGHTNING PROTECTION.

ALUMINUM-CLAD WIRE

A wire with a steel core for mechanical strength, and an outer jacket of aluminum for electrical conductivity, is called aluminum-clad wire. Its construction is shown. Copper is also sometimes used as the highly-conductive outer jacket (*see* COPPER-CLAD WIRE). Such wire is ideal for situations that require both excellent electrical conductivity and high tensile strength, such as longwire antennas.

Aluminum-clad wire comes in several sizes. The size is specified according to the diameter of the outside of the jacket.

At high frequencies, electrons tend to flow mostly at the outer edges of an electrical conductor (*see* SKIN EFFECT). At radio frequencies, this effect is quite pronounced, and this allows a thin aluminum jacket to carry almost all the current in an aluminum-clad wire. Therefore, the conductivity approaches that of solid aluminum. In addition to im-

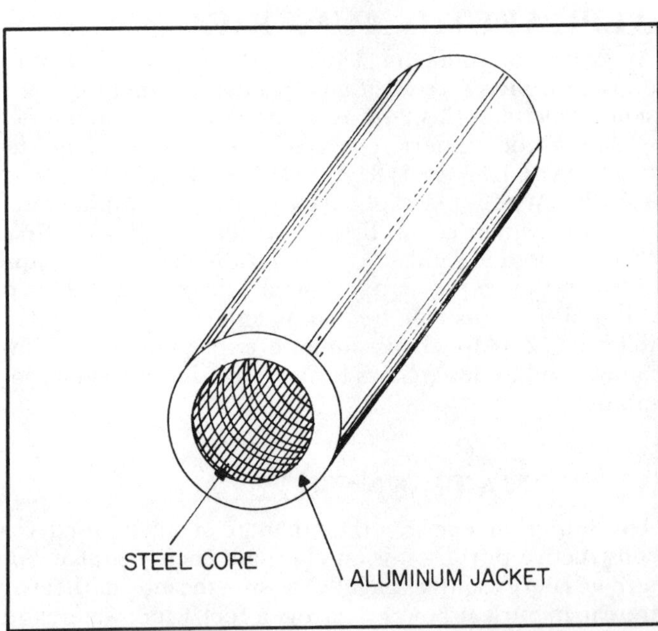

ALUMINUM-CLAD WIRE: Aluminum-clad wire has a steel core for strength and an aluminum coating or jacket for good electrical conductivity.

proving the conductivity of the wire, the aluminum jacket protects the steel core against rust.

AMATEUR RADIO

There are over 400,000 amateur radio operators, or "hams," in the United States, and several hundred thousand in the rest of the world. In the United States, Amateur Radio operators are licensed by the Federal Communications Commission. Hams converse, as shown in the photograph, over distances limited only by the size of the world. They also experiment with new devices and provide emergency backup communications during times of disaster. Hams have many different frequency bands, ranging from 1.8 MHz to the microwave part of the spectrum. Hams also have orbiting satellites (*see* ACTIVE COMMUNICATIONS SATELLITE).

To become an Amateur Radio Operator, it is necessary to pass an examination. Part of the examination requires demonstration of proficiency in the International Morse Code (*see* INTERNATIONAL MORSE CODE), and part of the test covers electronic theory. There are five different levels, or classes, of amateur operator license: the Novice, Technician, General, Advanced, and Extra classes. Each class carries certain frequency privileges and power and emission privileges. The Novice test is relatively simple; the Extra is considered quite sophisticated even by some engineers.

The Amateur Radio Service is not the same thing as the Citizen's Radio Service, or CB (*see* CITIZEN'S BAND). No test is necessary for the CB license. The Citizen's Band is confined to 40 channels in the 27-MHz frequency range. Power output is limited severely.

Amateur Radio operators have historically been pioneers in the development of new and improved communications techniques. They are also indispensable as a reservoir of highly trained radio operators who can provide assistance in times of disaster. Aside from its value to society, Amateur Radio is a source of enjoyment and relaxation for hams.

Some of the popular operating activities in which hams engage include contests (to see who can make the greatest number of contacts in a given amount of time), DXing (communicating with distant or tiny countries), rag chewing (just talking), and message handling. Hams use a special way of speaking, so you will probably find it difficult to understand them unless you are familiar with their "lingo." Many statements are abbreviated as Q signals, such as QRM, which means "There is interference to your signal." (*See* Q-SIGNAL.)

Amateur Radio clubs in many towns and counties provide a good way for an aspiring radio amateur to get more exposure to this hobby. Numerous publications exist, partly or wholly devoted to Amateur Radio. Many Amateur-Radio operators belong to such organizations and subscribe to one or more publications. *See also* FEDERAL COMMUNICATIONS COMMISSION.

AMATEUR RADIO: Amateur radio operators communicate with each other from stations as unobtrusive as this one, or occasionally from stations that occupy an entire building.

AMERICAN MORSE CODE

The American Morse Code is a system of dot and dash symbols, first used by Samuel Morse in telegraph communications. The American Morse Code is not widely used today. It has been largely replaced by the International Morse Code. Some telegraph operators still use American Morse.

The American Morse Code differs from the International Morse Code. The table shows the American Morse symbols, sometimes called "Railroad Morse." Some letters contain internal spaces. This causes confusion for operators familiar with International Morse Code. Some letters are also entirely different between the two codes. *See also* INTERNATIONAL MORSE CODE.

AMERICAN MORSE CODE: AMERICAN MORSE CODE SYMBOLS (SOMETIMES CALLED "RAILROAD MORSE").

Character	Symbol	Character	Symbol
A	•–	U	••–
B	–•••	V	•••–
C	•• •	W	•––
D	–••	X	•–••
E	•	Y	•• ••
F	•–•	Z	••• •
G	––•	1	•––•
H	••••	2	••–••
I	••	3	•••–•
J	–•–•	4	••••–
K	–•–	5	–––
L	——	6	••••••
M	––	7	––••
N	–•	8	–••••
O	• •	9	–••–
P	•••••	0	———
Q	••–•	PERIOD	••––••
R	• ••	COMMA	•–•–
S	•••	QUESTION MARK	–••–•
T	–		

AMERICAN WIRE GAUGE: AMERICAN WIRE GAUGE EQUIVALENTS IN MILLIMETERS.

AWG	Dia., mm	AWG	Dia., mm
1	7.35	21	0.723
2	6.54	22	0.644
3	5.83	23	0.573
4	5.19	24	0.511
5	4.62	25	0.455
6	4.12	26	0.405
7	3.67	27	0.361
8	3.26	28	0.321
9	2.91	29	0.286
10	2.59	30	0.255
11	2.31	31	0.227
12	2.05	32	0.202
13	1.83	33	0.180
14	1.63	34	0.160
15	1.45	35	0.143
16	1.29	36	0.127
17	1.15	37	0.113
18	1.02	38	0.101
19	0.912	39	0.090
20	0.812	40	0.080

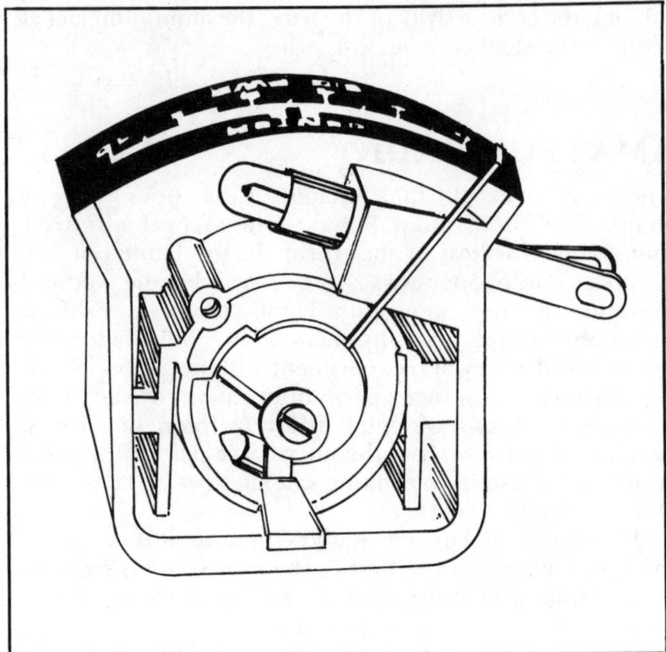

AMMETER: An ammeter, showing the coils and needle. This particular unit is used for registering signal strength on a radio receiver, and has a full-scale deflection current of 1 mA.

AMERICAN NATIONAL STANDARDS INSTITUTE, INC.

The American National Standards Institute, Inc. (ANSI), also stmetimes called the American Standards Association (ASA), is an organization that assures that the products of various manufacturers are compatible. Without such component standardization, building and servicing all kinds of equipment, non-electronic as well as electronic, would be much more difficult.

While we take for granted such things as the universal size of light-bulb sockets in the ordinary household lamp, or the uniformity of 117-volt, 60-Hz ac utility power, there are instances where standardization exists to a much lesser degree. A good example of this is RF cable installations, where many kinds of connectors are used. In some cases, standardization is not entirely practical. Better and better RF connectors evolved with time, but the best ones are rather difficult to install and sometimes more expensive. In general, however, standardization is a good thing because it makes things much more convenient.

AMERICAN RADIO RELAY LEAGUE

Amateur Radio operators have many different local, regional and national organizations. One such establishment, with members throughout the world, is the American Radio Relay League (ARRL).

In 1914, two Amateur Radio operators, Hiram Maxim and Clarence Tuska, established the ARRL for the purpose of increasing public awareness of Amateur Radio. Amateur Radio operators must continuously justify their existence. The headquarters of the American Radio Relay League is in Newington, Connecticut.

The word "relay" in the name of this organization has an interesting origin. In 1914, the worldwide, long-distance communication we know today had not yet been discovered. Transmitters and receivers were far less efficient. A communication range of 100 miles was often considered excellent. For the purpose of cross-country message handling, a "radio relay" was used. A message of special importance could be sent from New York to California quite rapidly by passing, or relaying, it among a chain of Amateur Radio stations.

Amateur Radio Operators were responsible, in part, for the discovery of long-distance communication on the short waves. But even today, the word "relay" still has meaning. Ham radio operators maintain an extensive nationwide network for the handling of messages. *See also* AMATEUR RADIO.

AMERICAN WIRE GAUGE

Metal wire is available in many different sizes or diameters. Wire is classified according to diameter by giving it a number. The designator for a given wire is known as the American Wire Gauge (AWG). In England, a slightly different system is used (*see* BRITISH STANDARD WIRE GAUGE). The numbers in the American Wire Gauge system range from 1 to 40, although larger and smaller gauges exist. The higher the AWG number, the thinner the wire.

The table shows the diameter vs AWG designator for AWG 1 through 40. The larger the AWG number for a given conductor metal, the smaller the current-carrying capacity becomes. The AWG designator does not include any coatings on the wire such as enamel, rubber, or plastic insulation. Only the metal part of the wire is taken into account.

AMMETER

An ammeter is a device for measuring electric current. The current passes through a set of coils, causing rotation of a central armature. An indicator needle attached to this armature shows the amount of deflection against a graduated scale (See the illustration). Ammeters may be designed to have a full-scale deflection as small as a few microamperes (*see also* AMPERE), up to several amperes.

To extend the range of an ammeter, allowing it to register very large currents, a resistor of precisely determined value is placed in parallel with the meter coils. This resistor diverts much, or most, of the current so that the meter actually reads only a fraction of the current.

Ammeters may be used as voltmeters by placing a resistor in series with the meter coils. Then, even a very high voltage will cause a small deflection of the needle (*see* VOLTMETER). The greatest accuracy is obtained when a sensitive ammeter is used with a large-value resistor. This minimizes the current drawn from the circuit. Ammeters should never be connected across a source of voltage without a series resistor, since damage to the meter mechanism may result.

Ammeters are available for measuring both ac and dc. Some ammeters register RF current. The devices must be specially designed for each of these applications. *See also* ALTERNATING CURRENT, DIRECT CURRENT, RF CURRENT.

AMPERE

The *ampere* is the unit of electric current. A flow of one coulomb per second, or 6.28×10^{18} electrons per second, past a given fixed point in an electrical conductor, is a curent of one ampere (see figure).

Various units smaller than the ampere are often used to measure electric current. A milliampere (mA) is one thousandth of an ampere, or a flow of 6.28×10^{15} electrons per second past a given fixed point. A microampere (μA) is one millionth of an ampere, or a flow of 6.28×10^{12} electrons per second. A nanoampere (nA) is a billionth of an ampere; it is the smallest unit of electric current you are likely to use. It represents a flow of 6.28×10^{9} electrons per second past a given fixed point.

A current of one ampere is produced by a voltage of one volt across a resistance of one ohm. This is Ohm's law (*see* OHM'S LAW). The ampere is applicable to measurement of alternating current or direct current. *See also* ALTERNATING CURRENT, DIRECT CURRENT.

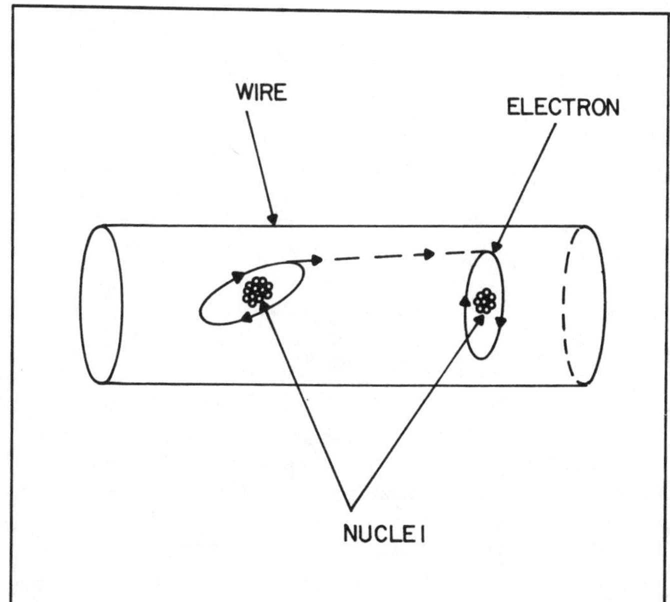

AMPERE: A current flow of one ampere represents 6.28×10^{18} electrons past any given point each second. Here, we see a somewhat exaggerated picture of one electron exchanging atoms.

AMPERE-HOUR METER

An *ampere-hour meter* is a device for measuring the total amount of electrical quantity (*see* COULOMB) passing a given point over a certain period of time. A current of one ampere for one hour represents 2.26×10^{22} electrons or 3600 coulombs. This is one ampere hour. An ampere hour can be represented by any combination of current and $\times 10_{22}$ electrons. See

Ampere-hour meters are generally used for measuring electrical energy. By calibrating an ampere-hour meter to register according to the voltage in the system, a watt-hour meter (*see* WATT-HOUR METER) is obtained. The meter that the power company installs at any business or residence to measure the total consumed energy is a watt-hour meter.

Ampere hours may be subdivided into smaller units; the milliampere hour is a thousandth of an ampere hour, for example, and a microampere-hour is a millionth of an ampere hour. These units represent the transfer of 2.26×10^{19} and 2.26×10^{16} electrons past a given point, respectively.

AMPERE'S LAW

The direction of an electric current is generally regarded by physicists and engineers as the direction of positive charge transfer. This is opposite to the direction of the movement of the electrons themselves, since electrons carry a negative charge. By convention, when speaking of the direction of current, it is considered to move from the positive to the negative terminal of a battery or power supply.

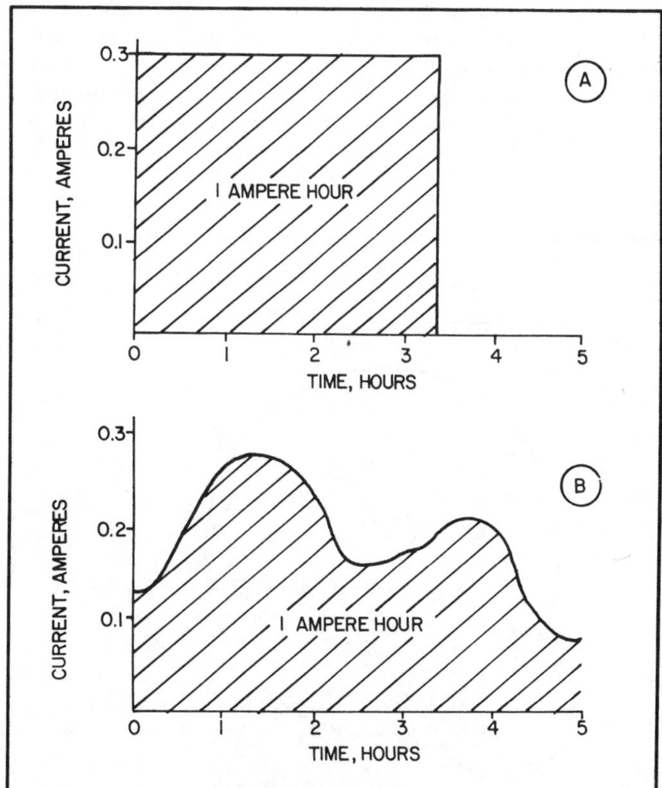

AMPERE-HOUR METER: One ampere hour might be represented as a steady current of 0.3 amperes for 3.33 hours (A), or as an irregular current for five hours (B). In any case, the area under the curve is the same, representing 2.26×10^{22} electrons.

According to Ampere's law, the magnetic field or flux lines generated by a current in a wire travel counterclockwise when the current is directed toward the observer (see illustration). This rule is sometimes also called the right-handed screw rule for magnetic-flux generation. A more universal rule for magnetic flux, applying to motors and generators, is called Fleming's Rule (*see* FLEMING'S RULE). As the right hand is held with the thumb pointed outward and the fingers curled, a current in the direction pointed by the thumb will generate a magnetic field in the circular direction pointed by the fingers.

AMPERE TURN

The ampere turn is a measure of magnetomotive force. One ampere turn is developed when a current of one ampere flows through a coil of one turn, or, in general, when a current of 1/n amperes flows through a coil of n turns.

One ampere turn is equal to 1.26 gilberts. The gilbert is the conventional unit of magnetomotive force. *See also* GILBERT, MAGNETOMOTIVE FORCE.

AMPLIDYNE

An amplidyne is a dc generator used for the special purpose of power amplification in servomechanisms (*see* SERVOMECHANISM). The amplidyne consists of a drive motor, powered by direct current, which causes rotation of the generator mechanism. The ac output of the generator is converted to dc by rectification, or by means of a commutator. See the block diagram of an amplidyne.

A control signal is applied to a control circuit, which changes the voltage or current applied to the field of the generator. This causes a large increase or decrease of output from the generator. Amplidynes once found extensive use in powering drive motors for pointing gun turrets aboard naval vessels, where varying slew rates were required to point guns at targets while the ship was in motion. *See also* DC AMPLIFIER.

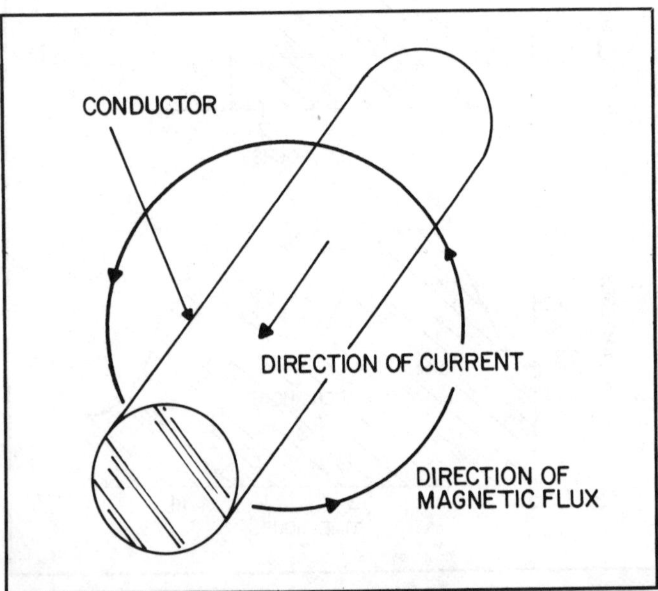

AMPERE'S LAW: Ampere's law states that the magnetic flux circulates in a counterclockwise direction as seen from a point where the positive current is approaching.

AMPLIFICATION

Amplification refers to any increase in the magnitude of a current, voltage, or wattage. Amplification makes it possible to transmit radio signals of tremendous power, sometimes over a million watts. Amplification also makes it possible to receive signals that are extremely weak. It allows the operation of such diverse instruments as light meters, public-address systems, and television receivers.

Usually, amplification involves increasing the magnitude of a change in a certain quantity. For example, a fluctuation from -1 volt to $+1$ volt, or 2 volts peak to peak, might be amplified so that the range becomes 0 to $+10$ volts, or five times greater. Alternating currents, when amplified, produce effective voltage gain (if the impedance is correct) and power gain.

Direct-current amplification is usually done with the intention of increasing the sensitivity of a meter or other measuring instrument. Alternating-current amplification is employed primarily in audio-frequency and radio-frequency applications, for the purpose of receiving or transmitting a signal. *See also* AMPLIFIER.

AMPLIFICATION FACTOR

Amplification factor is the ratio of the output amplitude in an amplifier to the input amplitude. The quantity is expressed for current, voltage or power, and is abbreviated by the Greek lower-case letter μ.

Usually, amplification factor is determined from the peak-to-peak voltage or current as shown. If the voltage amplification factor for an amplifier is a certain value, the current amplification factor need not be, and probably will not be, the same. Voltage or current gain is determined from the voltage or current amplification factor by the formula

$$\text{Gain (dB)} = 20 \log_{10} \mu$$

Power gain is related to the power-amplification fator by the equation

$$\text{Gain (dB)} = 10 \log_{10} \mu$$

See also DECIBEL, GAIN.

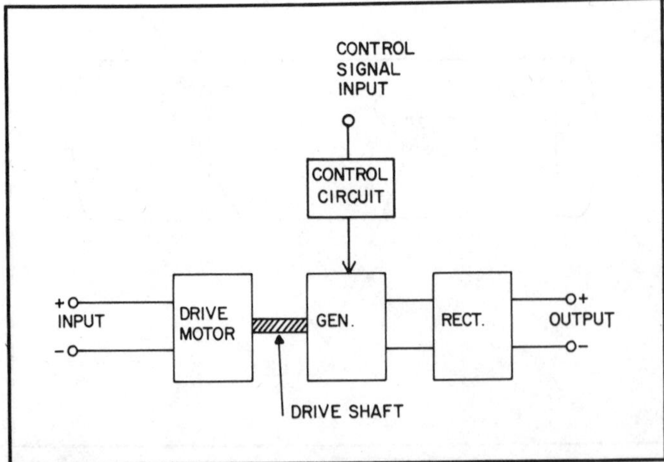

AMPLIDYNE: The amplidyne is a dc-powered dc generator.

AMPLIFICATION NOISE

All electronic circuits generate some noise. In an amplifier, the tube, transistor, or integrated circuit invariably generates some noise. This is called amplification noise. Amplification noise may be categorized as either thermal, electrical or mechanical.

The molecules of all substances, including the metal and other materials in an electronic circuit, are in constant random motion. The higher the temperature, the more active the molecules. This generates thermal noise in any amplifier.

As the electrons in a circuit hop from atom to atom, or impact against the metal anode of a vacuum tube, electrical noise is generated. The larger the amount of current flowing in a circuit, in general, the more electrical noise there will be.

Mechanical noise is produced by the vibration of the circuit components in an amplifier. Sturdy construction and, if needed, shock-absorbing devices, reduce the problem of mechanical noise.

While nothing can be done about the thermal noise at a given temperature, some equipment is cooled to extremely low temperatures to minimize thermal noise (*see* CRYOGENICS). Electrical noise is reduced by improved by the use of certain types of amplifying devices, such as the field-effect transistor. It is important that noise is kept as low as possible in the early stages of a multistage circuit, since any noise generated in one amplifier will be picked up and amplified, along with the desired signals, in succeeding stages. *See also* NOISE.

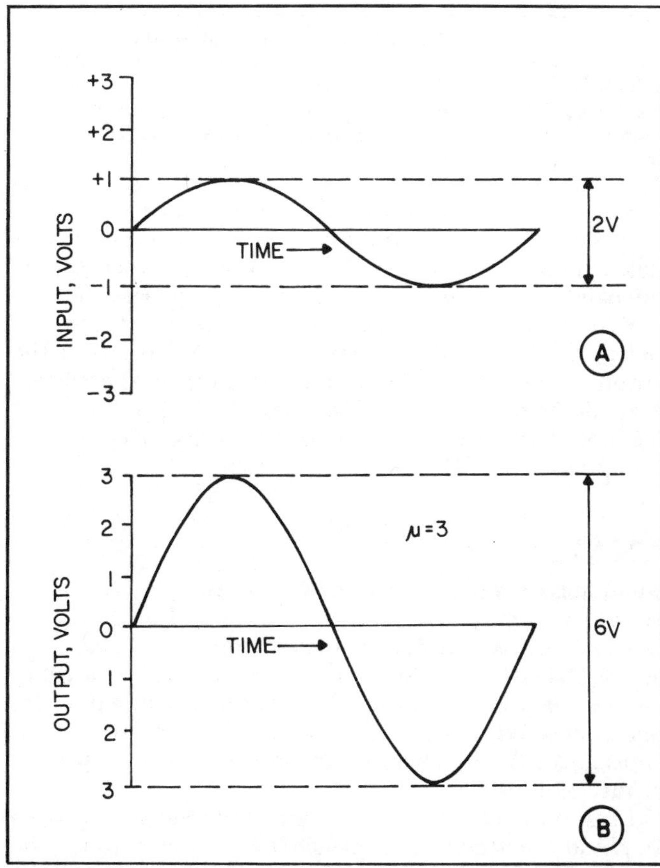

AMPLIFICATION FACTOR: At A, the peak-to-peak voltage is 2 volts, and at B, it is 6 volts. Therefore, in this case, the voltage amplification factor is 3.

AMPLIFIER

An amplifier is any circuit that increases the magnitude of a signal for constructive purposes, drawing power from an external source such as a battery or power supply.

Amplifiers may be built for the purpose of magnifying ac or dc, voltage, or power. Special amplifier circuits are designed for amplification of extremely weak signals; other kinds of amplifiers are built for high-power applications. All radio receivers and transmitters use amplifiers. A public-address or high-fidelity sound system uses power amplifiers at audio frequencies. (*See* POWER AMPLIFIER.)

The most common devices used in power amplifiers or weak-signal amplifiers are transistors, integrated circuits, and vacuum tubes. All function by causing a large change in output current or voltage when there is a small change in the input current or voltage. In doing this, the transistor, integrated circuit, or vacuum tube must have an external source of direct current for power.

A simple ac transformer is not an amplifier, even though it may provide a voltage or current increase (*see* GAIN). An amplifier increases the power, or energy, of a signal passing through its circuit; the transformer always has some power loss (*see* TRANSFORMER EFFICIENCY). Transformers are, however, often an integral part of an amplifier, because they provide the proper impedance matching for the amplifier input and output. *See also* TRANSFORMER, TRANSFORMER COUPLING.

AMPLITUDE

The strength of a signal is called its amplitude. Amplitude can be defined in terms of current, voltage, or power foı any given signal.

Knowing the root-mean-square (*see* ROOT MEAN SQUARE) current, I, and the root-mean-square voltage, E, for a particular ac signal, the power amplitude in watts is given by

$$P = EI$$

If we know the circuit impedance, Z, and either the current or voltage, then the power amplitude is

$$P = I^2Z = E^2/Z$$

Amplitude is usually described in reference to the strength of a radio-frequency signal, either at some intermediate point in a receiver or transmitter circuit, or at the output of a transmitter. Amplitude is measured using a wattmeter or oscilloscope. May also be measured using a spectrum analyzer (*see* SPECTRUM ANALYZER). On an oscilloscope, signals of increasing amplitude (*see* A, B, and C) appears as waveforms of greater and greater height, but of the same wavelength, assuming the frequency remains constant.

For weak signals at the antenna terminals of a receiver, the term strength is usually employed. Such signals are measured in microvolts. *See also* SENSITIVITY.

AMPLITUDE MODULATION

Amplitude modulation is the process of impressing information on a radio-frequency signal by varying its amplitude. The simplest example of amplitude modula-

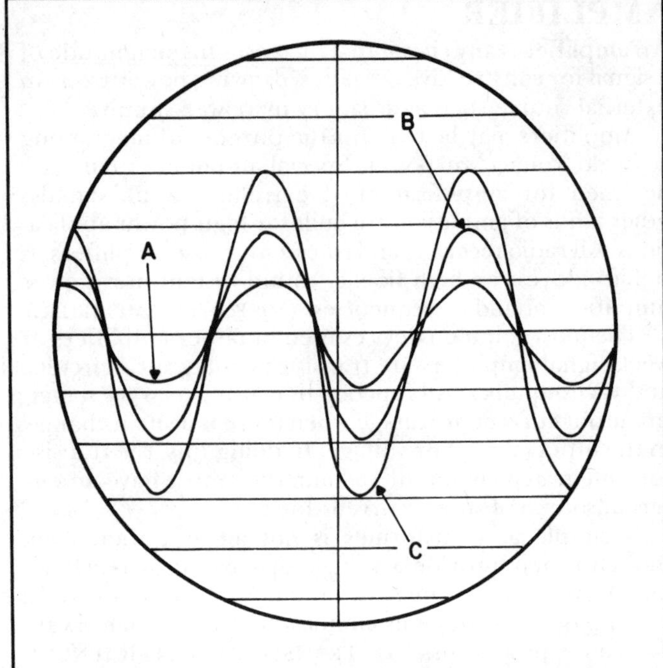

AMPLITUDE: Amplitude of a wave results in exaggerated height on an oscilloscope display, but does not affect the wavelength. Waves A, B, and C are in relative proportion 1:2:3 in terms of amplitude in this illustration.

tion (AM) is probably Morse-code transmissions, where the amplitude switches from zero to maximum.

Generally, amplitude modulation is done for the purpose of relaying messages by voice, television, facsimile, or other modes that are relatively sophisticated. The process is always the same: Audio or low frequencies are impressed upon a carrier wave of much higher frequency. A, B, and C illustrate the amplitude modulation of a carrier wave by a sine-wave or sinusoidal audio tone. The amplitude of the carrier is greatest on positive peaks of the sinusoidal tone, and smallest on negative peaks.

The modulation of an AM signal may be considerable, or it may be very small. The intensity or degrees of modulation is expressed as a percentage. This percentage may vary from zero to more than 100. An unmodulated carrier, as shown in the figure, has zero percent modulation by definition. If the negative peaks drop to zero amplitude, the signal is defined to have 100-percent modulation. At C, we see a signal with modulation of about 75 percent. If the modulation percentage exceeds 100, the negative peaks drop to zero amplitude and remain there for a part of the audio cycle. This is undesirable, because it causes distortion of the information reproduced by the receiver.

When a given radio-frequency signal is amplitude modulated, mixing occurs (*see* MIXER) between the modulation frequencies (f_M) and the carrier frequency (f_C). For sine-wave modulation such as shown in the figure, this mixing results in new radio-frequency signals, f_{LSB} and f_{USB}, given by

$$f_{LSB} = f_C - f_M$$

$$f_{USB} = f_C + f_M$$

These new frequencies are called sidebands. They are referred to as the upper sideband (USB) and lower sideband (LSB).

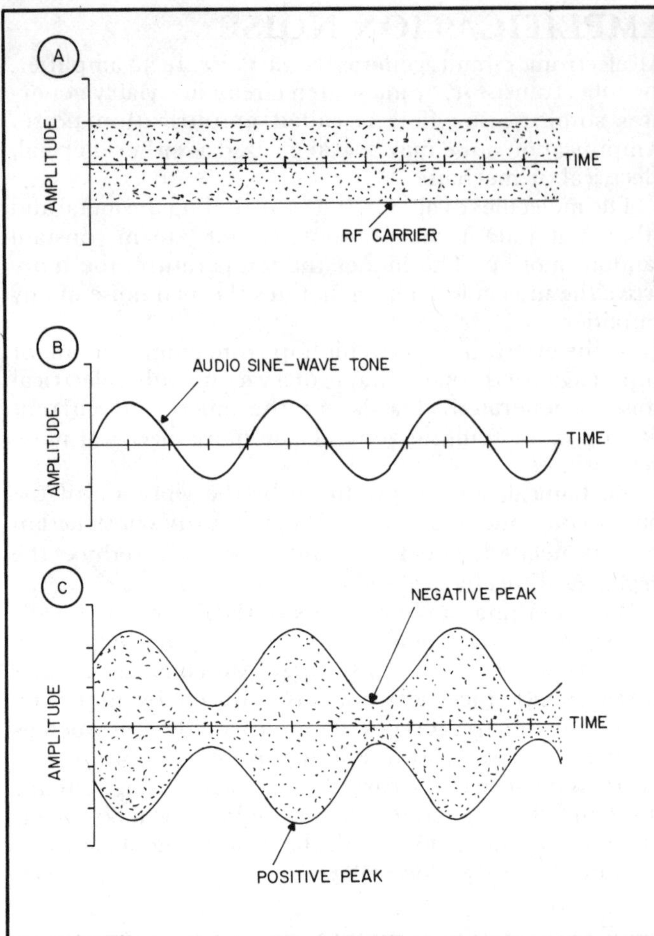

AMPLITUDE MODULATION: At A, an unmodulated carrier wave. (The waveform is too compressed on this scale to be visible.) At B, an audio waveform to be impressed on the carrier via amplitude modulation. At C, the AM signal. The greatest amplitude points are called positive peaks, and the smallest amplitude points are called negative peaks.

A special kind of amplitude modulation, commonly called single sideband (SSB), but properly named single-sideband, suppressed-carrier (SSSC), eliminates the carrier frequency (f_C) and also one of the sideband frequencies at the transmitter. Only one sideband is left at the output. This sideband is combined with a local oscillator signal at the receiver, resulting in a perfect reproduction of the modulating signal, provided the receiver frequency is correctly set. *See also* SINGLE SIDEBAND.

ANALOG

Quantities or representations that are variable over a continuous range are referred to as analog. In electronics, analog quantities are differentiated from digital quantities by the fact that analog variables can take an infinite number of values, but digital variables are limited to defined states. While a digital representation of a curve might appear like A, the analog representation gives a precise picture as in B.

Examples of analog quantities include the output of an amplitude-modulated, single-sideband transmitter. The amplitude of such a signal fluctuates over a continuous range from zero to the maximum, or peak, output. An example of a digital amplitude-modulated signal is the

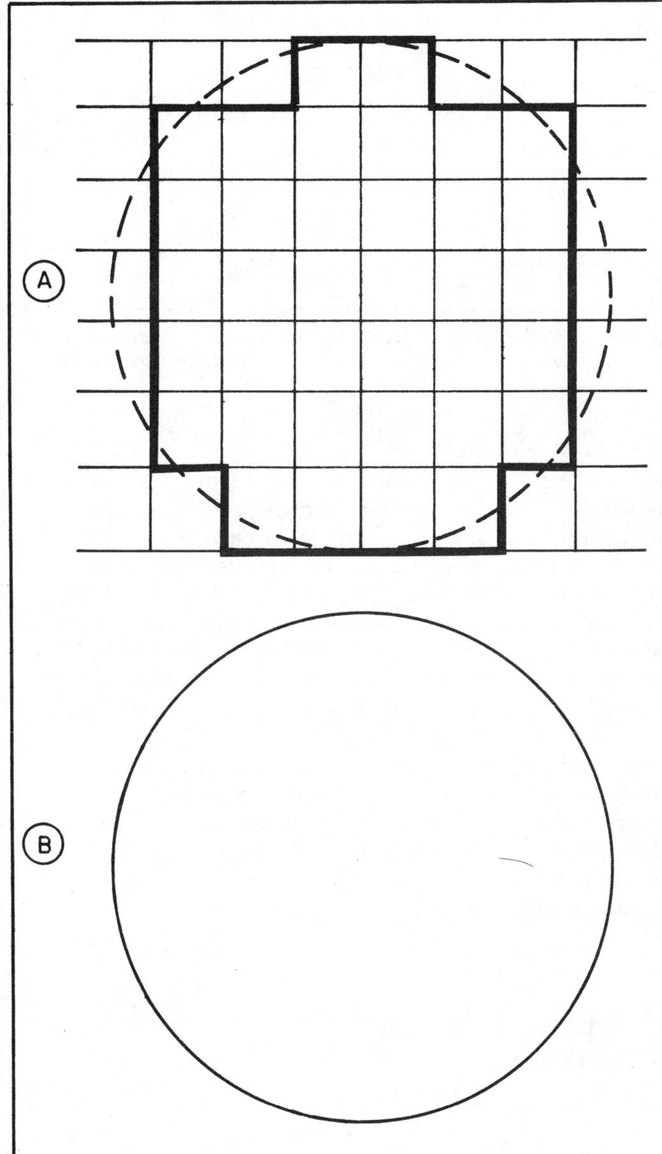

ANALOG: At A, digital representation of a circle. At B, analog representation of the same circle.

output of a cw transmitter sending Morse code. This signal has only two states: on and off.

While analog information usually provides more accuracy in reproducing a quantity, variable, or signal, digital information is transferred with greater efficiency. The difference is evident in the example just given. While voice inflections in an AM or SSB signal enable the transmission of emotions, a cw signal travels with better signal-to-noise ratio and hence greater efficiency. When a voice signal becomes unreadable because of interference or poor conditions, a cw signal is often still intelligible. *See also* DIGITAL.

ANALOG COMPUTER

An analog computer is a device that creates a mathematical analogy of a problem to be solved. The word analog, in fact, comes from the word analogy. Relations among variables are represented by certain mathematical curves. Sometimes these curves are simple, and sometimes they are very complicated. The analog computer approximates such curves according to its own repertoire of mathematical formulas. The digital computer, in contrast to this, approximates a function or relation in discrete bits, in the way the circle is approximated by A in ANALOG. The finer the mesh (the smaller the bits), the more accurate the representation (see B in ANALOG).

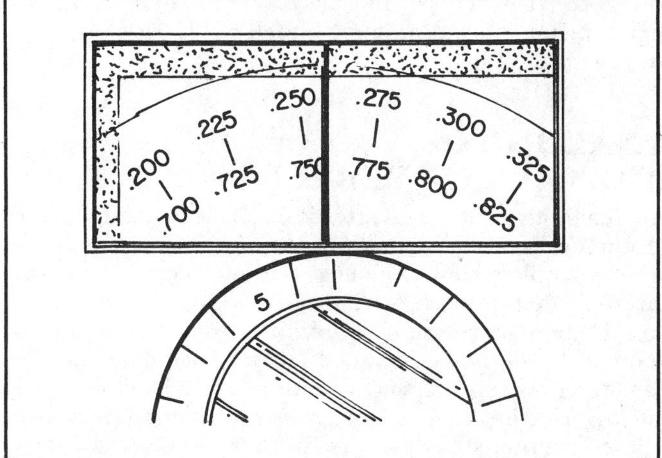

ANALOG CONTROL: An analog frequency readout.

Both analog and digital computers can be made to have varying degrees of accuracy. In certain cases, the analog computer will find an exact solution because the problem under scrutiny involves a relation identical to one of the storage functions in the computer. This happens less frequently with digital computers.

While the analog computer has some advantages with respect to accuracy in certain problems, the digital computer can be brought to a high degree of accuracy by repeated iterations of its processes, which effectively makes its mesh (such as shown by A in ANALOG) finer and finer. Digital information is, in general, more efficient than analog information. Almost all small calculators and computers today are digital. *See also* DIGITAL COMPUTER.

ANALOG CONTROL

A control or adjustment that is variable over a continuous range is an analog control. Examples of analog controls are the frequency dials on some communications transmitters and receivers (see illustration), volume controls, and RF gain controls. Analog controls are sometimes preferred over digital controls because the operator of the equipment can more easily visualize the entire range of a parameter in relation to a specific setting. Analog controls also allow adjustment to the exact desired value, whereas digital controls allow only an approximation.

ANALOG METERING

All of the common panel meters on electronic equipment are analog devices. Current, voltage, and power levels are monitored by devices that show a continuous range of a certain quantity, such as plate current or RF output in a communications transmitter (see the photograph).

In some situations, such as amplifier tuning indicators and signal-strength comparison tests, analog metering allows the operator or technician to visualize a continous range of possible values in relation to the actual reading. This is important when tuning circuits for optimum performance. It would be difficult to "peak" a signal on a digital meter, but on an analog meter it is easy. *See also* DIGITAL METERING.

ANALOG-TO-DIGITAL CONVERTER

An analog-to-digital converter is a circuit that converts a continuously variable quantity into a set of discrete values.

The simplest kind of analog-to-digital converter is the threshold detector (*see* THRESHOLD DETECTOR), which can be as elementary as a single semiconductor diode in association with another component such as a relay. As the voltage varies across a silicon diode, the diode either conducts (forward voltage 0.6 or greater) or is open (forward voltage less than 0.6, or reverse voltage). An analog quantity is thus converted to either of two digital states by the circuit, as shown in the illustration. If the relay is very sensitive, signals of 0.6 volt will close it, but anything of less amplitude will fail to close it. Thus the output will be 6 volts or zero volts, depending on the input. The low-voltage output condition is called the low state, 0, and the high-voltage output condition is called the high state, 1.

More sophisticated analog-to-digital converters have several, or many, different states. Slow-scan television is a good example of this (*see* SLOW-SCAN TELEVISION). While the actual image contains infinite shades of gray, from white to black, an excellent approximation is possible by limiting the signal information to just a few shades. This is done with an analog-to-digital converter.

The digital signal from an analog-to-digital converter is sometimes changed back to an analog signal at the receiving end of a digital-communications link. Thus an extremely accurate representation of the original signal is obtained. *See also* DIGITAL-TO-ANALOG CONVERTER.

AND GATE

An AND gate is a circuit that performs the logical operation "AND." An AND gate is schematically symbolized as shown (see A, B, and C). It may have two or more inputs.

Logic symbols 1, or high, at all the inputs of an AND gate will produce an output of 1 (high). But if any of the inputs are at logic 0, or low, then the output of the AND gate will be low. *See also* LOGIC GATE.

ANALOG METERING: An analog meter, showing plate current or relative RF output from a transmitter, depending on the function selected by a front-panel switch.

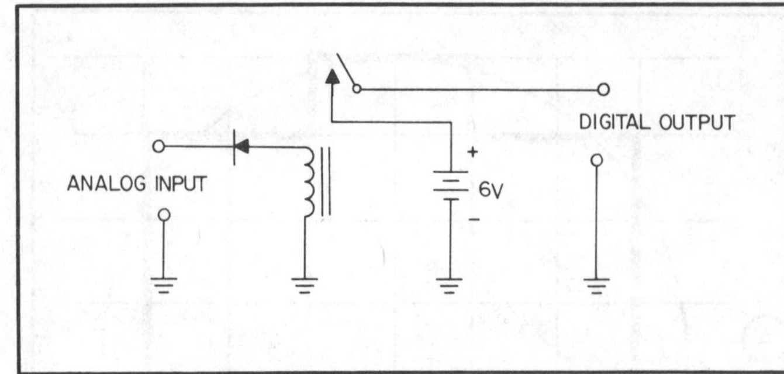

ANALOG-TO-DIGITAL CONVERTER: A simple analog-to-digital converter might consist of a diode and relay, such as is shown here.

ANDERSON BRIDGE

An Anderson bridge is a device for determining unknown capacitances or inductances. See the schematic diagram for an Anderson bridge designed to measure inductances.

For the proper operation of an Anderson bridge, it is necessary to have a frequency standard, a way to balance this standard with the known reactance and the unknown reactance, and an inductor to show when balance has been achieved. A galvanometer (*see* GALVANOMETER) is generally used as the indicator. It shows both positive and negative deflections from zero (null).

The Anderson bridge actually measures reactance, from which the inductance or capacitance is easily determined. Inductance bridges are calibrated in millihenrys or microhenrys. Capacitance bridges are calibrated in microfarads or picofarads. Some bridges are capable of measuring either inductance or capacitance. *See also* REACTANCE.

ANGLE OF BEAM

See BEAMWIDTH.

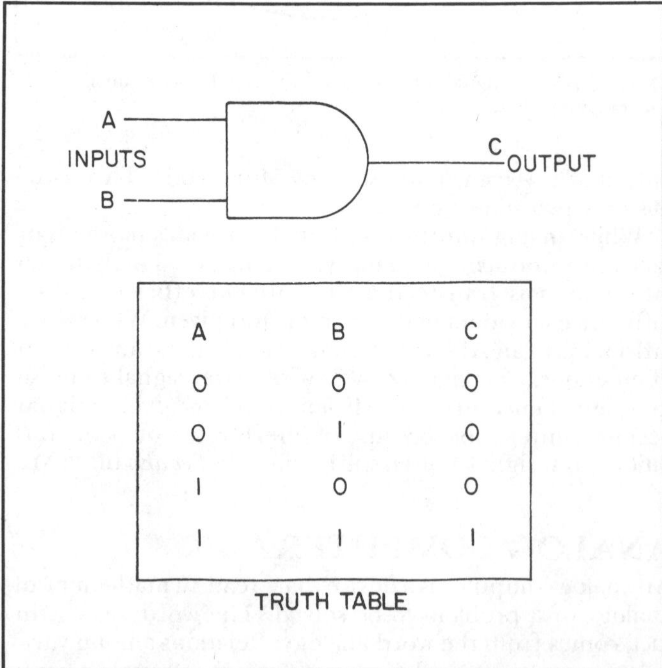

A	B	C
0	0	0
0	1	0
1	0	0
1	1	1

TRUTH TABLE

AND GATE: An AND gate with two inputs (A and B). The output (C) is 1 only when both inputs are 1, otherwise it is 0, as shown in the "truth" table.

ANGLE OF DEFLECTION

In a cathode-ray tube (*see* CATHODE-RAY TUBE), a narrow beam of electrons is sent through a set of electrically charged deflection plates to obtain the display. The angle of deflection of the beam is the number of degrees the beam is diverted from a straight path (see illustration). If the beam of electrons continues through the deflection plates in a perfectly straight line, the angle of deflection is zero.

In general, the greater the amplitude of the input signal to an oscilloscope, the greater the angle of deflection of the electron beam. The angle of deflection is directly proportional to the input voltage. Therefore, if an ac voltage of 2 volts peak-to-peak causes an angle of deflection of ±10 degrees, an ac voltage of 4 volts peak-to-peak will result in an angle of deflection of ±20 degrees. The angle of deflection in an oscilloscope is always quite small, so the displacement on the screen is essentially proportional to the angle of deflection. Some oscilloscopes are calibrated so that the angle of deflection increases in logarithmic proportion, rather than in direct proportion, to the input signal voltage. *See also* OSCILLOSCOPE.

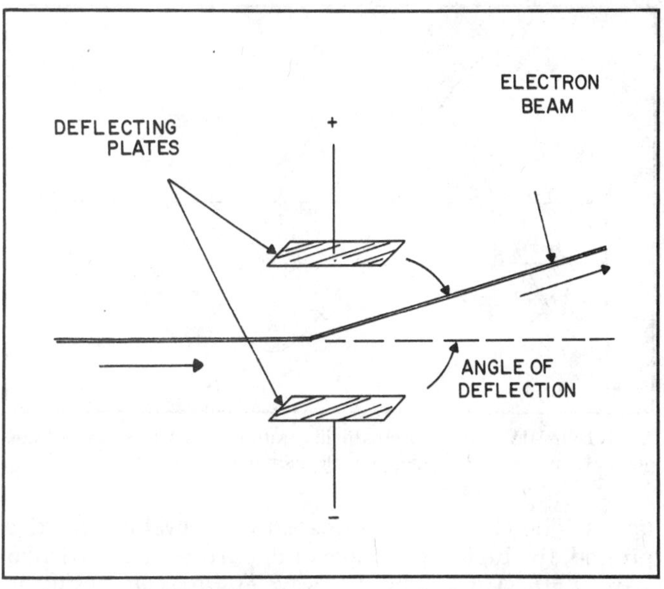

ANGLE OF DEFLECTION: The angle of deflection, in an oscilloscope, is the angle by which the electron beam changes direction after passing through the deflecting plates.

ANGLE OF DEPARTURE

The term angle of departure refers to the angle, relative to the horizon, with which a radio signal leaves a transmitting antenna.

At high frequencies, or about 3 to 30 MHz, the angle of departure from a horizontal antenna depends on the height of the antenna above effective ground. (*See* EFFECTIVE GROUND.) Sometimes a high angle of departure is desirable, such as when local communication is attempted at long wavelengths. At other times, a very low angle of departure is needed, such as for working over great dis-

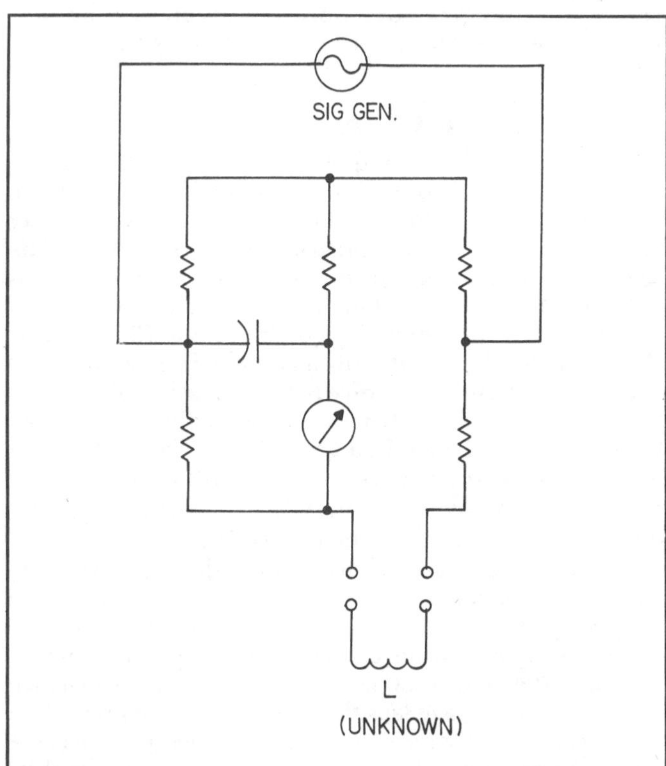

ANDERSON BRIDGE: An Anderson bridge for measuring an unknown inductance.

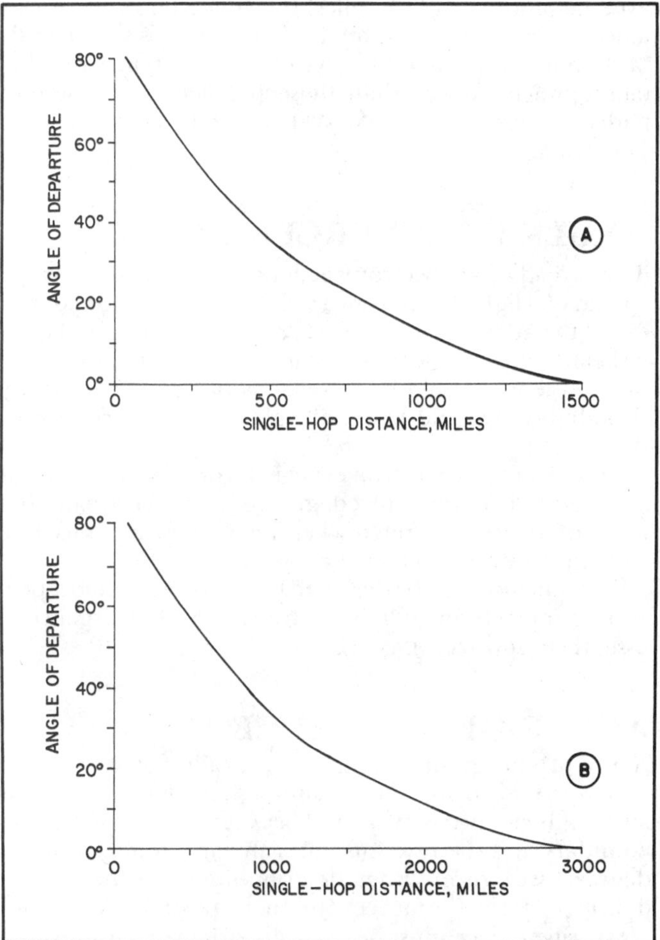

ANGLE OF DEPARTURE: The single-hop return distance, as a function of the angle of departure, for the E layer (A) and F layer (B) of the ionosphere under average conditions. Multi-hop distances are, of course, much greater. These graphs show the maximum possible distance assuming an angle of departure of zero degrees.

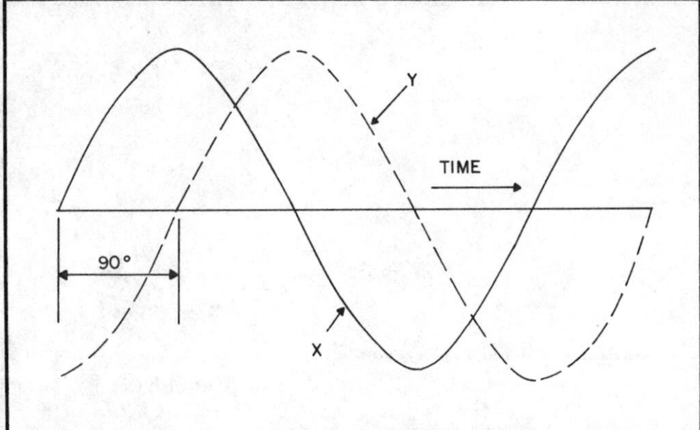

ANGLE OF DIVERGENCE: In this illustration, wave Y lags wave X by 90 degrees, since wave Y begins ¼ cycle later than wave X.

tances. The closer an antenna is to the level of effective ground, the higher the angle of departure at a particular wavelength. Any height of ¼ wavelength or less results in an angle of departure of 90 degrees, or directly upward.

The angle of departure is related to the distance at which a radio signal is returned to earth via the ionosphere. For E-layer propagation when the ionized layer is at an average height of 65 miles, the single-hop return distance as a function of the angle of departure is shown in A. For F-layer propagation when the ionized layer is at an average altitude of 200 miles, the single-hop return distance as a function of the angle of departure is shown in B. Multi-hop propagation allows communication over distances much greater than those possible via single-hop paths. *See also* E LAYER, F LAYER, PROPAGATION CHARACTERISTICS.

ANGLE OF DIVERGENCE

The spread of an electron beam in an oscilloscope, or the spread of a light beam from a collimating device or laser, is called the angle of divergence. A perfectly straight, parallel beam is impossible to realize in either case, but the angle of divergence should be as small as possible. A perfectly straight beam would have an angle of divergence of zero degrees.

In practice, the electron beam in a good oscilloscope has an angle of divergence of 2 degrees or less. The greater the angle of divergence, the thicker the oscilloscope trace line will appear on the screen (*see* OSCILLOSCOPE).

Laser devices can be designed to produce an almost perfectly parallel beam of light, with an angle of divergence of essentially zero (*see* LASER).

ANGLE OF INCIDENCE

A ray of energy impinging on a locally flat surface or region is said to have a certain angle of incidence. The angle of incidence is measured between the normal to the boundary and the ray, and thus can vary from zero to 90 degrees (see figure). When determining the angle of incidence of a ray, some confusion is possible, since the measuring device must be correctly oriented with respect to the ray and the plane. The true angle of incidence of a ray, R, relative to a plane, P, is obtained by first constructing a perpendicular line, L, passing through the plane at the same point, Q, as the ray. Then a ray, R′, is constructed

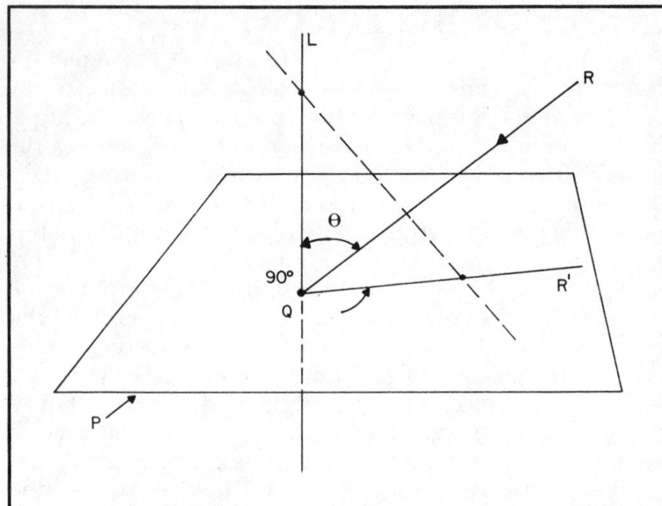

ANGLE OF INCIDENCE: The angle of incidence of a ray with respect to a plane.

within the plane such that all the points in L, R, and R′ lie along straight lines. That is, L, R, and R′ must be coplanar. The angle of incidence is then the angle between R and L. In the illustration, the angle of incidence is represented by the upper-case Greek letter Θ (theta).

When light strikes a flat reflecting surface such as a mirror, it is reflected at an angle equal to its angle of incidence. When a radio wave encounters a layer of the ionosphere, that wave is returned to earth at the same angle (roughly) as its angle of incidence. When sound energy strikes a wall, that energy is reflected at an angle equal to its angle of incidence. The term angle of incidence refers to impinging, or approaching, energy. *See also* ANGLE OF REFLECTION, ANGLE OF REFRACTION.

ANGLE OF LAG

Two waves of identical frequency and amplitude need not coincide with each other in terms of phase (*see* PHASE ANGLE). When they do not coincide, one wave is said to lag the other. The wave that begins its cycle earlier is called the leading wave, and the wave that begins its cycle later is called the lagging wave. One complete wave cycle is represented by 360 degrees. One-half cycle is 180 degrees. One wave may lag another by any angle from zero to 180 degrees as shown. If one wave lags another by more than a half cycle, we consider that wave to be leading the other by some angle less than 180 degrees. When the angle of lag between two waves is precisely 180 degrees, the waves are said to be in phase opposition. When the angle of lag is 90 degrees, the waves are in phase quadrature.

Angle of lag is an important quantity in alternating-current circuit theory. In a circuit containing resistance and no reactance, the voltage and current waves are exactly in phase. In an inductive reactance, the current cycle lags the voltage cycle by 90 degrees. In a capacitive reactance, the voltage lags the current by 90 degrees. In a circuit containing some resistance and some reactance, the waves are separated by some value between zero and 90 degrees. This occurs because reactances do not simply dissipate energy, as do resistances, but instead they store energy and release it later in the cycle. *See also* ANGLE OF LEAD, REACTANCE.

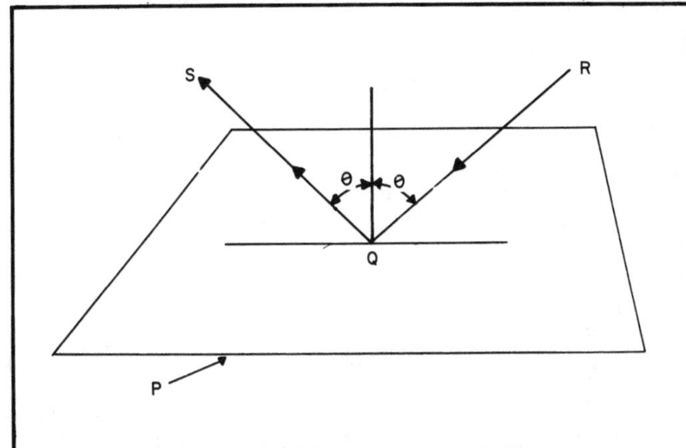

ANGLE OF REFLECTION: The angle of incidence and the angle of reflection are equal. Ray R is incident, and wave S is reflected. L is the normal to the surface at point Q.

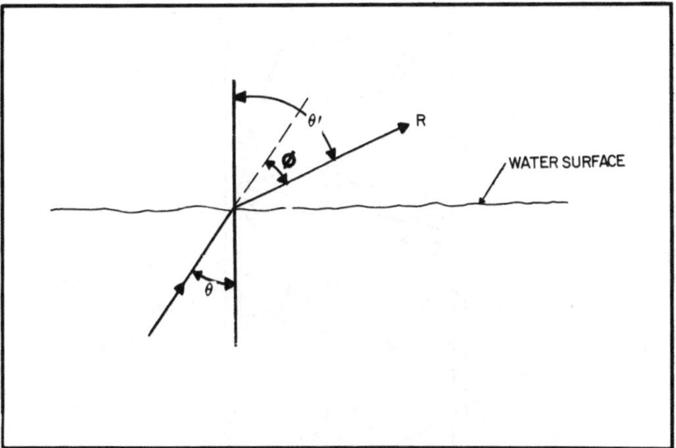

ANGLE OF REFRACTION: A ray of light leaving a pool of water changes its direction by an angle ϕ. The angle 90° − ϕ, is the angle of refraction. The angle of incidence is θ.

ANGLE OF LEAD

When two waves have the same frequency but are not in phase, we may say that one wave leads the other by a certain number of degrees. In the illustration in ANGLE OF LAG, wave Y is said to lag wave X (*see* ANGLE OF LAG). However, we can also say that wave X leads wave Y. This means exactly the same thing.

The angle of lead is specified as some value between zero and 180 degrees, as is the angle of lag. The leading wave is identified by the fact that its cycle begins earlier than that of the lagging wave, by an amount less than a half cycle. In most amplitude-vs-time illustrations, including that in ANGLE OF LAG, the lagging wave is displaced to the right and the leading wave is on the left.

If a wave leads another by some angle, ϕ, of more than 180 degrees, we consider it to lag the other by $360 - \phi$ degrees. This keeps confusion to a minimum. *See also* PHASE ANGLE, REACTANCE.

ANGLE OF RADIATION

See ANGLE OF DEPARTURE.

ANGLE OF REFLECTION

When a ray of energy strikes a flat object or barrier and is reflected, we speak of its departure angle as its angle of reflection. While this angle is also sometimes called the angle of departure, the latter term is generally reserved for energy generated at or near the flat surface or barrier, rather than energy reflected from it (*see* ANGLE OF DEPARTURE).

The angle of reflection is always the same as the angle of incidence for a locally flat, smooth surface (see figure). Further, the plane containing the incident and reflected rays is always perpendicular to the plane of the barrier at the point of reflection. *See also* ANGLE OF INCIDENCE

ANGLE OF REFRACTION

Energy passing from one medium to another is often refracted at the barrier (*see* REFRACTION). A good example of this is a visible light ray passing into or out of a pool of water as shown. Refraction can take place with sound waves, radio waves, infrared energy, ultraviolet energy, and X rays, as well as with visible light.

In the illustration, a ray of light, R, leaves a pool of water. At the surface, the ray changes its direction by a certain angle, ϕ. The angle of incidence of R with respect to the water surface is given by θ. Because of the refraction, ray R leaves the water surface with a new angle, θ', such that

$$\theta' = \theta + \phi$$

The angle of refraction is defined as θ'. The angle of direction change is ϕ.

When the angle of incidence θ is 0 degrees, the measure of ϕ will be zero. As θ is made larger, ϕ increases. At a certain point, the takeoff angle will become zero; then the angle of refraction is 90 degrees. That is, the ray will follow the water surface.Then if θ is made larger still, all of the light will be reflected back under the water, and none will pass through to the air.

For a light beam passing from air into water, the paths are identical to those for the water-to-air ray; however, the directions of the rays are reversed.

In the case of visible light, water has a higher index of refraction than air (*see* INDEX OF REFRACTION). The angle of refraction depends not only on the angle of incidence between two media, but also on their relative indices of refraction. A given substance often has a much different index of refraction for one type of energy, such as sound, than for another, such as light. The wavelength of a particular kind of energy also affects the index of refraction.

The ionosphere acts to bend shortwave radio signals because its index of refraction is greatly different, at high frequencies, than the index of refraction of air. *See also* PROPAGATION, PROPAGATION CHARACTERISTICS.

ANGLE STRUCTURE

An angle structure is a method of building a tower for mechanical strength. Braces are placed at angles with respect to the vertical support rods (see illustration). This provides greater rigidity and resistance to twisting, as compared to a tower without angle braces. This is especially important for towers that must support large antennas, since such antennas have a tendency to try to rotate in a high wind. *See also* TOWER.

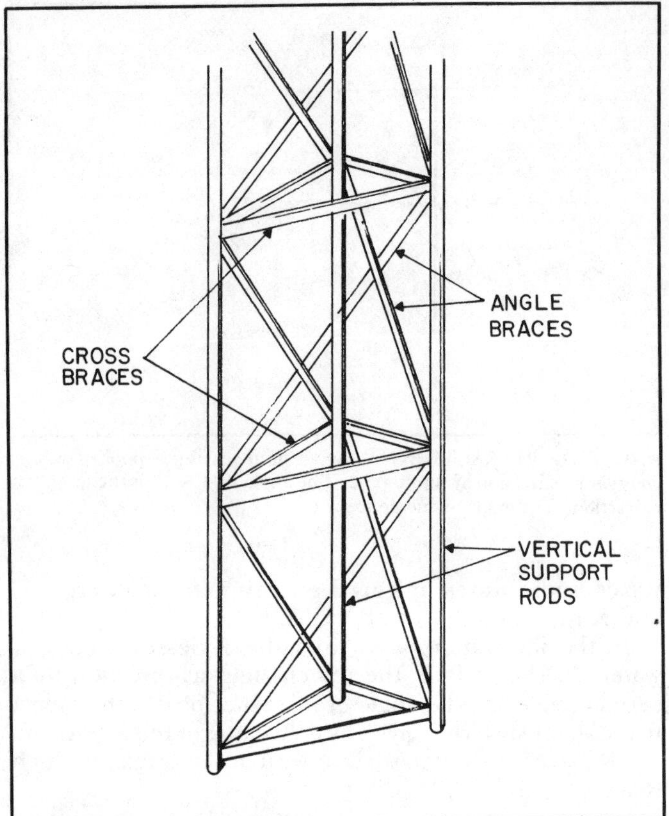

ANGLE STRUCTURE: Angle braces add rigidity to a tower.

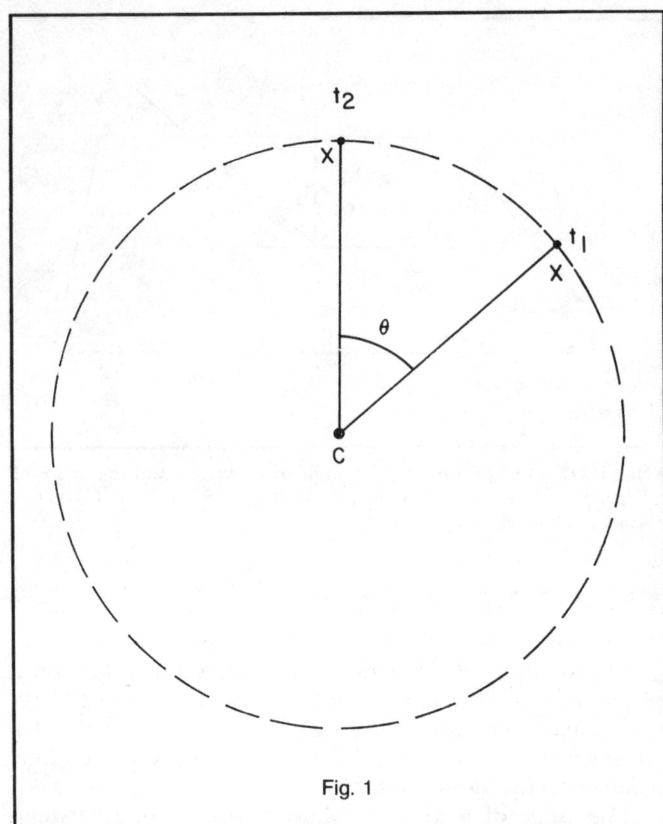

ANGULAR MOTION: Angular motion is simply circular motion. Angular displacement is measured in degrees or radians. Here, point X moves through an angular displacement θ from time t_1 to t_2.

ANGULAR FREQUENCY

Ordinarily, frequency is expressed in terms of cycles per second, or Hertz (*see* HERTZ). One complete cycle is broken into 360 degrees, representing one revolution around a circle. This same revolution may be expressed as 2π radians (*see* RADIAN).One radian represents approximately 57.3 degrees.

For some purposes, it is advantageous to express frequency in terms of degrees or radians per second, rather than in the conventional Hertz. For a frequency f_{Hz} in Hertz,

$$f_{d/s} = 360\ f_{Hz}$$

and

$$f_{r/s} = 2\pi f_{Hz}$$

where $f_{d/s}$ is the angular frequency in degrees per second, and $f_{r/s}$ is the angular frequency in radians per second.

In electronics applications, angular measures are seldom used to express frequency. *See also* ANGULAR MOTION.

ANGULAR MOTION

Angular motion is motion in a circle. It is sometimes also called rotational motion.

For a given object, X, orbiting a central point, C, as shown in Fig. 1, the angular displacement from time t_1 to time t_2 is given by the angle θ traversed by the line segment CX. The angular velocity is measured in degrees per second or radians per second (*see* RADIAN). The angular acceleration is the rate at which the angular velocity changes. It is expressed in degrees or radians-per-second-per-second.

Angular motion at constant velocity forms an exact mathematical representation of a sine wave. If the object X in the first illustration rotates with constant speed, completing one revolution per second around the central point C, its angular velocity is 360 degrees or 2π radians per second. If the circle is viewed edge-on, a back-and-forth oscillation of object X is observed. If the circle is then moved laterally (perpendicular to the plane in which it lies) as shown in Fig. 2 (A and B), this back-and-forth oscillation becomes a sine wave with a frequency of 1 Hz. This mathematical representation expresses how a sine wave is represented in terms of degrees: One cycle is given by one rotation of a point in a circle, or 360 degrees. A frequency of 1 Hz thus is equal to 360 degrees per second, or 2π radians per second. A half cycle is 180 degrees or π radians. A quarter cycle is 90 degrees or $\pi/2$ radians. *See also* SINE WAVE.

ANODE

In a vacuum tube or semiconductor diode, the anode is the electrode toward which the electrons flow. The anode of a vacuum tube is also known as the plate. The anode is always positively charged relative to the cathode (*see* CATHODE) under conditions of forward bias, and negatively charged relative to the cathode under conditions of reverse bias. Current therefore flows only when there is forward bias. (A small amount of current does flow with reverse bias, but it is usually negligible.) If the reverse voltage becomes excessive, there may be a sudden increase

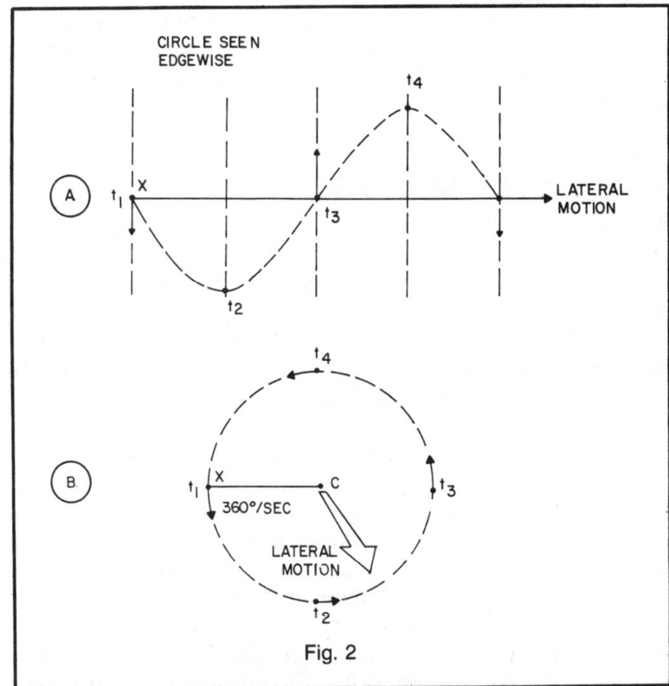

ANGULAR MOTION: Circular motion, projected through time in a direction perpendicular to the circle itself and viewed edgewise, is a perfect representation of sine-wave motion. At A, the circle seen edgewise; at B, seen face-on.

in current in the reverse direction (*see* ARCBACK) in a diode vacuum tube. A and B illustrate the normal conditions in a tube.

The term anode is sometimes used in reference to the positive terminal or electrode in a cell or battery. *See also* CELL, DIODE, TUBE.

ANTENNA

An antenna is a device for receiving or transmitting electromagnetic energy. In a receiving antenna, passing electromagnetic fields cause current to flow back and forth in the antenna conductors at the same frequency as the field oscillations. In a transmitting antenna, electrons flowing back and forth in the conductors generate electromagnetic fields that propagate far into space. An antenna is a special kind of transducer, then, that converts current to electromagnetic energy and vice-versa.

The simplest kind of antenna, in practice, is called the dipole. It consists of a length of electrical conductor, such as wire or metal tubing, ½ wavelength long and fed at the center. The dipole antenna is a resonant system at the frequency where it is ½ wavelength. The dipole also displays resonance at all harmonics of this frequency. The frequency at which a dipole represents ½ wavelength is called its fundamental frequency. (*See* RESONANCE.)

Any transmitting antenna will work well as a receiving antenna, at the same frequency. But the converse is not always true. Some small antennas, such as the ferrite-rod antenna, do not function as transmitting antennas.

The performance of a receiving or transmitting station is largely dependent on the antenna. If the antenna is badly designed or poorly located, station operation will suffer. A well-engineered antenna system can result in far more improvement to a communications circuit than the manipulation of any other single factor.

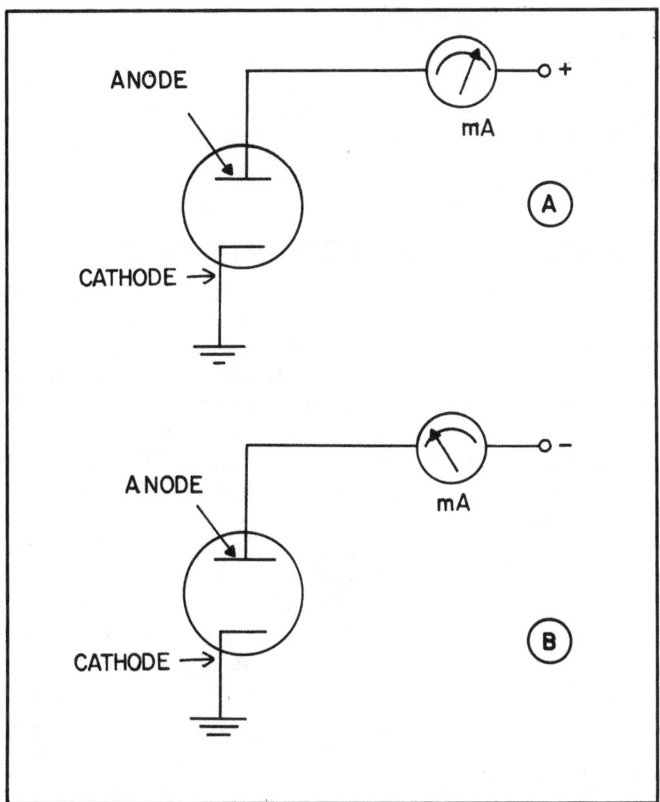

ANODE: When the anode is positive (A), current flows. When it is negative, current does not flow (B).

Various types of antennas are listed in this volume by name. Refer to the appropriate antenna name for information about specific kinds of antennas.

ANTENNA EFFICIENCY

Not all the electromagnetic field received by an antenna from its feed line is ultimately radiated into space. Some power is dissipated in the ground near the antenna in structures such as buildings and trees, the earth itself, and in the conducting material of the antenna. If P represents the total amount of available power at a transmitting antenna, P_R the amount of power eventually radiated into space, and P_L the power lost in surrounding objects and the antenna conductors, then

$$P_L + P_R = P, \text{ and}$$

$$\text{Efficiency (percent)} = 100\ (P_R/P)$$

The efficiency of a non-resonant antenna is difficult to determine in practice, but at resonance (*see* RESONANCE), when the antenna impedance, Z, is a known pure resistance, the efficiency can be found by the formula

$$\text{Efficiency (percent)} = 100\ (R/Z),$$

where R is the theoretical radiation resistance of the antenna at the operating frequency (*see* RADIATION RESISTANCE). The actual resistance, Z, is always larger than the theoretical radiation resistance R. The difference is the loss resistance, and there is always some loss.

Antenna efficiency is optimized by making the loss resistance as small as possible. Some means of reducing the loss resistance are the installation of a good RF-ground system (if the antenna is a type that needs a good RF ground), the use of low-loss components in tuning networks (if they are used), and locating the antenna itself as high above the ground, and as far from obstructions, as possible.

ANTENNA GROUND SYSTEM

Some types of antennas must operate against a ground system, or RF reference potential. Others do not need an RF ground. In general, unbalanced or asymmetrical antennas need a good RF ground, while balanced or symmetrical antennas do not. The ground-plane antenna (*see* GROUND-PLANE ANTENNA) requires an excellent RF ground in order to function efficiently. However, the half-wave dipole antenna (*see* DIPOLE ANTENNA) does not need an RF ground.

When designing an antenna ground system, it is important to realize that a good dc ground does not necessarily constitute a good RF ground. An elevated ground-plane antenna has a very effective RF ground that does not have to be connected physically to the earth in any way. A single thin wire hundreds of feet long may be terminated at a ground rod or the grounded side of a utility outlet and work very well as a dc ground, but it will not work well at radio frequencies.

The earth affects the characteristics of any antenna. The overall radiation resistance and impedance are affected by the height above ground. However, an RF ground system does not necessarily depend on the height of the antenna. Capacitive coupling to ground is usually sufficient, in the form of a counterpoise (*see* COUNTERPOISE).

A ground system for lightning protection must be a direct earth connection. Some antennas are grounded through inductors that do not conduct RF, but serve to discharge static buildup before a lightning stroke hits. *See also* LIGHTNING PROTECTION.

ANTENNA IMPEDANCE

Any antenna displays a defined impedance at its feed point at a particular frequency. Usually, this impedance changes as the frequency changes. Impedance at the feed point of an antenna consists of radiation resistance (*see* RADIATION RESISTANCE) and either capacitive or inductive reactance (*see* REACTANCE). Both the radiation resistance and the reactance are defined in ohms.

An antenna is said to be resonant at a particular frequency when the reactance is zero. Then, the impedance is equal to the radiation resistance in ohms. The frequency at which resonance occurs is called the resonant frequency of the antenna. Some antennas have only one resonant frequency. Others have many. The radiation resistance at the resonant frequency of an antenna depends on several factors, including the height above ground and the harmonic order, and whether the antenna is inductively or capacitively tuned. The radiation resistance can be as low as a fraction of an ohm, or as high as several thousand ohms.

When the operating frequency is made higher than the resonant frequency of an antenna, inductive reactance appears at the feed point. When the operating frequency is below the resonant point, capacitive reactance appears. To get a pure resistance, a reactance of the opposite kind from the type present must be connected in series with the antenna. *See also* ANTENNA RESONANT FREQUENCY, IMPEDANCE.

ANTENNA MATCHING

For optimum operation of an antenna and feed-line combination, the system should be at resonance. This is usually done by eliminating the reactance at the feed point, where the feed line joins the antenna radiator. In other words, the antenna itself is made resonant. The remaining radiation resistance is then transformed to a value that closely matches the characteristic impedance of the feed line.

If the reactance at the antenna feed point is inductive, series capacitors are added to cancel out the inductive reactance. If the reactance is capacitive, series coils are employed. An example of this is shown in A. The inductances are adjusted until the antenna is resonant. Both coils should have identical inductances to keep the system balanced (since this is a balanced antenna).

Once the antenna is resonant, only resistance remains. This value may not be equal to the characteristic impedance of the line. Generally, coaxial lines are designed to have a characteristic impedance, or Z_O, or 50 to 75 ohms, which closely approximates the radiation resistance of a half-wave dipole in free space. But, if the radiation resistance of the antenna is much different from the Z_O of the line, a transformer should be used to match the two parameters, as shown in B. This results in the greatest efficiency for the feed line. Without the transformer, standing waves on the line will cause some loss of signal. The amount of loss caused by standing waves is sometimes inconsequential, but sometimes it is large (*see* STANDING WAVE, STANDING-WAVE RATIO).

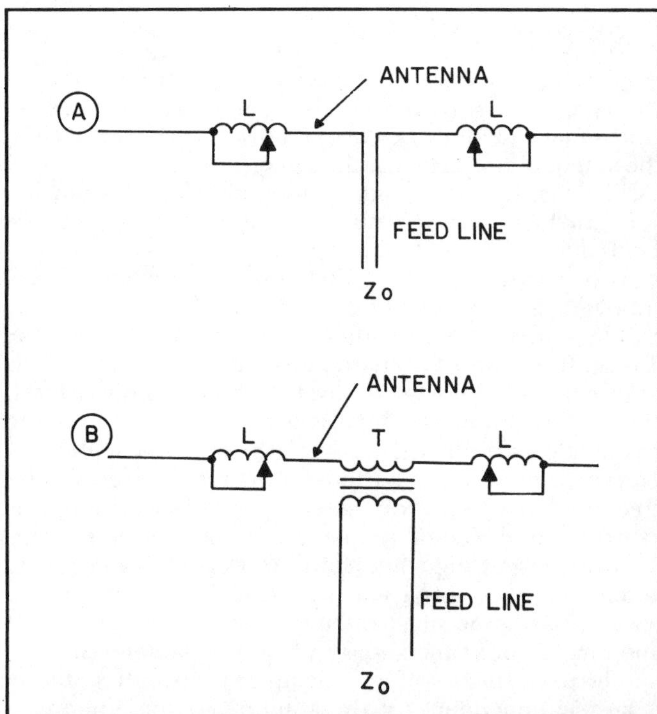

ANTENNA MATCHING: At A, the dipole is tuned to resonance by variable inductors. (This particular antenna is too short for the operating frequency, resulting in capacitive reactance.) At B, a transformer is used to match the radiation resistance of the dipole to the characteristic impedance, Z_o, of the feed line.

In some antenna systems, no attempt is made to obtain an impedance match at the feed point. Instead, a matching system (*see* TRANSMATCH) is used between the transmitter or receiver and the feed line. This allows operating convenience when the frequency is changed often. However, it does nothing to reduce the loss on the line caused by standing waves. *See also* TUNED FEEDERS.

ANTENNA PATTERN

The directional characteristics of any transmitting or receiving antenna, when graphed on a polar coordinate system, are called the antenna pattern. The simplest possible antenna pattern occurs when an isotropic antenna is used (*see* ISOTROPIC ANTENNA), although this is a theoretical ideal. It radiates equally well in all directions in three-dimensional space.

Antenna patterns are represented by diagrams such as those in A and B. The location of the antenna is assumed to be at the center of the coordinate system. The greater the radiation or reception capability of the antenna in a certain direction, the farther from the center the points on the chart are plotted. A dipole antenna, oriented horizontally so that its conductor runs in a north-south direction, has a horizontal-plane (H-plane) pattern similar to that in Fig. A. The elevation-plane (E-plane) pattern depends on the height of the antenna above effective ground at the viewing angle. With the dipole oriented so that its conductor runs perpendicular to the page, and the antenna ¼ wavelength above effective ground, the E-plane antenna pattern will resemble B.

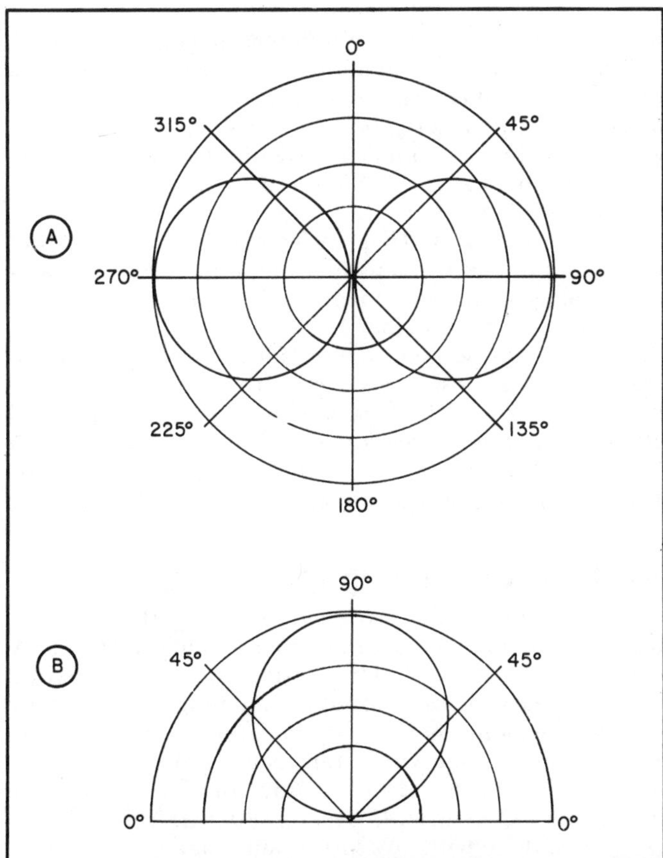

ANTENNA PATTERN: Antenna patterns for a dipole. At A, seen from above (horizontal or H plane); at B, seen from the end of the dipole (E plane), assuming a height of ¼ wavelength above effective ground.

The patterns in A and B are quite simple. Many antennas have patterns that are very complicated. For all antenna pattern graphs, the relative power gain (*see* ANTENNA POWER GAIN) relative to a dipole is plotted on the radial axis. The values thus range from 0 to 1 on a linear scale. Sometimes a logarithmic scale is used. If the antenna has directional gain, the pattern radius will exceed 1 in some directions. Examples of antennas with directional gain are the log periodic, longwire, quad, and Yagi. Some vertical antennas have gain in all horizontal directions. This occurs at the expense of gain in the E plane.

ANTENNA POLARIZATION

The polarization of an antenna is determined by the orientation of the electric lines of force in the electromagnetic field radiated or received by the antenna. Polarization may be linear, or it may be rotating (circular). Linear polarization can be vertical, horizontal, or somewhere in between. In circular polarization, the rotation can be either counterclockwise or clockwise (*see* CIRCULAR POLARIZATION).

For antennas with linear polarization, the orientation of the electric lines of flux is parallel with the radiating element, as shown in A and B. Therefore, a vertical element produces signals with vertical polarization, and a horizontal element produces horizontally polarized fields in directions broadside to the element.

In receiving applications, the polarization of an antenna is determined according to the same factors involved in transmitting. Thus, if an antenna is vertically polarized for transmission of electromagnetic waves, it is also vertically polarized for reception.

In free space, with no nearby reflecting objects to create phase interference, the circuit attenuation between a vertically polarized antenna and a horizontally polarized antenna, or between any two linearly polarized antennas at right angles, is approximately 30 dB compared to the attenuation between two antennas having the same polarization.

Polarization affects the propagation of electromagnetic energy to some extent. A vertical antenna works much

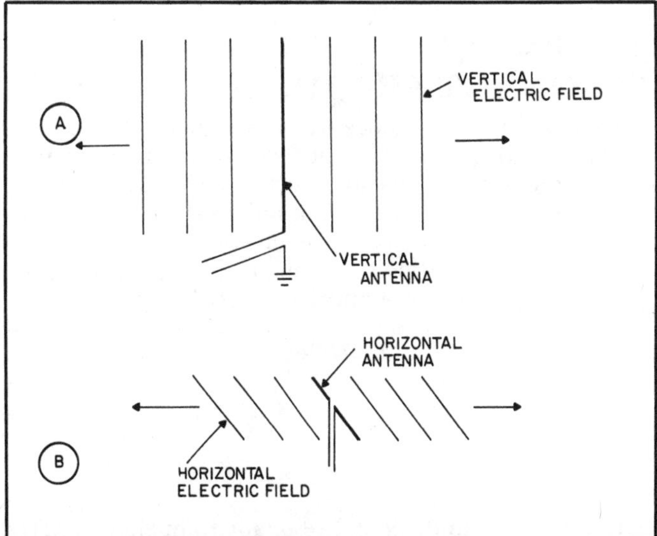

ANTENNA POLARIZATION: The polarization of an antenna is parallel to the conductor itself. At A, a vertical antenna; at B, a horizontal antenna.

better from transmission and reception of surface-wave fields (*see* SURFACE WAVE) than a horizontal antenna. For sky-wave propagation (*see* SKY WAVE), the polarization is not particularly important, since the ionosphere causes the polarization to be randomized at the receiving end of a circuit.

ANTENNA POWER GAIN

The power gain of an antenna is the ratio of the effective radiated power (*see* EFFECTIVE RADIATED POWER) to the actual RF power applied to the feed point. Power gain may also be expressed in decibels. If the effective radiated power is P watts and the applied power is P watts, than the power gain in decibels is

$$\text{Power Gain (dB)} = 10 \log_{10}(P_{ERP}/P)$$

Power gain is always measured in the favored direction of an antenna. The favored direction is the azimuth direction in which the antenna performs the best. For power gain to be defined, a reference antenna must be chosen with a gain assumed to be unity, or 0 dB. This reference antenna is usually a half-wave dipole in free space (*see* DIPOLE ANTENNA). Power gain figures taken with respect to a dipole are expressed in dBd. The reference antenna for power-gain measurements may also be an isotropic radiator (*see* ISOTROPIC ANTENNA), in which case the units of power gain are called dBi. For any given antenna, the power gains in dBd and dBi are different by approximately 2.15 dB:

$$\text{Power Gain (dBi)} = 2.15 + \text{Power Gain (dBd)}$$

Directional transmitting antennas can have power gains in excess of 20 dBd. At microwave frequencies, large dish antennas (*see* DISH ANTENNA, PARABOLOID ANTENNA) can be built with power gains of 30 dBd or more.

Power gain is the same for reception, with a particular antenna, as for transmission of signals. Therefore, when antennas with directional power gain are used at both ends of a communications circuit, the effective power gain over a pair of dipoles is the sum of the individual antenna power gains in dBd.

ANTENNA RESONANT FREQUENCY

An antenna is at resonance whenever the reactance at the feed point (the point where the feed line joins the antenna) is zero. This may occur at just one frequency, or it may take place at several frequencies. The impedance of an antenna near resonance is shown in the illustration.

For a half-wave dipole antenna in free space, the resonant frequency is given approximately by

$$f = 468/s_{ft}$$

or

$$f = 143/s_m,$$

where f is the fundamental resonant frequency in MHz, and s is the antenna length in feet (s_{ft}) or meters (s_m). For a quarter-wave vertical antenna operating against a perfect ground plane:

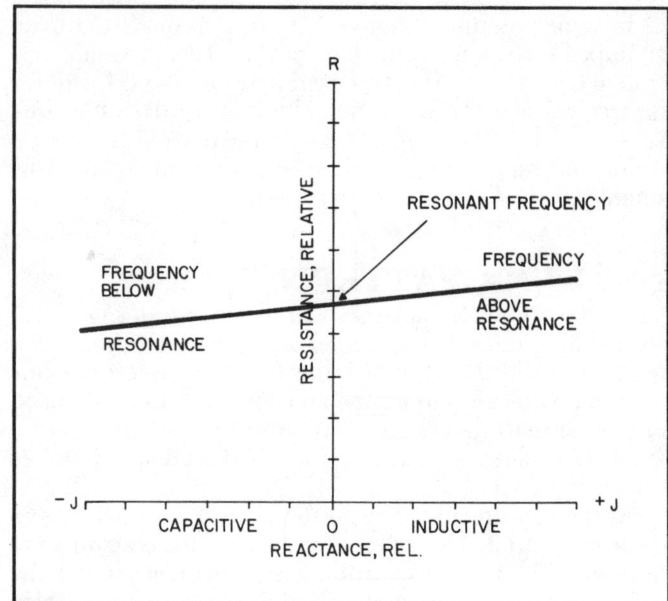

ANTENNA RESONANT FREQUENCY: At resonance, the impedance of an antenna is a pure resistance. The radiation resistance usually increases slightly as the frequency increases; this is shown by the slope of the line. Below resonance, the reactance is capacitive. Above resonance it is inductive.

$$f = 234/h_{ft},$$

or

$$f = 71/h_m,$$

where h is the antenna height in feet (h_{ft}) or meters (h_m).

The dipole antenna and quarter-wave vertical antenna display resonant conditions at all harmonics of their fundamental frequencies. Therefore, if a dipole or quarter-wave vertical is resonant at a particular frequency f, it will also be resonant at 2f, 3f, 4f, and so on. The impedance is not necessarily the same, however, at harmonics as it is at the fundamental. At frequencies corresponding to odd harmonics of the fundamental, the impedance is nearly the same as at the fundamental. At even harmonics, the impedance is much higher.

An antenna operating at resonance, where the radiation resistance is almost the same as the characteristic impedance, or Z_0, of the feed line, will perform with good efficiency, provided the ground system (if a ground system is needed) is efficient. *See also* CHARACTERISTIC IMPEDANCE, RADIATION RESISTANCE.

ANTENNA TUNING

Antenna tuning is the process of adjusting the resonant frequency of an antenna or antenna system (*see* ANTENNA RESONANT FREQUENCY). This is usually done by means of a tapped or variable inductor at the antenna feed point, or somewhere along the antenna radiator (see figure). It is also sometimes done with a transmatch at the transmitter, so that the feed line and antenna together form a resonant system (*see* TRANSMATCH, TUNED FEEDERS). Antenna tuning can be done with any sort of antenna.

An antenna made of telescoping sections of tubing is tuned exactly to the desired frequency by changing the amount of overlap at the tubing joints, thus changing the

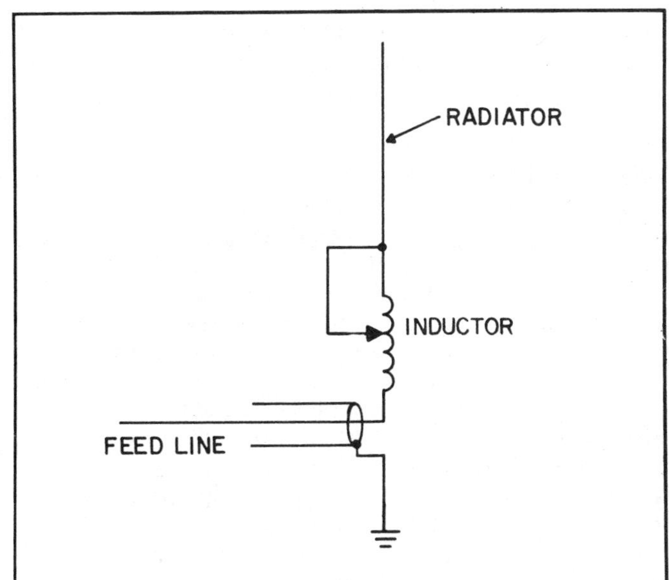

ANTENNA TUNING: A tuned short vertical antenna. The variable inductor is adjusted for resonance.

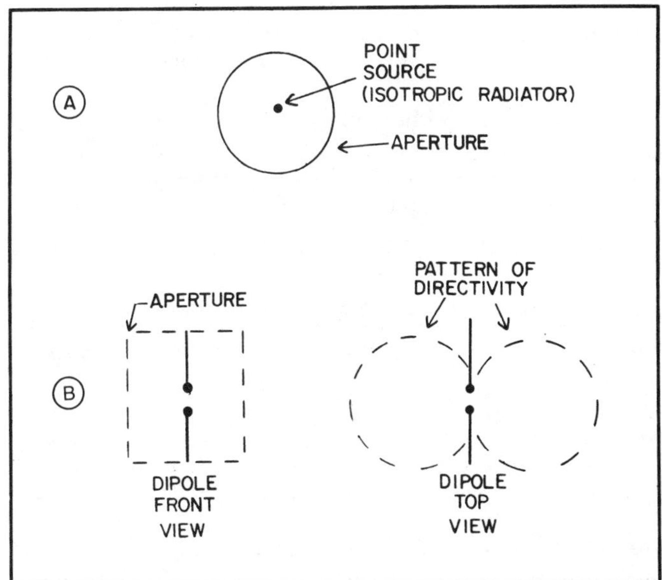

APERTURE: A picture of aperture in an isotropic radiator (A) and a half-wave dipole (B).

physical length of the radiating or parasitic elements. In a Yagi array, the director and reflector elements must be precisely tuned to obtain the greatest amount of forward power gain and front-to-back ratio. Phased arrays must be tuned to give the desired directional response. In some antennas, tuning is not critical, while in others it must be done precisely to obtain the rated specifications. In general, the higher the frequency, the more exacting are the tuning requirements.

ANTIHUNT DEVICE

An antihunt device is a circuit in an automobile direction-finding system (*see* DIRECTION FINDER). In such a device, the circuit sometimes overcorrects itself. The overcorrection in azimuth bearing causes another correction, which also exceeds the needed amount. This can happen over and over, resulting in a back-and-forth oscillation of the antenna. The indicator will read first to the left and then to the right of the desired target. An antihunt device prevents this endless oscillation of an automatic direction-finding system by damping the response. This lessens the extent of correction, so that overcorrection does not take place.

APERTURE

The area over which an antenna can effectively intercept an electromagnetic field is called its aperture. The true, or effective, aperture can range from an area that is larger than an antenna's physical size, as in the case of an array of wire antennas, to an area that is smaller than the physical size, as with horn or parabolic-reflector antennas. The aperture of an isotropic antenna is large compared to its size (the antenna is a point source, and the aperture is in the shape of a sphere), but the antenna has no gain because it lacks directivity (see A). The aperture of a dipole is not as large as that of an isotropic antenna, but a dipole exhibits some gain because of directivity (see B).

Antenna receiving gain is almost always expressed in decibels with reference to a half-wave dipole (dBd), or with respect to an isotropic antenna (dBi). The term aperture is seldom used. (*See* ANTENNA POWER GAIN.)

APPARENT POWER

In an alternating-current circuit containing reactance, the voltage and current reach their peaks at different times. That is, they are not exactly in phase. This complicates the determination of power. In a nonreactive circuit, we may consider

$$P = EI$$

where P is the power in watts, E is the RMS (*see* ROOT MEAN SQUARE) voltage in volts, and I is the RMS current in amperes. In a circuit with reactance, this expression is referred to as the apparent power. It is called apparent because it differs from the true power (*see* TRUE POWER) that would be dissipated in a resistor or resistive load. Only when the reactance is zero is the apparent power identical to the true power.

In a nonresonant or improperly matched antenna system, a wattmeter placed in the feed line will give an exaggerated reading. The wattmeter reads apparent power, which is the sum of the true transmitter output power and the reactive or reflected power (*see* REACTIVE POWER). To determine the true power, the reflected reading of a directional wattmeter is subtracted from the forward reading. The more severe the antenna mismatch, the greater the difference between the apparent and true power. As the antenna approaches a state of pure reactance, the apparent power approaches twice the true power in the feed line.

In the extreme, all of the apparent power in a circuit is reactive. This occurs when an alternating-current circuit consists of a pure reactance, such as a coil, capacitor, or short-circuited length of transmission line. *See also* REACTANCE.

APPLICATION

Computers are used for many different purposes. A solution may be sought to a large or complicated set of mathematical equations. A complicated integral might need to be evaluated. A computer might be used to store a list of things to do for each day of the year, by months, or it might be used as a word processor or electronic game. The purpose for which a computer is used in a given instance is its application.

Applications are either computational, as when solving a mathematical problem, or of a data-processing nature, as when storing a list of things to do. Electronic games might be considered to form a third category of computer application: recreational. *See also* COMPUTER.

APPROACH CONTROL

When aircraft are within 10 to 20 miles of their destination airport, they contact a radio guidance service called approach control. The control operator provides steering vectors and speed values to properly place the aircraft for entry into the traffic pattern at the airport and to provide safe spacing from the other aircraft in the vicinity. The service is conducted by voice communications in vhf and uhf radio, augmented by radar position displays in the control center. The actual landing may be done by visual reference, or, if weather conditions required, by instrument-landing procedures. After landing, the plane is guided on the airport grounds by a service called ground control. *See also* AIR TRAFFIC CONTROL.

ARC

An arc occurs when electricity flows through space. Lightning is a good example of an arc. When the potential difference between two objects becomes sufficiently large, the air (or other gas) ionizes between the objects, creating a path of relatively low resistance through which current flows.

An arc may be undesirable and destructive, such as a flashover across the contacts of a wafer switch. Or, an arc can be put to constructive use. A carbon-arc lamp is an extremely bright source of light, and is sometimes seen in large spotlights or searchlights where other kinds of lamps would be too expensive for the illumination needed (*see* ARC LAMP).

Undesirable or destructive arcing is prevented by keeping the voltage between two points below the value that will cause a flashover. An antenna lightning arrestor allows built-up static potential to discharge across a small gap (*see* AIR GAP, LIGHTNING ARRESTOR) before it gets so great that arcing occurs between components of the circuit and ground.

ARCBACK

Arcback is a flow of current in the reverse direction in a vacuum-tube rectifier. Ordinarily, electrons should flow from the cathode to the anode, but not vice-versa. This is how the tube rectifies alternating current.

If the reverse voltage becomes too large, electrons will start to flow from the anode to the cathode, and the tube will no longer rectify. The maximum reverse voltage that a tube can tolerate without arc back is called the peak inverse voltage (PIV), or peak reverse voltage (PRV) (*see* PEAK INVERSE VOLTAGE). In a semiconductor diode, the peak inverse voltage is sometimes called the avalanche voltage (*see* AVALANCHE VOLTAGE).

Rectifier vacuum tubes have been largely replaced by semiconductor diodes in recent years. Semiconductor diodes need no filament power supply, which itself consumes considerable power. They also generate far less heat, and are physically much smaller, than tubes. *See also* RECTIFIER, RECTIFIER TUBE.

ARC LAMP

An arc lamp is an extremely brilliant electric lamp that uses two closely spaced, pointed, carbon electrodes to generate an arc through the air (see illustration).

A voltage is first applied to the electrodes, which are initially spaced too far apart to allow arcing. The voltage may be either ac or dc. Then the electrodes are slowly brought closer together, until the air between them ionizes and arcing begins. A collimating lens or parabolic reflector serves to focus the light into a narrow beam.

Arc lamps may be relatively small or very large. Small carbon-arc lamps are sometimes used in theater movie projectors. Large carbon-arc lamps, with parabolic reflectors 6 to 10 feet in diameter, are used in promotional campaigns for some new businesses. These lamps are so brilliant that their light makes a visible spot on the clouds at night. Curiosity attracts people to the site. The lamps require a large portable generator, and consume thousands of watts of power. This same type of lamp was used in World War II to locate airplanes in the nighttime sky.

ARC RESISTANCE

Arc resistance is a measure of the durability of an insulating or dielectric material against arcing. When an arc occurs next to a dielectric material, a conductive path along the surface of the material will eventually be formed. The greater the arc resistance of the material, the longer it will be before this conductive path is formed. Such a conductive path, of course, ruins the insulating properties of the substance.

Arc resistance also refers to the electrical resistance of an arc, for example in an arc lamp (*see* ARC LAMP). With a certain interelectrode voltage, E, and current, I, the arc resistance is given by Ohm's law (*see* OHM'S LAW) as

$$R = E/I$$

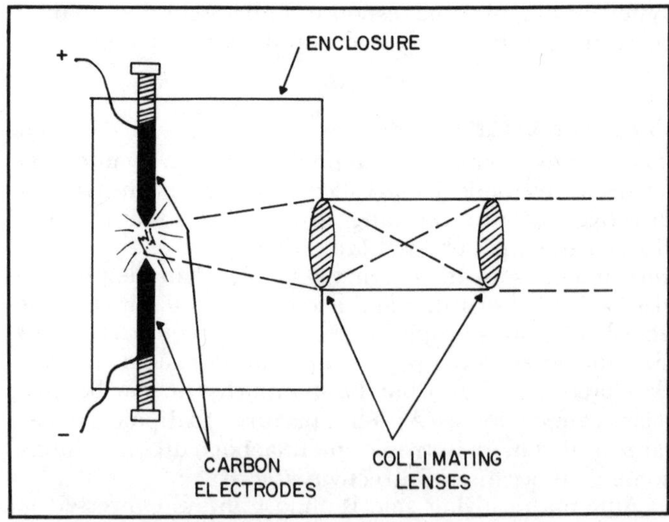

ARC LAMP: A carbon arc lamp. The carbon electrodes are brought together until an arc occurs. Such a lamp provides extremely brilliant light.

For alternating current, the voltage and current must be RMS values (*see* ROOT MEAN SQUARE) for this determination to be accurate.

AREA REDISTRIBUTION

An irregularly shaped pulse is sometimes difficult to measure in terms of duration; it may be difficult to pinpoint the moments of its beginning and ending. Area redistribution is a means of measuring the effective duration of an irregular pulse.

In A, an irregular pulse is shown as it might appear on an oscilloscope. The shaded region has a definite area, measurable in terms of amplitude and time (such as millivolt-seconds). It is virtually impossible to define the beginning and ending moments of this irregular pulse, and therefore its duration is vague. However, a rectangular pulse may be constructed containing the same peak amplitude and the same area as the irregular pulse (B). In a circuit, this pulse is constructed by using a rectangular pulse generator. The duration of the rectangular pulse that delivers the same amount of energy is considered to be the effective duration of the irregular pulse.

ARGUMENT

An argument is a variable in a function. The value of any function depends on the values of its arguments. In computer lingo, the term argument is used in the same context as the term variable in algebra.

A function, f, of arguments x, y, and z is denoted by f(x,y,z). Suppose that a function of three variables is specified by

$$f(x,y,z) = 3x - 2y + z$$

Then the values for f are computed for various combinations of argument values.

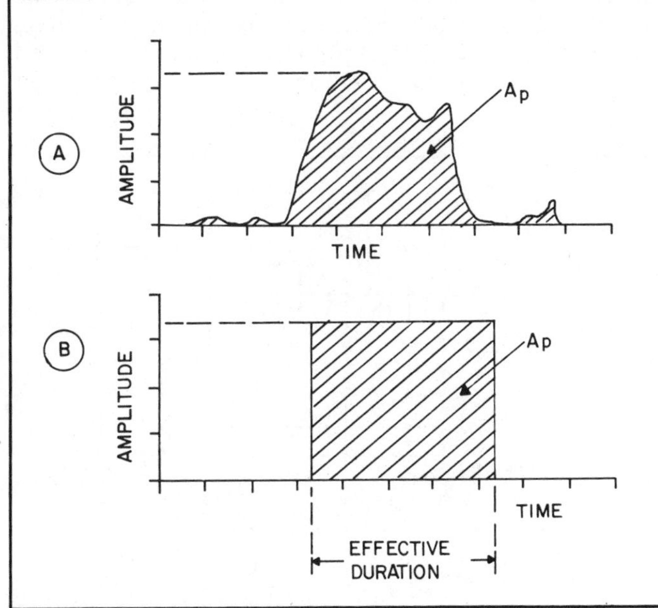

AREA REDISTRIBUTION: At A, an irregular pulse with indefinite starting and ending times. The area (A_P) under this curve is measured, and an identical rectangular pulse with the same amplitude (B) constructed. The effective duration time of the irregular pulse is then defined as the duration time of the rectangular pulse.

The set of possible values for the arguments of a function is called the domain of the function. We might have different domains for each argument. For example, the domain for x might be the set of integers, the domain for y might be all real numbers between, but not including, 0 and 1, and the domain for z might be the set of rational numbers. The range of a function is the set of all possible values resulting from combinations of the arguments. Functions can have only one argument, or hundreds of arguments. *See also* FUNCTION.

ARITHMETIC MEAN

The arithmetic mean of a set of numbers

$$\{x_1, x_2, x_3, \ldots, x_n\}$$

is a special function of those numbers, defined as

$$a(x_1, x_2, x_3, \ldots, x_n) = \frac{x_1 + x_2 + x_3 + \ldots + x_n}{n}$$

This function is also sometimes called the average of the set of numbers. The arithmetic mean is always larger than or equal to the smallest number in a set, and smaller than or equal to the largest number in the set. The arithmetic mean is not necessarily midway between the smallest and largest numbers in a given set. If x_{max} is the smallest number and x_{max} is the largest number in a set, then

$$m(x_1, x_2, x_3, \ldots, x_n) = \frac{x_{min} + x_{max}}{2}$$

is called the median of the set

$$\{x_1, x_2, x_3, \ldots, x_n\}.$$

Occasionally, numbers will be averaged by a different function than a $(x_1, x_2, x_3, \ldots, x_n)$, called the geometric mean. This does not usually result in the same value as the arithmetic mean. *See also* GEOMETRIC MEAN.

ARITHMETIC OPERATION

In an electronic computer, an operation is defined as a function of two arguments f(x,y), (*see* ARGUMENT, FUNCTION). An operation might be rather complicated, or very simple. The four operations

$$F_+ (x,y) = x + y$$
$$F_- (x,y) = x - y$$
$$F_\times (x,y) = xy$$
$$F_/ (x,y) = x/y$$

are easily recognizable as addition, subtraction, multiplication, and division. These are called arithmetic operations. Such functions as

$$f_L(x) = \log_{10}(x)$$
$$f_e(x) = e^x$$
$$f_{sin}(x) = \sin x$$

are not arithmetic operations. *See also* FUNCTION.

ARITHMETIC SYMMETRY

Arithmetic symmetry refers to the shape of a bandpass or band-refection filter response (*see* BANDPASS FILTER, BAND-REJECTION FILTER).

An example of arithmetic symmetry in a bandpass filter curve is shown in the drawing. The frequency scale (horizontal) is linear, so that each unit length represents the same number of Hertz in frequency. (In this case, each division represents 1 MHz). The vertical scale may either be linear, calibrated in volts or watts, or logarithmic, calibrated in decibels relative to a certain level. (Here, the vertical scale is calibrated in decibels relative to 1 milliwatt, or dBm. *See* DBM.)

The curve in the drawing is exactly symmetrical around the center frequency f_0 = 145 MHz; that is, the left-hand side of the response is a mirror image of the right-hand side. This is arithmetic symmetry.

In practice, arithmetic symmetry represents an ideal condition, and it can only be approximated. Modern technology has developed bandpass and band-rejection filters with almost perfect arithmetic symmetry with a variety of bandwidths and frequency ranges. *See also* CRYSTAL-LATTICE FILTER, MECHANICAL FILTER.

ARM-LIFT, BACK/CHEST-PRESSURE RESUSCITATION

The arm-lift, back-pressure and arm-lift, chest-pressure methods of artificial respiration (*see* ARTIFICIAL RESPIRATION) are of historical interest only. The mouth-to-mouth and mouth-to-nose methods (*see* MOUTH-TO-MOUTH RESUSCITATION, MOUTH-TO-NOSE RESUSCITATION) have replaced the manual methods. The manual forms are much less effective in restoring proper breathing to a victim, and they are more tiring to perform.

Artificial respiration is necessary if breathing has stopped. This can occur as a result of electric shock (*see* ELECTRIC SHOCK). Permanent brain damage may occur within four to six minutes of breathing failure unless some measures are taken to oxygenate the blood. If the heart has stopped as well as the breathing, additional measures are necessary, since air in the lungs is of no use when there is no circulation. *See also* CARDIOPULMONARY RESUSCITATION.

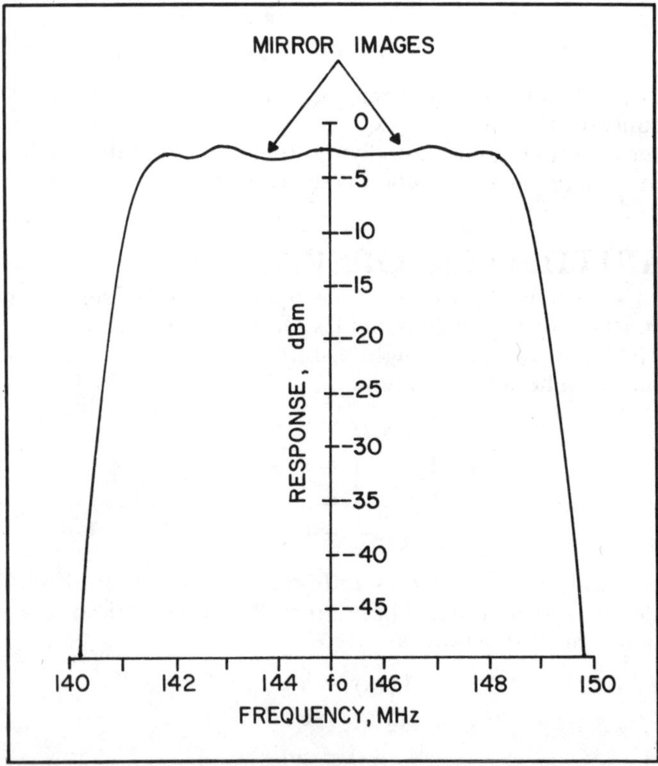

ARITHMETIC SYMMETRY: Arithmetic symmetry is the property of symmetry with respect to the center frequency for a bandpass or band-reject filter.

ARMSTRONG OSCILLATOR

An Armstrong oscillator is a circuit that produces oscillation by means of inductive feedback. See the simple circuit diagram of such an oscillator. A coil called the tickler is connected to the plate or collector, and is brought near the coil of the tuned circuit. The tickler is oriented to produce positive feedback. The amount of coupling between the tickler coil and the tuned circuit is adjusted so that stable oscillation takes place. The output is taken from the (tube) plate or (transistor) collector by means of a small capacitance, or by transformer coupling to the tuned circuit.

The frequency of the Armstrong oscillator is determined by the tuned-circuit resonant frequency. Usually, the capacitor is variable and the inductor is fixed, so the amount of feedback remains relatively constant as the frequency is changed. Armstrong oscillators are generally used in regenerative receivers (*see* REGENERATIVE DETECTOR). They are not often seen as variable oscillators in transmitters or superheterodyne receivers. Other types of oscillators are preferred for those applications. *See also* COLPITTS OSCILLATOR, ELECTRON-COUPLED OSCILLATOR, HARTLEY OSCILLATOR.

ARRAY

An array is an antenna system containing more than one element for the purpose of developing gain in some directions. Yagi antennas, for example, are sometimes stacked in pairs, one above the other, and fed in phase to obtain

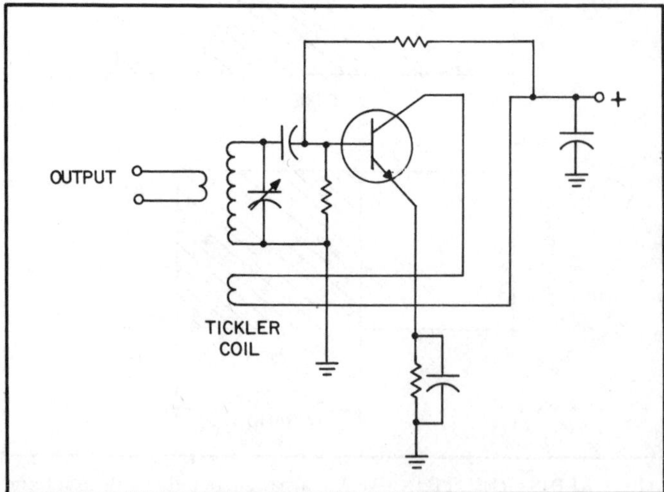

ARMSTRONG OSCILLATOR: An Armstrong oscillator. This kind of oscillator is usually employed as a regenerative detector in a simple receiver.

greater gain than would be possible in this manner as shown. The antennas must be fed in the proper phase to obtain the desired results.

An antenna array need not, of course, consist of Yagis. Any kind of antenna can form the basis for an array. A set of dipoles placed end-to-end is known as a collinear array (*see* COLLINEAR ANTENNA). The simplest collinear array is a half-wave dipole operated on its second-harmonic frequency, where it acts as a two-element combination of half-wave antennas.

The term array, in general, is used for any group of identical objects arranged in a logical pattern. A square or rectangular array is sometimes called a matrix (*see* MATRIX).

ARRESTOR

See LIGHTNING ARRESTOR.

ARRL

See AMERICAN RADIO RELAY LEAGUE.

ARSENIC

Arsenic is an element with atomic number 33 and atomic weight 75. It is significant in electronics because it is used as a semiconductor doping material in gallium-arsenide field-effect transistors (*see* GALLIUM-ARSENIDE FIELD-EFFECT TRANSISTOR). The atomic physicist abbreviates arsenic by the symbol *As*.

ARTICULATION

Articulation is a measure of the quality of a voice-communications circuit. It is given as a percentage of the speech units (syllables or words) understood by the listener. To test for articulation, a set of random words or numbers should be read by the transmitting operator. The words or numbers are chosen at random to avoid possible contextual interpolation by the receiving operator. This gives a true measure of the actual percentage of speech units received.

When plain text or sentences are transmitted, the receiving operator can understand a greater portion of the information because he can figure out some of the missing words or syllables by mental guesswork. The percentage of speech units received with plain-text transmission is called intelligibility (*see* INTELLIGIBILITY).

Articulation and intelligibility differ from fidelity (*see* FIDELITY). Perfect reproduction of the transmitted voice is not as important in a communications system as the accurate transfer of information. The best articulation generally occurs when the voice frequency components are restricted to approximately the range of 200 to 3000 Hz. Articulation can also be enhanced at times by the use of speech compressors or RF clipping (*see* SPEECH CLIPPING, SPEECH COMPRESSION).

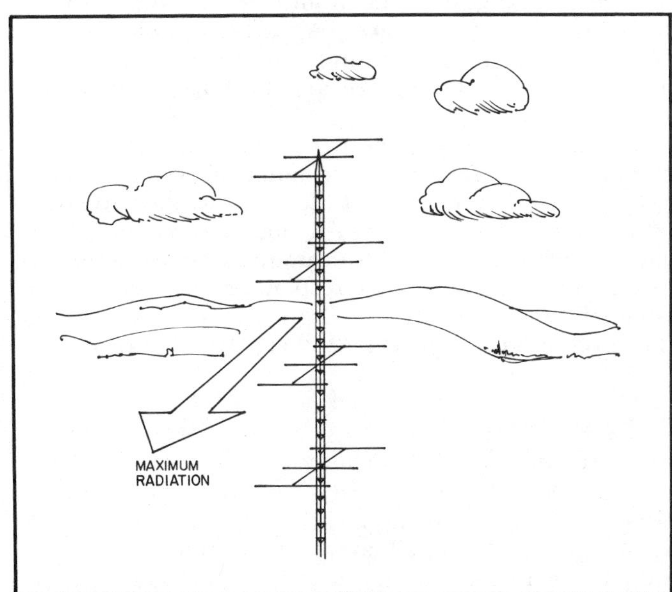

ARRAY: A stacked array of Yagi antennas. An array of four Yagi antennas, if ideally spaced, will produce approximately 6 dB gain over a single Yagi.

ARTIFICIAL EAR

The human ear has a certain frequency response and impedance to sound energy. To duplicate these properties of the ear, as well as its threshold sensitivity and relative perception of loudness, an artificial ear is constructed. Such a device is useful in the optimum design of high-fidelity recording and sound systems, and in the engineering of communications systems.

An artificial ear consists of a special microphone, enclosed in a box with acoustic properties similar to those of the human ear. An earphone or headset to be tested is placed over the device. Measuring instruments are then connected to the artificial ear to check the earphone output for frequency distribution, noise, and dynamic range as they would appear to a person. Sometimes a wide frequency response is desired from an earphone; sometimes the range is deliberately restricted. *See also* EARPHONE, FIDELITY, HIGH FIDELITY, INTELLIGIBILITY.

ARTIFICIAL GROUND

An artificial ground is an RF ground not directly connected to the earth. A good example of an artificial ground is the system of quarter-wave radials in an elevated ground-plane antenna (*see* GROUND-PLANE ANTENNA). A counterpoise (*see* COUNTERPOISE) is also a form of artificial ground.

In some situations, it is difficult or impossible to obtain a good earth ground for an antenna system. A piece of wire ¼ wavelength, or any odd multiple of ¼ wavelength, long at the operating frequency can operate as an artificial ground in such a case as shown. This arrangement does not form an ideal ground; the wire will radiate some energy, and thus is actually a part of the antenna. But an artificial ground is much better than no ground at all. Of course, some kind of dc ground should be used in addition to the RF ground to minimize the danger of electrical shock from built-up static on the antenna and from possible short circuits in the transmitting or receiving equipment. *See also* DC GROUND.

ARTIFICIAL INTELLIGENCE

Computers can, in some cases, perform not only calculation operations, data storage and processing, but also such

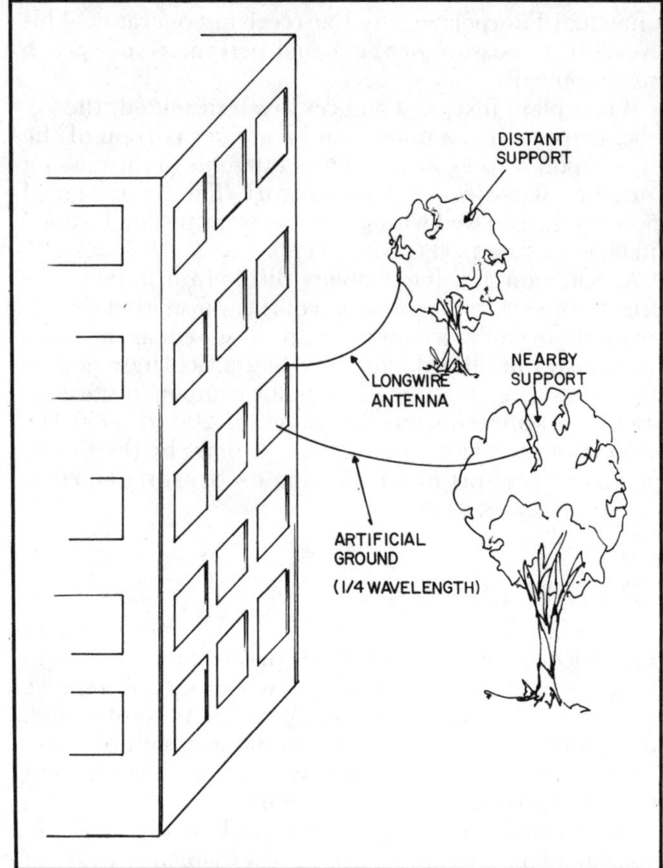

ARTIFICIAL GROUND: An artificial ground can be constructed by running a ¼-wavelength piece of wire in free space away from the operating location of a radio station. Here, an artificial ground is used for improving the performance of a shortwave or Amateur-Radio longwire antenna.

human functions as reasoning, learning, and decision making. Even some emotional response can be synthesized! A computer with such capabilities is said to have artificial intelligence. Technology is becoming so sophisticated that a highly complex conversation will soon be possible between computers and people. Of course, such "conversations" already take place in machine language between computers.

One of the greatest fantasies of science fiction is the creation of a computer more intelligent, and more ambitious, than man himself. Such a machine might reproduce itself by using robots to build more computers. The computer might obtain all the knowledge ever generated by man. Such a machine could perhaps build more and more memory into its own circuitry, increasing its own power almost without limit. Some scientists argue that it would be impossible for man to create a machine better than himself. This kind of thing, they argue, is as absurd as a perpetual-motion device. In the not-too-distant future, we may find out whether or not they are right!

Computers with a low level of artificial intelligence are available to the consumer in such forms as electronic games and computer chess. *See also* COMPUTER, ELECTRONIC GAME.

ARTIFICIAL RESPIRATION

A severe electric shock can cause unconsciousness, possibly cessation of breathing, and even heart failure. If the breathing has stopped but the heart is still pumping, artificial respiration can be used to revive a victim. If the heart has also failed, a technique called cardiopulmonary resuscitation must be used (*see* CARDIOPULMONARY RESUSCITATION, HEART FIBRILLATION).

Unless oxygenation of the blood is begun within four to six minutes of the cessation of breathing, permanent brain damage may occur. In 10 to 15 minutes, the victim will die. However, artificial respiration has, in some cases, kept people alive for hours until medical help arrived.

Artifical respiration is done by either of two methods: mouth-to-mouth and mouth-to-nose. Manual methods, using chest or back pressure and arm-lift maneuvers, are considered to be antiquated and inefficient. *See also* ELECTRIC SHOCK, MOUTH-TO-MOUTH RESUSCITATION, MOUTH-TO-NOSE RESUSCITATION.

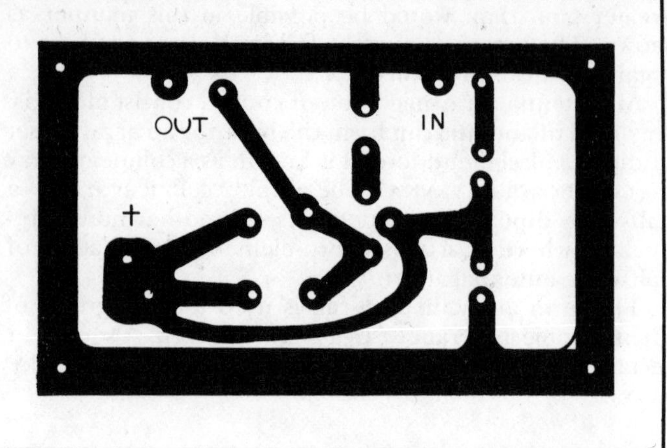

ARTWORK: Artwork for a simple amplifier printed-circuit board. This drawing is photographed and reduced for etching purposes.

ARTIFICIAL VOICE

An artificial voice is a speaker mounted in a special enclosure that has acoustic properties similar to a human mouth and head. Such a device duplicates the frequency content and dynamic range of a human voice by suitably modifying the sound from the speaker. An artificial voice is very useful in the design of high-fidelity recording microphones and in the engineering of communications systems.

Test instruments, connected to the microphone, determine its response to the voice. Some microphones must have a wide frequency response, such as those used to record music. Others have a deliberately narrowed response, such as those used for communications. Some microphones have a directional response pattern; others are essentially omnidirectional. *See also* ARTICULATION, FIDELITY, HIGH FIDELITY, INTELLIGIBILITY, MICROPHONE.

ARTWORK

In the construction of an integrated circuit (*see* INTEGRATED CIRCUIT), the pattern is first drawn to scale on a piece of glass or plastic film. If there are several layers to the integrated circuit, all layers are accurately drawn. A special camera then is used to reduce the pattern to its actual size for reproduction in the integrated circuit. In this way, a tremendous number of components can be squeezed into a very small space on a chip of semiconductor material. Transducers, resistors, capacitors, diodes, and wiring are all fabricated in this way.

A printed-circuit board (*see* PRINTED CIRCUIT) is made in a similar manner. Artwork is drawn on a piece of paper or film as shown, usually several times actual size. It is then photographed and reduced and put on a clear plastic film. A photographic process is employed to etch the wiring pattern onto a piece of copper-plated phenolic or glass-epoxy material. For this reason, this kind of artwork is called an etching pattern (*see* ETCHING).

Artwork is a tremendous cost saver in the fabrication of integrated circuits, because all wiring errors can be eliminated before the circuit is actually built. Sometimes the artwork for a particularly complicated circuit is drawn by a computer.

A-SCAN

On a cathode-ray tube display, time is usually represented by the horizontal scale, or abscissa, as the independent variable. The signal amplitude is usually shown on the vertical scale, or ordinate, as the dependent variable. A one-dimensional radar system, as shown, uses a display of this type to indicate the distances to various objects. Signals appear as vertical pulses or "pips." The farther to the right a pip appears, the greater its distance from the transmitter. The distance scale is linear, so that a correlation between scale divisions and kilometers or miles is easy to determine. *See also* RADAR.

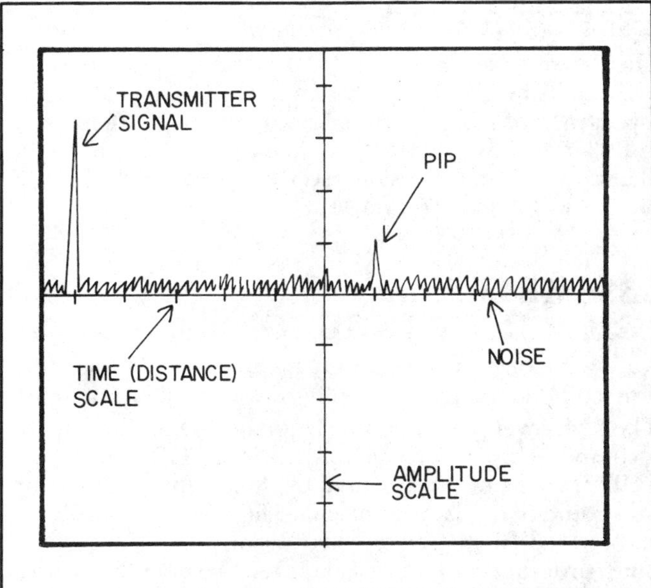

A-SCAN: A one-dimensional radar system using A-scan. The large pip on the left is the transmitter signal. A distant object appears as a pip to the right. The more distant the object, the farther to the right the echo appears.

ASCII: SYMBOLS FOR ASCII TELEPRINTER CODE.

First Four Signals	Last Three Signals 000	001	010	011	100	101	110	111
0000	NUL	DLE	SPC	0		P	/	p
0001	SOH	DC1	!	1	A	Q	a	q
0010	STX	DC2	"	2	B	R	b	r
0011	ETX	DC3	#	3	C	S	c	s
0100	EOT	DC4	$	4	D	T	d	t
0101	ENQ	NAK	%	5	E	U	e	u
0110	ACK	SYN	&	6	F	V	f	v
0111	BEL	ETB	'	7	G	W	g	w
1000	BS	CAN	(	8	H	X	h	x
1001	HT	EM	)	9	I	Y	i	y
1010	LF	SUB	*	:	J	Z	j	z
1011	VT	ESC	+	;	K	[	k	{
1100	FF	FS	,	<	L	/	l	/
1101	CR	GS	–	=	M	]	m	}
1110	SO	RS	.	>	N		n	~
1111	SI	US	|	?	O	–	o	DEL

ACK: Acknowledge
BEL: Bell
BS: Back space
CAN: Cancel
CR: Carriage return
DC1: Device control no. 1
DC2: Device control no. 2
DC3: Device control no. 3
DC4: Device control no. 4
DEL: Delete
DLE: Data link escape
ENQ: Enquiry
EM: End of medium
EOT: End of transmission
ESC: Escape
ETB: End of transmission block
ETX: End of text
FF: Form feed
FS: File separator
GS: Group separator
HT: Horizontal tab
LF: Line feed
NAK: Do not acknowledge
NUL: Null
RS: Record separator
SI: Shift in
SO: Shift out
SOH: Start of heading
SPC: Space
STX: Start of text
SUB: Substitute
SYN: Synchronous idle
US: Unit separator
VT: Vertical tab

Table 1

ASCII

American National Standard Code for Information Interchange (ASCII) is a seven-unit digital code for the transmission of teleprinter data. Letters, numerals, symbols, and control operations are represented. ASCII is designed primarily for computer applications, but is also used in some teletypewriter systems.

Each unit is either 0 or 1. In the binary number system, there are 2^7, or 128, possible representations. Table 1 gives the ASCII code symbols for the 128 characters.

The other commonly used teletype code is the Baudot code (*see* BAUDOT). The speed of transmission of ASCII or Baudot is called the baud rate (*see* BAUD RATE). If one unit pulse is s seconds in length, then the baud rate is defined as $1/s$. For example, a baud rate of 100 represents a pulse length of 0.01 second, or 10 ms. The speed of ASCII transmission in words per minute, or WPM (*see* WORDS PER MINUTE) is approximately the same as the baud rate. Commonly used ASCII data rates range from 110 to 19,200 baud, as shown in Table 2.

ASCII: SPEED RATES FOR THE ASCII CODE.

Baud Rate	Length of Pulse, ms	WPM
110	9.09	110
150	6.67	150
300	3.33	300
600	1.67	600
1200	0.833	1200
1800	0.556	1800
2400	0.417	2400
4800	0.208	4800
9600	0.104	9600
19,200	0.052	19,200

Table 2

ASPECT RATIO

The aspect ratio of a rectangular image is the ratio of its width to its height. For television in the United States, the aspect ratio of a picture frame is 4 to 3. That is, the picture is 1.33 times as wide as it is high. This ratio must be maintained in a television receiver or distortion of the picture will result. *See also* TELEVISION.

ASSEMBLER and ASSEMBLY LANGUAGE

An assembler is a computer program designed to produce a machine-language program (*see* MACHINE LANGUAGE). The high-level programming languages generally used by a computer operator, such as BASIC, FORTRAN, and COBOL, are not directly used by the computer; the computer understands only machine language. It would be extremely difficult and tedious to compose an entire computer program in machine language. The assembler serves to translate the higher-order operator language into machine language. This makes computer programming much easier and faster.

An assembler program must be written for the specific purpose of translating a given higher-order language into machine language. Assembly language is used to compose this program. Various higher-order languages require different assembler programs; BASIC is not the same, for example, as FORTRAN. Therefore the assembly language for each kind of higher-order language is unique. *See also* COMPUTER, HIGHER-ORDER LANGUAGE.)

ASSOCIATIVE STORAGE

Data may be stored in memory by assigning it an address (*see* ADDRESS) and recalling the information by calling the appropriate address. Alternatively, data can be stored according to part or all of its actual contents. This latter method of storing information is called associative storage.

Suppose, for example, that a schedule of events, by month, is stored according to the associative method. Then, to get a printout of the schedule for January, the input command might read "JAN," or "JANUARY," or "JAN92." Without associative storage, each month would require a separate address, such as A01 through A12 for a given year.

Associative storage has obvious advantages for situations in which a computer is operated by personnel having little or no training. Associative storage makes a computer seem more "friendly." *See also* MEMORY.

ASTABLE MULTIVIBRATOR

An astable circuit is a form of oscillator. The word astable means unstable. An astable multivibrator consists of two tubes or transistors arranged in such a way that the output of one is fed directly to the input of the other. Two identical resistance-capacitance networks determine the frequency at which oscillation will take place. The amplifying devices are connected in a common-cathode or common-emitter configuration as shown.

In the common-cathode or common-emitter circuit, the output of each tube or transistor is 180 degrees out of phase with the input. An oscillating pulse may begin, for example, at the base of Q1 in the illustration. It is inverted at the collector of Q1, and goes to the base of Q2. It is again inverted at the collector of Q2, and therefore returns to the base of Q1 in its original phase. This produces positive feedback, resulting in sustained oscillation.

The astable multivibrator is frequently used as an audio oscillator, but is not often seen in RF applications because its output is extremely rich in harmonic products. *See also* OSCILLATOR.

ASTIGMATISM

Astigmatism is the condition of an image being in focus in one plane and out of focus in another plane. In optics, this applies to lenses, and astigmatism can occur within the human eye. In electronics, it may apply to the electron beaming a cathode-ray tube (*see* CATHODE-RAY TUBE), or to dish and paraboloid antennas (*see* DISH ANTENNA, PARABOLOID ANTENNA).

With no horizontal or vertical signal input, the electron beam in an oscilloscope cathode-ray tube should come to focus as a sharp round spot (A). This should remain true when the beam is deflected to any part of the display screen. If the beam is out of focus, the spot will appear blurred and larger than normal. A focusing control is provided in most oscilloscopes to ensure that the beam is maintained in the sharpest possible focus.

If the vertical and horizontal focal points differ, however, astigmatism is produced and the beam will land on a spot with distorted shape (B). Focusing then becomes impossible; when the correct adjustment is reached in one plane, it will be incorrect in the other plane. Astigmatism degrades the accuracy of oscilloscope measurements. It causes "hazy" areas on a television picture screen, and blurred regions on a cathode-ray-tube alphanumeric display. Astigmatism can be corrected by careful alignment of the peripheral circuitry of a cathode-ray tube.

In a dish or paraboloid antenna, astigmatism widens the antenna pattern in one plane. This results in less effective concentration of the radiated energy, and therefore less power gain, than the rated values. This condition is caused

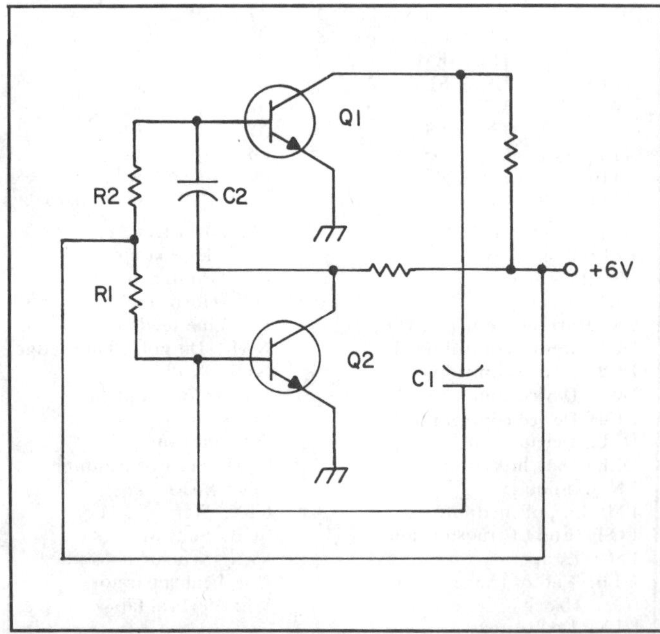

ASTABLE MULTIVIBRATOR: This circuit is a form of oscillator.

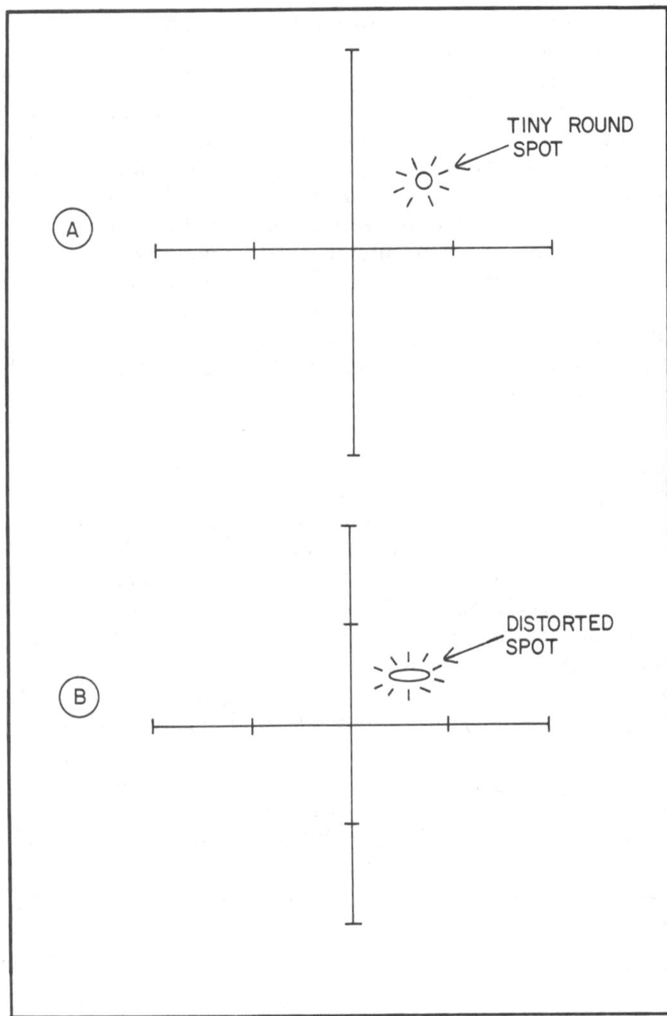

ASTIGMATISM: Astigmatism on a cathode-ray tube. The spot should appear round, as at A. An elongated spot, at B, indicates astigmatism.

by physical distortion of the antenna reflector. *See also* ANTENNA POWER GAIN.

ASYMMETRICAL CONDUCTIVITY

Ordinarily, a conductor carries direct current equally well throughout its cross section. Electrons flow just as easily on one side of a wire or length of tubing as on the opposite side. This is not generally true, however, for the earth. Current often follows an irregular path between two ground electrodes having a potential difference. This is asymmetrical conductivity of the ground.

Whenever an alternating current passes through a conductor such as metal wire or tubing, the flow of current should ideally be symmetrical about the center of the wire or tubing. Although skin effect (*see* SKIN EFFECT) will cause the conductivity at high frequencies to be better at the periphery than at the center, the nonuniformity is symmetrical about the center. If the conductivity is not symmetrical, the resistance will increase. In a carbon-composition resistor, asymmetrical conductivity is especially undesirable because it causes overheating of some parts of the resistor, even when the device is carrying less than its maximum rated current. *See also* CARBON RESISTOR.

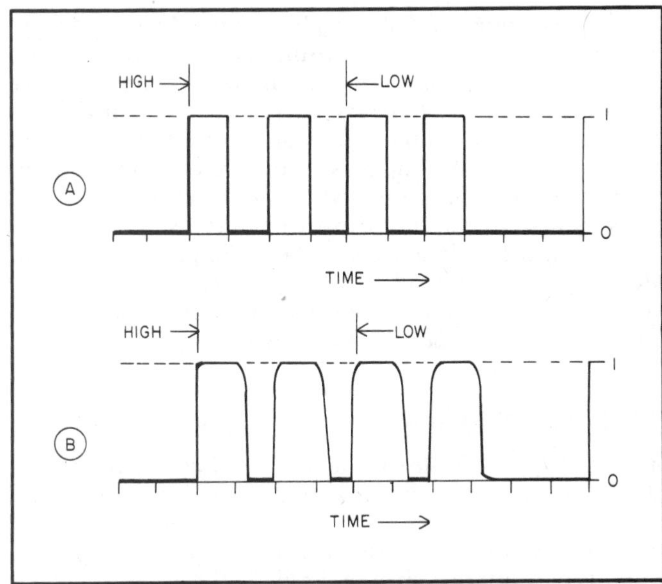

ASYMMETRICAL DISTORTION: Asymmetrical distortion. At A, the Morse letter H as it appears at the output of an electronic keyer. At B, the shaping circuits of the transmitter change the dot-to-space ratio.

ASYMMETRICAL DISTORTION

In a binary system of modulation, the high and low conditions (or 1 and 0 states) have defined lengths for each bit of information. The modulation is said to be distorted when these bits are not set to the proper duration. If the output bits of one state are too long or too short, compared with the signal input bits, the distortion is said to be asymmetrical.

A simple example of asymmetrical distortion often is found in a Morse-code signal. Morse code is a binary modulation system, with bit lengths corresponding to the duration of one dot. Ideally, a string of dots (such as the letter H in A and B) has high and low states of precisely equal length. An electronic keyer can produce signals of this nature, as illustrated in A. However, because of the shaping network in a CW transmitter (*see* SHAPING), the high state is often effectively prolonged, as shown in B, because the decay time is lengthened. While the rise, or attack, time is made slower by the shaping network, the change is much greater on the decay side in most cases. This creates asymmetrical distortion, since the dot-to-space ratio, 1-to-1 at the input, is greater than 1-to-1 at the transmitter output.

Asymmetrical distortion makes reception of a binary signal difficult, and less accurate, than would be the case for an undistorted signal. The effect may not be objectionable if it is small; in CW communications, a certain amount of shaping makes a signal more pleasant to the ear. Excessive asymmetrical distortion should be avoided. In teleprinter communications, asymmetrical distortion causes frequent printing errors. An oscilloscope can be used to check for proper adjustment of the high-to-low ratio.

ASYMMETRICAL SIDEBAND

See VESTIGIAL SIDEBAND.

ASYNCHRONOUS DATA

Asynchronous data is information not based on a defined time scale. An example of asynchronous data is manually sent Morse code, or CW. Machine-sent Morse code, in contrast, is synchronous. Synchronous transmission offers a better signal-to-noise ratio in communications systems among machines than asynchronous transmission. Nevertheless, the simple combination of a hand key and human ear for CW—the most primitive communications system—is commonly used when more sophisticated arrangements fail because of poor conditions!

Asynchronous data need not be as simple as hand-sent Morse code. A manually operated teletypewriter station and a voice system are other examples of asynchronous data transfer. *See also* SYNCHRONOUS DATA.

ATMOSPHERE

The atmosphere is the shroud of gases that surrounds our planet. Many other planets also have atmospheres; some do not.

Our atmosphere exerts an average pressure of 14.7 pounds per square inch at sea level. As the elevation above sea level increases, the pressure of the atmosphere drops, until it is practically zero at an altitude of 100 miles. Effects of the atmosphere, however, extend to altitudes of several hundred miles.

Scientists define our atmosphere in terms of three layers as shown. The lowest layer, the troposphere, is where all weather disturbances take place. It extends to a height of approximately 8 to 10 miles above sea level. The troposphere affects certain radio-frequency electromagnetic waves (*see* DUCT EFFECT, TROPOSPHERIC PROPAGATION). The stratosphere begins at the top of the troposphere and extends up to about 40 miles. No weather is ever seen in this layer, although circulation does occur. At altitudes from 40 to about 250 miles above the ground, several ionized layers of low-density gas are found. This region is known as the ionosphere. The layers of the ionosphere have a tremendous impact on the propagation of RF energy from dc into the vhf region. Without the ionosphere, radio communication as we know it would be much different. The long-distance shortwave propagation that we take for granted would not exist. *See also* D LAYER, E LAYER, F LAYER, IONOSPHERE, PROPAGATION, PROPAGATION CHARACTERISTICS.

ATOM

Elements can be broken down into smaller and smaller pieces, but there is a limit to how far this process can be continued before the identity of the element is lost. The smallest particle of any element is the atom.

Atoms are composed of positively charged, extremely dense centers, called nuclei, and negatively charged, orbiting electrons. The total positive charge in the nucleus of an atom is normally the same as the total negative charge of the surrounding electrons. The nucleus is made up of positively charged particles called protons, and electrically neutral particles called neutrons. One proton carries the same charge as one electron, but of opposite polarity. If

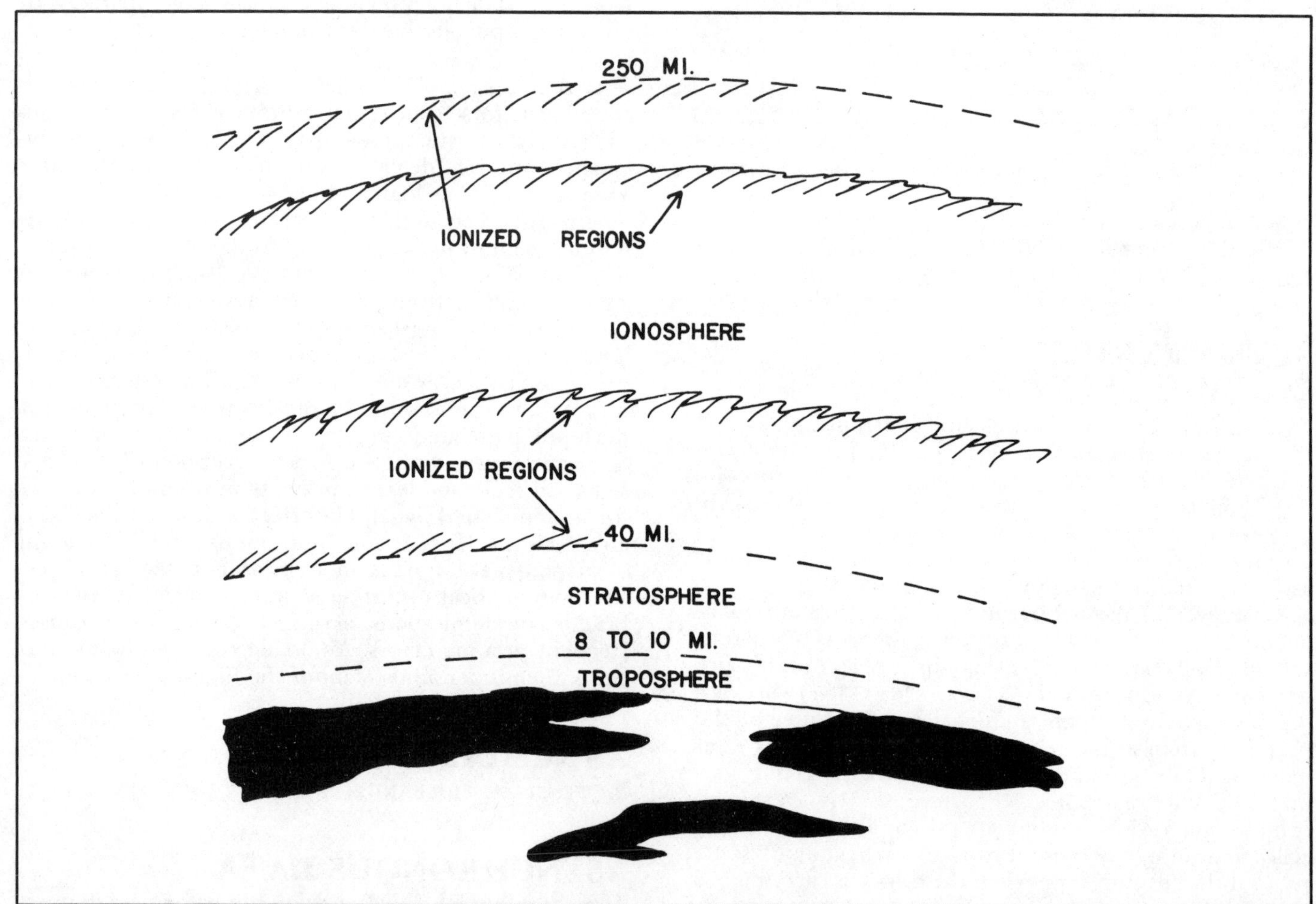

ATMOSPHERE: The atmosphere extends from the surface of the earth to approximately 250 miles altitude.

some electrons are added or taken away from an atom, the atom becomes an ion (*see* ION).

Electrons orbit the nucleus of an atom with great speed. They move so rapidly that, if an atom could be magnified to visible proportions, the electrons would appear as concentric spherical shells around the nuileus (see figure). The innermost shell of an atom can hold one or two electrons; it is called the K shell. The second shell can hold up to eight electrons, and is called the L shell. The third shell can hold up to 18 electrons, and is called the M shell. Heavy atoms may have more than 100 protons and 100 electrons, with several concentric shells. In general, the Nth shell of electrons in an atom can have as few as one, or as many as $2N^2$, electrons.

The negatively charged electron shells of adjacent atoms repel each other. This is why a chair, for example, does not merge with or sink into the floor. The repulsive force among atoms keeps objects from gravitationally collapsing upon themselves. *See also* ELECTRON, NEUTRON, PROTON.

ATOMIC CHARGE

When an atom contains more or less electrons than normal, it is called an ion (*see* ION). The atomic charge of a normal atom is zero. The atomic charge of an ion is positive if there is a shortage of electrons, and negative if there is a surplus of electrons. Some atoms ionize quite easily; others do not readily ionize.

The unit of atomic charge is the amount of electric charge carried by a single electron or proton; they carry equal but opposite charges. This is called an electron unit. One coulomb is a charge of 6.28×10^{18} electron units. Therefore, an electron unit is 1.59×10^{-19} coulomb (*see* COULOMB). An ion might have an atomic charge of +3 electron units or −2 electron units. The first case would indicate a deficiency of three electrons; the second case, an excess of two electrons. In general, the atomic charge number is always an integer, since the electron unit is the smallest possible quantity of charge. *See also* ELECTRON.

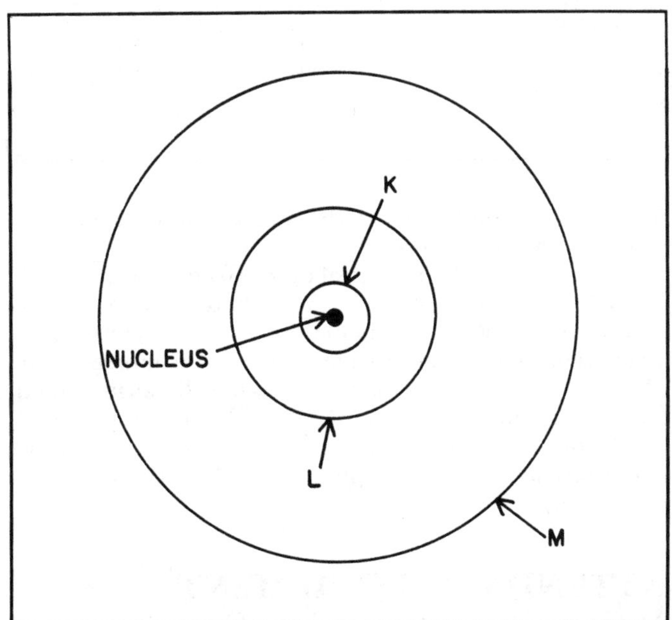

ATOM: Enlarged many times, an atom might look like this. The electrons orbit in spherical shells; here, the first three shells (K, L, M) are shown.

ATOMIC CLOCK

Time standards throughout the world today are based on atomic clocks. Such clocks also function as frequency standards. In the United States, time and frequency are broadcast by WWV in Fort Collins, Colorado, and by WWVH on the island of Kauai, Hawaii, by the National Bureau of Standards (*see* NATIONAL BUREAU OF STANDARDS, WWV/WWVH).

The element cesium is a frequently used atomic standard. The clock functions according to the resonant frequency of the vibration of this element. Such clocks are accurate to three parts in 10^{12}, or 3×10^{-10} percent, per year. Other atomic time-standard sources include the hydrogen maser and the rubidium gas cell.

Atomic clocks are generally set according to the mean solar day. This day is divided into 24 equal hours. Corrections are made periodically to compensate for irregularities in the rotational speed of the earth. *See also* COORDINATED UNIVERSAL TIME.

ATOMIC MASS UNIT

An atomic mass unit, or AMU, is defined as precisely 1/16 the mass of a neutral atom of O16, the most common isotope of oxygen (*see* ISOTOPE). One AMU is 1.66×10^{-24} gram, or approximately the mass of a single neutron or proton.

Atomic weights are expressed in atomic-mass units. The lightest element, hydrogen, has an atomic weight of 1 AMU. Lawrencium has an atomic weight of 257 AMU. *See also* ATOMIC WEIGHT.

ATOMIC NUMBER

The atomic number of an element is the number of protons in the nucleus of a single atom of that element. Hydrogen has a nucleus consisting of one proton alone, and thus has an atomic number of 1. Helium has a nucleus made of up of two protons and two neutrons; its atomic number is 2. There are elements corresponding to every possible atomic number up to over 100. The atom of uranium, with 92 protons, has the largest atomic number of any naturally occurring element. Man-made elements have greater atomic numbers.

The table lists atomic numbers of elements from 1 to 103. Atomic weights (*see* ATOMIC WEIGHT) are also given.

ATOMIC WEIGHT

The atomic weight of an element is the mass of that element in atomic mass units (*see* ATOMIC MASS UNIT). The atomic weight is approximately equal to the total number of protons and neutrons in the nucleus of an atom. An element of a given atomic number may have several different possible atomic weights, depending on the number of neutrons. These variations in the structure of the nucleus of an element are known as isotopes. For example, an atom of carbon (atomic number 6) usually has six neutrons in its nucleus, giving it an atomic weight of 12. However, some atoms of carbon have eight neutrons instead of six; this isotope is called carbon 14.

As the atomic number increases, the atomic weight generally increases as well. However, there are some instances in which atoms of higher atomic number have smaller atomic weight.

ATOMIC NUMBER, ATOMIC WEIGHT: ALPHABETICAL LIST OF ELEMENTS, WITH ATOMIC NUMBERS AND ATOMIC WEIGHTS. (ATOMIC WEIGHTS ARE ROUNDED TO NEAREST WHOLE.)

Element Name	Atomic Number	Atomic Weight	Element Name	Atomic Number	Atomic Weight
Actinium	89	227	Molybdenum	42	96
Aluminum	13	27	Neodymium	60	144
Americium	95	243	Neon	10	20
Antimony	51	122	Neptunium	93	237
Argon	18	40	Nickel	28	59
Arsenic	33	75	Niobium	41	93
Astatine	85	210	Nitrogen	7	14
Barium	56	137	Nobelium	102	254
Berkelium	97	247	Osmium	76	190
Beryllium	4	9	Oxygen	8	16
Bismuth	83	209	Palladium	46	106
Boron	5	11	Phosphorus	15	31
Bromine	35	80	Platinum	78	195
Cadmium	48	112	Plutonium	94	242
Calcium	20	40	Polonium	84	210
Californium	98	251	Potassium	19	39
Carbon	6	12	Praseodymium	59	141
Cerium	58	140	Promethium	61	145
Cesium	55	133	Proactinium	91	231
Chlorine	17	35	Radium	88	226
Chromium	24	52	Radon	86	222
Cobalt	27	59	Rhenium	75	186
Copper	29	64	Rhodium	45	103
Curium	96	247	Rubidium	37	85
Dysprosium	66	163	Ruthenium	44	101
Einsteinium	99	254	Samarium	62	150
Erbium	68	167	Scandium	21	45
Europium	63	152	Selenium	34	79
Fermium	100	257	Silicon	14	28
Fluorine	9	19	Silver	47	108
Francium	87	223	Sodium	11	23
Gadolinium	64	157	Strontium	38	88
Gallium	31	70	Sulfur	16	32
Germanium	32	73	Tantalum	73	181
Gold	79	197	Technetium	43	99
Hafnium	72	178	Tellurium	52	128
Helium	2	4	Terbium	65	159
Holmium	67	165	Thallium	81	204
Hydrogen	1	1	Thorium	90	232
Indium	49	115	Thulium	69	169
Iron	26	56	Tin	50	119
Krypton	36	84	Titanium	22	48
Lanthanum	57	139	Tungsten	74	184
Lawrencium	103	257	Uranium	92	238
Lead	82	207	Vanadium	23	51
Lithium	3	7	Xenon	54	131
Lutetium	71	175	Ytterbium	70	173
Magnesium	12	24	Yttrium	39	89
Manganese	25	55	Zinc	30	65
Mendelevium	101	256	Zirconium	40	91
Mercury	80	201			

An electron has almost no mass—about 5×10^{-4} AMU—and therefore electrons affect the atomic weight very little. The table in ATOMIC NUMBER shows the atomic weights of the most common isotopes of the elements having atomic numbers from 1 to 103. The atomic weights in the table have been rounded off to the nearest whole number.

ATTACK

The rise time for a pulse is sometimes called the attack or attack time. In music, the attack time of a note is the time required for the note to rise from zero amplitude to full loudness. The attack time for a control system, such as an automatic gain control (*see* AUTOMATIC GAIN CONTROL, AUTOMATIC LEVEL CONTROL), is the time needed for that system to fully compensate for a change in input parameters. The graphics show the attack time for a musical tone (A) and a dc pulse (B).

The attack time of a musical note affects its sound quality. A fast rise time sounds "hard;" and a slow rise time sounds "soft." With automatic gain or level control, an attack time that is too fast may cause overcompensation, while an attack time that is too slow will cause a loud popping sound at the beginning of each pulse or syllable.

The time required for a note or pulse to drop from full intensity back to zero amplitude is called the decay or release time. *See also* DECAY.

ATTENDANT EQUIPMENT

Attendant equipment consists of all accessory, or peripheral, apparatus needed for the operation of a device. In a television studio, for example, attendant equip-

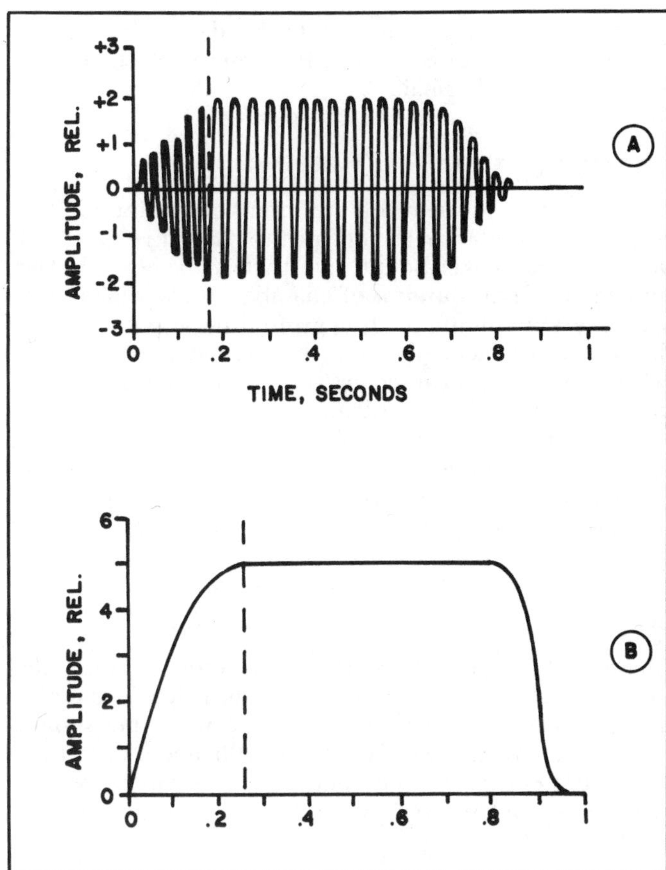

ATTACK: At A, of a musical note; at B, of a dc pulse.

ment includes such things as lights, props, background scenes, and remote cameras.

In a telephone switching system, an operator in the central office (*see* CENTRAL OFFICE SWITCHING SYSTEM) can receive or send a call, connect one line to another, and monitor or cut in on calls, using attendant equipment.

ATTENUATION

Attenuation is the decrease in amplitude of a signal between any two points in a circuit. It is usually expressed in decibels (*see* DECIBEL). Attenuation is the opposite of amplification (*see* AMPLIFICATION), and may be defined for voltage, current, or power. Occasionally the attenuation in a particular circuit is expressed as a ratio. For example, the reduction of a signal in amplitude from ±5 volts peak to ±1 volt peak is an attenuation factor of 5.

In general, the attenuation in voltage for an input E_{IN} and an output E_{OUT} is

$$\text{Attenuation (dB)} = 20 \log_{10} (E_{IN}/E_{OUT})$$

Similarly, the current attenuation for an input of I_{IN} and an output of I_{OUT} is

$$\text{Attenuation (dB)} = 20 \log_{10} (I_{IN}/I_{OUT})$$

For power, given an input of P_{IN} watts and an output of P_{OUT} watts, the attenuation in decibels is given by

$$\text{Attenuation (dB)} = 10 \log_{10} (P_{IN}/P_{OUT})$$

If the amplification factor (*see* AMPLIFICATION FACTOR) is X dB, then the attenuation is −X dB. That is, positive attenuation is the same as negative amplification, and negative attenuation is the same as positive amplification. *See also* ATTENUATOR.

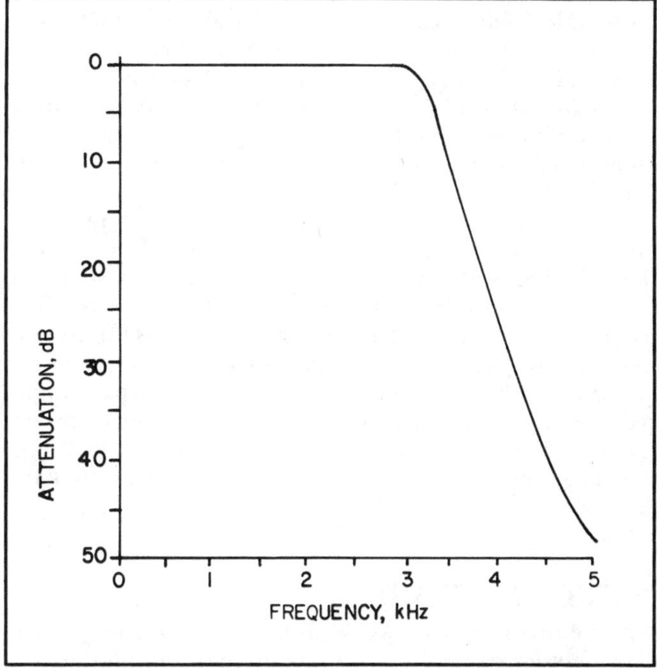

ATTENUATION: Attenuation vs frequency characteristic of a lowpass filter that might be used for communications purposes.

ATTENUATION DISTORTION

Attenuation distortion is an undesirable attenuation characteristic over a particular range of frequencies (*see* ATTENUATION VS FREQUENCY CHARACTERISTIC). This can occur in radio-frequency as well as audio-frequency applications. The lowpass response of the illustration is an advantage for a voice communications circuit, since most of the frequencies in the human voice fall below 3 kHz. However, for the transmission of music, the curve in the illustration would represent a circuit with objectionable attenuation distortion, because music contains audio frequencies as high as 20 kHz or more.

In sound systems, attenuation distortion depends to some extent on individual preferences. An equalizer (*see* EQUALIZER) is sometimes used to eliminate the attenuation distortion for a particular listener or application.

Attenuation distortion in a sound system can be the result of poor circuit design, use of improper speakers or headsets, use of poor-quality tape for recording, or excessively slow tape speed, and many other factors. *See also* HIGH FIDELITY.

ATTENUATION EQUALIZER

See EQUALIZER.

ATTENUATION VS FREQUENCY CHARACTERISTIC

The attenuation-vs-frequency characteristic of a circuit is the amount of loss through the circuit as a function of frequency. This function is generally shown with the amplitude, in decibels relative to a certain reference level, on the vertical scale and the frequency on the horizontal scale. The illustration in ATTENUATION DISTORTION shows the attenuation-vs-frequency characteristic curve for an audio lowpass filter. This filter is designed for use as an intelligibility enhancer for a voice-communications circuit.

Various devices are used for precise adjustment of the attenuation-vs-frequency characteristic in different situations. In high-fidelity recording equipment, the familiar equalizer or graphic equalizer (*see* EQUALIZER) is often used in place of simple bass and treble controls for obtaining exactly the desired audio response. In some RF amplifiers, it is advantageous to introduce a lowpass, highpass, or bandpass response to prevent oscillation or excessive radiation of unwanted energy. *See also* BANDPASS FILTER, BAND-REJECTION FILTER, HIGHPASS FILTER, LOWPASS FILTER.

ATTENUATOR

An attenuator is a network, usually passive rather than active, designed to cause a reduction in the amplitude of a signal. Such circuits are useful for sensitivity measurements and other calibration purposes. In a well-designed attenuator, the amount of attenuation is constant over the entire range of frequencies in the system to be checked. An attenuator introduces no reactance, and therefore no phase shift. Attenuators are built for a wide variety of input and output impedances. It is important that the impedances be properly matched, or the attenuator will not function properly.

Two simple passive attenuators made from noninductive resistors are shown in A and B. The circuit at A is called a pi-network attenuator, and the circuit at B is called a T-network attenuator. These circuits will function from the audio-frequency range well into the vhf spectrum. At uhf and above, however, the resistors begin to show inductive reactance because the wavelength is so short that the leads are quite long electrically. Then the attenuators will no longer perform their intended function.

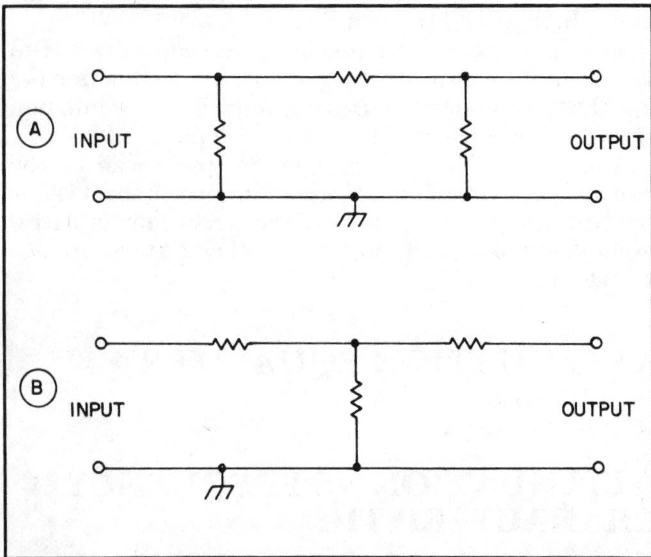

ATTENUATOR: Attenuators are often constructed from noninductive resistors. Here, two such devices are shown; at A, the pi network and at B, the T network. These attenuators are for use in unbalanced circuits.

Attenuators for uhf must be designed especially to suit the short wavelengths at those frequencies. Such circuits must be physically small.

AUDIBILITY

Audibility is a measure of the intensity of sound. It is expressed in decibels relative to the threshold of hearing—the weakest sound level that can be heard by the human ear. In a soundproof chamber, the audibility is 0 dB. A certain amount of background noise is created by the motion of blood through the capillaries of the ear. Some noise is also caused by the random motion of air molecules around the eardrum.

The table shows the audibility levels, in decibels, for some familiar sounds. The threshold of hearing for an average person is considered to be 10^{-16} watt per square centimeter. The audibility is given as

$$\begin{aligned}\text{Audibility (dB)} &= 10 \log_{10} (X/10^{-16})\\ &= 160 + 10 \log_{10} X,\end{aligned}$$

where X is the intensity of the given sound in watts per square centimeter. At 110 dB, tingling is felt in the ear. This is a level of 10^{-5} watt, or 10 microwatts, per square centimeter. At 120 to 130 dB, or 100 microwatts to 1 milliwatt per square centimeter, pain occurs. *See also* LOUDNESS, VOLUME, VOLUME UNIT.

AUDIO AMPLIFIER

An audio amplifier is an active device designed especially for the amplification of signals in the audio-frequency range.

Some audio amplifiers, such as microphone preamplifiers and radio-receiver audio stages, must work with very small signal input. They need not generate much output power. Other audio amplifiers, such as those found in high-fidelity or public-address systems, must develop large amounts of power output, sometimes thousands of watts.

In communications systems, the frequency response of an audio amplifier is restricted to a relatively narrow range. But in a high-fidelity system, a flat response is desired from perhaps 10 Hz to well above the range of human hearing.

Most audio amplifiers use transistors, although some use integrated-circuit operational amplifiers. It is rare to

AUDIBILITY: AUDIBILITY OF VARIOUS NOISES IN DECIBELS ABOVE THE THRESHOLD OF HEARING.

Sound	**Audibility, dB (approx.)**
Threshold of hearing	0
Whisper	10-20
Electric fan at 10 feet	30-40
Running water at 10 feet	40-60
Speech at 5 feet	60-70
Vacuum cleaner at 10 feet	70-80
Passing train at 50 feet	80-90
Jet at 1000 feet altitude	90-100
Loud discotheque	110-120
Air hammer at 5 feet	130-140

see vacuum tubes in an audio amplifier. *See also* AUDIO POWER, AUDIO RESPONSE, HIGH FIDELITY, OPERATIONAL AMPLIFIER.

AUDIO FREQUENCY

Alternating current in the range of approximately 10 to 20,000 Hz is called audio-frequency, or af, current. When passed through a transducer such as a speaker or headset, these currents produce audible sounds. A young person with excellent hearing can generally detect sine-wave tones from 10 to 20,000 Hz. An older person loses sensitivity to the higher frequencies and to the extremely low frequencies. By age 50 or 60, the limits of the hearing range are usually about 40 Hz to 10,000 Hz.

The range of audio frequencies is called the af spectrum. The drawing shows the af spectrum on a logarithmic scale from 10 Hz to 20,000 Hz. All sounds, from the simple tone of a sine-wave audio oscillator to the complex noises of speech, are combinations of frequencies in this range.

Radio communications allocations begin at 10 kHz, actually within the audio-frequency range. It is in fact possible at times to hear signals below 20 kHz by simply connecting a sensitive audio amplifier to an antenna.

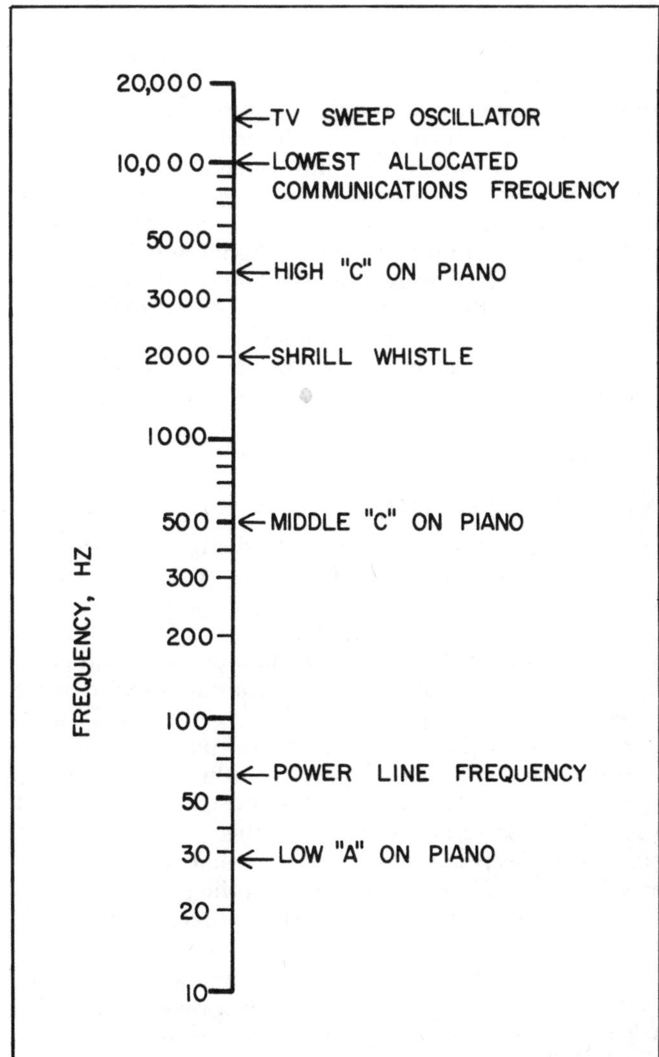

AUDIO FREQUENCY: The audio-frequency spectrum ranges from about 10 Hertz to 20 kHz.

AUDIO-FREQUENCY TRANSFORMER

An audio-frequency transformer is used for the purpose of matching impedances at audio frequencies. The output of an audio amplifier might have an impedance of 200 ohms, and the speaker or headset an impedance of only 8 ohms. The af transformer provides the proper termination for the amplifier. This assures the most efficient possible transfer of power.

The impedance-matching ratio of a transformer is proportional to the square of the turns ratio. Thus, if the primary winding has N_{PRI} turns and the secondary winding has N_{SEC} turns, the ratio of the primary impedance to the secondary impedance, Z_{PRI} to Z_{SEC}, is

$$Z_{PRI}/Z_{SEC} = (N_{PRI}/N_{SEC})^2$$

Also,

$$N_{PRI}/N_{SEC} = \sqrt{Z_{PRI}/Z_{SEC}}$$

Audio-frequency transformers are available in various power ratings and impedance-matching ratios. There are also different kinds of audio-frequency transformers to meet various frequency-response requirements. Such transformers are physically similar to ordinary alternating-current power transformers (see the illustration). They are wound on laminated or powdered-iron cores. *See also* TRANSFORMER.

AUDIO IMAGE

An audio image is a sound that comes from, or appears to come from, a certain point in space. Audio images are created by high-fidelity stereo systems because of the relative intensity and phase of the sound in the channels. By proper synthesis of amplitude and phase in the left-hand and right-hand channels, such sounds as a busy street, a passing train, or a concert orchestra are faithfully reproduced in a pair of speakers or stereo headphones.

Audio images always have an apparent distance and an apparent direction as shown. Sometimes there is difficulty in determining the distance to a sound image; a loud sound that is far away sounds very similar to a weaker sound nearby. There may be ambiguity in direction if two sounds from different places arrive at the ears with the same relative phase and amplitude. It is very hard, for example, to distinguish (with the eyes closed) among voices directly

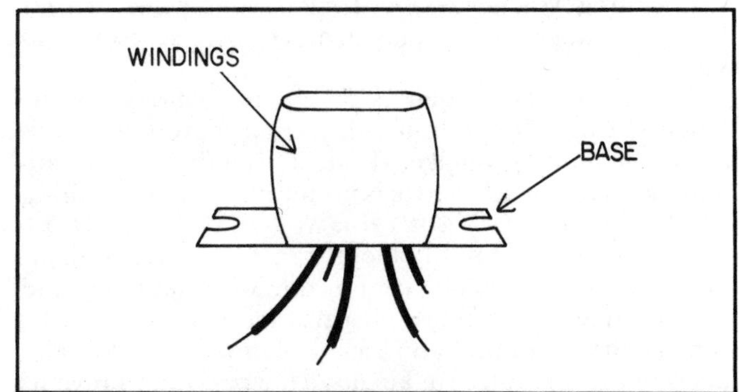

AUDIO-FREQUENCY TRANSFORMER: An audio-frequency transformer looks very much like an ac transformer.

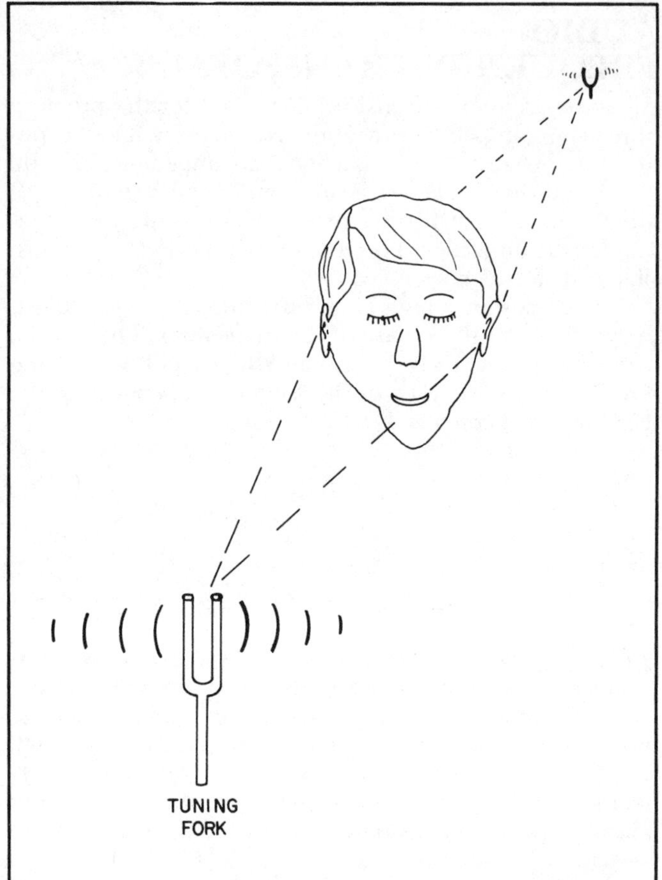

AUDIO IMAGE: An audio image has an apparent distance and an apparent direction. The apparent distance depends on the loudness of the sound, and the apparent direction depends on the relative phase of the sound at each ear.

in front, behind, above and below. All these directions are equidistant from both ears, and so the sound arrives in the same phase and at the same amplitude in all cases. *See also* STEREOPHONICS.

AUDIO LIMITER

An audio limiter is a device that prevents the amplitude of an audio-frequency signal from exceeding a certain value. All audio-frequency voltages below the limiting value are not affected. A limiter may be either active or passive.

The illustration shows a simple diode limiter. If germanium diodes are employed, the limiting voltage is approximately ±0.3 volts peak. If silicon diodes are used, the limiting voltage is about ±0.6 volts peak. Larger limiting voltages are obtained by connecting diodes in series.

Diode limiters, as well as most other kinds of audio limiters, cause considerable distortion at volume levels above the limiting voltage. If fidelity is important, an automatic gain or level control is preferable to audio limiting (*see* AUTOMATIC GAIN CONTROL, AUTOMATIC LEVEL CONTROL). In radio communications transmitters, audio limiters are sometimes used in conjunction with amplifiers and lowpass filters to increase the average talk power. The signal is first amplified, and then the limiter is employed to prevent overmodulation. Finally, a lowpass filter prevents excessive distortion products from being transmitted over the air. This is called speech clipping (*see* SPEECH CLIPPING). Under poor conditions, speech clipping can result in improved intelligibility and articulation (*see* ARTICULATION, INTELLIGIBILITY).

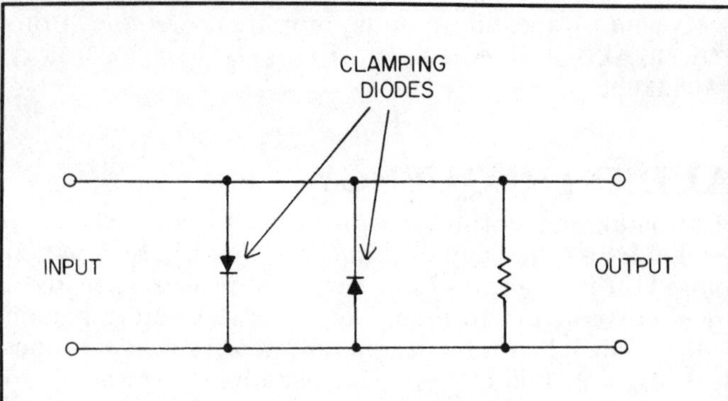

AUDIO LIMITER: A simple passive audio limiter might consists of two semiconductor diodes in reverse parallel.

AUDIO MIXER

An audio mixer is a circuit that combines two or more audio signals. It has an input terminal, or connection, for each input source. It also has an output signal terminal. Level controls are provided indepently for each input, and also for the output.

Combining audio signals is not a simple matter of joining wires. Unless the proper steps are taken to avoid interaction among the input signal sources, distortion and poor efficiency will result. The impedances at the input terminals of an audio mixer must remain constant, and at the required values, as the level controls are adjusted. The impedances must also remain constant at all frequencies and under conditions of varying signal amplitude. The output impedance must also remain constant. This requires careful design. Nonlinearity of the components must be minimized. The frequency response, or attenuation -vs-frequency characteristic, must be as flat as possible. Phase distortion must also be avoided. *See also* ATTENUATION VS FREQUENCY CHARACTERISTIC.

AUDIO POWER

Audio power is the power, in watts, dissipated in the output load of an audio amplifier. If the speakers or headphones of an audio system were replaced by nonreactive resistors of an identical impedance, all the audio power would be spent as heat. It could then be easily and accurately measured. Voltmeters are generally used in audio systems to measure power. They are calibrated in watts according to the impedance of the output.

Audio power can be expressed either as root-mean-square, or RMS, watts (*see* ROOT MEAN SQUARE) or as peak watts. For a pure sine-wave tone, the RMS power is 50 percent of the peak power. For a human voice, the RMS power is generally 15 to 25 percent of the peak power. For musical instruments, the ratio varies between about 20 and 50 percent.

Audio power ratings are important when choosing speakers for an audio system, since excessive audio power will damage a speaker. Some speakers are more efficient in converting audio-frequency currents into sound energy than others. The RMS power rating of an audio amplifier or speaker is the most common way of specifying its power

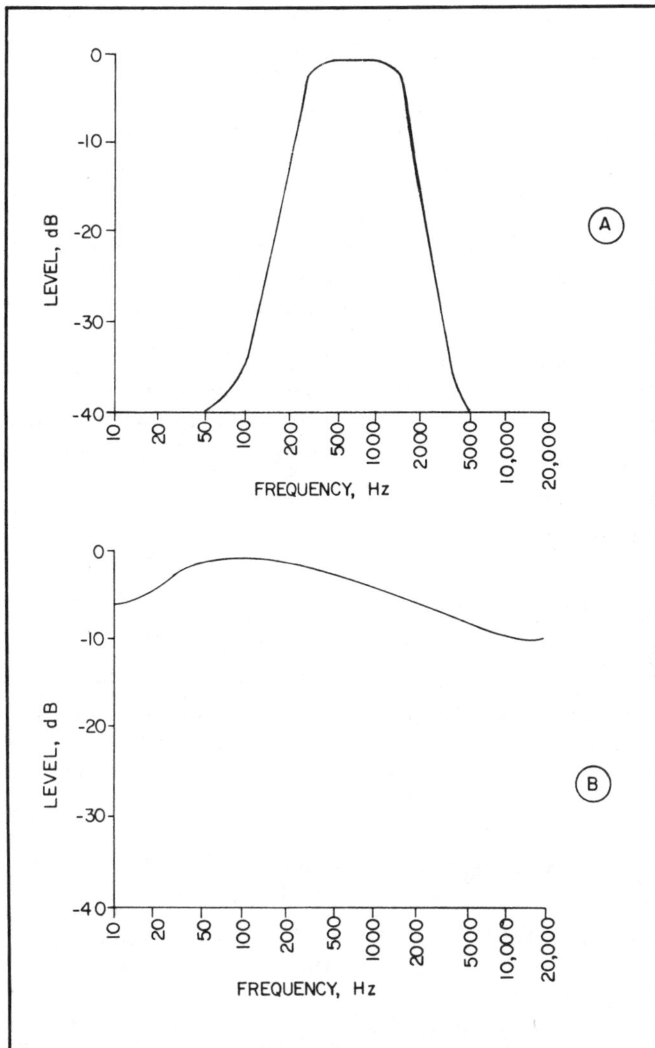

AUDIO RESPONSE: Two audio response curves. The response at A is ideal for voice communications, where fidelity is not as important as intelligibility. The response at B is better for reproducing music, where the range of frequencies is large.

capabilities, since RMS power is easier to measure than peak power. *See also* AUDIO AMPLIFIER, SPEAKER.

AUDIO RESPONSE

The audio response of a device is its sensitivity at all points in the audio-frequency (af) spectrum, expressed as a graph. Audio response is defined for devices such as speakers, headphones, and microphones. Sometimes an audio-response evaluation is performed on an entire af circuit, such as a tape recorder.

Whatever the device under scrutiny, the audio response is tailored to the intended application. For a voice communications system, the audio response of a microphone or speaker should be similar to that shown in A. In a high-fidelity application, however, this response would sound horrible! The more nearly flat response of B is essential for the faithful reproduction of music.

In filtering networks with no gain, the frequency response is sometimes called the attenuation-vs-frequency characteristic (*see* ATTENUATION VS FREQUENCY CHARACTERISTIC).

The audio response of a communications system is usually fixed. In a high-fidelity system, however, it is variable. A bass-treble, or tone, control may consist of one or two potentiometers to vary the low-frequency and high-frequency amplification. A more precise adjustment of the audio response in a high-fidelity system is accomplished by means of a device called an equalizer. This circuit allows control of several audio-frequency ranges independently. *See also* BASS, EQUALIZER, TONE CONTROL, TREBLE.

AUDIO SIGNAL GENERATOR

An audio signal generator, sometimes also called an audio function generator, is used in the testing, troubleshooting, and adjustment of audio-frequency circuits.

Modern audio signal generators can produce a variety of waveforms. Sine and square waves are the most common, but sawtooth and triangular waves can also be obtained with many of the devices (*see* SAWTOOTH WAVE, SINE WAVE, SQUARE WAVE, TRIANGULAR WAVE). The frequency range may extend well above and below the human hearing range. A typical audio signal generator costs a few hundred dollars.

By using an audio signal generator with output voltage independent of the frequency, audio-frequency response tests can be performed (*see* AUDIO RESPONSE). A special voltmeter for audio-frequency amplitude measurement, or an oscilloscope, is used to check the output. As the frequency is varied, the output amplitude can be plotted on a graph. The resulting curve clearly shows the audio response of the device under test.

A specialized form of audio signal generator/voltmeter is used to measure the amount of distortion produced in an audio amplifier. A pure sine-wave tone is supplied at the input of the amplifier and is notched out at the output, leaving only the harmonic components. The combined amplitude of these components is then measured. This kind of test instrument is called a distortion analyzer (*see* DISTORTION ANALYZER).

The most sophisticated instruments can generate many different waveforms, simulating various musical instruments. The Moog systhesizer is an example of such a device (*see* MOOG SYNTHESIZER). This type of audio signal generator is not normally used as a test instrument, but for making electronic music.

AUDIO TAPE

See RECORDING TAPE.

AUDIO TAPER

The human ear does not perceive sound intensity in a linear fashion. Instead, the apparent intensity, or audibility, of a sound increases according to the logarithm of the actual amplitude in watts per square centimeter. An increase of approximately 26 percent is the smallest detectable change for the human ear, and is called the decibel, abbreviated dB (*see* AUDIBILITY, DECIBEL). A threefold increase in sound power is 5 dB; a tenfold increase is 10 dB, and a hundredfold increase represents 20 dB.

The audio-taper volume control compensates for the nonlinearity of the human hearing apparatus. A special kind of potentiometer is used to give the impression of a linear increase in audibility as the volume is raised. Most radio receivers, tape players, and high-fidelity systems use audio-taper potentiometers in their af amplifier circuits. This kind of potentiometer is sometimes also called a logarithmic-taper, or log-taper, device.

For linear applications, a linear-taper potentiometer is needed. Use of a linear-taper potentiometer is undesirable in a nonlinear situation. The converse is also true; an audio-taper device does not function well in a linear circuit. *See also* LINEAR TAPER.

AUDIO-VISUAL COMMUNICATION

An audio-visual communication device uses both sight and sound to convey information. The most common example of this is ordinary television.

In an audio-visual communications system, it is important that the visible image be synchronized with the sound. This usually presents no problem unless the information is recorded.

A form of audio-visual communication requiring very little spectrum space is the slow-scan television (*see* SLOW-SCAN TELEVISION) and single-sideband (*see* SINGLE SIDEBAND) system. The picture is sent via one sideband, at the rate of one still frame every seven seconds; the voice is sent over the other sideband. This requires only about 6 kHz of spectrum, or the space taken by an average AM voice broadcast signal.

Audio-visual telephones are already available. The main problem with them is the amount of audio spectrum required for transmission of a detailed motion picture. Another problem is the fact that not everyone feels comfortable being seen as well as heard on the telephone! Another problem is cost. In time, technology should be able to solve some of these difficulties.

AURORA

A sudden eruption on the sun, called a solar flare, causes high-speed atomic particles to be thrown far into space from the surface of the sun. Some of these particles are drawn to the upper atmosphere over the north and south magnetic poles of our planet. This results in dramatic ionization of the thin atmosphere over the Arctic and Antarctic. On a moonless night, the glow of the ionized gas can be seen as the Aurora Borealis (Northern Lights) and Aurora Australis (Southern Lights).

The aurora are important in radio communication, because they affect the propagation of electromagnetic waves in the high-frequency and very-high-frequency ranges. Signals may be returned to earth at much higher frequencies than usual when the aurora are active. The unusual solar activity that causes the aurora, however, also can degrade normal ionospheric propagation. Sometimes the disturbance is so pronounced that even wire circuits are disrupted by the fluctuating magnetic fields. *See also* AURORAL PROPAGATION, IONOSPHERE, PROPAGATION, PROPAGATION CHARACTERISTICS.

AURORAL PROPAGATION

In the presence of unusual solar activity, the aurora (*see* AURORA) often return electromagnetic energy to earth. This is called auroral propagation. See illustration.

The aurora occur in the ionosphere, at altitudes of about 40 to 250 miles above the surface of the earth (*see* IONOSPHERE). Theoretically, auroral propagation is possible, when the aurora are active, between any two points on the ground from which the same part of the aurora can be seen. Auroral propagation seldom occurs when one end of the circuit is at a latitude less than 35 degrees north or south of the equator.

Auroral propagation is characterized by a rapid flutter, which makes voice signals such as AM, FM, or SSB unreadable most of the time. Morse code is generally most effective in auroral propagation, and even this mode may sometimes be heard but not understood because of the extreme distortion. The flutter is the result of severe selective fading and multipath distortion (*see* SELECTIVE FADING, MULTIPATH FADING). Auroral propagation can take place well into the very-high-frequency region, and is usually accompanied by moderate to severe deterioration in ionospheric conditions of the E and F layers. *See also* E LAYER, F LAYER, PROPAGATION, PROPAGATION CHARACTERISTICS.

AUTOALARM

An autoalarm is a receiving device tuned to 500 kHz, the international distress frequency. Whenever a signal is received on this frequency, the autoalarm is activated. The alarm produces an attention signal, and then the receiver is switched to a speaker so that the signal can be heard.

AUTODYNE

An autodyne is a combined oscillator and mixer using a vacuum tube. The oscillator may be crystal-controlled or variable in frequency. See the simple autodyne circuit diagram.

The cathode, control grid, and plate form the oscillator, in this case a Hartley type. The signal to be mixed with the oscillator frequency is applied to the second, or screen, grid.

The autodyne is often used as a detector in a direct-conversion receiver, or regenerative receiver (*see* DIRECT-CONVERSION RECEIVER, REGENERATIVE DETECTOR). The incoming signal beats with the oscillator signal, forming audio-frequency mixing products. This af is then amplified using ordinary audio amplifiers. The autodyne is sometimes used as a mixer when circuit simplicity is

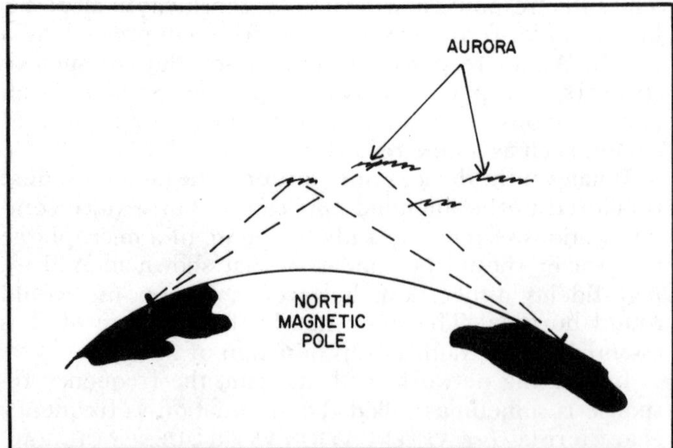

AURORAL PROPAGATION: In auroral propagation, electromagnetic energy is reflected from highly ionized auroral "curtains" above the north and south polar regions.

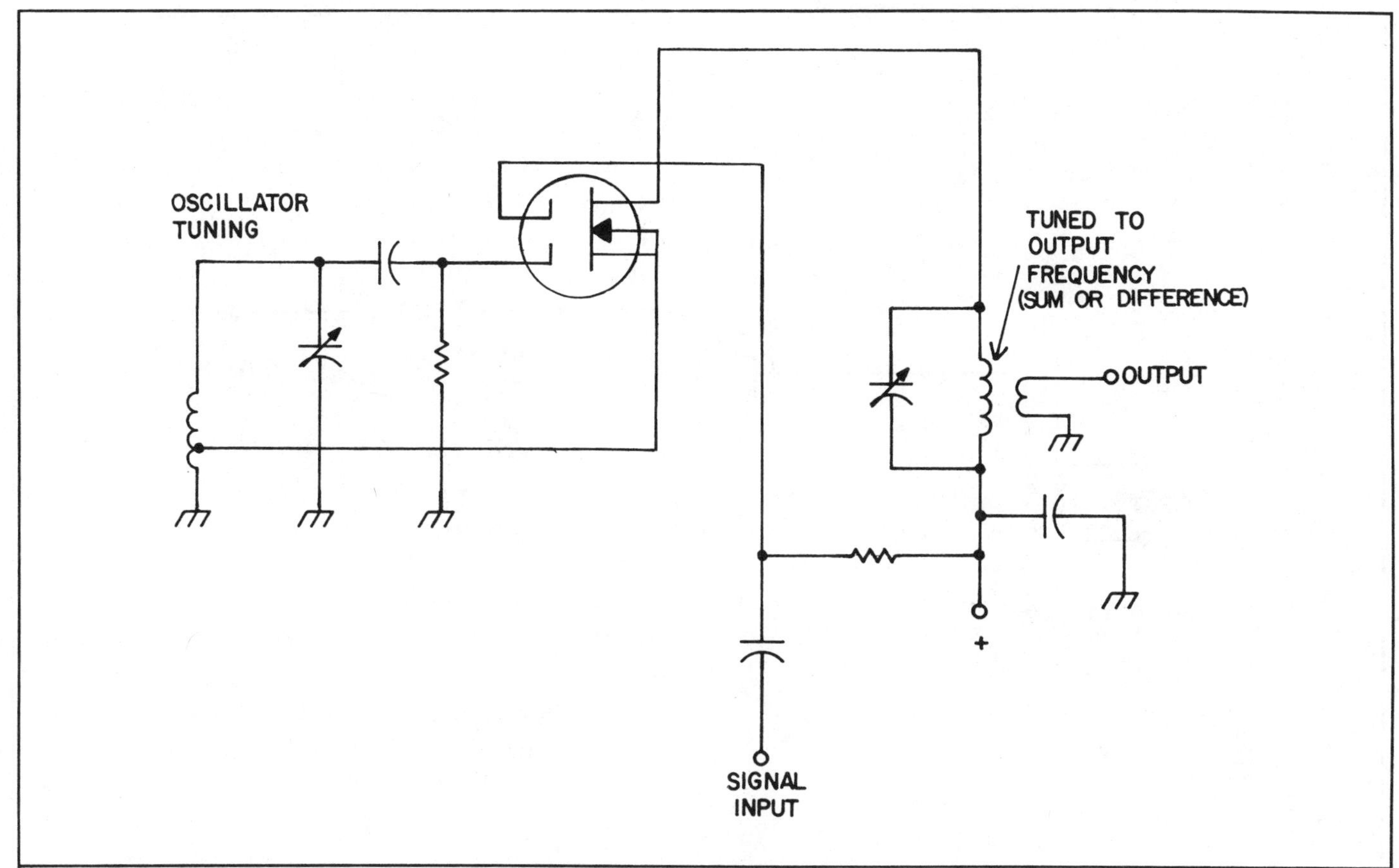

AUTODYNE: The autodyne is a combined oscillator and mixer in a single stage.

important. The autodyne replaces both the local oscillator and mixer stages in a receiver.

AUTOMATIC BIAS

Automatic bias is a method of obtaining the proper bias for a tube or transistor by means of a resistor, usually in the cathode or emitter circuit. A capacitor may also be used, in parallel with the resistor, to stabilize the bias voltage. The illustration shows an example of automatic bias. In this case, reverse bias is produced at the emitter-base junction of the transistor, facilitating the operation of a class-C amplifier.

Other means of developing bias include resistive voltage-dividing networks and special batteries or power supplies. *See also* BIAS, BIAS STABILIZATION, VOLTAGE DIVIDER.

AUTOMATIC BRIGHTNESS CONTROL

An automatic-brightness-control circuit is used in a television receiver to keep the relative brightness of the picture within certain limits. Excessively large changes in picture brightness are objectionable to the eye. Such fluctuations also affect the amount of contrast in the picture. Some combinations of brightness and contrast appear unnatural.

An automatic-brightness control operates in the same manner as automatic-gain and level controls (*see* AUTOMATIC GAIN CONTROL, AUTOMATIC LEVEL CONTROL). An increase in the brilliance of the scene causes the automatic brightness control to dim the picture; a decrease in brilliance causes the automatic brightness control to lighten the picture. The amount of compensation is adjusted so that some variation in picture brightness occurs, but to a lesser extent than the changes in the actual scene. Improperly adjusted automatic brightness control will result in unnatural television pictures. *See also* BRIGHTNESS.

AUTOMATIC CONTRAST CONTROL

Automatic contrast control serves to maintain the proper amount of contrast in a television picture receiver. The contrast of some scenes is excessive for good television reproduction, and the contrast in some scenes is insufficient. Changes in picture brightness affect the apparent contrast as seen by the human eye. The most natural contrast must be maintained for all levels of brightness.

The gain characteristics of the RF amplifiers in a television receiver determine the contrast. An automatic-contrast control functions by changing the gain of these stages. Automatic-contrast control is therefore a form of automatic gain control. *See* AUTOMATIC GAIN CONTROL, CONTRAST.

AUTOMATIC DIALING UNIT

A device for rapid dialing of telephone numbers is called an automatic-dialing unit (ADU) or automatic dialer.

Automatic-dialing units are made for either rotary or tone pushbutton telephone systems. The rotary type ADU stores several interrupt sequences, which are generated in the same manner as the interrupts produced by a rotary telephone dial. The tone ADU stores a sequence of tones

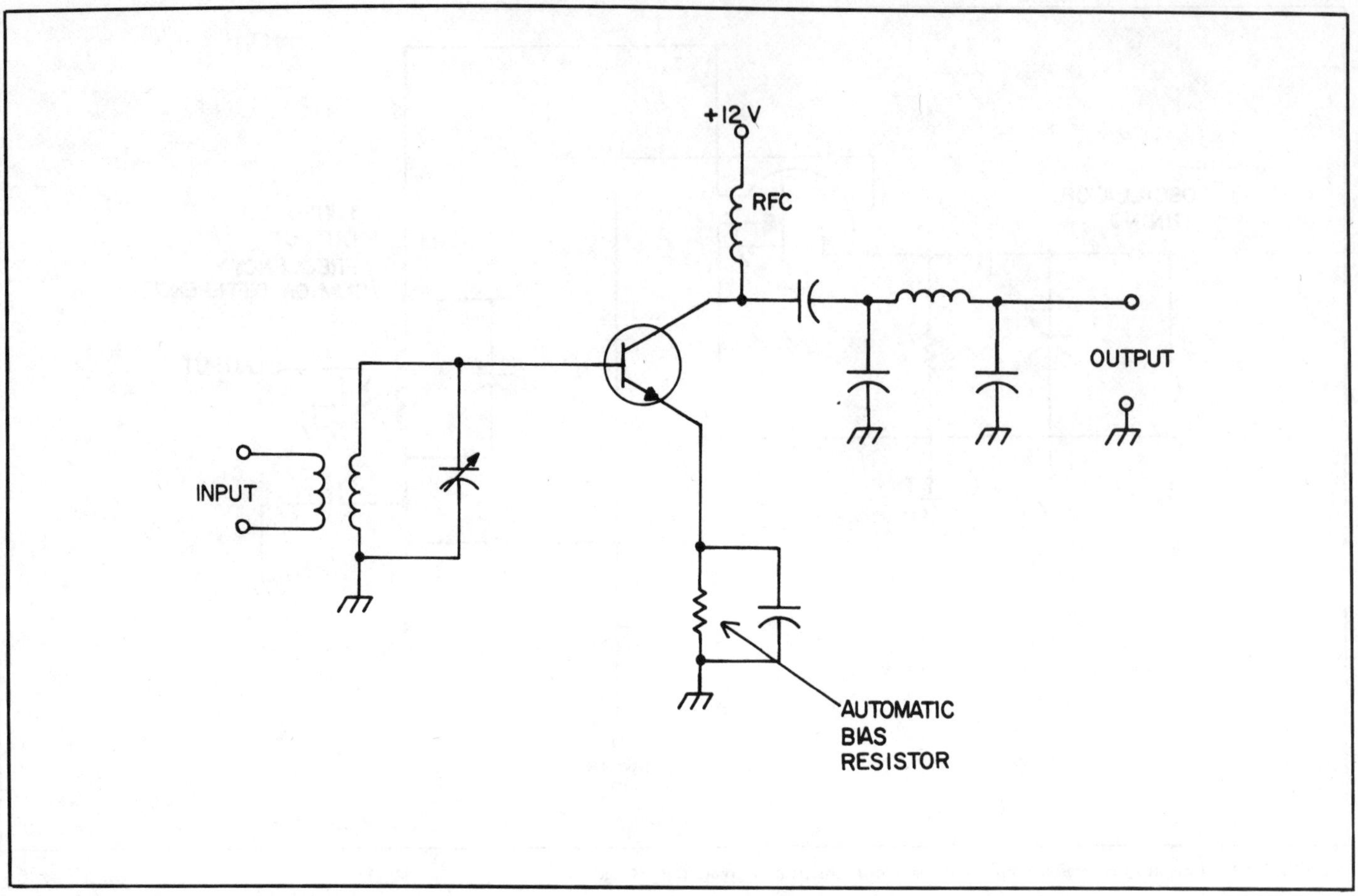

AUTOMATIC BIAS: Automatic bias is developed across a resistor. Here, the automatic bias circuit puts the transistor well past cutoff for operation inclass C.

corresponding to the desired telephone number (*see also* TOUCHTONE®).

The tone pushbutton automatic dialing unit is much faster than the rotary type. A seven-digit telephone number can be dialed via a tone-type ADU in less than one second. Both the rotary and tone-type automatic dialing units are available for consumer use in the home or business. Numbers are programmed into the ADU and recalled by pressing just one or two buttons. *See also* TELEPHONE.

AUTOMATIC DIRECTION FINDER

An automatic direction finder, or ADF, is a device that indicates the compass direction from which a radio signal is coming. A loop antenna is used to obtain the bearing. In conjunction with a small whip antenna, the loop gives an unidirectional indication.

A rotating device turns the loop antenna until the signal strength reaches a sharp minimum, or null. At this point the loop, connected to a sensing circuit, stops rotating automatically. The operator can simply read the compass bearing from the azimuth indicator in degrees.

The exact position of a transmitting station is obtained by taking readings with the automatic direction finder at two or more widely separated points as shown. On a detailed map, straight lines are drawn according to the bearings obtained at the observation points. The intersection point of these lines is the location of the transmitter. *See also* DIRECTION FINDER.

AUTOMATIC FREQUENCY CONTROL

Automatic frequency control is a method for keeping a receiver or transmitter on the desired operating frequency. Such devices are commonly used in FM stereo

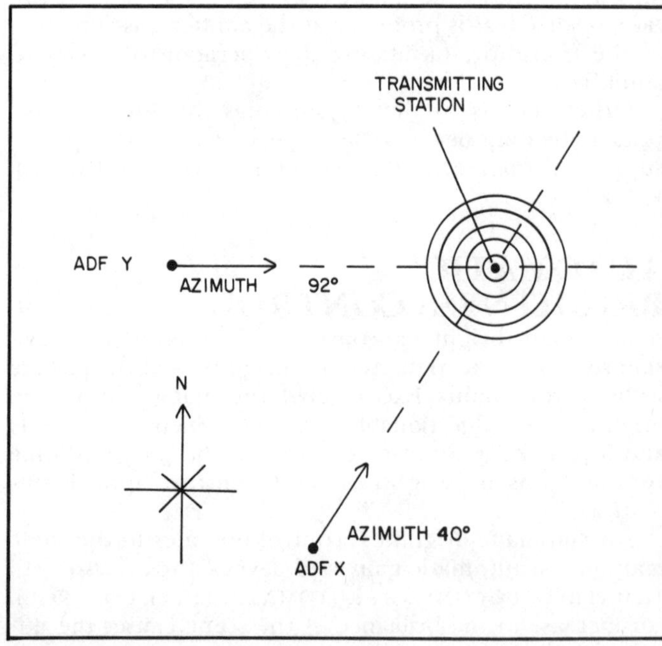

AUTOMATIC DIRECTION FINDER: An example of the use of automatic direction finders to locate a transmitter.

receivers, since tuning is critical with this type of signal, and even a small error can noticeably affect sound reproduction.

An automatic-frequency-control device senses any deviation from the correct frequency, and introduces a dc voltage across a varactor diode in the oscillator circuit to compensate for the drift. An increase in oscillator frequency causes the injection of a voltage to lower the frequency; a decrease in the oscillator frequency causes the injection of a voltage to raise the frequency.

A special kind of automatic frequency control is the phase-locked loop, abbreviated PLL. Such a circuit maintains the frequency of an oscillator within very narrow limits. *See also* PHASE-LOCKED LOOP.

AUTOMATIC GAIN CONTROL

In a communications receiver, it is desirable to keep the output essentially constant regardless of the strength of the incoming signal. This allows reception of strong as well as weak signals, without the need for volume adjustments. It also prevents "blasting," the effect produced by a loud signal as it suddenly begins to come in while the receiver volume is turned up.

Refer to the schematic diagram of a simplified automatic gain control, or AGC. An audio or RF amplifier stage can be used for AGC; usually, it is done in the intermediate-frequency stages of a superheterodyne receiver. Part of the amplifier output is rectified and filtered to obtain a dc voltage. This dc voltage is applied to the input circuit of that amplifier stage or a preceding stage. The greater the signal output, the greater the AGC voltages, which changes the amplifier bias to reduce its gain. By properly choosing the component values, the amplifier output can be kept at an almost constant level for a wide range of signal-input amplitudes. Automatic gain control is sometimes also called automatic level control or automatic volume control (*see* AUTOMATIC LEVEL CONTROL), although the term automatic level control is generally used to define a broader scope of applications.

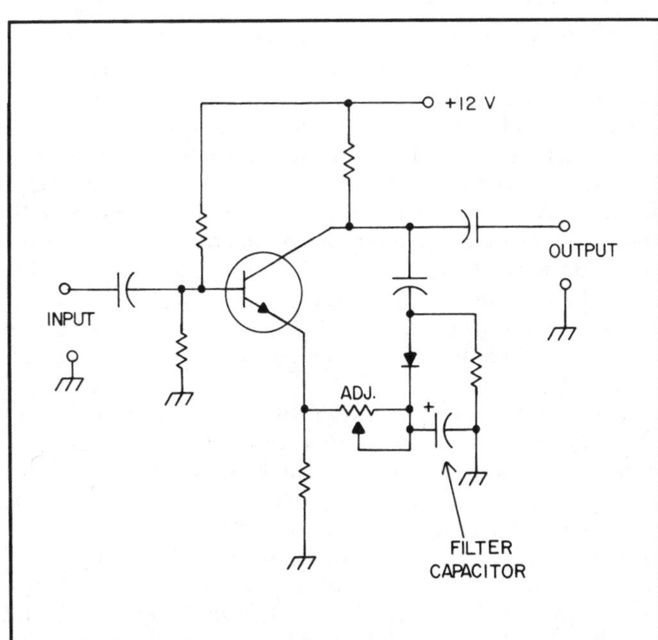

AUTOMATIC GAIN CONTROL: An automatic gain control circuit. Some of the output is rectified and filtered, and the resultant dc voltage is applied to the emitter to reduce the gain of the stage.

The attack time (*see* ATTACK) of an AGC system must be rapid, and the decay time (*see* DECAY) slower. Some receivers have AGC with variable decay or release time. The decay time is easily set to the desired value by suitably choosing the value of the filter capacitor in the dc circuit. The time constant of this decay circuit should be slow enough to prevent background noise from appearing between elements of the received signal, but rapid enough to keep up with fading effects. *See also* RECEIVER.

AUTOMATIC INTERCEPT

An automatic intercept device receives incoming messages and records them for later handling. A common example of such a circuit is a telephone answering machine. These machines are available to consumers in various forms, from simple to very sophisticated, for use in the home and business.

The automatic intercept answers the telephone after it rings, and plays a recorded announcement to the effect that the party is not presently available. The caller may then wait for a tone, whereupon the automatic intercept machine records a message of specified length, usually 30 to 45 seconds. Some automatic intercepts can record dozens of messages in this way.

AUTOMATIC LEVEL CONTROL

Automatic level control (ALC) is a generalized form of automatic gain control, or AGC (*see* AUTOMATIC GAIN CONTROL). While AGC is employed in communications receivers, ALC is found in a wide variety of devices, from tape recorders to AM, SSB, and FM transmitters. Both AGC and ALC function in the same way: Part of the output is rectified and filtered, and the resulting dc voltage is fed back to a preceding stage to control the amplification. This keeps the output essentially constant, even when the input amplitude changes greatly.

Automatic-level control is frequently used in voice transmitters to prevent flat topping of the modulated waveform (*see* FLAT TOPPING). Automatic-level control can also be used to boost the low-level modulation somewhat, since the amplifier may be made more sensitive without fear of overmodulation. A circuit designed to substantially increase the talk power in a voice transmitter is called a speech compressor or speech processor (*see* SPEECH COMPRESSION), and is a form of ALC in conjunction with an audio amplifier. Automatic-level control should not be confused with limiting or clipping (*see* AUDIO LIMITER, CLIPPER, SPEECH CLIPPING).

In a transmitter modulation circuit, ALC is sometimes called automatic-modulation control (*see* AUTOMATIC MODULATION CONTROL).

AUTOMATIC MESSAGE HANDLING

Messages are routed to the proper destination rapidly and efficiently by a system called automatic message handling. A message in this system is categorized by an electronic code, such as a combination of tone bursts or a subaudible tone. This eliminates the need for manual processing.

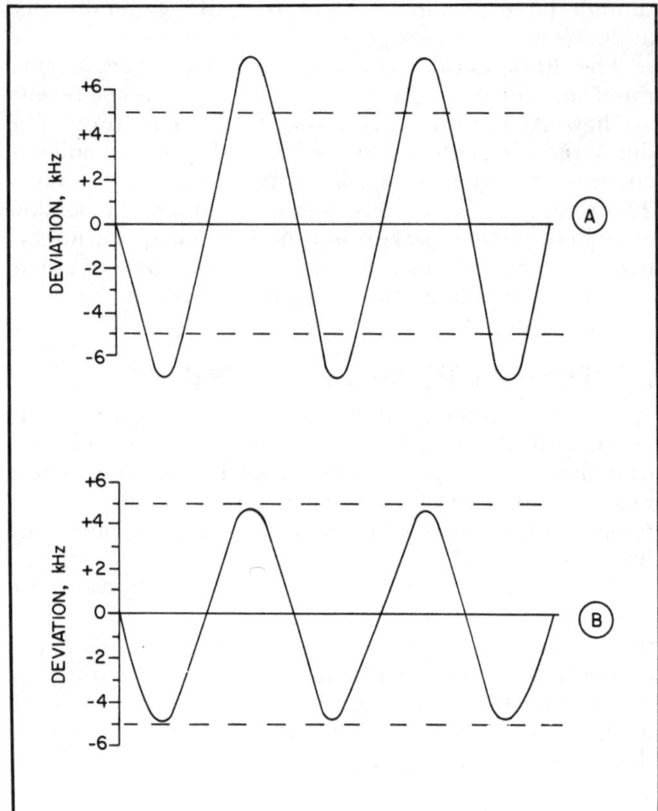

AUTOMATIC MODULATION CONTROL: An FM signal, as seen on a deviation monitor, without (A) and with (B) automatic modulation control.

The placing of a direct-dial, long-distance telephone call is an example of automatic message handling. The first digit, 1, is represented by a combination of two audio tones or a single pulse. It indicates to a computer that the caller intends to use the direct-dial system. The next three digits, also represented by either tones or pulses, define the geographical region to which the call is directed. The next three digits select the number prefix, further narrowing the possibilities. The final four digits direct the call to a particular business or residence.

Automatic message handling is done today by digital computers. This saves the time and cost of operator assistance. As a consequence of technological improvements such as this, great amounts of information are conveyed among an increasingly communication-conscious population.

AUTOMATIC MODULATION CONTROL

Automatic modulation control is a form of automatic-level control (*see* AUTOMATIC LEVEL CONTROL). In an AM, SSB, or FM transmitter, automatic modulation control prevents excessive modulation while allowing ample microphone gain.

In an AM or SSB transmitter, overmodulation causes a phenomenon known as splatter (*see* SPLATTER), which greatly increases the signal bandwidth and can cause interference to stations on nearby channels or frequencies. In an FM transmitter, overmodulation results in overdeviation. This makes the signal appear distorted in the receiver, and can also cause interference to stations on adjacent channels. The output of a sine-wave-modulated FM transmitter is shown both without (at A) and with (at B) automatic modulation control. In this particular system, the proper deviation is ± 5 kHz. In operation, while observing a deviation monitor, an operator would find it impossible to cause overdeviation, even by loud shouting into the microphone. The dotted lines in A and B show the proper modulation limits.

Without automatic modulation control, proper adjustment of a voice transmitter would be exceedingly critical. Therefore, nearly all voice transmitters use some form of automatic modulation control. *See also* AMPLITUDE MODULATION, FREQUENCY MODULATION, SINGLE SIDEBAND.

AUTOMATIC NOISE LIMITER

An automatic noise limiter, abbreviated ANL, is a circuit that clips impulse and static noise peaks while not affecting the desired signal. An automatic noise limiter sets the level of limiting, or clipping, according to the strength of the incoming signal. The AGC voltage (*see* AUTOMATIC GAIN CONTROL) in the receiver usually forms the reference for the limiter. The stronger the signal, the greater the limiting threshold. A manual noise limiter (*see* NOISE LIMITER) differs from the automatic noise limiter; the manual limiter must be set by the operator, and will cause distortion of received signals if not properly adjusted.

A noise limiter prevents noise from becoming stronger than the signals within the receiver passband, but the signal-to-noise ratio remains marginal in the presence of strong impulse noise. For improved reception with severe impulse noise, the noise blanker (*see* NOISE BLANKER) has been developed. Noise blankers are extremely effective against many kinds of manmade noise, but noise limiters are generally better for atmospheric static.

AUTOMATIC SCANNING RECEIVER

Some receivers have the capability to "search" for an occupied channel automatically, without the need for an operator. A channelized receiver with automatic-scanning capability can be programmed to constantly sweep through a given set of frequencies until a busy channel is found. This is called the busy-scan mode. Alternatively, the scanner can be set to search for an empty channel. This is called the vacant-scan mode. In some cases, a subaudible tone of a certain frequency may be necessary to cause the automatic scanning receiver to stop on a particular channel. This is called Private-Line®, or PL, operation (*see* PRIVATE LINE).

A common example of automatic scanning is found in the ordinary vhf/uhf scanner receiver (see photograph). This receiver operates only in the busy-scan mode. Channels are selected by installing crystals, or by programming the frequencies into a memory circuit. The receiver scans through all of the selected frequencies for approximately 0.1 second each, over and over, until a signal appears on one of the channels and opens the squelch. The open squelch then causes the scanner to stop, and the signal can be received. When the signal disappears, scanning resumes automatically after a specified time delay has elapsed. *See also* SCAN FUNCTIONS, SCANNING.

AUTOMATION

Whenever a machine takes the place of a living thing to perform a particular job, we have an example of automation. Automation is an integral part of our society and world.

Automation can be both good and bad. Machines can be substituted for people in dangerous or tedious jobs. This saves time for people, and also saves them from possible injury or death. But automation also sometimes means the loss of a job for someone. One machine may replace hundreds of people in certain instances.

The most important electronic breakthrough in automation has undoubtedly been the computer. This device saves literally millions of man-hours every day, performing arithmetic and mathematical calculations that would take years with a pencil and paper. *See also* COMPUTER.

AUTOMOTIVE ELECTRONICS

Automotive electronics deals with the engineering and implementation of electronic devices for use in cars and trucks.

All of the panel dials and lights in a modern vehicle are operated electronically. This allows more accurate monitoring than older mechanical methods. In some automobiles and trucks, vital functions are controlled by small computers (*see* MICROCOMPUTER). Such computers can be packed into just one or two integrated circuits.

Other electronic devices help ensure passenger safety in vehicles. An example of this is the interlock device that prevents a car from being started until all seat belts have been fastened. Another example, still not in widespread use, is an acceleration-sensitive air bag. An eventual possibility is a built-in breathalyzer to prevent drunk drivers from starting a car!

AUTOPATCH

An autopatch is a device for connecting a radio transceiver into the telephone lines by remote control. Autopatch is generally accomplished through repeaters (*see* REPEATER). The illustration is a schematic representation of an autopatch system.

The repeater is first accessed by transmitting a certain sequence of tone-coded digits. This is done with a telephone type keypad connected to the microphone input of the transmitter (*see* TOUCHTONE®). The autopatch system is then activated by another set of digits. At this point, the operator may dial the desired telephone number. With most Amateur-Radio autopatch repeaters, the repeater is always available for normal communication, but the autopatch must still be activated by a special sequence of digits.

Autopatch telephone conversations are not always the same as an ordinary phone hookup. Interruption of the radio operator is impossible if his receiver is disabled while he speaks. A simple form of autopatch system that works over a limited range, allowing duplex operation, is the increasingly popular cordless telephone. *See also* CORDLESS TELEPHONE.

AUTOPILOT

An autopilot is a device that automatically keeps an aircraft or guided missile on a predetermined course. This saves personnel time, on sophisticated aircraft, for other chores.

AUTOMATIC SCANNING RECEIVER: This is a vhf police-scanner radio.

On a commercial airliner, computer autopilots can actually fly the plane, with the exception of takeoffs, landings, and unplanned course changes.

The speed, orientation, and other variables for air flight are monitored by special instruments, and servomechanisms compensate for deviations from the proper parameters (*see* SERVOMECHANISM). The exact position of the aircraft is monitored by a radiolocation system (*see* RADIOLOCATION).

AUTOSYN

Autosyn is a trade name for a synchronous motor system developed by Bendix Corporation (*see* SYNCHRONOUS MOTOR). The Autosyn is used as a directional indicator for such things as rotatable antenna arrays. When the shaft of the sensor motor is moved, the shaft of the indicator motor follows. The Autosyn may be used as a remote-control device as well as an indicator.

A pair of Autosyn mechanisms can be used as an azimuth-elevation, or az-el, indicator for satellite or moonbounce antennas (*see* AZ-EL). The azimuth indicator is geared to a 1-to-1 ratio and calibrated from zero to 360 degrees. The elevation indicator is geared at a 1-to-4 ratio and calibrated from zero to 90 degrees. The illustration shows an az-el indicator system using Autosyn devices.

A machine that works in a similar way to the Autosyn is called the selsyn, used for antenna direction indicators. *See also* SELSYN, SERVOMECHANISM.

AUTOTRANSFORMER

The autotransformer is a special type of step-up or step-down transformer with only one winding. Usually, a transformer has two separate windings, called the primary and secondary, electrically isolated from each other as shown in A. The autotransformer, however, uses a single tapped winding.

For step-down purposes, the input to the autotransformer is connected across the entire winding, and the output is taken from only part of the winding as shown in B. For step-up purposes, the situation is reversed: the input is applied across part of the winding, and the output appears across the entire coil (C).

The autotransformer obviously contains less wire than a transformer with two separate windings. If electrical (dc) isolation is needed between the input and output, however, the autotransformer cannot be used.

The step-up or step-down voltage ratio of an autotransformer is determined according to the same formula as for an ordinary transformer (*see* TRANSFORMER). If N_{PRI} is the number of turns in the primary winding, and N_{SEC} is the number of turns in the secondary winding, then the voltage transformation ratio is

$$E_{PRI}/E_{SEC} = N_{PRI}/N_{SEC}$$

where E_{PRI} and E_{SEC} represent the ac voltages across the primary and secondary windings, respectively. The impedance-transfer ratio is

$$Z_{PRI}/Z_{SEC} = (N_{PRI}/N_{SEC})^2$$

where Z_{PRI} and Z_{SEC} are the impedances across the primary and secondary windings, respectively.

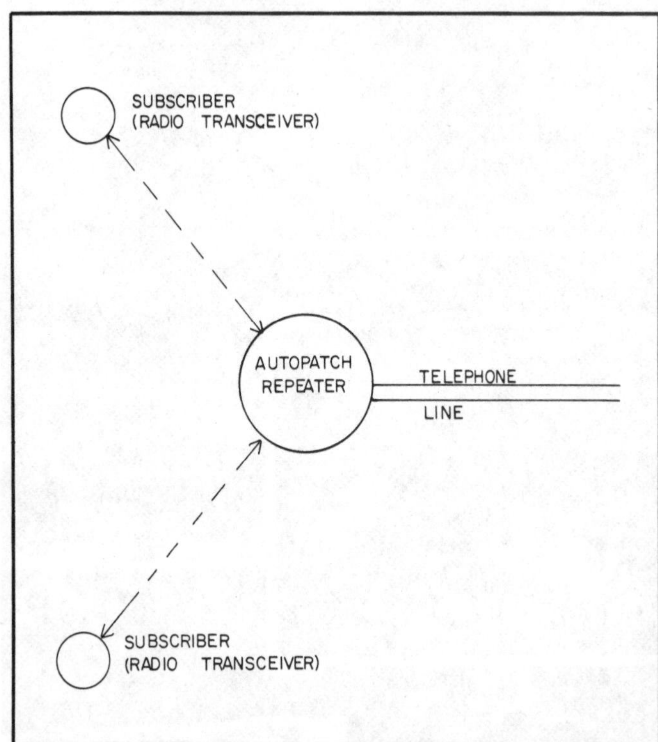

AUTOPATCH: An autopatch system allows the connection of radio transceivers to telephone lines.

AUXILIARY CIRCUIT

Any circuit other than the main, or primary circuit in an electrical or electronic system is called an auxiliary circuit. If the primary circuit malfunctions, the auxiliary circuit is used as a backup until the primary circuit has been repaired. For example, if a broadcast transmitter fails, a backup or auxiliary transmitter, usually having less power output, keeps the station on the air until the main transmitter has been repaired. *See also* AUXILIARY LINK, AUXILIARY MEMORY, AUXILIARY POWER.

AUXILIARY LINK

An auxiliary link is used when a primary communications system fails. This maintains communication, perhaps at a lower level of efficiency, until the main link can be reestablished.

For example, a communications link may be maintained by a satellite. If the satellite malfunctions, some other satellite might be used instead. If no other satellite is available, a high-frequency (shortwave) link can be sometimes be used. A wire or cable link, such as the telephone, might in some instances be the only auxiliary link available. *See also* ALTERNATE ROUTING.

AUXILIARY MEMORY

The capabilities of a computer are limited by the amount of memory, or storage, available. The computer can, however, access an auxiliary-memory device into which any amount of information can be placed.

Small computers have auxiliary memory devices such as disk drives or tape recorders (*see* DISK DRIVE, FLOPPY DISK). There is practically no limit to the number of floppy disks that can be kept near a computer! Programs and information can be stored on such disks. By simply placing the proper disk into the disk drive, the auxiliary memory is put to use.

A limited amount of auxiliary memory can be stored in integrated circuits, for immediate access without the need to physically insert a disk into a disk drive. *See also* RANDOM-ACCESS MEMORY, READ-ONLY MEMORY.

AUXILIARY POWER

When the main, or primary, power system for a circuit fails, auxiliary power must be used. A common example of an auxiliary, or backup, power supply for the home is a gasoline-powered generator.

Some electronic circuits, such as digital clocks or radio receivers that draw relatively little current, have built-in batteries that are kept charged by the main power source. A nickel-cadmium battery is often used for this purpose (*see* NICKEL-CADMIUM BATTERY). In the event of an interruption in the primary power source, the nickel-cadmium battery can take over for a short time. Switching between the main and auxiliary supplies is, in this case, done automatically. When the main power source is restored, the battery recharges, ready for another possible power failure.

The contents of a memory circuit are maintained in a microprocessor device, when disconnected from power, by a backup battery which is also a form of auxiliary power. *See also* MEMORY-BACKUP BATTERY.

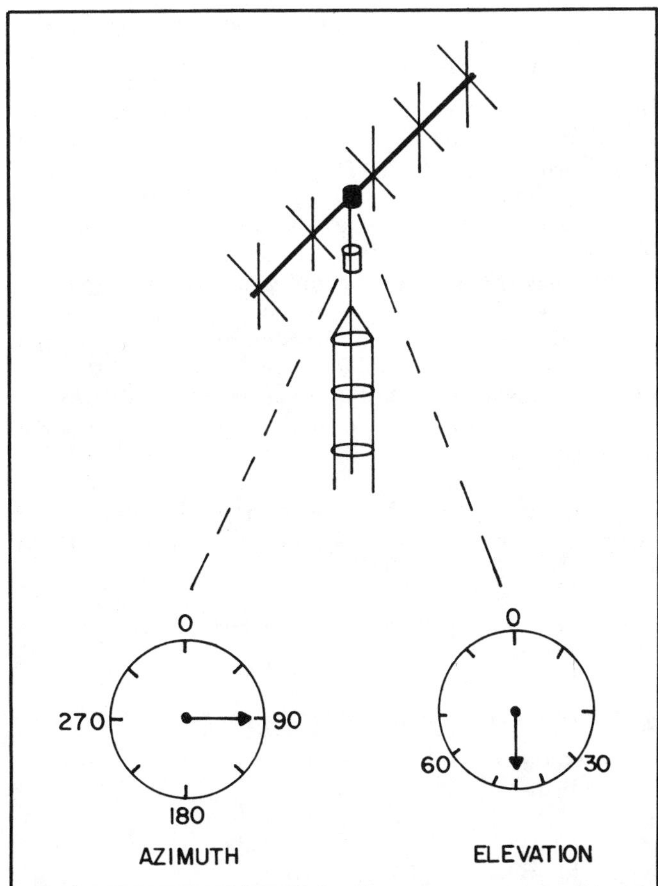

AUTOSYN: The Autosyn can be used with azimuth and elevation rotors to give precise bearings.

AVAILABLE POWER

When a transmission line is connected to an antenna, or a source of energy is connected to a load, power is dissipated in the load. This dissipation may occur as heat in a resistor, or as heat and visible light in a light bulb. It can also take the form of electromagnetic radiation, as in a radio antenna, or sound, as in a speaker or headset. There are many forms of energy dissipation. The amount of power dissipated in a load is measured in watts, and represents the rate of energy transfer.

If the load impedance is a conjugate match to the source or input impedance, then the transfer of energy from the source to the load occurs with 100 percent efficiency. However, if the source impedance and load impedance are not conjugate-matched, some of the energy from the source is not absorbed by the load. Instead, it is returned to the source as reactive power (*see* IMPEDANCE, IMPEDANCE MATCHING, REACTIVE POWER). The amount of power dissipated in the load is therefore not as great as it could be.

The available power from a generator, or source, is the amount of power dissipated in a conjugate-matched load. With a given generator, the true power dissipated by the load decreases as the match gets worse and worse. The available power remains the same no matter what the match. *See also* CONJUGATE IMPEDANCE, TRUE POWER.

AVAILABLE POWER GAIN

The available power gain of a circuit is the ratio of the available output power to the available input power. When impedances at the input and output of a circuit are properly matched (as they should be for most efficient operation), the available power gain is the same as the true power gain (*see* AVAILABLE POWER, TRUE POWER, TRUE POWER GAIN).

Available power gain is usually expressed in decibels, or dB (*see* DECIBEL). If P_{IN} is the available power input in

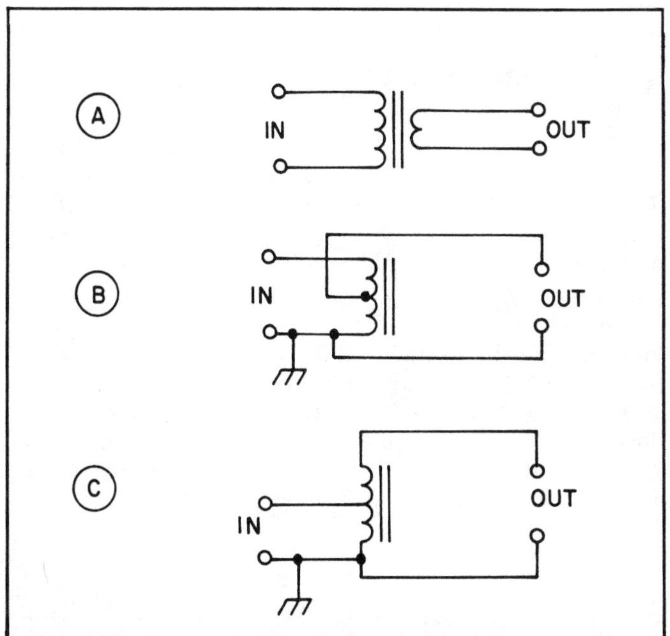

AUTOTRANSFORMER: At A, an ordinary transformer with two separate windings. At B, a step-down autotransformer. At C, a step-up autotransformer.

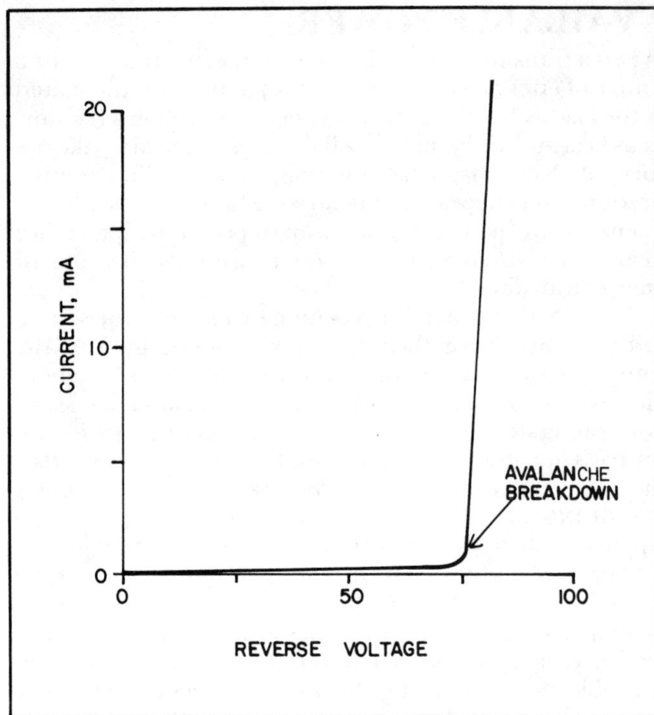

AVALANCHE BREAKDOWN: In a semiconductor diode, avalanche breakdown is typified by a rapid rise in current in the reverse direction.

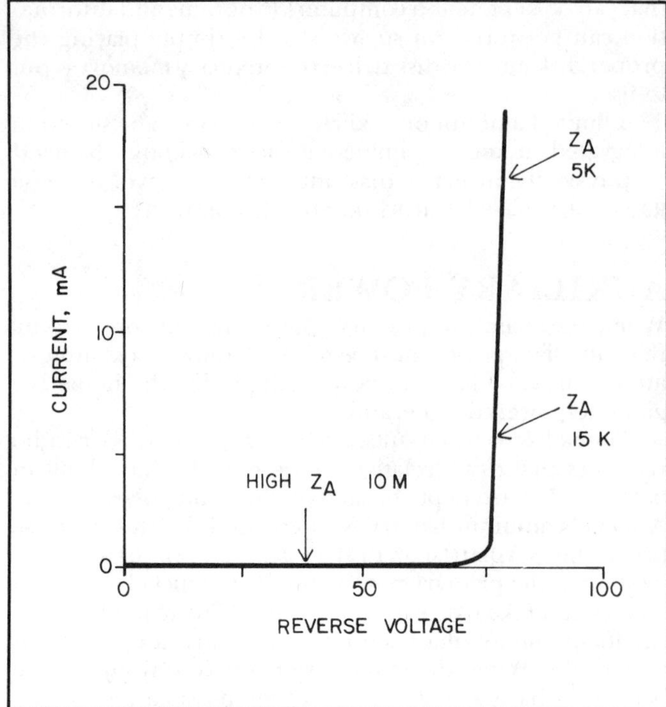

AVALANCHE IMPEDANCE: The avalanche impedance, Z_A, depends on the reverse voltage. At low reverse-voltage levels, Z_A is extremely high. Above the avalanche voltage, Z_A drops.

watts and P_{OUT} is the available power output in watts, then

$$\text{Available power gain (dB)} = 10 \log_{10} (P_{OUT}/P_{IN})$$

By comparing the available power gain with the true power gain of an amplifier, it is possible to tell whether or not a given circuit is performing as well as it should. The gain of an amplifier can be seriously degraded simply because impedances are not matched, making the true power gain much smaller than it could be if proper matching were arranged.

AVALANCHE

Avalanche occurs when an electric field becomes so strong that ionization occurs in a gas. One electron, accelerated by the field, collides with molecules of the gas and ionizes them (*see* ION). This sets many electrons free, each of which is accelerated by the field to strike, and ionize, more molecules. The eventual result is a conductive path through the gas.

The avalanche effect also occurs at semiconductor junctions, but in a somewhat different way. Ordinarily, a junction of N-type and P-type material (*see* N-TYPE SEMICONDUCTOR, P-TYPE SEMICONDUCTOR) will not conduct when reverse biased. However, if the reverse-bias voltage is sufficiently strong, conduction will take place. This is called avalanche breakdown. *See also* AVALANCHE BREAKDOWN.

AVALANCHE BREAKDOWN

When reverse voltage is applied to a P-N semiconductor junction, very little current will flow as long as the voltage is relatively small. But as the voltage is increased, the holes and electrons in the P-type and N-type semiconductor materials are accelerated to greater and greater speeds. Finally the electrons collide with atoms, forming a conductive path across the P-N junction. This is called avalanche breakdown. It does not permanently damage a semiconductor diode, but does render it useless for rectification or detection, since current flows both ways when avalanche breakdown occurs (*see* RECTIFICATION).

Avalanche breakdown in a particular diode may occur with just a few volts of reverse bias, or it may take hundreds of volts. Rectifier diodes do not undergo avalanche breakdown until the voltage becomes quite large (*see* AVALANCHE VOLTAGE). A special type of diode, called the zener diode (*see* ZENER DIODE), is designed to have a fairly low, and very precise, avalanche breakdown voltage. Such diodes are used as voltage regulators. Zener diodes are sometimes called avalanche diodes because of their low avalanche voltages.

The illustration shtws graphically the current flow through a diode as a function of reverse voltage. The current is so small, for small reverse voltages, that it may be considered to be zero for practical purposes. When the reverse voltage becomes sufficient to cause avalanche breakdown, however, the current rises rapidly.

AVALANCHE IMPEDANCE

When a diode has sufficient reverse bias to cause avalanche breakdown, the device appears to display a finite resistance. This resistance fluctuates with the amount of reverse voltage, and is called the avalanche resistance or avalanche impedance. It is given by

$$Z_A = E_R/I_R$$

for direct current, where Z_A is the avalanche impedance, E_R is the reverse voltage, and I_R is the reverse current. Units are ohms, volts and amperes respectively.

When E_R is smaller than the avalanche voltage (*see* AVALANCHE VOLTAGE), the value of Z_A is extremely large, because the current is small, as shown in the figure. When E_R exceeds the avalanche voltage, the value of Z_A drops sharply. As E_R is increased further and further, the value of Z_A continues to decrease.

The magnitude of Z_A for alternating current is practically infinite in the reverse direction, as long as the peak ac voltage never exceeds the avalanche voltage. When the peak ac reverse voltage rises to a value greater than the avalanche voltage, the impedance drops. The maximum voltage that a semiconductor diode can tolerate, for rectification purposes, without avalanche breakdown is called the peak inverse voltage (*see* PEAK INVERSE VOLTAGE).

The exact value of the avalanche impedance for alternating current is the value of resistor that would be necessary to allow the same flow of reverse current. The average Z_A for alternating current differs from the instantaneous value, which fluctuates as the voltage rises and falls. Avalanche is usually undesirable in alternating-current applications. *See also* INSTANTANEOUS EFFECTS.

AVALANCHE TRANSISTOR

An avalanche transistor is an NPN or PNP transistor designed to operate with a high level of reverse bias at the emitter-base junction. Normally, transistors are forward biased at the emitter-base junction except when cutoff conditions are desired when there is no signal input.

Avalanche transistors are seen in some switching applications. The emitter-base junction is reverse biased almost to the point where avalanche breakdown takes place. A small additional reverse voltage, supplied by the input signal, triggers avalanche breakdown of the junction and resultant conduction. Therefore, the entire transistor conducts, switching a large amount of current in a very short time. The extremely sharp "knee" of the reverse-voltage-vs-current curve facilitates this switching capability (see the illustrations in AVALANCHE BREAKDOWN and AVALANCHE IMPEDANCE). A small amount of input voltage can thus cause the switching of large values of current.

AVALANCHE VOLTAGE

The avalanche voltage of a P-N semiconductor junction is the amount of reverse voltage required to cause avalanche breakdown (*see* AVALANCHE, AVALANCHE BREAKDOWN). Normally, the N-type semiconductor is negative with respect to the P-type in forward bias, and the N-type semiconductor is positive with respect to the P-type in reverse bias.

In some diodes, the avalanche voltage is very low, as small as 6 volts. In other diodes, it may be hundreds of volts. When the avalanche voltage is reached, the current abruptly rises from near zero to a value that depends on the reverse bias (see the illustrations in AVALANCHE BREAKDOWN and AVALANCHE IMPEDANCE).

Diodes with high avalanche voltage ratings are used as rectifiers in dc power supplies. The avalanche voltage for rectifier diodes is called the peak inverse voltage or peak reverse voltage (*see* PEAK INVERSE VOLTAGE). In the design of a dc power supply, diodes with sufficiently high peak-inverse-voltage ratings must be chosen, so that avalanche breakdown does not occur.

Some diodes are deliberately designed so that an effect similar to avalanche breakdown occurs at a relatively low, and well defined, voltage. These diodes are called Zener diodes (*see* ZENER DIODE). They are used as voltage-regulating devices in low-voltage dc power supplies. *See also* DC POWER SUPPLY.

AVERAGE ABSOLUTE PULSE AMPLITUDE

A pulse of voltage, current, or power is often irregularly shaped. Sometimes the polarity of the pulse reverses one or more times. A rather complicated pulse is in A.

The average absolute pulse amplitude is a measure of the pulse intensity. In determining the average absolute pulse amplitude, the beginning and ending times, t_0 and t_1, must be known. Then, the absolute value of the pulse polarity is taken by inverting the negative part or parts of

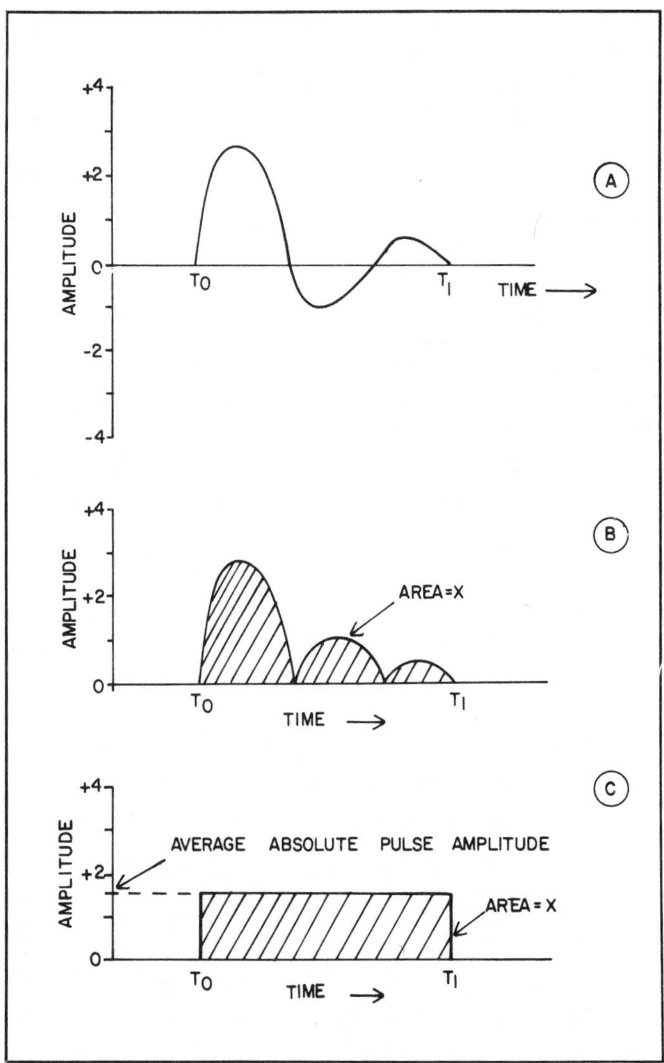

AVERAGE ABSOLUTE PULSE AMPLITUDE: Determination of average absolute pulse amplitude. At A, an irregular pulse, with beginning and ending times t_0 and t_1. At B, the negative part of the pulse is inverted and the area under the curve is shown by the shaded region. At C, a rectangle with the same area as the pulse is constructed, with identical beginning and ending times. The height of the rectangle is the average absolute pulse amplitude.

the pulse, forming a positive mirror image (B). Next, the total area under the pulse curve is found. It is shown by the shaded region. Finally, a rectangle is constructed, having the same beginning and ending times, t_0 and t_1, as the original pulse, and also the same area under the curve (C). The amplitude of this rectangular pulse is the average absolute pulse amplitude.

The procedure for determining the average absolute pulse amplitude should not be confused with area redistribution (*see* AREA REDISTRIBUTION). While area redistribution involves the effective duration of an irregular pulse, the average absolute-pulse amplitude is a measure of the effective strength, or intensity, of a pulse.

AVERAGE CURRENT

When the current flowing through a conductor is not constant, the average current is determined as the mathematical mean value of the instantaneous current at all points during one complete cycle. Most ammeters register average current. Some special devices register peak current.

Consider, for example, a class-B amplifier that has no collector current in the absence of an input signal. Then, under no-signal conditions, the average current at the collector is zero. When an input signal is applied, the collector current will flow during approximately 50 percent of the cycle as shown. In this particular illustration, the peak collector current (*see* PEAK CURRENT) is 70 mA. The average current is smaller—about 22 mA.

In the case of alternating current, the average current is usually zero, since the polarity is positive during half the cycle and negative during half the cycle, with peak values and waveforms identical in both the positive and negative directions. Of course, the effects of alternating current are very different from the effects of zero current! Average current is given little importance in ac circuits; the root-mean-square, or RMS, current is more often specified as the effective current in such instances (*see* ROOT MEAN SQUARE).

In a sine-wave half cycle, the average current is 0.637 times the peak current.

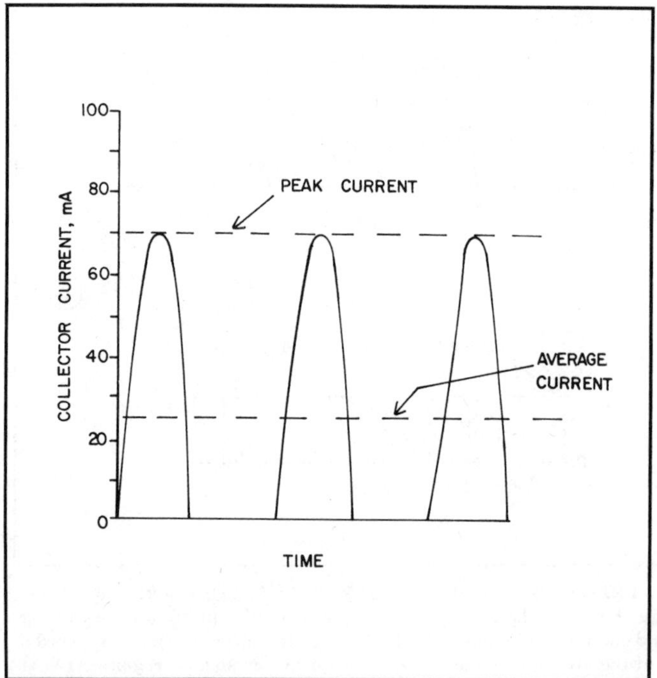

AVERAGE CURRENT: Average current and peak current in the output of a class-B amplifier.

AVERAGE LIFE

After excess carriers (electrons or holes) in a semiconductor N-type or P-type piece of material have been injected, they eventually combine with carriers of the opposite polarity. Holes are filled in by electrons, and electrons vacate atoms to create holes, restoring the material to its original condition. The time it takes for this process to be completed is called the average life. It is also sometimes called the lifetime or recombination time.

Excess electrons are introduced into an N-type semiconductor by a negative charge. Excess holes are put into P-type material by a positive charge (*see* HOLE, N-TYPE SEMICONDUCTOR, P-TYPE SEMICONDUCTOR).

The recombination of excess carriers, as a function of time, is exponential. It occurs very rapidly at first, and then more and more slowly, in a manner similar to the discharging of a capacitor (see illustration). In practice, the recombination process is effectively complete when most of the excess carriers have disappeared. This takes very little time. In silicon, it may be as little as a few nanoseconds (billionths of a second). In high-speed switching or vhf/uhf applications, it is important to have an extremely short average life. To make the recombination period as short as possible, impurity atoms are added to the semiconductor material. These impurity atoms, such as gold, act as recombination catalysts. *See also* DOPING, IMPURITY.

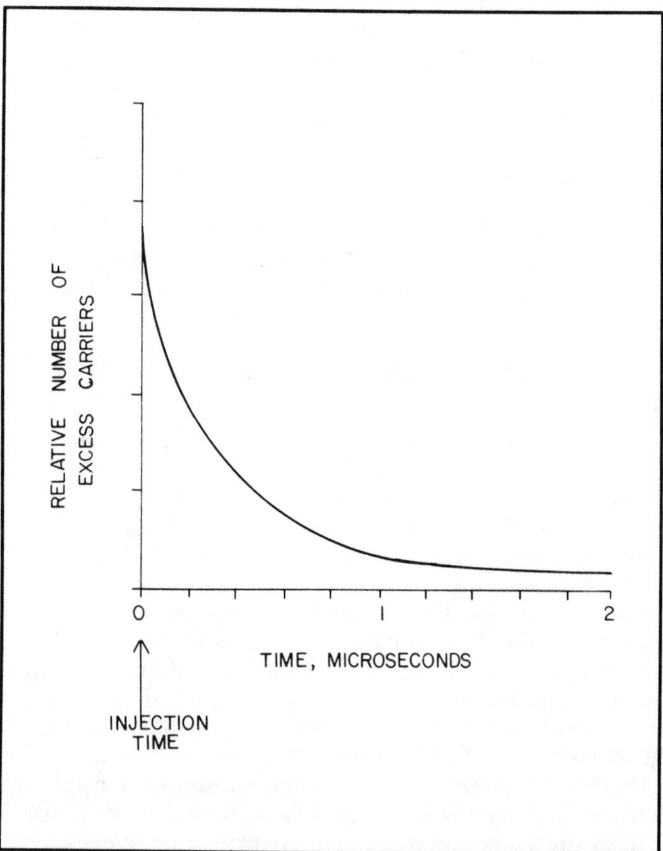

AVERAGE LIFE: Average life is the time required for excess injected carriers to almost totally recombine with carriers of the opposite polarity.

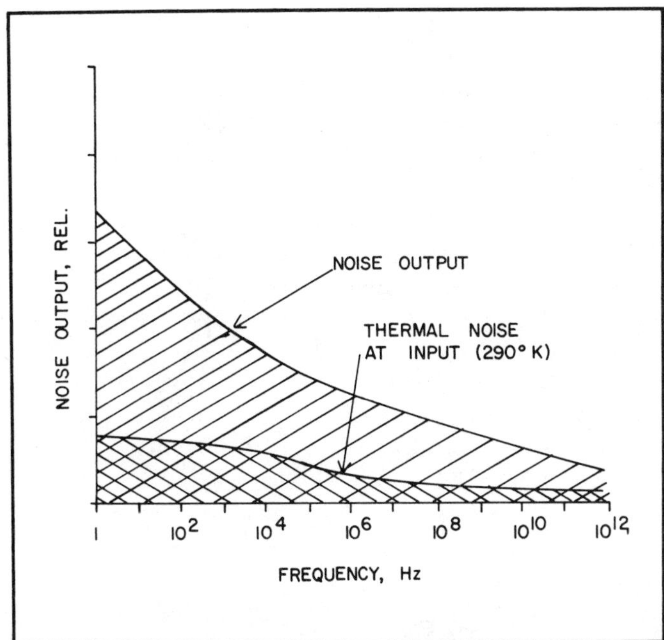

AVERAGE NOISE FIGURE: The average noise figure is the ratio of the total noise output to the thermal noise input in a circuit, at 290 degrees Kelvin, over the range of all possible frequencies from dc to infinity. Here, the range from 1 Hz to 10^{12} Hz is shown, which approximates the range of dc to infinity for practical purposes.

AVERAGE POWER: Average power in a modulation envelope. In this case, the envelope is that of a voice-modulated SSB signal.

AVERAGE NOISE FIGURE

The average noise figure of a device is its noise figure (*see* NOISE FIGURE) summed over all possible frequencies. This is theoretically dc to infinity. The illustration shows the noise input and output of a device from 1 Hz to 10^{12} Hz, which is a range of zero to infinity for most practical purposes. The input noise is attributable only to thermal agitation (*see* THERMAL NOISE), at a temperature of 290 Kelvin, or about 63 degrees Fahrenheit, for determination of the average noise figure.

The total output-noise power in the figure is represented by the area under the high curve. The total input thermal-noise power is represented by the area under the lower curve. Noise figures are generally expressed in decibels (*see* DECIBEL). The best possible noise figure is 0 dB, where the output-noise power is the same as the input-noise power. This is a theoretical ideal-never achieved in practice.

AVERAGE POWER

When the level of power fluctuates rapidly, such as in the modulation envelope of a single-side band transmitter (see the graph), the signal power may be defined in terms of peak power or average power. An ordinary wattmeter reads average power; special devices are needed to determine peak power.

In a given period of time, the area under the curve in the figure, if geometrically rearranged to form a perfect rectangle, will have a height corresponding to the average power. Between instants t_0 and t_1, chosen arbitrarily, the average power can be evaluated by considering the envelope as an irregular pulse (*see* AVERAGE ABSOLUTE PULSE AMPLITUDE).

The average power output of a circuit is never greater than the peak power output. If the power level is constant, as with an unmodulated carrier or frequency-modulated signal, the average and peak power are the same. If the amplitude changes, then the average power is less than the peak power. The output of a single-sideband signal modulated by a voice will have an average power level of about half the peak power level, although the exact ratio depends on the characteristics of the voice. *See also* PEAK ENVELOPE POWER, PEAK POWER.

AVERAGE VALUE

Any measurable quantity, such as voltage, current, power, temperature, or speed has an average value defined for a given period of time. Average values are important in determining the effects of a rapidly fluctuating variable.

To determine the average value of some variable, many instantaneous values (*see* INSTANTANEOUS EFFECTS) are mathematically combined to obtain the arithmetic mean (*see* ARITHMETIC MEAN). The more instantaneous, or sampling, values used, the more accurate the determination of the average value. For a sine-wave half cycle (such as one of the pulses in A or B), having a duration of 10 milliseconds (0.01 second) at an ac frequency of 50 Hz, instantaneous readings might be taken at intervals of 1 millisecond, then 100 microseconds, 10 microseconds, 1 microsecond, and so on. This would yield first 10, then 100, then 1000, and finally 10,000 sampling values. The average value determined from many sampling values is more accurate than the average value determined from just a few sampling values.

There is, of course, no limit to the number of sampling values that may be averaged in this way. The true average value, however, is defined as the arithmetic mean of all the instantaneous values in a given time interval. While the mathematical construction of this is rather complicated, a true average value always exists for a quantity evaluated

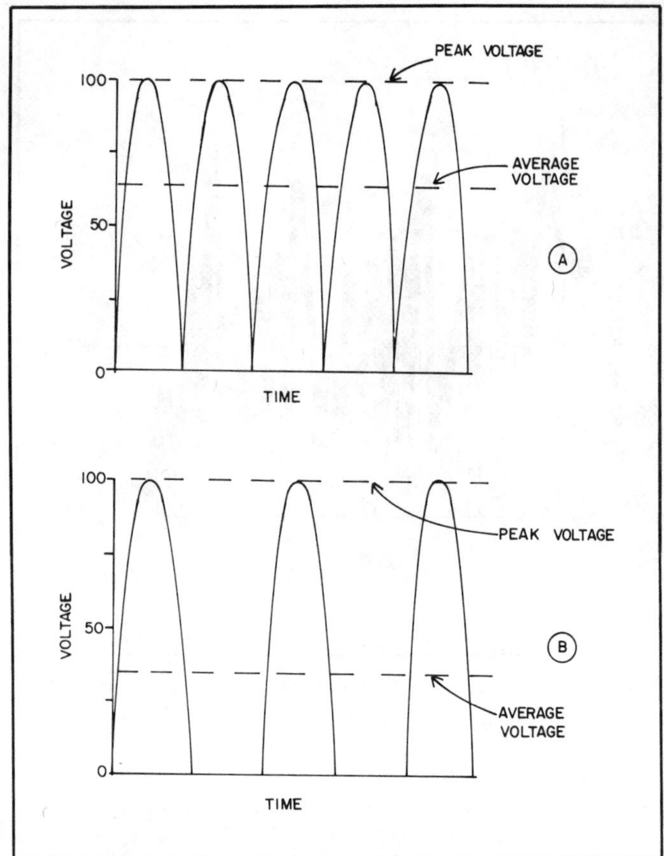

AVERAGE VALUE: At A, average vs peak voltage for a full-wave rectified ac power supply. At B, average vs peak voltage for a half-wave supply.

over a specified period of time.

An equivalent method of evaluating the average value for a variable quantity involves the construction of a rectangle, having a total area equal to the area under the curve, for a certain time interval. *See also* AVERAGE ABSOLUTE PULSE AMPLITUDE.

AVERAGE VOLTAGE

When the voltage in a circuit fluctuates, the average voltage is the mean value of the instantaneous voltage at all points during one complete cycle. Most voltmeters register average voltage when the frequency is greater than a few Hertz.

The output of a full-wave rectifier circuit is shown in A in AVERAGE VALUE. The peak voltage in this case is 100 volts (*see* PEAK VOLTAGE), but the average voltage is only about 63.7 volts, since, in one-half cycle of a sine wave, the average value is 0.637 times the peak value. In the case of a half-wave rectifier, where every other half cycle is cut off, the average voltage is just 31.9 volts when the peak voltage is 100 (B in AVERAGE VALUE).

For a source of alternating current, the average voltage is usually zero over a long period of time, since the polarity is positive during half the cycle and negative during the other half, and the peak values are identical in both directions. An ac voltage has, of course, very different effects from zero voltage, and therefore the average voltage is given little importance in ac circuits. The root-mean-square, or RMS, voltage is more often specified as the equivalent dc voltage. *See also* ROOT MEAN SQUARE.

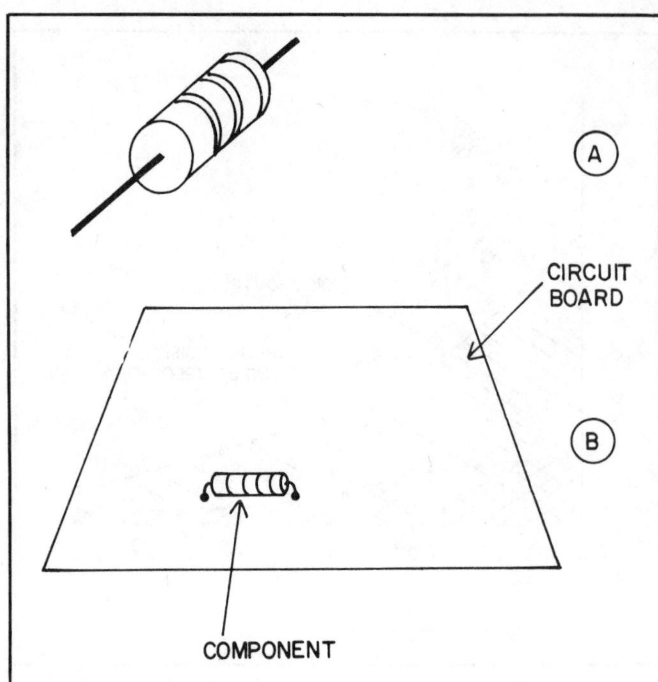

AXIAL LEADS: At A, general orientation. At B, component mounting on a circuit board. Axial leads allow the component to be placed first on the board.

AVIONICS

Avionics is the field of aviation electronics. Modern air transport depends heavily on electronic devices. Circuits are needed to monitor vital functions in an aircraft, determine the exact location of a plane, detect other planes in the vicinity, and maintain communications with ground-based stations (*see* AIR TRAFFIC CONTROL, RADIOLOCATION). There is little room for error in avionics; a single component failure could result in disaster. Therefore, avionics equipment must meet exacting standards.

A modern jet aircraft contains miles of wire and dozens of monitoring and controlling circuits. Modern aircraft also have computers on board that coordinate the operation of all electronic systems.

AVOGADRO CONSTANT

The Avogadro constant, often called Avogadro's number is approximately 6.022169×10^{23}, or 602,216,900,000,000,000,000,000. It is sometimes abbreviated *N*. This number of atoms or molecules is called one mole. The atomic weight (*see* ATOMIC WEIGHT) of a given element is the weight, in grams, of one mole of atoms of that element. One gram is therefore 6.022169×10^{23} atomic-mass units.

Avogadro's number is primarily of interest to chemists and physicists.

AWG

See AMERICAN WIRE GAUGE.

AXIAL LEADS

A component with axial leads has the leads protruding along a common, linear axis, as shown in A and B. Most

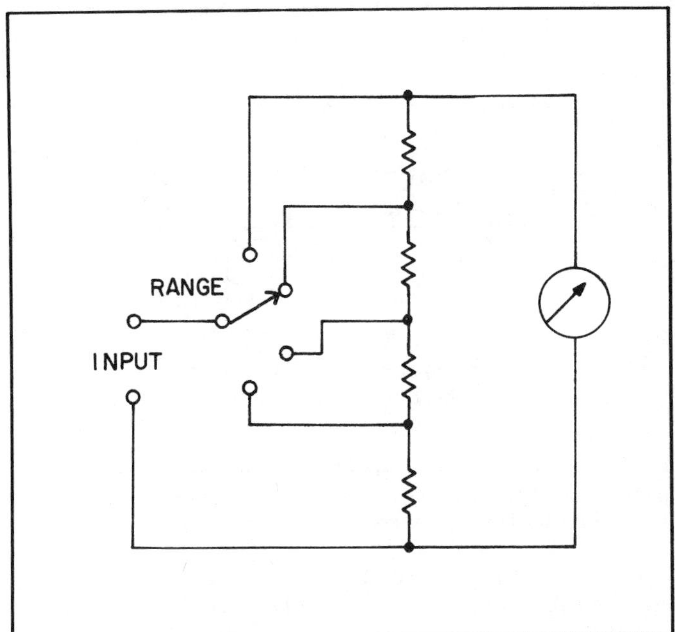

AYRTON SHUNT: The Ayrton shunt allows sensitivity adjustment of a meter, without changing the resistance across the meter itself.

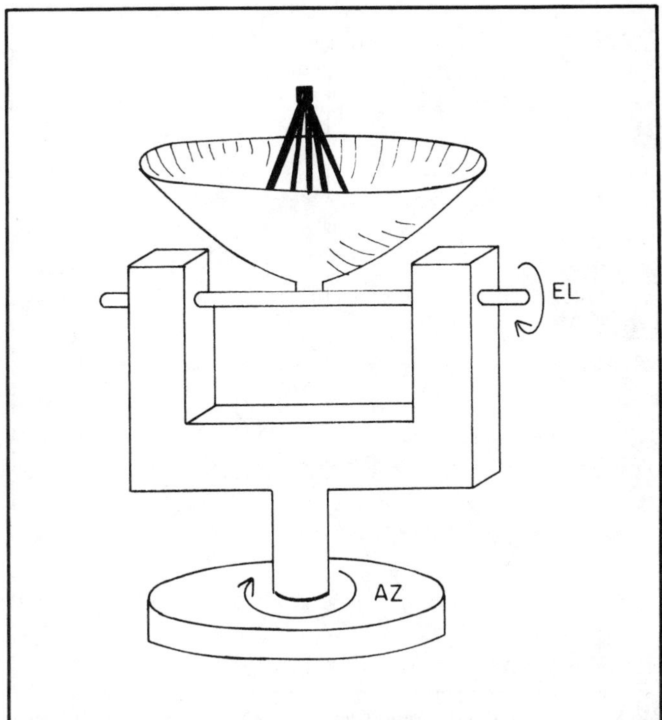

AZ-EL: An az-el mounting for a highly directional antenna.

resistors and semiconductor diodes, as well as many capacitors and inductors, have axial leads.

Axial leads give a component extra mechanical rigidity on a circuit board, since the component can be mounted flush with the board. Axial leads have less mutual inductance than parallel leads (*see* PARALLEL LEADS), because of their greater separation and their collinear orientation.

Axial leads are sometimes inconvenient when a component must be mounted within a small space on a printed-circuit board. In such situations, parallel leads are often preferred.

AXIS

The axis of an object is a set of points, usually a straight line, around which the object revolves or rotates, or could be revolved or rotated. For example, the axis of our planet is a straight line passi g through the geographic poles.

An axis may also be defined as a straight line about which an object is symmetical. The leads of a resistor, for example, are called axial leads (see the illustration for AXIAL LEADS), because the body of the component is symmetrical about the line containing the leads.

When one object orbits in a circle around another, the axis of revolution is a straight line passing through the center of the orbit and perpendicular to the plane containing the orbit.

The term axis is also occasionally applied to the scales of a graph. The horizontal axis is called the abscissa, representing the independent variable; the vertical axis is called the ordinate, representing the dependent variable.

AYRTON SHUNT

The sensitivity of a galvanometer is reduced by means of a shunt resistor. The addition of variable shunt resistors, however, sometimes affects the meter movement, or damping (*see* GALVANOMETER). A special type of shunt, called an Ayrton shunt, allows adjustment of the range of a galvanometer without affecting the damping. Such a shunt consists of a series combinatin of resistors. Several different galvanometer ranges can be selected by connecting the input current across variable portions of the total resistance (see illustration).

Because the resistance across the galvanometer coil is the same regardless of the range selected, the damping of the meter movement does not depend on the range chosen. The effective input resistance to the meter, however, does change. The more sensitive the meter range, the greater the resistance at the meter input terminals. A microammeter, for example, with a range of 0 to 100 microamperes, can be used as a milliammeter or ammeter without changing the damping when a suitable Ayrton shunt system is provided (*see* AMMETER). The common volt-ohm-milliammeter often utilizes an Ayrton shunt to facilitate operation in its various modes. *See also* VOLT-OHM-MILLIAMMETER.

AZ-EL

The term az-el refers to azimuth and elevation. Azimuth is the horizontal direction or compass bearing, and elevation is the angle relative to the horizon, for a directional device. Azimuth bearings range from zero (true north) clockwise through 90 degrees (east), 180 degrees (south), 270 degrees (west), to 360 degrees (true north). Elevation bearings range from zero (horizontal) to 90 degrees (straight up).

An az-el mounting is a two-way bearing in which the azimuth and elevation are independently adjustable. Satellite communication systems, and some telescopes, make use of such mountings (see the illustration). Some moonbounce systems also use this kind of mounting (*see* MOONBOUNCE).

If it is necessary to continuously follow a heavenly body such as the moon, a planet, the sun, or a star, the az-el

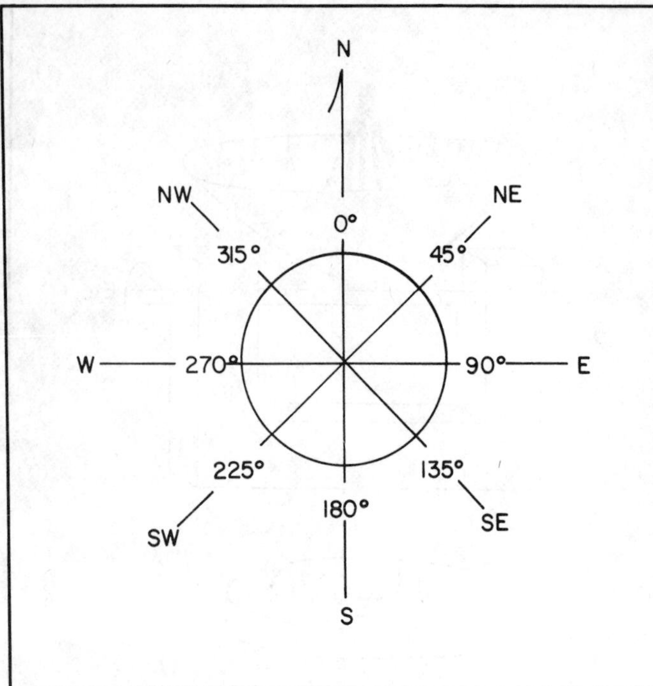

AZIMUTH: Azimuth bearings begin at true north (zero) and proceed clockwise, as viewed from above.

mounting requires continual adjustment of both the azimuth and elevation bearings. A special kind of mounting, called an equatorial mounting, consists of an az-el system oriented with its zenith toward the north celestial pole, or North Star. Then only the azimuth bearing requires adjustment to compensate for the rotation of the earth. This facilitates prolonged antenna orientation toward one celestial object. *See also* DECLINATION, RIGHT ASCENSION.

AZIMUTH

The term azimuth refers to the horizontal direction or compass bearing of an object or radio signal. True north is azimuth zero. The azimuth bearing is defined in degrees, measured clockwise from true north as shown in the illustration. Thus, east is azimuth 90 degrees, south is azimuth 180 degrees, and west is azimuth 270 degrees. Provided the elevation is less than 90 degrees, as is usually defined by convention, these azimuth bearings hold for any object in space.

Azimuth bearings are always defined as being at least zero, but less than 360 degrees, avoiding ambiguity. Measures smaller than zero degrees, and greater than or equal to 360 degrees, are not defined.
the sky, whether it is on the horizon or not.

AZIMUTH ALIGNMENT

In a tape recorder, the playback- and recording-head gaps should have their center lines oriented exactly parallel with each other, as shown in the illustration. These lines should also be exactly perpendicular to the recording tape. Azimuth alignment is correct if, and only if, both of these constraints are met.

If the recording and reproducing head gaps are not both perpendicular to the tape, or if they are not parallel with each other, the high-frequency response of the system will be impaired. This may not be important in voice re-

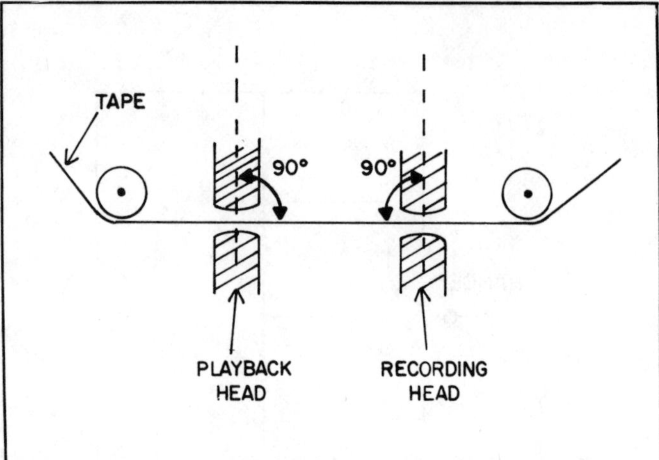

AZIMUTH ALIGNMENT: Azimuth alignment in a tape recorder is essential for the best sound reproduction.

cording, but the fidelity of music is seriously affected by such misalignment. (*See also* AUDIO RESPONSE, RECORDING HEAD.)

AZIMUTH BLANKING

As a radar antenna rotates, echoes are picked up from all directions, or azimuth bearings. Occasionally, it is necessary to eliminate the echoes from certain directions. This is done by simply shutting off the radar cathode-ray-tube screen while the antenna is pointed in the selected azimuth position or positions. A timing system accomplishes this; proper synchronization with the antenna maintains blanking in the desired directions. See illustration.

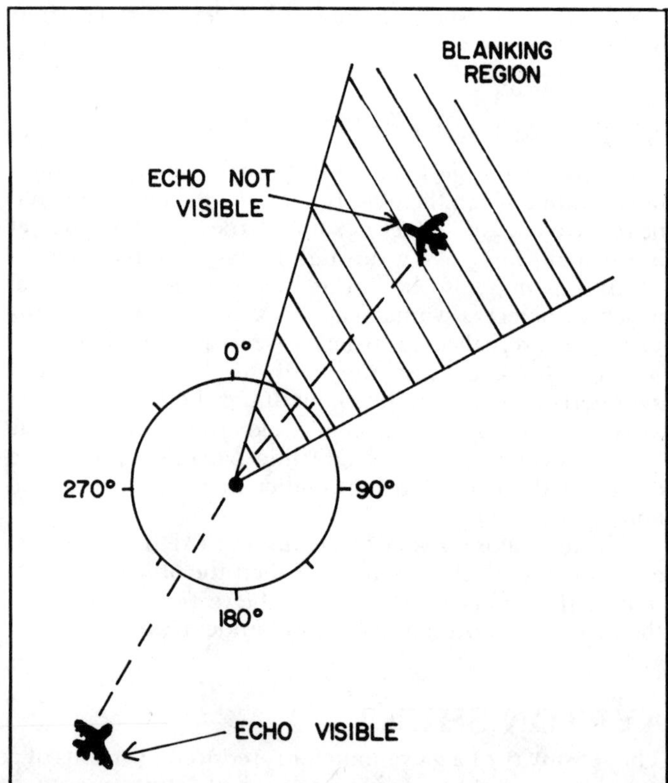

AZIMUTH BLANKING: Azimuth blanking is a means of eliminating echoes on radar from certain directions. Here, the blanking region extends from approximately azimuth 20 degrees to azimuth 60 degrees.

Azimuth blanking is used to eliminate unwanted echoes or distractions on a radar screen. *See also* AZIMUTH, RADAR.

AZIMUTH RESOLUTION

Azimuth resolution is the ability of a radar system to distinguish between objects at the same distance, but slightly different directions, from the antenna.

A and B illustrate the determination of azimuth resolution for a radar system. At A, the two aircraft are too close together to show up as separate echoes on the radar screen. But at B, the aircraft are sufficiently far apart for the radar to "see" them as two, rather than one.

There is a certain minimum angle that the objects must subtend with respect to the radar antenna in order for the radar to distinguish them as two separate objects rather than a single echo. This angle is the azimuth resolution of the system. It depends on many variables, including the radar wavelength, the size of the antenna, the pulse frequency, the speed at which the antenna rotates, and the resolution of the cathode-ray display tube.

Azimuth resolution is specified as an angle. Therefore, the farther away the objects are from the radar antenna, the greater the space between the objects must be in order for them to show up as separate echoes. For example, as two planes fly next to each other at constant spacing, directly away from the radar antenna, their echoes may appear distinct until they reach a certain distance. When they pass this distance, their echoes will appear to merge together because the angle between them has become smaller than the azimuth resolution of the system. *See also* AZIMUTH, RADAR, RESOLUTION.

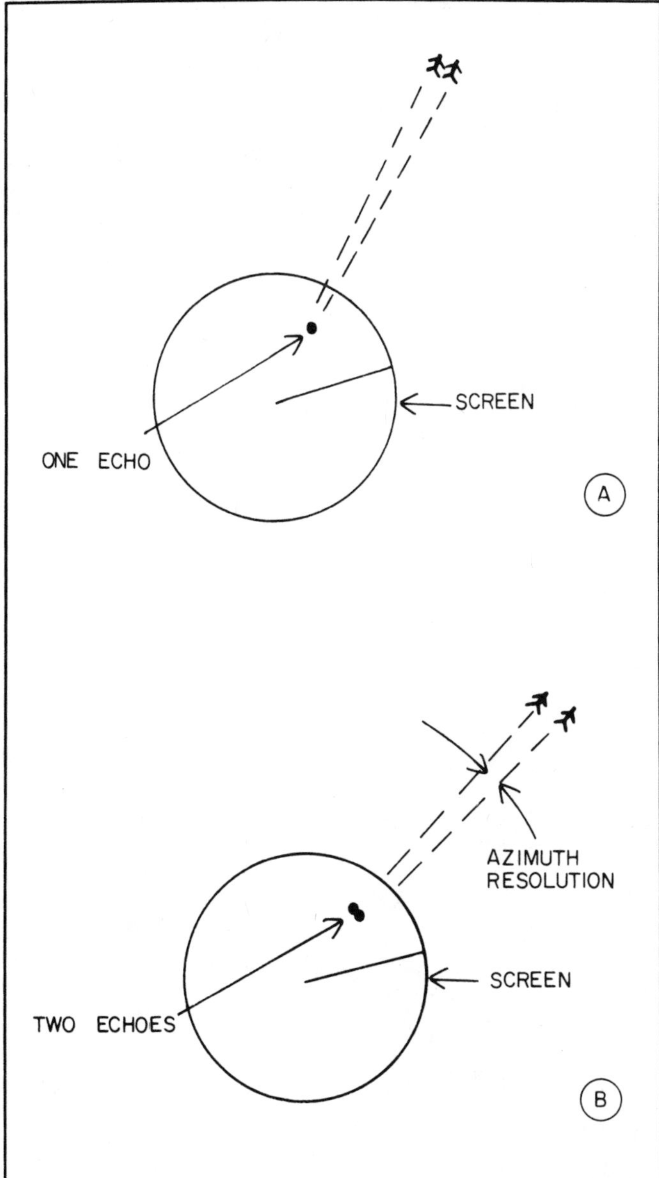

AZIMUTH RESOLUTION: At A, the distant objects subtend an angle too small for the radar to see the images separately. At B, however, the two planes are far enough apart in azimuth for the echoes to return and be seen as two spots on the screen. The azimuth resolution of the system is the smallest angle that allows a double echo when the objects are equidistant from the antenna.

AZUSA

Azusa is the name for a three-dimensional tracking system that operates in the frequency range of approximately 4 to 6 GHz (4×10^9 to 6×10^9 Hz).

The location of any object in three-dimensional space can be defined with respect to a reference point in terms of two angles and a radial distances (see the illustration). The two angles might be the azimuth and elevation (*see* AZIMUTH, ELEVATION), measured in degrees. The radial distance might be specified in meters. However, any units can be used for angle and distance measure.

Phase comparison is used in the Azusa system to locate

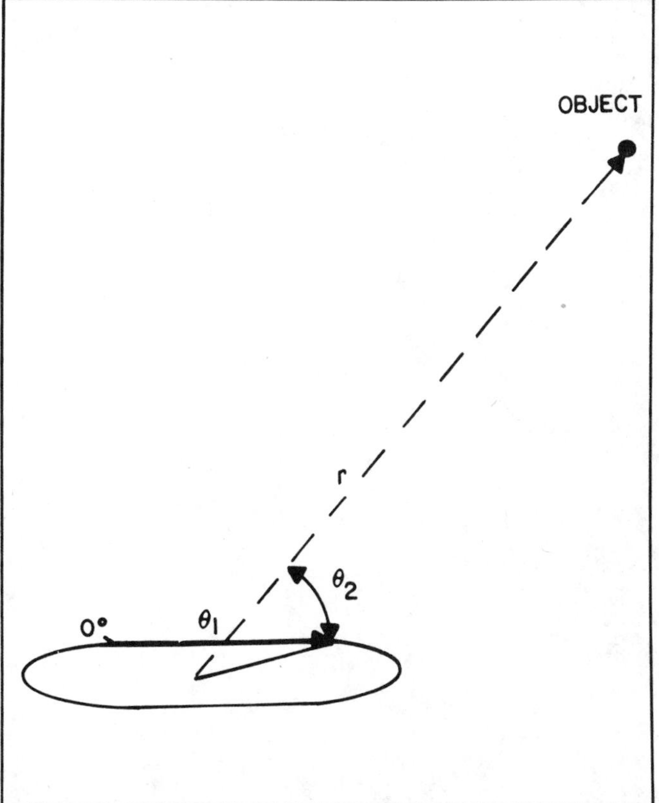

AZUSA: The Azusa system uses two direction angles, Θ_1 (azimuth in this drawing) and Θ_2 (elevation), and the slant range, r, to determine the exact location of an object.

an object in space. A short base line and two antennas are used. A computer provides two direction cosines and the radial distance, called the slant range.

The Azusa system was developed by Convair for use in measuring the positions of missiles immediately after launch. A transmitter in the missile provides a signal that facilitates precise determination of its position. *See also* GUIDED MISSILE.

B

BABBLE SIGNAL

A babble signal is a form of jamming device that is used for deception purposes.

Enemy signals are intercepted and recorded. Then they are retransmitted, at a later time, on the same frequency as the original signal. This is done with the intention of confusing the enemy. When previously recorded enemy signals are retransmitted on top of live signals, it becomes difficult to distinguish between the real transmission and the decoy. Special codes and directional signals can be used to overcome the confusion caused by babble signal. *See also* JAMMING.

BACK BIAS

Back bias is a voltage taken from one part of a multistage circuit and applied to a previous stage. This voltage is obtained by rectifying a portion of the signal output from an amplifier circuit. Back bias may be either regenerative or degenerative, or it can be used for control purposes.

A typical application of back bias is the automatic level control (*see* AUTOMATIC GAIN CONTROL, AUTOMATIC LEVEL CONTROL). A degenerative voltage is applied to either the input of the stage from which it is derived, or to a previous stage. This prevents significant changes in output amplitude under conditions of fluctuating input signal strength.

Back bias can be appied to either the anode or cathode of an active device. A negative voltage, for example, might be applied to the plate of a tube or the collector of an NPN transistor to reduce its gain. Alternatively, a positive voltage could be applied to the cathode of a tube or the emitter of an NPN transistor, if the cathode or emitter is elevated above dc ground. Still another form of degenerative back bias is the application of a negative voltage to the grid of a tube or the base of an NPN transistor.

Sometimes the term back bias is used to define reverse bias. This condition occurs when the anode of a diode, transistor, or tube is negative with respect to the cathode. *See also* REVERSE BIAS.

BACK DIODE

A back diode is a special kind of tunnel diode, operated in the reverse-bias mode. Tunnel diodes are used as oscillators and amplifiers in the uhf and microwave part of the electromagnetic spectrum (*see* TUNNEL DIODE).

The back diode has a negative-resistance region at low levels of reverse bias. This means that, within a certain range of voltage, an increase in reverse voltage will cause a decrease in the flow of current.

An ordinary semiconductor diode has a voltage-vs-current curve similar to that shown at A in the illustration. The back diode has a voltage-vs-current characteristic similar to that shown at B. The back diode is normally operated in the region indicated, at small values of reverse voltage (E_R). This results in oscillation at microwave frequencies.

BACK EMISSION

When a diode, triode, tetrode, or pentode tube is forward biased, electrons are emitted by the cathode and flow to the anode. Under conditions of reverse bias, when the anode or plate is negative with respect to the cathode, very little current flows. There is some emission of electrons by the plate under conditions of reverse bias. This is called back emission.

Back emission is generally very small, resulting in negligible current flow. If the reverse bias voltage is increased sufficiently, a sudden large increase in back emission occurs. This is called arcback (*see* ARCBACK). It is undesirable because it causes abnormal operation of a vacuum tube. All tubes have peak-inverse-voltage specifications that are well within the limitations of the device.

Back emission differs from secondary emission. Secondary emission results from the impact of high-speed electrons against the plate during normal operation. *See also* PRIMARY EMISSION, RECTIFIER, RECTIFIER TUBE, SECONDARY EMISSION.

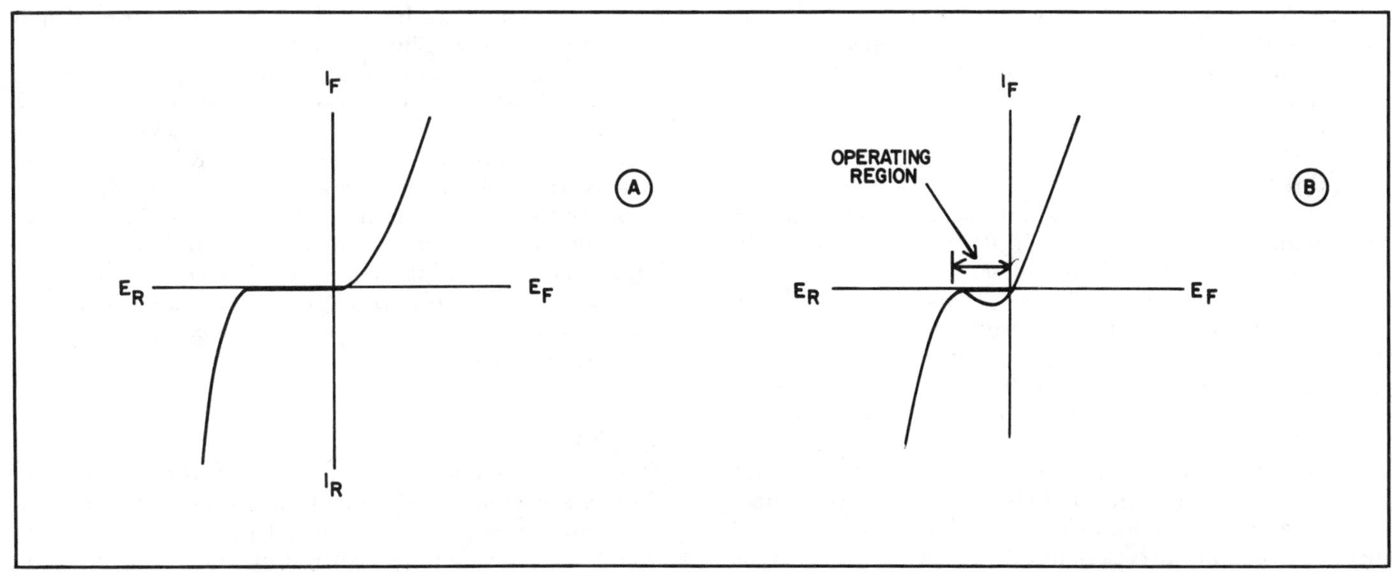

BACK DIODE: At A, voltage-vs-current plot for a typical semiconductor diode. At B, voltage-vs-current plot for a back diode.

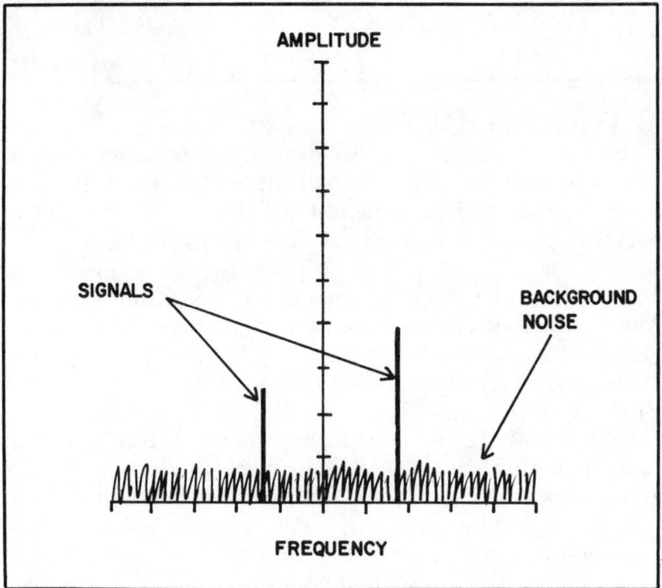

BACKGROUND NOISE: Background noise appears as "grass" on a spctrum analyzer.

BACKGROUND NOISE

In any amplifier, oscillator, or other device that draws current from a power supply, there is always some noise output in addition to the desired signals. This noise has no defined frequency, but instead consists of a random combination of impulses. This is called background noise. In a receiver, the background noise is heard as a steady hiss at the speaker. Some background noise also comes from external sources.

Background noise limits the sensitivity of a receiver and the dynamic range of some test instruments. Some thermal noise is generated as a result of molecular and atomic motion in all substances. Some noise is generated by the motion of electrons through conductors and semiconductors, from the impact of electrons against the plate and grids of vacuum tubes, and similar effects. A certain amount of background noise originates in outer space, emitted from distant stars and galaxies. Our own sun is a significant source of background noise at some frequencies.

On a spectrum analyzer, background noise and signals can be clearly identified (see illustration). Any signal that is weaker than the average level of the background noise will be undetectable. *See also* NOISE, SENSITIVITY, SIGNAL-TO-NOISE RATIO.

BACKLASH

Any rotary, or continuously adjustable, control will have a certain amount of mechanical imprecision. The accuracy of adjustment is limited, of course, by the ability of an operator to interpolate readings. (A good example of this is the ordinary receiver or transmitter frequency dial.) Some inaccuracy, however, is a consequence of the mechanical shortcomings of such things as gears, planetary drives, slide-contact devices, and the like. This is called *blacklash.*

Often, if a certain setting for a control is desired, that setting will depend on whether the control is turned clockwise or counterclockwise. For example, suppose it is necessary to tune precisely to the frequency of WWV at 15.0000 MHz. This is usually done in receivers by zero beating (*see* ZERO BEAT). When tuning upward from some frequency below 15 MHz, the dial might indicate 15.0002 MHz at zero beat. When tuning downward from some frequency above 15 MHz, the dial might read 14.9997 at zero beat. The difference, 15.0002 14.9997 or 0.0005 MHz, (0.5kHz), is the dial backlash.

Because backlash is a purely mechanical phenomenon, and not an electrical effect, it can be eliminated by the use of digital displays. The precision of the adjustment is limited only by the resolution of the display; backlash and human error are overcome. *See also* DIGITAL CONTROL DIGITAL DISPLAY.

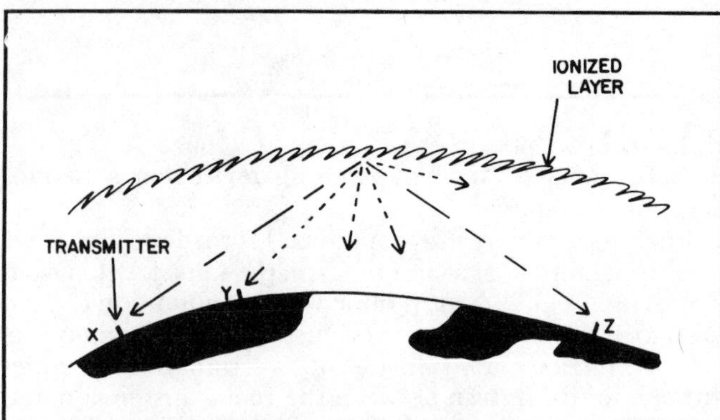

BACKSCATTER: Backscatter allows communication within the skip zone. Normally, station Y cannot hear station X. Station Z hears X by normal ionospheric propagation. When backscatter is strong enough, however, station Y can hear station X, even though Y is within the skip zone.

BACKSCATTER

Backscatter is a form of ionospheric propagation via the E and F layers (*see* E LAYER, F LAYER).

Generally, when a high-frequency electromagnetic field encounters an ionized layer in the upper atmosphere, the angle of return is roughly equal to the angle of incidence (*see* ANGLE OF INCIDENCE). However, a small amount of the field energy is scattered in all directions, as shown in the drawing. Some energy is scattered back toward the transmitting station, and into the skip zone if there is a skip zone (*see* SKIP ZONE). A receiving station within the skip zone, not normally able to hear the transmitting station, may hear this scattered energy if it is sufficiently strong. This is called backscatter.

Backscatter signals are often, if not usually, too weak to be heard. Under the right conditions, though, communication is possible at high frequencies via backscatter, over distances that are normally too short for E-layer or F-layer propagation. Backscatter signals are characterized by rapid fluttering and fading. This makes reception of amplitude-modulated or single-sideband signals almost impossible. The multipath nature of backscatter propagation (*see* MULTIPATH FADING) often makes even CW reception very poor. *See also* PROPAGATION, PROPAGATION CHARACTERISTICS.

BAFFLE

A baffle is a sound-shielding or sound-reflecting object. Baffles are commonly used in speakers to prevent the backward radiated sound waves from bouncing off the rear of the speaker enclosure and interfering with the desired forward radiation of sound energy (see illustra-

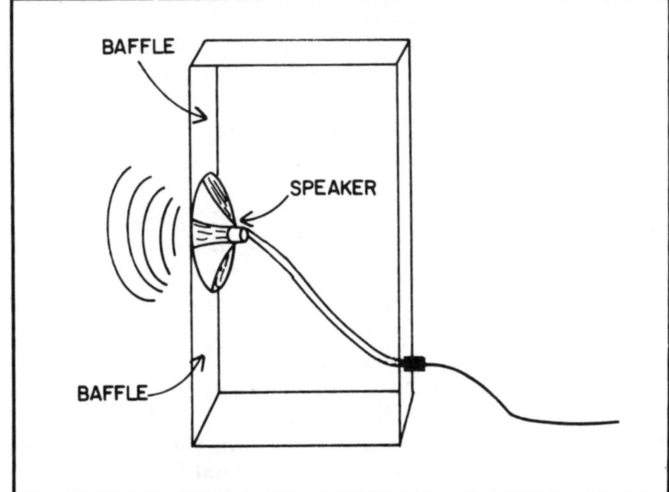

BAFFLE: A baffle at the front of a speaker enclosure prevents reflected sound from being heard from within the enclosure.

tion). Under certain conditions, a baffle is deliberately not used, or a hole is cut in it, so that a portion of the reflected sound energy is radiated in a forward direction. An example of this is the bass-reflex enclosure (*see* BASS-REFLEX ENCLOSURE).

Baffles are sometimes used in concert halls to direct sound from the stage to the entire audience in a uniform pattern. Such baffles may be hung in strange configurations from the ceiling of such an auditorium.

In an optical communications system, light baffles are sometimes ued to reduce the level of interfering background light picked up by the receiving device. Excessive ambient light can degrade the sensitivity and efficiency of such systems. *See also* OPTICAL COMMUNICATIONS.

BALANCE

The term balance is used to denote the most desirable set of circuit parameters in a variety of situations.

In a high-fidelity stereo sound system, balance refers to the proper ratio of sound intensity in the left and right channels. The correct balance is not necessarily the equalization of sound intensity in both channels; ideal balance is obtained when the sound reproduction most nearly duplicates the actual sound arriving at the pickup microphones. A balance control (*see* BALANCE CONTROL) is used in a stereo sound system to allow for correct adjustment of the balance.

In a radio-frequency transmission line consisting of two parallel conductors, balance is the condition in which the currents in the conductors flow with equal magnitude, but in opposite directions, everywhere along the line. A properly operating parallel-wire transmission line should always be kept in balance (*see* BALANCED TRANSMISSION LINE).

When two or more vacuum tubes or transistors are operated in push-pull or in parallel, it is desirable to balance the devices so that they both have essentialy the same gain characteristics. Supposedly identical tubes or transistors often display slightly different amounts of gain under the same conditions. If one tube or transistor draws more current than the other, a phenomenon known as current hogging can take place (*see* CURRENT HOGGING). Differences in input impedances, even if very small, can disrupt the efficiency and linearity of a circuit containing devices in push-pull or in parallel.

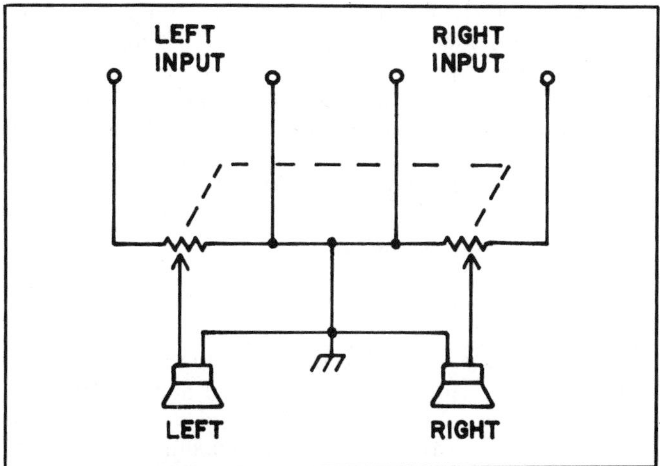

BALANCE CONTROL: A balance control consisting of two potentiometers in tandem.

In a single-sideband transmitter, carrier balance refers to the most nearly complete suppression of the carrier signal, leaving only the sidebands. An adjustment is usually provided in a balanced modulator to facilitate proper carrier balance (*see* BALANCED MODULATOR).

BALANCE CONTROL

A balance control is usually made up of one or two potentiometers in a high-fidelity stereo sound system, for the purpose of adjusting the balance between the left and right channels (*see* BALANCE). The schematic shows an example of a single stereo balance control consisting of two potentiometers in tandem. The amplifier gain is held constant in this example; the entire output of the left and right channel amplifiers appears across the resistances. This system may be used in low-power arrangements. In stereophonic systems with high-power amplifiers, the amplifier input is usually adjusted to control the gain.

The degree of balance necessary to achieve faithful reproduction of the original sound is usually, but not always, obtained when the left channel gain is the same as the right channel gain. The positioning of the speakers, and the characteristics of the room in which the stereo system is used, can affect the relative gain between the channels that results in the best balance. *See also* STEREOPHONICS.

BALANCED CIRCUIT

When an amplifier, oscillator, or modulator contains two elements operating on opposite-phase components of the signal cycle, with equal gain, the circuit is said to be balanced. A push-pull amplifier (see illustration) is one example of a balanced circuit. Each transistor operates on exactly half of the input cycle. Each transistor has the same gain as the other. The combination of their outputs therefore forms a complete, and undistorted, cycle.

Other examples of balanced circuits include the balanced modulator, push-pull amplifier/doubler, astable multivibrator, and parallel-transistor or parallel-tube amplifier. (*See* ASTABLE MULTIVIBRATOR, BALANCED MODULATOR, PARALLEL TRANSISTORS, PARALLEL TUBES.)

If the active components of a circuit, which is intended

to be balanced, have unequal gain characteristics or current drain, unbalance will be introduced. This results in degraded circuit performance. It can also lead to current hogging, and possible destruction of one or both of the amplifying or oscillating devices (*see* CURRENT HOGGING).

The direct input and output of a balanced circuit are characterized by the fact that neither lead is grounded. A transformer, however, allows an unbalanced input or load to be used with a balanced circuit.

BALANCED DETECTOR

A balanced detector is a special kind of demodulator for reception of frequency-modulated signals (FM). Any FM detector should, if possible, be sensitive only to variations in frequency, and not variations in amplitude, of the received signals. A schematic diagram of a simple circuit that accomplishes this, called a balanced detector, is shown in the schematic.

The tuned circuits, consisting of L1/C1 and L2/C2, are set to slightly different frequencies. By adjusting C1 so that the resonant frequency of L1/C1 is slightly above the center frequency of the signal, and by adjusting C2 so that the resonant frequency of L2/C2 is slightly below the center frequency of the signal, the circuit becomes a sort of frequency comparator. Whenever the signal frequency is at the center of the channel, midway between the resonant frequencies of L1/C1 and L2/C2, the transistors Q1 and Q2 produce equal outputs, and the output of the balanced detector is zero. When the signal frequency rises, the output of Q1 increases and the output of Q2 decreases. When the signal frequency falls, the output of Q2 increases and the output of Q1 decreases. The filtering capacitors, C_F, smooth out the RF component of the signals, leaving only audio frequencies at the output.

An amplitude-modulated signal will produce no output from the balanced detector. The outputs of Q1 and Q2 are always exactly equal for an AM signal at the center of the channel, although they may both change as the signal amplitude changes. The gain characteristics of Q1 and Q2, if properly chosen, will minimize the response of the balanced detector to AM signals not at the center of the channel. *See also* DISCRIMINATOR, FREQUENCY MODULATION, RATIO DETECTOR, SLOPE DETECTION.

BALANCED LINE

Any electrical line consisting of two conductors, each of which displays an equal impedance with respect to ground, is called a balanced line if the currents in the conductors are equal in magnitude and opposite in direction at all points along the line.

The advantage of a balanced line is that it effectively prevents external fields from affecting the signals it carries. The equal and opposite nature of the currents in a balanced line results in the cancellation of the electromagnetic fields set up by the currents everywhere in space except in the immediate vicinity of the line conductors.

Precautions must be taken to ensure that balance is maintained in a line that is supposed to be balanced. Otherwise, the line loss increases, and the line becomes susceptible to interference from external fields (in receiving applications) or radiation (in transmitting applications).

Some balanced lines use four, instead of two, parallel conductors arranged at the corners of a geometric square in the transverse plane. Diagonally opposite wires are connected together. This is called a four-wire line. It displays more stable balance than a two-wire line. *See also* FOUR-WIRE TRANSMISSION LINE, OPEN-WIRE LINE, TRANSMISSION LINE, TWIN LEAD.

BALANCED LOAD

A balanced load is a load that presents the same impedance, with respect to ground, at both ends or terminals. This means that the resistance, capacitance and inductance are identical at both terminals. A balanced load may be reversed—that is, the terminals transposed—without affecting circuit performance. A balanced load is required at the termination of a balanced transmission line, to ensure that the currents in the line will be equal and opposite (*see* BALANCED LINE, BALANCED TRANSMISSION LINE).

A good example of a balanced load is a center-fed length of conductor in free space, as shown at A in the illustration. Such an antenna, if fed at a right angle with a balanced transmission line, presents a balanced load to a transmitter at all frequencies in the electromagnetic spectrum. If the length or orientation of either side of this

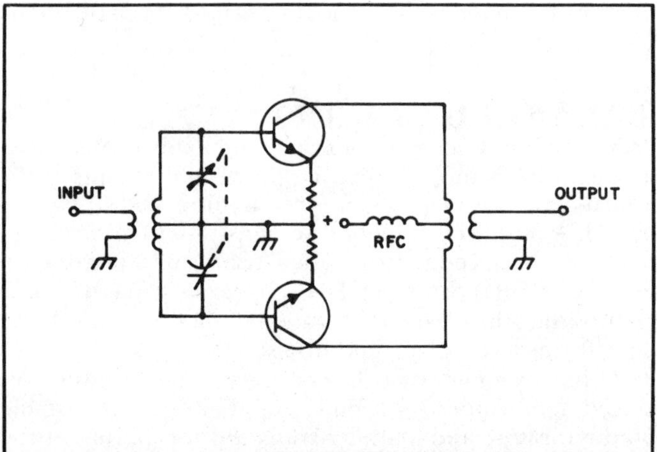

BALANCED CIRCUIT: The push-pull amplifier is an example of a balanced circuit.

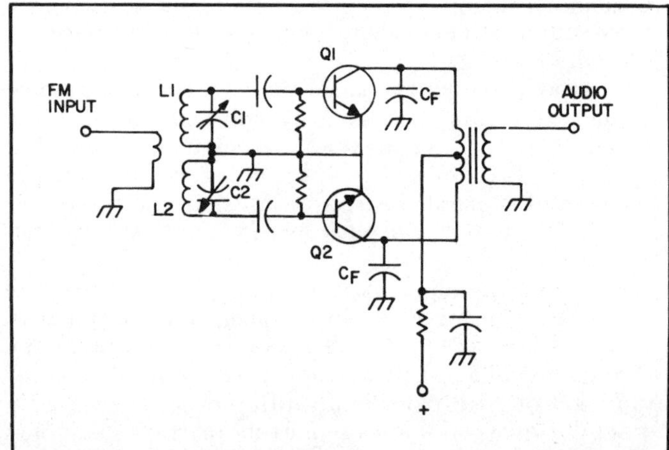

BALANCED DETECTOR: The two tuned circuits, L1/C1 and L2/C2, are tuned slightly above and below the center of the channel. Frequency deviations of the incoming signal thus cause an audio-frequency output.

antenna is changed, however, the balance of the system will be upset. One side of the antenna may, for example, be closer to some object (such as the ground) than the other side, as shown at B, or the transmission line might not be brought away from the conductor at a right angle, as at C. These things will ruin the balance of the load, as seen at the transmitter end of the feed line.

In practice, a perfectly balanced load can never truly be achieved in radio-frequency applications; it can only be approached. At audio frequencies, where the load does not radiate appreciable amounts of electromagnetic energy, a nearly balanced load is often achievable.

A commonly believed falsehood is that an impedance mismatch between the transmission line and the antenna upsets the system balance. Actually, a perfect impedance match between a line and the load is not essential for balance to exist. While an impedance mismatch does not affect system balance, poor balance can create an impedance mismatch where none would exist if the balance were maintained. *See also* IMPEDANCE MATCHING.

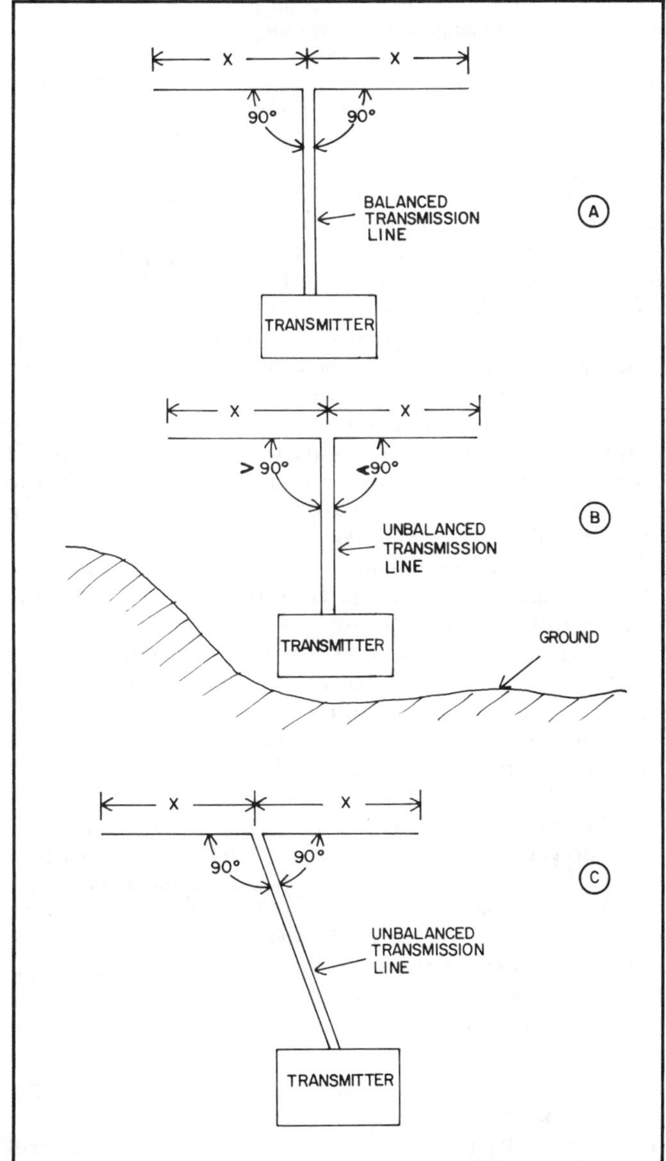

BALANCED LOAD: At A, a theoretically balanced load. At B, unbalance caused by proximity of ground on one side of the system. At C, unbalance caused by nonsymetrical positioning to the feed line with respect to the antenna.

BALANCED MODULATOR

To generate single-sideband signals, a balanced modulator is necessary to suppress the carrier energy (*see* SINGLE SIDEBAND). One such circuit is shown in simplified form (see drawing).

Transistors Q1 and Q2 act as ordinary amplitude modulators. While the audio input to Q1 is in phase with the audio input to Q2, however, their RF carrier inputs are out of phase. The collectors are connected in parallel and are thus in phase. The carrier is therefore cancelled in the output circuit, leaving only the sideband energy. The resulting signal is called a double-sideband, suppressed-carrier signal.

There are several ways to build balanced modulators, using both active devices (as shown here) and passive devices, such as diodes. All balanced modulators are designed to produce the double-sideband, suppressed-carrier type of signal. To get single-sideband emission, one of the sidebands from the balanced modulator is eliminated, either by filtering or by phase cancellation.

BALANCED OSCILLATOR

Any oscillator with center-tapped, grounded tank-circuit inductors is called a balanced oscillator. A simple balanced oscillator might consist of a tuned push-pull amplifier with inductive feedback, such as the circuit illustrated. The variable capacitors, C, determine the frequency of oscillation; the input circuit is kept in balance by using a split-stator (ganged) capacitor, ensuring identical values on either side of the tank circuit. Transistors Q1 and Q2 must have the same gain characteristics, as well, for good balance to be maintained.

Of course, the "push-pull oscillator" is not the only kind of balanced oscillator. The circuit shown will tend to cancel all even-ordered harmonics in the output circuit, just as a push-pull amplifier does. *See also* OSCILLATOR, PUSH-PULL AMPLIFIER, PUSH-PULL CONFIGURATION.

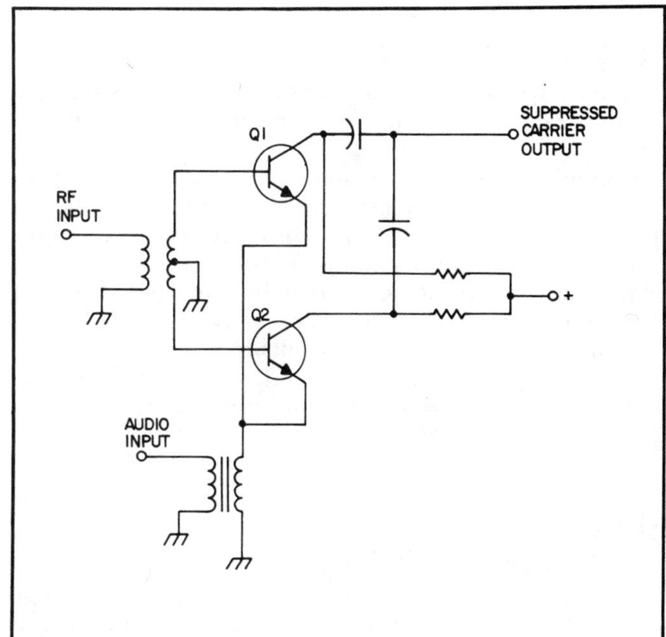

BALANCED MODULATOR: A balanced modulator cancels the carrier signal at the output, but leaves the sideband energy.

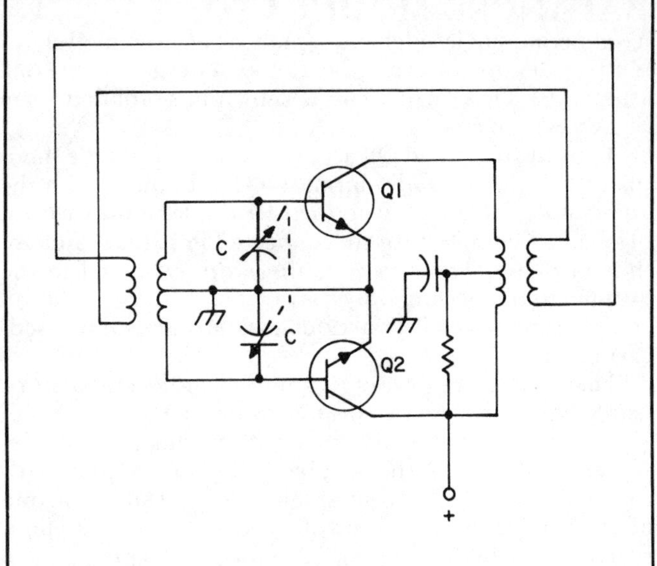

BALANCED OSCILLATOR: A balanced oscillator with inductive feedback.

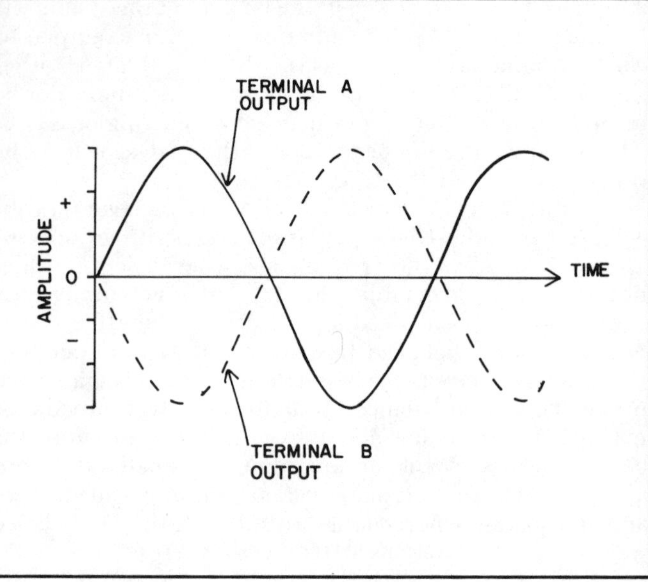

BALANCED OUTPUT: The signals at the two terminals of a balanced output are equal in amplitude but opposite in phase.

BALANCED OUTPUT

A balanced-output circuit is designed to be used with a balanced load and a balanced line. There are two output terminals in such a circuit; each terminal displays the same impedance with respect to ground. The phase of the output is opposite at either terminal with respect to the other (see illustration). The peak amplitudes at both terminals are identical.

A balanced output can be used as an unbalanced output (*see* UNBALANCED SYSTEM) by grounding one of the terminals. However, an unbalanced output cannot be used as a balanced output. A special transformer is needed for this purpose. In radio-frequency applications, such a transformer is called a balun, short for "balanced/unbalanced." *See also* BALUN.

BALANCED SET

When components such as tubes and transistors are connected in parallel or push-pull configurations, it is important that they have similar gain characteristics. Two semiconductor transistors might have identical manufacturer numbers, but unless a pair is chosen carefully, the probababilty is high that minor differences will exist in the actual current gain and amplification factors. Such dissimilarities can result in current hogging (*see* CURRENT HOGGING) or unwanted unbalances in the system.

When two or more components have been chosen on the basis of identical, or nearly identical, gain and load characteristics, the set of components is said to be balanced or matched.

BALANCED TRANSMISSION LINE

A balanced transmission line is a form of balanced line used in radio-frequency antenna transmitting and receiving systems. Such a line may have two or four parallel wires. A two-wire balanced transmission line operates as illustrated.

The currents in the parallel wires are equal in magnitude but opposite in direction (heavy arrows). These currents set up electric, or E, fields, shown by the dotted lines, and magnetic, or M, fields, shown by the solid circles. The E and M fields are perpendicular to each other everywhere in space, but they are of appreciable strength only in the immediate vicinity of the conductors.

Because the E and M fields are mutually perpendicular, an electromagnetic field is produced. This field propagates in a direction perpendicular to the E and M fields, or right along the conductors of the transmission line. This happens with a speed of almost 300,000,000 meters per second, or the velocity of light.

With a balanced transmission line such as that shown, it is important that the currents in the conductors be equal in amplitude and opposite in direction at all points along the line. This keeps the electromagnetic energy traveling along the line, and prevents it from being radiated into space. If the current balance is upset for any reason, some electromagnetic energy will be radiated away from the line. If an improperly balanced two-wire line is used for receiving, some signal pickup will occur along the line length. This adversely affects the performance of an antenna system, especially a directional one.

Balanced transmission lines usually have lower signal loss than unbalanced lines. This is because, in a balanced line, some of the dielectric is air, which has low loss, but in an unbalanced line, the dielectric often is entirely polyethylene, which has somewhat higher loss. Low loss is the chief advantage of open-wire balanced lines. Unbalanced lines are more convenient to install because they are less susceptible to the effects of nearby metallic objects. *See also* FOUR-WIRE TRANSMISSION LINE, OPEN-WIRE LINE, TRANSMISSION LINE, TWIN LEAD, UNBALANCED TRANSMISSION LINE.

BALLAST

The term ballast refers to current regulation or control under conditions of fluctuating voltage. Without ballast, a sudden, large increase in the voltage supplied to a circuit can result in malfunction or damage. In most electronic circuits today, voltage regulation in power supplies makes

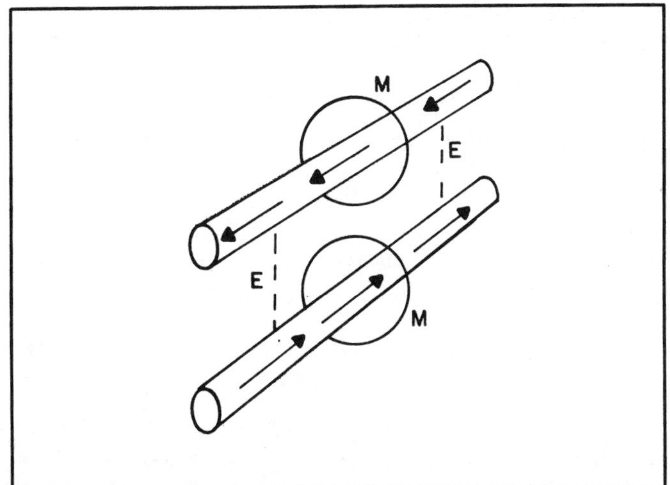

BALANCED TRANSMISSION LINE: The electric field (E) is represented by the dotted lines. The magnetic field (M) is represented by the circles. The currents in the conductors are shown by the arrows. The fields from the two conductors cancel each other everywhere in space, except very near the transmission line.

ballast devices unnecessary (*see* VOLTAGE REGULATION). Ballast devices are sometimes used with such things as portable ac generators, which have somewhat variable output voltages.

A ballast lamp is a bulb that displays increased resistance as the voltage goes up. This keeps the dissipated power nearly constant. If there is a sudden surge in the voltage, a ballast lamp will not burn out. If the voltage drops somewhat, the ballast lamp will maintain its brilliance.

A ballast tube maintains constant current flow in a circuit, when connected in series, under unstable voltage conditions. A ballast resistor operates on the same principle. The greater the input voltage, the larger the voltage drop across the resistor, and the flow of current thereby remains nearly constant.

BALUN

When a balanced load is connected to an unbalanced source of power, a balun is sometimes used. The word balun is a contraction of "balanced/unbalanced." A balun has an unbalanced input and a balanced output.

The simplest form of balun is an ordinary transformer. The isolation between the primary and secondary windings allows an unbalanced source of power to be connected to one end, and a balanced load to the other, as shown at A in the illustration. This condition can also be reversesd; a balanced source can be connected to the primary and an unbalanced load to the secondary.

At radio frequencies, baluns are sometimes constructed from lengths of coaxial cable. One example of such a device, which provides an impedance step-up ratio of 1 to 4, is shown at B. Coaxial baluns such as this are sometimes used for vhf.

Special transformers can be built to act as baluns over a wide range of frequencies. Two such devices are illustrated schematically. The balun at C provides a 1-to-1 impedance-transfer ratio. The balun at D provides a step-up ratio of 1 to 4. The coils are wound adjacent to each other on the same form. Toroidal powdered-iron or ferrite forms are often used for broadband balun transformers.

Although balun coils are quite frequently used when the

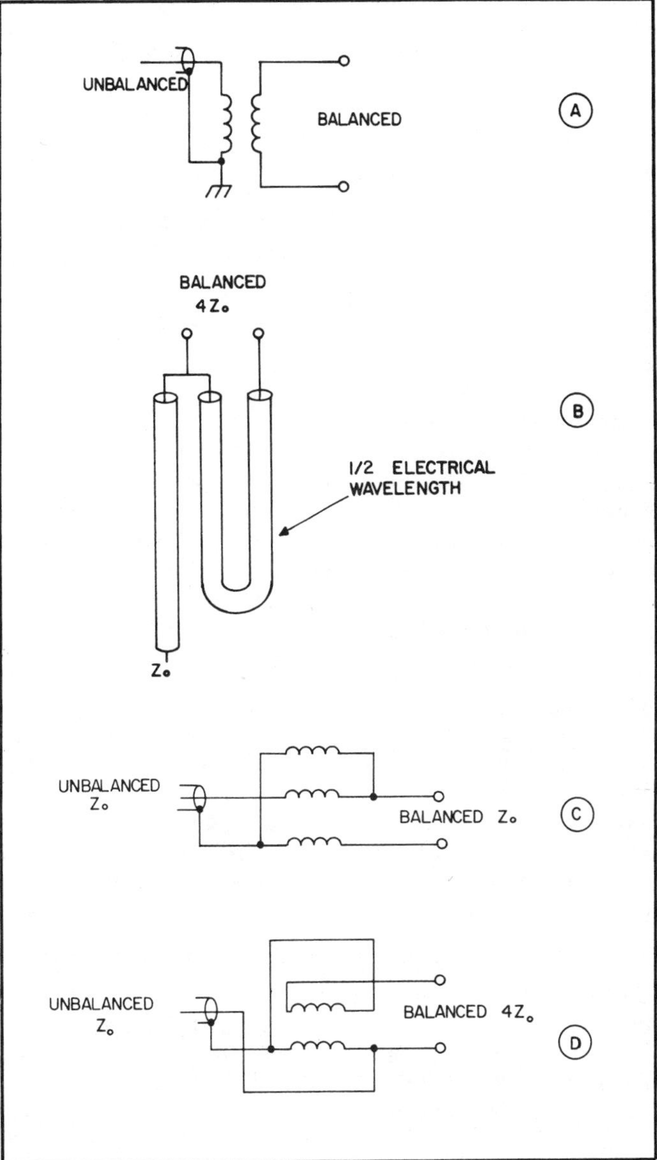

BALUN: Four types of balun transformers. At A, a simple transformer. At B, a coaxial balun with an impedance-transfer ratio (input to output) of 1 to 4. At C, a broadbanded balun with a 1-to-1 impedance-transfer ratio. At D, a broadbanded balun with an impedance-transfer ratio (input to output) of 1 to 4.

source of power is unbalanced and the load is balanced, the devices can also be used in the reverse situation—a balanced source and an unbalanced load. An example of the first type of application is a dipole or Yagi fed with coaxial cable. An example of the latter application is a vertical ground-plane antenna fed with balanced line. Baluns help maintain antenna or feed-line balance when it is important. *See also* BALANCED TRANSMISSION LINE, UNBALANCED TRANSMISSION LINE.

BANANA JACK AND PLUG

A convenient single-lead connector that slips easily in and out of its receptacle is called a banana connector. The banana plug looks something like a banana skin, and this is where it gets its name (see illustration).

Banana jacks are frequently found in screw terminals of low-voltage dc power supplies. The leads can be affixed using the screw terminals for more permanent use. If

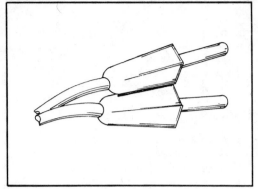

BANANA JACK AND PLUG: Banana plugs are convenient and neat.

BAND: RADIO-FREQUENCY BANDS.

Designation	Frequency	Wavelength
Very low (vlf)	10 kHz-30 kHz	30 km-10km
Low (lf)	30 kHz-300 kHz	10 km-1 km
Medium (mf)	300 kHz-3 MHz	1 km-100 m
High (hf)	3 MHz-30 MHz	100 m-10 m
Very high (vhf)	30 MHz-300 MHz	10 m-1 m
Ultra high (uhf)	300 MHz-3 GHz	1 m-100 mm
Super high (shf)	3 GHz-30 GHz	100 mm-10 mm
Extremely high (ehf)	30 GHz-300 GHz	10 mm-1 mm

frequent changing of leads is necessary, the banana jacks allow it with a minimum of trouble. Banana plugs are simply pushed in and pulled out. They lock with a moderate amount of friction. Banana connectors have fairly low dc loss and high current-handling capacity. They are not generally used at high voltages, however, because of the possibility of shock from exposed conductors.

BAND

Any range of electromagnetic frequencies, marked by lower and upper limits for certain purposes, is called a band. The AM broadcast band is from 535 to 1605 kHz. The FM broadcast band extends from 88 to 108 MHz. The 40-meter amateur band has a lower limit of 7.000 MHz and an upper limit of 7.300 MHz. There is no limit to how narrow or wide a band can be; it must only cover a certain range.

In radiocommunication, the electromagnetic spectrum is subdivided into bands according to frequency, as shown in the table. The very-low-frequency (vlf) band is only 20 kHz wide, but it covers a threefold range of frequencies. All of the higher bands cover a tenfold range of frequencies. For a more detailed breakdown of the bands in the radio-frequency spectrum, *see* FREQUENCY ALLOCATIONS, SUBBAND.

BANDPASS FILTER

Any resonant circuit, or combination of resonant circuits, designed to discriminate against all frequencies except a frequency f_0, or a band of frequencies between two limiting frequencies of f_0 and f_1 is called a bandpass filter. In a parallel circuit, a bandpass filter shows a high impedance at the desired frequency or frequencies, and a low impedance at unwanted frequencies. In a series configuration, the filter has a low impedance at the desired frequency or frequencies, and a high impedance at unwanted frequencies. Three types of bandpass filter circuits are illustrated.

Some bandpass filters are built with components other than actual coils and capacitors, but all such filters operate on the same principle. The crystal-lattice filter uses piezoelectric materials, usually quartz, to obtain a bandpass response (*see* CRYSTAL-LATTICE FILTER). A mechanical filter uses vibration resonances of certain substances (*see* MECHANICAL FILTER). A resonant antenna is itself a form of bandpass filter, since it allows efficient radiation at one frequency and neighboring frequencies, but discriminates against others. In optics, a simple color filter, discriminating against all light wavelengths except within a certain range, is a form of bandpass filter.

Bandpass filters are sometimes designed to have a very sharp, defined, resonant frequency. Sometimes the resonance is spread out over a fairly wide range. *See also* BANDPASS RESPONSE.

BANDPASS RESPONSE

The attenuation-vs-frequency characteristic of a bandpass filter is called the bandpass response (*see* ATTENUATION-VS-FREQUENCY CHARACTERISTIC). A bandpass filter may have a single, well-defined resonant frequency (A in the illustration), denoted by f_0; or, the response might be rectangular, having two well-defined limit frequencies f_0 and f_1, as shown at B (*see* RECTANGULAR RESPONSE). The

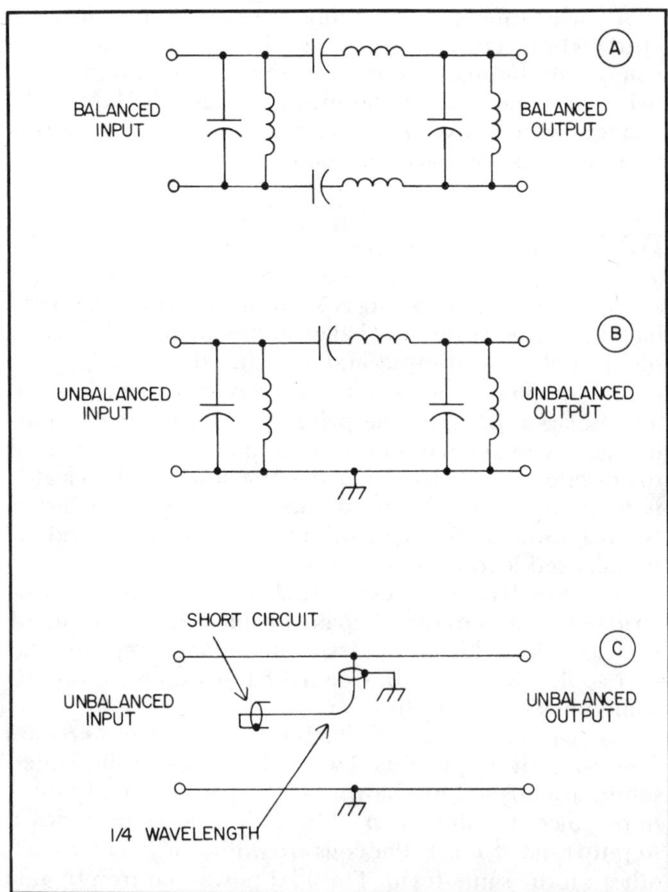

BANDPASS FILTER: At A, a balanced filter. At B, an unbalanced filter. At C, an unbalanced filter using a quarter-wave section of transmission line, suitable for use at vhf and uhf.

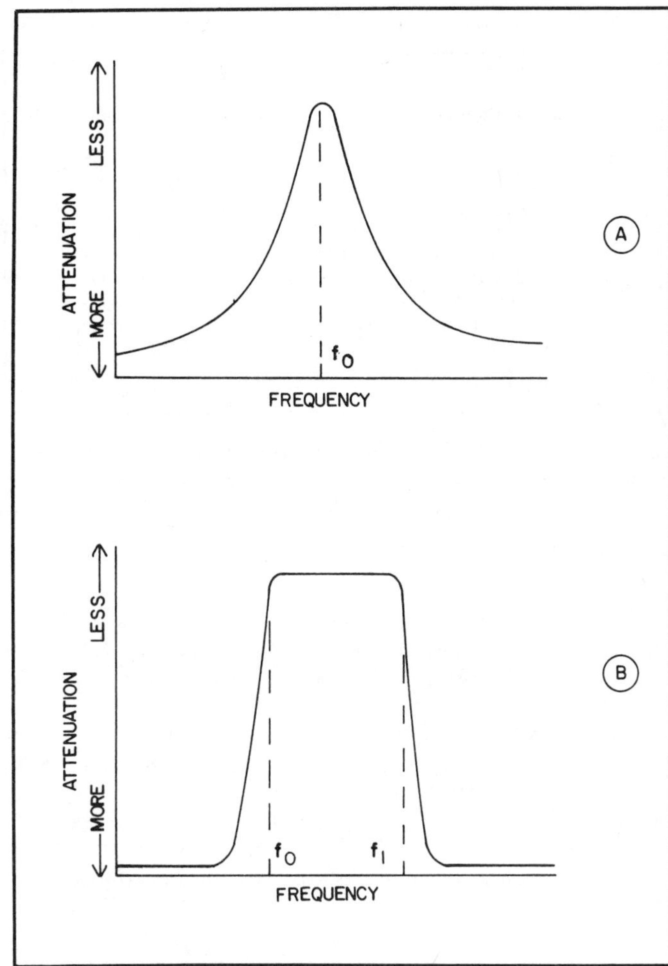

BANDPASS RESPONSE: At A, a single-frequency response. At B, wideband response.

bandwidth (*see* BANDWIDTH) might be just a few Hertz, such as with an audio filter designed for reception of Morse code. Or, the bandwidth might be several MHz, as in a helical filter designed for the front end of a vhf receiver.

A bandpass response is always characterized by high attenuation at all frequencies except within a particular range. The actual attenuation at desired frequencies is called the insertion loss (*see* INSERTION LOSS).

BAND-REJECTION FILTER

A band-rejection filter is a resonant circuit designed to pass energy at all frequencies, except within a certain range. The attenuation is greatest at the resonant frequency, f_0, or between two limiting frequencies, f_0 and f_1. Three examples of band-rejection filters are shown in the illustration. Note the similarity between the band-rejection and bandpass filters (*see* BANDPASS FILTER). The fundamental difference is that the band-rejection filter consists of parallel LC circuits connected in series with the signal path, or series LC circuits in parallel with the signal path; in bandpass filters, series-resonant circuits are connected in series, and parallel-resonance circuits in parallel.

Band-rejection filters need not necessarily be made up of actual coils and capacitors, but they usually are. Quartz crystals are sometimes used as band-rejection filters. Lengths of transmission line, either short-circuited or open, act as band-rejection filters for certain frequencies. A common example of a band-rejection filter is the notch

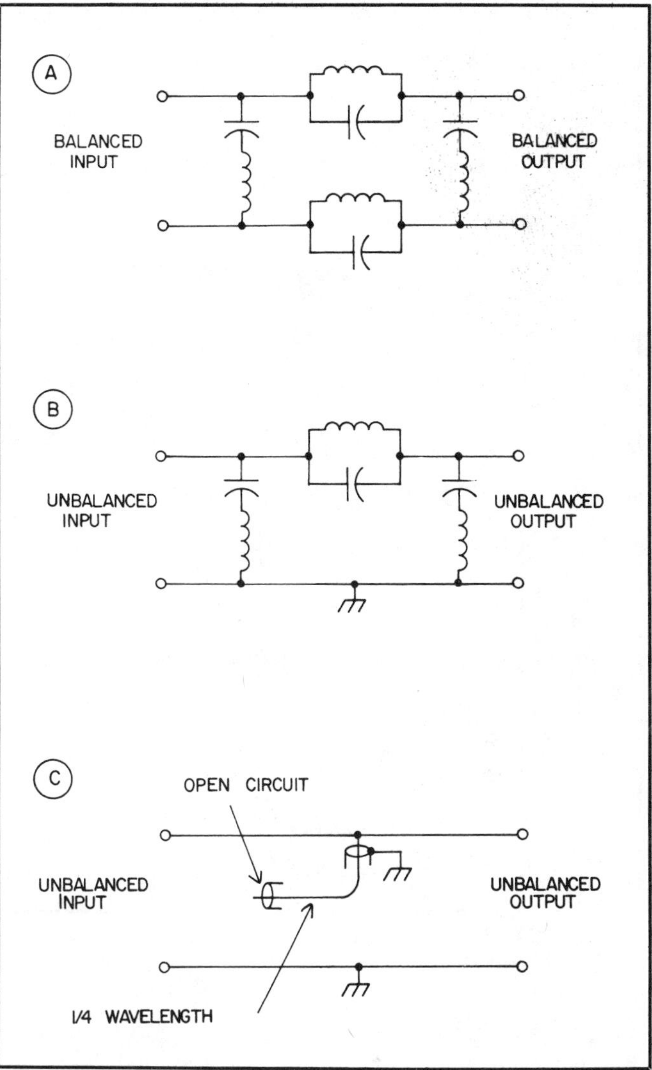

BAND-REJECTION FILTER: At A, a balanced filter. At B, an unbalanced filter. At C, an unbalanced filter using a quarter-wave section of transmission line, suitable for use at vhf and uhf.

filter found in some of the more sophisticated communications receivers. Another example is the antenna trap. Still another is the parasitic suppressor generally seen in the plate lead of a high-power RF amplifier. *See also* NOTCH FILTER, PARASITIC SUPPRESSOR, TRAP.

BAND-REJECTION RESPONSE

All band-rejection filters show an attenuation-vs-frequencies characteristic (*see* ATTENUATION-VS-FREQUENCY CHARACTERISTIC) marked by low loss at all frequencies except within a prescribed range. The illustration shows two types of band-rejection response. A sharp respone occurs at or near a single resonant frequency f_0. A rectangular response (*see* RECTANGULAR RESPONSE) is characterized by low attenuation below a limit f_0 and above a limit f_1, and high attenuation between these limiting frequencies.

Most band-rejection filters have a relatively sharp response. This is true of antenna traps and notch filters. Parasitic suppressors, used in high-frequency power amplifiers to prevent vhf parasitic oscillation, have a more broadband response. *See also* NOTCH FILTER, PARASITIC SUPPRESSOR, TRAP.

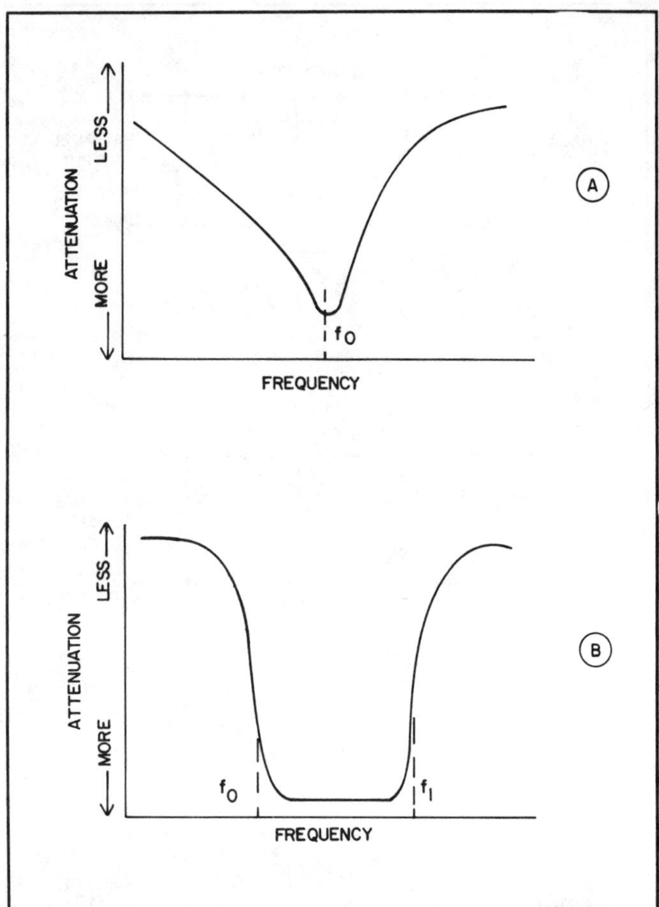

BAND-REJECTION RESPONSE: At A, a single-frequency response. At B, a wideband response.

BANDWIDTH

Bandwidth is a term used to define the amount of frequency space occupied by a signal, and required for effective transfer of the information to be carried by that signal. The term bandwidth is also sometimes used in reference to the nature of a bandpass or band-rejection filter response (*see* BANDPASS RESPONSE, BAND-REJECTION RESPONSE).

The bandwidths of several common types of signals are given by the table. The receivers for such signals must have a bandpass response at least as great as the signal bandwidth. If the receiver bandpass response is too narrow, the signal cannot be readily understood. In the extreme, where the receiver bandpass response is very narrow compared to the signal bandwidth, no information can be conveyed whatsoever.

While an excessively narrow bandpass is obviously undesirable for a signal having a given bandwidth, an unnecessarily wide receiver response is also bad. If the response is much wider than the bandwidth of the signal, a great deal of noise gets into the channel along with the desired signal. Other signals on nearby frequencies may also get into the channel and disrupt communications. So, therefore, the receiver bandpass response should always be tailored to the bandwidth of the signal received.

Bandwidth is sometimes used as a specification for the sharpness of a bandpass filter response. Two frequencies, f_0 and f_1, are found, at which the power attenuation is 3 dB with respect to the level at the center frequency of the filter. The bandwidth is the difference between these two frequencies. In general, if a signal occupies a certain bandwidth, then the receiver filter bandwidth (as determined from the 3-dB power attenuation points) should be the same for best reception.

BANDWIDTH: BANDWIDTHS OF SOME COMMON SIGNALS.

Emission Type	Typical Bandwidth
Morse code (CW)	10 Hz-250 Hz
Radioteletype (RTTY)	200 Hz-800 Hz
Single sideband (SSB)	3 kHz
Slow-scan television (SSTV)	3 kHz
Amplitude modulation, voice (AM)	6 kHz
Amplitude modulation, music (AM)	10 kHz
Frequency modulation, voice (FM)	10 kHz
Frequency modulation, music (FM)	100 kHz
Television (TV)	3 MHz-10 MHz

BANK

Whenever a group of similar or identical devices is connected together, the composite is called a bank. A bank of batteries, for example, is a set of batteries in series or parallel.

Banks are generally used to obtain more power, gain, or power-handling capability than would be possible using just one single component. A series bank of batteries provides greater voltage than one battery. A parallel bank of batteries can deliver more current, or the same amount of current for a longer period of time, as compared to a single battery.

BARDEEN, JOHN AND BRATTAIN, WALTER

John Bardeen and Walter Brattain are generally credited with the perfection of the first semiconductor triode (transistor). The three-element device allowed amplification and oscillation, as well as rectification. Bardeen and Brattain, working for the Bell Laboratories, made their discovery in the late 1940s.

Since the development of the first transistor, semiconductor technology has practically eliminated vacuum tubes in electronic equipment. Many different kinds of transistors are now in use. Bardeen and Brattain may be considered counterparts of Lee De Forest, who perfected the triode tube several decades before (*see* DE FOREST, LEE).

Nowadays, hundreds or even thousands of transistors can be fabricated on semiconductor chips just a few square millimeters in size. *see also* TRANSISTOR.

BAR GENERATOR

A bar generator is a device that produces regular pulses, in such a way that a pattern of bars is produced in a transmitted television picture signal. The bar pattern may be either vertical or horizontal. An example of a black-and-white bar pattern is shown in the illustration. A bar pattern can have just two bars, or several.

The pattern produced by a bar generator is used to check for vertical or horizontal linearity in a television transmitter and receiver. Vertical bars facilitate alignment of horizontal circuits. Horizontal bars allow adjustment for vertical linearity. In a color television system, colored

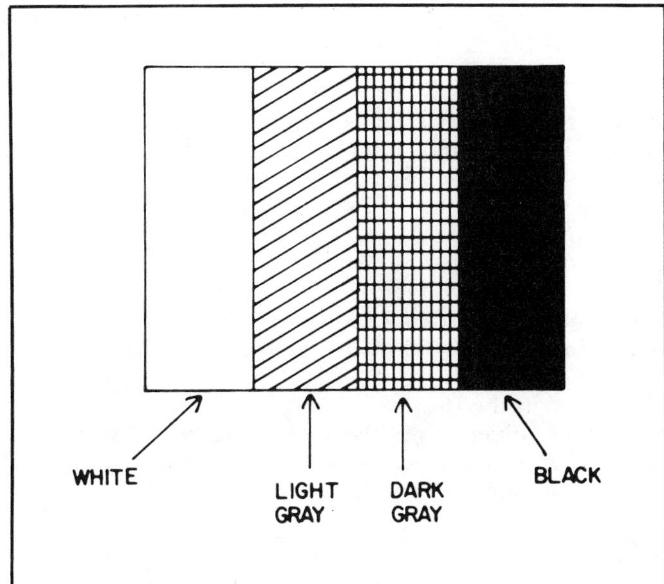

BAR GENERATOR: A black-and-white bar-generator pattern for adjusting horizontal linearity and focus.

bars are used to align the circuits for accurate color reproduction. In a black-and-white system, the various shades of gray are used to adjust the relative brightness and contrast. The lines of demarcation between the bars allow adjustment of the focus.

A bar pattern can be used for complete alignment of a television transmitting and receiving system. *See also* TELEVISION.

BARIUM

Barium is an element with atomic number 56 and atomic weight 137. This element is oxidized and used in the cathodes of vacuum tubes. A coating of barium oxide prolongs the life, and enhances the electron-emitting properties, of the cathode (*see* TUBE).

A substance called barium titanate, a compound of barium and titanium, is sometimes used in place of quartz in piezoelectric crystals. *See also* CRYSTAL, PIEZOELECTRIC EFFECT, QUARTZ.

BARKHAUSEN EFFECT

When the grid of an amplifier tube is capacitively coupled to a sufficient extent to the plate, oscillation at very-high or ultra-high frequencies sometimes takes place. This oscillation is called vhf or uhf parasitic oscillation (*see* PARASITIC OSCILLATION), since it is usually not desired.

Parasitic oscillation may occur without amplifier drive, in which case it can be detected by the presence of unusual levels of plate current while the exciter is off (*see* EXCITER). Parasitic oscillation can also take place when there is sufficient drive to cause the grid to go positive during part of the cycle. In this case, a spectrum analyzer is usually necessary to ascertain the existence of vhf or uhf parasitics.

Barkhausen effect is deliberately created in a device called a Barkhausen-Kurz oscillator. *See also* BARKHAUSEN-KURZ OSCILLATOR.

BARKHAUSEN-KURZ OSCILLATOR

A Barkhausen-Kurz oscillator is a special kind of oscillator for generating ultra-high-frequency energy. A tube is used with a grid voltage that is positive with respect to all the other electrodes, including the plate. Variations in the electric field between the grid and the plate of the tube cause electrons to oscillate in the interelectrode space. The frequency of oscillation is controlled by external tuned circuits.

Barkhausen oscillation can take place in a tube even when it is not wanted. In this form, such oscillation is called parasitic ocillation. Parasitic oscillation is detrimental to the performance of an amplifier, since it produces outputs on frequencies not intended. However, in the Barkhausen-Kurz oscillator, this form of oscillation is put to constructive use. *See also* OSCILLATION, OSCILLATOR, PARASITIC OSCILLATION.

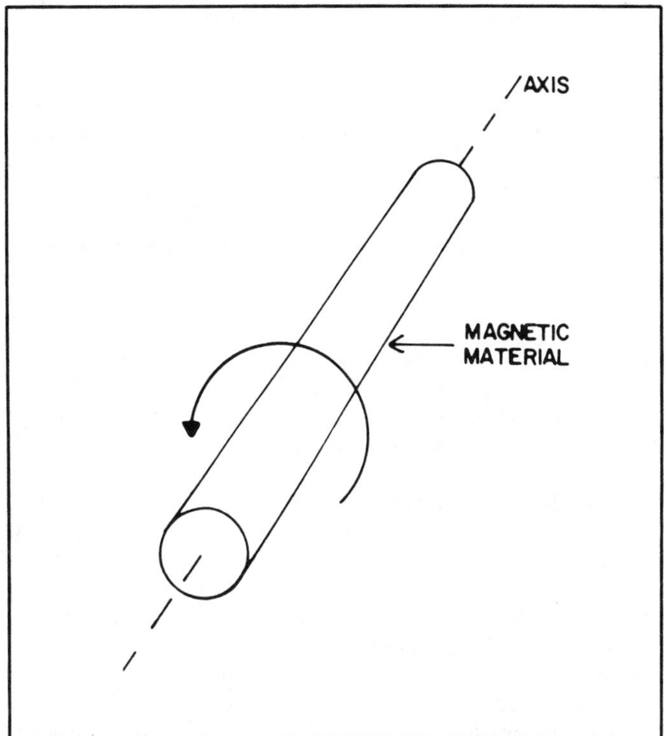

BARNETT EFFECT: Barnett effect results in the magnetization of a cylinder of magnetic material rotated about its longitudinal axis.

BARNETT EFFECT

When a cylinder of iron or other magnetic material is rotated about its axis, as shown in the drawing, a small amount of magnetization occurs. This is called the Barnett effect.

Magnetized metals differ from ordinary metals in that the atoms are aligned along a certain axis to a greater extent than would be the case as dictated by sheer probability. This happens only in metals containing iron and/or nickel. Magnetization causes attractive or repulsive forces.

Barnett effect results in permanent magnetization. *See also* MAGNETIC MATERIAL, MAGNETIZATION, PERMANENT MAGNET.

BARRETTER

A barretter is a form of ballast device (*see* BALLAST). A barretter tube consists of a helically wound length of iron wire inside a hydrogen-filled container. When connected in series at the output of a power supply or at the input to a circuit, the barretter tube keeps the current flow at a constant value for a wide range of voltages. Thus it acts as a current regulator for the circuit.

A resistor with a positive temperature coefficient (*see* TEMPERATURE COEFFICIENT), deliberately designed for a certain amount of resistance increase for each degree of temperature rise, is sometimes called a barretter or barretter resistor. Such a resistor should not be confused with a thermistor, which is a semiconductor device. *See also* THERMISTOR.

BARRIER CAPACITANCE

When a semiconductor P-N junction is reverse biased, so that it does not conduct, a capacitance exists between the P-type and N-type semiconductor materials. This is called barrier capacitance. It is also sometimes called the depletion-layer or junction capacitance.

Under conditions of reverse bias, a depletion region, or potential barrier, forms between the P-type and N-type materials (*see* DEPLETION LAYER). Positive ions are created in the N-type material, and negative ions are created in the P-type material. The greater the reverse bias, the wider the depletion layer, as shown in the illustration.

Barrier capacitance is a function of the width of the depletion layer in a given P-N junction. The capacitance is generally very small, on the order of a few picofarads. As the reverse-bias voltage increases, the width of the depletion layer increases. The barrier capacitance therefore gets smaller.

The barrier capacitance for a given reverse voltage depends on the cross-sectional area of the P-N junction, and also on the amount of doping and the kind of impurity used for doping (*see* DOPING, IMPURITY).

BARRIER-LAYER CELL

A barrier-layer cell is a photovoltaic cell (*see* PHOTOVOLTAIC CELL). When light strikes the surface of such a cell, a voltage appears across the output contacts.

Copper and cuprous oxide (a form of copper oxide) can be used in the construction of a barrier-layer cell. Other materials may also be used, such as selenium semiconductors. In any barrier-layer cell, photons striking the barrier create a voltage (see illustration).

Barrier-layer cells may someday provide an answer to the problem of deriving electrical energy from the sun. Presently, however, the use of such devices is, watt for watt, more expensive than conventional methods of generating power. *See also* SOLAR CELL, SOLAR POWER.

BARRIER VOLTAGE

Barrier voltage is the amount of forward bias necessary to cause conduction across a junction of two unlike materials. In a semiconductor P-N junction, the barrier voltage is approximately 0.3 volts for germanium, and 0.6 volts for silicon.

The forward-voltage-vs-current characteristic for a semiconductor diode looks like that shown in the graph. With small values of forward voltage, very little current flows. But when the voltage reaches the barrier voltage, a sharp rise in the flow of current is observed.

Barrier voltage is not the same thing as avalanche voltage, which takes place under conditions of reverse bias and with much higher voltages. *See also* AVALANCHE VOLTAGE.

BASE

A semiconductor transistor consists of either of two configurations, called PNP, illustrated at A, and NPN, at B. In the PNP device, a wafer of N-type semiconductor material is sandwiched in between two wafers of P-type material. In the NPN transistor, a wafer of P-type semiconductor material is sandwiched in between two wafers of N-type material (*see* N-TYPE SEMICONDUCTOR, P-TYPE SEMICONDUCTOR). The center wafer is called the *base* of the transistor, since the device is built geometrically around it.

The base of a transistor acts very much like the grid of a vacuum tube in practice. Theoretically, however, the base of a transistor operates on a different principle. The input signal in a transistor amplifier is applied between the base and the emitter. The output is taken between the emitter and collector, or between the collector and ground. *See also* COLLECTOR, EMITTER, TRANSISTOR.

BASEBAND

When a signal is amplitude-modulated or frequency-modulated, the range of modulating frequencies is called the baseband. For a human voice, intelligible transmissions are possible with baseband frequencies from about 200 to 3000 Hz. Experiments have been conducted to determine the minimum baseband needed for effective voice communication, and with analog methods, efficiency is impaired if the baseband is restricted to a range much

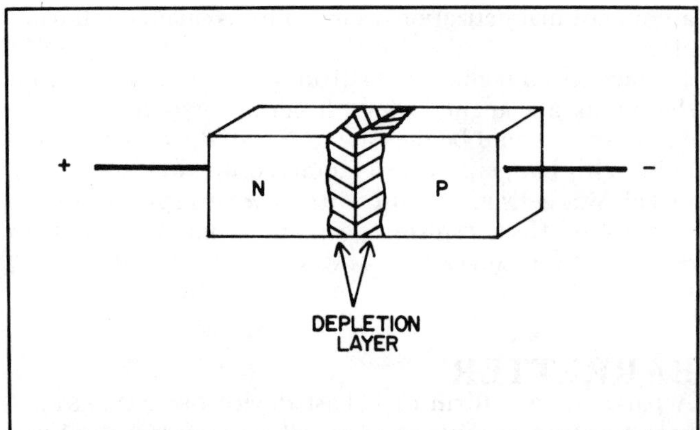

BARRIER CAPACITANCE: When a diode is reverse biased, the depletion layer acts as the dielectric to form a capacitor.

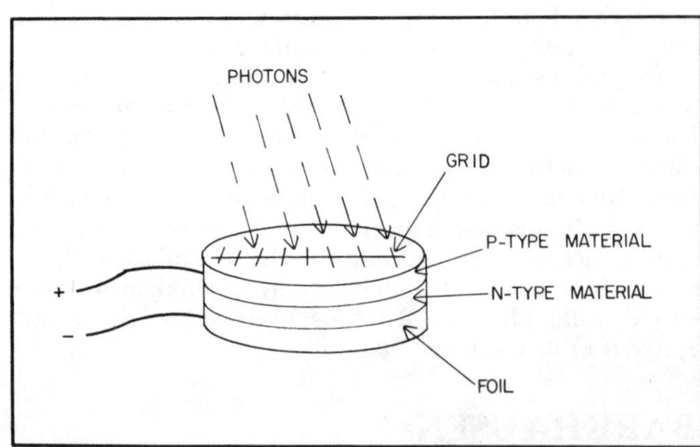

BARRIER-LAYER CELL: A barrier-layer cell made from semiconductor material.

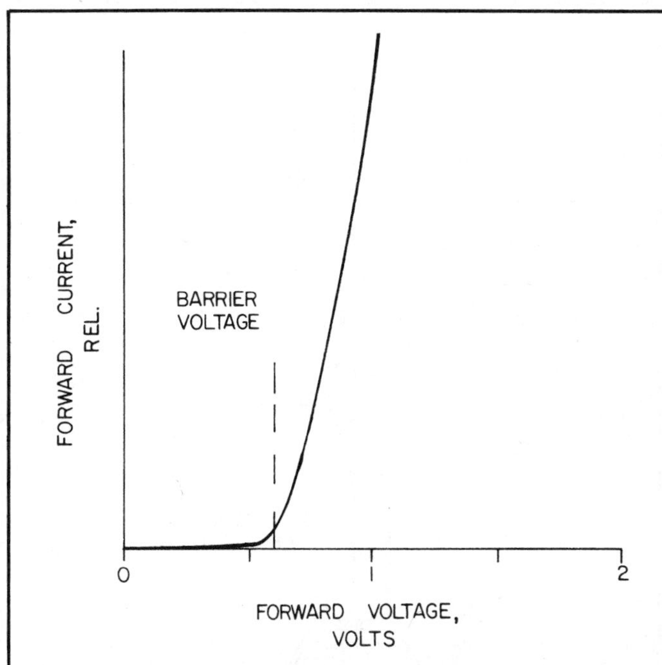

BARRIER VOLTAGE: The barrier voltage of a semiconductor diode is the forward voltage at which conduction begins.

smaller than 200 to 3000 Hz. One method of compressing baseband signals for voice communication that shows some promise is called narrow-band voice modulation (*see* NARROW-BAND VOICE MODULATION).

For good reproduction of music, the baseband should cover at least the range of frequencies from 20 to 5000 Hz, and preferably up to 15 kHz or more. The baseband in a standard AM broadcast signal is restricted to the range zero to 5 kHz, since the channel spacing is 10 kHz and a larger baseband would result in interference to stations on adjacent channels. In FM broadcasting, the baseband covers the entire range of normal human hearing. *See also* AUDIO FREQUENCY, HIGH FIDELITY.

BASE BIAS

The base bias of a transistor is the level of the base voltage with respect to the emitter or ground. Base bias can be derived from a battery, a resistive voltage-dividing network, or a resistor-capacitor combination. When the base is at the same voltage as the emitter of a transistor, the condition is called zero bias. Various kinds of base bias are shown in the illustration.

A transistor is normally cut off unless the forward bias at the emitter-base PN junction is greater than the barrier voltage (*see* BARRIER VOLTAGE). In a PNP transistor, forward base bias occurs when the base is negative with respect to the emitter. In an NPN device, forward bias means that the base is more positive than the emitter. To cause current to flow, this voltage difference must be at least 0.3 volts for a germanium transistor, and 0.6 volts for a silicon transistor.

For various amplification purposes, the base may be biased at class A, AB, B, or C. Class-A operation involves forward bias, as does class-AB operation. Class B usually results with zero bias. Class-C operation can be achieved either with zero bias or reverse bias, depending on the particular transistor. *See also* CLASS-A AMPLIFIER, CLASS-AB AMPLIFIER, CLASS-B AMPLIFIER, CLASS-C AMPLIFIER.

BASE INSULATOR

A base insulator is a piece of non-conducting material, such as glass or plastic, used to insulate the lower end of a vertical antenna from RF ground. The illustration shows

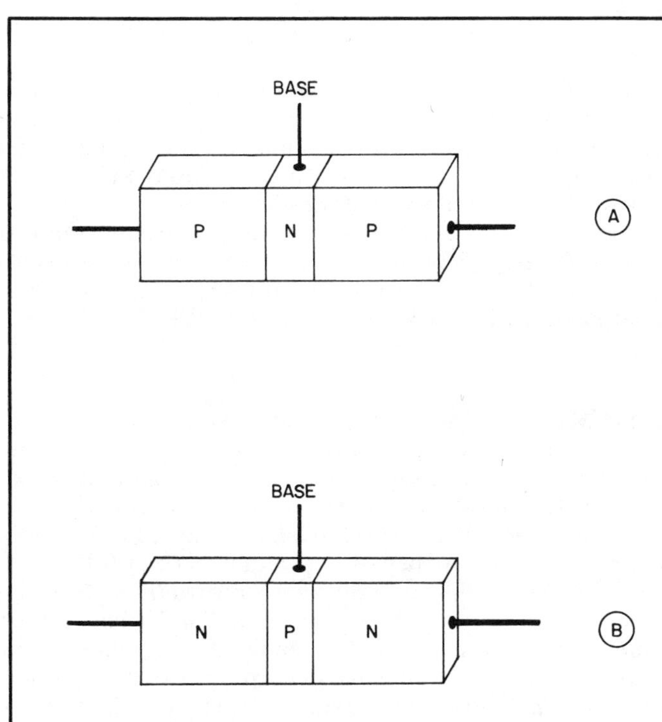

BASE: The base of a transistor is the central semiconductor. At A, a PNP transistor; at B, an NPN transistor.

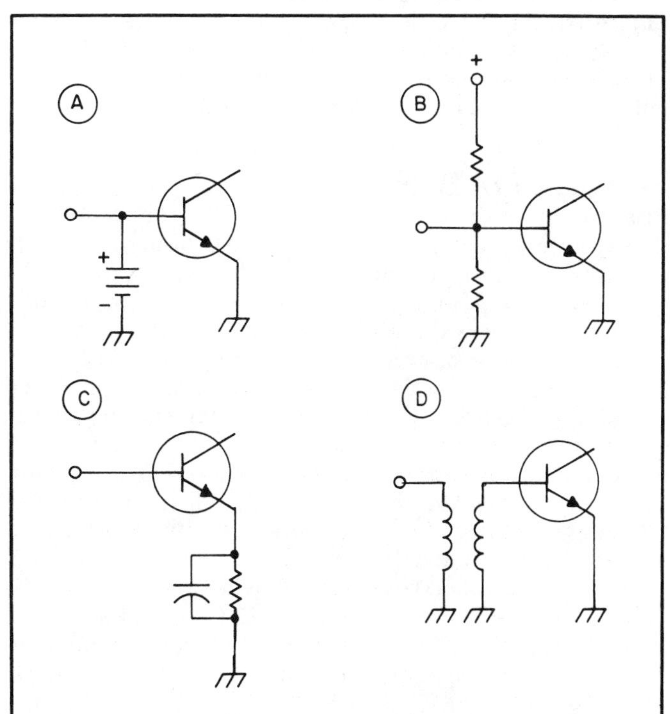

BASE BIAS: Four methods of achieving base bias. At A, a battery is connected between the base and ground. At B, a resistor voltage-divider network. At C, a resistor-capacitor combination in the emitter circuit. At D, zero bias.

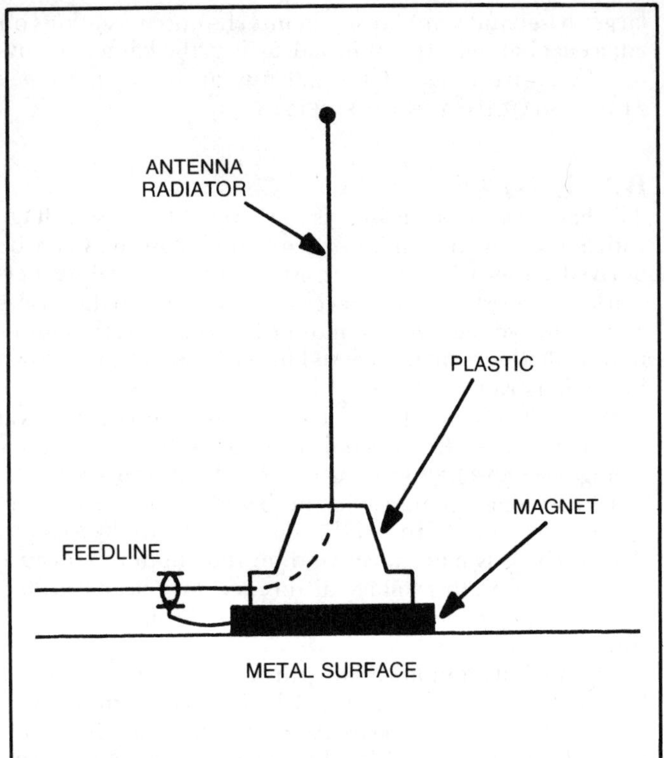

BASE INSULATOR: An antenna base insulator.

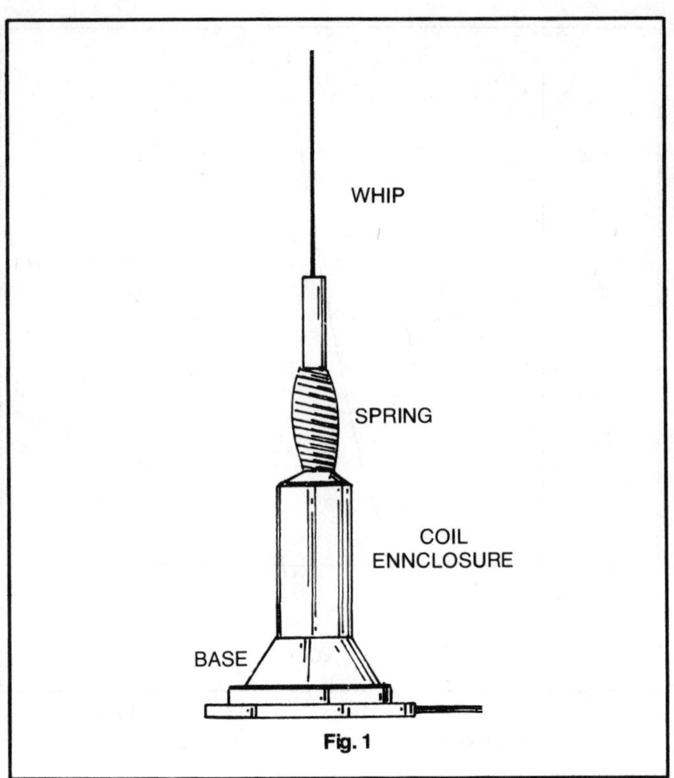

BASE LOADING: A base loading coil for a ⅝-wave, VHF mobile whip antenna, enclosed in a plastic cylinder.

an insulating base mount for a vhf mobile antenna.

A base insulator must be mechanically strong enough to support the weight of an antenna, and to endure stresses caused by wind. The insulator must be able to electrically withstand the voltage that appears between the base of the antenna and the RF ground at the highest level of power employed. When the base insulator is exposed to moisture, its dielectric properties should not change.

Some vertical antennas have no base insulator. Instead they are connected directly to the RF ground, and are fed at some point above the ground. *See also* SHUNT FEED.

BASE LOADING

Base loading is a term used to denote the connection of a series inductance or capacitance at the bottom of a vertical antenna radiator, for the purpose of changing the resonant frequency. Usually, a base-loading system consists of a coil which lowers the resonant frequency of the radiator for quarter-wave operation. Base loading is also frequently used to bring a ½-wave or ⅝-wave antenna to a matched condition. Figure 1 shows a base-loading coil, housed in protective plastic, for a ⅝-wave vhf mobile antenna.

For quarter-wave resonant operation with a radiator less than ¼ wavelength in height, some form of inductive loading is required to eliminate the capacitive reactance at the feed point. When the coil is placed at the antenna base, the scheme is called base loading. When the coil is placed above the base of the radiating element, it is center loading (*see* CENTER LOADING). Center loading requires more inductance, for the same frequency and radiator length, than base loading.

An 8-foot mobile whip antenna can be brought to quarter-wave resonance by means of base loading at all frequencies below its natural quarter-wave resonant frequency, which is approximately 29 MHz. Figure 2 gives the value of loading inductance L, in microhenrys (μH), as a function of the frequency in MHz, for base loading of an 8-foot radiator over perfectly conducting ground. While the RF ground in a mobile installation is not anything near perfect, the values shown are close enough to actual values to be of practical use.

BASIC

One of several different higher-order computer languages, BASIC is one of the most easily learned. It is easy to understand because the commands and functions are generally in plain English. BASIC is used primarily in mathematical and engineering situations.

For business purposes, the higher-order language called COBOL is preferred. In some scientific problems, FORTRAN is more efficient than BASIC. *See also* COBOL, FORTRAN, HIGHER-ORDER LANGUAGE.

BASS

The term BASS, pronounced "base," refers to low-frequency sound energy. On a piano or musical staff, any note below middle C is in the bass clef. All sounds above middle C are in the treble clef (*see* TREBLE). Middle C represents a frequency of approximately 261.6 Hz.

A tone control in a high-fidelity recording or receiving sound system can have separate bass and treble adjustments, or it may consist of a single knob or slide potentiometer. The amount of bass, compared to the amplitudes of other audio frequencies, affects the "presence" of the sound. Too little bass content causes a thin, diluted sound. Too much bass causes sound to seem muffled and heavy, and also creates annoying vibration. *See also* AUDIO RESPONSE, BASS RESPONSE, TONE CONTROL.

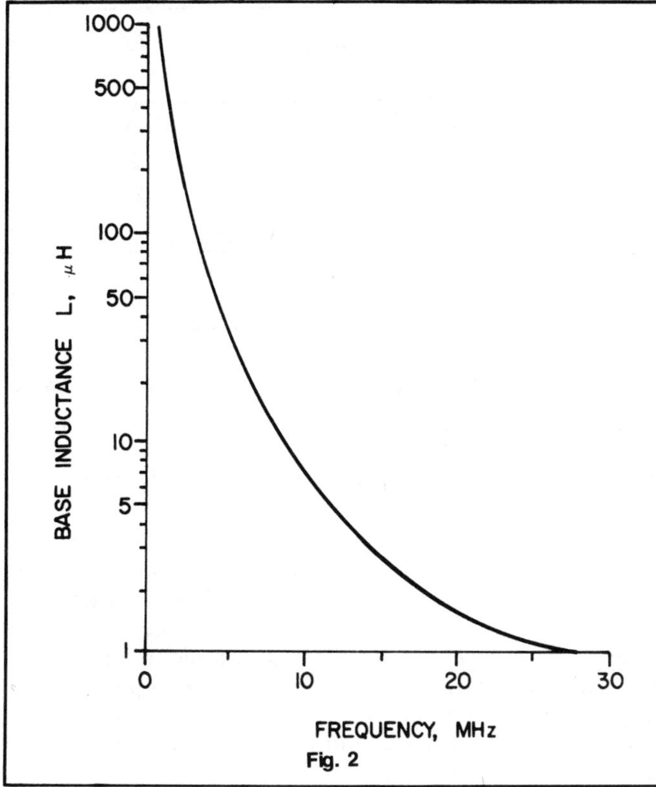

BASE LOADING: Inductance of a base-loading coil, as a function of frequency, for an 8-foot vertical radiator operated against theoretically perfect ground. In practice, where the ground conduction is marginal, the values may be somewhat larger or smaller.

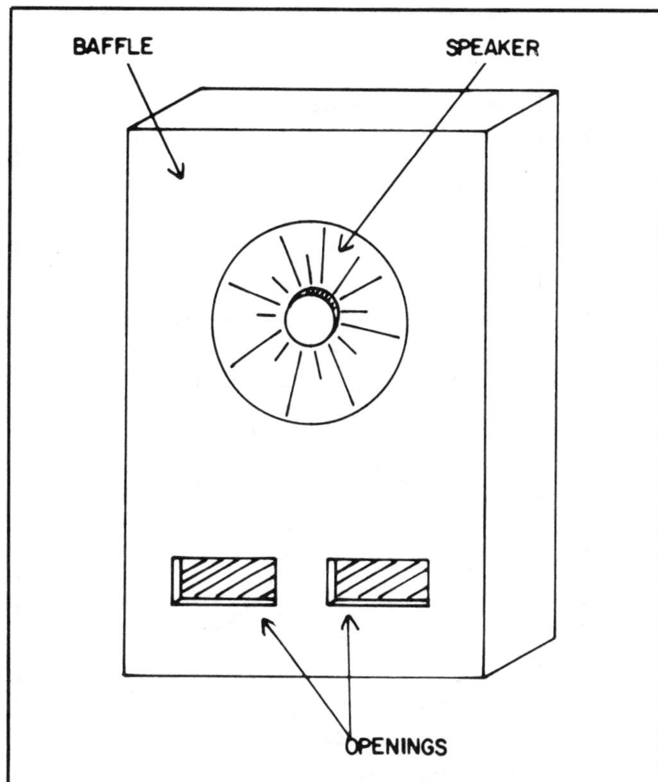

BASS-REFLEX ENCLOSURE: A bass-reflex enclosure uses baffle openings for reinforcement of low-frequency sound energy.

BASS-REFLEX ENCLOSURE

A type of speaker cabinet that reinforces the bass frequencies is called a bass-reflex enclosure. By providing an opening at the front of the speaker in the baffle (*see* BAFFLE), and by suitably choosing the internal dimensions of the speaker cabinet, bass frequencies are reflected from the rear of the enclosure and are reinforced in phase as they emerge through the opening or openings (see illustration).

A bass-reflex enclosure improves the bass response of physically small speakers. Generally, bass reproduction requires the use of large speaker cones that can follow the relatively slow rate of vibration of bass audio. A bass-reflex enclosure takes advantage of sound resonace, so that a smaller speaker can be used to achieve good bass quality in a high-fidelity system. *See also* BASS RESPONSE, WOOFER.

BASS RESPONSE

Bass response is the ability of a sound amplification or reproduction system to respond to bass audio-frequency energy. Bass response can be defined for microphones, speakers, radio receivers, record players, and recording tape—anything that involves the transmission of audio energy. In most high-fidelity sound systems, the bass response is adjustable by means of a tone control or a bass gain control (*see* TONE CONTROL).

Good bass response involves more than the simple ability of a sound system to reproduce bass audio. The sound will not be accurately reproduced if the audio response (*see* AUDIO RESPONSE) curve is not optimized. For example, a sound system might have excellent gain at low frequencies, with low distortion, but the gain must also be fairly uniform over the bass range of frequencies (A in the illustration). A nonuniform bass response will result in poor sound, as shown at B.

A device called an equalizer (*see* EQUALIZER) allows compensation for differences in sound systems. The bass and treble response, as well as the response at midrange and high frequencies, can be tailored for the best possible sound reproduction by using such a device.

BATCH PROCESSING

When a series of programs is run sequentially by a computer, the machine is said to be run in the batch-processing mode. This is usually done by large computers at high speed. Programs arrive from many different sources, or subscribers, in the form of electronic data or perhaps computer cards. Batch processing is often the most economical way to run a computer program, although it is inconvenient for program debugging, since the results are not known until the information is relayed back.

In contrast to batch processing, programs are sometimes run simultaneously. This is called multiprogramming or time sharing (*see* TIME SHARING).

BATHTUB CAPACITOR

A bathtub capacitor is a high-voltage, high-capacitance device that appears physically similar in shape to a bathtub (see illustration).Bathtub capacitors are generally oil-filled, and are used as filter capacitors in high-voltage dc power supplies.

Oil-filled capacitors are used less frequently today than

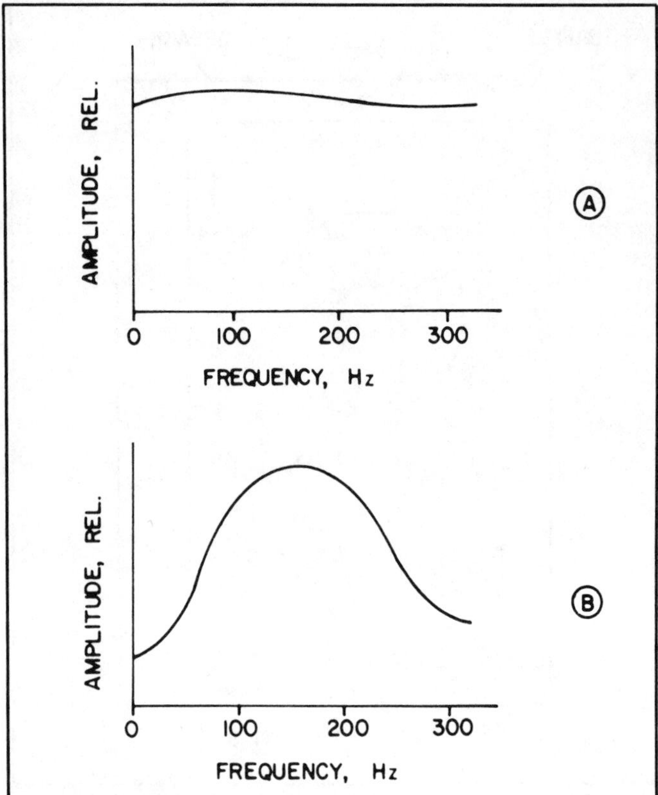

BASS RESPONSE: At A, a good bass response for music reproduction. At B, a poor response.

was the case a few years ago, since the solid-state circuits of modern electronic technology require low voltages. Ordinary electrolytic capacitors are generally suitable for modern applications. However, oil-filled capacitors can still be found in some high-power RF amplifiers, where vacuum tubes are still widely used. *See also* DC POWER SUPPLY, ELECTROLYTIC CAPACITOR.

BATTERY

A battery is a combination of cells in series (*see* CELL). There might be just two cells, or several dozen. Batteries provide greater dc voltage than a single cell.

A battery can be small in size, such as the 9-volt "transistor battery" shown in the photograph. A battery may be quite large, such as an automobile storage battery. In general, the larger the physical dimensions of a particular kind of battery, the more energy it will store.

At a given voltage, the energy capacity of a battery is measured in ampere hours or milliampere hours. The 9-volt miniature battery shown has a storage capacity of a few tens or hundreds of milliampere hours; an automobile storage battery has a storage capacity of 50 to 80 ampere hours. The actual energy in watt-hoursthat is stored in a battery is the product of its rated voltage and its ampere-hour capacity.

Some batteries are rechargeable while others are not. *See also* BATTERY CHARGING.

BATTERY CHARGING

Certain types of batteries, called storage batteries, can be recharged after their energy has been used up. A battery charger, consisting of a transformer and rectifier (and sometimes a milliammeter or ammeter to indicate charging current), is used to recharge a storage battery. The positive terminal of the charger is connected to the positive terminal of the battery for charging.

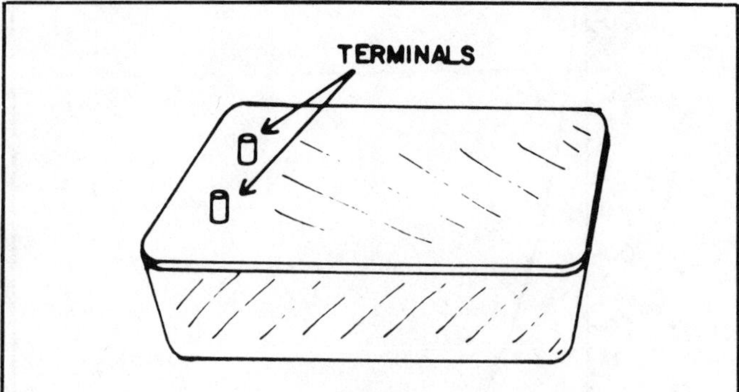

BATHTUB CAPACITOR: A bathtub capacitor looks something like a bathtub!

While a battery is charging, it draws the most current initially, and the level drops continuously until, when the battery is fully charged, the current is near zero. The total number of ampere hours indicated by a plot of current vs time would correspond, when charging is complete, to the capacity of the battery. Some chargers operate quickly, within a period of a few minutes, but charging usually takes several hours.

Electronic calculators and some radio equipment contain rechargeable nickel-cadmium batteries. The batteries are charged by means of a small current-regulated power supply that fits standard household electrical outlets. Nickel-cadmium batteries require four to six hours, typically, to reach a fully charged condition. *See also* NICKEL-CADMIUM BATTERY, STORAGE BATTERY.

BAUDOT

Baudot is a five-unit digital code for the transmission of teleprinter data. Letters, numerals, symbols, and a few control operations are represented. Baudot was one of the first codes used with mechanical printing devices. Sometimes this code is called the Murray code. In recent years, Baudot has been replaced in some applications by the ASCII code (*see* ASCII).

There are 2^5, or 32, possible combinations of binary pulses in the Baudot code, but this number is doubled by the control operations LTRS (lower case) and FIGS (upper case). In the Baudot code, only capital letters are sent. Upper-case characters consist mostly of symbols, numerals and punctuation. Table 1 shows the Baudot code symbols used in the United States and in most foreign countries. In some countries, the upper-case representations differ from those used in the United States. The International Consultative Committee for Telephone and Telegraph (CCITT) version of upper-case Baudot symbols is shown.

The Baudot code is still widely used by amateur radio operators. Baudot equipment is still occasionally seen in commercial systems, but the more efficient ASCII code is becoming the mode of choice, especially in computer applications, since ASCII has more symbol representations.

The most common Baudot data speeds range from 45.45 to 100 bauds, or about 60 to 133 words per minute (WPM), as shown in Table 2. *See also* BAUD RATE, WORDS PER MINUTE.

BATTERY: A battery can be as small as the miniature 9-volt unit shown here, or it can be made up of several large storage cells weighing hundreds of pounds.

BAUD RATE

Baud rate is a measure of the speed of transmission of a digital code. One baud consists of one element or pulse. The baud rate is simply the number of code elements transmitted per second. Sometimes the baud rate is stated simply as "baud."

Baud rates for teleprinter codes range from 45.45, the slowest Baudot speed, to 19,200 for the fastest ASCII speed. *See also* ASCII, BAUDOT.

BAYONET BASE AND SOCKET

Some equipment lamps use a special kind of base and socket called a bayonet (see illustration). Such sockets have a spring-loaded contact, and a slot into which the lamp base fits and is held in place. Some bayonet sockets have two contacts for dual-filament lamps, such as automobile tail indicator lights. The lamp can be placed in the socket only one way; it must first be rotated to the proper position.

The advantage of bayonet bases and sockets over screw-in type contacts is primarily that the bulb is less likely to come loose, and mechanical jamming is less frequent.

BAZOOKA BALUN

A bazooka, or bazooka balun, is a means of decoupling a coaxial transmission line from an antenna.

When coaxial feed lines are used with antenna radiators, the unbalanced nature of the line sometimes

BAUDOT: SYMBOLS FOR BAUDOT TELEPRINTER CODE.

Digits	Ltrs	US Figs	CCITT Figs	Digits	Ltrs	US Figs	CCITT Figs
00000	BLANK	BLANK	BLANK	10000	T	5	5
00001	E	3	3	10001	Z	″	+
00010	LF	LF	LF	10010	L	)	)
00011	A	--	--	10011	W	2	2
00100	SPACE	SPACE	SPACE	10100	H	#	£
00101	S	BELL	′	10101	Y	6	6
00110	I	8	8	10110	P	0	0
00111	U	7	7	10111	Q	1	1
01000	CR	CR	CR	11000	0	9	9
01001	D	$	WRU	11001	B	?	?
01010	R	4	4	11010	G	&	&
01011	J	′	BELL	11011	FIGS	FIGS	FIGS
01100	N	,	,	11100	M	•	•
01101	F	;	;	11101	X	/	/
01110	C	:	:	11110	V	;	=
01111	K	(	(	11111	LTRS	LTRS	LTRS

TABLE 1

BAUDOT: SPEED RATES FOR THE BAUDOT CODE.

Baud Rate	Length of Pulse, ms	WPM
45.45	22.0	60.6
45.45	22.0	61.3
45.45	22.0	65.0
50	20.0	66.7
56.92	17.6	75.9
56.92	17.6	76.7
74.20	13.5	99.0
74.20	13.5	100.0
100	10.0	133.3

TABLE 2

allows RF currents to flow along the outer conductor of the cable. This can create problems such as changes in the antenna directional pattern, and possibly radiation from the line itself. One method of choking off such currents is shown in the illustration. A metal cylinder or braided sleeve, measuring ¼ wavelength, is placed around the coaxial cable as shown. The end of the sleeve near the antenna is left free, and the other end is connected to the shield of the cable. For a given frequency, f, in MHz, the length of the bazooka in feet is given by

$$\text{Length (feet)} = 234/f$$

The bazooka is not a transformer, and consequently it cannot correct impedance mismatches. The bazooka, however, is useful for preventing radiation from a coaxial transmission line at high and very high frequencies. *See also* BALUN.

B BATTERY

The B battery is the source of dc power for the plate circuit of a tube. Generally, a B battery is a high-voltage, series combination of cells. Dozens of 1.5-volt cells were once combined in series for use in tube-operated portable equipment.

The B battery is connected to ground on the negative side, and to the plate circuit leads (and also the screen circuit leads, if the tube is a tetrode or pentode) on the positive side. Portable equipment is usually solid-state, so B batteries are seldom seen nowadays. Some older test equipment, however, may have vacuum-tube circuitry. Tube testers must have a B battery. When ac-operated power supplies came into use, the supply for the plate of the tube was called the "B + supply," continuing the terminology.

The high voltage of a B battery makes it a potential shock hazard. The internal resistance of large series combinations of cells is quite high, and this limits the amount of power that can be supplied by a B battery. *See also* TUBE.

B DISPLAY

A radar display that indicates the azimuth, or compass direction, of a target on a horizontal scale, and the range on a vertical scale, is called a B display. On this kind of display, targets show up as bright spots.

The B display differs from the standard radar display, where azimuth and range are shown on a circular screen so that targets appear in their true relative orientations. The illustration shows the appearance, on a B display, of a set of

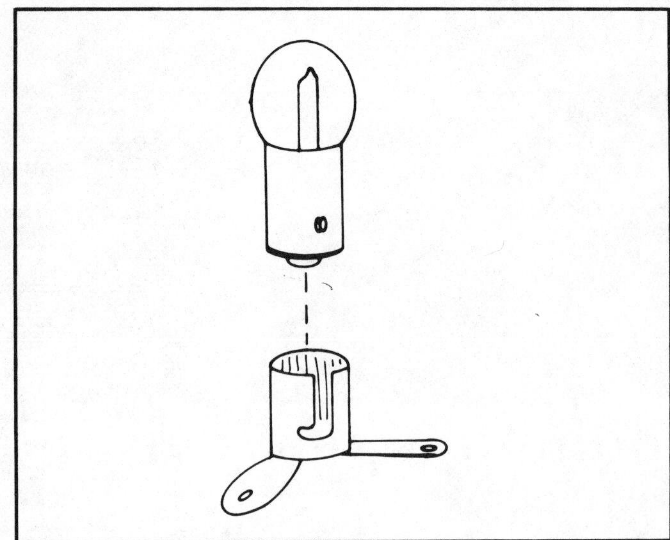

BAYONET BASE AND SOCKET: A bayonet base and socket allows quick installation and removal.

echoes arranged in a circle around the radar antenna, all at an equal range. A transformation of coordinates (*see* TRANSFORMATION) is necessary to get a visually correct display.

Motion, as well as relative position, is distorted on a B display. A target moving in a circle centered on the antenna would appear to move in a straight line horizontally across the screen. A target moving directly away from the station would appear to follow a path vertically up the screen of a B display. An object passing near the antenna, but actually moving in a straight path, would appear to describe a hyperbolic trace on a B display. *See also* RADAR.

BCD

See BINARY-CODED DECIMAL.

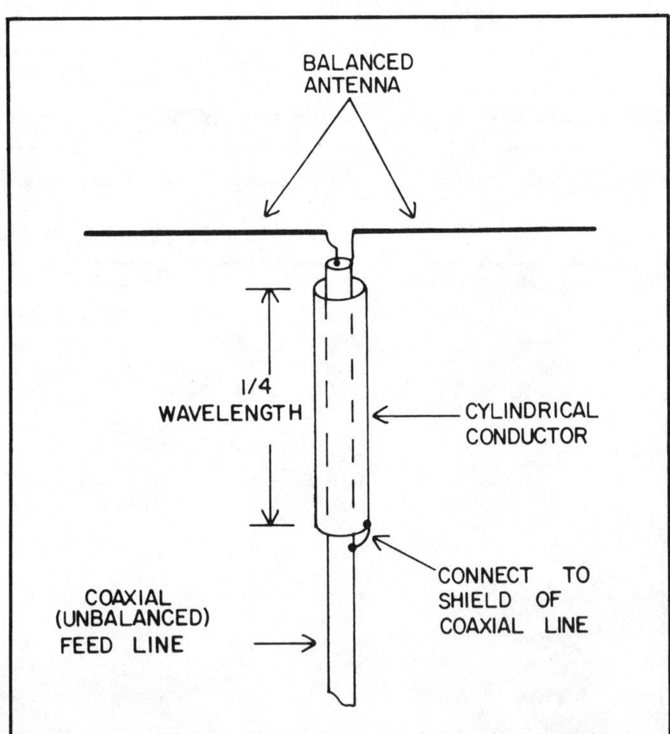

BAZOOKA BALUN: The bazooka balun chokes off RF currents flowing on the outside of a coaxial-cable feed line.

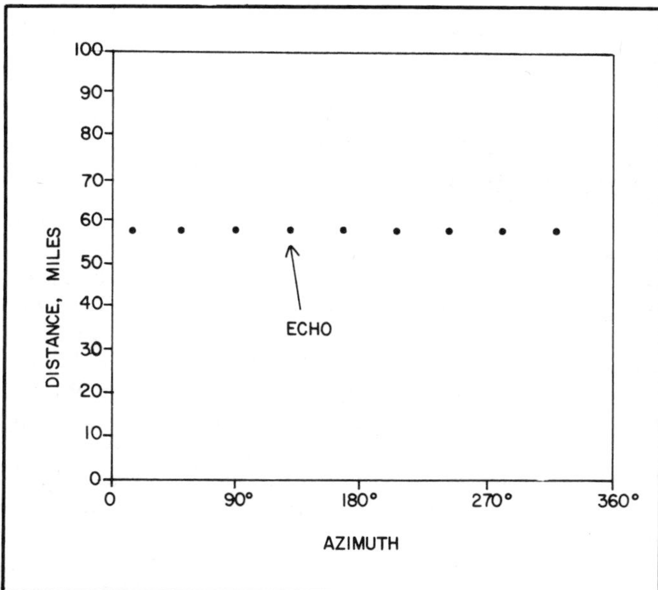

B DISPLAY: A radar B display showing echoes from objects in a circle centered on the antenna.

BEACON

A beacon is a transmitting station, usually having fairly low RF-power output, designed for the purpose of aiding in the monitoring of propagation conditions. The beacon sends a steady signal with frequent identification by call sign and geographical location. By listening on the beacon frequency, propagation conditions between the beacon and a receiving station are easily checked.

Beacons are used for radio location purposes, and for easy spotting of certain objects on radar equipment. Beacons are a valuable aid in radar tracking, especially if an object would be difficult to see or identify. *See also* PROPAGATION, RADAR.

BEAM ANTENNA

A beam antenna is any directional antenna with power gain for transmitting purposes. Generally, however, the term "beam" has come to mean a Yagi antenna. One such antenna is shown in the illustration.

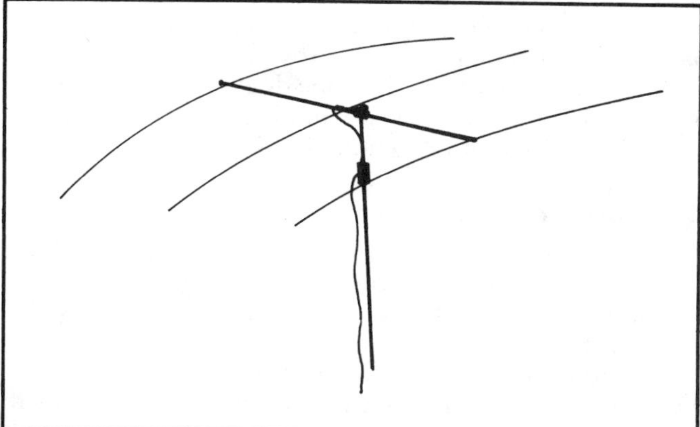

BEAM ANTENNA: A beam antenna for hf operation.

The effectiveness of a beam antenna is measured in terms of its forward gain, and also its front-to-back ratio (*see* FRONT-TO-BACK RATIO). In general, the greater the power gain, the greater the front-to-back ratio. However, there are instances where this simple correlation fails. Another means of evaluating the performance of a beam antenna is the beamwidth (*see* BEAMWIDTH).

Beam antennas are advantageous for both receiving and transmitting. The forward gain enhances the range, and good front-to-back ratio and narrow beamwidth aid in preventing interference to and from stations in directions other than that of the desired station. *See also* DIRECTOR, LOG-PERIODIC ANTENNA, QUAD ANTENNA, REFLECTOR, YAGI ANTENNA.

BEAM-POWER TUBE

A beam-power tube is a vacuum tube with a special beam-directing electrode that confines the electrons to certain directions (see drawing). The result is a concentrated beam, and this is where the tube gets its name. The beam-power tube is more efficient than an ordinary vacuum tube, especially when large amounts of power output are required, such as in RF transmitting equipment. The grids of a beam-power tube are usually aligned, which further aids in efficient operation. A beam-power tube may be a triode, tetrode, or pentode. They are usually pentodes, because a suppressor grid is needed with the high electron speeds. *See also* TUBE.

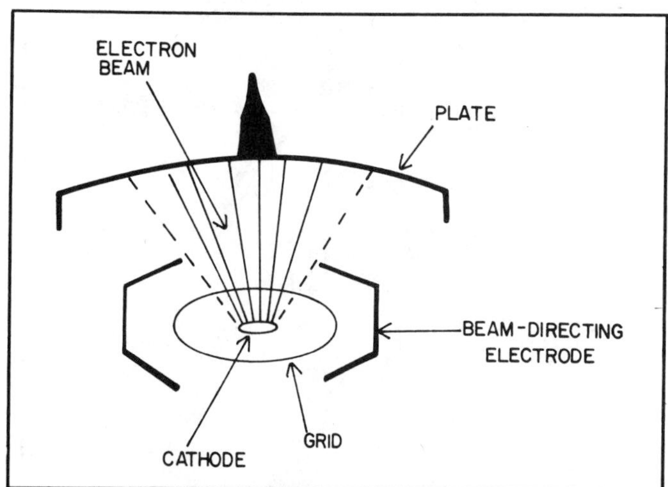

BEAM-POWER TUBE: The beam-power tube concentrates the electrons in a certain direction by means of a special electrode.

BEAMWIDTH

The beamwidth of a directional antenna or transducer is a measure of the degree to which its response or output is concentrated. Antenna beamwidth is usually specified in terms of horizontal direction, or azimuth. However, beamwidth may also be specified in the vertical plane.

To determine the beamwidth, the favored direction of the antenna (the direction in which it radiates the greatest amount of power) must be found, as in the illustration. Normally, less power is radiated to the left or right of, or above and below, this favored direction. As a field-strength meter is moved back and forth at a distance from the antenna, the two directions are found at which the power output from the antenna is 3 dB below the output in the favored direction. A power drop of 3 dB is a 50-percent reduction (*see* DECIBEL). Hence, these two directions, or bearings, are called the half-power points.

The beamwidth is the angle in degrees, in a specified plane (vertical or horizontal), between the half-power points as shown. Generally, the greater the power gain of an antenna, the narrower its beamwidth. A two-element Quad or Yagi antenna might have a beamwidth on the order of 60 to 80 degrees in the horizontal plane. A large parabolic dish antenna, at microwave frequencies, can have a beamwidth of less than one degree. The narrowest beamwidth of any device is found in the optical laser, where it is only a tiny fraction of a degree.

BEARING

The term bearing generally refers to a direction, such as azimuth or elevation, or a combination of directions. The compass bearing is the azimuth based on the north magnetic pole. A signal might be received from great distance, and its bearing found to be 90 degrees. This means the signal arrives from the east. Because of the spherical shape of the earth, however, the station where the signal originates will not usually have the same latitude in such a case. Bearings in terrestrial propagation define great circles (*see* GREAT CIRCLE). Charts of great circle bearings to various parts of the world, centered on certain locations, are available for communications purposes.

The term bearing is also used to mean a support for a rotating shaft. Bearings are found in motors, variable capacitors, potentiometers, and the like.

BEAT

When two waves are combined or superimposed, a beat occurs if the two frequencies are not the same. Waves beat together to create the appearance of either a change in amplitude (if the frequencies differ by just a few Hz) or new frequencies, called beat frequencies or heterodynes (if the original frequencies are far apart). Beat frequencies are the sum and difference of the original frequencies.

Sound waves often beat together when two musical notes are combined. Two tuning forks, having frequencies of, say, middle C (261.6 Hz) and D (293.7Hz) will beat together to create sound waves at 32.1 Hz and 555.3 Hz, the difference and sum frequencies, respectively. Beat frequencies are an important part of the sound of music, even though listeners are not usually aware of them.

Beat frequencies are important in the operation of mixers and superheterodyne receivers and transmitters. Heterodynes are responsible for the undesirable effects of cross modulation and intermodulation distortion. Heterodynes occur in any nonlinear circuit where two or more frequencies exist. *See also* CROSS MODULATION, HETERODYNE, INTERMODULATION, MIXER, SUPERHETERODYNE RECEIVER.

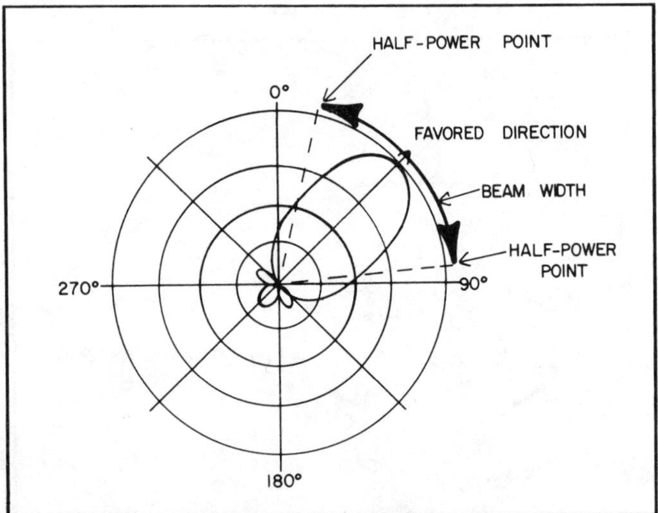

BEAM WIDTH: The beamwidth of an antenna is the angle in degrees between the half-power points in the azimuth plane.

BEAT-FREQUENCY OSCILLATOR

For detection of single-sideband (SSB), Morse code (CW), or radioteletype (RTTY) signals, an oscillator is necessary in the receiver. This oscillator, called a beat-frequency oscillator, is generally located in the intermediate-frequency stages. When tuned to the suppressed-carrier frequency of an SSB signal, or slightly away from the carrier frequencies of a CW or RTTY signal, the modulation information is retrieved. A detector that uses a beat-frequency oscillator is called a product detector (*see* PRODUCT DETECTOR). A block diagram of the use of a beat-frequency oscillator is shown in the illustration.

Without the beat-frequency oscillator (abbreviated BFO), CW and RTTY signals would be inaudible except for the clicks and thumps of the make and break moments. An SSB signal would sound muffled and unintelligible without a product detector. In some receivers, the frequency of the BFO is fixed at one side of the intermediate-frequency passband. The proper signal reproduction is obtained by adjusting the frequency of the receiver. In some receivers, the frequency of the BFO is tunable from one end of the intermediate-frequency passband to the other. In still other receivers, the passband itself is tunable and the BFO frequency remains constant.

A direct-conversion receiver has a BFO that is tunable over the entire reception range. When the BFO frequency gets sufficiently close to a signal, the beat frequency is audible. *See also* DIRECT-CONVERSION RECEIVER.

BELL, ALEXANDER GRAHAM

The inventor generally credited with the perfection of the wire telephone is Alexander Graham Bell (1847-1922). Bell was the first to successfully transmit voices via electrical impulses.

Bell believed that, since a voice is comprised of sounds that can be converted to electrical impulses, it ought to be possible to transmit those impulses over long distances and convert them back into recognizable sound. In order to do this, it was necessary to have a voice-to-impulse converter (microphone and battery), a two-wire electrical line, and an impulse-to-voice converter (speaker). Bell's apparatus was primitive by modern standards, but functioned well enough to allow communication via wire over moderate

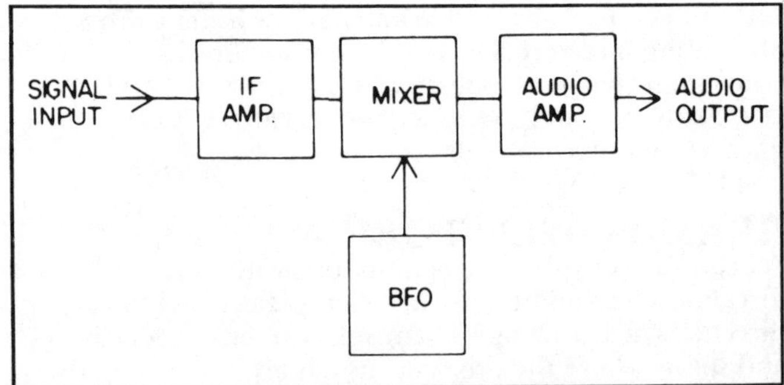

BEAT-FREQUENCY OSCILLATOR: A beat-frequency oscillator produces audio output with RF signal input by mixing.

distances.

Bell's actual discovery was made when he accidentally spilled some caustic solution onto himself and shouted for his assistant to come to his aid. His assistant heard the call through the telephone. *See also* TELEPHONE.

BELL TRANSFORMER

A bell transformer is a step-down transformer for use in alternating-current circuits. It has an iron core, and a voltage step-down ratio of approximately 6 to 1 or 12 to 1. Thus its output ranges from 10 to 20 volts with a 60-Hz, 120-volt primary.

Bell transformers can be used for the construction of any low-current ac or dc power supply. The transformer gets its name from the fact that it is frequently found in circuits for doorbells, alarm bells, and buzzers. *See also* TRANSFORMER.

BELL WIRE

Bell wire is a form of copper wire, usually solid rather than stranded, and soft-drawn rather than hard-drawn. Bell wire is covered with a cloth insulation; sometimes, however, any copper wire of American Wire Gauge (AWG) No. 18 to 22, with any kind of insulation, is called bell wire.

Bell wire is intended for low-voltage, low-current applications. It should not be used with 120-volt ac household appliances. Bell wire is often found in radio transmitters and receivers, where it is used as hookup wire.

Bell wire, being solid, has a tendency to break under stress. This is enhanced because of its relatively small size. Although bell wire can be used for receiving antennas and low-power transmitting antennas at high frequencies, larger wire is preferable since it does not stretch or break as easily as bell wire. *See also* HOOKUP WIRE, STRANDED WIRE, WIRE.

BENCHMARK

A benchmark is a simple program that is often used for testing microcomputers and microprocessors. In a given application, it may be possible to use any of several different chips, but one will perform better than the others. The benchmark program allows an engineer to determine which chip will be the best central processing unit (CPU) for the application.

A benchmark program simulates the actual operating requirements of the microcomputer or microprocessor under test. The program must be simple enough so that the engineer can follow it step by step manually. But the program must be complex enough to provide a reasonable simulation of the operating requirements. If the program is too simple, the wrong chip might be chosen; if it is too complicated, the test procedure will be tedious.

Benchmark programs are often the means by which a microcomputer or microprocessor chip is selected for use in a product such as a hand calculator or a radio transceiver. The most suitable device, as determined by the benchmark, is used as the CPU. *See also* CENTRAL PROCESSING UNIT, MICROCOMPUTER, MICROPROCESSOR.

BENT ANTENNA

A bent antenna is, as the name suggests, an antenna with a radiating element that is not straight. Usually, ideal antenna performance is obtained when the antenna radiator is straight. (A full-wave loop antenna, and various specialized types of receiving antennas, are exceptions.) Space limitations, though, sometimes make a straight antenna impossible to erect.

The illustration shows three examples of bent antennas. All of them are designed for a resonant frequency of 3.6 MHz. At this frequency, a quarter-wave radiator must be 65 feet long, and a half-wave radiator 130 feet long, for resonance. These lengths are derived from the standard formulas:

$$\text{Quarter wavelength (feet)} = 234/f$$

$$\text{Half wavelength (feet)} = 468/f$$

where f is the fundamental resonant frequency in MHz. These lengths are somewhat impractical at many locations, such as the standard-size city lot. Bent configurations allow the installation of an antenna in a place where a straight, full-size radiator will not fit.

Bent antennas are always a compromise. If properly chosen for utilization of available space, bent radiators can perform almost as well as straight ones. The feed-point impedance of a bent antenna is usually somewhat lower than that of a straight antenna, but the difference is seldom of great magnitude. *See also* WIRE ANTENNA.

BERYLLIUM OXIDE

Beryllium oxide (BeO) has an exceptional ability to con-

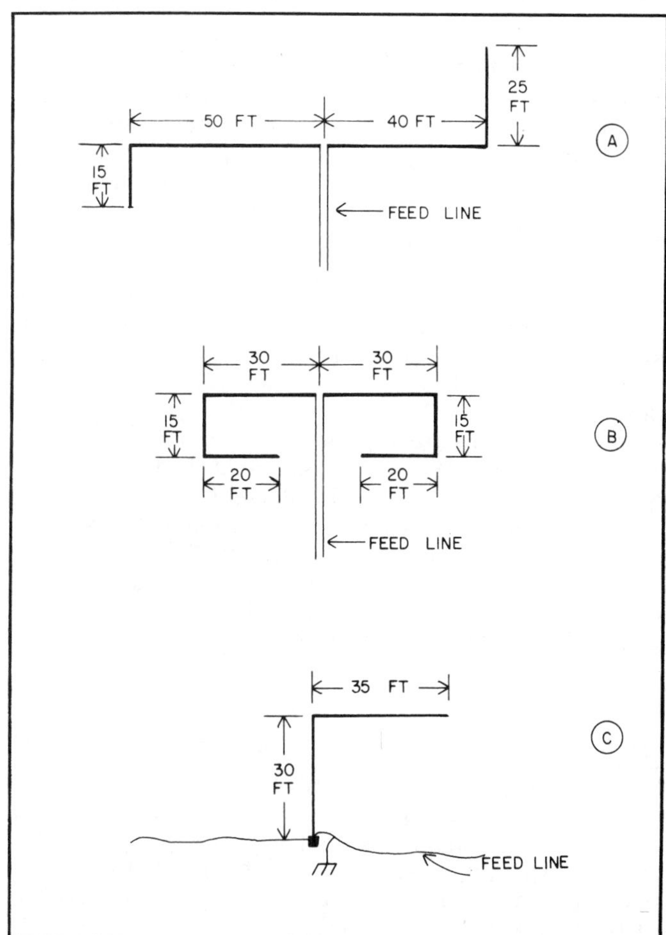

BENT ANTENNA: At A and B, two configurations for a center-fed dipole. At C, a bent vertical antenna.

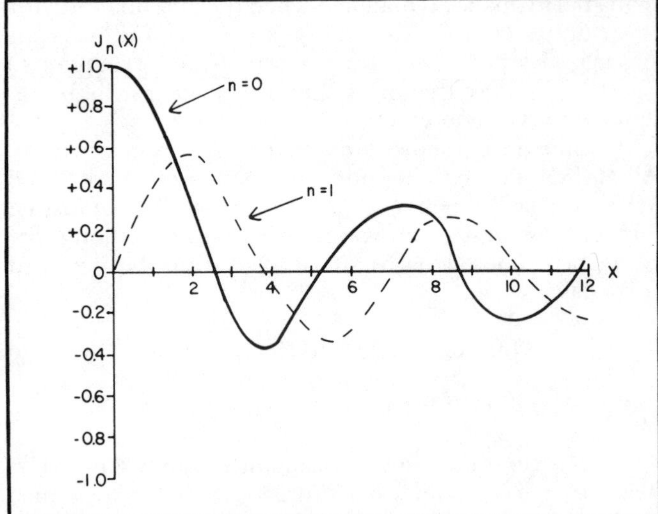

BESSEL FUNCTION: Bessel functions $J_0(x)$ and $J_1(x)$.

duct heat. It is an electrical insulator, but has heat-conducting characteristics similar to most metals.

Beryllium oxide is used in high-power amplifiers, of both tube and solid-state design, to dissipate the heat and maintain the amplifying device or devices at an acceptable temperature. Beryllium-oxide powder is extremely toxic when inhaled. Therefore, tubes and transistors that contain this substance as part of their heat-sink apparatus should be handled with great care. Beryllium oxide is also sometimes called beryllia.

BESSEL FUNCTION

A Bessel function is a solution to a certain kind of differential equation. Bessel functions are used in the determination of the effective bandwidth of a frequency-modulated signal. In FM, the distribution of sidebands above and below the carrier frequency depends on the deviation-to-modulating-frequency ratio, or modulation index (*see* MODULATION INDEX). This quantity affects the sideband amplitudes at various distances from the center frequency. Theoretically, FM sidebands extend infinitely to either side of the carrier. In practice, their amplitudes are significant only wcthin a certain range, given by a Bessel function.

Bessel functions are used by engineers in the design of bandpass filters. The optimum shape of the bandpass response can be determined with Bessel functions (see ARGUMENT) where n is a non-negatives integer called the order of the function. When graphed, Bessel functions look like damped sine waves. The drawing shows examples of the Bessel functions J_0 (x) and J_1 (x) for values of x between 0 and 12.

BETA

The term beta is used to define the current gain of a bipolar transistor when connected in a grounded-emitter circuit. A small change in the base current, I_B, causes a change in the collector current, I_C. The ratio of the change in I_C to the change in I_B is called the beta, and is abbreviated by the Greek letter β:

$$\beta = \Delta I_C / \Delta I_B$$

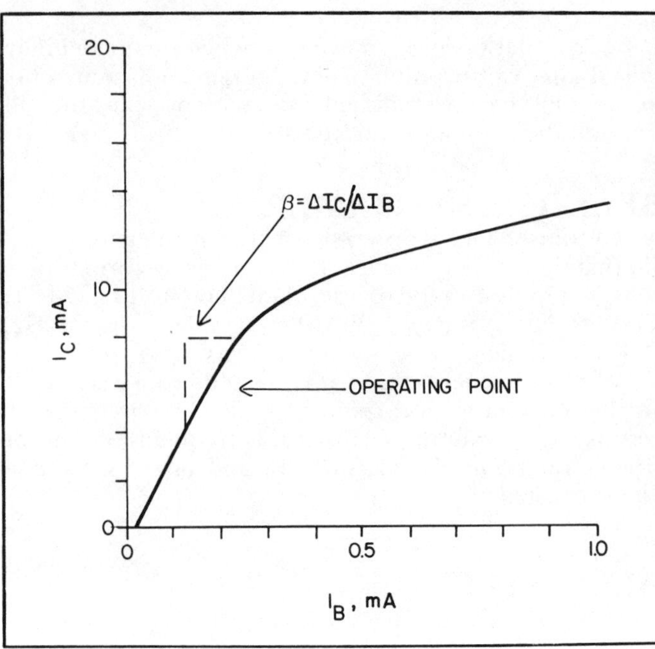

BETA: The beta of a transistor depends on the operating point chosen, and is a measure of amplification.

The collector voltage is kept constant when measuring the beta of a transistor.

The accompanying graph shows a typical I_C-vs-I_B curve for a transistor in a grounded-emitter configuration with constant collector voltage. The beta of the transistor is the slope of the curve at the selected operating point (*see* SLOPE). A line drawn tangent to the curve at the no-signal operating point, as shown, has a definite ratio ($\Delta I_C/\Delta I_B$) which is easily determined by inspection. *See also* ALPHA, TRANSISTOR.

BETA-ALPHA RELATION

The beta of a transistor is the base-to-collector current gain with the emitter at ground potential (*see* BETA). The alpha is the emitter-to-collector current gain with a grounded-base circuit (*see* ALPHA). The alpha of a transistor is always less than 1, and the beta is always greater than 1.

The beta of a transistor can be mathematically determined if the alpha is known according to the formula:

$$\beta = \alpha/(1-\alpha)$$

where the Greek letters α and β represent the alpha and beta, respectively. The following ratios also hold:

$$\alpha/\beta = 1 - \alpha = 1/(1 + \beta)$$
$$\beta/\alpha = 1/(1 - \alpha) = 1 + \beta$$

The alpha is mathematically derived from the beta by the equation:

$$\beta = \alpha\,(1 + \beta)$$

A typical bipolar transistor with an alpha of 0.95 has a beta of 19. As the beta approaches infinity, the alpha approaches 1, and vice-versa.

BETA CIRCUIT

Some amplifiers utilize feedback, either negative (for

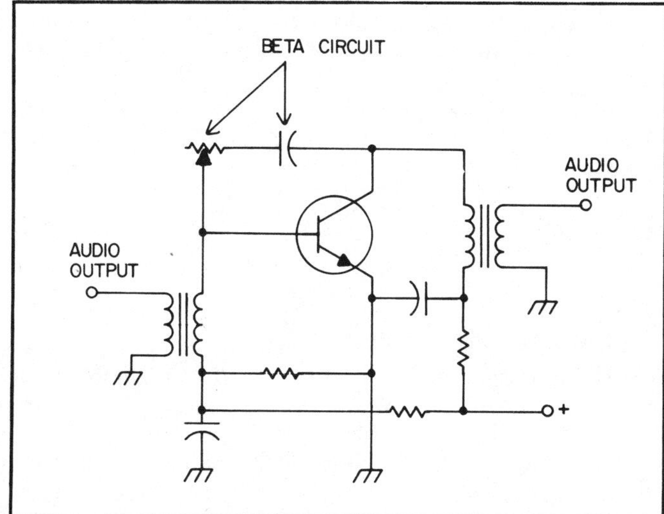

BETA CIRCUIT: The beta circuit is a feedback circuit.

stabilization) or positive (to enhance gain and selectivity). The part of the amplifier circuit responsible for the feedback is called a beta circuit. The schematic illustrates an amplifier with feedback, showing the beta circuit. In this particular case, the feedback is negative, intended to prevent oscillation.

In RF amplifiers, neutralization is a form of feedback, and the neutralizing capacitor forms the beta circuit. *See also* FEEDBACK AMPLIFIER, NEUTRALIZATION, NEUTRALIZING CAPACITOR.

BETA-CUTOFF FREQUENCY

As the operating frequency of an amplifier transistor is increased, the current amplification, or beta, decreases. In a common-emitter amplifier, the frequency at which the beta drops 3 decibels (dB) compared to its value at 1 kHz is called the beta-cutoff frequency. The beta-cutoff frequency is determined in the same manner as the alpha-cutoff frequency (*see* ALPHA-CUTOFF FREQUENCY). The only difference is that the emitter, rather than the base, is at ground potential.

Depending on the type of transistor, the beta-cutoff frequency might be a few MHz or several hundred MHz. The beta-cutoff frequency differs slightly from the alpha-cutoff frequency, since a common-emitter configuration is used instead of a common-base configuration. Both specifications are useful, however, in amplifier design with bipolar transistors at high and very-high radio frequencies. The graph illustrates the determination of beta-cutoff frequency.

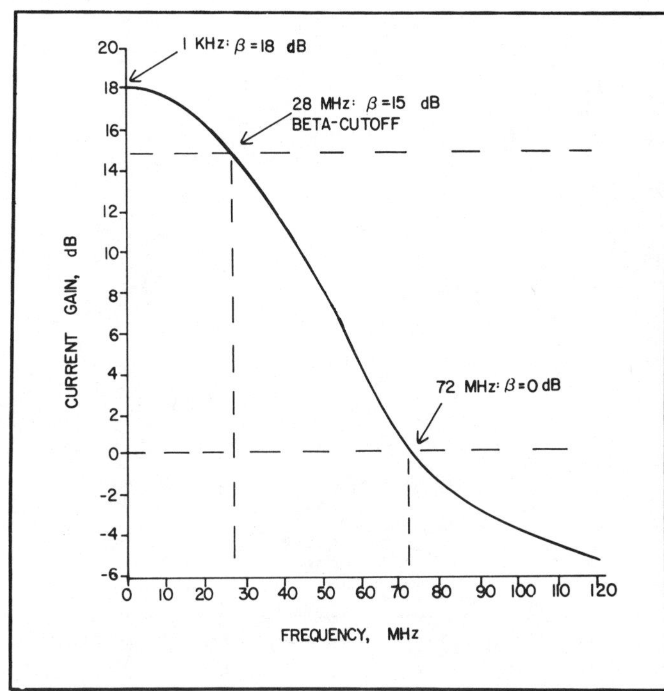

BETA-CUTOFF FREQUENCY: Determination of beta-cutoff frequency. When the beta of the transistor drops to 3 dB below its value at 1 kHz, the frequency (27 MHz in this case) is the beta-cutoff frequency.

BETA PARTICLE

A beta particle is a high-speed, electrically charged particle with the same mass and charge quantity as an electron. Beta particles are emitted by many radioactive substances, and are often given off by decaying atoms at nearly the speed of light. A barrage of beta particles is sometimes given the name beta rays.

Beta rays are, biologically, the least damaging of all forms of radiation. This is because the electron is the least massive of atomic particles. Neutrons, protons, and alpha particles, moving at high speed, do much greater damage to living cell tissue because of their greater mass.

Beta rays can consist of positively charged particles having the same mass as an electron. Such particles are called positrons, and are a form of anti-matter (*see* POSITRON). Beta rays consisting of electrons are sometimes called $\beta -$ radiation, and beta rays consisting of positrons are called $\beta +$ radiation. *See also* PARTICLE ACCELERATOR.

BETATRON

A betatron is a device for accelerating electrons to extremely high speeds. When electrons move at velocities approaching the speed of light, they become beta particles, a form of radioactive energy (*see* BETA PARTICLE). When the electrons strike a metal target, X rays are produced, which are used in medical and industrial applications.

The illustration is a pictorial diagram of a betatron. Electrons are injected into a toroidal, or doughnut-shaped, evacuated chamber. Powerful magnetic fields accelerate the electrons around and around, to impact against the target. These magnetic fields are carefully controlled to direct the electrons to the target with the maximum attainable speed. *See also* PARTICLE ACCELERATOR.

BEVERAGE ANTENNA

A Beverage antenna is a form of traveling-wave antenna, used for receiving at medium and high frequencies.

The Beverage antenna consists of a long, straight wire of at least several wavelengths, run close to the ground as shown in the illustration. A 300- to 600-ohm noninductive resistor is generally connected between the far end of the antenna and a ground rod. Such an antenna responds well to signals arriving from the direction in which the wire is pointed, but it responds poorly in all other directions. Traveling waves are set up along the wire as electromagnetic fields arrive from the favored direction.

A Beverage antenna is a form of terminated longwire

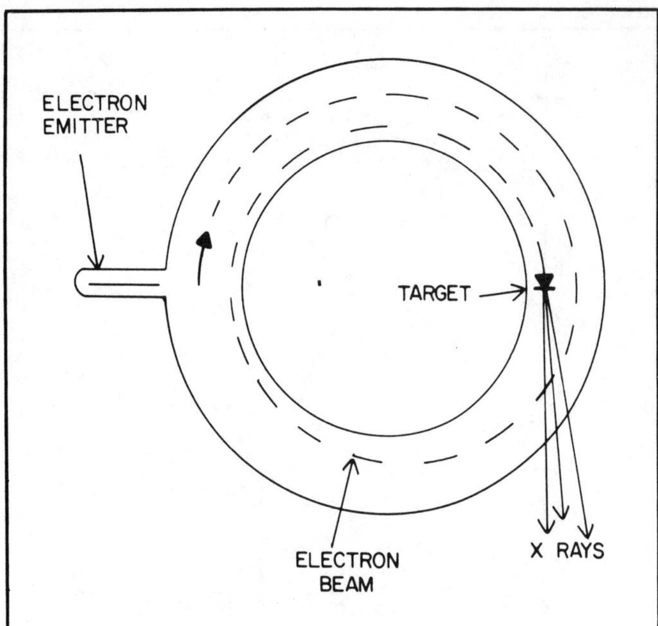

BETATRON: The betatron is an electron accelerator for producing X rays.

(*see* LONGWIRE ANTENNA). Because of its proximity to the ground, the Beverage is a poor antenna for transmitting. Its highly directional receiving characteristics, along with its relatively minimal noise pickup, make it a favorite among communications enthusiasts. The Beverage is frequently used by radio amateurs for listening on their 160-meter (1.8-MHz) and 80-meter (3.5-MHz) bands.

The resistor at the far end of the Beverage antenna should, ideally, be matched to the characteristic impedance of the wire for unidirectional operation. The resistor may be eliminated or short-circuited, resulting in bidirectional operation if desired. Without the terminating resistor, reception remains excellent in the favored direction of the illustration, but the response is also excellent in the opposite direction, making the antenna bidirectional.

BEZEL

A bezel is a holder for electronic metering devices, graticules, or lenses (see drawing). Bezels keep such devices firmly in place and protect them against dust and physical stress. They also provide an interesting or attractive appearance which, although not necessarily crucial in the test laboratory, is very important from a consumer point of view.

Some meter bezels have a small screw at the base for meter calibration. While the equipment is shut off, the screw is turned until the meter reads zero.

BFO

See BEAT-FREQUENCY OSCILLATOR.

B-H CURVE

A B-H curve is a graph that illustrates the magnetic properties of a substance. The magnetic field flux density, B, is plotted on the vertical scale as a function of the magnetizing force, H. Flux density is measured in units called gauss, and magnetizing force is measured in units called oersteds (*see* GAUSS, OERSTED). B-H curves are of importance in determining the suitability of a particular magnetic material for use in the cores of inductors and transformers. The graph shows an example of a B-H curve, also called a hysteresis loop.

When the magnetizing force is increasing, the flux density increases also. While the magnetizing force decreases, so does the flux density. The flux density does not change exactly in step with the magnetizing force in most core materials. This sluggishness of the magnetic response is called hysteresis. *See also* HYSTERESIS, HYSTERESIS LOOP, HYSTERESIS LOSS.

BIAS

Bias is a term generally used to refer to a potential difference, applied deliberately between two points for the purpose of controlling a circuit. In a vacuum tube, the bias is the voltage between the cathode and control grid. In a bipolar transistor, the bias is the voltage between the emitter and the base, or the emitter and the collector. In a field-effect transistor, the bias is the voltage between the source and the gate, or the source and the drain.

Certain bias conditions are used for specified purposes. In forward bias, the cathode of a vacuum tube is negative with respect to the grid or plate. In reverse bias, the opposite is true: The cathode is positive with respect to the grid or plate. In a semiconductor P-N junction, forward bias

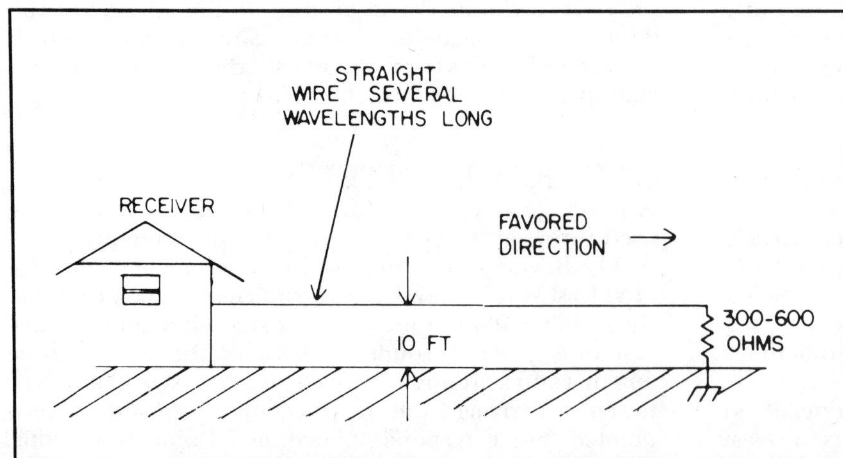

BEVERAGE ANTENNA: A Beverage antenna is a unidirectional receiving antenna sometimes used at medium and high frequencies.

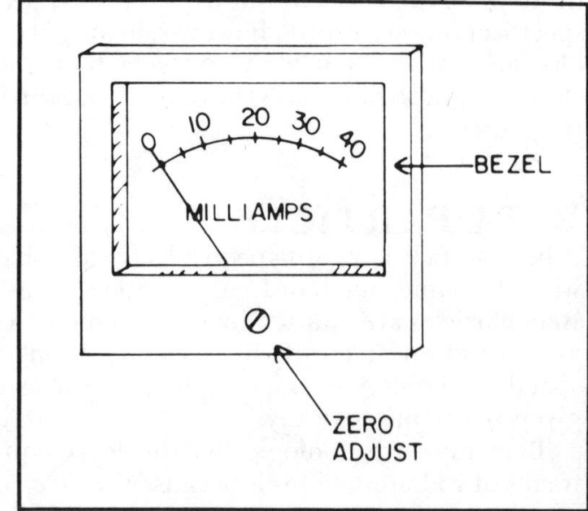

BEZEL: A meter bezel.

occurs when the P-type material is positive with respect to the N-type materials; in reverse bias, the P-type material is negative with respect to the N-type material. When two electrodes are at the same potential, they are said to be at zero bias. *See also* FORWARD BIAS, REVERSE BIAS, ZERO BIAS.

BIAS COMPENSATION

Bias compensation is a means of balancing the pickup arm of a record player. Ordinarily, the pickup arm has a tendency to pull inward toward the center of the record as the record rotates. If the stylus is too large for the grooves of the record (*see* STYLUS), it may actually slide inward across the surface.

By providing an outward-directed tension on the pickup arm, the tendency of the arm to slide toward the center can be reduced or eliminated. This bias compensation should be properly adjusted according to the speed of the record and the weight of the pickup arm. Bias compensation prevents the possibility of damage to a record from a sliding stylus, or from pressure of the stylus against the grooves of the record. This prolongs the life of the record and the stylus.

BIAS CURRENT

Bias current is the current between the emitter and the base of a bipolar transistor under conditions of no input signal. When the emitter-base junction is forward-biased, the bias current is normally a few microamperes or milliamperes. When the junction is at zero bias, the bias current is zero. When the junction is reverse-biased, the bias current is negligibly small.

Any particular transistor has an optimum bias-current operating point. When the bias voltage is properly chosen, the bias current is such that the distortion is minimal but the amplification sufficient. Bias-current specifications apply to class-A and class-AB amplifiers. In class-B and class-C amplifiers, the bias current is normally zero. *See also* BIAS, CLASS-A AMPLIFIER, CLASS-AB AMPLIFIER, FORWARD BIAS.

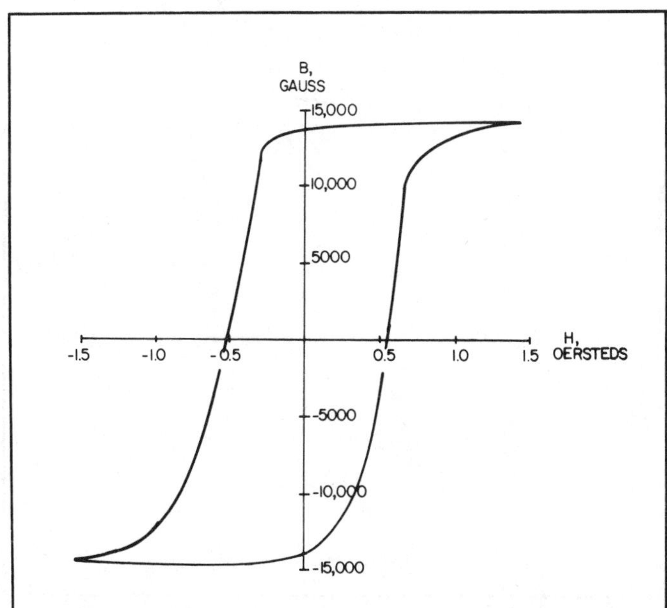

B-H CURVE: A B-H curve shows the lag between magnetizing force and flux density in a coil core for an alternating current.

BIAS DISTORTION

When a tube, transistor, or other amplifying device is operated with a bias resulting in nonlinearity, the distortion in the output signal is called bias distortion. Bias distortion is sometimes unimportant, such as in RF class-B and class-C amplifiers. In high-fidelity audio equipment, though, the bias must be chosen carefully to avoid this distortion as much as possible.

In the accompanying graphs, two operating points are shown. At A, the operating point is selected for linear operation, and there is essentially no distortion. At B, the operating point has been improperly selected; the bias current is too large. Therefore, the amplification factor is different on the positive part of the input cycle as compared with the negative part. At the output, the waveform is lopsided. This is bias distortion.

Bias distortion in an amplifier is not the same as the distortion resulting from excessive input-signal amplitude. While the two forms of distortion may produce similar output waveforms, excessive input amplitude, or overdrive, causes distortion no matter what the bias is set at. The drive level must thus be regulated to avoid distortion. *See also* BIAS, CLASS-A AMPLIFIER.

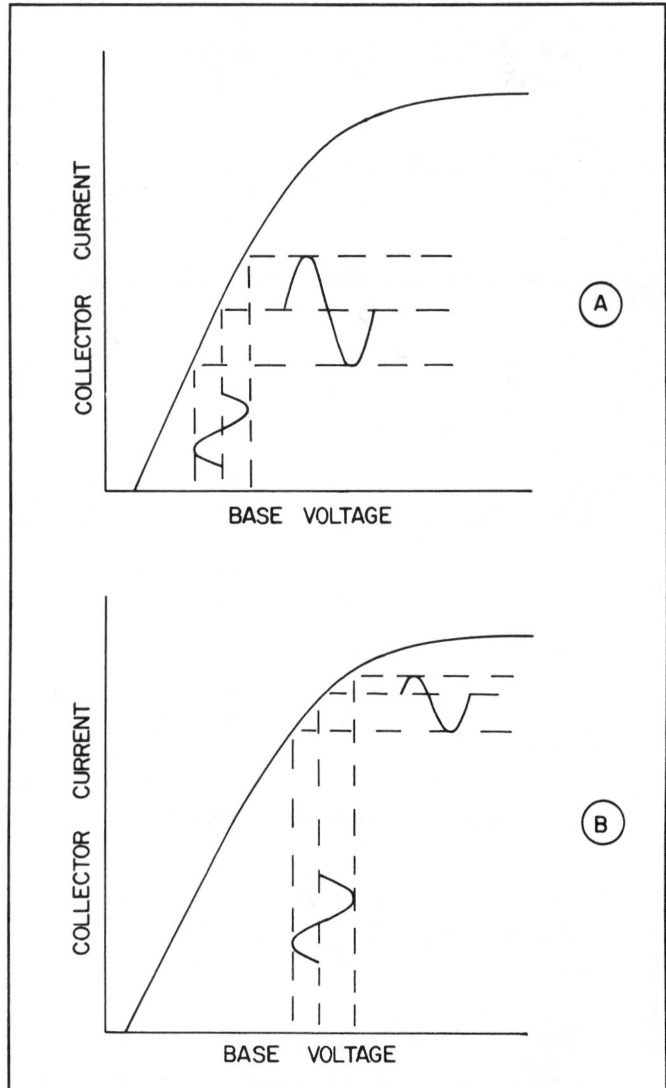

BIAS DISTORTION: Bias distortion occurs when the operating point of a transistor is improperly selected. At A, there is no bias distortion, but at B, the output waveform is not symmetrical because of improper bias.

BIAS STABILIZATION

Bias stabilization is any means of ensuring that the bias in a circuit will remain constant. In a transistor circuit, a common means of bias stabilization is the resistive voltage divider. By choosing the appropriate ratio of resistances and the correct magnitude of resistance values, the base voltage can be maintained within a precise range.

Without bias stabilization, the base of a transistor may become improperly biased because of temperature changes or because of variations in the strength of an input signal. In some cases, changes in bias can lead to thermal runaway, resulting in component overheating, loss of gain and linearity, and possibly even destruction of a transistor. *See also* BIAS, THERMAL RUN AWAY.

BICONICAL ANTENNA

A biconical antenna is a balanced broadband antenna that consists of two metal cones, arranged so that they meet at or near the vertices as shown in the drawing. The biconical antenna is fed at the point where the vertices meet. The exact feed-point impedance of a biconical antenna depends on the flare angle of the cones and the separation between their vertices.

A biconical antenna displays resonant properties at frequencies above that at which the height h of the cones is ¼ wavelength in free space. The highest operating frequency is several times the lowest operating frequency. A biconical antenna oriented vertically, as shown, emits and receives vertically polarized electromagnetic waves.

The biconical antenna is often used at vhf, but its size gets prohibitively large at lower frequencies. However, one of the cones can be replaced by a disk or ground plane for the purpose of reducing the physical dimensions of the antenna while retaining the broadband characteristics. If the top cone is replaced by a disk of a certain radius, the antenna becomes a discone. If the lower cone is replaced by a ground plane, the antenna becomes a conical monopole. Both the discone and the conical monopole are practical for use at frequencies as low as about 2 MHz. *See also* CONICAL MONOPOLE ANTENNA, DISCONE ANTENNA.

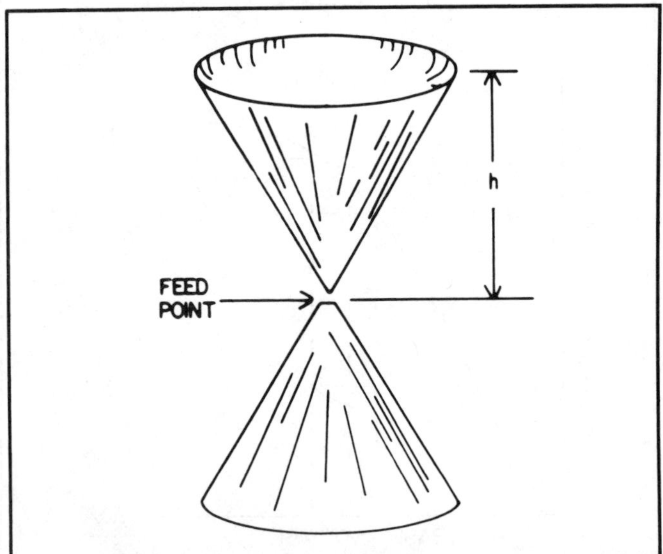

BICONICAL ANTENNA: A biconical antenna functions up to several octaves above the frequency where the cone height, h, is ¼ wavelength.

BIDIRECTIONAL PATTERN

Any transducer that performs in two opposite directions is said to have a bidirectional pattern. Such a transducer can be an antenna, microphone, speaker, or other device designed for converting one form of energy to another. A in the illustration shows a bidirectional response pattern in the azimuth (horizontal) plane.

Bidirectional patterns are often preferred over unidirectional patterns. Bidirectional antennas are quite common; the ordinary half-wave dipole antenna, for example, is bidirectional in the horizontal plane.

A bidirectional response pattern is usually symmetrical, as shown at A. That is, the response is equal in opposite directions, 180 degrees apart in the azimuth plane. For antennas, only the azimuth plane is used to define bidirectionality. But for microphones and speakers, the pattern must be symmetrical about a straight line in three dimensions, as shown at B, to be truly bidirectional. *See also* UNIDIRECTIONAL PATTERN.

BIFILAR TRANSFORMER

A bifilar transformer is a transformer in which the primary and secondary are wound directly adjacent to each other, as shown in the illustration. This provides the maximum possible amount of coupling between the windings. The mutual inductance of the windings of a bifilar transformer is nearly unity (*see* MUTUAL INDUCTANCE).

Bifilar transformers are especially useful when the greatest possible energy transfer is necessary. Capacitive coupling between the primary and secondary of a bifilar transformer, however, is great, and this allows the transfer of harmonic energy with essentially no attenuation. *See also* TRANSFORMER.

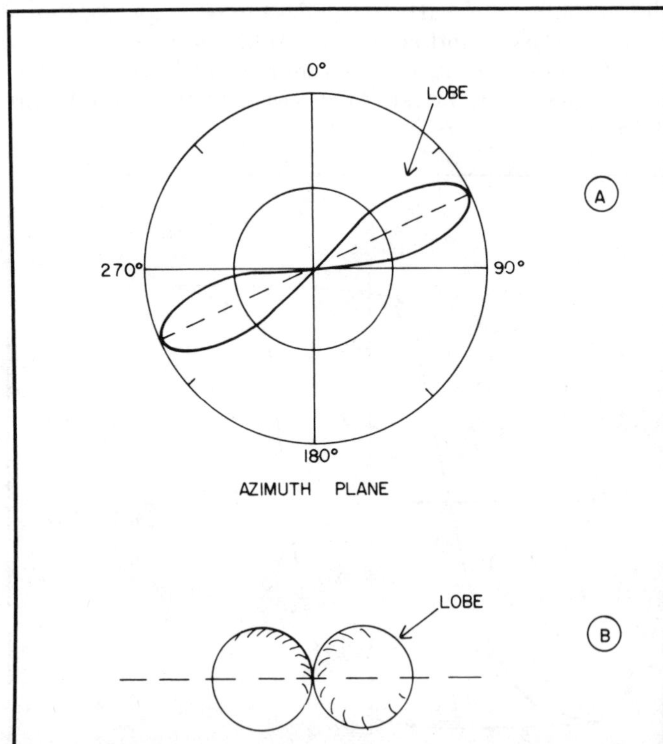

BIDIRECTIONAL PATTERN: At A, a bidirectional pattern in the azimuth plane, showing peaks at about 70 and 250 degrees. At B, a bidirectional pattern in three dimensions is symmetrical about a straight line.

BIFILAR WINDING

A wirewound resistor is made noninductive by a technique called bifilar winding. The resistance wire is first bent double. Then the wire is wound on the resistor form, beginning at the joined end of the double wire and continuing until all the wire has been put on the form. This results in cancellation of most of the inductance, since the currents in the adjacent windings flow in opposite directions. The field from one half of the winding thus cancels the field from the other half.

In general, a bifilar winding is any winding that consists of two wires wound adjacent to each other on the same form. A bifilar transformer (*see* BIFILAR TRANSFORMER) uses bifilar windings for maximum energy transfer. Some broadband balun transformers (*see* BALUN) use bifilar, trifilar, or even quadrifilar windings. *See also* QUADRIFILAR WINDING, TRIFILAR WINDING, WIREWOUND RESISTOR.

BILATERAL NETWORK

Any network or circuit with a voltage-vs-current curve that is symmetrical with respect to the origin is called a bilateral network. Another way of saying that the voltage-vs-current curve is symmetrical with respect to the origin is to say that no matter what the voltage, the current magnitude will remain the same if the voltage polarity is reversed.

Any network of resistors is a bilateral network. Certain combinations of other components are also bilateral circuits. The illustration shows a symmetrical voltage-vs-current curve, A, characteristic of a bilateral network. At B, a schematic diagram of one circuit, which gives the curve at A, is shown. There are many other circuits that will provide this same response.

Most tube and transistor circuits are not bilateral, since current ordinarily flows through such devices in only one direction.

BILLBOARD ANTENNA

A billboard antenna is a form of broadside array (*see* BROADSIDE ARRAY), consisting of a set of dipoles and a reflecting screen. The illustration shows the physical construction of a billboard antenna.

The reflecting screen is positioned in such a way that the electromagnetic waves at the resonant frequency are reinforced as they return in the favored direction. The more dipoles in a billboard antenna, the greater the forward gain. Because of the reflecting screen, a billboard antenna has an excellent front-to-back ratio (*see* FRONT-TO-BACK RATIO).

Billboard antennas are used at uhf and above. Below about 150 MHz or so, the physical size of a billboard antenna becomes prohibitively large for most installations.

BINARY-CODED DECIMAL

The binary-coded decimal (BCD) system of writing numbers assigns a 4-bit binary code to each numeral 0 through 9 in the base-10 system, as shown in the table. The 4-bit BCD code for a given digit is its representation in binary notation. Four places are always assigned.

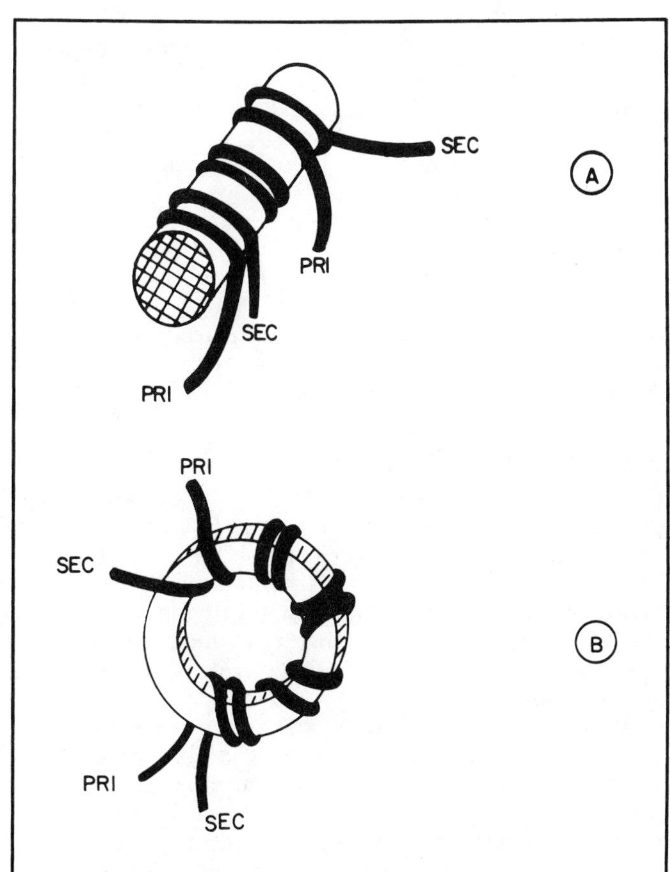

BIFILAR TRANSFORMER: At A, a solenoidal winding, and at B, a toroidal winding.

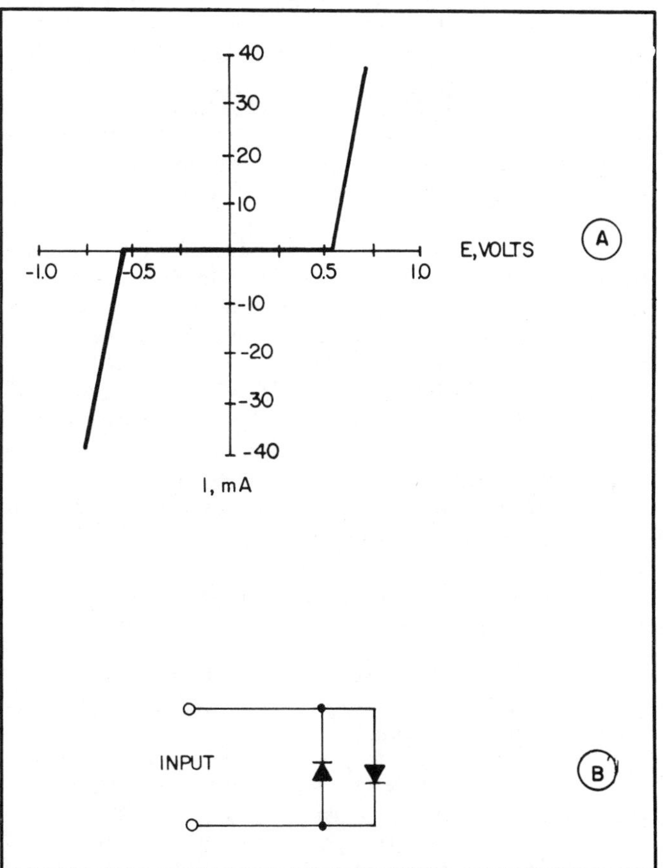

BILATERAL NETWORK: A bilateral network has a voltage-vs-current curve that is symmetrical about the origin. At A, a bilateral curve. At B, a circuit (one of many possible examples) that produces such a curve.

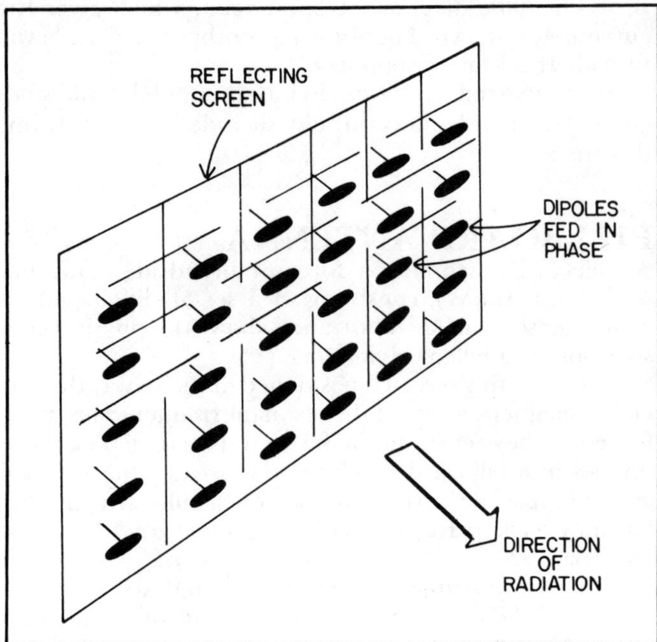

BILLBOARD ANTENNA: A billboard antenna consists of several, or many, dipoles driven in phase with a reflecting screen for use at short wavelengths.

Numbers larger than 9, having two or more digits, are expressed digit by digit in the BCD notation. For example, the number 189 in base-10 form is written as 0001 1000 1001 in BCD notation. The BCD representation of a number is not the same as its binary representation; in binary form, for example, 189 is written 10111101. *See also* BINARY-CODED NUMBER.

BINARY-CODED NUMBER

A binary-coded number is a number expressed in the binary system. The binary system of numbers is sometimes called base 2. Normally, we write numbers in base 10, called the decimal system.

In the base-10 system, the digit farthest to the right is the ones digit, and the digit immediately to its left is the tens digit. In general, if a given digit m represents $m \times 10^p$ in the decimal system, then the digit n, to the immediate left of m, represents $n \times 10^{p+1}$.

In base 2, the digit farthest to the right is the ones digit. Instead of 10 possible values for a digit, there are only two, 0 and 1. The digit to the left of the ones digit is the twos digit. Then comes the fours digit, the eights digit, and so on. In general, if a given digit m represents $m \times 2^p$ in the binary number system, then the digit n to its left represents $n \times 2^{p+1}$.

The illustration shows both the decimal and binary notations for the number 189. At A, in base 10:

$$189 = (9 \times 10^0) + (8 \times 10^1) + (1 \times 10^2)$$

At B, in binary notation:

$$10111101 = (1 \times 2^0) + (0 \times 2^1) + (1 \times 2^2) + (1 \times 2^3) + (1 \times 2^4) + (1 \times 2^5) + (0 \times 2^6) + (1 \times 2^7),$$

which is simply another way of saying that

$$189 = 1 + 4 + 8 + 16 + 32 + 128.$$

BINARY-CODED DECIMAL: BINARY-CODED DECIMAL REPRESENTATIONS OF THE DIGITS 0 THROUGH 9.

Digit in Base 10	BCD Notation
0	0000
1	0001
2	0010
3	0011
4	0100
5	0101
6	0110
7	0111
8	1000
9	1001

Binary-coded numbers are used by calculators and computers, because the 0 and 1 digits are easily handled as "off" and "on" conditions. Although a binary-coded number has more digits than its decimal counterpart, this presents no problem for a computer with capability of handling thousands or millions of bits. *See also* COMPUTER, DIGITAL, DIGITAL COMPUTER.

BINARY COUNTER

A binary counter is a device that counts pulses in the binary-coded number system. Each time a pulse arrives, the binary code stored by the counter increases by 1. A simple algorithm (*see* ALGORITHM) ensures that the counting procedure is correct.

Binary counters form the basis for most frequency counters (*see* FREQUENCY COUNTER). Such circuits actually count the number of cycles in a specified time period, such as 0.1 second or 1 second. Obviously, such counters must work very fast to measure frequencies approaching 1 GHz which is 10^9 Hz.

A simple divide-by-two circuit is sometimes called a binary counter or binary scaler. Such a circuit produces one output pulse for every two input pulses. A "T" flip-flop (*see* T FLIP-FLOP) is sometimes called a binary counter. Several divide-by-two circuits, when placed one after the other, facilitate digital division by any power of 2. *See also* DIGITAL CIRCUITRY.

BINDING POST

A binding post is a terminal used for a temporary, low-voltage connections. The photograph shows a typical set of binding posts on a dc power supply. Two thumb screws, made of metal and perhaps coated with plastic, are clamped onto the wire leads by hand turning. Sometimes the screw shafts have holes through which the wire leads can be placed to prevent the wire from slipping off the binding post.

Binding posts offer fairly good current-carrying capacity for low-voltage supplies, up to several amperes. Binding posts are not used with power supplies of more than about 30 volts, because of the shock hazard from exposed electrodes. Binding posts are not good for prolonged use outdoors, or for permanent installations. Soldered connections are preferable in such instances. Binding posts are frequently used with spade lugs, which fit neatly around the post (*see* SPADE LUG).

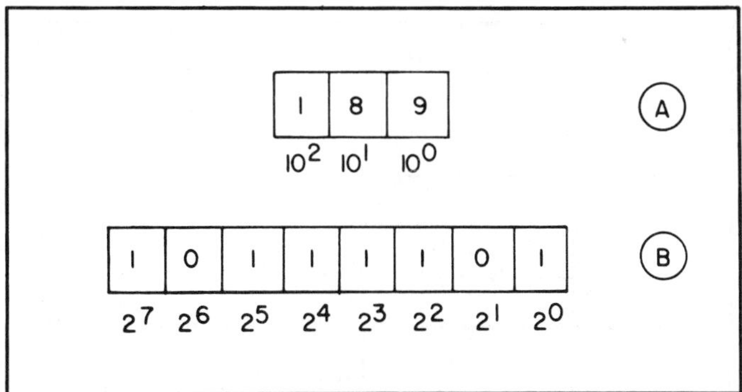

BINARY-CODED NUMBER: At A, a decimal number; at B, the same number written in binary-coded form.

BIOELECTRONICS

The application of electronics to biological problems is called the field of bioelectronics. This especially pertains to the development of electronic circuits to perform or regulate certain body functions. Probably the most common example of a bioelectronic device is the hearing aid, which consists of a microphone, audio amplifier, and earphone. Another frequently used bioelectronic device is the heart pacemaker.

As technology becomes more advanced, bioelectronic devices are being built that are physically smaller, and electronically better, than their predecessors.

Closely related to bioelectronics is the field of biomechanics, which involves the use of mechanical devices to perform, enhance, or regulate body functions. Electronically controlled biomechanical devices may somebody replace entire body organs such as the heart. *See also* BIONICS.

BIONICS

Bionics is a scientific field of study, devoted to the replacement of body organs by bioelectronic and biomechanical devices (*see* BIOELECTRONICS). While the development of a complete android, or "bionic man," is probably impossible, certain living organs can be replaced, at least temporarily, by bionic substitutes. One example of this is the kidney dialysis machine.

Bionics is closely related to robotics (*see* ROBOTICS). Robot arms and legs have been developed, and as the sophistication of bionic technology improves, it may someday be possible to build complete bionic arms and legs to replace injured limbs. It may also be possible to build bionic organs that actually perform better than their living counterparts.

BIPOLAR TRANSISTOR

A bipolar transistor is a semiconductor device consisting of P-type and N-type materials in a sandwich configuration, as shown in the illustration. The PNP bipolar transistor consists of a layer of N-type semiconductor in between two layers of P-type material; the NPN transistor is just the reverse of this (*see* N-TYPE SEMICONDUCTOR, P-TYPE SEMICONDUCTOR).

Bipolar transistors display relatively low input impedance, and variable output impedance. They are used as oscillators, amplifiers, and switches. Bipolar transistors are available in a variety of sizes and shapes. Some are built

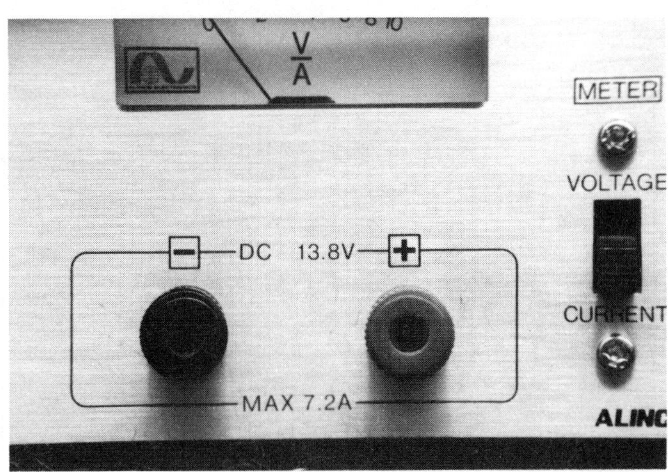

BINDING POST: Binding posts are commonly used with low-voltage power supplies.

especially for use as weak-signal amplifiers; others are built for high-power amplifiers, both audio and RF.

Generally, bipolar transistors are biased in a manner similar to that shown. The collector of a PNP device is always negative with respect to the emitter; the collector of an NPN device is always positive with respect to the emitter. The base bias for operation of a bipolar transistor depends on the class of operation desired (*see* BASE BIAS).

BIRMINGHAM WIRE GAUGE

The Birmingham Wire Gauge is a standard for measurement of the size, or diameter, of wire. Usually it is used for iron wire. Wire is classified according to diameter by giving it a number. The Birmingham Wire Gauge designators differ from the American and British Standard designators (*see* AMERICAN WIRE GAUGE, BRITISH STANDARD WIRE GAUGE), but the sizes are nearly the same. The higher the designator number, the thinner the wire.

The table shows the diameter vs Birmingham-Wire-Gauge number for designators 1 through 20. The Birmingham Wire Gauge designator does not include any coatings that might be on the wire, such as enamel, rubber,

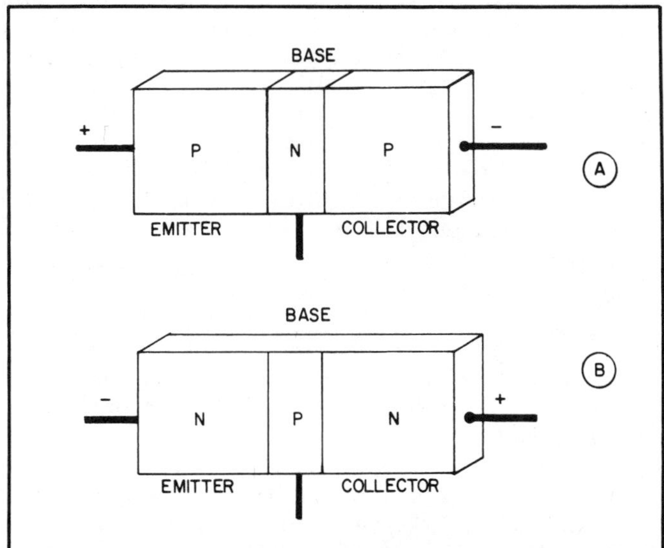

BIPOLAR TRANSISTOR: Bipolar transistor models. At A, PNP type; at B, NPN type.

BIRMINGHAM WIRE GAUGE: BIRMINGHAM WIRE GUAGE EQUIVALENTS IN MILLIMETERS.

BWG	Dia., mm	BWG	Dia., mm
1	7.62	11	3.05
2	7.21	12	2.77
3	6.58	13	2.41
4	6.05	14	2.11
5	5.59	15	1.83
6	5.16	16	1.65
7	4.57	17	1.47
8	4.19	18	1.25
9	3.76	19	1.07
10	3.40	20	0.889

or plastic insulation. Only the metal part of the wire is taken into account.

BISTABLE CIRCUIT

Any circuit that can attain either of two conditions, and maintain the same condition until a change command is received, is called a bistable circuit. The flip-flop (*see* FLIP-FLOP) is probably the most common type of bistable circuit.

A very simple kind of bistable circuit is the pushbutton switch, used with any device such as an electric light. When the button is pushed once, the lamp or appliance is switched on; pressing the button again turns the circuit off. Such switches can be either mechanical or electronic. The bistable circuit always maintains the same condition indefinitely unless a change-of-state command is received.

BISTABLE MULTIVIBRATOR

See FLIP-FLOP.

BIT

A bit is a binary digit. Each numeral place in a binary number represents one bit; for example, the binary number 11101 has five bits. A bit can be either 0 or 1. In the binary-coded decimal, or BCD, notation, each decimal digit is represented by four bits (*see* BINARY-CODED DECIMAL, BINARY-CODED NUMBER).

The 0 or 1 state of a bit can be denoted in various ways, such as low and high, off and on, black and white, or negative and positive (see illustration).

A group of bits is called a byte (*see* BYTE). Usually, a byte consists of eight bits, although the number varies depending upon the application.

BIT RATE

The bit rate of a binary-coded transmission is the number of bits sent per second. Digital computers process and exchange information at very high rates of speed, up to thousands of bits per second.

Computer data is often sent by means of a teleprinter code called ASCII (*see* ASCII). In teleprinter codes, a bit is called a baud, which can be either of two binary states. ASCII speeds range up to 19,200 bauds. *See also* BAUD.

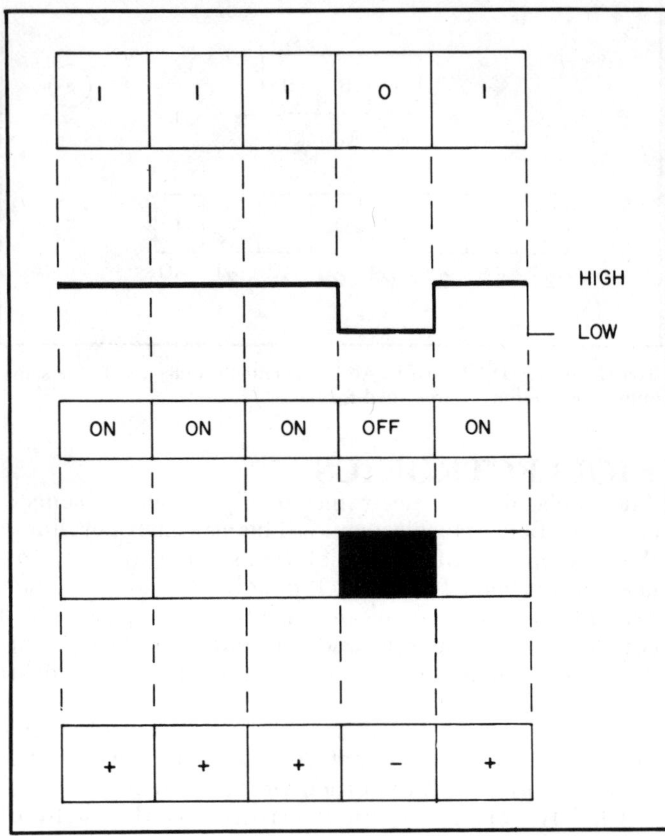

BIT: A bit is designated as either of two states, which can be expressed in a variety of ways.

BIT SLICE

A multi-chip microprocessor consists of a control section and one or more separate integrated circuits in which the register information is stored. Each register chip is called a bit slice (see drawing). Several bits slices, connected in parallel, complete the microprocessor along with the control chip.

A bit slice can contain any number of bits, usually a power of 2, such as 2, 4, 8, 16, and so on. A control IC with three 8-bit slices forms a 24-bit microprocessor. *See also* MICROPROCESSOR.

BLACK BODY

A black body is a theoretical object that emits or absorbs energy with complete efficiency at all wavelengths. Such an object cannot exist in reality, since no surface radiates or absorbs all energy. A black body would appear perfectly dark at all wavelengths of the electromagnetic spectrum.

When a black body becomes hot, it emits energy over a range of wavelengths that depends on its absolute temperature. This is called blackbody radiation. The higher the absolute temperature, the shorter the wavelength at which the maximum amount of radiation occurs. The graph shows an example of blackbody radiation at a temperature of 273 degrees Kelvin, which is 0 degrees Celsius or 32 degrees Fahrenheit. The peak wavelength of blackbody radiation for a given temperature, in degrees Kelvin, is given by Wien's displacement formula as

$$\lambda = 2898 / T,$$

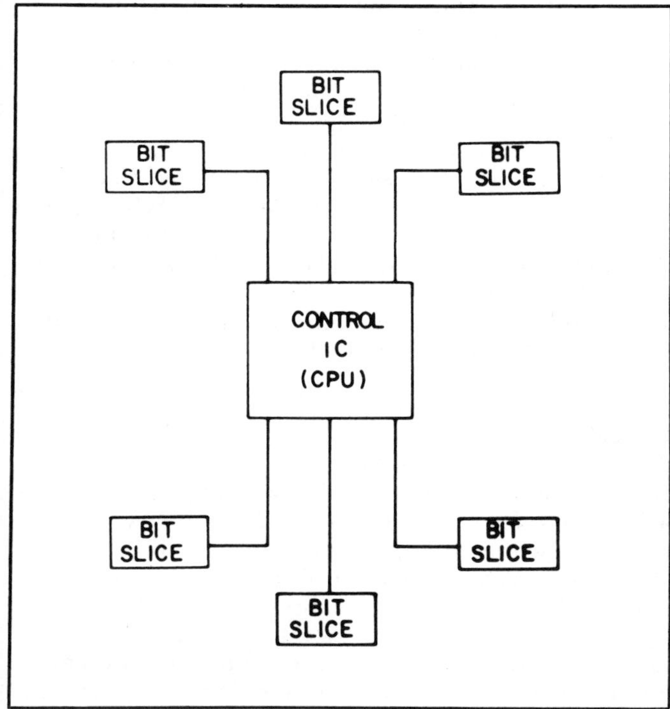

BIT SLICE: A bit slice is a section of memory connected to a control system.

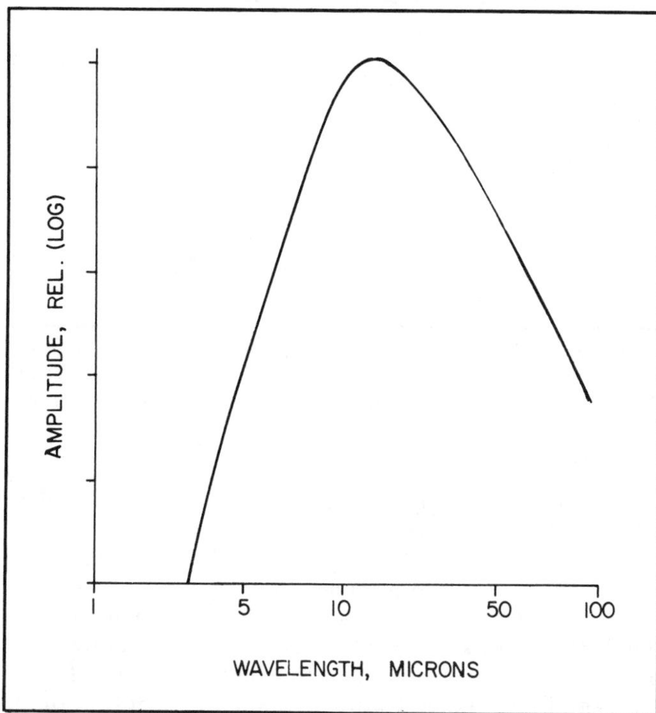

BLACK BODY: An example of black body-radiation, in this case for a temperature of 273 degrees Kelvin, or 32 degrees Fahrenheit.

where λ is the wavelength in microns (millionths of a meter) and T is the temperature in degrees Kelvin. *See also* KELVIN TEMPERATURE SCALE, WIEN'S DISPLACEMENT LAW.

BLACK BOX

Any circuit in which the internal details are unknown, or unimportant, is called a black box. Such a circuit becomes, in its intended application, a component itself. An example of a black box is an integrated circuit. While the internal details of various integrated circuits can be quite complicated, and very different, such a device appears as an empty box in a circuit diagram.

Two black boxes that behave in identical fashion under a certain set of circumstances may be used interchangeably even though they are internally different. An engineer does not have to be concerned with differences that are inconsequential in practice.

BLACK-LIGHT LAMP

See ULTRAVIOLET DEVICE.

BLACK TRANSMISSION

Black transmission is a form of amplitude-modulated facsimile signal (*see* FACSIMILE) in which the greatest copy density, or darkest shade, corresponds to the maximum amplitude of the signal. Black transmission is the opposite of white transmission, in which the brightest shade corresponds to the maximum signal amplitude (*see* WHITE TRANSMISSION).

In a frequency-modulated facsimile system, black transmission means that the darkest copy corresponds to the lowest transmitted frequency.

BLANKETING

Blanketing is a form of interference or jamming, in which a desired signal is obliterated by a more powerful undesired signal. A high-powered, frequency-modulated transmitter is capable of sending several kHz of signal bandwidth, rendering a sizable portion of the spectrum useless. Generally a complex combination of modulating tones is used. The multiplicity of FM sidebands extends more or less uniformly across a specified range of frequencies. *See also* JAMMING.

BLANKING

In television picture transmission, a blanking signal is a pulse that cuts off the receiver picture-tube during return traces. The blanking signal prevents the return trace from showing up on the screen, where it would interfere with the picture. Such a pulse is a square wave, with rise and decay times that are very short.

Blanking signals are sometimes used in radar equipment to shut off the display for a certain length of time with each antenna rotation. This makes the radar "blind" in certain directions, eliminating unwanted echoes. *See also* AZIMUTH BLANKING.

BLEEDER RESISTOR

A bleeder resistor is a fairly high-valued, low-wattage resistor connected across the output terminals of a power supply. Bleeder resistors serve two purposes: first, to aid in voltage regulation, and second, to reduce the shock hazard in high-voltage systems after power has been shut off. The schematic illustrates the connection of a bleeder resistor in a power supply.

The bleeder resistor prevents excessive voltages from developing across the power-supply filter capacitors when

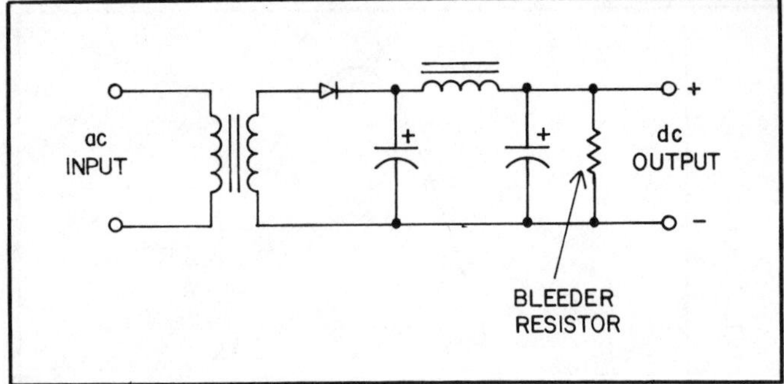

BLEEDER RESISTOR: Connection of a bleeder resistor in a power supply. The resistor regulates the voltage under no-load conditions, and speeds up the discharge of the capacitors when the load and input power are removed.

there is no load. Without such a current-draining device, the voltage at the supply output might build up to values substantially larger than the rated value.

In a high-voltage power supply, the filter capacitors will remain charged for some time after the power has been shut off, unless some means of discharging them is provided. Dangerous electrical shocks, from seemingly safe equipment, are made less likely by a bleeder resistor. However, the high-voltage terminal of such a supply should always be shorted to ground before servicing the equipment. This will ensure that no voltage is present. A metal rod with a well-insulated handle should be used to discharge any residual voltages.

BLIND ZONE

In radar,the term blind zone refers to a direction from which echoes cannot be received. The usual reason for the existence of blind zones is an object that gets in the way of the radar signals (see drawing). Such an obstruction shows up as a large, bright area on a radar display screen, unless azimuth blanking is used to eliminate it (*see* AZIMUTH BLANKING).

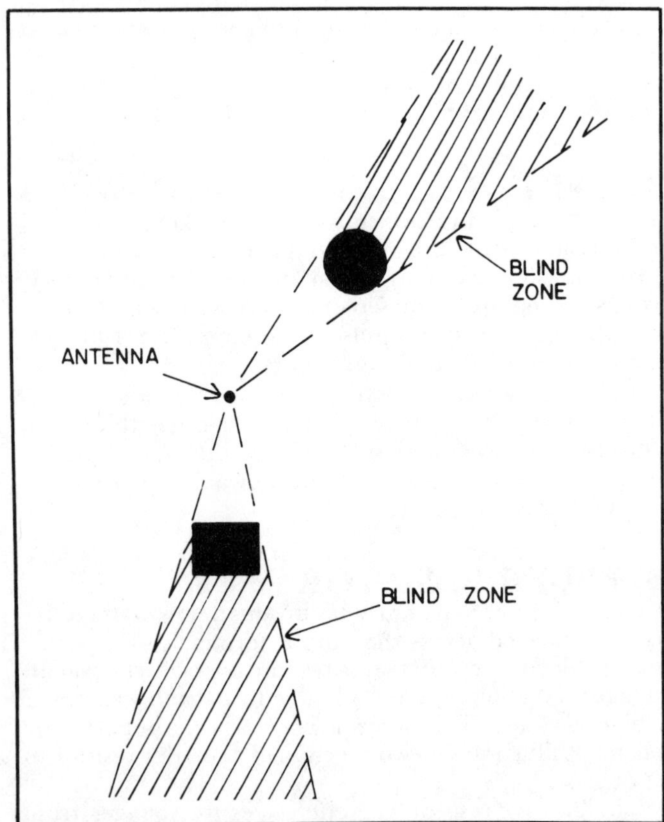

BLIND ZONE: Radar blind zones are caused by obstacles near the antenna.

In any communications system, a blind zone is an area or direction from which signals cannot be received, or to which messages cannot be transmitted. When communicating via the ionosphere, reception is often impossible within a certain distance called the skip zone (*see* SKIP ZONE). At vhf and above, a hill or building can create a blind zone. As the wavelength decreases, smaller and smaller obstructions cause blind zones.

BLIP

Blip is another word for radar echo. In any two-dimensional cathode-ray tube display, a bright spot, representing an echo, is sometimes called a blip (see illustration). In one-dimensional displays such as the A display (*see* "A" DISPLAY), a return is called a pip (*see* PIP0).

Radar blips can be the result of reflection of the signal from an object, or they may be caused by false reception, such as thunderstorm disturbances. Sometimes, deceptive transmissions are made to confuse radar by causing false blips. Tall buildings and other manmade structures cause a form of blip called ground clutter. *See also* AZIMUTH, RESOLUTION, RADAR.

BLOCK DIAGRAM

A circuit diagram showing the general configuration of a piece of electronic equipment, without showing the individual components, is called a block diagram. A simple superheterodyne receiver is block diagrammed in the illustration.

The path of the signal in a block diagram is generally from left to right and bottom to top. There are exceptions to this, but the signal flow, if not evident from the nature of the circuit, is often shown by arrows or heavy lines.

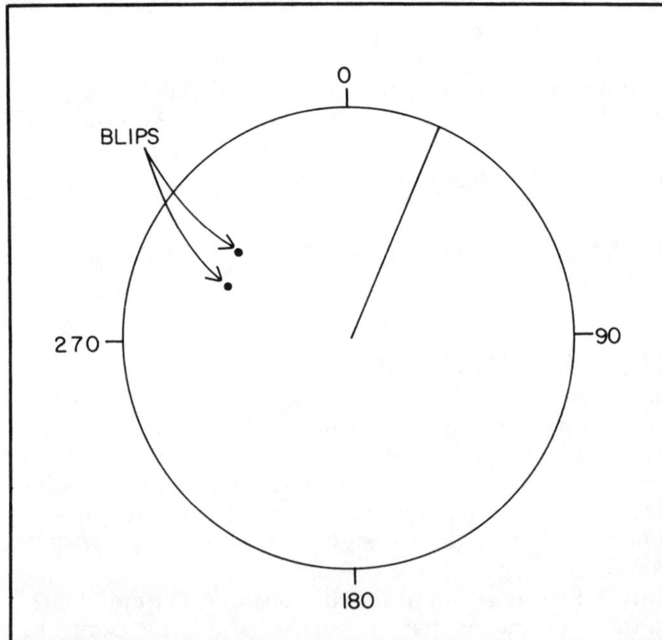

BLIP: Blips appear as spots on a radar screen.

In computer programming, block diagrams are used to indicate logical processes. This aids in developing programs. Such a block diagram is called a flowchart (*see* FLOWCHART). Block diagrams can also be used to represent algorithms (*see* ALGORITHM). Block diagrams are sometimes employed to denote the chain of command in a system. In such an illustration, the controller is at the top, and subordinate systems are underneath it.

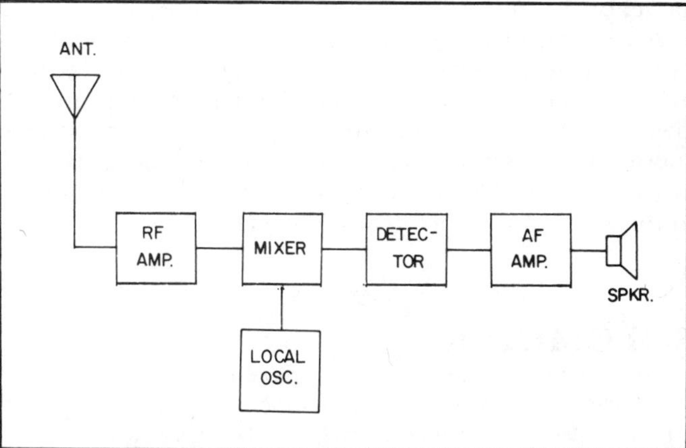

BLOCK DIAGRAM: A block diagram shows only discrete stages, and not individual components.

BLOCKING

Blocking is the application of a large negative bias voltage to the grid of a tube, the base of an NPN bipolar transistor, or the gate of an N-channel field-effect transistor, for the purpose of cutting off the device to prevent signal transfer. In a PNP bipolar transistor or a P-channel field-effect transistor, the blocking bias at the base or gate is positive.

Blocking is frequently used as a means of keying a code transmitter. The blocking voltage is applied to an amplifying stage while the key is up. When the key is down, the blocking bias is removed and the signal is transmitted. In transmitters, this is called base-block or gate-block keying, and is probably the most popular means of keying.

The term blocking is sometimes used to refer to the prevention of direct-current flow, while allowing alternating currents to pass. Such an electrical arrangement is necessary in many types of oscillators, and in capacitive coupling networks between amplifier stages. *See also* BLOCKING CAPACITOR.

BLOCKING CAPACITOR

When a capacitor is used for the purpose of blocking the flow of direct current, but allowing the passage of alternating current, the capacitor is called a blocking capacitor. Blocking capacitors facilitate the application of different dc bias voltages to two points in a circuit. Such a situation is common in multi-stage amplifiers. The schematic illustrates the connection of a blocking capacitor between two stages of a transistorized amplifier.

Blocking capacitors are usually fixed, rather than variable, and should be selected so that attenuation does not occur at any point in the operating-frequency range. Generally, the blocking capacitor should have a value that allows ample transfer of signal at the lowest operating

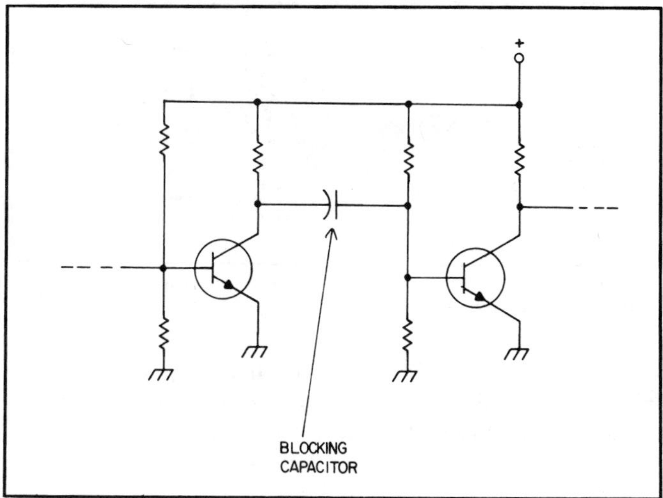

BLOCKING CAPACITOR: A blocking capacitor between two amplifier stages.

frequency, but the value should not be larger than the minimum to accomplish this. At vhf, the value of a blocking capacitor in a high-impedance circuit may be only a fraction of one picofarad. At audio frequencies in low-impedance circuits, values may range up to about 100 microfarads.

Blocking capacitors are used in the feedback circuits of some kinds of oscillators. The value of the feedback capacitor in an oscillator should be the smallest that will allow stable operation. *See also* CAPACITIVE COUPLING.

BLOCKING OSCILLATOR

A blocking oscillator is a form of relaxation oscillator (*see* RELAXATION OSCILLATION) that uses a feedback capacitor in conjunction with an inductance. The illustration shows a schematic diagram of one form of blocking oscillator. The feedback path is between the base of the transistor and the side of the output-transformer secondary that is in phase with the base.

The term blocking oscillator is used to refer to an oscillator that is switched on and off periodically by an external

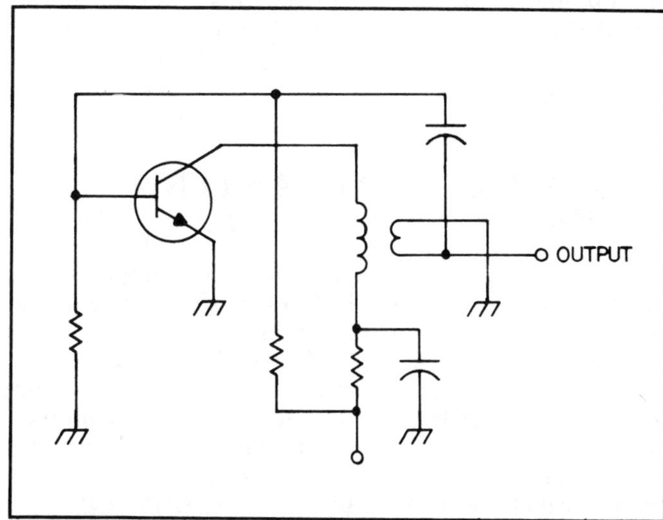

BLOCKING OSCILLATOR: Feedback is produced in phase by connecting the capacitor to the proper end of the output coil, and to the base of the transistor.

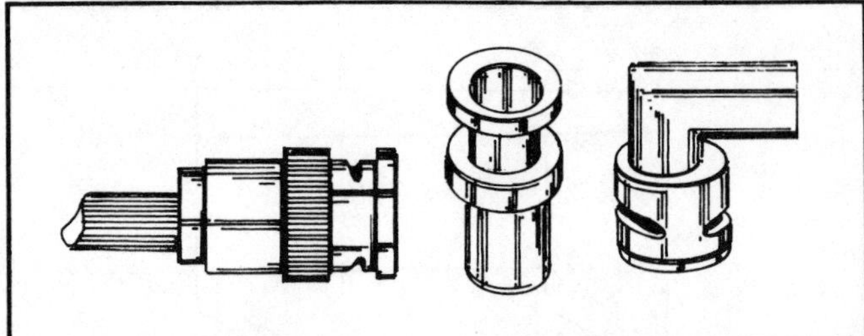

BNC CONNECTOR: The abbreviation stands for bayonet Neil-Concelman, the inventor.

source. Such an oscillator is also called a squegging oscillator. It transmits a series of pulses at the oscillator frequency. The pulses are used for testing purposes.

BLOWER

A device that facilitates air circulation for the purpose of cooling electronic equipment is called a blower. Blowers are especially common in high-power, tube-type RF amplifiers. The air circulation increases the dissipation rating of the tube.

Blowers can be found in a variety of forms. Some look like small fans. The squirrel-cage blower operates on a principle similar to the industrial exhaust fan. In modern solid-state amplifiers, conduction cooling is more commonly used than air cooling, since it is quieter, and there is no need for a special power supply to operate the blower motor. *See also* AIR COOLING, FAN.

BNC CONNECTOR

A small cable connector, used frequently in RF and test applications, is the BNC connector (see drawing). The BNC connector is designed to provide a constant impedance for 50-ohm coaxial cable connections. Some other types of connectors cause impedance discontinuities, which create losses at vhf and uhf.

The BNC connector has a quick-connect, quick-release feature similar to the bayonet socket (*see* BAYONET BASE AND SOCKET). BNC connectors are not intended for extremely high levels of power, nor for permanent installation outdoors. But they are extremely convenient when impedance matching must be good, or if frequent wiring changes are necessary. *See also* N-TYPE CONNECTOR, UHF CONNECTOR.

BOBTAIL CURTAIN ANTENNA

The bobtail curtain antenna is a vertically polarized, phased array for high-frequency operation. It is usually constructed from wire, although the vertical supports may be metal poles or towers. The drawing shows the general configuration of the bobtail curtain.

The horizontal sections are ½ wavelength electrically, and the three vertical sections, which radiate most of the energy, are ¼ wavelength. The bobtail curtain is a bidirectional antenna, with maximum radiation in line with the vertical radiators and minima in directions broadside to the vertical elements (*see* BIDIRECTIONAL PATTERN).

The bobtail curtain antenna is practical at frequencies as low as about 3 MHz, where two metal supporting poles about 78 feet high can be used for the outer vertical

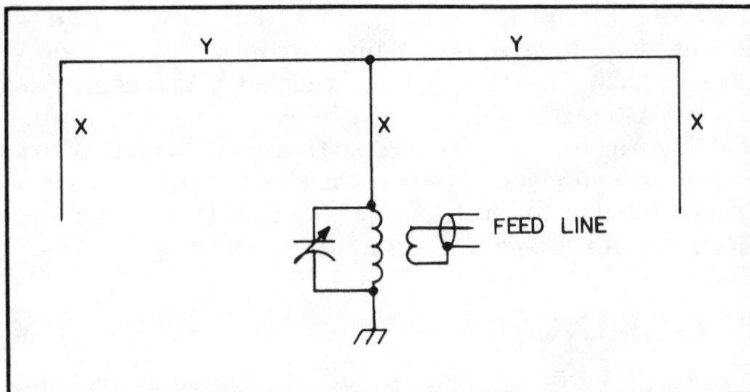

BOBTAIL CURTAIN ANTENNA: Sections X and ¼ wavelength and sections Y are ½ wavelength. The vertical portions radiate most of the energy, making this antenna similar to a group of three phased verticals.

radiators. The bobtail curtain has a low angle of radiation in the elevation plane.

BODY CAPACITANCE

When the human body is brought near an electronic circuit, some capacitance is introduced between the circuit wiring and the ground, unless the circuit is completely shielded. Usually, the effects of this body capacitance are not noticeable, since the value of capacitance is never more than a few picofarads. In certain tuned circuits such as ferrite-rod antennas or loop antennas, however, body capacitance can have a pronounced effect on the tuning, especially when the shunt capacitor is set for a small value. Usually, the hand is brought very near the shunt tuning capacitor as adjustments are made.

In general, the smaller the capacitances in a circuit, the more likely it is that body capacitance will be noticed. Shielding thus becomes very important at high frequencies and above.

BOHR ATOM

In 1913, Niels Bohr theorized that the atoms of all substances are composed of negatively charged particles, called electrons, in orbit about positively charged particles, called nuclei. The inward force, keeping the electrons from flying off into space, is described as the electrical attraction between the nucleus and the electron, since they are of opposite polarity. The drawing illustrates the Bohr model of the atom.

According to the Bohr theory, electrons can exist in various different levels of orbit. A higher orbit, farther from the nucleus, represents a higher energy state than a lower orbit close to the nucleus. If an electron gains energy, it moves into a higher orbit; a loss of energy results in a lower orbit. The modern theory of the atoms is similar to the Bohr model. *See also* ATOM, ELECTRON.

BOLOMETER

A bolometer is a device that changes its resistance in the presence of radiant energy. The most common example of a bolometer is the thermistor (*see* THERMISTOR).

The bolometer converts incident radiation into heat energy. The temperature change is then determined, and the amount of temperature change can be converted by a

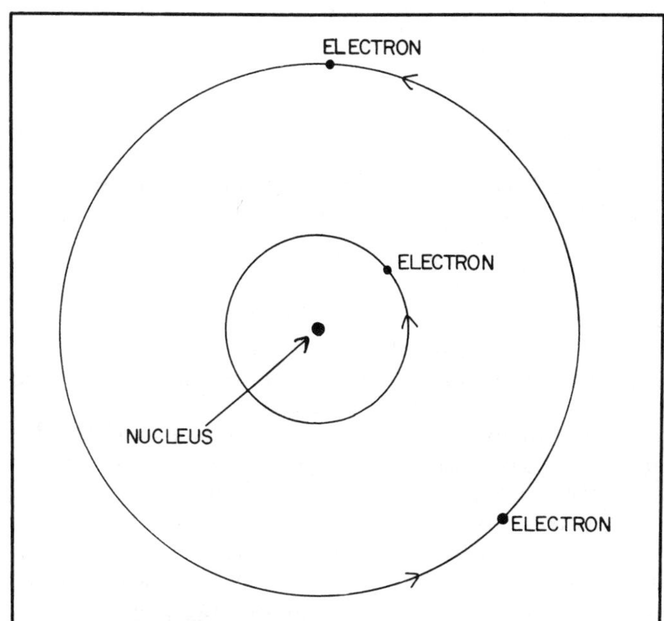

BOHR ATOM: The Bohr atom model postulates that negatively charged electrons orbit about a positive nucleus.

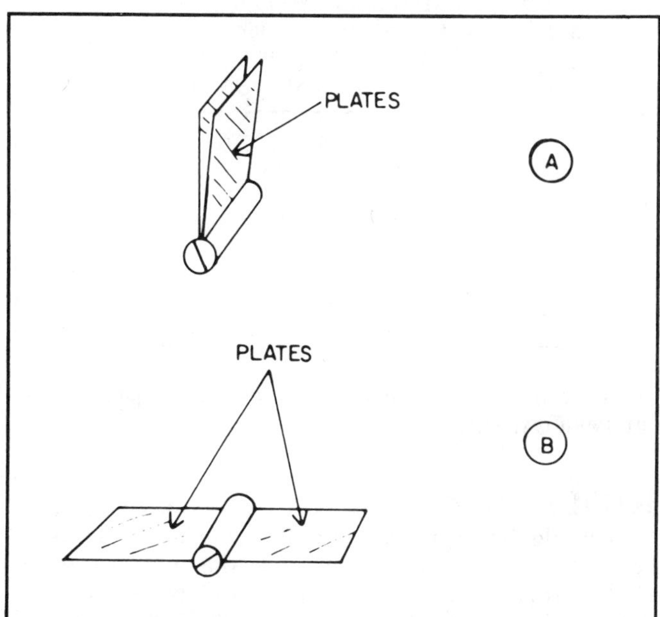

BOOK CAPACITOR: The book capacitor displays maximum capacitance, at A, when its plates are nearly parallel. Minimum capacitance occurs when the plates are coplanar, as shown at B.

known relation into a measure of the radiation intensity. The thermistor performs this function by itself; radiation intensity can be shown by a meter reading that depends on the resistance of the device with a constant supply voltage.

BOLTZMANN CONSTANT

The Boltzmann constant is a coefficient that defines the relationship between temperature and electron energy. The relation is linear, and is determined by this constant, which is abbreviated k. The higher the temperature, the greater the amount of energy in an electron.

The Boltzmann constant can be specified in a variety of different units. The two most common expressions are:

$$k = 1.38 \times 10^{-23} \text{ J/K}$$
$$k = 8.61 \times 10^{-5} \text{ eV/K}$$

where J/K represents joules per degree Kelvin, and eV/K represents electron volts per degree Kelvin. *See also* ELECTRON VOLT, JOULE, KELVIN TEMPERATURE SCALE.

BOMBARDMENT

In electronics, bombardment is the condition of subjecting an electrode, such as the plate of a vacuum tube, to high-speed impinging electrons. With sufficient bombardment, secondary emissions occur (*see* SECONDARY EMISSION). If a massive metal electrode such as lead or tungsten is bombarded by electrons with great velocity, X rays are produced.

In general, bombardment can refer to any situation in which a substance is exposed to high-speed atomic particles. By bombarding certain elements with accelerated protons or alpha particles, the nucleus structure is changed, creating new elements or isotopes.

The earth is always under bombardment from space by high-speed particles. Fortunately, most of these particles collide with atoms in the atmosphere before they can reach the ground. In this way, our atmosphere protects life from the deadly effects of this cosmic radiation. *See also* ALPHA PARTICLE, BETA PARTICLE, COSMIC RADIATION.

BONDING

Bonding is the joining of metals, especially the parts of a shielded enclosure, to prevent unwanted electromagnetic energy from entering or escaping a circuit. Bonding can be accomplished by soldering or welding, but a good electrical bond is usually obtained by simply clamping the two surfaces firmly together.

Examples of electrical bonding include ground clamps, screw-down power-amplifier enclosures, and tubing clamps for antenna construction. Bonding is used in the construction of automobiles to prevent potential differences from developing among the different metal parts. Sometimes, however, additional electrical bonding is necessary in mobile radio installations to prevent interference to the radio receiver from the vehicle electrical system.

Bonds between dissimilar metals are sometimes impossible to obtain by clamping, because the two metals tend to react with each other. In such cases, solder is often used to ensure that the electrical contact does not deterioriate.

BOOK CAPACITOR

A book capacitor is a small-value trimmer capacitor consisting of two or more plates. The capacitance is adjusted by changing the angle between the plates. The drawing illustrates a book capacitor; it gets this name from its physical resemblance to the covers of a book.

The capacitance is maximum when the two plates are parallel, or closed. The capacitance is minimum when the two plates are open so that they are coplanar. A book capacitor has a maximum value of only a few picofarads,

BOOLEAN ALGEBRA: BOOLEAN TRUTH TABLES. AT A, THE AND FUNCTION (MULTIPLICATION); AT B, THE NOT FUNCTION (COMPLEMENTATION); AT C, THE OR FUNCTION (ADDITION).

		A	B	C
X	Y	XY	X′	X+Y
0	0	0	1	0
0	1	0	1	1
1	0	0	0	1
1	1	1	0	1

Table 1

and a minimum value that may be less than 1 pF. *See also* TRIMMER CAPACITOR.

BOOLEAN ALGEBRA

Boolean algebra is a system of mathematical logic, using the functions AND, NOT, and OR. In the Boolean system, AND is represented by multiplication, NOT by complementation, and OR by addition. Thus X AND Y is written XY or X•Y, NOT X is written X′, and X OR Y is written X + Y. Table 1 shows the values of these functions, where true is represented by 1 and false by 0. Boolean functions are used in the design of digital logic circuits.

Using the Boolean representations for logical functions, some of the mathematical properties of multiplication, addition and complementation can be applied to form equations. The logical combinations on either side of the equation are equivalent. In some cases, the properties of the logical functions are not identical to their mathematical counterparts. As an example, multiplication is distributive with respect to addition in ordinary arithmetic, and it is true also in Boolean algebra. Thus:

$$X(Y+Z) = XY+XZ,$$

which means that the logical statements

X AND (Y OR Z)

and

(X AND Y) OR (X AND Z)

BOOLEAN ALGEBRA: SOME THEOREMS IN BOOLEAN ALGEBRA.

1. X+0 = X (additive identity)
2. X1 = X (multiplicative identity)
3. X+1 = 1
4. X0 = 0
5. X+X = X
6. XX = X
7. (X′) ′ = X (double negation)
8. X+X′ = 1
9. X′X = 0
10. X+Y = Y+X (commutativity of addition)
11. XY = YX (commutativity of multiplication)
12. X+XY = X
13. XY′+Y = X+Y
14. X+Y+Z = (X+Y)+Z = X+(Y+Z) (associativity of addition)
15. XYZ = (XY)Z = X(YZ) (associativity of multiplication)
16. X(Y+Z) = XY+XZ (distributivity)
17. (X+W) (Y+Z) = XY+XZ+WY+WZ (distributivity)

Table 2

are equivalent. The statement

$$XX = X$$

however, is quite alien to the ordinary kind of mathematical algebra.

Table 2 lists some of the most common theorems of Boolean algebra. *See also* LOGIC EQUATION, LOGIC FUNCTION.

BOOM

A boom is a horizontal support for the elements of a directive antenna such as a quad, log periodic, or Yagi (see drawing). The boom does not contribute to the radiation from such an antenna; it only supports the elements. The boom may be reinforced by supporting wires if it is long, or if there are many elements in the antenna.

A microphone is sometimes supported by a device called a boom, which is simply a long rod or pole that keeps the microphone in position with the base out of the way. Such microphones are often used in radio or television broadcast stations.

BOOST CHARGING

Boost charging is a high-current, short-term method of charging a storage battery. Boost charging is also sometimes called quick charging.

Generally, a storage battery must be charged for several hours at relatively slow rate in order to obtain a complete charge. However, a partial charge can be realized very quickly. Boost charging offers convenience when the available time is limited. However, slow charging should be done whenever possible.

BOOSTER

Any small, self-contained amplifier, for the purpose of improving the sensitivity of a receiver, is called a booster. Such amplifiers are popular in areas where television reception is poor. The booster is connected between the antenna lead-in wire and the television set. Amplifiers designed for enhancing the sensitivity of a microphone, or for providing extra audio power at the input of an audio-frequency amplifier, are sometimes called boosters.

In radio and television links between fixed stations, devices that intercept and retransmit the signals are sometimes called boosters. The active communications satellite is a sophisticated form of booster. *See also* ACTIVE COMMUNICATIONS SATELLITE, PREAMPLIFIER, REPEATER.

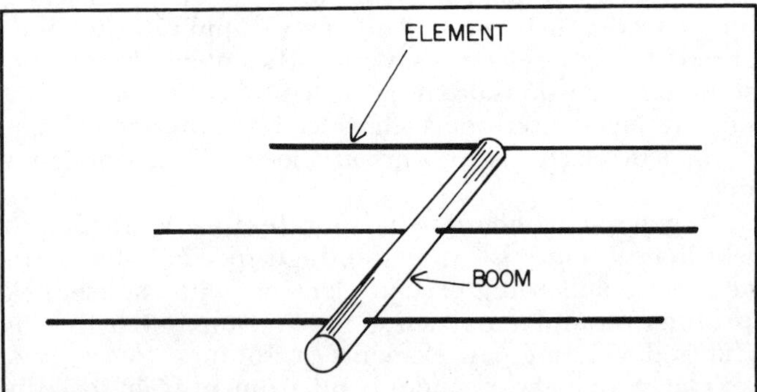

BOOM: A boom provides support for the elements of a Yagi antenna.

BOOTSTRAP CIRCUIT

A circuit, in which the output is taken from the source or emitter, is sometimes called a bootstrap circuit (see illustration), since the output voltage directly affects the bias. In this type of amplifier negative output pulses cause an increase in the negative voltage at the input, and positive output pulses cause a reduction in the negative voltage at the input.

The input of a bootstrap circuit is applied between the source and gate of an FET, or between the emitter and base of a transistor. This differentiates the bootstrap circuit from an ordinary grounded drain or grounded-collector amplifier, in which the input is applied between the gate and ground, or between the base and ground.

BOUND ELECTRON

An electron is said to be bound when it is under the influence of the nucleus of an atom, held in place by the electrical attraction between the positive charge of the nucleus and its own negative charge. A bound electron is sometimes stripped away from the nucleus easily; in other cases, the electron is very difficult to remove. The willingness of an electron to leave its orbit around an atomic nucleus depends on the position of the electron, and the number of other electrons in that orbit. Electrons that are stripped away in ordinary reactions among elements are called valence electrons.

When electrons are not under the influence of atomic nuclei, they are said to be free electrons. A free electron becomes bound when a nucleus captures it. Bound electrons are sometimes shared among more than one nucleus. *See also* FREE ELECTRON, SHARED ELECTRON, VALENCE ELECTRON.

BOWTIE ANTENNA

The bowtie antenna is a broadbanded antenna, often used at vhf and uhf. It consists of two triangular pieces of stiff wire, or two triangular flat metal plates, as shown in the drawing. The feed point is at the gap between the apexes of the triangles. There may be a reflecting screen to provide unidirectional operation, but this is not always the case.

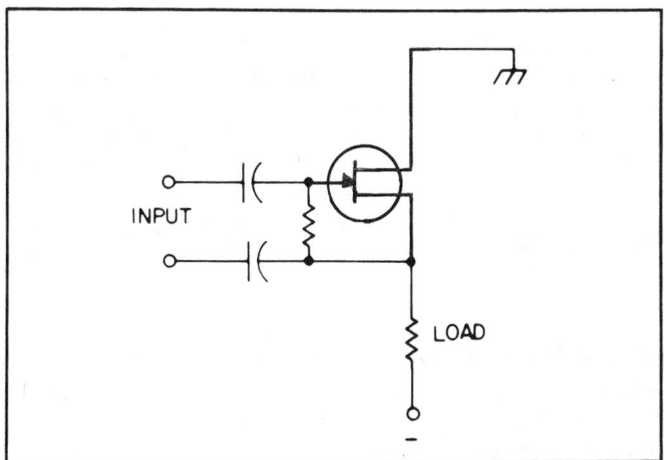

BOOTSTRAP CIRCUIT: The bootstrap circuit has an input potential that is affected by the instantaneous output voltage.

The bowtie antenna is a two-dimensional form of the biconical antenna (*see* BICONICAL ANTENNA). It obtains its broadband characteristics according to the same principles as the biconical antenna. The feed-point impedance depends on the apex angles of the two triangles. The polarization of the bowtie antenna is along a line running through the feed point and the centers of the triangle bases.

Bowtie antennas are occasionally used for transmitting and receiving at frequencies as low as about 20 MHz. Below that frequency, the size of the bowtie antenna becomes prohibitive.

B PLUS

The term B plus refers to the positive high-voltage dc supply used for the operation of a vacuum-tube circuit. Since tubes are becoming less and less common in modern electronics, the expression B plus is heard less often today than a few years ago, but sometimes is applied to the voltage source in transistor circuits.

B-plus voltages are frequently high enough to be dangerous. Some tubes require only a few volts, but most require hundreds of volts. Some tubes must be supplied with thousands of volts to operate properly. *See also* B BATTERY.

BRAID

Braid is a woven wire, made from many thin conductors. It is commonly used as the outer conductor, or shield, material in prefabricated coaxial cables (*see* COAXIAL CABLE). This allows physical flexibility and little chance of breakage. However, braid allows corrosion to occur more quickly than does solid tubing, since the surface area is not impervious to moisture and chemicals.

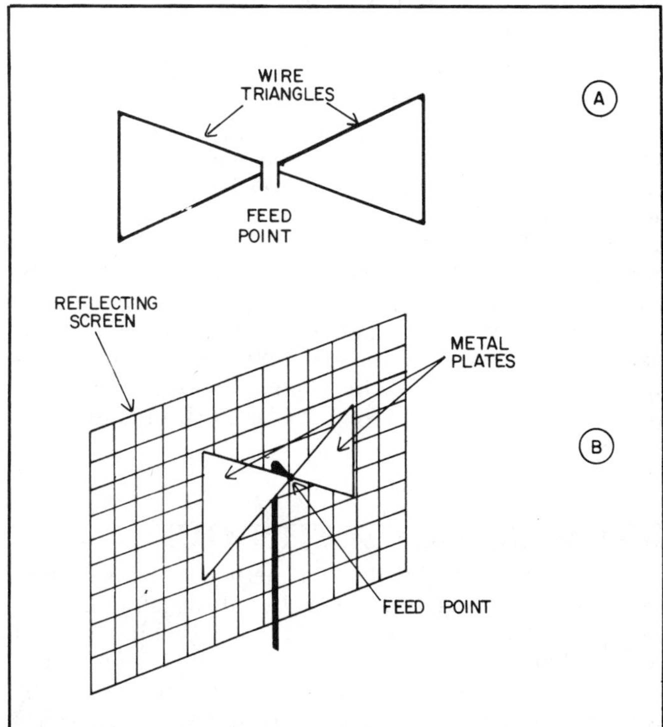

BOWTIE ANTENNA: Bowtie antennas. At A, a wire bowtie antenna, and at B, a bowtie antenna consisting of two triangular metal plates with a reflecting screen.

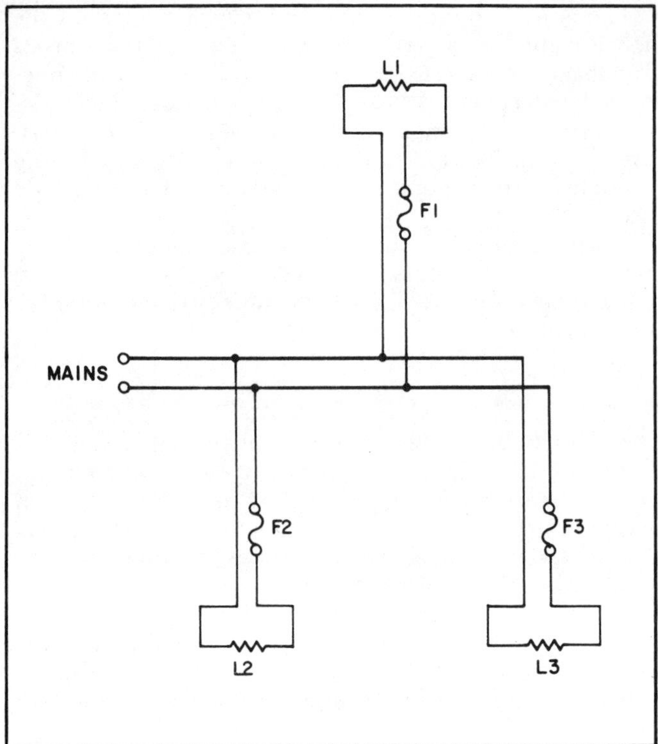

BRANCH CIRCUIT: Branch circuits are individually fused, and thus the failure of one branch does not affect the others.

Heavy braid is frequently used to electrically ground transmitters and receivers, and to bond station equipment together for safety purposes. Braid has a high current-carrying capacity, and very low inductance at radio frequencies. It is easy to install and adheres well to solder. In fact, a special type of braid called "Solder Wick" is used to remove solder from circuit boards for component replacement. *See also* SOLDER WICK.

BRANCH CIRCUIT

A branch circuit is a separately protected circuit for the operation of one appliance, or a few appliances. In a home wiring system, each branch circuit has its own separate fuse or breaker. The schematic illustrates branch circuits in a home or industrial electrical system. Because each branch has its own fuse or circuit breaker, a short circuit in one branch does not affect the other branches.

Branch circuits are always connected in parallel with the main circuit, since this ensures a constant voltage. Standard house power is supplied at 120 volts, plus or minus approximately 10 volts. Some appliances require twice this voltage. The standard ac frequency is 60 Hz in most of the world.

BREADBOARD TECHNIQUE

The breadboard technique is a method of constructing experimental circuits. A circuit board, usually made of phenolic or similar material, is supplied with a grid of holes. Components are mounted in the holes and wired together temporarily, either using hookup wire, or by copper plating on the underside of the board. A typical breadboard is shown in the photograph. Connections are easily removed or changed. The term comes from the early radio experimenter's practice of mounting components on

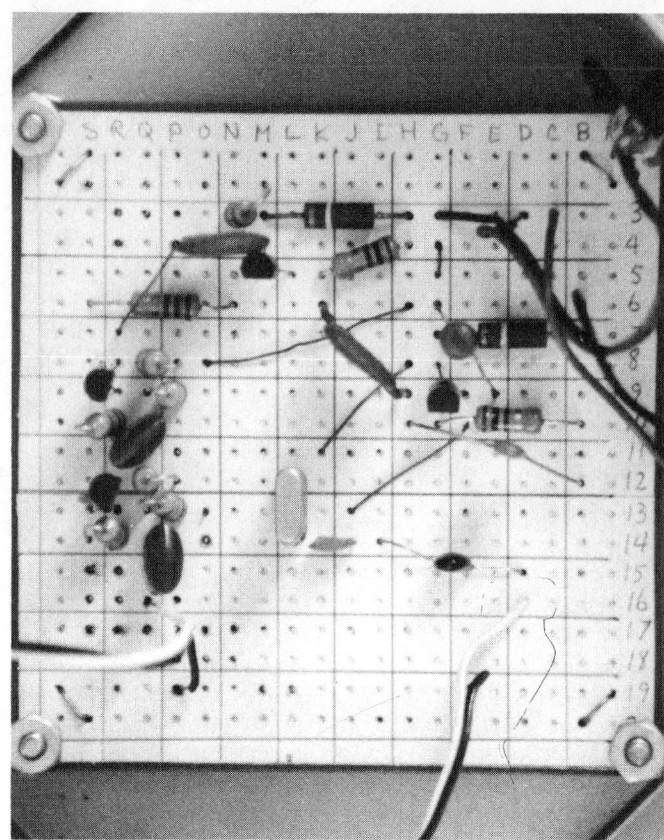

BREADBOARD TECHNIQUE: The breadboard technique uses a temporary circuit board, allowing components to be moved or rewired with ease.

a wooden board that was found in most homes and often used to mix bread dough.

Once the circuit has been perfected, a circuit board is designed and printed for permanent use. *See also* CIRCUIT BOARD, PRINTED CIRCUIT.

BREAK

When a receiving operator takes control of a communications circuit, a break is said to occur. The sending operator may say "break" to give the receiving operator a chance to ask a question or make a comment. The sending operator then pauses, or breaks, awaiting a possible reply.

When a third party with an important message interrupts a contact between two stations, the interruption is called a break. The third station indicates its intention to interrupt by saying "break" during a pause in the transmission of one of the other stations.

The term break is also used to refer to opening contacts in a circuit that is alternately opened and closed. For example, in Morse-code transmission, the transition between key-down and key-up conditions is called the break. The change from key-up to key-down condition is called make. *See also* MAKE.

BREAKDOWN

When the voltage across a space becomes sufficient to cause arcing, a breakdown occurs. In a gas, breakdown takes place as the result of ionization, which produces a conductive path. In a solid, breakdown actually damages the material permanently, and the resistance becomes much lower than usual.

In a semiconductor diode, the term breakdown is generally used in reference to a reverse-bias condition in which current flow occurs (*see* AVALANCHE BREAKDOWN). Ordinarily, the reverse resistance of a diode is extremely high, and the current flow is essentially zero. When breakdown occurs, the resistance abruptly lowers, and the current flow increases to a fairly large value.

Dielectric materials should be chosen so that breakdown will not take place under ordinary operating conditions. A sufficient margin of safety should be provided, to allow for the possibility of a sudden change in the voltage across the dielectric. Rectifier diodes should have sufficient peak-inverse-voltage ratings so that avalanche breakdown does not take place. Some semiconductor diodes, however, are deliberately designed to take advantage of their breakdown characteristics. *See also* ZENER DIODE.

BREAKDOWN VOLTAGE

The breakdown voltage of a dielectric material is the voltage at which the dielectric becomes a conductor because of arcing. In a solid dielectric, this usually results in permanent damage to the material. In a gaseous or liquid dielectric, the arcing causes ionization, but the breakdown damage is not permanent because of the fluid nature of the material.

In a reverse-biased diode, the breakdown voltage is the voltage at which the flow of current begins to rapidly increase (see illustration). Ordinarily, the current flow is very small in the reverse direction through a semiconductor diode; when the reverse voltage is smaller than the breakdown voltage, virtually no current flows. When the breakdown voltage is reached, a fairly large current flows. This is called avalanche breakdown. The transition between low and high current flow is abrupt. *See also* AVALANCHE BREAKDOWN.

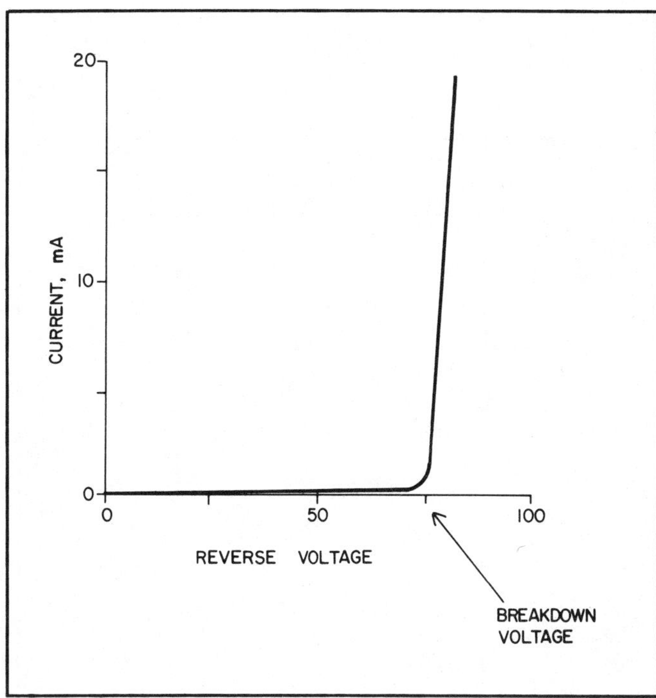

BREAKDOWN VOLTAGE: The breakdown voltage of a diode is the voltage at which it begins to appreciably conduct in the reverse direction.

BREAKER

A breaker is a device deliberately placed in series with an electrical circuit. If the current flow in the circuit becomes greater than a certain prescribed amount, then the breaker opens the circuit, and power is removed until the device is reset. Breakers are commonly used in modern household electrical wiring to prevent the danger of fire.

The term breaker is used in radio communication, especially on the Amateur and Citizen's bands. When a conversation between two or more operators is interrupted, the interrupting station is called the breaker. Since most Citizen's-band channels, and many Amateur channels, are occupied, anyone who enters a channel may be loosely called a breaker. *See also* AMATEUR RADIO, CIRCUIT BREAKER, CITIZEN'S BAND.

BREAK-IN OPERATION

Break-in operation is a form of radio communication in which a transmitting operator can hear signals at all times except during the actual emission of radiation by the transmitter. In break-in operation, when sending code (CW), the receiver is active between the individual dots and dashes, even if the speed is high. In single-sideband (SSB) operation, any pause allows reception when break-in is used. Break-in operation is fairly easy to achieve with low-power equipment, but when the transmitter output power is high, sophisticated techniques are needed to allow true break-in operation. Fast break-in operation is impractical with frequency-shift keying (FSK), amplitude modulation (AM), or frequency modulation (FM). However, semi break-in is sometimes used in these modes (*see* SEMI BREAK-IN OPERATION).

Break-in operation is not the same thing as duplex operation. In duplex, the transmitting operator is able to hear the other operator even while the transmitter is actually putting out power. This involves the use of two different frequencies. *See also* DUPLEX OPERATION.

BRIDGE CIRCUIT

A bridge circuit is a special form of network used in certain alternating-current circuit situations. A voltage is applied at one branch of the bridge circuit, and various components are connected to the other branches.

There are many kinds of bridge circuits. Some are used for signal generation, some for impedance matching, some for rectification or detection, some for mixing of signals, and some in measuring instruments.

An example of a bridge circuit is shown in the drawing. This bridge is used to measure inductance in terms of resistance and frequency. The balance of the circuit, indicated by a zero reading on the galvanometer, depends on the frequency of the applied voltage. Knowing the resistor values and the applied frequency, the unknown inductive reactance can be determined, and the value of inductance easily calculated from this.

BRIDGE RECTIFIER

A bridge rectifier is a form of full-wave rectifier circuit, consisting of four diodes arranged as shown in the illustration. The diodes may be either tube type or semiconductor type. Tube type diodes are seldom seen in modern rectifier circuits, however.

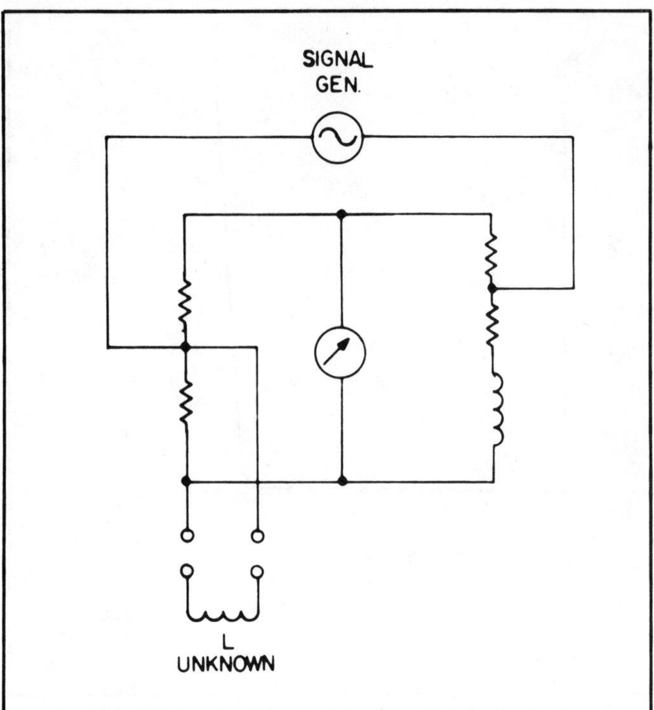

BRIDGE CIRCUIT: A bridge circuit for determining an unknown inductance.

During the part of the alternating-current cycle in which point X is negative and point Y is positive, electrons flow from point X through D3, then through the load R, and back through D2 to point Y. When point X is positive and point Y is negative, electrons flow from point Y through D4, then through the load R, and back through D1 to point X. With either polarity at X and Y, the electrons flow through R in the same direction.

Bridge rectifier circuits are commonly used in modern solid-state power supplies. Some integrated circuits are built especially for use as bridge rectifiers; they contain four semiconductor diodes in a bridge configuration, encased in a single package. Sometimes, circuits such as the one shown are used as detectors in radio receiving equipment. They may also be used as mixers. *See also* RECTIFIER CIRCUITS.

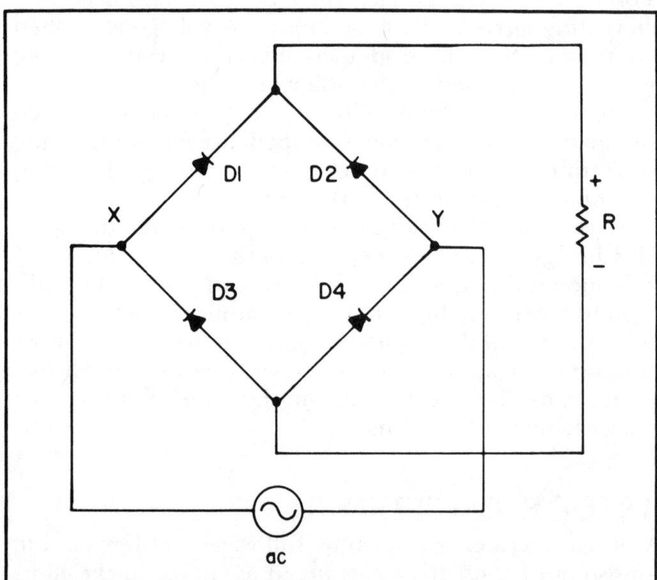

BRIDGE RECTIFIER: A bridge rectifier circuit consists of four diodes. The output is pulsating dc. The bridge circuit operates on both halves of the alternating-current cycle.

BRIDGED-T NETWORK

A bridged-T network is identical to a T network (*see* T NETWORK), with the exception that the series elements are shunted by an additional impedance. The shunting impedance may be a resistor, capacitor, or inductor. Generally, the shunting impedance is a variable resistance, used for the purpose of adjusting the sharpness of a filter response. The drawing shows such an arrangement. When the value of R is zero, the selectivity of the filter is minimum. When R is set at its greatest value, at least several times the impedance of the series elements, the selectivity of the filter is maximum.

BRIDGING CONNECTION

When a high-impedance circuit is connected across a component for the purpose of measurement, the addition of the high-impedance circuit should not noticeably affect the operation of the device under test. Such a connection is called a bridging connection. The ordinary voltmeter is used in this way, as are the test oscilloscope and the spectrum analyzer. Any loss caused by a bridging connection is called bridging loss. Some bridging loss is unavoidable, but the greater the impedance of the bridging circuit with respect to the circuit under test, the smaller the bridging loss.

The term bridging connection is often used interchangeably with the term shunt connection. *See also* SHUNT.

BRIGHTNESS

Brightness is the intensity of visible-light radiation or reflection from an object. In a cathode-ray tube, the brightness of the image depends on the intensity of the electron beam striking the phosphor material.

All television receivers have a brightness control which regulates the flow of electrons in the picture tube. Generally, one brightness control setting will give realistic pictures under all circumstances. An improperly adjusted brightness control creates unnatural pictures.

A bar pattern can be used to set the brightness control of a television receiver. In the photographs, relative brightness effects can be seen. At A, the brightness is insufficient; at B, it is properly adjusted; at C, it is excessive.

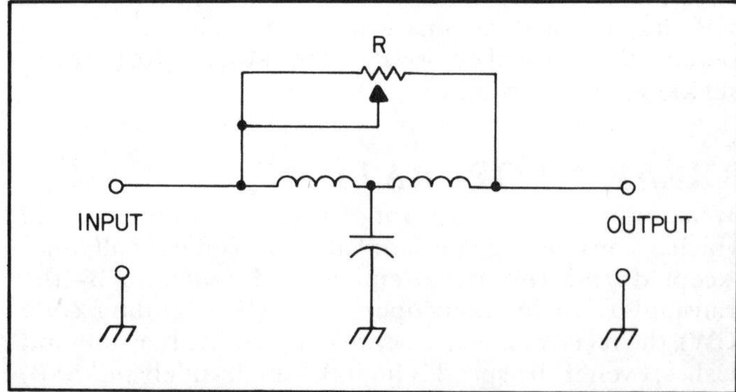

BRIDGED-T NETWORK: The bridged-T network contains a reactance across the series element of a T network. In this circuit, the variable resistor R affects the degree of selectivity of the T-network lowpass filter.

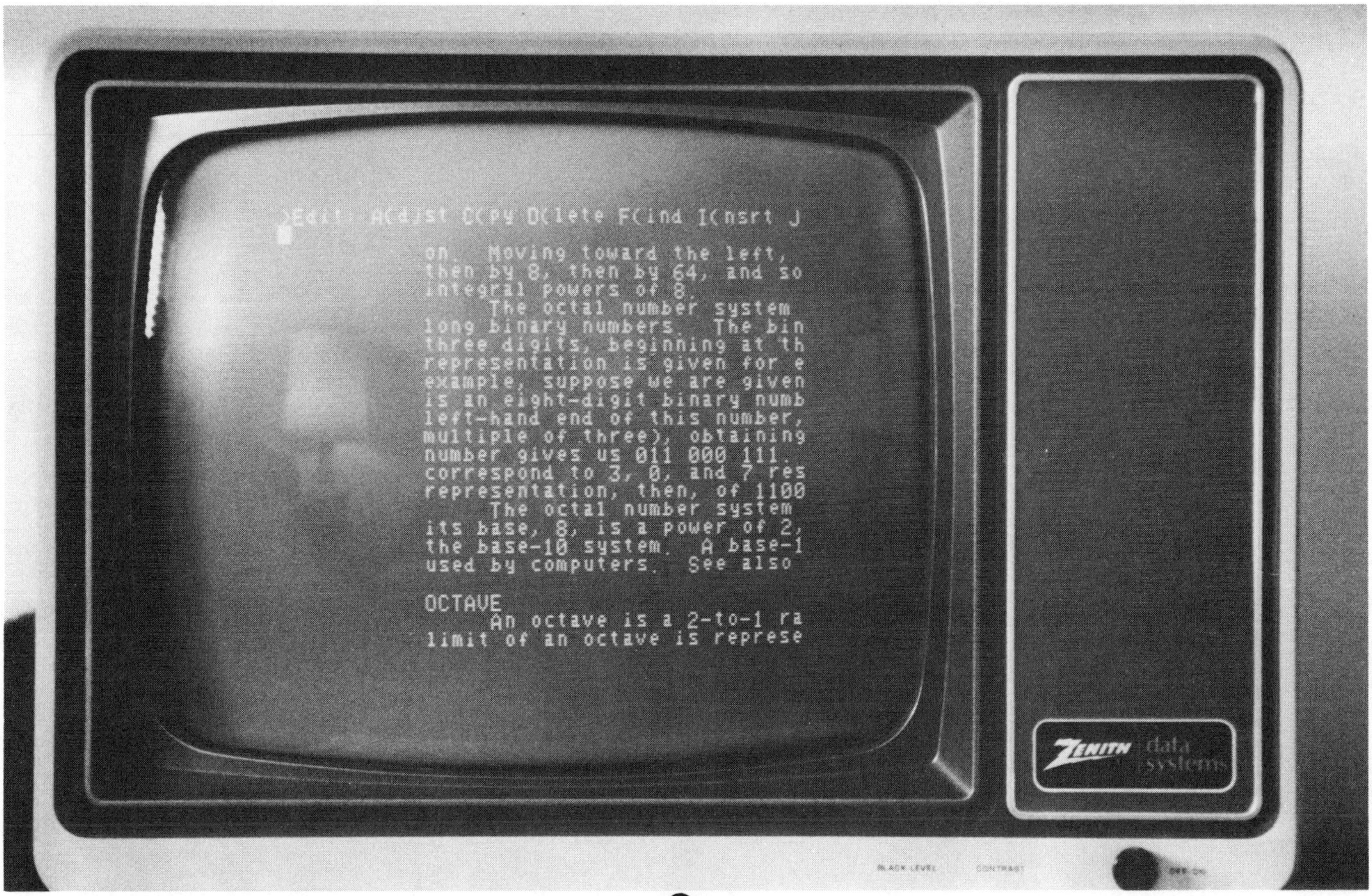

A

B

BRIGHTNESS: Brightness adjustment in a video monitor. At A, the brightness is insufficient. At B, it is acceptable.

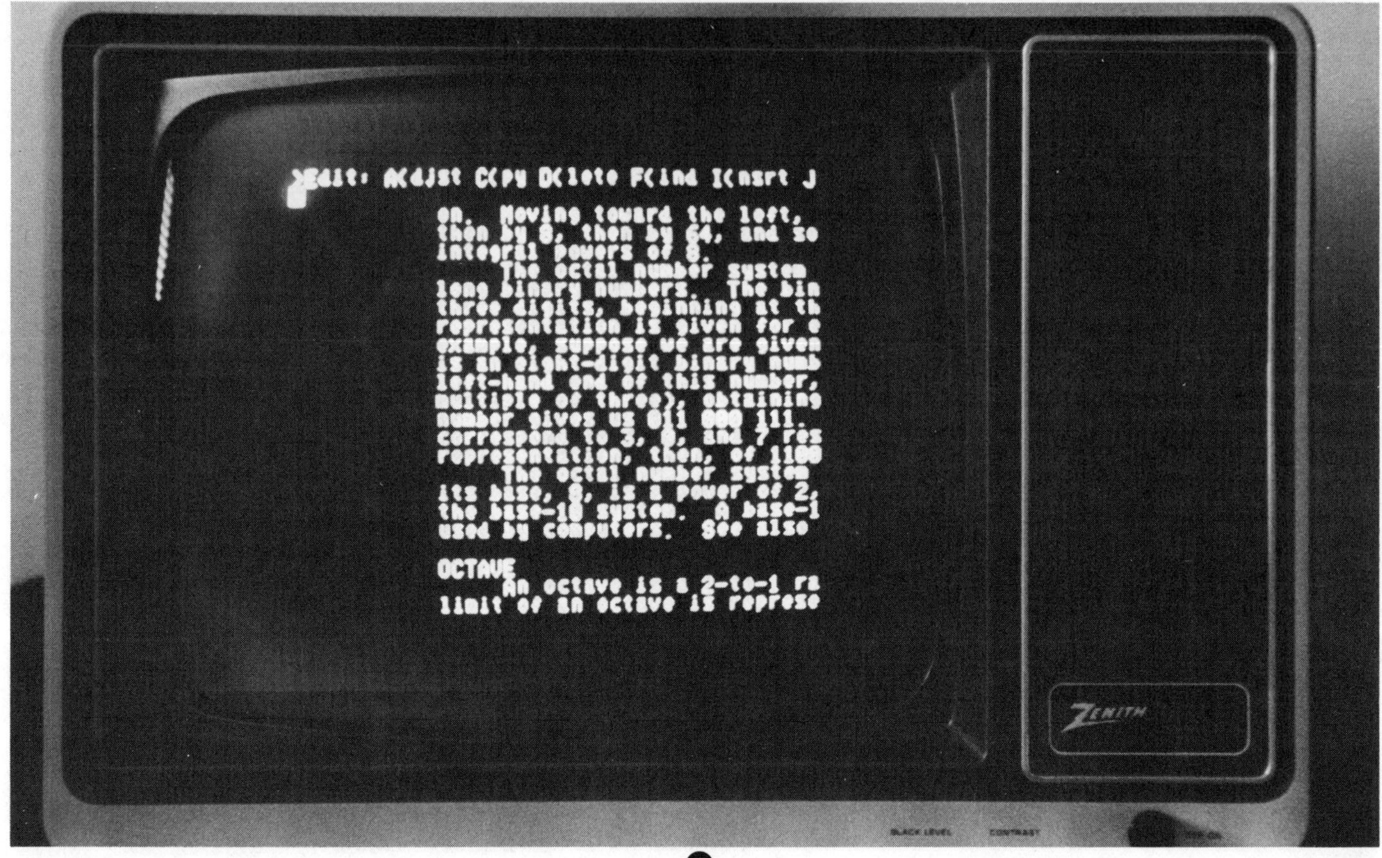

C

BRIGHTNESS: Brightness adjustment in a video monitor. At C, it is excessive. A bar pattern aids in proper brightness adjustment.

The brightness of a television picture is interrelated with the contrast to a certain extent. To keep the brightness and contrast of a television picture in proper proportion, automatic brightness and contrast controls are employed in the television receiver. *See also* AUTOMATIC BRIGHTNESS CONTROL, AUTOMATIC CONTRAST CONTROL, CONTRAST.

BRITISH STANDARD WIRE GAUGE

Metal wire is available in many different sizes or diameters. Wire is classified according to diameter by giving it a number. The designator most commonly used in the United States is called the American Wire Gauge, abbreviated AWG (*see* AMERICAN WIRE GAUGE). In some other countries, the British Standard Wire Gauge is used. The higher the number, the thinner the wire. The British Standard Wire Gauge sizes for designators 1 through 40 are shown in the table.

The larger the designator number for a given conductor metal, the smaller the current-carrying capacity becomes. The British Standard Wire Gauge designator does not take into account any coatings on the wire, such as enamel, rubber, or plastic insulation. Only the diameter of the metal itself is included in the measurement.

BRITISH THERMAL UNIT

The British Thermal Unit is a measure of energy transfer. It is abbreviated Btu. One Btu of energy will raise the temperature of one pound of pure water by one degree Fahrenheit. There are many other units of energy measure. The British Thermal Unit is commonly used to specify the cooling capacity of an air conditioner. *See also* ENERGY CONVERSION, ERG, FOOT POUND, JOULE, KILOWATT HOUR, WATT HOUR.

BROADBAND MODULATION

Broadband modulation is a relatively new concept in radiocommunications. Sometimes it is called wideband modulation.

The transfer of information by a radio signal requires a certain minimum amount of spectrum space. This minimum depends on the rate at which the information is conveyed. A very slow CW signal requires just 10 to 20 Hz of bandwidth. An ordinary voice signal requires about 2 to 3 kHz. High-speed teletype may require up to several kHz. The complex signal of a television picture needs several MHz of spectrum space. If two signals overlap at the receiving end of a communications circuit, there will be interference. Thus, a given amount of spectrum space has room for a limited number of signals.

In the past, attempts have been made to minimize the required bandwidth of radio signals to make the most possible room in the available amount of spectrum. However, by spreading the channel of a signal over a frequency range many times greater than its normal bandwidth, the density of power is made very small in all channels (see drawing).

A band might be filled to capacity by ordinary signals. But by spreading the energy of one more signal uniformly across the entire band, the interference to and from any fixed channel is negligible.

Various methods of obtaining broadband modulation

BRITISH STANDARD WIRE GAUGE: BRITISH STANDARD WIRE GAUGE EQUIVALENTS IN INCHES.

NBS SWG	Dia., in.	NBS SWG	Dia., in.
1	0.300	21	0.032
2	0.276	22	0.028
3	0.252	23	0.024
4	0.232	24	0.022
5	0.212	25	0.020
6	0.192	26	0.018
7	0.176	27	0.0164
8	0.160	28	0.0148
9	0.144	29	0.0136
10	0.128	30	0.0124
11	0.116	31	0.0116
12	0.104	32	0.0108
13	0.092	33	0.0100
14	0.080	34	0.0092
15	0.072	35	0.0084
16	0.064	36	0.0076
17	0.056	37	0.0068
18	0.048	38	0.0060
19	0.040	39	0.0052
20	0.036	40	0.0048

are being tested. One common way of achieving this kind of modulation is to constantly vary the frequency of an ordinary transmitter in a predetermined manner over a wide range. The receiver then simply follows the transmitter according to a certain algorithm.

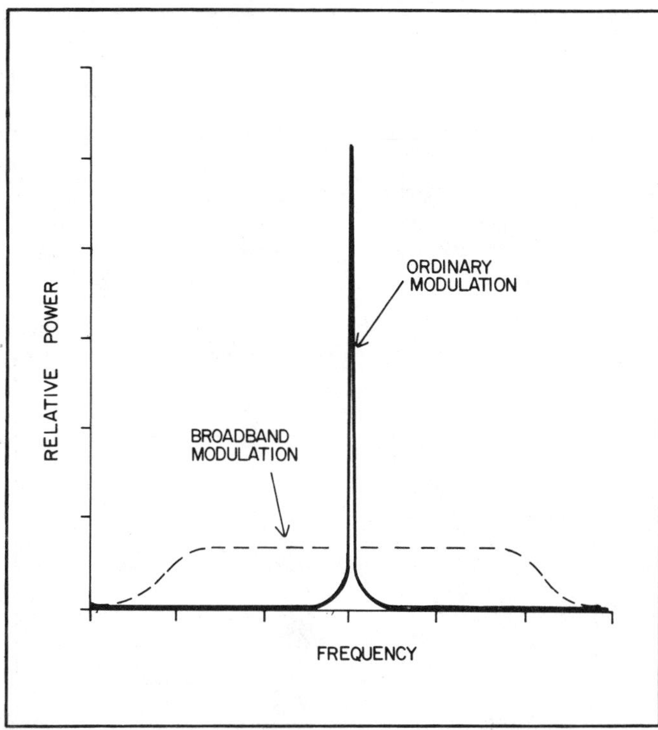

BROADBAND MODULATION: Broadband modulation contains a small concentration of energy within a given range of frequencies. Thus it is much different from ordinary modulation.

BROADCAST BAND

In the United States, there are several broadcast bands. The standard AM broadcast band is probably the most well-known, extending from 535 to 1605 kHz. The FM broadcast band occupies the space from 88 to 108 MHz. There are four television broadcast bands. The lower three of these are called the vhf TV broadcast band, and they contain channels 2 through 13. The fourth band is called the uhf TV broadcast band, and contains channels 14 through 69. All of these bands, along with the AM and FM radio broadcast bands, are listed in the table.

Several shortwave bands are allocated for international broadcasting because of the nature of propagation at high frequencies. These are listed also in the table. These bands are sometimes identified according to the approximate wavelength of the signals in meters. For example, the band 15.100 to 15.450 MHz is known as the 19-meter band. *See also* FREQUENCY ALLOCATIONS.

BROADCASTING

Broadcasting is one-way transmission, intended for reception by the general public. In the early days of radio, broadcasting began as a hobby pursuit. It was soon found to be an effective means of reaching thousands or millions of people simultaneously. Thus it has become a major pastime, and has proven profitable for corporations as well as broadcasting stations.

In the United States, broadcasting is done on several different frequency bands, using radio (both AM and FM), and television (*see* BROADCAST BAND). Some broadcasting stations are allowed only a few watts of RF power, while others use hundreds of thousands, or even millions, of watts. All broadcasting stations in the United States must be licensed by the Federal Communications Commission.

BROADSIDE ARRAY

A broadside array is a phased array of antennas (*see* PHASED ARRAY), arranged in such a way that the maximum radiation occurs in directions perpendicular to the plane containing the driven elements. This requires that all of the antennas be fed in phase, and special phasing harnesses are required to accomplish this. The illustration shows the geometric arrangement of a broadside array.

At frequencies as low as about 10 MHz, a broadside array may be constructed from as few as two driven antennas. At vhf and uhf, there might be several antennas. The antennas themselves may have just a single element, as shown in the figure, or they can consist of Yagi antennas, loops, or other systems with individual directive proper-

BROADCAST BAND: U.S. AND INTERNATIONAL BROADCAST BANDS.

A. United States Broadcast Bands

Frequency Range, MHz	Allocation
0.535-1.605	Standard AM
88.0-108.0	Standard FM
54.0-72.0	vhf Television
76.0-88.0	vhf Television
174.0-216.0	vhf Television
470.0-806.0	uhf Television

B. International Broadcast Bands

Frequency Range, MHz
5.950-6.200
9.500-9.775
11.700-11.975
15.100-15.450
17.700-17.900
21.450-21.750
25.600-26.100

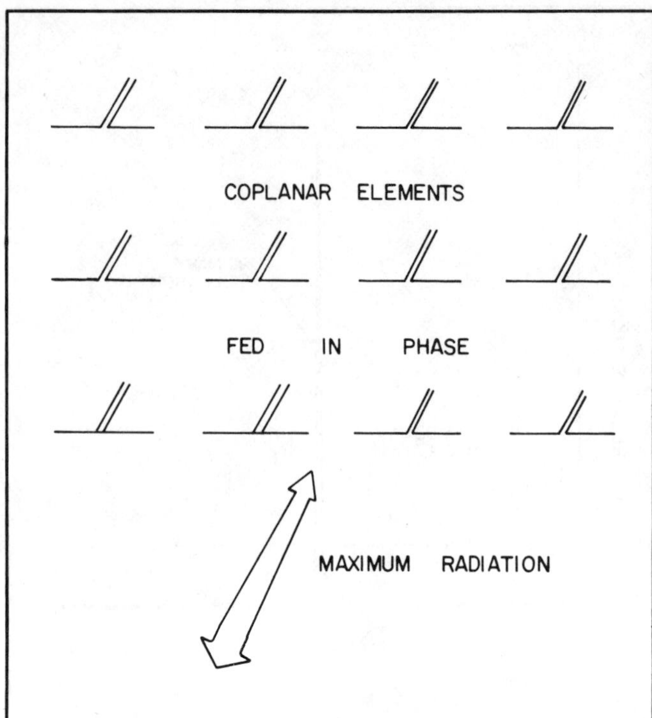

BROADSIDE ARRAY: A broadside array consists of a number of elements in a single plane, fed in phase. The maximum radiation occurs perpendicular to the plane containing the elements.

ties. If a reflecting screen is placed behind the array of dipoles in the drawing, the system becomes a billboard antenna (*see* BILLBOARD ANTENNA).

The directional properties of a broadside array depend on the number of elements, whether or not the elements have gain themselves, and on the spacing among the elements. *See also* DIRECTIONAL ANTENNA.

BROWN AND SHARP GAUGE

See AMERICAN WIRE GAUGE.

BUBBLE MEMORY

A bubble memory is a magnetic means of storing information. A magnetic field is used to store a binary bit of information (*see* BIT); this field is maintained until a command is received to change it. The location of a bit in a bubble memory is changed by applying external magnetic fields. *See also* MEMORY.

BUFFER

A buffer is a temporary storage circuit used with computers, word processors, and radioteletype or CW keyboard keyers. The buffer allows an operator to type far ahead of the transmitted information. Even though the operator may type erratically, the buffer smooths out the data as it is actually sent. A buffer circuit may either hold all of the information for later transmission, or it may transmit the information at a predetermined rate as it is fed in irregularly (see drawing).

The buffer is a first-in, first-out (FIFO) storage system. That is, if character X is fed into the buffer before character Y, then character X will be transmitted before Y. A buffer may have only a few characters of memory, or it might have thousands. *See also* FIRST-IN, FIRST-OUT.

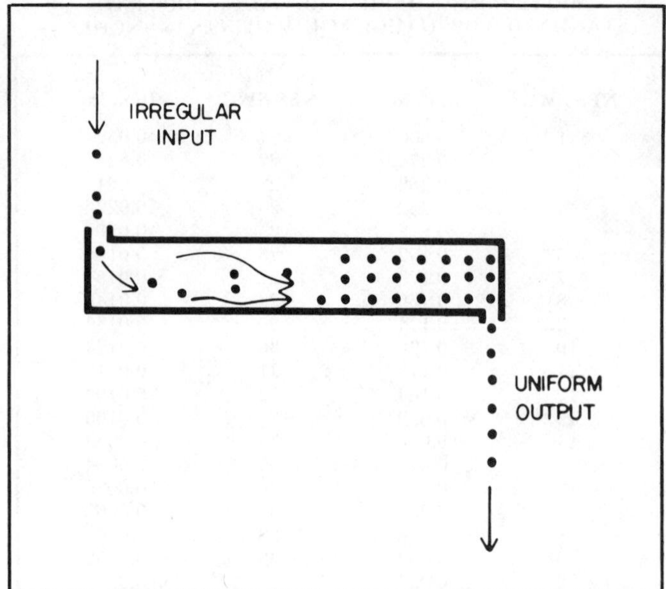

BUFFER: Pictorial representation of a buffer. Data may be fed into the buffer irregularly, but it is transmitted in perfect synchronization. The capacity of a buffer may be small, perhaps just a few characters; or it may be large, containing hundreds or even thousands of characters.

BUFFER STAGE

A buffer stage is a single-tube or single-transistor stage, used for the purpose of providing isolation between two other stages in a radio circuit. Buffer stages are commonly used following oscillators, especially keyed oscillators, to present a constant load impedance to the oscillator. As the amplifiers following an oscillator are tuned, their input impedances change, and this can affect the frequency of the oscillator unless a buffer stage is used. When an oscillator is keyed, the input impedance of the following amplifier stage may change; or, when an amplifier is keyed, its impedance can change. The buffer stage serves to isolate this impedance change from the oscillator, and thereby the oscillator frequency is stabilized.

The block diagram illustrates the location of a buffer stage in a CW transmitter. In this case, the oscillator is keyed. A buffer circuit generally has little or no gain.

BUG

The term "bug" is sometimes used in reference to a flaw in an electronic circuit or in a computer program. In a circuit, a "bug" is an imperfection in the general design, resulting in less-than-optimum operation. For example, a radio transmitter might have a problem with frequency stability. The cause of this kind of problem is sometimes difficult to find.

In a computer program, a "bug" results in inaccurate or incomplete output or execution. An example of the effect of this type of "bug" is a search-and-replace word-processing function that does not search an entire file.

A semiautomatic key, used by old-time radiotelegraphers and still used today by some radio amateurs, is often called a "bug." *See also* SEMIAUTOMATIC KEY.

BURN-IN

When a piece of equipment is brought into a test lab or service shop to be repaired, the malfunction does not always show up as soon as the device is connected to power.

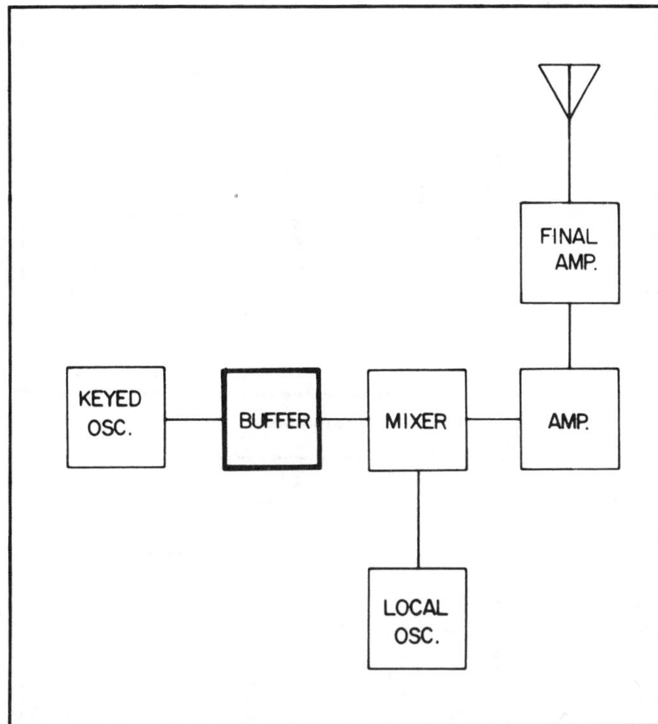

BUFFER STAGE: A buffer stage in a CW transmitter.

Often, the fault will be seen only occasionally, or the malfunction might not occur until the unit has been warmed or cooled to a certain temperature. Sometimes the humidity is a factor. Under these circumstances, a burn-in must be performed, and environmental parameters varied until the problem is found. The equipment is turned on and left on, sometimes for hours, or even days, until the malfunction is observed and can be diagnosed. Sometimes the problem is never seen. Such a situation is well known as a source of frustration for service technicians and customers in almost all electrical or mechanical fields.

At a factory, the burn-in is the initial testing phase of a unit. This ensures that the unit meets all specifications when it leaves the manufacturing plant. This procedure reduces the number of faulty new units that reach buyers. From a sales point of view, the importance of burn-in is obvious. *See also* INTERMITTENT FAILURE.

BUS

A communications path that handles several signals at once, from different originating points and to different destinations, is called a bus in computer application. The term bus is also used to refer to a common ground system for electrical or electronic equipment. A printed circuit board generally has a ground bus consisting of foil running around the perimeter of the board.

In radio stations, the equipment is grounded via a common strip or rod of metal called a bus bar. This ensures that the chassis of all units are at the same potential, and therefore the possibility of electrical shock to station personnel is minimized. The bus-bar grounding system also prevents the formation of ground loops, which can cause equipment malfunctions. *See also* GROUND LOOP.

BUSHING

A bushing is a device used to couple two sections of control shaft together longitudinally, as shown in the illustration. Bushings are used to lengthen the control shafts of rotary switches, variable capacitors, and potentiometers. Some bushings are made of insulating material such as porcelain or bakelite. Insulating bushings are used when it is necessary to keep a control shaft insulated from chassis ground, or when the effects of body capacitance must be kept at a minimum (*see* BODY CAPACITANCE).

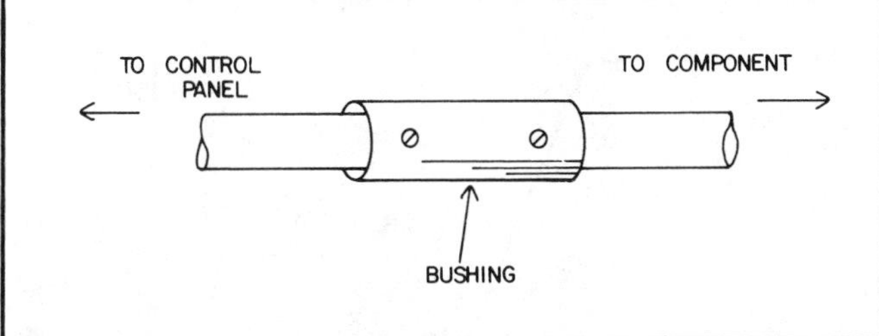

BUSHING: A bushing is a coupling between two sections of control shaft.

BUTTERFLY CAPACITOR

At very-high and ultra-high frequency ranges, a tuning device called a butterfly capacitor is often used in place of the conventional coil-and-capacitor tank circuit. A butterfly capacitor includes inductance as well as capacitance. The device gets its name from the fact that its plates resemble, physically, the opened wings of a butterfly.

The butterfly capacitor has a very high Q factor (*see* Q FACTOR). That is, it displays excellent selective characteristics, although the cavity resonator is better at vhf and uhf (*see* CAVITY RESONATOR). The drawing shows a butterfly capacitor. It appears very much like an ordinary trimmer capacitor.

BUTTERWORTH FILTER

A Butterworth filter is a special type of selective filter, designed to have a flat response in its passband and a uniform roll-off characteristic. A Butterworth filter may be designed for a lowpass, highpass, bandpass, or band-rejection response.

Figure 1 shows some ideal Butterworth responses for lowpass, highpass, and bandpass filters (A, B, and C). Note the absence of peaks in the passband. Schematic diagrams of sample lowpass (A), highpass (B), and bandpass (C) Butterworth filters are shown in Fig. 2. The input and load impedances must be correctly chosen for proper operation; the values of the filter resistors, inductors and

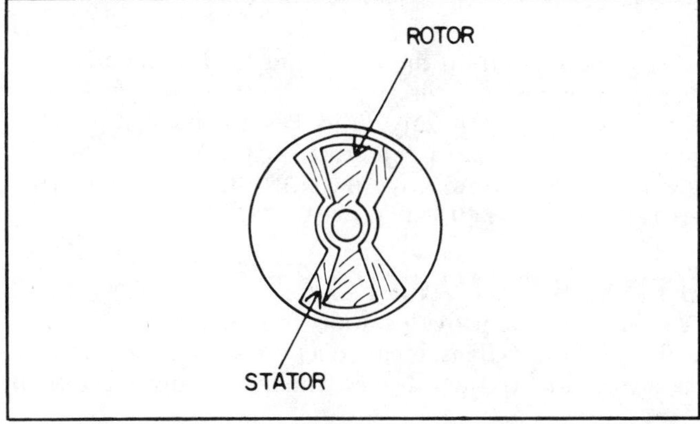

BUTTERFLY CAPACITOR: The butterfly capacitor looks very much like an ordinary trimmer capacitor.

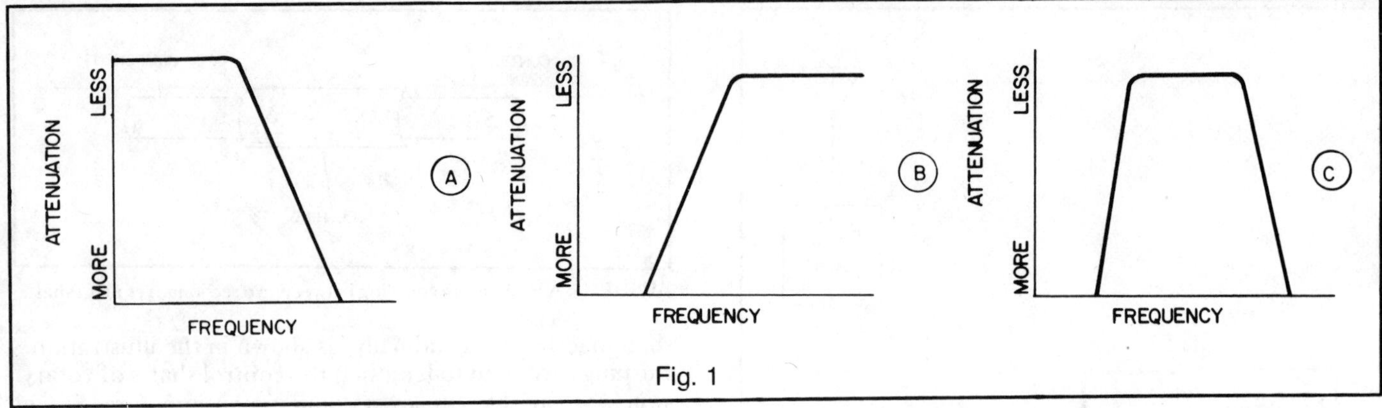

BUTTERWORTH FILTER: Ideal theoretical Butterworth filter responses. At A, lowpass; at B, highpass; at C, bandpass.

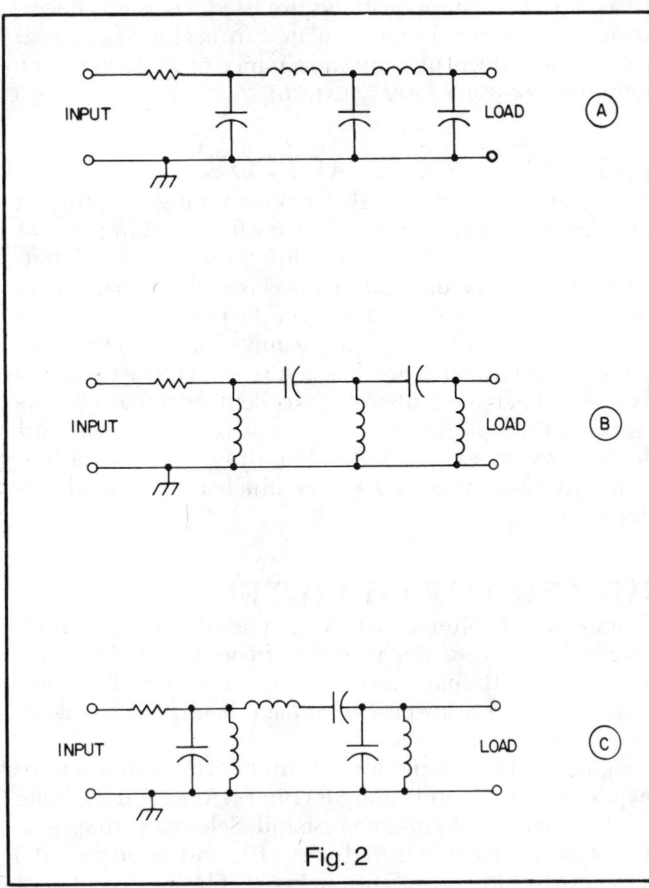

BUTTERWORTH FILTER: Examples of Butterworth filters. At A, lowpass; at B, highpass; at C, bandpass. The value of the resistor, as well as the reactances of the capacitors and inductors, depend on the input and output impedances for proper response.

capacitors depend on the input and load impedances.

A similar type of selective filter, called the Chebyshev filter, is also commonly used for lowpass, highpass, bandpass, and band-rejection applications. *See also* BANDPASS FILTER, BAND-REJECTION FILTER, CHEBYSHEV FILTER, HIGHPASS FILTER, LOWPASS FILTER.

BYPASS CAPACITOR

A capacitor that provides a low impedance for a signal, while allowing dc bias, is called a bypass capacitor. Bypass capacitors are frequently used in the emitter circuits of transistor amplifiers to provide stabilization. They are also used in the source circuits of field-effect transistors, and in the cathode circuits of tubes. The bypass capacitor places a circuit element at signal ground potential, although the dc potential may be several hundred volts.

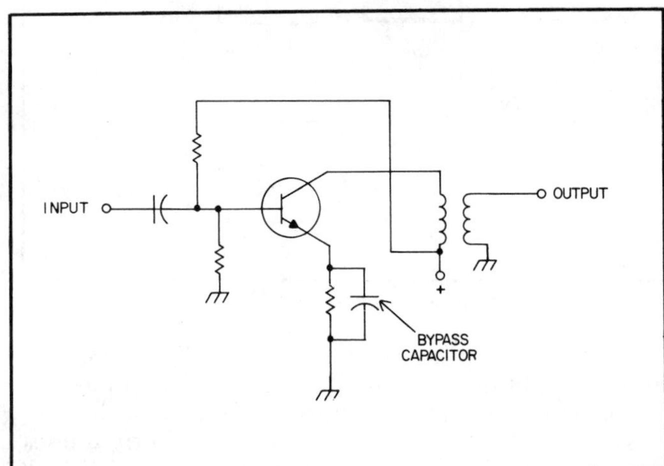

BYPASS CAPACITOR: A bypass capacitor in a simple amplifier circuit. This capacitor also provides stabilization and emitter bias.

The bypass capacitor is so named because the ac signal is provided with a low-impedance path around a high-impedance element. The schematic illustrates a typical bypass capacitor in the emitter circuit of a bipolar-transistor amplifier. A bypass capacitor is different from a blocking capacitor (*see* BLOCKING CAPACITOR); a bypass capacitor is usually connected in such a manner that it shorts the signal to ground, while a blocking capacitor is intended to conduct the signal from one part of a circuit to another while isolating the dc potentials.

BYTE

A byte is a group of bits (*see* BIT), processed in parallel. A byte may contain any number of bits, but the number is usually a power of 2, such as 2, 4, 8, 16, and so on. A computer processes each byte as one unit.

A byte is the smallest addressable unit of storage in a computer. It generally consists of eight data bits and one parity bit.

C

In electronics, the letter C is used to stand for capacitance. Capacitors in a circuit are generally given designators consisting of the letter C followed by a positive integer, such as C23, C47, and so on. The capital letter C is also used to stand for capacitance in equations.

The capital C is not only used for capacitance; it may also be used for coulombs, or for temperature in degrees Celsius. Whether 35C represents 35 coulombs or 35 degrees Celsius is usually evident from context. The small letter c is often used to stand for the speed of light, 3×10^8 meters per second.

CABLE

Any group of electrical conductors, bound together, is a *cable*. A cable might have just two conductors, such as the coaxial cable familiar to RF designers; or it might have many individual conductors (see illustration).

Cables are usually covered by a tough protective material such as rubber or polyethylene. This protects the conductors against the accumulation of moisture, and also against corrosion. Each individual conductor is, of course, covered with its own layer of insulation.

Cables are used extensively in communications and electronics. Most telephone circuits are completed via cable. In remote areas, cable is sometimes the only available means of obtaining television reception, except for satellite downlinks. *See also* CABLE COMMUNICATIONS, CABLE TELEVISION.

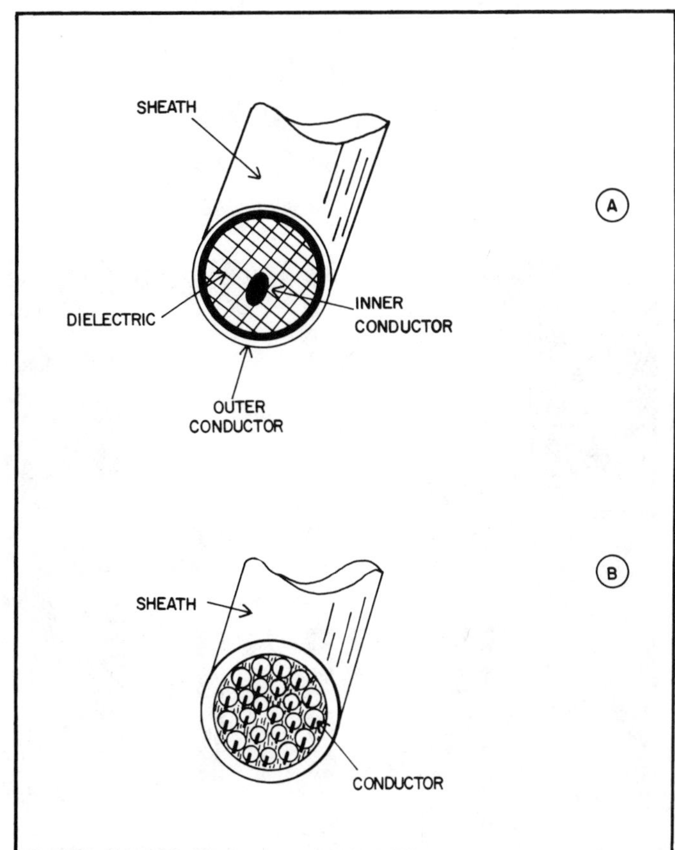

CABLE: Two kinds of cable. At A, a coaxial cable, with two concentric conductors. At B, a multi-conductor, unshielded cable.

CABLE COMMUNICATIONS

Any system of transferring information by wire is called cable communications. Cable communications, despite the inconvenience of having to install an electrical connection all the way from one end of a circuit to the other, offers reliable performance. Ionospheric conditions do not affect the performance of a shielded cable system; unshielded systems are only affected when a severe solar storm occurs, disrupting the magnetic field of the earth.

All cable systems have some loss because of the resistance of the conductors. Therefore, a signal transmitted in cable over long distances must be periodically amplified. Such amplifiers are called boosters. The useful range of a cable communications link is limited by the noise generated in the boosters, since each succeeding booster amplifies not only the signal, but the noise in the cable. The useful range is also limited by thermal noise, and the noise caused by the motion of electrons in the cable conductors. With modern techniques, though, world-wide cable links are possible.

Cable television and the telephone are the most common examples of cable communications systems. *See also* CABLE TELEVISION, TELEPHONE.

CABLE TELEVISION

In locations far from any television broadcast stations, cable television is popular. In recent years, cable television has replaced the conventional TV antenna in many cities. Cable television offers better picture quality than the individual antenna, and cable is also less susceptible to outside interference, such as automobile engines, airplanes, and the like.

Most cities have only two or three television broadcast stations, but with cable TV, all 12 vhf channels can be filled in every city in the United States. For example, a viewer in Miami might watch stations from Atlanta, New York, and Orlando, all via cable.

When television signals are transmitted by cable, the signals are sometimes heterodyned, or converted, to lower frequencies for more efficient transmission. All cable systems have increased loss as the frequency is raised. So, for long-distance cable transmission, uhf channels are generally heterodyned down to vhf. This is why a cable TV viewer might switch to channel 3 on a home TV set, and find it occupied by channel 17. Often, vhf channels are converted to lower vhf channels for the same reason.

CACHE MEMORY

A cache memory is a form of data storage used in computers for high-speed access. A small amount of data may be stored and quickly retrieved. The contents of a cache memory are stored for only a short time, and are easily changed or reprogrammed. Cache memory usually is contained in one or more integrated circuits. *See also* RANDOM-ACCESS MEMORY, SCRATCH-PAD MEMORY.

CADMIUM

Cadmium is an element with atomic number 48 and atomic weight 112. It is a metallic substance, often used as a plating material for steel to prevent corrosion. Cadmium adheres well to solder, and it is therefore a good choice for plating of chassis for the construction of electronic equipment.

Cadmium is used in the manufacture of photocells, which are devices that change resistance according to the amount of light that strikes them. Cadmium is also found in certain types of rechargeable cells and batteries, called nickel-cadmium batteries. *See also* CADMIUM PHOTOCELL, NICKEL-CADMIUM BATTERY.

CADMIUM PHOTOCELL

A cadmium photocell is a device that displays a variable resistance depending on the amount of light striking it (*see* PHOTOCELL). Most photocells are constructed from semiconductor materials. Two common types of semiconductor used in photocells are cadmium selenide and cadmium sulfide.

The cadmium-selenide photocell has a rapid response time, and is sensitive to light at the longer wavelengths. This makes it preferable for use with incandescent light. It also responds to infrared radiation.

The cadmium-sulfide photocell displays very large changes in resistance between dark and light conditions. Some such cells can be connected directly to 120-volt ac power, and used as light-actuated switches for common household appliances. *See also* LIGHT-ACTUATED SWITCH.

CAGE ANTENNA

A cage antenna is similar to a dipole anenna (*see* DIPOLE ANTENNA). Several half-wavelength, center-fed conductors are connected together at the feed point as shown in the illustration. The cage antenna is so named because it often physically resembles a cage. The multiplicity of conductors increases the effective bandwidth of the antenna. It also reduces losses caused by the resistance of the wire conductors.

A cage antenna with wires or rods arranged in the form of two cones, with apexes that meet at the feed point, is called a biconical antenna. Such an antenna offers a good impedance match for 50- or 75-ohm feed lines, over a frequency range of several octaves. *See also* BICONICAL ANTENNA.

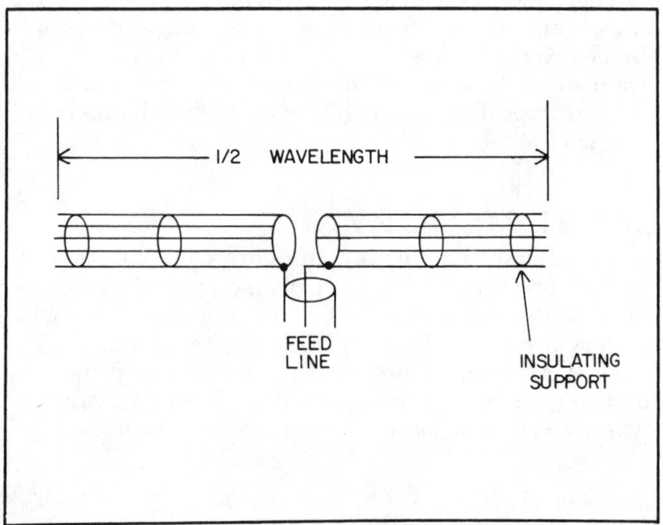

CAGE ANTENNA: The cage antenna has several parallel conductors.

CALCULATION

Calculation is a data-processing function. A combination of addition, subtraction, multiplication, and division forms a calculation. Other functions may also occasionally be used, such as the square and square root, sine, cosine, and so on. A calculation differs from an equation in that all quantities are known.

Generally, multiplication and division are performed before addition and subtraction, unless parentheses indicate otherwise. The square and square-root operations are performed before multiplication and division, unless otherwise specified. For example, the expression.

$$3 + 5 \times 4$$

is equal to $3 + (5 \times 4)$, or 23, and not $(3 + 5) \times 4$, which is 32.

Solving for unknown quantities, using equations, is called algebra. *See also* ALGEBRA.

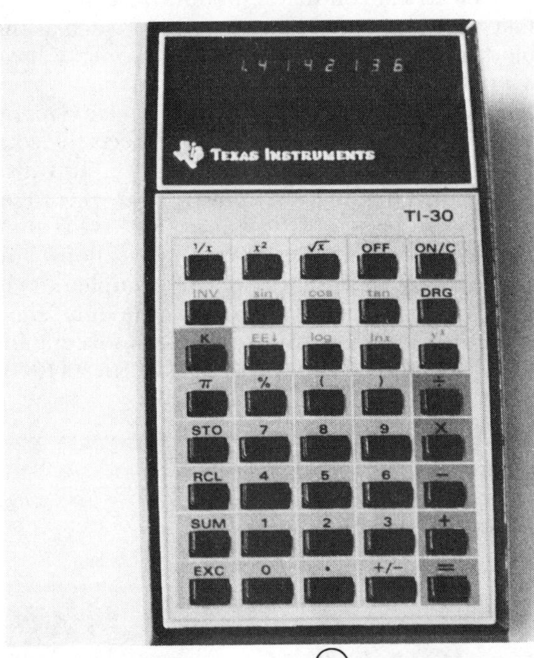

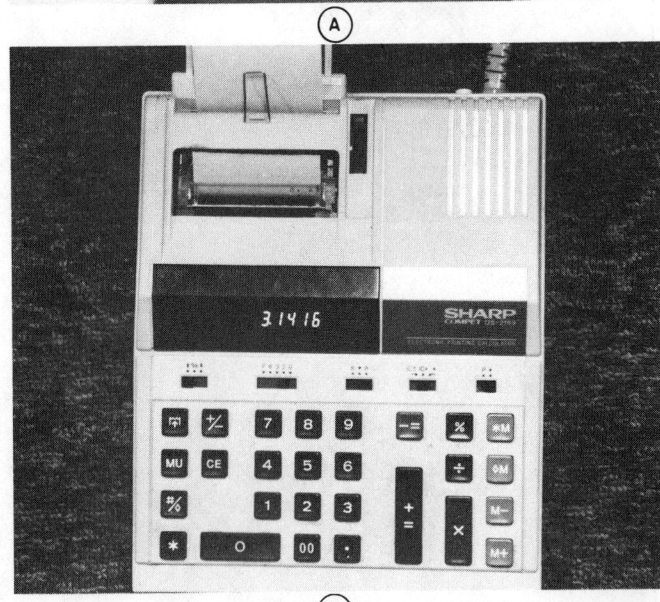

CALCULATOR: Two common calculators. At A, a pocket-size calculator for engineering and mathematical use. At B, a desk calculator, primarily used for bookkeeping.

CALCULATOR

An electronic calculator is a device that performs calculations using semiconductor devices. Calculators save a great deal of time. Some calculations that would require several minutes on paper can be performed in just a few seconds on an electronic calculator that costs only a few dollars. Calculators may be small enough to fit in a shirt pocket, such as the one at A in the illustration. This calculator is powered by a miniature 9-volt battery, and is about the size of a package of cigarettes.

Some calculators are built for desktop or office use, such as the one shown at B. This unit prints its calculations, if desired, on a small sheet of paper. Such calculators are sometimes used by retail stores in place of cash registers.

The calculator is less sophisticated than the computer. Computers can solve complicated algebraic equations that might require years of time to solve with a calculator. However, the line of distinction between the calculator and computer is not well-defined. Some calculators can be programmed to solve simple problems, just like a computer. *See also* COMPUTER.

CALCULUS

Calculus is a mathematical technique for precisely determining the rate of change of a quantity. Calculus is also used to determine the net average effect of a quantity over a period of time.

Differential calculus is concerned with the instantaneous rate of change of a variable (*see* INSTANTANEOUS EFFECT). The illustration shows a graph of collector current as a function of the base voltage for an NPN bipolar transistor. If a fairly accurate mathematical representation of this function can be found, a precise determination of the ratio of collector-current change to base-voltage change can be derived for any operating point. This ratio is the slope of a line tangent to the curve at the operating point. The rate of change of any variable is called the derivative.

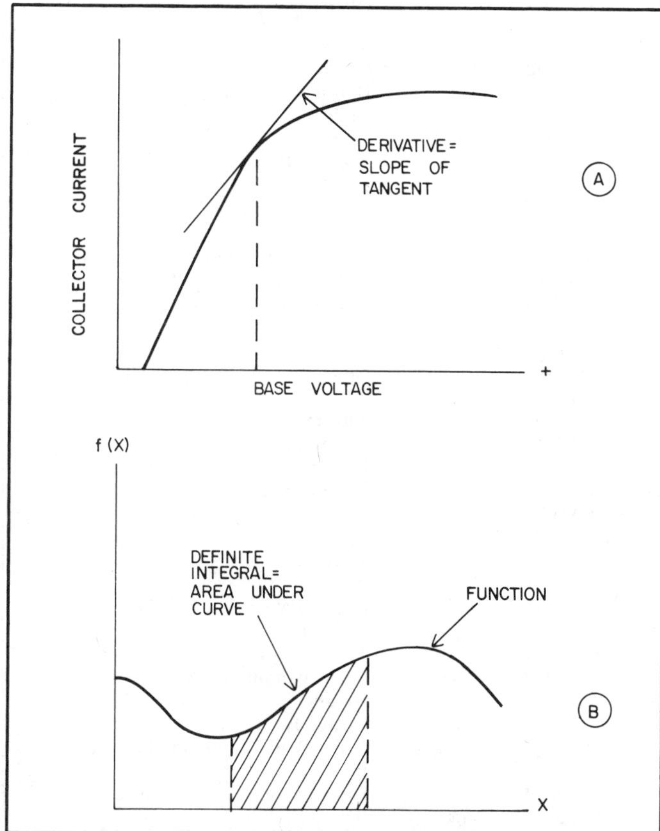

CALCULUS: At A, differential calculus gives the slope of a tangent line to a curve. This provides a means of determining the instantaneous rate of change of a quantity. At B, the area under a curve between two points is found by integral calculus.

In integral calculus, the area under the curve of a function is determined between two points on the abscissa or indepedent-variable axis. An example of this is illustrated at B. Regions above the abscissa (horizontal axis in the illustration) are considered to have positive area; regions below the abscissa have negative area. Integral calculus may be used to compute volumes in three dimensions, as well as areas. The area under a curve between two specified points is called a definite integral. The area function itself is called the indefinite integral.

Integration and differentiation are, in a certain sense, opposites. Usually, given a function f, the integral of the derivative yields f, and the derivative of the integral also yields f.

Special circuits, called differentiators and integrators, produce an output waveform that is the derivative or integral of the input waveform function. *See also* DIFFERENTIATOR CIRCUIT, INTEGRATOR CIRCUIT.

CALIBRATION

Calibration is the adjustment of a measuring instrument for an accurate indication. The frequency dials of radio receivers and transmitters, as well as the various dials and meters on test instruments, must be calibrated periodically to ensure their accuracy.

The precision of calibration is usually measured as a percentage, plus or minus the actual value. For example, the frequency calibration of a radio receiver may be specified as plus or minus 0.0001 per cent, written ±0.0001 percent. At a frequency of 1 MHz, this is a maximum error of one part in a million, or 1 Hz. At 10 MHz, it represents a possible error of 10 Hz either side of the indicated frequency.

Calibration requires the use of reference standards. Time and frequency are broadcast by radio statons WWV and WWVH in the United States (*see* WWV/WWVH), with extreme accuracy. Standard values for all units are determined by the National Bureau of Standards (*see* NATIONAL BUREAU OF STANDARDS).

CALL SIGN

Every licensed radio station in the world has a call sign. A call sign consists of a sequence of letters or letters and numbers. Commercial radio and television broadcast stations usually have call signs containing three or four letters. Other radio services assign alphanumeric call signs to their stations.

Call signs begin with a letter or number that depends on the country in which the station is licensed. In some cases, the country is subdivided into several zones, each with a distinctive call-sign format. The table lists international call-sign prefixes currently in use throughout the world. Identification requirements vary from country to country.

The nickname of a broadcast station is not the same as its call sign, and is usually not legal for purposes of identification. The Federal Communications Commission assigns call signs in the United States.

CALL SIGN: INTERNATIONAL CALL-SIGN PREFIXES, AS ASSIGED BY THE INTERNATIONAL TELECOMUNICATION UNION (ITU).

Prefix	Country	Prefix	Country
AAA-ALZ	United States	SNA-SRZ	Poland
AMA-AOZ	Spain	SSA-SSM	Egypt
APA-ASZ	Pakistan	SSN-STZ	Sudan
ATA-AWZ	India	SUA-SUZ	Egypt
AXA-AXZ	Australia	SVA-SZZ	Greece
AYA-AZZ	Argentina	TAA-TCZ	Turkey
BAA-BZZ	China	TDA-TDZ	Guatemala
CAA-CEZ	Chile	TEA-TEZ	Costa Rica
CFA-CKZ	Canada	TFA-TFZ	Iceland
CLA-CMZ	Cuba	TGA-TGZ	Guatemala
CNA-CNZ	Morocco	THA-THZ	France
COA-COZ	Cuba	TIA-TIZ	Costa Rica
CPA-CPZ	Bolivia	TJA-TJZ	Cameroon
CQA-CUZ	Portugal	TKA-TKZ	France
CVA-CXZ	Uruguay	TLA-TLZ	Central African Republic
CYA-CZZ	Canada	TMA-TMZ	France
DAA-DTZ	Germany	TNA-TNZ	Congo
DUA-DZZ	Philippines	TOZ-TQZ	France
EAA-EHZ	Spain	TRA-TRZ	Gabon
EIA-EJZ	Ireland	TSA-TSZ	Tunisia
EKA-EKZ	Union of Soviet Socialist Republics	TTA-TTZ	Chad
		TUA-TUZ	Ivory Coast
ELA-ELZ	Liberia	TVA-TXZ	France
EMA-EOZ	Union of Soviet Socialist Republics	TYA-TYZ	Benin
		TZA-TZZ	Mali
EPA-EQZ	Iran	UAA-UZZ	Union of Soviet Socialist Republics
ERA-ERZ	Union of Soviet Socialist Republics	VAA-VGZ	Canada
ETA-ETZ	Ethiopia	VHA-VNZ	Australia
EWA-EZZ	Union of Soviet Socialist Republics	VOA-VOZ	Canada
		VPA-VSZ	Great Britain and Northern Ireland
FAA-FZZ	France		
GAA-GZZ	Great Britain and Northern Ireland	VTA-VWZ	India
		VXA-VYZ	Canada
HAA-HAZ	Hungary	VZA-VZZ	Australia
HBA-HBZ	Switzerland	WAA-WZZ	United States
HCA-HDZ	Ecuador	XAA-XIZ	Mexico
HEA-HEZ	Switzerland	XJA-XOZ	Canada
HFA-HFZ	Poland	XPA-XPZ	Denmark
HGA-HGZ	Hungary	XQA-XRZ	Chile
HHA-HHZ	Haiti	XSA-XSZ	China
HIA-HIZ	Dominican Republic	XTA-XTZ	Voltaic Republic
HJA-HKZ	Colombia	XUA-XUZ	Cambodia
HLA-HMZ	South Korea	XVA-XVZ	Vietnam
HNA-HNZ	Iraq	XWA-XWZ	Laos
HOA-HPZ	Panama	XXA-XXZ	Portuguese Overseas Provinces
HQA-HRZ	Honduras	XYA-XZZ	Burma
HSA-HSZ	Thailand	YAA-YAZ	Afghanistan
HTA-HTZ	Nicaragua	YBA-YHZ	Indonesia
HUA-HUZ	El Salvador	YIA-YIZ	Iraq
HVA-HVZ	Vatican City	YJA-YJZ	New Hebrides
HWA-HYZ	France	YKA-YKZ	Syria
HZA-HZZ	Saudi Arabia	YMA-YMZ	Turkey
IAA-IZZ	Italy	YNA-YNZ	Nicaragua
JAA-JSZ	Japan	YOA-YRZ	Romania
JTA-JVZ	Mongolia	YSA-YSZ	El Salvador
JWA-JXZ	Norway	YTA-YUZ	Yugoslavia
JYA-JYZ	Jordan	YVA-YYZ	Venezuela
JZA-JZZ	Indonesia	YZA-YZZ	Yugoslavia
KAA-KZZ	United States	ZAA-ZAZ	Albania
OAA-OCZ	Peru	ZBA-ZJZ	Great Britain and Northern Ireland
ODA-ODZ	Lebanon		
OEA-OEZ	Austria	ZKA-ZMZ	New Zealand
OFA-OJZ	Finland	ZNA-ZOZ	Great Britain and Northern Ireland
OKA-OMZ	Czechoslovakia		
ONA-OTZ	Belgium	ZPA-ZPZ	Paraguay
OUA-OZZ	Denmark	ZQA-ZQZ	Great Britain and Northern Ireland
PAA-PIZ	Netherlands		
PJA-PJZ	Netherlands Antilles	ZRA-ZUZ	South African and Namibia
PKA-POZ	Indonesia	ZVA-ZZZ	Brazil
PPA-PYZ	Brazil	2AA-2ZZ	Great Britain and Northern Ireland
PZA-PZZ	Surinam		
QAA-QZZ	Q signals (no call signs)	3AA-3AZ	Monaco
SAA-SMZ	Sweden	3BA-3BZ	Mauritius

Prefix	Country
3CA-3CZ	Equatorial Guinea
3DA-3DM	Swaziland
3DN-3DZ	Fiji
3EA-3FZ	Panama
3GA-3GZ	Chile
3HA-3UZ	China
3VA-3VZ	Tunisia
3WA-3WZ	Vietnam
3XA-3XZ	Guinea
3YA-3YZ	Norway
3ZA-3ZZ	Poland
4AA-4CZ	Mexico
4DA-4IZ	Philippines
4JA-4LZ	Union of Soviet Socialist Republics
4MA-4MZ	Venezuela
4NA-4OZ	Yugoslavia
4PA-4SZ	Sri Lanka
4TA-4TZ	Peru
4UA-4UZ	United Nations
4VA-4VZ	Haiti
4WA-4WZ	North Yemen
4XA-4XZ	Israel
4YA-4YZ	International Civil Aviation Organization
4ZA-4ZZ	Israel
5AA-5AZ	Libya
5BA-5BZ	Cyprus
5CA-5GZ	Morocco
5HA-5IZ	Tanzania
5JA-5KZ	Colombia
5LA-5MZ	Liberia
5NA-5OZ	Nigeria
5PA-5QZ	Denmark
5RA-5SZ	Malagasy Republic
5TA-5TZ	Mauritania
5UA-5UZ	Niger
5VA-5VZ	Togo
5WA-5WZ	Western Samoa
5XA-5XZ	Uganda
5YA-5ZZ	Kenya
6BA-6BZ	Egypt
6CA-6CZ	Syria
6DA-6JZ	Mexico
6KA-6NZ	Korea
6OA-6OZ	Somali
6PA-6SZ	Pakistan
6TA-6UZ	Sudan
6VA-6WZ	Senegal
6XA-6XZ	Malagasy Republic
6YA-6YZ	Jamaica
6ZA-6ZZ	Liberia
7AA-7IZ	Indonesia
7JA-7NZ	Japan
7OA-7OZ	South Yemen
7PA-7PZ	Lesotho
7QA-7QZ	Malawi
7RA-7RZ	Algeria
7SA-7SZ	Sweden
7TA-7YZ	Algeria
7ZA-7ZZ	Saudi Arabia
8AA-8IZ	Indonesia
8JA-8NZ	Japan
8OA-8OZ	Botswana

Prefix	Country
8PA-8PZ	Barbados
8QA-8QZ	Maldive Island
8RA-8RZ	Guyana
8SA-8SZ	Sweden
8TA-8YZ	India
8ZA-8ZZ	Saudi Arabia
9AA-9ZZ	San Marino
9BA-9DZ	Iran
9EA-9FZ	Ethiopia
9GA-9GZ	Ghana
9HA-9HZ	Malta
9IA-9JZ	Zambia
9KA-9KZ	Kuwait
9LA-9LZ	Sierra Leone
9MA-9MZ	Malaysia
9NA-9NZ	Nepal
9OA-9TZ	Zaire
9UA-9UZ	Burundi
9VA-9VZ	Singapore
9WA-9WZ	Malaysia
9XA-9XZ	Rwanda
9YA-9ZZ	Trinidad and Tobago
A2A-A2Z	Botswana
A3A-A3Z	Tonga
A4A-A4Z	Oman
A5A-A5Z	Bhutan
A6A-A6Z	United Arab Emirates
A7A-A7Z	Qatar
A8A-A8Z	Liberia
A9A-A9Z	Bahrain
C2A-C2Z	Nauru
C3A-C3Z	Andorra
C4A-C4Z	Cyprus
C5A-C5Z	The Gambia
C6A-C6Z	Bahamas
C7A-C7Z	World Meteorological Organization
C8A-C9Z	Mozambique
D2A-D3Z	Angola
D4A-D4Z	Cape Verde
D5A-D5Z	Liberia
D6A-D6Z	State of Comoro
D7A-D9Z	South Korea
H2A-H2Z	Cyprus
H3A-H3Z	Panama
H4A-H4Z	Solomon Islands
H5A-H5Z	Bophuthatswana
J2A-J2Z	Djibouti
J3A-J3Z	Grenada
J4A-J4Z	Greece
J5A-J5Z	Guinea Bissau
L2A-L9Z	Argentina
P2A-P2Z	Papua New Guinea
P3A-P3Z	Cyprus
P4A-P4Z	Netherlands Antilles
P5A-P9Z	North Korea
S2A-S3Z	Bangladesh
S6A-S6Z	Singapore
S7A-S7Z	Seychelles
S8A-S8Z	Transkei
S9A-S9Z	St. Thomas and Principe
T2A-T2Z	Tuvalu
Y2A-Y9Z	East Germany

CALORIMETER

A calorimeter is a device for measuring heat. The calorimeter is used in some radio-frequency wattmeters. A special resistive load, which dissipates all the power applied to it as heat, is connected across the transmission line. The resistance of the load is precisely matched to the characteristic impedance of the transmission line, and the wattmeter serves as a dummy load. All the power from the transmitter is dissipated as heat in the load; the amount of heat is measured by the calorimeter, and this power is read on a meter as wattage. *See also* WATTMETER.

CAMERA CHAIN

In a television broadcasting system, the equipment that delivers the video information from the camera to the transmitter is called the camera chain. This includes not only the television camera itself, but its power supply, associated video monitors, and all connecting cables.

The signal from the camera chain is used to modulate the transmitting equipment so that the television station can send a picture at the proper frequency. *See also* COMPOSITE VIDEO SIGNAL.

CAMERA TUBE

A camera tube is a device that focuses a visible image onto a small screen, and scans this image with an electron beam to develop a composite video signal. The current in the electron beam varies, depending on the brightness of the image in each part of the screen. The response must be very rapid, since the scanning is done at a high rate of speed. The position of the electron beam, relative to the image, is synchronized between the camera tube and the receiver picture tube.

The camera tube most often used in television broadcast stations is called the image orthicon. A simpler and more compact camera tube, used primarily in industry and space applications, is called the vidicon. *See also* COMPOSITE VIDEO SIGNAL, IMAGE ORTHICON, VIDICON.

CANADIAN STANDARDS ASSOCIATION

The Canadian Standards Association, abbreviated CSA, is a group of volunteers that prepares specifications for engineering use by Canadian industry and government. Its equivalent in the United States is the Electronics Industries Association, insofar as electricity and electronics are concerned. *See also* ELECTRONICS INDUSTRIES ASSOCIATION.

CANDELA

The candela, abbreviated cd, is the standard unit of light-source intensity. It was formerly called the candle or candlepower; these latter two units have become obsolete.

One candela is defined as the luminous intensity of 0.0167 square centimeter of a blackbody radiator at a temperature of 2046 degrees Kelvin, the solidification temperature of platinum. A 40-watt electric bulb has a luminous intensity of about 3000 candela. An ordinary candle has a luminous intensity of about 1 candela.

One candela is equal to one lumen per steradian. *See also* LIGHT INTENSITY, LUMEN, STERADIAN.

CANDLE

See CANDELA.

CANDLEPOWER

The candlepower is an obsolete unit of light-source intensity, defined as the amount of light radiated horizontally by a common candle. The candela is now accepted as the standard unit of luminous intensity, since its definition is more precise. *See also* CANDELA.

CANNIBALIZATION

When a piece of equipment must be repaired in the field, and the only available components are contained in other equipment that is deemed less essential, those components may be robbed for use in important gear. Such robbing is called *cannibalization*. Theequipment from which parts are cannibalized is disabled until replacement parts can be obtained for it.

Cannibalization is a last-resort, emergency procedure, which technicians try to avoid. Cannibalization gets its somewhat humorous name from the fact that it involves a sacrifice of active equipment.

CAPACITANCE

Capacitance is the ability of a device to store an electric charge. Capacitance, represented by the capital letter C in equations, is measured in units called farads. One farad is equal to one coulomb of stored electric charge per volt of applied potential. Thus,

$$C \text{ (farads)} = Q/E$$

where Q is the quantity of charge in coulombs, and E is the applied voltage in volts.

The farad is, in practice, an extremely large unit of capacitance. Generally, capacitance is specified in microfarads, abbreviated uF, or in picofarads, abbreviated pF. These units are, respectively, a millionth and a trillionth of a farad:

$$1 \text{ uF} = 10^{-6} \text{ F}$$

$$1 \text{ pF} = 10^{-12} \text{ F}$$

Any two electrically isolated conductors will display a certain amount of capacitance. When two conductive materials are deliberately placed near each other to produce capacitance, the device is called a capacitor (*see* CAPACITOR).

Capacitances connected in parallel add together. For n capacitances C1, C2, . . ., Cn in parallel, then, the total capacitance C is:

$$C = C1 + C2 + \ldots + Cn$$

In series, capacitances add as follows:

$$C = \frac{1}{1/C1 + 1/C2 + \ldots + 1/Cn}$$

See also CAPACITIVE REACTANCE.

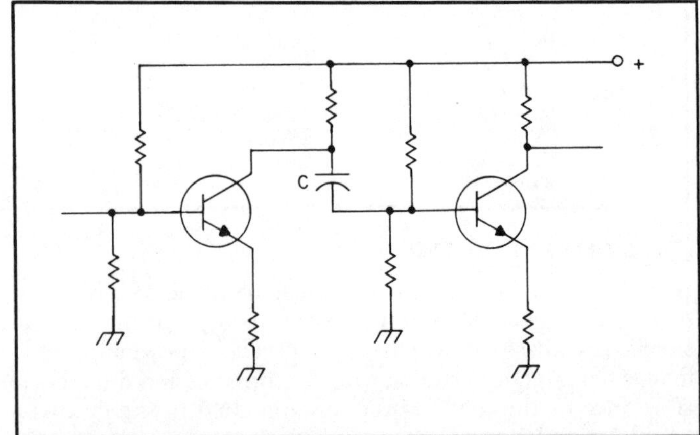

CAPACITIVE COUPLING: An example of capacitive coupling. Signal transfer occurs through C, which is chosen to allow efficient coupling.

CAPACITIVE COUPLING

Capacitive coupling is a means of electrostatic connection among the stages of amplification in a radio circuit. The output of one stage is coupled to the input of the succeeding stage by a capacitor, as illustrated in the schematic diagram. Capacitor C is the coupling capacitor.

Capacitive coupling allows the transfer of an alternating-current signal, but prevents the short-circuiting of a direct-current potential difference. Thus, the desired bias may be maintained on either side of the capacitor. A coupling capacitor is sometimes called a blocking capacitor (*see* BLOCKING CAPACITOR).

The chief disadvantage of capacitive coupling is that it is sometimes difficult, or impossible, to achieve a good impedance match between stages. This makes the power transfer inefficient. Also, capacitive coupling allows the transfer of all harmonic energy, as well as the fundamental or desired frequency, from stage to stage. Tuned circuits are necessary to reduce the transfer of this unwanted energy. Sometimes, the use of coupling transformers is better than the use of capacitors for interstage connection. *See also* TRANSFORMER COUPLING.

CAPACITIVE DIODE

See VARACTOR DIODE.

CAPACITIVE FEEDBACK

Capacitive feedback is a means of returning a portion of the output signal of a circuit back to the input, using a capacitor (see illustration). In most amplifier circuits, the output is 180 degrees out of phase with the input, so connecting a capacitor between the output and the input reduces the gain. If the output is fed back in phase, the gain of an amplifier increases, but the probability of oscillation is much greater.

Capacitive feedback may occur as a result of coupling between the input and output wiring of a circuit. This can produce undesirable effects, such as amplifier instability or parasitic oscillation. This kind of capacitive feedback is also possible within a tube or transistor, because of interelectrode capacitance. It is nullified by a process called neutralization (*see* NEUTRALIZATION).

Capacitive feedback is commonly employed in oscillator circuits such as the multivibrator, Hartley, Pierce, and Colpitts types. The Armstrong oscillator uses inductive feedback, via coupling between coils. *See also* INDUCTIVE FEEDBACK.

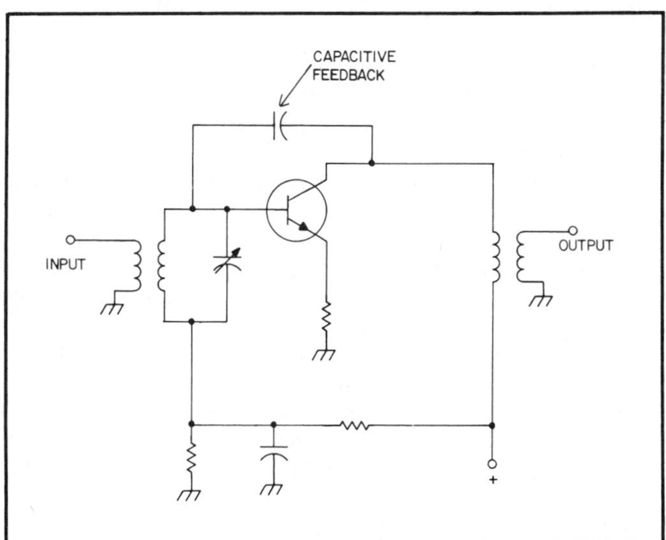

CAPACITIVE FEEDBACK: In this case, it is negative or degenerative feedback for the purpose of preventing oscillation in the amplifier.

CAPACITIVE LOADING

Capacitive loading can refer to either of two techniques for alternating the resonant frequency of an antenna system.

A vertical radiator, operated against ground, is normally resonant when it is ¼ wavelength high. However, a taller radiator can be brought to resonance by the installation of a capacitor or capacitors in series, as illustrated in the diagram (A). This is commonly called capacitive loading or tuning.

A vertical radiator shorter than ¼ wavelength may be resonated by the use of a device called a capacitance hat (B in the illustration). A capacitance hat is often used in conjunction with a loading coil. The capacitance hat results in an increase in the useful bandwidth of the antenna, as well as a shorter physical length at a given resonant frequency.

Both of the methods of capacitive loading described here are also used with dipole antennas. They may also be employed in the driven elements of Yagi antennas, and in various kinds of phased arrangements.

CAPACITIVE REACTANCE

Reactance is the opposition that a component offers to alternating current. Reactance does not behave as simply as dc resistance. The reactance of a component can be either positive or negative, and it varies in magnitude depending on the frequency of the alternating current. Negative reactance is called capacitive reactance, and posi-

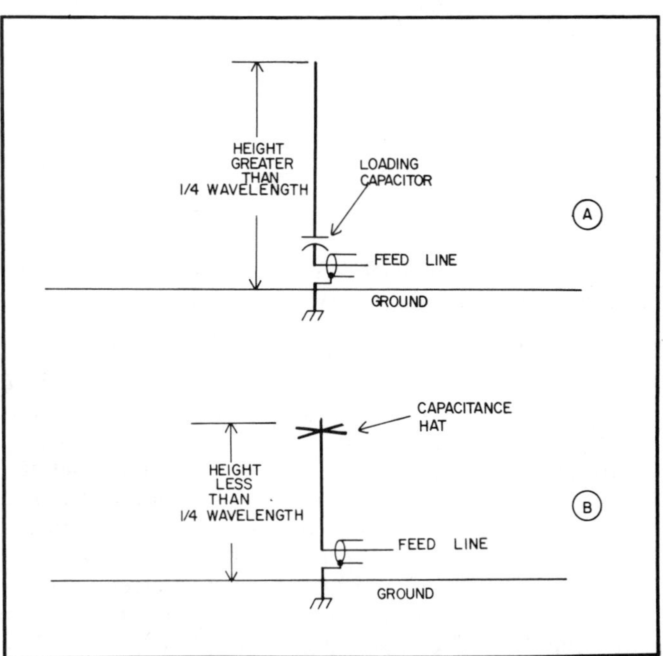

CAPACITIVE LOADING: Two types of capacitive loading. At A, a fixed capacitor connected in series with a radiating element raises the resonant frequency of the antenna. At B, a capacitance hat, consisting of a network of wires at the top of a radiator. This increases the bandwidth, and lowers the resonant frequency, of the antenna.

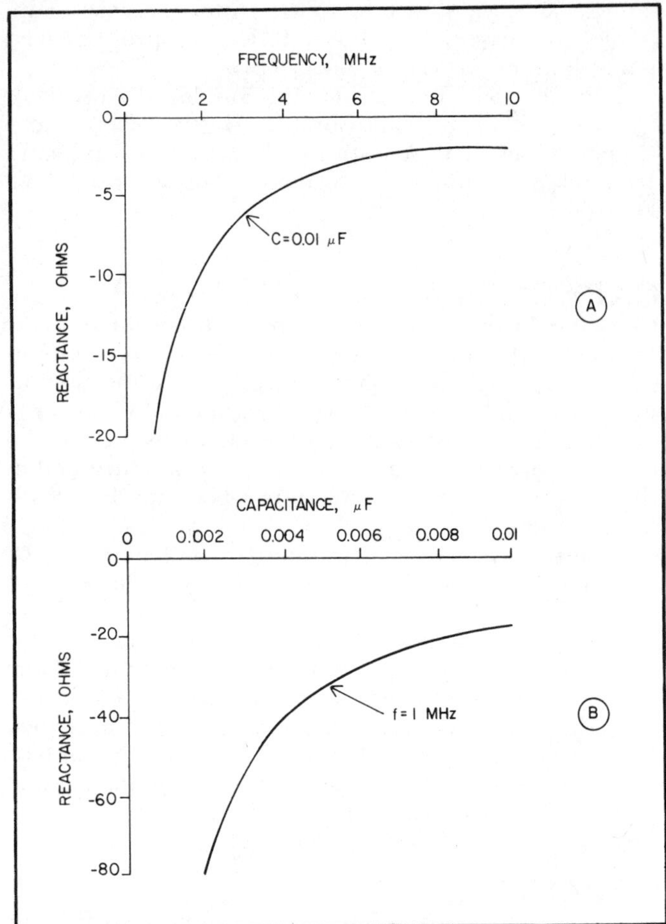

CAPACITIVE REACTANCE: At A, the reactance is shown for an 0.01-μF capacitor as a function of the frequency. At B, the reactance is shown as a function of capacitance at a frequency of 1MHz.

tive reactance is called inductive reactance. These choices of positive and negative are purely arbitrary, having been chosen by engineers for convenience.

The reactance of a capacitor of C farads, at a frequency of f Hertz, is given by:

$$X_C = \frac{-1}{2\pi fC}$$

This formula also holds for values of C in microfarads, and frequencies in megahertz. (When using farads, it is necessary to use Hertz; when using microfarads, it is necessary to use megahertz. Units cannot be mixed.)

The magnitude of capacitive reactance, for a given component, approaches zero as the frequency is raised. It becomes larger negatively as the frequency is lowered. This effect is illustrated by the graphs. At A, the reactance of a 0.01-μF capacitor is shown as a function of the frequency in MHz. At B, the reactance of various values of capacitance are shown for a frequency of 1 MHz.

Reactances, like dc resistances, are expressed in ohms. In a complex resistance-reactance circuit, the reactance is multiplied by the imaginary number $\sqrt{-1}$, which is mathematically represented by the letter j in engineering documentation. Thus a capacitive reactance of −20 ohms is specified as −20j; a combination of 10 ohms resistance and −20 ohms reactance is an impedance of 10 − 20j. *See also* IMPEDANCE, J OPERATOR.

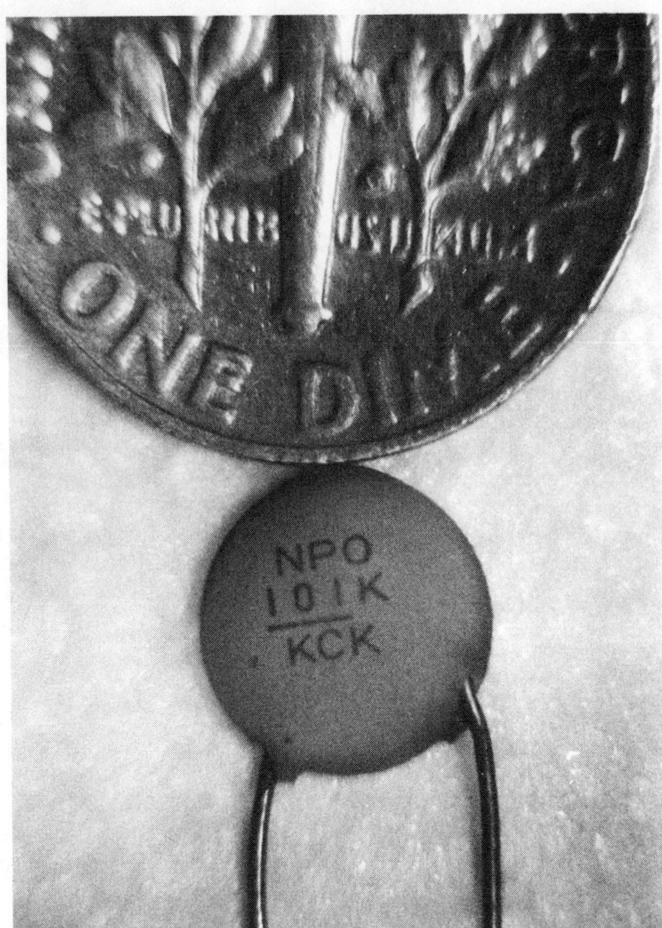

CAPACITOR: A small capacitor, frequently seen in solid-state communications equipment.

CAPACITOR

A capacitor is a device deliberately designed for providing a known amount of capacitance in a circuit. Capacitors are available in values ranging from less than 1 picofarad to hundreds of thousands of microfarads. It is rare to see a capacitor with a value of one farad. Capacitors are also specified according to their voltage-handling capability; these ratings must be carefully observed, to be certain the potential across a capacitor is not greater than the rated value. If excessive voltage, either from a signal or from a source of direct current, is applied to a capacitor, permanent damage can result.

Capacitors are often very tiny when their voltage ratings and values are small. A small capacitor is shown in the photograph. Some capacitors are very large, such as the electrolytic variety found in high-voltage power supplies, or the wide-spaced air variables used in high-power radio-frequency amplifiers.

Capacitors can be constructed in several different ways. Disk ceramic capacitors, often used at radio frequencies, consist of two metal plates separated by a thin layer of porcelain dielectric. Other types of capacitors include air variables, electrolytics, mica-dielectric types, mylar-dielectric capacitors, and many more. The most common kinds of capacitors are listed in the table, with their primary characteristics. For more detail concerning each type of capacitor, *see* AIR-VARIABLE CAPACITOR, CERAMIC CAPACITOR, ELECTROLYTIC CAPACITOR, MICA CAPACITOR, MYLAR CAPACITOR, PAPER CAPACITOR, POLYSTYRENE CAPACITOR, TANTALUM CAPACITOR, TRIMMER CAPACITOR.

CAPACITOR: COMMON TYPES OF CAPACITORS AND THEIR APPLICATIONS.

Capacitor Type	Approximate Frequency Range	Voltage Range
Air variable	lf, mf, hf, vhf, uhf	Med. to high
Ceramic	lf, mf, hf, vhf	Med. to high
Electrolytic	af, vlf	Low to med.
Mica	lf, mf, hf, vhf	Low to med.
Mylar	vlf, lf, mf, hf	Low to med.
Paper	vlf, lf, mf, hf	Low to med.
Polystyrene	af, vlf, lf, hf	Low
Tantalum	af, vlf	Low
Trimmer	mf, hf, vhf, uhf	Low to med.

Frequency Abbreviations

af: Audio frequency (0 to 20 kHz)
vlf: Very low frequency (10 to 30 kHz)
lf: Low frequency (30 to 300 kHz)
mf: Medium frequency (300 kHz to 3 MHz)
hf: High frequency (3 to 30 MHz)
vhf: Very high frequency (30 to 300 MHz)
uhf: Ultrahigh frequency (300 MHz to 3 GHz)

CAPACITOR MICROPHONE

See ELECTROSTATIC MICROPHONE.

CAPACITOR TUNING

Capacitor tuning is a means of adjusting the resonant frequency of a tuned circuit by varying the capacitance while leaving the inductance constant. When this is done, the resonant frequency varies according to the inverse of the square root of the capacitance. This means that if the capacitance is doubled, the resonant frequency drops to 0.707 of its previous value. If the capacitance is cut to one-quarter, the resonant frequency is doubled. Mathematically, the resonant frequency of a tuned inductance-capacitance circuit is given by the formula:

$$f = \frac{1}{2\pi\sqrt{LC}}$$

where f is the frequency in Hertz, L is the inductance in henrys, and C is the capacitance in farads. Alternatively, the units may be given as megahertz, microhenrys, and microfarads.

Capacitor tuning offers convenience when the values of capacitance are small enough to allow the use of air-variable units. If the capacitance required to obtain resonance is larger than about 0.001 μF, however, inductor tuning is generally preferable, since it allows a greater tuning range. *See also* INDUCTOR TUNING.

CAPACITY

Capacity may refer to any of several different properties of electronic components. Usually, the term capacity means the number of ampere hours that can be stored by a rechargeable battery, such as a storage battery. The term is also used to refer to the voltage or current rating of a component.

The total amount of data that a computer can handle, expressed in bytes, kilobytes, or megabytes, is sometimes called the capacity of the computer. In any memory circuit, the capacity is the number of bits or characters that can be stored.

The term capacity is sometimes used in place of capacitance. This is technically not a correct use of the term, but it usually causes no confusion as long as the meaning is evident from context. *See also* CAPACITANCE.

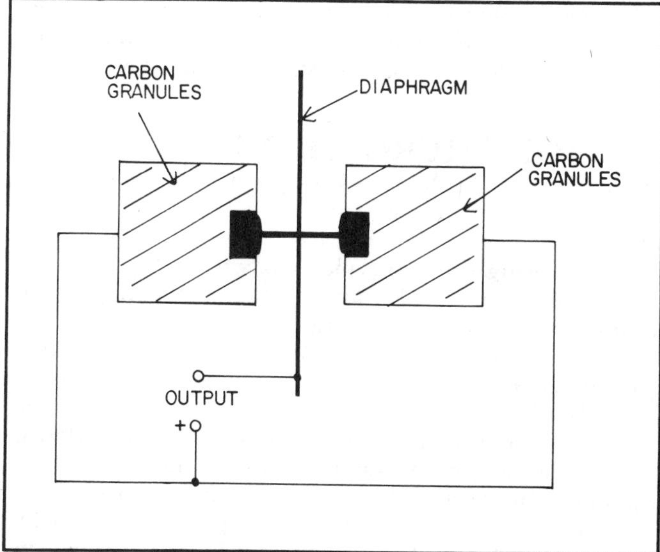

CARBON MICROPHONE: Sound waves, impinging on the diaphragm, cause vibration of the carbon granules and resulting changes in the conductivity of the device.

CAPTURE EFFECT

In a frequency-modulation receiver, a signal is either on or off, present or absent. Fadeouts occur instantly, rather than gradually as they do in other modes. When an FM signal gets too weak to be received, it comes in intermittently. This is called breakup.

If there are two FM stations on the same channel, a receiver will pick up only the stronger signal. This is because of the limiting effect of the FM detector, which sets the receiver gain according to the strongest signal present in the channel. The effect in FM receivers that causes the elimination of weaker signals in favor of stronger ones is called the capture effect.

Capture effect does not occur with amplitude-modulated, continuous-wave, or single-sideband receivers. In those modes, both weak and strong signals are audible in proportion, unless the difference is very great.

When two FM signals are approximately equal in strength, a receiver may alternate between one and the other, making both signals unintelligible.

CARBON

Carbon is an element with atomic number 6 and atomic weight 12. Carbon occurs naturally in two forms. The more common form is the black, nonmetallic substance familiar to most of us. This type of carbon is found, itself in two forms, a hard material (as seen in coal) and a softer material (called graphite). The least common form of carbon is the diamond.

The resistance of carbon is much greater than that of metal, and thus graphite is often used as the conducting material in the fabrication of noninductive resistors. Other substances are added to the graphite to vary the conductivity (*see* CARBON RESISTOR).

Carbon is used in some dry cells as the positive electrode. It is also used, in granular form, in certain kinds of microphones. *See also* CARBON MICROPHONE, ZINC CELL.

CARBON MICROPHONE

The carbon microphone is a device whose conductivity varies with the compression of carbon granules, under the influence of passing sound waves.

The drawing shows a typical carbon microphone. Vibration of air molecules against the metal diaphragm causes compression of the granules, which changes the resistance through the device. A direct-current potential, applied across the enclosure of carbon granules, is modulated by this resistance change. A transformer may be used at the microphone output to increase the impedance of the device as seen by the input circuit; the carbon microphone characteristically shows a low impedance.

Carbon microphones are quite susceptible to damage by excessive dc voltages. They have become almost obsolete in modern audio applications. Dynamic microphones are much more often seen today. *See also* DYNAMIC MICROPHONE.

CARBON RESISTOR

A carbon resistor is a low-wattage, noninductive resistor made from a mixture of carbon graphite and a nonconducting binder. The ratio of mixture determines the resistance, which can be as low as a fraction of an ohm, or as high as millions of ohms.

The carbon resistor is usually cylindrical in shape, such as is shown in the illustration. As the temperature rises, the resistance decreases. Thus a carbon resistor has a negative temperature coefficient. Carbon resistors are found in a variety of sizes; the most common are ¼ and ½ watt, but ⅛-watt, 1-watt, and 2-watt units are also used. Tolerances vary from less than 1 per cent to about 20 percent.

CARCINOTRON

The carcinotron is a backward-wave oscillator used to generate signals at extremely high frequencies. Electrons travel through a tube, interacting with electromagnetic fields induced externally. This produces oscillation. The carcinotron may be a linear tube or a circular tube. It operates in a manner similar to the magnetron (*see* MAGNETRON). The drawing is a pictorial diagram of a carcinotron. The electric and magnetic fields between the sole and the slow-wave structure cause oscillation in the electron beam as it travels down the tube.

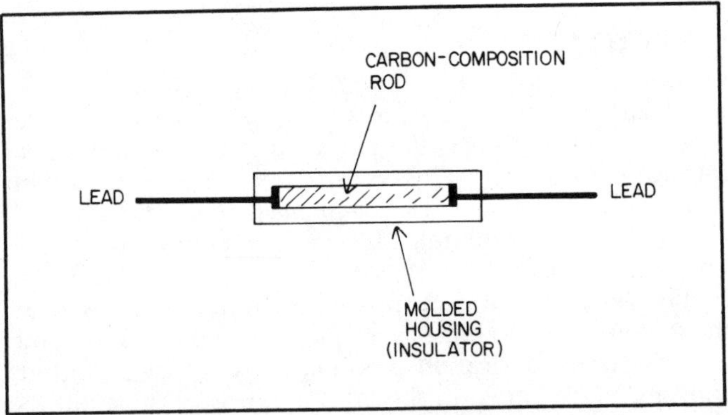

CARBON RESISTOR: Construction of a typical carbon resistor.

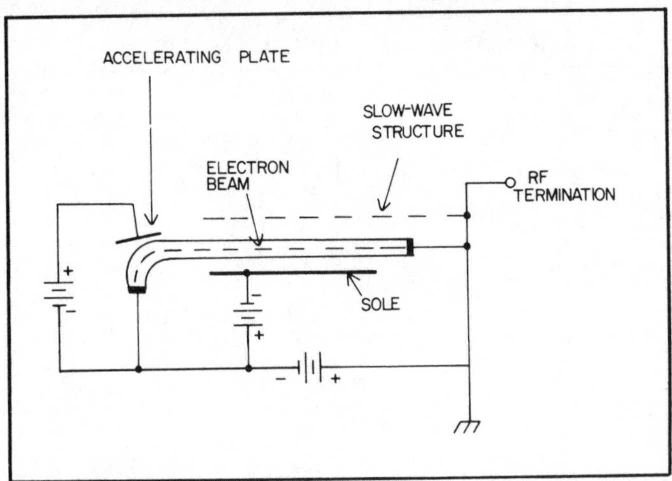

CARCINOTRON: The carcinotron generates microwave energy by the effects of electric and magnetic fields on an electron beam.

The carcinotron is capable of producing several hundred watts up to frequencies of more than 10 GHz. It will produce several milliwatts at frequencies greater than 300 GHz. The carcinotron tends to produce a large amount of noise along with the desired signal.

CARD

The card is an easily-replaced printed-circuit board. Electronic equipment often has modular construction, making in-the-field repairs much more convenient than in past days (*see* MODULAR CONSTRUCTION). The card with the faulty component or components is simply replaced on site with a properly-operating card. Then, the bad card is taken to a central repair facility, where it is repaired and made ready for use as a replacement in another unit.

The photograph shows a replaceable printed-circuit card. Some cards can be inserted and pulled out of a piece of equipment as easily as a household appliance is plugged in and unplugged. Others, such as the one shown here, require the use of simple tools such as a screwdriver and needle-nosed pliers for installation and removal. Interchangeable cards have greatly reduced the amount of test equipment needed by field technicians. At the central repair shop, computerized test instruments are sometimes used to diagnose and pinpoint the trouble in seconds.

Modular construction has improved the serviceability of electronic equipment, both in terms of turn-around time and effectiveness of repairs.

CARD: A replaceable printed-circuit card.

CARDIOID PATTERN

A cardioid pattern is an azimuth pattern representing the directional response of certain types of antennas, microphones, and speakers. The cardioid pattern is characterized by a sharp null in one direction, and a symmetrical response about the line running in the direction of the null. The illustration shows a typical cardioid pattern. It gets this name from the fact that it is heart-shaped.

Antennas with a cardioid directional response are frequently used in direction-finding apparatus. The extremely sharp null allows very accurate determination of the bearing.

A microphtne with a broad response and a null off the back is called a cardioid microphone. Such microphones are useful in public-address systems where the speakers are located around or over the audience. The poor response in that direction minimizes the chances of acoustic feedback.

CARDIOPULMONARY RESUSCITATION

Cardiopulmonary resuscitation, often abbreviated by the letters CPR, is a method of maintaining human life when the heart and/or lungs have stopped functioning.

Mouth-to-mouth resuscitation is combined with rhythmic pressure on the sternum to oxygenate and circulate the blood. For an adult, approximately 12 breaths and 60 heartbeats are administered each minute. For a small child, about 20 breaths and 90 heartbeats should be administered per minute.

Cardiopulmonary resuscitation has been effective as a life-saver in cases of electric shock, heart attacks, injuries, and-near drownings. Special courses are offered by the American Red Cross to familiarize people with CPR. Cardiopulmonary resuscitation can be performed by one or two persons. *See also* MOUTH-TO-MOUTH RESUSCITATION.

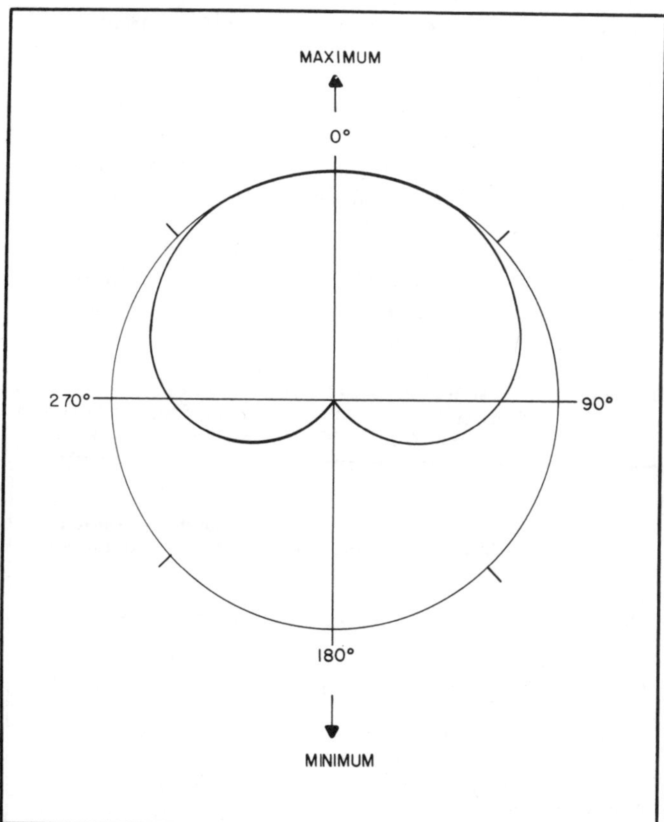

CARDIOID PATTERN: A cardioid pattern has a sharp null in one direction.

CARD READING

Card reading is a method of computer programming and data entry. A statement or short program is written by punching holes in a stiff paper card. The data is read by an electrical or optical system. Cards may be fed through the system, called a card reader, in rapid succession.

Certain programmable calculators use cards to execute an algorithm (*see* ALGORITHM). In computers that use batch processing (*see* BATCH PROCESSING), programs are written by placing one statement on each card. A complete program is run through the machine one card at a time. The card reader automatically restacks the cards in a collecting bin.

CARRIER

A carrier is an alternating-current wave of constant frequency, phase, and amplitude. By varying the frequency, phase, or amplitude of a carrier wave, information is transmitted. The term carrier is generally used in reference to the electromagnetic wave in a radio communications system; it may also refer to a current transmitted along a wire. An unmodulated carrier has a theoretical bandwidth of zero; all of the power is concentrated at a single frequency in the electromagnetic spectrum (see illustration). A modulated carrier has a nonzero, finite bandwidth (*see* BANDWIDTH).

The term carrier is also used for electrons in N-type semiconductor material, or for holes in P-type semiconductor material. In these cases, carriers are the presence or absence of extra electrons in atoms of the substance. *See also* ELECTRON, HOLE, N-TYPE SEMICONDUCTOR, P-TYPE SEMICONDUCTOR.

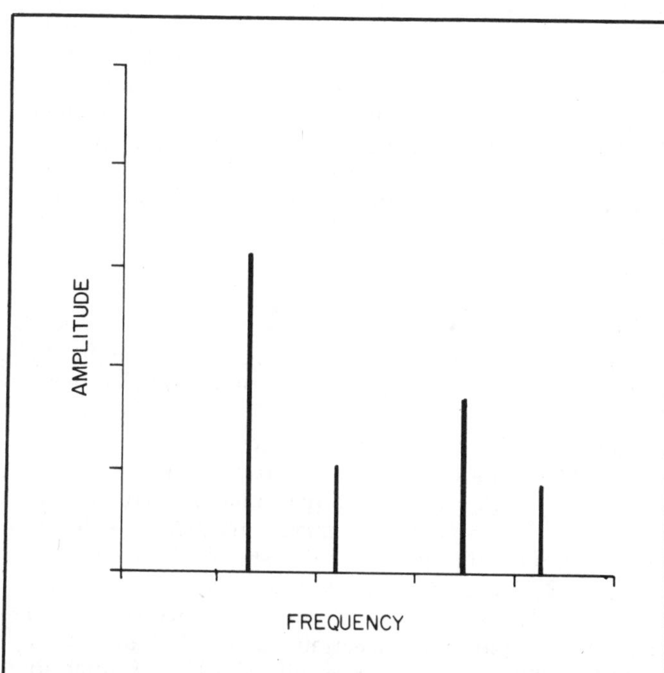

CARRIER: Spectral display of unmodulated carriers. They have a single frequency, and theoretically zero bandwidth.

CARRIER CURRENT

Carrier current is the magnitude of radio-frequency alternating currents in a transmitter carrier. Carrier current is generally measured along a transmission line. It may be expressed as a peak value, a peak-to-peak value, or a root-mean-square (RMS) value. The magnitude of the carrier current depends on the transmitter output power or carrier power, and on the impedance of the transmission line or antenna system. It also depends on the transmitter output power or carrier power, and on the impedance of the transmission line or antenna system. It also depends on the standing-wave ratio along the transmission line.

Carrier current can be measured with a radio-frequency ammeter. If the standing-wave ratio is 1 and the feed-line impedance, Z_O, is known, the RMS current, I, can be calculated from a wattmeter reading P watts according to the formula:

$$I = \sqrt{P/Z_O}$$

under conditions of zero transmitter modulation. *See also* CARRIER POWER, ROOT MEAN SQUARE.

CARRIER-CURRENT COMMUNICATION

Carrier current is a radio-frequency current deliberately sent over telephone or power lines. Many high-frequency carriers, each having a different frequency, can be individually modulated and sent simultaneously over a single wire. The signals are demodulated at the receiving end; a tuning circuit discriminates against unwanted frequencies, just as is the case in a radio communication circuit.

As the carrier frequency increases, a given line becomes more and more lossy. Therefore, there is a practical maximum to the number of signals that can be sent efficiently over one line. Carrier currents of different frequencies tend to beat together, forming mixing products on other frequencies (*see* MIXING PRODUCT). This effect can be minimized by making sure there are no nonlinear components in a circuit, but beating cannot be completely eliminated. The more carrier-current frequencies that are sent along one line, the greater the chances of mixing products causing noticeable interference on some channels.

CARRIER FREQUENCY

The carrier frequency of a signal is the average frequency of its carrier wave (*see* CARRIER). For continuous-wave (CW) and amplitude-modulated (AM) signals, the carrier frequency should be as constant as the state of the art will permit. The same is true for the suppressed-carrier frequency of a single-sideband (SSB) signal. For frequency-modulated (FM) signals, the carrier frequency is determined under conditions of no modulation. This frequency should always remain as stable as possible.

In frequency-shift keying (FSK), there are two signal frequencies, separated by a certain value called the carrier shift. The shift frequency is usually 170 Hz for amateur communications, and 425 Hz or 850 Hz for commercial circuits. The space-signal frequency is generally considered to be the carrier frequency in a frequency-shift-keyed circuit.

When specifying the frequency of any signal, the carrier frequency is always given.

CARRIER MOBILITY

In a semiconductor material, the electrons or holes are called carriers (*see* ELECTRON, HOLE). The average speed at which the carriers move, per unit of electric-field intensity, is called the carrier mobility of the semiconductor material. Electric-field units are specified in volts per centimeter or volts per meter. The mobility of holes is usually not the same as the mobility of electrons in a given material. However, the mobility depends on the amount of doping (*see* DOPING). At very high doping levels, the mobility decreases because of scattering of the carriers. As the electric field is made very intense, the mobility also decreases as the speed of carriers approaches a maximum, or limiting, velocity. This limiting velocity is approximately 10^4 to 10^5 meters per second, or 10^6 to 10^7 centimeters per second, in most materials.

The electron and hole mobility for four different semiconductor materials is given in the table. Units are specified in centimeters per second for an electric-field intensity of 1 volt per centimeter. (To determine the drift velocity for an electric-field intensity of E volts per centimeter, the given values must be multiplied by E.)

CARRIER POWER

Carrier power is a measure of the output power of a transmitter for continuous wave (CW), amplitude modulation (AM), frequency-shift keying (FSK), or frequency modulation (FM). Carrier power is measured by an RF wattmeter under conditions of zero modulation, and with the transmitter connected to a load equivalent to its normal rated operating load.

The output power of a CW transmitter is determined under key-down conditions. With FSK or FM, the output power does not change with modulation, but remains constant at all times. With AM, the output power fluctuates somewhat with modulation.

In single-sideband (SSB) or double-sideband, suppressed-carrier operation, there is no carrier and hence no carrier power. Instead, the output power of the transmit-

ELECTRON AND HOLE MOBILITY FOR DIFFERENT SEMICONDUCTOR MATERIALS, IN CENTIMETERS PER SECOND FOR A FIELD INTENSITY OF 1 VOLT PER CENTIMETER.

	Silicon	Germanium	Gallium Arsenide	Indium Antimonide
Electron Mobility	1350*	3900*	6800**	80,000**
Hole Mobility	480*	1900*	680*	4000**

*E. M. Conwell, "Properties of Silicon and Germanium II," *Proc. IRE,* Vol. 46, pp. 1281-1300.
**Bube, "Photoconductivity of Solids," John Wiley & Sons, New York.

ter is measured as the peak-envelope power. *See also* PEAK ENVELOPE POWER, RF POWER.

CARRIER SHIFT

Carrier shift is a method of transmitting a radioteletype signal. The two carrier conditions are called mark and space. While no information is being sent, a steady carrier is transmitted at the space frequency. When a letter, word, or message is sent, the carrier frequency is shifted in pulses by a certain prescribed number of Hertz. In amateur communications, the standard carrier shift is 170 Hz; in commercial and military communications, values of 425 Hz or 850 Hz are generally used, although non-standard carrier-shift values are sometimes employed.

The transmitter output power remains constant under both mark and space conditions. *See also* FREQUENCY-SHIFT KEYING.

CARRIER SUPPRESSION

In a single-sideband (SSB) transmitter, the carrier is eliminated along with one of the sidebands. The carrier is also eliminated in the double-sideband, suppressed-carrier (DSB) mode. The carrier is usually eliminated by phase cancellation in a special modulator called a balanced modulator.

The illustration shows spectral displays of typical amplitude- modulated AM, DSB, and SSB signals. The carrier can never be entirely eliminated, although it is generally at least 60 dB below the level of the peak audio energy. *See also* BALANCED MODULATOR, DOUBLE SIDEBAND, SINGLE SIDEBAND.

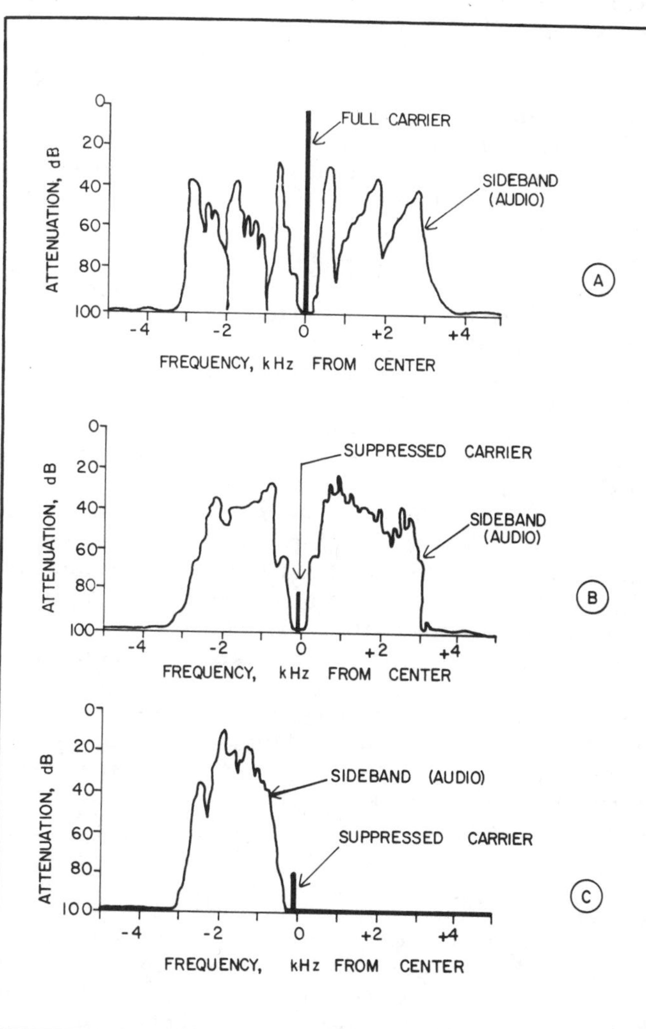

CARRIER SUPPRESSION: At A, an ordinary AM signal is shown as it is seen on a spectrum analyzer. At B, a double- sideband, suppressed-carrier signal; at C, a single-sideband signal. The carrier is eliminated by phase cancellation in the balanced modulator.

CARRIER SWING

In a frequency-modulation system, carrier swing is the total amount of frequency deviation of the signal (*see* DEVIATION). In a standard FM communications circuit, the maximum deviation is plus or minus 5 kHz with respect to the center of the channel; thus the carrier swing is 10 kHz (see illustration).

The term carrier swing is not often used in FM. More often, the deviation is specified. The modulation index is also occasionally specified (*see* MODULATION INDEX).

CARRIER TERMINAL

A carrier terminal is a transmitter and receiver circuit used in a carrier-current communication system (*see* CARRIER-CURRENT COMMUNICATION). The carrier terminal is actually a low-power, radio-frequency transmitter or set of transmitters, and a radio-frequency receiver or set of receivers. The drawing shows a block diagram of a single-frequency carrier terminal.

The transmitter and receiver frequencies are usually different to allow duplex operation. At the far end of the circuit, the transmitting and receiving frequencies are transposed. A carrier terminal is sometimes designed for single-sideband operation; in such a case, transmission may be via one sideband and reception via the other sideband. *See also* DUPLEX OPERATION.

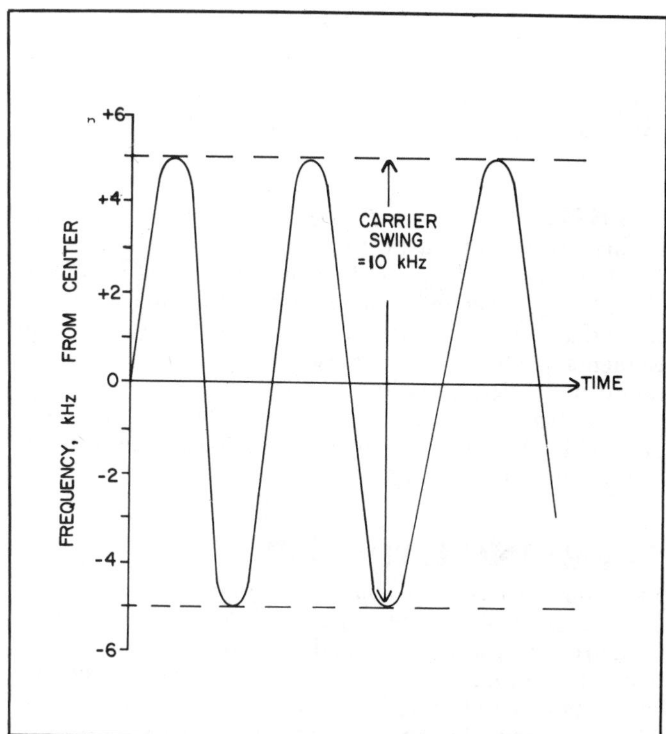

CARRIER SWING: Carrier swing is the frequency difference between the lowest and highest instantaneous frequency in an FM signal.

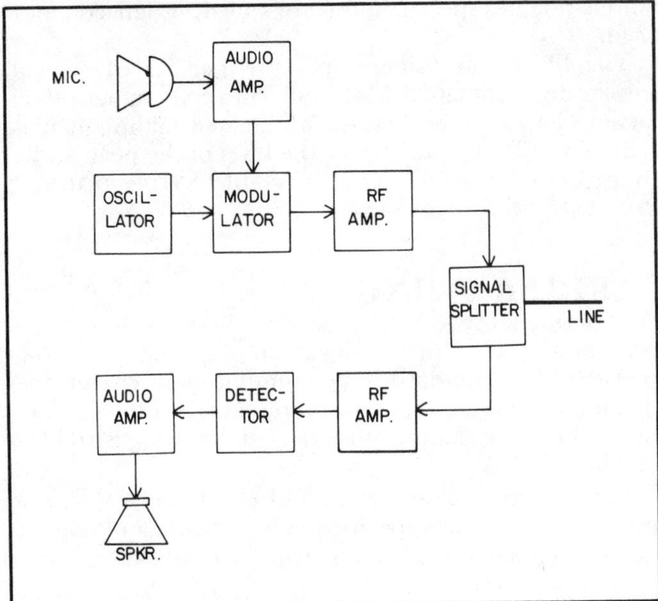

CARRIER TERMINAL: Block diagram of a carrier terminal.

CARRIER VOLTAGE

Carrier voltage is the alternating-current voltage of a transmitter radio-frequency carrier. It may be expressed as a peak voltage, a peak-to-peak voltage, or a root-mean-square (RMS) voltage; the magnitude of the carrier voltage depends on the transmitter output power (carrier power) and on the impedance of the feed line and antenna system. It also depends on the standing-wave ratio along the transmission line.

Carrier voltage can be measured with a radio-frequency voltmeter. If the standing-wave ratio is 1 and the feed-line impedance, Z_O is known, the RMS carrier voltage, E, can be calculated from an RF wattmeter reading P watts according to the formula:

$$E = \sqrt{PZ_O}$$

under conditions of zero transmitter modulation. *See also* CARRIER POWER, ROOT MEAN SQUARE.

CARRY

A carry is a mathematical operation in the addition of numbers. When the sum of a column of single digits exceeds 9 in the decimal system, or 1 in the binary system, some digits must be carried. This operation is taught in elementary school; in a digital computer, the carry operation is done electronically.

In subtraction, a carry is sometimes called a borrow. It is done when the difference of two digits is less than zero. *See also* COMPUTER.

CARRYING CAPACITY

Carrying capacity is the ability of a component to handle a given amount of current or power. Generally, the term is used to define the current-handling capability of wire. The maximum amount of current that can be safely carried by a wire depends on the type and diameter of the metal used. The table gives the carrying capacity for various sizes of copper wire in the American Wire Gauge (AWG) system, for continuous duty in open air.

CARRYING CAPACITY: CONTINUOUS CURRENT RATINGS, IN OPEN AIR, OF SOME AWG SIZES OF COPPER WIRE.

Size, AWG	Current, Amperes
8	73
10	55
12	41
14	32
16	22
18	16
20	11

When the carrying capacity of wire is exceeded, there is a possibility of fire because of overheating. If the carrying capacity of a wire is greatly exceeded, the wire becomes soft and subject to stretching or breakage. It may even melt. This sometimes happens when there is a short circuit and no fuse or circuit breaker. It often happens when lightning strikes a conductor.

CARTESIAN COORDINATES

The Cartesian coordinate system is probably the most commonly-used coordinate system in mathematics and engineering. It is sometimes called the rectangular coordinate system. The most familiar Cartesian system is the two-dimensional coordinate plane. The Cartesian system is named after the French mathematician Rene Descartes, who is credited with inventing it.

The Cartesian plane consists of two axes at right angles, calibrated in units. The units in a true Cartesian plane are marked off in linear fashion, so that a given distance always represents the same number of units along that axis. The scales do not necessarily have to be the same on both axes (see illustration). The axis representing the independent variable is called the abscissa, and the axis representing the dependent variable is called the ordinate. Usually, but not always, the abscissa is drawn horizontally and the ordinate is drawn vertically.

A variation of the Cartesian plane consists of nonlinear scales on one or both axes, as shown at B. This nonlinearity is usually in the form of a logarithmic scale.

Cartesian coordinate systems may be constructed in three or more dimensions. A typical three-dimensional Cartesian system is shown at C. All three axes are mutually perpendicular at their point of intersection. In four or more dimensions, the Cartesian system becomes impossible to visualize, although it can be represented mathematically.

Other types of coordinate systems exist, and are sometimes used. The most common of these is the polar system. Less common are curvilinear systems such as the Smith chart. *See also* CURVILINEAR COORDINATES, POLAR COORDINATES.

CARTRIDGE

A cartridge is an electromechanical transducer in a phonograph. It picks up the vibrations of the stylus, and converts them to electrical impulses, which are then amplified by the audio circuits in the phonograph. The illustration is a pictorial diagram of a typical cartridge.

The term cartridge is sometimes used in place of the term cassette. A cassette is an enclosed package containing magnetic tape. A cassette or cartridge of magnetic tape

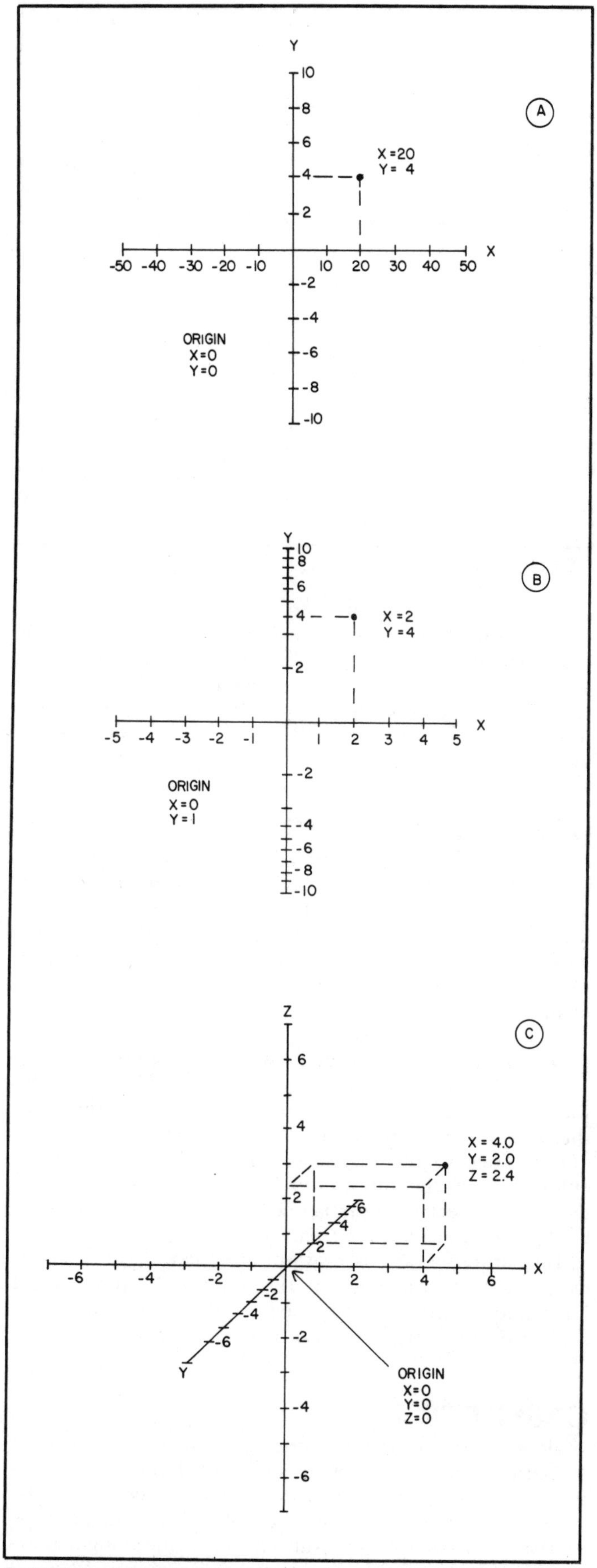

CARTESIAN COORDINATES: At A, a two-dimensional system, showing the point (x,y) = (20.4). At B, the y axis is logarithmic, and the point (x,y) = (2,4) is shown. At C, a three-dimensional Cartesian system, showing the point (x,y,z) = (4.0, 2.0, 2.4).

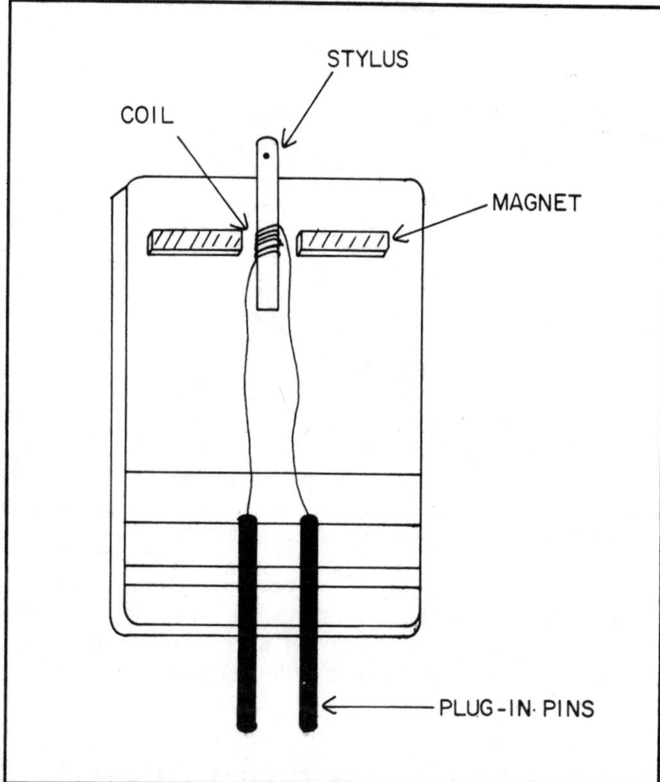

CARTRIDGE: A phonograph cartridge picks up sound vibrations from a record and transforms them into electrical impulses.

contains either an endless loop, or two spools that move in opposite directions when the device is inserted in either of two ways. *See also* CASSETTE.

CARTRIDGE FUSE

A cartridge fuse is a tubular fuse with metal end caps. The end caps make electrical contact with the socket. The illustration shows a cartridge fuse. These fuses are often used in radio apparatus, and in the peripheral equipment in a communications installation.

The cartridge fuse fits into a socket for easy replacement. When the fuse blows, it can clearly be seen; the glass appears darkened, or the internal wire obviously appears broken. Cartridge fuses are available in a variety of voltage and current ratings. *See also* FUSE.

CASCADE

When two or more amplifying stages are connected one after another, the arrangement is called a cascade circuit. Several amplifying stages can be cascaded to obtain much more signal gain than is possible with only one amplifying stage.

There is a practical limit to the number of amplifying stages that can be cascaded. The noise generated by the first stage, in addition to any noise present at the input of the first stage, will be amplified by succeeding stages along with the desired signal. When many stages are connected in cascade, the probability of feedback and resulting oscillation increases greatly.

Tuned circuits and stabilizing networks are used to maximize the number of amplifiers that can be connected in cascade. By shielding the circuitry of each stage, and by making sure that the coupling is not too tight, the chances

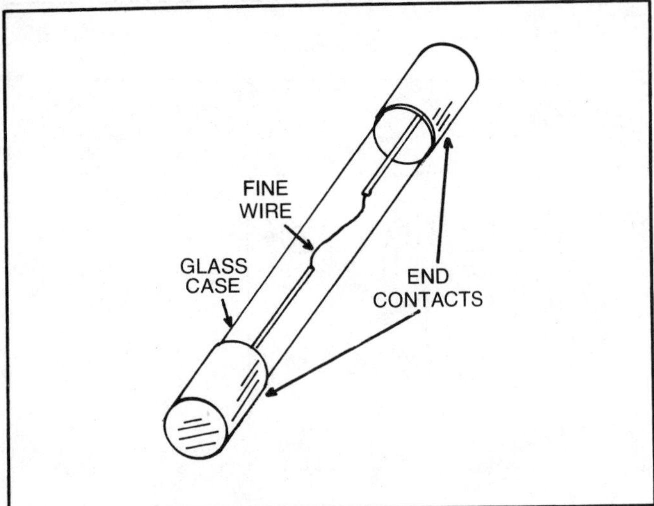

CARTRIDGE FUSE: Construction of a typical cartridge fuse.

of oscillation are minimized. Proper impedance matching ensures the best transfer of signals, with the least amount of noise. When several stages of amplification are connected in cascade, the entire unit is sometimes called a cascade amplifier.

CASCADE VOLTAGE DOUBLER

The cascade voltage doubler is a circuit for obtaining high dc voltages with an alternating-current input. It is a form of power supply, used in tube-type circuits with low to medium current requirements. A schematic diagram of a cascade voltage doubler is illustrated.

Capacitor C1 charges through diode D2 to the peak ac input potential during half of the cycle. During the other half of the cycle, it discharges through D1 to charge C2 to twice the peak ac input potential. Capacitor C2 also serves as a filtering capacitor to smooth out the ripple in the supply output.

Cascade voltage-doubler circuits have the advantage of a common input and output terminal, allowing unbalanced operation. *See also* VOLTAGE DOUBLER.

CASCODE

A circuit capable of obtaining high gain using tubes or transistors, with an excellent impedance match between two amplifying stages, is called a cascode amplifier.

The input stage is a grounded-cathode or grounded-emitter circuit. The plate or collector of the first stage is fed directly to the cathode or emitter of the second stage,

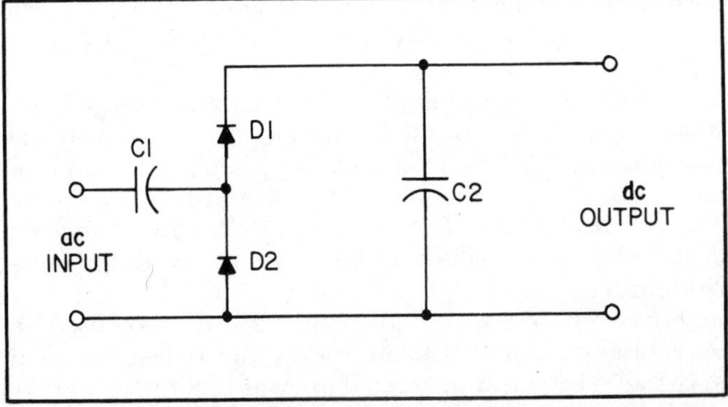

CASCADE VOLTAGE DOUBLER: Operation is discussed in text.

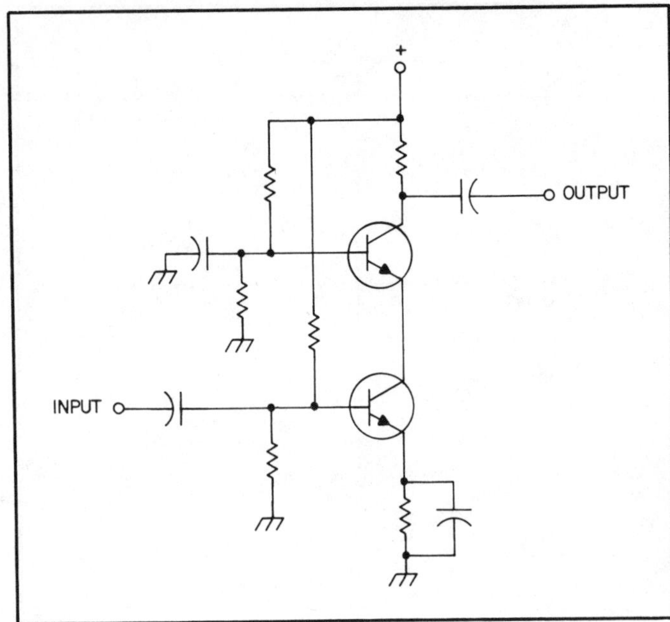

CASCODE: A cascode circuit, using NPN bipolar transistors.

which is a grounded-grid or grounded-base amplifier. The illustration shows a schematic diagram of a cascode amplifier using a pair of NPN bipolar transistors.

Because of the low-noise characteristics of the cascode circuit, it is often used in preamplifiers for the purpose of improving the sensitivity of HF and vhf receivers. *See also* PREAMPLIFIER.

CASSEGRAIN FEED

The Cassegrain arrangement is a method of feeding a parabolic or spherical dish antenna. It completely eliminates the need for mounting the feed apparatus at the focal point of the dish. In the conventional dish feed system, the waveguide terminates at just the right position, so that the RF energy emerges from the dish in a parallel beam (A in the illustration).

In the Cassegrain transmission system, the waveguide terminates at the base of the dish, which is physically more convenient from the standpoint of installation. The waveguide output is beamed to a small convex reflector at the focus of the dish, and the convex reflector directs the energy back to the main dish, shown at B. The beam emerges from the antenna in parallel rays. In reception, the rays of electromagnetic energy follow the same path, but in the opposite direction.

Dish antennas are used at uhf and microwave frequencies, when the diameter of the antenna, and the focal length, are many times the wavelength of the transmitted and received energy. *See also* DISH ANTENNA.

CASSETTE

A device that holds recording tape is called a cassette. Cassettes come in a variety of sizes and shapes for various applications. A typical cassette, used in a portable tape recorder, is shown in the photograph. This particular type of cassette plays for 30 minutes on each side; longer playing cassettes may allow recording for as much as 60 minutes per side. The internal construction of a cassette is also shown.

Some cassettes, such as the kind used in stereo tape

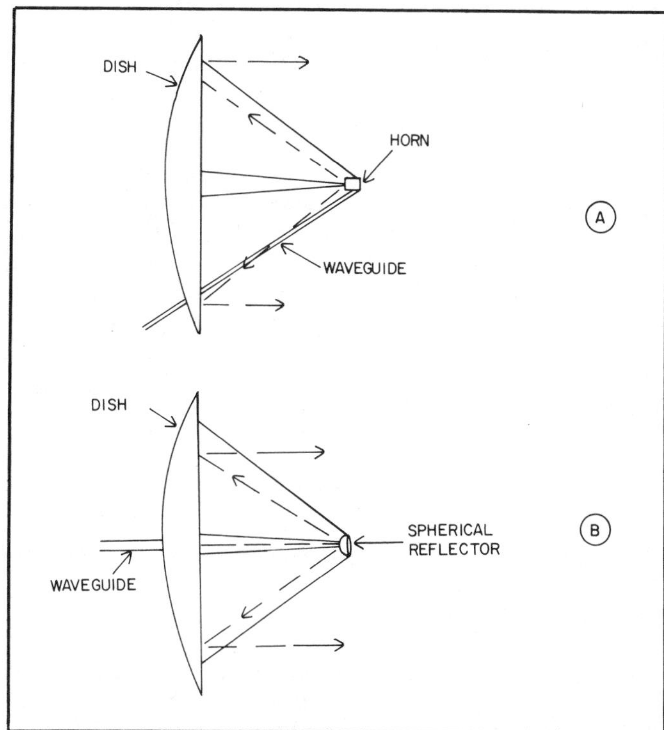

CASSEGRAIN FEED: At A, conventional dish-antenna feed. At B, Cassegrain feed.

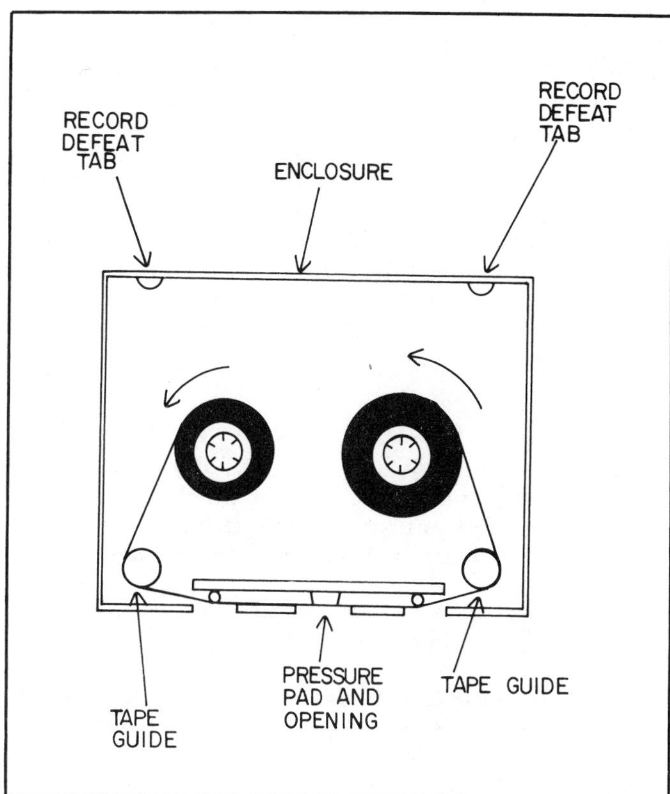

CASSETTE: Internal details of a typical audio cassette.

players, have endless loops of tape. Four recording paths, each with two channels for the left and the right half of the sound track, provide a total of eight recording tracks.

In recent years, cassette recorders have been increasingly used in video applications (*see* VIDEOTAPE RECORDER).

While cassettes offer convenience, some recording enthusiasts still prefer, and use, reel-to-reel recorders. *See also* RECORDING TAPE, REEL-TO-REEL TAPE RECORDER.

CATHODE

The cathode is the electron-emitting electrode in a vacuum tube. To enhance its emission properties, the cathode of a tube is heated to a high temperature by a wire called the heater (*see also* FILAMENT).

There are two types of cathode in common use. The directly-heated cathode consists of a specially treated filament, which itself emits electrons when connected to a potential that is negative with respect to the anode or plate of the tube. The indirectly-heated cathode is a cylinder of metal, separated from the heater but physically close to it.

Directly-heated cathodes require a direct current for heating in amplifier and oscillator applications, since alternating current would produce ripple in the output signal. The indirectly-heated cathode can make use of an alternating-current filament supply; the cathode acts as a shield against the ripple emission from the filament.

It is easy to tell from a schematic diagram whether a tube has a directly-heated cathode or an indirectly-heated cathode. The drawing shows schematic representations of both types.

In a bipolar transistor, the equivalent of the cathode is the emitter. In a field-effect transistor, the equivalent of the cathode is the source. *See also* EMITTER, SOURCE, TUBE.

CASSETTE: A tape-recorder cassette.

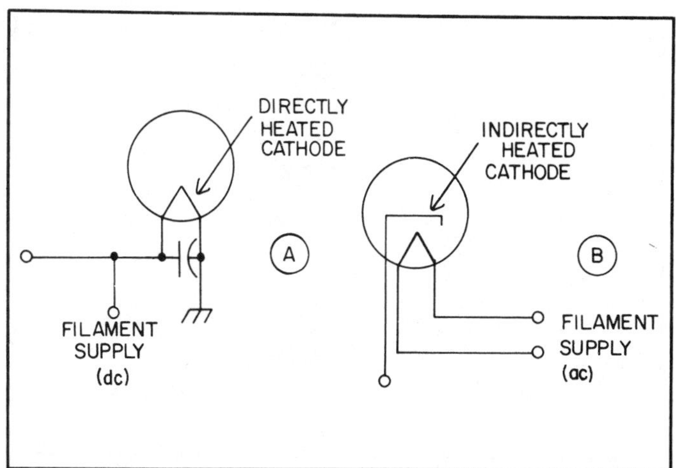

CATHODE: At A, schematic symbol for directly heated tube cathode. At B, symbol for indirectly heated cathode.

CATHODE BIAS

Cathode bias is the direct-current potential at which the cathode of a tube is placed with respect to ground, or with respect to the negative side of the power supply.

Cathode bias usually is obtained by means of a resistor shunted by a capacitor, as shown in the drawing. It may also be obtained by means of a separate power supply, or a semiconductor diode. Cathode bias provides stability and current limiting in a tube amplifier or oscillator. It also provides an easy means of making the dc grid voltage negative with respect to the cathode. Most tubes are operated in this way. *See also* TUBE.

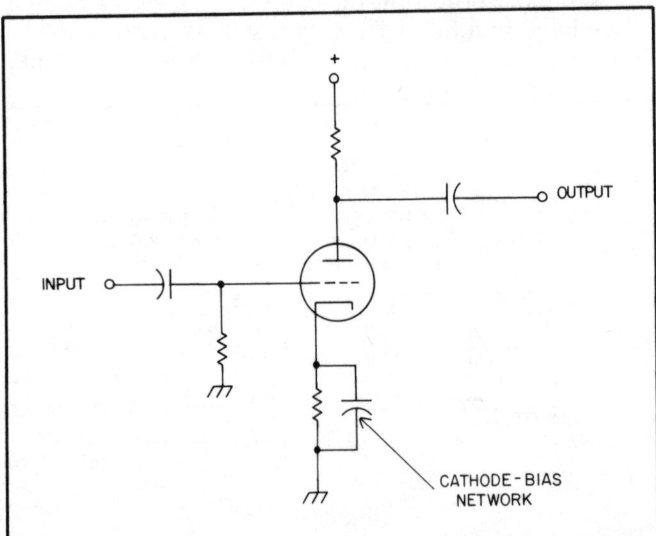

CATHODE BIAS: Cathode bias may be achieved by a resistance and capacitance, as shown here. Occasionally, a battery is used for cathode bias.

CATHODE COUPLING

Cathode coupling is a method of connecting one vacuum-tube circuit to another, or of providing the output from a tube amplifier to an external circuit. The amplifier output is taken from the cathode circuit, as shown in the illustration.

The bipolar-transistor equivalent of cathode coupling is emitter coupling. The field-effect-transistor equivalent is source coupling.

Cathode coupling provides a low output impedance. The circuit shown is called a cathode follower, and is used for the purpose of matching a high impedance to a low impedance. *See also* CATHODE FOLLOWER, EMITTER COUPLING, SOURCE COUPLING.

CATHODE FOLLOWER

The cathode follower is a tube circuit characterized by high input impedance, and gain less than unity. The output impedance is low. The cathode follower is often used as an impedance-matching circuit for broadband applications. A schematic diagram of a cathode follower is illustrated. The output is taken across part or all of the cathode resistance, depending on the output impedance desired.

In the cathode-follower circuit, the output voltage is in phase with the input voltage. Thus the cathode "follows" the input cycle, and this is how the circuit gets its name. The bipolar-transistor equivalent circuit is called the emitter follower, and the field-effect-transistor equivalent is called the source follower. These circuits have characteristics similar to the cathode follower. *See also* EMITTER FOLLOWER, SOURCE FOLLOWER.

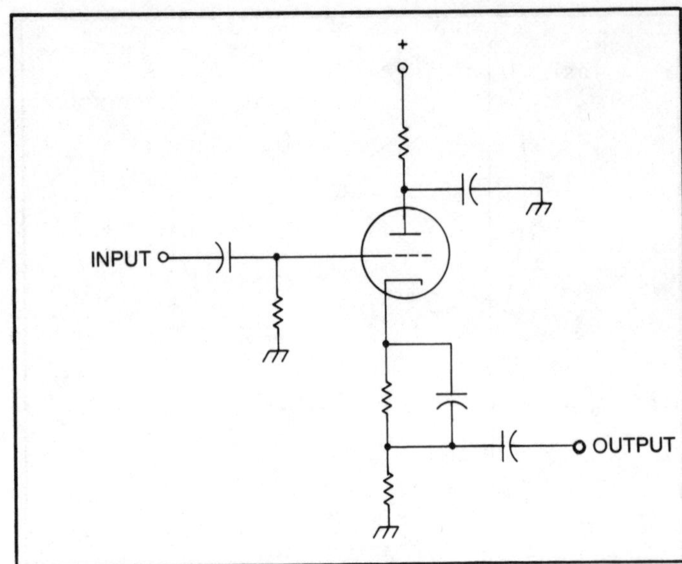

CATHODE FOLLOWER: A cathode follower circuit, using cathode coupling. The output is taken from the cathode circuit.

CATHODE KEYING

In older, tube-type code transmitters, keying was often accomplished by breaking the cathode circuit of one of the amplifying stages. Usually, this was done in a stage prior to the driver. The drawing shows an example of cathode keying. By cutting off the current-through the tube, signal transfer is prevented while the key is up.

The main difficulty with cathode keying is that a fairly high voltage is present at the key terminals. This can cause damage to manual telegraph keys because of sparking that takes place when contact is broken. It can also result in electrical shocks to operators of the equipment. Because of these drawbacks, cathode keying is rarely used in tube-type code transmitters today. A much more common method of keying a tube amplifier is the grid-block method (*see* GRID-BLOCK KEYING).

The equivalent of cathode keying in a bipolar-transistor circuit is emitter keying; in a field-effect transistor circuit,

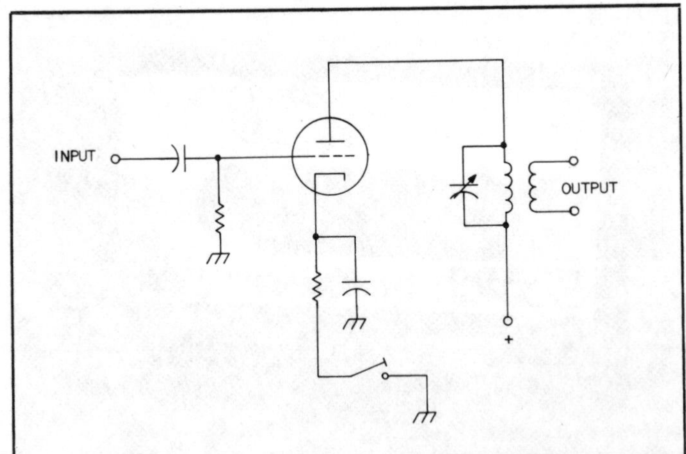

CATHODE KEYING: Cathode keying of an amplifier.

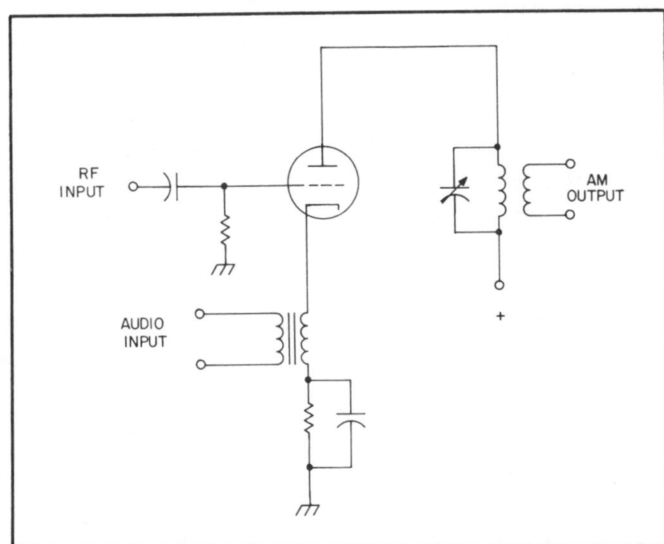

CATHODE MODULATION: In cathode modulation, the audio energy is applied at the cathode circuit and the radio-frequency signal is fed into the grid circuit.

it is source keying. Emitter and source keying are more practical than cathode keying, because of the lower voltages used in solid-state circuits. *See also* EMITTER KEYING, SOURCE KEYING.

CATHODE MODULATION

Cathode modulation is a method of obtaining amplitude modulation by applying audio-frequency energy to the cathode circuit of a vacuum tube. The radio-frequency energy is applied to the grid circuit, just as in an ordinary grounded-cathode amplifier. A schematic diagram of a circuit that uses cathode modulation to obtain AM is shown.

The audio-frequency energy applied to the cathode causes the gain of the amplifier to fluctuate. At positive peaks of the audio signal, the gain is minimum, and at negative peaks it is maximum. When the positive peaks cause the gain to drop to zero for just an instant, 100 percent modulation is achieved. Greater levels of audio input will cause overmodulation.

The bipolar-transistor equivalent of cathode modulation is emitter modulation. In a field-effect-transistor circuit, the equivalent of cathode modulation is source modulation. *See also* AMPLITUDE MODULATION, EMITTER MODULATION, SOURCE MODULATION.

CATHODE-RAY OSCILLOSCOPE

See OSCILLOSCOPE.

CATHODE-RAY TUBE

A cathode-ray tube is a device for obtaining a graphic display of an electronic function, such as a waveform or spectral display. A television picture tube is a cathode-ray tube, but in the electronics laboratory, the oscilloscope is the most familiar example of the use of a cathode-ray tube. Cathode-ray tubes are also used as video monitoring devices for computers and word processors.

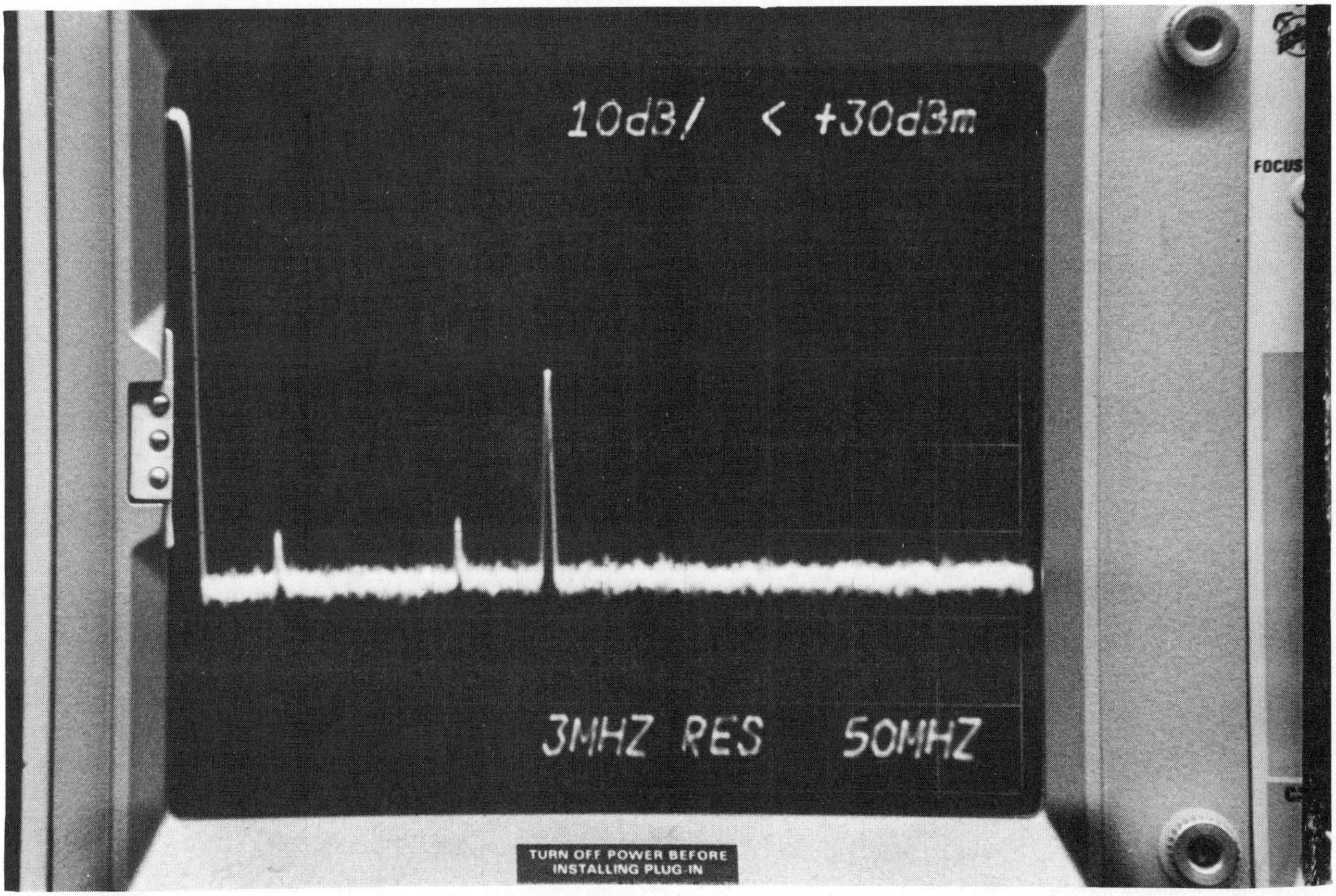

CATHODE-RAY TUBE: A cathode-ray-tube spectrum analyzer display.

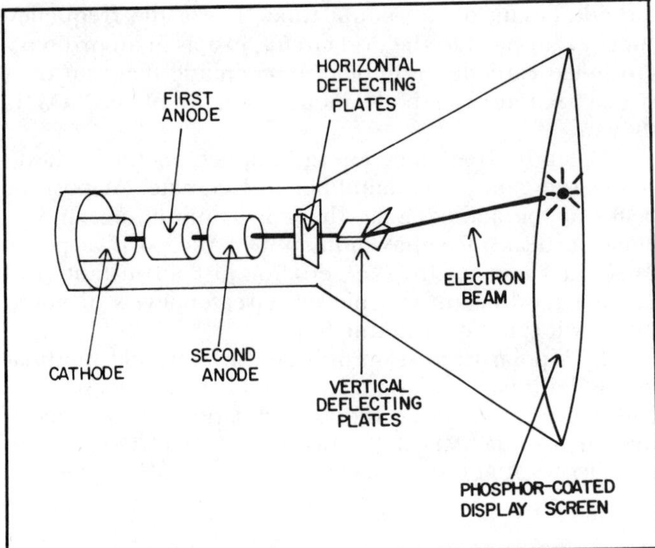

CATHODE-RAY TUBE: Internal details of a cathode-ray tube.

Oscilloscopes may be used in a variety of different ways. The spectrum-analyzer display, as shown in the first illustration, uses a cathode-ray tube to obtain a function of signal amplitude vs frequency. Other common oscillscope functions include amplitude vs time and frequency vs time. The cathode-ray-tube screen acts as a Cartesian plane (*see* CARTESIAN COORDINATES).

Also shown are the internal details and operation of a cathode-ray tube. The cathode, or electron gun, emits a stream of electrons. The first anode focuses the electrons into a narrow beam and accelerates them to greater speed. The second anode gives the electrons still more speed. The deflecting plates control the location at which the electron beam strikes the screen. The inside of the viewing screen is coated with phosphor material that glows when the electrons hit it.

The cathode-ray tube is capable of deflecting an electron beam at an extreme rate of speed. Some cathode-ray tubes can show waveforms at frequencies of hundreds of megahertz. *See also* OSCILLOSCOPE.

CAT'S WHISKER

Some semiconductor diodes have a tiny wire that makes contact with the semiconductor material. This can be seen in many kinds of glass-encased diodes. The illustration shows a cat's whisker. Some transistors also use this technique. Such diodes and transistors are called point-contact devices.

The earliest radio receivers used a crystal of galena or other semiconductor material, and a thin wire which had to be experimentally positioned for the best reception. A receiver consisting of only an antenna, tuned circuit, crystal detector, and headphones will actually operate in the presence of strong radio signals, without any amplification stages, and without the need for a source of power. Such a radio is called a crystal set. *See also* CRYSTAL DETECTOR, CRYSTAL SET.

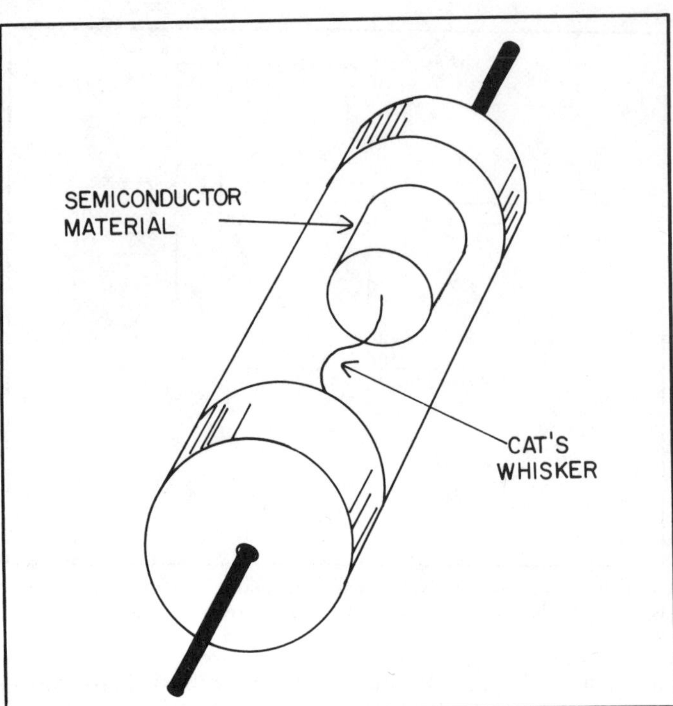

CAT'S WHISKER: A cat's whisker is a thin wire that makes contact with a wafer of semiconductor material to form a diode.

CAVITY FREQUENCY METER

A cavity frequency meter makes use of a cavity resonator (*see* CAVITY RESONATOR) to measure wavelengths at very high frequencies and above. An adjustable cavity resonator, and a radio-frequency voltmeter or ammeter, are connected to a pickup wire, or loosely coupled to the circuit under test. The cavity resonator is adjusted until a peak occurs in the meter reading. The free-space wavelength is easily measured and converted to the srequency according to the formula:

$$F = 300/\lambda$$

where f is the frequency in megahertz and λ is the wavelength in meters.

A cavity resonator is generally tuned to a length of ½ wavelength to obtain resonance in the cavity frequency meter. However, any multiple of ½ wavelength, such as 1, 1½, 2, and so on, will allow a resonant reading.

CAVITY RESONATOR

A cavity resonator is a metal enclosure, usually shaped like a cylinder or rectangular prism (see illustration). Cavity resonators operate as tuned circuits, and are practical for use at frequencies above about 200 MHz.

A cavity has an infinite number of resonant frequencies. When the length of the cavity is equal to any integral multiple of ½ wavelength, electromagnetic waves will be reinforced in the enclosure. Thus, a cavity resonator has a fundamental frequency and a theoretically infinite number of harmonic frequencies. Near the resonant frequency or any harmonic frequency, a cavity behaves like a parallel-tuned inductive-capacitive circuit. When the cavity is slightly too long, it shows inductive reactance, and when it is slightly too short, it displays capacitive reactance. At resonance, a cavity has a very high impedance and theoretically zero reactance. Cavity resonators are sometimes used to measure frequency (*see* CAVITY FREQUENCY METER).

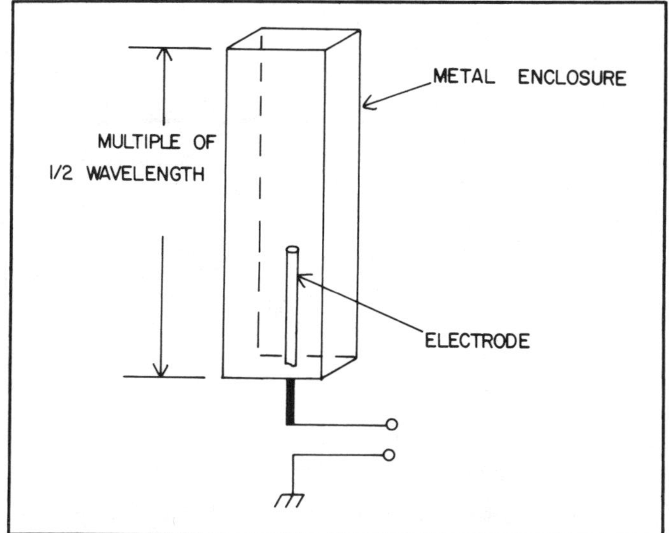

CAVITY RESONATOR: A cavity resonator, consisting of a section of rectangular waveguide.

The resonant frequency of a cavity is affected by the dielectric constant of the air inside. Temperatures and humidity variations therefore have some effect. If precise tuning is required, the temperature and humidity must be kept constant. Otherwise, a resonant cavity may drift off resonance because of changes in the environment.

A length of coaxial cable, short-circuited at both ends, is sometimes used as a cavity resonator. Such resonators can be used at lower frequencies than can rigid metal cavities, since the cable does not have to be straight. The velocity factor of the cable must be taken into account when designing cavities of this kind. *See also* VELOCITY FACTOR.

C BATTERY

A C battery is the grid-bias supply for a portable vacuum-tube device. Generally, the grid of a tube is negative with respect to the cathode; this negative voltage may be derived by means of a C battery, especially in class-B and class-C amplifiers (*see* CLASS-B AMPLIFIER, CLASS-C AMPLIFIER). The grid-bias supply in non-portable tube-operated equipment is called the C voltage or C-minus voltage.

Most modern portable equipment uses solid-state rather than tube-type devices. Thus, the C battery is not often seen. *See also* TUBE.

CELL

A cell is a voltage-producing device, usually operating via a chemical reaction. Two electrodes are placed in an electrolyte solution or paste, and a potential difference develops between the electrodes. Cells vary greatly in physical size. The photograph shows a common flashlight cell and a camera cell. Chemical cells usually develop about 1.5 volts. The ampere-hour capacity and current-delivering capacity depend mostly on the physical size of a cell.

When several cells are connected in series to obtain greater voltage, the combination is called a battery. Cells are often erroneously called batteries. The two most common types of dry-chemical cell are the alkaline and zinc-carbon cell (*see* ALKALINE CELL, ZINC CELL). Automotive batteries make use of lead-acid cells (*see* LEAD-ACID BATTERY). Certain kinds of semiconductor devices generate voltage when light strikes them; these devices are sometimes called cells (*see* PHOTOVOLTAIC CELL).

CELL: Commonly used cells.

CELLULAR TELEPHONE SYSTEM

A cellular telephone system, also sometimes called cellular radio, is a special form of mobile telephone that has been developed in recent years.

A cellular system consists of a network of repeaters, all connected to one or more central office switching systems (*see* CENTRAL OFFICE SWITCHING SYSTEM, REPEATER). The individual subscribers are provided with radio transceivers operating at very-high or ultra-high frequencies.

The network of repeaters is such that most places are always in range of at least one repeater; ideally, every geographic point in the country would be covered. As a subscriber drives a vehicle, operation is automatically switched from repeater to repeater, as shown in the illustration.

Eventually, most if not all telephone communication may be by cellular radio. Worldwide communication of high quality and low cost, using entirely wireless modes, might be achieved by the end of the twentieth century. *See also* MOBILE TELEPHONE, TELEPHONE.

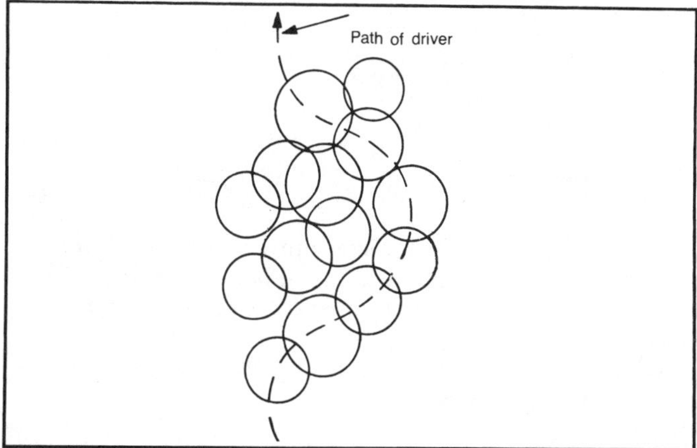

CELLULAR TELEPHONE SYSTEM: Repeater ranges are shown by circles and a hypothetical driving route by the dotted line. Ideally, the driver is always within range of at least one repeater.

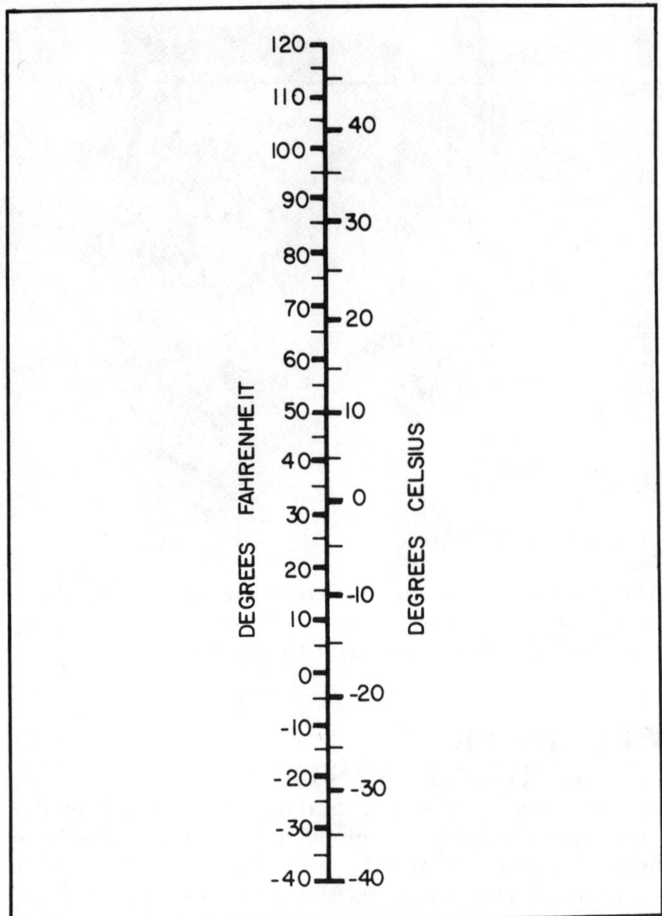

CELSIUS TEMPERATURE SCALE: Nomograph of Fahrenheit vs Celsius temperatures. At −40 degrees, both scales agree. Water freezes at 0°C and boils at 100°C.

CELSIUS TEMPERATURE SCALE

The Celsius temperature scale is a scale at which the freezing point of pure water at one atmosphere is assigned the value zero degrees, and the boiling point of pure water at one atmosphere is assigned the value 100 degrees. The Celsius scale was formerly called the Centigrade scale. The word Celsius is generally abbreviated by the capital letter C.

Temperatures in Celsius and Fahrenheit are related by the equations:

$$C = \frac{5}{9}(F - 32)$$

$$F = \frac{9}{5}C + 32$$

where C represents the Celsius temperature and F represents the Fahrenheit temperature. The nomograph aids in quick conversion between temperature readings in the two scales.

Celsius temperature is related to Kelvin, or absolute, temperature by the equation:

$$K = C + 273$$

where K represents the temperature in degrees Kelvin. A

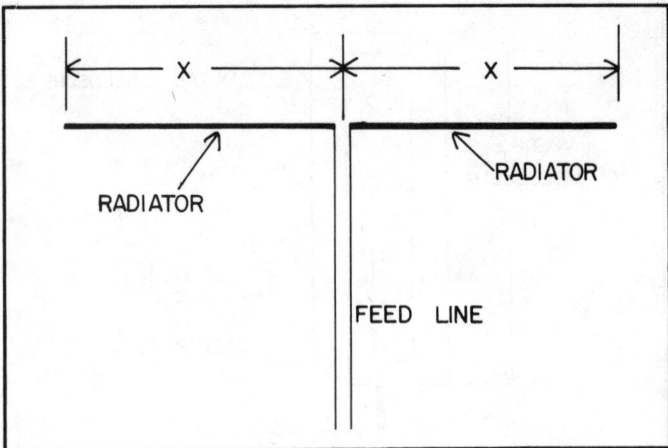

CENTER FEED: Center feed is geometrically symmetrical.

temperature of −273 degrees Celsius is called absolute zero, the coldest possible temperature. *See also* FAHRENHEIT TEMPERATURE SCALE, KELVIN TEMPERATURE SCALE.

CENTER FEED

When an antenna element, resonant or non-resonant, is fed at its physical center, the antenna is said to have center feed. Usually, such an antenna is a half-wave dipole, the driven element of a Yagi, or one of the elements of a phased array.

Center feed results in good electrical balance when a two-wire transmission line is used, (see drawing), whether the element is ½ wavelength or any other length, provided the two halves of the antenna are at nearly equal distances from surrounding objects such as trees, utility wires, and the ground.

For antenna elements measuring an odd multiple of ½ wavelength, center feed results in a purely resistive load impedance between approximately 50 and 100 ohms. For antenna elements measuring an even number of half wavelengths, center feed results in a purely resistive load

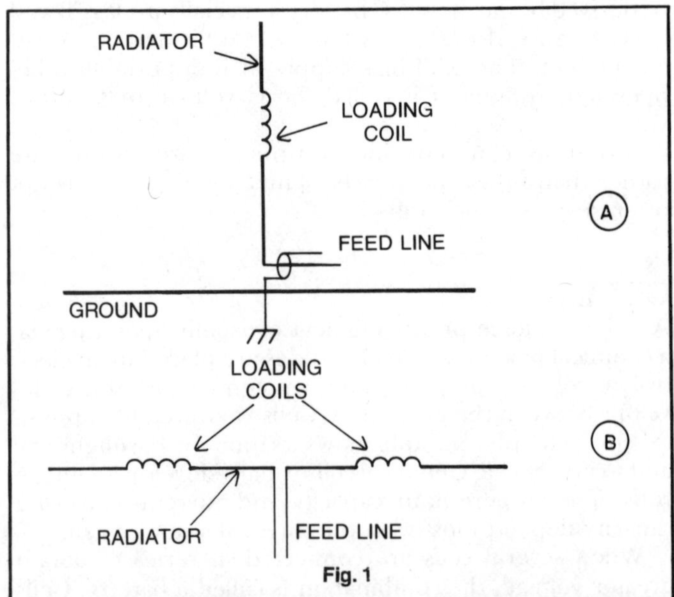

CENTER LOADING: Two versions of inductive center loading. At A, center loading of a short vertical radiator allows resonance at a height shorter than ¼ wavelength. At B, an antenna less than ½ wavelength is resonated with center feed.

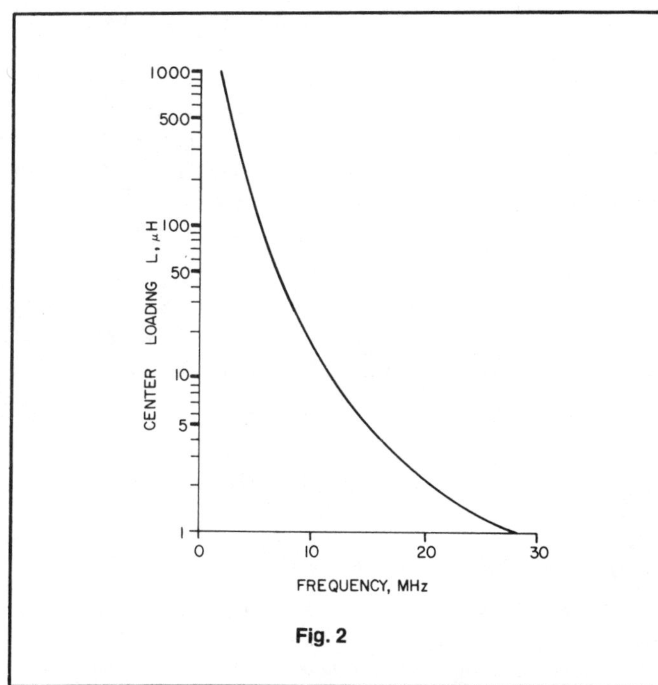

CENTER LOADING: Center-loading inductance in microhenrys vs frequency in megahertz for an 8-foot vertical whip operated against perfectly conducting ground.

impedance ranging from several hundred to several thousand ohms. If the element is not an integral multiple of ½ wavelength, reactance is present at the load in addition to resistance.

Antennas do not necessarily have to be center-fed to work well. *See also* END FEED, OFF-CENTER FEED.

CENTER LOADING

Center loading is a method of altering the resonant frequency of an antenna radiator. An inductance or capacitance is placed along the physical length of the radiator, roughly halfway between the feed point and the end. Figure 1 shows center loading of a vertical radiator fed against ground (A) and a balanced, horizontal radiator (B).

Inductances lower the resonant frequency of a radiator having a given physical length. Generally, for quarter-wave resonant operation with a radiator less than ¼ wavelength in height, some inductive loading is necessary to eliminate the capacitive reactance at the feed point. For quarter-wave resonant operation with a radiator between ¼ and ½ wavelength in height, a capacitor must be used to eliminate the inductive reactance at the feed point.

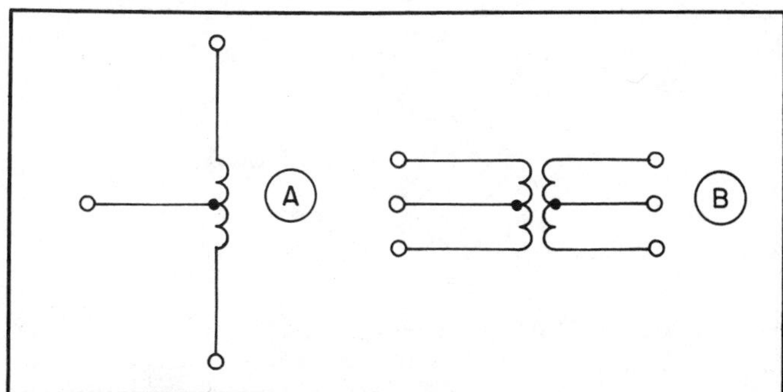

CENTER TAP: At A, in a single coil; at B, in a transformer.

An 8-foot mobile whip antenna can be brought to quarter-wave resonance by means of inductive center loading at all frequencies below its natural quarter-wave resonant frequency, which is about 29 MHz. Figure 2 gives the value of loading inductance L, in microhenrys (uH), as a function of the frequency in megahertz (MHz), for center loading of an 8-foot radiator over perfectly conducting ground. While the RF ground in a mobile installation is not anything near perfect, the values given are close enough to be of practical use.

When the inductor or capacitor in an antenna loading scheme is placed at the feed point, the system is called based loading. *See also* BASE LOADING.

CENTER TAP

A center tap is a terminal connected midway between the ends of a coil or transformer winding. The schematic symbols for center-tapped inductors and transformers are shown in the illustration.

In an inductor, a center tap provides an impedance match. A center-tapped inductor can be used as an autotransformer (*see* AUTOTRANSFORMER) at audio or radio frequencies.

A transformer with a center-tapped secondary winding is often used in power supplies to obtain full-wave operation with only two rectifier diodes. Audio transformers having center-tapped secondary windings are used to provide a balanced output to speakers. The center tap is grounded, and the ends of the winding are connected to a two-wire line. At radio frequencies, center-tapped output transformers also provide a means of obtaining a balanced feed system.

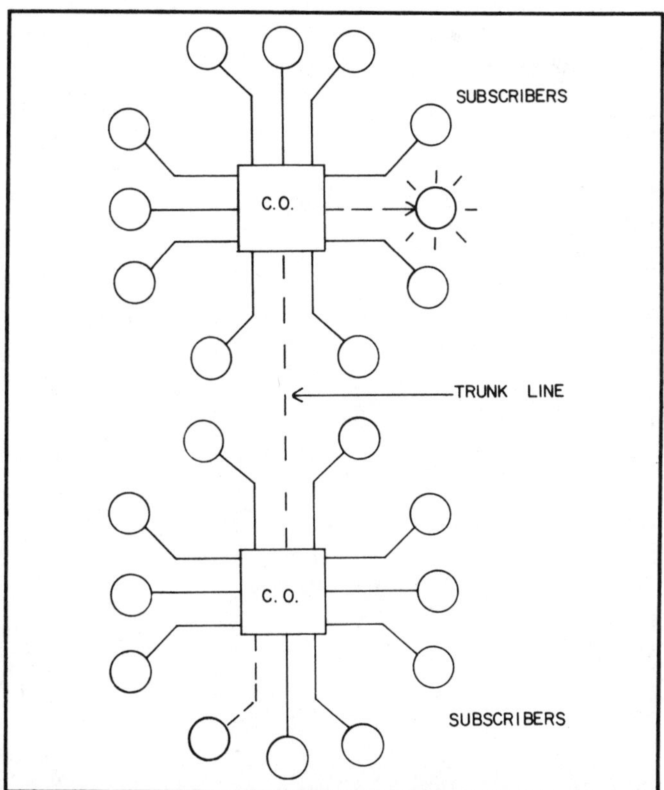

CENTRAL OFFICE SWITCHING SYSTEM: The central-office switching system connects the individual telephone subscribers to the trunk line.

CENTRAL OFFICE SWITCHING SYSTEM

In a telephone network, the station at which all subscriber lines come together is called the central office. There are, of course, many different possible interconnections; a large central office requires extensive switching apparatus to enable any customer to call any other at any time. Modern central-office switching systems are computer-controlled.

Each local area usually has its own central office. For long-distance communication, one central office is connected to another by means of a trunk line (*see* TRUNK LINE). A call thus originates on one subscriber line, goes to a central office, is routed along a trunk line to another central office (if the called line is connected to a different central office), and finally arrives at the distant subscriber line (see illustration). This entire process can be carried out in seconds without the aid of a human operator. *See also* TELEPHONE.

CENTRAL PROCESSING UNIT

The central processing unit, or central processor, is the part of a computer that coordinates the operation of all systems. The abbreviation, CPU, is often used when referring to the central processing unit.

Any computer-operated device has a CPU. Telephone switching networks, communications equipment, and many other electronic systems are coordinated by a CPU. Sometimes a microcomputer is called a CPU. *See also* COMPUTER, MICROCOMPUTER, MICROPROCESSOR.

CENTRIFUGAL FORCE

When an object is revolving around a central point, or changes its direction in any way, a force is introduced in a direction opposite to the direction of the acceleration. This force is sometimes called centrifugal force. In circular motion, the acceleration is directed inward, and thus the centrifugal force occurs outward, away from the center of revolution (see drawing). The greater the magnitude of the acceleration, the greater the centrifugal force.

The force F, in kilogram-meters per-second-per-second (kgm/s^2), is calculated according to the formula:

$$F = ma$$

where m is the mass of the object in kilograms, and a is the acceleration in meters per-second-per-second. The acceleration of a revolving body is determined by the formula:

$$a = 4\pi r^2/T^2$$

where r is the radius of revolution in meters, T is the time in seconds for one revolution at constant speed.

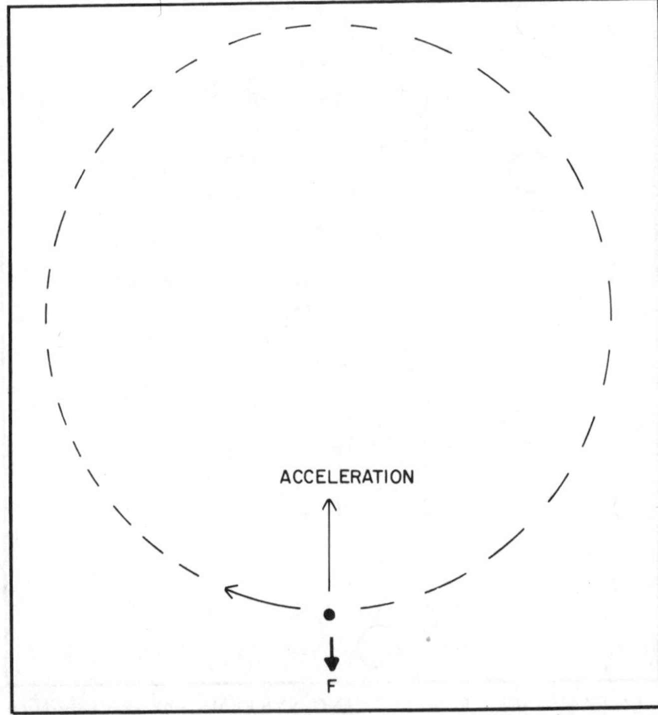

CENTRIFUGAL FORCE: Centrifugal force occurs in a direction opposite to the acceleration in circular motion.

CENTRIFUGAL SWITCH

A centrifugal switch is a switch that is actuated by centrifugal force. An acceleration switch is used for this purpose, since centrifugal force is actually acceleration force. When the acceleration reaches a certain intensity, the switch changes state. *See also* ACCELERATION SWITCH.

CERAMIC

Ceramic is a manufactured compound consisting of aluminum oxide, magnesium oxide, and other similar materials. It is a white, fairly lightweight, solid with a dull surface. Materials such as steatite and barium titanate are ceramics. A polarized crystal is sometimes called a ceramic crystal. The most familiar example of a ceramic material is porcelain.

Ceramics are used in a wide variety of electronic applications. Some kinds of capacitors employ a ceramic material as the dielectric. Certain inductors are wound on ceramic forms, since ceramic is an excellent insulator and is physically strong. Ceramics are used in the manufacture of certain types of microphones and phonographic cartridges. Some bandpass filters, intended for use at radio frequencies, have resonant ceramic crystals or disks. Ceramic materials are used in the manufacture of some kinds of vacuum tubes. *See also* CERAMIC CAPACITOR, CERAMIC FILTER, CERAMIC MICROPHONE, CERAMIC PICKUP, CERAMIC RESISTOR.

CERAMIC CAPACITOR

A ceramic capacitor is a device consisting of two metal plates, usually round in shape, attached to opposite faces

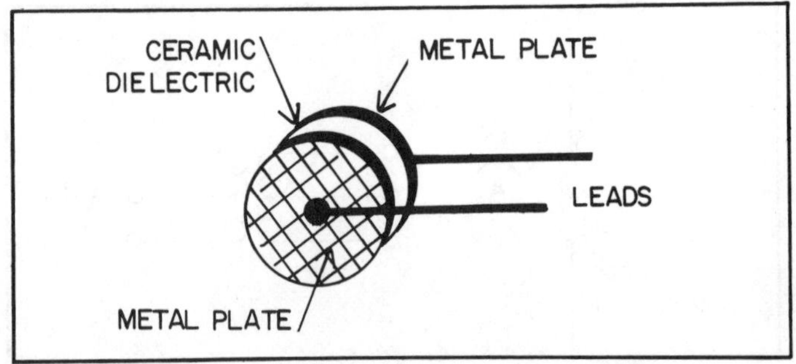

CERAMIC CAPACITOR: The protective coating is not shown.

of a ceramic disk, as shown in the illustration. The value of capacitance depends on the size of the metal plates and on the thickness of the ceramic dielectric material. Ceramic capacitors are generally available in sizes ranging from about 0.5 picofarad to 0.5 microfarad. Ceramic capacitors have voltage ratings from a few volts to several hundred volts.

The composition of the ceramic material determines the temperature coefficient of the capacitor (*see* TEMPERATURE COEFFICIENT). Ceramic capacitors are used from low frequencies up to several hundred megahertz. At higher frequencies, the ceramic material begins to get lossy, and this results in inefficient operation.

Ceramic capacitors are frequently used in high-frequency communications equipment. They are relatively inexpensive, and have a long operating life provided they are not subjected to excessive voltage. *See also* CERAMIC.

CERAMIC FILTER

A ceramic filter is a form of mechanical filter (*see* MECHANICAL FILTER) that makes use of piezoelectric ceramics to obtain a bandpass response. Ceramic disks are arranged as shown in the illustration. These disks resonate at the filter frequency. A ceramic filter is essentially the same as a crystal filter in terms of construction; the only difference is the composition of the disk material.

Ceramic filters are used to provide selectivity in the intermediate-frequency sections of transmitters and receivers. When the filters are properly terminated at their input and output sections, the response is nearly rectangular. *See also* BANDPASS RESPONSE, CRYSTAL-LATTICE FILTER, MECHANICAL FILTER, RECTANGULAR RESPONSE.

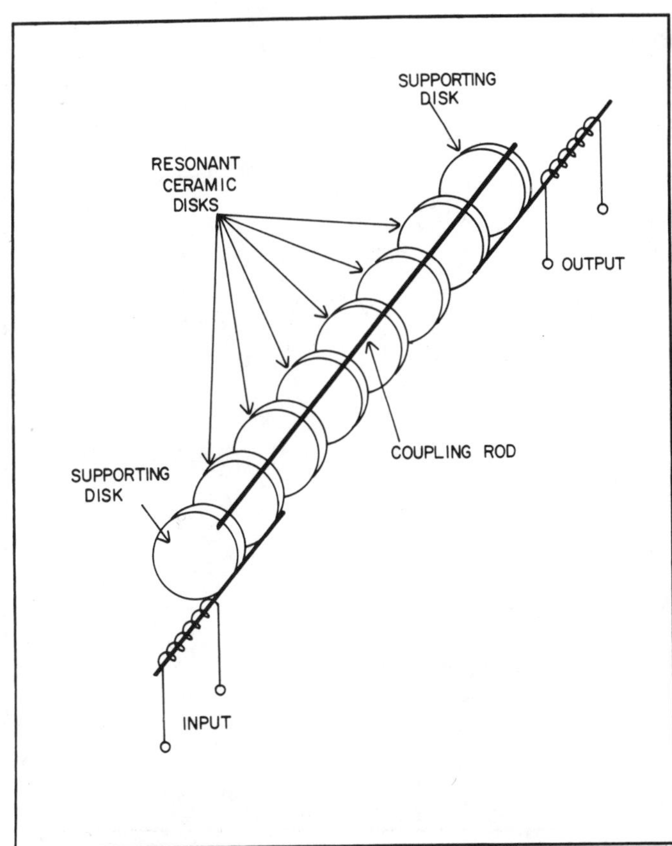

CERAMIC FILTER: Construction of a ceramic filter.

CERAMIC MICROPHONE

A ceramic microphone uses a ceramic cartridge to transform sound energy into electrical impulses. Its construction is similar to that of a crystal microphone (*see* CRYSTAL MICROPHONE). When subjected to the stresses of mechanical vibration, certain ceramic materials generate electrical impulses.

Ceramic and crystal microphones must be handled with care, since they are easily damaged by impact. Ceramic microphones display a high output impedance and excellent audio-frequency response. *See also* CERAMIC.

CERAMIC PICKUP

A ceramic pickup is a cartridge for phonograph use (*see* CARTRIDGE), in which a ceramic crystal is used to translate the vibrations of the stylus into electrical impulses. It operates according to the piezoelectric effect, in the same way that a ceramic microphone works. An artificially polarized ceramic crystal generates electrical impulses over a wide range of frequencies. A ceramic pickup is somewhat more durable than a conventional crystal type. The ceramic pickup has excellent sound-reproduction quality and excellent dynamic range. Sometimes a ceramic pickup is called a crystal pickup, in spite of the technical differences between the two. *See also* CERAMIC, CRYSTAL TRANSDUCER.

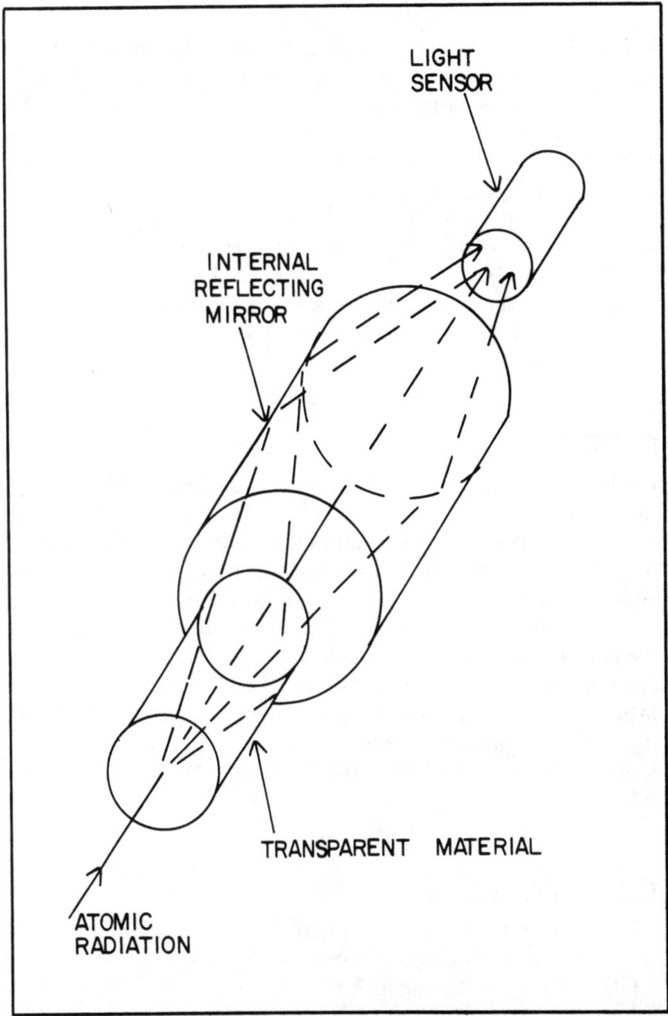

CERENKOV RADIATION: A Cerenkov counter. The atomic radiation produces visible light as it passes through the transparent material; the intensity of the light is then measured.

CERAMIC RESISTOR

A ceramic resistor is a device intended for limiting the current flowing in a circuit. It is made from carborundum, which is a compound of carbon and silicon. The value of resistance of a ceramic resistor decreases as the voltage across the component increases. Ceramic resistors can be obtained with either positive or negative temperature coefficients. A positive temperature coefficient means that the resistance increases with increasing temperature; a negative temperature coefficient means that the resistance decreases with increasing temperature.

Ceramic resistors generate less electrical noise, as a current passes through them, than ordinary carbon resistors generate. Ceramic resistors are available in fixed or variable form. They have a long service life.

CERENKOV RADIATION

When charged atomic particles, such as protons, electrons, or alpha particles, pass through a transparent substance at a speed greater than the speed of light in that material, Cerenkov radiation is produced. In a pool reactor, Cerenkov radiation appears as a bluish glow. A Russian scientist is responsible for orignally explaining this glow, and it is therefore named after him.

The Cerenkov effect is used to measure the intensity of atomic particle radiation in a device called a Cerenkov counter (see illustration). Atomic particles are passed through a transparent dielectric material in which the speed of the particles is greater than the speed of light. (This occurs because of the index of refraction of the material to visible light.) Cerenkov radiation, produced in the form of visible light, is measured by a photomultiplier or other light-sensitive device.

When electrons are accelerated through a hole in a dielectric material, and given just the right speed and modulation, a similar effect occurs. Radiation is produced, and it is called Cerenkov radiation. A device for producing this kind of Cerenkov radiation is called a Cerenkov rebatron radiator.

CESIUM

Cesium is an element. Its atomic number is 55, and its atomic weight is 133. The oscillation frequency of cesium is often used in atomic time standards. Cesium is also used as a "getter" in vacuum tubes, to eliminate any residual gases that might remain after the tube has been evacuated.

The nominal resonance frequency of a cesium-beam oscillator, used in atomic clocks, is 9192.631770 MHz. This frequency is extremely constant, as are all elemental oscillation frequencies, and it thus provides a good time base. The cesium standard was adopted at the 12th General Conference of Weights and Measures in 1964. *See also* ATOMIC CLOCK.

CGS SYSTEM

The centimeter-gram-second, or CGS, system is a standard basis for determining physical and electrical units. The CGS system is not used often, however; the meter-kilogram-second (MKS) system is much more often used.

All physical and electrical units are definable in terms of quantity, length, mass, time, and direction. Speed, for example, is measured in centimeters per second in the CGS system. Current is defined as coulombs (electrical quantity) per second. Voltage is defined in terms of mass, length, time, and quantity in a rather complicated form. Resistance is defined on the basis of current and voltage.

In the CGS system, electrical units are called ab units, such as the abampere, abvolt, and abohm. Electrostatic units in the CGS system are called stat units. *See also* AB UNITS, CURRENT MKS SYSTEM, STAT UNITS, VOLTAGE.

CHAIN RADAR

A tracking system consisting of several radar installations, each interconnected with the others, is called chain radar. As a target (such as a missile) follows its course, it is constantly within range of at least one radar station (see illustration). Thus, its course is monitored without interruption.

Chain radar systems facilitate continuous tracking of guided missiles, ensuring that they remain on course. A deviation from the prescribed course is immediately detected by one of the stations, and course corrections are thus made before the error becomes very great. *See also* RADAR.

CHAIN REACTION

When a chemical or physical reaction triggers more reactions, identical to the original reaction, the result is called a chain reaction. A familiar example of this is a set of

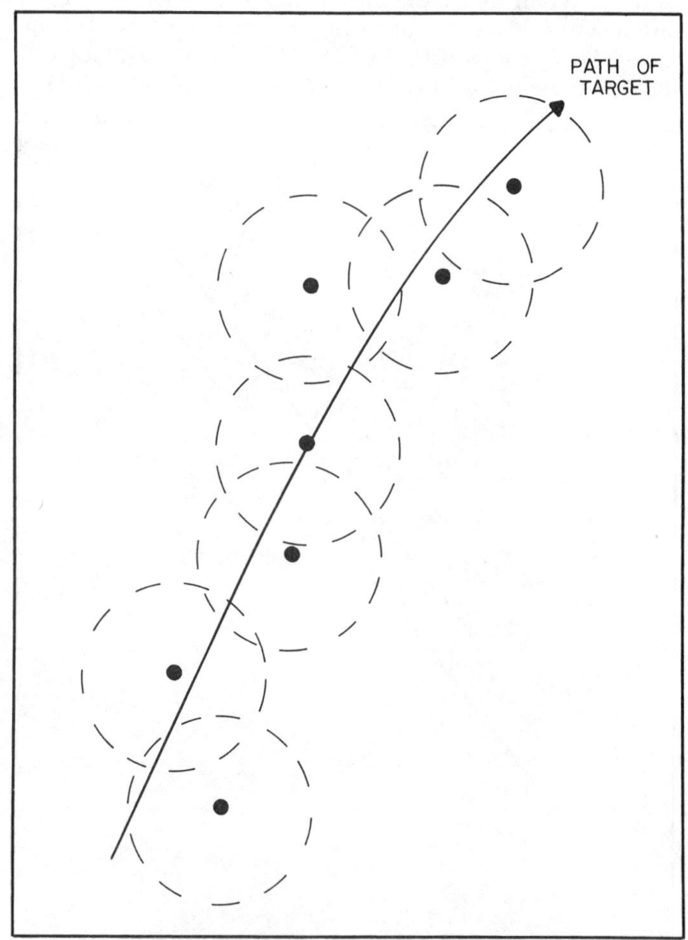

CHAIN RADAR: Overhead view of a chain radar system.

dominoes that falls, one by one, when the first is pushed. In nuclear physics, an atomic nucleus can be split by a high-speed neutron. When the nucleus splits, it throws off other high-speed neutrons, which in turn impact against and split other nuclei. The result is a self-sustaining reaction, or chain reaction. It can cause a tremendous release of energy. The atomic bomb operates on this principle.

In electronics, a form of chain reaction is called thermal runaway. The heating of a component can reduce its efficiency, causing further heating. Oscillation is another form of electronic chain reaction. *See also* THERMAL RUNAWAY.

CHANNEL

A channel is a particular band of frequencies to be occupied by one signal, or one 2-way conversation in a given mode. The term channel is also used to refer to the current path between the source and drain of a field-effect transistor (*see* CHANNEL EFFECT, FIELD-EFFECT TRANSISTOR).

Some frequency bands, such as the amplitude-modulation (AM) broadcast band and the frequency-modulation (FM) broadcast band, are allocated in channels by legal authority of the Federal Communications Commission in the United States. Some frequency bands, such as the Amateur-Radio FM band at 144 MHz, are allocated into channels by common agreement among the users. Some bands, such as the high-frequency Amateur bands, are not allocated into channels. Operation in such bands is done with variable-frequency oscillators.

Any signal requires a certain amount of bandwidth for efficient transfer of information. This bandwidth is called the channel, or channel width, of the signal. A typical A-M broadcast signal is 10 kHz wide, or 5 kHz above and below the carrier frequency at the center of the channel (see illustration).

Some signal channels are as narrow as 3 to 5 kHz; some are several megahertz wide, such as fast-scan commercial television channels. *See also* BANDWIDTH, FREQUENCY ALLOCATIONS.

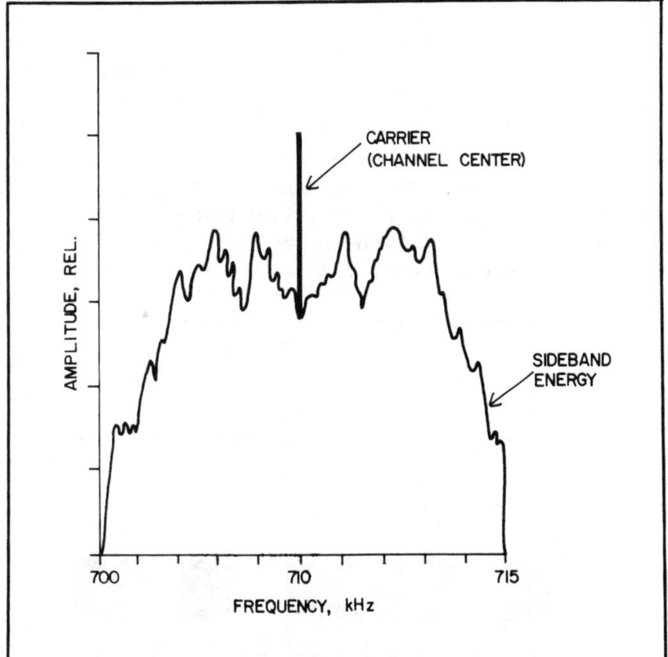

CHANNEL: A typical channel for an amplitude-modulated signal on the standard broadcast band. The center frequency is 710 kHz in this example.

CHANNEL ANALYSIS

When a signal is checked to ensure that all its components are within the proper assigned channel, the procedure is called channel analysis. Channel analysis requires a spectrum analyzer to obtain a visual display of signal amplitude as a function of frequency (*see* SPECTRUM ANALYZER).

The illustration shows an amplitude-modulated (AM) signal as it would appear on a spectrum-analyzer display. The normal bandwidth of an AM broadcast signal is plus or minus 5 kHz relative to the channel center. (A communications signal is often narrower than this, about plus or minus 3 kHz). At A, a properly operated AM transmitter produces energy entirely within the channel limits. At B, overmodulation causes excessive bandwidth. At C, an off-frequency signal results in out-of-band emission.

Channel analysis can reveal almost any problem with a modulated signal. But it takes some technical training to learn how different modes should appear on a spectrum analyzer.

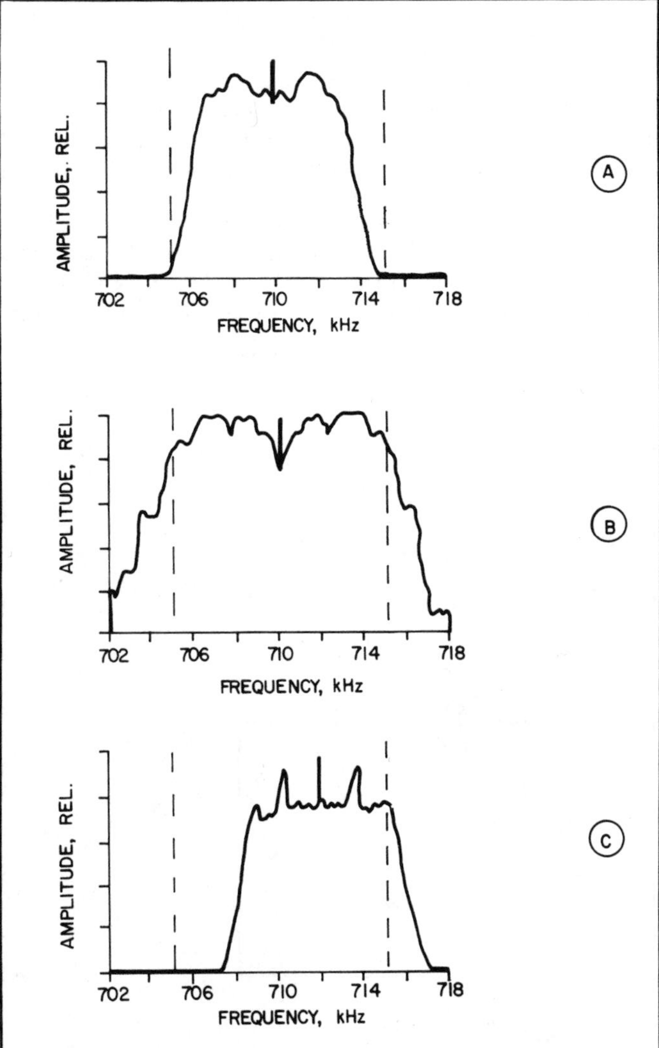

CHANNEL ANALYSIS: At A, a properly operating amplitude-modulated transmitter with a carrier frequency of 710 kHz. At B, excessive bandwidth. At C, the signal is off frequency.

CHANNEL CAPACITY

Any channel is capable of carrying some information at a certain rate of speed. A channel just a few Hertz wide can accommodate a code (CW) signal; a channel in the AM broadcast band can be used for the transmission of voices. In general, the greater the bandwidth of the channel (*see* BANDWIDTH), the more information, measured in characters or words, can be transmitted per unit time. The maximum rate at which information can be reliably sent over a particular channel is called the channel capacity.

Reliability is generally defined as an error rate not greater than a certain percentage of characters sent. As the speed of data transmission is increased, the error rate increases gradually at first. Then, as the data rate becomes too fast for the channel, the error percentage rises more and more rapidly. As the data speed is increased without limit, the data reception approaches a condition nearly equivalent to random characters. The precise error percentage representing the limit of reliability must be prescribed when defining channel capacity.

The channel capacity depends, to some extent, on the mode of transmission. Digital modes, such as frequency-shift keying, are generally more efficient than analog modes, such as voice transmission.

CHANNEL EFFECT

Channel effect is the tendency for current to leak from the emitter to the collector of a bipolar transistor. This is undesirable in bipolar transistors, but the field-effect transistor (FET) operates on this principle (*see* FIELD-EFFECT TRANSISTOR).

The illustration shows the channel of a field-effect transistor. A current flows from the source to the drain; its magnitude depends on the voltage at the gate electrodes. In this case, an N-channel field-effect transistor is illustrated. The P-channel unit operates on the same principle, but with reversed polarity.

The channel becomes narrower in the N-channel FET as the gate voltage becomes more and more negative. The channel width also depends on the voltage between the source and the drain.

The narrower the channel, the greater the effective resistance between the source and the drain. The channel effect as shown, for the ordinary field-effect transistor, is called the depletion mode. This is because the channel is normally conductive, and a charge on the gate cuts it off. Some metal-oxide field-effect transistors, or MOSFETs, operate in the enhancement mode, where there is no channel under conditions of zero gate voltage. In such devices, the gate voltage produces a channel by the same effect as the unwanted channel formation in a bipolar transistor. In the enhancement-mode MOSFET, however, the formation of a channel is desired, and is the basis for its operation. *See also* DEPLETION MODE, ENHANCEMENT MODE, METAL-OXIDE SEMICONDUCTOR FIELD-EFFECT TRANSISTOR.

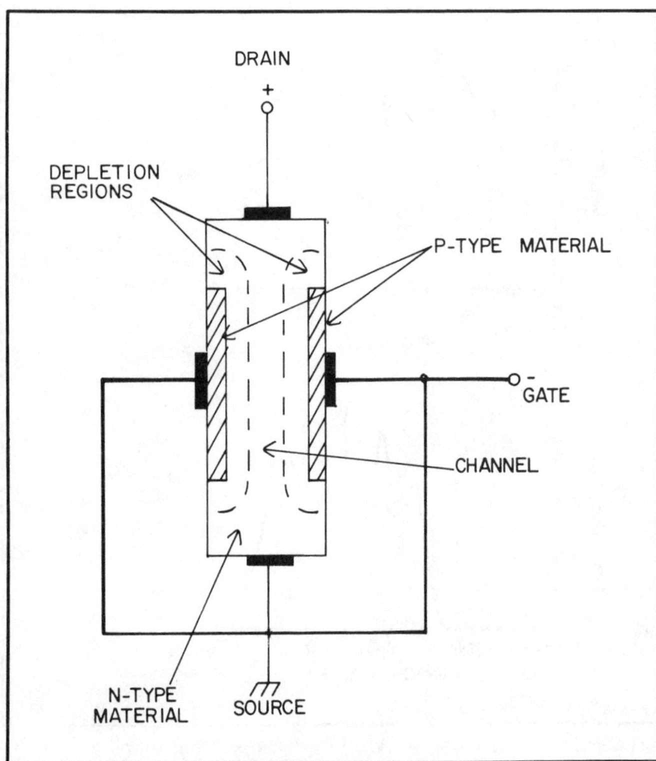

CHANNEL EFFECT: The channel effect is the basis for the operation of the field-effect transistor.

CHANNEL SEPARATION

In a channelized band, the frequency difference between adjacent channels is called the channel separation. The channel separation must always be at least as great as the channel bandwidth for the signals used; otherwise, interference will result. Occasionally, the channel separation is greater than the signal bandwidth, allowing a small margin of safety against adjacent-channel interference. This is the case in the frequency-modulation amateur band at 144 MHz. Sometimes the channel separation is insufficient to prevent interference between adjacent stations. This is a problem on some of the shortwave broadcast bands, where signals can often be found at a separation of 5 kHz although their bandwidth is 10 kHz.

The term channel separation is used to define the isolation, or attenuation, between the left and right channels of a stereo high-fidelity system. The figure is generally given in decibels. There is always some spillover between stereo channels, but in a good system, it should be possible to record two entirely different monaural signals on the two channels without significant mutual interference. Of course, this is seldom deliberately done; in recording, there is almost always some acoustical overlap between the left and right channels. However, the electrical separation between stereo channels should be as great as the state of the art will permit. *See also* STEREOPHONICS.

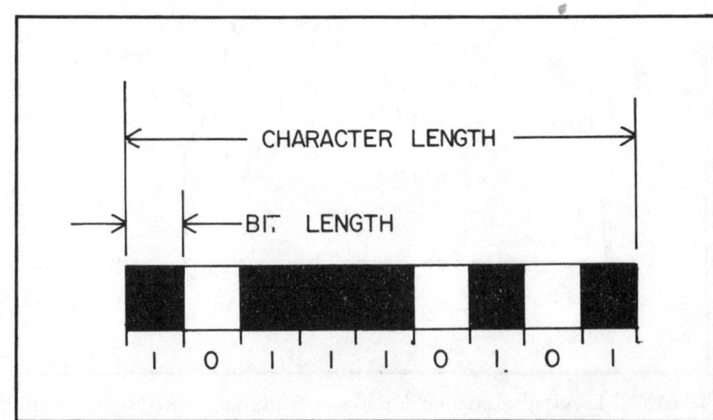

CHARACTER: The International Morse character L is nine bits in length, not including the following space.

CHARACTER

A character is an elementary symbol or unit of data. It may consist of a binary digit (bit), a decimal digit (0 through 9), a letter of the alphabet, a punctuation mark, or a nonstandard symbol. A space is technically considered a character as well. A character, in most codes, is represented by several bits. For example, in the International Morse code, the letter L is represented by the combination of bits 101110101, as shown in the illustration, where 0 represents the key-up condition, 1 represents the key-down condition, and the length of one dot is defined as one bit. (The space following the letter is not included as part of the letter.)

Words are made up of combinations of characters. In the English language, the length of a word may vary considerably. In some computer languages, a word is always the same length; in others, it varies. *See also* BIT, WORD.

CHARACTER GENERATOR

Any device that accepts alphabetic-numeric symbols and punctuation, and perhaps also nonstandard symbols, and converts them into a binary code, is a character generator. Examples of character generators include the Morse, Baudot, and ASCII keyboards, and the dot-matrix generator for printouts or cathode-ray-tube computer displays.

Character generators usually have adjustable speed, so that the data bits are transmitted at the desired rate. Buffer circuits ensure that characters are not lost if the operator types faster than the speed of data transmission. Sophisticated character generators are used as word processors. *See also* CHARACTER, WORD PROCESSING.

CHARACTERISTIC

Any specific property of a material or device is called a characteristic. A characteristic is definable and measurable, and can be expressed in electrical, electromagnetic, hydraulic, magnetic, mechanical, nuclear, or thermal terms.

Examples of characteristics include hardness, conductivity, temperature coefficient, and radiation resistance. These are, of course, just a few of many possible examples.

In logarithms, the characteristic is the portion of the logarithm to the left of the decimal point. For positive logarithms, the characteristic is the integral portion. For negative logarithms, the characteristic is the integral portion minus 1. For example, in the equation

$$\log_{10} 253 = 2.403$$

the characteristic is 2; in the equation

$$\log_{10} 0.253 = -0.5969$$

the characteristic is −1. *See also* LOGARITHM.

CHARACTERISTIC CURVE

A function or relation defining the interdependence of two quantities is called a characteristic curve. In electronics, characteristic curves are generally mentioned in reference to semiconductor or vacuum-tube triode, tetrode, or pentode devices.

A common example of a characteristic curve is illustrated. This curve defines the relation between the gate voltage (E_G) and the drain current (I_D) for an N-channel field-effect transistor, given a drain voltage of 3 volts.

Characteristic curves are used to find the best operating bias for an oscillator or amplifier. Various kinds of amplifiers operate at different points on the characteristic curve for a particular device. A linear amplifier should be biased at a point where the characteristic curve is nearly straight. A class-C amplifier is biased beyond the cutoff or pinchoff point. *See also* FIELD-EFFECT TRANSISTOR, TRANSISTOR, TUBE.

CHARACTERISTIC DISTORTION

Characteristic distortion is a fluctuation in the characteristic curve of a semiconductor device or vacuum tube. The characteristic curve is affected by any change of bias. A change of bias can be caused by the signal itself. Normally, the operation of an amplifier is fairly predictable from the direct-current bias and the characteristic curve of the device. The addition of an alternating-current signal, especially if it is high in voltage, can alter the bias. To get an accurate prediction of the operation of an amplifier in which there is high drive voltage, characteristic distortion must be taken into account. *See also* CHARACTERISTIC CURVE.

CHARACTERISTIC IMPEDANCE

The ratio of the signal voltage (E) to the signal current (I) in a transmission line depends on several things. Ideally, there should be no variations in current and voltage at different places along the line. When the load consists of a noninductive resistor of the correct ohmic value, the voltage-to-current (E/I) ratio Z_O will be constant all along the line, so that

$$Z_O = E/I$$

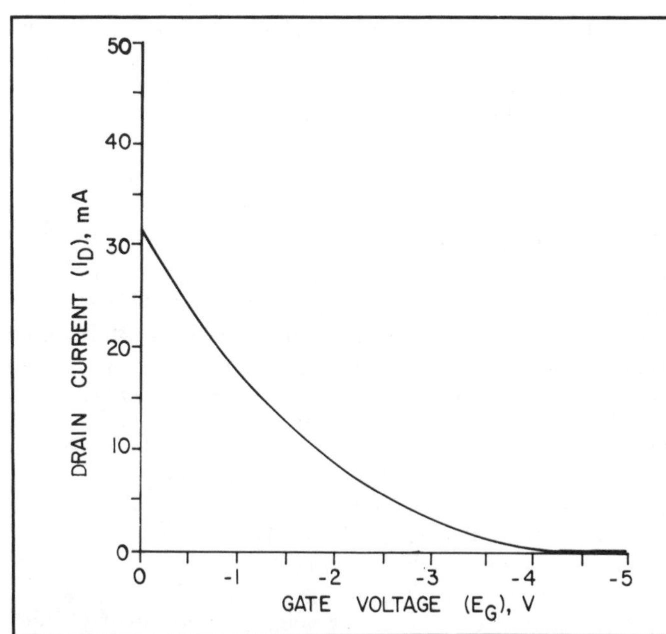

CHARACTERISTIC CURVE: An example of a characteristic curve for a field-effect transistor.

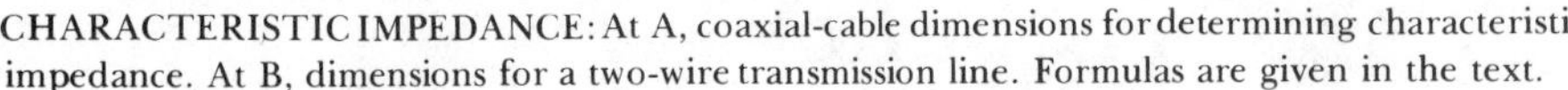

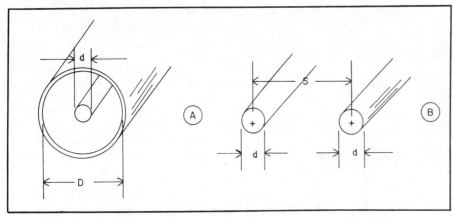

CHARACTERISTIC IMPEDANCE: At A, coaxial-cable dimensions for determining characteristic impedance. At B, dimensions for a two-wire transmission line. Formulas are given in the text.

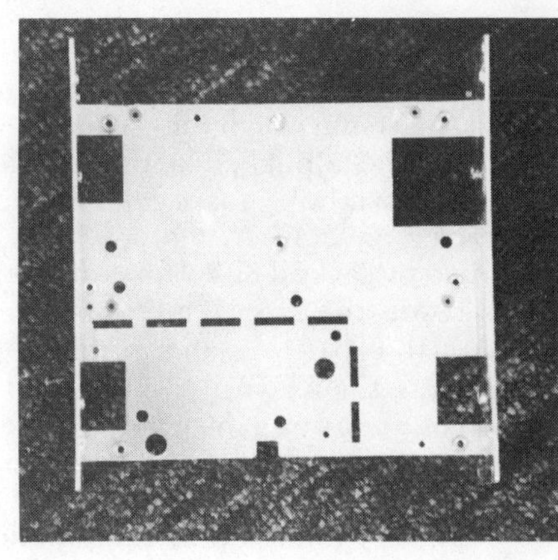

CHASSIS: A bare aluminum chassis.

The value Z_O under these circumstances is called the characteristic, or surge, impedance of the transmission line.

The characteristic impedance of a given line is a function of the diameter and spacing of the conductors used, and is also a function of the type of dielectric material. For air-dielectric coaxial line, in which the inside diameter of the outer conductor is D, and the outside diameter of the inner conductor is d (see A in the illustration), the characteristic impedance is:

$$Z_O = 138 \log_{10} (D/d)$$

For air-spaced two-wire line, in which the conductor diameter is d and the center-to-center conductor spacing is s (B in the illustration,) the formula is:

$$Z_O = 276 \log_{10} (2s/d)$$

The units for D, d, and s must, of course, be uniform in calculation, but any units may be used.

Generally, transmission lines do not have air dielectrics. Solid dielectrics, such as polyethylene, are often used, and this lowers the characteristic impedance for a given conductor size and spacing. For optimum operation of a transmission line, the load impedance should consist of a pure resistance R at the operating frequency, such that $R = Z_O$. *See also* TRANSMISSION LINE.

CHARACTRON

Charactron is the registered trade name for a special kind of cathode-ray tube made by Stromberg-Carlson Corporation. The Charactron has a high degree of resolution, so that alphabetic-numeric symbols can be displayed on the screen and easily read (*see* CATHODE-RAY TUBE).

The electron beam in the Charactron tube rapidly scans the screen in horizontal lines, in a manner similar to that of a television picture tube. The precise arrangement of light and dark gives, generally, light characters on a dark background. However, dark characters on a light background can also be obtained.

By properly adjusting the brightness and the contrast, the characters can be easily seen and identified. The Charactron is used in video displays for computers and word processors. A typical Charactron displays more than 20,000 characters per second, and can read up to 100,000 computer words per minute. *See also* VIDEO DISPLAY TERMINAL.

CHARGE

Charge is an electrostatic quantity, measured as a surplus or deficiency of electrons on a given object. When there is an excess of electrons, the charge is called negative. When there is a shortage of electrons, the charge is called positive. These choices are purely arbitrary, and do not represent any special qualities of electrons.

Charge is measured in units called coulombs (*see* COULOMB). A coulomb is the charge contained in 6.281×10^{18} electrons. Charge is usually represented by the letter Q in equations. The smallest possible unit of electrostatic charge is the amount of charge contained in one electron.

The quantity of charge per unit length, area, or volume is called the charge density on a conductor, surface, or object. Charge density is measured in coulombs per meter, coulombs per square meter, or coulombs per cubic meter. A charge may be carried or retained by an electron, proton, positron, anti-proton, atomic nucleus, or ion. *See also* ELECTRON, ION.

CHARGING

Charging is the process by which an electrostatic charge accumulates. Charging can occur with capacitors, inductors, or storage batteries of various types. In the capacitor, charge is stored in the form of an electric field. In an inductor, charge is stored in the form of a magnetic field. In a battery or cell, charge is stored in chemical form.

The rate of charging is measured in coulombs per second, which represents a certain current in amperes. This charging current can be measured with an ammeter. In a storage battery, charging occurs rapidly at first, and then more and more slowly as the storage capacity is reached. The same is true of capacitors and inductors. The charging current of a battery must be maintained at the proper levels for several hours, generally, in order to ensure optimum charging. *See also* NICKEL-CADMIUM BATTERY, STORAGE BATTERY.

CHASSIS

A chassis is a piece of aluminum, copper, or steel, on which radio components, circuit boards, and panels are

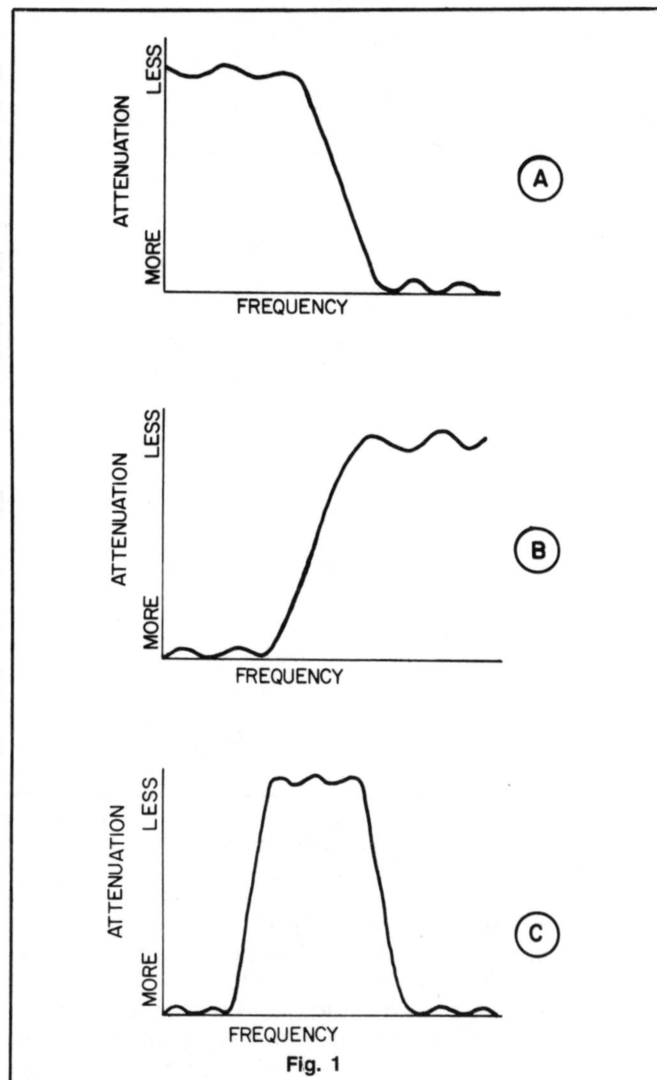

CHEBYSHEV FILTER: Chebyshev filter responses. At A, lowpass; at B, highpass; at C, bandpass.

mounted. The photograph shows a stripped chassis with various holes and bends in the metal for attachment of the other parts. The chassis shown here is about seven inches square; some chassis are larger or smaller than this, of course. The chassis adds mechanical rigidity to electronic equipment, and serves as a common ground.

Ground connections in electronic assemblies are often made to the chassis. The chassis must then, in turn, be provided with an external earth ground. A terminal on the chassis is used for this purpose. *See also* GROUND, GROUND CONNECTION.

CHATTER

When the voltage to a relay fluctuates, or is not quite sufficient to keep the contacts closed continuously, the contacts may open and close intermittently. This can occur at a high rate of speed, producing a characteristic sound that is called chatter.

Chatter can be deliberately produced by connecting the relay contacts in series with the coil power supply in such a way that oscillation is produced. As soon as the coil receives voltage, the contacts open and the voltage is interrupted. A spring then returns the contacts to the closed condition, and the cycle repeats itself. A resistance-capacitance combination determines the frequency of the oscillation. By varying the oscillation weight and speed, a simple buzzer can be constructed for use with code transmitters (*see* KEYER).

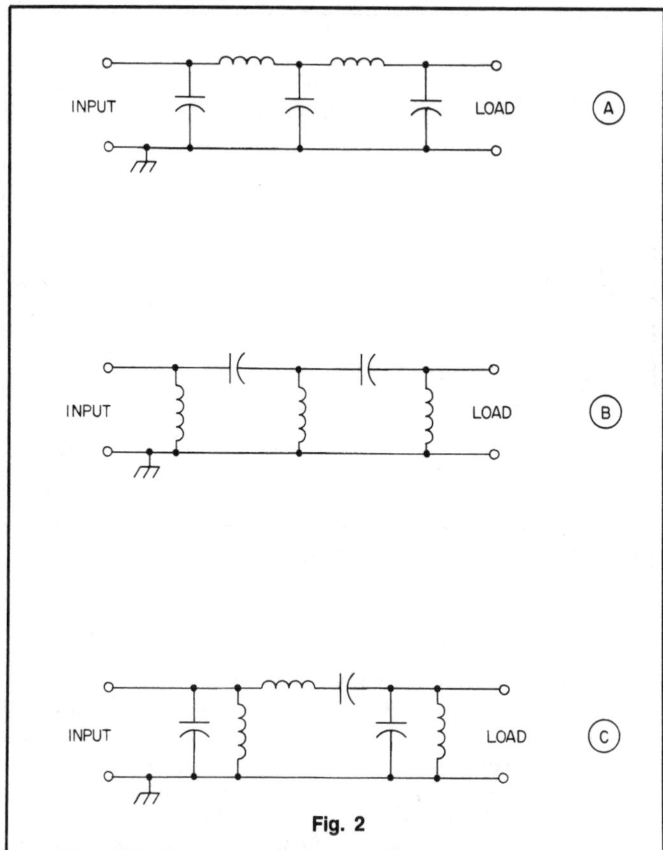

CHEBYSHEV FILTER: Typical Chebyshev filter circuits. At A, lowpass; at B, highpass; at C, bandpass.

Generally, chatter is an undesirable effect, and can totally disable a piece of electronic equipment. Solid-state switching is preferable to relay switching in many modern applications for this reason. Transistors and diodes do not chatter. *See also* RELAY.

CHEBYSHEV FILTER

(also spelled Tschebyscheff, Tschebyshev)

A Chebyshev filter is a special type of selective filter, having a nearly flat response within its passband, nearly complete attenuation outside the passband, and a sharp cutoff response. The extremely steep skirt selectivity of the Chebyshev filter is the primary advantage of it over other types of filters. The Chebyshev filter design is similar to that of the Butterworth filter (*see* BUTTERWORTH FILTER). Such filters may be designed for lowpass, highpass, and band-rejection applications, as well as for bandpass use.

Figure 1 shows, approximately, the Chebyshev responses for lowpass, highpass, and bandpass filters. Note the ripple; this ripple is usually of little or no consequence. Figure 2 shows schematic diagrams for simple Chebyshev lowpass, highpass, and bandpass filters. The input and load impedances must be properly chosen for the best filter response. The values of the inductors and capacitors depend on the input and load impedances, as well as the frequency response desired. *See also* BANDPASS FILTER, BAND-REJECTION FILTER, HIGHPASS FILTER, LOWPASS FILTER.

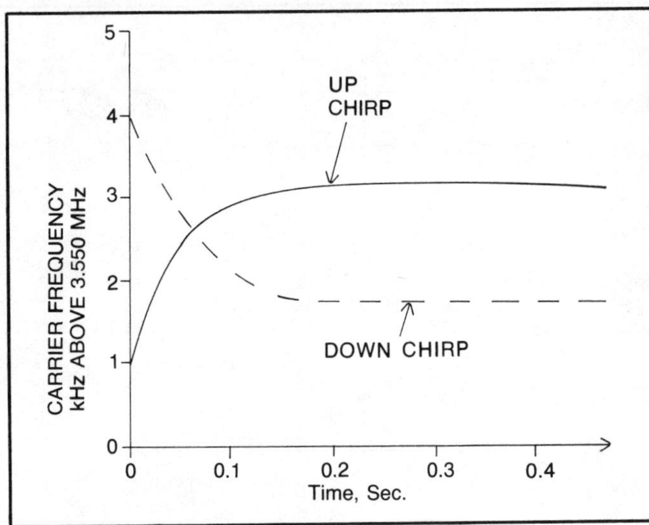

CHIRP: Chirp is a change in the frequency of a signal just after the key is depressed.

CHILD'S LAW

In a diode vacuum tube, the current varies with the 3/2 power (the square root of the cube) of the voltage, and inversely with the square of the distance between the electrodes. This relationship is called Child's Law. Mathematically, if E is the voltage between the cathode and the plate, I is the current through the device, d is the distance between the cathode and the plate, and k is a constant that depends on several factors, including the units chosen, then

$$I = kZ\left(\sqrt{E^3} / d^2\right)$$

See also TUBE.

CHIP

A chip is a piece of semiconductor material on which an integrated circuit is fabricated. The word chip is often used in place of the term integrated circuit. Chips are sliced from a section of semiconductor material called a wafer. Various processes are used in etching individual resistors, capacitors, inductors, diodes, and transistors onto the chip surface. A complete integrated circuit is supplied with external terminals called pins, and is encased in a hard package. *See also* INTEGRATED CIRCUIT.

CHIRP

Chirp is a change in the frequency of a keyed oscillator, resulting from variations in its load impedance or power supply voltage caused by keying. The frequency changes rapidly at first, and then more and more slowly as the load stabilizes. Usually, the frequency has stabilized within about 1 second. Chirp derives its name from the fact that it creates a bird-like sound in a CW receiver. The drawing illustrates chirp.

Chirp is undesirable because it increases the bandwidth of a continuous-wave signal. It also sounds bad. To reduce or eliminate chirp, the oscillator or oscillators must be provided with a load of constant impedance, and a stable power supply. This is more true of variable-frequency oscillators than crystal oscillators. A buffer stage (*see* BUFFER STAGE) helps to maintain a relatively constant impedance at the output of an oscillator. Another means of reducing chirp is to key an amplification stage of a transmitter, rather than the oscillator itself.

The term chirp is used in radar to refer to the widening or narrowing of pulses for the purpose of improving the signal-to-noise ratio. Narrow pulses are deliberately expanded at the transmitter, and reduced again as the signal is received. This results in a greater duty cycle for the transmitter, and greater average power output. It does not affect the range or resolution of the radar, but it improves the ability of the system to detect small targets. *See also* RADAR.

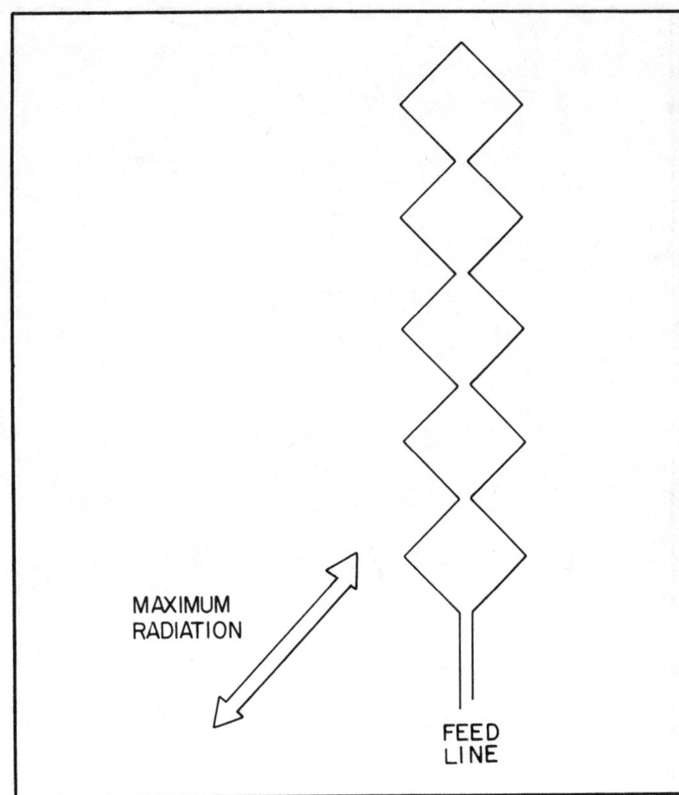

CHIREIX ANTENNA: A Chireix antenna consists of several loops, each with a circumference of one wavelength, fed end-to-end in phase.

CHIREIX ANTENNA

A Chireix antenna, sometimes also called a Chireix-Mesny antenna, is a phased array consisting of two or more square loops connected in series. The loops are coplanar, and have sides measuring ¼ wavelength at the resonant frequency. The radiation pattern is bidirectional, unless a reflecting screen or parasitic element is used. The illustration shows a five-element Chireix array.

The gain and radiation pattern of a Chireix antenna, operating at its fundamental frequency, depend on the number of elements. Chireix antennas are used at very-high frequencies and above, where they can be constructed from stiff wire or rigid tubing. *See also* PHASED ARRAY.

CHOKE

An inductor, used for the purpose of passing direct current while blocking an alternating-current signal, is called a choke. Typically, an inductor shows no reactance at dc, and increasing reactance at progressively higher ac fre-

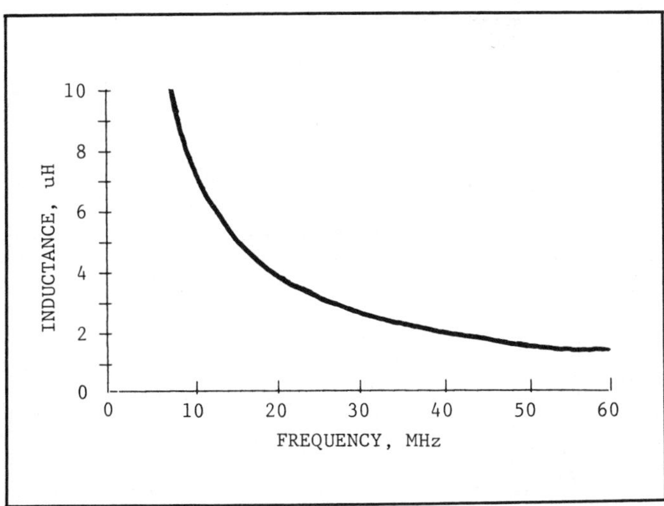

CHOKE: Inductance values for 500-ohm choke at various frequencies.

quencies. Some chokes are designed for the purpose of cutting off only radio frequencies. Some are designed to cut off audio as well as radio frequencies. Some chokes are designed to cut off essentially all alternating currents, even the 60-Hz utility current, and are used as power-supply filtering components. Together with large capacitances, such chokes make extremely effective power-supply filters.

Chokes may have air cores if they are intended for radio frequencies. Larger chokes use cores of powdered iron or ferrite. This increases the inductance of the winding for a given number of turns. Some chokes are wound on toroidal, or doughnut-shaped, cores. This provides a large increase in the inductance. Some chokes are wound on cores shaped like the core of an ac power transformer. These chokes can handle large amounts of current.

Chokes are useful when it is necessary to maintain a certain dc bias without short-circuiting the desired signal. *See also* INDUCTOR.

CHOKE-INPUT FILTER

A power-supply filter eliminates the pulsations from the direct-current output of a rectifier circuit. Often, a capacitor is the only filtering component. However, the addition of a large value of inductance, connected in series with the ungrounded side of the supply output, improves the quality of the output by further reducing the ripple (*see* RIPPLE). The drawing illustrates a choke-input power-supply filter.

The choke, L, is placed on the rectifier side of the filter capacitors. Capacitor C1 provides ripple suppression, and is a large-value electrolytic type. A smaller mylar or disk-ceramic capacitor, C2, effectively prevents radio-frequency energy from entering the supply through the power leads.

Choke-input filters are much bulkier and heavier than simple capacitor filters, especially when the supply must deliver a large amount of current at a high voltage. However, the choke-input filter improves the supply regulation and reduces the amount of ripple reaching the load. This is particularly desirable when the load draws a large amount of current, or when the most ripple-free direct current is needed. Sometimes, two or even three stages of choke/capacitor filtering are used. *See also* CHOKE, POWER SUPPLY.

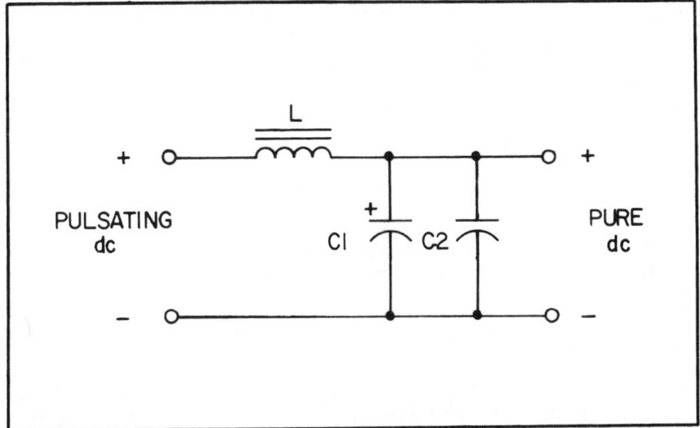

CHOKE-INPUT FILTER: A choke-input power supply filter offers better voltage regulation than a simple capacitor filter.

CHOPPER

A chopper is a device that interrupts a direct current, producing pulses of constant amplitude and frequency. A chopper consists of a low-frequency oscillator which opens and closes a high-current switching transistor. An example of a chopper circuit is shown in the illustration.

Choppers are used in a variety of ways. A direct-current voltage can be passed through a chopper, the resultant waveform then amplified or attenuated, and finally filtered to obtain a direct-current voltage different from the input. Choppers are sometimes also used as rectifiers; they can eliminate or invert one half of an alternating-current cycle. A rotating, perforated disk, used for the purpose of modulating a beam of light or a stream of atomic particles, is sometimes called a chopper.

A chopper is often used to obtain a high-voltage source of alternating current from a storage battery. This allows the operation of some household appliances from an automotive electrical system. *See also* CHOPPER POWER SUPPLY, VIBRATOR.

CHOPPER POWER SUPPLY

A circuit for obtaining a high-voltage source of alternating current from a direct-current source is called a chopper

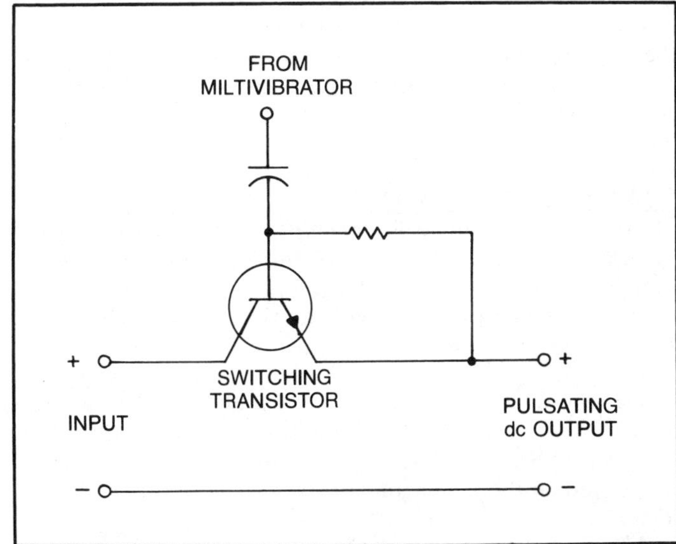

CHOPPER: A chopper circuit.

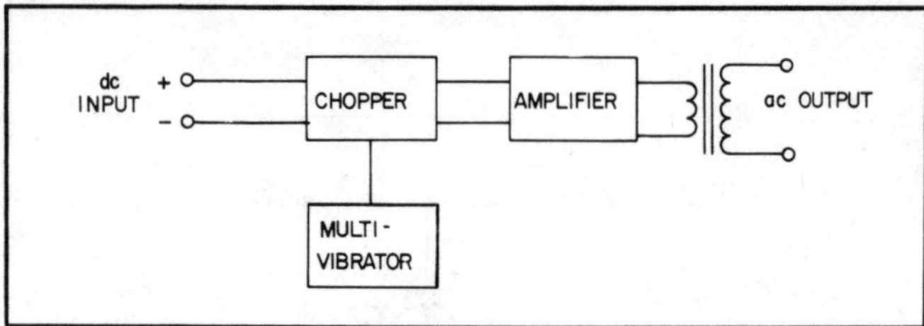

CHOPPER POWER SUPPLY: Block diagram of a chopper power supply.

power supply. Such devices were once used in the operation of mobile or portable equipment containing vacuum tubes. Chopper power supplies also facilitate the use of certain standard household appliances in boats and automobiles.

A block diagram of a chopper power supply is shown in the illustration. The chopper interrupts the direct-current power source at regular intervals, and the amplifier then increases the pulse amplitude. The transformer converts the pulsating direct current to alternating current. This supply thus produces power similar to that found in household utility outlets.

The output of a chopper power supply is generally not a good sine wave, since square-wave choppers are the most easy to manufacture. For this reason, many chopper power supplies are unsuitable for use with electric clocks and other appliances that need a nearly perfect, 60-Hz sine-wave supply.

If the transformer shown is replaced by a filtering network, a direct-current transformer or amplifier can be obtained. Such a transformer can provide hundreds of volts from a simple 12-volt, direct-current electrical system. *See also* CHOPPER, DC-TO DC CONVERTER.

CHROMA

Chroma is a combination of color hue and saturation (*see* HUE, SATURATION). White, black, and all intermediate shades of gray have no chroma.

Chroma is independent of the brightness of a color. The chroma is a subjective function of the dominant wavelength of a color, and the purity of the color. The chroma of red is different from the chroma of blue, since the wavelength is different. The chroma of a red laser, having essentially a single wavelength, is different from that of a red incandescent bulb, which emits wavelengths over a broad spectrum. Chroma is sometimes also called chromaticity.

CHROMINANCE

Chrominance is a measure of color. In color television, the chrominance is technically the difference between a given color and a standard color having the same brightness.

Chrominance primaries are the three colors which, in various combinations, can produce any color. The three chrominance primaries are red, green, and blue. Red and green combine to form yellow; red and blue combine to form yellow; red and blue combine to form magenta; green and blue combine to form cyan, or blue-green; and red, green, and blue in equal proportions combine to form white.

The red, green and blue chrominance components of a color television signal are transmitted independently. When they are combined at the receiver, a full-color picture results. A crystal-controlled oscillator generates a subcarrier at 3.57945 MHz, plus or minus 10 Hz, for the chrominance detector. *See also* COLOR TELEVISION.

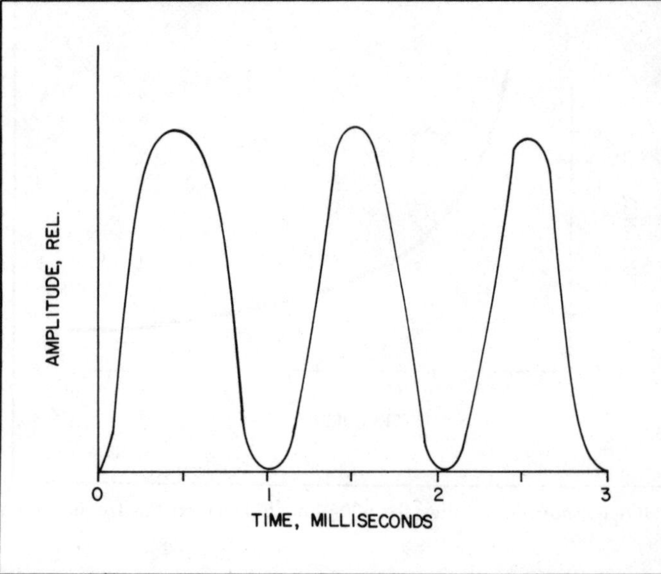

CHRONOGRAPH: A chronograph is any plot of a quantity, such as amplitude, versus time. Here, a chronograph of an amplitude-modulated signal is shown.

CHRONOGRAPH

A chronograph is a plot of any quantity as a function of time. A machine that records a quantity as a function of time is called a chronograph or chronograph recorder.

Chronographs are frequently used in electronics. The familiar sine-wave display is an example of a chronograph. Plots of temperature-versus-time, amplitude-versus-time, and other functions are frequently used. The accompanying drawing illustrates a chronograph of amplitude versus time for a sine-wave amplitude-modulated signal.

A chronograph is characterized by an accurate, and usually linear, time display along the horizontal axis. This allows precise determination of the period of a function. In the illustration, the period of the modulating waveform is 1 millisecond.

CHU

CHU is a time and frequency broadcasting station located in Canada. Its primary broadcasting frequency is 7.335 MHz. Time is broadcast each minute in English and French. Eastern Standard Time (EST) is announced; this is five hours behind Coordinated Universal Time (UTC). Brief tones are sent each second. The signal is amplitude-modulated.

In the United States, time and frequency stations are WWV and WWVH, maintained by the Nation Bureau of Standards. *See also* WWV/WWVH.

CIRCUIT

A circuit is a combination of electronic components, interconnected for a specific purpose. Examples include the oscillator, amplifier, detector, and power supply.

In general, any path between two or more points, capable of signal transfer, is a circuit. A telephone line, and the

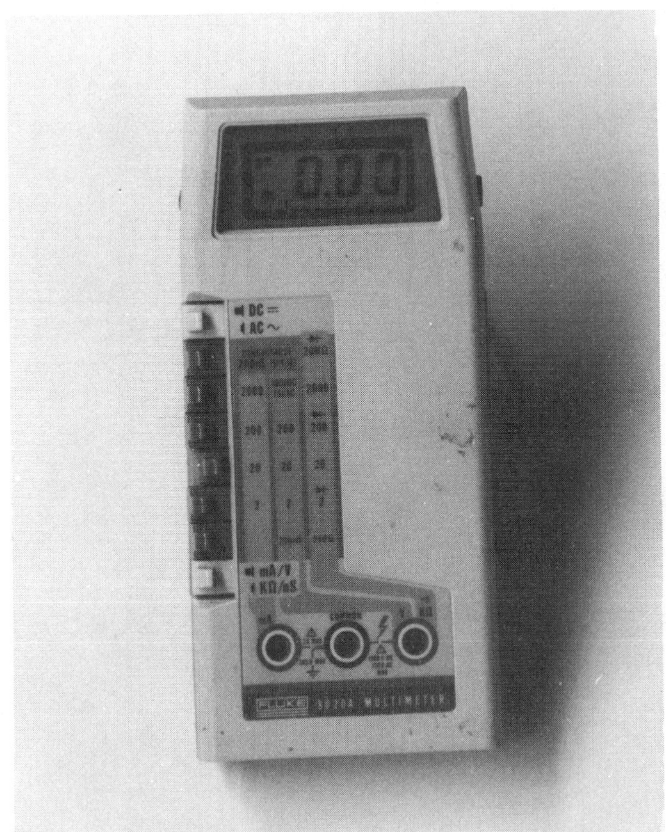

CIRCUIT ANALYZER: A simple test-laboratory analyzer.

terminals at either end, is a circuit; so is any electromagnetic path in a communications system.

Any electronic circuit can be represented by a schematic or block diagram. *See also* BLOCK DIAGRAM, SCHEMATIC DIAGRAM.

CIRCUIT ANALYZER

A device for measuring various electrical quantities, such as current, voltage, and resistance, is called a circuit analyzer. Such an instrument is also sometimes called a multimeter. Circuit analyzers may have either digital or analog displays. The photo shows a typical circuit analyzer, used in radio repair and testing.

An analog circuit analyzer generally consists of a sensitive ammeter with a full-scale reading of 30 to 50 microamperes. It has various shunt networks for measuring larger values of current. A battery produces the current needed for measuring resistance, and series resistors allow the measurement of voltage. A range-selector switch is provided for current, voltage, and resistance measurements.

A digital circuit analyzer gives an accurate, numerical reading without the need for visual interpolation, although a range selector is usually supplied for current, voltage, and resistance measurements.

Sometimes an analog device is preferable to a digital device; for example, a maximum or minimum quantity may be required with the adjustment of a certain control. Such an adjustment is more easily done with an analog device than with a digital device. If extreme accuracy is needed, a digital readout is better.

A special kind of circuit analyzer called an FET voltmeter is used in high-impedance circuits. This device allows measurement of circuit parameters with almost no effect in the operation. *See also* FET VOLTMETER.

CIRCUIT BOARD

A circuit board is a piece of phenolic, epoxy, or similar dielectric material, on which electronic components are mounted in the construction of a circuit. Circuit boards are used in nearly all modern electrical and electronic devices, having replaced the point-to-point method of construction in most applications.

Circuit boards may have copper foil on one or both sides, for the purpose of component interconnection. Some circuit boards, used in experimental construction, have no foil, or have a simple foil pattern on one side only. Circuit boards specifically fabricated for a specialized product often have very complicated foil patterns on both sides, or even between layers of dielectric. The circuit board shown is a simple experimenter's board with foil on the underside, connecting the holes together in groups of four.

Printed circuit boards, made for special purposes, are fabricated using a special photo-etching process. The board is initially plated completely with copper on one or both sides. A transparent piece of film is placed over the board, with a pattern preventing light from reaching certain parts of the copper foil. Then the entire board is placed in a special solution and exposed to bright light. The copper foil dissolves wherever light reaches it, and the result is a detailed circuit pattern. *See also* PRINTED CIRCUIT.

CIRCUIT BREAKER

A circuit breaker is a current-sensitive switch. It is placed in series with the power-supply line to a circuit. If the current in the line reaches a certain value, the breaker opens and removes power from the circuit. Breakers are easily reset when they open; this makes them more convenient than fuses, which must be physically removed and replaced when they blow. Circuit breakers are used to protect electronic circuits against damage when a malfunction causes them to draw excessive current. Circuit breakers are also used in utility wiring to minimize the danger of fire in the event of a short circuit.

There are many kinds of circuit breakers, for various voltages and limiting currents. Some breakers will open with just a few milliamperes of current, while others require hundreds of amperes. Some breakers open almost immediately if their limiting currents are reached or exceeded. Some breakers have a built-in delay. *See also* FUSE, SLOW-BLOW FUSE.

CIRCUIT CAPACITY

In a communications system, the number of channels that can be accommodated simultaneously without overload or mutual interference is called the circuit capacity. For example, in a 100-kHz-wide radio-frequency band, using amplitude-modulated signals that occupy 10 kHz of spectrum space apiece, the circuit capacity is 10 channels. It is impossible to increase the number of channels without sacrificing efficiency.

CIRCUIT BOARD: A circuit board for experimental work.

In an electrical system, the circuit capacity is usually specified as the number of amperes that the power supply can deliver without overloading. A typical household branch circuit has a capacity of approximately 10 to 30 ameres. A fuse or circuit breaker prevents overload. *See also* CIRCUIT BREAKER, FUSE.

CIRCUIT DIAGRAM

See SCHEMATIC DIAGRAM.

CIRCUIT EFFICIENCY

Circuit efficiency is the proportion of the power in a circuit that does the job intended for that circuit. Efficiency is usually expressed as a percentage. In an amplifier or oscillator, the efficiency is the ratio of the output power to the input power. For example, if an amplifier has a power input of 100 watts and a power output of 50 watts, its efficiency is 50/100, or 50 percent.

The circuit efficiency of an amplifier depends on the class of operation. Some Class-A amplifiers have a circuit efficiency as low as 20 to 30 percent. Class-B amplifiers usually have an efficiency of about 50 to 60 percent. Class-C amplifiers often have an efficiency rating of more than 80 percent.

An inefficient circuit generates more heat, for the same amount of input power, compared to an efficient one. This heat can, in some instances, destroy the amplifying or oscillating transistor or tube. Regardless of the class of operation, measures should always be taken to maximize the circuit efficiency. *See also* CLASS-A AMPLIFIER, CLASS-AB AMPLIFIER, CLASS-B AMPLIFIER, CLASS-C AMPLIFIER.

CIRCUIT NOISE

Whenever electrons move in a conductor or semiconductor, some electrical noise is generated. The random movement of molecules in any substance also creates noise. A circuit therefore always generates some noise in addition to the signal it produces or transfers. Noise generated within a piece of electronic equipment is called circuit noise.

In a telephone system, circuit noise is the noise at the input of the receiver. This noise comes from the system, and does not include any acoustical noise generated at the transmitter. Such noise is generated within the electrical circuits of the transmitter, and along the transmission lines and switching networks.

Circuit noise limits the sensitivity of any communications system. The less circuit noise produced, the better the signal-to-noise ratio (*see* SIGNAL-TO-NOISE RATIO). Therefore, all possible measures should be employed to minimize circuit noise when long-distance communication is desired. *See also* TELEPHONE.

CIRCUIT PROTECTION

Circuit protection is a means of preventing a circuit from drawing excessive current. This is done by means of either a circuit breaker, or by means of a fuse connected in series with the power-supply line (*see* CIRCUIT BREAKER, FUSE). The illustration shows a simple circuit-protection device connected in a power-supply line.

The current level at which the circuit breaker or fuse opens is usually about twice the normal operating current drain of the circuit. Circuit protection reduces the chances of extensive component damage in the event of a malfunction. It also reduces the possibility of fire caused by overheating of circuit components or wiring.

CIRCULAR ANTENNA

A circular antenna is a half-wave dipole bent into a circle.

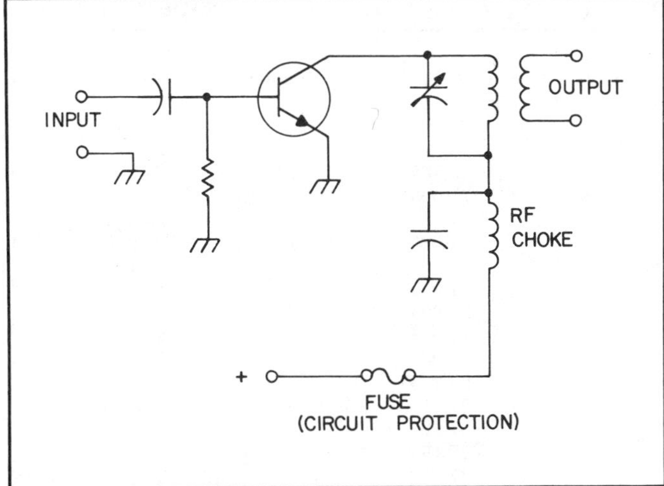

CIRCUIT PROTECTION: A fuse is a common method of circuit protection.

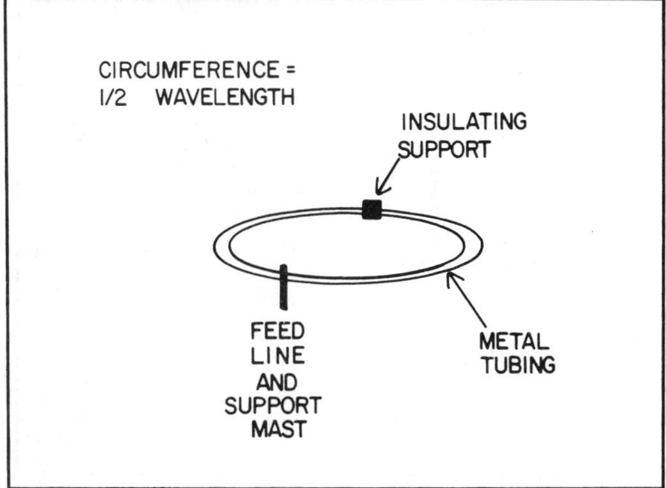

CIRCULAR ANTENNA: A circular antenna is a horizontal half-wave dipole bent into a circle. The ends are electrically insulated.

The ends are brought nearly together, but are not physically connected (see illustration). An insulating brace is often used to add rigidity to the structure. The circular antenna, when oriented in the horizontal plane, produces a nearly omnidirectional radiation pattern in all azimuth directions. Sometimes the circular antenna is called a halo.

Circular antennas, constructed from metal tubing, are sometimes used in mobile installations at frequencies above approximately 50 MHz. The horizontal polarization of the halo results in less fading or "picket fencing" than does vertical polarization. Several circular antennas may be stacked at ½-wavelength intervals to produce omnidirectional gain in the horizontal plane.

Circular antennas can be operated at odd multiples of the fundamental frequency, and a reasonably good impedance match will result when 50-ohm or 75-ohm feed lines are employed.

CIRCULAR POLARIZATION

The polarization of an electromagnetic wave is the orientation of its electric-field lines of flux. Polarization may be horizontal, vertical, or at a slant (*see* HORIZONTAL POLARIZATION, VERTICAL POLARIZATION). The polarization may also be rotating, either clockwise or counterclockwise. Uniformly rotating polarization is called circular polarization. The orientation of the electric-field lines of flux completes one rotation for every cycle of the wave, with constant angular speed.

Antennas for circular polarization are not, of course, themselves turned to produce the rotating electromagnetic field; the rotation is easily accomplished by electrical means. The illustration shows a typical antenna for generating waves with circular polarization. The antennas are fed 90 degrees out of phase by making feed-line stub X a quarter wavelength longer than stub Y. The signals from the two antennas thus add vectorially to create a rotating field. The direction, or sense, of the rotation can be reversed by adding ½ wavelength to either stub X or stub Y (but not both).

Circular polarization is compatible, with a 3-dB power loss, with either horizontal or vertical polarization, or with slanted linear polarization. When communicating with another station also using circular polarization, the senses must be in agreement. If a circularly polarized signal arrives with opposite sense from that of the receiving antenna, the attenuation is about 30 dB compared with matched rotational sense.

In uniform circular polarization, the vertical and horizontal signal components have equal magnitude. But this is not always the case. A more general form of rotating polarization, in which the components may have different magnitude, is called elliptical polarization. *See also* ELLIPTICAL POLARIZATION.

CIRCULAR SCANNING

See J DISPLAY.

CIRCULAR SWEEP

A circular sweep, or circular trace, is an oscilloscope trace that describes a circle. Such a sweep is obtained by applying two sine-wave signals of equal strength and frequency to

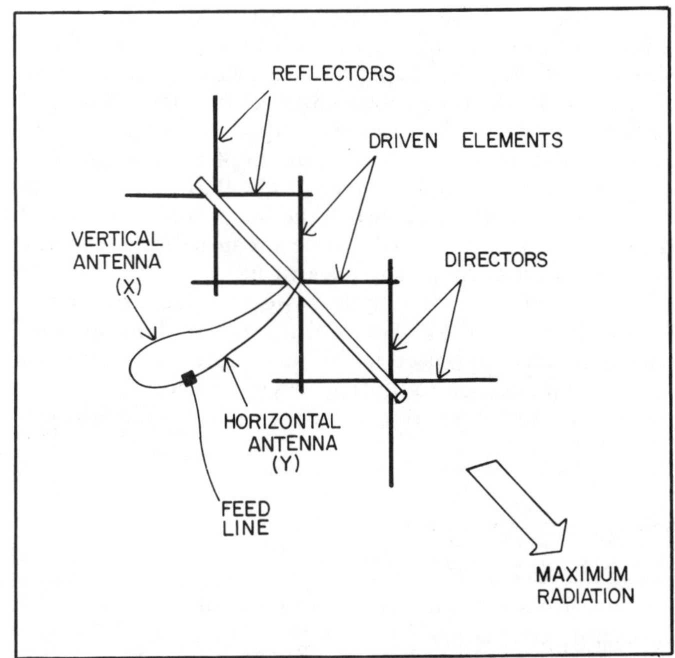

CIRCULAR POLARIZATION: An antenna for generating circular polarization. The feed line is split into two branches, X and Y; branch X is 90 electrical degrees longer than branch Y.

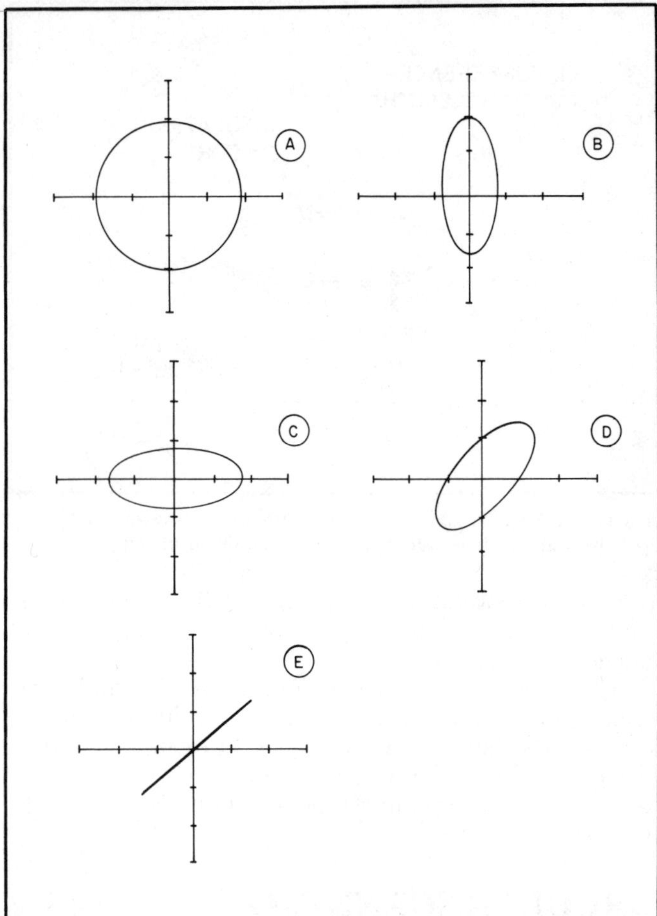

CIRCULAR SWEEP: At A, a circular sweep is the result of two signals of equal amplitude applied 90 degrees out of phase to the vertical and horizontal inputs of an oscilloscope. At B, the vertical signal is stronger; at C, the horizontal signal is stronger. At D, the two signals are 45 degrees out of phase. At E, the two signals are in phase.

the horizontal and vertical scope inputs, 90 degrees out of phase. Changes in the amplitude or phase of either signal flatten the circle into a horizontal or vertical ellipse (see illustration). A change in the frequency of one of the signals will give the display the appearance of a circle rotating in three dimensions. Changes in phase alone will result in a diagonally oriented ellipse; if the signals are either exactly in phase or exactly out of phase, the trace will appear as a line at a 45-degree angle. Two harmonically related signals, when applied at the horizontal and vertical deflecting plates of an oscilloscope, create a display called a Lissajous figure (*see* LISSAJOUS FIGURE).

A fixed circular or elliptical sweep indicates that the frequencies at both oscilloscope inputs are identical. The oscilloscope is often used in this way to precisely adjust an audio-frequency or radio-frequency oscillator exactly to match the frequency of a reference oscillator. *See also* OSCILLOSCOPE.

CIRCULATING TANK CURRENT

In a resonant circuit consisting of an inductor and capacitor, circulating currents flow back and forth between the two components. As the inductor discharges, the capacitor charges; as the capacitor discharges, the inductor charges.

CITIZEN'S BAND: CLASS-D CITIZEN'S BAND RADIO CHANNELS.

Channel	Frequency, MHz	Channel	Frequency, MHz
1	26.965	21	27.215
2	26.975	22	27.225
3	26.985	23	27.255
4	27.005	24	27.235
5	27.015	25	27.245
6	27.025	26	27.265
7	27.035	27	27.275
8	27.055	28	27.285
9	27.065	29	27.295
10	27.075	30	27.305
11	27.085	31	27.315
12	27.105	32	27.325
13	27.115	33	27.335
14	27.125	34	27.345
15	27.135	35	27.355
16	27.155	36	27.365
17	27.165	37	27.375
18	27.175	38	27.385
19	27.185	39	27.395
20	27.205	40	27.405

In class-B or Class-C amplifier, only part of the signal cycle is passed through the amplifying device. However, a resonant inductance-capacitance circuit, called a tank circuit, completes the cycle in the output by storing and releasing the energy at the resonant frequency. This storage of energy is where the tank circuit gets its name. *See also* CLASS-B AMPLIFIER, CLASS-C AMPLIFIER, TANK CIRCUIT.

CITIZEN'S BAND

The Citizen's Radio Service, or Citizen's Band, is a public, noncommercial radio service available to the general population of the United States. There are four classes of Citizen's Band radio, called Class A, Class B, Class C, and Class D. Class-A stations are licensed in the band 460 to 470 MHz, and are allowed 60 watts maximum input power. Class-B stations are allowed 5 watts of input power in the same band. Class-C stations are operated in the band 26.96 to 27.23 MHz, and on 27.255 MHz, as well as in the range 72 to 76 MHz, for radio-control purposes. Class-D stations are allowed to use 40 channels in the range 26.965 to 27.405 MHz for general communications. Class-D is by far the most popular of the Citizen's Radio classes. The table lists the 40 channels of the Class-D band.

Citizen's Band, or CB, offers a convenient communications system that can be used at home, in a car (see illustration), in a boat, on an airplane, or even while walking or hiking. The transmitter output power is legally limited on the Class-D band to 4 watts for amplitude modulation and 12 watts peak for single sideband. The communicating range is typically between 10 and 30 miles maximum. However, long-distance propagation is occasionally observed, since the 27-MHz band is susceptible to the effects of sunspot activity.

Citizen's Band enjoyed a great boom in the middle 1970s, when millions of Americans learned how easily they could obtain and operate simple two-way radio equipment. Hundreds of CB clubs exist on the local and national scale. An organization called Radio Emergency Associated Citizen's Teams, or REACT for short, monitors Channel 9, the officially designated emergency frequency, and provides assistance to motorists in trouble.

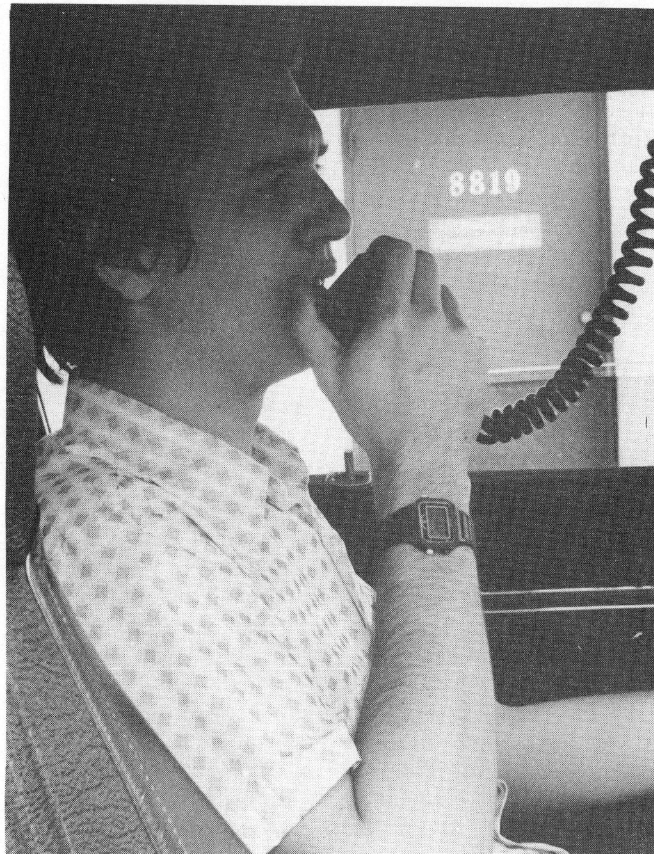

CITIZEN'S BAND: Citizen's Band radio is especially convenient for use in automobiles.

CLAMP

A clamp circuit is one that sets, or holds, the operating level of a component or another circuit. The illustration shows how a clamp tube can be used to hold (clamp) the plate dissipation of an RF amplifier to a safe level. While drive is applied to the amplifier, bias is developed across R_g, and the clamp tube is cut off, allowing the amplifier to operate normally. If drive fails or is removed, the bias becomes zero, and the clamp tube draws current through the screen resistor, R_s, which lowers the screen voltage to a point where the plate dissipation of the amplifier is reduced significantly. This system of protection was popular at one time with people who wanted to build a low-cost amplifier

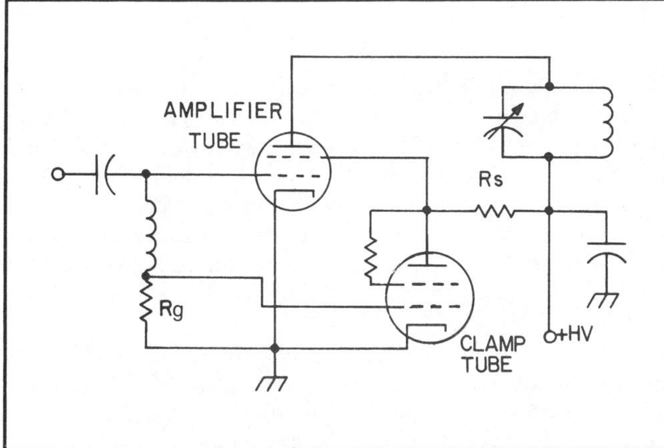

CLAMP: A tube is used to clamp the screen of an amplifier when excitation fails or is removed. Without grid bias, the clamp tube causes a voltage drop across the screen resistor, R_s thus lowering the dissipation of the amplifier tube to a safe level.

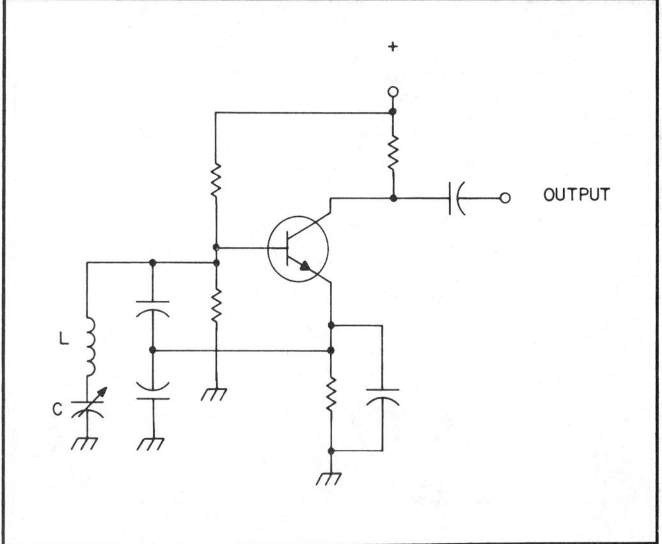

CLAPP OSCILLATOR: A Clapp oscillator using an NPN bipolar transistor.

because the expense of a protective fixed-bias supply was avoided. It was also used (with poor efficiency) as a modulator by impressing the audio voltage on the control grid of the clamp tube.

Another type of clamp circuit, also called a dc restorer, is used to restore the dc component to an ac waveform after the ac has been amplified. Early television receivers used one or more diodes to restore the dc level to the video information before it was applied to the cathode ray tube. This caused the black (or white) level of the picture to remain constant from one side of the screen to the other. Modern receivers have multipurpose integrated circuits that incorporate this feature automatically.

Waveform clippers or peak limiters using diodes have sometimes been called clamp circuits, although this terminology is not strictly correct. Their purpose is to limit the positive or negative excursions of an ac waveform. *See also* CLIPPER.

CLAPP OSCILLATOR

A specialized form of Colpitts oscillator, using a series-tuned tank circuit, is called a Clapp oscillator (*see* COLPITTS OSCILLATOR). The tank circuit is tapped using a capacitive voltage-divider network, but the voltage-divider capacitors are fixed. This is shown in the drawing. The frequency of the oscillator is adjusted either by varying the tank capacitance C, or by varying the tank inductance L. The output can be taken from a secondary winding coupled to L, or by means of capacitive coupling, as shown.

A Clapp oscillator is generally more stable than a Colpitts oscillator, since the variable capacitor is independent of the voltage-dividing network.

CLASS-A AMPLIFIER

A Class-A amplifier is a linear amplifier in which the plate, collector, or drain current flows for 100 percent of the input cycle. The tube or transistor is never driven to the cutoff point, and the input signal occurs over the linear part of the characteristic curve. The tube or transistor should be biased at the middle of the linear part of the characteristic curve for Class-A operation, as shown in the illustration.

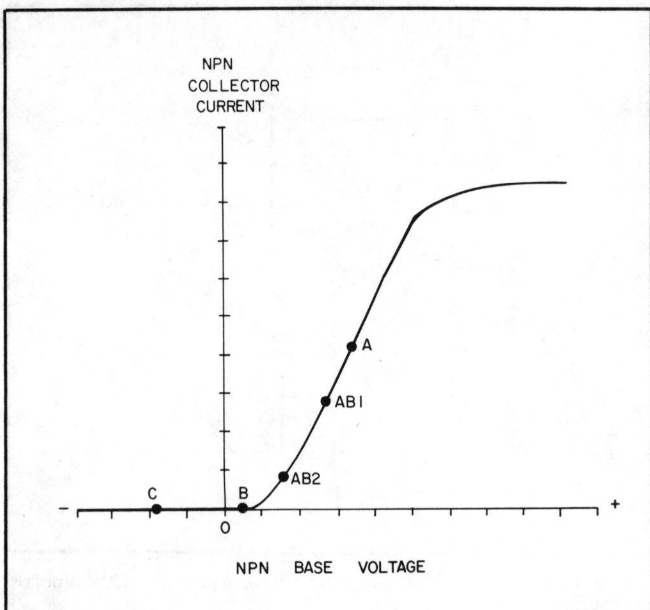

CLASS-A, CLASS-AB, CLASS-B, AND CLASS-C AMPLIFIER: Operating points on the characteristic curve of an NPN bipolar transistor for various classes of amplifier operation. The situations are similar for tubes and field-effect transistors.

Class-A amplifiers are often used in audio-frequency applications, where a minimum amount of waveform distortion is important. A Class-A amplifier may be either single-ended or push-pull (*see* PUSH-PULL AMPLIFIER). The efficiency of a Class-A amplifier is low, approximately 20 percent. Class-A amplifiers draw essentially no power from the input source; they are therefore good for use as receiver preamplifiers and front-end amplifiers. Radio-frequency power amplifiers are usually operated in Class AB, B, or C. *See also* AMPLIFIER, CHARACTERISTIC CURVE, CIRCUIT EFFICIENCY.

CLASS-AB AMPLIFIER

A Class-AB amplifier is an amplifier operated between the Class-A and Class-B bias conditions (*see* CLASS-A AMPLIFIER, CLASS-B AMPLIFIER). Plate, collector, or drain current flows for all or most of the input signal cycle. However, the tube or transistor is driven into the nonlinear portion of the characteristic curve, and therefore some waveform distortion occurs in the output.

There are two kinds of Class-AB amplifier. The Class-AB1 amplifier is not driven to cutoff, although the input signal drives the tube or transistor into the nonlinear region near cutoff. In a Class-AB2 amplifier, the tube or transistor is cut off during a small part of the cycle. See the illustration for approximate bias points for Class-AB1 and Class-AB2 operation of an NPN bipolar transistor.

The efficiency of a Class-AB amplifier is slightly higher than that of a Class-A amplifier, but not as good as that of a Class-B amplifier. A Class-AB1 amplifier draws very little power from the signal input source, but a Class-AB2 circuit draws significant power. Some distortion of the signal waveform occurs in both cases; thus Class-AB amplifiers are not used for high-fidelity applications. Class-AB circuits are sometimes used as radio-frequency power amplifiers. *See also* AMPLIFIER, CHARACTERISTIC CURVE.

CLASS-B AMPLIFIER

A Class-B amplifier is an amplifier operated at or near the cutoff point of the characteristic curve of a tube or transistor device. Plate, collector, or drain current flows for approximately 50 percent of the signal input cycle. During the other part of the cycle, the device is cut off. The illustration shows the direct-current bias point for Class-B operation of an NPN bipolar transistor.

Class-B amplifiers are used as radio-frequency amplifiers. In the push-pull configuration, Class-B circuits offer low distortion of the waveform, and good efficiency at audio frequencies. The efficiency of a Class-B amplifier is usually less than 50 percent. Class-B amplifiers draw considerable power from the source in the single-ended configuration. *See also* AMPLIFIER, CHARACTERISTIC CURVE, CIRCUIT EFFICIENCY.

CLASS-C AMPLIFIER

A Class-C amplifier is an amplifier operated beyond the cutoff point of the characteristic curve of a tube or transistor. The plate, collector, or drain current flows for less than half of the signal input cycle. During the remainder of the cycle, the device is cut off. The direct-current base bias for Class-C operation of an NPN bipolar transistor is shown in the illustration.

Class-C amplifiers are unsuitable for audio-frequency applications, since the output waveform is severely distorted. Class-C amplifiers are also unsuitable for weak-signal use, and for linear applications. However, Class-C amplifiers are often used in continuous-wave or frequency-modulated transmitters, where there is no fluctuation of carrier amplitude. The efficiency of the Class-C amplifier is high. In some cases it can approach 80 percent. Class-C amplifiers require a large amount of driving power to overcome the cutoff bias. *See also* AMPLIFIER, CHARACTERISTIC CURVE, CIRCUIT EFFICIENCY.

CLEAR

The term clear refers to the resetting or reinitialization of a circuit. All active memory contents of a microcomputer are erased by the clear operation. Auxiliary memory is retained when active circuits are cleared.

All electronic calculators have a clear function button. When this button is actuated, the calculation is discontinued and the display reverts to zero. By switching a calculator off and then back on, the clear function is done automatically.

CLEAR CHANNEL

A clear channel is an amplitude-modulation (AM) broadcast-band channel that renders service over a large area, and is protected within that area against interference. The maximum amount of power allowed a clear-channel station is 50 kW. This is the greatest amount of power permissible on the standard AM broadcast band in the United States. Some stations in other countries use considerably more power.

Clear-channel status is allocated to comparatively few stations. Much more common are the regional-channel and local-channel stations, with lower maximum power limits. *See also* LOCAL CHANNEL, REGIONAL CHANNEL.

CLICK FILTER

When a switch, relay, or key is opened and closed, a brief

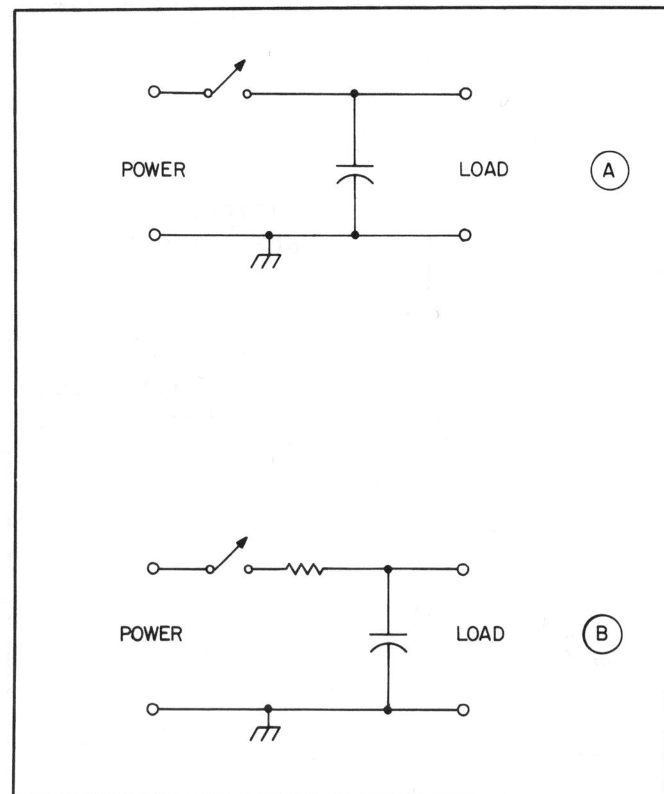

CLICK FILTER: Click filters. At A, a capacitor smooths out the decay portion of the code signal envelope. At B, a series resistor, in addition to the capacitor, smooths out both the rise and decay parts of the signal.

pulse of radio-frequency energy is emitted. This is especially true when the device carries a large amount of current. A capacitor connected across the device slows down the decay time from the closed to the open condition, where the click is most likely to occur. The illustration shows this arrangement at "A." A small resistor in series, combined with the capacitor in parallel, slows down the make time from the open to the closed state (at "B"). These devices are called click filters. Sometimes a choke is used in place of the resistor for circuits that draw high current.

In a code transmitter, a click filter is used to regulate the rise and decay times of the signal. Without such a filter, the rapid rise and decay of a signal can cause wideband pulses to be radiated at frequencies well above and below that of the carrier itself. This can result in serious interference to other stations. *See also* KEY CLICK, SHAPING.

CLIP LEAD

A clip lead is a short length of flexible wire, equipped at one or both ends with an alligator clip or similar temporary connector (*see* ALLIGATOR CLIP).

Clip leads come in many sizes. The photograph shows a typical clip lead used in a test laboratory for ground connections and temporary taps.

Clip leads are not suitable for permanent installations, especially outdoors, since corrosion occurs easily, and the connector may slip off. The current-carrying capacity of clip leads is quite limited.

CLIPPER

A clipper is a device that limits the peak amplitude of a signal at some value smaller than the peak value it normally

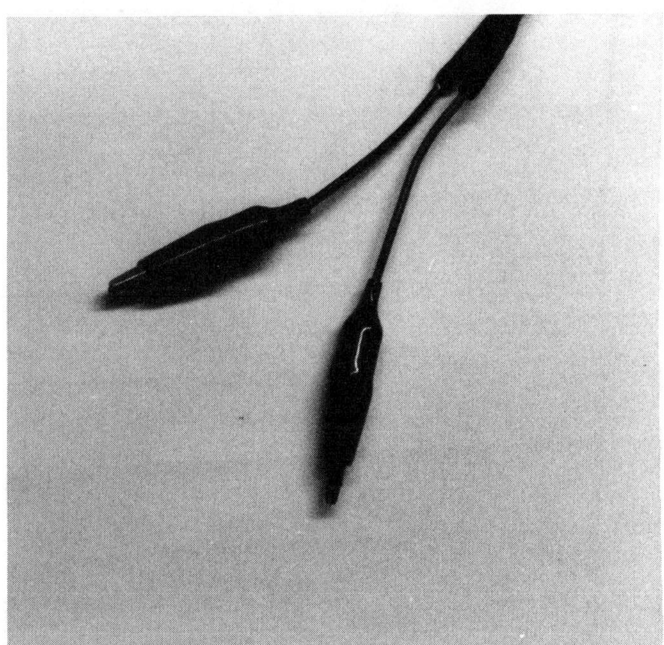

CLIP LEAD: Clip leads are convenient for temporary connections.

attains. This always results in distortion of the waveform or modulation envelope. Clippers are useful when it is necessary to limit the peak amplitude of a signal for any reason. Clippers are often used in single-sideband (SSB) transmitters to increase the average power in relation to the peak-envelope power.

The illustration shows unclipped and clipped sine-wave signals as they would appear on an oscilloscope display. In a process called radio-frequency speech clipping, an SSB signal is first amplified, then clipped, then put through a bandpass filter, and finally applied to the transmitter driver and final-amplifier stages. Clipping is sometimes used in code reception to limit the maximum audio output of the receiver, so that the automatic-gain-control (AGC) circuit can be disabled. Some code operators prefer this type of reception. *See also* CLAMPING, SPEECH CLIPPING.

CLOCK

A clock is a pulse generator that serves as a time-synchronizing standard for digital circuits. The clock sets the speed of operation of a microprocessor, microcomputer, or computer.

The clock produces a stream of electrical pulses with extreme regularity. Some clocks are synchronized with time standards. The speed may also be controlled by a resistance-capacitance network or by a piezoelectric crystal. The clock frequency is generally specified in pulses per second, or Hertz.

CLOSED CIRCUIT

Any complete circuit that allows the flow of current is called a closed circuit. All operating circuits are closed.

A transmission sent over a wire, cable, or fiber-optics medium, and not broadcast for general reception, is called a closed-circuit transmission. A telephone operates via a closed circuit (except, of course, for a radio telephone). Some closed-circuit radio and television systems are used as intercoms or security monitoring devices. *See also* CARRIER-CURRENT COMMUNICATION.

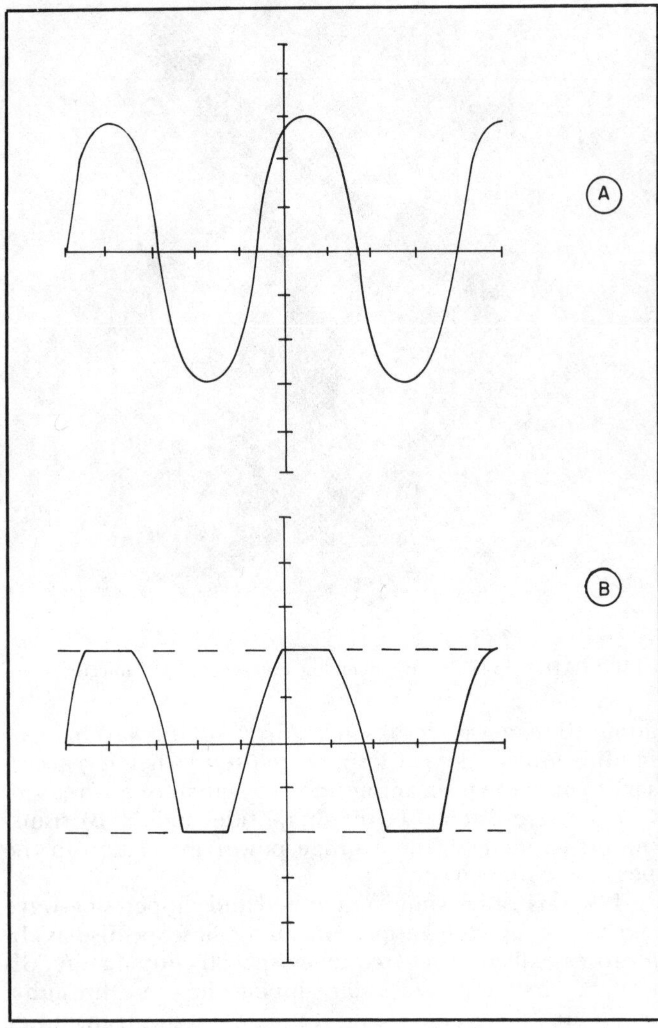

CLIPPER: Unclipped (A) and clipped (B) waveforms.

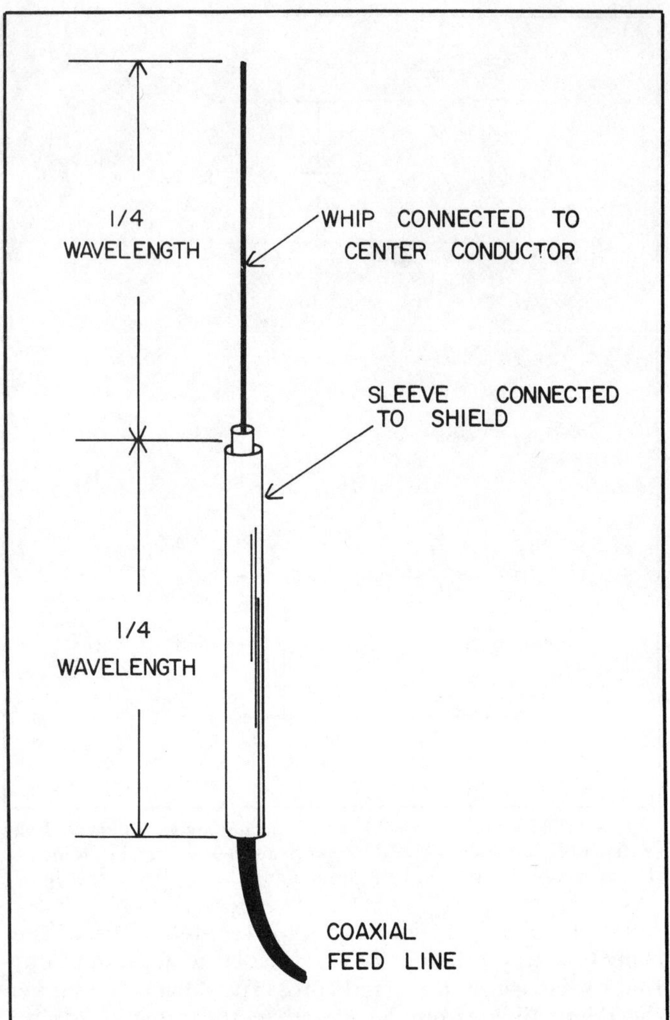

COAXIAL ANTENNA: Construction of a coaxial antenna.

COAXIAL ANTENNA

A coaxial antenna is a half-wave vertical dipole fed through one radiating element with coaxial cable. At the feed point, the cable center conductor is extended ¼ wavelength directly upward; the shield is folded back along the cable for ¼ wavelength. The illustration shows the construction of a coaxial antenna.

Coaxial antennas are frequently used on the 27-MHz Citizen's Band. Such antennas are, at that frequency, of practical dimensions—about 18 feet high—and the antenna provides a low angle of radiation. The radiation is vertically polarized.

Induced currents on the feed line are sometimes a problem with coaxial antennas. These currents can be choked off by winding the coaxial transmission line into a tight coil at the base of the antenna. The unobtrusive physical appearance of a Citizen's-Band coaxial antenna is shown in the photograph.

Coaxial antennas are practical at frequencies above about 5 MHz, where a vertical half-wave structure can be supported. *See also* DIPOLE ANTENNA, VERTICAL DIPOLE ANTENNA.

COAXIAL CABLE

Coaxial cable is a two-conductor cable consisting of a single center wire surrounded by a tubular metal shield. Most coaxial cables have a braided shield, insulated from the center conductor by polyethylene. Some coaxial cables have air dielectrics, and the center conductor is insulated from the shield by polyethylene beads or a spiral winding. The illustration shows the construction of a typical coaxial cable with a polyethylene dielectric.

Coaxial cable is commercially made in several diameters and characteristic-impedance values. A low-loss, well-shielded type of coaxial cable, with an outer conductor of solid metal tubing, is called hard line.

The table shows the characteristics of several commonly available varieties of prefabricated coaxial cable.

Coaxial cable is convenient to install, and it may be run next to metal objects, and even underground, without adversely affecting the loss performance. However, most coaxial cables have greater loss per unit length than two-conductor lines or waveguides. The characteristic impedance of coaxial lines is generally lower than that of two-wire lines. *See also* OPEN-WIRE LINE, TRANSMISSION LINE, WAVEGUIDE.

COAXIAL SWITCH

A coaxial switch is a multi-position switch designed for use with coaxial cable. The photograph shows a typical coaxial switch intended for radio-frequency use.

Coaxial switches must have adequate shielding. This necessitates that the enclosure be made of metal, such as aluminum. At very high frequencies and above, a coaxial switch must be designed to have a characteristic impedance

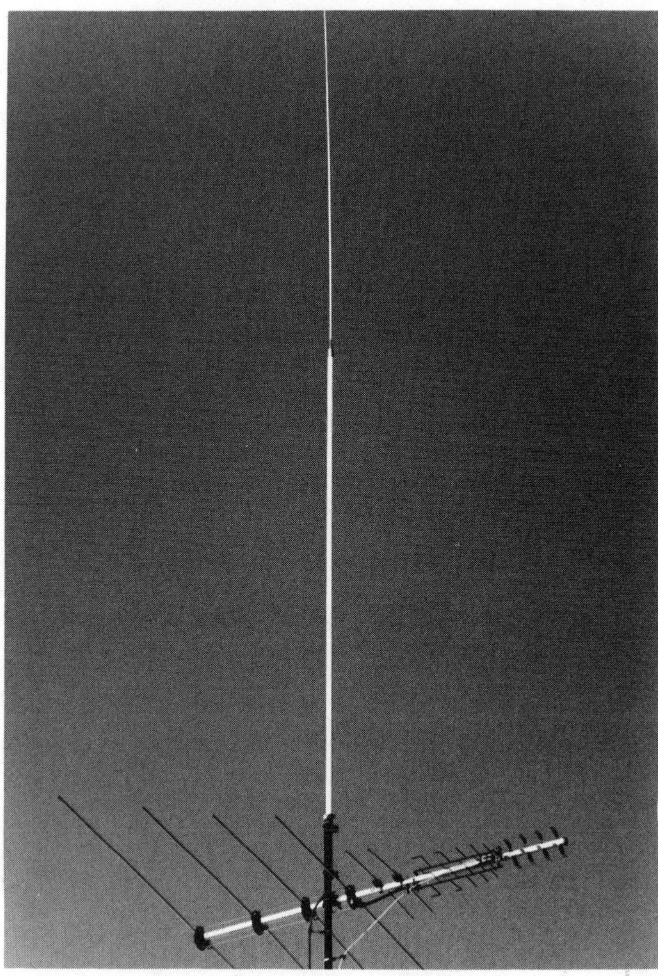
COAXIAL ANTENNA: A typical coaxial antenna. This unobtrusive radiator is designed for the 27-MHz Citizen's Band.

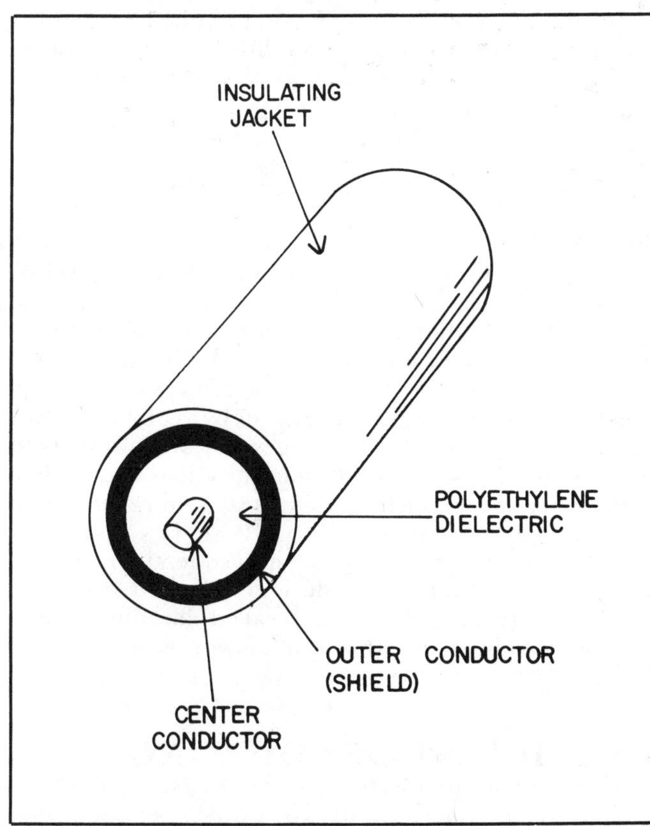

COAXIAL CABLE: Cross section of coaxial cable.

COAXIAL CABLE: CHARACTERISTICS OF PREFABRICATED COAXIAL TRANSMISSION LINES.

Type	Characteristic Impedance, Ohms	Velocity Factor	Outside Dia., in.	Picofarads per Foot
RG-8/U	52	0.66	0.41	29.5
RG-9/U	51	0.66	0.42	30.0
RG-11/U	75	0.66	0.41	20.6
RG-17/U	52	0.66	0.87	29.5
RG-58/U	54	0.66	0.20	28.5
RG-59/U	73	0.66	0.24	21.0
RG-174/U	50	0.66	0.10	30.8
Hard line 1/2-inch	50	0.81	0.50	25.0
	75	0.81	0.50	16.7
Hard line 3/4-inch	50	0.81	0.75	25.0
	75	0.81	0.75	16.7

identical to that of the transmission line in use; otherwise, impedance discontinuities may contribute to loss in the antenna system.

Some coaxial switches can be operated by remote control. This is especially convenient when there are several different antennas on a single tower, and no two of them have to be used at the same time. A single length of cable may then be used as the main feed line, and each antenna can be connected to a separate branch via the switch.

COAXIAL TANK CIRCUIT

A coaxial cable, cut to any multiple of ¼ electrical wavelength, can be used in place of an inductance-capacitance tuned circuit. If the length of the cable is an even multiple of ¼ wavelength, a low impedance is obtained by short-circuiting the far end, or a high impedance is obtained by opening the far end (A and B in the illustration). If the length of the cable is an odd multiple of ½ wavelength, a high impedance is obtained by short-circuiting the far end and a low impedance is obtained by opening the far end (C and D).

Coaxial tank circuits are used mostly at very-high and ultra-high frequencies, where ¼ or ½ wavelength is a short physical length. Coaxial tank circuits have excellent selectivity. Cavity resonators are also used as tuned circuits at very-high and ultra-high frequencies. *See also* CAVITY RESONATOR, COAXIAL WAVEMETER.

COAXIAL WAVEMETER

For measuring very-high, ultra-high, and microwave frequencies, a coaxial wavemeter is sometimes employed. This device consists of a rigid metal cylinder with an inner conductor along its central axis, and a sliding disk that shorts the cylinder and the inner conductor. The coaxial wavemeter is thus a variable-frequency coaxial tank circuit (see illustration).

By adjusting the position of the shorting disk, resonance can be obtained. Resonance is indicated by a dip or peak in an RF voltmeter or ammeter. The length of the resonant section is easily measured; this allows determination of the wavelength of the applied signal. The frequency is determined from the wavelength according to the formula

$$F = 300k/\lambda$$

where f is the frequency in megahertz and λ is the

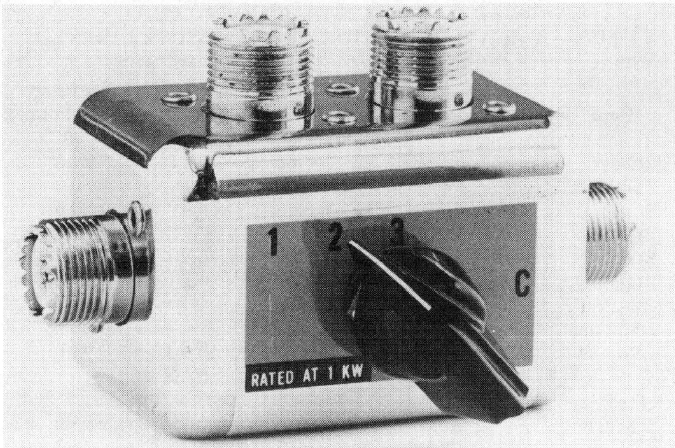

COAXIAL SWITCH: A coaxial switch, suitable for use in transmission lines at radio frequencies (courtesy of Gold Line Connector).

wavelength in meters; k is the velocity factor of the cable tank circuit, typically 0.95 for air dielectric. *See also* COAXIAL TANK CIRCUIT, VELOCITY FACTOR.

COBOL

COBOL is a computer language. The acronym stands for Common Business-Oriented Language. Commands and functions are expressed as words in the English language. COBOL is primarily intended, as its name implies, for business use; it is an internationally standardized computer language.

For problems in mathematics and physics, such as are encountered in scientific research, the languages BASIC and FORTRAN are generally preferred. *See also* BASIC, FORTRAN.

CODE

Any alternative representation of characters, words, or sentences in any language is a code. Some codes are binary, consisting of discrete bits in either an "on" or "off" state. The most common binary codes in use today for communications purposes are ASCII, BAUDOT, and the International Morse code (*see* ASCII, BAUDOT, INTERNATIONAL MORSE CODE). The "Q" and "10" signals, which are abbreviations for various statements, are codes (*see* Q-SIGNAL, TEN-CODE). Words in computer languages are a form of code.

Binary codes allow accurate and rapid transfer of information, since digital states provide a better signal-to-noise ratio than analog forms of modulation. The oldest telecommunications system, a combination of the Morse code and the human ear, is still used today when all other modes fail.

CODE TRANSMITTER

A code transmitter is the simplest kind of radio-frequency transmitter. It consists of an oscillator and one or more stages of amplification. One of the amplifiers is keyed, to turn the carrier on and off. The block diagram shows a simple code transmitter.

Sophisticated code transmitters may employ mixers for multiband operation. Many amplitude-modulated, frequency-modulated or single-sideband transmitters can function as code transmitters. An unmodulated carrier is simply keyed through the amplifying stages.

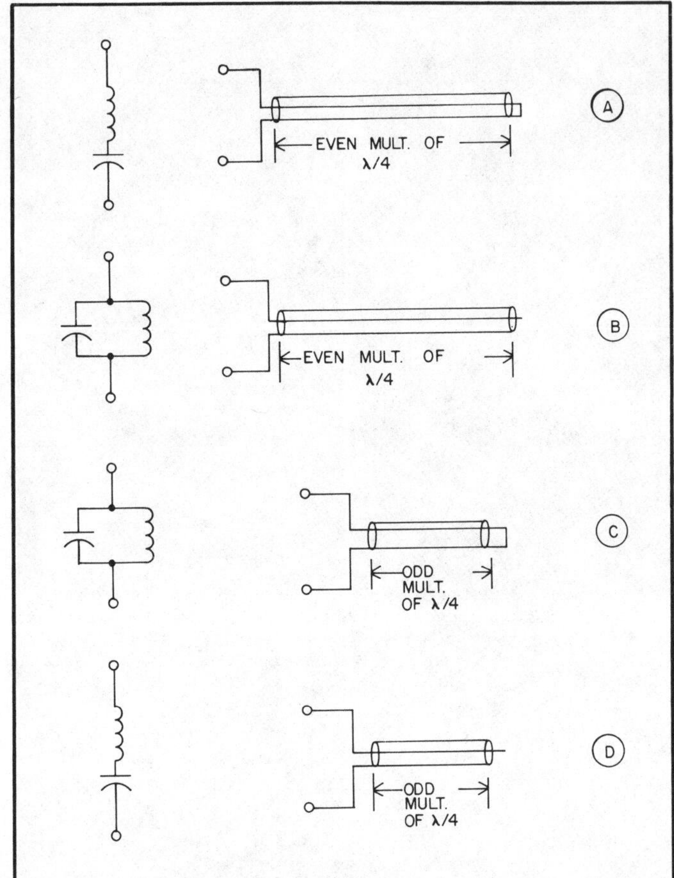

COAXIAL TANK CIRCUIT: At A and B, even multiples of one-quarter electrical wavelength; at C and D, odd multiples of one-quarter electrical wavelength. Equivalent inductance-capacitance circuits are shown.

Ideally, the output of a code transmitter is a pure, unmodulated sine wave at the operating frequency. Changes in amplitude under key-down conditions are undesirable. The rise and decay times of the carrier, as the transmitter is keyed, must be regulated to prevent key clicks. The frequency should be stable to prevent chirp. *See also* CHIRP, SHAPING.

CODING

The process of formulating a code is called coding. When preparing a code language, it is necessary to decide whether the smallest code element will represent a character, a word, or a sentence. The ASCII, BAUDOT, and Morse codes (*see* ASCII, BAUDOT, INTERNATIONAL MORSE CODE) represent each character by a combination of digital pulses. Computer languages use digital words to perform specific functions. Communications codes use a group of characters, such as QRX or 10-4, to represent an entire thought or sentence (*see* Q SIGNAL, TEN CODE).

When a language is translated into code, the process is called encoding. When a code is deciphered back into ordinary language, the process is called decoding. These functions may be done either manually or by machine. *See also* DECODING, ENCODING.

COEFFICIENT OF COUPLING

Two circuits may interact to a greater or lesser extent. The degree of interaction, or coupling, between two alternating-current circuits is expressed as a quantity

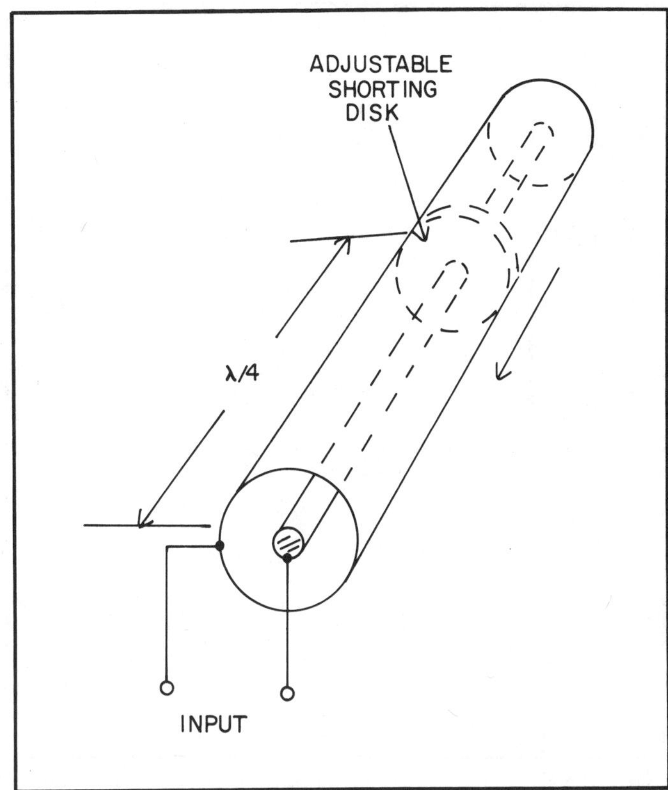

COAXIAL WAVEMETER: A coaxial wavemeter is an adjustable resonant coaxial tank circuit, connected across an indicating device.

called the coefficient of coupling, abbreviated in equations by the letter k. Usually, the coefficient of coupling is used in reference to inductors.

The coefficient of coupling, k, is related to the mutual inductance M and the values of two coils L1 and L2 according to the formula:

$$k = M/\sqrt{L1L2}$$

where the inductances are specified in henrys.

For impedances Z1 and Z2 in general, where they are of the same kind (predominantly capacitive or inductive):

$$k = M/\sqrt{Z1Z2}$$

where M is the mutual impedance. *See also* MUTUAL IMPEDANCE, MUTUAL INDUCTANCE.

COERCIVE FORCE

A magnetic material having some residual magnetism may be demagnetized by the application of a magnetic force opposing that of the existing field. For example, a permanent magnet, placed inside a coil carrying a direct current, can be demagnetized by that current if the field created by the current is opposite to that of the magnet. The amount of magnetizing force H, needed to demagnetize a certain object, is called coercive force.

A particular kind of magnetic material can be magnetized only to a certain extent, called the saturation induction. The amount of magnetizing force H, required to completely magnetize a given material, is sometimes called the coercive force. The ease with which a material is magnetized and demagnetized is called the coercivity of that material. *See also* COERCIVITY.

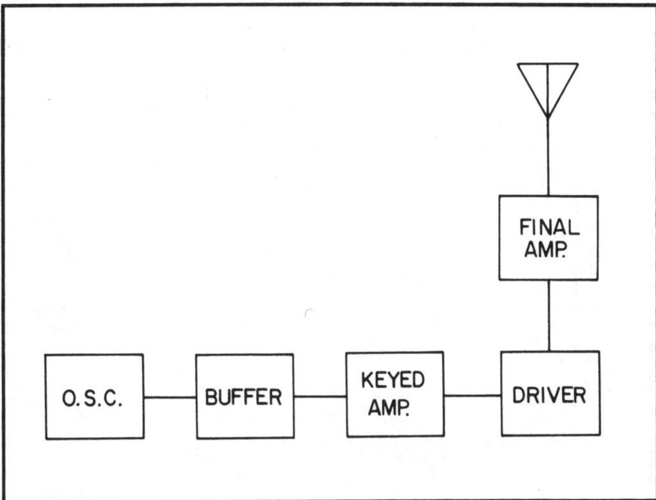

CODE TRANSMITTER: Block diagram of a simple code transmitter.

COERCIVITY

Some magnetic materials are fairly easy to magnetize and demagnetize; some materials are difficult to magnetize and demagnetize. The coercivity of a given material is an expression of the readiness with which it is magnetized and demagnetized.

In some applications, it is desirable to have a high degree of coercivity—that is, the material should be difficult to magnetize. Examples of such situations are the permanent magnet in a speaker, and a recording tape intended for the prolonged storage of information.

In some applications, a small amount of coercivity is desirable—that is, the material should be easy to magnetize and demagnetize. The core of an electromagnet has low coercivity. Some kinds of magnetic tape, intended for frequent erasing and re-recording, have low coercivity.

In general, the more coercive force required to demagnetize a material, the larger its coercivity. *See also* COERCIVE FORCE.

COHERENT LIGHT

Coherent light is light having a single frequency and phase. Most light, even if it appears to be monochromatic, consists of a certain range of wavelengths, and has random phase combinations. White light is made up of nearly equal radiation intensity at all visible frequencies; red light consists primarily of radiation at long visible wavelengths; green light is composed mostly of light in the middle of the visible frequency range. Figure 1 shows a spectral graph of sunlight passed through a red color filter (A). At B, we can see that the electromagnetic waves transmitted through the red color filter are in random phase, and variable frequency combinations.

The light transmitted by a helium-neon laser appears red, just as sunlight does through a red color filter. However, the laser light is emitted at just one wavelength (A in Fig. 2) and all the waves coming from the laser are in perfect phase alignment (B). Thus, the helium-neon laser emits coherent red light, while the red color filter transmits incoherent light.

Coherent light travels with greater efficiency—that is, lower attenuation per kilometer—than incoherent light. Using coherent light, a nearly parallel beam can be produced, and thus the energy is carried for tremendous distances with very little loss. Modulated-light communi-

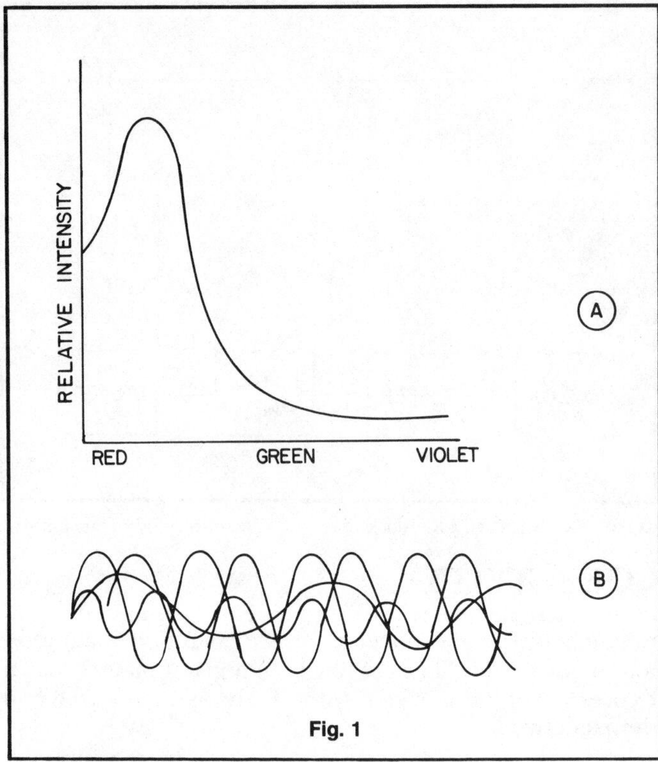

COHERENT LIGHT: At A, a spectral graph of incoherent light. At B, the various wavelengths and phase relationships of incoherent light (here greatly simplified).

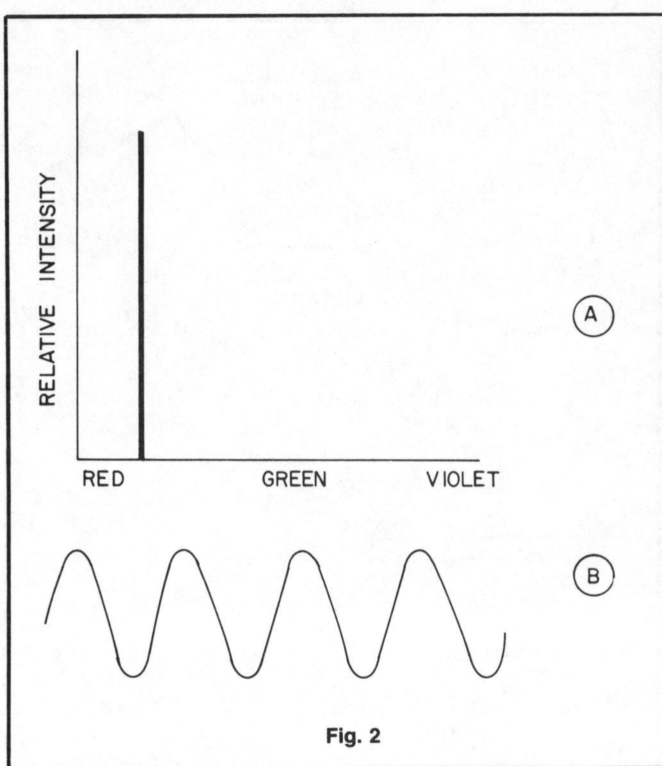

COHERENT LIGHT: At A, a spectral graph of coherent light. At B, the single wavelength emitted by a coherent-light generator.

cations systems generally use lasers, which produce coherent light. *See also* LASER, MODULATED LIGHT, OPTICAL COMMUNICATION.

COHERENT RADIATION

Coherent radiation is an electromagnetic field with a constant, single, frequency and phase. A continuous-wave radio-frequency signal is an example of coherent radiation. The static, or "sferics," produced by a thunderstorm, is an example of incoherent electromagnetic radio emission.

Energy is transferred more efficiently by coherent radiation than by incoherent radiation. The laser is an example of a visible-light device that produces coherent radiation. *See also* COHERENT LIGHT, LASER.

COIL

A coil is a helical winding of wire, usually intended to provide inductive reactance. The most common form of wire coil is the solenoidal winding (A in the drawing). The wire may be wound on an air core, or a core having magnetic permeability to increase the inductance for a given number of turns. Some coils are toroidally wound, as shown at B.

Coils are used in speakers, earphones, microphones, relays, and buzzers to set up or respond to a magnetic field. Coils are employed in transformers for the purpose of stepping a voltage up or down, or for the purpose of impedance matching. A coil wound on a ferrite rod can act as a receiving antenna at low, medium, and high frequencies. In electronic circuits, coils are generally used to provide inductance. *See also* COIL WINDING, INDUCTANCE, INDUCTOR.

COIL WINDING

When winding a coil to obtain a certain value of inductance, the dimensions of the coil, the number of turns, the type of core material, and the shape of the coil all play important roles.

Usually, if a powdered-iron or ferrite core materal is used for coil winding, data is furnished with the core as a guide to obtaining the desired value of inductance. For air-core solenoidal coils having only one layer of turns, the inductance L in microhenrys is given by the formula:

$$L = \frac{r^2N^2}{9R + 10M}$$

where R is the coil radius in inches, N is the number of turns, and m is the length of the winding in inches (see illustration). The inductance of a single-layer air-core solenoid thus increases with the square of the number of turns, and directly with the coil radius. For a coil with a given radius and number of turns, the greatest inductance is obtained when the length m is made as small as possible. *See also* INDUCTANCE, INDUCTOR.

COINCIDENCE CIRCUIT

A coincidence circuit is any digital circuit that requires a certain combination of input pulses in order to generate an output pulse. The input pulses must usually arrive within a designated period of time.

The most common form of coincidence circuit is a combination of AND gates (*see* AND GATE). For an output pulse to occur, all the inputs of an AND gate must be in the high state. Any complex logic circuit can be considered a coinci-

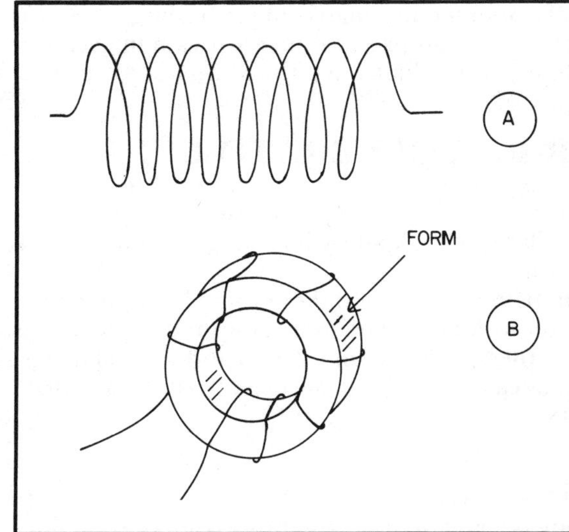

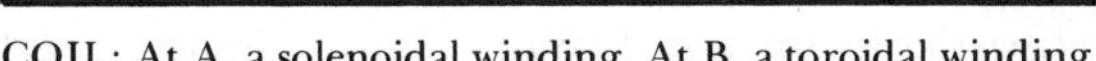

COIL: At A, a solenoidal winding. At B, a toroidal winding.

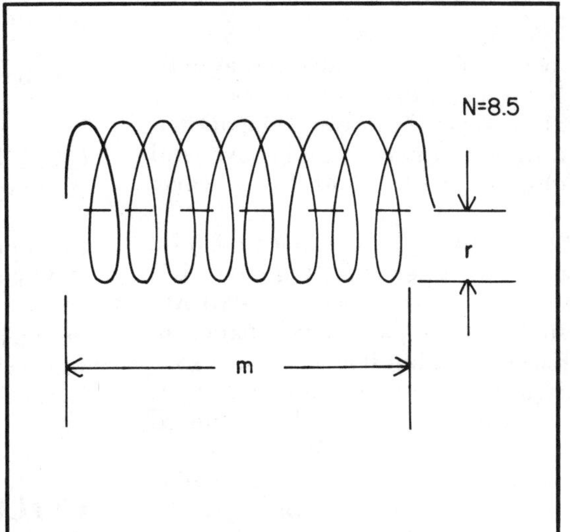

COIL WINDING: Single-layer, air-core, solenoidal coil dimensions for calculation of inductance.

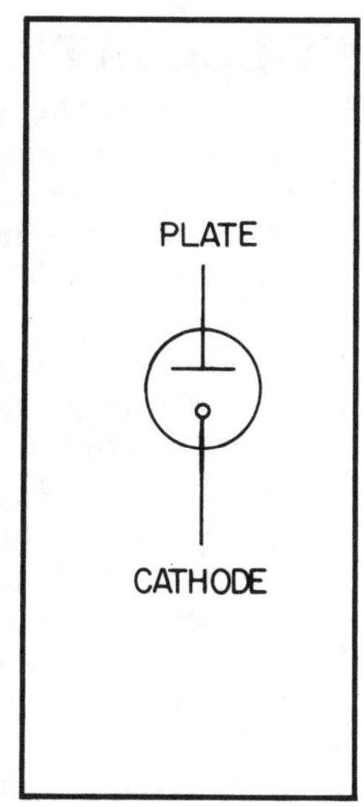

COLD CATHODE: Schematic symbol for a cold-cathode diode tube.

dence circuit, but the term is generally only used with reference to the logical operation AND.

COLD CATHODE

When a tube is operated without a filament, the cathode is said to be cold. Certain tubes are designed to operate with cold cathodes; these include photoelectric tubes, mercury-vapor rectifiers, and some voltage-regulator tubes. In a cold-cathode tube, the electrons are literally pulled from the cathode by a high positive anode potential.

The schematic symbol for a cold-cathode diode tube is shown in the drawing. *See also* MERCURY-VAPOR RECTIFIER, PHOTOTUBE.

COLLECTOR

The collector is the part of a semiconductor bipolar transistor into which carriers flow from the base under normal operating conditions. The base-collector junction is reverse-biased; in a PNP transistor the base is positive with respect to the collector, and in an NPN transistor the base is negative with respect to the collector.

The output from a transistor oscillator or amplifier is usually taken from the collector. The collector may be placed at ground potential in some situations, but it is usually biased with a direct-current power supply. The amount of power dissipated in the base-collector junction of a transistor must not be allowed to exceed the rated value, or the transistor will be destroyed. Resistors are often used to limit the current through the collector; such resistors are placed in series with either the emitter or collector lead. In some transistors, the collector is bonded to the outer case to facilitate heat conduction away from the base-collector junction.

The collector of a transistor corresponds roughly to the plate of a vacuum tube in circuit engineering applications, although the voltage is much smaller with the transistor than with the tube.

COLLECTOR CURRENT

In a bipolar transistor, the collector current is the average value of the direct current that flows in the collector lead. When there is no signal input, the collector current is a pure, constant direct current, determined by the bias at the base, the series resistance, and the collector voltage. The collector current for proper operation of a transistor varies considerably, depending on the application.

When a signal is applied to the base or emitter circuit of a transistor amplifier, the collector current fluctuates. But its average value, as indicated by an ammeter in the collector circuit (see drawing), may change only slightly. The collector current is the difference between the emitter current and the base current.

COLLECTOR RESISTANCE

The internal resistance of the base-collector junction of a bipolar transistor is called the collector resistance. This resistance may be specified either for direct current or for alternating current.

The direct-current collector resistance, R_{DC}, is given by:

$$R_{DC} = E/I$$

where E is the collector voltage and I is the collector current (see "A" in the illustration). This resistance varis with the supply voltage, the base bias, and any resistances in series with the emitter or collector.

The alternating-current resistance, R_{AC}, is given approximately by:

$$R_{AC} = \Delta E/\Delta I,$$

where ΔE and ΔI are the ranges of maximum-to-minimum collector voltage and current, as the fluctuating output current goes through its cycle. The illustration shows a method of approximately determining this dynamic resistance (B). The value of R_{AC} is affected by the same factors that influence R_{DC}. In addition, the class of operation has an effect, as does the magnitude of the input signal.

The alternating-current collector resistance is useful when designing a circuit for optimum impedance matching.

COLLIMATION

Collimation is the process of making electromagnetic waves parallel, or nearly parallel. This may theoretically be done at any frequency. In practice, however, collimation is done only at ultra-high radio frequencies and above, and at infrared, visible-light, ultraviolet, and X-ray wavelengths.

The dish antenna ("A" in drawing), when at least several wavelengths in diameter at the operating frequency, is a common example of a collimating device (*see* DISH ANTENNA). This dish usually has a paraboloid shape, although a spherical dish will work almost as well. Waves are emitted by a small horn at the focal point of the reflector. The waves then bounce off the reflector and are emitted from the assembly as a parallel beam. A parabolic or spherical mirror acts as a collimating device for visible light. Most flashlights and lanterns use this principle.

At infrared, visible-light, and ultraviolet wavelengths, collimation can be accomplished by means of a refracting lens, as shown at B. In a cathode-ray tube, the electron beam is collimated by the fields between sets of focusing electrodes. Electric fields may also be employed to collimate a beam of protons, alpha particles, or other charged atomic particles.

COLLINEAR ANTENNA

A set of half-wave radiators fed in phase and positioned in such a way that all the driven elements lie along one straight line, is called a collinear antenna. Such antennas may be either horizontal or vertical, as shown in the illustration. A set of stacked collinear antennas is sometimes called a collinear array.

Vertical collinear antennas are used in mobile and base installations at very-high and ultra-high frequencies, for the purpose of obtaining omnidirectional gain in the azimuth plane. A two-element collinear vertical provides about 3 dB power gain over a single dipole; a four-element collinear vertical gives about 6 dB gain over a vertical dipole.

When Yagi antennas are combined in collinear fashion, additional forward gain results, far greater than the gain from one Yagi alone. Sometimes twenty or more Yagi antennas are oriented in a matrix to form a large collinear array. Such arrays can produce forward gain in excess of 20 dB with respect to a half-wave dipole. *See also* STACKING.

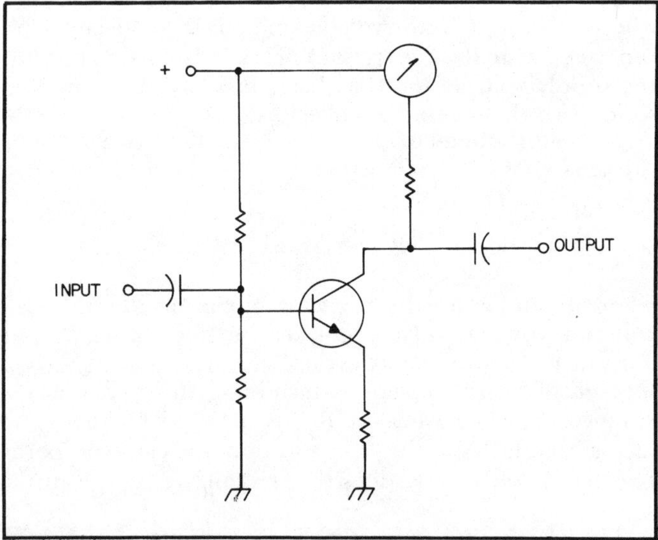

COLLECTOR CURRENT: Arrangement for measurement of the collector current in a bipolar-transistor amplifier.

COLOR-BAR GENERATOR

A color-bar generator is a device used in the testing and adjustment of a color television receiver. It operates in a manner similar to a black-and-white bar generator (*see* BAR GENERATOR).

The bar pattern may be either vertical or horizontal, and in various color combinations. Color reproduction, as well as brightness, contrast, horizontal and vertical linearity, and focus can be adjusted. *See also* COLOR TELEVISION, TELEVISION.

COLOR CODE

A color code is a means of representing component values and characteristics by means of colors. This scheme is used with almost all resistors (*see* RESISTOR COLOR CODE). Color codes are sometimes used on capacitors, inductors, transformers, and transistors.

When a cable has several different conductors, the individual wires are usually color-coded as a means of identifi-

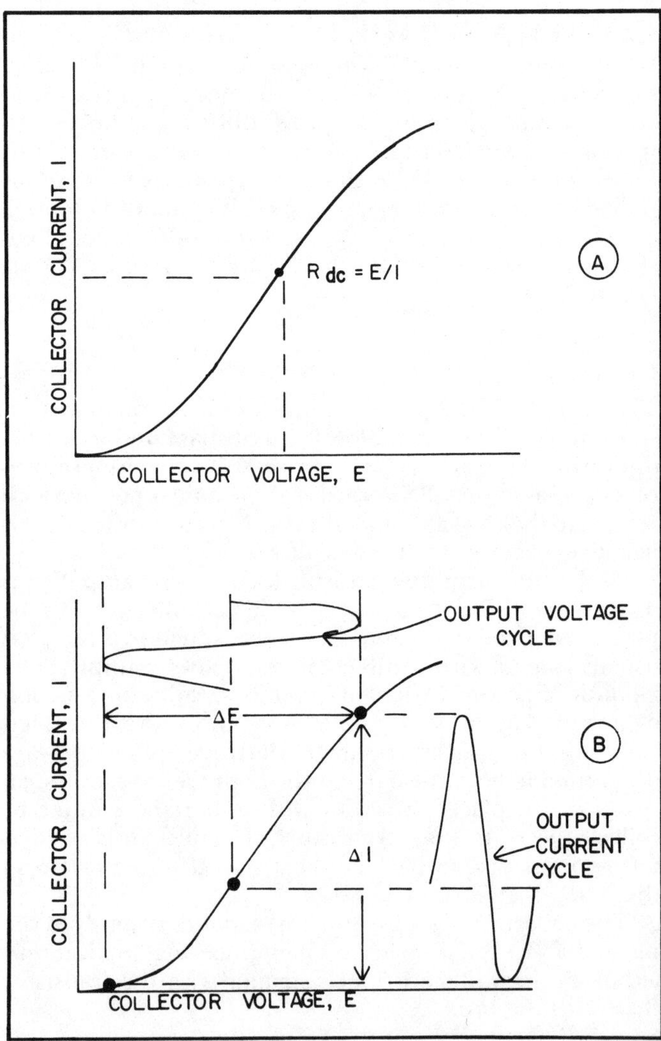

COLLECTOR RESISTANCE: At A, direct-current collector resistance is simply the quotient of the collector voltage and current. At B, the determination of alternating-current collector resistance is more complex. See text for discussion.

cation of conductors at opposite ends of the cable. In direct-current power leads, the color black often signifies ground or negative polarity, and red signifies the "hot" or positive lead. In house wiring, color coding may vary.

COLOR FIDELITY

The faithfulness or accuracy of the color in a color television is called color fidelity. Color fidelity depends on the proper intensity and linearity of the three primary colors—red, blue, and green—in the color signal from the transmitter. It also depends on proper alignment of the receiving equipment. Color fidelity may be degraded by interference.

The color fidelity of a television receiver is adjusted using a color-bar generator (*see* COLOR-BAR GENERATOR). Any scene will then appear nearly life-like. Fine adjustment of the color-intensity and color-tint functions are provided as front-panel controls on most television receivers. *See also* COLOR TELEVISION.

COLORIMETRY

The process of measuring color is called colorimetry. This is done with a device called a colorimeter, which compares any color to a reference standard. The three primary colors—red, blue, and green—are synthesized with constant chroma, but the luminance (brightness) of each color can be adjusted. By properly combining the three primary colors, the color under test can be matched. The relative components of the color are then indicated very precisely. Matching is done by visual comparison. *See also* CHROMA, HUE, SATURATION.

COLOR PICTURE SIGNAL

A color picture signal is a modulated radio-frequency signal that contains all the information needed to accurately reproduce a scene in full color. The channel width of a color-television picture signal is typically 6 MHz.

The illustration shows the modulation envelope of a single horizontal line of the video information in a color picture signal. The equalizing pulses, horizontal synchronization pulses, and vertical synchronization pulses are not shown. The horizontal blanking pulse turns off the picture-tube electron beam as it retraces from the end of one line to the beginning of the next line. This pulse is followed by a color-burst signal, which consists of eight or nine cycles at 3.579545 MHz. The phase of this burst provides the color information. The video information is then sent as the electron beam scans from left to right. There are 525 or 625 horizontal lines per frame.

The illustration at B shows the equalizing, horizontal synchronization, and vertical synchronization pulses of the picture signal. These pulses are sent after each complete frame. *See also* TELEVISION.

COLOR TELEVISION

Color television is the transmission and reception of motion pictures, in full color, by modulated electromagnetic waves. A special picture tube, combining the three primary colors—red, blue, and green—produces all possible colors.

A color television signal is somewhat more complex than a black-and-white signal (*see* COLOR PICTURE SIGNAL, PICTURE SIGNAL). A color pulse, with precise frequency

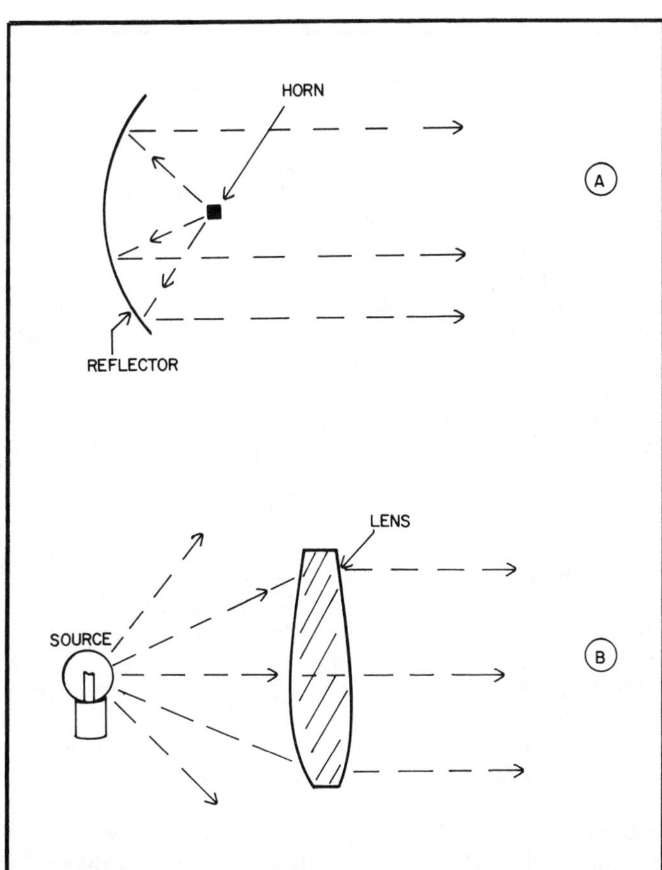

COLLIMATION: Collimation can be accomplished by means of a parabolic or spherical reflector (A), or, with infrared, visible light, and ultraviolet, by means of a refracting lens (B).

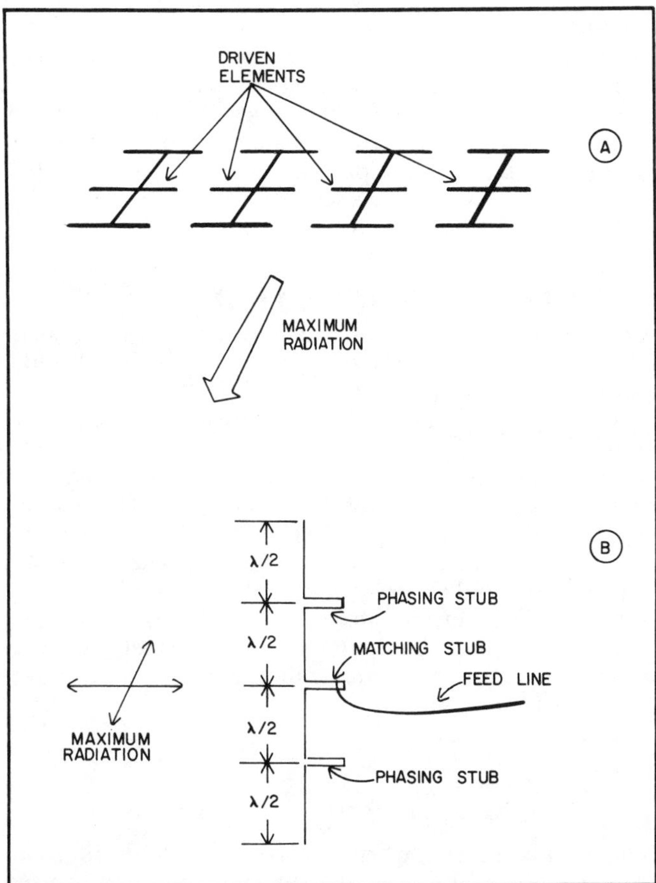

COLLINEAR ANTENNA: At A, a collinear antenna consisting of Yagis; at B, a vertical collinear antenna for omnidirectional operation.

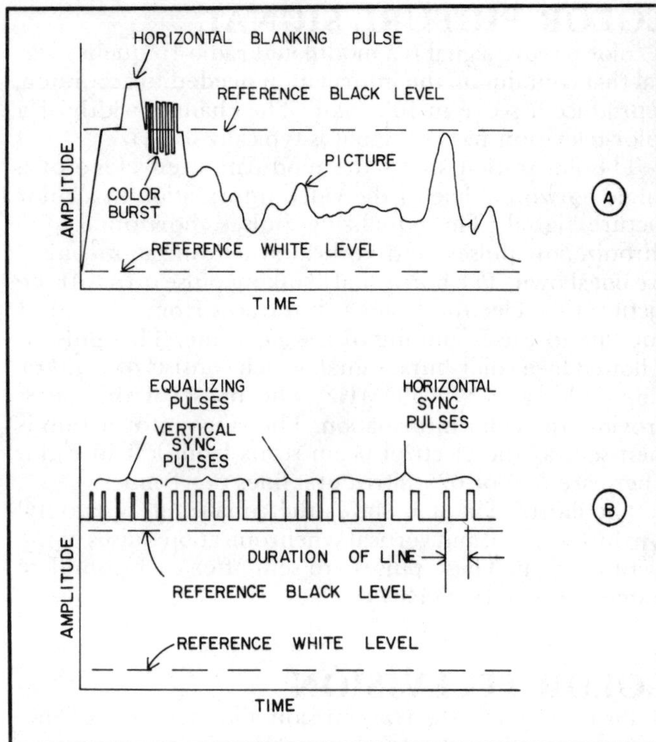

COLOR PICTURE SIGNAL: At A, the video information contained in one horizontal line of the picture; at B, the arrangement of equalizing pulses, vertical sync pulses, and horizontal sync pulses.

and phase, must be transmitted for accurate color reproduction. The standard frequency for this color burst is 3.579545 MHz.

The channels for standard television broadcasting may be used for either black-and-white or color transmissions. Color signals appear as normal black-and-white pictures on a television receiver not designed for color reception. When a black-and-white picture, such as an old movie, is broadcast via color television, the color function is disabled. *See also* TELEVISION.

COLPITTS OSCILLATOR

A Colpitts oscillator is an oscillator, usually of the variable-frequency type, characterized by capacitive feedback and a capacitive voltage-divider network. The illustration shows tube, transistor, and field-effect-transistor Colpitts circuits.

The operating frequency of the Colpitts oscillator is determined by the value of the inductance and the series combination of the two capacitors. Generally, the capacitors are variable. But alternatively, the capacitors may be fixed, and the frequency set by means of a variable inductor. The output can be taken from the circuit by inductive coupling, but better stability is usually obtained by capacitive or transformer coupling from the plate, collector, or drain circuit.

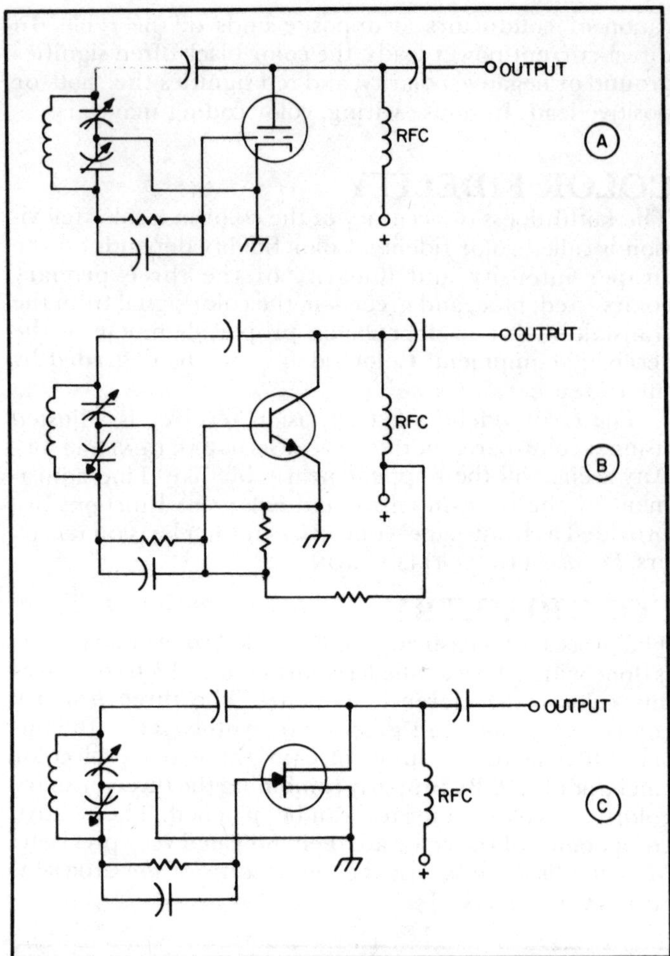

COLPITTS OSCILLATOR: Colpitts oscillator circuits, using a tube (A), a bipolar transistor (B), and a field-effect transistor (C).

COMMON CATHODE/ EMITTER/SOURCE

The common-cathode, common-emitter, and common-source circuits are probably the most frequently used amplifier arrangements with tubes, transistors, and field-effect transistors. The schematics show simple common-cathode, common-emitter, and common-source amplifier circuits. The cathode, emitter, or source is always operated at ground potential with respect to the signal; it need not necessarily be at ground potential for direct current. (The circuits shown are for illustrative purposes only and do not represent the only possible configurations.)

The common-cathode and common-source circuits have high input and output impedances. The common-emitter circuit is characterized by moderately high input impedance and high output impedance. In all three circuits, the input and output signals are 180 degrees out of phase. *See also* FIELD-EFFECT TRANSISTOR, TRANSISTOR, TUBE.

COMMON GRID/BASE/GATE

The common-grid, common-base, and common-gate circuits are amplifier or oscillator arrangements using tubes, transistors, and field-effect transistors respectively. These circuits have excellent stability as amplifiers. They are less likely to break into unwanted oscillation than the common-cathode, common-emitter, and common-source circuits (*see* COMMON CATHODE/EMITTER/SOURCE).

The grid, base, or gate is usually connected directly to ground; occasionally a direct-current bias may be applied and the grid, base, or gate shunted to signal ground with a bypass capacitor. The drawings illustrate typical common-grid, common-base, and common-gate amplifiers. (These diagrams do not, of course, represent the only

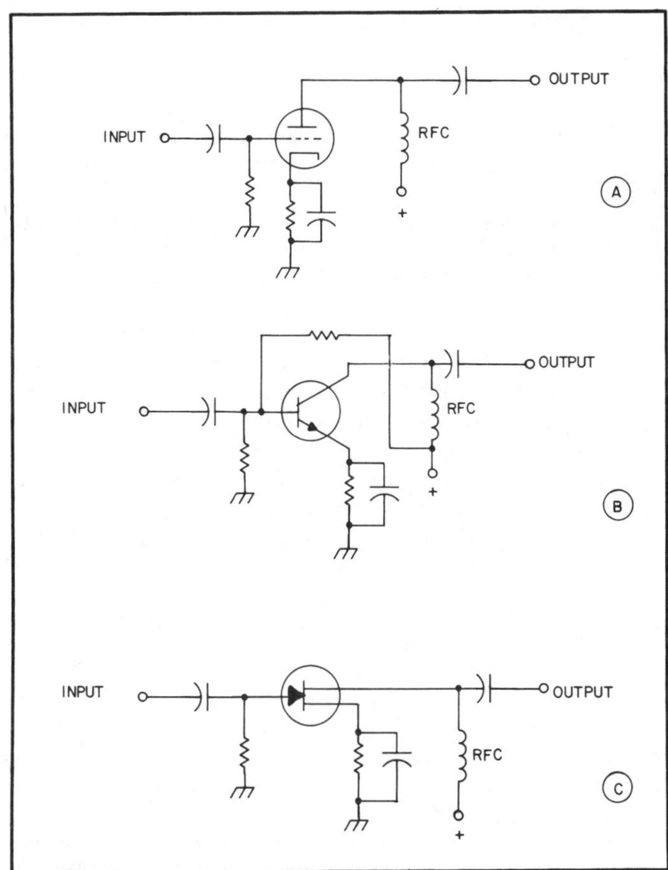

COMMON CATHODE/EMITTER/SOURCE: At A, a common-cathode amplifier; at B, a common-emitter amplifier; at C, a common-source amplifier.

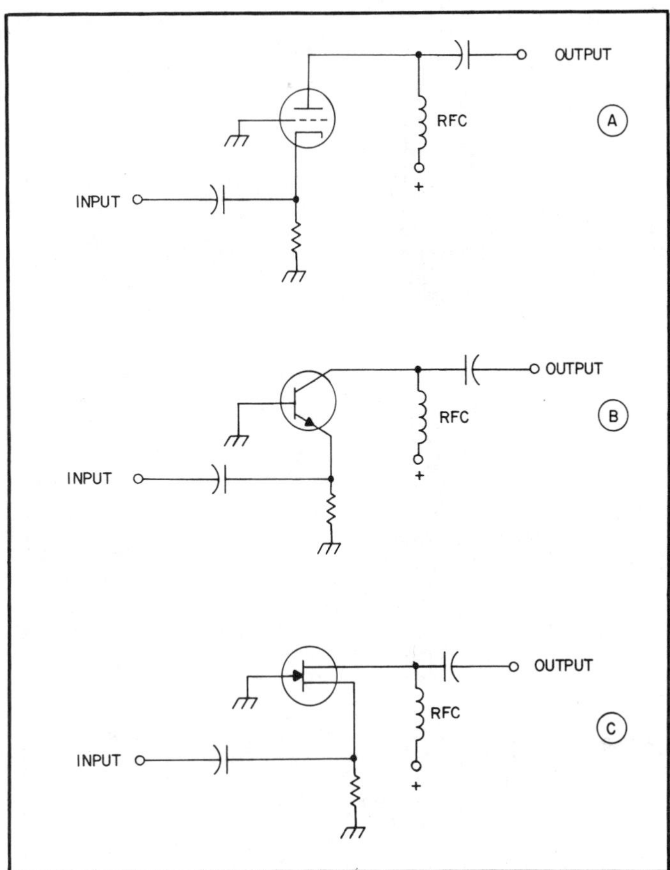

COMMON GRID/BASE/GATE: At A, a common-grid amplifier; at B, a common-base amplifier; at C, a common-gate amplifier.

possible configurations.)

The common-grid, common-base, and common-gate circuits display low input impedance. They require considerable driving power. The output impedance is high. The input and output waveforms are in phase. This kind of amplifier is often used as a power amplifier at radio frequencies. *See also* FIELD-EFFECT TRANSISTOR, TRANSISTOR, TUBE.

COMMON PLATE/ COLLECTOR/DRAIN

The common-plate, common-collector and common-drain circuits are generally used in applications where a high-impedance generator must be matched to a low-impedance load. The input impedances of the common-plate, common-collector, and common-drain circuits are high; the output impedances are low. The schematics show typical common-plate, common-collector, and common-drain circuits. (These circuits are not the only possible configurations.) They are sometimes called cathode-follower, emitter-follower, and source-follower circuits. The gain is always less than unity (*see* CATHODE FOLLOWER, EMITTER FOLLOWER, SOURCE FOLLOWER).

The plate, collector, or drain is sometimes grounded directly. However, this is not always done; biasing may be accomplished in a manner identical to that of the common-cathode, common-emitter, and common-source circuits (*see* COMMON CATHODE/EMITTER/SOURCE); the plate, collector, or drain is then placed at signal ground by means of a bypass capacitor, and the output is taken across a cathode, emitter, or source resistor or transformer. *See also* FIELD-EFFECT TRANSISTOR, TRANSISTOR, TUBE.

COMMUTATOR

A commutator is a mechanical device for obtaining a pulsating direct current from an alternating current. Commutators are used in motors and generators. A high-speed switch that reverses the circuit connections to a transducer, or rapidly exchanges them, is sometimes called a commutator.

In the direct-current motor, the commutator acts to reverse the direction of the current every half turn, so that the current in the coils always flows in one direction. As the motor shaft rotates, the commutator, attached to the shaft, connects the power supply to the motor coils. The method of obtaining contact is shown in the illustration.

In a direct-current generator, the commutator inverts every other half cycle of the output to obtain pulsating direct current rather than alternating current. The pulsations may be smoothed out using a capacitor. *See also* DC GENERATOR.

COMPACTRON

A compactron is a combination of several electron tubes within a single envelope (*see* TUBE). Common types of compactron include dual diodes, triodes, tetrodes, and pentodes. However, some compactrons contain three, four, or more individual tubes of different types.

The compactron is seldom seen in modern circuits. This is because tubes have been largely supplanted by solid-state

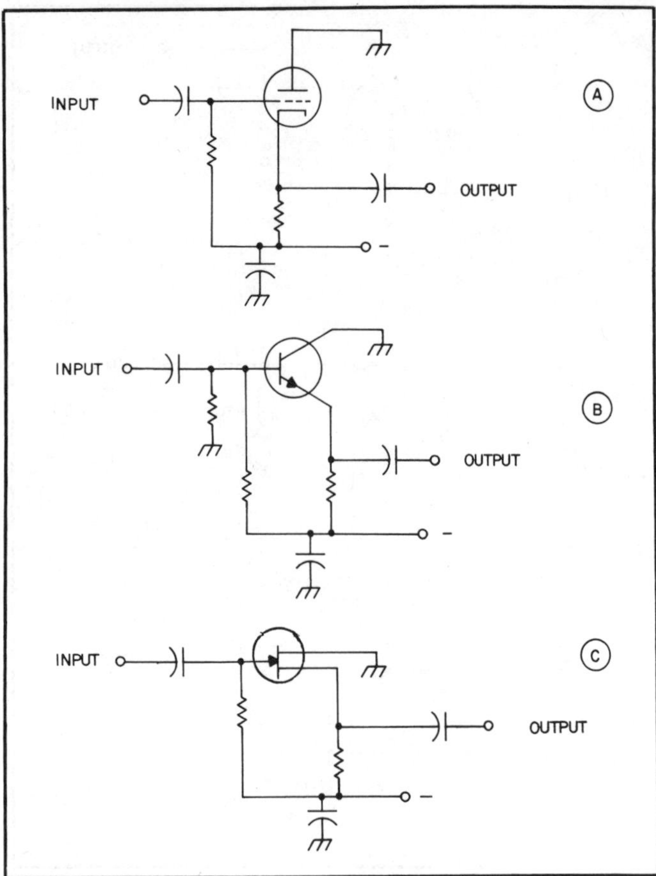

COMMON PLATE/COLLECTOR/DRAIN: At A, a common plate amplifier; at B, a common-collector amplifier; at C, a common-drain amplifier.

devices in all low-power applications. A single integrated circuit the size of a pencil eraser can perform the work of dozens of compactron tubes.

COMPANDOR

A compandor is a device used for the purpose of improving the efficiency of an analog communications system. The compandor consists of two separate circuits: an amplitude compressor, used at the transmitter, and an amplitude expander, used at the receiver.

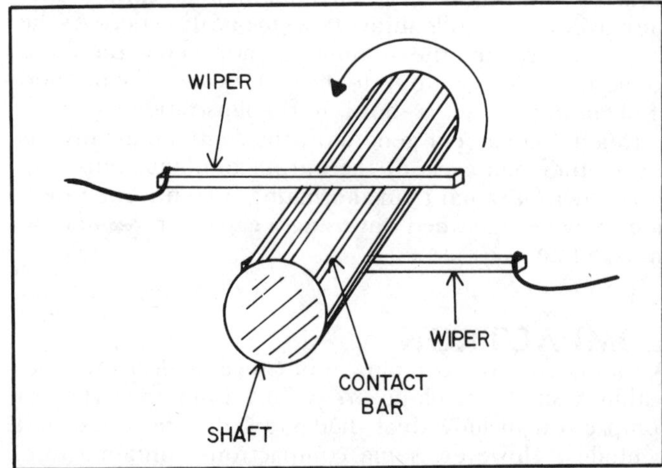

COMMUTATOR: A commutator makes contact between the retating shaft of a motor or generator, and a pair of wipers, to facilitate operation from, or generation of, direct current.

The amplitude compressor in a voice transmitter increases the level of the fainter portions of the envelope (see illustration). This increases the average power output of the transmitter, and increases the proportion of signal power that carries the voice information. This kind of compression can be done with amplitude-modulated, frequency-modulated, or single-sideband transmitters.

The amplitude expander follows the detector in the receiver circuit, and returns the voice to its natural dynamic range. Without the amplitude expander, the voice would be understandable, but less intelligible after pauses in speech. Speech expansion is shown at B.

Some devices are used to compress the modulating frequencies at the transmitter, and expand them at the receiver. This decreases the bandwidth of the transmitted signal. These devices are sometimes called frequency compandors. They are used in an experimental form of transmission called narrow-band voice modulation. *See also* NARROW-BAND VOICE MODULATION, SPEECH COMPRESSION.

COMPARATOR

A comparator is a circuit that evaluates two or more signals, and indicates whether or not the signals are matched in some particular way. Typically, the "yes" output (signals matched) is a high state, and the "no" output (signals different) is a low state. Comparators may test for amplitude, frequency, phase, voltage, current, waveform type, or numerical value. A number comparator has three outputs: greater than, equal to, and less than. A phase or frequency comparator may have an output voltage that varies depending on which quantity is leading or lagging, or larger or smaller.

In high-fidelity audio testing, it is often desirable to quickly switch back and forth between two systems for the purpose of comparing quality. A device that facilitates convenient switching is called a comparator.

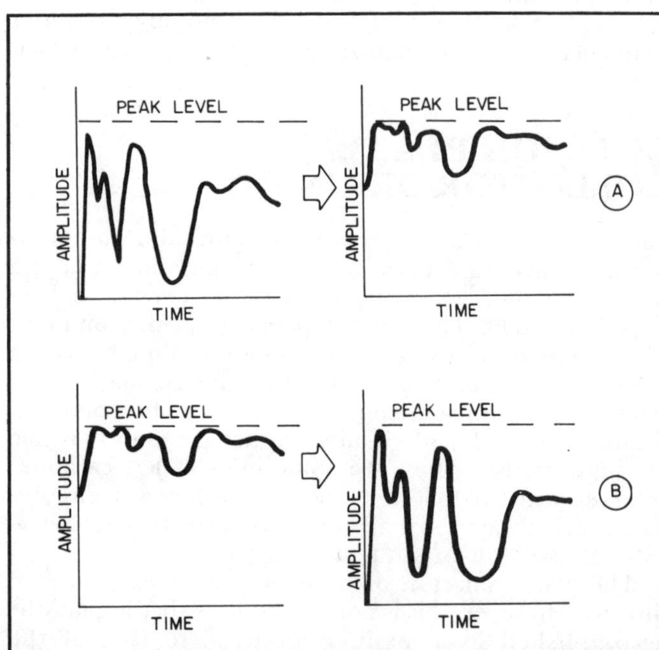

COMPANDOR: At A, the signal-amplitude compression at the transmitting end of a compandor circuit increases the average power in the signal. At B, the expanding circuit at the receiver restores the signal to its original sound.

COMPENSATION

Compensation is a method of neutralizing some undesirable characteristic of an electronic circuit. For example, a crystal might have a positive temperature coefficient: that is, its frequency might increase with increasing temperature. In an oscillator that must be stable under varying temperature conditions, this characteristic of the crystal can be neutralized by placing, in parallel with the crystal, a capacitor whose value increases with temperature and pulls the crystal frequency lower. If the positive temperature coefficient of the capacitor is just right, the result will be a frequency-stable oscillator.

In an operational-amplifier circuit, the frequency response must sometimes be modified in order to get stable operation. This may be done by means of external components, or it may be done internally. This is called compensation. *See also* OPERATIONAL AMPLIFIER.

COMPENSATION THEOREM

Any alternating-current impedance, produced by a combination of resistance, inductance, and capacitance, shows a certain phase relationship between current and voltage at a particular frequency. We usually think of an impedance as being produced by inductors, capacitors, and resistors (see illustration A). However, an equivalent circuit might be a solid-state device, a section of transmission line, or a signal generator.

If the equivalent circuit (black box) is inserted in place of the resistance-inductance-capacitance (RLC) network (illustration B), the overall operation of the rest of the circuit will remain unchanged. This fact is called the compensation theorem. *See also* BLACK BOX, IMPEDANCE.

COMPILER

In a digital computer, the compiler is the circuit that converts the higher-order language, such as BASIC, COBOL, or FORTRAN, into machine language. The operator understands and uses the higher-order language; the computer operates in machine language. A compiler is thus an electronic translator.

The compiler must itself be programmed to translate a given higher-order language into machine language, and vice versa. This program is called the assembler. Assembler programs are written in a language called assembly language. *See also* ASSEMBLER AND ASSEMBLY LANGUAGE, HIGHER-ORDER LANGUAGE, MACHINE LANGUAGE.

COMPLEMENT

The complement is a function of numbers—in particular, binary numbers. In the base-two system, where the only possible digits are 0 and 1, the complement of 0 is 1 and the complement of 1 is 0. In logic, the complement function is the same as negation.

When there are several digits in a binary number, the complement is obtained by simply reversing each digit. For example, the complement of 10101 is 01010. *See also* BINARY-CODED NUMBER.

COMPLEMENTARY COLORS

Two colors are called complementary when they combine to form white as seen by the human eye. Examples of primary colors are red and cyan, green and magenta, and yellow and blue. Of course, the two colors must be combined with the proper relative luminosity to appear white. Also, the levels of saturation of the two colors must be in the correct proportions.

When two colored pencils combine to form black, they are said to have complementary pigments. Pigment differs from color. The term color refers to radiant energy seen directly from a source, or reflected from a white object. Pigment is a reflecting characteristic of a surface. *See also* PRIMARY COLORS.

COMPLEMENTARY CONSTANT-CURRENT LOGIC

Complementary constant-current logic, abbreviated CCCL or C^3L, is a form of digital circuitry similar to transistor-transistor logic (TTL). It is a form of large-scale integration, used in the manufacture of digital integrated circuits. Complementary constant-current logic is characterized by high density.

The switching time in C^3L is very fast—just a few nanoseconds or billionths of a second—allowing more operations per unit time than other logic types. *See also* LARGE-SCALE INTEGRATION.

COMPLEMENTARY METAL-OXIDE SEMICONDUCTOR

Complementary metal-oxide semiconductor, abbreviated CMOS and pronounced "seamoss," is the name for a form of digital enhancement-mode switching device. Both N-

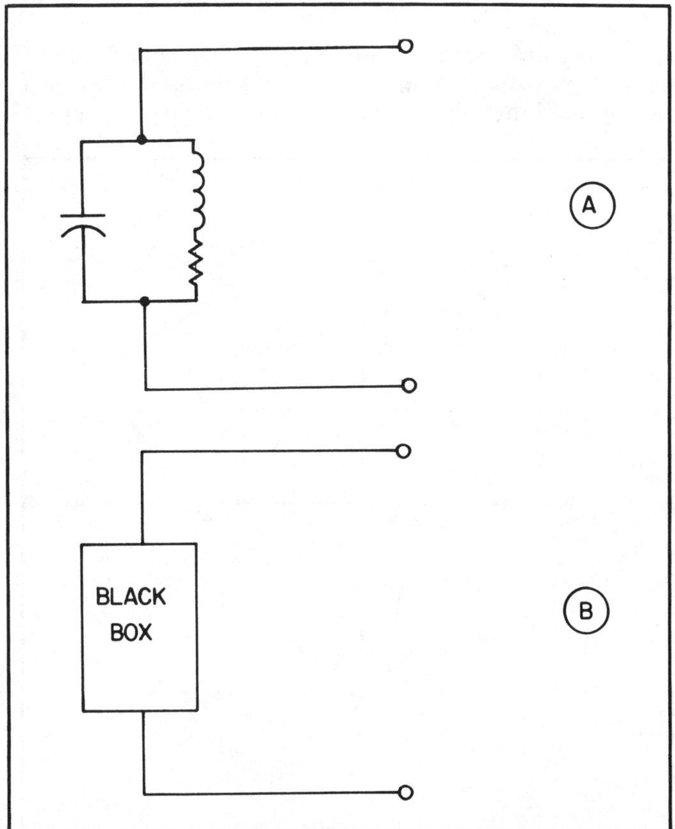

COMPENSATION THEOREM: A resistance-inductance-capacitance (RLC) circuit may be replaced by an equivalent circuit (black box) without affecting the external system.

channel and P-channel field-effect transistors are employed. A CMOS integrated circuit is fabricated on a chip of silicon.

The chief advantages of CMOS are its extremely low current consumption and its high speed. The main disadvantage of CMOS is its susceptibility to damage by static electricity. A CMOS integrated circuit should be stored with its pins embedded in a conducting foam material, available especially for this purpose. When building or servicing equipment containing CMOS devices, adequate measures must be taken to prevent static buildup. *See also* ENHANCEMENT MODE, METAL-OXIDE SEMICONDUCTOR FIELD-EFFECT TRANSISTOR.

COMPLEX NUMBER

A complex number is a quantity of the form a + bi, where a and b are real numbers, and $i = \sqrt{-1}$. The number a is called the real part of the complex quantity, and the number bi is called the imaginary part. In electronics, the number i is usually called j, and the complex number is written in the form R × jX (*see* J OPERATOR). Complex numbers are used by electrical engineers to represent impedances: the real part is the resistance and the imaginary part is the reactance. This representation is used because the mathematical properties of complex numbers are well suited to definition of impedance.

Complex numbers are represented on a two-dimensional Cartesian plane (*see* CARTESIAN COORDINATES). The real part, a, is assigned to the horizontal axis; the imaginary part, bi, is assigned to the vertical axis. The drawing shows the complex-number plane. Each point on this plane corresponds to one and only one complex number, and each complex number corresponds to one and only one point on the plane.

In electronics applications, only the right-hand side of the complex-number plane is ordinarily used. The top part of the right half of the plane, or first quadrant, represents impedances in which the reactance is inductive. The bottom part of the right half of the plane, or fourth quadrant, represents impedances in which the capacitance is the dominant form of reactance. *See also* IMPEDANCE, REACTANCE.

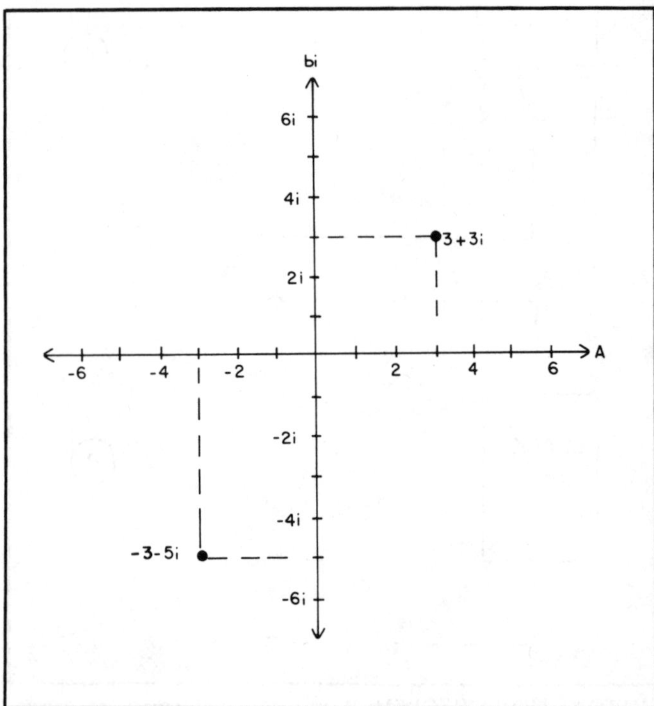

COMPLEX NUMBER: The complex-number plane, showing 3 + 3i and −3 −5i.

COMPLEX STEADY-STATE VIBRATION

When several, or many, sine waves are combined, the resulting waveform is called a complex steady-state vibration. The sine waves may have any frequency, phase, or amplitude. There can be as few as two sine waves, or infinitely many. The drawing illustrates the combination of two sine waves: a frequency and its second harmonic at equal amplitude.

When a sine wave is combined with some of its harmonic frequencies in the proper amplitude relationship, special waveforms result. The sawtooth and square waves are examples of infinite combinations of sine waves with their harmonics. Harmonic combinations are evaluated using Fourier series (*see* FOURIER SERIES).

An audio-frequency, complex, steady-state vibration, when applied to a speaker or headset, produces a sound called a complex tone. Essentially all musical instruments produce complex tones. A pure sine-wave tone is unpleasant to the ear. *See also* COMPLEX WAVEFORM, FOURIER SERIES, SAWTOOTH WAVE, SQUARE WAVE.

COMPLEX WAVEFORM

Any nonsinusoidal waveform is a complex waveform. The human voice produces a complex waveform. Almost all sounds, in fact, have complex waveforms. The most complex of all waveforms is called white noise, a random combination of sound frequencies and phases. If a complex waveform displays periodic properties, it is called a complex steady-state vibration (*see* COMPLEX STEADY-STATE VIBRATION).

All complex waveforms consist of combinations of sine waves. Even white noise can be resolved to sine-wave tones. A pure sine-wave tone has a rather irritating quality, especially at the midrange and high audio frequencies. *See also* SINE WAVE.

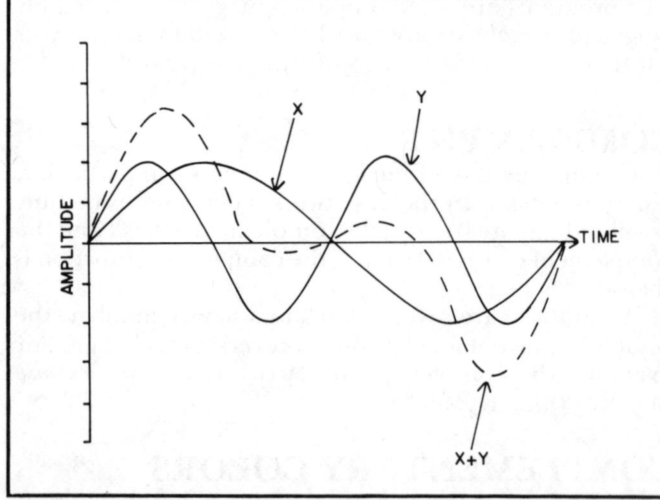

COMPLEX STEADY-STATE VIBRATION: Two sine waves, X and Y, add together to produce the complex steady-state wave X + Y. The combination wave is periodic, although not sinusoidal.

COMPLIANCE

Compliance is a measure of how well a phonograph stylus responds to vibration. Compliance is measured in terms of displacement per unit force, usually in centimeters or millimeters per dyne. In a speaker, compliance is the flexibility of the cone suspension. Compliance can be measured for either static or dynamic forces. Static compliance is the measure of displacement for a simple, direct force; dynamic compliance is the ease of displacement for an alternating force.

The greater the compliance of a stylus, the better the low-frequency response. The same is true of speakers; woofers should have a high degree of compliance for good reproduction of bass sounds.

COMPONENT

In any electrical or electronic system, a component is a lowest replaceable unit. A resistor, for example, is usually a component, although it may be contained inside an integrated circuit. A tube is a component, but the cathode of a tube is not.

In modular construction, the identity of the components depends on the point of view. To the field technician, a card or circuit board may be the lowest replaceable unit, and therefore it is a component. But in the shop, where the cards are repaired individually, the components are the resistors, capacitors, inductors, and other devices attached to the board.

A mathematical part of a quantity may sometimes be called a component. For example, an electromagnetic wave with slanted polarization has a vertical component and a horizontal component. The collector current in a transistorized amplifier circuit has a direct-current component (usually) and an alternating-current component.

COMPONENT LAYOUT

The arrangement of electronic components on a circuit board is called the component layout. The component layout is an important part of the design of a circuit. With modern printed-circuit boards, the placement of foil runs dictates the component layout to a certain extent. (Two foil runs must be electrically connected if they cross.) Poor circuit layout can increase the chances of unwanted feedback, either negative or positive. It can also cause inconvenience in servicing and adjustment of the equipment.

The illustration shows a circuit board on which the components have been carefully positioned for optimum efficiency and use of space. The component density is fairly uniform throughout the board; components are not sparse in one area and bunched together in another area. Components are clearly marked for easy identification when adjusting or repairing the equipment. Component layout becomes more and more important as the frequency gets higher. In some very-high-frequency and ultra-high-frequency circuits, the component layout alone can make the difference between a circuit that works and a circuit that doesn't work.

COMPOSITE VIDEO SIGNAL

A composite video signal is the modulating waveform of a television signal. It contains video, sync, and blanking information. When the composite video signal is used to modulate a radio-frequency carrier, the video information can be transmitted over long distances by electromagnetic propagation. By itself, the composite video signal is similar to the audio-frequency component of a voice signal, except that the bandwidth is greater.

Most video monitors, used with computers and communications terminals, operate directly from the compos-

COMPONENT LAYOUT: A well-engineered component layout.

ite video signal. Some monitors require a modulated, very-high-frequency picture signal, usually on one of the standard television broadcast channels. *See also* COLOR PICTURE SIGNAL, PICTURE SIGNAL.

COMPOUND

When two or more atoms combine via chemical reaction, the result is a compound. A compound consists of two atomic nuclei sharing electrons. For example, hydrogen and oxygen are elements; two atoms of hydrogen can be joined with one atom of oxygen to form the familiar compound H_2O: water.

Some combinations of elements join readily to form compounds. Some do not. Atoms whose outer shells are filled with electrons, such as helium and neon, are called non-reactive or inert elements because they do not ordinarily form compounds.

A compound differs from a mixture. In a mixture, there may be several elements and/or compounds, but no chemical reaction takes place among them. The atmosphere is a mixture of nitrogen, oxygen, carbon dioxide, and traces of many other gases. Some man-made pollutant gases combine to form compounds in the atmosphere.

Compounds are widely used in electronics as heat conductors, insulators, and dielectric materials. Compounds are also used in the manufacture of certain semiconductor devices, such as the metal-oxide semiconductor field-effect transistor.

COMPOUND MODULATION

When a modulated signal is itself used to modulate another carrier, the process is called compound modulation. An example of compound modulation is the impression of an amplitude-modulated signal of 100 kHz onto a microwave carrier. The main carrier may have many secondary signals impressed on it. The secondary signals do not all have to employ the same kind of modulation; for example, a code signal, an amplitude-modulated signal, a frequency-modulated signal and a single-sideband signal may all be impressed on a microwave carrier at the same time. The illustration shows compound modulation.

With compound modulation, the main signal frequency must be at least several times the frequency of the modulating signals. Otherwise, efficiency is poor and the signal-to-noise ratio is degraded.

Compound modulation is useful in telephone trunk lines, where a large number of conversations must be carried over a single circuit. Compound modulation is used in the transmission of signals via satellites. Compound modulation is used in fiber-optics systems, where many thousands of separate signals can be impressed on a single beam of light. *See also* ACTIVE COMMUNICATIONS SATELLITE, FIBER OPTICS, MODULATED LIGHT, TRUNK LINE.

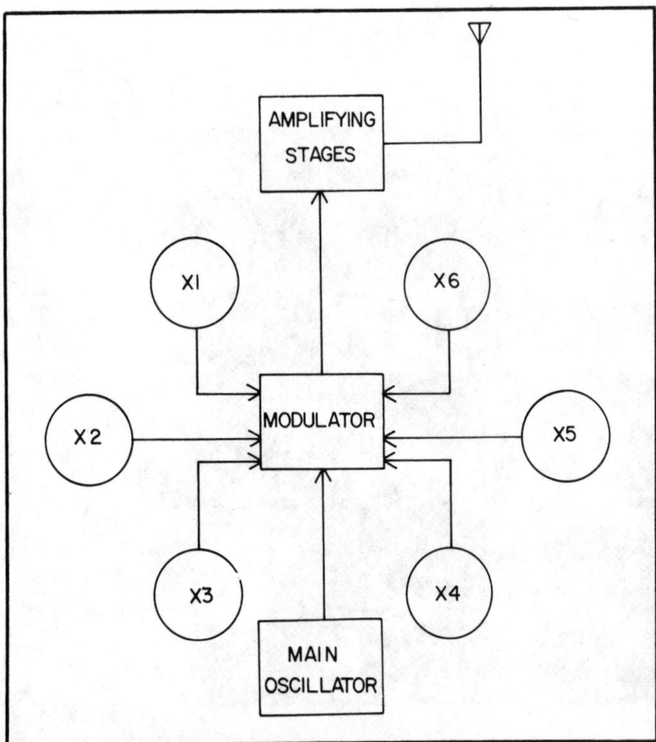

COMPOUND MODULATION: In compound modulation, several different signals (here shown as X1 through X6) are impressed on a main carrier.

COMPOUND CONNECTION

When two tubes or transistors are connected in parallel, or in any other configuration, the arrangement is sometimes called a compound connection. Examples of compound-connected devices include the Darlington pair, the push-pull configuration, certain rectifier circuits, and many different digital-logic circuits.

Compound connections are used in audio-frequency and radio-frequency circuits for the purpose of improving the efficiency or amplification factor. Compound connections may also be used to provide an impedance match.

COMPRESSION

Compression is the process of modifying the modulation envelope of a signal. This is done in the audio stages,

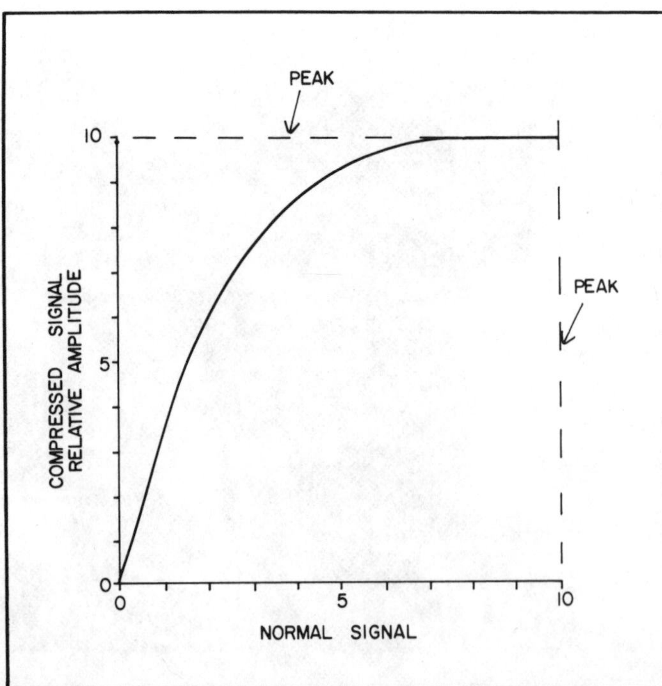

COMPRESSION: A compression curve. The amplification factor, which is the ratio between the compressed and normal signal amplitudes, decreases as the signal level increases.

before the modulator. The weaker components of the audio signal are amplified to a greater extent than the stronger components. The illustration shows a compression function, with the output amplitude as a function of the input signal amplitude. The peak, or maximum, amplitude of the compressed signal is the same as the amplitude of the normal signal. The weaker the instantaneous amplitude of the signal, the greater the amplification factor.

Compression is often used in communications systems to improve intelligibility under poor conditions. *See also* COMPRESSION CIRCUIT, SPEECH COMPRESSION.

COMPRESSION CIRCUIT

A compression circuit is an amplifier that displays variable gain, depending on the amplitude of the input signal. The lower the input signal level, the greater the amplification factor. A compression circuit operates in a manner similar to an automatic-level-control circuit (*see* AUTOMATIC LEVEL CONTROL).

One means of obtaining compression is to apply a rectified portion of the signal to the input circuit of the amplifying device, changing the bias as the signal amplitude fluctuates. This rectified bias should reduce the gain as the signal level increases. The time constant must be adjusted for the least amount of distortion. An example of this scheme is shown in the diagram.

There is a practical limit to the effectiveness of compression. Too much compression will cause the system to emphasize acoustical noise. In a voice communications transmitter, several decibels of effective gain can be realized using audio compression. *See also* SPEECH COMPRESSION.

COMPTON EFFECT

When photons (particles of radiant energy) strike electrons, the wavelength of the photons changes. This is because some of the photon energy is transferred to the electrons. The change in wavelength is called the Compton effect. The amount of change in the wavelength is a function of the scattering angle. Compton effect is generally observed with X rays.

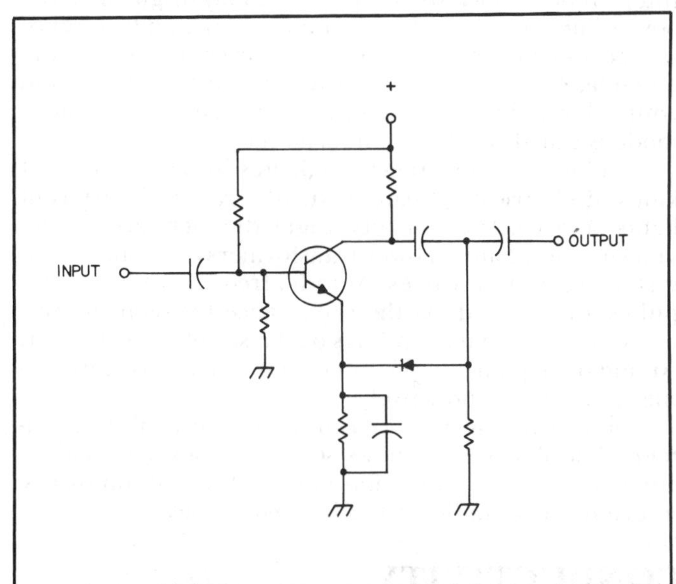

COMPRESSION CIRCUIT: An example of a compression circuit.

At a grazing angle of collision, very little energy is transferred to the electron, and the wavelength of the photon therefore increases only slightly (A). At a sharper angle of collision (B in the illustration), the photon loses more energy, and the change in wavelength is greater. At a nearly direct angle of collision, the wavelength change is the greatest, as shown at C.

Compton effect causes the wavelength of X rays to be spread out when the radiation passes through an obstruction. *See also* X RAY.

COMPUTER

A computer is a device that processes information. Computers have become commonplace in the lives of almost everyone in the industrialized countries. Computers are found in various forms in children's toys, radio communications equipment, desktop calculators, automobiles, and many other consumer devices. Computers make it relatively easy to store and retrieve almost any kind of information. They also make possible mathematical computations that would require years, or even centuries, for a single person to perform on paper.

Some microcomputers are smaller than a fingernail, contained on a single chip of silicon, and require almost no power for operation. Some computers occupy an entire room. The capacity of a computer to store information is measured in bytes, kilobytes, and megabytes. A home computer system, available for one or two thousand dollars, typically has from 16 to 128 kilobytes of storage. Larger computers may have many megabytes. *See also* BYTE, KILOBYTE, MEGABYTE.

COMPTON EFFECT: The Compton effect results in a change in the wavelength of a photon after it hits an electron. At A, a collision at a grazing angle produces little change in the photon energy. At B, a sharp angle causes significant change. At C, a direct hit results in the greatest change in wavelength.

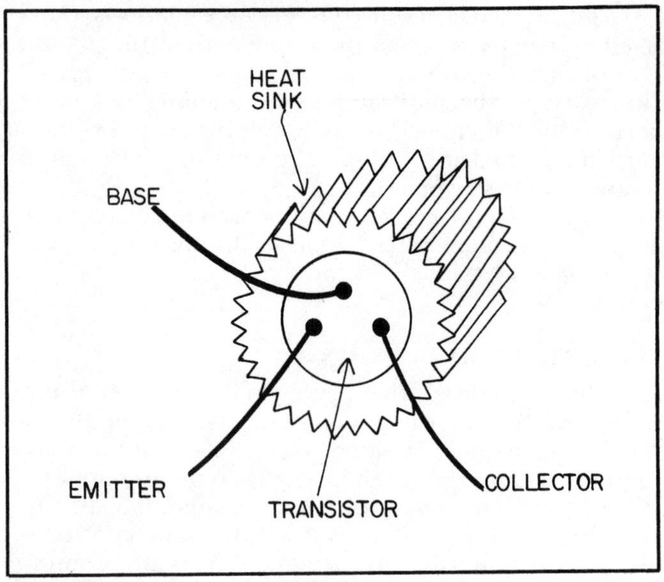

CONDUCTION COOLING: Conduction cooling of a bipolar transistor is often facilitated by a heat sink.

COMPUTER PROGRAMMING

In order to perform their functions, all computers must be programmed. Generally, a computer requires two different programs. The higher-order program allows the computer to interface, or communicate, with the operator. There are several different higher-order programming languages, such as BASIC, COBOL, FORTRAN, and others (*see* BASIC, COBOL, FORTRAN, HIGHER-ORDER LANGUAGE). The assembler program converts the higher-order language into binary machine language, and vice versa. The computer actually operates in machine language (*see* ASSEMBLER AND ASSEMBLY LANGUAGE, MACHINE LANGUAGE).

Preparation of assembler and higher-order programs is usually done by people. Some computers can generate their own programs for certain purposes. The program is a logical sequence of statements, telling the computer how to proceed.

CONDENSER

See CAPACITOR.

CONDOR

CONDOR is the name for a continuous-wave radionavigation and radiolocation system. A single ground station is used. The bearing, or azimuth-plane direction, is determined by direction-finding methods. The range, or distance, is determined by comparing the phase of the transmitted and reflected signals. The position of the distant object is indicated on a cathode-ray-tube display. *See also* RADIOLOCATION.

CONDUCTANCE

Conductance is the mathematical reciprocal of resistance. The unit of conductance is the mho—ohm spelled backwards. A resistance of 1 ohm is equal to a conductance of 1 mho. A resistance of 1 kilohm (1000 ohms) is equal to a conductance of 1 millimho; a resistance of 1 megohm (1,000,000 ohms) is equal to 1 micromho. Mathematically, if R is the resistance in ohms, then the conductance G in mhos is:

$$G = 1/R$$

Conductance is occasionally specified instead of resistance when the value of resistance is very low. The word mho is sometimes replaced by an upper-case, but inverted, Greek letter omega, such as 0.3℧. The mho is also called the siemens. *See also*MHO, RESISTANCE.

CONDUCTION COOLING

Conduction cooling is a method of increasing the dissipation capability of a tube or transistor. Heat is carried from the device by a thermally conductive material, such as beryllium oxide or silicone compound, to a large piece of metal. The metal, in turn, employs cooling fins to let the heat escape into space. The illustration shows a typical conduction-cooling arrangement for a transistor.

Conduction cooling is quieter than forced-air cooling, the other most common method of cooling. Blowers are not necessary for conduction cooling. However, care must be exercised to ensure that a good thermal bond is maintained between the cooling device and the tube or transistor. *See also* AIR COOLING, HEATSINK.

CONDUCTIVE MATERIAL

A conductive material is a substance that conducts electricity fairly well or very well. Such materials are called conductors. Most metals are good conductors; silver is the best. Strong liquid electrolytes are also good conductors.

Conduction occurs as a transfer of orbital electrons from atom to atom. When one atom accepts an extra electron, it usually gives up an electron, passing it on to a nearby atom. The best conductors do this readily. In a highly conductive material, a large current flows with the application of a relatively small voltage.

The most frequently used conductive material in electrical wiring is copper. Steel and aluminum are also employed, but less often. *See also* CONDUCTIVITY.

CONDUCTIVE INTERFERENCE

Interference to electronic equipment may originate in the power lines supplying that equipment. Such interference can be radiated by the lines and picked up by a radio antenna, or it may be conducted to the equipment and coupled via the power-supply transformer. The latter mode is called conductive interference.

Conductive interference originates in many different kinds of electrical appliances, such as razors, fluorescent lights, and electric blankets. Light dimmers are a well-known cause. Faulty power transformers can transmit interference on utility lines. At radio frequencies, these impulses are coupled via the capacitance between the primary and secondary windings of the supply transformer. At audio frequencies, the interference may pass through the transformer inductively.

Conductive interference can be very difficult to eliminate. The illustration shows some common alternating-current line filters for minimizing the effects of conductive interference at audio and radio frequencies.

CONDUCTIVITY

Conductivity is a measure of the ease with which a wire or

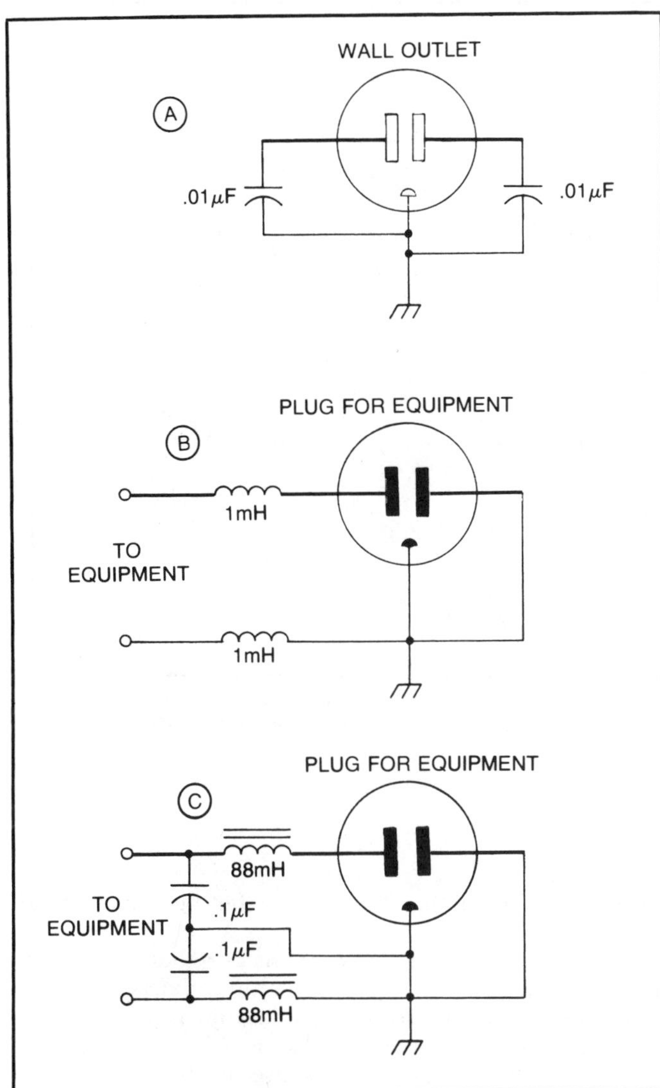

CONDUCTIVE INTERFERENCE: Three ways of reducing conductive interference. At A, bypass capacitors shunt stray RF energy to ground. At B, series choke coils prevent unwanted RF energy from reaching the equipment. At C, a combination of large-value chokes and capacitors for audio-frequency conductive interference.

material carries an electric current. Conductivity is expressed in mhos, millimhos, or micromhos per unit length. The table shows the conductivity of copper wire for American Wire Gauges 2 through 34.

Conductivity varies somewhat with temperature. The values in the table are for room temperature. Some materials get more conductive as the temperature rises; most get less conductive as the temperature rises.

The better the conductivity of a material, the more current will flow when a specific voltage is applied. The greater the conductivity, the smaller the voltage required to produce a given current through a length of conductor. Conductivity is the opposite of resistivity. *See also* RESISTIVITY.

CONE

A cone is a diaphragm responsible for the radiation of sound from a loudspeaker. The cone gets its name from the fact that it is conical in shape, as shown in the illustration. Speaker cones are usually made from thin cardboard material. The back-and-forth vibration of the cone, caused

CONDUCTIVITY: CONDUCTIVITY, IN MILLIMHOS PER METER, OF VARIOUS GAUGES OF COPPER WIRE.

American Wire Gauge	Millimhos per Meter
2	1,913,375
4	1,203,319
6	756,705
8	475,879
10	299,411
12	188,265
14	118,369
16	74,451
18	46,820
20	29,449
22	18,518
24	11,647
26	7,323
28	4,606
30	2,897
32	1,822
34	1,146

by the action of the speaker coil in the magnetic field, produces sound waves.

The speaker cone must be properly designed for optimum sound reproduction. In communications systems, a high-frequency cone response (about 3000 Hz) is undesirable. In a high-fidelity sound system, the response must extend to the limit of the human hearing range, or about 16,000 to 20,000 Hz. Special speaker combinations are necessary to achieve this frequency range. *See also* SPEAKER, TWEETER, WOOFER.

CONELRAD

CONELRAD stands for CONtrol of ELectromagnetic RADiation. In the event of an enemy attack on the United States by missiles or planes, all radio broadcast stations would be shut down with the exception of certain stations at 640 and 1240 kHz. On these frequencies, some designated stations would transmit intermittently, disseminating information relevant to the situation. This minimizes the possibility of using broadcast signals to locate cities.

In recent years, the term CONELRAD has become obsolete. It is part of a more comprehensive defense system called the Emergency Broadcast System (EBS). *See also* EMERGENCY BROADCAST SYSTEM.

CONE MARKER

A cone marker is a very-high-frequency or ultra-high-frequency beacon which radiates energy vertically, in a cone-shaped pattern. The vertex of the cone is located at the antenna, and the cone opens upward.

Cone markers are used by aircraft to locate their precise position. Knowing the exact location of two or more cone markers on the ground, and measuring the direction to each, the pilot of an aircraft can determine the location of the plane.

CONE OF SILENCE

A low-frequency radiolocation beacon radiates energy mainly in the horizontal plane. An inverted, vertical cone over the antenna contains very little signal, as shown in the

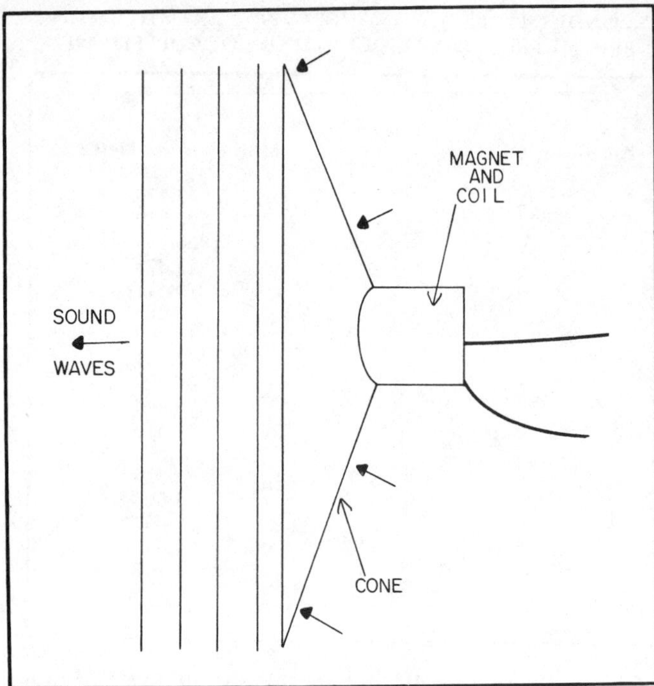

CONE: The speaker cone produces sound by vibrating against molecules of air.

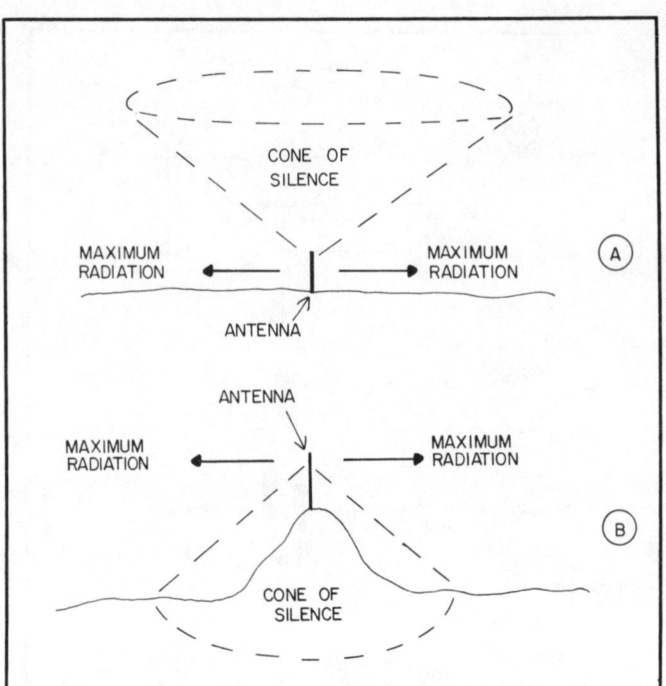

CONE OF SILENCE: At A, the cone of silence above a vertical antenna. At B, the cone of silence below a collinear vertical repeater antenna.

drawing. This zone is called the cone of silence.

Any omnidirectional antenna, designed for the radiation of signals primarily toward the horizon, has a cone of silence. The more horizontal gain, the broader the apex of this cone. In some repeater installations using high-gain collinear vertical antennas, located high above the level of average terrain, a cone of silence (poor repeater response) exists underneath the antenna. This zone may extend outward for several miles in all directions at ground level (B in the illustration). This effect often makes a nearby repeater useless. *See also* REPEATER.

CONFERENCE CALLING

When a telephone connection is made among three or more parties simultaneously, the mode is called conference calling. Each subscriber is then able to talk to any or all of the others. When one subscriber speaks, all the others can hear.

Conference calling requires special arrangements to match the impedance at the switching system. Several telephones, operating at the same time on one circuit, result in different loading conditions than a single telephone produces.

Conference calling is available for most home and business telephone subscribers in the United States today. Conference calling can be carried out over local or long-distance circuits. *See also* TELEPHONE.

CMOS

See COMPLEMENTARY METAL-OXIDE SEMICONDUCTOR.

CONICAL HORN

A conical horn is a radiating device used for acoustic or microwave electromagnetic energy. The cross-sectional area, A, at a given axial distance r from the apex of the horn, is given by:

$$A = kr^2$$

where k is a constant that depends on the apex angle of the horn.

A conical horn may have a cross section that is circular, square, rectangular, hexagonal, or any other geometric shape. The drawing illustrates a conical horn with a square cross section. Conical horns can be visually recognized by the fact that their sides, or faces, are linear. That is, the intersection of a plane passing through the radial axis with the face of the horn is always a straight line. Conical horns have a unidirectional radiation pattern.

CONICAL MONOPOLE ANTENNA

A conical monopole antenna is a form of biconical an-

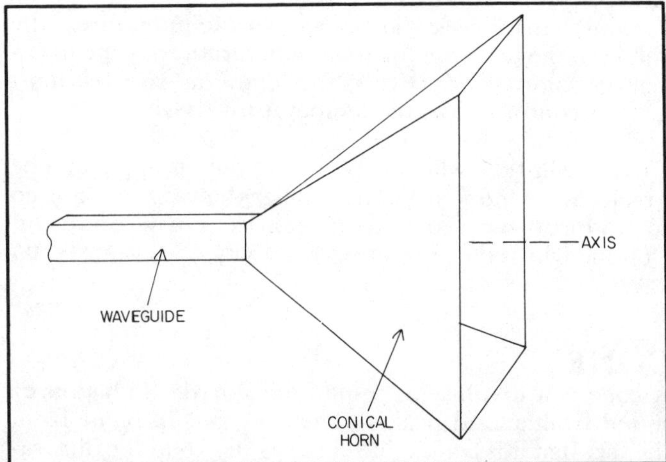

CONICAL HORN: A conical horn for microwave transmission.

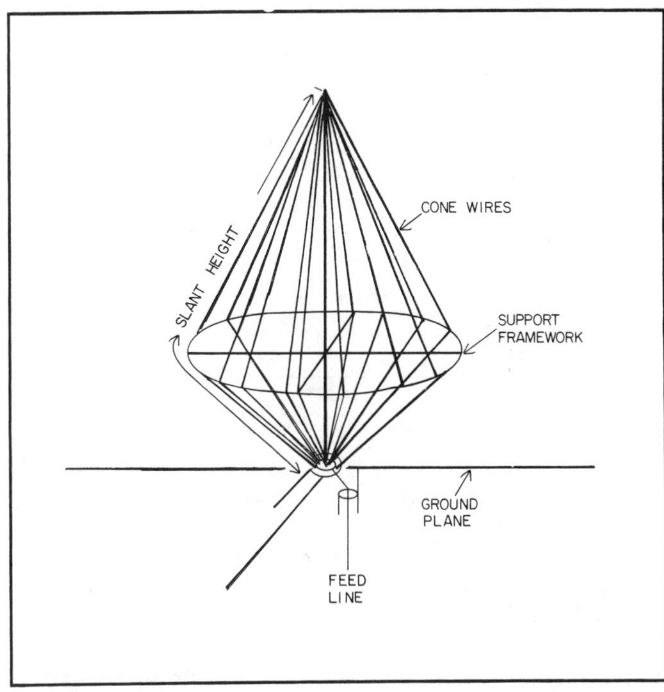

CONICAL MONOPOLE ANTENNA: The conical monopole antenna is usually folded inward at the top to make construction less cumbersome.

tenna, in which the lower cone has been replaced by a ground plane. The upper cone is usually bent inward at the top.

A conical monopole displays a wideband frequency response. The lowest operating frequency is determined by the size of the cone. At all frequencies up to several octaves above that at which the slant height (see illustration) of the cone is ¼ wavelength, the conical monopole provides a reasonably good match to a 50-ohm feed line.

Conical monopole antennas have vertical polarization, and are often constructed in the form of a wire cage as shown. This closely approximates a solid metal cone. A good ground system must be furnished for the conical monopole to work well. The conical monopole is an unbalanced antenna, and coaxial cable is best for the feed system. *See also* BICONICAL ANTENNA, DISCONE ANTENNA.

CONICAL SCANNING

Conical scanning is a method of determining the direction of a target in a radar system. The elevation as well as the azimuth can be found by this method. The radiation pattern of the antenna is made slightly lopsided, and the antenna is electrically or mechanically rotated, so that the lobe of maximum radiation rotates about the boresight (axis) of the radar dish.

When the target is exactly aligned with the boresight, the returned signal is constant in amplitude, as shown at A in the illustration. This is because the target is always the same angular distance from the lobe of maximum radiation. When the target is not aligned with the boresight, as shown at B, the returned signal is modulated at the frequency of antenna rotation. In this way, the antenna can be accurately pointed in the direction of the target. The conical-scanning method is very precise. *See also* RADAR.

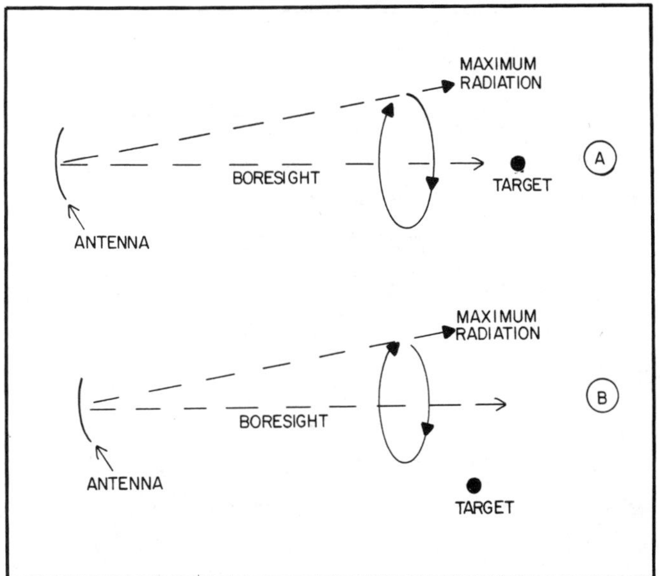

CONICAL SCANNING: At A, the target is always the same angular distance from the direction of maximum radiation. At B, the target is at varying angles from the direction of maximum radiation. As the antenna is rotated, the signal from the target at A stays constant in amplitude, while that at B fluctuates. This allows precise determination of the target position.

CONIC SECTION

A conic section is the geometric intersection of a cone and a plane. Conic sections include the circle, ellipse, parabola, and hyperbola (see drawing).

Conic sections can be generated, for illustrative purposes, using a bright flashlight and a large, flat surface such as a parking lot. When the flashlight is pointed straight down, the large area of relatively dim light (outside the bright central spot) is bounded by a circle. The flashlight must be oriented vertically, so that the bright central spot is directly below the bulb. In the Cartesian (x,y) plane, a circle is represented by the equation:

$$(x - a)^2 + (y - b)^2 = k$$

where a, b, and k are constants.

When the flashlight is pointed at an angle, in such a way that the perimeter of the area of dim light falls on the surface but is not a perfect circle, it forms an ellipse. In the Cartesian (x,y) plane, an ellipse is represented by the equation:

$$c(x - a)^2 + D(y - b)^2 = k$$

where a, b, c, D, and k are constants.

As the flashlight angle is raised, a point will be reached where the far edge of the perimeter of dim light no longer reaches the ground. At this angle, the outline of the dim region forms a parabola. In the Cartesian (x,y) plane, a parabola is represented by the formula:

$$(y - b)^2 = k(x - a)$$

where a, b, and k are constants.

If the angle of the flashlight is raised still further, the outline of the dim region becomes a hyperbola, until the dim region no longer reaches the ground at any point. In the Cartesian (x,y) plane, a hyperbola is represented by:

$$c(x - a)^2 - d(y - b)^2 = k$$

where a, b, c, d, and k are constants.

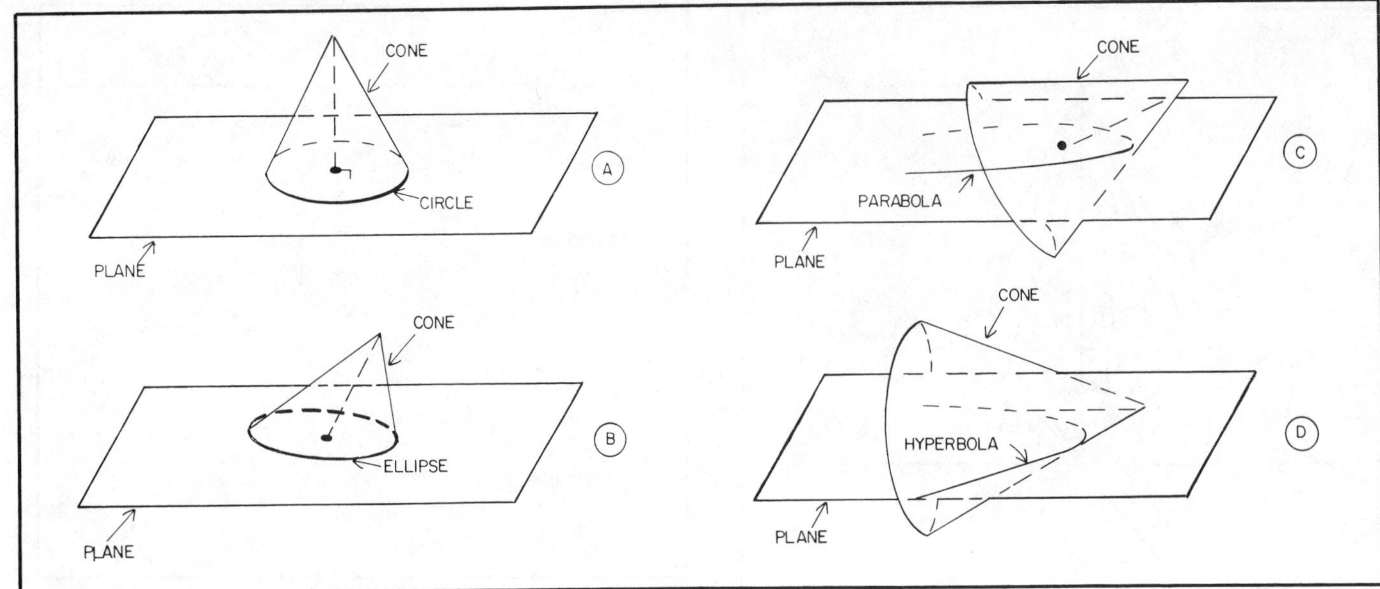

CONIC SECTION: Conic sections. At A, a circle; at B, an ellipse; at C, a parabola; at D, a hyperbola.

Conic sections are frequently seen as mathematical relations and functions. *See also* CARTESIAN COORDINATES.

CONJUGATE IMPEDANCE

An impedance contains resistance and reactance. The resistive component is denoted by a real number a, and the reactive component by an imaginary number jb (*see* COMPLEX NUMBER, IMPEDANCE, J OPERATOR, REACTANCE, RESISTANCE). The net impedance is denoted by R + jX. If X is positive, the reactance is inductive. If X is negative, the reactance is capacitive.

Given an impedance of R+jX, the conjugate impedance is R – jX. That is, the conjugate impedance contains the same value of resistance, but an equal and opposite reactance.

Conjugate impedances are important in the design of alternating-current circuits, and for the purpose of impedance matching between a generator and load.

CONNECTION

When two conducting materials are brought together, current can flow between them. If the conductors are deliberately joined, the point of contact is a connection. If they are accidentally joined, the point of contact is a short circuit.

Connections may be clamped, bolted, welded, or soldered. The greater the contact-surface area, the more current can flow through the connection, and the smaller the voltage drop for a given amount of current flow. Connections may be either temporary or permanent.

Permanent connections generally require sealing against the elements. In electronic circuits, this is done with solder. Outdoors, connections should be covered with a waterproof material, and also soldered, if possible. Certain combinations of dissimilar metals result in undesirable reactions between the metals. This can cause corrosion even if the connection is airtight. *See also* SOLDER, WIRE SPLICING.

CONNECTOR

A connector is a device intended for providing, and maintaining, a good electrical connection. Connectors may be either temporary or permanent. The alligator clip (*see* ALLIGATOR CLIP) is an example of a temporary connector. A coaxial-cable connector is suitable for permanent outdoor installation when the contacts are properly soldered to the cable conductors. There are many different kinds of connectors.

Choosing the proper connector for a given application is not difficult. Often, the choice is dictated by convention. In radio-frequency use, especially at very high frequencies and above, special connectors must be used to maintain impedance continuity. The BNC and N-type connectors are usually employed in coaxial-cable circuits at the higher frequencies (*see* BNC CONNECTOR, N-TYPE CONNECTOR). Single-wire connectors are simpler, and consist of such configurations as the binding post, spade lug, and various forms of plugs and clips. Refer to the appropriate connector name for specific information.

CONSERVATION OF ENERGY

The law of conservation of energy is an axiom of physics. When energy changes its form, the capacity for doing work remains the same. Energy cannot be created or destroyed within a system.

A battery may, for example, be used to run an electric motor (see A in illustration) or to illuminate a light bulb (as at B). If the battery is completely exhausted, the expended energy is the same in either case. With the motor, the chemical energy in the battery is converted mostly to mechanical energy. Some heat is produced as well because of losses in the conductors. With the light bulb, the chemical energy in the battery is converted to light and heat, which are forms of electromagnetic energy. *See also* ENERGY, WORK.

CONSONANCE

Consonance is a special manifestation of resonance. If two objects are near each other but not actually in physical contact, and both have identical or harmonically related resonant frequencies, then one may be activated by the other.

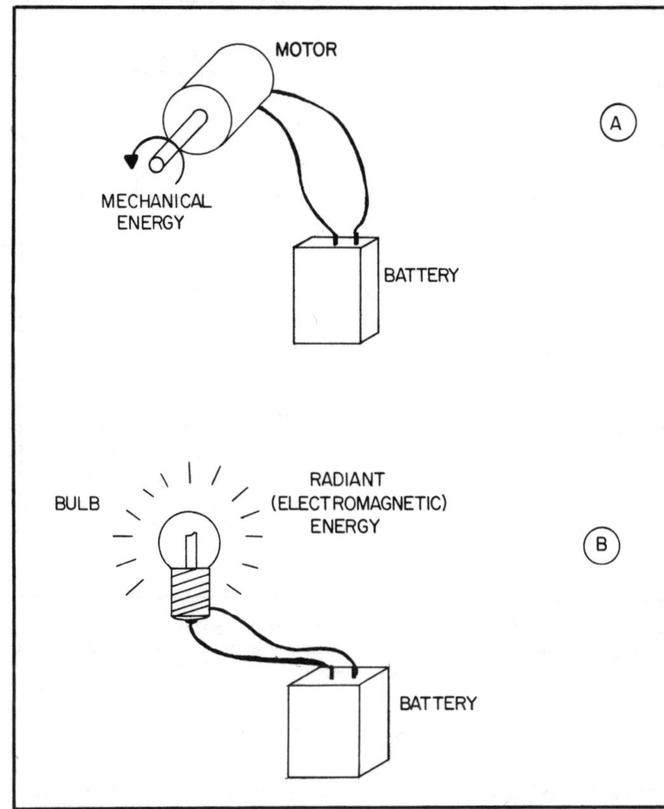

CONSERVATION OF ENERGY: At A, chemical energy is converted to mechanical energy. At B, chemical energy is converted to radiant energy. The law of conservation of energy states that the total amount of energy in the battery will be the same in either case.

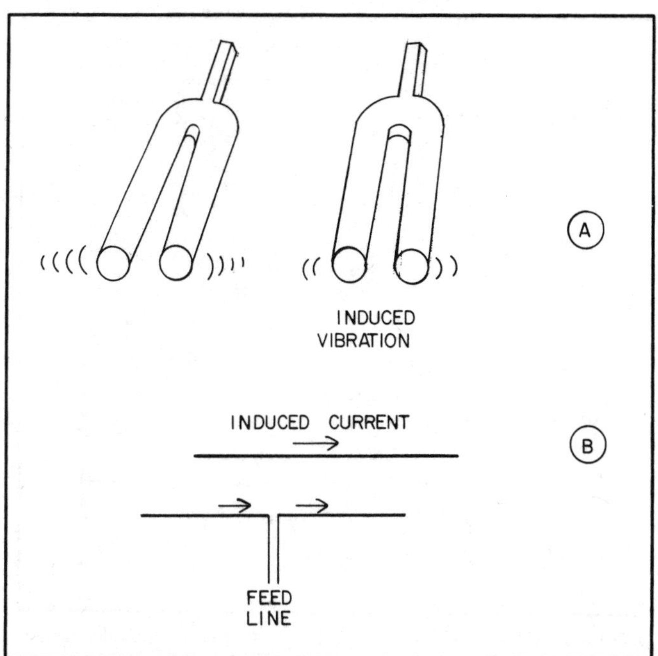

CONSONANCE: At A, acoustical consonance induces a tuning fork to oscillate. At B, electromagnetic consonance causes radio-frequency currents in a free length of conductor.

An example of acoustic consonance is shown by two tuning forks with identical fundamental frequencies. If one fork is struck and then brought near the other (see illustration at A), the second fork will begin vibrating. If the second fork has a fundamental frequency that is a harmonic of the frequency of the first fork, the same thing will happen; the second fork will then vibrate at its resonant frequency.

Electromagnetic consonance is produced when two antennas, both with the same resonant frequency, are brought in close proximity (B). If one antenna is fed with a radio-frequency signal at its resonant frequency, currents will be induced in the other antenna, and it, too, will radiate. Parasitic arrays operate on this principle. *See also* PARASITIC ARRAY, RESONANCE.

CONSTANT

In a mathematical equation, a constant is a fixed numerical value that does not change as the function is varied over its domain. In an equation with more than one variable, some of the variables may be held constant for certain mathematical evaluations. The familiar equation of Ohm's Law:

$$I = E/R$$

where I represents the current in amperes, E represents the voltage, and R represents the resistance in ohms, can be changed to the more general formula:

$$I^* = kE^*/R^*$$

where I* is the current in amperes, milliamperes, or microamperes, E* is the voltage in kilovolts, volts, millivolts, or microvolts, and R* is the resistance in ohms, kilohms, or megohms. The value of the constant, k, then depends on the particular units chosen. Once the units have been decided upon, k remains constant for all values of the function. For example, I* may be denoted in milliamperes, E* in volts, and R* in megohms; then $k = 0.001$, or 10^{-3}.

Certain constants are accepted as fundamental properties of the physical universe. Perhaps the best known of these is the speed of light, c, which is 3×10^8 meters per second in free space.

In mathematical equations, constants may be denoted by any symbol or letter. Letters from the first half of the alphabet, however, are generally used to represent constants. *See also* FUNCTION, VARIABLE.

CONSTANT-CURRENT MODULATION

Constant-current modulation, also called Heising modulation, is a form of plate or collector amplitude modulation. The plates or collectors of the radio-frequency and audio-frequency amplifiers are directly connected through a radio-frequency choke, as shown in the diagram.

The choke allows the audio signal to modulate the supply voltage of the radio-frequency stage. This produces amplitude modulation in the radio-frequency amplifier. The plate or collector current is maintained at a constant level because of the extremely high impedance of the choke at the signal frequency. Constant-current modulation is sometimes called choke-coupled modulation, because of the radio-frequency choke. *See also* AMPLITUDE MODULATION.

CONSTANT-CURRENT SOURCE

A constant-current source is a device that supplies the

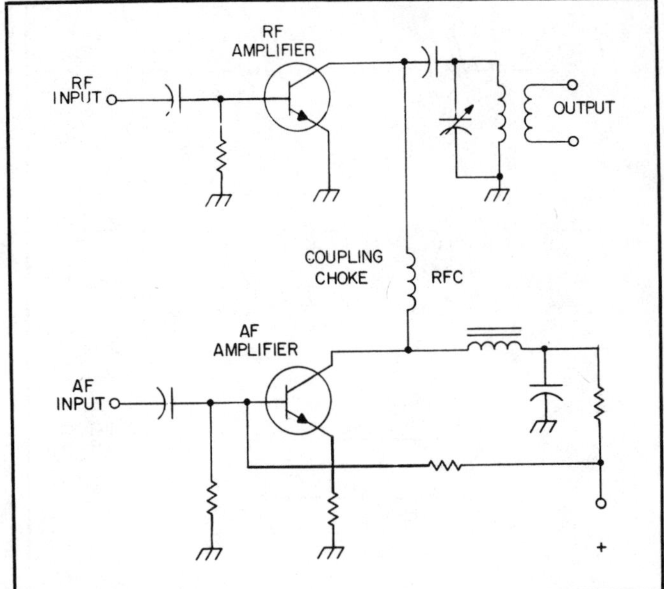

CONSTANT-CURRENT MODULATION: Constant-current modulation uses a radio-frequency choke to supply the audio energy to the modulating stage.

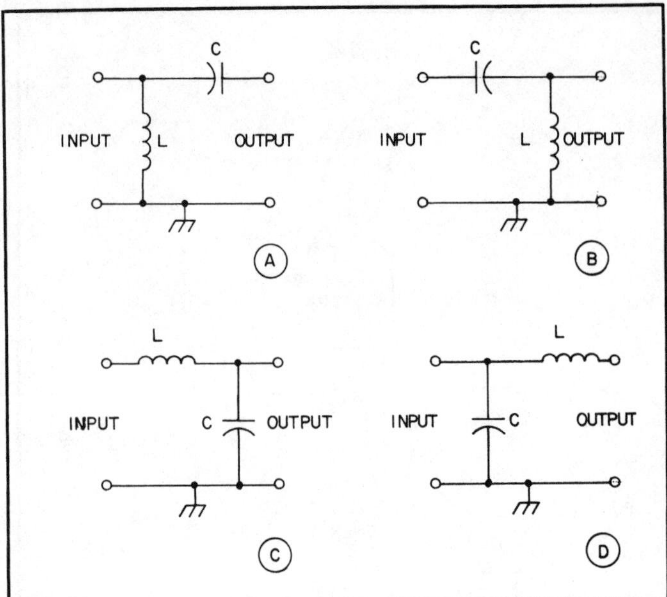

CONSTANT-K FILTER: Four types of constant-k filters. At A, highpass with choke input; at B, highpass with capacitor input; at C, low pass with choke input; at D, lowpass with capacitor input.

same amount of current, no matter what the load resistance. In a constant-current power supply, the voltage across the load varies in direct proportion to the load resistance.

As an illustration, suppose that a constant-current source delivers 100 milliamperes. If the load resistance is 10 ohms, then

$$E = IR = 0.1 \times 10 = 1 \text{ volt}$$

If the load resistance increases to 100 ohms, then

$$E = IR = 0.1 \times 100 = 10 \text{ volts}$$

The output voltage rises to its maximum value when there is no load—that is, an open circuit. For extremely high values of load resistance, the supply is not able to deliver its rated current. The load resistance must therefore be within certain limits for a constant-current source to operate properly.

CONSTANT-K FILTER

A constant-k filter is a series of L networks (*see* L NETWORK). There might be only one L section, or perhaps several. The L sections can be combined to form pi and T networks (*see* PI NETWORK, T NETWORK). The filter can be either a highpass or lowpass type (*see* HIGHPASS FILTER, LOWPASS FILTER). The diagrams illustrate various highpass and lowpass L-section filters.

The primary characteristic of a constant-k filter is the fact that the product of the series and parallel reactances is constant over a certain frequency range. Usually, only one kind of reactance is in series, and only one kind is in parallel. Capacitive reactance is given by the formula:

$$X_C = \frac{-1}{2 \pi fC}$$

where X_C is in ohms, f is in megahertz, and C is in microfarads. Inductive reactance is given by:

$$X_L = 2\pi fL$$

where X_L is in ohms, f is in megahertz, and L is in microhenries. The product of the series and parallel reactances in the illustration is thus:

$$k = X_L X_C = -\frac{2\pi fL}{2\pi fC} = -L/C$$

This value is a constant, and does not depend on the frequency. In practice, the inherent resistance of any circuit limits the frequency range over which it is useful.

CONSTANT-VELOCITY RECORDING

Constant-velocity recording is a method of recording phonograph disks. The amount of cutter movement is a function of the recording frequency. The higher the recording frequency, the smaller the excursion of the cutter.

When playing back a disk having a constant-velocity characteristic, the audio-frequency response of the equipment must compensate for the fact that the lower frequencies cause greater vibration amplitude of the stylus. The reproducing characteristic is shown in the illustration. The reference frequency is 1000 Hz. The higher the frequency, the greater the gain of the playback equipment. The sound reproduced by the phonograph thus sounds natural.

CONSTANT-VOLTAGE POWER SUPPLY

A constant-voltage power supply is a device that provides the same voltage, regardless of the load resistance. The lower the load resistance, the more current the supply delivers, to a certain point. All constant-voltage power supplies have a maximum current rating. As long as the load resistance is high enough to prevent excessive current

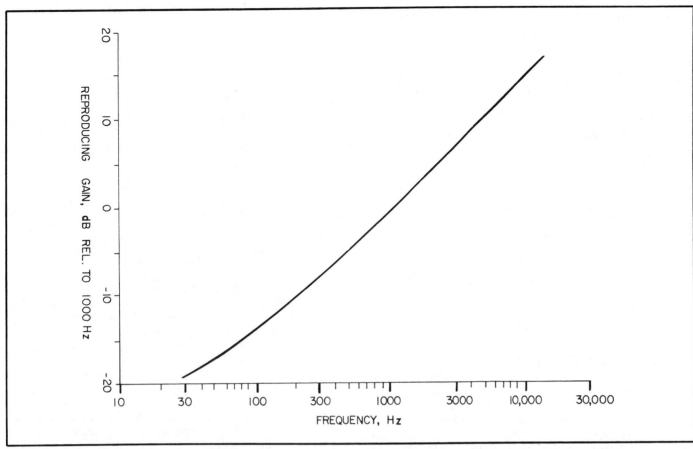

CONSTANT-VELOCITY RECORDING: Gain of the reproducing circuit, as a function of frequency, for constant-velocity recording.

drain, the voltage at the supply output remains nearly constant. Constant-voltage power supplies are also sometimes called regulated power supplies.

Voltage regulation can be accomplished in several different ways (*see* VOLTAGE REGULATION). Regulation is measured in percent. If E_1 is the open-circuit voltage at the output of a power supply and E_2 is the voltage under conditions of maximum rated current drain, then the regulation R in percent is

$$R = \frac{100\,(E_1 - E_2)}{E_1}$$

Most power supplies for radio equipment are intended to deliver a constant voltage for a wide range of load resistances, and are therefore constant-voltage power supplies. *See also* POWER SUPPLY.

CONTACT

A *contact* is a metal disk, prong, or cylinder, intended to make and break an electric circuit. Contacts are found in all electric switches, relays, and keying devices. A typical pair of relay contacts is shown in the drawing.

The surface area of the contact determines the amount of current that it can safely carry. Generally, convex contacts cannot carry as much current as flat contacts, since the contact area is smaller when the surfaces are curved. Contacts may be made and broken either by a sliding motion or by direct movement. Contacts must be kept clean and free from corrosion in order to maintain their current-handling capacity. Metals that oxidize easily do not make good contacts. Gold and silver are often used where reliability must be high. Some relay contacts are housed in evacuated glass or metal envelopes to keep them from corroding. *See also* REED RELAY, RELAY, SWITCH.

CONTACT ARC

Contact arc is a spark that occurs immediately following the break between two contacts carrying an electric current. Contact arc occurs to a greater extent when the current is large, as compared to when it is small. Contact arc is undesirable because it speeds up the corrosion of contact faces, eventually resulting in poor reliability and reduced current-carrying capacity.

Contact arc can be minimized by breaking the contacts with the greatest possible speed. Enclosing the contacts in a vacuum chamber prevents dielectric breakdown of the air, and thus eliminates oxidation of the contact metal. *See also* ARC.

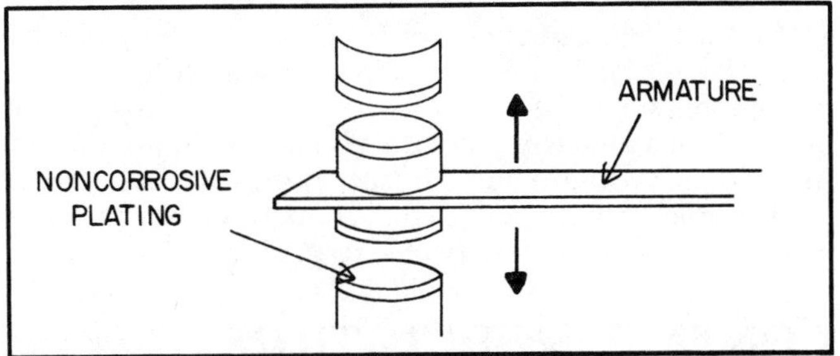

CONTACT: Relay contacts often are plated with a noncorrosive metal to increase reliability.

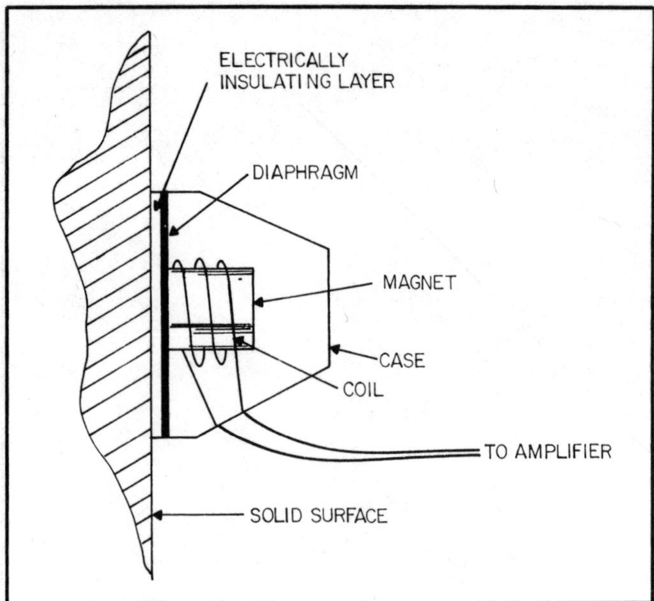

CONTACT MICROPHONE: Cutaway view of a simple contact microphone.

CONTACT BOUNCE

Contact bounce is an intermittent make-and-break action of relay or switch contacts as they close. Contact bounce is caused by the elasticity of the metal in the contacts and armature. Contact bounce is undesirable because it often results in contact arc (*see* CONTACT ARC) as the contacts intermittently break. Contact bounce can sometimes cause a circuit to malfunction because of modulation in the flow of current.

The effects of contact bounce can be minimized by placing a small capacitance across the contacts. This helps to smooth out the modulating effects on direct current. The magnetizing current in a relay coil should not be much greater than the minimum needed to reliably close the contacts. Adjustment of the armature spring tension is also helpful in some cases. *See also* RELAY.

CONTACT MICROPHONE

A microphone that picks up conducted sound energy, rather than sound propagated through the atmosphere, is called a contact microphone. A contact microphone works in the same way as an ordinary microphone (*see* MICROPHONE).The diaphragm, instead of being vibrated by air molecules, is placed directly against the surface from which sound is to be received. The illustration shows a simplified cutaway view of a dynamic type contact micrphone. An electrically insulating layer prevents a metallic surface from affecting the operation of the microphone.

Contact microphones can be used to listen through walls. People speaking in an adjacent room cause the walls to vibrate, since the air molecules exert pressure on the walls. Contact microphones can also pick up minute vibrations in the ground, caused by such things as trucks, trains, and people walking. Some contact microphones are especially designed for underwater recording.

CONTACT MODULATION

Contact modulation is the generation of square waves, using a relay, from a sine-wave, rectified sine-wave, or direct-current source. The input is applied to a fast-acting relay, as shown in the illustration. The relay contacts make and break at a certain threshold current, switching a source of direct current on and off intermittently to produce square waves.

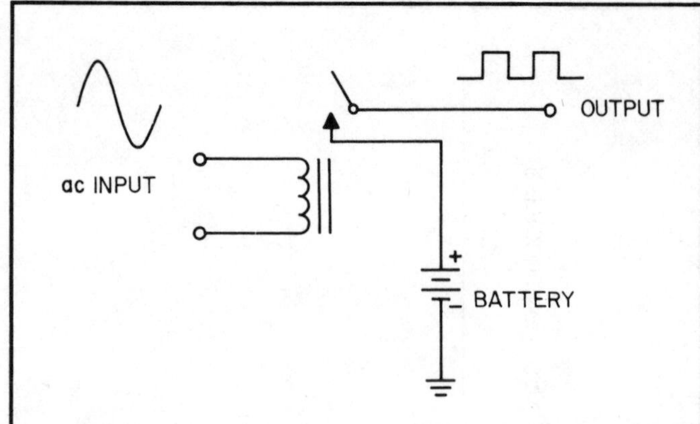

CONTACT MODULATION: Contact modulation uses a relay to convert alternating current from a sine wave to a square wave. Direct current may be interrupted by an oscillating relay to form square waves, which can then be transformed to the desired voltage.

Contact modulation is used in some chopper-type power supplies for mobile operation of equipment requiring high voltage. An oscillator circuit causes the relay to open and close, interrupting the battery voltage to form square waves. A step-up transformer is then used to obtain the necessary alternating-current voltage. Rectification produces high-voltage direct current. *See also* CHOPPER POWER SUPPLY.

CONTACT RATING

The contact rating of a switch or relay is the amount of current that the contacts can handle. The contact rating depends on the physical size of the contacts and the contact area. Usually, the contact rating is greatest for direct current, and decreases with increasing alternating-current frequency.

A switch or relay should always be operated well within its contact ratings. Otherwise, arcing will occur, and possibly even contact overheating with resulting damage. Contact ratings are sometimes specified in two ways—continuous and intermittent. The continuous rating is the maximum current that the contacts can handle indefinitely without interruption. The intermittent contact rating is the maximum current that the contacts can handle if they are closed half of the time. *See also* DUTY CYCLE.

CONTAMINATION

Contamination is the result of the jacket material of a coaxial cable bleeding through the outer braid and into the dielectric. As the cable ages, and is exposed to changes in temperature and humidity, contamination eventually takes place with certain kinds of prefabricated cables. The illustration shows this. The dielectric loss increases, the characteristic impedance of the line changes, and the velocity factor is altered. This can result in poor performance of an antenna system.

Contamination is more likely with some kinds of cable than with others. Many kinds of cable have non-contaminating jackets. Foamed polyethylene dielectric material is more susceptible to contamination than solid

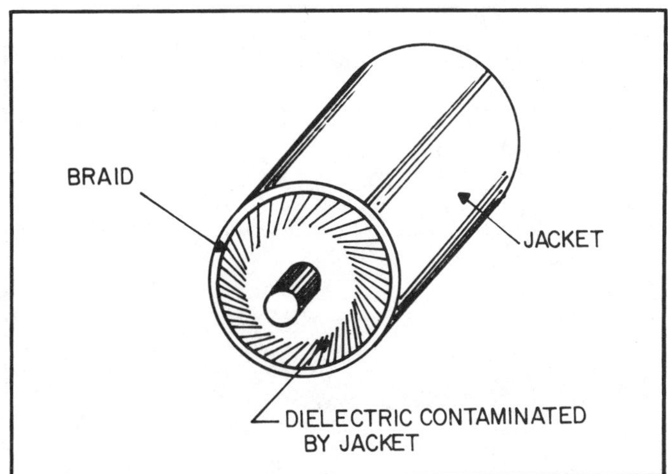

CONTAMINATION: Contamination of a coaxial-cable dielectric is caused by deterioration of the insulating jacket. It bleeds into the dielectric through the outer braid.

polyethylene. Hard line, which uses a solid cylindrical outer conductor, does not become contaminated, since the jacket material cannot penetrate to the dielectric.

The loss of any transmission line should be checked at the time of installation. The standing-wave ratio should also be checked. Any change in loss or standing-wave ratio may indicate that contamination has occurred. *See also* COAXIAL CABLE, STANDING-WAVE RATIO.

CONTINENTAL CODE

See INTERNATIONAL MORSE CODE.

CONTINUITY

In an electrical circuit, continuity is the existence of a closed circuit, allowing the flow of current. A simple device called a continuity tester is used to check for circuit continuity. The continuity tester consists of a power source and an indicating device, such as a battery and lamp as shown in the illustration. If the circuit under test is broken, or if the resistance is higher than it should be, the bulb will fail to light. For variable-resistance circuits, an ohmmeter can be used as a continuity tester (*see* OHMMETER).

In radio broadcasting, the written material used in the presentation of an on-the-air program is called continuity in the broadcaster's lingo.

CONTINUOUS-DUTY RATING

The continuous-duty rating of a device is the maximum amount of current, voltage, or power that it can handle or deliver with a 100-percent duty cycle (*see* DUTY CYCLE). This means that the device must operate constantly for an indefinite period of time.

A power supply, for example, may be specified as capable of delivering 6 amperes continuous duty. This means that if the necessary load is connected to the supply, such that 6 amperes of current is drawn, the supply can be left in operation for an unlimited amount of time, and it will continue to deliver 6 amperes.

Continuous-duty ratings are often given in watts for tubes, transistors, resistors, and other devices. Sometimes such devices can handle or deliver an amount of current, voltage, or power greater than the continuous-rated value

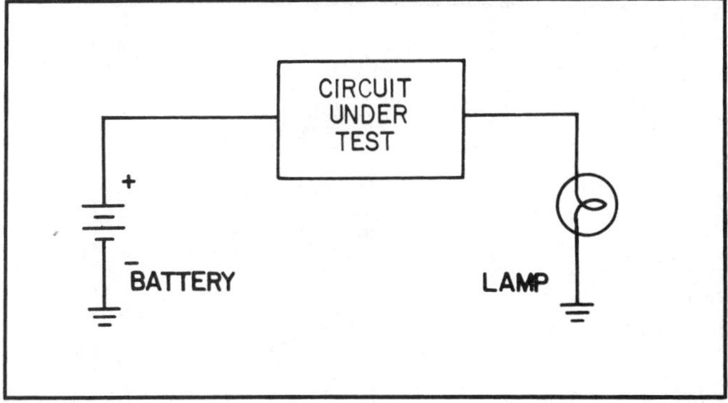

CONTINUITY: A simple continuity tester.

for short periods of time. *See also* INTERMITTENT-DUTY RATING.

CONTINUOUS FUNCTION

A continuous function is a mathematical function in which the dependent variable or variables change in an unbroken manner over the domains of the independent variables (*see* DEPENDENT VARIABLE, FUNCTION, INDEPENDENT VARIABLE).

Continuity at a point is rigorously defined as follows for a function f of one variable (see graph). For any point P = (x_o, y_o) such that $y_o = f(x_o)$, the function f is continuous at point P if, and only if:

$$\lim_{x \to x_o^-} f(x) = y_o$$

and

$$\lim_{x \to x_o^+} f(x) = y_o$$

This means that the function does not abruptly "jump" in value at P, moving either from left to right or from right to left.

If a function is continuous at all points in its domain, then it is called a continuous function. There are various mathematical ways to test a function to determine whether or not it is continuous. Examples of continuous functions are sine waves and transistor characteristic curves. Examples of discontinuous functions are square waves and sawtooth waves.

CONTINUOUS WAVE

A continuous wave is a sine wave that maintains a continuous amplitude of voltage peaks, or waveform peaks as long as the wave is being produced. The illustration shows a comparison between continuous wave (A) and damped wave (B). *See* DAMPED WAVE. In a damped wave, the waveform amplitude decreases with each cycle. Damped waves are produced by shock exciting an LC tuned circuit by suddenly applying and removing voltage, as in early spark-gap transmitters. With the advent of the vacuum tube, it became possible to continuously supply power to overcome the losses in an LC circuit. Therefore, a vacuum-tube/LC-oscillator circuit can sustain continuous waveform output for as long as the circuit has power applied.

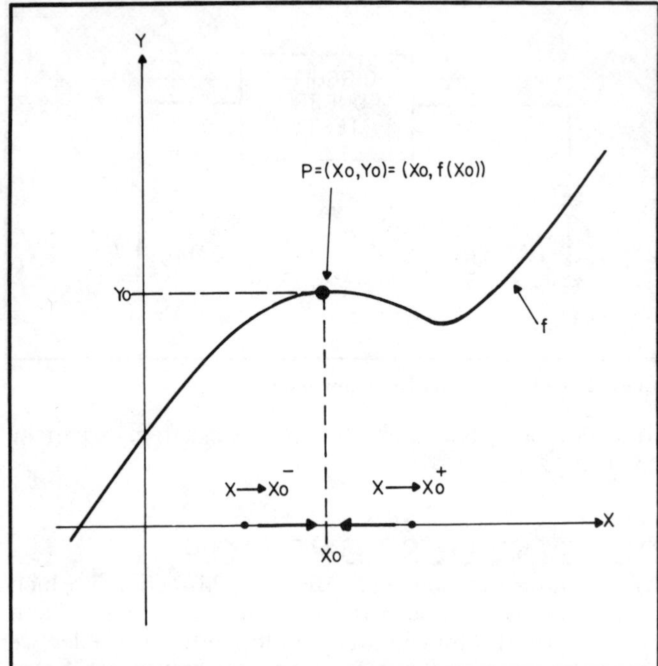

CONTINUOUS FUNCTION: A function is continuous at a point if and only if it does not "jump" in value at that point, approaching from either the right or from the left.

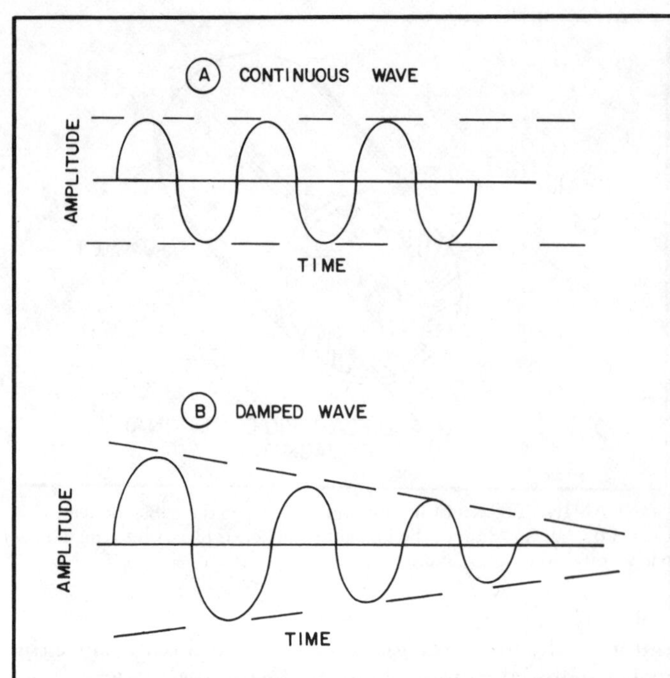

CONTINUOUS WAVE: Continuous waves maintain waveform or voltage amplitude as long as power is applied (A). The damped wave (B) decays to the vanishing point because of inherent resistance in the circuit.

Radio telegraphy today uses continuous-wave transmissions, broken into information bits in the form of dots and dashes. This mode of communication is often called cw (for continuous-wave communication), even though the distinction is no longer necessary because damped-wave (spark-gap) transmission is illegal. *See also* INTERNATIONAL MORSE CODE.

CONTRAN

CONTRAN is a high-order computer language that serves as its own assembler program (*see* ASSEMBLER AND ASSEMBLY LANGUAGE, HIGHER-ORDER LANGUAGE). In most programming situations, the compiler uses one program to translate the higher-order commands to machine language and vice versa; the operator uses the higher-order language to interface between himself and the compiler. CONTRAN eliminates the need for two programs. The translation between operator and machine language is performed in one step. *See also* COMPUTER PROGRAMMING.

CONTRAST

In a television picture signal, the contrast is the difference in magnitude between the white and black components. Contrast is affected by the brightness of the picture. It is important that the contrast be realistic for good picture reproduction.

Various levels of contrast are shown in the photographs. Too little contrast makes the picture appear washed-out and difficult to read (A). Proper contrast makes the words easy to discern (B). Too much contrast makes the words appear blurred, and can cause eyestrain (C).

Most television receivers have an external adjustment for the contrast. Once this control has been properly set, the contrast will remain correct, even for pictures of different brightness levels. *See also* AUTOMATIC CONTRAST CONTROL, PICTURE SIGNAL, TELEVISION.

CONTROL

The term control can mean any of several different things in electricity and electronics. Most often, the term control refers to an adjustable component, such as a potentiometer, switch, or variable capacitor, located for easy access in the operation of a piece of equipment. The channel selector on a television set, for example, is a control.

In computers, control is the maintenance of proper operation. This is accomplished by a central processing unit, or CPU (*see* CENTRAL PROCESSING UNIT). In general, control is the maintenance of operation at a certain level, or for a certain purpose, in a piece of electronic apparatus. Control can be either manual or automatic. In some modern communications and electronic devices, control is carried out by means of a microcomputer designed especially for that application. *See also* MICROCOMPUTER, MICROCOMPUTER CONTROL.

CONTROL GRID

The control grid is a part of a vacuum tube that acts in a manner similar to the base of a bipolar transistor or the gate of a field-effect transistor. Signal variations on the control grid of a vacuum tube result in changes in the plate current. A tube amplifies because a small change in the voltage between the grid and the cathode produces a large change in the current from the cathode to the plate. In a tube with more than one grid, the control grid is usually the grid located closest to the cathode. Some tubes have more than one control grid.

Generally, the control grid is biased at either the same voltage as the cathode, or negatively with respect to the cathode. The more negative the instantaneous grid voltage, the less plate current flows. When the grid bias becomes sufficiently negative to stop the plate current altogether, the condition is called cutoff or cutoff bias.

The drawings illustrate the schematic symbols for triode (A), tetrode (B), and pentrode (C) tubes, showing the control grid. *See also* BASE, GATE, GRID, GRID BIAS, TUBE.

A

B

CONTRAST: Contrast adjustment in a video monitor. At A, the contrast is insufficient; at B, it is acceptable; at C, it is excessive.

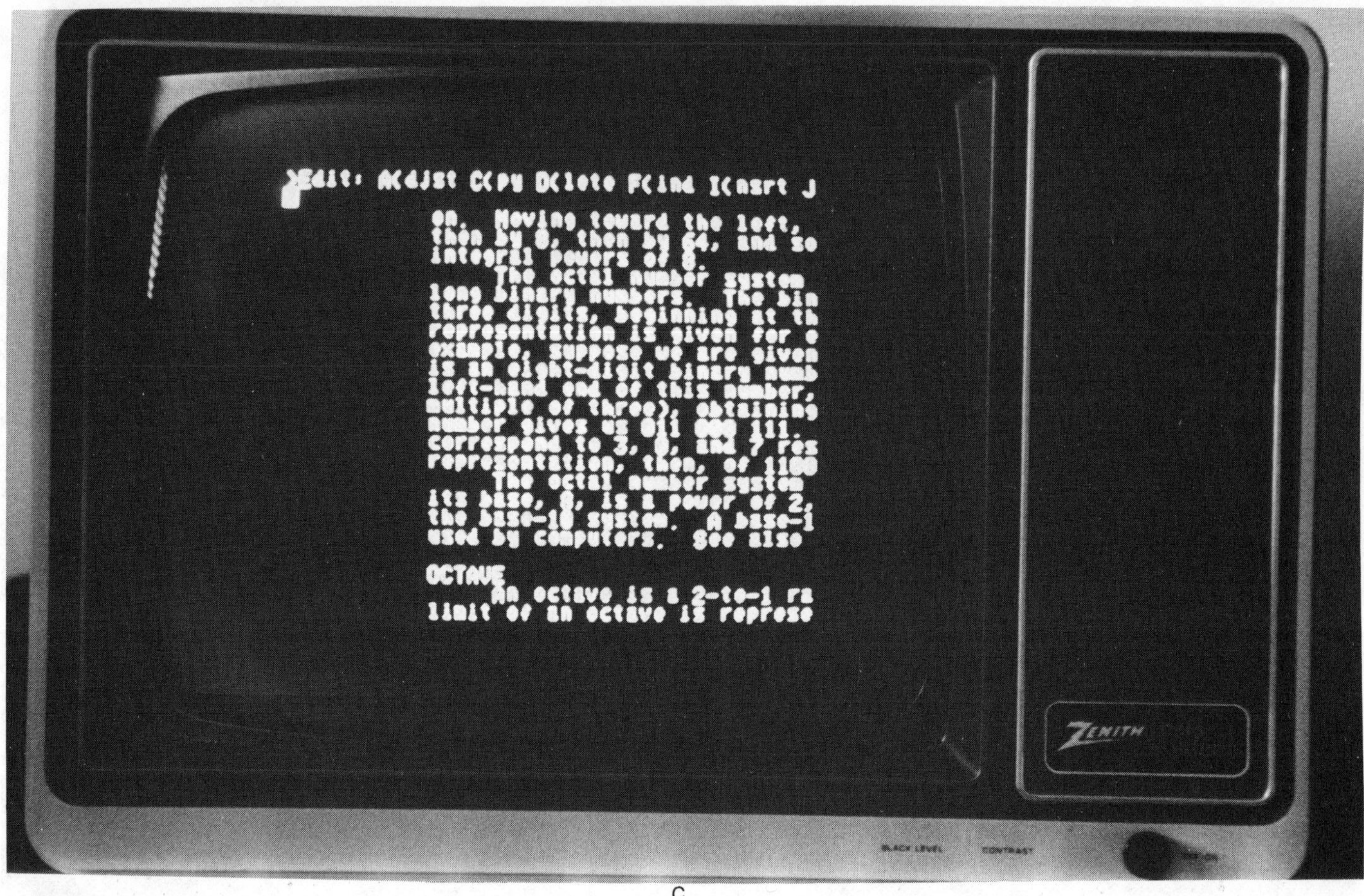

C

CONTROLLED-CARRIER TRANSMISSION

Controlled-carrier transmission or modulation is a special form of amplitude modulation (*see* AMPLITUDE MODULATION). It is sometimes called quiescent-carrier or variable-carrier transmission. The carrier is present only during modulation. In the absence of modulation, the carrier is suppressed. At intermediate levels of modulation, the carrier is present, but reaches its full amplitude only at modulation peaks.

The graphs illustrate the relationship between carrier amplitude and modulating-signal amplitude for ordinary amplitude modulation (A) and controlled-carrier modulation (B).

Controlled-carrier operation is more efficient than ordinary amplitude modulation because the transmitter does not have to dissipate energy during periods of zero modulation. The energy dissipation is also reduced when the modulation is less than 100 percent, which is almost all the time. *See also* CARRIER, MODULATION.

CONTROLLED RECTIFIER

See SILICON-CONTROLLED RECTIFIER.

CONTROLLER

A controller is a device or instrument that maintains a certain set of operating parameters for a piece of equipment. Sometimes, the term controller refers to a power-regulating device such as a light dimmer or motor speed adjustment, or a device that shuts off the equipment if it overheats or otherwise exceeds its safe ratings.

Controllers are used in servomechanisms to regulate their movement. As robot devices become more and more commonplace, and are being used to perform increasingly complex tasks, their controllers are getting more sophisticated. Microcomputers are often used in conjunction with controllers for electrical, electronic, and mechanical equipment. *See also* MICROCOMPUTER, MICROCOMPUTER CONTROL, ROBOTICS, SERVOMECHANISM.

CONTROL PANEL

A control panel forms the interface between a piece of equipment and its operator. A control panel may be very simple, such as that of a small radio receiver, or it may be quite complex, such as that of the laboratory test equipment shown in the photograph.

The arrangement of the controls is important from the standpoint of human engineering (*see* HUMAN ENGINEERING). An instrument that is easy to operate will, when all other things are equal, perform better than an instrument that is difficult to operate. Labeling is important. The meters and other display devices must be located so that they are easily seen. Functions must be carried out in a logical sequence. The arrangement of controls must also be compatible with an efficient component arrangement inside. *See also* COMPONENT LAYOUT.

CONTROL RECTIFIER

A control rectifier is a silicon device used for the purpose of switching large currents. A small control signal, applied to the rectifier, allows power regulation. Control rectifiers

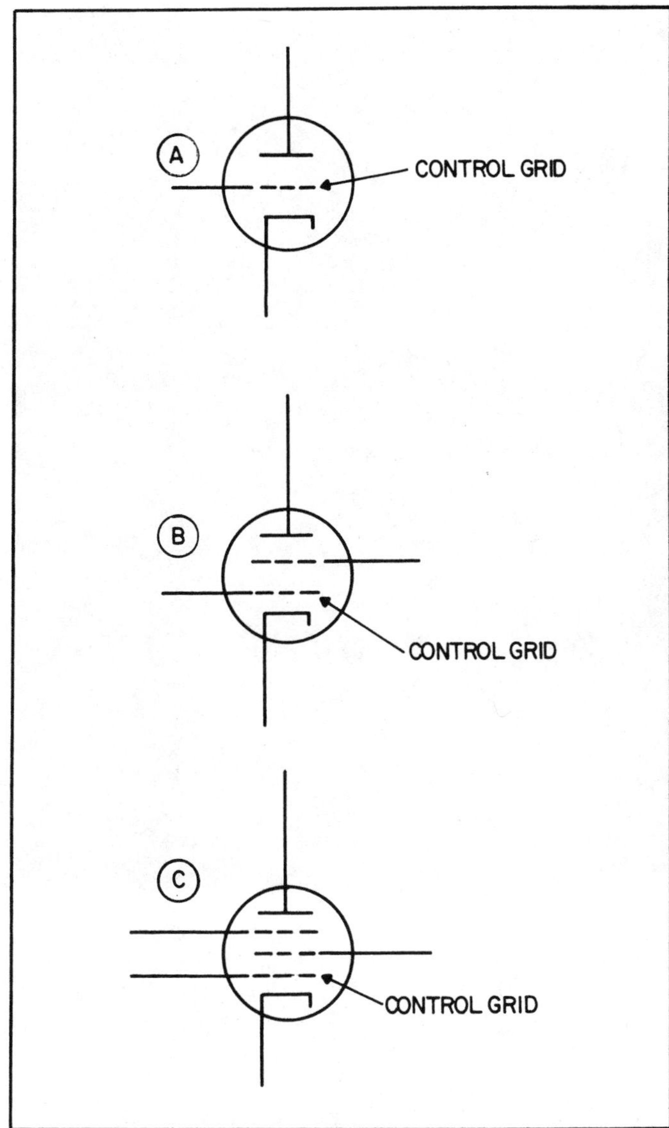

CONTROL GRID: Schematic symbols for the control grid of a triode (A), tetrode (B), and pentode (C).

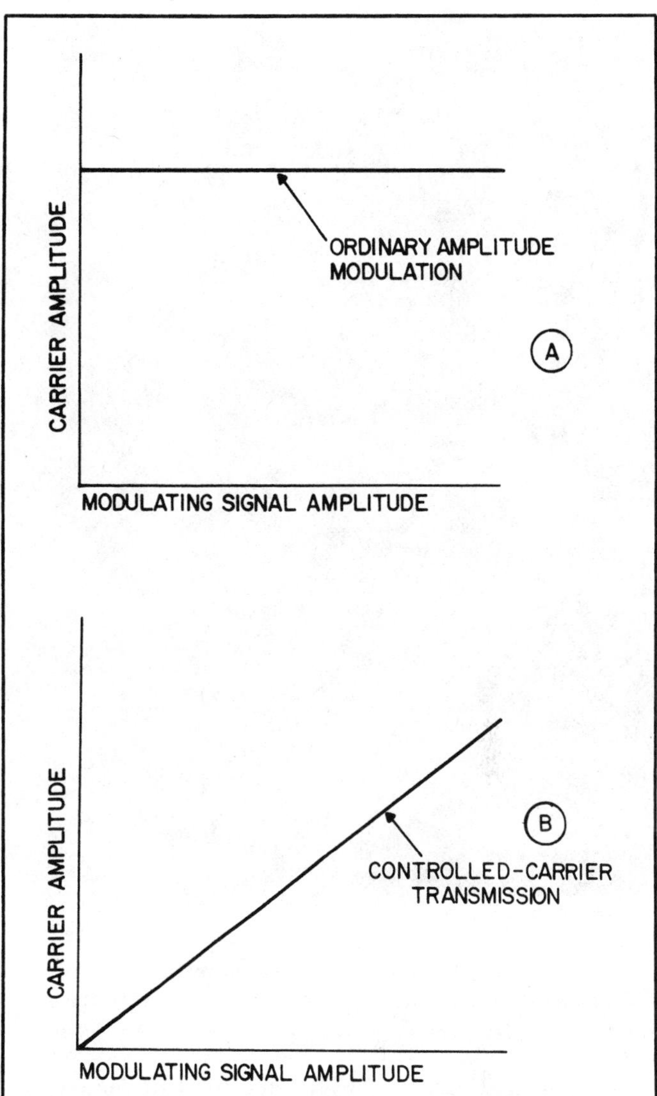

CONTROLLED-CARRIER TRANSMISSION: Carrier amplitude versus modulationg signal strength for ordinary amplitude modulation (A) and controlled-carrier transmission (B).

are more efficient, and also smaller, than vacuum tubes, relays, or rheostats for the adjustment of power.

Control rectifiers are used in such devices as light dimmers, motor-speed controls, remote switches, and current limiters. *See also* SILICON-CONTROLLED RECTIFIER.

CONTROL REGISTER

A control register stores information for the regulation of digital-computer operation. The control register consists of an integrated circuit or a set of integrated circuits. While one program statement is executed, the control register stores the next statement. Once a statement has been carried out by the computer, the statement in the control register is executed, and a new statement enters the register. *See also* COMPUTER, REGISTER.

CONTROL UNIT

In a digital computer, the control unit is the section that regulates the operation. The control unit constantly monitors the functioning of the other circuits, sending out instructions when necessary (*see* COMPUTER).

In a high-fidelity sound system, the preamplifier usually contains the volume, tone, and equalization controls. Such a preamplifier is called a control unit. It is much easier, electrically, to perform the control operations at a point where the sound signal amplitude is small, as compared to a point where it is large. A control unit receives the sound impulses from the tuner, turntable, microphone, tape player, or other transducer. The output of the control unit is connected to the power amplifier. *See also* HIGH FIDELITY.

CONVECTION COOLING

Convection cooling is a form of air cooling. Hot air rises, of course, and cool air falls. When a device such as a tube or transistor dissipates heat into the surrounding air, this well-known effect can be utilized to effectively cool the tube or transistor.

The illustration shows a typical arrangement for convection cooling of a vacuum tube. (The shape and size of the chamber may vary greatly.) Holes in the chassis under the tube allow cool air to enter the chamber as hot air rises through holes in the top of the chamber. Except at the top and bottom, the chamber is airtight. A continuous upward flow of air therefore takes place next to the tube.

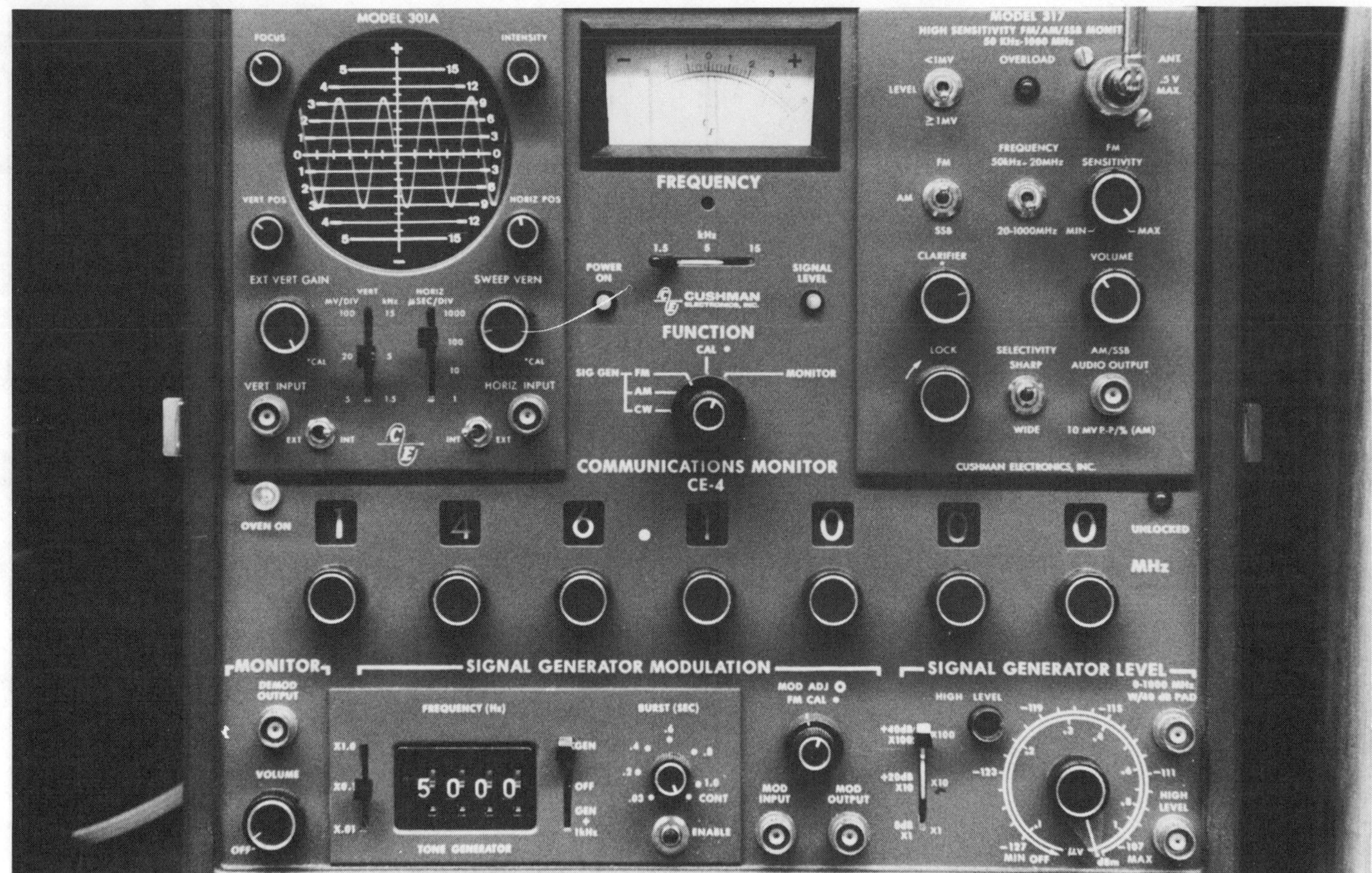

CONTROL PANEL: The control panel of this signal monitor/generator is complicated, but effectively laid out for easy use.

Convection cooling can be enhanced by the addition of a fan at the bottom of the chamber, to blow air upward. The fan may also be located at the top of the chamber to pull air through.

To a certain extent, convection is part of the cooling system in almost all high-power tube or solid-state amplifiers, A device called a heatsink, with vertical fins, facilitates heat dissipation by convection in almost all high-power semiconductor amplifiers. The heat is conducted to the heatsink, whereit is transferred to the surrounding environment. *See also* AIR COOLING, CONDUCTION COOLING, HEATSINK.

CONVERGENCE

In a cathode-ray tube with more than one electron beam, convergence is the condition in which all of the electron beams intersect in a single point, and this point must lie on the phosphor screen. Convergence is important in a color television picture tube, where three electron guns are employed, one for each of the primary colors (red, blue, and green). If convergence is not obtained at all points on a color picture screen, the colors will not line up properly. This results in strange colored borders around certain portions of objects in the picture. The drawing illustrates convergence of electron beams toward the phosphor screen of a cathode-ray tube.

As the electron beams scan across the tube, their point of convergence follows a two-dimensional surface in space, called the convergence surface. In a color television receiver, the three electron beams must not only always converge, but their convergence surface must correspond exactly with the phosphor surface of the picture tube. The controls in a color television receiver that provide precise alignment of the convergence surface are called the convergence and convergence-phase controls. *See also* COLOR TELEVISION, TELEVISION.

CONVERSATIONAL COMPUTER OPERATION

When a computer responds verbally to commands or questions from its operator, either giving the operator directions or asking for more information, the computer is said to be in the conversational mode. Conversational mode is common in large computers, but many microcomputers do not yet use it. However, a primitive form of the conversational mode is beginning to appear in some microcomputer-controlled consumer devices.

Conversational operation simplifies the procedure of communicating with a computer, and carrying out the desired tasks. Probably the most familiar example (to the public) of conversational computer operation is found in automatic bank tellers. The machine usually begins by saying something like:

"INSERT CARD FACE UP"

on an alphabetic-numeric display screen. Once the operator has put in his or her card, the machine responds with a statement such as:

"ENTER YOUR SECRET CODE"

The operator then presses the appropriate sequence of

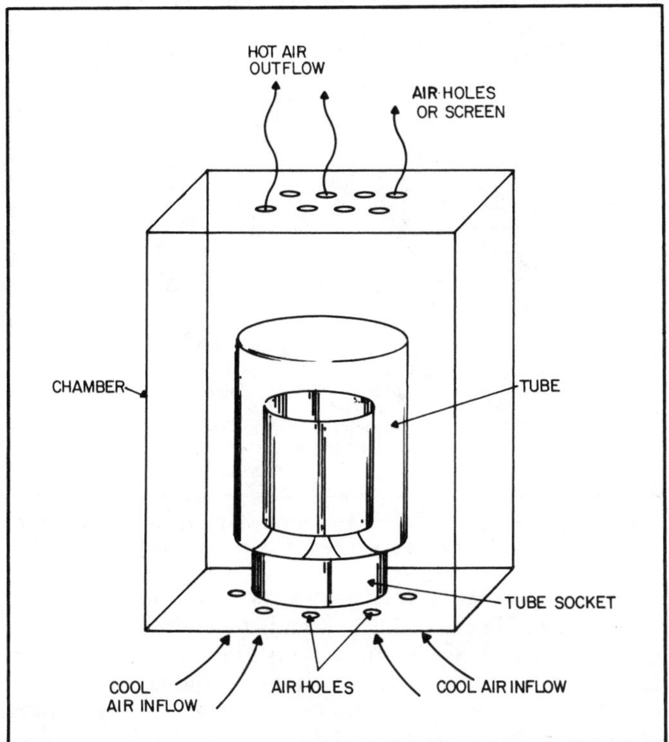

CONVECTION COOLING: Convection cooling of a vacuum tube makes use of the simple fact that hot air rises.

number buttons on the keyboard. The machine then says something like:

"SELECT TYPE OF TRANSACTION"

and the operator may then select from a variety of services. Operation continues in this manner until the process is complete. Almost anyone can operate a computer that is in the conversational mode. *See also* COMPUTER.

CONVERSION

Conversion is the process of changing frequency, voltage, current, or data-processing mode. A frequency converter changes the frequency of a signal without altering its bandwidth characteristics. A voltage converter changes the voltage of a power supply; such a converter may also change the supply from alternating to direct current or vice versa. A current converter changes the current supplied to a circuit.

In data processing, information can be changed from one form to another. Generally, this consists of altering the transmission mode from serial to parallel, or vice versa. *See also* PARALLEL DATA TRANSFER, SERIAL DATA TRANSFER.

CONVERSION EFFICIENCY

In a frequency converter, the ratio of the output power, voltage, or current to the input power, voltage, or current is called the conversion efficiency. For example, if a frequency converter is designed to change a 21-MHz signal to a 9-MHz signal, the conversion efficiency is the ratio of the 9-MHz signal power, voltage, or current to the 21-MHz signal power, voltage, or current. Conversion efficiency is always less than 100 percent.

In a rectifier, the conversion efficiency is defined as the ratio of the direct-current output power to the alternating-current input power. The conversion efficiency in a rectifier-and-filter power supply is always less than 100 percent because of losses in the rectifier devices and transformers. *See also* CONVERTER, POWER SUPPLY, RECTIFIER.

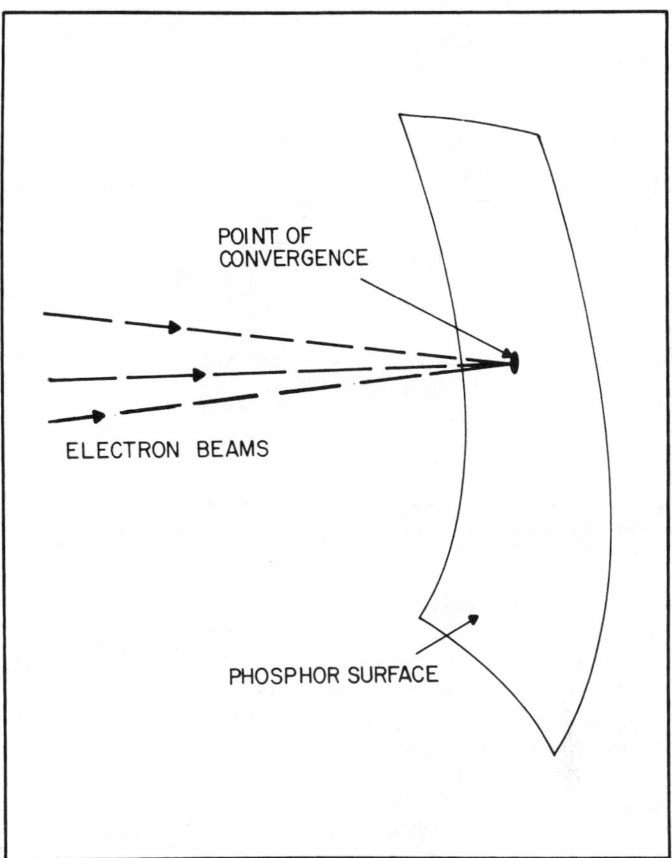

CONVERGENCE: Convergence of electron beams onto a phosphor screen.

CONVERSION GAIN AND LOSS

In a converter, the conversion gain or loss is the increase or decrease, respectively, of signal strength from input to output. This gain or loss may be determined in terms of power, voltage or current. It is specified in decibels.

A passive converter, such as a diode mixer, always has some conversion loss. However, converters with amplifiers may produce conversion gain.

If P1 is the input power to a converter and P2 is the output power, then the conversion gain is given in decibels by

$$\text{Power gain (dB)} = 10 \log_{10} (P2/P1)$$

If E1 is the input voltage and E2 is the output voltage, then

$$\text{Voltage gain (dB)} = 20 \log_{10} (E2/E1)$$

If I1 is the input current and I2 is the output current, then

$$\text{Current gain (dB)} = 20 \log_{10} (I2/I1)$$

When the output amplitude (P2, E2 or I2) is less than the input amplitude (P1, E1 or I1), the conversion gain will be negative. Then we have loss. We may say, for example, that a device has −3 dB conversion gain, or that it has 3 dB conversion loss. *See also* CONVERTER, GAIN, LOSS.

CONVERTER: A converter consists of an oscillator and a mixer.

CONVERTER

Any device that converts frequency, voltage, or current, or computer data from one form to another is called a converter. The term is most often used in reference to the frequency conversion in a superheterodyne receiver (*see* SUPERHETERODYNE RECEIVER).

In the superheterodyne, the signal frequency f is mixed with an unmodulated, pure carrier at a frequency f_0 to obtain a signal at the sum or difference frequency, f_1. Usually, the difference frequency is used; if $f > f_0$, then

$$f_1 = f - f_0$$

and if $f < f_0$, then

$$f_1 = f_0 - f$$

The new signal has exactly the same modulation characteristics as the original signal. The drawing illustrates frequency conversion.

Conversion is done for two reasons. First, a receiver must usually be capable of operation over a wide range of input signal frequencies. By mixing the signal frequency with the output of a variable-frequency oscillator, a signal of constant frequency can be obtained. This makes the tuning of successive amplifying stages much simpler. Second, a converter can be used to bring the frequency of a signal down to a value at which it is much easier to provide good selectivity and gain. This becomes increasingly important at very-high and ultra-high frequencies. Converters are often used for reception in this part of the electromagnetic spectrum. The converters are placed between the lower-frequency receiver and the antenna. *See also* MIXER.

COOK SYSTEM

An early method of obtaining stereophonic sound on a disk recorder was called the Cook system. Two separate grooves were cut next to each other in the disk. One groove contained the left-channel sound, and the other groove contained the right-channel sound. Two cutters were necessary in the recording process, and two styli were needed for playback. The Cook system is shown in the illustration.

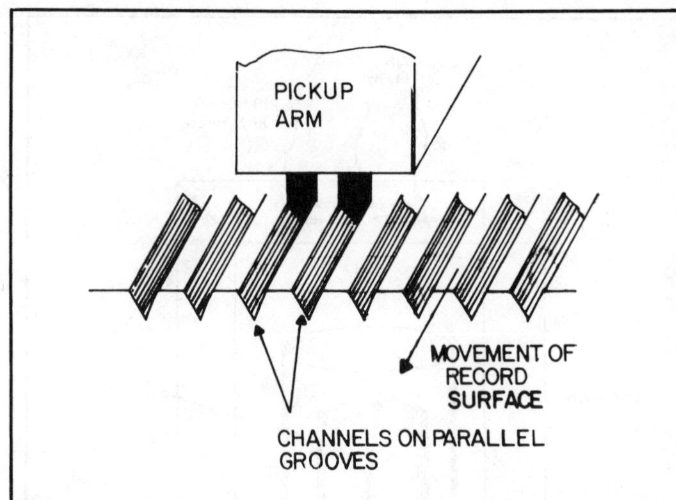

COOK SYSTEM: The Cook stereo disk recording system uses two needles, one for each channel, and separate grooves on the disk.

The Cook system reduced the playback time of a record by 50 percent, because the stylus advanced by two grooves per turn of the disk. Later, a system was devised that required only one groove for stereo sound reproduction on a disk. The impulses of the left channel were recorded on one face of the groove, and the right-channel sound was impressed on the other face of the same groove. By orienting the two faces at a 90-degree angle, excellent channel separation could be obtained. *See also* STEREO DISK RECORDING.

COOLING

In an electronic circuit, cooling is the process of removing undesirable heat to increase the power-handling capability. Cooling also reduces instability, because some component values change with temperature fluctuations. Cooling can take place in three ways: conduction, convection, and radiation. Quite often, a circuit is cooled by a combination of these three modes.

Vacuum tubes are sometimes cooled primarily by convection. However, most modern solid-state equipment uses conduction cooling. (*See* CONDUCTION COOLING, CONVECTION COOLING.) The transistor is mounted against a highly heat-conductive surface, which is thermally bonded to a large, finned piece of metal called a heatsink (*see* HEATSINK). The heatsink transfers heat to the environment via convection and radiation.

Cooling becomes especially important in power amplifiers. The tubes or transistors in such amplifiers generate considerable heat, because no amplifier is 100-percent efficient. Cooling allows the power input, and hence the output, to be maximized. Without effective cooling, the choice would be between a lower level of operating power and rapid destruction of the tubes or transistors.

COORDINATED UNIVERSAL TIME

Coordinated Universal Time, generally abbreviated UTC, is an astronomical time based on the Greenwich meridian (zero degrees longitude). The UTC day begins at 0000 hours and ends at 2400 hours. Midday is at 1200 hours.

The speed of revolution of the earth varies depending

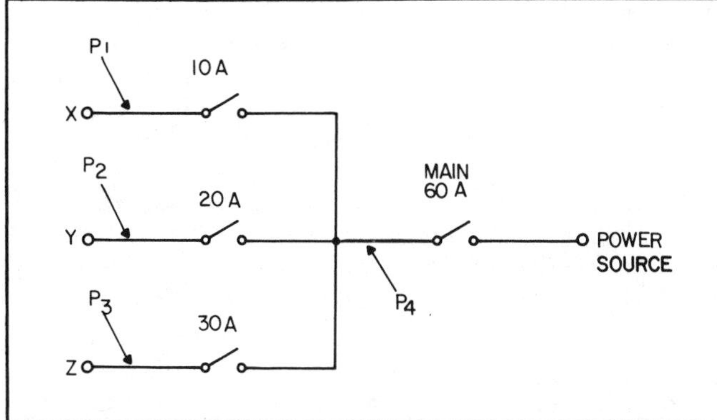

COORDINATION: In a coordinated system of circuit breakers, the branch breakers should trip first in the event of a short at point P_1, P_2, or P_3. The main breaker only should trip in the event of a short at point P_4.

on the time of the year. Coordinated Universal Time is based on the mean, or average, period of the synodic (sun-based) rotation. The earth is slightly behind UTC near June 1, and is slightly ahead near October 1. This variation is caused by tidal effects caused by the gravitational interaction between the earth and the sun.

Coordinated Universal Time is five hours ahead of Eastern Standard Time in the United States. It is six hours ahead of Central Standard Time, seven hours ahead of Mountain Standard Time, and eight hours ahead of Pacific Standard Time. These differences are one hour less during the months of daylight-saving time.

Coordinated Universal Time is broadcast by several time-and-frequency-standard radio stations throughout the world. In the United States, the principal stations are WWV in Colorado and WWVH in Hawaii. These stations are operated by the National Bureau of Standards. *See also* TIME ZONE, WWV/WWVH.

COORDINATE SYSTEM

A coordinate system is a means of precisely locating a point on a line, on a plane, in three-dimensional space, or in a space of more than three dimensions. The simplest system of coordinates is the number line of one dimension. In two dimensions, two coordinate numbers are necessary to uniquely define a point. In three dimensions, three numbers are needed. Coordinate values may be represented either by distances or by angles. In some advanced mathematical systems, four, five, or more dimensions are used.

Coordinate systems are often used in electricity and electronics. The familiar graph, showing one variable as a function of another, is an example of the Cartesian coordinate plane. A radar system shows the locations of objects on a two-dimensional display corresponding to a Cartesian or polar coordinate plane, depending on the type of radar display. The orientation of an antenna system is often determined by means of a coordinate system. *See also* CARTESIAN COORDINATES, CURVILINEAR COORDINATES, POLAR COORDINATES, SPHERICAL COORDINATES.

COORDINATION

When two circuit breakers are connected in series, and the breakers have different current ratings, it is advantageous to have the breaker with the lower current rating trip before the one with the higher rating in the event of a short circuit. The same is true of fuses. This is called coordination.

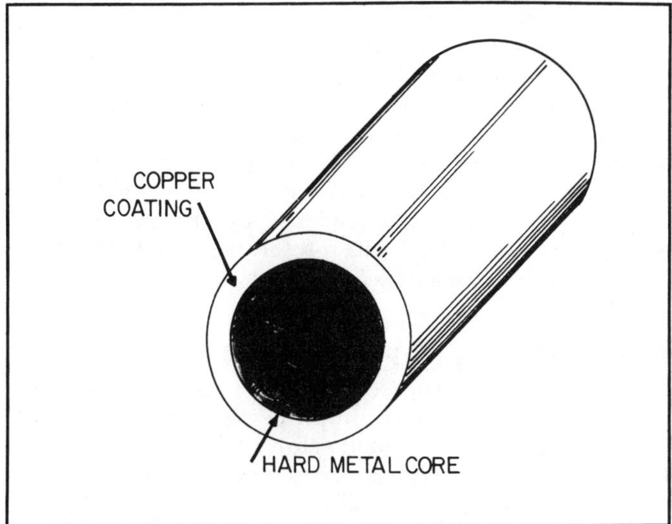

COPPER-CLAD WIRE: Copper-clad wire has a hard metal core and a coating of highly conductive copper.

The illustration shows a circuit with three branches, called X, Y, and Z. Branch X has a circuit breaker rated at 10 amperes. Branch Y is rated at 20 amperes, and branch Z is rated at 30 amperes. The entire circuit, consisting of the three branches in parallel, is rated at 60 amperes, which is the sum of the individual branch ratings.

If a short circuit occurs at point P1, P2, or P3, then the X, Y, or Z branch only should trip. If a short circuit occurs at point P4, then only the main breaker should trip. Coordination eliminates unnecessary inconvenience when resetting circuit breakers. In the illustration, coordination is achieved by making the X, Y, and Z branch breaker tripping times faster than the tripping time for the main breaker. *See also* CIRCUIT BREAKER.

COPPER

Copper is an element with atomic number 29 and atomic weight 64. In its pure form, copper appears as a dark, gold-like metal. It is flexible and malleable. Copper is used extensively in the manufacture of wire, since it is an excellent conductor of electric current. Copper is also a good conductor of heat. It is fairly resistant to corrosion.

Oxides of copper are used as rectifiers, especially in the manufacture of photovoltaic cells. Copper oxide exhibits the properties of a semiconductor. *See also* COPPER-CLAD WIRE, COPPER-OXIDE DIODE.

COPPER-CLAD WIRE

Copper-clad wire is a wire with a core of hard metal, such as iron or steel, and an outer coating of copper, (see illustration). Copper-clad wire is frequently used in radio-frequency antenna systems, where both tensile strength and good electrical conductivity are essential. The central core provides mechanical strength to resist stretching and breakage. The copper coating carries most of the current.

Although copper-clad wire has higher direct-current resistance than pure copper wire of the same diameter, because iron and steel are less effective electrical conductors than copper, the resistance at radio frequencies is almost as low as that of pure copper wire. This is because of the skin effect. At high alternating-current frequencies, most of the electron flow takes place near the outer part of a conductor (*see* SKIN EFFECT).

Copper-clad steel wire is more difficult to work with than soft-drawn or hard-drawn pure copper wire. When installing copper-clad steel wire, the wire must be unwound carefully to avoid kinks. The advantages of greater tensile strength, and less tendency to stretch, however, make copper-clad steel wire preferable to pure copper wire in a variety of applications. *See also* WIRE, WIRE ANTENNA.

COPPER-OXIDE DIODE

Copper oxide is a compound consisting of copper and oxygen. It has semiconductor properties, and can be used as a basis for a detector, modulator, rectifier, or photovoltaic cells. A copper-oxide diode is made by placing copper oxide against metallic copper.

Copper-oxide diodes are suitable only for low-voltage applications. In modern circuits, semiconductor diodes are almost always made using germanium, selenium, or silicon. Copper-oxide photovoltaic cells, consisting of copper-oxide diodes covered with clear, protective plastic, are still found in cameras and other devices. Such photovoltaic cells do not require an external source of power. *See also* DIODE, PHOTOVOLTAIC CELL.

COPY

The term copy is used in reference to reception in radiocommunication circuits, and sometimes also in landline circuits. The received message, whether written by a receiving operator or printed by a machine, is called the copy.

The expression "solid copy" means that everything has been received perfectly. It may also indicate to a transmitting operator that the signal is perfectly clear at the receiving end, with no significant interference, either manmade or otherwise. The ease or correctness with which a receiving operator or machine copies a signal is sometimes indicated by a readability number of 1 to 5. The readability designator 1 represents an unreadable signal; the designator 5 represents a perfectly readable signal. Numbers between these extremes represent intermediate levels of readability.

Radio operators use the term copy in a variety of ways. A telegraph operator "copies" the code. A printer "copies" a teletype signal. A voice operator "copies" a message. *See also* RST SYSTEM.

CORBINO EFFECT

When a disk carrying a radial current, either toward or away from the center, is subjected to a magnetic field whose lines of force run perpendicular to the disk, the flow of current is bent. The illustration shows this phenomenon. It is called the Corbino effect.

The Corbino effect allows a semiconductor disk, in a variable magnetic field, to act as a variable resistor. When there is no magnetic field present, the current flows in straight lines between the center of the disk and the edge, when one electrode is placed at the center and the other around the entire edge (see A). In the presence of a relatively weak magnetic field (B), the current path is bent slightly. Since the path is made longer from the center to the edge of the disk, the resistance increases. In an intense magnetic field (C), the current is forced to flow over a much longer path, since it is bent greatly. This increases the resistance by a large amount. A disk of semiconductor material, placed in a magnetic field for the purpose of controlling its center-to-edge resistance, is called a Corbino disk.

The Corbino effect takes place because electrons moving perpendicular to a magnetic field tend to be deflected in a direction orthogonal to both the field orientation and the direction of electron motion. *See also* HALL EFFECT.

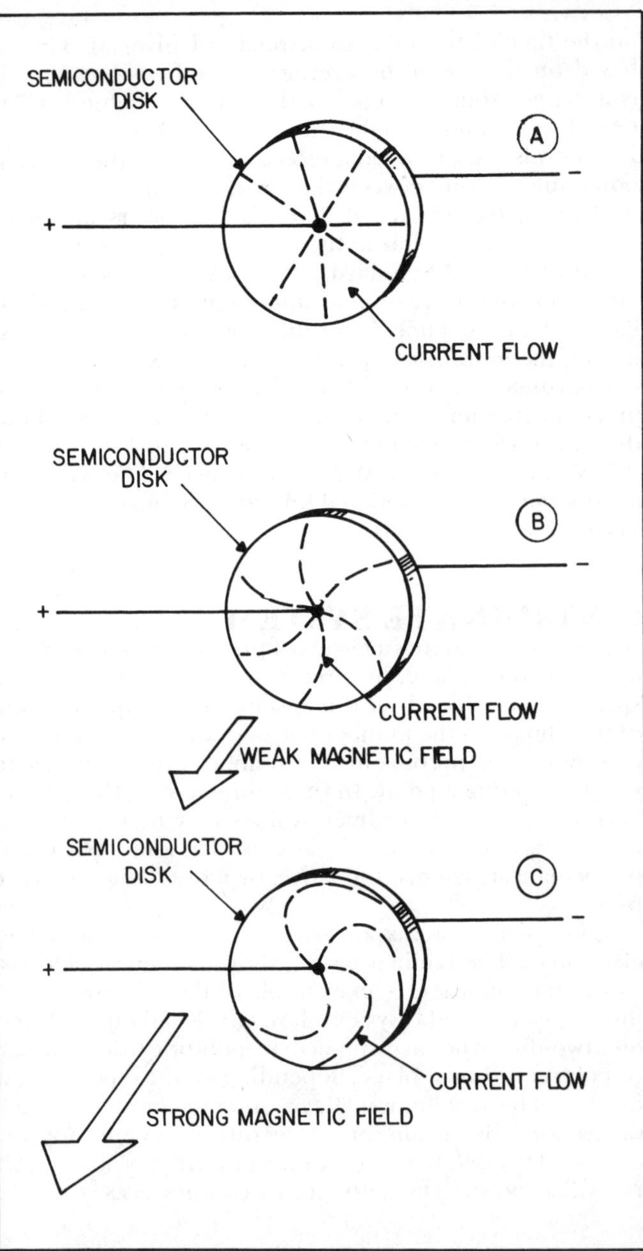

CORBINO EFFECT: The Corbino effect causes curvature of the current flow in a disk subjected to a magnetic field. At A, no field is present. At B, a weak orthogonal magnetic field causes slight bending of the electron or hole flow. At C, a strong field causes a large increase in the distance the current must flow through the disk. This increases the resistance between the center of the disk and the perimeter.

CORD

A cord is a simple, durable cable, having two or three conductors, and used for the purpose of transferring an electric current to an appliance or device. The most familiar types are two-conductor and three-conductor, 117-volt utility cords (see illustration).

Cords are generally not suitable for transferring alternating currents above the audio-frequency range. This is because of dielectric and conductor losses. Cables designed especially for radio frequencies should be used for such applications. *See also* CABLE.

CORDLESS TELEPHONE

A cordless telephone uses a radio connection between the receiver and the base, rather than the usual cord. Two small antennas, one at the main unit and the other at the receiver, allow use of the receiver as far away as 600 to 800 feet under ideal conditions. This can be extremely convenient, allowing calls to be made or received anywhere around the house (see illustration).

Cordless telephones are available in various configurations. With some units, calls can be received in a remote place, but outgoing calls must be dialed from the main base. With other units, both incoming and outgoing calls can be controlled entirely from the receiver.

Since cordless telephones are linked by low-power radio, they are subject to interference from some appliances and from stray electromagnetic impulses. However, the technology of the cordless telephone is improving rapidly. *See also* MOBILE TELEPHONE.

CORE

A core is a piece of ferrite or powdered-iron material that is placed inside a coil or transformer to increase its inductance. Cores for solenoidal inductors are cylindrical in shape. Cores for toroidal inductors resemble a doughnut. The drawing illustrates these core configurations.

The amount by which a ferromagnetic core increases the inductance of a coil depends on the magnetic permeability of the core material. Ferrite has a higher permeability, in most cases, than powdered iron.

The form on which a coil is wound, whether or not it is made of a material with high magnetic permeability, is sometimes called the core. Generally, however, the term core refers to a ferrite or powdered-iron substance placed within an inductor. *See also* FERRITE CORE, PERMEABILITY, POWDERED-IRON CORE, TRANSFORMER.

CORE LOSS

In a powdered-iron or ferrite inductor core, losses occur because of circulating currents and hysteresis effects (*see* EDDY-CURRENT LOSS, HYSTERESIS LOSS). All ferromagnetic core materials exhibit some loss.

Generally, the loss in a core material increases as the frequency increases. Ferrite cores usually have a lower usable-frequency limit than powdered-iron cores. In general, the higher the magnetic permeability of a core material, the lower the maximum frequency at which it can be used without objectionable loss.

Core losses tend to reduce the Q factor of a coil (*see* Q FACTOR). In ferrite-rod antennas, core losses limit the frequency range at which reception can be realized. In applications requiring large currents or voltages through or across an inductor, excessive core losses can result in overheating of the core material. The core may actually break under such conditions. *See also* FERRITE CORE, FERRITE-ROD ANTENNA, PERMEABILITY, POWDERED-IRON CORE, TRANSFORMER, TRANSFORMER EFFICIENCY.

CORE MEMORY

A core memory is a means of storing binary information in

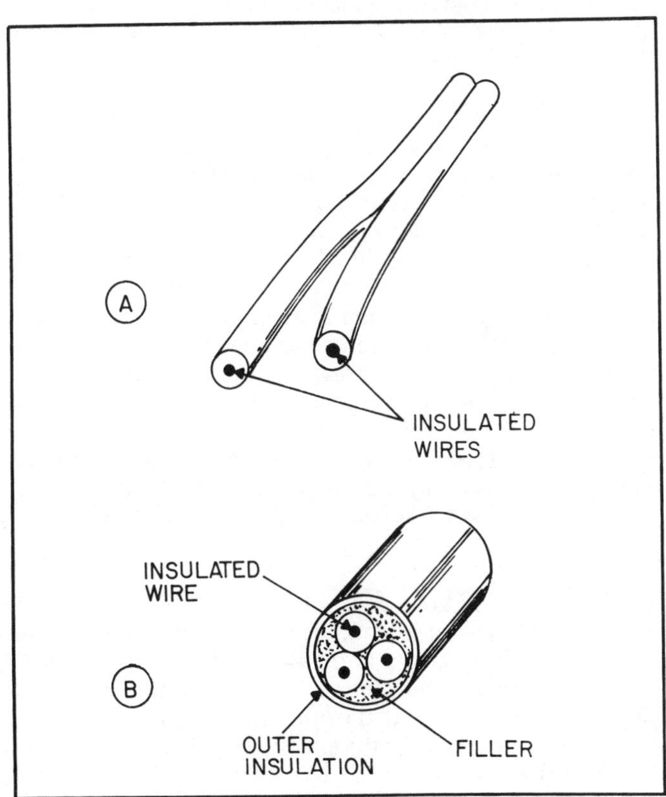

CORD: At A a two-conductor cord. At B a common three-wire utility cord.

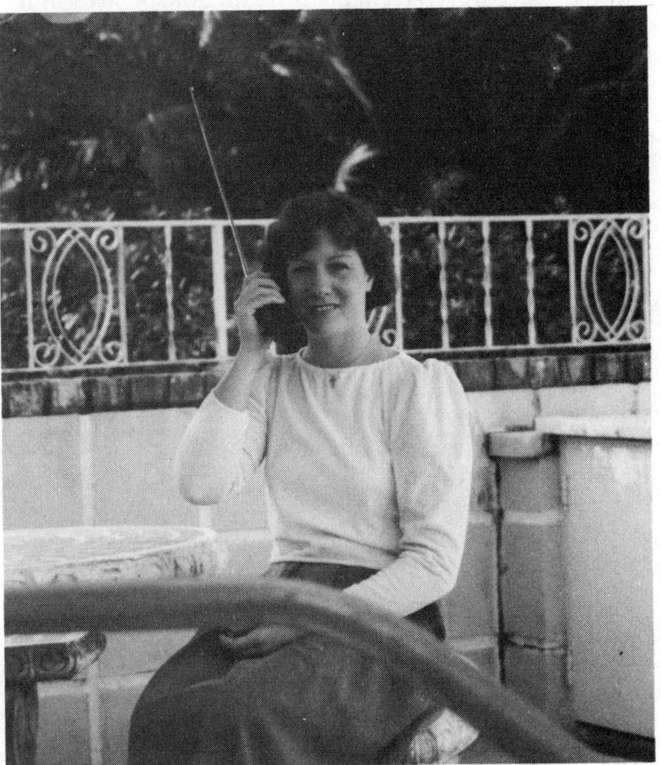

CORDLESS TELEPHONE: A cordless telephone can be used anywhere around the house.

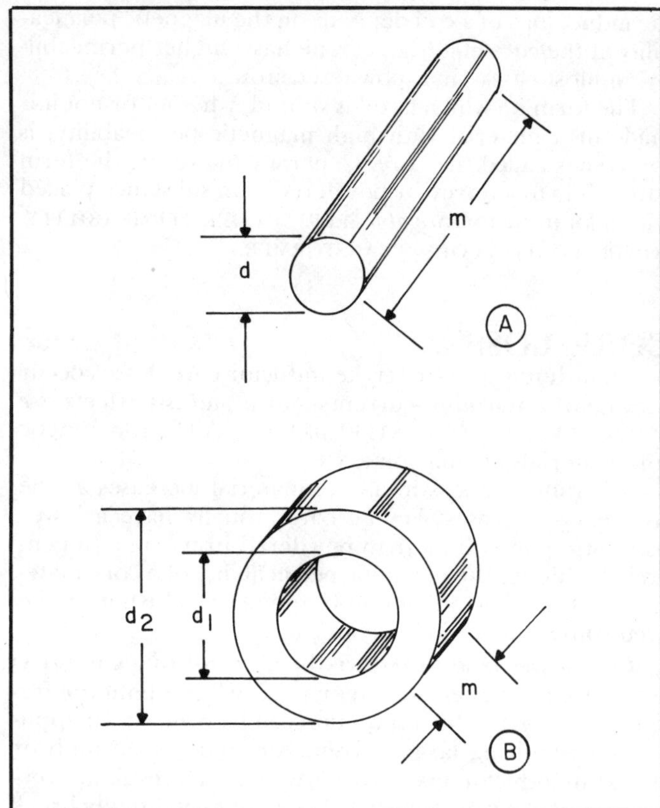

CORE: Inductor cores are usually made of ferrite or powdered iron. At A, a solenoidal core of diameter d and length m. At B, a toroidal core with inner diameter d_1, outer diameter d_2, and thickness m.

the form of magnetic fields. A group of small toroidal ferromagnetic cores comprises the memory. A magnetic field of a given polarity represents a digit 1; a field of the opposite polarity represents a digit 0. Wires are run through the cores to facilitate reversing the polarity of the magnetic flux in the cores.

Core memory is retained even when power is removed from the equipment. The cost of core memory is low, but the operating speed is lower than with some other forms of memory. *See also* MEMORY.

CORE SATURATION

As the magnetizing force applied to a ferrite or powdered-iron core is increased, the number of lines of flux through the core is also increased. However, there is a maximum limit to the number of flux lines a given core can accommodate. When the core is as magnetized as it can possibly get, it is said to be saturated. Some cores saturate more easily than others.

In inductors with ferrite or powdered-iron cores, saturation reduces the effective value of inductance for large currents. This is because the magnetic permeability of a saturated core is lessened. The change in inductance can have adverse effects on circuit performance in some situations. In a transformer, core saturation results in reduced efficiency and excessive heating of the core material. The current through an inductor or transformer should be kept below the level at which saturation occurs. *See also* CORE, FERRITE CORE, POWDERED-IRON CORE.

CORNER EFFECT

The response of a bandpass filter should ideally look like a

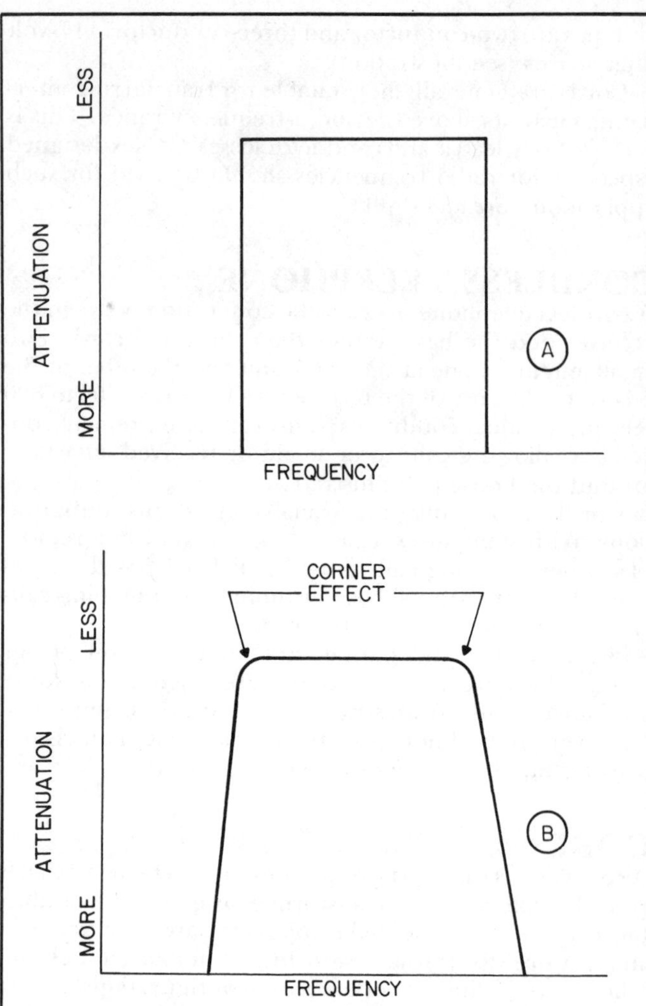

CORNER EFFECT: At A, an ideal rectangular filter response. At B, a typical filter response showing the rounding of the corners (corner effect).

perfect rectangle: zero attenuation within the passband, high attenuation outside the passband, and an instantaneous transition from zero attenuation to high attenuation at the limits of the passband. Lowpass, highpass, and band-rejection filters should have similar instantaneous transitions in the mathematically ideal case. The drawing illustrates this ideal response for a bandpass filter (A).

In practice, the rectangular response is never perfect. The corners of the rectangle—the transitions from zero to high attenuation—are not instantaneous, as shown at B. Immediately next to the passband, the corners appear rounded rather than sharp. This is called the corner effect, and it always occurs with selective filters. *See also* BANDPASS RESPONSE, BAND-REJECTION RESPONSE, HIGHPASS RESPONSE, LOWPASS RESPONSE, RECTANGULAR RESPONSE.

CORNER REFLECTOR

A corner reflector is a device for producing antenna gain at very-high frequencies and above. Two flat metal sheets or screens are attached together as shown in the illustration (A). Placed behind the radiating element of an antenna, the corner reflector produces forward gain. The beamwidth is quite broad (*see* BEAMWIDTH). The gain is the same for reception as for transmission at the same frequency. Corner-reflector antennas are often used for re-

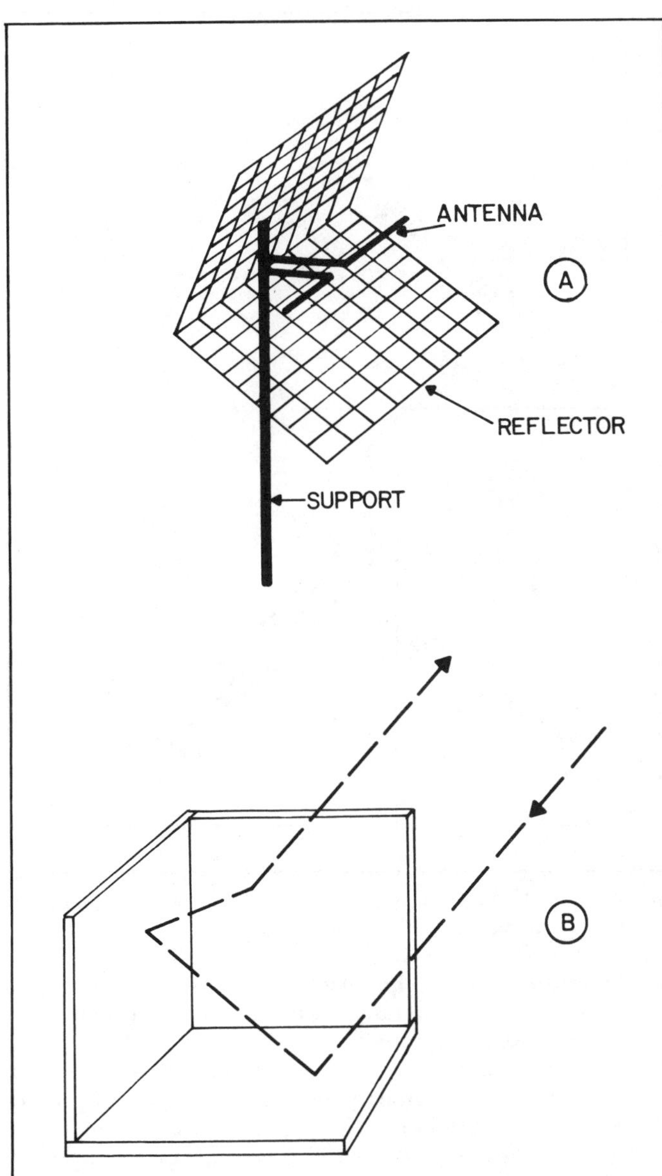

CORNER REFLECTOR: At A, a corner-reflector antenna often used for uhf reception. At B, a tri-corner reflector always returns a ray of energy in the same direction from which it comes.

ception of television broadcast signals in the ultra-high-frequency band.

A corner reflector may also consist of three flat metal surfaces or screens, attached together in a manner identical to the way two walls meet the floor or ceiling in a room (B). Such a device, if it is at least several wavelengths across, will always return electromagnetic energy in exactly the same direction from which it arrives. Because of this effect, corner reflectors of this type make excellent radar dummy targets. Such reflectors are sometimes called tri-corner reflectors.

CORONA

When the voltage on an electrical conductor, such as an antenna or high-voltage transmission line, exceeds a certain value, the air around the conductor begins to ionize. The result is a blue or purple glow called corona. This glow can sometimes be seen at the end or ends of an antenna at night, when a large amount of power is present. The ends of an antenna generally carry the largest voltages. Corona is more likely to occur when the humidity is high than when it is low, since moist air has a lower ionization potential than dry air.

Corona may occur inside a cable just before the dielectric material breaks down. Corona is sometimes observed between the plates of air-variable capacitors handling large voltages. A pointed object, such as the end of a whip antenna, is more likely to produce corona than will a flat or blunt surface. Some inductively loaded whip antennas have capacitance hats or large spheres at the ends to minimize corona.

Corona sometimes occurs as a result of high voltages caused by static electricity during thunderstorms. Such a display is occasionally seen at the tip of the mast of a sailing ship. Corona was observed by seafaring explorers hundreds of years ago. They called it Saint Elmo's fire. *See also* SAINT ELMO'S FIRE.

CORPUSCLE

The term corpuscle can refer to any sub-atomic particle. Electrons, neutrons, and protons are corpuscles. There are also other types of particles that may be called corpuscles.

Radiant energy has been found to occur in discrete packages. This has led to the corpuscular theory of electromagnetic radiation. The particle of electromagnetic energy is called the photon. The energy contained in a photon varies with the wavelength. The shorter the wavelength, the more energy is carried by a single corpuscle. Moving corpuscles have a definite mass and momentum. Some corpuscles also have spin. *See also* ELECTRON, NEUTRINO, NEUTRON, PHOTON, POSITRON, PROTON, QUARK.

CORRECTION

In an approximation of the value of a quantity, a correction or correction factor is a small increment, added to or subtracted from one approximation to obtain a better approximation.

Corrections are generally used when external factors influence the reading or result obtained with an instrument. For example, a frequency counter may be based on a crystal intended for operation at 20 degrees Celsius. If the temperature is above or below 20 degrees Celsius, a correction factor may be necessary, such as 2 Hertz for each degree Celsius.

Sometimes correction is performed by a circuit component especially designed for the purpose. In the above example, a capacitor with a precisely chosen temperature coefficient may be placed in parallel with the crystal to correct the drift caused by temperature changes. This kind of correction is usually called compensation. *See also* COMPENSATION.

CORRELATION

Correlation is a mathematical expression for the relationship between two quantities. Correlation may be positive, zero, or negative. A positive correlation indicates that an increase in one parameter is accompanied by an increase in the other. A negative correlation indicates that an increase in one parameter occurs with a decrease in the other. Zero correlation means that variations in the two parameters are unrelated.

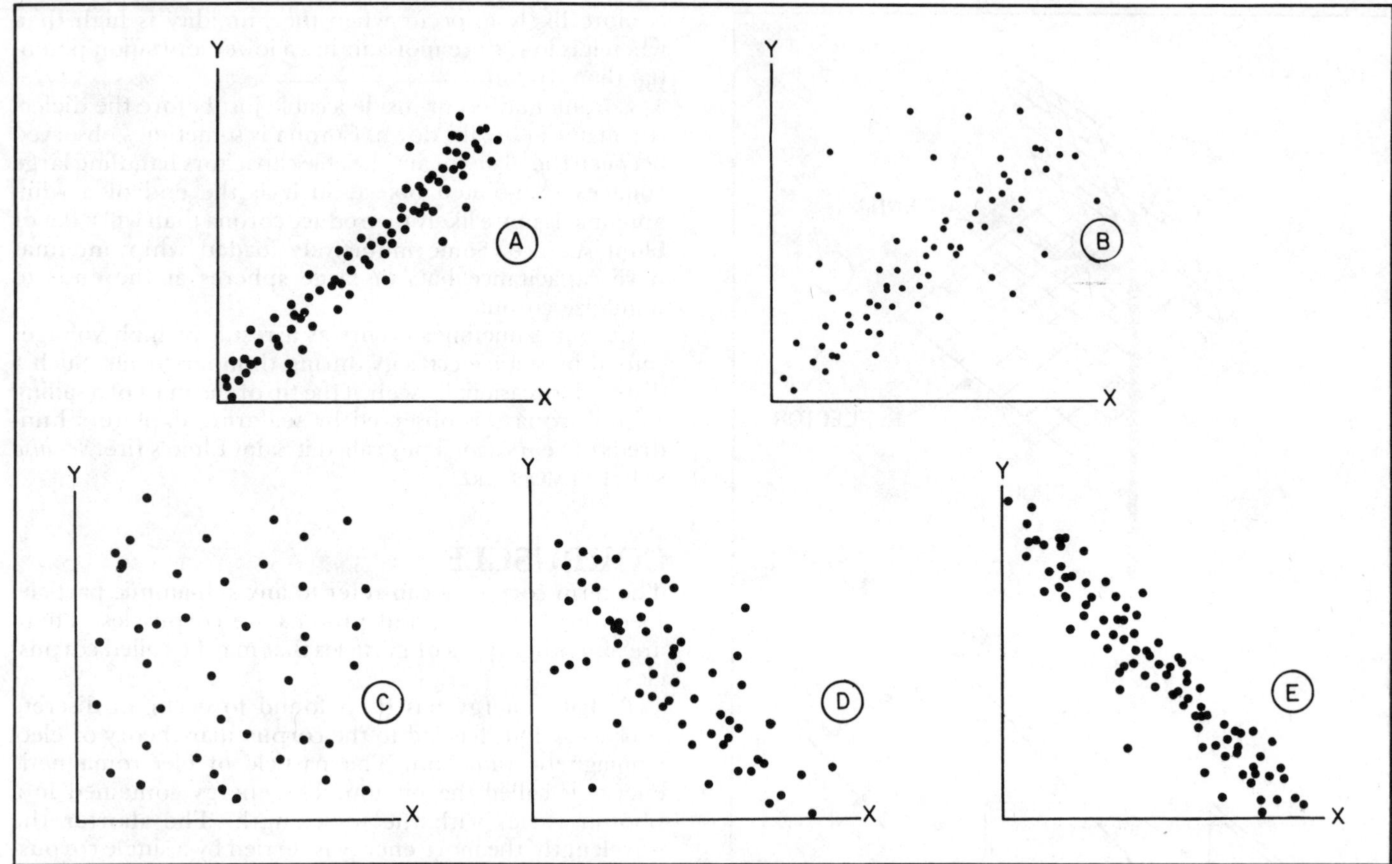

CORRELATION: Correlation is a measure of the relationship between two parameters, here shown as x and y. At A, strong positive correlation is shown; at B, weak position correlation; at C, zero correlation; at D, weak negative correlation; at E, strong negative correlation.

The coefficient of correlation between two variables is usually expressed as a number in the range between −1 and +1. The most negative correlation is given by −1, and the most positive by +1. We might say that the failure rate of a certain component (such as a transistor) is correlated with the temperature. The higher the temperature, for example, the more frequently the component fails. A statistical sampling can determine this correlation, and assign it a correlation coefficient.

The illustration shows examples of positive correlation, zero correlation, and negative correlation. The closer the correlation coefficient to +1, the more nearly the points lie along a straight line with slope of a positive value. This is shown at A and B. When two parameters are not correlated, or have zero correlation, the points are randomly scattered, as shown at C. When the parameters are negatively correlated, the points lie near a line with a negative slope, as shown at D and E. When the correlation coefficient is exactly −1 or +1, all the points lie along a perfectly straight line having either a negative or positive slope, respectively.

Correlation is important in determining the causes of various kinds of circuit malfunctions. Often, two quantities that might intuitively seem correlated are actually not significantly correlated. Sometimes, two quantities that do not intuitively seem correlated actually are.

CORRELATION DETECTION

Correlation detection is a form of detection involving the comparison of an input signal with an internally generated reference signal. The output of a correlation detector varies depending on the similarity of the input signal to the internally generated signal; maximum output occurs when the two signals are identical.

A phase comparator is an example of a correlation detector. In-phase signals produce maximum output; if the signals are not perfectly in phase, the output is reduced. A frequency comparator is another example of correlation detection. Two signals of the same frequency produce maximum output. The greater the difference between the input signal frequency and the internally generated signal frequency, the less the output. Other types of correlation detection are also possible. *See also* FREQUENCY COMPARATOR, PHASE COMPARATOR.

CORROSION

Corrosion is a chemical reaction between a metal and the atmosphere. The oxygen in the air is primarily responsible for corrosion. However, salt from the ocean, dissolved in water droplets in the air, can also cause corrosion. Chemicals created by manmade reactions, such as sulfur dioxide, can act to precipitate corrosion of metals.

Corrosion is characterized by deterioration of the surface of a metal. Sometimes, the oxide of a metal can act to protect the metal against further corrosion; an example of this is aluminum oxide. Sometimes, the oxide of a metal increases the vulnerability of the metal to further corrosion by increasing the surface area exposed to the air. An example of this is iron oxide, or rust.

Strong electrical currents can sometimes accelerate corrosion. An example of this is the rapid corrosion of the surface of a ground rod placed in acid soil. Conducted

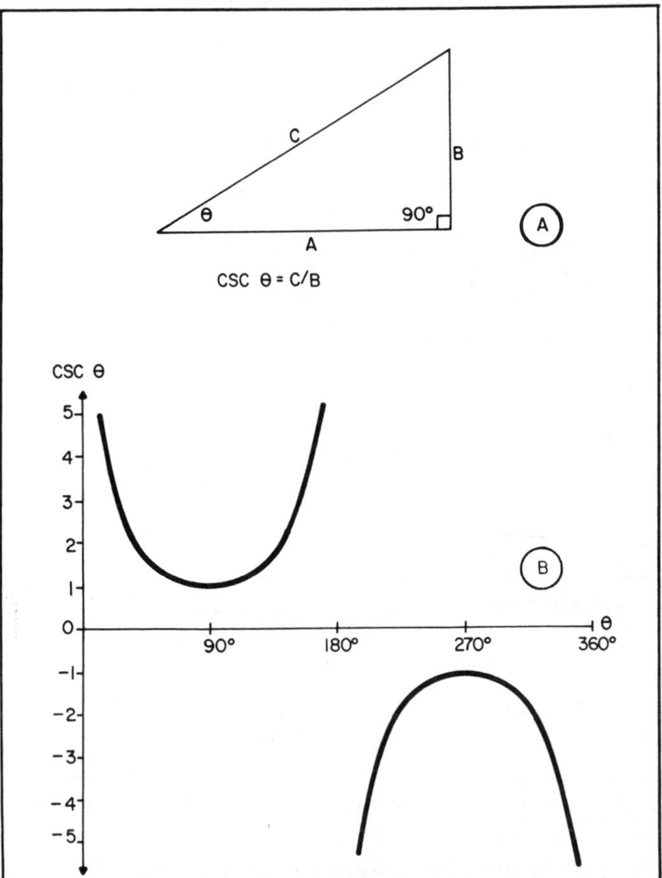

COSECANT: The cosecant function is the ratio ofthe length of the hypotenuse of a right triangle to the length of the opposite side, as shown at A. At B, an approximate graph of the cosecant function for angles between 0 and 360 degrees.

currents, through the electrolyte soil, cause rapid disintegration of the metal by electrolytic action. Reactions between certain chemicals placed in physical contact can also result in corrosion.

COSECANT

The cosecant function is a trigonometric function, equal to the reciprocal of the sine function (*see* SINE). In a right triangle A in the illustration, the cosecant of an angle θ between zero and 90 degrees is equal to the length of the hypotenuse divided by the length of the side opposite the angle. The smaller the angle, the greater the value of the cosecant function for angles θ between zero and 90 degrees. B illustrates the values of the cosecant function for angles ranging from zero to 360 degrees. The cosecant is undefined for $\theta = 0°$ and $\theta = 180°$.

In mathematical calculations, the cosecant function is abbreviated csc, and is given by the formula:

$$\csc \theta = 1/\sin \theta$$

where sin represents the sine function. *See also* TRIGONOMETRIC FUNCTION.

COSECANT-SQUARED ANTENNA

A special kind of radar antenna, designed to give echoes of the same intensity from targets at all distances, is called a cosecant-squared antenna. It gets this name from the shape of its vertical-plane radiation pattern. The radiation intensity varies according to the square of the cosecant of the elevation angle.

The least beam intensity is radiated directly upward, at an elevation angle of 90 degrees with respect to the horizon. The greatest beam intensity is radiated horizontally, at an elevation angle of zero degrees. The drawing illustrates this pattern. The intensity of radiation from a cosecant-squared antenna varies in such a way that the path attenuation is essentially the same over all signal paths to and from points in a given horizontal plane. *See also* RADAR.

COSINE

The cosine function is a trigonometric function. In a right triangle, the cosine is equal to the length of the adjacent side, or base, divided by the length of the hypotenuse, as shown at A in the illustration. In the unit circle $x^2 + y^2 = 1$, plotted on the Cartesian (x,y) plane, the cosine of the angle θ measured counterclockwise from the x axis is equal to x. This is shown at B. The cosine function is periodic, and begins with a value of 1 at the point $\theta = 0$. The shape of the cosine function is identical to that of the sine function (*see* SINE), except that the cosine function is displaced to the left by 90 degrees as shown at C.

In mathematical calculations, the cosine function is abbreviated cos. Values of cos θ for various angles θ are given in the table. *See also* TRIGONOMETRIC FUNCTION.

COSINE LAW

The cosine law is a rule for diffusion of electromagnetic energy reflected from, or transmitted through, a surface or medium (*see* DIFFUSION).

The energy intensity from a perfectly diffusing surface or medium is the most intense in a direction perpendicular to that surface. As the angle from the normal increases, the intensity drops until it is zero parallel to the surface (see illustration). The intensity, according to the cosine law, varies with the cosine of the angle θ relative to the normal. The intensity also varies with the sine of the angle ϕ relative to the surface, where $\phi = 90° - \theta$.

COSMIC NOISE

Cosmic noise is electromagnetic energy arriving from distant planets, stars, galaxies, and other celestial objects. Cosmic noise occurs at all wavelengths from the very-low-frequency radio band to the X-ray band and above. At the lower frequencies, the ionosphere of our planet prevents the noise from reaching the surface. At some higher frequencies, atmospheric absorption prevents the noise from reaching us.

Cosmic noise limits the sensitivity obtainable with receiving equipment, since this noise cannot be eliminated. Radio astronomers deliberately listen to cosmic noise in an effort to gain better understanding of our universe. To them, it is manmade noise rather than cosmic noise that limits the sensitivity of receiving equipment (*see* RADIO ASTRONOMY, RADIO TELESCOPE).

Cosmic noise is easy to mistake for tropospheric noise, but cosmic noise can be identified by the fact that it correlates with the plane of the galaxy. Perhaps the most intriguing form of cosmic noise, however, arrives with equal

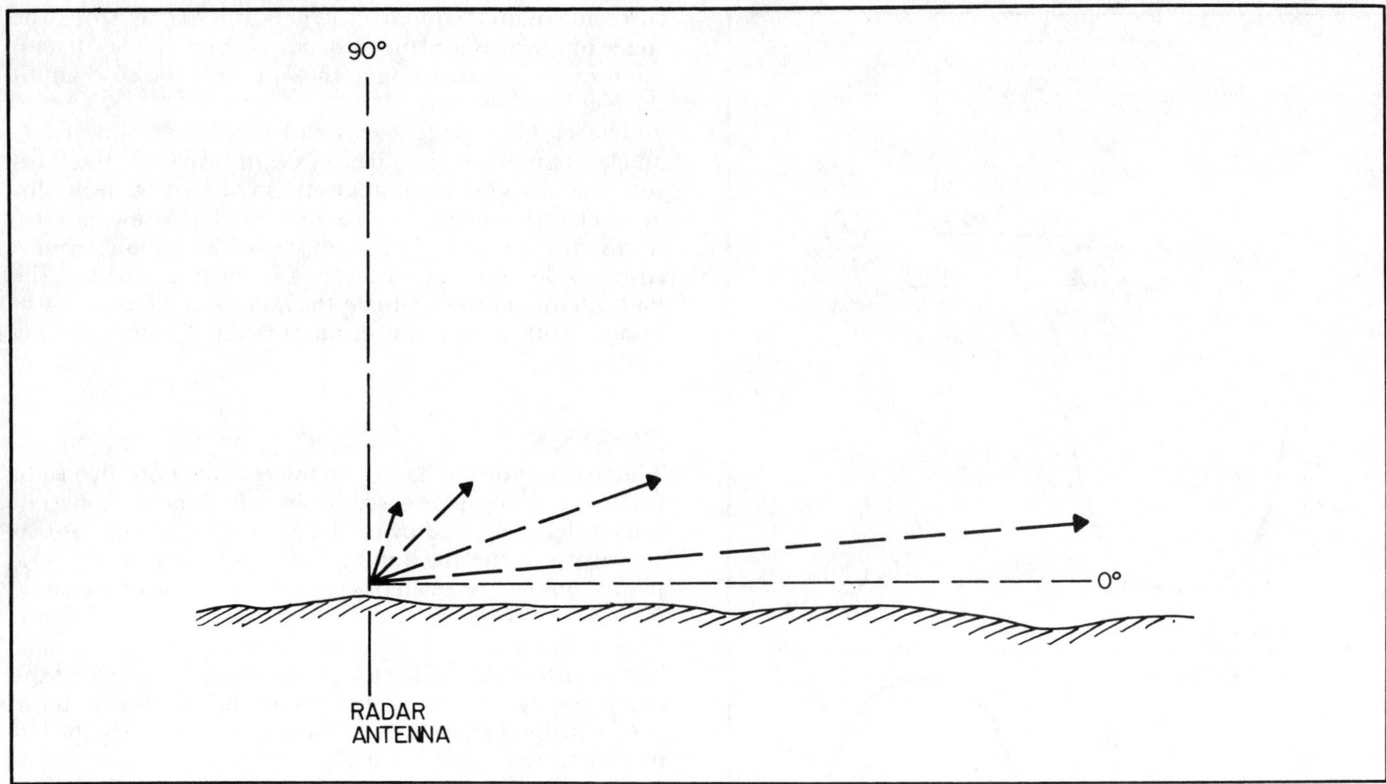

COSECANT-SQUARED ANTENNA: A cosecant-squared antenna radiates energy in proportion to the square of the cosecant of the elevation angle.

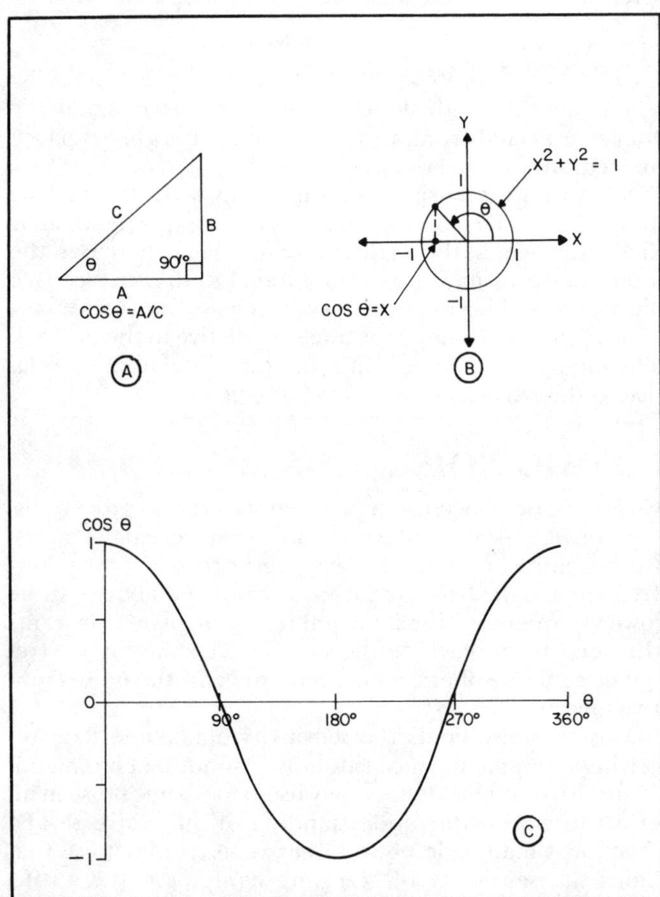

COSINE: The cosine function is the ratio of the length of the adjacent side of a right triangle to the length of the hypotenuse, as shown at A. At B, the unit-circle model for the cosine function; cos Θ = x (see text). At C, an approximate graph of the cosine function for angles between 0 and 360 degrees.

strength from all directions. In 1965, Arno Penzias and Robert Wilson of the Bell Laboratories observed faint cosmic noise that seemed to be coming from the entire universe. All other possible sources were ruled out. Astronomers have concluded that the noise originated with the fiery birth of our universe, in an event called the big bang. *See also* COSMOLOGY AND COSMOGONY.

COSMIC RADIATION

Cosmic radiation is a barrage of high-speed atomic particles from outer space. Such radiation may also be called cosmic rays or cosmic bombardment. Such radiation has high penetrating power, but the atmosphere of the earth absorbs much of it before it can reach the ground. Cosmic radiation may consist of alpha particles, beta particles, electrons, neutrons, protons, and other particles.

A Geiger counter or similar radiation detector can be used to show the presence of cosmic radiation. A device called a cloud chamber, which renders the paths of atomic particles visible, can also be used to prove the existence of cosmic radiation. Some cosmic rays originate in the sun. Some come from relatively nearby stars. Some originate in distant stars and galaxies in various stages of evolution. Sometimes, atomic particles travel millions, or even billions, of light years before ultimately creating a wisp in a cloud chamber, or a click in a Geiger counter. *See also* ALPHA PARTICLE, BETA PARTICLE, ELECTRON, NEUTRON, POSITRON, PROTON.

COSMOLOGY AND COSMOGONY

Cosmology is the study of the structure of the universe, and cosmogony is the science of the origin and evolution of our physical universe. Cosmology and cosmogony have

COSINE: VALUES OF COS θ FOR VALUES OF θ BETWEEN 0° AND 90°.

Θ, degrees	cos Θ	Θ, degrees	cos Θ
0	1.000	46	0.695
1	0.999	47	0.682
2	0.999	48	0.669
3	0.999	49	0.656
4	0.998	50	0.643
5	0.996	51	0.629
6	0.995	52	0.616
7	0.993	53	0.602
8	0.990	54	0.588
9	0.988	55	0.574
10	0.985	56	0.559
11	0.982	57	0.545
12	0.978	58	0.530
13	0.974	59	0.515
14	0.970	60	0.500
15	0.966	61	0.485
16	0.961	62	0.469
17	0.956	63	0.454
18	0.951	64	0.438
19	0.946	65	0.423
20	0.940	66	0.407
21	0.934	67	0.391
22	0.927	68	0.375
23	0.921	69	0.358
24	0.914	70	0.342
25	0.906	71	0.326
26	0.899	72	0.309
27	0.891	73	0.292
28	0.883	74	0.276
29	0.875	75	0.259
30	0.866	76	0.242
31	0.857	77	0.225
32	0.848	78	0.208
33	0.839	79	0.191
34	0.829	80	0.174
35	0.819	81	0.156
36	0.809	82	0.139
37	0.799	83	0.122
38	0.788	84	0.105
39	0.777	85	0.087
40	0.766	86	0.070
41	0.755	87	0.052
42	0.743	88	0.035
43	0.731	89	0.017
44	0.719	90	0.000
45	0.707		

existed for as long as men have looked up at the sky.

The earliest documented cosmological theories held that the earth was at the center of the universe, and that the stars were attached to a great sphere that turned around once every day. Later, it became evident that the motions of celestial objects were too complicated for this simple model to be correct. Astronomers of today put the earth in a relatively insignificant place, orbiting an ordinary star in a typical galaxy containing about 200 billion other stars. This galaxy, which we call the Milky Way, is one of millions or billions of galaxies.

The universe is believed to have originated in a fiery explosion, of incomprehensible violence, several billion years ago. Radio astronomers, with the aid of precision electronic equipment, have discovered strong evidence for this (*see* RADIO ASTRONOMY, RADIO TELESCOPE). The universe is apparently still expanding as a result of this initial event, called the big bang. It is not clear whether this expansion will continue forever, or whether the universe might eventually fall back in on itself.

COTANGENT

The cotangent function is a trigonometric function, equal to the reciprocal of the tangent function (*see* TANGENT). In a right triangle, the cotangent of an angle θ between zero and 90 degrees is equal to the length of the adjacent side

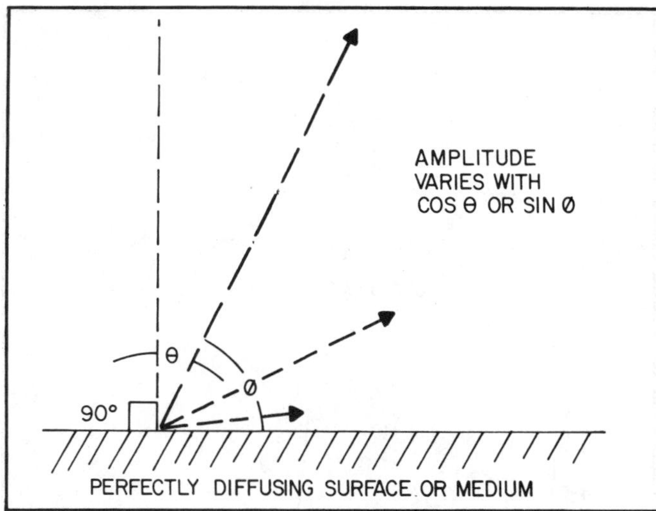

COSINE LAW: The intensity of transmitted or reflected energy from a perfectly diffusing medium or surface is proportional to the cosine of the angle θ, as shown.

divided by the length of the opposite side (A in the illustration). The smaller the angle, the larger this ratio. The values of the cotangent function for angles ranging from zero to 360 degrees are shown at B. The cotangent function is undefined at $\theta = 0°$ and at $\theta = 180°$.

In mathematical calculations, the cotangent function is abbreviated cot, and is given by the formula:

$$\cot \theta = 1/\tan \theta$$

where $\tan \theta$ represents the value of the tangent function at the angle θ. *See also* TRIGONOMETRIC FUNCTION.

COULOMB

The coulomb is the unit of electrical charge quantity. A coulomb of charge is contained in 6.28×10^{18} electrons. One electron thus carries 1.59×10^{-19} coulomb of negative electrical charge.

With a current of 1 ampere flowing in a conductor, exactly 1 coulomb of electrons (or other charge carriers) passes a fixed point in 1 second. The electron flow may take place in the form of actual electron transfer among atoms, or in the form of positive charge carriers called holes. A coulomb of positive charge indicates a deficiency of 6.28×10^{18} electrons on an object; a coulomb of negative charge indicates a surplus of 6.28×10^{18} electrons. *See also* ELECTRON, HOLE.

COULOMB, CHARLES AUGUSTIN DE

Charles Augustin de Coulomb (1736-1806), famous French physicist, is known primarily for his law of force for electrically charged objects (*see* COULOUMB'S LAW). Coulomb also conducted various experiments with electricity and magnetism and the manner in which they interact.

COULOMB'S LAW

The properties of electrostatic attraction and repulsion are given by a rule called Coulomb's law.

Given two charged objects X and Y (see illustration),

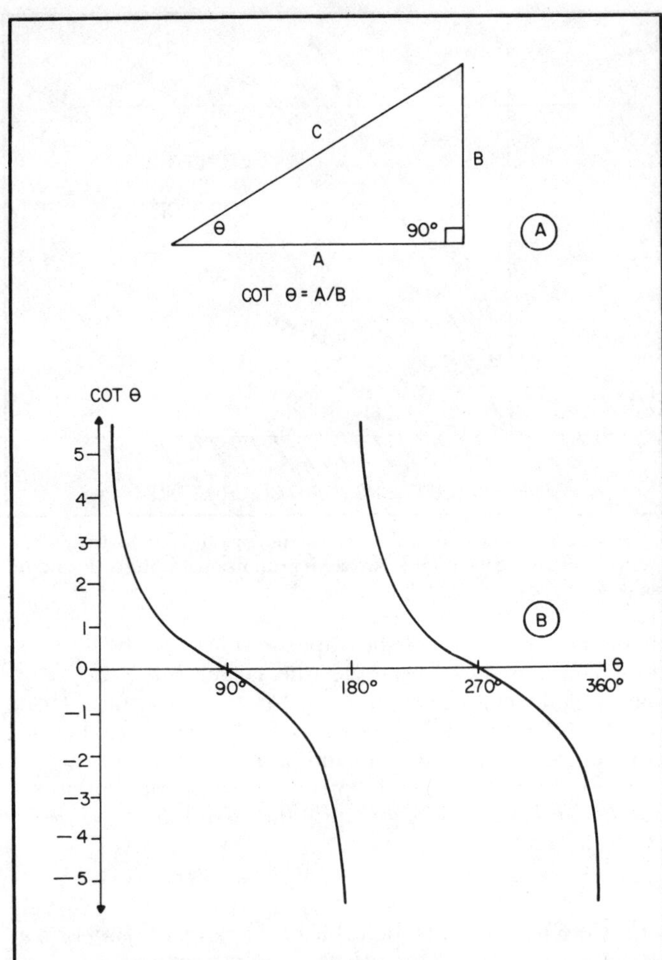

COTANGENT: The cotangent function is the ratio of the length of the adjacent side of a right triangle to the length of the opposite side, as shown at A. At B, an approximate graph of the contangent function for angles between 0 and 360 degrees.

separated by a charge-center to charge-center distance d containing charges Q_X and Q_Y, Coulomb's law states that the force F between the objects, caused by the electrostatic field is:

$$F = \frac{kQ_XQ_Y}{d^2}$$

where k is a constant that depends on the nature of the medium between the objects. The value of k is given by:

$$k = \frac{1}{4\pi\epsilon}$$

where ϵ is the permittivity of the medium between the objects (*see* PERMITTIVITY).

If the two charges Q_X and Q_Y are opposite, then the force F is an attraction. If Q_X and Q_Y are like charges, the force F is a repulsion. If positive charges are given positive values and negative charges are given negative values in the equation, then attraction is indicated by a negative force, and repulsion is indicated by a positive force.

COULOMETER

A coulometer is a device that measures a quantity of elec-

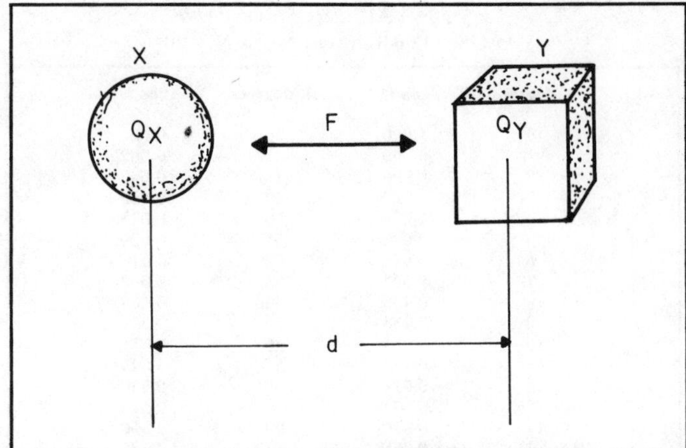

COULOMB'S LAW: The force F between two charged objects X and Y is proportional to the product of their charges Q_X and Q_Y, and inversely proportional to the square of the charge-center separation d.

tric charge. An electrolytic cell, capable of being charged and discharged makes an excellent coulometer.

When the charge is transferred to the electrolytic cell from an object, a certain amount of chemical action is produced. This chemical action is proportional in magnitude to the amount of charge. Knowing the relation between the charge and the extent of chemical action, the number of coulombs can be accurately determined. *See also* COULOMB.

COUNTER

A digital circuit that keeps track of the number of cycles or pulses entering it is called a counter. A counter consists of a set of flip-flops (*see* FLIP-FLOP) or equivalent circuits. Each time a pulse is received, the binary number stored by the counter increases by 1.

Counters may be used to keep track of the number of times a certain event takes place. Some counters measure the number of pulses within a specific interval of time, for the purpose of accurately determining the frequency of a signal. *See also* FREQUENCY COUNTER.

COUNTERMODULATION

Countermodulation is the bypassing of the cathode, emitter, or source resistor of the front end of a receiver, for the purpose of eliminating cross modulation (*see* CROSS MODULATION) in the circuit. The capacitor value is chosen so that the radio-frequency signal is shunted to ground, but the audio frequencies are not. The result is that audio-frequency signals are cancelled, or greatly reduced, by degenerative feedback. The desired radio-frequency signal is, however, easily passed through the amplifier.

The capacitor should have a reactance of less than one-fifth the resistor value at frequencies below 20 kHz, and it should have a reactance of least five times the resistor value at the signal frequency. Therefore, the capacitance depends on the value of the resistor. Countermodulation becomes less effective at low and very low frequencies. The schematic illustrates typical countermodulation bypass capacitor values as a function of the cathode, emitter, or source resistance.

COUNTERPOISE

A counterpoise is a means of obtaining a radio-frequency

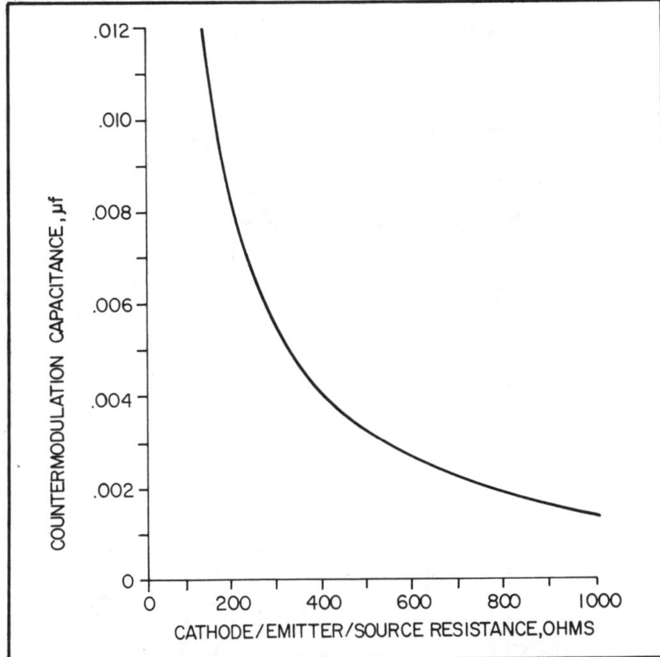

COUNTERMODULATION: Typical countermodulation capacitor values as a function of the cathode, emitter, or source resistance.

ground or ground plane without a direct earth-ground connection. A grid of wires is placed just above the actual surface to provide capacitive coupling to the ground. This greatly reduces ground loss at radio frequencies.

A simple counterpoise is shown in the illustration. Ideally, the radius of a counterpoise should be at least ¼ wavelength at the lowest operating frequency for a given system. The counterpoise is especially useful at locations where the soil conductivity is poor, rendering a direct ground connection ineffective. A counterpoise may be used in conjunction with a direct ground connection. *See also* GROUND CONNECTION, GROUND PLANE.

COUNTER TUBE

A counter tube is a gas-filled tube that registers the presence of high-speed, high-energy atomic particles. When radiation passes through the tube, the gas ionizes and produces conduction for an instant. The Geiger-Mueller counter, often called simply the Geiger counter, uses a counter tube. The illustration shows a cutaway view of a counter tube.

A brass wall is placed around the counter tube for counting gamma rays. A window can be opened for counting alpha particles, beta particles, and other less penetrating forms of radiation. Such a tube can generally detect up to about 2,000 particles per second. *See also* GEIGER COUNTER.

COUNTER VOLTAGE

When the current through a conductor is cut off, a reverse voltage, called a counter voltage, appears across the coil. This voltage can be very high if the current through the coil is high, and if the coil inductance is large. In some electric appliances containing motors, an interruption in current can present a serious shock hazard because the counter voltage can reach hundreds or even thousands of volts.

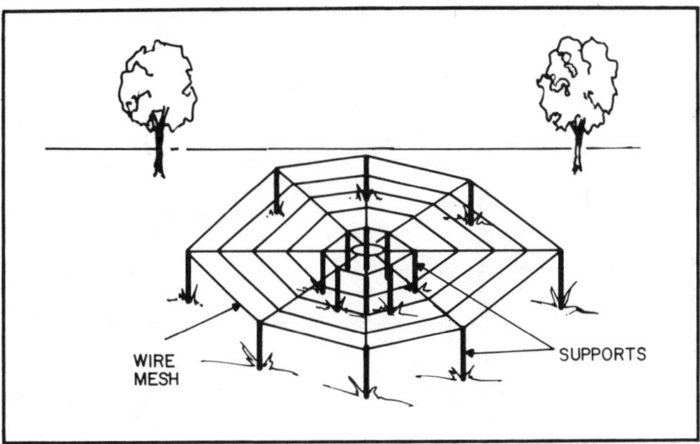

COUNTERPOISE: A counterpoise is a means of obtaining a ground at radio frequencies by means of capacitive coupling.

Counter voltage is employed in every automobile using spark plugs. An inductor called a spark coil stores the electric charge from the battery, which supplies 12 volts direct current, and discharges a pent-up electromotive force of thousands of volts. Counter voltage is also used in some electric fences.

COUPLED IMPEDANCE

When an oscillator or amplifier circuit is followed by another stage, or by a tuning network, the impedance that the oscillator or amplifier "sees" is called the coupled impedance. Ideally, the coupled impedance should contain only resistance, and no reactance. A device called a coupler or coupling network (*see* COUPLER) is sometimes used to eliminate stray reactances, especially in antenna systems.

The actual value of the resistive coupled impedance may range from less than 1 ohm to hundreds of thousands of ohms. Whatever its value, however, it should be matched to the output impedance of the amplifier or oscillator to which it is connected. The coupled impedance should also remain constant at all times. If it changes, the alteration can be passed back from stage to stage, affecting the gain of the circuit. If a change in impedance is passed all the way back to the oscillator stage, the frequency or phase of the signal will change. This can produce severe distortion of an amplitude-modulated or frequency-modulated signal. It results in chirp on a code signal. *See also* CHIRP.

COUPLER

A coupler is a device, usually consisting of inductors and/or

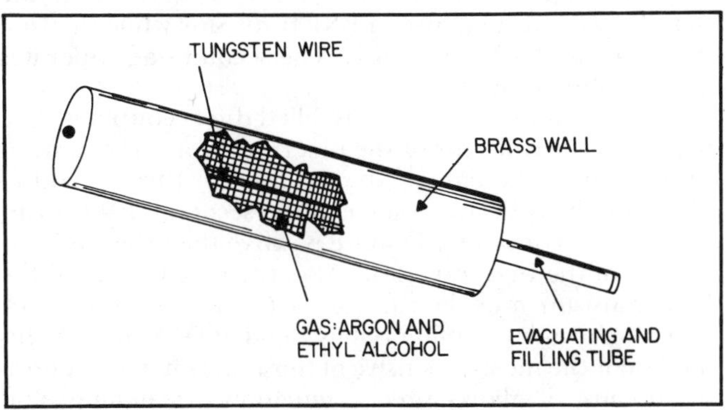

COUNTER TUBE: Cutaway view of a counter tube.

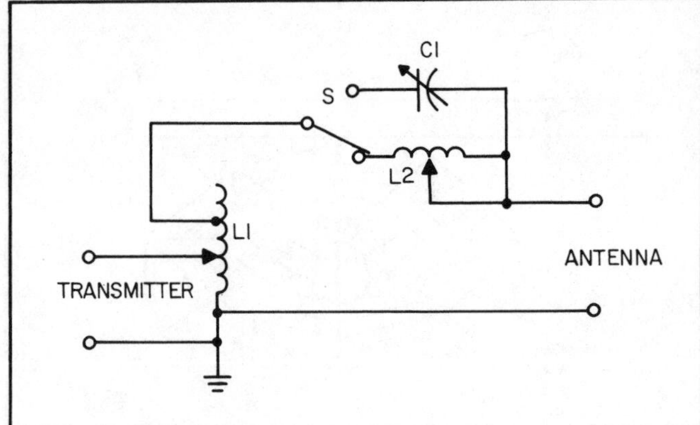

COUPLER: An antenna coupler for obtaining an impedance match with an antenna of unknown characteristics.

capacitors, for the purpose of facilitating the optimum transfer of power from an amplifier or oscillator to the next stage. A coupler may also be used between the output of a transmitter and an antenna. Some couplers have fixed components, and some are adjustable. A simple coupler circuit is illustrated in the diagram. It is intended for impedance matching between a radio-frequency transmitter and an antenna having an unknown impedance.

The resistive component of the impedance is matched by adjusting the tap on inductor L1. If the reactive component is capacitive, the switch S is positioned so that L2 appears in series with the antenna, and the value of L2 is set to exactly cancel the capacitive reactance. If the reactance is inductive, the switch S is positioned so that C1 appears in series with the antenna. Its capacitance is adjusted until the inductive reactance is balanced. *See also* ANTENNA MATCHING, ANTENNA TUNING.

COUPLING

Coupling is a means of transferring energy from one stage of a circuit to another. Coupling is also the transfer of energy from the output of a circuit to a load.

Interstage coupling, such as between an oscillator or mixer and an amplifier, can be done in a variety of ways. Four methods of coupling between two bipolar transistor stages are illustrated in the drawing.

The method at A is called capacitive coupling, because the signal is transferred through a capacitor. Capacitive coupling isolates the stages for direct current, so that their bias can be independently set.

The coupling scheme at B is called diode coupling. The diode passes signal energy in one direction, but isolates the stages for direct current in that direction. (Note that the second stage uses a PNP transistor while the first stage uses an NPN transistor.) The second stage operates in Class B or Class C.

The coupling scheme at C is called direct coupling. The voltage at the collector of the first transistor is the same as the voltage at the base of the second. For this method to function, the collector voltage of the second NPN transistor must be considerably more positive than the collector voltage of the first transistor. Also, the base voltage of the first transistor must be carefully set to avoid saturation.

Transformer coupling is illustrated at D. Although this method is the most expensive of those shown, it is preferable because it allows precise impedance matching, and offers good harmonic attenuation. Transformer coupling

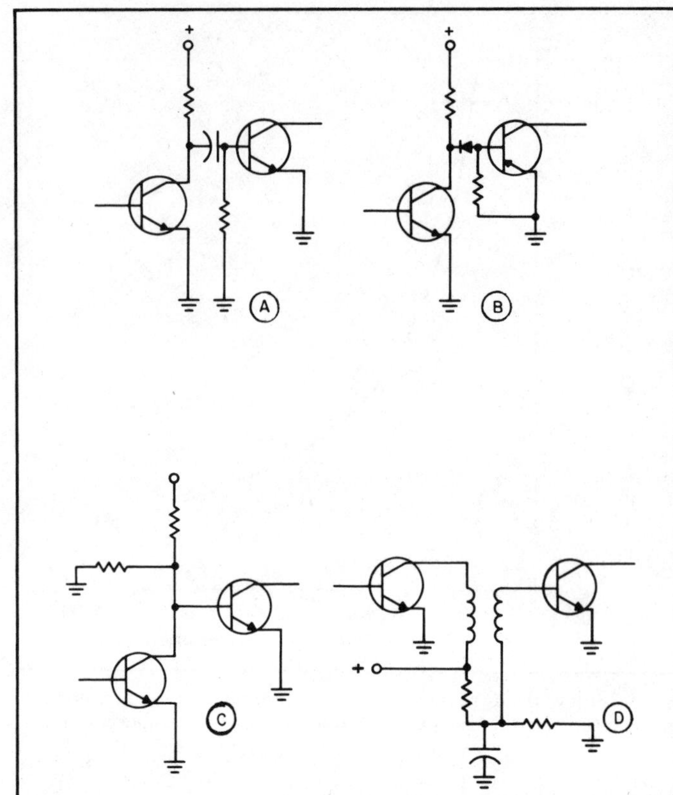

COUPLING: At A, capacitive coupling; at B, diode coupling; at C, direct coupling; at D, transformer coupling.

isolates the two stages for direct current, and allows the use of tuned circuits for improved efficiency. The phase can be reversed if desired. With a well-designed transformer-coupled circuit, electrostatic coupling is kept to a minimum. This improves the stability of the circuit.

The four methods of coupling shown are only a sampling of the many different arrangements possible. The most common method of interstage coupling is the capacitive method.

Coupling between a radio-frequency transmitter and its antenna is accomplished by means of a circuit called a coupling network, or coupler. *See also* COUPLER.

COVALENT BOND

A covalent bond occurs when two or more atoms share electrons. Some atoms give up electrons easily; these elements are said to have negative valence numbers. Some atoms readily accept additional electrons; they have positive valence numbers (*see* VALENCE NUMBER).

Atoms contain electrons in discrete orbits, called shells. The innermost shell is called the K shell, and may contain at most two electrons. The next shell is called the L shell, and may contain as many as eight electrons. The third shell, the M shell, can have up to 18 electrons. In general, the nth shell of an atom can contain from zero to $2n^2$ electrons. When an atom has just one or two electrons in its outermost shell, it gives up the electrons easily. When an atom has a shortage of one or two electrons in the outer shell, it readily accepts one or two more.

A covalent bond occurs among atoms having an equal surplus and shortage of electrons. For example, oxygen, with atomic number 8, has a deficiency of two electrons in its L shell. Hydrogen, with one electron, can either accept another to get two, or give one up to have none. Two hydrogen atoms can share their electrons with one atom of oxygen, creating the familiar compound H_2O. In this manner, all the atoms are "satisfied."

Some elements have all their shells filled completely. These atoms do not usually produce covalent bonds. Helium and neon are two examples of such elements.

COVERAGE

Coverage refers to the frequency range of a receiver or transmitter. The term coverage is also used to define the service area of a communications or broadcast station.

When specifying the frequency coverage of a transmitter or receiver, either the actual frequency range or the approximate wavelength range may be indicated. An Amateur-Radio receiver might, for example, be specified to cover 80 through 10 meters. This usually means that it operates only on the amateur bands designated in this range. A general-coverage receiver might be specified to work over the range 535 kHz to 30 MHz. This implies continuous coverage.

The coverage area of a broadcast station is determined by the level of output power and the directional characteristics of the antenna system. The Federal Communications Commission limits the coverage allowed to broadcasting stations in the United States. This prevents mutual interference among different stations on the same frequency. Stations may be designated to have local, regional, or national (clear-channel) coverage. *See also* CLEAR CHANNEL, LOCAL CHANNEL, REGIONAL CHANNEL.

CPU

See CENTRAL PROCESSING UNIT.

CQ

In radio communication, the term *CQ* is used to mean "calling anyone." It is used especially by Amateur-Radio operators (*see* AMATEUR RADIO). A radio operator of station W1GV, for example, may say, in a code transmission,

"CQ CQ CQ DE W1GV K"

which translates literally to "Calling anyone. Calling anyone. Calling anyone. This is W1GV. Go ahead."

Often, a directional or selective CQ call is used when an operator wants answers from only certain types of stations. For example, CQ DX indicates that the caller wants a station in a country different from his own. CQ MSN might mean "Calling all stations in the Minnesota Section Net."

In some radio services, calling CQ is considered boorish, or the mark of an inexperienced operator. This is the case in the 144-MHz amateur band, when using frequency modulation. It is also true on 27 MHz, in the Citizen's Band.

CREST FACTOR

The ratio of the peak amplitude to the root-mean-square amplitude of an alternating-current or pulsating direct-current waveform is called the crest factor. Sometimes it is called the amplitude factor. The crest factor depends on the shape of the wave.

In the case of a sine wave, the crest factor is equal to $\sqrt{2}$, or approximately 1.414 (see A in illustration). In the case of a square wave, the peak and root-mean-square amplitudes are equal, and therefore the crest factor is equal to 1, as shown at B.

In a complicated waveform, the crest factor may vary considerably, and may change with time. It is never less than 1, because the root-mean-square (RMS) voltage, current, or power is never greater than the peak voltage, current, or power. For a voice or music waveform, the crest factor is generally between 2 and 4. *See also* ROOT MEAN SQUARE.

CRITICAL ANGLE

When a beam of light or radio waves passes from one medium to another having a lower index of refraction, the energy may continue on into the second medium as shown at A in the illustration, or it might be reflected off the boundary and remain within the original medium, as shown at B. Whether refraction or reflection occurs depends on the angle of incidence (*see* ANGLE OF INCIDENCE, INDEX OF REFRACTION).

If the angle of incidence is very large, reflection will take place. Then the energy remains in the region having the larger index of refraction. If the angle of incidence is 0 degrees, the energy passes into the medium having the lower index of refraction, generally, and there is no change in its path. At some intermediate angle, called the critical angle, reflection just begins to occur as the angle of incidence is made larger and larger. The critical angle depends on the ratio of the indices of refraction of the two media, at the energy wavelength involved.

When radio waves encounter the E or F layers of the ionosphere (*see* E LAYER, F LAYER), the waves may be re-

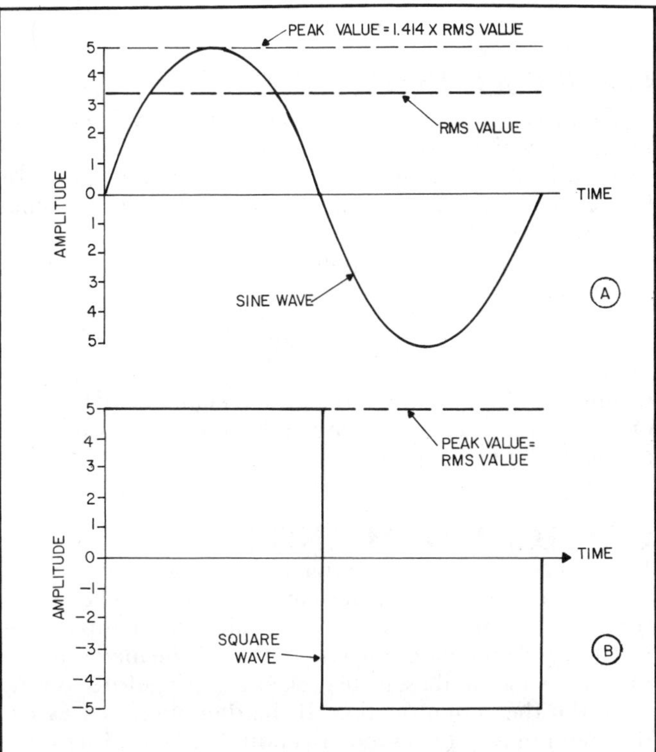

CREST FACTOR: The peak value is 1.414 times the RMS value for a sine wave (A), so the crest factor is 1.414. For a square wave (B), the crest factor is 1 because the peak and RMS values are equal.

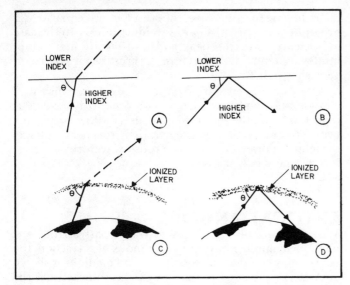

CRITICAL ANGLE: Refraction or internal reflection may take place when an energy beam goes from a medium of high refraction index to a medium of flow refraction index (A and B). When a radio signal meets the ionosphere, it may pass through into space or be returned to earth (C and D).

turned to the earth, or they might continue on into space (shown at C and D). The smallest angle of incidence, at which energy is returned to the earth is called the critical angle. The critical angle for radio waves depends on the density of the ionosphere, and on the wavelength of the signal. Sometimes, even energy arriving perpendicularly will be returned to the earth; in such a case, the critical angle is considered to be 0 degrees. Sometimes, electromagnetic energy is never returned to the earth by the ionosphere, no matter how great the angle of incidence. Then, the critical angle is undefined. *See also* PROPAGATION CHARACTERISTICS.

CRITICAL COUPLING

When two circuits are coupled, the optimum value of coupling (for which the best transfer of power takes place) is called critical coupling. If the coupling is made tighter or looser than the critical value, the power transfer becomes less efficient.

The coefficient of coupling, k, for critical coupling is given by:

$$k = 1/\sqrt{Q_1 Q_2}$$

where Q_1 is the Q factor of the primary circuit and Q_2 is the Q factor of the secondary circuit. *See also* COEFFICIENT OF COUPLING, Q FACTOR.

CRITICAL DAMPING

In an analog meter, the damping is the rapidity with which the needle reaches the actual current reading. The longer the time required for this, the greater the damping (*see* DAMPING). If the damping is insufficient, the meter needle may overshoot and oscillate back and forth before coming to rest at the actual reading. If the damping is excessive, the meter may not respond fast enough to be useful for the desired purpose.

Critical damping is the smallest amount of meter damping that can be realized without overshoot. This gives the most accurate and meaningful transient readings. *See also* D'ARSONVAL, METER MOVEMENT.

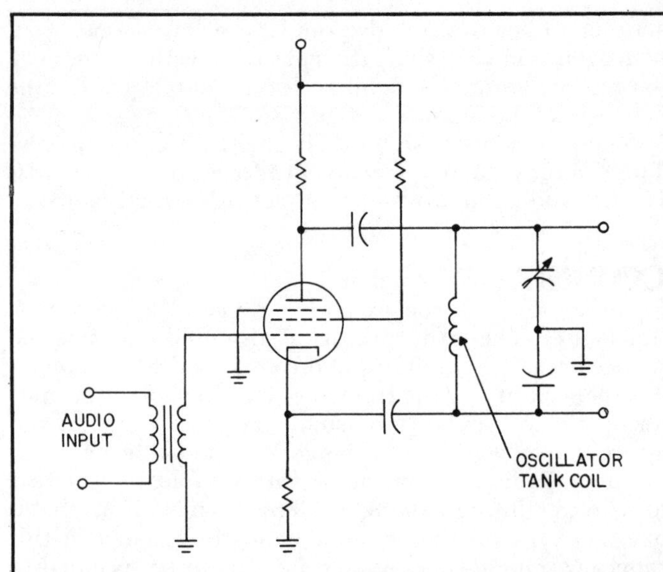

CROSBY CIRCUIT: A Crosby modulation circuit is a form of reactance modulator for obtaining frequency modulation.

CRITICAL FREQUENCY

At low and very low radio frequencies, all energy is returned to the earth by the ionosphere. This is true even if the angle of incidence of the radio signal with the ionized layer is 90 degrees (*see* ANGLE OF INCIDENCE).

As the frequency of a signal is raised, a point is reached where energy sent directly upward will escape into space. All signals impinging on the ionosphere at an angle of incidence smaller than 90 degrees will, however, still be returned to the earth. The frequency at which this occurs is called the critical frequency.

The critical frequency depends on the density of the ionized layers. This density changes with the time of day, the time of year, and the level of sunspot activity. The critical frequency for the ionospheric F layer is typically between about 3 and 5 MHz. *See also* PROPAGATION CHARACTERISTICS.

CROSBY CIRCUIT

The Crosby circuit is a method of obtaining frequency modulation. A reactance tube is connected across the tank-circuit coil or capacitor of the oscillator. The modulating signal is applied in series with the control-grid bias supply of the tube. This causes the reactance of the tube to fluctuate in accordance with the modulating signal. In turn, this changes the frequency of the oscillator. The Crosby frequency-modulation circuit is shown in the illustration.

Vacuum tubes are not generally used for obtaining frequency modulation nowadays. A much simpler way to get this kind of modulation is by means of a varactor diode. *See also* VARACTOR DIODE.

CROSS ANTENNA

A cross antenna consists of two or more horizontal anten-

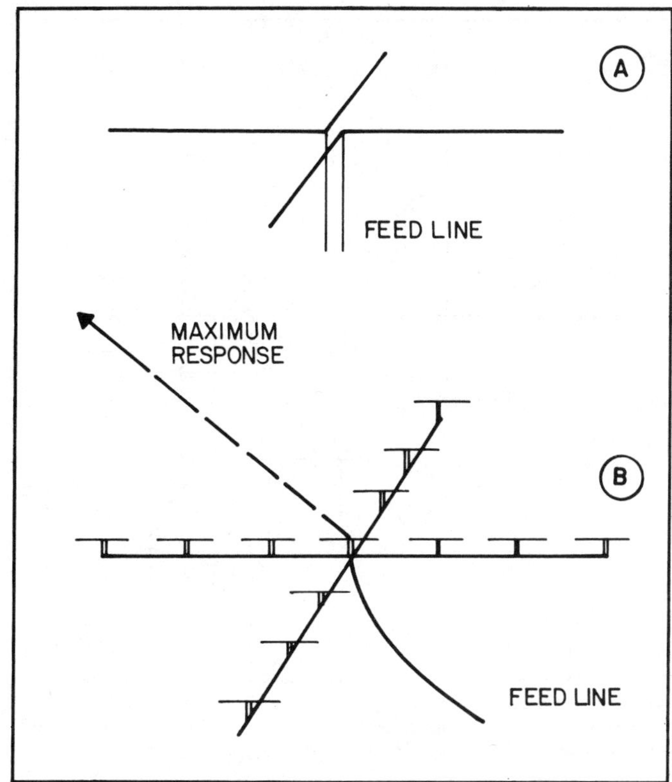

CROSS ANTENNA: Cross antennas can consist of just two dipoles, as shown at A, or many separate antennas, as shown at B.

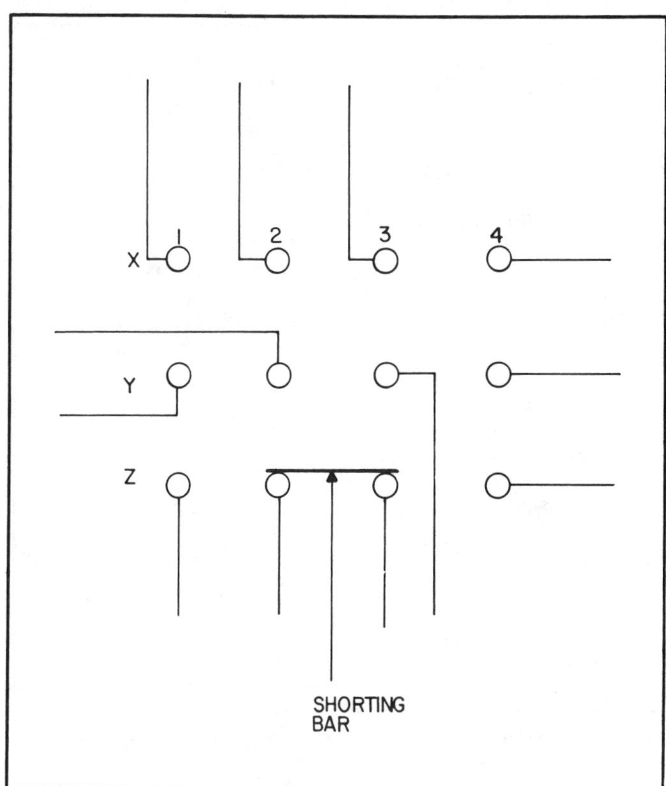

CROSSBAR SWITCH: A crossbar switch allows many different connection arrangements.

nas, connected to the same feed line and arranged so that they form a cross pattern. The drawing shows two kinds of cross antenna systems. The antennas may or may not be fed in phase.

A cross antenna consisting of two horizontal dipoles, as shown at A, produces a horizontally polarized signal. The radiation pattern is similar to that of an ordinary dipole, when the two dipoles are fed in phase.

A special form of cross antenna consists of an array of antennas arranged like a cross. They are fed in such a phase combination that a strong gain lobe occurs toward a certain part of the sky. There may be several, or perhaps many, of these antennas (see B). The azimuth and elevation of maximum radiation or reception can be adjusted by varying the phasing among the individual antennas. If its axes measure many wavelengths, this type of antenna can produce an extremely narrow beam. Therefore, such arrays are sometimes used with radio telescopes. *See also* MILLS CROSS.

CROSSBAR SWITCH

A crossbar switch is a special kind of switch that provides a large number of different connection arrangements. A set of contacts is arranged in a matrix. The matrix may be two-dimensional or three-dimensional. An example of a crossbar switch is shown in the illustration. The matrix may be square or rectangular in two dimensions; it may be shaped like a cube or a rectangular prism in three dimensions.

A shorting bar is used to select any adjacent pair of contacts that lie along a common axis. In this way, a large number of different combinations are possible with a relatively small number of switch contacts. The shorting bar is magnetically controlled.

CROSS BEARING

The cross-bearing system is a means of radionavigation (*see* RADIONAVIGATION). It allows the pilot of a boat or airplane to precisely determine his position.

To determine the location, two transmitting stations are used, and the locations of both must be known. Azimuth bearings are found for both stations using a direction finder. These bearings are then inverted by 180 degrees, obtaining the bearings the stations would observe for the boat or plane. Intersecting lines are drawn from the two stations according to these bearings. The point of intersection of the craft (see illustration).

CROSS-CONNECTED NEUTRALIZATION

In a push-pull amplifier (*see* PUSH-PULL AMPLIFIER), instability may cause unwanted oscillation. This is prevented in all radio-frequency amplifiers by means of a procedure called neutralization (*see* NEUTRALIZATION). The neutralization method most often used in a push-pull amplifier is called cross-connected neutralization, or simply cross neutralization.

In cross neutralization, two feedback capacitors are used. The output of one half of the amplifier is connected to the input of the other half, as shown in the diagram. The fed-back signals are out of phase with the input signals for undesired oscillation energy. When the capacitor values are correctly set, the probability of oscillation is greatly reduced. The values of the two neutralizing capacitors must always be identical to maintain circuit balance.

CROSS COUPLING

Cross coupling is a means of obtaining oscillation using two

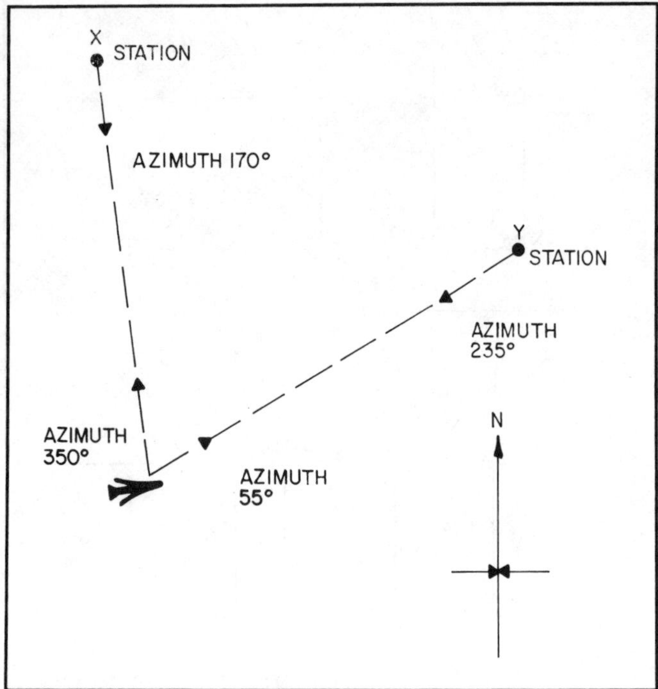

CROSS BEARING: Cross bearing is a method of radionavigation.

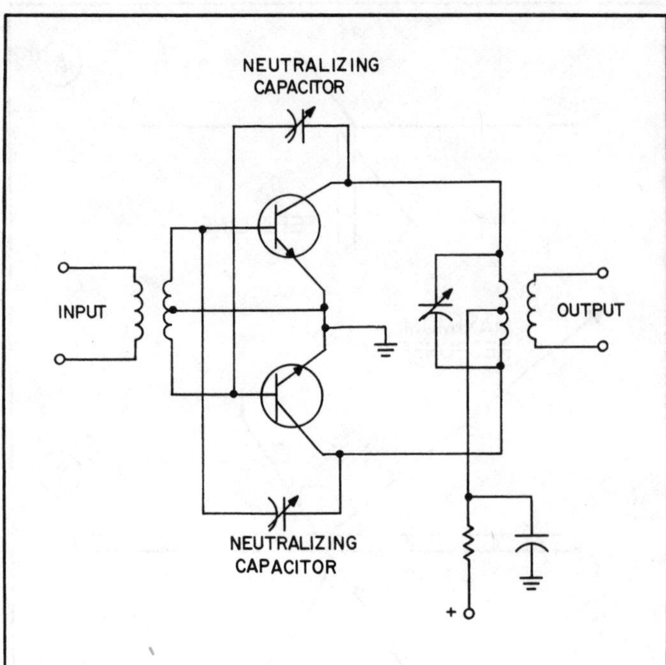

CROSS-CONNECTED NEUTRALIZATION: Cross-connected neutralization is the neutralization is the neutralization of a push-pull amplifier consisting of two identical capacitors.

stages of amplification. Normally, an amplifying stage produces a 180-degree phase shift in a signal passing through. Some amplifiers produce no phase shift. In either case, the signal will have its original phase after passing through two amplifying stages. By coupling the output of the second stage to the input of the first stage via a capacitor, oscillation can be obtained. The diagram illustrates simple cross coupling in a two-stage transistor circuit.

Cross coupling frequently occurs when it is not wanted. This is especially likely in multistage, high-gain amplifiers. The capacitance between the input and output wiring is often sufficient to produce oscillation because of positive feedback. This oscillation can be very hard to eliminate. The chances of oscillation resulting from unwanted cross coupling can be minimized by keeping all leads as short as possible. The use of coaxial cable is advantageous when lead lengths must be long. Individual shielded enclosures for each stage are sometimes necessary to prevent oscillation in such circuits. Neutralization sometimes works. *See also* NEUTRALIZATION.

CROSSED-NEEDLE METER

A crossed-needle meter is a device consisting of two pointer-type analog meters inside a single enclosure. The pointer movements are centered at different positions at the base of the meter. The meters are connected to different circuits, and their point of crossing illustrates some relationship between the two readings.

The illustration shows an example of a crossed-needle meter used for the purpose of showing forward power, reflected power, and the standing-wave ratio on a radio-frequency transmission line. One meter reads the forward power on its scale; the other meter reads reflected power on its scale. A third scale, consisting of lines on the meter face, indicates the standing-wave ratio (SWR), which is a function of forward and reflected power. The point of crossing between the two meter needles is observed against this scale, and the SWR can then be easily determined (*see* STANDING-WAVE RATIO).

Crossed-needle meters are more convenient in some cases than switched or separate meters. They allow the operator to constantly visualize the relationship between two parameters.

CROSSHATCH

When interference occurs to a television picture, a crosshatch pattern sometimes forms on the screen. A weak, unmodulated carrier close to the frequency of the television picture carrier will produce a diagonal set of parallel lines across the screen. If the interfering signal is modulated, sound bars may accompany the diagonal lines (*see* SOUND BAR). A crosshatch pattern does not necessarily wipe out the desired picture completely, although the

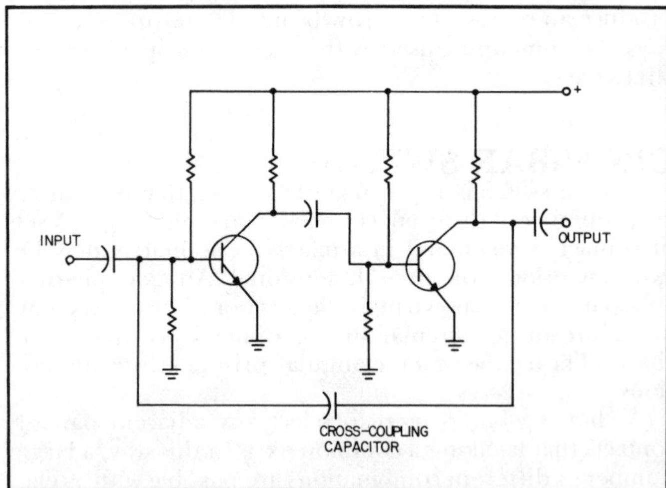

CROSS COUPLING: Cross coupling provides a means of making two successive amplifier stages oscillate. It may not be desired.

contrast often appears reduced, and the overall picture brightness may change.

The term crosshatch is also used to refer to a pattern for television testing. The crosshatch pattern consists of a set of horizontal and vertical lines, arranged like a grid, and transmitted into the television receiver. The horizontal lines are used to adjust the vertical linearity of the set. The vertical lines are used to adjust the horizontal linearity. *See also* HORIZONTAL LINEARITY, VERTICAL LINEARITY.

CROSS MODULATION

Cross modulation is a form of interference to radio and television receivers. Cross modulation is caused by the presence of a strong signal, and also by the existence of a nonlinear component in or near the receiver.

Cross modulation causes all desired signal carriers to appear modulated by the undesired signal. This modulation can usually be heard only if the undesired signal is amplitude-modulated, although a change in receiver gain might occur in the presence of extremely strong unmodulated signals. If the cross modulation is caused entirely by a nonlinearity within the receiver, it is sometimes called intermodulation (*see* INTERMODULATION).

Cross modulation can be prevented by attenuating the level of the undesired signal before it reaches nonlinear components, or drives normally linear components into nonlinear operation. If the nonlinearity is outside the receiver and antenna system, the cross modulation can be elminated only by locating the nonlinear junction and getting rid of it. Marginal electrical bonds between two wires, or between water pipes, or even between parts of a metal fence, can be responsible.

CROSSOVER NETWORK

A crossover network is a circuit designed to direct energy to different loads, depending on the frequency. Such a network generally consists of a splitter, if needed, and a highpass and lowpass filter. These components are interconnected as shown at A in the schematic.

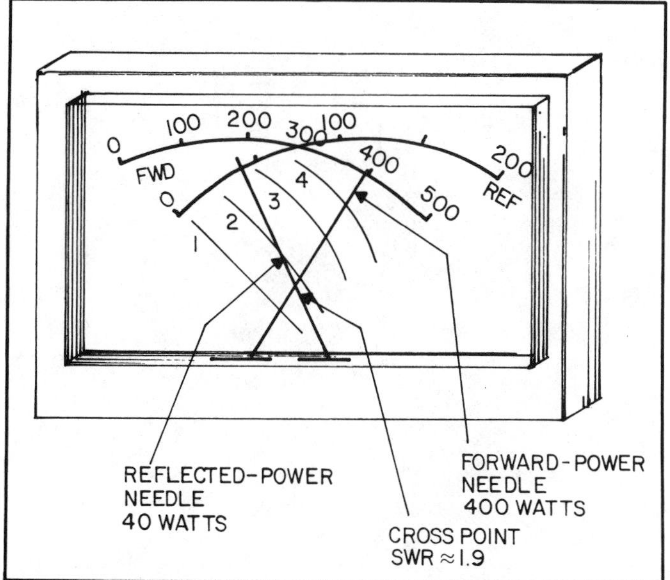

CROSSED-NEEDLE METER: A crossed-needle SWR meter. One scale shows the forward power, and the other scale shows the reflected power, A third scale indicates the standing-wave ratio as a function of the point where the two needles cross.

A simple crossover network, sometimes used in high-fidelity speaker systems, consists of two capacitors, as shown at B. One capacitor is connected in series with the tweeter line. This prevents low-frequency audio energy from reaching the tweeter. The other capacitor is connected in parallel with the woofer; this capacitor prevents high-frequency audio energy from reaching the woofer. There is no splitter in this circuit, since the energy is directly coupled to the speakers. The capacitor in series with the tweeter acts as a highpass filter, and the capacitor in parallel with the woofer acts as a lowpass filter.

The frequency at which both speakers receive an equal amount of energy is called the crossover frequency. Below the crossover point, the woofer gets more energy. Above the crossover frequency, the tweeter gets more energy. The crossover network should be designed so that the impedance of the whole system is nearly constant, as seen by the audio amplifier. It is also essential that the capacitor values be chosen correctly. *See also* HIGHPASS FILTER, LOWPASS FILTER, SPLITTER, TWEETER, WOOFER.

CROSS PRODUCT

The cross product or vector product, is a means of multiplying two vectors to obtain a third vector (*see* VECTOR). Given two vectors A and B, their cross product A × B is perpendicular to the plane containing A and B. The magnitude of the vector A × B is given by:

$$|A \times B| = |A|\,|B| \sin \Theta$$

where | A | is the length of A, | B | is the length of B, and Θ is the angle between A and B.

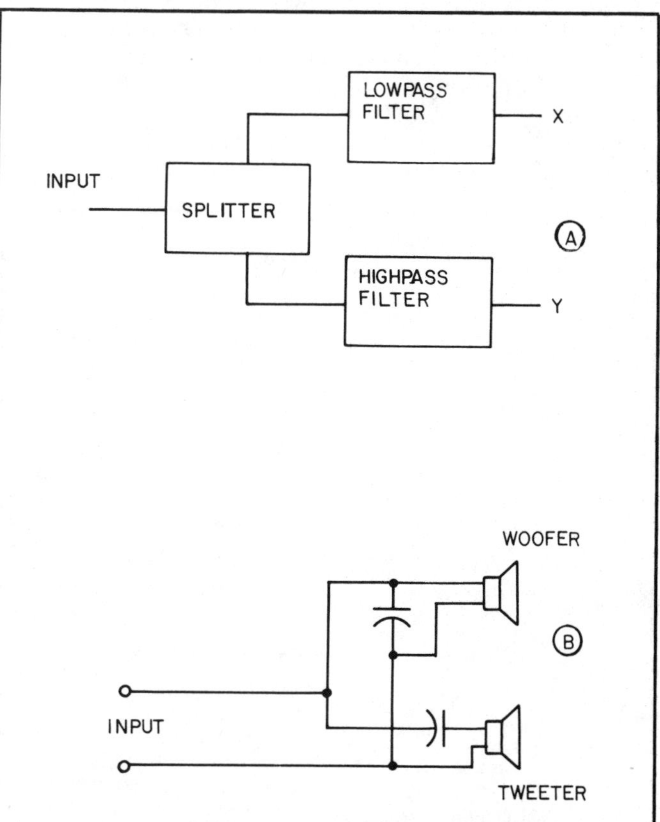

CROSSOVER NETWORK: At A, a crossover network using a splitter and two filters. At B, a simple crossover network, commonly used in high-fidelity speaker systems.

If the angle Θ is measured counterclockwise, as shown in the illustration, and vectors A and B lie in the plane of the page, then A × B points directly upward, out of the page, orthogonal to both vectors A and B. This constitutes a right-handed system; that is, if you curl your right-hand fingers in the direction of Θ, your thumb will point in the direction of the cross-product vector.

The cross product is useful in determining the magnitude and direction of an electromagnetic field, given known electric and magnetic field vectors. *See also* DOT PRODUCT, ELECTROMAGNETIC RADIATION.

CROSSTALK

Crosstalk is an undesired transfer of signals between circuits. Crosstalk is fairly common on telephone lines, especially in long-distance operation. Crosstalk may occur among different channels of a communications system.

Crosstalk rejection is the level of the crosstalk energy, in decibels, with respect to the desired signal. The crosstalk level is the amplitude of the crosstalk energy with respect to some reference value. Crosstalk loss is the effective degradation, in decibels, of the signal-to-noise ratio, caused by crosstalk in a communications circuit.

Crosstalk may be minimized by ensuring that a transmission line is properly shielded or balanced. In a carrier-current communications link (*see* CARRIER-CURRENT COMMUNICATION), crosstalk is minimized by the use of linear networks and the maintenance of good electrical connections throughout.

CRYOGENICS

Cryogenics is the science of the behavior of matter and energy at extremely low temperatures. The coldest possible temperature, called absolute zero, is the absence of all heat. This temperature is approximately −459.72 degrees Fahrenheit, or −273.16 degrees Celsius. The Kelvin temperature scale (*see* KELVIN TEMPERATURE SCALE) is based on absolute zero.

When the temperature of a conductor is brought to within a few degrees of absolute zero, the conductivity increases dramatically. If the temperature is cold enough, a current can be made to flow continuously in a closed loop of wire. This is called superconductivity.

Supercooling of a receiving antenna allows the use of amplifiers with greater gain than is possible without such cooling. Thermal noise is thereby reduced, and this improves the sensitivity of the receiver.

CRYSTAL

A crystal is a piece of piezoelectric material, used for the purpose of transforming mechanical vibrations into electrical impulses. Crystals are used in some microphones for this purpose (*see* CRYSTAL MICROPHONE, CRYSTAL TRANSDUCER).

Some crystal materials are used as detectors or mixers (*see* CRYSTAL DETECTOR, CRYSTAL SET).

Quartz crystals are widely used to generate radio-frequency energy. Such devices are typically housed in a metal can, as shown in the photograph. Two wire leads protrude from the base of the can. These leads are internally connected to the faces of the crystal, which consists of a thin wafer of quartz.

The frequency at which a quartz crystal vibrates depends on the manner in which the crystal is cut, and also on its thickness. The thinner the crystal, the higher the natural resonant frequency. The highest fundamental frequency of a common quartz crystal is in the neighborhood of 15 to 20 MHz; above this frequency range, harmonics must be employed to obtain radio-frequency energy.

Quartz crystals have excellent frequency stability. This is their main advantage. Crystals are much better than coil-and-capacitor tuned circuits in this respect. A crystal, however, cannot be tuned over a wide range of frequencies. Some crystals are used as selective filters because of their high Q factors. *See also* CRYSTAL CONTROL, CRYSTAL-LATTICE FILTER, CRYSTAL OSCILLATOR, Q FACTOR.

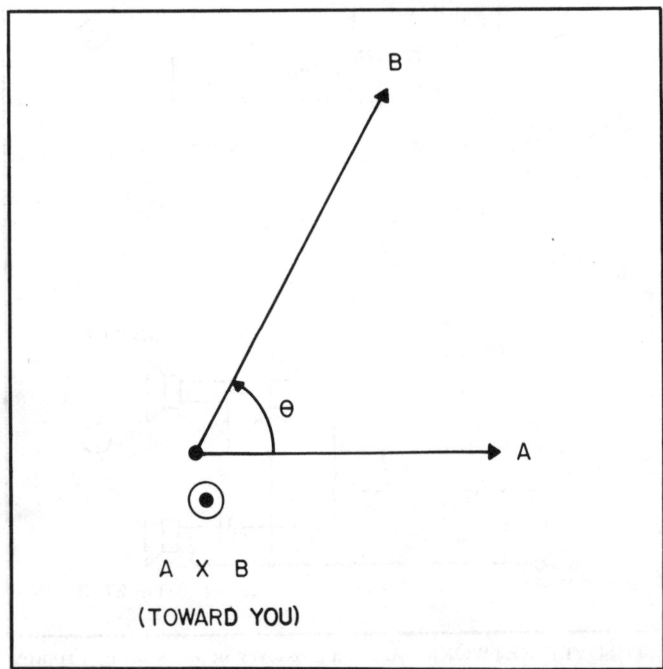

CROSS PRODUCT: The cross product of two vectors is orthogonal to them both. In this case, A × B comes out of the page, directly toward you.

CRYSTAL: A piezoelectric crystal may be smaller than a dime.

CRYSTAL CONTROL

Crystal control is a method of determining the frequency of an oscillator by means of a piezoelectric crystal. Such crystals, usually made of quartz (*see* CRYSTAL), have excellent oscillating frequency stability. Crystal control is much more stable than the coil-and-capacitor method.

Crystal oscillators may operate either at the fundamental frequency of the crystal, or at one of the harmonic frequencies. Crystals designed especially for operation at a harmonic frequency are called overtone crystals. Overtone crystals are almost always used at frequencies above 20 MHz, since a fundamental-frequency crystal would be too thin at these wavelengths, and might easily crack.

The operating frequency of a piezoelectric crystal can be increased slightly by placing an inductor in parallel with the crystal leads. A capacitor across a crystal will reduce the oscillating frequency. Generally, the amount of frequency adjustment possible with these methods is very small—approximately ±0.1 percent of the fundamental operating frequency. *See also* CRYSTAL OSCILLATOR.

CRYSTAL DETECTOR

In early radio receivers, a crystal detector was used to demodulate an amplitude-modulated signal. A piece of mineral, such as galena, was placed in contact with a fine piece of wire called a cat's whisker (*see* CAT'S WHISKER). This resulted in rectification. The cat's whisker often had to be moved around on the piece of mineral in order to get satisfactory reception.

Modern receivers employ silicon or germanium semiconductor diodes for the detection of amplitude-modulated signals. A bipolar or field-effect transistor, biased for Class-B operation, will act as a detector. *See also* DETECTION.

CRYSTAL-LATTICE FILTER

A crystal-lattice filter is a selective filter, usually of the bandpass type. It is similar in construction to a ceramic filter (*see* BANDPASS FILTER, CERAMIC FILTER) when housed in one container. Some crystal-lattice filters consist of several separate piezoelectric crystals. The crystals are cut to slightly different resonant frequencies to obtain the desired bandwidth and selectivity characteristics.

Crystal-lattice filters are found in the intermediate stages of superheterodyne receivers. They are also used in the filtering stages of single-sideband transmitters. Properly adjusted crystal-lattice filters have an excellent rectangular response, with steep skirts and high adjacent-channel attenuation. A simple crystal-lattice filter, using only two piezoelectric crystals at slightly different frequencies, is shown in the diagram. *See also* RECTANGULAR RESPONSE.

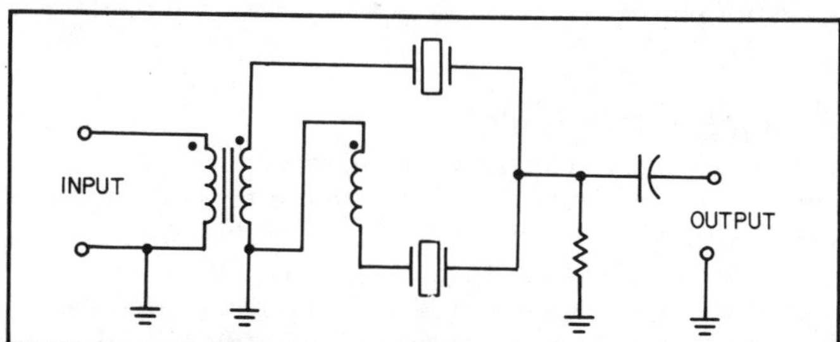

CRYSTAL-LATTICE FILTER: A two-crystal crystal-lattice filter. The crystals are resonant on slightly different frequencies.

CRYSTAL MICROPHONE

A crystal microphone is a device that uses a piezoelectric crystal to convert sound vibrations into electrical impulses. The impulses may then be amplified for use in public-address or communications circuits.

In the crystal microphone, which operates in a manner similar to the ceramic microphone (*see* CERAMIC MICROPHONE), vibrating air molecules set a metal diaphragm in motion. The diaphragm, connected physically to the crystal, puts mechanical stress on the piezoelectric substance. This, in turn, results in small currents at the same frequency or frequencies as the sound.

Crystal microphones are rather fragile. If the microphone is dropped, the crystal may break, and the microphone will be ruined. Crystal microphones have excellent fidelity characteristics. Dynamic microphones are also quite commonly used today. *See also* DYNAMIC MICROPHTNE.

CRYSTAL OSCILLATOR

A crystal oscillator is an oscillator in which the frequency is determined by a piezoelectric crystal. Crystal oscillators may be built using bipolar transistors, field-effect transistors, or vacuum tubes. The circuit generally consists of an amplifier with feedback, with the frequency of feedback governed by the crystal. Oscillation may take place at the fundamental frequency of the crystal, or at one of the harmonic frequencies.

The schematic illustrates three common types of crystal-oscillator circuits. A tuned output circuit provides harmonic attenuation, or it may be used to select one of the harmonic frequencies. If the oscillator is used at the fundamental frequency of the crystal, and if the amount of feedback is properly regulated, a tuned output circuit is not usually needed. However, the oscillator output will contain more energy at unwanted frequencies when a tuned circuit is not used. *See also* OSCILLATOR.

CRYSTAL OVEN

A crystal oven is a chamber in which the temperature is kept at an extremely constant level. Most piezoelectric oscillator crystals shift slightly in frequency as the temperature changes. Some crystals get higher in frequency with an increase in the temperature. This is called a positive temperature coefficient. Some crystals get lower in frequency when the temperature rises; this is called a negative temperature coefficient. A crystal oven is used to house crystals in circuits where extreme frequency accuracy is needed.

Crystal ovens employ thermostat mechanisms and small heating elements, just like ordinary ovens. The temperature is kept at a level just a little above the temperature in the room where the circuit is operated. Sometimes, several ovens are used, one inside the other, to obtain even more precise temperature regulation! Frequency-standard oscillators often use this kind of crystal oven.

CRYSTAL SET

An amplitude-modulation receiver, consisting only of a tuned circuit, a passive detector, and an earphone or

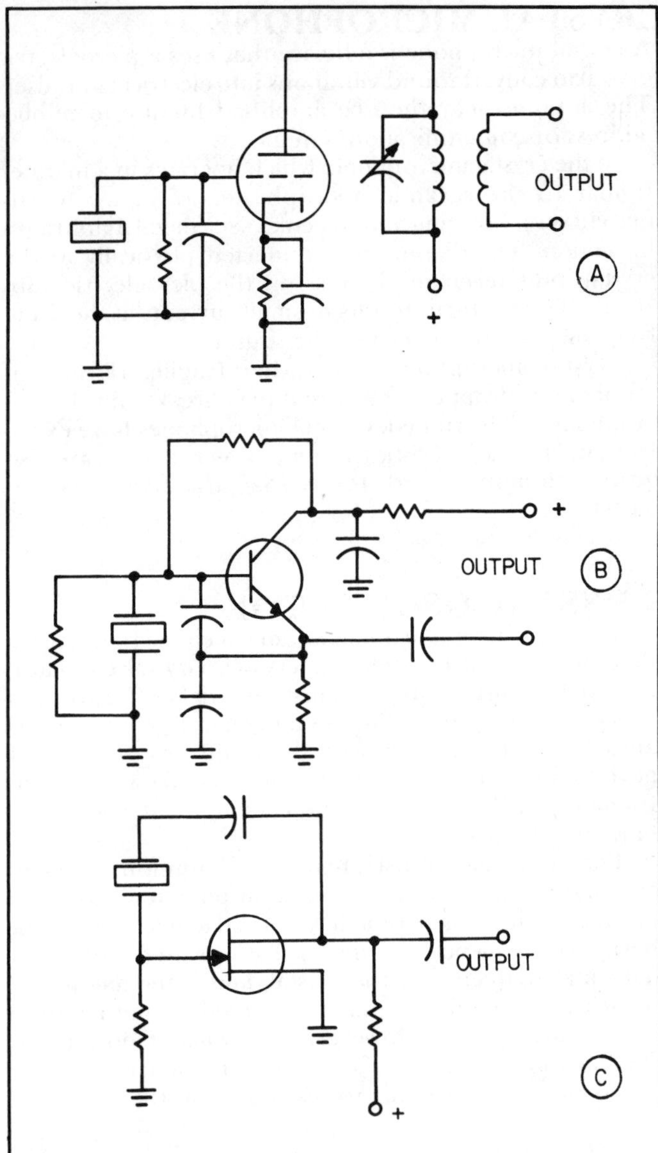

CRYSTAL OSCILLATOR: Three common kinds of crystal oscillator. At A, an oscillator using a triode vacuum tube. At B, an oscillator using a bipolar transistor. At C, a Pierce oscillator circuit using a field-effect transistor.

headset, with no amplification, is called a crystal set. The illustration shows a schematic diagram of such a receiver. Sometimes, one stage of audio amplification is added following the detector, and the resulting circuit is called a crystal receiver. A true crystal set, however, operates only from the power supplied by the received signal.

A crystal set is, obviously, not a very sensitive receiver, but fairly strong stations can be received with a very long, or a resonant, antenna. The selectivity of a crystal set is generally rather poor. Crystal sets were originally used for the reception of spark-generated radio signals, in the early days of wireless communication. The crystal set is not the most primitive possible type of radio receiver, however! In the presence of an extremely strong amplitude-modulated signal, a poor electrical junction in a metal tooth filling can sometimes pick up enough radio-frequency current to act as a detector. This allows the owner of the filling to hear the station, whether he wants to or not! Fortunately, such bizarre instances of crystal-set operation are extremely rare. *See also* DETECTION.

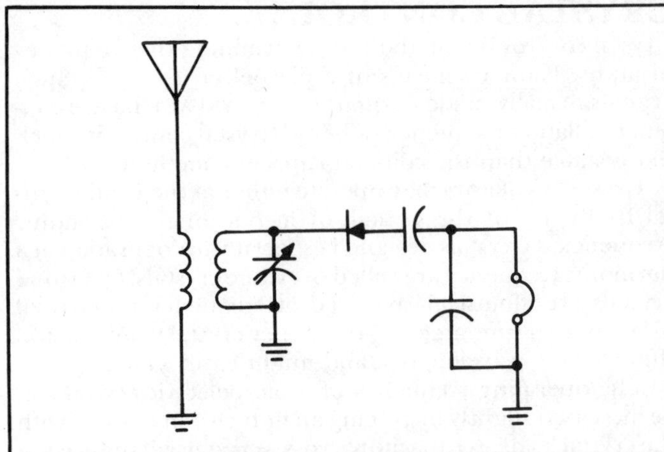

CRYSTAL SET: The crystal set is the height of simplicity. No power source, other than the received station itself, is used.

CRYSTAL TEST CIRCUIT

A crystal test circuit is a device for testing piezoelectric crystals for proper operation. Most crystal test circuits ascertain only that the crystal will oscillate on the correct frequency under the specified conditions. This allows faulty crystals to be easily identified.

More sophisticated crystal testing circuits are used to determine the temperature coefficient, crystal current, and other operating variables. The properties of a crystal or ceramic transducer are checked by a special type of crystal tester.

Semiconductor diodes are sometimes called crystals, although this is an antiquated and somewhat inaccurate term. A circuit for testing semiconductor diodes may be called a crystal tester. *See also* CRYSTAL.

CRYSTAL TRANSDUCER

Piezoelectric crystals are often used to convert mechanical vibrations into electrical impulses, and vice versa. This is called the piezoelectric effect (*see* PIEZOELECTRIC EFFECT). Crystals, either natural or synthetic, make excellent transducers for this reason. Such piezoelectric substances are used in microphones, phonograph cartridges, earphones, and buzzer devices. The acquisition tones now used on many devices with keyboards are generated by tiny piezoelectric crystals (*see* ACQUISITION TONE).

The illustration shows the operation of a simple crystal transducer. If alternating currents are applied to the holder plates, the crystal will vibrate, producing sound or ultrasound. Conversely, if sound or ultrasound impinges on the diaphragm, an alternating current will appear between the holder plates. Crystal transducers can operate at frequencies well above the range of human hearing. *See also* TRANSDUCER.

CURRENT

Current is a flow of electric charge carriers past a point, or from one point to another. The charge carriers may be electrons, holes, or ions (*see* ELECTRON, HOLE, ION). In some cases, atomic nuclei may carry charge.

Electric current is measured in units called amperes. A current of one ampere consists of the transfer of one coulomb of charge per second (*see* AMPERE, COULOMB). Current may be either alternating or direct. Current is

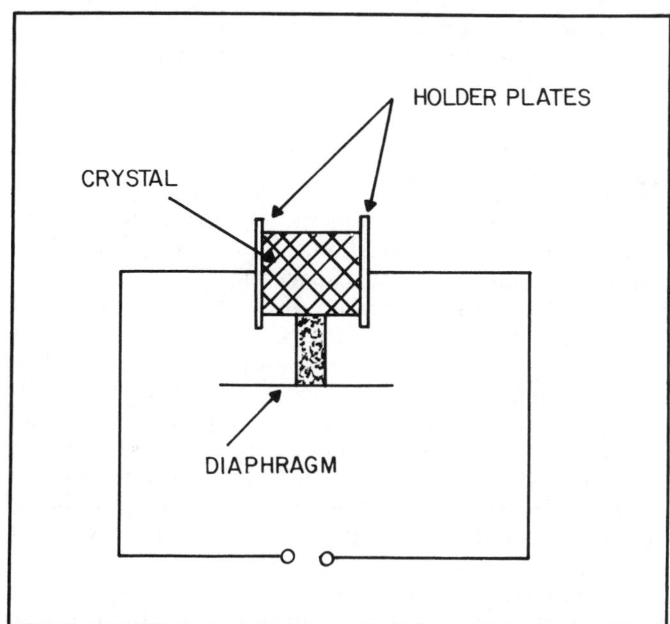

CRYSTAL TRANSDUCER: A crystal transducer operates via the piezoelectric effect.

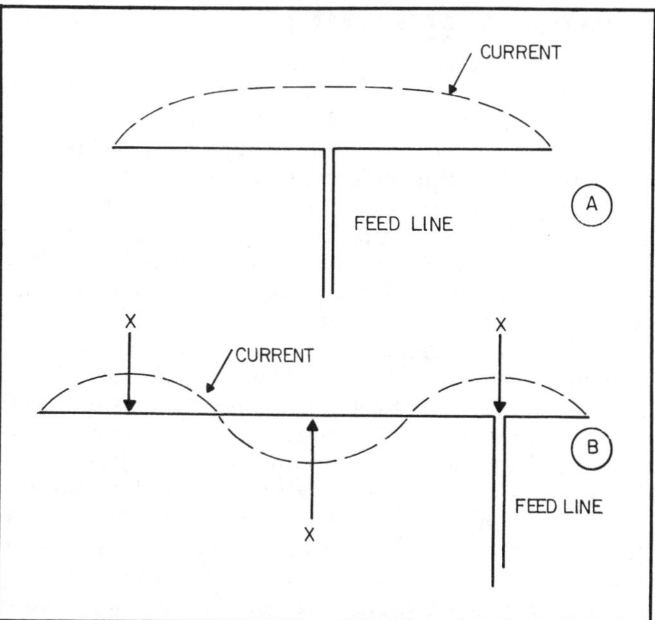

CURRENT FEED: At A, a center-fed, half-wave dipole antenna is always current-fed. At B, a 3/2-wavelength radiator may be current fed at any of the points marked X.

symbolized by the letter I in most equations involving electrical quantities.

The direction of current flow is theoretically the direction of the positive charge transfer. Thus, in a circuit containing a dry cell and a light bulb, for example, the current flows from the positive terminal of the cell, through the interconnecting wires, the bulb, and finally to the negative terminal. This is a matter of convention. The electron movement is actually in opposition to the current flow. Physicists use this interpretation of current flow purely as a mathematical convenience.

CURRENT AMPLIFICATION

Current amplification is the increase in the flow of current between the input and output of a circuit. It is also sometimes called current gain. In a transistor, the current-amplification characteristic is called the beta (*see* BETA).

Some amplifier circuits are designed to amplify current. Others are designed to amplify voltage. Still others are designed to amplify the power, which is the product of the current and the voltage. A current amplifier requires a certain amount of driving power to operate, since such a circuit draws current from its source. Current amplifiers generally have an output impedance that is lower than the input impedance, and therefore such circuits are often used in step-down matching applications. That is, they are used as impedance transformers.

Current amplification is measured in decibels. Mathematically, if I_{IN} is the input current and I_{OUT} is the output current, then:

$$\text{Current gain (dB)} = 20 \log_{10} (I_{OUT}/I_{IN})$$

See also DECIBEL, GAIN.

CURRENT-CARRYING CAPACITY

See CARRYING CAPACITY.

CURRENT DRAIN

The amount of current that a circuit draws from a generator, or other power supply, is called the current drain. The amount of current drain determines the size of the power supply needed for proper operation of a circuit.

If the current drain is too great for a power supply, the voltage output of the supply will drop. Ripple may occur in supplies designed to convert alternating current to direct current. With battery power, the battery life is shortened, the voltage drops, and the battery may overheat dangerously.

Current drain is measured in three ways. The peak drain is the largest value of current drawn by a circuit in normal operation. The average current drain is measured over a long period; the total drain in ampere hours is divided by the operating time in hours. Standby current drain is the amount of current used by a circuit during standby periods. Power supplies should always be chosen to handle the peak current drain without malfunctioning. *See also* POWER SUPPLY.

CURRENT FEED

Current feed is a method of connecting a transmission line to an antenna at a point on the antenna where the current is maximum. Such a point is called a current loop (*see* CURRENT LOOP). In a half-wavelength radiator, the current maximum occurs at the center, and therefore current feed is the same as center feed (A in the illustration). In an antenna longer than ½ wavelength, current maxima exist at odd multiples of ¼ wavelength from either end of the radiator. There may be several different points on an antenna radiator that are suitable for current feed (as at B).

The impedance of a current-fed antenna is relatively low. The resistive component varies between about 70 and 200 ohms in most cases. Current feed results in good electrical balance in a two-wire transmission line, provided the current in the antenna is reasonably symmetrical. *See also* VOLTAGE FEED.

CURRENT HOGGING

When two active components are connected together in parallel, or in push-pull configuration, one of them may draw most of the current. This situation, called current hogging, occurs because of improper balance between components. Current hogging can sometimes take place with poorly matched tubes or transistors connected in push-pull or parallel amplifiers.

Initially, one of the tubes or transistors exhibits a slightly lower resistance in the circuit. The result is that this component draws more current than its mate. If the resistive temperature coefficient of the device is negative, the tube or transistor carrying the larger current will show a lower and lower resistance as it heats. The result is a vicious circle: The more the component heats up, the lower its resistance becomes, and the more current it draws. Ultimately, one of the tubes or transistors does all the work in the circuit. This may shorten its operating life. It also disturbs the linearity of a push-pull circuit, and upsets the impedance match between the circuit and the load.

Current hogging may be prevented, or at least made unlikely, by placing small resistors in series with the emitter, source, or cathode leads of the amplifying devices. Careful selection of the devices, to ensure the most nearly identical operating characteristics, is also helpful.

CURRENT LIMITING

Current limiting is a process that prevents a circuit from drawing more than a certain predetermined amount of current. Most low-voltage, direct-current power supplies are equipped with current-limiting devices.

A current-limiting component exhibits essentially no resistance until the current, I, reaches the limiting value. When the load resistance R_L becomes smaller than the value at which the current I is at its maximum, the limiting component introduces an extra series resistance R_S. If the supply voltage is E volts, then

$$E = I (R_L + R_S)$$

when the limiting device is active. The resistance R_S increases as R_L decreases, so that

$$R_S = (E/I) - R_L$$

Current-limiting devices help to protect both the supply and the load from damage in the event of a malfunction. Transistors with large current-carrying capacity are used as limiting devices. Current limiting is sometimes called foldback.

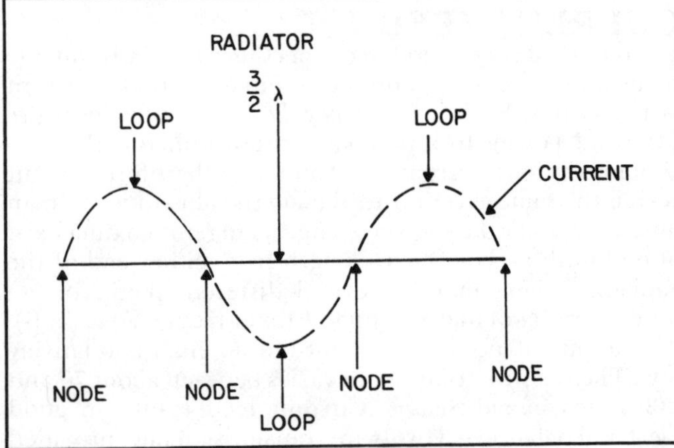

CURRENT LOOP AND CURRENT NODE: Locations of current loops and nodes on a radiator of 3/2 wavelength.

CURRENT LOOP

In an antenna radiating element, the current in the conductor depends on the location. At any free end, the current is negligible; the small capacitance allows only a tiny charging current to exist. At a distance of ¼ wavelength from a free end, the current reaches a maximum. This maximum is called a current loop. A ½-wavelength radiator has a single current loop at the center. A full-wavelength radiator has two current loops. In general, the number of current loops in a longwire antenna radiator is the same as the number of half wavelengths. The drawing illustrates the current distribution in an antenna radiator of 3/2 wavelength, showing the locations of the current loops.

Current loops may occur along a transmission line not terminated in an impedance identical to its characteristic impedance. These loops occur at multiples of ½ wavelength from the resonant antenna feed point when the antenna impedance is smaller than the feed-line characteristic impedance. The loops exist at odd multiples of ¼ wavelength from the feed point when the resonant antenna impedance is larger than the feed-line characteristic impedance. Ideally, the current on a transmission line should be the same everywhere, equal to the voltage divided by the characteristic impedance. *See also* CURRENT NODE, STANDING WAVE.

CURRENT NODE

A current node is a current minimum in an antenna radiator or transmission line. The current in an antenna depends, to some extent, on the location of the radiator. Current nodes occur at free ends of a radiator, and at distances of multiples of ½ wavelength from a free end. The illustration shows the locations of current nodes along a 3/2-wavelength antenna radiator. The number of current nodes is usually equal to 1 plus the number of half wavelengths in a radiator.

Current nodes may occur along a transmission line not terminated in an impedance equal to its characteristic impedance. These nodes occur at multiples of ½ wavelength from the resonant antenna feed point when the antenna impedance is larger than the feed-line characteristic impedance. They exist at odd multiples of ¼ wavelength from an antenna feed point when a resonant antenna impedance is smaller than the characteristic impedance of the line.

Current nodes are always spaced at intervals of ¼ wavelengths from current loops. Ideally, the current on a transmission line is the same everywhere, being equal to the voltage divided by the characteristic impedance. *See also* CURRENT LOOP, STANDING WAVE.

CURRENT REGULATION

Current regulation is the process of maintaining the current in a load at a constant value. This is done by means of a constant-current power supply (*see* CONSTANT-CURRENT SOURCE). A variable-resistance device is necessary to accomplish current regulation. When such a device is placed in series with the load, and the resistance increases in direct proportion to the supply voltage, the current remains constant. *See also* BARRETTER.

CURRENT SATURATION

As the bias between the input points of a tube or semiconductor device is varied in such a way that the output current increases, a point will eventually be reached at which the output current no longer increases. This condition is called current saturation. In a vacuum tube, the grid bias must usually be positive with respect to the cathode to obtain current saturation. In an NPN bipolar transistor, the base must be sufficiently positive with respect to the emitter (see illustration). In a PNP bipolar transistor, the base must be sufficiently negative with respect to the emitter. In a field-effect transistor, the parameters for current saturation are affected by the gate-source and drain-source voltages.

Saturation is usually not a desirable condition. It destroys the electrical ability of a tube or semiconductor device to amplify. Saturation sometimes occurs in digital switching transistors, where it may be induced deliberately in the high state to produce maximum conduction. *See also* SATURATION, SATURATION CURRENT, SATURATION CURVE.

CURRENT TRANSFORMER

A current transformer is a device for stepping current up or down. An ordinary voltage transformer functions as a current transformer, but in the opposite sense (*see* TRANSFORMER, VOLTAGE TRANSFORMER). The current step-up ratio of a transformer is the reciprocal of the voltage step-up ratio. If N_{PRI} is the number of turns in the primary winding and N_{SEC} is the number of turns in the secondary winding, then

$$I_{SEC}/I_{PRI} = N_{PRI}/N_{SEC}$$

where I_{PRI} and I_{SEC} are the currents in the primary and secondary, respectively. The impedance of the primary and secondary, given by Z_{PRI} and Z_{SEC}' are related to the currents by the equation

$$Z_{PRI}/Z_{SEC} = (I_{SEC}/I_{PRI})^2$$

These formulas assume a transformer efficiency of 100 percent. While this is an ideal theoretical condition, and it never actually occurs, the equations are usually accurate enough in practice. *See also* TRANSFORMER EFFICIENCY.

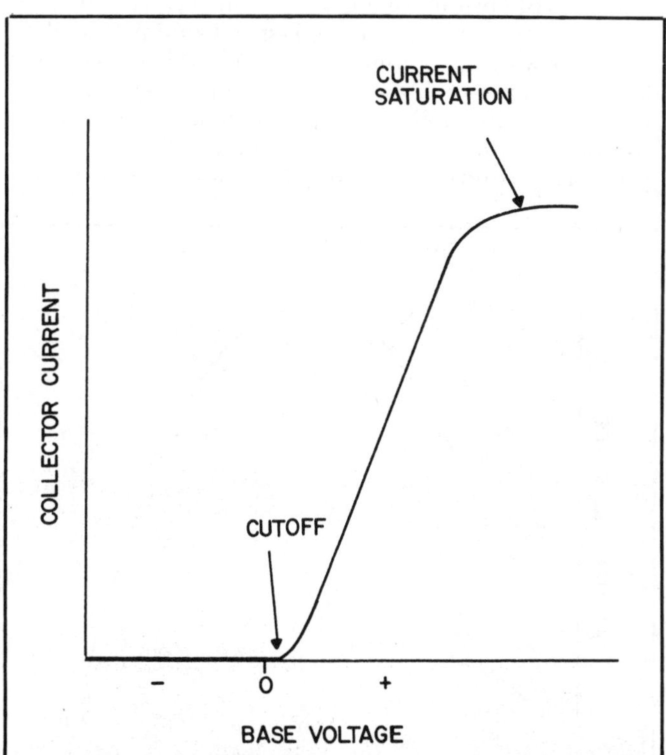

CURRENT SATURATION: As the current through a transistor increases, a point is reached at which it will increase no further. This condition is current saturation. Here, the condition is shown for an NPN bipolar transistor.

CURVE

A curve is a graphical illustration of a relation between two variables. In electronics, two-dimensional graphs are

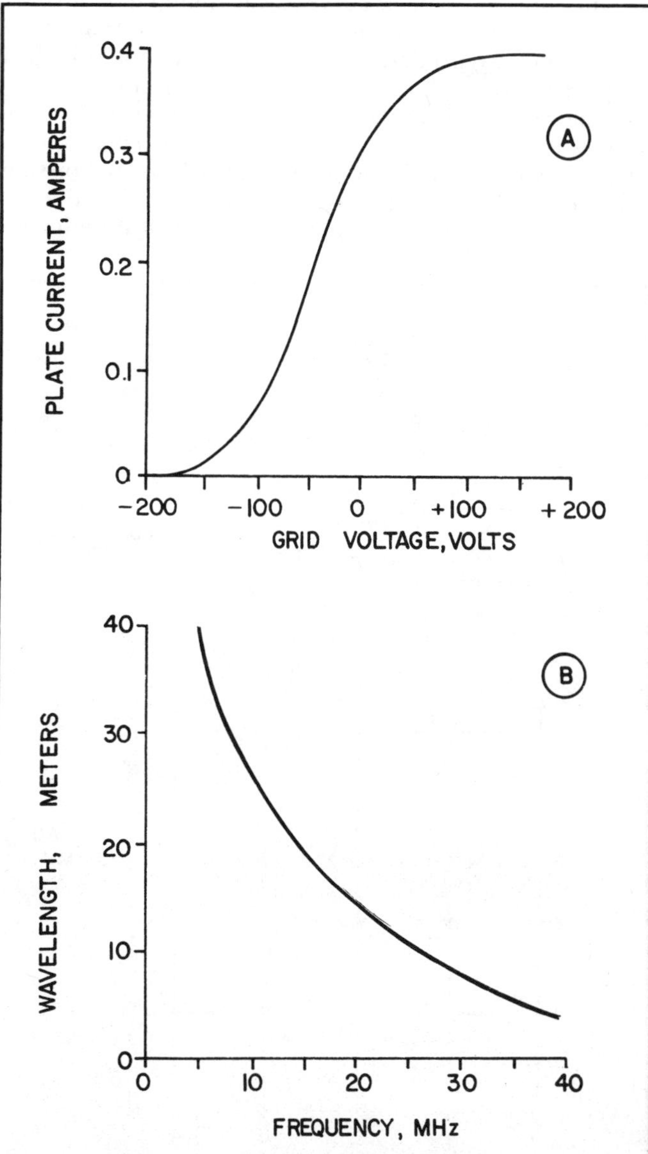

CURVE: At A, a curve showing grid voltage versus plate current. At B, a curve showing frequency versus wavelength for electromagnetic waves in free space.

commonly used to show the characteristics of circuits and devices. Generally, the Cartesian coordinate plane is used, although other coordinate systems are sometimes employed (*see* CARTESIAN COORDINATES, CURVILINEAR COORDINATES, POLAR COORDINATES).

The illustration shows two examples of curves. At A, the relation between the grid voltage and the plate current of a typical vacuum tube is shown. At B, the relationship between frequency and wavelength for electromagnetic radiation in free space is illustrated. In both of these cases, the curves represent a special kind of relation called a function, since there is never more than one value on the vertical (dependent) axis for any value on the horizontal (independent) axis. *See also* FUNCTION, RELATION.

CURVE TRACER

A curve tracer is a test circuit used to check the response of a component or circuit under conditions of variable input. A test signal is applied to the input of the component or circuit, and the output is monitored on an oscilloscope.

One common type of curve tracer is used to determine the characteristic curve of a transistor (*see* CHARACTERISTIC CURVE). A predetermined, direct-current voltage is applied between the emitter and the collector. Then, a variable voltage is applied to the base. The variable base voltage is also applied to the horizontal deflecting plates of an oscilloscope. The collector current is measured by sampling the voltage drop across a resistor in the collector circuit; this voltage is supplied to the vertical deflection plates of the oscilloscope. The result is a visual display of the base voltage versus collector-current curve.

Another common type of curve tracer uses a sweep generator (*see* SWEEP-FREQUENCY FILTER ANALYZER, SWEEP GENERATOR) and an oscilloscope. This provides a display of attenuation as a function of frequency for a tuned circuit.

Curve tracers allow comparison of actual circuit parameters with theoretical parameters. They are, therefore, invaluable in engineering and test applications.

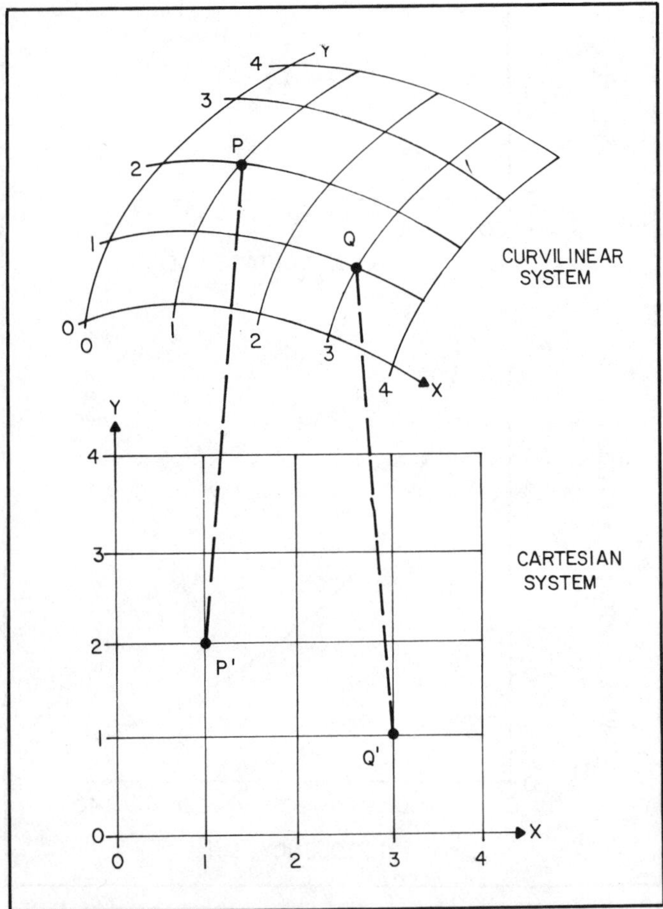

CURVILINEAR COORDINATES: Some curvilinear coordinates can be transformed to Cartesian coordinates, as shown here.

CURVILINEAR COORDINATES

When the coordinate lines in a graph system are not straight lines, the system is said to be curvilinear. The latitude-longitude lines on a globe are an example of a curvilinear coordinate system. Sometimes, a one-to-one mathematical transformation is possible between a curvilinear system and the more familiar Cartesian coordinate system (*see* CARTESIAN COORDINATES). Sometimes, such a transformation does not exist. The illustration shows a transformation between curvilinear and Cartesian coordinates in a local case.

Curvilinear coordinates are not generally used in electricity and electronics, except in some very advanced mathematical calculations. The exception to this rule is the polar coordinate system, used frequently in plotting antenna directional patterns and surface bearings. *See also* POLAR COORDINATES.

CUSTOM CALLING SERVICE

In a telephone system, a custom calling service is a function not usually performed. Such services can be obtained in most household or business systems today as an option. Examples of custom calling services include call forwarding, conference calling, and high-speed automatic dialing.

Call forwarding automatically dials a third party when desired. If a person is not at home, but knows of a number where he or she can be reached, the home phone may be programmed to transfer any incoming calls to that third number. Conference calling facilitates conversations

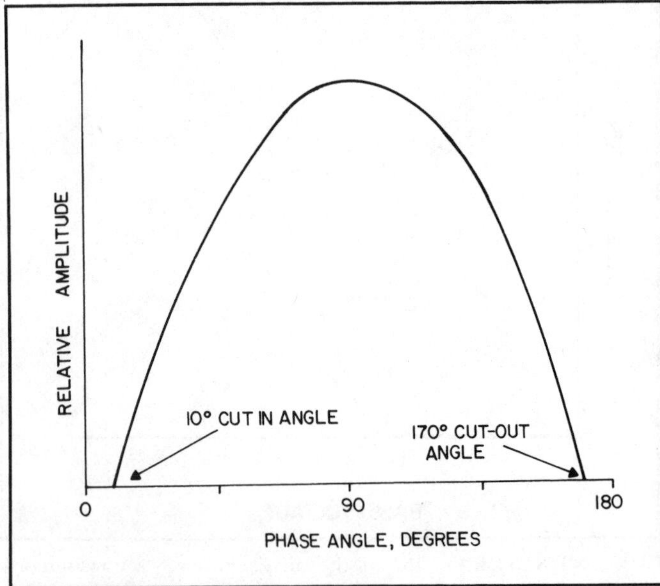

CUT-IN/CUT-OUT ANGLE: An example of cut-in and cut-out angles for a rectified wave.

among three or more parties. Each party may be at a different number in the telephone system. Automatic dialing allows certain selected numbers to be dialed at extremely high speed, simply by pressing one or two buttons.

As the telephone system becomes more and more advanced and sophisticated, more kinds of custom calling services are becoming available. These include such things as call screening. *See also* TELEPHONE.

CUT-IN/CUT-OUT ANGLE

A semiconductor diode requires between 0.3 volt and 0.6 volt of forward bias in order to conduct. In a rectifier circuit using semiconductor diodes, therefore, the conduction period is not quite one-half cycle. Instead, the conduction time is a little less than 180 degrees, as shown in the illustration.

The cut-in angle is the phase angle at which conduction begins. The cut-out angle is the phase angle at which conduction stops. In a 60-Hz rectifier circuit, if a phase angle of 0 degrees is represented by $t = 0$ second and a phase angle of 180 degrees is represented by $t = 1/120$ second or 8.33×10^{-3} second, then

$$\Theta_1 = 2.16 \times 10^4 \; t_1$$

and

$$\Theta_2 = 2.16 \times 10^4 \; t_2$$

where Θ_1 and Θ_2 are the cut-in and cut-out angles, respectively, and t_1 and t_2 are the cut-in and cut-out times, respectively.

The cut-in and cut-out angles become closer to 0 and 180 degrees as the voltage of a sine-wave, alternating-current waveform increases. In a square-wave rectifier circuit, the angles Θ_1 and Θ_2 are essentially equal to 0 and 180 degrees. *See also* RECTIFICATION.

CUTOFF

Cutoff is a condition in a tube or bipolar transistor in which the grid or base voltage prevents current from flowing through the device. In a field-effect transistor, the condition of current cutoff is usually called pinchoff (*see* PINCH-OFF).

In a vacuum tube, cutoff is achieved when the grid voltage is made sufficiently negative with respect to the cathode. In an NPN bipolar transistor, the base must generally be at either the same potential as the emitter, or more negative. In a PNP bipolar transistor, the base must usually be at either the same potential as the emitter, or more positive. In a field-effect transistor, pinchoff depends on the bias relationship among the source, gate, and drain. The drawing illustrates the cutoff point in the base-voltage-versus-collector-current curve of an NPN bipolar transistor.

The cutoff condition of an amplifying device is often used to increase the efficiency when linearity is not important, or when waveform distortion is of no consequence. This is the case in the Class-B and Class-C amplifier circuits (*see* CLASS-B AMPLIFIER, CLASS-C AMPLIFIER). A cut-off tube or transistor may also be used as a rectifier or detector.

The term cutoff is used also to refer to any point at which a certain parameter is exceeded in a circuit. For example, we may speak of the cutoff frequency of a lowpass filter, or the alpha-cutoff frequency of a transistor.

CUTOFF ATTENUATOR

A waveguide has a certain minimum operating frequency, below which it is not useful as a transmission line because it causes a large attenuation of a signal. The cutoff, or minimum usable, frequency of a waveguide depends on its cross-sectional dimensions (*see* WAVEGUIDE).

When a section of waveguide is deliberately inserted into a circuit, and its cutoff frequency is higher than the operating frequency of the circuit, the waveguide becomes an attenuator. This sort of device, used at very-high and ultra-high frequencies, is called a cutoff attenuator. The amount of attenuation depends on the difference between the operating frequency, f_o, and the cutoff frequency, f_c of the waveguide. As f_c-f_o becomes larger, so does the attenuation. The amount of attenuation also depends on the length of the section of waveguide. The longer the lossy section of waveguide, the greater the attenuation. *See also* ATTENUATOR.

CUTOFF FREQUENCY

In any circuit or device, the term cutoff frequency can refer to either a maximum usable frequency or a minimum usable frequency.

In a transistor, the gain drops as the frequency is increased. The cutoff in such a case is called the alpha-cutoff frequency or beta-cutoff frequency, depending on the application (*see* ALPHA-CUTOFF FREQUENCY, BETA-CUTOFF FREQUENCY).

As the frequency is raised in a lowpass filter, the frequency at which the voltage attenuation becomes 3 dB, relative to the level in the operating range, is called the cutoff frequency. In a highpass filter, as the frequency is lowered, the frequency at which the voltage attenuation becomes 3 dB, relative to the level within the operating range, is called the cutoff frequency. This is illustrated in the graph. A bandpass or band-rejection filter has two cutoff frequencies. (*See* BANDPASS RESPONSE, BAND-REJECTION RESPONSE, HIGHPASS RESPONSE, LOWPASS RESPONSE.)

Many different kinds of circuits exhibit cutoff frequencies, either at a maximum, minimum, or both. A broadband antenna has a maximum and a minimum cutoff frequency. All amplifiers have an upper cutoff frequency. Sections of coaxial or two-wire transmission line have an upper cutoff frequency; waveguides have a lower cutoff that is well defined. Usually, the specification for cutoff is 3 dB, representing 70.7 percent of the current or voltage in the normal operating range. However, other attenuation levels are sometimes specified for special purposes. *See also* ATTENUATION.

CUTOFF VOLTAGE

The cutoff voltage of a vacuum tube or transistor is the level of grid, base, or gate voltage at which cutoff occurs. In a field-effect transistor, the cutoff voltage is usually called the pinchoff voltage.

A bipolar transistor is normally cut-off. That is, when the base voltage is zero with respect to the emitter, the device does not conduct. Until approximately 0.3 to 0.6

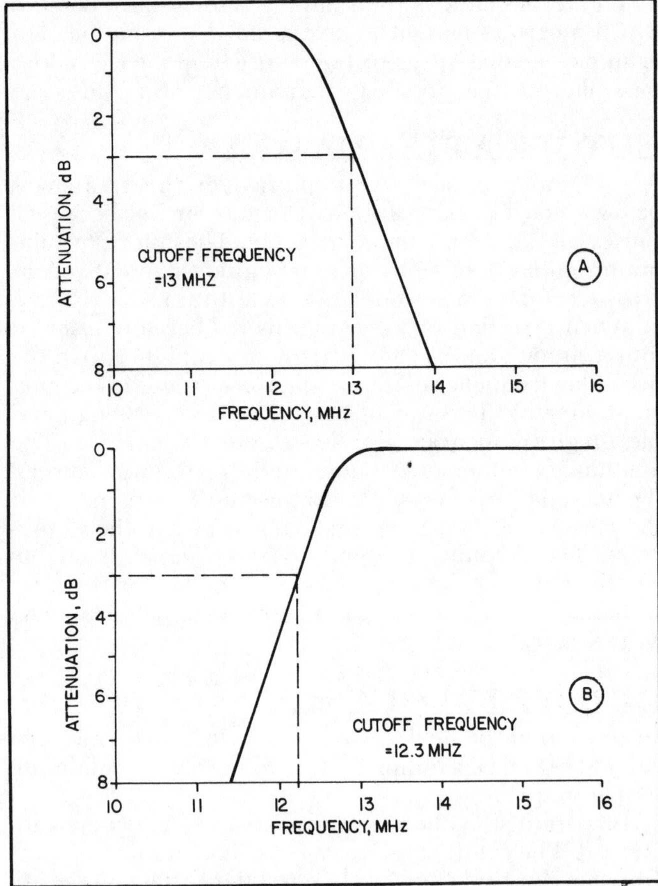

CUTOFF FREQUENCY: Attenuation-versus-frequency curves for a lowpass filter with a cutoff of 13.0 MHz (A) and a highpass filter with a cutoff of 12.3 MHz (B).

volts of base voltage is applied in the forward direction, the transistor remains cut-off. Above +0.3 to +0.6 volts, an NPN transistor will begin to conduct; below −0.3 to −0.6 volt, a PNP transistor begins to conduct. This is based on the assumption that the collector is properly biased—positive for an NPN device and negative for a PNP device.

In a vacuum tube, cutoff usually requires the presence of a large negative voltage on the control grid. This bias may be anywhere from a few to several hundred volts, depending on the tube type. *See also* CONTROL GRID, TRANSISTOR, TUBE.

CUTOUT

A cutout, or cutout device, is a circuit-breaking component such as a common breaker or fuse. When the current exceeds a certain predetermined level, the supply line is broken to protect the circuit from damage. This condition is sometimes called cutout. *See also* CIRCUIT BREAKER, FUSE.

CUT-OUT ANGLE

See CUT-IN/CUT-OUT ANGLE.

CW

See CONTINUOUS WAVE.

CYBERNETICS

All computers, and all living things, think by means of certain processes. Cybernetics is the study of these processes. Cybernetics is therefore the study of communication among, and control of, machines and organisms.

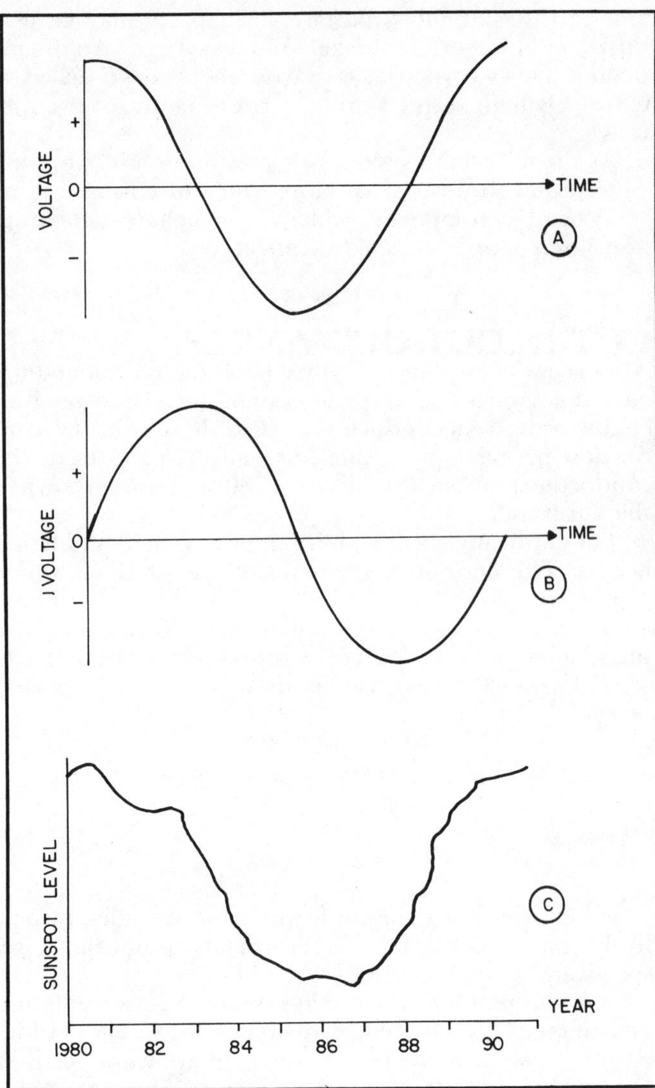

CYCLE: At A, a cycle of a sine wave measured from the positive peak to the next positive peak. At B, a cycle of a sine wave measured from the zero positive-going point to the next zero positive-going point. At C, a cycle of sunspot activity for the years 1980 to 1991 (approximately).

In the not-too-distant future, it will probably be possible for people and computers to think and reason together in an intelligent manner, and to actually converse. People may teach computers, and the computers may learn from people. The converse—eerie as it may sound—could also someday take place: People might gain new insights and ways of thinking from their computers! Computer teaching is already being done in some schools. Computers can replace teachers for some of the more mundane tasks.

The use, or abuse, of cybernetics raises some important moral and philosophical questions. For example, what effect will computer teaching have on the emotional development of a child? Will some people try to gain power over others by means of computerized mind control? Will computers themselves present a threat to man, perhaps becoming more intelligent than their creators? Science must prepare to deal with these and other possibilities which, just a few short years ago, were the subject of science-fiction books and movies, and even of jokes. *See also* ARTIFICIAL INTELLIGENCE.

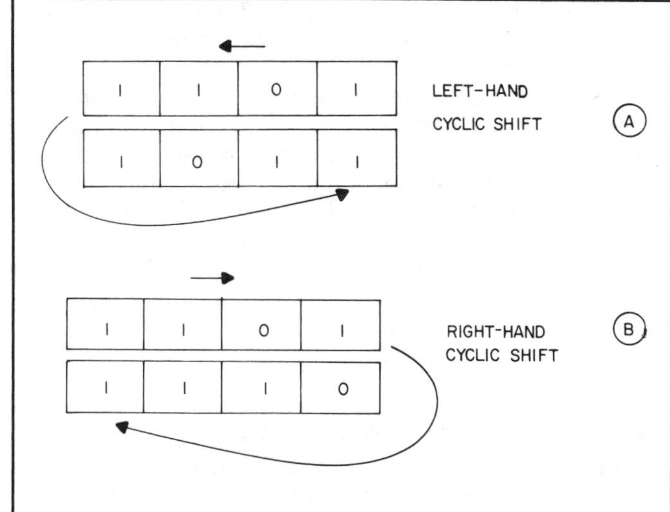

CYCLIC SHIFT: At A, a cyclic shift of one unit to the left. At B, a cyclic shift of one unit toward the right. The original binary number in each case is 1101.

CYCLE

In any periodic wave—that is, a waveform that repeats itself many times—a cycle is the part of a waveform between any point and its repetition. For example, in a sine wave, a cycle may be regarded as that part of the waveform between one positive peak and the next (A in the illustration), or between the point of zero, positive-going voltage and the next point of zero, positive-going voltage (as at B). It actually does not matter which point of reference is chosen to determine a cycle, as long as the waveform ends at the same place that it begins.

Cycles can be identified for any periodic waveform, such as the sunspot-variation curve shown at C. Here, although the waveshape varies slightly from one cycle to the next, a periodic variation is definitely present. In fact, the sunspot fluctuation, which repeats at intervals of approximately 11 years, is often called the sunspot cycle, since its recurrence is so predictable.

A cycle is routinely divided into 360 small, equal increments, called degrees. We may therefore identify small parts of a cycle, such as the 30-degree point and the 130-degree point. The difference between two points is called an angle of phase. Hence the angle between the 130-degree point and the 30-degree point is 100 degrees of phase. Engineers sometimes divide a cycle into radians. A radian is roughly equal to 57.3 degrees. There are exactly 2π radians in a complete cycle. *See also* PHASE ANGLE.

CYCLES PER SECOND

The term cycles per second is an obsolete expression for the frequency of a periodic wave. The commonly accepted electronic or electrical term nowadays is Hertz (*see* HERTZ). In older text and reference books, the frequency is still sometimes expressed in cycles, kilocycles, or megacycles per second, abbreviated cps, kc, and Mc, respectively.

CYCLIC IONSPHERIC VARIATION

The density of ionization in the upper atmosphere varies periodically with the time of day, the time of year, and the level of sunspot activity. Such fluctuations, which occur on a regular basis, are called cyclic ionospheric variations. Such variations affect the properties of radiocommunication on the medium and high frequencies.

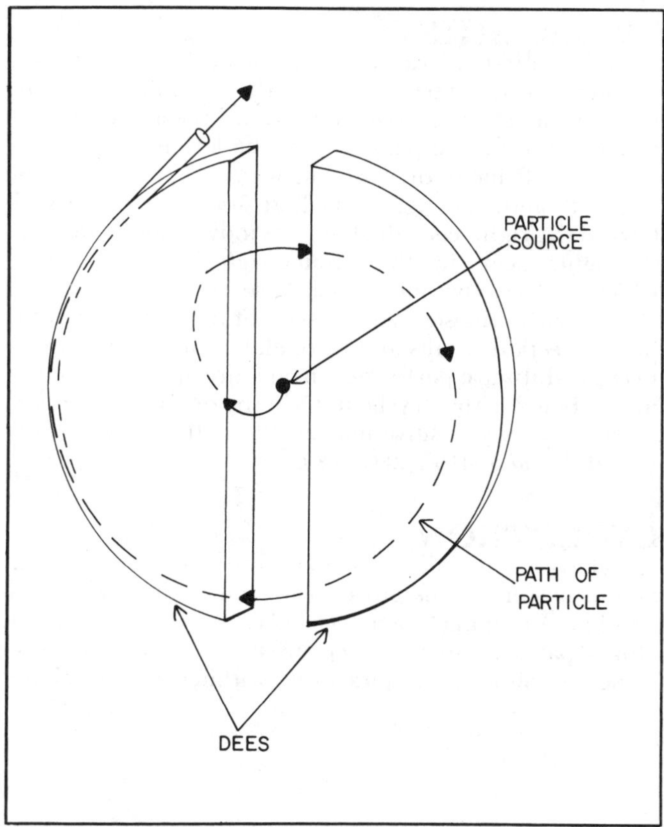

CYCLOTRON: A cyclotron forces charged particles into spiral paths of ever-increasing speed.

Generally, the density of ions is greater during the daylight hours than during the night. This is because the ultraviolet radiation from the sun causes the atoms in the upper atmosphere to ionize, at heights ranging from about 40 to 250 miles. This produces a daily cycle, which reaches its peak sometime after midday and reaches its minimum shortly before sunrise.

During the summer months, the level of ionization is usually greater, on the average, than during the rest of the year. During the winter months, the level of ionization is the least. Of course, when it is winter in the Northern Hemisphere, it is summer in the Southern Hemisphere and vice versa. The sun remains above the horizon for the longest time in the summer, allowing more atoms to become ionized. Also, the ultraviolet radiation is somewhat more intense in the summer, especially at lower levels in the atmosphere. The daily cycle is impressed upon this annual cycle.

The level of sunspot activity varies over a period of about 11 years. The years of maxima for this era are 1958, 1969, 1980, 1991, and 2002 A.D. The years of minima are 1964, 1975, 1986, 1997, and 2008 A.D. Ionospheric density is, on the average, greatest during the sunspot maxima and least during the minima. The annual and daily cyclic variations are impressed on the 11-year cycle. It is quite possible that even longer periodic sunspot variations occur, resulting in even longer ionospheric variations. We have not been able to measure the solar flux and sunspot numbers for a long enough time, however, to find such a cycle with certainty. *See also* PROPAGATION CHARACTERISTICS, SOLAR FLUX, SUNSPOT CYCLE.

CYCLIC SHIFT

A cyclic shift is a transfer of information in a storage register in one direction or the other, usually called the right or the left. In a cyclic shift toward the left, each digit or bit is moved one place toward the left, except for the extreme left-hand digit or bit, which replaces the one originally at the far right (A in illustration). In a cyclic shift toward the right, each digit or bit is moved one position to the right, except for the extreme right-hand digit or bit, which replaces the one on the far left, as at B.

In an n-bit register, a succession of n cyclic shifts in the same direction results in the original information. Also, if m cyclic shifts are performed in one direction, followed by or combined with m cyclic shifts in the opposite direction, where m is any positive integer, the initial storage is obtained. *See also* SHIFT REGISTER.

CYCLOTRON

A cyclotron is a particle accelerator used for the purpose of atom smashing. This process can change one element to another. A unipolar magnetic field is used to force charged atomic particles, such as electrons, ions, protons, or nuclei of heavier atoms, into spiral orbits within two metal chambers. The drawing shows a simplified diagram of a cyclotron. Because of their shape, resembling the capital letter D, the chambers are called "dees."

The charged particles are injected into the machine at the center. As the particles move outward in a spiral path that increases in radius, their angular frequency, or time to complete one revolution, remains constant or decreases only slightly. Therefore, their speed greatly increases. The beam of particles is ejected from the cyclotron at extreme speed, often a large fraction of the speed of light, so that relativistic effects occur. Such effects include changes in mass and spatial dimensions. *See also* PARTICLE ACCELERATOR.

CZOCHRALSKI METHOD

Semiconductor materials, such as silicon and germanium, are often obtained by "crystal growing." A tiny piece of the material, called a seed, is placed into a molten bath of the same substance. The seed crystal is slowly rotated in this bath, and the crystal increases in size over a period of time. This produces a large single crystal. This technique for producing large semiconductor crystals is called the Czrochralski method. *See also* GERMANIUM, SILICON.

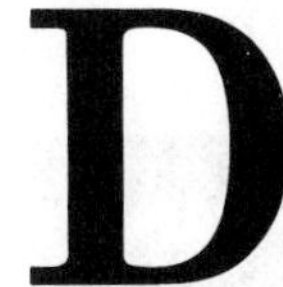

DAISY-WHEEL PRINTER

A daisy-wheel printer is a high-speed mechanical printing device used in electronic computers and typewriters. A circular type wheel, consisting of several dozen radial spokes, each with one character molded to its end face, rotates rapidly, then stops so that the proper character is at the top. Then, a hammer strikes the spoke from behind, pressing it against the back of the printing ribbon and onto the page. A typical daisy-wheel printer, used with an electronic typewriter, is shown at A. The printing rate of a daisy-wheel device varies, but is generally from 10 to 15 characters per second. The wheel itself is shown at B.

A daisy-wheel printer has the advantage of high reliability and relatively simple operation. Type wheels are easily interchanged when a different format or character style is desired. *See also* PRINTER.

DAMPED WAVE

A damped wave is an oscillation whose amplitude decays with time, as shown in the drawing. The damping may take place rapidly, in a few microseconds, to the point where the wave amplitude is essentially zero; the damping may also occur slowly, over a period of milliseconds or even several seconds. Damping can take place within the time of one cycle or less, or it can occur over a period of millions of cycles. Generally, the higher the Q factor in a circuit (*see* Q FACTOR), the more cycles occur before the signal amplitude deteriorates to essentially zero.

The decay in a damped wave occurs in the form of a logarithmic function, called a logarithmic decrement. *See also* CONTINUOUS WAVE, LOGARITHMIC DECREMENT.

DAMPING

Damping is the prevention of overshoot in an analog meter device, or the nature in which the needle comes to rest at a particular reading. Overshoot is generally undesirable, especially when meter readings change often. This is because overshoot creates confusion as to what the actual meter reading is. The greater the damping, the more slowly the meter needle responds to a change in current. The ideal amount of damping in a meter for a given application is called the critical damping (*see* CRITICAL DAMPING). If the damping is in excess of the critical damping, the meter will not respond fast enough to changes in the current through it.

Any absorption in a circuit, tending to reduce the amount of stored energy, is called damping. A resistor placed in a tuned circuit to lower the Q factor, for example, constitutes damping. This tends to reduce the chances of oscillation in a high-gain, tuned amplifier circuit. Mechanical resistance can be built into a transducer, such as an earphone or microphone, to limit the frequency response. This also is called damping. *See also* Q FACTOR.

DAMPING FACTOR

In a high-fidelity sound system, the actual output impedance of the amplifier may be much smaller than the impedance of the speaker. The ratio of the speaker impedance, which is usually 4, 8, or 16 ohms, to the amplifier output impedance, which is often less than 1 ohm, is called the damping factor.

The effect of this difference in impedance is to minimize the effects of speaker acoustic resonances. The sound output should not be affected by such resonances in a high-fidelity system. The frequency response should be as flat as is practicable. The damping factor in a high-fidelity system is somewhat dependent on the frequency of the audio energy. It also is a function of the extent of the negative feedback in the audio amplifier circuit. Damping factors in excess of 60 are quite common.

In a damped oscillation, the quotient of the logarithmic decrement and the oscillation period is sometimes called the damping factor. In a damped-wave circuit, where the coil inductance is given by L and the radio-frequency resistance is given by R, the damping factor, a, is defined as

$$a = \frac{R}{2L}$$

See also DAMPED WAVE, LOGARITHMIC DECREMENT.

DAMPING RESISTANCE

If the Q factor in a resonant circuit becomes too great (*see* Q FACTOR), an undesirable effect called ringing may occur. Ringing is especially objectionable in audio filters used in radioteletype demodulators and in code communications. To reduce the Q factor sufficiently, a resistor, called a damping resistor, is placed across a parallel-resonant circuit. Such a resistor may also be placed in series with a series-resonant circuit. In a parallel-resonant circuit, the Q factor decreases as the shunt resistance decreases. In a series-resonant circuit, the Q factor decreases as the series resistance increases. The lower the Q factor, the less the tendency for the resonant circuit to ring.

Damping resistance is an expression sometimes used in reference to a noninductive resistor placed across an analog meter to increase the damping. *See also* CRITICAL DAMPING, DAMPING.

DANIELL CELL

A Daniell cell is a wet cell that delivers approximately 1.1 volt. A zinc electrode, which develops a negative charge, and a copper electrode, which develops a positive charge, are placed in electrolyte solutions as illustrated in the drawing. The zinc electrode is put in a container with zinc sulfate. The copper electrode is put in a solution of copper sulfate. The container with the zi c-sulfate solution allows electrical conduction, but prevents the two electrolyte solutions from mixing.

Because the Daniell cell is a wet cell, it cannot be tipped or inverted. If this were to happen, the two electrolyte solutions would mix, and the cell operation would be upset. *See also* CELL.

DARAF

The daraf is the reciprocal unit of the farad. The word

A

B

DAISY-WHEEL PRINTER: At A, daisy-wheel printer, used with an electronic typewriter, At B, the wheel itself, showing the spokes on which the raised alphanumeric characters are molded.

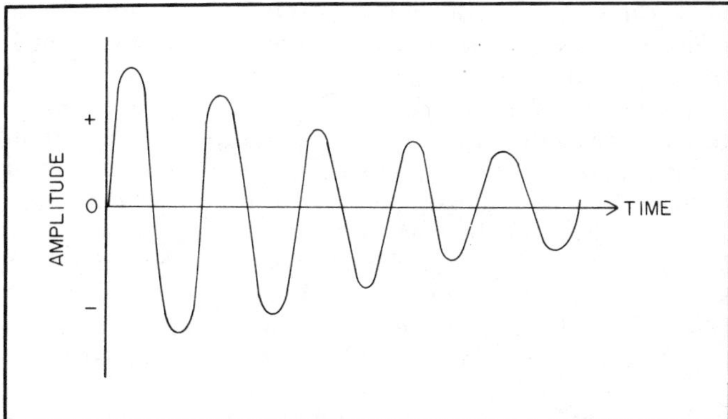

DAMPED WAVE: A damped wave decreases steadily in amplitude, according to a logarithmic function.

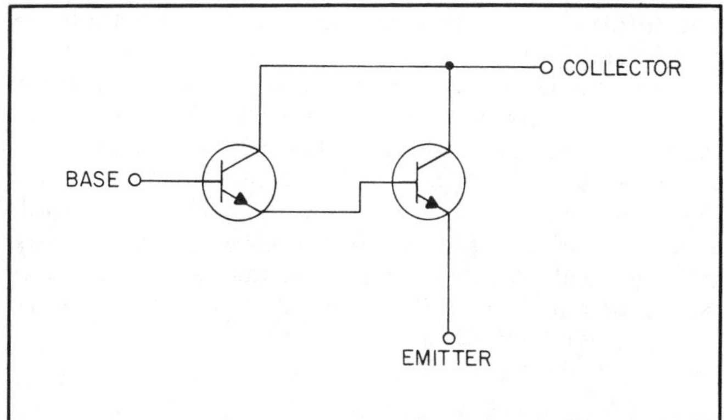

DARLINGTON AMPLIFIER: The Darlington amplifier consists of two bipolar transistors connected as shown here.

daraf is, in fact, farad spelled backwards! A value of 1 daraf is the reciprocal equivalent of 1 farad. The quantity 1/C, where C is capacitance, is called elastance (*see* ELASTANCE).

Generally, values of capacitance in practical circuits are much less than 1 farad. A capacitance of 1 microfarad, or 10^{-6} farad, corresponds to an elastance of 1 megadaraf, or 10^{6} daraf. A capacitance of 1 picofarad, or 10^{-12} farad, is an elastance of 1 teradaraf, or 10^{12} daraf.

DARLINGTON AMPLIFIER

A Darlington amplifier, or Darlington pair, is a form of compound connection between two bipolar transistors (*see* illustration). In the Darlington amplifier, the collectors of the transistors are connected together. The input is supplied to the base of the first transistor. The emitter of the first transistor is connected directly to the base of the second transistor. The emitter of the second transistor serves as the emitter for the pair. The output is generally taken from both collectors.

The amplification of a Darlington pair is equal to the product of the amplification factors of the individual transistors as connected in the system. This does not necessarily mean that a Darlington amplifier will produce far more gain than a single bipolar transistor in the same circuit. The impedances must be properly matched at the input and output to ensure optimum gain. Some Darlington pairs are available in a single case. Such devices are called Darlington transistors. The Darlington amplifier is sometimes called a double emitter follower or a beta multiplier. *See also* TRANSISTOR.

D'ARSONVAL

A D'Arsonval meter, or D'Arsonval movement, is a device used in analog monitoring. The current to be measured is passed through a coil that is attached to an indicating needle. The coil is operated within the field of a permanent magnet. As a current passes through the coil, a magnetic field is set up around the coil, and a torque appears between the permanent magnet field and the field of the coil. A spring allows the coil, and hence the pointer, to rotate only a certain angular distance; the greater the current, the stronger the torque, and the farther the coil turns. See drawing.

Generally, the coil in a D'Arsonval meter is mounted on jeweled bearings for maximum accuracy. D'Arsonval meters are widely employed as ammeters, milliammeters, and microammeters. With suitable peripheral circuitry, D'Arsonval meters are used as voltmeters and wattmeters as well, both for direct current and alternating current at all frequencies. *See also* AMMETER, ANALOG METERING.

DATA

The term data is generally used in place of the word "information" in electronic and computer applications.

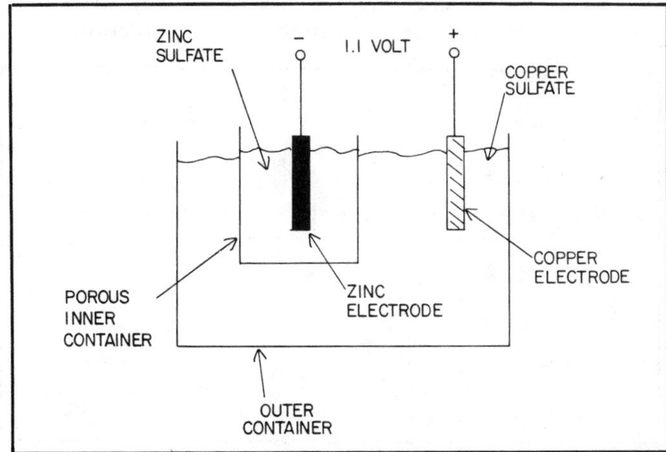

DANIELL CELL: A Daniell cell is a form of wet cell using zinc and copper electrodes.

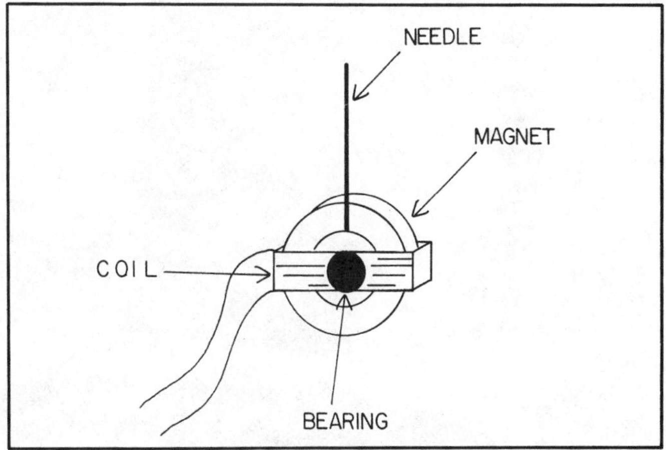

D'ARSONVAL: The D'Arsonval movement is the most common system in analog meters. Magnetic fields between the coil and the permanent magnet cause the needle to turn.

The information stored and handled by a computer is called data.

Any alphabetic-numeric sequence representing some identifiable quantity is regarded as data. This may be, for example, the notation showing that the high temperature on August 5 was 85 degrees Fahrenheit. Data differs from algebraic variables and constants (*see* CONSTANT, VARIABLE). Symbols giving instructions, or indicating mathematical operations are not data. In a digital computer, all data is stored and handled as sets of binary digits, either 0 or 1.

DATA-ACQUISITION SYSTEM

A data-acquisition system is a system intended for the purpose of gathering data for storage or manipulation. Such a system consists of a pickup device and a transmission device, which delivers the data to the memory or user.

Data may be stored in a variety of different ways. If the data is transmitted to a tape recorder, the signals can often be recorded directly from the transmission line. If the data is sent to a printer, the information can be conveniently stored on paper. If the data is sent to a computer, the data may be stored on magnetic disks, in core-memory devices, or in integrated circuits. The computer may, alternatively, act immediately on the data.

An example of data acquisition, perhaps in its most familiar form, is the business computer. Such a computer is shown in the illustration. The operator puts the data into the computer by means of a typewriter keyboard. Later, the information can be retrieved for use. *See also* DATA TRANSMISSION, MEMORY.

DATA COMMUNICATION

Data communication is the transfer of data in both directions between two points, or in all possible ways among three or more points. Each station must have a transmitting and receiving device.

Ideally, the data should arrive at a receiving station in exactly the same form as it is transmitted. Interference in the system occasionally changes the data as it is picked up by the receiver. However, some sophisticated data-communications systems can tell when something appears to be out of place; the receiving station will then ask the transmitting station to repeat that part of the data in which the apparent error has occurred.

Data communication is generally carried out in binary form, since this provides a better signal-to-noise ratio than analog methods. The alphabetic-numeric format of most data lends itself well to digital communications techniques. The impulses may be tranferred via radio, optical fibers, lasers, or landline (wire). Sometimes the data is coded, or scrambled, at the transmitting end and decoded at the receiving end. This reduces the chances of interception by unwanted parties. *See also* DATA, DATA TRANSMISSION.

DATA-ACQUISITION SYSTEM: A data-acquisition system often consists of a keyboard and video-display unit.

DATA CONVERSION

When data is changed from one form to another, the process is called data conversion. Data may, for example, be sent in serial form, bit by bit, and then be converted into parallel form for use by a certain circuit (*see* PARALLEL DATA TRANSFER, SERIAL DATA TRANSFER). Data may initially be in analog form, such as a voice or television-picture signal, and then be converted to digital form for transmission (see illustration). In the receiving circuit, the digital data may then be converted back into analog data (*see* ANALOG, ANALOG-TO-DIGITAL CONVERTER, DIGITAL, DIGITAL-TO-ANALOG CONVERTER).

Data conversion is performed for the purpose of improving the efficiency and/or accuracy of data transmission. For example, the digital equivalent of a voice signal is propagated with a better signal-to-noise ratio, in general, than the actual analog signal. It therefore makes sense to convert the signal to digital form when accuracy is of prime importance.

DATA PROCESSING

The work actually performed on data, once the data has been acquired (*see* DATA ACQUISITION SYSTEM), is called data processing. This may consist of calculations, organization of information, or both. Redundant or unnecessary data is eliminated, and the data is organized for the most efficient use.

Data processing requires a skilled operator as well as a computer. Programs must be written to provide efficient data processing. This has created many new jobs in recent years. Data processing is, in fact, current a major job market in the United States and other developed countries.

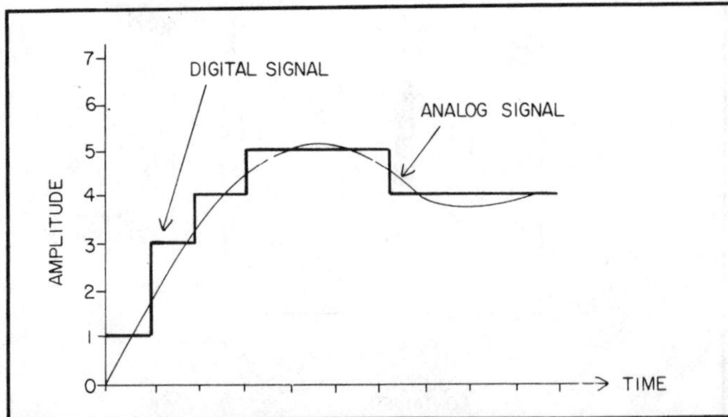

DATA CONVERSION: An analog signal, shown by the smooth curve, can be converted into digital form as shown here. The digital signal has eight different amplitude levels.

Data processing is used extensively by businesses to monitor cash flow, to maintain accurate, detailed, and organized records, and to maximize the efficiency of company operations. Most corporations now have a computer for this reason. A special computer language called COBOL, short for common business-oriented language, is designed especially for business data processing. Computers and data processing are also used extensively by government. *See also* COBOL.

DATA SIGNAL

Data signals are modulated waves, or pulse sequences, used for the purpose of transmitting data from one place to another. A data signal may be either analog or digital in nature (*see* ANALOG, DIGITAL). Analog data signals consist of a variable parameter, such as amplitude, frequency, or phase, that fluctuates over a continuous range. Theoretically, an analog data signal can have an infinite number of different states. A digital signal has only a few possible states, or levels. A television broadcast signal is an example of an analog data signal. A video-display terminal, showing only alphabetic-numeric characters, is an example of a device that uses a digital data signal. Other examples of digital data signals are the Morse code, and radioteletype frequency-shift keying.

Data signals may be transmitted in sequential form, that is, one after the other; or they may be sent in bunches. The former method of sending data is called serial transmission. The latter method is called parallel transmission. There are advantages and disadvantages to both methods. *See also* DATA TRANSMISSION, PARALLEL DATA TRANSFER, SERIAL DATA TRANSFER.

DATA TRANSDUCER

A data transducer is a device that converts non-electrical phenomena into electrical data for use by test instruments or computers. The non-electrical quantity may be light intensity or spectral color, sound, magnetic field intensity, radio wavelength, temperature, or any of various other parameters. In the aforementioned situations, the transducer may be a photocell, spectroscopic device, simple microphone, or frequency meter.

Perhaps the most common example of a data transducer is the optical price-code reader that has become increasingly employed in retail stores, especially grocery stores. This device reads the price from a series of parallel dark lines having variable spacing and thickness. This allows rapid, automatic adding of a bill. The illustration shows an example of this kind of price-code tag. An optical data transducer reads this pattern of lines, and enters the price into a small computer in the cash register.

DATA TRANSDUCER: A data transducer can quickly record the price of a piece of merchandise by optical means, from a pattern of lines such as this.

DATA TRANSMISSION

Data is sent from one place to another by means of a process called data transmission. Data may be transmitted via wires, beams of light, radio-frequency energy, sound, or simple direct-current impulses.

Data transmission is generally classified as either analog or digital. Analog data transmission fluctuates over a given range in terms of some parameter such as amplitude, frequency, or phase. Digital data has a finite number of discrete states, such as on/off or various precisely defined levels. Digital or analog data can be categorized as either serial or parallel. Serial data is transmitted sequentially; parallel data is sent several bits at a time.

There are advantages and disadvantages to analog and digital data-transmission modes. While analog data transmission allows a greater level of reproduction accuracy, and even helps carry meaning via inflection (such as tone of voice), digital transmission is generally faster and more efficient, especially under marginal conditions. Serial data transmission requires only one line or signal, but the transfer rate is comparatively slow. Parallel data transmission, although it necessitates the use of several lines or channels at once, is much faster. *See also* ANALOG, DATA, DIGITAL, PARALLEL DATA TRANSFER, SERIAL DATA TRANSFER.

dB

See DECIBEL.

dBa

The abbreviation dBa stands for adjusted decibels. Adjusted decibels are used to express relative levels of noise. First, a reference noise level is chosen, and assigned the value 0 dBa. All noise levels are then compared to this value. Noise levels lower than the reference level have negative values, such as −3 dBa. Noise levels greater than the reference level have positive values, such as +6 dBa.

The decibel is a means of expressing a ratio between two currents, power levels, or voltages. A reference level is therefore always necessary for the decibel to have meaning. *See also* DECIBEL.

dBd

The acronym dBd refers to the power gain of an antenna, in decibels, with respect to a half-wave dipole antenna. The dBd specification is the most common way of expressing antenna power gain (*see* DECIBEL, DIPOLE ANTENNA). The reference direction of the antenna under test is considered to be the direction in which it radiates the most power. The reference direction of the dipole is broadside to the antenna conductor.

Power gain in dBd is given by the formula:

$$\text{dBd} = 10 \log_{10} (P_a/P_d)$$

where P_a is the effective radiated power from the antenna in question with a transmitter output of P watts, and P_d is the effective radiated power from the dipole with a transmitter input of P watts.

An alternative method of measuring antenna power gain in dBd is possible using the actual field-strength values. If E_a is the field strength in microvolts per meter at a certain distance from the antenna in question, and E_d is the field strength at the same distance from a half-wave dipole getting the same amount of transmitter power, then

$$\text{dBd} = 20 \log_{10} (E_a/E_d)$$

See also ANTENNA POWER GAIN, dBi.

dBi

The acronym dBi refers to the power gain of an antenna, in decibels, relative to an isotropic antenna (*see* DECIBEL, ISOTROPIC ANTENNA). The direction is chosen in which the antenna under test radiates the best. An isotropic antenna, in theory, radiates equally well in all directions. The gain of any antenna in dBi is 2.15 dB greater than its gain in dBd; that is,

$$\text{dBi} = 2.15 + \text{dBd}$$

Power gain in dBi is given by the formula:

$$\text{dBi} = 10 \log_{10} (P_a/P_i)$$

where P_a is the effective radiated power from the antenna in question with a transmitter output of P watts, and P_i is the effective radiated power from the isotropic antenna with a transmitter output of P watts.

An alternative method of measuring antenna power gain in dBi is possible using actual field-strength values. If E_a is the field strength in microvolts per meter at a certain distance from the tested antenna, and E_i is the field strength from an isotropic antenna getting the same amount of power from the transmitter, then

$$\text{dBi} = 20 \log_{10} (E_a/E_i)$$

Actually, an isotropic antenna is not seen in practice. It is essentially impossible to construct a true isotropic antenna. Gain figures in dBi are sometimes used instead of dBd for various reasons. *See also* ANTENNA POWER GAIN, dBd.

dBm

The acronym dBm refers to the strength of a signal, in decibels, compared to 1 milliwatt, with a load impedance of 600 ohms. If the signal level is exactly 1 milliwatt, its level is 0 dBm. In general,

$$\text{dBm} = 10 \log_{10} P$$

where P is the signal level in milliwatts.

With a 600-ohm load, 0 dBm represents 0.775 volt, or 775 millivolts. With respect to voltage in a 600-ohm system, then,

$$\text{dBm} = 10 \log_{10} (E/775)$$

where E is the voltage in millivolts. A level of 0 dBm also represents a current of 1.29×10^{-3} ampere, or 1.29 milliamperes. With respect to current in a 600-ohm system:

$$\text{dBm} = 20 \log_{10} (I/1.29)$$

where I is the current in milliamperes. *See also* DECIBEL.

DC

See DIRECT CURRENT.

DC AMPLIFIER

Any device intended to increase the current, power, or voltage in a direct-current circuit is called a direct-current, or dc, amplifier. The most common type of dc amplifier is used for the purpose of increasing the sensitivity of a meter or other indicating device. Such an amplifier can be extremely simple, resembling an elementary alternating-current amplifier, as shown in the diagram.

Direct-current amplifiers may be used to amplify the voltage in an automatic-level-control circuit, for the purpose of accomplishing speech compression. In such an application, the time constant of the dc amplifier is critical. The microphone amplifiers are set to a high level of gain when there is no audio input; the greater the audio signal fed to the amplifiers, the more dc is supplied by the dc amplifiers acting on the automatic-level-control voltage. This dc voltage reduces the gain of the amplifying stages as the audio input increases. *See also* AMPLIFICATION, AMPLIFIER, AUTOMATIC LEVEL CONTROL, SPEECH COMPRESSION.

DC COMPONENT

All waveforms have a direct-current, or dc, component and an alternating-current, or ac, component. Sometimes one component is zero. For example, in pure dc, such as the output of a dry cell, the ac component is zero (see A in

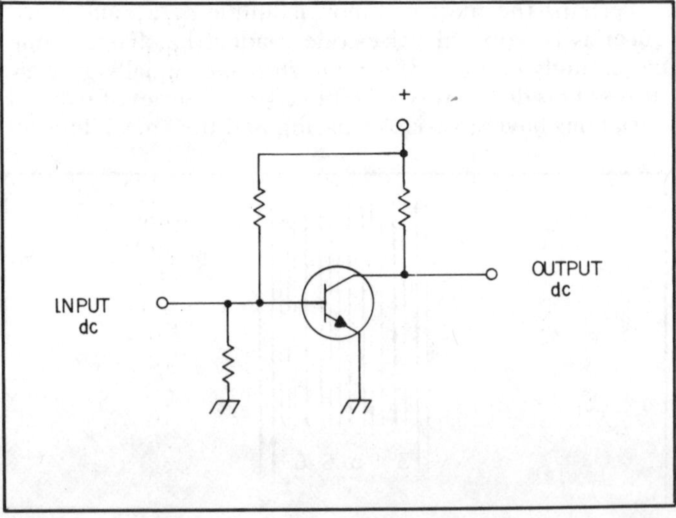

DC AMPLIFIER: A direct-current amplifier circuit may be very simple, resembling an ordinary alternating-current amplifier.

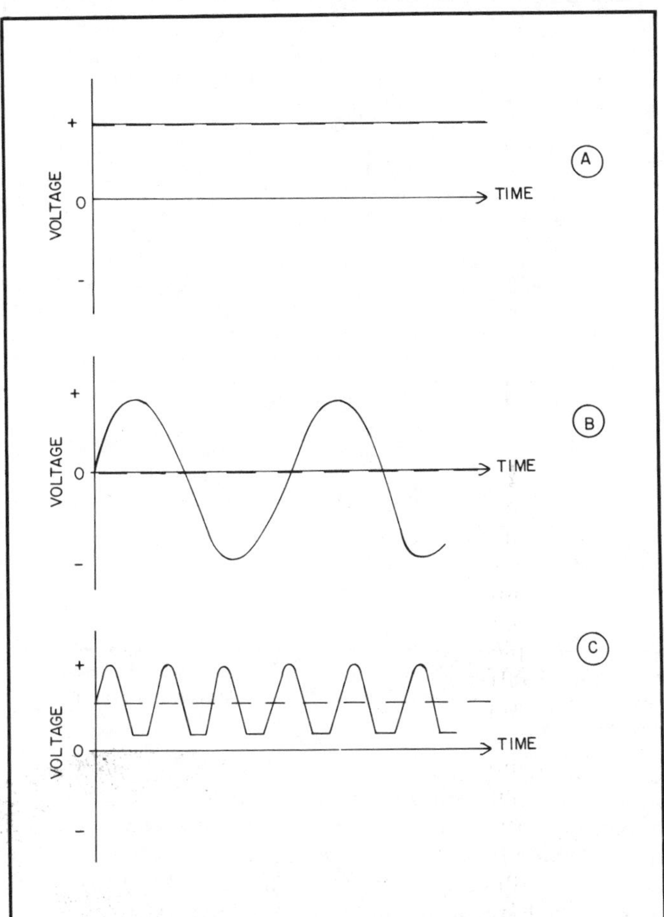

DC COMPONENT: At A, pure dc is exactly equivalent to its dc component. At B, pure ac has a dc component of zero. At C, pulsating dc, showing its dc component.

illustration). In a 60-Hz household outlet, the dc component is zero, because the average voltage from such a source is zero (B in illustration).

In a complex waveform, such as that illustrated at C, the dc component is the average value of the voltage. This average must be taken over a sufficient period of time. Some waveforms, such as the voltage at the collector of any amplifier circuit, have significant dc components. The dc component does not always change the practical characteristics of the signal, but the dc component must be eliminated to obtain satisfactory circuit operation in some situations. *See also* ALTERNATING CURRENT, DIRECT CURRENT.

DC ERASING HEAD

In a magnetic tape recorder, the dc erasing head is the head supplied with a pure direct current for the sole purpose of erasing all the magnetic impulses on the tape.

As the tape moves through the machine in the record mode, the dc erasing head is encountered first. All previously recorded information is removed prior to the tape arrival at the recording head. This ensures that the tape will contain the lowest possible amount of background noise, and that the old information will not interfere with the new. It is essential that the erasing head be supplied with unmodulated direct current. The dc erasing head is, of course, disabled in the playback mode. *See also* MAGNETIC RECORDING, MAGNETIC TAPE, TAPE RECORDER.

DC GENERATOR

A direct-current, or dc, generator is a source of direct current. A dc generator may be mechanical in nature, such as an alternating-current generator followed by a rectifier. A dc generator might consist of a chemical battery, a photovoltaic cell or thermocouple. The direct-current amplitude may remain constant, or it may fluctuate.

Direct-current generators are commonly used for a variety of purposes. The most common example is the dry cell, which generates electricity from a chemical reaction. Solar cells are another common type of dc generator. *See also* GENERATOR.

DC GROUND

A direct-current ground, or dc ground, is a dc short circuit (*see* DC SHORT CIRCUIT) to ground potential. Such a dc short circuit is provided by connecting a circuit point to the chassis of a piece of electronic equipment, either directly or through an inductor.

It is often desirable to place a component lead at dc ground potential, but still apply an alternating-current signal. An example of this is the zero-bias vacuum tube, in which the control grid is at dc ground, but which carries a large driving signal. Another example is the grounding of an antenna system through large inductors. This prevents the buildup of hazardous dc voltages on the antenna system, but does not interfere with the radiation or reception of radio-frequency signals.

DC POWER

Dc power is the rate at which energy is expended in a direct-current circuit. It is equal to the product of the direct-current voltage, E, and the current, I. This may be determined at one particular instant, or as an average value over a specified period of time. If R is the direct-current resistance in a circuit, and P is the direct-current power, then:

$$\begin{aligned} P &= EI \\ &= E^2/R \\ &= I^2R \end{aligned}$$

when units are given in volts, amperes, and watts for voltage, current, and power, respectively.

Direct-current energy is the average direct-current power multiplied by the time period of measurement. The standard unit of energy is the watt hour, although it may also be specified in watt seconds, watt minutes, kilowatt hours, or other variations. If W is the amount of energy expended in watt hours, then:

$$\begin{aligned} W &= Pt \\ &= EIt \\ &= E^2t/R \\ &= I^2Rt \end{aligned}$$

where t is the time in hours. *See also* ENERGY, POWER.

DC POWER SUPPLY

A direct-current, or dc, power supply is either a dc

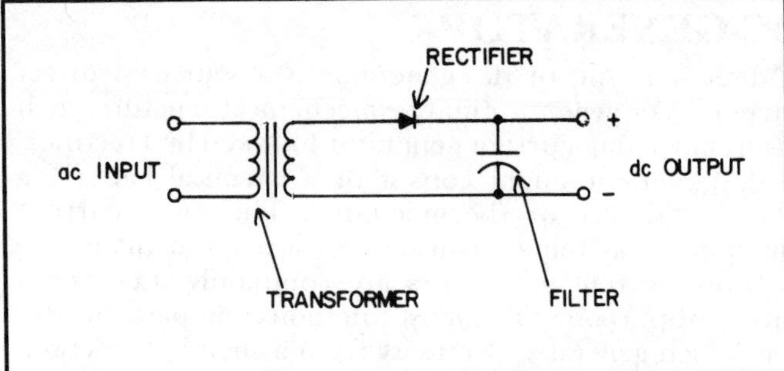

DC POWER SUPPLY: A simple half-wave dc power supply.

generator (*see* DC GENERATOR), or a device for converting alternating current to direct current for the purpose of operating an electronic circuit. Generally, the term dc power supply refers to the latter type of device.

A typical dc power supply circuit is shown in the schematic diagram. A transformer provides the desired voltage in alternating-current form. The semiconductor diode rectifies this alternating current. The capacitor smooths out The pulsations in the direct-current output of the rectifier.

The circuit shown is an extremely simple dc power supply. Many direct-current supplies have voltage-regulation devices, highly sophisticated rectifier circuits, current-limiting circuits, and other features. Some dc power supplies deliver only a few volts at a few milliamperes; others can deliver thousands of volts and/or hundreds of amperes. *See also* CURRENT LIMITING, RECTIFICATION, RECTIFIER CIRCUITS, TRANSFORMER, VOLTAGE REGULATION.

DC SHORT CIRCUIT

A direct-current, or dc, short circuit is a path that offers little or no resistance to direct current. A resistance may or may not be present in such a situation for alternating currents.

The simplest example of a direct-current short circuit is, of course, a length of electrical conductor. However, an inductor also provides a path for direct current. But an inductor, unlike a plain length of conductor, offers reactance to alternating-current energy. (*See* INDUCTIVE REACTANCE.) Coils are often used in electronic circuits to provide a dc short circuit while offering a high amount of resistance to alternating-current signals. Such coils are called chokes. When it is necessary to place a circuit point at a certain dc potential, without draining off the ac signal, a choke is used. Chokes also can be employed to eliminate the alternating-current component of a dc power-supply output. *See also* CHOKE, CHOKE-INPUT FILTER, DC POWER SUPPLY.

DC-TO-AC CONVERTER

A dc-to-ac (direct-current-to-alternating-current) converter is a form of power supply, often used for the purpose of obtaining household-type power from a battery or other source of low-voltage direct current. The chopper power supply makes use of a dc-to-ac converter (*see* CHOPPER POWER SUPPLY).

A dc-to-ac converter operates by modulating, or interrupting, a source of direct current. A relay or oscillating circuit is employed to accomplish this. The resulting modulated direct current is then passed through a transformer to eliminate the direct-current component, and to get the desired alternating-current voltage. A simple schematic diagram of a dc-to-ac converter is shown in the illustration.

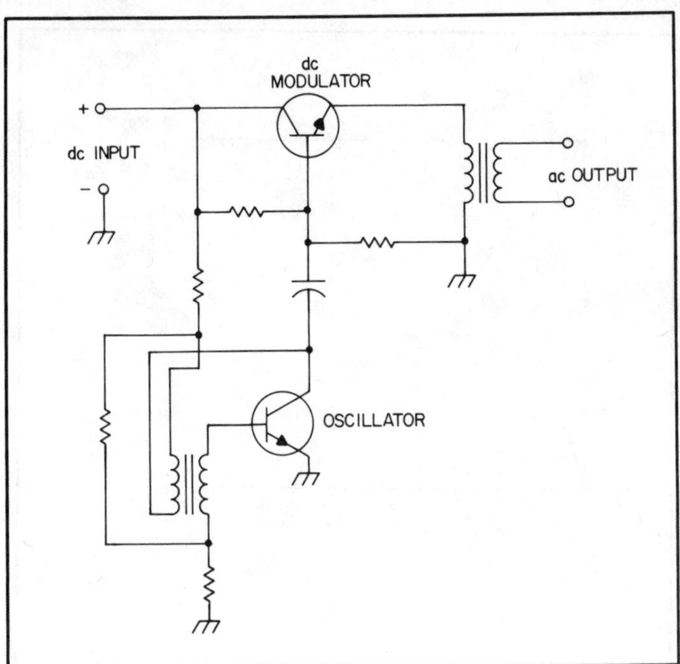

DC-TO-AC CONVERTER: A dc-to-ac converter consists of an oscillator and a chopper or modulator, followed by a transformer to eliminate the dc component.

When a dc-to-ac converter is designed especially to produce 120-volt, 60-Hz alternating current for the operation of household appliances, the device is called a power inverter. *See also* INVERTER.

DC-TO-DC CONVERTER

A dc-to-dc converter is a circuit that changes the voltage of a direct-current power supply. Such a device consists of a modulator, a transformer, a rectifier, and a filter, such as the circuit shown in the illustration. This circuit is similar to that of a dc-to-ac converter, except for the addition of the rectifier and filter (*see* DC-TO-AC CONVERTER).

A dc-to-dc converter may be used for either step-up or step-down purposes. Usually, such a circuit is used to obtain a high direct-current voltage from a comparatively low voltage. A fairly common type of dc-to-dc converter is used as a power supply for vacuum-tube equipment when the only available source of power is a 12-volt automotive battery or electrical system. The regulation of such a voltage step-up circuit depends on the ability of the battery or car alternator to handle large changes in the load current. A special regulator circuit is required if the voltage regulation must be precise.

A dc-to-dc converter is sometimes called a dc transformer. Low-power dc-to-dc converters can be built into a small integrated-circuit package.

DC TRANSDUCER

A dc, or direct-current, transducer is a device that converts direct-current energy into some other form, such as light,

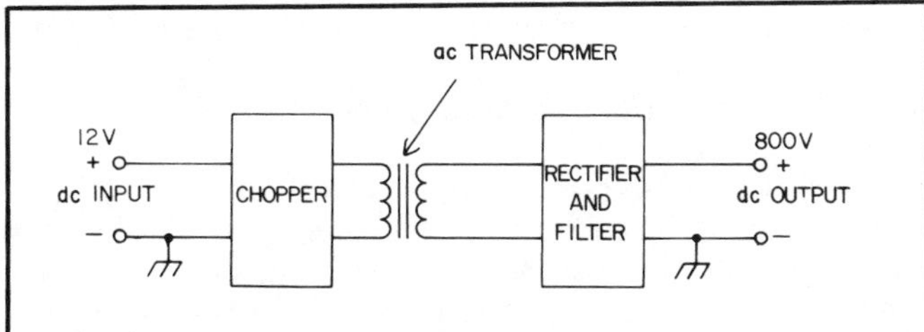

DC-TO-DC CONVERTER: A dc-to-dc converter block diagram. This circuit is essentially the same as a dc-to-ac converter, with the addition of a rectifier and filter.

heat, sound, or electromagnetic radiation. An ordinary incandescent bulb is a simple form of direct-current transducer. A typical buzzer is another example.

A device that converts a non-electrical form of energy into direct current is also called a dc transducer. A photovoltaic cell (*see* PHOTOVOLTAIC CELL) is a common type of dc transducer. A component that requires direct current for its operation, and modulates this direct current, is sometimes called a dc transducer. An example of this kind of device is the ordinary carbon microphone (*see* CARBON MICROPHONE).

DC TRANSMISSION

When electric power is sent from one point to another as a direct current, the method of transfer is called direct-current, or dc, transmission. Most power-transmission lines today carry alternating current. However, direct-current transmission, especially for long lines, is being used in many areas. Thomas Edison, the inventor of many electronic devices, advocated direct-current power transmission in all situations.

Alternating current is fairly easy to work with, since its voltage can be stepped up or down at will by the use of transformers. When power must be sent over long distances, high voltages are used, since higher voltages result in lower losses (*see* HIGH-TENSION POWER LINE). But at any voltage, direct current suffers less attenuation than alternating current, because transmission-line losses always increase with increasing frequency. The primary obstacle to dc transmission is the difficulty in converting extremely high voltages from alternating to direct form and vice-versa. There is another advantage to dc transmission, however; in a direct-current line, the electric and magnetic fields do not fluctuate. This greatly reduces the emission of electromagnetic energy, which can create problems near high-voltage power lines. The only electromagnetic radiation from a line carrying pure direct current would be the result of occasional transient spikes.

DEAD BAND

When no signals are received within a certain frequency range in the electromagnetic spectrum, that band of frequencies is said to be dead. A dead band can result from geomagnetic activity that disrupts the ionosphere of the earth; in fact, the term dead band is used only on those frequency bands that are affected by ionospheric propagation. During a severe geomagnetic storm, propagation becomes virtually impossible via the ionosphere. The illustration shows a hypothetical situation in which propagation deteriorates, making the band sound dead (*see* GEOMAGNETIC FIELD, IONOSPHERE).

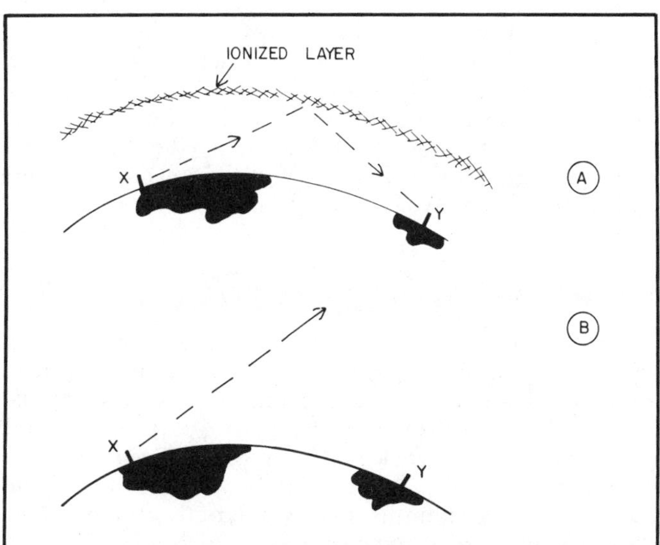

DEAD BAND: At A, the ionosphere provides communication between two points X and Y. When the ionization disappears, the signal can no longer be propagated, resulting in a dead band between the two points.

A dead band can be caused by the deterioration of the ionosphere with the setting of the sun, with low sunspot activity, and perhaps with coincidences of unknown origin.

A band may sometimes appear dead simply because no one is transmitting on it at a particular time. Amateur radio operators, when communicating for recreation, have sometimes listened to what they thought was a dead band, called CQ (*see* CQ), and found that the band was far from dead!

A band can go dead for just a few seconds or minutes, or it may remain unusable for hours or days. The range of frequencies can be as small as a few kilohertz, or may extend for several megahertz. *See also* PROPAGATION CHARACTERISTICS.

The term dead band is also used to describe the lack of response of a servomotor system through a part of its arc or range of operation. This lack of response can be caused by backlash in the gears or rotor of the servo, or by a lack of resolution of the position-sensing potentiometer (or other device) that feeds angular or position information to the servo system. The servo dead band can be expressed as degrees of arc, or as a percentage of total travel.

DEAD ROOM

A dead room is a chamber in which no sound is reflected from the walls, ceiling, or floor, and into which no sound is conducted or propagated from the outside. Such a chamber requires the use of sophisticated techniques for sound insulation and absorption. A dead room may be suspended inside a larger chamber by cables, and the gap evacuated. Dead rooms are used to conduct acoustic tests.

We seldom experience the absence of all sound. Background noise is, to some extent, always present in the form of fans, appliances, passing cars, and the like. Inside a dead room, some people have described the silence as oppressive, and even unbearable. A dead room allows the accurate measurement of the true threshold of human hearing at all frequencies.

A chamber that is completely shielded against electromagnetic fields is sometimes called a dead room. A shielded enclosure of this kind is used for testing and measurement at radio frequencies, when the rf noise floor must be kept as low as possible, and all interference eliminated. *See also* FARADAY CAGE, NOISE FLOOR.

DE BROGLIE'S EQUATION

All moving particles, and in fact all moving objects, display properties of wavelength. In the early part of this century, a physicist named de Broglie (pronounced de Broily) discovered that the wavelength of a moving particle is related to the mass of the particle and to its speed. The greater the mass, the greater the frequency of the wave action, and the shorter the wavelength. The wavelength also becomes shorter with increasing velocity.

De Broglie's equation is given by:

$$\lambda = \frac{6.54 \times 10^{-27}}{mv}$$

where λ is the wavelength in centimeters, m is the rest mass of the object in grams, and v is the speed of the object in centimeters per second.

De Broglie's equation is most often applied to moving electrons. Since the mass of an electron at rest is 9.03×10^{-28} grams, the equation for electrons can be simplified to:

$$\lambda = 7.24/v$$

where the wavelength is measured in centimeters per second.

The above formula applies only for nonrelativistic speeds. When the velocity of a particle is a sizeable fraction of the speed of light, c, which is 3×10^{11} centimeters per second, then the de Broglie formula must be modified to account for relativistic effects. It becomes:

$$\lambda = \frac{6.54 \times 10^{-27} \sqrt{1 - v^2 / c^2}}{mv}$$

where the factor $\sqrt{1 - v^2 / c^2}$ is the proportion by which time is dilated at speed v. The relativistic effect results in a greater particle mass, and therefore a shorter wavelength than would be expected without taking the effect into account.

De Broglie waves have been observed for electrons and other moving particles. Theoretically, de Broglie waves occur when any object is placed in motion—even as you walk down the street.

DEBUGGING

The process of perfecting the operation of an electronic circuit or computer program is called debugging. Literally, this means getting the bugs out! Usually, when an electronic circuit is first tested after it has been built exactly according to the plans, or when a computer program is run after it has been meticulously composed, a problem becomes evident. Sometimes, the circuit or program fails to work at all. Only rarely does the circuit or program work perfectly the first time it is tested. Therefore, debugging is almost always necessary.

The process of debugging may be very simple; it might, for instance, involve only a small change in the value of a component. Sometimes, the debugging process requires that the entire design process be started all over. Occasionally, the bugs are hard to find, and do not appear until the device has been put into mass production or the program has been published and extensively used. A debugging test must thus be very thorough so that the chances of production problems are minimized. *See also* GREMLIN.

DEBYE LENGTH

As an electron approaches a positive ion, the electron is at first not influenced by the ion. But at a certain distance, the electric fields of the ion and electron begin to interact, and the motion of the electron is influenced. This distance, which varies depending on the speed and direction of the electron, the positive charge of the ion, and the number of ions and electrons in the vicinity, is called the Debye length. The illustration shows a hypothetical example of electron capture by an ion, showing the Debye length. The electron may not actually be caught by the ion, but the path is always influenced within the Debye length.

As an electron is emitted from an atom, forming an ion, the Debye length is that distance from the atom at which the electron can be considered free rather than bound.

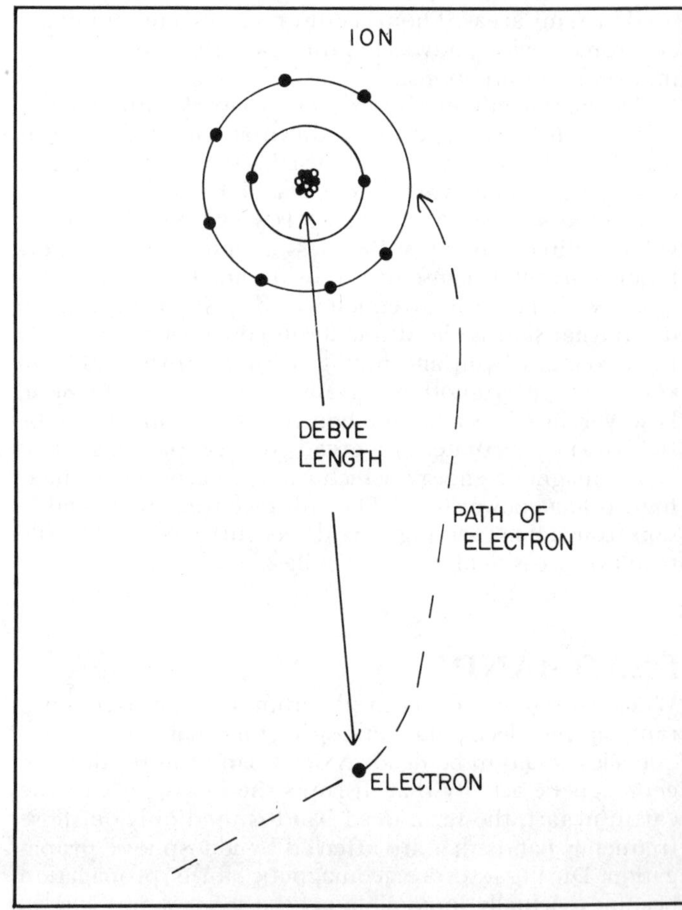

DEBYE LENGTH: The Debye length is the maximum distance at which an ion influences the path of an electron.

The Debye length may also be called the Debye radius, the Debye shielding distance, or the plasma length. *See also* BOUND ELECTRON, FREE ELECTRON, PLASMA.

DECADE

A range of any parameter, such that the value at one end of the range is ten times the value at the other end, is called a decade. The radio-frequency bands are arbitrarily designated as decades: 30 to 300 kHz is called the low-frequency band, 0.3 to 3 MHz is called the medium-frequency band, 3 to 30 MHz is called the high-frequency band, and so on.

There are an infinite number of decades between any quantity and the zero value for that parameter. For example, we may speak of frequency decades of 30 to 300 kHz, 3 to 30 kHz, 0.3 to 3 kHz, 30 to 300 Hz, and so on, without end, and we will never actually reach a frequency of zero. Of course, a parameter may be increased indefinitely, too, without end.

The decade method of expressing quantities is used by scientists and engineers quite often, because its logarithmic nature allows the evaluation of a larger range of quantities than is the case with a simple linear system. *See also* ORDER OF MAGNITUDE, SCIENTIFIC NOTATION.

DECADE BOX

A decade box is a device used for testing a circuit. A set of resistors, capacitors, or inductors is connected together via switches in such a way that values can be selected digit-by-digit in decade fashion.

The illustration shows a schematic diagram of a two-digit decade capacitance box. (Usually, more digits are provided, but for simplicity, this circuit shows only two.) Switch S1 selects any of ten capacitance values in microfarads: 0.00, 0.01, 0.02, 0.03, . . ., 0.09. Switch S2 also selects any of ten values of capacitance in microfarads, each ten times the values of capacitance in the circuit containing S1: 0.0, 0.1, 0.2, 0,3, . . ., 0.9. Therefore, there are 100 possible values of capacitance that can be selected by this system, ranging from 0.00 to 0.99 uF in increments of 0.01 uF.

Decade boxes are sometimes used to set the frequency of a digital radio receiver or transmitter. A signal generator and monitor may have a range of zero to 999.999999 MHz or higher, with frequencies selectable in increments as small as 1 Hz. This gives as many as 10^9 possible frequencies, with only nine independent selector switches. *See also* DECADE.

DECADE COUNTER

A counter that proceeds in decimal fashion, beginning at zero and going through 9, then to 10 and up to 99, and so on, is called a decade counter. A decade counter operates in the familiar base-10 number system. Counting begins with the ones digit (10^0), then the tens digit (10^1), and so on up to the 10^n digit. In a counter going to the 10^n digit, there are n+1 possible digits.

The display on a decade counter often, if not usually, contains a decimal point. Decades can proceed downward toward zero, as well as upward to ever-increasing levels. However, fractions of a pulse or cycle are never actually counted. Rather, the pulses or cycles are counted for a longer time when greater accuracy is needed. For example, a counting time of 10 seconds gives one additional digit of accuracy, as compared to a counting time of 1 second. This allows decimal parts of a cycle to be determined. *See also* DECADE, FREQUENCY COUNTER.

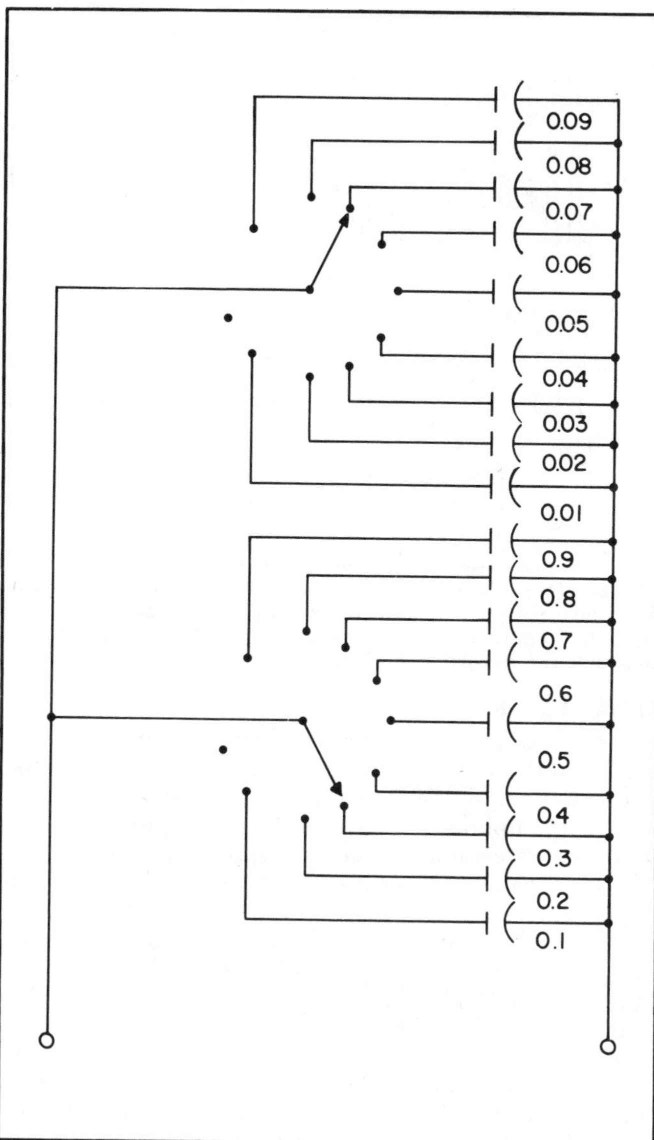

DECADE BOX: A simple decade capacitance box, with a range of 0.00 to 0.99 μF.

DECAY

The falling-off in amplitude of a pulse or waveform is called the decay. The decay of a pulse or waveform, although appearing to be instantaneous, is never actually so. A certain amount of time is always required for decay.

When you switch off a high-wattage incandescent bulb, for example, you can actually watch its brightness decay. But the decay in brilliance of a neon bulb or light-emitting diode is too rapid to be seen. Nevertheless, even a light-emitting diode has a finite brightness-decay time.

The graph illustrates the decay of the waveform in a continuous-wave (CW) transmitter following the transition from key-down to key-up condition. The decay curve is a logarithmic function. *See also* DAMPED WAVE, DECAY TIME, LOGARITHMIC DECREMENT, RISE TIME.

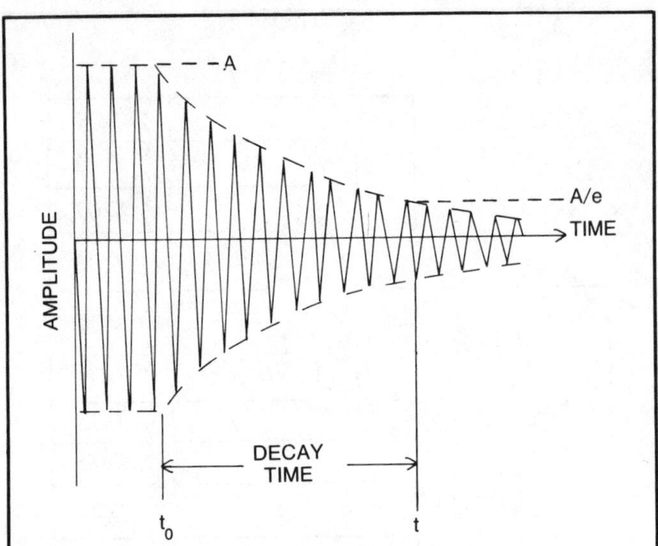

DECAY TIME: The decay time of signal is the time required for transition from maximum level to some predetermined amplitude. Here, the time interval starts at t_0 and ends at t. Initial amplitude is A and final amplitude is A/e, where e = 2.718.

DECAY TIME

The decay time of a pulse or waveform is the time required for the amplitude to decay to a certain percentage of the maximum amplitude. The time interval begins at the instant the amplitude starts to fall (see graph), and ends when the determined percentage has been attained.

The decay of a pulse or waveform proceeds in a logarithmic manner (*see* LOGARITHMIC DECREMENT). Therefore, in theory, the amplitude never reaches zero. In practice, of course, a point is always reached at which the pulse or wave amplitude can be considered to be zero. This point may be chosen for the determination of the decay time interval. In a capacitance-resistance circuit, the decay time is considered to be the time required for the charge voltage to drop to 37 percent of its maximum value. In this case the final amplitude is equal to the initial amplitude divided by e, where e is approximately 2.718. *See also* DECAY, TIME CONSTANT.

DECCA

Decca is the name given to a radionavigation system used by the British. The system operates in the range of 70 to 130 kHz, or approximately the same frequency as the American radionavigation system, loran C.

Decca is a phase-comparison system. Two or more fixed stations, with precisely coordinated phase, radiate the signals. An airplane pilot or boat captain locates his position by comparing the phases from the two stations, using a special integrator circuit designed for the purpose. When additional transmitting stations are used, several different readings can be obtained, and they can be averaged to get a close approximation of position. *See also* LORAN, RADIONAVIGATION.

DECIBEL

The decibel is a means of measuring relative levels of current, voltage, or power. A reference current I_0, or voltage E_0, or power P_0 must first be established. Then, the ratio of an arbitrary current I to the reference current I_0 is given by:

$$dB = 20 \log_{10} (I/I_0)$$

The ratio of an arbitrary voltage E to the reference voltage E_0 is given by:

$$dB = 20 \log_{10} (E/E_0)$$

A negative decibel figure indicates that I is smaller than I_0, or that E is smaller than E_0. A positive decibel value indicates that I is larger than I_0, or that E is larger than E_0. A tenfold increase in current or voltage, for example, is a change of +20 dB:

$$\begin{aligned} dB &= 20 \log_{10} (10E_0/E_0) \\ &= 20 \log_{10} (10) \\ &= 20 \times 1 = 20 \end{aligned}$$

For power, the ratio of an arbitrary wattage P to the reference wattage P_0 is given by:

$$dB = 10 \log_{10} (P/P_0)$$

As with current and voltage, a negative decibel value indicates that P is less than P_0; a positive value indicates that P is greater than P_0. A tenfold increase in power represents an increase of +10 dB:

$$\begin{aligned} dB &= 10 \log_{10} (10P/P_0) \\ &= 10 \log_{10} (10) \\ &= 10 \times 1 = 10 \end{aligned}$$

The reference value for power is sometimes set at 1 milliwatt, or 0.001 watt. Decibels measured relative to 1 milliwatt, across a pure resistive load of 600 ohms, are abbreviated dBm. Decibel figures are extensively used in electronics to indicate circuit gain, attenuator losses, and antenna power gain figures. *See also* ANTENNA POWER GAIN, dBa, dBd, dBi, dBm.

DECIBEL METER

A decibel meter is a meter that indicates the level of current, voltage, or power, in decibels, relative to some fixed reference value. The reference value may be arbitrary, or it may be some specific quantity, such as 1 milliwatt or 1 volt. In any case, the reference value corresponds to 0 dB on the meter scale. Levels greater than the reference level are assigned positive decibel values on the scale. Levels lower than the reference are assigned negative values. The illustration shows the scale of a typical decibel meter.

The "S" meters on many radio receivers, calibrated in S units and often in decibels as well, are forms of decibel meters. A reading of 20 dB over S9 indicates a signal voltage 10 times as great as the voltage required to produce a reading of S9. An S meter can be helpful in comparing the relative levels of signals received on the air. The reading of S9 corresponds to some reference voltage at the antenna terminals of the receiver, such as 10 microvolts. *See also* S METER, VOLUME-UNIT METER.

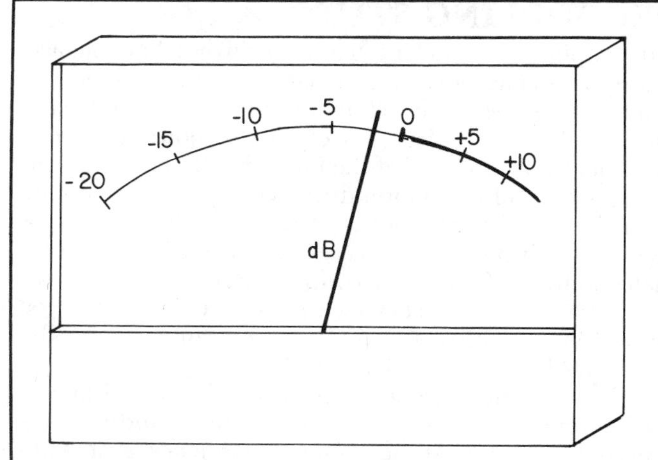

DECIBELMETER: A decibel meter is a logarthmic meter.

DECIMAL

The term decimal is used to refer to a base-10 number system. In this system, which is commonly used throughout the world, and which is the most familiar to us, numbers are represented by combinations of ten different digits in various decimal places.

The digit farthest to the right, but to the left of the decimal point, is multiplied by 10^0, or 1; the digit next to the left is multiplied by 10^1, or 10; the digit to the left of this is multiplied by 10^2, or 100. With each move to the left, the base value of the digit increases by a factor of 10, so that the nth digit to the left of the decimal point is multipled by 10^{n-1}.

The digit first to the right of the decimal point is multplied by 10^{-1}, or 1/10; the digit next to the right is multplied by 10^{-2}, or 1/100. This process continues, so that with each move to the right, the base value of the digit decreases by a factor of 10. Therefore, the nth digit to the right of the decimal point is multiplied by 10^{-n}.

Ultimately, the number represented by a decimal sequence is determined by adding the decimal values of all the digits. For example, the number 27.44 is equal to $2 \times 10^1 + 7 \times 10^0 + 4 \times 10^{-1} + 4 \times 10^{-2}$. We do not normally think of the value of 27.44 in this way, however.

While we use the decimal system in our everyday lives, digital circuits generally operate in a base-2 number system, where the only digits are 0 and 1, and where the values increment and decrement in powers of 2. *See also* BINARY-CODED NUMBER.

DECLINATION

Declination is the latitude of an object on the celestial sphere, measured with reference to the celestial equator. The celestial equator, having a declination of 0 degrees, runs through the sky in a huge imaginary circle around the earth. This circle lies in a plane passing through the equator of the earth. The north celestial pole, having a declination of 90 degrees north, lies directly above the north pole. The south celestial pole, having a declination of 90 degrees south, lies directly above the south pole in the sky. The locations of stars, planets, galaxies, and other cosmic objects are assigned a location in the sky according to the declination and right ascension (*see* RIGHT ASCENSION). An object in the sky with a given declination passes over any point on the earth with the corresponding latitude. See A in the illustration.

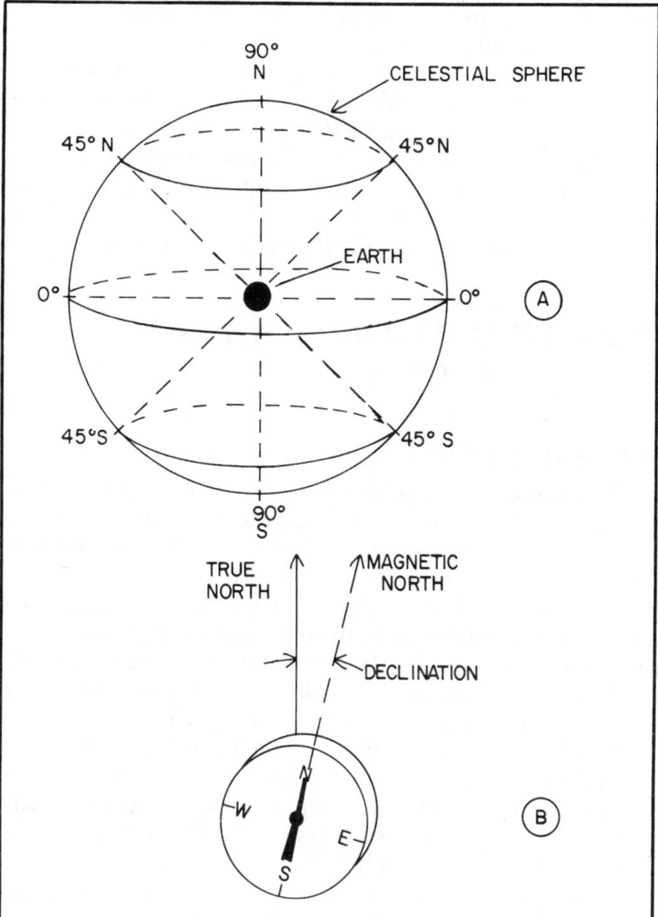

DECLINATION: At A, celestial declination is the latitude measure on the celestial sphere. At B, compass declination is the difference in degrees between true north and magnetic north.

The term declination is also sometimes used to refer to the difference between true north and magnetic north. Since the north magnetic pole is not located exactly at the north geographic pole, a magnetic compass will not always point toward true geographic north. The angular difference between a compass reading and true north is called declination. Declination (B in the illustration) depends on the location from which the compass reading is taken. The declination can be as small as 0 degrees, when the north magnetic pole is in line with the north geographic pole; or, it can be as great as 180 degrees, when the compass is located along a line connecting the two poles. *See also* INCLINATION.

DECODING

Decoding is the process of converting a message, received in code, into plain language. This is generally done by a machine, although in the case of the Morse code, a human operator often acts as the decoding medium.

Messages can be coded either for the purpose of efficiency and accuracy, such as with the Morse code or other codes, or for the purpose of keeping a message secret, as with voice scrambling or special abbreviations. Both types of code can be used at the same time. In this case, decoding requires two steps: one to convert the scrambling code to the plain text, and the other to convert the code itself to English or another language.

The conversion of a digital signal to an analog signal is sometimes called decoding. The opposite of the decoding process—the conversion of an analog signal to a digital signal, or the transformation of a plain-language message into coded form—is called encoding. Decoding is always done at the receiving end of a communications circuit. *See also* DIGITAL-TO-ANALOG CONVERTER, ENCODING.

DECOMPOSITION VOLTAGE

See BREAKDOWN VOLTAGE.

DECOUPLING

When undesired coupling effects must be minimized, a technique called decoupling is employed. For example, a multistage amplifier circuit will often oscillate because of feedback among the stages. This oscillation usually takes place at a frequency different from the operating frequency of the amplifier. In order to reduce this oscillation, or eliminate it entirely, the interstage coupling should be made as loose as possible, consistent with proper operation at the desired frequency. This is called decoupling.

Another form of decoupling consists of the placement of chokes and/or capacitors in the power-supply leads to each stage of a multistage amplifier (see diagram). This minimizes the chances of unwanted interstage coupling through the power supply.

When several different loads are connected to a single transmission line, such as in a multiband antenna system, resonant circuits are sometimes employed to effectively decouple all undesired loads from the line at the various operating frequencies. The trap antenna decouples a part of the radiator, to obtain resonance on two or more different frequencies. *See also* DECOUPLING STUB, TRAP, TRAP ANTENNA.

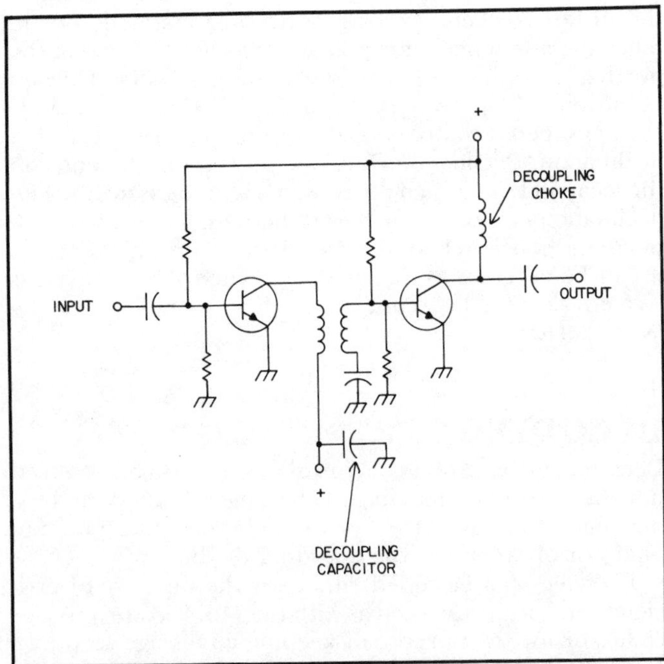

DECOUPLING: A decoupling capacitor and decoupling choke help to reduce the effects of feedback in a multistage amplifier.

DECOUPLING STUB

A decoupling stub is a length of transmission line that acts as a resonant circuit at a particular frequency, and is used in an antenna system in place of a trap (*see* TRAP). Such a stub usually consists of a ¼-wavelength section of transmission line, short-circuited at the far end. This arrangement acts as a parallel-resonant inductance-capacitance circuit.

At the resonant frequency of the stub, the impedance between the input terminals is extremely high. Therefore, such a stub can be used to decouple a circuit at the resonant frequency (*see* DECOUPLING). Such decoupling might be desired in a multiband antenna system, or to aid in the rejection of an unwanted signal.

A ¼-wavelength section of transmission line, open at the far end, will act as a series-resonant inductance-capacitance circuit. At the resonant frequency, such a device, also called a stub, has an extremely low impedance, essentially equivalent to a short circuit. This kind of stub can be extremely effective in rejecting signal energy at unwanted frequencies. By connecting a series-resonant stub across the antenna terminals, spurious responses or emissions are suppressed at the resonant frequency of the stub. The drawing illustrates the use of parallel-equivalent (A) and series-equivalent (B) stubs in antenna and feed-line systems.

Some stubs are ½ wavelength long, rather than ¼ wavelength. A short-circuited ½-wavelength stub acts as a series-resonant circuit, and an open-circuited ½-wavelength stub acts as a parallel-resonant circuit. All stubs show the same characteristics at odd harmonics of the fundamental frequency.

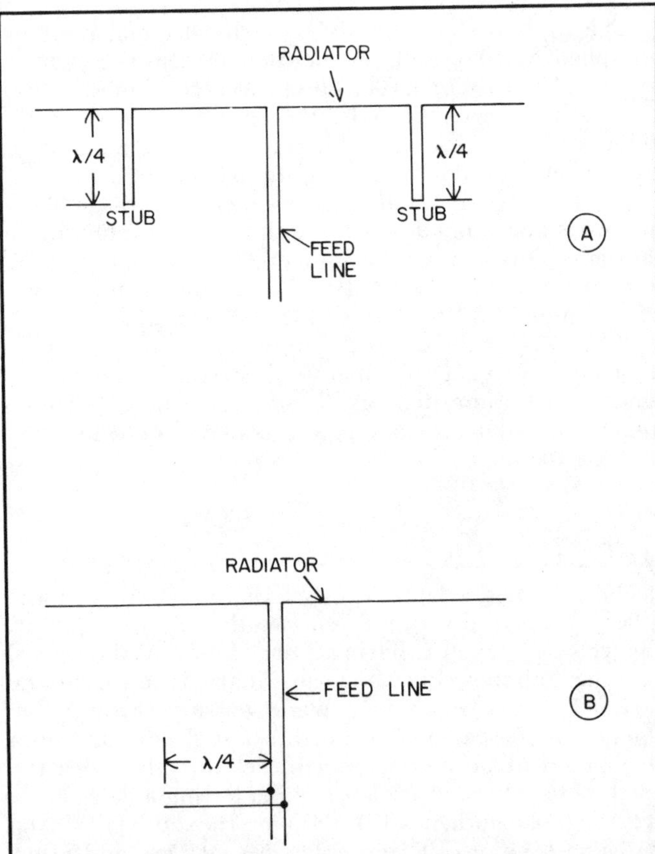

DECOUPLING STUB: Two kinds of decoupling stub. At A, a stub is used as a trap to provide resonance on two different frequencies. At B, a stub is used to attenuate a specific frequency in a transmission line.

DEEMPHASIS

Deemphasis is the deliberate introduction of a lowpass type response into the audio-frequency stages of a frequency-modulation receiver. This is done to offset the preemphasis introduced at the transmitter (*see* PREEMPHASIS). The schematic diagram shows a simple deemphasis network, "A." The graph ("B") illustrates, in a qualitative manner, the attenuation-versus-frequency characteristic of a typical deemphasis network.

By introducing preemphasis at the transmitter and deemphasis at the receiver in a frequency-modulation communications system, the signal-to-noise ratio at the upper end of the audio range is improved. This is because, as the transmitted-signal modulating frequency increases, the amplitude increases (because of preemphasis at the transmitter); when this amplitude is brought back to normal by deemphasis in the receiver, the noise is attenuated as well. *See also* FREQUENCY MODULATION, SIGNAL-TO-NOISE RATIO.

DEFIBRILLATION

An electric shock can cause the regular rhythm of the heartbeat to stop, and an uncoordinated twitching of the heart muscles occurs instead. This is called fibrillation (*see* ELECTRIC SHOCK, HEART FIBRILLATION). Unless the normal heart action is restored within a few minutes, death results because the blood supply to the body is cut off. A fibrillating heart does not effectively pump blood through the lungs to be oxygenated, nor can the blood get to the rest of the body.

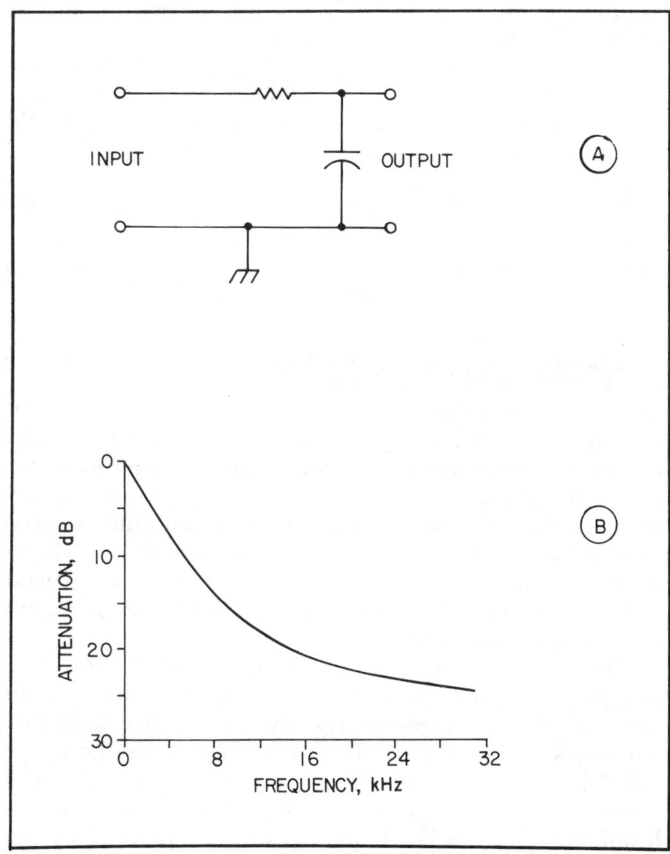

DEEMPHASIS: At A, a simple deemphasis circuit. At B, a sample representation of deemphasis, in terms of attenuation versus frequency.

The heart functions by means of electrical nerve impulses. A heart pacemaker (*see* PACEMAKER) actually regulates the heartbeat by transmitting electrical signals to the heart muscle. A device called a defibrillator also works via electrical impulses. The defibrillator produces an electric shock or series of shocks, which often gets the heart beating normally when it is in a state of fibrillation. Two metal electrodes are placed on the chest of the victim, in such a position that the current is sent through the heart.

If a defibrillator is not immediately available when a shock victim goes into fibrillation, the only alternative is to apply cardiopulmonary resuscitation until medical help arrives. *See also* CARDIOPULMONARY RESUSCITATION.

DEFINITE INTEGRAL

A definite integral is an integral evaluated between two defined points in the domain of a function (*see* INDEFINITE INTEGRAL). Geometrically, the definite integral of a one-variable function, evaluated between two points, is equal to the area under the curve of the function between those two points. The illustration shows an example of a definite integral, according to this geometric interpretation.

A definite integral, evaluated between two points a and b on the x axis of a function f (x), is written:

$$\int_a^b f(x)\,dx$$

The expression dx is the differential of x, and is necessary to make the quantity meaningful. The details of integration are covered in calculus textbooks.

An area that lies above the abscissa, or x axis, in the definite integral is considered positive. An area below the abscissa is considered negative. In the illustration, the value of the integral is positive, since the shaded region lies entirely above the x axis. *See also* CALCULUS.

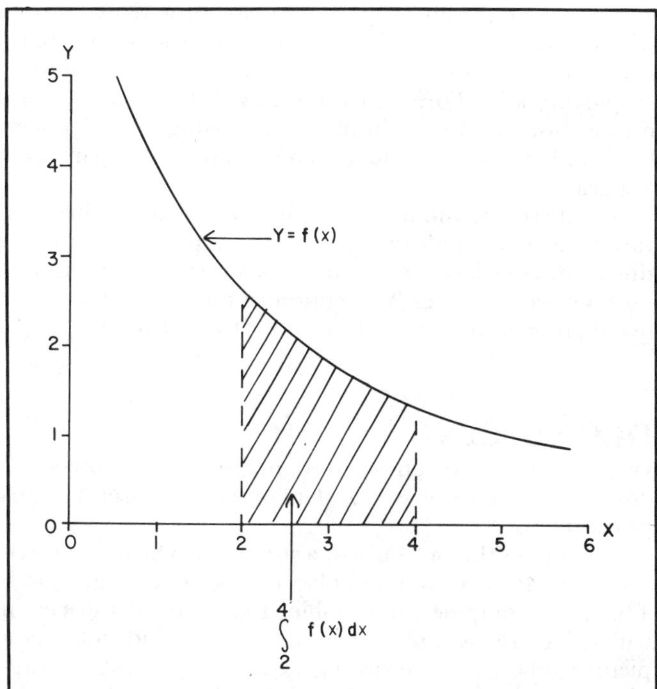

DEFINITE INTEGRAL: The definite integral of this function, evaluated between x = 2 and x = 4, is equal to the area of the shaded region.

DEFLECTION

Deflection is a deliberately induced change in the direction of an energy beam. The beam may consist of sound waves, radio waves, infrared radiation, visible light, ultraviolet radiation, X rays, or atomic-particle radiation such as a stream of electrons.

In a cathode-ray tube, electron beams are deflected to focus and direct the energy to a certain spot on a phosphor screen. In loudspeaker enclosures, deflecting devices are used to get the best possible fidelity. Deflectors called baffles are used for acoustic purposes in concert halls and auditoriums (*see* ACOUSTICS). Deflecting mirrors are employed in optical telescopes. Heat deflectors are sometimes used to improve energy efficiency in homes and buildings. *See also* BAFFLE, CATHODE-RAY TUBE, DEFLECTOR.

DEFLECTOR

A deflector is an electrode for the purpose of deflecting an energy beam. Generally, the term deflector is used to describe the plates in a cathode-ray tube. These plates are used to focus and direct the beam to the desired point on the phosphorescent screen.

In a loudspeaker, an attachment for spreading the sound waves over a broad angle is called a deflector. A loudspeaker deflector improves the fidelity of the sound by allowing the sound to be heard over a larger area.

DE FOREST, LEE

The invention of the triode vacuum tube is accredited to Lee De Forest, who successfully showed that a three-element evacuated device could be used to amplify signals and to produce oscillation at discrete radio frequencies.

De Forest's invention was revolutionary in its time, because it made possible the transmission of wireless signals within a narrow bandwidth. Previously, wideband spark transmission had been necessary. With an oscillator employing a De Forest vacuum tube, long-distance communication became commonplace using input-power levels much lower than had been required with spark-gap apparatus.

Triode tubes, and in fact all kinds of vacuum tubes, are rarely seen in modern electronic equipment. Semiconductor devices have replaced tubes in almost every application since De Forest's invention in the early part of the twentieth century. *See also* TRIODE TUBE, TUBE.

DEGAUSSING

Degaussing is a procedure for demagnetizing an object. A device called a degausser or demagnetizer is used for this purpose.

Sometimes the presence of a current in a nearby electrical conductor can cause an object to become magnetized. This effect may be undesirable. Examples of devices in which degaussing is sometimes required include television picture tubes, tape-recording heads, and relays. By applying an alternating current to produce an alternating magnetic field around the object, the magnetization can usually be eliminated. Another method of degaussing is the application of a steady magnetic field in opposition to the existing, unwanted field. *See also* DEMAGNETIZING FORCE.

DEGENERATION

See NEGATIVE FEEDBACK.

DEGREE

The degree is a unit of either temperature or angular measure. There are three common temperature scales, called the Celsius, Fahrenheit, and Kelvin scales (*see* CELSIUS TEMPERATURE SCALE, FAHRENHEIT TEMPERATURE SCALE, KELVIN TEMPERATURE SCALE).

In the Fahrenheit scale, which is most generally used in the United States, the freezing point of pure water at one atmosphere pressure is assigned the value 32 degrees. The boiling point of pure water at one atmosphere is 212 degrees on the Fahrenheit scale. In the Celsius temperature scale, water freezes at 0 degrees and boils at 100 degrees. Thus, a degree in the Celsius scale is representative of a greater change in temperature than a degree in the Fahrenheit scale. In the Kelvin, or absolute, temperature scale, the degrees are the same size as in the Celsius scale, except that 0 degrees Kelvin corresponds to absolute zero, which is the coldest possible temperature in the physical universe. In the respective temperature scales, readings are given in degrees F (Fahrenheit), degrees C (Celsius), or degrees K (Kelvin).

In angular measure, a degree represents 1/360 of a complete circle. The number 360 was chosen in ancient times, when scientists and astronomers noticed that the solar cycle repeated itself approximately once in 360 days. One day thus corresponds to a degree in the circle of the year. (It is fortunate indeed that the ancients were slightly off in their count; otherwise, we might be using degrees of measure of one part in 365.25!)

Phase shifts or differences are usually expressed in degrees, with one complete cycle represented by 360 degrees of phase. *See also* CYCLE, PHASE ANGLE, RADIAN.

DEHUMIDIFICATION

The operation of some electronic circuits is affected not only by the temperature, but also by the relative humidity. If the amount of water vapor in the air is too great, corrosion is accelerated, and condensation may occur. In an electrical circuit, condensation can cause unwanted electrical conduction between parts that should be isolated. Condensation can also cause the malfunction of high-speed switches. Dehumidification is the process of removing excess moisture from the air.

There are various ways to obtain dehumidification. The simplest method is to raise the temperature; for a given amount of water vapor in the air, the relative humidity decreases as the temperature rises. Dry crystals of calcium chloride or cobalt chloride, packed in a cloth sack, will absorb water vapor from the air, and will help to dehumidify an airtight chamber. Various dehumidifying sprays are also available. *See also* HUMIDITY.

DEIONIZATION

When an ionized substance becomes neutral, the change is called deionization. A collection of positive ions (*see* ION), for example, may be neutralized by the introduction of electrons or by the variation of parameters such as the temperature or the intensity of the electric field.

Generally, gases tend to ionize at high temperatures and/or high electrical potentials. In an ionized gas, as the temperature is reduced, the atoms will eventually become neutral, assuming that the intensity of the electric field is not too great. The temperature at which this neutralization takes place is called the deionization temperature. Given a constant temperature, as the intensity of the electric field is reduced, an ionized substance will become neutral at a certain point, called the deionization potential. *See also* IONIZATION.

DELAY

A delay is a specific time interval between a cause and its effect, or between one effect and another related effect. A delay always occurs between two related events. This delay may be inconsequential and perhaps unimportant, such as the time required for the voice impulses to traverse a local telephone circuit during a casual conversation, or the time between the transmission of an amateur-radio message and its reception in a distant receiver (see illustration). The delay may, however, be of great importance, or it might be deliberately introduced into a circuit. An example of this is the delay in broadcast programming, usually about 7 seconds, between the actual event and its transmission over the air.

The shortest possible delay between two circuit points is determined by the speed at which electric impulses can flow in that circuit. In an alternating-current circuit, delay is sometimes expressed as a fraction of a cycle, or in degrees of phase. *See also* DELAY CIRCUIT, DELAY LINE, PHASE ANGLE.

DELAY CIRCUIT

A delay circuit is a set of electronic components designed deliberately for the purpose of introducing a time or phase delay. Such a circuit might be a passive combination of resistors, inductors, and/or capacitors. A delay circuit may consist of a simple length of transmission line. Or, the device may be an active set of integrated circuits and peripheral components.

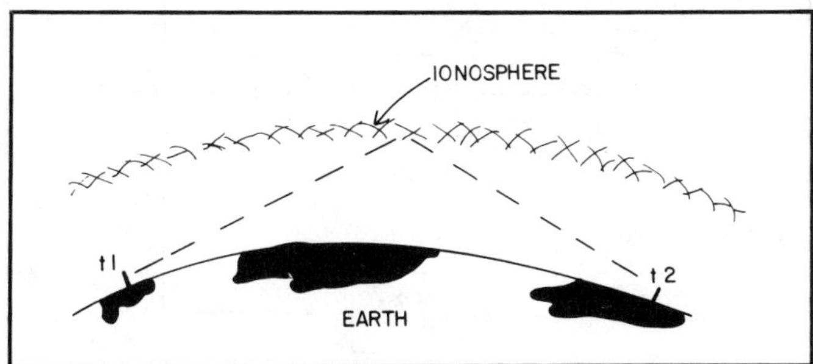

DELAY: The time delay $t_2 - t_1$ for ionospheric propagation between two radio stations is generally less than 100 milliseconds.

Delay circuits are used in a wide variety of applications. Broadcast stations delay their transmissions by approximately 7 seconds. This allows the signal to be cut off, if necessary, before it is transmitted over the air. Phase-delay circuits are extremely common. Certain kinds of switches and circuit breakers have a built-in time delay. Sometimes, delay is undesirable and hinders the performance of a circuit. *See also* DELAY, DELAY DISTORTION, DELAYED AUTOMATIC GAIN CONTROL, DELAYED MAKE/BREAK, DELAYED REPEATER, DELAYED TRANSMISSION, DELAY TIME, DELAY TIMER.

DELAY DISTORTION

In some electronic circuits, the propagation time varies with the signal frequency (*see* PROPAGATION TIME). When this happens, distortion takes place, because signal components having different frequencies arrive at the receiving end of the circuit at different times. This is delay distortion, which can happen in a radio communications circuit, in a telephone system, or even within a single piece of electronic equipment.

Generally, higher frequencies are propagated at a lower rate of speed than lower frequencies. This is shown hypothetically in the graph. If the propagation time is extremely short, the delay distortion will be inconsequential. But the longer the propagation time, the greater the chances of delay distortion. Delay distortion can be minimized by making the percentage difference between the lowest and highest frequencies in a signal as small as possible. A baseband signal, with components as low as 100 Hz and as high as 3000 Hz, for example, is more subject to delay distortion than a single-sideband signal with the same audio characteristics and a suppressed-carrier frequency of 1 MHz. In the former case, the percentage difference between the lowest and highest frequencies is very large, but in the latter case, it is extremely small.

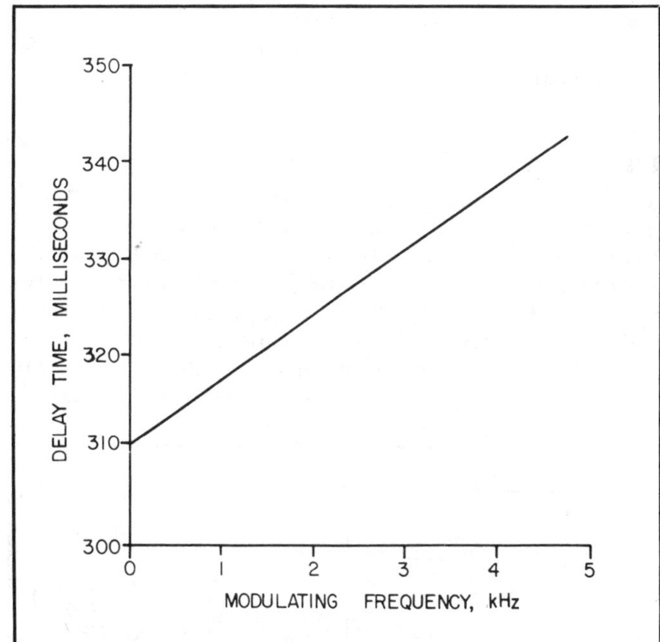

DELAY DISTORTION: An example of a delay-versus-frequency curve that might produce delay distortion.

DELAYED AUTOMATIC GAIN CONTROL

A delayed automatic-gain-control circuit is a special form of automatic-gain-control, or AGC, circuit (*see* AUTOMATIC GAIN CONTROL). It is used in many communications receivers.

In a delayed AGC circuit, signals below a certain threshold level are passed through the receiver with maximum gain. Only when the signal strength exceeds this threshold amplitude, does the AGC become active. Then, as the signal strength continues to increase, the AGC provides greater and greater attenuation.

The delayed AGC circuit allows better weak-signal reception than does an ordinary automatic-gain-control circuit.

DELAYED MAKE/BREAK

When a circuit is closed or opened a short while after the actuating switch or relay is energized or deenergized, the condition is called delayed make or break. For example, the contacts of a relay may close several milliseconds, or even several seconds or minutes, following application of current to the circuit. The contacts of a relay or other switching device may not open until some time after current has been removed. The former device is a delayed-make circuit; the latter is a delayed-break circuit.

Delayed-make and delayed-break devices are sometimes used in power supplies. For example, in a tube-type power amplifier, the filament voltage should be applied a few seconds or minutes before the plate voltage. A delayed-make circuit can be used in the plate supply, accomplishing this function automatically. A simple delayed-make/break relay arrangement, providing a delay of up to several seconds, is illustrated in the schematic diagram. The value of the capacitor determines the delay time. The larger the capacitance, the longer the delay. The resistor limits the current through the relay coil, and serves to lengthen the charging time of the capacitor. *See also* DELAY TIMER.

DELAYED REPEATER

A delayed repeater is a device that receives a signal and retransmits it later. The delay time between reception and retransmission may vary from several milliseconds to seconds or even minutes. Generally, a delayed repeater records the signal modulation envelope on magnetic tape, and plays the tape back into its transmitter.

A delayed repeater operates in essentially the same way as an ordinary repeater. The signal is received, demodulated, and retransmitted at a different frequency. An isolator circuit prevents interference between the receiver and the transmitter. The block diagram shows a simple delayed-repeater circuit. *See also* REPEATER.

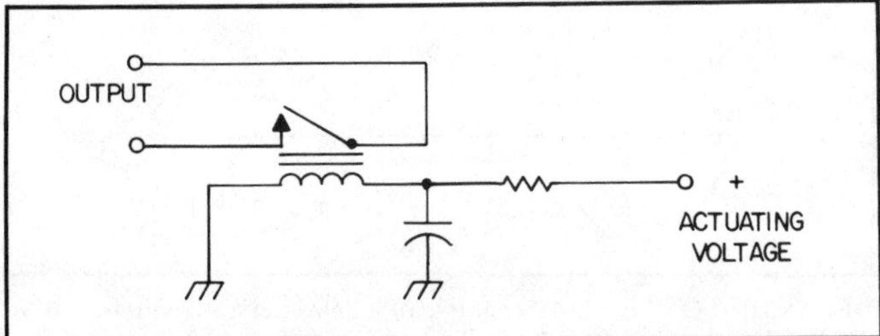

DELAYED MAKE/BREAK: A simple circuit for delayed make and break with a relay.

DELAYED TRANSMISSION

When a signal is transmitted after the event actually takes place, the transmission is called a delayed transmission. A broadcast station usually delays the transmission of a program for several seconds. This allows time for decision as to whether the signal should be cut off before it is sent over the air. This kind of delayed transmission is accomplished by recording the station program on magnetic tape, and playing it back into the transmitter after an elapsed time of about 7 seconds.

Another form of delayed transmission is the pre-recorded program. An audio or audio-video tape recorder is used to record a program. Then the program is edited to meet the time and other requirements of the station, and broadcast at a later time. Live broadcasts are becoming more and more rare since the advent of video tape recorders; pre-recorded programs are now the rule, rather than the exception.

DELAY LINE

A delay line is a circuit, often (but not necessarily) a length of transmission line, that provides a delay for a pulse or signal traveling through it. All transmission lines carry energy with a finite speed. For example, electromagnetic fields travel along a solid-polyethylene-dielectric coaxial cable at approximately 66 percent of the speed of light.

The illustration shows three different kinds of delay lines. At A, a length of coaxial cable is wound into a coil. Since the speed of propagation in the cable is about 2×10^8 meters per second, this line provides a delay of 5 nanoseconds (5×10^{-9} second) per meter. At B, a network of inductances and capacitances simulates a long length of transmission line. At C, an interior view of a helical delay line is illustrated. In this type of line, the center conductor is wound into a helix, like a spring, thus greatly increasing its length. This increases the propagation delay per unit length of the line. *See also* DELAY TIME.

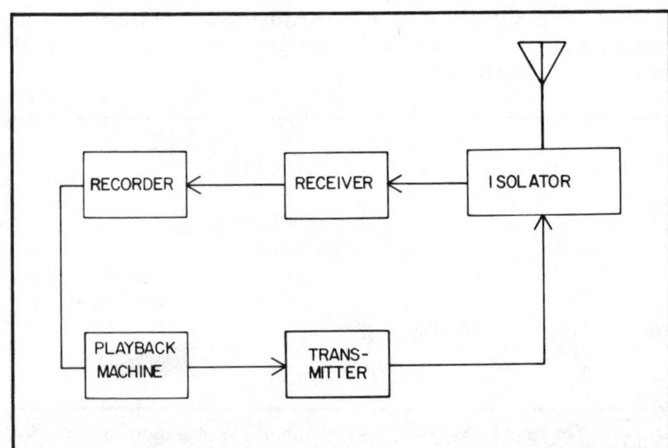

DELAYED REPEATER: A delayed repdater intercepts a signal, demodulates it, records the modulation envelope, and retransmits the signal on a different frequency at a later time.

DELAY TIME

The delay time is the length of time required for a pulse or signal to travel through a delay circuit or line, as compared to its travel over the same distance through free space. (*See* DELAY CIRCUIT, DELAY LINE.) Delay time is measured in seconds, milliseconds (10^{-3} second), microseconds (10^{-6} second), or nanoseconds (10^{-9} second).

Delay time is ordinarily measured with an oscilloscope. The illustration shows a simple arrangement for measuring the delay time through a circuit (A). A pulse or signal, supplied by the signal generator, is run through the delay circuit and also directly to the oscilloscope. This results in two pulses or waveforms on the oscilloscope screen. The frequency of the pulses or signal is varied, to be certain that the delay time observed is correct (it could be more than one cycle, misleading a technician if only one wavelength is used). The delay time is indicated by the separation of pulses or waves on the oscilloscope display (shown at B).

In a delay line with length m, in meters, and velocity factor v, in percent, the delay time t, in nanoseconds, is given by:

$$t = 333m/v$$

assuming the transmission and reception points are located negligibly close together for a signal traveling through free space. If the transmission and reception points are located n meters apart, then the delay time in nanoseconds is

$$t = 333m/v - 3.33n$$

See also PROPAGATION TIME.

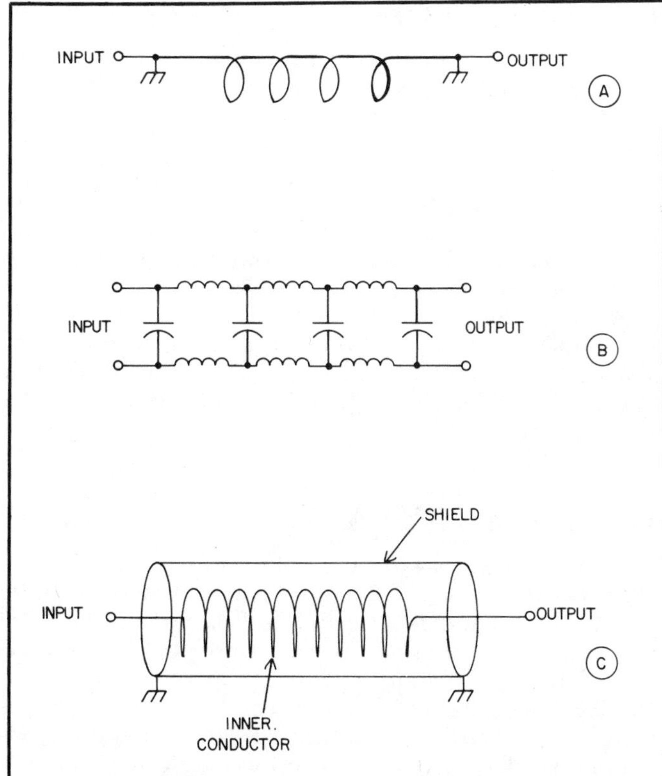

DELAY LINE: Three kinds of delay line. At A, a coiled length of coaxial cable; at B, an array of inductors and capacitors; at C, a coaxial line with a helical inner conductor.

DELAY TIMER

Any device that introduces a variable delay in the switching of a circuit is called a delay timer. Such a timer usually has a built-in, resettable clock. After the prescribed amount of time has elapsed, the switching is performed.

An example of the use of a delay timer is the power-up of a large vacuum-tube radio-frequency transmitter. The filament voltage is applied as the delay timer is first set to approximately 2 minutes. The plate voltage, however, does not get switched in until the timer has completed its cycle. Such a timer may be mechanical or electronic.

Delay timers are common in consumer items, such as microwave ovens. They are also used to actuate lamps or other devices, to scare off possible intruders when a home is unoccupied.

DELLINGER EFFECT

A sudden solar eruption causes an increase in the ionization density of the upper atmosphere of the earth. At high frequencies, especially those wavelengths known for long-distance daytime propagation, such ionization can cause an abrupt cessation of communications. This is called the Dellinger effect.

Normally, signals are propagated via the ionospheric E and F layers, as shown at A in the drawing. The lower layer, called the D layer, is not ordinarily ionized to a sufficient density to affect signals in long-distance propagation. However, a solar eruption causes a dramatic increase in the ionization density of all layers, including the D layer. This results in absorption of electromagnetic energy by the D

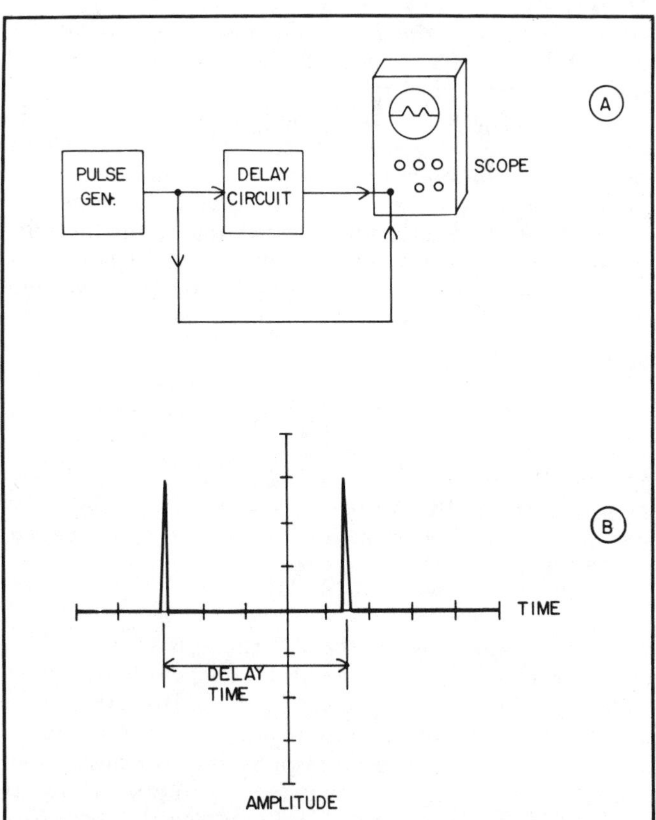

DELAY TIME: Measurement of delay time. At A, a simple arrangement for determining the delay through a circuit; at B, a hypothetical pulse display. The delay time is measured in terms of the sweep speed, in divisions per second, of the oscilloscope display.

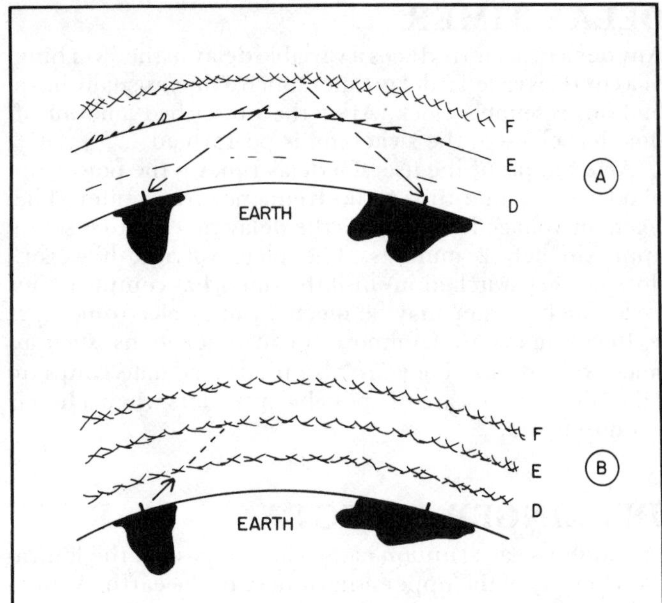

DELLINGER EFFECT: The Dellinger effect is the absorption of a signal by the ionospheric D layer, caused by a sudden solar eruption. At A, normal propagation. At B, the situation a few minutes after a solar disturbance. Note the increased density of the ionized layers.

layer, with a consequent disappearance of signals at distant points, as at B. *See also* D LAYER, PROPAGATION CHARACTERISTICS.

DELTA

Delta is a letter of the Greek alphabet. The capital letter delta, Δ, looks like a triangle. The lower-case letter delta, δ resembles our small letter d.

The lower-case letter δ is sometimes used in mathematical equations to represent a variable. The upper-case delta represents an increment in a variable, and therefore it is an operator. For example, at A in the graph, the slope m of the line is defined as the amount of change in the variable y divided by the amount of change in the variable x, as evaluated between two points (x_1, y_1) and (x_2, y_2). We may write this as

$$m = \frac{y_2 - y_1}{x_2 - x_1}$$

Often, however, mathematicians and engineers will call $y_2 - y_1$ by the name Δy (delta y), and $x_2 - x_1$ by the name Δx (delta x), and say that

$$m = \Delta y/\Delta x$$

The graph shown at A in the illustration is a straight line, and therefore $\Delta y/\Delta x$ is always the same, no matter what values are chosen for (x_1, y_1) and (x_2, y_2). However, in the graph at B, the points (x_1, y_1) and (x_2, y_2) affect the value of $\Delta y/\Delta x$. As the point (x_2, y_2) is moved closer and closer, along the curve, to the point (x_1, y_1), the value $\Delta y/\Delta x$ approaches a certain limiting value, called the derivative of the function at the point (x_1, y_1). This is the basis for differential calculus, an invaluable technique in mathematics for all phases of science. *See also* CALCULUS, DERIVATIVE, DIFFERENTIATION.

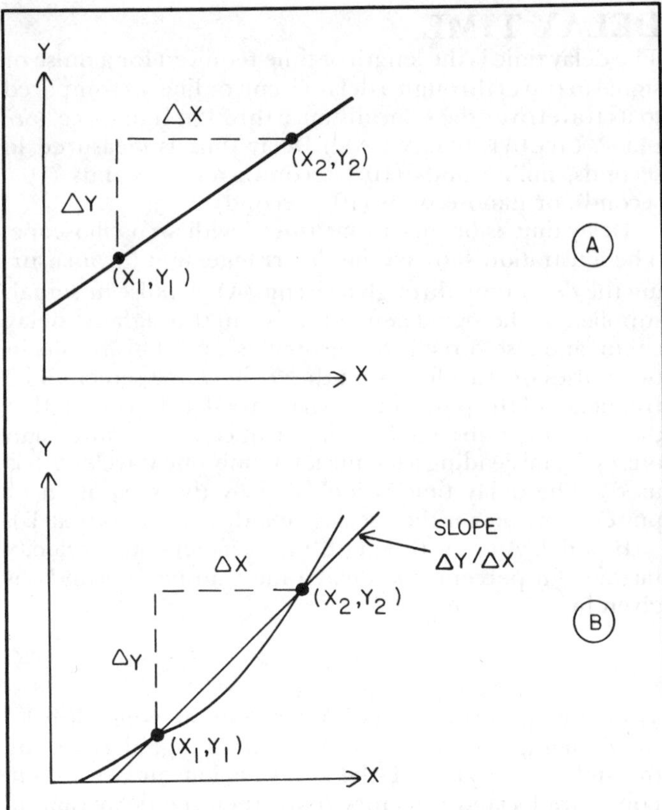

DELTA: At A, the slope $\Delta y/\Delta x$ of a straight line does not depend on the choice of points. At B, the value of $\Delta y/\Delta x$ changes as the point (x_2,y_2) moves toward (x_1,y_1). The limiting value of $\Delta y/\Delta x$ is the derivative of the function, at the point in question.

DELTA MATCH

A delta match is a method of matching the impedance of an antenna to the characteristic impedance of a transmission line. The delta-matching technique is used with balanced antennas and two-wire transmission lines.

The illustration shows a delta matching system. The length of the network, m, and the width or spacing between the connections, s, is adjusted until the standing-wave ratio on the feed line is at its lowest value. The length of the radiating element is ½ wavelength.

A variation of the delta match is called the T match. When the transmission line is unbalanced, such as is the case with a coaxial cable, a gamma match may be used for matching to a balanced radiating element. *See also* GAMMA MATCH, T MATCH.

DELTA MODULATION

Delta modulation is a form of pulse modulation (*see* PULSE MODULATION). In the delta-modulation scheme, the pulse spacing is constant, generated by a clock circuit. The pulse amplitude is also constant. But the pulse polarity can vary, being either positive or negative.

When the amplitude of the modulating waveform is increasing, positive unit pulses are sent. When the modulating waveform is decreasing in amplitude, negative unit pulses are sent. When the modulating waveform amplitude is not changing, the pulses alternate between positive and negative polarity (see illustration). Delta modulation gets its name from the fact that it follows the

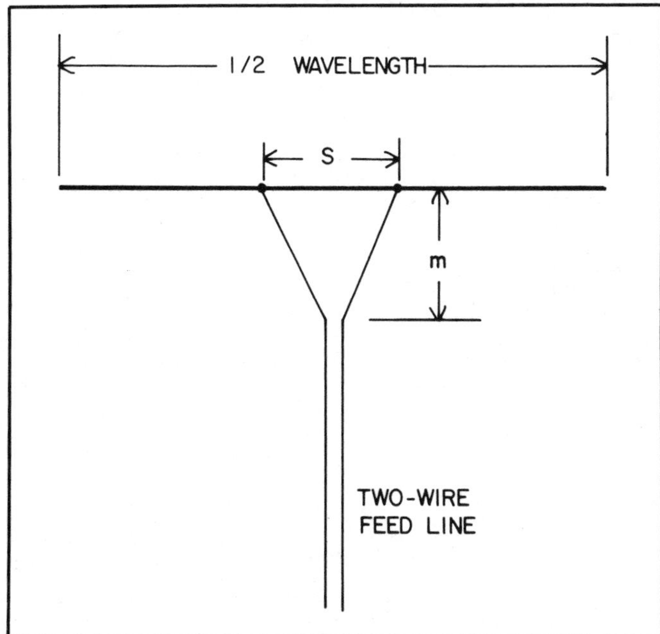

DELTA MATCH: A delta match is a means of matching the characteristic impedance of a two-wire feed line to a balanced radiating element.

difference, or derivative, of the modulating waveform.

In the delta-modulation detector, the pulses are integrated. This results in a waveform that closely resembles the original modulating waveform. A filter eliminates most of the distortion caused by sampling effects. The integrator circuit can consist of a series resistor and a parallel capacitor; the filter is usually of the lowpass variety. *See also* INTEGRATION.

DELTA WAVE

A delta wave is an impulse generated by the human brain. The frequency of delta waves is extremely low—less than 9 Hz. Usually, their frequency varies between approximately 0.2 and 4 Hz.

Delta waves occur in the human nervous system during the sleep cycle, especially during deep sleep. Brain waves, including delta waves, are observed by using a device called an electroencephalograph. *See also* ELECTROENCEPHALOGRAPH.

DEMAGNETIZING FORCE

A demagnetizing force is any effect that reduces the magnetism of an object. A certain amount of magnetic force, in opposition to the field of a magnetized object, will reduce the magnetism of the object to zero. High temperatures can cause demagnetization; so can a hard physical blow.

Objects are demagnetized by a device called a degausser. The degausser generates either an alternating magnetic field, or a steady magnetic field in opposition to the field of the object to be demagnetized (*see* DEGAUSSING).

In a powdered-iron or ferrite inductor or transformer core, the demagnetizing force required to reduce the core flux to zero is a function of the hysteresis characteristics of the material. This in turn depends on the magnetic permeability of the core material, and on the frequency

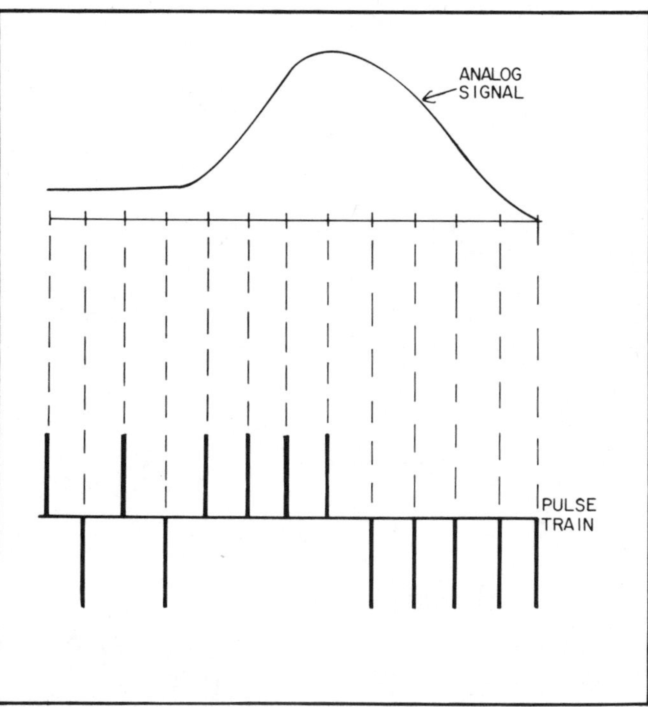

DELTA MODULATION: Delta modulation is the conversion of an analog waveform to digital pulses, according to the derivative of the waveform.

at which the core is used. *See also* HYSTERESIS, HYSTERESIS LOOP, PERMEABILITY.

DEMODULATION

See DETECTION.

DE MOIVRE'S EQUATION

The de Moivre equation, also called de Moivre's theorem, is a formula for determining the value of a complex number raised to a power.

Given a complex number a + jb, where $j = \sqrt{-1}$ and a and b are real numbers, let Θ be the angle subtended by the vector a + jb, relative to the real axis on a complex Cartesian plane, as illustrated at A. Also, let n be the power to which the complex number a + jb is raised. Then

$$(a + jb)^n = (\sqrt{a^2 + b^2})^n (\cos n\Theta + j \sin n\Theta)$$

where cos and sin represent the cosine and sine functions, respectively. The value of the exponent n may be any real number, not necessarily an integer.

As a simple example of the application of de Moivre's theorem, suppose we wish to evaluate the value of $(2 + j3)^2$. According to de Moivre's formula,

$$\begin{aligned}(2 + j3)^2 &= (\sqrt{2^2 + 3^2})^2(\cos 2\Theta = j \sin 2\Theta)\\ &= 13 \cos 2\Theta + 13j \sin 2\Theta\end{aligned}$$

We find Θ by recognizing, on the coordinate plane, that tan Θ = 3/2 = 1.5, as shown by B in the illustration. Therefore = Θ 56.31 degrees, and 2Θ = 112.62 degrees.

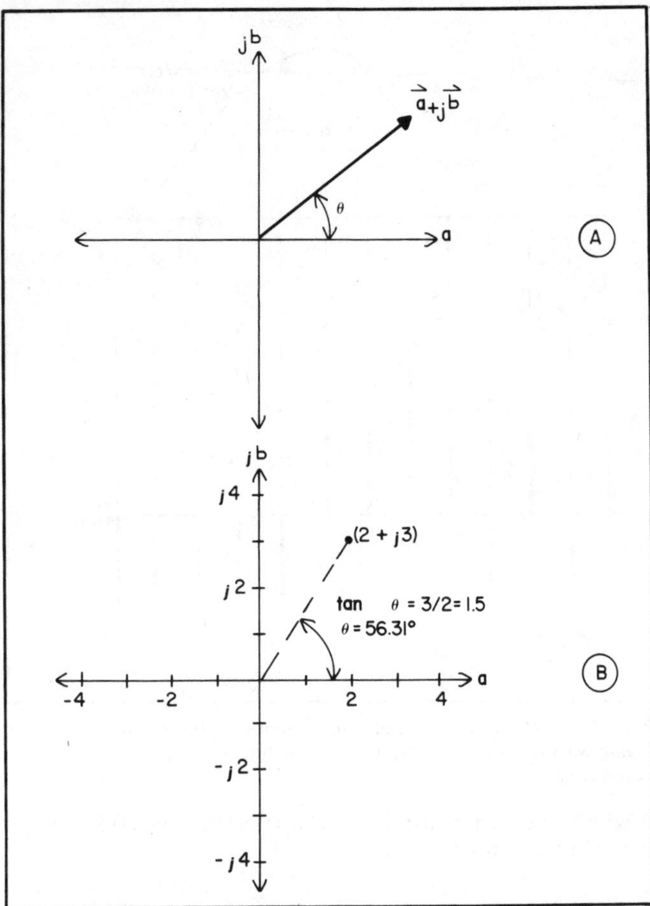

DE MOIVRE'S EQUATION: Any vector a + jb subtends an angle θ relative to the real axis, as shown at A. At B, for the complex number 2 + j3, it is geometrically apparent than tan $\theta = 3/2$; thus θ measures 56.31 degrees (approximately).

Then

$$\begin{aligned}(2 + j3)^2 &= 13 \cos(112.62°) + 13j \sin(112.62°)\\ &= 13 \times -0.3846 + 13j \times 0.9231\\ &= -5 + j12\end{aligned}$$

We can verify this by simply multiplying the complex number 2 + j3 by itself:

$$\begin{aligned}(2 + j3)(2 + j3) &= 4 + j6 + j6 + j^2 9\\ &= 4 + j12 - 9 = -5 + j12\end{aligned}$$

While it is obvious that we do not need de Moivre's theorem to evaluate a simple expression such as $(2 + j3)^2$, we would certainly need this equation to evaluate a complicated expression such as (2 + j3) exponent 8.3. *See also* CARTESIAN COORDINATES, COMPLEX NUMBER.

DE MORGAN'S THEOREM

De Morgan's theorem, also called the de Morgan laws, involves sets of logically equivalent statements. Let the logical operation OR be represented by multiplication and the operation AND be represented by addition, as in Boolean algebra (see BOOLEAN ALGEBRA). Let the NOT operation be represented by complementation, indicated by an apostrophe ('). Then de Morgan's laws are stated as:

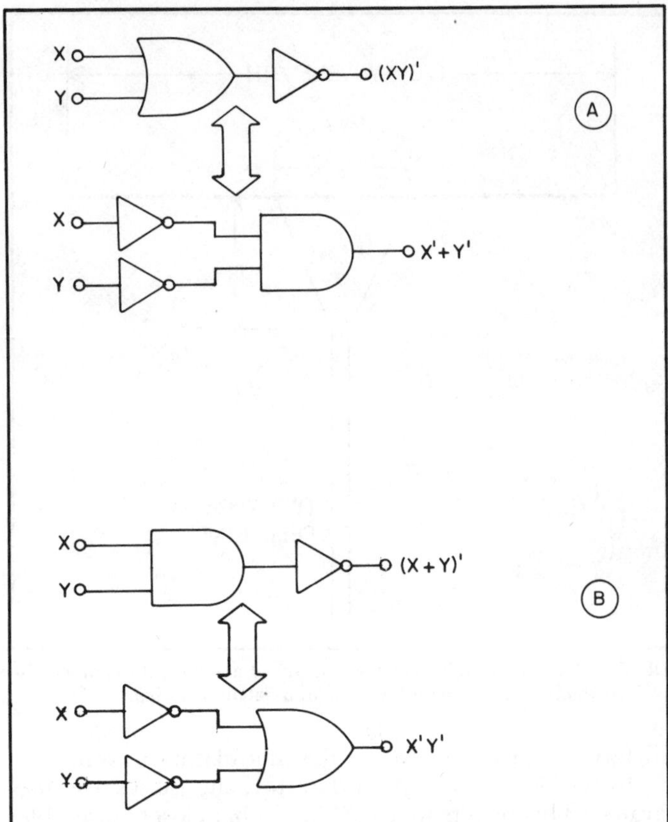

DE MORGAN'S THEOREM: Digital equivalents according to de Morgan's theorem. The two circuits shown at A are logically equivalent; the same is true of the two circuits at B.

$$(X + Y)' = X'Y'$$
$$(XY)' = X' + Y'$$

The illustration shows these rules in schematic form, showing logic gates, at A and B.

From these rules, it can be proven that the same laws hold for any number of logical statements $X_1, X_2, \ldots, X_n$

$$(X_1 + X_2 + \ldots + X_n)' = X_1'X_2' \ldots X_n'$$
$$(X_1 X_2 \ldots X_n)' = X_1' + X_2' + \ldots + X_n'$$

De Morgan's theorem is useful in the design of digital circuits, since a given logical operation may be easier to obtain in one form than in the other.

DEMULTIPLEXER

A demultiplexer is a circuit that separates multiplexed signals (*see* MULTIPLEX). Multiplexing is a method of transmitting several low-frequency signals on a single carrier having a higher frequency. Multiplexing may also be done on a time-sharing basis among many signals. A demultiplexer at the receiving end of a multiplex communications circuit allows interception of only the desired signal, and rejection of all the other signals.

The illustration shows a simplified block diagram of a demultiplexer. The main carrier is amplified and detected (*see* DETECTION), obtaining the subcarriers. A selective circuit then filters out all subcarriers except the desired one. A second detector demodulates the subcarrier, obtaining the desired signal. The main carrier modulation does not

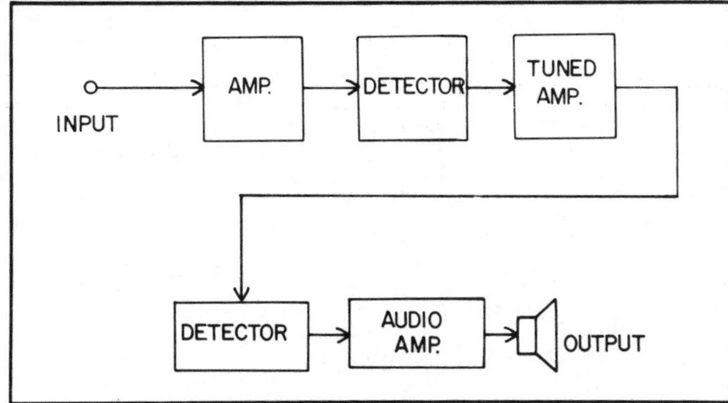

DEMULTIPLEXER: A block diagram of a demultiplexer.

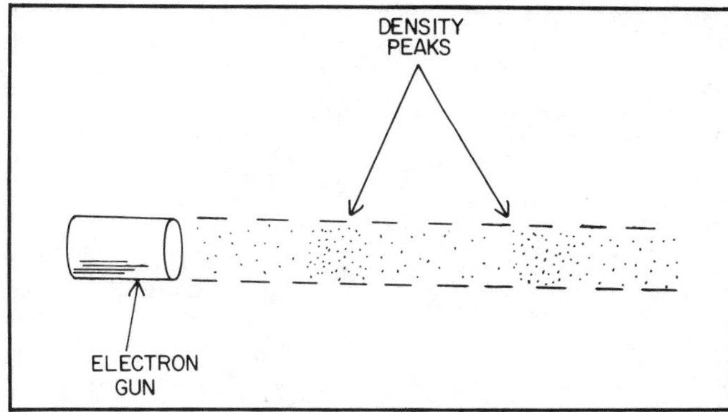

DENSITY MODULATION: In density modulation of a particle stream, the number of particles passing a point varies with time, but the speed of the particles is always constant.

have to be the same as the subcarrier modulation. Thus, different types of detectors may be needed for the two carriers. For example, the main carrier might be amplitude-modulated, while the subcarrier is frequency-modulated. *See also* MODULATION.

DENSITY MODULATION

Density modulation is the modulation of the intensity of a beam of particles, by varying only the number of particles. An example of density modulation is shown in the drawing. This type of modulation might be used in the electron stream of a cathode-ray tube. The greater the density of the electrons in the beam, the brighter the phosphorescent screen will glow when the electrons strike it. This is because, with increasing electron-beam density, more electrons strike the screen per unit time. The density is directly proportional to the number of particles passing any given point in a certain span of time.

In density modulation, all the particles of the beam move at constant speed, regardless of the intensity of the beam. When the speed of the electrons changes, the beam is said to be velocity-modulated. Density modulation may be combined with velocity modulation in some situations. *See also* VELOCITY MODULATION.

DEPENDENT VARIABLE

In a mathematical function y = f(x), the dependent variable is the variable that depends on the value of the other variable. In the case of y = f(x), y is the dependent variable, and x is called the independent variable (*see* INDEPENDENT VARIABLE).

In a Cartesian coordinate system, the dependent variable is usually plotted along the vertical axis, and the independent variable is usually plotted along the horizontal axis. The graph illustrates a plot of temperature versus the time of day for a hypothetical 24-hour period. In this graph, time is the independent variable, and is therefore plotted along the horizontal axis. The temperature is the dependent variable, and is plotted along the vertical axis. The horizontal axis is called the abscissa, and the vertical axis is called the ordinate.

A variable may depend on more than one independent factor. For example, the function z = f(w,x,y) is a function with three independent variables, namely w, x, and y; there is one dependent variable, namely z.

The coordinate system need not be Cartesian in order for the definitions of dependent and independent variables to apply. Any other type of system also has these variable designators. *See also* CARTESIAN COORDINATES, FUNCTION, VARIABLE.

DEPLETION LAYER

In a depletion-mode field-effect transistor, the region within the channel that does not conduct is called the depletion layer (*see* DEPLETION MODE, FIELD-EFFECT TRANSISTOR). A depletion layer forms at the junction of an N-type semiconductor and a P-type semiconductor whenever the junction is reverse-biased (*see* N-TYPE SEMICONDUCTOR, P-N JUNCTION, P-TYPE SEMICONDUCTOR).

A depletion layer is entirely devoid of charge carriers, and therefore no current can flow through the region. This is why a semiconductor diode will not conduct in the reverse direction. The depletion layer acts as a dielectric, and, in fact, a reverse-biased diode can be used as a capacitor.

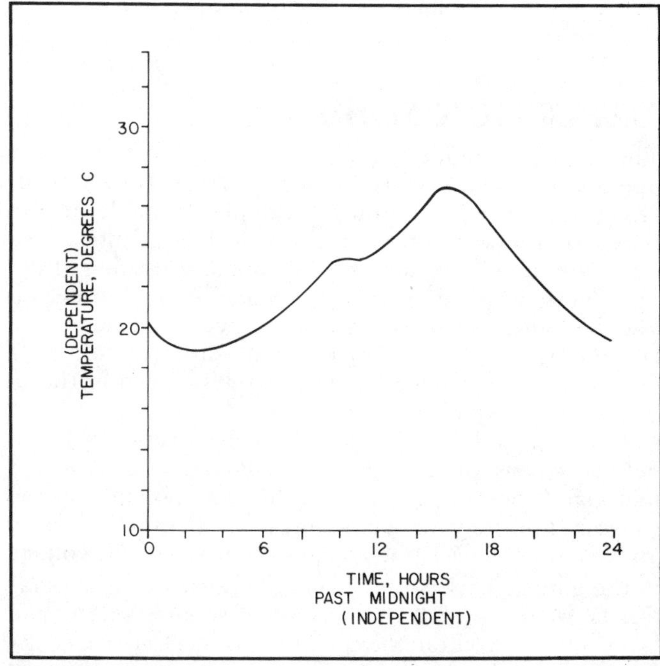

DEPENDENT VARIABLE: In this temperature-versus-time graph, the temperature is the dependent variable.

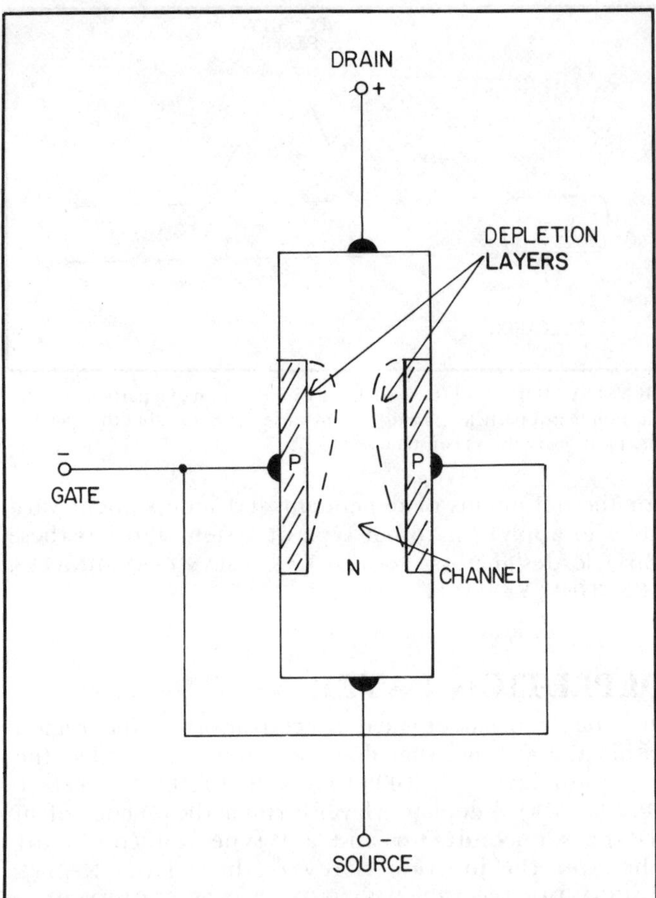

DEPLETION MODE: Pictorial illustration of the operation of a depletion-mode field-effect transistor. The conductivity depends on the width of the depletion layers.

The depletion layer reduces the cross-sectional area of the channel in a depletion-mode field-effect transistor when the gate-channel junction is reverse-biased. The illustration shows this. The greater the reverse-bias voltage, the larger the depletion layer becomes, until finally the channel is totally pinched off, and conduction ceases. *See also* PINCHOFF.

DEPLETION MODE

The depletion mode is a form of field-effect-transistor operation (*see* FIELD-EFFECT TRANSISTOR). In the field-effect transistor functioning as a depletion-mode device, the current flows through a region called the channel. Two electrodes on either side of the channel, consisting of the opposite type of semiconductor material and connected together and called the gate, are used to regulate the current through the device. The construction of a depletion-mode field-effect transistor is shown in the illustration.

In the depletion-mode field-effect transistor, the channel may consist of either N-type or P-type semiconductor material. When the gates are at the same potential as the source, a fairly large current flows through the channel. In an N-channel device, the application of a negative voltage to the gate constricts the channel because of the reverse bias between the gate-channel P-N junction. A depletion layer is formed within the channel (*see* DEPLETION LAYER). In a P-channel field-effect transistor, a positive voltage at the gate has this same effect. As the reverse bias is increased, the device will eventually stop conducting, because the depletion layer gets wider and wider. When the depletion layer extends all the way across the channel, the field-effect transistor is said to be pinched off.

Some field-effect transistors normally do not conduct, and require a forward bias for operation. These devices are called enhancement-mode field-effect transistors. *See also* DRAIN, ENHANCEMENT MODE, GATE, N-TYPE SEMICONDUCTOR, PINCHOFF, P-TYPE SEMICONDUCTOR, SOURCE.

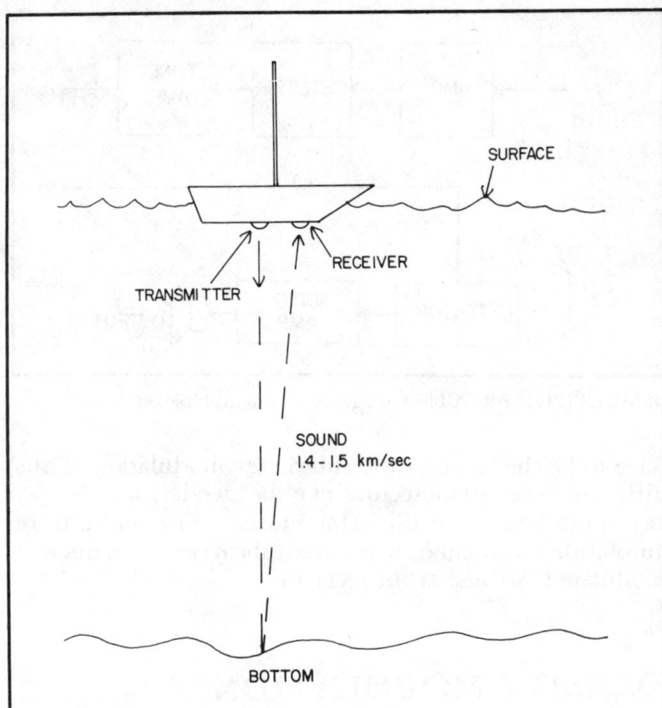

DEPTH FINDER: The depth finder operates by bouncing an acoustic pulse off the bottom of a body of water.

DEPOLARIZATION

Any electrolyte substance is subject to the formation of gases or other deposits around the electrodes in the presence of the flow of current. When this occurs, it is called polarization (*see* POLARIZATION). Polarization can prevent a dry cell or storage battery from operating properly. The deposits can act as electrical insulators, cutting off the flow of current by increasing the internal resistance.

Depolarization is the process of preventing, or at least minimizing, polarization in an electric cell. In dry cells, the compound manganese dioxide is used in the electrolyte material to retard polarization.

DEPTH FINDER

A depth finder, also called a fathometer, is a device for measuring the depth of a body of water at a given location. The depth finder operates by generating an underwater acoustic pulse, as shown in the drawing. The pulse travels outward from the generator in all directions under the water, but the maximum intensity is directed straight down. The pulse is reflected from the bottom of the body of water. The depth is determined by multiplying the speed of the pulse by the time required for the first echoes

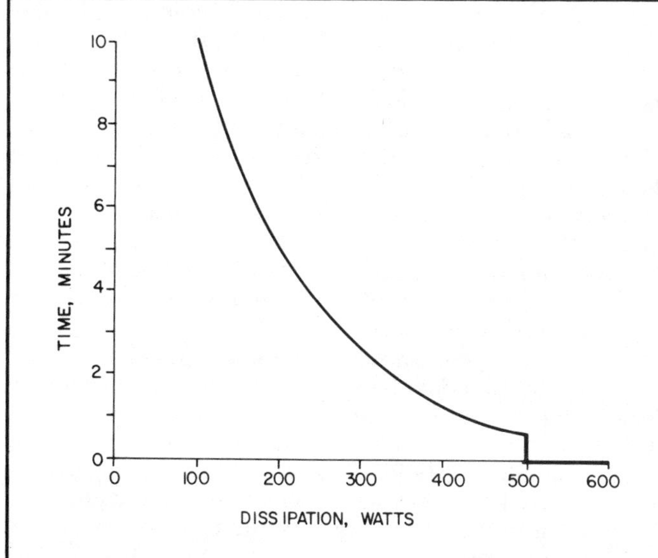

DERATING CURVE: A derating curve for a hypothetical dummy antenna. The unit can dissipate 100 watts indefinitely. Use at power levels above 500 watts is not recommended.

of the pulse by the time required for the first echoes to return. In fresh water, the sound pulses travel at approximately 1.4 kilometers per second; in salt water, the speed is about 1.5 kilometers per second. The depth determination is performed automatically within the receiver, giving a readout in feet, fathoms (units of 6 feet), or meters.

Depth finders are subject to some inaccuracy, since they are sometimes fooled by certain underwater objects or phenomena. However, depth finders have been extensively used to map the bottoms of lakes, rivers, and oceans. A special kind of underwater range finder, called the sonar, also operates on the acoustic principle. *See also* SONAR.

DERATING CURVE

A derating curve is a function that specifies the amount of dissipation a component can withstand safely. Such a function is plotted with time as the dependent variable, or ordinate, and dissipation or temperature as the independent variable, or abscissa (*see*DEPENDENT VARIABLE, INDEPENDENT VARIABLE).

The illustration shows an example of a derating curve for a dummy antenna, used for the purpose of testing radio transmitters off the air (*see* DUMMY ANTENNA). Such a device is usually made from a simple, noninductive resistor and a heatsink. The hypothetical dummy antenna in the illustration can withstand 100 watts of radio-frequency power dissipation for an indefinite length of time; the function is therefore asymptotic at the value 100 watts. As the amount of power increases above this value, however, the load cannot withstand the power dissipation continuously. The greater the power, the shorter the length of time allowed before overheating. Above 500 watts, the device should not be used at all; this fact is indicated on the derating curve by a drop to zero time.

DERIVATIVE

The derivative of a mathematical function is an expression of its rate of change, either at a certain instant, or in the

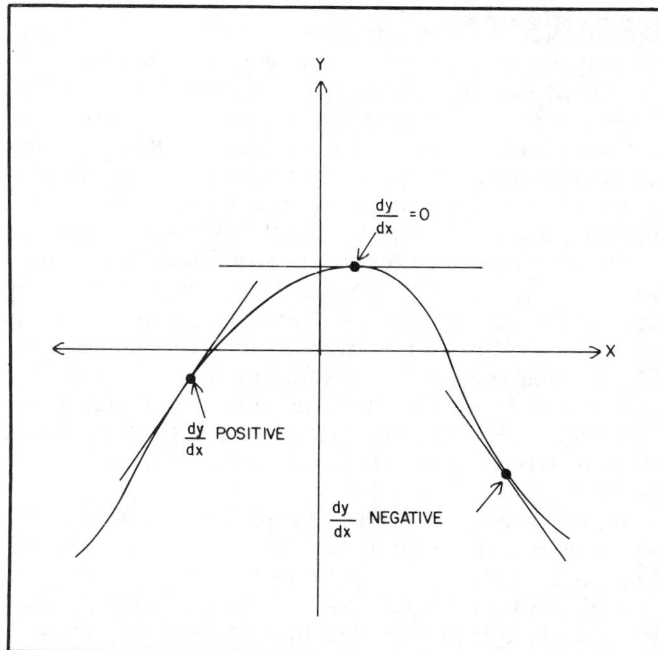

DERIVATIVE: The derivative of a function, at a point, is the slope of a line tangent to the function at that point.

general case. When the value of a function is constant, the rate of change, and therefore the derivative, is zero at all points. When the value of a function increases, the derivative is positive; the greater the rate of the increase, the larger the value of the derivative. When the value of a function decreases, the derivative is negative; greater rates of decrease result in derivatives that are more and more negative (*see* FUNCTION). The graph illustrates these three kinds of situations. The actual value of a derivative may be any real number at a specific point on a function. The derivative may, in some cases, be undefined at certain points of a function.

The derivative of a function may be the same at every point in the domain of a function, or it may change. The derivative of a function, in the general sense, is another function. For a given function $y = f(x)$, the derivative is denoted in a variety of ways, such as dy/dx, $df(x)/dx$, y', or $f'(x)$. The operation of finding a derivative for a function is called differentiation. Derivatives can be found in succession; such derivatives are called the second, third, fourth, and higher derivatives, generally written y'', y''', and so on, or $f''(x)$, $f'''(x)$, and so on.

The table shows the derivative functions for several of the most familiar mathematical functions. *See also* CALCULUS, DIFFERENTIATION, INDEFINITE INTEGRAL.

DERIVATIVE: DERIVATIVES OF SOME COMMON MATHEMATICAL FUNCTIONS. CONSTANTS ARE REPRESENTED BY THE LETTERS c AND n. THE VALUE OF e IS APPROXIMATELY 2.7183.

f(x)	df(x)/dx
$y = c$	$y' = 0$
$y = cx$	$y' = c$
$y = cx^2$	$y' = 2cx$
$y = cx^n$	$y' = ncx^{n-1}$
$y = \log_e(x)$	$y' = 1/x$
$y = e^x$	$y' = e^x$
$y = \sin x$	$y' = \cos x$
$y = \cos x$	$y' = -\sin x$
$y = \tan x$	$y' = \sec^2 x$

DESENSITIZATION

Desensitization is a reduction in the gain of an amplifier, particularly when the amplifier is employed in the radio-frequency stages of a receiver. Desensitization can be introduced deliberately; the radio-frequency (RF) gain control is used for the purpose of reducing the sensitivity of a receiver. Desensitization can occur when it is not wanted, and this sometimes presents a problem in radio reception.

The most frequent cause of unwanted desensitization of a radio-frequency receiver is an extremely strong signal near the operating frequency. Such a signal, after passing through the front-end tuning network, can drive the first RF-amplifier stage to the point where its gain is radically diminished. When this happens, the desired signal appears to fade out. If the interfering signal is strong enough, reception may become impossible near its frequency.

Some receivers are more susceptible to unwanted desensitization, or "desensing," than others. Excellent front-end selectivity, and the choice of amplifier designs that provide uniform gain for a wide variety of input signal levels, make this problem less likely. Sometimes, a tuned circuit in the receiver antenna feed line is helpful. *See also* FRONT END.

DESIGN

In electronics, design encompasses the physical and electrical arrangement of components and controls for the optimum performance of a piece of equipment. Good design helps to ensure reliability, conformance to required specifications, and ease and convenience of operation.

The design process begins when an engineer decides precisely what a circuit or device must do. Often, if not usually, there are two or more different designs that will accomplish the necessary task. Theoretical designs are drawn up in the form of schematic and wiring diagrams. Then, technicians build the different circuits, according to the diagrams and parameters provided by the engineer. Sometimes, a circuit fails to perform as expected; debugging is then required (*see* DEBUGGING). Finally, several different functioning models are completed, and the best one is selected and subjected to further refinement, until it is suitable for production.

Equipment design should be visually appealing as well as electrically functional, especially in consumer items. A person who develops electrical circuit designs is called a design engineer. There are various specialties within the design-engineering field, such as analog design, digital design, radio-frequency amplifier design, and so on. An important aspect of designing a circuit is the optimization of the usability of the equipment by people. This is called human engineering. *See also* HUMAN ENGINEERING.

DESKTOP COMPUTER

In recent years, computers have become smaller and smaller, as well as more affordable to the average consumer. Today, the desktop computer is a reality. An increasing number of homes and businesses have such computers, and they can be used for a wide variety of different purposes.

A typical small-business desktop computer, also suitable for home use, is shown in the photograph. The installation

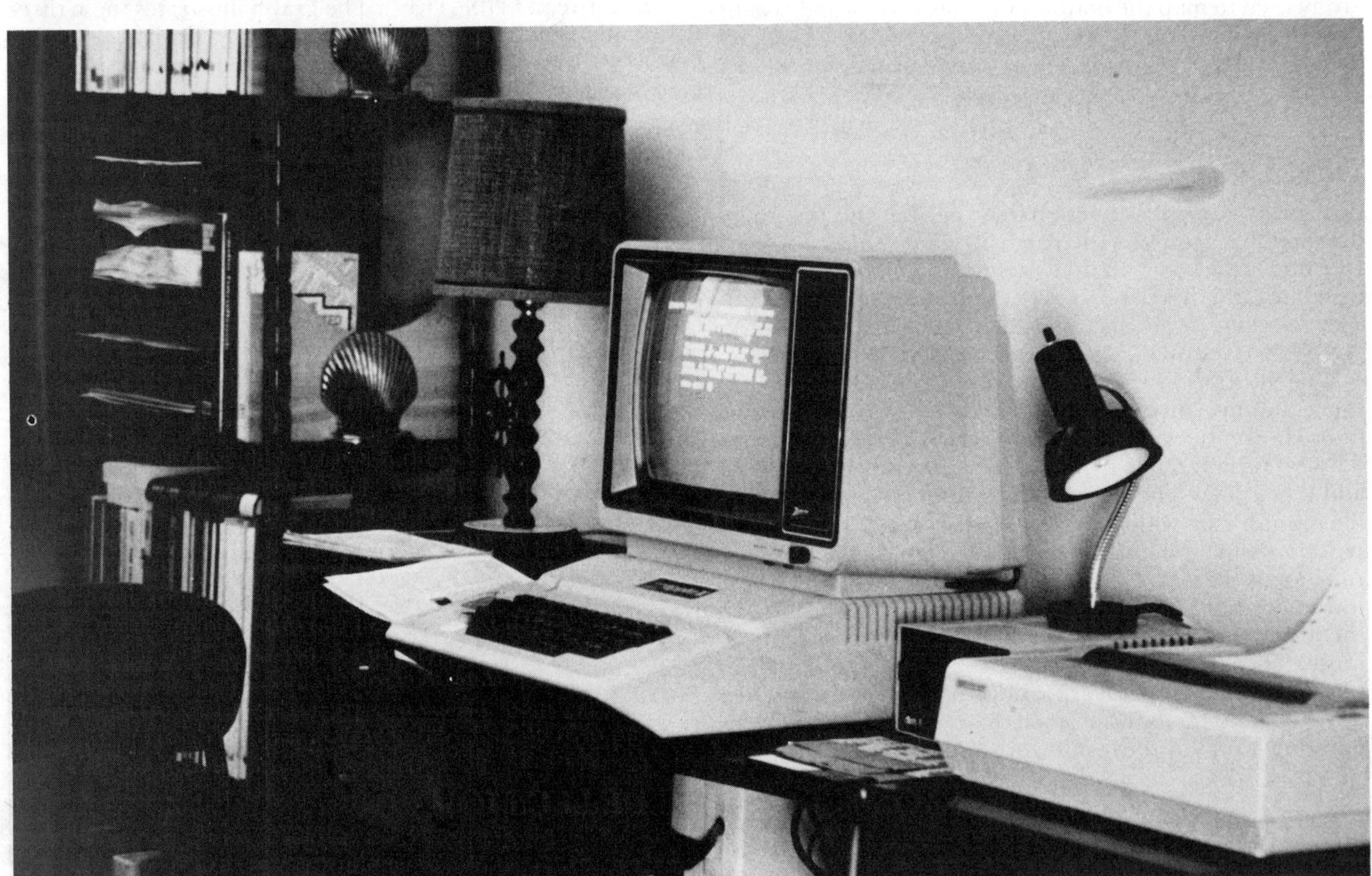

DESKTOP COMPUTER: Desktop computers are small enough to be used in any home or business.

consists of a video-display unit, keyboard, disk-drive system, and printer. The programs, or software, are provided on magnetic disks about the size of a small phonograph record. Some desktop computers can be programmed by impulses recorded on magnetic tape, and played on a typical cassette tape recorder.

Desktop computers can be used for record keeping, word processing, calculations, and the storage and transfer of information. They can also be used to automatically control servomechanisms and other machines. Of course, such computers are also used to play video games, and are increasingly used for educational purposes. *See also* COMPUTER.

DESOLDERING TECHNIQUE

When replacing a faulty component in an electronic circuit, it is usually necessary to desolder some connections. The method of desoldering employed in a given situation is called the desoldering technique.

With most printed-circuit boards, desoldering technique consists of the application of heat with a soldering iron, and the conduction of solder away from the connection by means of a braided wire called Solder Wick. The photograph shows this technique. The connection and the wick must both be heated to a temperature sufficient to melt the solder. Excessive heat, however, should be avoided, so that only the desired connection is desoldered, and so that the circuit board and nearby components are not damaged.

Many sophisticated desoldering devices are available. One popular device employs an air-suction nozzle, which literally swallows the solder by vacuum action as a soldering iron heats the connection. This sort of apparatus is especially useful when a large number of connections must be desoldered, since it works very fast. It is also useful in desoldering tiny connections, where there is little room for error.

For large connections, such as wire splices and solder-welded joints, it is sometimes better to remove the entire connection than to try to desolder it. *See also* SOLDER, SOLDERING GUN, SOLDERING IRON, SOLDERING TECHNIQUE, SOLDER WICK.

DETECTION

Detection is the extraction of the modulation from a radio-frequency signal. A circuit that performs this function is called a detector. Different forms of modulation require different detector circuits.

The simplest detector consists of a semiconductor diode, which passes current in only one direction. Such a detector is suitable for demodulation of amplitude-modulated (AM) signals. By cutting off one half of the signal, the modulation envelope is obtained. A Class-AB, B, or C amplifier can be used to perform this function, and also provide some gain. This is called envelope detection (*see* CLASS-AB AMPLIFIER, CLASS-B AMPLIFIER, CLASS-C AMPLIFIER, ENVELOPE DETECTOR).

There are several different ways of detecting a

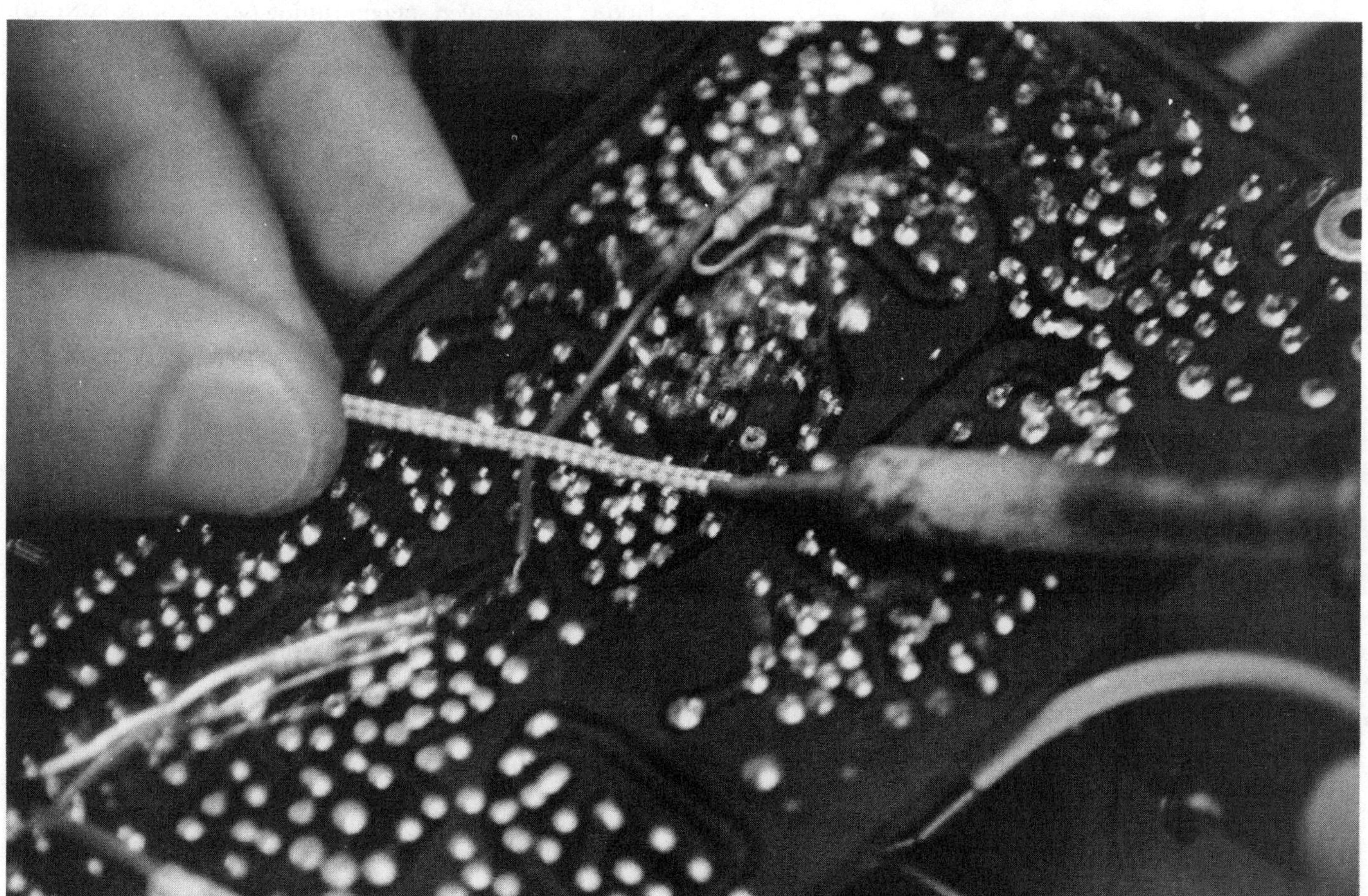

DESOLDERING TECHNIQUE: One desoldering method employs a braided wire called Solder Wick, which conducts the solder away from the connection for easy component removal.

frequency-modulated (FM) signal. These methods also apply to phase modulation. An ordinary AM receiver can sometimes be employed to receive FM by means of a technique called slope detection (*see* SLOPE DETECTION). Circuits that actually sense the frequency or phase fluctuations of a signal are called the discriminator and ratio detector (*see* DISCRIMINATOR, RATIO DETECTOR).

Single-sideband (SSB) and continuous-wave (CW) signals, as well as signals that employ frequency-shift keying (FSK), require a product detector for their demodulation (*see* PRODUCT DETECTOR). Such a circuit operates by mixing the received signal with the output of a local oscillator, resulting in an audio-frequency beat note.

Other kinds of detectors, such as digital-to-analog converters, are needed in specialized situations. The detector circuit in a radio receiver is generally placed after the radio-frequency or intermediate-frequency amplifying stages, and ahead of the audio-frequency amplifying stages. *See also* AMPLITUDE MODULATION, CONTINUOUS WAVE, FREQUENCY MODULATION, FREQUENCY-SHIFT KEYING, PHASE MODULATION, SINGLE SIDEBAND.

DETERMINANT

The determinant is a mathematical quantity, derived from the coefficients of a set of linear equations (*see* LINEAR ALGEBRA). A set of linear equations such as:

$$a_1x + a_2y = 0$$

$$b_1x + b_2y = 0$$

can be arranged in a matrix representation of coefficients, in the form:

$$\begin{pmatrix} a_1 & a_2 \\ b_1 & b_2 \end{pmatrix}$$

Similarly, a set of three linear equations in three variables, such as:

$$a_1x + a_2y + a_3z = 0$$

$$b_1x + b_2y + b_3z = 0$$

$$c_1x + c_2y + c_3z = 0$$

can be represented by the coefficient matrix:

$$\begin{pmatrix} a_1 & a_2 & a_3 \\ b_1 & b_2 & b_3 \\ c_1 & c_2 & c_3 \end{pmatrix}$$

This matrix representation can be made in any number of equations in any number of variables.

To have a determinant, a matrix must be square. That is, the number of rows must be the same as the number of columns. This means that the number of linear equations in the system must be identical to the number of unknowns. The determinant of a matrix having two rows and two columns is written

$$\begin{vmatrix} a_1 & a_2 \\ b_1 & b_2 \end{vmatrix} = a_1b_2 - a_2b_1$$

The determinant of a three-by-three matrix is

$$\begin{vmatrix} a_1 & a_2 & a_3 \\ b_1 & b_2 & b_3 \\ c_1 & c_2 & c_3 \end{vmatrix} = a_1b_2c_3 - a_1b_3c_2 - a_2b_1c_3 + a_2b_3c_1 + a_3b_1c_2 - a_3b_2c_1$$

Determinants of matrices larger than three-by-three are quite complicated, but are defined in linear-algebra textbooks. Such higher-order determinants exist for any matrix having the same number of rows as columns.

Determinants are useful in finding the solutions of large sets of linear equations. The calculation of a determinant is an algorithm well suited to computers, which can easily find the determinants of very large matrices, even having hundreds of rows and columns. *See also* MATRIX ALGEBRA.

DETUNING

Detuning is a procedure in which a resonant circuit is deliberately set to a frequency other than the operating frequency of a piece of equipment. Sometimes detuning is employed in the intermediate-frequency stages of a receiver to reduce the chances of unwanted interstage oscillation. This is called stagger tuning (*see* STAGGER TUNING).

Sometimes the preselector stage of a receiver is tuned to a higher or lower frequency than the intended one, for the purpose of reducing the gain in the presence of an extremely strong signal. A transmission line is usually made nonresonant, or detuned, to reduce its susceptibility to unwanted coupling with the antenna radiating element. *See also* PRESELECTOR.

DEUTERIUM

Deuterium is a form of the element hydrogen. Its atomic number is 1, and its atomic weight is 2.

Ordinary hydrogen consists of a single electron in orbit around a single proton (*see* HYDROGEN). This atom is the most common in the universe. Hydrogen makes up about 55 percent of the interstellar material in the cosmos. When a neutron is added to the nucleus of the simple hydrogen atom, deuterium is formed. Deuterium occurs in the fusion process, where hydrogen is converted into helium, liberating energy.

Initially, in the hydrogen fusion reaction, two protons combine and throw off a positive charged particle of anti-matter, known as a positron (*see* POSITRON). This results in a nucleus of deuterium, containing one proton and one neutron. This nucleus joins with another proton to form a helium-3 nucleus, which contains two protons and one neutron. Two nuclei of helium-3 then combine, giving off two protons, to form a nucleus of helium-4. Helium-4 is the second most abundant element in the universe, comprising about 44 percent of the interstellar matter. *See also* NUCLEAR FUSION.

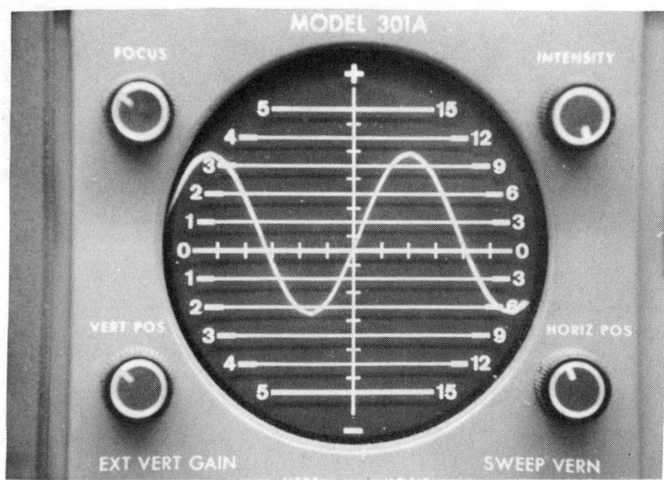

DEVIATION: A signal-monitor display of deviation of a frequency-modulated signal. The maximum frequency is at the top of the screen.

DEVIATION

In a frequency-modulated (FM) signal, deviation is the maximum amount by which the carrier frequency changes either side of the center frequency. The greater the amplitude of the modulating signal in an FM transmitter, the greater the deviation, up to a certain maximum. Deviation can be measured on a signal monitor, as shown in the photograph.

Deviation is generally expressed as a plus-or-minus (±) frequency figure. For example, the standard maximum deviation in a frequency-modulated communications system is ±5kHz. When the deviation is greater than the allowed maximum, an FM transmitter is said to be overdeviating. This often causes the received signal to sound distorted. Sometimes, the downward deviation is not the same as the upward deviation. This is the equivalent of a shift in the signal carrier center frequency with modulation, and it, too, can result in a distorted signal at the receiver. *See also* CARRIER SWING, FREQUENCY MODULATION, MODULATION INDEX.

DIAC

A diac is a semiconductor device that physically resembles a PNP bipolar transistor, except that there is no base connection. A diac acts as a bidirectional switch. The illustration shows the physical construction of a diac at A. The schematic symbol is shown at B.

The diac can be used as a variable-resistance device for alternating current. The output voltage can be regulated independently of the load resistance. In conjunction with a triac (*see* TRIAC), the diac can be employed as a motor-speed control, light dimmer, or other similar device. *See also* MOTOR-SPEED CONTROL.

DIAGRAM

A diagram is a drawing that depicts the organization of an electrical or electronic circuit. There are three basic kinds of diagrams, called the block diagram, the schematic diagram, and the wiring diagram.

A block diagram shows only the major circuits comprising a device. Rectangles or circles are used to represent each circuit, and are labeled according to their function.

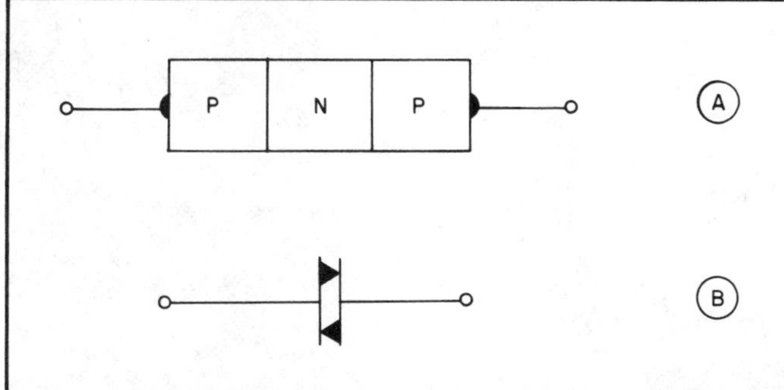

DIAC: At A, physical construction of a diac. At B, the schematic symbol.

The boxes or circles are interconnected by lines to show the general functional scheme. Arrows are sometimes used to show the path of a signal.

A schematic diagram shows all of the main components in a circuit, and how they are connected together. Component values are generally given, and also occasionally the reference designation. Wiring diagrams are still more complete; pin numbers for integrated circuits and tubes, color-code designators, and part numbers are shown for all components. This makes it easy for a technician to assemble the circuit, even if he or she does not know how the device works.

In schematic and wiring diagrams, certain standard symbols are employed to represent the various components. *See also* BLOCK DIAGRAM, SCHEMATIC DIAGRAM, SCHEMATIC SYMBOLS, WIRING DIAGRAM.

DIALING

In a telephone system, dialing is a means of selecting the telephone subscriber to be called. A sequence of seven digits is used for this purpose in the United States. Long-distance calls require a three-digit area-code prefix in addition to the seven-digit local designators. This entire sequence is preceded by still another digit to indicate direct dialing or operator assistance.

Telephone dialing devices are found in either of two forms: the rotary dial, and the Touchtone® system. The rotary dial transmits a series of pulses, with one pulse representing the number 1, two pulses representing the number 2, and so on up to ten pulses, representing the number 0. The pulses are sent at the rate of approximately ten per second, although other speeds are sometimes used. The Touchtone® system uses a 12-key or 16-key pad, with pairs of audible tones representing the digits 0 through 9, along with the symbols * and #. In the 16-key pad, four extra functions are included, designated A, B, C, and D. The Touchtone® system permits much more rapid dialing than the rotary system. The tone system also makes it easy to interface the telephone with radio communications equipment. *See also* TELEPHONE, TOUCHTONE®.

DIAL SYSTEM

Dial systems are the mechanical devices used for the purpose of setting the frequency of a radio transmitter or receiver to the desired value. Dial systems are also widely employed in such controls as preselectors, amplifier tuning networks, and antenna impedance-matching circuits.

A

B

DIAL SYSTEM: At A, a mechanical dial-drive system, using inductor permeability tuning. At B, graduations on the knob allow frequency determination to better than 1 kHz.

A dial system may be constructed in any of several ways. The simplest method is to mark the face of the equipment in a graduated scale, and to employ a control-shaft pointer knob. More sophisticated mechanisms have drive devices, rotating gears or disks, and a dial face with an adjustable indicator line. The photographs illustrate a precision dial mechanism with divisions of 1-kHz and interpolation accuracy to better than 0.5 kHz. At A, the mechanical device is shown, including the lighting system. At B, the front view is shown. The main dial is graduated in increments of 25 kHz, with a total span of 500 kHz. The knob skirt is calibrated in 1-kHz divisions. One turn of the knob represents exactly 25 kHz of frequency change.

The dial system shown has a linear readout. That is, a given angular knob rotation always corresponds to the same amount of frequency change. Some mechanisms are not linear. Tuning, in the case of the device illustrated, is accomplished by moving a powdered-iron core within an inductor. *See also* TUNING.

DIAMAGNETIC MATERIAL

Any material with a magnetic permeability less than 1 is called a diamagnetic material. The permeability of free space is defined as being equal to 1.

Examples of diamagnetic materials include bismuth, paraffin wax, silver, and wood. In all these cases, the magnetic permeability is only a little bit less than 1. *See also* FERROMAGNETIC MATERIAL, PARAMAGNETIC MATERIAL, PERMEABILITY.

DIAPHRAGM

A diaphragm is a thin disk, used for the purpose of converting sound waves into mechanical vibrations and vice versa. Diaphragms are used in speakers, headphones, and microphones.

In the speaker or headset, electrical impulses cause a coil to vibrate within a magnetic field. The diaphragm is attached to the coil, and moves along with the coil at the frequency of the electrical impulses. The result is sound waves, since the diaphragm imparts vibration to the air molecules immediately adjacent to it.

In a dynamic microphone, the reverse occurs. Vibrating air molecules set the diaphragm in motion at the same frequency as the impinging sound. This causes a coil, attached to the diphragm, to move back and forth within a magnetic field supplied by a permanent magnet or electromagnet. In a crystal or ceramic microphone, vibration is transferred to a piezoelectric material, which generates electrical impulses. In any microphone, the electrical currents have the same frequency characteristics as the sound hitting the diaphragm.

Some devices can operate as either a microphone or a speaker/headphone. *See also* CERAMIC MICROPHONE, CRYSTAL MICROPHONE, CRYSTAL TRANSDUCER, DYNAMIC LOUDSPEAKER, DYNAMIC MICROPHONE, HEADPHONE, SPEAKER, TRANSDUCER.

DIATHERMY

The process of heating body tissues by radio waves is called diathermy. A device that generates high-frequency electromagnetic fields for the purpose of diathermy is called a diathermy machine. Such machines are used by hospitals for physical therapy.

A diathermy machine heats the tissues deep under the skin, and is extremely effective in certain kinds of injuries. Diathermic heat is more penetrating than radiant or conducted heat, since it results from excitation of the body molecules by electromagnetic waves, and occurs with uniform intensity at all depths in living tissues. Diathermy machines usually operate in the high-frequency part of the electromagnetic spectrum, or between 3 and 30 MHz.

Diathermy machines can sometimes cause interference to high-frequency radio equipment, since the machines radiate at the same frequencies as shortwave receivers. This effect is called diathermic-machine interference or diathermy interference.

DICHOTOMIZING SEARCH

In a digital computer, a dichotomizing search is a method of locating an item, by number key, in a large set of items. Each item in the set is given a key. If there are 16 items, for example, they might be numbered 1 through 16.

The desired number key is first compared with a number halfway down the list. If the desired key is smaller than the halfway number, then the first half of the table is accepted, and the second half is rejected. If the desired key is larger than the halfway number, then the second half of the table is accepted, and the first half is rejected. The desired key is then compared with a number in the middle of the accepted portion of the table. On this basis, one half of this portion is accepted and the other half is rejected, just as in the first case. This process is repeated, selecting

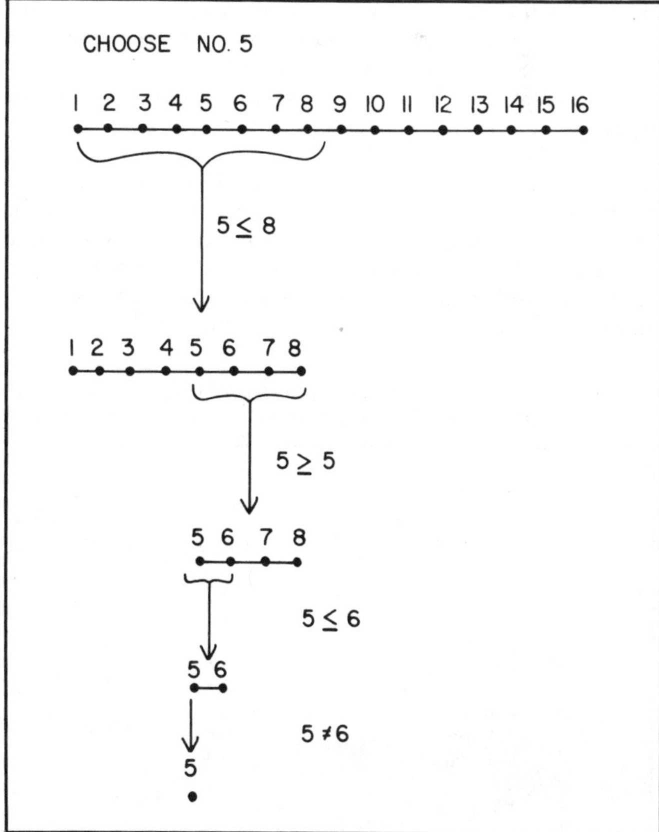

DICHOTOMIZING SEARCH: A dichotomizing search algorithm.

smaller and smaller parts of the table, until only one item remains. That key is then the desired key. The illustration shows the procedure of a dichotomizing search in a 16-item list.

The dichotomizing search is a form of algorithm often used in data processing. *See also* ALGORITHM, DATA PROCESSING.

DIELECTRIC

A dielectric is an electrical insulator, generally used for the purpose of manufacturing cables, capacitors, and coil forms. Dry, pure air is an excellent dielectric. Other examples of dielectrics include wood, paper, glass, and various rubbers and plastics. Distilled water is also a fairly good dielectric; water is known as a conductor only because it so often contains impurities that enhance its conductivity.

Dielectric materials are classified according to their ability to withstand electrical stress, and according to their ability to cause a charge to be retained when they are employed in a capacitor. Dielectric materials are also classified according to their loss characteristics. Generally, dielectric materials become lossier and less able to withstand electrical stress as the operating frequency is increased. *See also* DIELECTRIC ABSORPTION, DIELECTRIC AMPLIFIER, DIELECTRIC BREAKDOWN, DIELECTRIC CONSTANT, DIELECTRIC CURRENT, DIELECTRIC HEATING, DIELECTRIC LENS, DIELECTRIC LOSS, DIELECTRIC POLARIZATION.

DIELECTRIC ABSORPTION

After a dielectric material has been discharged, it may retain some of the electric charge originally placed across it. This effect is called dielectric absorption, because the material literally seems to absorb some electric charge. In capacitors, the dielectric absorption may make it necessary to discharge the component several times before the open-circuit voltage remains at zero.

Some dielectric materials absorb an electric charge to a greater extent than other substances. If a capacitor seems to have been completely discharged, it should still be checked for residual voltage before the assumption is made that there is no voltage across the component. Partially charged capacitors can present a dangerous shock hazard. *See also* CAPACITOR, DIELECTRIC.

DIELECTRIC AMPLIFIER

A dielectric amplifier is a voltage amplifier that functions because of variations in the dielectric constant of a special kind of capacitor (*see* DIELECTRIC CONSTANT). The input signal, applied to the capacitor, causes the dielectric constant of the device to fluctuate. This, in turn, changes its value. An alternating current, supplied by a local oscillator, is modulated by this fluctuation in the capacitance of the component. When this modulated current flows through a resistance of the correct value, a fluctuating voltage having a greater magnitude than the input voltage develops across the resistor. The output from the amplifier is taken at this point.

DIELECTRIC BREAKDOWN

When the voltage across a dielectric material becomes sufficiently high, the dielectric, normally an insulator, will begin to conduct. When this occurs in an air-dielectric cable, capacitor, or feed line, it is called arcing (*see* ARC). The breakdown voltage of a dielectric substance is measured in volts or kilovolts per unit length. Some dielectric materials can withstand much greater electrical stress than others. The amount of voltage, per unit length, that a dielectric material can withstand is called the dielectric strength.

In most materials, dielectric breakdown is the result of ionization. At a pressure of one atmosphere, air breaks down at a potential of about 2 to 4 kilovolts per millimeter. The value depends on the relative humidity and the amount of dust and other matter in the air; the greater the humidity or the amount of dust, the lower the breakdown voltage. In a solid dielectric material such as polyethylene, which has a dielectric strength of about 1.4 kilovolts per millimeter, permanent damage can result from excessive voltage. *See also* DIELECTRIC, DIELECTRIC HEATING, DIELECTRIC RATING.

DIELECTRIC CONSTANT

The dielectric constant of a material, usually abbreviated by the lower-case letter k, is a measure of the ability of a dielectric material to hold a charge. The dielectric constant is generally defined in terms of the capability of a material to increase capacitance. If an air-dielectric capacitor has a

DIELECTRIC CONSTANT, DIELECTRIC CURRENT, AND DIELECTRIC RATING: DIELECTRIC CHARACTERISTICS OF VARIOUS MATERIALS AT ROOM TEMPERATURE (APPROXIMATELY 25 DEGREES CELSIUS).

Material	Dielectric Constant 1 kHz	1 MHz	100 MHz	dc Resistivity, ohm-cm	Rating, kV/mm
Bakelite	4.7	4.4	4.0	10^{11}	0.1
Balsa wood	1.4	1.4	1.3	–	–
Epoxy resin	3.7	3.6	3.4	4×10^{7}	0.13
Fused quartz	3.8	3.8	3.8	10^{19}	0.1
Paper	3.3	3.0	2.8	–	0.07
Polyethylene	2.3	2.3	2.3	10^{17}	1.4
Polystyrene	2.6	2.6	2.6	10^{18}	0.2
Porcelain	5.4	5.1	5.0	–	–
Teflon	2.1	2.1	2.1	10^{17}	6
Water (pure)	78	78	78	10^{6}	–

value of C, then the same capacitor, with a dielectric substance of dielectric-constant value k, will have a capacitance of kC. The dielectric constant of air is thus defined as 1.

Various insulating materials have different dielectric constants. Some materials show a dielectric constant that changes considerably with frequency; usually the constant is smaller as the frequency increases. The table shows the dielectric constants of several materials at frequencies of 1 kHz, 1 MHz, and 100 MHz, at approximately room temperature. *See also* DIELECTRIC, PERMITTIVITY.

DIELECTRIC CURRENT

Dielectric materials are generally regarded as electrical insulators, but they are not perfect insulators. A small current flows through even the best quality dielectric substances. This current results in some heating, and therefore some loss (*see* DIELECTRIC HEATING, DIELECTRIC LOSS).

For direct current, the resistivity of a dielectric material is specified in ohm-millimeters or ohm-centimeters. This resistivity is always extremely high, assuming the dielectric material is not contaminated; it ranges from about 10^{12} to 10^{20} ohm-centimeters at room temperature. The table shows the direct-current resistivity of several types of dielectric materials. The greater the resistivity, the lower the dielectric current for a given voltage, provided the voltage is not so great that breakdown occurs. In the event of dielectric breakdown, a conductive path forms through the material, and the current increases dramatically. The substance then loses its dielectric properties. *See also* DIELECTRIC BREAKDOWN.

DIELECTRIC HEATING

Dielectric heating is the result of losses that occur in a dielectric material when it is subjected to an electric field (*see* DIELECTRIC LOSS). Dielectric materials heat up in direct proportion to the intensity of the electric field. The lossier a dielectric material, the hotter it will get in the presence of an electric field of a certain intensity.

Dielectric heating is sometimes deliberately used in the forming process of certain plastics. By subjecting the plastic to an intense radio-frequency electric field, the material becomes soft and pliable. In a prefabricated transmission line, dielectric heating can cause permanent damage. If the dielectric material in such a line, usually polyethylene, becomes soft, the conductors may move and cause a short circuit. Such dielectric heating might be the result of excessive transmitter power for the line in use. The large voltages caused by an excessive standing-wave ratio can also cause dielectric heating with consequent damage. *See also* STANDING-WAVE RATIO.

DIELECTRIC LENS

A dielectric lens is a device used for the purpose of focusing or collimating electromagnetic energy in the microwave frequency range. Such a lens operates in essentially the same manner as an optical lens. Any dielectric material may be used, although the dielectric loss should be as low as possible for maximum transmission of energy.

The drawing illustrates the principle on which a dielectric lens operates. Waves striking the lens at a 90-degree angle, that is, at the center, are not bent. When the waves do not strike the surface of the lens orthogonally, refraction occurs.

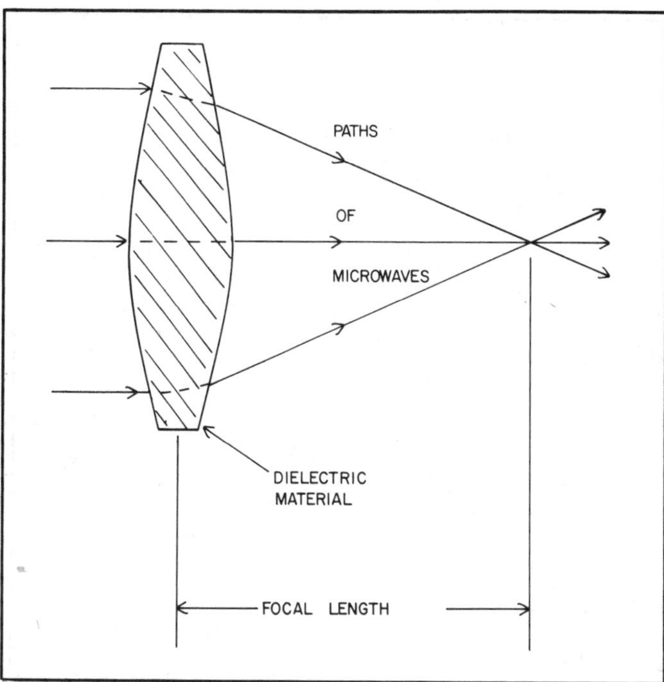

DIELECTRIC LENS: A dielectric lens focuses microwave energy, just as an optical lens focuses visible light. Here, parallel rays are brought to a focus by the dielectric refraction of the lens.

The focal length of a dielectric lens depends on the degree of curvature in the lens, and on the dielectric constant of the material from which the lens is made. The greater the curvature and/or the larger the dielectric constant, the shorter the focal length. *See also* FOCAL LENGTH, REFRACTION.

DIELECTRIC LOSS

Some dielectric materials transform very little of an electric field into heat. Some dielectrics transform a considerable amount of the field into heat. The dielectric loss in a particular material is an expression of the tendency of the material to become hot in the presence of a fluctuating electric field. Dielectric loss is usually expressed in terms of a quantity called the dissipation factor (*see* DISSIPATION FACTOR).

The dielectric loss of an insulating substance usually increases as the frequency of the fluctuating electric field is raised. Some of the best dielectric materials, such as polystyrene, have very low loss levels. Some materials, such as nylon, show generally lower losses as the frequency increases.

In transmission lines that carry radio-frequency energy, it is extremely important that the loss in the dielectric material be kept as low as possible. For this reason, many types of prefabricated transmission lines employ foamed dielectric material, rather than solid material. From the standpoint of low loss, air is one of the best dielectrics. Some transmission lines are sealed and filled with an inert gas such as helium to minimize the dielectric loss; spacers keep the inner conductor of such a coaxial line from short-circuiting to the braid. In some lines, such as the kind used in automobile radio antenna systems, the braid has a thin layer of polyethylene inside it, but most of the dielectric is air. *See also* DIELECTRIC.

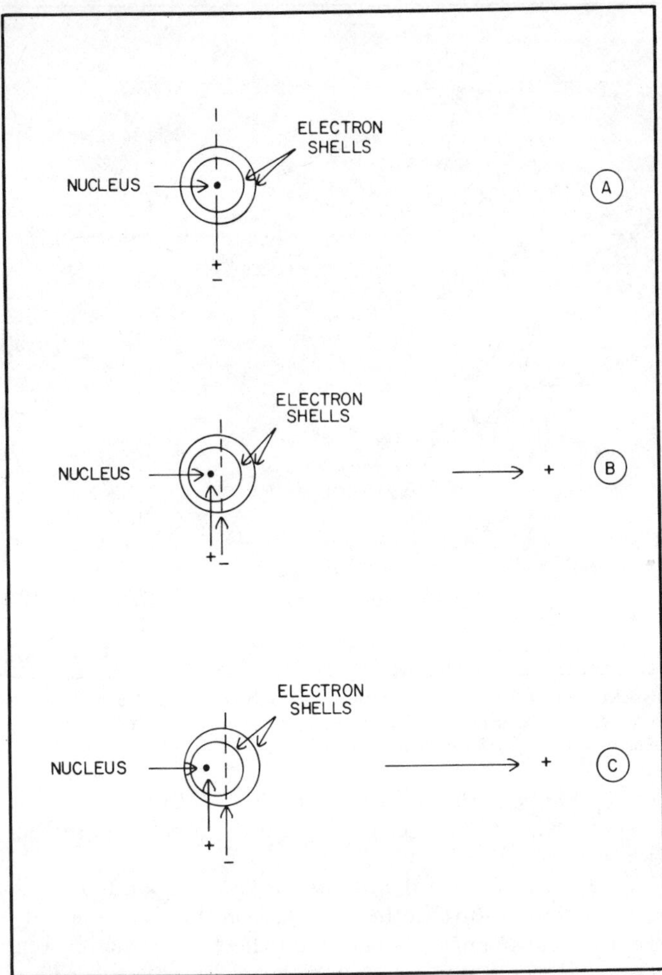

DIELECTRIC POLARIZATION: At A, the positive and negative charge centers of an atom of dielectric material are located in exactly the same place, with no electric field present. At B, a weak electric field causes some displacement. At C, a stronger field causes more displacement.

DIELECTRIC POLARIZATION

When an electric field is placed across a dielectric material, the location of the positive-charge center in each atom is slightly displaced relative to the negative-charge center. The greater the intensity of the surrounding electric field, the greater the charge displacement. This results in what is known as dielectric polarization (see illustration). At A, the atoms of the dielectric material are shown, with their charge-center locations, under conditions of no electric field. At B, the charge centers are illustrated under the influence of a weak electric field, with the positive part of the field toward the right in the diagram. At C, the atomic charge centers are shown under the influence of a strong electric field, again with the positive part of the field toward the right. When the polarity of the electric field is reversed, the dielectric polarization also reverses.

Dielectric polarization is caused by the forces of attraction between opposite charges, and repulsion between like charges. If the intensity of the electric field is sufficiently great, electrons are stripped from the atomic nuclei. This results in conduction, since the atoms then easily flow from one atom to the next, throughout the material. Sometimes this ionization causes permanent damage, such as in a solid dielectric. In the case of a gaseous dielectric material, such as air, the ionization does not cause permanent damage. *See also* DIELECTRIC, DIELECTRIC BREAKDOWN.

DIELECTRIC RATING

The dielectric rating of an insulating material refers to its ability to withstand electric fields without breaking down (*see* DIELECTRIC BREAKDOWN). The term may also be used to refer to the general characteristics of a dielectric material, such as the dielectric constant, the resistivity, and the loss. The dielectric charactersistics of several common substances were shown in the preceding table.

In a device that employs a dielectric material, such as a capacitor, the dielectric rating is usually the breakdown voltage of the entire device. This rating depends, in the case of a capacitor, on the thickness of the dielectric layer, as well as on the particular material used.

DIELECTRIC TESTING

Dielectric testing is a means of determining the properties of a dielectric substance, such as the dielectric constant, breakdown voltage, loss characteristics, and so on. These parameters often vary with the frequency, temperature, and sometimes with the relative humidity and atmospheric pressure.

To determine the dielectric constant of a material at a given electromagnetic frequency, the value of a capacitor is compared, using an air dielectric at a pressure of one atmosphere and zero humidty, with as much of the dust removed as possible. The frequency and temperature are varied and, if applicable, the humidity and air pressure.

The dielectric breakdown voltage is determined by increasing the intensity of the electric field until ionization, or arcing, occurs. The frequency and temperature are again varied and, if applicable, the humidity and air pressure, to determine the effects of these parameters.

The dielectric resistivity is measured using an extremely sensitive current meter and high voltages (but not so high as to cause dielectric breakdown). The layer of tested material is usually made very thin. Care must be exercised to ensure that the dielectric material does not ionize, even for an instant. Again, the temperature and frequency are varied and, if applicable, the humidity and air pressure.

Dielectric loss is measured in terms of the dissipation factor (*see* DISSIPATION FACTOR). All parameters are, as in the other cases, varied, to find out what effects they have on the dielectric loss.

Less commonly tested properties of dielectric substances include thermal expansion and contraction, the tendency to polarize in the presence of a direct-current electric field, and the tendency to absorb moisture. *See also* DIELECTRIC, DIELECTRIC BREAKDOWN, DIELECTRIC CONSTANT, DIELECTRIC CURRENT, DIELECTRIC LOSS, DIELECTRIC POLARIZATION.

DIETZHOLD NETWORK

A Dietzhold network is a shunt m-derived circuit, used for the purpose of tailoring the frequency response of a wideband radio-frequency power amplifier. The values of capacitance and inductance are chosen for optimum amplifier performance within a certain frequency range.

Examples of Dietzhold networks are illustrated in the schematic diagrams. At A, an unbalanced network, or half section, is shown. At B, a balanced Dietzhold network is illustrated. *See also* M-DERIVED FILTER.

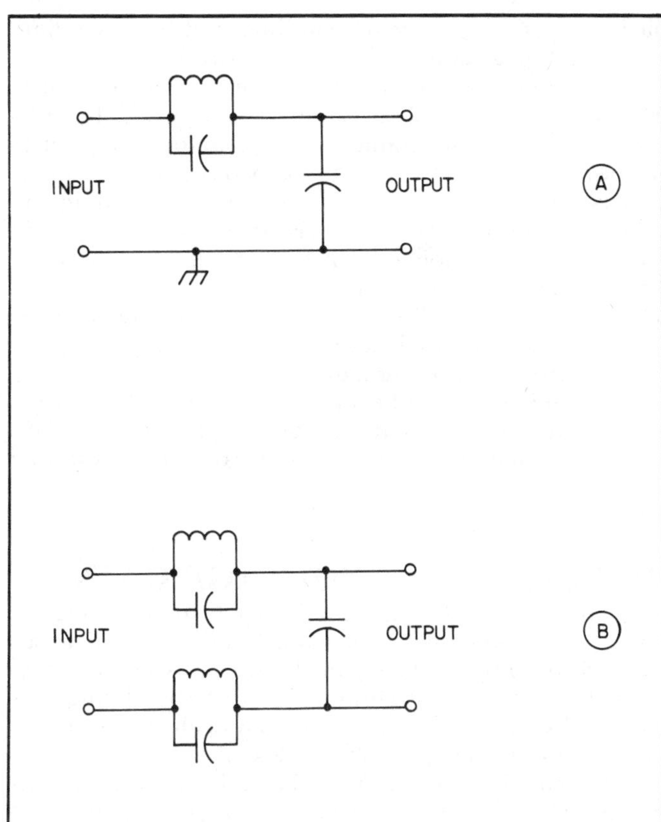

DIETZHOLD NETWORK: At A, unbalanced (half section); at B, balanced.

DIFFERENTIAL AMPLIFIER

A circuit that responds to the difference in the amplitude between two signals, and produces gain, is called a differential amplifier. Such an amplifier has two input terminals and two output terminals. Differential amplifiers are often found in integrated-circuit packages. The diagram illustrates a simple differential amplifier.

When two identical signals are applied to the input terminals of a differential amplifier, the output is zero. The greater the difference in the amplitudes of the signals, the greater the output amplitude. Differential amplifiers can be used as linear amplifiers, and have a broad operating-frequency range. Differential amplifiers can also be used as mixers, detectors, modulators, and frequency multipliers. *See also* DIFFERENTIAL VOLTAGE GAIN.

DIFFERENTIAL CAPACITOR

A form of variable capacitor having two sets of stator plates, and a single rotor-plate set, is called a differential capacitor. The photograph shows a differential air-variable capacitor, employed in an antenna-tuning network.

As the rotor is turned, the rotor plates move into one set of stators and out of the other. There are thus two variable capacitors in a differential capacitor. The values of these two capacitors are inversely related; as the capacitance of one unit increases, the capacitance of the other unit decreases. When one capacitor is at minimum value, the other is at the maximum. The rotor plates of the two capacitors are connected together and form the common terminal. The stator sets are brought out separately. *See also* AIR-VARIABLE CAPACITOR, VARIABLE CAPACITOR.

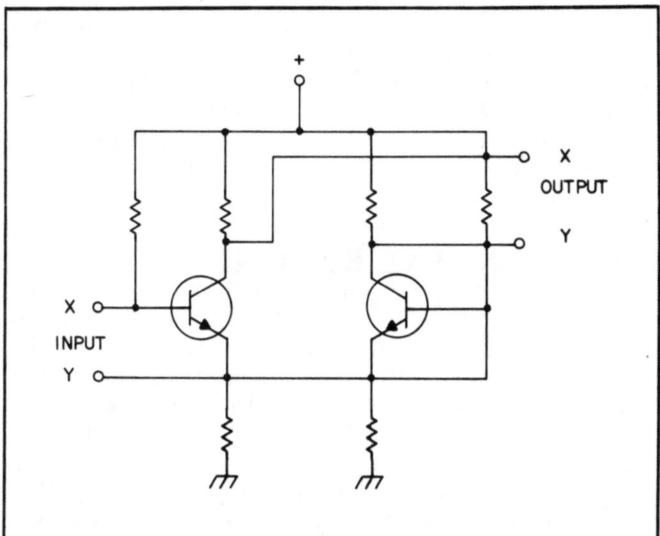

DIFFERENTIAL AMPLIFIER: A differential amplifier responds to the difference in amplitude of two input signals.

DIFFERENTIAL EQUATION

A differential equation is a mathematical equation in which some of the variables, or all of them, are derivatives or differentials (*see* DERIVATIVE, DIFFERENTIATION). Such equations can be categorized according to their form. Equations containing only first derivatives are called first-order differential equations. If the highest order derivative in a differential equation is the nth, then the equation is called an nth-order differential equation.

Differential equations can be fairly simple, or they can be quite complicated and hard to solve. Entire mathematics courses at the college level are devoted to these equations. The solution of a differential equation generally requires integration (*see* INDEFINITE INTEGRAL). Differential equations occur in the most advanced electronics engineering applications.

DIFFERENTIAL INSTRUMENT

A differential instrument or indicating device is a meter that shows the difference between two input signals. Two identical coils are connected to two sets of input terminals. The coils carry currents in opposite directions. When the

DIFFERENTIAL CAPACITOR: An air-variable differential capacitor.

currents are equal, their effects in the meter mechanism balance, and there is no deflection of the needle. When one current is greater in magnitude than the other, the net current flow is in one direction or the other.

Differential instruments may use amplifiers for better sensitivity. *See also* DIFFERENTIAL AMPLIFIER.

DIFFERENTIAL KEYING

Differential keying is a form of amplifier/oscillator keying in a code transmitter. While oscillator keying has the advantage of allowing full break-in operation, since the transmitter is off during the key-up intervals, chirp often results because of loading effects of the amplifier stages immediately following the keyed oscillator (*see* BREAK-IN, CHIRP, CODE TRANSMITTER).

An ingenious method of eliminating the chirp (that almost always accompanies oscillator keying) was devised several decades ago. In this technique, called differential keying, the amplifier is keyed along with the oscillator, but in a delayed fashion. When the key is pressed, the oscillator comes on first. Then, a few milliseconds later, the amplifier is switched on (see drawing). This gives the oscillator time to chirp before the signal is actually transmitted over the air! When the key is lifted, both the oscillator and the amplifier are switched off together; alternatively, the amplifier may be switched off a few milliseconds before the oscillator. *See also* KEYING.

DIFFERENTIAL TRANSDUCER

A differential transducer is a sensor device that has two input terminals and a single output terminal. The output is proportional to the difference between the input parameters. An example is a differential pressure transducer, which responds to the difference in mechanical pressure between two points. Any pair of transducers can be connected in a differential arrangement by using a differential amplifier (*see* DIFFERENTIAL AMPLIFIER).

When the two input parameters to a differential transducer are identical, the output is zero. The greater the difference between the two inputs, the greater the output will be. The most output occurs when one transducer section receives a large input and the other gets none. *See also* TRANSDUCER.

DIFFERENTIAL TRANSFORMER

A differential transformer is a special kind of transformer, having one or two primary windings and two secondary windings, and an adjustable powdered-iron or ferrite core. The windings are generally placed on a solenoidal form. The primaries, if dual, are connected in series. The secondary windings are connected in phase-opposing series fashion (*see* TRANSFORMER, TRANSFORMER PRIMARY, TRANSFORMER SECONDARY). The schematic diagram shows a differential transformer. As the core is moved in and out of the solenoidal form, the coupling ratio between winding pair X and winding pair Y (as illustrated) varies. This affects the amplitude and phase of the transformer output. When the core is at the center, so that the coupling between the two winding pairs is equal, the output of the transformer is zero. The farther off center the core is positioned, the greater the output amplitude. However, the output phase is reversed with the core nearer winding pair X, as compared to when it is closer to pair Y.

DIFFERENTIAL VOLTAGE GAIN

The gain of a differential amplifier, in terms of voltage, is known as differential voltage gain. Any differential device has a differential voltage-gain figure. It is measured in decibels (*see* DECIBEL). However, a differential amplifier usually displays a signal gain of significant proportions; passive devices generally show a loss.

Differential voltage gain is the ratio, in decibels, between a change in output voltage and a change in input voltage applied to either input terminal. *See also* DIFFERENTIAL AMPLIFIER.

DIFFERENTIATION

The mathematical determination of the derivative of a function is called differentiation. Differentiation may be performed either at a single point on a function, or as a general operation involving the entire function (*see* CALCULUS, DERIVATIVE).

Certain electronic circuits act to differentiate the waveform supplied to their inputs. The output waveform of such a circuit is the derivative of the input waveform, representing the rate of change of the input amplitude. If the amplitude is becoming more positive, then the output of the differentiator circuit is positive. If the amplitude is becoming more negative, then the output is negative. If the input amplitude to a differentiating circuit is constant, the output is zero.

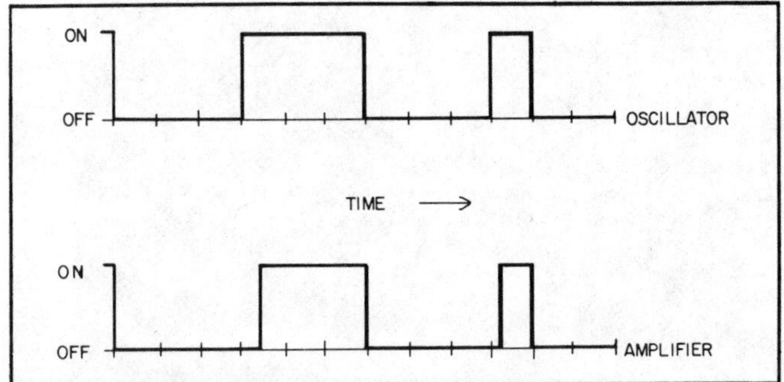

DIFFERENTIAL KEYING: In differential keying, the transmitter amplifier is keyed (switched on) slightly after the oscillator. This keeps oscillator chirp from being sent over the air.

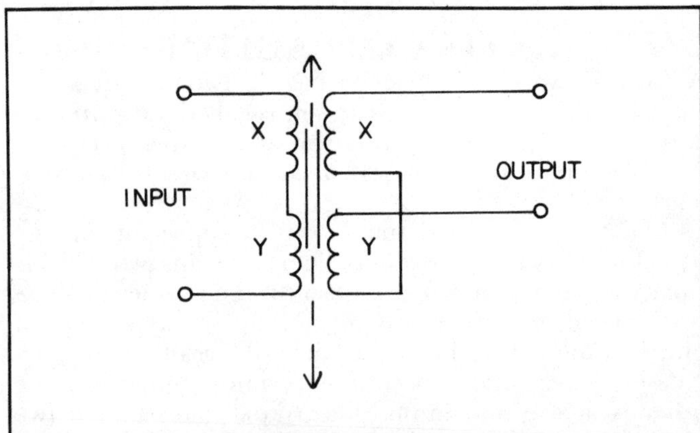

DIFFERENTIAL TRANSFORMER: A differential transformer allows adjustment of the output voltage and phase, by moving the core back and forth within the windings.

Differentiation is, both electrically and mathematically, the opposite of integration. When an integrator and a differentiator circuit are connected in cascade, the output waveform is usually identical to the input waveform. *See also* DIFFERENTIATOR CIRCUIT, INTEGRATION, INTEGRATOR CIRCUIT.

DIFFERENTIATOR CIRCUIT

An electronic circuit that generates the derivative, with respect to time, of a waveform is called a disferentiator circuit (*see* DERIVATIVE). When the input amplitude to a differentiator is constant—that is, a direct current—the output is thus zero. When the input amplitude increases at a constant rate, the output is a direct-current voltage. When the input is a fluctuating wave, such as a sine wave, the output varies according to the instantaneous derivative of the input. A perfectly sinusoidal waveform is shifted 90 degrees by a differentiator. Figure 1 shows the cases of constant input, uniformly increasing input, and sine-wave input, along with the resultant output waveforms.

A differentiator circuit can be extremely simple. Figure 2 shows two examples of differentiating circuits. The circuit at A is a series resistance-capacitance network; at B, an operational amplifier is used.

The differentiator circuit generally acts in an exactly opposite manner to an integrator circuit. This does not mean, however, that cascading an integrator and a differentiator will always result in the same signal at the output and input. Such is usually the case, but certain waveforms cannot be duplicated in this way.. *See also* INTEGRATION, INTEGRATOR CIRCUIT.

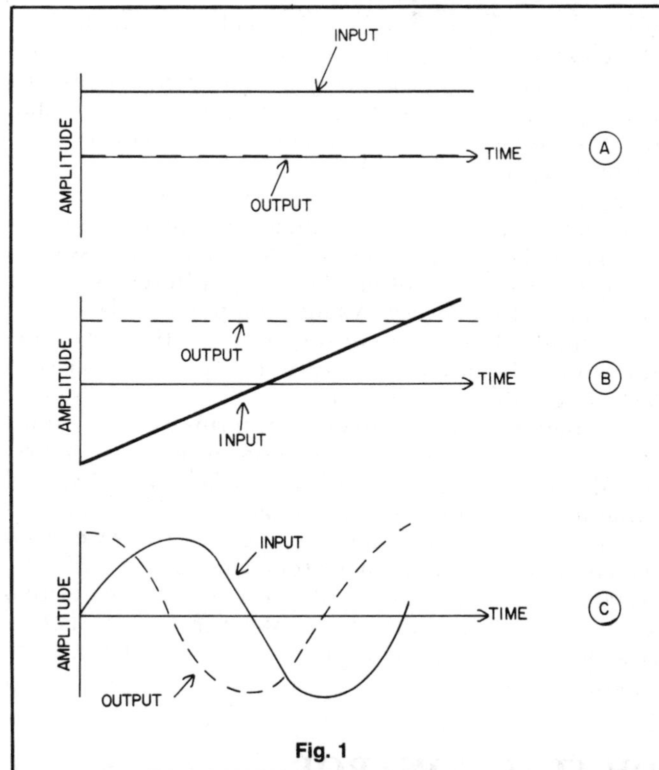

DIFFERENTIATION: Differentiation of waveforms. At A, a direct-current input gives zero output. At B, an increasing input gives a positive direct-current output. At C, a sine-wave input is shifted in phase by 90 degrees.

DIFFRACTION

Any effects that show wavelike properties, such as sound, radio-frequency energy, and visible light, have the ability to turn sharp corners and to pass around small obstructions. This property is called wave diffraction. The wavelike nature of visible light was discovered by observing interference patterns caused by diffraction.

Diffraction allows you to hear a friend speaking from around the corner of an obstruction such as a building, even when there are no nearby objects to reflect the sound. This effect, called razor-edge diffraction, is illustrated at A in the illustration. The corner of the obstruction acts as a second source of wave action. Electromagnetic fields having extremely long wavelength, comparable to or greater than the diameter of an obstruction, are propagated around that obstruction with very little attenuation. For example, low-frequency radio waves are easily transmitted around a concrete-and-steel building that is small compared to a wavelength, at B. As the wavelength of the energy becomes shorter, however, the obstruction causes more and more attenuation. At very high frequencies, a concrete-and-steel building casts a definite shadow in the wave train of electromagnetic signals.

A piece of clear plastic, having thousands of dark lines drawn on it, is called a diffraction grating. Light passing through such a grating is split into its constituent wavelengths, in very much the same way as light transmitted through a prism. This effect is the result of interference caused by diffraction through the many tiny slits. *See also* SPECTROSCOPE.

DIFFUSION

When one material spreads into, and permeates, another substance, the process is called diffusion. Many different kinds of semiconductor devices are made by diffusing one kind of substance into another. In diodes and transistors,

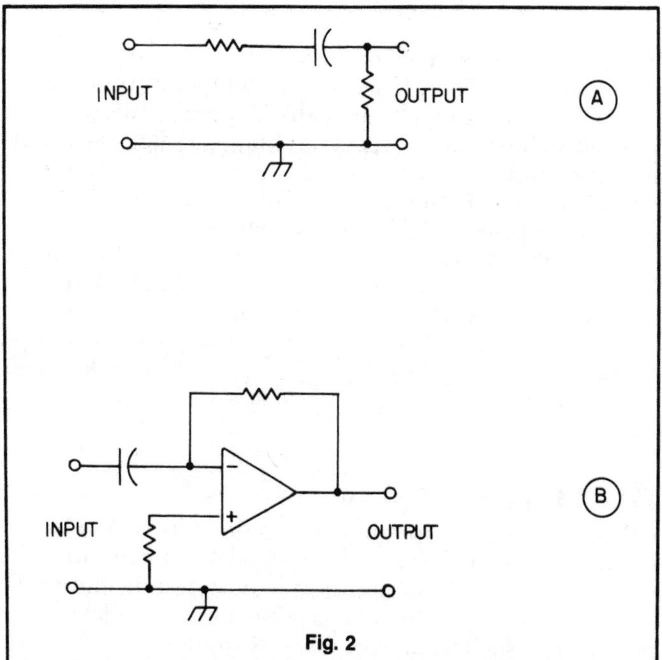

DIFFERENTIATOR CIRCUIT: Two forms of differentiator circuits. At A, a simple passive resistance-capacitance circuit is shown. At B, an active circuit using an operational amplifier.

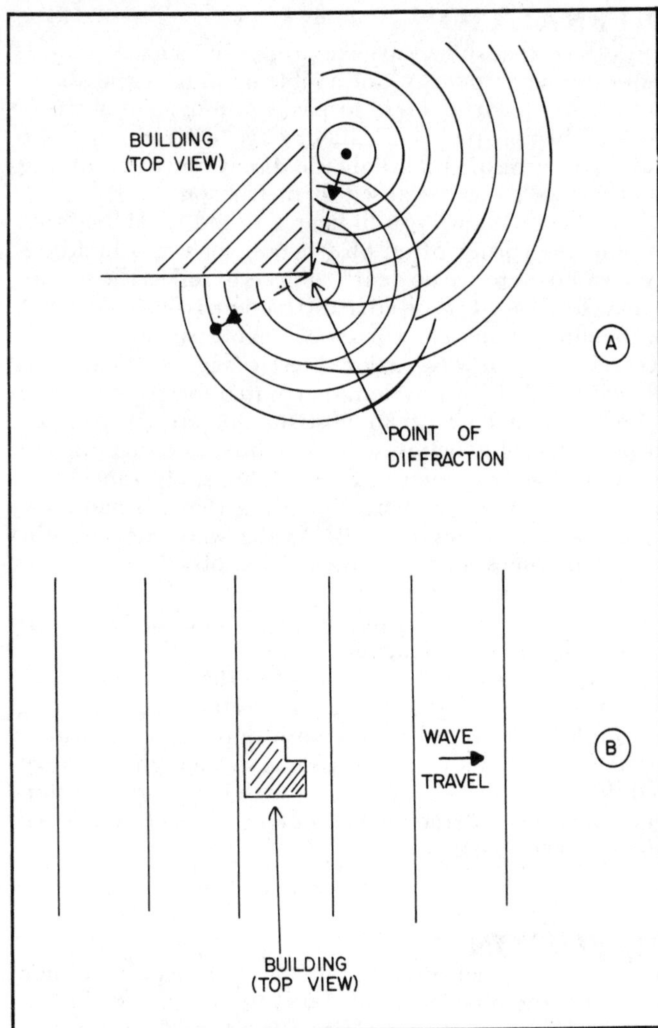

DIFFRACTION: Examples of diffraction. At A, sound diffracts around the corner of a building. At B, electromagnetic waves are diffracted around an obstruction that is small compared to a wavelength.

N-type and P-type semiconductors, as well as metal alloys and oxides, can be diffused at high temperatures to form various types of components with different properties. It is possible to build an integrated circuit on a silicon wafer by means of diffusion techniques.

The diffused metal-oxide semiconductor (DMOS) field-effect transistor is an extremely high-gain semiconductor device. It is made by the diffusion process. Diffused bipolar transistors can be made with extremely thin base regions, allowing the operating frequency to be higher than is the case with most devices. The gain is also optimized by this process. *See also* N-TYPE SEMICONDUCTOR, P-TYPE SEMICONDUCTOR.

DIGITAL

The term digital is used to define a quantity that exists at discrete states, or levels, rather than over a continuously variable range. Digital signals may have two or more different states. Usually, the total possible number of levels is a power of 2, such as 2, 4, 8, 16, and so on.

Probably the simplest kind of digital signal is the Morse code. The length of a bit in this code is called a dot length. More complex digital signals include frequency-shift keying, pulse modulation, and the like (*see* FREQUENCY-SHIFT KEYING, MORSE CODE, PULSE MODULATION).

Circuits that operate in a digital fashion are becoming the rule in all electronic applications. The digital mode is generally more efficient, more precise, and more straightforward than older analog techniques. Digital displays of current, voltage, time, frequency, and other parameters are extensively used because of easy readability and precision. Digital operations can be carried out with extreme speed. *See also* DIGITAL CIRCUITRY, DIGITAL DEVICE.

DIGITAL CIRCUITRY

Any circuit that operates in the digital mode is called digital circuitry (*see* DIGITAL). Examples of digital circuits include the flip-flop and the various forms of logic gates. Complicated digital devices are built up from simple fundamental units such as the inverter, AND gate, and OR gate. There is no limit, in theory, to the level of complexity attainable in digital circuit design. Physical space presents the only constraint. With increasingly sophisticated methods of miniaturization, more and more complex digital circuits are being packed into smaller and smaller packages. Entire computers can now be installed on a desktop for home or business use (*see* DESKTOP COMPUTER).

Digital circuitry is generally more efficient than older analog circuitry (*see* ANALOG), because of the finite number of possible digital states, compared to the infinite number of levels in an analog circuit. Many digital circuits can operate at greater speed than similar analog devices. *See also* AND GATE, FLIP-FLOP, INVERTER, OR GATE.

DIGITAL COMPUTER

A digital computer is a computer made up of digital circuitry (*see* DIGITAL CIRCUITRY). Such a device consists of logic circuits, which process signals, and memory circuits, which store information. Most computers today are digital, although some use analog designs (*see* ANALOG, ANALOG COMPUTER). Digital computers are used to compile and organize information, and to perform sophisticated mathematical calculations. A digital computer might, in fact, better be termed a digital information processor.

Digital computers often use many different kinds of input and output devices. A single computer can be used to serve hundreds of different subscribers at the same time. Sometimes, digital computers are accessed by remote control using radio or telephone.

All digital computers operate in terms of binary units called bits, and process information in terms of binary-coded numbers. This is called machine language. A special program translates the machine language into a higher-order language. The computer operator converses with the computer in the higher-order language. *See also* ASSEMBLER AND ASSEMBLY LANGUAGE, BASIC, BINARY-CODED NUMBER, COBOL, COMPUTER, COMPUTER PROGRAMMING, FORTRAN, HIGHER-ORDER LANGUAGE, MACHINE LANGUAGE.

DIGITAL CONTROL

Digital control is a method of adjusting a piece of equipment by digital means. Parameters frequently controlled by digital methods include the frequency of a radio receiver or transmitter, the programming of information into a memory channel, and the selection of a memory address. There are, of course, many other applications of digital control.

Digital control involves the selection of discrete states or levels, rather than the adjustment of a parameter over a continuous range. When the adjustment is continuous, the device is analog controlled (*see* ANALOG CONTROL). Digital controls are becoming increasingly common in all sorts of electronic devices.

Some modern electronic equipment is controlled almost entirely by digital circuits. But some functions are still regulated by analog means in most cases. The volume and tone controls in a high-fidelity stereo amplifier or receiver, for example, are often of analog design, although it is possible to make them digital. Undoubtedly, there are some aspects of machine control that will never be digitized. An example of this is the steering mechanism in an automobile. *See also* DIGITAL.

DIGITAL DEVICE

A digital device is a circuit, or a set of circuits, that operates in the digital mode (*see* DIGITAL, DIGITAL CIRCUITRY). Digital devices include most computers and calculators, and various kinds of test instruments. Some radio receivers and transmitters employ digital devices in their frequency-control systems.

Digital devices may be extremely simple, or highly complex. The simplest of all digital devices is the ordinary single-pole, single-throw switch. It is either on or off! Large networks of interconnected computers form extremely complicated digital devices. They may be made up of millions of individual switches. *See also* DIGITAL COMPUTER.

DIGITAL DISPLAY

A digital display shows a quantity, such as current, voltage, frequency, or resistance, in terms of numerals. This eliminates the reading errors that can occur with analog displays; there is no doubt about the indication. The photograph shows a typical digital frequency display. In this case, the display happens to be the readout of a calculator transceiver. Digital displays can use light-emitting diodes or liquid crystals.

While a digital display has obvious advantages over an analog display, there are some drawbacks in certain applications. For example, the operator of a piece of equipment that employs an analog display can get an intuitive feeling for how the indicated reading compares with other possible values of the parameter. It is visually apparent, for example, whether a receiver frequency is near the bottom, in the middle, or near the top of a frequency band, when an analog display is used. With a digital display, there is no such spatial reinforcement. *See also* ANALOG METERING, DIGITAL METERING, LIGHT-EMITTING DIODE, LIQUID-CRYSTAL DISPLAY, SEVEN-SEGMENT DISPLAY.

DIGITAL INTEGRATED CIRCUIT

A digital-logic circuit, built onto a chip or wafer of

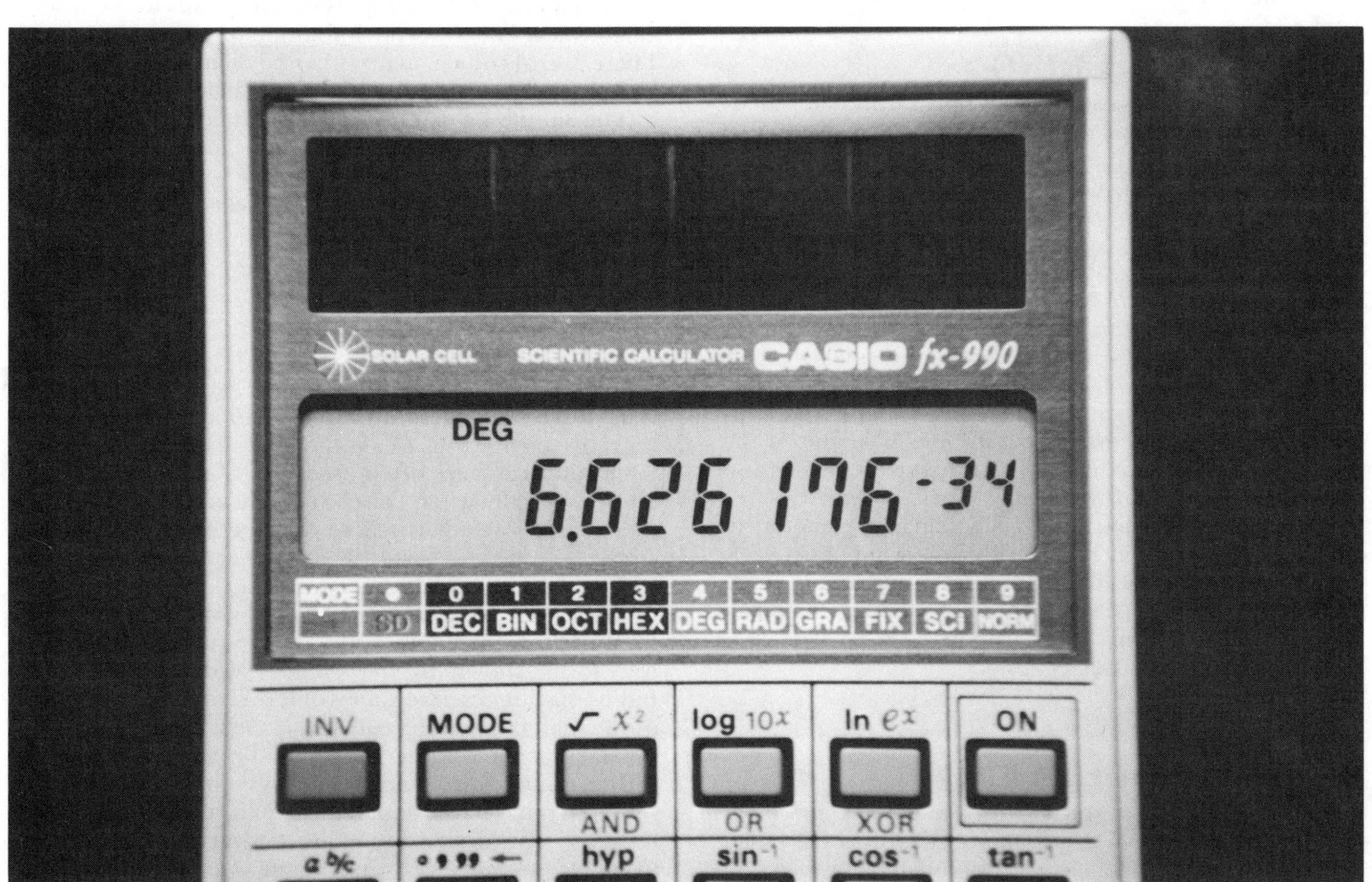

DIGITAL DISPLAY: A digital display indicates various functions and operating conditions precisely, with no interpolation error.

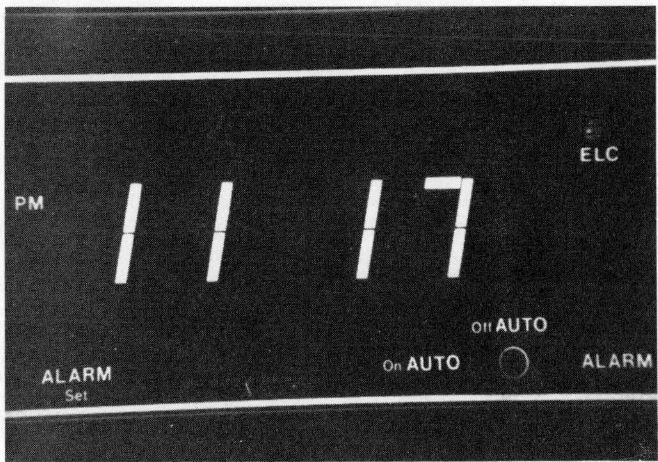

DIGITAL METERING: A digital meter may register time. Such is the case with this clock radio.

semiconductor material, is a digital integrated circuit. There are many different kinds of commercially manufactured digital integrated circuits. Sometimes, different manufacturers make digital integrated circuits that can be directly substituted for each other.

Some of the more sophisticated digital integrated circuits are the single-chip calculator, the microcomputer, the microprocessor, and the digital clock. Most digital integrated circuits were first developed using bipolar techniques. Now, metal-oxide-semiconductor design is also common (*see* METAL-OXIDE SEMICONDUCTOR).

A special kind of digital integrated circuit is the memory integrated circuit. This device has helped to make the handheld calculator and the desktop computer a reality. *See also* MEMORY.

DIGITAL METERING

An electrical or physical quantity may be monitored by means of a digital display of its value. This is called digital metering. In recent years, digital metering has become increasingly common, because of the proliferation, and the drop in the cost, of digital-circuit devices. The photograph shows one extremely common type of digital meter—a clock.

Digital methods are used in many metering applications, such as ammeters, ohmmeters, voltmeters, power meters, and frequency meters. Almost any analog meter can be replaced with a digital meter. Digital meters have the advantage of easy, quick readability. There is no margin for error because of inaccurate interpolation on the part of the person reading the meter. Digital meters also have the advantage of no moving parts.

Digital meters are less preferable than analog meters in certain situations. This is especially true when it is necessary to adjust a circuit for a minimum or maximum reading. Such a "dip" or "peak" is much more easily seen on an analog meter than on a digital meter. *See also* ANALOG METERING, DIGITAL DISPLAY.

DIGITAL MODULATION

Whenever a specific characteristic of a signal is varied for the purpose of conveying information, the process is called modulation (*see* MODULATION). If this is done by restricting the signal parameter to two or more discrete levels, or

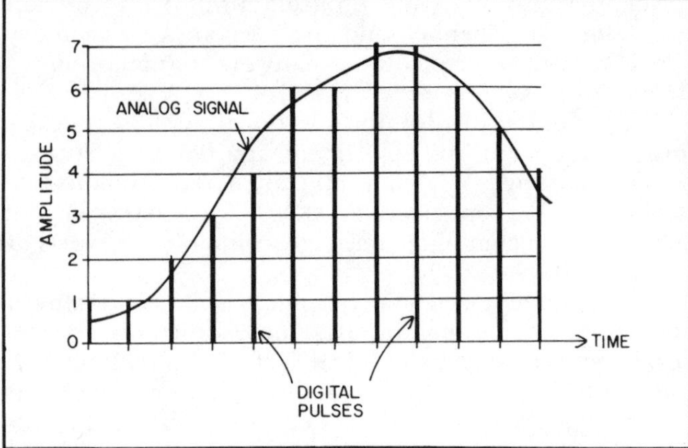

DIGITAL MODULATION: An analog waveform is approximated by digital pulses. In this example, the pulses may have eight different amplitudes.

states, the process is known as digital modulation. Digital modulation differs from analog modulation (*see* ANALOG), in which a parameter varies over a continuous range, and therefore has a theoretically infinite number of possible levels. The parameter in a digital-modulation system can be any signal characteristic, such as amplitude, frequency, or phase. In addition, pulses can be digitally modulated (*see* PULSE MODULATION).

The drawing illustrates a hypothetical digital pulse-amplitude modulation system. The analog waveform is shown by the solid line. A series of pulses, transmitted at uniform time intervals, approximates this waveform. The pulses can achieve any of eight different amplitude levels, as shown on the vertical scale. In this particular example, the amplitude of a given pulse corresponds to the instantaneous level nearest the amplitude of the analog signal. There are, of course, many other possible ways of getting a digital modulation output from an analog signal.

The simplest form of digital modulation is the Morse code. It has two possible states: on and off. These are generally called the key-down and key-up conditions, respectively. The common emission called frequency-shift keying, or FSK, is another simple form of digital modulation. It is used extensively in teletype. *See also* FREQUENCY-SHIFT KEYING, MORSE CODE.

DIGITAL-TO-ANALOG CONVERTER

A device that changes a digital signal to an analog signal (*see* ANALOG, DIGITAL) is called a digital-to-analog converter. Such converters are often used in digital-modulation systems, especially when voices are transmitted. The digital-to-analog converter is placed at the receiving end of the circuit.

The digital-to-analog converter works on the principle of averaging. A digital series of pulses, such as that shown in the previous illustration, approximates an analog waveform, shown by the solid curved line. When the pulses, having discrete values or levels, are fed to the input of the digital-to-analog converter, the analog waveform is obtained at the output.

In a digital voice-transmission system, the analog voice signal is converted to digital form in the transmitter. In the receiver, the digital-to-analog converter retrieves the original voice waveform. The digital signal often is propagated with better efficiency than the analog signal would be under the same conditions. This justifies the use of the digital mode. *See also* ANALOG-TO-DIGITAL CONVERTER, DIGITAL MODULATION, DIGITAL TRANSMISSION SYSTEM.

DIGITAL TRANSMISSION SYSTEM

Any system that transfers information by digital means is a digital transmission system. The simplest digital transmission system is a Morse code transmitter and receiver, along with the attendant operators. A teletype system uses digital transmission methods. Computers communicate by digital transmission.

Analog signals, such as voice and picture waveforms, can be transmitted by digital methods. At the transmitting station, a circuit called an analog-to-digital converter changes the signal to digital form. This signal is then transmitted, and the receiver employs a digital-to-analog converter to get the original analog signal back.

Digital transmission often provides a better signal-to-noise ratio, over a given communications link, than analog transmission. This results in better efficiency. *See also* ANALOG-TO-DIGITAL CONVERTER, DIGITAL MODULATION, DIGITAL-TO-ANALOG CONVERTER.

DIMENSIONLESS QUANTITY

A dimensionless quantity, also sometimes called a scalar, is a constant or variable having no properties of direction. When direction, or a component of direction, is implied, the quantity is a dimensional quantity. A set of dimensional quantities forms a vector (*see* VECTOR).

Examples of dimensionless quantities include the solutions of straightforward equations, such as $x^2 + 2x + 1 = 0$. Some constants are also dimensionless. Direct-current resistance is dimensionless, but the current or voltage are dimensional since they have polarity as well as magnitude. All real numbers can be considered dimensional, since they may be positive or negative. However, the absolute value of a real number is dimensionless. *See also* REAL NUMBER.

DIMMER

A dimmer is a device that controls the voltage supplied to a set of lights. Such a circuit may be a simple potentiometer, although a true dimmer usually provides a voltage that does not depend on the resistance of the load.

Dimmers are often found in household light switches. They can control the level of illumination by regulating the alternating-current voltage. Generally, the voltage can be set at any desired value from zero to 120 volts. Dimmers often use diacs or triacs to provide a voltage drop. The dimmer is similar in design to a motor-speed control. *See also* DIAC, MOTOR-SPEED CONTROL, TRIAC.

DIODE

A diode is a tube or semiconductor device, intended to pass current in only one direction. Diodes can be used for a wide variety of different purposes. The semiconductor diode is far more common than the tube diode in modern electronic circuits. The drawing at A shows the construction of a typical semiconductor diode; it consists of N-type semiconductor material, usually germanium or silicon, and P-type material. Electrons flow into the N-type material and out of the P terminal. The schematic symbol for a semiconductor diode is shown at B. Positive current flows in the direction of the arrow. Electron movement is contrary to the arrow. The positive terminal of a diode is called the anode, and the negative terminal is called the cathode, under conditions of forward bias (conduction).

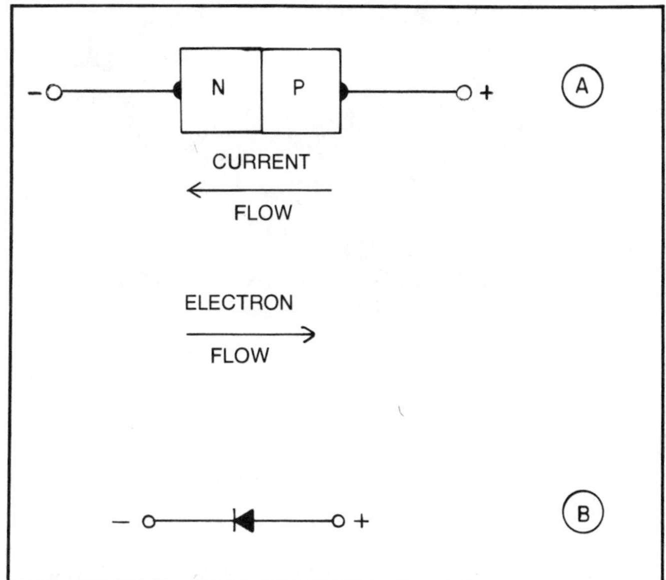

DIODE: At A, a pictorial illustration of the construction of a semiconductor diode. At B, the schematic symbol.

Semiconductor diodes can be very small, as shown in the photograph, and still handle hundreds or even thousands of volts at several amperes. The older tube type diodes are much bulkier and less efficient than the semiconductor diodes. Some of the tube type diodes require a separate power supply for the purpose of heating a filament. Semiconductor diodes are used for many different purposes in electronics. They can be used as amplifiers, frequency controllers, oscillators, voltage regulators, switches, mixers, and in many other types of circuits. *See also* DIODE ACTION, DIODE CAPACITANCE, DIODE CLIPPING, DIODE DETECTOR, DIODE FEEDBACK RECTIFIER, DIODE FIELD-STRENGTH METER, DIODE MIXER, DIODE OSCILLATOR, DIODE RECTIFIER, DIODE-TRANSISTOR LOGIC, DIODE TUBE, DIODE TYPES.

DIODE ACTION

Diode action is the property of an electronic component to pass current in only one direction. In a tube or semiconductor diode, the electrons can flow from the cathode to the anode, but not vice versa. Diode action occurs in all tubes and bipolar transistors, as well as in semiconductor diodes.

A voltage that allows current to flow through a diode is called forward bias. This occurs when the cathode is negative with respect to the anode. A voltage of the opposite polarity is called reverse bias. With most tubes and transistors, as well as with semiconductor diodes, a certain amount of forward bias voltage is necessary in order for current to flow. In a germanium diode, this bias is about

DIODE: A semiconductor diode may be extremely small, yet handle hundreds of volts or several amperes.

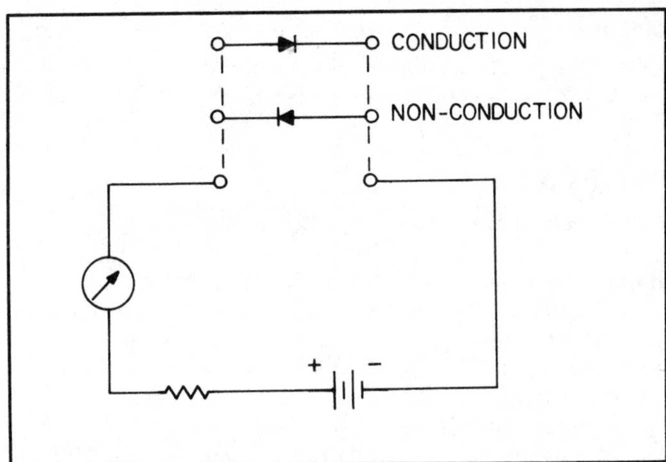

DIODE CHECKER: A simple diode checker consists of a battery, a meter, and a resistor.

0.3 volt; in a silicon diode, it is about 0.6 volt. In mercury-vapor rectifier tubes, it is about 15 volts. *See also* DIODE, P-N JUNCTION.

DIODE CAPACITANCE

When a diode is reverse-biased, so that the anode is negative with respect to the cathode, the device will not conduct. Under these conditions in a semiconductor diode, a depletion layer forms at the P-N junction (*see* DEPLETION LAYER, P-N JUNCTION). The greater the reverse-bias voltage, the wider the depletion region.

The depletion region in a semiconductor diode has such a high resistance that it acts as a dielectric material (*see* DIELECTRIC). Because the P and N materials both conduct, the reverse-biased diode acts as a capacitor, assuming the bias remains reversed during all parts of the cycle.

Some diodes are deliberately used as variable capacitors. These devices are called varactors (*see* VARACTOR DIODE). The capacitance of a reverse-biased diode limits the frequency at which it can effectively be used as a detector, since at sufficiently high frequencies the diode capacitance allows significant signal transfer in the reverse direction. Diode capacitance, when undesirable, is minimized by making the P-N junction area as small as possible.

A tube type diode also displays capacitance when reverse bias is applied to it. This is because of interelectrode effects. The capacitance of a reverse-biased tube diode does not depend to a great extent on the value of the voltage. *See also* DIODE TUBE.

DIODE CHECKER

A diode checker is a device intended for the purpose of testing semiconductor diodes. The simplest kind of diode checker consists of a battery or power supply, a resistor, and a milliammeter, as shown in the drawing. This device can easily be used to determine if current will flow in the forward direction (anode positive) and not in the reverse direction (cathode positive).

An ohmmeter makes a good diode checker. However, before using an ohmmeter for this purpose, check the polarity of the voltage present at the leads. In some volt-ohm-milliameters, the red lead presents a negative voltage with respect to the black lead when the instrument is in the ohmmeter mode.

More sophisticated diode checkers show whether or not a particular diode is within its rated specifications. However, for most purposes, the circuit shown, or an ohmmeter, is adequate. When a diode fails, the failure is generally catastrophic, such as an open or short circuit. *See also* DIODE.

DIODE CLIPPING

A diode clipper, or diode limiter, is a device that uses diodes for the purpose of limiting the amplitude of a signal. Generally, such a device consists of two semiconductor diodes connected in reverse parallel. One such scheme is shown at A in the illustration.

A silicon semiconductor diode has a forward voltage drop of about 0.6 volt. Thus, when two such diodes are placed in reverse parallel, the signal is limited to an amplitude of ±0.6 volt, or 1.2 volts peak-to-peak. If the signal amplitude is smaller than this value, the diodes have no effect, except for the small amount of parallel capacitance they present in the circuit. When the signal without the diodes would exceed 1.2 volts peak-to-peak, however, the diodes flatten the tops of the waveforms at +0.6 and −0.6 volts. This results in severe distortion. Thus, diode limiters are not generally useful in applications where complex waveforms are present, or in situations where fidelity is important.

Germanium diodes in the configuration shown would result in a limiting voltage of ±0.3 volt, or 0.6 volts peak-to-peak.

Clipping amplitudes greater than ±0.6 volt can be obtained by connecting two or more diodes in series in each branch of the clipping circuit. The diode clipper shown at B, using silicon diodes, limits the signal amplitude to ±1.8 volts, or 3.6 volts peak-to-peak. If germanium diodes are used, the clipping level is ±0.9 volt, or 1.8 volts peak-to-peak.

The diode clipper circuit has a maximum effective operating frequency. As the signal frequency is raised, the diodes begin to show capacitance when they are reverse-biased. Because one diode or the other is always reverse-biased in the configurations shown in the drawings, performance is degraded at excessively high frequencies. The maximum usable frequency depends on the type of diodes used, and on the impedance of the circuit. *See also* DIODE.

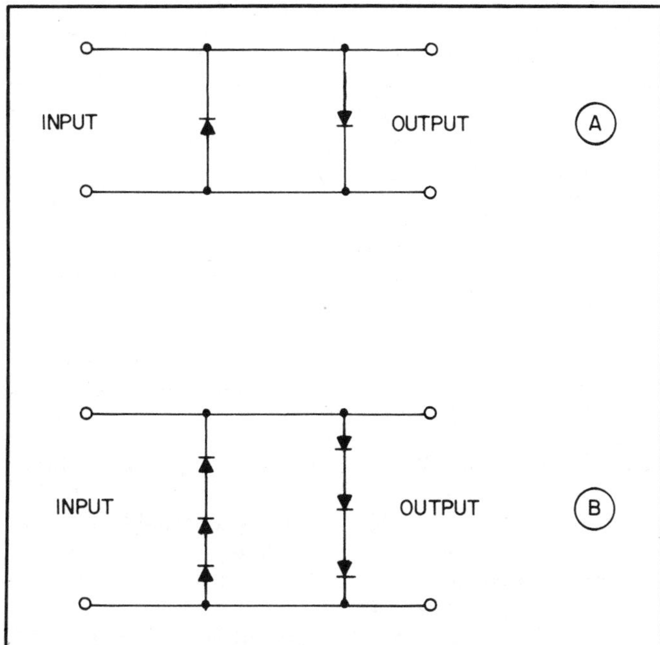

DIODE CLIPPING: Two examples of diode clipping. At A, two diodes in reverse parallel. At B, series diodes in reverse parallel allow a higher clipping threshold.

DIODE DETECTOR

A diode detector is an envelope-detector circuit (*see* ENVELOPE DETECTOR). Such a detector is generally used for the demodulation of an amplitude-modulated signal.

When the alternating-current signal is passed through a semiconductor diode, half of the wave cycle is cut off. This results in a pulsating direct-current signal of variable peak amplitude. The rate of pulsation corresponds to the signal carrier frequency, and the amplitude fluctuations are the result of the effects of the modulating information. A capacitor is used to filter out the carrier pulsations, in a manner similar to the filter capacitor in a power supply. The remaining waveform is the audio-frequency modulation envelope of the signal. This waveform contains a fluctuating direct-current component, the result of the rectified and filtered carrier. A transformer or series capacitor can be used to eliminate this. The resulting output is then identical to the original audio waveform at the transmitter.

The illustration is a simple schematic diagram of a diode detector. The waveforms at various points in the circuit are shown. *See also* DETECTION.

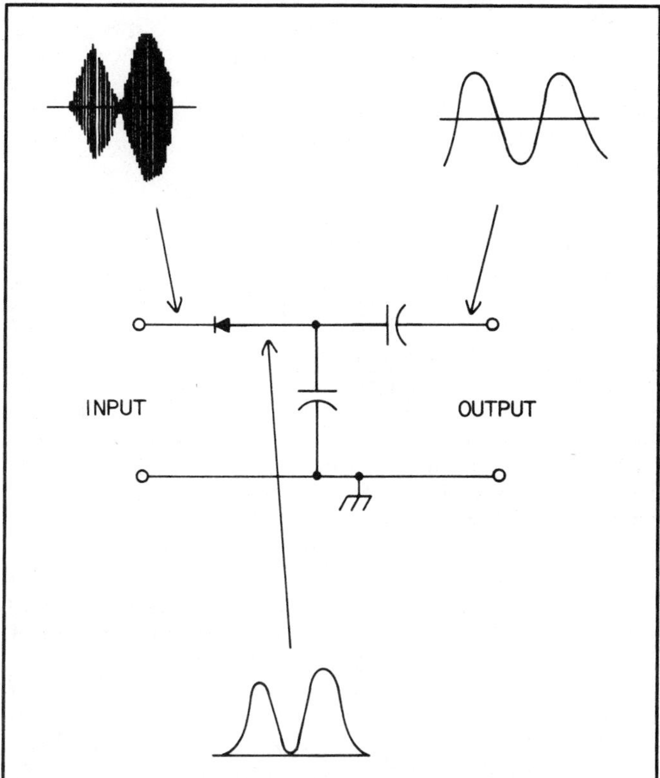

DIODE DETECTOR: A simple diode detector circuit, showing the waveforms at various points.

DIODE FEEDBACK RECTIFIER

A diode feedback rectifier is a device for obtaining a fluctuating voltage suitable for use in an automatic-gain-control circuit (*see* AUTOMATIC GAIN CONTROL, AUTOMATIC LEVEL CONTROL).

In a series of amplifying stages, a portion of the output from one of the later stages is rectified and filtered. This provides a direct-current voltage that varies in proportion to the strength of the signal. The voltage may be either positive or negative, depending on the direction in which the diode is connected. The polarity should be chosen so that the gain of a preceding stage is reduced when the voltage is applied to the base, gate, or grid of the earlier stage.

The effect of the diode feedback rectifier is to keep the output level constant, or almost constant, for a wide variety of signal amplitudes at the input. This maximizes sensitivity for weak signals, and reduces it for strong signals.

DIODE FIELD-STRENGTH METER

A field-strength meter that uses a semiconductor diode, for the purpose of obtaining a direct current to drive a microammeter, is called a diode field-strength meter. Such a field-strength meter, as shown in the schematic diagram, is the simplest kind of device possible for measuring relative levels of electromagnetic field strength. This kind of field-strength meter is not very sensitive. More sophisticated field-strength meters often have amplifiers and tuned circuits built in, and more nearly resemble radio receivers than simple meters.

The diode field-strength meter is easy to carry, and can be used to check a transmission line for proper shielding or balance. A handheld diode field-strength meter should show very little electromagnetic energy near a transmission line, but in the vicinity of the antenna radiator, a large reading is normally obtained. Some commercially made SWR (standing-wave-ratio) meters have a built-in diode field-strength meter, so that a small whip antenna can be used to monitor relative field strength. *See also* FIELD STRENGTH, FIELD-STRENGTH METER.

DIODE IMPEDANCE

The impedance of a diode is the vector sum of the resistance and reactance of the device in a particular circuit (*see* IMPEDANCE, REACTANCE, RESISTANCE). Both the resistance

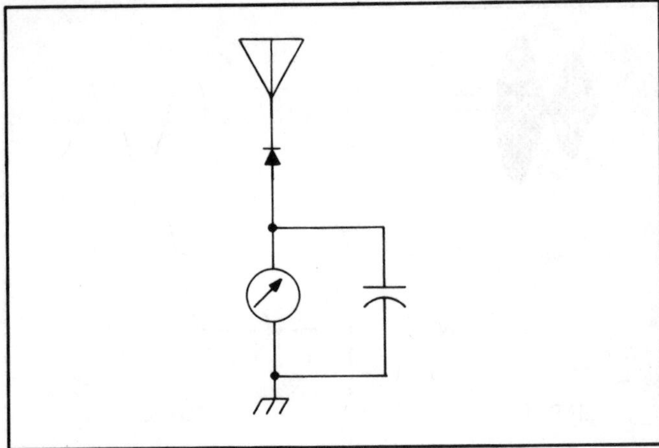

DIODE FIELD-STRENGTH METER: A diode field-strength meter simply rectifies the radio-frequency energy so that a direct-current meter can show the relative intensity.

and the capacitive reactance of a diode depend on the voltage across the device. The inductive reactance of a diode is essentially constant, and is primarily the result of inductance in the wire leads.

Generally, the resistance of a heavily forward-biased diode is extremely low, and the device in this case acts as a nearly perfect short circuit. When the forward bias is not strong, the resistance is higher, and the capacitive reactance is small. A reverse-biased diode has extremely high resistance. The capacitive reactance of a reverse-biased semiconductor diode is greater as the reverse-bias voltage rises. This is because the depletion region gets wider and wider (*see* DEPLETION LAYER, DIODE CAPACITANCE). The smaller the capacitance, the larger the capacitive reactance.

In a tube type diode, the capacitance under all bias conditions is essentially consistent, since it is the result of interelectrode effects, and the electrode sizes and spacing are always the same, no matter what the bias voltage. *See also* DIODE, DIODE TUBE.

DIODE MATRIX

A diode matrix is a form of high-speed, digital switching circuit, using semiconductor diodes in a large array. Two sets of wires, one shown horizontally on a circuit diagram and the other shown vertically, are interconnected at various points by semiconductor diodes. Diode matrices may be fairly small, such as in a simple counter. Or they may be huge, as in a digital computer.

Diode matrices are employed as decoders, memory circuits, and rotary switching circuits. *See also* DECODING, MEMORY, SWITCHING.

DIODE MIXER

A diode mixer is a circuit that makes use of the nonlinear characteristics of a diode, for the purpose of mixing signals (*see* MIXER). Whenever two signals having different frequencies are fed into a nonlinear circuit, the sum and difference frequencies are obtained at the output, in addition to the original frequencies.

The illustration is a schematic diagram of a typical, simple diode mixer circuit. Such a circuit has no gain, since it is passive. In fact, there is a certain amount of insertion

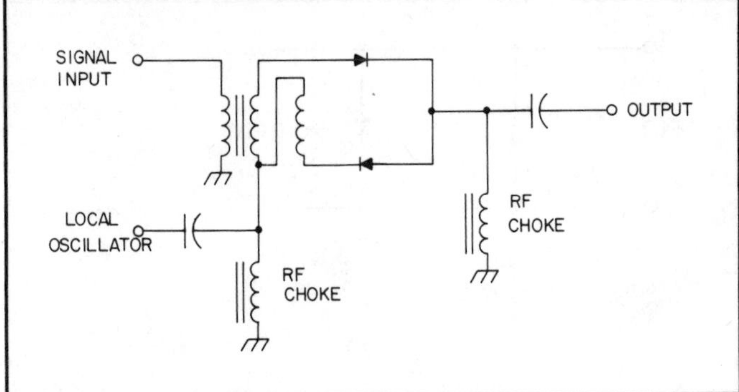

DIODE MIXER: A diode mixer uses the nonlinearity of the semiconductor diode to generate mixing products.

loss. However, amplification circuits can be used to boost the output to the desired level. Selective circuits reject all unwanted mixing products and harmonics, allowing only the desired frequency to pass. Diode mixers are often found in superheterodyne receivers, and also in transmitters. Frequency converters, used with receivers to provide operation on frequencies far above or below their normal range, sometimes employ diode mixers. These circuits can function well into the microwave spectrum. *See also* FREQUENCY CONVERSION, FREQUENCY CONVERTER, MIXING PRODUCT.

DIODE OSCILLATOR

Under the right conditions, certain types of semiconductor diodes will oscillate at ultra-high or microwave frequencies. A circuit that is deliberately designed to produce oscillation at these frequencies, using a diode as the active component, is called a diode oscillator. The drawing is a simple schematic diagram of a microwave oscillator that uses a semiconductor device called a Gunn diode. The Gunn diode has largely replaced the Klystron tube for ultra-high and microwave frequency oscillator applications.

The Gunn diode is mounted inside a resonant cavity. A direct-current bias is supplied to cause oscillation. The efficiency of the Gunn-diode oscillator is low, only a few percent. The frequency stability tends to be rather poor, since the slightest change in the bias voltage or temperature can cause a radical change in the oscillating frequency. The bias voltage must therefore be carefully regulated, and the temperature maintained at a level that is as nearly constant as possible. Automatic frequency control is sometimes used to improve the stability of the Gunn-diode oscillator (*see* AUTOMATIC FREQUENCY CONTROL). A phase-locked-loop device is also sometimes used (*see* PHASE-LOCKED LOOP).

The diode oscillator provides a maximum output of considerably less than 1 watt. Gunn-diode oscillators can be frequency-modulated by varying the bias voltage. A device called a tunnel diode, now essentially obsolete because of the superior characteristics of the Gunn diode, can also be used in a diode-oscillator circuit. *See also* GUNN DIODE, KLYSTRON, TUNNEL DIODE.

DIODE RATING

The rating of a diode refers to its ability to handle current.

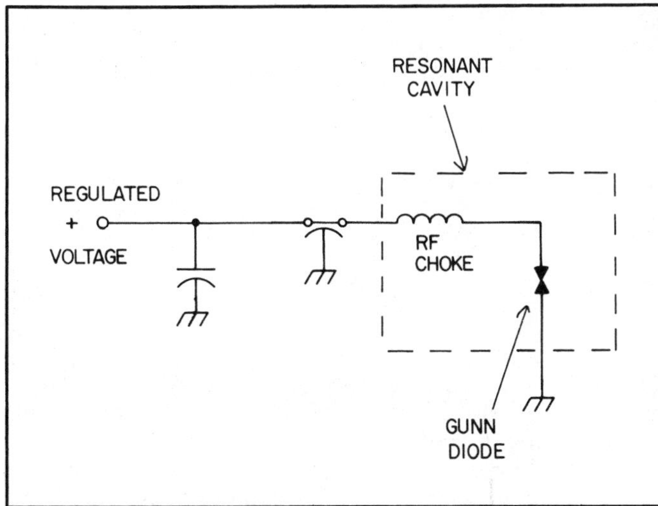

DIODE OSCILLATOR: A diode oscillator using a Gunn diode to generate microwave energy.

power, or voltage. Some semiconductor diodes are intended strictly for small-signal applications, and can handle only a few microamperes or milliamperes of current. Other semiconductor diodes are capable of handling peak-inverse voltages of hundreds or even thousands of volts, and currents of several amperes. These rugged diodes are found in power-supply rectifier circuits.

Diode ratings are generally specified in terms of the peak inverse voltage, or PIV, and the maximum forward current. Zener diodes, used mostly for voltage regulation, are rated in terms of the breakdown or avalanche voltage and the power-handling capacity. Other characteristics of diodes, that may be called ratings or specifications, are temperature effects, capacitance, voltage drop, and the current-voltage curve. *See also* AVALANCHE VOLTAGE, DIODE CAPACITANCE, DIODE IMPEDANCE, PEAK INVERSE VOLTAGE, ZENER DIODE.

DIODE-TRANSISTOR LOGIC

Diode-transistor logic, abbreviated DTL, is a form of digital-logic design in which a diode and transistor act to amplify and invert a pulse. Diode-transistor logic has a somewhat less rapid switching rate than most other bipolar logic families. The power-dissipation rating is medium to low. Diode-transistor logic is sometimes mixed with transistor-transistor logic (TTL) in a single circuit.

The drawing illustrates a simple DTL gate. This circuit can perform a NAND or NOR function, depending on whether positive or negative logic is employed. Diode-transistor logic gates are generally fabricated into an integrated-circuit package. *See also* DIRECT-COUPLED TRANSISTOR LOGIC, EMITTER-COUPLED LOGIC, HIGH-THRESHOLD LOGIC, INTEGRATED INJECTION LOGIC, METAL-OXIDE SEMICONDUCTOR LOGIC FAMILIES, NAND GATE, NEGATIVE LOGIC, NOR GATE, POSITIVE LOGIC, RESISTOR-CAPACITOR-TRANSISTOR LOGIC, RESISTOR-TRANSISTOR LOGIC, TRANSISTOR-TRANSISTOR LOGIC, TRIPLE-DIFFUSED EMITTER-FOLLOWER LOGIC.

DIODE TUBE

A diode tube is a vacuum tube with only two elements: a cathode and an anode. Diode tubes are generally used for the same purposes as semiconductor diodes; however,

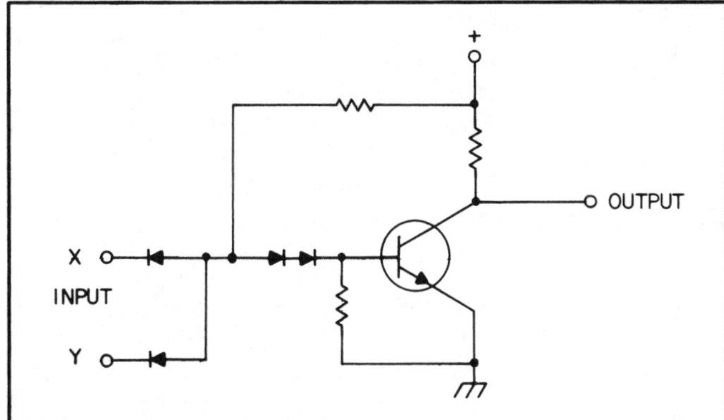

DIODE-TRANSISTOR LOGIC: Schematic diagram of diode-transistor logic.

diode tubes are essentially obsolete in modern circuits.

The illustration shows the schematic symbols for diode tubes. At A, a directly-heated-cathode tube is shown. At B, an indirectly-heated-cathode tube is shown. The tubes at A and B require a power supply for the purpose of heating the cathode. At C, a cold-cathode tube is shown.

Diode tubes are relatively inefficient, since they usually consume considerable power. Some older radio receivers, transmitters, and amplifiers may still be found that have diode rectifier tubes. *See also* TUBE.

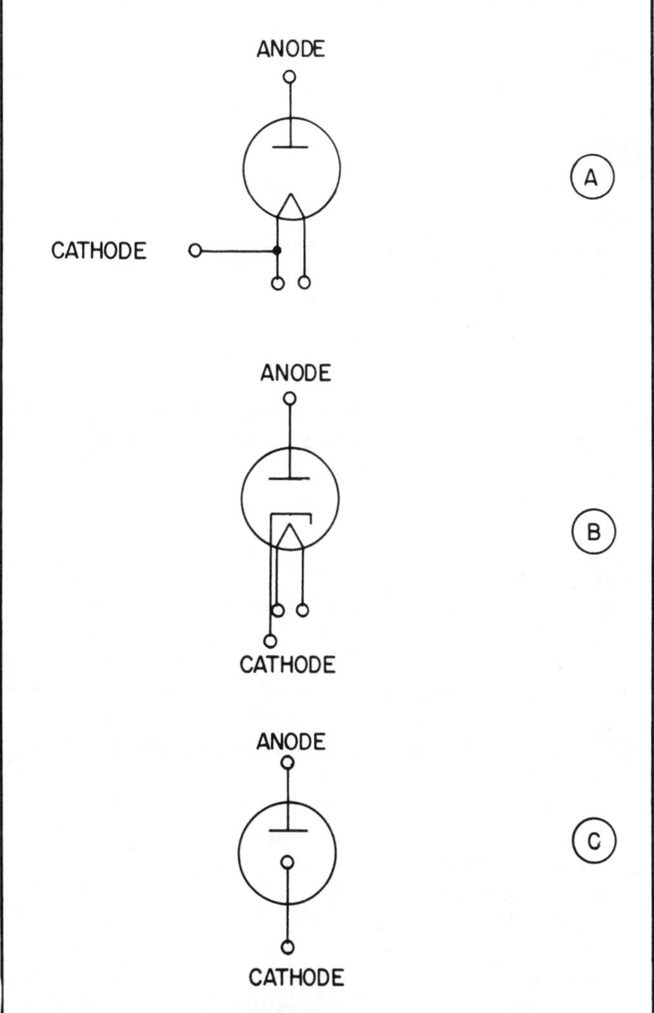

DIODE TUBE: At A, directly heated cathode; at B, indirectly heated cathode; at C, cold cathode.

DIODE TYPES

There are several different types of semiconductor diodes, each intended for a different purpose. The most obvious application of a diode is the conversion of alternating current to direct current. Detection and rectification use the ability of a diode to pass current in only one direction. But there are many other uses for these semiconductor devices.

Light-emitting diodes, called LEDs for short, produce visible light when forward-biased. Solar-electric diodes do just the opposite, and generate direct current from visible light. Zener diodes are used as voltage regulators and limiters. Gunn diodes and tunnel diodes can be employed as oscillators at ultra-high and microwave frequencies. Varactor diodes are used for amplifier tuning. A device called a PIN diode, which exhibits very low capacitance, is used as a high-speed switch at radio frequencies. Hot-carrier diodes are used as mixers and frequency multipliers. Frequency multiplication is also accomplished effectively using a step-recovery diode. The impact-avalanche-transit-time diode, or IMPATT diode, can act as an amplifying device.

Details of various diodes types and uses are discussed under the following headings: DIODE, DIODE ACTION, DIODE DETECTOR, DIODE FEEDBACK RECTIFIER, DIODE FIELD-STRENGTH METER, DIODE MATRIX, DIODE RECTIFIER, DIODE-TRANSISTOR LOGIC, GUNN DIODE, HOT-CARRIER DIODE, IMPATT DIODE, LIGHT-EMITTING DIODE, PIN DIODE, P-N JUNCTION, SOLAR CELL, STEP-RECOVERY DIODE, TUNNEL DIODE, VARACTOR DIODE, ZENER DIODE.

DIP

In electronics, the term dip usually refers to the adjustment of a certain parameter for a minimum value. A common example is the dipping of the plate current in a tube type radio-frequency amplifier. The dip indicates that the output circuit is tuned to resonance, or optimum condition. Antenna tuning networks are adjusted for a dip in the standing-wave ratio. A dip is also sometimes called a null (*see* NULL).

The dual-inline package, a familiar form of integrated circuit, is sometimes called a DIP for short. *See also* DUAL IN-LINE PACKAGE.

DIPLEX

When more than one receiver or transmitter are connected to a single antenna, the system is called a diplex or multiplex circuit. The diplexer allows two transmitters or receivers to be operated with the same antenna at the same time.

The most familiar example of a diplexer is a television feed-line splitter, which allows two television receivers to be operated simultaneously using the same antenna. Such a device must have impedance-matching circuits to equalize the load for each receiver. Simply connecting two or more receivers together by splicing the feed lines will result in ghosting because of reflected electromagnetic waves along the lines. Diplexers for transmitters operate in a similar manner to those for receivers.

Multiplex transmission is sometimes called diplex transmission when two signals are sent over a single carrier. Each of the two signals in a diplex transmission consists of a low-frequency, modulated carrier called a subcarrier. The main carrier, much higher in frequency than the subcarriers, is modulated by the subcarriers. *See also* MULTIPLEX.

DIP METER

See GRID-DIP METER.

DIPOLE

When an atom or molecule has a negative and positive charge that are separated in space, the atom or molecule is known as an electric dipole. A molecule or atom having separate north and south magnetic poles is called a magnetic dipole. An electric dipole is surrounded by an electric field, often represented by lines of flux as shown in the illustration at A. A magnetic dipole is also surrounded by lines of flux representing the magnetic field as at B.

Depending on the electric or magnetic field in the vicinity of a dipole, the atom or molecule tends to orient itself in a certain direction because of the attraction between opposite charges or magnetic poles, and because of the repulsion between like charges or poles. An oscillating field will cause the dipoles to move back and forth at a rate corresponding to the frequency of the oscillation. An oscillating dipole, likewise, produces an oscillating field. The forces of electric and magnetic attraction and repulsion are responsible for the propagation of radio signals through space. *See also* ELECTROMAGNETIC FIELD.

DIPOLE ANTENNA

The term dipole, or dipole antenna, is often used to de-

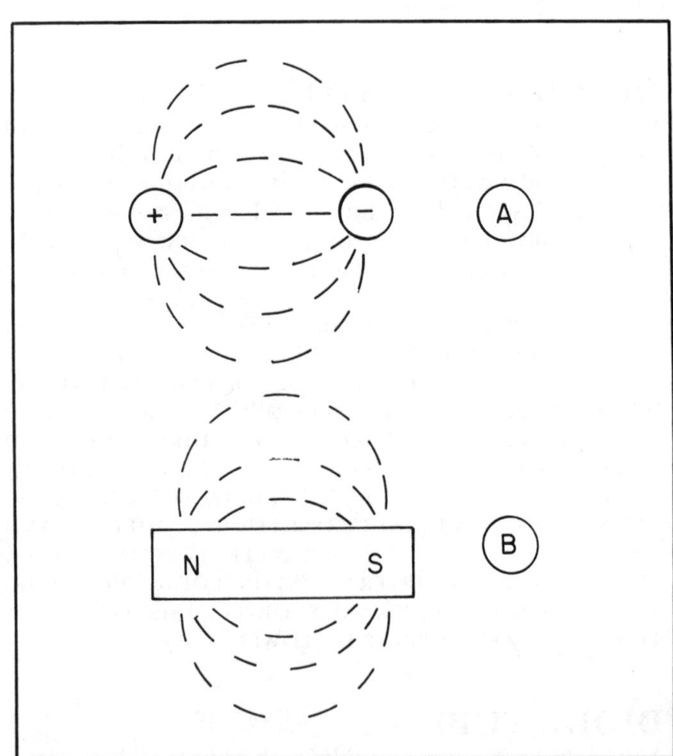

DIPOLE: At A, an electric dipole; at B, a magnetic dipole, showing the lines of flux.

scribe a half-wavelength radiator fed at the center with a two-wire or coaxial transmission line. Such an antenna may be oriented horizontally or vertically, or at a slant. The radiating element is usually straight; variations of the dipole go by other nicknames.

A half-wavelength conductor diplays resonant properties for electromagnetic energy. In free space—that is, when there are no objects near the radiator—the impedance at the center of a dipole is about 73 ohms, purely resistive. The impedance is a pure resistance at all harmonic frequencies. At odd harmonics, the value is about the same as at the fundamental frequency; but at even harmonics, it is very high. The radiation pattern of a dipole in free space is rather doughnut-shaped, as shown in the illustration. Maximum radiation occurs in directions perpendicular to the conductor.

Because of their relative simplicity, dipole antennas are quite popular among shortwave listeners and radio amateurs. This is especially true at frequencies below about 10 MHz, where more complicated antennas are often impractical. At higher frequencies, parasitic elements are often added to the dipole, creating power gain. Dipoles may also be fed in various multiple configurations to obtain power gain. *See also* PARASTIC ARRAY, PHASED ARRAY, VERTICAL DIPOLE ANTENNA, YAGI ANTENNA.

DIP SOLDERING TECHNIQUE

Dip soldering is a method of soldering an electrical connection, or coating a terminal or lead with solder, by dipping the entire connection, terminal, or lead into a container of molten solder. Excess solder is then removed. Solder melts at a fairly low temperature, so components can often be dipped without damage.

Printed-circuit boards are sometimes tinned, or coated with solder, by dipping the entire board into molten solder, or the face of the board to be tinned may be placed against the surface of the molten solder bath. This results in a coating of solder on all foil runs, making assembly easier and helping to protect the board against corrosion. Excess solder is easily removed from the non-foil parts of the board. The solder is an excellent conductor of electricity. *See also* SOLDER, SOLDERING TECHNIQUE.

DIRECT-CONVERSION RECEIVER

A direct-conversion receiver is a receiver whose intermediate frequency is actually the audio signal heard in the speaker or headset. The received signal is fed into a mixer, along with the output of a variable-frequency local oscillator. As the oscillator is tuned across the frequency of an unmodulated carrier, a high-pitched audio beat note is heard, which becomes lower until it vanishes at the zero-beat point. Then it rises in pitch again as the oscillator frequency gets farther and farther away from the signal frequency. The drawing shows a simple block diagram of a direct-conversion receiver.

For reception of code signals, the local oscillator is set slightly above or below the signal frequency. The audio tone will have a frequency equal to the difference between the oscillator and signal frequencies. For reception of amplitude-modulated or single-sideband signals, the oscillator should be set to zero beat with the carrier frequency of the incoming signal.

The direct-conversion receiver normally cannot provide the selectivity of a superheterodyne, since single-signal reception with the direct-conversion receiver is impossible. Audio filters are often used to provide some measure of selectivity. *See also* INTERMEDIATE FREQUENCY, SINGLE-SIGNAL RECEPTION, SUPERHETERODYNE RECEIVER.

DIRECT-COUPLED TRANSISTOR LOGIC

Direct-coupled transistor logic, abbreviated DCTL, is a bipolar logic family. It was the earliest logic design used in commercially manufactured integrated circuits. The DCTL logic scheme is fairly simple, as shown in the schematic.

Direct-coupled transistor logic has rather poor noise rejection characteristics. Current hogging can also cause some problems. Newer forms of direct-coupled transistor logic are available, and these designs have better operating characteristics than the original form. The signal voltages in DCTL are low. The switching speed and power-handling capabilities are about average, compared with other bipolar logic families. *See also* DIODE-TRANSISTOR LOGIC, EMITTER-COUPLED LOGIC, HIGH-THRESHOLD LOGIC, INTEGRATED INJECTION LOGIC, METAL-OXIDE-SEMICONDUCTOR LOGIC FAMILIES, RESISTOR-CAPACITOR-TRANSISTOR LOGIC, RESISTOR-TRANSISTOR LOGIC, TRANSISTOR-TRANSISTOR LOGIC, TRIPLE-DIFFUSED EMITTER-FOLLOWER LOGIC.

DIRECT COUPLING

Direct coupling is a form of circuit coupling, usually employed between stages of an amplifier. In direct coupling, the output from one stage is connected, by a direct

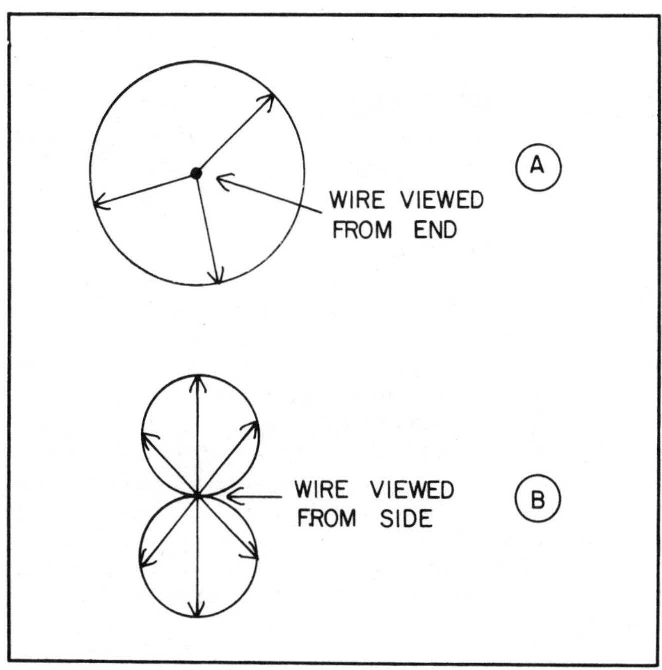

DIPOLE ANTENNA: The directional pattern of a dipole antenna is uniform in all directions at right angles to the wire (A), but is variable in any plane containing the wire (B).

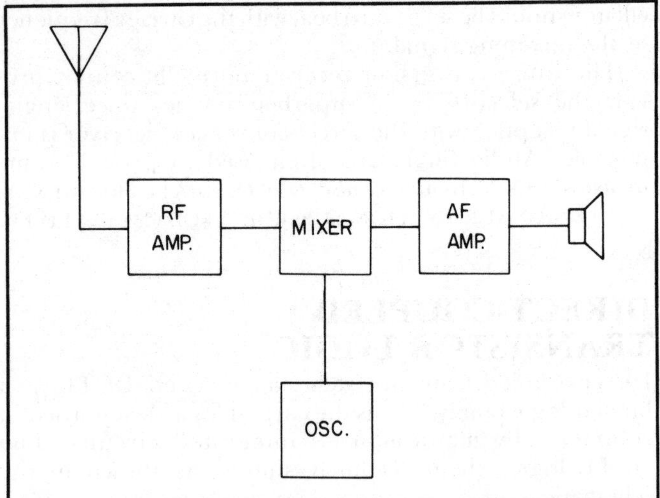

DIRECT-CONVERSION RECEIVER: A direct-conversion receiver mixes the incoming signal with the output of a local oscillator to obtain audio frequencies.

wire circuit, to the input of the next stage. This increases the gain of the pair over the gain of a single component, provided the bias voltages are correct. The Darlington amplifier is an example of direct coupling (*see* DARLINGTON AMPLIFIER). The schematic diagram showing direct-coupled transistor logic shows an example of direct coupling used to increase the sensitivity of a switching network.

Direct coupling is characterized by a wideband frequency response. This is because there are no intervening reactive components, such as capacitors or inductors, to act as tuned circuits. Therefore, direct-coupled amplifiers are more subject to noise than tuned amplifiers. Direct coupling transmits the alternating-current and direct-current components of a signal. Current hogging can sometimes be a problem, especially if adequate attention is not given to the maintenance of proper bias. *See also* CURRENT HOGGING.

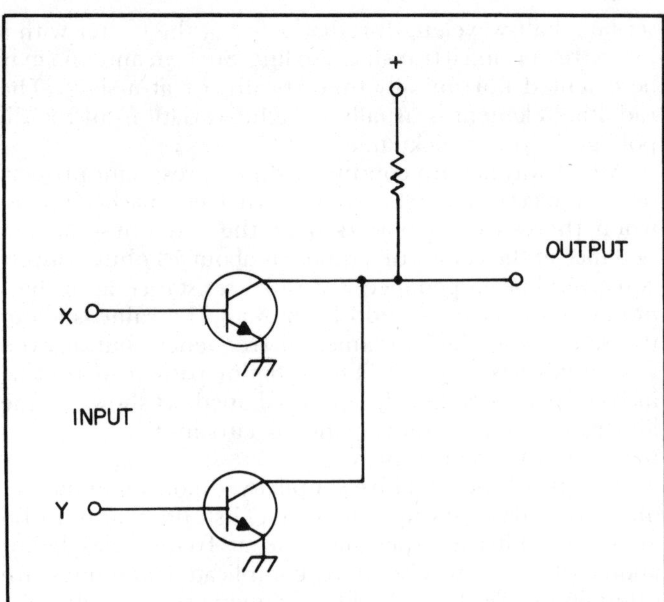

DIRECT-COUPLED TRANSISTOR LOGIC: Direct-coupled transistor logic uses direct coupling among transistors to form logic gates.

DIRECT CURRENT

A direct current is a current that always flows in the same direction. That is, the polarity never reverses. If the direction of the current ever changes, it is considered to be an alternating current (*see* ALTERNATING CURRENT).

Physicists consider the current in a circuit to flow from the positive pole to the negative pole. This is purely a convention. The movement of electrons in a direct-current circuit is therefore contrary to the theoretical direction of the current. In a P-type semiconductor material, however, the motion of the positive charge carriers (holes) is the same as the direction of the current.

Typical sources of direct current include most electronic power supplies, as well as batteries and cells. The intensity, or amplitude, of a direct current may fluctuate with time, and this fluctuation may be periodic. In this case, the current can be considered to have an alternating as well as a direct component, but the current itself is direct. *See also the listings under* DC.

DIRECT-DRIVE TUNING

When the tuning knob of a radio receiver or transmitter is mounted directly on the shaft of a variable capacitor, the tuning is said to be directly driven. A half turn of the tuning knob thus covers the entire range. Direct-drive tuning is found in small portable transistor radio receivers for the standard broadcast band. It is difficult to obtain precise tuning with a direct-drive control.

Most controls for the adjustment of frequency are indirectly driven. Thus, several turns of the control knob are needed to cover the entire range. This method of control is called vernier drive, and it may be employed in various other situations besides radio tuning. Cable-driven controls are also sometimes used to spread out the tuning range of a radio receiver or transmitter.

Most circuit-adjustment controls, such as volume and tone, are directly driven. However, vernier or cable drives may be used with any control to obtain precise adjustment. *See also* VERNIER.

DIRECTED NUMBER

Any number in which direction is implied, as well as magnitude, is called a directed number. The simplest examples of directed numbers are ordinary real numbers: They may be positive, negative, or zero. Positive numbers are usually assigned a direction toward the right or the top of a number scale. Negative numbers are generally assigned a direction toward the left or the bottom. Zero is the only real number that is not directed. The real-number line has one dimension, but more than one number line can be used to form multidimensional systems (*see* CARTESIAN COORDINATES, REAL NUMBER).

A common example of a two-dimensional, directed-number system is the set of complex numbers, formed by the real and imaginary number lines. Any number has a real directed component and an imaginary directed component in this system. Any direction and magnitude in the complex-number plane has exactly one complex correspondent, and any complex number is represented by exactly one vector in the plane. Directed numbers can have any number of dimensions; they are then usually referred to as vectors. *See also* COMPLEX NUMBER, VECTOR.

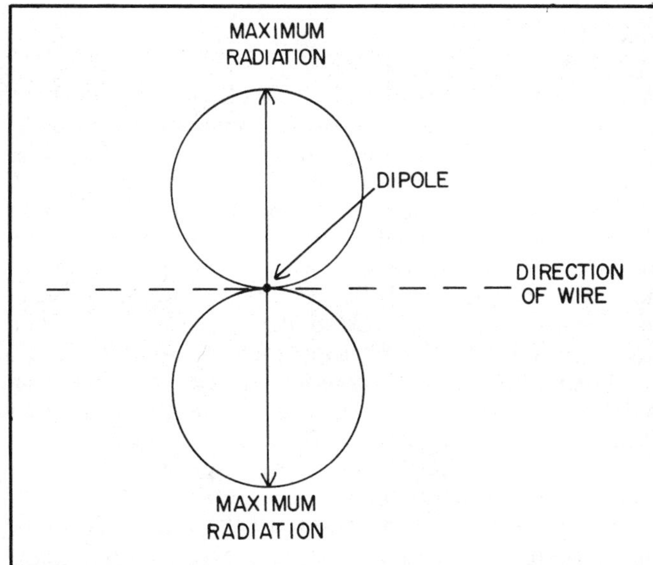

DIRECTIONAL ANTENNA: An example of a directional antenna, in this case a dipole. The radiation pattern is shown in a plane containing the radiating element.

DIRECTIONAL ANTENNA

A directional antenna is a receiving or transmitting antenna that is deliberately designed to be more effective in some directions than in others. For most radio-communications purposes, antenna directionality is considered to be important only in the azimuth, or horizontal, plane if communication is terrestrial. But for satellite applications, both the azimuth and altitude directional characteristics are important (*see* AZIMUTH, ELEVATION). Directional antennas are usually either bidirectional or unidirectional; that is, their maximum gain is either in two opposite directions or in one single direction (*see* BIDIRECTIONAL PATTERN, UNIDIRECTIONAL PATTERN). Some antennas have a large number of high-gain directions.

A vertical radiator, by itself, is omnidirectional in the azimuth plane. In the elevation plane, it shows maximum gain parallel to the ground, and minimum gain directly upward. A single horizontal radiator, such as a dipole antenna, produces more gain off the sides than off the ends, and therefore it is directional in the azimuth plane, as shown in the illustration. A dipole is considered a directional antenna, since it shows a bidirectional pattern.

Sophisticated types of directional antennas provide large amounts of signal gain in their favored directions. This gain and directionality can be obtained in a variety of ways. *See also* ANTENNA PATTERN, ANTENNA POWER GAIN, COLLINEAR ANTENNA, DIPOLE ANTENNA, DISH ANTENNA, PHASED ARRAY, QUAD ANTENNA, YAGI ANTENNA.

DIRECTIONAL FILTER

A directional filter is a frequency-selective filter or set of filters in a carrier-current communications system. Generally, the filter consists of a lowpass and/or highpass filter (*see* CARRIER-CURRENT COMMUNICATION, HIGHPASS FILTER, LOWPASS FILTER). The cutoff frequency of the filter is set in the middle of the band in which communication is carried out.

A directional-filter set allows communication in only one direction for the lower half of the band, and in only the other direction for the upper half of the band. The illustration is a block diagram of a carrier-current communications system, showing a directional-filter pair at each end of the circuit. At station X, the transmitter operates in the upper half of the band, at a higher frequency than the receiver, which is set for the lower half of the band. At station Y, this is exactly reversed: The transmitter operates in the lower part of the band, and the receiver in the upper part.

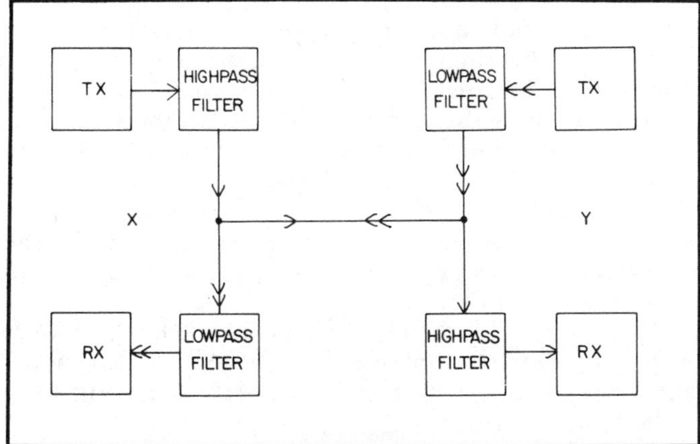

DIRECTIONAL FILTER: Directional-filter networks allow duplex communication via a single wire conductor.

The directional filters prevent the signal components of a transmitter from overloading the station receiver. With sufficient selectivity, full duplex operation is possible over a single current-carrying conductor. *See also* DUPLEX OPERATION.

DIRECTIONAL GAIN

Directional gain is a means of expressing the sound-radiating characteristics of a speaker. The term may also be applied to the acoustic-pickup characteristics of a microphone.

Generally, speakers do not radiate sound equally well in all directions. Instead, most of the sound energy is concentrated in a narrow cone, with its axis perpendicular to the speaker face. If a speaker could be designed that would radiate equally well in all directions, then its directional gain would theoretically be zero.

Suppose a hypothetical speaker, supplied with P watts of audio-frequency power, produces a sound-intensity level of p_x dynes per square centimeter, at a fixed distance of m meters, in all directions. The directional gain of this speaker is zero, since its sound radiation is the same in all directions. Imagine that this theoretical speaker is replaced by a real speaker, which radiates more effectively in some directions than in others. Suppose the real speaker has the same efficiency as the hypothetical one, and receives the same amount of power. If the sound pressure at a distance of m meters from this speaker, along the axis of its maximum sound output, is p dynes per square centimeter, then the directional gain of the real speaker, in decibels (dB), is:

$$\text{Directional gain (dB)} = 10 \log_{10}(p/p_x)$$

All real speakers have positive directional gain. When sound radiation is enhanced in one direction, it must be sacrificed in other directions. The converse of this is also true: A reduction in sound in some directions results in an increase in other directions, assuming the same total power output from the speaker.

Directional gain is expressed in decibels for microphones, too. But for a microphone, the directional gain is given in terms of pickup sensitivity, or the sound pressure required to produce a certain electrical output at the microphone terminals. For transducers in general, the directional gain is expressed as the directivity index (*see* DIRECTIVITY INDEX).

An omnidirectional microphone has zero directional gain. All other microphones have a positive directional gain. *See also* DIRECTIONAL MICROPHONE, OMNIDIRECTIONAL MICROPHONE.

DIRECTIONAL MICROPHONE

A directional microphone is a microphone designed to be more sensitive in some directions than in others. Usually, a directional microphone is unidirectional—that is, its maximum sensitivity occurs in only one direction.

Directional microphones are commonly used in communications systems to reduce the level of background noise that is picked up. This is important for intelligibility, especially in industrial environments where the ambient noise level is high. Directional microphones are also useful in public-address systems, since they minimize the amount of feedback from the speakers. Directional microphones can usually be recognized by their physical asymmetry.

Most directional microphones have what is called a cardioid response (*see* CARDIOID PATTERN). The maximum sensitivity of a cardioid microphone exists in a very broad lobe. The minimum sensitivity is sharply defined, and usually occurs directly opposite the direction of greatest sensitivity.

A microphone without directional properties is called an omnidirectional microphone. *See* OMNIDIRECTIONAL MICROPHONE.

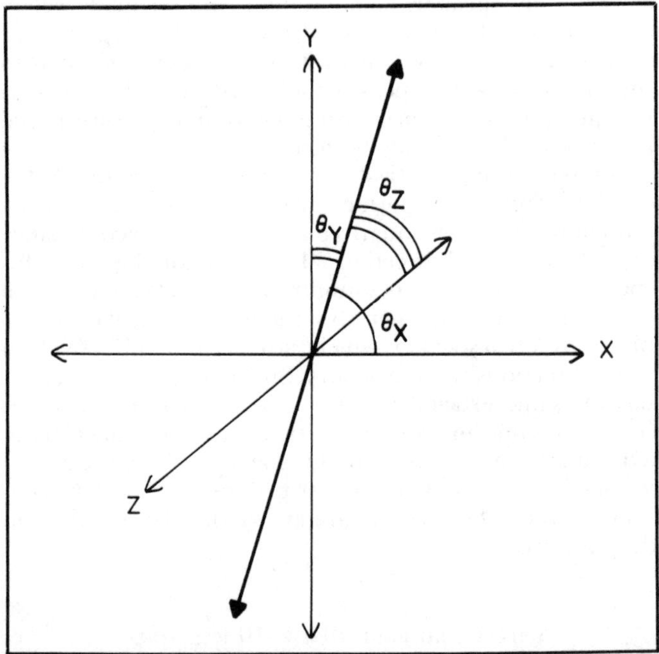

DIRECTION ANGLE AND COSINE: A straight line passing through the origin in a three-dimensional Cartesian system can be represented by the direction angles θ_x, θ_y, and θ_z, which represent the angles the line subtends with respect to the x, y, and z axes.

DIRECTION ANGLE AND COSINE

A direction angle is an angular component of the direction of an object in three or more dimensions, such as the bearing of one flying craft with respect to another. The most familiar examples of direction angles are azimuth and elevation. Direction angles specify a unique, straight line through the origin of a three-dimensional coordinate system; these angles are measured with respect to the x, y, and z axes independently (see illustration). Right ascension and declination are examples of direction angles. Latitude and longitude are also forms of direction angles.

Direction angles can provide the exact location of an object, with respect to a reference point, if the distance or range of the object is known. However, if the range is not known, the direction angles can only specify a particular straight geometric line in space.

In radar, the direction angle toward a target is the angle between the center of the antenna base line and a line connecting the base line with the target. The direction cosine is the cosine of the direction angle. A given direction angle can correspond to an infinite number of possible target locations. *See also* AZIMUTH, DECLINATION, ELEVATION, RADAR, RIGHT ASCENSION.

DIRECTION FINDER

A direction finder is a radio receiver with a precise signal-strength indicator, connected to a rotatable loop antenna

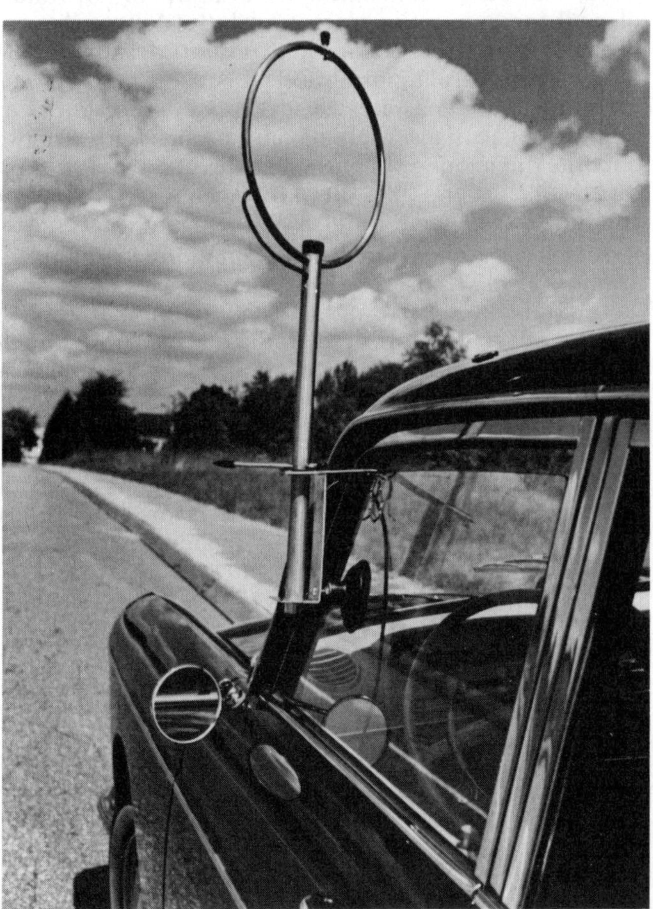

DIRECTION FINDER: A direction finder utilizes a loop antenna, which has a sharp bidirectional null (courtesy of Gold Line Connector).

having a defined directional response. Direction-finding equipment is often used for the purpose of locating a transmitter.

The receiver in a direction-finding system need not be especially sophisticated. It is the antenna design that is critical. A loop antenna is generally used, and it is often shielded against the electric component of the radio-wave front, so that it picks up only the magnetic field (see photograph). The circumference of the loop (*see* LOOP ANTENNA) should be less than about 0.1 wavelength at the operating frequency. Such an antenna displays a sharp null along a line passing through its center and perpendicular to the plane of the loop.

By rotating the loop antenna until a null is seen in the receiver, the line toward the transmitter can be found. This line provides an ambiguous bearing however, since it is not possible to tell whether the transmitter is at a certain bearing or 180 degrees opposite. Some direction finders have special antennas that eliminate this ambiguity. However, by taking readings from two different locations, the transmitter can be pinpointed by finding the intersection point of the two lines on a map.

Some direction finders work automatically. The antenna is rotated by a mechanical device until it points to the transmitter. *See also* AUTOMATIC DIRECTION FINDER.

DIRECTIVITY INDEX

The directivity index of a transducer is a measure of its directional properties. The directional index is similar to the directional gain of a speaker or microphone (*see* DIRECTIONAL GAIN); however, the mathematical definition is slightly different in terms of its expression.

For a sound-emitting transducer, let p_{av} be the average sound intensity in all directions from the device, at a constant radius m, assuming that the transducer receives an input power P. Then if p is the sound intensity on the acoustic axis, or favored direction, at a distance m from the device, the directivity index in decibels is given by:

$$\text{Directivity index (dB)} = 10 \log_{10}(p/p_{av})$$

For a pickup transducer, the same mathematical concept applies. However, it is in the reverse sense. If a given sound source, at a distance m from the transducer and having a certain intensity, produces an average voltage E_{av} at the transducer terminals as its orientation is varied in all possible directions, and the same source provides a voltage E when at a distance m from the transducer on the acoustic axis, then:

$$\text{Directivity index (dB)} = 10 \log_{10} (E/E_{av})$$

See also TRANSDUCER.

DIRECTIVITY PATTERN

See ANTENNA PATTERN.

DIRECT MEMORY ACCESS

In a digital computer, direct memory access is a means of obtaining information from the memory circuits without having to go through the central pocessing unit (CPU). The CPU is disabled while the memory circuits are being accessed. Direct memory access is abbreviated DMA.

Direct memory access saves time. It is much more efficient than getting memory information by routing through the central processing unit. There are several different methods of obtaining direct memory access. The process varies among different computer models. Direct memory access is used for the purpose of transferring memory data from one location to another, when it is not necessary to actually perform any operations on it. *See also* MEMORY.

DIRECTOR

A director is a form of parasitic element in an antenna system, designed for the purpose of generating power gain in certain directions. This increases the efficiency of a communications system, both by maximizing the effective radiated power of a transmitter, and by reducing the interference from unwanted directions in a receiving system. *See* ANTENNA POWER GAIN, PARASITIC ARRAY.

An example of the operation of a director is found in all Yagi- or quad-type antennas. A half-wave dipole antenna has a gain of 0 dBd (see dBd) in free space. This means, literally, that its gain with respect to a dipole is zero. The drawing illustrates the directive pattern of a half-wave dipole in free space at A.

If a length of conductor measuring approximately ½ wavelength, and not physically attached to anything, is brought near the half-wave dipole and parallel to it, the directivity pattern of the antenna changes radically. This was noticed by a Japanese engineer named Yagi, and the Yagi antenna is thus named after him. When the free element is a certain distance from the dipole, gain is produced in the direction of the free element, as shown at B.

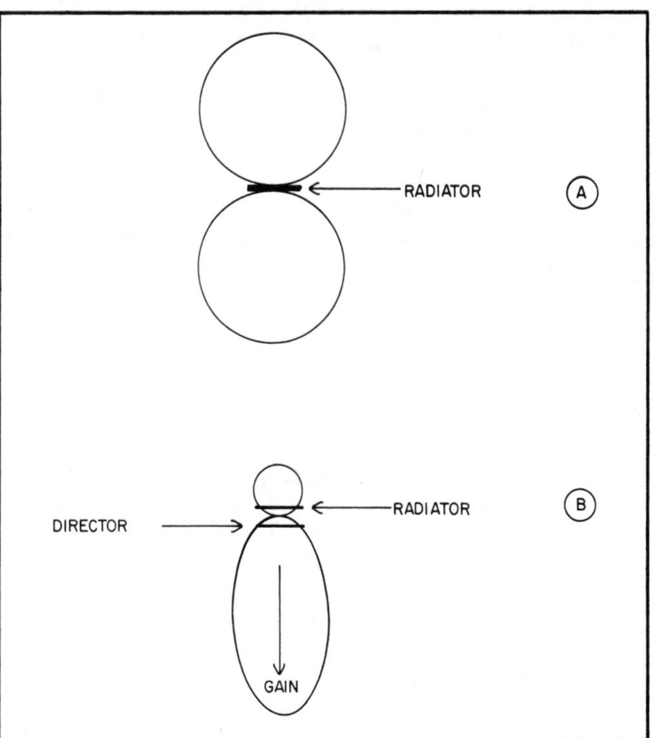

DIRECTOR: Effect of a director on the radiation pattern of a dipole antenna. At A, the pattern of the dipole alone is shown, in a plane containing the radiating element. At B, the pattern of the dipole in combination with the director is shown. This is an approximate representation.

The free element is then called a director. At certain other separations, the free element causes the gain to occur in the opposite direction, and then it is called a reflector (see REFLECTOR).

The most power gain that can be obtained using a dipole antenna and a single director is, theoretically, about 6 dBd. In practice it is closer to 5 dBd, because of ohmic losses in the antenna conductors. When two full-wavelength loops are brought in close proximity parallel to each other, with one loop driven (connected to the transmission line) and the other loop free, the same effect is observed. An antenna using loops in this manner is called a quad antenna.

The design of parasitic arrays is a very sophisticated art, and beyond the scope of this book. However, many excellent sources are available that discuss the operation and design of parasitic arrays. *See also* QUAD ANTENNA, YAGI ANTENNA.

DIRECT WAVE

In radio communications, the direct wave is the electromagnetic field that travels from the transmitting antenna to the receiving antenna along a straight line through space. The drawing illustrates the direct wave, which is also sometimes called the line-of-sight wave. Direct waves are responsible for part, but not all, of the signal propagation between two antennas when a line connecting the antennas lies entirely above the ground. The surface wave and the reflected wave also contribute to the overall signal at the receiving antenna in such a case. The combination of the direct wave, the reflected wave, and the surface wave is sometimes called the ground wave (*see* GROUND WAVE, REFLECTED WAVE, SURFACE WAVE). Depending on the relative phases of the direct, reflected, and surface waves, the received signal over a line-of-sight path may be very strong or practically nonexistent.

The range of communication via direct waves is, of course, limited to the line of sight. The higher the transmitting antenna, the larger the area covered by the direct wave. In mountainous areas, or in places where there are many obstructions such as buildings, it is advantageous to locate the antenna in the highest possible place. Direct waves are of little importance at low frequencies, medium frequencies, and high frequencies. But at very high frequencies and above, the direct wave is very important in propagation. *See also* LINE-OF-SIGHT COMMUNICATION.

DISCHARGE

When an electronic component that holds an electric charge, such as a storage battery, capacitor, or inductor, loses its charge, the process is called discharge. Charge is measured in coulombs, or units of 6.218×10^{18} electrons (*see* CHARGE).

Discharging may occur rapidly, or it may take place gradually. The discharge rate is the time required for a component to go from a fully charged state to a completely discharged condition. In a storage battery, the discharge rate is defined as the amount of current the battery can provide for a specified length of time. The discharging process occurs exponentially, as illustrated by the curve in the graph. With a given load resistance, the current is greatest at the beginning of the discharge process, and grows smaller and smaller with time.

Components such as capacitors and inductors can build up a charge over a long period of time, and then release the charge quickly. When this happens, large values of current or voltage can be produced. An automobile spark coil works on this principle. This property of charging and discharging can create a shock hazard, and precautions should be taken to ensure that a component has been completely discharged before any service work is performed. This is especially important with high-voltage power supplies.

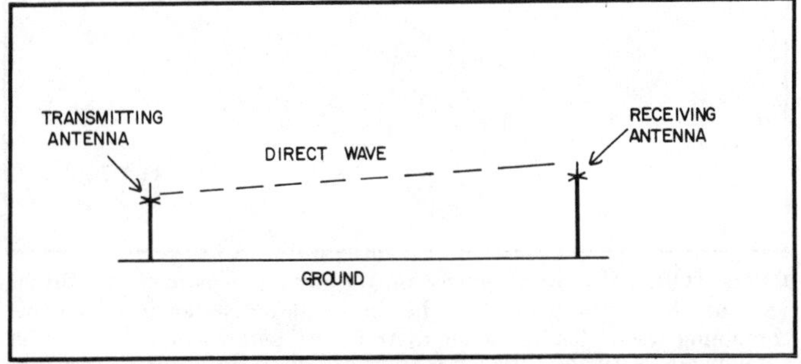

DIRECT WAVE: The direct wave travels entirely clear of the ground.

DISCONE ANTENNA

A discone antenna is a wideband antenna, resembling a biconical antenna, except that the upper conical section is replaced by a flat, round disk. A discone antenna is very similar to a conical monopole antenna as well (*see* BICONICAL ANTENNA, CONICAL MONOPOLE ANTENNA).

The discone, often used at very high frequencies, is fed at the point where the vertex of the cone joins the center of the disk, as shown in the illustration. The lowest operating frequency is determined by the height of the cone, h, and the radius of the disk, r. The value of h should be at least ¼ wavelength in free space, and the value of r should be at least 1/10 wavelength in free space. The discone presents a nearly constant, non-reactive load at all frequencies above the lower-limit frequency, for at least a range of several octaves. The exact value of the resistive impedance depends on the flare angle Θ of the cone. Typical values of Θ range between 25 and 40 degrees, resulting in impedances that present a good match for coaxial transmission lines.

A discone antenna is usually oriented so that the disk is horizontal, and on top of the cone. This produces a vertically polarized wave. The disk and the cone are made of

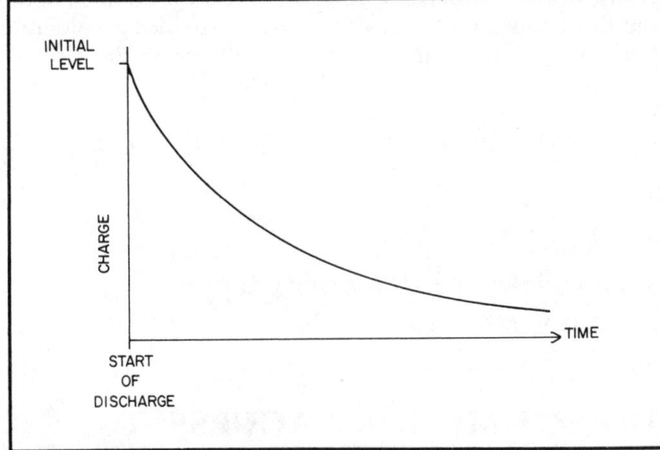

DISCHARGE: A typical discharge curve. The function is an exponential-decay curve.

sheet metal or fine wire mesh. The dimensions of a discone make it a practical choice at frequencies above 30 MHz; occasionally, it is used at frequencies as low as 3 MHz. The maximum radiation occurs approximately in the plane of the disk, or slightly below. The discone is omnidirectional in the azimuth plane.

DISCONNECT

The term disconnect refers to the separation of two previously connected circuits or components. In particular, the term is used to indicate the removal of power from a circuit, or the removal of a load from a source of power.

A device called a disconnect switch is employed in an electric motor to shut the power off when the motor draws too much current. This prevents damage to the moto windings, and reduces the chances of fire. A circuit breaker is a form of disconnect switch. *See also* CIRCUIT BREAKER.)

DISCONTINUITY

A discontinuity in an electrical circuit is a break, or open circuit, that prevents current from flowing. Discontinuity may occur in the power-supply line to a piece of equipment, resulting in failure. Sometimes a discontinuity occurs because of a faulty solder connection, or a break within a component. Such circuit breaks can be extremely difficult to find. Sometimes the discontinuity is intermittent, or affects the operation of a circuit only in certain modes or under certain conditions. This kind of problem is a well-known thing to experienced service technicians.

In a transmission line, a discontinuity is an abrupt change in the characteristic impedance (*see* CHARACTERISTIC IMPEDANCE). This may occur because of damage or deterioration in the line, a short or open circuit, or a poor splice. A transmission-line discontinuity may be introduced deliberately, by splicing two sections of line having different characteristic impedances. This technique is used for certain impedance-matching applications, and for distributing the power uniformly among two or more separate antennas.

In a discontinuous mathematical function, a discontinuity is a point in the domain of the function (along the axis of the independent variable) at which the function jumps from one value to another, or is undefined. Some functions have no discontinuities; these are called continuous functions. Some functions have many discontinuities, perhaps even an infinite number of them. *See also* CONTINUOUS FUNCTION, DISCONTINUOUS FUNCTION, FUNCTION.

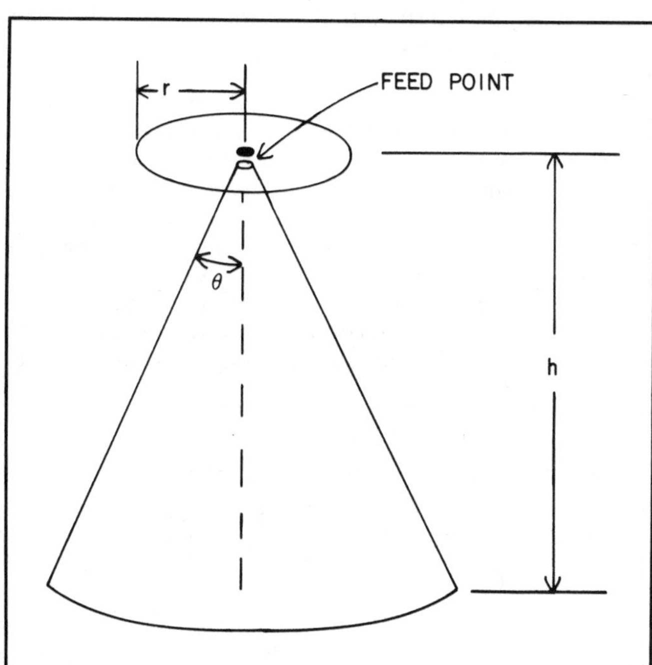

DISCONE ANTENNA: A discone antenna is fed at the intersection of the cone and the disk. The antenna is usually oriented such that the disk is on top of the cone, and lies in a hroizontal plane.

DISCONTINUOUS FUNCTION

A discontinuous function is a function that is not continuous at every point in its domain (*see* CONTINUOUS FUNCTION, FUNCTION). On a graph, such as the illustration, this appears as either a jump in value, or an undefined value. At A, the value of the function abruptly changes. At B and C, a point in the domain is not mapped into any value; that is, the value of the function is not defined.

Suppose x_0 is a point in the domain of the function $y = f(x)$, at which the function is not continuous. Then, the limiting value of the function as x approaches x_0 from the negative (left) side, and the limiting value of the function as x approaches x_0 from the positive (right) side, are not both equal to the value of the function $y_0 = f(x_0)$. This is expressed as an assertion that the following is not true:

$$\lim_{x \to x_0^-} f(x) = \lim_{x \to x_0^+} f(x) = y_0$$

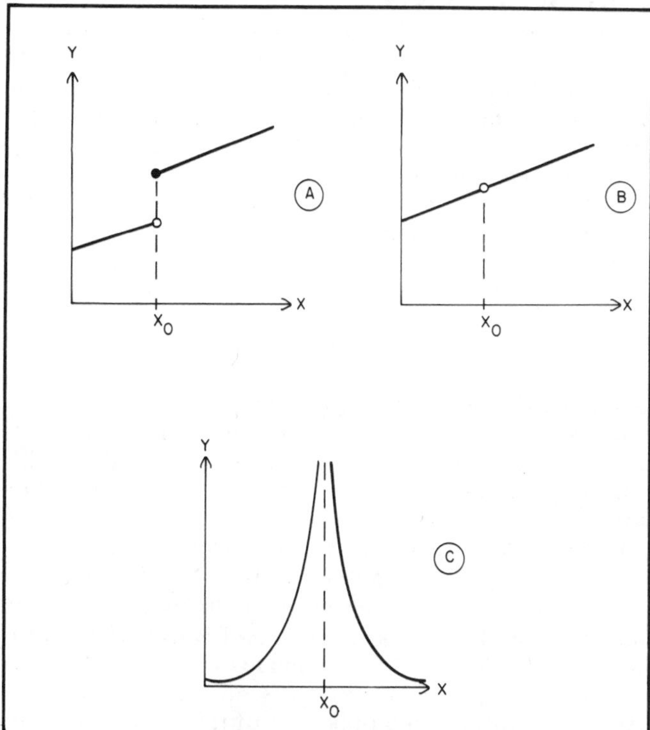

DISCONTINUOUS FUNCTION: Examples of discontinuous functions. At A, the value of the function jumps. At B and C, the value of the function is undefined at the point of discontinuity.

Examples of discontinuous functions are the tangent, cotangent, secant, and cosecant functions. All of these functions have undefined points. Another common example of a discontinuous function is the step function. Sometimes this is called the "U.S. Post Office" function. *See also* COSECANT, COTANGENT, SECANT, STEP FUNCTION, TANGENT.

DISCRETE COMPONENT

An electronic component such as a resistor, capacitor, inductor, or transistor is called a discrete component if it has been manufactured before its installation. In contrast to this, the resistors, capacitors, inductors, or transistors of an integrated circuit are not discrete; they are manufactured with the whole package, which may contain thousands of individual components.

In the early days of electronics, all circuits were built from discrete components. Only after the advent of solid-state technology, and especially miniaturization, did other designs emerge. At first, discrete components were assembled and sealed in a package called a compound circuit. But modern technology has provided the means for fabricating thousands of individual components on the surface of a semiconductor wafer.

Although discrete components are used less often than they were only a few years ago, there will always be a place for them. Such devices as fuses, circuit breakers, and switches must remain discrete. However, we will probably see fewer and fewer discrete components in electronic circuits in the coming years, as digital-control techniques become more refined. *See also* DIGITAL CONTROL, INTEGRATED CIRCUIT.

DISCRIMINANT

The discriminant is a mathematical expression in a quadratic, cubic, or high-order polynomial equation. The value of the discriminant determines the nature of the solutions to a polynomial equation. Discriminants are most often used in quadratic, or second-order, equations. In the quadratic equation generally expressed as:

$$ax^2 + bx + c = 0$$

the discriminant, D, is given by:

$$D = b^2 - 4ac$$

assuming that a, b, and c are real numbers. The roots, or solutions, of the quadratic equation are real and unequal if D is positive. If D is zero, the roots are real and identical, so the equation has but one solution. If D is negative, there are no real-number roots. In that case, the solutions are complex conjugates of the form $f + gi$ and $f - gi$, where $i = \sqrt{-1}$. *See* COMPLEX NUMBER, CONJUGATE IMPEDANCE.

In the cubic and higher-order equations, the discriminant is a complicated expression, and is usually found by means of a computer. Discriminants in cubic and higher-order equations are beyond the scope of this discussion. As an example of the application of the discriminant in a quadratic equation, suppose we are given the polynomial equation:

$$x^2 - 3x - 10 = 0$$

Then the discriminant, D, is:

$$D = 3^2 - (4 \times 1 \times -10)$$
$$= 9 + 40 = 49$$

This indicates that the solutions to the equation are real numbers, but unequal. This particular equation can be easily factored, giving:

$$(x - 5)(x + 2) = 0$$

and therefore the roots are obviously 5 and −2. *See also* QUADRATIC FORMULA.

DISCRIMINATOR

A detector often used in frequency-modulation receivers is called a discriminator (*see* FREQUENCY MODULATION). The discriminator circuit produces an output voltage that depends on the frequency of the incoming signal. In this way, the circuit detects the frequency-modulated waveform.

When a signal is at the center of the passband of the discriminator, the voltage at the output of the circuit is zero. If the signal frequency drops below the channel center, the output voltage becomes positive. The greater the deviation of the signal frequency below the channel center, the greater the positive voltage at the output of the discriminator. If the signal frequency rises above the channel center, the discriminator output voltage becomes negative; and, again, the voltage is proportional to the deviation of the signal frequency. The amplitude of the voltage at the output of the discriminator is linear, in proportion to the frequency of the signal. This ensures that the output is not distorted.

The illustration is a schematic diagram of a simple discriminator circuit suitable for use in a frequency-modulation receiver. A shift in the input signal frequency causes a phase shift in the voltages on either side of the transformer. When the signal is at the center of the channel, the voltages are equal and opposite, so that the net output is zero.

A discriminator circuit is somewhat sensitive to amplitude variations in the signal, as well as to changes in the frequency. Therefore, a limiter circuit is usually necessary when the discriminator is used in a frequency-modulation receiver. A circuit called a ratio detector, developed by RCA, is not sensitive to amplitude variations in the incoming signal. Thus, it acts as its own limiter. Immunity to amplitude variations is important in frequency-modulation reception, because it enhances the signal-to-noise ratio. *See also* LIMITER, RATIO DETECTOR.

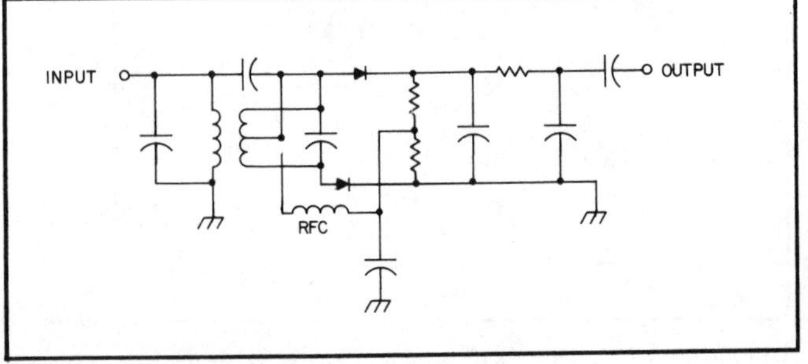

DISCRIMINATOR: A discriminator circuit, used for detecting a frequency-modulated signal.

DISH ANTENNA

A dish antenna is a high-gain antenna that is used for transmission and reception of ultra-high-frequency and microwave signals. The dish antenna consists of a driven element or other form of radiating device, and a large spherical or parabolic reflector, as shown in the illustration. The driven element is placed at the focal point of the reflector.

Signals arriving from a great distance, in parallel wavefronts, are reflected off the dish and brought together at the focus. Energy radiated by the driven element is reflected by the dish and sent out as parallel waves. The principle is exactly the same as that of a flashlight or latern reflector, except that radio waves are involved instead of visible light.

A dish antenna must be at least several wavelengths in diameter for proper operation. Otherwise, the waves tend to be diffracted around the edges of the dish reflector. The dish is thus an impractical choice of antenna, in most cases, for frequencies below the ultra-high range. The reflecting element of a dish antenna may be made of sheet metal, or it may be fabricated from a screen or wire mesh. In the latter case, the spacing between screen or mesh conductors must be a very small fraction of a wavelength in free space.

Dish antennas typically show very high gain. The larger a dish with respect to a wavelength, the greater the gain of the antenna. It is essential that a dish antenna be correctly shaped, and that the driven element be located at the focal point. Dish antennas are used in radar, and in satellite communications systems. Some television receiving antennas employ this configuration as well. *See* ANTENNA POWER GAIN, PARABOLOID ANTENNA.

DISINTEGRATION VOLTAGE

The voltage at the plate of a gas-filled tube must be within certain limits to ensure proper operation of the tube. The maximum limit for the plate-to-cathode voltage in such a tube is called the disintegration voltage. The minimum limit is called the ionization voltage (*see* IONIZATION VOLTAGE).

If the plate voltage becomes greater than the disintegration voltage, the cathode begins to lose its electron-emitting properties. The special coating on the cathode, providing a high electron emission, begins to get stripped away. Extreme plate voltages can thus permanently damage a gas tube. For the best tube life, it is important that the plate voltage be kept below the disintegration level. *See also* TUBE.

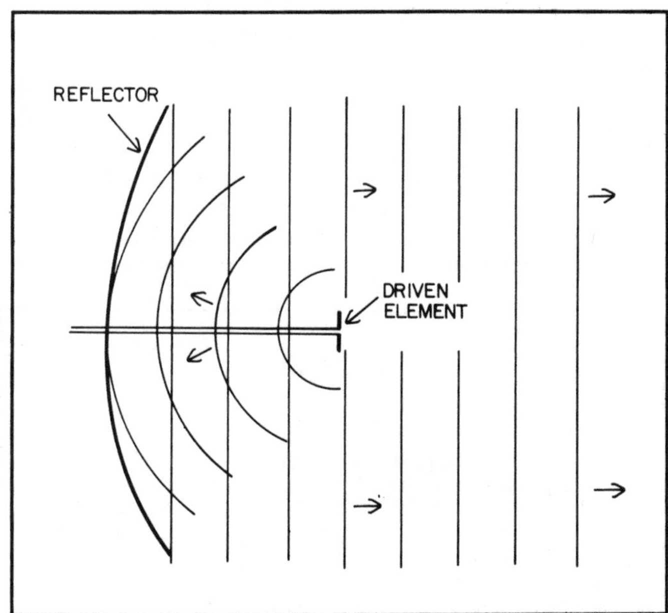

DISH ANTENNA: A dish antenna emits parallel wavefronts when used for transmitting. In reception, it focuses the waves at the driven element. Such antennas, if properly designed, can produce extremely high gain.

DISK CAPACITOR

A capacitor consisting of two round metal plates with leads attached, and with a layer of dielectric sandwiched between the metal plates, is called a disk capacitor, or sometimes disc capacitor. Disk capacitors are very common in radio-frequency electronic circuits, and are also sometimes seen in high-impedance audio applications. The leads of the capacitor protrude from the disk-shaped body, and are parallel to each other. The entire body of the capacitor is coated with a sealant to protect it against moisture and contamination.

The physical dimensions of a disk capacitor vary greatly. The capacitance depends on the size of the disks, the spacing between them, and the kind of dielectric material used. The working voltage depends mostly on the thickness of the dielectric. The photograph shows a typical disk capacitor, which is quite small in size, both physcally and electrically.

Capacitance values of disk capacitors generally range from less than 1 pF to about 1 uF. Voltage ratings are usually between about 10 and 1000 volts. The dielectric material is often a ceramic substance, which has very low loss. *See also* CERAMIC CAPACITOR.

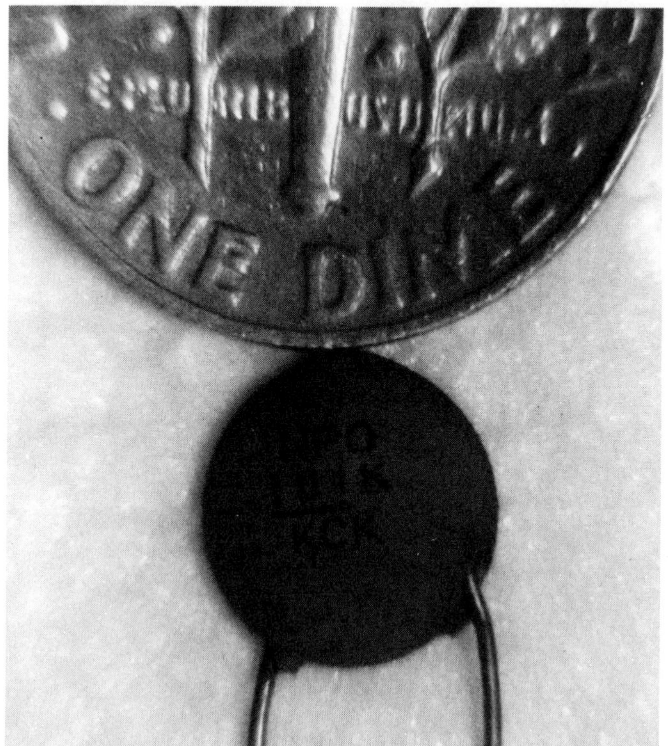

DISK CAPACITOR: A disk capacitor is easily recognizable.

DISK DRIVE

A disk drive is a device similar to a record player, except that the recording and playback is accomplished by magnetic means. This allows recording, in the same manner as with a tape recorder. Disk drives are extensively used in computers and word processors. An operator simply inserts the magnetic disk or diskette into a slot, where it is engaged by the drive. Information is easily stored or retrieved at any point on the disk. Disks are easily removed from the drive mechanism for convenient replacement.

Many small computers have built-in disk drives. Others do not have the drives already installed, but provisions are made for using them if desired. Many computers have two disk drives that operate in tandem. A single disk drive allows the storage of about 3,000,000 bits of information. The access time is very fast, only about ½ second in most cases, which makes the disk drive far superior to the magnetic tape recorder for the purpose of information storage and retrieval. *See also* DISKETTE.

DISKETTE

A diskette is a small magnetic disk, ranging in size from 2½ to 5¼ inches in diameter. Most diskettes come in a flat, square package for physical protection (see photograph). Data is recorded and stored on the diskette in concentric magnetic tracks, in a configuration similar to the grooves on a phonograph record. However, the principle is like the recording on a magnetic tape.

A magnetic tape system has excellent storage capacity, but disks and diskettes have a much higher storage density and a much shorter access time. Diskettes have about 200 tracks per radial inch; the resulting spiral is 300 to 500 feet long.

An especially common kind of diskette is sometimes called a floppy disk because it is physically flexible. The floppy disk (*see* FLOPPY DISK) is easy to store since it is very thin, and it is convenient to mail since it is not easily damaged and is lightweight. A typical floppy diskette can store 3,000,000 bits of data, and the serial access rate can be as high as 250 kHz. Thus, any information on the diskette can be accessed within about ½ second or less.

Although magnetic tape storage systems are less expensive than disks or diskettes, the access time in a tape system is considerably longer. *See also* DISK DRIVE, MAGNETIC RECORDING, MAGNETIC TAPE.

DISKETTE: A diskette is about the size of a 45-rpm phonograph record. Diskettes are usually enclosed in a square protective package.

DISK RECORDING

Disk recording is the process of recording sound onto a phonograph disk. The disk recorder cuts a spiral groove into the disk, starting at the outer edge and moving inward toward the center. The standard speed for disk recording is 33⅓ revolutions per minute. Some records are made at a speed of 45 revolutions per minute. The very oldest records were made at 78 revolutions per minute, but many modern phonographs do not have the capability to play these disks.

The sound vibrations in a disk recorder cause the cutter to vibrate back and forth. Most disks today are recorded in stereo, with two independent sound tracks. The cutter vibrates in two different planes at once. the planes are oriented at 90 degrees so that the vibrations in one can be completely independent of those in the other. The resulting groove contains two sound tracks on perpendicular faces (see drawing).

Since the angular speed of the disk is constant, the cutter travels over the surface of the disk more rapidly at the outer edge than near the center. Theoretically, then, the high-frequency reproduction is best at the outer periphery, and worst at the center. Most disks are made of a polyester material that provides excellent audio reproduction even near the center of the disk. *See also* HIGH FIDELITY, STEREOPHONICS.

DISPERSION

In some materials, electromagnetic waves are propagated with a velocity that depends on their wavelength. Ordinary glass exhibits this property with visible light. Red light goes through the glass faster than violet light. This effect is known as dispersion.

The most familiar example of dispersion is the rainbow produced when white light is passed through a prism (see illustration). The index of refraction is different for the different light wavelengths. Thus, violet light is bent more than red light (*see* INDEX OF REFRACTION).

When microwave energy strikes an obstruction that is not completely transparent to it, dispersion occurs. Gener-

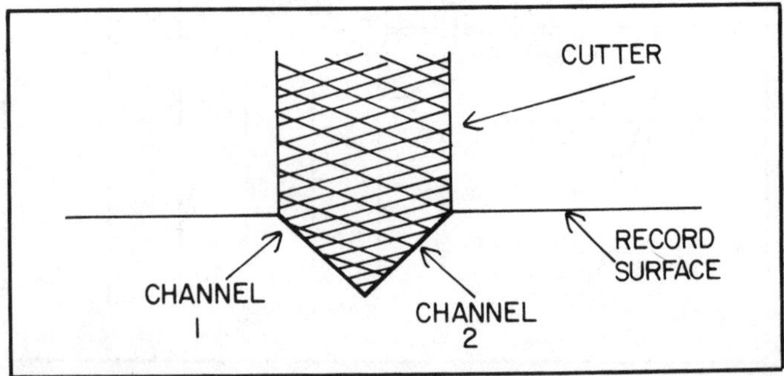

DISK RECORDING: In disk recording, the cutter produces a groove in the surface of the record. In stereo, shown here, two channels are cut, one at a right angle to the other.

ally, the lower frequencies are passed more rapidly than the higher frequencies. Thus the velocity factor depends on the wavelength (*see* VELOCITY FACTOR). This is not true for all dielectric materials, however.

In acoustics, dispersion refers to the effectiveness with which a speaker propagates the different sound wavelengths widely and uniformly. The better the speaker dispersion, the more realistic the sound reproduction over a wider angle. *See also* ACOUSTICS, SPEAKER.

DISPLACEMENT

Displacement is the movement of a particle or the change in position of a vector quantity (*see* VECTOR). Displacement is measured in linear units, such as centimeters or inches, or in angular units, such as degrees or radians.

The displacement of a particle or point in a multidimensional system can be defined in terms of the components parallel to each axis. In a polar coordinate system, the displacement may be defined in terms of a radial component and an angular component. Displacement might also be specified in terms of the overall distance the particle moves, and in a certain direction.

On the Cartesian (x,y) plane, a point may move from the origin (0,0) to the point (3,4). The x-axis component of the displacement is 3, and the y-axis component is 4. The overall distance traveled by the point is, in this case, 5 units, in a direction approximately 53 degrees counterclockwise from the x axis. In a polar system, this displacement would be given by $(\Theta,r) = (53°,5)$. In three dimensions, displacement becomes more difficult to define; in systems having many dimensions, the representations can get very complicated. *See also* CARTESIAN COORDINATES, POLAR COORDINATES.

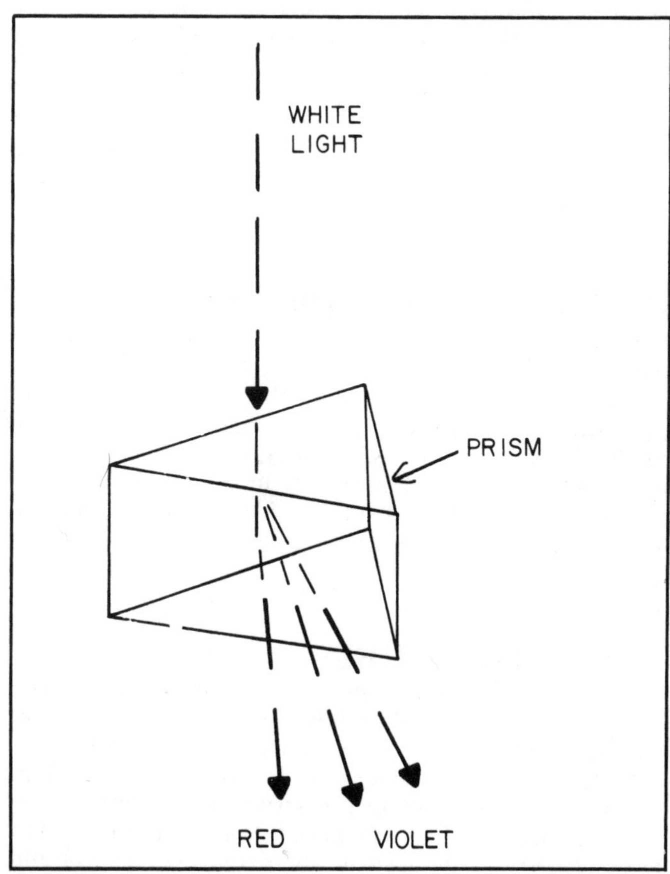

DISPERSION: White light is dispersed by a prism into its constituent wavelengths.

DISPLACEMENT CURRENT

When a voltage is applied to a capacitor, the capacitor begins to charge. A current flows into the capacitor as soon as the voltage is applied. At first this displacement current is quite large. But as time passes, with the continued application of the voltage, it grows smaller and smaller, approaching zero in an exponential manner.

The rate of the decline in the displacement current depends on the amount of capacitance in the circuit, and also on the amount of resistance. The larger the product of the capacitance and resistance, the slower the rate of decline of the displacement current, and the longer the time necessary for the capacitor to become fully charged. The magnitude of the displacement current at the initial moment the voltage is applied depends on the quotient of the capacitance and resistance. The larger the capacitance for a given resistance, the greater the initial displacement current. The larger the resistance for a given capacitance, the smaller the initial displacement current. Of course, the larger the charging voltage, the greater the initial displacement current.

In electromagnetic propagation, a change in the electric flux causes an effective flow of current. This current is called displacement current. The more rapid the change in the intensity of the electric field, the greater the value of the displacement current. The displacement current is 90 degrees out of phase with the electric-field cycle. The displacement current is perpendicular to the direction of wave propagation. *See also* ELECTROMAGNETIC FIELD, ELECTROMAGNETIC RADIATION.

DISPLACEMENT TRANSDUCER

A displacement transducer is a device that converts mechanical motion into electrical energy. A good example of a displacement transducer is an electric generator, in which a moving coil or magnet produces an electric current (*see* GENERATOR).

A displacement transducer usually operates on the principle that a changing magnetic field will produce a current in a conductor. A bar magnet within a solenoidal coil, for example, will cause a current pulse in the coil when moved axially within the coil.

Motion detectors may, in a sense, be considered displacement transducers. They usually operate by ultrasonic-wave interference. Any motion in the vicinity of such a device will cause a change in the wave pattern, and the device senses this. When this happens, an auxiliary circuit is switched on or off, such as a burglar alarm. *See also* ULTRASONIC MOTION DETECTOR.

DISPLAY

A display is a visual indication of the status of a piece of electronic equipment. Displays are also used in all metering devices. A display may be as simple as the frequency readout in a communications receiver or transmitter. Or, a display can be as complicated as the video monitors used with computers. The photograph illustrates a display used on an amateur-radio transceiver. It shows the frequency,

DISPLAY: An example of a display. In this case, it is entirely digital, and uses liquid crystals.

the signal strength, the transmitter relative output power, the memory channel in use, and other functions of the microcomputer. The physical layout of a display is important from the standpoint of operating efficiency and convenience.

The cathode-ray-tube screen of an oscilloscope or spectrum analyzer is a form of display. So is the face of a digital watch or timer, or the speedometer of an automobile. Displays may be either electronic or mechanical. Electronic displays can employ tubes, light-emitting diodes, or liquid crystals. An analog display shows a range of values in a continuous manner. A digital display shows either a set of numerals, or a bar indication. *See also* ANALOG, ANALOG METERING, DIGITAL, DIGITAL METERING, LIGHT-EMITTING DIODE, LIQUID-CRYSTAL DISPLAY.

DISPLAYED NUMERALS

In an electronic calculator, it is often possible to get greater accuracy than is indicated by the numeric display. This is because some calculators actually store more digits than they display. Not all calculators have this property, but those that do have the advantage of extra accuracy when it is needed or desired.

It is quite easy to test a calculator to see if it will store more digital information than it displays. One procedure is as follows. Suppose a particular calculator has an eight-numeral digital display. Determining $\sqrt{2}$ thus yields a display of 1.4142136, assuming the square-root function key is used. Subtracting 1 from this value will cause all of the digital information to move one place to the left. If the calculator has any "hidden" digits beyond the eight-digit display, the result will be:

$$\sqrt{2} - 1 = .41421356$$

but if the calculator does not have this property, the result will be either of the following:

$$\sqrt{2} - 1 = .41421360$$
$$\sqrt{2} - 1 = 0.4142136$$

Determining the extent of the extra available digits is a simple process. In this particular example, the display of .41421356 can be multiplied by 10, giving 4.1421356. Then, subtracting 4, we obtain:

$$((\sqrt{2} - 1) \times 10) - 4 = .14213562$$

if yet another "hidden" digit exists. But if it does not, the result will be either of the following:

$$((\sqrt{2} - 1) \times 10) - 4 = .14213560$$

$$((\sqrt{2} - 1) \times 10) - 4 = 0.1421356$$

Some calculators store three or four extra digits. This feature is seldom needed, but can be used to advantage if desired.

DISPLAY LOSS

Display loss is a term generally used with regard to receiver output monitoring devices, such as a spectrum monitor. A human operator, listening to the output of a receiver, will always be able to hear faint signals that do not show up on a spectrum monitor or other instrument (*see* SPECTRUM ANALYZER, SPECTRUM MONITOR). The ratio, expressed in decibels, between the minimum signal input power P_1 detected by an ideal instrument, and the minimum signal input power P_2 detected by a human operator using the same receiver, is called the display loss or visibility factor. Mathematically:

$$\text{Display loss (dB)} = 10 \log_{10}(P_1/P_2)$$

This is always a negative value. That is, the human operator is always better than the instrument. This is exemplified by the fact that a Morse-code copying machine will have difficulty in a marginal situation in which a human operator can get the message adequately. While code readers can "copy" signals at extremely high speed, given good propagation conditions and a good signal-to-noise ratio, a faint signal is often imperceptible to the machine even at a slow speed which the human operator can "copy" fairly well.

DISPOSABLE COMPONENT

Some circuit components are repairable, while others are not. Components that are not repairable, or are so inexpensive that it is cheaper to just throw them out and replace them, are called disposable components. Capacitors, diodes, integrated circuits, resistors, and transistors are all disposable. However, printed-circuit boards are often repairable, as are interconnecting cables, inductors, and the like. *See also* COMPONENT.

DISSECTOR TUBE

A dissector tube, also known as an image dissector, is a form of photomultiplier television camera tube (*see* PHOTOMULTIPLIER). Light is focused, by means of a lens, onto a translucent surface called a photocathode. This surface emits electrons in proportion to the light intensity. The electrons from the photocathode are directed to a barrier containing a small aperture. The vertical and horizontal deflection plates, supplied with synchronized scanning voltages, move the beam from the photocathode

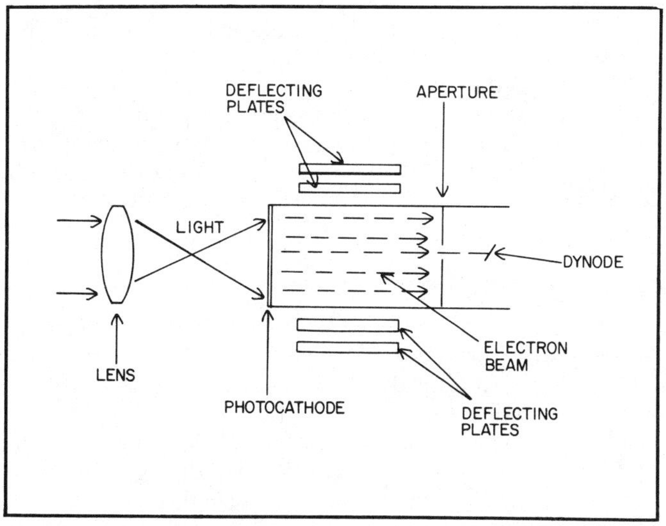

DISSECTOR TUBE: A dissector tube produces a modulated stream of electrons by rapidly scanning an electron image across a small hole (aperture). Dynodes allow large amplification factors.

across the aperture. Thus the aperture scans the entire image. The electron stream passing through the aperture is thus modulated depending on the light and dark nature of the image. A simplified diagram of a dissector tube is illustrated in the drawing.

After the electrons have passed through the aperture, they strike a dynode or series of dynodes. Each dynode emits several secondary electrons for each electron that strikes it (*see* DYNODE). In this way, the electron stream is intensified. Several dynodes in cascade can provide an extremely large amount of gain.

The resolving power, or image sharpness, of the dissector tube depends on the size of the aperture. The smaller the aperture, to a point, the sharper the image. However, there is a limit to how small the aperture can be, while still allowing enough electrons to pass, and avoiding diffraction interference patterns.

The image dissector tube, unlike other types of television camera tubes, produces very little dark noise. That is, there is essentially no output when the image is dark. This results in an excellent signal-to-noise ratio. *See also* TELEVISION.

DISSIPATION

Dissipation is an expression of power consumption, and is measured in watts. Power is defined as the rate of expenditure of energy; dissipation always takes place in a particular physical location.

When energy is dissipated, it may be converted into other forms, such as heat, light, sound, or electromagnetic fields. It never just disappears. The term dissipation is used especially with respect to consumption of power resulting in heat. A resistor, for example, dissipates power in this way. A tube or transistor converts some of its input power into heat; that power is said to be dissipated.

Generally, dissipated power is an undesired waste of power. Dissipated power does not contribute to the function of the circuit. Its effect can be detrimental; excessive dissipated power in a tube or transistor can destroy the device. Engineers use the term dissipation to refer to any form of power consumption. *See also* ENERGY, POWER, WATT.

DISSIPATION FACTOR

The dissipation factor of an insulating, or dielectric, material is the ratio of energy dissipated to energy stored in each cycle of an alternating electromagnetic field. This quantity, expressed as a number, is used as an indicator of the amount of loss in a dielectric material (*see* DIELECTRIC LOSS). The larger the dissipation factor, the more lossy the dielectric substance.

When the dissipation factor of a dielectric material is smaller than about 0.1, the dissipation factor is very nearly equal to the power factor, and the two may be considered the same for all practical purposes (*see* POWER FACTOR). The power factor is defined as the cosine of the angle by which the current leads the voltage. In general, the dissipation factor D is the tangent of the loss angle θ, which is the complement of the phase angle ϕ. Thus:

$$D = \theta = \tan (90° - \phi)$$

and, when $D < 0.1$,

$$D = \cos \phi$$

In a perfect, or lossless, dielectric material, the loss angle θ is zero. This means that the current leads the voltage by 90 degrees, as in a perfect capacitive reactance. In a conductor, the loss angle is 90 degrees. This means that the current and voltage are in phase. *See also* ANGLE OF LAG, ANGLE OF LEAD, LOSS ANGLE, PHASE ANGLE.

DISSIPATION RATING

The dissipation rating of a component is a specification of the amount of power it can safely consume, usually as heat loss. The dissipation rating is given in watts. Resistors are rated in this way; the most common values are ⅛, ¼, ½, and 1 watt, although much larger ratings are available. Zener diodes are also rated in terms of their power-dissipation capacity.

Transistors and vacuum tubes are rated according to their maximum safe collector, drain, or plate dissipation. In an amplifier or oscillator, the dissipated power P_D is the difference between the input power P_I and the output power P_O. That is:

$$P_D = P_I - P_O$$

The lower the efficiency, for a given amount of input power, the greater the amount of dissipation. It is extremely important that tubes and transistors be operated well within their dissipation ratings.

The dissipation rating for a tube or transistor is sometimes specified in two forms: the continuous-duty rating and the intermittent-duty rating. A device can often (but not always) handle more dissipation in intermittent service. For example, a tube can usually handle a higher amount of dissipation when used in a code transmitter, as compared to in an amplitude-modulated or frequency-modulated system in which the carrier is always present. *See also* DUTY CYCLE, POWER.

DISSONANCE

Dissonance is a term used in acoustics and music. It refers

to any unpleasant combination of tones.

Certain combinations of audio-frequency tones result in a pleasing sound. These combinations are called chords, and the pleasant quality is known as harmony. Whether a combination of audio notes is pleasing (harmonious) or displeasing (dissonant) is largely a matter of the perception of the listener. Someone from Beethoven's time would probably consider much of our music dissonant. We might, however, listen to the same sounds and call them harmonious. *See also* ACOUSTICS, HARMONY, MUSIC.

DISTANCE RESOLUTION

In a radar system, resolution is the ability of the receiver to identify two close, but distinct, targets individually. The minimum radial target separation for which this is possible is called the distance resolution or range resolution.

When two targets are extremely close to each other, they appear on the radar screen as a single echo, as shown at A in the illustration. But as they become farther and farther apart, a radial separation is reached where the targets appear separated on the screen, as shown at B. This minimum distance is the distance resolution.

The distance resolution of a radar system depends on the precision with which the receiver can measure a short time interval. This, in turn, depends on the shape of the radiated pulse, and the frequency of pulse emission. Echoes from a more distant target arrive later than the echoes from a nearby target.

Even if a radar system were able to measure arbitrarily small intervals of time, there would be a limit to the distance resolution. This is because of uncertainties in the propagation speed of the electromagnetic energy through the atmosphere. *See also* AZIMUTH RESOLUTION, RADAR.

DISTORTION

Distortion is a change in the shape of a waveform. Generally, the term distortion is used to refer to an undesired change in a wave, but distortion is sometimes introduced into a circuit deliberately.

The drawing illustrates a nearly perfect sine waveform, A, typical of radio-frequency transmitters. All of the energy is concentrated at a single wavelength. There is no harmonic energy in a perfect wave, but in practice there is always a little bit of distortion. Distortion results in the presence of the harmonics in various proportions (*see* HARMONIC), in addition to the fundamental frequency. Harmonics can be suppressed more than 80 dB in radio-frequency practice. This means that a harmonic is just one part in 100 million in amplitude, compared with the fundamental frequency. The sine wave in such a case is almost completely distortion-free.

A severely distorted waveform is shown at B. On a

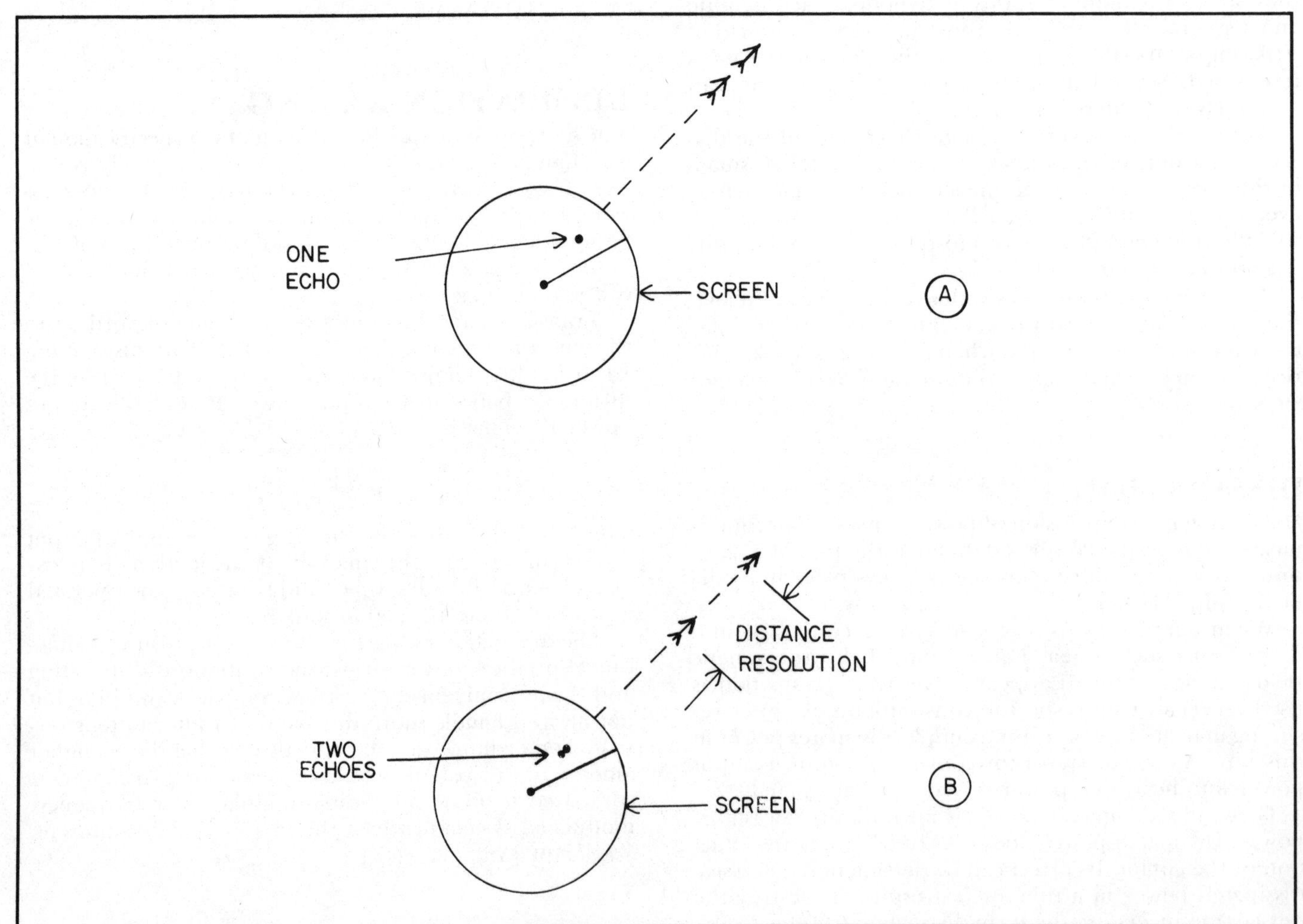

DISTANCE RESOLUTION: At A, the targets are too close to each other to be resolved; they appear at the same point on the radar display, even though they are separated in radial distance sufficiently for them to appear individually.

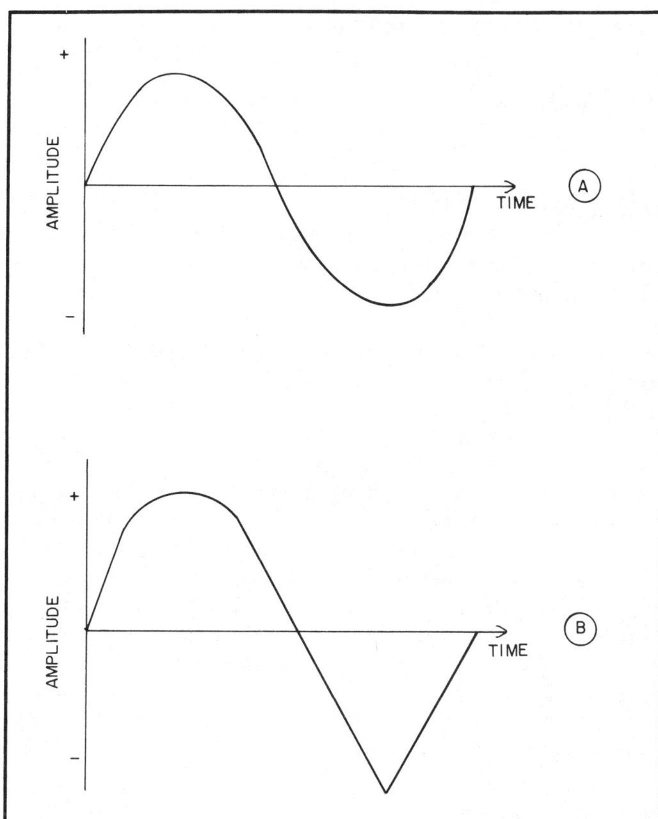

DISTORTION: At A, an undistorted sine wave. At B, distortion is seen, resulting in harmonic energy in addition to the fundamental frequency.

spectrum analyzer, distortion is visible as pips at harmonic frequencies in addition to the fundamental frequency (*see* SPECTRUM ANALYZER). The harmonics present, and their amplitudes, depend on the type and extent of the distortion.

In audio applications, distortion can occur with complex as well as simple waveforms. Sometimes the distortion can be so severe that a voice is unreadable or unrecognizable. Sound quality is measured in terms of fidelity, or accuracy of reproduction (*see* FIDELITY).

Distortion such as that illustrated is sometimes introduced into a circuit on purpose. Mixers and frequency multipliers require distortion in order to work properly. *See also* DISTORTION ANALYZER, FREQUENCY MULTIPLIER, MIXER.

DISTORTION ANALYZER

A distortion analyzer is a device, usually a meter, for measuring total harmonic distortion in an audio circuit. Such a meter has a special notch filter set for 1000 Hz. This filter can be used to remove the fundamental frequency of a standard audio test tone at the output of the amplifier. A voltmeter measures the audio signal remaining after notching out the fundamental frequency; this remaining energy consists entirely of harmonics, or distortion products. Total harmonic distortion is expressed as a percentage, measured with respect to the level of the output without the filter.

The photograph shows a typical distortion analyzer. This particular instrument can also measure the SINAD, or signal-to-noise-and-distortion, sensitivity of a communications receiver. The meter is extremely easy to use. *See also* SINAD, TOTAL HARMONIC DISTORTION.

DISTRESS FREQUENCY

In radio communication, a distress frequency is a frequency or channel specified for use in distress calling. Other calls may or may not be permitted.

In the aeronautical mobile service, radiotelegraphy distress calls are made at 500 kHz; radiotelephone distress frequencies are 2.182 MHz and 156.8 MHz. Survival craft use 243 MHz. The distress frequencies are the same in the maritime mobile service.

In the fixed and land mobile radio services, the disaster band is at 1.75 to 1.8 MHz. In the Citizen's-Band service, Channel 9, at 27.065 MHz, is the distress frequency. There is no specifically assigned distress frequency in the Amateur Radio Service. *See also* DISTRESS SIGNAL.

DISTRESS SIGNAL

A distress signal is a transmission indicating a life-threatening situation. The most familiar distress signals are Mayday and SOS. The word Mayday is used in

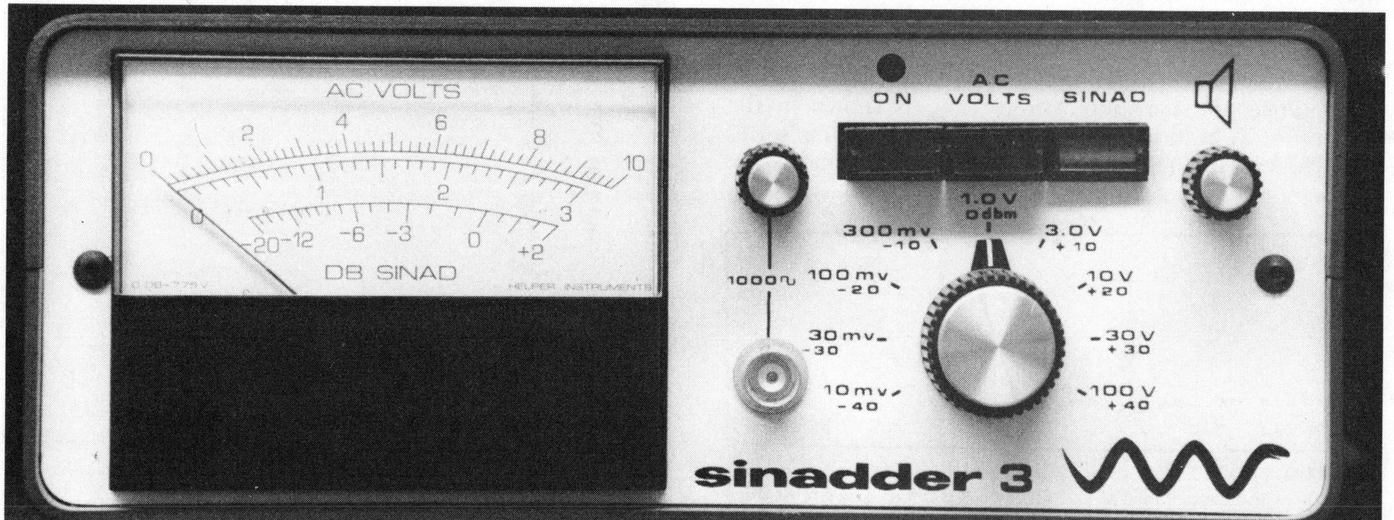

DISTORTION ANALYZER: A distortion analyzer consists of a simple meter and an audio filter, along with various controls for setting the sensitivity.

radiotelephony. The SOS signal is used in radiotelegraphy, and in the MORSE codes consists of three dots, three dashes, and three dots, usually sent in a continuous string (•••———•••).

In government radio services, standard frequencies are used for distress calling. *See also* DISTRESS FREQUENCY.

DISTRIBUTED AMPLIFIER

A distributed amplifier is a broadbanded amplifier sometimes employed as a preamplifier at very-high and ultra-high frequencies. Several tubes or transistors are connected in cascade, with delay lines between them (*see* DELAY LINE).

The more tubes or transistors in a distributed amplifier, the greater the gain. However, there is a practical limit to the number of amplifying stages that can be effectively cascaded in this way. Too many tubes or transistors will result in instability, possible oscillation, and an excessive noise level at the output of the circuit.

Since a distributed amplifier is not tuned, no adjustment is required within the operating range for which the circuit is designed. Distributed amplifiers are useful as preamplifiers for television receivers.

DISTRIBUTED ELEMENT

A constant parameter such as resistance, capacitance, or inductance is usually thought to exist in a discrete form, in a component especially designed to have certain electrical properties (*see* DISCRETE COMPONENT). However, there is always some resistance, capacitance, and inductance in even the most simple circuits. The equivalent resistance, capacitance, and inductance in the wiring of a circuit are called distributed elements or distributed constants.

The effects of distributed elements in circuit design and operation are not usually significant at low frequencies. However, in the very-high-frequency range and above, distributed elements are an important consideration. The lead length of a discrete component, at a sufficiently high frequency, becomes an appreciable fraction of the wavelength, and then the lead inductance and capacitance will affect the operation of a circuit. A resistor has distributed inductance and capacitance. An inductor has distributed resistance, on account of the ohmic loss in the conductor, and distributed capacitance, because of interaction between the windings.

In a transmission line, the inductance, resistance, and capacitance are distributed uniformly, as shown in the illustration. The inductance and resistance appear in series with the conductors. The capacitance appears across the conductors.

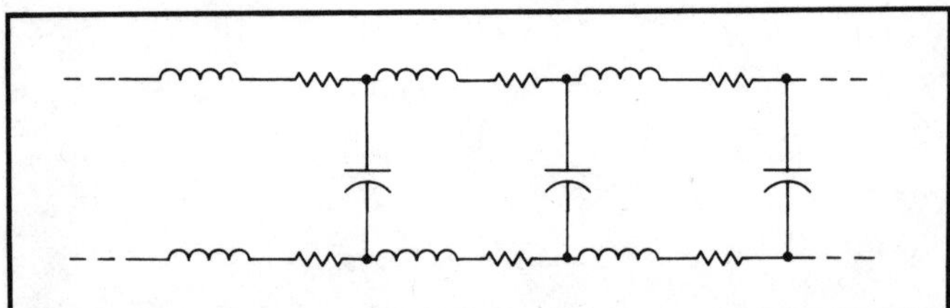

DISTRIBUTED ELEMENT: In a two-wire transmission line, resistance, inductance, and capacitance are uniformly distributed. The circuit thus appears, electrically, like a long string of series inductors and resistors, with parallel capacitors.

When discrete components are used in a circuit, or when the inductances, capacitances, or resistances are in defined locations, the elements are said to be lumped. *See also* LUMPED ELEMENT.

DISTRIBUTION

Distribution is a statistical property of a variable, and is important in probability analysis. The most familiar distribution is the standard or normal distribution, represented by a bell-shaped curve (*see* NORMAL DISTRIBUTION). However, there are many different kinds of distribution functions.

An example of a distribution function is seen in the results of an examination. Most of the scores in any test tend to be congregated within a certain range, such as is shown in the graph. The average score is called the mean; half of the area under the curve is to the left of the mean, and half is to the right. The score at which half of the examinations are worse and half are better is called the median. The degree to which the scores are spread out is expressed by the standard deviation. All distribution functions have a mean, a median, and a standard deviation. *See also* EXPONENTIAL DISTRIBUTION, MEAN, MEDIAN, POISSON DISTRIBUTION, RAYLEIGH DISTRIBUTION, STANDARD DEVIATION, STATISTICAL ANALYSIS.

DISTRIBUTION FRAME

In a telephone switching system, a distribution frame is the point at which all of the lines come together for interconnection. The distribution frame is part of the central office (*see* CENTRAL-OFFICE SWITCHING SYSTEM).

At the distribution frame, any line may be connected to any other line. In modern telephone circuits, this is done by a computer, so no human operator is needed. Many cross-connections can be made at once, and the combination can be changed at any time. The distribution frame

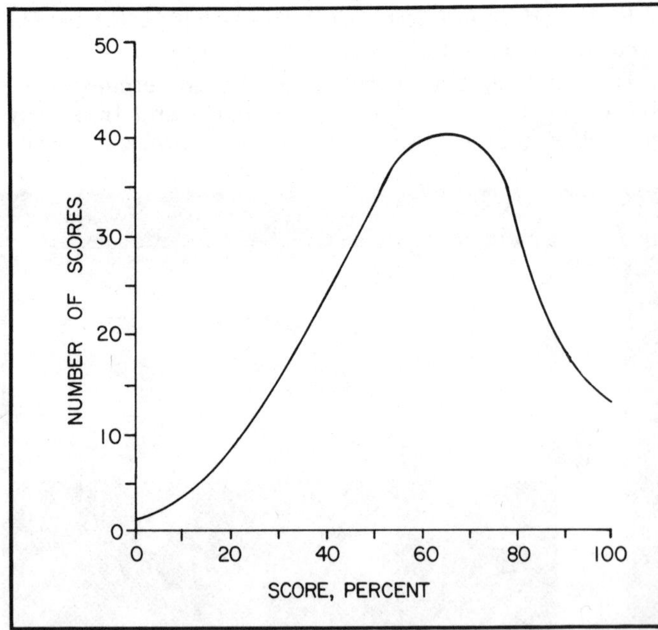

DISTRIBUTION: The results of a school test are a classic example of a statistical distribution.

should be large enough to handle calls at peak periods. *See also* TELEPHONE.

DISTRIBUTOR

A distributor is a device that switches current to two or more circuits on a time-sharing or rotating basis. Generally, a distributor consists of a rotating mechanical wiper, and a set of contacts arranged in a circle. As the wiper goes around, power is delivered momentarily to each circuit in succession.

The most common and familiar example of the application of a distributor is in an automotive electrical system. The distributor provides the current for the electric spark in each cylinder of the engine. The distributor is synchronized with the operation of the engine, so that each cylinder gets a burst of current at just the right moment. An electric commutator is sometimes called a distributor. Commutators are used in electric generators and motors to reverse the polarity of a voltage in synchronization with the rotation of the shaft. *See also* COMMUTATOR.

DISTRIBUTIVE LAW

The distributive law is a property of certain mathematical and logical operations. Given two operations, such as multiplication and addition, we say that multiplication is distributive with respect to addition if and only if:

$$x_1 (x_2 + x_3) = x_1x_2 + x_1x_3$$

for any values of x_1, x_2, and x_3. We are taught in elementary school that the above statement is in fact true. Multiplication is distributive over addition. But the reverse is not in general true. The following statement is false:

$$x_1 + x_2x_3 = (x_1 + x_2) (x_1 + x_3)$$

Therefore, addition is not distributive with respect to multiplication.

It is not difficult to prove that, if two operations such as multiplication and addition are distributive over two variables (as stated above), then they are distributive over any number of variables. That is, in the case of multiplication over addition:

$$x_1(x_2 + x_3 + \ldots + x_n) = x_1x_2 + x_1x_3 + \ldots + x_1x_n$$

In Boolean algebra, the AND operation is distributive with respect to the OR operation. This rule is expressed in the same way as the distributive law of multiplication over addition, when we let the AND operation be represented by multiplication and the OR operation be represented by addition. For any logical statements, X, Y, and Z, then:

$$X(Y + Z) = XY + XZ$$

See also BOOLEAN ALGEBRA.

DISTURBANCE

The term disturbance, in electronics, is generally used in reference to an unexpected or unwanted change in the propagation of radio signals. A disturbance in the geomagnetic field of the earth often causes a deterioration of ionospheric propagation at medium and high frequencies. Such a disturbance is caused by a solar flare, and is accompanied by the Aurora, or Northern and Southern lights. A severe geomagnetic disturbance may even affect telephone circuits.

An electrical storm is a form of disturbance, and the fields generated by such storms can cause severe local interference to radio circuits. Weather disturbances can also cause changes in the propagation at very-high and ultra-high frequencies. For more specific information, *see* GEOMAGNETIC STORM, LIGHTNING, SOLAR FLARE, SUDDEN IONOSPHERIC DISTURBANCE.

DIVERGENCE LOSS

A radiated energy beam generally diverges from the point of transmission. Sound, visible light, and all electromagnetic fields diverge in this manner. The exception is the laser beam (*see* LASER), which puts out a parallel beam of light.

Divergence results in a decrease in the intensity of the energy reaching a given amount of surface area, as a receiver is moved farther and farther from a transmitter. With most energy effects, the decrease follows the well-known inverse-square law (*see* INVERSE-SQUARE LAW).

Given an energy source with a constant output, such as a speaker (see illustration), the amount of energy striking a given area is inversely proportional to the square of the distance. Thus, if the distance doubles, the sound intensity is cut to one-fourth its previous level. If the distance becomes 10 times as great, the sound intensity drops to 1/100, or 0.01, of its previous value. Electromagnetic field strength, measured in microvolts per meter, decreases in proportion to the distance from the antenna. *See also* FIELD STRENGTH.

DIVERSITY RECEPTION

Diversity reception is a technique of reception that reduces the effects of fading in ionospheric communication. Two receivers are used, and they are tuned to the same signal. Their antennas are spaced several wavelengths apart. The outputs of the receivers are fed into a common audio amplifier, as shown in the illustration.

Signal fading usually occurs over very small areas, and at different rates, even at a separation of a few wavelengths in the receiving location, a signal may be very weak in one place and very strong in the other. Thus, when two antennas are used for receiving, and they are positioned sufficiently far apart, and they are connected to independent

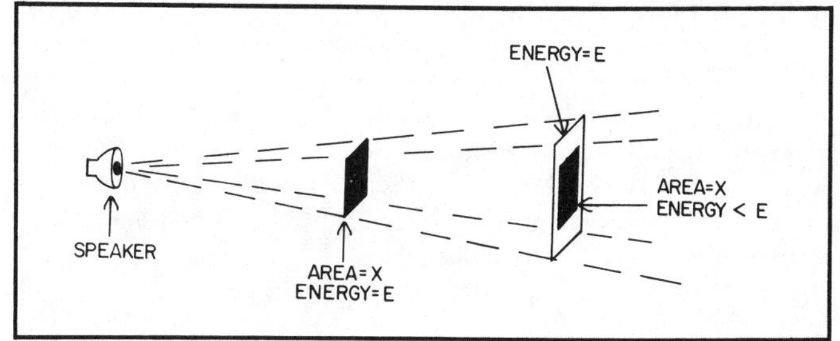

DIVERGENCE LOSS: Divergence loss results in a decrease in energy intensity with distance from the source.

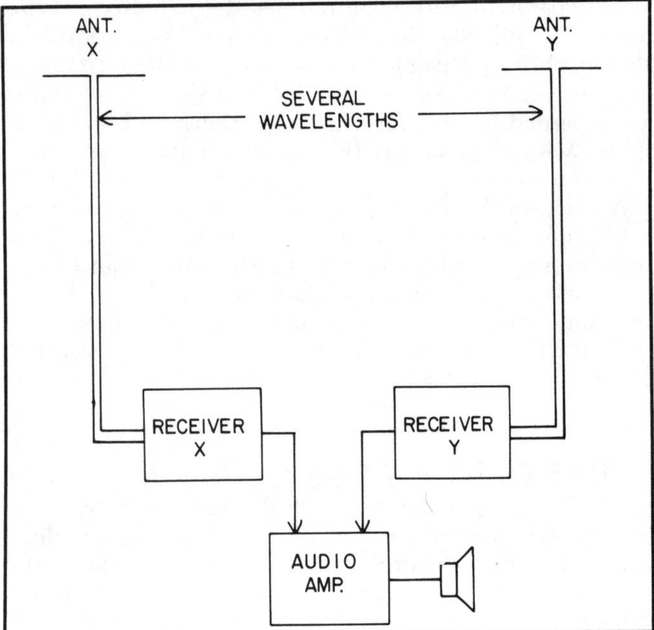

DIVERSITY RECEPTION: In diversity reception, two separate receivers are tuned to the same signal, using antennas spaced several wavelengths apart.

receivers, the chances are small that a fade will occur at both antenna locations simultaneously. At least one antenna almost always gets a good signal. The result of this, when the receiver outputs are combined, is the best of both situations! The fading is much less pronounced.

While diversity reception provides some measure of immunity to fading, the tuning procedure is critical and the equipment is expensive. More than two antennas and receivers can be used. In order to tune the receivers to exactly the same frequency, a common local oscillator should be used. Otherwise the audio outputs of the different receivers will not be in the proper phase, and the result may be a garbled signal, totally unreadable. *See also* FADING.

DIVIDER

A divider is a circuit that is used for the purpose of splitting voltages or currents. A common example of a divider is the resistive network often found in the biasing circuits of transistor oscillators and amplifiers.

In a calculator or computer, a divider is a circuit that performs mathematical division. A divider circuit is often used in a frequency counter to regulate the gate time. Divider circuits are used in digital systems to produce various pulse rates for frequency-control and measurement purposes. For example, a crystal calibrator having an oscillator frequency of 1 MHz may be used in conjunction with a divide-by-10 circuit, so that 100-kHz markers may be obtained. The divider can be switched in and out of the circuit so that the operator may select the markers he wants. A divide-by-100 circuit will allow the generation of 10-kHz markers from a 1-MHz signal. *See also* FREQUENCY DIVIDER, VOLTAGE DIVIDER.

D LAYER

The D layer is the lowest region of ionization in the upper atmosphere of our planet. The D layer generally exists only while the sun is above the visual horizon. During unusual solar activity, such as a solar flare, the D layer may become ionized during the hours of darkness. The D layer is about 35 to 60 miles in altitude, or 55 to 95 kilometers.

At very low frequencies, the D layer and the ground combine to act as a huge waveguide, making worldwide communication possible with large antennas and high-power transmitters (*see* WAVEGUIDE PROPAGATION). At the low and medium frequencies, the D layer becomes highly absorptive, limiting the effective daytime communication range to about 200 miles. At frequencies above about 7 to 10 MHz, the D layer begins to lose its absorptive qualities, and long-distance daytime communication is typical at frequencies as high as 30 MHz.

After sunset, since the D layer disappears, low-frequency and medium-frequency propagation changes, and long-distance communication becomes possible. This is why, for example, the standard amplitude-modulation (A-M) broadcast band behaves so differently at night than during the day. Signals passing through the D layer, or through the region normally occupied by the D layer, are propagated by the higher E and F layers. *See also* E LAYER, F LAYER, IONOSPHERE, PROPAGATION CHARACTERISTICS.

DMA

See DIRECT MEMORY ACCESS.

DMOS

See DIFFUSED METAL-OXIDE SEMICONDUCTOR.

DOHERTY AMPLIFIER

A Doherty amplifier is an amplitude-modulation amplifier consisting of two tubes or transistors used for different functions. One of the devices is biased to cutoff during unmodulated-signal periods, while the other acts as an ordinary amplifier. Enhancement of the modulation peaks is achieved by an impedance inverting line between the plate circuits of the two tubes, or between the collector circuits of the two transistors. The illustration shows a simplified schematic diagram of an FET type Doherty amplifier. This circuit is also sometimes called a Terman-Woodyard modulated amplifier.

Under conditions of zero modulation, Q1, called the carrier FET, acts as an ordinary radio-frequency amplifier. But Q2, called the peak FET, contributes no power during unmodulated periods. During negative modulation, when the instantaneous input power is less than the carrier power under conditions of zero modulation, the output of the carrier FET Q1 drops in proportion to the change in the input amplitude. At the negative peak, where the instantaneous amplitude may be as small as zero, the output of the carrier tube is minimum.

When a positive modulation peak occurs, the peak FET contributes some of the power to the output of the amplifier. The carrier-FET output power doubles at a 100-percent peak, compared to the output under conditions of no modulation. The peak FET also provides twice the output power of the carrier FET with zero modulation. Consequently, the output power of the amplifier is quad-

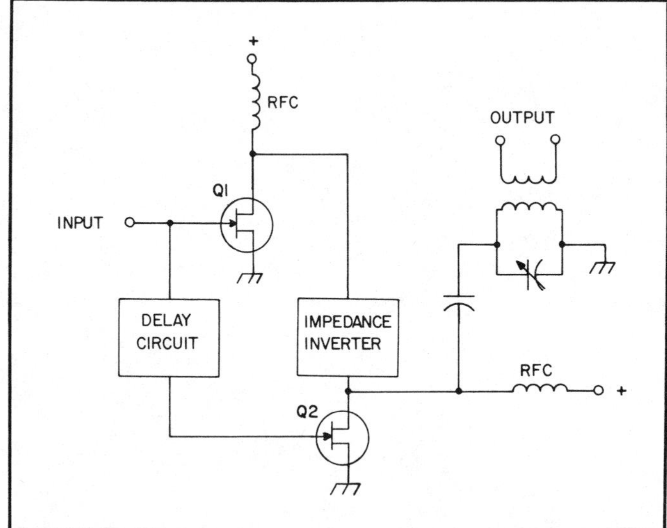

DOHERTY AMPLIFIER: A Doherty amplifier has two tubes or transistors. One amplifies the carrier signal, and the other amplifies the modulation peaks.

rupled over the unmodulated-carrier level, when a 100-percent peak comes along. This is normal for amplitude modulation at 100 percent. *See also* AMPLITUDE MODULATION.

DOLBY

Dolby is the name for a method of enhancing the signal-to-noise ratio in a magnetic recording system. The technique is similar to compression in radio-frequency communications systems (*see* COMPRESSION). The low-amplitude sound components, most likely to be masked by background hiss in the recording process, are boosted in volume in the Dolby method of recording. During playback, the low-level components are attenuated, thus obtaining the original sound.

Dolby techniques are sometimes categorized as Dolby A, the original Dolby recording method, and Dolby B, a simpler system for use by non-professionals. The more sophisticated Dolby A system has four independent noise-reduction circuits, which operate at the bass, midrange, treble, and high-frequency parts of the audible sound spectrum. The Dolby B system has only one noise-reduction frequency band, effective primarily on those audio frequencies at which the most tape hiss occurs.

For the best results, a Dolby tape should be played on a Dolby machine. This is because the dynamic range will be reduced otherwise. The Dolby system is standardized to ensure that the fidelity is optimum. *See also* MAGNETIC RECORDING, MAGNETIC TAPE, TAPE RECORDER.

DOMAIN

The domain of a mathematical function is the set of values on which the function operates. Usually, in Cartesian coordinate graphs, the domain is a subset of the values on the horizontal axis (*see* CARTESIAN COORDINATES).

A function may not be defined for all possible values of the independent variable. The function $f(x) = x^2$ is defined for all real numbers x, and thus its domain is the entire set of real numbers. But the function $g(x) = \log_{10}(x)$ is defined only for the positive real numbers. If x is zero or negative, then $\log_{10}(x)$ is not defined.

In some cases, the domain of the function depends on the extent of the number system we specify. The domain of the function $h(x) = \sqrt{x}$, for example, is restricted to the non-negative real numbers if we are willing to let h vary only over the set of reals. But if we let the value of h be any complex number (*see* COMPLEX NUMBER), then the domain of h becomes the entire set of real numbers. *See also* DEPENDENT VARIABLE, FUNCTION, INDEPENDENT VARIABLE, RANGE.

DOMAIN OF MAGNETISM

In any magnetic material, the domain is the region in which all of the magnetic dipoles are oriented in the same direction (*see* DIPOLE). When this occurs, the magnetic fields of the dipoles act in unison, and a large magnetic field is formed around the object, or in the vicinity of the domain.

An iron nail is normally not magnetized; its dipoles are oriented at random, and not all in the same direction, so the magnetic fields do not add together. However, a part of the nail may become magnetized under the influence of an electric current or external magnetic field. The part of the nail that becomes magnetized is called the magnetic domain. The domain may extend the whole length of the nail, or only through part of it. *See also* MAGNETIZATION.

DON'T-CARE STATE

In a binary logic function or operation, some states do not matter; they are unimportant insofar as the outcome is concerned. This occurs when the function is defined for some, but not all, of the logical states. In such a case, those states that do not affect conditions, one way or the other, are called don't-care states.

As an example, suppose we wish to represent the decimal digits 0 through 9 in binary form. Then we need four binary places, and we obtain:

0 = 0000
1 = 0001
2 = 0010
3 = 0011
4 = 0100
5 = 0101
6 = 0110
7 = 0111
8 = 1000
9 = 1001

Actually, then, there are six binary numbers that are not used: 1010, 1011, 1100, 1101, 1110, and 1111. These, of course, correspond to the decimal numbers 10 through 15. If the left-hand binary digit is a 1, then we know the decimal representation must be either 8 or 9. We might say that the binary numbers 1000, 1010, 1100, and 1110 correspond to the decimal value 8, and the binary numbers 1001, 1011, 1101, and 1111 correspond to the decimal number 9. It doesn't matter, then, what the two middle digits are when the far left digit is 1. In that situation, the two middle binary digits are don't-care states. *See also* BINARY-CODED NUMBER.

DOORKNOB TUBE

At ultra-high and microwave frequencies, a special kind of

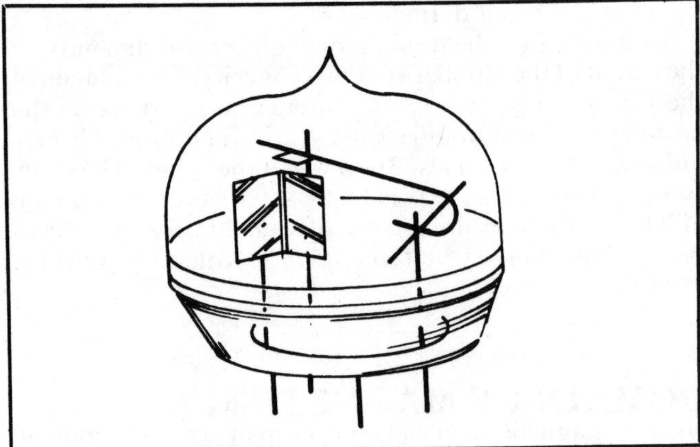

DOORKNOB TUBE: A doorknob tube derives its name from its physical shape.

vacuum tube can be used for oscillation and amplification purposes. Because of its shape, this tube is called a doorknob tube.

The design of the doorknob tube allows an extremely rapid transit time for the electrons flowing from the cathode to the plate. This is important at the ultra-high and microwave frequencies, since the transit time limits the maximum usable frequency of a tube. The doorknob tube exhibits low loss at the upper range of the radio spectrum. The illustration shows a pictorial diagram of a typical doorknob tube. It is similar in design to the acorn tube. *See also* ACORN TUBE, TUBE.

DOPING

Doping is the addition of impurity materials to semiconductor substances. Doping alters the manner in which such substances conduct currents. This makes the semiconductor, such as germanium or silicon, into an N-type or P-type substance.

When a doping impurity containing an excess of electrons is added to a semiconductor material, the impurity is called a donor impurity. This results in an N-type semiconductor (*see* N-TYPE SEMICONDUCTOR). The conducting particles in such a substance are the excess electrons, passed along from atom to atom.

When an impurity containing a shortage of electrons is added to a semiconductor material, it is called an acceptor impurity. The addition of acceptor impurities results in a P-type semiconductor material (*see* P-TYPE SEMICONDUCTOR). The charge carriers in this kind of substance are called holes, which are atoms lacking electrons. The flow of holes is from positive to negative in a P-type material (*see* HOLE).

Typical donor elements, used in the manufacture of N-type semiconductors, include antimony, arsenic, phosphorus, and bismuth. Acceptor elements, for P-type materials, include boron, aluminum, gallium, and indium. The N-type and P-type materials resulting from the addition of these elements to silicon, germanium, and other semiconductor materials, are the basis for all of solid-state electronics. *See also* GERMANIUM, SEMICONDUCTOR, SILICON.

DOPPLER EFFECT

An emitter of wave energy, such as sound or elec-

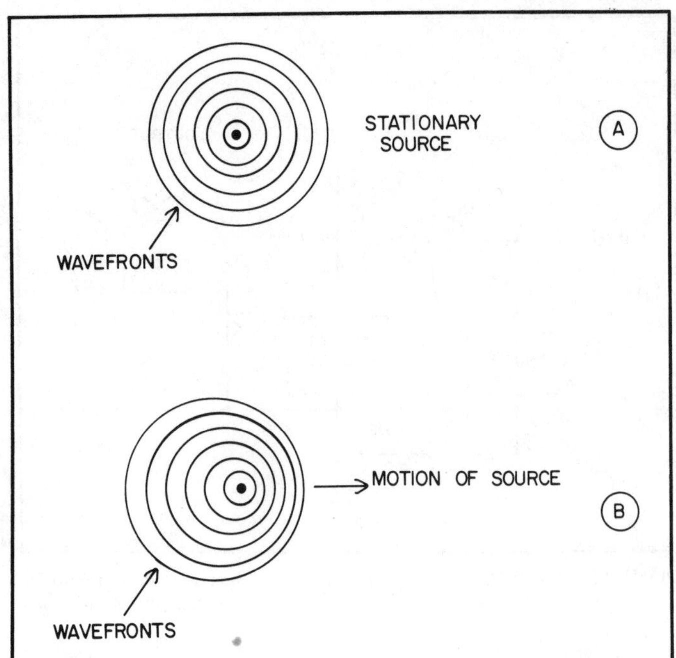

DOPPLER EFFECT: The Doppler effect results in eccentricity of the wavefronts emanated from a source of energy. At A, the waves from a stationary source appear concentric. At B, the waves from a source moving with respect to the observer appear to be compressed in the direction of motion.

tromagnetic fields, shows a frequency that depends on the radial speed of the source with respect to an observer. If the source is moving closer to the observer, the apparent frequency increases. If the source moves farther from the observer with time, the apparent frequency decreases. For a given source of wave energy, different observers may thus measure different emission frequencies, depending on the motion of each observer with respect to the source. The illustration shows the Doppler effect for a moving wave source. Only the radial component of the motion with respect to the observer results in a Doppler shift.

For sound, where the speed of propagation in air at sea level is about 1100 feet per second, the apparent frequency f^* of a source having an actual frequency f is:

$$f^* = f(1 + v/1100)$$

where v is the approach velocity in feet per second. (The approach velocity is considered negative if the observer is moving away from the source.)

For electromagnetic radiation, where the speed of propagation is c, the apparent frequency f^* of the emission from a source having an actual frequency f is:

$$f^* = f(1 + v/c)$$

where v is the approach velocity in the same units as c. (Again, the approach velocity is considered negative if the source and the observer are moving farther apart.)

The above formula for electromagnetic Doppler shifts holds only for velocities up to about ±0.1c, or 10 percent of the speed of light. For greater speeds, the relativistic correction factor must be added, giving:

$$f^* = f(1 + v/c)\sqrt{1 - v^2/c^2}$$

The wavelength becomes shorter as the frequency gets

higher, and longer as the frequency gets lower. Thus, if the actual wavelength of the source is λ, the apparent wavelength λ* is given by:

$$\lambda^* = \frac{\lambda}{(1 + v/c)\sqrt{1 - v^2/c^2}}$$

If there is no radial change in the separation of two objects, there will be no Doppler effect observed between them, although, if the tangential velocity is great enough, there may be a lowering of the apparent frequency because of relativistic effects.

DOPPLER RADAR

A form of radar that allows measurement of the radial velocity of an object, by means of the Doppler effect, is called a Doppler radar. Electromagnetic bursts are transmitted in the direction of the moving object, and are received after they have been reflected from the object. An approaching object will cause an increase in the frequency of the signal, and the greater the speed, the greater the increase in the frequency. A retreating object will cause a decrease in the frequency of the signal, in direct proportion to the radial speed.

Police radar systems operate on the Doppler principle. Automobiles are, in general, excellent reflectors of microwave energy. An automatic circuit in the radar receiver calculates the speed of an oncoming car. A microcomputer may be used for this purpose. The speed is displayed numerically on a two-digit light-emitting diode device. If the oncoming car is directly approaching, so that the speed is entirely accountable by Doppler effects, the Doppler radar system is accurate to better than 1 mile per hour. *See also* DOPPLER EFFECT.

DOSIMETRY

A dosimeter is a device used for measuring the amount of atomic radiation to which a person or object has been exposed over a certain period of time. The standard unit of atomic radioactivity is the roentgen (*see* ROENTGEN). Most people get from 10 to 20 roentgens of atomic radiation within their lifetimes. Some of this comes from medical X rays; some comes from the environment in the form of natural and manmade background radiation.

A dosimeter consists of a Geiger counter or similar rate counter, in conjunction with an integrator circuit that adds up the number of counts in a certain period of time. Some dosimeters measure primarily alpha rays; some measure primarily the beta and gamma forms of radiation. It is also possible to measure the radiation resulting from high-speed bombardment by neutrons.

Exposure to about 100 roentgens of radiation, over a period of a few hours or days, will cause radiation sickness. Symptoms include fever, loss of appetite, possible loss of hair, and a sunburn-like redness of the skin. A dose of more than about 300 roentgens in a short time will usually cause death. *See also* ALPHA PARTICLE, BETA PARTICLE, COUNTER TUBE, GAMMA RAY, GEIGER COUNTER, INTEGRATION, X RAY.

DOT GENERATOR

In a color television receiver, it is important that the three color electron beams converge at all points on the phosphor screen. Red, green, and blue beams meet to form white. If the alignment is not correct, the picture will appear distorted in color. A dot generator is a device used by television technicians to aid in the adjustment of the beam convergence. The pattern produced by the dot generator contains a number of white circular regions distributed over the picture frame.

If the convergence alignment of a color television receiver is perfect, all of the dots will appear white and in good focus. But if the convergence alignment is faulty, the red, green, and blue electron beams will not land together at all points of the screen. Some of the white dots will then have colored borders. If the convergence alignment is very bad, the red, green, and blue color beams might land so far away from each other that a picture is hardly recognizable in color. Alignment with the pattern produced by the dot generator is comparatively easy. With an actual picture, it would be very difficult. *See also* CONVERGENCE, PICTURE TUBE, TELEVISION.

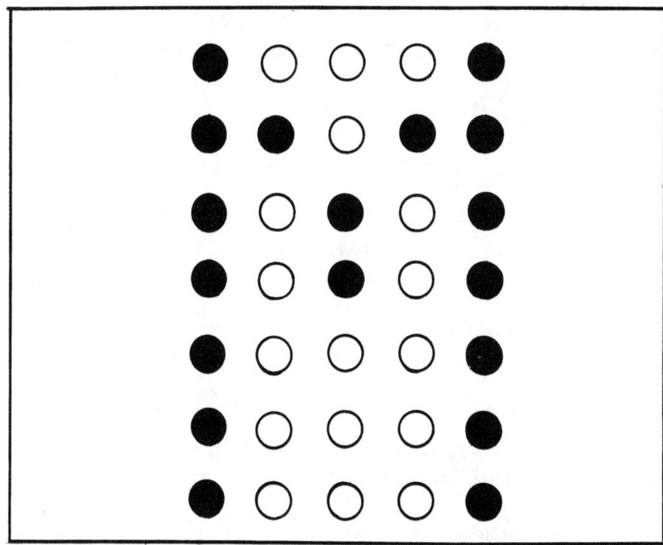

DOT-MATRIX PRINTER: An example of a dot-matrix printing of the capital letter M.

DOT-MATRIX PRINTER

A dot-matrix printer is a high-speed printing device often used in computer and communications terminals. A group of styli, usually in an array of five wide by seven high, hits a pressure-sensitive paper or transfers the ink from a ribbon, forming a letter, numeral, or symbol. Such a matrix is fine enough to resolve all of the common characters used in most printing systems. Upper-case and lower-case letters can be adequately printed to allow fairly easy reading. Some dot-matrix printers use a finer array of styli, providing still better resolution. The illustration shows the principle of the dot-matrix printer, showing the capital letter M.

Dot-matrix printing can be easily recognized by the fact that the lines in all the characters are broken up into small dots. A good dot-matrix printer can produce text that is as easy to read as typewritten material. *See also* PRINTER.

DOT PRODUCT

The dot product is a mathematical method of multiplying

vector quantities (*see* VECTOR). The result is a real number or scalar, rather than a vector. For this reason, the dot product is sometimes called the scalar product. Given two vectors A and B, the dot product A • B is given by:

$$A \cdot B = |A||B| \cos \Theta$$

where | A | is the length of A, | B | is the length of B, and Θ is the angle between the two vectors.

Another means of expressing the dot product in Cartesian coordinates is as follows: Let $A = (x_1,y_1,z_1)$ and $B = (x_2,y_2,z_2)$ in the Cartesian three-space. Then:

$$A \cdot B = x_1x_2 + y_1y_2 + z_1z_2$$

where x_1, y_1, and z_1 represent the end point of the vector A originating at (0,0,0), and x_2, y_2, and z_2 represent the end point of B originating at (0,0,0).

The dot product is commutative. That is, for any two vectors A and B:

$$A \cdot B = B \cdot A$$

The dot product is also distributive with respect to vector addition. Therefore:

$$A \cdot (B + C) = A \cdot B + A \cdot C$$

See also CARTESIAN COORDINATES, CROSS PRODUCT.

DOUBLE BALANCED MIXER

A double balanced mixer is a mixer circuit that operates in a manner similar to a balanced modulator (*see* BALANCED MODULATOR, MIXER, MODULATOR). The input energy in a double balanced mixer does not appear at the output. All of the output ports and input ports are completely isolated, so that the coupling between the input oscillators is negligible, and the coupling between the input and output is also negligible. The double balanced mixer is different from the single balanced mixer; in the latter circuit, there is some degree of interaction between the input signal ports (*see* SINGLE BALANCED MIXER).

Hot-carrier diodes are generally used in modern balanced-mixer circuits. They can handle large signal amplitudes without distortion, and they generate very little noise. They are also effective at frequencies up to several gigahertz. Since these circuits, such as the one shown in the diagram, are passive rather than active, they show some conversion loss. This loss is typically about 6 dB. However, the loss is easily overcome by the use of amplifiers following the mixer. *See also* HOT-CARRIER DIODE.

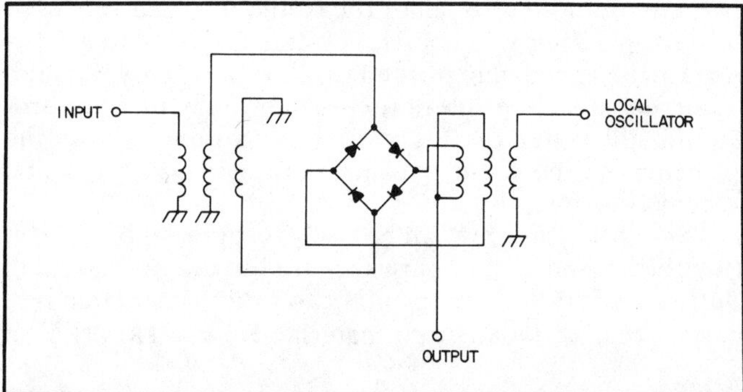

DOUBLE BALANCED MIXER: A double balanced mixer circuit isolates the input, local oscillator, and output ports.

DOUBLE-BASE JUNCTION TRANSISTOR

See TETRODE TRANSISTOR.

DOUBLE CIRCUIT TUNING

When the input circuit and the output circuit of an amplifier are independently tunable, the system is said to have double circuit tuning. In an oscillator, double tuning involves separate tank circuits in the base and collector portions for a bipolar transistor, the gate and plate portions for a vacuum tube. In a coupling transformer, double circuit tuning refers to the presence of a tuning capacitor across the secondary winding as well as the primary winding.

Double circuit tuning provides, as we should expect, a greater degree of selectivity and harmonic suppression than the use of only one tuned circuit per stage. In a multistage amplifier, however, double circuit tuning increases the possibility of oscillation at or near the operating frequency. This tendency can be reduced by tuning each circuit to a slightly different frequency, thereby reducing the Q factor of the entire amplifier chain. In such a situation, the circuit is said to be stagger-tuned. *See also* STAGGER TUNING.

DOUBLE-CONVERSION RECEIVER

A double-conversion receiver, also known as a dual-conversion receiver, is a superheterodyne receiver that has two different intermediate frequencies. The incoming signal is first heterodyned to a fixed frequency, called the first intermediate frequency. Selective circuits are used at this point. In a single-conversion receiver (*see* SINGLE-CONVERSION RECEIVER) the first intermediate frequency

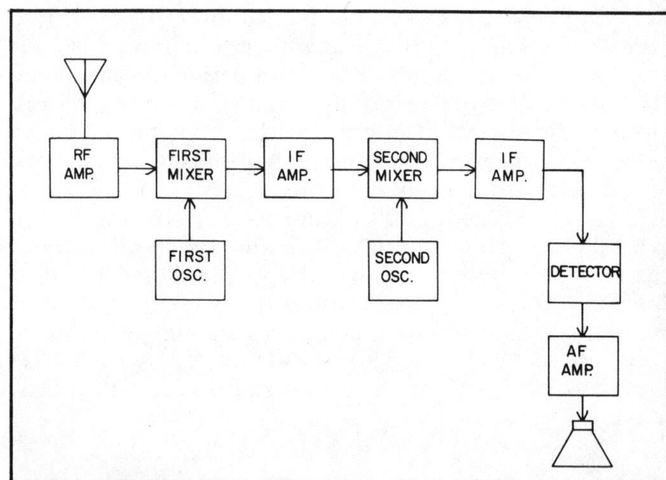

DOUBLE-CONVERSION RECEIVER: A double-conversion receiver has two mixers and two intermediate frequencies. The second intermediate frequency is generally much lower than the first.

signal is fed to the detector and subsequent audio amplifiers. However, in a double-conversion receiver, the first intermediate frequency is heterodyned to a second, much lower frequency, called the second intermediate frequency. The illustration is a block diagram of a double-conversion receiver. The second intermediate frequency may be as low as 50 to 60 kHz.

The low-frequency signal from the second mixer stage is very easy to work with from the standpoint of obtaining high gain and selectivity. Such low frequencies are not practical in a single-conversion receiver because the local-oscillator frequency would be too close to the actual signal frequency; this would cause very poor image rejection. A double-conversion receiver, while offering superior selectivity, usually has more "birdies" or local-oscillator heterodynes. But, judicious selection of the intermediate frequencies will minimize this problem. *See also* INTERMEDIATE FREQUENCY, MIXER, SUPERHETERODYNE RECEIVER.

DOUBLE INTEGRAL

A double integral is a form of mathematical integration. Double integrals are sometimes used to determine the area enclosed by two or more geometric functions in two dimensions. In some cases, the area bounded by a set of functions cannot be found by the usual methods of integration, and thus double or even triple integration may be necessary. Double integrals are used to determine the volumes of three-dimensional figures bounded by multivariable functions.

In very complicated cases in several dimensions, integration may have to be performed several times. The mathematical techniques for doing this are quite sophisticated.

The procedures for multiple integration are beyond the scope of this volume, but textbooks in intermediate and advanced calculus are good sources of information on the subject. In electrical engineering, higher-level calculus is often used, and any serious engineer should be at least casually acquainted with double and multiple integration. *See also* CALCULUS, DEFINITE INTEGRAL, INDEFINITE INTEGRAL.

DOUBLE POLE

In a switching arrangement, the term double pole refers to the simultaneous switching of more than one circuit. Double-pole switches are commonly used in many electronics and radio circuits. They are especially useful when two circuits must be switched on or off at once. For example, the plate voltages to the driver and final amplifier stages of a vacuum-tube transmitter are generally of different values, and must be switched separately. A single switch, having two poles, is much more practical than two single-pole switches.

In band-switching applications in radio receivers and transmitters, several poles of switching are sometimes used. There may be three, four, or more poles in the same switch. Such multiple-pole switches are usually ganged rotary switches. *See also* MULTIPLE POLE, ROTARY SWITCH.

DOUBLER

A doubler is an amplifier circuit designed to produce an output at the second harmonic of the input frequency. Such a circuit is therefore a frequency multiplier (*see* FREQUENCY MULTIPLIER). The input circuit may be untuned, or it may be tuned to the fundamental frequency. The output circuit is tuned to the second-harmonic frequency. The active element in the circuit—that is, the transistor or tube—is deliberately biased to result in nonlinear operation. This means it must be biased for Class AB, B, or C. Alternatively, a nonlinear passive element, such as a diode, may be used. The nonlinearity of the device produces a signal rich in harmonic energy.

A special type of circuit that lends itself well to application as a doubler is the push-push circuit. In this arrangement, which resembles an active full-wave rectifier, two transistors or tubes are connected with their inputs in phase opposition, and their outputs in parallel. This tends to cancel the fundamental frequency and all odd harmonics in the output circuit. But the even harmonics are reinforced. *See also* PUSH-PUSH CONFIGURATION.

DOUBLE SIDEBAND

Whenever a carrier is amplitude-modulated, sidebands are produced above and below the carrier frequency. These sidebands represent the sum and difference frequencies of the carrier and the modulating signal (*see* SIDEBAND). A typical amplitude-modulated signal, as it might appear on a spectrum analyzer, is shown at A in the illustration. This is called a double-sideband, full-carrier

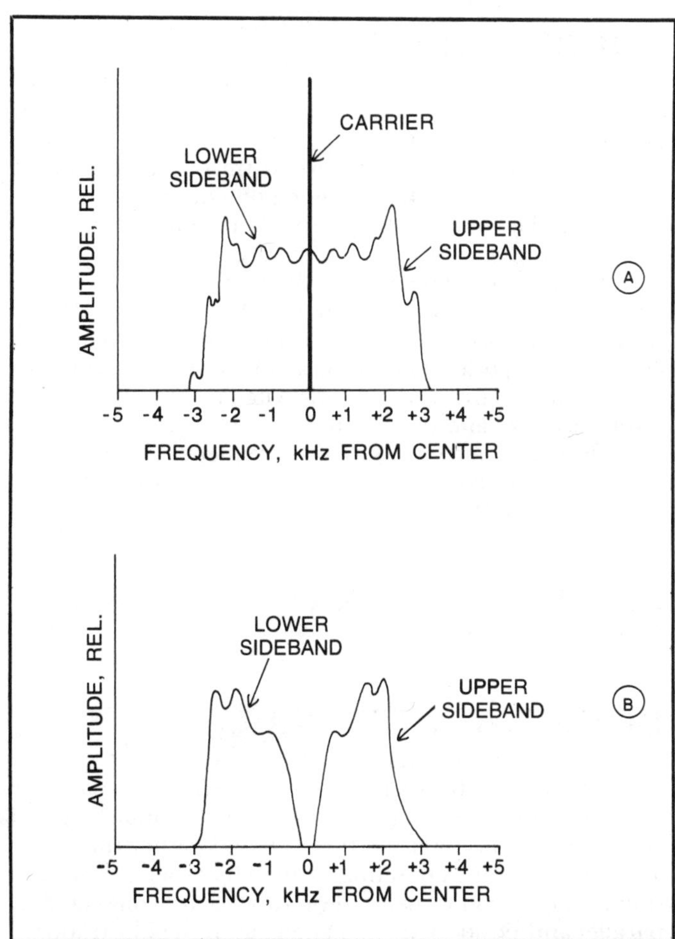

DOUBLE SIDEBAND: At A, an ordinary amplitude-modulated signal as it would appear on the display of a spectrum analyzer. At B, a double-sideband, suppressed-carrier signal.

signal; both sidebands and the carrier are plainly visible on the display.

If a balanced modulator (*see* BALANCED MODULATOR) is used for the purpose of obtaining the amplitude-modulated signal, the output lacks a carrier, as shown at B. The sidebands are left intact. Actually, of course, the carrier is not completely eliminated, but it is greatly attenuated, typically by 40 to 60 dB or more. Such a signal is called a double-sideband, suppressed-carrier signal. To obtain a normal amplitude-modulated signal at the receiver, a local oscillator is used to re-insert the missing carrier. The double-sideband, suppressed-carrier signal makes more efficient use of the audio energy than the full-carrier signal. This is because, in the suppressed-carrier signal, all of the transmitter output power goes into the sidebands, which carry the intelligence. In the full-carrier signal, no less than two-thirds of the transmitter power is used by the carrier, which alone carries no information at all.

A double-sideband, suppressed-carrier signal is difficult to receive, because the local oscillator must be on exactly the same frequency as the suppressed carrier. The slightest error will cause beating between the two sidebands, and the result will be a totally unreadable signal. Therefore, this kind of modulation is not often found in radiocommunication situations; single-sideband emission is much more common. *See also* AMPLITUDE MODULATION, SINGLE SIDEBAND.

DOUBLE THROW

A double-throw switch is used to connect one line to either of two other lines. For example, it might be desired to provide two different plate voltages to a tube type amplifier for operation at two different power-input levels. In such a case, the common point of a double-throw switch can be connected to the plate circuit, and the two terminals to the different power supplies.

Some switches have several different throw positions. These are called multiple-throw switches; a rotary switch is the most common configuration. There may be 15 or even 20 different positions in such a rotary switch. Multiple-throw switches are used in band-switching applications for radio receivers and transmitters. They are also employed in multi-function controls, digital frequency controls, and in a wide variety of other situations. *See also* MULTIPLE THROW, ROTARY SWITCH.

DOUBLET ANTENNA

See DIPOLE ANTENNA.

DOUBLE-V ANTENNA

A double-V, or fan, antenna is a form of dipole antenna. In any antenna, the bandwidth increases when the diameter of the radiating element is made larger. A double-V takes advantage of this by using two elements rather than one, and geometrically separating them so that they behave as a single, very broad radiator. Two dipoles are connected in parallel and positioned at an angle, as in the illustration.

The maximum radiation from a double-V antenna, and the maximum gain for receiving, occur in all directions perpendicular to a line bisecting the angle between the

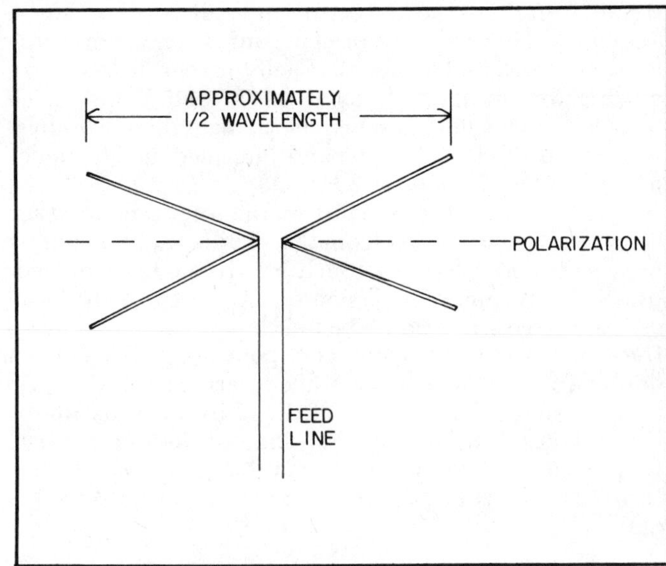

DOUBLE-V ANTENNA: The double-V antenna is essentially two half-wave dipoles connected in parallel.

dipoles. In the horizontal plane, this is broadside to the plane containing the elements. The polarization is parallel to the line bisecting the apex angles of each pair of jointed conductors. The impedance at the feed point is somewhat higher than that of an ordinary dipole antenna; it is generally on the order of 200 to 300 ohms. This provides a good match for the common types of prefabricated twin-lead line.

Double-V antennas may be used for either transmitting or receiving, and are effective at any frequency on which a dipole antenna is useful. They are quite often used as receiving antennas for the standard frequency-modulation (FM) broadcast band at 88 to 108 MHz, because of their broad frequency-response characteristics. *See also* DIPOLE ANTENNA.

DOUBLE-ZEPP ANTENNA

A full-wavelength, straight conductor, fed at the center, is sometimes called a double-zepp antenna. Such an antenna is actually a two-element collinear antenna, and shows a small power gain over a half-wave dipole in free space. The directional pattern of the double-zepp antenna is similar to that of the dipole antenna. Maximum radiation occurs perpendicular to the element, but the lobes are somewhat more oblong than for a dipole (see illustration). The polarization is in directions parallel to the wire.

The impedance at the feed point of a double-zepp antenna is a pure resistance, and the value is quite high. If wire is used in the construction of the antenna, the feed-point resistance may be as high as several thousand ohms. If tubing is used or the antenna is placed close to the ground or other obstructions, the value is lower, ranging from perhaps 300 to 1000 ohms.

Double-zepp antennas are used by radio amateurs, especially at the lower frequency bands such as 1.8 MHz and 3.5 MHz. The double zepp can, however, be used to advantage at any frequency. The versatility of this antenna is enhanced by the fact that it is resonant on all harmonic frequencies, with about the same feed-point impedance as on the fundamental frequency. It will function as a half-wave dipole at an operating frequency of half the funda-

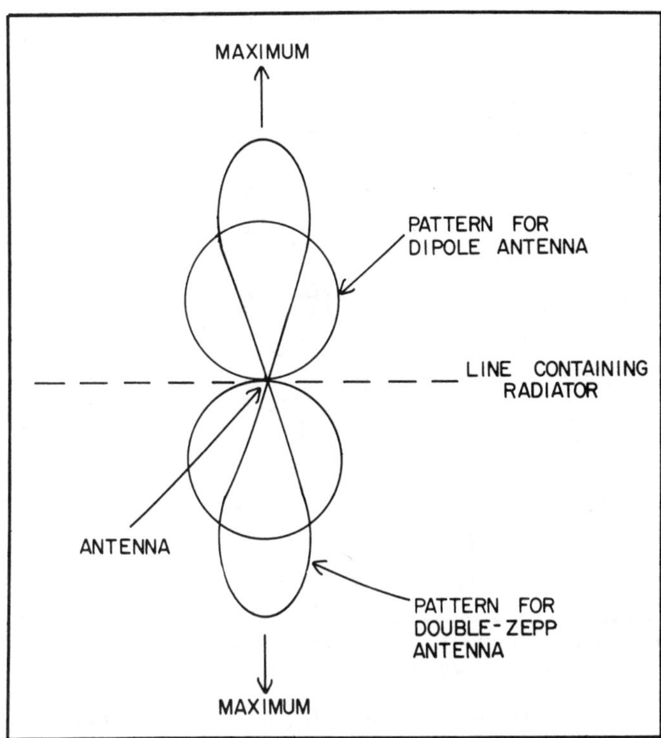

DOUBLE-ZEPP ANTENNA: The directional pattern of the double-zepp antenna is similar to that of a dipole antenna, except that the lobes are narrower and the double-zepp antenna produces some gain.

mental. Generally, the double-zepp antenna is fed with a low-loss transmission line such as open wire, so that the loss caused by standing waves will not be significant. *See also* DIPOLE ANTENNA, EXTENDED DOUBLE-ZEPP ANTENNA, ZEPPELIN ANTENNA.

DOWN CONDUCTOR

In an antenna installation, a down conductor is a length of heavy wire or tubing, run from the top of a supporting mast to a ground rod. The purpose of the down conductor is to provide a low-impedance path for the discharge of lightning strokes. Down conductors are especially important for wooden masts, since a lightning stroke can set such a structure on fire. Down conductors are sometimes used to protect trees against lightning.

If an antenna mast or lightning rod is located on the roof of a house or other building, or in some location where it is not possible to obtain a straight run to the ground rod, the down conductor should be positioned in the shortest possible path between the mast or rod and the ground. Ground rods should be driven at least 8 feet into the soil, and at least a few feet away from underground objects such as sewer pipes, gas mains, and building foundations. *See also* LIGHTNING, LIGHTNING PROTECTION, LIGHTNING ROD.

DOWN CONVERSION

Down conversion refers to the heterodyning of an input signal with the output of a local oscillator, resulting in an intermediate frequency that is lower than the incoming signal frequency. Technically, any situation in which this occurs is down conversion. But usually, the term is used with reference to converters designed especially for reception of very-high, ultra-high, or microwave signals with communications equipment designed for much lower frequencies than that contemplated.

The available amount of spectrum space at ultra-high and microwave frequencies is extremely great, covering many hundreds of megahertz, and the operator of a receiver using a down converter may find the selectivity too great! For example, with a communications receiver designed to tune the high-frequency band in 1 MHz ranges, the 10,000 to 10,500-MHz band would appear, following down-conversion, to occupy 500 such ranges. Each of these 500 ranges would require the receiver to be tuned across its entire bandspread. One signal might then require a long time to find!

This same spreading-out property, however, promises no lack of available spectrum space in the future, especially with the advent of microwave satellites. *See also* MIXER, UP CONVERSION.

DOWNLINK

The term downlink is used to refer to the band on which an active communications satellite transmits its signals back to the earth. The downlink frequency is much different from the uplink frequency, on which the signal is sent up from the earth to the satellite. The different uplink and downlink frequencies allow the satellite to act as a repeater, retransmitting the signal immediately.

Satellite uplink and downlink frequencies are usually in the very-high or ultra-high frequency range. They may even be in the microwave range. The downlink transmitting antenna on a satellite must produce a fairly wide-angle beam, so that all of the desired receiving stations get some of the signal. *See also* ACTIVE COMMUNICATIONS SATELLITE, REPEATER, UPLINK.

DOWNTIME

Downtime is the time during which an electronic circuit or system is not operational. This may be because of routine servicing, or because of a catastrophe such as a power failure. The term downtime is used especially with respect to computers and other highly sophisticated electronic systems.

Obviously, the objective of any user of any system is to minimize the amount of downtime. Periodic shutdowns for maintenance are generally preferable, and result in less overall downtime, than waiting for a malfunction and then repairing it. The problem may not occur when it is convenient!

DOWNWARD LEADER

See LEADER.

DOWNWARD MODULATION

If the average power of an amplitude-modulated transmitter decreases when the operator speaks into the microphone, the modulation is said to be downward. Generally, in downward modulation, the instantaneous carrier amplitude is never greater than the amplitude under zero-modulation conditions. That is, the positive modulation peaks never exceed the level of the unmodulated

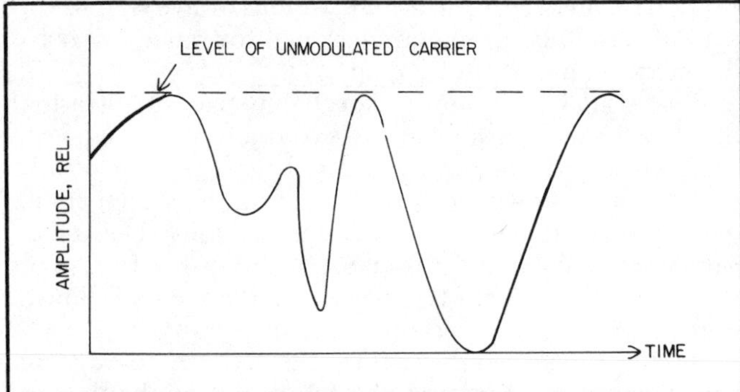

DOWNWARD MODULATION: In downward modulation, the signal never exceeds the amplitude with zero modulation.

carrier. Negative modulation peaks may go as low as the zero amplitude level, just as is the case with ordinary amplitude modulation, but clipping should not take place. The drawing illustrates a typical amplitude-versus-time function for a downward-modulated signal.

In theory, downward modulation is less efficient than ordinary amplitude modulation, since more transmitter power is required in proportion to the intelligence transmitted. That is, less effective use is made of the energy in the transmitter. Downward modulation is not often used in communications systems, since the large amplitude of the unmodulated carrier is a waste of power. *See also* AMPLITUDE MODULATION, UPWARD MODULATION.

DRAIN

The term drain is used to refer to the current or power drain from a source. For example, the current drain of a 12-ohm resistor connected to a 12-volt power supply is 1 ampere.

In a field-effect transistor, the drain is the electrode from which the output is generally taken. The drain of a field-effect transistor is at one end of the channel, and is the equivalent of the plate of a vacuum tube or the collector of a bipolar transistor. In an N-channel field-effect transistor; the drain is biased positive with respect to the gate and the source. In the P-channel device, the drain is negative with respect to the gate and source. *See also* FIELD-EFFECT TRANSISTOR.

DRAIN-COUPLED MULTIVIBRATOR

A drain-coupled multivibrator is a form of oscillator that uses two field-effect transistors. The circuit is analogous to a tube type multivibrator (*see* MULTIVIBRATOR). The drain of each field-effect transistor is coupled to the gate of the other through a blocking capacitor, as shown in the diagram.

The drain-coupled multivibrator is, in effect, a pair of field-effect-transistor amplifiers, connected in cascade with positive feedback from the output to the input. Since each stage inverts the phase of the signal, the output of the two-stage amplifier is in phase with the input, and oscillation results. The values of the resistors and capacitors determine the oscillating frequency of the drain-coupled multivibrator. Alternatively, tuned circuits may be used in place of, or in series with, the drain resistors. This will create a resonant situation in the feedback circuit, confining the feedback to a single frequency.

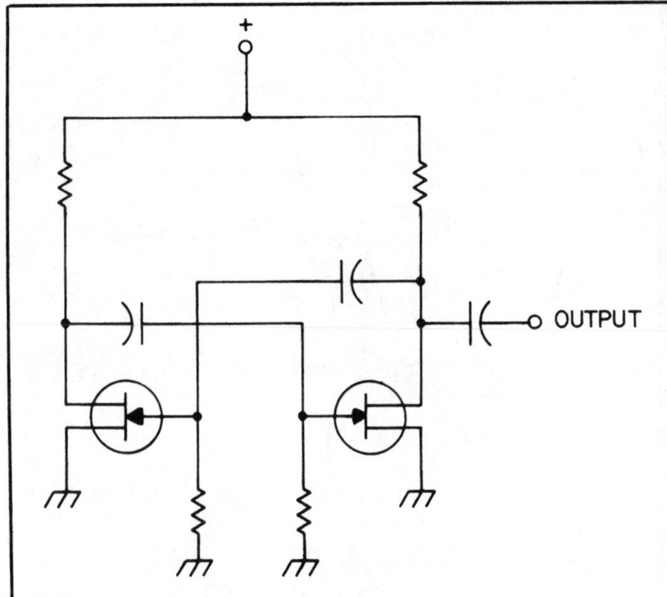

DRAIN-COUPLED MULTIVIBRATOR: The drain-coupled multivibrator is a form of oscillator.

DRIFT

In a conductor or semiconductor, drift is the movement of charge carriers. The term is especially used in conjunction with the flow of current in a semiconductor substance. In an N-type semiconductor material, the charge carriers are extra electrons among the atoms of the substance. In a P-type material, the charge carriers are atoms deficient in electrons; these are called holes (*see* ELECTRON, HOLE).

Drift velocity is often mentioned with respect to a semiconductor device. Drift velocity is given in centimeters per second or meters per second. In most materials, an increase in the applied voltage results in an increase in the drift velocity, but only up to a certain maximum. The greater the drift velocity for a given material, the greater the current per unit cross-sectional area.

The ease with which the charge carriers in a semiconductor are set in motion is called the drift mobility. Generally, the drift mobility for holes is less than that for electrons. The drift mobility determines, in a transistor or field-effect transistor, the maximum frequency at which the device will produce gain in an amplifier circuit. This frequency also depends on the thickness of the base or channel region.

The term drift is sometimes used in electronics to refer to an unwanted change in a parameter. This is especially true of frequency in an oscillator. The oscillator frequency of a radio transmitter may change, in either an upward or a downward direction, because of temperature variations among the components. Frequency drift is not nearly as serious a problem today as it was before the advent of solid-state technology. Transistors and integrated circuits generate almost no heat in an oscillator circuit; vacuum tubes in previous years were notorious for this. The phase-locked-loop oscillator is an extremely drift-free form of oscillator, which uses a crystal and various frequency-dividing integrated circuits. *See also* PHASE-LOCKED LOOP.

DRIFT-FIELD TRANSISTOR

A drift-field transistor is a special form of bipolar transistor, intended as an amplifier or oscillator at very high frequencies. The base region of the device is made very thin by an alloy-diffusion process. The thinner the base region, the shorter the time required for charge carriers to get from one side of the base to the other, and therefore the higher the frequency at which the device can be effectively used.

In the base region of the drift-field transistor, the impurity concentration is greater toward the emitter side than toward the collector side. This causes the drift mobility of the charge carriers to increase toward the collector side (*see* DRIFT). The charge carriers are thus accelerated ad they move from the emmitter-base junction to the base-collector junction. The result is a very high maximum-usable frequency. *See also* TRANSISTOR.

DRIVE

Drive is the application of power to a circuit for amplification, dissipation, or radiation. In particular, the term drive refers to the current, voltage, or power applied to the input of the final amplifier of a radio transmitter. The greater the drive, the greater the amplifier output. But excessive drive can cause undesirable effects such as harmonic generation and signal distortion.

Some amplifiers, such as the Class-A and Class-AB amplifier, require very little driving power. The Class-A amplifier theoretically needs no driving power at all; it runs solely off of the voltage supplied to its input. Some amplifiers, such as the Class-C radio-frequency amplifier, must have a large amount of driving power to function properly. (*See* CLASS-A AMPLIFIER, CLASS-AB AMPLIFIER, CLASS-B AMPLIFIER, CLASS-C AMPLIFIER.)

In an antenna, the element that is connected directly to the feed line, and therefore receives power from the transmitter directly, is called the driven element. An antenna may have one or more such elements (*see* DRIVEN ELEMENT).

The term drive is sometimes used to refer to the mechanical device in a disk recorder. The device used to operate a rotating control, such as a tuning dial, may also be called a drive. *See also* DIRECT-DRIVE TUNING, DISK DRIVE, VERNIER.

DRIVEN ELEMENT

In an antenna with parasitic elements (*see* PARASITIC ARRAY), those elements connected directly to the transmission line are called driven elements. In most parasitic arrays, there is one driven element, one reflector, and one or more directors (*see* DIRECTOR, REFLECTOR).

When several parasitic antennas are operated together, such as in a collinear or stacked array, there are several driven elements. Each driven element receives a portion of the output power of the transmitter. Generally, the power is divided equally among all of the driven elements. In some phased arrays, all of the elements are driven.

The driven element in a parasitic array is always resonant at the operating frequency. The parasitic elements are usually (but not always) slightly off resonance; the directors are generally tuned to a higher frequency than that of the driven element, and the reflector is generally set to a lower frequency. The impedance of the driven element, at the feed point, is a pure resistance when the antenna is operated at its resonant frequency. When parasitic elements are near the driven element, the impedance of the driven element is low compared to that of a dipole in free space.

For the purpose of providing an impedance match between a driven element and a transmission line, the driven element may be folded or bent into various configurations. Among the most common matching systems are the delta, gamma, and T networks. Sometimes the driven element is a folded dipole rather than a single conductor. *See also* DELTA MATCH, FOLDED DIPOLE ANTENNA, GAMMA MATCH, QUAD ANTENNA, T MATCH, YAGI ANTENNA.

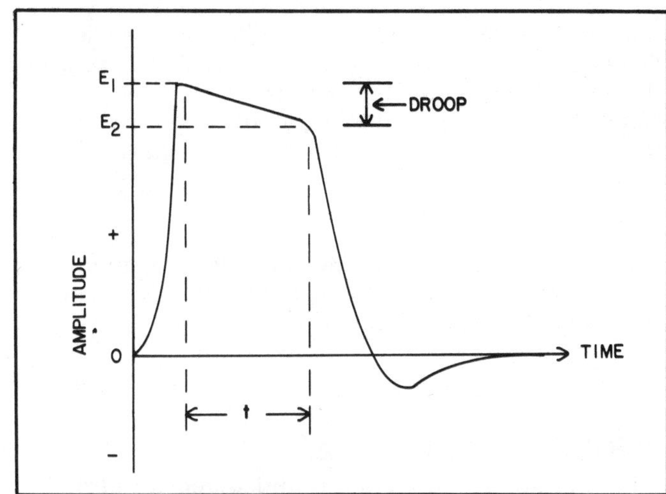

DROOP: Droop is a measure of imperfection in a square pulse.

DRIVER

A driver is an amplifier in a radio transmitter, designed for the purpose of providing power to the final amplifier. The driver stage must be designed to provide the right amount of power into a certain load impedance. Too little drive will result in reduced output from the final amplifier, and reduced efficiency of the final amplifier. Excessive drive can cause harmonic radiation and distortion of the signal modulation envelope.

The driver stage receives a signal at the operating frequency and provides a regulated output with a minimum of tuning. Most or all of the transmitter tuning adjustments are performed in the final amplifier stage. *See also* DRIVE, FINAL AMPLIFIER.

DROOP

Droop is a term used to define the shape of an electric pulse. Droop is generally specified as a percentage of the maximum amplitude of a pulse.

The drawing illustrates an electric pulse as it might appear on an oscilloscope display screen. The maximum amplitude of the pulse is indicated by E_1. After the initial rise to maximum amplitude, the pulse should ideally remain at that amplitude for its duration, and then rapidly fall back to the zero or minimum level. However, in practice, this ideal is seldom attained. The amplitude of the pulse decreases until it drops to zero, but the drop is often followed by overshoot (*see* OVERSHOOT), or a reversal of the polarity. The pulse duration at maximum amplitude is shown by t. At time t, the amplitude of the pulse is E_2, somewhat less than E_1. The droop, in percent, is given by:

$$D = 100(E_1 - E_2)/E_1$$

and represents the amount by which the maximum pulse amplitude drops during the high state. In a perfectly rectangular pulse, there is no droop, and therefore it is 0 percent.

DROOPING RADIAL

The radial system in a ground-plane antenna (*see* GROUND-PLANE ANTENNA) is usually horizontal, and the radiating element is vertical. A quarter-wave vertical radiator in this configuration gives a resistive impedance of about 37 ohms. This represents a fairly good match to 50-ohm coaxial-cable transmission line.

A perfect match to 50-ohm cable, using a ground-plane antenna, may be obtained by slanting the radials downward at an angle of about 45 degrees (see illustration). The greater the slant with respect to the horizontal plane, the larger the feed-point impedance becomes. When the radials are oriented straight down, the antenna impedance is about 73 ohms, identical with that of a dipole in free space. In fact, such an antenna is called a vertical dipole.

Ground-plane antennas with drooping radials are often used at frequencies above approximately 27 MHz, or the Class-D Citizen's Band. At these frequencies, rigid tubing can be used for the radials as well as for the vertical radiator. In some amateur-radio installations, ground-plane antennas are used at frequencies as low as 1.8 MHz, and the radials serve the function of guy wires. *See also* COAXIAL ANTENNA, VERTICAL DIPOLE ANTENNA.

DROP-IN AND DROP-OUT

In a magnetic recording of digital information, such as a computer program or data store, bits are sometimes falsely recorded or played back. The generation of extra bits is called drop-in. Occasionally, a bit is lost in the recording or playback process; the loss of a bit is called drop-out.

Drop-in and drop-out can be caused by dust or other foreign material on the surface of a magnetic tape. Drop-in and drop-out can also be caused by defects in the tape, poorly regulated or incorrect tape speed, electromagnetic interference, or hum picked up from nearby utility lines. To prevent drop-in and drop-out, the tape must be kept clean. Only the best quality of tape should be used, and extremely thin tape should be avoided since it can stretch. When tapes are stored, the temperature and humidity should be kept within reasonable limits. The recorder should be cleaned frequently. Interconnecting cables must be well shielded, and a good electrical ground should be provided for the entire system. The tape recorder should be well made, so that the playback speed matches the recording speed, and so that the speed is uniform. *See also* MAGNETIC RECORDING.

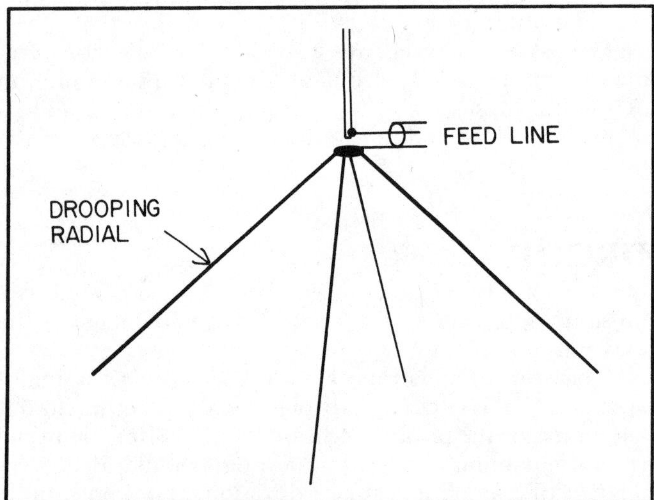

DROOPING RADIAL: A ground-plane antenna with drooping radials has a higher feed-point impedance than a ground-plane antenna with horizontal radials.

DROPPING RESISTOR

A dropping resistor is a resistor installed in a circuit for the purpose of lowering the voltage supplied to a load. The amount by which the voltage is dropped depends on the power-supply output voltage, and on the load resistance.

Dropping resistors should be used only when the load resistance remains constant. Otherwise, the voltage regulation will be extremely poor. For example, a dropping resistor may be used to operate a 12-volt incandescent bulb, which normally draws a current of 100 mA, from 120-volt alternating-current utility mains, as shown in the diagram. In such a case, the normal operating resistance of the bulb is:

$$R_{BULB} = 12/0.1 = 120 \text{ ohms}$$

In a 120-volt circuit, the total resistance R_{LOAD} necessary to provide a current of 100 mA is:

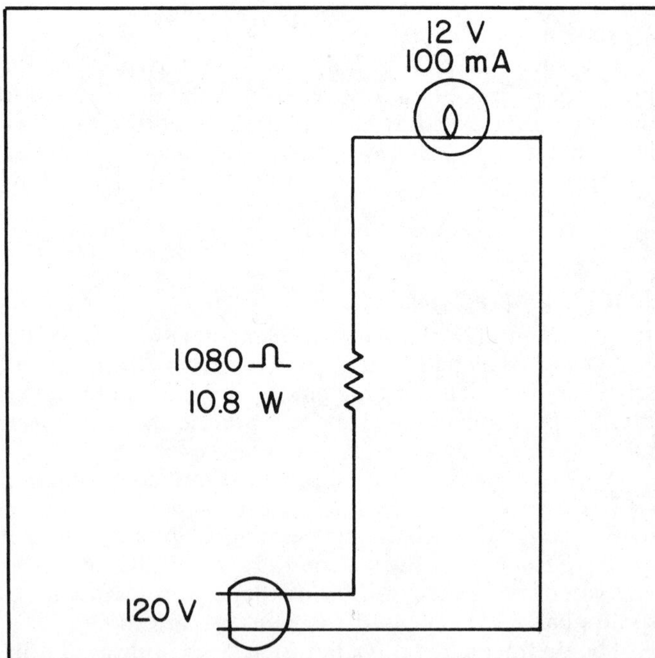

DROPPING RESISTOR: A dropping resistor can be used to run a 12-volt bulb from 120-volt utility mains.

$$R_{LOAD} = 120/0.1 = 1200 \text{ ohms}$$

Thus the resistor R should have a value of 1200 − 120 ohms, or 1080 ohms. Under these conditions in the circuit shown, the bulb will light normally, provided the dropping resistor can handle a power level of at least

$$P = (0.1)^2 \times 1080 = 10.8 \text{ watts}$$

To illustrate the poor regulation of this arrangement under conditions of varying load resistance, suppose a second 12-volt bulb, identical to the first, is connected in parallel with the first. The result would be a decrease in the current through both bulbs, because of the lowered resistance of the bulb combination. In fact, the current would drop to about half its original value in either bulb. To provide the bulbs with 12 volts, the value of the dropping resistor would have to be cut in half.

Dropping resistors are used only in circuits having a constant load resistance. They are also quite wasteful of power. In the above example, with the single 12-volt bulb and the 1080-ohm resistor, the bulb consumes just 1.2 watts, but the resistor takes 10.8 watts!

DRY CELL

A dry cell is a voltage-generating dry chemical device. It consists of two electrodes separated by a thick electrolyte paste. The cells commonly used in flashlights and portable radio receivers are dry cells. They are available in hardware, grocery, and drug stores everywhere.

Dry cells exist in a variety of different physical sizes, ranging from the tiny camera cells, smaller than a dime, to large cylindrical cells measuring more than a foot high. In general, the milliampere-hour capacity of a dry cell is proportional to its physical size. The amount of current that a cell can deliver at a given time is also proportional to its physical size.

Dry cells are commercially made primarily in two different forms: the alkaline cell and the zinc cell. There are other forms. of dry cells. *See also* ALKALINE CELL, CELL, NICKEL-CADMIUM BATTERY, SILVER-OXIDE CELL, SILVER-ZINC CELL, ZINC CELL.

DRY-REED SWITCH

The dry-reed switch is a form of relay. Two metal strips, called reeds, are sealed in a cylindrical glass tube. Some dry-reed switches contain several poles, and may be either single-throw or double-throw (*see* DOUBLE-THROW, SINGLE-THROW).

One of the strips in the dry-reed switch is made of a magnetic material. Therefore, in the presence of a magnetic field, it is moved from one position to the other. Some single-throw dry-reed switches are normally open, and some are normally closed.

Dry-reed switches are noteworthy for their high speed of operation. However, their current-carrying capacity is rather limited because of their generally small physical size. They cannot be subjected to very high voltages, or arcing will take place across the contacts. Dry-reed switches are favored for use in radio-frequency switching applications because of the low capacitance between the contacts, resulting from the small physical dimensions of the devices.

DTL

See DIODE-TRANSISTOR LOGIC.

DUAL-BEAM OSCILLOSCOPE

A dual-beam oscilloscope is an oscilloscope with two electron guns rather than one. This allows the simultaneous viewing of two different signals. Two independent traces appear on the oscilloscope screen, one above the other, as shown in the illustration.

A dual-beam oscilloscope is extremely useful for comparing two different signals for frequency or phase. Such an oscilloscope can be used to great advantage for checking the gain and linearity of an amplifier. One beam can be used to monitor the amplifier input, and the other can be used to monitor the amplifier output. Dual-beam oscilloscopes are also used in the troubleshooting and design of digital circuits. *See also* OSCILLOSCOPE.

DUAL COMPONENT

When two individual components are combined into a single package, the total package is called a dual component. The components may be identical, or they may be substantially different.

The most common dual components are dual diodes and dual capacitors, often used in full-wave power supplies. Dual transistors are sometimes seen as well; one such component consists of two transistors connected in a Darlington pair and housed in a single case.

Sometimes several components are contained in a single

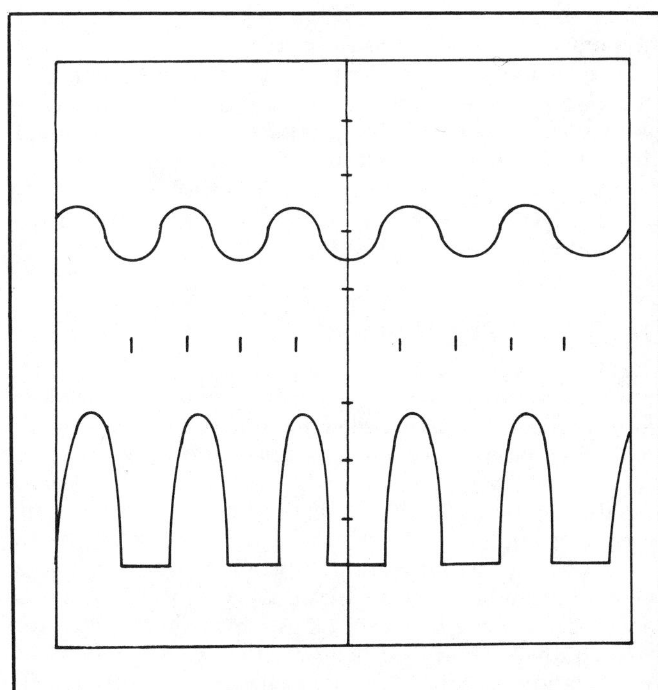

DUAL-BEAM OSCILLOSCOPE: A dual-beam oscilloscope can show two traces at once. Here, the upper trace shows the input to a Class-B amplifier, and the lower trace shows the output waveform.

DUAL IN-LINE PACKAGE: A dual in-line package is common with integrated circuits.

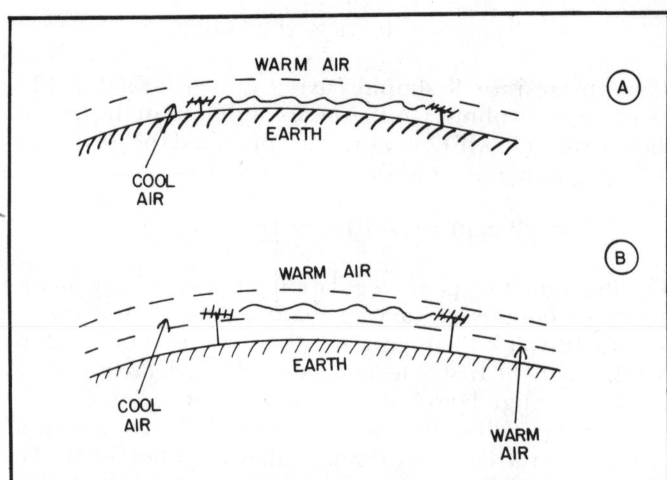

DUCT EFFECT: The duct effect allows propagation of signals through a mass of cool air near the ground, as shown at A, or through a layer of cool air sandwiched between layers of warmer air, as shown at B.

package. An example of this is the bridge rectifier circuit, which contains four semiconductor diodes in a single housing. In the case of the integrated circuit, of course, hundreds or even thousands of components are fabricated on a semiconductor wafer and placed into a single package. *See also* INTEGRATED CIRCUIT.

DUAL-DIVERSITY RECEPTION

See DIVERSITY RECEPTION.

DUAL IN-LINE PACKAGE

The dual in-line package is a common type of housing for integrated circuits. A flat, rectangular box containing the chip is fitted with lugs along either side, as shown in the photo. There may be just a few pins on each side, or there may be 15, 20, or even 30. The number of lugs is generally the same on either side of the device.

Dual in-line packages are convenient for installation in sockets. Sometimes they are soldered directly to printed-circuit boards. The short length of the lugs, and their small physical diameter, result in very low interelectrode capacitances. A dual in-line package device is sometimes called a DIP. *See also* INTEGRATED CIRCUIT, SINGLE IN-LINE PACKAGE.

DUCT EFFECT

The duct effect is a form of propagation that takes place with very-high-frequency and ultra-high-frequency electromagnetic waves. Also called ducting, this form of propagation is entirely confined to the lower atmosphere of our planet.

A duct forms in the troposphere when a layer of cool air becomes trapped underneath a layer of warmer air, as shown at A in the illustration, or when a layer of cool air becomes sandwiched between two layers of warmer air, as shown at B. These kinds of atmospheric phenomena are fairly common along and near weather fronts in the temperate latitudes. They also take place frequently above water surfaces during the daylight hours, and over land surfaces at night. Cool air, being more dense than warmer air at the same humidity level, exhibits a higher index of refraction for radio waves of certain frequencies. Total internal reflection therefore takes place inside the region of cooler air, in much the same way as light waves are trapped inside an optical fiber or under the surface of a body of water (*see* INDEX OF REFRACTION, TOTAL INTERNAL REFLECTION).

For the duct effect to provide communications, both the transmitting and receiving antennas must be located within the same duct, and this duct must be present continuously between the two locations. Sometimes, a duct is only a few feet wide. Over cool water, a duct may extend just a short distance above the surface, although it may persist for hundreds or even thousands of miles in a horizontal direction. Ducting allows over-the-horizon communication of exceptional quality on the very-high and ultra-high frequencies. Sometimes, ranges of over 1,000 miles can be obtained by this mode of propagation. *See also* TROPOSPHERIC PROPAGATION.

DUMMY ANTENNA

A dummy antenna is a resistor, having a value equal to the characteristic impedance of the feed line, and a power-dissipation rating at least as great as the transmitter output power. There is no inductance in the resistor. A dummy antenna is used for the purpose of testing transmitters off the air, so that no interference will be caused on the frequencies at which testing is carried out. A dummy antenna can also be useful for testing communications receivers for internal noise generation, when no outside noise sources can be allowed to enter the receiver antenna terminals.

The resistor in a dummy antenna is always noninductive. This ensures that the feed line can be terminated in a perfect impedance match. The sizes of dummy loads can range from very small to very large. The photograph shows a tiny dummy antenna designed for use with a Citizen's-Band transceiver having only 4 watts of output power. Some dummy loads consist of large resistors immersed in oil-coolant containers. They can handle a kilowatt or more of power on a continuous basis.

Dummy antennas must be well-shielded to prevent accidental signal radiation. This is especially important at the higher power levels.

DUMMY ANTENNA: A dummy antenna may be tiny, such as the 4-watt device shown here, or it may be very large, capable of handling thousands of watts (courtesy of Gold Line Connector).

DUMP

The term dump is used in computer practice to refer to the transfer or loss of memory. Memory bits may be moved from one storage location to another, to make room in the first memory for more data. In the second memory, sometimes called a hard memory, the data is not affected by programming changes or power failures.

Some computer programs use dump points in their execution at periodic intervals. This transfers the data from the active, or soft, memory into a permanent, or hard, memory. During a pause in the execution of a program, the computer may backtrack to the most recent dump point, and use the data there to continue the program. This process is called dump-and-restart. *See also* MEMORY.

DUPLEXER

A duplexer is a device in a communications system that allows duplex operation. This means that the two operators can interrupt each other at any time, even while the other operator is actually transmitting. Duplex operation is usually carried out using two different frequencies. Notch filters between the transmitter and receiver at each station prevent overloading of the receiver front end. A repeater uses a duplexer to allow the simultaneous retransmission of received signals on a different frequency (*see* NOTCH FILTER, REPEATER). The illustration shows a block diagram of a duplex communications system, including the duplexer.

In a radar installation, a duplexer is a device that automatically switches the antenna from the receiver to the transmitter whenever the transmitter puts out a pulse. This prevents the transmitter from damaging or overloading the receiver. In this application, the duplexer acts as a high-speed, radio-frequency-actuated transmit-receive switch. The receiver is not actually able to operate while the transmitter sends out the pulse. *See also* TRANSMIT-RECEIVE SWITCH.

DUPLEX OPERATION

Duplex operation is a form of radiocommunication in

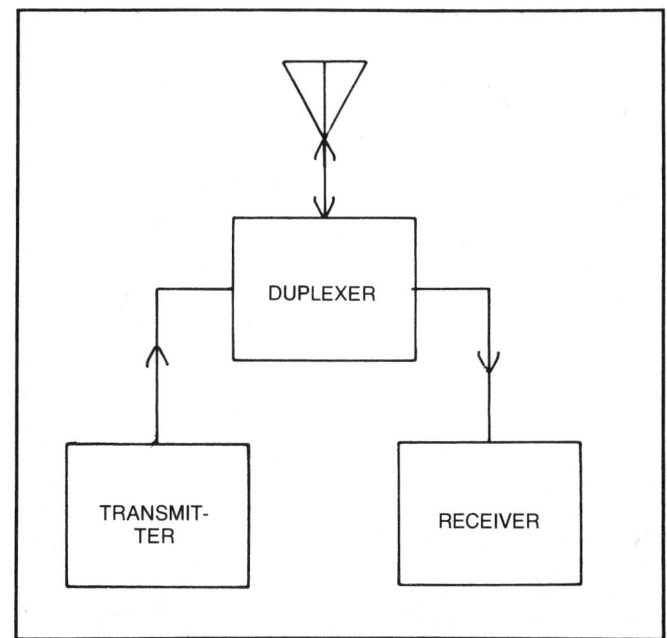

DUPLEXER: A duplexer allows the simultaneous operation of a receiver and transmitter, on different frequencies, using the same antenna.

which the transmitters and receivers of both stations are operated simultaneously and continuously. Each station operator may interrupt the other at any time, even while the other is actually transmitting. Therefore, a normal conversation is possible, similar to that over a telephone. There are no "blind" transmissions, and therefore there is no need for exchange signals such as "over."

In order to achieve full duplex operation, it is necessary that the transmitting and receiving frequencies be different. Two stations in duplex communication, say station X and station Y, must have opposite transmitting and receiving frequencies. For example, station X may transmit on 146.01 MHz and receive on 146.61 MHz; then station Y must transmit on 146.61 MHz and receive on 146.01 MHz. A duplexer (*see* DUPLEXER) is usually required to prevent the station receiver from being desensitized or overloaded by the transmitter.

In radiotelegraphy, a form of nearly full duplex operation is sometimes called break-in operation. A fast transmit-receive switch blocks the receiver during a transmitted dot or dash, but allows extremely rapid recovery. If the receiving operator desires to interrupt the transmitting operator, a tap on the key will be heard by the sending operator within a few milliseconds. Break-in operation is not true duplex, however, since the sending operator cannot actually hear the receiving operator interrupt during a transmitted pulse. *See also* BREAK-IN OPERATION.

DURATION TIME

Duration time is the length of an electric pulse, measured from the turn-on time to the turn-off time. If the pulse is rectangular, then the duration time is quite easy to determine, as shown in the drawing. But electric pulses are usually not perfectly rectangular. They often have rounded corners, droop, or other irregularities. Then, the duration of the pulse may be expressed in several different ways. Generally, the duration time of a non-rectangular pulse is considered to be the theoretical duration time of an

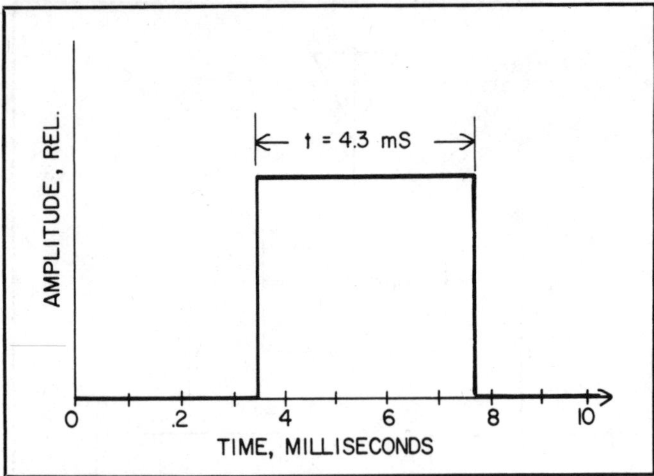

DURATION TIME: The duration time of a pulse is the time from its beginning to its end, in terms of the high condition.

equivalent rectangular pulse. The equivalent pulse may be generated by area redistribution, or by amplitude averaging (*see* AREA REDISTRIBUTION, AVERAGE ABSOLUTE PULSE AMPLITUDE).

The average amplitude of a pulse train depends on the duration and the frequency of the pulses. *See also* DUTY FACTOR.

DUST PRECIPITATOR

An electrostatic device for removing dust particles from the air is called a dust precipitator. Such devices are sometimes used in homes and businesses as air cleaners and purifiers. Dust precipitators are usually located at the air-intake point of a heating or air-condition system.

The air to be cleaned is passed through a grid of highly charged metal plates. The plates are alternately charged in terms of electrical polarity. The voltage is set so high that arcing almost occurs. When a dust particle passes between the plates, it acquires an electric charge, and is accelerated toward one of the plates. The particle then sticks to the plate, and is therefore removed from the air.

Occasionally, if a very large particle passes through the dust precipitator, an arc occurs, and the particle is vaporized. This causes a loud snapping noise.

DUTY CYCLE

The duty cycle is a measure of the proportion of time during which a circuit is operated. Suppose, for example, that during a given time period t_0, a circuit is in operation for a total length of time t, measured in the same units as t_0. Of course, t cannot be larger than t_0; if $t = t_0$, then the duty cycle is 100 percent. In general, the duty cycle is given by:

$$\text{Duty cycle (percent)} = 100t/t_0$$

When determining the duty cycle, it is important that the evaluation time t_0 be long enough. The evaluation time must be sufficiently lengthy so that an accurate representation is obtained. For example, a circuit may be on for one minute, and then off for one minute, and then on again for one minute, in a repeating cycle. In such a case, an evaluation time of 10 seconds is obviously too short; t_0 must be at least several minutes for a reasonably accurate estimate of the duty cycle.

Power-dissipating components such as resistors, transistors, or vacuum tubes, can often handle more power as the duty cycle decreases. A transmitter transistor may have a dissipation rating of 100 watts for continuous duty; but at a 50-percent duty cycle, the rating may increase to 150 or 200 watts. The duty cycle is assumed to be reasonable under such circumstances, such as a maximum of 15 seconds on and 15 seconds off. *See also* DISSIPATION RATING.

DUTY FACTOR

The duty factor of a system is the ratio of average power to peak power, expressed as a percentage. The duty factor is similar to the duty cycle, except that the system does not necessarily have to shut down completely during the low period (*see* DUTY CYCLE). In a continuous-wave transmitter, the duty factor and the duty cycle are the same. Letting P_{max} be the peak power and P_{av} be the average power, then:

$$\text{Duty factor (percent)} = 100P_{av}/P_{max}$$

In a pulse train, or string of pulses, the duty factor is the product of the pulse duration and the frequency. As a percentage, letting the duration time be t and the frequency in Hertz be f, then:

$$\text{Duty factor (percent)} = 100tf$$

In a string of square or rectangular pulses, the duty factor is the same as the duty cycle. *See also* DURATION TIME.

DX

DX is a term used mainly by shortwave broadcast enthusiasts and Amateur Radio operators. Literally, DX means distance. However, the term DX also carries a connotation of foreign lands—especially unusual places, even if they are not very far away. The actual distance of a place, to a serious DXer, is not as important as the rarity of the location.

For example, Japan is quite far away from Miami, Florida; the coast of Africa is closer in terms of great-circle mileage. But most radio amateurs would consider a contact in, say, Togo (a tiny country on the western coast of Africa) much more like DX than a contact in Japan, simply because Japan has many amateur radio operators and Togo has very few.

Sometimes, DXers travel to a remote place which has no Amateur Radio stations at all, so that they might provide DX contacts to their friends back in the U.S. This is called a "DXpedition." *See also* AMATEUR RADIO.

DYNAMIC CHARACTERISTIC

Electronic components carrying alternating currents behave differently from those carrying only pure direct currents. Transistors and tubes are especially significant in this respect. The characteristic curves of such devices (*see* CHARACTERISTIC CURVE) are generally determined by testing with direct currents only; such curves are therefore

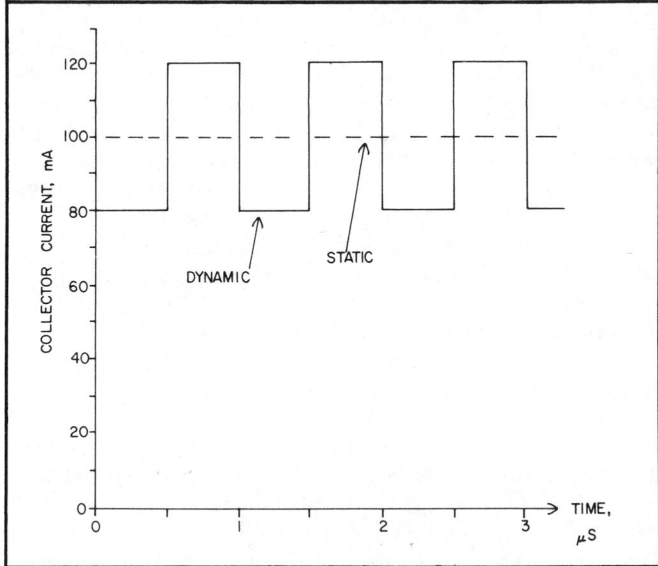

DYNAMIC CHARACTERISTIC: The dynamic characteristic takes the voltage, current, or power fluctuations into account, when a device is operated with an alternating-current or pulsating direct-current signal.

static characteristics, meaning that they have been determined with steady signals. In actual operation of most amplifiers and oscillators, of course, the signal is not constant, but fluctuating. This may cause the characteristics to be slightly different. Real-operation conditions are reflected in the dynamic characteristics of a component or circuit.

An example of a dynamic characteristic versus a static characteristic is shown in the illustration. The dotted line illustrates the static collector current of a hypothetical NPN transistor amplifier under conditions of zero input signal. When a driving signal is applied, the collector current fluctuates. As the base voltage becomes more negative during one half of the input cycle, the collector current falls. As the base voltage becomes more positive during the other half of the input cycle, the collector current rises.

We may simply say that the static collector current, as evidenced by the graph, is 100 mA when there is no input signal. We must be more specific in talking about the dynamic collector current when there is a signal at the input. The collector current in actual operation fluctuates, as can be seen on the vertical scale, between 80 mA and 120 mA. It fluctuates at a frequency of 1 MHz, as can be ascertained from the horizontal time scale. (Each cycle requires 1 μS.) Finally, it can be seen that the fluctuation is a square wave. In this particular example, the average dynamic collector current is the same as the static collector current: 100 mA. This is an example of a Class-A amplifier.

In general, the dynamic characteristics are the actual operating parameters of circuits and components. These parameters are not always the same as the static characteristics, which are usually theoretical or are derived under rigid test conditions. *See also* STATIC CHARACTERISTIC.

DYNAMIC CONTACT RESISTANCE

In a relay, the resistance caused by the contacts depends on several factors. Such factors include the pressure with

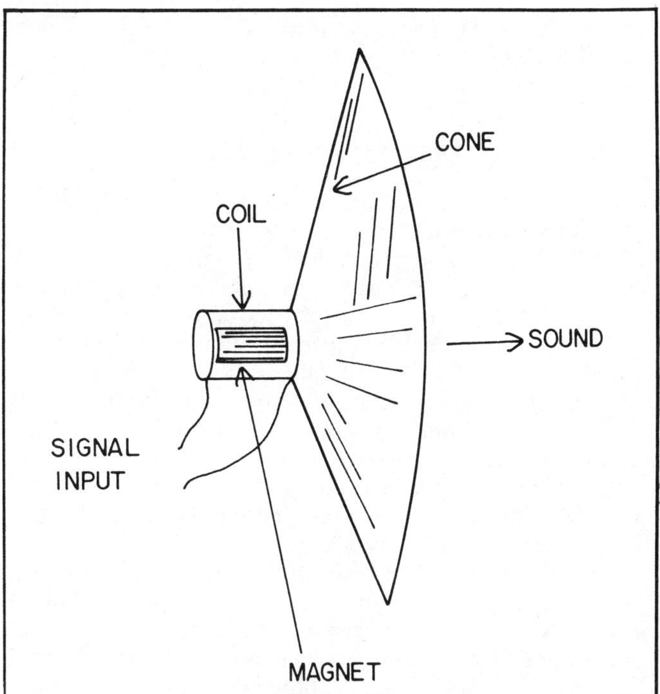

DYNAMIC LOUDSPEAKER: A simplified pictorial rendition of a dynamic loudspeaker.

which the contacts press against each other, as well as their surface area and the amount of corrosion. The greater the pressure, the lower the contact resistance. The effective resistance of a current-carrying pair of relay contacts may vary because of mechanical vibration between the contacts. The actual contact resistance, measured with the relay in an actual circuit in normal operating circumstances, is the dynamic contact resistance.

Normally, the dynamic resistance of any pair of relay contacts should be as low as possible. This can be ensured by keeping the contacts clean, and by making certain that the relay coil gets the right amount of current, so that the contacts will open and close properly. Too much coil current will cause contact bounce; too little coil current will result in insufficient pressure between the contacts. *See also* RELAY.

DYNAMIC LOUDSPEAKER

A dynamic loudspeaker is a speaker that uses magnetic fields and electric currents to produce sound vibrations.

In the dynamic loudspeaker, a permanent magnet supplies the necessary magnetic field. A solenoidal coil carries the audio-frequency alternating or pulsating direct current. As the current in the coil fluctuates, the changing magnetic field, ireated by the current, interacts with the field around the permanent magnet. The resulting forces cause sound to be propagated.

The coil in the dynamic loudspeaker, called the voice coil, is mounted in such a way that it can move back and forth as the magnetically induced forces act on it. A large paper cone is attached to the coil, and moves along with it, as shown in the simplified illustration. The moving cone imparts some of the energy to the surrounding air molecules, resulting in acoustical waves. Most speakers today are of this type.

A device similar to the dynamic loudspeaker can be used

to convert sound energy into audio-frequency electric currents. *See also* DYNAMIC MICROPHONE, SPEAKER.

DYNAMIC MICROPHONE

A dynamic microphone is a device that utilizes magnetic fields and mechanical movements to create alternating electric currents. The principle of operation is exactly the same as that of the dynamic loudspeaker, except that it operates in the reverse sense (*see* DYNAMIC LOUDSPEAKER).

Vibrating air molecules cause mechanical movement of a diaphragm, as shown in the illustration. A coil, attached rigidly to the diaphragm, moves along with it in the field of a permanent magnet. As a result of this movement, currents are induced in the coil. These currents are in exact synchronization with the sound vibrations impinging against the diaphragm. Amplifiers may be used to boost the currents to any desired level.

Dynamic microphones are one of the most common types of microphones used today. They are physically and electrically rugged, and last a long time. The magnet should not be subjected to extreme heat or severe physical blows. Other types of microphones are sometimes used, such as the ceramic and crystal microphones. *See also* CERAMIC MICROPHONE, CRYSTAL MICROPHONE, MICROPHONE.

DYNAMIC PICKUP

A dynamic pickup is a phonograph transducer that operates on the same principle as the dynamic microphone or loudspeaker (*see* DYNAMIC LOUDSPEAKER, DYNAMIC MICROPHONE). A stylus vibrates as it follows the groove of the phonograph disk. A small coil, attached to the stylus, moves along with it in the field of a permanent magnet. This results in audio-frequency currents in the coil.

Dynamic pickups are seen quite often in high-fidelity applications. The ceramic and crystal type pickups are also fairly common. The dynamic pickup has the advantage of being electrically and physically rugged, although it should not be subjected to excessive heat or severe mechanical abuse, since this may weaken the magnet. *See also* CERAMIC PICKUP, CRYSTAL TRANSDUCER.

DYNAMIC RANGE

In an audio sound system or a communications receiver, the dynamic range is often specified as an indicator of the signal variations that the system can accept, and reproduce, without objectionable distortion.

In high-fidelity sound systems, there is a weakest signal level that can be reproduced without being masked by the internal noise of the amplifiers. There is also a strongest signal level that can be reproduced without exceeding a certain amount of total harmonic distortion in the output. Usually, the maximum allowable level of total harmonic distortion is 10 percent (*see* TOTAL HARMONIC DISTORTION). If P_{max} represents the maximum level of input power that can be reproduced without appreciable distortion, and P_{min} represents the smallest audible input power, then the dynamic range in decibels is given by:

$$\text{Dynamic range (dB)} = 10 \log_{10}(P_{max}/P_{min})$$

In a communications receiver, the dynamic range refers to the amplitude ratio between the strongest signal that can be handled without appreciable distortion in the front end, and the weakest detectable signal. The formula for the dynamic range in a communications receiver is exactly analogous to that for audio equipment. Most receivers today have adequate sensitivity, so this is not a primary limiting factor in the dynamic range of a receiver. Much more important is the ability of a receiver to tolerate strong input signals without developing nonlinearity. Such nonlinearity in the front end, or initial radio-frequency amplification stages, can result in problems with intermodulation and desensitization. *See also* DESENSITIZATION, FRONT END, INTERMODULATION.

DYNAMIC REGULATION

In a voltage-regulated power supply, dynamic regulation refers to the ability of the supply to handle large, sudden changes in the input voltage with a minimum of change in the output voltage or current. The most common kind of sudden change in the input voltage to a power supply is called a transient. A transient can be the result of a lightning stroke or arc in a power transmission line (*see* TRANSIENT).

Poor dynamic regulation in a power supply can result in sudden changes in the output voltage. This can cause damage to modern solid-state equipment. *See also* POWER SUPPLY, VOLTAGE REGULATION.

DYNAMIC RESISTANCE

According to Ohm's law, we think of the resistance between two points as the quotient of the voltage and the current (*see* OHM'S LAW). This quantity is the static resistance in a circuit. In operation, the effective resistance or impedance of a component may not be the same as the static resistance or impedance. The actual, or effective, operating resistance is called the dynamic resistance.

Suppose a component is subjected to a changing voltage, which results in a changing current. The illustration shows two possible resulting curves. At A, the current increases linearly with the voltage. At B, the current does not change in direct proportion to the voltage.

A curve such as the one at B might result from a gradual change in the voltage across a component, with a resulting heat change that causes a fluctuation in the resistance. The dynamic resistance over a certain range is found by dividing the change in the voltage, ΔE (delta E), by the change in

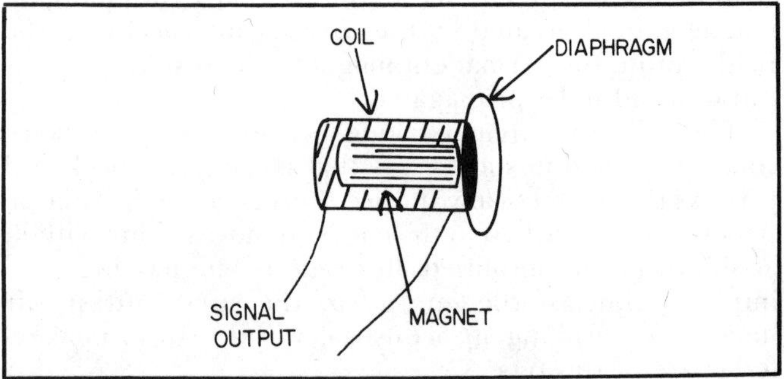

DYNAMIC MICROPHONE: A simplified pictorial rendition of a dynamic microphone.

the current, ΔI, between two specific points along the curve. Thus, at A, the dynamic resistance is:

$$\Delta E/\Delta I = 10/0.5 = 20 \text{ ohms}$$

and it does not matter what points on the curve we choose. The dynamic resistance is always the same in the illustration at A. But at B, the dynamic resistance changes depending on the points selected. For the two points shown, the dynamic resistance is

$$\Delta E/\Delta I = 10/0.3 = 33 \text{ ohms}$$

but this is the unique situation for these points. *See also* RESISTANCE.

DYNAMIC SPATIAL RECONSTRUCTOR

The dynamic spatial reconstructor is an X-ray machine that displays body organs in three-dimensional detail. The machine was developed at the Mayo Clinic in Rochester, Minnesota.

With normal X-ray pictures, the image produced on the film or screen is two-dimensional. The exact location of a tumor or other abnormality cannot be determined without taking at least two or three pictures from different angles. The dynamic spatial reconstructor gives a three-dimensional representation in motion, which makes it much easier to determine the location of an abnormality in the body and to make a proper diagnosis. *See also* X-RAY.

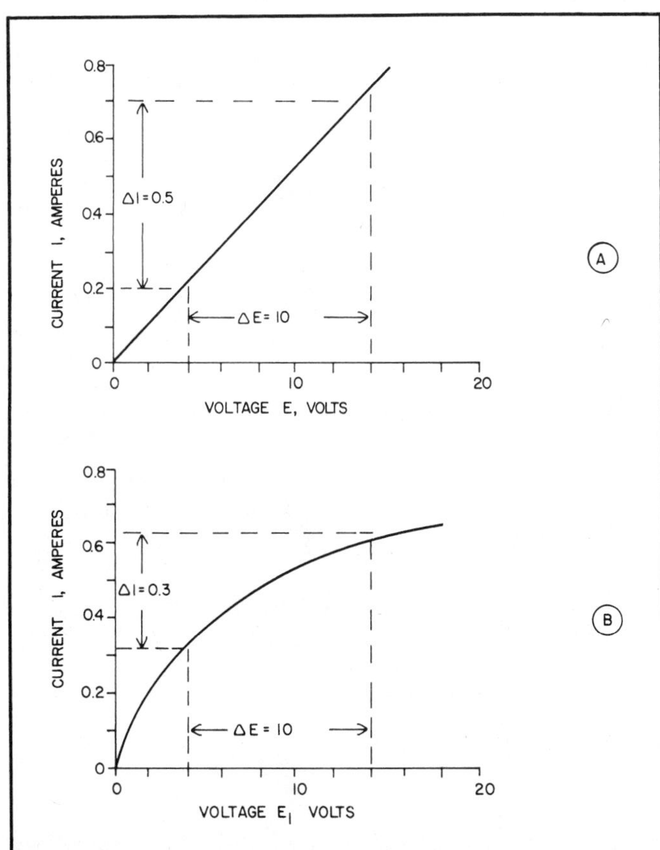

DYNAMIC RESISTANCE: Dynamic resistance is determined as the difference in voltage divided by the difference in current over a certain range of operation. At A, dynamic resistance does not depend on the range chosen. But at B, the dynamic resistance changes with the range chosen.

DYNAMO

See GENERATOR.

DYNAMOMETER

See ELECTRODYNAMOMETER.

DYNAMOTOR

A dynamotor is a small motor-generator, contained in a single housing. The motor, usually operated from a low voltage, turns a generator that produces a high voltage. Dynamotors usually have directly-coupled shafts.

The generation of a high voltage at an appreciable current level requires great turning power on the part of the motor in a dynamotor device. Therefore, the low-voltage motor draws a large amount of current. A dynamotor can be used to produce either alternating or direct current from a low-voltage, direct-current source. Such a dynamotor serves as a power inverter when the output voltage is 120 with a 60-Hz frequency.

In the construction of a dynamotor, it is important that the motor speed respond rapidly to changes in the load on the generator. Otherwise the voltage regulation will be very poor. A sudden drop in the load resistance puts extra turning resistance on the generator shaft, and this happens immediately; the motor needs a short time to compensate. *See also* GENERATOR, INVERTER, MOTOR.

DYNATRON EFFECT

In a vacuum tube, when the direct-current screen voltage exceeds the plate voltage, a negative-resistance region develops in the plate circuit. This means that, within a certain range, an increase in the plate voltage will cause a decrease in the plate current. Normally, of course, increasing plate voltages result in increases in the plate current. The negative-resistance effect is called the dynatron effect.

The dynatron effect takes place because the screen grid attracts secondary electrons from the plate under conditions of large positive screen voltage. Normally, the secondary electrons would return to the more positive plate electrode. When a tube has a screen voltage larger than the plate voltage, the plate is deprived of many of these secondary electrons, and therefore the plate current is reduced.

The dynatron effect can be used to advantage in a circuit known as a dynatron oscillator. The dynatron effect can also cause an amplifier circuit to oscillate, if the plate voltage is reduced without a corresponding reduction in the screen-grid voltage. Such oscillation can produce spurious signals, outside the band intended. Such oscillator effects are called parasitics. *See also* PARASITIC OSCILLATION, TUBE.

DYNODE

A dynode is an electrode often used in a photomultiplier or

dissector tube to amplify a stream of electrons. When the dynode is hit by an electron, it emits several secondary electrons. Approximately five to seven secondary electrons are given off by the dynode for each electron striking it.

Dynodes in photomultiplier tubes are slanted in such a way that an electron must strike a series of the dynodes in succession. The electron hits first one dynode, where several secondary electrons are emitted along with it. The electrons then hit a second dynode, where one incident electron again produces several secondary electrons. It does not take many dynodes to get very large gain figures in this way. For example, four dynodes, each of which emits five secondary electrons in addition to the one that strikes it, will result in a gain of 6^4 or 1296. This is 62 dB in terms of current!

Dynodes in a photomultiplier or dissector tube are placed around the inner circumference, and are oriented at the proper angles so that an impinging electron strikes them one after the other. With the electrons in the tube, as with light against a mirror, the angle of incidence is equal to the angle of reflection. *See also* DISSECTOR TUBE, PHOTOMULTIPLIER.

EARPHONE

A small speaker, designed to be worn in or over the ear, is called an earphone. Earphones are sometimes used in communications for privacy or for quiet reception. With an earphone worn on one ear, the other ear is free, allowing the operator to hear external sounds. The photograph shows a typical lightweight earphone. Such a small earphone is sometimes called an earpiece.

A pair of earphones is often mounted on a headband and worn over both ears. The combination is then called a headphone or headset. Such a combination of earphones essentially obstructs the listener's sensitivity to external noises.

Earphones are available in impedances as low as 4 ohms and as high as several thousand ohms. An earphone consumes far less power than a speaker, and so the earphone can be useful when it is necessary to conserve power, as in battery operation. Some earphones operate on the dynamic principle; others are crystal or ceramic transducers. *See also* CRYSTAL TRANSDUCER, DYNAMIC LOUDSPEAKER, HEADPHONE.

EARPHONE: An earphone may be very small. This unit will fit comfortably inside the ear for private listening.

EARTH CONDUCTIVITY

The earth is a fair conductor of electric current. The conductivity of the soil is, however, highly variable depending on the geographic location. In some places, the soil may be considered an excellent conductor; in other locations it is a very poor conductor. In general, the better the soil for farming, the better it will be in terms of electrical conductivity. Wet, black soil is a very good conductor, while rocky or sandy dry soil is poor. A brief rain shower can drastically alter the soil conductivity in some locations.

The conductivity of the earth is generally measured in millimhos per meter. A mho (also known as a Siemens) is the unit of conductance, and is expressed as the reciprocal of the resistance. A millimho is equivalent to a resistance of 1,000 ohms. A conductance of 1,000 millimhos is equivalent to a resistance of 1 ohm. In the United States, the earth conductivity ranges from approximately 1 millimho per meter to more than 30 millimhos per meter. All earth is, by comparison to sea water, a very poor conductor; salt water in the ocean has an average conductivity of about 5,000 millimhos per meter.

In radiocommunications, the earth conductivity is important from the standpoint of obtaining a good electrical ground. The earth conductivity is also important for surface-wave propagation at frequencies below about 20 to 30 MHz. The better the conductivity of the earth, the better the radio-station ground system will be for a given installation, and the better the surface-wave propagation. Good ground conductivity is also advantageous in electrical installations, and for the grounding of equipment for protection against lightning. *See also* GROUND CONNECTION, SURFACE WAVE.

EARTH CURRENT

Electric wires under or near the ground, carrying alternating currents, cause currents to be induced in the ground. This occurs because the ground is a fairly good conductor, and the electromagnetic field around the wire causes movement of the charge carriers in any nearby conductor. High-tension lines can induce considerable current in the ground near the wires. Radio broadcast stations cause radio-frequency currents to flow in the ground in the vicinity of their antennas.

Currents flow in the earth as a natural part of the environment. Two ground rods, driven several feet apart, will often have a small potential difference, and this can be measured with an ordinary alternating-current voltmeter. The intensity of the ground current in the vicinity of a conductor carrying alternating current depends on the conductivity of the ground in the area, and on the current intensity in the wire and the distance of the wire above the ground.

The earth conductivity is quite variable in different geographic locations. In rocky or sandy soil, the conductivity is poor and the earth currents can be expected to be small. In black soil, especially if wet, ground currents can be very large. *See also* EARTH CONDUCTIVITY.

E BEND

In a waveguide, an E bend is a change in the direction of the waveguide, made parallel to the plane of electric-field lines of flux. It is called an E bend for exactly this reason: It follows the E-field component of the electromagnetic wave.

The bending of radio waves by the ionospheric E layer is sometimes called E bend or E skip, although both of these terms are somewhat inaccurate from a technical standpoint. Signal propagation via the E layer is fairly common at frequencies below about 50 MHz, and may sometimes occur well into the very-high-frequency spectrum. *See also* E LAYER, WAVEGUIDE.

EBS

See EMERGENCY BROADCAST SYSTEM.

ECCENTRICITY

When a dial or control is not properly centered, it is said to be eccentric. Eccentricity in a control results in difficulty of adjustment, since the position of the knob with respect to the calibrated scale is not consistent as the setting is changed. Eccentricity is a mechanical, and not an electrical, problem. In well-designed, high-quality equipment, eccentricity will not be found.

In mathematics, eccentricity is a measure of the degree to which an ellipse differs from a circle, or an ellipsoid deviates from a sphere. A circle and sphere have zero eccentricity. If c represents the distance from the center of an ellipse or ellipsoid to either of the foci, and a is the larger (major) radius, then the eccentricity is given by c/a.

In a phonograph disk, eccentricity refers to a condition wherein the spiral groove is not centered on the disk. This may happen because the hole is off center, or because the grooves themselves are off center. Eccentricity in a phonograph disk results in warbling of the sound when it is played back, as the stylus first speeds up and then slows down in its travel through the length of the groove. If the eccentricity is severe, this warbling will be very noticeable and objectionable. *See also* DISK RECORDING.

ECG

See ELECTROCARDIOGRAM.

ECHO

Whenever energy is sent outward from a source, is reflected off of a distant object, and then returns in detectable magnitude, the return signal is called an echo. The most familiar echoes are, of course, the sound echoes that we all hear when we shout in the direction of a distant, acoustically reflective object. In radar, the term echo is used to refer to any return of the transmitted pulses, creating a blip in the screen (*see* RADAR).

Echoes can sometimes be heard on radio signals in the high-frequency part of the electromagnetic spectrum, where ionospheric propagation is prevalent. A few milliseconds after receiving a strong signal pulse, a weaker echo may be heard. This occurs because the signal has propagated via two paths: a short path and a long path (see illustration). The echo delay time is the difference in the time required for the signal to propagate by means of the two paths. *See* LONG-PATH PROPAGATION, SHORT-PATH PROPAGATION.

Echoes are occasionally detected in wire transmission circuits. This can happen, for example, in a long-distance telephone connection. Such electrical echoes result from impedance discontinuities, or "bumps," along the line. These discontinuities cause some of the incident signal to be reflected back toward the transmitting station. A device called an echo eliminator is used in telephone circuits to suppress such reflected signals. Sometimes, however, some of the echo gets through and can be heard by the person talking on the telephone. *See also* TELEPHONE.

ECHO BOX

An echo box is an apparatus designed for the purpose of testing and calibrating a radar set.

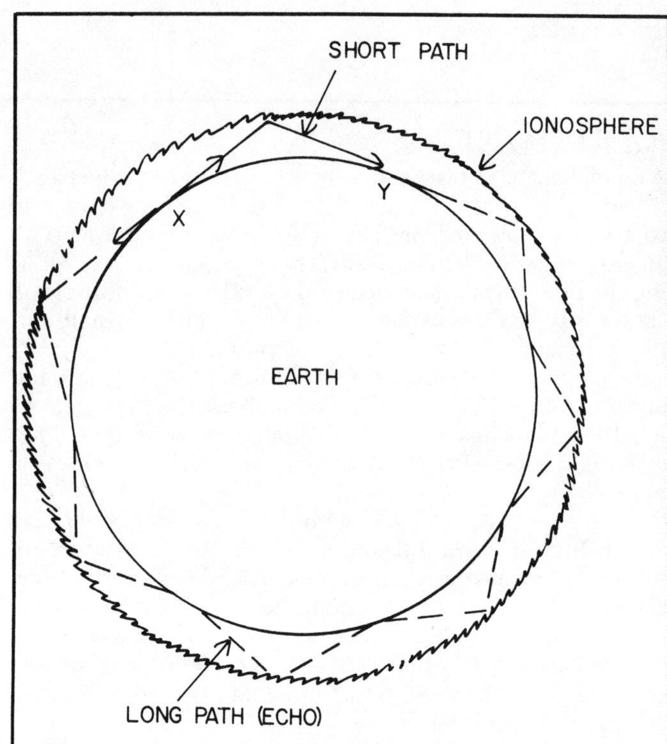

ECHO: A propagation echo can result from the delayed reception of a long-path signal, following the reception of the short-path signal.

The transmitted pulse is fed into the echo box, instead of to the radar antenna. The echo box then produces a delayed return pulse, the intensity of which gradually falls at a certain rate. The decaying echo signal is fed to the radar receiver input. Eventually the pulse decays to the point where the radar receiver can no longer detect it.

If the amplitude-versus-time function of the output signal is known, the receiver sensitivity can be determined according to the length of time for which the receiver is able to detect the signal. When the echo can no longer be seen on the radar screen, the level of the signal is checked, and this represents the weakest signal that the radar can "see." *See also* RADAR.

ECHO INTERFERENCE

When a high-frequency radio signal is propagated via the ionosphere, an echo sometimes occurs (*see* ECHO). This may be the result of long-path propagation in addition to the normal short-path propagation. Or, echo interference may be the result of propagation of the signal over several more-or-less direct paths of variable distance.

When echo interference is severe, it can render a signal almost unintelligible, because phase modulation is superimposed on the envelope of the signal. This phenomenon is especially noticeable with auroral propagation, when the signal path may change at a very rapid rate.

In a long-distance telephone circuit, the echo is sometimes of sufficient amplitude to create interference at the transmitting station. This can distract the person speaking, since he hears his own words with a slight delay in the earphones, and it can actually be difficult to maintain a train of thought! Echo eliminators are commonly used in telephone systems to prevent this kind of interference, which results from reflection of signals from impedance discontinuities along the line.

In a radar system, echo interference is caused by the presence of unwanted "false" targets such as buildings and thunderstorms. During the second world war, it was discovered that thundershowers and rain squalls produced echoes on radar sets; since then, the radar has been a valuable tool in the determination of short-term weather forecasts and warnings. *See also* LONG-PATH PROPAGATION, RADAR, TELEPHONE.

ECLIPSE EFFECT

The conditions of the ionospheric D, E, and F layers vary greatly with the time of day. During a solar eclipse, when the moon passes between the earth and the sun, nighttime conditions can take place for a brief time within the shadow of the moon. This effect on the ionized layers of the upper atmosphere is called the eclipse effect.

The D layer, the lowest layer of the ionosphere, seems to be affected to a greater extent during a solar eclipse than the E or F layers, which are at greater altitudes, and ionize more slowly. The ionization of the D layer depends on the immediate level of ultraviolet radiation from the sun, but the E and F layers have some "lag" time. An eclipse in the sun causes a dramatic decrease in the ionization level of the D layer, but the E and F layers are generally not affected to a great extent. The reduction in the ionization density of the D layer causes improved propagation at some frequencies, because the absorption created by the D layer is not as great during an eclipse, as under normal circumstances during daylight hours.

Investigations are continuing into the effects of solar eclipses on the behavior of the ionosphere. This may help us learn more about the nature of the ionosphere, and how it propagates radio signals. Of course, since solar eclipses are rare at any given location, they are not of great practical concern in ionospheric propagation. *See also* D LAYER, E LAYER, F LAYER, PROPAGATION CHARACTERISTICS.

E CORE

An E core is a form of transformer core, made of ferromagnetic material such as powdered iron, and shaped like a capital letter E. The drawing illustrates a typical E core. Wire coils are wound on the horizontal bars of the letter E; up to three different windings are possible, so that a transformer may have two secondary windings giving different voltages and impedances with respect to a single primary winding.

The interaction among the windings on an E core are, of course, enhanced by the action of the ferromagnetic material, which has a high permeability for magnetic fields. Most of the magnetic flux is contained within the E-shaped core, and very little exists in the surrounding air. *See also* TRANSFORMER.

EDDY-CURRENT LOSS

In any ferromagnetic material, the presence of an alternating magnetic field produces currents that flow in circular or elliptical paths, around and around. These currents, because of their resemblance to vortex motions, are called eddy currents.

Eddy currents are responsible for some degradation in

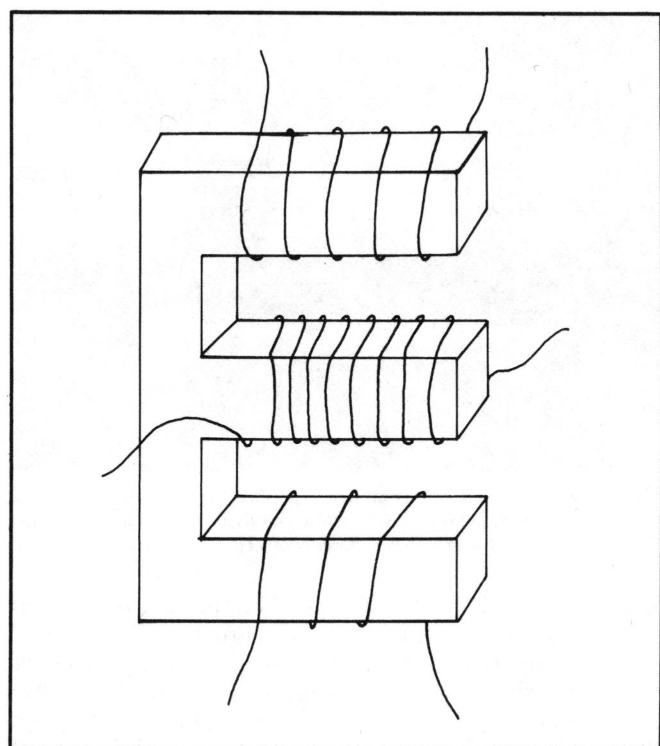

E CORE: An E-core is shaped like a capital letter E, and may have two or three windings.

the efficiency of any transformer that uses an iron core. To minimize the effects of eddy currents, transformers sometimes have laminated cores, consisting of a number of thin, flat pieces of ferromagnetic material glued together but insulated from each other. This chokes off the flow of eddy currents to a large extent, because circular paths are cut by the insulating layers. *See also* TRANSFORMER.

EDGE CONNECTOR

An edge connector is a rigid plug-and-socket device used for multiple-conductor interconnections. Edge connectors provide a reliable means of contact when many conductors must be employed between circuits. Some edge connectors contain dozens of pins.

The male edge connector is usually fabricated from printed-circuit material, such as phenolic, and may actually be part of a printed circuit containing other components. The female part of the connector contains spring pins, which make contact with the thin foil strips on the male edge connector. Several connections can be made on either side of the connector board. A typical edge-connector arrangement is shown in the photograph.

In modern, highly sophisticated solid-state equipment, edge connectors are commonly used. They are sometimes provided on circuit boards, in digital devices such as communications monitors, for the purpose of attaching a troubleshooting computer system. *See also* PRINTED CIRCUIT.

EDGE EFFECT

In a capacitor, most of the electric field is confined between the two plates. However, near the edges of the surfaces, some of the electric lines of flux extend outward from the

EDGE CONNECTOR: An edge connector is generally fabricated on a piece of printed-circuit board, which inserts into a spring-loaded receptacle. Such a connector may have many contacts.

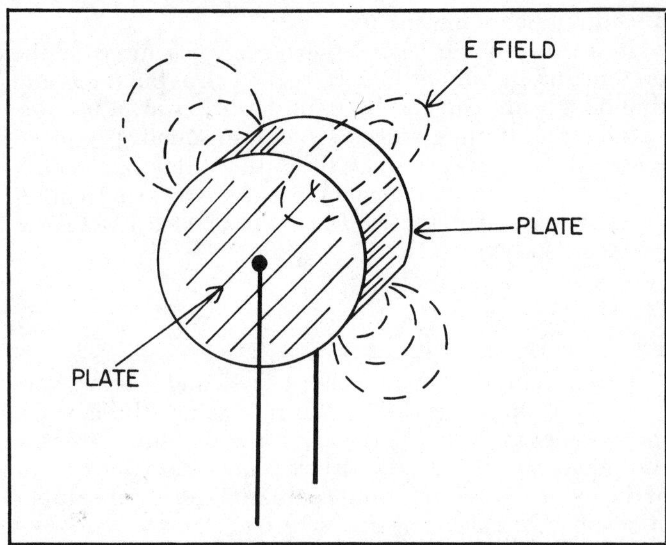

EDGE EFFECT: The edge effect results in the presence of lines of electric flux outside the space between the plates of a capacitor. Here, a disk capacitor is shown. Most of the electric field is contained between the plates, but some leakage occurs, and this can result in unwanted coupling with nearby components.

space between the plates. This is called the edge effect, and is shown pictorially for a disk capacitor in the drawing.

The edge effect is most pronounced with air-dielectric capacitors, since the surrounding air has the same dielectric constant as the medium between the plates. However, the edge effect also occurs to some extent in other types of capacitors. It is least pronounced in electrolytic capacitors, which usually have shielded enclosures. The edge effect can result in unwanted capacitive coupling among components in a circuit, and for this reason, the component layout of a circuit is critical at frequencies where the smallest intercomponent capacitance may cause feedback or other undesirable effects. *See also* CAPACITOR.

EDISON EFFECT

When the filament of an incandescent bulb is heated by means of passing an electric current through it, it gives off electrons. This is called the Edision effect, because it was Thomas Edision, in his early experiments with light bulbs, who first discovered it. The electrons caused the bulbs to get black on the inside. Edision used positively charged electrodes in the evacuated bulbs to attract these electrons, resulting in a flow of current, but preventing the blackening.

Edison's discovery that a positive electrode would eliminate this blackening inside an incandescent bulb eventually led to the perfection of the vacuum tube, which uses a positively charged plate to attract electrons from a hot filament. *See also* TUBE.

EDISON, THOMAS A.

A prolific inventor and experimentalist, Thomas A. Edison (1847-1931) is probably best known for his invention of the incandescent lamp. Edison was also responsible for improving methods of power transmission and wire telegraphy.

Edison spent about 15 to 18 hours a day in his laboratory, abandoning almost everything else in his life to work on his projects. Perfecting the light bulb took a tremendous amount of effort, and there were many failures before success was finally achieved.

Edison recommended direct current for long-distance transmission of electricity. Most of his colleagues disagreed with him because alternating current allows the use of voltage transformers while direct current does not. As our energy demands increase, necessitating the use of longer and longer power-transmission lines, however, some scientists are beginning to reconsider this issue. Power loss is less, per unit distance, for direct current as compared with alternating current. Over long distances this difference could offset other factors.

A memorial to Thomas Edison can be found in Fort Myers, Florida, where he did most of his work in his later years. *See also* AC POWER TRANSMISSION, HIGH-TENSION POWER LINE, INCANDESCENT LAMP.

EEG

See ELECTROENCEPHALOGRAM.

EFFECTIVE BANDWIDTH

All bandpass filters allow some signal to pass, and reject energy at frequencies outside the passband range (*see* BANDPASS RESPONSE). The bandwidth of a filter of this kind is usually determined according to the 3-dB power-attenuation points. However, there is another method of specifying the bandpass characteristic of a bandpass filter. This is the effective-bandwidth method.

Suppose a hypothetical bandpass filter, having a perfectly rectangular response (*see* RECTANGULAR RESPONSE), is centered at the same frequency f_0 as an actual bandpass filter. This situation is illustrated in the drawing. Suppose that the bandwidth of the rectangular filter can be varied at will, all the way from zero to a value much greater than that of the real filter, but always remaining centered at f_0. At some value of bandwidth of the rectangular filter, the characteristics will be identical to that of the actual filter, with respect to the amount of energy transferred. This rectangular bandwidth is the effective bandwidth of the real filter.

The effective bandwidth of a filter depends on the

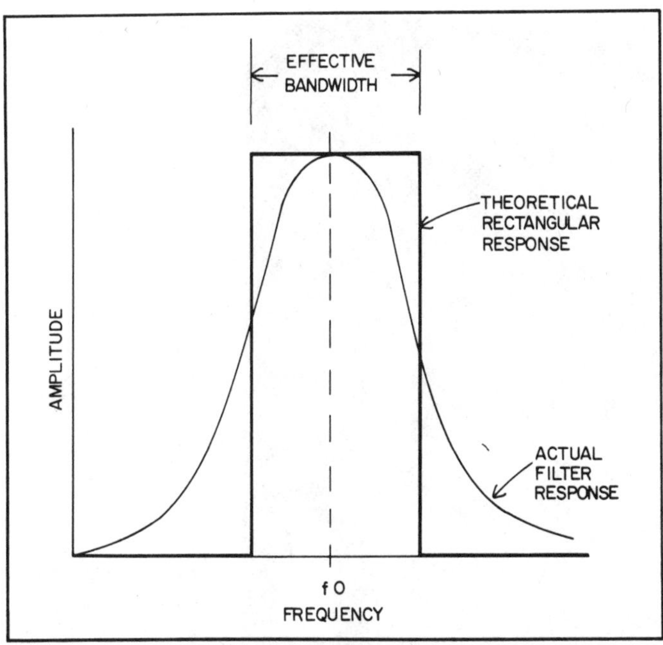

EFFECTIVE BANDWIDTH: The effective bandwidth of a bandpass filter is determined by finding a rectangular response that provides the same theoretical energy-transfer ratio.

steepness of the response, and also on the nature of the response within the passband. *See also* BANDPASS FILTER.

EFFECTIVE GROUND

In radiocommunications practice, below a frequency of about 100 MHz, the ground plays a significant role in the behavior of antenna systems, both for receiving and for transmitting. This is true whether or not an antenna is physically connected to an earth ground. The earth, since it is a fairly good conductor at most locations, absorbs some radio-frequency energy and reflects some. The lower the frequency, in general, the better the conductivity of the soil in a particular location (*see* EARTH CONDUCTIVITY). Some soils are good conductors at virtually all frequencies below 100 MHz; some soils are poor conductors even for direct current.

Because reflected electromagnetic wave combines with the direct wave for any antenna system, the height of the antenna above the ground is an important consideration, since this height determines the elevation angles at which the direct and reflected waves will act together, and also the elevation angles at which they will oppose each other. The earth surface as we see it is not the surface from which the electromagnetic waves are reflected. The effective ground plane, in most cases, lies several feet below the actual earth surface. The poorer the earth conductivity, the farther below the surface lies the effective ground plane at a certain electromagnetic frequency.

Irregularity in the terrain near an antenna, whether natural or man-made, affects the level of the effective ground. In areas with numerous buildings, utility wires, fences, and other obstructions, the determination of the effective ground plane is practically impossible from a theoretical standpoint. Antenna installations in such environments usually proceed on a trial-and-error basis, insofar as the optimum height and position are concerned. *See also* DIRECT WAVE, REFLECTED WAVE.

EFFECTIVE RADIATED POWER

When a transmitting antenna has power gain, the effect in the favored direction is equivalent to an increase in the transmitter power, compared to the situation in which an antenna with no gain is used. To obtain a doubling of the effective radiated power in a certain direction, for example, either of two things may be done: The transmitter output power can actually be doubled, or the antenna power gain can be increased in that direction by 3 dB. Effective radiated power takes into account both the antenna gain and the transmitter output power. It is also affected by losses in the transmission line and impedance matching devices.

With an antenna that shows zero gain with respect to a dipole (0 dBd), the effective radiated power or ERP is the same as the power P reaching the antenna feed point. If the antenna has a power gain of A_p dB, then:

$$A_p = 10 \log_{10}(\mathrm{ERP}/P)$$

and the ERP is thus given by:

$$\mathrm{ERP} = P \times 10^{(A_p/10)}$$

The greater the forward gain of the antenna, the greater the effective radiated power.

Effective radiated power is usually determined by using a half-wave dipole as the reference antenna. However, occasionally the effective radiated power may be determined with respect to an isotropic radiator (*see* DIPOLE ANTENNA, ISOTROPIC ANTENNA).

Effective radiated power is an important criterion in the determination of the effectiveness of a transmitting station. The ERP may be many times the transmitter output power. The ERP specification is more often used at very-high frequencies and above, than at high frequencies and below. *See also* ANTENNA POWER GAIN, dBd,dBi, DECIBEL.

EFFECTIVE VALUE

A component, pulse, or other electrical quantity may have, in practice, a value that appears different from the actual value. This actual value is called the effective value of the quantity.

An example of the difference between an actual value and an effective value can be found when a resistor is used at different alternating-current frequencies. This is especially true of wirewound resistors. The actual value, at direct-current and very low alternating-current frequencies, might be 10 ohms. As the frequency is raised, however, a point is reached at which the inductance of the windings introduces an additional resistance in the circuit. At a frequency of, say, 10 MHz, the effective resistance might be 20 ohms, consisting of 10 ohms pure resistance and 10 ohms of inductive reactance.

In alternating-current theory, the root-mean-square value of a waveform is sometimes called the effective value. *See also* ROOT MEAN SQUARE.

EFFICIENCY

Efficiency is the ratio, usually expressed as a percentage, between the output power from a device and the input power to that device. All kinds of electronic apparatus, such as speakers, amplifiers, transformers, antennas, and light bulbs, can be evaluated in terms of efficiency. Generally, only the output in the desired form is taken as the actual output power; dissipated power in unwanted forms is considered to have been wasted. The efficiency of any device is less than 100 percent; while some circuits or devices approach 100-percent efficiency, this ideal is never actually achieved.

Given that the input power to a device is P_{IN} watts and the useful output power is P_{OUT} watts, then the efficiency in percent is given by:

$$\text{Efficiency (percent)} = 100P_{OUT}/P_{IN}$$

Some devices, such as air-core transformers, are very efficient. Others, such as speakers and incandescent lamps, are very inefficient, and deliver only a small proportion of their input power in the desired output form. Amplifiers are often evaluated in terms of circuit efficiency; the input power is the product of the collector, drain, or plate current and voltage, and the output power is the actual wattage of audio-frequency or radio-frequency signal available. *See also* CIRCUIT EFFICIENCY.

EFFICIENCY MODULATION

Amplitude modulation of a carrier signal can be obtained in many different ways (*see* AMPLITUDE MODULATION). One rarely used method is called efficiency modulation. In this modulation scheme, the efficiency of a radio-frequency amplifier is varied by an audio input signal, and this causes changes in the level of the carrier output.

One technique for obtaining efficiency modulation is described as follows. An unmodulated carrier is supplied to the input of a radio-frequency amplifier stage. The output circuit is tuned so that it is slightly off resonance. The applied audio signal, acting via a varactor diode, affects the resonant frequency of the output tank circuit, alternately bringing it closer and farther away with respect to the resonant frequency of operation. As the circuit approaches resonance, the output of the amplifier increases because of improved efficiency; as the circuit deviates from resonance, the output drops because of poorer efficiency. The resulting amplitude variations are in step with the audio signal, and amplitude modulation is the result.

Efficiency modulation is not often used in amplitude-modulated transmitters. More practical, and efficient, methods are available. *See also* MODULATION METHODS.

EIA

See ELECTRONIC INDUSTRIES ASSOCIATION.

EIGHT-TRACK RECORDING

Eight-track recording is a common method of recording stereo audio signals on a magnetic tape. The common "cassette album" uses eight-track tape, and it is only about one-quarter of an inch wide (see photograph). In this kind of cassette, an endless loop of tape, which may have a single-revolution playing time of from 8 to 30 minutes, contains four separate full-stereo recordings. Each recording contains two separate magnetic tracks on the tape, one for the left-hand channel and one for the right-hand channel.

EIGHT-TRACK RECORDING: An eight-track cassette, used in the recording and reproduction of stereo music, contains a tape only about ¼ inch in width, but has eight parallel magnetic recordings for a revolution time of from 8 to 30 minutes. This photograph shows the tape end of such a cassette.

At the end of each revolution of the tape, an electromechanical device causes the tape player to switch to the next pair of tracks. Thus the total playing time of the tape is four times the playing time of a single revolution.

Excellent channel separation is possible with modern eight-track tapes. The tapes can be stored almost indefinitely, provided the temperature and humidity are kept within reasonable limits. *See also* MAGNETIC RECORDING, MAGNETIC TAPE.

EINSTEIN, ALBERT

Almost everyone knows the name of Albert Einstein (1879-1955) because of his theory of relativity. Einstein made his famous discoveries mostly by pure mathematics and imagination while working in the Swiss patent office during the early years of the twentieth century.

Einstein believed that all natural forces such as gravitation, electricity, magnetism, and nuclear-particle attraction are the result of a single characteristic of the universe. This theory, as yet not conclusively proven, is known as the unified-field theory.

EINSTEIN EQUATION

The Einstein equation, $E = mc^2$, is the famous equation of relativity theory that exemplifies the duality of matter and energy. According to Albert Einstein, who developed the theory of relativity in the early part of the twentieth century, matter and energy are simply different manifestations of the same thing. This is true no matter what kind of elements or energy forms are involved.

According to the Einstein equation, a very small amount of mass is equivalent to a large amount of energy. For example, suppose we are given a mass of 1 kilogram, or about 2.2 pounds; imagine we are somehow able to directly transform this mass into energy with 100-percent efficiency. Then the energy, in joules (*see* JOULE), would be:

$$E = mc^2 = 1 \times (3 \times 10^8)^2 = 9 \times 10^{16} \text{ joules}$$

or 90 quadrillion joules! A joule is a watt second, and so the above amount of energy is equivalent to 2.5×10^{10} kilowatt hours. With this amount of energy, a city of 3 million people consuming 1 gigawatt (1 billion watts) of power on the average, could be fully supplied for nearly 3 years. And all from a piece of matter having only 1 kilogram of mass!

Nuclear reactions convert matter into energy, but the amounts are tiny. In nature, the sun and all stars convert matter into energy by the fusion of hydrogen into helium. But a direct matter-to-energy conversion, illustrated in the bizarre example above, has not yet been found. *See also* ENERGY, POWER, RELATIVITY THEORY.

EKG

See ELECTROCARDIOGRAM.

ELASTANCE

Elastance is a measure of the opposition a capacitor offers to being charged or discharged. The unit of elastance is called the daraf, which is farad spelled backwards. Elastance is the reciprocal of capacitance. That is, if C is the capacitance in farads, then:

$$\text{Elastance (darafs)} = 1/C$$

A value of capacitance of 1 picofarad is equivalent to an elastance of 1 teradaraf, or 10^{12} darafs. A capacitance of 1 nanofarad is an elastance of 1 gigadaraf (10^9 darafs); a value of capacitance of 1 microfarad is 1 megadaraf.

The daraf is an extremely small unit of elastance, seldom observed in actual capacitors. The elastance of 1 daraf is the value at which 1 volt of applied electric charge will produce 1 coulomb of displacement (*see* COULOMB). In most actual capacitors, 1 volt of charge produces no more than a few thousandths of a coulomb of charge displacement. *See also* DARAF.

ELASTIC WAVE

An elastic wave is a mechanically transferred wave disturbance. Such a wave requires a medium that can be compressed or physically moved in some way, such as air, water, or a taut rope. The most familiar example of an elastic wave is the longitudinal wave of sound through the air.

Elastic waves can be propagated in either of two substantially different ways. The longitudinal wave, illustrated at A in the drawing, travels because of alternate compression and expansion of the medium along the axis of propagation. Sound in the atmosphere is, as has been mentioned, a longitudinal disturbance. The transverse

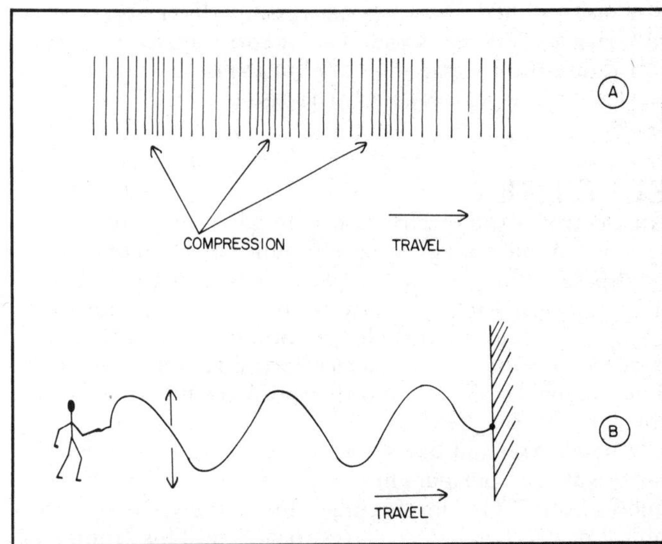

ELASTIC WAVE: Elastic waves may be either longitudinal, as shown at A, or transverse, as shown at B.

wave is another mode of elastic propagation, and can be illustrated by shaking a rope or string that is pulled tight, as at B. The transverse wave disturbance occurs in directions perpendicular to the direction of wave travel.

Until the early part of the twentieth century, scientists believed that all wave disturbances are elastic. It was thought that this is true of electromagnetic waves, and that the whole universe is permeated with a tenuous, massless, elastic substance, that carries light and other forms of radiant wave energy. This theory was eventually discarded by Albert Einstein, who revolutionized all of physics. *See also* ETHER THEORY.

E LAYER

The E layer of the atmosphere is an ionized region that lies at an altitude of about 60 to 70 miles, or 100 to 115 kilometers, above sea level. The E layer is frequently responsible for medium-range communication on the low-frequency through very-high-frequency radio bands. The ionized atoms of the atmosphere at that altitude affect electromagnetic waves at wavelengths as short as about 1 meter.

The ionization density of the E layer is affected greatly by the height of the sun above the visual horizon. Ultraviolet radiation and X rays from the sun are primarily responsible for the existence of the E layer. The maximum ionization density generally occurs around midday, or slightly after. The minimum ionization density takes place during the early morning hours, after the longest period of darkness. The E layer almost disappears shortly after sundown under normal conditions.

Occasionally, a solar flare will cause a geomagnetic disturbance, and the E layer will ionize at night in small areas or "clouds." The resulting propagation is called sporadic-E propagation, and is well known among communications people who use the upper portion of the high-frequency band and the lower portion of the very-high-frequency band. Often, the range of communication is sporadic-E mode can extend for over 1,000 miles. However, the range is not as great as with conventional F-layer propagation (*see* SPORADIC-E PROPAGATION).

At frequencies above about 150 MHz, or a wavelength

of 2 meters, the E layer becomes essentially transparent to radio waves. Then, instead of being returned to earth, they continue on into space. *See also* D LAYER, F LAYER, IONOSPHERE, PROPAGATION CHARACTERISTICS.

ELECTRET

An electret is the electrostatic equivalent of a permanent magnet. While a magnet has permanently aligned magnetic dipoles, the electret has permanently aligned electric dipoles (*see* DIPOLE). An electret is generally made of an insulating material, or dielectric. Some varieties of waxes, ceramic substances, and plastics are used to make electrets. The material is heated and then cooled while in a strong, constant electric field.

The electret can be used in a microphone, in much the same way as a permanent magnet is used in the dynamic microphone. The impinging sound waves cause a diaphragm, attached to an electret, to vibrate. This produces a fluctuating electric field, which produces an audio-frequency voltage at the output terminals. *See also* DYNAMIC MICROPHONE.

ELECTRICAL ANGLE

A sine-wave alternating current or voltage can be represented by a counterclockwise rotating vector in a Cartesian coordinate plane. The origin of the vector is at the point (0,0), or the center of the plane, and the end of the vector is at a point having a constant distance from the origin.

In such a representation, the amplitude of the current or voltage is indicated by the length of the vector. The greater the amplitude, the longer the vector, and the farther its end point is from the point (0,0). The frequency of the sine wave is represented by the rotational speed of the vector; one cycle is given by a full-circle, or 360-degree, rotation. This model is illustrated in the drawing. The y axis, or ordinate, represents the amplitude scale, and the instantaneous amplitude of the wave is therefore given by the y value of the end point of the vector (*see* CARTESIAN COORDINATES, VECTOR).

The electrical angle θ is the angle, in degrees or radians, that the vector subtends relative to the positive x axis at any given instant. The wave cycle begins at 0 degrees, with the vector pointing along the x axis in the positive direction. The amplitude, or y component, at this time is zero. At an electrical angle of 90 degrees, the vector points along the y axis in a positive direction. This is the maximum positive point of the wave cycle. At 180 degrees, the vector points along the x axis in a negative direction, and the amplitude is zero once again. At 270 degrees, the vector points along the y axis in a negative direction, and this is the maximum negative amplitude of the wave.

While the y component of the electrical-angle vector gives the instantaneous amplitude, the x component gives the instantaneous rate of change of the signal amplitude. The speed of vector rotation, in radians per second, is called the angular frequency of the wave, and is equal to 2π (about 6.28) times the frequency in Hertz. *See also* PHASE ANGLE, SINE WAVE.

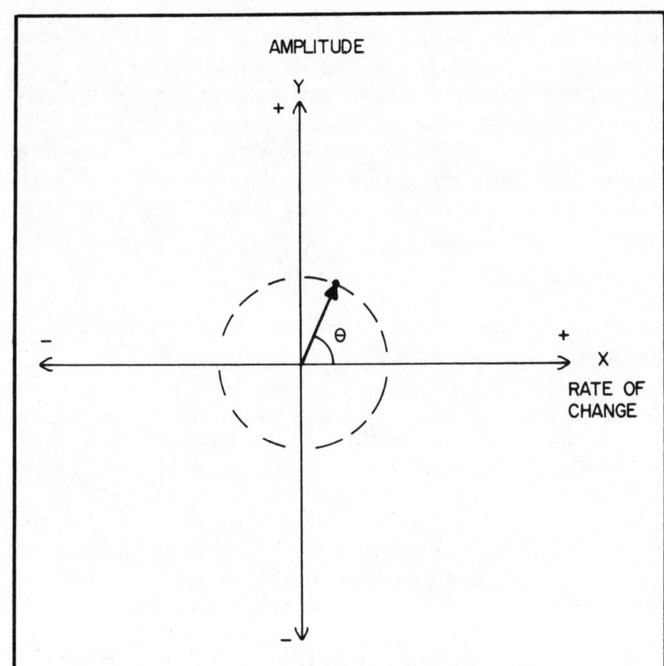

ELECTRICAL ANGLE: The electrical angle of a sine wave can be shown by means of a vector in the Cartesian coordinate system. The y axis shows the instantaneous amplitude, and the x axis shows the rate of change. The strength of the disturbance is shown by the length of the vector, and the angle θ is the electrical angle at a given instant.

ELECTRICAL BANDSPREAD

In a radio receiver, electrical bandspread refers to bandspread obtained by adjustment of a separate inductor or capacitor, in parallel with the main-tuning inductor or capacitor. The main-tuning control in a shortwave receiver may cover several megahertz in one revolution. Bandspread is considered electrical if it is derived using a separate component; simply gearing down the main-tuning control constitutes mechanical bandspread.

An electrical-bandspread control can be made to have any desired degree of resolution, consistent with the stability of the frequency of the receiver. Generally, one turn of a bandspread knob covers between 10 and 100 kHz of spectrum in a high-frequency receiver, with 25 kHz probably being about average.

ELECTRICAL CONDUCTION

Electrical conduction is the movement of charge carriers through a material. Substances may be good conductors, seminconductors, or insulators. Even the best conductor offers some resistance, and even the best insulator conducts to a tiny extent. For all practical purposes, electrical conduction is of use only in good conductors and semiconductors.

Conduction normally occurs by the exchange of excess electrons between atoms, or by the exchange of electron deficiencies between atoms. The electron carries a negative charge, and thus the flow of electrons in a circuit is from the negative to the positive. An electron deficiency, called a hole, carries a positive charge, and the flow of holes is therefore from the positive to the negative.

Electrical conduction is measured in terms of current. The unit of current is the ampere; 1 ampere of current is the conduction of 1 coulomb of electrons per second past a given point in a circuit. *See also* AMPERE, COULOMB, CURRENT, ELECTRON, HOLE.

ELECTRICAL DISTANCE

Electrical distance is the distance, measured in wavelengths, required for an electromagnetic wave to travel between two points. The electrical distance depends on the frequency of the electromagnetic wave, and on the velocity factor of the medium in which the wave travels.

In free space, an electrical wavelength at 1 MHz is about 300 meters. In general, in free space:

$$\lambda = 300/f$$

where λ is the length of a wavelength in meters and f is the frequency in megahertz. If v is the velocity factor, given as a fraction rather than a percent, then:

$$\lambda = 300\ v/f$$

Electrical distances are often specified for antenna radiators and transmission-line stubs. Such distances are always given in wavelengths.

ELECTRICAL ENGINEERING

Electrical engineering is the study of the application of physical and mathematical skills to the solving of electrical problems. An electrical engineer is a trained professional in this field. College degrees in electrical engineering include the Associate's degree (ASEE), the Bachelor's degree (BSEE), the Master's degree (MSEE), and the Doctor of Philosophy degree (Ph. D.).

Electrical engineering has become synonymous with electronic engineering in recent years. All electrical engineers are concerned with the design and development of innovative electronic devices. Electrical engineers are becoming more and more valuable. This trend can be expected to continue as the electronic and computer age becomes more established.

Electrical engineering has numerous sub-specialites, and the study of the subject can go on in great detail for many years in any of these specialized areas. Such subspecialties include antenna design, digital design, and radio-frequency design at all parts of the electromagnetic spectrum.

ELECTRICAL INTERLOCK

An electrical interlock is a switch designed for the purpose of removing power from a circuit when the enclosing cabinet is opened. Interlocks are especially important in high-voltage circuits, since they protect service personnel from the possibility of electric shock.

An interlock switch is usually attached to the cabinet door or other opening, in such a way that access cannot be gained without removing the high voltage from the interior circuits.

Interlock switches may fail, and therefore they should not be blindly relied upon to protect the lives of service personnel. Interlock switches should, of course, never be defeated, unless it is absolutely necessary. Even then, this should be done only by technicians familiar with the equipment. *See also* ELECTRIC SHOCK.

ELECTRICAL WAVELENGTH

In an antenna or transmission line carrying radio-frequency signals, the electrical wavelength is the distance between one part of the cycle and the next identical part. The electrical wavelength depends on the velocity factor of the antenna or transmission line, and also on the frequency of the signal.

In free space, where a signal has a frequency f in megahertz, the wavelength λ is given by the equations:

$$\lambda\ (\text{meters}) = 300/f$$
$$\lambda\ (\text{feet}) = 984/f$$

Along a thin conductor, the electrical wavelength at a given frequency is somewhat shorter than in free space. The velocity factor (*see* VELOCITY FACTOR) in such a situation is about 95 percent, or 0.95, therefore:

$$\lambda\ (\text{meters}) = 285/f$$
$$\lambda\ (\text{feet} = 935/f$$

along a wire conductor.

In general, in a conductor or transmission line with a velocity factor v, given as a fraction:

$$\lambda\ (\text{meters}) = 300v/f$$
$$\lambda\ (\text{feet}) = 984v/f$$

The electrical wavelength of a signal in a transmission line is always less than the wavelength in free space. When designing impedance-matching networks or resonant circuits using lengths of transmission line, it is imperative that the line be cut to the proper length in terms of the electrical wavelength, taking the velocity factor into account. *See also* ELECTRICAL DISTANCE.

ELECTRIC CHARGE

See CHARGE.

ELECTRIC CONSTANT

The electrical permittivity of free space is sometimes called the electric constant, represented by ϵ_0. This value is about 8.854 picofarads per meter. Essentially all dielectric materials have an electrical permittivity greater than that of a vacuum.

The electric constant is the basis for the determination of dielectric constants for atl insulating materials. *See also* DIELECTRIC CONSTANT, PERMITTIVITY.

ELECTRIC COUPLING

Electric coupling exists between, or among, any objects that show mutual capacitance. Electric coupling may be desirable, or it may be unwanted.

When two objects are electrically coupled, the charged particles on both objects exert a mutual attractive force or a mutual repulsive force. Like charges repel, and opposite charges attract. The plates of a capacitor provide a good illustration of the effects of electric coupling. A negative charge on one plate produces a positive charge on the other plate, by repelling the electrons in the other plate and literally pushing them from it. A positive charge on one plate produces a negative charge on the other, by attracting extra electrons to that plate. *See also* CHARGE, MAGNETIC COUPLING.

ELECTRIC EYE

An electric eye is a device that optically senses the presence of an obstruction, and operates an electric circuit or mechanical device in response to the obstruction. In particular, an electric eye is designed to sense the presence of a person. Such a device usually has a light-beam generator and sensor, along with switching circuits. The illustration is a simplified diagram of an electric eye.

Normally, the beam from the light-generating device is picked up by the sensor in a continuous and unchanging manner. But, when something passes between the light generator and the sensor, even for a very brief moment, the light beam is cut off, and this produces a pulse in the output of the sensor. The pulse actuates a flip-flop or other circuit, and this change in current can be employed to open or close a door, set off an alarm, or perform some other function.

Electric eyes are extensively used in burglar-alarm systems. They are also commonly employed in automatic doors. Some elevator systems employ these devices to keep the door open for as long as people continue to enter or leave the elevator.

Electric-eye devices do not necessarily have to use visible light. Some such devices use infrared or ultraviolet light, so the beams cannot be seen with unaided eyes.

ELECTRIC FIELD

Any electrically charged body sets up an electric field in its vicinity. The electric field produces demonstrable effects on other charged objects. The electric field around a charged object is represented, by the physicist and the engineer, as lines of flux. A single charged object produces radial lines of flux as at A in the illustration. The direction of the field is considered to be from the positive pole outward, or inward toward the negative pole.

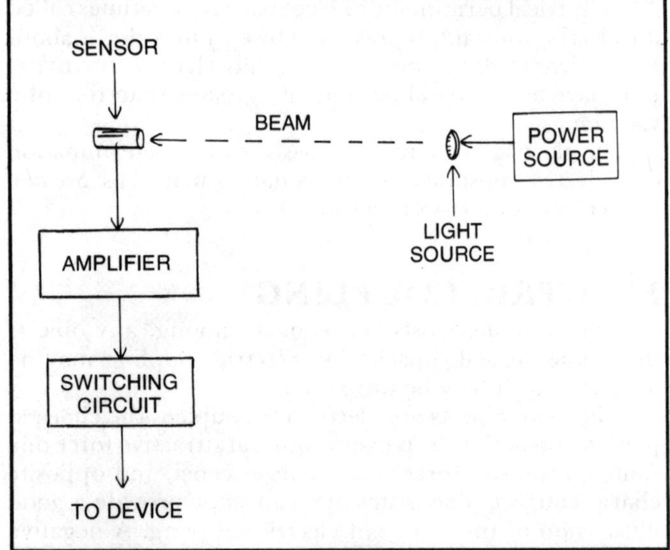

ELECTRIC EYE: A block diagram of an electric-eye system, showing the peripheral components.

When two charged objects are brought near each other, their electric fields interact. If the charges are both positive or both negative, a repulsion occurs between their electric fields, as shown at B. If the charges are opposite, attraction takes place, as at C.

The intensity of an electric field in space is measured in volts per meter. Two opposite charges with a potential difference of 1 volt, separated by 1 meter, produce a field of 1 volt per meter. The greater the electric charge on an object, the greater the intensity of the electric field surrounding the object, and the greater the force that is exerted on other charged objects in the vicinity. *See also* CHARGE, DIPOLE, ELECTRIC COUPLING, ELECTRIC FLUX.

ELECTRIC FLUX

Electric flux is the presence of electric lines of flux in space or in a dielectric substance. The greater the intensity of the electric field, the more lines of flux present in a given space, and therefore the greater the electric flux density.

Of course, the "lines" of an electric field are not actual lines that can be seen. They are simply imaginary, with each "line" representing a certain amount of electric charge. Electric flux density is measured in coulombs per square meter.

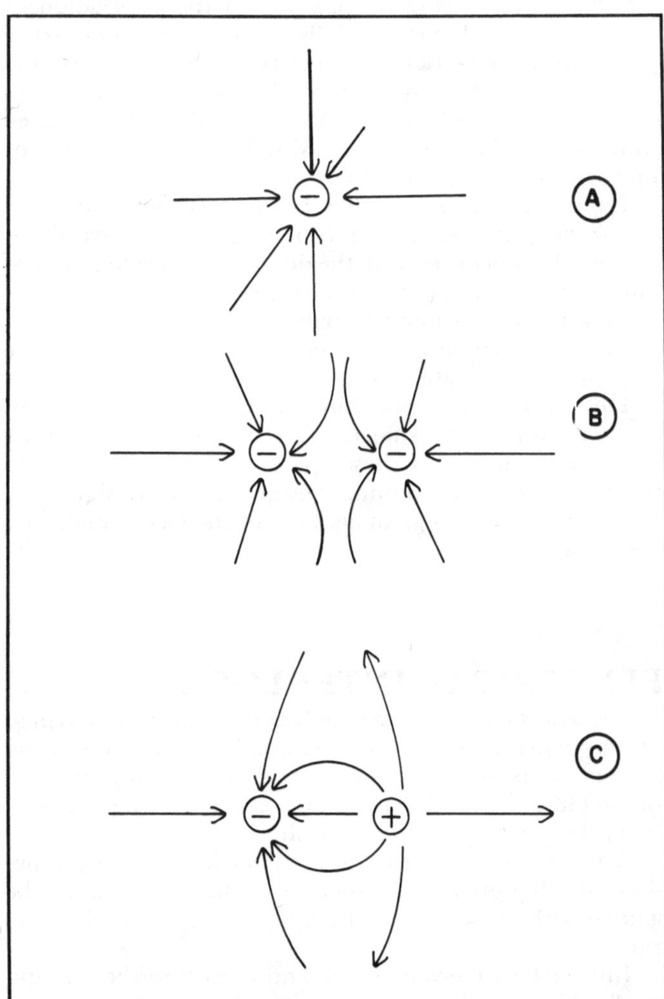

ELECTRIC FLUX: At A, a single charge produces radial lines of electric flux. At B, two like charges adjacent to each other produce a repulsive force, and the lines of flux are as shown. At C, two opposite electric charges near each other produce an attractive force, with flux lines as shown.

The electric flux density falls off, or decreases, with increasing distance from a charged object, according to the law of inverse squares (*see* INVERSE-SQUARE LAW). If the distance from a charged object is doubled, the electric flux density is reduced to one-quarter its previous value. This can be envisioned by imagining spheres centered on the charged particle shown at A in the previous illustration. No matter what the radius of the sphere, all of the electric flux around the charged particle must pass through the surface of the sphere. As the radius of the sphere is increased, the surface area of the sphere grows according to the square of the radius. Therefore, a region of the sphere with a certain area, such as 1 square meter, becomes smaller in proportion to the square of the radius of the sphere; fewer electric lines of flux will cut through the region. *See also* CHARGE, ELECTRIC FIELD.

ELECTRICITY

Electricity is a branch of science, and in particular, of physics. The study of electricity is involved with the use and behavior of electric current, power, and voltage. This applies especially to power distribution and utilization.

Any electric current or voltage may be called electricity. For example, the 120-volt electrical wires in any home or building contain electricity. Electricity can be manifested as a potential difference between two objects, or as the flow of electrons (or other charged particles) past a certain point.

ELECTRIC SHOCK

Electric shock is the harmful flow of electric current through living tissue. The effects of such current can be very damaging, and may even cause death. All living cells contain some electrolyte matter, and therefore they are capable of carrying electric current. Harmful currents are produced by varying amounts of voltage, depending on the resistance of the circuit through the living tissue.

In the human body, the most susceptible organ, as far as life-threatening effects are concerned, is the heart. A current of 100 to 300 mA for a short time can cause the heart muscle to stop its rhythmic beating and begin to twitch in an uncoordinated way. This is called heart fibrillation. An electric shock is much more dangerous when the heart is part of the circuit, as compared to when the heart is not in the circuit.

Under certain conditions, even a low voltage can produce a lethal electric shock. When working on equpment carrying more than perhaps 12 volts, therefore, strict precautions must be taken to minimize the chances of electric shock. The 120-volt utility lines produce an extreme hazard, since this voltage is more than sufficient to cause death.

If electric shock occurs, the power source should be cut off immediately. No attempt should be made to rescue a person who is in contact with live wires. Cardiopulmonary resuscitation (CPR) or mouth-to-mouth resuscitation may be necessary for a victim of electric shock. A medical professional should be called immediately. *See also* CARDIOPULMONARY RESUSCITATION, HEART FIBRILLATION, MOUTH-TO-MOUTH RESUSCITATION.

ELECTRIC WIRING

Electric wiring is the arrangement of the wire conductors in an electrical or electronic system. Wiring can be categorized in two ways: power-distribution wiring and electronic-circuit wiring.

All wiring should be capable of handling the current in the circuit, without generating excessive heat. This necessitates that the wire size be sufficiently large. Wires carrying high voltages should be so marked, so that technical personnel are aware of the potential risk, and they should have sufficient insulation to prevent arcing to other conductors or parts of the circuit.

Sometimes it is necessary to shield a wire to prevent the radiation or pickup of electromagnetic fields. Coaxial cable is generally used in such situations (*see* COAXIAL CABLE). In an electronic circuit, wire leads should be color-coded or otherwise labeled, for easy identification and tracing by service personnel.

In modern printed circuits, the intercomponent wiring consists largely of foil runs on the circuit boards themselves. This eliminates much of the "loose" wiring in a circuit, and greatly simplifies the design and construction of electronic equipment. Wiring between modules or circuit boards often consists of multiconductor cables, with plugs installed at either end. *See also* PRINTED CIRCUIT.

ELECTRIFICATION

Any object that carries an electric charge is said to be electrified, and the process of putting a charge on a body is called electrification. A familiar example of electrification takes place in any dry climate when hard-soled shoes are worn on a carpeted floor! The static charge that results can cause an arc of, in some cases, up to several millimeters.

When an object becomes electrified, it carries either a surplus of electrons (a negative charge) or a deficiency of electrons (a positive charge). When two objects with opposite charge come into contact, electrons flow from the negatively charged body to the positively charged body, until the potential difference between the objects is zero.

The term electrification is sometimes used to mean the availability of electric service in a given area, or the presence of voltage in power lines. Thus a community or wire may be considered "electrified." *See also* CHARGE.

ELECTROCARDIOGRAM

An electrocardiogram is a recording of the electrical impulses produced by the human heart. The machine used to record these impulses is called an electrocardiograph. The electrocardiogram is a medical diagnostic aid.

To record an electrocardiogram, a set of electrodes is attached to the skin on the chest of the patient. These electrodes pick up the weak electrical currents that are generated by the body to control the heart muscle. After amplification, the impulses are fed to a pen recorder. They may also be viewed on an oscilloscope screen.

Healthy people have a characteristic electrocardiogram. In various disorders of the heart, different abnormalities of the electrocardigram are found. This helps doctors determine the nature of the disease. *See also* MEDICAL ELECTRONICS.

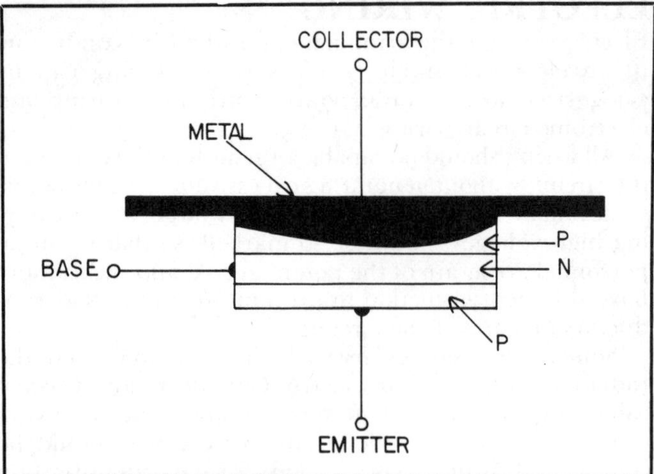

ELECTROCHEMICAL DIFFUSED-COLLECTOR TRANSISTOR: An electrochemical diffused-collector transistor has a metal portion that is diffused into the collector material. This metal serves as a heat sink and is connected to the case of the device directly.

ELECTROCAUTERY

An electrocautery is a device used in medicine for the purpose of cauterizing, or controlled burning, of tissue. An application of cauterization might consist of the removal of a small skin lesion. The electrocautery may also be called an electric needle.

A small wire, made of platinum or other metal with a high melting temperature and a high resistance to corrosion, can be connected to a source of high-frequency electric current. At radio frequencies, electric currents cause heating of living cells, and this can be used as an electrocautery device. An insulated probe protects the hands of the operator. *See also* MEDICAL ELECTRONICS.

ELECTROCHEMICAL DIFFUSED-COLLECTOR TRANSISTOR

All transistors generate heat in operation, especially at the base-collector junction. In transistors designed to act as power amplifiers at audio or radio frequencies, heatsinks are often employed to prevent the base-collector junction from heating to destructive temperatures. The electrochemical diffused-collector transistor is a device designed especially for the effective removal of heat from the base-collector region.

The internal structure of the electrochemical diffused-collector transistor is illustrated in the drawing. It is a PNP type transistor. Much of the P-type semiconductor material has been etched away by metal in an electrochemical diffusion process. The piece of metal is actually the case, or enclosure, for the transistor, and it may be bolted to a heatsink through a thin sheet of mica or other electrically insulating, heat-conducting material. *See also* DIFFUSION, HEATSINK, TRANSISTOR.

ELECTROCHEMICAL EQUIVALENT

The electrochemical equivalent is a constant that expresses the rate at which a metal is deposited in the electroplating process (*see* ELECTROPLATING). The constant can be given in a variety of different ways. It is often specified in coulombs required to deposit one gram of the metal onto the electrode. The reciprocal of this quantity may also be specified.

The value of the electrochemical equivalent varies for different metals. The table lists the electrochemical equivalents in coulombs per gram for various metals. Electrochemical equivalents can be given for nonmetallic elements as well, although in electronics the value is of concern only for the metals used in electroplating. The larger the value of the electrochemical equivalent given in the table, the greater the amount of total electric charge that is necessary to cause a given mass of the metal to be deposited.

ELECTROCHEMICAL EQUIVALENT: ELECTROCHEMICAL EQUIVALENTS, IN COULOMBS PER GRAM, FOR SOME COMMON METALS.

Element	Electrochemical Equivalent
Aluminum	1.1×10^4
Chromium	5.6×10^3
Copper	3.0×10^3
Gold	1.5×10^3
Iron	5.2×10^3
Lead	1.9×10^3
Magnesium	7.9×10^3
Nickel	3.3×10^3
Platinum	2.0×10^3
Silver	8.9×10^2
Tin	3.3×10^3
Zinc	3.0×10^3

ELECTROCHEMICAL TRANSDUCER

Any device that converts electrical energy into chemical energy, or vice versa, is called an electrochemical transducer. Such devices are extremely common in modern electronics. The most familiar examples of electrochemical transducers are the dry cell and the storage battery. In such devices, chemical energy is transformed into electric current. In rechargeable cells and batteries, such as the storage battery, electric current is also transformed into chemical energy during the process of charging.

Cells and batteries are not the only electrochemical transducers. A simple device that measures the earth conductivity is an electrochemical transducer. The same sort of device can also be used to measure the electrolyte concentration in a solution. *See also* ELECTROCHEMISTRY.

ELECTROCHEMISTRY

Electrochemistry is a branch of chemistry relating to electrical phenomena. All elements, compounds, and mixtures have certain definable chemical properties, as well as electrical characteristics, such as charge, conductivity, and valence number. Some elements are useful as dielectrics, some as conducting media, and some as semiconductors. Electrochemistry enables researchers to more easily find the best substances for these different electrical purposes.

A striking example of the importance of electrochemistry in modern electronics has been its role in solid-state technology. Electrochemical processes are extremely useful in the fabrication of certain kinds of transistors and integrated circuits. Electrochemistry has also given consumers the dry cell and the storage battery, on which we all rely. *See also* ALKALINE CELL, STORAGE BATTERY, ZINC CELL.

ELECTROCHRONOMETER

See ELECTRONIC CLOCK.

ELECTROCUTION

See ELECTRIC SHOCK.

ELECTRODE

An electrode is any object to which a voltage is deliberately applied. Electrodes are used in such things as vacuum tubes, electrolysis devices, cells, and other applications. An electrode is generally made of a conducting material, such as metal or carbon.

In a device that has both a positive and a negative electrode, the positive electrode is called the anode, and the negative electrode is called the cathode. The cathode in a vacuum tube emits electrons, and the anode, sometimes called the plate, collects them.

All electrodes display properties of capacitance, conductance, and impedance. These characteristics, in individual cases, depend on many things, such as the size and shape of the electrodes, their mutual separation, and the medium between them. *See also* ELECTRODE CAPACITANCE, ELECTRODE CONDUCTANCE, ELECTRODE IMPEDANCE, TUBE.

ELECTRODE CAPACITANCE

Electrode capacitance is the effective capacitance that an electrode shows relative to the ground, the surrounding environment, or other electrodes in a device. All electrodes have a certain amount of capacitance, which depends on the surface area of the electrode, the distances and surface areas of nearby objects, and the medium separating the electrode from the nearby objects.

In a vacuum tube, the interelectrode capacitance, or capacitance among the cathode, grids, and plate, is an important parameter in the design of circuits at very high frequencies and above. The interelectrode capacitance in a tube can, at such frequencies, affect the resonant frequency of associated tank circuits. The interelectrode capacitance also places an upper limit on the frequency at which the tube can be effectively used to produce gain. Generally, the interelectrode capacitance in a vacuum tube is on the order of just a few picofarads, or trillionths of a farad. *See also* ELECTRODE, TUBE.

ELECTRODE CONDUCTANCE

The conductance of an electrode is the ease with which it emits or picks up electrons. Electrode conductance is measured in units called mhos or Siemens, which are the standard units of conductance, and are equivalent to the reciprocal of the ohmic resistance.

The conductance of an electrode or group of electrodes depends on many factors, including the physical size of the electrodes, the material from which they are made, the coating (if any) on electron-emitting electrodes, and the medium among the electrodes. In a vacuum tube, the interelectrode conductance is the ease with which electrons flow from the cathode to the plate. This depends, of course, on the grid bias, as well as on the characteristics of the tube itself. *See also* ELECTRODE, TUBE.

ELECTRODE DARK CURRENT

In a photomultiplier or television camera tube, the electrode dark current is the level of current that flows in the absence of light input to the tube. Dark current is important as a measure of the noise generated by a television camera tube; the smaller the dark current, the better.

The electrode dark current in most photomultiplier tubes is very small. The anode dark current originates partially from the photocathode, and partly from the other components within the tube. Some of the dark current also originates because of thermal disturbances. The electrode dark current results in what is called dark noise, as the electrons strike the anode. Dark noise limits the sensitivity of a television picture camera tube. *See also* PHOTOMULTIPLIER.

ELECTRODE IMPEDANCE

Electrode impedance is the resistance encountered by a current as it flows through an electrode. In the case of direct current, electrode impedance is simply the reciprocal of the electrode conductance (*see* ELECTRODE CONDUCTANCE). It is measured, therefore, in ohms.

For alternating current, the electrode impedance consists of both resistive and reactive components. The electrode impedance is determined in the same manner as ordinary impedance, by adding, vectorially, the resistance and the reactance (*see* IMPEDANCE, REACTANCE, RESISTANCE).

Electrode impedance in a particular situation depends on several factors, including the surface area of the electrode, the material from which it is made, the surrounding medium, and, in a vacuum tube, the bias and the grid voltage. The frequency of the applied alternating-current signal also affects the electrode impedance in some cases, since the capacitive and inductive reactances change as the frequency changes. *See also* ELECTRODE CAPACITANCE, TUBE.

ELECTRODYNAMICS

Electrodynamics is a branch of physics and electricity, concerned with the interaction between electrical and mechanical effects. The attraction of opposite charges and the repulsion of like charges are examples of electrodynamic phenomena. The forces and the electric charges are directly related. The forces of magnetism are also electrodynamic forces, since a magnetic field can be generated by a current in a wire.

Electric motors and generators are examples of electrodynamic devices. They operate via magnetic forces, generated by electric currents or resulting in electric cur-

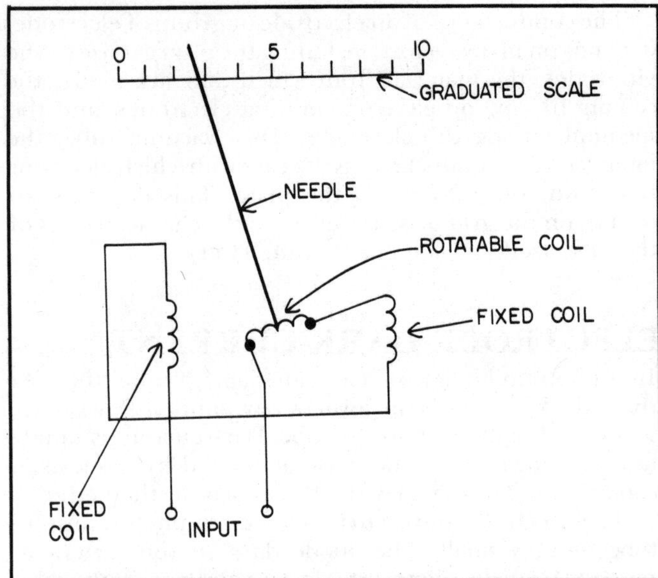

ELECTRODYNAMOMETER: An electrodynamometer operates by means of interaction between the magnetic fields produced by fixed and moving coils.

rents. Such mechanical devices as indicating meters are also electromechanical instruments. *See also* D'ARSONVAL, ELECTRODYNAMOMETER, GENERATOR, MOTOR.

ELECTRODYNAMOMETER

An electrodynamometer, also sometimes called a dynamometer, is a form of analog metering device. The principle of operation is similar to that of the D'Arsonval movement, but the magnetic field is supplied by an electromagnet, rather than by a permanent magnet. The current being measured supplies the magnetic-field current for the electromagnet.

The drawing illustrates, schematically, the operation of the electrodynamometer. Two stationary coils are connected in series with a moving coil. The moving coil is attached to the indicating needle, and is mounted on bearings. The moving coil is held at the meter zero position, under conditions of no current, by a set of springs.

When a current is applied to the coils, magnetic fields are produced around both the stationary coils and the movable coil. The resulting magnetic forces cause the movable coil to rotate on its bearings. The larger the current, the more the coil is allowed to rotate by the springs.

Electrodynamometers can be used in any situation where a D'Arsonval meter is used, although the D'Arsonval meter is generally more sensitive when it is necessary to measure very small currents. *See also* D'ARSONVAL.

ELECTROENCEPHALOGRAPH

An electroencephalograph is a device that records brain waves. Brain impulses are electrical in nature, and they can be picked up by electrodes attached at various points on the scalp. An amplifier and a pen recorder provide a permanent record of the brain waves. The waves may also be viewed on a cathode-ray oscilloscope, and photographed.

The electroencephalograph is an important device in medical technology, since it aids in the diagnosis of nervous-system abnormalities. Brain waves, as seen on the electroencephalogram, show characteristic patterns during sleep, alertness, anger, and other nervous states. *See also* MEDICAL ELECTRONICS.

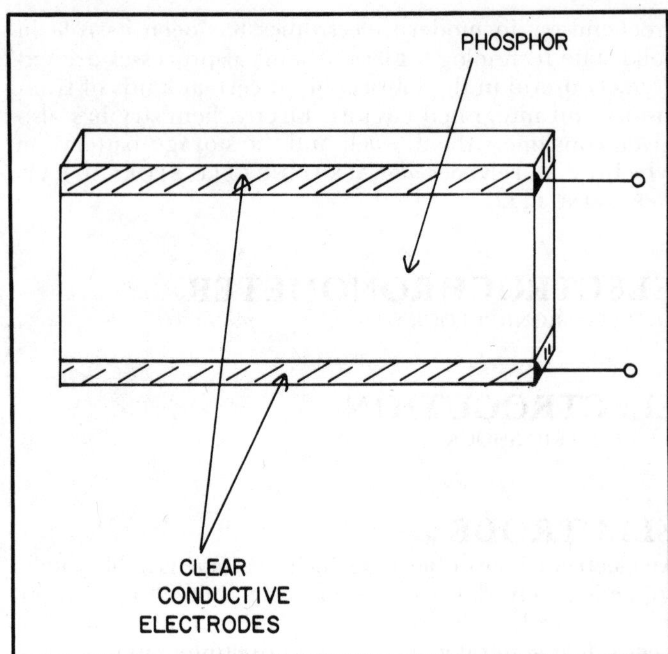

ELECTROLUMINESCENT CELL: Cross-sectional view of an electroluminescent cell.

ELECTROKINETICS

Electrokinetics is a branch of physics and electricity concerned with the behavior of moving charged particles, and the behavior of moving substances in electrical contact. Examples of electrokinetic effects are the electrolysis of salt water, electrophoresis, and the electrical operation of a dry cell or storage battery. *See also* ELECTROLYSIS, ELECTROLYTIC CAPACITOR, ELECTROPHORESIS.

ELECTROLUMINESCENT CELL

An electroluminescent cell is a device that emits visible light when an alternating current is passed through it. Generally, an electroluminescent cell consists of two transparent conducting electrodes, with a layer of phosphor in between, as shown in the simple diagram. When a voltage is applied between the two transparent electrodes, a current passes through the phosphor, and this causes the phosphor to emit visible light. Since the electrodes are clear, the light can easily be seen through them.

Small electroluminescent cells are sometimes found in stores, and they plug directly into the 120-volt utility outlet to act as night lights. Their soft glow is somewhat intriguing.

A gas-filled bulb is a form of electroluminescent cell. The glowing material may be neon, mercury vapor, sodium vapor, or some other gas. A light-emitting diode is also a form of electroluminescent cell, and it operates by means of semiconductor reactions. *See also* LIGHT-EMITTING DIODE, MERCURY-VAPOR LAMP, NEON LAMP, PHOSPHOR, SODIUM-VAPOR LAMP.

ELECTROLYTIC CAPACITOR: A typical electrolytic capacitor for low-voltage applications.

ELECTROLYSIS

Electrolysis is a process in which a compound is separated into its constituent elements, through the application of an electric current. Electrolysis can occur in a solid, a liquid, or a gas. Most commonly, we think of electrolysis in liquids or semi-solid materials.

All compounds consist of elements, some of which have positive valence numbers, and some of which have negative valence numbers (*see* VALENCE NUMBER). Atoms with positive valence numbers are attracted to negatively charged electrodes, and atoms with negative valence numbers are attracted to positively charged electrodes. An example of this effect can be seen on any automobile battery terminals, assuming the battery has been in use for at least several months. The positive electrode of the battery attracts oxygen, which has a valence number of −2. Since oxygen is a highly reactive element, the terminal metal often becomes noticeably corroded because of reactions with the oxygen.

Electrolysis can be used for a variety of different purposes. Electroplating is an example of the action of electrolysis (*see* ELECTROPLATING). A simple home experiment can be performed to illustrate electrolysis in a liquid. Place two electrodes in a salt-water solution, and apply a potential difference of 12 volts (obtained with dry cells) between the electrodes. Use a clear drinking glass to hold the solution. (Do not, under any circumstances, immerse your fingers in the solution while the voltage is applied!) Bubbles will appear at the electrodes. Hydrogen gas will be formed at the negative electrode, and oxygen will be at the positive electrode. These are from the water. Although you will not be able to see it, sodium from the salt will accumulate at the negative electrode. It reacts with the water immediately to form hydrogen gas. Chlorine gas will be formed, to a limited extent, at the positive electrode, along with the oxygen from the water.

Electrolysis results not only in the deposition of elements because of an applied electric current; electrolysis can also result in the generation of electricity when the appropriate electrodes are placed into an electrolyte. *See also* ALKALINE CELL, ELECTROLYTE, LEAD-ACID BATTERY, STORAGE BATTERY, ZINC CELL.

ELECTROLYTE

An electrolyte is any substance which, in a solution carrying an electric current, will separate into its constituent elements. Common table salt (sodium chloride) is a good example of an electrolyte (*see* ELECTROLYSIS). So is common baking soda (sodium bicarbonate). Most acids and bases are electrolytes. Electrolyte solutions always conduct electricity.

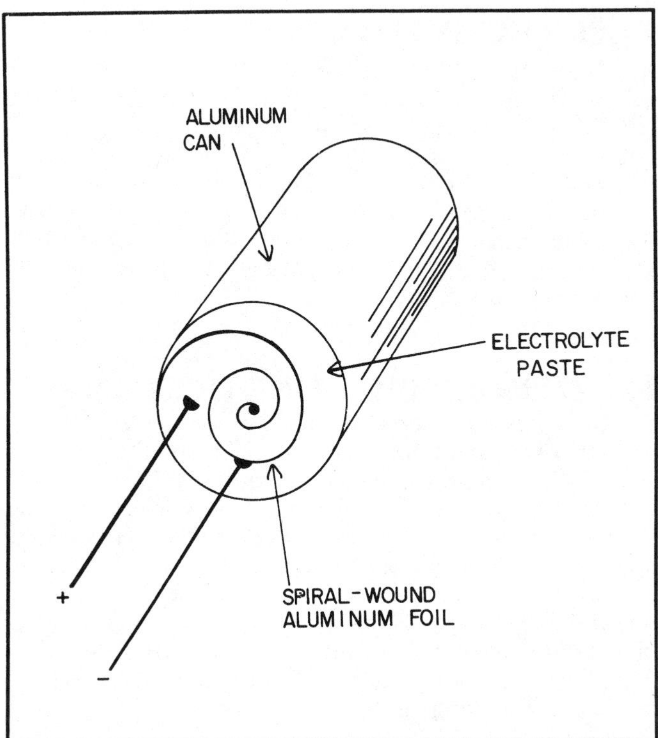

ELECTROLYTIC CAPACITOR: Cross-sectional view of an electrolytic capacitor. The dielectric consists of a thin layer of aluminum oxide on the surface of the aluminum foil.

Electrolyte substances are important in the manufacture of electric cells and batteries. The action of an electrolyte solution on a pair of electrodes results in the generation of an electric potential between the electrodes. Chemical energy is thus transformed into electrical energy. *See also* ALKALINE CELL, LEAD-ACID BATTERY, STORAGE BATTERY, ZINC CELL.

ELECTROLYTIC CAPACITOR

An electrolytic capacitor is a device that makes use of an electrochemical reaction to generate a large capacitance within a small volume. An electrolyte paste is contained in a cylindrical aluminum can. One of the leads is connected to the aluminum can, and the other to the electrolyte paste. The photograph shows a typical electrolytic capacitor, and the drawing illustrates a cutaway view inside such a capacitor.

The electrolysis process within the electrolytic capacitor causes a thin, nonconducting layer of aluminum oxide to form on the surface of the aluminum-foil sheet. This dielectric coating provides a very small physical separation between the conductive electrolyte and the aluminum. Therefore, the capacitance between the two subtances is very large per unit surface area. Electrolytic capacitors can have values as high as hundreds of thousands of microfarads. Other types of capacitors are generally much smaller in value. A disk-ceramic capacitor, for example, rarely has a value of more than 1 μF.

Unlike other kinds of capacitors, electrolytic devices are sensitive to polarity. A positive and a negative lead are clearly marked on all commercially manufactured electrolytic capacitors. The capacitor must be connected so that the voltage across the capacitor has the polarity shown by the markings. *See also* CAPACITANCE, CAPACITOR.

ELECTROLYTIC CELL

An electrolytic cell is any device containing an electrolytic substance and at least two electrodes. All electrochemical cells fall into this category, as do electrolytic capacitors and electrolytic resistors.

Electrolytic cells may be used for either of two purposes: to generate direct currents, or to operate from or with direct currents. *See also* ELECTROLYSIS, ELECTROLYTE, ELECTROLYTIC CAPACITOR, ELECTROLYTIC RECTIFIER, ELECTROLYTIC RESISTOR.

ELECTROLYTIC CONDUCTION

Electrolytic conduction is the passage of current through an electrolytic solution. Any electrolytic solution tends to ionize when a voltage is applied to electrodes immersed in it. Negative ions are attracted to the positive electrode, and they give up their excess electrons there. Positive ions are attracted to the negative electrode, and at that point, they acquire electrons. If the ion is a gas, it can be seen bubbling out of the electrolyte solution. If the ion is a solid, a noticeable accumulation may eventually form on the electrode. The ionization will continue for as long as current is applied, or as long as the ions remain available in the solution.

The flow of current in an electrolyte is different from the movement of electrons in a metal wire. The current is, nevertheless, measurable in terms of unit charges per second, and the current can be evaluated by placing an ammeter in series with the electrolyte device. *See also* ELECTROLYSIS, ELECTROLYTE.

ELECTROLYTIC RECTIFIER

An electrolytic rectifier is a device that allows current to flow in only one direction because of electrolysis properties of electrodes and electrolyte solutions. There are many different combinations of electrolyte metals and electrolyte solutions that can be used for the purpose of making an electrolytic rectifier. One example is an aluminum-and-lead electrode pair in a solution of sodium bicarbonate, or ordinary baking soda. This device resembles a common storage battery.

The operation of an electrolytic rectifier may be thought of as the result of adding a direct-current circuit, as shown in the illustration, to an alternating-current circuit. One half-cycle of the alternating current causes energy to be stored in the electrolytic cell. A direct current is therefore produced, in series with the alternating current. The series addition of a sufficient number of direct-current cells will cause the alternating current to become a pulsating direct current. For this to happen, the direct current must be at least equal to the peak value of the alternating current. *See also* RECTIFICATION.

ELECTROLYTIC RESISTOR

An electrolytic resistor is a device that consists of two wire leads immersed in a solution such as salt water. The resistance between the leads can be varied by adjusting the concentration of the electrolyte in the solution. For example, if salt water is used, more salt can be added to decrease the resistance. This applies, of course, only up to the saturation point of the solution.

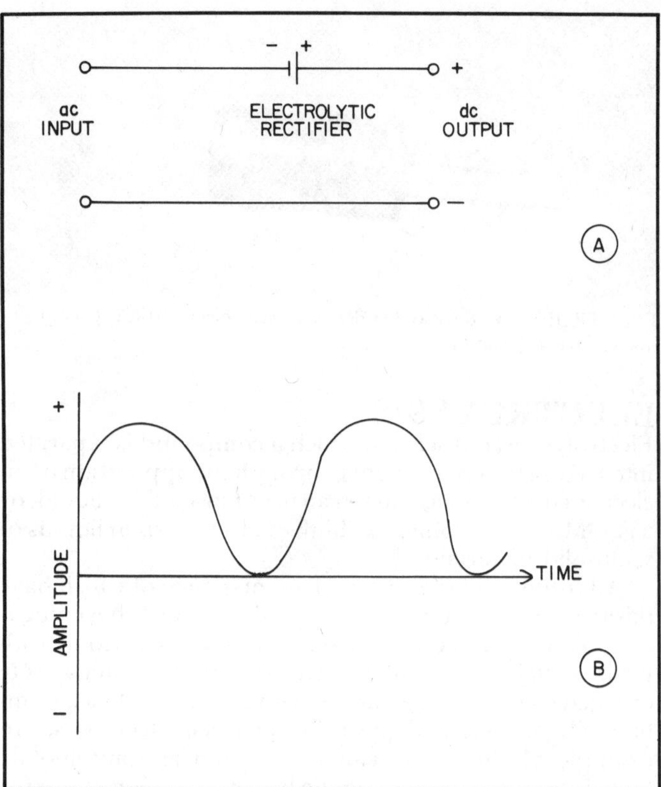

ELECTROYTIC RESISTOR: At A, schematic diagram of an electrolytic rectifier circuit. At B, the output waveform, showing the effect of the addition of a direct-current component to an alternating-current cycle.

Electrolytic resistors will function in direct-current or alternating-current circuits. They are intended for emergency use only. Salt water or baking-soda water, along with a cup and some hookup wire, are the only necessary materials for the construction of an electrolytic resistor! But these resistors tend to change value with time, and also with the applied voltage. The oxygen in the water, and the chlorine or carbon in the electrolyte, will attack the positive electrode and cause corrosion. This changes its effective resistance. *See also* ELECTROLYTE, ELECTROLYTIC CONDUCTION.

ELECTROMAGNET

An electromagnet is a form of temporary magnet that makes use of the effects of electric currents to produce a strong magnetic field. Electromagnets display the same properties, when in operation, as ordinary permanent magnets. Some dynamic devices and metering instruments use electromagnets in place of permanent magnets (*see* DYNAMIC LOUDSPEAKER, DYNAMIC MICROPHONE, ELECTRODYNAMOMETER).

An electromagnet is constructed by placing an iron or steel rod inside of a solenoidal coil, as shown in the drawing. When a direct current is passed through the coil winding, a magnetic field is produced, and the rod behaves exactly like a permanent magnet until the current is shut off. If an alternating current is applied to the coil, an alternating magnetic field is produced. This will attract magnetic metals, but will not develop a constant north or south magnetic pole in the iron or steel rod within the coil.

The intensity of the magnetic field generated by an

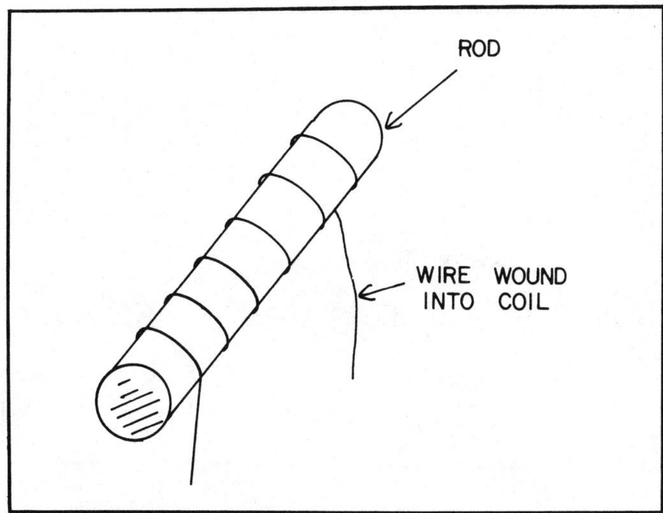

ELECTROMAGNET: An electromagnet consists of a coil of wire on a solenoidal magnetic core.

electromagnet depends on the number of turns in the coil, the kind of metal in the rod, and the current through the coil windings. *See also* MAGNETIC FIELD.

ELECTROMAGNETIC CONSTANT

In a vacuum, the speed of propagation of electric, electromagnetic, and magnetic fields is a constant of about 186,282 miles pers second, or 299,792 kilometers per second. This constant is sometimes abbreviated by the lowercase letter c.

The electromagnetic constant is always the same in a vacuum, no matter from what viewpoint it is measured. A passenger on a rapidly moving space ship would observe the same value as a person standing on the earth, or anywhere else. This fact was established by the physicist Albert Einstein in the early part of the twentieth century, and became part of his famous theory of special relativity. Before that time, scientists thought that the speed of light, and the speed of other electromagnetic effects, might depend on the motion of the observer. The special theory of relativity postulated that the speed of light is always the same, no matter from what point of view it is measured. This dramatically affected the course of physics. *See also* ELECTRIC FIELD, ELECTROMAGNETIC FIELD, MAGNETIC FIELD, RELATIVITY THEORY.

ELECTROMAGNETIC DEFLECTION

Electromagnetic deflection is the tendency of a beam of charged particles to follow a curved path in the presence of a magnetic field. This takes place because any moving particle itself produces a magnetic field, and the two fields interact to produce forces on the particles.

In a television picture tube, as well as in some cathode-ray oscilloscope tubes, electromagnetic deflection is used to guide the electron beam, emitted by the electron gun or guns, across the phosphor screen. Electric fields can also be used for this purpose.

Electromagnetic deflection can be demonstrated in a simple home experiement. Bring a powerful magnet, either a permanent magnet or an electromagnet, near the phosphor screen of a television receiver picture tube. Tune the receiver to a channel that comes in clearly. The magnet will produce a noticeable distortion in the picture. *See also* ELECTRIC FIELD, MAGNETIC FIELD.

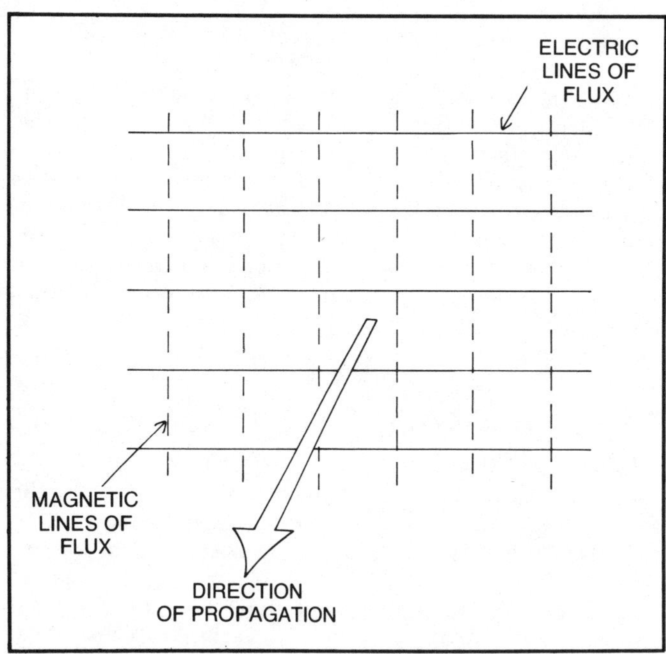

ELECTROMAGNETIC FIELD: In an electromagnetic field, the lines of electric flux are perpendicular to the lines of magnetic flux; the direction of travel is perpendicular to both sets of flux lines.

ELECTROMAGNETIC FIELD

An electromagnetic field is a combination of electric and magnetic fields. The electric lines of force are perpendicular to the magnetic lines of force at all points in space. An electromagnetic field can be either static or alternating. An alternating electromagnetic field propagates in a direction perpendicular to both the electric and magnetic components, as shown in the diagram.

Propagating electromagnetic fields are produced whenever charged particles are subjected to acceleration. The most common example of acceleration of charged particles is the movement of electrons in a conductor carrying an alternating current. The constantly changing velocity of the electrons causes the generation of a fluctuating magnetic field, which results in an electromagnetic field that propagates outward from the conductor.

The frequency of an alternating electromagnetic field is the same as the frequency of the alternating current in the conductor. The frequency can be as low as a fraction of a cycle per second, or Hertz; it can be as high as many trillions or even quadrillions of Hertz. Visible light is a form of electromagnetic-field disturbance. So are infrared rays, ultraviolet rays, X rays, and gamma rays. Of course, radio waves are an electromagnetic phenomenon.

All propagating electromagnetic fields display properties of wavelength λ as well as frequency f. In a vacuum, where the speed of electromagnetic propagation is about 300 million (3×10^8) meters per second, the wavelength in meters is given by:

$$\lambda = 300/f$$

where f is specified in megahertz. The higher the frequency, the shorter the wavelength.

Electromagnetic fields were first discovered by physicists during the nineteenth century. They observed that the fluctuating fields had a peculiar way of exerting their effects over great distances. Today, the results of this discovery are all around us, in the form of a vast and complex network of wireless communication systems, having frequencies ranging over several orders or magnitude. *See also* ELECTROMAGNETIC RADIATION, ELECTROMAGNETIC SPECTRUM.

ELECTROMAGNETIC FOCUSING

Electromagnetic focusing is a form of electromagnetic deflection, used in cathode-ray tubes (*see* ELECTROMAGNETIC DEFLECTION). A stream of electrons is brought to a sharp focus at the phosphor surface of the tube, by means of a coil or set of coils that carry electric currents. The coils produce magnetic fields, which direct the electrons so that they all land on the phosphor screen at the same point.

Electromagnetic focusing provides optimum resolution in the picture of a television receiver. It also operates in any device using a cathode-ray tube, such as an oscilloscope, electron microscope, or electron telescope. *See also* CATHODE-RAY TUBE.

ELECTROMAGNETIC INDUCTION

When an alternating current flows in a conductor, a nearby conductor that is not physically connected to the current-carrying wire will show some electron movement at the same frequency. This phenomenon is known as electromagnetic induction, and is a result of the effects of the electromagnetic field set up by a conductor having an alternating current. The varying electric and magnetic fields produced by acceleration of charged particles cause forces to be exerted on nearby charged particles.

All radio communication is possible because of electromagnetic induction acting over long distances. The effect, originally observed by the discoverers of the propagation as happening over small distances, actually takes place over an unlimited distance. The effect travels through space with the speed of light, or about 186,282 miles per second. Such a disturbance takes the form of mutually perpendicular electric and magnetic fields. *See also* ELECTROMAGNETIC FIELD.

ELECTROMAGNETIC INTERFERENCE

When an electrical or electronic circuit malfunctions in the presence of an electromagnetic field, the interference is called electromagnetic interference (abbreviated EMI). Usually, the electromagnetic field must be relatively strong for this kind of interference to occur; but sometimes it has been known to happen even in the presence of a fairly weak field. Then, of course, more than one field at a time may be responsible for the interference.

A very common form of electromagnetic interference is the alternating-current "hum" picked up by inadequately shielded microphone or interconnecting cables in audio-amplifier systems. The interfering field is the 60-Hz field produced by utility wires. The cure for this form of EMI is the proper shielding of the appropriate wires and, if necessary, bypassing of leads or the insertion of parallel choke coils.

Another common form of electromagnetic interference is seen in home entertainment equipment such as high-fidelity stereo systems. The electromagnetic fields from nearby radio broadcast stations, or other transmitting stations, are intercepted by dynamic pickups or wire leads, and are rectified by the amplifier circuits. If the nearby transmitter is amplitude-modulated, the modulating signal will actually be heard through the speakers of the system. If the signal is unmodulated, a buzz or change in volume will be observed. As with 60-Hz electromagnetic interference, the shielding of wire cables is helpful. Bypassing with small capacitors may be necessary as well; small radio-frequency chokes can be installed in series with audio leads to provide improvement in some cases. But radio-frequency EMI can be very difficult to eliminate. It can occur even if the radio transmitter is operation properly, since it is the purpose of the transmitter to generate the electromagnetic field. *See also* ELECTROMAGNETIC FIELD.

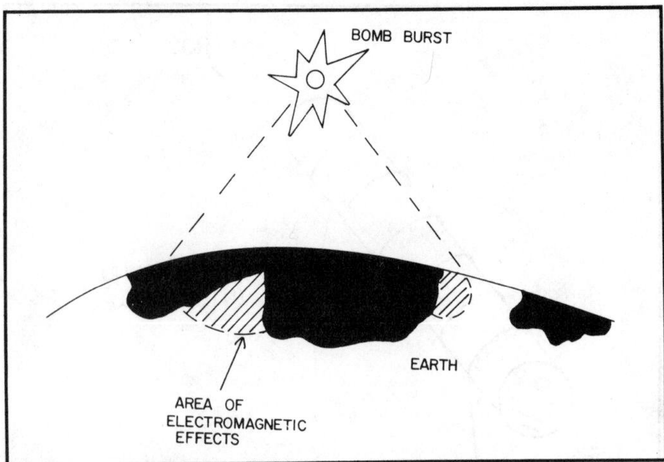

ELECTROMAGNETIC PULSE: The electromagnetic pulse from an atomic bomb, exploded high above the ground, could cause damaging effects over a large area.

ELECTROMAGNETIC PULSE

An electromagnetic pulse is a sudden burst of electromagnetic energy, caused by a single, abrupt change in the speed or position of a group of charged particles. An electromagnetic pulse does not generally have a well-defined frequency or wavelength, but instead it exists over practically the entire electromagnetic spectrum. This may include the radio wavelengths, infrared, visible light, ultraviolet, X rays, and even gamma rays. Electromagnetic pulses can be generated by arcing, and on a radio receiver they sound like popping or static bursts. Lightning produces an electromagnetic pulse of bandwidth extending from the very-low-frequency range to the ultraviolet range.

An electromagnetic pulse can contain a fantastic amount of power for a short time. Lightning discharges have been known to induce current and voltage spikes in nearby electrical conductors, of such magnitude that

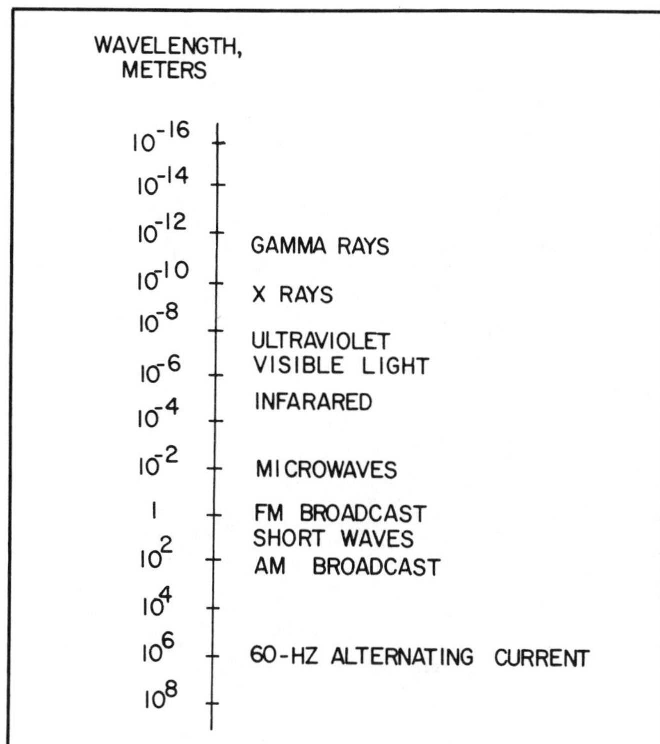

ELECTROMAGNETIC SPECTRUM: The electromagnetic spectrum, showing the wavelength in meters. This logarithmic representation shows the range of the most common wavelenghts; theoretically, the spectrum extends infinitely above and below the illustration.

equipment is destroyed and fires are started.

The detonation of an atomic bomb creates a strong electromagnetic pulse. The explosion of a multi-megaton bomb at a very high altitude, while not creating a devastating shock wave or heat blast at the surface of the earth, could cause the generation of a damaging electromagnetic pulse over an area of many thousands of square miles (see illustration). It is possible that a major nuclear attack might be preceded by the explosion of bombs aboard satellites. The resulting electromagnetic pulses could induce damaging voltages and currents in radio antennas, telephone wires, and power transmission lines over a vast geographic area. For this reason, all communications systems should have some form of protection against the effects of such an electromagnetic pulse. One possible means of defense is provided by equipment for direct-hit lightning protection. *See also* ELECTROMAGNETIC FIELD, LIGHTNING PROTECTION.

ELECTROMAGNETIC RADIATION

Electromagnetic radiation is the propagation of an electromagnetic field through space (*see* ELECTROMAGNETIC FIELD). All forms of radiant energy, such as infrared, visible light, ultraviolet, and X rays, are forms of electromagnetic radiation.

The wavelength is the property of electromagnetic radiation that determines the form in which it is manifested. Radio waves may be as long as thousands of meters, or as short as a few millimeters, or anywhere in between these two extremes. Other forms of electromagnetic radiation have very short wavelengths. The shortest known gamma rays measure about 10^{-15} meter, or 10^{-5} Angstrom unit. This is so small that even the most powerful microscope could not resolve it. Theoretically, no limits exist for how long or how short an electromagnetic wave can be.

In a vacuum, electromagnetic radiation travels at the speed of light, or about 186,282 miles per second or 299,792 kilometers per second. In media other than a perfect vacuum, the speed of propagation is slower. The precise velocity depends on the substance, and also on the wavelength of the electromagnetic radiation. Some materials are opaque to electromagnetic radiation at certain wavelengths. *See also* ELECTROMAGNETIC CONSTANT, ELECTROMAGNETIC FIELD, ELECTROMAGNETIC SPECTRUM.

ELECTROMAGNETIC SHIELDING

Electromagnetic shielding is a means of keeping an electromagnetic field from entering or leaving a certain area. The most common form of electromagnetic shieleing consists of a grounded enclosure, made of sheet metal or perforated metal. A screen may also be used.

An electromagnetic shield is important in many different kinds of electronic systems. Shielding prevents unwanted electromagnetic coupling betweeen circuits. If a shielded enclosure is not made of solid metal, the openings should be very small compared with the wavelength of the electromagnetic field. Perforated shields allow ventilation for cooling.

Coaxial cable provides an example of an electromagnetic shield that is is employed for the purpose of preventing radiation from a transmission line. By keeping the electromagnetic field inside the line, the shield guides the wave along the line. The shield continuity or effectiveness is a measure of the degree to which such a cable confines the electromagnetic field. This parameter depends on the frequency of the field. The higher the frequency for a given cable, the less the effective shielding. *See also* COAXIAL CABLE, ELECTROMAGNETIC FIELD, SHIELDING EFFECTIVENESS.

ELECTROMAGNETIC SPECTRUM

The electromagnetic spectrum represents the entire range of frequencies or wavelengths of electromagnetic energy. In theory, the electromagnetic spectrum has no lower or upper limit; the wavelength can be any size.

Physicists and engineers use a logarithmic scale, such as the one shown in the drawing, to illustrate the electromagnetic spectrum. The wavelength is shown in meters in this example. Some of the more familiar "landmarks" in the electromagnetic spectrum are indicated. The shorter the wavelength, the more energy is contained in a single photon (*see* PHOTON).

Electromagnetic radiation has properties that vary immensely with the wavelength. Some radio waves are bent or reflected by the ionized layers in the upper atmosphere of the earth, while others are not affected. In general, radio waves shorter than about 2 meters will pass through the ionosphere into space. The air itself is practically opaque at some electromagnetic wavelengths; some of the infrared range cannot penetrate the atmosphere, and the short ultraviolet rays, X rays, and gamma rays are also blocked. (It is fortunate for life on the earth that this is so!) The shortest known gamma rays have a length of about 10^{-15} meter. But, theoretically, shorter rays can exist.

The wavelength λ of an electromagnetic wave is related to the frequency f by the simple equation:

$$\lambda = 300/f$$

where λ is in meters and f is in megahertz. If λ is to be given in feet, then the relation becomes:

$$\lambda = 984/f$$

See also ELECTROMAGNETIC FIELD, ELECTROMAGNETIC RADIATION, GAMMA RAY, INFRARED, LIGHT, RADIO WAVE, ULTRAVIOLET, X RAY.

ELECTROMAGNETIC SWITCH

An electromagnetic switch is a device that allows remote-control switching. A current, either direct or alternating, is supplied to a coil of wire with a ferromagnetic, solenoidal core, creating an electromagnet (*see* ELECTROMAGNET). The field attracts an armature toward or away from a fixed set of contacts, opening and closing various circuits.

Electromagnetic switches, also often called relays, are used in a variety of applications for switching at different speeds, and with frequencies ranging from direct current up to several hundred megahertz. *See also* RELAY.

ELECTROMAGNETIC THEORY OF LIGHT

Visible light behaves, in some ways, like a barrage of particles called photons (*see* PHOTON). In other respects, light behaves like an electromagnetic-wave disturbance. The wave model is sometimes called the electromagnetic theory of light.

The wavelength of electromagnetic disturbances in the range of visible light is extremely short. At the longest, it is about 0.75 microns (a micron is a millionth of a meter). The shortest visible light waves are about 0.39 microns in length. This visible range varies somewhat from person to person, just as the range of audible frequencies is greater in some people than in others. The longest waves look red to the eye, and as the waves get shorter, the colors progress through orange, yellow, green, blue, indigo, and violet. The corresponding frequencies range from about 400 terahertz to 770 terahertz. (A terahertz, abbreviated THz, is a trillion hertz, or a million megahertz.)

All radiant energy behaves, in some ways, like visible light in terms of electromagnetic properties. At wavelengths greater than the longest visible red, the disturbance is called infrared. At wavelengths shorter than the violet, the waves are called ultraviolet. *See also* LIGHT, PARTICLE THEORY OF LIGHT.

ELECTROMAGNETIC TUBE

See CATHODE-RAY TUBE.

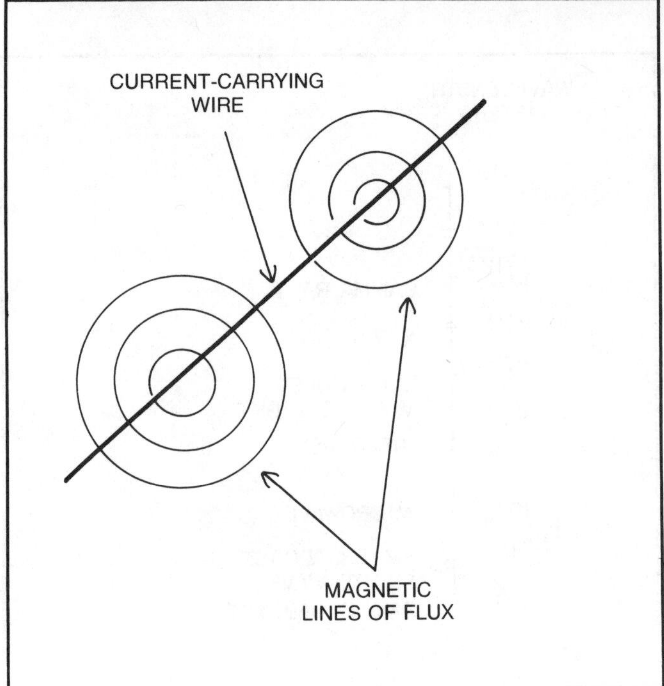

ELECTROMAGNETISM: A current-carrying wire produces circular magnetic lines of flux.

ELECTROMAGNETISM

When a charged particle, or a stream of charged particles, is set in motion, a magnetic field is produced. The lines of magnetic force occur in directions perpendicular to the motion of the charged particles. This effect is called electromagnetism.

Electromagnetism is well illustrated by the flow of current in a straight wire conductor. The magnetic lines of flux lie in a plane or planes orthogonal to the wire. They appear as concentric circles around the wire axis, as illustrated in the drawing. The intensity of the magnetic field, or the number of flux lines in a given amount of area, is proportional to the intensity of the current in the wire.

If an alternating current is flowing in the wire, the intensity of the magnetic field varies, and these fluctuations result in a spatial charge differential or electric field. A changing, or oscillating, magnetic field therefore contains an electric-field component as well. The electric lines of flux occur parallel to the wire, or perpendicular to the magnetic lines of force. The combination of the two fields has an ability to propagate long distances through space, and is called an electromagnetic field. *See also* ELECTROMAGNET, ELECTROMAGNETIC FIELD, ELECTROMAGNETIC RADIATION.

ELECTROMECHANICAL AMPLIFIER

An electromechanical amplifier is a device that converts electrical energy into mechanical energy, and then back into electrical energy at a higher level of current, voltage, or power.

A dynamotor is an example of an electromechanical amplifier designed to convert a low voltage to a higher voltage (*see* DYNAMOTOR). Such devices are sometimes used in power supplies. A chopper device is another means of getting the same effect (*see* CHOPPER POWER SUPPLY).

Electroacoustic amplifiers are another form of electromechanical amplifier. One sort of electracoustic

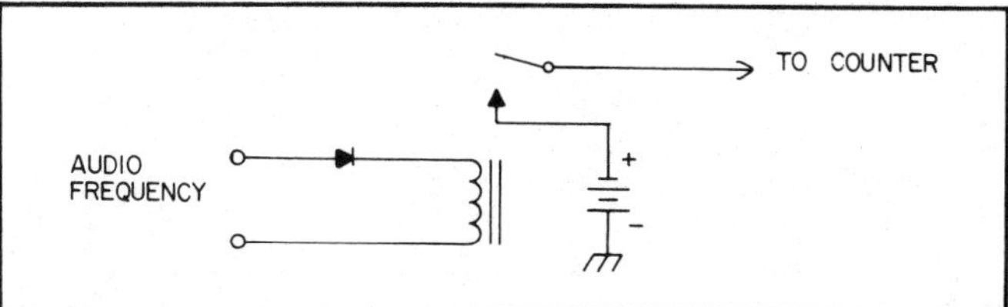

ELECTROMECHANICAL FREQUENCY METER: This device is suitable for frequency measurement at relatively low frequencies.

amplifier, called the carbon-button amplifier, consists of an earphone with its diaphragm attached to a device similar to a carbon microphone (*see* CARBON MICROPHONE). A direct current is passed through the button containing the carbon granules. The acoustic vibrations from the earphone modulate the direct current. The resulting modulated current has greater amplitude than the original signal applied to the earphone. *See also* AMPLIFICATION.

ELECTROMECHANICAL CHOPPER CIRCUIT

See CHOPPER POWER SUPPLY, CONTACT MODULATION.

ELECTROMECHANICAL FILTER

See MECHANICAL FILTER.

ELECTROMECHANICAL FREQUENCY METER

An electromechanical frequency meter is a device used to measure low and midrange audio frequencies. The device operates by converting the sound or audio-frequency vibrations into mechanical motion. The frequency of the mechanical motion is easily measured.

A fast-acting relay, with its coil connected to the audio-frequency signal source through a half-wave rectifier circuit, generates pulses in the direct-current power supply connected through the contacts (see illustration). These pulses occur at the rate of one pulse per audio-frequency cycle. The pulses are square, since the relay contact is either all the way open or all the way closed. The pulses are counted by an electronic circuit with a gate time of 1 second. The resulting count is the frequency in Hertz.

For greater accuracy, a longer gate time may be used. If the gate time is 10 seconds, for example, the pulse count can be divided by 10 to get the frequency in Hertz; this provides an additional digit of accuracy. For more frequent (but less accurate) readings, a gate time of 0.1 second may be used. *See also* FREQUENCY COUNTER.

ELECTROMECHANICAL OSCILLOSCOPE

An electromechanical oscilloscope is a device that converts alternating current or pulsating direct current into mechanical motion in order to display the waveform. Such an oscilloscope usually utilizes a meter movement, a light-beam generator, a reflecting mirror attached to the meter movement, and a translucent or reflecting display screen.

When low-frequency alternating or pulsating direct current is passed through the meter coil, the meter movement oscillates back and forth in synchronization with the current variations. The mirror causes the light beam to be projected onto the screen in an oscillating manner. Another mirror provides a scanning movement of the beam. The result is a synchronized waveform display on the screen. The human eye can only perceive about 20 flashes per second of visible light; if the sweep frequency of the device is greater than 20 Hz, therefore, the display appears continuous to the eye.

Electromechanical oscilloscopes are useful only at rather low frequencies. This is because the meter movement cannot respond to high-frequency oscillations or fluctuations of current. For higher frequencies, the cathode-ray oscilloscope is much more often used than any other form of display device. *See also* OSCILLOSCOPE.

ELECTROMECHANICAL RECTIFIER

An electromechanical rectifier is a fast-acting relay device, which alternately opens and closes in the presence of an alternating current. During one half of the cycle, the contacts are open, and that half of the cycle is cut off. During the other half of the cycle, the contacts are closed, and that half is passed. For example, if the contacts are open during the negative half of the cycle, the resulting current is positive. This is a pulsating direct current, exactly the same as would be produced by a half-wave diode rectifier circuit.

Another form of electromechanical rectifier is the commutator. This device operates as a shaft rotates, alternating the polarity in step with the frequency of the current. This results in full-wave rectification. Commutators are used in motors to allow them to operate from direct current; they are also employed to produce direct current from generators. *See also* COMMUTATOR, GENERATOR, MOTOR.

ELECTROMECHANICAL TRANSDUCER

Any device that converts electrical energy into mechanical energy, or vice versa, is called an electromechanical transducer. Examples of such transducers include common motors and generators. The transformation is usually the result of electromagnetic effects, in which a moving charge creates magnetic forces, or a moving object having a magnetic field creates an electric current (*see* ELECTROMAGNETISM).

Devices that convert sound, or acoustic, energy into electric energy are sometimes considered electromechanical transducers. Speakers, earphones, and microphones are examples of such devices. They may also be called electroacoustic transducers. The mechanical movement of sound can be converted into electrical energy, or vice versa, in a number of different ways, including the electromagnetic, electrostatic, and piezoelectric methods.

A vacuum-tube triode having a mechanically variable plate-to-cathode spacing is sometimes called an electromechanical transducer. Such a tube has a variable amplification factor, depending on the amount of physical pressure applied to an external stylus. The device is sensitive to very rapid changes in stylus pressure. The electromechanical transducer tube can be used for a variety of purposes, such as the determination of irregularity of a surface. *See also* TRANSDUCER, TUBE.

ELECTROMETER

An electrometer is a highly sensitive device, incorporating a meter and an amplifier, used for measuring small voltages. The input resistance is extremely high, and may be as large as several quadrillion ohms (a quadrillion is 10^{15}, or a million billion). The electrometer, because of its high input resistance, draws essentially no current from the source under measurement. This allows the device to be used for measuring electrostatic charges.

An electrometer uses a special vacuum tube called an electrometer amplifier or electrometer tube (*see* TUBE). The interelectrode resistance is very high, and the amplifier circuit is designed in a manner similar to that of a Class-A audio or radio-frequency amplifier. The noise generation is minimal; this is important from the standpoint of obtaining the maximum possible sensitivity. A tiny change in the input current, in some cases as small as 1 picoampere (a millionth of a millionth of an ampere), can be detected.

The electrometer is employed whenever it is necessary that the current drain be negligible. An electrometer can also be used to measure electrostatic voltages in which the amount of charge (quantity) is actually small. *See also* FET VOLTMETER.

ELECTROMOTIVE FORCE

Electromotive force is the force that causes movement of electrons in a conductor. The greater the electromotive force, the greater the tendency of electrons to move. Other charge carriers can also be moved by electromotive force; in some types of semiconductor, the deficiency of an electron in an atom can be responsible for conduction. Such a deficiency is called a hole; electromotive force can move holes as well as electrons. *See also* ELECTRON, HOLE, VOLTAGE.

ELECTRON

An electron is a sub-atomic particle that carries a unit negative electric charge. Electrons may be free in space, or they may be under the influence of the nuclei of atoms. An electron is tiny indeed; it is hard to imagine its size. A single electron is to a particle of dust as the particle of dust is to the whole planet earth! The mass of an electron is 9.11×10^{-31} kilogram; therefore, a gram of them would have a count of 1.1×10^{27}, or about 1 trillion quadrillion. A single electron carries a charge of 1.59×10^{-19} coulomb; thus a coulomb of charge represents 6.28×10^{18} electron charges (*see* COULOMB).

Electrons are believed to move around the nuclei of atoms, in such a way that their average position is represented by a sphere at a certain distance from the nucleus. The force of electrostatic repulsion prevents gravity from causing atoms to fall into one another. It is primarily the electrons that are responsible for this repulsion.

In an electric conductor, the flow of electrons occurs as a passing-on of the particles from one atom to another. It does not take place as a simple flow, like water in a hose. Some atoms have a tendency to pick up extra electrons, and some have a tendency to lose electrons. Sometimes a flow of current occurs because of a deficiency of electrons, and sometimes it occurs because of an excess of electrons among the constituent atoms of a particular substance. The deficiency of an electron in an atom is called a hole. Hole conduction is important in the operation of semiconductor diodes and transistors. *See also* HOLE.

ELECTRON AVALANCHE

See AVALANCHE.

ELECTRON-BEAM GENERATOR

An electron-beam generator, also sometimes called an electron gun, is an electrode designed for the purpose of producing a stream of electrons in a vacuum tube. An electron-beam generator is used in cathode-ray tubes, such as those used in television receivers or oscilloscopes.

The electron-beam generator is made up of a heated cathode, a control electrode, accelerating electrodes, and focusing electrodes. The cathode emits electrons because of the negative voltage applied to it. The control electrode regulates the intensity of the electron emission from the cathode, and is used to control the brightness of the image. The accelerating electrodes give the electron beam more speed. The focusing electrodes direct the electrons into a narrow, defined beam.

After being emitted from an electron-beam generator, a beam of electrons is generally passed through a set of deflecting electrodes. These control the position at which the beam lands on the phosphor screen. Voltages applied to the deflecting electrodes cause the beam to scan across the screen; the intensity or position of the beam may be further modified by signal information. *See also* CATHODE-RAY TUBE.

ELECTRON-BEAM TUBE

An electron-beam tube is a vacuum tube that uses an electron-beam generator rather than a typical cathode. The beam-power tube, the klystron tube, the magnetron tube, the cathode-ray tube, and the photomultiplier tube are examples of electron-beam tubes.

The electron-beam tube, in contrast to the conventional vacuum tube, operates using a focused and directed beam of electrons. This can create amplification or oscillation in the same way as with conventional tubes, but according to special parameters not available with other kinds of tubes. Electric and magnetic fields are used in electron-beam tubes to control the paths of the beams. In an ordinary vacuum tube, the intensity of the electron beam is regulated but its directional properties are not. *See also* BEAM-POWER TUBE, CATHODE-RAY TUBE, KLYSTRON, MAGNETRON, PHOTOMULTIPLIER, TUBE.

ELECTRON-COUPLED OSCILLATOR

An electron-coupled oscillator is a special form of

vacuum-tube oscillator that is less susceptible to loading effects than is a conventional oscillator circuit. A change in the load impedance of an oscillator can cause frequency changes, and can sometimes even result in cessation of oscillation. The electron-coupled oscillator is designed to overcome these effects. A changing load impedance to an electron-coupled oscillator does not result in significant changes in the frequency, and almost never causes the oscillator to stop functioning.

The illustration is a schematic diagram of a Hartley type electron-coupled oscillator. The screen grid acts as the plate for the oscillator circuit, while the output is taken from the actual plate of the tube. An intervening suppressor grid may or may not be used. The stream of electrons arriving at the plate of the tube is modulated by the oscillator. The coupling to the oscillator output is provided only by the electron beam as it traverses the inside of the tube from the control grid through the screen grid to the plate. The interelectrode capacitance is generally very small within a vacuum tube, and therefore a large change in the output impedance does not significantly load down the oscillator. The electron-coupled oscillator acts as its own buffer stage, and this often eliminates the need for such a stage following the oscillator. The disadvantage of the electron-coupled oscillator is that a vacuum tube must be used to obtain the necessary results. *See also* OSCILLATOR.

ELECTRON EMISSION

Electron emission occurs whenever an object gives off electrons to the surrounding medium. This occurs in all vacuum tubes, and also in incandescent and fluorescent light bulbs. It also takes place in a variety of other devices. Some materials such as barium oxide or strontium oxide, are especially well-suited to electron emission. Electrodes are sometimes coated with such because this enhances their capability to give off electrons when heated and subjected to a negative voltage.

Electron emission can occur because of thermionic effects, where the temperature of a negatively charged electrode is raised to the point that electrons break free from the forces that normally hold them to their constituent atoms. The cathodes of most vacuum tubes act in this way. They may be directly heated or indirectly heated (*See* CATHODE.)

Electron emission can be secondary in nature, resulting from the impact of high-speed electrons against a metal surface. This sometimes takes place at the plate of a vacuum tube, and in that case it can be detrimental. (The suppressor grid of a pentode tube keeps the secondary electrons from escaping from the plate.) A dynode operates on the principle of secondary emission to amplify an electron beam (*see* DYNODE).

If an electric charge is applied to the surface of an object, and the voltage is great enough, electrons will be thrown off. Also, if a nearby positive voltage is sufficiently strong, electrons will be pulled from an object. This is called field emission. Cold-cathode tubes operate on this principle (*see* COLD CATHODE).

In a photoelectric tube, light striking a barrier called a photocathode results in the emission of electrons. This is called photoemission. The intensity of photoemission depends on the brightness of the light, and also on its wavelength (*see* PHOTOCATHODE, PHOTOMULTIPLIER).

The electrons emitted from an electrode are always the same elementary particles, carrying a unit charge and having the same mass, no matter what the cause of their emission. *See also* ELECTRON, TUBE.

ELECTRON-HOLE PAIR

In an atom of a semiconductor material, the conduction band and the valence band are separated by an energy gap. An electron in the conduction band is free to move to another atom or escape entirely. But an electron in the valence band is held to the atom by the positive charge of the protons in the nucleus (*see* VALENCE BAND). An electron may move from the valence band to the conduction band of an atom if it receives a certain amount of energy. This leaves an electron vacancy, called a hole, in the valence band of the atom, as shown in the illustration. The

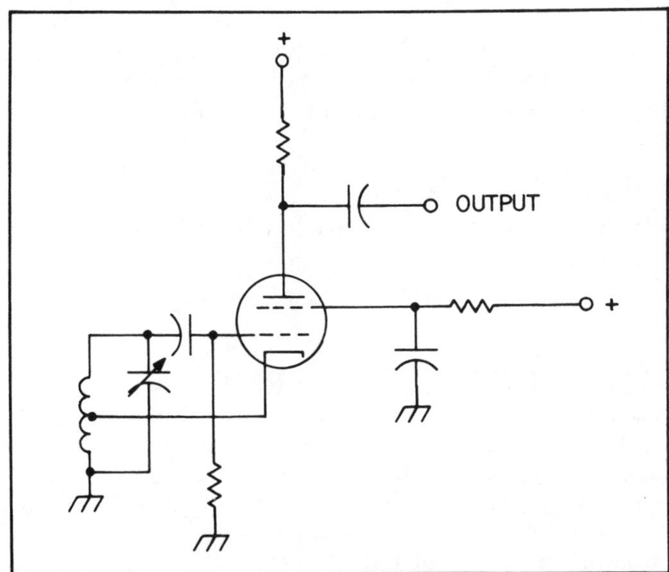

ELECTRON-COUPLED OSCILLATOR: In the electron-coupled oscillator, the screen grid acts as the plate for the oscillator, while the output is taken from the actual plate of the tube.

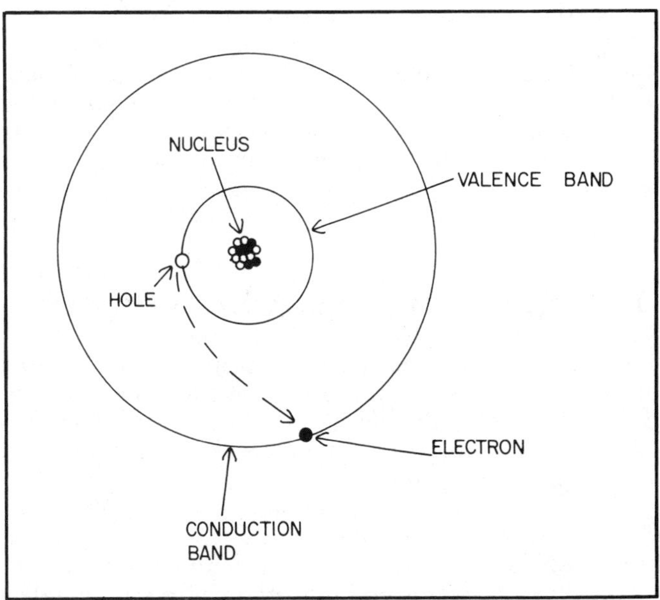

ELECTRON-HOLE PAIR: When an electron moves from the valence band to the conduction band, a hole remains in the valence band.

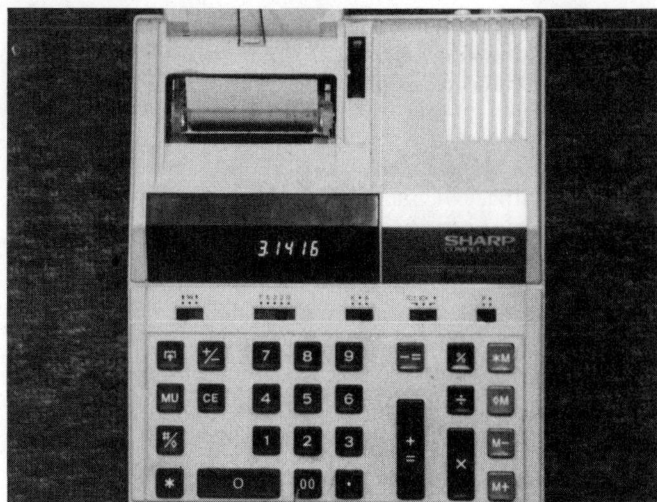

ELECTRONIC CALCULATOR: A typical desktop electronic calculator, used for bookkeeping purposes.

ELECTRONIC CLOCK: Electronic clocks can be small enough to fit inside a pen, and yet they can keep time to within 1 second per month.

electron and the hole form what is called an electron-hole pair.

The current in certain semiconductor materials, called N-type semiconductors, flows because of the extra electrons that are passed from atom to atom. In other substances, called P-type semiconductors, the current flows as the result of a transfer of electron deficiencies, or holes, from atom to atom. The electron carries a negative charge, and the deficiency, or hole, carries a positive charge. Therefore, the flow of electrons is from the negative to the positive, but the flow of holes is in the other direction, from the positive to the negative. *See also* ELECTRON, HOLE, N-TYPE SEMICONDUCTOR, P-TYPE SEMICONDUCTOR.

ELECTRONIC BUZZER

An electronic buzzer is an oscillator for audio frequencies, with the output connected to a transducer such as a speaker. When power is applied to the device, it oscillates and produces a sound.

Electronic buzzers are used in a wide variety of applications. Many digital calculators use such oscillators, with tiny piezoelectric transducers, for the psychological effect they produce. Electronic clocks and watches often have such buzzers to be used as alarms. In computer games, the complex tones are generated by electronic buzzers having variable or multiple frequency outputs. *See also* OSCILLATOR.

ELECTRONIC CALCULATOR

An electronic calculator is a machine, using transistors and integrated circuits, that performs arithmetic operations such as addition, subtraction, multiplication, and division. Some electronic calculators can also perform more advanced operations, such as the square-root, logarithm, and trigonometric functions.

Electronic calculators come in a variety of different designs, some for home use, some for business use, and some for various kinds of scientific applications. The desktop calculator shown in the photograph is intended primarily for business and financial use. It contains a small printer that records, if desired, all of the operations on a sheet of narrow paper, in much the same way as a cash register. In fact, modern cash registers are actually electronic calculators of this kind.

Sophisticated electronic calculators are available that can be programmed to perform complex mathematical operations many times. Such calculators are one step below the microcomputer. *See also* COMPUTER, MICROCOMPUTER.

ELECTRONIC CLOCK

An electronic clock is a timepiece of any size, the activity of which is governed by electronic circuitry. Electronic clocks have almost completely replaced the older mechanical and electrical devices. Electronic clocks are more physically rugged and more accurate than the clocks of the past. Electronic clocks can be made very tiny, and are sometimes found in calculators or even in pens (see photograph).

A mechanical clock, with a motor governed by an electronic oscillator, is sometimes called an electronic clock. The oscillator uses a frequency standard such as a tuning fork or quartz crystal. Some electronic clocks even have a built-in radio receiver, tuned to one of the time-standard frequencies such as the WWV/WWVH frequencies of 5, 10, 15, or 20 MHz and others. A tone, transmitted hourly by such stations, automatically resets such clocks (see WWV/WWVH).

A timing circuit, used in a digital system, is sometimes called an electronic clock. Such a circuit generates pulses at precise intervals to govern the operation of microcomputers and other digital equipment.

ELECTRONIC CONTROL

Electronic control is the operation or adjustment of a device by electronic, rather than mechanical, means. As mechanical methods of control become less and less common, the electronic method is becoming more widely used.

A good example of electronic control is the frequency adjustment in a digital synthesized radio receiver. The 1-MHz, 100-kHz, 10-kHz, and 0.1-kHz digits are selected independently by means of pushbutton switches. Alternatively, in some units, the frequency can be incremented rapidly up or down by pressing and holding a button. This type of control circuit has taken the place of older mechanical dial devices in many modern radio and electronic circuits.

In general, the science of controlling things by electronic means can be called the science of electronic control. It is no longer limited to the control of machines. A heart pacemaker, for example, is an electronic device that controls the rate of beating of a human heart.

ELECTRONIC COUNTER

See COUNTER.

ELECTRONIC ENGINEERING

See ELECTRICAL ENGINEERING.

ELECTRONIC FLASH TUBE

See FLASH TUBE.

ELECTRONIC GAME

With the availability of computers for home use, electronic games have become commonplace in recent years. An electronic game uses electronic circuits and devices to create a certain scenario, which may then be controlled by one or more operators. A video monitor, resembling a television receiver, is almost always used in an electronic game. Some electronic games can be connected directly to the antenna terminals of a home television receiver.

In the late 1970s and early 1980s, rapid advances in computer technology resulted in the proliferation of electronic games in such places as amusement centers and bars, as well as in the home. Some electronic games have educational value, and allow the operators to actively participate in a learning process. Other electronic games are intended primarily for entertainment. The variety of different scenarios obtainable with such computer games is almost unlimited. Computer-game type cassettes and diskettes are sold in electronics and hobby stores. *See also* COMPUTER.

ELECTRONIC INDUSTRIES ASSOCIATION

The Electronic Industries Association, or EIA, is an agency in the United States that sets standards for electronic components and test procedures. The EIA is also responsible for the setting of performance standards for various types of electronic equipment. Standardization is advantageous both to the manufacturers (since it broadens their markets) and to the users of electronic equipment (since it makes it easier for them to get replacement parts and peripheral devices).

The Electronic Industries Association works with other national and international standards agencies. The American National Standard Institute (ANSI) is one such association. The International Electrotechnical Commission (IEC), headquartered in Geneva, Switzerland, is responsible for electronic standards worldwide.

ELECTRONIC MUSIC

Audio-frequency electronic oscillators are capable of producing complex as well as simple waveforms. All musical instruments produce rather complex, but identifiable and reproducible, waveforms. Recent technology has made it possible to duplicate the sound of any musical instrument by means of audio-frequency oscillators and transducers. A device called a Moog synthesizer, used by many modern bands as part of their instrumentation, can produce many different kinds of sound. This is an example of electronic music.

Electronically amplified or modified music is so commonplace today that it is difficult to find any other kind of music! Even classical music is somewhat electronic when it is reproduced through a high-fidelity system. A live orchestra is not electronic; neither is a marching band. But popular music bands always use some electronic means to create their sounds.

Very complicated musical themes can be created using a synthesizer and a computer. Many sound tracks can be recorded, one on top of the other, reproducing the sound of a whole band or orchestra. Some experimenters are working with the idea of having computers actually compose music. *See also* MOOG SYNTHESIZER.

ELECTRONIC TIMER

See ELECTRONIC CLOCK.

ELECTRONIC WARFARE

Electronic warfare is the use of electronic equipment and techniques to aid the successful use of conventional weapons. Some examples would be the use of jamming signals to render enemy radar ineffective, or using sophisticated signals to cause the enemy radar to report false targets or targets that are not where they are expected. Signals emitted by either a ground station or an aircraft can confuse an enemy missile, causing it to miss the intended target. Missiles can be programmed to follow an enemy radar beam to its source, thus destroying the radar site. Communications circuits can be disrupted by noise transmitters, or penetrated by false signals, thus making information transfer more difficult.

Every transmitter, and every type of emission, has its own characteristics, called a "signature." Identifying these sources and their role in any situation is also part of electronic warfare. The technique of counteracting the enemy's use of electronic warfare is termed "electronic countermeasures," or ECM. Both fields are becoming increasingly important, and complex, in today's military preparedness. The search for new and better methods of EW and ECM often leads to discoveries useful to many in business and consumer electronics.

ELECTRONIC WATCH

See ELECTRONIC CLOCK.

ELECTRON MICROSCOPE

An electron microscope is a form of microscope that uses electron beams rather than visible light to obtain an image of a small object. The electron microscope allows extreme

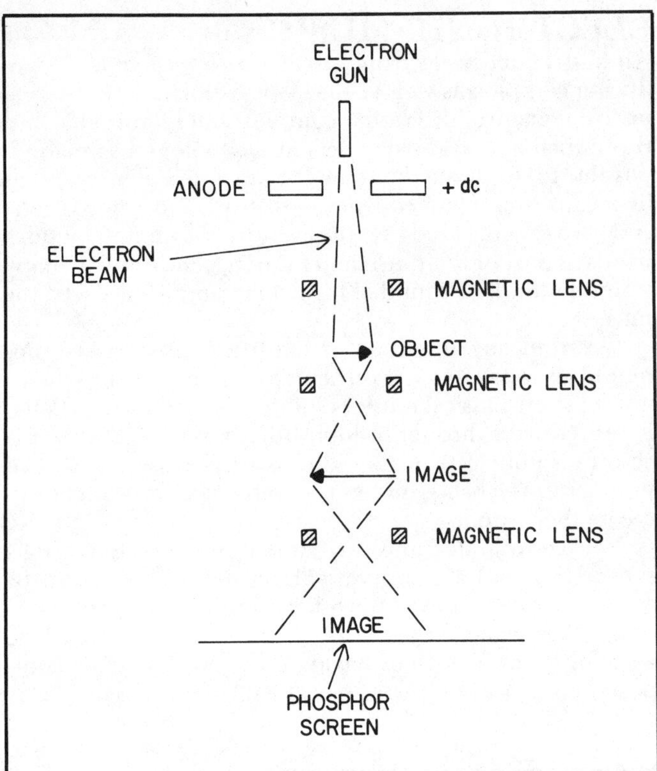

ELECTRON MICROSCOPE: Principle of operation of an electron microscope. A beam of electrons is used instead of visible light. The image is formed on a phosphor screen.

magnification, much greater than any optical microscope. The electron beam is focused and refracted by means of magnetic fields. The magnetic deflectors, sometimes called magnetic lenses, produce a magnified image of the object under observation, when the object is placed at the proper point in the electron beam. The illustration shows a simplified diagram of an electron microscope.

With the electron microscope, the image of the object is viewed on a phosphor screen, which resembles a television picture screen. Alternatively, the image may be photographed on a special kind of electron-sensitive film. The magnification factor can be as large as 300,000 to 400,000. Under ideal conditions, this means that a grain of sand could be rendered to have an effective diameter of over 100 feet.

ELECTRON-MULTIPLIER TUBE

An electron-multiplier tube is a device that amplifies a beam of electrons. Such tubes operate on the principle of secondary emission (*see* SECONDARY EMISSION). The electron beam is reflected off of a series of plates, each of which throws off several electrons for each one that strikes it. With several such plates, called dynodes, in succession, very large current-gain factors can be realized. The gain may exceed 60 dB.

Electron-multiplier tubes are commonly used in a device called a photomultiplier, which is in turn often used in sensitive television cameras and infrared detectors. *See also* DYNODE, PHOTOMULTIPLIER.

ELECTRON ORBIT

In an atom, the electrons orbit the nucleus in defined ways.

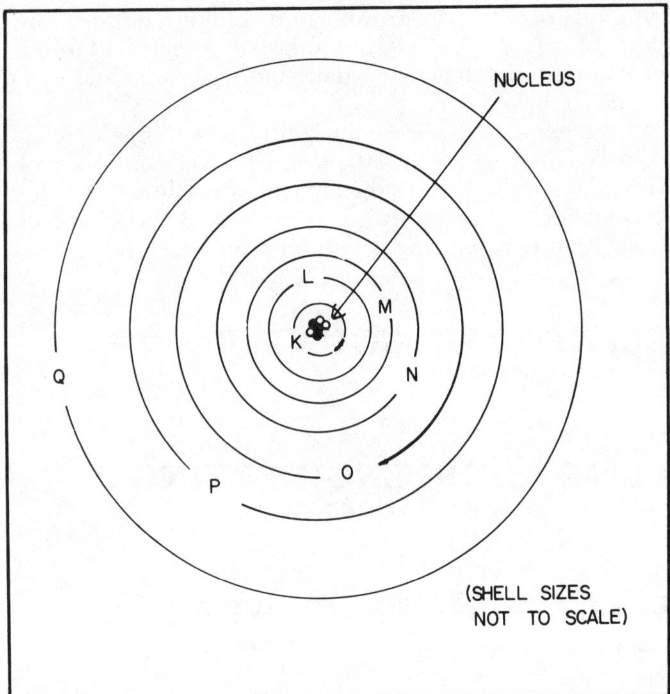

ELECTRON ORBIT: Electron orbits around an atomic nucleus can exist only at discrete distances from the nucleus. (This drawing is not to scale.)

The positions of the electrons are sometimes called orbits or shells, but actually, the orbit or shell of an electron represents the average path that it follows around the nucleus. Electrons orbit only at precise distances from the nucleus, in terms of their average positions. The greater the radius of an orbit, the greater the energy contained in the individual electron.

The defined orbits or shells for the electrons in an atom are called, beginning with the innermost shell, the K, L, M, N, O, P, and Q shells. These are roughly diagrammed, although not to scale, in the illustration. The drawing has been dimensionally reduced, so that the shells, which are actually spherical in shape, appear as concentric circles around the nucleus. The K shell is the least energetic, and can contain at most two electrons. The L shell, slightly more energetic, can have up to eight electrons. The M shell may have as many as 18 electrons; the N shell, 32; the O shell, 50; the P shell, 72; and the Q shell, 98. The outermost shells of an atom are never filled. The inner ones tend to fill up first, and the number of electrons in an atom is limited.

An electron may change its orbit from one shell to another. If an electron absorbs energy, it will move into a more distant shell, provided it gets enough extra energy. By moving inward to a shell closer to the nucleus, an electron gives off energy. When an electron absorbs energy, it leaves a definite trace in a spectral dispersion, called an absorption line. When it gives off energy, it creates an emission line. Astronomers have used absorption and emission lines to detect the presence of various elements on the planets, in the sun, in other stars, and in other galaxies. *See also* ELECTRON.

ELECTRON TELESCOPE

An electron telescope is a device that obtains at least part of its magnification by electronic means. It is similar to a closed-circuit television system.

An objective lens focuses the incoming light onto a photocathode in a television camera tube. The resulting electron beam is spread out, and then focused onto a phosphor screen. The divergence, or spreading, of the electron beam results in magnification of the image cast by the objective lens onto the photocathode. The electron telescope is similar in certain ways to the electron microscope. The photocathode takes the place of the object to be magnified in the microscope. *See also* ELECTRON MICROSCOPE.

ELECTRON TUBE

See TUBE.

ELECTRON VOLT

The electron volt is a unit of electrical energy. An electron, or any object having a unit charge, moving through a potential difference of 1 volt, has an energy of 1 electron volt (abbreviated eV). An electron volt is a very small amount of energy. It is only about 1.6×10^{-19} joule (*see* ENERGY, JOULE). Common units of energy in this sytem are the kilo-electron volt (keV), mega-electron volt (MeV), and giga-electron volt (GeV or BeV); these units represent, respectively, a thousand, a million, and a billion electron volts.

The electron volt is generally specified to represent the energy contained in fast-moving atomic particles. This may include electrons, protons, alpha particles, or heavier nuclei. Photons may also be assigned an energy in electron volts; the amount of energy in a photon of 1.24×10^{-6} meters, or 12,400 Angstrom units in the infrared spectrum, is equal to 1 eV. *See also* ELECTRON, ENERGY.

ELECTROPHORESIS

Electrophoresis is a method of coating a metal with an insulating material. A suspension is prepared, consisting of dielectric particles in a liquid. Two electrodes are immersed into the suspension, and a voltage is applied between the electrodes. The drawing illustrates the basic arrangement (A).

When the current flows in the suspension, particles of the dielectric material are attracted to, and desposited on, the anode or positively charged electrode. The resulting coating of insulation adheres very well. Electrophoresis is a superior means of applying insulation to certain metals.

Electrophoresis effects are useful in the manufacture of certain kinds of electronic displays. Charged particles, suspended in a colored liquid, will migrate toward the top or the bottom of a flat container, depending on the polarity of an applied electric field. The container is extremely thin, so that the particles can move from one face to the other in a short time. The liquid in which the particles are suspended is colored or cloudy so that the particles will be seen only when they are at the top of the container. Such an arrangement, sometimes called an electrophoretic display, is shown at B. The familiar liquid-crystal display operates according to a principle much like this. *See also* LIQUID-CRYSTAL DISPLAY.

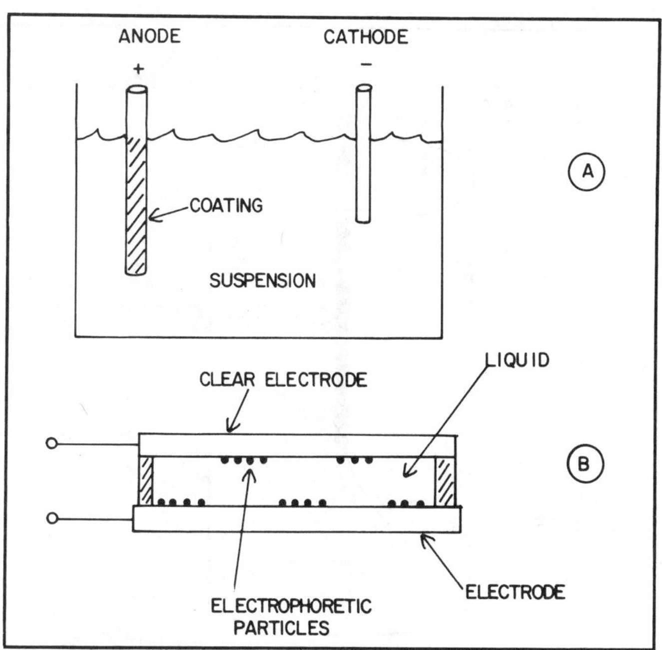

ELECTROPHORESIS: At A, electrophoresis process is used to coat an electrode with an insulating substance. At B, the electrophoresis principle allows the operation of a display.

ELECTROPLATING

Electroplating is a method of coating one substance with another substance by means of electrolysis. This is quite often done with metals, to reduce the rate at which they corrode; a less corrosive metal plating is applied to the surface of a metal that tends to corrode easily. This is especially important in marine environments, since the chlorine ions from salt-water vapor can cause extreme problems with corrosion in metals. Tropical climates, because of their high humidity and temperature, are also very hard on some metals.

Electroplating allows two dissimilar metals, which would normally react with each other, to be put in electrical contact. Both metals are electroplated with the same coating metal. For example, copper and iron may both be electroplated with tin. This results in a good electrical contact that is immune to interaction between dissimilar metals.

Aluminum, steel, and zinc should not be used in an outdoor environment, especially where corrosion is a problem, without first being electroplated with a less reactive metal. *See also* ELECTROCHEMICAL EQUIVALENT, ELECTROLYSIS.

ELECTROSCOPE

An electroscope is a device that is used for detecting an electrostatic charge. Such a device is quite simple, as illustrated in the drawing, consisting of a metal rod with a ball at one end and a pair of gold-foil leaves at the other end. The gold leaves are usually contained in a sealed glass jar with an insulating cover. The air in the jar is kept clean and dry to minimize discharging effects.

When an electrostatic charge, consisting of either a surplus of electrons (negative charge) or a shortage of electrons (positive charge) is placed on the electroscope rod, the gold-foil leaves will stand apart from each other.

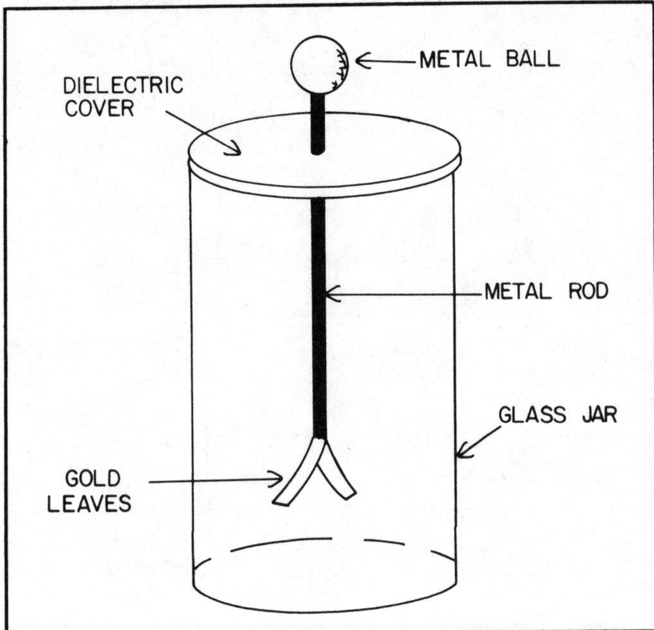

ELECTROSCOPE: An electroscope is used to detect the presence of electrostatic charges. A charge, either positive or negative, causes the foil leaves to stand apart.

The metal ball prevents the charge from escaping through the rod and into the surrounding air.

An electric field will cause the leaves of the electroscope to stand apart; this makes the electroscope useful for detecting the presence of such a field. The field causes electrons to be driven either onto or away from the gold-foil leaves. *See also* CHARGE, ELECTRIC FIELD.

ELECTROSTATIC DEFLECTION

When a beam of electrons or other charged particles is passed through an electric field, the direction of the beam is altered. Since a beam of electrons is composed of particles having negative charge, the beam will be accelerated away from the negative pole and toward the positive pole of an electric field, as shown in the illustration.

The phenomenon of electrostatic deflection is used in the oscilloscope cathode-ray tube. A pair of deflecting plates causes the electron beam in such a tube to be deflected; if an alternating field is applied, the electron beam swings back and forth or up and down. This causes movement of the spot of the waveforms applied to the deflecting plates.

Electrostatic deflection is important in the operation of many different kinds of devices. Some television picture tubes make use of this effect, although electromagnetism is more often used in this kind of device. *See also* CATHODE-RAY TUBE. ELECTROMAGNETIC DEFLECTION, OSCILLOSCOPE.

ELECTROSTATIC FIELD

See ELECTRIC FIELD.

ELECTROSTATIC FLUX

See ELECTRIC FLUX.

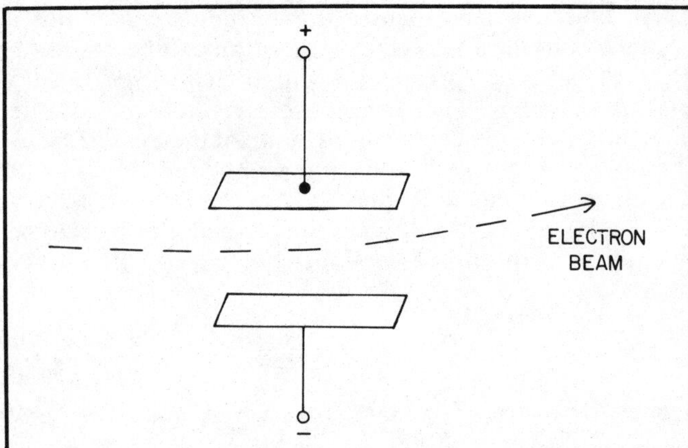

ELECTROSTATIC DEFLECTION: Electrostatic deflection causes the path of an electron beam to be bent away from the negative electrode and toward the positive electrode.

ELECTROSTATIC FORCE

Electrostatic force is the attraction between opposite electric charges and the repulsion between like electric charges. The electrostatic force depends on the amount of charge on an object or objects, and the amount of distance separating the objects. It also depends on the medium between the objects.

Electrostatic forces cause movement of the electrons in substances subjected to electric fields. Electrostatic forces are also responsible for the action of the deflecting plates in a cathode-ray oscilloscope, and in some other devices. The flow of current in a conductor or semiconductor is made possible by electrostatic forces. The force is generally measured in newtons per coulomb of electric charge. *See also* CHARGE.

ELECTROSTATIC GENERATOR

Any device that generates an electrostatic charge is called an electrostatic generator. A common form of electrostatic generator is the Van de Graaff generator (*see* VAN DE GRAAFF GENERATOR).

When you walk with hard-soled shoes on a carpeted floor, especially in dry or cool weather, you become a sort of electrostatic generator. A charge is built up on your body with respect to the surrounding objects because of friction between the shoe soles and the carpet material. This potential difference can become quite large, sometimes as much as several thousand volts. All electrostatic generators operate on principles similar to the scuffing of shoes on a carpet. The largest Van de Graaff generators can generate millions of volts in this way, and the resulting spark discharge can jump across an entire room.

Of course, the largest known electrostatic generators are the cumulonimbus clouds that we see in thunderstorms. We have all witnessed the discharges that result from the generation of charge differences in such clouds. *See also* LIGHTNING.

ELECTROSTATIC HYSTERESIS

In a dielectric substance, the polarization of the electric field within the material does not always follow the polarization of the external field under alternating-current conditions. A small amount of delay, or lag, occurs, because

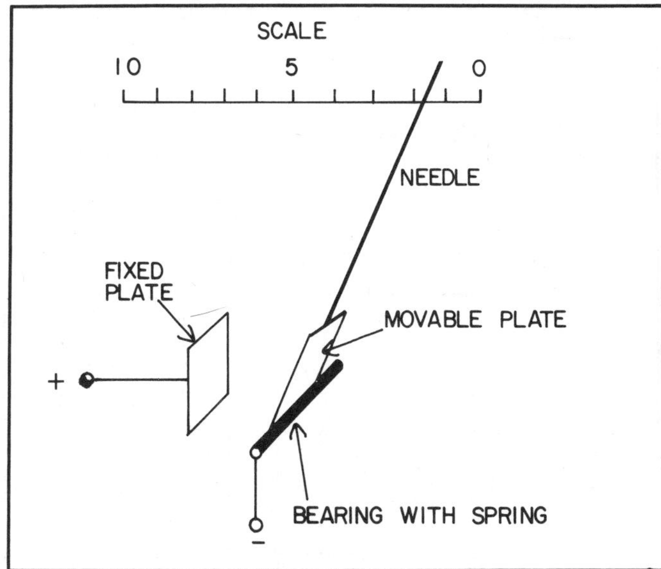

ELECTROSTATIC INSTRUMENT: An electrostatic voltmeter operates by means of the attraction between opposite poles. Such a meter can, with appropriate peripheral devices, be used to measure current or power, as well as voltage.

any dielectric substance requires some time to charge and discharge. This effect is called electrostatic hysteresis.

Electrostatic hysteresis is not usually significant, in most dieletrics, at the lower radio frequencies. As the frequency is increased into the very high and ultra high ranges, however, the effects of electrostatic hysteresis become noticeable. Different dielectric materials exhibit different electrostatic hysteresis properties. Ferroelectric substances are especially noted for their large amount of electrostatic hysteresis. *See also* DIELECTRIC ABSORPTION, FERROELECTRICITY, HYSTERESIS.

ELECTROSTATIC INDUCTION

When any object is placed in an electric field, a separation of charge occurs on that object. Electrons move toward the positive pole of the electric field, and away from the negative pole, within the object. This creates a potential difference between different regions in the object.

An example of electrostatic induction is provided by bringing a charged object near the ball of an electroscope (*see* ELECTROSCOPE). If the object is negatively charged, the gold-foil leaves of the electroscope will receive a surplus of electrons. This is because the electrons will be driven away from the ball. If the object brought near the ball is positively charged, electrons will be attracted to the ball, and away from the gold-foil leaves. In either case, the leaves will stand apart, indicating that they have a charge, even though no actual contact has been made between the charged object and the electroscope. *See also* CHARGE.

ELECTROSTATIC INTSTRUMENT

An electrostatic instrument is a form of metering device that operates via electrostatic forces only. Generally, a stationary, fixed metal plate is mounted near a rotatable plate, creating a sort of air-variable capacitor. The indicating needle is attached to the rotatable plate. A simplified diagram of an electrostatic instrument is shown in the drawing.

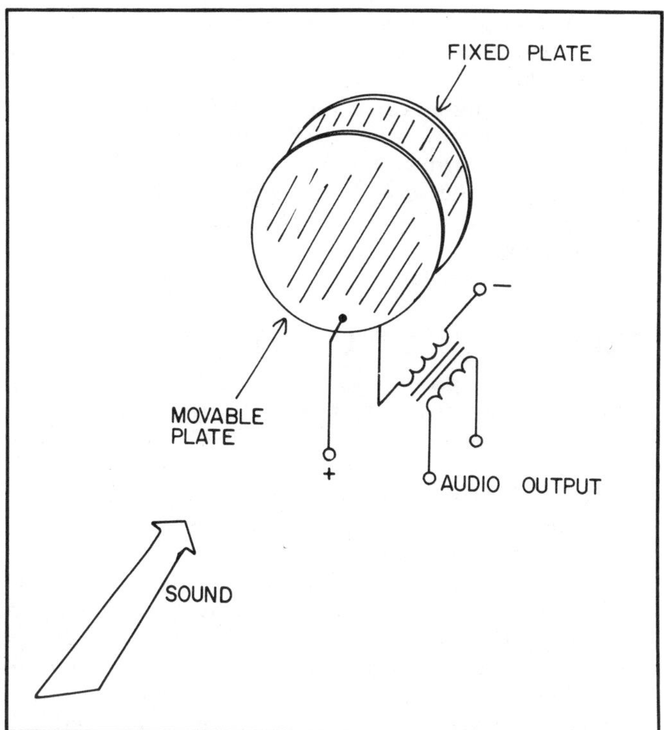

ELECTROSTATIC MICROPHONE: An electrostatic microphone resembles a variable capacitor. As the capacitance between the plates changes because of impinging sound waves, a current is produced at the output terminals.

When a voltage is applied between the two metal plates, they are attracted to each other because of their opposite polarity (*see* ELECTROSTATIC FORCE). A spring, attached to the rotatable plate, works against this force to regulate the amount of rotation. The greater the voltage between the plates, the greater the force of attraction, and the farther the rotatable plate will turn against the tension of the spring.

An electrostatic voltmeter draws very little current. Only the tiny leakage current through the air between the plates contributes to current drain. Thus, the electrostatic voltmeter has an extremely high impedance. When the appropriate peripheral circuitry is added, the electrostatic voltmeter can be used as an ammeter or wattmeter. *See also* CHARGE.

ELECTROSTATIC MICROPHONE

An electrostatic microphone is a microphone that produces a variable capacitance as sound waves strike its diaphragm. For this reason, such a microphone is sometimes called a capacitor microphone or condenser microphone.

The drawing illustrates, very roughly, the construction of an electrostatic microphone. The capacitor has one fixed plate, which is rigid. The other plate comprises the diaphragm of the device. A direct-current voltage is applied between the plates. When an acoustic disturbance causes the diaphragm to vibrate, the instantaneous separation between the plates is affected. This causes the capacitance between the plates to fluctuate, exactly in step with the sound waves impinging on the diaphragm.

A variable capacitance, applied across a source of direct-current voltage, results in a certain amount of charging and discharging. In the electrostatic microphone, the changing capacitance between the diaphragm and the fixed plates causes current to flow because of the changing amount of charge. This current can be amplified, and the result is an excellent reproduction of the sound waves in the form of electrical impulses. *See also* MICROPHONE.

ELECTROSTATIC POTENTIAL

Any electric field produces a voltage gradient, measured in volts per meter of distance. Two points in space, separated by a given distance within the influence of an electric field, will have a voltage difference. This is called an electrostatic potential. The voltage between the two electrodes, from which the field results, is the largest possible electrostatic potential within the field.

An electrostatic potential between two objects produces an electrostatic force, and this force is directly proportional to the potential difference for a given constant separation distance. For a given amount of electrostatic potential between two objects, the force drops off as the distance increases. Electrostatic potential is measured by a device that draws very little current. This is necessary, because although the voltages may be quite large, the total amount of electric charge may be rather small. *See also* ELECTROSTATIC FORCE, ELECTROSTATIC INSTRUMENT.

ELECTROSTATIC PRECIPITATION

Electrostatic precipitation is an effect produced on particles in a gas or liquid by an electric field. In a sufficiently intense electric field, particles become ionized, or charged. The charge causes the particles to be attracted toward one of the field-generating electrodes. Positive ions are attracted to the negative electrode, and negative ions are attracted to the positive electrode.

Electrostatic precipitation is utilized to advantage in the dust precipitator, which is a device for cleaning the air. The effect of electrostatic precipitation is also useful in a process called electrophoresis, in which metals may be coated with insulating material by applying a charge to a pair of electrodes immersed in a suspension. *See also* DUST PRECIPITATOR, ELECTROPHORESIS.

ELECTROSTATICS

Electrostatics is a branch of physics. The science of electrostatics involves the behavior of electric charges at rest, when no current is flowing. Such charges produce forces among objects on which they exist. Other objects or particles in the vicinity are affected by these forces. Electrostatics differs from electrodynamics, which is concerned with the behavior of moving charges. *See also* CHARGE, ELECTRIC FIELD, ELECTRIC FLUX, ELECTRODYNAMICS, ELECTROSTATIC FORCE.

ELECTROSTATIC SPEAKER

An electrostatic speaker is a transducer that converts audio-frequency electric currents into sound waves by means of electrostatic forces. The speaker consists of two fixed metal plates and a thin, flexible plate in between. The illustration is a simplified diagram of the construction of an electrostatic speaker. The fixed plates are perforated to allow the transmission of acoustical energy.

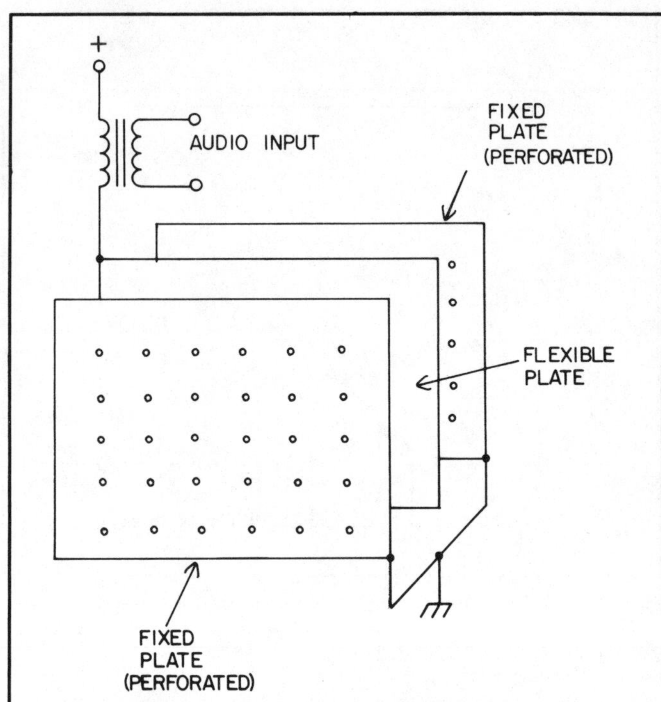

ELECTROSTATIC SPEAKER: The electrostatic speaker operates by means of electrical attraction and repulsion between a set of rigid plates and a movable, flexible plate.

When an audio-frequency signal exists at the center-tapped secondary of the transformer, the electric field between the fixed plates is made to fluctuate. A direct-current voltage, applied between the fixed plates and the flexible plate, causes forces to be exerted on the flexible plate. When there is no audio-frequency input, these forces are equal and opposite. But when a signal is applied to the speaker, the voltage between the fixed plates varies, and the center diaphragm moves back and forth.

Electrostatic speakers have almost no depth,and are very light in weight. For good low-frequency response, however, electrostatic speakers must be large. Typical electrostatic speakers for use with high-fidelity equipment range from 2 to 3 feet square. *See also* SPEAKER.

ELECTROSTATIC SHIELDING

Electrostatic shielding, also commonly known as Faraday shielding, is a means of blocking the effects of an electric field, while allowing the passage of magnetic fields. Electrostatic shielding reduces the amount of capacitive coupling between two objects to practically zero. However, inductive coupling is not affected.

An electrostatic shield consists of a wire mesh, a screen, or sometimes a plate of non-magnetic metal such as aluminim or copper, grounded at a single point. This causes the mesh, screen, or plate to acquire a constant electric charge, but little or no current can flow in it. Such a shield, placed between the primary and secondary windings of an air-core transformer as shown at A in the illus-

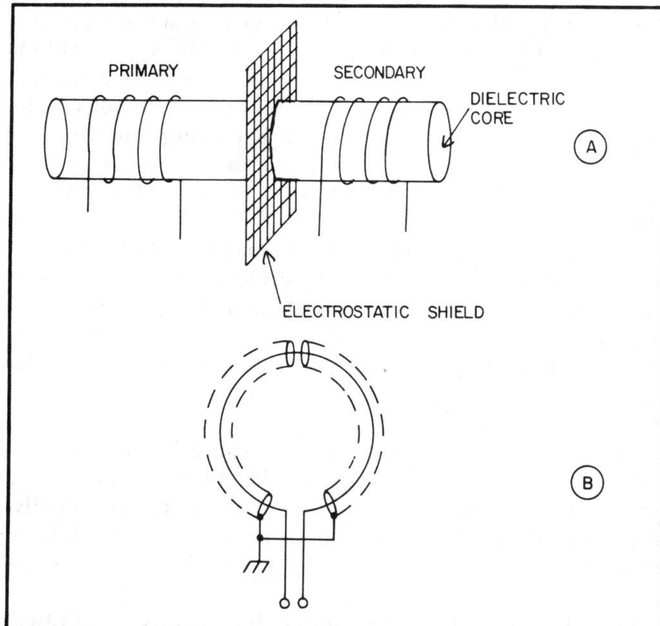

ELECTROSTATIC SHIELDING: Electrostatic shielding can be used in a radio-frequency transformer, as shown at A, to reduce capacitive coupling between the coils. At B, a loop antenna can be electrostatically shielded to improve directional effects and reduce noise pickup.

tration, practically eliminates capacitive interaction between the windings. Electrostatic shields are often used in the tuned circuits of radio-frequency equipment, when transformer coupling is employed. The shield, by eliminating the electrostatic coupling between stages, reduces the amount of unwanted harmonic and spurious-signal energy that is transferred from stage to stage (*see* TRANSFORMER COUPLING).

An electrostatic shield is sometimes used in the construction of a direction-finding loop antenna, as shown at B. A section of nonmagnetic tubing or braid is placed around, and insulated from, the loop itself. The shield is grounded at the feed point, but is broken at the top, preventing the circulation of current but maintaining a fixed electric charge. This allows the magnetic component of an electromagnetic wave to get to the loop, but the electric component is kept out. This enhances the directional characteristics of the antenna. It also may improve the signal-to-noise ratio. *See also* LOOP ANTENNA.

ELECTROSTATIC VOLTMETER

See ELECTROSTATIC INSTRUMENT.

ELECTROSTRICTION

When an electric field is applied to a dielectric material, an inward force is created on the material. This is called electrostriction.

The charge within a dielectric causes attraction between one face of the dielectric and the other. The electrodes, having opposite polarity, are also attracted to each other. If the intensity of the electric field becomes too great, the dielectric may physically crack under the stress of electrostriction. This sometimes happens in disk capacitors and other ceramic devices when they are subjected to excessive voltages. *See also* ELECTROSTATIC FORCE.

ELEMENT

An element is one of the fundamental building blocks of matter. According to contemporary theories of matter, all elements are made up of atoms. The atoms consist of a nucleus, having some positively charged particles and some neutral particles, and the electrons, which are negatively charged and orbit around the nucleus. The positively charged particles in the nucleus are called protons. The number of protons in an atom is normally the same, or about the same, as the number of electrons, so that the atom has no net charge. The neutral particles are called neutrons.

An atom may have only one proton and one electron; this is the simplest atom, hydrogen. Hydrogen is the most abundant element in the universe. Some atoms have more than 100 protons and electrons. The elements with many protons and electrons are called heavy elements, and they tend to be unstable, breaking up into elements with fewer protons and electrons.

The table under the listing ATOMIC NUMBER is an alphabetical list of the known elements, according to their atomic number (number of protons) and atomic weight (relative mass). In general, the greater the atomic number, the greater the atomic weight, although there are occasional exceptions to this rule (*see* ATOM, ATOMIC WEIGHT).

Elements may exist in the form of a gas, a liquid, or a solid. Elements often combine to form compounds (*see* COMPOUND). Elements and compounds may exist in the same medium without being chemically attached; such a combination is called a mixture. Elements, compounds, and mixtures make up all of the matter in the universe.

The term element is sometimes used to define a component in an electrical or electronic system. This is true especially of antennas. The antenna element that receives the energy directly from the feed line is called the driven element. Some antennas have elements that are not connected to the feed line directly; these are called parasitic elements. *See also* DRIVEN ELEMENT, ELECTRON, NEUTRON, PARASITIC ELEMENT, PROTON.

ELEMENT SPACING

In an antenna having more than one element, such as a parasitic or phased array (*see* PARASITIC ARRAY, PHASED ARRAY), the element spacing is the free-space distance, in wavelengths, between two specified antenna elements. This might be the director spacing with respect to the driven element or another director; it might be the reflector spacing with respect to the driven element; it might be the spacing between two driven elements. In an antenna with several elements, the spacing between adjacent elements may differ depending on which elements are specified.

For a given frequency in megahertz, the spacing in wavelengths is given by the simple formula:

$$s = df/300$$

if d is in meters. If d is in feet, then:

$$s = df/984$$

At shorter wavelengths, d may be given in centimeters or

inches. If d is in centimeters, then:

$$s = df/30,000$$

and if d is in inches, then:

$$s = df/11,800$$

See also DIRECTOR, DRIVEN ELEMENT, QUAD ANTENNA, YAGI ANTENNA.

ELEVATION

Elevation is the angle, in degrees, that an object in the sky subtends with respect to the horizon. It is also specified as the vertical deviation of the major lobe of an antenna from the horizontal. The smallest possible angle of elevation is 0 degrees, which represents the horizontal; the largest possible angle of elevation is 90 degrees, which represents the zenith.

Elevation is also sometimes called altitude. It is one of two coordinates needed for uniquely determining the direction of an object in the sky. The other coordinate, called azimuth, represents the angle measured clockwise around the horizon from true north (*see* AZIMUTH).

Astronomers use a variation of the azimuth/elevation system. The coordinates, rather than being fixed with respect to the horizon of the earth at a particular location, are fixed with respect to the heavens. The equivalent of azimuth is called right ascension; the equivalent of elevation is called declination. The declination in this system may vary between +90 degrees (toward the north celestial pole) and −90 degrees (toward the south celestial pole). *See also* DECLINATION, RIGHT ASCENSION.

ELLIPTICAL POLARIZATION

The polarization of an electromagnetic wave is the orientation of the electric lines of flux in the wave. Often, this orientation remains constant; but sometimes it is deliberately made to rotate as the wave propagates through space. If the orientation of the electric lines of flux changes as the signal is propagated from the transmitting antenna, the signal is said to have elliptical polarization. (*See* HORIZONTAL POLARIZATION, POLARIZATION, VERTICAL POLARIZATION.)

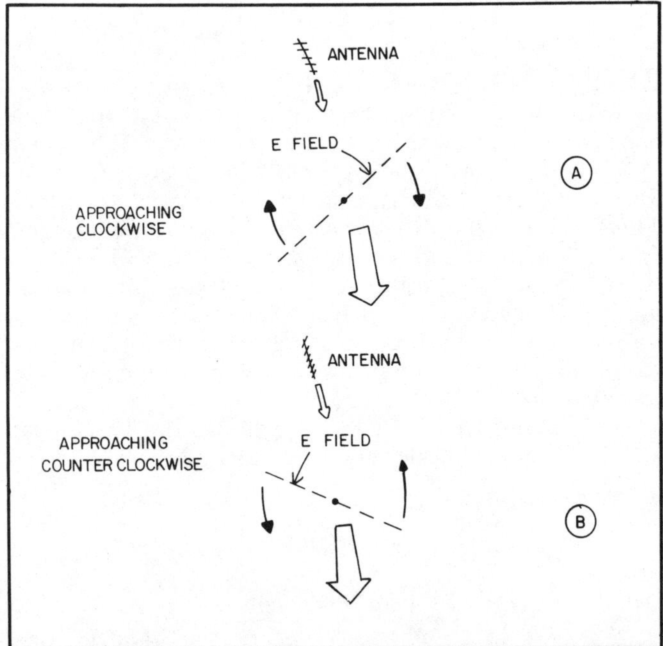

ELLIPTICAL POLARIZATION: At A, clockwise elliptical polarization; at B, counterclockwise elliptical polarization. Only the electric (E) field line is shown; the magnetic field line is perpendicular to the electric field line. Both rotate in the same sense.

An elliptically polarized electromagnetic field may rotate either clockwise or counterclockwise as it moves through space. This is shown in the illustration. The intensity of the signal may or may not remain constant as the wave rotates. If the intensity does remain constant as the wave rotates, the polarization is said to be circular (*see* CIRCULAR POLARIZATION).

Elliptical polarization is useful because it allows the reception of signals having unpredictable or changing polarization, with a minimum of fading and loss. Ideally, the transmitting and receiving antennas should both have elliptical polarization, although signals with linear polarization can be received with an elliptically polarized antenna. If the transmitted signal has opposite elliptical polarization from the receiving antenna, however, there is substantial loss.

Elliptical polarization is generally used at ground stations for satellite communication. In receiving, the use of an elliptically polarized antenna reduies the fading caused by changing satellite orientation. In transmitting, the use of elliptical polarization ensures that the satellite will always receive a good signal for retransmission. *See also* LINEAR POLARIZATION.

EMERGENCY BROADCAST SYSTEM

In the event of a national emergency, especially the threat of a large-scale nuclear war, all normal broadcasting would cease. An attention tone would first be transmitted over all stations. This tone would be followed by specific instructions concerning the frequencies to which to tune in a given area. This system is called the Emergency Broadcast System, or EBS. All broadcasting stations in the United States are required by the Federal government to conduct periodic tests of the Emergency Broadcast System.

The reason for changing the frequencies and locations of broadcasting transmitters, in case of nuclear attack, is that enemy missiles might use the signals from broadcast stations to zero in on large cities. The Emergency Broadcast System is an updated version of the older system called Control of Electromagnetic Radiation, or CONELRAD. *See also* CONELRAD.

EMERGENCY COMMUNICATIONS

Emergency communications is the highest-priority form of communications. Messages having emergency status always take precedence over other kinds of messages. An emergency or distress signal implies a threat to life. The emergency designation is to be used only in such a situation. In some communications services, certain frequencies are set aside for use in emergency communications (*see* DISTRESS FREQUENCY, DISTRESS SIGNAL).

An emergency communications system should be periodically tested, to be sure that it is working perfectly. Such a communications system must have an independent source of power (*see* EMERGENCY POWER SUPPLY). Portable and mobile equipment can be valuable in a large-scale emergency, but fixed stations are generally capable of more radio-frequency output power.

Distress signals in radiotelegraphy are easily recognizable; the standard signal is the SOS (. . . — — — . . .). In radiotelephony, the word "Mayday" is used. An emergency message should be sent slowly and clearly. If interference is encountered, the continuous-wave mode is best. The type A1 emission, or continuous-wave Morse code, is still the best choice of emission for adverse conditions. Code transmitters are simple and can sometimes be built from receivers. *See also* CODE TRANSMITTER.

EMERGENCY POWER SUPPLY

An emergency power supply is an independent, self-contained power source that can be used in the event of a failure in the normal utility power. Such an emergency supply is usually either a set of rechargeable batteries, or a fossil-fuel generator. The primary requirements for an emergency power supply are:

- It must supply the necessary voltages for the operation of the equipment to be used.
- It must be capable of delivering sufficient current to run the equipment to be used.
- It must be capable of operating for an extended period of time.
- It must be accessible immediately when needed.

A gasoline generator is an excellent emergency power supply. Such a generator can supply 120 volts at about the standard line frequency, up to power levels of several kilowatts. Department stores often carry small gasoline-powered generators for portable use. A supply of gasoline is needed for the operation of these generators, and the gasoline should be stored safely to prevent explosions.

A set of storage batteries, such as 12-volt automotive batteries, can also be used as a source of emergency power. Many radio receivers and transmitters, as well as some specialized appliances, will run directly from 12 volts direct current. Power inverters can be used to get 120 volts alternating current from such a source. A solar recharging system can be used to keep the batteries operational. *See also* GENERATOR, INVERTER, SOLAR POWER, STORAGE BATTERY.

EMISSION CLASS

The emission class of a radio-frequency transmitter is a standard designation of the type of modulation used. Such designators consist of a letter followed by a number; sometimes another letter comes after the number. The letter A stands for amplitude modulation; the letter F stands for frequency modulation; the letter P stands for pulse modulation. Emissions are designated, for each form of modulation, by numbers from 0 to 9. The table lists the standard emission classes, with short descriptions of the characteristics of each class.

The bandwidth of a signal varies greatly with the emission class used. A signal with no modulation, called A0 or F0, takes almost no spectrum space. A television signal such as A5 might take several megahertz of space. The bandwidth of a particular type of emission may be different in one communications service, as compared with another service. For example, the standard F3 signal in the frequency-modulation broadcast band is much wider than the same kind of signal in the two-way communications service. Some kinds of emission are prohibited on certain frequencies. *See also* AMPLITUDE MODULATION, BANDWIDTH, FREQUENCY MODULATION, MODULATION, PULSE MODULATION.

EMISSION CLASS: STANDARD EMISSION CLASSIFICATIONS.

Emission	Description
A0	Unmodulated, pure carrier
A1	Carrier telegraphy
A2	Amplitude-modulated telegraphy
A3	Amplitude-modulated telephony
A3A	Single-sideband reduced-carrier telephony
A3J	Single-sideband suppressed-carrier telephony
A4	Amplitude-modulated facsimile
A5	Amplitude-modulated television
A7	Multi-channel voice-frequency amplitude-modulated telegraphy
A9	Two independent sidebands
F0	Unmodulated, pure carrier
F1	Frequency-shift keying
F2	Frequency-modulated telegraphy
F3	Frequency-modulated telephony
F4	Frequency-modulated facsimile
F5	Frequency-modulated television
F6	Four-frequency diplex telegraphy
F9	Frequency-modulated emission other than above
P0	Unmodulated-pulse carrier
P1	Keyed pulsed carrier
P2	Pulse-modulated telegraphy
P3	Pulse-modulated telephony
P9	Pulse-modulated emission other than above

EMISSION-TYPE TUBE TESTER

An emission tube tester is a device that measures the electron emission of the cathode of a vacuum tube. Such an instrument is used to determine whether or not a given tube is producing the electron emission it should. Different kinds of tubes have different emission characteristics, and sophisticated tube testers have several different sockets with adjustable voltage for the plate and filament. These voltages must be properly set before an accurate indication is possible.

The emission tube tester places a vacuum tube in a diode configuration by connecting all the grids to the plate, and supplying this connection with the positive voltage for the plate. The amount of current drain then indicates the extent of the cathode emission. A meter, either graduated in numbers from 0 to 10 (or some other range) or having regions marked "bad" and "good," is used to show the amount of current drawn by the tube. Some tube testers have an indicator that shows whether there is gas in the tube.

If a tube tester says that a tube is bad, it probably is bad. Short circuits among the grids and plate will not show up in the emission tester, however, since all of these electrodes are short-circuited when the device is used. *See also* CATHODE, FILAMENT, TUBE.

EMISSIVITY

Emissivity is the ease with which a substance emits or absorbs heat energy. This property is important in the choice of materials for, and in the design of, heatsinks (*see* HEATSINK), since such devices give up much of their heat by means of radiation. Emissivity is a measure of the brightness of an object in the infrared part of the electromagnetic spectrum. The darker the object appears in the infrared, the better the emissivity. Simply painting an object black does not guarantee that the emissivity will be excellent, since visually black paint may not be as black in the infrared region.

The capability of a material to dissipate power depends on the temperature; the higher the temperature, in general, the more power can be dissipated by a given material. Emissivity depends on the shape of an object; irregular surfaces are better for heat radiation than smooth surfaces. A sphere is the poorest possible choice for a heat emitter; in heatsink design, finned surfaces are employed to maximize the ratio of the surface area to the volume. Different substances have different emissivity characteristics under identical conditions. Graphite is one of the most heat-emissive known substances. *See also* FINNED SURFACE, INFRARED.

EMITTER

In a semiconductor bipolar transistor, the emitter is the region from which the current carriers are injected. The emitter of a transistor is somewhat analogous to the cathode of a vacuum tube, or the source of a field-effect transistor. But the actual operating principles of the bipolar-transistor emitter are different from those of the cathode or source of a tube or field-effect device.

The emitter of a bipolar transistor may be made from either an N-type semiconductor wafer (in an NPN device) or a P-type semiconductor wafer (in a PNP device). The drawing shows the schematic symbols for both of these kinds of bipolar transistors, illustrating the position of the emitter. The emitter lead is indicated by the presence of an arrowhead (*see* N-TYPE SEMICONDUCTOR, P-TYPE SEMICONDUCTOR).

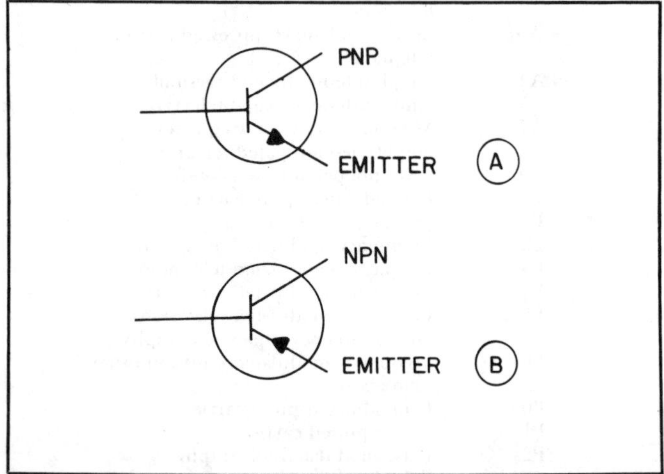

EMITTER: The emitter of an NPN transistor is shown by the arrowhead at A, pointing outward. The emitter of a PNP transistor is shown by the arrowhead at B, pointing inward.

The input of a transistor amplifier may be supplied to the emitter circuit or to the base circuit. If it is supplied to the emitter circuit, the impedance is low. The output of a transistor amplifier may be taken from either the collector or the emitter circuit; if the output is taken from the emitter circuit, the impedance is low. In either the emitter-input or emitter-output case, an amplifier is said to be emitter-coupled. Generally, the input of a transistor amplifier is supplied to the base, and the output is taken from the collector. The emitter is usually kept at signal ground potential. *See also* BASE, COLLECTOR, EMITTER-BASE JUNCTION, EMITTER-COUPLED LOGIC, EMITTER COUPLING, EMITTER CURRENT, EMITTER DEGENERATION, EMITTER FOLLOWER, EMITTER KEYING, EMITTER MODULATION, EMITTER RESISTANCE, EMITTER STABILIZATION, EMITTER VOLTAGE, TRANSISTOR.

EMITTER-BASE JUNCTION

The emitter-base junction of a bipolar transistor is the boundary between the emitter semiconductor and the base semiconductor. It is always a P-N junction (*see* P-N JUNCTION). In an NPN transistor, the emitter is made from N-type material and the base is made from P-type; in the PNP transistor it is the other way around.

In most transistor oscillators and low-level amplifiers, the emitter-base junction is forward-biased. In the NPN device, therefore, the base is usually somewhat positive with respect to the emitter, and in the PNP device, somewhat negative. The emitter-base junction of a transistor will conduct only when it is forward-biased. In this respect, it behaves exactly like a diode. For the current to pass to the collector circuit, the emitter-base junction must be forward-biased. But a small change in this bias will result in a large change in the collector current. The input signal to a bipolar-transistor circuit is always applied at the emitter-base junction.

In Class-B and Class-C amplifiers, the emitter-base junction does not conduct under conditions of zero input signal. Thus, it conducts for only part of the signal cycle. Sometimes a reverse bias is deliberately applied, and the the collector current flows for just a tiny part of input cycle. When no current flows in the emitter-base junction because of a lack of forward bias, the transistor is said to be cut off. *See also* CLASS-B AMPLIFIER, CLASS-C AMPLIFIER, TRANSISTOR.

EMITTER-COUPLED LOGIC

Emitter-coupled logic, abbreviated ECL, is a bipolar form of logic. Most digital switching circuits in the bipolar family utilize saturated transistors; they are either fully conductive or completely cut off. But in emitter-coupled logic, this is not the case. The drawing illustrates a typical emitter-coupled logic gate. The emitter-coupled logic circuit acts as a comparator between two different current levels.

The input impedance with ECL is very high. The switching speed very rapid. The output impedance is relatively low. The principal disadvantages of emitter-coupled logic are the large number of components required, and the susceptibility of the circuit to noise. The noise susceptibility of ECL arises from the fact that the transistors, not operated at saturation, tend to act as amplifiers of analog signals. *See also* DIODE-TRANSISTOR LOGIC, DIRECT-COUPLED TRANSISTOR LOGIC, HIGH-THRESHOLD LOGIC, INTEGRATED INJECTION LOGIC, METAL-OXIDE SEMICONDUCTOR LOGIC FAMILIES, RESISTOR-CAPACITOR-TRANSISTOR LOGIC, RESISTOR-TRANSISTOR LOGIC, TRANSISTOR-TRANSISTOR LOGIC, TRIPLE-DIFFUSED EMITTER-FOLLOWER LOGIC.

EMITTER COUPLING

When the output of a bipolar-transistor amplifier is taken from the emitter circuit, or when the input to a transistor amplifier is applied in series with the emitter, the device is said to have emitter coupling. Emitter coupling may be capacitive, or it may make use of transformers. The illustration shows an example of emitter coupling in a two-stage, NPN-transistor amplifier. Capacitive coupling is employed in this case. The output of the first stage is taken from the emitter.

Emitter coupling results in a low input or output impedance, depending on whether the coupling is in the input or output of the amplifier stage. Emitter input coupling is used with grounded-base amplifiers. Such circuits are occasionally used as radio-frequency power amplifiers. Emitter output coupling is used when the following stage requires a low driving impedance. A circuit utilizing emitter output coupling is sometimes called an emitter follower. *See also* EMITTER FOLLOWER.

EMITTER CURRENT

In a bipolar-transistor circuit, emitter current is the rate of charge-carrier transfer through the emitter. Emitter current is generally measured by means of an ammeter connected in series with the emitter lead of a bipolar-transistor amplifier circuit.

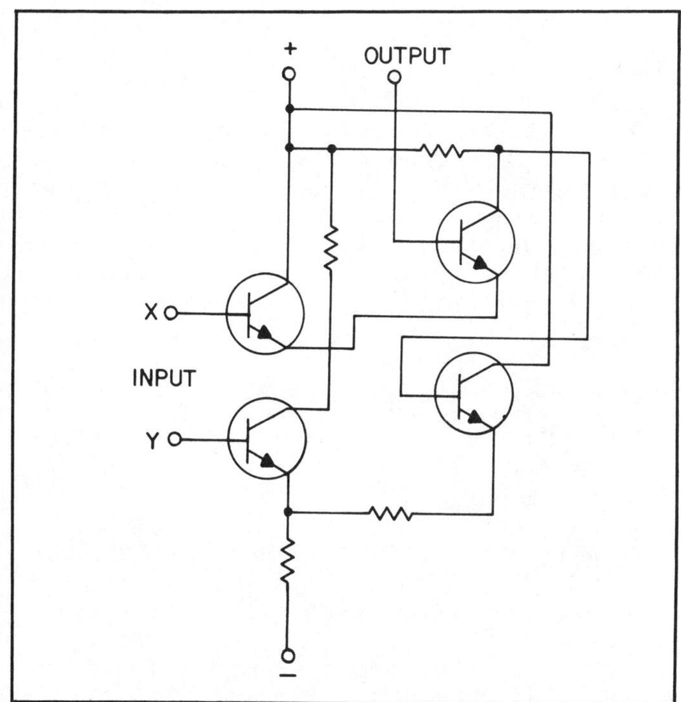

EMITTER-COUPLED LOGIC: Four bipolar transistors are used for analog switching.

Most of the emitter current flows through a transistor and appears at the collector circuit. A small amount of the emitter current flows in the base circuit. When the emitter-base junction is reverse-biased, no emitter current flows, and the transistor is cut off. As the emitter-base junction is forward-biased to larger and larger voltages, the emitter current increases, as does the base current and the collector current. The emitter current fluctuates along with the signal input at the emitter-base junction of a transistor amplifier. *See also* EMITTER-BASE JUNCTION.

EMITTER DEGENERATION

Emitter degeneration is a simple means of obtaining degenerative, or negative, feedback in a common-emitter transistor amplifier (*see* COMMON CATHODE/EMITTER/SOURCE). A series resistor is installed in the emitter lead of a transistor amplifier stage; no capacitor is connected across it.

When an alternating-current input signal is applied to the base of an amplifier having emitter degeneration, the emitter voltage fluctuates along with the input signal. This happens in such a way that the gain of the stage is reduced. The distortion level is also greatly reduced. The amplifier is able to tolerate larger fluctuations in the amplitude of the input signal when emitter degeneration is used; the output is virtually distortion-free by comparison with an amplifier that does not use emitter degeneration.

Emitter degeneration is often used in high-fidelity audio amplifiers. Distortion must be kept to a minimum in such circuits. Emitter degeneration acts to stabilize an amplifier circuit, protecting the transistor against destruction by such effects as current hogging or thermal runaway. *See also* EMITTER CURRENT, EMITTER STABILIZATION, EMITTER VOLTAGE.

EMITTER FOLLOWER

An emitter follower is an amplifier circuit in which the output is taken from the emitter circuit of a bipolar transistor. The output impedance of the emitter follower is low. The voltage gain is always less than 1; that is to say, the output signal voltage is smaller than the input signal voltage. The emitter-follower circuit is generally used for the purpose of impedance matching. The amplifier compo-

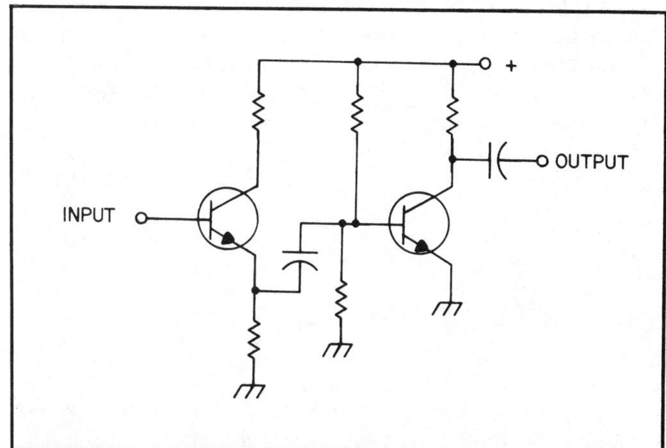

EMITTER COUPLING: The first stage output is taken from the emitter circuit. The first stage is called an emitter follower.

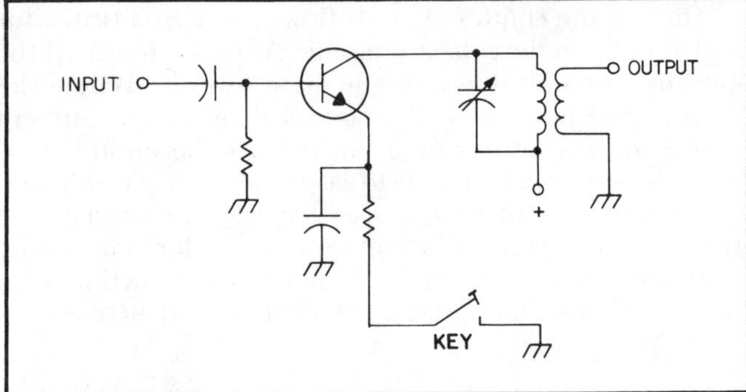

EMITTER KEYING: Emitter keying is used to accomplish code transmission. The resistor and capacitor provide shaping, or controlled rise and fall time, in the emitter current.

nents are often less expensive, and offer greater bandwidth, than transformers.

The previous illustration shows a two-stage, NPN-transistor amplifier circuit, in which the first stage is an emitter follower. The output of the emitter follower is in phase with the input. The impedance of the output circuit depends on the particular characteristics of the transistor used, and also on the value of the emitter resistor. Sometimes, two series-connected emitter resistors are employed. The output is taken, in such a case, from between the two resistors. Alternatively, a transformer output may be used. Emitter-follower circuits are useful because they offer wideband impedance matching. *See also* EMITTER COUPLING.

EMITTER KEYING

In an oscillator or amplifier, emitter keying is the interruption of the emitter circuit of a bipolar-transistor stage for the purpose of obtaining code transmission. Emitter keying completely shuts off an oscillator in the key-up condition. In an amplifier, only a negligible amount of signal leakage occurs when the key is up. The illustration shows emitter keying in a radio-frequency amplifier stage. Emitter keying is an entirely satisfactory method of keying in bipolar circuits although, in high-power amplifiers, the switched current may be very large. Usually, emitter keying is done at low-level amplifier circuits. In a code transmitter, emitter keying is often performed on two or more amplifier stages simultaneously. A shaping circuit, illustrated as a series resistor and a parallel capacitor, provides reduction of key clicks at the instants of make and break in code transmission. The series resistor, in conjunction with the parallel capacitor, slows down the emitter voltage drop when the key is closed. This allows a controlled signal rise time. When the key is released, the capacitor discharges through the resistor and the emitter-base junction, slowing down the decay time of the signal. *See also* KEY CLICK, SHAPING.

EMITTER MODULATION

Emitter modulation is a method of obtaining amplitude modulation in a radio-frequency amplifier. The radio-frequency input signal is applied to the base of a bipolar transistor in most cases, although it may be applied in series with the emitter. The audio signal is always applied in series with the emitter. The illustration shows a typical

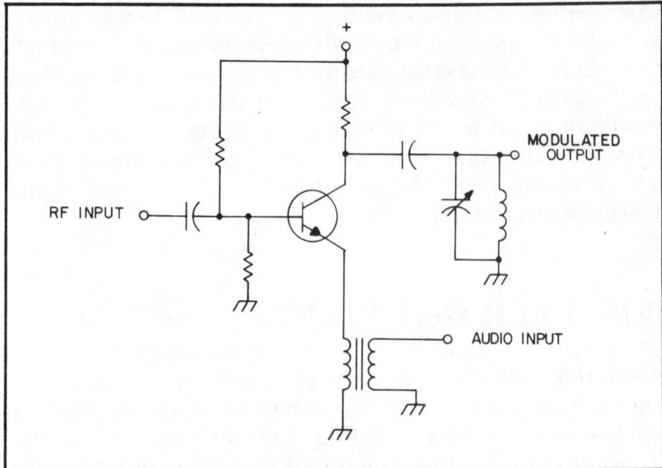

EMITTER MODULATION: In emitter modulation, the radio-frequency signal is applied to the base and the audio-frequency signal is applied in series with the emitter.

emitter-modulated radio-frequency amplifier circuit.

As the audio-frequency signal at the emitter swings negative in the NPN circuit, the current through the transistor increases. As the audio voltage swings positive, the current through the transistor decreases. When the audio voltage reaches its positive peak, the amplitude of the output signal drops to zero under conditions of 100-percent amplitude modulation.

Emitter modulation offers the advantage of requiring very little audio input power for full modulation. *See also* AMPLITUDE MODULATION, MODULATION.

EMITTER RESISTANCE

The emitter resistance of a bipolar transistor is the effective resistance of the emitter in a given circuit. The emitter resistance depends on the bias voltages at the base and collector of a transistor. It also depends on the amplitude of the input signal, and on the characteristics of the particular transistor used. The emitter resistance can be controlled, to a certain extent, by inserting a resistor in series with the emitter lead in a transistor circuit.

The external resistor in the emitter lead of a bipolar transistor oscillator or amplifier is, itself, sometimes called the emitter resistance. Such a resistor is used for impedance-matching purposes, for biasing, for current limiting, or for stabilization. The emitter resistance may have a capacitor connected in shunt. *See also* EMITTER CURRENT, EMITTER STABILIZATION, EMITTER VOLTAGE.

EMITTER STABILIZATION

Emitter stabilization is a means of preventing undesirable effects that sometimes occur in transistor amplifier circuits as the temperature varies. Emitter stabilization is used in common-emitter transistor amplifiers to prevent the phenomenon known as thermal runaway (*see* THERMAL RUNAWAY). A resistor, connected in series with the emitter lead of the transistor, accomplishes the objective of emitter stabilization.

If the collector current increases because of a temperature rise in the transistor, and there is no resistor in series with the emitter lead, the increased current will cause further heating of the transistor, which may in turn cause more current to be drawn in the collector circuit. This results in a vicious circle. It can end with the destruction of the transistor base-collector junction. If an emitter-stabilization resistor, having the proper value, is installed in the circuit, the collector-current runaway will not take place. Any increase in the collector current will then cause an increase in the voltage drop across the resistor; this will cause the bias at the emitter-base junction to change in such a way as to decrease the current through the transistor. The emitter-stabilization resistor thus regulates the current in the collector circuit. A decrease in the current through the transistor will, conversely, cause a decrease in the voltage drop across the emitter resistor, and a tendency for the collector current to rise.

Emitter stabilization is especially important in transistor amplifiers using two or more bipolar transistors in a parallel or push-pull common-emitter configuration. *See also* COMMON CATHODE/EMITTER/SOURCE, EMITTER CURRENT, EMITTER VOLTAGE.

EMITTER VOLTAGE

In bipolar transistor circuits, the emitter voltage is the direct-current potential difference between the emitter and ground. If the emitter is connected directly to the chassis ground, then the emitter voltage is zero. But if a series resistor is used, then the emitter voltage will not be zero. In such a case, if an NPN transistor is employed, the emitter voltage will be positive; if a PNP transistor is used, it will be negative. The emitter voltage increases as the current through the transistor increases, and also as the value of the series resistance is made larger. In the common-collector configuration, when the collector is directly connected to chassis ground, the emitter voltage is negative in the case of an NPN circuit, and positive in the case of a PNP circuit. Letting I be the emitter current in a transistor circuit, and R the value of the series resistor in the emitter circuit, specified in amperes and ohms respectively, then the emitter voltage V_e is given in volts by:

$$V_e = I_e R_e$$

A capacitor, placed across the emitter resistor, keeps the emitter voltage constant under conditions of variable input signal. If no such capacitor is employed, the emitter voltage will fluctuate along with the input cycle. *See also* EMITTER CURRENT, EMITTER DEGENERATION, EMITTER STABILIZATION.

EMPIRICAL DESIGN

Empirical design is a method of circuit engineering in which experience and intuition are used more than theoretical techniques. The empirical-design process can be described as a trial-and-error method of arriving at an optimum circuit design.

Empirical design is employed by the professional as well as by the hobbyist. Engineers are all familiar with the phenomenon in which a circuit performs well "on paper," but inadequately in practice!

Sophisticated circuits are generally designed theoretically before they are put together for testing and debugging. But more simple circuits, such as oscillators and amplifiers, can often be thrown together and then tailored to perfection by empirical methods. Such an approach often saves considerable time in these situations. *See also* DESIGN.

ENABLE

The term enable means the initialization of a circuit. To enable a device is to switch it on, or make it ready for operation. This can be done by means of a simple switch, or by a triggering pulse. The initializing command or pulse is itself called the enable command or pulse. The term enable is used mainly in computer and digital circuit applications.

When a magnetic core is induced to change polarity by an electronic signal, the signal is called an enable pulse. This signal allows a binary bit to be written into, or removed from, the magnetic memory. *See also* MAGNETIC CORE, MEMORY.

ENAMELED WIRE

When a wire is insulated by a thin coating of paint, it is called enameled wire. Enameled wire is available in sizes ranging from smaller than AWG 40 to larger than AWG 10 (*see* AMERICAN WIRE GAUGE). Enameled wire is often preferable to ordinary insulated wire, because the enamel does not add significantly to the wire diameter.

Enameled wire is commonly used in coil windings, when many turns must be put in a limited space. Examples of such situations include audio-frequency chokes, transformers, and coupling or tuning coils. Some radio-frequency chokes also use enameled wire. Whenever miniaturization is important, enameled wire will probably be found. Because the enamel is thin, however, this type of wire is not suitable for high-voltage use. *See also* WIRE.

ENCAPSULATION

When a group of discrete components is enclosed or embedded in a rigid material such as wax or plastic, the circuit is said to be encapsulated. Encapsulation protects the components against physical damage. Encapsulation also reduces the effects of mechanical vibration on the operation of a circuit.

The photograph shows a group of components on a printed-circuit board. The components are encapsulated in wax. This particular set of components is a voltage-controlled oscillator in a frequency-modulated transceiver. The transceiver is designed for mobile and portable operation, and it is therefore subjected to much mechanical vibration. The wax, by absorbing the shock, reduces the effects of vibration on the voltage-controlled oscillator. Without the wax, microphonics would be a problem. The transmitted and/or received signals might be noticeably modulated whenever the transceiver was moved or bumped. *See also* MICROPHONICS.

ENCLOSURE

An enclosure is the cabinet or box in which a circuit is housed. Enclosures can have almost any size and shape, and they may be made from almost any material, although

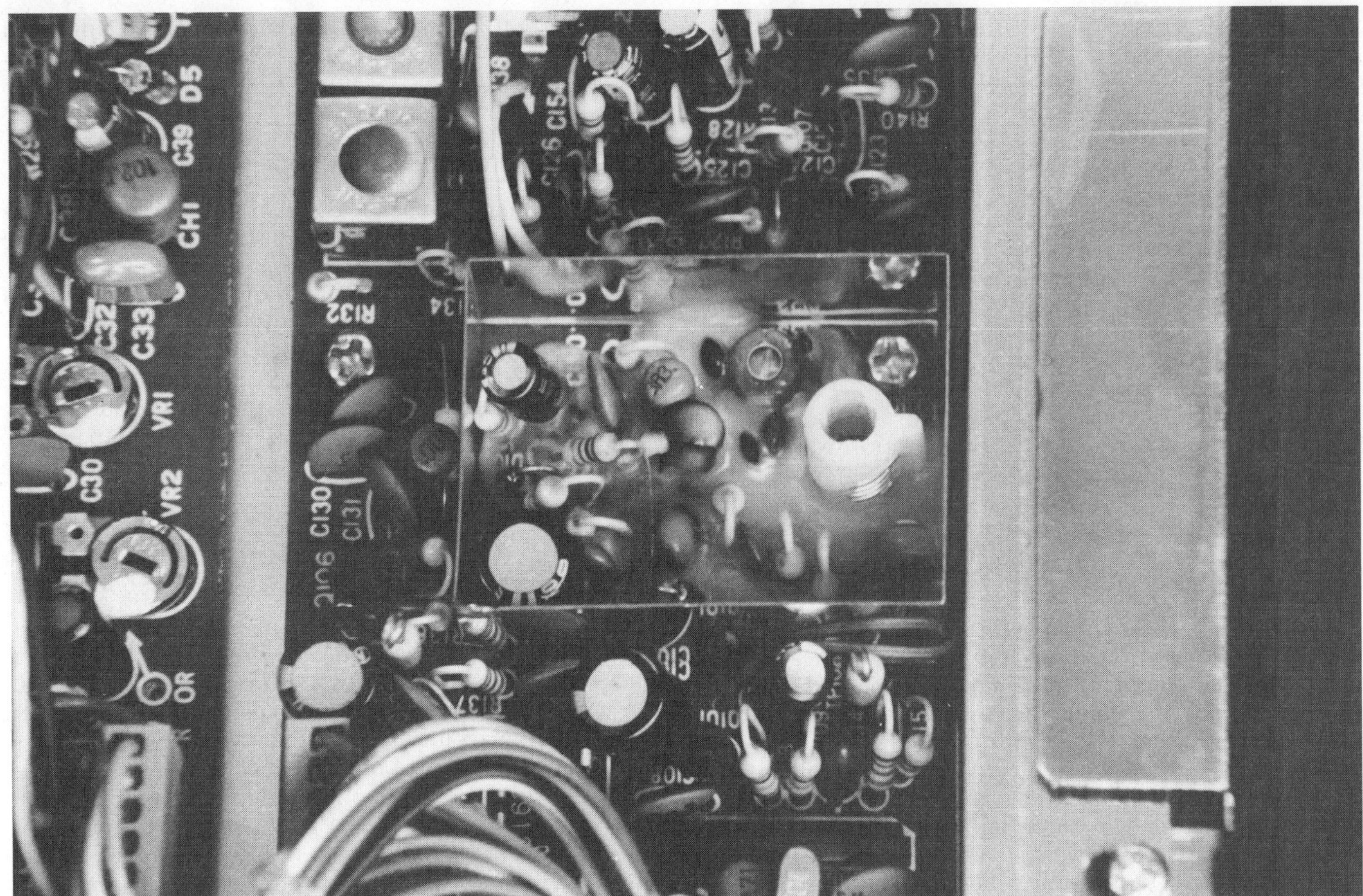

ENCAPSULATION: This voltage-controlled oscillator circuit(at center) is encapsulated with wax to minimize the effects of mechanical vibration.

they are generally made of metal or plastic. The choice of a proper enclosure is extremely important in the design of any piece of electronic equipment. The enclosure must be made of metal if electromagnetic shielding is needed. It must be large enough to allow room for ventilation if any of the components generate heat in operation. The front-panel arrangement must be functional and reasonably attractive. In consumer items, the overall appearance of the enclosure is very important, and can greatly affect the sales of an electronic device.

ENCLOSURE: Radio-frequency power amplifiers are often placed in shielded enclosures to prevent interaction with other parts of a transmitter. When the top cabinet is in place, the shielding completely surrounds the amplifier components.

Within the main enclosure of a piece of apparatus, there are often other, smaller enclosures. The photograph shows a metal enclosure for the final-amplifier components of a high-frequency transmitter. When the top part of the outer cabinet (which has been removed for the photograph) is replaced, the final-amplifier components are shielded from the surrounding circuitry. This reduces coupling between the final amplifier and other parts of the transmitter; such coupling might otherwise result in oscillation. *See also* ELECTROMAGNETIC SHIELDING.

ENCODER

An encoder is a device that translates a common language into a code. Such a device often has a keyboard resembling that of a typewriter. When the operator presses a particular key, the machine produces the code element corresponding to that character. The code may have any of many different forms. Some of the most common are the ASCII, BAUDOT, and Morse codes (*see* ASCII, BAUDOT, CODE, MORSE CODE).

An analog-to-digital converter may be called an encoder if a communications system uses digital transmission. In general, any device that interfaces a human operator with a transmission medium, for the purpose of sending signals, can be called an encoder. At the receiving end of the circuit, a decoder performs exactly the opposite function, interfacing the transmission medium with another human operator or machine. *See also* ANALOG-TO-DIGITAL CONVERTER, DECODING, DIGITAL-TO-ANALOG CONVERTER, ENCODING.

ENCODING

Encoding is the process of translating a common spoken or written language, such as English, or Japanese, into code. This process may be performed by a mechanical or electronic machine called an encoder (*see* ENCODER), or it may, in some cases, be done by a human operator. A Teletype machine is an electronic encoder (some of the older machines are mechanical). A radiotelegraph operator is a human encoder and decoder!

Under certain circumstances, a message may be encoded two or more times. For example, when a radiotelegrapher sends a Q signal (*see* Q SIGNAL), he performs two encoding operations. The letters QRN, for example, mean "I am bothered by static interference." The radiotelegrapher first mentally encodes that statement as the sequence of letters QRN; he then sends the letters QRN in Morse code. Double encoding is very common in communications practice.

Most codes are digital in nature. The spoken voice is an analog phenomenon. In digital voice transmission, the process of encoding is usually performed by an analog-to-digital converter. *See also* ANALOG-TO-DIGITAL CONVERTER, CODE, DECODING.

END EFFECT

When a solenoidal inductor carries an alternating current, some of the magnetic lines of flux extend outside the immediate vicinity of the coil. Although the greatest concentration of magnetic flux lines is inside the coil itself, especially if the coil has a ferromagnetic core such as powdered iron, the external flux can result in coupling with other components in a circuit. Most of the external magnetic flux appears off the ends of a solenoidal coil; hence it is sometimes called the end effect. In recent years, toroidal inductor cores have become more commonly used, because toroidal coils do not exhibit the end effect (*see* TOROID).

In an antenna radiator, the impedance at any free end is theoretically infinite; that is, no current should flow. In practice, a small charging current exists at a free end of any antenna, because of capacitive coupling to the environment. This is called the end effect. The amount of end capacitance, and therefore the value of the charging current, depends on many factors, including the conductor diameter, the presence of nearby objects, and even the relative humidity of the air surrounding the antenna. The end effect reduces the impedance at the feed point of an end-fed antenna or a center-fed antenna measuring a multiple of one wavelength. *See also* END FEED.

END FEED

When the feed point of an antenna radiator is at one end of the radiator, the arrangement is called end feed. End feed always constitutes voltage feed, since the end of an antenna radiating element is always at a current node (*see* VOLTAGE FEED). End feed is usually accomplished by means of a two-wire balanced transmission line. One of the feed-line conductors is connected to the radiator, and the other end is left free, as shown in the illustration.

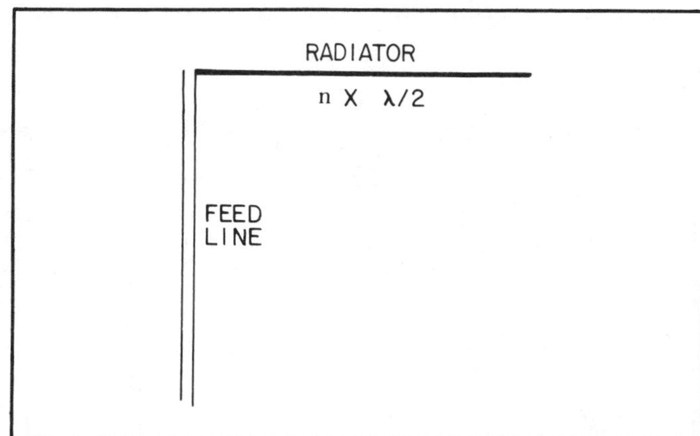

END FEED: An end-fed antenna always measures a multiple of ½ wavelength. The feed line, usually a two-wire line, may be any length.

In an end-fed antenna system, the radiator must have an electrical length that is an integral multiple of ½ wavelength. If f is the frequency in megahertz, then ½ wavelength, denoted by L, is given approximately by the formulas:

$$L \text{ (meters)} = 143/f$$

$$L \text{ (feet)} = 468/f$$

The length of the radiating element is quite critical when end feed is used. If the antenna is slightly too long or too short, the current nodes in the transmission line will not occur in the same place. this will result in unbalance between the two conductors of the line, with consequent radiation from the line. Even if the antenna radiator is exactly a multiple of ½ electrical wavelength, the end effect will cause the terminated end of the line to have a somewhat different impedance than the unterminated end (*see* END EFFECT). This, again, results in unbalance between the currents in the two line conductors. The zeppelin, or zepp, antenna is the most common example of an end-fed system. End feed is often much more convenient, from an installation standpoint, than other methods of antenna feed. This is particularly true at the lower frequencies. The slight amount of radiation from the feed line in a zepp antenna can be tolerated under most conditions. *See also* ZEPPELIN ANTENNA.

END-FIRE ANTENNA

An end-fire antenna is a bidirectional or unidirectional antenna in which the greatest amount of radiation takes place off the ends. An end-fire antenna consists of two or more parallel driven elements if it is a phased array (*see* PHASED ARRAY); all of the elements lie in a single plane. A parasitic array is sometimes considered an end-fire antenna, although the term is used primarily for phased systems (*see* PARASITIC ARRAY).

A typical end-fire array might consist of two parallel half-wave dipole antennas, driven 180 degrees out of phase and spaced at a separation of ½ wavelength in free space. This results in a bidirectional radiation pattern, as shown in the illustration. In the phasing system, the two branches of transmission line are different in length by 180 electrical degrees; when cutting the sections of line, the velocity factor of the line must be taken into account (*see* VELOCITY FACTOR).

End-fire antennas show some power gain, in their favored directions, compared to a single half-wave dipole

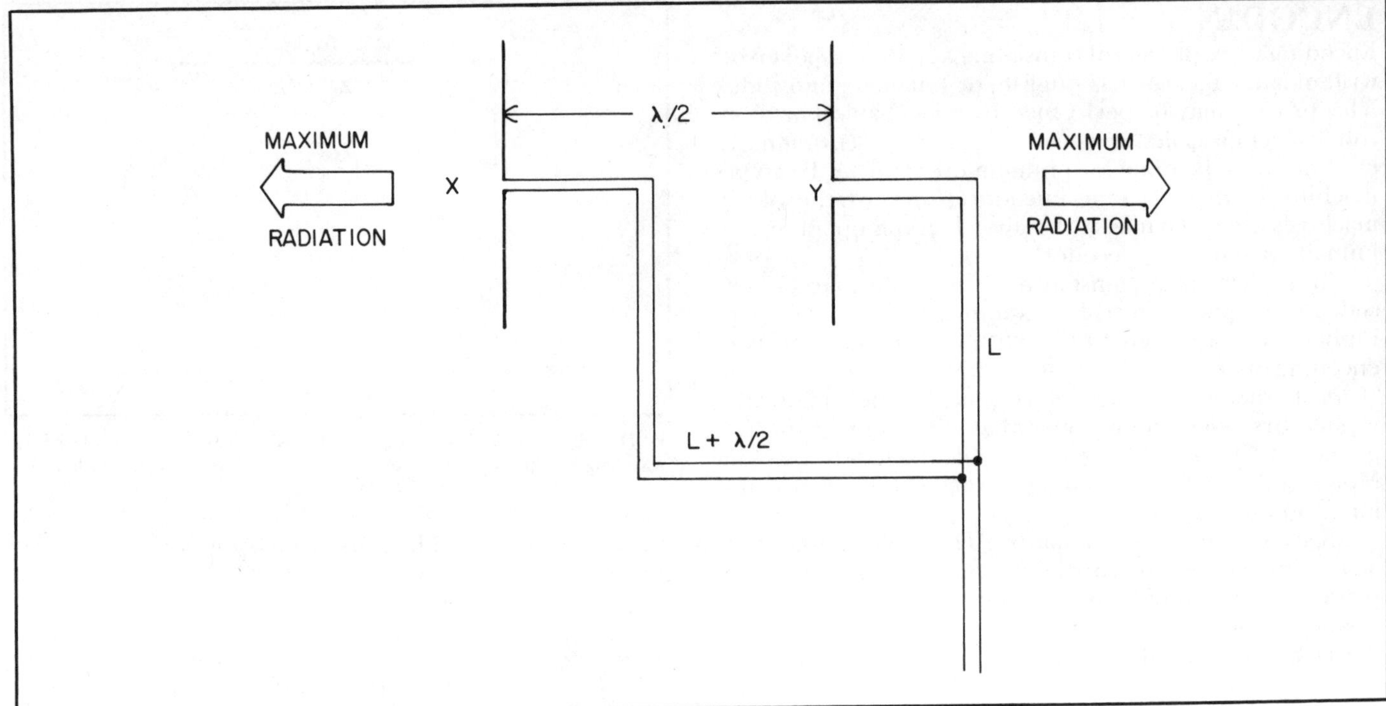

END-FIRE ANTENNA: An example of an end-fire antenna. The length L may be any value. One section of the phasing system is ½ wavelength longer than the other in this case. The two elements are thus fed in opposing phase.

antenna. The larger the number of elements, with optimum phasing and spacing, the greater the power gain of the end-fire antenna. *See also* ANTENNA POWER GAIN, DIPOLE ANTENNA.

ENDORADIOSONDE

An endoradiosonde is a tiny transmitting device that can be swallowed, and thereby placed in the stomach and intestines. A transducer monitors the physiological conditions as the device makes its way through the digestive tract. The radio transmitter, modulated by the output of the transducer, sends a low-power signal to outside instruments. The endoradiosonde is enclosed in a capsule that is impervious to the powerful digestive chemicals that are found in the body.

The endoradiosonde is a valuable diagnostic aid in the treatment of some types of digestive disorders. *See also* MEDICAL ELECTRONICS.

END-PLATE MAGNETRON

An end-plate magnetron is a special kind of magnetron tube (*see* MAGNETRON). In the ordinary magnetron, the electrons are accelerated into spiral paths by means of a magnetic field, as at A in the illustration. The greater the intensity of the magnetic field, the more energy is imparted to the electrons, up to a certain practical maximum.

With the addition of an electric field between two plates at the ends of the cylindrical chamber, the electrons are accelerated longitudinally as well as in an outward spiral. This is illustrated at B. The radial-velocity component remains the same, and is the result of the magnetic field. The longitudinal component, imparted by the electric field between the two end plates, thus gives the electrons extra speed. With the end plates, greater electron speed can be realized than without the plates, and this increases the amplitude of the tube output. *See also* ELECTRIC FIELD, ELECTRON, ELECTROSTATIC FORCE.

ENERGY

Energy is the capacity for doing work. Energy is manifested in many forms, including chemical, electrical, thermal, electromagnetic, mechanical, and nuclear. Energy is expended at a rate that can be specified in various ways; the standard unit of energy is the joule (*see* JOULE). The rate of energy expenditure is known as power (*see* POWER); the unit of power is the watt, which is a rate of energy expenditure of 1 joule per second. The units

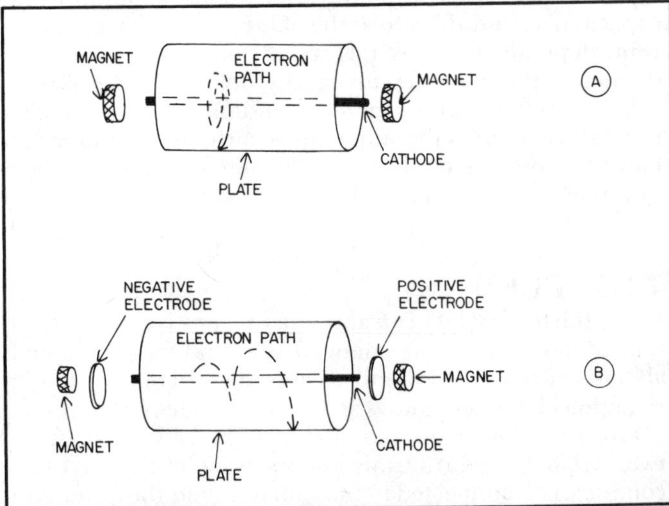

END-PLATE MAGNETRON: At A, a schematic diagram of a magnetron. The electrons travel in spiral paths. At B, the addition of electrodes, or end plates, with opposite electric polarity, causes the electrons to move longitudinally as well as spirally outward.

specified for energy depend on the application. The British thermal unit is a measure of thermal energy; the foot pound, of mechanical energy, and so on (*see* BRITISH THERMAL UNIT, ELECTRON VOLT, ENERGY UNITS, ERG, FOOT POUND, KILOWATT HOUR, WATT HOUR).

Energy is becoming an increasingly familiar term in the modern world. The supply of energy, in readily usable form, is not unlimited, and in some places there have been spot shortages of this indispensable commodity. Raw energy from the universe is, for all practical purposes, infinite in supply, but people must learn how to harness it, and this will require a great research effort. When controlled, energy provides us with light, transportation and, indirectly, food. Unharnessed energy leaves us at its mercy. It can even be destructive. *See also* ENERGY CONVERSION, ENERGY CRISIS, ENERGY DENSITY, ENERGY EFFICIENCY, ENERGY LEVEL, ENERGY STORAGE, ENERGY UNITS.

ENERGY CONVERSION

Energy occurs in many different forms. Energy conversion is the process of changing energy from one form to another. For example, mechanical energy may be changed into electrical energy; this is done by a generator system. Visible-light energy can be converted to electrical energy by means of a solar cell. Chemical energy can be converted to electrical energy by means of a battery. The number of possible conversion modes is practically unlimited.

Energy conversion always takes place with an efficiency of less than 100 percent. No device has ever been invented that will produce the same output energy, in desired form, as it receives in another form. Some devices are almost 100 percent efficient; others are much less than perfect. An incandescent light bulb, which converts electrical energy into visible-light energy, has a very poor efficiency; such a bulb is better for generating heat than light! The fluorescent light bulb is a much more efficient energy converter.

In general, if P_{IN} is the input power to an energy converter, and P_{OUT} is the output from the converter in the desired form, then the conversion efficiency in percent is given by:

$$\text{Efficiency (percent)} = 100P_{OUT}/P_{IN}$$

Energy is measured in different units, depending on the application, and on the form in which the energy occurs. The universal unit of energy is the joule. *See also* BRITISH THERMAL UNIT, ELECTRON VOLT, ENERGY UNITS, ERG, FOOT POUND, JOULE, KILOWATT HOUR, WATT HOUR.

ENERGY CRISIS

The energy available for use by civilization has, until recently, seemed inexhaustible. The actual presence of unharnessed energy is almost infinite, but civilization must perfect new methods for utilizing this energy. As the older methods become inadequate to support the ever-increasing world population and technological sophistication, newer methods will have to be developed and perfected. This changeover will not be easy, and some of the symptoms of the inadequacy of the older energy-utilization methods have already been felt. Unless new methods are found, the world will eventually face what has been called an energy crisis.

The first warnings of the crisis became apparent in the 1970s, when a shortage of fossil fuels occurred in the industrialized nations. The direct cause of that crisis was trade-related and did not actually represent depletion of the natural supply of fossil fuels. But the event prompted scientists to call attention to the fact that the world supply of fossil fuels is finite, and some day it will run out.

Some energy alternatives currently being researched include nuclear fission, nuclear fusion, solar power, heat from the earth's interior, wind energy, and water energy. Hopefully, civilization will meet the challenge of the energy crisis before the fossil fuels are depleted. *See also* ENERGY, POWER.

ENERGY DENSITY

In an energy-producing cell, the energy density is the ability of the cell to produce a given amount of power for a given length of time, for a given amount of cell mass. Theoretically, the energy density is expressed as the total energy in joules divided by the cell mass in kilograms. Obviously, it is advantageous to have an energy density that is as high as possible.

Energy density in a cell may also be represented as the total energy in joules divided by the cell volume in cubic meters. Again, it is desirable to maximize the energy density as expressed in terms of cell volume.

Technology has greatly improved the energy density of the common chemical cell over the past several decades. Ordinary dry cells, available in grocery, drug, and hardware stores, have excellent energy density. Some such cells are smaller than a dime, and yet will run a digital watch for months. *See also* ENERGY.

ENERGY EFFICIENCY

The energy efficiency of a device is the ratio between the energy output of the device, in the desired form, and the amount of energy it receives. This value is always smaller than 100 percent, but it can vary greatly. The exact value depends on many things, including the type of device and the form of energy desired in the output.

In recent years, with ever-growing concern about the availability of energy supplies, energy efficiency has become a more and more important consideration in the manufacture of consumer and industrial appliances and machines. The greater the energy efficiency of a device, the less energy it requires to perform a given task. This results in more work for a fixed amount of energy. As an example of the application of energy efficiency, incandescent bulbs are being replaced by fluorescent bulbs in homes and businesses. The incandescent bulb wastes much energy producing heat, when the desired form of energy is visible light. Fluorescent bulbs produce more light, and less heat, for a given amount of input energy. In automotive engineering, the concept of energy efficiency is well known, and cars with greater gasoline mileage per gallon are constantly being developed. Energy efficiency depends, to some extent, on the proper adjustment and operation of a device. Therefore, apparatus that is out of alignment should be realigned for maximum energy efficiency, and operation should be conducted in an energy-efficient manner. The maximization of energy efficiency today will give civilization the greatest possible amount of time to perfect new methods of energy utilization. *See also* ENERGY, ENERGY CRISIS.

ENERGY LEVEL

In an atom of any element, the electrons orbit the nucleus in discrete average positions. These positions are called shells. The farther a shell is situated from the nucleus, or center, of the atom, the greater the energy contained in an electron orbiting in that shell. The innermost shell of an atom is called the K shell. Outside of the K shell are the L, M, N, O, P, and Q shells, with progressively increasing energy levels.

If an electron, orbiting in a given shell, absorbs a certain amount of energy, then it will move to a more distant shell. Conversely, if an electron moves to a closer orbital shell, it will emit energy. If an electron absorbs enough energy, it will escape from the nucleus of the atom and become what is called a free electron. Extremely hot substances have an abundance of free electrons, since they contain a large amount of thermal energy. Such a substance is called a plasma.

The effects of electron energy levels are responsible for the absorption and emission lines in the spectra of substances. The different elements have unique patterns of emission and absorption lines, and this enables astronomers to determine the composition of distant stars and planets. Electrons are believed to behave in the same manner everywhere in the universe. *See also* ELECTRON ORBIT.

ENERGY STORAGE

In certain situations, it is desirable to store energy for possible future use. Energy may be stored in chemical form, as in the common cell or battery. It may be stored in electrical form, as in a capacitor. It may be stored in magnetic form, as in an inductor. Energy may also be stored as heat.

Whenever energy is stored for use at a later time, there is some loss. The amount of useful energy recovered is never as great as the amount of energy initially stored. However, in most energy-storage devices, the loss is quite small within a reasonable period of time. But the longer the elapsed time between the initial storage period and the utilization period, the greater the loss.

The most common vehicle for energy storage in electronics and communications is the rechargeable battery. There are many kinds of rechargeable cells and batteries; the most common types are the lead-acid battery and the nickel-cadmium battery. *See also* LEAD-ACID BATTERY, NICKEL-CADMIUM BATTERY, STORAGE BATTERY.

ENERGY UNITS

Energy is manifested in many different forms, and used for different purposes. Energy may be measured in any of several different systems of units, depending on the application. The fundamental unit of energy is the joule, or watt second (*see* JOULE). The other units are used as a matter of convention, and they each bear a relationship to the joule that can be given in terms of a numerical constant, or conversion factor. The table shows the conversion factors for the most common energy units: the British thermal unit (BTU), the electron volt, the erg, the foot pound, the kilowatt hour, and the watt hour.

ENERGY UNITS: CONVERSION FACTORS FOR VARIOUS ENERGY UNITS.

Unit	To convert to joules, multiply by	Conversely, multiply by
British thermal unit (Btu)	1.055×10^{3}	9.480×10^{-4}
electron volt	1.602×10^{-19}	6.242×10^{18}
erg	10^{-7}	10^{7}
foot pound	1.356	7.376×10^{-1}
kilowatt hour	3.600×10^{6}	2.778×10^{-7}
watt hour	3.600×10^{3}	2.778×10^{-4}

The unit chosen for a given application depends mostly on the consensus of the users. While joules represent a perfectly satisfactory way to express energy in any situation the quantities are sometimes rather unwieldy and may appear ridiculous. For example, an average residential dwelling might consume 1,000 kilowatt hours of electricity in a month; this is 3.6 billion joules. It makes more sense to use a kilowatt-hour meter in this application, and not a joule meter! *See also* BRITISH THERMAL UNIT, ELECTRON VOLT, ENERGY, ERG, FOOT POUND, KILOWATT HOUR, WATT HOUR.

ENHANCEMENT MODE

The enhancement mode is an operating characteristic of certain kinds of semiconductor devices, notably the metal-oxide-semiconductor field-effect transistor (MOSFET). Some of the MOSFET devices have no channel under conditions of zero gate bias. In other words, they are normally pinched off. These MOSFET devices are called enhancement-mode devices, because a voltage must be applied in order for conduction to take place. The channel is enhanced, rather than depleted, with increasing bias. The other type of field-effect transistor is normally conducting, and the application of gate bias reduces the conduction. This is the depletion mode (*see* DEPLETION MODE).

In an N-channel enhancement-mode device, a negative voltage at the gate electrode causes a channel to be formed. In the P-channel enhancement-mode device, a positive voltage at the gate is necessary for a channel to be produced. The larger the voltage, assuming the polarity is correct, the wider the channel gets, and the better the conductivity from the source to the drain. This occurs, however, only up to a certain maximum. The enhancement-mode effect is shown schematically in the drawing.

The enhancement-mode MOSFET has an extremely low capacitance between the gate and the channel. This is also true of the depletion-mode MOSFET. This characteristic allows the MOSFET to be used at the very-high-frequency range and above, with good efficiency. The enhancement-mode device, since it is normally cut off, works well as a Class-B or Class-C amplifier, although it can also be used in the Class-A and Class-AB configurations. *See also* METAL-OXIDE SEMICONDUCTOR FIELD-EFFECT TRANSISTOR.

ENTLADUNGSSTRAHLEN

Electric arcing results in the emission of ultraviolet as well

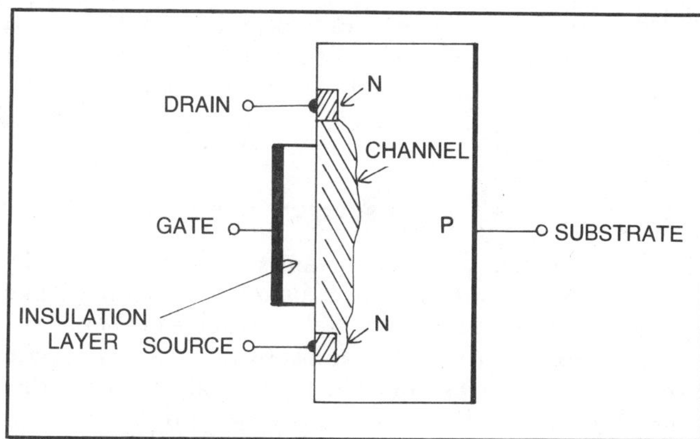

ENHANCEMENT MODE: Cross-sectional view of an enhancement-mode device. The channel, formed when suitable bias is applied to the gate, comprises the conductive channel.

as visible radiation. The term entladungsstrahlen is a German expression that means discharge radiation and refers in particular to the ultraviolet rays put out by an electric arc. This radiation occurs at very short ultraviolet wavelengths. Generally, the wavelength of discharge radiation is between 400 and 900 Angstrom units, or approximately one tenth the wavelength of visible light.

An electric arc should never be observed without some form of eye protection. Ultraviolet rays are damaging to the retina of the eye. Ordinary glasses, if the lenses are actually made of glass, are satisfactory, because glass is opaque to ultraviolet rays. Anyone involved in arc welding, or in any other occupation in which it is necessary to look at an electric arc, should have some form of eye protection, not only against the sparks, but against the ultraviolet rays. *See also* ARC, ULTRAVIOLET.

ENTROPY

Energy always moves from a region of greater concentration to areas of lesser concentration. For example, a hot object radiates its heat into the surrounding environment; a cold object absorbs heat from the environment. The tendency for energy to equalize its level everywhere in the universe is called entropy. Just as water seeks to find the lowest level of elevation, energy seeks to become uniformly distributed throughout the universe.

According to theroretical physics, the process of entropy is, in the general universe, irreversible. This fact has caused some scientists to theorize that the ultimate fate of the cosmos is a cold, dark end. But in recent years, there have been some suggestions that this theory might not be correct. Nevertheless, on a small scale, entropy does hold as a physical principle.

In information theory, a function called the entropy function is used to determine the amount of intelligence that is contained, and the details are beyond the scope of this book. A text on information theory is recommended for those interested in pursuing the details of the entropy function. *See also* INFORMATION THEORY.

ENVELOPE

The envelope of a modulated signal is the modulated waveform. Generally, the term envelope is used for amplitude-modulated or single-sideband signals. An imaginary line, connecting the peaks of the radio-frequency carrier wave, illustrates the envelope of a signal. A typical envelope display for a single-sideband signal is shown in the illustration.

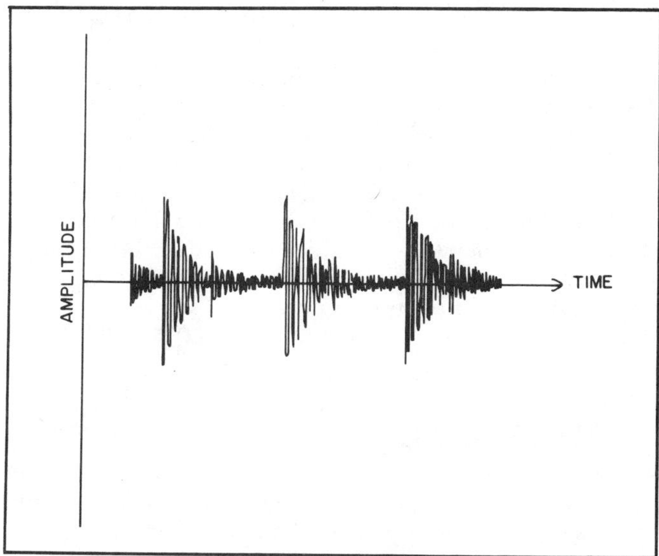

ENVELOPE: A single-sideband modulation envelope, as seen on an oscilloscope.

The appearance of a signal envelope is an indicator of how well a transmitter is functioning. Different kinds of distortion appear as different abnormalities in the modulation envelope of a transmitted signal. An experienced engineer can immediately tell, by looking at the radio-frequency modulation envelope of a transmitter supplied with a test-modulation signal, whether or not severe distortion is present. The most serious kind of distortion in an amplitude-modulated wave is called peak clipping or flat topping. Such a condition causes the generation of sidebands at frequencies far removed from that of the carrier. This can result in interference to stations on frequencies outside the normal channel required by such a transmitter (*see* AMPLITUDE MODULATION, SINGLE SIDEBAND).

The glass enclosure in which a vacuum tube is housed is sometimes called the tube envelope. With the decreasing use of vacuum tubes in electronic circuits, however, this use of the term is getting less and less common. *See also* TUBE.

ENVELOPE DETECTOR

An envelope detector is a form of demodulator in which changes of signal amplitude are converted into audio-frequency impulses. The most simple form of envelope detector is the simple half-wave rectifier circuit for radio frequencies. A semiconductor diode is adequate to perform the function of an envelope detector.

While an envelope detector is suitable for demodulating an amplitude-modulated signal having a full carrier, it is not satisfactory for the demodulation of a single-sideband, suppressed-carrier signal. When an envelope detector is used in an attempt to demodulate a suppressed-carrier signal, the resulting sound at the receiver output is unintelligible, and has been called "monkey chatter." This sound is familiar to those involved in communications. In order to properly demodulate a single-sideband, suppressed-carrier signal, a beat-frequency oscillator is necessary. This replaces the missing carrier wave in the receiver.

An envelope detector may be used to demodulate a frequency-modulated signal by means of a technique called slope detection. The receiver is tuned slightly away from the carrier frequency of the signal, so that the frequency variations bring the signal in and out of the receiver passband. This produces amplitude fluctuations in the receiver; when these are demodulated by the envelope detector, the result is amplitude variations identical to those of an amplitude-modulated signal. *See also* AMPLITUDE MODULATION, DETECTION, SLOPE DETECTION.

EPISCOTISTER

A device for mechanically modulating a light beam is called an episcotister. This device generally employs reflecting mirrors or disks to interrupt the light beam at an audio-frequency rate. In this manner, signals may be transmitted in Morse code, and picked up by a light-sensitive electronic device. Bright sources of light, which cannot ordinarily be modulated at an audio-frequency rate, are sometimes used with an episcotister to obtain tone modulation at audio frequencies.

The illustration shows the operation of a rotating-disk episcotister. The disk has a uniformly spaced pattern of slots or holes around its perimeter, and as the disk turns, the slots pass and block the light. The light from the bulb must be brought to a sharp focal point by means of mirrors or lenses, and the disk slots eclipse the light at this focal point. Collimating mirrors or lenses are then used to obtain an intense, parallel beam of modulated light. The frequency of the modulation depends on the rate of rotation of the disk, and on the number of slots in the disk. In practice, frequencies up to several kilohertz are possible.

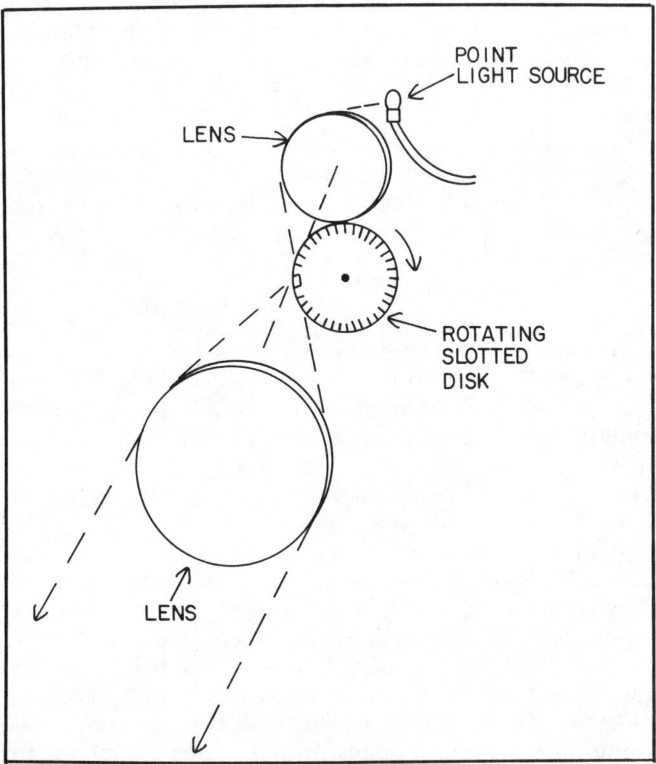

EPISCOTISTER: The rotating-disk episcotister produces a tone-modulated light beam.

Of course, a voice signal cannot be transmitted by means of an epsicotister. More sophisticated devices are needed for voice modulation of a light beam. *See also* MODULATED LIGHT.

EPITAXIAL TRANSISTOR

An epitaxial transistor is a transistor that is made by "growing" semiconductor layers on top of each other. The process of semiconductor-crystal growth is called epitaxy.

In the epitaxy process, a silicon substrate is used to begin the growth. This substrate, which eventually serves as the collector of the transistor, is immersed in a prepared gas or liquid solution. The base is later diffused into this epitaxial layer (*see* DIFFUSION). The emitter is alloyed to the top of the epitaxial layer.

The epitaxial transistor is characterized by a very thin base region, which is important for effective operation at high frequencies. It is also possible to get a relatively large surface area at the base-collector junction, and this is important for the effective dissipation of heat. The drawing shows a cross-sectional view of a typical epitaxial transistor. Sometimes it is called an epitaxial mesa transistor. *See also* TRANSISTOR.

E PLANE

The E plane, or electric-field plane, is a theoretical plane in space surrounding an antenna. There are an infinite number of E planes surrounding any antenna radiator; they all contain the radiating element itself. The principal E plane is the plane containing both the radiating element and the axis of maximum radiation. Thus, for a Yagi antenna, the E plane is oriented horizontally; for a quarter-wave vertical element without parasitics, there are infinitely many E planes, all oriented vertically.

The magnetic lines of flux in an electromagnetic field cut through the E plane at right angles at every point in space. The direction of field propagation always lies within the E plane, and points directly away from the radiating element. In a parallel -wire transmission line, the E plane that plane containing both of the feed-line conductors in a given location; the E plane follows all bends that are made in the line. In a coaxial transmission line, the E planes all contain the line of the center conductor. In waveguides, the orientation of the E plane depends on the manner in which the waveguide is fed. *See also* ELECTRIC FIELD, ELECTROMAGNETIC FIELD.

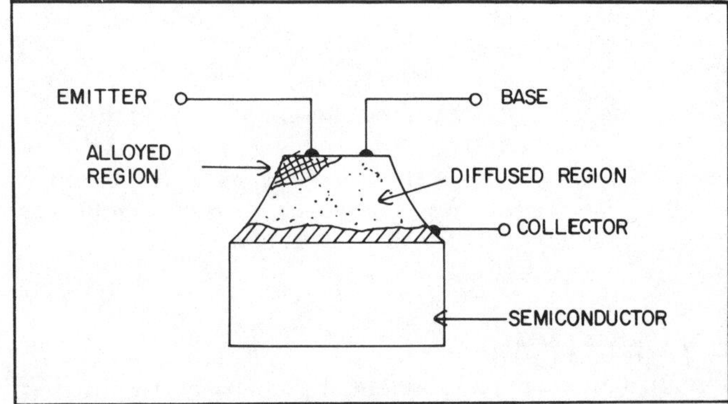

EPITAXIAL TRANSISTOR: An epitaxial mesa transistor is built up on a semiconductor base.

EPROM

The term EPROM, pronounced e-prom, is an abbreviation for electricially programmable read-only memory. This type of memory is erased and reprogrammed by an electric signal. *See also* PROGRAMMABLE READ-ONLY MEMORY.

EQUALIZER

An equalizer is a high-fidelity device that is used to tailor the frequency response of an audio system. The simplest equalizers are the base-treble controls on all stereo amplifiers, phonographs, and tape players. More sophisticated equalizers are available for those who desire the best possible sound from their systems.

Different speakers generally sound different when used with the same amplifier; also, a given set of speakers will often have different sounds when used with different audio systems. This is well known among hi-fi enthusiasts. By using a precision equalizer, these discrepancies can be avoided.

The more complex equalizers generally have several sliding-bar potentiometers located side-by-side, as illustrated. Each potentiometer controls the gain at a different range of frequencies. There may be eight or more such potentiometers. The maximum gain for each frequency range is toward the top, and the low-to-high frequency range goes from left to right. Thus the adjustment of the potentiometers gives the impression of a graphic display. Such devices are, for this reason, sometimes called graphic equalizers. *See also* HIGH FIDELITY.

EQUATION

An equation is a mathematical expression that says two quantities are equal. An equation always has a left side and a right side. Sometimes equations may be strung in run-on fashion. Equations may contain numbers or unknowns (variables). Equations may also contain functions.

In engineering, equations are the primary mathematical tools by which all computations are made. Equations can occur in a variety of forms. The simplest is the linear equation, so called because its graph in the Cartesian coordinate system (*see* CARTESIAN COORDINATES) is a straight line. The linear equation takes the form:

$$ax + by = 0$$

Nonlinear equations can exist in many forms. The quadratic equation is an equation in which one expression is squared; a cubic equation has an expression raised to the third power. Other examples of nonlinear equations are those containing trigonometric functions, logarithmic functions, exponential functions, and roots.

The appearance of equations in a text often intimidates readers with little or no mathematical background, but in general, even the most complicated equations are based on fairly simple principles. Without equations as a means of expressing concepts, the engineer would hardly be able to carry on. *See also* FUNCTION.

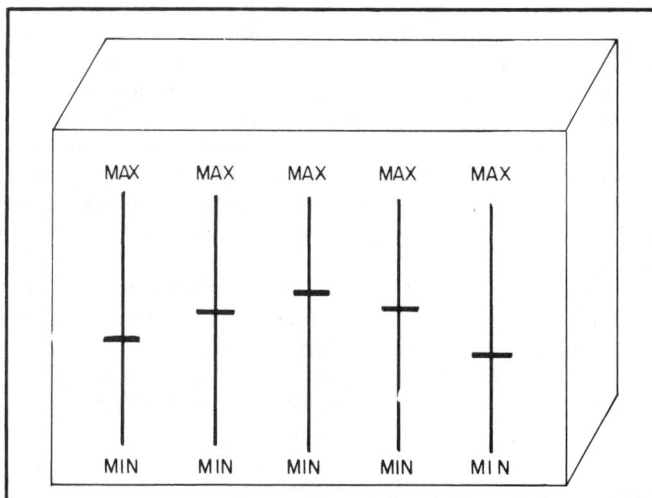

EQUALIZER: An equalizer for use in high-fidelity sound reproducing. Each sliding-bar potentiometer controls the gain at a different range of frequencies. This device is shown with the greatest gain in midrange, and the least gain at the bass (left) and treble (right).

EQUIVALENT CIRCUITS

When two different circuits behave the same way under the same conditions, the circuits are said to be equivalent. There are many examples of, and uses for, equivalent circuits. One good example is the dummy antenna (*see* DUMMY ANTENNA). Such a device consists of a pure resistance, and when connected to a transmission line having a characteristic impedance equal to this resistance, the dummy load behaves exactly like an antenna with a purely resistive impedance of the same value.

Equivalent circuits are often used in the test laboratory and on the technician's bench, for the purpose of simulating actual operating conditions when designing and repairing electronic devices. The equivalent circuit is often much simpler than the actual circuit. The equivalent must have the same current, voltage, impedance, and phase relationship as the actual circuit. The equivalent circuit must also be capable of handling the same amount of power as the actual circuit; in the dummy-antenna example given above, a ⅛-watt resistor would obviously not suffice for use with a transmitter delivering 25 watts of RF output!

An important aspect of technological advancement is the development of simpler and simpler equivalent circuits. This has led, for example, to the handheld calculator, which, for a price of a few dollars, replaces the large tube type devices of just a few decades ago. It has also led to the development of more and more complex computers, as a larger number of components can be fit into the same space. *See also* MINIATURIZATION.

ERASE

The erase process is the removal of information or data from a certain medium. Some media are erasable while others are not. Examples of erasable media include magnetic tape and disks, and various integrated-circuit memories.

In some devices, it is necessary to erase the stored data before new data can be recorded. In other instances,the new data is automatically recorded, without the need to erase or re-initialize the medium.

Erasing may be done accidentally as well as intentionally, and care should be exercised, especially with magnetic storage, to prevent this. Magnetic tapes and disks should be kept far away from magnetic fields. They should not be placed near current-carrying wires, coils, or devices containing permanent magnets, such as speakers! *See also* MAGNETIC RECORDING.

ERG

The erg is a unit of energy. It is generally abbreviated by the small letter e. A force of 1 dyne, acting through a distance of 1 centimeter, is an energy expenditure of 1 erg. It has been said that 1 erg is approximately the amount of energy required for a mosquito to take off after it has bitten you!

The erg is not often used as a unit of energy in electronics. Much more often, the watt hour or kilowatt hour is specified. The universal unit of energy is the joule (*see* JOULE). A joule is 10 million ergs. A kilowatt hour is 3.6 million joules, or 36 trillion ergs! *See also* ENERGY UNITS.

ERROR

Whenever a quantity is measured, there is always a difference between the instrument indication and the actual value. This difference, expressed as a percentage, is called the error or instrument error. Mathematically, if x is the actual value of a parameter and y is the value indicated by a measuring instrument, then the error is given by:

$$\text{Error (percent)} = 100(y - x)/x$$

if y is larger than x, and:

$$\text{Error (percent)} = 100(x - y)/x$$

if x is larger than y. Not all of the error in instrumentation is the fault of the instrument. There is a limit to how well the person reading an instrument can interpolate values. For example, the needle of a meter might appear to be 0.3 of the way from one division to the next, as far as the person can tell, but in reality it is 0.38 of the way from one division to the next. Digital meters, of course, eliminate this problem, but the resolution of a digital device is limited by the number of digits in the readout. The term error is often used in reference to an improper command to a computer. This will result in an undesired response from the computer, or will cause the computer to inform the operator of the error. *See also* INTERPOLATION.

ERROR ACCUMULATION

Errors in measurement (*see* ERROR) tend to add together when several measurements are made in succession. This is called error accumulation. If the maximum possible error in a given measurement is x units, for example, and the measurement is repeated n times, then the maximum possible error becomes nx units. The percentage of maximum error, however, does not change.

An example of error accumulation is the measurement of a long distance, such as the length of a piece of antenna wire, with a short measuring stick such as a ruler. It is not convenient, of course, to measure long distances with rulers, but sometimes this is all that is available! Suppose the actual length of a piece of wire is 100 feet, and this length is measured with a ruler, resulting in a maximum error of 0.01 foot. The ruler must be placed against the wire 100 times, and this results in a maximum possible error of 100 × 0.01 foot, or 1 foot. The maximum possible error per measurement is 1 percent, or 0.01/1; the maximum possible overall error is also 1 percent, or 1/100.

When making measurements, it is best, if possible, to avoid iterating the procedure, such as in the above example, because this multiplies the total error. *See also* ERROR.

ERROR CORRECTION

Error correction is a computer operating system in which certain types of errors are automatically corrected. An example of such a routine is a program that maintains a large dictionary of common English words. The operator of a word-processing machine may make mistakes in typing or spelling, but when the error-correction program is run, the computer automatically corrects all typographical errors, except in words not contained in its dictionary. With modern computers, extremely large dictionary files are realizable, and a program using such a file can perform almost all of the editing functions. This saves the word-processor operator the trouble of having to peruse the text for minor errors.

In measurement of parameters such as voltage, current, power, or frequency, error correction is used when an instrument is known to be inaccurate. For example, the operator of a radio transmitter may know that the station wattmeter reads 3 percent too high; this may have been determined by using a standard wattmeter borrowed from a laboratory. In such a case, if the transmitter output is supposed to be 100 watts, the operator will know that this will be shown on his instrument by a reading of 103 watts. *See also* ERROR.

ERROR-SENSING CIRCUIT

An error-sensing circuit is a means of regulating the output current, power, or voltage of a device. In a power supply, for example, an error-sensing circuit might be used to keep the output voltage at a constant value; such a circuit requires a standard reference with which to compare the output of the supply. If the power-supply voltage increases, the error-sensing circuit produces a signal to reduce the voltage. If the power-supply output voltage decreases, the error-sensing circuit produces a signal to raise it. An error-sensing circuit is therefore a form of amplitude comparator.

An error circuit does not necessarily have to provide regulation for the circuit to which it is connected; some error-sensing circuits merely alert the equipment personnel that something is wrong, by causing a bell to ring or a warning light to flash. *See also* ERROR SIGNAL.

ERROR SIGNAL

An error signal is the signal put out by an error-sensing circuit (*see* ERROR-SENSING CIRCUIT). The error signal is actuated whenever the output parameter of a circuit or device differs from a standard reference value. An error signal may be used simply for the purpose of informing personnel of the discrepancy, or the signal may be used to initiate corrective measures.

An example of an error signal is the rectified voltage in an automatic-level-control circuit (*see* AUTOMATIC LEVEL CONTROL). The output of the amplifier chain in such a circuit is held to a constant value. If the signal tends to increase in amplitude, the error signal reduces the gain of the system.

Error signals are used to regulate the operation of servo systems. Each mechanical movement is monitored by an error-sensing circuit; error signals are generated whenever the system deviates from the prescribed operating conditions, and the signals cause corrective action. *See also* SERVOMECHANISM, SERVO SYSTEM.

ESAKI DIODE

See TUNNEL DIODE.

ESNAULT-PELTERIE FORMULA

The Esnault-Pelterie formula is a formula for determining the inductance of a single-layer, solenoidal, air-core coil based on its physical dimensions. The inductance increases roughly in proportion to the coil radius for a given number of turns. The inductance increases with the square of the number of turns for a given coil radius. If the number of turns and the radius are held constant, the inductance decreases as the solenoid is made longer.

Letting r be the coil radius in inches, N the number of turns, and m the length of the coil in inches, then the inductance L in microhenrys is given by:

$$L = r^2N^2/(9r + 10m)$$

See also INDUCTANCE.

ETCHING

Etching is a process by which printed-circuit boards are made. Initially, all circuit boards are fabricated of phenolic or glass-epoxy material, plated on one or both sides with copper.

The circuit layout is drawn on a piece of paper, and perfected for optimum use of the available space. This is called the etching pattern. An example of such a pattern is

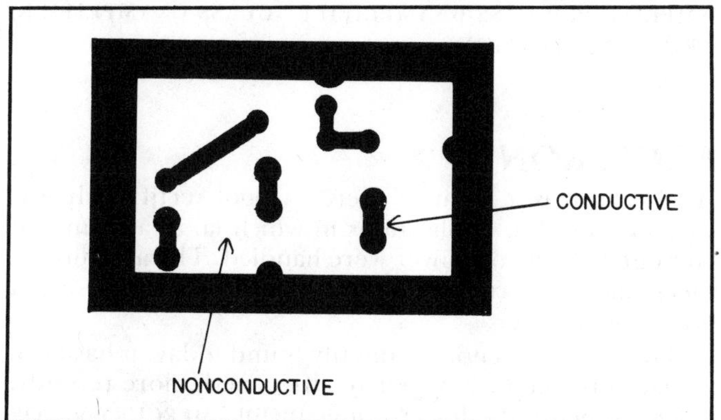

ETCHING: A simple circuit-board etching pattern.

shown in the illustration. The conductive paths are dark, and the portions of the board to be etched are white. The etching pattern is photographed and reduced or enlarged to the size of the circuit board. The photograph is printed on clear mylar.

The mylar sheet is placed over the copper plating of the board, and then the board and mylar are immersed in an etching solution. Ammonium persulfate crystals are added to water, in a ratio of about one part ammonium persulfate to two parts pure water. To this, a small amount of mercuric chloride is added. Alternatively, the etching solution can consist of one part ferric chloride and two parts pure water. When the circuit board, immersed in the etching solution, is exposed to bright light, the etching solution reacts with the copper. The etching pattern on the mylar prevents the copper from disappearing underneath the black areas. The circuit board should be continuously moved to speed up the etching process, and it should be kept at a temperature somewhat above normal room temperature. The etchant solution, ready-made, is commercially available. The skin should never be allowed to come into contact with etching solutions. *See also* PRINTED CIRCUIT.

ETHER THEORY

Isaac Newton originally theorized that light was composed of particles, called corpuscles or photons (*see* PHOTON). In the nineteenth century, however, physicists noticed that light had some of the properties of a wave disturbance, and it was discovered that light is a form of electromagnetic wave.

The wave theory of light gave rise to the idea that light must be propagated through space in a manner similar to the propagation of sound through air or water. This theory necessitated that there exist a substance of some kind everywhere in space, to carry the waves. This substance was called the luminiferous ether, and the theory that light is propagated through this substance was called the ether theory.

When scientists tried to determine the motion of the lumeniferous ether with respect to the earth, they found that it appeared to be stationary relative to our planet. This resulted in an apparent contradiction, since it implied that the earth was at rest with respect to the whole universe! Albert Einstein finally discarded the ether theory, since there was no evidence to support it.

Today, we realize that electromagnetic fields travel through a complete vacuum, without any intervening medium. Radio waves, infrared, visible light, ultraviolet, X rays, and gamma rays do not propagate through space in the same way that sound travels through the air. The speed of electromagnetic waves is, unlike the speed of sound, independent of the point of view from which it is measured, and this proves that the ether theory cannot be valid. *See also* ELECTROMAGNETIC FIELD, ELECTROMAGNETIC RADIATION, ELECTROMAGNETIC SPECTRUM, ELECTROMAGNETIC THEORY OF LIGHT.

EVEN-ORDER HARMONIC

An even-order harmonic is any even multiple of the fundamental frequency of a signal. For example, if the fundamental frequency is 1 MHz, then the even-order harmonics occur at frequencies of 2 MHz, 4 MHz, 6 MHz, and so on.

Certain conditions favor the generation of even-order harmonics in a circuit, while other circumstances tend to cancel out such harmonics. The push-push circuit accentuates even-order harmonics (*see* PUSH-PUSH CONFIGURATION), and is often used as a frequency doubler. The push-pull circuit (*see* PUSH-PULL CONFIGURATION) tends to cancel out the even-order harmonics.

A half-wave dipole antenna generally discriminates against the even-order harmonics. The impedance at the feed point of a half-wave dipole is very high at the even-order harmonic frequencies; while it may range from 50 to 150 ohms at the fundamental and odd harmonics, it can be as large as several thousand ohms at the even harmonics. *See also* ODD-ORDER HARMONIC.

EXALTED-CARRIER RECEPTION

Selective fading causes severe distortion on amplitude-modulated signals, especially when the carrier frequency is attenuated significantly. When that happens, and the sidebands are much stronger than the carrier for a moment, the signal becomes nearly unintelligible (*see* SELECTIVE FADING). Exalted-carrier reception is a method of reducing the effects of selective distortion on the signal carrier.

In the exalted-carrier system, the signal is split into two branches in the intermediate-frequency section of the receiver. One branch amplifies the entire signal in the usual manner. The other branch contains a narrow-band filter that allows only the carrier signal to pass. The carrier signal is amplified greatly in the stages immediately following the narrow-band filter, so that when the carrier is recombined with the signal at a later stage, the amplitude of the carrier is much greater than that of the sidebands. The drawing illustrates a block diagram of this principle.

If the carrier amplitude is exaggerated, or exalted, the signal readability is not adversely affected. Distortion occurs only when the amplitude of the carrier is not great enough. By overamplyifying the carrier by means of the technique illustrated, it is always at a sufficient amplitude, even during a fade, to allow good signal readability. Thus the adverse effects of fading are reduced. An even better system, and much more often used today, is single sideband, in which the entire carrier is supplied at the receiver. *See also* SINGLE SIDEBAND.

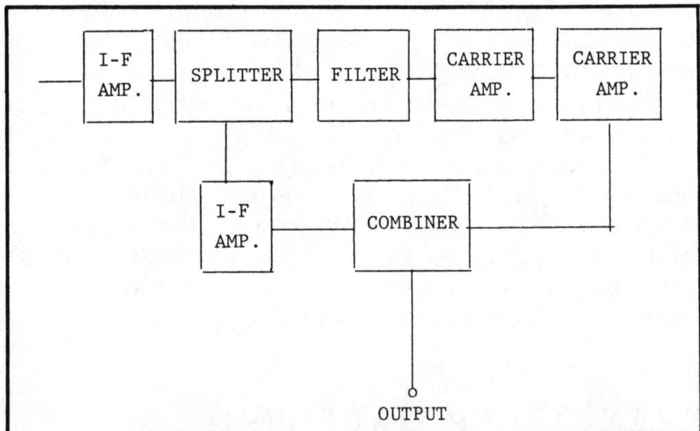

EXALTED-CARRIER RECEPTION: In exalted-carrier reception, the carrier is amplified separately from the rest of the signal in the intermediate-frequency stages of a receiver. When the carrier is reinserted, the effects of fading are reduced.

EXCEPT GATE

See EXCLUSIVE-OR GATE.

EXCITATION

Excitation is the driving power, current, or voltage to an amplifier circuit. The term is generally used in reference to radio-frequency power amplifiers. Excitation is sometimes called drive or driving power.

Class-A amplifiers theoretically require only a driving voltage, but no excitation power. In practice, a small amount of excitation power is required in the Class-A amplifier. In the Class-AB and Class-B amplifiers, some excitation is required to obtain proper operation. In the Class-C amplifier, a large amount of excitation power is needed in order to obtain satisfactory operation.

The circuit that supplies the excitation to a power amplifier is called the driver or exciter. When a transmitter is used in conjunction with an external power amplifier, the transmitter itself is called an exciter. *See also* CLASS-A AMPLIFIER, CLASS-AB AMPLIFIER, CLASS-B AMPLIFIER, CLASS-C AMPLIFIER, DRIVE, DRIVER, EXCITER.

EXCITER

An exciter is a circuit that supplies the driving power to an amplifier. Usually, the term exciter is applied to radio-frequency equipment. The exciter must be capable of delivering sufficient power to operate the amplifier circuit properly. With Class-A amplifiers, this power level may be almost negligibly small, although considerable peak-to-peak voltage may be necessary. With Class-AB and Class-B amplifiers, the power level varies depending on the exact bias level. With Class-C amplifiers, a large amount of exciter power is needed for satisfactory operation.

The exciter generally has a tuned output, so that optimum impedance matching can be obtained between it and the power amplifier. The exciter may be run by itself without the power amplifier, provided the exciter output tuning circuits are capable of providing a good direct match to the antenna system. When a transmitter is used in conjunction with an external power amplifier, the transmitter is sometimes called the exciter. The exciter may also be called the driver. *See also* CLASS-A AMPLIFIER, CLASS-AB AMPLIFIER, CLASS-B AMPLIFIER, CLASS-C AMPLIFIER, DRIVER, EXCITATION.

EXCITRON

An excitron is a form of mercury-pool rectifier. It was formerly used in applications in which large amounts of current, voltage, or power were handled. The arc must be mechanically started, by tilting the tube or by using a magnetic plunger.

The excitron is not commonly found today; it has been replaced by mercury-vapor rectifiers and, more recently, semiconductor diodes. *See also* DIODE, MERCURY-VAPOR RECTIFIER, RECTIFICATION.

EXCLUSIVE-OR GATE

The OR function may be either inclusive or exclusive; normally, unless otherwise specified, it is considered inclusive. The normal OR function is true if either or both of the inputs are true, and false only if both of the inputs are false. The exclusive OR function is true only if the inputs are opposite; if both inputs are true or both are false, the exclusive-OR function is false.

The illustration shows the schematic symbol of the exclusive-OR logic gate, along with a truth table of the function. The logic symbol 1, or high, indicates true, and 0, or low, indicates false. *See also* LOGIC GATE, OR GATE.

EXPANDER

An expander is a circuit that increases the amplitude variations of a signal. Expansion is the opposite, electrically, of compression (*see* COMPRESSION, COMPRESSION CIRCUIT). At low signal levels, the expander has little effect; the amplification factor increases as the input-signal amplitude increases.

An amplitude expander is used at the receiving end of a circuit in which compression is employed at the transmitter. By compressing the amplitude variations at the transmitting end of a circuit, and expanding the amplitude at the receiver, the signal-to-noise ratio is improved, and communications efficiency thereby is made better. This is because the weaker parts of the signal are boosted before propagation via the airwaves; when the weak components are attenuated at the receiver, the noise is attenuated also. A system that makes use of compression at the transmitter and expansion at the receiver is sometimes called an amplitude compandor. *See also* COMPANDOR.

EXPANSION CHAMBER

An expansion chamber, also called a cloud chamber, is a device used for viewing the paths of high-speed subatomic particles.

A glass housing, with very humid air inside, is depressurized by means of a piston until the temperature falls below the dew point. This causes tiny water droplets to form on ions within the chamber. When a radiactive particle such as a high-speed proton, neutron, or alpha particle enters the chamber, the air is ionized along the path of the particle. The water vapor condenses on the ions and forms a visible cloud trail.

High-speed atomic particles are continually arriving from outer space. Virtually all of this cosmic radiation is blocked by the atmosphere of our planet, and does not reach the ground. However, the collision of a cosmic particle with an air molecule results in the emission of another proton, neutron, or alpha particle, called a secondary particle. These are observed on the ground in cloud chambers or with radiation counters. The cloud chamber offers the advantage of showing the direction in the sky from which a subatomic particle arrives. Cloud chambers in space may be used for the purpose of locating celestial sources of cosmic radiation. *See also* COSMIC RADIATION.

EXPERIMENTAL RADIO SERVICE

New techniques of communication are always being tested. This is done in a service called the experimental service, or experimental radio service. Some experimentation is done by radio amateurs. But the amateurs cannot always get permission from the Federal Communications Commission to test new methods of communication. The experimental radio service is designated especially for this function.

The electromagnetic spectrum between 10 kHz and about 3,000 GHz may have uses as yet undiscovered. By means of experimentation, these applications can be found. Recent developments include the perfection of the microwave oven, which has replaced conventional ovens in many homes throughout the United States and other industrialized countries.

EXPLORING COIL

An exploring coil, sometimes called a sniffer, is a form of

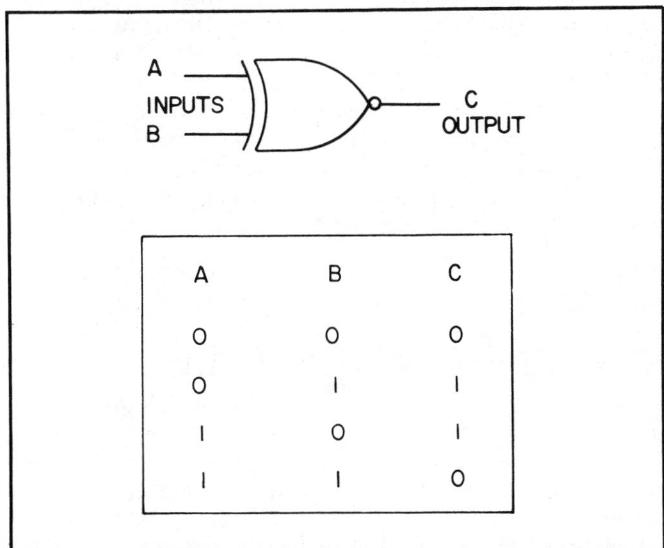

A	B	C
0	0	0
0	1	1
1	0	1
1	1	0

EXCLUSIVE-OR GATE: The exclusive-OR gate, showing a truth table for various input combinations.

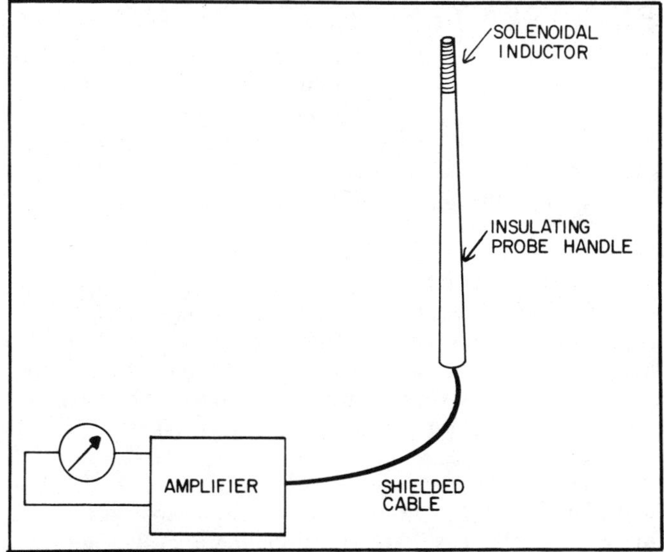

EXPLORING COIL: A simple diagram of an exploring coil, used for detecting magnetic fields.

radio-frequency or magnetic-field detector. The device consists basically of a small inductor at the end of a probe, and a shielded or balanced cable connection to an amplifier. The amplifier output is connected to an indicating device such as a meter, light-emitting diode, or oscilloscope. The illustration is a simple diagram of an exploring-coil device.

When a magnetic field is present in the vicinity of the exploring coil, an indication is obtained whenever the coil is moved across the magnetic lines of flux. If a radio-frequency field is present, an indication will be obtained even if the coil is held still. An exploring coil can be used to determine if the shielding is adequate in a particular circuit. The exploring coil may also be used to obtain a small signal for monitoring purposes, without affecting the load impedance of the circuit under observation. *See also* EXPLORING ELECTRODE, RADIO-FREQUENCY PROBE.

EXPLORING ELECTRODE

An exploring electrode is a device similar to an exploring coil. The device consists of a small pickup electrode, connected to an amplifier circuit. The output of the amplifier circuit is connected to an indicating meter, light-emitting diode, or oscilloscope.

Either an exploring coil or an exploring electrode may be used to detect radio-frequency fields. However, for some applications, the electrode is preferable to the coil, and in other situations, the coil is more effective. An exploring electrode, like the exploring coil, causes little or no change in the load impedance of a circuit under test. The coupling mode of the exploring electrode is capacitive, or electrostatic; that of the exploring coil is inductive, or magnetic.

An exploring electrode generally presents an extremely high impedance to the input of the amplifier. Shielding of the cable from the probe to the amplifier is an absolute necessity, and the radio-frequency ground must be excellent. Otherwise, stray electromagnetic fields will be picked up by the cable and amplified, resulting in severe interference. *See also* EXPLORING COIL, RADIO-FREQUENCY PROBE.

EXPONENTIAL DISTRIBUTION

The exponential distribution is a form of statistical distribution, used for determining the probability that an event will occur within a specified time interval. The longer the time interval, the greater the chances of an event occurring within that interval. In general, the exponential distribution function is given by:

$$P = fe^{-ft}$$

where P is the probability, f is the frequency of the occurrence, t is the length of the time interval, and e is approximately 2.718. A qualitative example of the use of an exponential distribution is as follows. Suppose that a thunderstorm is in progress, and lightning strikes occur, resulting in static pulses in a radio receiver. If you listen to the receiver for just a few seconds, it is not likely that you will hear a pulse of noise; but the longer you listen, the greater the chances that you will hear static. As the listening time gets longer and longer, the probability of your hearing a static pulse approaches 100 percent. Using the exponential function, knowing the mean frequency f of the lightning pulses, the chances of your hearing a pulse can be calculated on the basis of the listening time t. This probability must be determined by mathematical integration. *See also* STATISTICAL ANALYSIS.

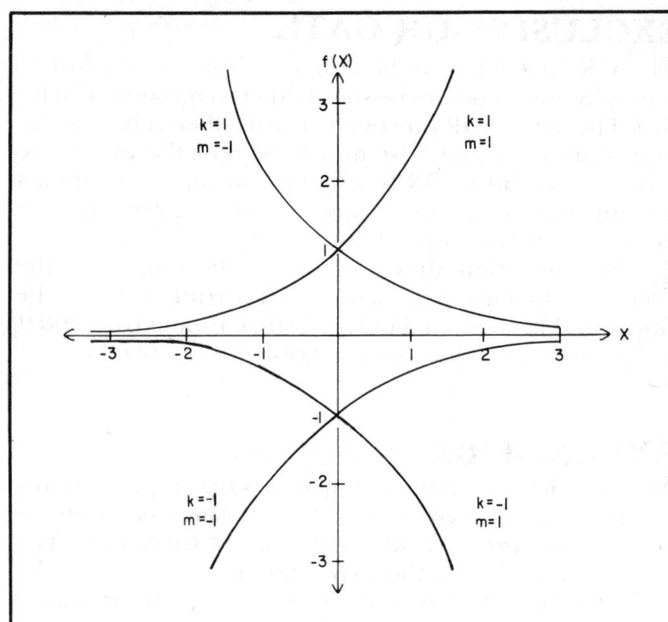

EXPONENTIAL FUNCTION: Four basic exponential functions.

EXPONENTIAL FUNCTION

An exponential function is a function that describes the rise and decay curves of many different phenomena. The general form of the exponential function is:

$$f(x) = ke^{mx}$$

where k and m are constants and e is approximately 2.718.

If the value of the exponential constant, m, is positive, then the function is positive and increasing (if k is positive) or negative and decreasing (if k is negative). If m is negative, then the value of the function is positive and decreasing (if k is positive) or negative and increasing (if k is negative). These four situations are illustrated in the drawing for the cases (k,m) = (1,1), (k,m) = (−1,1), (k,m) = (1,−1), and (k,m) = (−1,−1).

A charging or discharging capacitor behaves according to an exponential function. So does a damped oscillation. Radioactive substances decay according to an exponential function. The probability functions of certain events, with respect to time, are exponential. *See also* EXPONENTIAL DISTRIBUTION, FUNCTION.

EXTENDED-CUTOFF TUBE

An extended-cutoff tube, also called a variable-mu tube, is a form of vacuum tube in which cutoff cannot be obtained by any value of negative grid bias. The gain of the tube depends on the bias; the greater the negative grid bias, the lower the gain. The grid of an extended-cutoff tube is constructed in a somewhat different manner than is the grid of an ordinary tube. The spacing of the wires is closer at the center of the grid than at the top or bottom.

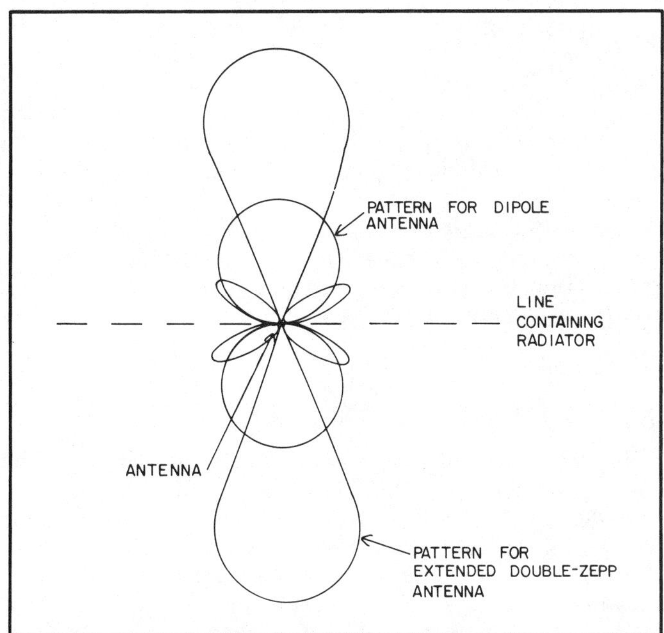

EXTENDED DOUBLE-ZEPP ANTENNA: The extended double-zepp antenna has a power gain of about 3 dB over a dipole.

Extended-cutoff tubes are especially useful in automatic-level-control applications, since a rectified negative voltage can be applied to the grid to reduce the gain to a greater extent than would be possible with an ordinary vacuum tube. In modern circuits, however, solid-state devices can perform this function as well as, or better than, the extended-cutoff tube. One solid-state device that exhibits variable gain isthe dual-gate metal-oxide-semiconductor field-effect transistor (MOSFET). *See also* AUTOMATIC LEVEL CONTROL, METAL-OXIDE-SEMICONDUCTOR FIELD-EFFECT TRANSISTOR, TUBE.

EXTENDED DOUBLE-ZEPP ANTENNA

The extended double-zepp antenna is similar to the double-zepp antenna, except that it is slightly longer. The optimum length of the extended-double zepp antenna is about 1.3 wavelengths, compared with 1 wavelength for the double zepp. This extension of the antenna provides approximately 1 dB additional gain; while the power gain of the double zepp is 2 dB over a dipole, the extended version provides 3 dB power gain over a dipole.

The radiation pattern of the extended double zepp is very much like that of the double zepp, except that the lobes are slightly broader, and there are four minor lobes in addition to the two major lobes. The drawing illustrates the directional pattern of the extended double zepp, compared with that of a half-wave dipole, as viewed from directly above. The maximum radiation occurs in directions perpendicular to the axis of the antenna conductor. The minor lobes are at angles of about 35 degrees with respect to the conductor.

The extended double-zepp antenna is popular among amateur radio operators, especially on the lower-frequency bands such as 1.8 and 3.5 MHz. For proper operation, the entire antenna should be run in a single straight line, and should be reasonably clear of obstructions. The feed-point impedance of the antenna contains some reactance as well as resistance, because it is not fed at a current or voltage loop. low-loss, parallel-wire transmission line should be used to feed the antenna. *See also* DOUBLE-ZEPP ANTENNA.

EXTERNAL FEEDBACK

When feedback, either positive or negative, occurs because of the external connections to a circuit, the phenomenon is called external feedback. External feedback can take place because of improper shielding of the input and/or output leads of a device. For example, in an audio-amplifier system, poor shielding of the microphone and speaker leads can result in coupling between the input and output of the circuit. This can, in turn, cause oscillation.

External feedback is generally not desired. If feedback is wanted, it is usually incorporated into a circuit, so that it can be regulated for optimum operation. If external feedback becomes a problem in a circuit, it can usually be eliminated by grounding the system, shielding all input, output, and power-supply leads, and bypassing the power-supply leads. Under certain circumstances, in radio-frequency apparatus, accidental resonances in external leads can cause feedback that is difficult to eliminate. In such cases, it is often necessary to change the lengths of one or more external leads. *See also* FEEDBACK

EXTRACTOR

An extractor is a device that is used for the purpose of removing a signal from a circuit for monitoring purposes, or for sampling a constituent of a signal. For example, a radio-frequency probe (*see* RADIO-FREQUENCY PROBE) may be used to extract a small amount of voltage from the tank circuit of a radio-frequency amplifier, for monitoring on an oscilloscope. The modulation can then be monitored, in addition to other signal parameters.

A device for removing integrated circuits from a printed-circuit board is sometimes called an extractor. This device generally consists of a multi-pronged soldering iron, which heats up all of the pins of the device simultaneously. An absorbing braid or suction device removes the solder from the connections. Then, the integrated circuit can be easily pulled from the board. *See also* SOLDER WICK.

EXTRAPOLATION

When data is available within a certain range, an estimate of values outside that range is sometimes made by a technique called extrapolation. Extrapolation can be done by "educated guesswork," although computers provide the best results. Some functions can be extrapolated, while others cannot.

An example of extrapolation is shown in the graph. An attenuation-versus-frequency curve is illustrated for a hypothetical bandpass filter. The center frequency is 3.600 MHz; data are provided for the range 3.595 to 3.605 MHz. If we wish to estimate the attenuation at a frequency of, say, 3.607 MHz, we must extrapolate. This is done simply by extending the function over the range 3.605 to 3.607 MHz, based on the nature of the curve within the range given. Assuming the function contains no unnatural ir-

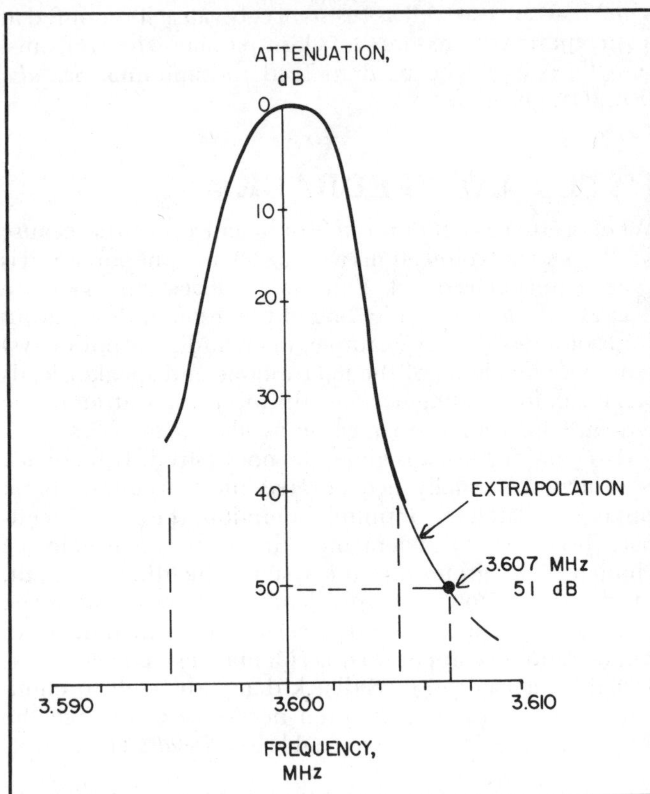

EXTRAPOLATION: An example of extrapolation. At 3.607 MHz, the attenuation is probably about 51 dB.

regularities, a good estimate can be obtained for the attenuation at 3.607 MHz.

Random-number functions are impossible to extrapolate. Other very complicated functions can be difficult or impossible to extrapolate with reasonable accuracy. The accuracy of extrapolation diminishes as the independent variable gets farther from the domain of values given. Thus, while we may get a very good idea of the attenuation of the filter circuit (see illustration) at 3.607 MHz, we cannot readily ascertain the attenuation at 3.650 MHz, or any other frequency far removed from the bandpass range, on the basis of the data given. *See also* FUNCTION, INTERPOLATION.

EXTREMELY HIGH FREQUENCY

The extremely-high-frequency range of the radio spectrum is the range from 30 GHz to 300 GHz. The wavelengths corresponding to these limit frequencies are 10 mm to 1 mm, and thus the extremely-high-frequency waves are called millimetric waves. Most of the frequencies in the millimetric range are allocated by the International Telecommunication Union, headquartered in Geneva, Switzerland.

Electromagnetic waves in the extremely-high-frequency range are not affected by the ionosphere of the earth. Signals in this range pass through the ionosphere as if it were not there. Millimetric waves behave very much like infrared rays, visible light, and ultraviolet. They can be focused by parabolic reflectors of modest size. Because of their tendency to propagate in straight lines unaffected by the atmosphere of the earth, millimetric waves are used largely in satellite communications. The frequencies allow modulation by wideband signals. *See also* ELECTROMAGNETIC SPECTRUM, FREQUENCY ALLOCATIONS .

EXTREMELY LOW FREQUENCY

The extremely-low-frequency band is the range of the electromagnetic spectrum from 30 Hz to 300 Hz. The wavelengths corresponding to these frequencies are 10,000 km to 1,000 km, or 10 million meters to 1 million meters. For this reason the waves are called megametric.

The ionosphere is an almost perfect reflector for megametric waves. Signals in this range cannot reach us from space; they are all reflected back by the ionized layers. The wavelength is many times the height of the ionosphere above the ground, and for this reason, severe attenuation occurs when megametric waves are propagated over the surface of the earth. Antennas for the extremely-low-frequency range are ridiculously long; for example, at 30 Hz, a half-wave dipole must be approximately 3,100 miles in length!

Megametric waves may be used to generate ground currents, in which the entire planet may become resonant. Research is continuing in this area. Megametric waves penetrate underwater, and therefore are useful in communicating with submarines. However, only very narrow-band modulation is possible. *See also* ELECTROMAGNETIC SPECTRUM, FREQUENCY ALLOCATIONS.

EXTRINSIC SEMICONDUCTOR

An extrinsic semiconductor is a semiconductor to which an impurity has been deliberately added. The pure form of the semiconductor is called intrinsic. Impurities must generally be added to semiconductor materials to make them suitable for use in solid-state devices such as transistors and diodes. The process of adding impurities to semiconductors is called doping (*see* DOPING).

Generally, the greater the amount of impurity added to a material such as silicon or germanium, the better the conductivity becomes. Certain impurities result in charge transfer via electrons; this kind of substance is called an N-type semiconductor. Other impurity elements form a substance in which holes carry the charge; these materials are called P-type semiconductors. *See also* ELECTRON, HOLE, INTRINSIC SEMICONDUCTOR, N-TYPE SEMICONDUCTOR, P-TYPE SEMICONDUCTOR.

FACSIMILE

Facsimile is a mode of modulation by which still pictures are transmitted. This code is usually specified by the designator A4 for amplitude modulation and F4 for frequency modulation (*see* EMISSION CLASS).

Facsimile may be transmitted by means of radio or wire. A picture or page of writing may be transmitted. A common use for facsimile is sending satellite weather pictures to earth-based stations. A recording device prints the picture on a piece of paper. Since facsimile involves the transmission of a still picture rather than a moving picture, the bandwidth is less than that for television transmission.

The number of facsimile pictures that can be sent in a given time depends on the bandwidth of the signal. The greater the number of pictures per unit time, the greater the bandwidth. *See also* TELEVISION.

FADING

Fading is a phenomenon that occurs with radio-frequency signals propagated via the ionosphere. Fading takes place because of changes in the conditions of the ionosphere between a transmitting and receiving station, at frequencies where ionospheric propagation is common.

Fading may occur in two different ways. One way in which fading takes place is the combination of waves of variable phase at the receiving station in a communications circuit. The ionosphere is somewhat unstable, and as its height above the earth fluctuates, the phase of a signal reflected from it also changes. Several signals may propagate over different paths simultaneously, and the resulting phase combination at a receiver can fluctuate greatly within a short time interval.

Fading may also occur because of the ionosphere between the transmitting and receiving station. This kind of fading is apt to take place near the maximum usable frequency for the ionosphere at a given time (*see* MAXIMUM USABLE FREQUENCY). A signal may be perfectly readable and then suddenly disappear.

Fading sometimes takes place because of multipath propagation, in which a signal is reflected from the ionosphere at several different points en route to the receiving station (*see* MULTIPATH FADING). When the ionosphere changes orientation rapidly, the fading may occur first at one frequency, and then at another, progressing across the passband of a signal from top to bottom or vice versa. This is called selective fading. The narrower the bandwidth of a signal, the less susceptible it is to selective fading. Since selective fading is common, narrowband signals are preferred over wideband signals in the frequency range where ionospheric propagation is predominant. *See also* SELECTIVE FADING.

FAHRENHEIT TEMPERATURE SCALE

The Fahrenheit temperature scale is used in the United States, and some other industrialized countries, mainly for weather reporting and forecasting. The scale is centered at the freezing point of a saturated salt-water solution; this temperature is assigned the value zero. Pure water freezes at 32 degrees above zero. Pure water boils at a temperature that depends on the atomspheric pressure, but at sea level under normal (standard) pressure, water boils at a Fahrenheit temperature of 212 degrees. Fahrenheit readings are usually followed by the capital letter F.

Scientists generally do not use the Fahrenheit scale. Instead, they employ the Celsius scale. A temperature increment of 1 degree Celsius is equivalent to 9/5, or 1.8, degree Fahrenheit. The absolute, or Kelvin, scale is also sometimes used, especially by astronomers and physicists. The Kelvin degree represents the same amount of temperature change as the Celsius degree, except that the Kelvin scale is centered at the lowest possible temperature while the Celsius scale is centered at the freezing point of pure water. *See also* CELSIUS TEMPERATURE SCALE, KELVIN TEMPERATURE SCALE.

FAILSAFE

Failsafe is a term that refers to a form of failure protection in electronic circuits and systems. A failsafe circuit is so designed that, in the event of malfunction of one or several malfunctions, the circuit will remain safe for operation; no further damage will occur, and no hazard will be presented to personnel using the equipment. A failsafe system is a system that places a backup circuit in operation if a circuit fails. This allows continued operation. A warning signal alerts the personnel that a circuit has failed, and this expedites repair.

In a computer, failsafe is sometimes called failsoft. A malfunction in a failsoft computer will result in some degradation of efficency, but will not cause a total shutdown. The computer alerts the user that a problem is present, giving its location in the system.

No system is 100-percent reliable, but failsafe circuits and systems reduce the amount of downtime in the event of component failures. *See also* DOWNTIME, FAILURE RATE.

FAILURE RATE

Failure rate is an expression of the frequency with which a component, circuit, or system fails. Generally, failure rate is specified in terms of the average number of failures per unit time. Of course, the longer the time interval chosen for the specification, the greater the probability that a given component, circuit, or system will malfunction.

An example of component failure rate is as follows. Suppose that, out of 100 diodes, three of them malfunction within a period of 1 year. We assume that all of the diodes are operational at the start of the 1-year period. Then the failure rate is 3 percent per year for that particular application of the diode. The probability is 3 percent that a given diode will fail within 1 year.

When determining the failure rate for a component, circuit or system, it is customary to test many units at the same time. This increases the number of observed failures, and allows a better probability estimate than would be possible with just one test unit.

As the time interval approaches zero in the failure-rate specification, the failure rate itself does not approach zero. Some components, circuits, or systems fail the instant they are switched on. Any failure occurs at a defined instant. The probability that a component will fail at any given moment is called the hazard rate. *See also* HAZARD RATE, QUALITY CONTROL, RELIABILITY.

FALL TIME

See DECAY TIME.

FAMILY

Any set of similar things, such as components, devices, or mathematical functions, is called a family. For example, in digital logic, there are several families of gate design, including the bipolar and complementary-metal-oxide semiconductor families.

The illustration shows what is called a family of curves. This kind of graph is actually a two-dimensional representation of a three-dimensional function. There are two independent variables; one is shown (in this example) along the horizontal axis, and the other is represented in incremental form by the different curves. The horizontal axis shows collector voltage, and the different curves are illustrated for various values of base voltage for a bipolar transistor. The dependent variable, shown on the vertical axis, is the collector current.

The constituents of a family are always related in some way. Totally different items cannot be members of the same family. However, there are various ways of specifying the nature of the relationship among members of a family. A bipolar transistor and a field-effect transistor are not both members of the family of bipolar devices, but they are both members of the family of solid-state devices. A synonym for "family" might be the word "category."

FAN

A fan is a form of air-cooling device. It is a blower that uses blades inclined at an angle. Fans are employed in electronic circuits that generate large amounts of heat, especially vacuum-tube power amplifiers. Fans are sometimes also used for cooling solid-state equipment such as computers, since excessive temperatures can cause an increase in the frequency of errors in digital apparatus.

The illustration shows a small fan used in a computer system to keep the integrated circuits at a reasonable temperature. With the fan, the computer can be operated at a higher room temperature with a lower error rate. If the fan is not used, the system will malfunction at a lower ambient temperature. This can be a problem in warm climates. *See also* AIR COOLING, BLOWER.

FAN-IN AND FAN-OUT

Fan-in and fan-out are terms used to describe the connection of several terminals to a common circuit. When a circuit has more than one input terminal, the inputs are said to be fanned in. When a circuit has more than one output terminal, the outputs are said to be fanned out.

In a digital computer, fan-in is the number of inputs that can be accommodated by a particular logic circuit, and fan-out is the number of outputs that can be fed by a particular logic circuit. For example, an AND gate may have any number of inputs; for the output to be high, all of the inputs must be high. There may be several outputs, as well, each feeding a subsequent circuit.

FARAD

A farad is the unit of capacitance. When the voltage across a capacitor changes at a rate of 1 volt per second, and a current flow of 1 ampere results, the capacitor has a value of 1 farad. A capacitance of 1 farad also results in 1 volt of potential difference for a charge of 1 coulomb.

The farad is, in practice, an extremely large unit of capacitance. Rarely does a capacitor have a value of 1 farad. Capacitance is generally measured in microfarads, or millionths of a farad (abbreviated uF); for small capacitors, the picofarad, or trillionth of a farad (abbreviated pF) is often specified. Sometimes the nanofarad, or billionth of a farad (abbreviated nF) is indicated.

At radio frequencies, typical capacitances range from 1 pF to perhaps 1,000 pF in tuned circuits, and from 0.001 uF to about 1 uF for bypassing purposes. At audio frequencies, typical capacitances are in the range 0.1 uF to 100 uF. In power-supply filtering applications, capacitances may range up to 10,000 uF or more. *See also* CAPACITANCE, CAPACITOR.

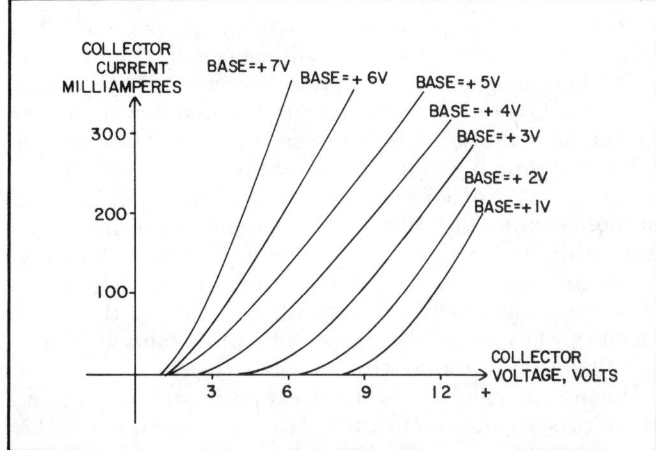

FAMILY: A family of curves illustrating the operation of an NPN bipolar transistor.

FAN: A small fan used to cool the circuitry of a home computer.

FARADAY

The faraday is a unit of electrical quantity or charge. A charge of 1 faraday represents the amount of electrical quantity required to free 1 gram atomic weight of a univalent element in the electrolysis process (*see* ELECTROLYSIS, GRAM ATOMIC WEIGHT). A charge of 1 faraday represents about 96,500 coulombs, or 6.06×10^{23} electrons.

Generally, the faraday is not specified in electronics applications. The coulomb is much more often used to represent charge quantities. *See also* COULOMB.

FARADAY CAGE

A Faraday cage is an enclosure that prevents electric fields from entering or leaving. However, magnetic fields can pass through. The Faraday cage consists of a wire screen or mesh, or a solid non-magnetic metal, broken so that there is not a complete path for current flow. The Faraday cage is similar to the electrostatic shield employed in some radio-frequency transformers to reduce the transfer of harmonic energy (*see* ELECTROSTATIC SHIELDING).

Faraday cages are used in experimentation and testing, when the presence of electric fields is not desired. An entire room may be shielded. Sometimes a complete electromagnetic shield is called a Faraday cage, although this is technically a misnomer. *See also* ELECTROMAGNETIC SHIELDING.

FARADAY EFFECT

When radio waves are propagated through the ionosphere, their polarization changes because of the effects of the magnetic field of the earth. This is called the Faraday effect. Because of the Faraday effect, sky-wave signals arrive with random, and fluctuating, polarization. It makes very little difference what the polarization of the receiving antenna may be; signals originating in a vertically polarized antenna may arrive with horizontal polarization some of the time, vertical polarization at other times, and slanted polarization at still other times.

The Faraday effect results in signal fading over ionospheric circuits. There are other causes of fading, as well. The fading caused by changes in signal polarization can be reduced by the use of a circularly polarized receiving antenna (*see* CIRCULAR POLARIZATION).

When light is passed through certain substances in the presence of a strong magnetic field, the plane of polarization of the light is made to rotate. This is a form of Faraday effect. In order for this to occur, the magnetic lines of force must be parallel to the direction of propagation of the light. *See also* POLARIZATION.

FARADAY, MICHAEL

One of the pioneers in the research of the nature of electric and magnetic fields, Michael Faraday (1791-1867), demonstrated that such fields produce forces on objects. While these forces appear similar, they result from different causes. Faraday also conducted experiments showing that electric currents produce magnetic fields and that changing magnetic fields induce current in electrical conductors. This knowledge ultimately led to the discovery of electromagnetic fields that propagate over vast distances. *See also* FARADAY'S LAWS.

FARADAY SHIELDING

See ELECTROSTATIC SHIELDING, FARADAY CAGE.

FARADAY'S LAWS

Faraday's laws are concerned with two different phenomena, and are therefore placed into two categories.

The law of electromagnetic induction is sometimes called Faraday's law. This is the familiar principle of the generation of a current in a wire that moves in a magnetic field. Actually, the wire may be stationary and the field may be moving; but as long as the wire cuts across magnetic lines of flux, a current is induced. The current is directly proportional to the rate at which the wire crosses the lines of force of the magnetic field. The faster the motion, the greater the current; the more intense the field, the greater the current. It is this effect, in part, that makes radio communication possible (*see* ELECTROMAGNETIC INDUCTION).

The other principle form of Faraday's law is involved with electrolytic cells. In any electrolytic cell, the greater the amount of charge passed through, the greater the mass of substance deposited on the electrodes. The actual mass deposited for a given amount of electric charge depends on the electrochemical equivalent of the substance. *See also* ELECTROCHEMICAL EQUIVALENT, ELECTROLYSIS.

FAR FIELD

The far field of an antenna is the electromagnetic field at a great distance from the antenna. The far field has essentially straight lines of electric and magnetic flux, and the lines of electric flux are perpendicular to the lines of magnetic flux. The wavefronts are essentially flat planes.

The far field of an antenna has a polarization that depends on several factors. The polarization of the transmitting antenna dictates the polarization of the far field under conditions in which there is no Faraday effect (*see* FARADAY EFFECT). The power density of the far field diminishes with the square of the distance from the antenna. The field strength, in microvolts per meter, diminishes in direct proportion to the distance from the antenna.

The far field of an antenna is sometimes called the Fraunhofer region. The far field begins at a distance that depends on many factors, including the wavelength and the size of the antenna. The signal normally picked up by a receiving antenna is the far field, except at extremely long wavelengths. *See also* ELECTROMAGNETIC FIELD, NEAR FIELD, TRANSITION ZONE.

FATIGUE

As a circuit or component gets older, it becomes less reliable. This aging process occurs rapidly with some components or circuits, and slowly with others. When a component or circuit needs to be replaced or repaired because of the simple effects of aging, the condition is called fatigue.

Fatigue usually occurs only in power-dissipating components such as vacuum tubes and transistors. To a lesser extent, fatigue takes place in all active devices. Passive components such as coils and capacitors generally do not suffer from fatigue unless they are forced to handle currents or voltages that put them under strain. That should not happen in a well-designed circuit.

Equipment should be checked periodically for possible

fatigue. A drop in the power output of an amplifier, for example, may go unnoticed if it is gradual and no periodic tests are conducted. A change in the standing-wave ratio on a transmission line, especially a gradual drop, often indicates feed-line aging, with a resulting loss increase. *See also* AGING.

FAULT

Any discontinuity or problem in a circuit is sometimes called a fault. The fault is always localized in nature, such as a short-circuited capacitor or a burned-out diode. With proper tools and equipment, the fault can always be found and corrected.

A fault will often result in a leakage current, known as a fault current, that is not present under proper operating conditions. This current can, in some cases, cause destruction of additional components. For example, a partial short circuit in a coupling capacitor may change the bias at the next stage sufficiently to cause excessive current in that stage, and consequent damage to its components.

Troubleshooting is sometimes called fault finding, and an instrument used by a technician is called a fault finder. *See also* TROUBLESHOOTING.

FEASIBILITY

A feasibility study is an analysis of the desirability of using a computer to handle certain problems, particularly in a business. In some cases, a computer would not pay for itself, but in other situations it would be profitable for the company to purchase a computer, install it, and train personnel to use it.

Ironically, a feasibility study often requires the use of a computer! Factors that affect the feasibility of using a computer in a business include the volume of work, the department involved, and the nature of the work. Computers can perform much of the mundane, boring paperwork that is necessary for the efficient operation of a company. But a computer would be less suited to innovation and creative work, although the computer can be an invaluable aid to engineers in solving complex problems.

The feasibility study must provide the optimum solution for efficiency, both in terms of cost and time. The decision must be made as to the size of a computer, if it is determined that one should be used. *See also* COMPUTER.

FEDERAL COMMUNICATIONS COMMISSION

The Federal Communications Commission, or FCC, is an agency of the United States Government. The FCC is responsible for the allocation of frequencies for radiocommunications and broadcasting within the United States. The FCC is also responsible for the enforcement of the laws concerning telecommunications.

The FCC issues various kinds of licenses for communications and broadcasting personnel. Some licenses can be obtained only by passing an examination. The FCC composes and administers these examinations. Other licenses require only the filing of an application.

When the FCC deems it necessary to create a new rule, it first publishes a notice of proposed rulemaking. Then, within a certain specified time period, interested or concerned parties are allowed to make comments on the subject. The FCC, after considering the comments, makes its decision. The power of the FCC is limited, however, by the Congress.

Each country determines its own frequency allocations and system of radiocommunications laws. The International Telecommunication Union provides coherence among the many nations of the world, for the purpose of optimum utilization of the limited space in the spectrum. *See also* INTERNATIONAL TELECOMMUNICATION UNION.

FEED

Feed is the application of current, power, or voltage to a circuit. The term feed is used particularly with reference to antenna systems. There are basically three electrical methods of antenna feed: current feed, voltage feed, and reactive feed.

A current-fed antenna has its transmission line connected where the current in the radiating element is greatest. This occurs at odd multiples of ¼ wavelength from free ends of the radiator. A voltage-fed antenna has the transmission line connected at points where the current is minimum; such points occur at even multiples of ¼ wavelength from free ends, and also at the free ends. Both current feed and voltage feed are characterized by the absence of reactance; only resistance is present. In current feed, the resistance is relatively low; in voltage feed, it is high.

When an antenna is not fed at a current or voltage loop, the feed is considered reactive. This may be the case when an antenna is not resonant at the operating frequency, or when it is fed at a point not located at an integral multiple of ¼ wavelength from a free end.

Feed may be classified in other ways, such as according to the geometric position of the point where the transmission line joins the radiating element. In end feed, the line is connected to the end of the element; in center feed, to the center; in off-center feed, somewhat to either side of the center. A vertical radiator may be fed at the base, or part of the way up from the base to the top. A special method of feed for a vertical radiator is called shunt feed. *See also* CENTER FEED, CURRENT FEED, END FEED, OFF-CENTER FEED, SHUNT FEED, TRANSMISSION LINE, VOLTAGE FEED.

FEEDBACK

When part of the output from a circuit is returned to the input, the situation is known as feedback. Sometimes feedback is deliberately introduced into a circuit; sometimes it is not wanted. Feedback is called positive when the signal arriving back at the input is in phase with the original input signal. Feedback is called negative (or inverse) when the signal arriving back at the input is 180 degrees out of phase with respect to the original input signal. Positive feedback often results in oscillation, although it can enhance the gain and selectivity of an amplifier if it is not excessive. Negative feedback reduces the gain of an amplifier stage, makes oscillation less likely, and enhances linearity.

Feedback can result from a deliberate arrangement of components. All oscillators use positive feedback. Unwanted feedback can result from coupling among external wires or equipment. This external coupling may be capacitive, inductive, or perhaps acoustic. We have all heard the acoustic feedback in an improperly operated public-address system.

The number of amplifying stages that can be cascaded,

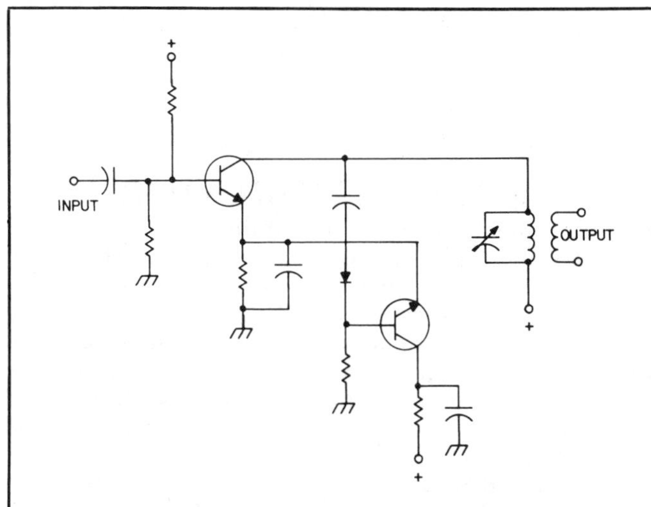

FEEDBACK AMPLIFIER: A feedback amplifier, used for the purpose of automatic level control.

for the purpose of obtaining high gain, is limited partly by the effects of feedback. The more stages connected in cascade, the higher the probability that some positive feedback will occur, with consequent oscillation. *See also* ACOUSTIC FEEDBACK, NEGATIVE FEEDBACK, OSCILLATION.

FEEDBACK AMPLIFIER

A feedback amplifier is a circuit, placed in the feedback path of another circuit, to increase the amplitude of the feedback signal. The phase may also be inverted. A feedback amplifier is used when the feedback signal would not otherwise be strong enough to obtain the desired operation.

An example of a feedback amplifier is shown in the illustration. In this particular case, the feedback signal is direct current obtained by rectifying the signal output. The greater the amplitude of the signal at the output, the larger the direct-current voltage applied to the input. The direct-current voltage reduces the gain of the amplifier. This example illustrates amplified automatic level control (*see* AUTOMATIC LEVEL CONTROL).

The term feedback amplifier is sometimes used to describe an amplifier that uses feedback to obtain a certain desired characteristic. Negative feedback, for example, is often used to reduce the possibility of unwanted oscillations in a radio-frequency amplifier. Combinations of negative and positive feedback can be used to modify the frequency response of an amplifier. This is especially true of operational-amplifier circuits. *See also* FEEDBACK, OPERATIONAL AMPLIFIER.

FEEDBACK CONTROL

Whenever feedback is deliberately introduced into a circuit for any reason, some means must be provided to control the amount of feedback. Otherwise, the desired circuit operation may not be obtained. A feedback control may consist of a simple potentiometer, used to adjust the amount of feedback signal arriving at the input of the circuit. More sophisticated methods of feedback control consist of self-regulating circuits, such as the feedback amplifier employed in the amplified automatic-level-control circuit (*see* AUTOMATIC LEVEL CONTROL, FEEDBACK AMPLIFIER).

The elimination of undesired feedback in a circuit is sometimes called feedback control. The output of a circuit can be partially coupled back to the input by the capacitance or inductance between the input and output peripheral wiring. In a public-address system, acoustic feedback can sometimes take place as the microphone picks up sound from the speakers. Coupling between the input and output should be minimized. This might be done by means of bypass capacitors, series chokes, or phase-inverting circuits. *See also* FEEDBACK.

FEEDBACK RATIO

The feedback ratio, usually represented by the Greek letter beta (β), is a measure of the amount of feedback in a feedback amplifier. If e_o is the output voltage of an amplifier with no load, and e_f is the feedback voltage, then the feedback ratio is given simply as:

$$\beta = e_f/e_o$$

When the feedback ratio is to be expressed as a percentage, it is generally represented by n, and

$$n = 100e_f/e_o$$

This quantity is sometimes called the feedback percentage. The feedback ratio is never greater than 1, and the feedback percentage is never greater than 100.

If A is the open-loop gain of an amplifier, and β is the feedback ratio, then the feedback factor m is the quantity given by:

$$m = 1 - \beta A$$

See also FEEDBACK.

FEEDER CABLE

Any cable that carries signal information from one place to another is a feeder cable. A coaxial transmission line for a radio-frequency antenna system is sometimes called a feeder cable. Usually, the term feeder cable is used to describe a communication cable running from a central station to a distribution station. From the distribution station, secondary cables, sometimes also called distribution cables, run to local stations and subscribers.

The coaxial cable used to carry cable-television signals is called a feeder cable. Amplification circuits are used at intervals in the feeder cable to keep the signal at a sufficient level for reception by television sets. *See also* CABLE COMMUNICATIONS, CABLE TELEVISION.

FEED LINE

A feed line is a transmission line used for the purpose of transferring an electromagnetic field from a transmitter to an antenna. The point where the feed line joins the antenna is called the feed point.

Feed lines may take a variety of different forms. The most common type of feed line in use today is the coaxial line, which is prefabricated and usually has a solid or foamed polyethylene dielectric material. Two-wire line, also called open-wire or parallel-wire line, is often used in the feed system of a television receiving antenna. Homemade open-wire line is sometimes used by radio amateurs. Such a feed line has very low loss, and can tolerate high power levels and large impedance mismatches.

A feed line should be chosen on the basis of low loss, cost effectiveness, and power-handling capacity. Other important factors include convenience of installation, the antenna impedance, and antenna-system balance. The best feed line for a particular antenna depends on all these factors. *See also* COAXIAL CABLE, FOUR-WIRE TRANSMISSION LINE, HARD LINE, OPEN-WIRE LINE, TWIN-LEAD.

FEEDTHROUGH CAPACITOR

A feedthrough capacitor is a device that is used for passing a lead through a chassis, while simultaneously bypassing the lead to chassis ground. The center (axial) part of the feedthrough capacitor is connected to the lead; the outer part of the component is connected to chassis ground. The illustration shows a feedthrough capacitor.

Feedthrough capacitors are useful in power-supply leads, when it is desirable to prevent radio-frequency energy from being transferred into or out of the equipment along these leads. Feedthrough capacitors are also sometimes used in the speaker or microphone leads of an audio amplifier circuit.

Feedthrough capacitors are available in many different physical and electrical sizes. The proper value of capacitance should be chosen for the application desired. *See also* BYPASS CAPACITOR, CAPACITOR.

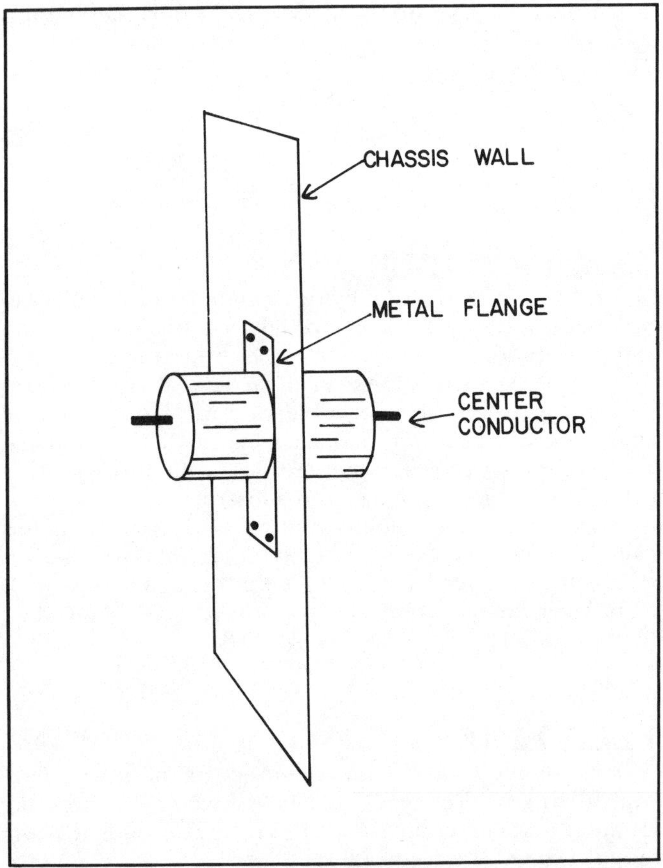

FEEDTHROUGH CAPACITOR: Drawing of a feedthrough capacitor. The metal flange of the capacitor is connected to chassis ground.

FEEDTHROUGH INSULATOR

A feedthrough insulator is used for the purpose of passing a lead through a metal chassis while maintaining complete electrical insulation from the chassis. Feedthrough insulators are also sometimes used when it is necessary to pass leads through a partial conductor or nonconductor with a minimum amount of dielectric loss. Feedthrough insulators generally consist of a threaded shaft with two conical sections of porcelain or glass held in place by nuts on the shaft.

Feedthrough insulators are available in a variety of different sizes. The voltage across a feedthrough insulator may, in some applications, reach substantial values. Therefore, a sufficiently large insulator should always be used. In the antenna-tuning network shown, the feedthough insulators may be subjected to several thousand volts under certain conditions. The feedthrough insulators should have the lowest possible loss at the highest frequency of the equipment in which the are used. The insulators should also have the smallest possible amount of capacitance. *See also* INSULATOR.

FELICI MUTUAL-INDUCTANCE BALANCE

The Felici mutual-inductance balance is a device used for the purpose of determining the mutual inductance, or degree of coupling, between two coils. The Felici balance is used primarily at lower frequencies. At higher frequencies, the capacitance among the windings of the inductors results in difficulty of measurement.

The accompanying schematic diagram shows a simple Felici balance. A standard mutual inductor, which is variable, is connected together with the pair of inductors for which the mutual inductance is to be determined. The reference transformer is adjusted until a null is observed on the indicator. This indicator may be a meter or, if the signal generator produces audio-frequency energy, a headset or speaker. When the null is seen, it indicates that the reference inductor has been set to have the same mutual inductance as the unknown. The mutual inductance may then be read directly from a calibrated scale. *See also* MUTUAL INDUCTANCE.

FEMALE

Female is an electrical term used to describe a jack. A female plug is recessed, and the conductors are not exposed. An example of a female plug is the common utility wall outlet.

The female jack is usually, but not always, mounted in a fixed position on the panel of a piece of electronic apparatus. The male plug (*see* MALE) is generally attached to the cord or cable and fits into the female plug. Connectors may have only one conductor, or they may have dozens of conductors.

FEMTO

See PREFIX MULTIPLIERS.

FERNICO

Fernico is a material used in the manufacture of metal-envelope vacuum tubes. Such tubes are not commonly used today, having been largely replaced by semiconductor devices.

The pins at the base of a metal-envelope tube are passed through glass beads to insulate them from the envelope. The base of the tube is made from an alloy of iron, cobalt, and nickel, called Fernico. This alloy keeps losses between the pins at a minimum, while still providing the desired shielding of the tube interior from the external environment. *See also* TUBE.

FERRIC OXIDE

Ferric oxide is a form of iron oxide. Its chemical formula is Fe_2O_3. Ferric oxide is reddish in color, and we know it as rust.

Ferric oxide has magnetic properties, and also is easily powdered and applied to surfaces. Ferric oxide is ideally suited to the manufacture of recording tapes, and is commonly used for this purpose. The material is ground into a very fine dust, and is applied to the surface of the tape, which is usually made from mylar or a similar flexible material. Impulses from the recording heads cause the particles to become magnetized; with sufficiently fine ferric-oxide powder, frequencies in excess of 10 kHz can be reproduced at a tape speed of 3¾ inches per second. This represents more than 2,600 complete cycles per inch of tape. *See also* MAGNETIC RECORDING, MAGNETIC TAPE.

FERRITE

Ferrite is a substance designed for the purpose of increasing the magnetic permeability in an inductor core. Ferrite has a higher degree of permeability than ordinary powdered iron. Ferrite materials are classified as either soft or permanent. Ferrite acts as as electrical insulator, resembling a ceramic substance with respect to current conductivity. The magnetic conductivity, however, is excellent, and the eddy-current losses are very small.

Ferrite materials are available in a great variety of shapes and permeability values. The tiny receiving antennas used in small transistor radios consist of a coil wound on a solenoidal ferrite core. Such antennas have excellent sensitivity at frequencies below the shortwave range. Ferrite is employed in toroidal inductor cores to obtain large values of inductance with a relatively small amount of wire. Ferrite is also used to make pot cores, which allows even larger inductances to be obtained. Typical permeability values for ferrites range from about 40 to more than 2,000. Ordinary powdered-iron cores generally have much smaller permeability values. Ferrite is well suited to low-frequency and medium-frequency applications in which loss must be kept to a minimum. *See also* FERRITE BEAD, FERRITE CORE, FERRITE-ROD ANTENNA, PERMEABILITY, POT CORE, POWDERED-IRON CORE, TOROID.

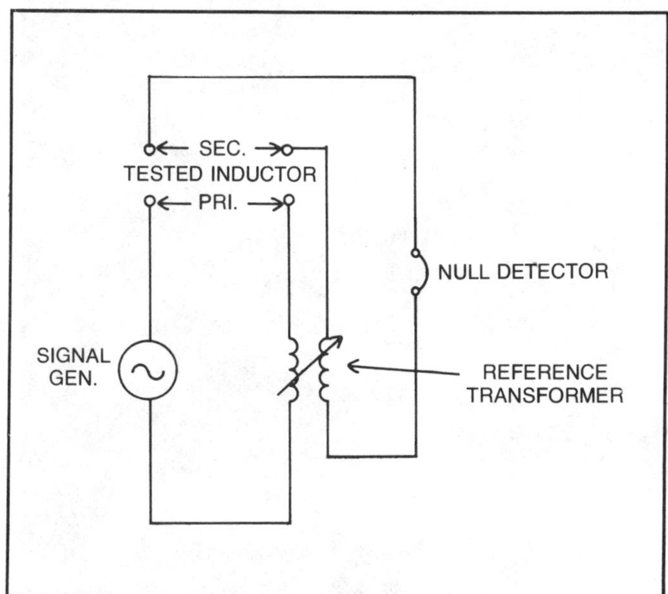

FELICI MUTUAL-INDUCTANCE BALANCE: The Felici mutual-inductance balance is used for determining the degree of coupling between the windings of a transformer.

FERRITE BEAD

A ferrite bead is a small, toroidal piece of ferrite. Ferrite beads are often used for the purpose of choking off radio-frequency currents on wire leads and cables. The bead is simply slipped around the wire or cable, as illustrated. This introduces an inductance for high-frequency alternating currents without affecting the direct currents and low-frequency alternating currents.

Ferrite beads are especially useful for choking off antenna currents on coaxial transmission lines in the very-high-frequency range and above. Antenna currents, or radio-frequency currents flowing on the outside of a coaxial feed line, can cause problems with transmitting equipment. One or more ferrite beads along the feed line are effective in choking off these unwanted currents. But the desired transmission of electromagnetic fields inside the cable is not affected.

Ferrite beads are used in certain digital memory systems

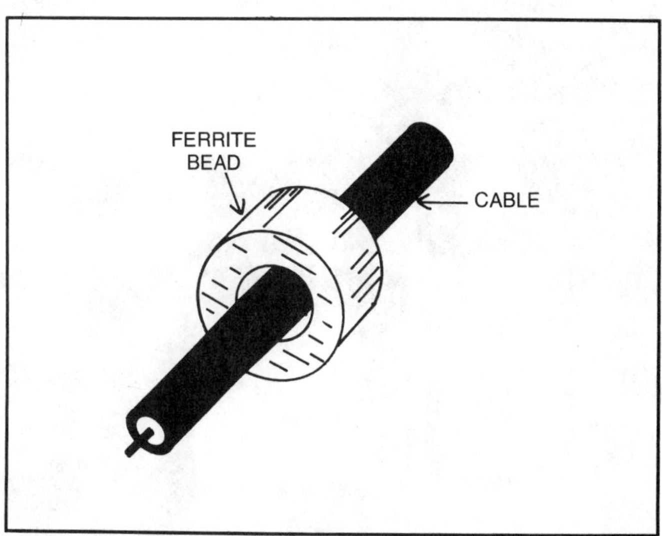

FERRITE BEAD: A ferrite bead, placed around a coaxial cable, chokes off radio-frequency currents flowing on the outer conductor of the cable.

to store information by means of magnetic fields. The polarity of the magnetic field may be either clockwise or counterclockwise. The polarity will remain the same until a current is sent through a conductor passing through the bead. If the current is such as to reverse the polarity of the magnetic field stored in the ferrite, then the polarity will remain reversed until another surge of current, in the opposite direction, changes the field orientation again. *See also* FERRITE, MAGNETIC CORE, TOROID.

FERRITE CORE

A ferrite core is an inductor core made from ferrite material. Ferrite cores are extensively used in electronic applications at low, medium, and high frequencies, for the purpose of maximizing the inductance of a coil while minimizing the number of turns. This results in low-loss coils.

The most common ferrite-core configurations are the solenoidal type and the toroidal type. The toroid core has gained popularity in recent years because the entire magnetic flux is contained within the core material. This eliminates coupling to external components (*see* TOROID).

At audio frequencies and very low frequencies, the pot core is often used. This type of core actually surrounds the coil, whereas in other inductors the coil surrounds the core. Pot cores allow inductance values in excess of 1 henry to be realized using a moderate amount of wire. Such coils, because of the low loss in the ferrite material, exhibit extremely high Q factors, and are thus useful in applications where extreme selectivity is needed (*see* POT CORE).

At higher frequencies, ferrite becomes lossy, and powdered-iron cores are preferred. *See also* FERRITE, POWDERED-IRON CORE.

FERRITE-ROD ANTENNA

For receiving applications at low, medium, and high frequencies, up to approximately 20 MHz, a ferrite-rod antenna is sometimes used. This type of antenna consists of a coil wound on a solenoidal ferrite core. A series or parallel capacitor, in conjunction with the coil forms a tuned circuit. The operating frequency is determined by the resonant frequency of the inductance-capacitance combination.

Ferrite-rod antennas display directional characteristics similar to the dipole antenna (*see* DIPOLE ANTENNA). The sensitivity is maximum off the sides of the coil, and a sharp null occurs off the ends. This has little effect with sky-wave signals, which tend to arrive from varying directions. However, the null can be used to advantage to eliminate interference from local signals and from man-made sources of noise. The ferrite-rod antenna is physically very small (see photograph). This makes it easy to orient the rod

FERRITE-ROD ANTENNA: Ferrite-rod antennas are found in all small transistorized amplitude-modulation broadcast receivers. The rod appears near the center of the photograph.

in any direction. Some ferrite-rod antennas are affixed to az-el mountings (*see* AZ-EL).

Ferrite-rod antennas are not generally used at vhf and uhf, because the required inductance is so small that the coil would have very few turns, and capacitive effects would predominate. A further limitation is imposed by the fact that ferrite materials are seldom suitable for use above about 50 MHz. They become lossy because of hysteresis effects. *See also* FERRITE.

FERROELECTRIC CAPACITOR

The ferroelectric capacitor is a form of ceramic capacitor in which the dielectric is a ferroelectric substance (*see* FERROELECTRICITY). Ferroelectric materials show excellent dielectric properties and have very large dielectric constants. Ferroelectric capacitors therefore have large values of capacitance for a relatively small demand on physical space.

Because of hysteresis effects, the value of a ferroelectric capacitor depends on the voltage across it. In general, the greater the voltage across a ferroelectric capacitor, the smaller its value. Ferroelectric capacitors tend to drop somewhat in value as they age, because the polarization of the ferroelectric material deteriorates, reducing the dielectric constant.

Ferroelectric capacitors tend to be self-resonant at frequencies ranging from a few megahertz to the very-high frequency region. This is because of inherent inductance in the leads and plates. Such capacitors are used mostly for coupling and bypassing purposes, where a constant value of capacitance is not important. They are generally not suitable for use much above the high-frequency region. *See also* CAPACITOR, DIELECTRIC.

FERROELECTRICITY

Ferroelectricity is the polarization of electric dipoles in certain insulating materials (*see* DIPOLE). It is very similar to magnetism, in which magnetic dipoles become polarized. A ferroelectric material is sometimes called an electret (*see* ELECTRET). Ferroelectric substances are used to make certain kinds of microphones and speakers, as well as capacitors with large values.

Examples of ferroelectric substances include barium titanate, barium strontium titanate, potassium dihydrogen phosphate, Rochelle salts, and triglycine sulfate. These materials are essentially ceramics. Certain waxes and plastics can be heated and then cooled in a strong electric field, and they then become ferroelectric substances. *See also* DIELECTRIC, DIELECTRIC POLARIZATION, FERROELECTRIC CAPACITOR.

FERROMAGNETIC MATERIAL

A ferromagnetic material is a substance that has a high magnetic permeability. The most common ferromagnetic materials used in electronics are ferrite and powdered iron. Such substances greatly increase the inductance of a coil when used as the core material.

Ferromagnetic materials may consist of various combinations of iron fragments and other substances. Such materials as ferrite and powdered iron are highly resistive to the direct application of electric current, but will carry magnetic fields with very little loss. The permeability of a ferromagnetic material is a measure of the degree to which the magnetic lines of force are concentrated within the substance (*see* PERMEABILITY).

Ferromagnetic materials are formed into a variety of different shapes for different applications. *See also* FERRITE, FERRITE-ROD ANTENNA, POT CORE, POWDERED-IRON CORE, TOROID.

FET

See FIELD-EFFECT TRANSISTOR.

FET VOLTMETER

An FET voltmeter is a device similar to a vacuum-tube voltmeter (*see* VACUUM-TUBE VOLTMETER). The input impedance is extremely high, so the device does not draw much current from the source. An FET voltmeter uses a field-effect-transistor amplifier circuit to achieve the high input impedance.

The illustration shows a simplified schematic diagram of an FET voltmeter. The ranges are selected by varying the gain of the amplifying stage, or by connecting (by means of a switch) various resistances in parallel with the microammeter.

The FET voltmeter can also be used to measure current and resistance; this can be done simply by using the meter and power supply in conjunction with various resistors, or the field-effect transistor may be employed to act as an amplifier to increase the sensitivity. Such a combination meter is called an FET volt ohm-milliammeter, or FET VOM. *See also* FIELD-EFFECT TRANSISTOR.

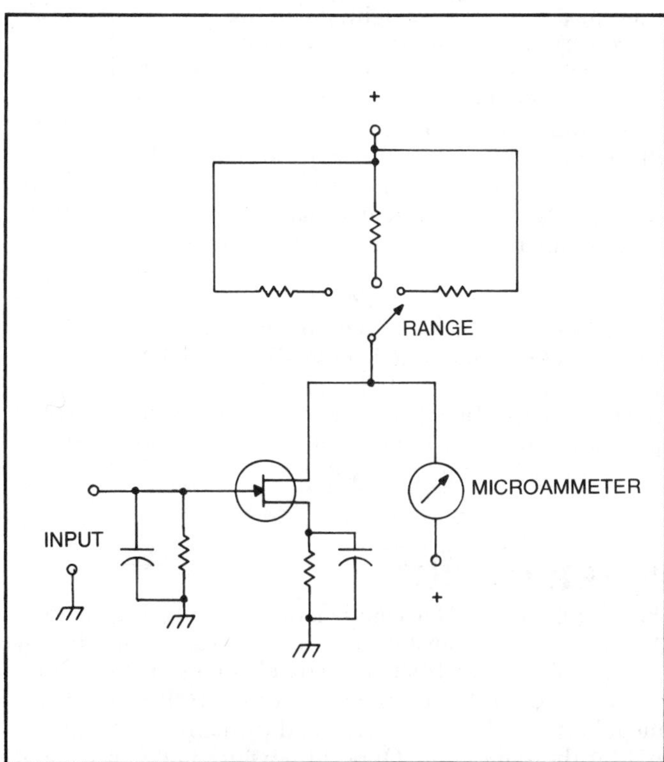

FET VOLTMETER: The FET voltmeter provides a high degree of sensitivity, and the current drain is very low.

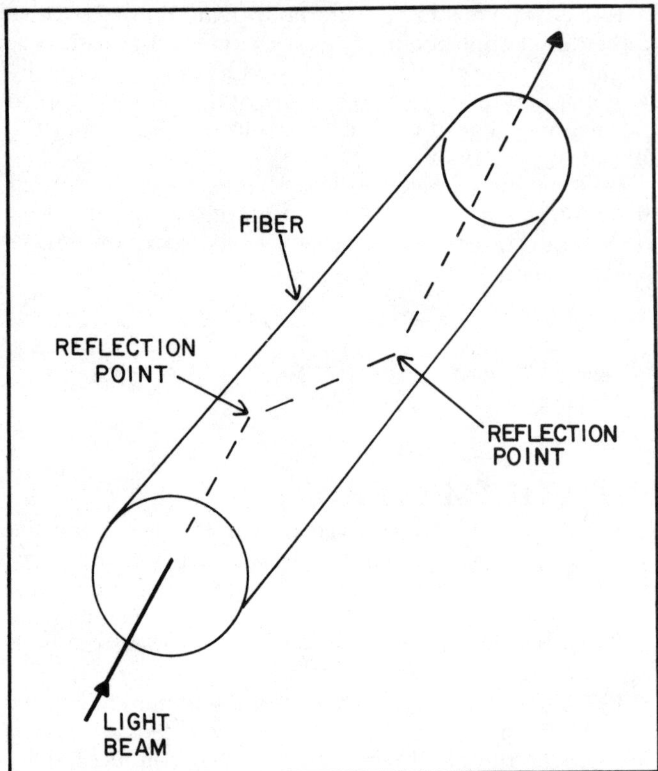

FIBER: Light beams propagate through optical fibers because the index of refraction in the fiber material is higher than the index of refraction in the surrounding medium.

FIBER

A fiber is a thin, transparent strand of material, usually glass or plastic, that is used to carry light beams. A fiber confines the light energy inside its walls because of high internal reflection. The index of refraction of the fiber substance is, for visible light, much higher than the index of refraction for air; therefore, light rays are reflected from the inner walls as they propagate lengthwise along the fiber (see illustration).

A bundle of optical fibers may be used to transfer many different light beams at the same time. A single light beam can be simultaneously modulated by hundreds, or even thousands, of independent signals. Fiber cables of current design can carry millions of different conversations at once!

A large bundle of optical filaments can carry a picture. The image is focused onto one end of the bundle of fibers, and is remagnified at the other end. This principle makes it possible to look deep inside the human body. A device called a fiberscope can be swallowed by a patient and slowly fed through the entire gastrointestinal system. *See also* FIBER OPTICS, FIBERSCOPE.

FIBER OPTICS

Fiber optics is a modern method of conveying information. Light beams are transferred from one place to another by an optical fiber or fibers (*see* FIBER). Light beams can be modulated at frequencies as high as hundreds of megahertz, and can be modulated by many different signals at the same time. Optical fibers are inexpensive and efficient. They allow the transmission of data at extremely high speeds. Optical fibers are compact and lightweight, and are essentially immune to electromagnetic interference. Modern optical fibers may soon replace ordinary wire conductors in long-distance, high-volume communication cables.

A fiber-optics communications system requires a light source that can be easily modulated, a light receptor that can follow the modulation of the light, and a system of amplifiers. Certain kinds of diodes are excellent sources of light for this purpose; some such diodes emit coherent light, which suffers less attenuation than ordinary light. Gallium-arsenide (GaAs) electroluminescent diodes are widely used in fiber-optics communications systems. Phototransistors, silicon solar cells, and similar devices can be employed to intercept the light.

Modulation of a light beam for fiber-optics communications can be accomplished electrically or externally. The electrical method essentially resembles amplitude modulation, in which the intensity of the light source is varied periodically with the signal information. External modulation is accomplished by passing the light or reflecting it from, a device with variable transmittivity. Magnetic fields, is sufficiently strong and properly oriented, can be used to modulate the polarization of a light beam. *See also* MODULATED LIGHT, OPTICAL COMMUNICATION.

FIBERSCOPE

A fiberscope is a flexible bundle of optical fibers, with a focusing lens at each end, used for the purpose of viewing areas not normally observable. The image to be viewed is focused onto one end of the fiber bundle, and is magnified at the other end. The illustration shows how the principle by which the fiberscope operates.

The resolution of the fiberscope depends on the size of the lens at the receptor end, and also on the number and fineness of the fibers in the bundle. Magnification can be obtained by using a highly convex lens at the receptor end of the bundle. Each fiber carries a small part of the image. Of course, the fibers must be oriented the same way at both ends of the bundle if the image at the display end is to be a faithful reproduction of the image at the receptor end.

The fiberscope is useful for viewing inside the body. A small light can be placed at the receptor end, and the fiber bundle can then be swallowed. This allows medical doctors to observe, for example, the stomach lining. *See also*FIBER, FIBER OPTICS, MEDICAL ELECTRONICS.

FIBRILLATION

See HEART FIBRILLATION.

FIDELITY

Fidelity is the faithfulness with which a sound, or other signal, is reproduced. In general, perfect fidelity represents a complete lack of distortion in the signal waveform at all frequencies of modulation. Fidelity is particularly important in the reproduction of music, and in modern stereo high-fidelity equipment, great attention is given to minimizing the distortion.

Fidelity is not very important in communications practice. Intelligibility is of greater concern in this application. Maximum fidelity does not always occur along with

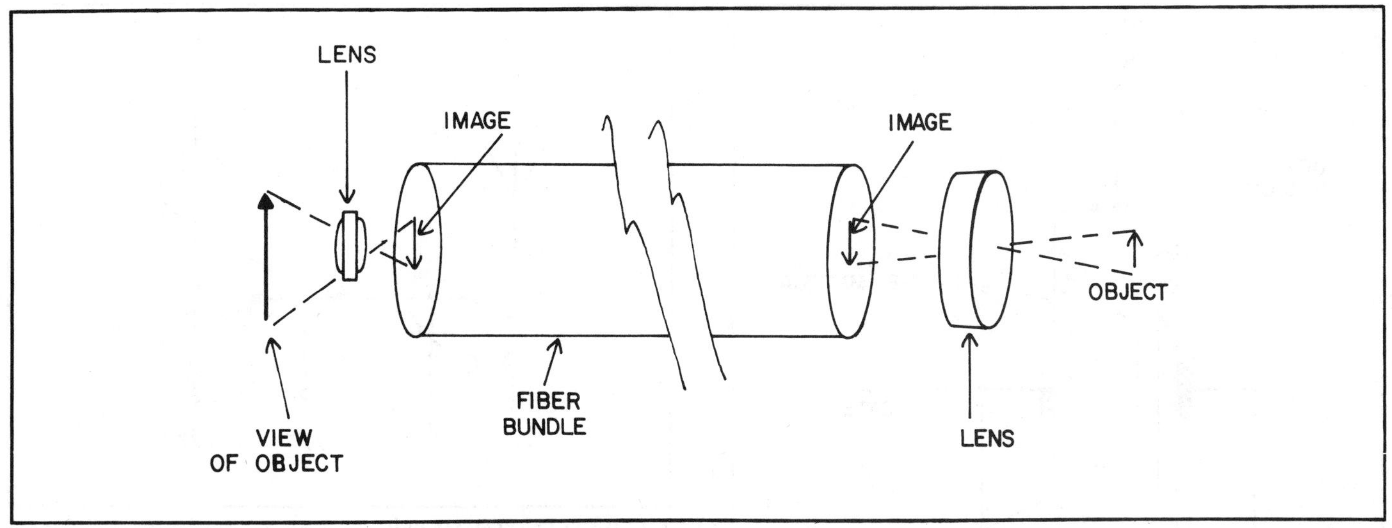

FIBERSCOPE: The fiberscope transmits an image in discrete segments or bits, by means of parallel optical fibers.

maximum intelligibility. Although the human voice contains components of very low and very high audio frequencies, the range from about 300 Hz to 3 kHz is sufficient for communications purposes, and thus the audio-frequency response in a voice communications circuit is usually limited to this range.

For the best possible fidelity, an audio amplifier must usually be operated in Class A or Class B push-pull. The amplifier must be linear over the range of amplitudes to be reproduced. This requires careful engineering. *See also* CLASS-A AMPLIFIER, HIGH FIDELITY.

FIELD

Many effects occur between or among objects separated by empty space. Examples of this are electricity, magnetism, and gravitation. The means by which an effect is propagated with no apparent medium to carry it is called a field. In electronics, the most significant fields are the electric field, the electromagnetic field, and the magnetic field.

The effects of fields travel through empty space at a speed of approximately 186,282 miles per second, or 299,792 kilometers per second. This is the familiar speed of light. In material substances, the speed is slower than that in a vacuum, and depends on several factors.

Radio waves, infrared, visible light, ultraviolet, X rays, and gamma rays are all electromagnetic-field effects. The whole range of electromagnetic-field effects is called the electromagnetic spectrum (*see* ELECTROMAGNETIC SPECTRUM).

In television, half of the image, or 262.2 lines, is called a field (*see* TELEVISION). Computer-record subdivisions are also sometimes called fields. Generally, the term field is used in reference to the effects of electricity and magnetism. *See also* ELECTRIC FIELD, ELECTROMAGNETIC FIELD, MAGNETIC FIELD.

FIELD-EFFECT TRANSISTOR

A field-effect transistor, or FET, is a semiconductor device used as an amplifier, oscillator, or switch. The field-effect transistor operates on the basis of the interaction of an electric field with semiconductor material.

The drawing illustrates a cross-sectional representation of a field-effect transistor. The source is at ground potential, and the drain, at the opposite end of the N-type wafer, is connected to a positive voltage. The path from the source to the drain is called the channel. In this case, the channel consists of an N-type semiionductor wafer, and thus the device is called an N-channel FET. Electrons flow from the source to the drain, through the channel.

The gate electrode consists of two sections of P-type semiconductor material in the N-channel FET. Alternatively, the P-type semiconductor may be wrapped around the channel. When a negative voltage is applied to the gate electrode, a depletion region, or area of nonconduction, forms in the channel, as shown. The larger the negative voltage, the bigger the depletion region becomes. This narrows the channel, and increases the resistance from the source to the drain. In this way, the FET acts as a sort of current valve. An alternating-current signal can be applied to the gate, and the current through the FET will fluctuate with much greater amplitude than the original signal. The result is amplification.

If the negative voltage at the gate electrode becomes large enough, the channel will be entirely severed by the depletion region. This condition is called pinchoff and is somewhat analogous to cutoff in a tube or bipolar transistor.

If the N-type and P-type semiconductors are reversed, the FET will operate in much the same way, except that the channel conducts via holes rather than electrons. Such a device is called a P-channel FET. The voltage at the gate electrode must be positive in order to narrow the channel. The drain is normally made negative with respect to the grounded source in this type of FET.

There are numerous other kinds of field-effect transistors. The type shown here is sometimes called a junction field-effect transistor (JFET). The other common type of field-effect transistor is the metal-oxide semiconductor FET, or MOSFET.

Some field-effect transistors operate by enhancement, rather than depletion, of the channel. These are called enhancement-mode devices; the example shown here is of a depletion-mode FET. (*See* DEPLETION MODE, ENHANCEMENT MODE.)

All field-effect transistors are characterized by high input impedance. The gate draws almost no current, and this makes the FET useful in weak-signal amplifier appli-

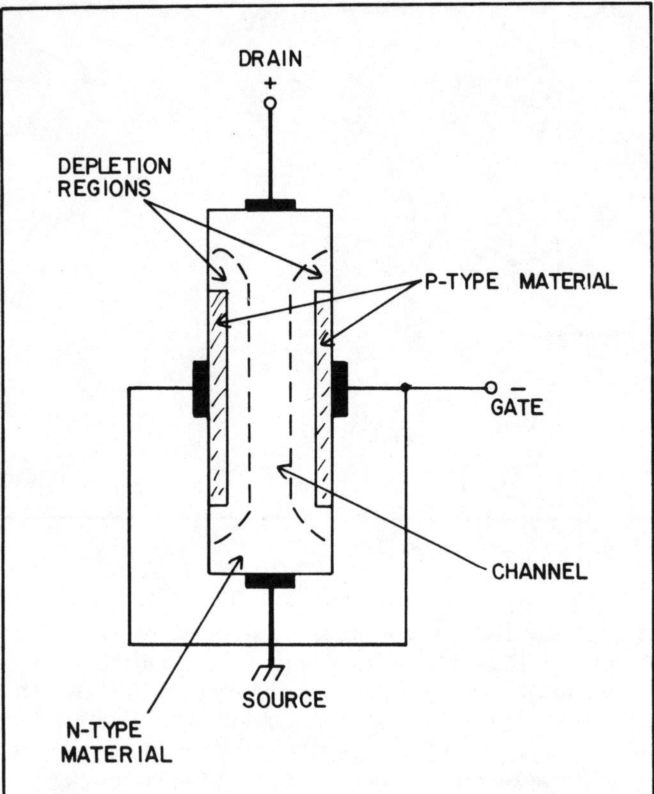

FIELD-EFFECT TRANSISTOR: Cross-sectional diagram of a junction field-effect transistor.

cations. Some field-effect transistors are useful as power amplifiers. *See also* METAL-OXIDE SEMICONDUCTOR FIELD-EFFECT TRANSISTOR, VERTICAL METAL-OXIDE SEMICONDUCTOR FIELD-EFFECT TRANSISTOR.

FIELD STRENGTH

Field strength is a measure of the intensity of an electric, electromagnetic, or magnetic field. The strength of an electric field is measured in volts per meter. The strength of a magnetic field is measured in gauss (*see* GAUSS). The intensity of an electromagnetic field is generally measured in volts per meter, as registered by a field-strength device. The intensity of an electromagnetic field may also be measured in watts per square meter.

The electromagnetic field strength from a transmitting antenna, as measured in volts, millivolts, or microvolts per meter, is proportional to the current in the antenna and to the effective length of the antenna. The field strength is inversely proportional to the wavelength and distance from the antenna. The field strength in watts, milliwatts, or microwatts per square meter varies with the power applied to the antenna and inversely with the square of the distance. The field strength always depends on the direction from the antenna as well, and is influenced by such things as phasing systems and parasitic elements. *See also* ELECTROMAGNETIC FIELD.

FIELD-STRENGTH METER

A field-strength meter is a device designed for the purpose of measuring the intensity of an electromagnetic field. Such a meter may be extremely simple and broadbanded, or highly complex, incorporating amplifiers and tuned

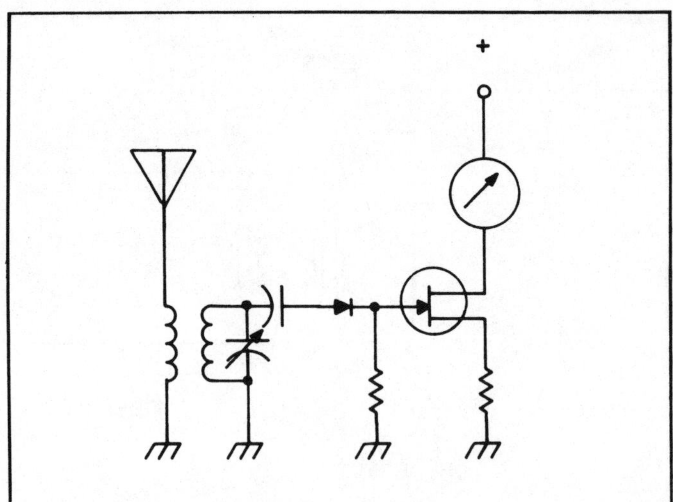

FIELD-STRENGTH METER: A simple field-strength meter incorporating an FET amplifier and a tuned circuit.

circuits. The field-strength meter usually provides an indication in volts, millivolts, or microvolts per meter.

The simplest type of field-strength meter consists of a microammeter, a semiconductor diode, and a short length of wire that serves as the pickup. Such a device is uncalibrated, but is useful for getting an idea of the level of radio-frequency energy in a particular location. This kind of field-strength meter may, for example, be used to test a transmission line in an amateur-radio station.

A more sophisticated field-strength meter may use an amplification circuit to measure weak fields. Generally, the amplifier uses a tuned circuit to avoid confusion resulting from signals on frequencies other than that desired. The drawing illustrates a simplified schematic diagram of such a field-strength meter.

The most complex field-strength meters are built into radio receivers. Accurately calibrated S meters can serve as field-strength meters in advanced receivers. Most S meters in common receivers are calibrated, but are not precise enough for actual measurements of electromagnetic field strength. The precision field-strength meter is employed in antenna testing and design, primarily for measurements of gain and efficiency. *See also* S METER.

FIFO

See FIRST-IN/FIRST-OUT.

FIGURE OF MERIT

The figure of merit is a measure of the quality of a capacitor or inductor. It is specified as the ratio of the reactance to the resistance. The greater the resistance in proportion to the reactance, the larger the loss in the device, and the lower the figure of merit. In a tuned circuit, the figure of merit affects the selectivity, and is known as the Q factor.

Generally, the larger the value of an inductor or capacitor, the poorer the figure of merit becomes. This is because large-value capacitors have more dielectric material to cause loss, and/or use lossier dielectrics. The larger the value of an inductor, in general, the more wire is needed, and/or the lossier the core material tends to be. There are some exceptions to this. *See also* Q FACTOR.

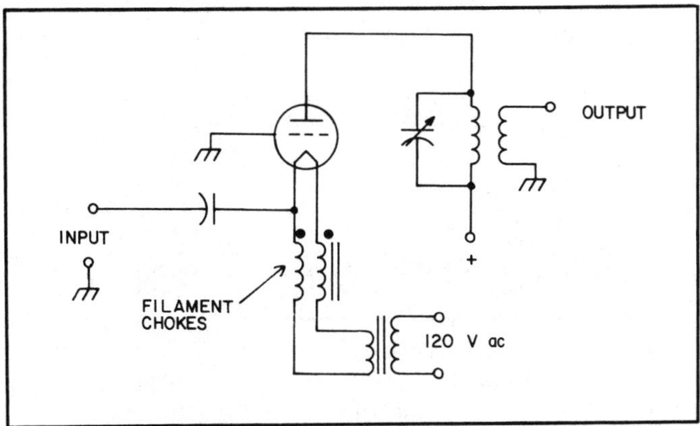

FILAMENT CHOKE: Filament chokes are used in directly heated, grounded-grid tube amplifiers. The result is a low-impedance path for the filament current, but a high impedance is presented to radio-frequency currents.

FILAMENT

A filament is a thin piece of wire that emits heat and/or light when an electric current is passed through it. In an incandescent bulb, the filament glows white hot, and is intended mainly to produce light. In a vacuum tube, the filament is intended to generate heat, which drives electrons off the cathode.

Filaments in vacuum tubes take two basic forms. The directly heated cathode consists of a filament, through which a current is passed, and this filament also serves as the cathode, connected to a negative source of voltage for tube operation. The directly heated tube has a thin, cylindrical cathode around the filament; the filament heats the cathode, but is not directly connected to it.

Tube filaments may require as little as 1 volt for their operation. Some filaments require hundreds of volts. In small tubes, the filaments can hardly be seen to glow while the device is operating. In large power-amplifier tubes, the filaments glow as brightly as small incandescent bulbs. *See also* CATHODE, TUBE.

FILAMENT CHOKE

A filament choke is a heavy-duty, radio-frequency choke, used in the filament leads of vacuum-tube power amplifiers. Such chokes are necessary in the directly heated, grounded-grid type amplifier, which receives its drive at the cathode circuit. Without the filament chokes in such a circuit, the RF driving power would be short-circuited to ground. The drawing illustrates the connection of filament chokes in a grounded-grid, zero-bias power amplifier.

Filament chokes must be capable of carrying large currents, since they are connected in series with the filaments of large tubes. A filament-choke pair may be wound on a single core, which is made of powdered iron or ferrite. The reactance of the filament choke must be large enough to prevent significant drive power from being short-circuited to ground. *See also* FILAMENT, COMMON GRID/BASE/GATE.

FILAMENT SATURATION

In a vacuum tube, given a constant level of plate voltage and a constant grid bias, the plate current increases as the filament voltage increases. This is because the higher the filament voltage gets, the more heat is developed, and the better the filament acts to drive electrons from the cathode.

There is a limit, however, to the plate current as the filament voltage is increased. Beyond a certain point, further increases in the filament voltage will not result in any increase in the plate current. When the filament is operated at this point, the condition is called filament saturation. It is pointless to have a filament voltage larger than the saturation value; this will unnecessarily heat the filament and shorten its life.

The filament-saturation voltage depends, to a certain extent, on the grid bias and the plate voltage. When a tube is cut off because of large negative grid bias, no amount of filament voltage will result in plate current. The same is true, of course, if the plate voltage is zero. If the plate voltage becomes large enough, the tube will begin to conduct even in the absence of voltage at the filament. *See also* FILAMENT, TUBE.

FILAMENT TRANSFORMER

In a vacuum-tube circuit, the filament transformer provides the filament or filaments with the needed voltage. Most tubes use either 6 or 12 volts for the operation of their filaments; this voltage need not usually be rectified or filtered. Some power-supply transformers have several secondary windings, some of which are designed for the filaments.

The secondary winding or windings of a filament transformer must be capable of handling the current drawn by all of the tubes in a circuit. The tube filaments are connected in parallel, and consequently the current requirement is proportional to the number of tubes used with a given winding. If too many tubes are connected across a given filament transformer, the transformer will overheat. This will result in reduced voltage and current for all of the filaments, and may cause damage to the transformer itself.

Filament transformers and power supplies are not needed in solid-state devices, and thus such transformers are rarely seen today, except in large power-amplifier circuits. *See also* FILAMENT, TUBE.

FILE

A file is a data store. All of the elements in a file are generally related according to format or application. Data files may be modified easily; it is a simple matter to add, delete, or change a file.

An example of a file is a section of text in a word-processing device. The file is stored on a magnetic disk. There is a limit to how large the file can be; it can hold information only up to a certain maximum. Anything consisting of the standard alphabetic, numeric, and punctuation symbols, along with spaces in any arrangement, can be placed in the file by the operator of the system. The file can be deleted, changed, or appended at any time. The file is given a name, and this name is used to retrieve the file and to distinguish it from others on the same disk or tape.

In a larger sense, a file set is a collection of data, consisting of smaller, individual files. For example, a file may be kept of tax-deductible expenditures for a given year; the combination of all the annual files, over a long period of time, is the file set. *See also* DATA PROCESSING, WORD PROCESSING.

FILTER

A filter is a passive or active circuit designed primarily for the purpose of modifying a signal or source of power. A passive filter requires no power for its operation, and always has a certain amount of loss. An active filter requires its own power supply, but may provide gain.

In a power supply, the filter is the network of capacitors, resistors, and inductors that eliminates the fluctuations in the direct current from the rectifier. A good power supply filter will produce a smooth direct-current output under variable load conditions, with good regulation.

Filters are used in communications practice to eliminate energy at some frequencies while allowing energy at other frequencies to pass with little or no attenuation. Such filters, if passive, are constructed using capacitors, inductors, and sometimes resistors. Active filters generally utilize operational amplifiers (*see* OPERATIONAL AMPLIFIER). Mechanical filters make use of the resonant properties of certain substances. Crystal and ceramic filters use piezoelectric materials to obtain the desired frequency response.

Filters can be categorized, in terms of their frequency-response characteristics, into four groups. The bandpass filter allows signals between two predetermined frequencies to pass, but attenuates all other frequencies. The band-rejection or bandstop filter eliminates all energy between two frequencies, and allows signals outside the limit frequencies to pass. The highpass filter allows only the signals above a certain frequency to pass. The lowpass filter allows only the signals below a certain frequency to pass. The actual response characteristics of all four types of filter are highly variable, and depend on many different factors. Selective filters are employed in both transmitting and receiving applications. Some are designed for use at radio frequencies, while others operate at audio frequencies. *See also* ACTIVE FILTER, BANDPASS FILTER, BAND-REJECTION FILTER, CERAMIC FILTER, CRYSTAL-LATTICE FILTER, FILTER ATTENUATION, FILTER CAPACITOR, FILTER CUTOFF, FILTER PASSBAND, FILTER STOPBAND, HIGHPASS FILTER, LOWPASS FILTER, MECHANICAL FILTER, POWER SUPPLY.

FILTER ATTENUATION

Filter attenuation is the loss caused by a selective filter at a certain frequency. If the specified frequency is within the passband of the filter, the attenuation is known as insertion loss.

Outside the passband of a selective filter, the attenuation depends on many things, including the distance of the frequency from the passband and the sharpness of the filter response. The ultimate attenuation of the filter is the greatest amount of attenuation obtained at any frequency. This normally occurs well outside the passband of the filter.

Filter attenuation is specified in decibels. If E is the root-mean-square signal voltage at a given frequency with the filter, and E_o is the root-mean-square signal voltage with the filter short-circuited, then the attenuation of the filter at that frequency is:

$$\text{Attenuation (dB)} = 20 \log_{10} (E_o/E)$$

When an active filter is used, the filter may produce gain. In that instance, the attenuation is negative. The filter gain is given by:

$$\text{Gain (dB)} = 20 \log_{10} (E/E_o)$$

See also FILTER.

FILTER CAPACITOR

A filter capacitor is used in a power supply to smooth out the ripples in the direct-current output of the rectifier circuit. Such a capacitor is usually quite large in value, ranging from a few microfarads in high-voltage, low-current power supplies to several thousand microfarads in low-voltage, high-current power supplies. Filter capacitors are often used in conjunction with other components such as inductors and resistors.

The filter capacitor operates because it holds the charge from the output-voltage peaks of the power supply. The smaller the load resistance, the greater the amount of capacitance required in order to make this happen to a sufficient extent.

Filter capacitors in high-voltage power supplies can hold their charge even after the equipment has been shut off. Resistors of a fairly large value should be placed in parallel with the filter capacitors in such a supply, so that the shock hazard is reduced. Before servicing any high-voltage equipment, the filter capacitors should be discharged with a shorting stick several times. *See also* POWER SUPPLY.

FILTER CUTOFF

In a selective filter, the cutoff frequency, or filter cutoff, is that frequency or frequencies at which the signal output voltage is 6 dB below the level in the passband. A bandpass or band-rejection filter normally has two cutoff frequencies. A high-pass or lowpass filter has just one cutoff point.

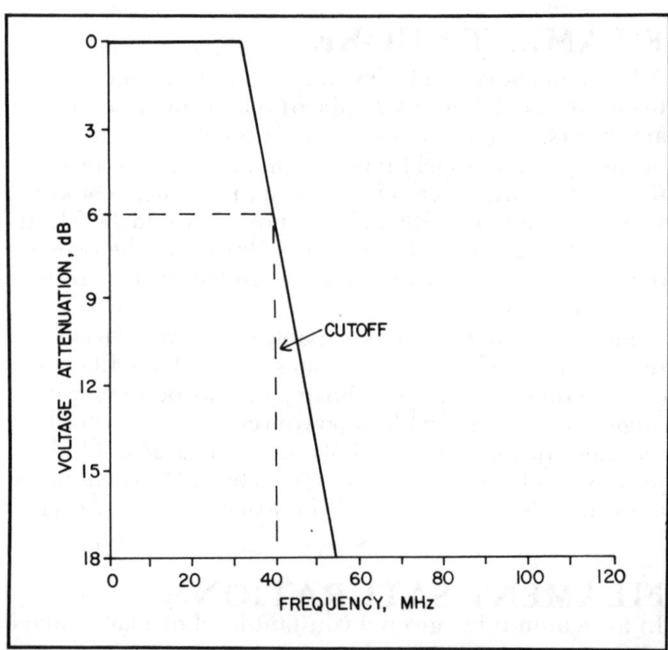

FILTER CUTOFF: The cutoff frequency of a filter is that point at which the voltage is attenuated by 6 dB relative to the attenuation in the passband.

The filter cutoff frequency is an important characteristic in choosing a filter for a particular application. But it is not the only significant factor. The sharpness of the response is also important, as is the general shape of the response within the passband.

The drawing illustrates the response of a typical lowpass filter, such as might be used in a high-frequency radio transmitter to reduce emissions at very-high and ultra-high frequencies. The cutoff frequency is clearly shown, marking the point at which the voltage drops by 6 dB and the power drops by 3 dB. The cutoff frequency depends on the impedance of the transmission line and antenna system. A filter should slways be terminated properly both at its input and output. *See also* BANDPASS FILTER, BAND-REJECTION FILTER, FILTER, HIGHPASS FILTER, LOWPASS FILTER.

FILTER PASSBAND

The passband of a selective filter is the range of frequencies over which the attenuation is less than a certain value. Usually, this value is specified as 6 dB for voltage or current, and 3 dB for power.

The passband of a filter depends on the kind of filter. If L represents the lower filter cutoff (*see* FILTER CUTOFF) and U represents the upper cutoff, then the passband of a bandpass filter is given according to

$$L < x < U,$$

where x represents frequencies in the passband. For a band-rejection filter:

$$x < L \text{ or } x > U$$

That is, the passband consists of all frequencies outside the limit frequencies.

For a highpass filter, if L is the cutoff, then:

$$x > L$$

and for a lowpass filter, if U represents the cutoff, then

$$x < U$$

This means that, for the highpass filter, the passband consists of all frequencies above the cutoff; for a lowpass filter, the passband consists of all frequencies below the cutoff. *See also* BANDPASS FILTER, BAND-REJECTION FILTER, FILTER, HIGHPASS FILTER, LOWPASS FILTER.

FILTER STOPBAND

The stopband of a selective filter consists of those frequencies not inside the filter passband. Generally, this means those frequencies for which the filter causes a voltage or current attenuation of 6 dB or more, or a power attenuation of 3 dB or more (*see* FILTER PASSBAND).

For a bandpass filter, the stopband consists of two groups of frequencies, one below or equal to the lower cutoff L, and the other above or equal to the upper cutoff U. For a band-rejection filter, the stopband consists of all frequencies between and including L and U. For a highpass filter, the stopband is that range of frequencies less than or equal to the cutoff. For a lowpass filter, the stopband is the range of frequencies higher than or equal to the cutoff. *See also* BANDPASS FILTER, BAND-REJECTION FILTER, FILTER, HIGHPASS FILTER, LOWPASS FILTER.

FILTER TUBE

A filter tube is a vacuum tube that was once employed as a filtering device in a direct-current power supply. The tube was used in place of a filter choke, and was extremely effective in eliminating residual ripple in the output of the supply. A transistor is now used for this purpose, assuming it can handle the voltage.

The illustration shows a power-supply output circuit in which a filter tube is used. The unfiltered direct current, which contains some ripple, is applied to the control grid of the tube. The tube amplifies this ripple signal, and simultaneously inverts its phase, since any common-cathode tube amplifier produces a 180-degree phase shift. Provided the gain of the amplifier is just right, the resulting ripple signal at the plate will cancel the ripple from the supply output when the two signals are combined. Filter tubes are almost never used today; they have been replaced by semiconductor devices in virtually all power supplies. *See also* FILTER, POWER SUPPLY.

FINAL AMPLIFIER

A final amplifier is the last stage of amplification in any system containing several amplifying circuits. The final amplifier drives the load. Generally, the term final amplifier is used in reference to radio-frequency transmitting equipment.

In a transmitter, the final amplifier is often tuned in the output circuit, to allow precise adjustment of the resonant frequency and impedance-matching characteristics. However, in recent years, untuned or broadbanded final amplifiers have become increasingly common. These amplifiers are simple to use, since they require little or no adjustment for proper operation; however, they cannot compensate for wide fluctuations in the load impedance.

An amplifier attached to the output of a transmitter for

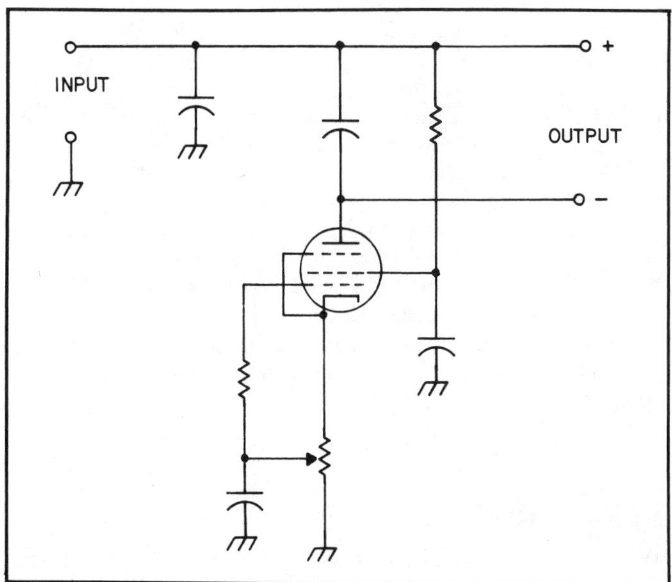

FILTER TUBE: A filter tube in the output of a rectifier circuit.

the purpose of increasing the power level is sometimes called a final amplifier. Such an amplifier usually has its own tuned circuits, and if the transmitter has a tuned output, this means that two sets of adjustments must be made.

The final-amplifier tube or transistor is itself often called the final. It is important that this device be operated according to its specifications. Otherwise, excessive harmonic output can occur, and distortion may be introduced into the modulation envelope. Finals are often operated quite close to their maximum power-dissipation limits, and in some cases it is necessary to observe a duty cycle of less than 100 percent. *See also* TRANSMITTER.

FINE TUNING

Fine tuning is the precise adjustment of the frequency of a radio transmitter or receiver. Fine tuning can be accomplished in many different ways, both electrical and mechanical. In a general-coverage communications receiver, the bandspread control is used for fine tuning. In some communications transceivers, a "clarifier" control accomplishes fine tuning.

Mechanical fine-tuning controls are not often seen in modern equipment, but may occasionally be encountered in older receivers. Such a fine-tuning control uses knobs, connected to the main-tuning shaft through gears, to obtain a spread-out control.

Electrical fine tuning can be obtained using small variable capacitors, inductors, or potentiometers. The fine-tuning control is normally connected in parallel with the main-tuning control, and has a much smaller minimum-to-maximum range.

In modern digital transmitters and receivers, fine-tuning controls are essentially obsolete. Each individual digit can be selected independently, and this obviates the need for a fine-tuning control. For example, to obtain a frequency of 14.2311 MHz each of the six digits may be selected independently in some digital communications equipment. *See also* ELECTRICAL BANDSPREAD, TUNING, VERNIER.

FINNED SURFACE

In the design of heatsinks (*see* HEATSINK), it is desirable to maximize the surface area from which heat can be carried away by convection and radiation. This has resulted in the use of finned surfaces. The photograph shows a typical heatsink, with the characteristic finned surface.

The finned surface of a well-designed heatsink allows vertical air flow to enhance convection. Air warmed by the heatsink can rise upward through the spaces between the fins. Cool air can flow inward from the bottom.

The finned surface has a much larger surface area than a flat surface, and this increases the radiation of heat. Typical finned surfaces may have 10 times the surface area of a flat surface occupying roughly the same amount of space.

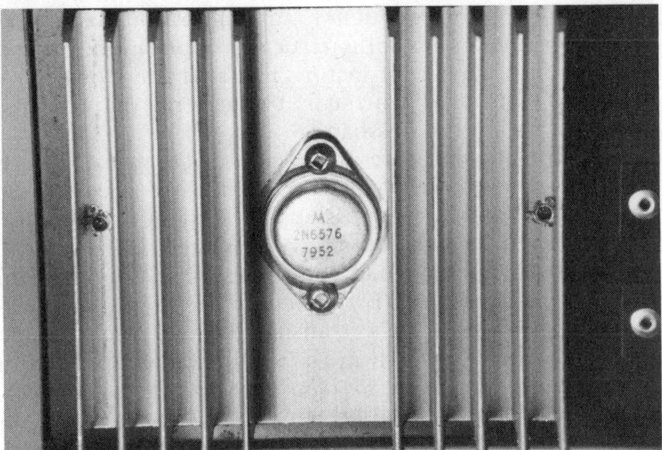

FINNED SURFACE: Finned surfaces allow efficient radiation of heat, and are therefore used in the design of heatsinks. This heatsink is used with a high-current power supply.

FIRE PROTECTION

Wherever electrical or electronic equipment is used, the danger of fire exists. Electrical fires may start because of overheating of wires or components. Fires can also begin as the result of sparks or electrical arcing.

Prevention of fire is far preferable to dealing with a fire already started. Care should be exercised in the installation of electrical wiring, to be certain that the conductors can safely handle the current they will be required to carry (*see* CARRYING CAPACITY). Connections should be made so that the chances of sparking or short circuits are minimized. Wall outlets should not be overloaded; electrical circuits should be protected with conservatively rated fuses or breakers.

In the construction of electronic equipment, again, the wiring should be capable of handling the current. Components should be rated such that they will not overheat and catch fire; this is especially true of resistors. Electronic circuits should always be fused to protect them from damage in the event of a short circuit.

No matter what precautions are taken against fire, it is always possible that a fire will start. For electrical fires, special extinguishers are necessary. Baking-soda or carbon-dioxide types are best. Water extinguishers are not sufficient because the water can result in further short circuiting and can increase the chances of electric shock to personnel.

The proper fire-protection measures can save thousands or millions of dollars, and make electrical and electronic equipment safe to operate. Local fire departments are excellent sources of further information regarding fire protection. Standards for fire protection are established by the American National Standards Institute, Inc.

FIRING ANGLE

The firing angle is an expression of the phase angle at which a thyratron or silicon-controlled rectifier fires (*see* SILICON-CONTROLLED RECTIFIER, THYRATRON). The firing angle is denoted by the lower-case Greek letter alpha (α). The firing angle is measured in degrees or radians; it represents the point on the control-voltage cycle at which the device is activated.

In a magnetic amplifier, the firing angle is denoted by the lower-case Greek letter phi (ϕ). As the input-voltage vector rotates, the core of the magnetic amplifier is driven into saturation at a certain point (*see* MAGNETIC AMPLIFIER). This point, measured as a phase angle in degrees or radians, is called the firing angle. *See also* PHASE ANGLE.

FIRMWARE

Firmware is a form of microcomputer or computer programming, or software (*see* SOFTWARE). Firmware is programmed into a circuit permanently, however; to change the programming, it is necessary to replace one or more components in the system. The read-only memory, or ROM, is an example of firmware.

In many different kinds of consumer and industrial devices, microcomputers are found in which firmware programming is used. Simple control systems are especially well suited to firmware programming. For example, a radio transceiver may have certain control functions performed by a microcomputer which has been programmed in a certain way. It is impossible to change the programming of the device because it uses firmware. *See also* COMPUTER, MICROCOMPUTER.

FIRST AID

When a person has been injured or becomes suddenly ill, the immediate attention he or she needs is called first aid. Medical assistance is not always available soon enough. Proper administration of first aid can make the difference, in some situations, between life and death, or between short-term disability and long-term disability. Proper first aid can also reduce the duration of a hospital stay.

Personnel who work with electronic equipment run the risk of electric shock. Even comparatively low voltages can sometimes cause a serious shock; many low-voltage power supplies are operated from 120-volt alternating current sources. First aid for electric-shock victims must be directed at the possible stoppage of breathing and/or failure of the heart.

In the event of injury requiring first aid, proper training is invaluable. The American National Red Cross publishes a booklet about first aid; courses are widely available for training purposes. *See also* CARDIOPULMONARY RESUSCITATION, ELECTRIC SHOCK, MOUTH-TO-MOUTH RESUSCITATION.

FIRST HARMONIC

Whenever frequency multipliers are used in a circuit, harmonics are produced. The input signal to a frequency multiplier is called the first harmonic. Higher-order harmonics appear at integral multiples of the first harmonic.

A crystal calibrator, often used in communications receivers, provides a good example of deliberate harmonic generation. A 1-MHz crystal may be used in the oscillator circuit. A nonlinear device follows, resulting in harmonic output; the second harmonic is 2 MHz, the third harmonic is 3 MHz, and so on. The 1-MHz output signal is the first harmonic.

Some crystal calibrators use frequency dividers. For example, a divide-by-10 circuit may be switched in between the oscillator and the nonlinear device in the above example, producing markers at multiples of 100 kHz. The first harmonic of the divider then becomes 100 kHz, the second harmonic becomes 200 kHz, and so on. The first harmonic can be identified because all of the other signals are integral frequency multiples of it. This is true only of the first harmonic.

The first harmonic is sometimes called the fundamental frequency. In circuits where harmonic output is not desired, the term first harmonic is generally not used. *See also* FREQUENCY MULTIPLIER, FUNDAMENTAL FREQUENCY, HARMONIC.

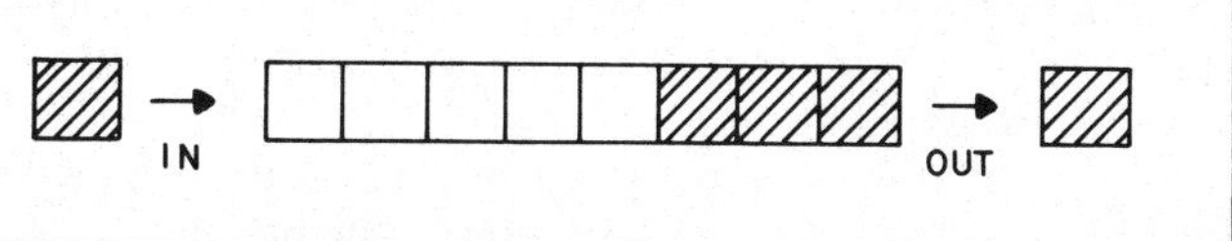

FIRST-IN/FIRST-OUT: Pictorial rendition of the operation of a first-in/first-out buffer. Dark boxes represent characters; this buffer has an 8-character capacity.

FIRST-IN/FIRST-OUT

A first-in/first-out circuit, abbreviated FIFO, is a form of read-write memory. The buffer circuits used in such devices as electronic typewriters, word processors, and computer terminals are examples of first-in/first-out memory stores.

The operation of a first-in/first-out buffer is evident from its name. The illustration shows an FIFO circuit with eight characters of storage. If certain characters are fed into the input at an irregular rate of speed, the buffer eliminates some of the irregularity without changing the order in which the characters are transmitted. Suppose, for example, that the operator of a teletype machine can type 60 words per minute, but not at a smooth, constant rate of speed. If the system speed is 60 words per minute, a FIFO buffer of sufficient capacity can be used to produce continuous, regular output characters. The FIFO circuit transmits the characters in exactly the same order in which they are received.

Not all buffers operate on the FIFO principle. Sometimes it is desirable to have a first-in/last-out buffer; as its name implies, this type of buffer inverts the order of the characters it receives. *See also* BUFFER, MEMORY.

FIRST IONIZATION POTENTIAL

Atoms are normally neutral, having the same number of orbiting electrons as nuclear protons. However, under certain conditions, electrons may be removed or added to atoms. This causes the atoms to acquire electric charge, positive in the event electrons are removed, and negative if they are added. A charged atom is called an ion. The process in which an atom acquires a charge is called ionization.

Ionization can be produced by extremely high temperatures or by the presence of a strong electric field. It takes energy to strip electrons from neutral atoms. The amount of energy, or work, required to strip a single electron from an isolated atom is called the first ionization potential. It is normally specified in electron volts.

Some substances are more easily ionized than others, and thus the first ionization potential varies with different elements. *See also* ELECTRON, ION.

FIRST LAW OF THERMODYNAMICS

The First Law of Thermodynamics is an expression of the equivalence between different forms of energy. Specifically, heat can be converted into mechanical work, and vice versa. For example, mechanical work of 4.183 joules results in the generation of 1 calorie of heat energy.

The conversion of heat into mechanical energy, and thence into electrical energy, is one of the problems facing our population today. Geothermal heat, originating at the center of the earth, may someday be used to provide some of the energy needs in a world with increasing energy problems. *See also* ENERGY, GEOTHERMAL ENERGY, SECOND LAW OF THERMODYNAMICS, THERMODYNAMICS, THIRD LAW OF THERMODYNAMICS.

FISHBONE ANTENNA

A fishbone antenna is a form of wideband antenna having a unidirectional end-fire response. The antenna gets its name from its physical appearance (see illustration). The fishbone antenna is used primarily for receiving applications at low, medium, and high frequencies.

Several collector antennas are coupled by means of capacitors to a transmission line. The spacing of the antennas is not critical, although they are normally separated by about 0.1 to 1 wavelength. The terminating resistor provides a unidirectional response; if the resistor is removed, the response becomes bidirectional, along the direction in which the feed line runs.

The fishbone antenna is not rotatable, but it can be installed fairly close to the ground. The antenna is also sometimes called a herringbone antenna. Such an antenna can provide greatly improved reception, when compared with a dipole or ground-plane antenna, under conditions of severe interference.

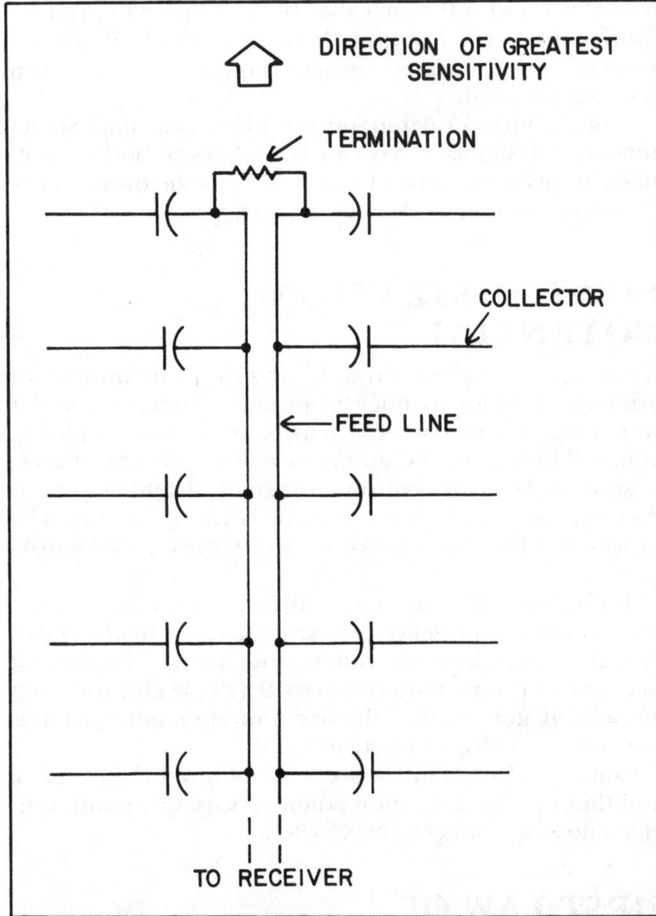

FISHBONE ANTENNA: The fishbone antenna is a directional receiving antenna. If the terminating resistor is removed, the pattern becomes bidirectional; the antenna will then respond to signals not only in the indicated direction, but also in the opposite direction.

FISSION

See NUCLEAR FISSION.

FIXED BIAS

When the bias at the base, gate, or grid of an amplifying or oscillating transistor, field-effect transistor, or tube is unchanging with variations in the input signal, the bias is called fixed bias. Fixed bias may be supplied by resistive voltage-divider networks, or by an independent power supply. Fixed bias is often used in amplifying circuits for both audio and radio frequencies.

The schematics illustrate two ways of getting fixed bias with a bipolar-transistor amplifier. At A, a voltage divider is shown. The values of the resistors determine the voltage at the base of the transistor. At B, an independent power supply is employed. The circuit at A is a Class-A amplifier, since the base is biased to draw current even in the absence of signal; the resistors should be chosen for operation in the middle of the linear part of the collector-current curve. At B, a Class-C amplifier is shown; the independent power supply is used to bias the transistor beyond the cutoff point.

When the bias is not fixed, the characteristics of an amplifier will change, along with whatever parameter is responsible for the changes in bias. An automatic-level-control circuit, for example, employs variable bias to change the gain of an amplifier. *See also* AUTOMATIC BIAS, AUTOMATIC LEVEL CONTROL.

FIXED DECIMAL

When the decimal point in a numeric display does not

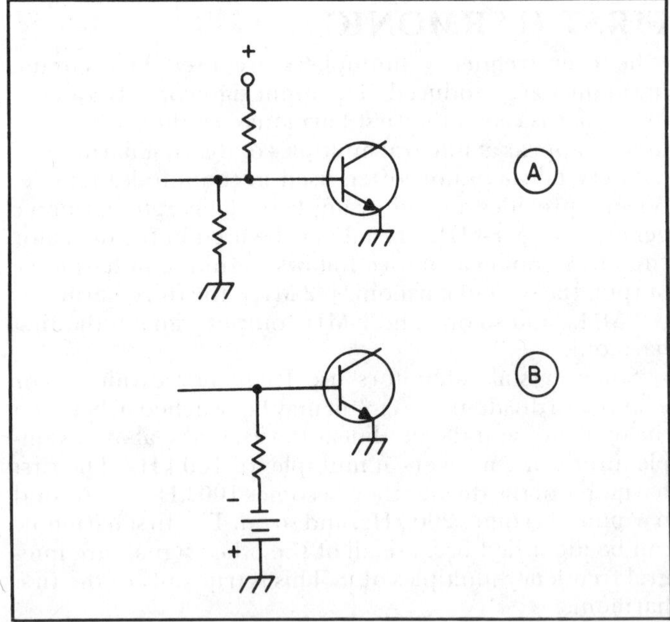

FIXED BIAS: Two methods of obtaining fixed bias. At A, a voltage divider; at B, a separate power supply.

move regardless of the operations performed, the display is said to be in the fixed-decimal mode. Fixed-decimal systems are commonly found in frequency-measuring devices and certain electronic calculators. The fixed-decimal display is always used in bookkeeping, for example; the decimal point is two places from the right in such applications.

While the fixed-decimal mode is useful in some situations, it is undesirable in others. For scientific calculations, a floating-decimal display is preferred. This allows the display of extremely small or large numbers without sacrificing accuracy. If a bookkeeping type calculator were used for scientific operations, small numbers would be displayed with reduced accuracy. The value 0.015 would be rounded to 0.02; any value smaller than 0.005 would be displayed as 0. *See also* FLOATING DECIMAL.

FIXED FREQUENCY

A fixed-frequency device is an oscillator, receiver, or transmitter designed to operate on only one frequency. Such devices are usually crystal-controlled (*see* CRYSTAL CONTROL).

Fixed-frequency communication offers the advantage of instant contact; no search is necessary at the receiving station in order to locate the frequency of the transmitter. However, if this advantage is to be realized, the frequencies must be accurately matched. The slightest error can cause reduction in communications efficiency, or even total system failure. To maintain the operating frequency, a standard source must be used, such as WWV or WWVH. Phase-locked-loop circuits can, in conjunction with such frequency standards, keep the transmitter and receiver frequencies matched to a high degree of precision. *See also* PHASE-LOCKED LOOP, VARIABLE FREQUENCY, WWV/WWVH.

FIXED RADIO SERVICE

The Fixed Radio Service is a communications network operated by the U.S. Government. Similar services exist in many other countries. The Fixed Service is intended, as its name implies, for communication between fixed locations. Such links include circuits at all frequencies, from the very-low to the ultra-high and microwave ranges.

The Fixed Service is assigned many different frequency bands on a world-wide basis. Some bands are assigned differently in various countries. Power limitations vary depending on the frequency band and the country. In the United States, the frequency allocations are administered by the Federal Communications Commission. *See also* FEDERAL COMMUNICATIONS COMMISSION.

FLAGPOLE ANTENNA

A flagpole antenna is a vertical antenna in which the ground system is either underground or oriented directly downward. Such an antenna gets its name because it has the appearance of a flagpole; it is simply a vertical rod without appendages. A coaxial antenna is an example of a flagpole antenna (*see* COAXIAL ANTENNA). Flagpole antennas are characterized by their unobtrusiveness. The polarization is vertical, and the radiation pattern is omnidirectional in the horizontal plane. Flagpole antennas can be inductively loaded for use at medium and high frequencies, while maintaining a height of only a few feet.

Some inventive radio amateurs and CB operators have used actual flagpoles as vertical antennas. There are many different ways of feeding a flagpole to make it work as a radiating element in the medium-frequency and high-frequency ranges. Probably the most common feed method is shunt feed. The feed point is connected part of the way from the base of the vertical mast to the top. Another method is to insulate the base of the flagpole from the ground, and feed it at that point. *See also* BASE LOADING, SHUNT FEED.

FLAME MICROPHONE

A flame microphone is a device that converts sound waves into electrical energy by means of the action of a flame. Two electrodes are placed in the flame, and a finite electrical resistance develops between them. A voltage is applied between the electrodes, and a transformer is coupled in series with the electrode leads, as shown in the illustration.

When sound waves pass through the flame, its electrical resistance fluctuates along with the acoustic vibrations. This modulates the direct current passing through the transformer. The audio-frequency currents are taken from the transformer secondary, where they can then be amplified.

Flame microphones are not commonly used in communications practice, for understandable reasons. *See also* MICROPHONE.

FLASH ARC

When a vacuum tube is subjected to excessive plate voltage, arcing may occur between the cathode and the plate. The result is a near short circuit in the plate power supply, and a bright flash from the tube. This is called a flash arc; it results in improper operation of the tube, and may cause permanent damage.

Vacuum tubes should be operated at a level of plate voltage considerably below the point at which flash arcing occurs. Proper operating voltages are given in vacuum-tube data tables. The flash-arc voltage of a tube depends on the physical size and separation of the cathode and the plate. It also depends, to a certain extent, on the grid bias. *See also* ARC, TUBE.

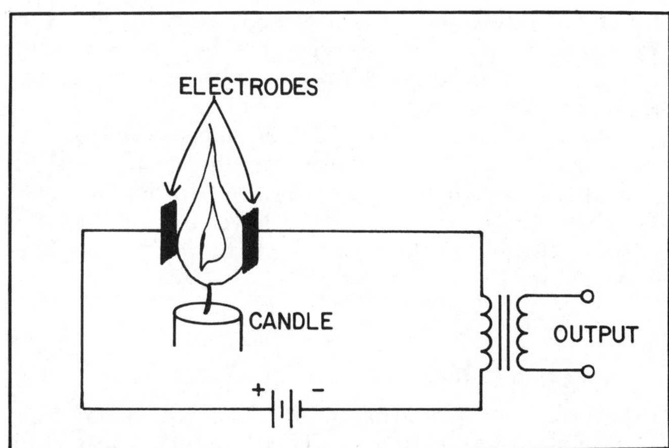

FLAME MICROPHONE: The flame resistance varies, producing audio output.

FLASHBACK

In a gas type rectifier tube, conduction normally takes place in only the forward direction. Electrons normally flow from the cathode to the plate, but not in the opposite direction. When the voltage across a rectifier tube becomes high enough in the reverse direction (cathode positive), however, the gas in the tube ionizes. This results in some reverse current. This effect is called flashback. The minimum peak-inverse voltage at which flashback occurs is called the flashback voltage. Rectifier tubes become less efficient when the flashback voltage is exceeded, because of reverse conduction.

Flashback occurs only during that part of the alternating-current cycle at which the reverse voltage is in excess of the ionization voltage in the tube. As the voltage becomes greater and greater, reverse conduction occurs over a larger portion of the alternating-current cycle. Rectifier tubes are not commonly used in modern electronic equipment, but semiconductor diodes have the same sort of reverse-voltage limitation. *See also* AVALANCHE BREAKDOWN, IONIZATION VOLTAGE, PEAK INVERSE VOLTAGE, TUBE.

FLASHOVER

Flashover is a term used to describe the arcing that takes place when the voltage across an insulator becomes excessive. Flashover is undesirable because it results in conduction over the ionized path of the arc, where conduction is not wanted. Flashover can cause fire. It can also cause permanent damage to the insulating material.

Flashover can take place in radio-frequency circuits when a severe impedance mismatch exists. If the mismatch is of enough magnitude, extreme voltages can appear at various points along a transmission line, or across feed insulators. High voltages may also appear across switch contacts. The resulting flashover causes a change in the load impedance as seen by the transmitter; this can cause damage to the final amplifier of the transmitter. *See also* ARC, FLASHOVER VOLTAGE.

FLASHOVER VOLTAGE

The flashover voltage is the minimum voltage, under certain conditions, at which flashover or arcing occurs. The flashover voltage in a given situation depends on several things, including the space between the electrodes that carry opposite polarity, the dielectric substance separating the electrodes, the temperature and, in some instances, the humidity.

Dielectric materials should not be subjected to voltages too near the flashover voltage. Flashover can result in permanent damage to some materials. In other cases, the flashover voltage becomes much lower once an ionized path has formed in the insulating material. *See also* ARC, FLASHOVER.

FLASH TUBE

A flash tube is a device that produces a brief burst of light because of ionization of a gas. The flash tube consists of a glass tube filled with a gas at low pressure, and two electrodes (see illustration). The tube may be coiled to increase the length while minimizing the occupied volume; this provides a brighter flash.

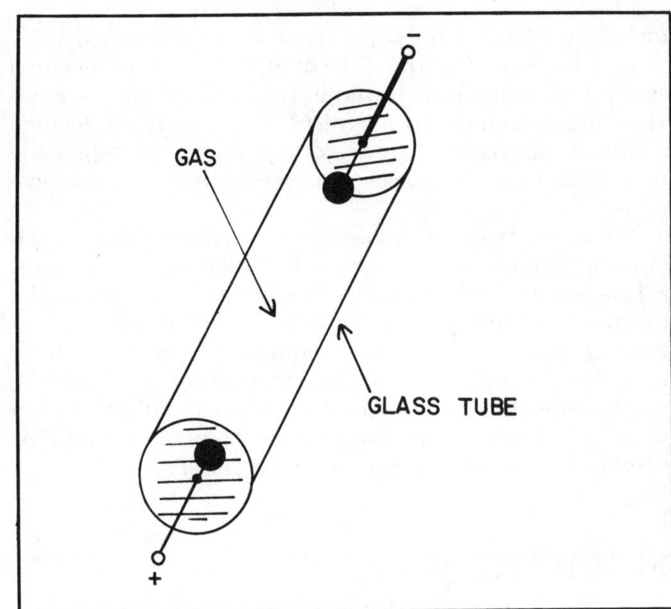

FLASH TUBE: Pictorial drawing of a flash tube. A high voltage between the electrodes causes the gas to ionize and emit visible light.

When a sufficiently high voltage is applied between the electrodes for a moment, a flash arc occurs. The flash tube can be operated many times. Strobe lights commonly employ flash tubes. Such tubes are frequently used in photography as well. *See also* FLASH ARC.

FLAT PACK

A flat pack is a form of housing for integrated circuits. The flat pack is characterized by physical thinness; the pins protude straight outward from the main package, in the plane of the housing. The photograph shows a typical example of a flat-pack integrated circuit. The entire package is smaller than a penny.

Flat-pack integrated circuits are used in situations where space is restricted. The devices are normally soldered directly to the circuit boards. Extreme care is necessary in soldering and desoldering flat-pack devices, because they are easily damaged, and solder bridges may form between adjacent pins because of the small spacing. The integrated-circuit flat pack shown contains a microcomputer. *See also* INTEGRATED CIRCUIT.

FLAT RESPONSE

When a transducer or filter displays uniform gain or attenuation over a wide range of frequencies, the device is said to have a flat response. Normally, the response is considered flat if the gain or attenuation is more or less constant throughout the operating range. At frequencies above or below the operating range, the response is not important.

A flat response is desirable for speakers, headphones, and microphones in high-fidelity equipment. Such devices should have uniform response throughout the audio-frequency spectrum, or from about 20 Hz to 20 kHz. In

FLAT PACK: A flat-pack integrated circuit is used when space is at a premium.

communications practice, the response is normally flat over a much smaller range of frequencies. The illustration shows a flat response, such as is found in high-fidelity transducers, and a narrower response, such as is typical of communications equipment.

Controls in a high-fidelity recording or reproducing system can be employed to tailor the response to the liking of a particular listener. In the recording process, the amplifier response is not always flat; the gain may be greater at some frequencies than at others. This is corrected in the playback process. *See also* FREQUENCY RESPONSE.

FLAT-TOP ANTENNA

Any antenna system in which the radiating element is horizontal can be called a flat-top antenna. Examples of flat-top antennas are the dipole antenna, the double-zepp antenna, and the extended double-zepp antenna. (*See* DIPOLE ANTENNA, DOUBLE-ZEPP ANTENNA, EXTENDED DOUBLE-ZEPP ANTENNA.)

A vertical antenna may be connected to a horizontal radiating wire, as shown in the illustration, for the purpose of lowering the resonant frequency. The resulting antenna is sometimes called a flat-top antenna. Most of the radiation in such an antenna takes place from the vertical portion. The top section provides additional capacitance, and carries a relatively high voltage. The flat-top antenna radiates a signal that has predominantly vertical polarization.

FLAT TOPPING

Flat topping is a form of distortion that sometimes occurs on an audio-frequency waveform or a modulation envelope. Flat topping takes place because of a severe nonlinearity in an amplifying circuit. Flat topping is undesirable because it results in the generation of harmonic energy, and degrades the quality of a signal.

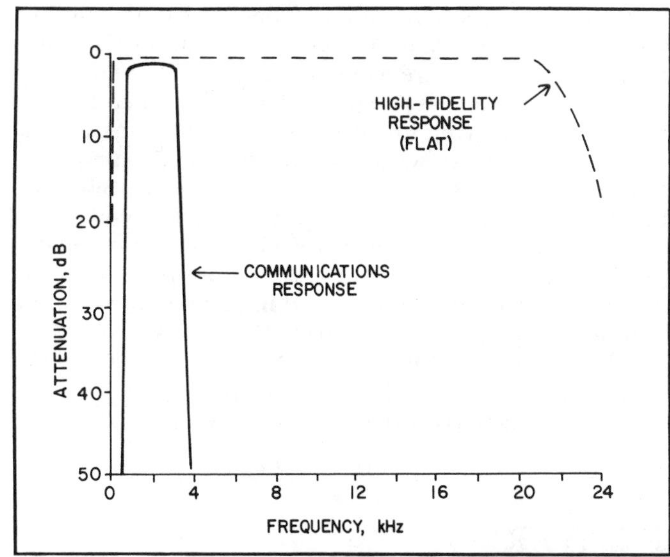

FLAT RESPONSE: The dotted line shows the flat audio response typical of high-fidelity recording/reproducing systems. The solid line shows the restricted response usually found in communications equipment.

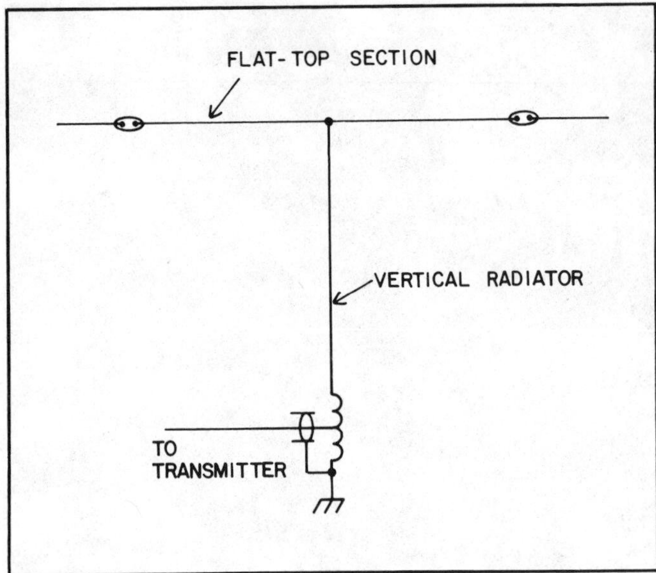

FLAT-TOP ANTENNA: An example of a flat-top antenna.

Flat topping in a modulation envelope of a single-sideband transmitter is shown in the illustration. The peaks of the signal are cut off, and appear flat. Flat topping on a single-sideband or amplitude-modulated signal results in excessive bandwidth because of the harmonic distortion. This can cause interference to stations on frequencies near the distorted signal. This type of distortion is sometimes called splatter.

Flat topping is usually caused by improper bias in an amplifying stage, or by excessive drive. If the bias is incorrect, a tube or transistor may saturate easily. Saturation can also be caused by too much driving voltage or power. *See also* DISTORTION.

FLAT TRANSMISSION LINE

When a transmission line contains no reflected electromagnetic field, but only the forward field, the line is called flat. A flat line is terminated in an impedance with a resistive component equal to the characteristic impedance of the line; there is no reactance. A flat line has no standing waves, and the current and voltage remain in the same proportion everywhere along the length of the line.

In communications practice, a flat feed line may or may not be worth striving for. A transmission line operates at maximum efficiency when it is flat; standing waves cause some additional loss. At very high frequencies and above, a flat line is desirable because the additional loss tends to be greater than at lower frequencies. Flatness is more important in coaxial lines than in two-wire transmissions lines.

When a feed line is terminated in an impedance other than its characteristic impedance, the currents and voltages become nonuniform. The current and voltage change with the location in the line. *See also* CHARACTERISTIC IMPEDANCE, STANDING WAVE, STANDING WAVE RATIO.

F LAYER

The F layer is a region of ionization in the upper atmosphere of the earth. The altitude of the F layer ranges from about 100 to 260 miles, or 160 to 420 kilometers. The F

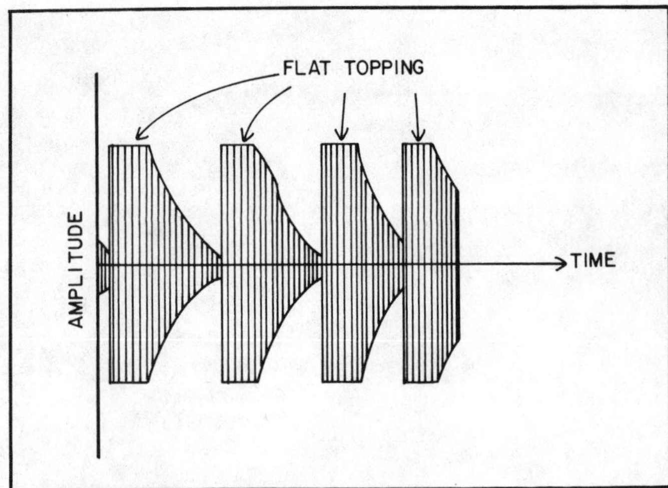

FLAT TOPPING: On an oscilloscope, flat topping of a single-sideband envelope might look like this. Note the clipped peaks.

layer of the ionosphere is responsible for most long-distance radio propagation at medium and high frequencies.

The F layer actually consists of two separate ionized regions during the daylight hours. The lower layer is called the F1 layer and the upper layer the F2 region. During the night, the F1 layer usually disappears. The F1 layer rarely affects radio signals, although under certain conditions it may intercept and return them.

The F layer attains its maximum ionization during the afternoon hours. But the effect of the daily cycle is not as pronounced in the F layer as in the D and E layers. Atoms in the F layer remain ionized for a longer time after sunset. During times of maximum sunspot activity, the F layer often remains ionized all night long.

Because the F layer is the highest of the ionized regions in the atmosphere, the propagation distance is longer via F-layer circuits than via E-layer circuits. The single-hop distance for signals returned by the F2 layer is about 500 miles for a radiation angle of 45 degrees. For a radiation angle of 10 degrees, the single-hop distance may be as great as 2,000 miles. For horizontal waves the single-hop F2-layer distance is from 2,500 to, 3,000 miles. For signals to propagate over longer distances, two or more hops are necessary.

The maximum frequency at which the F layer will return signals depends on the level of sunspot activity. During sunspot maxima, the F layer may occasionally return signals at frequencies as high as about 80 to 100 MHz. During the sunspot minimum, the maximum usable frequency can drop to less than 10 MHz. *See also* D LAYER, E LAYER, IONOSPHERE, PROPAGATION CHARACTERISTICS.

FLEMING'S RULES

Fleming's rules are simple means of remembering the relationship among electric fields, magnetic fields, voltages, currents, and forces.

If the fingers of the right hand are curled and the thumb is pointed outward as shown in A, then a current flowing in the direction of the thumb will cause a magnetic field to flow in a circle, the sense of which is indicated by the fingers. This is called the right-hand rule for the magnetic flux generated by an electric current.

If the thumb, first finger, and second finger are

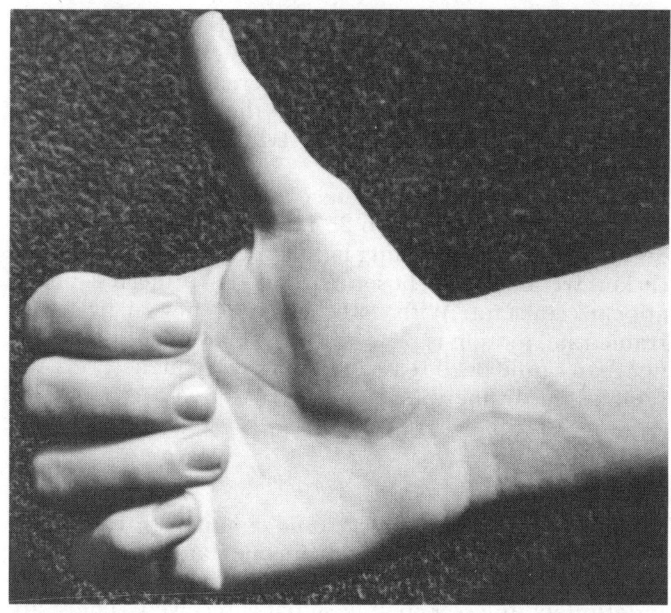

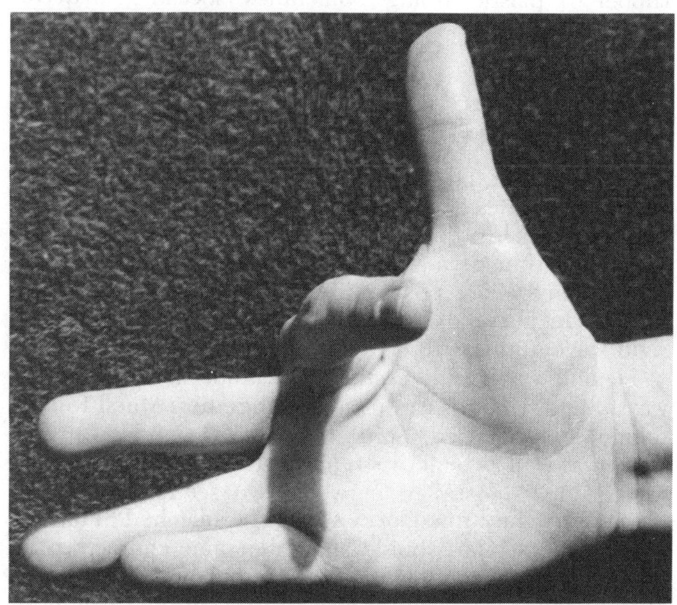

FLEMING'S RULES: At A, the right-hand rule for magnetic fields generated by electric currents. At B, the right-hand rule for a generator. At C, the left-hand rule for a motor.

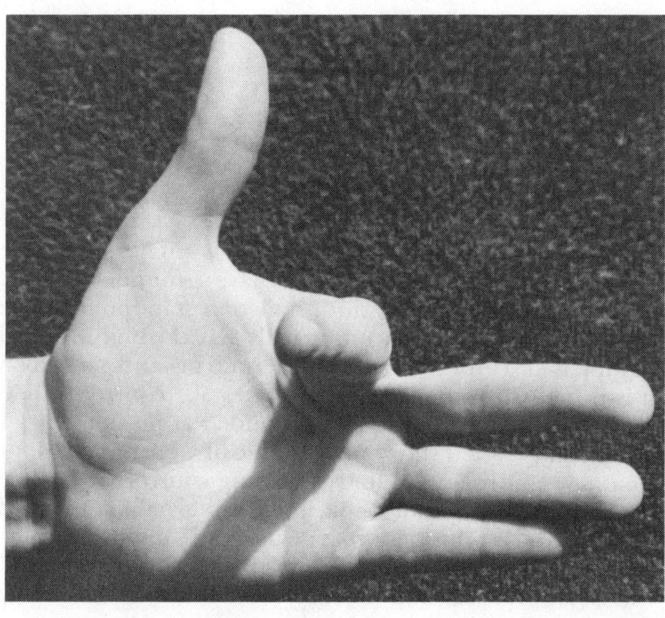

oriented at right angles to each other as illustrated in B and C, then the right hand will show the relationship among the direction of wire motion, the magnetic field, and the current in an electric generator (B), and the left hand will indicate the relationship among the direction of wire motion, the magnetic field, and the current in an electric motor (C). The thumb shows the direction of wire motion. The index finger shows the direction of the magnetic field from north to south. The middle finger shows the direction of electric current from positive to negative. *See also* GENERATOR, MOTOR.

FLEXIBLE COUPLING

A flexible coupling is a mechanical coupling between two rotary shafts. The flexibility allows the shafts to be positioned so that they are not perfectly aligned with each other, while still transferring rotation from one shaft to the other. Flexible couplings are used in a variety of different situations.

Flexible couplings may be made of metal springs or rubber or plastic tubing. Sometimes, special gear-drive systems are used. *See also* BUSHING.

FLEXIBLE WAVEGUIDE

A flexible waveguide is a section of waveguide used to join rigid waveguides that are not in precise orientation with respect to each other. Flexible-waveguide sections make the installation of a waveguide transmission line much easier, since exact positioning is not required. Flexible waveguides also allow for the expansion and contraction of rigid waveguides with changes in temperature.

Flexible-waveguide sections are made in a variety of ways. Metal ribbons can be joined together. Metal foil or tubing may be used. Some flexible waveguides can be twisted as well as bent or stretched. If a section of flexible waveguide is installed between two sections of rigid waveguide, the impedances should be matched. That is, the flexible section should show the same characteristic impedance as the rigid waveguide. The joints themselves must be made properly, as well, to prevent the formation of impedance "bumps." *See also* WAVEGUIDE.

FLICKER EFFECT

The flicker effect is a phenomenon resulting in noise in a vacuum tube. Flicker effect occurs mostly at the lower frequencies; the noise level decreases as the frequency increases.

The physics of flicker effect are not well known. The noise is thought to occur because of irregularity in the cathode surface or coating. The amount of current the tube carries, the plate and grid voltages, and the temperature may all affect the level of the noise. Flicker effect is only one source of noise in vacuum tubes; shot noise and thermal agitation are also significant. *See also* SHOT NOISE, THERMAL NOISE, TUBE.

FLICKER FREQUENCY

The flicker frequency in a motion-picture projection system is the number of times the screen is illuminated each second. Normally, the flicker frequency is twice the number of frames per second. The screen is blanked as the frame is positioned, and again while the frame is projected. The standard frame rate is 24 Hz, and the flicker frequency is 48 Hz.

The human eye can detect a light-modulation frequency of only about 15 to 25 Hz. When the frequency is greater than this, a flashing light appears continuous. The flicker frequency is chosen so that the projected image appears constant. With a flicker frequency of twice the frame rate, motion is reproduced in a more realistic manner than would be the case if the flicker frequency were the same as the frame rate.

FLIP-FLOP

A flip-flop is a simple electronic circuit with two stable states. The circuit is changed from one state to the other by a pulse or other signal. The flip-flop maintains its state indefinitely unless a change signal is received. There are several different kinds of flip-flop circuits.

The D type flip-flop operates in a delayed manner, from the pulse immediately preceding the current pulse.

The J-K flip-flop has two inputs, commonly called the J and K inputs. If the J input receives a high pulse, the output is set to the high state; if the K input receives a high pulse, the output is set low. If both inputs receive high pulses, the output changes its state either from low to high or vice versa.

The R-S flip-flop has two inputs, called the R and S inputs. A high pulse at the R input sets the output low; a high pulse on the S input sets the output high. The circuit is not affected by high pulses at both inputs.

The R-S-T flip-flop has three inputs called R, S, and T. The R-S-T flip-flop operates exactly as the R-S flip-flop works, except that a high pulse at the T input causes the circuit to change states.

A T flip-flop has only one input. Each time a high pulse appears at the T input, the output state is reversed.

Flip-flop circuits are interconnected to form the familiar logic gates, which in turn comprise all digital apparatus. *See also* LOGIC GATE.

FLOAT CHARGE

See TRICKLE CHARGE.

FLOATING CONTROL

Most control components, such as switches, variable capacitors, and potentiometers, have grounded shafts. This is convenient from an installation standpoint, since the shafts of controls are normally fed through a metal front panel. However, there are certain instances in which it is not possible to ground the shaft. The control is then said to be floating, since it is grounded at no point.

The illustration shows an example of a grounded (non-floating) control, at A, and a floating control, at B. The rotor plates of a variable capacitor are internally connected to the shaft; for this reason, it is customary to put the rotor plates at ground potential. This is the case at A. But at B, this is not practical, and the frame and shaft of the variable

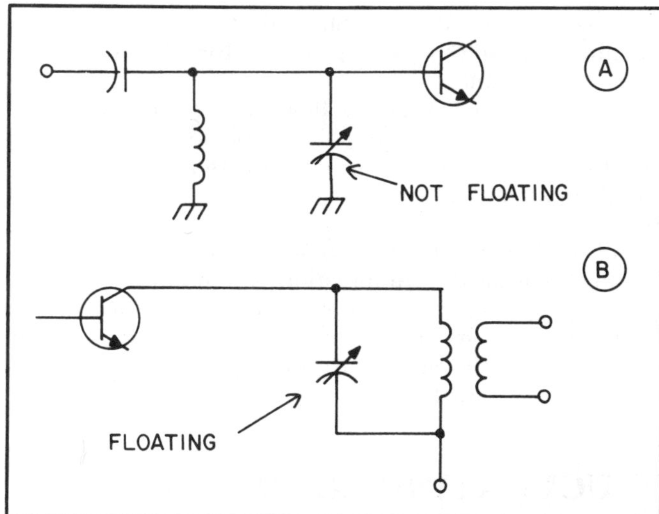

FLOATING CONTROL: At A, the rotor plates of the capacitor are grounded; at B, they are floating (not grounded).

capacitor must be floated. To accomplish this, the capacitor is mounted on insulators above the chassis, and the shaft must be insulated from the panel and knob or shaft.

Floating controls are more subject to the effects of body capacitance than grounded controls. Therefore, floating controls should be used only when there is no alternative. *See also* BODY CAPACITANCE.

FLOATING DECIMAL

A floating-decimal or floating-point display is a numeric display often found in calculators. All calculators have a fixed number of digits in the display, and accuracy is maximized by the use of a movable decimal point.

A ten-digit display with a fixed decimal point, two places from the far right, can accurately display quantities from 0.01 to 99,999,999.99. However, when the decimal point is movable, the display can accurately render quantities as small as 0.0000000001, or as large as 9,999,999,999. Because of the larger range using a floating-decimal display, this scheme is preferred in scientific work.

The range of a display can be further increased by using scientific notation. Many handheld calculators have this feature, and it is actuated whenever the display can no longer render a quantity in the simple decimal form. *See also* FIXED DECIMAL, SCIENTIFIC NOTATION.

FLOATING PARAPHASE INVERTER

A floating paraphase inverter is a circuit devised for inverting the phase of a signal. Two transistors, field-effect transistors, or vacuum tubes are used in the circuit. The schematic diagram shows a floating paraphase inverter.

A signal applied to the input is inverted in phase at output X, because of the phase-reversing effect of the bipolar transistor amplifier Q1. A portion of the signal from output X is fed to the base of transistor Q2. This signal is obtained by means of a resistive network, so that it is equal in amplitude to the input signal. Amplifier Q2 inverts the phase of this signal, delivering it to output Y. Consequently, the waves at the two outputs are out of phase, and equally elevated above the common terminal.

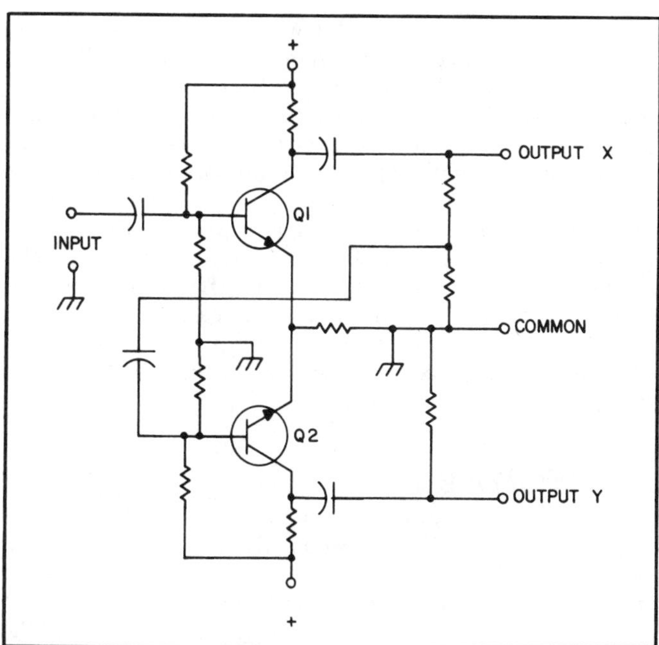

FLOATING PARAPHASE INVERTER: Outputs X and Y are in phase opposition.

Either output may be used to obtain signals in phase opposition, or they may be combined to form a floating pair of terminals. *See also* PARAPHASE INVERTER.

FLOATING POINT

In an electronic circuit, a point is called "floating" if it is ungrounded and not directly connected to the power supply terminal. In the illustration for FLOATING PARAPHASE INVERTER, for example, the output terminals X and Y are floating.

A floating decimal is sometimes called a floating point. *See also* FLOATING DECIMAL.

FLOOD GUN

A flood gun is an electron gun employed in a storage cathode-ray tube. The flood gun operates in conjunction with another electron gun, called the writing gun, to store a display for a period of time (*see* ELECTRON-BEAM GENERATOR).The writing gun emits a thin beam of electrons, which strikes the phosphor screen and forms the image. The beam is modulated by means of deflecting plates, exactly as in the conventional oscilloscope. A storage mesh holds the image. The flood gun is used to illuminate the phosphor screen to view the stored image. Low-energy electrons from the flood gun can pass through the mesh only where the image has been stored. When the image is to be erased, the flood gun is used to clear the mesh. *See also* OSCILLOSCOPE, STORAGE OSCILLOSCOPE, WRITING GUN.

FLOPPY DISK

A floppy disk is a flexible disk made from a material similar to that used for magnetic tape. The recording and storage scheme is the same as that for magnetic tape, except that the tracks are in the form of concentric rings around the disk.

The floppy disk allows much faster storage and retrieval

of information than is possible with magnetic tape. This is because the magnetic head can move back and forth across the disk surface. The head need never move a greater distance than half the outer circumference of the disk—just a few inches. In contrast, the distance between two bits of data on a magnetic tape may be hundreds or thousands of feet.

Floppy disks are used in computers and word-processing systems. A large amount of information can be stored on a single disk that is smaller than a phonograph record. The information is easily accessed, replaced, or altered. Home computers use small floppy disks known as diskettes. *See also* DISKETTE, MAGNETIC RECORDING.

FLOWCHART

A flowchart is a diagram that depicts a logical algorithm or sequence of steps. The flowchart looks very much like a block diagram. Boxes indicate conditions, and arrows show procedural steps. Flowcharts are often used to develop computer programs. Such charts can also be used to document troubleshooting processes for all kinds of electronic equipment.

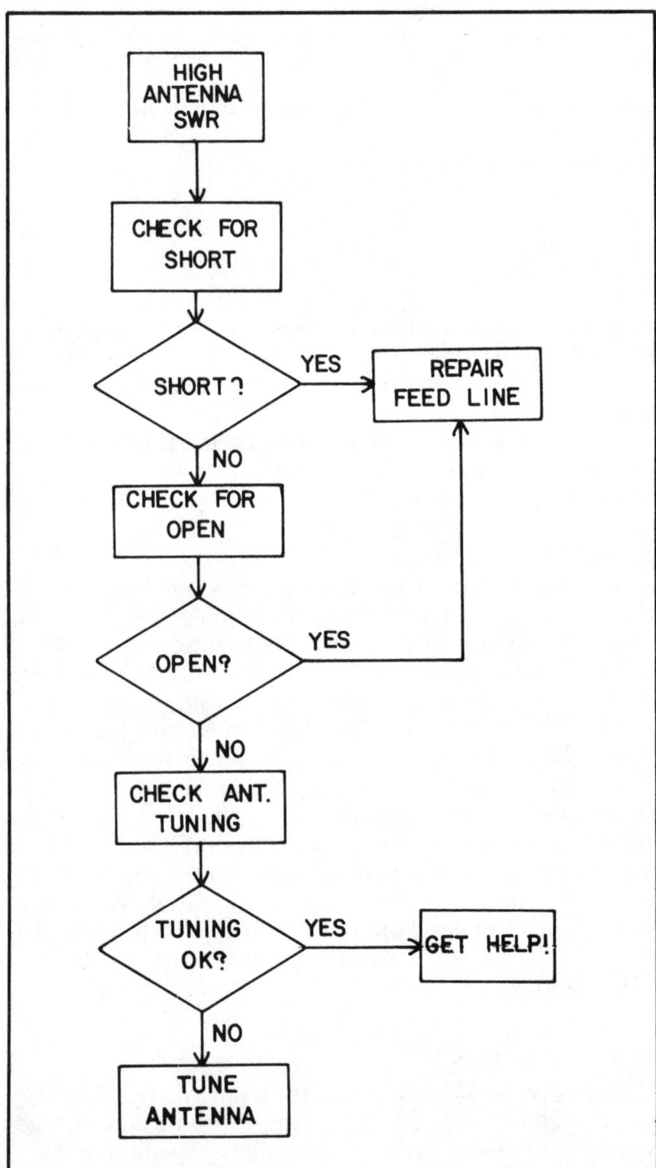

FLOWCHART: A flowchart, indicating the procedure to be followed in repaircng an antenna system with an excessive standing-wave ratio (SWR).

The symbology used in flowcharts is not well standardized. An example of a simple troubleshooting flowchart is shown in the illustration. Decision steps are indicated by diamonds. Various conditions are indicated by boxes.

A flowchart must represent a complete logical process. No matter what the combination of conditions and decisions, a conclusion must always be reached. Sometimes the conclusion consists simply of an instruction to return to an earlier stage in the algorithm, but infinite loops should not exist. *See also* ALGORITHM.

FLUCTUATING LOAD

The resistance or impedance of a circuit load may not remain constant. Some loads maintain an essentially constant impedance as a function of time; an example of such a load is the antenna system in a radio-frequency transmitting station. Some loads show variable resistance or impedance. An example of this is an ordinary household electrical system. The load impedance of such a circuit depends on the number and kinds of appliances that are used.

A generator subjected to a fluctuating load must be capable of delivering the proper voltages and currents to the load, no matter what the state of the load at a given instant. The generator must be able to respond rapidly to changes in the load impedance. Under certain extreme conditions, the generator may not be capable of dealing with a change in load resistance; this sometimes happens, for example, when all the residents of a large city operate air conditioners at the same time.

The load impedance that a transmitter oscillator circuit "sees," at the input of the following amplifier stage, often fluctuates with the keying of the oscillator. A buffer stage between the oscillator and the first amplifier is used to reduce the magnitude of the load fluctuations. Oscillators tend to change frequency when the load impedance changes; the buffer stage therefore improves the stability of the oscillator. *See also* BUFFER STAGE, IMPEDANCE.

FLUID LOGIC

Fluid logic, also sometimes called fluid-flow or fluidic logic, is a method of obtaining digital-logic circuits by means of fluid valves. There are no electronic elements or moving parts. The fluid may be either a liquid or a gas.

Fluid logic systems can operate under conditions impossible with ordinary electronic logic circuits. Fluid logic is essentially immune to electromagnetic interference. However, fluid logic operates at a rather slow speed, limiting the number of operations that can be performed per unit time. Logic gates, flip-flops, inverters, and amplifiers can all be built using fluidic methods.

FLUORESCENCE

When a substance emits visible light because it has absorbed energy, the phenomenon is called fluorescence. Fluorescence can occur in gases, liquids, or solids. Most gases will exhibit this property when bombarded with, or forced to carry, an electron beam of sufficient intensity .

Fluorescent materials are found in many different electronic devices. The cathode-ray tube, for example, has a fluorescent material, as a coating on the inside surface of the screen, to display images produced by electron beams. The phosphor coating on such screens is made up of fine crystals that emit visible light of various colors. Some examples of fluorescent substances employed in cathode-ray tubes are zinc sulfide, zinc fluoride, and zinc oxide.

Fluorescence provides an efficient means of producing light for household or business purposes. *See also* FLUORESCENT TUBE.

FLUORESCENT TUBE

The fluorescent tube, or fluorescent lamp, is an electrical device used for illumination purposes. It has a high degree of efficiency. The fluorescent tube generates more visible light, and less heat, than incandescent bulbs of the same rating.

The illustration is a simple cross-sectional diagram of a fluorescent tube. A glass tube is first evacuated, and then filled with a combination of argon gas and mercury. When an electric current flows through the gas, the mercury is vaporized. This results in ultraviolet radiation, which falls on the phosphor coating on the inside of the tube. The phosphor produces white light when excited by the radiation.

Fluorescent tubes are available in many different sizes and wattage ratings. The smallest commercially available fluorescent lamps use about 15 watts at 120 volts, and measure just a few inches in length. The largest such tubes are over 6 feet long and use about 100 watts. *See also* FLUORESCENCE.

FLUORINE

Fluorine is an element, symbolized by the capital letter F. Its atomic number is 9, and the atomic weight is 19. Fluorine is a member of the halogen family of elements (*see* HALOGEN).

Fluorine is used in some cathode-ray-tube coatings as a constituent of phosphors, which glow when bombarded by high-energy electrons, ultraviolet radiation, or X rays.

FLUOROSCOPE

A fluoroscope is a device used to obtain an internal view of the human body by means of X rays. Certain phosphors will fluoresce when bombarded with radiation; the fluoroscope uses this effect to allow an in-motion view of certain parts of the body.

A source of X rays is placed a short distance from a large screen coated with barium platinocyanide or other X-ray-sensitive phosphor. When a part of the body is placed between the X-ray source and the screen, a shadow is cast on the screen (see illustration). The soft tissues of the body are partially transparent to X rays, but to a variable extent, and medical professionals are able to recognize certain diseases or tumors by looking at the X-ray display. The image of the body on a fluoroscope screen has a bizarre appearance to the uninitiated.

The fluoroscope is being replaced in modern medicine by other devices, which subject patients to less radiation whenever possible. *See also* MEDICAL ELECTRONICS.

FLUTTER

Flutter is the term used to describe a warbling, or change in pitch, of the sound in an audio recording/reproducing system. Slow flutter is sometimes called wow, because of the way it sounds. Flutter can occur in any recording or reproducing process in which the vibrations are stored in a fixed medium.

Flutter may result from a periodic change in the speed of the recording medium, relative to the transducer. An example of this is obtained by placing a record onto a turntable off-center. (This can be easily done with standard 45-rpm records because of the large center hole.) The stylus speeds up and slows down as the disk rotates. This is often called wow.

Flutter may also occur because of nonuniform motor speed in either a disk system or a magnetic system. This

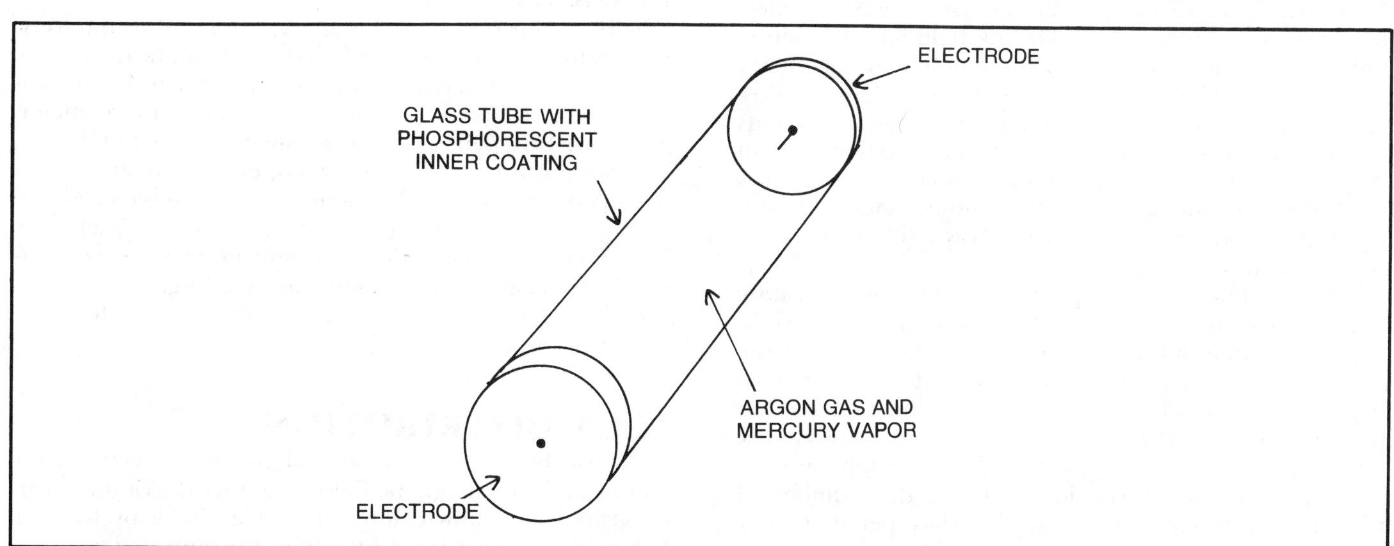

FLUORESCENT TUBE: The fluorescent tube produces ultraviolet light because of the ionization of gases. The ultraviolet light strikes the phosphor inside the glass tube, resulting in visible light.

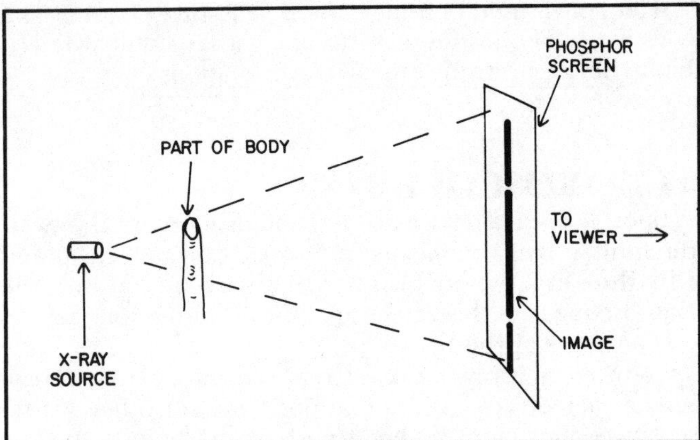

FLUOROSCOPE: A fluoroscope allows the observation of internal details of the body, while the body is in motion.

type of flutter is generally much more rapid than the periodic type. Random flutter may be so minor that it cannot be noticed on voice recordings, but it becomes evident when music is recorded and reproduced. Some tape recorders are intended for voice reproduction only, and the reasons for this are apparent when an attempt is made to reproduce music.

When the local oscillator in a superheterodyne receiver is unstable, the received signal appears to fluctuate in frequency. Sometimes the oscillator is affected by mechanical vibration, resulting in rapid and dramatic changes in frequency. The sound of the best note resulting from this instability is called flutter.

The flutter in a recording/reproducing system, as well as that in a superheterodyne receiver, may result from engineering deficiencies, or they may take place because a recording and reproducing device is used for an application other than that intended. Flutter problems can be corrected, if necessary, by replacing the faulty component or device with one having greater precision. *See also* DISK RECORDING, MAGNETIC RECORDING, SUPERHETERODYNE RECEIVER, WOW.

FLUX

Flux is a measure of the intensity of an electric, electromagnetic, or magnetic field. Any field has an orientation, or direction, which can be denoted by imaginary "lines of flux." In an electric field, the lines of flux extend from one charge center to the other in various paths through space. In a magnetic field, the lines extend between the two poles. In an electromagnetic field, the electric lines of flux are generally denoted. (*See* ELECTRIC FIELD, ELECTRIC FLUX, ELECTROMAGNETIC FIELD, MAGNETIC FIELD, MAGNETIC FLUX.)

The flux lines of any field have a certain concentration per unit of surface area through which they pass at a right angle. This field intensity varies with the strength of the charges and/or magnetic poles, and with the distance from the poles. The greatest concentration of flux is near either pole, and at points lying on a line connecting the two poles. For radiant energy such as infrared, visible light, and ultraviolet, the flux is considered to be the number of photons that strike an orthogonal surface per unit time. Alternatively, the flux may be expressed in power per unit area, such as watts per square meter. *See also* FLUX DENSITY, FLUX DISTRIBUTION.

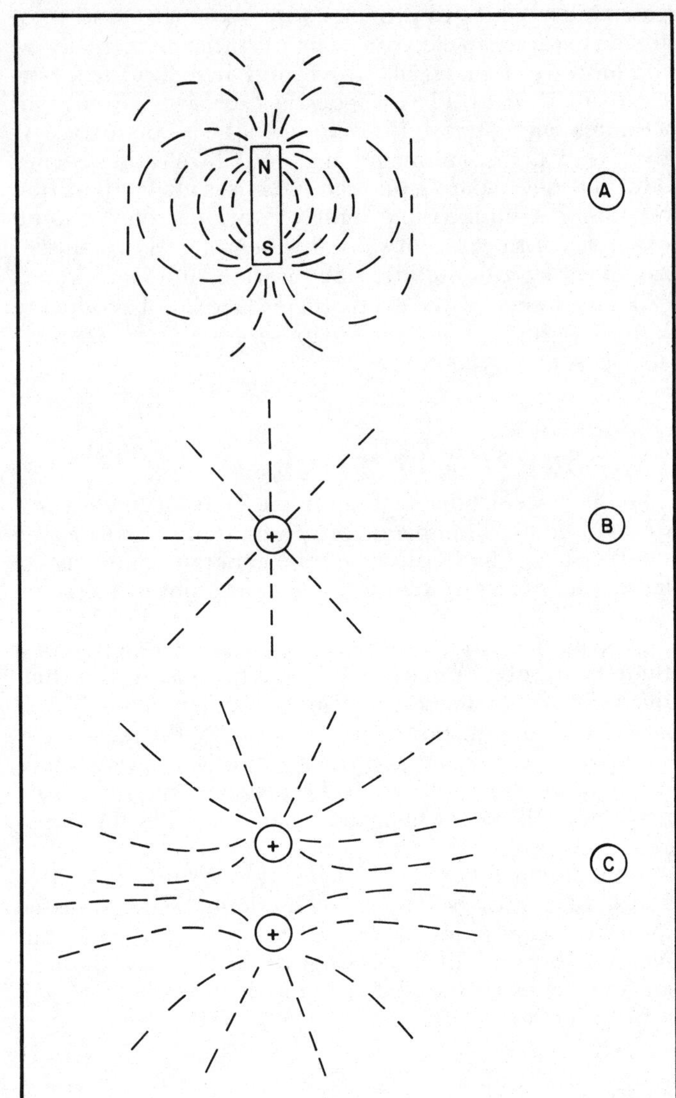

FLUX DISTRIBUTION: At A, the flux distribution around a magnetic dipole. At B, the flux distribution around an isolated electric charge. At C, the flux distribution around a pair of like charges.

FLUX DENSITY

The flux density of a field is an expression of its intensity. Generally, flux density is specified for magnetic fields. The standard unit of magnetic flux density is the tesla, which is equivalent to 1 volt second (weber) per square meter. Magnetic flux density is abbreviated by the letter B.

Another unit commonly used to express magnetic flux density is the gauss. A flux density of 1 gauss is considered to represent a line of force per square centimeter. The "lines" are imaginary, and represent an arbitrary concentration of magnetic field. There are 10,000 gauss per tesla. *See also* FLUX, GAUSS, MAGNETIC FIELD, MAGNETIC FLUX. TESLA, WEBER.

FLUX DISTRIBUTION

Flux distribution is the general shape of an electric, electromagnetic, or magnetic field. The flux distribution can be graphically shown, for a particular field, by lines of force. The flux tends to be the most concentrated near the poles of an electric or magnetic dipole.

The flux distribution around a bar magnet can be viv-

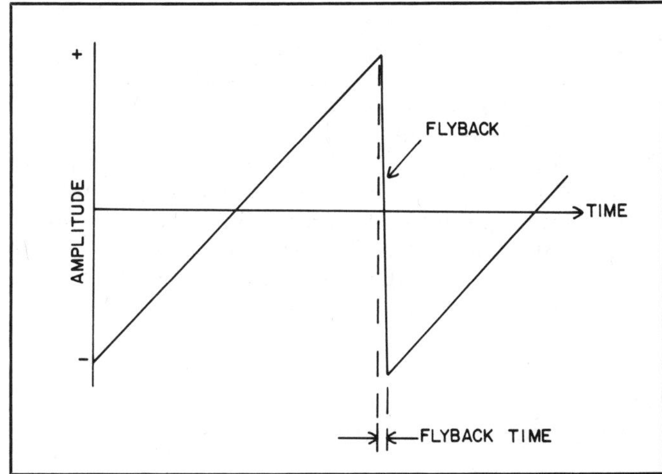

FLYBACK: Flyback is the return of a sawtooth pulse to its minimum, or starting, voltage.

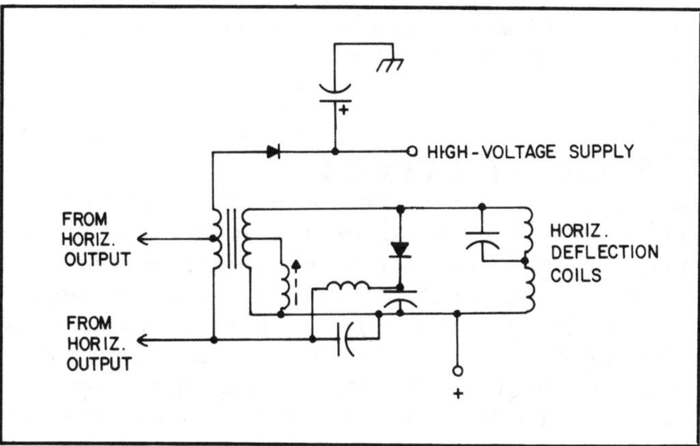

FLYBACK POWER SUPPLY: Also called a kickback supply.

idly illustrated by the use of iron filings and a sheet of paper. The paper is placed over the bar magnet, and the filings are sprinkled onto the paper. The filings align themselves in such a way as to render the field lines visible (see illustration at A). The general distribution of the magnetic field is characteristic.

The flux distribution around an isolated electric charge is shown at B. The lines of force extend radially outward, having their greatest concentration near the charge center. If two opposite electric charges are placed near each other, the flux is distributed in the same manner as the magnetic field shown at A; but if the charges are of the same polarity, the flux distribution takes on a shape similar to that shown at C.

The greatest electric or magnetic flux is always near the poles, and along a line connecting two poles. *See also* DIPOLE, ELECTRIC FLUX, FLUX, MAGNETIC FLUX.

FLUXMETER

A fluxmeter is a device for measuring the intensity of a magnetic field. Such a meter is also called a gaussmeter.

The fluxmeter generally consists of a wire coil, which can be moved back and forth across the flux lines of the magnetic field. This causes alternating currents to flow in the coil, and the resulting current is indicated by a meter. The greater the intensity of the magnetic field for a given coil speed, the greater the amplitude of the alternating currents in the coil. Knowing the rate of acceleration of the coil, and the amplitude of the currents induced in it, the flux density can be determined.

Fluxmeters are usually calibrated in gauss. A flux density of 1 gauss is the equivalent of a line of flux per square centimeter. *See also* FLUX, GAUSS, MAGNETIC FLUX.

FLYBACK

Flyback is a term that denotes the rapid fall of a current or voltage that has been increasing for a period of time. The flyback time is practically instantaneous under ideal conditions; in practice, however, it is finite. The illustration shows flyback in a sawtooth waveform.

Sawtooth waves are used in such devices as oscilloscopes and television receivers to sweep the electron beam rapidly back after a line has been scanned. The flyback sweep is much more rapid than the forward sweep; the display is blanked during the flyback to prevent any possible interference with the image created by the forward sweep. In an oscilloscope or television picture tube, the flyback is from right to left, while the visible sweep is from left to right.

When a coil or capacitor is rapidly discharged, the voltage or current can be very large for a short time. This can cause electric shock, arcing, and possible damage to components of a circuit. This kind of flyback is called kickback. The spark generator in an automobile engine, and in any other internal combustion engine (except a diesel engine), operates on this principle. *See also* DISCHARGE, FLYBACK POWER SUPPLY, PICTURE SIGNAL, PICTURE TUBE.

FLYBACK POWER SUPPLY

In a television picture receiver, the horizontal scanning is controlled by a device called a flyback power supply. It is also often called a kickback power supply. The flyback supply provides a high-voltage sawtooth wave, which is fed to the horizontal deflecting coils in the picture tube. A typical flyback power supply is shown in the illustration.

The output waveform of the flyback power supply is such that the deflection of the electron beam occurs rather slowly from left to right, but rapidly from right to left. The picture modulation is impressed on the electron beam during the left-to-right sweep. The beam is blanked out during the return sweep, to prevent interference with the picture. A similar power supply is employed in cathode-ray oscilloscope, but the left-to-right (forward) scanning speed is variable rather than constant. *See also* FLYBACK, PICTURE SIGNAL, PICTURE TUBE.

FLY'S-EYE LENS

In the manufacture of microminiature integrated circuits, photographic processes are often used. A small image of the circuit is cast onto a semiconductor wafer by means of a lens, and a photochemical reaction etches the circuit into the semiconductor material.

A fly's-eye lens allows many identical circuits to be fabricated at the same time from a single image. The lens gets its name from the fact that, in appearance, it resembles the eye of a fly! There are numerous small lenses arranged in such a way that the light from one image gets cast onto many focal points at once. All of the images are identical, and can be of any desired size, depending on the focal lengths of the separate convex lenses.

FLYWHEEL EFFECT

In any tuned circuit containing inductance and capacitance, oscillations tend to continue at the resonant frequency of the circuit, even after the energy has been removed. The higher the Q factor, or selectivity, of the circuit, the longer the decay period. The Q factor increases as the amount of resistance in the tuned circuit decreases. This phenomenon, wherein resonant circuits tend to "ring," is called flywheel effect, and it occurs because the inductor and capacitor in a tuned circuit store energy.

The flywheel effect makes it possible to operate a Class-AB, Class-B, or Class-C radio-frequency amplifier, with minimal distortion in the shape of the signal wave. The output waveforms of such amplifiers not having tuned output circuits would be nonsinusoidal. The Class-B and Class-C output waveforms would be greatly distorted, and would contain great amounts of harmonic energy. But, because of the flywheel effect, the tuned output circuit requires only brief pulses, occurring at the resonant frequency, to produce a nearly pure sine-wave output (see illustration).

The flywheel effect does not make an inherently nonlinear amplifier, such as a Class-C amplifier, operate in linear fashion. The modulation envelope of a single-sideband signal, for example, will be badly distorted if the signal is passed through a Class-C amplifier circuit. The shape of the carrier wave is, however, maintained even in a Class-C amplifier, because of the flywheel effect of the tuned output circuit. *See also* CLASS-AB AMPLIFIER, CLASS-B AMPLIFIER, CLASS-C AMPLIFIER, Q FACTOR, TUNED CIRCUIT.

FLYWHEEL TUNING

The tuning controls of some radio communications receivers use heavy weights to obtain a large amount of angular momentum. This weight may take the form of a thick metal disk attached to the shaft of the control. This is known as flywheel tuning.

Flywheel tuning offers some advantages. It creates a smooth "feel" to the tuning dial of a receiver. By spinning a dial with flywheel tuning, several revolutions of the knob can be rapidly executed; the dial just keeps on going because of the momentum. Many operators like flywheel tuning. However, in some situations, flywheel tuning is not desirable. In mobile or portable operation, the flywheel can cause misadjustment of the tuning control in the event of mechanical vibration. And, just as some operators like the "feel" of flywheel tuning, some operators dislike it. Flywheel tuning is most often found in older communications and short-wave broadcast receivers, especially those with slide-rule tuning dials. *See also* SLIDE-RULE TUNING, TUNING, TUNING DIAL.

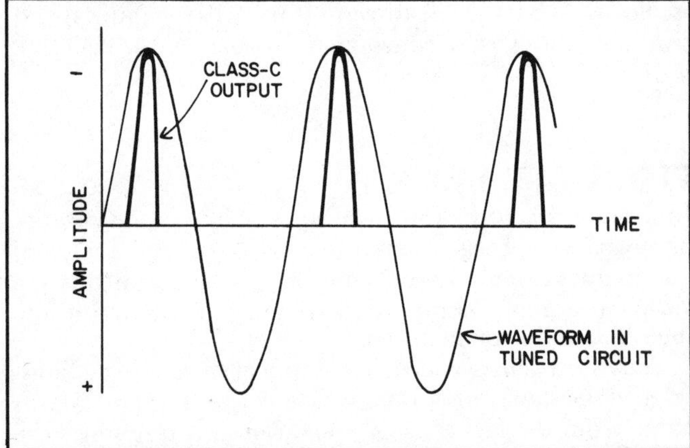

FLYWHEEL EFFECT: The flywheel effect completes the alternating-current cycle in the output of a Class-B or Class-C amplifier.

FM

See FREQUENCY MODULATION.

FM STEREO

Most frequency-modulation broadcast stations use stereo transmission. Some frequency-modulation (FM) broadcast receivers are capable of reproducing the stereo sound, and others are not. For this reason, FM stereo broadcasting must be accomplished in a special way, so it is compatible with both types of receivers. Modulated subcarriers (*see* SUBCARRIER) are used to obtain FM transmission of two independent channels.

An FM stereo signal consists of a main audio channel and two stereophonic channels. The main-channel audio consists of the combined audio signals from the left and right stereo channels; this signal frequency modulates the main carrier between 15 Hz and 15 kHz. This allows reception of the broadcast by receivers not having stereo capability. The left and right stereo channels are impressed on the signal by means of subcarriers. The pilot subcarrier frequency modulates the main carrier at 19 kHz. At twice this frequency, or 38 kHz, the stereophonic subcarrier is amplitude-modulated by the difference signal between the left and right channels. The pilot subcarrier serves to maintain the frequency of the stereophonic subcarrier in the receiver.

The channel-difference sidebands occupy a modulating frequency range from 23 to 53 kHz, centered at the frequency of the stereophonic subcarrier. A subsidiary-

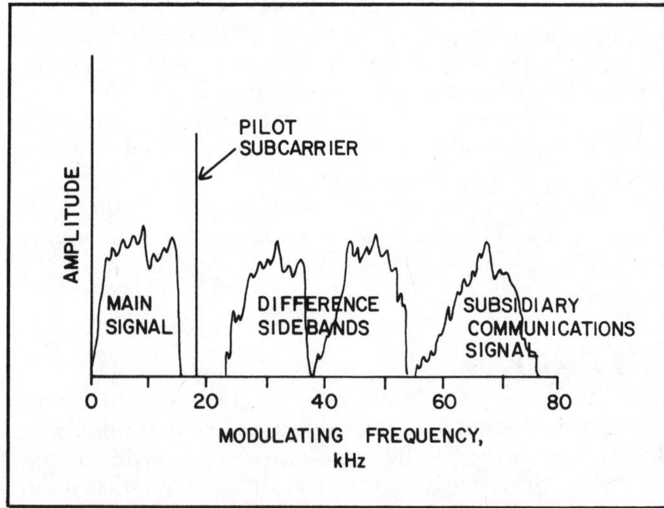

FM STEREO: Spectral illustration of the modulation band in a frequency-modulation stereo broadcast signal. The subsidiary communications signal may or may not be used.

communications signal may be impressed on the main carrier in the modulating-frequency band from 53 to 75 kHz; the maximum modulating frequency allowed by law is 75 kHz. The different sidebands contain all of the information necessary to reproduce a complete stereo signal in the FM receiver. The illustration shows the modulation-frequency band of a standard FM broadcast signal.

Stereo broadcasting has enjoyed great popularity since it was first introduced. Stereo receivers can be found in most homes and automobiles today. Stereo FM receivers can now be manufactured almost as cheaply as amplitude-modulation (AM) receivers. *See also* FREQUENCY MODULATION, SUBSIDIARY COMMUNICATIONS AUTHORIZATION.

FOCAL LENGTH

The focal length of a lens or reflector is the distance from its center to its focal point. Parallel rays of energy arriving at the lens or reflector from a great distance converge, and form an image, at the focal point. Conversely, a point source of energy at the focal point of a lens or reflector will produce parallel rays.

The drawing illustrates the focal lengths for typical reflectors and lenses. At A, the focal point of a spherical reflector is approximately half way to its center; the focal length is thus about half the radius. For a parabolic reflector having a curve according to the function $y = x^2$, the focal point is at $(x,y) = (0,0.25)$; the focal length is 0.25 unit. At B, the focal point of a convex lens is illustrated.

The focal length of a lens or reflector determines the size of the image resulting from impinging parallel rays. The greater the focal length, the larger the image. For a given source of energy, the greater the focal length, the more nearly parallel the transmitted rays. *See also* PARABOLOID ANTENNA.

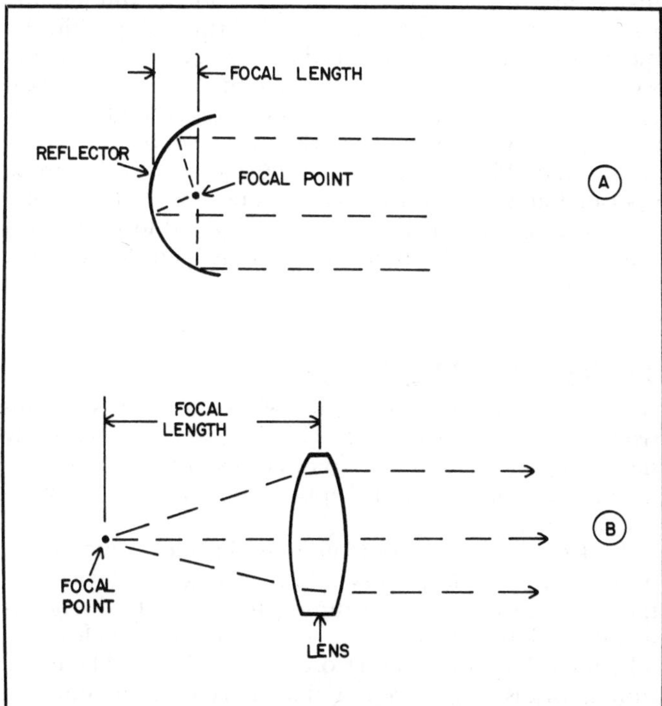

FOCAL LENGTH: At A, the focal point and focal length of a reflector; at B, the focal point and length of a convex lens.

FOCUS CONTROL

In a television receiver or cathode-ray oscilloscope, it is desirable to have the electron beam be as narrow as possible. This is accomplished by means of focusing electrodes. The voltage applied to these electrodes controls the deflection of electrons so that they are brought to a point as they land on the phosphor screen.

Most television receivers have internal focusing potentiometers, and the controls are preset. Once properly adjusted, they seldom require further attention. In the oscilloscope, the focus control is usually located on the front panel.

Focusing can be accomplished by electromagnetic as well as electrostatic deflection. Some cathode-ray tubes use coils, which produce magnetic fields, to direct and focus the electron beam. *See also* ELECTROMAGNETIC DEFLECTION, ELECTROMAGNETIC FOCUSING, ELECTROSTATIC DEFLECTION, PICTURE TUBE.

FOLDBACK

See CURRENT LIMITING.

FOLDED DIPOLE ANTENNA

A folded dipole antenna is a half-wavelength, center-fed antenna constructed of parallel wires in which the outer ends are connected together. The folded-dipole antenna may be thought of as a "squashed" full-wave loop (see illustration).

The folded dipole has exactly the same gain and radiation pattern, in free space, as a dipole antenna. However, the feed-point impedance of the folded dipole is four times that of the ordinary dipole. Instead of approximately 73 ohms, the folded dipole presents a resistive impedance of almost 300 ohms. This makes the folded dipole desirable for use with high-impedance, parallel-wire transmission lines. It also can be used to obtain a good match with 75-ohm coaxial cable when four antennas are connected in phase, or with 50-ohm coaxial cable when six antennas are connected in phase.

Folded dipoles are often found in vertical collinear antennas, such as are used in repeaters at the very high frequencies. Folded dipoles have somewhat greater bandwidth than ordinary dipoles, and this makes them useful for reception in the frequency-modulation (FM)

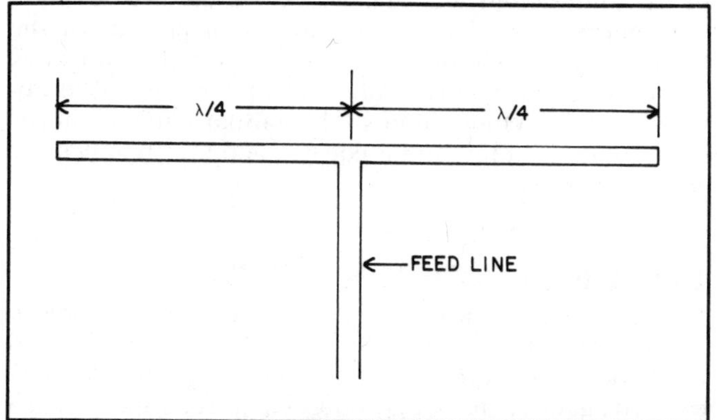

FOLDED DIPOLE ANTENNA: The folded dipole antenna is the same size, and has the same radiation pattern, as an ordinary dipole antenna. But the feed-point impedance is four times as large.

broadcast band, between 88 and 108 MHz. *See also* DIPOLE ANTENNA.

FOLLOWER

A follower is a circuit in which the output signal is in phase with, or follows, the input signal. Follower circuits always have a voltage gain of less than 1. The output impedance is lower than the input impedance. Follower circuits are typically used as broadbanded impedance-matching circuits; they are often cheaper and more efficient than transformers.

The follower circuit is characterized by a grounded collector if a bipolar transistor is used, a grounded drain if a field-effect transistor is used, and a grounded plate if a vacuum tube is used. The vacuum-tube follower is called a cathode follower; the bipolar circuit is called an emitter follower; the field-effect-transistor circuit is called a source follower. *See also* CATHODE FOLLOWER, EMITTER FOLLOWER, SOURCE FOLLOWER.

FOOT CANDLE

The foot candle is a unit of visible-light intensity. A light source with a luminance of 1 candela (*see* CANDELA) produces 1 foot candle of power on a spherically curved surface having an area of 1 square foot, and placed so that every point is 1 foot from the source. A light source having an intensity of 1 candela will produce about 12.6 foot candles of power on any spherical surface completely surrounding it. The foot candle is abbreviated fc. *See also* ILLUMINANCE.

FOOT LAMBERT

The foot lambert is a unit of luminance. A surface that emits or reflects 1 lumen per square foot is considered to have a brightness of 1 foot lambert. A luminance of 1 foot lambert is the equivalent of 0.32 candela per square foot. The abbreviation for foot lambert is fL. *See also* CANDELA, LUMEN, LUMINANCE.

FOOT POUND

The foot pound is a unit of energy or work. When a weight of 1 pound is lifted against gravity for a distance of 1 foot, the energy expended is equal to 1 foot pound. In the metric system, the unit most commonly specified for work is the kilogram meter; 1 kilogram meter is equal to approximately 7.3 foot pounds. The standard unit of energy is the joule; 1 foot pound is equal to about 1.36 joules. *See also* ENERGY, WORK.

FORCE

Force is that effect which, when exerted against an object, induces motion or causes acceleration. Force is measured in a variety of different units. In general, a certain force F, exerted against a mass m, produces an acceleration a, such that $F = ma$. A force of 1 kilogram meter per second per second (1 kgm/s^2) is called 1 newton. A force of 1 gram centimeter per second per second (1 $gmcm/s^2$) is called 1 dyne. The newton represents 100,000 dynes.

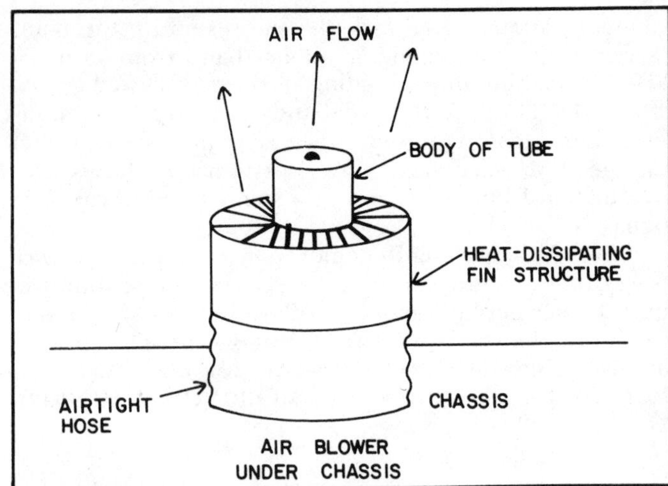

FORCED-AIR COOLING: An example of forced-air cooling of a high-power vacuum tube.

When a force acts on an object, the resulting acceleration may consist of a change in the speed of the object, or a change in the direction of motion, or both. The acceleration is always in the same direction as the force. *See also* ACCELERATION.

FORCED-AIR COOLING

Forced-air cooling is a method of maintaining the temperature of electronic equipment by passing air through an enclosure. Forced-air cooling is accomplished by means of electric fans. Generally, the term forced-air cooling is used in reference to localized cooling of a particular component or components, especially vacuum tubes in high-power amplifiers.

Certain vacuum tubes are provided with heat-dissipating fins, positioned radially inside a cylinder surrounding the plate. The air is guided through the fins by means of a hose (see illustration). Large glass-envelope vacuum tubes are sometimes placed inside cylindrical chambers called chimneys. The air is forced up the chimney, and as the air passes over the surface of the tube, heat is removed. The forced-air cooling allows the tubes to be operated at a higher level of input power. If the cooling system fails, some means must be provided to shut down the amplifier, or the tubes will be damaged. *See also* AIR COOLING.

FORK OSCILLATOR

A fork oscillator is an oscillator whose frequency is controlled by a tuning fork. Tuning forks characteristically have very stable resonant frequencies. This is especially true if the temperature is kept constant, avoiding expansion and contraction.

Fork oscillators are sometimes used in electronic clocks, in place of piezoelectric crystals. A mechanical device sustains the vibration of the tuning fork, and frequency is maintained by the physical dimensions of the fork. A schematic diagram of a fork oscillator is shown. The metal tuning fork is magnetized, so that alternating currents are induced in the coils.

Fork oscillators are generally used at low audio frequencies. The size of a tuning fork for higher-frequency

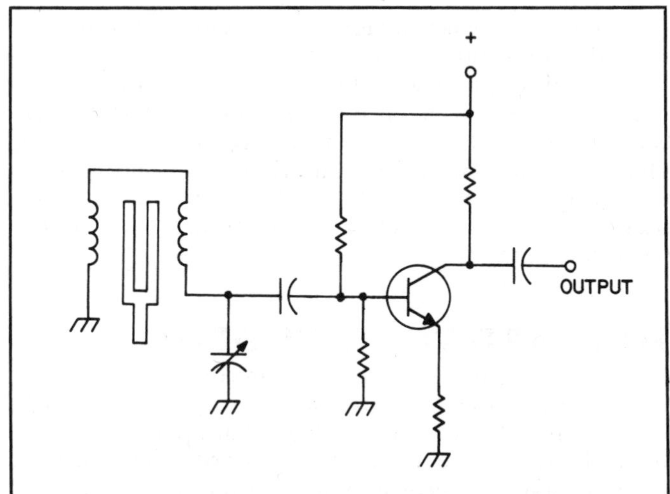

FORK OSCILLATOR: A fork oscillator amplifies the alternating currents generated by a magnetized tuning fork.

oscillations is too small to be practical. *See also* TUNING FORK.

FORM FACTOR

The form factor is a function of the diameter-to-length ratio of a solenoidal coil. It is used for the purpose of calculating the inductance. The value of the form factor, as a function of the ratio of diameter to length, is given by the graph. Given the value of the form factor F, the number of turns N, and the diameter of the coil d in inches, the coil inductance L in microhenrys is given by the formula:

$$L = FN^2d$$

In a tuned-circuit resonant response, the shape factor is sometimes called the form factor (*see* SHAPE FACTOR).

In an alternating-current wave, the ratio of the root-mean-square, or RMS, value to the average value of a half cycle is sometimes called the form factor (*see* ROOT MEAN SQUARE).

FORMULA

A formula is a mathematical expression that gives the value of a desired quantity as a function of other quantities. A formula contains variable quantities and constant quantities. The left side of a formula shows the desired quantity, and the right side is more or less complicated, depending on the nature of the function.

Formulas are invaluable tools for the engineer. Every formula contains information that can be graphed, and often formulas are shown in graphical form rather than mathematical form for simplicity. Although the appearance of formulas in a discussion may give the illusion of complexity, most formulas greatly simplify a scientific presentation. *See also* EQUATION, FUNCTION.

FORTRAN

FORTRAN is a form of higher-order computer language, designed for scientific and engineering applications. The word FORTRAN is a contraction of FORmula TRANSlation. The language was originally developed by International Business Machines (IBM).

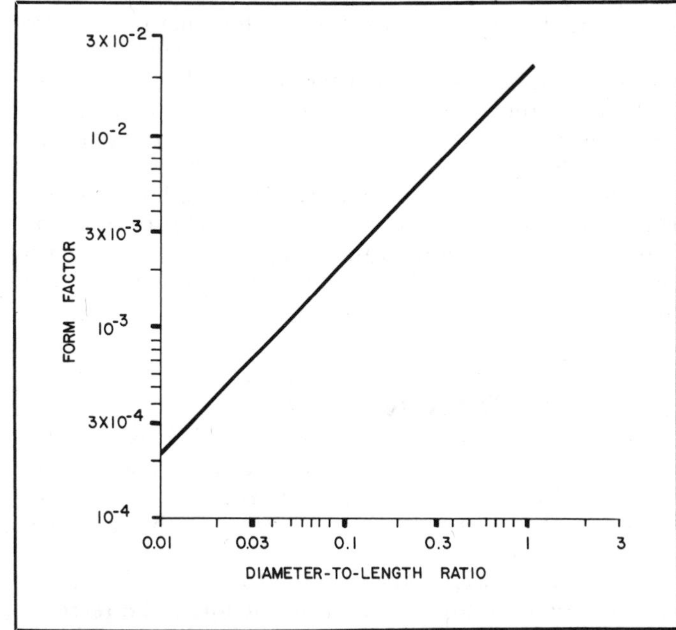

FORM FACTOR: The form factor, for determining the inductance of a solenoidal coil, as a function of the diameter-to-length ratio of the coil.

FORTRAN is very similar to the higher-order language BASIC. It is easy to learn and understand; functions are in the English language, combined with mathematical formulas.

For business purposes, the language COBOL is preferred. In purely mathematical problems, BASIC is often easier to use than FORTRAN. *See also* BASIC, COBOL, HIGHER-ORDER LANGUAGE.

FORWARD BIAS

In any diode or semiconductor junction, forward bias is a voltage applied so that current flows. A negative voltage is applied to the cathode, and a positive voltage to the anode. In some diode or semiconductor devices, a certain amount of forward bias is necessary before current will flow.

The emitter-base junction of a bipolar transistor is normally forward-biased. In the NPN device, the voltage applied to the base is usually positive with respect to the voltage at the emitter; in the PNP device, the base voltage is negative with respect to the emitter voltage. This causes a constant current to flow through the junction. The exception is the case of the Class-C amplifier, where the junction is reverse-biased to put it at cutoff.

Any semiconductor diode will conduct, for all practical purposes, only when it is forward-biased. Under conditions of reverse bias, there is no current flow. This effect makes rectification and detection possible. *See also* REVERSE BIAS.

FORWARD BREAKOVER VOLTAGE

In a silicon-controlled rectifier, a certain amount of voltage must be applied to the control, or gate, terminal in order for the device to conduct in the forward direction. However, if the forward bias is great enough (*see* FORWARD

BIAS), the device will conduct no matter what the voltage on the control electrode. The minimum forward voltage at which this happens is called the forward breakover voltage. This voltage is extremely high in the absence of a signal at the control electrode; the greater the signal voltage, the lower the forward breakover voltage.

After the control signal has been removed, the silicon-controlled rectifier continues to conduct. However, if forward bias is removed and then reapplied, the device will not conduct again until the control signal returns. *See also* SILICON-CONTROLLED RECTIFIER.

FORWARD CURRENT

In a semiconductor junction, the forward current is the value of the current that flows in the presence of a forward bias (*see* FORWARD BIAS). Generally, the greater the forward bias, the greater the forward current, although there are exceptions.

In any diode, a certain amount of forward bias is necessary to cause the flow of current in the forward direction. This minimum forward voltage varies from about 0.3 volt in the case of a germanium diode to 15 volts for a mercury-vapor device. The function of forward current versus forward bias voltage is called the forward characteristic of a diode, transistor, or tube. *See also* REVERSE BIAS.

FORWARD RESISTANCE

The forward resistance of a diode device, or any device that typically conducts in only one direction, is the value of the forward voltage divided by the forward current. If E is the forward voltage in volts and I is the forward current in amperes, then the forward resistance R, in ohms, is:

$$R = E/I$$

This is simply an expression of Ohm's law (*see* OHM'S LAW).

The forward resistance of a P-N junction or other unidirectionally conducting device depends on the value of the forward voltage. Until the forward voltage reaches the minimum value required for current to flow, the resistance is extremely high, and in some cases may be considered practically infinite. In some devices, the resistance increases with increasing forward voltage; in other situations it decreases. In most semiconductor diode devices, the forward resistance is very high until a forward voltage of 0.3 to 0.6 volt is obtained. Then the resistance abruptly drops to a low value, and with further increases in the forward voltage, the resistance approaches a minimum limit. *See also* FORWARD BIAS, FORWARD CURRENT.

FORWARD SCATTER

When a radio wave strikes the ionospheric E and F layers, it may be returned to earth. This kind of propagation most often takes place at frequencies below about 30 MHz, although it may occasionally be observed at frequencies well into the very-high part of the spectrum. When the electromagnetic field encounters the ionosphere, most of the energy is returned at an angle equal to the angle of incidence, much like the reflection of light from a mirror. However, some scattering does occur. Forward scatter is the scattering of electromagnetic waves in directions away from the transmitter.

Forward scatter results in the propagation of signals over several paths at once. This contributes to fading, because the signal reaches the receiving antenna in varying phase combinations. Some scattering occurs in the backward direction, toward the transmitter; this is called backscatter. *See also* BACKSCATTER, IONOSPHERE, PROPAGATION CHARACTERISTICS.

FORWARD VOLTAGE DROP

In a rectifying device such as a semiconductor diode or vacuum tube, the forward voltage drop is the voltage, under conditions of forward bias, that appears between the cathode and the anode. The forward voltage drop of most rectifying devices remains constant at all values of forward bias. For a germanium semiconductor P-N junction, the forward voltage drop is about 0.3 volt. For silicon semiconductor devices it is about 0.6 volt. For vacuum-tube rectifiers, the forward voltage drop varies depending on the type of tube. Mercury-vapor tube rectifiers have a voltage drop of about 15 volts with forward bias.

If the forward bias is less than the forward voltage drop for a particular device, the component will normally not conduct. When the voltage reaches or exceeds the forward voltage drop, the component abruptly begins to conduct. *See also* FORWARD BIAS, FORWARD CURRENT, FORWARD RESISTANCE.

FOSTER-SEELEY DISCRIMINATOR

The Foster-Seeley discriminator is a form of detector circuit for reception of frequency-modulated signals. A center-tapped transformer, with tuned primary and secondary windings, is connected to a pair of diodes. The center tap of the secondary winding is coupled to the primary. See the schematic diagram of a simple Foster-Seeley type discriminator. The circuit is recognized by the single, center-tapped secondary winding.

As the frequency of the input signal varies, the voltage across the secondary fluctuates in phase. The diodes produce audio-frequency variations from these changes in phase. The Foster-Seeley discriminator circuit is sensitive to changes in the signal amplitude as well as phase and frequency. Therefore, the circuit must have a limiter stage

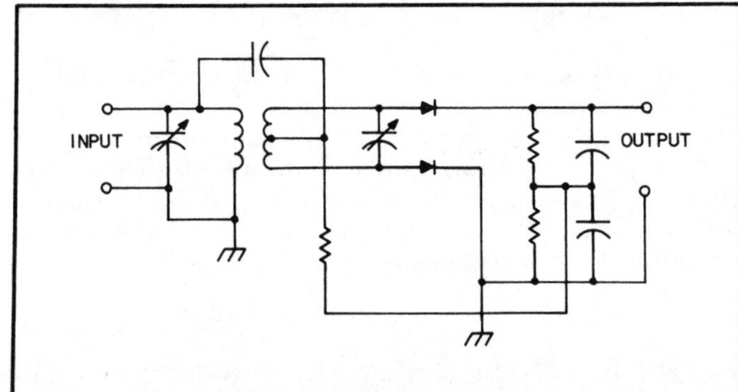

FOSTER-SEELEY DISCRIMINATOR: The Foster-Seeley discriminator is a common circuit for detecting frequency-modulated signals.

preceding it for best results. *See also* DISCRIMINATOR, FREQUENCY MODULATION, RATIO DETECTOR.

FOURIER SERIES

Any periodic function can be expressed in terms of an infinite series called a Fourier series. The Fourier series is a sum of sine and cosine functions. Any periodic wave, such as a sawtooth wave or square wave, consists of a certain combination of fundamental and harmonic frequencies. The Fourier series is a means of denoting this combination.

The General form of a Fourier series is given by:

$$f(x) = (A_o/2) + \sum_{n=1}^{\infty} (A_n \cos(nx) + B_n \sin(nx))$$

where A_o, A_n, and B_n are constants. The more terms are evaluated in the series, the more accurate the representation of the actual function. A Fourier series is an infinite series, however, so any finite sum represents an approximation of the actual function.

Fourier series can be extremely complex, and computers are often used in the evaluation of a waveform by Fourier analysis. *See also* FOURIER TRANSFORM.

FOURIER TRANSFORM

The Fourier transform is a mathematical method of obtaining the power spectrum, as a function of frequency, for a given signal waveform. This is done by determining the Fourier series representing the particular waveform, and is a mathematical approximation.

Fourier transformations allow the evaluation of nonperiodic as well as periodic waveforms. Any waveform can be mathematically approximated over a finite interval, provided there are a finite number of maxima and minima in that interval. The mathematical techniques in Fourier analysis are quite sophisticated, and a detailed discussion is beyond the scope of this book. Engineering texts are good sources of information on the subject of Fourier series, Fourier analysis, and Fourier transforms. *See also* FOURIER SERIES.

FOUR-LAYER SEMICONDUCTOR

A four-layer semiconductor, also called a four-layer diode, is a special form of semiconductor device containing four semiconductor wafers with three P-N junctions. The four-layer diode is sometimes called a Shockley diode or PNPN device. Four-layer diodes are commonly used as switching devices.

In the reverse-bias condition, a four-layer diode behaves in the manner similar to an ordinary diode. The resistance is high, and very little current flows. In the forward direction, as the voltage increases, the current remains small until a certain point is reached. Then the current becomes large abruptly. Once this critical point has been reached, the forward voltage can be reduced almost to zero, and the current will continue to rise. The voltage-versus-current relation for a typical four-layer diode is illustrated by the graph. Four-layer diodes are switched on by means of a brief forward pulse called a trigger pulse. To be switched off again, the voltage must be removed or reversed. The switching time is quite rapid, in some cases just a few nanoseconds (billionths of a second). The time required for the device to switch on is shorter than that required for it to switch off. Other PNPN or four-layer devices include the silicon-controlled rectifier and the triac. *See also* SILICON-CONTROLLED RECTIFIER, TRIAC.

FOUR-LAYER TRANSISTOR

A four-layer transistor is a form of four-layer semiconductor of the PNPN form, having three terminals. Such devices are generally used for switching purposes. Examples include the silicon-controlled rectifier and the thyristor.

Four-layer transistors have rapid switching times, sometimes as short as a few nanoseconds (billionths of a second). The devices are frequently used in such circuits as light dimmers and motor-speed controls. *See also* FOUR-LAYER SEMICONDUCTOR, SILICON-CONTROLLED RECTIFIER, THYRISTOR, TRIAC.

FOUR-TRACK RECORDING

Four-track recording is a method of recording stereo audio signals on magnetic tape. A reel-to-reel or cassette arrangement may be used. Each recording requires two tracks on the tape, one for the left channel and the other for the right channel. A four-track tape can therefore contain two separate stereo recordings or four separate monaural recordings. The four magnetic tracks run parallel to each other along the tape. When stereo is used, tracks 1 and 3 in a reel-to-reel tape are recorded in one direction, and tracks 2 and 4 in the other direction.

Most cassettes in modern stereo recording and reproducing systems are of the eight-track variety, which can handle four different stereo recordings. *See also* EIGHT-TRACK RECORDING, MAGNETIC RECORDING, MAGNETIC TAPE.

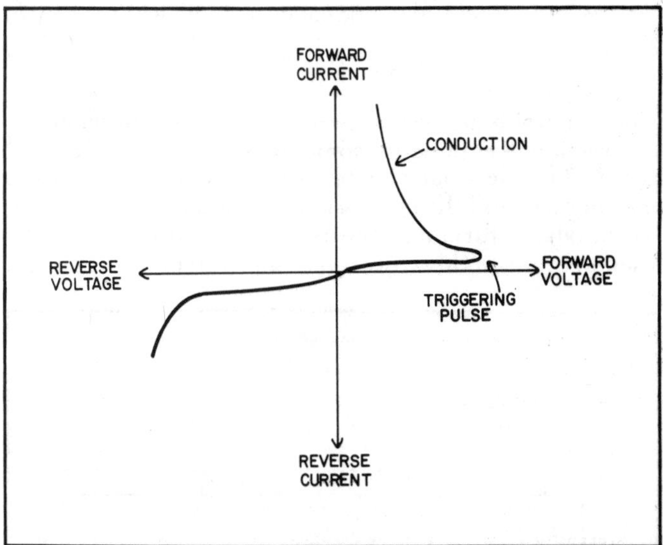

FOUR-LAYER SEMICONDUCTOR: Current-versus-voltage relation for a four-layer semiconductor device. The triggering pulse causes the resistance to drop, and it stays low until the voltage is reversed or removed.

FOUR-WIRE TRANSMISSION LINE

A four-wire transmission line is a special form of balanced line, used to carry radio-frequency energy from a transmitter to an antenna, or from an antenna to a receiver. Four conductors are run parallel to each other, in such a way that they form the edges of a long square prism (see illustration). Each pair of diagonally opposite conductors is connected together at either end of the line.

Four-wire transmission line may be used in place of parallel-wire line in almost any installation. For a given conductor size and spacing, four-wire line has a somewhat lower characteristic impedance, since the four-wire line is essentially two two-wire lines in parallel. The four-wire line is not as susceptible to the influences of nearby objects as is the two-wire line; more of the electromagnetic-field energy is contained within the line. Thus, the four-wire line has inherently better balance, and is less likely to radiate or pick up unwanted noise. Four-wire line is somewhat more difficult to install, and is more expensive, than two-wire lines. Four-wire lines are not commonly used. *See also* BALANCED TRANSMISSION LINE, OPEN-WIRE LINE, TWIN-LEAD.

FOX MESSAGE

The fox message is a familiar test signal to operators of teletype equipment, especially radioteletype. The short message contains every letter of the alphabet and all ten numeric digits. The most common fox message is:

THE QUICK BROWN FOX JUMPS OVER THE LAZY DOG 0123456789

There are variations. *See also* TEST MESSAGE.

FRACTIONAL EXPONENT

In some mathematical expressions, fractional exponents are seen. Although they may look formidable, they are actually not difficult to evaluate. A fractional exponent is any exponent that is not a whole number.

The general rule for fractional exponents is:

$$x^{a/b} = \sqrt[b]{x^a},$$

where a and b are integers and x is any real number.

Fractional exponents sometimes appear in decimal form. The decimal can be converted to a fractional number a/b, and the expression then evaluated according to the above formula. Alternatively, logarithms may be used to evaluate the expression, since for any positive real number x and any real number y,

$$\log_{10}(x^y) = y \log_{10} x$$

See also LOGARITHM.

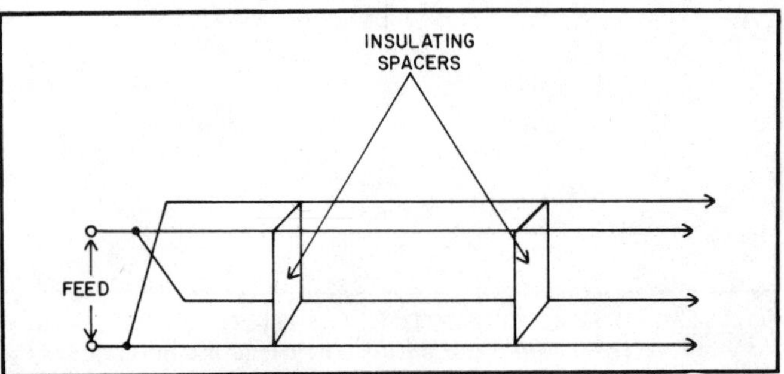

FOUR-WIRE TRANSMISSION LINE: This is a balanced line.

FRAME

A frame is a single, complete television picture. The picture is generally repeated 25 times per second. The electron beam scans 625 horizontal lines per frame, with a horizontal-to-vertical size ratio of 4 to 3. Scanning takes place from left to right along each horizontal line, and from top to bottom for the lines in a complete frame.

There are some variations, in different parts of the world, in the number of scanning lines in a frame and in the overall bandwidth of the picture signal. However, each frame is always a still picture. The human eye can perceive only about 15 to 25 still pictures per second; if the pictures are repeated more often than this, they combine to form an apparent moving picture. Television picture frames are repeated so often that they combine to form a moving image. *See also* TELEVISION.

FRANKLIN ANTENNA

See COLLINEAR ANTENNA.

FRANKLIN, BENJAMIN

One of the most well-known inventors in American history is Benjamin Franklin. An entrepreneur and adventurer, Franklin (1706-1798) contributed to our knowledge of electricity by showing that lightning is caused by the same natural phenomenon as static electricity such as is produced by rubbing silk against glass or by shuffling on a carpet.

Franklin conducted his experiment by flying a kite in a thundershower. Sparks flew from a key that dangled from the string. His experiment was extremely dangerous, and Franklin was fortunate that he was not killed. Others have tried the experiment and have died from burns or electrocution. No one should fly a kite anywhere near a thundershower.

Franklin was instrumental in the invention and implementation of the lightning rod, a device for protecting property against damage by lightning strokes. *See* LIGHTNING, LIGHTNING PROTECTION, LIGHTNING ROD.

FRANKLINE

Frankline is a name that has been suggested as a unit of electric charge. A charge of 1 frankline is that amount of electric potential that exerts a force of 1 dyne on an equal charge of 1 centimeter in empty space.

The frankline is not commonly used as a unit of charge—the standard unit is the coulomb. *See also* COULOMB.

FRANKLIN OSCILLATOR

A Franklin oscillator is a form of variable-frequency oscillator circuit, using transistors, field-effect transistors, or tubes. A resonant circuit is connected in the base, gate, or grid circuit of one device. Feedback is provided by a second stage, which serves to amplify the signal and invert its phase. The output of the amplifying stage is coupled back to the base, gate, or grid of the first device.

In the common-emitter, common-source, or common-cathode configurations, active devices show a 180-degree phase difference between their input and output circuits. In the Franklin oscillator, since there are two devices, the output is in phase with the input. Coupling between the two devices is usually accomplished by means of capacitors. However, tuned circuits can be used. The output of the oscillator may be taken from any point in the circuit. The illustration is a schematic diagram of a simple Franklin oscillator. *See also* OSCILLATION, OSCILLATOR.

FREE ELECTRON

When an electron is not held in an orbit around the nucleus of an atom, the electron is called a free electron. In order to become independent of the influence of atomic nuclei, electrons must receive or contain a certain minimum amount of energy. The less energy an electron possesses, the more tightly it is held in orbit around the nucleus of an atom. This is because the positive charge of the nucleus attracts the negative charge of the electron.

A substance in which most, or all, of the electrons are free is called a plasma. When a large electric voltage is placed across a substance, some of the electrons may be stripped from the nuclei and become free for a short time before falling under the influence of another nucleus. This can also happen when a substance is greatly heated, or bombarded with intense electromagnetic radiation such as ultraviolet light or X rays. When an electron is not free, it is called a bound electron. *See also* BOUND ELECTRON, ELECTRON, IONIZATION, PLASMA.

FREE-RUNNING MULTIVIBRATOR

See ASTABLE MULTIVIBRATOR.

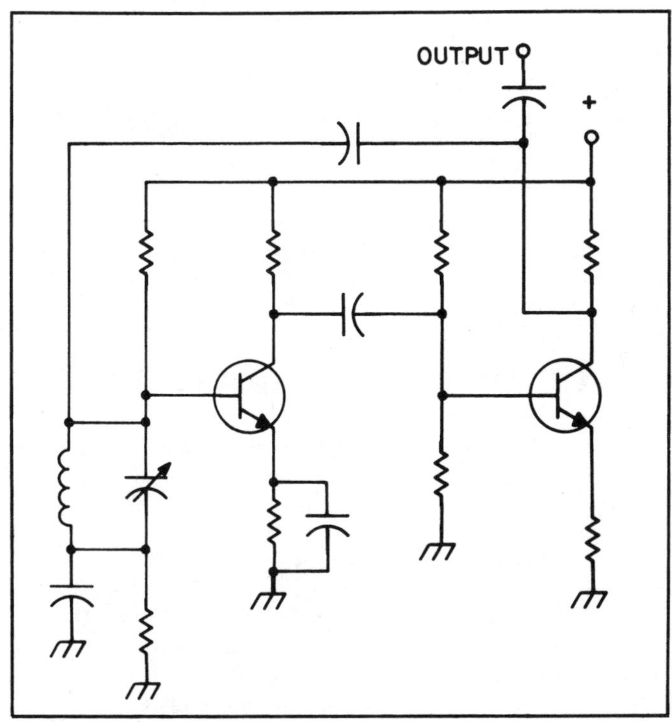

FRANKLIN OSCILLATOR: A Franklin oscillator uses a phase-inverting amplifier in the feedback circuit.

FREE SPACE

Free space is a term used to denote an environment in which there are essentially no objects or substances that affect the propagation of an electromagnetic field. Perfect free space exists nowhere in the universe; the closest thing to true free space is probably found in interstellar space. Free space has a magnetic permeability of 1 and a dielectric constant of 1. Electromagnetic fields travel through free space at approximately 186,282 miles per second.

When designing and testing antenna systems, the free-space performance is evaluated for the purpose of determining the directional characteristics, impedance, and power gain. A special testing environment provides conditions approaching those of true free space. An antenna installation always deviates somewhat from the theoretical free-space ideal. However, antennas placed several wavelengths above the ground, and away from objects such as trees, utility wires, and buildings, operate in nearly the same fashion as if they were in free space. *See also* FREE-SPACE LOSS, FREE-SPACE PATTERN.

FREE-SPACE LOSS

As an electromagnetic field is propagated away from its source, the field becomes less intense as the distance increases. In free space, this happens in a predictable and constant manner, regardless of the wavelength.

The field strength of an electromagnetic field, in volts per meter, varies inversely with distance. If F is the field strength expressed in volts per meter and d is the distance from the source, then:

$$F = k/d$$

where k is a constant that depends on the intensity of the source.

The field strength as measured in watts per square meter varies inversely with the square of the distance from the source. If G is the field strength in watts per square meter and d is the distance from the source, then:

$$G = m/d^2$$

where m is, again, a constant that depends on the source intensity (though not the same constant as k).

In free space, the field strength drops by 6 dB when the distance from the source is doubled. This is true no matter how the field strength is measured in the far field. *See also* FAR FIELD, FIELD STRENGTH, FREE SPACE, INVERSE-SQUARE LAW.

FREE-SPACE PATTERN

The free-space pattern is the radiation pattern exhibited by an antenna in free space. Free-space patterns are usually specified when referring to the directional characteristics and power gain of an antenna. Under actual operating conditions, in which an antenna is surrounded by other objects, the free-space pattern is modified. If the antenna is close to the ground, or is operated in the midst of trees, buildings, and utility wires within a few wavelengths of the radiator, the antenna pattern may differ substantially from the free-space pattern.

An antenna that radiates equally well in all possible directions is called an isotropic antenna (*see* ISOTROPIC ANTENNA). The free-space pattern of the isotropic antenna is a sphere, with the antenna at the center (see

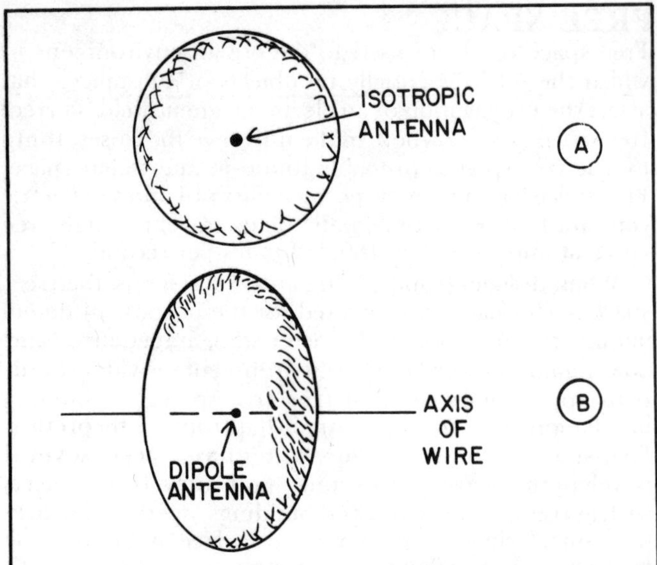

FREE-SPACE PATTERN:Free-space radiation patterns. At A, the pattern for an isotropic radiator is spherical. At B, the radiation pattern for a dipole is doughnut-shaped.

illustration at A). A dipole antenna (*see* DIPOLE ANTENNA) radiates equally well in all directions perpendicular to the axis of the radiator, but the field strength drops as the direction approaches the axis of the wire. Along the axis of the wire, there is no radiation from the dipole. The free-space pattern of a dipole is torus-shaped, as at B. The dipole and the isotropic antenna are used as references for antenna power-gain measurements. *See also* ANTENNA PATTERN, ANTENNA POWER GAIN,, dBd, dBi.

FREON

Freon is a synthetic gas, consisting of carbon, chlorine, and fluorine. Its chemnical formula is CCl_2F_2. Freon is commonly used as a coolant in such devices as air conditioners and refrigerators. However, in rarefied form, freon can serve as an excellent dielectric substance.

Freon has about 2½ times the dielectric strength of air. This means that, at atmospheric pressure, freon can withstand about 2½ times as much voltage as dry, pure air. When large voltages must exist within a limited space, freon can be useful for insulation purposes. *See also* DIELECTRIC, DIELECTRIC BREAKDOWN.

FREQUENCY

For any periodic disturbance, the frequency is the rate at which the cycle repeats. Frequency is generally measured in cycles per second, or Hertz, abbreviated Hz. Rapid oscillation frequencies are specified in kilohertz (kHz), megahertz (MHz), gigahertz (GHz), and terahertz (THz). A frequency of 1 kHz is 1,000 Hz; 1 MHz = 1,000 kHz; 1 GHz = 1,000 MHz; and 1 THz = 1,000 GHz. If the time required for 1 cycle is denoted by p, in seconds, then the frequency f is equal to 1/p.

Wave disturbances can have frequencies ranging from less than 1 Hz to many trillions of terahertz. The audio-frequency range, at which the human ear can detect acoustic energy, ranges from about 20 Hz to 20 kHz. The radio-frequency spectrum of electromagnetic energy ranges up to hundreds or thousands of gigahertz. The frequency of electromagnetic radiation is related to the wavelength according to the formula:

$$\lambda = 300v/f$$

where λ is the wavelength in meters, v is the velocity factor of the medium in which the wave travels, and f is the frequency in megahertz. *See also* ELECTROMAGNETIC SPECTRUM, VELOCITY FACTOR, WAVELENGTH.

FREQUENCY ALLOCATIONS

Frequency allocations are frequencies, or bands of frequencies, assigned in the electromagnetic spectrum by a government or other authority for use by specific personnel and for specific purposes.

Frequency assignments are made on a worldwide basis by the International Telecommunication Union (ITU), headquartered in Geneva, Switzerland. The lowest allocated electromagnetic frequency is 10 kHz. The highest allocated frequency is 275 GHz at the time of this writing; however, the maximum frequency can be expected to rise as technology improves at the short wavelengths. In the United States, frequency assignments are made by the Federal Communications Commission (FCC). A list of United States allocations can be obtained from the Superintendent of Documents, U.S. Government Printing Office, Washington, DC 20402.

FREQUENCY CALIBRATOR

A frequency calibrator is a device that generates unmodulated-carrier markers at precise, known frequencies. Frequency calibrators generally consist of crystal-controlled oscillators. The crystal may be housed in a special heat-controlled chamber for maximum stability; phase-locked-loop circuits may also be employed in conjunction with standard broadcast frequencies.

Frequency calibrators are a virtual necessity in all receivers. Without a properly adjusted calibrator, the operator of a receiver cannot be certain of the accuracy of the frequency dial or readout. Calibrators are especially important in transmitting stations, because out-of-band transmission can cause harmful interference to other stations.

The frequency calibrator in a receiver should be checked periodically against a standard frequency source such as WWV or WWVH. The calibrator should be adjusted, if it produces markers at even-megahertz points, to zero beat with the carrier frequency of the standard source. *See also* CRYSTAL CONTROL, CRYSTAL OVEN, PHASE-LOCKED LOOP, WWV/WWVH.

FREQUENCY COMPARATOR

A frequency comparator is a circuit that detects the difference between the frequencies of two input signals. The output of a frequency comparator may be in video or audio form. Basically, all frequency comparators operate on the principle of beating. The beat frequency, or difference between the two input frequencies, determines the output.

A simple frequency comparator circuit is illustrated in the schematic diagram. When the frequencies of the two input signals are identical, there is no output. When the frequencies are different, a beat note is produced, in proportion to the difference between the input-signal frequencies. This output signal is fed to an audio amplifier. The audio beat note appears at the output of the amplifier,

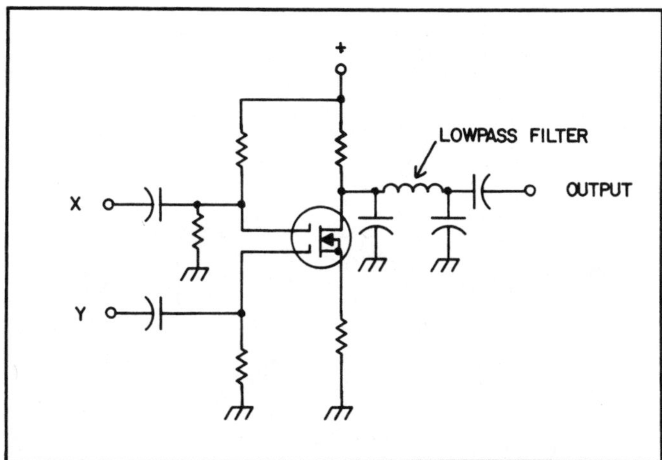

FREQUENCY COMPARATOR: A simple frequency comparator. This circuit operates as a mixer, producing the difference signal in the output. The lowpass filter has a cutoff frequency well below either of the two input-signal frequencies.

where it may be fed to a speaker, oscilloscope, or frequency counter.

Frequency comparators are used in such devices as phase-locked-loop circuits, where two signals must be precisely matched in frequency. The output of the frequency comparator is used to keep the frequencies identical, by means of controlling circuits. *See also* PHASE-LOCKED LOOP.

FREQUENCY COMPENSATION

When the frequency response of a circuit must be tailored to have a certain characteristic, frequency-compensating circuits are often used. The simplest form of frequency-compensating circuit is a capacitor or inductor, in conjunction with a resistor. More sophisticated frequency-compensating circuits use operational amplifiers or tuned circuits.

The bass and treble controls in a stereo hi-fi system are a form of frequency compensation. The desired response can be obtained by adjusting these controls. In frequency-modulation communications systems, a form of frequency compensation called preemphasis is often used at the transmitter, with deemphasis at the receiver, for improved intelligibility. Some communications receivers have an audio tone control, to compensate for the speaker response. *See also* DEEMPHASIS, PREEMPHASIS, TONE CONTROL.

FREQUENCY CONTROL

A frequency control is a knob, switch, button, or other device, that is used to set the frequency of a receiver, transmitter, transceiver, or oscillator.

The most common method of frequency control is the tuning knob; the frequency can be adjusted over a continuous range by rotating the knob. This is analog frequency control. The knob is connected either to a variable capacitor or a variable inductor, which in turn sets the frequency of an oscillator-tuned circuit.

In recent years, digital frequency control has become increasingly common. There are several different configurations of digital frequency control, but this method of frequency adjustment is always characterized by discrete frequency intervals instead of a continuous range. One method of digital frequency control uses a rotary knob, very similar in appearance to the analog tuning knob. However, when the knob is turned, it produces a series of discrete changes in frequency, at defined intervals such as 10 Hz. Another form of digital control uses spring-loaded toggle switches or buttons; by holding down the switch lever or button, the frequency increments at a certain rate in discrete steps, such as 1 kHz per second in 10-Hz jumps. The third common form of digital frequency control uses several buttons, each of which sets one digit of the desired frequency. Some communications devices have more than one of these digital tuning methods, so that the operator may choose the particular frequency-control mode he wants. The electrical control of the frequency is usually accomplished by means of varactor diodes in the oscillator tuned circuit, and/or synthesized oscillators.

Both analog and digital frequency-control methods have advantages and disadvantages. Often, the optimum method is determined largely by the preference of the individual operator. *See also* ANALOG CONTROL, DIGITAL CONTROL.

FREQUENCY CONVERSION

In some situations, it is necessary to change the carrier frequency of a signal, without changing the modulation characteristics. When this is deliberately done, it is called frequency conversion. Frequency conversion is an integral part of the superheterodyne receiver, and it is found in many communications transmitters as well.

Frequency conversion is accomplished by means of heterodyning. The signal is combined with an unmodulated carrier in a circuit called a mixer (*see* MIXER). If the input signal has a frequency f, and the desired output frequency is h, then the frequency of the unmodulated oscillator g must be such that either $f - g = h$, $g - f = h$, or $f + g = h$. The output frequency must, in other words, be equal to either the sum or the difference of the input frequencies.

Frequency conversion is not the same thing as frequency multiplication. With certain types of modulation, the bandwidth changes when the frequency of the carrier is multiplied. But with true frequency conversion, the signal bandwidth is never affected. Frequency conversion is desirable in many applications, because it simplifies the design of the selective circuits in a receiver or transmitter. *See also* FREQUENCY CONVERTER, FREQUENCY MULTIPLIER, SUPERHETERODYNE RECEIVER.

FREQUENCY CONVERTER

A frequency converter is a mixer, designed for use with a communications receiver to allow operation on frequencies not within the normal range of the receiver. If the operating frequency is above the receiver range, the converter is called a down converter. If the operating frequency is below the receiver range, the converter is called an up converter. Converters are often used with high-frequency, general-coverage receivers to allow reception at very-low, low, very-high, and ultra-high frequencies.

Converters are relatively simple circuits. With a good communications receiver, capable of operating in the range of 3 to 30 MHz, frequency converters can be used to obtain excellent sensitivity and selectivity throughout the radio spectrum. The use of a converter with an existing receiver is often not only superior to a separate receiver, but far less expensive.

Some converters, designed to operate with transceivers, allow two-way communication on frequencies outside the range of the transceiver. This kind of device has two separate converters, and the entire circuit is called a transverter. *See also* DOWN CONVERSION, FREQUENCY CONVERSION, MIXER, UP CONVERSION.

FREQUENCY COUNTER

A frequency counter is a device that measures the frequency of a periodic wave by actually counting the pulses in a given interval of time.

The usual frequency-counter circuit consists of a gate, which begins and ends each counting cycle at defined intervals. The accuracy of the frequency measurement is a direct function of the length of the gate time; the longer the time base, the better the accuracy. At higher frequencies, the accuracy is limited by the number of digits in the display. Of course, an accurate reference frequency is necessary in order to have a precision counter; crystals are used for this purpose. The reference oscillator frequency may be synchronized with a time standard by means of a phase-locked loop.

Frequency counters are available that allow accurate measurements of signal frequencies into the gigahertz range. Frequency counters are invaluable in the test laboratory, since they are easy to use and read. Typical frequency counters have readouts that show six to ten significant digits. *See also* FREQUENCY MEASUREMENT.

FREQUENCY DEVIATION

See DEVIATION.

FREQUENCY DIVIDER

A frequency divider is a circuit whose output frequency is some fraction of the input frequency. All frequency dividers are digital circuits. While frequency multiplication can be accomplished by means of simple nonlinear analog devices, frequency division requires counting devices. Fractional components of a given frequency are sometimes called subharmonics; a wave does not naturally contain subharmonics.

Frequency dividers can be built up from bistable multivibrators. Dividers, in conjunction with frequency multipliers, can be combined to produce almost any rational-number multiple of a given frequency. For example, four divide-by-two circuits can be combined with a tripler, resulting in multiplication of the input frequency by 3/16. *See also* FREQUENCY MULTIPLIER.

FREQUENCY-DIVISION MULTIPLEX

One method of sending several signals simultaneously over a single channel is called frequency-division multiplex. An available channel is subdivided into smaller segments, all of equal size, and these segments are called subchannels. Each subchannel carries a separate signal. Frequency-division multiplexing can be used in wire transmission circuits or radio-frequency links.

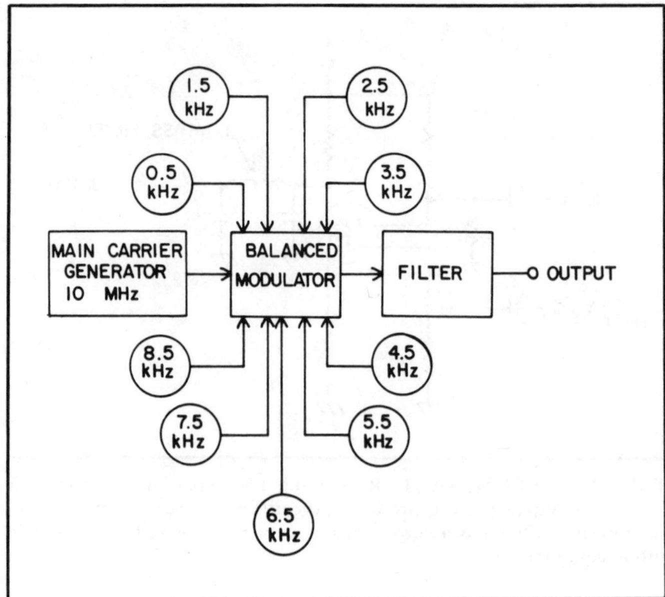

FREQUENCY-DIVISION MULTIPLEX: Schematic representation of the generation of a frequency-division-multiplex signal.

One method of obtaining frequency-division multiplexing is shown in the illustration. The available channel is the frequency band from 10,000 MHz to 10.009 MHz, or a space of 9 kHz. This space is divided into nine subchannels, each 1 kHz wide. Nine different medium-speed carriers are impressed on the main carrier of 10.000 MHz. This is accomplished by using audio tones of 500, 1500, 2500, 3500, 4500, 5500, 6500, 7500, and 8500 Hz, all of which modulate the main carrier to produce a single-sideband signal. The main carrier is suppressed by means of a balanced modulator. Each audio tone may be keyed independently in a digital code such as Morse, BAUDOT, or ASCII. The tones must be fed into the modulator circuit using appropriate isolation methods to avoid the generation of mixing products. The resulting output signals appear at 1-kHz intervals from 10.0005 MHz to 10.0085 MHz.

The above example is given only for the purpose of illustrating the concept of frequency-division multiplex. There are many different ways of obtaining this form of emission. Frequency-division multiplex necessitates a sacrifice of data-transmission speed in each subchannel, as compared with the speed obtainable if the entire channel were used. For example, each signal in the above illustration must be no wider than 1 kHz; this limits the data-transmission speed to 1/9 the value possible if the whole 9-kHz channel were employed. Frequency-division multiplex is a form of parallel data transfer. *See also* MULTIPLEX, PARALLEL DATA TRANSFER.

FREQUENCY MEASUREMENT

There are many different ways of determining the frequency of a signal. Basically, these methods can be categorized in two ways: direct measurement and indirect measurement.

Direct frequency measurement is done by means of frequency counters. The number of alternations, or cycles, within a given time period is actually counted by means of digital circuits. The time interval need not necessarily be exactly 1 second. The longer the counting interval, the greater the accuracy of the measurement.

Indirect frequency measurement is done with frequency meters or calibrated receivers. The simplest frequency meter consists of an adjustable tuned circuit and an indicator that shows when the device is set for resonance at a particular frequency. The approximate frequency may then be read from a calibrated scale. If a receiver is available, the signal can be tuned in and the frequency read, with fair to good accuracy, from the tuning dial. If a calibrated signal generator is available, it may be set, by means of an indicating device, to zero beat with the signal under measurement. *See also* FREQUENCY, FREQUENCY METER.

FREQUENCY METER

A frequency meter is a device that is used for measuring the frequency of a periodic occurrence. There are several different kinds of frequency meters. An increasingly common method of frequency measurement is the direct counting of the cycles within a given interval of time; this is done with a frequency counter. Frequency counters provide extremely accurate indications of the frequency of a periodic wave. In some cases the resolution is better than ten significant digits.

The absorption wavemeter uses a tuned circuit in conjunction with an indicating meter. The capacitor or inductor in the tuned circuit is adjustable, and has a calibrated scale showing the resonant frequency over the tuning range. When the device is placed near the signal source, the meter shows the energy transferred to the tuned circuit; the coupling is maximum, resulting in a peak reading of the meter, when the tuned circuit is set to the same frequency as the signal source.

The gate-dip meter operates on a principle similar to the absorption wavemeter, except that the gate-dip meter contains a variable-frequency oscillator. An indicating meter shows when the oscillator is set to the same frequency as a tuned circuit under test. Gate-dip meters are commonly used to determine unknown resonant frequencies in tuned amplifiers and antenna systems.

The heterodyne frequency meter contains a variable-frequency oscillator, a mixer, and an indicator such as a meter. The oscillator frequency is adjusted until zero beat is reached with the signal source. This zero beat is shown by a dip in the meter indication.

The cavity frequency meter is often used to determine resonant frequencies at the very-high, ultra-high, and microwave parts of the spectrum. The principle of operation is similar to that of the absorption wavemeter, except that an adjustable resonant cavity is used rather than a tuned circuit.

The lecher-wire system of frequency measurement employs a tunable section of transmission line to determine the wavelength of an electromagnetic signal in the very-high, ultra-high, and microwave regions. The system consists of two parallel wires with a movable shorting bar. The shorting bar is set until the system reaches resonance, as shown by an indicating meter.

Spectrum analyzers are sometimes used for the purpose of frequency measurement. General-coverage receivers can be used for determining the frequency of a signal. An oscilloscope can be used for very approximate measurements. *See also* ABSORPTION WAVEMETER, CAVITY FREQUENCY METER, FREQUENCY COUNTER, HETERODYNE FREQUENCY METER, LECHER WIRE, OSCILLOSCOPE, SPECTRUM ANALYZER.

FREQUENCY MODULATION

Information may be transmitted over a carrier wave in many different ways. Frequency modulation, abbreviated FM, is a common method. The frequency of the carrier wave is made to fluctuate in accordance with the modulating waveform.

Frequency modulation is generally accomplished by a device called a reactance modulator. The audio-frequency signal is applied to the tuned circuit of an oscillator across a varactor diode, which introduces a fluctuating capacitance into the circuit (*see* REACTANCE MODULATOR, VARACTOR DIODE). The amount by which the signal frequency changes is called the deviation (*see* DEVIATION). In communications practice, the maximum deviation is typically plus or minus 5 kHz with respect to the unmodulated-carrier frequency.

Frequency modulation results in the generation of sidebands, very similar to those in an amplitude-modulated system. However, as the deviation is increased, sidebands appear at greater and greater distances from the main carrier. The amplitude of the main carrier also depends on the amount of deviation. The amplitudes of the sidebands, which appear at integral multiples of the modulating-signal frequency above and below the carrier, as well as the amplitude of the carrier itself, are a function of the ratio of the deviation to the modulating frequency. The function is rather complicated, but in general, the greater the deviation, the greater the bandwidth of the signal. The ratio of the maximum frequency deviation to the highest modulating frequency is called the modulation index (*see* MODULATION INDEX).

When a frequency-modulated signal is put through a frequency multiplier, the deviation is multiplied by the same factor as the carrier frequency. This results in an increase in the bandwidth of the signal. This effect does not occur with mixing. These facts must be kept in mind when designing a frequency-modulation transmitter.

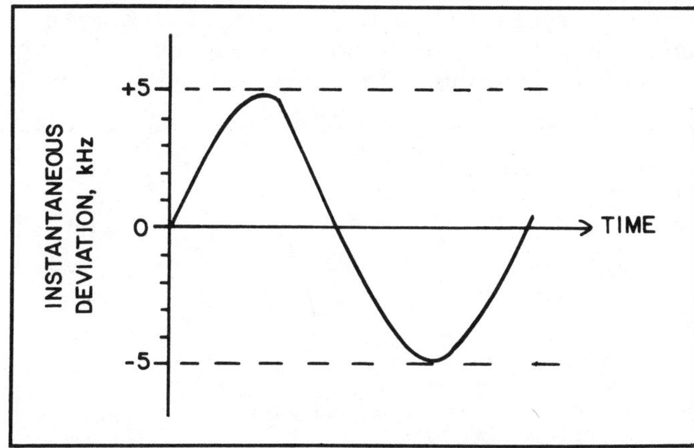

FREQUENCY MODULATION: The frequency-versus-time function for a communications FM signal with sine-wave modulation.

Reception of FM signals requires the use of a special kind of detector, called a discriminator. The discriminator, in conjunction with a limiter, is sensitive only to variations in the signal frequency, and is essentially immune to variations in amplitude. The ratio detector, another form of FM detector, needs no limiter circuit (*see* DISCRIMINATOR, RATIO DETECTOR).

A frequency-modulated signal should remain constant in amplitude, and should change only in frequency. The illustration shows the function of frequency versus time for a sine-wave frequency-modulated communications signal. Note that the function of frequency variation is a linear representation of the modulating waveform. That is, the rate of frequency change is directly proportional to the rate of change in the voltage of the modulating signal.

The phase of a signal is inherently related to its frequency. With frequency modulation, changes in phase are introduced. With changes in phase, instantaneous variations also occur in the signal frequency. Phase modulation can be used in place of frequency modulation in a communications system, with only minor differences. *See also* PHASE MODULATION.

FREQUENCY MULTIPLIER

A frequency multiplier is a device that produces an integral multiple, or many integral multiples, of a given input signal. A frequency multiplier circuit may also be called a harmonic generator.

The design of a frequency multiplier is straightforward. In order to cause the generation of harmonic energy, a nonlinear element must be placed in the path of the signal. A semiconductor diode is ideal for this purpose. (However, the type of diode must be chosen so that the capacitance is not too great.) Then, the signal waveform is distorted, and harmonics are produced. Following the nonlinear element, a tuned circuit may be used to select the particular harmonic desired. The tuned circuit is set to resonate at the harmonic frequency. The illustration is a simple schematic diagram of a frequency multiplier. There are other methods of obtaining frequency multiplication; odd multiples may be produced by means of a push-pull circuit, and even multiples by a push-push circuit (*see* PUSH-PULL CONFIGURATION, PUSH-PUSH CONFIGURATION).

Frequency multiplication does not change the characteristics of an amplitude-modulated or continuous-wave signal. However, the bandwidth of a single-sideband, frequency-shift-keyed, or frequency-modulated signal is multiplied by the same factor as the frequency. With frequency-shift keying or frequency modulation, it is possible to compensate for this effect, but with a single-sideband signal, severe distortion is introduced. Frequency multiplication increases the tuning range and the tuning rate of a variable-frequency oscillator by the factor of multiplication. Any signal instability, such as chirp or drift, is also multiplied. For this reason, frequency multiplication is generally limited to a factor of about four or less. *See also* DOUBLER, TRIPLER.

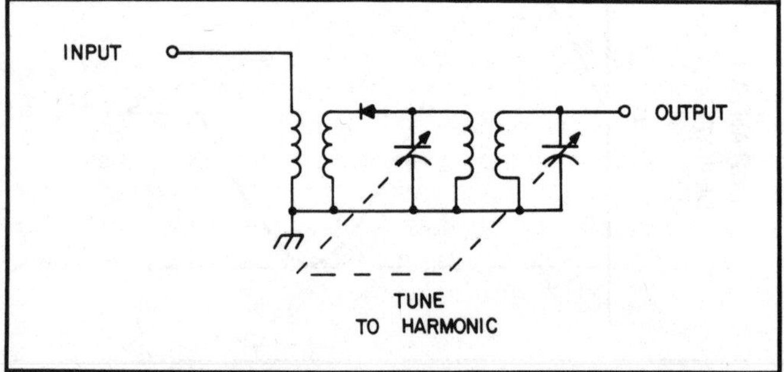

FREQUENCY MULTIPLIER: This circuit uses a single diode.

FREQUENCY OFFSET

In a communications transceiver, the frequency offset is the difference between the transmitter frequency and the receiver frequency. If the transmitter and the receiver are set to the same frequency, then the offset is zero, and the mode of operation is called simplex.

In continuous-wave (code) communications, the two stations usually set their transmitters to exactly the same frequency. This necessitates that their receivers be offset by several hundred Hertz, so that audible tones appear at the speakers. The receiver offset is obtained by a control called a clarifier or a receiver-incremental-tuning (RIT) control.

In transceivers designed for operation with repeaters, the transmitter frequency is significantly different from the receiver frequency. The offset may be several megahertz at the ultra-high-frequency range. This offset is needed because the repeater cannot transmit on the same frequency at which it receives. If the transmitter frequency of the transceiver is higher than the receiver frequency, the offset is called positive; if the transmitter frequency is lower, the offset is considered negative. For example, a typical amateur transceiver, operating on the 2-meter band, might be set to receive on 146.94 MHz and transmit on 146.34 MHz. The offset is thus −600 kHz. The repeater, of course, has an input or receiving frequency of 146.34 MHz, and an output or transmitting frequency of 146.94 MHz. *See also* RECEIVER INCREMENTAL TUNING, REPEATER.

FREQUENCY RESPONSE

Frequency response is a term used to define the perfor-

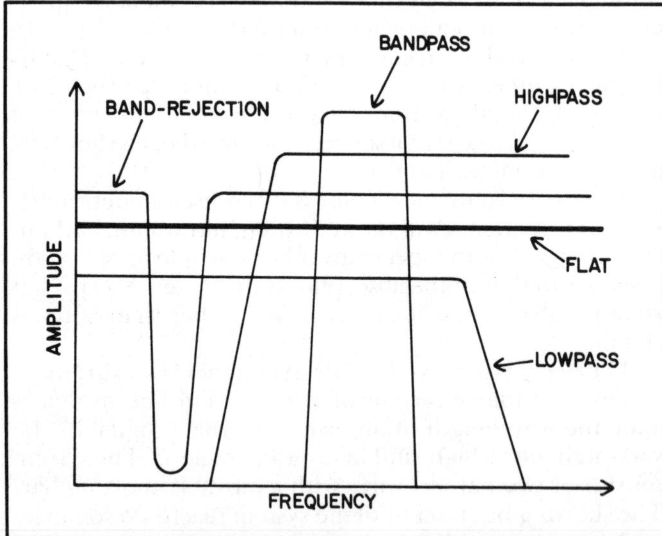

FREQUENCY RESPONSE: Five kinds of frequency response. (The zero-attenuation levels are different for clarity of illustration.)

mance or behavior of a filter, antenna system, microphone, speaker, or headphone. There are basically five kinds of frequency response: flat, bandpass, band-rejection, highpass, and lowpass. The attenuation-versus-frequency functions, in simple generalized form, for each type of response are shown in the illustration.

The frequency responses of some devices are more complicated than the elementary functions illustrated. There may be several "peaks" and "valleys" in the attenuation-versus-frequency functions of some systems. An example of this is the audio high-fidelity system employing a graphic equalizer (*see* EQUALIZER). The equalizer allows precise adjustment of the response.

Devices with various kinds of frequency response are useful in particular electronics applications. Bandpass filters are employed in radio-frequency transmitters and receivers. Lowpass filters are used to minimize the harmonic output from a transmitter. Highpass filters may be used to reduce the susceptibility of television receivers to interference by lower-frequency signals. Band-rejection filters are often used to attenuate a strong, undesired signal in the front end of a receiver. *See also* BANDPASS RESPONSE, BAND-REJECTION RESPONSE, FLAT RESPONSE, HIGHPASS RESPONSE, LOWPASS RESPONSE.

FREQUENCY SHIFT

A frequency shift is a change in the frequency of a signal, either intentional or unintentional. The term shift implies a sudden, or rapid, change in frequency.

The frequency of a transmitter or receiver may suddenly shift because of a change in the capacitance or inductance of the oscillator tuned circuit. The amount of the shift is given by the difference between the signal frequencies before and after the change. An increase in frequency represents a positive shift; a decrease is a negative shift.

Frequency shift is deliberately introduced into a carrier wave in all forms of frequency-modulation transmission. *See also* FREQUENCY MODULATION, FREQUENCY-SHIFT KEYING.

FREQUENCY-SHIFT KEYING

Frequency-shift keying is a digital mode of transmission, commonly used in radioteletype applications. In a continuous-wave or code transmission, the carrier is present only while the key is down, and the transmitter is turned off while the key is up. In frequency-shift keying, however, the carrier is present all the time. When the key is down, the carrier frequency changes by a predetermined amount. The carrier frequency under key-up conditions is called the space frequency. The carrier frequency under key-down conditions is called the mark frequency (see illustration).

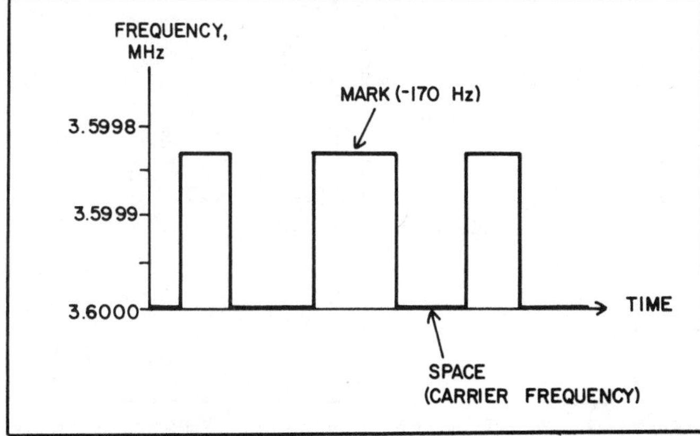

FREQUENCY-SHIFT KEYING: An example of frequency-shift keying, as a function of frequency versus time. The space frequency is considered to be the carrier frequency. The mark frequency is down-shifted, in this case, by 170 Hz.

In high-frequency radioteletype communications, the space frequency is usually higher than the mark frequency by a value ranging from 100 Hz to 900 Hz. The most common values of frequency shift are 170 Hz, 425 Hz, and 850 Hz. In recent years, because of the improvement in selective-filter technology, the narrower shift values have become increasingly common.

Frequency-shift keying is considered a form of frequency modulation, and is designated type F1 emission. Frequency-shift keying can be obtained in two ways. The first, and obvious, method is the introduction of reactance into the tuned circuit of the oscillator in a transmitter. The other method uses an audio tone generator with variable frequency, the output of which is fed into the microphone input of a single-sideband transmitter. If this method is used, precautions must be taken to ensure that there is no noise in the injected signal, for such noise will appear on the transmitted signal.

In some radioteletype links, especially at the very-high frequencies and above, audio frequency-shift keying is used. This is called F2 emission. An audio tone generator with variable frequency is coupled to the microphone input of an amplitude-modulated or frequency-modulated transmitter. The tone frequencies are between about 1000 Hz and 3000 Hz. This mode of frequency-shift keying requires considerably greater bandwidth than F1 emission. Its primary advantage is that the receiver frequency setting is not critical, whereas with F1 emission, even a slight error in the receiver setting will seriously degrade the reception.

Frequency-shift keying provides a greater degree of transmission accuracy than is possible with an ordinary continuous-wave system. This is because the key-up condition, as well as the key-down condition, is positively indicated. The frequency-shift-keyed system is therefore less susceptible to interference from atmospheric or man-made sources. *See also* RADIOTELETYPE.

FREQUENCY SWING

See CARRIER SWING.

FREQUENCY SYNTHESIZER

A frequency synthesizer is a device that generates highly precise frequency signals by means of a single crystal oscillator in conjunction with frequency dividers and multipliers. Frequency synthesizers are often used in modern communications equipment in place of the older variable-frequency oscillators.

Frequency synthesizers are digital devices, and they are characterized by discrete steps in frequency rather than adjustability over a continuous range. The steps may be rather large or very small, depending upon the application in which the device is to be used. The crystal oscillator may be set to zero beat with a standard frequency source, greatly increasing the accuracy of the frequency output. Phase-locked-loop circuits are used for the purpose of maintaining the accuracy of the frequency output of a synthesizer. *See also* PHASE-LOCKED LOOP.

FREQUENCY TOLERANCE

Frequency tolerance is a means of expressing the accuracy of a signal generator. The more stable an oscillator, the narrower the frequency tolerance. Tolerances are always expressed as a percentage of the fundamental quantity under consideration; in this case, it is the frequency of the oscillator.

Suppose, for example, that a 1-MHz oscillator has a frequency tolerance of plus or minus 1 percent. Since 1 percent of 1 MHz is 10 kHz, this means that the oscillator can be expected to be within 10 kHz of its specified frequency of 1 MHz. This is a range of 0.990 to 1.010 MHz.

Frequency-tolerance specifications are often given for communications receivers and transmitters. The tolerances may be given as a percentage, or as an actual maximum expected frequency error. The frequency tolerance required for one communications service may be much different from that needed for another service. *See also* TOLERANCE.

FRONT END

Front end is another name for the first radio-frequency amplifier stage in a receiver. The front end is one of the most important parts of any receiver, because the sensitivity of the front end dictates the sensitivity of the entire receiver. A low-noise, high-gain front end provides a good signal for the rest of the receiver circuitry. A poor front end results in a high amount of noise or distortion.

The front end of a receiver becomes more important as the frequency gets higher. At low and medium frequencies, there is a great deal of atmospheric noise, and the design of a front-end circuit is rather simple. But at frequencies above 20 or 30 MHz, the amount of atmospheric noise becomes small, and the main factor that limits the sensitivity of a receiver is the noise generated within the receiver itself. Noise generated in the front end of a receiver is amplified by all of the succeeding stages, so it is most important that the front end generate very little noise.

In recent years, improved solid-state devices, especially field-effect transistors, have paved the way for excellent receiver front-end designs. As a result of this, receiver sensitivity today is vastly superior to that of just a few years ago.

Another major consideration in receiver front-end design is low distortion. The front-end amplifier must be as linear as possible; the greater the degree of nonlinearity, the more susceptible the front end is to the generation of mixing products. Mixing products result from the heterodyning, or beating, of two or more signals in the circuits of a receiver; this causes false signals to appear in the output of the receiver.

The front end of a receiver should be capable of handling very strong as well as very weak signals, without nonlinear operation. The ability of a front end to maintain its linear characteristics in the presence of strong signals is called the dynamic range of the circuit. If a receiver has a front end that cannot handle strong signals, desensitization will occur when a strong signal is present at the antenna terminals. Mixing products may also appear because the front end is driven into nonlinear operation. *See also* DESENSITIZATION, INTERMODULATION, NOISE FIGURE, SENSITIVITY.

FRONT-TO-BACK RATIO

Front-to-back ratio is one expression of the directivity of an antenna system. The term applies only to antennas that are unidirectional, or directive in predominantly one direction. Such antennas include (among others) the Yagi, the quad, and the parabolic-dish antenna.

Mathematically, the front-to-back ratio is the ratio, in decibels, of the field strength in the favored direction to the field strength exactly opposite the favored direction, given an equal distance (see illustration). In terms of power, if P is the power gain of an antenna in the forward direction and Q is the power gain in the reverse direction, both expressed in decibels with respect to a dipole or isotropic radiator, then the front-to-back ratio is given simply by $P - Q$.

The other primary consideration in antenna directive behavior is the forward gain, or power gain. Power gain is more important in terms of transmitting; the front-to-back ratio is of more interest in terms of receiving directivity. *See also* ANTENNA POWER GAIN, dBd, dBi.

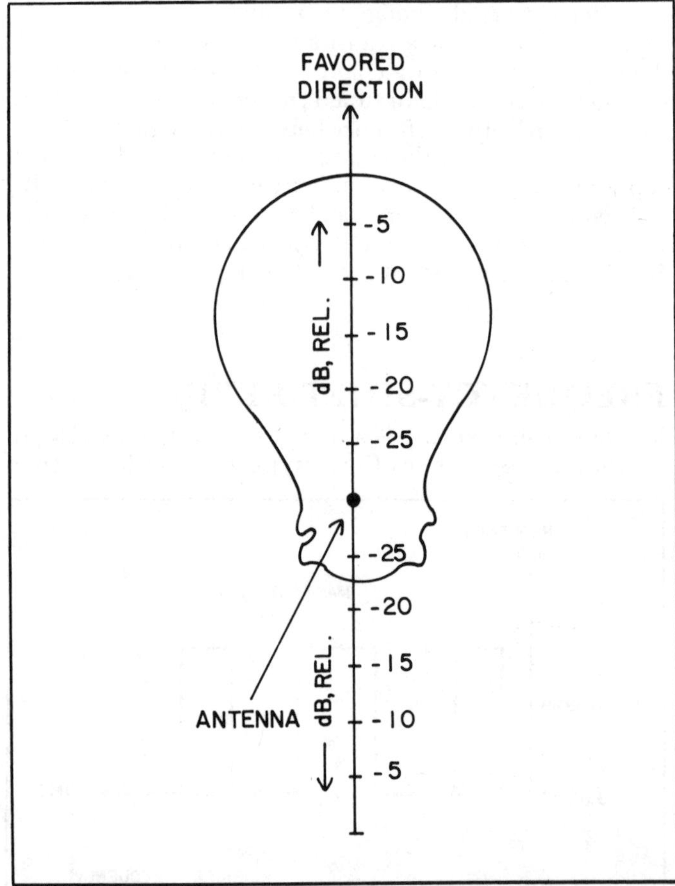

FRONT-TO-BACK RATIO: An example of the radiation pattern of a directional antenna. The relative gain in decibels is shown, with the maximum gain set at 0 dB. The gain in the direction opposite the favored direction is −23 dB, indicating that the front-to-back ratio of this antenna is 23 dB.

F SCAN

In radar, an F scan is a form of display used for the purpose of aiming. The F scan looks very much like the view through a gun sight, except that it is electronic rather than visual. The target appears at, or near, the center of the display. The horizontal and vertical scales represent aiming error. When the aim is accurate, the target blip is at the center of the display. The illustration shows a typical F-scan display.

The F scan may be used not only with radar devices, but with visual apparatus, infrared detectors, and even ultraviolet or X-ray devices. Aiming can be accomplished by automatic means; sensing circuits can orient the apparatus so that the blip appears at the center of the F-scan display. *See also* RADAR.

FUCHS ANTENNA

A Fuchs antenna is the simplest possible form of antenna. It consists simply of a length of wire, one end of which is attached to the transmitter output terminals, and the other end of which is left free. An antenna coupler is used if the antenna is not an odd multiple of ¼ wavelength at the operating frequency.

The Fuchs antenna is also called a random-wire antenna, and is used primarily in portable or emergency installations in the high and very-high frequency ranges. The main advantage of the Fuchs antenna is that it is simple to install, and works relatively well. There is no transmission line, and thus there is no loss in the line. However, since the radiating part of the antenna is immediately adjacent to the transmitter, electromagnetic interference can sometimes be a problem when this kind of antenna is used. A good ground system is essential when the Fuchs antenna is employed. If possible, the antenna feed point should be removed at least a few feet from the transmitter, to minimize the possibility of electromagnetic-interference effects. *See also* END FEED.

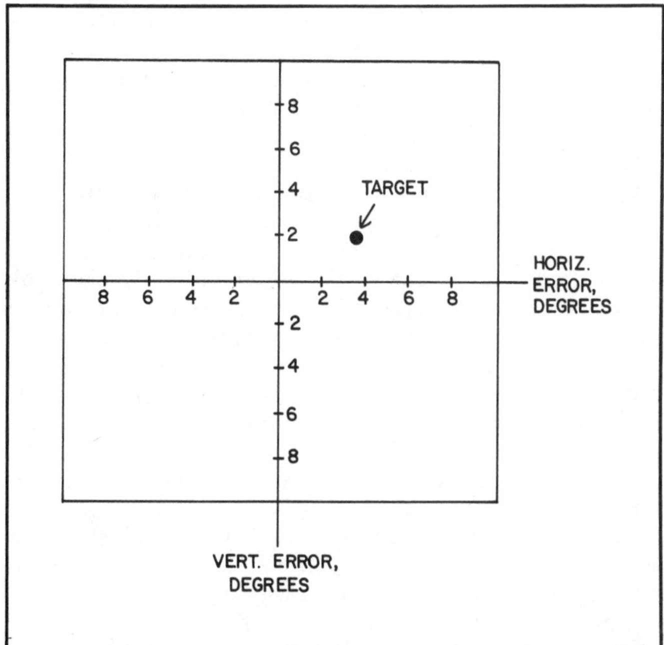

F SCAN: An F-scan radar display resembles the view through a gun sight.

FULL-SCALE ERROR

The full-scale error is a means of determining the accuracy of an analog metering device. Often, the accuracy specification of such a meter is given in terms of a percentage of the full-scale reading.

As an example, suppose that a milliammeter, with a range of 0 to 1 mA, is stated as having a maximum error of plus or minus 10 percent of full scale. This means that, if the meter reads 1 mA (or full scale), the actual current might be as small as 0.9 mA or as large as 1.1 mA. If the meter reads half scale, or 0.5 mA, the actual current might be as small as 0.4 mA or as large as 0.6 mA. In other words, the maximum possible error is always plus or minus 0.1 mA, regardless of the meter reading.

Meters are specified for full-scale readings because that is one point where all errors can conveniently be summed and compensated. Errors introduced by resistive components in the circuit are constant percentages, regardless of the current flowing through them. Errors introduced by the meter movement are not necessarily linear; the springs that return the pointer to its zero position produce nonlinear forces which can cause an error at midscale to be significantly different from the error at full scale. Calibration of all the errors in a meter at many points on the scale would be costly and time-consuming. *See also* ANALOG METERING.

FULL-WAVE RECTIFIER

A full-wave rectifier is a circuit that converts alternating current into pulsating direct current. The full-wave rectifier operates on both halves of the cycle. One half of the wave cycle is inverted, while the other is left as it is. The input and output waveforms of a full-wave rectifier in the case of a sine-wave alternating-current power source are illustrated at A.

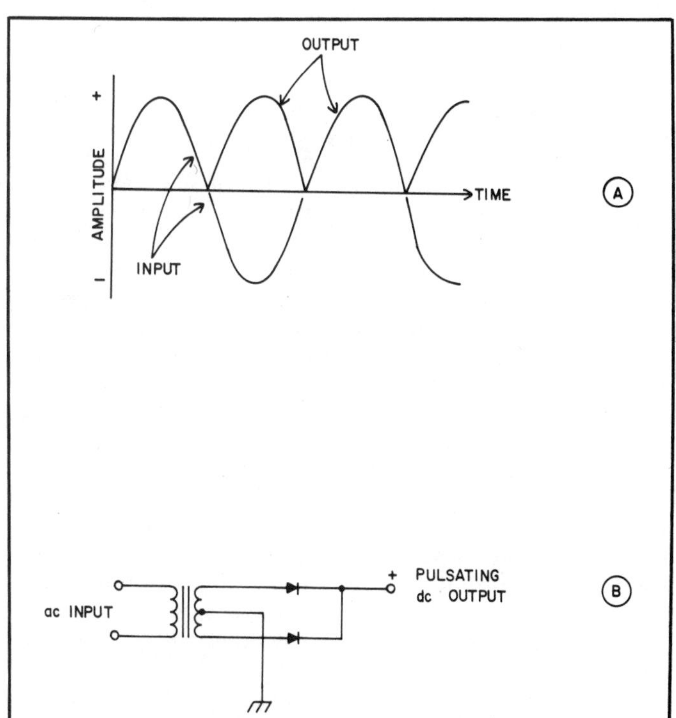

FULL-WAVE RECTIFIER: At A, the alternating-current input and the pulsating direct-current output of a full-wave rectifier circuit. At B, a typical full-wave rectifier circuit.

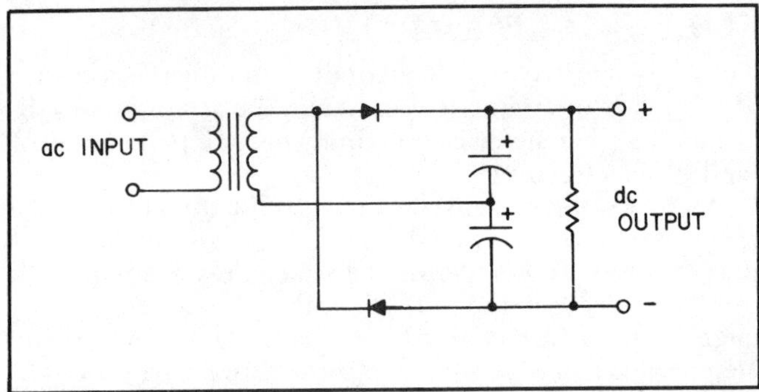

FULL-WAVE VOLTAGE DOUBLER: A full-wave voltage-doubler power supply. The output voltage is twice the voltage appearing across either capacitor. This circuit is essentially two half-wave rectifiers operating in series.

A typical full-wave rectifier circuit is shown at B. The transformer secondary winding is center-tapped, and the center tap is connected to ground. The opposite sides of the secondary winding are out of phase. Two diodes, usually semiconductor types, allow the current to flow either into or out of either end of the winding, depending on whether the rectifier is designed to supply a positive or a negative voltage. A form of full-wave rectifier that does not require a center-tapped secondary winding, but necessitates the use of four diodes rather than two, is called a bridge rectifier (*see* BRIDGE RECTIFIER).

The primary advantage of a full-wave rectifier circuit over a half-wave rectifier circuit is that the output of a full-wave circuit is easier to filter. The voltage regulation also tends to be better. The ripple frequency at the output of a full-wave rectifier circuit over a half-wave rectifier circuit is that the output of a full-wave circuit is easier to filter. The voltage regulation also tends to be better. The ripple frequency at the output of a full-wave rectifier circuit is twice the frequency of the alternating-current input. *See also* HALF-WAVE RECTIFIER, RECTIFICATION.

FULL-WAVE VOLTAGE DOUBLER

A full-wave voltage doubler is a form of full-wave power supply. The direct-current output voltage is approximately twice the alternating-current input voltage. Voltage-doubler power supplies are useful because smaller transformers can be used to obtain a given voltage, as compared with ordinary half-wave of full-wave supplies. Medium-power and low-power vacuum-tube circuits sometimes have voltage-doubler supplies.

See the schematic diagram of a full-wave voltage doubler. During one half of the cycle, the top capacitor charges to the peak value of the alternating-current input voltage; during the other half of the cycle, the other capacitor charges to the peak of the alternating-current input voltage, but in the opposite direction. Both capacitors hold their charge throughout the input cycle, resulting in the effective connection of two half-wave, direction-current power supplies in series, with twice the voltage of either supply.

Full-wave voltage doublers are not generally suitable for applications in which large amounts of current are drawn. The regulation and filtering tends to be rather poor in such cases. *See also* HALF-WAVE RECTIFIER, RECTIFICATION.

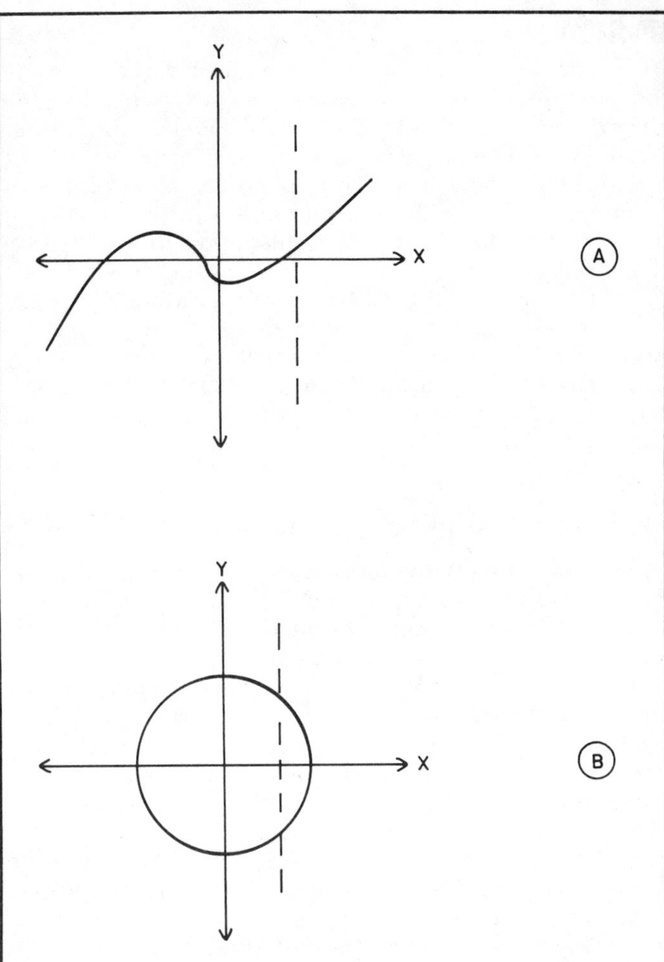

FUNCTION: At A, a mathematical function has at most one y value for each x value. At B, a relation that is not a function. Some values of x have two corresponding y values. A function may have, and often does have, x values for which there are no y values.

FUNCTION

A function is a form of mathematical relation. The function assigns particular values to a variable or set of variables. The value of the function is called the dependent variable because it depends on the other variable or variables. The variables that affect the value of the function are called the independent variables. A function may be considered an operation, which assigns, or "maps," the constituents of one set into those of another set.

For a relation to be a function, a special property must be satisfied: Each value of the independent variable or variables must correspond to at most one value of the dependent variable. In other words, each point in the domain must have no more than one constituent in the range. The illustration shows a function of one variable, with the independent variable on the horizontal or x axis, and the dependent variable on the vertical or y axis (A). Note that it is possible to draw a vertical line anywhere in this graph, and never does it intersect the function at more than one point. A relation that is not a function is shown at B; some of the values of the independent variable correspond to more than one value of the dependent variable.

Functions are generally written as a letter of the alphabet, followed by the independent variables in parentheses. For example, we might say that $f(x) = 2x + 3$, or $g(x,y,z) = x + y + z$. Sometimes, functions are written in simplified form, such as $y = 2x + 3$ or $x + y + z = w$. *See also* DEPENDENT VARIABLE, DOMAIN, INDEPENDENT VARIABLE, RANGE, RELATION.

FUNCTION GENERATOR

A function generator is a signal generator that can produce various different waveforms. All electrical waveforms can be expressed as mathematical functions of time; for example, the instantaneous amplitude of a sine wave can be expressed in the form $f(t) = a \sin(bt)$, where a is a constant that determines the peak amplitude, and b is a constant that determines the frequency. Square waves, sawtooth waves, and all other periodic disturbances can be expressed as mathematical functions of time, although the functions are quite complicated in some cases (*see* FUNCTION).

Most function generators can produce sine waves, sawtooth waves, and square waves. Some can also produce sequences of pulses. More sophisticated function generators can create a large variety of different waveforms are used for testing purposes in the design, troubleshooting, and alignment of electronic apparatus such as audio amplifiers and digital circuits.

For audio frequencies, a special sort of function generator is used by numerous modern musical groups. This is the Moog synthesizer. The Moog can duplicate the sound of almost any musical instrument, as well as create entirely unique sounds. *See also* MOOG SYNTHESIZER, SAWTOOTH WAVE, SINE WAVE, SQUARE WAVE.

FUNCTION TABLE

A function table is a list of the values of a mathematical function at discrete points within its domain. Functions can be represented by means of graphs, or written in mathematical form, or listed as a table. Each representation has its advantages and disadvantages. The mathematical expression of a function always allows the determination of its precise value, but it is necessary to repeat the calculation whenever a new value is needed. A graph allows instant determination of the value of the function for any value of the independent variables, but there is always some error in reading the graph. A table gives very accurate representations for discrete values in the domain, but it is often necessary to interpolate to obtain values not in the table.

A simple representation of the function $f(x) = 3x + 4$ is given in the table. Since thid particular function is a linear function, the tabular representations can be interpolated to obtain exact values for the function. In the case of a nonlinear function, interpolation does not yield exact values, but the accuracy is good if the representaions are at closely spaced intervals.

Function tables are frequently seen in electronics, because they provide convenient and accurate means for determining the values of functions at a glance. *See also* FUNCTION.

FUNCTION TABLE: TABULAR REPRESENTATION FOR $f(x) = 3x + 4$, OVER THE DOMAIN OF VALUES 0 TO 4.

x	f(x)
0.0	4.0
0.5	5.5
1.0	7.0
1.5	8.5
2.0	10.0
2.5	11.5
3.0	13.0
3.5	14.5
4.0	16.0

FUNDAMENTAL FREQUENCY

The fundamental frequency of an oscillator, antenna, or tuned circuit is the primary frequency at which the device is resonant. All waveforms or resonant circuits have a fundamental frequency, and integral multiples of the fundamental frequency, called harmonics. A theoretically pure sine wave contains energy at only its fundamental frequency. But in practice, sine waves are always slightly impure or distorted, and contain some harmonic energy. Complex waves, such as the sawtooth or square wave, often contain large amounts of harmonic energy in addition to the fundamental.

The amplitude of the fundamental frequency is used as a reference standard for the determination of harmonic suppression. The degree of suppression of a particular harmonic is expressed in decibels with respect to the level of the fundamental. The fundamental frequency need not necessarily be the frequency at which the most energy is concentrated; a harmonic may have greater amplitude. The fundamental also might not be the desired frequency in a particular application; harmonics are often deliberately generated and amplified. *See also* HARMONIC, HARMONIC SUPPRESSION.

FUNDAMENTAL SUPPRESSION

In the measurement of the total harmonic distortion in an amplifier circuit, the fundamental frequency must be eliminated, so that only the harmonic energy remains. This is called fundamental suppression. Fundamental suppression is generally done by means of a trap circuit, or band-rejection filter, centered at the fundamental frequency (*see* TOTAL HARMONIC DISTORTION).

In a frequency multiplier, the fundamental frequency is suppressed at the output while one or more harmonics are amplified. In such circuits, the fundamental frequency is not wanted. Band-rejection filters may be used in the output of the frequency multiplier to attenuate the fundamental signal. Highpass filters may also be used. Bandpass filters, centered at the frequency of the desired harmonic, also provide fundamental suppression. *See also* FREQUENCY MULTIPLIER, HARMONIC.

FUSE

A fuse is a device used in power supplies and power sources for electronic equipment. The fuse protects the components of the power supply, as well as the components of the equipment, against possible damage in the event of a malfunction of either the supply or the equipment. A fuse is a simple device, consisting of a thin piece of wire in an enclosure. The thickness of the wire determines the current level at which the fuse will open when the fuse is connected in series with the power-supply leads.

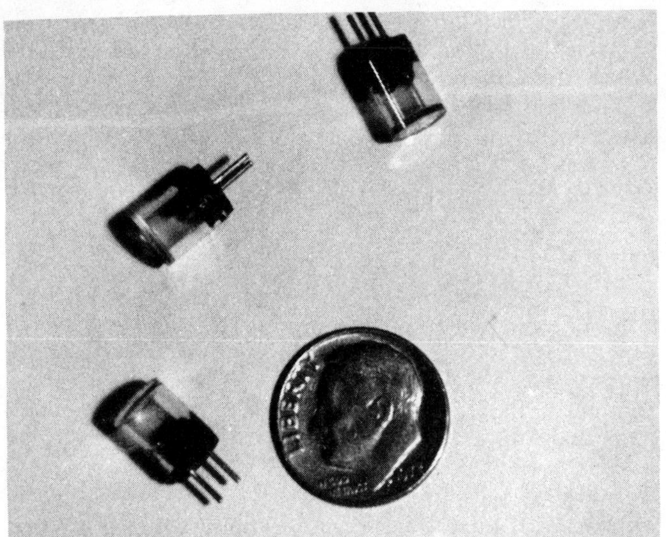

FUSE: Miniature fuses, used for protecting the sensitive input circuits of a communications monitor.

Fuses are commonly found in household electrical systems. Electronic devices such as communications receivers and transmitters have their own independent fuses. A fuse may be as tiny as those shown in the photograph—about the size of a pencil eraser—or they may be extremely large and bulky.

Fuses can be obtained with a variety of time constants. Some fuses blow out almost instantly when the current exceeds the rated value. Others have a built-in delay, which allows a brief period of excessive current without blowing the fuse. The latter type of fuse is known as a slow-blow fuse.

The main disadvantage of fuses is that they must be replaced when they blow out. This is not always convenient. Extra fuses must be kept on hand in case of an accidental short circuit or other temporary malfunction that causes a fuse to blow. Circuit breakers, which may be reset an indefinite number of times without replacement, are becoming more and more common. But there are some situations where fuses are still used because they offer better circuit protection. *See* CIRCUIT BREAKER.

FUSION

See NUCLEAR FUSION.

GAGE

See GAUGE.

GAIN

Gain is a term that describes the extent of an increase in current, voltage, or power. The gain of an amplifier circuit is an expression of the ratio between the amplitudes of the input signal and the output signal. In an antenna system, gain is a measure of the effective radiated power compared to the effective radiated power of some reference antenna (*see* ANTENNA POWER GAIN). Gain is almost always given in decibels.

In terms of voltage, if E_{IN} is the input voltage and E_{OUT} is the output voltage in an amplifier circuit, then the gain G, in decibels, is given by:

$$G = 20 \log_{10} (E_{OUT}/E_{IN})$$

The current gain of an amplifier having an input current of I_{IN} and an output current of I_{OUT} is given in decibels by:

$$G = 20 \log_{10} (I_{OUT}/I_{IN})$$

In a power amplifier, where P_{IN} is the input power and P_{OUT} is the output power, the power gain G, in decibels, is:

$$G = 10 \log_{10} (P_{OUT}/P_{IN})$$

When the gain of a particular circuit is negative, there is insertion loss. Some types of circuits, such as the follower, exhibit insertion loss. For typical radio-frequency amplifiers, the gain ranges from about 15 to 25 dB per stage of amplification. Sometimes the gain may be as high as 30 dB per stage, but this is rare. *See also* AMPLIFICATION FACTOR, DECIBEL, INSERTION LOSS.

GAIN CONTROL

A gain control is an adjustable control, such as a potentiometer, that changes the gain of an amplifier circuit. The volume control in every radio receiver or audio device is a form of gain control. Communications equipment generally has an audio-frequency gain control and a radio-frequency gain control.

A gain control is fairly simple to incorporate into an amplifier circuit. A potentiometer may be used to provide varying amounts of drive to an amplifier circuit; this should be done at a low power level. The illustration shows a typical means of obtaining variable gain in an audio amplifier. The position of the potentiometer determines the proportion of the input that is applied to the amplifier circuit.

A gain control should not affect the linearity of the amplifier circuit in which it is placed. Therefore, changing the bias on the transistor or tube in an amplifier is not an acceptable way of obtaining gain control. The gain control should be chosen so that succeeding stages cannot be overdriven if the gain is set at maximum. The gain control should have sufficient range so that, when the gain is set at minimum, the output of the amplifier chain is reduced essentially to zero. *See also* GAIN.

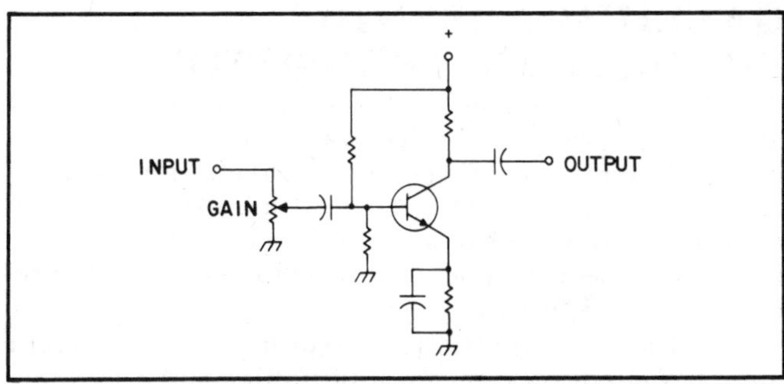

GAIN CONTROL: A gain control in the input of an audio amplifier, allowing adjustment from zero to full volume. The potentiometer setting does not affect the bias of the amplifier circuit, but provides a constant input impedance.

GALACTIC NOISE

All celestial objects radiate a certain amount of energy at radio wavelengths. This is especially true of stars and galaxies, and the diffuse matter among the stars.

Galactic noise was first observed by Karl Jansky, a physicist working for the Bell Telephone Laboratories, in the 1930s. It was an accidental discovery. Jansky was investigating the nature of atmospheric noise at a wavelength of about 15 m, or 20 MHz. Jansky's antenna consisted of a rotatable array, not much larger than the Yagi antennas commonly used by radio amateurs at the wavelength of 15 m.

Most of the noise from our galaxy comes from the direction of the galactic center. This is not surprising, since that is where most of the stars are concentrated. Galactic noise contributes, along with noise from the sun, the planet Jupiter, and a few other celestial objects, to most of the cosmic radio noise arriving at the surface of the earth. Other galaxies radiate noise, but since the external galaxies are much farther away from us than the center of our own galaxy, sophisticated equipment is needed to detect the noise from them. *See also* RADIO ASTRONOMY, RADIO TELESCOPE.

GALENA

Galena is a compound consisting of lead and sulfur, with a chemical formula PbS. In nature, galena occurs in the form of silver-gray crystals. The crystals have a roughly cubical shape.

Galena was used in the earliest semiconductor detectors. By placing a fine wire, called a cat's whisker, at the right point of the surface of a crystal of galena, rectification could be obtained at radio frequencies. The galena detector was a rather finicky sort of device, and experimentation was usually necessary in order to obtain good performance.

It is an interesting experiment to build a simple crystal set, using only a variable capacitor, a home-wound coil, an earphone, an antenna wire, and a piece of galena with a cat's whisker! Such a primitive receiver can actually pick up strong radio signals in the AM broadcast band, with only the antenna current as a power source. *See also* CAT'S WHISKER, CRYSTAL SET.

GALLIUM-ARSENIDE FIELD-EFFECT TRANSISTOR

The field-effect transistor is well known as a low-noise amplifying device. Field-effect transistors are available in many different forms. One type of field effect transistor is fabricated from gallium-arsenide (GaAs) semiconductor material, and is called a gallium-arsenide field-effect transistor. The nickname for this device is GaAsFET, pronounced just as it looks.

The GaAsFET is ideal for use in high-gain, low-noise amplifier circuits, especially at ultra-high and microwave frequencies. Gallium arsenide (an N-type material) has higher carrier mobility than the germanium or silicon usually found in semiconductor devices. Carrier mobility is the speed with which the charge carriers—either electrons or holes—move within a substance with a given electric-field intensity (*see* CARRIER MOBILITY). The GaAsFET is a depletion-mode field-effect transistor. That is, it normally conducts, and the application of bias to the gate electrode reduces the conductivity of the channel.

The GaAsFET can be used in receiver preamplifiers, converters, and in power amplifiers up to several watts. They are generally N-channel devices. *See also* DEPLETION MODE, FIELD-EFFECT TRANSISTOR.

GALVANISM

The production of electric current by chemical action is called galvanism. The word comes from the name of an eighteenth-century scientist, Luigi Galvani. The principles of galvanism are used in common energy cells such as the dry cell and the storage battery. Galvanism also causes certain metals to corrode, or react with other substances to form compounds.

When two dissimilar metals are brought into contact with each other in the presence of an electrolyte, they may act like a small cell, and a voltage difference will be produced between the metals. Salt in the air, or even slightly impure water vapor, can act as the electrolyte. The metals gradually corrode because of the chemical reaction; this is called galvanic corrosion.

Electrolytic action can be used to coat iron or steel with a thin layer of zinc. This is called galvanizing. Galvanized metal is more resistant to corrosion than bare metal, although in a marine or tropical environment, even galvanized iron or steel will eventually corrode. *See also* CELL, ELECTROLYSIS.

GALVANOMETER

A galvanometer is a sensitive device for detecting the presence of, and measuring, electric currents. The galvanometer is similar to an ammeter, but the needle of the galvanometer normally rests in the center position. Current flowing in one direction results in a deflection of the needle to the right; current in the other direction causes the needle to move to the left (see illustration).

The galvanometer is especially useful as a null indicator in the operation of various bridge circuits, since currents in both directions can be detected. An ordinary meter in such a situation will read zero when a reverse current is present, thus making an accurate null indication impossible to obtain. *See also* AMMETER.

GAMMA MATCH

A gamma match is a device for coupling a coaxial transmission line to a balanced antenna element, such as a half-wave dipole. The radiating element consists of a single conductor. The shield part of the coaxial cable is connected to the radiating element at the center. The center conductor is attached to a rod or wire running parallel to the radiator, toward one end. The illustration shows the gamma match. The device gets its name from its resemblance in shape to the upper-case Greek letter gamma.

The impedance-matching ratio of the gamma match depends on the physical dimensions: the spacing between the matching rod and the radiator, the length of the matching rod, and the relative diameters of the radiator and the matching rod. The gamma match allows the low feed-point impedance of a Yagi antenna driven element to be precisely matched with the 50-ohm or 75-ohm charac-

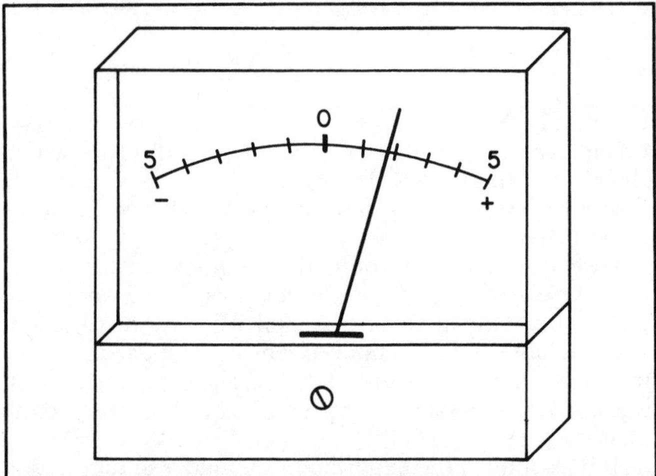

GALVANOMETER: The galvanometer registers not only current intensity, but current direction. This characteristic is useful in bridge circuits, since it allows precise adjustment for a null reading.

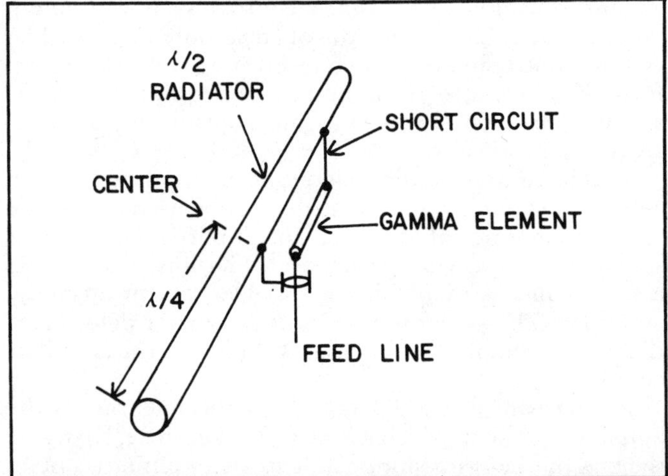

GAMMA MATCH: A gamma match is used to provide a precise impedance match between a coaxial feed line and a straight, half-wave radiating element. The impedance-transfer ratio is variable, and depends on the dimensions of the paralleled gamma element.

teristic impedance of a coaxial feed line. The gamma match also acts as a balun, which makes the balanced radiator compatible with the unbalanced feed line. Gamma matching is often seen in multielement parasitic arrays, especially rotatable arrays. *See also* BALUN, IMPEDANCE MATCHING, YAGI ANTENNA.

GAMMA RAY

Gamma ray is a term used to describe electromagnetic energy at extremely short wavelengths. Gamma rays have the shortest wavelengths of any known form of radiant energy. The longest gamma rays are perhaps 0.1 Angstrom unit, or 10^{-11} meter, in length. At longer wavelengths, the rays are called X rays. The demarcation between the X-ray and gamma-ray ranges is, however, not universally standardized. The shortest gamma rays have wavelengths on the order of 10^{-5} Angstrom units, although there is theoretically no limit to how short the rays can be.

Gamma rays are emitted by certain radioactive substances. They are also generated in atomic reactions, such as in a particle accelerator. High-speed atomic particles from space, called cosmic particles, strike the atoms of the upper atmosphere and produce other particles along with gamma rays.

Gamma radiation is extremely penetrating. The most energetic gamma rays can easily pass through concrete. For shielding against gamma rays, lead is usually necessary. Atomic reactors are placed deep under water to prevent the gamma radiation from escaping. Gamma rays cause damage to living tissue. An overexposure to gamma radiation can result in radiation sickness and even death. Gamma radiation is measured in units called roentgens. *See also* ALPHA PARTICLE, BETA PARTICLE, COUNTER TUBE, DOSIMETRY, GEIGER COUNTER, ROENTGEN, X RAY.

GANG CAPACITOR

Variable capacitors are sometimes connected together in parallel, on a common shaft. Such a combination capacitor is called a gang capacitor. The individual capacitors rotate together. The sections may or may not have the same values of capacitance.

Gang capacitors are used in circuits having more than one adjustable resonant stage, and a requirement for tuning each resonant stage along with the others. This is often the case in communications equipment. The photograph shows a typical gang capacitor. This unit is in the final-amplifier section of a communications transmitter.

The rotor plates of the gang capacitor are usually all connected to the frame of the unit, and the frame is in turn put at common ground. Gang capacitors with electrically separate sets of rotor plates are rare. *See also* AIR-VARIABLE CAPACITOR, GANGED CONTROLS.

GANGED CONTROLS

When several adjustable components, such as potentiometers, switches, variable capacitors, or variable inductors are connected in tandem, the controls are said to be ganged. Ganged controls are frequently found in electronic devices, because it is often necessary to adjust several circuits at the same time.

An example of the use of a ganged control is the volume

GANG CAPACITOR: This three-gang capacitor forms the tuning control for an RF amplifier.

adjustment in a stereo high-fidelity system. There are two channels, and usually a single volume control. This requires a two-gang potentiometer. (Some systems have a separate volume control for each channel.) Ganged rotary switches are almost always found in multiband communications receivers and transmitters, because of the need to switch several stages of amplification simultaneously. When several tuned circuits must be adjusted together, a gang of variable capacitors is employed. *See also* GANG CAPACITOR.

GAP

See AIR GAP.

GAS TUBE

A gas tube is an electron tube in which a gas is deliberately introduced. When the voltage across the electrodes of a gas-filled tube reaches a certain value, the gas ionizes, and the tube conducts. Below this voltage, the tube does not conduct. This results in a constant internal voltage drop, which makes the gas tube useful as a rectifier and voltage regulator.

Gas-filled tubes can be classified as mercury-vapor rectifiers, voltage-regulator tubes, and thyratron tubes. The mercury-vapor rectifier is capable of handling high voltages and currents. Xenon gas may sometimes be used in place of mercury vapor to obtain satisfactory operation over a wider temperature range.

Voltage-regulator tubes make use of the constant internal voltage drop in the gas-filled tube. The larger the current through the tube, the lower the resistance. The tube is connected in parallel with the source of voltage to be regulated.

The thyratron tube employs a grid between the cathode and the plate, for the purpose of controlling the voltage at which the gas will ionize. Once the gas in the thyratron has ionized, the voltage on the grid no longer affects the conduction of the tube, and it continues to conduct.

In modern electronic circuits, gas tubes are not generally found. This is because solid-state devices have largely replaced tubes. *See also* MERCURY-VAPOR RECTIFIER, RECTIFIER TUBE, THYRATRON, TUBE, VOLTAGE REGULATION.

GATE

The gate of a field-effect transistor is the electrode that allows control of the current through the channel of the device. The gate is analogous to the base of a bipolar transistor, or the grid of a vacuum tube. In an N-channel junction field-effect transistor (JFET), the gate voltage is usually negative with respect to the source voltage. In the P-channel device, the gate is biased positively.

In the JFET, the gate consists of two sections of P-type semiconductor material in the N-channel device, placed on either side of the channel. Alternatively, the gate may have the configuration of a cylinder of P-type material around the channel. In the P-channel device, of course, the gate is fabricated from N-type semiconductor material.

The greater the reverse bias applied to the gate, the less the channel of the JFET conducts, until, at a certain value of reverse bias, the channel does not conduct at all. This condition is called pinchoff. The pinchoff voltage depends on the voltage between the source and the drain of the field-effect transistor.

In the insulated-gate (metal-oxide-semiconductor) field-effect transistor or MOSFET, the gate operates differently. There are two basic modes of operation, known as depletion mode and enhancement mode (*see* DEPLETION MODE, ENHANCEMENT MODE, METAL-OXIDE-SEMICONDUCTOR FIELD-EFFECT TRANSISTOR). A MOSFET may have more than one gate.

Small fluctuations in the gate voltage result in large changes in the current through the field-effect transistor. This is the reason that amplification is possible with such devices. *See also* FIELD-EFFECT TRANSISTOR.

In digital logic, the term gate refers to a switching device with one or more inputs, and one output. The gate performs a logical function, depending on the states of the inputs. The three basic logic gates are the inverter, or NOT gate, the AND gate, and the OR gate. *See also* LOGIC GATE.

GATE CIRCUIT

In a field-effect-transistor amplifier, the gate circuit is the combination of components associated with the gate of the device. Usually, the input to a field-effect-transistor amplifier is applied to the gate circuit.

In digital electronics, a gate circuit is a combination of logic gates, designed for the purpose of performing a certain function. Many different logical operations can be carried out by combining the simple NOT, AND, and OR gates. In fact, all logical switching arrangements can theoretically be realized by various combinations of logic gates. *See also* GATE, LOGIC GATE.

GATED-BEAM DETECTOR

See QUADRATURE DETECTOR.

GATED-BEAM TUBE

A gated-beam tube is a vacuum tube used in a gated-beam or quadrature detector (*see* QUADRATURE DETECTOR). The gated-beam has three grids, and is thus a pentode. As the electrons travel from the cathode to the plate, they first encounter the signal or limiter grid. The frequency-modulated signal is applied to this grid. The accelerator grid comes next; it imparts extra speed to the electrons. Finally, the electrons encounter the quadrature grid. Here, a signal is applied that is 0 degrees out of phase with

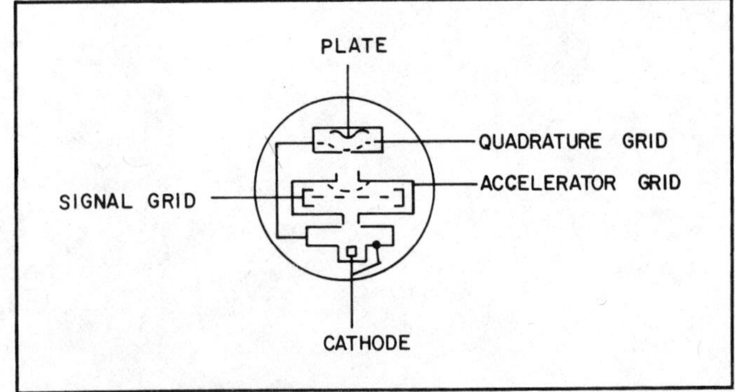

GATED-BEAM TUBE: The gated-beam tube, used in a frequency-modulation quadrature detector.

the wave at the signal grid.

The drawing illustrates the internal details of the gated-beam tube. The tube is used for detection of frequency-modulated or phase-modulated signals. *See also* FREQUENCY MODULATION, PHASE MODULATION.

GATE-DIP METER

A gate-dip meter is a field-effect-transistor device which operates on the same principle as a grid-dip meter. The gate-dip and grid-dip meters are used for the purpose of determining the resonant frequency of a tuned circuit or antenna.

The gate-dip meter has the obvious advantage of requiring less power than a grid-dip meter. There is no filament, and the necessary operating voltage is much smaller. The characteristics of the gate-dip meter are, however, identical to those of the grid-dip meter in practical applications. *See also* GRID-DIP METER.

GATE-PROTECTED MOSFET

A gate-protected MOSFET is a type of metal-oxide semiconductor field-effect transistor, or MOSFET. The gate electrode in a MOSFET is susceptible to breakdown in the presence of static electricity. This can be such a problem that handling the devices can be risky. The gate-protected MOSFET has built-in Zener diodes connected in reverse parallel, which prevent the accumulation of charges of over a few volts. *See also* METAL-OXIDE SEMICONDUCTOR FIELD-EFFECT TRANSISTOR.

GAUGE

A gauge is a device that measures a certain quantity, such as current, voltage, power, or frequency. All meters are examples of gauges. Gauges usually are found in the form of a meter type device.

The diameter of an electric wire is sometimes specified in terms of the gauge, which is a number assigned to indicate the approximate size of a conductor. In the United States, the American Wire Gauge (AWG) designator is most commonly used for this purpose. There are numerous other wire gauges. The term gauge is also sometimes used in reference to the thickness of a piece of sheet metal.

GAUSS

The gauss is a unit of magnetic flux density. A field strength of 1 gauss is considered to be 1 line per square centimeter of surface perpendicular to the flux lines in a magnetic field. The tesla is the other major unit of magnetic flux density; 1 tesla is equal to 10,000 gauss. *See also* MAGNETIC FLUX, TESLA.

GAUSSIAN DISTRIBUTION

See NORMAL DISTRIBUTION.

GAUSSIAN FUNCTION

A Gaussian function is a mathematical function used in the design of filters. A Gaussian filter passes a pulse with essentially no overshoot, but the rise time is extremely rapid.

In order to pass a square pulse with maximum effectiveness, a filter must have a nearly flat response, with skirts of optimum steepness. *See also* BANDPASS FILTER, BANDPASS RESPONSE, BESSEL FUNCTION.

GAUSS, KARL FRIEDERICH

One of the most famous pioneers in electrostatic theory, Karl Friederich Gauss (1777-1855) is best known for his proof of the inverse-square law for electric fields. Gauss also experimented and theorized on the relationships between electricity and magnetism. He was one of the top mathematicians of his time. *See also* GAUSS' THEOREM.

GAUSS'S THEOREM

Gauss's Theorem is an expression for determining the intensity of an electric field, depending on the quantity of charge.

For any closed surface in the presence of an electric field, the electric flux passing through that surface is directly proportional to the enclosed quantity of charge. *See also* ELECTRIC FIELD.

GEIGER COUNTER

The Geiger counter is an instrument for measuring the intensity of high-energy radiation, such as alpha particles, beta particles, X rays, and gamma rays.

The heart of the Geiger counter is the counter tube (*See* COUNTER TUBE). When a high-energy particle or photon enters the tube, the gas in the tube is ionized, causing a brief pulse of current to flow. This current pulse is amplified by means of a transistorized circuit. Then the pulse is fed to a digital counter, an audio amplifier, or a meter. A typical Geiger counter can register up to approximately 2,000 pulses per second. *See also* ALPHA PARTICLE, BETA PARTICLE, GAMMA RAY, X RAY.

GENERATOR

A generator is a source of signal in an electronic circuit. Such a device may be an oscillator, or it may be an electromechanical circuit. Signal generators are widely used in electronic design, testing, and troubleshooting (*See* SIGNAL GENERATOR).

A device for producing alternating-current electricity, by means of a rotating coil and magnetic field, is called a generator. Whenever a conductor moves within a magnetic field, so that the conductor cuts across magnetic lines of force, current is induced in the conductor. A generator may consist of either a rotating magnet inside a coil of wire, or a rotating coil of wire inside a magnet. The illustration shows the principle of an alternating-current generator. The shaft is driven by a motor powered by gasoline or some other fossil fuel; alternatively, steam turbines may be used.

Small portable gasoline-powered generators, capable of delivering from about 1 kW to 5 kW, can be purchased in department stores. Larger generators can supply enough electricity to run an entire house or building. The largest electric generators, found in power plants, are as large as a

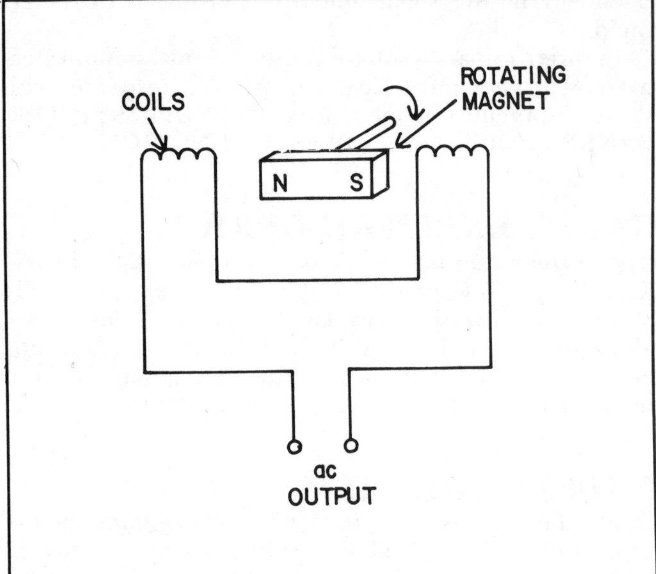

GENERATOR: A simple generator can consist of a horseshoe magnet, rotated within a coil of wire.

small building, produce millions of watts, and can provide sufficient electricity for a community.

A generator, like a motor, is an electromechanical transducer. The construction of an electric generator is, in fact, almost identical to that of a motor; some motors can operate as generators when their shafts are turned by an external force. *See also* MOTOR.

GEOMAGNETIC FIELD

The geomagnetic field is a magnetic field surrounding our planet. The earth is magnetized like a huge bar magnet. The north magnetic pole is located near the north geographic pole, and the south magnetic pole near the south geographic pole. The magnetic lines of force extend far out into space (see illustration). The earth is not the only planet that has a magnetic field. Other planets, especially Jupiter, as well as stars, such as the sun, are known to have magnetic fields.

The magnetic field of the earth is responsible for the behavior of the magnetic compass. The needle of a compass is actually a small bar magnet, suspended on a bearing that allows free rotation. One pole of the bar magnet points south, and the other pole points north, in most locations on the surface of the earth. There is a slight error because of the difference in location between the magnetic and geographic poles.

The geomagnetic field attracts charged particles emitted by the sun, and causes them to be concentrated near the magnetic poles. When there is a solar flare, resulting in large amounts of charged-particle emission from the sun, the upper atmosphere glows in the vicinity of the magnetic poles. This is known as the Aurora. The appearance of the Aurora heralds a disturbance in the magnetic environment of the earth. This is called a geomagnetic storm. *See also* AURORA, GEOMAGNETIC STORM, MAGNETIC FIELD.

GEOMAGNETIC STORM

A geomagnetic storm is a disturbance, or fluctuation, in the magnetic field of the earth. Such a "storm" is caused by

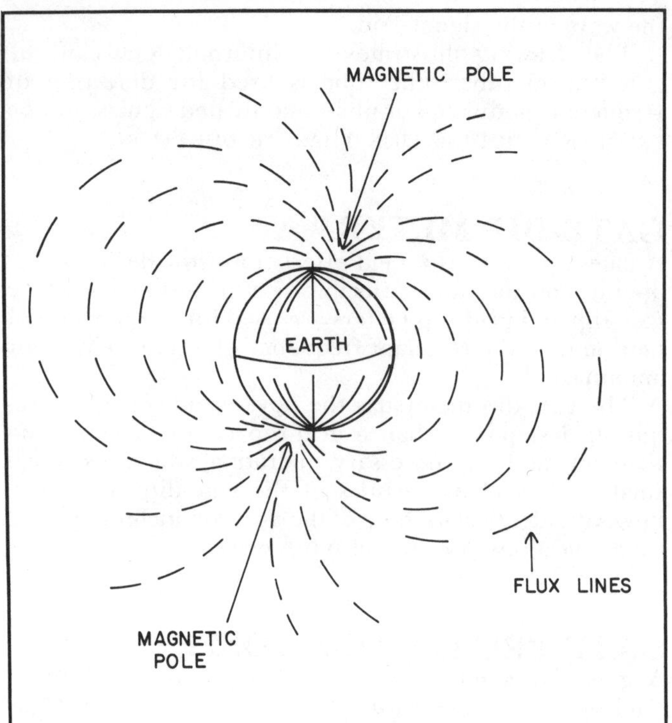

GEOMAGNETIC FIELD: The geomagnetic lines of force resemble the magnetic flux in the vicinity of a bar magnet.

charged particles from the sun.

A solar flare causes the emission of high-speed protons, electrons, and alpha particles, all of which carry an electric charge. Since these charges are in motion, they have an effective electric current, which is influenced by the magnetic field of the earth. These particles also create their own magnetic field as they move.

A geomagnetic storm changes the character of the ionosphere of the earth, with a profound impact on radio communications at the low, medium, and high frequencies. Often, communications circuits will be completely severed because of changes in the ionization in the upper atmosphere. Instead of returning signals to the earth, the ionosphere may absorb electromagnetic energy, or may allow it to pass into space.

To a certain extent, we can predict a geomagnetic storm. A solar flare can be seen with optical telescopes, and it is from one hour to several hours before the charged particles arrive at the earth. During the night, the Aurora can sometimes be seen during a geomagnetic storm. If a disturbance is especially severe, telephone circuits may be disrupted, as well as other wire services. *See also* AURORA, GEOMAGNETIC FIELD, IONOSPHERE, SOLAR FLARE.

GEOMETRIC MEAN

The geometric mean is a form of mathematical average of two or more numbers. In general, if there are n different numbers to be averaged according to the geometric mean, they are multiplied together, and then the nth root is taken.

The geometric mean is not the same as the arithmetic mean. When we think of the average value of a set of numbers, we normally think of the arithmetic mean. In a theoretical sense, however, the geometric mean is a more useful representation of the average value of a set of numbers in certain situations. *See also* ARITHMETIC MEAN.

GEOTHERMAL ENERGY

The supply of fossil-fuel energy, on which the industrialized nations of the world rely, will not last forever. Research has been continuing, in the last few years, toward the goal of reducing dependency on fossil fuels. Geothermal energy may provide an alternative.

The earth has a hot, semiliquid core. It is not necessary to descend very far into the crust of our planet before this fact becomes evident; great energy has been contained within the earth ever since it was formed. This is strikingly evident when volcanoes erupt. By digging shafts several thousand feet into the ground, the heat from within the earth might be harnessed for use by man. One method that has been suggested is to pump water or steam into a deep shaft, where it will be heated until its pressure increases. Then the steam could be used to drive turbines connected to electric generators.

Scientists have theorized that the use of this heat can help to solve long-term energy problems. The heat of the earth is called geothermal energy. The supply of geothermal energy is vast indeed, and perhaps it can be tapped with very little or no effect on the environment.

Other alternative energy sources include hydroelectric, nuclear, solar, and wind energy. *See also* HYDROELECTRIC POWER PLANT, NUCLEAR ENERGY, NUCLEAR FISSION, NUCLEAR FUSION, SOLAR POWER, WIND POWER.

GERMANIUM

Germanium is an element, with atomic number 32 and atomic weight 73. Germanium is a semiconductor material. In its pure form, it is an insulator; its conductivity depends on the amount of impurity elements added. It is used in the manufacture of some diodes, photocells, and transistors.

Germanium has been largely replaced by silicon in modern semiconductor devices. This is because germanium is sensitive to heat, and germanium devices can be destroyed by the soldering process used in the manufacture of electronic circuitry.

Germanium diodes are still occasionally found in electronic devices. The germanium diode has a forward voltage drop of approximately 0.3, compared with 0.6 for the silicon diode. Germanium diodes are employed primarily in radio-frequency applications. *See also* DIODE.

GETTER

A getter is a device used in the process of evacuating an electron tube. After the tube has been pumped out and sealed, a small amount of gas remains, since no evacuation mechanism works perfectly. A small electrode, the getter, is charged with a radio-frequency voltage. This causes the remaining gas in the tube to react with the electrode, and the vacuum within the tube becomes less contaminated.

The getter serves no purpose once the evacuation is complete. However, it can still be seen inside the tube. *See also* TUBE.

GHOST

A ghost is a false image in a television receiver, caused by reflection of a signal from some object just prior to its arrival at the antenna. The ghost appears in a position different from the actual picture because of the delay in propagation of the ghost signal with respect to the actual received signal.

Ghosts usually appear as slightly displaced images on a television screen. In especially severe cases, the ghost may be almost as prominent as the actual picture, which makes it difficult or impossible to view the television signal. Ghosting can often be minimized simply by reorienting an indoor television antenna, or rotating an outdoor antenna. In some cases, ghosts can be difficult to eliminate. The object responsible for the signal reflection must then be moved far away from the antenna. In certain cases, this is not practical.

With modern cable television, ghosting is seldom a problem. The signals are entirely confined to the cable, and propagation effects are unimportant. Ghosting can take place in a cable system, however, if impedance "bumps" are present in the line near the receiver. *See also* TELEVISION.

GIGA

Giga is a prefix meaning 1 billion (1,000,000,000). This prefix is attached to quantities to indicate the multiplication factor of 1 billion. For example, 1 gigohm is 1 billion ohms; 1 gigaelectronvolt is 1 billion electron volts; 1 gigahertz is 1 billion Hertz; and so on.

Giga is one of many different prefix multipliers commonly used in electronics. *See also* PREFIX MULTIPLIERS.

GILBERT

The gilbert is a unit of magnetomotive force, named after the scientist William Gilbert. The standard symbol for the gilbert is the upper-case Greek letter sigma (Σ).

In a coil, a magnetomotive force of 1 gilbert is equivalent to 1.26 ampere turns. In general, if N is the number of turns in a coil and I is the current, in amperes, that flows through the coil, then the magnetomotive force Σ in gilberts is equal to:

$$\Sigma = 1.26\ NI$$

See also MAGNETOMOTIVE FORCE.

GLASS FIBER

See FIBER, FIBER OPTICS.

GLASS INSULATOR

Glass is an excellent dielectric, or insulating, material, and has good mechanical strength as well. Thus, glass is sometimes used in the manufacture of insulators for electrical wiring and antenna installations.

Glass insulators are available in many different sizes and shapes. The surface of the insulator is generally ribbed to increase the amount of surface distance between the two ends. Glass insulators are somewhat less common than porcelain insulators. *See also* INSULATOR.

G-LINE

See SURFACE-WAVE TRANSMISSION LINE.

GLOW DISCHARGE

When an electric current is passed through a rarefied gas, the gas glows with a characteristic spectrum. Different gases appear different in color. This emission is called glow discharge.

When the glow-discharge spectrum is separated by means of a spectroscope, the energy can be observed to occur at discrete wavelengths, rather than over a continuous range. The discrete energy wavelengths are called emission lines. Each gas has its own set of emission lines, like a fingerprint that identifies it. Astronomers can recognize the emission lines of gases in stars, planets, and in interstellar space.

When a voltage is applied to a gas, causing the flow of current, electrons absorb energy from the voltage source. Some of the electrons fall back again to lower energy levels, and release discrete bursts of radiation in electromagnetic form. This can happen only in specific quantities, resulting in emission at defined wavelengths. Some emission lines are stronger than others; this is because some energy quantities are more common than others as the electrons fall back.

As seen with the eye, a mercury-vapor glow discharge looks bluish or bluish white. Sodium vapor has a candle-flame yellow color. Other gases have various colors from red to violet. Some of the emission lines are invisible, occurring at the infrared, ultraviolet, and even the X-ray or radio wavelengths. *See also* ELECTROMAGNETIC SPECTRUM.

GLOW LAMP

Any device that makes use of glow discharge for the purpose of generating visible light is called a glow lamp. Glow lamps are extremely common in modern applications. They last longer than incandescent lamps, and produce more light and less heat, resulting in greater energy efficiency.

Some glow lamps are comparatively dim, such as the neon lamps often seen in radio-equipment panels or in night lights. Other glow lamps can be very bright, such as the mercury-vapor and sodium-vapor street lights now in widespread use. Some glow lamps, such as the common fluorescent tube, produce ultraviolet radiation by the excitation of internal gases; this ultraviolet then strikes a phosphor coating on the inside of the lamp envelope to produce visible light. *See also* GLOW DISCHARGE.

GOLD

Gold is a metallic element, with atomic number 79 and atomic weight 197. We recognize this yellowish metal as an international money standard. It is also useful in some electronic applications.

Gold plating is occasionally used to reduce the electrical resistance of a set of switch or relay contacts. Gold is highly malleable, and can withstand the physical stress of repeated contact openings and closings. Also, gold is highly resistant to corrosion. Gold-plated contacts maintain their reliability over a long period of time without the need for frequent cleaning. Gold is a good conductor of electricity.

Gold is used in the manufacture of certain semiconductor diodes. A fine gold wire is bonded to a tiny wafer of germanium. This device is called a gold-bonded diode. Gold may be used as an impurity substance in the doping of semiconductor transistors. *See also* DOPING.

GOLDSCHMIDT ALTERNATOR

Radio-frequency energy can be generated by means of an alternator, similar to an electric generator, in conjunction with frequency multipliers and resonant circuits. Such a device, used in the early days of radio, is called a Goldschmidt alternator.

Alternators can directly produce energy only at relatively low radio frequencies, such as 100 Hz. But there is practically no limit to the factor by which radio-frequency energy can be multiplied. By using nonlinear componenets and resonant circuits, the Goldschmidt alternator is capable of producing energy well into the high-frequency range. In modern radio circuits, such generators are not used; vacuum tubes and, more recently, semiconductor devices are much preferred as oscillators and amplifiers. *See also* GENERATOR.

GONIOMETER

A goniometer is a direction-finding device commonly used in radiolocation and radionavigation. Many direction finders have rotatable antennas, with certain characteristic directional patterns. The goniometer uses a mechanically fixed antenna, and the directional response is varied electrically.

A simple pair of phased vertical antennas can serve as a goniometer (see illustration). The two antennas may be spaced at any distance between ¼ and ½ wavelength. Depending on the phase in which the antennas are fed, the null can occur in any compass direction. The null may be unidirectional in some situations, and bidirectional in other cases. The precise direction is determined by knowing the phase relationship of the two feed systems. *See also* DIRECTION FINDER, RADIOLOCATION, RADIONAVIGATION.

GRACEFUL DEGRADATION

When a portion of a computer system malfunctions, it is desirable to have the computer continue operating, even though the efficiency is impaired. The failure of a single diode should not result in the shutdown of an entire system. This property is called graceful degradation.

In the event of a subsystem malfunction, a sophisticated computer can use other circuits to temporarily accomplish the tasks of the failed portions of the system. The attendant personnel are notified by the computer that a malfunction has occurred, so that it may be corrected as soon as possible. Graceful degradation is sometimes called failsafe or failsoft. *See also* COMPUTER, FAILSAFE.

GRADED FILTER

A graded filter is a form of power-supply output filter, which provides various degrees of ripple elimination. Some circuits can tolerate more ripple than others. Those circuits that can operate with higher ripple levels are connected to earlier points in the filter sequence (see illustration). Some circuits need less power-supply filtering be-

GONIOMETER: Phase relationships between two antennas produce the directional response in the goniometer. These illustrations show quarter-wave verticals viewed from directly above, and separated by a quarter wavelength. At A, the antennas are in phase. At B, the antenna on the right is phased 90 degrees ahead of the antenna on the left. At C, the antenna on the right is phased 90 degrees behind the antenna on the left.

cause they draw very little current; these circuits are also connected to the earlier points in the graded filter.

The graded filter offers the advantage of superior ripple elimination and voltage regulation for those circuits that require the purest direct current. The current drain is reduced at the farthest point in the graded filter. *See also* POWER SUPPLY.

GRADED JUNCTION

A graded semiconductor junction is a P-N junction that is grown, by means of epitaxial methods, in a carefully controlled manner.

Graded-junction devices have reverse-bias characteristics that differ from ordinary semiconductor devices. The capacitance across the junction of a reverse-biased, graded-junction diode decreases more rapidly, with increasing reverse voltage, than the capacitance across an ordinary P-N junction. *See also* P-N JUNCTION.

GRAM ATOMIC VOLUME

The gram atomic volume of an element is the volume, usually expressed in cubic centimeters, occupied by 6.02×10^{23} atoms of the element in the solid state. Gram atomic volume is an expression of the relative size of an atom of a particular substance.

The gram atomic volume of an element does not, strangely enough, depend on the atomic number. The more complicated atoms do not necessarily take up more space, in the solid state, than the simpler atoms. For example, solid hydrogen, the simplest atom in the universe, has a gram atomic volume of 13 cubic centimeters. But gold,

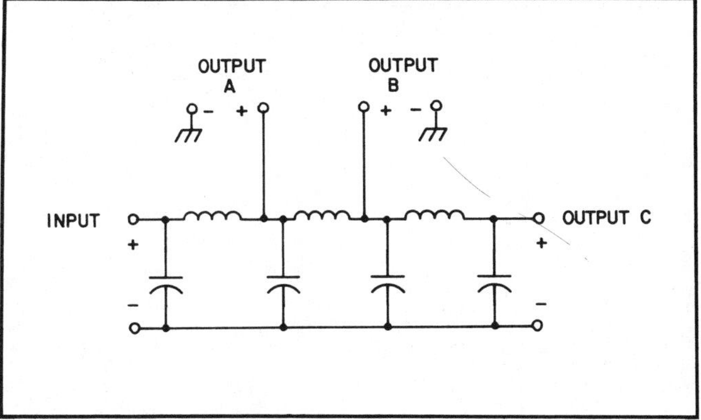

GRADED FILTER: A typical graded filter. Output A provides the least filtering, output B provides more, and C gives the greatest amount of filtering.

with an atomic number of 79, has a gram atomic volume of only 10 cubic centimeters. *See also* ATOM.

GRAM ATOMIC WEIGHT

The gram atomic weight of an element is the mass, in grams, of 1 mole (6.02×10^{23} atoms) of that element. The gram atomic weight is usually expressed as the atomic weight in element tables. Thus, when we say that the atomic weight of gold is 197, we know that 1 mole of gold will mass 197 grams.

In general, the gram atomic weight increases with the atomic number of an element, but there are some exceptions. Some elements can occur in different forms, which have slightly different gram atomic weights.

Scientists usually speak of "1 gram atomic weight" of an element, meaning a given mass of that element, corresponding in grams to its atomic weight. *See also* ATOM, AUTOMIC WEIGHT.

GRAPH

A graph is a means of illustrating a relation between two or more variable quantities. Graphs can be constructed in a variety of different ways, but the most common method is the Cartesian system (*see* CARTESIAN COORDINATES).

In the common graph, the controlled or independent variable is usually represented on the horizontal axis, and the dependent variable is represented on the vertical axis. The axis may or may not be calibrated in linear form.

The illustration shows a simple graph, depicting the current-versus-voltage function for a typical semiconductor diode. The independent variable is the voltage across the P-N junction. The dependent variable is the current through the device. *See also* CURVILINEAR COORDINATES, DEPENDENT VARIABLE, FUNCTION, INDEPENDENT VARIABLE, POLAR COORDINATES.

GRAPHICAL ANALYSIS

Graphical analysis is a method of evaluating the interaction among different quantities. A graphical representation, such as that in the previous illustration, is used as the basis for graphical analysis.

Suppose, for example, that we wish to know the avalanche voltage, the forward breakover voltage, and the

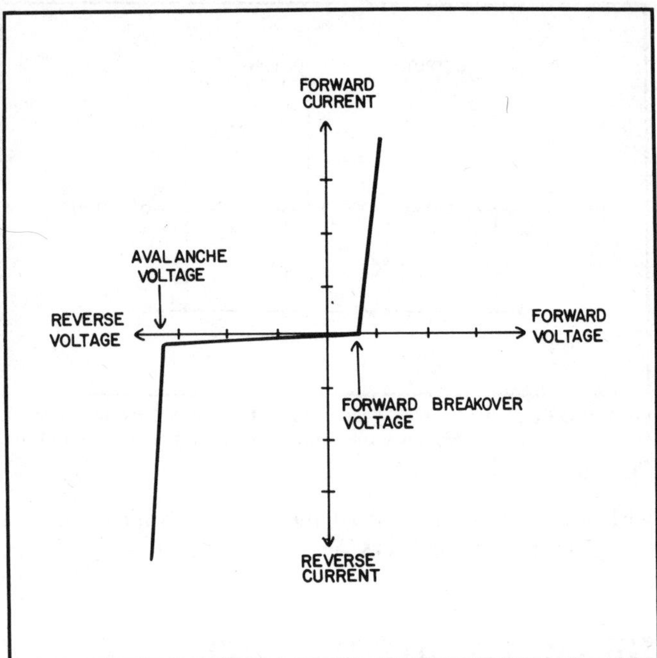

GRAPH: A graph, in this case showing the characteristics of a typical semiconductor diode.

rate of current increase with increasing forward voltage for a particular semiconductor diode. We can plot its current-versus-voltage curve on a Cartesian graph. The illustration is an example of the current-versus-voltage curve of a typical diode. The avalanche voltage can be clearly seen as the reverse voltage at which the current abruptly rises. The forward breakover voltage can also be clearly seen, as the forward voltage at which the current abruptly rises. The rate of current increase with increasing voltage, in the forward direction, can be determined by placing a straight edge tangent to the graph at the desired point. The slope of the line indicates the rate of change.

Graphical analysis is used in many different phases of electronic engineering. Antenna performance, transmission-line design, and the design of oscillators, amplifiers, and filters can all benefit from the use of graphical analysis. Modern computers can perform graphical analysis much more quickly and accurately than is possible with paper and pencil. *See also* GRAPH.

GRAPHIC EQUALIZER

See EQUALIZER.

GRAPHICS

Graphics is a general term that represents data in the form of charts and graphs (*see* GRAPH). Graphics facilitates the presentation and interpretation of complex information.

In a computer system, graphics is the capability of the computer to draw pictures. The pictures may, in a sophisticated computer, include depth and color. Computer graphics can provide enhanced representations of physical objects, bringing out certain features to be observed. The simplest graphics involve two-dimensional Cartesian plots. The most complex graphics may include such things as computer-enhanced color radar pictures and perspective drawings of intricate objects. Motion can sometimes be depicted in computer-graphics displays.

Graphics capability is commonplace today, and is available in many home computers as well as the more sophisticated business computers. Computer games are probably the most familiar form of home-computer graphics. Graphics systems almost always make use of video display monitors, which resemble television sets. *See also* VIDEO DISPLAY TERMINAL.

GRAPHITE

Graphite is a form of carbon, atomic number 12. Graphite is a soft, grayish-black substance. It is used in many different kinds of electronic components, especially resistors and the plates of certain vacuum tubes. Graphite is used in the manufacture of common pencils.

Graphite, because of its high tolerance for heat, is often combined with nonconducting material, such as clay, to make resistors. Most small noninductive resistors are of the graphite variety.

In power-amplifier vacuum tubes, the plates dissipate large amounts of heat. Graphite can withstand heat better than the metals normally used in vacuum-tube plates; graphite plates are found in thicknesses as great as several millimeters. The use of graphite plates increases the tube life. *See also* CARBON RESISTOR, TUBE.

GRATING

A grating, also called a diffraction grating, is a transparent piece of glass or plastic with many fine, straight, parallel dark lines. The diffraction grating acts like a prism for visible light. When a beam of light is passed through the grating, the beam is split into a spectrum of constituent wavelengths from red to violet. Some diffraction gratings allow the passage of infrared and ultraviolet wavelengths.

Gratings can be used at radio frequencies as well as at visible wavelengths. A set of parallel metal bars, or slots in a metal sheet, produce an interference pattern when radio waves pass through. If the bars or slots are spaced within certain limits, the radio waves are spread out into a spectrum. Frequency resolution can be obtained in this way; a movable detector can be set anywhere along the spectral distribution to choose a particular frequency.

Diffraction gratings are used in optical spectroscopes and other devices. *See also* DIFFRACTION, SPECTROSCOPE.

GREAT CIRCLE

A great circle is a path between two points over the surface of a sphere, such that the path forms part of a circle whose center is the same as the center of the sphere. A great-circle path lies in a plane that passes through the geometric center of the sphere (see illustration). On the earth, great-circle paths represent the shortest distance between two points. Great-circle paths are sometimes called geodesic paths. Examples of great circles on the earth include all longitude lines and the equator; but the latitude lines are not great circles.

When radio waves propagate around the world, they usually follow paths that are nearly great circles. For this reason, in high-frequency radio communications, it is often desirable to know the great-circle bearing toward a particular place. This is especially important if a direc-

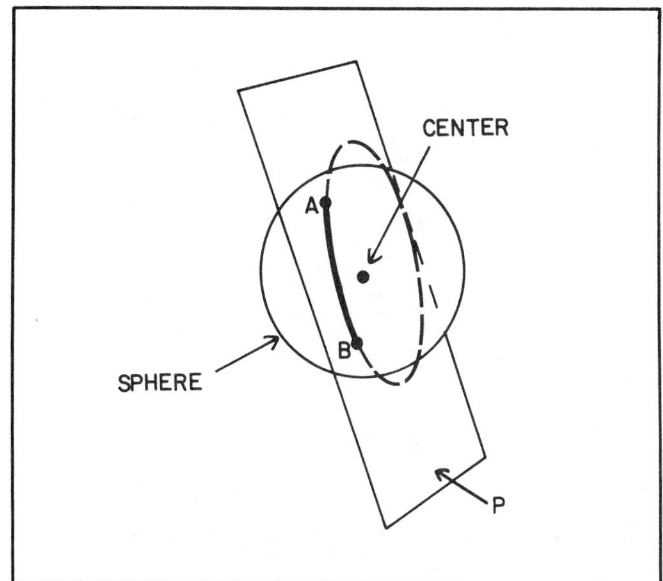

GREAT CIRCLE: On a spherical surface, a great circle is centered at the center of the sphere. Here, the shortest distance on the sphere between points A and B lies in the plane P.

tional antenna system is in use. The great-circle bearing can be determined by means of a great-circle map of the world. Such a map shows the geography of the earth as seen from a viewpoint that is centered on your location.

Great-circle bearings are often surprising. For example, South Africa lies almost directly east, along a great-circle path, from the center of the United States. India lies north, over the pole. Radio waves occasionally travel the long way around the world, rather than the short way; in such cases, the antenna should be pointed west for South Africa and south for India, from a receiving location in the center of the United States. *See also* LONG-PATH PROPAGATION, SHORT-PATH PROPAGATION.

GREENWICH MEAN TIME

Greenwich mean time, abbreviated GMT, is the time in the zone along the meridian of zero degrees, passing through Greenwich, England. Greenwich mean time is based on the position of this meridian with respect to the sun. At 1200 GMT, the Greenwich meridian lines up precisely with the sun. At 0600 and 1800 GMT, the Greenwich meridian is 90 degrees (¼ rotation) from the sun. At 0000 GMT, the Greenwich meridian lies exactly opposite the sun. The 24-hour time scale is always specified for GMT.

Greenwich mean time used to be the standard for time throughout the world. In recent years, however, GMT has been replaced with coordinated universal time (UTC). The two are the same for all practical purposes. *See also* COORDINATED UNIVERSAL TIME.

GREMLIN

Gremlin is a term sometimes used by technicians to refer to a continuing problem that is difficult or impossible to track down. The expression originated among aircraft personnel. A gremlin is a little devil, wandering around the apparatus, on which the problem can be blamed!

Gremlins are a frustratingly common phenomenon in electronics. Every technician or engineer knows how gremlins behave. *See also* TROUBLESHOOTING.

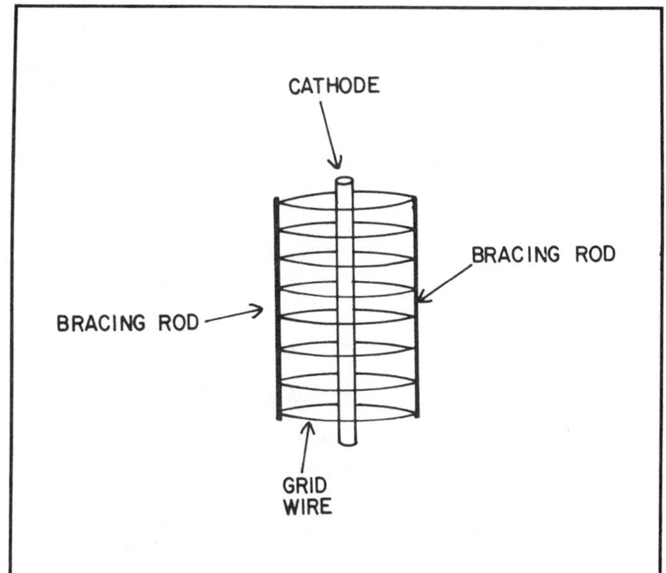

GRID: The grid in a tube surrounds the cathode.

GRID

A grid is an element in a vacuum tube, placed between the cathode, which emits electrons, and the plate, which collects electrons. The grid resembles a screen, and surrounds the cathode in the shape of a cylinder or oblate cylinder (see illustration). A vacuum tube may have two, three, or more concentric grids.

The function of the grid is to control the current through the tube. When the grid is at the same potential as the cathode, a certain amount of current will flow from the cathode to the plate. As the grid voltage is made more negative, less current flows through the tube. If the grid is made positive, it draws some current away from the path between the cathode and the plate. When a rapidly varying signal voltage is applied to the grid of a tube, the current in the plate circuit fluctuates along with the signal. When this current is put through a load, the resulting voltage fluctuations are often much larger than those of the original input signal; this is how amplification is obtained.

The innermost grid in a tube is called the control grid. The second grid is called the screen; the third grid is called the suppressor. All have different functions. The input signal is usually applied to the control grid. *See also* CONTROL GRID, SCREEN GRID, SUPPRESSOR GRID, TUBE.

GRID BIAS

In a vacuum tube, the grid bias is the direct-current voltage applied to the control grid. The grid bias is usually measured with respect to the cathode. The grid bias may be obtained in a variety of different ways. Three different methods are shown in the illustration. At A, a resistor and capacitor are used to elevate the cathode circuit above direct-current ground, generating negative grid bias. At B, a separate power supply is used for obtaining the negative grid bias. At C, the grid bias is zero.

Most vacuum tubes require a negative grid bias, the

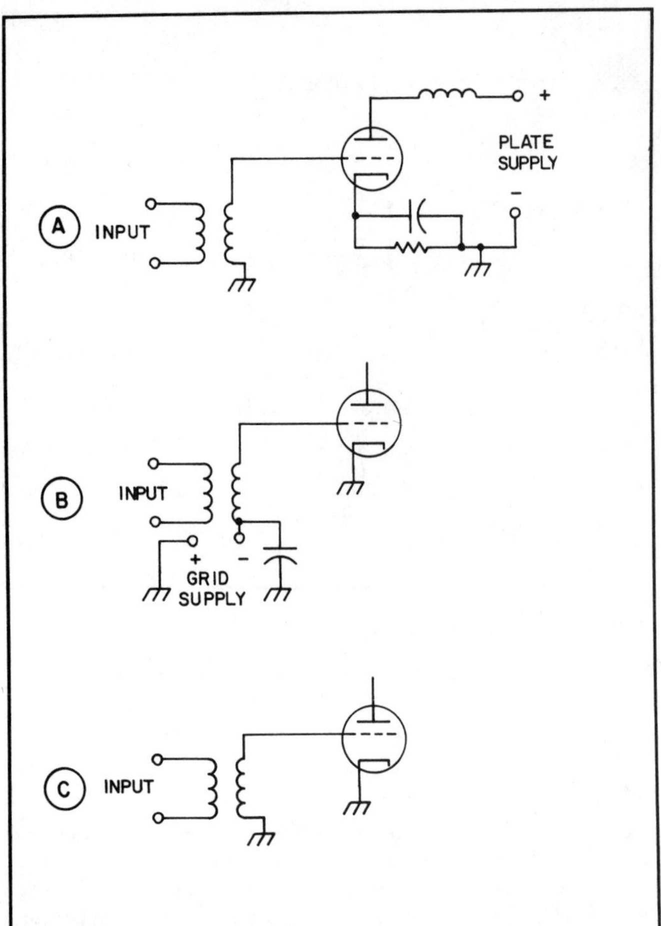

GRID BIAS: Three methods of obtaining grid bias. At A, the cathode is made slightly positive by means of a resistor and capacitor. At B, a separate negative power supply is used. At C, a zero-bias tube needs no grid bias.

exact value depending on the characteristics of the tube and the application for which the tube is to be used. Some tubes operate with zero bias.

As the grid bias is made progressively more negative, the plate current decreases until, at a certain point, the tube stops conducting altogether. This value of grid bias is called the cutoff bias, and it depends on the plate voltage as well as the particular type of tube used. If the grid bias is made positive, the grid will draw current away from the plate circuit. An input signal may drive the control grid positive for part of the cycle, but deliberate positive bias is almost never placed on the control grid of a tube.

The level of grid bias must be properly set for a tube to function according to its specifications in a given application. The required grid bias is different among Class-A, Class-AB, Class-B, and Class-C amplifiers. *See also* CLASS-A AMPLIFIER, CLASS-AB AMPLIFIER, CLASS-B AMPLIFIER, CLASS-C AMPLIFIER, GRID, TUBE.

GRID-BLOCK KEYING

Grid-block keying, also sometimes called blocked-grid keying, is a form of amplifier keying. A large negative voltage is applied to the grid of one of the low-power amplifier stages of a transmitter during the key-up condition. This cuts the tube off, preventing the input signal from reaching the output. The illustration is a schematic diagram of a typical tube amplifier using grid-block keying.

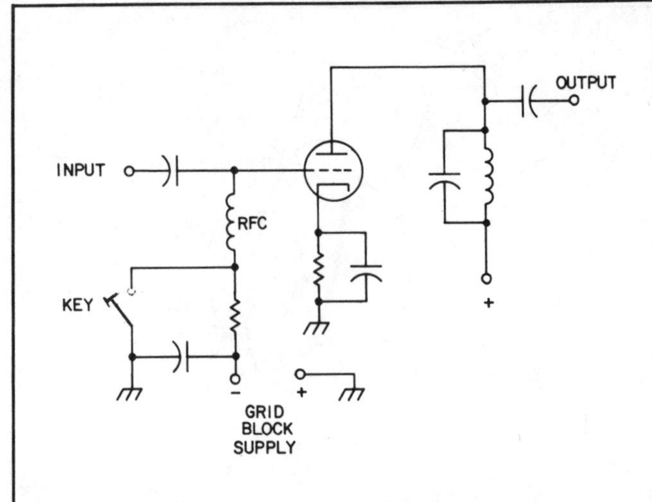

GRID-BLOCK KEYING: An example of grid-block keying.

When the key is pressed, the negative cutoff voltage is short-circuited, and the grid attains normal operating bias. Then, the tube amplifies normally, and the input signal is allowed to reach the following stages.

Grid-block keying offers the advantage of relatively low-current switching, and moderate voltages. Grid-block keying also allows the keying of a transmitter with minimal impedance change at the oscillator stage. This helps to prevent chirping. *See also* GRID BIAS, KEYING.

GRID-DIP METER

A grid-dip meter is a device that is used for determining the resonant frequency of a tuned circuit or antenna. The heart of the grid-dip meter is a vacuum-tube oscillator. The tuned circuit of the oscillator contains a large, solenoidal inductor which allows coupling to external circuits.

As the frequency of the oscillator is tuned back and forth, with the coil near the resonant circuit to be tested, a sharp dip occurs in the grid current, indicated by the meter, when the resonant frequency of the external circuit is reached. The external circuit may have several resonant frequencies; at each, the grid current dips sharply. The resonant frequency is read from a calibrated dial on the grid-dip oscillator.

Grid-dip meters usually have several different plug-in coils, so that the frequency range is maximized and the calibration is fairly accurate. The illustration is a schematic diagram of a typical grid-dip meter.

A field-effect transistor may be used in place of the vacuum tube in a grid-dip meter. The field-effect transistor does not need a source of power for a filament, and is considerably lighter and more compact. This kind of meter is called a gate-dip meter. *See also* GATE-DIP METER.

GRID-LEAK DETECTOR

Grid-leak detection is a form of detection used in some vacuum-tube amplifier circuits. It results in envelope detection, of the kind normally used for amplitude-modulated signals. The grid-leak detector makes use of

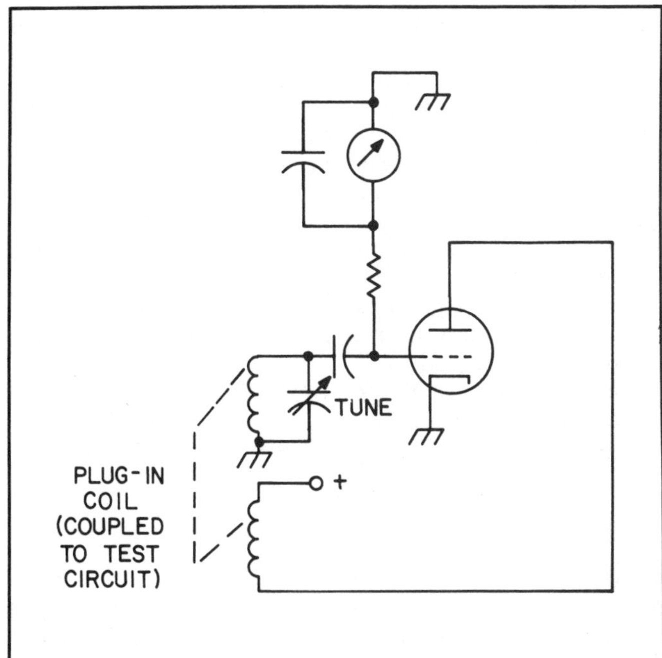

GRID-DIP METER: A simple grid-dip meter.

the diode action between the cathode and the control grid of a vacuum tube.

The diagram illustrates a typical grid-leak detector. The circuit is very simple. A resistor, with a high value, is placed in series with the control grid and bypassed with a small capacitor. A large direct-current voltage drop develops across the resistor in the presence of an input signal.

In modern receivers, tubes are generally not used, and semiconductor-diode circuits can easily perform the action of envelope detection for amplitude-modulated signals. Therefore, grid-leak detectors are rarely seen today, except in very old tube type radio receivers. *See also* DETECTION, ENVELOPE DETECTOR.

GRID MODULATION

Grid modulation is a method of obtaining amplitude modulation in a vacuum-tube amplifier circuit. Grid modulation requires very little audio power to achieve 100-percent modulation.

The drawing is a schematic diagram of an amplifier with grid modulation. The audio signal is simply applied to the control grid, along with the carrier signal. Alternatively, the audio signal may be applied to the screen grid of a tetrode or pentode tube. The output is amplitude-modulated.

The principal disadvantage of grid modulation is that all of the amplifying stages following the modulated stage must be linear. This means that they must be Class-A, Class-AB, or Class-B amplifiers. The efficient Class-C amplifier will cause distortion of an amplitude-modulated signal.

In some amplitude-modulated transmitters, a Class-C final amplifier stage is used, and the audio signal is applied to the plate circuit of this stage. A large amount of audio signal power is necessary in order to achieve 100-percent modulation in this form of circuit. This method is called plate modulation. *See also* AMPLITUDE MODULATION, PLATE MODULATION.

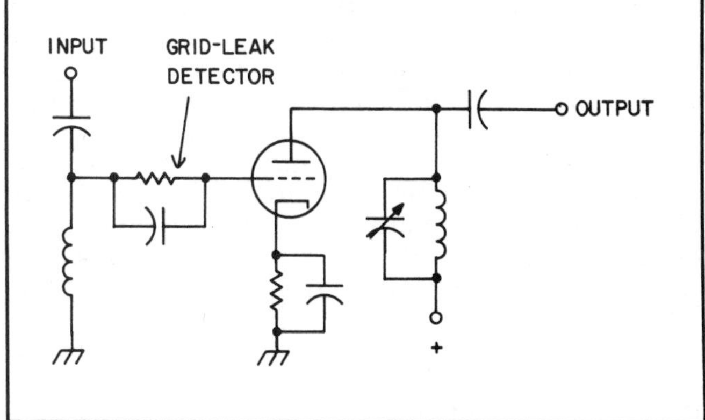

GRID-LEAK DETECTOR: The grid-leak detector uses a resistor and capacitor to provide envelope demodulation.

GROUND

Ground is the term generally used to describe the common connection in an electrical or electronic circuit. The common connection is usually at the same potential for all circuits in a system.

The common connection for electronic circuits is almost always ultimately routed to the earth. The earth is a fair to good conductor of electricity, depending on the characteristics of the soil. The earth itself provides an excellent common connection where the conductivity is good.

For an earth ground, the best connection is found in areas where salt comprises part of the soil. Such places include beach areas. Dark, moist soil also provides a good ground. *See also* EARTH CONDUCTIVITY.

GROUND ABSORPTION

Electromagnetic energy propagates, to a certain extent, along the surface of the earth. This occurs mostly at the very low, low, and medium frequencies. It can also take place at the high frequencies, but to a lesser degree. In this situation, the earth actually forms part of the circuit by which the wave travels; this mode is called surface-wave propagation (*see* SURFACE WAVE).

The ground is not a perfect conductor, and therefore a surface-wave circuit tends to be rather lossy. The better the earth conductivity in a given location, the better the surface-wave propagation, and the lower the ground ab-

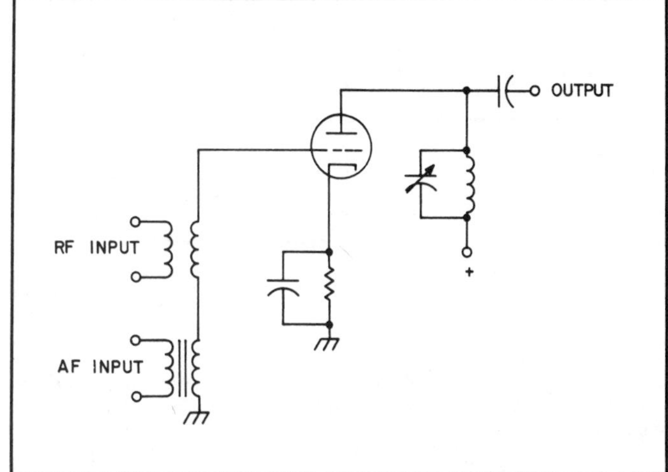

GRID MODULATION: An example of a grid-modulation circuit.

sorption. Salt water forms the best surface over which electromagnetic waves can travel. Dark, moist soil is also good. Rocky, sandy, or dry soil is relatively poor for the propagation of electromagnetic waves, and the absorption is high.

No matter what the earth conductivity, the ground absorption always increases as the frequency increases. At very high frequencies and above, the ground absorption is so high that signals do not propagate along the surface, but instead travel directly through space. *See also* EARTH CONDUCTIVITY.

GROUND BUS

A ground bus is a thick metal conductor, such as a strap, braid, or section of tubing, to which all of the common connections in a system are made. The ground bus is routed to an earth ground by the most direct path possible.

The advantage of using a ground bus in an electronic system is that it avoids the formation of ground loops (*see* GROUND LOOP). All of the individual components of the system are separately connected to the bus, as shown in the schematic diagram.

The size of the conductor used for a ground bus depends on the number of pieces of equipment in the system, and on the total current that the system carries. Typical ground-bus conductors are made from copper tubing having a diameter of ¼ to ½-inch. For radio-frequency applications, the larger conductor sizes are preferable; also, the distance to the ground connection should be made as short as possible in such cases. *See also* GROUND.

GROUND CLUTTER

See GROUND RETURN.

GROUND CONDUCTIVITY

See EARTH CONDUCTIVITY.

GROUND CONNECTION

A ground connection is the electrical contact between the common point of an electrical or electronic system and the earth. A ground connection should have the least possible resistance. The most effective ground connection consists of one or more ground rods driven into the soil to a depth of at least 6 to 8 feet. Alternatively, a ground connection can be made to a cold-water pipe, although the continuity of the pipe should be checked before assuming the connection will be good. In modern homes, pipes often have plastic joints; the entire pipe may even be made of plastic.

The ground connection is of prime importance to the operation of unbalanced radio-frequency systems. In balanced radio-frequency systems, the ground connection is less critical, although from a safety standpoint, it should always be maintained. All components of a system should be connected together, preferably via a ground bus leading to an earth ground, to prevent possible potential differences that can cause electric shock. *See also* GROUND, GROUND BUS.

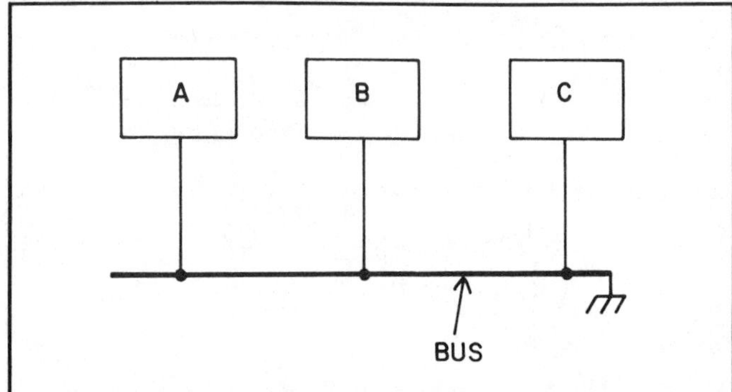

GROUND BUS: The ground bus is a convenient and effective method of obtaining a trouble-free common connection for systems with several separate units.

GROUND EFFECT

The directional pattern of an antenna system, especially at the very low, low, medium, and high frequencies, is modified by the presence of the surface of the earth underneath the antenna. This is called ground effect. It is more pronounced in the vertical, or elevation, plane than in the horizontal plane.

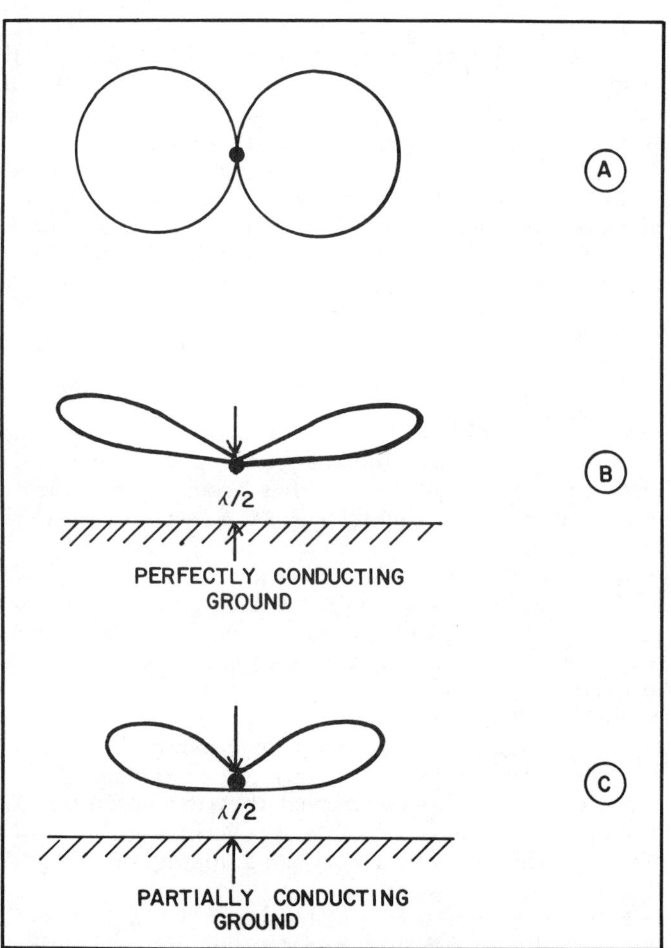

GROUND EFFECT: The ground effect on a half-wave, horizontal dipole antenna, viewed from the axis of the wire. At A, the radiation pattern in the absence of ground; at B, the radiation pattern for an antenna height of ½ wavelength over perfectly conducting ground; at C, a typical radiation pattern for an antenna height of ½ wavelength over normal ground.

The effective surface of the earth usually lies somewhat below the actual surface. The difference depends on the earth conductivity, and on the presence of conducting objects such as buildings, trees, and utility wires. The effective ground surface reflects radio waves to a certain extent. At a great distance, the reflected wave and the direct wave add together in variable phase.

The drawing illustrates an example of the effect of perfectly conducting ground and partially conducting ground on the vertical-plane radiation pattern or a half-wave dipole antenna. At A, the free-space pattern of the dipole is shown, in a situation where no ground surface is present. At B, the effect of a perfectly conducting ground surface ½ wavelength below the antenna is illustrated. At C, the effect of a partially conducting ground surface (typical of the actual earth), ½ wavelength below the antenna, is shown.

Ground effects have a large influence on the takeoff angle of electromagnetic waves from an antenna. This, in turn, affects the optimum distance at which communications is realized. In general, the higher a horizontally polarized antenna is positioned above the ground, the better the long-distance communication will be. For a vertically polarized antenna with a good electrical ground system, the height of the radiator above the ground is of lesser importance.

Ground effect takes place in the same manner for reception, with a given antenna system, as for transmission at the same frequency. *See also* ANGLE OF DEPARTURE, EARTH CONDUCTIVITY, EFFECTIVE GROUND, GROUND WAVE.

GROUND FAULT

A ground fault is an unwanted interruption in the ground connection in an electrical system. This can result in loss of power to electrical equipment, because there is no circuit via which the current can return. Excessive current may flow in the ground conductor of an ac electrical system, and dangerous voltages may be present at points that should be at ground potential.

A ground fault is usually not difficult to locate, but it may not immediately manifest itself. Only one piece of equipment may be involved, and the ground fault will not be noticed until that piece of equipment is operated. Ground faults can sometimes result in dangerous electrical shock.

Ground connections should, if possible, be soldered into place, so that corrosion will not result in ground faults. The number of joints in a ground circuit should be kept to a minimum. *See also* GROUND, GROUND CONNECTION.

GROUND LOOP

A ground loop occurs when two pieces of equipment are connected to a common ground bus, and are also connected together by means of separate wires or cables. Ground loops are quite common in all kinds of electrical and electronic apparatus, and normally they do not present a problem. However, in certain situations they can result in susceptibility to electromagnetic interference, such as hum pickup or interaction with nearby transmitters.

In the event of undesirable hum pickup or other electromagnetic interference that defies attempts at troubleshooting, ground loops should be suspected. Ground loops may be extremely difficult to locate. However, removing interconnecting ground wires one at a time, and checking for interference each time, will eventually pinpoint the source of the trouble. *See also* ELECTROMAGNETIC INTERFERENCE, GROUND, GROUND BUS.

GROUND PLANE

A ground plane is an artificial radio-frequency ground, constructed from wires or other conductors, and intended to approximate a perfectly conducting ground. Ground planes are often used with vertical antennas to reduce the losses caused by ground currents. All amplitude-modulation broadcast stations use vertical antennas with ground-plane systems.

A ground plane can be constructed by running radial wires outward from the base of a quarter-wave vertical antenna. The conductors may be buried under the soil or placed just above the surface of the earth. They may be insulated or bare. The more radials that are used, the more effective the ground plane. Also, the quality of the ground plane improves with increasing radial length. An optimum system consists of at least 100 radials of ½-wavelength or longer, arranged at equal angular intervals. However, smaller radial systems provide considerable improvement in antenna performance, compared with no radials at all. The radials are connected to a driven ground rod at the base of the antenna.

In some high-frequency antenna systems, the ground plane is elevated above the actual surface of the earth. When the ground plane is ¼ wavelength or more above the surface, only three or four quarter-wave radials are necessary in order to obtain a nearly perfect ground plane. The ground-plane antenna operates on this principle. *See also* GROUND-PLANE ANTENNA.

GROUND-PLANE ANTENNA

A ground-plane antenna is a vertical radiator operated against a system of quarter-wave radials and elevated at least a quarter wavelength above the effective ground. The

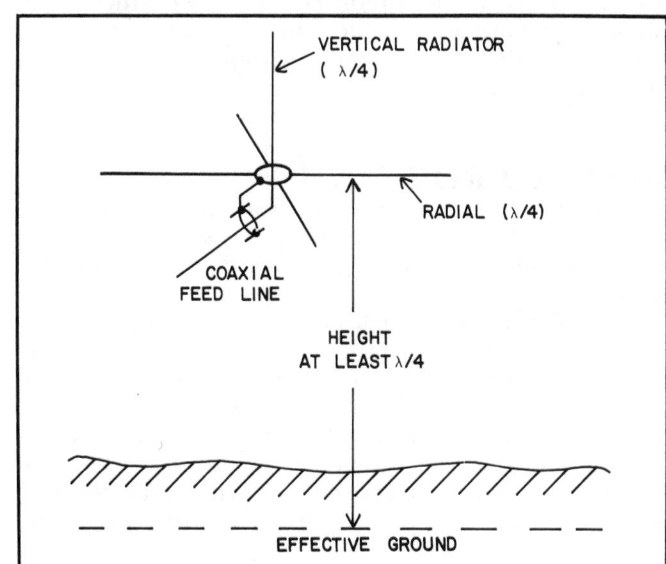

GROUND-PLANE ANTENNA: The ground-plane antenna requires only three or four quarter-wave radials, provided the feed point is at least a quarter wavelength above effective ground.

radiator itself may be any length, but should be tuned to resonance at the desired operating frequency.

When a ground plane is elevated at least 90 electrical degrees above the effective ground surface, only three or four radials are necessary in order to obtain an almost lossless system. The radials are usually run outward from the base of the antenna at an angle that may vary from 0 degrees to 45 degrees with respect to the horizon. The drawing illustrates a typical ground-plane antenna.

A ground-plane antenna is an unbalanced system, and should be fed with coaxial cable. A balun may be used, however, to allow the use of a balanced feed line. The feed-point impedance of a ground-plane antenna having a quarter-wave radiator is about 37 ohms if the radials are horizontal; this impedance increases as the radials are drooped, reaching about 50 ohms at an angle of 45 degrees (*see* DROOPING RADIAL). The radials may be run directly downward, in the form of a quarter-wave tube concentric with the feed line. Then the feed-point impedance is approximately 73 ohms. This configuration is known as a coaxial antenna. *See also* COAXIAL ANTENNA, EFFECTIVE GROUND, GROUND PLANE, RADIAL.

GROUND RESISTANCE

See EARTH CONDUCTIVITY.

GROUND RETURN

In direct-current transmission, as well as some alternating-current circuits, one leg of the connection is provided by the earth ground. This part of the circuit is called the ground return. The ground return allows the use of only one conductor, rather than two conductors, in a transmission system. A good ground connection is essential at each end of the circuit if the ground return is to be effective.

In radar, objects on the ground cause false echoes. This effect is especially prevalent in populated areas, where the numerous manmade structures "fool" the radar. This is called the ground-return effect; it is also sometimes called ground clutter. Ground-return effects make it difficult to track a target at a low altitude. This phenomenon can also confuse weather forecasters attempting to track thunderstorms by radar. *See also* RADAR.

GROUND ROD

A ground rod is a solid metal rod, usually made of copper-plated steel, used for the purpose of obtaining a good ground connection for electrical or electronic apparatus.

Ground rods are available in a wide variety of diameters and lengths. The most effective ground is obtained by driving one or more rods at least 8 feet into the soil, well away from the foundations of buildings. Smaller ground rods are sometimes adequate for very low-current or low-power installations in locations where the soil conductivity is excellent. Ground rods can be obtained at most electrical equipment stores. *See also* GROUND, GROUND CONNECTION.

GROUND WAVE

In radio communication, the ground wave is that part of the electromagnetic field that is propagated parallel to the surface of the earth. The ground wave actually consists of three distinct components: the direct or line-of-sight wave, the reflected wave, and the surface wave. Each of these three components contributes to the received ground-wave signal.

The direct wave travels in a straight line from the transmitting antenna to the receiving antenna. At most radio frequencies, the electromagnetic fields pass through objects such as trees and frame houses with little attenuation. Concrete-and-steel structures cause some loss in the direct wave at higher frequencies. Obstructions such as hills, mountains, or the curvature of the earth cut off the direct wave completely.

A radio signal can be reflected from the earth or from certain structures such as concrete-and-steel buildings. The reflected wave combines with the direct wave (if any) at the receiving antenna. Sometimes the two are exactly out of phase, in which case the received signal is extremely weak. This effect occurs mostly in the very-high-frequency range and above.

The surface wave travels along the earth, and occurs only with vertically polartized energy at the very low, low, medium, and high frequencies. Above 30 MHz, there is essentially no surface wave. At the very low and low frequencies, the surface wave propagates for hundreds or even thousands of miles. Sometimes the surface wave is called the ground wave, in ignorance of the fact that the direct and reflected waves can also contribute to the ground wave in very short-range communications. The actual ground-wave signal is the phase combination of all three components at the receiving antenna. *See also* DIRECT WAVE, REFLECTED WAVE, SURFACE WAVE.

GROVE CELL

A Grove cell is an electrochemical cell. It makes use of liquid electrolytes and metal electrodes to generate a direct-current voltage.

A solution of nitric acid is contained in a porous cup, which is immersed in a solution of sulfuric acid held in a larger container. The porous cup allows electric currents to pass, but keeps the two solutions from mixing.

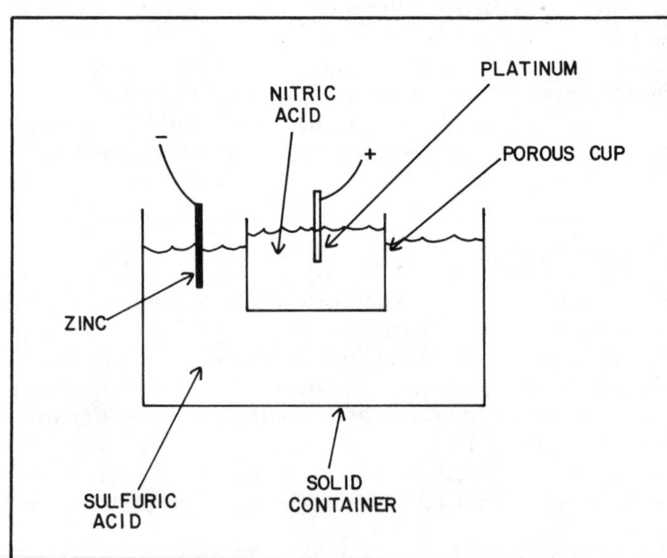

GROVE CELL: The Grove electrochemical cell.

A platinum electrode is immersed in the nitric-acid solution, forming the positive terminal. A zinc electrode is immersed in the sulfuric-acid solution, forming the negative electrode. The drawing illustrates the construction of the Grove cell. *See also* CELL.

GUARD BAND

In a channelized communications system, a guard band or guard zone is a small part of the spectrum allocated for the purpose of minimizing interference between stations on adjacent channels. It is desirable to have some unused frequency space between channels, so that the sidebands of one signal will not cause interference in the passband of a receiver tuned to the next channel.

The width of the guard band should be sufficient to allow for the imperfections in the bandpass filters of the transmitter and receiver circuits. However, excessive guard-band space is wasteful of the spectrum. It is sometimes a matter of trial-and-error to determine the optimum guard-band width in a communications system. An example of this is the 2-meter frequency-modulation amateur band. The original band plan used channels spaced at 15-kHz intervals between 146 and 148 MHz. The deviation of the signals in this band is plus or minus 5 kHz, or a total theoretical bandwidth of 10 kHz. However, adjacent-channel interference was often observed in this band. Later, when the band was extended below 146 MHz, the channels were established at 20-kHz increments. This has reduced adjacent-channel interference.

In some bands, notably the international shortwave broadcast band, there is often no guard zone at all. Stations in this part of the spectrum usually have a bandwidth of 10 kHz, but may be spaced only 5 kHz apart, resulting in severe adjacent-channel interference. *See also* ADJACENT-CHANNEL INTERFERENCE.

GUIDED MISSILE

A guided missile is a missile that uses electronic devices or an inertial reference system to direct it to its target. Guided missiles may be of the air-to-air, ground-to-air, air-to-ground, or ground-to-ground variety.

Any of two methods can be employed to direct the missile to its destination. The missile may contain a homing device, or it may be controlled from the launch point by means of radio transmitters, or it may employ a set of gyroscopes that serve as an inertial guidance system.

Most modern missiles use guidance systems. Such systems are often computer-controlled; the trajectory of the missile is precisely determined by a computer, which performs all of the steering operations. The missile may sense sound waves, infrared radiation, radio emissions, or signals from a radiolocation beacon in order to guide itself to a target. *See also* HOMING DEVICE.

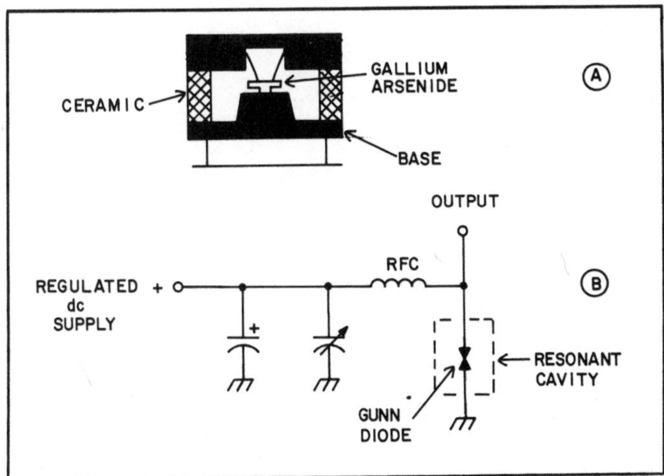

GUNN DIODE: At A, cutaway view of the inside of a Gunn diode. At B, schematic diagram of a simple Gunn-diode oscillator.

GUNN DIODE

A Gunn diode is a form of semiconductor device that operates as an oscillator in the ultra-high-frequency and microwave parts of the electromagnetic spectrum. The Gunn diode has replaced the Klystron tube in many situations.

The Gunn diode is mounted in a resonant enclosure, as shown at A in the illustration. A direct-current voltage is applied to the device and, if this voltage is large enough, the Gunn diode will oscillate. A schematic diagram of an oscillator using a Gunn diode is shown at B.

Gunn diodes are not very efficient. Only a small fraction of the consumed power actually results in radio-frequency power. Gunn diodes tend to be highly sensitive to changes in temperature and bias voltage. The frequency may vary considerably even with a small change in the ambient temperature; for this reason, the temperature must be carefully regulated. The change in frequency with voltage can be useful for frequency-modulation purposes, but voltage regulation of some sort is essential. Most Gunn diodes require about 12 volts for proper operation. Oscillation can take place at frequencies in excess of 20 GHz. The output power from a Gunn-diode oscillator can, with proper bias and circuit design, be more than 0.1 W. *See also* DIODE OSCILLATOR.

GUYING

Guying is the installation of supporting wires for tall structures. Some antenna supports cannot stand up without guy wires. Others need guying only in the event of high winds.

Guy wires are usually installed in sets of three or four, and may be attached to the antenna support structure at one or more levels. If three sets of guy wires are used, they

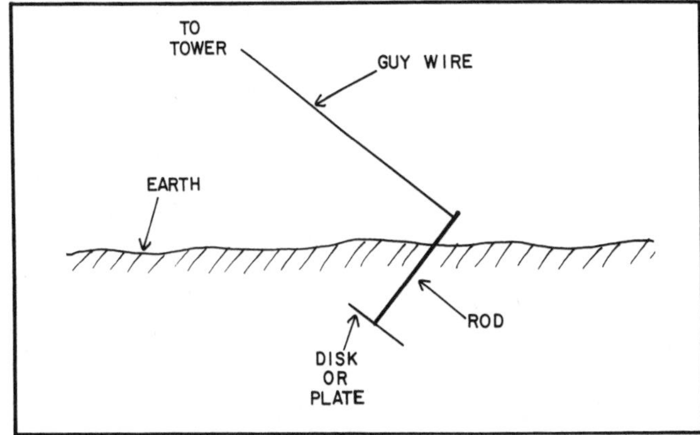

GUYING: A guy anchor should be installed so that it will not be pulled out of the earth by forces on the guy wire.

should be brought out from the support structure at 120-degree angles (as viewed from above); if sets of four guys are used, they should be positioned at 90-degree angles.

The number of sets of guy wires needed for a particular support structure depends on its height, its rigidity, and the expected maximum wind loading. Wind loading depends on the diameter of the structure, the kind of antenna array it supports, and the wind velocity. The maximum expected wind velocity varies considerably depending on geographic location. In general, the worst areas are those exposed to hurricanes arriving from the ocean.

Guy wires should generally not be brought down from a tower at an angle of less than 30 degrees with respect to the tower in the vertical plane. The steeper the angle at which a guy wire is installed, the greater the strain on the wire and on the guy anchor.

Guy anchors should be installed so that they are not easily pulled from the ground. The illustration shows a typical guy anchor. With large towers, a separate guy anchor should be used for each guy wire. In smaller towers having multiple guy sets, a wire from each set may be attached to a single, common guy anchor.

The guying of antenna support structures should not be undertaken by an amateur. It is advisable that a professional be consulted, so that the optimum system may be installed in terms of safety, cost, and longevity. Guesswork can result in disaster. *See also* TOWER, WIND LOADING.

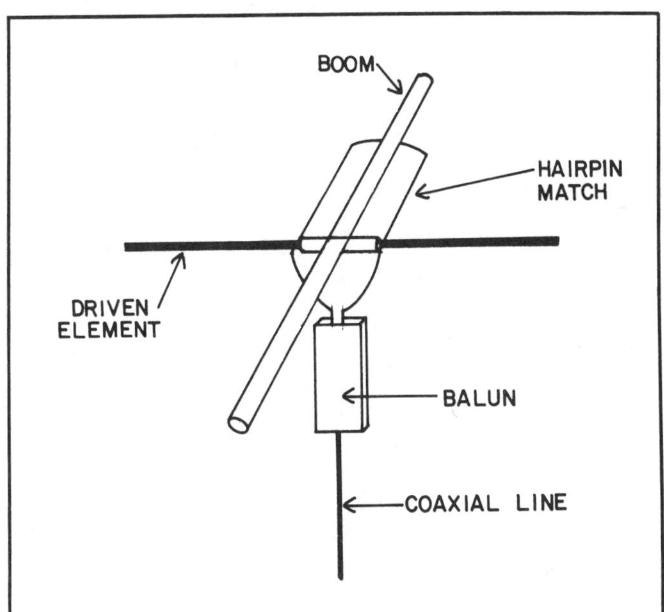

HAIRPIN MATCH: An example of a hairpin match installed in the driven element of a Yagi antenna.

HAIRPIN MATCH

A hairpin match is a means of matching a half-wave, center-fed radiator to a transmission line. The hairpin match is especially useful in the driven element of a Yagi antenna, where the feed-point impedance is lowered by the proximity of parasitic elements.

The hairpin match requires that the radiating element be split at the center. It also requires the use of a balanced feed system; if a coaxial line is used, a balun is needed at the feed point. The hairpin match consists of a section of parallel-wire transmission line, short-circuited at the far end (see illustration). The length of the section is somewhat less than ¼ wavelength, and thus it appears as an inductance.

The adjustment of the hairpin match is quite simple; a sliding bar can be used to vary the length of the section. The section itself should be perpendicular to the driven element; this generally necessitates mounting it along the boom of the Yagi antenna. The use of a hairpin match causes a slight lowering of the resonant frequency of a radiating element. Therefore, the radiator must be shortened by a few percent to maintain operation at the desired frequency. When the length of the hairpin section and the length of the radiating element are just right, a nearly perfect match can be obtained with either 50-ohm or 75-ohm coaxial feed lines. *See also* IMPEDANCE MATCHING, YAGI ANTENNA.

HALF-ADDER

A half-adder is a digital logic circuit with two input terminals and two output terminals. The output terminals are called the sum and carry outputs.

The sum output of a half-adder circuit is the exclusive-OR function of the two inputs. That is, the sum output is 0 when the inputs are the same and 1 when they are different. The carry output is the AND function of the two inputs: It is 1 only when both inputs are 1. (See the truth table.)

HALF-ADDER: SUM AND CARRY OUTPUTS AS A FUNCTION OF LOGIC INPUT.

INPUTS X	INPUTS Y	SUM OUTPUT	CARRY OUTPUT
0	0	0	0
0	1	1	0
1	0	1	0
1	1	0	1

The half-adder circuit differs from the adder in that the half-adder will not consider carry bits from previous stages. There are several different combinations of logic gates that can function as half adders. *See also* ADDER, BINARY-CODED NUMBER.

HALF-BRIDGE

A half-bridge is a form of rectifier circuit, similar to the bridge rectifier except that two of the diodes are replaced by resistors (see illustration). The two resistors have equal values. The voltages at either end of the series combination are always equal and opposite. Thus, at the center point, the voltage is zero, and this point is generally grounded. In the schematic, a positive voltage appears at the output. By reversing the diodes, a negative-voltage supply is obtained.

The efficiency of the half-bridge circuit is less than that of the conventional bridge rectifier. This is because the resistors tend to dissipate some power as heat as the current flows through them. The half-bridge operates over the entire alternating-current input cycle, and is therefore a form of full-wave rectifier. *See also* BRIDGE RECTIFIER.

HALF CYCLE

In any alternating-current system, a complete cycle occurs between any two identical points on the waveform. These

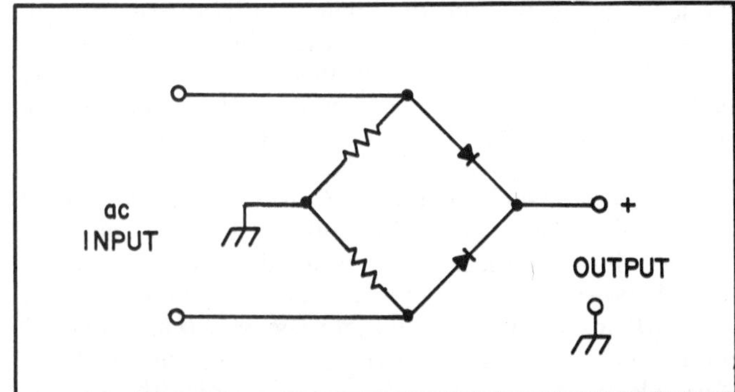

HALF-BRIDGE: The half-bridge rectifier circuit.

points can be chosen arbitrarily; they may be positive peaks, zero-voltage points, negative peaks, or any other point. The complete cycle requires a certain length of time, P, which is called the period of the wave.

A half cycle is simply any part of the waveform that occurs during a time interval of P/2, if the period is P. Usually, the term half-cycle is used in reference to either the negative or positive portion of a sine-wave alternating-current waveform. A half-cycle represents 180 electrical degrees. *See also* ALTERNATING CURRENT, CYCLE, ELECTRICAL ANGLE.

HALF LIFE

The half life of a radioactive substance is the time it takes for the substance to decay to a point where its radiation intensity is 50 percent of its initial value. Half life is a convenient expression of the speed with which a radioactive material disintegrates.

The half life of any particular substance is always the same. It does not depend on the intensity of the radiation. If a substance has a half life of 10 years, for example, then its radioactivity will e half as great after 10 years as at the time of first measurement; after 20 years it will be ¼ as great; after 30 years, ⅛ as great; and so on. This will be true no matter what the time of the initial measurement.

The half-life figures of the constituents of radioactive fallout, which would cover the earth following a global thermonuclear war, gives us a terrifying glimpse of the possible effects such a war would have on our planet. The half lives of some of these substances are measured in years. *See also* RADIOACTIVITY.

HALF-POWER POINTS

The sharpness of an antenna directive pattern, or of the selective response of a bandpass filter, is often specified in terms of the half-power points. In the case of a directive antenna, the variable parameter is compass direction (see A in the illustration). In the case of a bandpass filter, the variable parameter is frequency, as at B.

In antenna systems, the reference power level is the effective radiated power in the favored direction. This may occur in more than one direction. In most parasitic arrays, the favored direction is a single compass point; in the case of the half-wave dipole, the favored direction is two compass points; an unterminated longwire generally has favored directions at four compass points. The effective radiated power at the half-power points is 3 dB below the level in the favored direction. The field strength, in volts per meter, is 0.707 times the field strength in the favored direction. The angle, in degrees, between the half-power points of a single lobe is called the beamwidth. *See also* BEAMWIDTH, FIELD STRENGTH.

For a bandpass filter, the half-power points are the frequencies at which the power from the filter drops 3 dB below the power output at the center of the passband. The bandwidth of the filter is sometimes specified in terms of the half-power points, but more often it is given in terms of the 6-dB attenuation points and the 60-dB attenuation points. This combined figure gives an indication of the skirt selectivity as well as the actual bandwidth. *See also* BANDPASS FILTER, BANDPASS RESPONSE, SHAPE FACTOR, SKIRT SELECTIVITY.

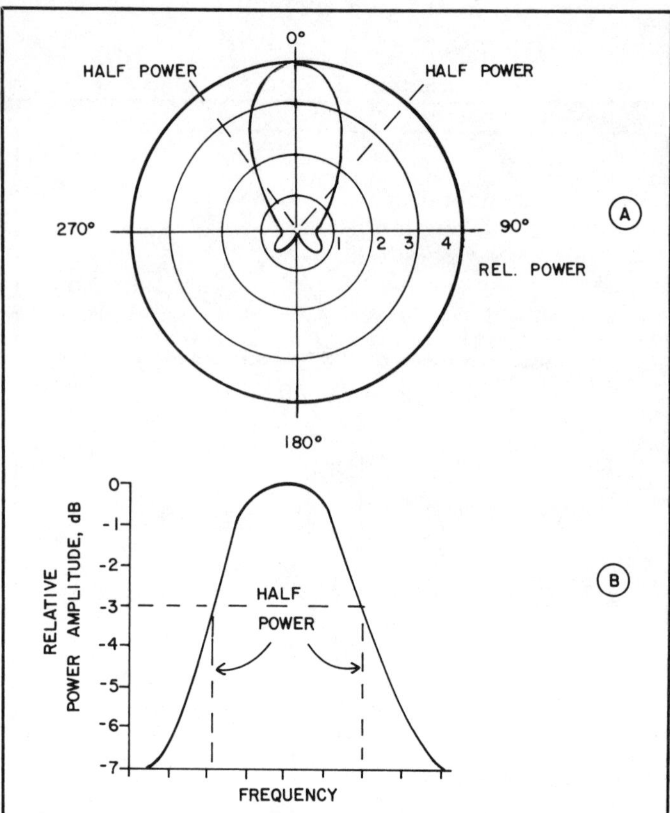

HALF-POWER POINTS: At A, the horizontal-plane radiation pattern of a typical directional antenna, showing the half-power points. At B, a bandpass frequency response showing half-power points.

HALF STEP

A half step is a change in the frequency of an audio tone, or a difference in the frequencies of two audio tones, equivalent to 6 percent of the frequency of the lower-pitched tone. Any two immediately adjacent keys on a piano are a half step apart in frequency. In most musical arrangements, every tone frequency is at a defined half step with respect to the other tones. However, in some cultures, smaller intervals can be found in musical tones. *See also* MUSIC.

HALF-WAVE ANTENNA

A half-wave antenna is a radiating element that measures an electrical half wavelength in free space. Such an antenna may have a physical length anywhere from practically zero to almost the physical dimensions of a half wavelength in free space. A dipole antenna is the simplest example of a half-wave antenna.

In theory, a half wavelength in free space is given in feet according to the equation:

$$L = 492/f$$

where L is the linear distance and f is the frequency in megahertz. A half wavelength in meters is given by:

$$L = 150/f$$

In practice, an additional factor must be added to the above equations, because electromagnetic fields travel somewhat more slowly along the conductors of an antenna than they do in free space. For ordinary wire, the results as obtained above are multiplied by about 0.95. For tubing or large-diameter conductors, the factor is slightly smaller, and may range down to about 0.90 (*see* VELOCITY FACTOR).

A half-wave antenna may be made much shorter than a physical half wavelength. This is accomplished by inserting inductances in series with the radiator. The antenna may be made much longer than the physical half wavelength by inserting capacitances in series with the radiator. *See also* CAPACITIVE LOADING, ELECTRICAL WAVELENGTH, INDUCTIVE LOADING.

HALF-WAVE DIPOLE

See DIPOLE ANTENNA.

HALF-WAVE RECTIFIER

A half-wave rectifier is the simplest form of rectifier circuit. It consists of nothing more than a diode in series with one line of an alternating-current power source, and a transformer (if necessary) to obtain the desired voltage. A is a schematic diagram of a half-wave rectifier circuit designed to supply a positive output voltage. The diode may be reversed to provide a negative output voltage.

The output of the half-wave rectifier contains only one half of the alternating-current input cycle, as shown at B. The other half is simply blocked. The half-wave rectifier gets its name from the fact that it operates on only one half of the input cycle.

Half-wave rectifiers have the advantage of being extremely simple in terms of design. An unbalanced alternating-current input source can be used; in some instances, no transformer is needed and the direct-current voltage can be derived straight from a wall outlet. However, the half-wave rectifier has rather poor voltage-regulation characteristics. The pulsating output is more difficult to filter than that of the full-wave rectifier. Half-wave rectifier circuits are often used in situations where the current drain is low and the voltage regulation need not be especially precise. *See also* BRIDGE RECTIFIER, FULL-WAVE RECTIFIER.

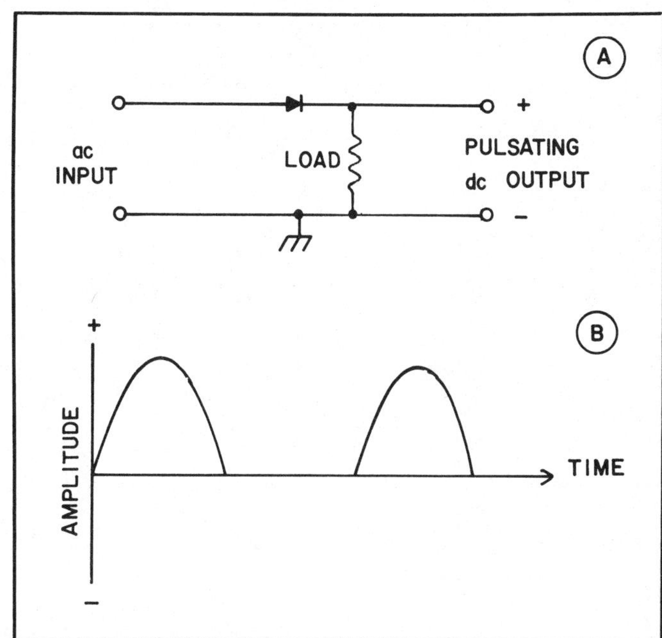

HALF-WAVE RECTIFIER: At A, a half-wave rectifier circuit. At B, the waveform resulting from the application of a sine-wave alternating current to the input of a half-wave rectifier.

HALF-WAVE TRANSMISSION LINE

A half-wave transmission line is a section of electromagnetic feed line that measures an electrical half wavelength. The physical length L of such a transmission-line section, for a frequency f in megahertz, is given in feet by:

$$L = 492v/f$$

where v is the velocity factor of the line (*see* VELOCITY FACTOR). The length L in meters is:

$$L = 150v/f$$

A half-wave section of transmission line has certain properties that make it useful as a tuned circuit. The impedance at one end of such a line is exactly the same, neglecting line loss, as the impedance at the other end. This is true, however, only within a narrow range of frequencies, centered at the half-wave resonant frequency. If the far end of a half-wave transmission line is an open circuit, the line behaves as a parallel-resonant tuned circuit. If the far end is short-circuited, the section behaves as a series-resonant tuned circuit.

At frequencies above approximately 20 MHz, where half-wave sections of transmission line have reasonable length, these sections are often used in place of coils and capacitors as resonant circuits. If the line loss is reasonably low, the Q factor of the half-wave transmission line is very high. The half-wave transmission line may be either balanced or unbalanced, depending on the circuit in which it is used. *See also* Q FACTOR.

HALL CONSTANT

In a current-carrying conductor, the transverse electric field, the magnetic field, and the current density are mutually dependent. Given a transverse electric-field intensity of e, a current density of i, and a magnetic-field strength m, then:

$$e/(im) = k$$

where k is a constant. This is called the Hall constant.

The Hall constant is an expression of the Hall effect, in which a magnetic field produces a transverse voltage in a current-carrying conductor. *See also* HALL EFFECT.

HALL EFFECT

When a current-carrying electrical conductor is placed in a magnetic field, a voltage sometimes develops between one side of the conductor and the other. For this to happen, the magnetic lies of force must be perpendicular, or nearly perpendicular, to the line containing the conductor. The voltage then appears at right angles to the magnetic lines of

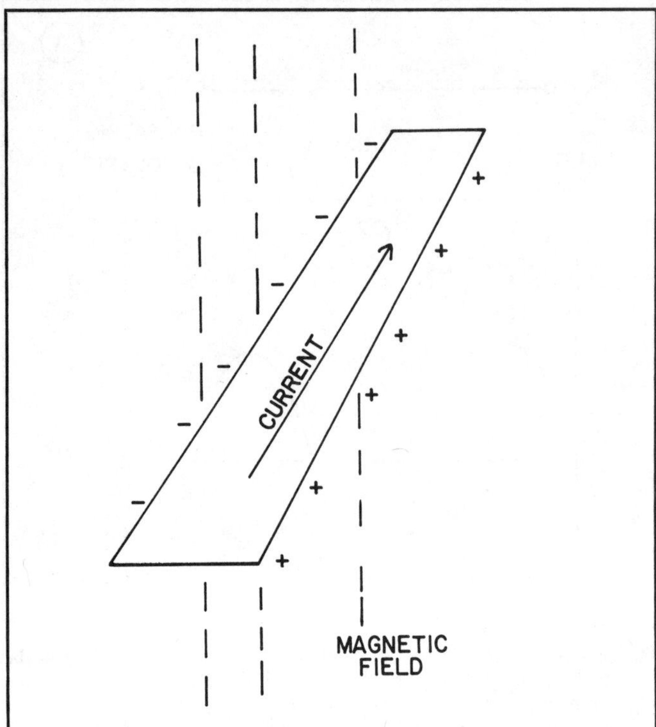

HALL EFFECT: The Hall effect in a metal strip. A potential difference develops between opposite edges of the strip.

force. If the conductor is a strip, and the magnetic lines of force are perpendicular to the strip, then the voltage will appear between opposite edges of the strip (see illustration). This is known as the Hall effect. The electric-field intensity e, generated by the Hall effect, is given by the formula:

$$e = kim$$

where i is the current in the conductor, m is the magnetic-field strength, and k is a constant called the Hall constant.

In some metals, the voltage displays a polarity opposite to that in other metals under the same conditions. The polarity depends on the atomic structure of the metal. The Hall effect occurs in semiconductor materials as well as in metals; a device called a Hall generator makes use of the Hall effect in semiconductors for the purpose of measuring magnetic-field strength. The voltage is quite small unless the current and the magnetic field are extremely intense. However, the voltage can produce current in an external circuit. *See also* HALL GENERATOR, HALL MOBILITY.

HALL GENERATOR

A Hall generator is a device that makes use of the Hall effect for the purpose of generating a direct-current voltage in the presence of a magnetic field. A semiconductor wafer is used. A battery, or other source of direct current, is connected to opposite ends of the wafer. A voltmeter is connected to the adjacent sides of the wafer, as illustrated in the drawing. Some semiconductor materials are much better than others for use in the Hall generator. The greater the Hall mobility (*see* HALL MOBILITY), the better; indium antimonide is especially effective.

When a magnetic field exists in the vicinity of the wafer,

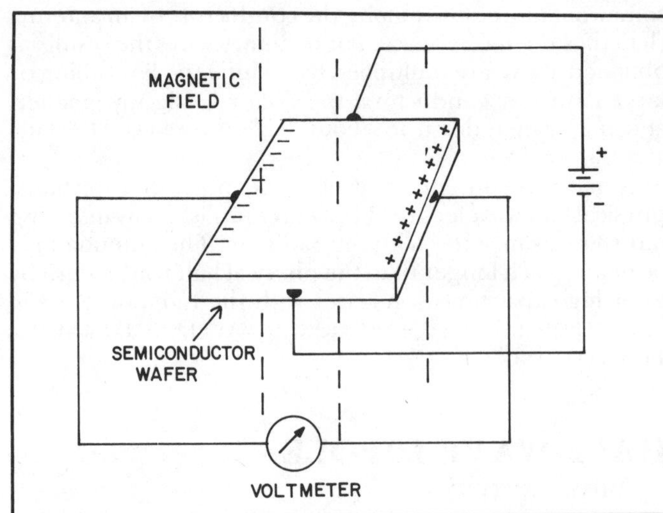

HALL GENERATOR: A Hall generator can be used to measure the strength of a magnetic field.

such that the magnetic lines of force are perpendicular to the plane of the wafer, a voltage is induced and will register on the voltmeter. The value of the voltage is quite small, but is proportional to the current in the wafer and to the intensity of the magnetic field. The Hall generator is therefore useful for measuring the strength of a magnetic field. *See also* HALL EFFECT.

HALL MOBILITY

The Hall mobility is an expression of the extent to which the Hall effect takes place in a semiconductor material. For a given magnetic-field intensity and current value, the voltage generated by the Hall effect is greater when the Hall mobility is higher.

The Hall mobility is given by the product of the Hall constant and the conductivity for a given material. In general, the greater the carrier mobility in a semiconductor, the greater the Hall mobility. *See also* HALL EFFECT, HALL GENERATOR.

HALO ANTENNA

A halo antenna is a special form of horizontal half-wave antenna. Basically, the halo consists of a dipole whose

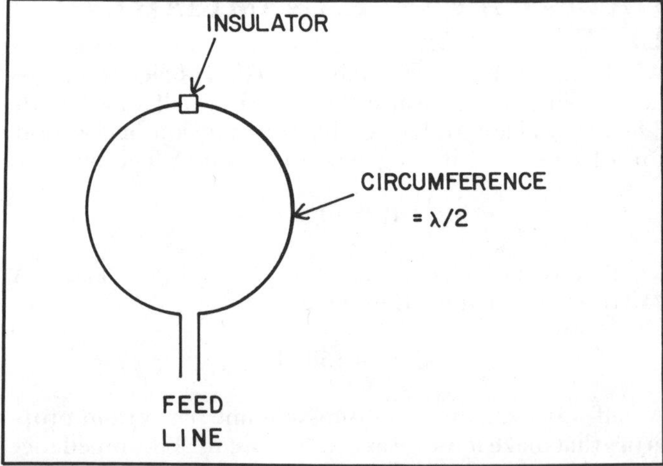

HALO ANTENNA: The halo antenna is simply a half-wave dipole, bent into a circle.

elements have been bent into a circle, so that the circumference of the circle is ½ electrical wavelength. The ends of the dipole are insulated from each other at the opposite side of the circle from the feed point (see illustration).

The halo antenna is used mostly in the very-high-frequency part of the radio spectrum. At 30 MHz, for example, the circumference of a halo antenna is only about 15 feet, so that the diameter is less than 5 feet. Copper or aluminum tubing is practical for the construction of a halo in the very-high-frequency range.

Halo antennas exhibit a nearly omnidirectional radiation pattern in the horizontal plane. The polarization is horizontal. Halo antennas are often vertically stacked to obtain omnidirectional gain in the horizontal plane. *See also* DIPOLE ANTENNA.

HALOGEN

A halogen is an element that is a member of group VII-A of the periodic table. The nonmetallic halogens are astatine, bromine, chlorine, fluorine, and iodine. These elements are highly reactive, and have similar characteristics.

Halogen gases, such as chlorine, are sometimes introduced into radiation counter tubes to reduce the recovery time between counts. This enables the tube to accurately register higher levels of radiation. *See also* COUNTER TUBE.

HAM

Amateur radio operators are sometimes called hams, and amateur radio itself is sometimes called ham radio. The exact origin of the term ham is uncertain, although in the old days of telegraphy, the expression was used to refer to poor operators! It is believed that the amateurs embraced the name in a light-hearted way, and the term has stuck ever since.

Today, ham radio is a respected hobby, undertaken by hundreds of thousands of people. *See also* AMATEUR RADIO.

HANDSET

A handset is a telephone type communications device, containing a microphone and an earphone in a configuration that can be easily held against the side of the head. Some portable radio transceivers use handsets, but the most familiar handset is that found in the common household telephone (see the photograph).

The telephone handset may have the dialing apparatus contained within itself, as shown, or the dialer may be in the base of the telephone. *See also* TELEPHONE.

HANDSET: Handsets are used in almost every telephone.

HANDSHAKING

In a digital communications system, accuracy can be improved by synchronizing the transmitter and receiver precisely before the beginning of data transfer. This is called handshaking. The process may be repeated at intervals to maintain synchronization.

Handshaking is becoming more and more universal in electronic communications, because it dramatically improves the signal-to-noise ratio. In general, the more frequently the handshaking operation is done in the process of signal transmission, the better the signal-to-noise ratio, although there is a point of diminishing returns.

In some digital systems, an independent reference standard, such as a time and frequency station, is used to synchronize the digital signals between the transmitter and receiver. This is called coherent or synchronized digital communications. *See also* SYNCHRONIZED COMMUNICATIONS.

HARD LINE

Hard line is a form of coaxial cable with 100 percent shielding continuity. The perfect shield is obtained because the outer conductor of the hard line is made from metal tubing. Aluminum tubing is commonly used for the manufacture of hard line; copper or galvanized metal may also occasionally be used.

Hard line is not as flexible as coaxial cable with a braided-wire outer conductor. Hard line is therefore somewhat more difficult to install. The dielectric material may be solid, or it may consist of beaded sections; it is often made from polyethylene; sometimes Teflon is used.

Hard lines are known for their durability and relatively low loss. They are available in a number of sizes and charcteristic-impedance ratings. The smallest common hard line is about ⅜ inch in diameter; the largest commonly available is about 1 inch across. *See also* COAXIAL CABLE.

HARDWARE

In computer engineering, hardware is a term used to describe the actual circuitry that makes up the device. The wiring, circuit boards, integrated circuits, diodes, transistors, and resistors comprise the hardware. So do the operator interface devices, such as display terminals, printers, and keyboards. The computer programs, in contrast, are called software or firmware, depending on whether or not they are easily modified (*see* FIRMWARE, SOFTWARE).

A person who designs computer hardware is called a hardware engineer. The hardware engineer is an electrical engineer. *See also* ELECTRICAL ENGINEERING.

HARMONIC

Any signal contains energy at multiples of its frequency, in addition to energy at the desired frequency. The lowest frequency component of a signal is called the fundamental frequency; all integral multiples are called harmonic frequencies, or simply harmonics.

In theory, a pure sine wave contains energy at only one frequency, and has no harmonic energy. In practice, this ideal is never achieved. All signals contain some energy at harmonic frequencies, in addition to the energy at the fundamental frequency. The signal having a frequency of twice the fundamental is called the second harmonic. The signal having a frequency of three times the fundamental is called the third harmonic, and so on.

Wave distortion always results in the generation of harmonic energy. While the nearly perfect sine wave has very little harmonic energy, the sawtooth wave, square wave, and other distorted periodic oscillations contain large amounts of energy at the harmonic frequencies. Whenever a sine wave is passed through a nonlinear circuit, harmonic energy is produced. A circuit designed to deliberately create harmonics is called a harmonic generator or frequency multiplier.

Harmonic output from radio transmitters is undesirable, and the designers and operators of such equipment often go to great lengths to minimize this energy as much as possible. *See also* FREQUENCY MULTIPLIER, FUNDAMENTAL FREQUENCY, HARMONIC SUPPRESSION.

HARMONIC GENERATOR

See FREQUENCY MULTIPLIER.

HARMONIC SUPPRESSION

Harmonic suppression is an expression of the degree to which harmonic energy is attenuated, with respect to the fundamental frequency, in the output of a radio transmitter. Harmonic suppression is also a term used to denote the process of minimizing harmonic energy in the output of a signal generator, especially a radio transmitter. Harmonics are undesirable in the output of such equipment because they cause interference to other services and operations.

If a given harmonic signal has a power level of Q watts in the output circuit of a transmitter, while the fundamental-frequency output is P watts, then the harmonic suppression S in decibels is given by:

$$S = 10 \log_{10} (P/Q)$$

Harmonic suppression may be accomplished in three ways. The most frequently used method is the insertion of one or more tuned bandpass filters in the output of the transmitter. The filter frequency is centered at the operating frequency of the transmitter. This provides additional harmonic attenuation in the output of the final amplifier.

The second method of obtaining harmonic suppression is the insertion of a lowpass filter in the transmitter output. The cutoff frequency of the filter should be the lowest that will result in negligible attenuation at the fundamental frequency.

The third method of obtaining harmonic suppression is the use of band-rejection filters, also sometimes called traps. The trap offers harmonic attenuation at only one frequency; the bandpass and lowpass filters provide rejection of all harmonics above the fundamental. However, the trap circuit often gives better results at the design frequency.

Of course, if an amplifier is intended to operate as a linear amplifier, the grid bias and grid drive should be maintained at the proper levels to ensure minimal generation of harmonic energy. The Class-C amplifier causes more harmonics to be produced than other types of amplifiers. *See also* BANDPASS FILTER, BAND-REJECTION FILTER, HARMONIC, LOWPASS FILTER.

HARMONY

Harmony is the combination of musical tones such that the resulting sound is pleasing to the ear. The kinds of audio tone combinations that are harmonious, instead of dissonant, vary greatly among different people and different cultures.

Harmony is important in the composition of music. Tone combinations produce mixing products, in addition to the tones themselves. This contributes to the overall harmony. Some harmony is extremely complex, consisting of different tone waveforms as well as frequencies. There may be dozens of individual musical instruments contributing to a single chord. *See also* MUSIC.

HARP ANTENNA

A harp antenna is a special form of broadband unbalanced ground-plane antenna. The radiating portion of the an-

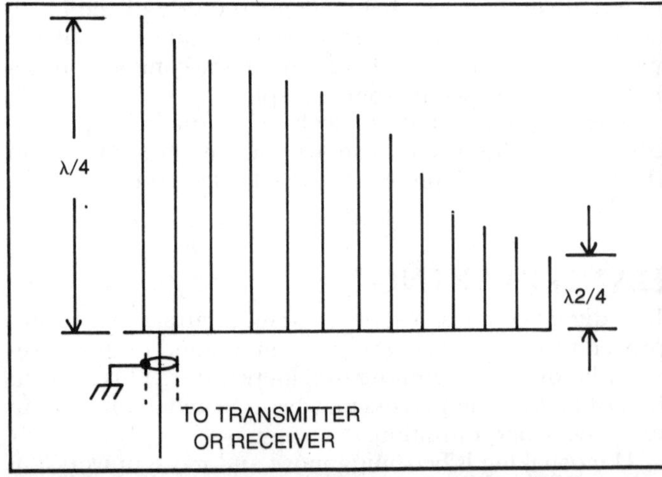

HARP ANTENNA: The harp antenna consists of a series of quarter-wave vertical radiators of graduated length. Here, λ/4 represents the longest wavelength, and λ2/4 represents the shortest wavelength, at which the antenna will function.

tenna is vertically polarized, and consists of a number of different vertical conductors of various lengths. The conductors extend upward from a common feed line, giving the antenna the appearance of a harp (see illustration).

Generally, the lowest operating frequency of the harp antenna is determined by the length of the longest radiator. The highest operating frequency is determined by the length of the shortest radiator. The radiating elements are an electrical quarter wavelength at resonance. Harp antennas are most practical at the very-high frequencies, where the element lengths are easily manageable, allowing construction from metal rods or tubing. The harp antenna requires a good ground plane. The polarization is vertical, and the radiation pattern is essentially omnidirectional in the azimuth plane. *See also* GROUND-PLANE ANTENNA.

HARTLEY OSCILLATOR

The Hartley oscillator is a form of variable-frequency oscillator. The operating frequency is determined by a parallel combination of inductance and capacitance. The feedback system is provided by a tap in the coil of the tank circuit. The illustration shows schematic diagrams for Hartley oscillators using a vacuum tube, a bipolar transistor, and a field-effect transistor.

The cathode, emitter, or source of the amplifying device is always connected to the coil tap in the Hartley configuration. This is how it can be recognized. The output is usually taken from the plate, collector, or drain circuit of the oscillator. However, the output may also be obtained by means of a loosely coupled coil near the tank coil. *See also* OSCILLATOR.

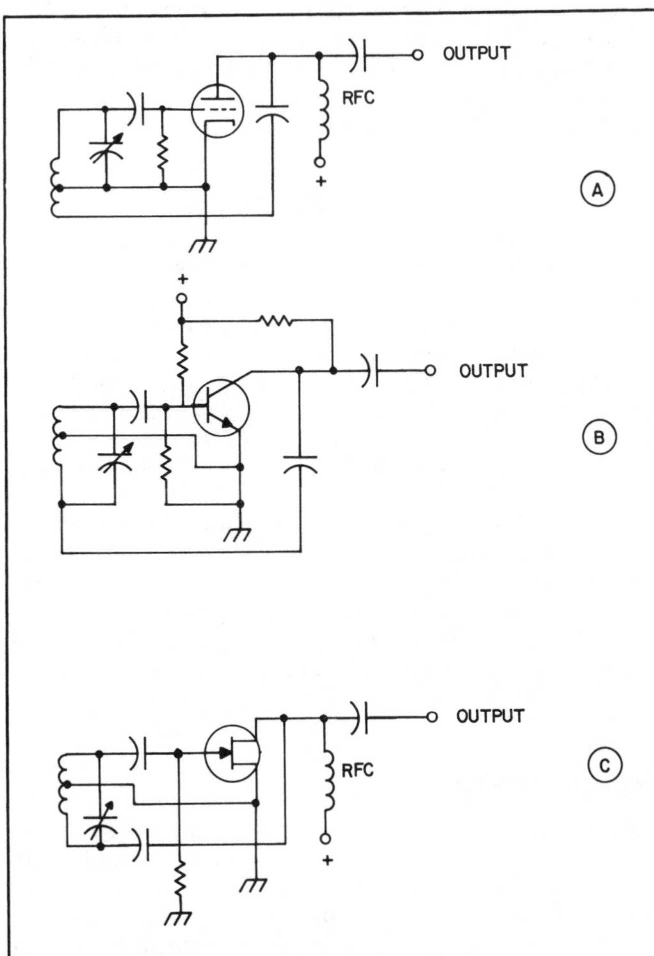

HARTLEY OSCILLATOR: Three forms of Hartley oscillator. At A, a triode vacuum-tube circuit; at B, an NPN bipolar-transistor circuit; at C, an N-channel field-effect-transistor circuit.

HASH NOISE

Hash, or hash noise, is a form of electrical noise generated by gas and mercury-vapor tubes. It is generally broadband in nature, and may also be called white noise. Hash noise is also generated to a certain extent by semiconductor rectifiers.

Hash noise may be a problem with sensitive receiving circuits, since some of the noise occurs at radio frequencies. The hash noise can be suppressed by housing the power supply components in a shielded enclosure, and placing radio-frequency chokes and bypass capacitors in the power-supply leads at the points where they leave the enclosure. *See also* WHITE NOISE.

HAY BRIDGE

The Hay bridge is a circuit designed for the purpose of measuring the value of an unknown inductance. The Hay bridge also gives an indication of the Q factor of the inductor under test. A typical Hay bridge is illustrated by the schematic diagram.

Two balance controls, consisting of potentiometers, are used in the bridge. One control is calibrated in terms of inductance, in millihenrys or microhenrys. The other is calibrated in terms of the Q factor, giving a measure of the reactance-to-resistance ratio of the tested coil. Correct adjustment is indicated by a null meter or headset.

The Hay bridge is generally used for the measurement

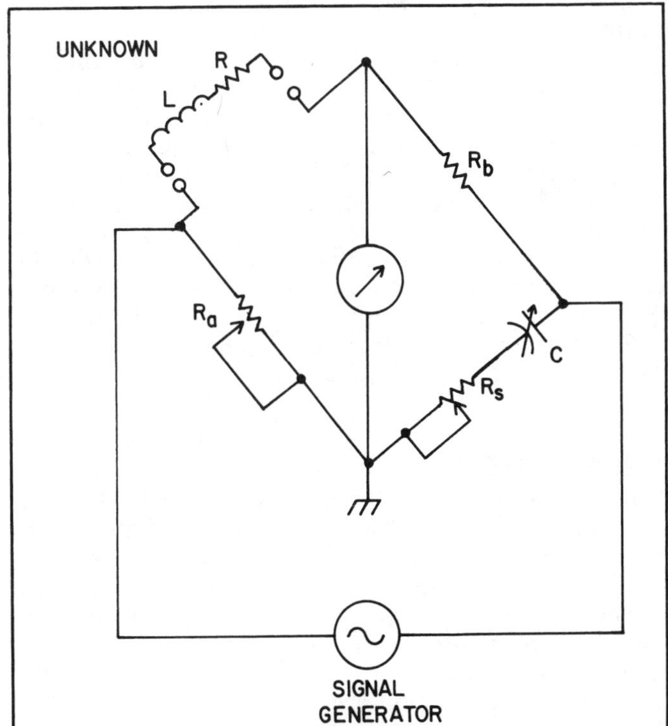

HAY BRIDGE: The Hay bridge is used for measuring relatively large inductances.

of relatively large inductances. The signal generator is thus an audio device in most cases. At very-high frequencies, the inductance of the component leads affects the accuracy of the bridge. If the resistances, as shown in the diagram, are given in ohms, and the capacitance is specified in farads, then the inductance of the coil in henrys is given by:

$$L = R_a R_b C/(1 + \omega^2 C^2 R_s^2)$$

where ω is the angular frequency in radians per second. The angular frequency is approximately equal to 6.3 times the frequency in hertz.

The Q factor of the unknown combination L and R is:

$$Q = 1/(\omega C R_s)$$

where ω is, again, the angular frequency in radians per second. *See also* INDUCTANCE, Q FACTOR.

HAZARD RATE

The hazard rate of an electronic circuit or component is the instantaneous failure rate. It is an expression of the probability that a device will fail immediately when it is put to use.

As the interval of time approaches zero, the failure rate of a component or circuit approaches the hazard rate. Some components do fail immediately when put to use, and therefore the hazard rate is never zero. *See also* FAILURE RATE.

HEAD

The term head is used in a variety of electronic applications. Generally, the head is a critical operating part of a device, especially a transducer such as a microphone, speaker, headphone, tape-recording circuit, or phonograph.

In a tape recorder, the magnetic device that impresses the information on the tape, or picks up the impulses from the tape, is called the head. The tape head is a dynamic device. Changing currents in a recording head produce fluctuating magnetic fields. Variable magnetic fields in the vicinity of a reproducing (playback) head result in electric currents.

It is important that the head in a tape recorder be properly aligned for optimum recording and reproduction with the magnetic tape. The gap should be perpendicular to the plane of the recording tape. *See also* MAGNETIC RECORDING.

HEADPHONE

A headphone is a pair of earphones, designed to be worn over the head so that both ears are covered. Headphones provide excellent intelligibility in communications, because external sounds are minimized.

Some headphones operate with both earphones connected in parallel or in series. Such devices are called monaural headphones. Some headphones have two independent earphones, and are used for listening to high-fidelity stereo (see photograph).

Headphones are available in impedances ranging from 4 ohms to more than 2,000 ohms. Some headphones have a wide-range frequency response; these are ideal for high-fidelity applications. Other types of headphones have peaked responses, intended for communications. The proper impedance and frequency response should be chosen for the desired situation. Most headphones are of the dynamic type, and resemble pairs of miniature speakers. *See also* EARPHONE, SPEAKER.

HEADPHONE: A typical headphone. This unit has cushioned earpads to help muffle external noises.

HEART FIBRILLATION

When a current of about 100 to 200 milliamperes passes through the heart muscle, the normal beating sometimes stops, and the muscles contract in a random manner. This condition is called heart fibrillation. Low blood potassium and chronic heart disease can also cause fibrillation.

Heart fibrillation is an extremely dangerous condition. Unless normal heart activity is restored, the body cells cannot receive the oxygen they need to survive. Brain damage occurs within a few minutes. Death follows shortly thereafter.

Cardiopulmonary resuscitation is sometimes effective in keeping a fibrillation victim alive until medical help arrives. An ambulance or physician should be called immediately if heart fibrillation is suspected. Medical professionals can often get the heart beating normally again by means of a device called a defibrillator. *See also* CARDIOPULMONARY RESUSCITATION, ELECTRIC SHOCK.

HEAT SINK

A device that helps to remove excess heat from electronic devices, by means of conduction, convection, and radiation, is called a heat sink. The use of a heat sink increases the power-dissipating capability of a transistor, tube, integrated circuit, or other active device. Some heat sinks are small, and fit around a single miniature component (see photograph). Other heat sinks are large, and may be used with several different components at once.

HEATSINK: A heatsink may be small and designed for a single component, as shown here; or it may be large and bulky, and used with several components.

The heat sink is made of metal, usually aluminum or iron, and is an excellent conductor of heat. The heat is thus carried away from the component by means of thermal conduction. The surface of the heat sink is deliberately designed to facilitate radiation and convection. This is usually accomplished by means of a finned surface. Heat sinks are often painted black to enhance the radiation.

When a heat sink is used, it is necessary that the component be thermally bonded to the device. Otherwise, the benefits will be lost. Silicone compound is generally used for thermal bonding. If it is necessary to electrically insulate the component from the heat sink, a flat piece of mica may be placed between the component and the heat sink. *See also* FINNED SURFACE.

HEAVISIDE LAYER

See KENNELLY-HEAVISIDE LAYER.

HECTO

Hecto is a prefix multiplier, corresponding to the quantity 100. This prefix is seldom used in electronics; generally, the prefixes in scientific applications are multiples of 1,000. The terms hectowatt (100 watts), hectovolt (100 volts), or hectohertz (100 hertz) are almost unheard of. *See also* PREFIX MULTIPLIERS.

HEIGHT ABOVE AVERAGE TERRAIN

At very-high frequencies and above, antenna performance is dependent on the height. The higher the antenna, the better the performance in most cases. In locations where the terrain is irregular, it is often difficult to determine the effective height of an antenna for communications purposes. Thus, the height-above-average-terrain (HAAT) figure has been devised.

In order to accurately determine the HAAT of an antenna, a topographical map is necessary. The location of the antenna must be found on the map. A circle is then drawn around the point corresponding to the location of the antenna. This circle should have a radius large enough to encompass at least several hills and valleys in the vicinity of the antenna. Within the circle, the average elevation of the terrain is calculated. This process involves choosing points within the circle in a grid pattern, adding up all of the elevation figures, and then dividing by the number of points in the pattern.

Once the average elevation, A, is found in the vicinity of the antenna, the elevation B of the antenna must be found by adding the height of the supporting structure to the ground-elevation figure at the antenna location. The HAAT is then equal to B−A.

The Federal Communications Commission imposes certain limitations, in some services, on antenna HAAT. This is especially true for repeater antennas. *See also* REPEATER.

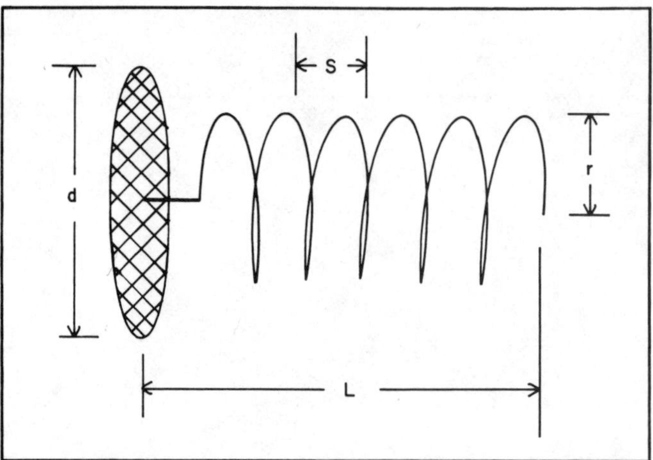

HELICAL ANTENNA: A helical antenna may be used to generate gain at short wavelengths. Here, d is the reflector diameter, L is the length of the antenna, s is the spacing between turns, and r is the radius of the helix.

HELICAL ANTENNA

A helical antenna is a form of circularly polarized, high-gain antenna used mostly in the ultra-high-frequency part of the radio spectrum. Circular polarization offers certain advantages at these frequencies (*see* CIRCULAR POLARIZATION).

The drawing illustrates a typical helical antenna. The reflecting device may consist of sheet metal or screen, in a disk configuration with a diameter d of at least 0.8 wavelength at the lowest operating frequency. The radius r of the helix should be approximately 0.15 wavelength at the center of the intended operating frequency range. The longitudinal spacing between turns of the helix, given by s, should be approximately ¼ wavelength in the center of the operating frequency range. The overall length of the helix, shown by L, may vary, but should be at least 1 wavelength at the lowest operating frequency. The longer the helix, the greater the forward power gain. Gain figures in excess of 15 dB can be realized with a single, moderate-sized helical antenna; when several such antennas are phased, the gain increases accordingly. Bays of two or four helical antennas are quite common.

The helical antenna illustrated will show a useful operating bandwidth equal to about half the value of the center frequency. An antenna centered at 400 MHz will function between approximately 300 and 500 MHz, for example. The helical antenna is normally fed with coaxial cable. The outer conductor should be connected to the reflecting screen or sheet, and the center conductor should be connected to the helix. The feed-point impedance is about 100 to 150 ohms throughout the useful operating frequency range.

Helical antennas are ideally suited to satellite communications, since the circular polarization of the transmitted and received signals reduces the amount of fading as the satellite orientation changes. The sense of the circular polarization may be made either clockwise or counterclockwise, depending on the sense of the helix. *See also* SATELLITE COMMUNICATIONS.

HELICAL FILTER

A helical filter, or helical resonator, is a quarter-wave device often used at very-high and ultra-high frequencies as a bandpass filter. A shielded enclosure contains a coil, one end of which is connected to the enclosure and the other

HELICAL FILTER: A typical helical filter.

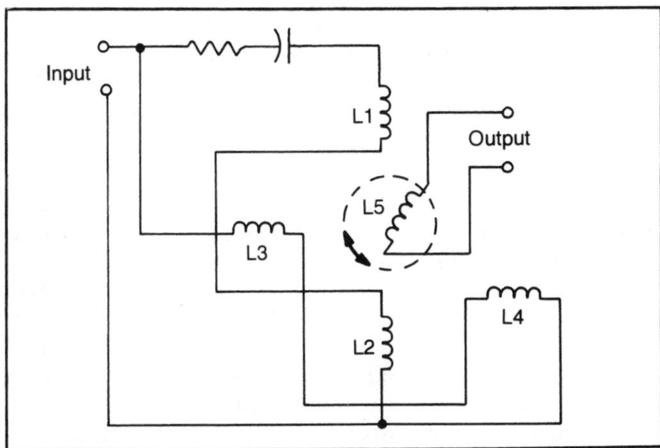

HELMHOLTZ COIL: The Helmholtz coil provides continuously variable adjustment of altering-current phase. Coil L5 is rotatable; coils L1 through L4 are fixed.

end of which is either left free or connected to a variable capacitor.

The photograph shows a three-section helical filter designed for operation in the band 142 MHz to 150 MHz. The three filter sections are tuned to slightly different frequencies, resulting in very little attenuation throughout the passband, but steep skirts and high attenuation outside the passband. Helical filters provide a high Q factor, which is important in the reduction of out-of-band interference. *See also* BANDPASS FILTER, Q FACTOR.

HELMHOLTZ COIL

A Helmholtz coil is an inductive device that provides continuously variable phase shift for an alternating-current signal. Two primary windings are oriented at right angles and split into two sections, as shown in the illustration. The currents in the two primary coils differ by 90 degrees; this phase shift is provided by a resistor and capacitor. The secondary coil is mounted on rotatable bearings. As the secondary coil is turned through one complete rotation, the phase of the signal at the output terminals changes continuously from 0 to 360 degrees. Any desired signal phase may be chosen by setting the coil to the proper position.

The Helmholtz coil works because the fields from the primary windings add together as vectors. The magnitudes of the two component vectors change in the secondary coil as the secondary coil is turned. The Helmholtz coil is frequency sensitive; that is, it will work at only one frequency. To change the operating frequency, the values of the resistor and capacitor must be changed to provide a 90-degree phase difference between the two primary coils. *See also* PHASE ANGLE.

HENRY

The henry, abbreviated H, is the standard unit of inductance. In a circuit in which the current is changing at a constant rate of 1 ampere per second, an inductance of 1 henry results in the generation of 1 volt of potential difference across an inductor.

The henry is an extremely large unit of inductance. It is rare to find a coil with a value of 1 H. Therefore, inductance values are generally given in millihenrys (mH), microhenrys (uH), or nanohenrys (nH). An inductance of 1 mH is equal to 0.001 H; 1 uH is 0.001 mH; 1 nH is 0.001 uH. *See also* INDUCTANCE.

HEPTODE

A heptode is a vacuum tube having seven internal elements. In addition to the plate and cathode, there are usually five grids in the heptode. Alternatively, there may be one or more control electrodes for such purposes as accelerating the electron beam.

Heptodes are seldom seen, but they are sometimes found in mixers and modulators. However, in modern electronic equipment, transistors have largely replaced vacuum tubes. The heptode is a rarity among rarities. *See also* TUBE.

HERMAPHRODITIC CONNECTOR

A hermaphroditic connector is an electrical plug that mates with another plug exactly like itself. Such a connector has an equal number of male and female contacts.

Hermaphroditic connectors can be put together in only one way. This makes them useful in polarized circuits such as direct-current power supplies. *See also* CONNECTOR.

HERMETIC SEAL

Some electronic components are susceptible to damage or malfunction from moisture in the air. This is especially true of piezoelectric crystals and certain semiconductor devices. Such components are often enclosed in hermetically sealed cases. A hermetic seal is simply an airtight, durable seal. The common oscillator crystal, housed in a metal can, is hermetically sealed.

A hermetic seal must be long-lasting and physically rugged. Many different kinds of glues and cements are acceptable for hermetic sealing. Sometimes the interior of a hermetically sealed enclosure is filled with an inert gas such as helium, to further retard the deterioration of the component or components inside.

HERTZ

Hertz is the standard unit of frequency. It is abbreviated Hz. A frequency of 1 complete cycle per second is a frequency of 1 Hz. The term hertz became widespread in the late 1960s, and is now used instead of the term cycles per second.

In radiocommunication, signals are typically thousands, millions, or billions of hertz in frequency. A frequency of 1,000 Hz is called 1 kilohertz (kHz); a frequency of 1,000 kHz is 1 megahertz (MHz); a frequency of 1,000 MHz is 1 gigahertz (GHz). Sometimes the terahertz (THz) is used as a measure of frequency; 1 THz = 1,000 GHz.

The angular frequency in radians per second is equal to approximately 6.3 times the frequency in hertz. *See also* FREQUENCY.

HERTZ ANTENNA

A Hertz antenna is any horizontal, half-wavelength antenna. The feed point may be at the center, at either end, or at some intermediate point. The Hertz antenna operates independently of the ground, and is therefore a balanced antenna.

Examples of the Hertz antenna include the dipole and the zeppelin antenna. The driven element of a Yagi antenna is often of the Hertz configuration. *See also* DIPOLE ANTENNA, YAGI ANTENNA, ZEPPELIN ANTENNA.

HERTZ, HEINRICH RUDOLPH

The experimental demonstration of electromagnetic-wave effects is generally credited to physicist Heinrich Rudolph Hertz (1857-1894). Hertz showed that electromagnetic waves propagate in the same way as light waves. This eventually led to the wave theory of light and the discovery of the electromagnetic spectrum. The standard unit of frequency, previously known as the cycle per second, is now called the Hertz. *See also* ELECTROMAGNETIC SPECTRUM, HERTZ.

HERTZ OSCILLATOR

The Hertz oscillator was one of the first radio-frequency generating devices. It was developed by Hertz to demonstrate the action of electromagnetic induction in the latter part of the nineteenth century.

The Hertz oscillator contains a spark gap which is operated from a high-voltage power source. The high voltage is derived from a coil called an induction coil. Resonant inductance-capacitance circuits, placed in series with the supply voltage, confine the noise from the spark to a narrow band of radio frequencies. Radiation takes place from two metal plates, which act as a large capacitor.

The earliest radio transmitters made use of the spark principle to generate their radio-frequency energy. The output of such a transmitter took place over a small band of frequencies, rather than at a single discrete frequency as is the case with modern continuous-wave transmitters. In place of the metal plates, a tuned antenna system was connected to the output of the oscillator. The Hertz oscillator is not a true oscillator, but instead is a resonant device that concentrates the noise from a spark generator to a narrow band of frequencies. *See also* OSCILLATION, SPARK TRANSMITTER.

HETERODYNE

The term heterodyne can refer to either of two things. A heterodyne is a mixing product resulting from the combination of two signals in a nonlinear component or circuit. Mixing is sometimes called heterodyning.

When two waves having frequencies f and g are combined in a nonlinear component or circuit, the original frequencies appear at the output along with energy at two new frequencies. These frequencies are the sum and difference frequencies, $f + g$ and $f - g$. They are sometimes called heterodyne frequencies. A tuned circuit can be used to choose either the sum or the difference frequency. In the mixer, $f + g$ and $f - g$ are usually much different. In the heterodyne or product detector, they may be very close together.

Heterodyning occurs to a certain extent whenever two signals are present in the same medium. This is because no circuit is perfectly linear, and some distortion always takes place. A diode, capable of effectively handling frequencies f and g, or a transistor biased to cutoff, are ideal for heterodyning purposes. *See also* FREQUENCY CONVERSION, FREQUENCY CONVERTER, HETERODYNE DETECTOR, MIXER.

HETERODYNE DETECTOR

A heterodyne detector is a detector that operates by beating the signal from a local oscillator against the received signal. This form of detector is more often called a product detector, and is required for the reception of continuous-wave, frequency-shift-keyed, and single-sideband signals.

The incoming signal information is extracted by heterodyning, resulting in audible difference frequencies. In the case of a continuous-wave signal, the difference may be as small as about 100 Hz or as large as about 3 kHz; the same is true with frequency-shift keying. For single-sideband reception, the local oscillator frequency should correspond to the frequency of the suppressed carrier.

A heterodyne detector is used in many superheterodyne receivers and all direct-conversion receivers. *See also* PRODUCT DETECTOR.

HETERODYNE FREQUENCY METER

A heterodyne frequency meter is a device similar to a direct-conversion receiver, used for the purpose of measuring unknown frequencies. A calibrated local oscillator is used, along with a mixer and amplifier, as shown in the diagram.

The signal of unknown frequency is fed into the mixer along with the output of the local oscillator. It is helpful to have some idea of the frequency of the signal beforehand, because harmonics of the local oscillator will produce false readings. As the local oscillator is tuned, or heterodynes occur in the output. These heterodynes occur at frequencies f, f/2, f/3, f/4, and so on, as read on the local-oscillator calibrated scale. The correct reading is f, the highest frequency. Readings may also be obtained at 2f, 3f, 4f, and so on, corresponding to harmonics of the input signal; but these components are actually present, while the lower-frequency indications are false signals. (Some heterodyne frequency meters have tuned input circuits, which track along with the local oscillator fundamental frequency, reducing harmonic effects.) The correct reading usually gives the loudest heterodyne The local oscillator should be adjusted for zero beat before the final reading is taken.

All receivers with product detectors can be used as heterodyne frequency meters within their operating ranges, provided the dial is calibrated with reasonable accuracy. It is helpful to have a crystal calibrator in such a

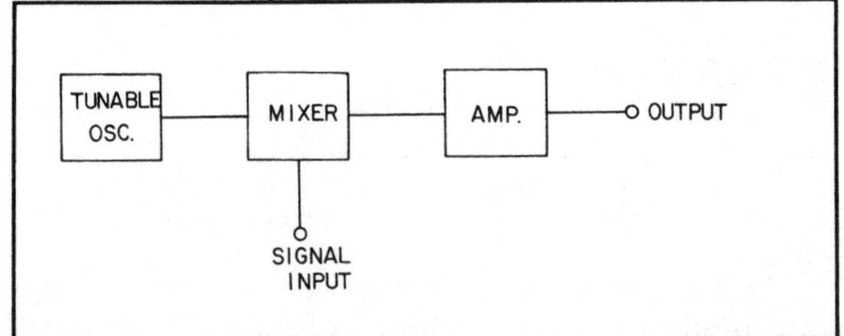

HETERODYNE FREQUENCY METER: Block diagram of a simple heterodyne frequency meter.

receiver, and this calibrator should be adjusted against a frequency standard, such as WWV or WWVH. *See also* FREQUENCY MEASUREMENT, HETERODYNE.

HETERODYNE REPEATER

When a signal is amplified and retransmitted, the frequency of the signal must usually be changed. Otherwise, feedback is likely to result in oscillation between the receiver and transmitter circuits. Virtually all repeaters use frequency converters, and they are therefore called heterodyne repeaters.

The input signal to the heterodyne repeater might have a frequency f. This signal is heterodyned with the output of a local oscillator at frequency g, producing mixing products at frequencies f + g and f − g. Generally, the difference frequency is chosen. A tuned bandpass filter, centered at frequency f − g, serves to eliminate other signals. An amplifier tuned to frequency f − g provides the output. At the input of the repeater, a notch or band-rejection filter may be used, centered at frequency f − g, to prevent desensitization of the receiver by the transmitter.

In general, the greater the difference between the input and output frequencies of a heterodyne repeater, the more easily the circuit is designed, and the less difficulty is encountered with receiver desensitization. *See also* FREQUENCY CONVERSION, REPEATER.

HEXODE

A hexode is a vacuum tube having six internal elements. In addition to the plate and the cathode, there are usually four grids in the hexode. Alternatively, there may be one or more control electrodes for such purposes as accelerating the electron beam.

Hexodes are seldom seen, but they are sometimes found in mixers, modulators, and television camera and picture devices. However, in modern electronic equipment, transistors have largely replaced vacuum tubes. The hexode is therefore virtually an anachronism, except in television applications. *See also* TUBE.

HEXADECIMAL NUMBER SYSTEM

The hexadecimal number system is a base-16 system. Hexadecimal notation is commonly used by computers, because 16 is a power of 2. The numbers 0 through 9 are the same in hexadecimal notation as they are in the familiar decimal number system. However, the values 10 through 15 are represented by single "digits," usually the letters A through F.

Addition, subtraction, multiplication, and division are somewhat different in hexadecimal notation. For example, in decimal language, we say that 5 + 7 = 12; but in hexadecimal, 5 + 7 = C. The carry operation does not occur until a sum is greater than F. Thus, 8 + 7 = F, and 8 + 8 = 10. *See also* NUMBER SYSTEMS.

HIGHER-ORDER LANGUAGE

A higher-order or high-level computer language is the interface between the machine and the human operator. The computer itself "thinks" in binary or machine language, and the construction of programs in this language is tedious. The higher-order language is often quite similar to ordinary language. A translation program, called an assembler, converts the machine language back into terms the operator can readily understand.

Examples of higher-order languages are BASIC (used for mathematical and scientific calculations), COBOL (used in business applications), FORTRAN (used mostly by scientists and engineers), and the languages of computer games. *See also* ASSEMBLER AND ASSEMBLY LANGUAGE, BASIC, COBOL, FORTRAN.

HIGH FIDELITY

High fidelity is a term used to describe audio systems engineered for highly accurate reproduction of the original sound. The typical high-fidelity recording and reproducing system is sensitive to sounds at all audio frequencies, and even beyond.

In the recording process, two channels are generally used, called the left and right channels. The microphones have a wide frequency response. It is important that the recording tape be capable of accurately holding impulses at frequencies of at least 20 kHz.

In the reproducing process, the audio amplifiers must be capable of linear operation over a range of frequencies from at least 20 Hz to 20 kHz. The speakers or headphones are an important part of the high-fidelity reproducing system. Communications type speakers or headphones are not satisfactory because of their peaked audio response.

In recent years, great progress has been made in the field of high-fidelity recording and reproducing. This has been largely because of the advent of solid-state technology. As a result of this, many households in the industrialized countries have high-fidelity tape playback systems, phonographs, and radio receivers. *See also* FM STEREO, MAGNETIC RECORDING, STEREO DISK RECORDING, STEREOPHONICS.

HIGH FREQUENCY

High frequency is the designator applied to the range of frequencies from 3 MHz to 30 MHz. This corresponds to a wavelength between 100 meters and 10 meters. The high frequencies include all of the so-called shortwave bands. Ionospheric propagation is of great importance at the high frequencies.

At the lower end of the high-frequency part of the electromagnetic spectrum, the ionosphere almost always reflects energy back to the earth. Conditions are highly variable within the high-frequency range, and depend largely on the sunspot cycle, the time of day, and the time of year. At the upper end of the high-frequency range, near 30 MHz, the ionosphere is often transparent to radio signals. The F layer is the primary ionospheric layer responsible for long-distance communication in the high-frequency spectrum. *See also* F LAYER, IONOSPHERE, PROPAGATION CHARACTERISTICS.

HIGHPASS FILTER

A highpass filter is a combination of capacitance, inductance, and/or resistance, intended to produce large

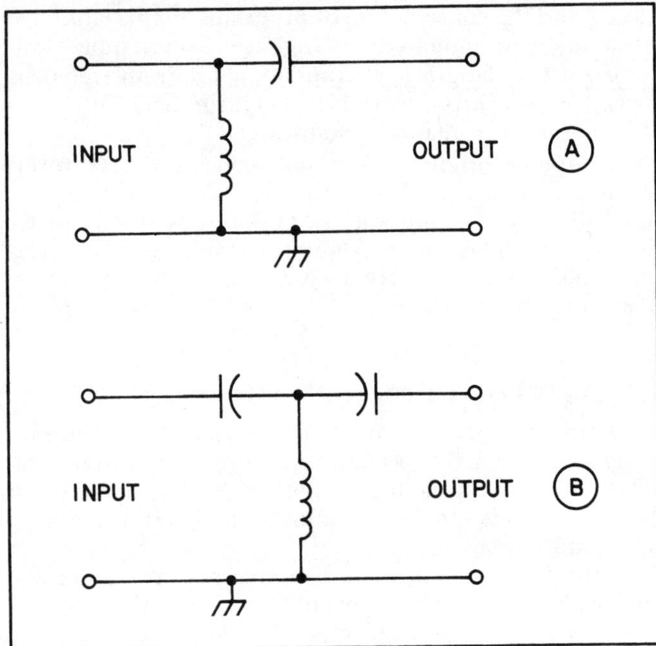

HIGHPASS FILTER: Two common forms of unbalanced highpass filter. At A, the L network; at B, the T network.

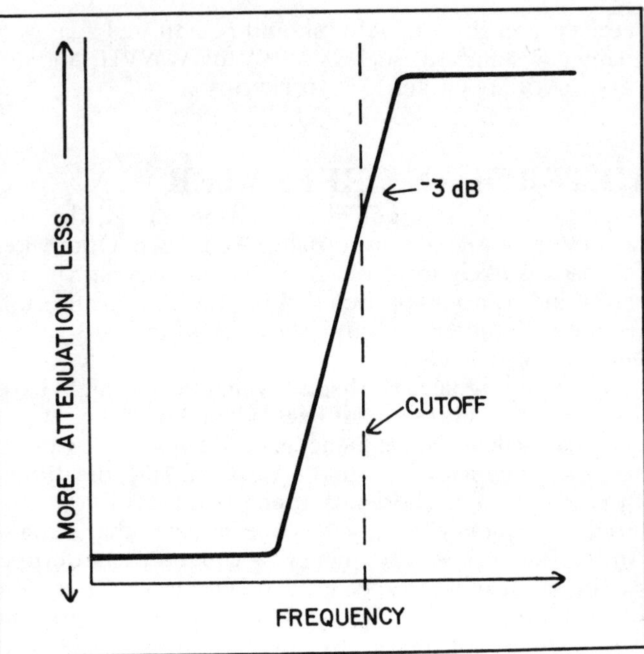

HIGHPASS RESPONSE: The ideal highpass response is characterized by uniform attenuation at frequencies well above and below the cutoff. The transition is smooth.

amounts of attenuation below a certain frequency and little or no attenuation above that frequency. The frequency at which the transition occurs is called the cutoff frequency (*see* CUTOFF FREQUENCY). At the cutoff frequency, the voltage attenuation is 3 dB with respect to the minimum attenuation. Above the cutoff frequency, the voltage attenuation is less than 3 dB. Below the cutoff, the voltage attenuation is more than 3 dB.

The simplest highpass filters consist of a parallel inductor or a series capacitor. More sophisticated highpass filters have a combination of parallel inductors and series capacitors, such as the filters shown in the illustration. The filter at A is called an L-section filter; that at B is called a T-section filter. These names are derived from the geometric shapes of the filters as they appear in the schematic diagram.

Resistors are sometimes substituted for the inductors in a highpass filter. This is especially true if active devices are used, in which case many filter sections can be cascaded.

Highpass filters are used in a wide variety of situations in electronic apparatus. One common use for the highpass filter is at the input of a television receiver.The cutoff frequency of such a filter is about 40 MHz. The installation of such a filter reduces the susceptibility of the television receiver to interference from sources at lower frequencies. *See also* HIGHPASS RESPONSE.

HIGHPASS RESPONSE

A highpass response is an attenuation-versus-frequency curve that shows greater attentuation at lower frequencies than at higher frequencies. The sharpness of the response may vary considerably. Usually, a highpass response is characterized by a high degree of attenuation up to a certain frequency, where the attenuation rapidly decreases. Finally the attenuation levels off at near zero insertion loss. The highpass response is typical of highpass filters.

The cutoff frequency of a highpass response is that frequency at which the insertion loss is 3 dB with respect to the minimum loss. The ultimate attenuation is the level of attenuation well below the cutoff frequency, where the signal is virtually blocked. The ideal highpass response should look like the attenuation-versus-frequency curve in the illustration. The curve is smooth, and the insertion loss is essentially zero everywhere well above the cutoff frequency. *See also* CUTOFF FREQUENCY, HIGHPASS FILTER, INSERTION LOSS.

HIGH Q

High Q is a term used to describe a filter circuit with a great deal of selectivity. The term is usually applied to bandpass and band-rejection filters.

High-Q filters are desirable in situations where the response must be confined to a single signal, or to a very narrow range of frequencies. However, this characteristic is not always wanted. In general, high-Q circuits require low-loss inductors and capacitors, and the resistance should be as low as possible. *See also* Q FACTOR.

HIGH-TENSION POWER LINE

A high-tension power line is an alternating-current power line in which the voltage is extremely high. Typical voltage values in commercial high-tension lines are in the range 10,000 to 100,000 volts. Some high-tension lines have voltages of up to a million.

High-tension lines are a more efficient means of transferring power than low-voltage lines. This is because, for a given amount of power, higher voltage means lower current — and it is the current that produces ohmic losses in electrical conductors. When the current is as low as possible, the so-called I^2R losses are minimized.

The main problem with high-tension lines is that the voltages are so high that large insulators and massive towers are required to prevent destructive arcing. This is primarily an expense factor. However, the losses are so

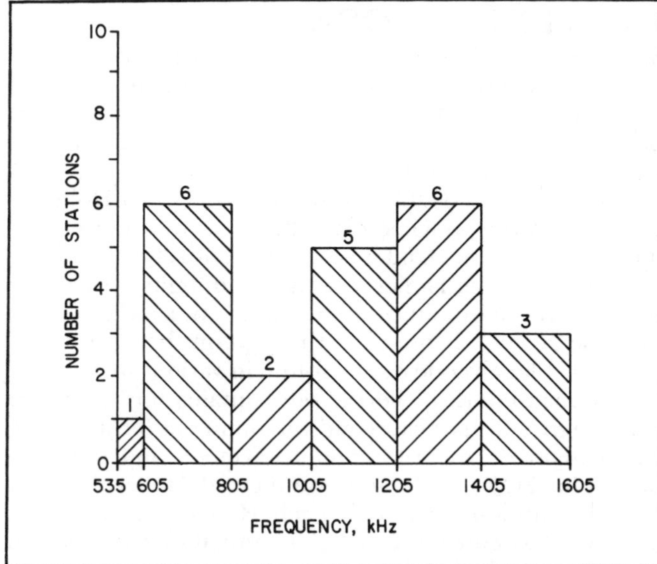

HISTOGRAM: A histogram showing the number of received broadcast stations in various segments of the standard amplitude-modulation broadcast band. (The lowest segment, since it covers a narrower range of frequencies than the others, gives a disproportionately small value.)

much lower than with equivalent low-voltage lines, that this expense is more than offset by improvements in efficiency.

Most high-tension power lines operate at 60 Hz, the same frequency as the alternating current in conventional utility outlets. However, some thought has been given to direct-current power transmission over long distances, since power loss is reduced with direct current as compared with alternating current. *See also* AC POWER TRANSMISSION.

HIGH-THRESHOLD LOGIC

High-threshold logic is a form of bipolar digital logic, often abbreviated HTL. High-threshold logic is exactly the same as diode-transistor logic (DTL), except that Zener diodes are added in series with each input line. The Zener diodes greatly reduce the noise susceptibility of DTL.

High-threshold logic devices require rather large operating voltages, and the power dissipation is therefore higher than with DTL. But, in situations where noise must be minimized, HTL gives superior performance. *See also* DIODE-TRANSISTOR LOGIC.

HISS

A form of audio noise, in which the amplitude is peaked near the midrange or treble regions, is often called hiss. Hiss is always heard in active audio-frequency circuits. The sound of hiss is familiar to anyone who has worked with high-fidelity equipment, public-address systems, or radio receivers.

In audio equipment, the hiss is generated as the result of random electron or hole movement in components. The hiss generated in the early stages is amplified by subsequent stages. In a radio receiver, hiss can be generated in the intermediate-frequency stages and in the front end, mixers, and oscillators. Some hiss even originates outside the receiver, in the form of thermal noise in the antenna conductors and atmosphere. *See also* WHITE NOISE.

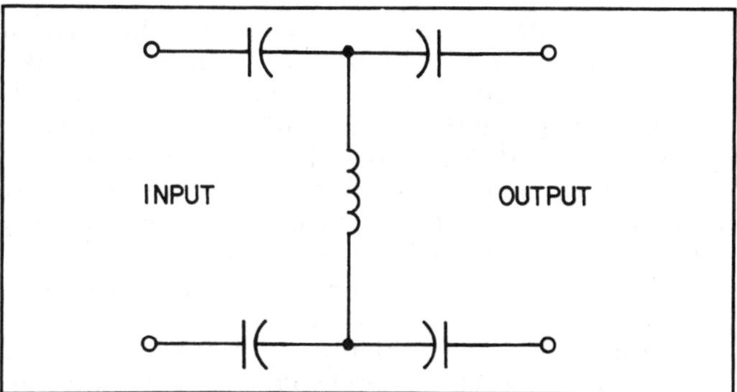

H NETWORK: An H-network highpass filter.

HISTOGRAM

A histogram is a form of graphical representation. The histogram shows probabilities of various events, as a function of some parameter. The histogram can be recognized by the presence of vertical rectangles of varying heights. Such graphical illustrations are commonly used to show statistical distributions.

The illustration is a histogram showing the distribution of received radio signals within the standard AM broadcast band at a hypothetical location. The signals are first found, by means of a radio receiver, and then categorized according to carrier frequency. Note, however, that discrete frequencies are not given; instead, the band is divided into a number of equal-sized ranges. The result is an indication of the relative signal density over the band. This histogram will vary in appearance over the course of a day, since reception in the AM broadcast band is much different during the darkness hours than during daylight.

The histogram is an extremely simple and effective means of illustrating certain statistical functions. However, it is relatively imprecise. The example shown gives us an approximate idea of where to tune the receiver to find the most signals; but it does not show the exact frequencies of the signals. *See also* CARTESIAN COORDINATES, GRAPH.

H NETWORK

An H network is a form of filter section, sometimes seen in balanced circuits. The H network gets its name from the fact that the schematic-diagram component arrangement looks like the capital letter H turned sideways (see illustration).

The H configuration may be used in the construction of bandpass, band-rejection, highpass, and lowpass filters. The H configuration is a popular arrangement for attenuator design; noninductive resistors are employed. Several H networks may be cascaded to obtain better filter or attenuator performance.

The H network is not the only configuration for filters and attenuators. *See also* L NETWORK, PI NETWORK, T NETWORK.

HOLDING CURRENT

Holding current is the minimum amount of current that will keep a switching device actuated. The term applies to relays, and also to various tube and solid-state devices, especially the silicon-controlled rectifier and thyristor.

In any switching device, a certain amount of current

must flow before a change of state occurs. This turn-on current is always larger than the holding current. In relays, the difference between the turn-on and holding currents is usually rather small; the initial current must therefore be nearly maintained. But in the silicon-controlled rectifier and thyristor, the current may be reduced considerably following turn-on, and the device will remain actuated. Relays, silicon-controlled rectifiers, and thyristors with various ratios of turn-on current on holding current are available.

Silicon-controlled rectifiers and thyristors with low holding-current levels are preferable when the load impedance is high. This keeps the device in the actuated state under conditions of variable current. A larger holding-current value may be wanted when the load impedance is low, so that the device can be switched off easily. *See also* RELAY, SILICON-CONTROLLED RECTIFIER, THYRISTOR.

HOLE

A hole is a carrier of electric charge. In certain semiconductor materials, the charge carriers are predominantly holes rather than electrons. A hole is an atom with one electron missing. Holes have positive charge, while electrons have negative charge.

An electric current in a P-type semiconductor material flows as the result of hole movement. The electron-deficient atoms do not themselves move; the electrons migrate from atom to atom, creating a chain of vacancies which moves from the positive to the negative (see illustration). The holes in a P-type material behave in much the same way as the electrons in an N-type substance or ordinary wire conductor.

While an electron is an identifiable particle of matter, a hole is not. The concept of the hole makes it easier to explain the operation of a semiconductor device in some cases. *See also* ELECTRON, ELECTRON-HOLE PAIR, N-TYPE SEMICONDUCTOR, P-N JUNCTION, P-TYPE SEMICONDUCTOR.

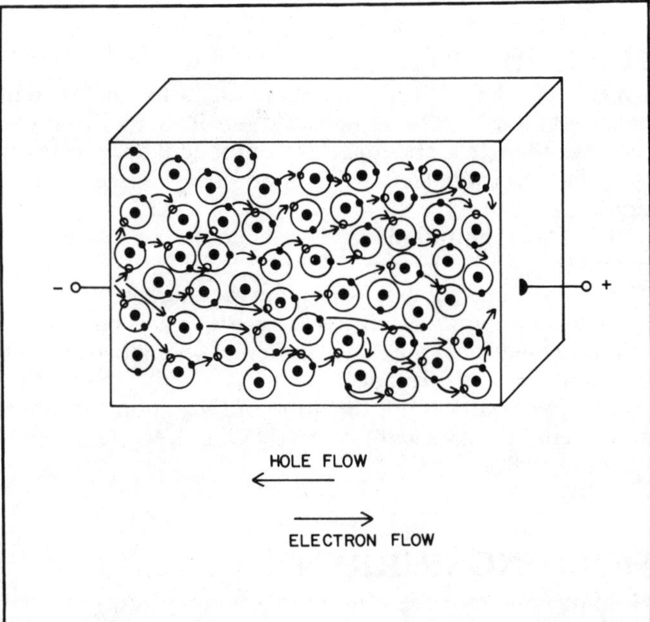

HOLE: A greatly simplified rendition of hole flow and electron flow in a P-type semiconductor. Atomic nuclei are shown as heavy black dots. Electrons are represented by small black dots, and electron vacancies (holes) by small open dots. The large circles are electron orbits.

HOLLERITH CODE

Many digital computers make use of punched cards. The standard rectangular computer card is easy to recognize; it measures 3¼ inches by 7⅜ inches, and is made of stiff paper. Holes in the card are arranged according to a code known as the Hollerith code.

There are 12 possible positions for data in a Hollerith character. Most punched cards have 12 rows and 80 columns. Each column contains one Hollerith character. The table indicates the Hollerith letters and digits, along with the more common punctuation marks. An x indicates that the hole position is punched through on the card.

The Hollerith code can be found in various forms. The basic alphanumeric Hollerith code contains only the letters A through Z (upper case) and the numerals 0 through 9. The extended Hollerith contains symbols and punctuation. The FORTRAN and commercial Hollerith codes contain additional symbols. Each Hollerith card, which normally has 80 characters, can contain one full line of data. Card readers can process the cards at extreme speed. *See also* CARD READING.

HOLOGRAPHY

Holography is a means of reproducing three-dimensional images. The three-dimensional photograph is called a hologram. Holography can reproduce the effects of mo-

HOLLERITH CODE: PUNCHED LOCATIONS ARE MARKED BY AN X. IN THIS TABLE, ROWS ARE SHOWN VERTICALLY AND COLUMNS ARE SHOWN HORIZONTALLY.

Character	12	11	0	1	2	3	4	5	6	7	8	9
A	x			x								
B	x				x							
C	x					x						
D	x						x					
E	x							x				
F	x								x			
G	x									x		
H	x										x	
I	x											x
J		x		x								
K		x			x							
L		x				x						
M		x					x					
N		x						x				
O		x							x			
P		x								x		
Q		x									x	
R		x										x
S			x		x							
T			x			x						
U			x				x					
V			x					x				
W			x						x			
X			x							x		
Y			x								x	
Z			x									x
0			x									
1				x								
2					x							
3						x						
4							x					
5								x				
6									x			
7										x		
8											x	
9												x
.	x					x					x	
,			x			x					x	

tion as well as the three-dimensional details of an object or scene. The hologram can also show color.

The hologram is recorded as an interference pattern on photographic film. Under ordinary light, the hologram looks meaningless. A source of coherent light, such as a laser, is required in order to view the image.

Viewing a hologram for the first time is a strange experience. Models are often used to demonstrate the versatility of the hologram; typical images are a chess board or a forest. When the viewer moves, different parts of the scene become visible.

HOMING DEVICE

Some automatic guidance systems make use of homing devices. The homing device requires a signal source for reference. The signal source may be a transmitter that sends out a beacon signal, which the homing device locates and follows. Or, the signal source may be an enemy transmitter, a source of sound waves, or a source of other emission such as infrared.

All homing devices have receivers or detectors, which sense the signal or emission to which a vehicle is guided. A directional receptor determines whether the vehicle is or is not on the correct course. Sophisticated homing devices use computers to make the necessary course corrections, so that the vehicle continues toward the target.

Homing devices are especially useful in the guided missile. The infrared radiation from populated areas may be used to guide ballistic missiles to their targets. The signals from radio and television broadcasting towers can be picked up and used for guidance. Even the visible light, generated in cities during the nighttime hours, can be used. *See also* DIRECTION FINDER, GUIDED MISSILE.

HOOK TRANSISTOR

A hook transistor is a four-layer semiconductor device, also called a PNPN transistor. The outer P-type and N-type semiconductor layers serve as the emitter and collector; the inner N-type layer serves as the base.

The hook transistor contains an extra P-type layer between the base and the collector. This extra layer results in increased gain at the higher frequencies. The extra P-N junction enhances the flow of charge carriers. It also tends to reduce the capacitance in the base-collector junction, allowing the device to operate effectively at higher frequencies than is possible with a conventional PNP or NPN bipolar transistor. *See also* TRANSISTOR.

HOOKUP WIRE

Hookup wire is a flexible, insulated form of copper wire, employed in electronic circuits for the purpose of circuit-board and component interconnection. The stranded form of wire is more resistant to breakage, and adheres more readily to solder, than solid wire. Some hookup wire is pre-tinned, and has a silvery color, while other forms are made from plain copper.

Hookup wire is available in a wide variety of sizes and colors. The smallest hookup wire is approximately No. 26 size; the largest is about No. 14. (These sizes are AWG, or American Wire Gauge.) *See also* AMERICAN WIRE GAUGE, WIRE.

HORIZON

The horizon is the apparent edge of the earth, as viewed from a particular height above the surface at a particular electromagnetic frequency. The radio horizon is generally farther away than the visual horizon. The radio horizon is of considerable importance in communication at very-high frequencies and above, because most communication in this part of the spectrum takes place via so-called direct or line-of-sight propagation.

The distance to the horizon depends on several factors. In general, the higher the viewing point, the greater the distance to the horizon, regardless of frequency. The distance is also dependent on the frequency; at some wavelengths the atmosphere displays refractive effects to a greater extent than at other wavelengths. In calculating the effective distance to the radio horizon, it must be realized that the values obtained are theoretical, and are based on smooth terrain. In areas having many large hills or mountains, the actual distance to the radio horizon can vary greatly from the theoretical values.

Let h be the height of the viewpoint in feet over smooth earth, and let d be the distance to the horizon in miles. Then, at visual wavelengths,

$$d = \sqrt{1.53\,h}$$

The infrared horizon is essentially the same as the visual horizon. The ultraviolet and gamma-ray horizons are also the same, for all practical purposes, as the visual horizon. However, as the wavelength becomes much longer than the infrared, the horizon begins to lengthen. This is equivalent to an effective increase in the radius of the earth. For radio waves in the very-high and ultra-high frequency bands, a good approximation for the distance to the horizon is given by:

$$d = \sqrt{2h}$$

where d is again specified in miles and h is specified in feet.

For a complete radio circuit, with a transmitting antenna at height g feet and a receiving antenna at height h feet, the effective line-of-sight path can be calculated by:

$$d = \sqrt{2g} + \sqrt{2h}$$

See also DIRECT WAVE, LINE-OF-SIGHT COMMUNICATION.

HORIZONTAL LINEARITY

Horizontal linearity is one expression of the degree to which a television picture is an accurate reproduction of a scene. Most television receivers have an internal adjustment for calibrating the horizontal linearity. In a few receivers, the control is external.

The horizontal linearity of a television receiver should be adjusted only by a technician with the proper test equipment. Improper horizontal linearity results in a distorted picture. If the horizontal linearity is not properly set, objects that appear to have a certain width in one part of the screen will seem to be wider or narrower at another location on the screen.

All television picture signals are broadcast with a linear horizontal scan. Improper horizontal linearity is usually the fault of the receiver. *See also* TELEVISION.

HORIZONTAL POLARIZATION

When the electric lines of force of an electromagnetic wave are oriented horizontally, the field is said to be horizontally polarized. In communications, horizontal polarization has certain advantages and disadvantages at various wavelengths.

At the low and very-low frequencies, horizontal polarization is not often used. This is because the surface wave, an important factor in propagation at these frequencies, is more effectively transferred when the electric field is oriented vertically. Most standard AM broadcast stations, operating in the medium-frequency range, also employ vertical rather than horizontal polarization.

In the high-frequency part of the electromagnetic spectrum, horizontal polarization becomes practical. The polarization is always parallel to the orientation of the radiating antenna element; horizontal wire antennas are simple to install above about 3 MHz. The surface wave is of lesser importance at high frequencies than at low and very-low frequencies; the sky wave is the primary mode of propagation above 3 MHz. This becomes increasingly true as the wavelength gets shorter. Horizontal polarization is just as effective as vertical polarization in the sky-wave mode.

In the very-high and ultra-high frequency range, either vertical or horizontal polarization may be used. Horizontal polarization generally provides better noise immunity and less fading than vertical polarization in this part of the spectrum. *See also* CIRCULAR POLARIZATION, POLARIZATION, VERTICAL POLARIZATION.

HORIZONTAL SYNCHRONIZATION

In television communications, the picture signals must be synchronized at the transmitter and receiver. The electron beam in the television picture tube scans from left to right and top to bottom, in just the same way as you read the page of a book. At any given instant of time, in a properly operating television system, the electron beam in the receiver picture tube is in exactly the same relative position as the scanning beam in the camera tube. This requires synchronization of the horizontal as well as the vertical position of the beams.

When the horizontal synchronization in a television system is lost, the picture becomes totally unrecognizable. This is illustrated by misadjustment of the horizontal-hold control in any television set. Even a small synchronization error results in severe "tearing" of the picture. The transmitted television signal contains horizontal-synchronization pulses at the end of every line. This tells the receiver to move the electron beam from the end of one line to the beginning of the next line. Vertical synchronization pulses tell the receiver that one complete picture, called a frame, is complete, and that it is time to begin the next frame. *See also* PICTURE SIGNAL, TELEVISION, VERTICAL SYNCHRONIZATION.

HORN ANTENNA

A horn antenna is a device used for transmission and reception of signals at ultra-high and microwave frequencies. There are several different configurations of the horn antenna, but they all look similar. The illustration is a pictorial drawing of a commonly used horn antenna.

The horn antenna provides a undirectional radiation pattern, with the favored direction coincident with the opening of the horn. Horn antennas are characterized by a lowest usable frequency. The feed system generally consists of a waveguide, which joins the horn at its narrowest point.

Horn antennas are often used in feed systems of large dish antennas. The horn is pointed toward the center of the dish, in the opposite direction from the favored direction of the dish. When the horn is positioned at the focal point of the dish, extremely high gain and narrow-beam radiation are realized. *See also* DISH ANTENNA.

HORN SPEAKER

A horn speaker is a directional speaker used at midrange and high audio frequencies. Horn speakers are generally used as part of a multiple-speaker system, including woofers for the bass range, midrange speakers, and tweeters for the high range.

The horn speaker has a fairly narrow frequency response. The sound is concentrated in a fairly sharp beam. Baffles are often used in the horn structure to help distribute the sound in the horizontal plane. The illustration shows a typical horn speaker with internal baffles.

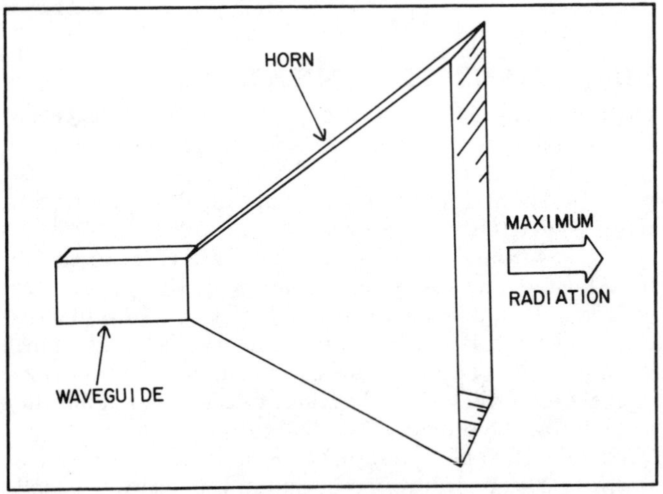

HORN ANTENNA: A typical horn antenna for microwave use.

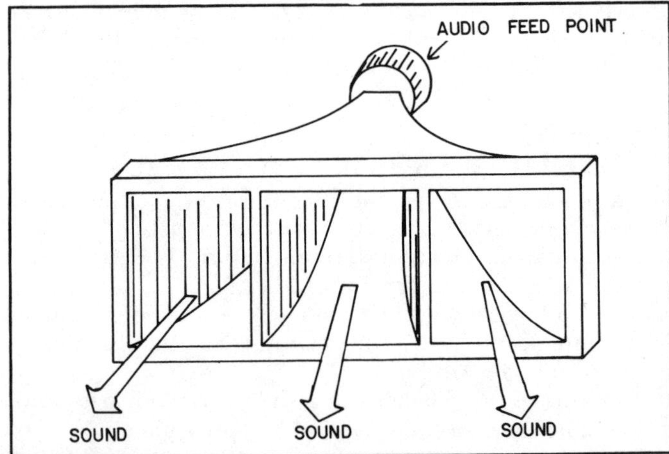

HORN SPEAKER: A horn speaker with internal baffles for sound distribution.

Multiple-speaker systems, using horns in the midrange and high parts of the audio spectrum, are among the most sophisticated sound-transmission devices. They are often used in outdoor concerts, where a high sound level is wanted.

Horn speakers are sometimes used by themselves in public-address systems. For good results in this application, however, a large horn must be used, or the low-frequency components of the voice will be severely attenuated. *See also* SPEAKER.

HORSEPOWER

The horsepower is a unit of power, generally used in reference to mechanical devices such as motors and internal-combustion engines. The horsepower is equal to 746 watts. The watt is therefore equal to 0.00134 horsepower.

In electronic applications, the horsepower is seldom used. The watt is the preferred unit. *See also* POWER.

HOT-CARRIER DIODE

The hot-carrier diode, sometimes called the HCD for short, is a special kind of semiconductor diode. The HCD is used in various circuits, especially mixers and detectors at very-high and ultra-high frequencies. The HCD will work at much higher frequencies than most other semiconductor diodes.

The hot-carrier diode generates relatively little noise, and exhibits a high breakdown or avalanche voltage. The reverse current is minimal. The internal capacitance is low, since the diode consists of a tiny point contact (see illustration). The wire is gold-plated to minimize corrosion effects. An N-type silicon wafer is used as the semiconductor. *See also* DIODE.

HOT SPOT

When a small region in the plate of a vacuum tube becomes much hotter than the material in its vicinity, the region of higher temperature is called a hot spot. Hot spots can often be seen as a vacuum tube operates; the plate will actually glow red in a round or elliptical spot. This condition is undesirable in some types of tubes, while it is perfectly normal in other types. *See also* PLATE DISSIPATION, TUBE.

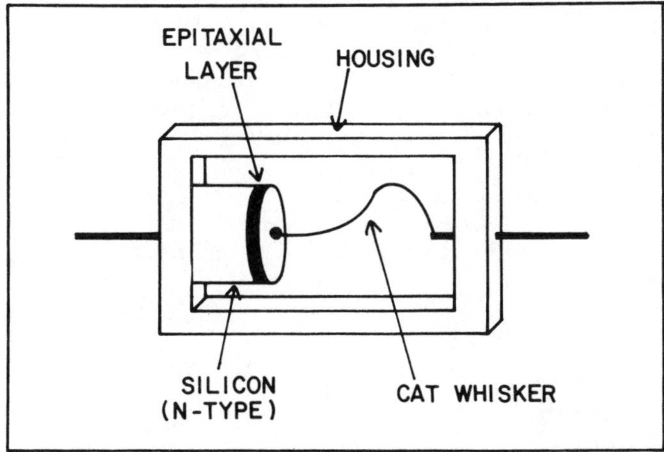

HOT-CARRIER DIODE: Cross-sectional drawing of a hot-carrier diode.

In a communications system, a hot spot is a region in which reception is much better than in the immediate surroundings. This effect can be caused by phase addition of several components of a signal as a result of geography in the area. It can also be caused by unusually good earth conductivity. Sometimes the cause is unknown. A receiving hot spot is not necessarily a good transmitting location. *See also* EARTH CONDUCTIVITY, PROPAGATION CHARACTERISTICS.

HOT-WIRE METER

A hot-wire meter is a device that makes use of the thermal expansion characteristics of a metal wire for the purpose of measuring current. When a current flows through the tightly stretched wire, the wire expands. This causes a pointer, attached to the wire, to move across a graduated scale. Hot-wire current-measuring devices can be used as ammeters, voltmeters, or wattmeters. Hot-wire ammeters can be attached to a variety of different devices to indicate such parameters as wind speed, motor speed, or rate of fluid flow.

The hot-wire meter is not particularly sensitive, but can register large amounts of current without damage. The damping, or rate at which the meter responds to changes in the current passing through it, is slow. This is an advantage in situations where rapid fluctuations are of little interest, but is not desirable when precise indications are required for rapidly changing parameters. The hot-wire meter can measure alternating current just as well as it can measure direct current. For this reason, hot-wire ammeters are often used for the determination of radio-frequency current in antenna transmission lines. *See also* AMMETER.

HUE

Hue is a term that is essentially synonymous with color. The hue of an object or light beam is dependent on the wavelength of the light. Hue is determined by the wavelength at which the light intensity is the greatest.

Colors may have much different levels of saturation, even though they have the same hue. This can be illustrated with a color television receiver. The color-intensity knob controls the saturation. The tint knob controls the hue. *See also* CHROMA, SATURATION.

HUM

Hum is the presence of 60-Hz or 120-Hz modulation in an electronic circuit. Hum may occur in the carrier of a radio transmitter, in a radio-frequency receiving system, or in an audio system.

When the filtering is inadequate in the output of a power supply, hum is often introduced into the circuits operating from the supply. With half-wave power supplies, the hum has a frequency of 60 Hz; with full-wave supplies, the frequency is 120 Hz. Both of these frequencies are within the range of human hearing, and produce objectionable modulation.

Hum may be picked up by means of inductive or capacitive coupling to nearby utility wires. This hum is always at a frequency of 60 Hz. Poorly shielded amplifier-input wiring is a major cause of hum in audio circuits. Hum may also be picked up by the magnetic heads of a tape recorder. Improper shielding or balance in the output leads of an audio amplifier can also cause hum modulation to be introduced into the circuit.

Power-supply hum can be remedied by the installation of additional filtering chokes and/or capacitors. Hum from utility wiring can be minimized by the use of excellent shielding in the input leads to an amplifier, and by proper shielding or balance in the output. A good ground system, without ground loops, can be helpful as well. *See also* ELECTROMAGNETIC SHIELDING, POWER SUPPLY.

HUMAN ENGINEERING

The art of designing electronic equipment to suit the needs of the operator is called human engineering. Even if a circuit is well-designed electrically, it will be difficult to use if proper attention has not been given to human engineering.

Human engineering involves judicious positioning of controls and indicators. Meters and displays should be easy to read, and controls convenient to adjust. The number of controls should be sufficient to accomplish the desired functions, but too many adjustments make it difficult to use the apparatus.

The art of human engineering also involves the proper choice of electrical characteristics. A receiver with a product detector, but no envelope detector, can be used to receive amplitude-modulated signals, but tuning is difficult. Many telegraph operators prefer a receiver in which the automatic level control can be switched off. The tuning rate should be slow enough so that signals are easy to pinpoint, but it should be fast enough so that the operator does not have to spend a lot of time spinning the dial to get from one frequency to another. The above examples refer to radio receivers, but human engineering is important in the design of computer terminals, test equipment, transmitters, and all other electronic devices.

As the field of electronics gets more and more advanced, human engineering should play a greater role. *See also* DESIGN, ELECTRICAL ENGINEERING.

HUMIDITY

Humidity is the presence of water vapor in the air. The humidity is usually specified in terms of a ratio, called the relative humidity. This is the amount of water vapor actually in the air, compared to the amount of water vapor the air is capable of holding without condensation. The higher the temperature, the more water vapor can be contained in the air.

Relative humidity is determined by the use of two thermometers, one with a dry bulb and the other with its bulb surrounded by a wet cloth or wick. Evaporation from the wick causes the wet-bulb reading to be lower than the dry-bulb reading. For a given temperature, the water from the wet bulb evaporates more and more rapidly as the humidity gets lower. The two readings are found on a table, and the relative humidity is indicated at the point corresponding to both readings.

In northern latitudes during winter, the indoor relative humidity can become very low. This is because the cold outside air cannot hold much moisture, and when the cold air is heated, its capacity for holding moisture goes up while the actual amount of moisture stays the same.

In some mountainous, cold locations, the relative humidity can drop to less than 5 percent. The resulting static electricity can cause damage to electronic circuits when they are handled. This is especially true of metal-oxide semiconductor (MOS) devices. In marine or tropical regions, the humidity may rise to the point where condensation takes place on circuit boards and components. This can degrade performance and accelerate corrosion.

The specifications for a piece of electronic equipment often contain humidity limitations. The primary reason for this is the danger to MOS devices when it is very dry, and the possibility of condensation, with resultant component damage, when it is wet. *See also* SPECIFICATIONS.

HUNTING

Hunting is the result of overcompensation in an electronic circuit. Hunting is particularly common in direction-finding apparatus and improperly adjusted phase-locked-loop devices.

Any circuit that is designed to lock on some signal is subject to hunting if it is set in such a way that overcompensation occurs. The circuit will then oscillate back and forth on either side of the desired direction, frequency, or other parameter. The oscillation may be fast or slow. It may eventually stop, and the circuit achieve the desired condition. But if the misadjustment is especially severe, the hunting may continue indefinitely.

Hunting can be eliminated by proper alignment of a circuit. Additional damping usually gets rid of this problem. *See also* DIRECTION FINDER, PHASE-LOCKED LOOP.

HYBRID DEVICES

A component or circuit that uses two or more different technological forms of design is sometimes called a hybrid device. For example, a radio transmitter may utilize bipolar transistors and field-effect transistors in every stage except the final amplifier, which uses a vacuum tube. Such a transmitter would be considered to have hybrid design, since it makes use of both solid-state and vacuum-tube technology.

Whether or not a circuit is to be called a hybrid device depends, in part, on the distinction that is drawn among various technological methods. An integrated circuit that uses both microminiature discrete components and integrated components is often called a hybrid integrated circuit, even though all of the components are solid-state. Computers that use both analog and digital techniques are often called hybrid computers. A switch that makes use of relays and transistors might be called a hybrid switching circuit.

Hybrid devices often contain the best of various different forms of design, resulting in a circuit that is more efficient than would be possible if only one form of design is used.

HYDROELECTRIC POWER PLANT

Moving water can produce tremendous force, and this force can be harnessed to produce electric power. Electricity derived from moving or falling water is called hydroelectric power.

A hydroelectric power plant requires that a dam be built in a river. The dam creates a situation where the water must fall from a higher elevation to a lower elevation. The water produces great force as it falls, and this force is used to drive a set of electric generators. The greater the electric power supplied by a generator, the more turning force is needed. Thus the dam must be at least a certain height, and must contain at least a certain amount of water, to generate a given amount of power.

Hydroelectric power plants produce essentially no pollution, and they can be operated continuously as long as the water flows at the necessary rate. It is possible to place several hydroelectric power plants along a single river, provided they are spaced far enough apart. The water that accumulates behind the dam can be used as a lake or reservoir.

Hydroelectric power plants can be built only in certain locations. Of course, such a plant must be on a river. The terrain must be fairly hilly or mountainous, so that the reservoir will not flood a large land area. High-tension power lines must be used to supply electricity to people who do not live near the generators. In a period of prolonged drought, the river may almost dry up, and the output of the plant will decrease. Despite these disadvantages, however, hydroelectric power is increasingly important as a replacement for fossil fuels. *See also* ENERGY CRISIS, GENERATOR, POWER PLANT.

HYDROGEN

Hydrogen is the simplest element in the universe. It is also the most abundant; more than half of all matter is thought to be hydrogen. In its most common form, the hydrogen atom consists of one electron and one proton; thus its atomic number is 1, and its atomic weight is 1. Some hydrogen atoms have one neutron in the nucleus, resulting in an atomic weight of 2. This variant of hydrogen is called deuterium. It is possible for two neutrons to exist in the nucleus of a hydrogen atom; this results in tritium.

Hydrogen is a gas at room temperature. When combined with oxygen, hydrogen forms water. Hydrogen is extremely flammable, and can be dangerous. If a spark or flame ignites a large amount of hydrogen, it combines violently with oxygen in the atmosphere. The sole end product of this reaction is water.

Hydrogen holds some promise as a replacement for natural gas as a source of energy. Hydrogen burns more efficiently than any other flammable gas. It also results in no pollution. Hydrogen comprises the fuel for the nuclear-fusion reactor, which has not yet been perfected. The use of hydrogen on a large scale will require improved methods of purification and extraction. *See also* DEUTERIUM, NATURAL GAS, NUCLEAR FUSION.

HYPERBOLA

A hyperbola is a form of conic section, representing the intersection of a plane and a double cone. The plane must be parallel to the axis of the double cone (see illustration at A). The cone may have any apex angle of at least 0 degrees but less than 90 degrees with respect to the axis. The plane may be any finite, nonzero distance from the cone axis.

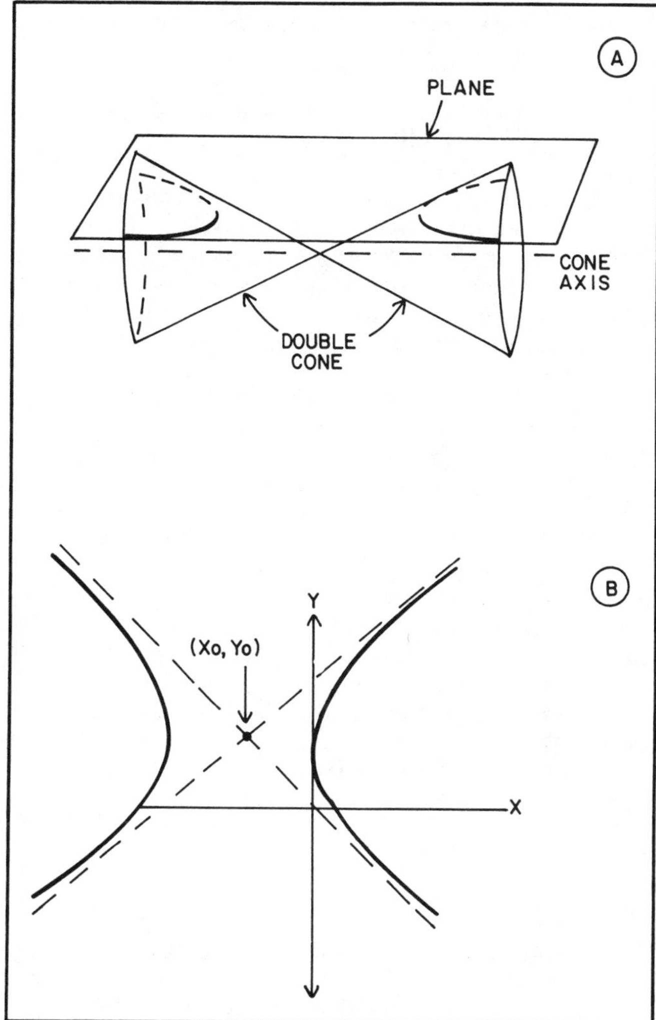

HYPERBOLA: A hyperbola (heavy lines) consists of the intersection of a double cone and a plane, as shown at A. At B, the Cartesian representation of the hyperbola.

The general form of the equation for a hyperbola in the Cartesian coordinate system is:

$$(x - x_o)^2 / a^2 - (y - y_o)^2 / b^2 = 1$$

where a and b are constants, and (x_o, y_o) represents the center of the hyperbola, as shown at B.

The hyperbola with the simplest possible equation, $x^2 - y^2 = 1$, is sometimes called the unit hyperbola. It is centered at the origin, or the point (o,o). The unit hyperbola gives rise to a special form of trigonometric function, called the hyperbolic trigonometric function, which is invaluable in the solution of differential equations. *See also* CARTESIAN COORDINATES, CONIC SECTION, HYPERBOLIC TRIGONOMETRIC FUNCTIONS.

HYPERBOLIC TRIGONOMETRIC FUNCTIONS

The ordinary trigonometric functions are based on the unit circle, having the equation $x^2 + y^2 = 1$. A less well-known set of trigonometric functions exists, based on the unit hyperbola $x^2 - y^2 = 1$. These functions are called the hyperbolic trigonometric functions.

All of the hyperbolic functions can be derived from the

hyperbolic sine and cosine (abbreviated sinh and cosh) which are defined as follows:

$$\sinh x = \tfrac{1}{2} (e^{x} - e^{-x})$$
$$\cosh x = \tfrac{1}{2} (e^{x} + e^{-x})$$

where x is any real number, and e is approximately 2.718.

Four other hyperbolic functions arise from sinh and cosh; these are the hyperbolic tangent (tanh), the hyperbolic contangent (coth), hyperbolic secant (sech), and hyperbolic cosecant (csch). These four functions bear the same relation to sinh and cosh as is the case with circular functions:

$$\tanh x = \sinh x / \cosh x = (e^{x} - e^{-x})/(e^{x} + e^{-x})$$
$$\coth x = \cosh x / \sinh x = (e^{x} + e^{-x})/(e^{x} - e^{-x})$$
$$\operatorname{sech} x = 1 / \cosh x = 2/(e^{x} + e^{-x})$$
$$\operatorname{csch} x = 1 / \sinh x = 2/(e^{x} - e^{-x})$$

The hyperbolic functions display many properties similar to the circular functions. The functions sinh and cosh are their own second derivatives. The derivative of the sinh function is the cosh function, and vice versa. Hyperbolic functions are useful in the solution of certain differential equations. This, in turn, makes the hyperbolic functions invaluable in electronics design and analysis of such things as attenuators and filters. *See also* COSECANT, COSINE, COTANGENT, DIFFERENTIAL EQUATION, SECANT, SINE, TANGENT.

HYSTERESIS

Hysteresis is the tendency for certain electronic devices to act "sluggish." This effect is especially important in the cores of transformers, because it limits the rate at which the core can be magnetized and demagnetized. This, in turn, limits the alternating-current frequency at which a transformer will operate efficiently.

Hysteresis effects occur in many different devices, not just in transformer cores. For example, the thermostat in a heater or air conditioner must have a certain amount of hysteresis, or "sluggishness." If there were no hysteresis in such a device, the heater or air conditioner would cycle on and off at a rapid rate, once the temperature reached the thermostat setting. With too much hysteresis, the temperature would fluctuate far above and below the thermostat setting. The correct amount of hysteresis in a thermostat results in a nearly constant temperature without excessive cycling of the heater or air conditioner.

Hysteresis is an important consideration in the design of certain electronic controls, such as a receiver squelch. Too little hysteresis results in squelch popping. Too much hysteresis makes it difficult to obtain a precise squelch setting. Certain digital control devices have built-in hysteresis. *See also* B-H CURVE, HYSTERESIS LOOP, HYSTERESIS LOSS.

HYSTERESIS LOOP

A hysteresis loop is a graphical representation of the effects of hysteresis. Such a graph may also be called a hysteresis curve, or box-shaped curve. The B-H curve is the graphical representation of magnetization versus magnetic force (*see* B-H CURVE), and thus is a special form of hysteresis loop.

The drawing illustrates a hysteresis loop for a typical

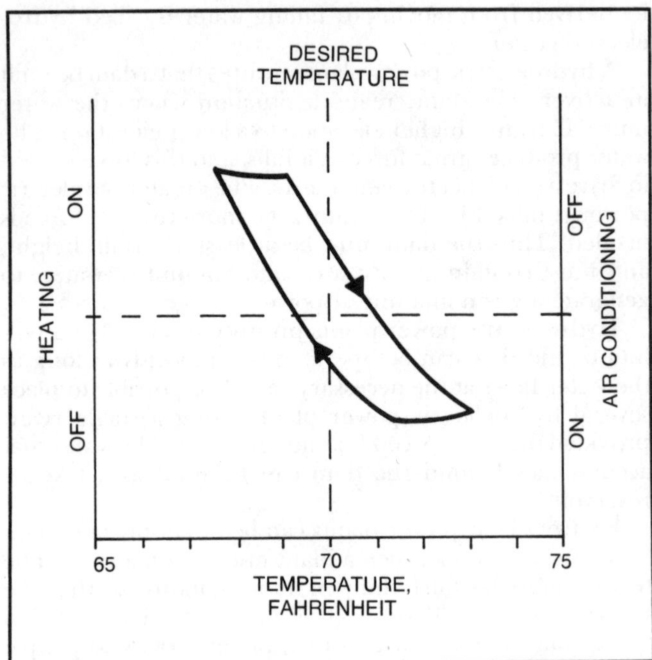

HYSTERESIS LOOP: A thermostat hysteresis loop. This particular thermostat is set at 70 degrees Fahrenheit, but allows the temperature to fluctuate between about 68 and 72 degrees Fahrenheit.

thermostat mechanism, used in conjunction with a heater or air conditioner. The temperature, in degrees Fahrenheit, is shown on the horizontal scale. The on and off conditions are shown on the vertical scale. For heating, the on condition is above the dotted line; for air conditioning, the on condition is below the dotted line. Time is depicted by tracing clockwise around the loop.

The hysteresis loop is useful for evaluating the performance of any device that exhibits hysteretic properties. In general, the wider the hysteresis loop, the greater the effect. *See also* HYSTERESIS, HYSTERESIS LOSS.

HYSTERESIS LOSS

The hysteretic properties of magnetic materials cause losses in transformers having ferromagnetic cores. Hysteresis loss occurs as a result of the tendency of ferromagnetic materials to resist rapid changes in magnetization. Hysteresis and eddy currents are the two major causes of loss in transformer core materials.

The amount of hysteresis loss in a core material depends on two factors. In general, the higher the alternating-current frequency of the magnetizing force, the greater the hysteresis loss for any given material. The way in which the material is manufactured affects the amount of hysteresis loss at a specific frequency.

Solid or laminated iron displays considerable hysteresis loss. For this reason, transformers having such cores are useful at frequencies up to only about 15 kHz. Ferrite has much less hysteresis loss, and can be used at frequencies as high as 20 to 40 MHz, depending on the particular mix. Powdered-iron cores have the least hysteresis loss of any ferromagnetic material, and are used well into the vhs range. Of course, air has practically no hysteresis loss, and is preferred in many radio-frequency applications for this reason. *See also* EDDY-CURRENT LOSS, FERRITE, FERRITE CORE, FERROMAGNETIC MATERIAL, HYSTERESIS, LAMINATED CORE, POWDERED-IRON CORE.

I

IC

See INTEGRATED CIRCUIT.

ICBM

See INTERCONTINENTAL BALLISTIC MISSILE.

ICE LOADING

Ice loading is the additional stress placed on an antenna or power line, and the associated supporting structures, by accumulation of ice. When a structure is designed and installed in an area subject to ice storms, allowances must be made for ice loading. Otherwise, the antennas, power lines, or towers may actually collapse under the stress.

The most susceptible areas of the United States, in terms of the probability and frequency of ice storms, are the northeast and midwest regions. The northwest and southeast are somewhat less likely to have ice storms. The desert southwest and the extreme southern United States seldom have such storms.

The presence of ice on a structure increases the wind loading dramatically. This effect, too, must be taken into account when power lines or antenna systems are designed or installed in a particular geographic region. *See also* WIND LOADING.

ICONOSCOPE

An iconoscope is a form of television camera tube. The operation of the iconoscope is similar to that of the vidicon. See the pictorial diagram of an iconoscope tube.

The visible image is focused onto a flat photocathode by means of a convex lens. The photocathode consists of a fine grid of light-sensitive capacitors, called a mosaic. Each capacitor is supplied with a charging voltage. The light from the image causes each capacitor to discharge as the electron beam, supplied by the electron gun, scans the mosaic. The amount of discharging depends on the intensity of the light at that particular location: With no light, the capacitors remain almost fully charged, and under bright light, they almost completely discharge. As the electron beam strikes each capacitor in turn, an output pulse occurs. The intensity of the pulse is proportional to the brightness of the light striking the capacitor.

The iconoscope, like the vidicon, is somewhat less sensitive, and has a slightly longer image lag time, than the image orthicon. *See also* CAMERA TUBE, IMAGE ORTHICON, VIDICON.

IDEAL

An ideal circuit or object is theoretically perfect. For example, an ideal diode has zero resistance in the forward direction and infinite resistance in the reverse direction. Ideal ground conducts with zero resistance. An ideal capacitor has no leakage. Obviously, ideal components cannot actually exist in practice, but they serve as a standard for comparison among real components.

In theoretical calculations, it is often necessary to use ideal components initially, and then insert a correction factor at a later time to account for imperfections in the actual circuit. Some electronic circuits actually function because of the imperfections of a component, and an ideal component would not work. An example of this is the

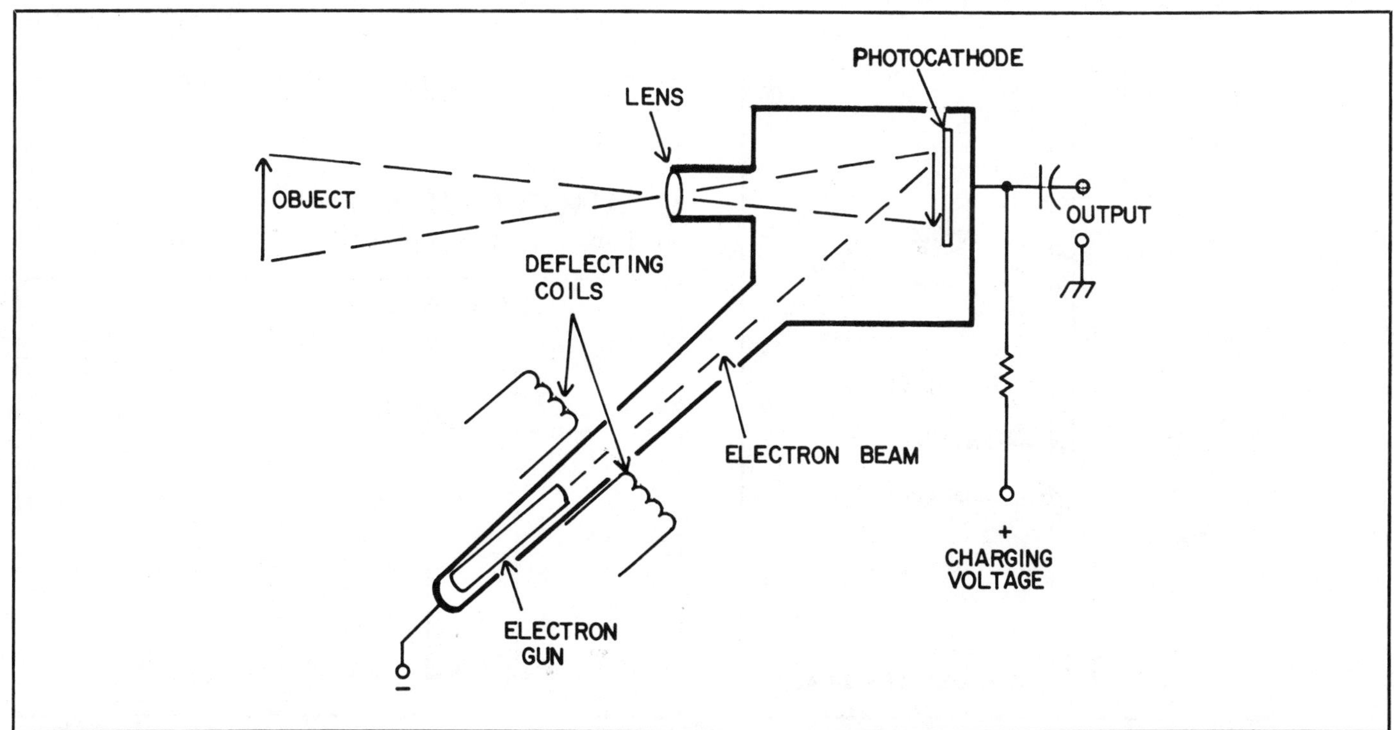

ICONOSCOPE: Pictorial diagram of an iconoscope tube.

series installation of a forward-conducting semiconductor diode to obtain a small but constant drop in the voltage from a direct-current power supply. An ideal diode would have no voltage drop in the forward direction, and would be useless for such an application.

IEEE

See INSTITUTE OF ELECTRICAL AND ELECTRONIC ENGINEERS.

IF

See INTERMEDIATE FREQUENCY.

IGNITION NOISE

Ignition noise is a wideband form of impulse noise, generated by the electric arc in the spark plugs of an internal combustion engine. Ignition noise is radiated from many different kinds of devices, such as automobiles and trucks, lawn mowers, and gasoline-engine-driven generators.

Ignition noise can usually be reduced or eliminated in a receiver by means of a noise blanker. The pulses of ignition noise are of very short duration, although their peak intensity may be considerable. Ignition noise is a common problem for mobile radio operators, especially if communication in the high-frequency range is contemplated. Ignition noise can be worsened by radiation from the distributor to wiring in a truck or automobile. Sometimes, special spark plugs, called resistance plugs, can be installed in place of ordinary spark plugs, and the ignition noise will be reduced. An automotive specialist will know whether or not this is feasible in a given case. An excellent ground connection is imperative in the mobile installation. Mobile receivers for high-frequency work should have effective, built-in noise blankers.

Ignition noise is not the only source of trouble for the mobile radio operator. Noise can be generated by the friction of the tires against a pavement. High-tension power lines often radiate noise at radio frequencies. *See also* IMPULSE NOISE, NOISE, NOISE BLANKER, NOISE LIMITER, POWER-LINE NOISE.

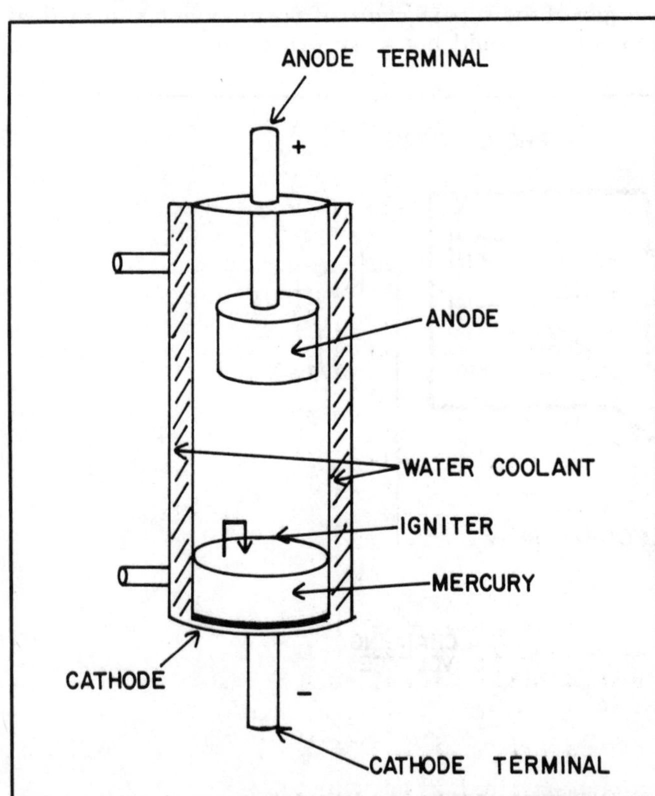

IGNITRON: Cutaway view of an ignitron tube.

IGNITRON

The ignitron is a form of high-voltage, high-current rectifier. The device consists of a mercury-pool cathode and a single anode. Forward conduction occurs as a result of arcing during the part of the cycle in which the anode is positive with respect to the cathode.

During the forward-voltage part of the alternating-current cycle, an igniting electrode dips into the mercury and initiates the arc. The ignitron generates large amounts of heat, and therefore it must be cooled in some manner. Water, pumped around the outside of the tube, serves this purpose well. The illustration is a cutaway view of an ignitron tube. It is apparent from this diagram that the ignitron can only operate when it is right-side-up.

The ignitron is characterized by the ability to withstand temporary currents or voltages far in excess of its normal ratings. *See also* RECTIFICATION.

ILLUMINANCE

Illuminance is an expression of the amount of light that strikes a surface. Illuminance is measured in lumens per square centimeter (phots), lumens per square meter (lux), or lumens per square foot (foot candles).

For a given light source, the illuminance impinging on a surface, oriented at a right angle to the incident radiation, is inversely proportional to the square of the distance of the object from the source. For example, if the distance is doubled, the illuminance becomes just ¼ of its previous value. This is the familiar law of inverse squares. *See also* FOOT CANDLE, INVERSE-SQUARE LAW, LUX, PHOT.

ILLUMINATION METER

An illumination meter is a device for measuring the inten-

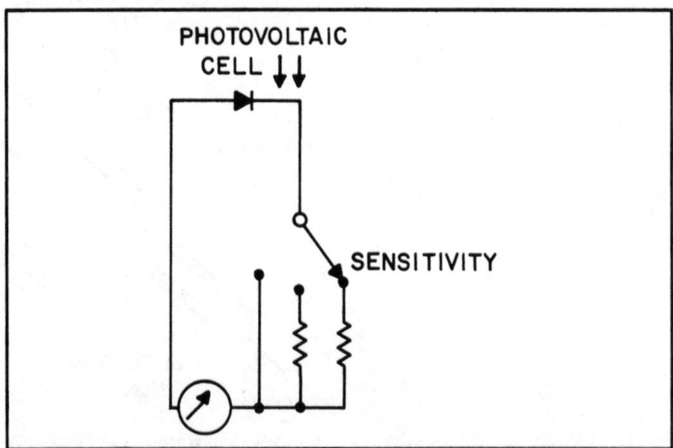

ILLUMINATION METER: Sensitivity is easily varied by means of a switchable set of resistors.

sity of light in a given place. Illumination meters are used by photographers and television camera operators. Illumination meters are calibrated in terms of illuminance.

The schematic diagram shows a simple illumination meter. The device consists of a solar cell and a microammeter or milliammeter. The greater the illuminance on the solar cell, the more current it generates, and the higher the meter reading.

Illumination meters are wavelength-sensitive. That is, they respond more readily to light of certain wavelengths than to light at other wavelengths. For best results, the wavelength response of an illumination meter should be similar to that of the film used in photographic apparatus. Otherwise, overexposure may result in some situations and underexposure will occur in other cases. *See also* ILLUMINANCE.

IMAGE CONVERTER

An image converter is a device that changes an electrical, acoustic, infrared, ultraviolet, X-ray, or gamma-ray image into a visual image. Image converters are employed by medical personnel in the form of ultrasonic scanners and X-ray machines. An infrared image converter makes it possible to "see" through some obstacles that are opaque to visible light. Astronomers use infrared, ultraviolet, X-ray, and gamma-ray image converters to view celestial objects at invisible wavelengths. The electron microscope is a form of image converter.

Image converters may consist of special photographic films, sensitive to radiation at wavelengths outside the normal visible spectrum; these devices normally do not allow the observation of motion. More sophisticated image converters, such as the fluoroscope or snooperscope, enable personnel to see motion. Image converters may be interconnected with computers for such diverse purposes as missile guidance and medical diagnosis. *See also* ELECTRON MICROSCOPE, ELECTRON TELESCOPE, FLUOROSCOPE, GAMMA RAY, INFRARED, SNOOPERSCOPE, ULTRASONIC SCANNER, ULTRAVIOLET, X RAY.

IMAGE FREQUENCY

In a superheterodyne receiver, frequency conversion results in a constant intermediate frequency. This simplifies the problem of obtaining good selectivity and gain over a wide range of input-signal frequencies. However, it is possible for input signals on two different frequencies to result in output at the intermediate frequency. One of these signal frequencies is the desired one, and the other is undesired. The undesired response frequency is called the image frequency.

The illustration is a simple block diagram of a superheterodyne receiver having an intermediate frequency of 455 kHz. This is a common value for the intermediate frequency of a superheterodyne circuit. The local oscillator is tuned to a frequency 455 kHz higher than the desired signal; for example, if reception is wanted at 10.000 MHz, the local oscillator must be set to 10.455 MHz. However, an input signal 455 kHz higher than the local-oscillator frequency, at 10.910 MHz, will also mix with the local oscillator to produce an output at 455 kHz. If signals are simultaneously present at 10.000 and 10.910 MHz, then, they may interfere with each other. We call 10.910 MHz, in this example, the image frequency.

The image frequency in a superheterodyne receiver always differs from the signal frequency by twice the value of the intermediate frequency. A selective circuit in the front end of the receiver must be used to attenuate signals arriving at the image frequency, while passing those arriving at the desired frequency. The lower the intermediate frequency in a superheterodyne receiver, the closer the image frequency is to the desired frequency, and the more difficult it becomes to attenuate the image signal. For this reason, modern superheterodyne receivers use a high intermediate frequency, such as 9 MHz, so that the image and desired frequencies are far apart. The high intermediate frequency may later be heterodyned to a low frequency, making it easier to obtain sharp selectivity. *See also* DOUBLE-CONVERSION RECEIVER, IMAGE REJECTION, SINGLE-CONVERSION RECEIVER, SUPERHETERODYNE RECEIVER.

IMAGE IMPEDANCE

An image impedance is an impedance that is "seen" at one end of a network, when the other end is connected to a load or generator having a defined impedance. The image impedance may be the same as the terminating impedance at the other end of the network, but this is not always the case.

As an example, suppose that a pure resistive impedance Z_1 is connected to one end of a quarter-wave transmission line having a characteristic impedance Z_0. Then the image impedance Z_2 at the opposite end of the line is related to the other impedances according to the formula:

$$Z_0 = \sqrt{Z_1 Z_2}$$

If the transmission line measures a half electrical wavelength, however, Z_1 and Z_2 will always be equal.

The relationships become more complicated when reactance is present in the terminating impedance. For a quarter-wave line, capacitive reactance at one end is transformed into an inductive reactance at the opposite end, and vice versa. For a half-wave line, the terminating and image impedances are equal even when reactance is present.

There are many different kinds of networks that produce various relationships between terminating and image impedances. *See also* IMPEDANCE, REACTANCE.

IMAGE INTENSIFIER

An image intensifier is a device that makes a visual or electronic image brighter. Such a device can be used to

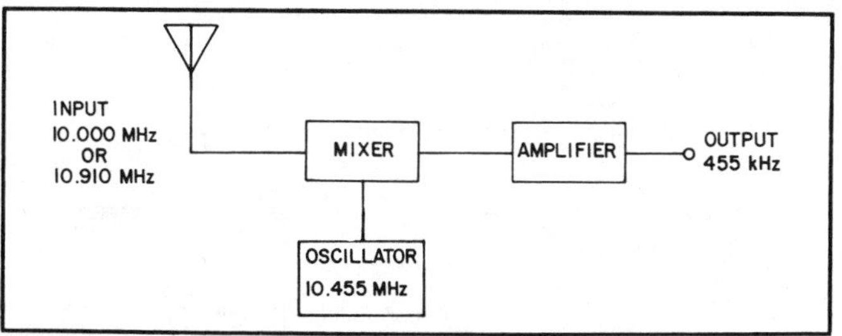

IMAGE FREQUENCY: In a simple superheterodyne circuit such as this, two input-signal frequencies can result in output at the intermediate frequency. Here, the desired signal is at 10.000 MHz, and the image is at 10.910 MHz.

literally see in the dark. Image intensifiers also can be used to help people with visual handicaps to see better in dim light.

The image intensifier generally consists of a television camera and picture tube, thus forming a self-contained closed-circuit television system. An amplifier increases the brightness of the picture signal from the camera tube. *See also* CAMERA TUBE, PICTURE TUBE.

IMAGE ORTHICON

The camera tube most often used for live commercial television broadcasting is called the image orthicon. The image orthicon is a highly sensitive camera tube. It responds rapidly to motion and changes in light intensity. A pictorial diagram of an image orthicon camera tube is shown in the illustration.

Light is focused by means of a convex lens onto a translucent plate called the photocathode. The photocathode emits electrons according to the intensity of the light in various locations. A target electrode attracts these electrons, and accelerator grids give the electrons additional speed. When the electrons from the photocathode, called photoelectrons, strike this target electrode, secondary electrons are produced. Several secondary electrons are emitted from the target for each impinging photoelectron. This is part of the reason for the high sensitivity of the image orthicon.

A fine electron beam, generated by an electron gun, scans the target electrode. (The scanning rate and pattern correspond with the scanning in television receivers.) Some of these electrons are reflected back toward the electron gun by the secondary emissions from the target. Areas of the target having greater secondary-electron emission result in an intense return beam; areas of the target that emit few secondary electrons result in lower returned-beam intensity. Therefore, the returned beam is modulated as it scans the target electrode. A receptor electrode picks up the returned beam.

The main disadvantage of the image orthicon is its rather high noise output. However, in situations where there is not much light, or where a fast response time is required, the image orthicon is superior to other types of camera tubes. *See also* CAMERA TUBE, ICONOSCOPE, VIDICON.

IMAGE REJECTION

In a superheterodyne receiver, image rejection is the amount by which the image signal is attenuated with respect to a signal on the desired frequency. Image rejection is always specified in decibels.

In general, the higher the intermediate frequency of a receiver, the better the image rejection. *See also* IMAGE FREQUENCY, SUPERHETERODYNE RECEIVER.

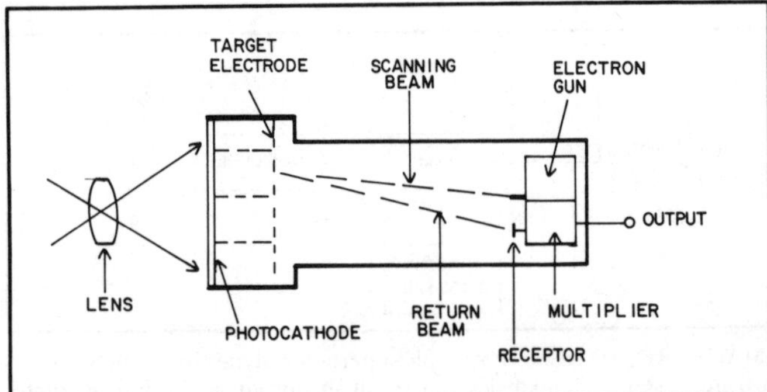

IMAGE ORTHICON: Cross-sectional diagram of an image orthicon.

IMPATT DIODE

The IMPATT, or impact-avalanche-transit-time, diode is a form of semiconductor device that is useful as an amplifier of microwave radio-frequency energy. Some IMPATT diodes can provide output power in excess of 1 watt, providing considerable power gain in transmitters using Gunn-diode oscillators.

In future years, improved devices of this kind can be expected. *See also* GUNN DIODE.

IMPEDANCE

Impedance is resistance to alternating-current flow. The behavior of alternating-current circuits is somewhat more complicated than the behavior of direct-current circuits. In an alternating-current system, the impedance depends not only on the resistance, but also on the reactance. Reactance in turn varies with frequency. Resistances are always positive. Reactances can be positive or negative (*See* REACTANCE.)

The illustration illustrates the complex-plane system of defining impedances. The resistance component is plotted along the horizontal axis toward the right. The positive, or

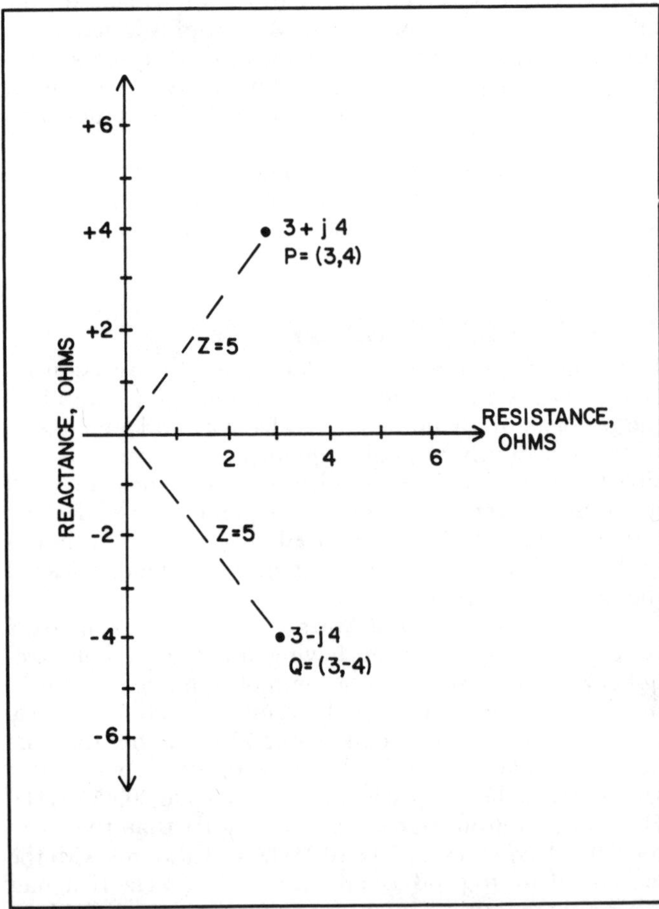

IMPEDANCE: The complex impedance plane, showing two combinations of resistance and reactance.

inductive, reactance is plotted on the vertical axis, going upward. The negative, or capacitive, reactance is plotted along the vertical axis, going downward. Both the resistance and the reactance are specified in ohms. The reactance component is sometimes multiplied by $\sqrt{-1}$, abbreviated j. Each point in the coordinate plane therefore corresponds to exactly one impedance, and any impedance can be defined as a unique point on the plane.

Suppose an inductor presents 3 ohms of resistance and 4 ohms of reactance. Then the impedance is defined on the plane by the point P = (3,4). Engineers write this as 3 + j4. Suppose a capacitor offers 3 ohms of resistance and 4 ohms of reactance. Then the impedance is defined by the point Q = (3,−4); engineers would write 3 − j4.

Sometimes the impedance is defined simply as the length of the line from the origin, or (0,0), to the point on the plane defining the impedance. This length is given by:

$$Z = \sqrt{R^2 + X^2}$$

where R is the resistance and X is the reactance, both given in ohms. This method of defining impedance does not tell the complete story, however, because Z does not contain information concerning the components. In the illustration, points P and Q both correspond to Z = 5 ohms, but the two impedances are vastly different.

Various kinds of circuits have different impedance characteristics. Antennas, tuned circuits, and transmission lines show impedances at radio frequencies. A transmission line also displays a property called the characteristic impedance, which depends on its physical dimensions and construction.

Impedances in series add vectorially together. In the example, if the capacitor and inductor are connected in series, the impedance becomes:

$$P + Q = (3 + j4) + (3 - j4) = 6 + j0,$$

indicating 6 ohms of resistance and no reactance.

Impedances in parallel add like resistances in parallel, providing the complex representations are used. Thus, in a parallel configuration,

$$\begin{aligned} P + Q &= PQ / (P + Q) \\ &= (3 + j4)\,(3 - j4) / (3 + j4) + (3 - j4) \\ &= 4.17 + j0, \end{aligned}$$

indicating 4.17 ohms of resistance and no reactance. The absense of reactance in the above circuits indicates that they are resonant circuits. *See also* ADMITTANCE, CONJUGATE IMPEDANCE, PARALLEL RESONANCE, RESONANCE, SERIES RESONANCE.

IMPEDANCE BRIDGE

An impedance bridge is a device used for the purpose of determining the resistive and reactive components of an unknown impedance. The impedance of any reactive circuit changes with the frequency, so the bridge can function only at a specific frequency. Impedance bridges are often used by radio-frequency engineers to determine the impedances of tuned circuits and antenna systems.

The impedance bridge generally has a null indicator and two adjustments. One adjustment indicates the resistance, and the other indicates the reactance. *See also* IMPEDANCE.

IMPEDANCE MATCHING

An alternating-current circuit always functions best when the impedance of a power source is the same as the impedance of the load to which power is delivered. This does not happen by chance; often, the two impedances must be made the same by means of special transformers or networks. This process is called impedance matching. Impedance matching is important in audio-frequency as well as radio-frequency applications.

In a high-fidelity system, audio transformers ensure that the output impedance of an amplifier is the same as that of the speakers; this value is generally standardized at 8 ohms. Radio-frequency transmitting equipment is usually designed to operate into a 50-ohm, nonreactive load, although some systems have output tuning circuits that allow for small resistance fluctuations and/or small amounts of reactance in the load.

Radio-frequency engineers probably face the most difficult impedance-matching tasks. At very-high, ultra-high, and microwave frequencies, poorly matched impedances can result in large amounts of signal loss. If the generator and load impedances are not identical, some of the electromagnetic field is reflected from the load back toward the source. In any alternating-current system, the load will accept all of the power only when the impedances of the load and source are identical. *See also* IMPEDANCE, REACTANCE, RESISTANCE.

IMPEDANCE TRANSFORMER

An impedance transformer is used to change one pure resistive impedance to another. Impedance transformers are sometimes used at audio frequencies to match the output impedance of an amplifier to the input impedance of a set of speakers. In radio-frequency work, impedance transformers comprise part of an impedance-matching system, generally used for optimizing the antenna-system for a transmitter.

Suppose that the impedance of a purely resistive (that is, nonreactive) load is given by Z_s, and this load is connected to the secondary winding of a transformer having a primary-to-secondary turns ratio of T. This situation is shown in the illustration. Then the impedance Z_p appearing across the primary winding, neglecting transformer losses, is given by:

$$Z_p = Z_s\,T^2$$

An impedance transformer must be designed for the

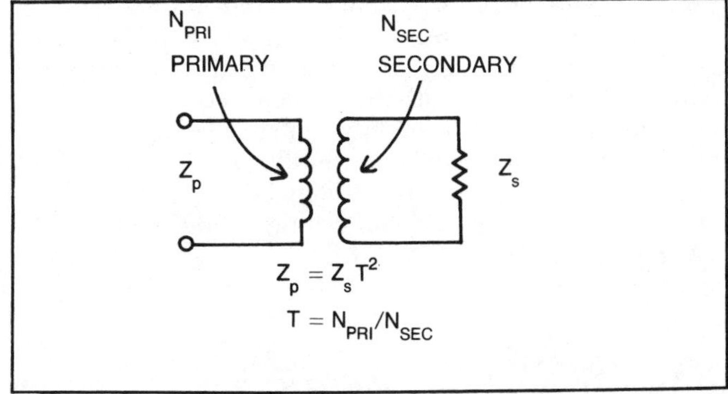

IMPEDANCE TRANSFORMER: The impedance-transfer ratio is equal to the square of the turns ratio.

proper range of frequencies. This generally means that the reactances of the windings must be comparable to the source and load reactances at the frequency in use. *See also* IMPEDANCE, IMPEDANCE MATCHING, TRANSFORMER.

IMPULSE

An impulse is a sudden surge in voltage or current. An impulse that appears at a utility outlet is sometimes called a transient or transient spike.

Since impulses are of very short duration, they contain high-frequency components. This can cause radio-frequency emissions from electrical wiring and appliances, resulting in interference to nearby radio receivers. *See also* IMPULSE NOISE.

IMPULSE GENERATOR

Any device that creates electrical impulses can be called an impulse generator. The term is usually used to refer to a device that is intentionally used to produce impulses.

There are many different kinds of impulse generators. The most common types employ inductance and/or capacitance to store energy and release it abruptly. The spark coil in an internal combustion engine is a good example of an impulse generator. *See also* IMPULSE, IMPULSE NOISE.

IMPULSE NOISE

Any sudden, high-amplitude voltage pulse will cause radio-frequency energy to be generated. In electrical systems without shielded wiring, the result can be electromagnetic interference. Impulse noise is generated by internal combustion engines because of the spark-producing voltage pulses. This form of noise is called ignition noise (*see* IGNITION NOISE). Impulse noise can also be produced by household appliances such as vacuum cleaners, hair dryers, electric blankets, thermostat mechanisms, and fluorescent-light starters.

Impulse noise is usually the most severe at very-low and low radio frequencies. However, serious interference can often occur in the medium and high frequency ranges. Impulse noise is seldom a problem above about 30 MHz, except in extreme cases. Impulse noise may be picked up by high-fidelity audio systems, resulting in interference to phonograph devices and tape-deck playback equipment.

Impulse noise in a radio receiver can be reduced by the use of a good ground system. Ground loops should be avoided. A noise blanker or noise limiter can be helpful. A receiver should be set for the narrowest response bandwidth consistent with the mode of reception. Impulse noise in a high-fidelity sound system can be more difficult to eliminate. An excellent ground connection, without ground loops, is imperative. It may be necessary to shield all speaker leads and interconnecting wiring. *See also* ELECTROMAGNETIC INTERFERENCE.

IMPURITY

An impurity is a substance that is added to a semiconductor material. The type and amount of impurity determines whether the semiconductor will conduit by means of electron transfer or hole transfer. The process of adding an impurity is called doping (*see* DOPING).

An impurity that results in the creation of an N-type material is called a donor impurity. Donor substances include arsenic, phosphorus, and bismuth. The N-type material conducts by passing excess electrons from atom to atom. The electron mobility depends on the amount and kind of donor impurity that is added to the silicon or germanium semiconductor. This, in turn, affects the characteristics of the diode, transistor, or field-effect transistor that is made from the material. *See also* N-TYPE SEMICONDUCTOR.

An impurity that results in the formation of a P-type semiconductor is called an acceptor impurity. Acceptors include boron, aluminum, gallium, and indium. Such substances are electron-deficient. The P-type material conducts by passing electron deficiencies, or holes, from atom to atom. Hole mobility is generally lower than electron mobility. However, the hole mobility is determined, as is the electron speed in the N-type material, by the kind and amount of impurity. The velocity of electrons or holes in a semiconductor, for a given electric-field intensity, is called the carrier mobility. *See also* CARRIER MOBILITY, GERMANIUM, HOLE, P-TYPE SEMICONDUCTOR, SILICON.

INCANDESCENT LAMP

An incandescent lamp is a device that produces visible light by means of a current in a resistive wire. The resistive wire is called the filament. With the application of sufficient voltage, the filament glows white hot. The filament is enclosed in an evacuated chamber to prevent oxidation. The photograph shows a typical incandescent lamp with a coiled tungsten filament. Incandescent lamps are available for use with power supplies ranging from less than 1 volts to several hundred volts.

INCANDESCENT LAMP: A small high-intensity bulb that operates from 117-volts.

Incandescent lamps are relatively inefficient. They produce more heat than light. For this reason, such bulbs are being used less often nowadays than in the past; energy conservation is of great concern. Fluorescent lamps are considerably more efficient than incandescent lamps. In electronic equipment, some dial lights still employ incandescent lamps, although the trend is toward light-emitting-diode displays. *See also* FLUORESCENT TUBE, GLOW LAMP.

INCLINATION

Inclination is the angle, measured in degrees with respect to the horizon, that the axis of a permanent magnet will point if allowed to rotate freely in the vertical plane. The magnet tends to align itself parallel to the geomagnetic lines of flux in its vicinity.

At the geomagnetic equator, midway between the two geomagnetic poles, the inclination of a magnet is zero. (The geomagnetic equator does not precisely coincide with the zero-latitude, geographic equator, since the magnetic poles are not located exactly at the geographic poles.) As a magnet is moved away from the geomagnetic equator, the inclination increases. *See also* DECLINATION, GEOMAGNETIC FIELD.

INCLUSIVE-OR GATE

See OR GATE.

INCOHERENT RADIATION

Incoherent electromagnetic radiation occurs when there are many different frequency and/or phase components in the wave. An example of incoherent radiation is the light from any ordinary bulb, such as an incandescent lamp or fluorescent tube.

Most radiation, including radio-frequency noise, infrared, visible light, ultraviolet rays, X rays, and gamma rays, is incoherent. This is because such emissions result from a random sort of disturbance. When radiation is concentrated at a single wavelength, and all of the wavefronts are lined up in phase, the radiation is said to be coherent. A radio transmitter generates coherent radiation. So does a maser or laser. *See also* COHERENT RADIATION.

INCREMENT

An increment is a change in the value of a variable, such as current, voltage, power, or frequency. The Greek uppercase letter delta (Δ) is often used to represent an incremental quantity; for example, Δf represents a defined change in frequency from one value to another.

Increments are important in the dynamic evaluation of some circuits. The slope of a line in the Cartesian (x,y) plane, for example, is given by $\Delta y/\Delta x$. This indicates the rate of change in y with respect to the rate of change in x. On a curve, the value of $\Delta y/\Delta x$ depends on the part of the graph in which the evaluation is made, and also on the size of the increments chosen. As the increments get smaller and smaller, the value of $\Delta y/\Delta x$ approaches a limiting value, known as the derivative of the function at a point. *See also* DERIVATIVE.

INDEFINITE INTEGRAL: INDEFINITE INTEGRALS OF SOME COMMON MATHEMATICAL FUNCTIONS. CONSTANTS ARE REPRESENTED BY THE LETTERS C, K, AND N. THE VALUE OF e IS APPROXIMATELY 2.7183.

f(x)	**$\int f(x)\,dx$**
$y = k$	$\int y\,dx = kx + c$
$y = kx$	$\int y\,dx = (k/2)x^2 + c$
$y = kx^2$	$\int y\,dx = (k/3)x^3 + c$
$y = kx^n$	$\int y\,dx = ((k/(n+1))x^{n+1} + c$
$y = \log_e(x)$	$\int y\,dx = x \log_e(x) - x + c$
$y = e^x$	$\int y\,dx = e^x + c$
$y = \sin(x)$	$\int y\,dx = -\cos(x) + c$
$y = \cos(x)$	$\int y\,dx = \sin(x) + c$
$y = \tan(x)$	$\int y\,dx = -\log_e(\cos(x)) + c$

INDEFINITE INTEGRAL

The area under the curve of a function is found by a calculus technique called integration. The indefinite integral is an expression that can be evaluated for any two points in the domain of the function, so that any defined portion of the area under the curve may be calculated.

Given a function f(x), the indefinite integral is written in the form:

$$\int f(x)\,dx$$

where dx is the differential of x, representing the limiting increment in the value of x. (For purposes of calculation, the differential is not of importance, except that it should always be written when specifying an indefinite integral.)

The table shows some common indefinite integrals. More comprehensive tables are available in a variety of mathematics data books. Note that all of the indefinite integral functions contain a constant, denoted by c. This constant is present because indefinite integration is just the opposite of differentiation, and any constant term, when differentiated, becomes zero. In the table, the functions after the integral signs are the derivatives of their indefinite-integral functions. Definite integrals are found by first establishing the limits, x_1 and x_2, in the domain, within which the area under the curve is to be evaluated. The value of the integral function is then calculated for x_1 and x_2. The definite integral is the difference between these two numerical values. *See also* CALCULUS, DEFINITE INTEGRAL, DERIVATIVE, FUNCTION.

INDEPENDENT VARIABLE

In any relation or function, an independent variable is a value on which other variables depend. There may be one independent variable, or there may be several. In the Cartesian coordinate plane, wherein functions of one variable are represented, the independent variable is almost always plotted along the horizontal axis. In a polar coordinate system, the independent variable is usually plotted as an angle progressing counterclockwise (*see* CARTESIAN COORDINATES, POLAR COORDINATES).

The difference between the independent and dependent variables can be illustrated by imagining a graph of temperature versus time of day. The time is the independent variable on which the temperature depends. *See also* DEPENDENT VARIABLE, FUNCTION.

INDEX OF MODULATION

See MODULATION INDEX.

INDEX OF REFRACTION

The speed of electromagnetic-wave propagation is approximately 186,000 miles (300,000 kilometers) per second in empty space. However, in various other media, the speed is slower. For a given substance at a specified electromagnetic wavelength, the index of refraction, n, is defined as:

$$n = 299792/v$$

where v is the speed of propagation in the given medium, measured in kilometers per second.

For most substances, the index of refraction changes with the wavelength. For example, the index of refraction for glass is greater at the violet end of the visible spectrum than at the red end. This is called dispersion (*see* DISPERSION).

Refractive effects are evident not only at the visible wavelengths, but also at radio frequencies. In certain substances such as polyethylene, radio waves travel much more slowly than in free space. The value 1/n in such situations is known as the velocity factor. *See also* REFRACTION, VELOCITY FACTOR.

INDICATOR

An indicator is a device that displays the status of a circuit function, such as current, voltage, power, on/off condition, or frequency. Indicators usually take the form of analog or digital meters (*see* ANALOG METERING, DIGITAL METERING). However, light-emitting devices are also sometimes used as indicators.

Indicators are characterized by a direct reading of the quantity to be measured. No calculations are necessary. If the measured parameter changes value or state, the indicator follows the change. *See also* METER.

INDIRECTLY HEATED CATHODE

In a vacuum tube, the cathode may consist of a tubular metal electrode, within which a heating filament is inserted. This type of cathode is called an indirectly heated cathode, since the heating current does not actually pass through the cathode itself.

Indirectly heated tubes have certain advantages over the directly heated type. The cathode can be electrically separated from the filament. This makes it easy to control the bias between the grid and the cathode of the tube; a parallel resistor-capacitor combination, connected in series with the cathode lead, provides negative grid bias without the need for a separate power supply. The input signal can be easily applied to an indirectly heated cathode. The main disadvantage of the indirectly heated cathode is that the tube requires time to "warm up." *See also* CATHODE.

INDIUM

Indium is an element with atomic number 49 and atomic weight 115. In the manufacture of semiconductor materials, indium is used as an impurity or dopant. Indium is an acceptor impurity. This means that it is electron-deficient. When indium is added to a germanium or silicon semiconductor, a P-type substance is the result. P-type semiconductors conduct via holes. *See also* DOPING, HOLE, IMPURITY, P-TYPE SEMICONDUCTOR.

INDOOR ANTENNA

For radio and television reception or transmission, an outdoor antenna is always preferable to an indoor antenna. However, in certain situations it is not possible to install an outdoor antenna.

Indoor antennas are always a compromise, especially at the very-low, low, medium, and high frequencies. Electrical wiring interferes with wave propagation at these frequencies. In a concrete-and-steel structure, the shielding effect is even more pronounced than in a frame building. Indoor antennas are generally more subject to manmade interference from electrical appliances. If an indoor antenna is used for transmitting purposes, the chances of electromagnetic interference are greatly increased, compared with the use of an outdoor antenna. The efficiency of an indoor transmitting antenna is generally lower than that of an identical outdoor antenna.

Indoor antennas have some advantages over outdoor antennas. An indoor antenna is less susceptible to induced voltages resulting from nearby lightning strikes. Indoor antennas do not corrode as rapidly as outdoor antennas, and maintenance is simpler. At very-high and ultra-high frequencies, indoor antennas can perform very well if they are located high above the ground.

In general, if an indoor antenna must be used, the antenna should be kept reasonably clear of metallic obstructions. It should be placed as high as possible. At frequencies below about 30 MHz, the antenna should be cut to resonance or made as long as possible.

INDUCED EFFECTS

Voltages and currents can be induced in a material in a variety of ways. When a conductor is moved through a magnetic field, current flows in the conductor. An object in an electric field exhibits a potential difference, in certain instances, between one region and another. Alternating currents in one conductor can induce similar currents in nearby conductors.

Induced effects are extremely important in many applications. The transformer works because a current in the primary winding induces a current in the secondary. Radio communication is possible because of the induced effects of alternating currents at high frequencies. *See also* ELECTRIC FIELD, ELECTROMAGNETIC FIELD, MAGNETIC FIELD.

INDUCTANCE

Inductance is the ability of a device to store energy in the form of a magnetic field. Inductance is represented by the capital letter L in mathematical equations. The unit of inductance is the henry. One henry is the amount of inductance necessary to generate 1 volt with a current that changes at the rate of 1 ampere per second. Mathematically:

$$L \text{ (henrys)} = E/(dI/dt)$$

where E is the induced voltage and dI/dt is the rate of current change in amperes per second.

In practice, the henry is an extremely large unit of inductance. Usually, inductance is specified in millihenrys (mH), microhenrys (uH), or nanohenrys (nH). These units are, respectively, a thousandth, a millionth, and a billionth of 1 henry:

$$1 \text{ mH} = 10^{-3} \text{ H}$$

$$1 \text{ uH} = 10^{-3} \text{ mH} = 10^{-6} \text{ H}$$

$$1 \text{ nH} = 10^{-3} \text{ uH} = 10^{-9} \text{ H}$$

Any length of electrical conductor displays a certain amount of inductance. When a length of conductor is deliberately coiled to produce inductance, the device is called an inductor (*See* INDUCTOR).

Inductances in series add together. For n inductances L1, L2, . . ., Ln in series, then, the total inductance L is:

$$L = L1 + L2 + \ldots + Ln$$

In parallel, inductances add according to the equation:

$$L = 1/(1/L1 + 1/L2 + \ldots + 1/Ln)$$

See also INDUCTIVE REACTANCE.

INDUCTANCE MEASUREMENT

Inductance is usually measured by means of a device incorporating known capacitances that cause resonance effects at measurable frequencies. In a tuned circuit, an unknown inductance can be combined with a known capacitance, and the resonant frequency determined by means of a signal generator and indicating device. The illustration shows this arrangement.

Once the resonant frequency f has been found for a known capacitance C and an unknown inductance L, the value of L can be determined according to the following equation:

$$L = 1/(39.5f^2C)$$

where L is given in microhenrys, f in megahertz, and C in microfarads. Alternatively, L may be found in henrys if f is given in hertz and C is given in farads.

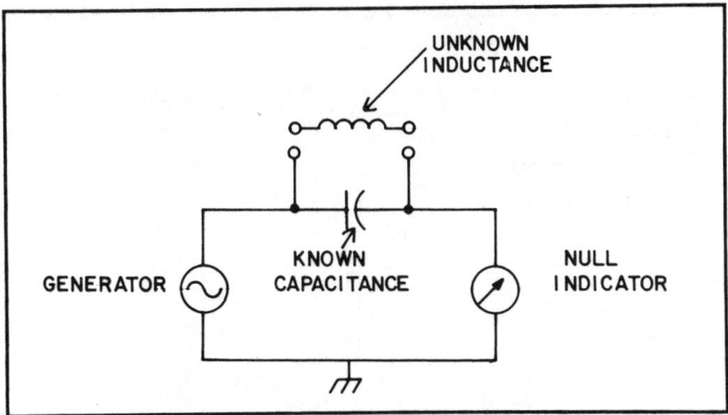

INDUCTANCE MEASUREMENT: A simple method of determining the value of an unknown inductance. The signal-generator frequency is adjusted until a dip, or null, is observed on the indicator.

A more sophisticated device for determining unknown inductances is called the Hay bridge. This circuit allows not only the inductance, but also the Q factor, to be found for a particular inductor. *See also* HAY BRIDGE, INDUCTANCE, Q FACTOR, RESONANCE.

INDUCTION

See INDUCED EFFECTS.

INDUCTION COIL

An induction coil is a device that generates very high ac voltages. The induction coil operates on the transformer principle (see illustration).

The primary winding of an alternating-current step-up transformer is connected to a source of direct current through an interrupter, such as a vibrator or chopper. The interrupter produces a series of pulses in the primary winding. The secondary winding exhibits a large alternating-current voltage. If the secondary-to-primary turns ratio of the transformer is large enough, this potential may reach several thousand volts. The maximum obtainable value is limited by the efficiency of the transformer. *See also* CHOPPER POWER SUPPLY, TRANSFORMER.

INDUCTION HEATING

Induction heating is a method of heating a metal by causing induced current to flow. The induction heater consists of a large coil, known as a work coil, to which a high-power radio-frequency signal is applied. The metal object is placed inside this coil. If the radio-frequency signal is of the correct wavelength, and the power level is high enough, the metal object will become very hot.

Induction heating works because the fluctuating magnetic field, within the coil, causes circulating currents in the metal; the ohmic losses in the metal result in the generation of heat. Induction heating makes use of magnetic effects in the same way that electric fields cause dielectrics to become hot. *See also* DIELECTRIC HEATING.

INDUCTION LOSS

When a conductor carries radio-frequency energy, it is important that the conductor either be shielded, or that it be kept away from metallic objects. This is because the inductive coupling to nearby metallic objects can cause losses in the conductor.

Induction loss can occur when an antenna radiator is placed too close to utility wiring, a metal roof, or other conducting medium. Circulating currents occur, resulting

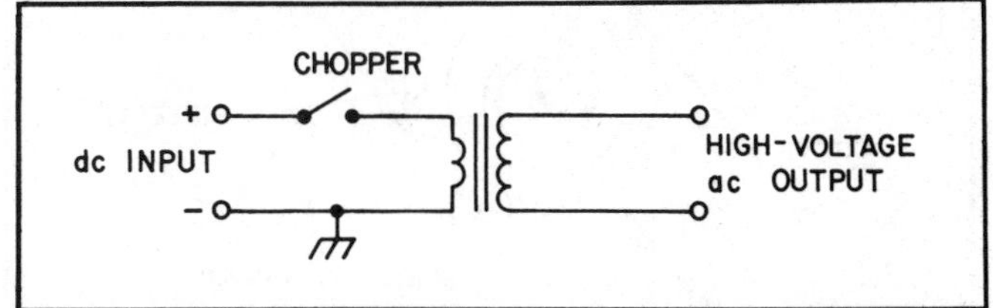

INDUCTION COIL: An induction coil is used to generate extremely high alternating-current voltages.

in heating of the nearby material. A parallel-wire transmission line should be kept at least several inches away from metal objects such as wires, towers, or siding, so that the possibility of induction loss is minimized. Induction losses increase the overall loss in an antenna system.

INDUCTIVE CAPACITOR

An inductive capacitor is a capacitor with built-in inductance. Such capacitors are useful as self-contained tuned circuits. Inductive capacitors are generally wound in a spiral pattern, with two concentric plates separated by a rigid dielectric material (see illustration). The amount of capacitance depends on the area and separation of the metal plates, and also on the nature of the dielectric. The amount of inductance depends largely on the number of turns in the spiral structure.

In the very-high-frequency range and above, all capacitors begin to show significant inductive reactance, simply because of the finite length of the component leads. In the ultra-high and microwave parts of the spectrum, the wiring inductance of all components must be taken into account when circuits are designed and constructed. *See also* LEAD INDUCTANCE.

INDUCTIVE COUPLING

See TRANSFORMER COUPLING.

INDUCTIVE FEEDBACK

Inductive feedback results from inductive coupling between the input and output circuits of an amplifier. The feedback may be intentional or unintentional; it can be positive (regenerative) or negative (degenerative), depending on the number of stages and the circuit configuration. Inductive feedback can take place at audio or radio frequencies.

Some oscillators use inductive feedback for the purpose of signal generation. An example of such an oscillator is the Armstrong oscillator, in which a feedback coil from the output circuit is deliberately placed near the tank coil (*see* ARMSTRONG OSCILLATOR).

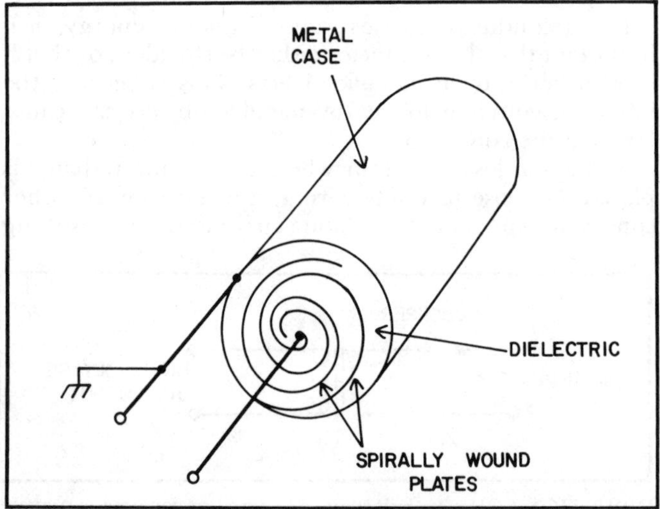

INDUCTIVE CAPACITOR: An example of an inductive capacitor. There are other varieties.

Unwanted inductive feedback is generally more of a problem in the very-high-frequency range and above, as compared to its likelihood at lower frequencies. This is because the inductive reactance of intercomponent wiring becomes greater as the frequency increases, and this in turn increases the inductive coupling among different parts of a circuit. Inductive feedback of a regenerative nature can result in spurious oscillations in a receiver or transmitter. The spurious signals in a transmitter may be radiated along with the desired signal, causing interference to other stations. In audio-frequency equipment, inductive feedback can sometimes occur if unshielded speaker leads are placed near the input cables of the amplifier. The result, if the feedback is regenerative, will be hum or an audio-frequency tone. *See also* FEEDBACK.

INDUCTIVE LOADING

In radio-frequency transmitting installations, it is often necessary to lengthen an antenna electrically, without making it physically longer. This is especially true at the very-low, low, and medium frequencies. Inductive loading is the most common method of accomplishing this.

In an unbalanced antenna, such as a quarter-wave resonant vertical working against ground, the inductor may be placed at any location less than approximately 90 percent of the way to the top of the radiator. The most common positions for the inductor are at the base and near the center. With a given physical radiator length, larger and larger inductances result in lower and lower resonant frequencies. There is no theoretical limit to how low the resonant frequency of an inductively loaded quarter-wave vertical can be made. In practice, however, the losses become prohibitive when the physical length of the radiator is less than about 0.05 wavelength.

In a balanced antenna system, such as a half-wave resonant dipole, identical inductors are placed in each half of the antenna, in a symmetrical arrangement with respect to the feed line (see illustration). For a given physical antenna length, larger and larger inductances result in lower and lower resonant frequencies. As with the quarter-wave vertical antenna, there is no theoretical lower limit to the resonant frequency that can be realized with a given physical antenna length. However, the losses become prohibitive when the physical length becomes too short. If the dipole is placed well above the ground and away from obstructions, and if the conductors and inductors are

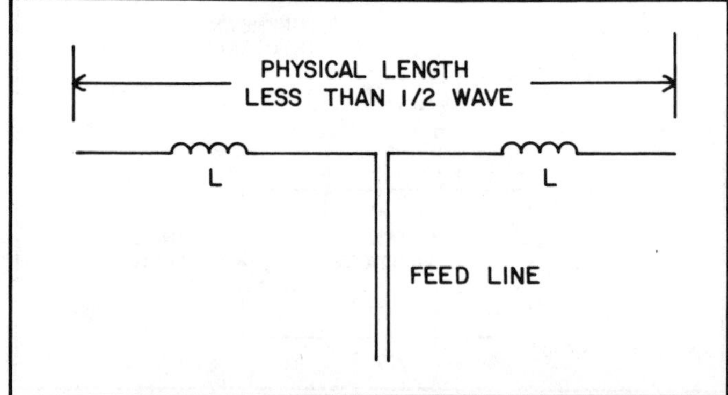

INDUCTIVE LOADING: Inductive loading of a dipole antenna. The two coils have identical inductance L, and are positioned symmetrically with respect to the feed point.

made from material with the least possible ohmic loss, good performance can be had with antennas as short as 0.05 free-space wavelength.

Inductively loaded antennas always show a low radiation resistance, because the free-space physical length of the radiator is so short. As the length of an antenna becomes shorter than 0.1 wavelength, the radiation resistance drops rapidly. This makes it difficult to achieve good transmitting efficiency with very short radiators. *See also* BASE LOADING, CENTER LOADING, RADIATION RESISTANCE.

INDUCTIVE REACTANCE

Reactance is the opposition that a component offers to alternating current. Reactance does not behave as simply as resistance. The reactance of a component can be either positive or negative, and it varies in magnitude depending on the frequency of the alternating current. Positive reactance is called inductive reactance; negative reactance is called capacitive reactance. These choices of positive and negative are purely arbitrary, and are used as a matter of mathematical convenience.

The reactance of an inductor of L henrys, at a frequency of f hertz, is given by:

$$X_L = 2\pi fL,$$

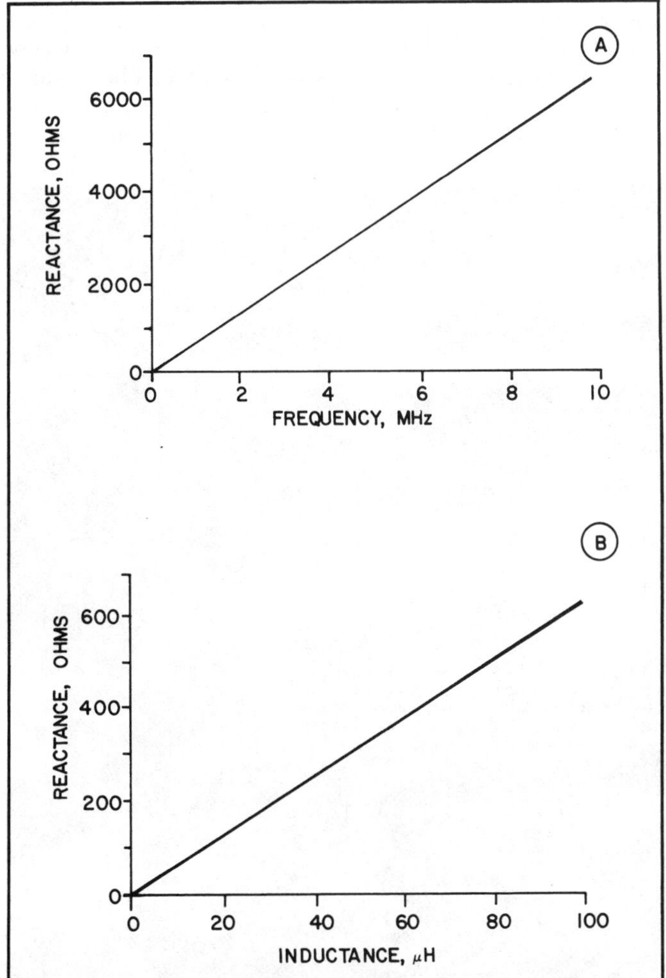

INDUCTIVE REACTANCE: At A, the reactance of a 100-uH inductor as a function of frequency. At B, the reactance of various values of inductances at a frequency of 1 MHz.

where X_L is specified in ohms. If the frequency f is given in kilohertz and the inductance L in millihenrys, the formula also applies. The frequency may be given in megahertz and the inductance in microhenrys, as well. (Units should not be mixed.)

The magnitude of the inductive reactance, for a given component, approaches zero as the frequency is lowered. The inductive reactance gets larger without limit as the frequency is raised. This effect is illustrated in the graph. At A, the reactance of a 100-uH inductor is shown as a function of the frequency in megahertz. At B, the reactance of various values of inductance are shown for a constant frequency of 1 MHz.

Reactances, like direct-current resistances, are always expressed in ohms. In a complex circuit containing resistance and reactance, the reactance is multiplied by $\sqrt{-1}$, which is mathematically represented by the letter j in engineering documentation. Thus an inductive reactance of +20 ohms is specified as +j20; a combination of 10 ohms resistance and +20 ohms inductive reactance is a complex impedance of 10 + j20. *See also* IMPEDANCE, J OPERATOR.

INDUCTOR

An inductor is an electronic component designed especially for the purpose of providing a controlled amount of inductance. Inductors generally consist of a length of wire wound into a solenoidal or toroidal shape (see illustration). The inductance may be increased by placing a substance with high magnetic permeability within the coil. Such materials include iron, powdered iron, and ferrite. Commercially made inductors have values ranging from less than 1 uH to about 10 H. Small inductors are used in radio-frequency tuned circuits and as radio-frequency chokes. Larger inductors are employed at audio frequen-

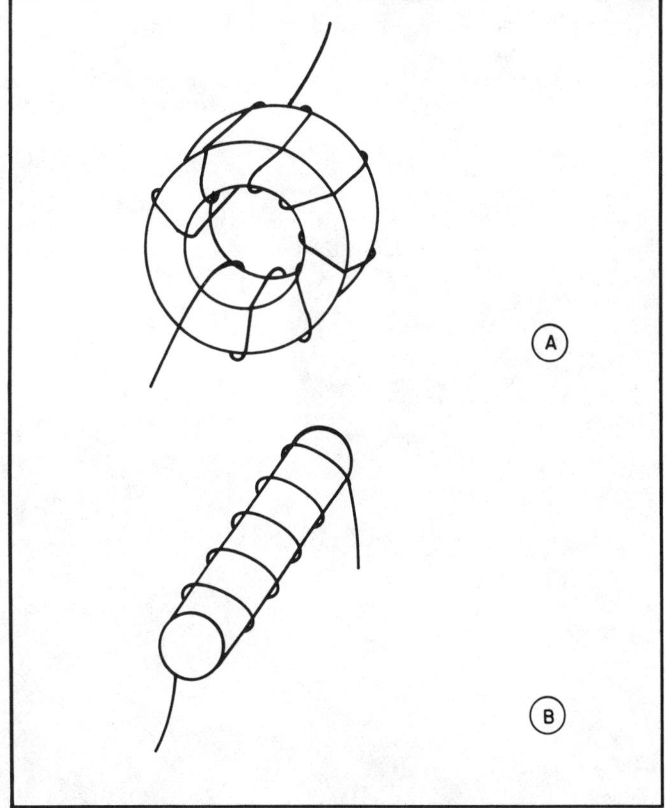

INDUCTORS: Toroid (A) and solenoid (B).

cies; the largest inductors are used as filter chokes in power supplies.

Coil-shaped inductors are used in tuned circuits from audio frequencies to the ultra-high radio-frequency region. In the ultra-high-frequency and microwave parts of the spectrum, short lengths of transmission line can serve as inductors. Any length of line shorter than ¼ electrical wavelength, and short-circuited at the far end, acts as an inductor. The same is true of a section of line between ¼ and ½-wavelength, with an open circuit at the far end.

A perfect inductor shows only inductive reactance, and no resistance. Such a component can exist only in theory; all real inductors have some ohmic loss as well as reactance. *See also* INDUCTANCE, INDUCTIVE REACTANCE.

INDUCTOR TUNING

Inductor tuning is a means of adjusting the resonant frequency of a tuned circuit by varying the inductance while leaving the capacitance constant. When this is done, the resonant frequency varies according to the inverse square root of the inductance. This means that if the inductance is doubled, the resonant frequency drops to 0.707 of its previous value. If the inductance is cut to one-quarter, the resonant frequency is doubled. Mathematically, the resonant frequency of a tuned inductance-capacitance circuit is given by the formula:

$$f = 1/(2\pi \sqrt{LC})$$

where f is the frequency in hertz, L is the inductance in henrys, and C is the capacitance in farads. Alternatively, the units may be given as megahertz, microhenrys, and microfarads.

Inductor tuning is somewhat more difficult to obtain than capacitor tuning, from a mechanical standpoint. However, inductor tuning usually offers a more linear frequency readout than capacitor tuning. If the capacitance required to obtain resonance is larger than the values normally provided for variable capacitors, inductor tuning is preferable. For the purpose of inductor tuning, a rotary device may be used, as shown in the photograph. A movable core, consisting of powdered iron or ferrite, may also be used. The latter form of inductor tuning is sometimes called permeability tuning. *See also* CAPACITOR TUNING.

INDUSTRIAL ELECTRONICS

Industrial electronics is the field of application of electronics in industry. In recent years, industrial electronics has become increasingly important in the manufacturing and quality-control areas. Computers have played a major role in the advance of industrial electronics. But the applications of electronics in industry are widespread, and include other fields, such as communications and automotive systems.

In a modern manufacturing plant, electronic devices now perform many of the mundane tasks previously done by people. Some people believe that this has done injury to the job market. However, computer operators and designers are needed more than ever before, for the purpose of perfecting electronic devices in industry. The advance of electronics has brought about a greatly increased consumer demand for electronic home-entertainment de-

INDUCTOR TUNING: Inductor tuning can be accomplished by means of a rotary coil, such as the one in this antenna tuner.

vices. These factors point to an increase in the demand for industrial personnel. *See also* INDUSTRIAL RADIO AND TELEVISION.

INDUSTRIAL RADIO AND TELEVISION

Electromagnetic communications serves an important role in industry. Two-way radio communication is invaluable in the dispatching of personnel and in the overall control of mobile networks. The telephone "beeper," which alerts personnel of incoming calls, is used by thousands of small businesses in the United States and other industrialized countries today. Business two-way radio systems generally operate in the very-high-frequency and ultra-high-frequency parts of the spectrum, and are licensed by government agencies. The Citizen's band, at approximately 27 MHz, is widely used for commercial purposes in many countries.

Closed-circuit television systems are often used in retail stores, banks, and other places in which theft is a continuing problem. Wide-angle lenses allow constant monitoring of the premises, and video-tape recorders can be used to photograph suspicious persons. In the extreme, tiny radio transmitters called "bugs" are sometimes used to eavesdrop on conversations relating to business. A remote receiver, often equipped with a recorder, picks up the transmission. *See also* CITIZEN'S BAND, TWO-WAY RADIO.

INFORMATION

Information is a term used to describe many kinds of data. In general, any message, whether analog or digital, contains information. The amount of information in an analog message is difficult to measure, but can be roughly estimated in terms of the size of an equivalent digital message. The size of a digital message is specified in bits, bytes, or words.

In general, for the transmission of a given amount of information, a certain minimum amount of bandwidth is required for a certain amount of time. The wider the bandwidth allotted, the more rapidly a given quantity of information can be conveyed. The longer the time allowed, the narrower the bandwidth can be. *See also* BIT, BYTE, DATA, WORD.

INFORMATION THEORY

Information theory is a form of statistical communication theory. The development of information theory is usually credited to C. E. Shannon. Shannon believed that a signal could be processed at the transmitting end of a circuit, as well as at the receiving end, for the purpose of improved communications reliability under adverse conditions.

Information theory allows an engineer to calculate the best possible reliability that can be expected for a communications circuit under specified conditions. For any given noise level, it is possible to transmit information with any desired degree of accuracy. The rate of data transfer is the primary variable. The slower the rate at which data is sent, the greater the accuracy.

In recent years, information theory has been refined to help engineers not only to determine the reliability of a particular system, but to design the optimum system for a certain set of conditions. Information theory makes extensive use of probability and calculus techniques.

INFRADYNE RECEIVER

An infradyne receiver is a form of superheterodyne receiver. Most superheterodyne receivers produce an intermediate-frequency signal that has a frequency equal to the difference between the input and local-oscillator frequencies. The infradyne receiver, however, produces an intermediate frequency that is equal to the sum of the input and local-oscillator frequencies. The term infradyne arises from the fact that both the input and local oscillator signals are lower than the intermediate frequency.

If f is the input-signal frequency to an infradyne receiver and g is the frequency of the local oscillator, heterodynes are produced at frequencies corresponding to $f - g$ (if f is larger than g) or $g - f$ (if f is less than g); a signal is also produced at $f + g$. In the infradyne, $f + g$ is the intermediate frequency.

The infradyne receiver is characterized by the fact that the signal always appears right side up. That is, the frequency components are not inverted. In a superheterodyne receiver using the difference frequencies, the signal sometimes appears upside down. The main disadvantage of the infradyne is that harmonics of the local-oscillator signal may fall within the input-tuning range of the receiver. *See also* HETERODYNE, SUPERHETERODYNE RECEIVER.

INFRARED

Infrared is a form of electromagnetic radiation. The wavelength of infrared is somewhat longer than the wavelength of visible light. The shortest infrared wavelengths occur in the range just outside the visible spectrum. The longest infrared wavelengths border on the microwave part of the electromagnetic spectrum.

Infrared radiation is classified in four ways. The near infrared extends from a wavelength of 750 nanometers (nm) to 1,500 nm. The middle infrared extends from approximately 1,500 nm to 6,000 nm. The far infrared extends from 6,000 nm to 40,000 nm, and the far-far infrared encompasses wavelengths as great as 1 millimeter (mm).

We usually think of infrared as "heat radiation," although in a technical sense this is not entirely correct. Infrared radiation feels warm as it strikes the skin. This is because the radiant energy is transformed into heat as it is absorbed.

Infrared radiation is produced in large quantities by incandescent lamps. All stars in the heavens produce some infrared radiation; this is especially true of the red stars. Infrared wavelengths can be used for communications purposes, although the atmosphere is opaque to the energy at certain wavelengths. Water in the air causes severe attenuation between about 4,500 and 8,000 nm; carbon dioxide interferes with infrared transmission between about 14,000 and 16,000 nm. There are various other infrared-opaque bands.

Special forms of light-emitting diode (LED) produce emissions in the infrared region. Infrared photographs are often used for the purpose of evaluating the heat loss from a house or building. *See also* ELECTROMAGNETIC SPECTRUM, INFRARED DEVICE.

INFRARED DEVICE

Any device that converts infrared energy into some other form of energy, or vice-versa, is called an infrared device. The most common infrared devices are those which generate infrared emissions from electricity, and those which generate electrical impulses from infrared energy.

Infrared devices are used in communications systems, intrusion alarm systems, and tracking systems. Certain semiconductor substances act as photocells or photovoltaic cells at infrared wavelengths. Thermocouples can be used to sense infrared radiation at the longer wavelengths. Some diodes, such as gallium-arsenide types, generate infrared energy in the near range, or about 780 to 1000 nanometers (nm), when a current is passed through them. The neodymium (Nd) laser is used to generate a high-energy, coherent infrared beam at 1060 nm.

The most sophisticated infrared devices include photographic apparatus, facilitating detailed still pictures in the infrared range, and the snooperscope, which actually allows the detection of motion. Objects frequently emit infrared radiation even when they do not emit visible light. Thus, infrared devices can be used to literally "see in the dark." *See also* INFRARED, OPTICAL COMMUNICATIONS, SNOOPERSCOPE.

INFRASOUND

Infrasound is an acoustic disturbance that occurs at frequencies below the human hearing range. The lower limit of the audio-frequency range is about 16 to 20 Hz. The infrasonic range extends downward in frequency from this point.

Infrasound is characterized by long acoustic wavelengths. For this reason, an infrasonic disturbance can be propagated around many obstructions that would interfere with the transfer of ordinary sound. Infrasound can, if sufficiently intense, cause damage to physical objects because of resonance effects.

In air, sound waves travel at about 1100 feet, or 335 meters, per second. The same is true of infrasound. The wavelengths of infrasonic disturbances are generally greater than 55 feet, or 17 meters. Some disturbances may contain components at wavelengths of over 100 meters.

INHIBIT

In a digital system such as a computer, an inhibit command is a signal or pulse that prevents or delays a certain operation. If the inhibit line is in the low or zero-voltage state, certain operations do not take place, while they will occur if the inhibit line is high. In negative logic, the situation is reversed; a low inhibit line allows the operation to proceed, while the high state prevents it.

Inhibit gates are used in memory circuits for the purpose of preventing a change of state. This keeps the contents of the memory intact and inalterable. Until the inhibit command is removed, the memory cannot be changed. *See also* MEMORY.

INITIALIZATION

Initialization is a process in which all of the lines in a microcomputer are set to the low, or zero-voltage, condition prior to operation. Most microcomputers are initialized every time power is removed and reapplied. In some cases, certain memory stores are retained by memory-backup power supplies. Removal of the power source results in initialization.

It is sometimes necessary to initialize a microcomputer device during the course of normal operation. This is especially true of the complementary metal-oxide semiconductor, or CMOS, integrated circuit. Such devices are characterized by extremely low leakage current, and "false signals" can be maintained for extended periods, causing what appears to be a malfunction.

In some microcomputer devices, initialization consists of a resident routine for the purpose of aiding in the programming of the system. Such a routine is called the initial instruction. *See also* MICROCOMPUTER.

INITIAL SURGE

When power is first applied to a device that draws a large amount of current, an initial surge may occur in the circuit current. The surge is often considerably higher than the normal operating current load of the device. This effect is exemplified by the momentary dimming of the lights in a house when the refrigerator, air conditioner, or electric heater first starts. The current surge is normally not dangerous, but it can occasionally result in a blown fuse or tripped circuit breaker. The same effect occurs when power is applied to radio equipment, although it is not as dramatic. Slow-blow fuses are employed in devices that cause an initial surge (*see* SLOW-BLOW FUSE).

In the event of an electric power failure, an initial voltage surge may occur when electricity is restored. This surge can, in some instances, cause damage to appliances. This kind of initial surge is most likely to occur if the power failure covers a widespread area. In the event of a power failure, it is wise to shut off devices that draw large amounts of current. The same precautions should be taken with appliances that might be damaged by voltages slightly in excess of the normal 110 to 120 volts.

The initial surge is not the same thing, and is not caused by the same conditions, as a transient. Transients reach much higher peak amplitudes. *See also* TRANSIENT.

INJECTION LASER

The injection laser is a form of light-emitting diode (LED), with a relatively large and flat P-N junction. The injection laser emits coherent light, provided the applied current is sufficient. If the current is below a certain level, the injection laser behaves much like an ordinary LED, but when the so-called threshold current is reached, the charge carriers recombine in such a manner that laser action occurs.

Most injection-laser devices are fabricated from gallium arsenide (GaAs), and have a primary emission wavelength of about 905 nanometers (nm). This is in the infrared range. Other types of injection lasers are available for different wavelengths. Among these, the GaAsP and GaP injection lasers produce outputs at approximately 660 nm and 550 nm, respectively; these wavelengths are in the visible-light range. The maximum peak power, in short pulses, can be as great as 100 watts, but the pulse duration must be very short. Injection lasers produce emission at exceedingly narrow bandwidth. This is characteristic of

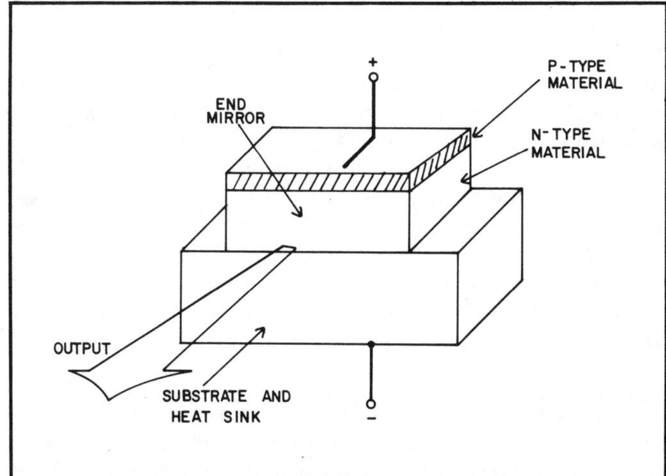

INJECTION LASER: An injection-laser diode.

laser devices. The drawing illustrates the construction of a typical injection-laser diode.

Injection lasers are used mainly in optical communications systems. The coherent light output from the injection laser suffers less attenuation than incoherent light as it passes through the atmosphere. *See also* OPTICAL COMMUNICATIONS.

INPUT

Input is the application of a signal to a processing circuit. An amplifier circuit, for example, receives its input in the form of a small signal; the output is a larger signal. A logic gate may have several signal inputs. The input terminals are themselves sometimes called inputs.

The input signal in a given situation must conform to certain requirements. The amplitude must be within certain limits, and the impedance of the signal source must match the impedance presented by the input terminals of the circuit. *See also* INPUT CAPACITANCE, INPUT DEVICE, INPUT IMPEDANCE, INPUT RESISTANCE.

INPUT CAPACITANCE

The input terminals of any amplifier, logic gate, or other circuit always contain a certain amount of parallel capacitance. The capacitance appears between the two terminals in a balanced circuit, and between the active terminal and ground in an unbalanced circuit.

The amount of input capacitance must, in most circuits, be smaller than a certain value for proper circuit operation. In general, the higher the frequency, the smaller the maximum tolerable input capacitance. Also, the higher the input impedance of a circuit (*see* INPUT IMPEDANCE), the smaller the maximum tolerable input capacitance. Generally, the input capacitance should not exceed 10 percent of the resistive input impedance. When the input capacitance becomes too high, the input impedance changes so much that some of the incident signal is reflected back toward the source.

In some circuits, the input capacitance may be very large. This is the case, for example, in the output filtering network of a power supply when a capacitor is used at the input. This condition is normal. *See also* INPUT RESISTANCE.

INPUT DEVICE

In a computer system, an input device consists of any peripheral equipment that serves the function of providing information to the computer. Examples of input devices include the keyboard in a terminal unit, a tape recorder or disk drive, and a telephone or radio interface. Card readers are also input devices. *See also* CARD READING, DISK DRIVE, TERMINAL.

In any electronic system, a circuit through which an input is applied is called an input device. Examples are the power transformer, a signal transformer, or a coupling network. *See also* INPUT.

INPUT IMPEDANCE

The input impedance of a circuit is the complex impedance that appears at the input terminals. Input impedance normally contains resistance, but no reactance, under ideal operating conditions. However, in certain situations, large amounts of reactance may be present. This occurs as either an inductance or a capacitance.

Input impedances are often matched to the output impedance of the driving device. For example, an audio amplifier with a 600-ohm resistive input impedance should be used with a microphone having an impedance of close to 600 ohms. An half-wave dipole antenna with a pure resistive input impedance of 73 ohms should be fed with a transmission line having a characteristic impedance of nearly this value; the transmitter output circuits should also have an impedance of nearly this value.

If reactance is present in the input impedance of a device, some of the electromagnetic energy is reflected back toward the source. This can happen in some circuits, to a limited extent, with little or no adverse effect. In other circuits, such a condition must be avoided. *See also* IMPEDANCE, IMPEDANCE MATCHING, OUTPUT IMPEDANCE, REACTANCE.

INPUT-OUTPUT PORT

In computer systems, an input-output port or device is a peripheral unit that serves two purposes: to provide data to the computer, and to transfer data from the computer. Most input devices, such as a tape recorder, disk drive, and terminal unit, can also serve as output ports. Punched cards, in some computers, serve as input-output ports. The modem, which provides an interface with a telephone or radio circuit, is a common form of input-output device (*see* MODEM).

An input-output port can serve any of three different functions. It may operate as an interface between the computer and a human operator; a video-display terminal and keyboard function in this manner. The input-output device may serve as a read-write memory medium. The disk drive is an example of this. The input-output device may also act as an interface between one computer and another. The modem is generally used for this purpose. *See also* CARD READING, DISK DRIVE, INTERFACE TERMINAL.

INPUT RESISTANCE

In a direct-current circuit, a certain resistance is always present at the input terminals. This is called the input resistance, and is equivalent to the input voltage divided by the input current.

The input resistance of a direct-current circuit may remain constant, regardless of the applied voltage. This is generally the case with passive direct-current networks that contain only resistors. Often, however, the input resistance changes with the amount of applied voltage. This is generally true with passive circuits containing nonlinear elements such as semiconductor diodes. Active devices, such as audio amplifiers and radio equipment, usually have an input resistance that changes, not only with the level of applied voltage, but with the level of the input signal.

In an alternating-current circuit, the input resistance is the resistive component of the input impedance. *See also* INPUT IMPEDANCE.

INPUT TUNING

When the input circuit of an amplifier contains a tuned inductance-capacitance circuit, the amplifier is said to have input tuning. Not all amplifiers require input tuning. The illustration shows three radio-frequency amplifier circuits. The amplifier at A has a pi-network input-tuning circuit. The amplifier shown at B uses a resonant inductance-capacitance network at its input. The two configurations are used for different purposes. At C, an amplifier is illustrated that does not make use of input tuning.

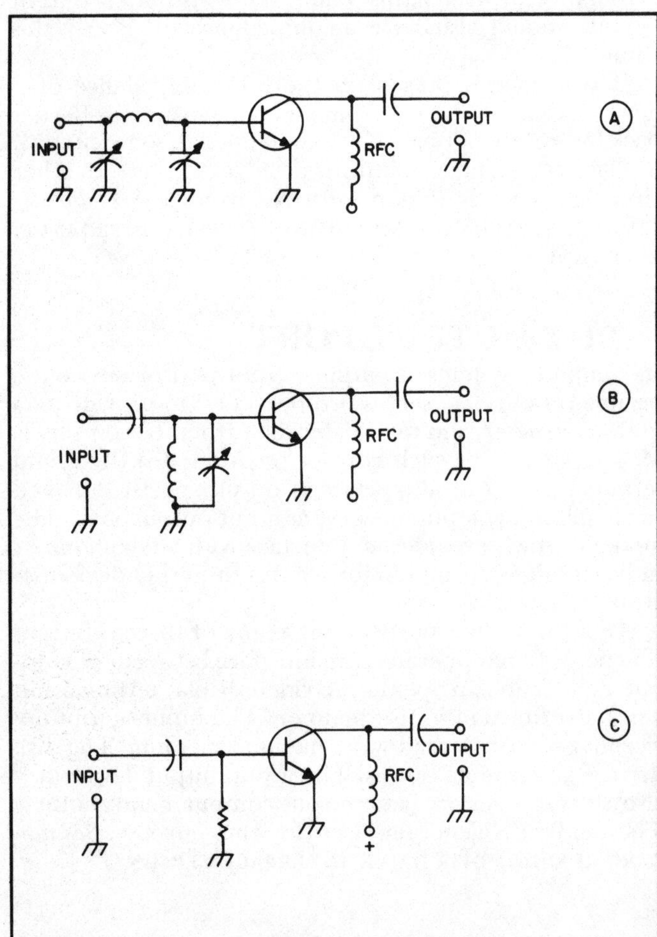

INPUT TUNING: At A, a simple radio-frequency amplifier with a pi-network input-tuning circuit. At B, an amplifier with a parallel-resonant input-tuning circuit. At C, an amplifier with no input tuning.

Input tuning has certain advantages and disadvantages. The primary advantage is realized in situations where the amplifier input impedance differs from the output impedance of the driving stage, or when additional selectivity is desired. The pi-network circuit (A) allows unequal impedances to be matched. The parallel-resonant network (B) provides high input selectivity. The main disadvantage of input tuning is that it increases the complexity of adjustment of the amplifier circuit. A change in the frequency of the driving signal usually necessitates retuning of the input networks. *See also* IMPEDANCE MATCHING, PI NETWORK.

INSERTION GAIN

Insertion gain is an expression of signal amplitude with and without the use of some intermediate circuit. Insertion gain is expressed in decibels.

If the voltage amplitude at the output of a device is E_1 without the intermediary circuit and E_2 with the intermediary circuit, then the insertion gain G is given in decibels by:

$$G\ (\text{dB}) = 20 \log_{10}(E_2/E_1)$$

For current, the formula is similar; if I_1 represents the output current without the intermediary circuit and I_2 is the output current with the circuit in place, then the gain G is:

$$G\ (\text{dB}) = 20 \log_{10}(I_2/I_1)$$

For power, if P_1 is the power without the intermediary circuit and P_2 is the power with the circuit in place, then the insertion gain G is:

$$G\ (\text{dB}) = 10 \log_{10}(P_2/P_1)$$

It is evident from the above equations that if the insertion of the intermediary circuit causes a reduction in voltage, current, or power, then the insertion gain for that parameter is negative. In such cases, the intermediary circuit is said to have insertion loss (*see* INSERTION LOSS). Passive circuits always have negative insertion gain. Active circuits may display positive or negative insertion gain, depending on whether voltage, current, or power is evaluated.

Amplifier circuits may show insertion gain for some parameters, but a loss for other parameters. It is necessary to define insertion gain in terms of current, voltage, or power, if the expression is to have meaning. *See also* DECIBEL, GAIN.

INSERTION LOSS

Insertion loss is a reduction in the output voltage, current, or power of a system, resulting from the addition of an intermediary network such as a filter or attenuator. Insertion loss is generally specified in decibels.

For a selective filter, the insertion loss is normally specified for the frequency, or band of frequencies, at which the attenuation is the least. Most selective filters have low insertion loss at the frequencies to be passed, and high loss at other frequencies.

If the output voltage of a circuit is E_1 before the insertion of an intermediary network and E_2 after the insertion of the network, the insertion loss L is given by:

$$L\ (dB) = 20 \log_{10}(E_1/E_2)$$

For current, if I_1 is the output current before insertion of the device and I_2 is the current with the device installed, then the loss in decibels is given by:

$$L\ (dB) = 20 \log_{10}(I_1/I_2)$$

For power, if P_1 is the output wattage before the insertion of the device and P_2 is the wattage afterward, then:

$$L\ (dB) = 10 \log_{10}(P_1/P_2)$$

Passive networks always have a certain amount of insertion loss, but an efficient device normally exhibits less than 1 dB of loss at the frequencies to be passed. Active networks may have negative insertion loss. In general, if the insertion loss in decibels is −x, then the insertion gain in decibels is x. That is, negative loss is the same as positive gain, and vice-versa. *See also* DECIBEL, INSERTION GAIN, LOSS.

INSTANTANEOUS EFFECT

Whenever a parameter fluctuates, rather than maintaining a constant magnitude or value, instantaneous effects become important. Amplitude and frequency are the most common parameters, in electronic circuits, that change value. However, resistance and impedance also change rapidly under some conditions.

The instantaneous magnitude of a parameter is the value at a given instant. This is, of course, a function of time. The familiar sine-wave alternating-current waveform provides a good example of the difference between an instantaneous value and the average and effective values. The illustration shows the voltage-versus-time function for the output of a typical household electrical outlet. The instantaneous voltage can attain any value between −165 volts and +165 volts. The precise voltage changes with time, completing one cycle every 16.67 milliseconds. The average voltage is actually zero, but this is obviously not the effective voltage! The root-mean-square (RMS) voltage, which is usually considered the effective voltage, is approximately 117 volts.

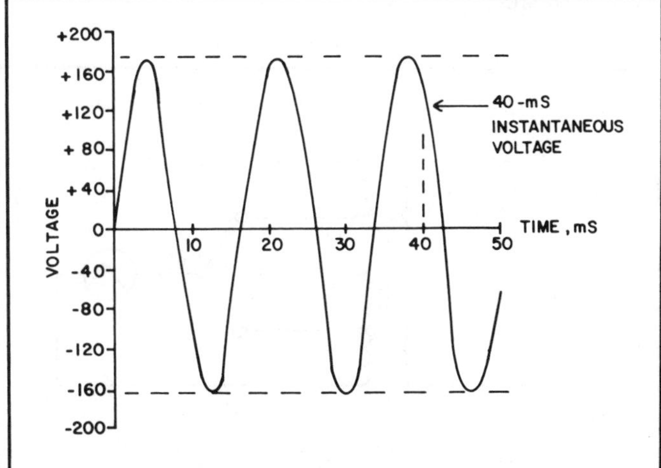

INSTANTANEOUS EFFECT: The voltage at a utility outlet would look like this on an oscilloscope display. Any point on the waveform shows the instantaneous voltage (vertical axis) at some moment (horizontal axis).

Instantaneous effects are sometimes of little concern, but in other cases they can be very important. A sudden surge in the voltage from a household utility line might last for only a few millionths of a second, and thus not affect the average or RMS voltages significantly. However, the instantaneous voltage may reach several hundred volts during that brief time, and this can be damaging to some kinds of electronic equipment. Instantaneous effects can be measured only with devices capable of displaying the details of a waveform. The oscilloscope is the most common such instrument. *See also* AVERAGE VALUE, EFFECTIVE VALUE, OSCILLOSCOPE, PEAK VALUE, ROOT MEAN SQUARE.

INSTITUTE OF ELECTRICAL AND ELECTRONIC ENGINEERS

The Institute of Electrical and Electronic Engineers, abbreviated IEEE, is a professional organization in the United States and some other countries. The purpose of the association is to advance the state of the art in electricity and electronics. Originally, the IEEE consisted of two separate groups called the American Institute of Electrical Engineers (AIEE) and the Institute of Radio Engineers (IRE).

The IEEE sets standards for construction and performance of many kinds of electrical and electronic equipment. Members work in conjunction with other organizations, such as the Electronic Industries Association. *See also* ELECTRONIC INDUSTRIES ASSOCIATION.

INSTRUCTION

An instruction is a set of data bits that directs a computer to follow certain procedures. A computer program, for example, is a string of instructions. The execution of the instructions results in the computer performing the complete operations desired. A program may contain in excess of 1,000,000 individual instructions.

A computer is capable of performing only a finite number of different kinds of instruction; the complete set of instructions that the machine can execute is called the instruction repertoire. The total number of individual instructions that a machine can handle, in a specific program, depends on the memory capacity. Most modern computers have an instruction repertoire of fewer than 100 different identifiable commands, although the number of possible combinations in a program is enormous.

An instruction normally consists of a code signal and a memory address. The code signal denotes the particular instruction, such as "multiply" or "divide." The memory address tells the machine where to get and store the information. Instructions may be fed into a computer by means of a disk drive, a magnetic tape, a set of punched cards, a modem, or by a human operator via a keyboard and video display unit. *See also* COMPUTER, DATA.

INSTRUMENT

An instrument is a device for measuring some parameter. The instrument may simply display the quantity for direct viewing by a human operator, or the information may be recorded or transmitted to a remote location. In general, any device that registers a variable quantity is an instrument. Examples are the voltmeter, ammeter, wattmeter, speedometer, oscilloscope, and many other devices. Instruments may be either analog (capable of registering a continuous range of values) or digital (showing only discrete values). Both types of instrument have advantages and disadvantages in various applications.

Some instruments are more accurate than others. The accuracy of an analog instrument always depends on two factors: the electrical or mechanical precision of the device, and the ease with which the indication is read or interpolated by the operator. A digital instrument eliminates operator errors by registering the value as a numeral. Mechanical difficulties are overcome in digital instruments by the use of electronic displays, which have no moving parts. The chief limiting factor in the accuracy of a digital instrument is the electrical calibration. The fineness or coarseness of the indicated steps is also important.

Instruments form the interface between a human operator and a piece of equipment. Well-designed instrumentation is imperative if an electronic or mechanical device is to be used effectively. *See also* ANALOG METERING, DIGITAL METERING, HUMAN ENGINEERING.

INSTRUMENT AMPLIFIER

An instrument amplifier is an input amplifier for an electronic instrument. The amplifier is used for the purpose of increasing the sensitivity of the instrument. Such amplifiers can be used with direct-current or alternating-current instruments. Some devices that employ amplifiers include the active field-strength meter, the oscilloscope, and the FET voltmeter.

There is a limit to the sensitivity that can be obtained with any instrument, no matter how high the gain of the amplifier. This is because all amplifiers generate some noise, and a small amount of background noise occurs as a result of currents and thermal effects. Quantities of magnitude smaller than the level of the noise, therefore, cannot be reliably measured even when instrument amplifiers are used.

Instrument amplifiers should be housed in shielded enclosures, and bypass capacitors or appropriate chokes should be used, so that the instrument will not be affected by signals other than those to be measured. The power supply must be checked periodically. Calibration should also be regularly checked against a known standard when an instrument amplifier is used. *See also* INSTRUMENT.

INSTRUMENT ERROR

All indicating devices have some degree of inaccuracy; none display the actual value of a parameter. The difference between the actual value and the instrument reading is called the instrument error.

In an analog instrument, the amount of error depends on the electrical and mechanical calibration of the device, and also on the ease with which the operator can interpolate among the divisions on the scale. The error is usually specified, for a given instrument, as a maximum percentage at full scale (*see* FULL-SCALE ERROR).

In a digital instrument, operator error is not a factor, since the parameter is displayed as a numeral. The limitations on the accuracy of a digital device are imposed by the electrical precision of the indicator circuit, and by the resolution of the display. A digital meter having a resolution of three decimal places for a given range (such as 0.00 to 9.99 volts) cannot provide as much accuracy as a meter with a resolution of four decimal places (such as 0.000 to 9.999 volts). The instrument error becomes just a tenth as great when one digit is added to the display, assuming the electrical precision is sufficient.

If the value of a parameter changes rapidly, the instrument error may increase dramatically. This occurs, for example, in the collector-current metering of a single-sideband amplifier. The meter cannot follow the rapid modulation of the voice signal. The needle appears to jump up during voice peaks, but it does not accurately register the actual instantaneous current. A digital meter would be entirely useless in this kind of application; the display would present a constantly changing array of digits. For optimum indication of the instantaneous current values in such a situation, an oscilloscope would be necessary. *See also* ANALOG METERING, DIGITAL METERING.

INSTRUMENT TRANSFORMER

In an instrument intended for measuring alternating currents, an instrument transformer is sometimes used to provide a variety of metering ranges. Transformers may be used for alternating-current ammeters or voltmeters. Such transformers usually have several tap positions, connected to a rotary switch for choosing the desired range (see illustration).

In a voltmeter, the range is inversely proportional to the number of turns in the secondary winding of the transformer. For example, if the secondary has n turns and the maximum voltage on the scale is x, then switching to 10n turns will change the full-scale reading to 0.1x. In an ammeter, the situation is reversed; the full-scale range is directly proportional to the number of turns. Thus, if n turns provide a reading of x amperes at full scale, 10n turns will result in a full-scale reading of 10x. These relations are based on the assumptions that the input of the meter is provided at the primary winding of the transformer, and that the number of turns in the primary winding remains the same under all conditions. *See also* TRANSFORMER.

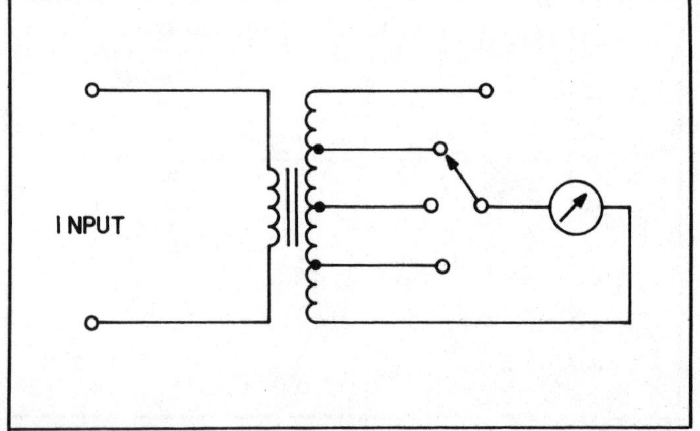

INSTRUMENT TRANSFORMER: A switchable instrument transformer allows the selection of various full-scale indicating ranges.

INSULATED CONDUCTOR

An insulated conductor is a wire, length of tubing, or other electrically conducting substance that is covered with a layer of non-conducting material. Insulated conductors are used in electrical wiring to prevent short circuits, and to reduce the danger of shock or fire. Insulated conductors are also sometimes used in radio-frequency antenna systems.

The insulation covering a conductor must be thick enough to prevent accidental arcing. Wire is available with a variety of insulating materials and thicknesses. Enamel is the thinnest insulation. Cloth is sometimes used because of its durability, although it tends to lose some of its insulating properties if it gets wet. Rubber and plastic are commonly used in insulation on household electrical wires. *See also* ENAMELED WIRE, INSULATING MATERIAL.

INSULATING MATERIAL

Electrical insulating materials are used for a variety of purposes. In general, insulating substances are used in situations where conduction must be prevented. Insulating materials are classified according to their ability to withstand heat. In order of ascending temperature rating, these categories are designed as follows: O (90 degrees Celsius), A (105 degrees Celsius), B (130 degrees Celsius), F (155 degrees Celsius), H (180 degrees Celsius), C (220 degrees Celsius), and over Class C (more than 220 degrees Celsius).

Class-O materials include paper and cloth. Class-A insulating substances include impregnated paper and cloth. Class-B, Class-F, Class-H, and Class-C materials are substances such as glass fiber, asbestos, and mica. All insulators over Class C consist of inorganic materials such as specially treated glass, porcelain, and quartz. Teflon is able to withstand temperatures of more than 250 degrees Celsius, and it is therefore rated over Class C. Enamel is usually a Class-O material. Vinyl, often used for insulating electrical wiring, is generally a Class-A substance, although silicone enamel may rate as high as Class H.

The class of insulating material needed for a given application depends on the amount of current expected to flow in the conductors, and on the conductor size. If the class of insulation is too low, fire may occur. If the class is unnecessarily high, the expense is unwarranted.

Insulating materials are also rated according to dielectric constant, dielectric strength, tensile (breaking) strength, and leakage resistance. *See also* DIELECTRIC, DIELECTRIC CONSTANT, DIELECTRIC RATING, INSULATION RESISTANCE.

INSULATION RESISTANCE

The resistance of an insulating material is a measure of the ability of the substance to prevent the undesired flow of electric currents. In general, the higher the insulation resistance, the more effective the material for purposes of preventing current flow.

Insulation resistance is generally measured for direct currents. This value, given in ohm-centimeters (ohm-cm), may change for radio-frequency alternating currents. Among the substances with the highest known insulation resistance are polystyrene, at 10^{18} ohm-cm, fused quartz, at 10^{19} ohm-cm, and teflon and polyethylene, which have direct-current resistivity of 10^{17} ohm-cm. Vinyl substances have insulation resistances ranging from 10^{14} to 10^{16} ohm-cm. Rubbers rate from about 10^{12} to 10^{15} ohm-cm. *See also* DIELECTRIC, DIELECTRIC RATING.

INSULATOR

An insulating material is sometimes called an insulator (*see* INSULATING MATERIAL). However, the term insulator generally refers to a device specifically designed to prevent current from flowing between or among nearby electrical conductors. An insulator is a rigid object, usually made from porcelain or glass. Insulators are commonly used in radio-frequency antenna systems and in utility lines.

Insulators are found in a tremendous range of sizes and shapes. Two common types are shown (A and B). At A is a small porcelain insulator, which is intended for use in receiving antenna systems and in transmitting systems at power levels up to about 1 kW. At B are large insulators, which are typical of high-tension power-line structures. The small insulator is about 2 inches long, while the large ones are from 2 feet to 6 feet in length. The extremely high-voltage cross-country power lines, with voltages in the hundreds of thousands, have insulators up to several meters long.

INTEGRAL

See DEFINITE INTEGRAL, INDEFINITE INTEGRAL.

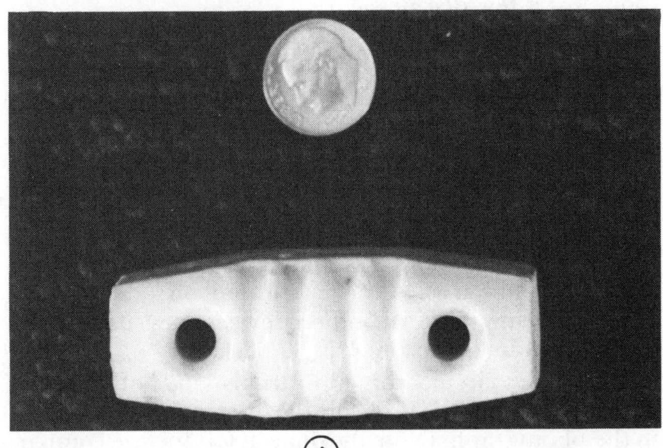

A

B

INSULATOR: At A, a small porcelain insulator. At B, large power-line insulators.

INTEGRATED CIRCUIT: A typical integrated circuit is tiny, yet it may perform a complex function. Note the dime for size comparsion.

INTEGRATED CIRCUIT

Modern solid-state electronics has produced the integrated circuit, or IC. With the advent of the semiconductor age, it became feasible to build an entire electronic circuit, including resistors, capacitors, inductors, diodes, and transistors, onto a single chip of silicon or germanium. Other substances, notably metal oxides, are also used in fabricating integrated circuits. The result is a package that may be smaller than a pencil eraser, yet it can contain the complexity of a system that would have taken up an entire room just a few decades ago.

Integrated circuits are classified in a variety of ways. Analog, or linear, ICs operate over a continuous range, and include such devices as operational amplifiers. Digital ICs are employed mostly in computers, electronic counters, frequency synthesizers, and digital instruments. Bipolar integrated circuits employ bipolar-tansistor technology. Complementary-metal-oxide-semiconductor (CMOS) ICs are becoming increasingly common because of their extremely low current demand. Monolithic integrated circuits are entirely fabricated on one piece of semiconductor material; other devices may use two or more individual chips in a single package.

Integrated circuits come in various kinds of packages. The most common is probably the dual in-line type, shown in the photograph. (Note the dime at left for size comparison.) Other housing methods are the single in-line package, with pins along one edge, and the flat pack, with pins along two, three, or four edges (*see* DUAL IN-LINE PACKAGE, FLAT PACK, SINGLE IN-LINE PACKAGE).

A theoretical limit probably exists for the degree of miniaturization that can be achieved. Some experts believe that limit has nearly been reached today; others think that much further miniaturization is still possible. *See also* SOLID-STATE ELECTRONICS.

INTEGRATED INJECTION LOGIC

A bipolar form of logic, using both NPN and PNP transistors, is called integrated injection logic (abbreviated I^2L). This type of logic circuit allows the smallest physical size of any member of the bipolar logic family. The illustration is a simple schematic diagram of an I^2L gate.

The manufacture of I^2L is simple. No resistors are needed; a single pair of transistors completes the circuit for one gate. The operating speed is very rapid, approaching the speed of transistor-transistor logic. The current requirement of I^2L is quite small, resulting in low power consumption. *See also* DIODE-TRANSISTOR LOGIC, DIRECT-COUPLED TRANSISTOR LOGIC, EMITTER-COUPLED LOGIC, HIGH-THRESHOLD LOGIC, METAL-OXIDE-SEMICONDUCTOR LOGIC FAMILIES, RESISTOR-CAPACITOR-TRANSISTOR LOGIC, RESISTOR-TRANSISTOR LOGIC, TRANSISTOR-TRANSISTOR LOGIC, TRIPLE-DIFFUSED EMITTER-FOLLOWER LOGIC.

INTEGRATION

Integration is a mathematical or electrical process, by which the cumulative value of a function is constantly added up. Integration is the opposite of differentation (*see* DIFFERENTIATION), in both the mathematical and the electrical sense. Graphically integration is represented by the area under a curve.

The graph shows the integration process, beginning from a designated point. Time is depicted along the horizontal axis, and amplitude along the vertical axis. With each half cycle, the total area under the curve fluctuates, resulting in a 90-degree phase shift in this example. *See also* DEFINITE INTEGRAL, INDEFINITE INTEGRAL, INTEGRATOR CIRCUIT.

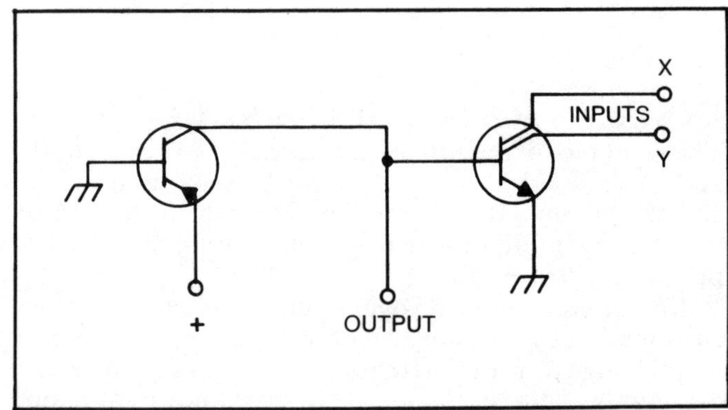

INTEGRATED INJECTION LOGIC: Schematic diagram of an integrated-injection-logic gate.

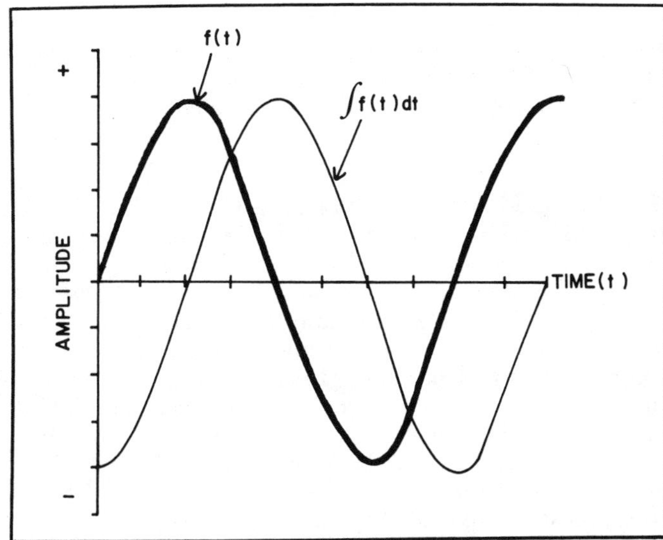

INTEGRATION: Electrical integration of a sine wave. The heavy wave corresponds to the function of time, f(t), to be integrated. The lighter line is the result of integration, and is the output of an integrator circuit with sine-wave input.

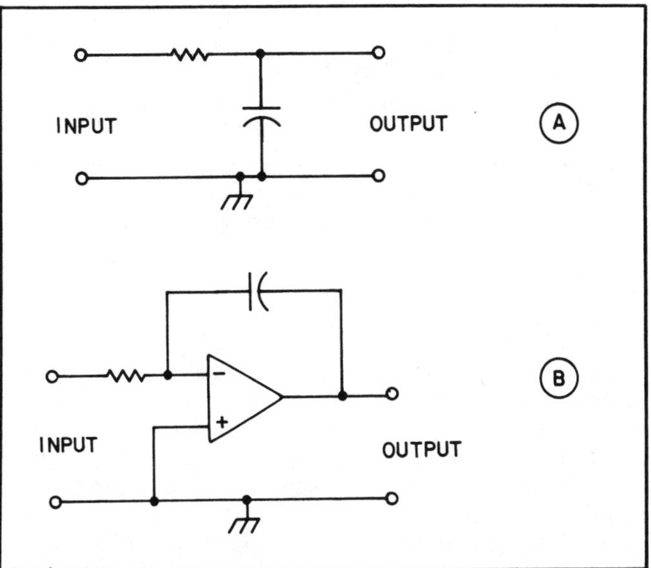

INTEGRATOR CIRCUIT: Two integrator circuits. At A, a simple passive resistance-capacitance network; at B, a circuit that uses an operational amplifier.

INTEGRATOR CIRCUIT

An electronic circuit that generates the integral, with respect to time, of a waveform is called an integrator circuit. When the input amplitude to an integrator is constant—that is, a direct current—the output constantly increases. If the input is zero, the output is zero. When the input amplitude fluctuates alternately toward the positive and the negative, the output reflects the integral of the waveform (see previous illustration).

An integrator circuit can be extremely simple. The diagrams shows two examples of circuits designed for electrical integration. At A, a resistance-capacitance network is shown. At B, an operational amplifier is used.

The integrator circuit generally acts in a manner exactly opposite to a differentiator. This does not mean, however, that cascading the two forms of circuit will always result in an output that is the same as the input. This is usually the case, but there are certain waveforms that are not duplicated when passed through an integrator and differentiator in cascade. *See also* DIFFERENTIATION, DIFFERENTIATOR CIRCUIT, INTEGRATION.

INTELLIGIBILITY

Intelligibility is a measure of the ease with which a signal is accurately received. It is measured as a percentage of correctly received single syllables in a plain-text transmission. The intelligibility may be affected by the context of the transmission, since the receiving operator can sometimes guess what is coming.

The percentage of correctly received syllables in a random transmission, in which the receiving operator cannot guess what is coming, is called articulation (*see* ARTICULATION).

Good intelligibility is imperative in a communications circuit. Normally, a band of audio frequencies ranging from about 300 to 3000 Hz is sufficient to convey a voice message with good intelligibility. Intelligibility is not highly correlated with fidelity; excellent fidelity requires a much wider band of audio frequencies, and this may actually reduce the intelligibility. Intelligibility is directly related to the signal-to-noise ratio in a voice circuit. *See also* FIDELITY, SIGNAL-TO-NOISE RATIO.

INTENSITY

Intensity is a general term that refers to the magnitude of a signal, sound, or beam of radiation. The magnitude of the electron beam in a cathode-ray tube is often called the intensity. Most television receivers and oscilloscopes are equipped with intensity controls for varying the brightness of the picture.

The intensity of a signal may be rapidly modulated to convey information. When a radio wave is thus modulated, the mode is known as amplitude modulation. If an electron beam is modulated, such as in the common television picture tube, the mode is called intensity modulation. It may also be called Z-axis modulation. *See also* AMPLITUDE.

INTENSITY MEASUREMENT

Methods of measuring signal intensity depend on the kind of signal: acoustic, radio, infrared, visible, ultraviolet, X-ray, gamma-ray, or electron-beam.

Acoustic disturbances, or sound waves, may be defined in terms of intensity by means of a microphone, audio amplifier, rectifier, and an indicating ammeter or voltmeter. Sound intensity is generally measured in decibels above the threshold of hearing (*see* DECIBEL).

The intensity of radio waves is generally measured by means of a field-strength meter. The most common units are microvolts per meter. The field-strength meter consists of an antenna, a rectifier, perhaps an amplifier, and a direct-current meter (*see* FIELD-STRENGTH METER).

Visible-light intensity is measured by means of a photocell or photovoltaic cell, a power supply if needed, and a meter. Light intensity is usually specified in candela (*see* CANDELA).

The intensity of infrared or ultraviolet radiation may be measured in a manner similar to the method used for the measurement of visible-light intensity. A thermocouple may be used for measurement of infrared intensity, since radiation at such wavelengths causes heating of objects. For X rays and gamma rays, dosimeters are used (*see* DOSIMETRY).

The intensity of an electron beam is usually specified in terms of the equivalent current in amperes. In any electron-emitting device, this current can be easily measured. *See also* ELECTRON-BEAM GENERATOR.

INTERACTION

When a change in a parameter affects the value of some other parameters, interaction is said to occur. Interaction takes place in many different forms in electrical and electronic devices. For example, the setting of the loading control in a transmitter affects the optimum setting of the output tuning control; the two controls therefore interact. In a shortwave general coverage radio receiver, the main-tuning control and the bandspread control often interact. Interaction may be desired, or it may be unwanted.

When one effect (such as the presence of a fluctuating magnetic field) produces another (such as an alternating current), interaction is said to take place. This form of interaction is very common, and occurs in a variety of different situations.

Two or more circuits that operate in a state of mutual dependence, such as a group of interconnected computers, comprise an interactive system. Information is transmitted from one point to another in such a system; the functioning of one member affects the functioning of the other members. This mode of operation is often called the interactive or conversational mode, in which two-way data exchanges take place.

INTERCARRIER RECEIVER

An intercarrier receiver is a form of television receiver. All of the signal components, including the picture information, sound signal, and synchronization pulses, are heterodyned to the intermediate frequency. The intermediate-frequency stages amplify the signal. The signal is then fed to the detector.

In the detector stage, the picture signal, sound signal, and synchronizing pulses are separated and fed to the picture tube modulator, speaker, and picture tube deflection plates. The illustration is a block diagram of an intercarrier television receiver. *See also* TELEVISION.

INTERCHANGEABILITY

Interchangeability is the degree of ease with which one component type may be replaced by another type, while providing acceptable operation for the whole system. Standardization of electrical and electronic components makes interchangeability feasible in many cases. The American National Standards Institute and the Electronic Industries Association set standards for component interchangeability in commercial electronics in the United States. International standardization is carried out by a number of organizations run by the International Electrotechnical Commission.

In circuits made up of discrete components, interchangeability is often possible among many items. This is especially true of vacuum tubes and transistors. Other components, too, are interchangeable; for example, a mylar capacitor may sometimes be used in place of a ceramic capacitor. A 2-watt, 47-ohm resistor may, of course, replace a 1-watt, 47-ohm resistor.

In modern electronic systems, the trend is toward integrated-circuit construction, in which complex circuits are housed in discrete packages. Interchangeability is often impossible in such systems, but the troubleshooting and repair process is greatly simplified. *See also* INTEGRATED CIRCUIT.

INTERCONNECTION

Interconnection is the system of wiring, radio links, or other media by which various parts of a system communicate. The interconnection scheme in a system may be as simple as a one-way link; it may be very complicated. Computers are often interconnected for the purpose of information transfer, or for obtaining additional memory capacity. This is generally accomplished by telephone. A home computer, for example, can "tap" into a central computer. The industrialized world is extensively criss-crossed by electronic interconnections of all kinds.

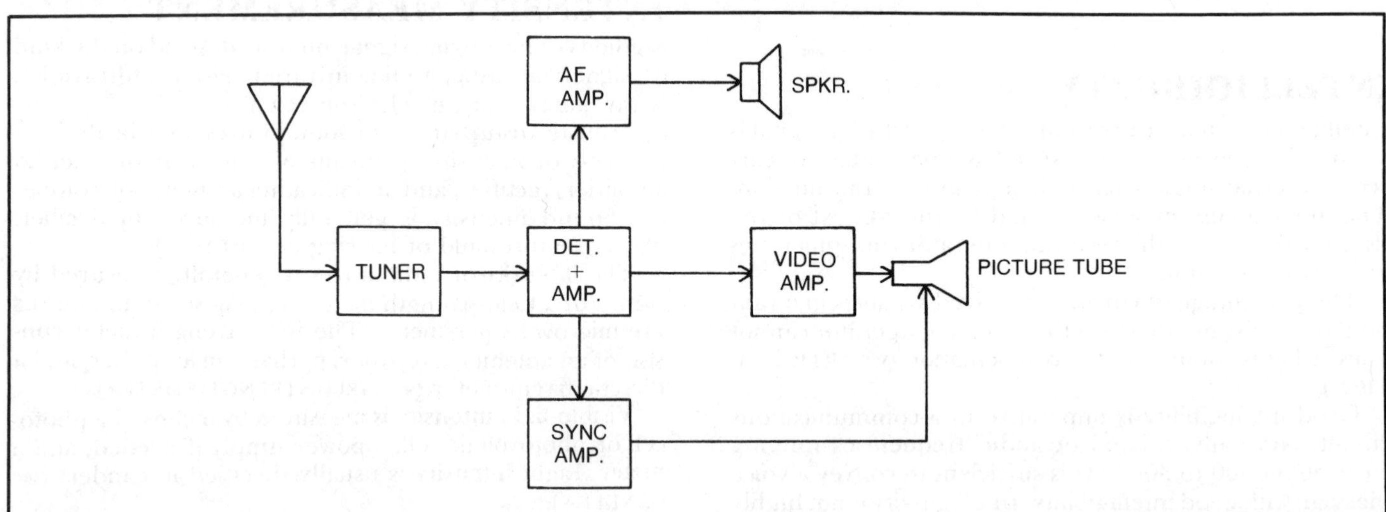

INTERCARRIER RECEIVER: Block diagram of an intercarrier television receiver.

In an electronic circuit consisting of discrete components such as transistors, resistors, capacitors, and inductors, the wiring scheme is sometimes called component interconnection. This interconnection is shown by means of the familiar schematic diagram. Groups of integrated circuits, too, can be interconnected to form complex electronic systems. Integrated-circuit interconnection diagram look much different from the diagrams for discrete-component circuits. (*see* SCHEMATIC DIAGRAM).

Two or more power generators may be interconnected to provide more energy than a single generator. This kind of interconnection requires that the generators be operated in phase, so that the alternating currents reinforce each other. Otherwise the efficiency of the system is greatly impaired. *See also* POWER PLANT.

INTERCONNECTION DIAGRAM

See SCHEMATIC DIAGRAM.

INTERCONTINENTAL BALLISTIC MISSILE

A rocket, armed with a nuclear warhead and capable of traveling thousands of miles, is called an intercontinental ballistic missile. Such missiles first came into common existence in the 1950s and 1960s. The earliest intercontinental ballistic missile was the German V-2 rocket, which contained a conventional warhead and was capable of traveling a few hundred miles. This rocket is famous for the destruction it caused to England in World War II.

The intercontinental ballistic missile has made it possible to carry out a thermonuclear war within about one hour. A missile may be launched from any part of the world, travel through space, and land at any other location, within 60 minutes. Such missiles are similar to the first rockets with which men were propelled into orbit around earth. In fact, missiles can be placed into orbit.

Intercontinental ballistic missiles are guided by means of computer programs, or they may use homing devices to zero in on their targets. Inertial guidance systems are also used. The United States and the Soviet Union possess most of the intercontinental ballistic missiles in the world. *See also* GUIDED MISSILE, HOMING DEVICE.

INTERELECTRODE CAPACITANCE

When electrodes are placed in close proximity, they show a certain amount of mutual capacitance. For any two electrodes, this capacitance varies inversely with the separation, and directly according to the size of either electrode. Interelectrode capacitance is not very important at low and medium frequencies, but as the frequency increases, the interelectrode capacitance in a system becomes more significant.

Interelectrode capacitance is specified for tubes and transistors. In a vacuum tube, the grid and plate electrodes display mutual capacitance, and this is the primary limiting factor on the frequency range in which the device can produce gain. The reverse-biased base-collector junction of a transistor behaves in a similar manner. In the field-effect transistor, capacitance exists between the channel and the gate or gates. The interelectrode capacitance in a typical tube, transistor, or field-effect transistor is a few picofarads. *See also* ELECTRODE CAPACITANCE.

INTERFACE

In an electronic system, an interface is a point at which data is transferred from one component to another. An interface generally involves conversion of data so that compatibility is achieved for both components. Thus we may speak of a computer-to-operator interface (a terminal) or a radio-to-telephone interface (autopatch). The circuit that performs the data conversion is often called an interface.

An interface may perform a relatively simple task, such as the wiring of audio-frequency energy into a transmitter. The data conversion may be exceedingly complicated, such as the translation of visual data and keyboard commands into machine language for the operation of a computer. In computer systems, there are many interface points; a home computer, for example, requires a video display unit, a tape recorder or disk drive, and often a printer and telephone modem. *See also* DATA, DATA CONVERSION, MODEM.

INTERFERENCE

Interference is the presence of unwanted signals or noise that increases the difficulty of radio reception. There are three kinds of interference: natural noise, manmade noise, and manmade signals.

Natural electromagnetic noise originates in outer space and in the atmosphere of the earth. There is nothing man can do to reduce the actual level of this noise. Atmospheric noise tends to be greatest at the lower frequencies and in the vicinity of thundershowers. Noise from outer space is prevented from reaching the surface of the earth in the very-low, low, and medium-frequency ranges because of the shielding effect of the ionosphere. However, at higher frequencies, such noise can sometimes be significant in communications applications. In general, the narrower the signal bandwidth in a communications link, the greater the immunity to natural noise (*see* NOISE).

Manmade noise is caused mostly by electrical wiring and appliances, and by internal-combustion engines and motors. Manmade noise is most intense at the very-low frequencies. As the electromagnetic frequency increases, the level of manmade noise goes down. As with natural noise, the immunity of a system depends on the bandwidth. In some cases, special circuits can reduce the level of manmade noise (*see* IGNITION NOISE, IMPULSE NOISE, NOISE BLANKER, NOISE LIMITER).

Interference from manmade communications transmitters may be further categorized as either unintentional or intentional. The radio-frequency spectrum is heavily used in modern industrialized nations, and a certain amount of accidental interference is to be expected. The narrower the average signal bandwidth, in a given band of frequencies, the lower the probability of accidental interference. Intentional interference is called jamming, and is often used in wartime for the purpose of cutting enemy communications links. (*See* INTERFERENCE FILTER, INTERFERENCE REDUCTION.)

When two or more electromagnetic waves combine in variable phase, producing areas of stronger and weaker field intensity, the interaction is sometimes called interference. *See also* INTERFERENCE PATTERN, INTERFEROMETER.

INTERFERENCE FILTER

An interference filter is a device, used in a radio receiver, for the purpose of reducing or eliminating interference (*see* INTERFERENCE). There are several different types of interference filters.

For natural noise, such as the noise originating in the atmosphere, a noise limiter may provide some reduction in the level of interference (*see* NOISE LIMITER). Narrow-band transmission and reception is preferable to wide-band communications in the presence of high levels of natural noise. The noise limiter and narrow-band selective filter can thus act as noise filters.

Manmade noise may be reduced or eliminated by the same means as natural noise. However, a noise blanker (*see* NOISE BLANKER) is often effective against manmade ignition or impulse noise, whereas it is usually not effective against natural noise.

Interference from manmade signals is best reduced by means of highly selective bandpass filters, in conjunction with narrow-band transmission. A device called a notch filter can be used to attenuate fixed-frequency, unmodulated carriers within the communications passband (*see* NOTCH FILTER). Broadband, deliberate interference, or jamming, may be almost impossible to overcome by means of filtering. In such cases, a directional antenna system is usually needed. *See also* DIRECTIONAL ANTENNA.

INTERFERENCE PATTERN

When two electromagnetic fields interact, regions of phase reinforcement and phase opposition are created. The result is a pattern of more and less intense emission. This pattern is known as an interference pattern.

The simplest interference patterns are the directional radiation and response patterns of antenna systems. The configuration of the pattern depends on the spacing of the antennas, measured in wavelengths, and on the relative phase of the signals at the different antennas (*see* PHASED ARRAY).

When visible light is passed through a pair of narrow slits, an interference pattern occurs. This was noticed by physicists during the nineteenth century, and it led to the wave theory of light (see illustration).

Interference patterns are often noticed with acoustic (sound) waves. Such patterns must be considered when designing a high-fidelity arrangement. The interference pattern from a pair of stereo speakers, for example, cause "loud" and "dead" spots in certain instances, unless the physical layout is carefully planned. *See also* PHASE ANGLE, PHASE OPPOSITION, PHASE REINFORCEMENT.

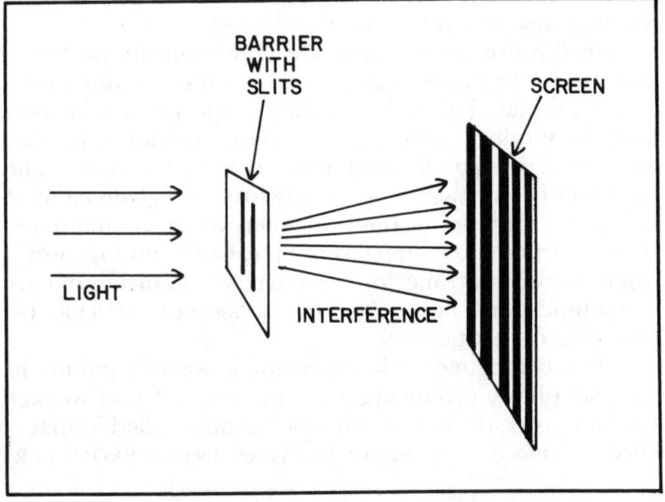

INTERFERENCE PATTERN: When monochromatic visible light passes through an opaque barrier with two thin, closely spaced slits, a pattern of light and dark bands is produced on a screen.

INTERFERENCE REDUCTION

The reduction of interference is important in a communications system. The lower the interference level, the better the signal-to-noise ratio in a communications link. Interference reduction can be effected at the transmitting station, by the use of the narrowest possible emission bandwidth. With this single exception, however, all interference reduction must be accomplished at the receiving end of a circuit.

Interference can be reduced in a receiver by means of various filtering devices (*see* INTERFERENCE FILTER). Directional antenna systems (*see* DIRECTIONAL ANTENNA) can also be used. If the interference originates at a single point in space, and this point remains fixed or moves slowly, a loop antenna (*see* LOOP ANTENNA) can be used to "null out" the interference. If the interference comes from a broad general direction different from the direction of the desired signal, a unidirectional antenna such as a quad, Yagi, or dish may be used.

In certain situations, interference reduction can be extremely difficult. This is especially true in cases of deliberate interference, or jamming (*see* JAMMING). A new technique of communication, called spread-spectrum operation, can sometimes be used to advantage to reduce interference from jamming. *See also* SPREAD-SPECTRUM TECHNIQUES.

INTERFEROMETER

The interferometer is a form of radio telescope in which two antennas are used in a phased configuration. The interferometer provides much greater resolving power, using two small antennas, than is possible with a single radio antenna. The technique was pioneered by two radio astronomers, Martin Ryle of England and J. L. Pawsey of Australia.

A single dish antenna, having a diameter of many wavelengths, can yield good resolution in radio astronomy. But two such antennas, separated by a great distance, allows far greater resolution. The sensitivity of the interferometer is not very much higher than that of a single-antenna system, but in radio astronomy, resolving power is often more important than sensitivity.

The illustration shows a typical directional pattern for a single large dish antenna (A). At B is the response of a typical interferometer system, using two identical antennas spaced at a great distance. Each fine line in B represents a direction in which the incoming wavefronts add in phase at both antennas. The gaps between the lines represent directions in which the incoming wavefronts arrive in phase opposition. The interferometer is sensitive, therefore, in many directions. The radio astronomer, knowing the exact relationship between the two signals, can determine or select the exact direction of reception.

The greater the spacing between the antennas in the interferometer, the better the resolution. If the spacing in meters between the aerials is L and the wavelength in meters is λ, then the angular separation a, in degrees, between the lobes of the directional pattern is given by:

$$a = 57.3\lambda/L$$

Of course, the shorter the wavelength, the greater the resolving power for a given antenna separation. Some interferometer systems have resolution of a few seconds of arc at wavelengths of several centimeters. *See also* RADIO ASTRONOMY, RADIO TELESCOPE.

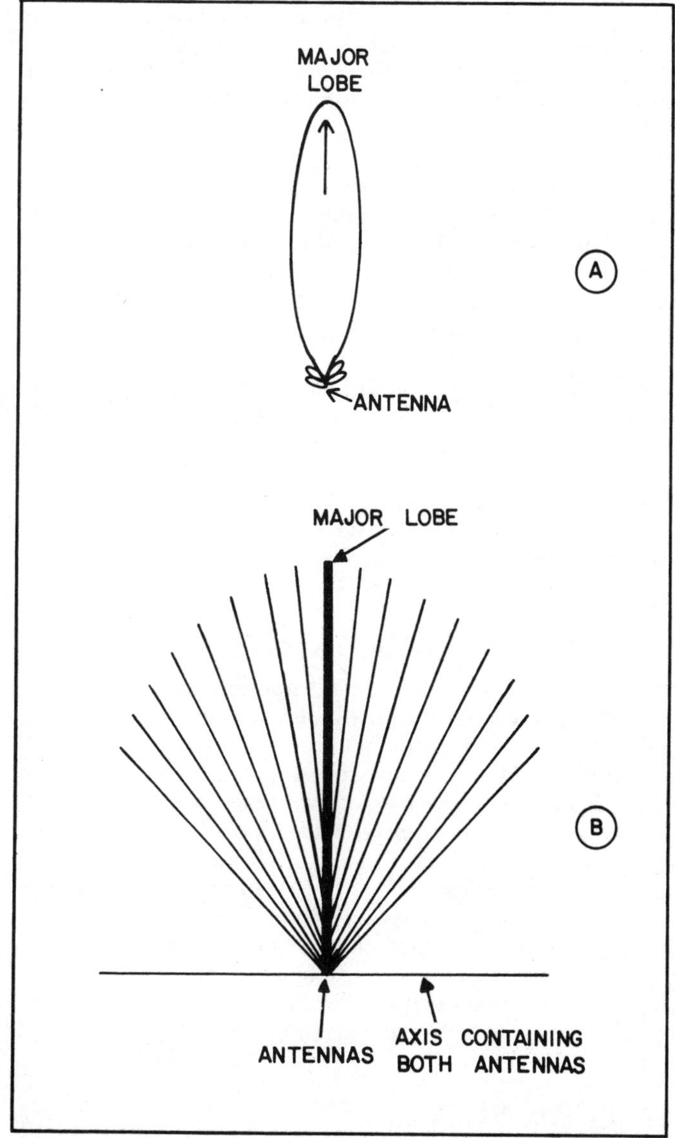

INTERFEROMETER: At A, typical directional response for a single large dish antenna, such as is used in a radio telescope. At B, response pattern of an interferometer. (In practice, the pattern usually has lobes that are much finer than those shown here.)

INTERMEDIATE FREQUENCY

In a superheterodyne receiver, the intermediate frequency, abbreviated i-f, is the ouput frequency from the mixer stage (*see* MIXER, SUPERHETERODYNE RECEIVER). The intermediate frequency of the superheterodyne is usually a fixed frequency. This makes it easy to obtain high gain and excellent selectivity, since the i-f amplifier stages can be tuned precisely for optimum performance at a single frequency (*see* INTERMEDIATE-FREQUENCY AMPLIFIER).

Some superheterodyne receivers employ one intermediate frequency; these are called single-conversion receivers. Other receivers have two intermediate frequencies. The incoming signal is heterodyned to a fixed frequency called the first i-f, which is in turn heterodyned in a later stage to the second i-f. The second i-f is generally a low frequency, which facilitates high selectivity. This type of receiver is called a double-conversion or dual-conversion receiver. Some receivers even have three intermediate frequencies; these are called triple-conversion receivers.

A high intermediate frequency, such as several megahertz, is preferable to a low i-f for purposes of image rejection. However, a low i-f is better for obtaining sharp selectivity. This is why double-conversion receivers are common: They provide the advantages of both a high first i-f and a low second i-f. *See also* DOUBLE-CONVERSION RECEIVER, IMAGE REJECTION, SINGLE-CONVERSION RECEIVER.

INTERMEDIATE-FREQUENCY AMPLIFIER

An intermediate-frequency amplifier is a fixed radio-frequency amplifier commonly used in superheterodyne receivers. Such amplifiers generally are cascaded two or more in a row, with tuned-transformer coupling (see A). The intermediate-frequency (i-f) amplifiers follow the mixer stage, and precede the detector stage. Double-conversion receivers (*see* DOUBLE-CONVERSION RECEIVER)

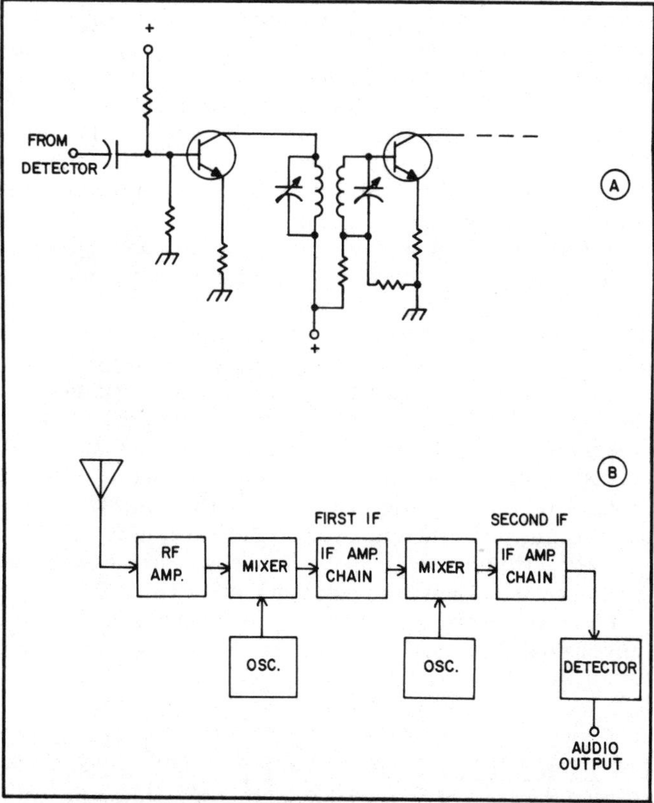

INTERMEDIATE-FREQUENCY AMPLIFIER: At A, schematic diagram of the tuned-transformer coupling generally used between stages of the intermediate-frequency chain in a super-heterodyne receiver. At B, a simplified block diagram of a double-conversion receiver, showing the first and second i-f chains.

have two sets of i-f amplifiers; the first set follows the first mixer and precedes the second mixer, and the second set follows the second mixer and precedes the detector, as at B.

The intermediate-frequency amplifier chain serves two main purposes: to provide high gain, and to provide excellent selectivity. Gain and selectivity are much easier to obtain with amplifiers that operate at a single frequency, as compared with tuned radio-frequency amplifiers. *See also* INTERMEDIATE FREQUENCY, SUPERHETERODYNE RECEIVER.

INTERMITTENT-DUTY RATING

The intermittent-duty rating of a component is its ability to handle current, voltage, or power with a duty cycle of less than 100 percent (*see* DUTY CYCLE). Usually, intermittent duty is specified as a duty cycle of 50 percent. However, the length of a single active period must also be specified. The intermittent-duty rating of a device is almost always greater than the continuous-duty rating (*see* CONTINUOUS-DUTY RATING).

Some electronic devices are intended primarily for intermittent service, while others are designed for continuous service. Sometimes, specifications will be given for both continuous and intermittent service.

INTERMITTENT FAILURE

When a system, circuit, or component malfunctions only occasionally, the malfunction is called an intermittent failure, or simply an intermittent.

Intermittents are well known to every experienced service technician. Troubleshooting is difficult if the problem is not always evident. In some cases, the failure refuses to manifest itself under test conditions, but only occurs in actual service, where repair is inconvenient or impossible. The failure may occur for only a fraction of 1 percent of the operating time, or it may take place for more than 99 percent of the time.

Intermittent failures can be induced by mechanical vibration, thermal shock, excessive humidity, or operation at excessive voltage, current, or power levels. Sometimes there is no apparent cause for the failure. The repair process consists of locating and replacing the defective components, or improving the operating environment to conform with equipment specifications. Components must be wiggled around, or the unit bumped and shaken. It may be necessary to use cooling compounds such as "freeze mist," localized heating devices, and the like. The troubleshooting process can be very time-consuming. When it appears as though the problem has been corrected, extensive testing is needed to ensure that this is actually the case.

The most harrowing intermittent problems are sometimes called gremlins—meaning little devils—by the technician. *See also* GREMLIN.

INTERMODULATION

In a receiver, an undesired signal sometimes interacts with the desired signal. This usually, but not always, occurs in the first radio-frequency amplifier stages, or front end. The desired signal appears to be modulated by the undesired signal, as well as having its own modulation. This effect is known as intermodulation. Intermodulation is sometimes mistaken for interference from another signal; occasionally, interference is mistaken for intermodulation. Intermodulation may be comparatively mild, or it may be so severe that reception becomes impossible.

Intermodulation can take place for any or all of several reasons. Inadequate selectivity in the receiver input can result in unwanted signals passing into the front-end transistor. Nonlinear operation of the front-end transistors increases the chances of intermodulation. An excessively strong signal may drive the front end into nonlinear operation, even though it is properly designed. A poor electrical connection in the antenna system may cause intermodulation in a receiver used with that antenna system. In the extreme, a poor electrical connection near the antenna, but physically external to it, can produce intermodulation effects. The poor connection literally forms a diode mixer.

Intermodulation distortion (abbreviated IMD) is, of course, never desirable, and modern receivers are designed to keep this form of distortion to a minimum. Manufacturers and designers speak of a specification called intermodulation spurious-response attenuation. This is usually expressed in decibels. Two signals are applied simultaneously to the antenna terminals of the receiver under test. The amplitude ratio is prescribed according to preset specifications. The strength of the unwanted signal required to produce a given amount of intermodulation is then compared with the strength of the desired signal, and the ratio is calculated in decibels.

Alternatively, the strength of the unwanted signal may be held constant, and the extent of the distortion measured as a modulation percentage in the desired signal. This percentage is called the intermodulation-distortion percentage.

Standards for intermodulation spurious-response attenuation are set in the United States by the Electronic Industries Associaton. When testing a receiver for intermodulation response, the procedures and standards of this association should be consulted. The susceptibility or immunity of a receiver to IMD is primarily dependent on the design of its front end. *See also* FRONT END.

INTERNAL IMPEDANCE

The internal impedance of a component or device is the impedance between its terminals, at a given frequency, without external components. Internal impedance consists of direct-current or ohmic resistance and either positive (inductive) or negative (capacitive) reactance. Internal impedance is an important consideration when using test instruments such as meters and oscilloscopes.

The internal impedance of current-measuring devices should be as low as possible. When voltage is to be measured, the internal impedance should be high. In general, the internal impedance of a test instrument should be such that the performance of the circuit is not altered when the instrument is used. *See also* IMPEDANCE.

INTERNAL NOISE

In a radio receiver, noise is an important consideration, since it limits the sensitivity of the circuit. Noise may originate outside the receiver, in the form of atmospheric and cosmic noise (*see* COSMIC NOISE, SFERICS). Noise is also produced by the movement of atoms as a result of temperature (*see* THERMAL NOISE). Some noise is produced by the active components within the receiver itself. This is known as internal noise.

At very-low, low, and medium frequencies, external noise is quite intense, and internal receiver noise normally does not play a significant part in limiting the sensitivity. In the high-frequency spectrum, external noise becomes less intense, and in the very-high, ultra-high, and microwave parts of the spectrum, internal noise is much more important than external noise. Low-noise receiver design is of the greatest concern, therefore, at frequencies above about 30 MHz.

Modern semiconductor technology allows the design and manufacture of low-noise receivers having sensitivity unheard-of just a few years ago. In some scientific applications, noise is further reduced by supercooling, or operation of equipment at extremely low temperatures. *See also* NOISE, SENSITIVITY. SIGNAL-TO-NOISE RATIO, SUPERCONDUCTIVITY.

INTERNATIONAL MORSE CODE

The International Morse Code is a system of dot and dash symbols commonly used by radiotelegraph operators throughout the world. The International Morse Code is much more often used than the American Morse Code, although a few telegraph operators still use the American Morse.

The International Morse Code differs from the American Morse Code; the table shows the International Morse symbols. The dot represents one unit length. The dash represents three units and the spaces between dots and dashes is one unit. The space between letters is three units, and between words, seven units. The International Morse Code is less confusing than the American Morse, because the latter code contains odd spaces and unit lengths in some instances. *See also* AMERICAN MORSE CODE.

INTERNATIONAL SYSTEM OF UNITS

The International System of Units, abbreviated SI, is based on the meter, kilogram, second, ampere, degree Kelvin, candela, and mole (Avogadro constant). The system is sometimes called the meter-kilogram-second (MKS) or meter-kilogram-second-ampere (MKSA) system of units. All physical units can be derived in terms of these standard units. The SI, or Systeme International d'Unites, was agreed upon in the Treaty of the Meter in 1960.

Prior to 1948, the international system of units was based on the centimeter, gram, second, and degree Kelvin. The older international-unit system was discarded by international treaty on January 1, 1948. The most recent agreement was made in 1960. Sometimes a distinction is made between "absolute" and "international" electrical quantities. The difference is primarily a matter of the standard quantities from which the units are derived. *See also* AMPERE, AVOGADRO CONSTANT, CANDELA, KELVIN TEMPERATURE SCALE, KILOGRAM, METER, SECOND.

INTERNATIONAL MORSE CODE:
INTERNATIONAL MORSE CODE SYMBOLS.

Character	Symbol	Character	Symbol
A	•–	U	••–
B	–•••	V	•••–
C	–•–•	W	•––
D	–••	X	–••–
E	•	Y	–•––
F	••–•	Z	––••
G	––•	1	•––––
H	••••	2	••–––
I	••	3	•••––
J	•–––	4	••••–
K	–•–	5	•••••
L	•–••	6	–••••
M	––	7	––•••
N	–•	8	–––••
O	–––	9	––––•
P	•––•	0	–––––
Q	––•–	PERIOD	•–•–•–
R	•–•	COMMA	––••––
S	•••	QUESTION MARK	••––••
T	–		

INTERNATIONAL TELECOMMUNICATION UNION

The International Telecommunication Union, abbreviated ITU, is an international organization that sets worldwide standards for electromagnetic communication. The ITU governs the frequency allocations and call-sign prefixes for all regions of the world. This helps maximize the efficiency with which the electromagnetic spectrum is utilized. The ITU has established a set of Radio Regulations. These regulations may be obtained by contacting the Secretary General, International Telecommunication Union, Place des Nations, CH-1211, Geneva 20, Switzerland.

In the United States, telecommunications regulations are determined by the Federal Communications Commission, although their rules are based on those given by the ITU. *See also* FEDERAL COMMUNICATIONS COMMISSION.

INTERPOLATION

Interpolation is a mathematical process by which intermediate values are determined. Interpolation is a com-

INTERPOLATION:
VALUES OF THE SINE FUNCTION FOR ANGLES BETWEEN 30 AND 40 DEGREES. FUNCTION VALUES ARE SPLIT INTO TENTHS BETWEEN 35 AND 36 DEGREES FOR ILLUSTRATIVE PURPOSES (SEE TEXT).

Angle θ	sin θ
30°	0.500
31°	0.515
32°	0.530
33°	0.545
34°	0.559
35°	0.574
35.1°	0.5754
35.2°	0.5768
35.3°	0.5782
35.4°	0.5796
35.5°	0.581
35.6°	0.5824
35.7°	0.5838
35.8°	0.5852
35.9°	0.5866
36°	0.588
37°	0.602
38°	0.616
39°	0.629
40°	0.643

monly used technique in all fields of engineering. It is especially useful in tabular function listings.

The table is an example of a listing of values for the sine function. Values are specified for each angular degree. But suppose we wish to determine the sine of 35.5 degrees, the value of which is not shown in the table. This value may be obtained simply by averaging the values of sin(35°) and sin(36°):

$$\sin(35.5^\circ) = (\sin(35^\circ) + \sin(36^\circ))/2$$
$$= (0.574 + 0.588)/2 = 0.581$$

While this value is not exact, since the sine function is not a perfectly linear function, the value can be considered accurate for practical purposes.

If the value of sin(35.7°) is to be found according to the table, the interpolation process is somewhat more complicated. The interval between sin(35°) and sin(36°) must then be divided into ten equal parts. Thus, sin(35.7°) is the value that is 7/10 of the way from sin(35°) to sin (36°).

More sophisticated methods of interpolation are sometimes used when great accuracy is needed. Such interpolation is usually carried out via computer.

Interpolation is performed mentally in a variety of instrument-reading situations. The photograph shows a spectrum-analyzer display, showing three pips or signals. Interpolation is necessary to determine both the frequencies and amplitude levels of the signals. Each large division on the horizontal scale corresponds to 50 MHz. Each large division on the vertical scale corresponds to 10 dB. Two technicians might disagree slightly on the precise readings.

Visual interpolation is sometimes easy, and sometimes very difficult, depending on the instrument used. Analog devices usually require some visual interpolation, while digital devices do not. *See also* ANALOG METERING, DIGITAL METERING.

INTERPRETER

An interpreter is a complex computer program designed to translate ordinary-language commands into the computer's machine language.

The interpreter translates each command and executes it immediately. In this way the interpreter differs from the compiler, which "collects" all the commands before translating and executing them.

In practice, an interpreter is somewhat more convenient to use than a compiler. The most popular interpreter high-level language is BASIC. *See also* BASIC, COMPILER.

INTERRUPT

An interrupt is a computer instruction that tells the machine to stop a program and perform some other, more important task. The interrupted program is called the background job. The background job has low priority, but may be very time-consuming. The more important task is

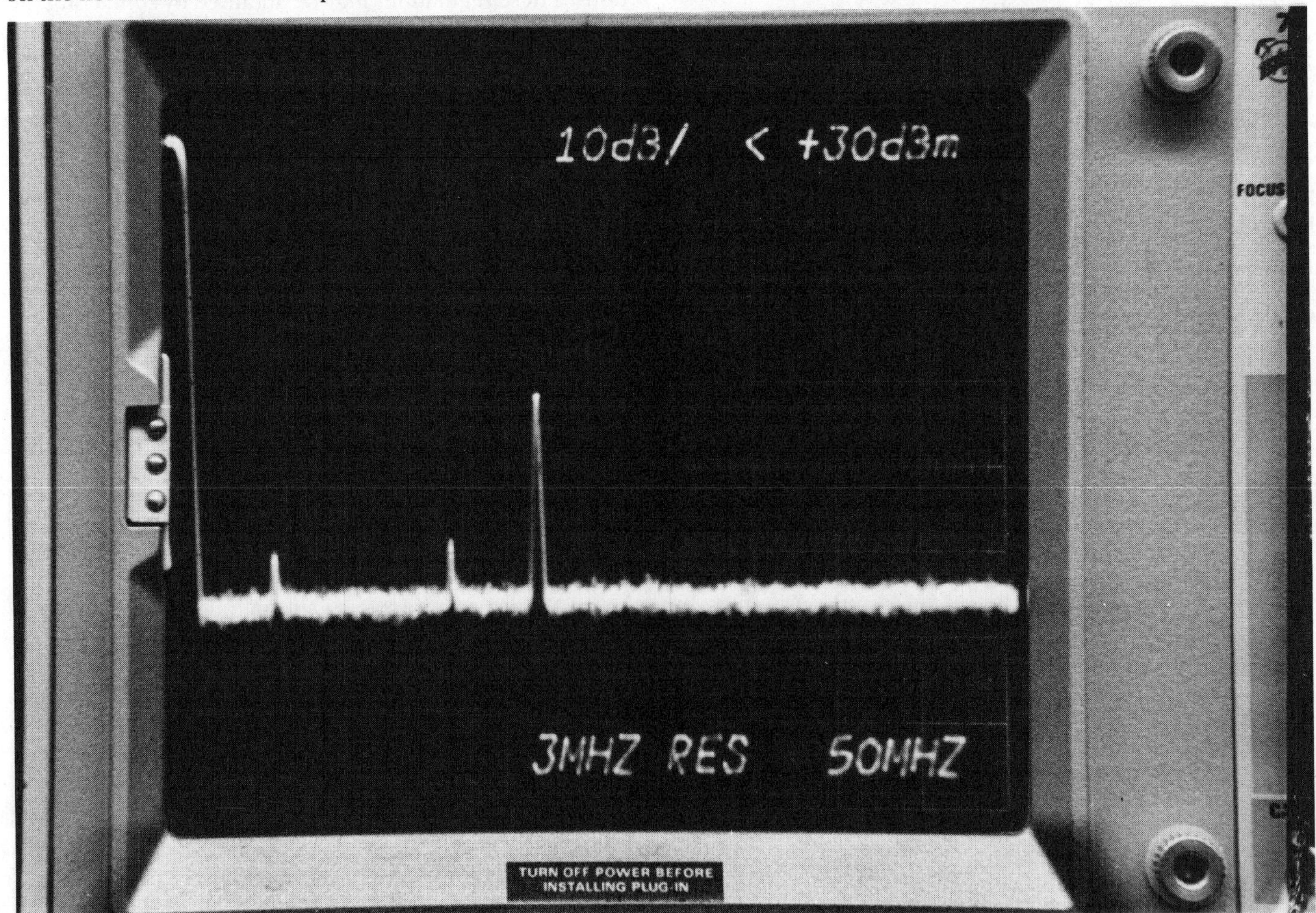

INTERPOLATION: Interpolation is often necessary in order to read an instrument display. The spectrum-analyzer display is set for 50 MHz per large horizontal division and 10 dB per large vertical division. Signals appear at approximately 60 MHz, 65 MHz, and 220 MHz. Their respective amplitudes, relative to the top line, are approximately −50 dB, −48 , and −31 dB.

called the foreground job; it is usually of short duration but high priority. A background job may be interrupted several times for different foreground jobs.

A computer program may occasionally be interrupted when it becomes evident that an error has been made. For example, a computer operator may notice that a program contains an infinite loop. Rather than let the computer run aimlessly for an indefinite period of time, the operator interrupts the execution of the program by issuing an interrupt command.

INTERRUPTER

An interrupter is a device that intermittently opens a circuit. The interrupter can be used for a variety of purposes, such as generation of pulsating direct current from pure direct current, half-wave rectification of alternating current, or keying of a continuous-wave transmitter.

Interrupters generally fall into two categories: electrical and mechanical. Electrical interrupters are such devices as switching diodes and transistors and silicon-controlled rectifiers. Mechanical interrupters include switches and relays. In a direct-current to alternating-current power inverter, the interrupter is sometimes called a chopper. *See also* CHOPPER.

INTERSTAGE COUPLING

Interstage coupling is the means by which a signal is transferred from one circuit to another. The circuits may be oscillators, amplifiers, mixers, or detectors. There are several different methods of interstage coupling.

Capacitive coupling is commonly used at all frequencies, from the low audio range into the ultra-high-frequency radio spectrum. A series capacitor passes the signal while allowing different direct-current voltages to be applied to the stages. (*See* CAPACITIVE COUPLING)

Direct coupling consists of a short circuit between two stages. This method of coupling does not allow for different direct-current voltages between stages. Direct coupling exhibits a constant attenuation level over a wide range of frequencies. (*See* DIRECT COUPLING.)

Transformer coupling makes use of a two-winding audio-frequency or radio-frequency transformer for the purpose of passing a signal. Either or both of the windings may contain a parallel capacitor for tuning purposes. An electrostatic shield may be placed between the windings to reduce the stray capacitance. Transformer coupling results in a narrow range of operating frequencies when tuning is used. Direct-current isolation is excellent. (*See* TRANSFORMER COUPLING.)

INTERTIE

See INTERCONNECTION.

INTRINSIC SEMICONDUCTOR

A pure semiconductor material, containing no added impurities, is called an intrinsic semiconductor. Such substances are almost insulators; they do not conduct electricity very well. Usually, impurities must be added for the purpose of manufacturing semiconductor devices. Then the semiconductor material becomes extrinsic. *See also* DOPING, ELECTRON, EXTRINSIC SEMICONDUCTOR, HOLE, N-TYPE SEMICONDUCTOR, P-TYPE SEMICONDUCTOR.

INVERSE FUNCTION

A mathematical function is an operation that assigns the members of one set to the members of another set according to certain rules. A relation is a function if, but only if, a given member from the domain has at most one constituent in the range (*see* DOMAIN, RANGE). Examples of functions are the sine, cosine, and tangent operations. But there are infinitely many different mathematical functions.

An inverse relation or function is the operation that carries out exactly the reverse of the original function. If f is a given function, the inverse is written f^{-1}. The inverse of a function may not be a function unless its domain or range are restricted. This is the case with all of the trigonometric functions, as well as many others.

In general, if a function and its inverse are performed in succession, the original value is obtained. That is, for any x in the domain of a function f:

$$f^{-1}(f(x)) = x, \text{ and}$$
$$f(f^{-1}(x)) = x$$

See also FUNCTION, RELATION.

INVERSE-SQUARE LAW

The inverse-square law is a physical principle that defines the manner in which three-dimensional radiation is propagated. This principle applies to radio waves, infrared radiation, visible light, ultraviolet, X rays, and gamma rays. It applies to barrages of particles, such as high-speed protons or neutrons. It also applies to sound waves. The inverse-square law is used for the purpose of determining, or predicting, the power intensity, per unit area, of a diverging effect at a given distance from a point source. The inverse-square law does not apply to parallel energy beams such as the maser or laser. The rule is also invalid when a parameter is measured in terms other than intensity per unit area.

To illustrate the derivation of the inverse-square principle, suppose we have a light bulb that emits 100 watts of power. If we completely surround this bulb by a sphere we know that 100 watts of power is striking the inside surface of the sphere. This is true no matter what the radius of the sphere, since any photon from the bulb must eventually encounter the sphere. Recall the formula for the surface area of the sphere:

$$A = 4\pi r^2$$

where A is the area, r is the radius, and π is approximately equal to 3.14. The energy intensity per unit area depends on the size of the sphere. If we double the radius of the sphere, we quadruple its surface area, and this results in just ¼ as much light for each square meter, square inch, or square foot of the sphere. If we triple the radius, the surface area increases by a factor of 9, resulting in 1/9 the energy density. In general, given a power level of P watts per square centimeter at a distance d from a source of energy, the power Q at distance xd will be:

$$Q = P/x^2$$

where x may be any positive real number.

The inverse-square law only applies to quantities that are measured over a specific surface area. The rule does not apply to parameters that are measured along a line. Therefore, the field strength from an antenna in watts per square meter does obey the inverse-square law, but the field strength in microvolts per meter does not obey the principle. Some effects, such as magnetic-field intensity surrounding an inductor, also do not obey the inverse-square law, since flux density is determined partly in terms of three-dimensional volume, and is not a simple radiant effect.

INVERSE VOLTAGE

The inverse voltage across a rectifier is the extent to which the cathode becomes positive with respect to the anode. Inverse voltage results in nonconduction of a rectifier device, unless the potential becomes so great that avalanche or flashover occurs. In a semiconductor device, inverse voltage is sometimes called reverse voltage.

All rectifier devices are rated according to the maximum peak value of inverse voltage they can withstand. If this rating is exceeded, avalanche may take place. *See also* AVALANCHE, PEAK INVERSE VOLTAGE.

INVERSION

Normally, the temperature of the lower atmosphere drops with increasing altitude. This is primarly because the surface of the earth is warmed by radiation from the sun, and heat is transferred less and less effectively into higher and higher parts of the atmosphere. The increased density of the air near the surface results in a greater index of refraction for radio waves near the earth; the index gradually decreases with altitude. This effect (see A) results is a form of propagation called tropospheric bending (*see* INDEX OF REFRACTION, TROPOSPHERIC PROPAGATION).

Occasionally, a layer of cold air exists near the surface, and the index of refraction greatly increases. At low levels, the air temperature increases with increasing altitude under such conditions. This is called a temperature inversion, or simply inversion. The cold air is more dense than the warm air, and this results in a more pronounced difference in the index of refraction at various altitudes. At the boundary between the cold air and the overlaid warm air, the radio-wave index of refraction drops abruptly, as at B.

Temperature inversions often occur during the early morning hours, after the land has cooled. Over water, an inversion may persist for 24 hours a day, although this

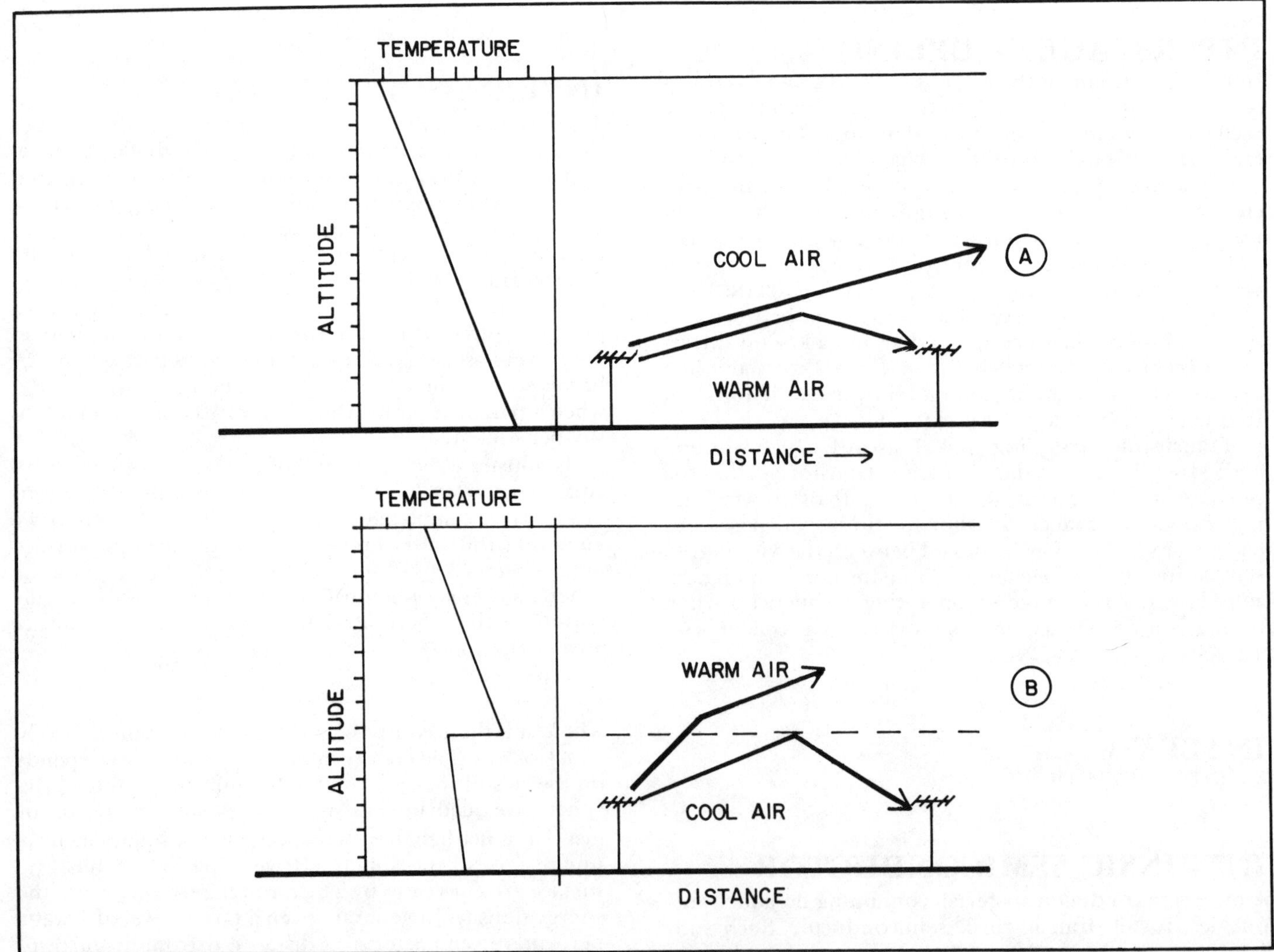

INVERSION: At A, normal atmospheric conditions, showing the effect on radio waves. At B, a temperature inversion causes waves to be reflected at the boundary plane, provided the angle of incidence is small.

depends on the water temperature. Weather fronts are commonly associated with inversions. Inversions produce a form of tropospheric propagation known as ducting, in which a radio wave is actually trapped within a cold layer of air. Inversions are also associated with severe air pollution in large cities. *See also* DUCT EFFECT.

INVERTED-L ANTENNA

An inverted-L antenna is a form of receiving and transmitting antenna sometimes used at medium and high frequencies. The antenna gets its name from the fact that it is shaped like an upside-down letter L.

The inverted-L antenna may be either end-fed or center-fed. The illustration at A shows a quarter-wave, end-fed inverted-L antenna. This antenna requires an excellent ground system. The height of the vertical section is normally about 1/10 to 1/8 wavelength, with the rest of the length made up in the horizontal part. This antenna is often employed in situations where a full-size quarter-wave vertical structure is impractical. The feed-point impedance is approximately 30 to 40 ohms at resonance.

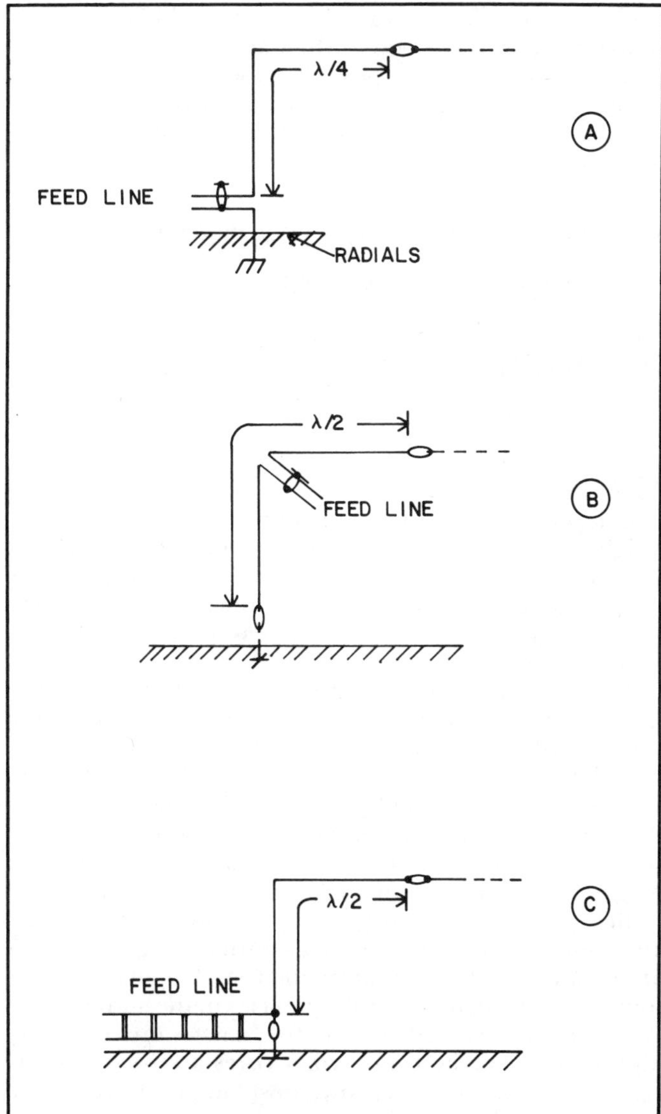

INVERTED-L ANTENNA: Three forms of inverted-L antenna. At A, a quarter-wave base-fed radiator; at B, a half-wave center-fed radiator; at C, a half-wave end-fed radiator.

The arrangement at B is a center-fed, half-wave inverted-L antenna. This arrangement is nearly the electrical equivalent of a dipole antenna (*see* DIPOLE ANTENNA). A good ground is not essential for proper functioning of this antenna. The feed point is elevated, and the feed line runs down at about a 45-degree angle. The physical lengths of the antenna legs, each ¼ electrical wavelength, may be reduced by inductive loading (*see* INDUCTIVE LOADING). The feed-point impedance is about 50 ohms, but lower if inductive loading is used.

At C, an end-fed, half-wave inverted-L antenna is illustrated. This antenna is fed in the same manner as a zeppelin antenna (*see* ZEPPELIN ANTENNA). It does not require a particularly good ground system for proper operation, although a radial network enhances performance. The feed-point impedance of this antenna is quite high, while in the other arrangements it is relatively low. For this reason, open-wire line or an impedance-matching network must be used.

The inverted-L antenna produces both vertical and horizontal polarization. At medium and high frequencies, this allows good performance both for DX (distance) and local work. *See also* HORIZONTAL POLARIZATION, VERTICAL POLARIZATION.

INVERTER

The term inverter is used to denote either of two different kinds of electronic circuits. A logical inverter is a NOT gate, while a power inverter converts low-voltage direct current into 120-volt alternating current.

A logical inverter has one input and one output. If the input is high, the output is low; if the input is low, the output is high. The illustration shows the symbol for the logical inverter, along with a truth table showing its characteristics.

A power inverter is essentially a chopper power supply. An interrupting device produces a pulsating direct-current output from the power source, which is usually a 12-volt battery. This pulsating direct current is regulated in frequency, so that it is as close as possible to 60 Hz. A step-up transformer then provides the 120-volt output needed to operate low-power household appliances. *See also* CHOPPER POWER SUPPLY.

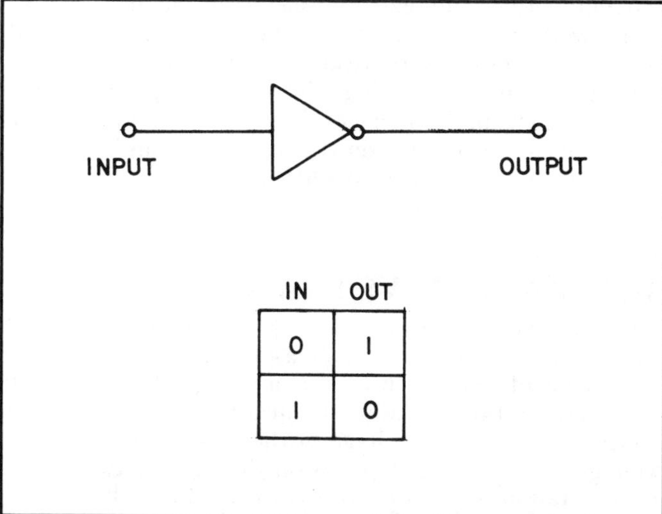

IN	OUT
0	1
1	0

INVERTER: An inverter simply changes the state of an input line from high (1) to low (0) or vice-versa.

INVERTING AMPLIFIER

An inverting amplifier is any amplifier that produces a 180-degree phase shift in the process of amplification. Virtually all common-cathode, common-emitter, and common-source amplifiers are of this variety (*see* COMMON CATHODE/EMITTER/SOURCE). In general, the grounded-grid, grounded-base, and grounded-gate configurations are not inverting amplifiers (*see* COMMON GRID/BASE/GATE).

A pushpull amplifier may be inverting or non-inverting, depending on the transformer output arrangement.

Inverting amplifiers are less susceptible to oscillation than non-inverting amplifiers, since any stray coupling between input and output results in degenerative, rather than regenerative, feedback.

ION

An ion is an atom from which one or more electrons have been removed or added. Normally, atoms are electrically neutral; this is the state in which they "prefer" to be. But some atoms give up or accept electrons readily.

Ions are created by a process called ionization. *See also* IONIZATION.

IONIZATION

An atom normally has the same number of negatively charged electrons and positively charged protons. Since both particles have equal charge quantities, most atoms are electrically neutral. Sometimes, however, electrons are lost or captured by an atom, so that the charge is not neutral. The non-neutral atom is called an ion, and any process or event that creates ions is known as ionization.

Ionization occurs when substances are heated to high temperatures, or when large voltages are impressed across them. Lightning is the result of ionization of the air. An electric spark is caused by a large buildup of charges, resulting in forces on the electrons in the intervening medium. These forces pull the electrons away from individual atoms. Ionized atoms generally conduct electric currents with greater ease than electrically neutral atoms.

Ionized gases are prevalent in the upper atmosphere of our planet. These layers are known as the ionosphere. The ionosphere has refracting effects on electromagnetic waves at certain frequencies. This makes long-distance communication possible on the shortwave bands. *See also* IONIZATION VOLTAGE, IONOSPHERE.

IONIZATION VOLTAGE

In a gas type electron tube, a large voltage will cause ionization. As the voltage increases between the cathode and plate of such a tube, a point is reached at which conduction begins because of ionization. This potential is called the ionization voltage of the tube. The ionization voltage of a gas tube depends on the kind of gas and its concentration. Gas tubes were once used as voltage regulators and rectifiers. *See also* GAS TUBE.

IONOSPHERE

The atmosphere of our planet becomes less and less dense with increasing altitude. Because of this, the amount of ultraviolet and X-ray energy received from the sun gets greater at increased altitude. At certain levels, the gases in the atmosphere become ionized by solar radiation. These regions comprise the ionosphere of the earth.

Ionization in the upper atmosphere occurs mainly at three levels, called layers. The lowest region is called the D layer. The D layer exists at altitudes ranging from 35 to 60 miles, and is ordinarily present only on the daylight side of the planet. The E layer, at about 60 to 70 miles above the surface, also exists mainly during the day, although night-time ionization is sometimes observed. The upper layer is called the F layer. This region sometimes splits into two regions, known as the F1 (lower) and F2 (upper) zones. The F layer may be found at altitudes as low as 100 miles and as high as 260 miles.

The ionosphere has a significant effect on the propagation of electromagnetic waves in the very-low, low, medium, and high-frequency bands. Some effect is observed in the very-high-frequency part of the spectrum. At frequencies above about 100 MHz, the ionosphere has essentially no effect on radio waves. Ionized layers cause absorption and refraction of the waves. This makes long-distance communication possible on some frequencies. At the longer wavelengths, energy from space cannot reach the surface of the earth because of the ionosphere. *See also* D LAYER, E LAYER, F LAYER.

IONOSPHERIC PROPAGATION

See PROPAGATION CHARACTERISTICS.

I/O PORT

See INPUT-OUTPUT PORT.

IRON

Iron is an element with atomic number of 26 and atomic weight 56. Iron is best known for its magnetic properties. Iron is found in great abundance on our planet, especially deep under the surface. The core of the earth is believed to consist largely of iron. This is the reason for the existence of a magnetic field surrounding the planet.

Iron is used in many applications in electricity and electronics. The atoms of iron tend to align themselves in the presence of a magnetic field. If the field is strong enough, the alignment persists even after the field is removed. In this case, the iron is said to be permanently magnetized. Permanent magnets are used in dynamic microphones, speakers, earphones, and other transducers. Permanent magnets are also used in analog metering devices.

Because of its high magnetic permeability, iron makes an excellent core material for transformers. Various forms of iron are used at different frequencies. Laminated iron is found in transformers for frequencies in the audio range. Powdered iron is used at radio frequencies up to the very high region. Ferrite transformer cores are used at radio frequencies in the very-low, low, medium, and high ranges. *See also* FERRITE, FERROMAGNETIC MATERIAL, LAMINATED CORE, PERMEABILITY, POWDERED-IRON CORE.

IRON CORE

See FERRITE CORE, LAMINATED CORE, POWDERED-IRON CORE.

IRON-VANE METER

An iron-vane meter is a device used for measuring alternating currents. The ordinary direct-current meter will not respond to alternating current, because such a meter produces directional readings, according to the current polarity.

The iron-vane instrument contains a movable iron vane, attached to a pointer. The pointer is held at the zero position on a calibrated scale by a spring under conditions of zero current. A stationary iron vane is positioned next to the movable vane as shown in the illustration. Both vanes are placed inside a coil of wire, the ends of which form the meter terminals.

When current is supplied to the coil, a magnetic field is produced in the vicinity of the iron vanes. The field causes the vanes to become temporarily magnetized. The vanes are positioned in such a way that like magnetic poles form near each other on each vane end; the strength of the magnetism is proportional to the current in the coil. Since like magnetic poles repel, the movable vane is deflected away from the stationary vane. The spring allows the pointer to move a specific distance, depending on the force of repulsion. This effect occurs for alternating or direct currents in the coil. The meter deflection is therefore proportional to the current.

The iron-vane instrument is, by itself, an ammeter. However, it can be used as a voltmeter or wattmeter with the addition of appropriate peripheral components. The scale of the iron-vane meter is usually nonlinear. The meter is accurate only at power-line frequencies because of the coil inductance and losses in the iron vanes. Thus, the iron-vane meter is used only for measurement of currents and voltages in ordinary utility devices. It is not suitable for use at audio or radio frequencies.

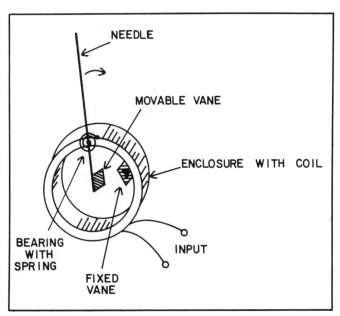

IRON-VANE METER: The iron-vane meter can be used to measure either alternating or direct currents.

IRRADIANCE

Irradiance is a measure of the intensity of radiant energy that falls on a given surface area. Irradiance is generally measured in watts per square centimeter (W/cm^2) or watts per square meter (W/m^2). Irradiance can be measured at any electromagnetic wavelength.

In the visible-light range, irradiance is called illuminance. *See also* ILLUMINANCE.

IRRATIONAL NUMBER

An irrational number is a quantity that cannot be expressed as the quotient of two integers. In decimal form, an irrational number consists of an infinite sequence of digits, with no repeating block of digits.

Irrational numbers are much more common in mathematics than rational numbers. In fact, the number of irrational quantities is infinitely greater than the number of rational quantities. Irrational numbers include such factors as pi (approximately 3.14), e (approximately 2.72), and many others. *See also* RATIONAL NUMBER, REAL NUMBER.

I SCAN

An I scan is a form of radar display in which the target distance is indicated, but the altitude or direction is not given. In the I scan, the target appears as a circle having a definite radius. The larger the radius of the circle, the greater the distance to the target (see illustration). The I scan is a mono-dimensional form of display. *See also* RADAR.

ISOLATION

Isolation is any means of electrical, acoustical, or electromagnetic separation between components, circuits, or systems. Isolation is important from the standpoint of circuit independence in certain instances.

In direct-current and low-frequency alternating-current applications, isolation is often called insulation. In radio-frequency applications, isolation is called shielding.

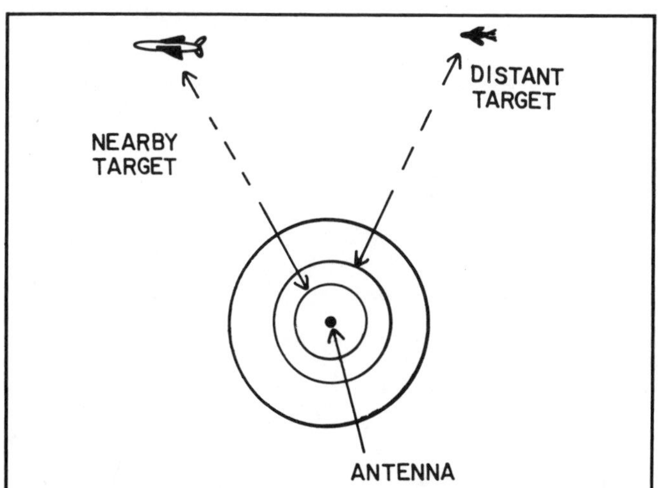

I SCAN: I-scan radar. Echoes appear as concentric circles.

ISOTOPE

The nuclei of all atoms are identified according to the number of protons they contain. For example, an atom with six protons is a carbon atom; an atom with eight protons is an oxygen atom. All atoms also contain neutrons in their nuclei, with the exception of simple hydrogen. The number of neutrons may vary, however, even while the number of protons remains the same.

The number of protons in an element, known as the atomic number, uniquely determines the identity of the element. The number of neutrons may vary, so that a given element can occur in more than one natural form. The different forms of an element, resulting from nuclei having different numbers of neutrons, are called isotopes of the element.

Hydrogen, while normally containing no neutrons, may have one or two neutrons in addition to the proton in its nucleus. If there is one neutron, the isotope is called deuterium. If there are two neutrons, the isotope is called tritium. Carbon, with atomic number 6, generally occurs with six neutrons, so that the atomic weight is 12. However, some carbon atoms have eight neutrons, and the atomic number of such atoms is 14. Carbon 14 is a radioactive substance, because it decays fairly quickly. Radioactivity is characteristic of many different isotopes.

In general, the larger the atomic number of an element, the greater the number of isotopes that can be formed. The most common naturally occurring isotopes are the most stable. Some isotopes are made by man for use in atomic energy. *See also* ATOM, ATOMIC NUMBER, ATOMIC WEIGHT.

ISOTROPIC ANTENNA

An isotropic antenna is a device that radiates electromagnetic energy equally well in all directions. Such an antenna is a theoretical construct, and does not actually exist. However, the isotropic-antenna concept is occasionally used for antenna-gain comparisons.

The power gain of an isotropic antenna is about −2.15 dB with respect to a half-wave dipole in free space. That is, the field strength from a half-wave dipole antenna, in its favored direction, is approximately 2.15 dB greater than the field strength from an isotropic antenna at the same distance and at the same frequency.

The radiation pattern of the isotropic antenna, in three dimensions, appears as a perfect sphere, since the device works equally well in all directions. In any given plane, the radiation pattern of the isotropic antenna is a perfect circle, centered at the antenna. *See also* ANTENNA PATTERN, ANTENNA POWER GAIN, dBd, dBi.

ITERATION

Iteration is a process of repeating a mathematical operation over and over, for the purpose of making a calculation. Iteration is used for such purposes as determining square roots of numbers, and for addition of infinite series.

Some mathematical calculations involve approximations, the accuracy of which depend on the number of iteratons. Generally, long iterative processes are carried out by means of computers. The more repetitions that are performed, the greater the accuracy of the final result.

ITU

See INTERNATIONAL TELECOMMUNICATION UNION.

J

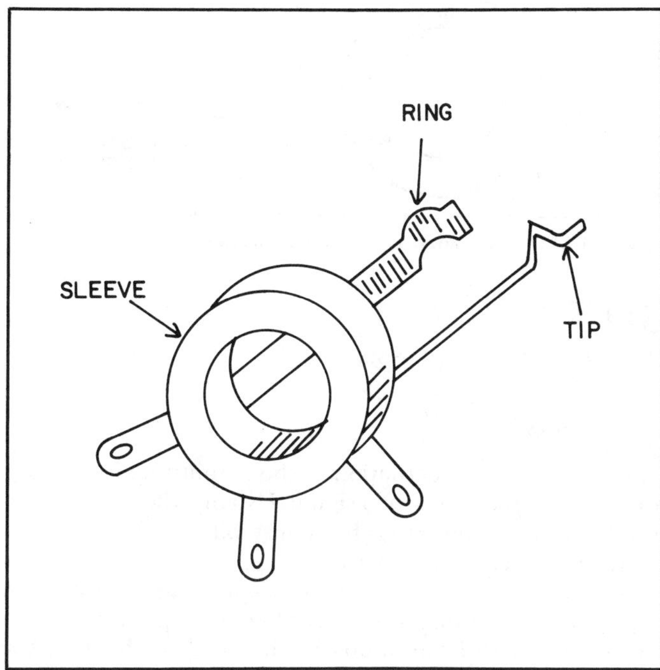

JACK: The jack is a female, or recessed, connector. This is a view of a 3-connector, ¼-inch phone jack.

JACK

A jack is a receptacle for a plug. The jack is sometimes called the female connector. An example of a jack is shown in the illustration.

Usually, the jack is mounted on the equipment panel in a fixed position, and the plug is mounted on the end of a cord. However, in some cases this is reversed. *See also* PLUG.

JACKET

A jacket is the coating on a multi-conductor cable. The term is used especially with coaxial cables. The coating on an electronic component, such as a capacitor, is also sometimes called a jacket.

The jacket is generally made of insulating material, rather than conducting material. This isolates the component from surrounding objects. It also protects the component or cable against corrosion.

JACK PANEL

A jack panel is a flat, usually metallic, panel in which a number of jacks are mounted (*see* JACK). Each jack is connected to a specific circuit in a system. An operator may insert one or more plugs into various jacks in the jack panel for the purpose of connecting various circuits.

Jack panels were common in the early days of wire telephone systems. Telephone operators made connections by inserting plugs into various jacks on a jack panel. Today, most telephone interconnections are performed by automatic switching devices, controlled by computers. *See also* TELEPHONE.

JAM ENTRY

When a logic-gate input is deliberately supplied with a high or low voltage by force, the entry is called a jam entry or jam input. The jam entry overrides whatever logic state may be present on the line initially.

Jam entry is often used for the purpose of intializing, or presetting, a logic circuit or microcomputer. Once the desired state has been achieved, the circuit is free to operate normally. *See also* INITIALIZATION.

JAMMING

Jamming is deliberate interference to radar or communication signals. Jamming techniques can vary from the most elementary to the most complex.

The simplest form of jamming is the intentional transmission of an unmodulated carrier on the frequency of a two-way radio link. This form of interference is prohibited by law in the United States, but it is sometimes done anyway.

Jamming is a common practice among countries that broadcast international radio transmissions on the shortwave bands. A highly modulated or overmodulated, high-power signal is transmitted on the frequency to be jammed. A jamming signal can be recognized by its characteristic whining or buzzing sound as heard on an AM receiver.

In radar, jamming is used for the purpose of masking a target. This may be done by the transmission of a synchronized set of pulses at or near the frequency of the radar transmitter. The radar receiver then "sees" false targets or large bright spots that cannot be identified. Jamming, when skillfully done, can be almost impossible to overcome. *See also* INTERFERENCE.

JANSKY, KARL

The first primitive radio astronomy was performed in the early part of this century by engineer Karl Jansky. The discovery of noise from the center of the galaxy was in fact accidental. Jansky was conducting expeiments to determine the directional characteristics of atmospheric noise (sferics) at a wavelength of about 15 meters when he found an amplitude periodicity of 23 hours and 56 minutes.

Jansky's conclusion that the noise originated at the center of our galaxy was not taken seriously until years later, when astronomers began building radio telescopes. *See also* RADIO ASTRONOMY, RADIO TELESCOPE.

J ANTENNA

A J antenna is a form of end-fed antenna. It is electrically similar to the zeppelin antenna, since it is an end-fed, half-wave antenna (*see* ZEPPELIN ANTENNA). The J antenna is sometimes known as a J pole. The antenna gets these names from its physical shape.

The J antenna consists of a half-wave, vertical radiator, with a quarter-wave section of transmission line that serves as a matching stub (see illustration). If the radiator is made

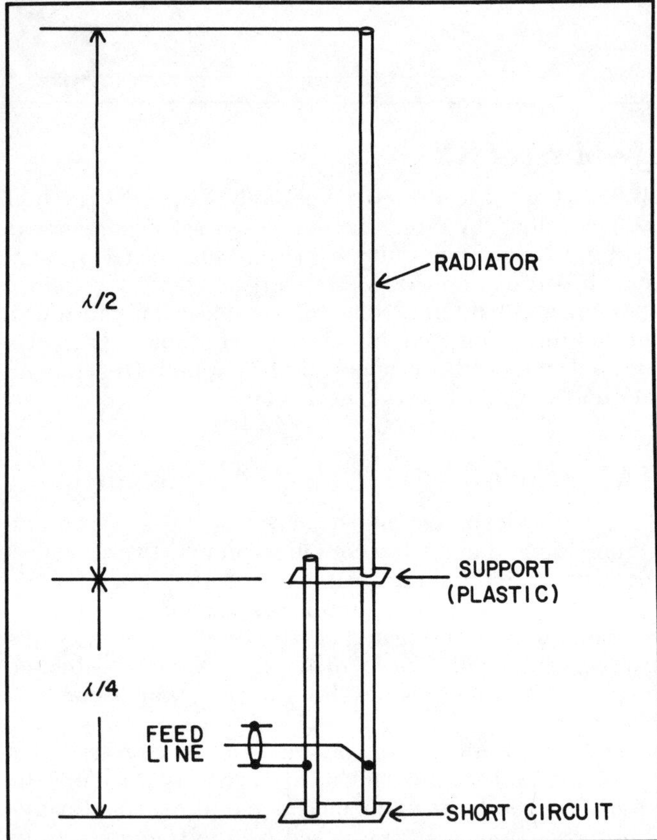

J ANTENNA: The J antenna consists of a half-wave vertical radiator and a quarter-wave matching stub.

from a length of metal rod or tubing, the matching section may be incorporated into the physical construction of the antenna.

The matching section is short-circuited at the bottom. The antenna is fed with coaxial cable at a point between the shorted end of the stub and the antenna end. The feed-point impedance is a pure resistance at all points on the matching stub; the value of this pure resistance is zero at the shorted end and very high—several hundred ohms or more—at the antenna end. If 50-ohm coaxial cable is used, the best impedance match is obtained when the feed point is about 1/6 of the way from the shorted end to the antenna end. It actually does not matter which coaxial conductor goes to the longer element, but in most cases the center conductor is placed there.

The J antenna has an omnidirectional radiation pattern in the horizontal plane. The J antenna is used primarily at very-high and ultra-high frequencies. The primary advantage of this antenna is that it does not require a ground system.

J DISPLAY

In radar, a J display, or J scan, is a display with a circular time base. The display appears as a circle with the transmitter pulse at the top. The scanning proceeds clockwise at a rapid rate. Echoes appear as radial pips. The greater the distance, or range, to the target, the farther clockwise the pip appears (see illustration).

The J display is a one-dimensional display. That is, by itself it can show only the distance to a target. *See also* RADAR.

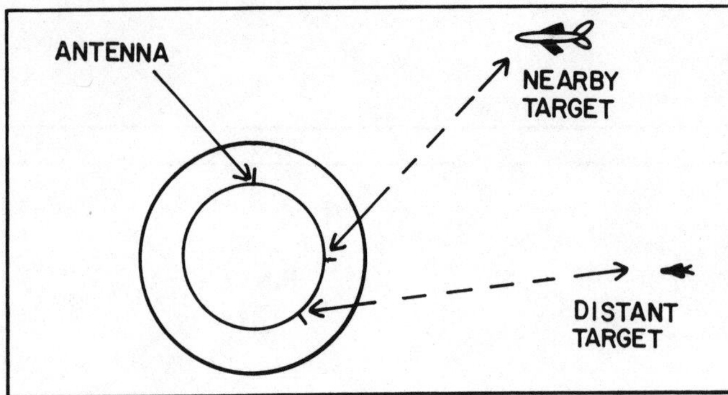

J DISPLAY: J-scan radar. Echoes appear as radial pips.

JFET

See FIELD-EFFECT TRANSISTOR.

JITTER

A jitter is a rapid fluctuation in the amplitude, frequency, or other characteristic of a signal. Usually, the term jitter is used in video systems to refer to instability in the displayed picture on a cathode-ray tube.

Jitter in a cathode-ray-tube display can result from changes in the voltages on the deflecting plates. It can also occur because of variations in the current through the deflecting coils, or because of changes in the speed of the electrons passing through the tube. Mechanical vibration, an interfering signal, or internal noise can also cause jitter. Sometimes the synchronizing circuits fail to "lock" onto the signal in the display, causing vertical or horizontal jitter. *See also* CATHODE-RAY TUBE, OSCILLOSCOPE, TELEVISION.

J-K FLIP-FLOP

See FLIP-FLOP.

JOHNSON COUNTER

See RING COUNTER.

JOHNSON NOISE

See THERMAL NOISE.

JOINT

See JUNCTION.

JOINT COMMUNICATIONS

When a communications medium or facility is used by more than one service, the arrangement is called joint communications, Many frequency bands in the electromagnetic spectrum are shared by more than one service. In such situations, one service generally has priority over the others. The service having the top priority is called the primary service, and the others are called secondary services.

The International Telecommunication Union is responsible for assignment of frequency bands throughout the world. In the United States, the Federal Communications Commission makes frequency allocations. *See also* FEDERAL COMMUNICATIONS COMMISSION, INTERNATIONAL TELECOMMUNICATION UNION.

JONES PLUG

A Jones plug is a special form of connector often used for providing power to electronic equipment. The female Jones plug has two or more recessed contacts, oriented in a defined pattern of vertical and horizontal (see A). The male Jones plug has protruding contacts that mate with those of the female plug (B). The contacts are arranged so that the two plugs can be put together in only one way.

Jones plugs are available in numerous configurations and sizes, for different power-supply applications. Jones plugs are normally intended for indoor use only, unless they are sealed with a coating of waterproof material such as silicone. *See also* CONNECTOR.

J OPERATOR

The j operator is an imaginary number, mathematically defined as the square root of −1. Mathematicians denote this quantity by the lowercase letter i; however, electrical and electronic engineers prefer the lowercase letter j because i may be used to represent current.

The j operator is used to represent reactance. An inductive reactance of X ohms, where X is any positive real number, is written jX. Sometimes it may be written Xj. A capacitive reactance of − X ohms is written −jX or −Xj. The set of possible reactances is the set of imaginary numbers, or real-number multiples of j.

A complex impedances is represented by the sum of a pure resistance and an inductive (positive) or capacitive (negative) reactance. If the resistive component is R, then the impedance Z takes the form R + jX or R − jX.

Complex impedance can be added together, or even multiplied, in this form. The important thing to remember is that when j is multiplied by itself, the result is −1. *See also* CAPACITIVE REACTANCE, IMPEDANCE, INDUCTIVE REACTANCE.

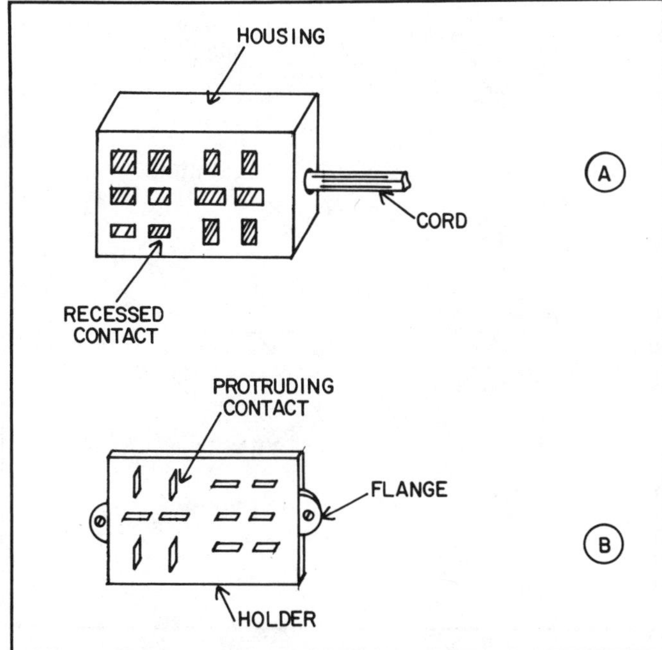

JONES PLUG: Jones plugs. At A, female; at B, male.

JOSEPHSON EFFECT

When two superconducting materials are brought close together, a current flows across the gap between the tips of the two objects. Superconductivity is characterized by extremely low ohmic resistance in a substance (*see* SUPERCONDUCTIVITY).

When the current crosses the gap between the superconducting objects, high-frequency electromagnetic energy is generated. This was noticed by Brian Josephson, and thus it is called the Josephson effect.

JOULE

The joule is the standard unit of energy or work. When a force of 1 newton is applied over a distance of 1 meter, the amount of work done is 1 joule. The joule is equivalent in energy to 0.24 calories; that is, 1 joule of energy will raise the temperature of 1 gram of water by 0.24 degrees Celsius. A kilowatt hour is equivalent to 3.6 million joules. A power level of 1 watt is a rate of expenditure of 1 joule per second. *See also* ENERGY.

JOULE'S LAW

When a current flows through a resistance, heat is produced. This heat is called joule heat or joule effect. The amount of heat produced is proportional to the power dissipated. The power P, in watts, dissipated in a circuit carrying I amperes, and having resistance of R ohms, is given by the formula:

$$P = I^2R$$

Joule's law recognizes this, by stating that the amount of heat generated in a constant-resistance circuit is proportional to the square of the current.

JOYSTICK

A joystick is a control device that functions in two dimensions: the up-down dimension and the left-right dimension. A joystick consists of a movable lever or handle and a ball-bearing unit (see illustration). The joystick gets its name from its physical resemblance to the joystick of an

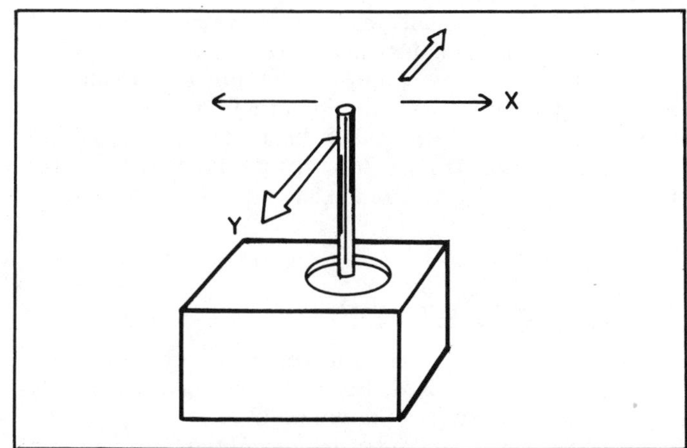

JOYSTICK: The joystick can be manipulated in two dimensions.

airplane. Some joysticks have a rotational control dimension, in addition to the up-down and right-left functions. The lever is rotated either clockwise or counterclockwise. This makes it possible to control three different parameters.

The joystick is used extensively in modern electronic computer games, for the purpose of manipulating objects on the display screen. The joystick is also employed for entering coordinates in the X, Y, and Z input registers of a computer. The X and Y coordinates are set by moving the joystick from side to side and up and down. The Z coordinate is set by twisting. *See also* GRAPHICS.

JUNCTION

A junction is the point, line, or plane at which two or more different components or substances are joined. A simple electrical connection may be called a junction. In a waveguide, a junction is a fitting used for the purpose of joining one section to another. In a transmission line, a junction is the splice between two sections of the feed system.

A semiconductor junction is a plane surface at which P-type and N-type materials meet. This is called a P-N junction. Such a junction forms a one-way barrier for current; that is, it has diode action. *See also* P-N JUNCTION.

JUNCTION CAPACITANCE

When a semiconductor P-N junction is reverse-biased (*see* P-N JUNCTION), the resistance is high. However, there is some capacitance, known as junction capacitance.

The junction capacitance of a given diode, under conditions of reverse bias, depends on several factors. The surface area of the junction is important; in general, the larger the area, the greater the junction capacitance. The value of the reverse voltage affects the junction capacitance in some cases. This is especially true in the varactor diode, which acts as a voltage-controlled, variable capacitor (*see* VARACTOR DIODE).

The capacitance of a reverse-biased semiconductor junction determines the maximum frequency at which a device can be used. The smaller the junction capacitance, the higher the limiting frequency. At very-high and ultra-high frequencies, therefore, it is necessary that semiconductor devices have small values of junction capacitance. Rectifier diodes have large junction capacitance. Gallium-arsenide devices usually have small values of junction capacitance; the same is true of metal-oxide semiconductor components. The point-contact junction exhibits a smaller capacitance than the P-N junction. *See also* DIODE CAPACITANCE, GALLIUM-ARSENIDE FIELD-EFFECT TRANSISTOR, METAL-OXIDE SEMICONDUCTOR, METAL-OXIDE SEMICONDUCTOR FIELD-EFFECT TRANSISTOR, POINT-CONTACT JUNCTION.

JUNCTION DIODE

A junction diode is a common type of semiconductor device, manufactured by diffusing P-type material into N-type material, thereby obtaining a P-N junction (*see* P-N JUNCTION). Junction diodes are usually made from germanium or silicon, with added dopants.

The junction diode conducts when the N-type material is negatively charged with respect to the P-type material. This negative charge must be at least 0.3 volts if germanium is used as the semiconductor, or 0.6 volts if silicon is used. If the potential difference is less than this threshold value, or if the P-type material is negative with respect to the N-type material, the junction diode will exhibit a very large direct-current resistance, infinite for all practical purposes.

Junction diodes are commonly used in rectifier and detector circuits. They may also be used as mixers, modulators, converters, and frequency-control devices.

A silicon junction diode can withstand higher temperatures than a germanium device. However, germanium has certain advantages because of its lower threshold. *See also* DIODE, DIODE CAPACITANCE, JUNCTION CAPACITANCE, POINT-CONTACT JUNCTION.

JUNCTION FIELD-EFFECT TRANSISTOR

See FIELD-EFFECT TRANSISTOR.

JUNCTION TRANSISTOR

A junction transistor is a form of bipolar tranistor. The junction transistor is formed by joining N-type and P-type semiconductor materials. The emitter-base and base-collector junctions are thus semiconductor P-N junctions.

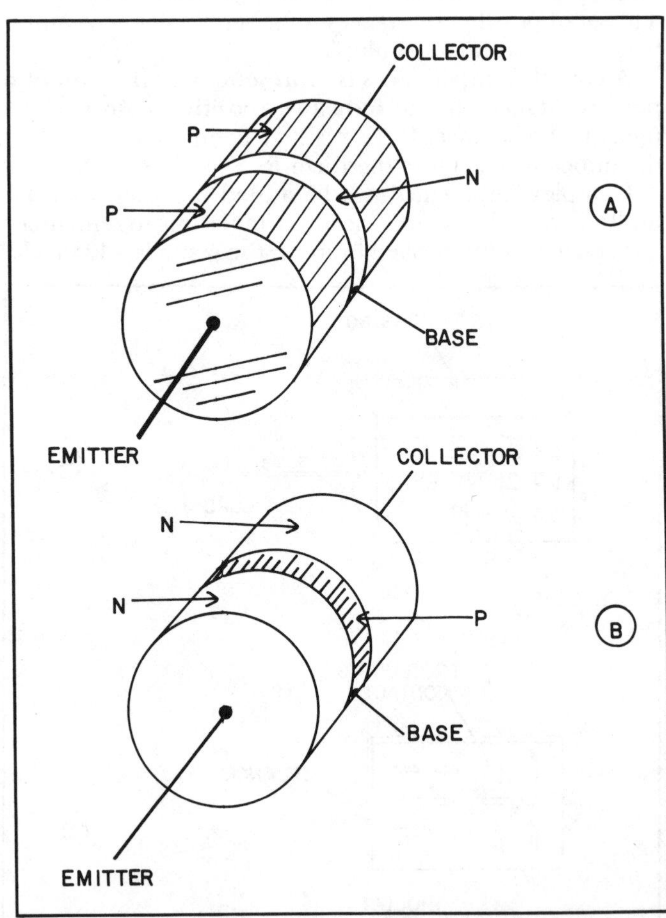

JUNCTION TRANSISTOR: The two common types of junction transistor. At A, a PNP bipolar device. At B, an NPN bipolar device.

The illustration shows the two common kinds of junction transistor. The junction transistor is characterized by relatively high junction capacitance. This limits the frequency at which the device can be used. *See also* JUNCTION CAPACITANCE, POINT-CONTACT JUNCTION, P-N JUNCTION, POINT-CONTACT TRANSISTOR.

JUSTIFICATION

Justification is a term sometimes used in electronic word-processing applications. Justification literally means the vertical alignment of lines. Most word-processing systems use left-margin justification. This means that the lines are vertically aligned on the left side. Some word processors have the option of right-margin justification. This function requires that extra spaces be inserted between words in some lines, so that each line is exactly the same physical length. Typewriters produce copy without right-margin justification; books are usually typeset in a format with right-margin justification. *See also* WORD PROCESSING.

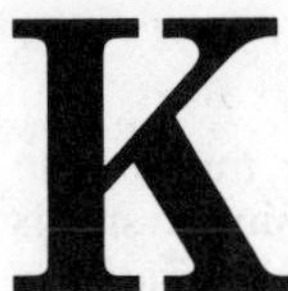

KARNAUGH MAP

A Karnaugh map is a truth table, arranged in such a way that a logical expression is broken down into its constituents. Some logic functions are complicated, and may be obtained by putting logic gates in different combinations. The Karnaugh map shows which of these combinations is the simplest. This minimizes the number of individual gates to be used.

There are several different methods that are used for reducing a logic equation. The fundamental rules of Boolean algebra are often employed for simplifying equations. Truth tables may also be used. *See also* BOOLEAN ALGEBRA, TRUTH TABLE.

KELVIN ABSOLUTE ELECTROMETER

The Kelvin absolute electrometer is a device for measuring electrostatic voltages. The meter consists of two stationary metal plates, with a movable plate between them. The movable plate is attached to a pointer, and is held at the zero position by a set of return springs. The illustration is a simplified diagram of the device.

When an alternating-current or direct-current voltage is applied to the input terminals of the electrometer, the movable plate is deflected away from the stationary plates. The extent of the movement is determined by the voltage; the greater the input voltage, the farther the plate moves before equilibrium is reached with the return-spring tension. The pointer moves along a graduated scale.

The Kelvin absolute electrometer has extremely high input impedance, and therefore draws essentially no current from the circuit under test. The device does not use an amplifying device; this is why it is called an absolute electrometer. *See also* ELECTROMETER.

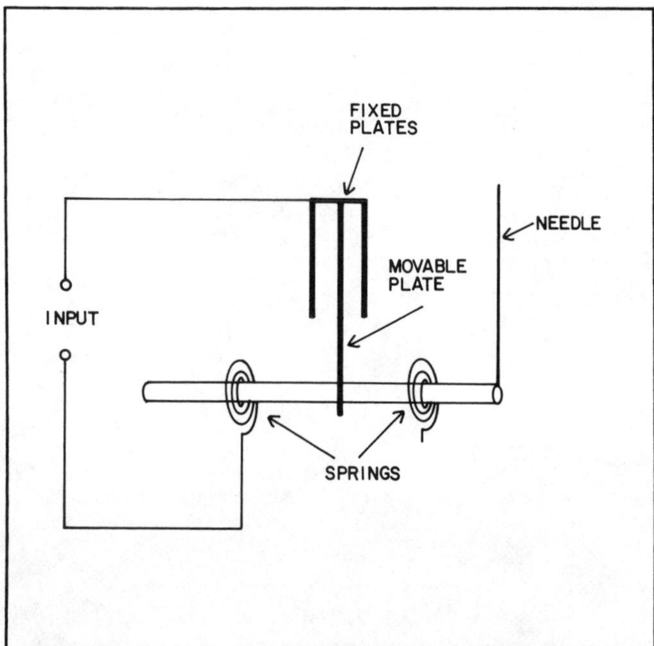

KELVIN ABSOLUTE ELECTROMETER: The Kelvin absolute electrometer can be used to measure either ac or dc potential.

KELVIN BALANCE

The Kelvin balance is a device for measuring alternating and direct electric currents. The device uses a mechanical balance, similar to the balance found in a chemistry laboratory. Coils are attached to the arm of the balance. The coils are interconnected in series, so that a current at the input terminals will cause magnetic fields to form around the coils. The coils are arranged so that one pair attracts while the other repels (see illustration).

When there is no current at the input of the Kelvin balance, the indicator shows equilibrium. As a small alternating or direct current is applied to the coils, the needle is deflected. The greater the current, the more the needle is deflected. A sliding weight is moved along the balance arm until the needle returns to the center position. The balance arm has a calibrated scale, from which the value of the current is determined according to the position of the weight at equilibrium.

The Kelvin balance is an extremely sensitive device. However, it has a very slow response to changes in the current. If the current fluctuates, the weight must be repositioned. *See also* AMMETER.

KELVIN DOUBLE BRIDGE

The measurement of small resistance requires special techniques. One device, specifically designed for the measurement of low resistance, is the Kelvin double bridge (see illustration).

The unknown resistance, R, is connected as shown in the circuit with the other known resistances. The ratios R1/R2 and R3/R4 are known, and they are identical. The value of R5, a precisely calibrated, low-resistance potentiometer, is adjusted for a zero reading on the meter.

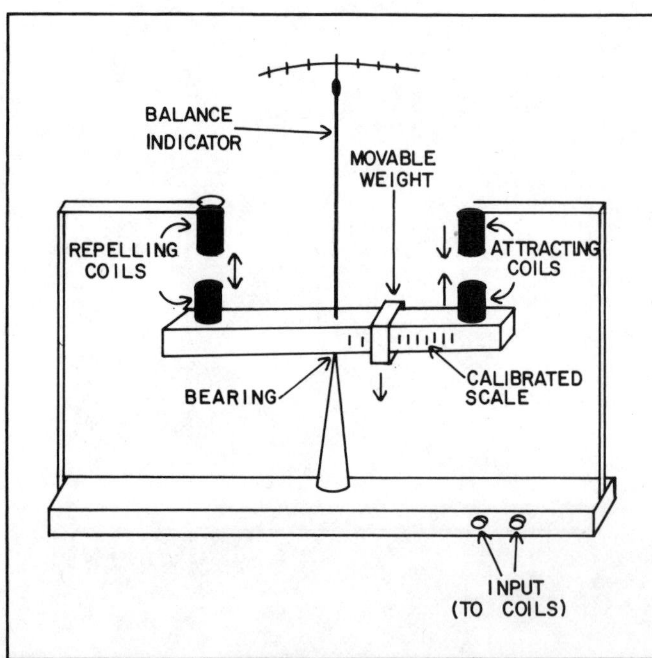

KELVIN BALANCE: The Kelvin balance is an extremely sensitive current meter. It can be used to measure either alternating or direct current.

When balance is obtained, the ratio R/R5 is equal to the ratios R1/R2 and R3/R4. The value R is calculated from the other values by the formulas:

$$R = (R1/R2)R5, \text{ or}$$

$$R = (R3/R4)R5$$

The Kelvin double bridge is sometimes called the Thomson bridge.

KELVIN TEMPERATURE SCALE

The Kelvin temperature scale, also called the absolute temperature scale, is based on the coldest possible temperature. All readings on the Kelvin scale are positive readings; 0 degrees Kelvin is known as absolute zero, and represents the total absence of thermal energy.

The Kelvin degree is the same size as the Celsius degree. If C represents the temperature in degrees Celsius and K represents the temperature in degrees Kelvin, then the readings are approximately related according to the simple formula:

$$K = C + 273$$

For conversion from Fahrenheit (F) to Kelvin, the following formulas apply:

$$K = 0.555F + 255$$
$$F = 1.8K - 459$$

See also CELSIUS TEMPERATURE SCALE, FAHRENHEIT TEMPERATURE SCALE.

KENNELLY-HEAVISIDE LAYER

In the upper part of the atmosphere of the earth, rarefied gases become ionized because of radiation from the sun. This region begins at about 35 miles altitude and extends upward to heights exceeding 250 miles. This part of the atmosphere is called the ionosphere (*see* IONOSPHERE).

In the early days of radio, it was noticed that some parts of the ionosphere cause electromagnetic energy to be returned to the earth. The ionospheric layer responsible for this was called the Kennelly-Heaviside layer, after the scientists A. E. Kennelly and O. Heaviside. Today, this name has become obsolete, and we speak of the three primary ionospheric zones as the D, E, and F layers. *See also* D LAYER, E LAYER, F LAYER.

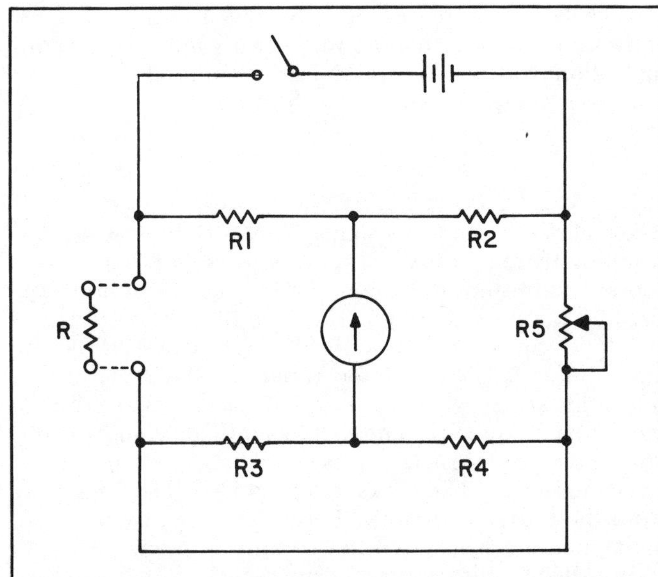

KELVIN DOUBLE BRIDGE: The Kelvin double bridge is used to determine very small values of resistance. The values R1 through R4 are known and fixed. The value of R5 is known and variable. The unknown resistance is R. A galvanometer indicates balance.

KEY

A key is a simple switch used for the purpose of sending radiotelegraph code. It is sometimes called a straight key or brass pounder. The key consists of a spring-mounted lever with a knob, which is held by the fingers (see illustration). Characters of the Morse code are formed by pushing the lever down at intervals.

The radiotelegraph key was the earliest device for modulating a transmitter. Today, more sophisticated keying devices are used. The speed at which an operator can generate code via a simple key is limited. Some experts can send at speeds of 35 to 40 words per minute with a straight key, but most people find it difficult to exceed about 20 words per minute. *See also* CODE TRANSMITTER, KEYER.

KEYBOARD

A keyboard is a set of momentary switches, generally used for the purpose of entering data into a calculator or computer. Keyboards may also be used to actuate Morse-code generators or teletype machines.

A keyboard may be any size, and labeled in any manner. The photograph (A) shows the keyboard of a typewriter. The arrangement of letters, numerals, and common punctuation is standardized. The less common symbols

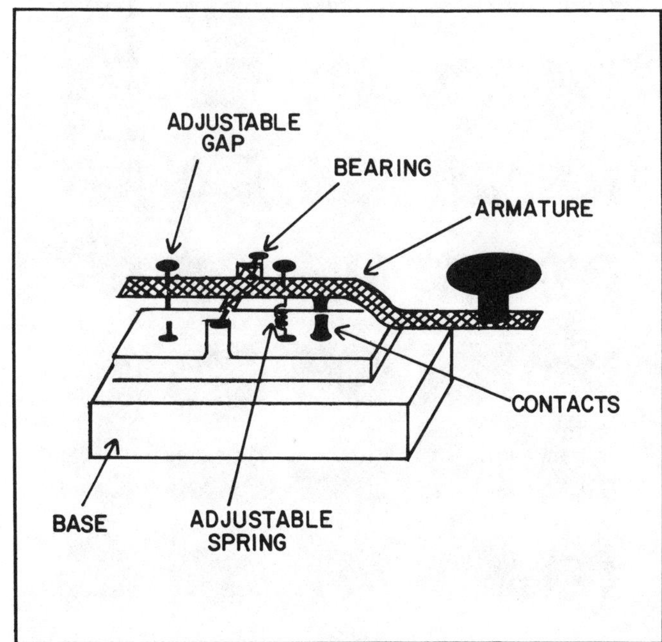

KEY: A telegraph key.

KEYBOARD: At A, a typewriter keyboard. Letters and numerals are in standard positions. Symbols and control keys may be in different locations in different machines. At B, a 40-key calculator keyboard. The arrangement of numbers is standard.

may be arranged in different ways in different keyboard units. This keyboard has been called the QWERTY keyboard, because those are the first letters that appear in sequence at upper left. Early typewriter and teletype keyboards were mechanical devices. Today, they are fabricated from electronic switches, and often have large buffers, allowing the operator to type ahead of the actual data transmission.

A typical handheld-calculator keyboard is shown at B. This keyboard is a matrix of eight rows and five columns, or 40 keys total. Calculators vary greatly in keyboard configuration, with the exception of the numerals, which are usually arranged in a standard format.

KEY CLICK

A key click is a spurious signal from a radio transmitter, resulting from excessively rapid rise time or fall time. In a properly adjusted continuous-wave radio transmitter, the rise and fall time should be finite, so that the modulation envelope appears similar to the illustration at A. If the rise time and/or fall time are too rapid, as shown at B or C, sidebands are generated at frequencies far removed from the signal frequency. These sidebands have short duration, and take place at the "make" or "break" instants. Such sidebands sound like clicks to an operator listening away from the signal frequency. These clicks can cause objectionable interference to other communications.

In order to ensure that a transmitter will not produce key clicks, it is necessary to provide a shaping network. Such a network is usually comprised of series resistors and parallel capacitors in the keying circuit. *See also* CODE TRANSMITTER, SHAPING.

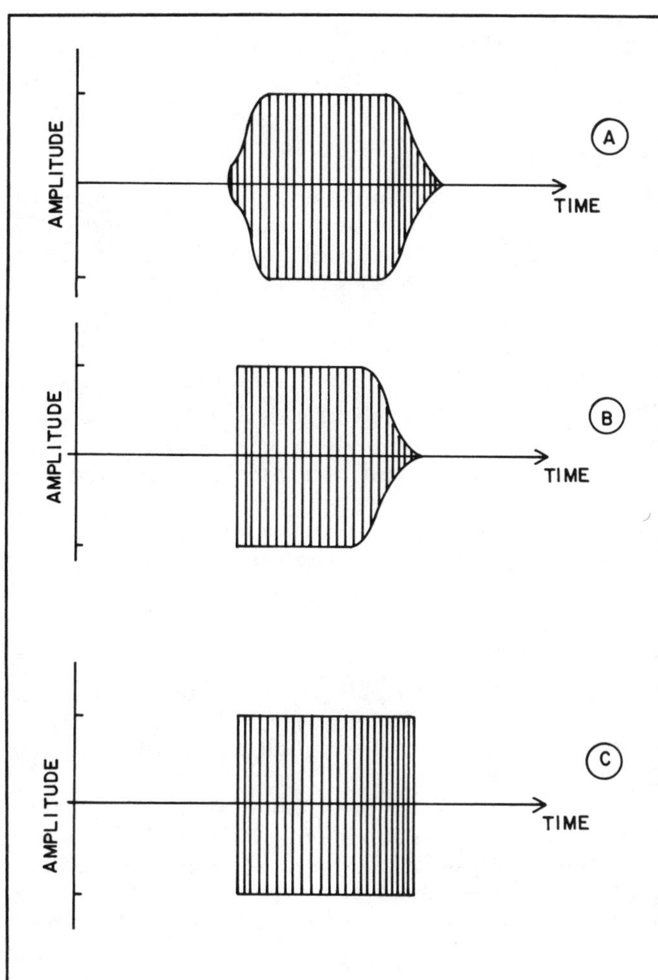

KEY CLICK: At A, the output of a properly operating code transmitter is shown as it would appear on an oscilloscope screen. At B, excessively fast rise time; at C, excessively fast rise and decay times. The signals at B and C will have key clicks.

KEYER

A keyer is a device for switching a continuous-wave (code) transmitter on and off, or for shifting its frequency, to produce Morse code. Keyers are available in several different basic configurations. Commercial units can be found in an almost countless selection of styles, some with sophisticated memory circuits and operating aids.

The basic electronic keyer is actuated by a single-pole, double-throw, center-off, lever switch that can be moved from side to side. This lever switch, called a paddle, resembles the semiautomatic key or "bug" (*see* SEMIAUTOMATIC KEY). The paddle is normally at center, or off. When the lever is pushed to the left, the keyer produces a string of dashes. When the lever is pressed to the right, the keyer produces a string of dots. The dots and dashes are perfectly timed and formed by the keyer circuit. The actual duration of the dot is ⅓ the duration of the dash. The keyer inserts a space of one dot length after each dot or dash. When these spaces are included in the element length, the dash is twice as long as the dot. The dot-to-space ratio is usually 1 to 1, but can be adjusted in some keyers, via a weight control (*see* WEIGHT).

Some keyer devices contain two single-pole, single-throw, momentary-contact switches arranged in such a way that either or both can be actuated. Pressing one switch produces dashes, and pressing the other results in the generation of dots. Pressing both switches together produces an alternating string of dots and dashes. This is called squeeze keying, because of the physical arrangement of the paddles.

The most sophisticated keyers are actuated by means of a keyboard. With the keyboard keyer, the speed of transmission is limited only by the speed at which the operator can type. With a conventonal paddle keyer or squeeze keyer, the most proficient operators can send code at speeds of 60 to 70 words per minute. But with a keyboard keyer, speeds of more than 100 words per minute are possible. *See also* KEY, KEYING.

KEYING

Keying is the means by which a code transmitter is modulated. Keying is accomplished either by switching the carrier on and off, or by changing its frequency. The latter method is called frequency-shift keying (*see* FREQUENCY-SHIFT KEYING).

The code transmitter may be keyed at the oscillator, or at any of the amplifiers following the oscillator. If a heterodyne circuit is used, any of the local oscillators may be keyed. Oscillator keying makes it possible for the operator to listen in between dots and dashes. This is called break-in operation (*see* BREAK-IN OPERATION). Amplifier keying usually results in the transmission of a local signal when the key is up; this disables the receiver unless excellent isolation is provided in the transmitter circuit.

The most common method of keying in a transmitter is called gate block keying. In the bipolar circuit, the equivalent is called base-block keying. *See also* CODE TRANSMITTER, KEY CLICK, SHAPING.

KEYPUNCH

A keypunch is a device that facilitates the recording of data on cards. Modern word processors are sometimes called keypunch machines, and the operators are known as keypunch operators (although this is not technically correct).

A device for recording data on perforated paper tape, using a keyboard and a device called a reperforator, is sometimes called a keypunch. This machine was once common in teletype applications. The paper tape was run through a reading device for data entry. In modern computer and word-processing applications, the video-display unit, with magnetic disks, is more common than the reperforator. *See also* VIDEO DISPLAY TERMINAL, WORD PROCESSING.

KEYSTONING

Keystoning is a form of television picture distortion in which the horizontal gain is not uniform at various vertical picture levels. This form of distortion occurs when the picture tube circuitry is out of alignment (see illustration).

Keystoning can be detected by directing the camera toward a square object. The distortion will cause the object to appear trapezoidal—that is, narrower at the top than at the bottom, or vice-versa. *See also* TELEVISION.

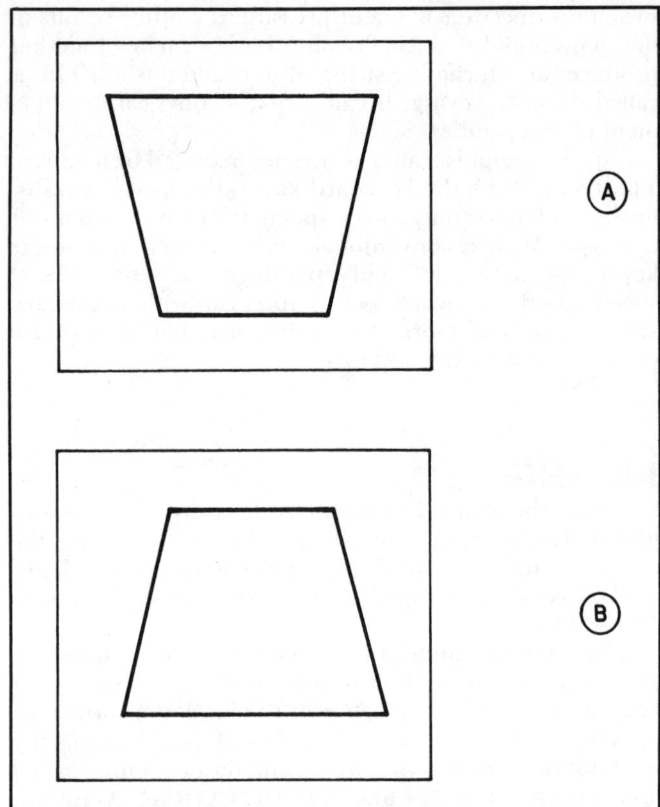

KEYSTONING: At A, the horizontal gain is greater at the top than at the bottom of the picture. At B, the reverse situation is shown. The object should be perfectly square.

KICKBACK

When the current through an inductor is suddenly interrupted, a voltage surge, called kickback, occurs. The polarity of this voltage is opposite to the polarity of the original current. The surge may be quite large; the peak voltage depends on the coil inductance and the original current.

Kickback voltage can, in some instances, reach lethal values. Whenever large inductances carry large amounts of current, the components of the circuit must be handled carefully because of the kickback shock hazard. The electric-fence generator makes use of the kickback phenomenon. So does the spark generator in an automobile.

KILO

Kilo is a prefix multiplier. The addition of "kilo-" before a designator denotes a quantity 1,000 (one thousand) times as large as the designator without the prefix. *See also* KILOGRAM, KILOHERTZ, KILOWATT, KILOWATT HOUR, PREFIX MULTIPLIERS.

KILOBYTE

The kilobyte is a commonly used unit by which information quantity is expressed. A kilobyte is equivalent to 1024, or 2^{10}, bytes.

Most personal computers have a memory capacity of several tens of kilobytes. The term kilobyte is often abbreviated K, so we might speak of a computer having 32K or 64K bytes of memory. *See also* BYTE, MEGABYTE.

KILOCYCLE

See KILOHERTZ.

KILOGRAM

The kilogram is the standard international unit of mass. At the surface of the earth, a mass of 1 kilogram weights 2.205 pounds. The abbreviation for kilogram is kg. A kilogram is 1,000 grams. *See also* INTERNATIONAL SYSTEM OF UNITS.

KILOHERTZ

The kilohertz, abbreviated kHz, is a unit of frequency, equal to 1,000 hertz, or 1,000 cycles per second.

Audio frequencies, and radio frequencies in the very low, low, and medium ranges, are often specified in kilohertz. The human ear can detect sounds at frequencies as high as 16 to 20 kHz. *See also* FREQUENCY.

KILOWATT

The kilowatt, abbreviated kW, is a unit of power, or rate of energy expenditure. A power level of 1 kW represents 1,000 watts. In terms of mechanical power, 1 kW is about 1.34 horsepower.

The kilowatt is a fairly large unit of power. An average incandescent bulb uses only about 0.1 kW. Radio broadcast transmitters produce radio-frequency signals ranging in power from less than 1 kW to over 100 kW. An average residence uses from about 1 kW to 10 kW. *See also* POWER, WATT.

KILOWATT HOUR

The kilowatt hour, abbreviated kWh, is a unit of energy. An energy expenditure of 1 kWh represents an average power dissipation of P kilowatts, drawn for t hours, such that $Pt = 1$.

The kilowatt hour is the standard unit in which electric energy consumption is measured in businesses and residences. An average household consumes about 500 to 1,500 kWh each month. *See also* ENERGY.

KINESCOPE

See PICTURE TUBE.

KINESCOPE RECORDER

A kinescope recorder is a camera that takes motion pictures directly from the screen of a television picture tube or kinescope. An ordinary movie camera will not work as a kinescope recorder, because the rate at which the camera takes the pictures must correspond to the scanning rate of the television signal. An ordinary movie camera, used in place of a kinescope recorder, produces a picture with light and dark bands that appear to move vertically across the picture frame.

A normal television picture has a repetition rate of 60 Hz in the United States. This means that 60 complete frames are sent each second. The kinescope, therefore, must have a frame rate that is an integral fraction of this rate: 60 Hz, 30 Hz, 20 Hz, and so on. For good visual reproduction, a frame rate of at least 20 Hz is necessary.

The same principles that apply to film recording from the television receiver also apply to transmission of motion pictures by television. *See also* PICTURE SIGNAL, TELEVISION.

KINETIC ENERGY

Kinetic energy is a form of energy manifested as the motion of objects or particles. The most common form of kinetic energy is thermal energy, in which atoms and molecules constantly move about and collide with each other. Kinetic energy also takes place in the form of organized motion, such as the movement of an automobile.

Kinetic energy may be measured in a variety of units. The standard unit for quantifying energy is the joule. *See also* ENERGY, JOULE.

KIRCHHOFF, GUSTAV ROBERT

The physicist Gustav Robert Kirchhoff (1824-1887) is best known for his research of the nature of electric conduction. Kirchhoff noticed certain things about electric currents and voltages. His laws (*see* KIRCHHOFF'S LAWS) are taught in all beginning-level electronics courses today.

KIRCHHOFF'S LAWS

The physicist Gustav R. Kirchhoff is credited with two fundamental rules for behavior of currents in networks. These rules are calledd Kirchhoff's First Law and Kirchhoff's Second Law.

According to Kirchhoff's First Law, the total current flowing into any point in a direct-current circuit is the same as the total current flowing out of that point. This is true no matter what the number of branches intersecting at the point (see A).

Kirchhoff's Second Law states that the sum of all the voltage drops arond a circuit is equal to zero (see B). This sum must include the voltage of the generator, and polarity must be taken into account: equal negative and positive voltages total zero.

Using Kirchhoff's Laws, the behavior of networks, from the simplest to the most complex, can be evaluated.

A lesser known of Kirchhoff's discoveries concerns blackbody radiation. From this principle, the Wien Displacement Law for blackbody radiation can be derived. *See also* BLACK BODY, WIEN'S DISPLACEMENT LAW.

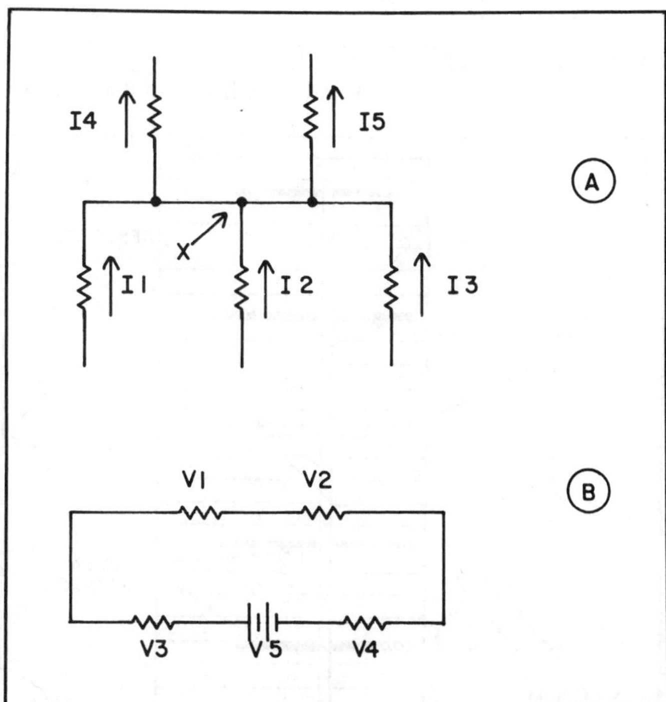

KIRCHHOFF'S LAWS: Example of Kirchhoff's laws. At A, the current I1 + I2 + I3, flowing into point X, is equal to the current I4 + I5 flowing away from point X. At B, the sum of the voltages V1 + V2 + V3 + V4 is equal but opposite in polarity to V5. That is, V1 + V2 + V3 + V4 + V5 = 0.

KLYSTRON

The Klystron is a form of electron tube used for generation and amplification of microwave electromagnetic energy. The Klystron is a linear-beam tube; it incorporates an electron gun, one or more cavities, and apparatus for modulating the beam produced by the electron gun. There are several different types of Klystron tube. The most frequently used are the two-cavity Klystron, the multi-cavity Klystron, and the reflex Klystron.

The illustration at A is a simplified pictorial diagram of the two-cavity Klystron, which is used as an amplifier or oscillator at moderate power levels. In the first cavity, the electron beam is velocity-modulated, resulting in variations in the density of the electron beam in the second cavity. The output is taken from the second cavity. The signal fluctuations are greater in the second cavity than at the input point. This results in amplification.

The multi-cavity Klystron contains one or more inter-

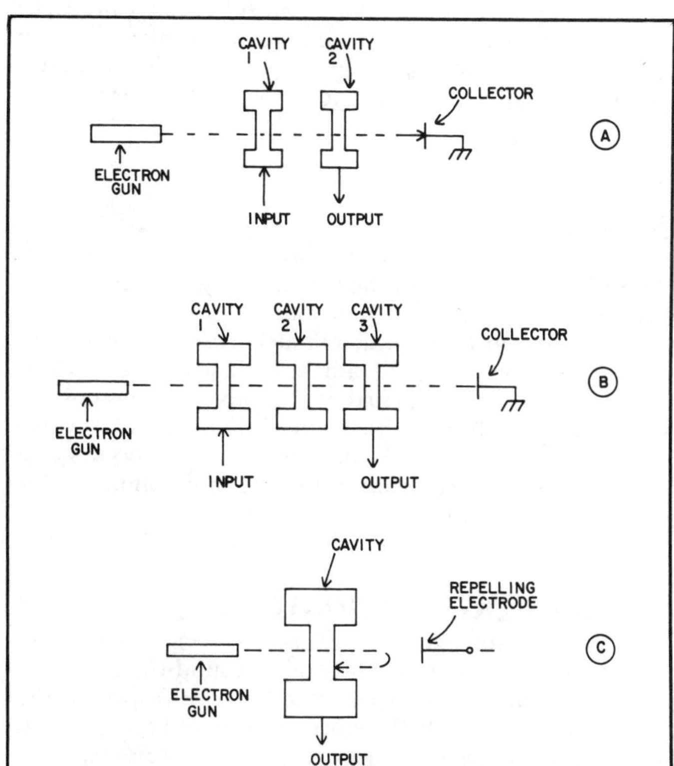

KLYSTRON: Three types of Klystron tube. At A, the two-cavity device; at B, a three-cavity Klystron; at C, the reflex Klystron.

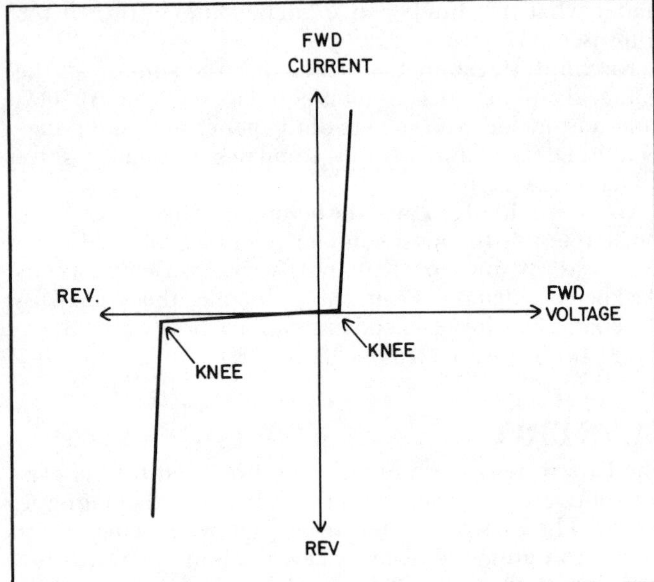

KNEE: Knees in the current-versus-voltage curve for a semiconductor diode.

mediate cavities between the input and output cavities, as at B. The intermediate cavity or cavities produce enhanced modulation of the electron beam, and therefore the amplification factor is greater. Multi-cavity Klystrons are used at high power levels, in some cases more than 1 million watts.

The reflex Klystron (C) contains only one cavity. A retarding field causes the electron beam to reverse direction; this phase reversal allows large amounts of energy to be drawn from the electrons. The reflex Klystron is used as an oscillator at low and medium power levels. It can also be used as a frequency-modulating device. Reflex-Klystron oscillators can produce a few watts of radio-frequency power in the microwave range. *See also* TRAVELING-WAVE TUBE.

KNEE

A sharp bend in a response curve is somtimes called a knee. This is especially true of the current-versus-voltage curve for a semiconductor P-N junction.

As the forward voltage across such a junction increases, the current remains small until the voltage reaches 0.3 V to 0.6 V. At this point, the current abruptly rises; this point is the knee (see illustration). Under conditions of reverse voltage, the response is similar, but the knee occurs at a much greater voltage than in the forward condition. *See also* DIODE, P-N JUNCTION.

KNIFE-EDGE DIFFRACTION

When an electromagnetic wave encounters a barrier with a sharp edge, the energy propagates around the edge to a certain extent. The sharper the edge with respect to the size of the wavelength, the more pronounced this effect. It is called knife-edge diffraction.

In general, the lower the electromagnetic frequency, the less the knife-edge diffraction loss in any given case. For this reason, knife-edge diffraction is common at the very low, low, and medium frequencies; it becomes less prevalent at high and very-high frequencies. In the presence of natural obstacles such as hills and mountains, knife-edge diffraction is not common at ultra-high and microwave frequencies. *See also* DIFFRACTION.

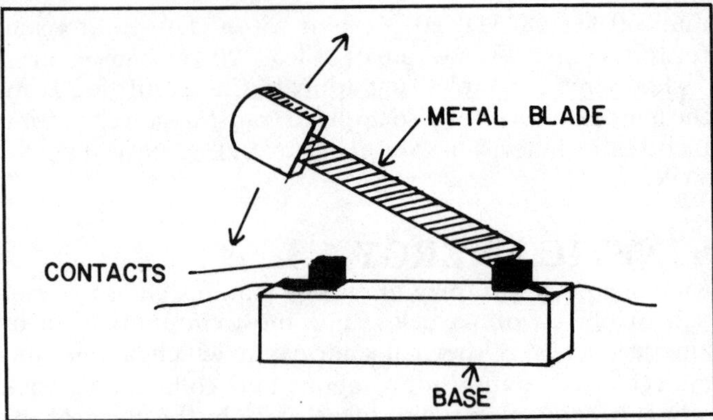

KNIFE SWITCH: A single-pole, single-throw knife switch.

KNIFE SWITCH

A knife switch is a mechanical switch capable of handling very large voltages and currents. The switch gets its name from its resemblance to a knife. A movable blade slides in and out of a pair of contacts (see illustration). Knife switches are found in sizes ranging from about 1 inch to over 1 foot in length. There may be just one blade, or there may be two or more connected in parallel. The switch may have one or two throw positions.

The principle advantage of the knife switch is its ability to withstand extreme voltages without arcing. This makes it valuable for lightning protection, and in high-voltage applications. However, the knife switch usually has exposed contacts, and this presents a shock hazard. *See also* SWITCH.

KOOMAN ANTENNA

A Kooman antenna is a high-gain, undirecitonal antenna

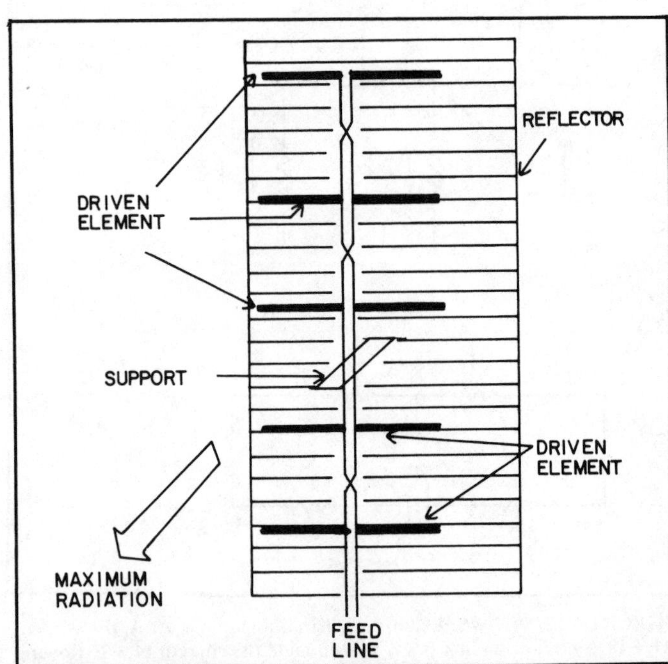

KOOMAN ANTENNA: The Kooman antenna consists of a stacked array of full-wave radiating elements, in conjunction with a reflector.

often used at ultra-high and microwave frequencies. The antenna uses a reflector in conjunction with collinear and broadside characteristics (see illustration).

Several full-wave, center-fed conductors are oriented horizontally and stacked above each other. They are separated by intervals of ½ electrical wavelength (as measured along the phasing line). The phasing line consists of two parallel conductors. The conductors are transposed at each succeeding driven element. Thus, all of the driven elements are fed in the same phase. The reflecting network may consist of a wire mesh or another set of conductors. *See also* BROADSIDE ARRAY, COLLINEAR ANTENNA.

KOZANOWSKI OSCILLATOR

A vacuum-tube oscillator, using positively biased grids and intended for operation at ultra-high and microwave frequencies, is called a Kozanowski oscillator. The device uses two tubes with the grids connected in parallel and the plates and cathodes in a resonant-circuit configuration. The resonance is achieved by means of parallel wires, the lengths of which are tuned to the operating frequency (see illustration).

Certain tubes tend to oscillate when their grids are biased with a positive voltage. This oscillation can be a nuisance in high-frequency amplifier circuits. The Kozanowski oscillator makes constructive use of it. *See also* OSCILLATION, PARASITIC OSCILLATION.

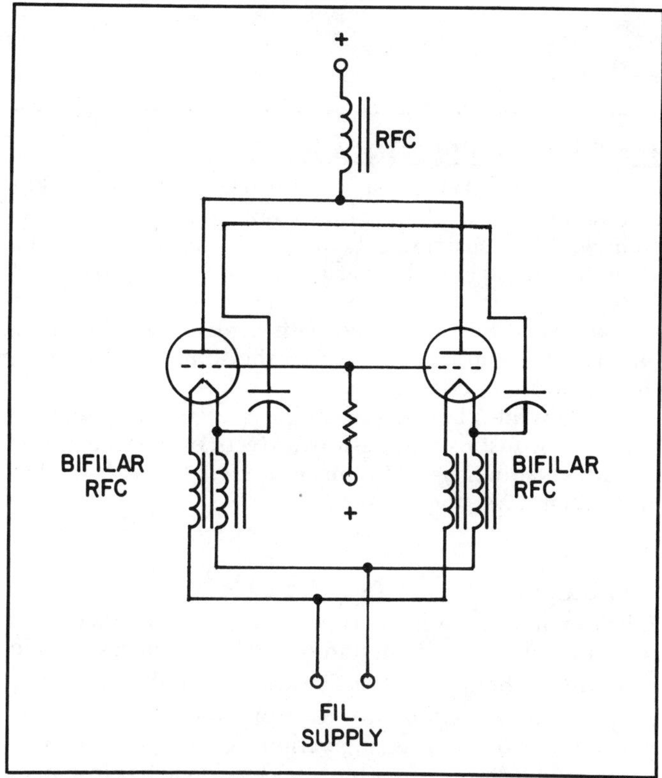

KOZANOWSKI OSCILLATOR: The Kozanowski oscillator uses two tubes, with resonant lines in the cathode/plate circuits.

L

LABYRINTH SPEAKER

A labyrinth speaker is a special form of enclosed speaker containing a network of sound-absorbing chambers. The chambers are positioned behind the speaker cone. The chambers are oriented in various ways, for the purpose of reducing internal resonances. The inside walls of the speaker enclosure are lined with acoustically absorbent material, preventing reflection of the sound waves within the enclosure.

The labyrinth loudspaker displays a flat response, with very little acoustic standing-wave effects, over a wide range of audio frequencies. This makes it ideal for high-fidelity applications. *See also* SPEAKER.

LADDER ATTENUATOR

A ladder attenuator is a network of resistors, connected in such a way that the input and output impedances remain constant as the attenuation is varied. The drawing illustrates a ladder attenuator equipped with a multiple-position switch for selecting various levels of attenuation. Some ladder attenuators have selector switches in banks, facilitating operation at any desired decibel value.

The ladder attenuator is often used in the test laboratory for comparing the amplitudes of different signals when calibrated oscilloscopes or spectrum analyzers are not sufficiently accurate. Ladder attenuators are also used in situations where the test signal is too strong for the input circuit of the test equipment. Ladder attenuators are characterized by negligible reactance at frequencies well into the ultra-high range, provided the resistor leads are very short. *See also* ATTENUATION, ATTENUATOR.

LADDER NETWORK

A ladder network is, in general, any cascaded series of L networks or H networks (*see* H NETWORK, L NETWORK). The L-network sequence is used in unbalanced systems; a series of H networks is used in balanced systems. The ladder attenuator is a network of resistors (*see* LADDER ATTENUATOR). However, capacitors and inductors may also be used in ladder networks.

Ladder networks, consisting of inductances and capacitances, are used as lowpass and highpass filters in radio-frequency applications. By cascading several L-section or H-section filters, greater attenuation is obtained outside the passband, and a sharper cutoff frequency is achieved. Some power-supply filter sections consist of ladder networks, using series inductors and/or resistors, and parallel capacitors. *See also* HIGHPASS FILTER, LOWPASS FILTER.

LAGGING PHASE

When two alternating-current waveforms, having identical frequency, do not precisely coincide, the waveforms are said to be out of phase. The difference in phase can be as great as ½-cycle.

In a display of amplitude versus time, two waveforms with different phase, but identical frequency, might appear as shown in the graph. One wave occurs less than ½-cycle later than the other. The later wave is called the lagging wave. If the waveforms are exactly ½-cycle apart in phase, the two signals are said to be in phase opposition. If they are precisely aligned in phase, they are reinforcing. *See also* ANGLE OF LAG, PHASE ANGLE.

LAMBERT

The lambert is a unit of luminance in the centimeter-gram-second (CGS) system. The lambert is not often specified; the standard international unit is candela per square meter. A luminance of 1 lambert is equal to 3,183 candela per square meter. Conversely, 1 candela per square meter is equivalent to a luminance of 0.000314 lambert. *See also* CANDELA, LUMINANCE.

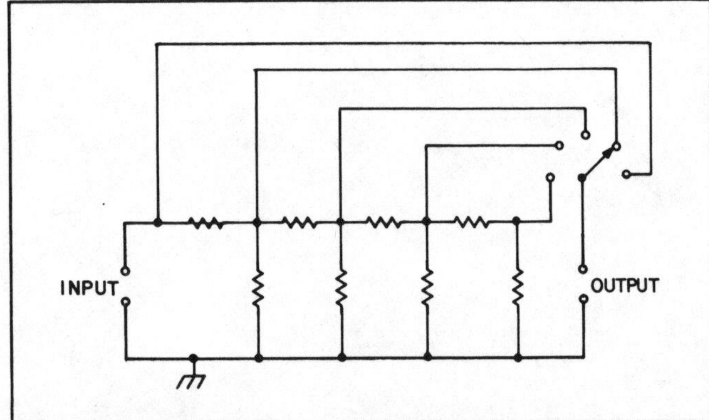

LADDER ATTENUATOR: General representation of five-position ladder attenuator.

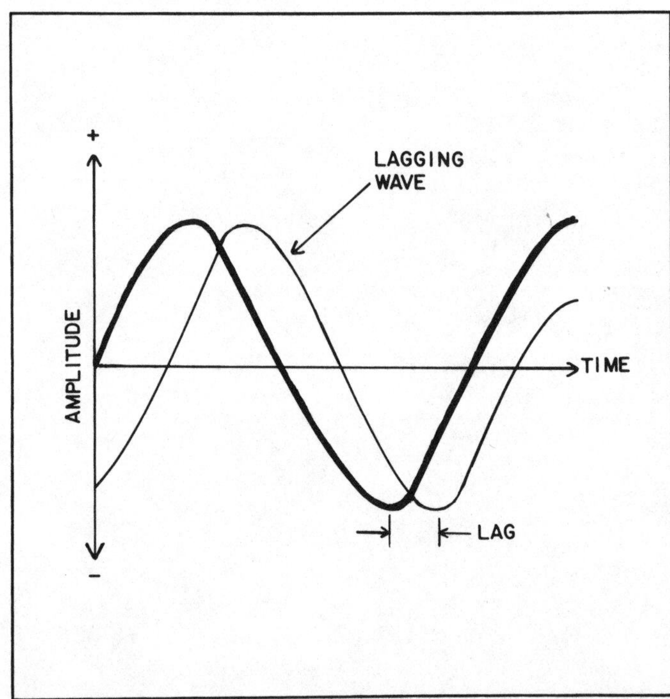

LAGGING PHASE: Two waves may have identical frequencies, but one wave (the lagging wave) occurs later than the other wave.

LAMBERT'S LAW OF ILLUMINATION

The inverse-square law states that the illuminance of a surface is inversely proportional to the square of the distance between the surface and the light source. This is strictly true, however, only if the angle between the surface and the light beam is the same at all times. Lambert's law is a modification of the inverse-square law for visible light, incorporating an additional factor to compensate for possible changes in this angle.

If a light beam falls on a surface with illuminance of P watts per square meter at a 0-degree angle of incidence, and then the surface is deflected by a certain angle ϕ from that position, the illuminance will be multiplied by the cosine of that angle; that is, it will become equal to p cos ϕ. This multiplication factor is always less than 1, and thus the light intensity always decreases as a surface is deflected from the orthogonal position. *See also* ANGLE OF INCIDENCE, INVERSE-SQUARE LAW.

LAMINATED CORE

A laminated core is a form of transformer core, generally used at power-line and audio frequencies. The transformer core, rather than consisting of a solid piece of iron, is made up of several thin cross-sectional sheets, glued together with an insulating adhesive (see illustration).

The laminated core reduces losses caused by eddy currents in the transformer core. If the core is solid iron, circulating currents develop in the material. These currents do not contribute to the operation of the transformer, but simply heat up the core and result in loss of efficiency. The laminated core chokes off these eddy currents (*see* EDDY-CURRENT LOSS). At radio frequencies, laminated cores become too lossy for practical transformer use. For this reason, ferrite and powdered iron are preferred at these frequencies. *See also* FERRITE CORE, POWDERED-IRON CORE.

LAND-MOBILE RADIO SERVICE

A two-way radio station in a traveling land vehicle, such as a car or bus, engaged in non-amateur communication, is called a land-mobile station. Any fixed station, used for the purpose of non-amateur communication with the operator of a land vehicle, is also a land-mobile station.

The land-mobile radio service is allocated a number of frequency bands in the radio spectrum. Many of these bands are in the very-high and ultra-high ranges. The allocations vary somewhat from country to country.

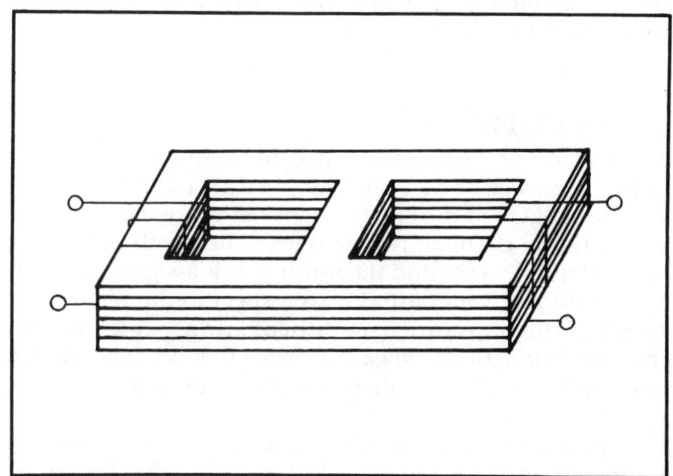

LAMINATED CORE: A laminated transformer core consists of several thin layers of iron, held together by insulating glue.

LANGUAGE

In computer applications, a language is a system of representing data by means of digital bits, or by characters such as letters and numerals.

Languages fall into three basic categories. Machine language is the actual digital information that is processed by the computer. This language appears almost meaningless to the casual observer. The operator of the computer must converse in a higher-order language, such as BASIC, COBOL, or FORTRAN. These languages use actual words and phrases, as well as standard numerals and symbols. The higher-order language is converted into the machine language by the compiler program. This program must be written in an intermediate langauge known as assembly language.

Different computer languages are best suited to different applications. For example, BASIC and FORTRAN are intended mainly for mathematical and scientific applications, while COBOL is used primarily by business. *See also* ASSEMBLER AND ASSEMBLY LANGUAGE, BASIC, COBOL, FORTRAN, HIGHER-ORDER LANGUAGE, MACHINE LANGUAGE.

L ANTENNA

Any antenna that is bent at a right angle can be called an L antenna. The term is used mainly in reference to medium-frequency and high-frequency wire antennas. The antenna can have any orientation. If the L appears upside-down, a common configuration, the radiator is called an inverted-L antenna (*see* INVERTED-L ANTENNA).

The various forms of L antenna are commonly used in locations where space is limited. Such situations often arise for the shortwave listener and amateur-radio operator. The L antenna is characterized by a single bend, at approximately the center of the radiating element. The bend subtends an angle of about 90 degrees.

LARGE-SCALE INTEGRATION

Large-scale integration is the process by which integrated-circuits, containing more than 100 logic elements per chip, are fabricated. Large-scale integration, or LSI, has advanced solid-state electronics to the point where complex systems can be housed in tiny packages.

Both bipolar and metal-oxide-semiconductor (MOS) technology have been adapted to LSI. The electronic wrist watch, single-chip calcualtor, and microcomputer are some of the most recent developments.

There is evidently a physical limit to the amount of miniaturization that is possible in semiconductor technology. The emphasis in future years will probably be directed toward lower cost and improved efficiency and reliability, as well as toward miniaturization. *See also* INTEGRATED CIRCUIT, SOLID-STATE ELECTRONICS.

LASER

The laser, also called the optical maser, is a device that generates coherent electromagnetic radiation in, or near, the visible part of the spectrum. There are several different methods for obtaining coherent light. Coherent light is characterized by a narrow beam and the alignment of all wavefronts in the disturbance (*see* COHERENT LIGHT, COHERENT RADIATION).

Laser action occurs in many different materials, and a complete description of all possible means of constructing a laser is beyond the scope of this volume. The laser process can occur in numerous different ways. Lasers can be described in terms of the following categories: chemical, gas, liquid, metal-vapor, semiconductor, and solid-state. An additional category, consisting of unusual laser systems, can be added to this list.

An early laser was constructed from a ruby rod. For this reason it is called a ruby laser. The illustration is a simplified diagram of the ruby laser. A flashlamp surrounds the ruby rod, which has mirrors at each end. The rear mirror is a total reflector, but the front mirror allows about 4 to 6 percent of the light to pass through (that is, it reflects 94 to 96 percent of the light). A resonant condition occurs within the ruby rod. If the gain of the system is greater than the loss, oscillation occurs, resulting in a coherent-light output at about 700 nm (7×10^{-7}m). This falls in the red part of the visible spectrum.

A common form of commercially available laser is comprised of helium and neon gas. This laser also produces a red visible beam. Some helium-neon lasers are available for as little as 150 to 200 dollars. The laser diode, or injection laser, is a simple type of device that is universally available at a low cost (*see* INJECTION LASER).

Lasers are used in an almost incredible variety of different applications. Low-power lasers can be employed for short-range visible-light communications, since laser light can be modulated, and it suffers far less angular divergence than ordinary light (*see* OPTICAL COMMUNICATIONS). Lasers have been used for such diverse purposes as measurement of distances, measurement of velocities, and medical surgery. High-power lasers can be used for heating and welding, and they can be deadly weapons. *See also* MASER.

LASER DIODE

See INJECTION LASER.

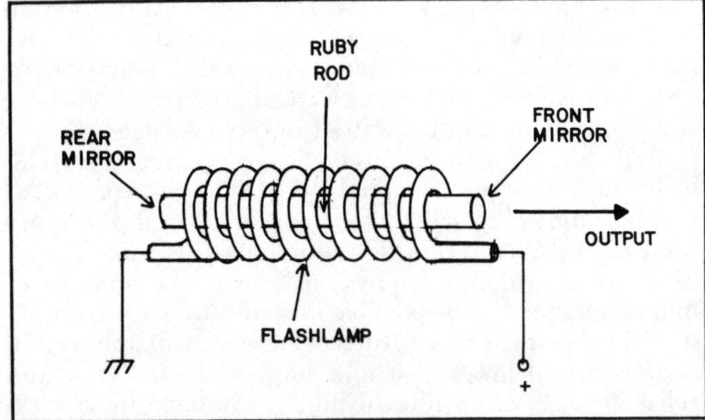

LASER: Simplified diagram of a ruby laser.

LATCH

A latch is a digital circuit intended for the purpose of maintaining a particular condition. A latch consists of a feedback loop. The latch prevents changes from high to low, or vice versa, resulting from external causes.

A simple circuit for storing a logic element is called a latch. A flip-flop can be used for this purpose; a cross-coupled pair of logic gates can also be used. *See also* LOGIC GATE.

LATCHUP

Latchup is an undesirable and abnormal operating condition in which a transistor or logic circuit becomes disabled because of the application of an excessive voltage or current. Latchup results in a circuit malfunction, even though there is nothing physically wrong with any of the circuit components.

In a switching circuit, a transistor will sometimes fall into the avalanche region at the base-collector junction, because of excessive reverse bias, and it will not return to normal until the voltage is entirely removed. This is called P-N-junction latchup or transistor latchup. It sometimes occurs in regulated power supplies.

In a digital circuit, latchup can occur in numerous ways. The application of an excessive or improper voltage, at some point in the circuit, is usually responsible. Microcomputer circuits can sometimes operate improperly because of a stray voltage at some point. In such cases, reinitialization is usually needed to bring the device back to normal (*see* REINITIALIZATION).

In an amplifier circuit, unwanted oscillation is sometimes called latchup because it disables the amplifier. This can occur in improperly shielded audio-frequency or radio-frequency systems. *See also* FEEDBACK, PARASTIC OSCILLATION.

LATENCY

In a digital computer system, the response to a command may be very rapid, but it is not instantaneous. The delay time is called the latent time, and the condition of the computer during this time is called latency. Of course, the latent time should be as short as possible when high operating speed is needed. In a serial storage system, latency is defined as the difference between access time and word time. *See also* ACCESS TIME.

LATITUDE

Latitude is one of two angular coordinates that uniquely define the location of a point on the surface of the earth.

The equator of the earth, or any sphere, is a great circle with latitude defined as 0 degrees. The north pole is assigned latitude +90, and the south pole is assigned latitude −90. Latitude is measured with respect to the plane containing the equator (see illustration). Therefore, intermediate-latitude lines are circles that become smaller and smaller as the latitude gets higher, either positively or negatively.

The latitude line +23.5 is known as the Tropic of Cancer. The latitude line −23.5 is called the Tropic of Capricorn. The arctic and antarctic circles are at +66.5 and −66.5 degrees, respectively.

For a point to be determined unambiguously, another coordinate is needed. That coordinate is, of course, the longitude. *See also* LONGITUDE.

LATITUDE EFFECT

The magnetic field surrounding the earth, known as the geomagnetic field, has an effect on charged atomic particles passing through the upper atmosphere. Charged particles, traveling at high speed through space, have an effective electric current, and consequently they produce magnetic fields. This is true of electrons, protons, alpha particles, and other atomic nuclei. When such particles come near the geomagnetic field, they are diverted toward the north and south magnetic poles by the interacting magnetic forces.

Because of this effect, more charged particles are observed near the poles than near the equator. This is called the latitude effect. The higher the latitude, either north or south, the more cosmic particles are observed at the surface of the earth at any given time. *See also* COSMIC RADIATION, GEOMAGNETIC FIELD.

LATTICE

In crystalline substances, the atoms are arranged in an orderly pattern. This pattern is the same wherever the substance is found. The piezoelectric crystal is an examle of a substance with such a structure, called a lattice structure. The term lattice is used to describe the arrangement of components in certain electronic circuits. *See also* CRYSTAL-LATTICE FILTER, LATTICE FILTER.

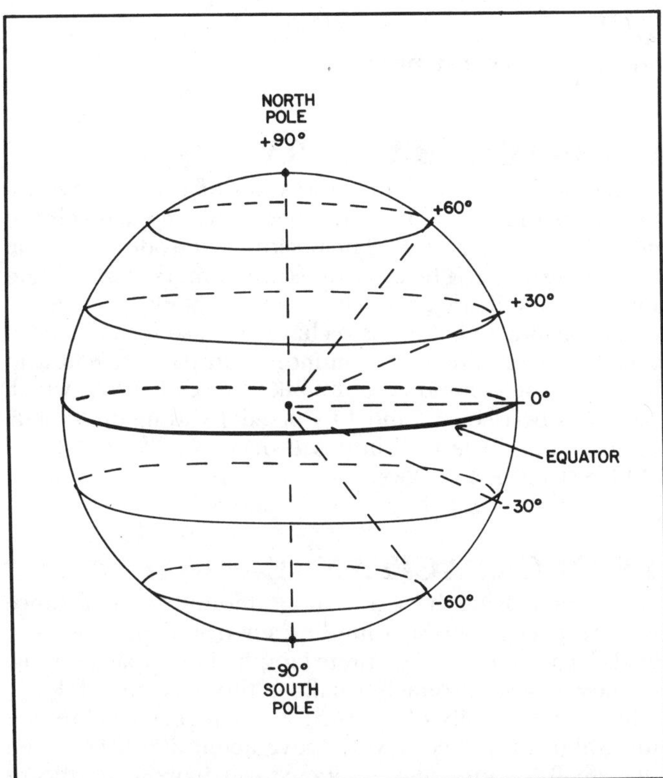

LATITUDE: Latitude circles exist in parallel planes. Their angle designator is determined according to their angular displacement from the equator. Latitudes range from −90 (the south pole) to +90 (the north pole). Here, the latitude circles are shown for −60, −30, 0, +30, and +60 degrees, in addition to the poles.

LATTICE FILTER

A lattice filter is a selective filter in which the components are arranged in a lattice configuration. The illustration at A shows one general arrangement of impedances in a lattice filter. (Variations are possible.) The details of the impedance can vary greatly; in the simplest form it may be simply an inductor or capacitor. However, the impedances in a lattice filter are usually bandpass, or resonant, circuits, such as that shown at B. The impedances may also be piezoelectric crystals.

Lattice filters provide a better bandpass response than is obtainable with a single selective circuit. *See also* BANDPASS RESPONSE, CRYSTAL-LATTICE FILTER.

LAVALIER MICROPHONE

A lavalier microphone is a small, voice-sensitive microphone that is suspended from the user's neck. The microphone hangs near the breastbone and is pointed upward toward the mouth.

The lavalier microphone has an obvious advantage over handheld or boom microphones in situations where the hands must be left free and the user must be able to move around. Lavalier microphones are used in television interviewing. They are also used when it is necessary to keep the microphone hidden; this is the case, for example, for security personnel. Lavalier microphones must be sensitive because they are not normally placed very close to the mouth. Some lavalier microphones have built-in, miniature transmitters, eliminating the need for a long cord. *See also* MICROPHONE, WIRELESS MICROPHONE.

LAW OF AVERAGES

The law of averages is a theorem of probability and statistics. It states that, given a large sampling of events, the probability values obtained by experiment will closely approximate the theoretical values.

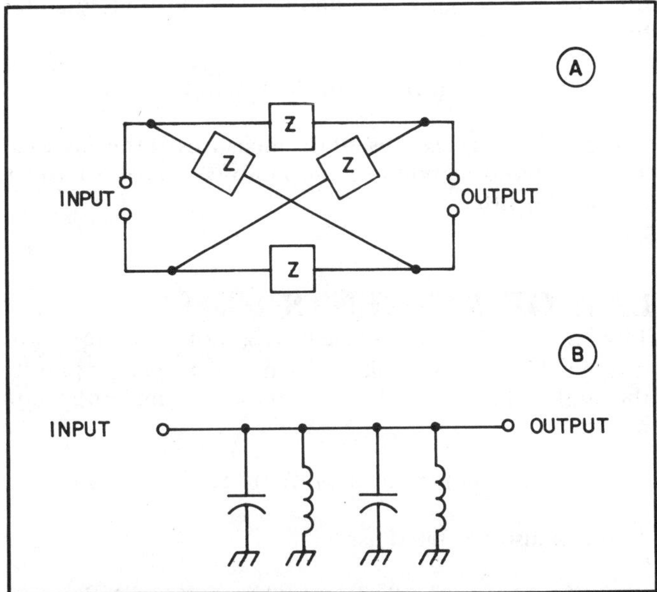

LATTICE FILTER: At A, one possible configuration for a lattice filter, showing the arrangement of impedances. At B, one possible circuit configuration for the bandpass impedances in the lattice filter. (Other arrangements are feasible in both cases.)

An example of the law of averages is illustrated by the repeated tossing of a coin. The probability of getting "heads" on any given toss is 0.5, or 50 percent. If the coin is tossed just eight times, it is quite possible that this theoretical value will not be realized. The coin might show "heads" all eight times, or not at all. However, if the coin is tossed 800 times, the theoretical value will be more closely approached. The greater the number of tosses, the more nearly the experimental and theoretical values will agree. As the number of tosses increases without limit, the proportion of "heads" will approach 0.5.

The law of averages is sometimes misinterpreted. If a coin is tossed "heads" eight times in a row, the probability of "heads" on the ninth toss does not decrease; it is still 0.5, as for any toss. The notion that an event is "due" because of a disproportionate series of past events is, for a random sampling, a false notion. The situation may be different if cause and effect is involved.

LAW OF COSINES

The law of cosines is a rule of trigonometry. Given any triangle ABC (see illustration) having sides of length a, b, and c, then:

$$c^2 = a^2 + b^2 - 2ab \cos C$$

where C is the angle subtended by sides a and b. *See also* COSINE, LAW OF SINES, LAW OF TANGENTS.

LAW OF INVERSE SQUARES

See INVERSE-SQUARE LAW.

LAW OF SINES

The law of sines is a rule of trigonometry. Given any triangle ABC (see next illustration) having sides opposite the angles of lengths a, b, and c, the following equation holds:

$$a/\sin A = b/\sin B = c/\sin C$$

where A, B, and C represent the measures of the angles at the corresponding points. *See also* LAW OF COSINES, LAW OF TANGENTS, SINE.

LAW OF TANGENTS

The law of tangents is a rule of trigonometry. Given any triangle ABC (see next illustration) having sides opposite the angles of lengths a, b, and c, the following proportion holds:

$$(a-b)/(a+b) = \tan((A-B)/2) \,/\, \tan((A+B)/2)$$

This can also be stated as:

$$(a-b) \tan((A+B)/2) = (a+b) \tan((A-B)/2)$$

In both equations, A and B represent the measures of the angles at points A and B in the triangle. *See also* LAW OF COSINES, LAW OF SINES, TANGENT.

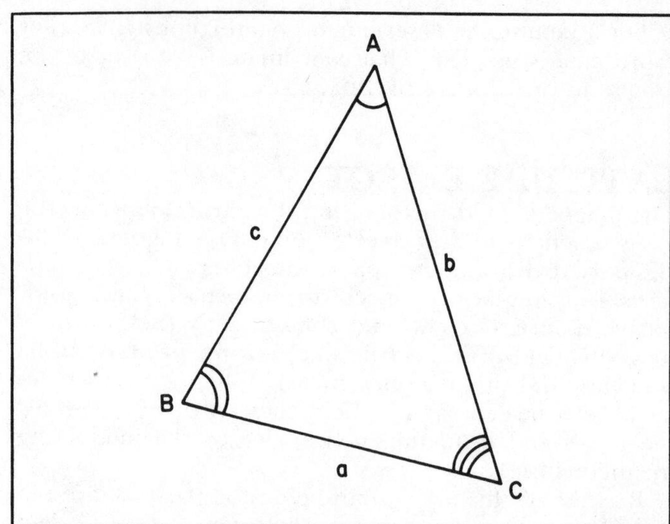

LAW OF COSINES: Triangle for illustration of the laws of cosines, sines, and tangents. The vertices are designated by points A, B, and C. These designators also represent the measures of the angles at these vertices. Sides opposite A, B, and C are denoted by a, b, and c; these letters also stand for the lengths of the three sides.

LC CIRCUIT

An LC circuit is a circuit consisting of a combination of inductances and capacitances. Such circuits may be series or parallel resonant, or they may take the form of an H network, L network, or pi network. The response may be of the bandpass, band-rejection, highpass, or lowpass type. *See also* BANDPASS FILTER, BAND-REJECTION FILTER, HIGHPASS FILTER, LOWPASS FILTER. The illustration on next page shows some examples of LC circuits.

LCD

See LIQUID-CRYSTAL DISPLAY.

LEAD-ACID BATTERY

A lead-acid battery is a combination of cells, forming a storage battery. The negative electrodes in each cell are made from spongy lead. The positive electrodes are made of lead peroxide. The electrodes are immersed in a dilute solution of sulfuric acid, which acts as an electrolyte.

The lead-acid battery is rechargeable, so it can be used over and over. The most common example of a lead-acid battery is the common automobile storage battery, which produces between 12 and 14 volts, depending on its state of charge and the load imposed on it. *See also* BATTERY, CELL, STORAGE BATTERY.

LEAD CAPACITANCE

Component leads have a certain amount of capacitance with respect to the surrounding environment, since the leads have a measurable, finite length. This capacitance is, in most cases, extremely small—a tiny fraction of 1 pF. This is not usually of concern at frequencies below the ultra-high range. However, above about 300 MHz, even this small amount of capacitance can have a significant effect on the performance of a circuit.

In the design of ultra-high-frequency and microwave circuits, component leads should be as short as possible to minimize the capacitance. Lead inductance can also influ-

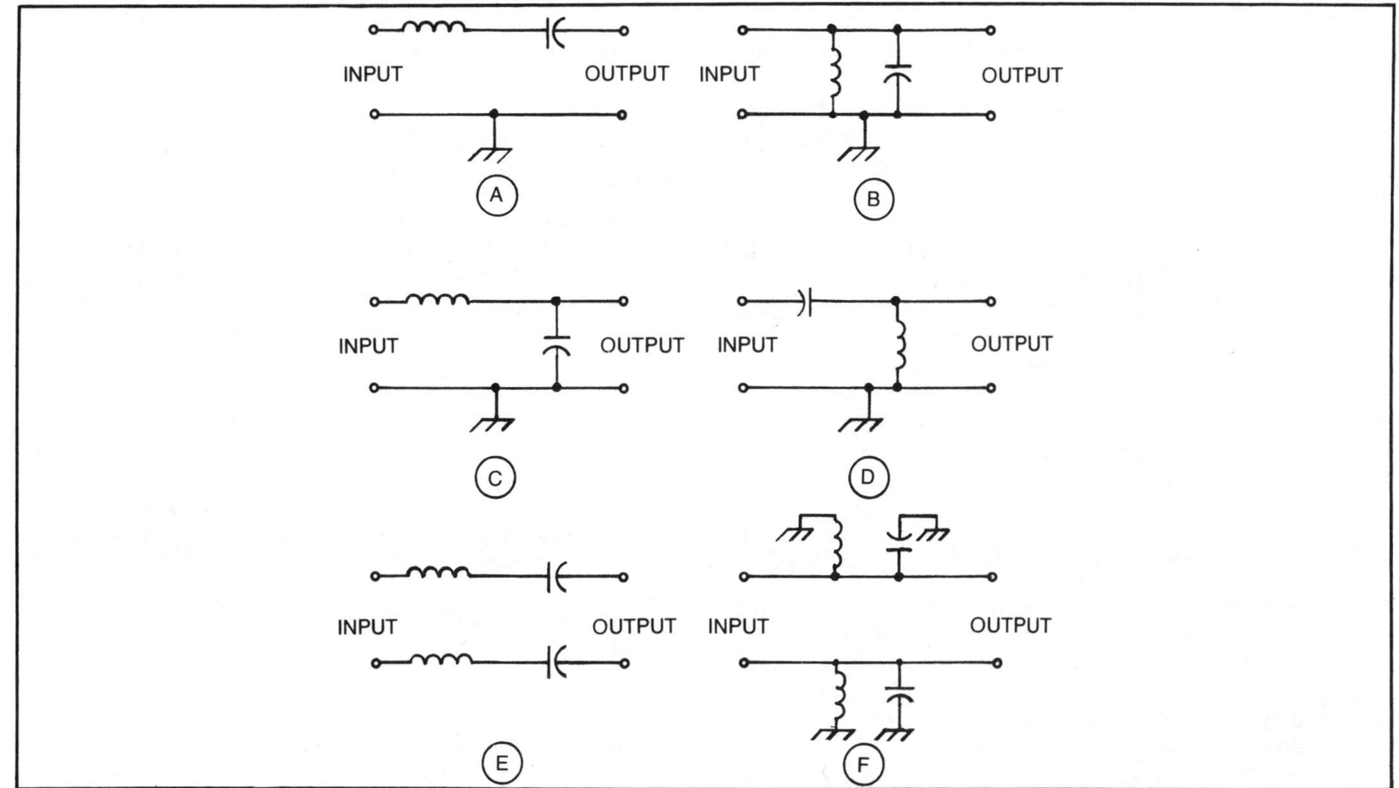

LC CIRCUIT: Examples of inductance-capacitance circuits. At A, unbalanced series-resonant; at B, unbalanced parallel-resonant; at C, unbalanced lowpass L-section; at D, unbalanced highpass L-section; at E, balanced series-resonant; at F, balanced parallel-resonant.

ence circuit operation at these frequencies.

Long leads, such as test leads for voltmeters and oscilloscopes, have much larger capacitance than component leads. Typical coaxial cable has a capacitance of 15 to 30 pF per foot. Parallel-wire leads exhibit somewhat lower lead capacitance than coaxial cable, although it can still be significant at high and very-high frequencies. *See also* LEAD INDUCTANCE.

LEADER

When a large charge difference builds up between a cloud and the earth, or between different parts of a cloud, lightning occurs. The lightning stroke begins with the leader.

The leader consists of a probing flow of electrons, which seek the path of lowest resistance between the two charged clouds. This is a sort of trial-and-error process, and many suboptimal paths are found, resulting in "dead ends." The leader may proceed between two clouds, from a cloud to the ground, or from the ground to a cloud.

Once the leader has nearly reached the other charge pole, a second leader, having positive charge, reaches out toward it. The place where the two leaders meet is called the point of strike, and is usually 30 to 100 feet from the positive charge pole.

Once the leaders have found the path of lowest resistance, the full discharge takes place, in the form of the bright, sometimes terrifying spark that we call lightning. *See also* LIGHTNING.

LEADER TAPE

A leader tape is a section of blank magnetic or paper tape that precedes the portion of the tape containing the recorded data. In a magnetic tape, the leader usually consists of clear plastic, spliced to the main tape. In a paper tape, the leader is generally a blank section of tape, without any punched holes.

The purpose of the leader tape is simply to facilitate loading or winding of the tape for recording or playback. *See also* MAGNETIC TAPE, PAPER TAPE.

LEAD-IN

An antenna feed line is sometimes called a lead-in. Generally, the term lead-in is used to refer to a single-wire feed line in a shortwave receiving antenna. A single-wire line displays a characteristic impedance of between 600 and 1,000 ohms under ordinary circumstances (*see* CHARACTERISTIC IMPEDANCE). Little or no attempt is made to match the characteristic impedance of a single-wire lead-in to the impedance of the antenna itself.

The lead-in, if it consists of a single wire, contributes somewhat to the received signal. However, most of the signal is picked up by the main antenna. If the lead-in consists of a coaxial cable or a parallel-wire line, and the lead-in is properly terminated, the lead-in does not contribute to reception or transmission of signals. *See also* FEED LINE.

LEAD INDUCTANCE

Component leads, since they consist of wire and have measurable length, invariably display a certain amount of inductance. This inductance is normally very small—on the order of a few nanohenrys (billionths of a henry) for the average resistor or capacitor with 1-inch leads.

The lead inductance of electronic components is not generally of great concern at frequencies below the ultra-high range. However, above 300 MHz, the effects can become significant, because at these frequencies the lead lengths become a substantial fraction of a wavelength. At ultra-high and microwave frequencies, therefore, leads must be cut very short to minimize the effects of lead inductance.

Long leads, such as test leads for voltmeters and oscilloscopes, have much larger inductance than component leads. A typical cable or parallel-wire test lead has sufficient inductance to affect circuit operation at very-high frequencies. In conjunction with the lead capacitance, resonance can occur at various frequencies in the very-high or ultra-high range, and this will have an effect on measurements at those frequencies. *See also* LEAD CAPACITANCE.

LEADING PHASE

When two alternating-current waveforms, having the same frequency, do not precisely coincide, the waveforms are said to be out of phase. The difference in phase can be as great as ½-cycle.

In a display of amplitude versus time, two waveforms with different phase, but identical frequency, might appear as in the graph. One wave occurs less than ½-cycle earlier than the other. The earlier wave is called the leading wave. If the waveforms are exactly ½-cycle apart in phase, the two signals are said to be in phase opposition. If they are precisely aligned in phase, they are reinforcing. *See also* ANGLE OF LEAD, PHASE ANGLE.

LEAKAGE CURRENT

In a reverse-biased semiconductor junction, the current flow is ideally, or theoretically, zero. However, some current flows in practice, even when the reverse voltage is well below the avalanche value. This small current is called the leakage current. *See* P-N JUNCTION.

All dielectric materials conduct to a certain extent, although the resistance is high. Even a dry, inert gas allows a tiny current to flow when a potential difference exists between two points. This current is called the dielectric leakage current. *See also* DIELECTRIC CURRENT.

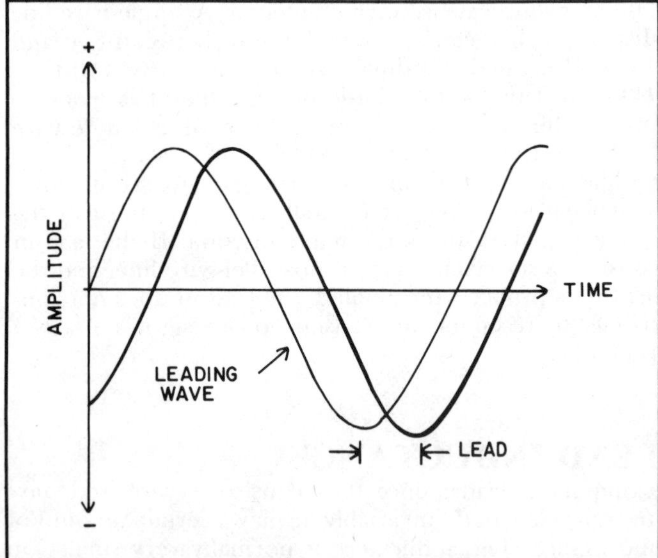

LEADING PHASE: Two waves may have identical frequencies, but one wave (the leading wave) occurs earlier than the other wave.

LEAKAGE FLUX

In a transformer, the coupling between the primary and the secondary windings is the result of a magnetic field that passes through both windings. In some transformers, not all of the magnetic lines of flux pass through both windings. Some of the flux generated by the current in the primary does not pass through the secondary. This is called the leakage flux.

In general, the closer the primary and secondary windings are located to each other, the smaller the leakage flux. An iron-core transformer has a smaller leakage flux than an air-core transformer of the same configuration. The toroidal transformer has the least leakage flux of any transformer. The leakage flux is inversely related to the coefficient of coupling between two inductive windings. *See also* LEAKAGE INDUCTANCE, MAGNETIC FLUX, TRANSFORMER.

LEAKAGE INDUCTANCE

In a transformer, the mutual inductance between the primary and secondary windings is often less than 1 because of the existence of leakage flux (*see* LEAKAGE FLUX). As a result of this, the primary and secondary windings contain inductance that does not contribute to the transformer action. This inductance, known as leakage inductance, is effectively in series with the actual primary and secondary windings.

A transformer can be considered to consist of a perfect component, having primary and secondary windings with no leakage flux; there is also a series self-inductive component in both the primary and secondary circuits (see illustration). These series inductances should be as small as possible, so that the transformer will have negligible losses in the windings. *See also* TRANSFORMER.

LEAKAGE REACTANCE

See LEAKAGE INDUCTANCE.

LEAKAGE RESISTANCE

In theory, a capacitor should not conduct any current when a direct-current voltage is placed across it, except for

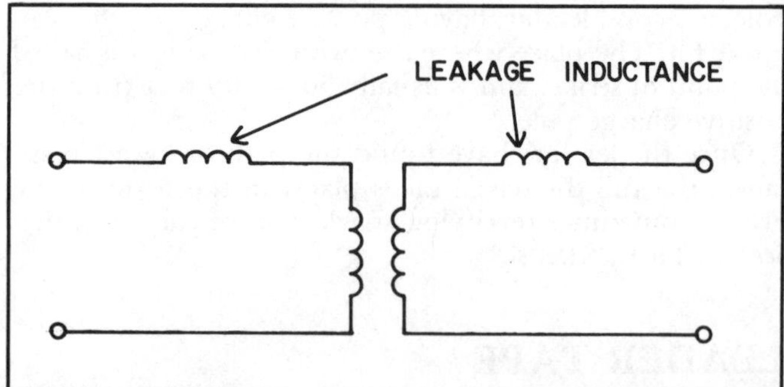

LEAKAGE INDUCTANCE: Leakage inductance appears in series with the windings of a transformer.

the initial charging current. However, once the capacitor has become fully charged, a small current continues to flow. The leakage resistance of the capacitor is defined as the voltage divided by the current.

Air-dielectric capacitors generally have the highest leakage resistance. Ceramic and mylar capacitors also have very high leakage resistance. Electrolytic capacitors have somewhat lower leakage resistance. Typically, capacitors have leakage resistances of billions or trillions of ohms or more. *See also* DIELECTRIC CURRENT, LEAKAGE CURRENT.

LECHER WIRES

Lecher wires are a form of selective circuit used at ultra-high frequencies for the purpose of measuring the frequency of an electromagnetic wave. The assembly consists of a pair of parallel wires or rods, mounted on an insulating framework. A movable bar allows adjustment of the length of the section of parallel conductors (see illustration).

The electromagnetic energy is coupled into the system by means of an inductor at the shorted end of the parallel-conductor section. An indicating device, such as a meter or neon lamp, is inductively coupled to the other end. Both the input and output couplings must be loose, so that the resonant properties of the parallel-conductor system are not affected.

The position of the movable shorting bar is varied until a peak is indicated on the meter or lamp. This shows that the system of wires is resonant at the input frequency, or at some multiple of the input frequency. The shortest resonant length indicates the fundamental frequency of the input signal. A calibrated scale, adjacent to the movable bar, shows the approximate frequency or wavelength of the input signal. The fundamental frequency also corresponds to the distance between adjacent resonant positions of the bar.

LECLANCHE CELL

The Leclanche cell is a special form of dry cell. A carbon electrode is placed in a manganese-dioxide paste. This paste is held in a porous container that allows electric current to flow but prevents the manganese dioxide from escaping. The porous container is surrounded by a solution of a monium chloride. The whole assembly is contained in a zinc enclosure (see illustration). The carbon electrode forms the positive pole of the cell. The zinc electrode forms the negative pole. The cell produces about 1.5 volts direct current.

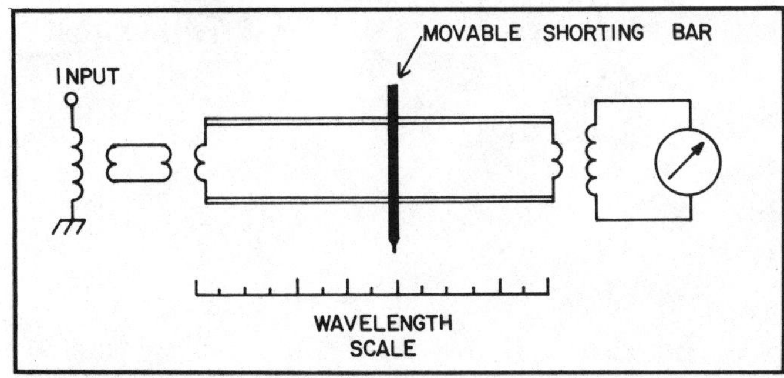

LECHER WIRES: Lecher wires are used for measuring frequency or wavelength. The movable bar is set for resonance according to the indicator.

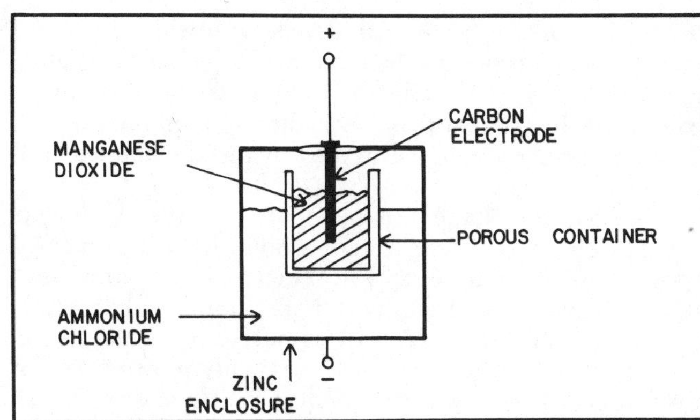

LECLANCHE CELL: The Leclanche cell consists of carbon and zinc electrodes immersed in manganese dioxide and ammonium chloride.

An ordinary dry cell is sometimes called a Leclanche cell. There are, however, some differences between the ordinary dry cell and the true Leclanche cell. *See also* DRY CELL.

LED

See LIGHT-EMITTING DIODE.

LENZ'S LAW

When a current is induced by a changing magnetic field, or by the motion of a conductor across the lines of flux of a magnetic field, that current itself generates a magnetic field. According to Lenz's Law, the induced current causes a magnetic force that acts against the motion.

An example of Lenz's Law can be observed in the operation of an electric generator. When the output of the generator is connected to a load, so that current flows in the generator coils, considerable turning force (torque) is necessary to make the generator work. The lower the load resistance, the greater the current in the coils, and the greater the magnetic force that opposes the coil rotation. This is why a powerful engine or turbine is needed to operate a large generator. *See also* MAGNETIC FIELD.

LEYDEN JAR

A Leyden jar is a form of capacitor. It was first used to demonstrate that electric charges can be stored. The

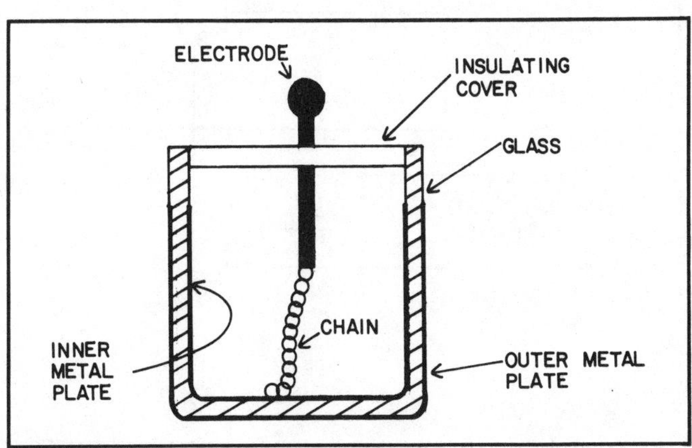

LEYDEN JAR: The Leyden jar is a primitive capacitor capable of storing many thousands of volts.

Leyden jar consists of an ordinary glass jar with aluminum foil or plating on the inside and outside surfaces (see illustration). A rod, generally made of brass or aluminum, extends through the top of the jar, and a chain provides an electrical connection between the rod and the inner metallic surface.

When a high voltage is placed on the rod, the Leyden jar becomes charged. It can retain a potential difference of more than 100,000 volts. The Leyden jar is sometimes used in classroom demonstrations, along with a static generator such as the Van de Graaff device. When the Leyden jar is discharged, the spark may jump a gap of several inches. Home experiments with these devices can be dangerous. *See also* VAN DE GRAAFF GENERATOR.

LIE DETECTOR

See POLYGRAPH.

LIGHT

Light is visible electromagnetic radiation, in the wavelength range of about 750 nanometers (nm) to 390 nm. (A nanometer is equal to a billionth of a meter.) The longest wavelengths appear red to the human eye, and the colors change as the wavelength gets shorter, progressing through orange, yellow, green, blue, indigo, and violet.

The earliest theory of light held that it is a barrage of particles. This is known as the corpuscular theory of light, and it is still accepted today, although we now know that light also has electromagnetic-wave properties. The light particle is called a photon. The shorter the wavelength of light, the more energy is contained in a single photon.

In a vacuum, light travels at a speed of about 186,282 miles per second (299,792 kilometers per second). This is the same speed with which all electromagnetic fields propagate. In some materials, light travels more slowly than this.

Light can be amplitude-modulated and polarization-modulated, for the purpose of line-of-sight communications. This is commonly done in optical-fiber systems (*see* FIBER, FIBER OPTICS, OPTICAL COMMUNICATIONS).

Light can be generated by thermal reactions or by other means, such as ionization and lasing. Ordinary white or colored light is called incoherent light, since the waves arrive in random phase. Laser light is called coherent light, because all of the waves are in phase. *See also* COHERENT LIGHT, ELECTROMAGNETIC THEORY OF LIGHT, INCOHERENT RADIATION, LIGHT INTENSITY, PARTICLE THEORY OF LIGHT, PHOTON.

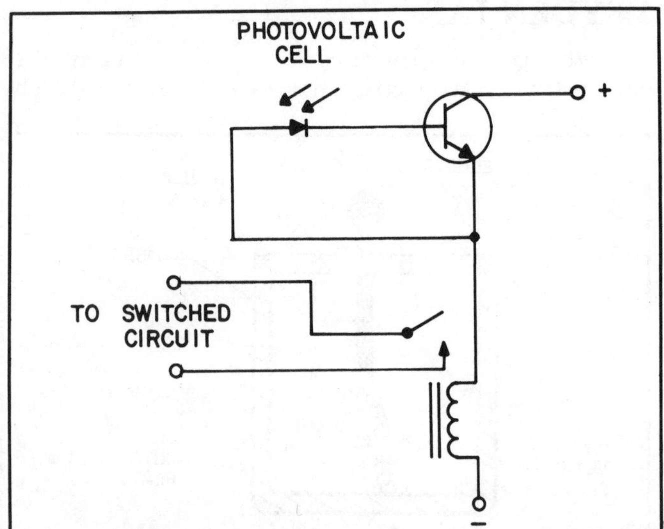

LIGHT-ACTIVATED SWITCH: A light-activated switch using a photovoltaic cell (solar cell), a transistor, and a relay.

LIGHT-ACTIVATED SILICON-CONTROLLED RECTIFIER

A light-activated silicon-controlled rectifier, or LASCR, is a switching device activated by visible light. Impinging light waves perform the same function as the gate current in the ordinary silicon-controlled rectifier (SCR).

Ordinarily, the LASCR does not conduct. However, when the incident light reaches a certain intensity, the device conducts. The amount of light necessary for conduction is determined by the characteristics of the device, and also by the value of an externally applied bias. *See also* SILICON-CONTROLLED RECTIFIER.

LIGHT-ACTIVATED SWITCH

A light-activated switch is any device that opens or closes a circuit in the presence of irradiation by visible light. Such a device may consist of a phototransistor, or a solar cell and switching circuit (see illustration).

Light-actuated switches are used for a variety of purposes. Some types of light-actuated switches close a circuit when light impinges on them; some open a normally closed circuit. An example of the latter type of light-actuated switch is a device often found in homes for the purpose of deterring burglars. When it gets dark, the switch turns on one or more lights. The same sort of light-actuated switch is used by some municipalities to turn on street lights at dusk. *See also* LIGHT-ACTIVATED SILICON-CONTROLLED RECTIFIER, PHOTOTRANSISTOR, PHOTOVOLTAIC CELL

LIGHT-EMITTING DIODE

A light-emitting diode, technically called an electroluminescent diode, is a device that emits infrared, visible light,

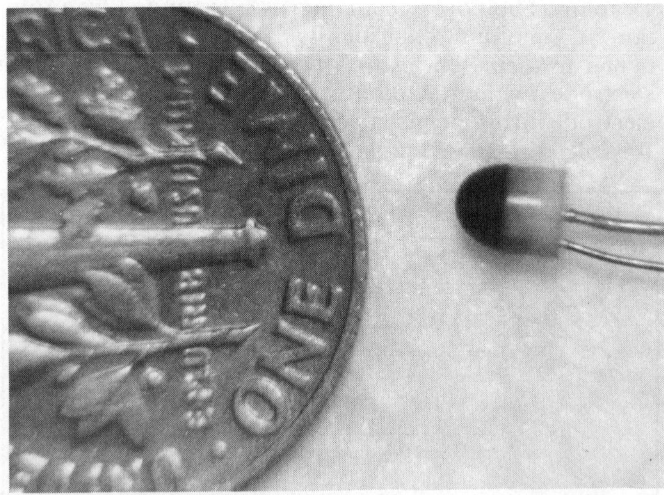

LIGHT-EMITTING DIODE: Light-emitting diodes may be very small, such as the unit shown here, or they may be fairly large, such as in an alphanumeric display.

or ultraviolet radiation when it is supplied with a certain amount of forward voltage. Light-emitting diodes (abbreviated LEDs) are used in many different kinds of electronic circuits, especially indicators and digital displays. Many different colors are available; the most common are red, yellow, and green.

The LED is generally fabricated from a direct band-gap semiconductor material such as gallium arsenide (GaAs). Many different kinds of commercially manufactured LEDs are available. The typical LED is mounted inside an epoxy material that allows maximum transmission of light at the emission frequency. Sometimes a lens-type enclosure is used to enhance the radiation or give it desired directional characteristics. The photograph shows a typical light-emitting diode, next to a dime for size comparison.

An LED with a flat and uniform junction, capable of emitting coherent light, is the injection laser. Such a device requires a certain minimum forward current in order to produce coherent-light output. These devices are sometimes called laser diodes or laser LEDs. *See also* INJECTION LASER.

LIGHT INTENSITY

There are numerous different units for measuring light intensity . The most common unit is the candela, formerly called the candlepower (*see* CANDELA). This is a unit of radiant-light energy, equivalent to 1 lumen per steradian. A steradian is a unit of 3-dimensional angular measure (*see* STERADIAN).

Surface luminance, or the intensity of light radiated by a large surface, is usually specified in candela per square meter. Sometimes the lambert is used (*see* LAMBERT, LUMINANCE). Surface luminance is given for large radiant surfaces such as the sun or a photoluminescent device.

Illuminance, or the amount of light energy falling on a surface, is usually measured in lumens per square meter. The illuminance of a surface depends on the intensity of the light source, the distance of the surface from the source, and the angle that the surface subtends with respect to the rays from the light source (*see* ILLUMINANCE, LUMEN).

LIGHT METER

A light meter is a device for measuring the relative intensity of radiant or ambient light. Light meters are commonly used with photographic apparatus, either as built-in devices with cameras, or as external units. Sometimes the light meter is called an exposure meter or illumination meter. The most common form of light meter employs a light-sensitive semiconductor device, such as a phototransistor, photodiode, or photovoltaic cell, in conjunction with a milliammeter or microammeter. *See also* ILLUMINATION METER.

LIGHTNING

Lightning is a massive electrical discharge that often occurs in rain storms. Lightning may also take place in the charged clouds of dust storms, volcanoes, and other disturbances.

The dangers of lightning are generally underestimated by the public. In an average year, lightning and tornadoes can be expected to kill about the same number of people. In the United States, lightning strikes are most frequent in central Florida. This region has been called the lightning belt. Lightning is also common in the mountains of the southwest.

Thunderstorm lightning can be manifested as a movement of electrons from a cloud to another cloud, from a cloud to the ground, from a cloud into free air, or from the ground to a cloud. The most common kind of lightning is the cloud-to-cloud stroke.

A lightning discharge begins as a hesitant, tentative movement of electrons away from the center of negative charge. This discharge, which moves in jumps or steps, is called the leader (*see* LEADER). The leader "searches" for the path of least resistance between the negative and positive concentrations of charge. In the process, many "dead ends" are found, but within a few seconds the path is complete. When the leader approaches the positive charge pole, a second leader, originating at the center of positive charge, meets it. The two leaders intersect at a point approximately 30 to 100 feet from the positive charge center. This point is called the point of strike.

When the leaders have established an ionized path between charge poles, a substantial electron flow occurs, very rapidly, from the negative charge center to the positive charge center. This constitutes the visible flash (see photograph). The magnitude of the discharge current is usually over 100,000 amperes. There may be several strokes in rapid sequence. The air within the discharge path is heated to extreme temperatures and expands, producing thunder.

LIGHTNING: Lightning is actually a gigantic spark, spanning distances up to several miles. (Courtesy of U.S. Department of Commerce, NOAA.)

If lightning strikes a dwelling that is not properly protected, fire may result as the discharge produces dangerous sparks. When lightning strikes a tree, the trunk may explode or catch on fire. If lightning hits a power line, the resulting current surge can damage or destroy transformers and household appliances. Radio communications antennas are vulnerable to lightning; a nearby strike can damage transmitters and receivers connected to an antenna, and a direct hit often results in substantial destruction unless proper protection is provided.

Lightning seems to occur most often around the edges of thundershowers, and less frequently in the center of a storm cell. However, lightning may strike at a considerable distance from the cloud. Sometimes, the positive charge center at the top of a thundercloud causes a gigantic ground-to-cloud discharge, carrying millions of amperes and spanning a distance of several miles. This is known as a superbolt. *See also* LIGHTNING ARRESTOR, LIGHTNING PROTECTION, LIGHTNING ROD.

LIGHTNING ARRESTOR

A lightning arrestor is a device that allows discharge of high voltages on an antenna feed line. Lightning arrestors generally consist of one or more spark gaps, adjusted so that a voltage in excess of a certain value will be routed to ground. The illustration at A shows a cutaway view of a coaxial lightning arrestor. A lightning arrestor for balanced line can be home-constructed as shown at B.

The gap is adjusted so that the maximum transmitter power does not produce a discharge at any operating frequency. This is a trial-and-error process. If the gap is located at a current node on a transmission line having a high standing-wave ratio, the voltage may be sufficient to cause arcing even if the contacts are fairly far apart. The closest possible spacing should be used.

Lightning arrestors provide some reduction in the chances of a direct lightning strike. This is because the charge on an antenna is neutralized, via arcing to ground, before it can build up to large values. However, a lightning arrestor does not guarantee that a direct hit will not occur. Radio equipment should not be used during thunderstorms unless it is absolutely necessary. The safest precaution is to disconnect the feed line from the radio equipment, and connect the line to a well-grounded point, whenever the radio equipment is not in use. *See also* LIGHTNING, LIGHTNING PROTECTION.

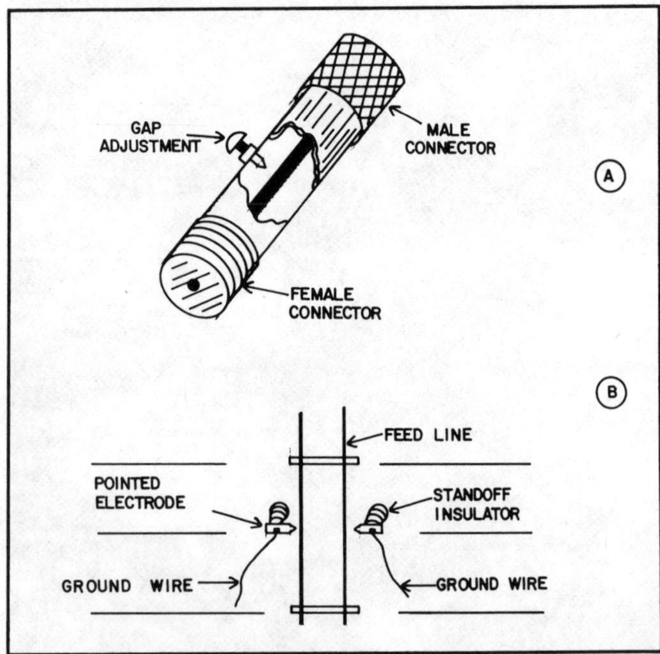

LIGHTNING ARRESTOR: At A, a cutaway view of a coaxial lightning arrestor. At B, a homemade lightning arrestor for balanced line. The standoff insulators are mounted on the wall or siding of the building.

LIGHTNING PROTECTION

Lightning is more dangerous than many people realize. Dozens of people are killed or injured by lightning every year. Most of these people are outside when they are struck, although lightning can sometimes hit a person who is indoors.

During a thunderstorm, electrical appliances should not be used. This is especially true of radio or television equipment connected to outdoor antennas. The telephone should not be used. The shower and bathtub should be avoided. Unnecessary major appliances, such as air conditioners, should be switched off.

If you are outdoors during a thunderstorm, try to get inside. If that is impossible, try to find a ditch or ravine or other low-lying area. If a substantial charge begins to accumulate near you, your hair may actually stand on end. This indicates that you may be struck. In such instances you should squat with your feet close together and your head down. This will minimize the potential difference among different parts of your body.

Houses and other buildings can be protected against lightning by means of lightning rods and grounded metal roofing. Power lines are generally protected by a grounded wire running above, and parallel with, the electric wires. Trees can be protected against damage by installation of lightning rods in the top branches. For radio equipment, the best precaution is to disconnect and ground the antenna feed line, and to unplug the equipment itself. Lightning arrestors can reduce damage from nearby lightning strikes, and prevent dangerous charge buildup on an antenna, but such devices are of little help in the event of a direct hit. *See also* LIGHTNING, LIGHTNING ARRESTOR, LIGHTNING ROD.

LIGHTNING ROD

A lightning rod is a grounded metal electrode, placed on the roof of a building for the purpose of lightning protec-

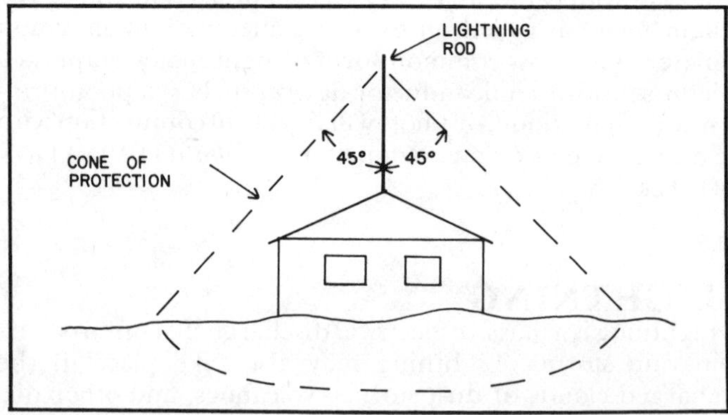

LIGHTNING ROD: A lightning rod provides a cone of protection with an apex angle of approximately 45 degrees.

tion. Lightning rods are commercially available. They may also be home-constructed.

Generally, a lightning rod gives protection within a cone-shaped region under the rod. The apex of the cone is located at the top of the rod. The apex angle of the cone, with respect to the vertical axis of the rod, is about 45 degrees (see illustration). If the ground is level, a lightning rod can therefore be expected to provide safety within a circular region whose radius is equal to the height of the rod.

According to Benjamin Franklin, who is given partial credit for inventing the lightning rod, the tip of the rod should be tapered to a sharp point. This is, in fact, the way most lightning rods are built today. King George III of England believed that lightning rods should have blunt or rounded tips. Recently, some evidence has been gathered to suggest that the blunt tip is more effective than the pointed tip. Discharges occur more readily from pointed objects than from blunt objects. Lightning is therefore more likely to strike a blunt rod than a pointed one. There is some disagreement in the scientific community as to which kind of rod is better.

A lightning rod must be properly grounded. Heavy wire should be used, preferably AWG No. 4 or larger. The ground rod should be driven at least 8 feet into the ground, and it should be located several feet from the foundation of the building. Multiple rods may be used. The ground wire should take the shortest possible path from the base of the lightning rod to the ground. The electrical bonds between the lightning rod and the wire, and between the wire and the ground rod, must be as substantial as possible. The ground rod must be located high enough so that the cone of protection covers the whole building. *See also* LIGHTNING, LIGHTNING PROTECTION.

LIGHT PEN

A light pen is a device used with video-display units for the purpose of creating graphic images. The light pen makes it possible to literally draw pictures on a television screen.

The image on a television screen is created by an electron beam that scans rapidly from left to right, in horizontal lines beginning at the top of the screen and moving downward. It is similar to the path your eyes follow as you read a book. When the light-sensitive tip of the light pen is placed over a certain spot on the screen, a pulse is produced as the electron beam scans past that point. This pulse is fed to a microcomputer, which causes the spot to change state. A letter X may, for example, be converted to a blank (space) or to a solid square. The light pen may also be used to produce certain functions or effects, as in a video game.

As the tip of the light pen is moved around on the screen, successive character blocks are changed from one state to the other, and/or the designated function is applied at the points scanned. The resolution is limited by the number of character blocks in the entire screen. *See also* GRAPHICS, VIDEO GAME.

LIMIT

A limit is a defined value for a mathematical function or relation. Limits are used in the theoretical definitions for derivatives and integrals in calculus. They are also used in many other mathematical situations.

For a limit to be defined for a function $y = f(x)$, a point x_0 must first be chosen for the independent variable x. The function may or may not be defined at this point. However, two limits are possible. As the value x approaches the point x_0 from the left or negative side, we write:

$$\lim_{x \to x_0^-} f(x) = y_1$$

and as x approaches x_0 from the right or positive side, we say that:

$$\lim_{x \to x_0^+} f(x) = y_2$$

The values y_1 and y_2 may both be undefined; one may be defined but not the other; they may both be defined but unequal; they may both be defined and equal. Sometimes we see the expression:

$$\lim_{x \to +\infty} f(x)$$

which denotes the limit of f(x) as x increases without bound ("approaches positive infinity"). We may also sometimes encounter the expression:

$$\lim_{x \to -\infty} f(x)$$

which denotes the limit of f(x) as x decreases without bound ("approaches negative infinity"). *See also* FUNCTION, RELATION.

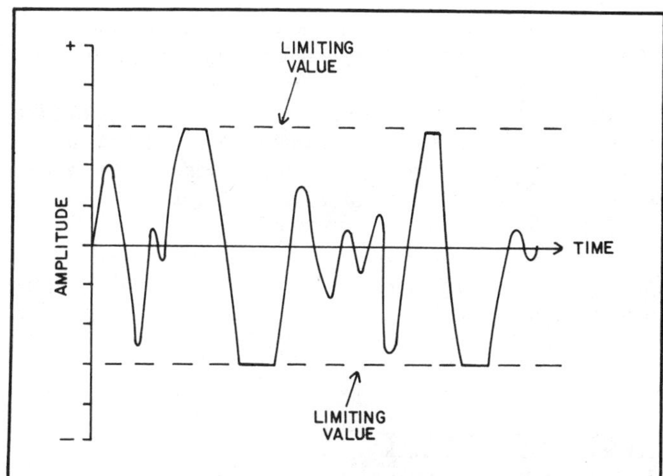

LIMITER: A limiter prevents a signal from exceeding a certain amplitude. In this drawing, the solid line indicates the output signal after limiting.

LIMITER

A limiter is a device that prevents a signal voltage from exceeding a certain peak value. When the peak signal voltage is less than the limiting value, the limiter has no effect. But if the peak input-signal voltage exceeds the limiting value, the limiter clips off the tops of the waveform at the peak-limiting value (see illustration).

Limiters are used in frequency-modulation receivers for the purpose of reducing the response to variations in signal amplitude. The limiting threshold is set very low, so that even a rather weak signal will exceed the limiting voltage. Then, changes in amplitude are eliminated. This is why frequency-modulation receivers are less susceptible than amplitude-modulation receivers to impulse noise and atmospheric static. The limiter stage is placed immediately befor the discriminator stage. *See also* DISCRIMINATOR, FREQUENCY MODULATION.

In low-frequency, medium-frequency, and high-frequency communications receivers, audio-peak limiters are sometimes used to improve the signal-to-noise ratio under adverse conditions. *See also* NOISE LIMITER.

LIMIT SWITCH

A limit switch is a device that opens or closes a circuit when a certain parameter, such as current, power, voltage, or illumination, reaches a specified value, either from below or above. A common circuit breaker is a form of limit switch that is actuated by excessive current in a circuit. A voice-operated relay is another form of limit switch. There are many other examples.

Limit switches may be electromechanical or electrical. The circuit breaker and voice-actuated relay are electromechanical devices. Solid-state limit switches can be fabricated using diodes, transistors, and silicon-controlled rectifiers. Some limit switches are actuated when the parameter exceeds a certain value. Others are actuated when the parameter falls below a specified cutoff level.

LINE

A line is a wire, or set of wires, over which currents or electromagnetic fields are propagated. A power line, telephone line, or radio transmission line may each be simply called a line. *See also* FEED LINE, POWER LINE, TRANSMISSION LINE.

Electric and magnetic fields are defined in terms of flux lines. The lines of flux represent the theoretical direction of the field. Each line represents a certain quantity of electric or magnetic flux. *See* ELECTRIC FLUX, FLUX, MAGNETIC FLUX.

LINEAR ALGEBRA

Linear algebra is a branch of mathematics dealing with the solving of sets of multivariable linear functions (*see* LINEAR FUNCTION). Linear algebra is widely used in engineering applications. One of the most significant applications is the optimization of a procedure involving numerous factors that vary in a linear manner. This is called linear programming.

A set of linear functions can only be solved uniquely when the number of equations is equal to, or larger than, the number of variables in each function. Sometimes a set of linear equations has no solutions; sometimes it has more than one solution.

Sets of linear equations can be solved by any of three methods: substitution, addition, and by using matrices. The matrix method is used by computers to solve sets of hundreds, or even thousands, of linear equations in hundreds or thousands of variables. *See also* MATRIX ALGEBRA.

LINEAR AMPLIFIER

A linear amplifier is a circuit that provides current gain, power gain, or voltage gain, such that the instantaneous output amplitude is a constant multiple of the instantaneous input amplitude. The linear amplifier ideally produces no distortion in the waveform or envelope of a signal.

All high-fidelity audio amplifiers are linear amplifiers. In radio-frequency work, linear amplifiers may produce waveform distortion but not envelope distortion. An amplitude-modulated or single-sideband signal must be amplified with a linear circuit. A linear radio-frequency amplifier may be operated in Class A, Class AB, or Class B. However, a Class-C circuit cannot be used as a linear amplifier. *See also* CLASS-A AMPLIFIER, CLASS-AB AMPLIFIER, CLASS-B AMPLIFIER, ENVELOPE.

LINEAR EQUATION

See LINEAR FUNCTION.

LINEAR FUNCTION

A linear function is a mathematical function of one or more variables such that no variable is raised to a power. That is, there are no exponents in the expression. A one-variable linear function can be written in the form:

$$f(x) = mx + b$$

where m and b are real-number constants. The value m is called the slope of the function (*see* SLOPE), and b is the point at which the graph of the function crosses the dependent-variable axis, or ordinate (see illustration).

The linear function gets its name from the fact that, graphically, the representation is always free from curvature in the Cartesian coordinate system. A one-variable linear function appears as a straight line. If the linear function has two independent variables, the graph appears as a flat plane in space.

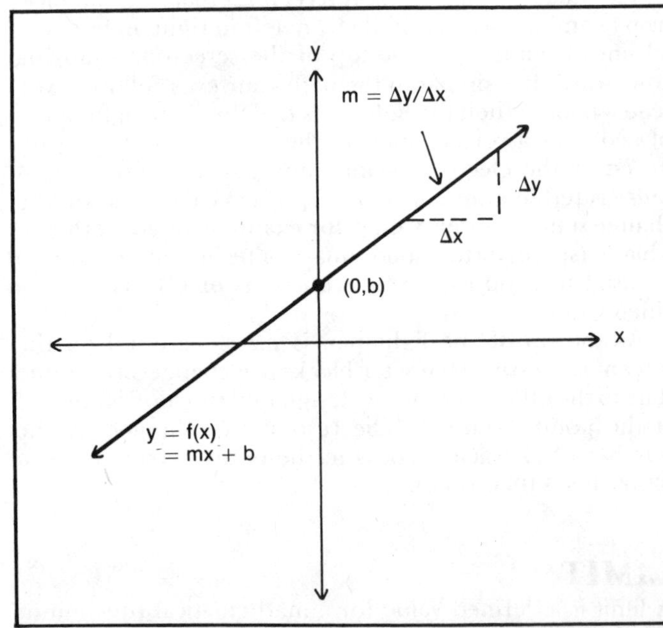

LINEAR FUNCTION: A graph of the line y = mx + b, where m is the slope and b is the y-intercept value.

In electronic applications, a linear function is identical to a linear response (*see* LINEAR RESPONSE). A linear electrical function is a special form of linear mathematical function. If y represents the output current or voltage and x represents the input current or voltage, then a linear electrical function takes the form

$$y = kx,$$

where k is a real-number constant. *See also* FUNCTION, NONLINEAR FUNCTION.

LINEAR INTEGRATED CIRCUIT

A linear integrated circuit is a solid-state analog device. Linear integrated circuits are characterized by a theoretically infinite number of possible operating states; in contrast, the digital integrated circuit usually has just two possible states. Linear integrated circuits are used as amplifiers, oscillators, and regulators.

Within a certain input range, the input and output voltages of the linear integrated circuit are directly proportional to each other. If the instantaneous output is plotted on the Cartesian plane as a function of the instantaneous input, the graph appears as a straight line. *See also* DIGITAL INTEGRATED CIRCUIT, INTEGRATED CIRCUIT, LINEAR AMPLIFIER.

LINEARITY

Linearity is an expression of the resemblance between the input and output signals of a circuit. In general, the better the linearity, the less distortion is generated in a device.

In a strict mathematical sense, linearity is the condition in which the instantaneous input and output signal amplitudes are related by a constant factor. In audio-frequency applications, the input and output waveforms must be such that all points are related by this constant. In radio-frequency work, a circuit may be considered linear as long as the input and output envelope amplitudes are related by a constant. The actual radio-frequency waveform may be distorted. *See also* LINEAR AMPLIFIER.

LINEAR POLARIZATION

An electromagnetic wave is said to have linear polarization when the electric field maintains a constant orientation. Most radio-frequency signals have linear polarization as they leave an antenna. Any antenna consisting of a single, fixed radiator, with or without parasitic elements, produces a signal with linear polarization. A linearly polarized wave may be oriented horizontally, diagonally, or vertically.

When a radio signal is returned to the earth from the ionosphere, the direction of the electric-field lines may no longer be constant, because of phasing effects in the ionospheric layers. The polarization of a radio signal may be deliberately made to rotate as it is emitted from the antenna. This is called elliptical or circular polarization. *See also* CIRCULAR POLARIZATION, ELLIPTICAL POLARIZATION, HORIZONTAL POLARIZATION, VERTICAL POLARIZATION.

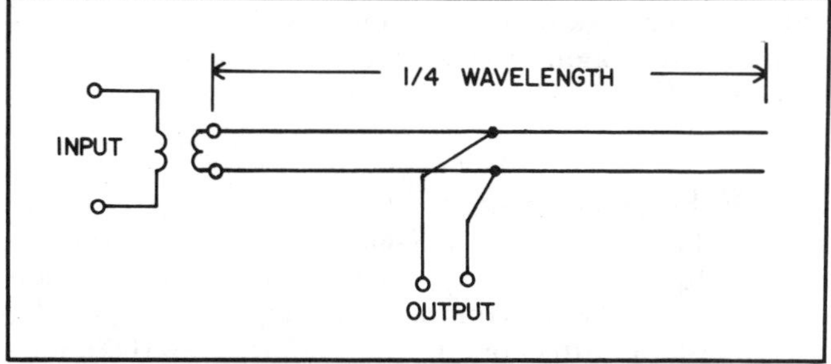

LINEAR TRANSFORMER: A simple linear transformer consists of a quarter-wave section of transmission line.

LINEAR RESPONSE

A transducer is said to have a linear response if the input and output current or voltage are related in a certain way. Specifically, the response is linear if the graph of the output versus input appears as a straight line passing through the origin point (0,0).

A linear response can be defined in algebraic terms. If an input current or voltage of x_1 results in an output current or voltage of y_1, and if an input current or voltage of x_2 results in an output current or voltage of y_2, then the response is linear if and only if $y_1 = kx_1$, $y_2 = kx_2$ and $y_1 + y_2 = k(x_1 + x_2)$.

Aa transducer may exhibit a linear response over a certain range of input current or voltage. Often though, a transducer is linear only within a certain range; outside of that range it becomes nonlinear. *See also* LINEARITY.

LINEAR TAPER

Some potentiometers have a resistance that varies in linear proportion to the rotation. Such potentiometers are said to have a linear taper. If a linear-taper potentiometer is rotated through a certain angle in any part of its range, the change in resistance is always the same.

Linear-taper potentiometers are used in many electronic circuits for alignment or adjustment purposes. For volume control, however, the linear-taper potentiometer is not generally used, because people perceive sound in a logarithmic manner. For volume control, audio-taper potentiometers are preferred. *See also* AUDIO TAPER, POTENTIOMETER.

LINEAR TRANSFORMER

A linear transformer is a radio-frequency transformer, sometimes used at very-high frequencies and above for the purpose of obtaining an impedance match.

The linear transformer consists of a quarter-wave section of transmission line, short-circuited at one end and open at the other end (see illustration). The line may be of the parallel-conductor type or the coaxial type. Power is applied via a small link at the short-circuited end of the line. The output is taken from some point along the length of the transmission line.

The output impedance is near zero at the short-circuited end of the linear transformer, and is extremely large (theoretically infinite) at the open end. At intermediate points, the impedance is a pure resistance whose value depends on the distance from the short-circuited end.

The linear transformer is used, in various configurations, in many kinds of antenna systems for radio-frequency reception and transmission. *See also* IMPEDANCE MATCHING.

LINE BALANCE

A parallel-conductor transmission line must be balanced in order to function properly. Line balance is achieved when the currents in the two conductors are equal in magnitude, but opposite in direction, at every point along the line.

In an antenna system, line balance may be difficult to obtain because of interaction between the line and the radiating part of the antenna. In a center-fed dipole antenna, the open-wire or twin-lead feed line should be oriented at a right angle to the radiating element. The two halves of the radiating element must be exactly the same electrical length, and they must present the same impedance at the feed point. In an antenna system, poor line balance results in radiation from the feed line. *See also* BALANCED TRANSMISSION LINE.

LINE FAULT

A line fault is an open or short circuit in a transmission line, resulting in partial or complete loss of signal or power at the output end of the line.

An open circuit in a line can be located by shorting the terminating, or output, end of the line, and measuring the circuit continuity at various points with an ohmmeter or other device (see A in illustration). Measurements should be taken first at the input end of the line; subsequent checks should be made at points closer and closer to the output end. The resistance will appear infinite, or very large, until the testing apparatus is moved past the line fault. Then the resistance will abruptly drop.

A short circuit in a line can be found without breaking the circuit if a high-frequency signal is applied at the input, as shown at B. An indicating device, which may consist of a meter with a radio-frequency diode in series, is moved back and forth along the line. The meter produces readings that fluctuate, depending on the distance from the input, until the short is passed. Beyond the short, the meter reading drops to nearly zero and remains there as the meter is moved further.

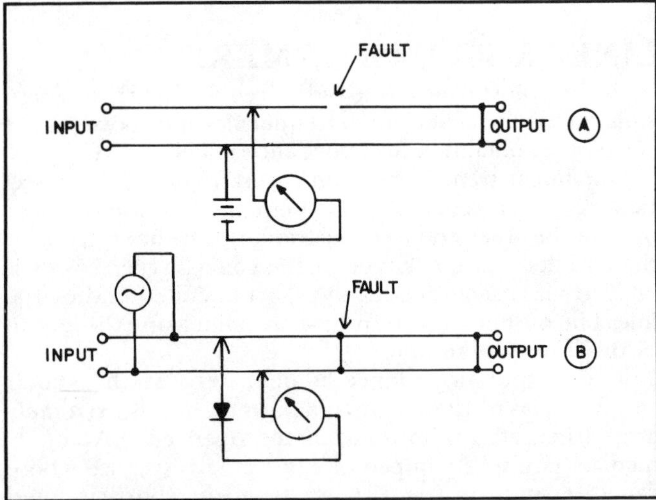

LINE FAULT: At A, a method of checking for an open-circuit line fault. At B, a technique for finding a short-circuit line fault.

LINE FILTER

A line filter is a device that may be placed in the alternating-current power-supply cord of an electronic device. Line filters generally consist of series inductors and/or parallel capacitors (see illustration).

A line filter is useful for eliminating transient spikes in utility-line voltage. Such spikes may cause an electronic device to malfunction; in some cases a spike can result in permanent damage. Line filters are also helpful in reducing electromagnetic noise from alternating-current power lines. *See also* POWER-LINE NOISE, TRANSIENT.

LINE LOSS

Line loss is the dissipation of power in, or radiation of power from, a transmission line. Given a line-input power of P1 watts, and an output power of P2 watts, the loss power is the difference P1-P2.

Line loss can result from any or all of several different factors. The conductors of any transmission line have some resistance, known as ohmic loss. The dielectric material between the conductors of a transmission line produces some loss. Conductor losses and dielectric losses are manifested in the form of heat. If the line balance is poor, power is lost by radiation from the line.

Line loss is generally specified in decibels per unit length. For a particular transmission line, the loss usually increases as the frequency increases. The decibel value, or proportion, of line loss does not depend on the power level of the applied signal, as long as the power does not exceed the maximum rating for the line. In a radio-frequency feed line, the loss increases as the standing-wave ratio increases. *See also* DECIBEL, STANDING-WAVE-RATIO LOSS.

LINE-OF-SIGHT COMMUNICATION

Radio communication by means of the direct wave is sometimes called line-of-sight communication. The range of line-of-sight communication depends on the height of the transmitting and receiving antennas above the ground, and on the nature of the terrain between the two antennas.

Line-of-sight communication is the primary mode at microwave frequencies. While the range is obviously limited in this mode, propagation is virtually unaffected by external parameters, such as ionospheric or tropospheric disturbances. Line-of-sight communication range is limited to the radio horizon. *See also* DIRECT WAVE, HORIZON.

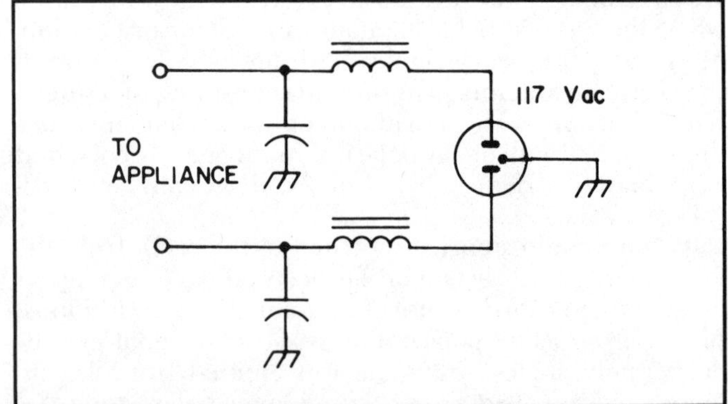

LINE FILTER: A typical alternating-current utility-line filter.

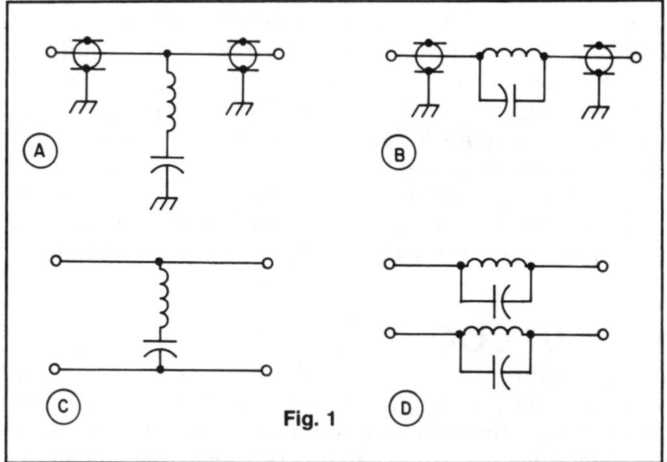

LINE TRAP: Line traps. At A, series-resonant unbalanced; at B, parallel-resonant unbalanced; at C, series-resonant balanced; at D, parallel-resonant balanced. The traps have a band-rejection response.

LINE-OF-SIGHT WAVE

See DIRECT WAVE.

LINE PRINTER

A line printer is a device that prints the output of a computer on paper, line by line. The normal line lengths are 80 and 132 characters.

Line printers are similar to teletype receiving devices. Such printers may operate on the dot-matrix principle, or they may use ball or daisy-wheel devices. The dot-matrix printer is fastest. *See also* DAISY-WHEEL PRINTER, DOT-MATRIX PRINTER, PRINTER.

LINE REGULATION

See VOLTAGE REGULATION.

LINES OF FLUX

Electric or magnetic fields are sometimes theoretically described as consisting of lines of flux. The lines of flux in a field run in the general direction of the field effect. The field is considered to originate at one end of each line of force, and to terminate at the other end of the same line. The lines of flux converge at the electric charge centers or magnetic poles.

Lines of flux do not physically exist. They simply represent a certain quantity of electric or magnetic field flux. *See also* ELECTRIC FLUX, FLUX, MAGNETIC FLUX.

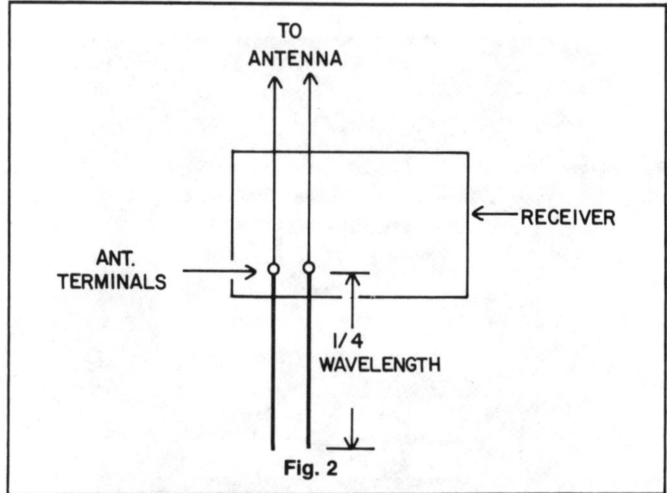

LINE TRAP: A quarter-wave, balanced radio-frequency line trap, fabricated from a transmission line.

LINE TRAP

Any band-rejection filter, employed in a transmission line for the purpose of notching out signals at a certain frequency or frequencies, is called a line trap. Line traps may be installed in various different ways. Figure 1 shows series-resonant and parallel-resonant line traps, installed in coaxial and parallel-wire feed lines.

A line trap may be used to suppress the harmonic output of a radio transmitter. The trap is simply tuned to the harmonic frequency. In a receiving antenna system, a line trap is sometimes used to reduce the level of a strong local signal that would otherwise overload the front end of the receiver.

A quarter-wavelength section of transmission line, either coaxial or parallel-wire, can be used as a series-resonant trap. One end of the quarter-wavelength section is connected to the receiver or transmitter antenna terminals along with the feed line, and the other end is simply left open (see Fig. 2). This kind of trap is sometimes called a line trap, since it is actually constructed from transmission line. At the resonant frequency, this device appears as a short circuit across the input terminals. *See also* TRAP.

LINE TUNING

Line tuning is a method of tank-circuit construction sometimes used at ultra-high and microwave frequencies. A pair of parallel conductors, or a length of coaxial line, is used as a parallel-resonant circuit. This is accomplished by short circuiting one end of a quarter-wavelength section of line (see illustration). The other end of the line then attains the characteristics of a parallel-resonant, inductance-capacitance circuit.

The resonant line may be tuned by means of a movable shorting bar, or the frequency may be fixed. A tuned line has better frequency stability than an inductor-capacitor arrangement at ultra-high and microwave frequencies. However, the quarter-wave tuned line is resonant at odd harmonic frequencies as well as at the fundamental frequency.

A quarter-wavelength section of line may be left open at the far end, resulting in the equivalent of a series-resonant inductance-capacitance circuit. Half-wavelength sections

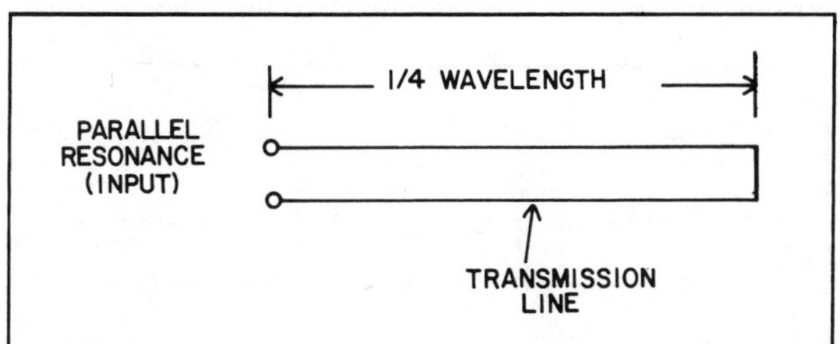

LINE TUNING: Line tuning is accomplished by means of a resonant length of transmission line.

of line are also sometimes used for tuning purposes. A shorted half-wavelength line acts as a series-resonant circuit; if the far end is open, the half-wavelength line acts as a parallel-resonant tank.

A tuned line may be used to eliminate undesired signals in an antenna system. Such a device is called a line trap. *See also* LINE TRAP.

LINK

A link is a connection, via wire, radio, light beams, or other medium, between two circuits within a larger system. A radio or television broadcasting station may use a link between the studio and the transmitter. A computer may be linked to other computers for the purpose of transferring data or to obtain more memory. In general, any one-way or two-way communications path may be called a link.

When two circuits are coupled loosely by means of a pair of inductive transformers, the arrangement is called a link, or link coupling. *See also* LINK COUPLING.

LINK COUPLING

Link coupling is a means of inductive coupling. The output of one circuit is connected to the primary winding of a step-down transformer. The secondary winding of the transformer is directly connected to the primary winding of a step-up transformer. The secondary of the step-up transformer provides the input signal for the next stage or circuit. The illustration is a schematic diagram illustrating link coupling.

The smaller, intermediate windings in a link-coupling arrangement usually have just one or two turns. As a result, there is very little capacitance between the two stages. This minimizes the transfer of unwanted harmonic energy. Link coupling also minimizes loading effects in the output of the first stage.

Test instruments are sometimes connected to a circuit by means of link coupling. Because the capacitance is so small, the presence of the test instrument does not significantly affect the circuit under test. This is important in radio-frequency work, especially at the very-high frequencies and above.

Link coupling can be used for the purpose of generating feedback in an amplifier system. The feedback may be positive, resulting in oscillation; negative feedback can be employed for the purpose of neutralization. *See also* LINK FEEDBACK.

LINK FEEDBACK

Link feedback is the use of link coupling between the output and input of an amplifier circuit. The collector, drain, or plate transformer winding is inductively coupled, by means of a link, to the base, gate, or grid winding.

Link feedback can be either positive (regenerative) or negative (degenerative). Positive feedback results in oscillation if the coupling is tight enough. Positive link feedback can be used to make a regenerative receiver. Negative link feedback is sometimes used in radio-frequency power amplifiers for the purpose of neutralization. *See also* LINK COUPLING, NEUTRALIZATION, REGENERATIVE DETECTOR.

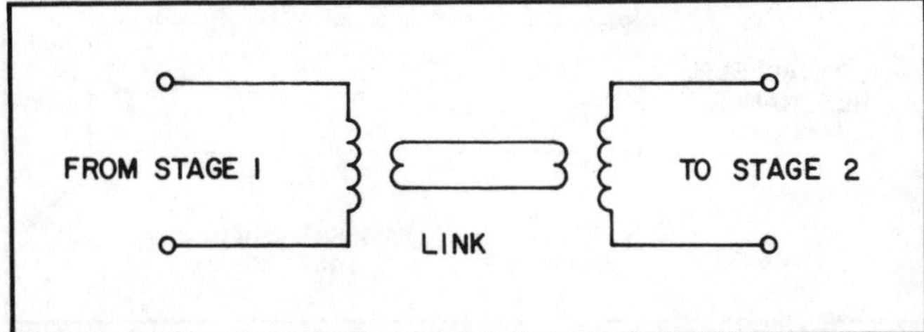

LINK COUPLING: Link coupling provides minimal capacitive interaction between radio-frequency stages.

LIQUID COOLING

Liquid cooling is a method of removing excess heat from an active device such as a transistor, field-effect transistor, or tube. A confined liquid is pumped around the outside of the device. Heat is removed from the amplifying device by means of conduction. After passing around the active component, the liquid is forced through a radiator, where the liquid is cooled for recirculation.

Liquid cooling is sometimes used in high-power broadcast transmitters. At low and moderate power levels, liquid cooling is uncommon. *See also* AIR COOLING, CONDUCTION COOLING, CONVECTION COOLING, COOLING, FORCED-AIR COOLING.

LIQUID-CRYSTAL DISPLAY

A liquid-crystal display, abbreviated LCD, is a form of alphanumeric or digital display. The LCD has become increasingly common in recent years because of its extremely low current requirements. Liquid-crystal displays are employed in such devices as electronic clocks and watches, calculators, and radio test equipment.

Most liquid-crystal displays consist of one or more seven-segment groups of bars. Other configurations may also be included for specialized purposes, such as indication of operating mode. The photograph shows a handheld radio transceiver with an LCD frequency/function readout.

A typical LCD is constructed as shown in the drawing. Each segment is connected to one contact on the edge of the board. The front and back plates are treated so that the light passing through them is polarized. When the liquid-crystal material changes state, its polarization changes. A dark region is produced when the polarization of the

LIQUID-CRYSTAL DISPLAY: A typical liquid-crystal display, used in a handheld portable transceiver.

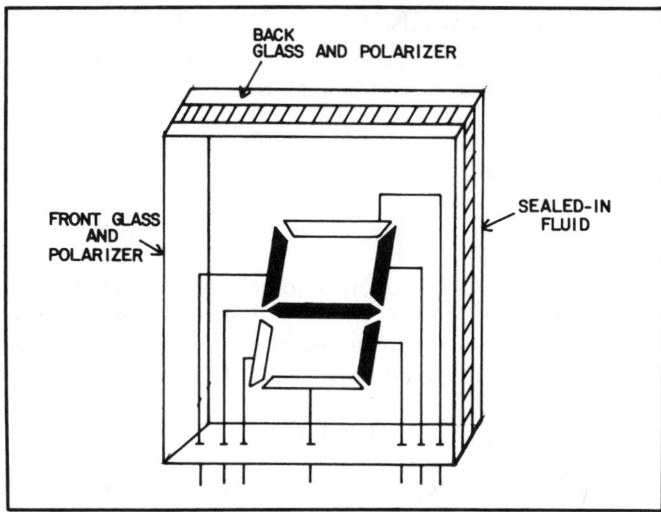

LIQUID-CRYSTAL DISPLAY: Pictorial drawing of a liquid-crystal seven-segment character.

liquid-crystal material occurs at a right angle to the polarization of the viewing glass.

Liquid-crystal displays are available in many different sizes. Some LCDs have a backlighting arrangement, so they can be easily read in complete darkness. Other LCDs rely on an external source of light. *See also* NEMATIC CRYSTAL.

LISSAJOUS FIGURE

A Lissajous figure is a pattern that appears on the screen of an oscilloscope when the horizontal and vertical signal frequencies are integral multiples of some base frequency.

When the horizontal and vertical signals are equal in amplitude and frequency and differ in phase by 90 degrees, a circle appears on the display. When the vertical signal has twice the frequency of the horizontal signal, a sideways figure-8 pattern is observed. If the vertical signal has half the frequency of the horizontal signal, an upright figure-8 pattern is traced. Different frequency combinations produce different patterns. The shapes of the ellipses vary depending on relative amplitude and/or phase angle. Some common Lissajous figures are shown in the illustration.

Lissajous figures are useful for precise adjustment of variable-frequency oscillators against a reference oscillator with a known audio frequency. Using an oscilloscope and a reference oscillator, Lissajous figures simplify such procedures as the adjustment of Private-Line® (PL) oscillators and Touchtone® encoders. A fixed pattern indicates a harmonic zero-beat condition. *See also* OSCILLOSCOPE, ZERO BEAT.

LITER

The liter, abbreviated by the lower-case letter l, is the standard metric unit of volume. A liter is the volume occupied by 1 kilogram of pure water at a temperature of 4 degrees Celsius. A cube measuring 10 centimeters on each edge occupies a volume of exactly 1 liter. A volume of 1 cubic centimeter represents 0.001 liter; this unit is called the milliliter (ml).

A volume of 1 liter is approximately 33.7 fluid ounces, or a little more than a liquid quart.

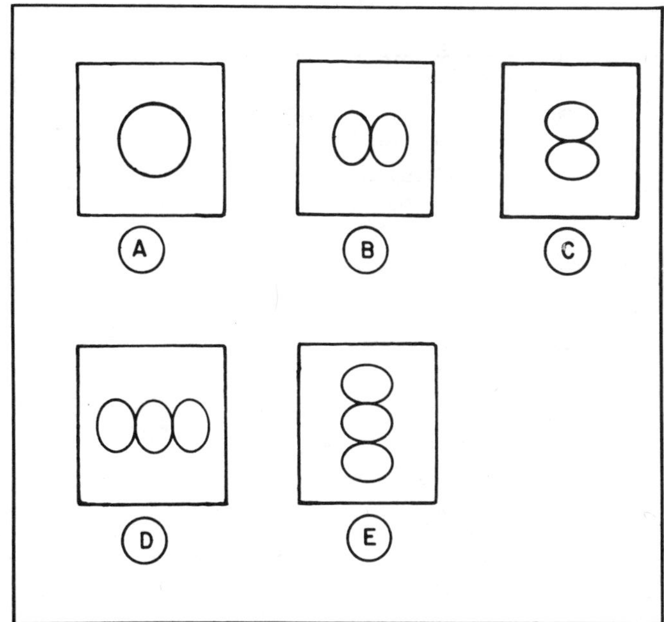

LISSAJOUS FIGURE: Simple Lissajous figures, for horizontal frequencies h and vertical frequencies v. At A, v = h; at B, v = 2h; at C, v = h/2; at D, v = 3h; at E, v = h/3.

LITHIUM CELL

A lithium cell is an electrochemical cell, in which the positive electrode is fabricated from lithium. The lithium cell produces approximately 2.8 volts.

Lithium cells are noted for their excellent shelf life, and are ideally suited for low-current applications over long periods of time. Lithium cells are becoming more and more common in electronic clocks and watches, and as memory-backup power supplies for microcomputer devices. *See also* CELL.

LITZ WIRE

At radio frequencies, current tends to flow mostly near the outside surface of a conductor. This is called skin effect (*see* SKIN EFFECT). For direct current and low-frequency alternating current, the conductivity of a wire is proportional to the cross-sectional area, or the square of the wire diameter. However, at high frequencies, the conductivity is directly proportional to the diameter of the wire. At high frequencies, therefore, it makes better sense to maximize the conductor surface area, rather than to simply maximize the cross-sectional area of the wire. Litz wire is designed with this principle in mind.

Litz wire consists of several individual, enameled conductors, interwoven in a special way. The wire is fabricated so that all inner strands come to the outside at regular intervals, and all outer strands come to the center at equal intervals. Litz wire is basically a stranded wire in which the conductors are insulated from each other.

Litz wire exhibits low losses at radio frequencies, since the conducting surface area is much greater than that of an ordinary solid wire of the same diameter.

L NETWORK

An L network is a form of filter section that is used in unbalanced circuits. The L network gets its name from the

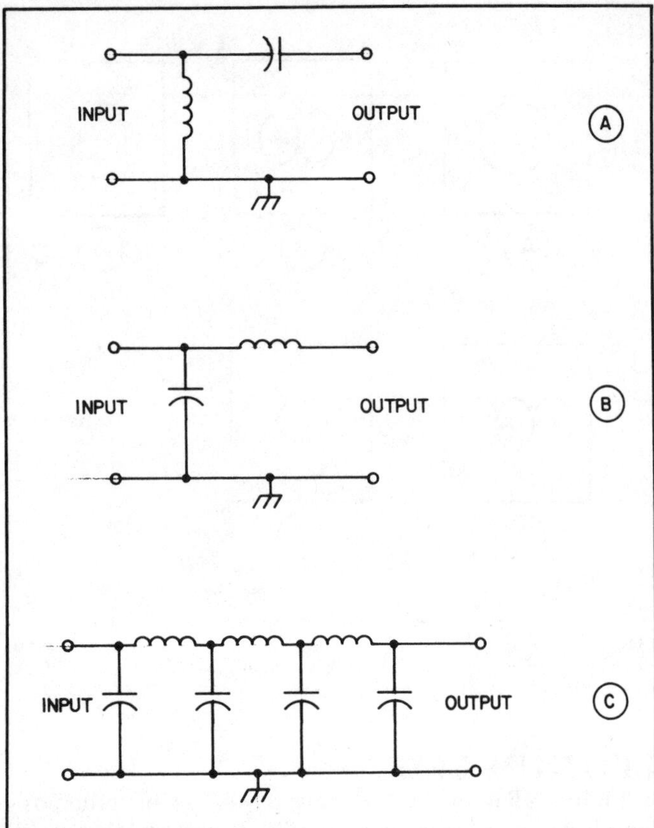

L NETWORK: At A, a highpass network; at B, a lowpass network; at C, a cascaded lowpass network.

fact that the schematic-diagram component arrangement resembles the capital letter L (see illustration).

The L configuration may be used in the construction of highpass or lowpass filters. The series component may appear either before or after the parallel component. In the illustration, a typical highpass L network is shown at A; a lowpass network is shown at B. Several L networks may be cascaded to obtain a sharper cutoff response, as shown at C.

The L network is one of several different configurations for selective filters. *See also* H NETWORK, PI NETWORK, T NETWORK.

LO

See LOCAL OSCILLATOR.

LOAD

A load is any circuit that dissipates, radiates, or otherwise makes use of power. A direct-current load exhibits a definite resistance. This resistance may vary with the amount of power applied. An alternating-current load exhibits resistance, and may also show capacitive or inductive reactance. Examples of loads include audio speakers, electrical appliances, and radio-frequency transmitting antennas.

For optimum transfer of power from a circuit to its load, the load impedance should be the same as the output impedance of the circuit. *See also* IMPEDANCE MATCHING, LOAD IMPEDANCE, LOADING.

LOAD CURRENT

The amount of current flowing in a load is called the load current. For a direct-current load, the current I is given by the formula:

$$I = E/R$$

where E is the voltage across the load, and R is the resistance of the load.

For an alternating-current load, the current depends on several different factors. In general, given a load impedance Z, and a root-mean-square voltage E across the load (see ROOT MEAN SQUARE), the current I is given by:

$$I = E/Z$$

The load current depends on the power supplied to the load and on the load impedance. *See also* LOAD, LOAD IMPEDANCE.

LOADED ANTENNA

The natural resonant frequency of an antenna can be changed by placing reactance in series with the radiating element. An antenna that has such a reactance in its radiating element is called a loaded antenna.

The most common type of loaded antenna has a resonant frequency that has been lowered by the series installation of an inductor. This type of loaded antenna is often used for mobile operation at medium and high frequencies. A capacitance hat, or loading disk, may be placed at the end or ends of such an antenna to increase the bandwidth.

The series connection of capacitors raises the resonant frequency of an antenna element. This kind of loaded antenna is sometimes used at very-high and ultra-high frequencies. *See also* CAPACITIVE LOADING, INDUCTIVE LOADING.

LOADED LINE

The characteristic impedance of a transmission line depends on the distributed capacitance and inductance. Any transmission line can be theoretically represented as a set of series inductors and parallel capacitors, although in reality these reactances are not discrete components.

Sometimes, it is necessary to change the characteristics of a transmission line at a certain location. This is done by the installation of capacitors and/or inductors in the line. The result of adding reactance to a transmission line is a change in the resonant frequency of a line. A transmission line in which reactance is deliberately added is called a loaded line. A line may be loaded for the purpose of matching the characteristic impedance to the output impedance of a radio transmitter. Line loading may also be used to match the characteristic impedance to the impedance of an antenna or other load. *See also* CHARACTERISTIC IMPEDANCE, FEED LINE, IMPEDANCE MATCHING, LOADING.

LOAD IMPEDANCE

Any power-dissipating or power-radiating load has an impedance. Part of the impedance is a pure resistance, and is given in ohms. Reactance may also be present; it is also specified in ohms. A direct-current load, or a resonant alternating-current load, has resistance but no reactance. A nonresonant alternating-current load may have capacitive or inductive reactance.

The load impedance should always be matched to the impedance of the power-delivering circuit. Normally, this means that the load contains no reactance (that is, the load is resonant), and the value of the load resistance is equal to the characteristic impedance of the feed line. If the load impedance is not the same as the characteristic impedance of the line, special circuits can be installed at the feed point to change the effective load impedance.

If the load impedance differs from the feed-system impedance, the load will not absorb all of the delivered power. Some of the electromagnetic field will be reflected back toward the source. This results in an increased amount of power loss in the system. The magnitude of this loss increase may or may not be significant. *See also* IMPEDANCE, IMPEDANCE MATCHING, REFLECTED POWER.

LOADING

The deliberate insertion of reactance into a circuit is called loading. Generally, this is done for the purpose of obtaining resonance.

If a load contains capacitive reactance, then an equal amount of inductive reactance must be added to obtain resonance. Conversely, if a load contains inductive reactance, an equal amount of capacitive reactance must be added.

A common example of loading is the insertion of an inductor in series with a physically short antenna. A short antenna has capacitive reactance, and the loading inductor provides an equal and opposite reactance. The larger the value of the loading inductance, the lower the resonant frequency becomes. A capacitor may be inserted in series with a physically long antenna to obtain resonance. *See also* CAPACITIVE LOADING, INDUCTIVE LOADING, LOADED ANTENNA, LOADED LINE, RESONANCE.

LOADING CAPACITOR

See CAPACITIVE LOADING.

LOADING COIL

See INDUCTIVE LOADING.

LOADING DISK

A loading disk, also known as a capacitance hat or top hat, is a device attached at or near the free end or ends of an antenna (see illustration). The loading disk serves two functions. It allows the antenna to be physically shorter, for a given resonant frequency, than an antenna without the disk; the loading disk also results in a more broadbanded response.

Loading disks are often used in conjunction with inductive loading to reduce the voltage at the end of the antenna and to raise the radiation resistance. *See also* INDUCTIVE LOADING.

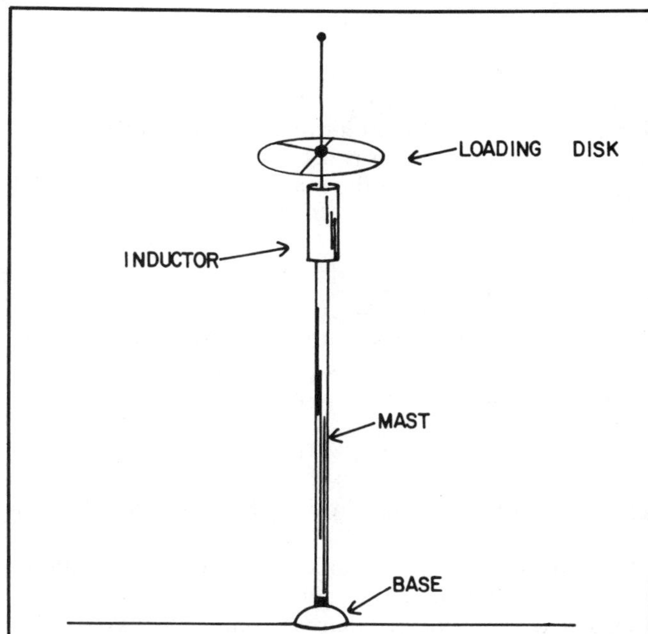

LOADING DISK: An antenna loading disk, in this case constructed from stiff wire conductors.

LOADING INDUCTOR

See INDUCTIVE LOADING.

LOAD LOSS

When power is applied to a load, the load ideally dissipates all of the power in the manner intended. For example, a light bulb should produce visible light but no heat; a radio antenna should radiate all of the energy it receives, and waste none as heat. However, in practice, no load is perfect. If P watts are delivered to a load, and the load dissipates Q watts in the manner intended, then the value P - Q is called the load loss. Load loss may be expressed in decibels by the formula:

$$\text{Loss (dB)} = 10 \log_{10} (P/Q)$$

Some loads exhibit large losses. The incandescent bulb is an example of a lossy load. Other loads have very little loss; a half-wave dipole antenna in free space is an example of an efficient load. *See also* LOAD.

LOAD POWER

Load power can be defined in two ways. The load input power is the amount of power delivered to a load. The load output power is the amount of power dissipated, in the desired form, by the load. The load output power is always less than the load input power; their difference is called the load loss (*see* LOAD LOSS).

If the load impedance is a pure resistance, then the load input power P can be determined in terms of the root-mean-square load current I and the root-mean-square load voltage E by the formula:

$$P = EI$$

If the purely resistive load impedance is R ohms, then:

$$P = E^2/R = I^2R$$

When reactance is present in a load, the load input power differs from the incident power as determined by the current and voltage. This is because some power appears in imaginary form across the reactance. *See also* APPARENT POWER, REACTIVE POWER, TRUE POWER.

LOAD VOLTAGE

The voltage applied across a load is called the load voltage. For a direct-current load, the load voltage E can be determined according to the formula:

$$E = IR$$

where I is the current through the load and R is the resistance of the load.

For an alternating-current load, the voltage depends on several different things. In general, given a load impedance Z and a root-mean-square current I through the load (*see* ROOT MEAN SQUARE), the root-mean-square voltage E is:

$$E = IZ$$

The load voltage depends on the power supplied to the load, and also on the load impedance. *See also* LOAD, LOAD IMPEDANCE.

LOBE

In the radiation pattern of an antenna, a lobe is a local angular maximum. An antenna pattern may have just one lobe, or it may have several lobes. Different lobes may have different magnitudes (see illustration). The strongest lobe is called the main or major lobe. The weaker lobes are called secondary or minor lobes. *See also* ANTENNA PATTERN.

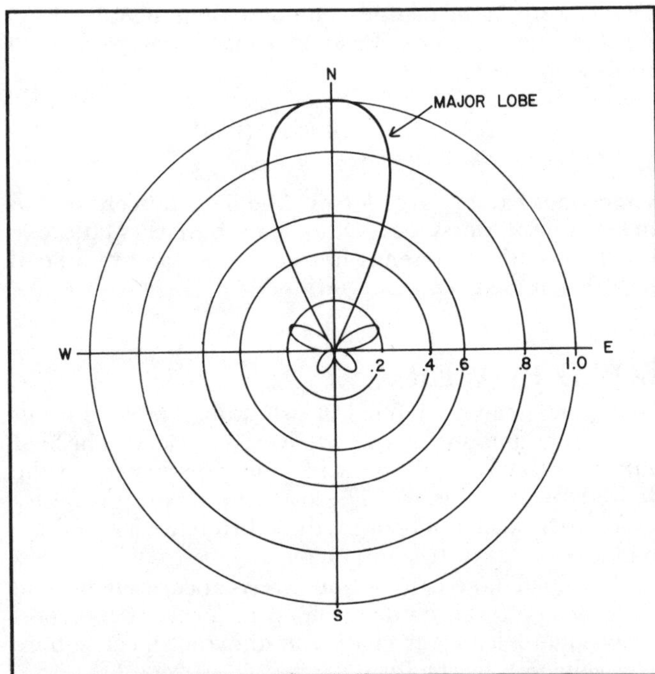

LOBE: An antenna pattern may have several lobes. The strongest lobe, in this case oriented toward the north, is called the major lobe or main lobe. The antenna is at the center of the diagram, and the vantage point is from directly above the antenna. Field strength in a particular direction is indicated by the distance of the curve from the center in that direction.

Lobes occur in the directional responses of various kinds of transducers, such as microphones and speakers. *See also* DIRECTIONAL MICROPHONE.

LOCAL ACTION

In an electrochemical cell, impurities in an electrode may result in a potential difference between different parts of that electrode. This effect is called local action. Local action results in degradation of the performance of an electrochemical cell or battery.

When certain impurities exist in a part of an electrode, that region of the electrode may act as an anode, or positive terminal, even though the electrode itself forms the cathode of the cell. Conversely, an impurity in the anode may result in a local negative charge.

Local action can be minimized by using the purest possible materials for the electrodes in a cell. *See also* CELL.

LOCAL CHANNEL

In standard amplitude-modulation broadcasting, a local channel is a short-range channel. A station operating on a local channel is sometimes called a Class-IV station. The local-channel station has the least priority of any standard-broadcast station.

Local-channel stations are limited to a power level of 250 watts to 1,000 watts during the daylight hours (sunrise to sunset). During the night, the maximum permissible power level is 250 watts. These limitations apply in the United States, and are set by the Federal Communications Commission. *See also* CLEAR CHANNEL, REGIONAL CHANNEL.

LOCAL CONTROL

When a device, such as a computer or radio transmitter, is controlled on location, the method is known as local control. Local control allows direct monitoring and adjustment of equipment. Local control requires that an engineer or operator be present at the site. *See also* REMOTE CONTROL.

LOCAL LOOP

In a teletype system, the local loop is an arrangment that allows paper tapes to be run through the machine for checking. In modern, computerized systems, the information is usually stored in a solid-state memory bank rather than on paper tape. By running the data through the machine locally, the station operator can correct errors in the text before they are transmitted. The local loop does not involve connection to an external line or circuit; it is contained entirely at one place. *See also* RADIOTELETYPE, TELETYPE.

LOCAL OSCILLATOR

A local oscillator is a radio-frequency oscillator, the output

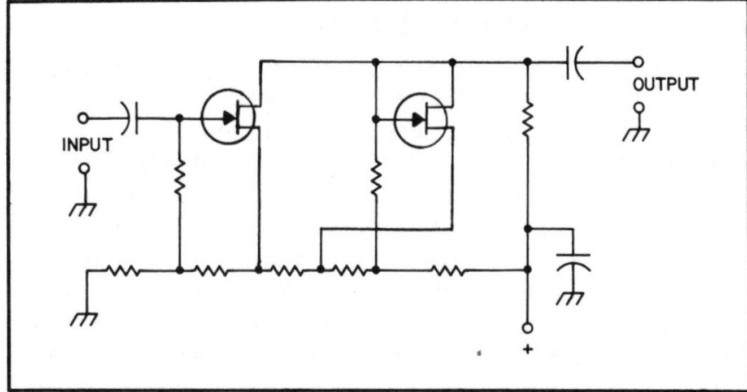

LOFTIN-WHITE AMPLIFIER: An FET Loftin-White audio amplifier.

of which is not intended for direct transmission over the air. All of the oscillators in a superheterodyne receiver are local oscillators. In transmitters, local-oscillator signals may be combined in mixers, amplified, perhaps multiplied, and then sent over the air.

Local-oscillator design requires excellent shielding and isolation, so that the output will not cause interference to nearby radio receivers. *See also* OSCILLATOR, SUPERHETERODYNE RECEIVER.

LOFTIN-WHITE AMPLIFIER

A two-transistor audio amplifier circuit, in which the output of the first stage is directly coupled to the input of the second stage, is called a Loftin-White amplifier. Either bipolar or field-effect transistors may be used.

The Loftin-White circuit is characterized by a multielement resistive voltage divider that provides bias for the emitters or sources of both active devices and for the collector or drain of the first device (see illustration).

LOG

A log is a record of the operation of a radio station. Logs generally include such information as the date and time of a transmission, the nature of the transmission, the mode of emission, the power input or output of the transmitter, the frequency of transmission, and the message traffic handled. In some situations, logs must be kept by law. Logging requirements are determined, in the United States, by the Federal Communications Commission.

The mathematical logarithm operation is sometimes abbreviated as "log." *See also* LOGARITHM.

LOGARITHM

A logarithm is a mathematical operation. The logarithm of a number x, in base b, is the power to which the number b must be raised to obtain the value x. If $y = \log_b x$, then:

$$b^y = x$$

Logarithms are usually specified either in base e, where e is about 2.718, or in base 10. The base-e logarithm is sometimes called the natural logarithm. The natural logarithm of a number x is written ln x. The base-10

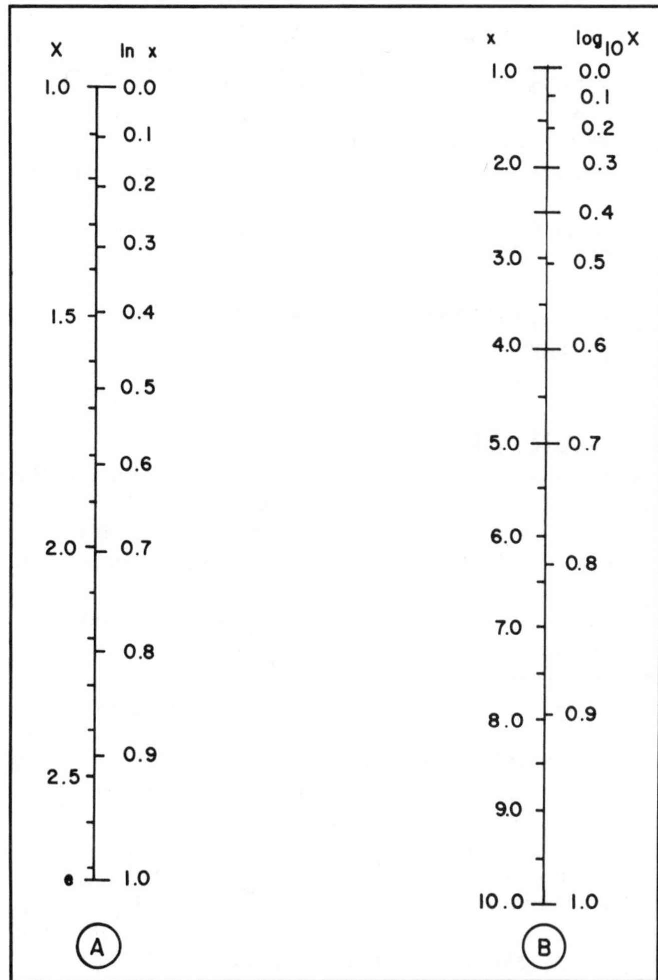

LOGARITHM: At A, natural logarithms for values between 1 and e, where e is approximately 2.718. At B, base-10 logarithms for values between 1 and 10.

logarithm of a number x is written as $\log_{10}$ x, or sometimes simply log x.

The logarithm function is defined only for positive real numbers. Non-positive numbers do not have logarithms. The illustration at A is a nomograph illustrating natural logarithms. A nomograph illustrating base-10 logarithms is shown at B. Logarithmic functions are included in many modern pocket calculators.

Logarithms are important in scientific use, because they allow a wide range of values to be compressed into a relatively small graphic or linear space. Some graph scales are calibrated logarithmically instead of in linear fashion, providing superior rendition of certain mathematical functions. The amplitude scales of some test instruments are calibrated logarithmically. *See also* LOGARITHMIC FUNCTION, LOGARITHMIC SCALE.

LOGARITHMIC FUNCTION

Any mathematical function that involves a logarithm, in conjunction with constants, is called a logarithmic function. The general form of a logarithmic function is given by either of the following two forms:

$$f(x) = c \log_b kx$$

$$g(x) = c/(\log_b kx)$$

where f or g represent the function operator, c and k are real-number constants, and b is a positive real number representing the base of the logarithm.

Logarithmic functions appear in many different electronic applications. *See also* FUNCTION, LOGARITHM, LOGARITHMIC DECREMENT, LOGARITHMIC SCALE.

LOGARITHMIC GRAPH

See LOGARITHMIC SCALE.

LOGARITHMIC SCALE

When a coordinate scale is calibrated according to the logarithm of the actual scale displacement, the scale is called a logarithmic scale. Such scaling is universally done according to the base-10 logarithm of the distance.

Logarithmic scales may be used in a Cartesian coordinate system for one axis or both axes (see illustration). The coordinate system at A is sometimes called a semi-logarithmic graph; the coordinate system at B is known as a logarithmic or log-log graph (*see* CARTESIAN COORDINATES). In a polar graph, the radial coordinate axis may be calibrated in a logarithmic fashion (*see* POLAR COORDINATES). Logarithmic scales are also sometimes used in nomographs (*see* NOMOGRAPH).

Logarithmic scales are especially useful in plotting amplitudes when the range is expected to vary over many orders of magnitude. Any decibel scale is a logarithmic scale; some meters and oscilloscope displays are calibrated in this way. *See also* DECIBEL, LOGARITHM.

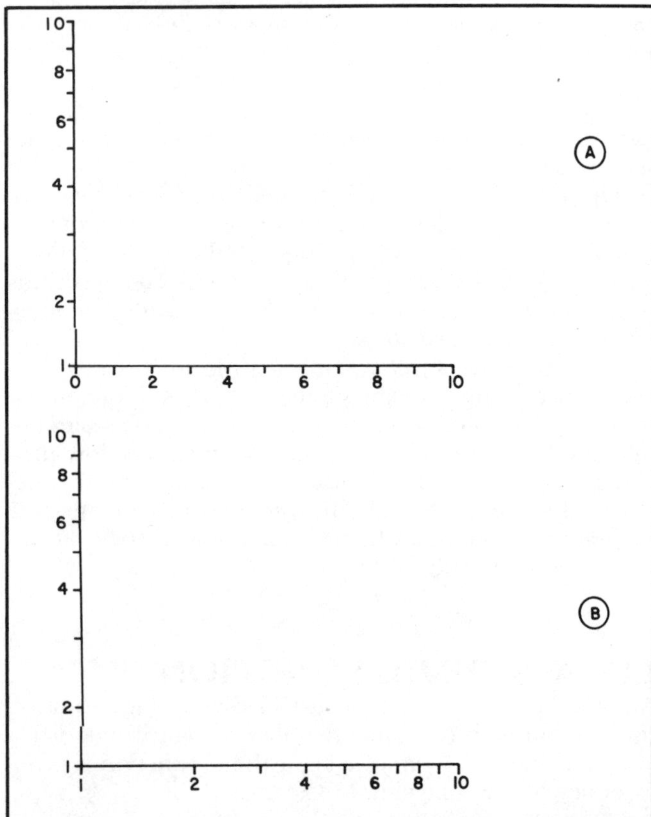

LOGARITHM IC SCALE: At A, semi-logarithmic; at B, logarithmic or log-log.

LOGARITHMIC TAPER

Some controls have a value or function that varies with the logarithm of the angular displacement, rather than directly with the angular displacement. Such controls are said to have a logarithmic taper. Logarithmic-taper potentiometers are fairly common in audio-frequency applications and in brightness controls.

The human senses perceive disturbances approximately according to the logarithm of their intensity. Therefore, a linear-taper control is often not suitable for adjustment of variables perceived by the senses, such as loudness and brightness. A logarithmic-taper device compensates for human perception, making the control "seem" linear. *See also* AUDIO TAPER, LINEAR TAPER.

LOGIC

Logic is a term used in electronics to describe the operation of computers or switching devices. A circuit input or output may have either of two logic states, known as high and low. Sometimes these conditions are called 1 and 0, or true and false, respectively. A logic device usually has one or more inputs, and one output. The device performs a logical or Boolean function on the input or inputs, in order to obtain the output (*see* BOOLEAN ALGEBRA).

Digital logic may be categorized as either positive or negative. In positive logic, the high condition is represented by the more positive voltage, and the low condition by the more negative voltage at a given circuit point. The more positive voltage represents the condition of truth (1); the more negative voltage represents the condition of falsity (0). In negative logic, these conditions are reversed; the more positive voltage represents logical falsity (0) and the more negative voltage represents logical truth (1). *See* NEGATIVE LOGIC, POSITIVE LOGIC.

In mathematics, logic is the field of study dealing with reasoning. Mathematical logic involves not only the elementary functions of Boolean algebra and truth tables, but delves into highly sophisticated theorems and applications. *See also* LOGIC EQUATION, LOGIC FUNCTION, MATHEMATICAL LOGIC, TRUTH TABLE.

LOGICAL EQUIVALENCE

Two logical expressions are said to be equivalent when their truth or falsity is always the same. Given two logically equivalent expressions A and B, A is true if B is true and vice-versa.

Logical equivalence is represented by means of a logic equation. *See also* LOGIC EQUATION.

LOGIC CIRCUIT

A combination of logic gates is called a logic circuit. Logic circuits vary from the simplest (a single inverter, or NOT gate) to extremely complex arrays.

A logic circuit is designed to perform a certain logic function. That is, each combination of inputs always produces the same output or outputs. Logic circuits are often fabricated onto integrated circuits; sometimes a single integrated circuit contains several different logic circuits. *See also* LOGIC FUNCTION, LOGIC GATE.

LOGIC DIAGRAM

A logic diagram is a schematic diagram of a logic circuit. The complete logic diagram shows the interconnection of individual gates. The complexity of the logic diagram depends on the function that the circuit performs.

The illustration is an example of a logic diagram. This circuit performs the logic function shown by the accompanying truth table. There are other possible logic circuits that will accomplish this same logic function, and the diagrams of those circuits are much different from the one illustrated. *See also* LOGIC CIRCUIT, LOGIC FUNCTION, LOGIC GATE.

LOGIC EQUATION

A logic equation is a symbolic statement showing two logically equivalent Boolean expressions on either side of an equal sign (*see* BOOLEAN ALGEBRA). The logic equation A = B, where A and B are Boolean expressions, means literally "A is true if B is true, and B is true if A is true." Mathematicians would say, "A if and only if B."

Logic equations are important in electronics. A particular logic circuit, intended to accomplish a given logic function, can often be assembled from logic gates in several different ways. One of these arrangements is usually simpler than all of the others, and is therefore most desirable from an engineering standpoint.

Many logic equations are known to be true, and are called theorems of Boolean algebra. By applying these theorems, the digital engineer can find the simplest possible logic circuit for a given function. *See also* LOGIC CIRCUIT, LOGIC FUNCTION.

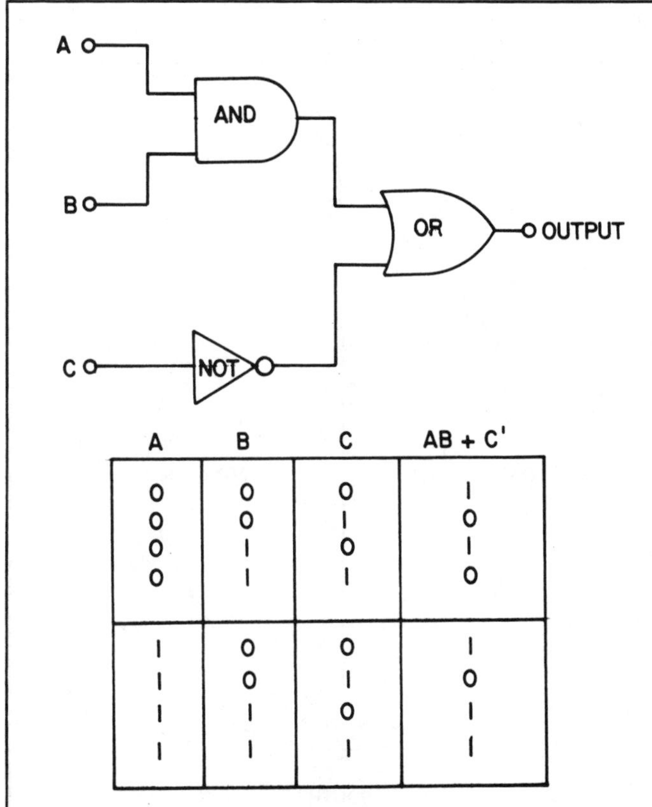

A	B	C	AB + C'
0	0	0	1
0	0	1	0
0	1	0	1
0	1	1	0
1	0	0	1
1	0	1	0
1	1	0	1
1	1	1	1

LOGIC DIAGRAM: A logic diagram, showing a circuit with an AND gate, an OR gate, and an inverter, obtaining the Boolean function AB + C'.

LOGIC FUNCTION

A logic function is an operation, consisting of one or more input variables and one output variable. The logic function is a simple form of mathematical function (*see* FUNCTION), in which the input variables may achieve either of two states in various combinations. A logic function may be written in the form:

$$F(x_1, x_2, x_3, \ldots, x_n) = y$$

where x_1 through x_n represent the input variables and y represents the output. The values of the variables may be either 0 (representing falsity) or 1 (representing truth). If there are n input variables, then there are 2^n possible input combinations.

Every logic circuit performs a particular logic function. Logic functions may be written in Boolean form. Most logic functions can be represented in several different Boolean forms. Logic functions can also be written as a truth table, or a listing of all possible input combinations along with the corresponding output states. Logic functions can also be written in the form of a schematic diagram of the logic circuit, showing the interconnection of gates. *See also* BOOLEAN ALGEBRA, LOGIC CIRCUIT, LOGIC DIAGRAM, LOGIC EQUATION, LOGIC GATE, TRUTH TABLE.

LOGIC GATE

A logic gate is a simple logic circuit. The gate performs a single AND, NOT, or OR operation, perhaps preceded by a NOT operation.

Logic gates are shown schematically in the illustration.

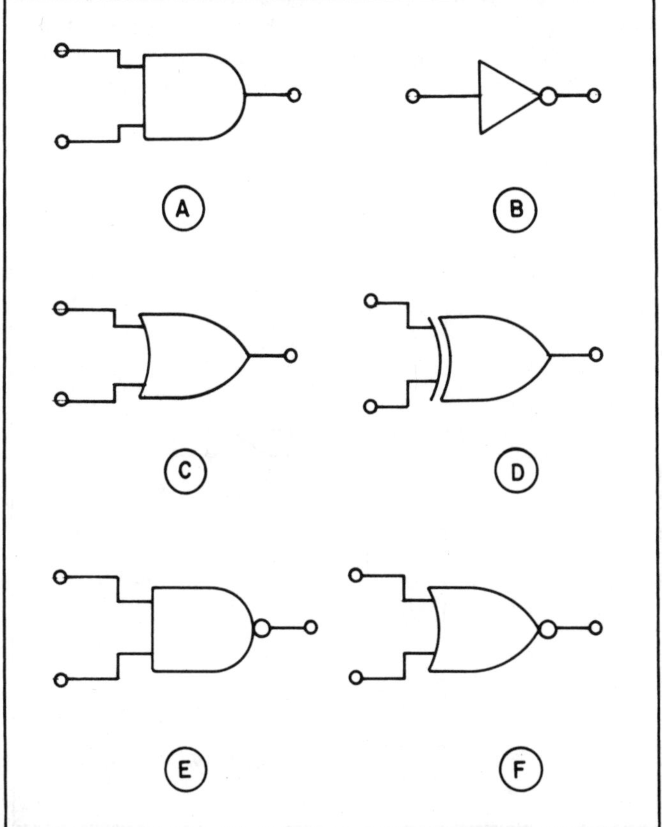

LOGIC GATE: A: AND. B: NOT. C: OR. D: Exclusive OR. E: NAND. F: NOR.

The AND gate, at A, produces a high output only if all of the inputs are high. The NOT gate or inverter (B) simply changes the state of the input. The OR gate, at C, produces a high output if at least one of the inputs is high. The exclusive-OR gate, shown at D, produces a high output if and only if the two inputs are different. The AND gate may be followed by an inverter, resulting in a device called a NAND gate (E). The OR gate may be followed by an inverter to form a NOR gate (F).

Logic gates are combined to perform complicated logical operations. This is the basis for all of digital technology. *See also* AND GATE, EXCLUSIVE-OR GATE, INVERTER, LOGIC CIRCUIT, LOGIC FUNCITON, NAND GATE, NOR GATE, OR GATE.

LOG-PERIODIC ANTENNA

A log-periodic antenna, also known as a log-periodic dipole array (LPDA), is a special form of unidirectional, broadband antenna. The log-periodic antenna is sometimes used in the high-frequency and very-high-frequency parts of the radio spectrum, for transmitting or receiving.

The log-periodic antenna consists of a special arrangement of driven dipoles, connected to a common transmission line. The illustration is a schematic diagram of the general form of a log-periodic antenna. The design parameters are beyond the scope of this discussion, since they vary depending on the gain and bandwidth desired. In general, the elements become shorter nearer the feed point (forward direction) and longer toward the back of the antenna. Note that the element-interconnecting line is twisted 180 degrees between any two adjacent elements.

The log-periodic antenna exhibits a fairly constant input impedance over a wide range of frequencies. Typically, the antenna is useful over a frequency spread of about 2 to 1. The forward gain of the log-periodic antenna is comparable to that of a two-element or three-element Yagi. The gain may be increased slightly by slanting the elements forward. The gain is also proportional to the number of elements in the antenna.

The log-periodic antenna is especially useful in situations where a continuous range of frequencies must be covered. This is the case, for example, in television receiving. Therefore, many (if not most) commercially manufactured very-high-frequency television receiving antennas are variations of the log-periodic design.

LOG-PERIODIC ANTENNA: General configuration of a log-periodic dipole array.

LONGITUDE

Longitude is one of two coordinates used to uniquely define a point on the surface of the earth. Longitude values range from −180 degrees to +180 degrees. The longitude line corresponding to 0 degrees runs through Greenwich, England. This line is called the Greenwich Meridian. Points west of this line are assigned negative longtude values; points east of the line are given positive values.

All longitude lines on the globe are great circles (*see* GREAT CIRCLE), intersecting at the north and south geographic poles, as shown in the illustration. The two halves of each longitude line represent different longitude coordinates. If the eastern-hemisphere half of a given circle represents x degrees east, then the western-hemisphere half of the same circle represents 180 − x degrees west.

The other coordinate commonly used for locating points on a sphere is called latitude. *See also* LATITUDE.

LONGITUDINAL WAVE

A wave disturbance in which the displacement is parallel to the direction of travel is called a longitudinal wave. The most common example of a longitudinal wave is the sound wave in air.

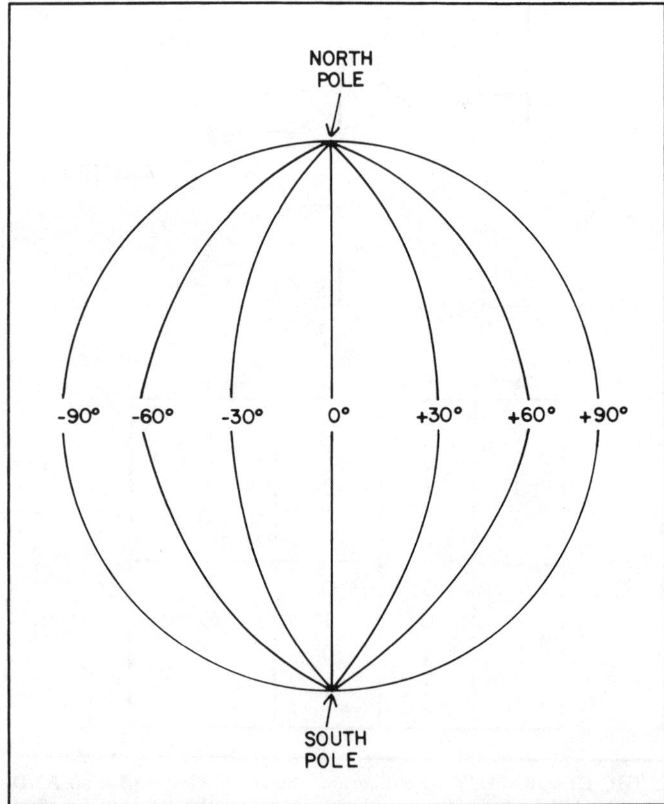

LONGITUDE: Longitude lines, as viewed from the plane of the equator.

All longitudinal waves are propagated via the movement of particles, such as air molecules. Some waves are propagated by lateral, or transverse, motion of particles. Others are propagated by electromagnetic effects. *See also* ELECTROMAGNETIC FIELD, TRANSVERSE WAVE.

LONG-PATH PROPAGATION

At some radio frequencies, worldwide communication is possible because of the effects of the ionosphere. Radio waves travel over, or near, great-circle paths across the globe (*see* GREAT CIRCLE). There are two possible paths over which the radio waves may propagate between two points on the surface of the earth. Both paths lie along the same great circle, but one path is generally much longer than the other.

When the radio waves travel the long way around the world to get from one point to the other, the effect is called long-path propagation. Long-path propagation is more common between widely separated points than between nearby points. Long-path propagation always occurs via the F layer, or highest layer, of the ionosphere. Whether the signal will take the long path or the short path depends on the ionospheric conditions at the time. In some cases a signal will propagate via both paths. For two points located exactly opposite each other on the globe, all paths are the same length, and the signal may travel over many different routes at the same time.

Under some conditions, long-path propagation occurs all the way around the planet. When this happens, the signal from a nearby transmitter may seem to echo: The signal is heard first directly, and then a fraction of a second later as the long-path wave arrives. *See also* F LAYER, IONOSPHERE, PROPAGATION CHARACTERISTICS, SHORT-PATH PROPAGATION.

LONGWIRE ANTENNA

A wire antenna measuring 1 wavelength or more, and fed at a current loop or at one end, is called a longwire antenna. Longwire antennas are sometimes used for receiving and transmitting at medium and high frequencies.

Longwire antennas offer some power gain over the half-wave dipole antenna. The longer the wire, the greater the power gain. For an unterminated longwire antenna measuring several wavelengths, the directional pattern resembles A in the illustration. As the wire is made longer, the main lobes get more nearly in line with the antenna, and their amplitudes increase. As the wire is made shorter, the main lobes get farther from the axis of the antenna, and their amplitudes decrease. If the longwire is terminated at the far end (opposite the feed point), half of the pattern disappears, as at B.

Longwire antennas have certain advantages. They offer considerable gain and low-angle radiation, provided they are made long enough. The graph (C) shows the theoretical power gain, with respect to a dipole, that can be realized with a longwire antenna. The gain is a function of the length of the antenna. Longwire antennas are inexpensive and easy to install, provided there is sufficient real estate. The longwire antenna must be as straight as possible for proper operation.

There are two main disadvantages to the longwire antenna. First, it cannot conveniently be rotated to change the direction in which maximum gain occurs. Second, a great deal of space is needed, especially in the medium-frequency spectrum and the longer high-frequency wavelengths. For example, a 10-wavelength longwire antenna measures about 1,340 feet at 7 MHz. This is more than a quarter of a mile. *See also* WIRE ANTENNA.

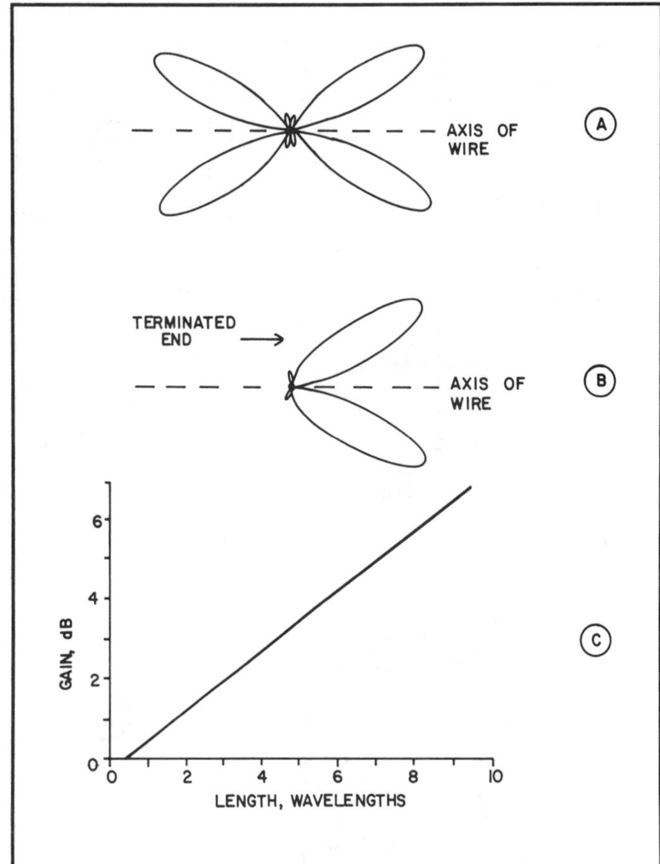

LONGWIRE ANTENNA: At A, the directional pattern for an unterminated longwire antenna. At B, the directional pattern for a terminated longwire. At C, the maximum theoretical gain, as a function of length in wavelengths, for longwire antennas.

LOOP

In electronics, the term loop can have any of several different meanings.

A local current or voltage maximum on an antenna or transmission line is called a loop (*see* CURRENT LOOP, VOLTAGE LOOP).

A closed signal path in an amplifier, oscillator, or other circuit may be called a loop. Examples are the feedback loop in an oscillator, and the loop in an operational-amplifier circuit (*see* FEEDBACK).

In a computer program, a loop is a portion of the program through which the computer cycles many times. The loop ends when some condition is satisfied.

A single-turn coil of wire is called a loop. A large coil, having one or more turns and used for the purpose of receiving or transmitting radio signals, is called a loop or loop antenna. *See also* LOOP ANTENNA.

LOOP ANTENNA

Any receiving or transmitting antenna, consisting of one or more turns of wire forming a direct-current short cir-

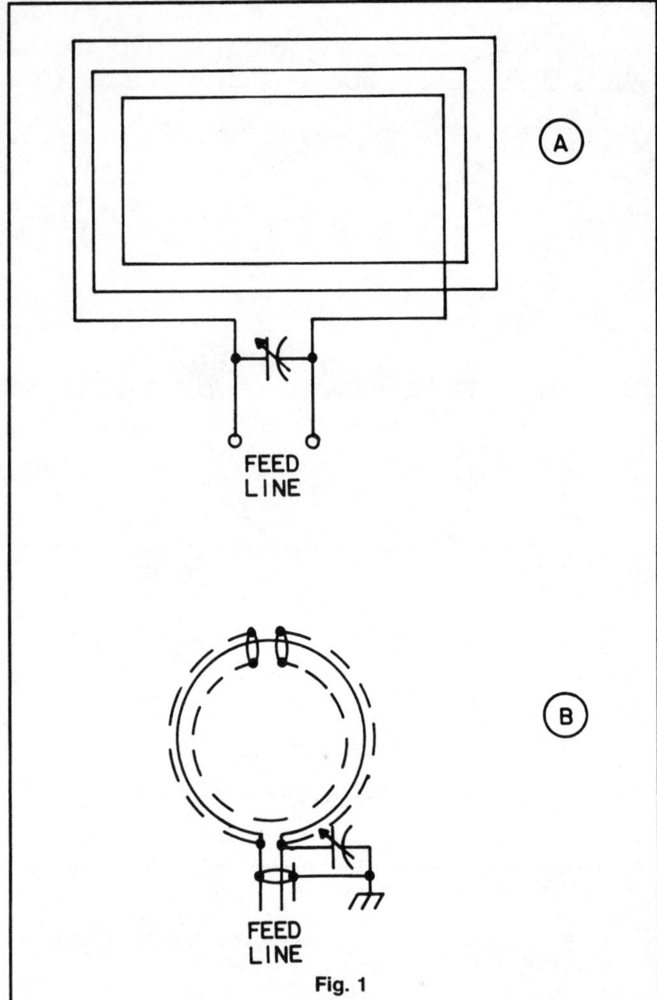

LOOP ANTENNA: At A, an unshielded loop with several turns. At B, a shielded loop with one turn.

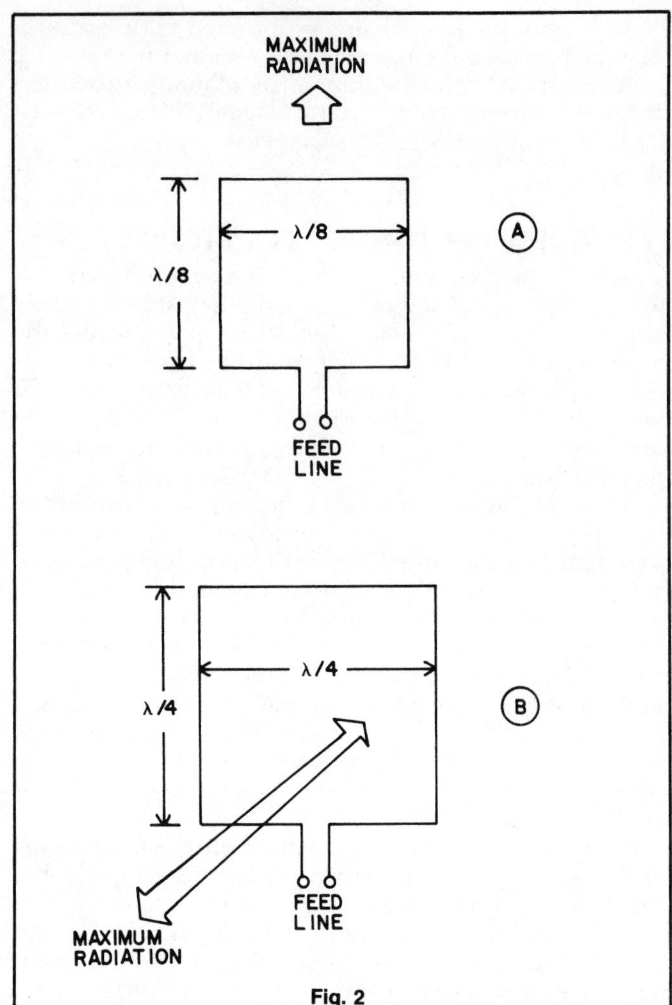

LOOP ANTENNA: At A, a large loop with a circumference of 0.5 wavelength. At B, a large loop with a circumference of 1 wavelength.

cuit, is called a loop antenna. Loop antennas can be categorized as either small or large.

Small loops have a circumference of less than 0.1 wavelength at the highest operating frequency. Such antennas are suitable for receiving, and exhibit a sharp null along the loop axis. The small loop may contain just one turn of wire, or it may contain many turns. The loop may be electrostatically shielded to improve the directional characteristics. Figure 1 shows a multi-turn unshielded loop antenna (A). A single-turn shielded loop antenna is shown at B. The ferrite-rod antenna is a form of small loop in which a ferromagnetic core is used to enhance the signal pickup (*see* FERRITE-ROD ANTENNA). Small loops are used for direction finding and for eliminating manmade noise or strong local interfering signals. Small loops are not suitable for transmitting, because of their extremely low radiation resistance.

Large loops have a circumference of either 0.5 wavelength or 1 wavelength at the operating frequency. The half-wavelength loop (A in Fig. 2) presents a high impedance at the feed point, and the maximum radiation occurs in the plane of the loop. The full-wavelength loop (B) presents an impedance of about 50 ohms at the feed point, and the maximum radiation occurs perpendicular to the plane of the loop. A large loop can be used either for transmitting or receiving. The half-wavelength loop exhibits a slight power loss relative to a dipole, but the full-wavelength loop shows a gain of about 2 dBd. The full-wavelength loop forms the driven element for the popular quad antenna. *See also* QUAD ANTENNA.

LOOPSTICK ANTENNA

See FERRITE-ROD ANTENNA.

LOOSE COUPLING

When two inductors have a small amount of mutual inductance, they are said to be loosely coupled. Loose coupling provides relatively little signal transfer.

Loose coupling is desirable when it is necessary to pick off a small amount of signal energy for monitoring purposes, without disturbing the operation of the circuit under test. *See also* COEFFICIENT OF COUPLING, MUTUAL INDUCTANCE.

LORAC

A special type of radionavigation system, used by the Seismograph Service Corporation, is called LORAC. Two transmitters are used. Both are continuous-wave transmitters, spaced far apart. The heterodyne, or beat, frequency derived from the two transmitters is obtained by means of a receiver at a third fixed location. This heterodyne frequency is retransmitted at the third location, resulting in a total of three transmitted signals on different frequencies.

The captain of a ship or aircraft determines his position by means of a receiver that picks up all three signals and compares their phase. LORAC is somewhat different from the other commonly used long-range radionavigation system, known as loran. *See also* LORAN, RADIOLOCATION, RADIONAVIGATION.

LORAN

The term loran is a contraction of "LOng-RAnge Navigation." Loran is used by ships and aircraft for determination of position, and for navigation.

There is one system of loran in common use today. The system, called loran C, uses pulse transmission at a rate of about 20 cycles per second. Two transmitters are used, and they are spaced at a distance of 300 miles (see illustration). The operating frequency is 100 kHz, which corresponds to a wavelength of 3 km. Location is found by comparing the time delay in the arrival of the signal from the more distant transmitter, relative to the arrival time of the signal from the nearer transmitter. The lines of constant time difference are hyperbolic, as shown.

An older system of loran, which has now been discontinued, was known as loran A. This system operated in the range 1.85 to 1.95 MHz, or a wavelength of about 160 m. Loran C is favored since the ground-wave propagation range is greater, resulting in better reliability. *See also* RADIOLOCATION, RADIONAVIGATION.

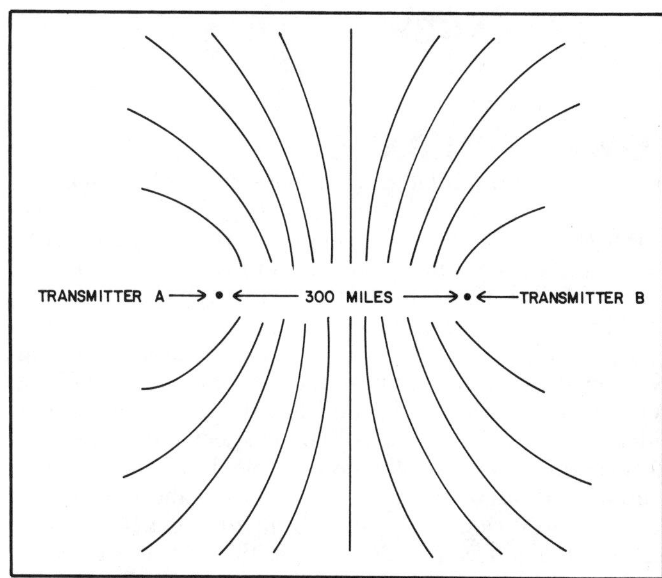

LORAN: The loran radionavigation system is a hyperbolic system. Sets of points having constant propagation delay times form hyperbolae on the surface of the earth, as shown here.

LOSS

Loss is a term that describes the extent of a decrease in current, voltage, or power. Most passive circuits exhibit some loss. The loss is an expression of the ratio between the input and output signal amplitudes in a circuit. Loss is usually specified in decibels.

Loss in a circuit often results in the generation of heat. The more efficient a circuit, the lower the loss, and the smaller the amount of generated heat. Sometimes, loss is deliberately inserted into a circuit or transmission line. A device designed for such a purpose is called an attenuator (*see* ATTENUATOR).

If a circuit has a gain of x dB, then the loss is −x dB. Conversely, if a circuit has a loss of x dB, then the gain is −x dB. Loss is just the opposite of gain. *See also* DECIBEL, GAIN, INSERTION LOSS.

LOSS ANGLE

In a dielectric material, the loss angle is the complement of the phase angle. If the phase angle is given by ϕ and the loss angle is given by θ, then

$$\theta = 90 - \phi$$

where both angles are specified in degrees.

In a perfect, or lossless, dielectric material, the current leads the voltage by 90 degrees, and thus $\phi = 90$. The loss angle is therefore zero. As the dielectric becomes lossy, the phase angle gets smaller, and the loss angle gets larger. This is because resistance is present in a lossy dielectric, in addition to capacitance. The tangent of the loss angle is called the dissipation factor. *See also* ANGLE OF LAG, ANGLE OF LEAD, DISSIPATION FACTOR, PHASE ANGLE.

LOSSLESS LINE

A theoretically perfect transmission line is called a lossless line. Such a line does not exist in practice, but it is useful, for the purposes of some mathematical calculations, to assume that a transmission line has no loss.

In a lossless line, no power is dissipated as heat. The available power at the line output is the same as the input power. If the line is terminated in a pure resistance exactly equal to its characteristic impedance, the current and voltage along the lossless line are uniform at all points.

In a lossless line, an impedance mismatch at the terminating end would result in no additional power loss. But in a real feed line, a mismatch results in some additional loss because of the standing waves. *See also* FEED LINE, STANDING-WAVE-RATIO LOSS.

LOSS TANGENT

See LOSS ANGLE, DISSIPATION FACTOR.

LOUDNESS

Loudness is an expression of relative apparent intensity of sound. Loudness is sometimes called volume.

The human ear perceives loudness approximately in terms of the logarithm of the actual intensity of the disturbance. Loudness is often expressed in decibels (*see* DECIBEL). An increase in loudness of 2 dB represents the smallest detectable change in sound level that a listener can detect if he is expecting the change. The threshold of hearing is assigned the value 0 dB.

In an audio circuit, such as a radio receiver output or high-fidelity system, the control that affects the sound intensity is often called the loudness control. *See also* SOUND.

LOUDSPEAKER

See SPEAKER.

LOWER SIDEBAND

An amplitude-modulated signal carries the information in the form of sidebands, or energy at frequencies just above and below the carrier frequency (*see* AMPLITUDE MODULATION, SIDEBAND). The lower sideband is the group of frequencies immediately below the carrier frequency. Lower sideband is often abbreviated LSB.

The LSB frequencies result from difference mixing between the carrier signal and the modulating signal. For a typical voice amplitude-modulated signal, the lower sideband occupies about 3 kHz of spectrum space (see illustration). For reproduction of television video information, the sideband may be several megahertz wide.

The lower sideband contains all of the modulating-signal intelligence. For this reason, the rest of the signal (carrier and upper sideband) can be eliminated. This results in a single-sideband (SSB) signal on the lower sideband. *See also* SINGLE SIDEBAND, UPPER SIDEBAND.

LOWEST USABLE FREQUENCY

For any two points separated by more than a few miles, electromagnetic communication is possible only at certain frequencies. At very low frequencies, communication is almost always possible between any two locations in the world, provided the power level is high enough. Usually, however, communication is also possible at some much higher band of frequencies, in the medium or high range.

Given some arbitrary medium or high frequency f, at which communication is possible between two specific points, suppose this frequency is made lower until communication is no longer possible. This cutoff frequency might be called f_L. It is known as the lowest usable high frequency, or simply the lowest usable frequency. It is sometimes abbreviated LUF. The LUF changes as the locations of the transmitting or receiving stations are changed. The LUF is a function of the ionospheric conditions, which fluctuate with the time of day, the season of the year, and the environment in the sun. *See also* MAXIMUM USABLE FREQUENCY, MAXIMUM USABLE LOW FREQUENCY, PROPAGATION CHARACTERISTICS.

LOW FREQUENCY

The range of electromagnetic frequencies extending from 30 kHz to 300 kHz is sometimes called the low-frequency band. The wavelengths are between 1 km and 10 km. For this reason, low-frequency waves are sometimes called kilometric waves.

The ionospheric E and F layers return all low-frequency signals to the earth, even if they are sent straight upward. Low-frequency signals from space cannot reach the surface of our planet because the ionosphere blocks them.

Low-frequency waves, if vertically polarized, propagate well along the surface of the earth. This becomes more and more true as the frequency gets lower. Low-frequency signals may, therefore, be used someday by space travelers on planets that have no ionosphere. Even on the moon, which has no ionosphere to return radio waves back to the ground, low-frequency waves might be useful for over-the-horizon communication via the surface wave. *See also* SURFACE WAVE.

LOWPASS FILTER

A lowpass filter is a combination of capacitance, inductance, and/or resistance, intended to produce large amounts of attenuation above a certain frequency and little or no attenuation below that frequency. The frequency at which the transition occurs is called the cutoff frequency (*see* CUTOFF FREQUENCY). At the cutoff frequency, the attenuation is 3 dB with respect to the minimum attenuation. Below the cutoff frequency, the attenuation is less than 3 dB. Above the cutoff, the attenuation is more than 3 dB.

The simplest lowpass filters consist of a series inductor or a parallel capacitor. More sophisticated lowpass filters have a combination of series inductors and parallel capacitors, such as the examples shown in the illustration. The filter at A is called an L-section lowpass filter; the circuit at B is a pi-section lowpass filter. These names are

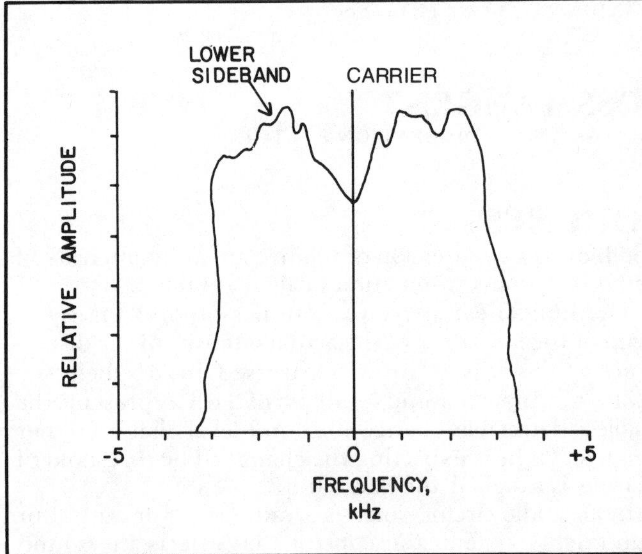

LOWER SIDEBAND: An amplitude-modulated signal as seen on a spectrum-analyzer display, showing the lower sideband.

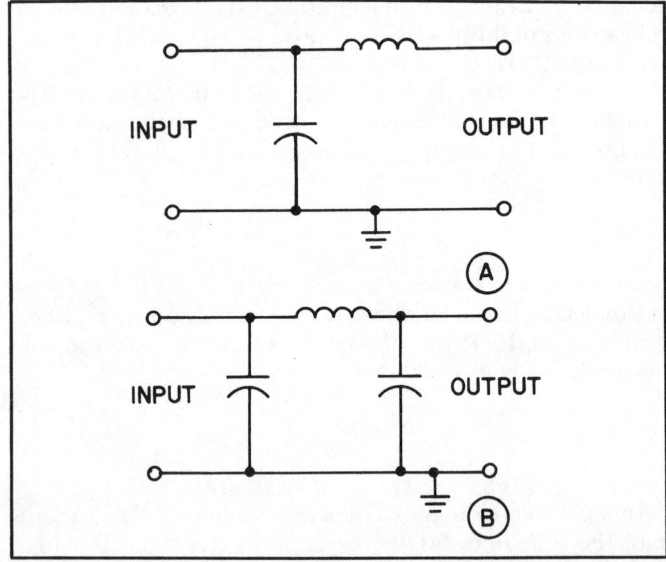

LOWPASS FILTER: At A, a lowpass L-section filter. At B, a lowpass pi-section filter. Both are for use in unbalanced circuits.

derived from the geometric arrangement of the components as they appear in the diagram.

Resistors are sometimes substituted for the inductors in a lowpass filter. This is especially true when active devices are used, in which case many filter stages can be cascaded.

Lowpass filters are used in many different applications in radio-frequency electronics. One common use of a lowpass filter is at the output of a high-frequency transmitter. The cutoff frequency is about 40 MHz. When such a lowpass filter is installed in the transmission line between a transmitter and antenna, very-high-frequency harmonics are greatly attenuated. This reduces the probability of television interference. In a narrow-band transmitter, lowpass filters are often built-in for reduction of harmonic output. *See also* LOWPASS RESPONSE.

LOWPASS RESPONSE

A lowpass response is an attenuation-versus-frequency curve that shows greater attenuation at higher frequencies than at lower frequencies. The sharpness of the response may vary considerably. Usually, a lowpass response is characterized by a low degree of attenuation up to a certain frequency; above that point, the attenuation rapidly increases. Finally the attenuation levels off at a large value. Below the cutoff frequency, the attenuation is practically zero. The lowpass response is typical of lowpass filters.

The cutoff frequency of a lowpass response is that frequency at which the insertion loss is 3 dB with respect to the minimum loss. The ultimate attenuation is the level of attenuation well above the cutoff frequency, where the signal is virtually blocked. The ideal lowpass response should look like the attenuation-versus-frequency curve in the illustration. The curve is smooth, and the insertion loss is essentially zero everywhere well below the cutoff frequency. *See also* CUTOFF FREQUENCY, INSERTION LOSS, LOWPASS FILTER.

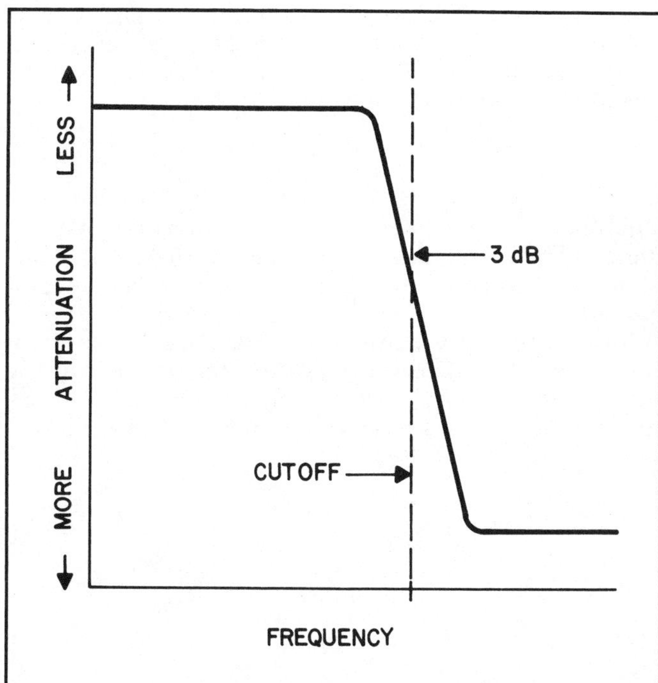

LOWPASS RESPONSE: A typical lowpass response. The cutoff frequency is that frequency where the attenuation is 3 dB with respect to the minimum insertion loss.

LSB

See LOWER SIDEBAND.

LSI

See LARGE-SCALE INTEGRATION.

LUF

See LOWEST USABLE FREQUENCY.

LUG

A lug is a terminal attached to the end of a wire for convenient connection to a screw or binding post. Lugs provide a better electrical connection than simply bending the wire around the screw or binding post. Some lugs can be soldered to the end of a wire; others are clamped on without soldering.

The terminals on a tie strip, to which wires are soldered in point-to-point electronic assemblies, are sometimes called lugs. *See also* TIE STRIP.

LUMEN

The lumen is the standard international unit of light flux. Given a light source that produces 1 candela, a flux of 1 lumen is emitted within a solid angle measuring 1 steradian (*see* STERADIAN). If a light source is surrounded by a sphere with a radius of 1 meter, and the source has a luminous intensity of 1 candela, then 1 lumen of flux crosses each square meter of the surface of the sphere.

All lamps are rated according to their output in lumens. A typical 25-watt incandescent bulb emits about 220 lumens of visible light within the wavelength range of 390 to 750 nanometers. *See also* CANDELA, LUMINOUS FLUX, LUMINOUS INTENSITY.

LUMINAIRE

A complete studio lighting assembly is called a luminaire. The luminaire consists of a set of lamps, interconnecting cords, a power supply, and all of the peripheral hardware needed to mount and position the lamps.

A luminaire is used for the purpose of setting up a temporary television studio, movie set, or photographic set.

LUMINANCE

Luminance is a measure of the luminous flux passing through a given surface area. Luminance is also known as luminous sterance. In the standard international system of units, luminance is measured in candela per square meter. Luminance may also be measured in candela per square centimeter, candela per square foot, and other combinations. Luminance is sometimes called brightness.

Given a light source of fixed intensity, the luminance decreases as the distance from the source increases. If the distance is doubled, the luminance drops to one-fourth its previous value. In general, the luminance varies inversely with the square of the distance from the light source. *See also* CANDELA, LUMEN, LUMINOUS FLUX, LUMINOUS INTENSITY.

LUMINESCENCE

When a substance absorbs energy in a non-visible form, and retransmits the energy as light, that material is said to be luminescent. Examples of luminescent substances include the phosphor on a cathode-ray-tube screen, certain semiconductors, and various other chemicals. Luminescence may result from bombardment by electrons or other subatomic particles, from the presence of an electric current, or from irradiation by electromagnetic fields such as infrared, ultraviolet, or X rays.

Luminescence always involves the conversion of energy from one form to another. Electrical energy is converted to light by means of an electroluminescent cell. Invisible radiation is converted into light energy by phosphorescent substances. *See also* ELECTROLUMINESCENT CELL, PHOSPHORESCENCE.

LUMINIFEROUS ETHER

See ETHER THEORY.

LUMINOUS FLUX

Luminous flux is the rate at which light energy is emitted or transmitted. The unit of luminous flux is the lumen (*see* LUMEN).

In general, the greater the luminous flux produced by a light source, the brighter the source appears from a given distance. Luminous flux is determined within the optical range only. Optical wavelengths are considered to be between approximately 390 and 750 nanometers (nm).

When luminous flux is measured within a given solid angle, it is called luminous intensity. *See also* LUMINOUS INTENSITY.

LUMINOUS INTENSITY

Luminous intensity is a measure of the amount of luminous flux emitted within a certain solid angle. Luminous intensity is generally measured in units called candela.

An intensity of 1 candela is equal to a flux of 1 lumen within a solid angle of 1 steradian. A solid angle of 1 steradian represents $1/(4\pi)$ of the total three-dimensional emission from an object. Therefore, a light source that produces 1 candela has a total luminous flux of about 4π lumens. *See also* CANDELA, LUMEN, LUMINOUS FLUX, STERADIAN.

LUMINOUS STERANCE

See LUMINANCE.

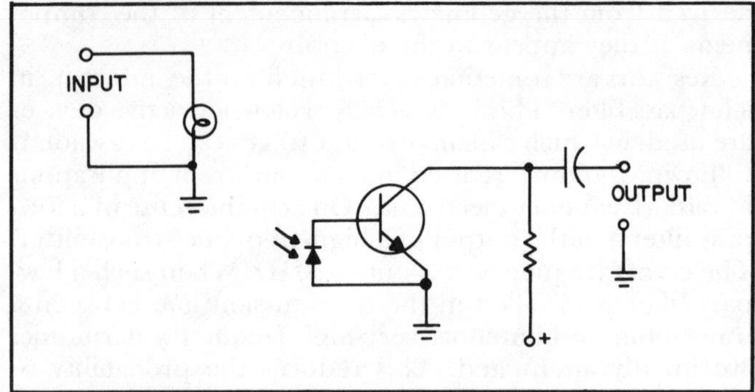

LUMISTOR: A lumistor is an optical-coupling device.

LUMISTOR

A lumistor is an optical coupling circuit. The signal to be transferred is applied to a light-emitting device. This device may be a light bulb for signals at very low frequencies. A light-emitting diode produces better results at higher frequencies. A laser may also be used. A photocell, photovoltaic cell, or other light-sensitive device picks up the modulated light and delivers it to the load. The illustration shows a simplified schematic diagram of a lumistor.

The lumistor makes it possible to transfer a signal with essentially no "pulling," or effects on the output impedance of the signal generator. A change in the load impedance has no effect on the operating characteristics of the signal generator when optical coupling is used. Optical coupling also facilitates the transfer of signals between circuits at substantially different potentials. *See also* OPTICAL COUPLING.

LUMPED ELEMENT

When a reactance, resistance, or voltage source appears in a discrete location, the component is called a lumped element. All ordinary capacitors, inductors, and resistors are lumped elements. However, a transmission line is not a lumped element because it exhibits inductance and capacitance that are distributed over a region. *See also* DISCRETE COMPONENT, DISTRIBUTED ELEMENT.

LUX

The lux is the unit of visible-light illuminance. When 1 lumen of light falls on 1 square meter of surface area, the illuminance is equal to 1 lux. There are other units of illuminance.

The illuminace at a given surface depends on the intensity of the light source, the distance of the source from the surface, and the angle that the surface subtends with respect to the rays from the source. *See also* ILLUMINANCE.

MACHINE LANGUAGE

Machine language is the mode in which a computer actually operates. Small groups of binary digits, or bits, comprise instructions for the computer. These groups of bits are called microinstructions. One or more microinstructions together form a macroinstruction (*see* MACROINSTRUCTION, MICROINSTRUCTION).

Computers are usually not operated or programmed in machine language. Instead an assembler or compiler determines the machine language from a higher-order language more easily understood by people. There are several different higher-order languages for various computer applications.

When a new computer is checked, or if a new higher-order language is to be developed, it may be necessary to work directly with machine language. In some program debugging situations, and in computer maintenance and repair work, machine language must be understood by the programmer or technician. *See also* ASSEMBLER AND ASSEMBLY LANGUAGE, HIGHER-ORDER LANGUAGE.

MACRO

Macro is a prefix that refers to a collection of things, or to an extremely large group. Macro means large-scale.

Sometimes a macroinstruction or macroprogram is called a macro for short. *See also* MACROINSTRUCTION, MACROPROGRAM.

MACROINSTRUCTION

A macroinstruction is a machine-language computer instruction made up of a group of microinstructions (*see* MICROINSTRUCTION). A macroinstruction corresponds to a specific, unique command in a higher-order language. The translation from command to macroinstruction is done by the compiler. *See also* MACHINE LANGUAGE, MICROINSTRUCTION.

MACROPROGRAM

A macroprogram is a computer program written in the language used by the operator. When we think of a computer program, we generally think of the macroprogram, which takes the form of a series of statements in higher-order language.

A macroprogram may be written in assembly language, for the purpose of translating between machine language and the user language. Such a program is sometimes called a macroassembler or macroassembly program. *See also* ASSEMBLER AND ASSEMBLY LANGUAGE, HIGHER-ORDER LANGUAGE.

MAGNESIUM

Magnesium is an element with atomic number 12 and atomic weight 24. In its natural form, magnesium is a metal, and it appears much like aluminum. It reacts readily with oxygen or chlorine, as well as certain other elements.

Magnesium compounds are used as phosphor substances in cathode-ray tubes. Magnesium fluoride and magnesium silicate produce a red-orange glow when bombarded by electrons. Magnesium tungstate fluoresces with a bluish color when exposed to an electron beam.

MAGNET

A magnet is a device that produces a flux field because of the effects of molecular alignment, or because of electric currents. The molecules of iron or nickel, when aligned uniformly, produce a continuous magnetic field, and these substances may be used as permanent magnets. A coil of wire with an iron core will produce a strong magnetic field when a current passes through the coil, but the field disappears when the current stops flowing. Such a device is called an electromagnet or temporary magnet.

Magnets are used for many purposes in electronics. The dynamic microphone or speaker uses a magnet to convert mechanical forces into electrical impulses, or vice-versa. Generators and motors use magnets. Indicating devices, such as meters, often use magnets to obtain needle deflection. The list of devices using magnets is almost endless. *See also* ELECTROMAGNET, MAGNET COIL, MAGNETIC FIELD, MAGNETIC POLARIZATION, MAGNETIC POLE, MAGNETIZATION, PERMANENT MAGNET.

MAGNET COIL

An electric current always produces a magnetic field. The magnetic lines of flux occur in directions perpendicular to the flow of current. Thus if a wire is wound into a helical or circular coil, a strong magnetic field is produced, with a configuration similar to that of the magnetic field surrounding a bar magnet (see illustration). A coil, deliberately wound around an iron rod for the purpose of creating a magnetic field, is called a magnet coil or magnetic coil.

All electromagnets contain magnet coils. When a magnet coil carries a large amount of current, the magnetic field becomes extremely strong, and the electromagnet can pick up large iron or steel objects. If the magnet coil is

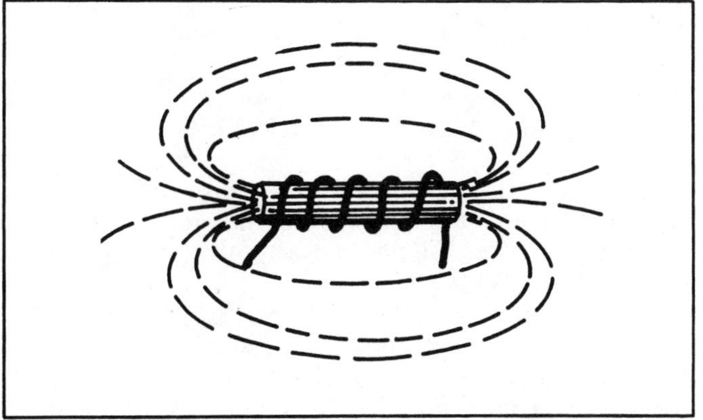

MAGNET COIL: A magnet coil, when carrying current, produces a magnetic field in the same configuration as a bar magnet.

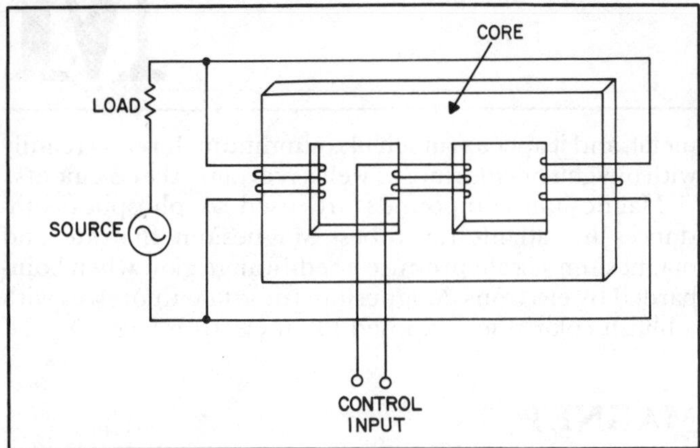

MAGNETIC AMPLIFIER: A magnetic amplifier is a form of variable-impedance transformer.

supplied with an alternating current, the magnetic field collapses and reverses polarity in step with the change in the direction of the current.

Some meters, motors, generators, and other devices use magnet coils, usually for the purpose of converting electrical energy to mechanical energy or vice-versa. *See also* ELECTROMAGNET.

MAGNETIC AMPLIFIER

A magnetic amplifier is a device that is used for the purpose of modulating or changing the voltage across the load in an alternating-current circuit. The magnetic amplifier consists of an iron-core transformer, with an extra winding to which a control signal can be applied.

The illustration is a schematic diagram of a simple magnetic amplifier. Two windings appear in series with the alternating-current power supply and the load. These windings are connected in opposing phase, and together they are called the output winding.

A third coil is wound around the center column of the transformer core. This coil is called the input coil. When a direct current is applied to the input coil, the impedance of the output coil changes. This occurs because of the saturable-reactor principle (*see* SATURABLE REACTOR). A small change in the current through the input coil results in a large fluctuation in the output-coil impedance. Therefore, amplification occurs. Magnetic amplifiers can, if properly designed, produce considerable power gain.

MAGNETIC BEARING

The magnetic bearing, or magnetic heading, refers to the azimuth as determined by a magnetic compass. Magnetic bearings are measured in degrees clockwise from magnetic north.

The magnetic bearing is usually a little different from the geographic bearing, since the geographic and geomagnetic poles are not in exactly the same places. *See also* GEOMAGNETIC FIELD.

The shafts of some meters, gyroscopes, and other rotating devices may employ bearings using magnetic fields to minimize friction. Such bearings are called magnetic bearings. The shaft is literally suspended in the air because of the repulsive magnetic force between like poles of permanent magnets or electromagnets. The friction of such a bearing, when operated in a vacuum, is very nearly zero. *See also* MAGNETIC FORCE.

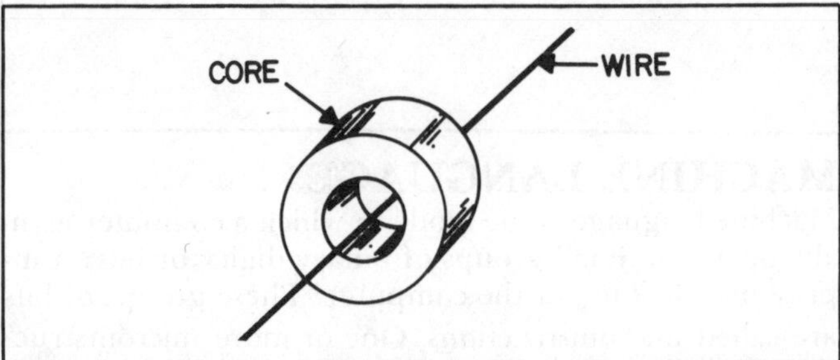

MAGNETIC CORE: A magnetic core consists of a toroidal powdered-iron or ferrite core surrounding a length of wire. A pulse of current in the wire causes a magnetic field to be stored in the core.

MAGNETIC BLOWOUT

An electric arc can be extinguished by a strong magnetic field. A device intended for this purpose is called a magnetic blowout.

The magnetic blowout consists of a permanent magnet that can be brought close to the arc. The magnetic field causes the arc to be deflected. This occurs because of the interaction between the fields surrounding the magnet and the moving electrons in the arc. When sufficient lateral deflection is produced, the arc is "blown out." Sometimes an arc will blow itself out as a result of the magnetic field produced by its own current. *See also* ARC.

MAGNETIC CIRCUIT

A magnetic circuit is a complete path for magnetic flux lines (*see* MAGNETIC FLUX). The lines of flux are considered to originate at the north pole of a magnetic dipole, and to terminate at the south pole. A magnetic circuit has certain similarities to an electric circuit.

The magnetic equivalent of voltage is known as magnetomotive force, and is measured in gilberts. The magnetic equivalent of current is flux density. The magnetic equivalent of resistance is reluctance. These three parameters behave, in a magnetic circuit, in the same way as voltage, current, and resistance behave in an electric circuit. *See also* FLUX DENSITY, MAGNETOMOTIVE FORCE, RELUCTANCE.

MAGNETIC CORE

A magnetic core is a ferromagnetic object, such as an iron rod or powdered-iron toroid core, used for the purpose of concentrating or storing a magnetic field. Magnetic cores are used in many different electrical and electronic devices, such as the electromagnet, the transformer, and the relay.

A special form of magnetic core is used to store digital information in some computers. A toroidal ferromagnetic bead surrounds a wire or set or wires, as shown in the illustration. When a current pulse is sent through the wire, a magnetic field is produced, and the toroid becomes magnetized in either the clockwise or counterclockwise sense (depending on the direction of the current pulse). The toroid remains magnetized in that sense indefinitely, until another current pulse is sent through the wire in the opposite direction. One sense represents the binary digit 0 (false); the other sense represents 1 (true). Magnetic-core storage is retained even in the total absence of power in a computer circuit. Several magnetic cores, when combined, allow the storage of large binary numbers.

MAGNETIC COUPLING

When a magnetic field influences a nearby object or objects, the effect is known as magnetic coupling. Examples of magnetic coupling include the deflection of a compass needle by the earth's magnetic field, the interaction among the windings of a motor or generator, and the operation of a transformer. There are many other examples. Magnetic coupling can take place as a direct force, such as attraction or repulsion, between magnetic fields. Magnetic coupling can also occur because of motion among magnetic objects, or because of changes in the intensity of a magnetic field.

A fluctuating current produces a changing magnetic field, which in turn produces changing currents in nearby objects. This is especially true when coils of wire are placed along a common axis. This form of magnetic coupling is called electromagnetic induction (*see* ELECTROMAGNETIC INDUCTION).

Magnetic coupling is enhanced by the presence of a ferromagnetic substance between two objects. At extremely low frequencies, solid iron is an effective magnetic medium. At very low frequencies, laminated iron or ferrite is preferable. At higher frequencies, ferrite and powdered iron are used.

The degree of magnetic coupling between two wires or coils is measured as mutual inductance. *See also* MUTUAL INDUCTANCE.

MAGNETIC DEFLECTION

Magnetic force, like any force, can cause objects to be accelerated. This acceleration may take place as a change in the speed of an object, a change in the direction of its motion, or both.

Any moving charged object generates a magnetic field.

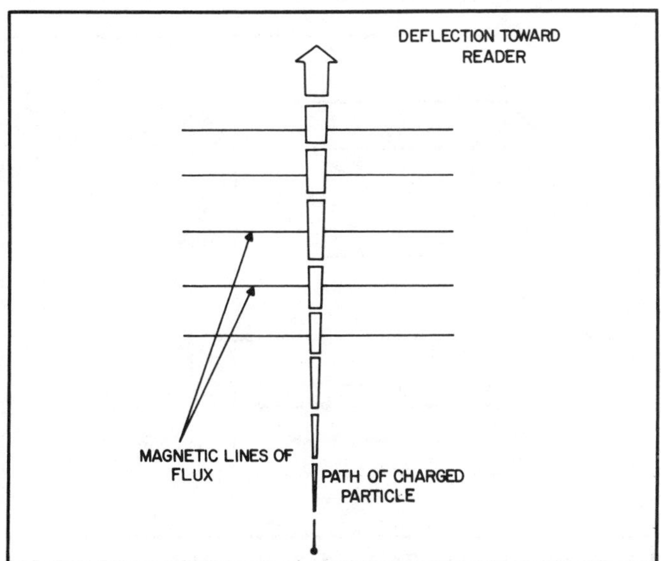

MAGNETIC DEFLECTION: A magnetic field experts a lateral force on charged particles. For the field shown here, the path of the particle is deflected outward from the page, directly toward the reader.

The most common example of this is the magnetic field that surrounds a wire carrying an electric current. However, charged particles traveling through space also produce magnetic fields. When such moving charged particles encounter an external magnetic field, the particles are accelerated or deflected (see illustration).

Charged particles from the sun are deflected by the earth's magnetic field. During an intense solar storm, when many charged particles are emitted from the sun, the deflection causes such a concentration of particles near the poles that a visible glow—aurora—occurs.

The principle of magnetic deflection is used in the common indicating meter, the cathode-ray tube, and many other electronic devices. *See also* MAGNETIC FORCE.

MAGNETIC EQUATOR

The magnetic equator is a line around the earth, midway between the north and south magnetic poles. Every point on the magnetic equator is equidistant from either magnetic pole. At the magnetic equator, a compass needle experiences no dip; the geomagnetic lines of flux are parallel to the surface of the earth.

The magnetic equator does not exactly coincide with the geographic equator, since the magnetic poles do not lie exactly at the geographic poles. *See also* GEOMAGNETIC FIELD.

MAGNETIC FEEDBACK

See INDUCTIVE FEEDBACK.

MAGNETIC FIELD

A magnetic field is a region in which magnetic forces can occur. A magnetic field is produced by a magnetic dipole or by the motion of electrically charged particles.

Physicists consider magnetic fields to be made up of flux lines. The intensity of the magnetic field is determined according to the number of flux lines per unit surface area (*see* FLUX DENSITY). While the magnetic lines of flux do not physically exist in any tangible form, they do have a direction that can be determined at any point in space. Magnetic fields can have various shapes, as shown in the illustration.

A magnetic field is considered to originate at a north magnetic pole, and to terminate at a south magnetic pole. A pair of magnetic poles, surrounded by a magnetic field, is called a magnetic dipole. *See also* MAGNETIC FLUX, MAGNETIC FORCE, MAGNETIC POLE.

MAGNETIC FIELD INTENSITY

See FLUX DENSITY.

MAGNETIC FLUX

Magnetic lines of flux are sometimes referred to as magnetic flux. Magnetic flux exists wherever an electrically charged particle is moving. Magnetic flux surrounds certain materials with aligned or polarized molecules. Magnetic flux lines are considered to emerge from a magnetic north pole and to enter a magnetic south pole.

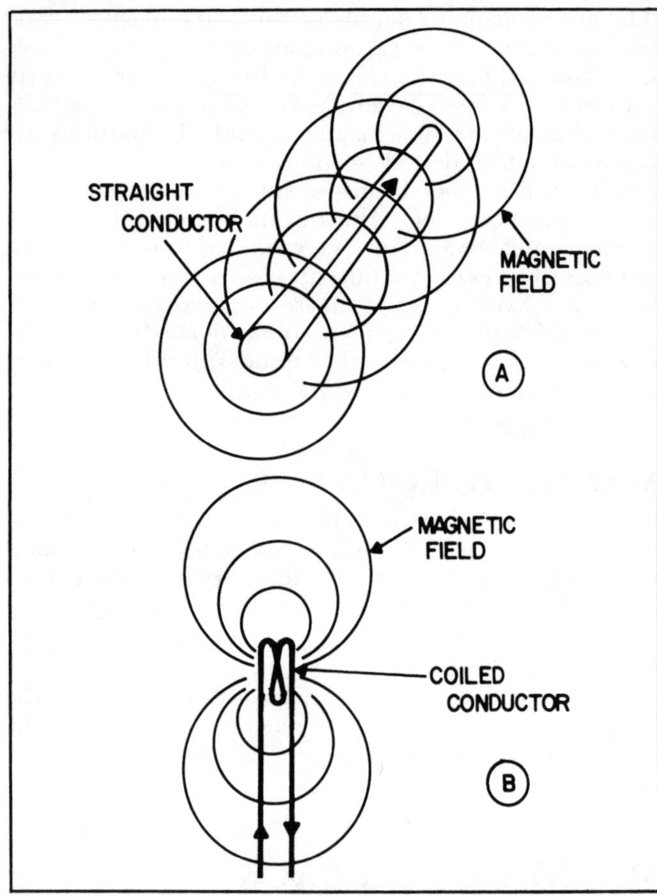

MAGNETIC FIELD: At A, the magnetic field surrounding a straight conductor. At B, the magnetic field surrounding a coiled conductor.

Every flux line is, however, a continuous, closed loop. This is true no matter what is responsible for the existence of the magnetic field.

The more intense a magnetic field, the greater the number of flux lines crossing a given two-dimensional region in space. A line of flux is actually a certain quantity of magnetic flux, usually 1 weber or 1 maxwell. The strength of a field may be measured in webers per square meter. Another commonly used unit of flux density is the line of flux per square centimeter; this unit is known as the gauss. *See also* FLUX, FLUX DENSITY, FLUX DISTRIBUTION, GAUSS, MAGNETIC FIELD, MAXWELL, WEBER.

MAGNETIC FORCE

Magnetic fields produce forces that can be directly measured. Magnetic forces occur between magnetic poles and objects made with ferrous material. Magnetic forces also occur between any two magnetic poles, and between current-carrying wires.

When two like magnetic poles are brought near each other, repulsion occurs. When two unlike poles are brought near each other, attraction occurs. The magnitude of the force depends on the strengths of the magnetic fields surrounding the poles, and on the distance separating the poles. The force is proportional to the product of the flux densities, and inversely proportional to the cube of the distance between the poles.

Magnetic force causes deflection in the paths of moving charged particles. The force occurs in a direction perpendicular to both the magnetic field and the path traveled by the particle (*see* MAGNETIC DEFLECTION).

Magnetic forces can, if properly harnessed, produce levitation effects. An experimental train, capable of traveling at more than 300 miles per hour, uses magnetic forces to provide a low-friction track and to produce acceleration. Magnetic forces are responsible for the operation of many electronic devices, including transducers, cathode-ray tubes, and meters. *See also* MAGNETIC FIELD, MAGNETOSTRICTION.

MAGNETIC HEADING

See MAGNETIC BEARING.

MAGNETIC INDUCTION

See ELECTROMAGNETIC INDUCTION.

MAGNETIC MATERIAL

A magnetic material is any substance whose molecules become aligned, or polarized, in the presence of a magnetic field. Such materials include iron and nickel, and various mixtures derived from them.

Some magnetic materials can be come permanently magnetized because their molecules remain aligned even after the external field has been removed. Such objects are then called permanent magnets (*see* MAGNET, PERMANENT MAGNET). Permanent magnets are used in transducers and meters in a variety of situations.

At frequencies below several hundred megahertz, transformer cores are routinely made from magnetic materials. These substances are generally known as ferrite, laminated iron, and powdered iron. *See also* FERRITE CORE, LAMINATED CORE, POWDERED-IRON CORE, TRANSFORMER.

MAGNETIC PERMEABILITY

See PERMEABILITY.

MAGNETIC POLARIZATION

In a ferrous substance, such as iron, nickel, or steel, the molecules produce magnetic fields. Each molecule forms a

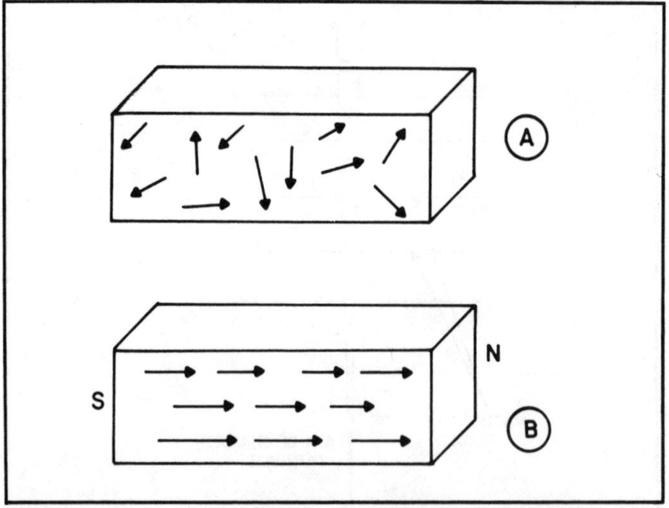

MAGNETIC POLARIZATION: At A, the dipoles (molecules) in an unmagnetized material are oriented at random. At B, a magnetizing force causes alignment of the dipoles, resulting in a magnetic field surrounding the piece of material.

magnetic dipole, containing a north and south magnetic pole (*see* MAGNETIC POLE).

The molecules in a ferrous substance are normally aligned in random fashion, so that the net effect of their magnetic fields is zero (A in the illustration). However, in the presence of an external magnetic field, the molecules become more nearly aligned. If the field is very strong, the molecules line up in nearly the same orientation. This greatly increases the strength of the magnetic field within, and near, the substance. When the field is removed, the molecules of some ferrous materials remain aligned, forming a permanent magnet (B).

Magnetic memory devices operate on the principle of magnetic polarization. A current through a wire produces a magnetic field. This field influences the properties of some ferrous substance, such as powdered iron, causing the substance to become polarized. The magnetic memory retains its polarization until reverse current pulse is sent through the wire (*see* MAGNETIC CORE, MAGNETIC RECORDING).

The core of the earth is believed to be composed mostly of iron. The molecules in the interior of our planet are aligned in such a way that a permanent magnetic field exists around the earth. The axis of this magnetic field nearly coincides with the axis of rotation of the earth. Some other planets, such as Jupiter, also exhibit this form of magnetic polarization. *See also* GEOMAGNETIC FIELD, PERMANENT MAGNET.

MAGNETIC POLE

A magnetic pole is a point, or localized region, at which magnetic flux lines converge. Magnetic poles are called either north or south. This is basically a matter of convention. In the magnetic compass, one end of the magnetized needle points toward the north magnetic pole; this is called the north pole of the magnet. Similarly, the end of a magnet that orients itself toward the south is called the south pole. Magnetic flux lines, according to theory, leave north poles and enter south poles.

It is believed that magnetic poles must always exist in pairs. This conclusion arises from one of the fundamental laws of Maxwell. Magnetic flux lines, according to Maxwell, always take the form of closed loops. Thus the magnetic lines of flux must arise as a result of dipoles—pairs of north and south magnetic poles (*see* DIPOLE). Magnetic monopoles may be possible, and much interesting theoretical research has been done in an attempt to find an isolated magnetic north or south pole. However, such a phenomenon would appear contrary to the laws of Maxwell. Some scientists think they may have found magnetic monopoles. *See also* MAGNETIC FIELD, MAXWELL'S LAWS.

MAGNETIC RECORDING

Magnetic recording is a method of storing and retrieving data by means of magnetic tape or disks. The magnetic medium consists of powdered iron oxide on a flexible plastic base. A magnetic transducer, called the head, generates a magnetic field that causes polarization of the molecules in the iron oxide. *See* DISKETTE, FLOPPY DISK, MAGNETIC TAPE.

In the magnetic-disk recording system, the head follows a spiral path on the disk. The main advantage of magnetic-disk recording is rapid access to any part of the information stored on the disk. Magnetic disks are used for storage of computer data, but they are becoming more common for video recording as well.

Magnetic-tape recording makes use of a long strip of magnetic material, usually 0.5 mil, 1.0 mil, or 1.5 mil thick and ⅛ inch, ¼ inch, or 2 inches wide. In magnetic sound recording, the information is placed along straight paths called tracks. A tape may have one, two, four, or eight tracks, each with a different recording impressed on it. For sound applications, the standard tape sizes are ⅛ inch or ¼ inch wide, and the most common speeds are 1-⅞, 3-¾, or 7-½ inches per second. In video recording, the tape speed is 15 inches per second; the video information is recorded on oblique tracks with a frequency-modulated carrier, and the sound is recorded on one or two straight tracks.

Magnetic recording techniques have improved since they were first introduced decades ago. Today, magnetic recordings make it possible to store vast amounts of information for long periods of time. *See also* TAPE RECORDER, VIDEO DISK RECORDING, VIDEO TAPE RECORDING.

MAGNETIC STORM

See GEOMAGNETIC STORM.

MAGNETIC TAPE

Magnetic tape is a form of information-storage medium, commonly used for storing sound, video, and digital data. Magnetic tape is available in several different thicknesses and widths for different applications.

Magnetic tape consists of millions of fine particles of iron oxide, attached to a plastic or mylar base. A magnetic field, produced by the recording head, causes polarization of the iron-oxide particles. As the field fluctuates in intensity, the polarization of the iron-oxide particles also varies (see illustration). When the tape is played back, the magnetic fields surrounding the individual iron-oxide

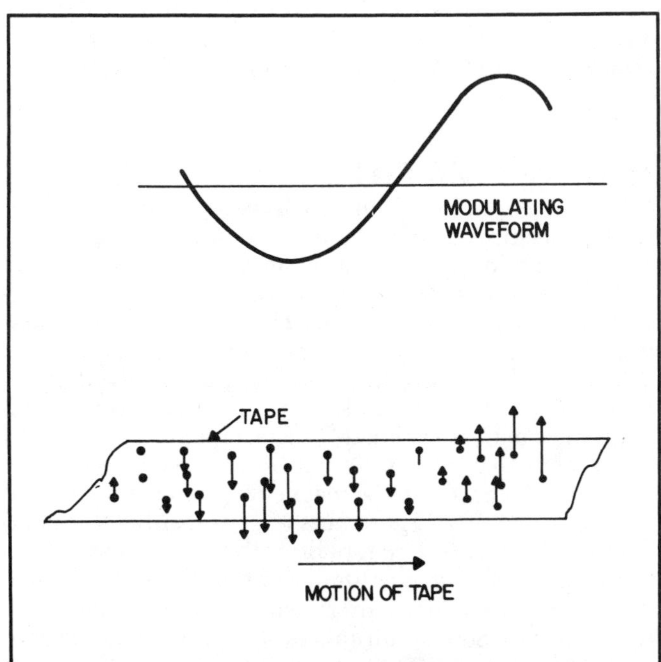

MAGNETIC TAPE: Magnetic dipoles in a magnetic tape become aligned by the magnetizing force of the applied signal.

particles produce current fluctuations in the playback or pickup head.

For sound and computer-data recording, magnetic tape is available either in cassette form (*see* CASSETTE) or wound on reels (*see* REEL-TO-REEL TAPE RECORDER). The thickness may be 0.5 mil, 1.0 mil, or 1.5 mil. (A mil is 0.001 inch.) The thicker tapes have better resistance to stretching, although the recording time, for a given length of tape, is proportionately shorter than with thin tape. In the cassette, the tape may be either ⅛ inch or ¼ inch wide. Reel-to-reel tape is generally ¼-inch wide. Wider tape is often used, however, in recording studios and in video applications. The tape may have up to 24 individual recording tracks. The most common speeds are 1-⅞, 3-¾, and 7-½ inches per second. For voice recording, the slower speeds are adequate, but for music, higher speeds such as 15 inches per second are preferable for enhanced sensitivity to high-frequency sound.

Video tape ranges from 1/2 inch to 2 inches wide, and the standard recording speed is 15 inches per second. Even this speed is not directly sufficient for the reproduction of the high frequencies encountered in video, but the recording is done in a slanted or oblique direction, resulting in much higher effective speeds (*see* VIDEO TAPE RECORDING).

Magnetic tape provides a convenient and compact medium for long-term storage of information. However, certain precautions should be observed. The tape must be kept clean, and free from grease. Magnetic tapes should be kept at a reasonable temperature and humidity and they should not be subjected to magnetic fields. *See also* MAGNETIC RECORDING, TAPE RECORDER.

MAGNETISM

Magnetism is a term that is used to describe any magnetic effects. Magnetic attraction and repulsion, interaction with moving charged particles, and polarization of molecules may all be loosely categorized as magnetism. *See also* MAGNETIC DEFLECTION, MAGNETIC FIELD, MAGNETIC FORCE, MAGNETIC MATERIAL, MAGNETIC POLARIZATION.

MAGNETIZATION

When certain materials are subjected to magnetic fields, their molecules become aligned (*see* MAGNETIC POLARIZATION), resulting in intensification of the magnetic field within the substance. When this occurs, the substance is said to be magnetized. Magnetization may be temporary, disappearing as soon as the magnetizing force is removed; or it may be permanent, persisting indefinitely after removal of the external field. Transformer cores exhibit temporary magnetization. Magnetic tape is an example of a substance that attains permanent magnetization.

Some substances can be magnetized and demagnetized rapidly, while other substances react sluggishly to changing magnetic fields. The rapidity with which a material can be magnetized, demagnetized, and remagnetized is called the degree of magnetic hysteresis for the substance. (*See* HYSTERESIS.) Some substances are easily magnetized, while others are difficult to magnetize. The ease with which a material can be magnetized is called its permeability. (*See* PERMEABILITY.)

MAGNETOHYDRODYNAMICS

Electrically charged or conductive substances interact with magnetic fields. For example, a moving charged particle is deflected by a magnetic field. Similarly, a conductive fluid, flowing in any medium, is influenced by the presence of a magnetic field. The study of the behavior of charged or conductive gases and fluids, under the influence of magnetic fields, is known as magnetohydrodynamics.

Magnetohydrodynamics has many applications. For example, a device called a magnetohydrodynamic generator allows the production of electricity from a hot gas. The gas is passed through a magnetic field, resulting in precipitation of electrons. A circulating magnetic field can be used to impart spin to a conductive fluid, producing inertia; this device is called a magnetohydrodynamic gyroscope. *See also* MAGNETIC DEFLECTION.

MAGNETOMOTIVE FORCE

Magnetomotive force is sometimes called magnetic potential or magnetic pressure. Magnetomotive force, in a magnetic circuit, is analogous to electromotive force in an electric circuit.

The most common unit of magnetomotive force is the gilbert. Another often-used unit is the ampere turn. A current of 1 ampere, passing through a one-turn coil of wire, sets up a magnetomotive force of 1 ampere turn or 1.26 gilberts. In general, the magnetomotive force M generated by a coil of N turns, carrying a current of I amperes, is given in ampere turns by the formula:

$$M = NI$$

The larger the magnetomotive force in a region of space, the greater the magnetic effects, all other factors being constant. *See also* AMPERE TURN, GILBERT.

MAGNETORESISTANCE

Certain semiconductor substances, such as bismuth and indium antimonide, exhibit variable resistance depending on the intensity of a surrounding magnetic field. Even ordinary wire conductors can be influenced somewhat, if the magnetic field is sufficiently strong.

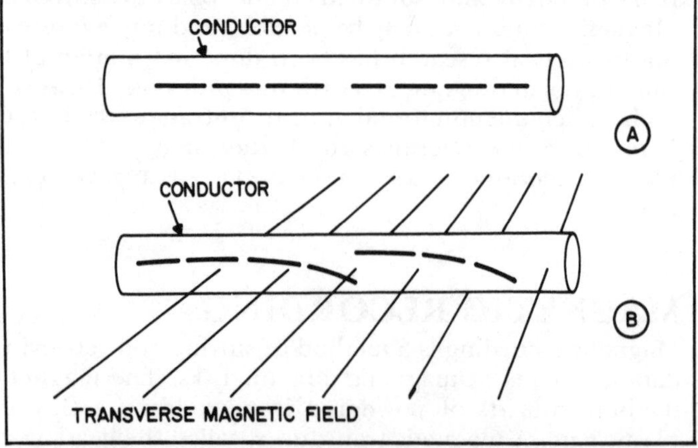

MAGNETORESISTANCE: At A, current flows parallel to the conductor walls when no external magnetic field is present. At B, an external magnetic field, if applied at an angle to the conductor axis, causes deflection of the charge carriers in the conductor. This increases the resistance through the conductor.

Magnetoresistance occurs because of magnetic deflection of the paths of charge carriers. If no magnetic field is present, charge carriers move in paths that, on the average, are parallel to a line connecting the centers of opposite charge poles as at A in the illustration. If a magnetic field is introduced, however, the paths of the charge carriers are deflected, and their paths tend to curve toward one side of the conducting medium, as at B. This increases the effective distance through which the charge carriers must travel to get from one charge pole to the other. The effect is very much like a current in a river, which makes it necessary for a swimmer to work harder to reach a specified point on the opposite shore. *See also* MAGNETIC DEFLECTION.

MAGNETOSPHERE

The magnetic field surrounding the earth produces dramatic effects on charged particles in space. The geomagnetic lines of flux extend far above the surface of our planet, and high-speed atomic particles constantly arrive from the sun and the distant stars. When these charged particles enter the earth's magnetic field, they are deflected toward the poles.

The magnetosphere is the boundary, roughly spherical in shape and surrounding the earth, that represents the maximum extent of the magnetic-deflection effects of the geomagnetic field on charged particles from space (see illustration). *See also* GEOMAGNETIC FIELD.

MAGNETOSTRICTION

Magnetostriction is the tendency for certain substances to expand or contract under the influence of a magnetic field. Magnetostriction, in a magnetic field, is analogous to electrostriction in an electric field (*see* ELECTROSTRICTION).

Substances that exhibit magnetostrictive properties include Alfer (an alloy of aluminum and iron), nickel, and Permalloy (an alloy of iron and nickel). Nickel exhibits negative magnetostriction; that is, it is compressed by a magnetic field. Alfer and Permalloy exhibit positive magnetostriction; they expand as the surrounding magnetic field becomes stronger. Other magnetostrictive substances include various forms of iron oxide, nichrome, powdered iron, and ferrite.

If a magnetostrictive substance is subjected to a sufficiently strong magnetic field, the material may actually break under the strain. This occasionally happens with ferrite and powdered-iron inductor cores when the surrounding coils carry current far in excess of rated values.

Magnetostrictive materials can be employed in some of the same applications as piezoelectric crystals. For example, a magnetostriction oscillator works by means of resonance in a rod of magnetostrictive material. Transducers can also be made with magnetostrictive materials. *See also* MAGNETOSTRICTION OSCILLATOR, PIEZOELECTRIC EFFECT.

MAGNETOSTRICTION OSCILLATOR

A magnetostriction oscillator is similar to a crystal oscillator. Instead of a piezoelectric crystal, the magnetostriction oscillator makes use of a rod of magnetostrictive material. The rod serves two purposes: It vibrates at the frequency of oscillation, and it acts as the core for the feedback transformer (see illustration).

The fundamental oscillating frequency is determined by the physical length and diameter of the rod, whether it exhibits positive or negative magnetostriction, and the extent of the magnetostrictive effects. *See also* MAGNETOSTRICTION.

MAGNETRON

The magnetron is a form of traveling-wave tube used as an oscillator at ultra-high and microwave frequencies.

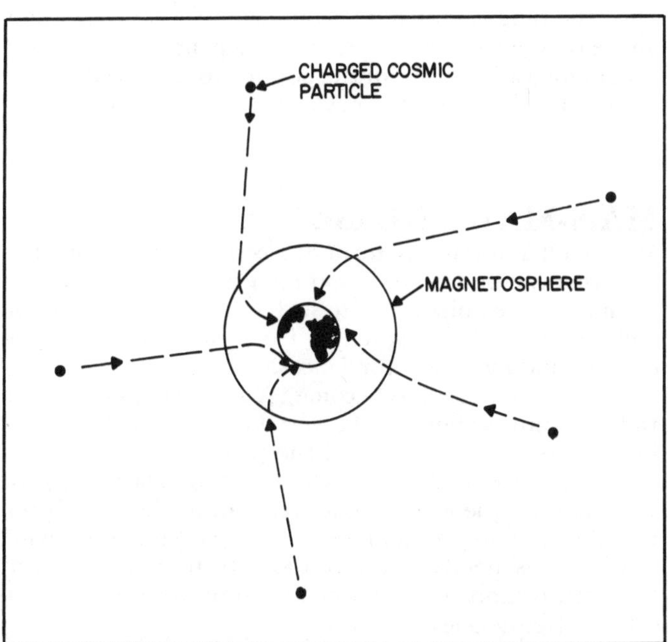

MAGNETOSPHERE: The magnetosphere of the earth is the outer boundary of the region within which charged particles are affected by the geomagnetic field.

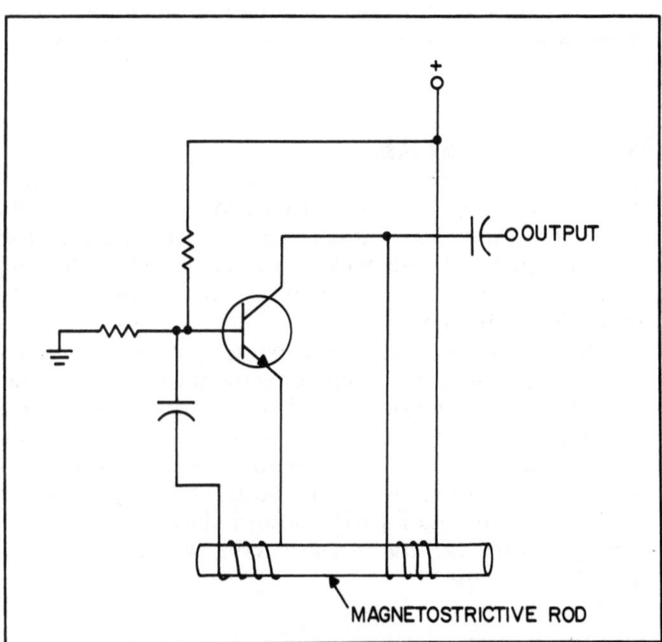

MAGNETOSTRUCTION OSCILLATOR: This circuit uses a bipolar transistor.

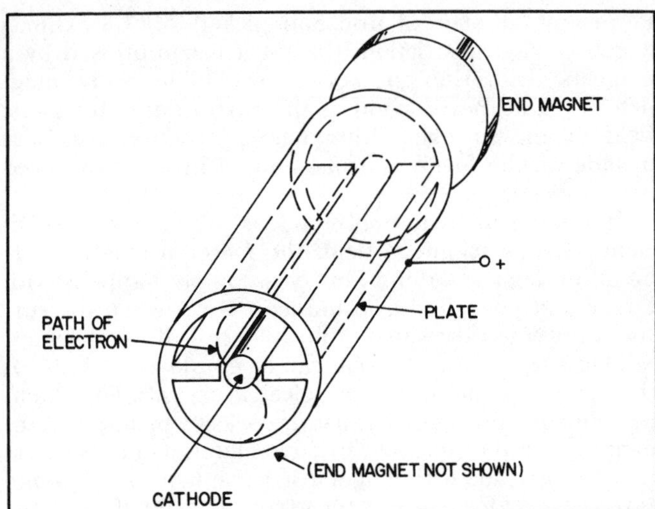

MAGNETRON: The magnetron produces oscillation because of the influence of a magnetic field on the motion of electrons.

Most magnetrons contain a central cathode and a surrounding plate, as shown in the illustration. The plate is usually divided into two or more sections by radial barriers called cavities. The r-f output is taken from a waveguide opening in the anode.

The cathode is connected to the negative terminal of a high-voltage source, and the anode is connected to the positive terminal. This causes electrons to flow radially outward from the cathode to the anode. A magnetic field is applied in a longitudinal direction; this causes the electrons to travel outward in spiral, rather than straight, paths. The electrons tend to travel in bunches because of the interaction between the electric and magnetic fields. This results in oscillation with the frequency being somewhat stabilized by the cavities.

Magnetrons can produce continuous power outputs of more than 1 kW at a frequency of 1 GHz. The output drops as the frequency increases; at 10 GHz, a magnetron can produce about 10 to 20 watts of continuous radio-frequency output. For pulse modulation, the peak-power figures are much higher. *See also* TRAVELING-WAVE TUBE.

MAGNET WIRE

Magnet wire is a term sometimes applied to solid-copper, insulated wire for low-current use. Magnet wire gets its name from the fact that it is popular in winding coils for electromagnets. Magnet wire may have plastic, rubber, or enamel insulation, and ranges in size from approximately AWG No. 14 to AWG No. 40.

Magnet wire is rigid; when bent, it stays bent. This makes magnet wire especially convenient for coil winding. Magnet wire is sometimes used for shortwave receiving antennas, and for transmitting antennas at low power levels. However, magnet wire tends to stretch when it is unsupported in a long span, because the wire is made of soft copper. For circuit wiring, stranded wire is preferable to magnet wire. *See also* HOOKUP WIRE, WIRE.

MAINFRAME

See CENTRAL PROCESSING UNIT.

MAJORITY CARRIER

All semiconductors carry current in two ways. Electrons transfer negative charge from the negative pole to the positive pole; holes carry positive charge from the positive pole to the negative pole (*see* ELECTRON, HOLE). Electrons and holes are known as charge carriers.

In an N-type semiconductor material, electrons are the dominant form of charge carrier. In a P-type material, holes are the more common charge carrier. The dominant carrier is called the majority carrier. *See also* MINORITY CARRIER, N-TYPE SEMICONDUCTOR, P-TYPE SEMICONDUCTOR.

MAJOR LOBE

See LOBE.

MAKE

The action of closing a circuit is often called make. The term is generally used with relays, or with keying circuits in a code transmitter.

In the code transmitter, make represents the length of time required, after the key is closed, for the output signal to rise from zero to maximum amplitude. This is a rapid process, normally taking a few milliseconds. However, it is not instantaneous. The rate at which the carrier amplitude rises is extremely important; if it is too fast, key clicks will occur, but if it is too slow, the signal will sound "soft" and may even be difficult to read. *See also* CODE TRANSMITTER, KEY CLICK, SHAPING .

MALE

Male is the term that denotes any connector designed for insertion into a receptacle. Examples of male connectors include the common lamp plug, the phone plug, and the edge connectors found on some printed-circuit boards.

A male connector can be recognized by the fact that, when the plug is pulled, the conductors are exposed. For this reason, male connectors are usually used for the load in a circuit, rather than for the power supply, to reduce the chances for short circuits or electric shock. *See also* FEMALE.

MAN-MADE NOISE

Man-made noise is any form of electromagnetic interference that can be traced to non-natural causes. In particular, man-made noise refers to such interference as ignition and impulse noise, originating from internal-combustion engines and electrical appliances.

Man-made noise has become an increasing problem in radiocommunications at the very-low, low, medium, and high frequencies, because of the proliferation of household appliances that generate electromagnetic energy. At very-high frequencies and above, man-made noise is not usually a serious problem, except when extreme receiving sensitivity is needed. The radio astronomer, however, faces frustration from man-made noise sources at practically all frequencies.

Man-made noise, unlike natural noise, generally comes from a constant direction at a given frequency. This makes it possible, in most cases, to cancel man-made noise by means of highly directional receiving antennas such as the ferrite rod or small loop, although the apparent direction of the noise may change with frequency. Noise blankers and limiters can also help in the reduction of interference from man-made noise. *See also* ELECTROMAGNETIC INTERFERENCE, FERRITE-ROD ANTENNA, IGNITION NOISE, IMPULSE NOISE, LOOP ANTENNA, NOISE, NOISE BLANKER, NOISE LIMITER.

MARCONI ANTENNA

Any antenna measuring ¼ electrical wavelength at the operating frequency, and fed at one end, is called a Marconi antenna. One of the earliest communications antennas was of the Marconi type, used at low frequencies, and consisting of an end-fed wire with a short vertical section and a long horizontal span (see illustration). The most common modern form of Marconi antenna is the ground-mounted, quarter-wave vertical antenna. However, a simple length of wire, connected directly to a transmitter, is also sometimes used by radio amateurs and shortwave listeners today, much as it was in the first days of wireless communication.

The Marconi antenna is simple to install, but it must be operated against a good radio-frequency ground. When this is done, the Marconi antenna can be extremely effective. The grounding system may consist of a set of radial wires, a counterpoise, or a simple ground-rod or water-pipe connection. The feed-point impedance of the Marconi antenna is a pure resistance, and ranges from about 20 to 50 ohms, depending on the surrounding environment and the precise shape of the radiating element. The Marconi antenna may be operated at any odd harmonic of the fundamental frequency. *See also* GROUND-PLANE ANTENNA, MARCONI EFFECT, VERTICAL ANTENNA.

MARCONI EFFECT

A center-fed or end-fed antenna may show resonance at a frequency determined by the length of the feed line and radiator combined. Such resonance usually occurs at a frequency at which the combined length of the feed line and radiator is ¼ wavelength, or some odd multiple thereof. Marconi effect can result in unwanted radiation from an antenna feed line, as the whole system behaves like a Marconi antenna. This can occur with antenna systems using coaxial feeders, as well as with systems using open-wire line. The currents flow on the outside of the coaxial outer conductor, or in phase along both conductors of open-wire line.

At low, medium, and high frequencies, antenna systems should generally not be operated at frequencies where Marconi effect might take place. For a particular frequency, it is wise to choose the feed line length to avoid Marconi resonances. *See also* MARCONI ANTENNA.

MARCONI, GUGLIELMO

A pioneer in wireless communications, Guglielmo Marconi (1874-1937) is best known for his work in long-distance transmission of electromagnetic waves. Marconi also invented and patented a horizontal aerial that displayed directional properties (*see* MARCONI ANTENNA).

In his later years, Marconi experimented with shortwave wireless communication. His work ultimately led to the vast network of high-frequency stations that still exists today. In 1909, Marconi won the Nobel prize for physics.

MARITIME-MOBILE RADIO SERVICE

A two-way radio station in an ocean-going vessel, such as a small boat or a large ship, engaged in non-amateur communication, is called a maritime-mobile station. Any fixed station, used for the purpose of non-amateur communication with the operator of an ocean-going vessel, is also part of the maritime-mobile service.

The maritime-mobile radio service is allocated a number of frequency bands in the radio spectrum. The bands range from the very low frequencies through the ultra-high frequencies; precise allocations vary somewhat from country to country.

MARK/SPACE

In a digital communications system in which there are two

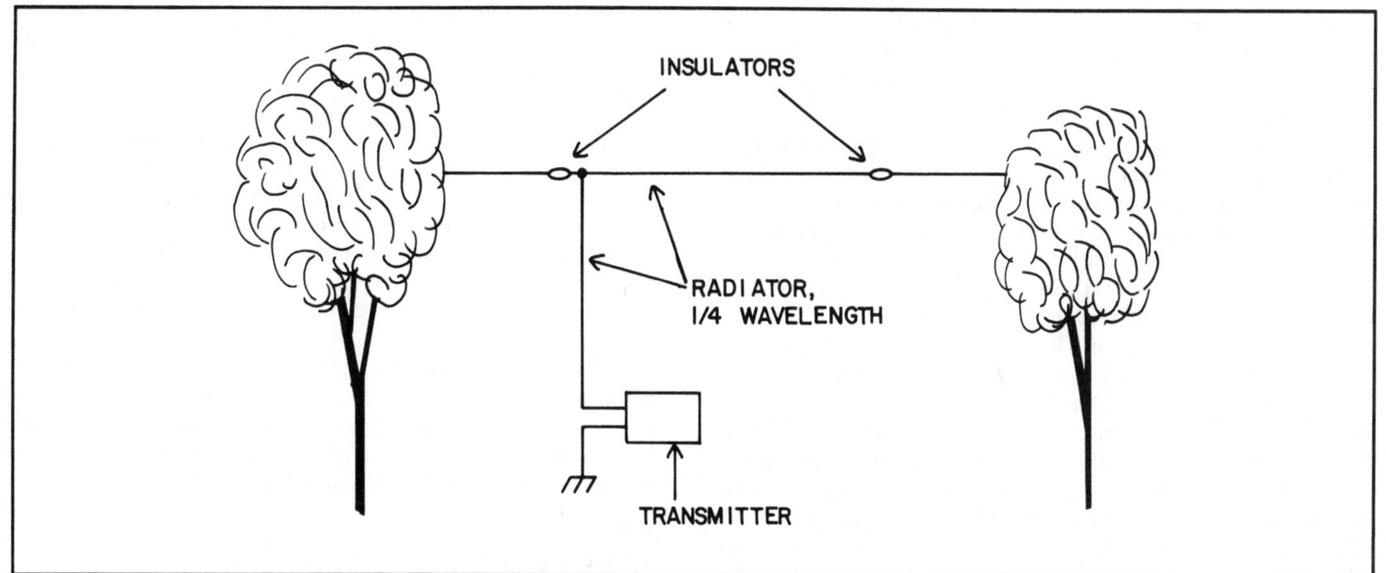

MARCONI ANTENNA: A Marconi antenna is a quarter-wavelength radiator operated against ground.

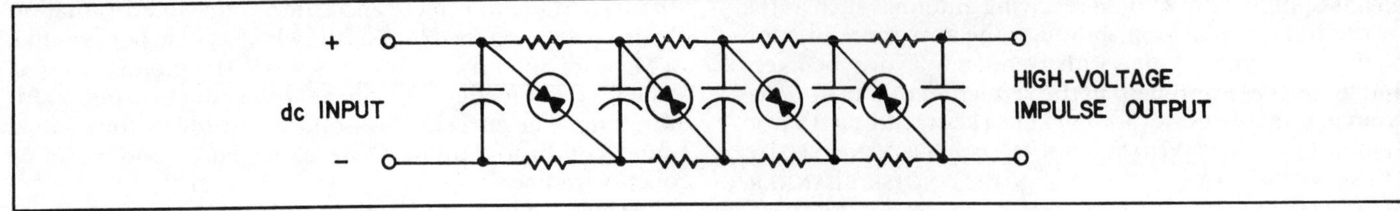

MARX GENERATOR: A Marx generator produces high-voltage pulsed output from a low-voltage direct-current input.

possible states, such as a Morse-code system or radioteletype system, the ON condition is usually called mark, and the OFF condition is called space. Frequency-shift keying is often used in radioteletype; the lower frequency represents mark and the higher frequency represents space. *See also* FREQUENCY-SHIFT KEYING.

MARX GENERATOR

A Marx generator is a form of high-voltage, direct-current impulse generator. The principle of operation is somewhat analogous to a voltage-multiplier power supply. A source of direct current is connected to a set of capacitors that are arranged in parallel with series resistances, as shown in the illustration. A set of spark gaps is connected to the network of capacitors in such a way that the gaps are subjected to the series combination of the capacitor voltages. The voltage appearing across the spark gaps is therefore much greater than the voltage across the individual capacitors.

When the capacitors reach a certain critical charge value, they discharge to create a spark across all of the gaps at once. This process occurs repeatedly as the capacitors alternately charge and discharge. The sparks produce short pulses of extremely high voltage at the output terminals.

MASER

The maser is a special form of amplifier for microwave energy. The term itself is an acronym for Microwave Amplification by Stimulated Emission of Radiation. The maser output is the result of quantum resonances in various substances.

When an electron moves from a high-energy orbit to an orbit having less energy, a photon is emitted (*see* ELECTRON ORBIT). For a particular electron transition or quantum jump, the emitted photon always has the same amount of energy, and therefore the same frequency. By stimulating a substance to produce many quantum jumps from a given high-energy level to a given low-energy level, an extremely stable signal is produced. This is the principle of the maser oscillator. Ammonia and hydrogen gas are used in maser oscillators as frequency standards. Rubidium gas may also be used. The output-frequency accuracy is within a few billionths of 1 percent in the gas maser.

Solid materials, such as ruby, can be used to obtain maser resonances. This kind of maser is called a solid-state maser, and it can be used as an amplifier or oscillator. When an external signal is applied at one of the quantum resonance frequencies, amplification is produced. The solid-state maser must be cooled to very low temperatures for proper operation; the optimum temperature is nearly absolute zero. While satisfactory maser operation can sometimes be had at temperatures as high as that of dry ice, the most common method of cooling is the use of liquid helium or liquid nitrogen. These liquid gases bring the temperature down to just a few degrees above absolute zero.

The traveling-wave maser is a common form of device for use at ultra-high and microwave frequencies. The cavity maser is another fairly simple device. Both the traveling-wave and cavity masers use the resonant properties of materials cut to precise dimensions. The traveling-wave and cavity masers are solid-state devices, and the temperature must be reduced to a few degrees Kelvin for operation. The gain of a properly operating traveling-wave or cavity type maser amplifier may be as great as 30 to 40 dB.

Maser amplifiers are invaluable in radio astromony, communications, and radar. Some masers operate in the infrared, visible-light, and even ultraviolet parts of the electromagnetic spectrum. Visible-light masers are called optical masers or, more commonly, lasers. *See also* LASER.

MASS

Mass is a measure of the amount of matter present in an object. Mass is expressed in kilograms (kg) in the standard international system of units. The kilogram is the amount of mass which, when subjected to a force of 1 newton in free space, undergoes an acceleration of 1 meter per second per second.

Mass is often confused with weight. However, weight depends on the intensity of a gravitational field in which the mass exists; mass is independent of the environment in which an object is found. The pound (lb) is an expression of weight; on the surface of our planet, a mass of 1 kg weighs about 2.2 lb. But on the moon, the same mass weighs just 0.36 lb, and on Jupiter, it weighs 5.5 lb.

MAST

A mast is a supporting structure for an antenna. A mast is not part of the radiating system, but serves only to hold the antenna in place. Generally, a mast consists of a vertical pole, from 1 inch to several inches in diameter. A mast may be self-supporting, as at A in the illustration, or it may require guying, as at B. A short mast is often attached to a rotator atop a tower (as at C) when directional antenna systems are used.

When using a mast to support an antenna, it is important that the mast have sufficient diameter to withstand twisting forces in high winds. It is also important to be sure that unsupported mast sections are strong enough to stand up against the wind resistance. *See also* TOWER.

MASTER

Master is a term that refers to the main, or primary, item in

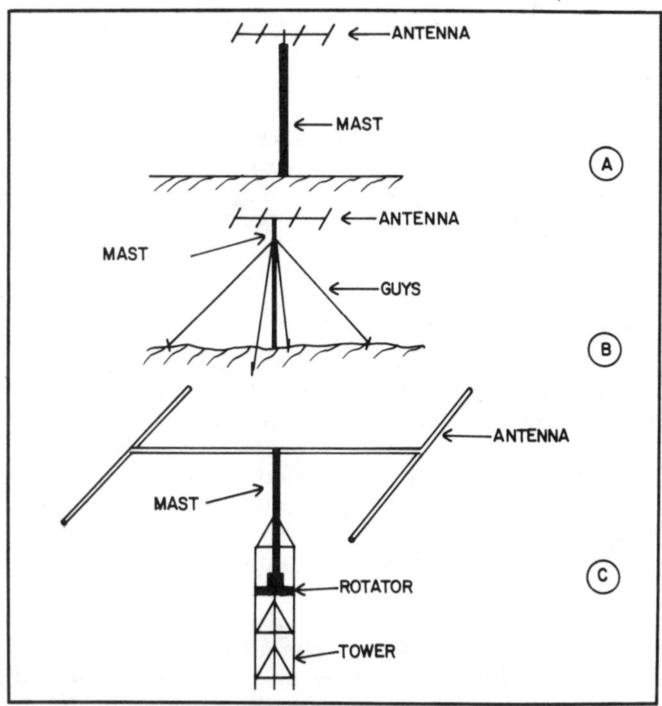

MAST: At A, an unsupported antenna mast. At B, a guyed mast. At C, a tower-extension mast with rotator.

a group of items. For example, the main or original copy of a recording is sometimes called the master. A computer file that is retained for a long time, and is not significantly altered, is called a master file. The central or primary control console for an electronic system is called the master console.

A system-controlling device is sometimes called a master. For example, in a large communications network, the main control point is called the master control center. In some situations, the devices regulated by the master are called slaves. *See also* SLAVE.

MATCHED IMPEDANCE

See IMPEDANCE MATCHING.

MATCHED LINE

A matched line is a feed line or transmission line, terminated by a nonreactive load whose impedance is identical to the characteristic impedance of the line. In a matched line, there is no reflected power, and there are no standing waves. A matched condition is desirable for any line, because the line loss is minimized.

Given a feed-line characteristic impedance of Z_o and a purely resistive load impedance R, where $Z_o = R$, the current and voltage appear in the same proportion at all points along the line. If E is the voltage at any point on the line, and I is the current at that same point, the following holds:

$$E/I = Z_o = R$$

This uniformity of voltage and current exists only in a matched line.

Since any feed line has some loss, the voltage and current are somewhat higher near the transmitter than near

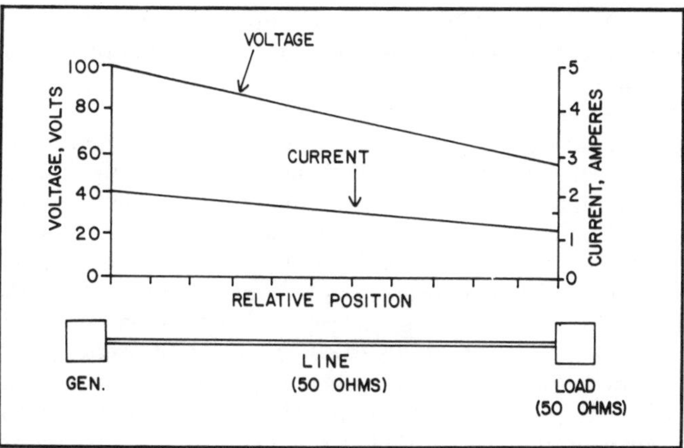

MATCHED LINE: In a matched transmission line, the voltage-to-current ratio is the same at every point along the line, even though the actual voltage and current values may vary because of line loss.

the load, as shown in the illustration. However, the ratio E/I is always the same. *See also* CHARACTERISTIC IMPEDANCE, IMPEDANCE MATCHING, MISMATCHED LINE, REFLECTED POWER, STANDING WAVE, STANDING-WAVE RATIO.

MATCHING

When two or more active devices, such as field-effect transistors, bipolar transistors, or vacuum tubes, are operated in parallel, push-pull, or push-push, it is desirable to ensure that they have operating characteristics that are identical. For a given type of transistor or tube, there are often considerable differences in operating parameters among different sample units. If the components are chosen at random, current hogging or other undesirable effects may occur. Component matching eliminates such potential problems. *See also* CURRENT HOGGING, PUSH-PULL CONFIGURATION, PUSH-PUSH CONFIGURATION.

The term matching is sometimes used to describe the process or condition of matched impedances. *See also* IMPEDANCE MATCHING.

MATCHTONE

A matchtone is a device that is used for the purpose of monitoring the transmission of code signals. The matchtone device is connected to the output of a transmitter to make the signals audible.

The matchtone operates via the radio-frequency energy from a strong signal. The radio-frequency energy is rectified to produce direct-current pulses. These pulses cause a changing bias in an audio-frequency oscillator, so that the key-down condition results in an audible-tone output.

MATHEMATICAL LOGIC

Mathematical logic is the science of reasoning, approached from a rigorous standpoint. Mathematical logic is sometimes called symbolic logic. There are several different recognized forms of mathematical logic; the simplest is called sentential logic. In sentential logic, complete statements or sentences are the smallest units used. A more complex form of mathematical logic is known as first-order logic; sentences are broken down into smaller units. Second-order logic is still more detailed. Specialized forms of mathematical logic exist; an example is a form of logic in which sentences are infinitely long! The more complicated forms of logic are of interest primarily to the theoretician, and not (yet) to the engineer.

Digital logic, of interest to the electronic engineer, takes the form of elementary sentential logic. Sentential logic is fundamentally simple, but the equations and combinations can become quite involved. The mathematician studies not only the truth and falsity of individual logical statements, but generalized theorems concerning the nature of logic itself. Mathematical logic begins with set theory, and progresses through the study of complex logical languages. A full description of mathematical logic is utterly impossible within the scope of this volume. However, many textbooks, mostly at the college level, are available for the interested reader. *See also* BOOLEAN ALGEBRA, LOGIC, LOGIC EQUATION, LOGIC FUNCTION.

MATRIX

A matrix is a high-speed switching array, consisting of wires and diodes in a certain orderly arrangement. The matrix itself is an array of wires; some of the wires are interconnected via diodes (*see* DIODE MATRIX).

In linear algebra, a matrix is a rectangular array of numbers that can represent a set of simultaneous linear equations. The coefficients of the equations are represented in an orderly fashion, eliminating the variables. This facilitates algebraic operations with a minimum of confusion. For example, the set of equations:

$$2x - 4y + 2z = 0$$

$$x + 7y + 6z = 0$$

$$-2x - 4y - 3z = 0$$

may be represented by the matrix:

$$\begin{pmatrix} -2 & -4 & 2 & 0 \\ 1 & 7 & 6 & 0 \\ -6 & -4 & -3 & 0 \end{pmatrix}$$

Matrices are useful in the solving of sets of linear equations, especially by computer. *See also* MATRIX ALGEBRA.

MATRIX ALGEBRA

Matrix algebra is a means of solving sets of simultaneous linear equations. Basically, matrix algebra consists of various operations on matrices, in a manner similar to the familiar addition and multiplication of numbers. Given two matrices, their sum or product may be found by fairly simple operations.

Matrix addition is commutative. That is, for any two matrices A and B, it is always true that $A + B = B + A$. This is not, however, true in general for multiplication of matrices. The distributive and associative laws both hold. Many theorems exist in matrix algebra, and a general review of these theorems is beyond the scope of this book. A linear algebra text should be consulted for details.

In the solution of a set of linear equations, three elementary row operations are used. These operations can be performed in any order. Two rows may be exchanged; any row may be multiplied by a nonzero number; any row may be replaced by the sum of its values and x times the values of another row, where x is a nonzero number. These operations are rapidly carried out by computer in solving sets of linear equations containing hundreds or even thousands of variables. *See also* LINEAR ALGEBRA.

MAXIM, HIRAM P.

A co-founder of the American Radio Relay League, Hiram P. Maxim was one of the first radio amateurs to send messages over long distances by relaying them from one station to another.

In the early part of the twentieth century, radio signals were transmitted by spark generators at low and medium frequencies. The maximum possible range was normally less than 100 miles. Efficient handling of messages, however, allowed rapid dissemination of information from one end of the country to the other. Radio amateurs still send messages in this way, free of charge, using nets (*see* NET).

Maxim had to fight to keep amateur radio from being outlawed in the years following World War I. He succeeded, at great cost to himself, and the American Radio Relay League today has over 150,000 active members. *See also* AMATEUR RADIO, AMERICAN RADIO RELAY LEAGUE.

MAXIMUM POWER TRANSFER

When the power from a source is optimally transferred to a load, maximum power transfer is said to occur. Maximum power transfer requires that the load impedance be matched to the source impedance.

When the load and source impedances differ, some of the electromagnetic field is reflected back to the source. This causes loss in the system. *See also* IMPEDANCE MATCHING, STANDING-WAVE-RATIO LOSS.

MAXIMUM USABLE FREQUENCY

For any two points separated by more than a few miles, electromagnetic communication is possible only at certain frequencies. At very low frequencies, using high transmitter power, worldwide communication is almost always possible. In general, there is a band of higher frequencies at which communication is also possible, and the power requirements are moderate.

Given some arbitrary medium or high frequency f, at which communication is possible between two specific points, suppose the frequency is raised until communication is no longer possible. The cutoff frequency might be called f_u. This cutoff point is called the maximum usable frequency, or MUF, for the two locations in question.

The MUF depends on the locations and separation distance of the transmitter and receiver. The MUF also varies with the time of day, the season of the year, and the sunspot activity. All of these factors affect the ionosphere. For a given signal path, the propagation generally improves as the frequency is increased toward the MUF; above the MUF the communication abruptly deteriorates. When a communications circuit is operated just below the

MUF, and the MUF suddenly drops, a rapid and total fadeout occurs. This effect is familiar to users of the high-frequency radio spectrum. *See also* LOWEST USABLE FREQUENCY, MAXIMUM USABLE LOW FREQUENCY, PROPAGATION CHARACTERISTICS.

MAXIMUM USABLE LOW FREQUENCY

In the very-low-frequency radio band, worldwide communication is almost always possible with high-power transmitters. The ionosphere returns all signals to the earth at these frequencies. Surface-wave propagation alone can provide long-range communications at very-low frequencies.

Given a circuit at some very-low frequency f, between two defined points on the earth, suppose the frequency is gradually raised. If the transmitter and receiver are more than a few miles apart, the path loss generally increases as the frequency is increased. This happens for two reasons: The ionospheric D layer becomes less reflective and more absorptive, and the surface-wave loss increases (*see* D LAYER, SURFACE WAVE). Under some conditions, a frequency f_u exists at which communication deteriorates to the point of uselessness. This frequency may be in the low or medium range, and is known as the maximum usable low frequency (MULF).

The MULF is not the same as the maximum usable frequency. If the frequency is raised above the MULF, communication will often become possible again at a frequency called the lowest usable high frequency (LUF). As the frequency is raised still more, propagation ultimately deteriorates at the maximum usable frequency (MUF). *See also* LOWEST USABLE FREQUENCY, MAXIMUM USABLE FREQUENCY, PROPAGATION CHARACTERISTICS.

MAXWELL

The maxwell is a unit of magnetic flux, representing one line of flux in the centimeter-gram-second system of units. Magnetic flux density may be measured in maxwells per square centimeter.

The weber is the more common, standard international unit of magnetic flux. The weber is equivalent to 10^8 maxwells. *See also* MAGNETIC FLUX, WEBER.

MAXWELL BRIDGE

The Maxwell bridge is a device used for measuring unknown inductances. The internal resistance of a coil can also be determined. A typical Maxwell bridge circuit is illustrated by the schematic diagram.

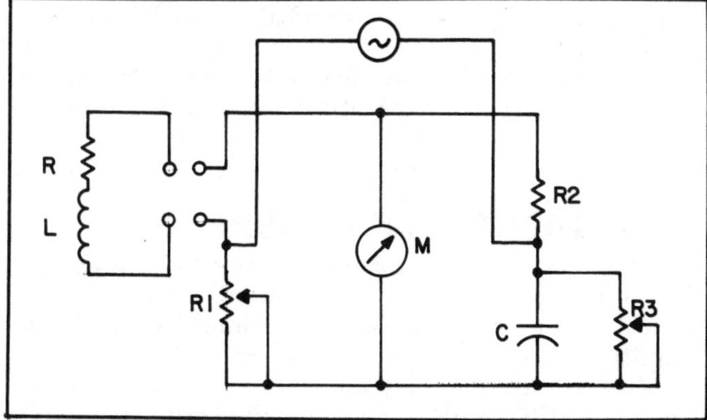

MAXWELL BRIDGE: A Maxwell bridge for inductance determination.

The unknown inductance, L, and the unknown series resistance, R, are determined by manipulating the variable resistors R1 and R3. Balance is indicated by a null reading on the indicator meter. When balance has been reached, the value of L in henrys is found by:

$$L = C(R1R2)$$

where C is given in farads and R1 and R2 are given in ohms. The series resistance, R, is:

$$R = R2^2/R3$$

where all resistances are in ohms. *See also* INDUCTANCE.

MAXWELL, JAMES CLERK

A Scottish scientist and professor of experimental physics, James Clerk Maxwell (1831-1879) was involved with many aspects of physics. His main contribution to electricity and electronics was his extensive work in electromagnetic theory and practice. Maxwell showed that visible light is an electromagnetic wave. He also developed his famous equations concerning electric and magnetic fields and their behavior. *See also* MAXWELL'S EQUATIONS.

MAXWELL'S EQUATIONS

Maxwell's equations are a set of four mathematical relations that describe the behavior of electromagnetic fields. Verbally stated, the rules are:

(1) The amount of work necessary to move a unit magnetic pole completely around any closed path is equal to the total current linking the path. The total current consists of conduction and displacement currents.
(2) The electromotive force induced in any nonmoving, closed loop is proportional to the rate of change of magnetic flux within the loop.
(3) For an electrically charged object, the total electric flux surrounding the object is precisely equal to the charge quantity.
(4) Magnetic lines of flux are always closed loops. That is, magnetic lines of flux have no points of beginning or ending.

A detailed discussion of Maxwell's equations is beyond the scope of this book. The integral forms of Maxwell's equations, corresponding to the four rules stated above, are given below for reference, but a physics textbook should be consulted for further detail. The following notations are used:

I	Conduction current is a closed path
I^1	Displacement current linking a path
H	Magnetic-field vector
E	Electric-field vector
B	Magneic-flux-density vector
D	Electric-flux-density vector
df/dt	Rate of change of electric flux
dg/dt	Rate of change of magnetic flux

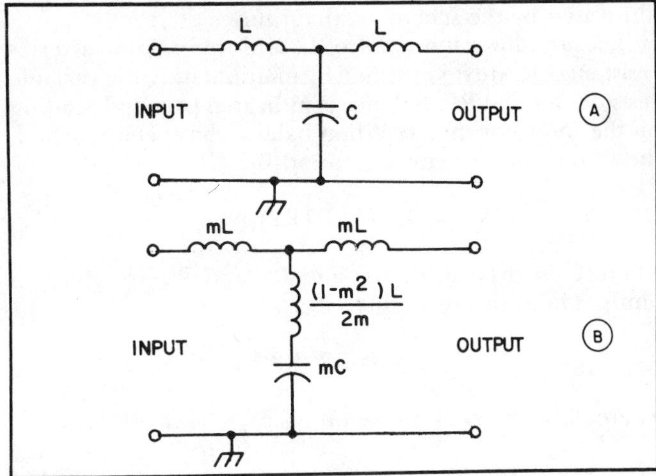

M-DERIVED FILTER: Examples of T-section low pass filters: At A, typical constant-k, andat B, m-derived.

dp	Path-length increment vector
dS	Vector element normal to a surface representing a small portion of the surface
Q	Charge quantity on a surface

Using these notations, the four rules can be stated as:

(1) $\int H \bullet dp = I + I^1 = I + df/dt$

(2) $\int E \bullet dp = -dg/dt$

(3) $\int D \bullet dS = Q$

(4) $\int B \bullet dS = 0$

See also ELECTRIC FIELD, ELECTROMAGNETISM, MAGNETIC FIELD.

MAYDAY

See DISTRESS SIGNAL.

McPROUD TEST

The McProud test is a simple means of checking the tracking efficiency of a phonograph. The phonograph is set for 45 revolutions per minute (RPM). A standard 45-RPM disk is placed on the turntable without the usual center adapter, and the disk is pushed all the way to one side to produce the greatest possible back-and-forth swing as the turntable rotates. The pickup arm is placed on the disk surface. Satisfactory tracking efficiency requires that the stylus remain in the groove continuously, with no slippage, in spite of the fact that the resulting sound is warbled.

M-DERIVED FILTER

An m-derived filter is a variation of a constant-k inductance-capacitance filter (*see* CONSTANT-K FILTER). The m-derived filter gets its name from the fact that the values of inductance and capacitance are multiplied by a common factor m. The illustration at A shows a typical T-section, constant-k, lowpass filter for unbalanced line.

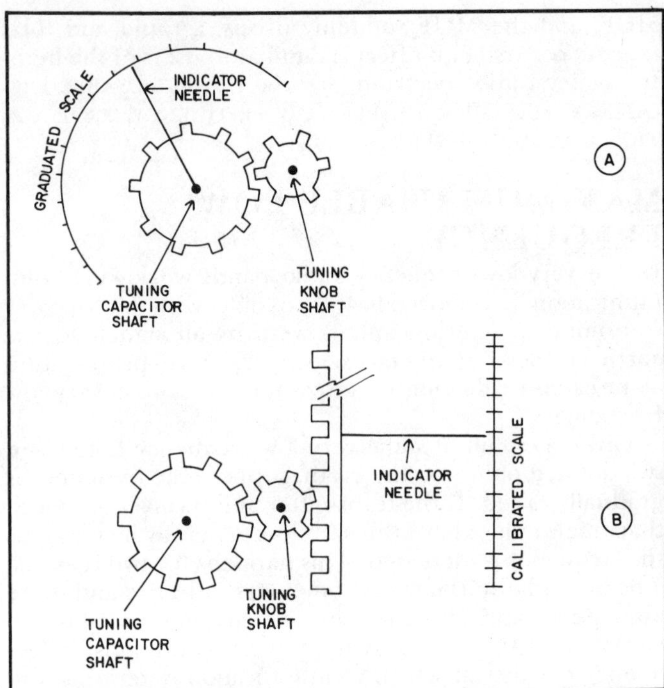

MECHANICAL BANDSPREAD: Two methods of obtaining mechanical bandspread. At A, a rotary indicator is used. At B, a rack-and-pinion arrangement provides a slide-rule dial readout.

The illustration at B shows an m-derived filter with component values altered by the factor m. An additional inductor is placed in series with the capacitor in the m-derived filter.

The value of the factor m is always between 0 and 1. The optimum value for m in a given situation depends on the type of response and the cutoff frequency desired. A properly designed m-derived filter has a sharper cutoff than a constant-k filter for a given frequency. *See also* HIGHPASS FILTER, HIGHPASS RESPONSE, LOWPASS FILTER, LOWPASS RESPONSE.

MEAN

The average of a set of values is sometimes called the mean. In statistics, the mean value of a distribution may involve the average of a large number, or even an infinite number, of individual samples. If there are an infinite number of values, the distribution is represented by a continuous function, called an expected-value function. The mean of such a distribution is found by integrating the expected values over the entire set of real numbers.

The mean for a finite set of values may be calculated either arithmetically or geometrically. Usually, the arithmetic mean is used. The values are added together, and the sum is then divided by the number of values in the set. *See also* ARITHMETIC MEAN, GEOMETRIC MEAN.

MECHANICAL BANDSPREAD

In a radio receiver or transmitter, mechanical bandspread is a way of reducing the tuning rate by means of gear reduction or other mechanical techniques. Mechanical bandspread is sometimes called vernier.

Most standard-broadcast receivers use mechanical bandspread. This makes it easier to tune in the signals. The drawing illustrates two common methods for obtaining mechanical bandspread. At A, a pair of gears is used, with the tuning knob connected to the smaller gear. At B, a rack-and-pinion arrangement is shown. The extent of the bandspreading is determined by the gear ratio.

In sophisticated communications receivers, electrical bandspread is generally preferred over mechanical bandspread because the precision obtainable by mechanical methods is limited. *See also* ELECTRICAL BANDSPREAD.

MECHANICAL DIALER

See ROTARY DIALER.

MECHANICAL FILTER

A mechanical filter is a form of bandpass filter. It is sometimes called an ultrasonic filter.

The mechanical filter operates on a principle similar to the ceramic or crystal filter. Two transducers are employed, one for the input and the other for the output. The input signal is converted into electromechanical vibration by the input transducer. The resulting vibrations travel through a set of resonant disks to the output transducer. The output transducer converts the vibrations back into an electrical signal by means of magnetostriction.

The mechanical filter offers some advantages over electrical filters at low and medium frequencies. Mechanical elements are of reasonable size above about 75 kHz, allowing a filter to be enclosed in a small package. Above perhaps 750 kHz, mechanical filters become difficult to construct because the elements become too small. No adjustment is needed, since the resonant disks are solid and of fixed size. The resonant characteristics are extremely pronounced, often far superior to electrical filters of the same physical size. Mechanical filters can be designed to have a nearly flat response within the passband, and very steep skirts with excellent ultimate attenuation (*see* SKIRT SELECTIVITY). The mechanical filter is a good choice for bandpass applications at intermediate frequencies from 75 to 750 kHz.

The elements of a mechanical filter have more than one resonant frequency. Therefore, it is necessary to provide some external means of attenuation at spurious frequencies. Mechanical filters are somewhat sensitive to physical shock, and care must be exercised to ensure that such filters are not subjected to excessive vibration.

The construction of a mechanical filter is very similar to the construction of a ceramic filter. Instead of ceramic disks, metal disks, usually made of nickel, are used. *See also* CERAMIC FILTER.

MECHANICAL RECTIFIER

Any device that converts alternating current to direct current by mechanical means is called a mechanical rectifier. The simplest form of mechanical rectifier is the commutator. This device acts by synchronizing the make-and-break of a rotating contact with the alternating-current cycle. Conduction occurs only when current flows in the desired direction. Current flowing in the opposite direction is either cut off or reversed. *See also* COMMUTATOR, RECTIFICATION.

MECHANICAL SCANNING

In early television apparatus, the scene was scanned by mechanical means rather than by electrical means. This method of scanning can still be used today, although it is not found in television.

The mechanical scanner consists of a moving light receptor that focuses on various parts of an image. The image is broken into one or more horizontal lines.

Mechanical scanning can be used in a low-frequency oscilloscope. In this device, a thin pencil of light is reflected from a mirror and cast onto a display screen. A second mirror causes the light-beam image to rapidly move across the display screen in a horizontal path. Incoming signals cause vertical deflection of the beam.

MEDIAN

Median is a statistical expression. In general, in a given set of values or samples, the median value is that value for which an equal number of samples fall above and below.

If the number of values in a set is finite and odd, the median is the middle value. For example, if the values are 1, 3, 4, 5, 7, 8, and 9, the median is 5: There are three values above 5 and three below 5. If the number of values is even, then the median is generally the arithmetic mean of the two center values. For example, if the values are 2, 3, 7, and 8, then the median is equal to the arithmetic mean of 3 and 7. That value is 5. (Note that if the number of elements in a sample set is finite and even, then the median is not in the set.)

In a statistical distribution having an infinite number of values, the median must be found by integration. The median is that value for which half of the area under the curve lies to the left, and half of the area under the curve lies to the right (see illustration). Thus, if the total area under the curve is 1, that is:

$$\int_{-\infty}^{\infty} p(x)\,dx = 1$$

where p is the continuous probability distribution, then the

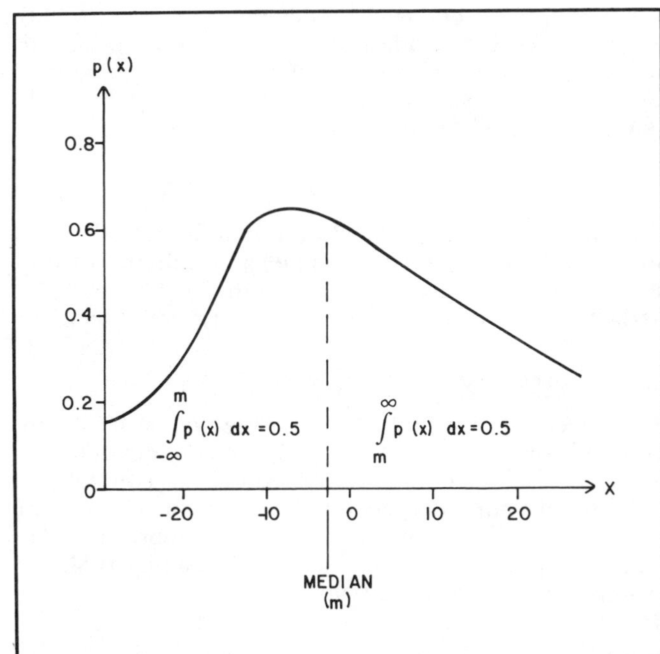

MEDIAN: The median of a continuous probability function is the value for which the area under the curve is equal on both sides.

median is the value m such that

$$_{-\infty}\int^{m} p(x)\,dx = {}_{m}\int^{\infty} p(x)\,dx = 0.5$$

The median is sometimes the same, for a given set of values, as the arithmetic mean or average. But this is not true in general. *See also* ARITHMETIC MEAN.

MEDICAL ELECTRONICS

Electronic devices can be used to diagnose illness or to treat illness. Examples of diagnostic electronic equipment include the ultrasonic scanner, the electrocardiograph, and the electgroencephalograph. Equipment used for treatment includes radio-frequency generators, the heart pacemaker, and even microcomputers for causing controlled muscle movement in paralyzed people. These examples are but a few of the applications of electronic equipment in medicine.

More specific information about various aspects of medical electronics is given in the following articles:ARM-LIFT, BACK/CHEST-PRESSURE RESUSCITATION, ARTIFICIAL RESPIRATION, BIOELECTRONICS, BIONICS, CARDIOPULMONARY RESUSCITATION,CYBERNETICS, DEFIBRILLATION, DELTA WAVE, DIATHMERY, DOSIMETRY, ELECTRIC SHOCK, ELECTROCARDIOGRAM, ELECTROCAUTERY, ELECTROCEPHALOGRAPH, ENDORADIOSONDE, FIBERSCOPE, FLUOROSCOPE, HEART FIBRILLATION, MICROCOMPUTER CONTROL, MOUTH-TO-MOUTH RESUSCITATION, MOUTH-TO-NOISE RESUSCITATION, NUCLEAR MAGNETIC RESONANCE, PACEMAKER, PSYCHOACOUSTIC, RADIOLOGY, ROENTGEN, SHOCK HAZARD, SUBLIMINAL COMMUNICATIONS, ULTRASONIC HOLOGRAPHY, X-RAY TUBE.

MEDIUM-SCALE INTEGRATION

Medium-scale integration is the process by which integrated circuits, containing up to 100 individual gates per chip, are fabricated. Medium-scale integration, or MSI, allows considerable miniaturization of electronic circuits, but not to the extent of large-scale integration or LSI.

Both bipolar and metal-oxide semiconductor (MOS) technology have been adapted to MSI. Various linear and digital circuits use MSI circuitry. When extreme miniaturization is necessary, LSI is more often used. *See also* INTEGRATED CIRCUIT, SOLID-STATE ELECTRONICS.

MEGA

A prefix multiplier, attached to a quantity to indicate 1 million (10^6), is mega. For example, a megahertz is 1 million hertz, and a megavolt is 1 million volts. *See also* MEGAHERTZ, MEGAWATT, MEGOHM, PREFIX MULTIPLIERS.

MEGABYTE

The megabyte is a unit of information equal to 2^{20}, or 1,048,576 bytes. The megabyte is a common unit of memory used in minicomputers and larger digital computers.

Personal computers generally have memory of less than one megabyte. The term megabyte is often abbreviated M; we might speak of a memory capable of holding 10M, for example. *See also* BYTE, KILOBYTE.

MEGACYCLE

See MEGAHERTZ.

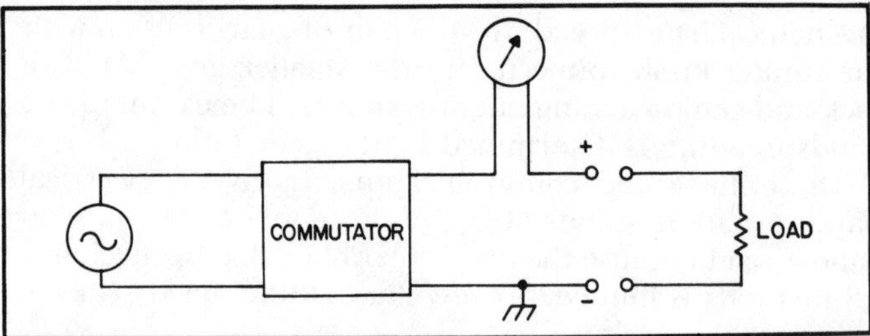

MEGGER: A megger is used for measuring extremely high values of resistance.

MEGAHERTZ

The megahertz, abbreviated MHz, is a unit of frequency, equal to 1 million hertz, or 1 million cycles per second.

Radio frequencies in the medium, high, very-high, and ultra-high ranges are often specified in megahertz. A frequency of 1 MHz is at approximately the middle of the standard AM broadcast band in the United States. *See also* FREQUENCY.

MEGAWATT

The megawatt, abbreviated MW, is a unit of power, or rate of energy expenditure. A power level of 1 MW represents 1 million watts. In terms of mechanical power, 1 MW is about 1,340 horsepower.

The megawatt is a very large unit of power. A small town of 5,000 residents uses approximately 1 MW of power on an average day. Some radio broadcast transmitters have an input power level approaching 1 MW. *See also* POWER, WATT.

MEGGER

A megger is a device that is used to measure extremely high resistances. The resistances of electrical insulators range from several megohms (millions of ohms) up to billions or trillions of ohms, or even higher. For the measurement of such resistances, a high-voltage source is needed.

The illustration is a schematic diagram of a typical megger device. The high-voltage generator is turned by hand. The voltage is dangerous, and presents an electric shock hazard.

The megger may be used for continuity testing and short-circuit location, as well as for the measurement of high direct-current resistances. *See also* OHMMETER.

MEGOHM

A megohm is a unit of resistance equivalent to 1 million ohms. With a potential difference of 1 volt, a current of 1 microampere (1 millionth of an ampere) flows through a resistance of 1 megohm.

Some common resistors have values that range as high as a few hundred megohms. Typical insulating materials have resistances of millions of megohms or more. *See also* OHM, RESISTANCE.

MEMORY

Memory is the retention of digital information for future

use. There are several types of memory devices. Some memory is retained by electrical means, and some memory is retained by magnetic means. Some memory circuits are intended for immediate access; others are designed to store information for periods of days, weeks, or even months.

Various types of integrated circuits are used for storing or recalling memory information. A read-only memory (ROM) contains data that can be accessed for various purposes, but data cannot be re-written into the ROM. A random-access memory (RAM) can be programmed and re-programmed, as desired. Memory capacity is measured in bytes (*see* BYTE). In a computer, the memory capacity is an important consideration; the greater the memory-storage room, the more versatile the computer. Typical home computers have memory capacity ranging up to several hundred kilobytes; a kilobyte is 1,024 bytes. An industrial computer may have room for up to many megabytes.

Memory may be retained in ferrite cores, in the form of directional magnetic fields (*see* MAGNETIC CORE). Such a memory device can retain information for prolonged periods, although it is much bulkier than the integrated-circuit form of memory.

Some memory devices require a source of power in order to be retained for long periods. Such a source of power is called a memory-backup battery. *See also* MEMORY BACKUP BATTERY, RANDOM-ACCESS MEMORY, READ-ONLY MEMORY.

MEMORY-BACKUP BATTERY

A memory backup battery is a set of electrochemical cells used for the purpose of retaining electrical data in integrated circuits. The memory backup battery may be rechargeable, and connected to the main power source, or it may be independent of the main power source.

Memory backup batteries are used in microcomputer-controlled devices in which the main source of power is removed for long periods of time. For example, a transceiver may contain programmed operating frequencies; the memory backup battery keeps these frequencies in storage even if the unit is not in use. The memory backup battery in a microcomputer should be capable of supplying sufficient energy to retain the data for at least a month. *See also* MEMORY.

MERCURY BATTERY

See MERCURY CELL.

MERCURY CELL

A mercury cell is a form of electrochemical cell that generates about 1.4 volts. The mercury cell is enclosed in a steel housing. The cathode is made from mercuric oxide; this is how the cell gets its name. The anode is made of zinc. The electrolyte is made up of potassium hydroxide and zinc oxide; hence the mercury cell is an alkaline cell.

The mercury cell is characterized by a relatively long life for its size and mass. Also, the mercury cell tends to produce a very constant voltage throughout its life. A series combination of mercury cells is called a mercury battery. *See also* CELL.

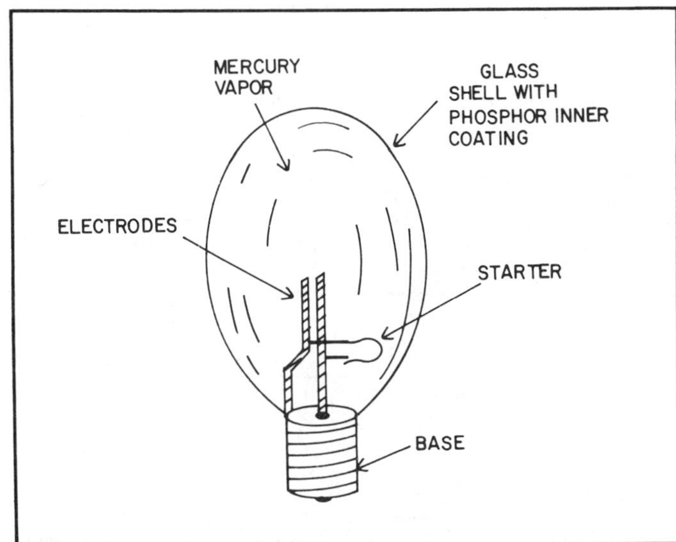

MERCURY-VAPOR LAMP: A form of electroluminescent device.

MERCURY-VAPOR LAMP

A mercury-vapor lamp is a form of electroluminescent lamp. Mercury vapor, when ionized, produces intense visible and ultraviolet light at several discrete wavelengths. Mercury-vapor lamps are commonly used as street lights in cities. The mercury-vapor lamp is much more efficient than the common incandescent lamp.

The typical mercury-vapor lamp (see drawing) consists of a pair of electrodes, a small amount of mercury, and a phosphor coating on the inner surface of the lamp shell. The ionization of the vaporized mercury results in considerable ultraviolet radiation; when this radiation strikes the phosphor, visible light is produced. Mercury-vapor lamps can be recognized by their intense blue-white light.

MERCURY-VAPOR RECTIFIER

The mercury-vapor rectifier is a cold-cathode, tube type rectifier used in high-voltage power supplies. While solid-state rectifier diodes have largely replaced tube type rectifiers in modern electronic power supplies, mercury-vapor tubes are still sometimes found.

The mercury-vapor tube operates because of ionization of a small amount of mercury at a high temperature and a low pressure. The voltage drop across the mercury-vapor tube electrodes is about 15 volts, regardless of the supply voltage. Mercury-vapor rectifiers are usually paired in a full-wave, center-tap configuration. Four tubes can be used as a bridge rectifier. *See also* RECTIFICATION, RECTIFIER TUBE.

MESH

A mesh is a collection of circuit branches with the following three properties: (1) The collection forms a closed circuit; (2) Every branch point in the circuit is incident to exactly two branches; and (3) No other branches are enclosed by the collection. While a circuit loop may contain smaller loops, such as in the example at A in the illustration, a mesh is a "smallest possible loop," as at B.

The mesh provides a means of evaluating the currents and voltages in various parts of a complex circuit. Kirchhoff's laws can be applied to all of the individual meshes in a circuit. *See also* KIRCHHOFF'S LAWS, MESH ANALYSIS.

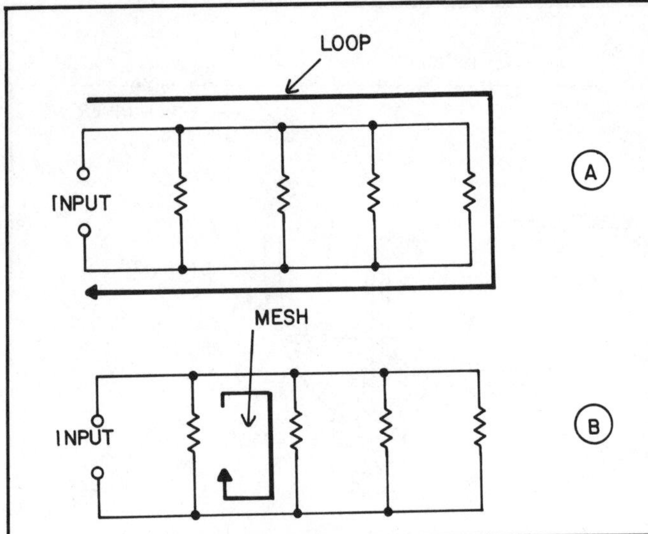

MESH: At A, a loop. At B, a mesh.

MESH ANALYSIS

Mesh analysis is a method of evaluating the operation or characteristics of an electronic circuit containing one or more closed loops. The current, voltage, and resistance through or between various parts of the circuit can be determined according to Kirchhoff's laws. Kirchhoff's laws form the basis for mesh analysis.

Mesh analysis is relatively simple for direct-current circuits, although the equations become rather cumbersome for circuits having many meshes. Mesh analysis is mathematically sophisticated in the alternating-current case, where reactance must be considered as well as resistance.

A full description of the techniques of mesh analysis is beyond the scope of this book. For details, a circuit-theory text should be consulted. *See also* KIRCHHOFF'S LAWS, MESH.

MESNY CIRCUIT

A Mesny circuit, or Mesny oscillator, is a tuned-input, tuned-output, push-pull circuit used to produce signals at ultra-high and microwave frequencies. A pair of transistors is connected together as shown in the illustration. A feedback loop causes oscillation.

The Mesny circuit uses quarter-wave matching sections to obtain resonance. A pair of parallel wires, with a movable shorting bar, is connected in the input circuit. A similar pair of wires is connected in the output. The positions of the shorting bars determine the oscillating frequency of the circuit. Normally, the input and output tuned circuits are set for resonance at the same frequency. The output is generally taken from the source or emitter circuit to minimize loading effects.

METAL

Metal is a type of material element with certain properties. All metals are characterized by good electrical conductivity. This is because the electrons of a metal are easily stripped from their orbits around the atomic nuclei. Electrical conduction occurs as the electrons move from atom to atom. Metals also are generally good conductors of heat.

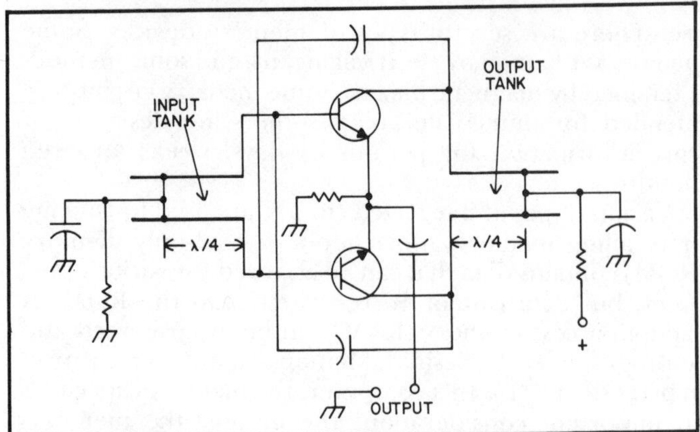

MESNY CIRCUIT: The Mesny oscillator uses quarter-wave matching sections to obtain oscillation at short wavelengths.

Common metals include aluminum, copper, and iron. All metals except mercury are solids at room temperature. Metals display various different melting temperatures; while mercury melts at about −39 degrees Celsius, tungsten remains in solid form at temperatures up to about 3,400 degrees Celsius.

Metals and metal compounds are extensively used in electronics. The most familiar application of metals is in the manufacture of wires and other electrical conductors. Metals are used as electrodes in vacuum tubes, as transformer cores, as equipment housings, and in countless other ways. Metal oxides are used in the manufacture of semiconductor devices. *See also* CONDUCTIVE MATERIAL, CONDUCTIVITY, METAL-OXIDE SEMICONDUCTOR.

METAL-OXIDE SEMICONDUCTOR

The oxides of certain metals exhibit insulating properties. In recent years, so-called metal-oxide-semiconductor (MOS) devices have come into widespread use.

MOS materials include such compounds as aluminum oxide and silicon dioxide. Metal-oxide semiconductor devices are noted for their low power requirements. MOS integrated circuits have high component density and high operating speed. Metal oxides are also used in the manufacture of certain field-effect transistors.

All MOS devices are subject to damage by the discharge of static electricity. Therefore, care must be exercised when working with MOS components. MOS integrated circuits and transistors should be stored with the leads inserted into a conducting foam, so that large potential differences cannot develop. When building, testing, and servicing electronic equipment in which MOS devices are present, the body and all test equipment should be kept at direct-current ground potential. *See also* METAL-OXIDE-SEMICONDUCTOR FIELD-EFFECT TRANSISTOR, METAL-OXIDE-SEMICONDUCTOR LOGIC FAMILIES.

METAL-OXIDE-SEMICONDUCTOR FIELD-EFFECT TRANSISTOR

An active component commonly used today is the metal-oxide-semiconductor field-effect transistor, or MOSFET. A typical MOSFET cross section is illustrated at A in the illustration. The schematic symbol for a MOSFET is shown

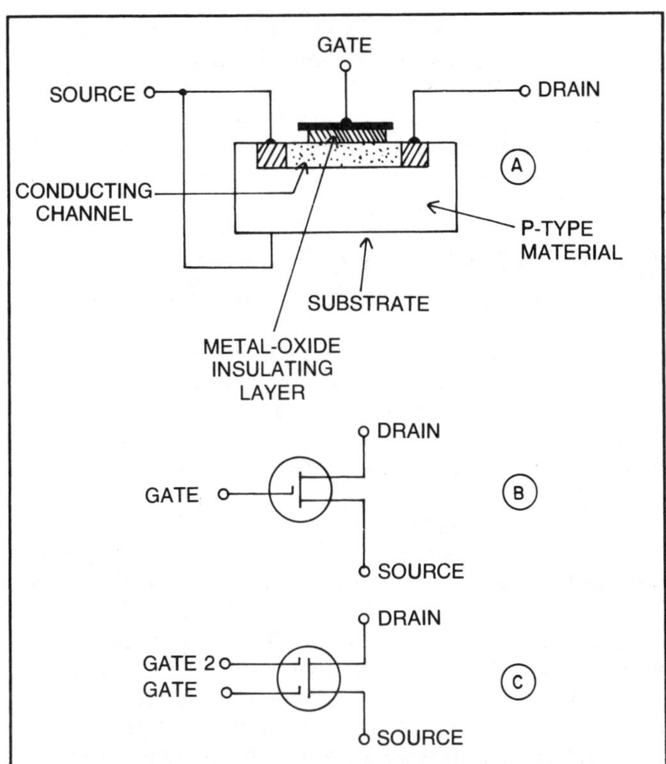

METAL-OXIDE-SEMICONDUCTOR FIELD-EFFECT TRANSISTOR: At A, a simplified cross-sectional diagram of a metal-oxide-semiconductor field-effect transistor (MOSFET). This is an N-channel, enhancement-mode device. At B, schematic symbol for MOSFET; this symbol can represent either an N-channel or a P-channel device. At C, schematic symbol for dual-gate MOSFET (either N-channel or P-channel).

at B. The MOSFET can be recognized because the gate electrode is insulated from the channel by a thin layer of metal oxide. Because the gate is insulated from the channel, the MOSFET is sometimes called an insulated-gate field-effect transistor, or IGFET.

The MOSFET device has extremely high input impedance. It is typically billions or trillions of ohms. Thus, the MOSFET requires essentially no driving power. MOSFETs are commonly used in high-gain receiver amplifier circuits. A single-gate MOSFET amplifier stage can produce better than 15 dB gain at a frequency of 100 MHz. Some MOSFETs have two gates, and are known as dual-gate MOSFETs. They are useful as mixers and modulators, as well as high-gain amplifiers. The schematic symbol for a dual-gate MOSFET is shown at C.

Some MOSFETs are normally conducting, or saturated, under conditions of zero gate bias. They are therefore depletion-mode devices. However, enhancement-mode MOSFETs are common. MOSFETs may have a channel consisting of either N-type material or P-type material. *See also* DEPLETION MODE, ENHANCEMENT MODE, FIELD-EFFECT TRANSISTOR.

METAL-OXIDE-SEMICONDUCTOR LOGIC FAMILIES

Metal-oxide semiconductor technology lends itself well to the fabrication of digital integrated circuits. In recent years, several metal-oxide-semiconductor (MOS) logic families have been developed. Miniaturization is more easily obtained with MOS technology than with bipolar technology, and the power requirements are lower.

Metal-oxide-semiconductor logic families include complementary (CMOS), N-channel (NMOS), and silicon-on-sapphire (SOS).

CMOS utilizes both N-type and P-type materials on the same substrate. CMOS is characterized by extremely low power requirements and relative insensitivity to external noise (*see* COMPLEMENTARY METAL-OXIDE SEMICONDUCTOR). The operating speed is relatively high.

The NMOS technology is commonly used today. It has high operating speed and the advantage of simplicity. P-channel (PMOS) metal-oxide-semiconductor logic is occasionally used, although it operates at a slower speed than NMOS or bipolar logic devices.

SOS provides the fastest operation of currently available MOS logic families. The operating speed of SOS approaches that of transistor-transistor-logic bipolar circuits. SOS processing is more expensive than CMOS, NMOS, or PMOS.

Metal-oxide-semiconductor logic families are especially useful in high-density memory applications. Many microcomputer chips make use of MOS technology today. *See also* METAL-OXIDE SEMICONDUCTOR.

METEOR SCATTER

When a meteor from space enters the upper part of the atmosphere, an ionized trail is produced because of the heat of friction. Such an ionized region reflects electromagnetic waves at certain wavelengths. This phenomenon, known as meteor scatter, can result in over-the-horizon propagation of radio signals (see illustration).

A single meteor generally produces a trail that persists from about 1 second to several seconds, depending on the size of the meteor, its speed, and the angle at which it enters the atmosphere. This amount of time is not sufficient for the transmission of very much information, but during a meteor shower, ionization is often almost continuous. Meteor-scatter propagation has been observed well into the very-high-frequency range of the radio spectrum.

Meteor-scatter propagation is mainly of interest to experimenters and radio amateurs; it will probably never be used routinely in radio communication. Meteor-scatter propagation occurs over distances ranging from just be-

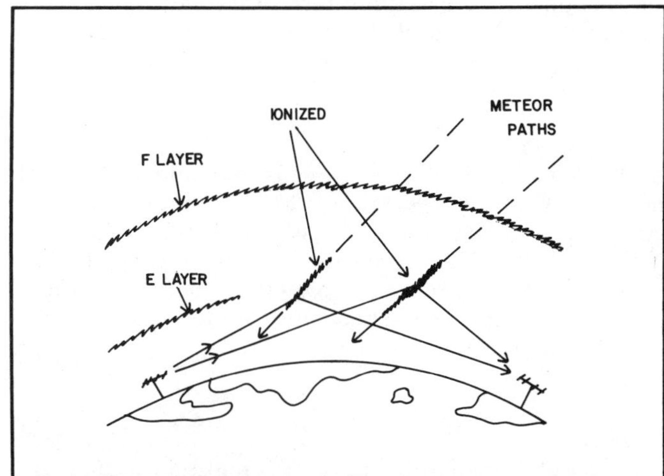

METEOR SCATTER: Meteor-scatter propagation usually occurs in or near the ionospheric E layer, allowing over-the-horizon communication for brief periods.

yond the horizon up to about 1,500 miles, depending on the height of the ionized trail and the relative locations of the trail, the transmitting station, and the receiving station.

METER

A meter is a device, either electrical or electromechanical, used for the purpose of measuring an electrical quantity. The most familiar kind of meter is the moving-needle device. However, digital meters are becoming increasingly common. *See also* ANALOG METERING, DIGITAL METERING.

The basic unit of displacement in the metric system is the meter. Originally, the meter was defined as 1 ten-millionth of the distance from the north pole to the equator of the earth. That is, there were supposed to be 10,000,000 meters in that distance. Now, the meter is more precisely defined as the standard international unit of length. A distance of 1 meter represents 1,650,763.73 wavelengths, in a vacuum, of the radiation corresponding to the transition of electrons between the levels $2p_{10}$ (in the L shell) and $5d_5$ (in the 0 shell) of the atom of krypton 86. The meter is approximately 39.37 inches, or a little more than an English yard. *See also* METRIC SYSTEM.

METRIC SYSTEM

The metric system is a decimal system for measurement of length, area, and volume. Because it is a decimal system, the metric system has gained widespread use in many countries throughout the world. Scientists routinely use the metric system. The principal unit of length is the meter (*see* METER). The decimeter is 0.1 meter; the centimeter is 0.01 meter; the millimeter is 0.001 meter; the kilometer is 1,000 meters.

The decimal nature of the metric system makes it much easier to use than the older English system. However, the English system is still used for non-scientific applications in the United States. *See also* MKS SYSTEM, STANDARD INTERNATIONAL SYSTEM OF UNITS.

MHO

The mho is a unit of electrical conductance. In recent years, the mho has been called the siemens in the standard international system of units. Electrical conductance is the mathematical reciprocal of resistance. Given a resistance of R ohms, the conductance S in siemens is simply

$$S = 1/R$$

A resistance of 1 ohm represents a conductance of 1 siemens. When the resistance is 1,000 ohms, the conductance is 0.001 siemens or 1 millisiemens. When the resistance is 1,000,000 ohms, the conductance is 1 microsiemens. The siemens is also used as the unit of admittance in alternating-current circuits. *See also* ADMITTANCE, CONDUCTANCE.

MICA

Mica is a silicate material that occurs naturally in the crust of the earth. Physically, mica appears as a transparent, sheet-like substance similar to plastic or cellophane. Mica has excellent dielectric properties.

Mica conducts heat quite well, and it is therefore useful in transistor heat-sink applications when it is necessary to electrically insulate the case of the transistor from the heat sink. Mica is commonly used in the manufacture of low-value to medium-value capacitors for use at moderately high voltages. *See also* MICA CAPACITOR.

MICA CAPACITOR

Mica capacitors are characterized by low loss, high dielectric strength, and good stability under variable voltage and temperature conditions. Mica capacitors are widely used in radio-frequency circuits, especially in the construction of selective inductance-capacitance filters. Mica capacitors are available in a wide range of values, from a few picofarads up to a few thousand picofarads. Voltage ratings range from about 100 volts to 35 kilovolts. Mica capacitors are non-polarized.

Several different methods of construction are employed in the manufacture of mica capacitors. The simplest method is to clamp a pair of foil electrodes to either side of a thin sheet of mica. Other methods include the bonding of metals (especially silver) to the mica sheet, the attachment of concentric ring-shaped electrodes to the mica, and the stacking of plated mica sheets. Mica capacitors are encased in resin to protect them from the environment. *See also* CAPACITOR.

MICRO

Micro is a prefix multiplier that means 0.000001, or 10^{-6}. For example, 1 microfarad is 10^{-6} farad, and 1 microvolt is 10^{-6} volt. *See also* PREFIX MULTIPLIERS.

The term micro is sometimes used to refer to something very small. For example, a microcircuit is a small, or miniaturized, electronic circuit; a microcomputer is a miniature computer. More examples can be found in the following several terms.

MICROALLOY TRANSISTOR

A microalloy transistor is a form of bipolar semiconductor transistor. A thin wafer of semiconductor material forms the base of the transistor. The emitter and collector are formed by alloying a small amount of impurity material at two points on opposite faces of the wafer.

The microalloy transistor is characterized by relatively low junction capacitance. Microalloy transistors are available in either NPN or PNP types. They are not commonly used today. *See also* TRANSISTOR.

MICROCIRCUIT

A microcircuit is a highly miniaturized electronic circuit. In modern electronics, the microcircuit is found in the form of densely packed components, all etched onto a single semiconductor wafer. The microcircuit is usually called an integrated circuit. *See also* INTEGRATED CIRCUIT.

MICROCODE

See MICROINSTRUCTION.

MICROCOMPUTER

A microcomputer is a small computer, with the central processing unit (CPU) enclosed in a single integrated-circuit package. Today, microcomputer CPU chips are available in sizes ranging down to ¼ inch on a side. The microcomputer CPU is sometimes called a microprocessor.

Microcomputers vary in sophistication and memory storage capacity, depending on the intended use. Some personal microcomputers are available for less than $100 as of this writing. Such devices employ liquid-crystal displays and have typewriter-style keyboards. Larger microcomputers are used by more serious computer hobbyists and by small businesses. Such microcomputers typically cost from several hundred to several thousand dollars. *See also* COMPUTER, MICROCOMPUTER CONTROL, MICROPROCESSOR.

MICROCOMPUTER CONTROL

Microcomputers are often used for the purpose of regulating the operation of electrical and electromechanical devices. This is known as microcomputer control. Microcomputer control makes it possible to perform complex tasks with a minimum of difficulty.

An example of a microcomputer-controlled device is the programmable scanning radio receiver. Microcomputer control is also widely used in such devices as robots, automobiles, and aircraft. For example, a microcomputer can be programmed to switch on an oven, heat the food to a prescribed temperature for a certain length of time, and then switch the oven off again. Microcomputers can be used to control automobile engines to enhance efficiency and gasoline mileage. Microcomputers can navigate and fly airplanes. The list of potential applications is almost uncountable.

One of the most recent, and exciting, applications of microcomputer control is in the field of medical electronics. Microcomputers can be programmed to provide electrical impulses to control erratically functioning body organs, to move the muscles of paralyzed persons, and for various other purposes.

MICROELECTRONICS

Microelectronics is the application of miniaturization techniques to the design and construction of electronic equipment. Microelectronics has progressed to the point that a circuit once requiring an entire room, and several thousand watts of power, can now be housed in a package smaller than a dime and requiring only a few milliwatts or microwatts. The biggest breakthrough in microelectronics has been the integrated circuit. *See also* INTEGRATED CIRCUIT.

MICROINSTRUCTION

In computer machine language, a bit pattern comprising an elementary command is called a microinstruction. The microinstruction is the smallest form of machine-language instruction. Several microinstructions can be combined to form a macroinstruction, which is a common and identifiable computer instruction. *See also* MACHINE LANGUAGE, MACROINSTRUCTION.

MICROMINIATURIZATION

The techniques of fabricating electronic components by etching of semiconductors is known as microminiaturization. Modern methods of microminiaturization have, within a few decades, reduced the physical size of electronic equipment by a factor of several million. Power requirements have decreased by almost the same proportion. A small computer can now be fit into a case no larger than a wallet.

The most advanced form of microminiaturization is found in integrated-circuit chips. The most compact chips contain hundreds or even thousands of individual logic gates on a semiconductor wafer having a surface area of a fraction of a square inch. Very-large-scale integration (VLSI) chips contain from 100 to 1,000 gates on a single wafer. There is some research being done in the field of ultra-large-scale integration (ULSI), in which up to 10,000 gates might be placed on a single chip.

Although there must be a limit to how small an electronic circuit can be—electrons are tiny, but are nevertheless bigger than geometric points—we have not yet reached this limit. Some scientists think we are still nowhere near it. *See also* LARGE-SCALE INTEGRATION, VERY-LARGE-SCALE INTEGRATION, ULTRA-LARGE-SCALE INTEGRATION.

MICROPHONE

A microphone is an electroacoustic transducer, designed to produce alternating-current electrical impulses from sound waves. This can be done in many ways, and thus there are many different types of microphones.

One of the earliest microphones was the carbon-granule microphone. This device required an external source of direct current. The sound waves caused changes in the resistance of a carbon-granule container, resulting in modulation of the current (*see* CARBON MICROPHONE). This kind of microphone is seldom used today.

More recently, the piezoelectric effect has been used to generate electric currents from passing sound waves. Various substances, when subjected to vibration, produce weak electric impulses that can be amplified. The ceramic microphone operates on this principle (*see* CERAMIC MICROPHONE).

A more rugged type of microphone operates by means of electromagnetic effects. A diaphragm, set in motion by passing sound waves, causes a coil and magnet to move with respect to each other. The fluctuating magnetic field thus results in alternating currents through the coil. This device is called a dynamic microphone (*see* DYNAMIC MICROPHONE).

There are other, less common forms of microphones, used for special purposes. Electrostatic and optical devices, for example, can be used as microphones in certain situations.

Microphones are available in a wide variety of sizes, input impedances, and with various frequency-response characteristics. Some microphones are omnidirectional, while others have a cardioid or unidirectional response. The optimum choice of a microphone is important in any audio-frequency system. Specialized microphones are made for communications, high-fidelity, and public-address applications.

MICROPHONICS

Mechanical vibration can cause unwanted modulation of a radio-frequency oscillator circuit, or in an audio- or radio-frequency amplifier circuit. This can occur in a transmitter, resulting in over-the-air noise in addition to the desired modulation; it can take place in a receiver, causing apparent noise on a signal. Such unwanted modulation is known as microphonics.

In fixed-station equipment, microphonics are not usually a problem, since mechanical vibration is not severe. However, in mobile applications, equipment must be designed for minimum susceptibility to microphonics. Microphonics can, if severe, result in out-of-band modulation in a transmitter. It can also make a signal almost unintelligible.

For immunity to microphonics, equipment circuit boards must be firmly anchored to the chassis. Component leads must be kept short. Especially sensitive circuits, such as oscillators, may have to be encased in wax or some other shock-absorbing substance.

MICROPROCESSOR

The central processing unit of a microcomputer is sometimes called a microprocessor. There is some confusion between the terms microcomputer and microprocessor, and the two are often used in place of each other (*see* MICROCOMPUTER).

Technically, a microcomputer consists of the microprocessor integrated circuit and perhaps one or more peripheral integrated circuits. The peripheral circuits may be integrated onto the same chip as the central processing unit, or they may be separate. The peripheral integrated circuits contain memory and programming instructions. *See also* CENTRAL PROCESSING UNIT.

MICROPROGRAM

A microprogram is a set of microinstructions, at the machine-language level, that results in the execution of a specific function, independent of the functions of the program being run. The microprograms implement routine operating functions in a computer. Microprograms may be permanently placed in a computer or microcomputer in the form of firmware. *See also* FIRMWARE.

MICROSTRIP

Microstrip is a form of unbalanced transmission line. A flat conductor is bonded to a ground-plane strip by means of a dielectric material (see illustration). This results in an image conductor, parallel to the flat wire conductor, but on the other side of the ground-plane strip. The effective current in the image conductor is equal in magnitude to the current in the actual conductor, but flows in the opposite direction. Thus, very little radiation occurs (in theory) from a microstrip transmission line.

Microstrip lines are sometimes used at ultra-high and microwave frequencies. Microstrip lines have somewhat lower loss than coaxial lines, but radiate less than most open-wire lines at the short wavelengths. The characteristic impedance of the microstrip line depends on the width of the flat wire conductor, the spacing between the flat wire conductor and the ground place, and on the type of dielectric material used.

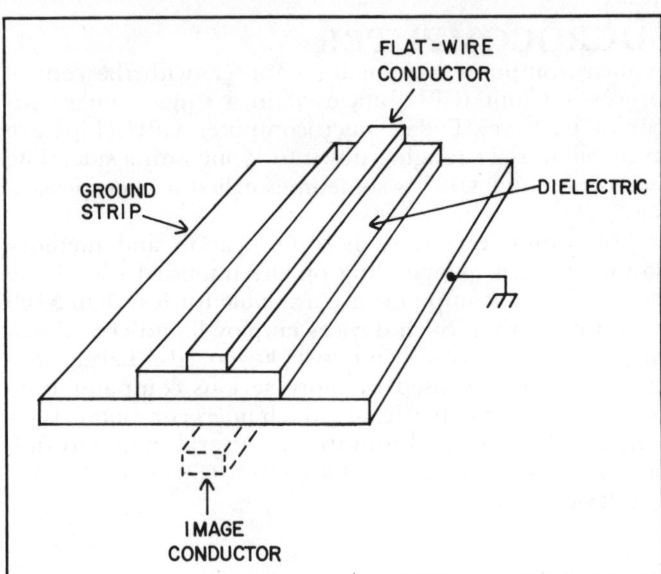

MICROSTRIP: This type of unbalanced transmission line is used at very short wavelengths.

MICROWAVE

Microwaves are defined as that part of the electromagnetic spectrum at which the wavelength falls between about 1 millimeter and 30 centimeters. The microwave frequencies range from approximately 1 GHz to 300 GHz.

Microwaves are very short electromagnetic radio waves, but they have a longer wavelength than infrared energy. Microwaves travel in essentially straight lines through the atmosphere, and are not affected by the ionized layers.

Microwave frequencies are useful for short-range, high-reliability radio and television links. In a radio or television broadcasting system, the studio is usually at a different location than the transmitter; a microwave link connects them. Satellite communication and control is generally accomplished at microwave frequencies. The microwave region contains a vast amount of spectrum space, and can therefore hold many wideband signals.

Microwave radiation can cause heating of certain materials. This heating can be dangerous to human beings when the microwave radiation is intense. When working with microwave equipment, care must be exercised to avoid exposure to the rays. The heating of organic tissue by microwaves can be put to constructive use, however; this idea led to the invention of the microwave oven. *See also* MICROWAVE OVEN.

MICROWAVE OVEN

A microwave oven is a device used for cooking food by means of microwave heating. Organic molecules become agitated when subjected to microwave energy, and the temperature therefore rises. The more intense the microwave energy, the greater the rise in temperature for a given organic substance and a given electromagnetic frequency.

Microwave ovens make it possible to cook food in a fraction of the time required by a conventional oven. Since

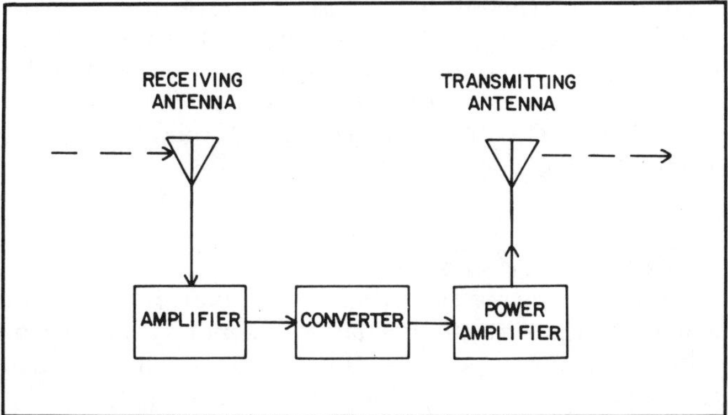

MICROWAVE REPEATER: Block diagram of a microwave repeater.

the microwave energy penetrates the food with virtually no attenuation, the food is uniformly heated, resulting in more even cooking. Caution must be exercised when using a microwave oven, because any metallic objects inside the oven will result in microwave interference patterns, with resulting uneven cooking. *See also* MICROWAVE.

MICROWAVE REPEATER

A microwave repeater is a receiver/transmitter combination used for relaying signals at microwave frequencies. Basically, the microwave repeater works in the same way as a repeater at any lower frequency. The signal is intercepted by a horn or dish antenna, amplified, converted to another frequency, and retransmitted (see illustration).

Microwave repeaters are used in long-distance overland communications links. With the aid of such repeaters, microwave links supplant wire-transmission systems. *See also* REPEATER.

MIDRANGE

The middle portion of the audio-frequency spectrum is sometimes called the midrange. Although the lower and upper limits of the midrange are not specifically defined, midrange may be thought of as the minimum band of frequencies necessary for the understanding of a human voice: about 300 Hz to 3 kHz.

In a high-fidelity system, a speaker must respond well to the midrange frequencies as well as the low and high frequencies. However, the midrange should be in the proper proportion with respect to the lower and higher frequencies. Many high-fidelity speaker combinations have separate transducers for the low, midrange, and high frequencies. *See also* BASS, TREBLE.

MIL

A mil is a small unit of linear measure, equal to 0.001 inch or approximately 0.0254 millimeter.

The mil is not often used nowadays. Scientists and engineers generally prefer to use metric units of distance, especially in the small sense. Cross-sectional area is sometimes still specified in terms of circular mils, however. An area of 1 circular mil is equal to the area of a circle whose diameter is 1 mil. Therefore, 1 circular mil is about 0.785 square mil, or 0.000507 square millimeter.

MILLER EFFECT

A transistor, field-effect transistor, or tube exhibits variable input capacitance under conditions of changing direct-current input bias. In general, an active amplifying device shows smaller input capacitance when it is biased at or beyond cutoff, as compared with bias below the cutoff point. This change in capacitance is called the Miller effect.

The Miller effect can result in rapidly fluctuating input impedance to a stage driven into the cutoff region during the alternating-current input cycle. For example, in a Class-B amplifier circuit, the input capacitance normally increases with increasing drive. This can result in nonlinearity unless the circuit is designed with the Miller effect in mind. *See also* INPUT IMPEDANCE.

MILLER OSCILLATOR

A Miller oscillator is a special form of crystal oscillator. A transistor, field-effect transistor, or tube may be used as the active element in the circuit. The illustration shows Miller oscillators using an NPN bipolar transistor (A), an N-channel field-effect transistor (B), and a tirode vacuum tube (C).

The Miller oscillator can be recognized by the presence of the crystal between the base, gate, or grid and ground. The output contains a tuned circuit. The internal capacitance of the active device provides the necessary feedback for oscillation. The Miller oscillator is sometimes called a conventional crystal oscillator. *See also* CRYSTAL OSCILLATOR, OSCILLATION.

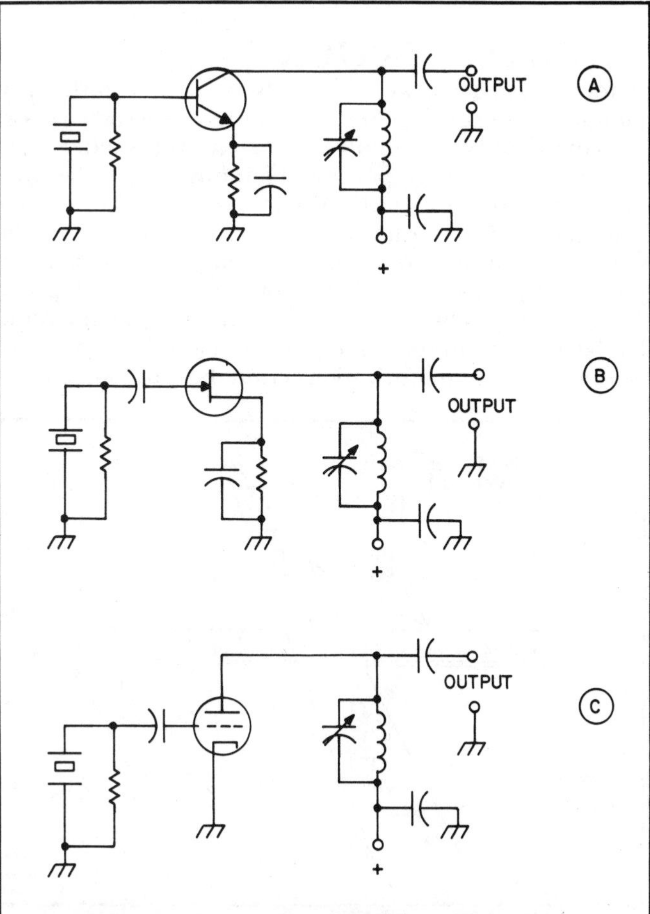

MILLER OSCILLATOR: A bipolar-transistor circuit is shown at A, a field-effect-transistor circuit at B, and a vacuum-tube circuit at C.

MILLI

Milli is a prefix multiplier meaning 0.001 or 1/1,000. When this prefix is placed in front of a quantity, the magnitude is decreased by a factor of 1,000. Thus, for example, 1 millimeter is equal to 0.001 meter, 1 milliampere is equal to 0.001 ampere, and so on. *See also* PREFIX MULTIPLIERS.

MILLS CROSS

The Mills cross is a form of antenna used mostly by radio astronomers. Two phased arrays are positioned at an angle, as illustrated. The antenna is named after the radio astronomer B. Y. Mills, who originally called the antenna the "super cross."

Mills perfected his design at the University of Sydney in Australia. His antenna consists of two cylindrical parabolic reflectors, each 1 mile long, and oriented at right angles to each other. By changing the phase relationship between the two linear arrays, it is possible to steer the main lobe to a certain extent.

By combining two linear arrays as shown, it is possible to obtain much greater resolution than with either linear array by itself. In fact, the Mills cross, with its two mile-long arrays, has resolution almost equivalent to that of a single dish 1 mile in diameter. At a wavelength of 2.7 meters (about 112 MHz), the main-lobe beamwidth is just 10 minutes of arc, or 0.167 degree. At higher frequencies, the beamwidth is even smaller. *See also* RADIO ASTRONOMY, RADIO TELESCOPE.

MINIATURIZATION

A few years ago, it would have been absurd to think of building a pocket-sized microcomputer; now such devices are commonplace. Radio receivers and transmitters that once required a large portion of a room can now be combined in a desktop unit. Miniaturization of electronic equipment accomplishes two things: first, it reduces the size of a given piece of apparatus, and second, it makes it possible to put more into a given amount of physical space.

Solid-state technology, espcially the integrated circuit, has been the main reason for the rapid advances that have been made in miniaturization. Printed circuit boards and modular construction, along with better use of space, have also been responsible. Miniaturization has not only resulted in greater component and circuit density, but it has also greatly reduced the power requirements and improved electrical efficiency. *See also* INTEGRATED CIRCUIT, SOLID-STATE ELECTRONICS.

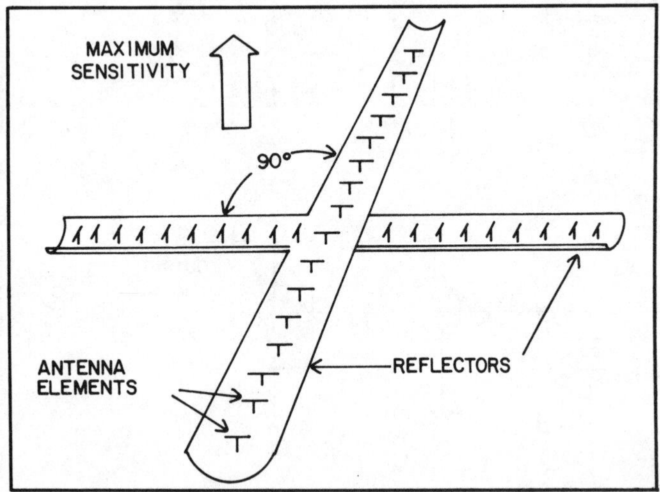

MILLS CROSS: The Mills Cross antenna is constructed by combining two fan-beam antennas at right angles.

MINICOMPUTER

A minicomputer is a small digital computer. It is similar to a microcomputer, but has larger memory capacity and higher speed of operation. The demarcation between a minicomputer and a microcomputer is not precisely defined. We might call a 128-kilobyte computer a micro and a 32-megabyte computer a mini. As the technology advances, however, we may someday speak of a 32-megabyte computer as a micro and a 512-megabyte computer as a mini!

Minicomputers are used by medium-size and large businesses. For small-business and personal applications, microcomputers are more often used. *See also* MICROCOMPUTER.

MINORITY CARRIER

Semiconductor materials carry electric currents in two ways. Electrons transfer negative charge from the negative pole to the positive pole; holes carry positive charge from the positive pole to the negative pole (*see* ELECTRON, HOLE).

Electrons and holes are known as charge carriers. In an N-type semiconductor material, most of the charge is carried by electrons, and relatively little by holes. Thus, holes are called the minority carrier in N-type material. Conversely, in P-type material, the electrons are the minority carriers. *See also* MAJORITY CARRIER, N-TYPE SEMICONDUCTOR, P-TYPE SEMICONDUCTOR.

MINOR LOBE

See LOBE.

MISMATCHED LINE

When the terminating impedance, or load impedance, of a transmission line is not a pure resistance with a value equal to the characteristic impedance of the line, the line is said to be mismatched. A line is always mismatched if the load contains reactance; a purely resistive load of the wrong value can also present a mismatch to a transmission line.

A mismatched line may degrade the performance of a system significantly, but this is not always the case. In a mismatched line, the voltage and current do not exist in uniform proportion. The current and voltage, instead, occur in a pattern of maxima and minima along the line (see illustration). Current minima occur at the same points as voltage maxima, and current maxima correspond to voltage minima. This pattern of current and voltage is known as a standing-wave pattern. Standing waves result in increased conductor and dielectric heating in a transmission line, and this causes more power to be dissipated as heat, reducing the efficiency of the system.

Small mismatches in a transmission line can often be tolerated with little or no adverse effects. However, large mismatches can cause problems such as reduced generator efficiency, increased line loss, and overheating of the line. The degree of the mismatch is called the standing-wave ratio. *See also* STANDING-WAVE RATIO, STANDING-WAVE-RATIO LOSS.

MIXER

A mixer is a device that combines two signals of different frequencies to produce a third signal, the frequency of which is either the sum or difference of the input frequencies. Usually, the difference frequency is used as the mixer output frequency. Mixers are widely used in superheterodyne receivers, and in radio transmitters (*see* SUPERHETERODYNE RECEIVER).

A mixer requires some kind of nonlinear circuit element in order to function properly. The nonlinear element may be a diode or combination of diodes; such a circuit is shown in the illustration at A. The diode mixer is a passive circuit, since it does not require an external source of power. However, there is some insertion loss when such a mixer is used.

Mixers can have gain; at B, a bipolar-transistor mixer is shown. At C, a dual-gate metal-oxide-semiconductor field-effect transistor (MOSFET) is used. At D, a tetrode vacuum tube is used.

The mixer operates on a principle similar to that of an amplitude modulator. However, in the mixer, both signals are usually in the radio-frequency spectrum, while in a modulator, one signal is much lower in frequency than the other (*see* MODULATOR). The output of the mixer circuit is tuned to the sum or difference frequency, as desired.

Mixing may occur in a circuit even when it is not desired. This is especially likely in high-gain receiver front-end amplifiers. Mixing can also occur between a desired signal and a parasitic oscillation in an amplifier or oscillator. Mixing sometimes takes place in semiconductor devices when two or more strong signals are present. Such mixing is called intermodulation, and the resulting unwanted signals are called mixing products.

A device used for combining two or more audio signals in a broadcast or recording studio is known as a mixer. In this type of device, nonlinear operation, with the consequent generation of harmonics and distortion products, is not desired. The circuit must therefore be as linear as possible.

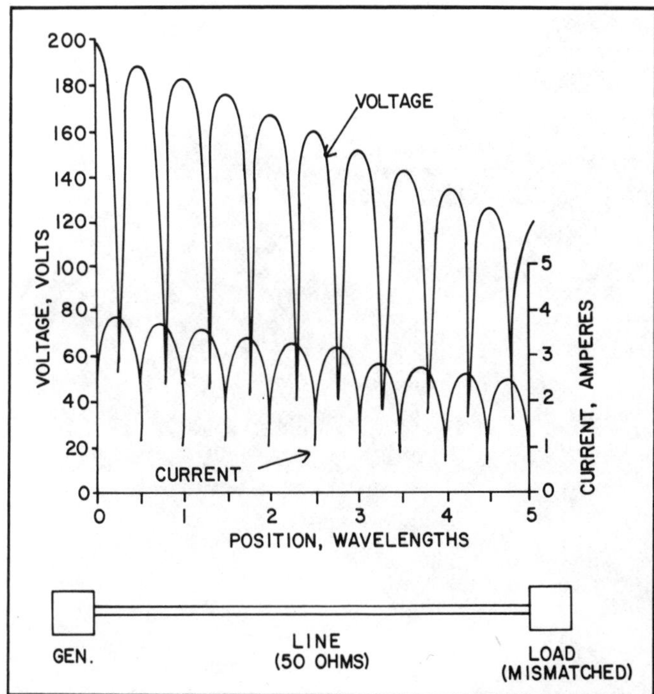

MISMATCHED LINE: A mismatched transmission line has irregular voltage and current levels along its length. The current and voltage levels are generally lower toward the load because of line loss.

MIXING PRODUCT

A mixing product is a signal that results from mixing of two other signals. The output of a mixer, for example, is a mixing product (*see* MIXER). Mixing products may occur either as a result of intentional mixing, as in the mixer, or as a result of unintentional mixing, in amplifiers, oscillators, or filters.

Whenever two signals with frequencies f and g are combined in a nonlinear component, mixing products occur at frequencies f − g and f + g. If there are three or more signals, many different mixing products will exist. Mixing

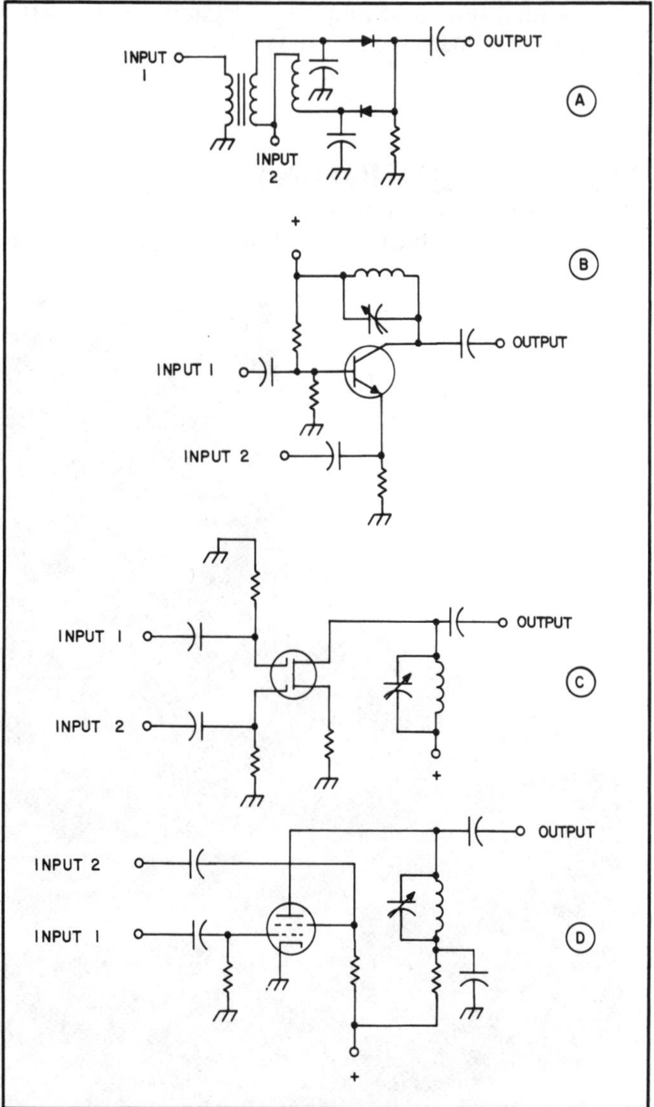

MIXER: At A, a passive mixer circuit. At B, an active mixer circuit using a bipolar transistor. At C, a mixer employing a dual-gate metal-oxide-semiconductor field-effect transistor (MOSFET). At D, a mixer using a tetrode vacuum tube.

products may themselves combine with the original signals or other mixing products, producing further signals known as higher-order mixing products. It is not difficult to see how, when numerous signals are combined in a nonlinear environment, the situation can get very complex!

Mixing products can result in interference to radio receivers. Two external signals, at frequencies f and g, may mix in the receiver front end and result in interference at frequencies f − g and f + g. This is called intermodulation. Mixing products generated within a receiver, from signals originating in the local oscillators, are sometimes called birdies. *See also* INTERMODULATION.

MKS SYSTEM

The system of units most commonly used by scientists today is called the meter-kilogram-second (MKS) system. The MKS system has recently been expanded to include the ampere (for current), kelvin (for temperature), candela (for light intensity), and mole (for quantity of substance). This expanded system is called the standard international system of units, or SI system. *See also* KILOGRAM, METER, SECOND, STANDARD INTERNATIONAL SYSTEM OF UNITS.

MOBILE EQUIPMENT

Mobile electronic equipment is any apparatus that is operated in a moving vehicle such as a car, truck, train, boat, or airplane. Mobile equipment includes all attendant devices, including antennas, microphones, power supplies, and so on.

The design and operation of mobile electronic equipment is much simpler today than it was just a few years ago. The main reason for the improvement is the development of solid-state technology. Most mobile power supplies provide low-voltage direct-current power; 13.8 volts is by far the most common. A tube type circuit requires a complex power inverter for operation from such a supply, but solid-state equipment can usually be operated directly from the mobile supply.

Mobile equipment is generally more compact than fixed-station apparatus. In addition, mobile equipment must be designed to withstand larger changes in temperature and humidity, as well as severe mechanical vibration.

Most two-way mobile communications installations consist of a vertically polarized omnidirectional antenna and a transceiver. A typical mobile Citizen's-Band transceiver is shown in the photograph. The transceiver is a transmitter-receiver combination housed in a single cabinet (*see* TRANSCEIVER). This saves significant space and cost. Most mobile operation is done at very-high and ultra-high radio frequencies, where the antennas are manageable in size. However, some mobile operation is carried out at low, medium, and high frequencies, using inductively loaded antennas.

Equipment designed for portable operation can also be used in mobile applications. Generally, however, portable radio transmitters and receivers are less powerful and versatile than apparatus designed particularly for mobile use. *See also* PORTABLE EQUIPMENT.

MOBILE EQUIPMENT: A compact mobile radio transceiver. (Courtesy of Radio Shack, a division of Tandy Corporation.)

MOBILE TELEPHONE

A mobile telephone is a radio transceiver designed for access to an autopatch system (*see* AUTOPATCH). Mobile telephones originally did not allow continuous two-way conversation; neither party could interrupt the other. However, some mobile telephones now allow true duplex operation.

Mobile telephones, since they involve radio transmissions, cannot legally be used in some countries without a license from the government. Mobile and portable telephones generally operate in the very-high-frequency and ultra-high-frequency radio bands, or between about 30 MHz and 3 GHz. This provides reliable operation within a radius of several miles from the home station or repeater system.

Mobile telephones are of obvious value to businesses, and can be useful in emergency situations as well.

A special form of portable telephone consists of a small handheld unit and a base station, with an operating range of several hundred feet. This device, which can be used by anyone in most countries without a government license, is called a cordless or wireless telephone. *See also* CORDLESS TELEPHONE.

MOBILITY

See CARRIER MOBILITY.

MODE OF EMISSION

See EMISSION CLASS.

MODEM

The term modem is a contraction of the words MOdulator and DEModulator. A modem is a two-way interfacing device that can perform either modulation or demodulation. Modems are extensively used in computer communications, for interfacing the digital signals of the computer with a transmission medium such as a radio or telephone circuit.

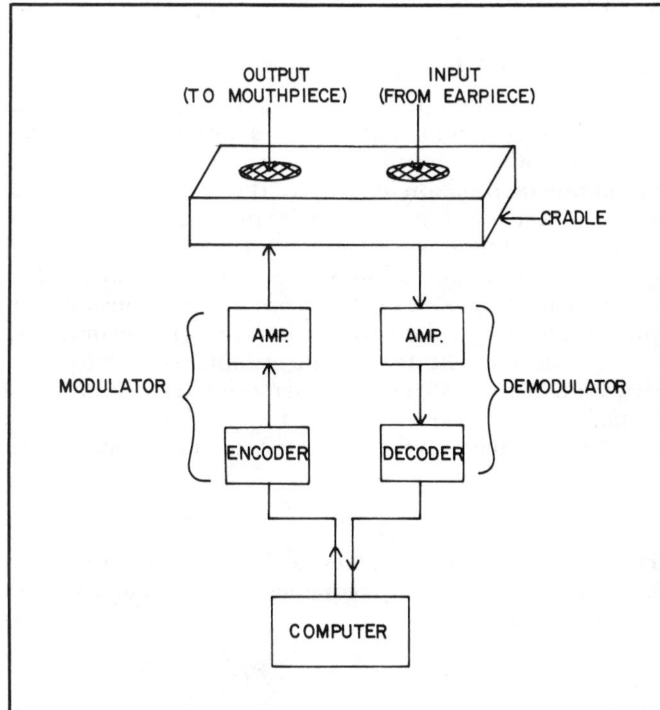

MODEM: Basic construction of a telephone modem.

The illustration is a block diagram of a modem suitable for interfacing a home or business computer with an ordinary telephone. The modulator converts the digital signals from the computer into audio tones. The output is similar to that of an ordinary audio-frequency-shift keyer. The demodulator converts the incoming tones back into digital signals. The audio tones fall within the band of approximately 300 Hz to 3 kHz, so they can be efficiently transmitted over a telephone circuit or narrow-band radio transmitter.

There are many different kinds of devices that perfrom modulation and demodulation, and can therefore be called modems. In general, any signal converter that consists of modulator and demodulator is a modem. *See also* DETECTION, MODULATION.

MODULAR CONSTRUCTION

A few decades ago, electronic equipment was constructed in a much different way than it is today. Components were mounted on tie strips, and wiring was done in point-to-point fashion (*see* POINT-TO-POINT WIRING). This kind of wiring is still used in some high-power radio transmitters, but in recent years, modular construction has become the rule. Solid-state technology has been largely responsible for this change; circuits are far more compact, and there is much less shock hazard, than was the case in the vacuum-tube days.

In the modular method of construction, individual circuit boards are used. Each circuit board contains the components for a certain part or parts of the system. The circuit boards are entirely removable, usually with a simple tool resembling a pliers. Edge connectors facilitate easy replacement. The edge connectors are wired together for interconnection between circuit boards.

Modular construction has greatly simplified the maintenance and servicing of complicated apparatus. In-the-field repair consists of nothing more than the identification, removal, and replacement of the faulty module. The faulty circuit board is then sent to a central facility, where it can be fixed using highly sophisticated equipment. Once the faulty board has been repaired, it is ready to serve as a replacement module in another system when the need arises.

MODULATED LIGHT

Electromagnetic energy of any frequency can theoretically be modulated for the purpose of transmitting intelligence. The only constraint is that the frequency of the carrier be at least several times the highest modulating frequency. Modulated light has recently become a significant method of transmitting information. The frequency of visible light radiation is exceedingly high, and therefore modulated light allows the transfer of a great amount of information on a single beam.

A simple modulated-light communications system can be built for a few dollars, using components available in hardware and electronics retail stores. The schematic diagram shows such a system. It is capable of operating

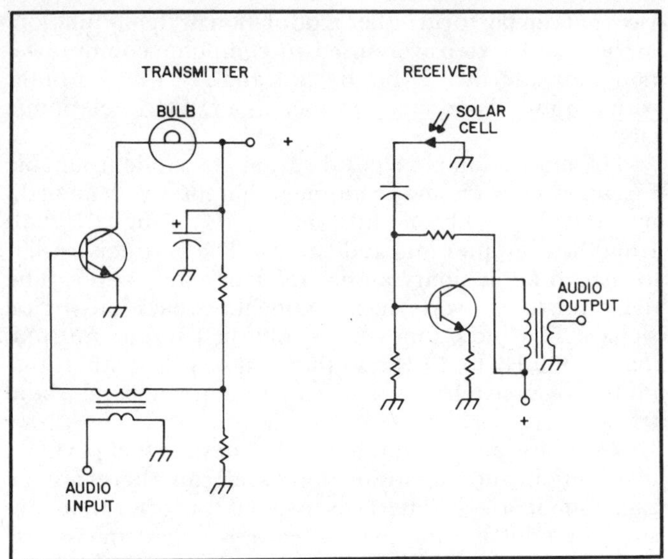

MODULATED LIGHT: A simple modulated-light transmitter and receiver can be built with easily obtained parts.

over a range of several feet. More sophisticated modulated-light systems use laser beams for greater range. Optical fibers may also be used (*see* FIBER, FIBER OPTICS).

The system shown is an amplitude-modulation system. There are other ways of modulating a light beam. Polarization modulation is one such alternative. Lasers make pulse modulation a viable method of transmitting information over light beams. Position modulation is also possible, especially with narrow-beam sources such as lasers. Modulation of the actual frequency of a light beam is difficult to obtain directly, but phase modulation can be achieved with coherent-light sources.

Modulated-light transmission through the atmosphere is, of course, limited by weather conditions. However, cloudless weather is frequent enough in some parts of the world to make modulated-light satellite communications and control feasible. *See also* OPTICAL COMMUNICATIONS.

MODULATION

When some characteristic of an electromagnetic wave is deliberately changed or manipulated for the purpose of transmitting information, the energy is said to be modulated. Modulation can be accomplished with any form of electromagnetic energy. There are many different forms of modulation.

Amplitude modulation was the first method of transmitting complex information via electromagnetic waves. The simplest form of amplitude modulation is Morse-code transmission. Voices and other analog signals can be easily impressed onto a carrier wave. Analog amplitude modulation can take several forms. *See* AMPLITUDE MODULATION, DOUBLE SIDEBAND, SINGLE SIDEBAND.

Frequency modulation is another common method of conveying intelligence via electromagnetic waves. Phase modulation works in a similar way. *See* FREQUENCY MODULATION, PHASE MODULATION.

A more recent development has been the technique of pulse modulation. There are several ways of modulating signal pulses; the amplitude, frequency, duration, or position of a pulse may be modified to achieve modulation. *See* PULSE MODULATION.

To recover the intelligence in a modulated signal, some means is necessary to separate the modulating signal from the carrier wave at the receiving end of a communications circuit. This process is called detection. Different forms of modulation require different processes for signal detection. *See* DETECTION, DISCRIMINATOR, ENVELOPE DETECTOR, HETERODYNE DETECTOR, PRODUCT DETECTOR.

In general, if a characteristic of an electromagnetic wave can be made to fluctuate rapidly enough to convey intelligence at the desired rate, then that parameter offers a means of modulation. The more data that must be transmitted in a given amount of time, however, the more difficult it becomes to efficiently modulate an electromagnetic wave in a particular manner. *See also* MODULATION METHODS, MODULATOR.

MODULATION COEFFICIENT

The modulation coefficient is a specification of the extent of amplitude modulation of an electromagnetic wave. The modulation coefficient is abbreviated by the small letter m, and it can range from a value $m = 0$ (for an unmodulated carrier) to $m = 1$ (for 100-percent or maximum distortion-free modulation).

Let E_c be the peak-to-peak voltage of the unmodulated carrier wave. Let E_m be the maximum peak-to-peak voltage of the modulated carrier. Then:

$$m = (E_m - E_c)/E_c$$

Theoretically, m can be larger than 1, but this represents overmodulation, and distortion inevitably occurs under such conditions. *See also* AMPLITUDE MODULATION, MODULATION PERCENTAGE.

MODULATION ENVELOPE

See ENVELOPE.

MODULATION INDEX

In a frequency-modulated transmitter, the modulation index is a specification that indicates the extent of modulation. Frequency modulation differs from all forms of amplitude modulation in terms of the way it must be measured. There is a limit to the extent to which a signal can be usefully amplitude-modulated, but the only limitation on frequency deviation is imposed by the fact that the frequency cannot be lower than zero at any given instant. For practical purposes, such a limitation is nearly meaningless.

Suppose that the maximum instantaneous frequency deviation (*see* DEVIATION) of a carrier is f kHz. Suppose the instantaneous audio modulation frequency is g kHz. Let us call the modulation index by the letter m. Then:

$$m = f/g$$

For example, if the frequency deviation is plus or minus 5 kHz and the modulating frequency is 3 kHz, then $f = 5$ and $g = 3$; therefore:

$$m = 5/3 = 1.67$$

The modulation index is a useful means of measuring frequency modulation when a pure sine-wave tone and constant deviation are used. However, with a voice signal, this is not the case, and a more general form of the modulation index must be used. This specification is called the deviation ratio. The deviation ratio d is given by:

$$t = f/g$$

where f is the maximum instantaneous deviation and g is the highest audio modulating frequency, both specified in kilohertz. *See also* FREQUENCY MODULATION.

MODULATION METHODS

There are numerous ways of modulating an electromagnetic wave for the purpose of conveying intelligence. The most common modulated parameters are the amplitude (AM), the frequency (FM), and the phase (PM). A series of pulses may also be modulated in terms of pulse duration, pulse amplitude, or pulse position. (Pulse modulation is, like phase modulation, abbreviated PM).

The best type of modulation for a given situation depends on the amount of information to be conveyed, the desired accuracy, the propagation conditions, the amount of interference, the alloted bandwidth, and many other considerations. Amplitude modulation is most common in the very-low, low, medium, and high-frequency spectra. Frequency modulation and phase modulation are generally used in the very-high and ultra-high ranges. Pulse modulation is often found in the ultra-high and microwave parts of the radio spectrum. *See also* AMPLITUDE MODULATION, FREQUENCY MODULATION, PHASE MODULATION, PULSE MODULATION.

MODULATION PERCENTAGE

In an amplitude-modulated signal, the modulation percentage is a measure of the extent to which the carrier wave is modulated. A percentage of zero refers to the absence of amplitude modulation. A percentage of 100 represents the maximum modulating-signal amplitude that can be accommodated without envelope distortion.

Suppose the unmodulated-carrier voltage of a signal is E_c volts, and that the peak amplitude (with maximum modulation) is E_m volts. Then the percentage of modulation, m, is given by

$$m = 100\,(E_m - E_c)/E_c$$

Under conditions of 100-percent amplitude modulation, the peak power is four times the unmodulated power, regardless of the waveform of the modulating signal. The extent to which the average power increases over the unmodulated-carrier power, for a given modulation percentage, depends on the waveform. The average-power increase with 100-percent sine-wave modulation is approximately 50 percent. For a human voice, the average power increases by only about half that amount. Generally, amplitude modulation levels of more than 100 percent are not used. This is because negative-peak clipping occurs under such conditions, resulting in distortion of the signal and unnecessary bandwidth. *See also* AMPLITUDE MODULATION.

MODULATOR

A modulator is a circuit that combines information with a radio-frequency carrier for the purpose of transmission over the airwaves. Different forms of modulation require different kinds of modulator circuits.

The simplest modulator is the telegraph key for producing code transmissions. Code is a form of amplitude modulation. More complex amplitude modulation is produced by a circuit such as that shown in the illustration. This circuit is essentially an amplifier with variable gain. The gain is controlled by the input signal, in this case an audio signal from the microphone and audio amplifier. All amplitude modulators work according to this basic principle, although the details vary. *See* AMPLITUDE MODULATION.

A special sort of amplitude modulator is designed to eliminate the carrier wave, leaving only the sideband energy. Such a circuit is called a balanced modulator. *See* BALANCED MODULATOR, DOUBLE SIDEBAND, SINGLE SIDEBAND.

Frequency or phase modulation requires the introduction of a variable reactance into an oscillator circuit. This can be done in various ways. The variable reactance is controlled by the modulating signal, and causes the phase and/or resonant frequency of the oscillator to fluctuate. *See* FREQUENCY MODULATION, PHASE MODULATION, REACTANCE MODULATOR.

Pulse modulation is obtained by circuits that cause changes in the amplitude, timing, or duration of high-powered radio-frequency pulses. The circuit details depend on the type of pulse modulation used. *See* PULSE MODULATION.

MODULATOR-DEMODULATOR

See MODEM.

MODULUS

The term modulus can have any of several different meanings in electricity and electronics.

The absolute value or magnitude of an impedance is sometimes called the modulus. An impedance generally

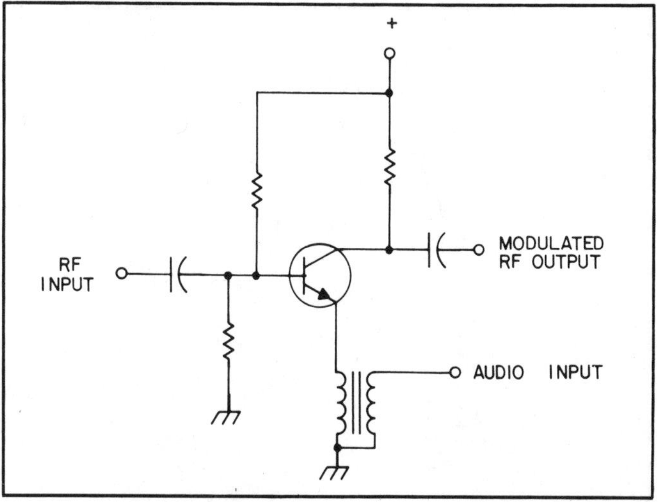

MODULATOR: A simple modulator circuit, suitable for low-level amplitude modulation.

consists of a resistive component and a reactive component (*see* IMPEDANCE), both specified in ohms. Suppose the resistance is denoted by R and the reactance by X; then the modulus impedance Z is given by the equation:

$$Z = \sqrt{R^2 + X^2}$$

In computer practice, the number of counter states per cycle is called the modulus. This value is usually greater than or equal to 1.

The degree to which a substance or device exhibits a certain property is called the modulus. For example, the elasticity of a material is expressed in terms of the elasticity modulus. In this application, the modulus might also be called the coefficient.

When a logarithm is specified in one base, and it is necessary to convert the logarithm to another base, a constant multiplier is used. This multiplier is called the modulus. For two given logarithm bases a and b, the modulus does not depend on the particular number for which the logarithm is found. As an example, the natural logarithm of any number can be obtained from the base-10 logarithm by multiplying by 2.303. The constant 2.303 is called the modulus. *See also* LOGARITHM.

MOLE

The mole is a unit of quantity, generally used by chemists and physicists. A mole is about 6.02×10^{23} molecules. A mole is like any other unit of quantity, such as a dozen (12) or a gross (144).

For any element, 1 mole of atoms has a certain mass in grams. This mass is known as the atomic weight for that substance. For example, the atomic weight of the most common isotope of oxygen is approximately 16. Therefore, 1 mole of oxygen atoms has a mass of about 16 grams. *See also* ATOMIC WEIGHT.

MOLECULE

A molecule is a group of atoms representing a substance as it occurs in nature. Every substance is made of molecules, and the molecules are always moving. Some molecules consist of just one atom. Others have thousands or even millions of atoms. All molecules are too small to see with ordinary optical microscopes, although some molecules have been observed with the electron microscope.

An atom of oxygen consists of a nucleus having eight protons and eight neutrons. In the atmosphere, oxygen atoms tend to group themselves in pairs.Sometimes they clump together in groups of three. A pair or triplet of oxygen atoms comprises an oxygen molecule in our air. Hydrogen atoms also tend to pair off; thus a normal hydrogen molecule consists of two hydrogen atoms. Oxygen and hydrogen atoms particularly like to group together; one oxygen atom joins with two hydrogen atoms (see illustration) to form the familiar water molecule. The foregoing examples are simple. Some molecules consist of complicated combinations of several different kinds of atoms. *See also* ATOM.

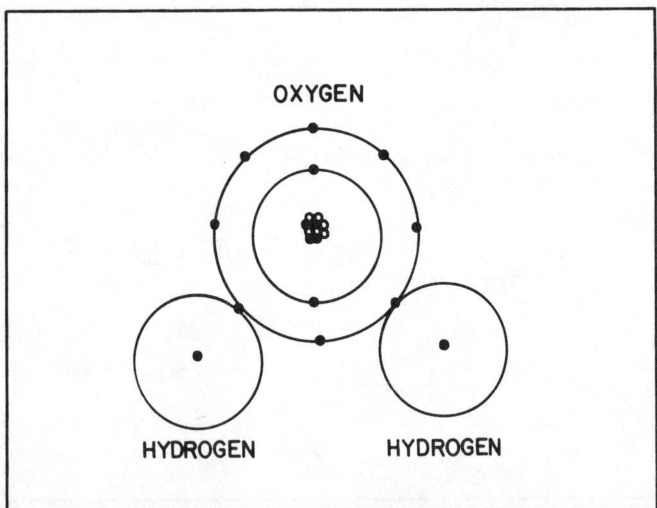

MOLECULE: The water molecule consists of two hydrogen atoms sharing electrons with a single atom of oxygen.

MOLYBDENUM

Molybdenum is an element with atomic number 42 and atomic weight 96. Molybdenum occurs in nature as a metallic substance.

Molybdenum is used in the electrodes of certain vacuum tubes. Molybdenum alloys are used as ferromagnetic cores for audio coils. The permeability of molybdenum permalloy is comparable to that of powdered iron or ferrite.

MOMENT

Moment is a mathematical expression involving rotation around a point. Moment is determined by the product of a parameter and a radial distance. For example, suppose a force F is applied to a lever of length d, as shown in the illustration. Then the product Fd, where F is given in newtons and d is given in meters, is called the moment of force.

In physics, moment is defined in a variety of ways. For example, the moment of inertia is a measure of the inertia of a rotating or revolving object. For a magnet, the product of the pole strength and the distance between the poles is known as the magnetic moment.

MONIMATCH

See REFLECTOMETER.

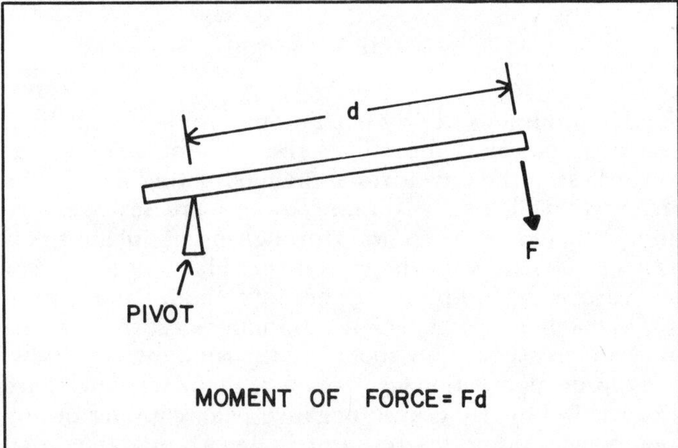

MOMENT: The moment of force is an angular quantity.

MONITOR

A monitor is an instrument that allows a signal to be analyzed. The oscilloscope is a common type of signal monitor (*see* OSCILLOSCOPE). However, more specialized types of monitors are used for such purposes as analyzing the modulating waveform in an amplitude-modulated or frequency modulated signal (see illustration) or evaluating the spectral output of a radio transmitter.

In a video display terminal, a monitor is used to view the text or data in alphanumeric form. Such a monitor consists of a cathode-ray tube and associated circuitry, and appears like a television set. *See also* VIDEO DISPLAY TERMINAL.

MONKEY CHATTER

When a single-sideband signal is not tuned in properly, the voice is impossible to understand. This can occur because the receiver and transmitter are not on precisely the same frequency. It can also happen when a receiver is set for one sideband and the transmitted signal is on the opposite sideband. The sound of the signal under such conditions is called monkey chatter. In some cases the signal really sounds like a monkey. *See also* SINGLE SIDEBAND.

MONOCHROMATICITY

Monochromaticity is the presence of light at only one wavelength, or within a single narrow band of wavelengths. (The absence of color, such as in a black-and-white television picture, is sometimes called monochromaticity, but that is a technical misnomer.) A monochromatic light source has a difinite color (hue), depending on the wavelength of the emission. The saturation may vary.

A laser emits highly monochromatic light, since the output is essentially at only one wavelength. A color filter makes things appear monochromatic, although the transmitted light actually occupies a band of wavelengths. The greater the level of saturation of light, the more perfectly monochromatic it is. The precise level of saturation at which a light source is considered monochromatic is, however, not precisely defined. *See also* HUE, SATURATION.

MONOLITHIC CIRCUIT

See INTEGRATED CIRCUIT.

MONOPOLE

A single unit electric charge, or group of electric charges, is called a monopole if it is isolated in space, and not accompanied by an equal and opposite charge center. Unlike a dipole, a monopole has a nonzero net charge.

The flux lines immediately surrounding an electric monopole are almost perfectly straight, and radiate outward in all directions from the charge center. The lines become curved near other electric charges. In theory, the electric lines of flux around a monopole extend indefinitely.

It is possible that magnetic monopoles may exist, but scientists have not yet found one. A north magnetic pole is always accompanied, it appears, by a south magnetic pole. The existence of a magnetic monopole apparently contradicts Maxwell's laws. Physicists continue, however, to search for the elusive magnetic monopole. *See also* DIPOLE.

MONOPOLE ANTENNA

Any unbalanced antenna, measuring ¼ physical wavelength or less in free space, is known as a monopole antenna. The monopole antenna is always quarter-wave resonant. The monopole is always fed at one end, and requires a ground plane for proper operation. The drawing illustrates the basic concept of the monopole antenna.

A monopole antenna is normally fed with unbalanced transmission line, such as coaxial cable. The feed-point

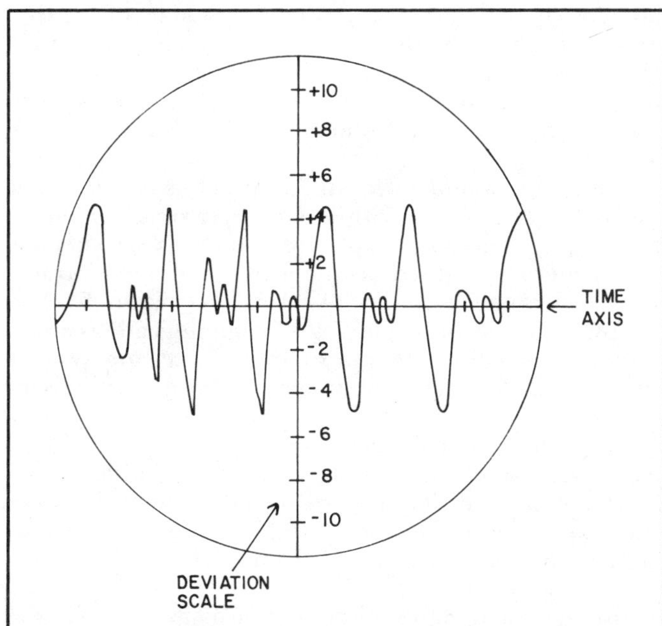

MONITOR: A signal-monitor screen, showing the modulating waveform from a frequency-modulated transmitter.

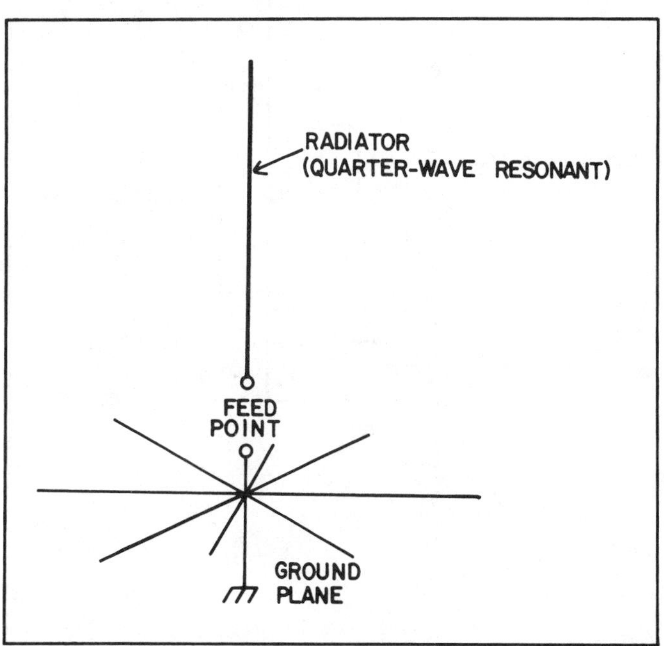

MONOPOLE ANTENNA: Basic design scheme for the monopole antenna. The ground plane may be artificial, such as a set of radials, or it may be the actual earth.

impedance varies, depending on the design. For a straight quarter-wave monopole over perfectly conducting ground, the feed-point impedance is approximately 37 ohms.

There are many variations of the monopole antenna. *See also* CONICAL MONOPOLE ANTENNA, DISCONE ANTENNA, MARCONI ANTENNA, VERTICAL ANTENNA.

MONOSTABLE MULTIVIBRATOR

A monostable multivibrator is a circuit with only one stable condition. The circuit can be removed from this condition temporarily, but it always returns to that condition after a certain period of time. The monostable multivibrator is sometimes called a one-shot multivibrator.

The illustration is a simple schematic diagram of a monostable multivibrator. Normally, the output is high, at the level of the supply voltage (+5 V). When a positive triggering pulse is applied to the input, the output goes low (0 V) for a length of time that depends on the values of the timing resistor R and the timing capacitor C. If R is given in ohms and C is given in microfarads, then the pulse duration T, in microseconds, can be found by the equation:

$$T = 0.69RC$$

After the pulse duration time T has elapsed, the monostable multivibrator returns to the high state.

Monostable multivibrators are used as pulse generators, timing-wave generators, and sweep generators for cathode-ray-tube devices. *See also* MULTIVIBRATOR.

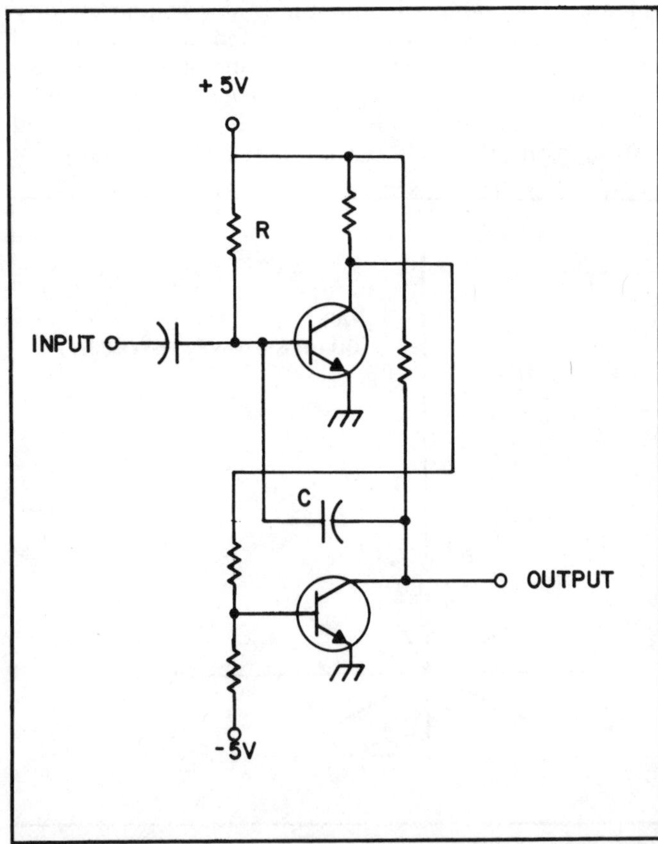

MONOSTABLE MULTIVIBRATOR: A monostable, or one-shot multivibrator circuit.

MONTE CARLO METHOD

The Monte Carlo method is a mathematical process for performing statistical calculations. The Monte Carlo method simplified some statistical problems by shortening the computation process.

An event with a given probability can be simulated by using random numbers. This can be done with any desired degree of accuracy. For example, the chances are 25 percent of getting "heads" twice for two tosses of a coin. If the coin is tossed 1 million times, how often might we expect to get "heads" exactly twice in a row? Rather than tossing the coin 1 million times, a computer can be used to simulate the coin tosses by drawing pairs of random numbers.

MOOG SYNTHESIZER

The Moog synthesizer is an electronic waveform generator that can produce almost any kind of periodic audio-frequency signal. A single audio-frequency tone can have thousands of different sound qualities, depending on the details of the waveform. A clarinet sounds different from a flute or a trumpet. The Moog can simulate the sound of practically any musical instrument, and can also produce sounds unlike those made by any known musical instrument. To the uninitiated, a Moog looks like an incredibly complicated jumble of wires, switches, and plugs, all surrounding a piano-like keyboard. Actually, the Moog is nothing more than a group of audio signal generators.

The Moog became popular among musicians, especially rock-and-roll and popular bands, in the 1960s. It is still extensively used today, and even its most exotic sounds are becoming familiar to our ears. A Moog synthesizer is expensive, and, like any musical instrument, its use must be learned through practice. Some musicians say that a Moog is one of the most difficult instruments to master.

MOONBOUNCE

The moon reflects radio signals to a certain extent. This has allowed accurate measurement, by radar, of the distance to the moon. With large antennas and high-powered transmitters, communications can be carried out by means of signals reflected from the moon. This mode of communication is called earth-moon-earth (EME) or moonbounce.

Achieving moonbounce communications requires that the tremendous EME path loss be overcome. Antennas must be physically large. A phased system of Yagi antennas must generally be used, along with a high-power transmitter. Feed-line losses must be kept to a minimum. The antenna must be fully steerable if communications are to be maintained for a length of time. Receiving systems must employ low-noise preamplifiers for maximum sensitivity.

Moonbounce signals have a rapid fluttering sound. This occurs because of random phase effects as the signals arrive from many different parts of the moon at the same time; the phase combinations are constantly changing because of the rotation of the earth and the wobbling (libration) of the moon on its axis.

Moonbounce communication is usually done at the very-high or ultra-high frequencies. The most popular amateur bands for EME are at 144 MHz (2 m) and 432 MHz (¾ m). Morse code is generally used, but voice (single-sideband) and slow-scan television signals have been used successfully under ideal conditions. There is even an annual EME contest for radio amateurs, sponsored by the American Radio Relay League.

MORSE CODE

The Morse code is a binary means of sending and receiving messages. It is a binary code because it has just two possible states: on (key-down) and off (key-up).

There are two different Morse codes in use by English-speaking operators today. The more commonly used code is called the international or continental code. A few telegraph operators use the American Morse code. (*See* AMERICAN MORSE CODE, INTERNATIONAL MORSE CODE.) The code characters vary somewhat in other languages.

Modern communications devices today can function under weak-signal conditions that would frustrate a human operator. But when human operators are involved, the Morse code has always been the most reliable means of getting a message through severe interference. This is because the bandwidth of a Morse-code radio signal is extremely narrow, and it is comparatively easy for the human ear to distinquish between the background noise and a code signal.

MORSE, SAMUEL F. B.

An artist and inventor, Samuel F. B. Morse (1791-1872) is known for his work with the transmission of electrical impulses via wire. Morse invented a code, consisting of off/on pulses of various duration, for sending messages by means of electric current (*see* AMERICAN MORSE CODE, INTERNATIONAL MORSE CODE).

Morse put the first working telegraph line between Washington, DC and Baltimore, Maryland and, in 1844, send the first message ever conveyed at electronic speeds: "What hath God Wrought!"

MOS

See METAL-OXIDE SEMICONDUCTOR.

MOSAIC

See PHOTOMOSAIC.

MOSFET

See METAL-OXIDE-SEMICONDUCTOR FIELD-EFFECT TRANSISTOR.

MOTION DETECTOR

See ULTRASONIC MOTION DETECTOR.

MOTOR

A motor is a device that converts electrical energy into mechanical energy. Motors can operate from alternating or direct current, and can run at almost any speed. Motors range in size from the tiny devices in a wristwatch to huge, powerful machines that can pull a train at over 100 miles per hour.

All motors operate by means of electromagnetic effects. Electric current flows through a set of coils, producing powerful magnetic fields. The attraction of opposite magnetic poles, and the repulsion of like magnetic poles, results in a rotating force. The greater the current flowing in the coils, the greater the rotating force. When the motor is connected to a load, the amount of force required to turn the motor shaft increases. The more the force, the greater the current flow becomes, and the greater the amount of energy drawn from the power source.

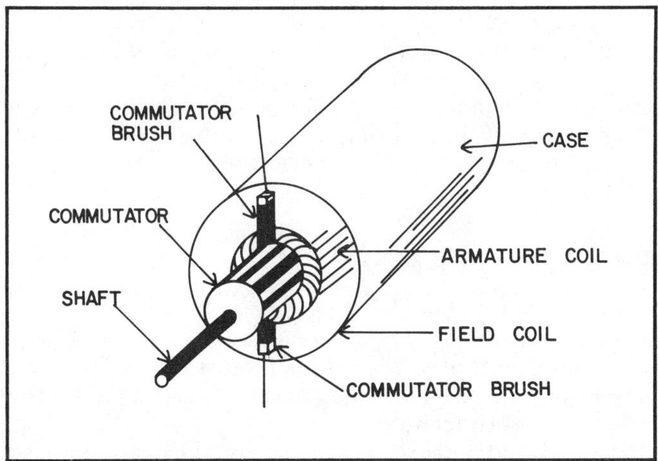

MOTOR: Cutaway view of a typical electric motor.

The illustration is a cutaway diagram of a typical electric motor. One set of coils rotates with the motor shaft. This is called the armature coil. The other set of coils is fixed, and is called the field coil. The commutator reverses the current with each half-rotation of the motor, so that the force between the coils maintains the same angular direction.

The electric motor operates on the same principle as an electric generator. In fact, some motors can actually be used as generators. *See also* GENERATOR.

MOTORBOATING

Excessive feedback in an audio amplifier can result in a fluttering or popping sound that resembles the sound of a motorboat. Such oscillation usually results from undesirable coupling between the output and the input of an amplifier or chain of amplifiers. This can occur as a result of capacitive or inductive coupling in the wiring. It can also occur via the power supply. Because of the characteristic sound, this low-frequency oscillation is called motorboating.

Motorboating can be eliminated by minimizing the coupling between the output and the input of an amplifier system. Interstage connecting wires should be balanced or shielded, and should be as short as possible. A filtering device may have to be placed in series with the leads to the power supply. Such a filter usually consists of a large-value choke and one or two capacitors. *See also* FEEDBACK.

MOTOR-SPEED CONTROL

A motor-speed-control device is a circuit designed for alternating-current power control. Such a circuit can be used to control the amount of power used by various appliances or electric lights. The motor-speed control provides a continuously adjustable output voltage, ranging from zero to the full input value. The output voltage is independent of the load resistance, as long as that resistance is not less than a certain specified minimum.

The triac is the most common device used in the construction of a motor-speed control. The illustration is a schematic diagram of a simple triac motor-speed control. *See also* SILICON-CONTROLLED RECTIFIER, TRIAC.

MOUTH-TO-MOUTH RESUSCITATION

When a person has stopped breathing because of electric shock or other injury, resuscitation must be applied. The most commonly used method of artificial respiration is the mouth-to-mouth technique.

If an injured person is not breathing, the heart may also have stopped. The pulse should be checked immediately. If the heart has stopped, cardiopulmonary resuscitation (CPR) must be used. If the heart is beating, mouth-to-mouth resuscitation alone is sufficient.

The person's mouth should be checked for debris, and then the head should be tilted back. The nose should be pinched shut. To perform the resuscitation, air is exhaled from the rescuer, by mouth, into the mouth of the victim. The victim is then allowed to exhale. This process should be repeated about 12 times per minute for an adult victim, and 20 times per minute for a child.

A full and comprehensive discussion of resuscitation techniques is beyond the scope of this book. This essay should not, therefore, be considered as instruction. The American Red Cross publishes books and provides courses in cardiopulmonary resuscitation and mouth-to-mouth resuscitation. These books and courses should be used for training purposes. Everyone should be trained in cardiopulmonary resuscitation and mouth-to-mouth resuscitation. Information may be obtained from local chapters of the American Red Cross. *See also* CARDIOPULMONARY RESUSCITATION.

MOUTH-TO-NOSE RESUSCITATION

The mouth-to-nose method of resuscitation is basically the same as the mouth-to-mouth method. The only difference is that the rescuer places his or her mouth over the nose of the victim.

Mouth-to-nose resuscitation is preferred by some people in the event the victim has vomited. *See* MOUTH-TO-MOUTH RESUSCITATION.

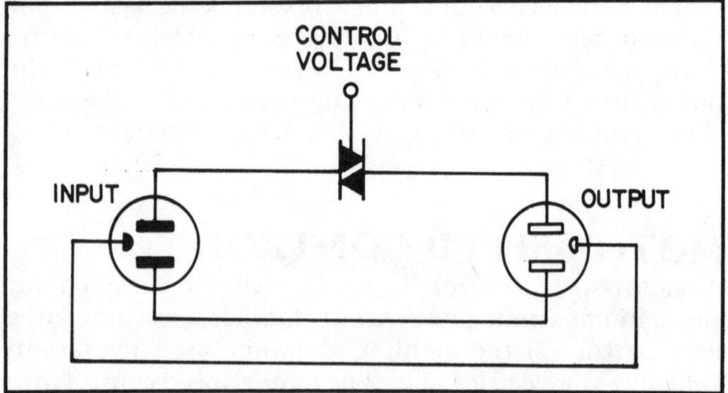

MOTOR-SPEED CONTROL: A simple motor-speed control, using a triac.

MOVING-COIL METER

See D'ARSONVAL.

MOVING-COIL LOUDSPEAKER

See DYNAMIC LOUDSPEAKER.

MOVING-COIL MICROPHONE

See DYNAMIC MICROPHONE.

MOVING-COIL PICKUP

See DYNAMIC PICKUP.

MSI

See MEDIUM-SCALE INTEGRATION.

MU

Mu is a letter of the Greek alphabet. Its symbol is written like a small English letter u with a "tail" (μ). The English u is often used in place of the actual symbol μ.

Mu is used as a prefix multiplier meaning micro (*see* MICRO). The symbol μ is also used to indicate amplification factor, permeability, inductivity, magnetic moment, and molecular conductivity.

MUF

See MAXIMUM USABLE FREQUENCY.

MULTIBAND ANTENNA

A multiband antenna is an antenna that is designed for operation on more than one frequency. A half-wave dipole antenna is a multiband antenna; it is resonant at all odd

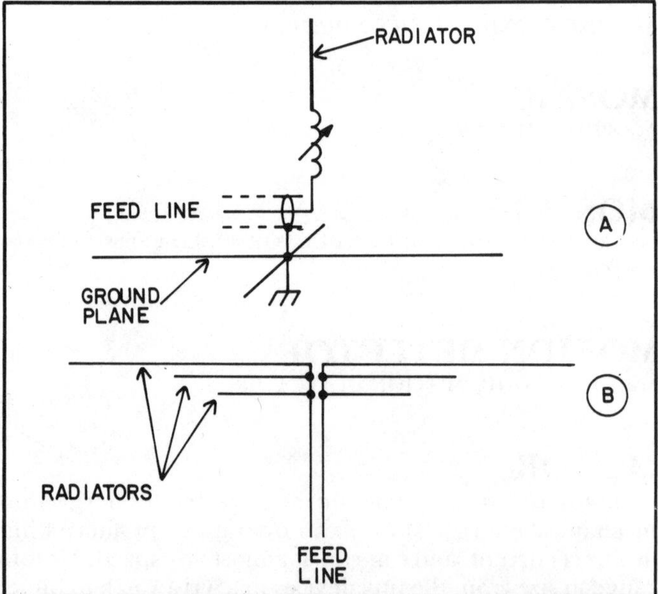

MULTIBAND ANTENNA: Two methods of obtaining multiband antenna. At A, a short radiator can be adjusted for resonance via a variable inductor. At B, several half-wave antennas, one for each frequency desired, are connected in parallel at a common feed point.

multiples of the fundamental resonant frequency. An end-fed half-wave antenna can be operated at any multiple of the fundamental frequency. All antennas have a theoretically infinite number of resonant frequencies; not all of these frequencies, however, are useful.

Multiband antennas can be designed deliberately for operation on specific frequencies. Traps are commonly used for achieving multiband operation (*see* TRAP, TRAP ANTENNA). A variable inductor can be used for changing the resonant frequency of an antenna, as shown at A in the illustration (*see* INDUCTIVE LOADING). A tuning network may be used to adjust the resonant frequency of an antenna/feeder system (*see* TUNED FEEDERS). Several different antennas can be connected in parallel to a single feed system to achieve multiband operation as at B.

Multiband antennas offer convenience; it is simple to switch from one frequency band to another. However, harmonic-resonant multiband antennas can radiate unwanted harmonic signals. For this reason, when a harmonic-resonant antenna is used, care must be exercised to ensure that the transmitter harmonic output is sufficiently attenuated in the final-amplifier and output circuits. *See also* HARMONIC, RESONANCE.

MULTIELEMENT ANTENNA

Some antennas employ multiple elements for the purpose of obtaining a directional response. An antenna element consists of a length of conductor. A conductor may be directly connected to the feed line; then it is called an active or driven element. A conductor may be physically separate from the feed line; then it is called a passive or parasitic element.

Multielement antennas can be broadly classified as either parasitic arrays or phased arrays. In the parasitic array, passive elements are placed near a single driven element (*see* DIRECTOR, DRIVEN ELEMENT, PARASITIC ARRAY, REFLECTOR). In the phased array, two or more elements are driven together (*see* PHASED ARRAY).

Multielement antennas are used mostly at high frequencies and above, although some large broadcast installations make use of phased vertical arrays at low and medium frequencies. The main advantages of a multielement antenna are power gain and directivity, which enhance both the transmitting and receiving capability of a communications station. The design and construction of a multielement antenna is, however, more critical than for a single-element antenna.

MULTILEVEL TRANSMISSION

Multilevel transmission is a form of digital transmission, in which some signal parameter has three or more discrete values. The number of possible levels must, however, be finite. Therefore ordinary analog modulation is not considered multilevel transmission.

Multilevel transmission can be used to digitally transmit a complex waveform, provided the element (bit) duration is short enough, and provided the number of levels is large enough. The complex waveform shown at A in the illustration is converted to a coarse three-level amplitude-

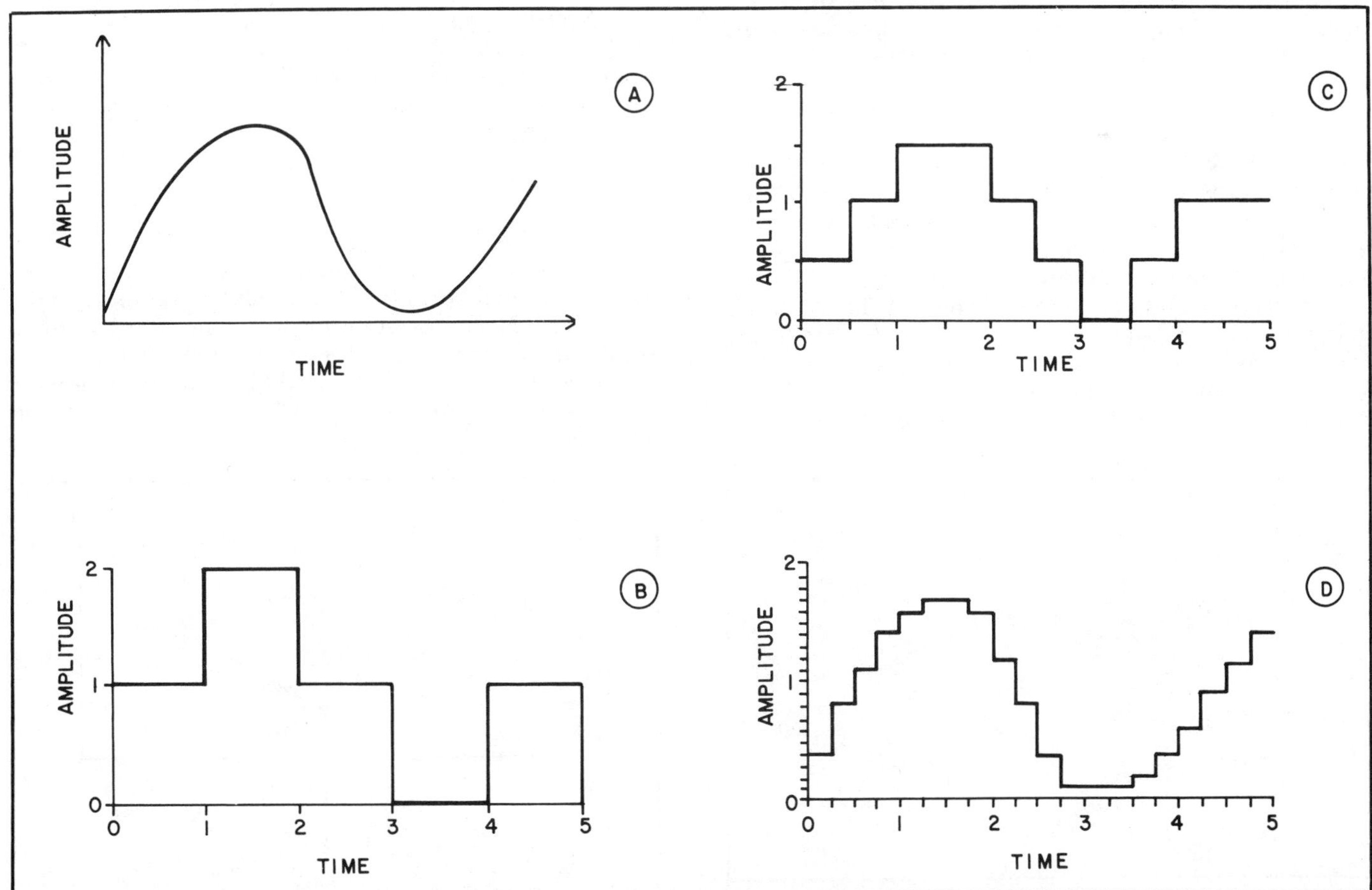

MULTILEVEL TRANSMISSION: At A, an irregular analog waveform. At B, a three-level digital representation of the waveform; it is a relatively poor approximation. At C and D, finer digital representations provide better approximations.

modulated signal at B. Finer multilevel conversion signals are illustrated at C and D.

The amplitude is not the only parameter that can be varied in a multilevel signal. In a slow-scan television signal, for example, the frequency may attain any of several discrete values, resulting in various shades of brightness. Multilevel transmission may be employed with any form of modulation. The primary advantage of multilevel transmission is its narrow bandwidth compared with ordinary analog modulation.

MULTIPATH FADING

Multipath fading is a form of fading that occurs primarily at medium and high frequencies, and results from the random phase combination of received ionospheric signals arriving along more than one path at the same time.

The ionosphere is not a smooth reflector of electromagnetic energy. Instead, the ionization occurs in irregular patches and with variable density. The result is that signals may be transmitted between two points via several different paths simultaneously (see illustration). These several paths do not all have the same overall length, and therefore the signals combine in random phase at the receiver. As the ionosphere undulates, the various path lengths constantly vary, and so does the overall phase combination at the receiver.

Multipath fading effects can be reduced by using two receivers for the same signal, with antennas located at least several wavelengths apart. The probability that a severe fade will occur at both receivers, simultaneously, is less than the probability that a fade will take place at a single receiver. *See also* DIVERSITY RECEPTION, FADING.

MULTIPLE POLE

A multiple-pole relay or switch is designed for switching two or more circuits at once. Multiple-pole relays and switches may be of the single-throw or multiple-throw variety (*see* MULTIPLE THROW).

Schematic symbols for multiple-pole switches or relays generally show the same circuit designator for each pole, followed by A, B, C, and so on (see illustration). The various poles may be connected in the diagram by a dotted line. When one pole changes state, the others all change state.

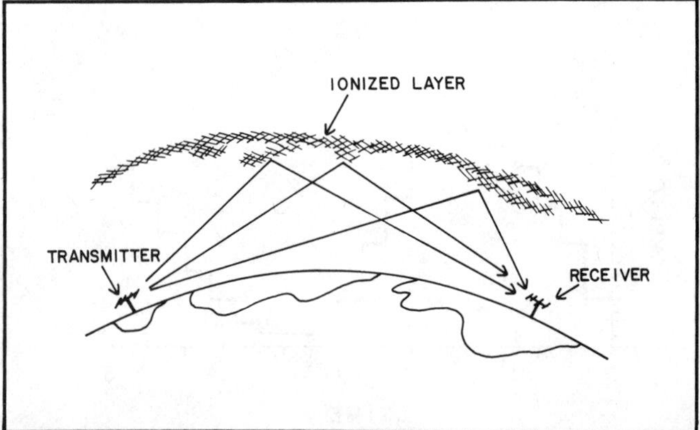

MULTIPATH FADING: Multipath fading occurs when several waves, propagated by different paths (solid lines), combine in varhing phase at the receiving antenna.

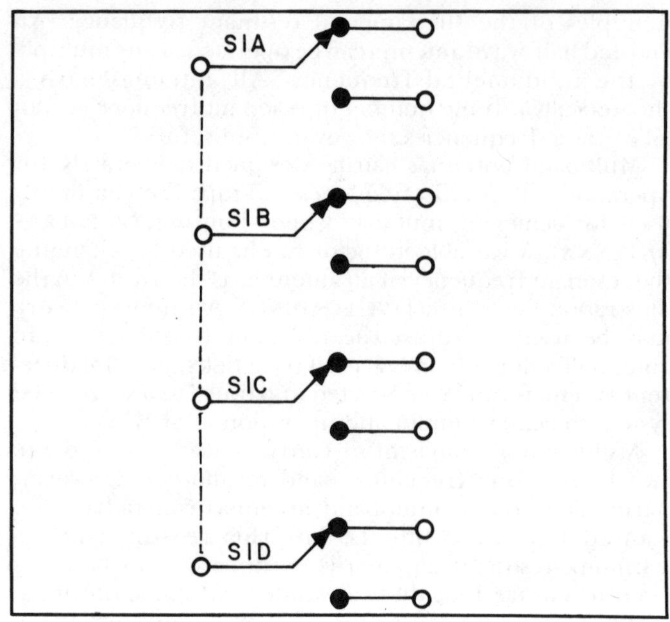

MULTIPLE POLE: A multiple-pole switch. Here, the switch is designated S1, and it has four poles.

MULTIPLE THROW

A multiple-throw relay or switch is a device that can connect a given conductor to two or more other conductors. A relay normally has at most two throw positions because of mechanical constraints, but a switch may have several throw positions. Multiple-throw relays and switches may be ganged to form a multiple-pole, multiple-throw device (*see* MULTIPLE POLE).

The illustration shows the schematic symbol for a five-position multiple-throw switch. Generally, a switch with three or more throw positions is a wafer switch. *See also* WAFER SWITCH.

MULTIPLEX

Multiplex refers to the simultaneous transmission of two or more messages over the same medium or channel at the same time. Multiplex transmission may be achieved in various different ways, but the most common methods are frequency-division multiplex and time-division multiplex (*see* FREQUENCY-DIVISION MULTIPLEX, TIME-DIVISION MULTIPLEX).

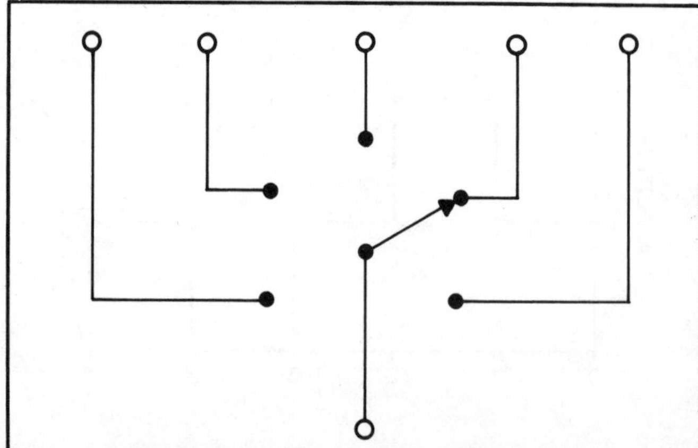

MULTIPLE THROW: A five-position, multiple-throw switch.

Multiplex transmission requires a special encoder at the transmitting end and a special decoder at the receiving end. The medium must be such that interference does not occur among the various channels transmitted.

A form of digital signal transmission, used especially with display devices, is called multiplex. When it is necessary to provide a voltage to many different display components at the same time, the data may be cumbersome to transmit in parallel form. The data may, instead, be sent in serial form, using time-division multiplex. For example, a four-digit light-emitting-diode display can be illuminated by "rotating" the applied voltage from left to right. If this is done rapidly enough, the eye cannot tell that each digit is illuminated for only 25 percent of the time. *See also* PARALLEL DATA TRANSFER, SERIAL DATA TRANSFER.

MULTIPLICATION

Multiplication is an arithmetic operation. When a number x is multipled by an integer n, the number x is added to itself n times. In electronics, parameters such as frequency or voltage may be multiplied by means of specialized circuits. The multiplication factor is always an integer. *See also* FREQUENCY MULTIPLIER, VOLTAGE MULTIPLIER.

MULTIPLIER CIRCUIT

See FREQUENCY MULTIPLIER, VOLTAGE MULTIPLIER.

MULTIPLIER PREFIX

See PREFIX MULTIPLIERS.

MULTIVIBRATOR

A multivibrator is a form of relaxation oscillator. A multivibrator circuit consists of a pair of inverting amplifiers. The amplifiers are connected in series, and a direct feedback loop runs from the output of the second stage to the input of the first. A block diagram of the basic scheme is shown in the illustration. Multivibrators are extensively used in digital circuit design.

There are three main types of multivibrator. The astable multivibrator is a free-running oscillator. The bistable multivibrator, often called a flip-flop, may attain either of two stable conditions. The monostable multivibrator maintains a single condition except when a triggering pulse is applied; then the state changes for a predetermined length of time. *See also* ASTABLE MULTIVIBRATOR, FLIP-FLOP, MONOSTABLE MULTIVIBRATOR.

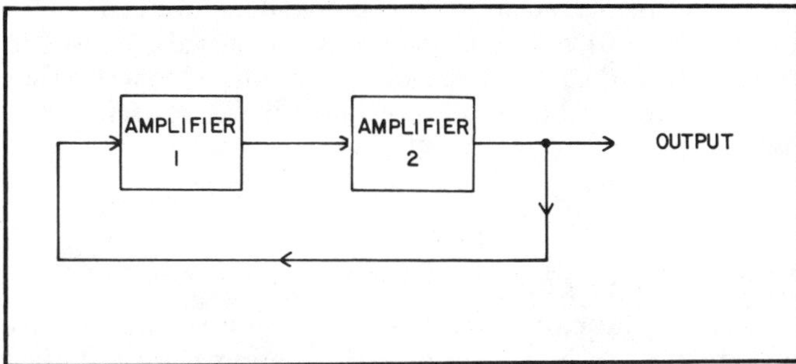

MULTIVIBRATOR: A multivibrator basically consists of two amplifier circuits in cascade, with positive feedback.

MUMETAL

Mumetal is a ferromagnetic substance commonly used as a magnetic shield in cathode-ray tubes.

Mumetal consists of a mixture of iron, nickel, copper, and chromium. The magnetic permeability is very high. The direct-current resistivity is moderately low, resulting in good magnetic shielding characteristics. *See also* FERROMAGNETIC MATERIAL.

MURRAY LOOP TEST

The Murray loop test is a method of determining the distance from a given point in a telephone line to a ground fault. For this to be possible, an extremely accurate means of resistance measurement is needed. The device used for this purpose is called a Murray bridge, a special form of Wheatstone bridge.

The drawing illustrates the connection of the Murray bridge to a telephone line. Resistor R3 is adjusted until the indicator shows a null reading. The resistance along the line, R, is then given by:

$$R = R2R3/R1$$

where R1 and R2 are fixed resistances.

Knowing the resistance of the line per unit length in either conductor (for example, x ohms per meter), the distance d, in meters, to the ground fault can be found by simple calculation:

$$d = R/(2x)$$

See also WHEATSTONE BRIDGE.

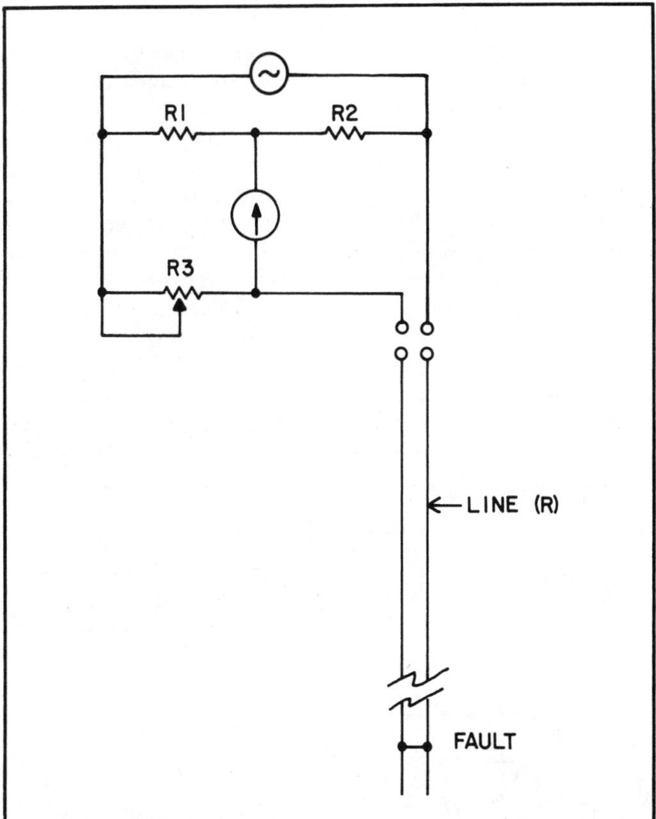

MURRAY LOOP TEST: The Murray loop test is a method for determining the location of a telephone-line fault.

MUSIC

Music is a combination of audio tones that is regarded as pleasing to the ear. The earliest music consisted of speaking in a monotone voice. Then, it was discovered that certain tone sequences had a pleasant sound. Devices were invented to make more and more complex tone combinations, with many different types of waveforms. Today, music is incredibly diverse and sophisticated.

For music to be reproduced faithfully, precision audio amplifier circuits are needed. Most of today's music is electronically reproduced or processed by high-fidelity circuits before anyone hears it. *See also* HIGH FIDELITY, MOOG SYNTHESIZER, STEREOPHONICS.

MUSIC POWER

Music power is an expression of the power output obtained from a high-fidelity audio amplifier. The music power is the root-mean-square (RMS) power output of the amplifier. It varies considerably with time.

Some sounds in a musical selection produce relatively low power output from an amplifier. There may even be moments of silence, when the music power is zero. Other sounds produce large momentary power output. Music power varies not only with the actual volume of the music, but also with the waveforms of the instruments. *See also* ROOT MEAN SQUARE.

MUTING CIRCUIT

A muting circuit is a device that cuts off a television or radio receiver while the power is on. This prevents signals from being heard, but is allows instantaneous reactivation of the circuit when desired by the operator.

A muting circuit generally consists of a switching transistor or relay in series with one of the stages of the receiver. The muted stage is usually one of the radio-frequency or intermediate-frequency amplifiers in the receiver chain. However, an audio-frequency circuit may be muted if an extremely fast recovery is not needed. The schematic diagram shows a simple muting system using a transistor switch. A negative voltage at the muting terminal opens the emitter circuit of the amplifier. Other types of muting circuits also exist; some are actuated by a positive voltage, others by an open circuit at the input terminals, and still others by a short circuit.

Muting circuits are used for break-in operation for Morse-code communication. Muting circuits are also used to prevent receiver overload or acoustic feedback in two-way voice radio systems. They can also be found in television channel switching circuits. *See also* BREAK-IN OPERATION.

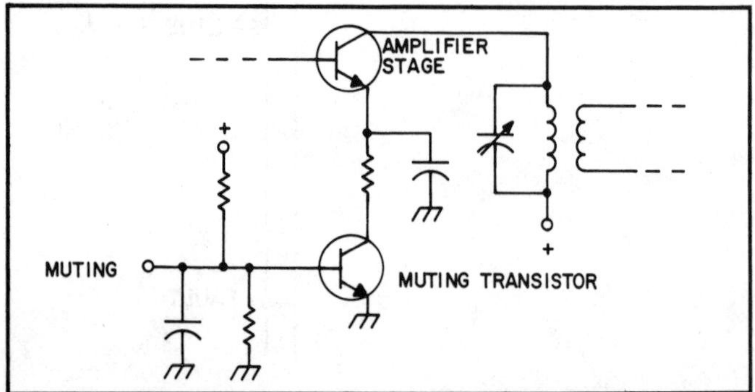

MUTING CIRCUIT: A simple muting circuit for cutting off a radio-frequency amplifier stage.

MUTUAL CAPACITANCE

Any two conductors, no matter where they are located or how they are oriented, have a certain amount of capacitance with respect to each other. That is, given a device capable of measuring a sufficiently tiny value of capacitance, any set of two conductors can be shown to act as a capacitor. The existence of capacitance among electrical conductors is called mutual capacitance. The mutual capacitance between two conductors increases as the surface area of either conductor is made larger; the mutual capacitance decreases as the distance between conductors becomes greater.

Mutual capacitance is often too small to be measured, and is of no consequence. However, in radio-frequency circuit wiring, mutual capacitance is sometimes sufficient to cause feedback and other problems. This is especially true at very-high and ultra-high frequencies. Radio circuits at these frequencies are designed to minimize the mutual capacitance among wires and components.

Mutual capacitance exists among the elements within certain electronic components, particularly diodes, relays, switches, transistors, and vacuum tubes. In radio-frequency circuit design, this capacitance is often significant, and must be taken into account in the design process. *See also* ELECTRODE CAPACITANCE, INTERELECTRODE CAPACITANCE.

MUTUAL CONDUCTANCE

See TRANSCONDUCTANCE.

MUTUAL IMPEDANCE

Two electrical conductors invariably display a certain amount of mutual capacitance and mutual inductance (*see* MUTUAL CAPACITANCE, MUTUAL INDUCTANCE). The combination of mutual capacitance and inductance presents a complex impedance, known as mutual impedance, that varies with frequency. Because of this, a resonant circuit is formed. The resonant frequency depends on the amount of mutual capacitance and inductance. Generally, the natural resonant frequency of two conductors or electrodes is very high—on the order of hundreds or even thousands of megahertz.

The impedance between or among the electrodes of certain components can cause oscillations. These oscillations are harnessed in certain types of vacuum tubes for the purpose of generating microwave signals. However, the oscillations can occur in a vacuum tube even when they are not wanted. Then they are called parasitic oscillations. *See also* PARASITIC OSCILLATION.

MUTUAL INDUCTANCE

Any two conductors have a certain amount of inductance with respect to each other. A fluctuating magnetic field around one conductor will invariably cause an alternating current in the other. The degree of inductive coupling between conductors or inductors is called mutual inductance.

Mutual inductance, represented by the letter M and expressed in henrys, is an expression of the degree of coupling between two coils. The mutual inductance is mathematically related to the coefficient of coupling, which may range from 0 (no coupling) to 1 (maximum possible coupling). If the coefficient of coupling is given by k and the inductances of the two coils, in henrys, are given by L1 and L2, then:

$$M = k \sqrt{L1L2}$$

For example, suppose that k = 0.5, and the coils have inductances of 0.5 henry and 2 henrys. Then:

$$M = 0.5 \sqrt{0.5 \times 2} = 0.5$$

In practice, since most coils have inductances that are a small fraction of 1 henry, the value of M is quite small.

For two inductors having values L1 and L2 (given in henrys) and connected in series with the flux linkages in the same direction, the total inductance L, in henrys, is:

$$L = L1 + L2 + 2M$$

If the flux linkages are in opposition, then:

$$L = L1 + L2 - 2M$$

See also COEFFICIENT OF COUPLING, INDUCTANCE.

MYLAR CAPACITOR

Mylar, a form of plastic, has excellent dielectric properties, and it is therefore well suited for making capacitors. Mylar capacitors are common in audio-frequency and radio-frequency circuits.

A mylar capacitor is relatively impervious to the effects of moisture, since plastic does not absorb water. Therefore, mylar capacitors are an excellent choice in marine or tropical environments. Mylar capacitors are relatively inexpensive. Typical values range from several picofarads to several microfarads. The voltage rating is low to moderate. Mylar capacitors are suitable for use at high frequencies. But at very-high frequencies and above, the dielectric loss is somewhat increased. The maximum frequency at which a mylar capacitor can efficiently operate depends on the particular type of plastic used. *See also* CAPACITOR.

NAND GATE

A NAND gate is a logical AND gate, followed by an inverter (*see* AND GATE, INVERTER). The term NAND is a contraction of NOT AND. The illustration shows the schematic symbol for the NAND gate, along with a truth table indicating the output as a function of the input.

When both or all of the inputs of the NAND gate are high, the output is low. Otherwise, the output is high.

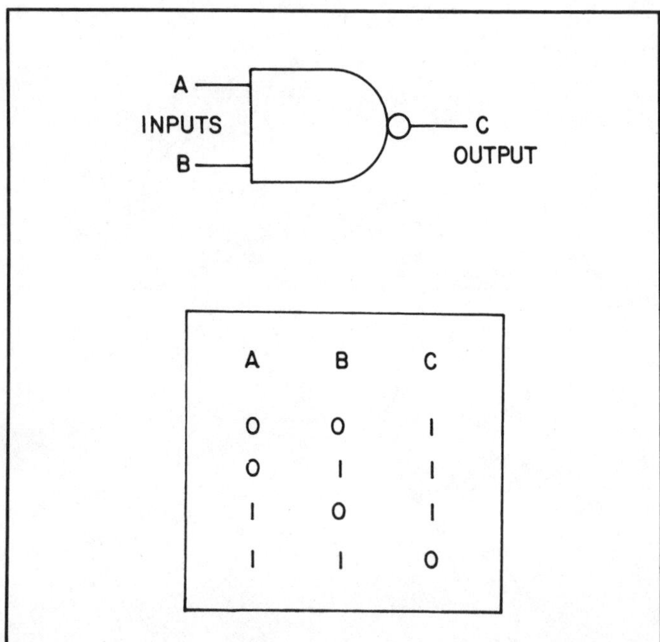

A	B	C
0	0	1
0	1	1
1	0	1
1	1	0

NAND GATE: The NAND gate is equivalent to an AND gate followed by an inverter.

NANO

Nano is a prefix multiplier, representing 1 billionth (10^{-9}). For example, 1 nanofarad is equal to 10^{-9} farad, or 0.001 microfarad; 1 nanowatt is equal to 10^{-9} watt. The abbreviation for nano is the small letter n. *See also* PREFIX MULTIPLIERS.

NAPIER

See NEPER.

NARROW-BAND VOICE MODULATION

Narrow-band voice modulation, abbreviated NBVM, is a technique for reducing the bandwidth necessary for the transmission of a human voice. Normally, this bandwidth is about 2.5 to 3 kHz; a bandpass of approximately 300 Hz to 3 kHz is required. Using NBVM, the bandwidth can be substantially reduced.

The sounds of the human voice do not occupy the entire frequency range between 300 Hz and 3 kHz; rather, the sound tends to be concentrated within certain bands called formants. For a voice to be clear and understandable, three formants must be passed. The lower formant occurs at a range of about 300 Hz to 600 Hz. The upper two formants exist within the range 1.5 to 3 kHz. Between 600 Hz and 1.5 kHz, very little sound occurs. In NBVM, this unused range is eliminated, reducing the overall signal bandwidth by about 900 Hz. Narrow-band voice modulation is thus a frequency-companding process.

Narrow-band voice modulation is achieved by mixing at the baseband, or audio-frequency, level. The resulting audio signal sounds "scrambled." The audio output of the NBVM compressor can be fed to the microphone input of any voice transmitter. At the receiver, the signal is restored to its original form by frequency expansion. *See also* COMPANDOR.

NATIONAL BUREAU OF STANDARDS

The National Bureau of Standards (NBS) is an agency in the United States that maintains values for physical constants in the standard international system of units.

Among shortwave radio listeners and radio amateurs, the National Bureau of Standards is best known for its continuous standard time and frequency broadcasts. These signals are sent by WWV in Fort Collins, Colorado, and WWVH on the island of Kauai, Hawaii. *See also* STANDARD INTERNATIONAL SYSTEM OF UNITS, WWV/WWVH.

NATIONAL ELECTRIC CODE

The National Electric Code (NEC) is a set of recommendations for safety in electric wiring. The NEC is prepared by the National Fire Protection Association and the Institute of Electrical and Electronic Engineers, and is purely advisory. However, the NEC is enforced by some local governments.

The NEC is primarily concerned with types of insulation, and suitable wire sizes and conductor materials, and for various levels of voltage and current. Maximum allowable current values are given for different kinds and sizes of wire and cable. Temperature derating curves are given.

The NEC also makes recommendations regarding the wiring of electronic equipment, espenially if high voltage and/or current levels are used. Detailed information can be obtained from the American National Standards Institute, New York, NY.

NATURAL FREQUENCY

The natural frequency of a circuit, antenna, or set of electrical conductors is the lowest frequency at which the system is resonant for electromagnetic energy. In acoustics, physical objects have a lowest resonant sound frequency; this is sometimes called the natural frequency.

Certain systems have a tendency to oscillate at the natural frequency. For example, a radio-frequency power amplifier, if not neutralized, may develop parasitic oscillations at the natural frequency of interelectrode impedance.

In an antenna system, the radiator and feed line together have a natural frequency that differs from the resonant frequency of the antenna by itself. When tuned feeders are used to obtain broadband operation, the feed line may radiate if the system is used at its natural frequency. *See also* PARASITIC OSCILLATION, RESONANCE, TUNED FEEDERS.

NATURAL GAS

Natural gas is a flammable gas that is used for a variety of industrial purposes. The most common ingredient of natural gas is methane. It is a fossil fuel.

Natural gas is important in electricity because of its usefulness as a fuel for electric generators. *See also* GENERATOR.

NATURAL LOGARITHM

See LOGARITHM.

NATURAL NUMBER

In mathematics, a natural number is a positive integer, or whole number. The set of natural numbers is an infinite set. Natural numbers are also known as counting numbers.

The number e, the base for natural logarithms, is sometimes called the natural number. It is an irrational number; its value in decimal form, to ten significant digits, is 2.718281828. The natural number is the sum of the infinite series:

$$e = 1 + 1/1! + 1/2! + 1/3! + \ldots$$

where n! is equal to n factorial:

$$n! = n(n-1)(n-2)(n-3) \ldots (n-(n-1))$$

See also LOGARITHM.

NAUTICAL MILE

The nautical mile is a unit of distance used by mariners. It is slightly greater than the statute mile. A nautical mile is equal to 1.151 statute miles, or 1.85 kilometers.

Nautical speed is expressed in units called knots. A knot is one nautical mile per hour. Wind speed is sometimes given in knots; to obtain the speed in statute miles per hour, divide by 1.151.

NAVIGATION EQUIPMENT

See RADIOLOCATION, RADIONAVIGATION.

NEAR FIELD

The near field of radiation from an antenna is the electromagnetic field in the immediate vicinity of the antenna. The electric and magnetic lines of flux in the near field are curved because of the proximity of the radiating element (A in the illustration). The wavefronts are not flat, but instead are convex.

For a paraboloidal or dish antenna, the near field is the region within which the radiation occurs mostly within a cylindrical region having the same diameter as the dish, as at B. The near field of radiation from a dish antenna is sometimes called the Fresnel zone. The distance to which the near field extends depends on the diameter of the antenna, the antenna aperture, and the wavelength.

In radio communications, the near field is generally of little importance. The signal normally picked up by the receiving antenna results not from the near field, but from the far field or Fraunhofer region. *See also* FAR FIELD.

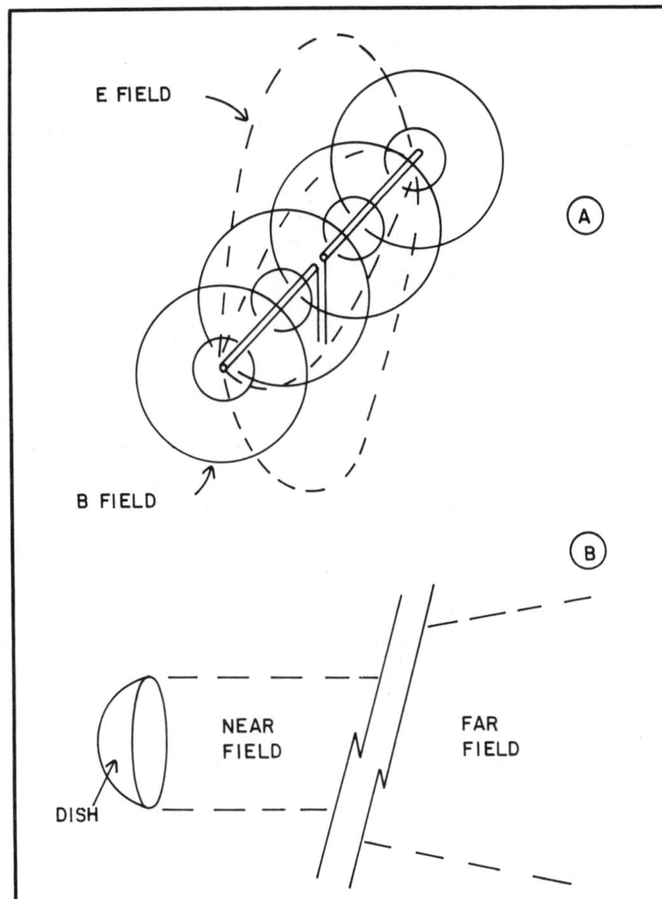

NEAR FIELD: At A, the near field surrounding a half-wave dipole. The electric field (dotted lines) and magnetic field (solid lines) have curved wavefronts. At B, the near field and far field in front of a dish antenna.

NECESSARY BANDWIDTH

The necessary bandwidth of a signal is the minimum bandwidth required for transmission of the data with reasonable accuracy. The degree of accuracy is somewhat arbitrary; it is impossible to obtain perfection. In general, the greater the signal bandwidth, the better the accuracy. As the bandwidth is increased beyond a certain point, however, the improvement is small and spectrum space is wasted. Typical values of necessary bandwidth for various emission types are:

For a Morse-code signal, the necessary bandwidth, in hertz, is generally considered to be about 2.4 times the speed in words per minute. Thus, for example, a Morse transmission at 20 words per minute (a typical speed) is 48 Hz. The receiver bandpass should be at least 48 Hz for accurate reception of a Morse signal at 20 words per minute.

For BAUDOT and ASCII codes, the necessary bandwidth also depends on the speed. Generally, the necessary bandwidth, in hertz, for such signals is about 2.4 times the speed in words per minute, or 3.2 times the speed in bauds. As with Morse transmission, this is a somewhat subjective value, and does not represent an absolute standard.

For a single-sideband transmission, the necessary bandwidth is about 2.5 kHz. However, with narrow-band modulation techniques, this value can be reduced considerably (*See* NARROW-BAND VOICE MODULATION). A slow-scan television signal requires about 2.5 kHz.

A normal amplitude-modulated or frequency-modulated voice signal requires approximately 5 to 6 kHz of spectrum space for reasonable intelligibility. Some frequency-modulated signals, however, are spread over a much larger space to improve the fidelity. The transmission of high-fidelity music requires a minimum of 40 kHz of spectrum space with amplitude or frequency modulation.

Video signals and high-speed data transmissions have large necessary bandwidth. The typical television broadcast channel is 6 MHz wide.

See also BANDWIDTH, BAUD RATE, SPEED OF TRANSMISSION.

NEGATION

Negation is a logical NOT operation. In Boolean algebra, negation is called complementation. Electronically, negation is performed by a NOT gate or inverter. *See also* BOOLEAN ALGEBRA, INVERTER.

NEGATIVE CHARGE

Negative charge, or negative electrification, is the result of an excess of electrons on a body. Friction between objects can result in an accumulation of electrons on one object (a negative charge) at the expense of electrons on the other object. When an atom has more electrons than protons, the atom is considered to be negatively charged.

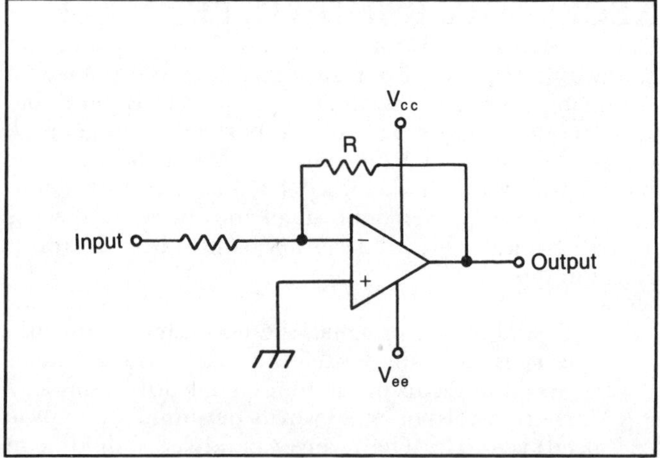

NEGATIVE FEEDBACK: Amplifier with variable negative feedback. This is an operational-amplifier circuit.

The smallest unit of negative charge is carried by a single electron. Conversely, the smallest unit of positive charge is carried by the proton. The terms negative and positive are arbitrary. *See also* CHARGE, POSITIVE CHARGE.

NEGATIVE FEEDBACK

When the output of an amplifier circuit is fed back to the input in phase opposition, the feedback is said to be negative. Negative feedback is used for a variety of purposes in electronic circuits.

Some amplifiers tend to break into oscillation easily. This can be prevented by a neutralizing circuit, which is a form of negative-feedback arrangement (*see* NEUTRALIZATION).

Negative feedback is used in some audio amplifiers to improve stability and fidelity. Generally, negative feedback can be provided by simply installing a resistor of the proper value in series with the cathode, emitter, or source of the active amplifying device. Negative feedback may also be obtained by directly applying some of the output signal to the input of the amplifier, 180 degrees out of phase.

Negative feedback is routinely used in operational-amplifier circuits to control the gain and improve the bandwidth. When no negative feedback is used, the operational amplifier is said to be operating in the open-loop condition (*see* OPEN LOOP, OPERATIONAL AMPLIFIER). Negative feedback is provided by placing a resistor in the feedback path, as shown in the illustration. The amount of negative feedback depends on the value of the resistor between the output and the input, denoted by R in the illustration. The smaller the value of R, the greater the negative feedback, and the smaller the gain of the circuit. *See also* FEEDBACK.

NEGATIVE LOGIC

The way in which logical signals are defined is a matter of convention. Normally, the logic 1 is the more positive of the voltage levels and the logic 0 is the more negative of the voltage levels. That is, logic 1 is high and logic 0 is low. When the voltages are reversed, the logic is said to be negative or inverted.

Either positive or negative logic will provide satisfactory operation of a digital device. *See also* LOGIC.

NEGATIVE RESISTANCE

Normally, the current through a device increases as the applied voltage is made larger. This is true of most active and passive components. However, certain components exhibit a different characteristic within a certain range: The current decreases as the applied voltage is made larger. This is called a negative-resistance characteristic. The illustration shows an example of a negative-resistance characteristic.

Negative resistance occurs in certain diodes. Some transistors and vacuum tubes also exhibit this property. Negative resistance causes oscillation when the applied voltage is within a certain range. This oscillation usually occurs at ultra-high or microwave frequencies. The magnetron and the tunnel diode operate on this principle. *See also* MAGNETRON, NEGATIVE TRANSCONDUCTANCE, TUNNEL DIODE.

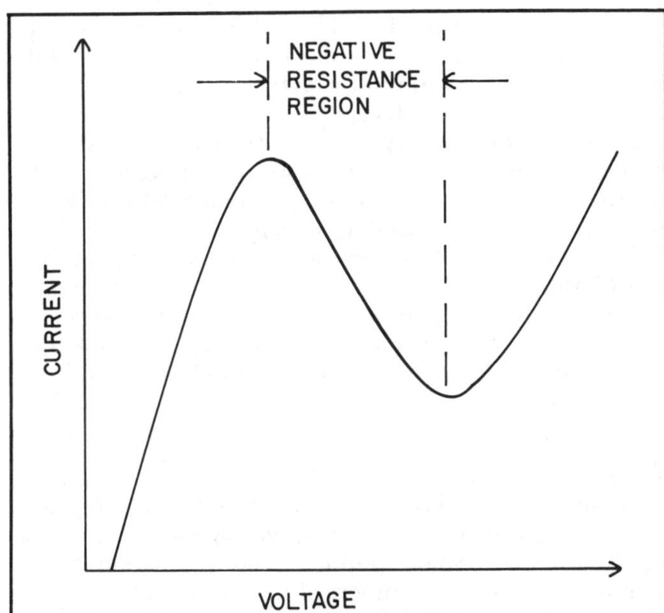

NEGATIVE RESISTANCE: Negative resistance occurs when the current level drops with increasing voltage.

NEGATIVE TRANSCONDUCTANCE

In a pentode vacuum tube, the transconductance between the suppressor grid and the screen grid is normally positive. However, when the screen grid is supplied with a more positive voltage than the plate, the suppressor-screen transconductance becomes negative within a certain range. That is, if the negative voltage of the suppressor is increased, the screen current increases. This occurs because of the repulsion of electrons by the suppressor. The more negative the suppressor, the more electrons it repels back toward the screen, and thus the more electrons are picked up by the screen.

Negative transconductance often results in oscillation at ultra-high or microwave frequencies. The plate voltage should always be higher than the screen voltage in a tetrode or pentode radio-frequency amplifier. Otherwise, parasitics are likely. Negative transconductance can, however, be introduced into a tetrode or pentode deliberately, for the purpose of generating signals. The dynatron and transitron oscillators are examples of circuits in which negative transconductance is used to generate signals. *See also* DYNATRON EFFECT, PARASITIC OSCILLATION, TRANSCONDUCTANCE, TUBE.

NEMATIC CRYSTAL

A nematic crystal, or nematic fluid, is an organic liquid with long molecules. Nematic fluid is widely used in digital displays. The liquid is normally transparent. When an electric field of sufficient intensity is introduced, the liquid becomes opaque until the field is removed. Then the liquid immediately becomes transparent again.

Displays using nematic crystals have become increasingly popular in calculators, watches, and other electronic devices. The nematic crystal draws practically no current, since it is the electric field alone that causes the change of state. This is an advantage in battery-powered devices, where the current drain must be minimized. *See also* LIQUID-CRYSTAL DISPLAY.

NEON

Neon is an element with atomic number 10 and atomic weight 20. At room temperature, neon is a gas. Neon is non-reactive, since the inner two electron shells are completely filled.

Neon is used in glow-discharge lamps. When neon is ionized by a high voltage, some of the emission lines fall in the visible-light spectrum. *See also* NEON LAMP.

NEON LAMP

A neon lamp is a device consisting of two electrodes in a sealed glass envelope containing neon gas. When a voltage is introduced between the electrodes, the neon gas becomes ionized, and the lamp glows. Neon lamps have a characteristic red-orange color.

Neon lamps are used in indicating devices, displays, oscillators, and voltage regulators. *See also* NEON-LAMP OSCILLATOR.

NEON-LAMP OSCILLATOR

A neon lamp, a capacitor, and a resistor can be interconnected to form an audio-frequency oscillator. Such an oscillator produces a sawtooth wave. The frequency is determined by the values of the parallel capacitance and the series resistance. The illustration is a schematic diagram of a neon-lamp oscillator with adjustable frequency.

When power is applied to the neon-lamp oscillator, the capacitor charges at a rate determined by the value of the resistor. When the capacitor is fully charged, the bulb ionizes and depletes the charge in the capacitor. The voltage across the bulb then falls below the minimum level necessary to cause ionization; the bulb is extinguished, and a new cycle begins. The process repeats at an audio-frequency rate.

The neon-lamp oscillator is simple and inexpensive, and is commonly used as an audio-frequency oscillator in applications where the waveform and frequency are not critical. The deionization time of the neon lamp prevents oscillation at frequencies above about 5 kHz.

NEPER

The neper, abbreviated Np, is a unit that is sometimes used for expressing a ratio of currents, voltages, or wattages. The neper may also be called the napier. The neper is similar to the decibel. In fact, nepers and decibels are related by a simple constant:

$$1 \text{ Np} = 8.686 \text{ dB}$$

$$1 \text{ dB} = 0.1151 \text{ Np}$$

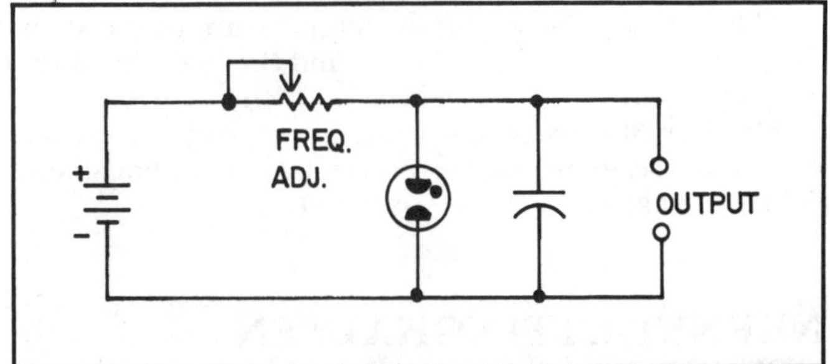

NEON-LAMP OSCILLATOR: A simple neon-lamp audio oscillator.

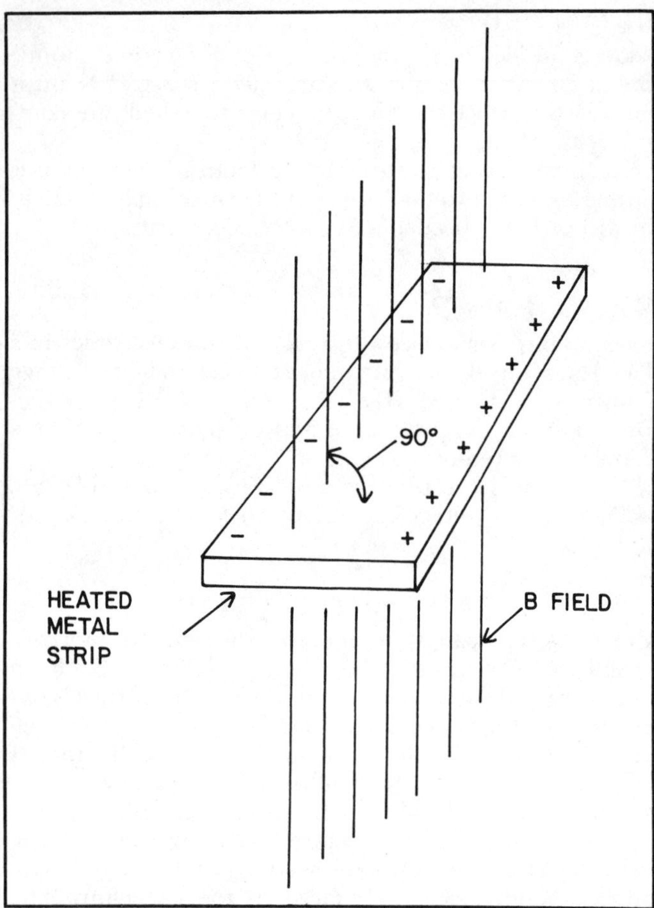

NERNST EFFECT: A voltage is produced between the edges of a heated metal strip in a magnetic (B) field.

A gain or loss is determined in nepers according to the formula:

$$Np = \ln \sqrt{x1/x2},$$

where x1 and x2 are the quantities to be compared and ln represents the natural (base-e) logarithm. *See also* DECIBEL, LOGARITHM.

NERNST EFFECT

When a strip of metal is nonuniformly heated and placed in a magnetic field, with the magnetic lines of flux perpendicular to the plane of the strip, a potential difference develops between the edges of the strip. This is called the Nernst effect. The drawing illustrates the relative orientation of the metal strip, the magnetic flux, and the generated voltage.

The voltage depends on the temperature of the strip, the particular kind of metal used, and the intensity of the magnetic field.

The Nernst effect can operate in reverse: An applied voltage can result in heating of a crystal strip in a magnetic field. *See* NERNST-ETTINGHAUSEN EFFECT.

NERNST-ETTINGHAUSEN EFFECT

When certain substances are placed in a magnetic field and current is passed through them, heating occurs. The current flow must be perpendicular to the magnetic lines of flux. This thermomagnetic effect is known as the Nernst-Ettinghausen effect.

The extent of the temperature differential depends on the level of the current, the intensity of the magnetic field, and the particular crystal used. The temperature difference occurs along an axis that is perpendicular to the flow of the current, and is also perpendicular to the magnetic lines of flux.

The Nernst-Ettinghausen effect can operate in reverse: A heated metal strip in a magnetic field develops a potential difference between its edges. *See* NERNST EFFECT.

NET

A communications network, intended for the relaying and delivering of information or messages by radio, is called a net. Nets are especially popular among amateur radio operators. However, commercial nets also exist.

Most amateur-radio nets are "traffic" nets, and their operators handle a wide variety of messages. Other amateur-radio nets are "conversational" nets, intended for special-interest discussion such as antenna theory and construction. Still other nets operate for specific purposes, such as the relaying of information in a severe-weather situation.

The term net is widely used as a synonym for the term network (*see* NETWORK).

A resultant or effective quantity is called a net quantity. For example, a circuit may have currents flowing in opposite directions at the same time; the net current is the effective current is one direction. A circuit may have two or more voltage sources, some of which buck each other; the net voltage is the sum of the individual voltages, taking polarity into account.

NETWORK

A network is a set of electronic components, interconnected for a specific purpose. In particular, a network is a circuit intended for insertion in a power-supply line or transmission line. Some networks are quite simple; some are exceedingly complicated.

Networks may be classified as either active or passive. An active network contains at least one component that requires an external source of power for operation. A passive network requires no external source of power; it is entirely made up of passive components.

Networks are often identified according to the way in which the components are interconnected. For example, a pi network consists of two parallel components, one on either side of a series-connected component. A T network consists of two series components, one on either side of a parallel-connected component. *See also* H NETWORK, L NETWORK, PI NETWORK, T NETWORK.

NEUTRAL CHARGE

When the total charge on an object is zero—the positive balances the negative—the object is said to be electrically neutral. A neutral charge results in a zero net electric field.

Atoms normally have neutral charge, since the number of protons is the same as the number of orbiting electrons.

However, atoms may gain or lose orbiting electrons, becoming charged. *See also* NEGATIVE CHARGE, POSITIVE CHARGE.

NEUTRALIZATION

Radio-frequency amplifiers have a tendency to oscillate. This is especially true of power amplifiers. This unwanted oscillation is called parasitic oscillation (*see* PARASITIC OSCILLATION), and may occur at any frequency, regardless of the frequency at which the amplifier is operating. Neutralization is a method of reducing the likelihood of such oscillation.

The neutralizing circuit is a negative-feedback circuit. A small amount of the output signal is fed back, in phase opposition, to the input. This may be done in a variety of different ways; the most common method is the insertion of a small, variable capacitor in the circuit. The drawing illustrates two methods of neutralization in a simple bipolar-transistor power amplifier. The method at A is called collector neutralization, since the capacitor is connected to the bottom of the collector tank circuit. The method at B is called base neutralization, because the capacitor is connected to the bottom of the base tank circuit. The negative feedback occurs through the variable capacitor. The capacitance required for optimum neutralization depends on the operating frequency and impedance of the amplifier. Normally the neutralizing capacitance is a few picofarads.

The neutralizing capacitor is first adjusted by removing the driving power from the input of the amplifier. A sensitive broadband wattmeter, or other output-sensing indicator, is connected to the amplifier output. The capacitor is adjusted for minimum output indication. *See also* NEGATIVE FEEDBACK, NEUTRALIZING CAPACITOR, POWER AMPLIFIER.

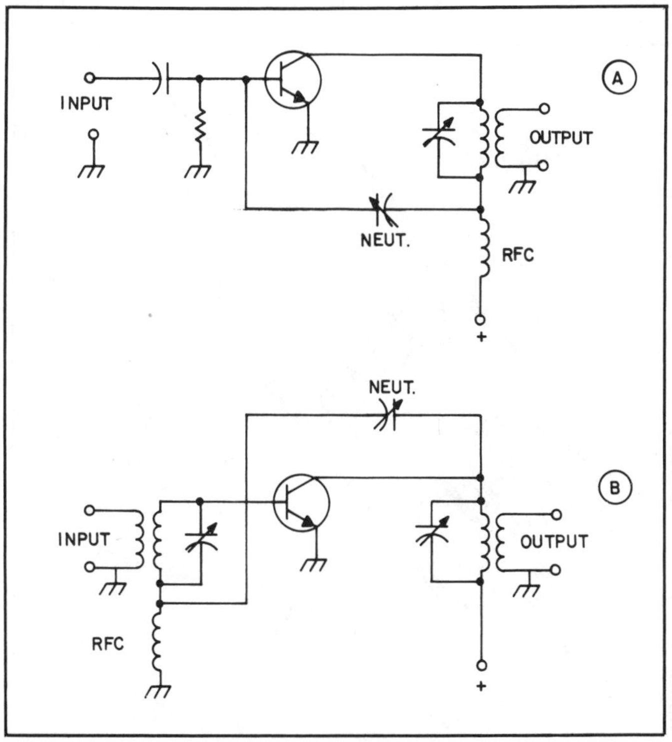

NEUTRALIZATION: Two methods of neutralizing a radio-frequency power amplifier. At A, collector neutralization. At B, base neutralization.

NEUTRALIZING CAPACITOR

A neutralizing capacitor is a small-valued, variable capacitor used for providing negative feedback in a radio-frequency power amplifier. The negative feedback can be varied by adjusting the value of the capacitance.

The capacitor method is the most common means of neutralizing a power amplifier. However, other negative-feedback circuits may also be used. *See also* NEGATIVE FEEDBACK, NEUTRALIZATION, POWER AMPLIFIER.

NEUTRINO

A neutrino is a subatomic particle that is emitted in the nuclear-fusion process. Neutrinos are generated in the hot centers of stars. The particles have theoretically zero mass, although they posses "spin."

Neutrinos can pass through planets, stars, and interstellar matter almost unimpeded. This makes it difficult to find them. A large container, filled with carbon tetrachloride and buried deep within the earth, is used for detection. When a neutrino interacts with a molecule of carbon tetrachloride, an argon nucleus is formed. Neutrinos have been detected, but in less quantity than was originally theorized. Some scientists believe that the hydrogen fusion process in the sun goes on and off, like a thermostat, over millions of years—and right now, it's off.

NEUTRON

The neutron is a subatomic particle with zero electric charge. Neutrons are found in the centers (nuclei) of all atoms, with the exception of the hydrogen atom. The neutron has a slightly larger mass than the proton. Neutrons are unbelievably dense: a teaspoonful of them would weigh tons.

Neutrons are given off by certain radioactive materials as they decay. Neutron radiation, in large quantities, is dangerous to life. Some neutron radiation arrives at the earth from outer space.

Under certain conditions, a neutron will split into a proton and an electron. In other cases, a proton and electron can combine to form a neutron. *See also* ATOM, ELECTRON, PROTON.

NEWTON

The newton is the standard international unit of force. The unit gets its name from Sir Isaac Newton, the physicist who revolutionized much of scientific thinking in his time.

A force of 1 newton, applied to a mass of 1 kilogram, will result in an acceleration of 1 meter per second per second. Newtons are sometimes called kilogram meters per second squared (kgm/s^2) for this reason.

A force of 1 newton is the equivalent of 100,000 dynes. A force of 1 dyne results in an acceleration of 1 centimeter per second per second, against a mass of 1 gram. *See also* FORCE.

NEWTON'S LAWS OF MOTION

In the seventeenth century, Sir Isaac Newton changed the course of physics by formulating his three laws of motion. From experimentation, Newton deduced these facts concerning the behavior of objects and forces:

1. Any body at rest, or moving at a uniform velocity, tends to maintain that condition of motion until an external force is applied.

2. When a force is applied to a mass, the mass accelerates. The acceleration occurs in direct proportion to the force, and in inverse proportion to the mass of the body. The acceleration takes place in the same direction in which the force is applied.

3. For every acting force, there is an equal reacting force that occurs in the opposite direction.

Newton's laws of motion are still the basis for classical mechanics today. These laws apply to charged particles moving at non-relativistic speeds. As the speed of a particle becomes very high, however, Newton's laws no longer accurately explain its behavior. *See also* RELATIVISTIC SPEED, RELATIVITY THEORY.

NICHROME

Nichrome is an alloy of two common metals, nickel and chromium. Nichrome has high resistivity and a high melting point. It is used for electrical heating elements, and also in the fabrication of thin-film and wirewound resistors.

Nichrome is sometimes also called nickel-chromium. The word Nichrome was originally coined by Driver-Harris Company. *See also* WIREWOUND RESISTOR.

NICKEL

Nickel is an element with atomic number 28 and atomic weight 59. Nickel is used for many different electrical and electronic purposes.

An alloy of nickel and chromium exhibits a high resistance and a high melting temperature. This alloy, called Nichrome, is used in the manufacture of heating elements and resistors (*see* NICHROME).

Nickel is sometimes used as a plating material for steel. Nickel-plated steel has a dull, grayish-white appearance when unpolished; when polished it appears dark gray. Nickel plating retards the corrosion of steel, but does not entirely prevent it (*see* ELECTROPLATING).

Nickel is used in the manufacture of electrodes in some vacuum tubes. Pressed or impregnated nickel is found in the cathodes of such tubes. Nickel cathodes exhibit high electron emission at relatively low temperatures. However, the tungsten cathode can withstand higher temperature levels (*see* CATHODE, TUBE).

Nickel is used in the manufacture of some ferromagnetic transformer-core and inductor-core materials. The characteristics of such cores vary greatly, depending on the method of manufacture (*see* FERROMAGNETIC MATERIAL).

A compound of nickel, hydrogen, and oxygen is used in the manufacture of a form of rechargeable cell, the nickel-cadmium cell. Such cells, and batteries made from them, are widely used in electronic devices (*see* NICKEL-CADMIUM BATTERY).

Nickel is a magnetostrictive material. In the presence of an increasing magnetic field, nickel contracts. For this reason, nickel is often used in the manufacture of magnetostrictive transducers (*see* MAGNETOSTRICTION, MAGNETOSTRICTION OSCILLATOR).

NICKEL-CADMIUM BATTERY

A nickel-cadmium battery, often abbreviated Ni-Cd or NICAD, is a battery made with cadmium anodes, nickel-hydroxide cathodes, and an electrolyte of potassium hydroxide. A single nickel-cadmium cell is shown in the diagram. A single Ni-Cd cell provides approximately 1.2 volts direct current.

Nickel-cadmium cells and batteries are rechargeable, and thus they can be used repeatedly. For this reason, they have become popular in such items as calculators, radio receivers, and low-powered portable radio transceivers.

A typical, 500-milliampere-hour Ni-Cd battery is about the size of eight AA cells. Such a battery requires about three to five hours to be fully charged. When the battery is used, it should not be allowed to become totally discharged, or its polarity may be reversed and the battery will be ruined. However, Ni-Cd batteries should not be repeatedly charged under partial-discharge conditions either. Most electronic devices operating from these batteries have a warning indicator that shows when it is time to recharge. *See also* BATTERY, CELL, CHARGING, RECHARGEABLE CELL.

NITROGEN

Nitrogen is an element with atomic number 7 and atomic weight 14. At room temperature, nitrogen is a gas. The atmosphere of our planet is made up of 78 percent nitrogen.

Nitrogen is a relatively non-reactive gas, although it can form certain compounds under the right conditions. Since nitrogen is so abundant, it is relatively inexpensive to obtain in pure form. One simple method of getting a 99-

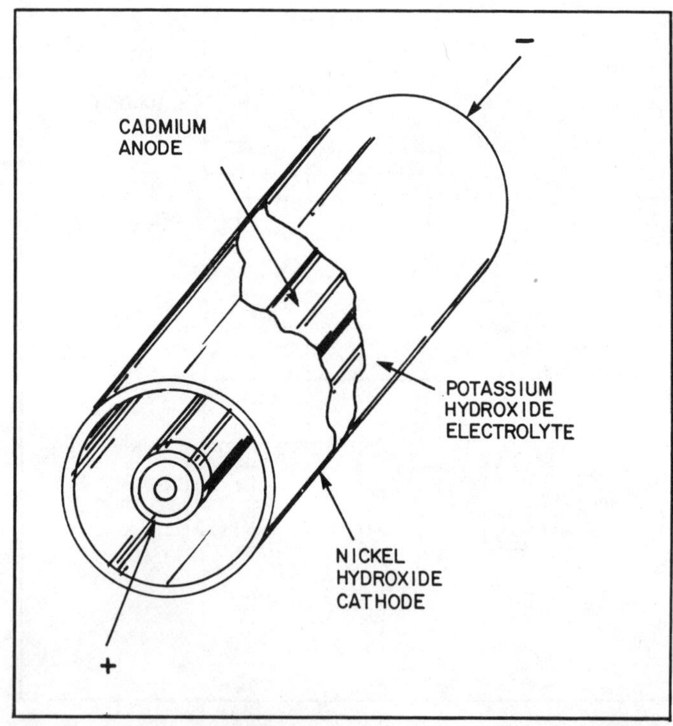

NICKEL-CADMIUM BATTERY: Cutaway view of a nickel-cadmium cell.

percent pure sample of nitrogen is to burn the oxygen out of the air in an enclosed chamber.

Nitrogen is used in the fabrication of certain sealed transmission lines, since it is a noncorrosive gas. An air-dielectric coaxial cable, for example, may be evacuated and then refilled with pure, dry nitrogen gas. This improves the dielectric qualities, and prolongs the life of the transmission line.

NIXIE TUBE

A Nixie® tube, also called a readout tube or readout lamp, is a vacuum tube containing several cathodes. The cathodes are arranged in such a way that various combinations result in alphanumeric characters. Decimal points may also be included. The tube is filled with a gas, usually neon, that glows when a voltage is applied between the anode and one or more of the cathodes. The illustration is a diagram of a typical seven-segment Nixie tube.

Each of the cathodes is connected to a separate pin at the base of the tube. The anode has its own pin. When voltages are applied to various cathodes, the corresponding segments are illuminated with a reddish-orange or pink color.

Nixie tubes are characterized by rapid response and high brightness. They were once common in digital displays, such as frequency-readout indicators. Nixie tubes can still be found today, but in recent years they have been largely replaced by the more compact light-emitting diodes and liquid-crystal displays. *See also* LIGHT-EMITTING DIODE, LIQUID-CRYSTAL DISPLAY.

NMOS

See METAL-OXIDE-SEMICONDUCTOR LOGIC FAMILIES.

NOBLE GAS

A noble gas is an inert, or non-reactive, gas. The noble gases include argon, helium, krypton, neon, and xenon. Noble gases are generally used in glow-discharge devices, such as indicator lamps and Nixie tubes. *See also* GLOW DISCHARGE, GLOW LAMP, NIXIE TUBE.

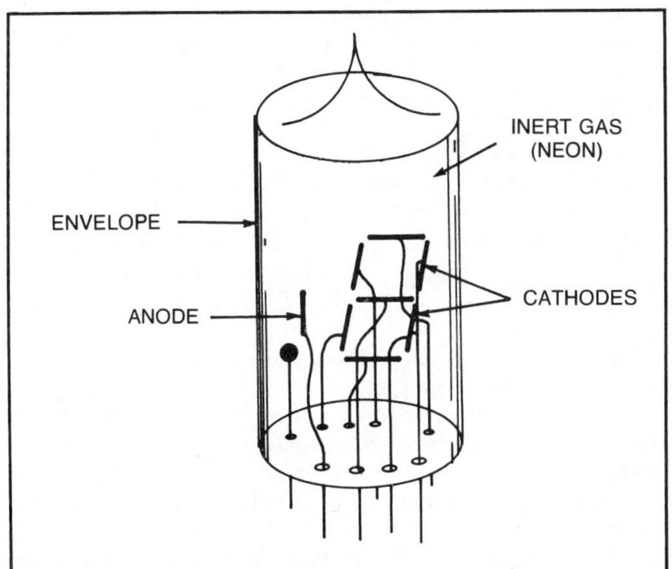

NIXIE TUBE: A typical Nixie® tube with a seven-segment display and a decimal point.

NODE

A node is a local minimum in a variable quantity. In a transmission line or antenna radiator, for example, we may speak of current nodes or voltage nodes. A current node is a point along a transmission line or antenna at which the current reaches a local minimum; at such a point, the voltage is usually at a local maximum. A voltage node is a point at which the voltage reaches a local minimum, and the current is usually at a local maximum. Nodes of a given kind are separated by electrical multiples of ½ wavelength. The opposite of a node is a loop (*see* CURRENT NODE, LOOP, VOLTAGE NODE).

In a circuit with two or more branches, a node is any circuit point that is common to at least two different branches. At a node, the current inflow is always the same as the current outflow, according to Kirchhoff's laws (*see* KIRCHHOFF'S LAWS).

NOISE

Noise is a broadbanded electromagnetic field, generated by various environmental effects and artificial devices. Noise can be categorized as either natural or manmade.

Natural noise may be either thermal or electrical in origin. All objects radiate noise as a result of their thermal energy content. The higher the temperature, the shorter the average wavelength of the noise. This is known as black-body radiation (*see* BLACK BODY, WIEN'S DISPLACEMENT LAW). Black-body radiation constantly bombards the earth from outer space. It also originates in all objects on the earth, and in the earth itself.

Cosmic disturbances, solar flares, and the movement of ions in the upper atmosphere all contribute to the natural electromagnetic noise present at the surface of the earth. Sferics, or noise generated by lightning, is a source of natural noise (*see* SFERICS).

Electromagnetic noise is produced by many different manmade devices. In general, any circuit or appliance that produces electric arcing will produce noise. Such devices include fluorescent lights, heating devices, automobiles, electric motors, thermostats, and many other appliances.

The level of electromagnetic noise affects the ease with which radio communications can be carried out. The higher the noise level, the stronger a signal must be if it is to be received. The signal-to-noise ratio (*see* SIGNAL-TO-NOISE RATIO) can be maximized in a variety of different ways. The narrower the bandwidth of the transmitted signal, and the narrower the passband of the receiver, the better the signal-to-noise ratio at a given frequency. This improvement occurs, however, at the expense of data-transmission speed capability. Circuits such as noise blankers and limiters are sometimes helpful in improving the signal-to-noise ratio. Noise-reducing antennas can also be used to advantage in some cases. But there is a limit to how much the noise level can be reduced; a certain amount of noise will always exist. *See also* NOISE BLANKER, NOISE FILTER, NOISE FLOOR, NOISE LIMITER, NOISE-REDUCING ANTENNAp10.

NOISE BANDWIDTH

See NOISE EQUIVALENT BANDWIDTH.

NOISE BLANKER

A noise blanker is a form of noise-reducing circuit com-

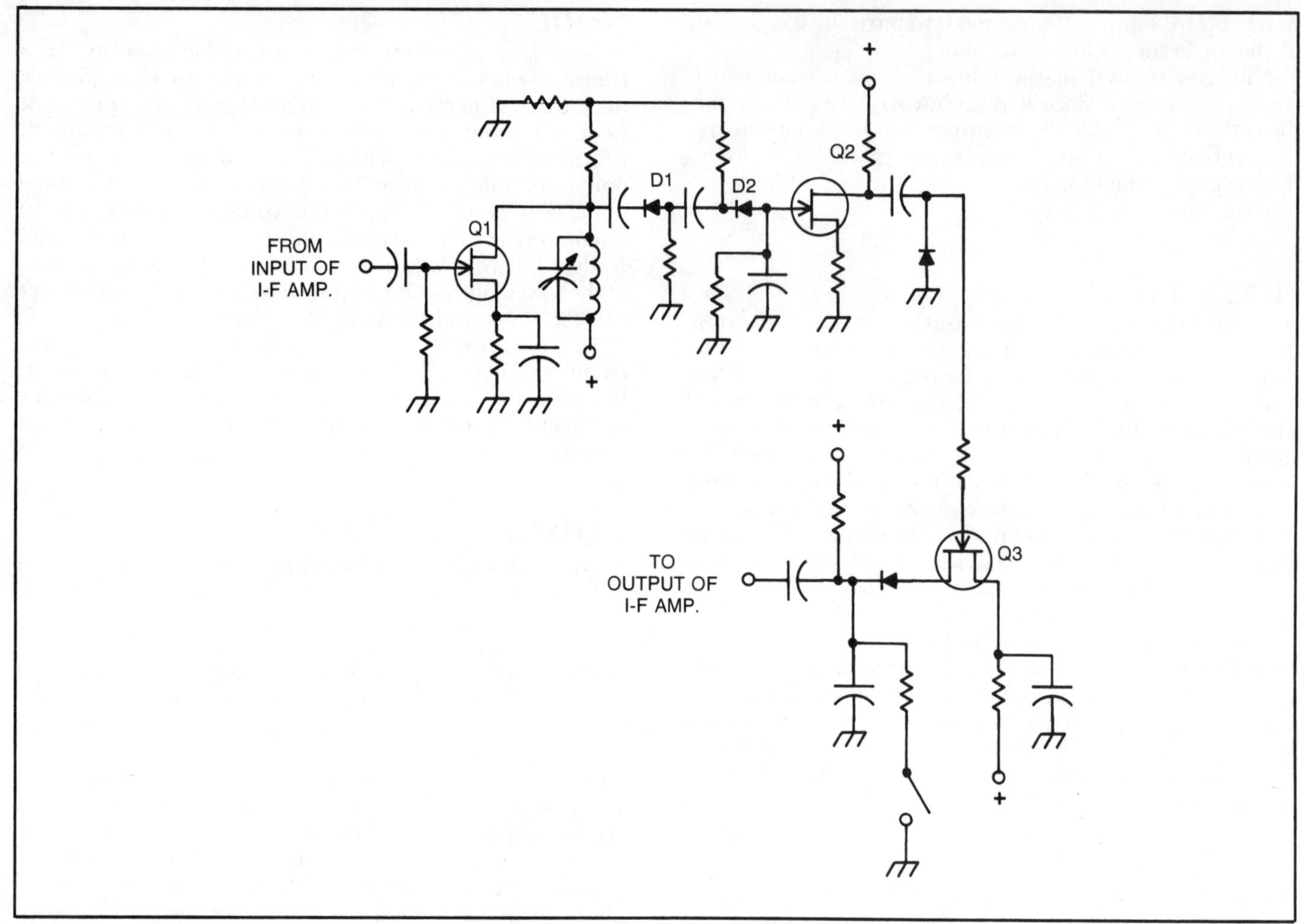

NOISE BLANKER: A noise-blanker circuit for insertion in the intermediate-frequency chain of a superheterodyne receiver. The functions of Q1, Q2, Q3, D1 and D2 are discussed in the text.

monly used in radio receivers at very low, low, medium, and high frequencies.

The noise blanker operates by discriminating between the short, intense noise bursts and the more uniform characteristics of a desired signal. The noise blanker is usually installed in one of the intermediate-frequency stages of a superheterodyne receiver, or just prior to the detector stage in a direct-conversion receiver.

Under low-noise conditions, the noise blanker has no effect on the signals passing through the amplifier stage. But when a sudden, high-amplitude, short-duration pulse occurs, the noise blanker cuts off the stage. The illustration is a schematic diagram of a simple noise-blanker circuit that may be used in an intermediate-frequency amplifier.

The circuit operates as follows. The first stage, Q1, amplifies the incoming signals. The two diodes D1 and D2 rectify the signal, producing negative-going pulses at the gate of the second stage, Q2. Under normal conditions Q2 conducts, but when a noise pulse occurs, the negative voltage spike cuts off this stage and prevents any signal or noise from reaching the input of Q3. The purpose of the stage Q3 is to prevent positive overshoot following noise pulses. Such overshoot might otherwise render the circui only partially effective. Very short noise pulses will not be heard at the output of the receiver when this circuit is used. Pulses of longer duration (a few milliseconds) will result in "holes" in received signals, but this is much less annoying than the actual noise.

Noise blankers are most effective against impulse type noise (*see* IMPULSE NOISE), in which the interfering pulses are intense but of very short duration. Impulse noise is generated mostly by manmade sources. Noise blankers do not work as well against the more random noise resulting from thundershowers or cosmic disturbances. A circuit called a noise limiter is preferred in cases where natural noise is a problem. *See also* NOISE, NOISE LIMITER, SFERICS.

NOISE-EQUIVALENT BANDWIDTH

Any noise source has a spectral distribution. Some noise is very broadbanded, and some noise occurs with a fairly well-defined peak in the spectral distribution (see illustration, A and B). The total noise can be expressed in terms of the area under the curve of amplitude versus frequency.

Let f be the frequency at which the power density of the noise is greatest in the spectral distribution in the graph at B. Let a represent the power density of the noise at the frequnecy f. We can construct a rectangle with height a, centered at f, that has the same enclosed area P as the curve encloses. This is illustrated at C. The width of this rectangle, N, is a frequency span $f_2 - f_1$, such that $f_2 - f = f - f_1$. The value N is the noise equivalent bandwidth at the frequency f.

Noise equivalent bandwidth is often specified for de-

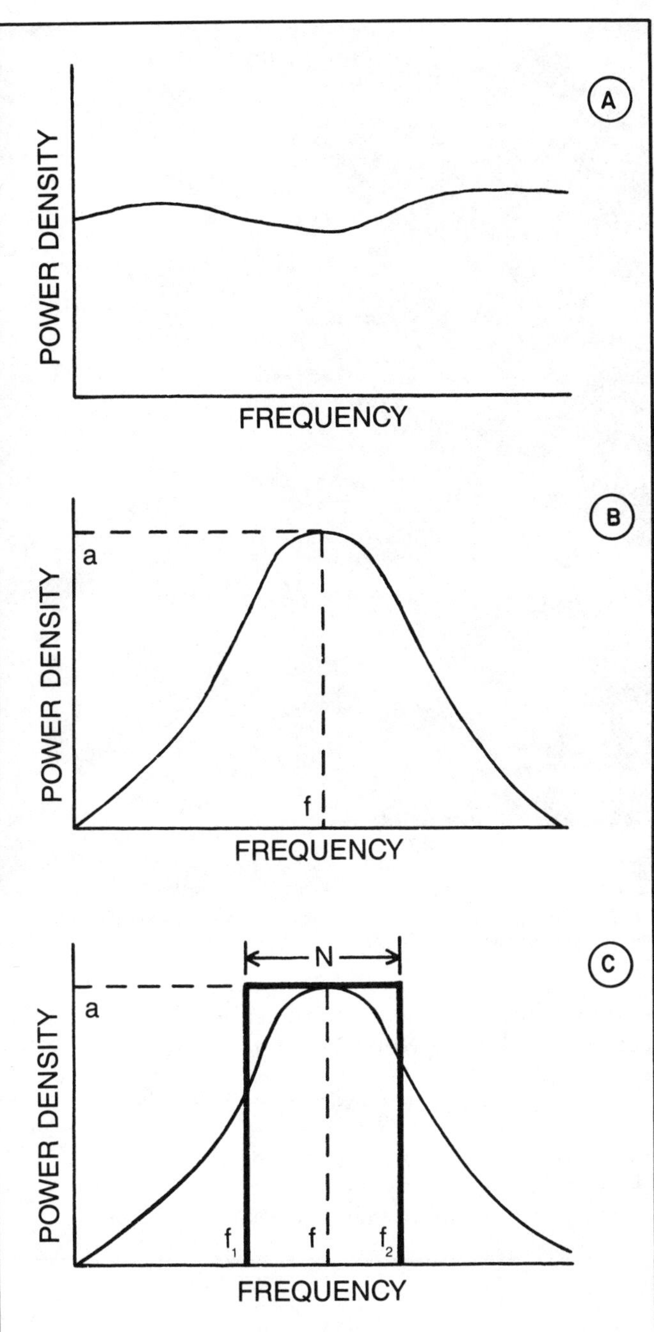

NOISE EQUIVALENT BANDWIDTH: At A, a qualitative example of noise with a broadband spectral distribution. At B, an example of noise having a defined frequency f at which the power density is maximum. A C, computation of the noise equivalent bandwidth at frequency f.

vices such as filters, bolometers, and thermistors. The noise equivalent bandwidth is an expression of the frequency characteristics of the devices. *See also* NOISE.

NOISE EQUIVALENT TEMPERATURE

A black body produces a characteristic spectral noise distribution according to its absolute temperature. The higher the temperature, the higher the peak frequency of the noise output (*see* BLACK BODY). The correlation is so precise that the temperature of the black body can be accurately determined by finding its spectral noise distribution.

The noise equivalent temperature of a noise source is the temperature that corresponds to its spectral energy distribution. Most noise sources emit energy that resembles black-body radiation for relatively cold objects. *See also* WIEN'S DISPLACEMENT LAW.

NOISE FIGURE

The noise figure is a specification of the performance of an amplifier or receiver. Noise figure, expressed in decibels, is an indication of the degree to which a circuit deviates from the theoretical ideal.

Suppose that the noise output of an ideal network results in a signal-to-noise ratio:

$$R1 = P/Q1$$

where P is the signal power and Q1 is the noise power. Let the noise power in the actual circuit be given by Q2, resulting in an actual signal-to-noise ratio:

$$R2 = P/Q2$$

Then the noise figure, N, is given in decibels by the equation:

$$N\ (dB) = 10 \log_{10} (R1/R2)$$

$$= 10 \log_{10} (Q2/Q1)$$

A perfect circuit would have a noise figure of 0 dB. This value is never achieved in practice. Values of a few decibels are typical. The noise figure is important at very-high frequencies and above, where relatively little noise occurs in the external environment, and the internal noise is the primary factor that limits sensitivity. *See also* SENSITIVITY, SIGNAL-TO-NOISE RATIO.

NOISE FILTER

A noise filter is a passive circuit, usually consisting of capacitance and/or inductance, that is inserted in series with the alternating-current power cord of an electronic device. The noise filter allows the 60-Hz alternating current to pass with essentially no attenuation, but higher-frequency noise components are suppressed.

A typical noise filter is simply a lowpass filter, consisting of a capacitor or capacitors in parallel with the power leads, and an inductor or inductors in series (see the illustration). The values are chosen for a cutoff frequency just above 60 Hz.

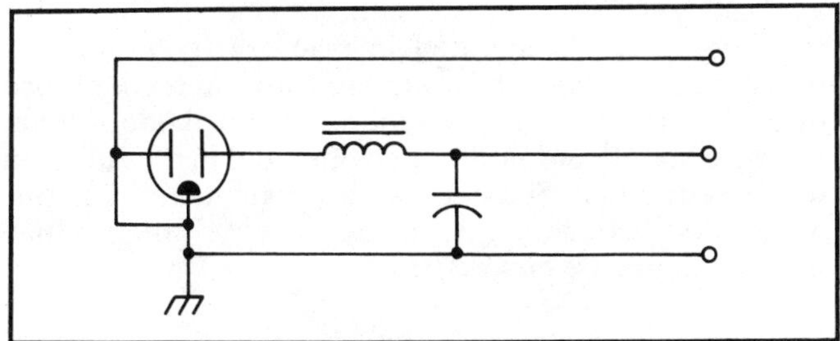

NOISE FILTER: A lowpass type line noise attenuator.

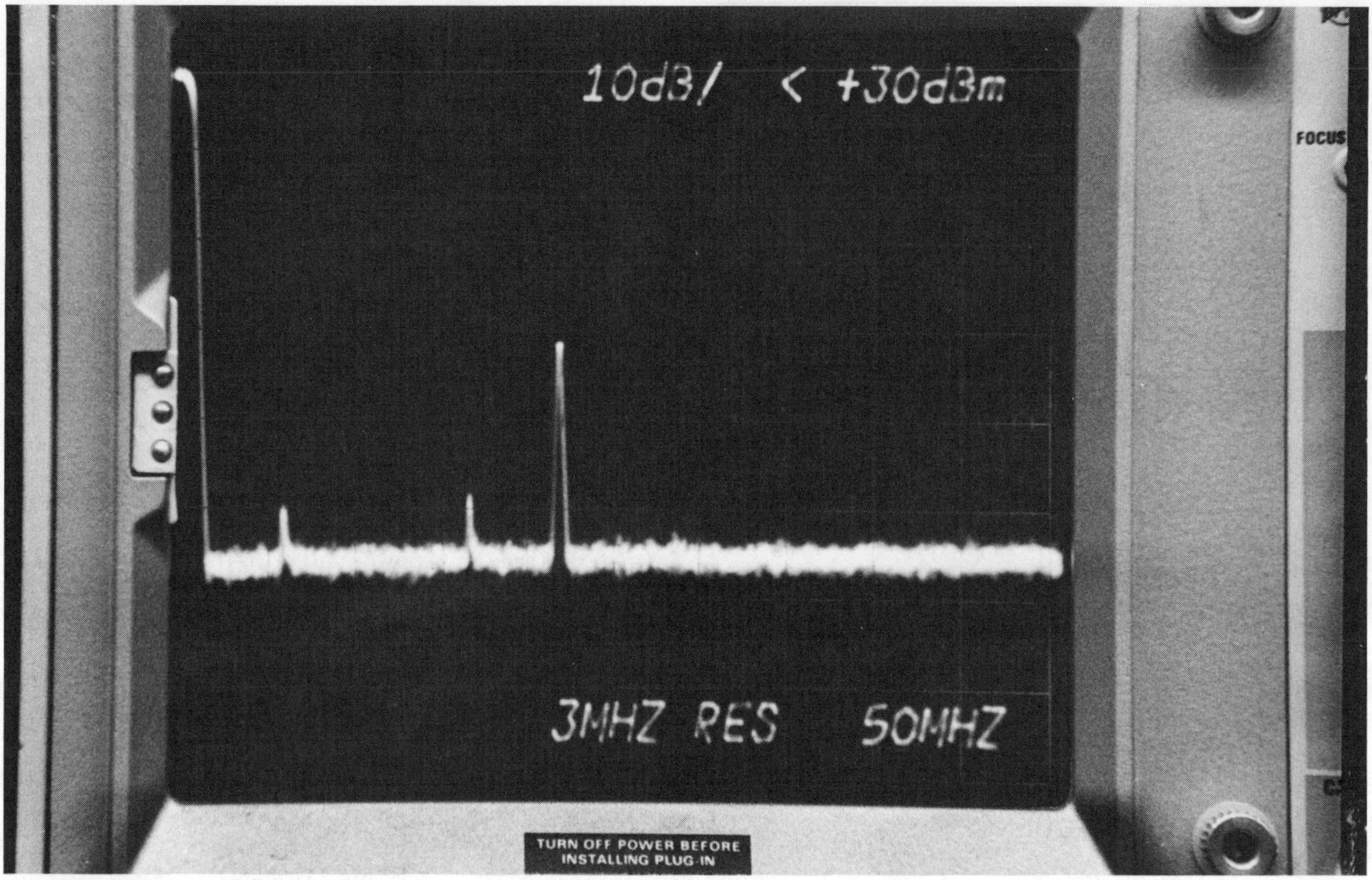

NOISE FLOOR: The noise floor of this spectrum-analyzer display is approximately 55 dB below the level of the local-oscillator signal. (The local oscillator signal is the tallest pip, at the extreme left.) Signals can be seen above the level of the noise floor.

NOISE FLOOR

The level of background noise, relative to some reference signal, is called the noise floor. Signals normally can be detected if their levels are above the noise floor; signals below the noise floor cannot be detected. In a radio receiver, the level of the noise floor may be expressed as the noise figure (*see* NOISE FIGURE).

In a spectrum-analyzer display, the noise-floor level determines the sensitivity and the dynamic range of the instrument. The noise floor is generally specified in decibels with respect to the local-oscillator signal (see photograph). A typical spectrum analyzer has a noise floor of −50 dB to −70 dB. *See also* SPECTRUM ANALYZER.

NOISE GENERATOR

Any electronic circuit that is designed to produce electromagnetic noise is called a noise generator. Noise generators are used in a variety of testing and alignment applications, especially with radio receivers.

Noise may be generated in many different ways. A diode tube, operated at saturation (full conduction), produces broadband noise. A semiconductor diode will also produce broadband noise when operated in the fully conducting condition. Some diodes generate noise when they are reverse-biased. A current-carrying resistor produces thermal noise. *See also* NOISE.

NOISE LIMITER

A noise limiter is a circuit, often used in radio receivers, that prevents externally generated noise from exceeding a certain amplitude. Noise limiters are sometimes called noise clippers.

A noise limiter may consist of a pair of clipping diodes with variable bias for control of the clipping level (see illustration). The bias is adjusted until clipping occurs at the signal amplitude. Noise pulses then cannot exceed the signal amplitude. This makes it possible to receive a signal

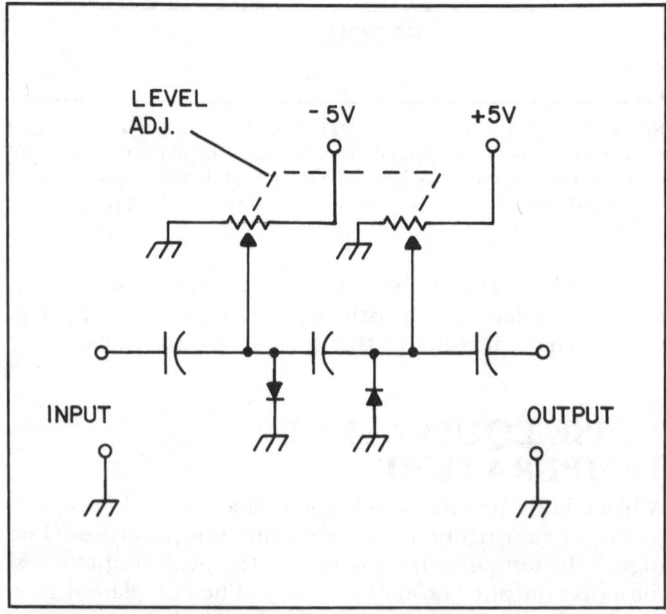

NOISE LIMITER: A variable-threshold noise-limiter circuit.

that would otherwise be drowned out by the noise. The noise limiter is generally installed between two intermediate-frequency stages of a superheterodyne receiver. In a direct-conversion receiver, the best place for the noise limiter is just prior to the detector stage.

A noise limiter may use a circuit that sets the clipping level automatically, according to the strength of an incoming signal. This is a useful feature, since it relieves the receiver operator of the necessity to continually readjust the clipping level as the signal fades. Such a circuit is called an automatic noise limiter (*see* AUTOMATIC NOISE LIMITER).

Noise limiters are effective against all types of natural and manmade noise, including noise not affected by noise-blanking circuits. However, the noise limiter cannot totally eliminate the noise. *See also* NOISE BLANKER.

NOISE PULSE

A noise pulse is a short burst of electromagnetic energy. Often, the instantaneous amplitude of a noise pulse rises to a much higher level than the amplitude of a received signal. However, since the duration of a noise pulse is short, the total electromagnetic energy is usually small.

Noise pulses are produced especially by electric arcing. The noise pulse may produce energy over a wide band of frequencies, from the very-low to the ultra-high, at wavelengths from kilometers to millimeters. *See also* IMPULSE NOISE, NOISE.

NOISE QUIETING

Noise quieting is a decrease in the level of internal noise in a frequency-modulation receiver, as a result of an incoming signal. With the squelch open (receiver unsquelched) and no signal, a frequency-modulated receiver emits a loud hissing noise. This is internally generated noise. When a weak signal is received, the noise level decreases. As the signal gets stronger, the level of the noise continues to decrease until, when the signal is very strong, there is almost no hiss from the receiver.

The noise quieting phenomenon provides a means of measuring the sensitivity of a frequency-modulation receiver. The level of the noise at the speaker terminals is first measured under no-signal conditions. Then a signal is introduced, by means of a calibrated signal generator. The signal level is increased, without modulation, until the noise voltage drops by 20 dB at the speaker terminals. The signal level, in microvolts at the antenna terminals, is then determined. Typical unmodulated signal levels for 20-dB noise quieting are in the range of 1 uV or less at very-high and ultra-high frequencies, with receivers using modern solid-state amplifiers.

The noise-quieting method is one of two common ways of determining the sensitivity of a frequency-modulation receiver. The other method is called the signal-to-noise-and-distortion, or SINAD, method. *See also* SINAD, SINAD METER, SQUELCH.

NOISE-REDUCING ANTENNA

Electromagnetic-noise pickup can be minimized by the use of various specialized types of receiving antennas. The most effective method of reducing noise in an antenna system is the phase-cancellation or nulling method. With antennas designed for transmission as well as reception, noise reduction is generally more difficult, and the results less spectacular.

Any receiving antenna with a sharp directional null can be used to reduce the level of received manmade noise. A ferrite-rod or loop antenna is especially useful for this purpose (*see* FERRITE-ROD ANTENNA, LOOP ANTENNA). Both of these antennas have sharp directional nulls along their axes. When the null is oriented in the focal direction of a noise source, the noise level drops. When tuned to resonance by means of a variable capacitor, the bandwidth of a ferrite-rod or loop antenna is very narrow, and this serves to further increase the immunity of the antenna system to broadband noise. The feed line must be properly balanced or shielded, however, if the benefits of such antennas are to be realized.

The noise rejection of a small loop antenna can be enhanced by the addition of a Faraday shield surrounding the loop element. The Faraday shield discriminates against the electrostatic component of a signal, while allowing the magnetic component to penetrate unaffected. Many noise signals are locally propagated by predominantly electrostatic coupling.

The overall noise susceptibility of a receiving/transmitting antenna system can be minimized by careful balancing or shielding of the feed line, maintaining an adequate radio-frequency ground, and maximizing the antenna Q factor (that is, minimizing the bandwidth). In general, at the medium and high frequencies, a horizontally polarized antenna is less susceptible to manmade noise than is a vertically polarized antenna.

NOISE SUPPRESSOR

See NOISE BLANKER, NOISE LIMITER.

NOISE TEMPERATURE

See NOISE EQUIVALENT TEMPERATURE.

NO-LOAD CURRENT

When an amplifier is disconnected from its load, some current still flows in the collector, drain, or plate circuit. This current is usually smaller than the current under normal load conditions. The no-load current depends on the class of amplification and the full-load output impedance for which the circuit is designed.

When the secondary winding of an alternating-current power transformer is disconnected from its load, a small amount of current still flows. This current occurs because of the inductance of the secondary winding; it is called no-load current. The value of the no-load current depends on the voltage supplied to the primary winding of the transformer, and also on the primary-to-secondary turns ratio. *See also* TRANSFORMER.

NO-LOAD VOLTAGE

When the load is removed from a power supply, the voltage rises at the output terminals. The voltage at the output terminals of a power supply, with the load disconnected, is called the no-load voltage.

The difference between the no-load and full-load voltages of a power supply depends on the amount of current drawn by the load. The greater the current demanded by the load, the greater the difference between the no-load voltage and the full-load voltage.

The percentage difference between the no-load and full-load voltages of a power supply is sometimes called the regulation, or the regulation factor, of the supply. In some instances, it is important that the output voltage change very little with large fluctuations in the load current. In other cases, a large voltage change can be tolerated. *See also* POWER SUPPLY, REGULATED POWER SUPPLY, VOLTAGE REGULATION.

NOMINAL VALUE

All components have named, or specified, values. For example, a capacitor may have a specified value of 0.1 μF; a microphone may have a named impedance of 600 ohms. An amplifier may have a specified power-output rating of 25 watts. A piezoelectric crystal may have a frequency rating of 3.650 MHz. The named, or specified, value is called the nominal value.

The nominal value for a component or circuit is the average for a large sample of manufactured items, but individual units will vary somewhat either way. Nominal values are given without reference to the tolerance, or deviation that an individual sample may exhibit from the nominal value. *See also* TOLERANCE.

NOMOGRAPH

A nomograph is a simple, straightforward means of illustrating the relationship between the independent and dependent variables in a mathematical function. A nomograph can be called a nomogram, which is also correct.

The illustration is an example of a nomograph of the relationship between the frequency of an electromagnetic wave, in megahertz, and the free-space wavelength, in meters. This is a simple, linear nomograph. More complex nomographs require the use of a straight-edge for determining the values of functions of two variables. *See also* FUNCTION.

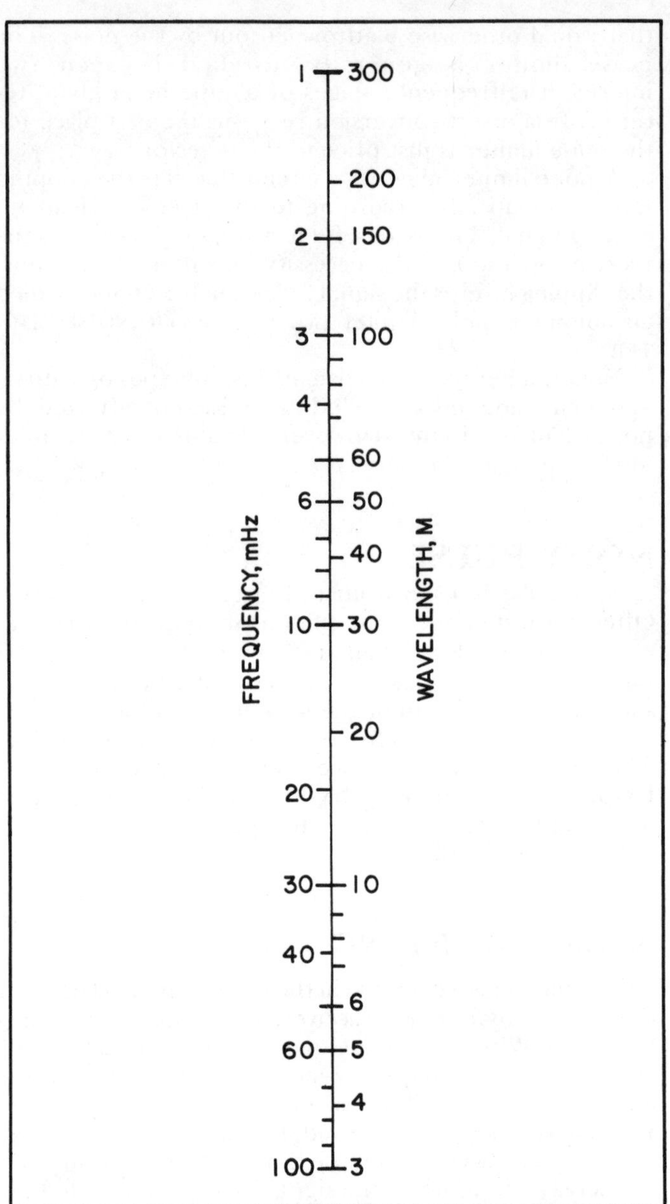

NOMOGRAPH: A simple nomograph, showing the relation between frequency and wavelength between 1 MHz and 100 MHz.

NONLINEAR CIRCUIT

A nonlinear circuit is a circuit having a nonlinear relationship between its input and output. That is, if the output is graphed against the input, the function is not a straight line.

Sometimes a circuit will behave in a linear manner within a certain range of input current or voltage, but will be nonlinear outside that range. An example is a radio-frequency power amplifier designed to act as a linear amplifier. It will behave in a linear manner only as long as the drive (input power) is not excessive. But if the amplifier is overdriven, it will no longer act as a linear amplifier. A similar effect is observed in other linear circuits.

Nonlinearity results in increased harmonic and intermodulation distortion in a circuit. In fact, harmonic generators and mixers make use of this characteristic.

Nonlinear circuits are extensively used as radio-frequency power amplifiers. Class-AB, Class-B, and Class-C amplifiers all cause some distortion of the input waveform. Class-AB and Class-B amplifiers do not, however, cause envelope distortion of an amplitude-modulated signal. *See also* CLASS-A AMPLIFIER, CLASS-AB AMPLIFIER, CLASS-B AMPLIFIER, CLASS-C AMPLIFIER, LINEAR AMPLIFIER, LINEARITY, NONLINEARITY.

NONLINEAR FUNCTION

A nonlinear function is a mathematical function that is not a linear function. A nonlinear function of one variable, for example, may have any appearance except a straight line in the Cartesian plane (see illustration). A nonlinear function of two variables has any appearance except that of a flat plane in Cartesian three-space.

Any function in which an exponent appears is generally nonlinear. Logarithmic, trigonometric, and many other kinds of functions are nonlinear. *See also* CARTESIAN COORDINATES, FUNCTION, LINEAR FUNCTION.

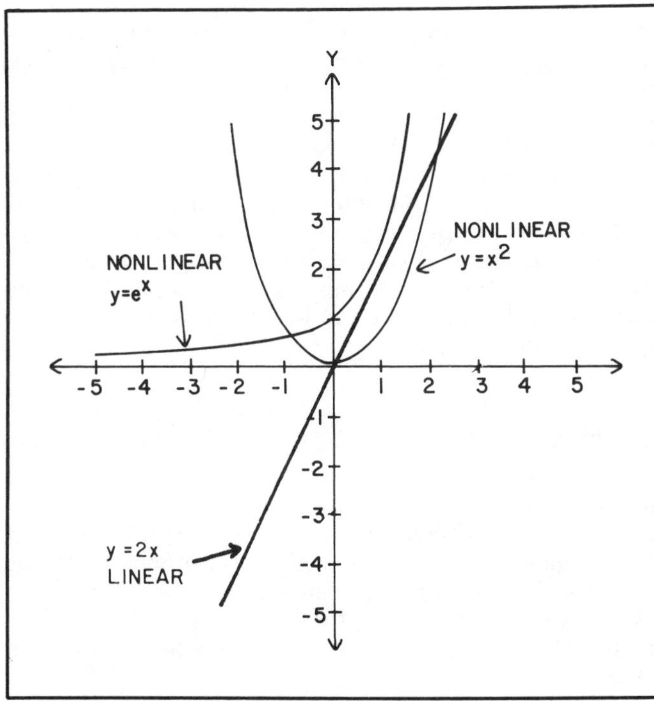

NONLINEAR FUNCTION: Two nonlinear functions are illustrated in this graph by curved lines. The straight line shows a linear function.

NONLINEARITY

Any departure of a circuit from linear operation is called a nonlinearity. Nonlinearity always results in distortion of a signal. This may be deliberately introduced, or it may be undesirable.

Nonlinearity can be caused by improper bias voltage in a transistor, field-effect transistor, or tube. Nonlinearity can also be caused by excessive drive at the input and/or underloading at the output of an amplifier. In a push-pull or parallel amplifier configuration, nonlinearity can result from improper balance between the two transistors, field-effect transistors, or tubes.

In a linear radio-frequency amplifier, nonlinearity of the modulation envelope is undesirable, although nonlinearity of the carrier waveform is often produced. Nonlinearity of the modulation envelope results in increased bandwidth because of harmonic and intermodulation products. *See also* LINEAR AMPLIFIER, LINEARITY, NONLINEAR CIRCUIT.

NONPOLARIZED COMPONENT

A component is considered nonpolarized if it can be inserted either way in a circuit with the same results. Examples of nonpolarized components are resistors, inductors, and many types of capacitors.

If the leads of a component cannot be interchanged or reversed without adversely affecting circuit performance, the component considered is polarized. *See also* POLARIZED COMPONENT.

NONPOLARIZED ELECTROLYTIC CAPACITOR

Most electrolytic capacitors are polarized components; that is, they must be installed in the correct direction with regard to the dc polarity of the circuit. However, certain types of electrolytic capacitors are deliberately designed so that they can be installed without concern for polarity. Such capacitors are called nonpolarized electrolytic capacitors. They are useful in applications where a large amount of capacitance is needed, but there is no direct-current bias to form the oxide layer in a normal electrolytic capacitor.

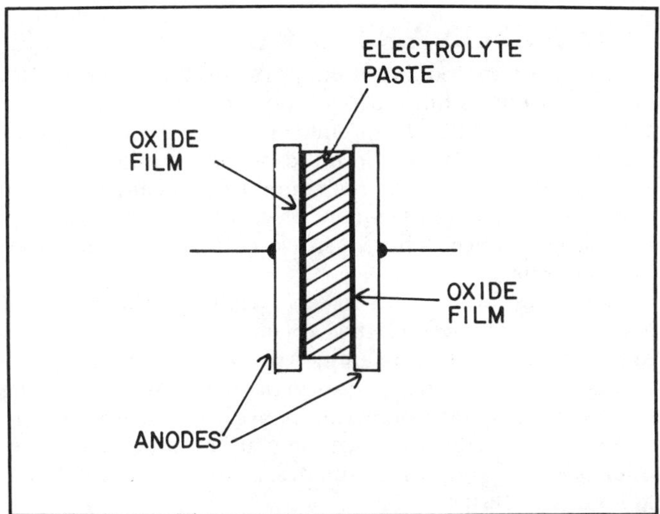

NONPOLARIZED ELECTROLYTIC CAPACITOR: The nonpolarized electrolytic capacitor can be used with dc circuits of either polarity, or with an ac circuit.

The schematic diagram shows a nonpolarized electrolytic capacitor. Both terminals are called anodes. The electrolyte is placed between the anodes. An oxide film forms around each anode, insulating it from the electrolyte material. In an alternating-current circuit, each anode receives its bias during half of the cycle. *See also* ELECTROLYTIC CAPACITOR.

NONREACTIVE CIRCUIT

In theory, all circuits contain at least a tiny amount of inductive or capacitive reactance because of effects among the various leads and electrodes. However, this reactance may be so small that it can be neglected in some instances. A circuit is nonreactive if it contains no effective inductive or capacitive reactance, as seen from its input terminals.

A circuit may be nonreactive at just one frequency. An example of such a circuit is the common resonant circuit. At the resonant frequency, the inductive and capacitive reactances cancel, and only resistance is left (*see* RESONANCE).

A circuit may be nonreactive at one frequency and an infinite number of harmonic frequencies. For example, a quarter-wave section of transmission line, having a fundamental frequency f, is also resonant at frequencies 2f, 3f, 4f, and so on. At all of these frequencies, the line measures an integral multiple of a quarter wavelength. A quarter-wave radiating element exhibits the same property.

Some circuits are essentially nonreactive over a wide range of frequencies. This type of circuit generally contains no reactive components, but only resistances. At all frequencies below a certain maximum frequency, the circuit can be considered nonreactive; but above this subjective maximum, component leads introduce significant reactance. *See also* REACTANCE, REACTIVE CIRCUIT.

NONRESONANT CIRCUIT

A circuit is nonresonant if either of the following is true: (1) The circuit is nonreactive over a continuous range of frequencies, or (2) The circuit contains capacitive and inductive reactances that do not cancel each other.

Resistive networks do not contain significant reactance over a wide range of frequencies, but such circuits do not exhibit resonance. Thus, resistive networks are nonresonant circuits.

Bandpass filters, band-rejection filters, highpass filters, lowpass filters, piezoelectric crystals, transmission lines, and antennas are all examples of circuits that are nonresonant at most frequencies. This is because either the capacitive reactance or the inductive reactance predominates. The reactances cancel only at certain discrete frequencies or bands of frequencies. *See also* RESONANCE, RESONANT CIRCUIT.

NONSATURATED LOGIC

Most digital-logic circuits use semiconductors that are either saturated (fully conducting) or cut off (fully open). However, some types of logic circuits operate under conditions of partial conduction in one or both states. Examples of such logic designs include the emitter-coupled, integrated injection, and triple-diffused emitter-follower varieties (*see* EMITTER-COUPLED LOGIC, INTEGRATED INJECTION LOGIC, TRIPLE-DIFFUSED EMITTER-FOLLOWER LOGIC).

Nonsaturated logic has certain advantages. The most notable advantage is its higher switching speed, compared with most saturated logic forms. However, a nonsaturated transistor is somewhat more susceptible to noise than is a saturated transistor. Recent technological improvements have reduced the noise susceptibility of nonsaturated bipolar logic devices. *See also* SATURATED LOGIC.

NONSINUSOIDAL WAVEFORM

A waveform is nonsinusoidal if its shape cannot be precisely represented by the sine function. That is, a nonsinusoidal waveform is any waveform that is not a sine wave (*see* SINE WAVE). The illustration shows several periodic nonsinusoidal waves, all with the same fundamental period.

Most naturally occurring waveforms are nonsinusoidal. All nonsinusoidal waveforms contain energy at more than one frequency, although a definite fundamental period is usually ascertainable. Most musical instruments, as well as the human voice, produce nonsinusoidal waves.

Certain forms of periodic nonsinusoidal waveforms are used for test purposes. Of these, the most common are the sawtooth and square waves. *See also* SAWTOOTH WAVE, SQUARE WAVE.

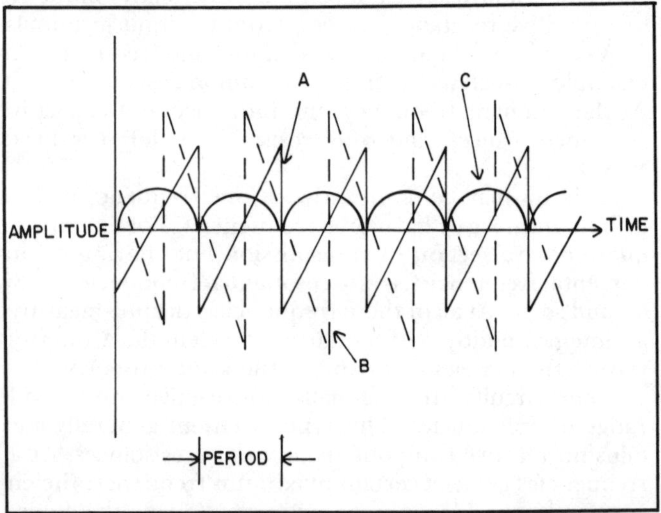

NONSINUSOIDAL WAVEFORM: Three nonsinusoidal waveforms with identical periods: sawtooth (A), reversed sawtooth with distortion (B), and sine-square-root (C).

NOR GATE

A NOR gate is an inclusive-OR gate followed by an inverter. The expression NOR derives from NOT-OR. The NOR gate outputs are exactly reversed from those of the OR gate. That is, when both or all inputs are 0, the output is 1; otherwise, the output is 0. The illustration shows the schematic symbol for the NOR gate, along with a truth table showing the logical NOR function. *See also* OR GATE.

NORMAL DISTRIBUTION

The normal distribution, also called the Gaussian distribution or bell-shaped curve, is a probability function. The normal distribution is often found in nature when random events occur. The maximum probability density of the normal distribution occurs in the center of the range (see illustration). The total area under the curve is 1.

There are infinitely many possible variations of the normal distribution, depending on the value of the standard deviation (s). The smaller the standard deviation, the more the distribution is concentrated toward the center. The larger the standard deviation, the more flat the function becomes. The functions illustrated show normal distributions for s = 0.5, s = 1, and s = 2. When s = 0, the function is entirely concentrated at the center of the range or x axis. As the value of s becomes very large, the function approaches the x axis, and the central maximum becomes less and less defined.

Assuming that the mean of the normal probability-

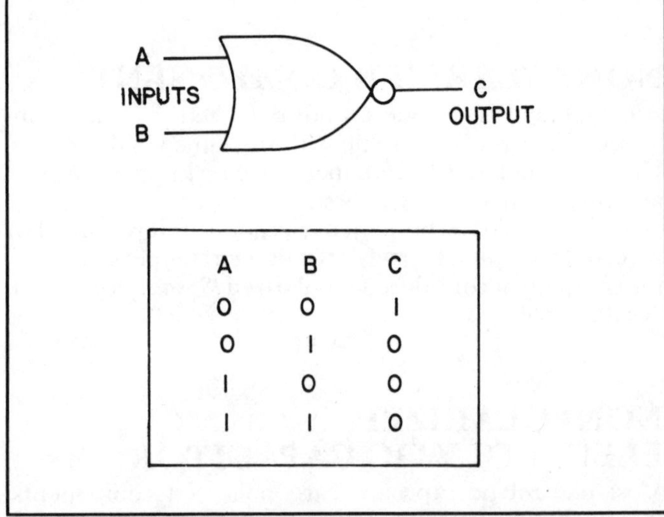

A	B	C
0	0	1
0	1	0
1	0	0
1	1	0

NOR GATE: Schematic symbol for NOR gate, with truth table.

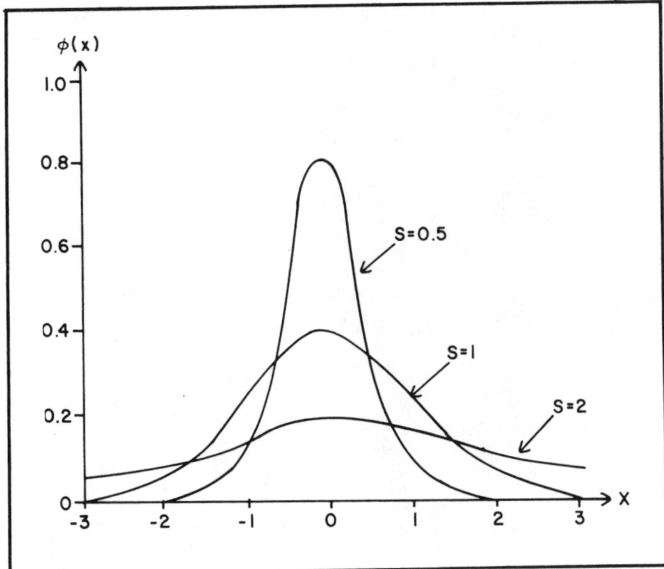

NORMAL DISTRIBUTION: Three examples of normal distributions for different values of standard deviation (s).

density function is 0 and the standard deviation is given by s, then the density function ϕ (x) can be represented by

$$\phi\ (x) = (0.4/s)e^{(-x^2/2s^2)}$$

where e is approximately equal to 2.718. *See also* STATISTICAL ANALYSIS, STANDARD DEVIATION.

NORTON'S THEOREM

Norton's theorem is a principle involving the properties of circuits with more than one generator and/or more than one constant impedance.

Given a circuit with two or more constant-current, constant-frequency generators and/or two or more constant impedances, there exists an equivalent circuit, consisting of exactly one constant-current, constant-frequency generator and exactly one constant impedance. If the more complicated circuit is replaced by the simpler circuit, identical conditions will prevail. *See also* COMPENSATION THEOREM, RECIPROCITY THEOREM, SUPERPOSITION THEOREM, THEVENIN'S THEOREM.

NOT AND GATE

See NAND GATE.

NOTCH FILTER

A notch filter is a narrowband-rejection filter. Notch filters are found in many superheterodyne receivers. The notch filter is extremely convenient for reducing interference caused by strong, unmodulated carriers within the passband of a receiver.

Notch-filter circuits are generally inserted in one of the intermediate-frequency stages of a superheterodyne receiver, where the bandpass frequency is constant. There are several different kinds of notch-filter circuit. One of the simplest is a trap configuration, inserted in series with the signal path, see A in illustration. The notch frequency is adjustable, so that the deep null (B) can be tuned to any frequency within the receiver passband.

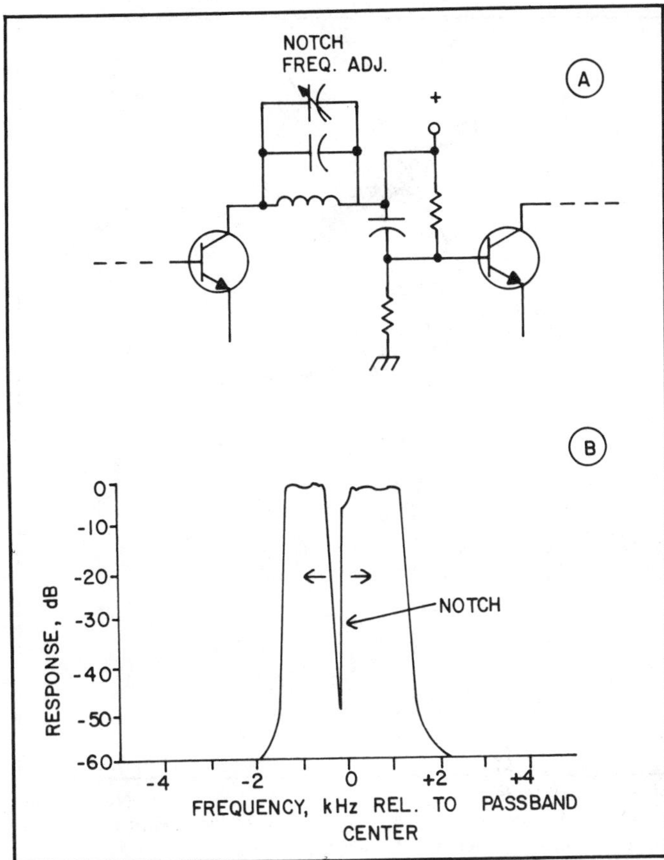

NOTCH FILTER: At A, a simple notch-filter circuit. At B, spectral diagram of typical passband response of a single-sideband receiver, with notch about 50 dB deep.

A properly designed notch filter can produce attenuation well in excess of 40 dB. *See also* BAND-REJECTION FILTER.

NOT GATE

See INVERTER.

N-P JUNCTION

See P-N JUNCTION.

NPN TRANSISTOR

An NPN transistor is a three-layer, bipolar semiconductor device. A layer of P-type material is sandwiched in between two layers of N-type material (see illustration A). One N-type layer is thinner than the other; this layer is the emitter. The P-type layer in the center is the base; the thicker N-type layer is the collector.

The NPN transistor is normally biased with the base positive with respect to the emitter. The collector is positive with respect to the base. This biasing configuration is shown at B.

For Class-B or Class-C amplifier operation, the base

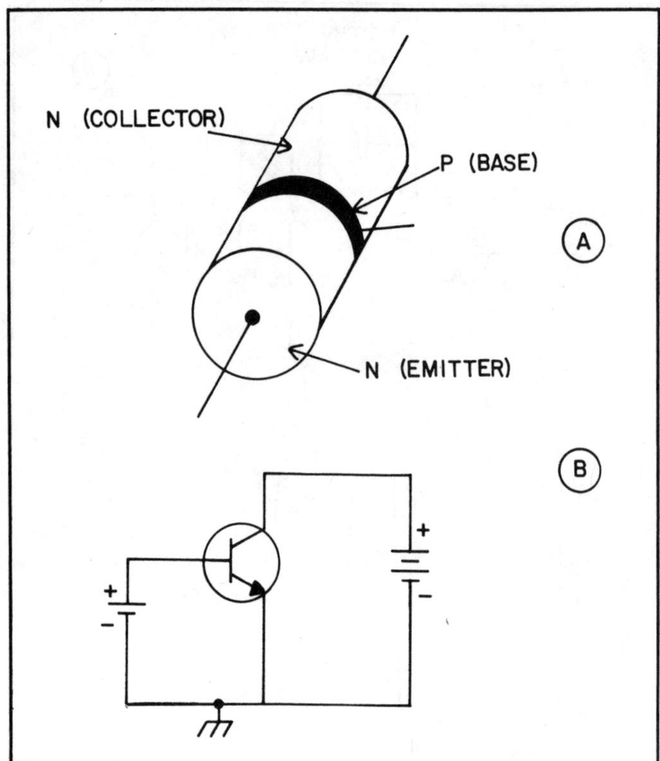

NPN TRANSISTOR: At A, pictorial representation of an NPN bipolar transistor. At B, the basic biasing scheme. The collector-emitter voltage is larger than the base-emitter voltage.

may be biased at the same voltage as the emitter, or at a more negative voltage (*see* CLASS-B AMPLIFIER, CLASS-C AMPLIFIER).

The NPN transistor uses a biasing polarity that is opposite to that of the PNP transistor. *See also* PNP TRANSISTOR.

NTSC COLOR TELEVISION

See COLOR TELEVISION, TELEVISION.

N-TYPE CONNECTOR

The N-type connector is a connector similar to the PL-259, used with coaxial cables in radio-frequency applications. The N-type connector is known for its low loss and constant impedance. These features are important at very-high and ultra-high frequencies, and therefore the N-type connector is preferred at these frequencies.

N-type connectors are available in either male or female form (see photograph) and in other configurations such as male-to-male and female-to-female. Adaptors are available for coupling an N-type connector to a cable or chassis having a different type of connector.

N-TYPE SEMICONDUCTOR

An N-type material is a semiconductor substance that conducts mostly by electron transfer. N-type germanium or silicon is fabricated by adding certain impurity elements to the semiconductor. These impurity elements are called donors, because they contain an excess of electrons. The N-type semiconductor gets its name from the fact that it conducts via negative charge carriers (electrons).

N-TYPE CONNECTOR: Male (left) and female (right).

The carrier mobility in an N-type semiconductor is greater than the carrier mobility in a P-type material, since electrons are transferred more rapidly, from atom to atom, than are holes, for a given applied electric field.

N-type material, combined with P-type material, is the basis for manufacture of most forms of semiconductor diodes, transistors, field-effect transistors, integrated circuits, and other devices. *See also* CARRIER MOBILITY, DOPING, ELECTRON, HOLE, P-N JUNCTION, B-TYPE SEMICONDUCTOR.

NUCLEAR ENERGY

In nature, heavier elements tend to break down, over a long period of time, into smaller elements. This process is called nuclear fission (*see* NUCLEAR FISSION). In the fission process, energy is liberated.

In the early and middle part of the twentieth century, scientists discovered ways to accelerate this process, developing first the atomic bomb, and later controlled atomic reactors. Finally, the nuclear fusion process was discovered (*see* NUCLEAR FUSION), and the hydrogen bomb was invented. A nuclear-fusion reactor has not yet been perfected.

Nuclear energy has both constructive and destructive applications. The world is demanding more and more energy, and as a result of this demand, nuclear reactors have been built to supply electricity. This has reduced dependence on the limited supply of fossil fuels such as coal, natural gas, and oil.

The fission reactor produces nuclear waste, and a controversy has developed over how to get rid of the waste. Some groups believe the waste presents a serious hazard to mankind, and that there is no safe way to dispose of it. Others believe the danger is relatively small. *See also* ENERGY.

NUCLEAR FISSION

Nuclear fission is a process by which the nuclei of heavy elements break down to form lighter elements. Fission takes place slowly in nature. Uranium 238, a material found in the earth's crust, is a substance that is slowly deteriorating to form lighter elements. As this happens, the uranium gives off energy, mostly in the form of high-energy electromagnetic radiation (*see* RADIOACTIVITY, URANIUM).

Certain isotopes of lighter elements also undergo nuclear fission because they are unstable (*see* ISOTOPE). Carbon 14 is an example of such an element. Normally, the carbon nucleus contains six protons and six neutrons; a nucleus of carbon 14 has eight neutrons. The radioactivity of carbon 14 is used to date ancient materials such as fossils.

Using isotopes of the heavier elements, such as uranium 235, a method was discovered for causing rapid nuclear fission. Large amounts of energy can be liberated in a short time when a sufficient mass of such a substance is brought together. This is how the first atomic bombs were developed. Later, such reactions were harnessed to produce energy for constructive purposes. *See also* NUCLEAR ENERGY.

NUCLEAR FUSION

Nuclear fusion is the joining of two or more nuclei of light elements, resulting in a heavier nucleus. The most common form of nuclear fusion is the joining of hydrogen nuclei to form helium nuclei (see illustration). This reaction is believed to be responsible for most of the energy radiated by stars, including our sun.

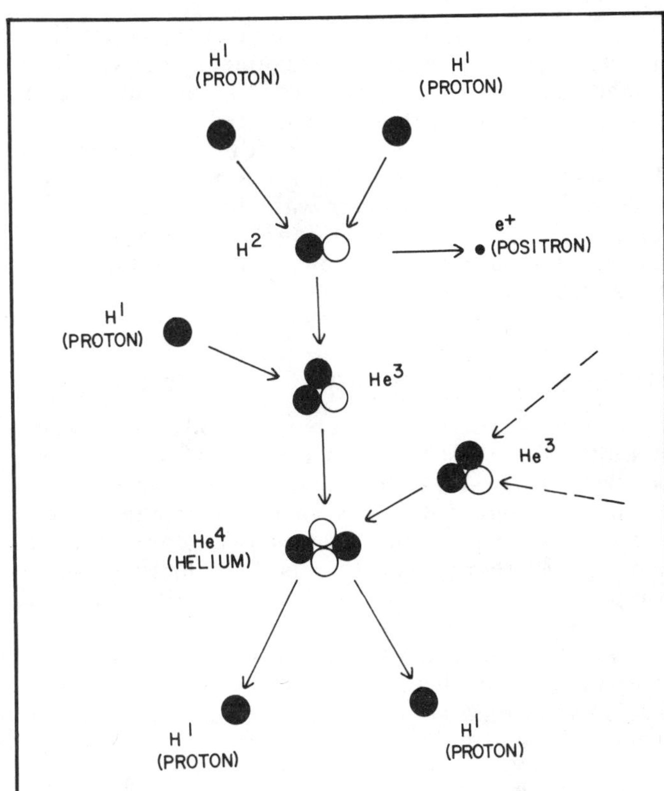

NUCLEAR FUSION: The hydrogen-to-helium fusion reaction. Six hydrogen nuclei combine to form one helium nucleus, plus one positron, two leftover hydrogen nuclei, and energy. This representation shows the process as it is believed to occur at the centers of stars.

Nuclear fusion results in the liberation of tremendous amounts of energy. For the fusion process to begin, the temperature must be millions of degrees. The temperature at the center of the sun is believed to be about 25 million degrees Fahrenheit. The centers of some stars are much hotter.

When the supply of hydrogen in the center of a star begins to run out, helium atoms begin joining to form carbon. This requires a still hotter temperature. Carbon atoms join to form other, heavier elements. Some physicists and astronomers believe that all of the existing elements were formed inside ancient stars by the fusion process. This includes the matter in our bodies.

Nuclear fusion has been harnessed by man for destructive purposes. The hydrogen bomb works in the same way that our sun shines. A fission bomb (*see* NUCLEAR FISSION) is first detonated, producing temperatures high enough to cause hydrogen fusion. A single hydrogen bomb can devastate several hundred square miles of the earth's surface. A typical hydrogen bomb produces as much energy as 20 million tons of TNT.

The fusion process has not yet been tamed for use in a nuclear reactor, but many scientists hope for such a development some day. The main problem has been attaining the high temperatures necessary for hydrogen fusion, while confining the plasma. *See also* NUCLEAR ENERGY.

NUCLEAR MAGNETIC RESONANCE

In the presence of a strong, fluctuating magnetic field, the nuclei of certain atoms will oscillate back and forth. This occurs because of electromagnetic forces between the charged nucleus and the magnetic field. The phenomenon is known as nuclear magnetic resonance.

Nuclear magnetic resonance is being used for medical diagnostic purposes. The patient's body is exposed to a strong, radio-frequency electromagnetic field. This causes the nuclei of the atoms in the body to oscillate, and consequently to emit their own radio waves. These radio waves can be received and processed via computer, producing detailed cross-sectional pictures of the interior of the body. The technique also allows physicians to evaluate the functioning of body organs. *See also* MEDICAL ELECTRONICS.

NULL

A null is a condition of zero output from an alternating-current circuit, resulting from phase cancellation or signal balance.

An alternating-current bridge circuit is adjusted for a zero-output, or null, reading in order to determine the values of unknown capacitances, inductances, or impedances. The radiation pattern from an antenna system may exhibit zero field strength in certain directions; these directions are called nulls. *See also* PHASE BALANCE.

NUMBER LINE

A number line is a geometric representation of numerical values. The number line is infinitely long, extending forever in both directions from the value zero. Each point on the line corresponds to exactly one real number, and each real number has exactly one matching point on the line. The integer values are usually labeled at equal intervals. However, this need not be the case; some number lines use logarithmic or otherwise nonlinear intervals. The distance between the assigned points for adjacent integers is chosen according to the range desired.

Number lines are used in coordinate systems and alignment charts. For example, two number lines, placed at right angles to each other with their zero points in common, form a two-dimensional coordinate system known as the Cartesian plane. *See also* CARTESIAN COORDINATES, NOMOGRAPH.

NUMBER SYSTEMS

A number system is a scheme or method of evaluating quantity. The most common number system used by people is the base-10, or decimal system, in which there are ten possible digits. Machines, such as computers, use a different number system, called the binary or base-2 system, in which there are only two possible digits. For some computer applications, the octal (base-8) or hexadecimal (base-16) systems are used.

A number system may be set up using any integral base. In a base-n system, there are n possible digits. The number places are assigned, according to powers of n. A number is written in the form:

$$\ldots m_2 m_1 m_0 . m_{-1} m_{-2} m_{-3} \ldots$$

having the value:

$$\ldots + m_2 n^2 + m_1 n^1 + m_0 n^0 + m_{-1} n^{-1} + m_{-2} n^{-2} + m_{-3} n^{-3} + \ldots$$

where m may be any integer between, and including, 0 and n−1. *See* BINARY-CODED NUMBER, HEXADECIMAL NUMBER SYSTEM, OCTAL NUMBER SYSTEM.

Different sets of decimal (base −10) numbers exist. The simplest set is the set of natural numbers or positive integers. The set of rational numbers is obtained by forming all possible quotients of the positive and negative integers. The set of real numbers is even larger; some of these cannot be written out in decimal form. A two-dimensional number system, known as the complex numbers, is used for certain impedance-related calculations. *See also* COMPLEX NUMBER, RATIONAL NUMBER, REAL NUMBER.

NUMERICAL CONSTANT

See CONSTANT.

NUVISTOR TUBE

Nuvistor is a trade name of RCA. The Nuvistor tube is a small metal-and-ceramic receiving vacuum tube. The Nuvistor is noted for its high gain, especially at very-high and ultra-high frequencies.

In the Nuvistor, the filament is at the center, and the cathode, grid or grids, and plate surround the filament in the form of concentric cylinders. The electrodes are closely spaced, so the tube is useful at very short wavelengths. *See also* TUBE.

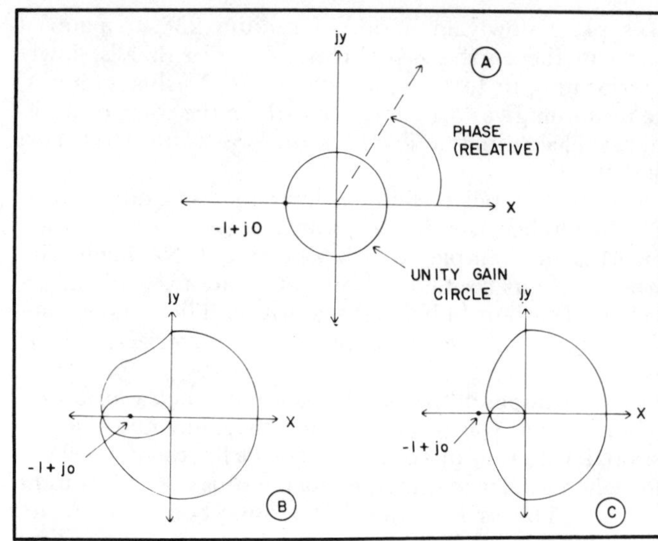

NYQUIST DIAGRAM AND NYQUIST STABILITY CRITERION: At A, the Nyquist-diagram plane. At B, an example of a Nyquist plot for an unstable feedback amplifier. At C, a Nyquist plot for a stable feedback amplifier.

NYQUIST DIAGRAM

A Nyquist diagram, also called a Nyquist plot, is a complex-number graph used for evaluating the performance of a feedback amplifier. The Nyquist-diagram plane is a Cartesian system. The real-number axis (or x axis) is horizontal, and the imaginary-number axis (or y axis) is vertical.

In the Nyquist diagram, open-loop gain is represented by circles centered at the origin. The circle with radius 1 is called the unity-gain circle. Points inside this circle represent amplifier gain less than 1; points outside this circle correspond to gain figures greater than 1. The output phase is represented by radial lines passing through the origin. This is shown in A of the illustration. The Nyquist diagram is made by plotting the real and imaginary components of the amplifier output in the complex plane, for all possible input frequencies. Two examples of such plots are shown at B and C.

The Nyquist diagram is used to determine whether or not a feedback amplifier is stable. *See* NYQUIST STABILITY CRITERION.

NYQUIST STABILITY CRITERION

To determine if a single-loop feedback amplifier is stable the Nyquist diagram can be used. This is known as the Nyquist stability criterion.

The Nyquist plot is obtained for all possible input frequencies. If the Nyquist plot contains a closed loop surrounding the point −1 + j0, then the system is unstable. If the plot contains no such loop and the open-loop configuration is stable, then the amplifier is stable.

The illustration shows the Nyquist plane at A. At B, an example of a Nyquist plot for an unstable single-loop feedback amplifier is shown. At C, a plot is illustrated for a stable single-loop feedback amplifier. *See also* NYQUIST DIAGRAM.

O

OCTAL NUMBER SYSTEM

The octal number system is a base-8 system. Only the digits 0 through 7 are used. When counting in base 8, the number following 7 is 10; the number following 77 is 100, and so on. Moving toward the left, the digits are multiplied by 1, then by 8, then by 64, and so on, upward in successively larger integral powers of 8.

The octal number system is sometimes used as shorthand for long binary numbers. The binary number is split into groups of three digits, beginning at the extreme right, and then the octal representation is given for each three-digit binary number. For example, suppose we are given the binary number 11000111. This is an eight-digit binary number. We first add a zero to the left-hand end of this number, making it nine digits long (a multiple of three), obtaining 011000111. Breaking up this number gives us 011 000 111. These three binary numbers correspond to 3, 0, and 7 respectively. The octal representation, then, of 11000111 is 307.

The octal number system is used by some computers, since its base, 8, is a power of 2, making it easier to work with than the base-10 system. A base-16 number system is also sometimes used by computers. *See also* HEXADECIMAL NUMBER SYSTEM.

OCTAVE

An octave is a 2-to-1 range of frequencies. If the lower limit of an octave is represented in hertz by f, then the upper limit is represented by 2f, or the second harmonic of f. If the upper limit of an octave is given in hertz by f, then the lower limit is f/2.

The term octave is sometimes used when specifying the gain-versus-frequency characteristics of an operational amplifier. A circuit may, for example, have a rolloff characteristic (decrease in gain) of 6 dB per octave. This means that if the gain is x dB at frequency f, then the gain is x − 6 dB at frequency 2f, x − 12 dB at frequency 4f, x − 18 dB at frequency 8f, and so on.

For audio frequencies, the octave can be recognized by ear. This phenomenon is familiar to the musician; from a given tone frequency f, it is easy to imagine the notes having frequencies 2f and 4f. It is also easy to imagine the notes having frequencies f/2 and f/4. *See also* HARMONIC.

ODD-ORDER HARMONIC

An odd-order harmonic is any odd multiple of the fundamental frequency of a signal. For example, if the fundamental frequency is 1 MHz, then the odd-order harmonics occur at frequencies of 3 MHz, 5 MHz, 7 MHz, and so on.

Certain conditions favor the generation of odd-order harmonics in a circuit, while other circumstances tend to cancel out the odd harmonics. The push-pull circuit accentuates the odd harmonics (*see* PUSH-PULL CONFIGURATION). Such circuits are often used as frequency triplers. The push-push circuit (*see* PUSH-PUSH CONFIGURATION) tends to cancel out the odd harmonics in the output.

A half-wave dipole antenna will operate at all odd-harmonic frequencies, assuming it is a full-size antenna (not inductively loaded). The impedance at the center of such an antenna is purely resistive at all odd-harmonic frequencies. At the odd harmonics, the dipole antenna presents a slightly higher feed-point resistance than at the fundamental frequency, but the difference is not usually large enough to cause an appreciable mismatch when the antenna is operated at the odd harmonics. *See also* EVEN-ORDER HARMONIC.

OERSTED

The oersted is a unit of magnetic-field intensity in the centimeter-gram-second (CGS) system, expressed in terms of magnetomotive force per unit length. A field of 1 oersted is equivalent to a field of 1 gilbert, or 0.796 ampere turns, per centimeter. *See also* AMPERE TURN, GILBERT, MAGNETOMOTIVE FORCE.

OERSTED, HANS CHRISTIAN

A Danish scientist named Hans Christian Oersted (1777-1851) is credited with first discovering that a magnetized rod is influenced when brought near a wire carrying an electric current. Oersted found that a current produces a magnetic field with lines of flux in a plane perpendicular to the direction of the current flow. Oersted is sometimes credited with founding the science of electromagnetism.

OFF-CENTER FEED

Off-center feed is a method of applying power to an antenna at a current or voltage loop not at the center or end of the radiating element. Off-center feed is sometimes called windom feed, but the windom is only one form of off-center-fed antenna (*see* WINDOM ANTENNA).

For off-center feed to be possible, an antenna must be at least 1 wavelength long. A 1-wavelength antenna has current loops at distances of ¼ wavelength from either end. A 1-wavelength radiator may thus be off-center-fed with coaxial cable, as shown in the illustration.

In general, for a longwire antenna measuring an integral multiple of ½ wavelength, a coaxial feed line may be connected ¼ wavelength from one end, and a reasonably good impedance match will be obtained. If the longwire is terminated, the feed point should be ¼ wavelength from

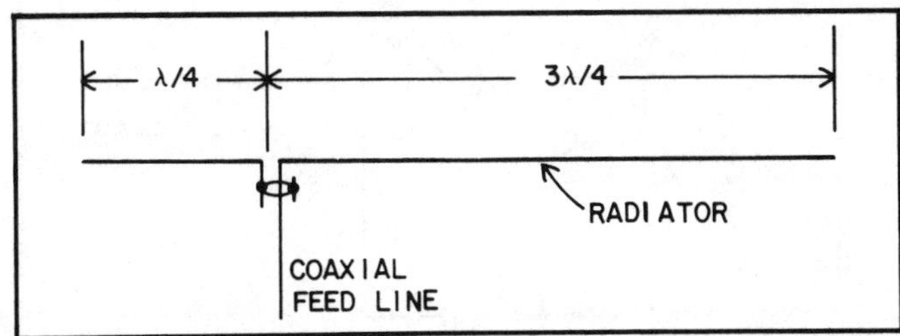

OFF-CENTER FEED: Off-center current feed for a full-wavelength radiator.

the unterminated end. A low-loss, high-impedance, balanced feed line may be connected at any multiple of ¼ wavelength from the unterminated end of such an antenna. *See also* CENTER FEED, END FEED, LONGWIRE ANTENNA.

OHM

The ohm is the standard unit of resistance, reactance, and impedance. The symbol is the Greek capital letter omega (Ω). A resistance of 1 ohm will conduct 1 ampere of current when a voltage of 1 volt is placed across it. *See also* IMPEDANCE, REACTANCE, RESISTANCE.

OHM, GEORG SIMON

Georg Simon Ohm (1787-1854), a German scientist and professor, discovered a mathematical relationship among the potential difference, flow of current, and resistance in a direct-current circuit. This relationship is quite simple, and is now called Ohm's Law. *See also* OHM'S LAW.

OHMIC LOSS

Ohmic loss is the loss that occurs in a conductor because of resistance. Ohmic loss is determined by the type of material used to carry a current, and by the size of the conductor. In general, the larger the conductor diameter for a given material, the lower the ohmic loss. Ohmic loss results in heating of a conductor.

For alternating currents, the ohmic loss of a conductor generally increases as the frequency increases. This occurs because of a phenomenon known as skin effect, in which high-frequency alternating currents tend to flow mostly near the surface of a conductor (*see* SKIN EFFECT).

In a transmission line, ohmic loss is just one form of loss. Additional losses can occur in the dielectric material. Loss can also be caused by unwanted radiation of electromagnetic energy, and by impedance mismatches. *See also* DIELECTRIC LOSS, RADIATION LOSS, STANDING-WAVE-RATIO LOSS.

OHMMETER

An ohmmeter is a simple device used for measuring direct-current resistance. A source of known voltage, usually a battery, is connected in series with a switchable known resistance, a milliammeter, and a pair of test leads (see illustration).

Most ohmmeters have several ranges, labeled according to the magnitude of the resistances in terms of the scale indication. The scale is calibrated from 0 to "infinity" in a nonlinear manner. The range switch, by inserting various resistances, allows measurement of fairly large or fairly small resistances.

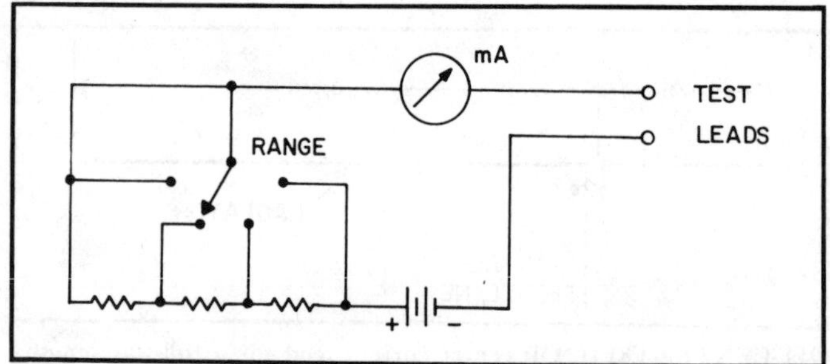

OHMMETER: Schematic diagram of a simple ohmmeter.

Most ohmmeters can accurately measure any resistance between about 1 ohm and several hundred megohms. Specialized ohmmeters are needed for measuring extremely small or large resistances. *See also* MEGGER, WHEATSONE BRIDGE.

OHM'S LAW

Ohm's law is a simple relation between the current, voltage, and resistance in a circuit. The current, voltage, and resistance in a direct-current situation are interdependent. If two of the quantities are known, the third can be found by a simple equation.

Letting I represent the current in amperes, E represent the voltage in volts, and R represent the resistance in ohms, the following relation holds:

$$I = E/R$$

The formula may also be restated as:

$$E = IR, \text{ or}$$

$$R = E/I$$

In an alternating-current circuit, pure reactance can be substituted for the resistance in Ohm's law. If both reactance and resistance are present, however, the situation becomes more complex. *See also* CURRENT, RESISTANCE, VOLTAGE.

OHMS PER VOLT

A voltmeter is sometimes rated for sensitivity according to a specification known as ohms-per-volt. The ohms-per-volt rating is an indication of the extent to which the voltmeter will affect the operation of a high-impedance circuit.

To determine the ohms-per-volt rating of a voltmeter, the resistance through the meter must be measured at a certain voltage. Normally this is the full-scale voltage. This resistance can be found by connecting the voltmeter in series with a milliammeter or microammeter, and then placing a voltage across the combination such that the meter reads full scale. The ohms-per-volt (R/E) rating is then equal to 1 divided by the milliammeter or microammeter reading expressed in amperes.

For example, suppose that a voltmeter with a full-scale reading of 100 V is connected in series with a microammeter. A source of voltage is connected across this combination, and the voltage adjusted until the voltmeter reads full scale. The meter reads 10μA. Then the ohms-per-volt rating of the meter is 1/0.00001, or 100,000, ohms per volt.

It is desirable that a voltmeter have a very high ohms-per-volt rating. The higher this figure, the smaller the current drain produced by the voltmeter when it is connected into a circuit. In a high-impedance circuit, this is extremely important. Some voltmeters are designed so that they exhibit extremely high ohms-per-volt figures. These meters employ field-effect transistors having extremely high input impedances. Older meters of this type use vacuum tubes, although this technology is now obsolete. *See also* FET VOLTMETER, VACUUM-TUBE VOLTMETER, VOLTMETER.

OMNIDIRECTIONAL ANTENNA

An omnidirectional antenna is an antenna that radiates with equal field strength in all directions. Such an antenna does not exist in reality. However, some antennas do radiate with equal field strength within a given plane. Most vertical antennas are omnidirectional in the horizontal plane (*see* VERTICAL ANTENNA). The vertical-plane directional pattern of such an antenna exhibits maximum field strength at or near the horizontal plane.

An isotropic antenna is a true omnidirectional antenna, but it exists only in theory. An isotropic antenna radiates with equal field strength in all directions in three dimensions. *See also* ISOTROPIC ANTENNA.

OMNIDIRECTIONAL MICROPHONE

An omnidirectional microphone is a microphone that responds equally well to sound from all directions. Omnidirectional microphones often have spherical heads (*see* illustration).

Omnidirectional microphones are used when it is necessary to pick up background noises as well as the voice or music of the subject. Omnidirectional microphones are not generally desirable for communications applications. *See also* DIRECTIONAL MICROPHONE.

ONE-SHOT MULTIVIBRATOR

See MONOSTABLE MULTIVIBRATOR.

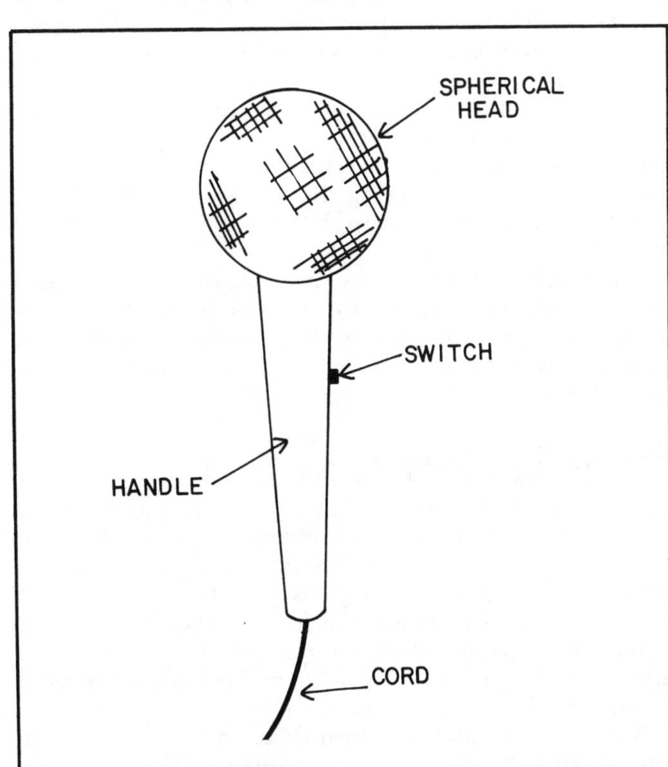

OMNIDIRECTIONAL MICROPHONE: The omnidirctional microphone can generally be recognized by its spherical head.

OPEN CIRCUIT

A circuit is open when current cannot flow. An open circuit may occur as a simple result of the power supply being shut off, or it may be because of a malfunction.

In an open circuit, the current is zero, and the resistance is theoretically infinite. *See also* CLOSED CIRCUIT.

OPEN LOOP

In an amplifier, the condition of zero negative feedback is called the open-loop condition. With an open loop, an amplifier is at maximum gain.

In an operational-amplifier circuit, the open-loop gain is sometimes specified as a measure of the amplification factor of a particular integrated circuit. The input and output impedances are generally specified for the open-loop configuration. *See also* OPERATIONAL AMPLIFIER.

OPEN-WIRE LINE

Open-wire line is a form of radio-frequency feed line. Open-wire line is sometimes called parallel-wire line. Open-wire feed lines are commercially manufactured with characteristic impedances of approximately 300 and 450 ohms. Homemade open-wire lines can be made to have characteristic impedances as high as 600 ohms.

The characteristic impedance of an air-dielectric balanced line depends on the size of the conductors, and also on the spacing between the conductors. If the radius of the conductors is given by r and the center-to-center spacing by s, then the characteristic impedance Z_O is:

$$Z_O = 276 \log_{10} (s/r)$$

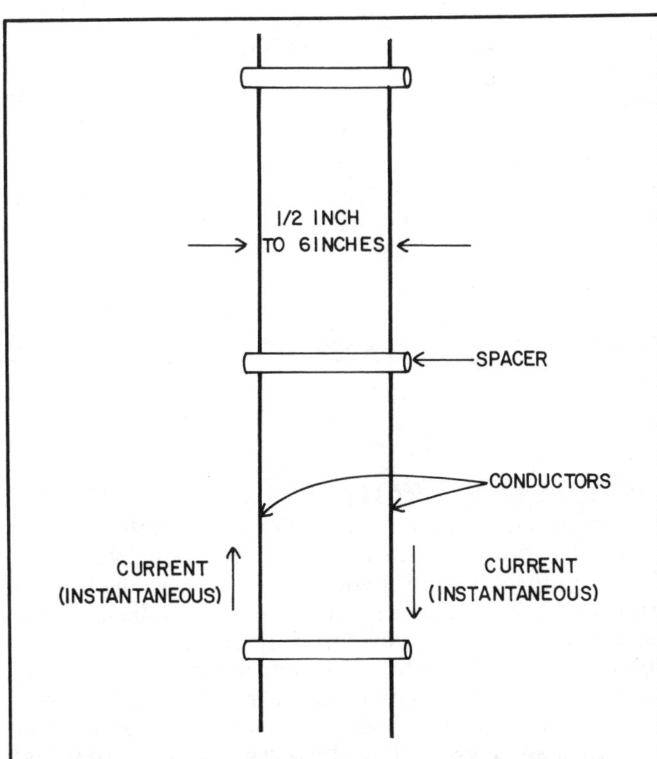

OPEN-WIRE LINE: Construction of open-wire transmission line. The instantaneous current is equal in magnitude in both conductors, but flows in opposing directions.

provided, of course, that r and s are specified in the same units. The presence of spacers tends to lower this value slightly. Also, if the line is placed near objects with a higher dielectric constant than air (and that includes almost everything), the characteristic impedance will be lowered slightly (*see* CHARACTERISTIC IMPEDANCE).

The illustration shows a typical open-wire line. The wires are spaced at a distance of from ½ inch to 6 inches in most cases. The spacers are placed at regular intervals to keep the wires at a roughly constant separation. Open-wire line is a balanced transmission line.

Open-wire feed line is characterized by low loss and high power-handling capacity, even in the presence of significant mismatches between the line and the antenna. Since open-wire line is an inherently balanced line, it should be used with a balanced load. A tuning network, having a balanced output, is needed at the transmitter or receiver (*see* BALANCED LOAD, BALANCED TRANSMISSION LINE).

Open-wire line is ideal for television reception in fringe areas, since it has lower loss, per unit length, than the standard twin-lead lines. *See also* TWIN-LEAD.

OPERATING ANGLE

In an amplifier, the operating angle is the number of degrees, for each cycle, during which current flows in the collector, drain, or plate circuit. The operating angle varies, depending on the class of amplifier operation.

In a Class-A amplifier, the output current flows during the entire cycle. Therefore, in such an amplifier, the operating angle is 360 degrees. In a Class-AB amplifier, the current flows for less than the entire cycle, but for more than half of the cycle; hence the operating angle is larger than 180 degrees but less than 360 degrees. In a single-ended Class-B amplifier, the operating angle is about 180 degrees. In a Class-C amplifier, the current flows for much less than half of the cycle, and the operating angle is thus smaller than 180 degrees.

The operating angle varies with the relative base/collector, gate/drain, or grid/plate bias of an amplifier. The operating angle is also affected by the driving voltage. *See also* CLASS-A AMPLIFIER, CLASS-AB AMPLIFIER, CLASS-B AMPLIFIER, CLASS-C AMPLIFIER, OPERATING POINT.

OPERATING BIAS

See OPERATING POINT.

OPERATING POINT

The operating point of an amplifier circuit is the point at which the direct-current bias is applied along the curve depicting the collector current versus base voltage (for a transistor), the drain current versus gate voltage (for a field-effect transistor), or the plate current versus grid voltage (for a tube). The choice of operating point determines the class of operation for the amplifier. *See* CLASS-A AMPLIFIER, CLASS-AB AMPLIFIER, CLASS-B AMPLIFIER, CLASS-C AMPLIFIER, FIELD-EFFECT TRANSISTOR, TRANSISTOR, TUBE.

Certain transducers and semiconductor devices require direct-current bias for proper operation. When the signal is received or applied, the instantaneous current fluctuates above and below the no-signal current. The no-signal current is called the operating current or operating point. The correct choice of operating point ensures that the device will function according to its ratings.

OPERATIONAL AMPLIFIER

An operational amplifier is an integrated-circuit semiconductor amplifier having high stability and linear characteristics. The ideal operational amplifier, under optimum conditions, would theoretically have infinite input impedance, zero output impedance, infinite gain, and infinite bandwidth. Of course, these ideals cannot be realized in practice, but many operational amplifiers or "op amps" have very high input impedance (they draw almost no current and hence almost no power), very low output impedance, extremely high gain (more than 100 dB in some cases), and large bandwidth (up to several megahertz).

A typical operational amplifier consists of an arrangement of resistors, diodes, and transistors, in a configuration such as shown in A of the schematic diagram. There are two inputs called the inverting and non-inverting inputs. The inverting input provides a 180-degree phase shift at the output. A negative-feedback loop, consisting of a resistor and/or capacitor between the inverting input and the output, is used to control the gain of the amplifier. The non-inverting input is in phase with the output.

Two power-supply terminals are provided; they are usually called Vcc (the collector terminal) and Vee (the emitter terminal). In the NPN-type operational amplifier shown, Vcc is positive (about +5 V to +15 V) and Vee is negative (about −5 V to −15 V). A single-polarity supply may be used in conjunction with voltage-divider networks, or a two-pole supply may be used.

The op amp is used in a variety of electronic devices, including analog-to-digital converters, averaging amplifiers, differentiators, direct-current amplifiers, integrators, multivibrators, oscillators, and sweep generators. The op amp is represented in schematic diagrams by a trangle showing the inputs, the output, and (sometimes) the power-supply terminals, as illustrated at B.

Operational amplifiers are available in most electronic stores in integrated-circuit form. Some integrated-circuit packages contain two or more operational amplifiers within a single housing. Op amps are inexpensive and easy to work with. *See also* INTEGRATED CIRCUIT.

OPTICAL COMMUNICATION

Optical communication is the application of light beams for sending and receiving messages. In recent decades, optical communication has become increasingly important. Early sailors actually used a form of optical communication—flags and flashing lights—but modern systems are capable of transferring millions of times more information than were the seamen who originally used visible light for sending and receiving signals.

Modern optical communications systems use modulated-light sources and receivers sensitive to rapidly varying light intensity (*see* MODULATED LIGHT). Lasers and light-emitting diodes are the favored method of generating modulated light, since they have rapid response time

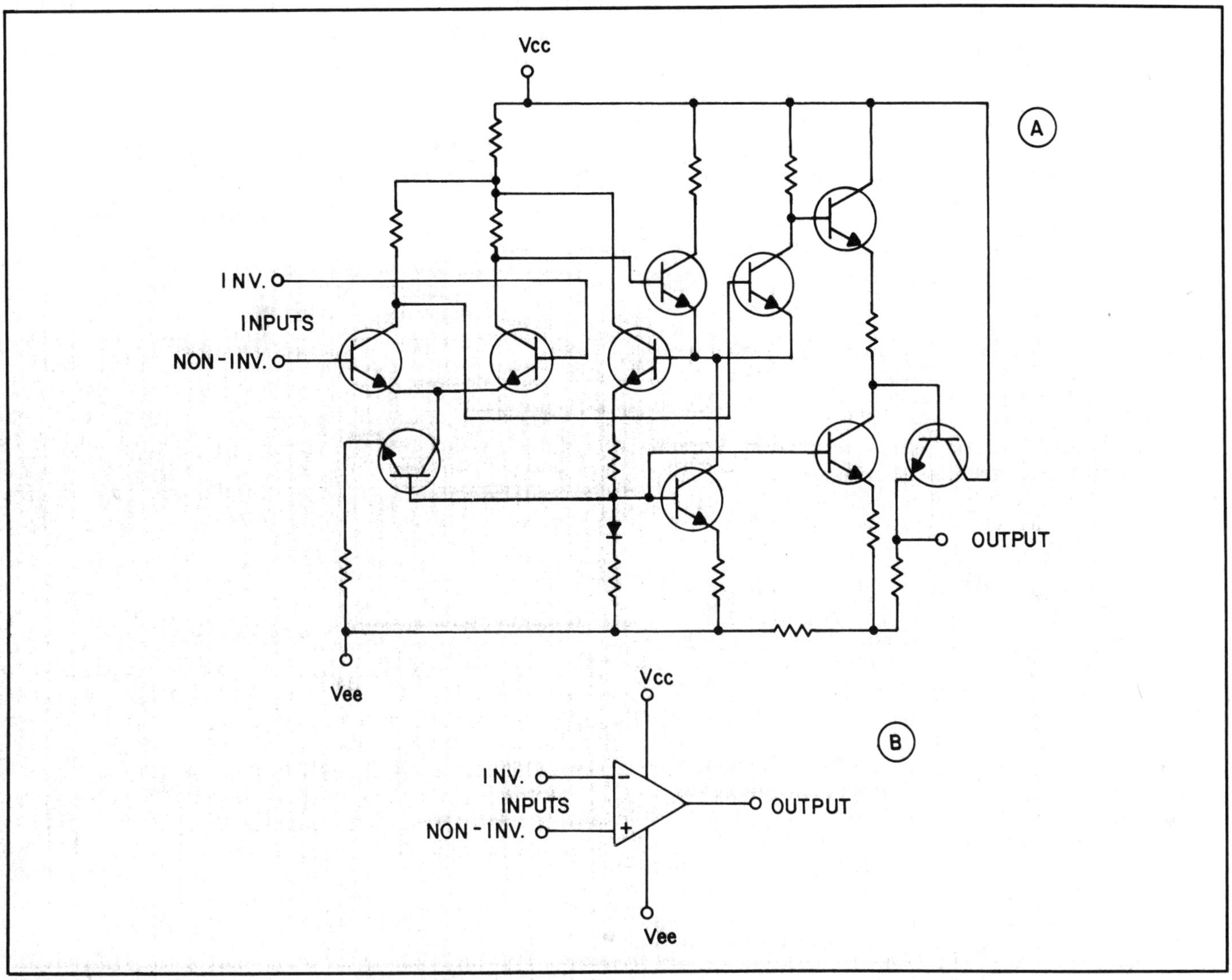

OPERATIONAL AMPLIFIER: At A, a schematic diagram of an operational amplifier. At B, the schematic symbol typically used to represent an operational amplifier.

and emit light within a narrow range of wavelengths (*see* LASER, LIGHT-EMITTING DIODE).

Communication can be obtained in the infrared and ultraviolet regions of the spectrum, as well as in the visible range, provided that the wavelength is chosen for low atmospheric attenuation. Infrared and ultraviolet communications can be considered, for practical purposes, as optical communication (*see* INFRARED, ULTRAVIOLET).

Optical communication can be carried out directly through the atmosphere, in much the same way as microwave links are maintained. However, optical propagation is susceptible to the weather. Rain, snow, and fog can obscure the light and prevent the transfer of data. The transmittivity of clear air varies with the wavelength, as well. The atmosphere is fairly transparent at visible wavelengths. Blue light tends to be scattered more than red light. Considerable attenuation occurs over significant portions of the infrared and ultraviolet spectra.

Glass or plastic fibers can transmit light for long distances, in much the same way that wires carry electricity. A single beam of light can carry far more information than can an electric current, however, and so-called optical fibers are becoming more common for use in place of wire systems (*see* FIBER, FIBER OPTICS).

OPTICAL COUPLING

Optical coupling is a method of transferring a signal from one circuit to another, by converting it to modulated light and then back to its original electrical form.

The signal is first applied to a light-emitting diode or other light-producing device. The resulting modulated-light signal propagates across a small gap, where it is intercepted by a photocell, phototransistor, or other light-sensitive device and converted back into electrical impulses. The basic principle is shown in the illustration. Some optical-coupling devices contain the light transmitter and the light receptor in a single package with a glass or plastic transmission medium.

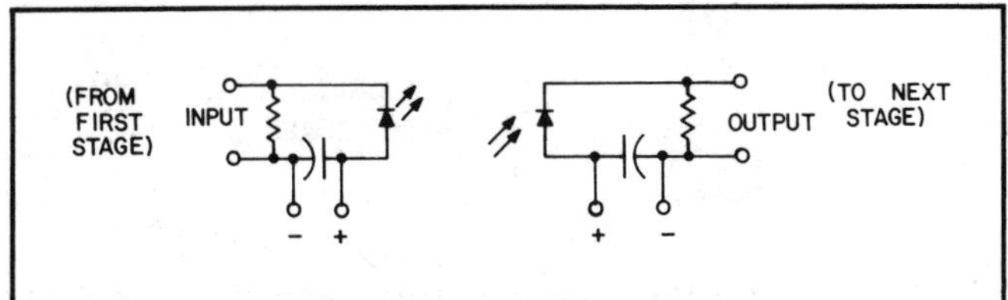

OPTICAL COUPLING: Basic technique for optical coupling. The light-emitting diode and the photodiode (or other light-sensitive device) are placed close together in an opaque housing.

Optical coupling is used when it is necessary to prevent impedance interaction between two successive stages in a multistage circuit. Normally, changes in the input signal to an amplifier cause the input impedance to vary, and this in turn affects the preceding stage. Optical coupling avoids this problem by providing essentially complete isolation. Optical coupling also facilitates coupling between circuits at substantially different potentials.

Optical-coupling devices can be assembled in various different forms for different purposes. One common optical coupler is called a lumistor. *See also* LIGHT-EMITTING DIODE, LUMISTOR, MODULATED LIGHT, PHOTOCELL, PHOTODIODE, PHOTOFET, PHOTORESISTOR, PHOTOTRANSISTOR, PHOTOTUBE, PHOTOVOLTAIC CELL.

OPTICAL FIBER

See FIBER, FIBER OPTICS, MODULATED LIGHT.

OPTICAL LEVER

An optical lever is a device that is sometimes used for increasing the sensitivity of a metering device. A beam of light, reflected from a mirror on a rotating bearing, is directed toward a screen some distance away from the mirror (see illustration). A small change in the orientation of the mirror results in a large deflection of the distant image.

Optical levers were used in some early galvanometers and microammeters for the purpose of obtaining extreme sensitivity. However, optical levers present certain mechanical problems, such as sensitivity to vibration and air currents. The apparatus is somewhat cumbersome, since it requires considerable physical space. Recent developments in solid-state technology have made it possible to achieve metering sensitivity approaching that of the optical lever, but with greater physical stability.

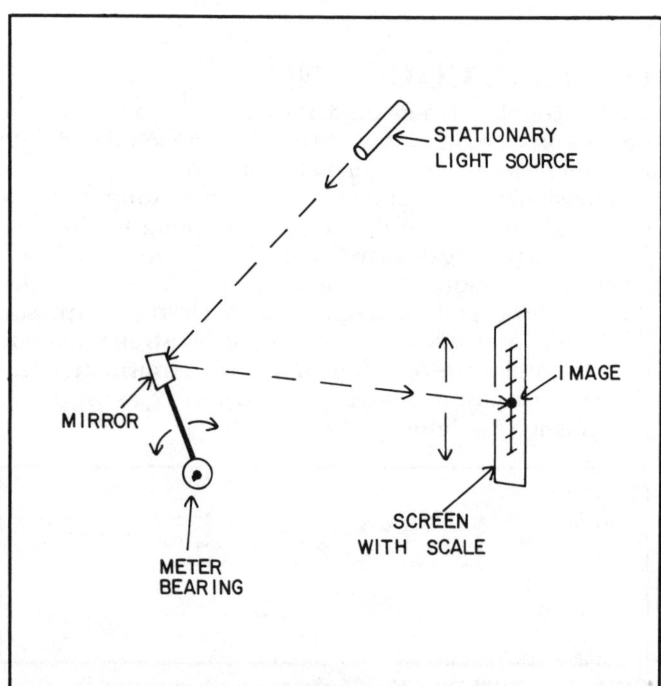

OPTICAL LEVER: The optical lever has a long, massless "needle"—a light beam—for enhanced metering sensitivity.

OPTICAL MASER

See LASER, MASER.

OPTICAL SOUND RECORDING AND REPRODUCTION

Movie-film sound tracks are recorded by means of a modulated-light source (*see* MODULATED LIGHT). As the film moves through the recorder, the modulated-light source causes the film to be exposed in a linear pattern that follows the sound vibrations. The sound track appears, on the film, as a band having variable density or width. The band is located to one side of the picture-frame sequence.

When the film is run through the projector, the sound is retrieved by a form of optical coupling (*see* OPTICAL COUPLING). A constant-brightness light source is placed on one side of the film, and the beam of light is passed through the sound band. A photocell on the other side of the film picks up the modulated light and converts it into the original audio impulses.

OPTICAL TRANSMISSION

Optical transmission is the generation and propagation of a modulated-light signal for communications purposes. Optical transmission can be achieved in various different ways.

The most common method of optical transmission is the direct amplitude modulation of a light source. Light-emitting diodes and lasers are generally best for this, because these devices have a rapid response rate. However, even an incandescent bulb can be modulated to a certain extent at audio frequencies.

Other methods of optical transmission include polarization modulation, pulse modulation, and multiplexing. *See* MODULATED LIGHT, OPTICAL COMMUNICATION.

The relative ease with which visible light propagates through a medium is sometimes called the optical transmission characteristic, or optical transmittivity, of the medium. The attenuation varies with the wavelength of the light.

OPTICS

Optics is a branch of physics, concerned with the behavior of visible light. Optics is significant in electronics, because of its importance in the design and construction of optical communications systems.

A set of optical devices is sometimes referred to as optics. For example, the hardware comprising a fiber-communications system may be called optics. *See also* FIBER OPTICS, MODULATED LIGHT, OPTICAL COMMUNICATION.

OPTIMIZATION

Optimization is a process or operation, by which a set of parameters is adjusted for the most efficient, the maximum, or the minimum result.

In electronics, the process of optimization is a part of all engineering practice. Once a circuit design has been made to work, it is usually possible to improve the performance considerably. This may involve the changing of component values or the changing of component types. Optimization is partly determinable by calculation, but often it is a trial-and-error procedure.

In mathematics, an optimization problem usually involves the maximizing or minimizing of a function of several variables. This can be a relatively easy procedure, but often it is so complicated that a computer is needed. *See also* FUNCTION.

OPTIMUM ANGLE OF RADIATION

When a radio signal is returned to earth by the ionosphere, the angle of incidence is approximately equal to the angle of reflection. Although there exist some irregularities in the ionosphere, this rule can be considered valid on the average. For a given distance between the transmitting and receiving station, and a given ionized-layer altitude, the optimum angle of radiation for single-hop propagation can be determined. The greater the distance between the stations, the lower the optimum angle; the higher the ionized-layer altitude, the higher the optimum angle. An example is shown in Fig. 1.

At the high frequencies, ionospheric propagation usually occurs from the F1 layer, at an altitude of about 100 miles. For single-hop propagation via the F1 layer, the optimum angle varies with distance as shown at A in Fig. 2. Propagation from the F2 layer occurs over greater distances because of the greater ionization altitude of about 200 miles, as at B. The F1 layer is responsible for most high-frequency propagation during the daylight hours. The F2 layer usually returns the signals at night (*see* F LAYER).

Under certain conditions at high frequencies, but more commonly at very-high frequencies, the E layer, at an altitude of about 60 miles, returns signals to the earth. The resulting relation between the optimum angle and the station separation is shown at C (*see* E LAYER, SPORADIC-E PROPAGATION).

The angle of departure from an antenna can be controlled to a certain extent at the high frequencies. The higher the antenna above the ground, the lower the angle at which maximum radiation takes place.

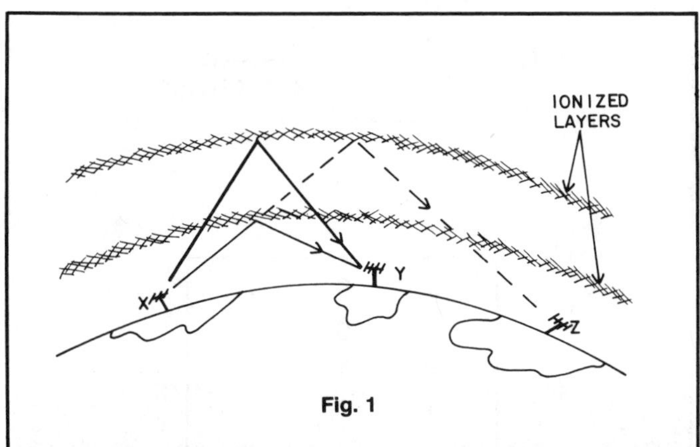

OPTIMUM ANGLE OF RADIATION: Propagation distance depends on the angle of radiation, and also on the altitude of the ionized layer. Here, three stations are shown; X is the transmitting station and Y and Z are receiving stations. Note the differences in the optimum angle of radiation.

When the distance between the transmitting and receiving station is greater than the maximum single-hop distance, multi-hop propagation occurs. In such cases, the optimum angle of radiation is less well defined. In general, the best results for multi-hop paths are obtained when the angle of departure is as low as possible. This means that, for long-distance communication, the transmitting antenna should be placed as far above the ground as can be managed. *See also* ANGLE OF DEPARTURE, PROPAGATION CHARACTERISTICS.

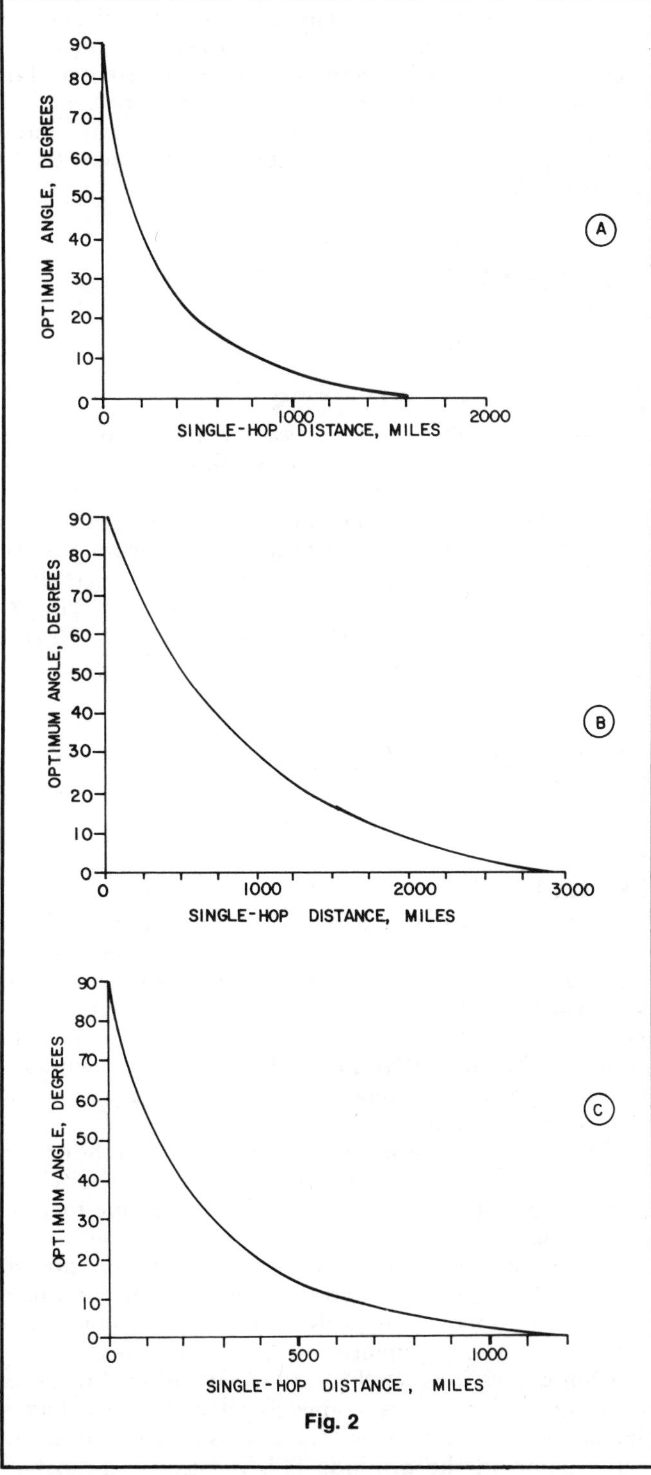

OPTIMUM ANGLE OF RADIATION: Optimum angle of radiation, as a function of propagation distance, for (A) the F1 layer, (B) the F2 layer, and (C) the E layer.

OPTIMUM WORKING FREQUENCY

For any high-frequency communications circuit, there exists an optimum working frequency, at which the most reliable communications are obtained. The optimum working frequency lies between the lowest usable frequency and the maximum usable frequency (*see* LOWEST USABLE FREQUENCY, MAXIMUM USABLE FREQUENCY).

The optimum working frequency depends on the distance between the transmitting and receiving stations, the time of day, the time of year, and the level of sunspot activity. In general, the optimum working frequency is increased as the station separation becomes greater. The optimum working frequency is usually greater during the day than at night; it is higher in summer than in winter, and is highest during periods of peak sunspot activity. *See also* PROPAGATION CHARACTERISTICS.

OPTOELECTRONIC COUPLING

See OPTICAL COUPLING.

OPTOELECTRONICS

Optoelectronics is a field of science common to optics as well as electronics. Optoelectronics has become increasingly important in recent years, with the advent of lasers, optical fibers, and optical communications in general. *See also* FIBER, FIBER OPTICS, LASER, LASER DIODE, LIGHT-EMITTING DIODE, OPTICAL COMMUNICATION, OPTICAL COUPLING, OPTICAL TRANSMISSION, PHOTOCELL, PHOTODIODE, PHOTOFET, PHOTORESISTOR, PHOTOTRANSISTOR, PHOTOTUBE, PHOTOVOLTAIC CELL.

OPTOISOLATOR

See LUMISTOR, OPTICAL COUPLING.

ORBIT

An orbit is the path that one body follows around another body, or around the center of force between two bodies. Electrons follow orbits around the nuclei of atoms (*see* ELECTRON ORBIT). Satellites follow orbits around the earth; the earth orbits the sun, and the sun orbits the center of the Milky Way galaxy. In every case, the inward and outward forces exactly balance, resulting in a condition of equilibrium.

An orbit may be either circular or elliptical. If the orbit is a circle, the center of force is at the center of the circle. If the orbit is elliptical, the center of force is at one focus of the ellipse.

If the outward force exceeds the inward force for any reason, the orbit is broken and the path of the object becomes a parabola or hyperbola. The orbiting object then escapes from the influence of the central object.

Communications satellites are usually placed in special orbits around the earth, known as geostationary orbits. A satellite in a geostationary orbit always remains above the same point on the ground. This is because the orbital period is 23 hours and 56 minutes—exactly the same as the sidereal period of rotation of the planet. *See also* ACTIVE COMMUNICATIONS SATELLITE.

ORDER OF MAGNITUDE

The order of magnitude of a number is an expression of its relative size, on a logarithmic scale. Given a specific number, its order of magnitude is found by first determining its base-10 logarithm (*see* LOGARITHM). Then the resulting number is rounded off to the next lowest integer. (If the logarithm happens to be an exact integral value, it is left alone.)

For example, the order of magnitude of 3.355 is 0, since the base-10 logarithm is 0.5257, and truncating this number yields 0. The order of magnitude of 33.55 is 1; the order of magnitude of 335.5 is 2, and so on. Working down instead of up, the order of magnitude of 0.3355 is -1; the order of magnitude of 0.03355 is -2, and so on.

In scientific notation, the order of magnitude is the value of the exponent. Scientists often speak of one quantity as being "several orders of magnitude" smaller or larger than another. A quintillion, for example, is 12 orders of magnitude larger than a million, which is in turn 12 orders of magnitude larger than one millionth. *See also* SCIENTIFIC NOTATION.

OR GATE

An OR gate is a form of digital logic gate, with two or more inputs and one output. The OR gate performs the Boolean inclusive-OR function. That is, if all inputs are 0 (false), the output is 0; if any or all inputs are 1 (true), the output is 1.

The illustration shows the schematic symbol for an OR gate, along with a truth table for a two-input device.

OSCILLATION

Oscillation is any repeating, periodic effect, such as the swinging of a pendulum, the back-and-forth motion of a spring, or the continual back-and-forth movement of electrons in an electrical conductor. Oscillation may be short-lived, dying down with time, or it may continue indefinitely, if an outside source of energy is provided to keep it going.

In electronics, oscillation generally refers to a

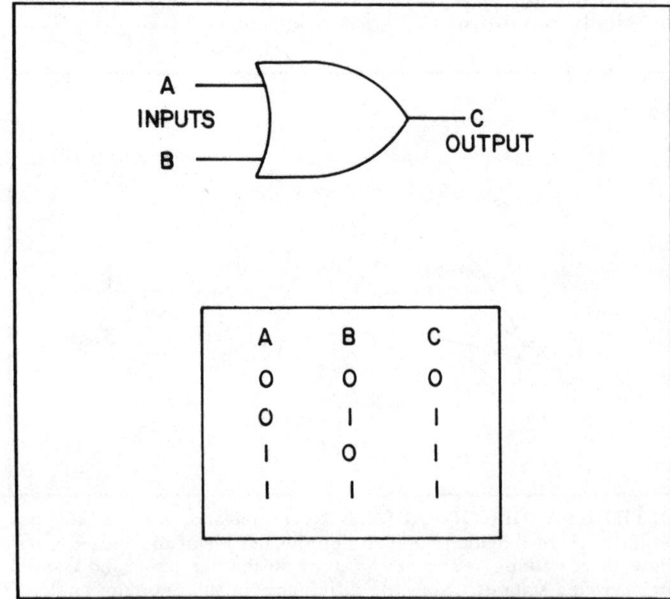

A	B	C
0	0	0
0	1	1
1	0	1
1	1	1

OR GATE: Schematic symbol for an OR gate, with truth table.

phenomenon caused by positive feedback (*see* FEEDBACK). Circuits can be made to oscillate deliberately at a precise frequency. Sometimes, oscillation occurs when it is not desired. This is likely to happen when an amplifier circuit has too much gain or is improperly designed for the application intended.

Oscillation is always the result of periodic storage and release of energy. In the pendulum, energy is stored as the weight rises, and is released as the weight falls. In the spring, energy is stored with compression and released with expansion. In an inductance-capacitance circuit, energy is alternately transferred between the inductive and capacitive reactances. *See also* OSCILLATOR.

OSCILLATOR

An oscillator is a circuit designed specifically to produce electric oscillation. All oscillators use the feedback principle. Although there are many different types of oscillator circuits, they all consist basically of an amplifier in which some of the output is applied, in phase, to the input.

An oscillator requires an active amplifying device, such as a transistor, field-effect transistor, or tube. An external source of power is also needed. If the oscillator is to produce alternating currents at a specific frequency, some form of resonant circuit, such as an inductance-capacitance network or a piezoelectric crystal, is necessary. Specialized devices, such as the Gunn diode and the Klystron tube, can produce oscillation because of negative-resistance effects (*see* NEGATIVE RESISTANCE, NEGATIVE TRANSCONDUCTANCE).

Specific oscillator circuits are shown under the following listings: ARMSTRONG OSCILLATOR, COLPITTS OSCILLATOR, CRYSTAL OSCILLATOR, GUNN DIODE, HARTLEY OSCILLATOR, PIERCE OSCILLATOR.

OSCILLATOR KEYING

Oscillator keying is a method of obtaining code transmission by interrupting the action of an oscillator circuit. Oscillator keying may be accomplished in various ways; the most common method is simply to break the cathode, emitter, or source circuit of the oscillator (see the schematic diagram).

Oscillator keying offers the advantage of completely shutting down the transmitter when the key is opened. This allows reception between code elements. However, oscillator keying results in chirp unless the circuit is carefully designed. *See also* CHIRP, KEYING.

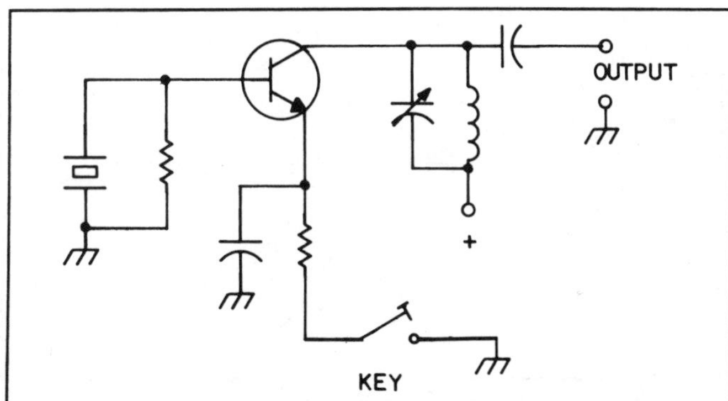

OSCILLATOR KEYING: An example of oscillator keying.

OSCILLOGRAPH

An oscillograph is a device similar to an oscilloscope (*see* OSCILLOSCOPE). Originally, an oscilloscope was called an oscillograph, but today the two terms have different meanings.

The oscillograph makes a permanent record of a waveform, while the oscilloscope simply displays the waveform on a cathode-ray tube.

For extremely low-frequency oscillations or periodic fluctuations in the magnitude of a quantity, the oscillograph may utilize a pen recorder (*see* PEN RECORDER). For higher frequencies, an oscillograph usually consists of a camera attached to an oscilloscope. The resulting photograph of the oscilloscope trace is called an oscillogram. *See also* OSCILLOSCOPE CAMERA.

OSCILLOSCOPE

An oscilloscope is an electronic instrument that displays the waveforms of electronic signals. Oscilloscopes are used extensively in the design, construction, alignment, and servicing of electronic equipment. A typical test-laboratory oscilloscope is shown in the photograph.

Oscilloscopes operate by causing an electron beam to sweep rapidly from left to right across the phosphor screen of a cathode-ray tube (*see* CATHODE-RAY TUBE). The sweep rate is variable over a wide range, and is specified in seconds, milliseconds, or microseconds per division. (A division is approximately 1 centimeter, as marked on the screen.)

A signal is applied to the vertical deflection plates of the cathode-ray tube, causing the electron beam to move up and down as the signal voltage varies. The vertical sensitivity of the oscilloscope is adjustable from a few volts per division down to a few millivolts or even microvolts per division. The result is a graphic display on the screen, showing the signal waveform, the amplitude, and the frequency.

Some oscilloscopes have input terminals connected to the horizontal deflection plates, as well as to the vertical deflection plates. This allows comparison of the relative phase and/or frequency of two signals, using patterns called Lissajous figures (*see* LISSAJOUS FIGURE).

Sophisticated oscilloscopes are available for specialized purposes. The dual-trace oscilloscope allows simultaneous observation of two waveforms side-by-side, for comparing frequency, phase, pulse duration, and other parameters. Persistence-type displays are available for "fixing" a waveform on the screen for several minutes. The maximum display frequency of an oscilloscope may be several hundred kilohertz up to several gigahertz, depending on the application.

Oscilloscopes are incorporated into various other test devices, such as the signal monitor and spectrum analyzer. *See also* MONITOR, OSCILLOGRAPH, SPECTRUM ANALYZER.

OSCILLOSCOPE CAMERA

An oscilloscope camera is a special camera designed for taking photographs of oscilloscope displays. This provides a permanent record of the display. Oscilloscope cameras are often used for comparing data obtained at different times in the design of electronic equipment.

An ordinary camera, mounted on a tripod, may be used

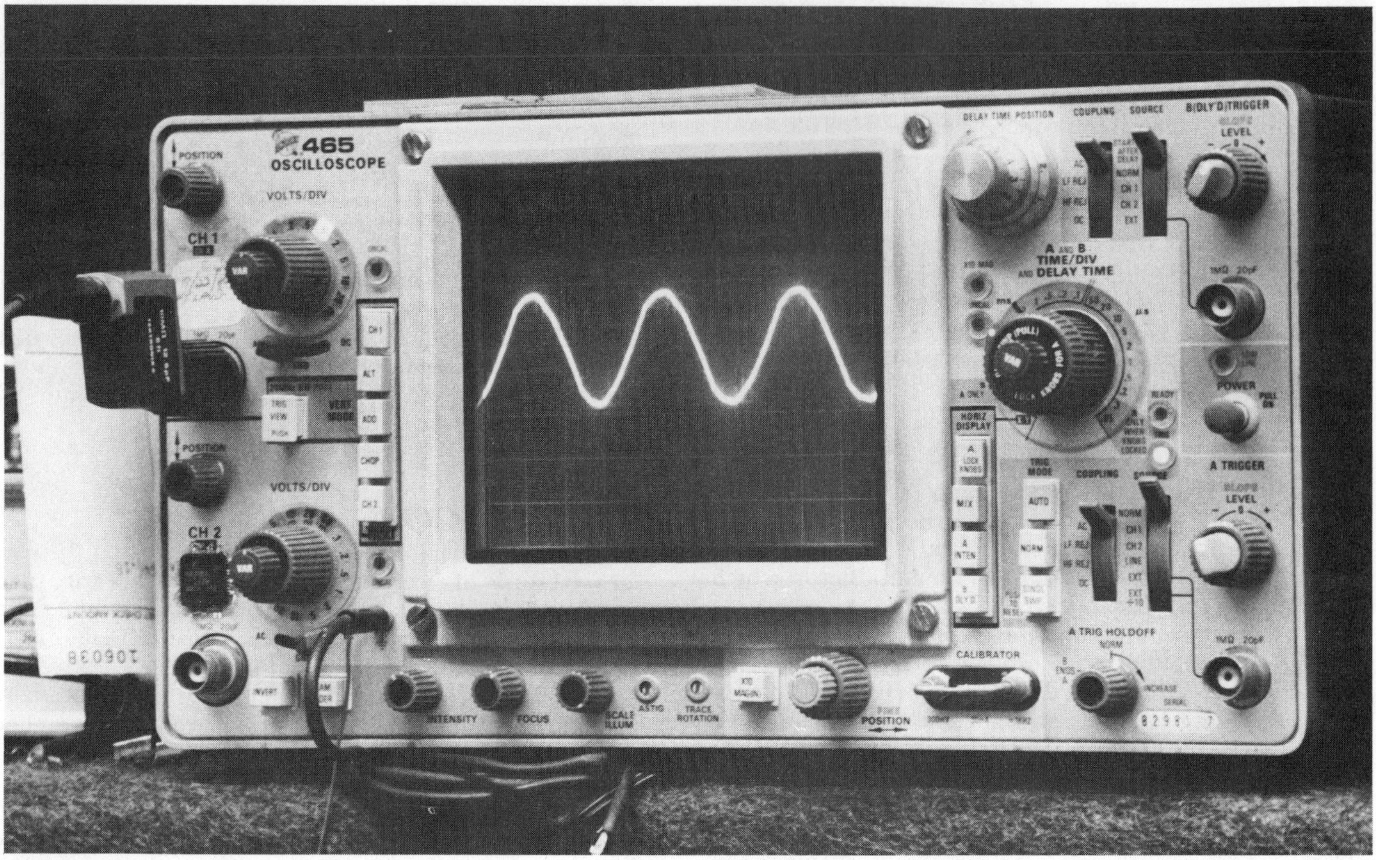

OSCILLOSCOPE: A typical test-laboratory oscilloscope.

as an oscilloscope camera. It is necessary that the shutter speed be slow enough to allow at least one complete trace across the oscilloscope screen. Otherwise, only part of the waveform will be shown in the resulting photograph.

Specially designed cameras are made for the purpose of photographing oscilloscope displays. Such a camera is fitted with an opaque hood for keeping out stray light. The hood is attached to the screen frame of the oscilloscope. Many oscilloscopes are provided with mounting screws or flanges around the screen, for the attachment of an oscilloscope camera. *See also* OSCILLOGRAPH, OSCILLOSCOPE.

OUNCE

The ounce is a unit of weight. In the gravitational field of the earth, an ounce is the amount of weight corresponding to a mass of 28.35 grams. An ounce is 1/16 (0.0625) pound.

The fluid ounce is a unit of measurement of volume. A fluid ounce is 1/16 (0.0625) pint, or 1/32 (0.03125) quart. A fluid ounce is equivalent to 0.02957 liter, or 29.57 cubic centimeters.

OUTDOOR ANTENNA

An outdoor antenna is any antenna that is placed outside a house or building. In general, outdoor antennas are preferable to indoor antennas, since there are fewer obstructions for the transmission or reception of signals in most outdoor locations as compared with most indoor locations.

An outdoor antenna is almost always much better, in terms of transmitting performance, than an indoor antenna. However, indoor antennas can sometimes be reasonably effective, particularly when the structure is a frame house without metal reinforcements or metal siding. Narrowband antennas, equipped with low-noise preamplifiers, can be successfully used for certain receiving purposes even when they are not located outdoors.

Indoor antennas usually perform poorly in steel-frame buildings, because the frame acts as a shield against radio-frequency energy. *See also* INDOOR ANTENNA.

OUT OF PHASE

Two signals are considered to be out of phase when their phase difference is 180 degrees. That is, they must be ½ cycle different in phase.

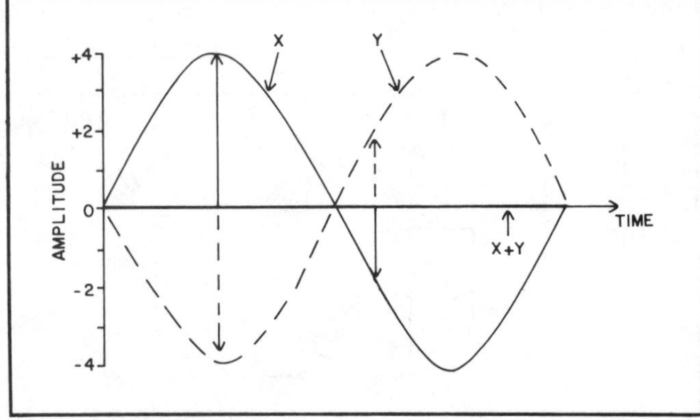

OUT OF PHASE: Sine waves of opposite phase and equal amplitude cancel, resulting in zero signal.

For certain waveforms, such as the sine wave and the square wave, two out-of-phase signals of equal amplitude totally cancel (see illustration). When this happens, the net signal is zero, because the instantaneous signal amplitudes are equal and opposite. *See also* PHASE, PHASE ANGLE.

OUTPUT

Output is the signal produced by a circuit, or processed by a circuit. An oscillator generates its own output. An amplifier circuit produces its output from a signal of lesser intensity, applied to the input. A circuit may have just one set of output terminals, or it may have several. The output terminals themselves are sometimes called outputs.

The output signal in a given situation must conform to certain requirements. The amplitude may have to be within specified limits. A load of a certain impedance may have to be connected to the output terminals of a circuit for proper operation. *See also* OUTPUT IMPEDANCE.

OUTPUT DEVICE

In a computer system, an output device consists of any peripheral equipment that transfers data out of the computer. Examples of output devices include a video display terminal, a printer, card-punching devices, and the modulator sections of modems. *See also* MODEM, PRINTER, TERMINAL, VIDEO DISPLAY TERMINAL.

In any electronic system, a circuit from which an output signal is obtained may be called an output device. Examples of such output devices are a power transformer, a signal transformer, and a coupling network. *See also* OUTPUT.

OUTPUT IMPEDANCE

The output impedance of a circuit is the load impedance which, when connected to the output terminals, results in optimum transfer of power. The output impedance of a circuit is normally a pure resistance. If a load contains reactance, that reactance must, for best operation, be tuned out by means of an equal and opposite reactance connected in series with the load.

The resistance of a load should always be matched to the output resistance of the driving circuit. Removing the reactance may, by itself, not be sufficient. A transformer is required for resistance matching. For example, if an audio amplifier is designed to work into an 8-ohm load, a step-up transformer must be used for best results with 16-ohm speakers, and a step-down transformer should be used with 4-ohm speakers.

When the output impedance of a device differs from the load impedance, some of the electromagnetic field is reflected back from the load toward the source. This may or may not cause significant deterioration in the performance of the driving circuit. In the above audio-amplifier example, the transformer can, in some cases, be removed with only a slight reduction in system efficiency. However, in many instances, the impedance match must be nearly perfect. *See also* IMPEDANCE, IMPEDANCE MATCHING, INPUT IMPEDANCE, REACTANCE.

OUTPUT RESISTANCE

See OUTPUT IMPEDANCE.

OVERCURRENT PROTECTION

Overcurrent protection is any means of preventing excessive current from flowing through a circuit. The most common forms of overcurrent-protection devices are the circuit breaker and the fuse (*see* CIRCUIT BREAKER, FUSE).

Some power supplies have built-in overcurrent protection. If the load resistance drops to a value that would cause excessive current to flow, the power supply automatically inserts an effective resistance in series with the load, preventing the current from rising above the maximum rated value. When the load resistance returns to normal, the power supply continues to deliver the rated current without the need for resetting a circuit breaker or replacing a blown fuse. *See also* CURRENT LIMITING, CURRENT REGULATION.

OVERDAMPING

In an analog meter, the damping is the rapidity with which the needle reaches the actual current reading. Critical damping is the smallest amount of damping that does not result in needle overshoot (*see* CRITICAL DAMPING, DAMPING). If the damping is greater than the critical value, a meter is said to be overdamped.

When a meter is overdamped, it responds slowly to changes in the current. This is desirable in some applications. For example, the value of a parameter may fluctuate rapidly to some extent, while the average value remains constant. An overdamped meter is preferable in such a case.

OVERDRIVE

Overdrive occurs when excessive input is applied to a

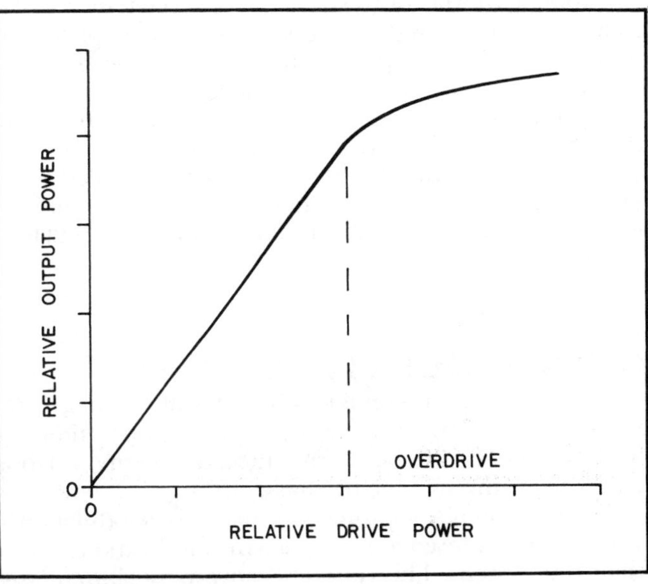

OVERDRIVE: Overdrive occurs as the output of an amplifier increases less rapidly with greater and greater input.

power amplifier. Overdrive is an undesirable condition, because it results in distortion of the envelope of a modulated signal. It also causes the generation of excessive harmonics. In some cases, overdrive can damage a transistor or tube.

In a properly operating power amplifier, the output power increases, up to a certain point, as the driving power is increased. Beyond this point, the output power increases more gradually. It is in this range that overdrive occurs (see illustration). *See also* DRIVE.

OVERFLOW

The capacity of any memory device is limited. When the amount of data is too great for storage in a memory, an overflow condition occurs. This can be observed in handheld calculators, minicomputers, and large computers.

Overflow can result in various responses from a calculator or computer. Generally, the excess data is ignored or compensated for in some way; or, an "error" signal may be generated. For example, a handheld calculator may have room for 10 digits. Keying in 3.6666666666 (an 11-digit number) would then result in a rounded-off display of 3.666666667, or perhaps a truncated display of 3.666666666. An attempt to divide by zero results, with most calculators, in an "error" signal.

Sophisticated computers will inform an operator if a memory overflow occurs. *See also* MEMORY.

OVERLOADING

Overloading is the result of an attempt to draw too much current from a source of power, or to apply excessive voltage to a load.

A familiar example of overloading occurs when the frying pan, the oven, the coffee pot, and an air conditioner are all turned on at once—and the fuse or circuit breaker opens the circuit! Another, hopefully less familiar, example is the inadvertent connection of a 117-volt appliance to a 234-volt power source.

Still another example of overloading is the improper adjustment of the output matching circuit in a radio transmitter, resulting in excessive collector or plate current in the final-amplifier stage, and reduced radio-frequency output power and efficiency.

In a radio receiver, an incoming signal may be so strong that the front end is driven into nonlinear operation or desensitization. This condition is sometimes called overloading. It results in false signals, intermodulation, or a general reduction in sensitivity. *See also* DESENSITIZATION, FRONT END.

OVERMODULATION

When a radio transmitter is modulated to an extent greater than that specified or required for normal operation, the transmitter is said to be overmodulated. Overmodulation can occur with any type of emission.

In an amplitude-modulated signal, overmodulation is generally considered to exist when the modulation is more than 100 percent. This causes negative-peak clipping (see A in the illustration). If especially severe, it may also cause flat-topping of the positive envelope peaks, as at B. The signal bandwidth becomes excessive, and the audio is distorted. *See* AMPLITUDE MODULATION, SINGLE SIDEBAND.

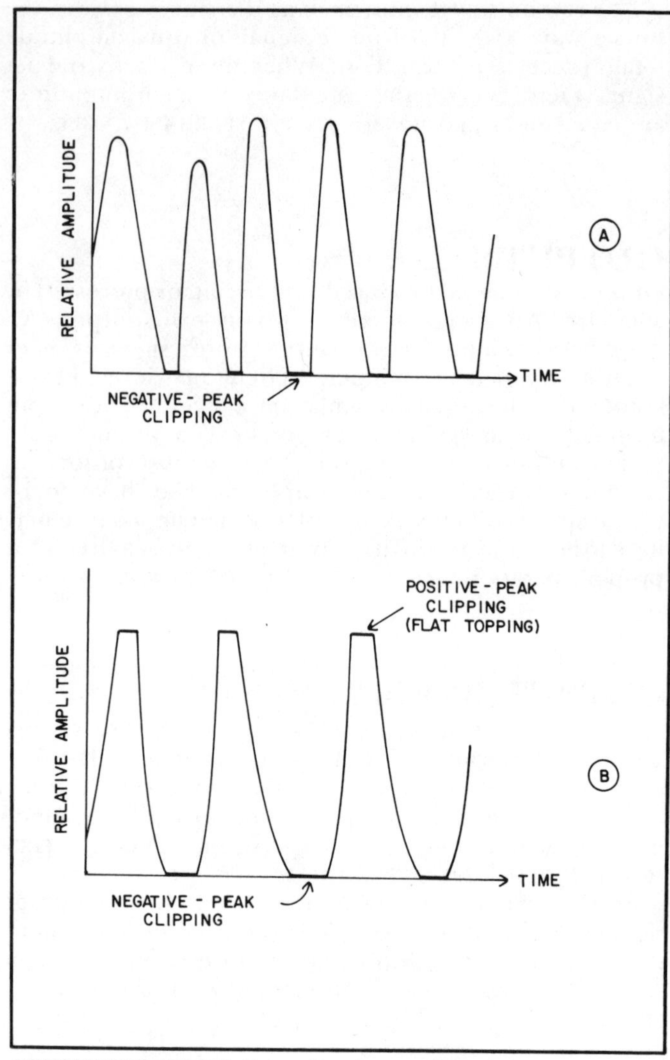

OVERMODULATION: Examples of overmodulation. At A, negative-peak clipping is shown. At B, clipping occurs on the positive as well as the negative envelope peaks.

In a continuous-wave signal, overmodulation can occur because of an excessively rapid rate of rise or decay in amplitude. The result is excessive bandwidth in the form of key clicks. *See* CONTINUOUS WAVE, KEY CLICK.

In a frequency-modulated signal, overmodulation exists when the deviation is greater than the rated system deviation. *See also* FREQUENCY MODULATION.

OVERSHOOT

In an analog metering device, insufficient damping results in a condition called overshoot (*see* CRITICAL DAMPING, DAMPING). When the current changes dramatically, the meter needle goes past the actual new reading, and then returns to the correct position. In severe cases of overshoot, the meter needle may oscillate back and forth several times before coming to rest at the proper position.

In a pulse waveform, a condition called overshoot often occurs in the rise or decay (see illustration). This may or may not affect the operation of a circuit. The overshoot can usually be prevented by lengthening the rise and/or decay time. *See also* DROOP.

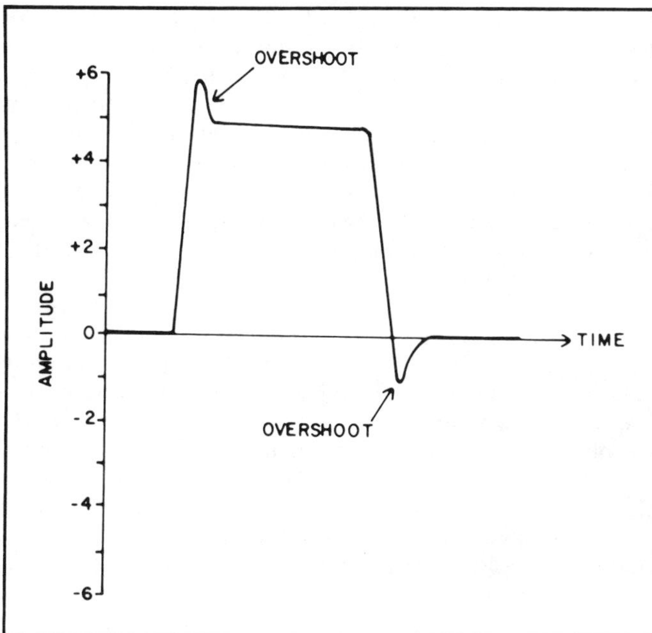

OVERSHOOT: Overshoot can occur on the rise and/or decay of a pulse.

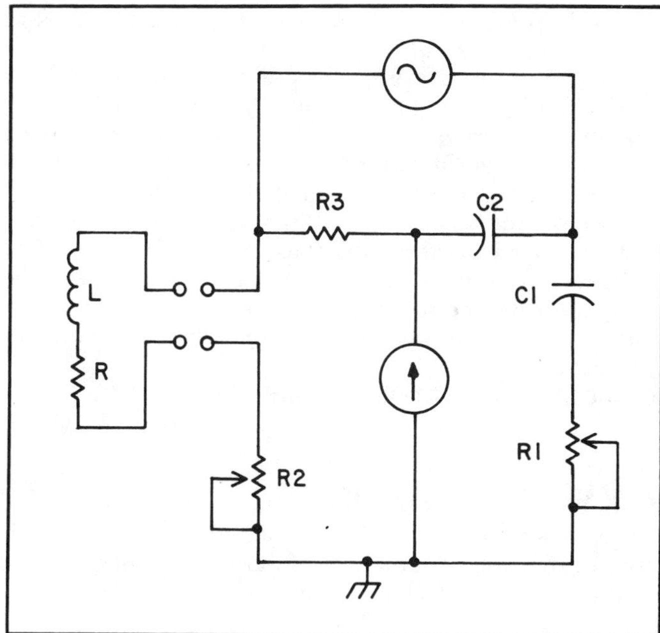

OWEN BRIDGE: The Owen bridge can measure inductances over a wide range. It can also determine the effective series resistance of a coil.

OVERVOLTAGE PROTECTION

In a regulated power supply, the regulation system may fail. If this happens, and the power supply does not have an overvoltage-protection circuit, the voltage will rise dramatically. In some cases it may increase to more than twice the normal value. This can cause damage to equipment connected to the supply.

Most overvoltage-protection circuits consist of a Zener diode for sensing the output voltage, and a silicon-controlled rectifier for short-circuiting the output terminals when overvoltage occurs. This causes the regulator circuit to switch off the supply, or the fuse may blow. A schematic diagram of this type of overvoltage-protection circuit is shown.

There are other methods of obtaining overvoltage protection besides that shown in the illustration. One common alternative method is called remote sensing. The remote-sensing circuit consists of a device that has a certain threshold voltage, shutting down the supply when the threshold is exceeded. A bipolar transistor can be used for this purpose. The threshold voltage depends on the base bias. Under normal conditions the transistor, which is connected in series with the supply output, conducts. If the voltage becomes excessive, the transistor opens the circuit. *See also* REGULATED POWER SUPPLY.

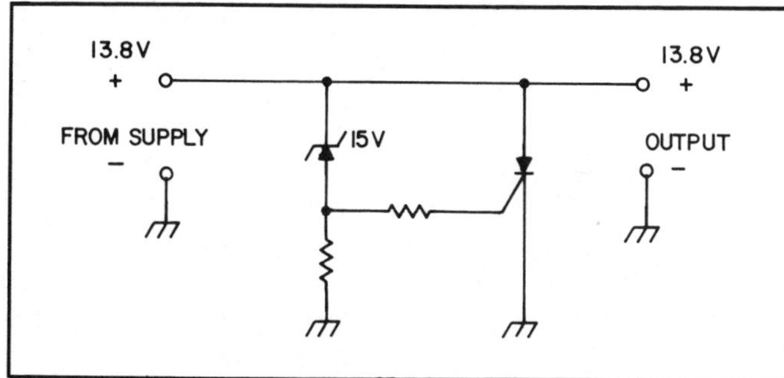

OVERVOLTAGE PROTECTION: An example of overvoltage protection. The zener diode conducts at 15 volts, causing the silicon-controlled rectifier to shut down the supply.

OWEN BRIDGE

The Owen bridge is a device for measuring unknown values of inductance. The Owen bridge is noted for its extremely wide range—it can determine any inductance from less than 0.1 nH to more than 100 H. This covers essentially all possible inductance values.

The illustration is a schematic diagram of an Owen bridge. The unknown inductance, L, has an effective series resistance R. Potentiometers R1 and R2 are adjusted for a null reading on the indicator. The generator produces an audio-frequency signal. The indicator may be a speaker, a headset, or a meter with a rectifier diode connected in series. Capacitors C1 and C2, and resistor R3, are fixed in value.

When a balance indication is obtained, the unknown inductance can be determined according to the formula:

$$L = C2R1R3$$

and the series resistance can be determined by the formula:

$$R = C2R3/C1 - R2$$

See also INDUCTANCE.

OXIDATION

Oxidation is a form of corrosion. A substance becomes oxidized when it combines with oxygen. Many different materials undergo oxidation when exposed to the air. The oxidation process is similar to combustion, but oxidation occurs more slowly.

Most metals will oxidize if exposed to the atmosphere for long periods. Some metals oxidize more rapidly than others. Iron, for example, rusts easily, and deteriorates much more quickly than chromium. For all materials, the oxidation process is hastened by high temperature and high humidity. This is why metals corrode so rapidly in tropical or subtropical regions. Electrical connections should be soldered and, if possible, insulated, if they are to be located outdoors.

Oxygen is not the only chemical that can cause corrosion. Chlorine is highly corrosive. Salt (sodium chloride) greatly exacerbates corrosion, because most metals react readily with chlorine. Marine environments are notorious for causing rapid deterioration of electrical and mechanical equipment because of the "salt air" (*see* CORROSION).

Oxidation is used to advantage in the manufacture of certain kinds of electronic components. An example is the copper-oxide rectifier; another is the electrolytic capacitor. An oxide coating on some materials provides protection against further corrosion. *See also* COPPER-OXIDE DIODE, ELECTROLYTIC CAPACITOR.

OXYGEN AND OZONE

Oxygen is an element, with atomic number 8 and atomic weight 16. Oxygen is the second most abundant gas in the atmosphere of our planet; the air is about 21 percent oxygen. Oxygen is necessary for the survival of living things, but in electronics, it is better known for its ability to cause corrosion of metals (*see* OXIDATION).

In the air, oxygen molecules tend to be grouped in pairs. However, sometimes oxygen molecules combine in groups of three. The result is called ozone. Ozone is formed by electric sparks. Ozone is also created when intense ultraviolet light passes through oxygen. Ozone has a characteristic metallic odor, somewhat like that of chlorine bleach.

In the upper atmosphere of the earth, much of the oxygen occurs in the form of ozone because of the intense ultraviolet rays from the sun. The ozone is, fortunately, relatively opaque to ultraviolet. The atmosphere thus creates its own protective layer for life on earth. If the ozone layer were to disappear, much life on our planet would perish because of the effects of short-wave ultraviolet light reaching the ground. In recent years, some scientists have expressed concern that the activities of mankind might reduce, or destroy, this ozone layer. *See also* ULTRAVIOLET.

P

PA

See POWER AMPLIFIER.

PACEMAKER

A pacemaker is an electronic device that regulates the heartbeat of a person whose heart cannot properly regulate itself. Pacemakers can be implanted in, or attached to, the body of a patient. Pacemakers operate from small batteries.

Pacemakers work by generating electrical impulses at regular intervals. The impulses are transmitted to the nerves that control the heart muscles. Each impulse causes the heart to contract. Without pacemakers, many heart patients would die; their hearts would beat irregularly or go into fibrillation. *See also* DEFIBRILLATION, HEART FIBRILLATION.

PAD

A pad is an attenuator network that displays no reactance, and exhibits constant input and output impedances. Pads are often used between radio transmitters and external linear amplifiers to regulate the amplifier drive. Pads are also used in some receiver front ends to reduce or prevent overloading in the presence of extremely strong signals. Such pads can be switched in or out of the front end as desired. Pads are sometimes used for impedance matching.

A typical attenuator pad configuration is shown in the illustration. The resistors are noninductive, and must be capable of dissipating the necessary amount of power. In a receiver front end, small resistors (¼-watt or ⅛-watt) are satisfactory. For external power-amplifier applications, the resistors may have to dissipate as much as 100 watts or more. *See also* ATTENUATOR.

PADDER CAPACITOR

A padder capacitor is a small, variable capacitor used for the purpose of adjusting the frequency of an oscillator circuit. The padder capacitor is placed in parallel with the tuning capacitor in the tank circuit of the oscillator (see A). The value of the padder capacitor is much smaller than that of the main tuning capacitor. Most padders have a maximum capacitance of only a few picofarads. The padder allows precise calibration of the tuning dial or frequency readout of a receiver or transmitter.

When a ganged variable capacitor is used to tune more than one resonant circuit at a time, padder capacitors may be placed across each section of the ganged capacitor. The padders are adjusted until proper tracking is obtained (*see* GANG CAPACITOR, TRACKING). Some gang capacitors have built-in padders.

A typical padder capacitor is shown at B. The adjustment is made with a plastic device resembling a miniature screwdriver. The adjusting tool must have an insulating shaft to prevent detuning from hand-capacitance effects.

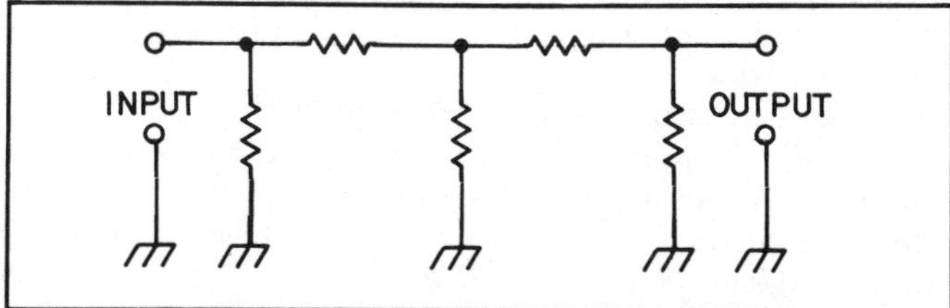

PAD: An attenuator pad. This pad is for unbalanced lines. The resistance values depend on the characteristic impedance of the line in which the pad is installed.

PADDING

Padding is the process of precisely adjusting a circuit by means of small, series or parallel components. An example is the use of a padder capacitor for adjustment of oscillator frequency (*see* PADDER CAPACITOR).

In computer practice, padding is a process in which "blanks" or other meaningless data characters are added to a file to make the file a certain standard size. For example, the meaningful data in a computer file may comprise 12 kilobytes, but the standard file size might be 16 kilobytes. The file is padded by adding 4 kilobytes of "blanks." The "blanks" can be added anywhere in the file; the most common location is at the beginning or end.

When an attenuator pad is inserted into a circuit, the process is sometimes called padding (*see* PAD).

PAGE PRINTER

See PRINTER.

PANEL

See CONTROL PANEL, JACK PANEL.

PANORAMIC RECEIVER

A panoramic receiver is a device that allows continuous visual monitoring of a specified band of frequencies. Most receivers can be adapted for panoramic reception by connecting a specialized form of spectrum analyzer into the first intermediate-frequency chain, just after the first (see A in illustration).

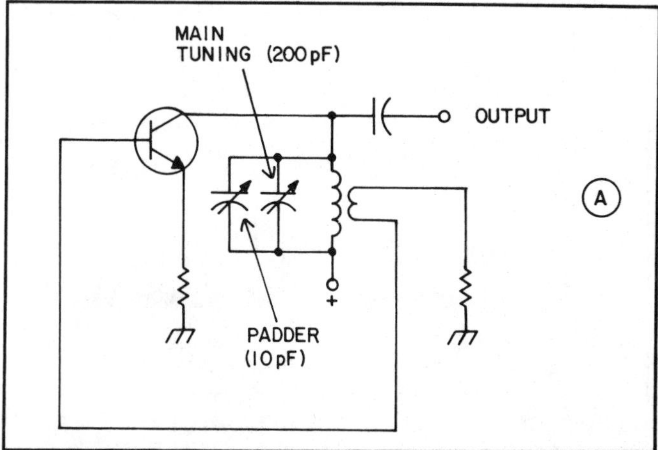

PADDER CAPACITOR: At A, schematic diagram of an oscillator with a padder capacitor in the tuning network.

(B)

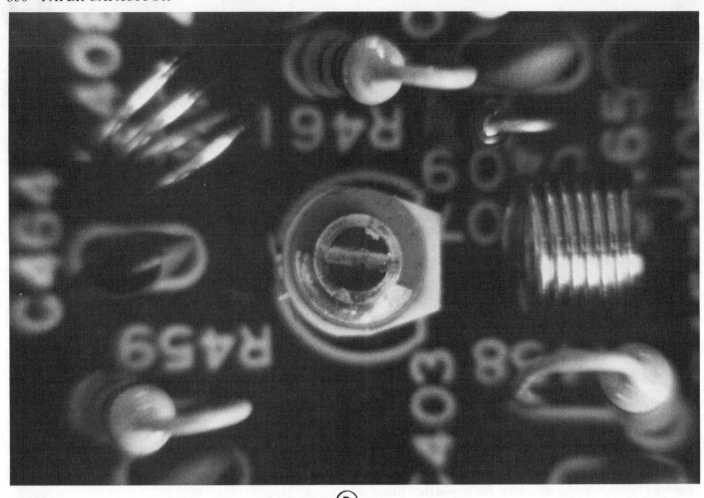

PADDER CAPACITOR: At B, a photograph of a padder capacitor (shown at center).

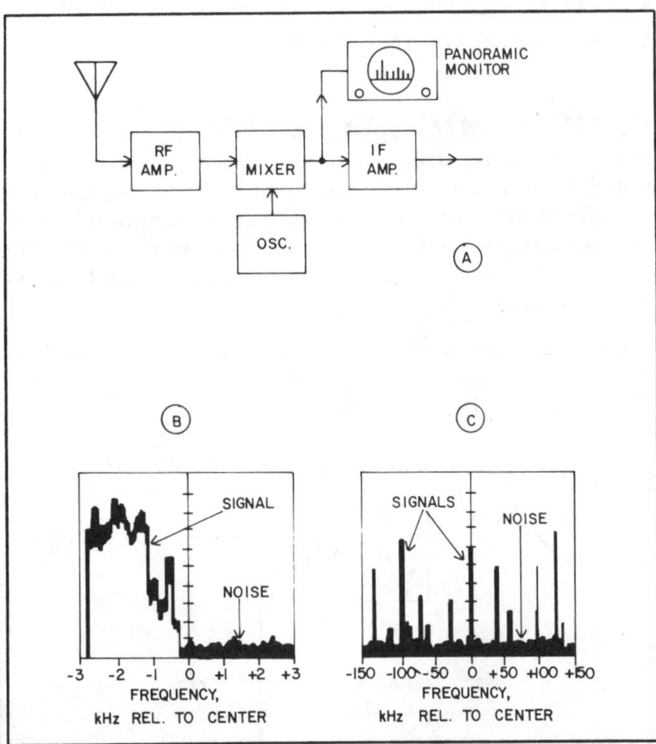

PANORAMIC RECEIVER: At A, a block diagram shows the installation of a panoramic monitor in a typical superheterodyne receiver. At B, a narrowband panoramic display, showing a single'sideband signal. At C, a wideband panoramic display.

A monitor screen displays the signals as vertical pips along a horizontal axis. The signal amplitude is indicated by the height of the pip. The position of the pip along the horizontal axis indicates its frequency. The frequency at which the receiver is tuned appears at the center of the horizontal scale. The number of kilohertz per horizontal division can be set for spectral analysis of a single signal or a narrow frequency band (B) or the scale may be set for observation of a wide range of frequencies (C). The maximum possible range is limited by the selectivity characteristics of the radio-frequency receiver stages. *See also* SPECTRUM ANALYZER.

PAPER CAPACITOR

A paper capacitor is a capacitor that uses a thin film of paper as the dielectric material. The paper is soaked with wax or some other water-resistant material. Paper capacitors are usually cylindrical in shape. The paper is placed between strips of foil, and the entire assembly is rolled up to form the capacitor (see illustration). Values range from about 1000 pF to 0.5 μF. Paper capacitors are non-polarized.

Paper capacitors were once common from the very-low to the high frequencies. Paper capacitors are not generally suitable for use at very high frequencies and above, because paper dielectrics become lossy at these frequencies. Paper capacitors were used at low to moderate voltage levels. They are rarely employed today. *See also* CAPACITOR.

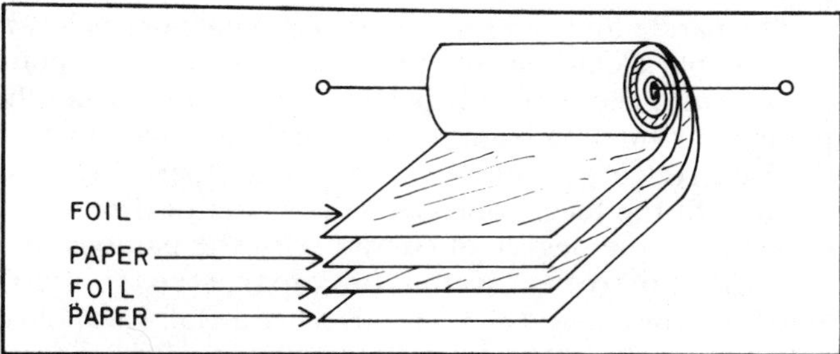

PAPER CAPACITOR: Construction of a paper capacitor.

PAPER TAPE

Paper tape is a permanent data-storage medium. Paper tape has been largely replaced by electronic memory devices in modern equipment, but occasionally it can still be found.

Each character is punched into a paper tape as a set of small holes in a row perpendicular to the edges of the tape. This is done by a device called a reperforator. The physical arrangement in which the holes are punched is shown at A in the illustration. The eight-channel hole code, according to Electronic Industries Association (EIA) Standard RS-244, is shown for the basic symbols A through Z, 0 through 9, and various punctuation and function characters. Holes correspond to the high (or 1) condition; blanks correspond to the low (or 0) condition. The eight-level code is now more commonly used than the old five-level code (B).

Paper tape can be read mechanically, at a rate of approximately 60 characters per second. Electronic devices, employing optical coupling, can read the tape much faster. Paper tape has the advantage of being a permanent form of memory. It cannot be erased by accidental removal of power or exposure to magnetic fields. However, paper tape tends to be bulky for storage of large amounts of data. The disk or magnetic-tape storage media are generally used for large files. *See also* DISKETTE, FLOPPY DISK, MAGNETIC TAPE.

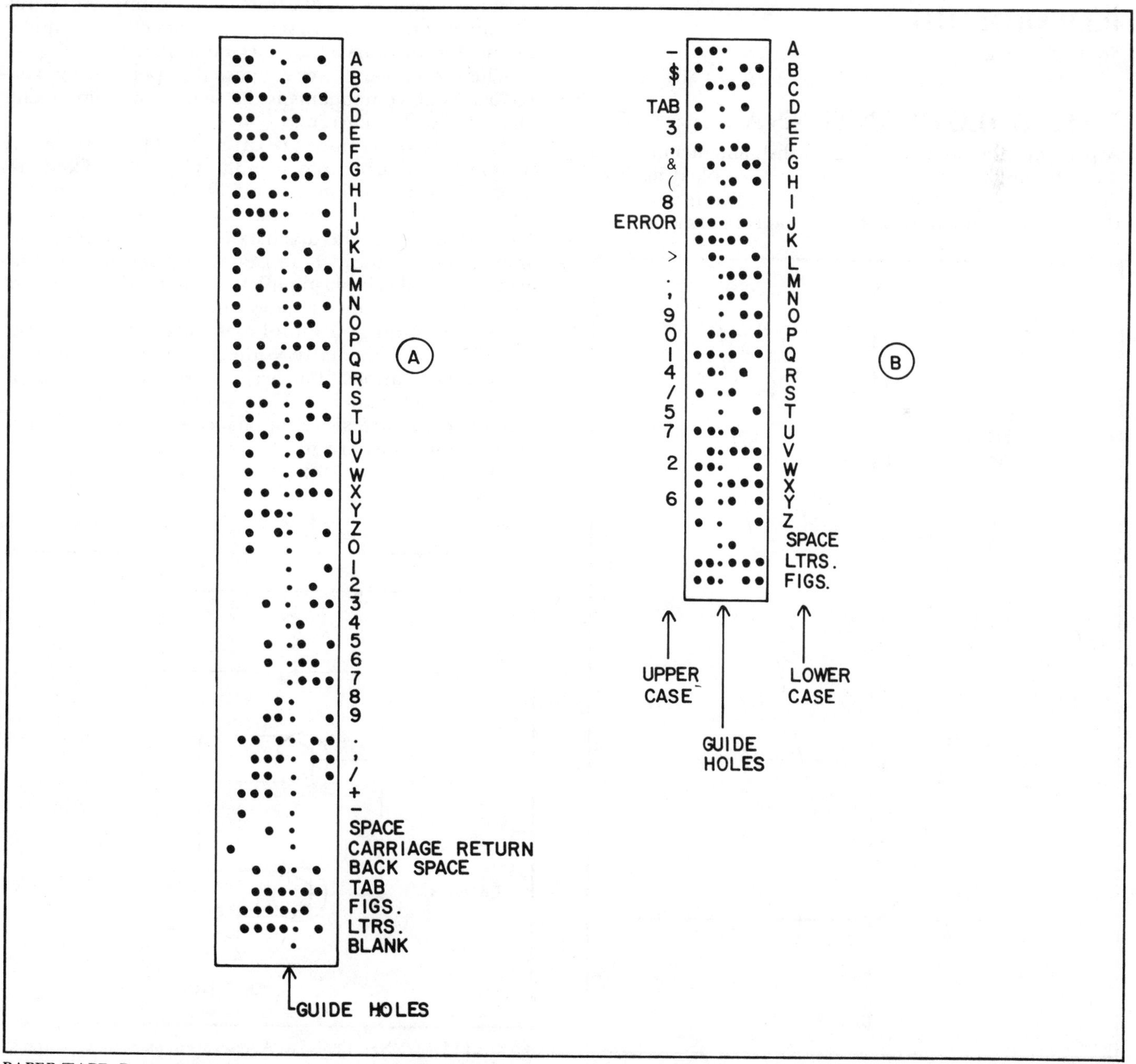

PAPER TAPE: Paper-tape codes. At A, standard eight-level code; at B, standard five-level code.

PARABOLA

A parabola is a conic section in two dimensions. The parabola, in the Cartesian system of coordinates, is a function of the form:

$$y = ax^2 + bx + c$$

where a, b, and c are constants, x is the independent variable, and y is the dependent variable. Any parabolic shape, in terms of constant units, can be defined by selecting appropriate values for a, b, and c. The drawing illustrates the simple parabola $y = x^2$.

If a parabola is rotated about its axis, the result, in three dimensions, is called a paraboloid. In the example shown, the axis of the parabola is the y axis. Dish antennas are often of paraboloidal form. The paraboloid acts as a collimator for electromagnetic waves. *See also* DISH ANTENNA, PARABOLOID ANTENNA.

PARABOLOID

See PARABOLA.

PARABOLOID ANTENNA

A paraboloid antenna is a form of dish antenna in which the reflecting surface is a geometric paraboloid in three dimensions. A paraboloid is the surface that results from the rotation of a parabola about its axis (*see* PARABOLA).

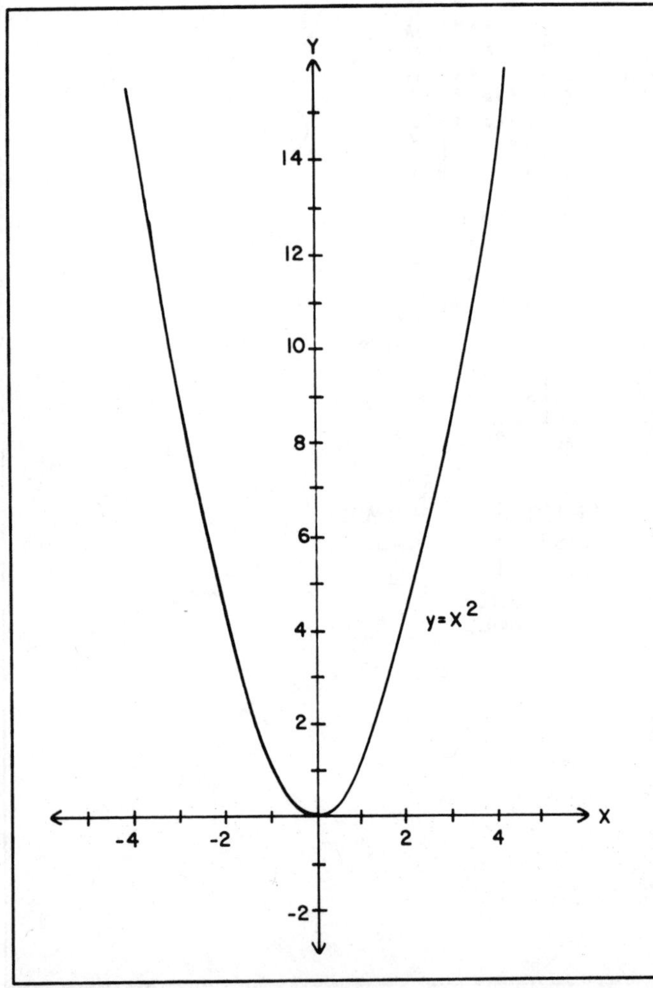

PARABOLA: An example of a parabola in Cartesian coordinates.

The paraboloid antenna has a focal point, at which rays arriving parallel to the antenna axis converge. This point is, in theory, geometrically perfect. This distinguishes the paraboloid antenna from the spherical antenna, in which the rays converge almost, but not quite, perfectly. The distance of the focal point from the center of the surface depends on the degree of curvature of the paraboloid.

Paraboloid reflectors may be constructed from wire mesh, or they may be made of sheet metal. Paraboloid antennas are generally used at ultra-high frequencies and above. Wire mesh is satisfactory at the longer wavelengths, but at very short wavelengths, sheet metal is preferable. The driven element is located at the focal point of the paraboloid. *See also* DISH ANTENNA, SPHERICAL ANTENNA).

PARALLEL CONNECTION

Components are said to be connected in parallel when corresponding leads of each component are connected together. The drawing illustrates some examples of parallel connections; at A, resistors are connected in parallel, and at B, transistors are connected in parallel.

When components are connected in parallel, the voltage across all components is the same. The components share the load of current.

Parallel connections are often used to increase the current-handling or power-handling capability of a circuit. A parallel combination of ten 500-ohm, 1-watt resistors can handle 10 watts as a 50-ohm dummy load, for example, which is ten times the dissipation rating for a single 50-ohm, 1-watt resistor. Transistors or tubes may be connected in parallel in an amplifier circuit, for the purpose of increasing the obtainable power output.

In a common household electric circuit, all of the appliances are normally connected in parallel in each branch. The various branches, in turn, are connected in parallel at the circuit box. The parallel connection assures that the voltage across each appliance will not change when other appliances are used. *See also* PARALLEL LEADS, PARALLEL TRANSISTORS, PARALLEL TUBES, SERIES CONNECTION.

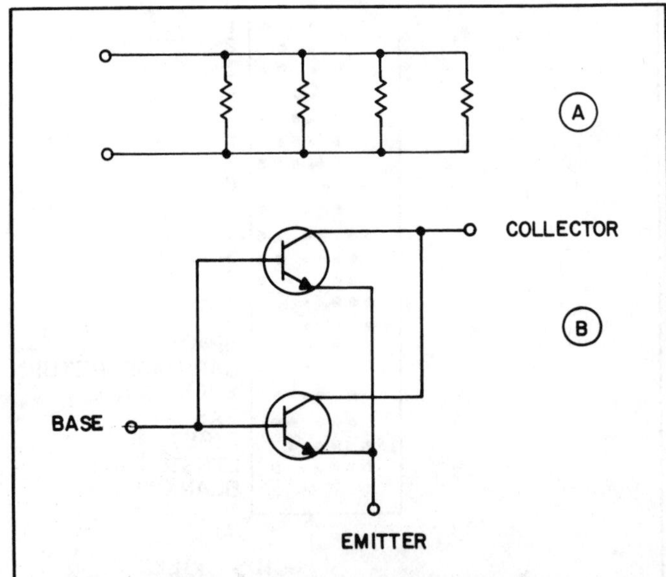

PARALLEL CONNECTION: At A, resistors in parallel. At B, transistors in parallel.

PARALLEL DATA TRANSFER

Information may be transmitted bit by bit, along a single line, or it may be transmitted simultaneously along two or more lines. The first method is known as serial data transfer (*see* SERIAL DATA TRANSFER). The latter method is called parallel data transer. This term arises from the fact that the transmission lines each carry portions of the same data.

In computer practice, parallel data transfer refers to the transmission of all bits in a word at the same time, over individual parallel lines; however, different words are generally sent one after the other.

Parallel data transfer has the advantage of being more rapid than serial transfer. However, more lines are required, in proportion to the factor by which the speed is increased. It takes ten lines, for example, to cut the data transmission time from a serial value of 60 seconds to a parallel value of 6 seconds.

PARALLEL LEADS

When two or more wire leads are used together in a circuit, with their ends joined, the leads are said to be parallel. Parallel leads are sometimes employed in radio-frequency circuits to reduce the inductance per unit length. Parallel leads also improve the conductivity at radio frequencies over that of a single conductor, by reducing losses caused by skin effect *See also* LITZ WIRE, SKIN EFFECT.

PARALLEL RESONANCE

When an inductor and capacitor are connected in parallel, the combination will exhibit resonance at a certain frequency (*see* RESONANCE, RESONANT FREQUENCY). This condition occurs when the inductive and capacitive reactances are equal and opposite, thereby cancelling each other and leaving a pure resistance.

The resistive impedance that appears across a parallel-resonant circuit is extremely high. In theory, assuming zero loss in the inductor and capacitor, the impedance at resonance is infinite. In practice, since there is always some loss in the components, the impedance is finite, but it may be on the order of hundreds of kilohms. The exact value depends on the Q factor, which in turn depends on the resistive losses in the components and interconnecting wiring (*see* IMPEDANCE, Q FACTOR). Below the resonant frequency, the circuit appears capacitively reactive; above the resonant frequency it is inductively reactive (see graph). Resistance is also present under non resonant conditions, making the impedance complex.

Parallel-resonant circuits are commonly used in the output configurations of radio-frequency oscillators and amplifiers. They may also be seen in some input circuits and coupling-transformer arrangements.

A section of transmission line may be used as a parallel-resonant circuit. A quarter-wave section, short-circuited at the far end, behaves like a parallel, and resonant, inductance and capacitance. A half-wave section, open at the far end, also exhibits this property (*see* LINE TRAP). Such resonant circuits can be used in place of inductance-capacitance (LC) circuits. This is often done at very-high and ultra-high frequencies, where the lengths of quarter-wave or half-wave sections are reasonable.

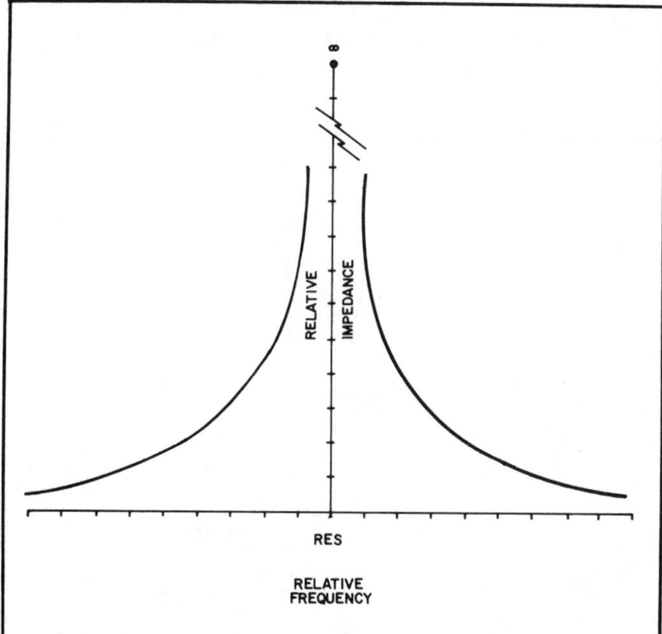

PARALLEL RESONANCE: A parallel-resonant circuit displays a pure resistance of a very high value.

PARALLEL TRANSISTORS

Two or more bipolar or field-effect transistors may be connected in parallel to increase the power-handling capability. All of the emitters or sources, bases or gates, and collectors or drains, are connected together. All of the transistors are of identical manufacture.

Ideally, the power-handling capability of a parallel combination of n transistors is n times that of a single transistor of the same kind. When two or three transistors are connected in parallel, this is essentially the case. However, in practice, certain limitations exist; it is not feasible to tie many transistors together, expecting an effective high-power amplifier to be the result. It is always better, if possible, to use one high-power transistor rather than serveral small ones.

The main reason that massive parallel combinations of bipolar transistors do not work well is that the input impedance is reduced in such a configuration, and most bipolar power transistors exhibit low imput impedances to begin with. If ten transistors are connected in parallel, for example, the input impedance becomes just one-tenth of the

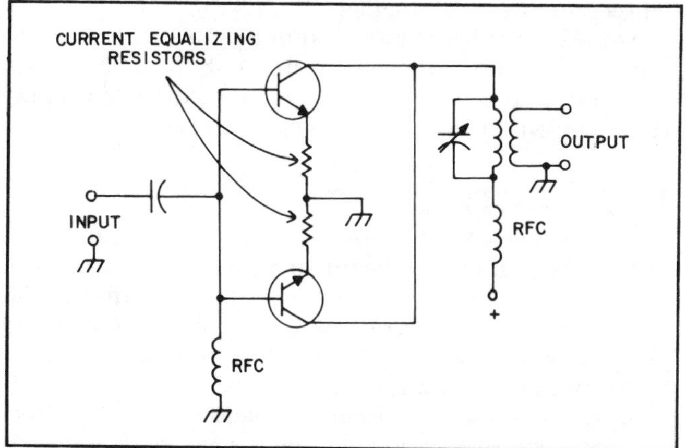

PARALLEL TRANSISTORS: Practical amplifier circuit using parallel transistors. The emitter resistors prevent current hogging.

value for a single transistor. This makes it necessary to use a cumbersome and lossy input transformer. The input capacitance also increases in a parallel combination, as compared with a single unit.

In any parallel combination of transistors, it is advantageous to place small resistors in the emitter or source leads of each of the units (see illustration). This prevents the phenomenon called current hogging, in which one transistor does all the work while the other one does practically no work (*see* CURRENT HOGGING).

Parallel combinations of two or three transistors can be used with excellent results in a power amplifier, if care is exercised in the design of the circuit. Many well-designed, commercially manufactured audio-frequency and radio-frequency power amplifiers use parallel-transistor circuits. *See also* FIELD-EFFECT TRANSISTOR, PARALLEL TUBES, POWER AMPLIFIER, TRANSISTOR.

PARALLEL TUBES

Vacuum tubes may be connected in parallel for the purpose of power amplification. The cathodes, grids, and plates are all tied together. The resulting circuit is analogous to a parallel-transistor amplifier (*see* PARALLEL TRANSISTORS).

Parallel-tube amplifiers enjoy the same advantages, and suffer from the same difficulties, as parallel-transistor amplifiers. Since tubes normally exhibit a higher impedance than transistors, it is sometime possible to use six or eight tubes in parallel. Larger combinations, however, result in design difficulties.

Parallel-tube amplifiers are used in some broadcast stations, and in some medium-power communications systems such as amateur-radio transmitters, with excellent performance and reliability. *See also* TUBE.

PARALLEL-WIRE LINE

See OPEN-WIRE LINE.

PARAMAGNETIC MATERIAL

A substance with a magnetic permeability greater than 1, but less than that of a ferromagnetic material, is called paramagnetic. Paramagnetic materials tend to maintain the same permeability even when the intensity of the magnetic field fluctuates. This is not characteristic of the ferromagnetics. Both ferromagnetic and paramagnetic materials exhibit changes in permeability with changes in temperature and frequency.

Some examples of paramagnetic materials are compounds of aluminum, beryllium, cobalt, magnesium, manganese, and nickel. *See also* DIAMAGNETIC MATERIAL, FERROMAGNETIC MATERIAL, PERMEABILITY.

PARAMETER

The term parameter is used to describe any factor that influences some other factor or factors. The term is, in some cases, used as a synonym for the term variable. Examples of parameters in electronics include such things as radiation resistance, the value of the feedback resistor in an operational-amplifier circuit, and the value of the capacitor in a resonant circuit. Radiation resistance affects the operation of a radio antenna; the feedback resistance affects the gain of an operational amplifier; the capacitance in a resonant circuit affects the resonance frequency. These are just a few of many possible examples of parameters.

In mathematics, a parameter is a variable that affects or relates two or more other variables. Time is often specified as a parameter in physics or mathematics. We might have a set of equations such as

$$x = 3t + 4$$
$$y = -6t - 6$$
$$z = 2t - 2$$

The parameter is t, and the values of x, y and z depend on t in this case. Sets of equations are often put into parametric form for convenience. *See also* EQUATION, FUNCTION, RELATION.

PARAMETRIC AMPLIFIER

A parametric amplifier is a form of radio-frequency amplifier that operates from a high-frequency alternating-current source, rather than the usual direct-current source. Some characteristic of the circuit, such as reactance or impedance, is made to vary with time at the power-supply frequency. Parametric amplification is commonly used with electron-beam devices at microwave frequencies. The traveling-wave tube is an example of such a device (*see* TRAVELING-WAVE TUBE).

Parametric amplifiers are characterized by their low noise figures, which may, in some instances, be less than 3 dB. The gain is about the same as that of other types of amplifiers; typically it ranges from 15 to 20 dB.

PARAMETRIC EQUATIONS

See PARAMETER.

PARAPHASE INVERTER

A paraphase inverter is a form of circuit that provides two outputs, differing in phase by 180 degrees. A transistor, field-effect transistor, or tube may be employed.

The illustration shows an example of a paraphase inverter using a bipolar transistor. Two output connections are provided; one is at the collector and the other is at the emitter. The input is applied to the base.

The emitter output signal is in phase with the input signal, but the collector output signal is 180 degrees out of

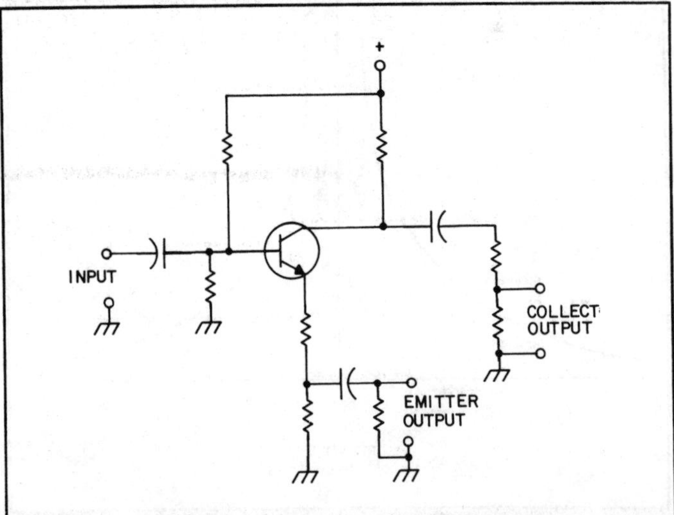

PARAPHASE INVERTER: A paraphase inverter using a bipolar transistor.

phase with the input signal. Since the collector output impedance is normally higher than the emitter output impedance, a voltage-dividing network may be used in the collector output if it is necessary for the two outputs to show equal impedances. If this is done, the circuit has gain of less than unity for both signals, and the output impedances are lower than the input impedance.

The paraphase inverter offers a simple and inexpensive way to obtain two signals in opposing phase.

PARASITIC ARRAY

An antenna with one or more parasitic elements is called a parasitic array. Parasitic arrays are commonly used at the high, very-high, and ultra-high frequencies for obtaining antenna power gain.

Common examples of parasitic arrays are the quad and the Yagi. Other configurations also exist. *See also* PARASITIC ELEMENT, QUAD ANTENNA, YAGI ANTENNA.

PARASITIC ELEMENT

In an antenna, a parasitic element is an element that is not directly connected to the feed line. Parasitic elements are used for the purpose of obtaining directional power gain. Generally, parasitic elements can be classified as either directors or reflectors. Directors and reflectors work in opposite ways.

Parasitic elements operate by electromagnetic coupling to the driven element. The principle of parasitic-element operation was first discovered by a Japanese engineer named Yagi; the Yagi antenna is named after him. Yagi found that elements parallel to a radiating element, at a specific distance from it, and of a certain length, caused the radiation pattern to show gain in one direction and loss in the opposite direction. Nowadays, Yagi's principle is used at high frequencies by amateur and commercial radio operators.

At high frequencies, parasitic elements are often used in directional antennas. The most common of these are the quad and the Yagi. Such antennas are known as parasitic arrays. They exhibit a unidirectional pattern. *See also* DIRECTOR, QUAD ANTENNA, REFLECTOR, YAGI ANTENNA.

PARASITIC OSCILLATION

In a radio-frequency power amplifier, oscillation sometimes takes place. This oscillation is undesirable, and it may occur at frequencies much different from the actual operating frequency. Such oscillations are called parasitic oscillations, or simply parasitics.

Parasitics can be eliminated by providing a certain amount of negative feedback in a power amplifier. This is accomplished by a neutralization circuit. A parasitic suppressor may also be used to choke off parasitics in some instances. *See also* NEUTRALIZATION, PARASITIC SUPPRESSOR.

PARASITIC SUPPRESSOR

A radio-frequency amplifier sometimes oscillates at a frequency far removed from the operating frequency. Such oscillation is called parasitic oscillation (*see* PARASITIC OSCILLATION).

When parasitic oscillations are at a frequency much higher or lower than the operating frequency, the unwanted oscillation can sometimes be choked off by means of parasitic suppressors. This method is especially common for eliminating parasitics in medium-frequency and high-frequency power amplifiers. The parasitic choke consists of a resistor and a small coil connected in parallel, and placed in series with the collector, drain, or plate lead of the amplifier (see illustration). The resistor is typically 50 to 150 ohms; it is a noninductive carbon type. The coil consists of three to five turns of wire, wound on the resistor. In an amplifier having two or more transistors or tubes in parallel, suppressors are installed at each device individually.

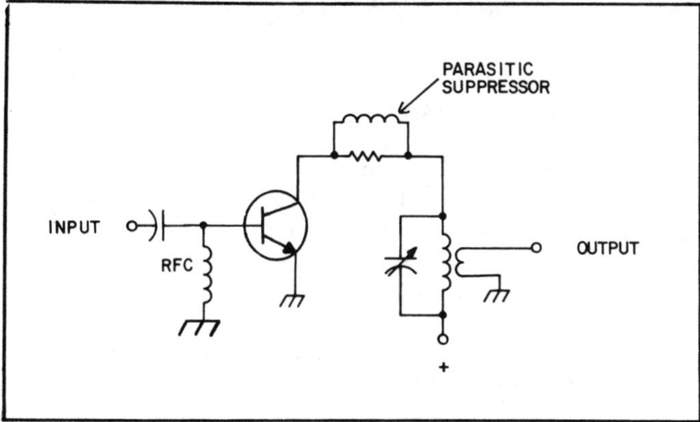

PARASITIC SUPPRESSOR: A parasitic suppressor can be used to choke off unwanted very-high-frequency oscillations in a power amplifier.

Low-frequency parasitics are not affected by the type of suppressor shown. To eliminate low-frequency parastics in a radio-frequency power amplifier, neutralization is usually necessary. *See also* NEUTRALIZATION.

PARITY

Parity is an expression that indicates whether the sum of the binary digits in a code word is even or odd. Accordingly, if the sum is even, the parity is called even; if the sum is odd, the parity is called odd. In certain situations, it is necessary to have the digits in all code words add up to an even number; in other cases the sum must always be odd.

Suppose, for example, that a code consists of five-digit words. Some examples might be 01001, 11000, and 11101 if the parity is even; in the case of odd parity, some code words might be 10000, 10101, and 00111.

If the sum of the digits in a word is incorrect (odd when it should be even, or vice-versa), an extra bit may be added to the word to correct the discrepancy. Some codes have an extra bit that may be 0 (if the parity of the word is correct) or 1 (if not).

In the above example, we might add a sixth digit at the end of each word as a parity bit. Suppose that the parity is even. Then the first three words, given above, would have a 0 added at the end, and they would become, respectively, 010010, 110000, and 111010. The second three words, however, would require the addition of a 1, becoming 100001, 101011, and 001111. With the addition of the parity bit, the sum of the digits in all of the words is made to be even.

In a data-processing or transmission system, a parity check is sometimes made as a test for accuracy. If the parity is wrong, the state of the parity bit can be reversed, or the receiver may "ask" the transmitter to repeat the word.

PARTIAL DERIVATIVE

A partial derivative is a mathematical derivative in a function of two or more variables. A partial derivative is determined with respect to one of the independent variables of the function; the rest of the independent variables are treated as constants. The partial derivatives are generally different with respect to different variables. For example, suppose we are given the function:

$$f(x,y,z) = x + y^2 + z^3$$

We determine the partial derivative with respect to x (written $\partial f/\partial x$ by holding y and z constant, obtaining:

$$\partial f/\partial x = 1$$

The partial derivative with respect to y is determined by holding x and z constant:

$$\partial f/\partial y = 2y$$

and the partial derivative with respect to z is found by treating x and y as constants:

$$\partial f/\partial z = 3z^2$$

Partial derivatives are used to evaluate the rates of change of different parameters in mathematically complicated situations. *See also* DERIVATIVE, FUNCTION.

PARTICLE ACCELERATOR

A particle accelerator is a device that is used for splitting atomic nuclei. Charged particles, such as protons, electrons, and helium nuclei (alpha particles), can be made to move at high speeds by powerful electric fields. When such particles, which have extreme density, strike the nuclei of other atoms, fission may occur (*see* NUCLEAR FISSION), resulting in new elements.

Some particle accelerators have a straight-line configuration. Such devices consist of a long tube, in which atomic particles are accelerated by electric fields. The tube may be hundreds or even thousands of meters in length. Such a device is called a linear accelerator.

Other particle accelerators direct the atomic nuclei in circular paths. The most common such accelerator is the cyclotron (*see* CYCLOTRON). Both the linear accelerator and the cyclotron cause charged particles to move so fast that, when they strike atomic nuclei, the nuclei are literally blasted to pieces. Thus new elements and isotopes are formed.

Particle accelerators are used by experimental physicists for numerous purposes, including the fabrication of various isotopes of one element, or the conversion of an element to another element. *See also* ATOM, ELEMENT, ISOTOPE.

PARTICLE THEORY OF LIGHT

Visible light, and all electromagnetic radiation, exhibits certain properties that suggest it consists of a barrage of particles. Isaac Newton, the famous English physicist of the seventeenth century, first formally proposed that light might consist of particles or corpuscles. Such particles are also thought to exist by modern scientists. They are called photons (*see* PHOTON).

Light occurs in discrete packets of miniscule size. Although these packets cannot be seen, they have measurable mass, and they exert pressure on physical objects. The particles travel at the recognized speed of light—approximately 186,282 miles per second, or 299,792 kilometers per second.

Although light, and other electromagnetic radiation including radio waves, infrared, ultraviolet, X rays, and gamma rays, exhibits behavior suggestive of particles, the radiation also behaves something like a wave disturbance. This duality is still a source of some bewilderment. It seems that each photon carries with it a certain quantity of wave energy; for this reason, the particle theory is sometimes called the quantum theory. *See also* ELECTROMAGNETIC THEORY OF LIGHT.

PARTITION NOISE

In a pentode or tetrode tube with a positively charged screen grid, some of the cathode current is directed into the screen circuit, although most is directed into the plate circuit. This diversion of electrons occurs in a definite proportion, depending on the structure of the tube and the relative screen and plate currents.

The diversion of the electrons, forming two paths within the tube itself, causes wideband noise. This noise is called partition noise. Because of this, pentode and tetrode tubes are generally "noisier" than triode tubes, which have no screen grids. This disadvantage is offset by the fact that the pentode and tetrode produce greater gain than the triode. *See also* PENTODE TUBE, PLATE, SCREEN GRID, TETRODE TUBE, TRIODE TUBE, TUBE.

PASSBAND

In a selective circuit, the passband is the band of frequencies at which the attenuation is less than a certain value with respect to the minimum attenuation. This value is usually specified as 3 dB, representing half the power at the maximum-gain frequency.

In a superheterodyne receiver having a narrow bandpass filter, the passband may be as small as a few hundred hertz for continuous-wave signals, or as wide as 5 to 10 kHz for amplitude-modulated signals. In a receiver designed for reception of frequency-modulated signals, the passband may be as wide as 100 kHz or more. In a television receiver the passband is about 6 MHz in width.

If the passband is too wide for a given emission mode, excessive interference is received, reducing the sensitivity of the receiver. If the passband is too narrow, the full signal cannot be received, and again, the sensitivity is degraded. There exists an optimum passband for every situation.

For continuous-wave signals of moderate speed, the passband can be as narrow as 50 to 100 Hz. For single-sideband and slow-scan television, 2 to 3 kHz is best. For amplitude-modulation, 6 kHz is standard for voice signals, about 10 kHz for music signals, and 6 MHz for fast-scan television. For frequency modulation, phase modulation, and pulse modulation, the optimum passband depends on the deviation, phase change, or pulse rate. *See also* BANDPASS FILTER, BANDPASS RESPONSE.

PASSIVE COMPONENT

In any electronic circuit, some components require no external source of power for their operation. Such components are called passive. Passive components include semiconductor diodes (in most cases), resistors, capacitors, inductors, and the like. Components that require an external source of power, and/or produce gain in a circuit, are called active (*see* ACTIVE COMPONENT).

Some components can act as either active or passive devices, depending on their application. The varactor diode is an example of such a device. *See also* VARACTOR DIODE.

PASSIVE ELEMENT

See PARASITIC ELEMENT.

PASSIVE FILTER

A passive filter is a form of selective filter that does not require an external source of power for its operation. Passive filters may be of the bandpass, band-rejection, highpass, or lowpass variety (*see* BANDPASS FILTER, BAND-REJECTION FILTER, HIGHPASS FILTER, LOWPASS FILTER). Passive filters are commonly made using inductors, resistors, and/or capacitors. Crystal and mechanical filters are also passive (*see* CRYSTAL LATTICE FILTER, MECHANICAL FILTER).

Passive filters are commonly used in radio-frequency circuits. At audio frequencies, active filters, utilizing integrated-circuit operational amplifiers, are often seen. *See also* ACTIVE FILTER.

PASSIVE SATELLITE

A satellite intended for over-the-horizon radio communication may operate in two different ways. The most common type of satellite is the active satellite, which picks up the signals, converts them to another band of frequencies, and retransmits them (*see* ACTIVE COMMUNICATIONS SATELLITE, MOONBOUNCE). However, passive satellites can also be used. Passive satellites simply reflect incident electromagnetic signals, without the need for any electronic apparatus.

In the 1960s, the Echo satellites were put into orbit around the earth. These were passive satellites. These devices were large, spherical, and metallic. They reflected radio signals in much the same way as the ground surface does. Another, more familiar but less often-used, passive satellite is our own moon. The main advantage of passive satellites is that they carry no equipment that might malfunction. The disadvantage of the passive satellite is that the reflected signals tend to be very weak. *See also* MOONBOUNCE.

PATCH CORD

A patch cord is a length of conductor or cable, usually equipped with plugs on both ends, and used for the purpose of making temporary connections between circuits. Patch cords are useful in the test lab. Patch cords were once used by telephone operators. Today, computers have largely replaced operators with patch cords in the telephone industry—although patch cords and panels are still sometimes used in the smaller municipalities.

Patch cords can be configured in an immense variety of different ways, because of the proliferation of connector types used in the electronics industry. Patch cords with different connectors on either end can serve as adaptors (*see* ADAPTOR, CONNECTOR). The general experience of electronic technicians seems to be that there are never enough patch cords, or the existing cords are too short or lack the proper connectors. A good laboratory should be thoroughly stocked with a complete repertoire of these cords. The cords should also be periodically checked to ensure that the electrical connections are intact.

PATENT

A patent is a legal guarantee of exclusive rights to manufacture a product. A patent for an original invention can be obtained in the United States by prosecuting the details with the Patent Office in Washington, DC.

In order to obtain a patent for a device, such as an electronic circuit, the inventor must show novelty and originality. Improvements in previously existing devices are sometimes eligible for patent protection. For information concerning patents in the United States, the prospective inventor should contact a reputable patent attorney or the U.S. Government Printing Office, Washington, DC 20402.

PEAK

In a waveform or other changing parameter, a peak is an instantaneous or local maximum or minimum. The value of the parameter falls off on either side of a maximum peak (see A in illustration), and rises on either side of a minimum peak (B).

The term peak is used in many different situations. A local maximum or minimum in a mathematical function may be called a peak. In the response curve of a bandpass filter, the frequency of least attenuation is often called the

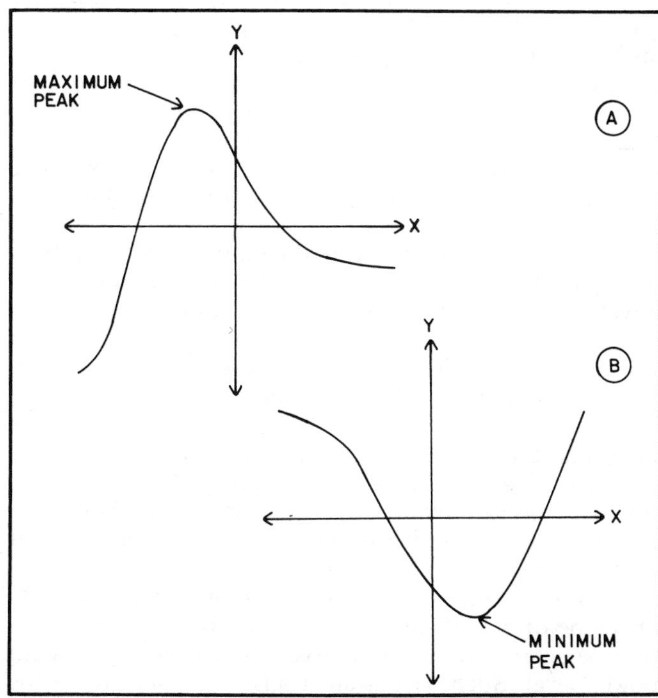

PEAK: Peaks in mathematical functions. At A, a local maximum; at B, a local minimum.

peak frequency or peak point. In the adjustment of device, the peak setting is the control position that results in the maximum value for a certain parameter. An example of this is the adjustment of the plate-tuning capacitor for maximum output power in a radio-frequency transmitter.

A waveform peak usually has infinitesimal, or nearly zero, duration. (In some cases a peak may be sustained for a certain length of time; the square wave, for example, has peaks that last for one-half cycle.) A peak may be either positive or negative in terms of polarity or direction of current flow. *See also* PEAK CURRENT, PEAK ENVELOPE POWER, PEAK POWER, PEAK-READING METER, PEAK VALUE, PEAK VOLTAGE.

PEAK CURRENT

Peak current is the maximum instantaneous value reached by the current in an alternating or pulsating waveform.

In a symmetrical alternating-current sine wave, the peak current is the same in either direction, and is reached once per cycle in either direction. In nonsymmetrical waveforms, the peak current may be greater in one direction than in the other. The peak current may be sustained for a definite length of time during each cycle, but more often the peak current is attained for only an infinitesimal time.

The peak current in a sinusoidal waveform is 1.414 times the root-mean-square (RMS) current, assuming there is no direct-current component associated with the signal. For other waveforms, the peak-to-RMS current ratio is different. *See also* ROOT MEAN SQUARE.

PEAK ENVELOPE POWER

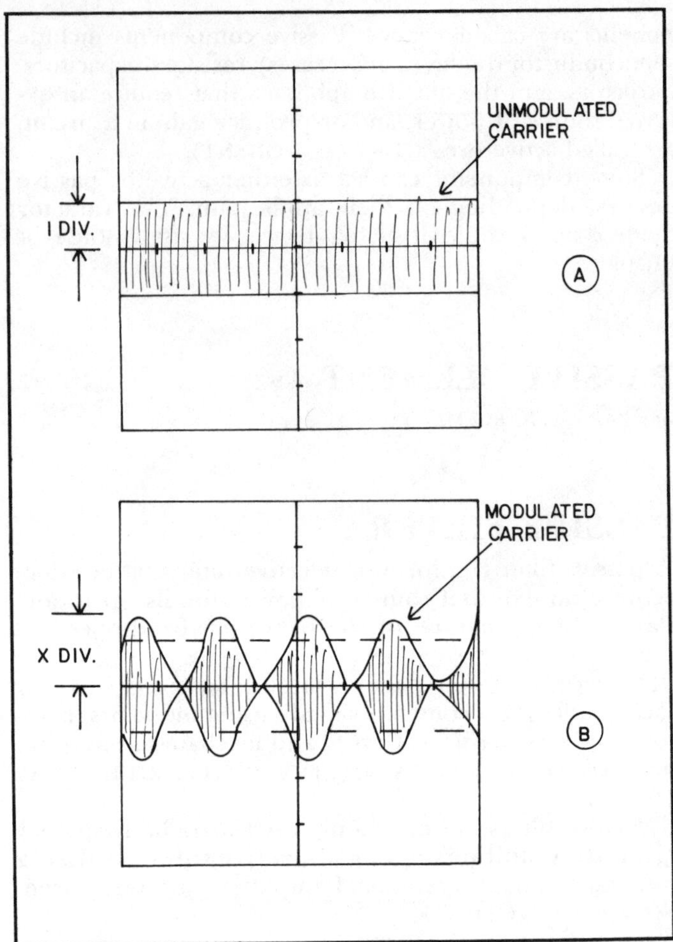

PEAK ENVELOPE POWER: Determination of peak envelope power in an amplitude-modulated signal. At A, zero modulation; at B, full modulation.

In an amplitude-modulated or single-sideband signal, the radio-frequency carrier output power of the transmitter varies with time. The input power also fluctuates. The maximum instantaneous value reached by the carrier power is called the peak envelope power. For any form of amplitude modulation, the peak envelope power is always greater than the average power. The peak envelope power differs from the actual peak waveform power. The peak waveform power is determined according to the product of the instantaneous waveform voltage and current; the peak envelope power is the maximum root-mean-square waveform power (*see* PEAK POWER). Nevertheless, the terms peak envelope power and peak power are often used interchangeably.

In the case of frequency modulation or phase modulation, the carrier amplitude does not change, and the peak envelope power is the same as the average power. In a continuous-wave (CW) Morse-code signal, the peak envelope power is the key-down average power, which is approximately twice the long-term average power.

The peak envelope power is not shown by ordinary metering devices when signals are voice-modulated. This is because the actual transmitter input and output power fluctuate too rapidly for any meter needle to follow. Special meters exist for measurement of peak envelope power in such cases (*see* PEAK-READING METER).

An oscilloscope can be used to show the modulation envelope of a transmitter; the peak envelope power can then be determined by comparison with a known, constant power level. Such a constant transmitter output can be obtained in a conventional amplitude-modulated transmitter by cutting off the modulating audio. In a single-sideband transmitter, a sine-wave tone can be supplied to the audio input, at a level sufficient to maximize the transmitter output without envelope distortion.

To determine the peak-envelope output power, a wattmeter is first used to read the continuous output as obtained in the above manner. The constant-amplitude signal is then displayed on an oscilloscope, and the sensitivity of the scope adjusted until the display indicates plus or minus 1 division (see A in illustration). Then normal modulation is applied. The peaks of the envelope can be seen, reaching a level of plus or minus x divisions, where x is greater than 1, as at B.

The oscilloscope indicates voltage, not power; thus, the ratio of the peak envelope power P to the constant-carrier power Q is:

$$P/Q = x^2$$

and the peak envelope power is given by:

$$P = x^2Q$$

Peak envelope power is often specified for single-sideband signals. The ratio of the peak envelope power to the average power in a single-sideband emission depends on the modulating waveform. In a single-sideband signal, the peak-envelope power for a human voice signal is approximately two to three times the average power. *See also* ENVELOPE, SINGLE SIDEBAND.

PEAK INVERSE VOLTAGE

The peak inverse voltage in a circuit is the maximum instantaneous voltage that occurs with a polarity opposite from the polarity of normal conduction. The term is often used with respect to rectifier diodes. The peak inverse voltage across a diode is the maximum instantaneous negative anode voltage. This voltage is sometimes called the peak reverse voltage.

Rectifier diodes are rated according to their maximum peak-inverse-voltage or peak-reverse-voltage (PIV or PRV) tolerance. If the inverse voltage significantly exceeds the PIV rating for a diode, avalanche breakdown occurs and the diode conducts in the reverse direction. Certain diodes, called Zener diodes, are deliberately designed to be operated near, or at, their avalanche points. *See also* AVALANCHE, AVALANCHE BREAKDOWN, RECTIFIER DIODE, RECTIFIER TUBE, ZENER DIODE.

PEAK LIMITING

In a fluctuating or alternating waveform, it is sometimes desirable to prohibit the peak value from exceeding a certain level (*see* PEAK VALUE). This is called peak limiting. An example of peak limiting in an alternating-current waveform is shown in the illustration.

Peak limiting can be accomplished in a variety of different ways. A pair of back-to-back semiconductor diodes results in peak limiting of low-voltage alternating-current signals; this technique is often used in the audio stages of communications receivers to prevent "blasting" (*see* AUDIO LIMITER).

In a modulated transmitter, peak limiting can be used to increase the ratio of average power to peak-envelope power. This technique is sometimes also called radio-frequency clipping. Peak limiting of an amplitude-modulated signal can be done indirectly by modifying the audio waveform; this is called speech clipping. *See also* CLIPPER, RF CLIPPING, SPEECH CLIPPING.

PEAK POWER

The peak power of an alternating-current signal is the maximum instantaneous power. The peak power is equivalent to the peak voltage multiplied by the peak current when the phase angle is zero (no reactance).

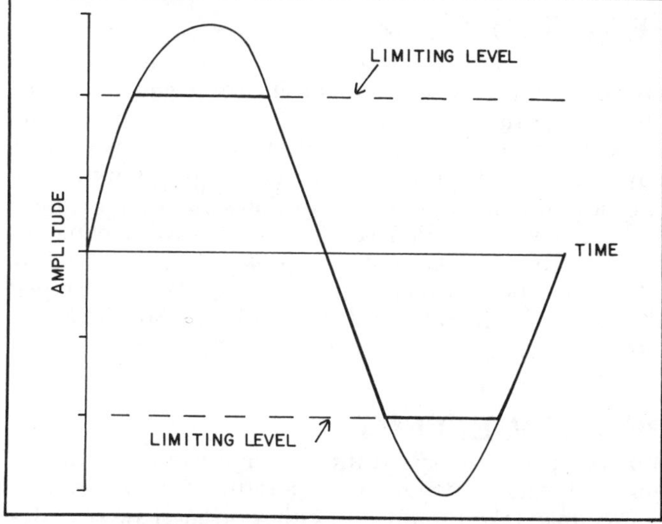

PEAK LIMITING: The amplitude cannot exceed the limiting level.

In a sinusoidal alternating-current signal, the peak power, P_p, can be determined from the root-mean-square (RMS) voltage, E_{RMS}, and the RMS current, I_{RMS}, according to the formula:

$$P_p = 2E_{RMS}\ I_{RMS}$$

where P_p is given in watts, E_{RMS} is given in volts, and I_{RMS} is given in amperes. The peak power, in terms of the peak voltage, E_p, and the peak current, I_p, is:

$$P_p = E_p I_p$$

In a modulated radio-frequency signal, the peak envelope power is sometimes called the peak power. *See also* PEAK ENVELOPE POWER.

PEAK-READING METER

A peak-reading meter is a special device intended for determining peak values. Most meters register average values for signals that fluctuate rapidly; this is simply because the meter needle or display time-constant cannot follow the variations.

Peak-reading meters can be designed for measurement of current, power, or voltage. All such devices function in basically the same way, using voltmeters or ammeters with specialized peripheral circuitry.

A typical peak-reading meter operates as follows, and as illustrated by the simplified schematic diagram. A detecting circuit, having an extremely fast time constant, senses the peak voltage across a resistance. (A semiconductor diode can be used for this purpose.) A capacitor is charged by the rectified voltage. The circuit is wired so that the capacitor charges very quickly but discharges slowly. The voltage across the capacitor is monitored by a voltmeter having extremely high internal resistance. The peak current, power, or voltage can be determined from this voltmeter reading in terms of the resistance in the circuit. The meter needle, upon application of signal, jumps up to the peak value and remains there for several seconds.

Peak-reading meters are employed for convenient determination of peak values without an oscilloscope. A peak-reading wattmeter may, for example, be used to measure the peak envelope power from a single-sideband transmitter. *See also* PEAK CURRENT, PEAK ENVELOPE POWER, PEAK POWER, PEAK VALUE, PEAK VOLTAGE.

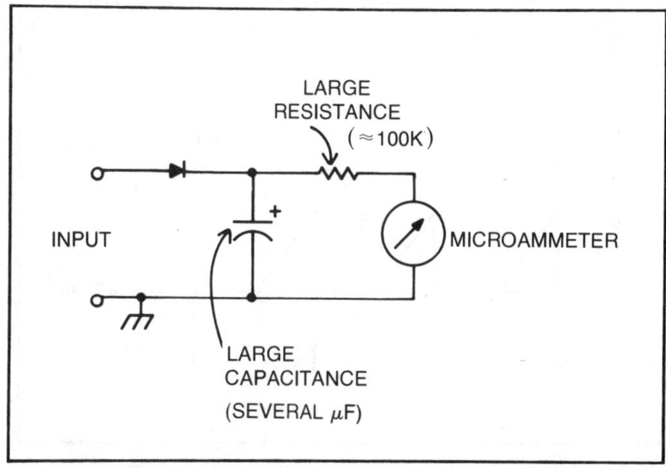

PEAK-READING METER: A simple design for a peak-reading meter.

PEAK-TO-PEAK VALUE

In a fluctuating or alternating waveform, the peak-to-peak value is the difference between the maximum positive instantaneous value and the maximum negative instantaneous value. Peak-to-peak is sometimes abbreviated as p-p or pk-pk.

The drawing illustrates peak-to-peak values for an alternating-current sine wave (A), a rectified sine wave (B), and a sawtooth wave (C). The peak-to-peak value of a waveform is a measure of signal intensity at a given frequency. Signal voltage is often given in terms of the peak-to-peak value. Note that changes in the direct-current component of a signal have no effect on the peak-to-peak value.

Peak-to-peak values for any waveform are readily observed on an oscilloscope display. This is possible even if the frequency is too high to allow viewing of the waveform, or if waveform irregularity makes it difficult to determine the average value.

For an alternating-current sinusoidal waveform, the peak-to-peak value is equal to twice the peak value, and 2.828 times the root-mean-square value. *See also* PEAK VALUE, ROOT MEAN SQUARE.

PEAK VALUE

The peak value of a fluctuating or alternating waveform is the maximum instantaneous value. Some waveforms reach peaks in two directions of polarity or current flow (*see* PEAK CURRENT, PEAK VOLTAGE). These peaks may be equal in absolute magnitude, but this is not necessarily so. The peak value may last for a definite length of time; this is the case for a square wave. The peak value of a sine wave lasts for a theoretically infinitesimal (zero) time. The illustration at A illustrates peak values for a sine-wave alternating-current waveform. The peak value for the same waveform with the addition of a direct-current component is shown at B.

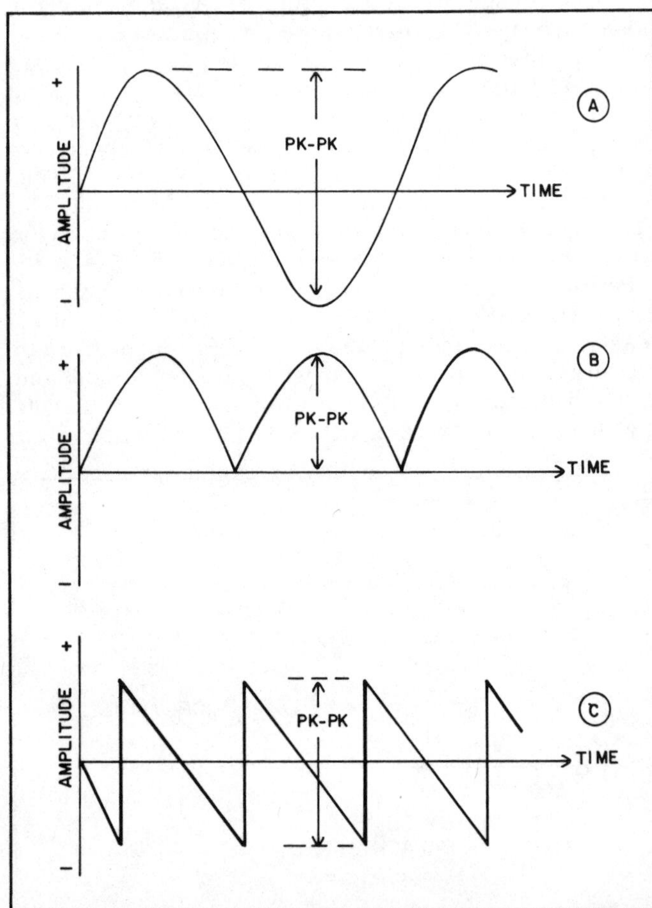

PEAK-TO-PEAK VALUE: Peak-to-peak amplitudes for a sine wave (A), a rectified sine wave (B), and a sawtooth wave (C).

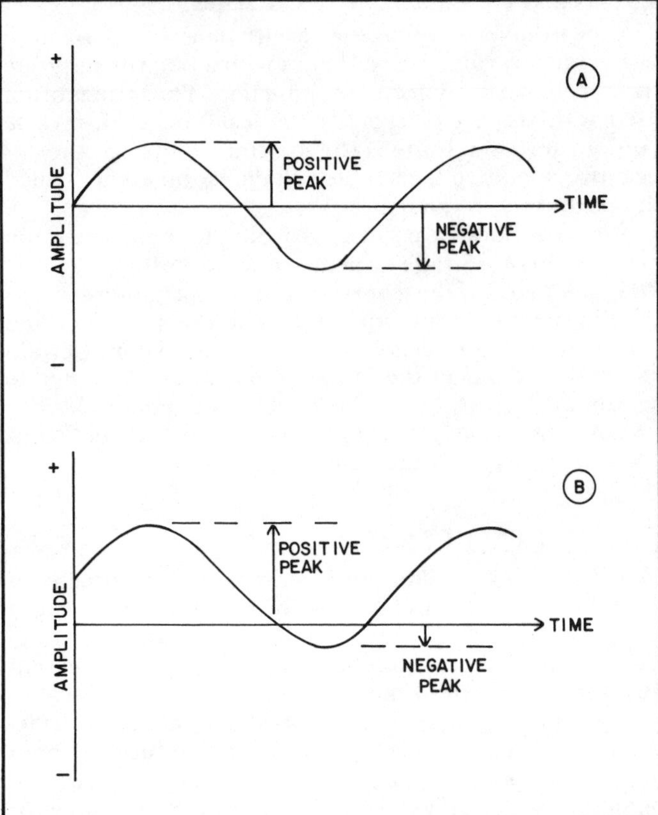

PEAK VALUE: The waveform at A has the same peak-to-peak amplitude as the waveform at B, but the positive and negative absolute peak values differ. A dc component is impressed on the wave at B.

Peak values of power may be specified in either of two ways. The maximum instantaneous waveform power is the product of the peak voltage and the peak current (*see* PEAK POWER). The peak envelope power is the maximum carrier power in an amplitude-modulated or single-sideband signal (*see* PEAK ENVELOPE POWER).

PEAK VOLTAGE

The peak voltage of a fluctuating or alternating waveform is the instantaneous maximum positive or negative voltage. The peak positive voltage is often equal in magnitude to the peak negative voltage, but not always. The peak voltage may be sustained for a certain length of time, but in most cases it is attained only for an infinitesimally short time.

For a sinusoidal alternating-current wave, having no direct-current component, the peak voltage is equal to 1.414 times the root-mean-square voltage. For other types of waveforms, the relationship is different. *See also* ROOT MEAN SQUARE.

PELTIER EFFECT

When a current is passed through a junction of two dissimilar metals, a change in temperature may occur. The change may take the form of either an increase or a decrease in the temperature, depending on the direction of the current through the junction. If a current in one direction causes heating, then a current in the other direction produces cooling. This effect is known as the Peltier effect. Peltier effect might be thought of as a sort of reverse thermocouple effect (*see* THERMOCOUPLE).

The amount of thermal energy transferred, for a given current, depends on the metals. The thermal energy that is emitted or absorbed with the passage of current through a junction of dissimilar metals is called the Peltier heat. The quotient of the Peltier heat and the current is called the Peltier coefficient.

PENETRATING FREQUENCY

See CRITICAL FREQUENCY.

PEN RECORDER

A pen recorder is a device for graphically plotting the magnitude of a parameter as a function of time. The device is a form of analog meter. Pen recorders are used for long-term monitoring of a variable quantity, or whenever a permanent record is needed. Pen recorders are used for plotting such functions as heartbeat, temperature, barometric pressure, and wind speed versus time.

The pen recorder works very much like an ordinary analog meter. The needle, to which a pen is attached, moves back and forth according to the magnitude of the measured parameter. A paper is guided under the pen at a certain speed, corresponding to the required number of seconds, minutes, or hours per division. The paper is usually attached to a rotating drum. The illustration shows a pen recorder.

A pen recorder cannot be used to plot functions that vary with extreme rapidity, because the metering device cannot respond fast enough. For detailed permanent records of rapidly fluctuating or complicated waveforms, an oscilloscope, equipped with a camera, is preferable. *See also* ANALOG, METERING, OSCILLOSCOPE, OSCILLOSCOPE CAMERA.

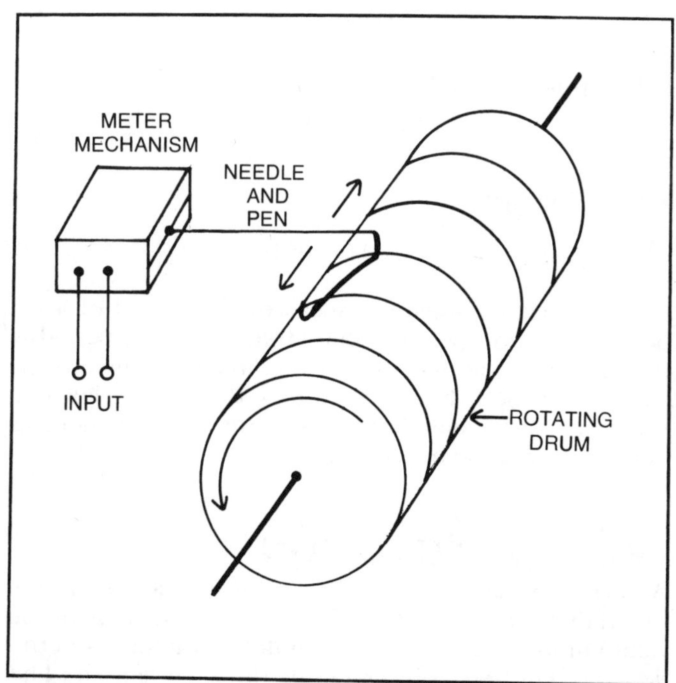

PEN RECORDER: Principle of the pen recorder.

PENTAGRID CONVERTER

A pentagrid converter is a vacuum tube having five different grids. The tube is, as its name implies, generally used in converter circuits (*see* CONVERTER). In modern electronic converters, transistors, both bipolar type and field-effect type, have largely replaced pentagrid converters and other tubes. Dual-gate metal-oxide semiconductor field-effect transistors (MOSFETs) are common in converter circuits today.

The pentagrid converter has a cathode (negative electrode) and anode (positive electrode), like an ordinary tube. It also has a control grid, screen grid, and suppressor grid, as does the pentode tube (*see* PENTODE TUBE). In addition to these electrodes, the pentagrid converter has two signal-injection grids. The two mixing signals are applied to these injection grids.

The main advantage of the pentagrid converter is that it produces higher gain than tetrode or pentode tubes when used in a converter. *See also* TUBE.

PENTODE TUBE

A pentode tube is a vacuum tube having five elements. In addition to the cathode and the plate or anode, there are three grids. The three grids serve different purposes.

The control grid operates as the input-signal electrode in most pentode-tube circuits (*see* CONTROL GRID). The screen grid serves the same function as in the tetrode tube (*see* SCREEN GRID, TETRODE TUBE). The third grid, the suppressor, increases the gain of the pentode over that usually obtainable with the tetrode. It does this by repelling secondary electrons from the plate, directing them back to the plate so that they can contribute to the signal output (*see* SECONDARY EMISSION, SUPPRESSOR GRID).

The control grid in the pentode tube is usually biased negatively with respect to the cathode. The screen grid is biased at a voltage more positive than the cathode but less positive than the plate. The suppressor grid is usually biased at the same voltage as the cathode. In fact, many pentode tubes have an internal short circuit between the cathode and the suppressor grid.

The illustration is a schematic symbol for the pentode tube, showing a typical biasing arrangement. The three

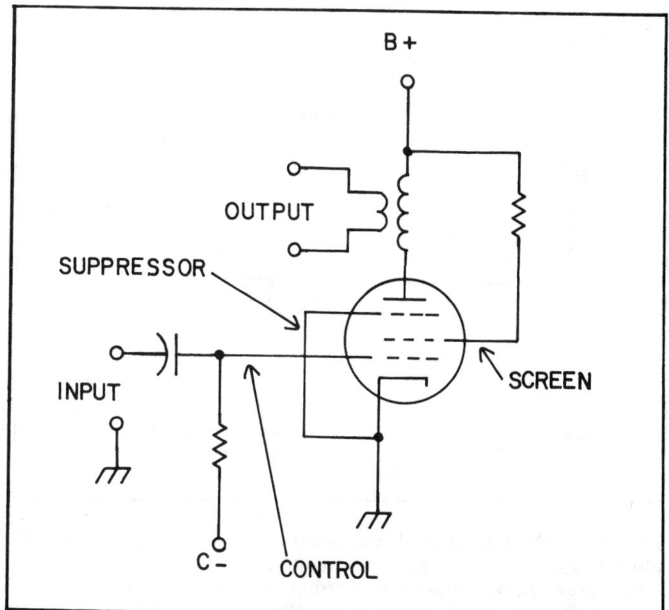

PENTODE TUBE: Typical biasing arrangement.

grids are, in the physical construction of the pentode, roughly concentric. The control grid is closest to the cathode. The screen grid is immediately outside the control grid. The suppressor is closest to the plate. *See also* TUBE.

PENZIAS, ARNO

While working with colleague Robert Wilson at the Bell Laboratories in 1965, Arno Penzias made one of the most important discoveries in modern cosmology.

Penzias and Wilson accidentally discovered radio waves coming from all points in space. This radiation was found by cosmologists to be the remnants of a brilliant flash that is believed to have taken place about 20 billion years ago. That flash, it is thought, was an explosion that brought the universe into being.

PEP

See PEAK ENVELOPE POWER.

PERCENTAGE OF MODULATION

See MODULATION PERCENTAGE.

PERFORMANCE CURVE

A graph, representing the characteristics of a component

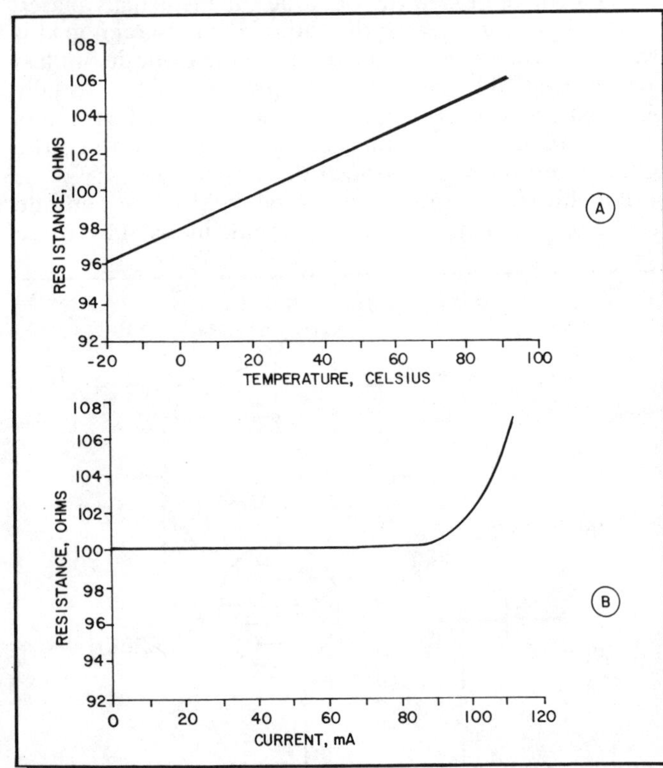

PERFORMANCE CURVE: Performance curves for a hypothetical 100-ohm, ½-watt resistor. At A, the resistance is shown as a function of the temperature. At B, the function of resistance versus current through the resistor. Destruction occurs at about 120 mA. Serious performance degradation begins at about 90 mA.

under variable conditions, is called a performance curve. A single component usually has several different performance curves, one for each independent variable for which the device is to be evaluated. Typical independent variables that affect component performance include the voltage across the component, the current through it, the frequency of the signal passing through or applied to the component, the temperature, and many other factors. The dependent variable may be the gain, the component value, the emitted power, or some other performance-related factor.

The illustration shows two examples of performance curves for a resistor having a positive temperature coefficient (that is, the resistance increases as the temperature rises). At A, the low-current resistance is shown as a function of the ambient temperature. At B, the resistance is plotted as a function of the current at higher values. The current heats the resistor, increasing the resistance. It is possible, using the two performance curves shown, to determine the temperature to which the resistor will be heated for a given amount of current. It is also quite obvious how much current the resistor can handle without burning out!

Performance curves can be plotted for almost any electronic component, whether active or passive. Such curves are invaluable to engineers in the choice of components for a particular circuit application or design.

Performance curves are generally plotted in the two-dimensional Cartesian coordinate system. However, other coordinate systems are sometimes used for special purposes. *See also* CARTESIAN COORDINATES, COORDINATE SYSTEM.

PERIOD

A period is the least length of time required for an alternating or pulsating waveform to complete one cycle. For example, in a 60-Hz electric system, the period is 1/60 second. In general, if f is the frequency in hertz, then the period T in seconds is:

$$T = 1/f$$

If f is specified in kilohertz, then:

$$T = 0.001/f$$

and for f in megahertz:

$$T = 0.000001/f$$

Mathematicians sometimes speak of the period of a repeating function (*see* PERIODIC FUNCTION); the period is the length of an interval, on the independent-variable axis, over which the function goes through exactly one repetition. An example of this is the sine function; its period is 2π, or about 6.28.

PERIODIC FUNCTION

A periodic function is a mathematical function (*see* FUNCTION) that repeats itself. Such a function attains all of the values in its range within certain defined intervals in the domain. The intervals are all of the same length. The interval length is called the period of the function. An

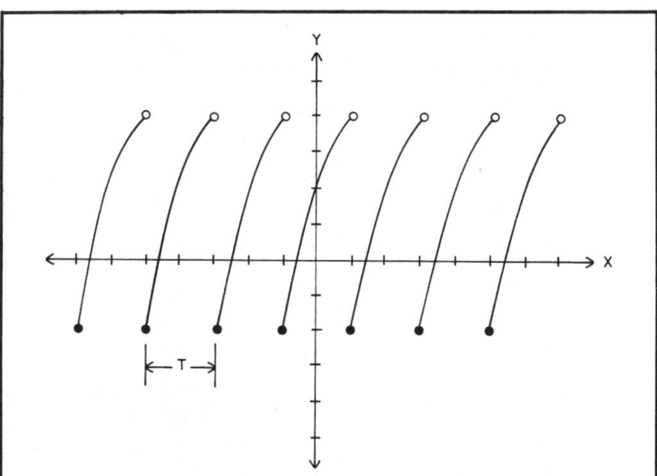

PERIODIC FUNCTION: The period is T.

arbitrary example of a periodic function is shown in the illustration.

In electronics, periodic functions are used in many different situations. The most familiar periodic function is the sine function and multiples of it. The cosecant, cosine, cotangent, secant, and tangent functions, and their multiples, are periodic. *See* COSECANT, COSINE, COTANGENT, SECANT, SINE, TANGENT.

Periodic functions may, in some instances, be very complicated. For example, several different multiples of the sine function may be superimposed on each other. This is the case in wave disturbances containing harmonic energy or having multiple frequency components. The period of such a function is sometimes hard to determine. However, it can always be defined. *See also* PERIOD.

PERIODIC TABLE OF THE ELEMENTS

The material elements are often listed in a special arrangement known as the periodic table. Chemists and physicists use the periodic table because it is illustrative of the groups of elements having similar properties. There are seven different periods. They are numbered, not surprisingly, 1 through 7.

The elements are also classified in 16 groups according to certain characteristics. The groups are designated by Roman numerals and English letters, as follows: I-A, II-A, III-B, IV-B, V-B, VI-B, VII-B, VIII, I-B, II-B, III-A, IV-A, V-A, VI-A, VII-A, and ZERO. This is the order of

PERIODIC TABLE OF THE ELEMENTS: ABBREVIATIONS OF THE CHEMICAL ELEMENTS.

Ac	Actinium	Gd	Gadolinium	Pm	Promethium
Ag	Silver	Ge	Germanium	Po	Polonium
Al	Aluminum	H	Hydrogen	Pr	Praseodymium
Am	Americium	He	Helium	Pt	Platinum
Ar	Argon	Hf	Hafnium	Pu	Plutonium
As	Arsenic	Hg	Mercury	Ra	Radium
At	Astatine	Ho	Holmium	Rb	Rubidium
Au	Gold	I	Iodine	Re	Rhenium
B	Boron	In	Indium	Rh	Rhodium
Ba	Barium	Ir	Iridium	Rn	Radon
Be	Beryllium	K	Potassium	Ru	Ruthenium
Bi	Bismuth	Kr	Krypton	S	Sulfur
Bk	Berkelium	La	Lanthanum	Sb	Antimony
Br	Bromine	Li	Lithium	Sc	Scandium
C	Carbon	Lu	Lutetium	Se	Selenium
Ca	Calcium	Lw	Lawrencium	Si	Silicon
Cb	Columbium	Md	Mendelevium	Sm	Samarium
Cd	Cadmium	Mg	Magnesium	Sn	Tin
Ce	Cerium	Mn	Manganese	Sr	Strontium
Cf	Californium	Mo	Molybdenum	Ta	Tantalum
Cm	Curium	N	Nitrogen	Tb	Terbium
Co	Cobalt	Na	Sodium	Tc	Technetium
Cr	Chromium	Nb	Niobium	Te	Tellurium
Cs	Cesium	Nd	Neodymium	Th	Thorium
Cu	Copper	Ne	Neon	Ti	Titanium
Dy	Dysprosium	Ni	Nickel	Tl	Thallium
Er	Erbium	No	Nobelium	Tm	Thulium
Es	Einsteinium	Np	Neptunium	U	Uranium
Eu	Europium	O	Oxygen	V	Vanadium
F	Fluorine	Os	Osmium	W	Tungsten
Fe	Iron	P	Phosphorus	Xe	Xenon
Fm	Fermium	Pa	Proactinium	Y	Yttrium
Fr	Francium	Pb	Lead	Yb	Ytterbium
Ga	Gallium	Pd	Palladium	Zn	Zinc
				Zr	Zirconium

TABLE 2

PERIODIC TABLE OF THE ELEMENTS: LONG FORM.

I A	II A	III B	IV B	V B	VI B	VII B	VIII			I B	II B	III A	IV A	V A	VI A	VII A	0	
H 1																H 1	He 2	1
Li 3	Be 4											B 5	C 6	N 7	O 8	F 9	Ne 10	2
Na 11	Mg 12											Al 13	Si 14	P 15	S 16	Cl 17	Ar 18	3
K 19	Ca 20	Sc 21	Ti 22	V 23	Cr 24	Mn 25	Fe 26	Co 27	Ni 28	Cu 29	Zn 30	Ga 31	Ge 32	As 33	Se 34	Br 35	Kr 36	4
Rb 37	Sr 38	Y 39	Zr 40	Nb 41	Mo 42	Tc 43	Ru 44	Rh 45	Pd 46	Ag 47	Cd 48	In 49	Sn 50	Sb 51	Te 52	I 53	Xe 54	5
Cs 55	Ba 56	La 57	Hf 72	Ta 73	W 74	Re 75	Os 76	Ir 77	Pt 78	Au 79	Hg 80	Tl 81	Pb 82	Bi 83	Po 84	At 85	Rn 86	6
Fr 87	Ra 88	Ac 89																7
			Ce 58	Pr 59	Nd 60	Pm 61	Sm 62	Eu 63	Gd 64	Tb 65	Dy 66	Ho 67	Er 68	Tm 69	Yb 70	Lu 71		Lanthanide
			Th 90	Pa 91	U 92	Np 93	Pu 94	Am 95	Cm 96	Bk 97	Cf 98	Es 99	Fm 100	Md 101	No 102	Lw 103		Actinide

TABLE 1

the groups as they appear across the periodic table from left to right.

Table 1 shows the form of the periodic table that is probably the most common. It is called the long form. Atomic numbers are given with the element abbreviations. Table 2 lists the chemical abbreviations, in alphabetical order, followed by the full name of the element. *See also* ATOMIC NUMBER, ATOMIC WEIGHT.

PERMANENT MAGNET

Certain metals, notably iron and nickel, are affected by magnetic fields in a unique way. The magnetic dipoles within such substances can, in the presence of a sufficiently intense magnetizing force, become permanently aligned. This results in a so-called permanent magnet—a piece of material that is constantly surrounded by a magnetic field.

A piece of iron or steel can be permanently magnetized by stroking it with another permanent magnet. This action causes the magnetic dipoles, which are normally aligned at random, to be oriented more or less in a common direction. Their effects then average out to create a magnetic field around the object. (Normally, the effects of the dipoles average out to a zero magnetic field.)

Permanent magnets are used in speakers, microphones, meters, and certain types of transducers. In the laboratory or the factory, tools such as screwdrivers are sometimes weakly magnetized for convenience in dismantling or assembling electronic equipment. Tiny permanent magnets are the medium by which a magnetic tape stores its information. *See also* D'ARSONVAL, DYNAMIC LOUDSPEAKER, DYNAMIC MICROPHONE, DYNAMIC PICKUP, MAGNETIC FIELD, MAGNETIC RECORDING, MAGNETIZATION.

PERMANENT-MAGNET METER

See D'ARSONVAL.

PERMANENT-MAGNET MICROPHONE

See DYNAMIC MICROPHONE.

PERMANENT-MAGNET SPEAKER

See DYNAMIC LOUDSPEAKER.

PERMANENT-MAGNET PICKUP

See DYNAMIC PICKUP.

PERMEABILITY

Certain materials affect the concentration of the lines of force in a magnetic field (*see* MAGNETIC FIELD, MAGNETIC FLUX). Some substances cause the lines of flux to move farther apart, resulting in a decrease in the intensity of the field compared with its intensity in a vacuum. Other substances cause the density of the magnetic flux to increase.

The permeability of a substance is an expression of the extent to which that material affects the density of the magnetic flux. Free space (a vacuum) is assigned the permeability value 1. Materials that reduce the flux density, known as diamagnetic materials, have permeability less than 1. Substances that cause an increase in the flux density are known as ferromagnetic and paramagnetic materials. They have permeability factors greater than 1. The paramagnetic materials have values slightly larger than 1; the ferromagnetics have values much greater. (*See* DIAMAGNETIC MATERIAL, FERROMAGNETIC MATERIAL, PARAMAGNETIC MATERIAL.)

PERMEABILITY: PERMEABILITY FACTORS OF SOME COMMON SUBSTANCES.

Substance	Permeability (Approx.)
Aluminum	Slightly more than 1
Bismuth	Slightly less than 1
Cobalt	60-70
Ferrite	100-3000
Free Space	1
Iron	60-100
Iron, refined	3000-8000
Nickel	50-60
Permalloy	3000-30,000
Silver	Slightly less than 1
Steel	300-600
Super-permalloys	100,000-1,000,000
Wax	Slightly less than 1
Wood, dry	Slightly less than 1

The table lists several common types of materials, along with their approximate permeability values at room temperature. The permeability of a given substance may change substantially with fluctuations in the temperature. The permeability factors of some materials, notably the ferromagnetics, may vary depending on the intensity of the magnetic field. For these reasons, the values in the table should not be taken as precise. The table reflects the fact that no substances have permeability values much smaller than 1, although some have values far in excess of 1. Certain nickel-iron alloys exhibit permeability factors of more than 1,000,000.

Substances with high magnetic-permeability values are used primarily for the purpose of increasing the inductances of coils. When a high-permeability core is inserted into a coil, the inductance is multiplied by the permeability factor. This effect is useful in the design of transformers and chokes at all frequencies. *See also* CHOKE, FERRITE CORE, LAMINATED CORE, POWDERED-IRON CORE, TRANSFORMER.

PERMEABILITY TUNING

In a resonant radio-frequency circuit, tuning may be accomplished by varying the value of either the inductor or the capacitor. The value of the inductor can be adjusted by changing the number of turns, or by changing the magnetic permeability of the core. The latter method of inductor tuning is known as permeability tuning.

Permeability tuning is accomplished by moving a powdered-iron core in and out of the coil (see illustration). A threaded shaft is attached to the core, and this shaft is rotated to allow precise positioning of the core. The farther into the coil the core is set, the larger the inductance, and the lower the resonant frequency becomes. Conversely, as the core gets farther outside the coil, the

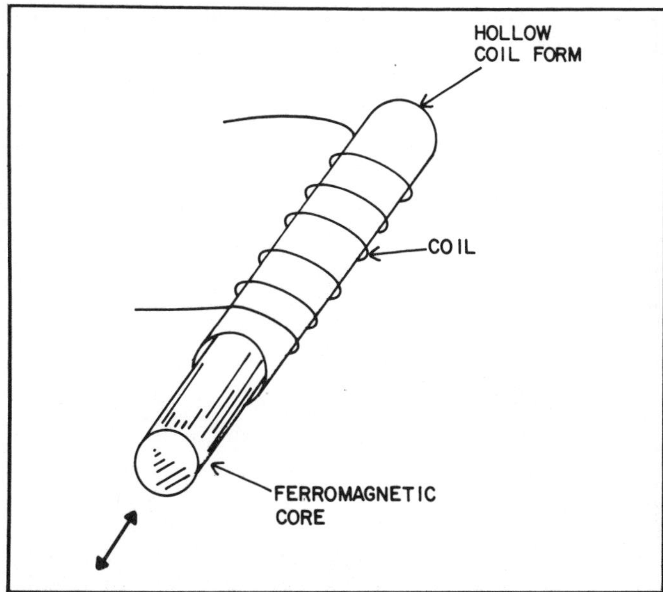

PERMEABILITY TUNING: Principle of permeability tuning. The ferromagnetic core is moved into the coil to increase the inductance, and is pulled out of the coil to reduce the inductance.

inductance gets smaller, and the resonant frequency becomes higher.

The main advantage of permeability tuning, as compared with capacitor tuning, is that the control adjustment is more linear. This allows the use of a dial calibrated in linear increments. Permeability tuning also has the advantage that the capacitor in the tuned circuit may be fixed, rather than variable. This reduces the effects of external capacitance, and allows the use of a capacitor having optimum temperature coefficient. *See also* CAPACITOR TUNING, INDUCTOR TUNING.

PERMEANCE

In a magnetic circuit, permeance is an expression of the ease with which a magnetic field is conducted. Permeance is the reciprocal of reluctance. Permeance in a magnetic circuit is analogous to conductance in an electric circuit. Permeance is generally measured in webers per ampere. *See also* RELUCTANCE.

PERMITTIVITY

Permittivity is an expression of the absolute dielectric properties of a material or medium. The dielectric constant is determined from the permittivity.

Permittivity is represented by the lowercase Greek letter epsilon (ϵ). The permittivity of free space is represented by the symbol ϵ_o. The quantity is usually expressed in farads per meter; the permittivity of free space, in these units, is:

$$\epsilon_o = 8.85 \times 10^{-12} \text{ farad per meter}$$

Given a substance with permittivity ϵ farad per meter, the dielectric constant, k, is determined according to the formula

$$k = \epsilon/\epsilon_o$$

$$= \epsilon/(8.85 \times 10^{-12})$$

See also DIELECTRIC, DIELECTRIC CONSTANT.

PERMUTATION

A set of data characters or bits may be arranged in various different ways. The possible arrangements, or ordering schemes, are called permutations.

Given n different objects or characters, the number of possible permutations, P, is equal to n! (n factorial), or:

$$P = n(n-1)(n-2) \ldots (1)$$

For example, the letters A, B, C, D, and E may be arranged in $P = 5 \times 4 \times 3 \times 2 \times 1 = 120$ different ways. The set of all letters in the alphabet may be arranged in 26! different ways; this is about 4.03×10^{26} permutations.

PERSISTENCE TRACE

Some oscilloscopes are equipped with a device that "freezes" the trace on the screen for prolonged viewing. This feature is called a persistence trace. A single sweep of the electron beam is stored in a memory circuit, and the waveform is shown on the screen for as long as the operator wants.

Persistence trace oscilloscopes is especially useful when analyzing waveforms of extremely low frequency. Without such a feature, the display appears as nothing but a spot that jumps as it moves, slowly, from left to right across the screen. The persistence trace reveals the details. The persistence trace may also be used with an oscilloscope camera when the sweep frequency is relatively low. *See also* OSCILLOSCOPE, OSCILLOSCOPE CAMERA.

pH

The pH is an expression of the relative acidity or alkalinity of a liquid. The pH is represented by a value between 0 and 14. Numbers smaller than 7 represent acidity, and numbers greater than 7 represent alkalinity. A liquid is neutral when its pH value is exactly 7.

Theoretically, pH is the negative logarithm of the hydrogen-ion concentration, measured in gram equivalents per liter. The pH is significant as a measure of electrolyte concentration in electrochemical devices.

PHANTASTRON

A special form of monostable multivibrator, often used as a sweep generator, is called a phantastron. The phantastron circuit uses a pentode tube and a semiconductor or tube type diode. The drawing illustrates a phantastron circuit.

Normally, the control grid is biased slightly positive with respect to the cathode. The suppressor grid prevents plate current from flowing. To generate the output signal, a positive pulse is applied to the suppressor. This causes the tube to briefly conduct. This condition is unstable, and the tube returns to the non-conducting condition within a short time. The length of time during which the tube conducts can be adjusted by varying the bias on the control grid. *See also* MONOSTABLE MULTIVIBRATOR.

PHASE

Phase is a relative quantity, describing the time relationship between or among waves having identical frequency.

The complete wave cycle is divided into 360 equal parts, called degrees of phase. The time difference between two waves can then be expressed in terms of these degrees. Phase is sometimes also expressed in radians. One radian corresponds to about 57.3 degrees of phase.

One wave may occur sooner than the other by as much as 180 degrees. The earlier wave is called the leading wave. A disturbance may occur as much as 180 degrees later than its counterpart; the later wave is called the lagging wave (*see* LAGGING PHASE, LEADING PHASE). When waves are exactly 180 degrees different in phase, they are said to be perfectly out of phase, or in phase opposition.

Two signals having equal frequency add together vectorially, depending on their phase difference. This is illustrated for two signals of equal frequency but different amplitude. At A, the phase difference is zero; that is, the signals are exactly in phase. The resulting amplitude is the sum of the amplitudes of the two signals. At B, the signals differ in phase by 45 degrees. The resulting amplitude is smaller than that at A, as can be seen by the parallelogram method for adding vectors. At C, the signals are 90 degrees out of phase, with the resulting amplitude still smaller. At D, the signals are 135 degrees out of phase, yielding an even smaller composite signal; and at E, they are in phase opposition, resulting in the minimum possible amplitude.

The length of time corresponding to one degree of phase depends on the frequency of the signal. If the frequency in hertz is given by f, then the time t, in seconds, corresponding to one degree of phase is:

$$t = 1/(360f) = 0.00278/f$$

For specific information about various aspects of phase, see the articles immediately following.

PHASE ANGLE

Phase angle is an expression, given in degrees, for the relative difference in phase between two signals (*see* PHASE). Phase angle is often given to indicate the difference in phase between the current and the voltage in a circuit containing reactance (*see* LAGGING PHASE, LEADING PHASE).

In a dielectric material, the phase angle is the extent to which the current leads the voltage. The phase angle, indicated by ϕ, is equal to the complement of the loss angle θ. That is:

$$\phi = 90 - \theta$$

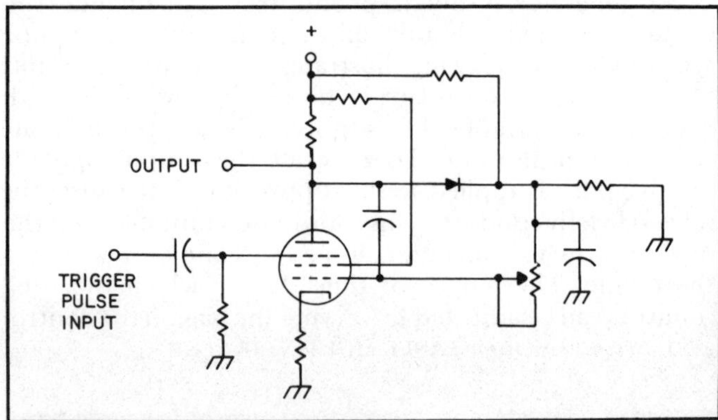

PHANTASTRON: Simplified schematic diagram of a vacuum-tube phantastron circuit.

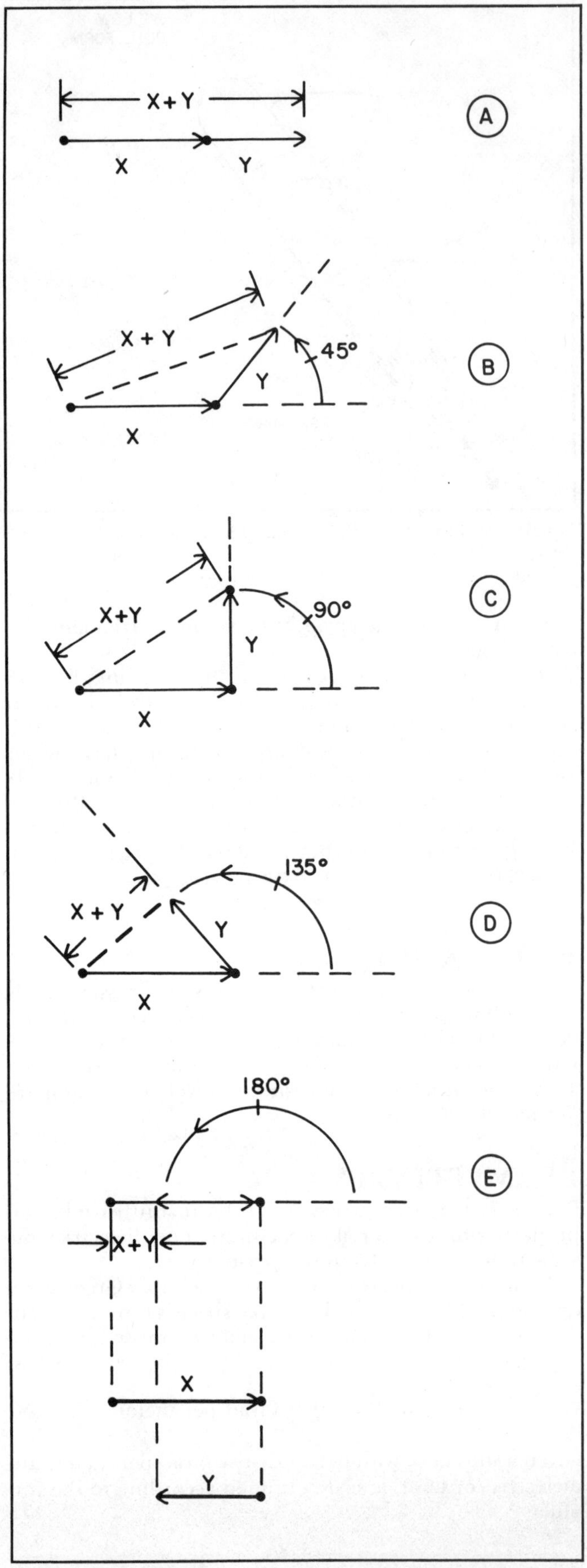

PHASE: Vector diagrams showing the addition of two hypothetical signals with various phase relationships. At A, signals X and Y are in phase. At B, they differ by 45 degrees; at C, by 90 degrees; at D, by 135 degrees; at E, by 180 degrees.

The phase angle may vary from 0 to 90 degrees. The larger the phase angle, the lower the dielectric loss (*see* DISSIPATION FACTOR).

Phase angle also indicates the lossiness of an inductor. In a perfect inductive reactance, the current lags the voltage by 90 degrees. The phase angle ϕ may, again, vary from 0 to 90 degrees; the larger the phase angle, the lower the loss in the inductor.

A phase angle of 0 degrees indicates that the current and voltage are in phase. This occurs only when there is no reactance in the circuit. *See also* CAPACITIVE REACTANCE, IMPEDANCE, INDUCTIVE REACTANCE.

PHASE BALANCE

Phase balance is an expression of the relative symmetry of a square wave. The phase-balance angle is defined as the difference between 180 degrees and the measured phase angle between the centers of the positive and negative pulses.

The drawing illustrates phase balance. The period, T, of the waveform, is the length of one complete cycle, from the midpoint of one positive pulse to the midpoint of the following positive pulse. This interval is divided into 360 degrees. The midpoint of the negative pulse is then located; ideally, it should be at the 180-degree point. If the midpoint of the negative pulse occurs at x degrees, then the phase-balance angle is 180 − x.

In a square wave with zero transition time, the phase-balance angle is 0 degrees, since x = 180. This is not the case in the illustration; in this example, the phase-balance angle is about +10 degrees, since the midpoint of the negative pulse is at 170 degrees.

In a phased array, the condition of signal cancellation, resulting in a directional null, is sometimes called phase balance. As seen from a distant point along such a null, the currents in the various phased elements oppose each other in phase, resulting in a field strength of zero. *See* PHASED ARRAY.

Phase opposition may occasionally be spoken of as phase balance. This is the condition wherein two signals of identical frequency differ in phase by 180 degrees. *See also* PHASE, PHASE OPPOSITION.

PHASE CANCELLATION

See PHASE OPPOSITION.

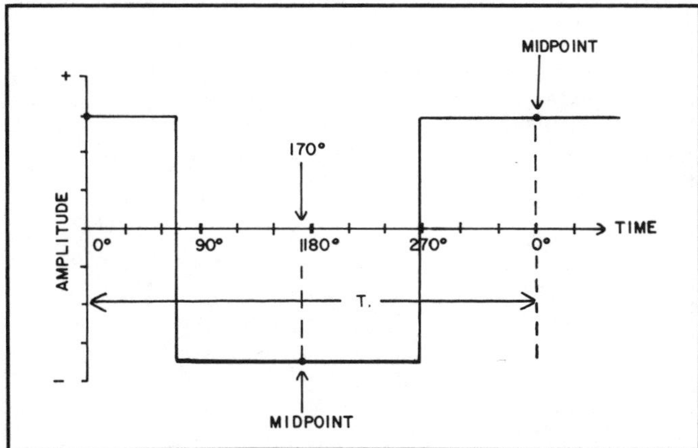

PHASE BALANCE: Phase balance in a square wave. Here, the phase balance is +10 degrees.

PHASE COMPARATOR

A phase comparator is a part of a phase-locked-loop circuit (*see* PHASE-LOCKED LOOP). The phase comparator does exactly what its name implies: It compares two signals in terms of their phase. The output voltage of the phase comparator depends on whether the signals are in phase or not. If the signals are not in phase, the voltage from the comparator causes the phase of an oscillator signal to be shifted back and forth until its signal is exactly in phase with the signal from a reference oscillator.

In some types of radiolocation and radionavigation systems, the relative phase of two signals is used to determine the position of a ship or airplane. The relative phase is determined by the phase comparator. The output of the phase comparator is fed to a computer, which determines the position. *See also* RADIOLOCATION, RADIONAVIGATION.

PHASED ARRAY

A phased array is an antenna having two or more driven elements. The elements are fed with a certain relative phase, and they are spaced at a certain distance, resulting in a directivity pattern that exhibits gain in some directions and little or no radiation in other directions.

Phased arrays may be very simple, consisting of only two elements. Two examples of simple pairs of phased dipoles are shown in the illustration. At A, the two dipoles are spaced one-quarter wavelength apart in free space, and they are fed 90 degrees out of phase. The result is that the signals from the two antennas add in phase in one direction, and cancel in the opposite direction, as shown by the arrows. In this particular case, the radiation pattern is unidirectional. However, phased arrays may have directivity patterns with two, three, or even four different optimum directions. A bidirectional pattern can be obtained, for example, by spacing the dipoles at one wavelength, and feeding them in phase, as shown at B.

More complicated phased arrays are sometimes used by radio transmitting stations. Several vertical radiators, arranged in a specified pattern and fed with signals of specified phase, produce a designated directional pattern. This is done to avoid interference with other broadcast stations on the same channel.

Phased arrays may have fixed directional patterns, or they may have rotatable or steerable patterns. The pair of phased dipoles (A) may, if the wavelength is short enough to allow construction from metal tubing, be mounted on a rotator for 360-degree directional adjustability. With phased vertical antennas, the relative signal phase can be varied, and the directional pattern thereby adjusted to a certain extent. *See also* PHASE.

PHASE DIFFERENCE

See PHASE, PHASE ANGLE.

PHASE DISTORTION

When the output of an amplifier fluctuates in phase, even though the input does not, the circuit is said to introduce phase distortion into the signal.

Phase distortion does not always cause problems in radio equipment. When phase distortion occurs, the effect is similar to phase modulation or frequency modulation of the signal. The signal may have a buzzing or fluttering

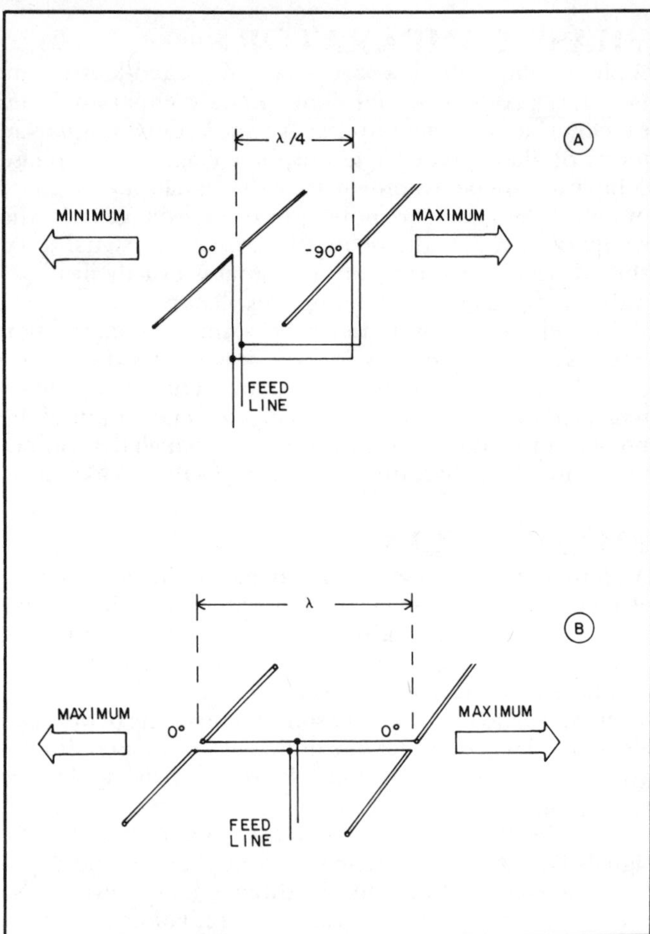

PHASED ARRAY: Examples of phased dipole arrays. At A, a unidirectional pattern is produced. At B, a bidirectional pattern is obtained. (Patterns are in the horizontal plane.)

sound as heard in a frequency-modulation receiver. In an amplitude-modulation or code receiver, phase distortion is usually not noticeable.

Phase distortion sometimes occurs when a phase-locked-loop circuit malfunctions. If the circuit is not able to lock properly, the frequency and/or phase of the signal will continually shift as the device "attempts" to stabilize. In a phase-modulation or frequency-modulation transmitter or receiver, phase distortion is extremely undesirable, since it interferes with the transmitted information. *See also* PHASE, PHASE-LOCKED LOOP.

PHASED VERTICALS

See PHASED ARRAY.

PHASE INVERTER

A phase inverter is a circuit that literally flips the waveform upside down. Phase inversion may be accomplished by most ordinary single-ended amplifier circuits. Transformers can also be used.

Phase inversion differs from a phase delay of 180 degrees. A certain delay time will result in 180 degrees of phase delay at some frequency f, and the equivalent of 180-degree delay at all odd harmonics 3f, 5f, 7f, and so on. But a phase inverter will operate effectively at all frequencies.

The paraphase inverter circuit, having a single-ended input and a push-pull output, is sometimes called a phase inverter. This circuit produces two output signals, in phase opposition with respect to each other. Such a circuit may be used to drive a push-pull or push-push amplifier, without the need for an input transformer. *See also* PARAPHASE INVERTER, PHASE.

PHASE LAG

See ANGLE OF LAG.

PHASE LEAD

See ANGLE OF LEAD.

PHASE-LOCKED LOOP

A phase-locked loop is a circuit that produces a signal with variable frequency. The phase-locked loop is extensively used in radio equipment, both for transmitting and receiving.

The phase-locked loop, abbreviated PLL, has a voltage-controlled oscillator, the frequency of which can be varied by means of a varactor diode. The voltage-controlled oscillator is set for a frequency close to the intermediate frequency. A phase comparator causes the voltage-controlled oscillator to stabilize at exactly the intermediate frequency. If the voltage-controlled-oscillator frequency departs from the intermediate frequency, the phase comparator produces a voltage that brings the oscillator back to the correct frequency (*see* PHASE COMPARATOR). If the signal frequency changes, the voltage-controlled oscillator follows along, provided that the change is not too rapid. The drawing shows a block diagram of a phase-locked-loop circuit.

A PLL requires a certain amount of time to compensate for changes in the frequency of the incoming signal. If the incoming-signal frequency is modulated at more than a

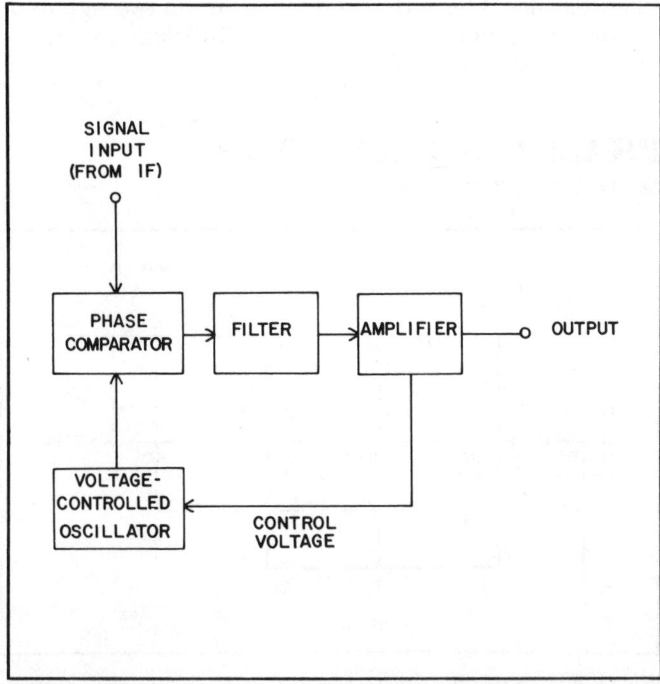

PHASE-LOCKED LOOP: Block diagram of a phase-locked loop circuit.

few hertz, the PLL will not lock continuously, but it will produce a fluctuating error voltage. The PLL can therefore be used for the purpose of detecting frequency-modulated or phase-modulated signals.

Modern PLL circuits can be found housed in single integrated-circuit packages. Phase-locked-loop circuits are used in most frequency-modulation communications equipment at the high, very-high, and ultra-high frequencies today. *See also* FREQUENCY MODULATION, PHASE MODULATION, VOLTAGE-CONTROLLED OSCILLATOR.

PHASE MODULATION

Phase modulation is a method of conveying information via radio-frequency carrier waves. The instantaneous phase of the carrier is shifted in accordance with the modulating waveform.

Phase modulation is very similar, in practice, to frequency modulation. This is because any change in the instantaneous phase of a carrier also results in an instantaneous fluctuation in the frequency, and vice-versa.

In phase modulation, the extent of the phase shift is directly proportional to the amplitude of the modulating signal. The rapidity of the phase shift is directly proportional to both the amplitude and the frequency of the modulating signal. This differentiates phase modulation from frequency modulation; the result is a difference in the frequency-response characteristics.

Many frequency-modulated transmitters actually employ phase modulation of one of the amplifier stages. When this is done, the high frequencies appear exaggerated at the receiver unless an audio-frequency lowpass filter is used at the transmitter. The output of such a filter must decrease in direct proportion to the modulating frequency. When this modification has been made at the transmitter, it is impossible to distinguish between phase modulation and true frequency modulation at the receiver. *See also* FREQUENCY MODULATION, PHASE.

PHASE OPPOSITION

When two signals are exactly 180 degrees different in phase, they are said to be in phase opposition.

Phase opposition results in a net amplitude that is the absolute value of the difference in the amplitudes of two signals having identical frequency. The phase of the resultant signal is the same as the phase of the stronger of the two constituent signals (see illustration). If the two signals have the same amplitude, they combine in phase opposition to produce no signal. This effect is called phase cancellation.

Two signals may be in phase opposition for either of two reasons: One signal may be delayed with respect to the other by one-half cycle, or one signal may be inverted (upside down) with respect to the other signal. A phase delay of 180 degrees can be obtained with a half-wavelength delay line. Phase inversion can be accomplished with an amplifier circuit or a transformer. *See also* PHASE INVERTER.

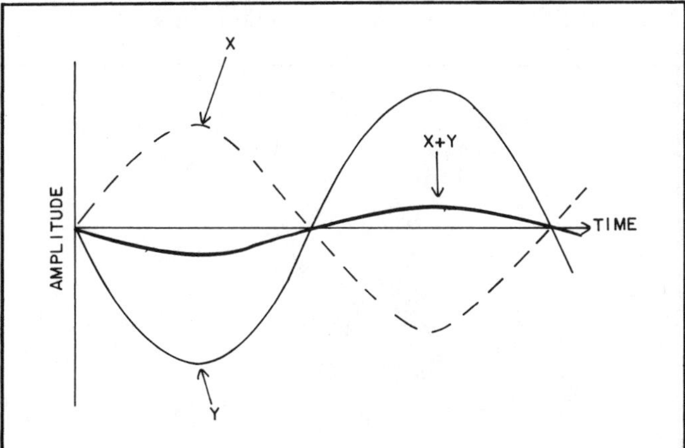

PHASE OPPOSITION: The sum of the two out-of-phase signals, X and Y, has phase corresponding to the stronger signal (in this case Y).

PHASE REINFORCEMENT

Two signals of identical frequency are said to be in phase reinforcement when they have the same phase.

Waves in phase reinforcement add in amplitude arithmetically. That is, the amplitude of the resultant signal is equal to the arithmetic sum of the amplitudes of the constituent signals (see illustration). *See also* PHASE.

PHASE SHIFT

A phase shift is a change in the phase of a signal. Phase shift may occur over a small fraction of one cycle, or it may occur over a span of many cycles. Phase shift is normally measured in degrees. One degree of phase shift corresponds to 1/360 (0.00278) cycle.

Certain electronic circuits shift the phase of an input signal. Phase shift is normally expressed as either positive or negative in such cases. A negative phase shift refers to a delay of up to one-half cycle (see A in the illustration). A positive phase shift is a delay of more than one-half cycle, and up to one full cycle, as at B. Generally, a phase delay of one full cycle is equivalent to zero phase shift.

A phase shift of more than 360 degrees can generally be expressed as equivalent to some value of shift smaller than 360 degrees. For example, if a signal is delayed by 500 degrees, the result is effectively equivalent to a delay of 500 − 360, or 140, degrees.

Phase shift can be used to convey information via radio-frequency carrier waves. In such a system, the instantaneous phase of a carrier wave is made to follow the waveform of some input signal, such as a human voice. This is called phase modulation. *See also* PHASE, PHASE ANGLE, PHASE MODULATION.

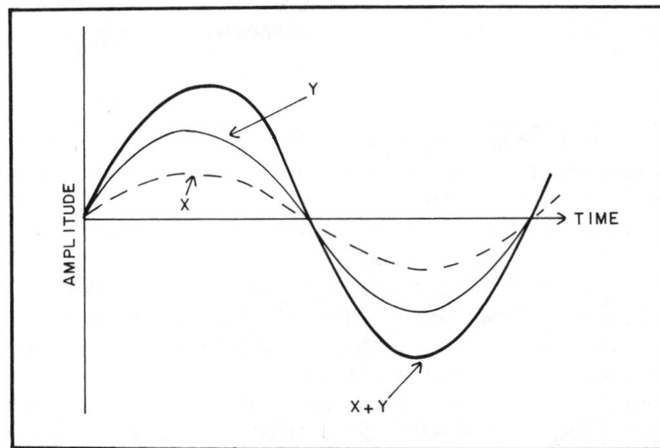

PHASE REINFORCEMENT: The two in-phase signals, X and Y, add arithmetically.

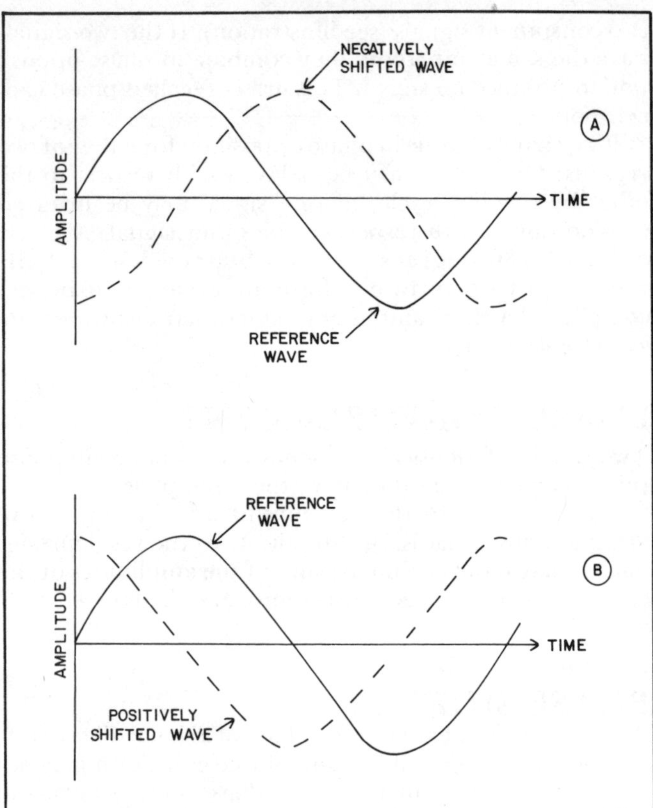

PHASE SHIFT: At A, 90-degree (¼-cycle) negative shift is shown. At B, 90-degree positive shift is shown.

PHASE-SHIFT DISCRIMINATOR

See FOSTER-SEELEY DISCRIMINATOR.

PHASE-SHIFT KEYING

Phase-shift keying is a method of transmitting digital information. It is similar to frequency-shift keying (*see* FREQUENCY-SHIFT KEYING), except that the phase, not the frequency, is shifted.

In phase-shift keying, the carrier is transmitted at a constant amplitude. During the space, or key-up, condition, the wave is at one phase; during the mark, or key-down, condition, the phase is altered by a predetermined amount.

Phase-shift keying may be used in place of frequency-shift keying in certain applications. The primary advantage of phase-shift keying is that it can be accomplished in an amplifier stage, whereas frequency-shift keying cannot. *See also* PHASE.

PHASING

Phasing is the technique by which a phased array is made to have a certain directional characteristic (*see* PHASED ARRAY). A device that accomplishes this phasing is sometimes called a phasing harness.

Depending on the relative phase of the elements in a phased antenna, and on their spacing, the radiation pattern may have one, two, three, or more major lobes. Phasing harnesses usually consist of simple delay lines. A transmission line, measuring an electrical quarter wavelength, produces a delay of 90 degrees. A half-wave section of line causes the signal to be shifted in phase by 180 degrees. *See also* PHASE.

PHASOR

See COMPLEX NUMBER.

PHONE

See RADIOTELEPHONE, TELEPHONE.

PHONE JACK

A phone jack is a female connector, having two or three conductors, and usually measuring ⅛ inch to ¼ inch in diameter. Phone jacks are commonly used in communications equipment, especially in audio-frequency applications where impedance continuity is not critical.

Phone jacks have an outer, grounded terminal, called the sleeve. The innermost terminal is called the tip. There may be an intermediate terminal, called the ring. Some phone jacks have an auxiliary set of terminals, used for breaking an external circuit when a plug is inserted. The drawing is an illustration of a typical phone jack. The phone jack mates with a phone plug of the same size and the same number of conductors.

The phone jack is usually mounted in the chassis of a piece of apparatus, while the plug, or male connector, is mounted on the cord. Phone jacks are widely available in electronics retail stores. *See also* PHONE PLUG.

PHONE PLUG

A phone plug is a male connector that mates with a phone jack (*see* PHONE JACK). Phone plugs are connected to two-conductor or three-conductor cords; accordingly, a plug may have two or three conductors. Phone plugs may be ⅛ inch or ¼ inch in diameter.

The outer portion is called the sleeve, and is generally grounded. If coaxial cable is used, the sleeve is connected to the shield. The inner contact is called the tip. In a three-channel phone plug, the middle contact is called the ring. The illustration shows a typical, two-conductor, ¼-inch phone plug.

Phone plugs are used mostly in audio applications, where impedance continuity is not critical. Such plugs can be found in most electronics retail stores for a very low price.

PHONETIC ALPHABET

The phonetic alphabet is a set of 26 words, one for each letter of the alphabet, used by radiotelephone operators

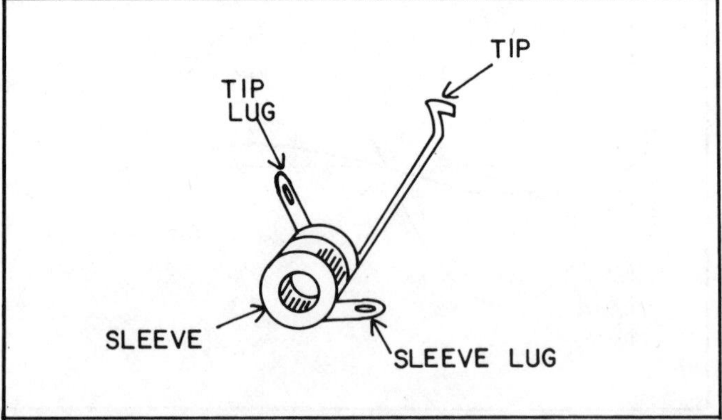

PHONE JACK: A two-conductor phone jack.

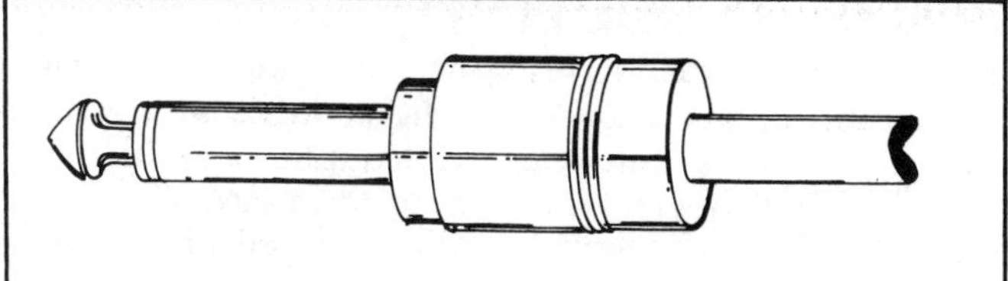

PHONE PLUG: A standard, ¼-inch, two-conductor phone plug.

for the purpose of clarifying messages under marginal conditions. The words are deliberately chosen so that they are not easily confused with other words in the list. The table shows the most commonly used English phonetic alphabet.

When it is necessary to spell out a word in a radiotelephone communication, the operator will say, for example, "City of Miami. I spell: Mike, India, Alpha, Mike, India."

Phonetics should be used only when necessary. Otherwise, the receiving operator may be confused by them. Phonetics should not be used to spell out common words (such as prepositions, verbs, and the like).

PHONETIC ALPHABET PHONETIC ALPHABET RECOMMENDED BY THE INTERNATIONAL TELECOMMUNICATION UNION (ITUz.

Letter	Word	Letter	Word
A	Alpha	N	November
B	Bravo	O	Oscar
C	Charlie	P	Papa
D	Delta	Q	Quebec
E	Echo	R	Romeo
F	Foxtrot	S	Sierra
G	Golf	T	Tango
H	Hotel	U	Uniform
I	India	V	Victor
J	Juliet	W	Whiskey
K	Kilo	X	X ray
L	Lima	Y	Yankee
M	Mike	Z	Zulu

PHONO JACK

A phono jack is a female connector, usually chassis-mounted, and used for audio-frequency and radio-frequency applications. The phono jack measures approximately ¼ inch in diameter. It resembles the phone jack (*see* PHONE JACK).

Phono jacks are convenient and easy to use. They have fairly good impedance continuity, and are sutiable for low-power and medium-power applications up through the high frequencies. They begin to get somewhat lossy at very-high and ultra-high frequencies. The illustration is a drawing of a typical female phono jack. It mates with the phono plug. *See also* PHONO PLUG.

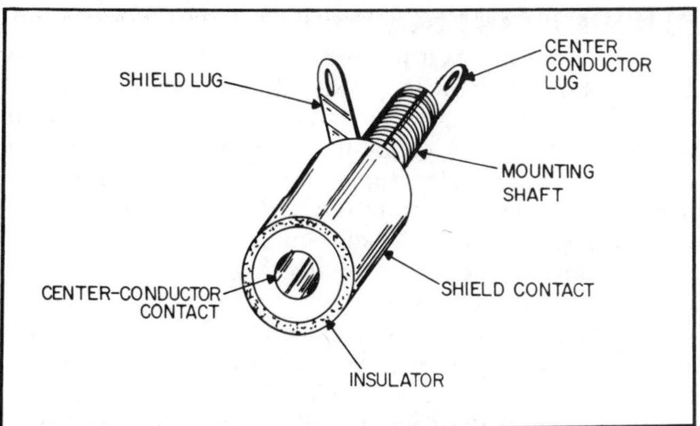

PHONO JACK: The outside diameter of the shield contact is ¼ inch.

PHONO PLUG

A phono plug is a ¼-inch male connector that is often used in audio-frequency applications. It is also suitable for use at radio frequencies up to approximately 30 MHz, in low and medium-power systems.

Phono plugs have a coaxial configuration (see illustration). They can be installed on all types of coaxial cable up to, and including, RG-59/U. Phono plugs exhibit better impedance continuity than phone plugs and jacks, but when an excellent match is needed, other types of connectors, such as PL-259/SO-239 or N type, are preferable. *See also* PHONO JACK.

PHOSPHOR

A phosphor material is a substance that glows when bombarded by a beam of electrons or other high-speed subatomic particles. Phosphor materials will also usually glow when exposed to ultraviolet rays, X rays, or gamma rays.

Phosphor materials are available in a wide variety of different glow colors. A common color is yellow-green, which is used in oscilloscopes, radar, and video-display terminals. White phosphors are used in black-and-white television receivers. Color television receivers utilize tricolor phosphors—red, blue, and green. Phosphors can be found in almost any color.

Various different phosphors have different persistence characteristics. The persistence is a measure of how long the phosphor will continue to glow after the radiation has been removed. The optimum degree of persistence depends on the application. For example, in television reception, a short persistence is needed; in radar, a longer persistence is desirable.

Common phosphor materials include zinc, silicon, and potassium compounds. *See also* CATHODE-RAY TUBE.

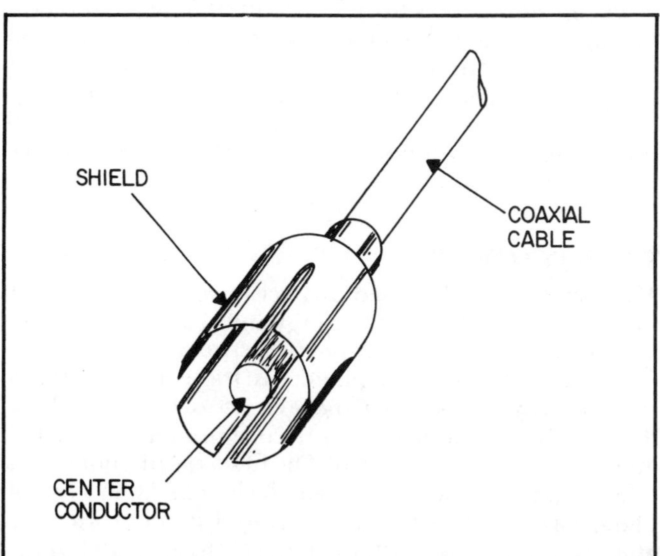

PHONO PLUG: The inside diameter of the shield section is ¼ inch, so that it fits snugly onto the phono jack.

PHOSPHORESCENCE

Phosphorescence is a property that some substances exhibit, causing them to glow under bombardment by high-speed atomic particles or short-wave radiation such as ultraviolet, X rays, or gamma rays. Some phosphorescent materials stop glowing almost immediately after the energizing radiation is removed. Others continue to glow for seconds, minutes, or even hours.

Phosphorescence occurs because high-energy radiation imparts additional energy to the atoms in the substance. This causes the electrons to move into higher orbits (*see* ELECTRON ORBIT). When the energizing radiation is removed, the electrons fall back to their original lower-energy orbits more or less rapidly, giving off photons of visible light in the process. *See also* PHOSPHOR.

PHOT

The phot is the centimeter-gram-second unit of illuminance. The more frequently used unit is the lux, which is equivalent to 1 lumen per square meter. The phot is equal to 1 lumen per square centimeter.

Since there are 10,000 square centimeters in one square meter, 1 phot represents 10,000 lux. *See also* ILLUMINANCE, LUMEN.

PHOTOCATHODE

In a phototube, the photocathode is a light-sensitive electrode. The photocathode emits electrons when light of sufficient intensity falls on it. The electrons emitted by the photocathode, as a result of impinging light, are called photoelectrons. In the case of monochromatic light, the number of photoelectrons emitted is directly proportional to the energy in the arriving light. (For non-monochromatic light, this is not true.)

The photocathode of a camera tube or phototube may respond to a wide spectrum of wavelengths. Typically, the response is maximum at one particular wavelength, and the sensitivity falls off at longer and shorter wavelengths. The peak wavelength can be adjusted, in practice, by the use of optical filters.

Photocathode sensitivity is measured in microamperes or milliamperes per lumen. Various materials result in different sensitivity levels. For incandescent light (tungsten filament), typical sensitivity values range from less than 5 microamperes per lumen to more than 100 microamperes per lumen. *See also* CAMERA TUBE, PHOTOMOSAIC, PHOTOTUBE.

PHOTOCELL

The term photocell is used to describe any of a variety of devices that convert light energy into electrical energy. Photocells are often called photoelectric cells.

There are two kinds of photocells: those that generate a current all by themselves in the presence of light, and those that simply cause a change in effective resistance when the intensity of the light is varied. The first type of photocell is called a photovoltaic cell or solar cell. The latter type of photocell can be found in a variety of different forms. *See also* PHOTOCONDUCTIVITY, PHOTODIODE, PHOTOFET, PHOTORESISTOR, PHOTOTRANSISTOR, PHOTOTUBE, PHOTOVOLTAIC CELL, SOLAR CELL.

PHOTOCONDUCTIVITY

Certain materials exhibit a resistance, or conductivity, that varies in the presence of visible light, infrared, or ultraviolet. Such substances are called photoconductive materials. The property of changing resistance in accordance with impinging light intensity is called photoconductivity.

In general, a photoconductive substance has a certain finite resistance when there is no visible light falling on it. As the intensity of the visible light increases, the resistance decreases. There is a limit, however, to the extent that the resistance will continue to decrease as the light gets brighter and brighter (see illustration).

Photoconductivity occurs in almost all materials to a certain extent, but it is much more pronounced in semiconductors. When light energy strikes a photoconductive material, the charge-carrier mobility increases. Thus, current can be more easily made to flow when a voltage is applied. The more photons are absorbed by the material for a given electromagnetic wavelength, the more easily the material conducts an electric current.

Photoconductive materials are used in the manufacture of photoelectric cells. Some examples of photoconductive substances are germanium, silicon, and the sulfides of various other elements. *See also* PHOTODIODE, PHOTOFET, PHOTORESISTOR, PHOTOTRANSISTOR.

PHOTODIODE

A photodiode is, as its name suggests, a diode that exhibits a variable effective resistance depending on the intensity of visible light that lands on its P-N junction.

Photodiodes are quite common. In fact, most ordinary diodes having P-N junctions are photodiodes (*see* DIODE, P-N JUNCTION). The reason that not all diodes can be used as photodiodes is simply that the housing is opaque, or that the P-N junction is situated between two opaque pieces of semiconductor material. Thus, no light can reach the junction.

A photodiode is specially designed so that its P-N junction is readily exposed to visible light. This necessitates the use of a large suface area for the junction. A transparent enclosure is also needed. The conductivity of a photodiode

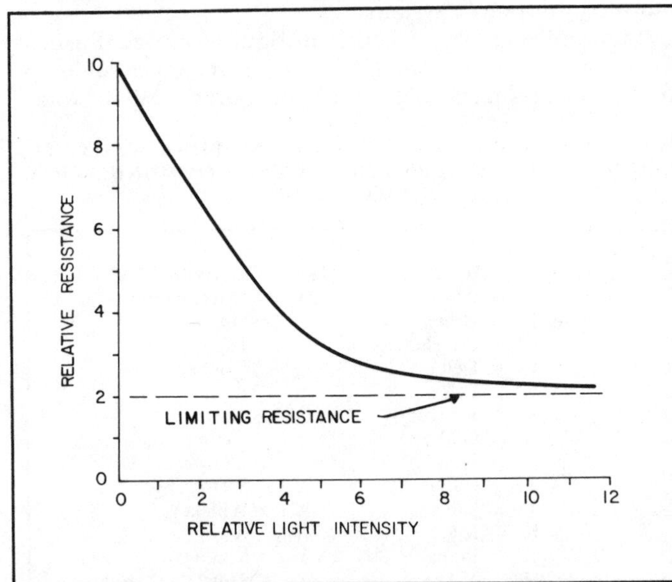

PHOTOCONDUCTIVITY: A typical photoconductivity curve.

generally increases in the reverse direction as the impinging light gets brighter, but there is a limit to the drop in forward resistance. This limiting value is called the saturation value.

A photodiode is connected in series with the circuit to be controlled, and is reverse-biased. Care must be exercised to ensure that the photodiode does not carry too much current. The diagram illustrates a typical circuit incorporating a photodiode for the purpose of opening and closing a relay. *See also* PHOTOTRANSISTOR.

PHOTOELECTRIC CELL

See PHOTOCELL.

PHOTOELECTRIC EFFECT

See PHOTOCELL, PHOOCONDUCTIVITY, PHOTODIODE, PHOTOET, PHOTORESISTOR, PHOTOTRANSISTOR, PHOTOTUBE, PHOTOVOLTAIC CELL, SOLAR CELL.

PHOTOELECTRON

See PHOTOCATHODE.

PHOTOEMISSION

See PHOTOCATHODE.

PHOTOFET

A photoFET is a field-effect transistor that exhibits photoconductive properties (*see* FIELD-EFFECT TRANSISTOR, PHOTOCONDUCTIVITY).

Most ordinary field-effect transistors would be suitable for use as photoFETs, except that their housings and construction are opaque, and it is therefore impossible for any light to reach the P-N junction inside the device.

The photoFET must be constructed in such a way that its P-N junction, forming the boundary between the gate and the channel, has the largest possible area. The P-N junction must be situated so that light can fall on it. The enclosure for the device must be transparent. PhotoFETs differ from other kinds of photocells primarily because of their higher impedance.

PHOTOMETRY

Photometry is the science of measurement of visible-light intensity. There are many ways to electronically measure the brightness of a light source, but photometry is especially concerned with the intensity of light as the human eye sees it.

The human eye is more sensitive to light in the yellow and green part of the spectrum than it is at other wavelengths. Most people can see electromagnetic radiation between about 390 and 750 nanometers wavelength (a nanometer is 10^{-9} meter), or about 3900 to 7500 Angstroms (an Angstrom unit corresponds to 0.1 nanometer of 10^{-10} meter). The longer wavelengths appear as a progressively deepening red; the shorter wavelengths as fading violet. Some people can see energy at wavelengths a little bit shorter than 390 nanometers, and/or a little longer than 750 nanometers. The human-eye response curve is approximately shown in the illustration.

Different types of photoconductive materials exhibit different functions of sensitivity versus wavelength. In photometry, it is advantageous to use photoconductive substances with response closely resembling that of the human eye. A typical light-measuring device consists of a photocell, a source of voltage, perhaps a variable resistance, and a milliameter or microammeter, all connected in series. The meter scale must be calibrated appropriately before the instrument can be used. *See also* LIGHT METER, PHOTOCELL, PHOTOCONDUCTIVITY.

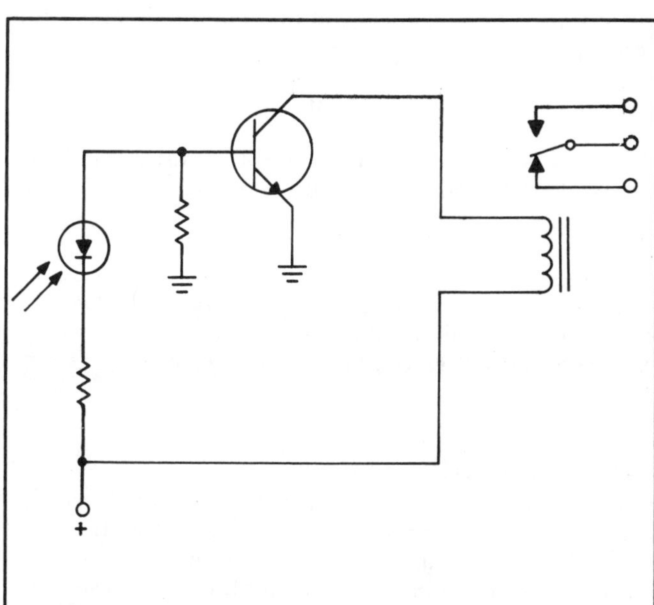

PHOTODIODE: Schematic diagram of a simple relay-actuating circuit using a photodiode. The photodiode is reverse-biased. Light causes it to conduct, switching the transistor on and providing current to the relay coil.

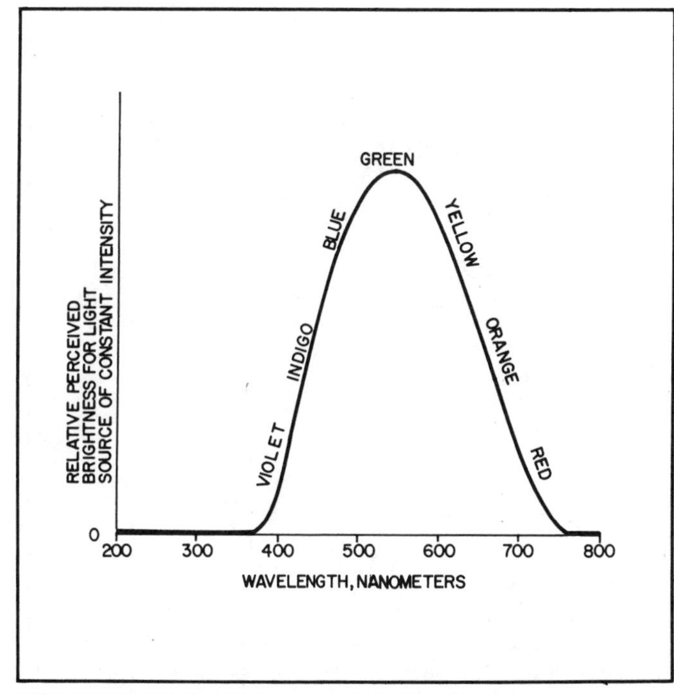

PHOTOMETRY: Sensitivity curve of the human eye.

PHOTOMOSAIC

In a camera tube, the photocathode is divided into a grid of small, photosensitive dots (*see* CAMERA TUBE, PHOTOCATHODE). This grid of dots is called the photomosaic.

Each small spot on the photomosaic recieves a certain amount of visible-light energy, depending on the image received. A lens focuses the image onto the flat photomosaic. An electron beam then scans the photomosaic, and the beam is modulated according to the amount of light falling on each spot (see illustration). *See also* COMPOSITE VIDEO SIGNAL, IMAGE ORTHICON, VIDICON.

PHOTOMULTIPLIER

A photomultiplier is a vacuum-tube device that generates a variable current depending on the intensity of the light that strikes it. The photomultiplier gets its name from the fact that it literally multiplies its own output, thereby obtaining extremely high sensitivity. Photomultipliers are used for measurement of light intensity at low levels.

The photomultiplier consists of a photocathode (*see* PHOTOCATHODE), which emits electrons in proportion to the intensity of the light impingng on it. These electrons are focused into a beam, and this beam strikes an electrode called a dynode (*see* DYNODE). the dynode emits several secondary electrons for each electron that strikes it. Generally, the dynode produces four or five secondary electrons for each primary electron (*see* SECONDARY EMISSION). The resulting, intense electron beam is collected by the anode.

A photomultiplier tube may have several dynodes, resulting in a very large amount of gain. The amount by which the sensitivity can be improved, by cascading one dynode after another, is limited by the amount of background electron emission from the photocathode. This background emission is called the dark noise, or photocathode dark noise.

The illustration shows the basic principle of operation of the photomultiplier. *See also* PHOTOTUBE.

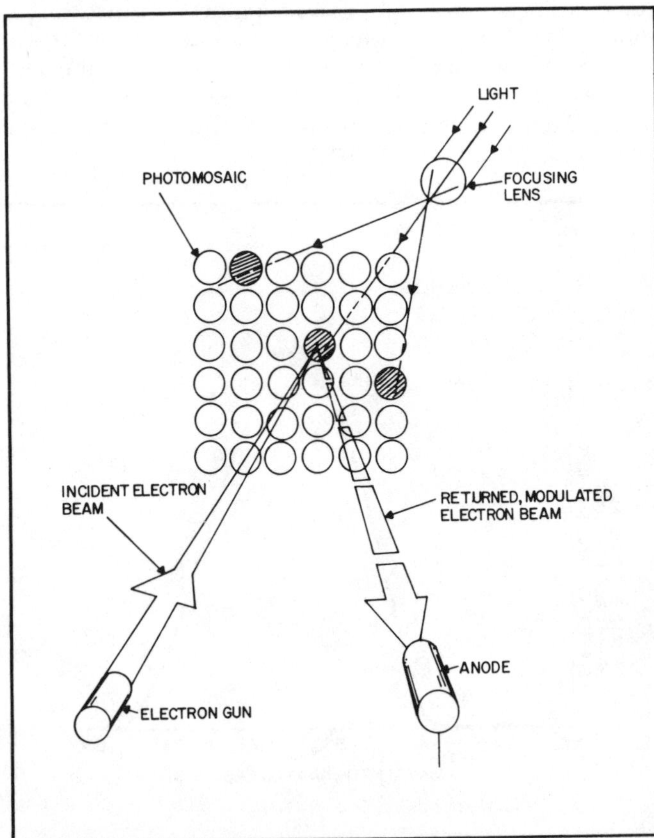

PHOTOMOSAIC: Operation of the photomosaic, resulting in modulation of an electron beam.

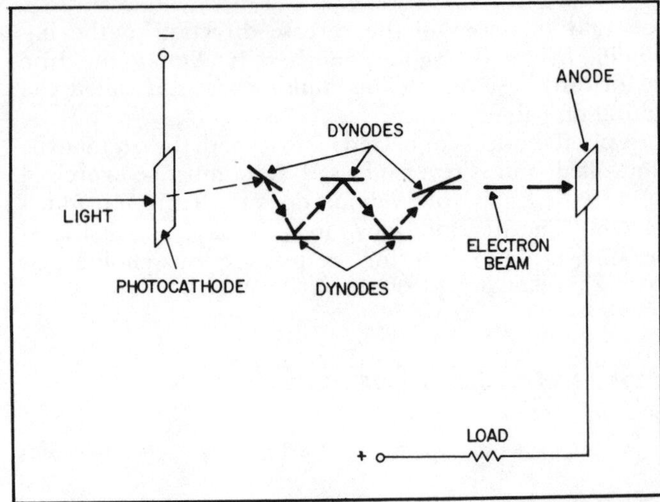

PHOTOMULTIPLIER: Operation of the photomultiplier tube. The electron beam (dotted line) becomes more intense as it bounces off each dynode.

PHOTON

A photon is a particle of electromagnetic radiation. We normally think of a photon as a packet of visible-light energy, but actually all electromagnetic radiation is made up of particles.

Isaac Newton was the first scientist to formulate, in detail, a particle theory of light. Modern scientists can observe both particle-like and wave-like properties of electromagnetic radiation. The energy contained in a single photon depends on the wavelength.

If e represents the energy of one photon, in ergs, and the frequency in hertz if f, then:

$$e = hf$$

where h is Planck's constant, 6.62×10^{-27} erg-seconds. If the wavelength is given by λ, in meters, then:

$$e = (3 \times 10^8)h/\lambda$$

Photons can be observed to exert pressure on objects, as a barrage of particles would. *See also* ELECTROMAGNETIC THEORY OF LIGHT, PARTICLE THEORY OF LIGHT.

PHOTORESISTOR

A photoresistor is a device that exhibits a variable resistance, depending on the amount of light that strikes it. Normally, the resistance is a certain finite value in total darkness; as the light intensity increases, the resistance decreases. But this only occurs up to a certain point. If the intensity of the light increases further than the maximum limit of the device, the resistance does not drop any further. *See also* PHOTOCONDUCTIVITY.

PHOTOTRANSISTOR

A phototransistor is a bipolar transistor that varies in effective resistance as the intensity of radiant light changes (*see* PHOTOCONDUCTIVITY).

Transistors are normally enclosed in opaque packages, for the purpose of eliminating noise caused by the action of ambient light in the collector-base junction. Phototransistors have transparent or translucent packages, so that light can reach this junction. Phototransistors are constructed in such a way that the base-collector P-N junction can receive the maximum possible amount of light.

The most common phototransistors are of the NPN type, and are manufactured from silicon. Most phototransistors do not have terminals at the base electrode, but only at the emitter and the collector. Phototransistors are sometimes biased by means of a light-emitting diode near the device; but base bias is not necessary and, in fact, may even be detrimental to the performance of a phototransistor.

Phototransistors are used in much the same way as photodiodes and photoresistors. The effective conductivity increases as the light intensity increases. This occurs, however, only up to a certain point, known as the saturation point. *See also* PHOTODIODE, PHOTORESISTOR.

PHOTOTUBE

A phototube is an electron tube that exhibits variable resistance, depending on the amount of light that strikes its cathode. Phototubes are generally sensitive to infrared, ultraviolet, X rays, and gamma rays as well as to visible light.

The greater the intensity of the radiation striking the cathode of the phototube, the greater the number of electrons that are emitted. Thus, the resistance of the tube goes down as the intensity of the light increases. The tube shows a certain finite resistance when there is no light; the value of the resistance eventually reaches a defined minimum as the light gets brighter.

Phototubes were once used in a variety of control devices, such as light-actuatd switches. However, in recent years, solid-state devices have largely replaced phototubes. The solid-state photoelectric devices require far less voltage than the phototube. Solid-state photoelectric devices also have no need for a filament power supply. However, phototubes are still favored in applications where a high voltage must be switched or regulated by visible light, infrared, ultraviolet, X rays, or gamma rays.

Television camera tubes are sometimes called phototubes. Such tubes, like ordinary phototubes, are light-sensitive. However, the camera tube is more complex than the conventional phototube. *See also* CAMERA TUBE, PHOTOCONDUCTIVITY.

PHOTOVOLTAIC CELL

A photovoltaic cell is a semiconductor device that generates a direct current when it is exposed to visible light. Photovoltaic cells generally consist of a P-N junction having a large surface area, and a transparent housing to allow light to enter easily (see A in illustration).

Most photovoltaic cells are made from silicon. Some are made from selenium. Photovoltaic cells require no external bias; they generate their electricity all by themselves.

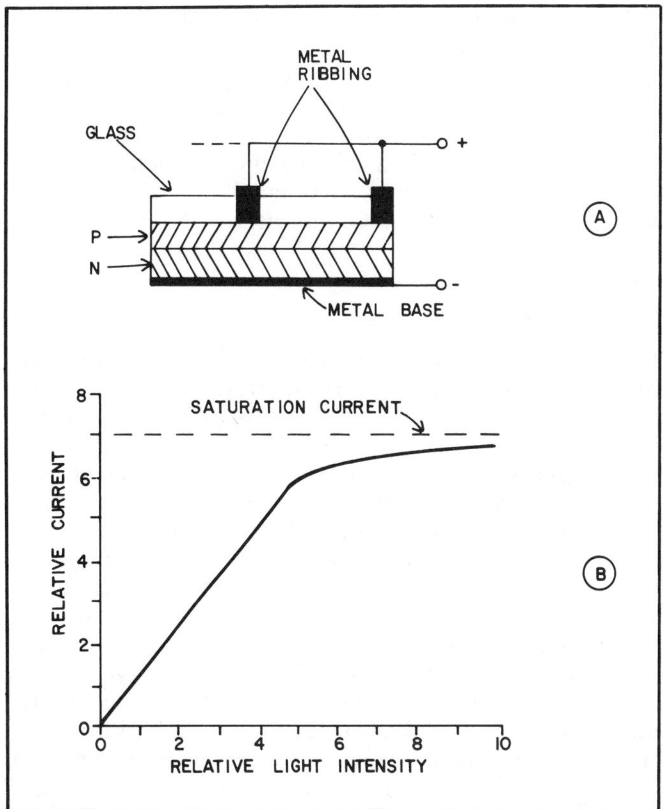

PHOTOVOLTAIC CELL: At A, construction of a typical silicon photovoltaic cell. At B, current-versus-brightness characteristic of the photovoltaic cell.

When light (and, in some cases, infrared or ultraviolet energy) impinges on the P-N junction, electron-hole pairs are produced. The intensity of the current from the photovoltaic cell, under constant load conditions, varies in linear proportion to the brightness of the light striking the device, up to a certain point. Beyond that point, the increase is more gradual, and it finally levels off at a maximum current called the saturation current, as at B.

The ratio of the available output power to the light power striking a photovoltaic cell is called the efficiency, or the conversion efficiency, of the cell. Most photovoltaic cells have relatively low conversion efficiency, about 10 to 15 percent. Advances in the state of the art are expected to improve this figure in the future.

Photovoltaic cells are used in many different electronic devices, including modulated-light systems and solar-power generators (*see* MODULATED LIGHT, SOLAR POWER).

One important application of the photovoltaic cell is its power-generating capability. Photovoltaic cells can, in conjunction with storage batteries, provide continuous electric power from sunlight. This is especially true in the lower latitudes, and at the higher elevations, where sunshine is relatively abundant. At present, solar power, obtained from photovoltaic cells, is still relatively expensive, but its cost is decreasing. *See also* SOLAR CELL, SOLAR POWER.

PI π

Pi is an irrational number, representing the ratio of the circumference of a circle to its diameter. Pi is symbolized by the Greek lower-case letter π. The value of π, to ten significant digits, is 3.141592654.

The number π occurs in many different mathematical, physical, and enineering equations. Some mathematicians have calculated the value of π, using computers, to hundreds of thousands of decimal places.

An unbalanced filter, containing two parallel elements on either side of a single series element, is sometimes called a pi network, because of its resemblance (in schematic diagrams, at least) to the Greek letter π. *See also* PI NETWORK.

PICKUP

Any kind of sensing device may be called a pickup. This includes sensors for electromagnetic energy, sound energy, motion, acceleration, vibration, and other parameters.

Pickups operate in various different ways. Some devices exhibit a variable resistance with changes in the intensity of the quantity to be measured. Such pickups include photoconductive devices, carbon microphones, and various sorts of switches. Some pickups generate a fluctuating direct current; photovoltaic cells are an example of such a device. Some pickups are transducers; this includes most microphones and other sensors of acoustic energy.

In high-fidelity sound applications, the device at the end of the phonograph arm is known as the pickup. Pickups generally operate either by means of a magnet and coil, or by piezoelectric action. The needle, or stylus, follows the groove in the disk as the disk rotates. Irregularities in the groove cause the stylus to vibrate. These vibrations coincide with the sound that was recorded on the disk. The motion of the stylus is translated into electrical impulses by the pickup device. *See also* CERAMIC PICKUP, DYNAMIC PICKUP.

PICO

Pico is a prefix multiplier meaning one trillionth, or 10^{-12}. The abbreviation for pico is the small letter p.

Some electrical units, such as the farad, are extremely large in practice. At the high, very-high, and ultra-high radio frequencies, capacitances are often specified in picofarads (pF). A capacitance of 1 pF is equivalent to a millionth of a microfarad (1 pF = 10^{-6} μF). *See also* PREFIX MULTIPLIERS.

PICTURE SIGNAL

A picture signal, sometimes called a video signal, refers to any of a variety of different methods of conveying visual images by modulating electromagnetic waves. The picture signal most familiar to us is the television signal.

A television signal consists of a composite video signal that modulates a radio-frequency carrier (*see* COMPOSITE VIDEO SIGNAL, TELEVISION). Some television signals convey color as well as brightness information; others convey only brightness.

The standard fast-scan television signal in the United States is defined by the Federal Communications Commission. The channel bandwidth is nominally 6 MHz. The picture and sound carriers are separated by 4.5 MHz within this passband. The picture carrier is near the bottom of the passband, and the sound carrier is almost at the very top. The picture carrier is amplitude-modulated, while the sound carrier is frequency-modulated. The illustration is a spectral representation of the typical fast-scan television signal. An amplitude-versus-time diagram, showing the modulating components of a picture signal, is illustrated under the listing COLOR PICTURE SIGNAL. The black-and-white signal is the same as the color signal, except that the color-information components are absent.

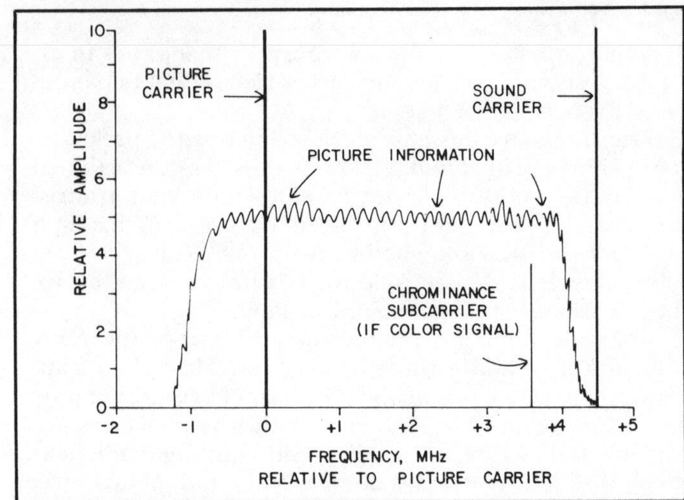

PICTURE SIGNAL: Fast-scan, amplitude-modulated television picture signal.

The number of scanning lines in a fast-scan picture signal varies from country to country. In the United States, 525 lines are used, with a horizontal scanning frequency of 15.734264 MHz for color signals and 15.75 MHz for black-and-white signals. The picture scanning occurs from left to right and from top to bottom, exactly as you would read a single-column book. The frame-scanning rate is 59.94 Hz for color and 60 Hz for black-and-white.

In recent years, technology has improved with respect to television picture transmission, and some thought has been given to increasing the number of scanned lines per frame. This would provide a picture with greater resolution than is now possible. Since the advent of cable television transmission, which has almost eliminated reception problems associated with airwave propagation, high-resolution television has become entirely practical.

Fast-scan television is not the only means of transmitting a picture. Radio amateurs commonly use a method of transmission called slow scan (*see* SLOW-SCAN TELEVISION). The resolution of slow scan is not as good as that of fast scan, and motion information is not conveyed, but the bandwidth is much narrower. A method of picture transmission that gives excellent resolution, but requires much more time per frame, is called facsimile or fax. This mode is extensively used by weather satellites (*see* FACSIMILE, WEATHER SATELLITE).

PICTURE TUBE

A picture tube is a special form of cathode-ray tube, used for the purpose of receiving fast-scan television signals (*see* CATHODE-RAY TUBE). Television picture tubes can be found in an enormous variety of sizes; some are smaller than a low-wattage incandescent light bulb, while others are larger than a person and weigh hundreds of pounds. The two main types of picture tubes are the black-and-white tube and the color tube.

The picture tube, also known as the kinescope, operates as shown in the illustration. An electron gun or guns, located at the back of the tube, produces a stream of

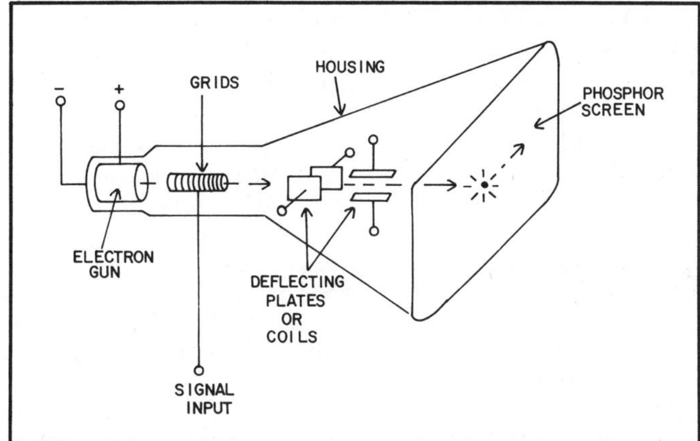

PICTURE TUBE: Operation of the television picture tube. This drawing shows a black-and-white tube.

electrons. The gun consists of a cathode and a set of focusing anodes (*see* ELECTRON-BEAM GENERATOR). Deflecting electrodes cause the beam to scan the screen in the characteristic 525-line format, with a frame-repetition rate of 59.94 Hz for color and 60 Hz for black-and-white.

The signal modulation is applied to the control grid or grids, resulting in fluctuations in the intensity of the electron beam as it scans the lines. As the beam strikes the phosphor screen, this modulation produces lighter and darker regions. From a distance, we see a picture.

In black-and-white television, there is only one electron gun, one set of deflecting plates, and one set of modulating grids. This is because only the brightness of the signal varies. The phosphor material glows white when the electron beam strikes it.

In color television, the phosphor screen consists of an array of dots or vertical phosphors that glow red, blue, and green in various brightness combinations. The color picture tube contains three electron guns, three sets of deflecting coils, and three sets of modulating grids: one each for the red, blue, and green components. *See also* COLOR TELEVISION, TELEVISION.

PIERCE OSCILLATOR

The Pierce oscillator is a form of crystal oscillator. The Pierce configuration may be recognized by the connection of the crystal between the grid and plate of a vacuum-tube circuit (see A in illustration), between the base and collector of a bipolar-transistor circuit (B) or between the gate and drain of a field-effect-transistor circuit (C).

The Pierce circuit is used extensively in radio-frequency applications. The main advantage of the Pierce oscillator is that the crystal acts as its own tuned circuit. This eliminates the need for an adjustable inductance-capacitance tank circuit in the output. *See also* CRYSTAL OSCILLATOR.

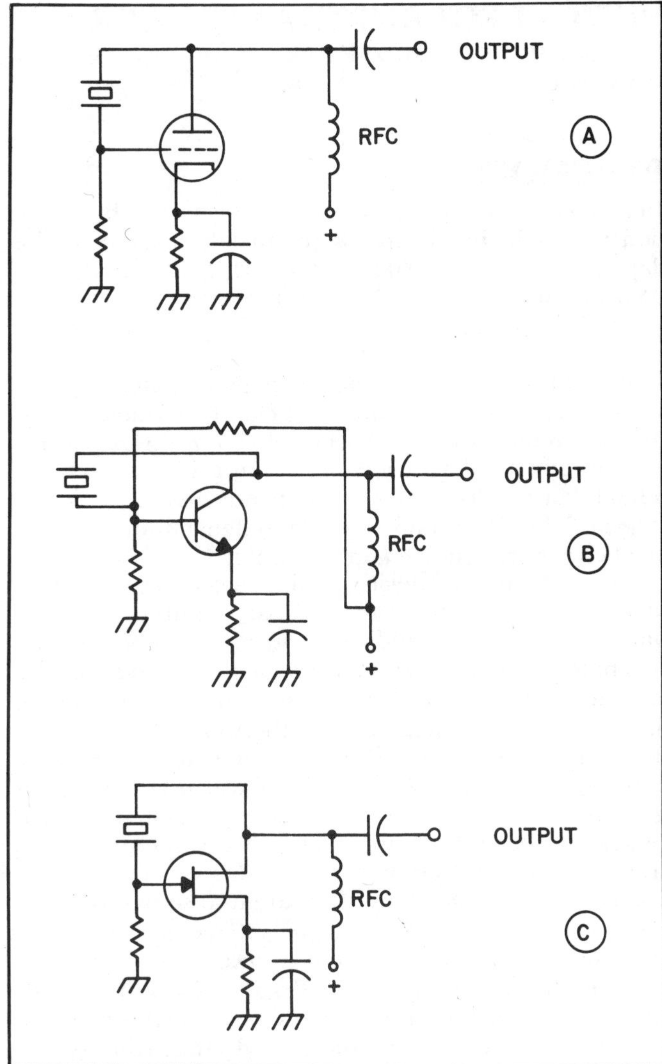

PIERCE OSCILLATOR: At A, a vacuum-tube circuit; at B, a bipolar-transistor circuit; at C, a field-effect-transistor circuit.

PIEZOELECTRIC EFFECT

Certain crystalline or ceramic substances can act as transducers at audio and radio frequencies. When subjected to mechanical stress, these materials produce electric currents; when subjected to an electric voltage, the substances will vibrate. This effect is known as the piezoelectric effect. Piezoelectric substances include such materials as quartz, Rochelle salts, and various artificial solids.

Piezoelectric devices are employed at audio frequencies as pickups, microphones, earphones, and beepers or buzzers. At radio frequencies, piezoelectric effect makes it possible to use crystals and ceramics as oscillators and tuned circuits. *See also* CERAMIC, CERAMIC FILTER, CERAMIC MICROPHONE, CERAMIC PICKUP, CRYSTAL, CRYSTAL CONTROL, CRYSTAL-LATTICE FILTER, CRYSTAL OSCILLATOR, CRYSTAL MICROPHONE, CRYSTAL TRANSDUCER.

PIEZOELECTRIC FILTER

See CERAMIC FILTER, CRYSTAL-LATTICE FILTER.

PIEZOELECTRICITY

See PIEZOELECTRIC EFFECT.

PIEZOELECTRIC MICROPHONE

See CERAMIC MICROPHONE, CRYSTAL MICROPHONE.

PIEZOELECTRIC PICKUP

See CERAMIC PICKUP, CRYSTAL TRANSDUCER.

PIEZOELECTRIC TRANSDUCER

See CERAMIC MICROPHONE, CERAMIC PICKUP, CRYSTAL MICROPHONE, CRYSTAL TRANSDUCER.

PINCHOFF

In a junction field-effect transistor, pinchoff is the condition in which the channel is completely severed by the depletion region (*see* FIELD-EFFECT TRANSISTOR, METAL-OXIDE-SEMICONDUCTOR FIELD-EFFECT TRANSISTOR). Pinchoff results in minimum conductivity from the source to the drain.

Pinchoff is the result of a large negative gate-to-source voltage in an N-channel junction field-effect transistor. In the P-channel device, pinchoff results from a large positive gate-to-source voltage. The minimum gate-to-source potential that results in pinchoff depends on the particular type of field-effect transistor. It also depends on the voltage between the source and the drain.

In the N-channel device, as the gate voltage becomes more negative, the drain voltage required to cause pinchoff becomes less and less positive. This is because the potential difference between the gate electrode and the channel is greater near the drain than near the source, and this effect is exaggerated as the drain-to-source voltage is increased. If the drain-to-source voltage is made sufficiently large, in fact, pinchoff will occur will zero gate-to-source bias. The graph shows the gate-to-source pinchoff voltages for a hypothetical N-channel fild-effect transistor. The different curves represent the pinchoff voltages for various values of the drain-to-source voltage.

In Class-A and Class-AB amplifiers, a field-effect transistor is normally biased at a lower value than that required to produce pinchoff. For operation in Class B, the device is biased at, or slightly beyond, the pinchoff voltage. In Class-C operation, a field-effect transistor is biased considerably beyond pinchoff (*see* CLASS-A AMPLIFIER, CLASS-AB AMPLIFIER, CLASS-B AMPLIFIER, CLASS-C AMPLIFIER).

The field-effect transistor depicted is normally conductive through the channel; application of larger and larger bias voltages will eventually result in pinchoff. This is known as depletion-mode operation. Some types of field-effect transistors are normally nonconductive under conditions of zero bias. For conduction to occur in this kind of device, a certain gate-to-source bias must be applied. These field-effect transistors are called enhancement-mode devices. *See also* DEPLETION MODE, ENHANCEMENT MODE.

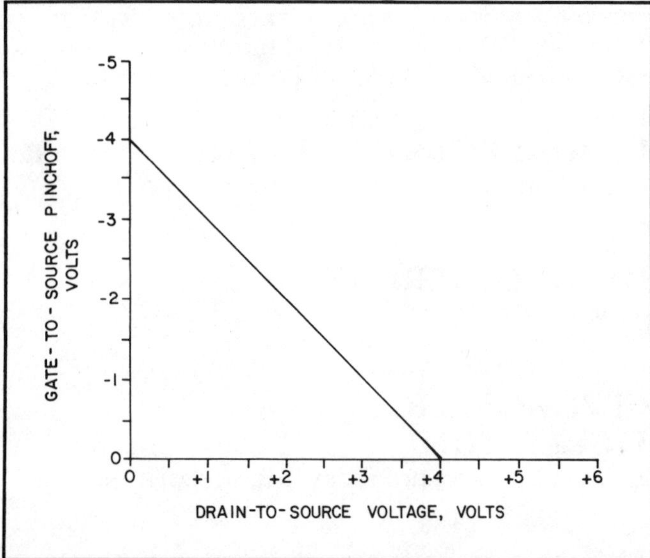

PINCHOFF: Gate-to-source pinchoff voltage, as a function of the drain-to-source voltage, for a hypothetical field-effect transistor.

PINCHOFF VOLTAGE

In a depletion-mode field-effect transistor, the pinchoff voltage is the smallest gate-to-source bias voltage that results in complete blocking of the channel by the depletion region.

The pinchoff voltage depends on the kind of field-effect device, and also on the voltage between the drain and the source. *See* PINCHOFF.

PINCUSHION DISTORTION

In a television system, pincushion distortion refers to a type of distortion in which both sides, and the top and bottom, of the picture sag inward toward the center (see illustration). Pincushion distortion is the result of improper alignment of a television receiver. It often occurs because of mutual effects between the horizontal and vertical scanning circuits in the picture tube. *See also* PICTURE TUBE, TELEVISION.

PIN DIODE

A special form of diode, having an intermediate layer of intrinsic semiconductor material between the P-type and N-type layers, is known as a PIN diode. The term PIN is derived from the terms P-type, intrinsic, and N-type.

The PIN diode has relatively low inherent capacitance. This makes it useful as a switching diode at high, very-high, and ultra-high radio frequencies. The illustration at A shows a simple radio-frequency switching circuit using a PIN diode. The PIN diode can also be successfully used as a rectifier at these frequencies.

The PIN device is manufactured by diffusion process, in which heavily doped P-type and N-type semiconductor materials are combined. The intrinsic layer may be quite

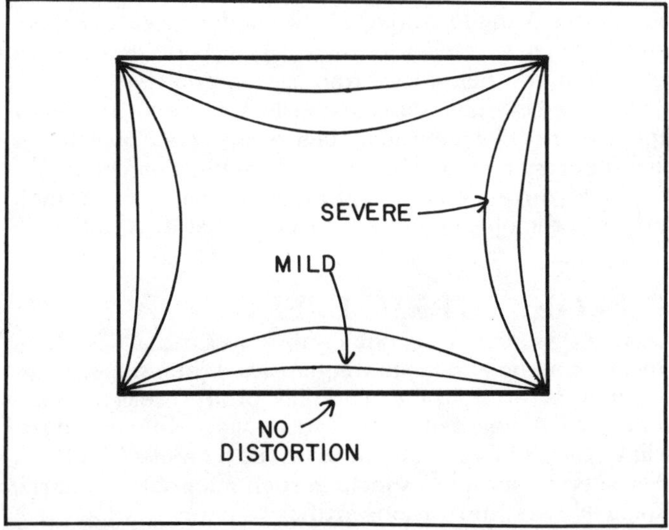

PINCUSHION DISTORTION: Pincushion distortion causes the borders of a picture to bow inward.

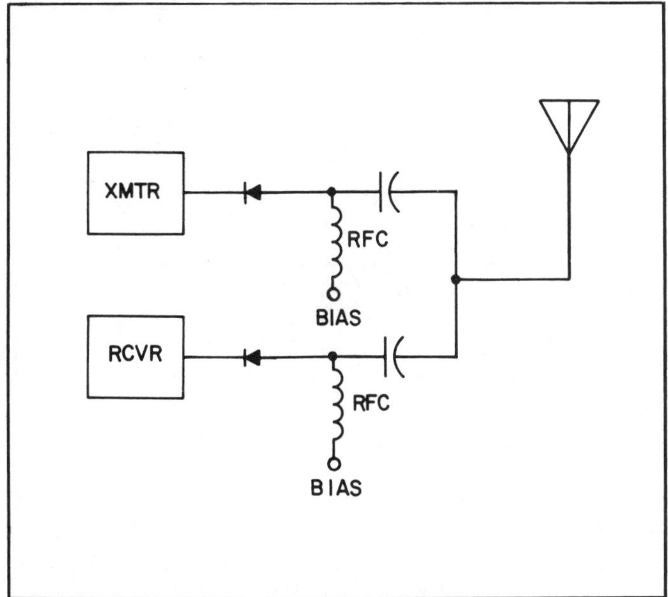

PIN DIODE: A PIN-diode circuit used for switching an antenna between a transmitter and receiver.

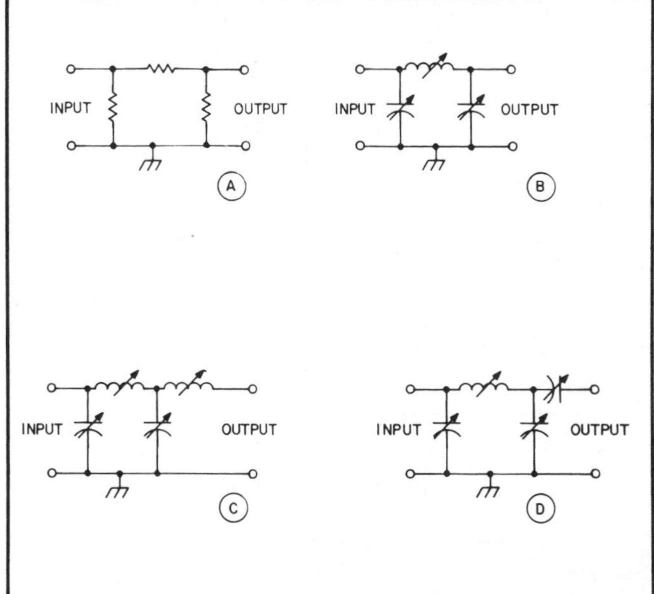

PI NETWORK: At A, an attenuator comprised of noninductive resistors; at B, an impedance-matching circuit for use at radio frequencies; at C, a radio-frequency pi-L network; at D, a radio-frequency pi-C network.

thin or relatively thick; the thickness affects the resulting performance of the diode. The I layer acts as a conductor when the diode is forward-biased. Under conditions of reverse bias, the I layer behaves somewhat like a dielectric, having very low loss and low dielectric constant. *See also* DIODE CAPACITANCE, INTRINSIC SEMICONDUCTOR.

PI NETWORK

A pi network is an unbalanced circuit, used especially in the construction of attenuators, impedance-matching circuits, and various forms of filters. Two parallel components are connected on either side of a single series component. The drawing at A illustrates a pi-network circuit consisting of non-inductive resistors. An inductance-capacitance impedance-matching circuit is shown at B. The pi network gets its name from its resemblance, at least in diagram form, to the Greek letter pi (π).

In a pi-network impedance-matching circuit, the capacitor nearer the generator or receiver is called the tuning capacitor, and the capacitor nearer the load or antenna is called the loading capacitor. By properly adjusting the values of the two capacitors and the inductor, matching between non-reactive impedances can be obtained over a fairly wide range in practice.

Pi-network impedance-matching circuits are extensively used in the output circuits of tube-type and transistor-type radio-frequency power amplifiers. In such circuits, the plate or collector impedance is often several times the actual load impedance. The loading-capacitor value is set so that, when the tuning capacitor is adjusted, the "dip" in the plate or collector current results in the optimum operation of the final-amplifier tube or transistor.

In certain applications, a series inductor may be added to a pi-network impedance-matching circuit. This allows the cancellation of capacitive reactances in the load. Such a circuit (C) is called a pi-L network. Similarly, if inductive reactance exists in the load, a capacitor may be included in series with the network. This is known as a pi-C circuit (D). *See also* IMPEDANCE MATCHING.

PI-C NETWORK

See PI NETWORK.

PI-L NETWORK

See PI NETWORK.

PINK NOISE

Pink noise is a form of wideband acoustic noise in which the amplitude is inversely proportional to the frequency. The lower the frequency, the greater the noise amplitude; the higher the frequency, the smaller the amplitude within the range of human hearing. Pink noise sounds like a low-pitched, or muffled, hiss.

Pink noise is relaxing to the ear and the brain. Natural sources of pink noise include wind through tree branches, light rain, and the roar of surf on an ocean beach. Pink noise can be electronically generated from white noise by means of an audio lowpass filter. Pink noise gets its name from the fact that it consists primarily of lower audio frequencies, just as pink light consists mainly of longer visible wavelengths. *See also* WHITE NOISE.

PIP

A visual signal on an electronic display is called a pip. Such a signal may be a radar echo, an indication on a spectrum analyzer, or some similar phenomenon. In radar, the pip is often called a blip. *See* RADAR, SPECTRUM ANALYZER.

PITCH

Pitch is an expression for the perceived frequency of an acoustic disturbance. The shorter the sound wavelength, or the greater the frequency, the higher the pitch. Some sounds have pitch that is easy to determine; an example is

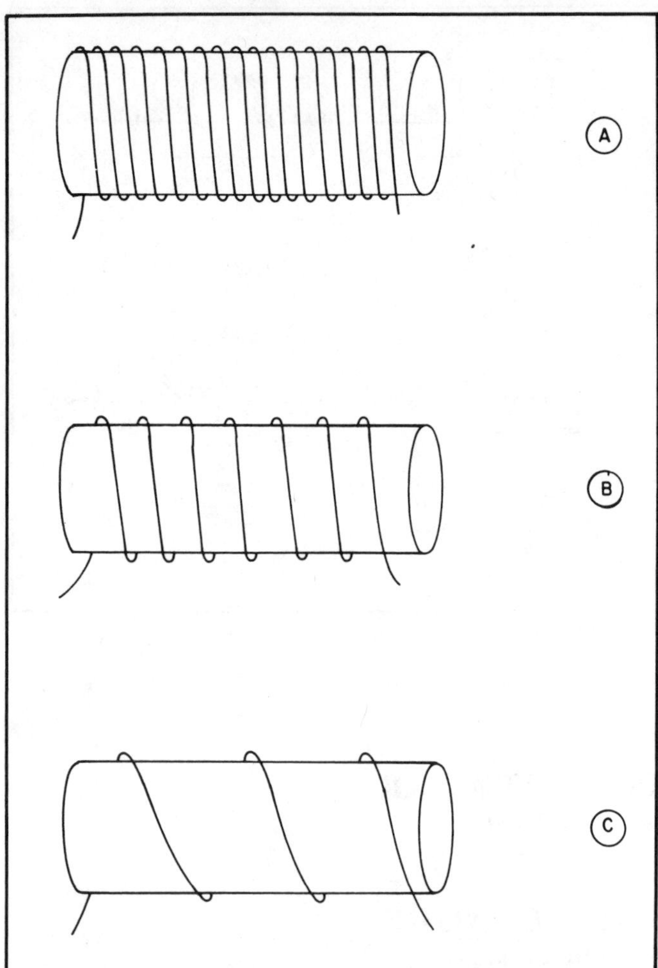

PITCH: Three examples of coil pitch. At A, the pitch is maximum, and decreases at B; the least pitch is at C.

the note from a musical instrument. Other sounds have less well-defined pitch, such as a jet-airplane engine.

On a phonograph disk, the radial distance between grooves is known as pitch. The finer the pitch, the more closely the grooves are spaced. The pitch in a single phonograph disk may vary considerably from one part of the disk to another.

In an inductor winding, the linear separation between adjacent turns is called the pitch (see illustration). The larger the number of turns per linear inch of the coil, the finer the pitch. Given a coil having N turns in a layer measuring m inches long, the pitch P is numerically expressed as:

$$P = N/m$$

turns per inch. For a coil having N turns, the inductance increases as the pitch increases. *See also* COIL WINDING.

PLANAR ARRAY

A planar array is a form of phased antenna system, in which three or more coplanar elements are fed with a certain phase relationship to obtain directional gain. A planar array may consist of linear elements or looped elements; it may or may not have a reflecting device or parasitic elements. The billboard antenna, and various forms of broadside arrays, are forms of planar arrays (*see* BILLBOARD ANTENNA, BROADSIDE ARRAY, PHASED ARRAY).

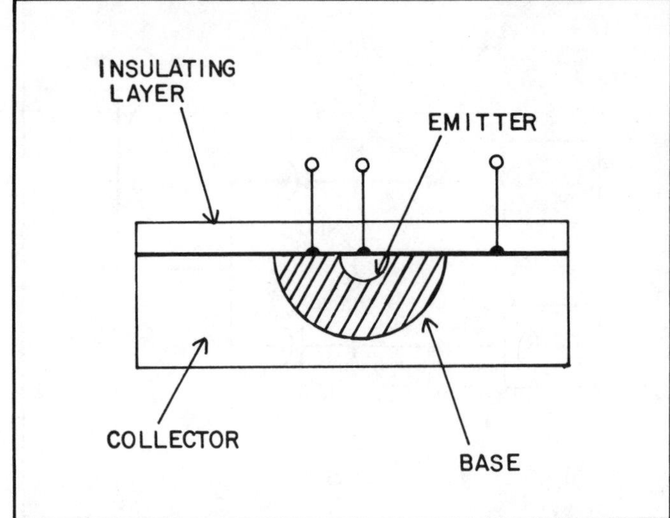

PLANAR TRANSISTOR: Cross-sectional drawing of a planar transistor. This drawing is somewhat distorted from the true dimensions for clarity.

Planar arrays may exhibit the major lobe in a direction or directions perpendicular to the plane containing the elements, but this need not necessarily be the case. Some planar arrays, such as the Mills Cross or Christiansen antenna, have steerable main lobes. *See also* MILLS CROSS.

PLANAR TRANSISTOR

A planar transistor is a special form of bipolar transistor. It may be of either the NPN or the PNP type. The planar transistor gets its name from the fact that all three elements—the emitter, the base, and the collector—are fabricated from a single layer of material.

The illustration shows a cross-section of the construction of a planar transistor. The emitter is at the center, the base surrounds the emitter, and the collector is at the outside and bottom. The entire structure is covered with a flat film of silicon dioxide or other insulating material. This protects the transistor.

The collector has by far the greatest volume. The base-collector junction has much larger surface area, because of the concentric construction, than the emitter-base junction. This gives the planar transistor the ability to dissipate relatively large amounts of heat at the base-collector junction. Planar design is well suited to the fabrication of power transistors. *See also* TRANSISTOR.

PLANCK'S CONSTANT

Radiant energy exhibits both particle-like and wave-like properties. All electromagnetic energy consists of discrete packets, or "corpuscles," known as photons. The photon represents the smallest possible amount of energy that can exist at a given wavelength (*see* PHOTON). Electromagnetic energy cannot be divided into smaller components.

The shorter the wavelength (or the higher the frequency) of an electromagnetic disturbance, the more energy is contained in each photon. Conversely, the more energy in a photon, the shorter the wavelength and the higher the frequency. The energy and the wavelength are related according to a precise linear function:

$$e = hf$$

where e is the energy in the photon, f is the frequency of the electromagnetic wave, and h is a constant known as Planck's constant. The constant gets its name from the physicist Max Planck, one of several great scientists whose discoveries revolutionized particle physics around the turn of the century.

The constant has approximately the value:

$$h = 6.62 \times 10^{-27} \text{ erg-seconds}$$

$$= 6.62 \times 10^{-34} \text{ joule-seconds}$$

Thus, if e is to be specified in ergs, the first value is used, and if e is given in joules, the second value is used. The frequency, f, is given in hertz. *See also* ELECTROMAGNETIC THEORY OF LIGHT, PARTICLE THEORY OF LIGHT.

PLANE

A plane is a theoretical object of geometry. The plane is usually thought of as a flat surface, such as a floor or wall. In geometry, any three points in space lie in exactly one plane. Any two intersecting lines determine a unique plane.

Planes are often specified for certain purposes, especially in antenna theory and applications. For example, we may speak of the radiation pattern of an antenna in the horizontal plane; this means the pattern in the plane tangent to the surface of the earth at the location of the antenna. We may speak of a vertical-plane pattern; this refers to the pattern in one of a set of planes passing through the antenna, perpendicular to the horizontal plane. The specific vertical plane is determined according to the azimuth, or compass direction.

In three-dimensional Cartesian space, a plane is determined by a linear equation of the form:

$$ax + by + cz + d = 0$$

where a, b, c, and d are constants, and x, y, and z are the coordinates of the Cartesian three-space. *See also* CARTESIAN COORDINATES.

PLANE OF POLARIZATION

See POLARIZATION.

PLANE REFLECTOR

A plane reflector is a passive antenna reflector that is commonly used for transmitting and receiving at ultrahigh and microwave frequencies. The reflector gets its name from the fact that it is perfectly flat and electrically continuous. A plane reflector is positioned ¼ wavelength behind a set of dipole antennas, producing approximately 3 dB of power gain.

The operation of the plane reflector is shown in the illustration. The direct, or forward, radiation from the antenna is not affected. The radiation in the opposite direction encounters the metal plane, where the wave is reversed in phase by 180 degrees and reflected back in the forward direction. Since the plane is ¼ wavelength away from the radiating element, the total additional path distance for the reflected wave is ½ wavelength, or 180 degrees, with respect to the direct wave. The extra distance and the phase reversal, together, result in phase reinforcement between the direct wave and the reflected wave.

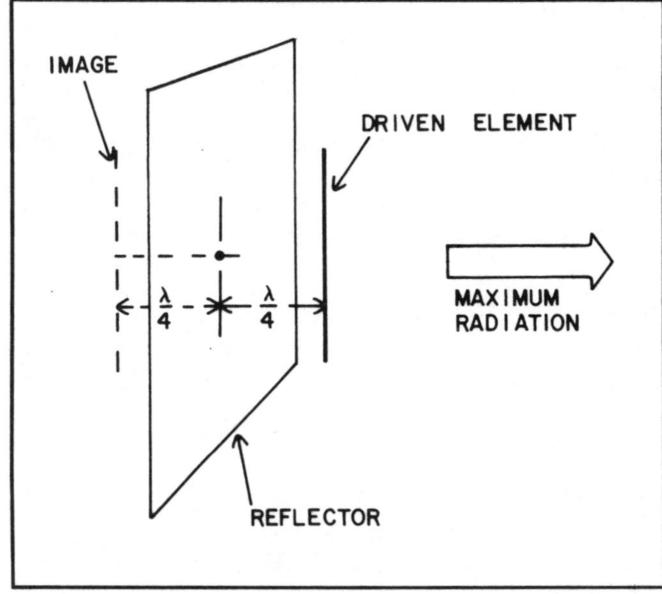

PLANAR REFLECTOR: Operation of a planar reflector. The gain is about 3 dB, representing a doubling of the effective radiated power.

Plane reflectors may be made of wire mesh, screen, a set of parallel bars, or a solid sheet. If wire screen or mesh is used, the spacing between the wires or bars should be less than about 0.05 wavelength for best results.

The plane reflector can be used with any number of radiating elements. A large array of dipoles with a flat reflector is called a billboard antenna. Plane reflectors can be used with helical or loop antennas as well as with dipole elements. In all cases, the extra power gain is 3 dB, corresponding to a doubling of the effective radiated power compared to an identical antenna without the reflector. All antennas that incorporate plane reflectors are a form of planar array. *See also* BILLBOARD ANTENNA, PLANAR ARRAY.

PLASMA

When a gas is heated to an extremely high temperature, or is subjected to an intense electric field, the electrons are stripped away from their normal orbits around the atomic nuclei. When this happens, the gas, which is ordinarily an excellent insulator, begins to conduct. The electrons are easily passed from nucleus to nucleus. This state of matter is called a plasma.

A plasma is affected by electric and magnetic fields in a manner different from a gas in its usual state. Since the plasma is an effective conductor, it can be confined or deflected by electric and magnetic fields. *See also* ELECTRON, ELECTRON ORBIT.

PLATE

The anode of a vacuum tube is often called the plate. The plate is supplied with a positive voltage, so that it attracts electrons from the cathode. The plate of a tube is analogous to the collector of a bipolar transistor, and to the drain of a field-effect transistor.

The plate of a tube may be shaped like a cylinder, a

rectangular prism, or some other configuration. In some vacuum tubes, the plate is ribbed or finned to increase its surface area. This, in turn, increases its ability to dissipate power.

Tube plates are subject to heating to a greater or lesser extent, depending on the type of tube and the circuit in which it is used. Some tubes have plates made from carbon or a metal capable of withstanding extreme temperatures. *See* PLATE CURRENT, PLATE DISSIPATION, PLATE MODULATION, PLATE POWER INPUT, PLATE RESISTANCE, PLATE VOLTAGE, TUBE.

The term plate is sometimes used to describe the electrodes in a capacitor, or the anode of an electrochemical cell. *See also* ANODE, CAPACITOR, CELL, ELECTRODE.

PLATE CURRENT

In a vacuum tube, the plate current is the amount of current that flows in the plate circuit. Plate current is sometimes abbreviated I_P.

The plate current in a vacuum-tube circuit depends on a number of factors. Plate current is usually minimum at resonance. It generally increases as the plate voltage is increased. The plate current decreases as the control-grid bias voltage gets more and more negative. When the grid bias reaches the point at which plate current is reduced to zero, the tube is said to be cut off.

Vacuum tubes are generally rated in terms of maximum allowable plate current. If the plate current becomes excessive, the plate dissipation may rise to such a level that damage occurs (*see* PLATE DISSIPATION).

The instantaneous plate current fluctuates in accordance with the variations in the instantaneous control-grid voltage, caused by the input signal. This is how a tube amplifies. *See also* TUBE.

PLATE DISSIPATION

In a vacuum tube amplifier or oscillator circuit, there is a difference between the plate power input (*see* PLATE POWER INPUT) and the circuit power output. This power difference is largely spent as heat in the plate of the tube. The power that is used up in heating the plate of the tube is the plate dissipation.

In general, for a given amount of plate power input, the plate dissipation increases as the efficiency decreases. The Class-A amplifier results in a plate dissipation of more than 50 percent of the total plate power input; the Class-C amplifier may result in a plate dissipation of as little as 20 percent of the total plate power input (*see* CLASS-A AMPLIFIER, CLASS-AB AMPLIFIER, CLASS-B AMPLIFIER, CLASS-C AMPLIFIER). The above figures are based on the assumption that the amplifier is properly adjusted. Improper adjustment will substantially increase the plate dissipation. This may occur, for example, when a radio-frequency power amplifier is not tuned to resonance in the output circuit (*see* POWER AMPLIFIER).

Vacuum tubes are rated according to the maximum amount of power they can safely dissipate without the likelihood of damage. The continuous rating is smaller than the intermittent-duty rating (*see* DUTY CYCLE). Cooling systems may, in some cases, dramatically increase the plate-dissipation rating of a tube (*see* COOLING).

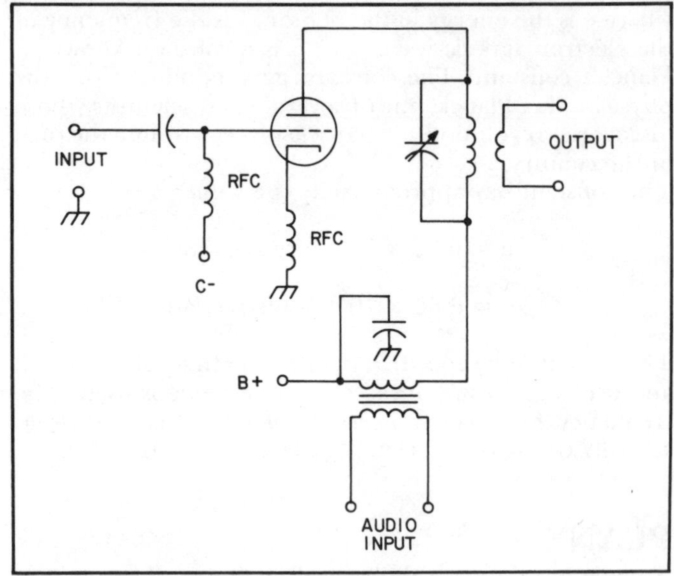

PLATE MODULATION: Plate modulation of a radio-frequency Class-C power amplifier.

PLATE MODULATION

Plate modulation is a method that is sometimes used in tube type power-amplifier circuits for the purpose of obtaining an amplitude-modulated signal.

Plate modulation is normally applied at the final-amplifier stage of a high-power radio-frequency transmitter. An audio transformer, connected in series with the plate circuit of the tube, is used to couple the audio-frequency energy into the amplifier. The plate transformer is inserted on the power-supply side of the output tank circuit (see illustration).

To achieve 100-percent modulation, the audio-frequency power must be at least half the plate power input of the amplifier under zero-modulation conditions. That is, the audio power must be ⅓ (33 percent) of the amplifier plate input power when modulated. In a high-power transmitter, this obviously requires the use of a substantial audio-frequency amplification chain, along with a bulky plate transformer. However, the advantage of plate modulation is that it may be used with Class-C radio-frequency final amplifiers. This results in greater efficiency in that circuit. *See also* AMPLITUDE MODULATION, PLATE POWER INPUT.

PLATE POWER INPUT

The plate power input to a tube circuit is the product of the direct-current plate voltage and the plate current. Plate power input is the sum of the output power and the power dissipated as heat in the tank circuit, conductors, and the plate of the tube itself.

The plate power input to an amplifier is always greater than the output. The ratio of the output to the plate power input is called the efficiency. In Class-A amplifiers, the plate power input may be more than twice the output. In Class-C amplifiers, the output is almost as large as the input (*see* CLASS-A AMPLIFIER, CLASS-AB AMPLIFIER, CLASS-B AMPLIFIER, CLASS-C AMPLIFIER).

In the determination of plate power input, the plate voltage is usually measured with a direct-current voltmeter, and the plate current is measured with a direct-current ammeter.

PLATE RESISTANCE

The plate resistance of a vacuum tube is an expression of the internal resistance of the device. When there is no reactance in the output circuit—that is, it is resonant—the output impedance of the tube is equal to the plate resistance. Plate resistance is abbreviated R_P.

Most vacuum tubes have plate-resistance ratings of hundreds, or even thousands, of ohms. This usually necessitates the use of an impedance transformer between the plate circuit of the tube and the load. For example, an antenna system may show a resistive impedance of 50 ohms, while the final-amplifier tube of a transmitter has a plate resistance of 2,500 ohms. This would require a transformer with an impedance-transfer ratio of 25:1; the turns ratio in such a case would be 5:1.

The plate resistance, R_P, of a tube under non-reactive conditions is the quotient of the plate voltage, E_P, and the plate current, I_P:

$$R_P = E_P/I_P$$

This value is sometimes called the static plate resistance. When a signal is applied, the instantaneous plate resistance may differ from the static value. Under such conditions, the instantaneous, or dynamic, plate resistance is equal to the rate of change in the plate voltage divided by the rate of change in the plate current:

$$R_P = dE_P/dI_P$$

See also PLATE CURRENT, PLATE VOLTAGE.

PLATE VOLTAGE

The plate voltage of a vacuum tube is the potential difference between the plate and the cathode, or between the plate and chassis ground. Plate voltages are always positive. The plate voltage may range from a few volts for small receiving tubes to thousands of volts for high-power transmitting tubes.

Plate voltage is abbreviated E_P. The plate voltage of a tube is equal to the power-supply voltage, minus the voltage drop across any series resistance in the plate circuit. If the plate-cathode voltage is not identical with the voltage between the plate and ground, the cathode-circuit resistance must be included in calculation of the plate-cathode voltage.

PLATINUM

Platinum is an element with atomic number 78 and atomic weight 195. In its pure form, platinum is a metallic substance, similar to silver in appearance.

Platinum is an excellent conductor of electricity, and is highly resistant to corrosion. Thus it is sometimes used as the electrode or electrodes in chemical and physical experiments. Platinum may also be used for coating switch and relay contacts. Platinum is, however, expensive. It is considered one of the precious metals.

PLL

See PHASE-LOCKED LOOP.

PLOT

See GRAPH.

PLUG

Any male connector can be called a plug. Examples of plugs include the standard ¼-inch phone plug, lamp-cord plugs, phono plugs, Jones plugs, and many others. A plug mates with a female jack of the same size and number of conductors. Sometimes the female socket or jack, together with the male connector, is called a plug.

Plugs always have their conductors exposed when they are disconnected. Therefore, plugs should not be installed in the part of the circuit normally carrying the voltage from the power supply. This precaution will minimize the shock hazard, and prevent possible inadvertent short circuits. *See also* BANANA JACK AND PLUG, JONES PLUG, MALE, PHONE PLUG, PHONO PLUG, SOCKET.

PLUG-IN COMPONENT

A plug-in component is any component that can be removed for easy replacement. Some transistors are of the plug-in type. Many integrated circuits are installed in sockets, and thus they are plug-in components.

In certain types of equipment, especially test apparatus, plug-in components are used to allow the adjustment of certain parameters. For example, a common type of radio-frequency wattmeter has a socket in which various measuring elements can be inserted. The elements are plug-in components. One element may be used for a full-scale indication of 25 watts in the frequency range 25 MHz to 200 MHz, for example; another element will provide a full-scale indication of 500 watts at 200 MHz to 500 MHz. The element is chosen for maximum meter readability in the appropriate frequency range.

In older radio receivers, bandswitching was done by means of plug-in inductors. For the low frequencies, an inductor having many turns was used. This plug-in inductor could be removed, and a smaller one put in its place, for reception of medium and high frequencies. This technique of bandchanging is not used nowadays. Rotary switches have proven considerably more convenient for the purpose.

PLUNGER-TYPE METER

A plunger-type meter is a current-measuring device using a coil and a magnetic rod called a plunger. The plunger moves in and out of the coil against the tension of a spring. A needle is attached to the end of the plunger, and also to a rotating bearing (see illustration).

Under conditions of zero current in the coil, the rod is mostly outside of the coil. When a current flows through the coil, the magnetic field attracts the plunger, pulling it into the coil until the inward force is balanced by the tension of the spring. The more current that flows in the coil, the stronger the magnetic attraction becomes, and the farther the rod is pulled. The needle therefore moves upscale in proportion to the intensity of the current.

Plunger-type meters are useful in the measurement of alternating currents, since the direction of the current in the coil does not affect the magnetic force against the rod.

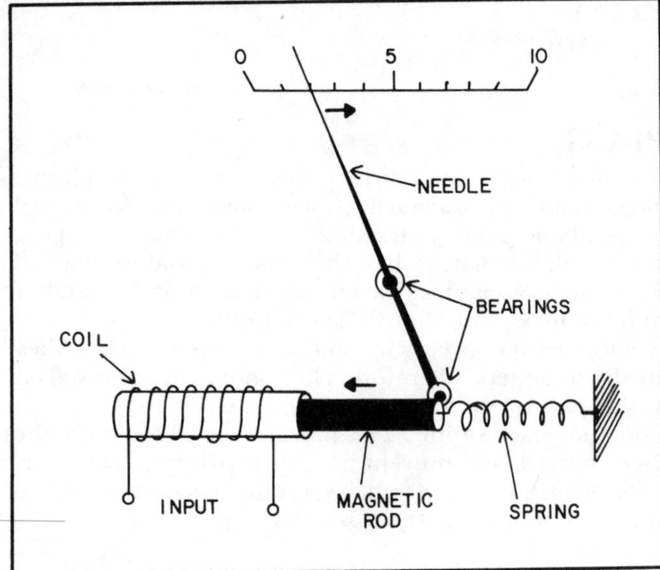

PLUNGER-TYPE METER: Operation of the plunger-type meter.

The rod itself is not a permanent magnet, and thus it is attracted by either a magnetic north pole or a magnetic south pole. *See also* METER.

PLUTONIUM

Plutonium is an element with atomic number 94 and atomic weight of about 242 (although it varies somewhat, since there are many different isotopes). Plutonium does not occur in nature; it is too heavy. All of the plutonium known to exist has been made by man.

Plutonium is a byproduct of nuclear fission. The element is radioactive. Plutonium is a waste product of nuclear-fission reactors, but it can be used to manufacture atomic bombs.

PL-259

See UHF CONNECTOR.

PMOS

See METAL-OXIDE-SEMICONDUCTOR LOGIC FAMILIES.

P-N JUNCTION

A P-N junction is the boundary between layers of P-type and N-type semiconductor materials. A P-N junction tends to conduct when the N-type material is negative with respect to the P-type material, but it exhibits high resistance when the polarity is opposite.

A P-N junction is said to be forward-biased when the N-type material is negative with respect to the P-type material (see A in illustration). The junction conducts well in this mode. The P-type material, having a deficiency of electrons, receives electrons from the N-type material (*see* N-TYPE SEMICONDUCTOR, P-TYPE SEMICONDUCTOR).

A P-N junction is said to be reverse-biased when the N-type material is positive with respect to the P-type material as at B. In this condition, electrons are pulled from the

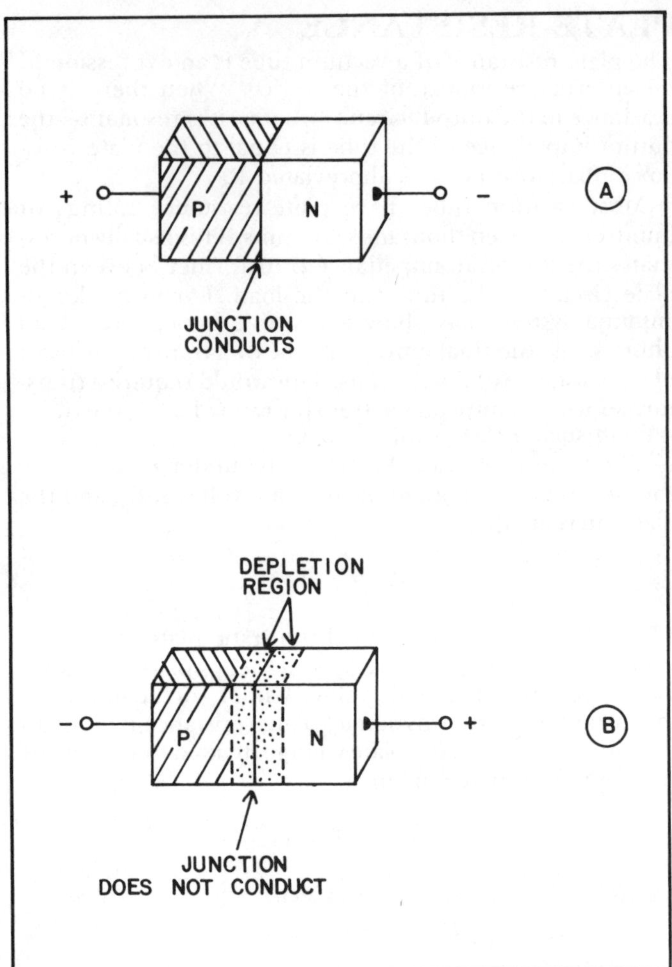

P-N JUNCTION: The P-N junction under conditions of forward bias (A) and reverse bias (B). When the junction is reverse-biased, the depletion region essentially prevents conduction.

P-type material, which already has a deficiency of electrons; and electrons accumulate in the N-type material, which already has an excess of electrons. The result is a zone that has extremely high resistance. This region occurs immediately on either side of the boundary between the two different semiconductor layers. Therefore, the P-N junction does not conduct well in the reverse direction; the resistance may be millions of megohms or more. If the reverse bias is increased to larger and larger values, however, the P-N junction suddenly begins to conduct at a certain bias level, known as the avalanche voltage (*see* AVALANCHE, AVALANCHE BREAKDOWN).

Since a P-N junction conducts well in one direction but not in the other, devices consisting of P-N junctions are useful as detectors and rectifiers (*see* DETECTION, DIODE ACTION, RECTIFICATION). The P-N junction of a semiconductor device may be capable of accumulating and depleting the insulating layer very rapidly, perhaps on the order of millions or billions of times per second. In such instances, the resulting diode can be used for detection at megahertz and even gigahertz frequencies. Other P-N-junction devices can follow alternating-current frequencies of only a few tens or hundreds of kilohertz. The ability of a P-N-junction diode to handle high frequencies depends largely on the surface area of the junction. It also depends on the thickness of the depletion layer that results from a given amount of reverse voltage. *See also* DIODE CAPACITANCE.

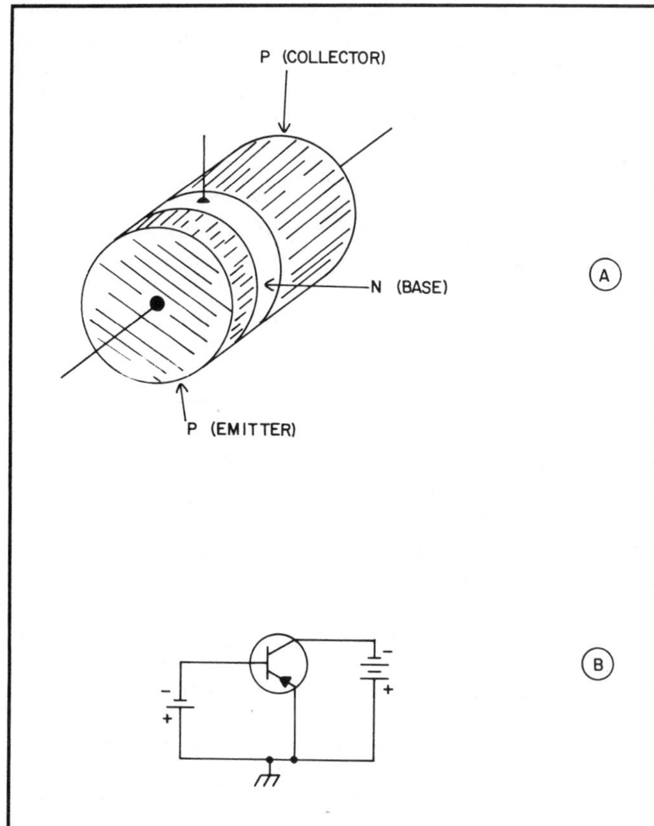

PNP TRANSISTOR: At A, pictorial illustration of a PNP bipolar transistor. At B, The basic biasing scheme. The collector-emitter voltage is larger than the base-emitter voltage.

PNP TRANSISTOR

A PNP transistor is a three-layer, bipolar semiconductor device. A layer of N-type semiconductor material is sandwiched in between two layers of P-type material (see A in illustration). One of the P-type layers is thicker than the other. The thicker P-type layer is designated the collector of the transistor, and the thinner layer is the emitter. The N-type layer in the center is the base of the transistor.

The PNP transistor is normally biased with the base negative with respect to the emitter. The collector is biased negatively with respect to the base. This biasing configuration is illustrated schematically at B.

For Class-B or Class-C amplifier operation, the base may be biased at the same level as the emitter, or perhaps even positively with respect to the emitter. This places the transistor at, or beyond, cutoff (*see* CLASS-B AMPLIFIER, CLASS-C AMPLIFIER).

The PNP transistor uses a biasing polarity that is exactly opposite to that of the NPN transistor. *See also* NPN TRANSISTOR.

PNPN TRANSISTOR

See FOUR-LAYER TRANSISTOR.

POINT

A point is a geometric spot or position in space. In theory, a point has zero dimensions. That is, it has no height, no width, and no depth. In real life, of course, points do not concretely exist in this form, but a tiny spot in space may be called a point for certain practical purposes.

In geometry, the point is considered one of three elementary or fundamental objects, and its definition is not rigorously given (the other two elementary objects are the line and the plane).

Mathematicians, physicists, engineers, and other scientists often speak of points on a number line (*see* NUMBER LINE). Each numerical value corresponds to a unique point on the line. Also, each point on the line corresponds to one number. Such points have zero length.

Points are often specified in graphs of mathematical functions. For example, in the function:

$$y = f(x) = 3x + 4$$

we may speak of the point $(x,y) = (2,10)$, which lies on the graph, or the point $(x,y) = (2,3)$, which does not lie on the graph. Most functions have an infinite number of points that lie on their graphs; a few functions have only a finite number of points. *See also* FUNCTION, GRAPH.

In practice, an arbitrarily (but not infinitely) small spot in space is sometimes called a point. For example, we may speak of a certain point in an electronic circuit, or a point source of light.

POINT-CONTACT JUNCTION

A point-contact junction is a form of semiconductor junction. A fine wire, called a cat's whisker, is placed in contact with a piece of semiconductor material, such as germanium or silicon (see illustration). The semiconductor material forms the anode of the junction, and the cat's whisker forms the cathode. The point-contact junction conducts well when the anode is positive with respect to the cathode, but the device conducts poorly when the anode is negative with respect to the cathode. This one-way conductivity results in diode action.

The point-contact method is used in the fabrication of various types of diodes and transistors. A point-contact junction generally has much lower capacitance than a P-N junction, because the surface area is much smaller. This

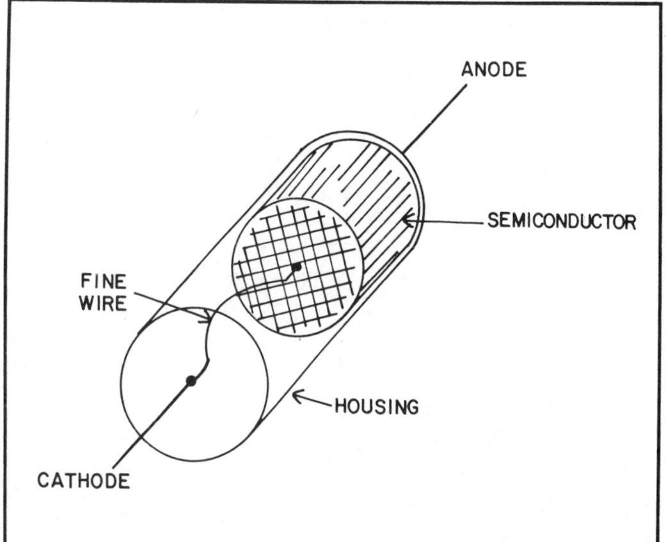

POINT-CONTACT JUNCTION: Pictorial drawing of a diode with a point-contact junction.

makes the point-contact junction useful at higher frequencies. However, the point-contact junction cannot carry much current. Therefore, point-contact devices are limited to low-power applications. *See also* CAT'S WHISKER, P-N JUNCTION, POINT-CONTACT TRANSISTOR.

POINT-CONTACT TRANSISTOR

Two point-contact junctions can be combined to form a transistor (*see* POINT-CONTACT JUNCTION). A piece of germanium or silicon serves as the base, and two fine wires (cat's whiskers) are placed in contact with the wafer to form the emitter and collector (see illustration). The result is a device that behaves like an NPN bipolar transistor.

Point-contact transistors were the earliest form of semiconductor transistor. Such devices are still occasionally used, although junction-type transistors are now more common. The point-contact transistor works well at very-high and even ultra-high frequencies, because the point-contact junctions exhibit low capacitance. However, point-contact transistors cannot carry much current, and are therefore used only in receivers or local-oscillator circuits. *See also* TRANSISTOR.

POINT SOURCE

A point source of radiation is, theoretically, an infinitely tiny spot in space from which the energy emanates. Of course, there is no such thing as a real point source of radiation, but for certain practical purposes, it is useful to consider point sources. An energy source appears more and more like a point source as it is observed from greater and greater distances.

The intensity of electromagnetic radiation from a point source, in terms of power per unit surface area, decreases with increasing distance, according to the inverse-square law (*see* INVERSE-SQUARE LAW). When the source is not a perfect point (and in practice it never is), the inverse-square law is not precisely correct. However, at large distances, most sources of electromagnetic radiation can be considered point sources. Thus, the inverse-square law is reasonably accurate at great distances from a source of radiation, regardless of the wavelength.

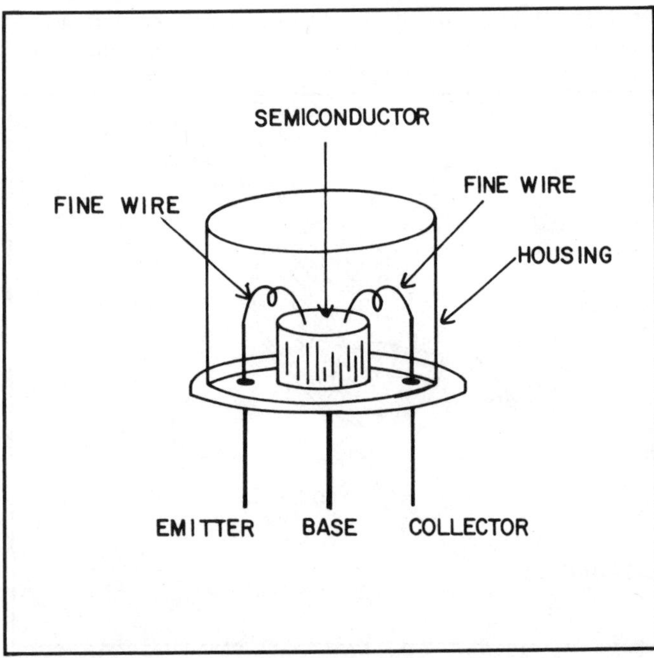

POINT-CONTACT TRANSISTOR: Pictorial drawing of a point-contact transistor.

POINT-TO-POINT WIRING

Point-to-point wiring is a form of circuit wiring used mostly in vacuum-tube circuits. Point-to-point wiring is not often seen in modern solid-state circuits; printed-circuit boards are much more common nowadays (*see* PRINTED CIRCUIT). Point-to-point wiring consists of individual hookup-wire links, soldered among various points in a circuit. The beginning and ending points of each wire usually consist of tie-strip terminals. Tie strips are also known as terminal strips (*see* TIE STRIP).

In some applications, especially at frequencies below about 30 MHz when high radio-frequency power levels are involved, point-to-point wiring is still used. This is because individual wire conductors can withstand higher current levels than the foil runs of printed circuits. The main advantage of point-to-point wiring is its ability to handle large amounts of current. The main disadvantage of point-to-point wiring is that it requires more physical space than circuit-board wiring.

A modern form of point-to-point wiring, having greater compactness and convenience than the old-fashioned form, is called wire wrapping. *See also* WIRE WRAPPING.

POISSON DISTRIBUTION

Poisson distribution is a form of normal distribution used for determining the average number of random events that occur within a given interval of time. The events may be of any sort, such as lightning flashes during a storm or the number of radio stations transmitting a certain commercial message.

During a specified time interval t, the probability P that an event will take place n times is given by the formula:

$$P = ((ft)^n/n!)e^{-ft}$$

or, in terms of percentage,

$$P\ (\text{percent}) = 100((ft)^n/n!)e^{-ft}$$

where f is the average frequency of the event. The values f and t must be based on the same time units; for example, if f is given in events per hour, then t must be specified in hours.

As an example of the application of the Poisson-distribution formula, suppose that a certain building is hit by lightning on the average of five times every ten years. Then the frequency f, in strikes per years, is 0.5. What is the probability that the building will be hit five times in a single year? Using the formula:

$$\begin{aligned} P &= ((0.5 \times 1)^5/5!)e^{-0.5} \\ &= (0.03125/120) \times 0.6065 \\ &= 0.0001579 \end{aligned}$$

To obtain the value as a percentage, we multiply by 100. The chance that the building will be struck five times in one year is therefore only 0.01579 percent. *See also* NORMAL DISTRIBUTION, STATISTICAL ANALYSIS.

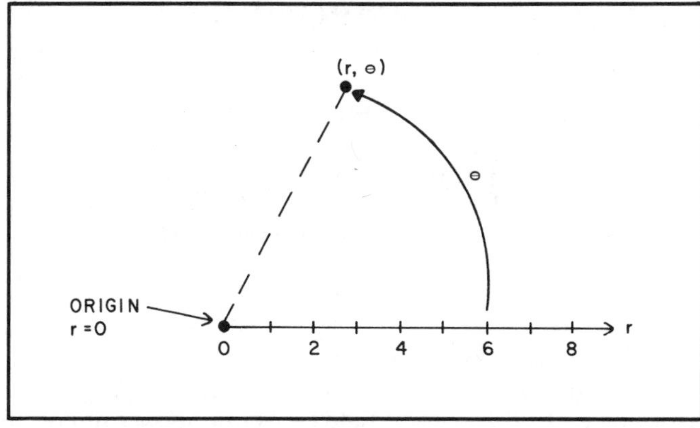

POLAR COORDINATES: Polar coordinates are often used to uniquely locate, and define, the points in a geometric plane.

POLAR COORDINATES

Polar coordinates refers to a geometric system for locating points on a plane. The polar coordinate system is often used for graphing mathematical functions. Polar coordinates are almost always used for plotting antenna directional patterns (*see* ANTENNA PATTERN).

The polar system is based on a central point. Other points lie a specific radial distance r from the center, and have a direction that is indicated in terms of an angle θ. The value of r is always nonnegative in practice, but in theory it may be positive or negative. The value of θ, in practice, is always at least zero but less than 360 degrees. In theory, however, θ may attain any value, positive or negative, representing multiple rotations about the center. The angle θ is measured counterclockwise from a reference axis that coincides with the positive x axis of the Cartesian system (*see* CARTESIAN COORDINATES). The drawing illustrates the polar coordinate system.

A given equation does not have the same graph in polar coordinates as in rectangular coordinates. It is not possible to simply substitute r for x and θ for y, expecting to convert a function from rectangular form to polar form. The conversion from rectangular to polar coordinates is done according to the formulas:

$$r = \sqrt{x^2 + y^2}$$

$$\theta = \arctan (y/x)$$

Conversion from polar to rectangular coordinates is done by means of the formulas:

$$x = r \cos \theta$$

$$y = r \sin \theta$$

Relations that appear simple when expressed in rectangular coordinates may become quite complicated when converted to polar coordinates. Similarly, simple polar equations are often horrendous in rectangular form. An illustrative example (but certainly not the most complicated) is the circle. In rectangular coordinates, the equation is difficult to work with. A circle of radius k, centered at the origin, appears in rectangular form as:

$$x^2 + y^2 = k^2$$

but in polar form the equation is simply:

$$r = k$$

See also COORDINATE SYSTEM.

POLARITY

Polarity refers to the relative voltage between two points in a circuit. One pole, or voltage point, is called positive, and the other pole is called negative. The polarity affects the direction in which the current flows in the circuit.

Physicists consider current to flow from the positive pole to the negative pole. The actual movement of electrons is from the negative pole to the positive pole, since electrons carry a negative charge.

In magnetism, polarity refers to the orientation of the north and south magnetic poles. *See also* DIPOLE, POLE.

POLARIZATION

Polarization is an expression of the orientation of the lines of flux in an electric, electromagnetic, or magnetic field. Polarization is of primary interest in electromagnetic effects. The polarization of an electromagnetic field is considered to be the orientation of the electric flux lines.

An electromagnetic field may be horizontally polarized, vertically polarized, or slantwise polarized (*see* HORIZONTAL POLARIZATION, VERTICAL POLARIZATION). The polarization is generally parallel with the active element of an antenna; thus, a vertical antenna radiates and receives fields with vertical polarization, and a horizontal antenna radiates and receives fields having horizontal polarization.

Electromagnetic fields may have polarization that continually changes. This kind of polarization can be produced in a variety of ways (*see* CIRCULAR POLARIZATION, ELLIPTICAL POLARIZATION).

Polarization effects occur at all wavelengths, from the very low frequencies to the gamma-ray spectrum. Polarization effects are quite noticeable in the visible-light range. Light having horizontal polarization, for example, will reflect well from a horizontal surface (such as a pool of water), while vertically polarized light reflects poorly off the same surface. Simple experiments, conducted with polarized sunglasses, illustrate the effects of visible-light polarization. *See also* ELECTRIC FIELD, ELECTRIC FLUX, ELECTROMAGNETIC FIELD.

In an electrochemical cell or battery, the electrolysis process sometimes results in the formation of an insulating layer of gas on one of the plates or sets of plates. This effect, called polarization, causes an increase in the internal resistance of the cell or battery. This, in turn, limits the amount of current that the device can deliver. If the polarization is severe, the cell or battery may become totally useless. *See also* CELL.

POLARIZATION MODULATION

Information can be impressed on an electromagnetic wave by causing rapid changes in the polarization of the wave (*see* POLARIZATION). This method of modulation is known as polarization modulation. Polarization modulation can theoretically be achieved at any wavelength from the very low frequencies through the gamma-ray spectrum. However, polarization modulation is most often used at microwave and visible-light wavelengths.

An example of an apparatus for obtaining polarization modulation is shown in the illustration at A. Two antennas are oriented at right angles with respect to each other. Under zero-modulation conditions, one antenna, say the vertical one, receives all of the signal, while the other antenna gets none. When modulation is applied, some of the signal is applied to the horizontal antenna at the expense of the signal applied to the vertical antenna. The total signal power reaching both antennas remains constant. Under maximum instantaneous modulation conditions, the horizontal antenna gets all of the power and the vertical antenna gets none. This causes the polarization of the transmitted signal to shift by values up to 90 degrees, in accordance with the amplitude of the modulating waveform (B).

When the signal from the antenna shown at A encounters a receiving antenna having linear polarization, the instantaneous signal strength fluctuates in sync with the modulating waveform. This will occur regardless of the orientation of the receiving antenna. For best results, the receiving antenna should be parallel with one of the transmitting-antenna elements.

The polarization of visible light can be modulated in a variety of different ways. A pair of light sources, equipped with mutually perpendicular polarizing filters, can be amplitude-modulated in a manner similar to that described above. The receiving apparatus must then incorporate a polarizing filter for retrieval of the information.

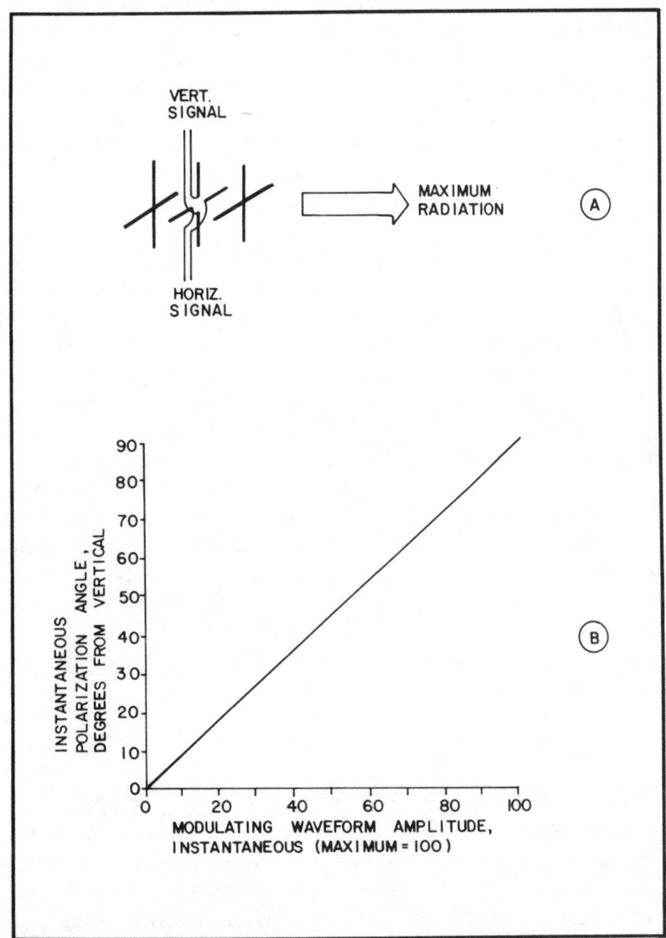

POLARIZATION MODULATION: At A, one method of obtaining polarization modulation. Two perpendicular Yagi antennas are used. At B, the signal polarization is shown as a function of the instantaneous modulating-waveform amplitude, resulting from the arrangement at A.

POLARIZED COMPONENT

A polarized component is an electronic component that must be installed with its leads in a certain orientation. If a polarized component is installed incorrectly, the circuit will not work properly.

Examples of polarized components include diodes, certain types of capacitors, and, of course, all direct-current power sources. *See also* NONPOLARIZED COMPONENT.

POLARIZED LIGHT

Most visible light is randomly polarized. That is, the orientation of the electric-field component does not occur in any particular direction. When the electric lines of flux are all parallel, the light is polarized.

Polarized light can be produced by passing any light through a special filter. Some sunglasses have lenses that polarize the light passing through them. Such glasses are transparent to vertically polarized light, but almost opaque to horizontally polarized light. This greatly reduces the glare from light reflected from horizontal surfaces. *See also* POLARIZATION.

POLE

A pole is a point toward which the flux lines of an electric or magnetic field converge, or from which the flux lines diverge. Electric poles may be either positive, representing a deficiency of electrons, or negative, representing an excess of electrons. Magetic poles may be either north or south.

Electric poles can exist alone—that is, a single positive or negative pole may be isolated—but magnetic poles are always paired. (*See* DIPOLE, MONOPOLE).

In a power supply, the positive terminal is sometimes called the positive pole, and the negative terminal the negative pole. The manner in which the supply is connected to the load, or circuit, is called the polarity. *See also* POLARITY.

POLISH NOTATION

Polish notation is a method of writing mathematical expressions in which the operators precede or follow the numbers or variables. This scheme was invented by the Polish mathematician and logician, Lukasiewicz.

The main advantage of Polish notation is the fact that it eliminates the need for parentheses in mathematical and logical expressions. Polish notation appears somewhat confusing to the uninitiated, but it is fairly easy to learn. Polish notation is sometimes used in writing expressions in Boolean algebra and mathematical logic. For example, we normally write X OR Y as X+Y in Boolean form; but in the Polish form it is written +XY.

Some small calculators utilize a data-entry method that is similar to Polish notation. Such calculators are sometimes called Polish calculators. In a calculator using the Polish method, an ENTER key is used in conjunction with the operation keys. For example, suppose that you want to determine the value of the expression:

$$y = (2 + 3) \times 4$$

With a conventional calculator, you might press keys in the following order:

(2 + 3) × 4 =

but with a Polish calculator, the sequence would be

2 ENTER 3 + ENTER 4 ×

See also BOOLEAN ALGEBRA.

POLYESTER

Polyester is a plastic material, the chemical name of which is polyethylene glycol terephthalate. Polyester has high tensile strength, is extremely flexible, and is resistant to changes in temperature and humidity.

A polyester film is added to many kinds of magnetic recording tapes. The plastic is placed on the back (non-recording) side of the tape. Tapes are available in various thicknesses for different uses. The polyester greatly increases the longevity of the recording tape. *See also* MAGNETIC TAPE.

POLYETHYLENE

Polyethylene is a form of plastic. The full name of the substance is polymerized ethylene. This material is noted for its resistance to moisture, its low dielectric loss, and its high dielectric strength. Polyetheylene is a tough, fairly flexible, semi-transparent plastic.

Polyethylene is widely used in electronics as an insulator, especially in coaxial cables and twin-lead transmission lines. The polyethylene may be used either in solid form or in foamed form. Solid polyethylene lasts longer than the foamed variety, and also is less susceptible to contamination. Foamed polyethylene has somewhat lower dielectric loss, and a lower dielectric constant. *See also* COAXIAL CABLE, TWIN-LEAD.

POLYGRAPH

The polygraph is better known as a lie detector. It is an established medical fact that stress causes changes in certain body functions such as blood pressure, breathing rate, and skin conductivity. The word polygraph literally means "a recorder of several things," and that describes the machine quite well. A typical polygraph uses pen recorders that monitor several body functions at the same time, on parallel tracks (*see* PEN RECORDER).

The use of the polygraph is fairly familiar to most people. An interrogator asks the subject numerous questions, keeping watch on the pen recorder. The resulting record contains several curves, one for each body function. Each question corresponds to a certain interval on the graph. Large fluctuations in the body functions indicate stress. This is assumed to mean that the subject has not told the truth for that particular question or set of questions.

Polygraph tests are sometimes used in criminal investigations when there is a shortage of other evidence. Some employers require job applicants to take polygraph tests as a part of their evaluation.

POLYPHASE ANTENNA

See TURNSTILE ANTENNA.

POLYPHASE CURRENT

See THREE-PHASE ALTERNATING CURRENT.

POLYPHASE RECTIFIER

A polyphase rectifier is a device used to convert three-phase alternating current to direct current (*see* THREE-PHASE ALTERNATING CURRENT). Polyphase rectifiers operate according to essentially the same principles as single-phase rectifiers. All of the polyphase circuits make use of diodes; either solid-state or tube type diodes can be used.

For three-phase rectification, the most common circuits are illustrated by the diagrams. At A, the three-phase bridge circuit is shown; at B, the three-phase double-wye circuit is shown; at C, the three-phase star circuit is shown. The circuits at A and B provide twice the output ripple frequency compared with the circuit at C. Thus, the outputs of the bridge and double-wye circuits are easier to filter.

The main advantage of polyphase rectifiers over single-phase rectifiers is the fact that the ripple frequency is higher, resulting in more pure direct current after filtering.

POLYSTYRENE

Polystyrene is a plastic material, noted for its low dielectric loss at radio frequencies. The dielectric constant is ap-

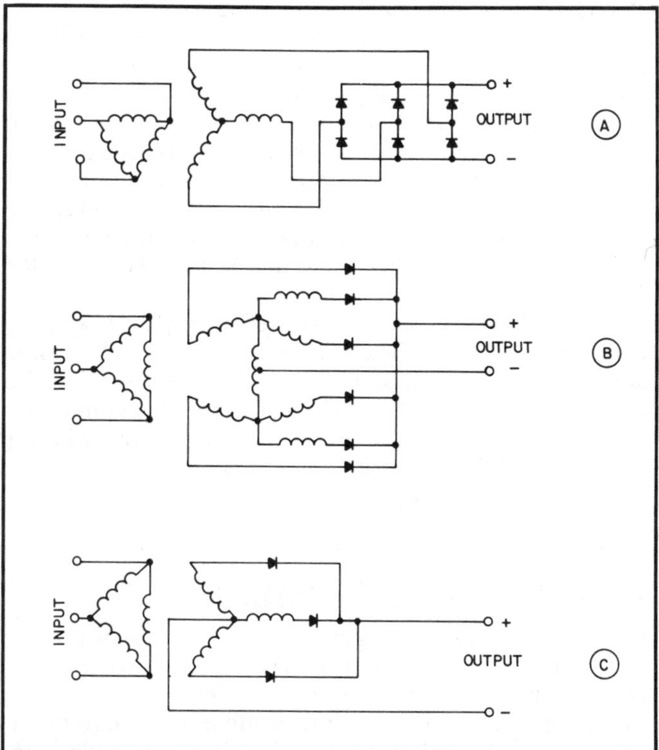

POLYPHASE RECTIFIER: Three common types of polyphase rectifier circuits. At A, three-phase bridge; at B, three-phase double-wye; at C, three-phase star.

proximately 2.6, regardless of the frequency. The direct-current resistivity is very high, on the order of 10^{18} ohm-centimeters. The power factor is very low at all frequencies. Polystyrene exhibits very little change in characteristics with large fluctuations in temperature and humidity. The material appears as a clear plastic in its pure form.

Polystyrene is widely used in the manufacture of capacitors for use in radio-frequency oscillators and amplifiers at low and medium power levels. *See also* POLYSTYRENE CAPACITOR.

POLYSTYRENE CAPACITOR

Polystyrene is well suited as a dielectric material for the manufacture of capacitors. The electrical characteristics are excellent at radio frequencies (*see* POLYSTYRENE). Polystyrene capacitors are easy to manufacture, and for this reason, they are relatively inexpensive.

Polystyrene capacitors are noted for their stability under variable temperature conditions, and for this reason, they are preferred for use in variable-frequency oscillators at the low, medium, and high frequencies. Polystyrene capacitors can be made to have very low tolerance values (high precision). Polystyrene capacitors are physically rugged and moisture-resistant. However, they are not tolerant of oils and petroleum-based solvents.

While polystyrene itself has low dielectric loss up through the ultra-high-frequency range, the electrode resistivity limits the maximum frequency at which polystyrene capacitors can be used. Polystyrene capacitors are used in low-voltage applications.

PORCELAIN

Porcelain is a ceramic material having excellent dielectric properties. Porcelain appears as a hard, white, dull substance when not glazed; it is shiny when glazed. Porcelain has high tensile strength, but it is brittle.

Porcelain is widely used in the manufacture of insulators for direct current as well as radio-frequency alternating current; it can withstand very high temperatures and voltages. The dielectric constant of porcelain decreases somewhat as the frequency goes up; while it is about 5.4 at 1 kHz, it decreases to 5.1 at 1 MHz and 5.0 at 100 MHz.

The dielectric loss of dry porcelain is quite low. Some trimmer capacitors utilize porcelain as the dielectric. Many air-variable capacitors, especially those designed for transmitting, incorporate porcelain for insulating purposes. Porcelain is also occasionally used as a coil-winding form, especially in power amplifiers. *See also* CERAMIC, CERAMIC CAPACITOR.

PORTABLE EQUIPMENT

Portable communications equipment is any apparatus that is designed for operation in remote locations. Portable equipment is battery-operated, and can be set up and dismantled in a minimum time. Some portable equipment can be operated while being carried; an example is the handy-talkie or walkie-talkie. The photograph shows a common type of handheld transceiver for very-high-frequency use.

PORTABLE EQUIPMENT: A portable transceiver. This unit is designed for use on the 2-meter (144-MHz) amateur band.

With the improvement in miniaturization and semiconductor technology, portable equipment can have complex and sophisticated features. A few decades ago, the mention of a microcomputer-controlled walkie-talkie would have elicited ridicule. Today, such devices are commonplace.

Portable equipment must be compact and light in weight. For this reason, such equipment always has modest power requirements, since a high-power battery is invariably bulky and heavy. The unit shown produces 3 watts of r-f power output.

In addition to being small and light, portable equipment must be designed to withstand physical abuse such as vibration, temperature and humidity extremes, and prolonged use without complicated servicing.

Portable equipment is sometimes used in mobile applications. For example, the handheld unit shown can be connected to a mobile antenna and used in a car or truck. *See also* MOBILE EQUIPMENT.

PORTABLE TELEPHONE

A portable telephone is a radio transceiver designed for access to an autopatch system (*see* AUTOPATCH). Portable telephones resemble handsets with antennas.

There are two types of portable telephones. The short-range variety is called a cordless or wireless telephone (*see* CORDLESS TELEPHONE). It is designed for use within a few hundred feet of a base unit. The long-range type is similar to a walkie-talkie, and its range is limited only by the access range of the autopatch repeater and transceiver (typically 10 to 20 miles). The amateur-radio transceiver in the previous illustration is equipped with an autopatch encoder, so it can be used as a portable telephone. *See also* MOBILE TELEPHONE.

POSITIVE CHARGE

Positive charge, or positive electrification, is the result of a deficiency of electrons in the atoms of which an object is composed. Friction between two objects can cause an electron imbalance, resulting in positive charge on one of the bodies. When an atom has fewer electrons than protons that atom is said to be positively charged.

The smallest unit of positive charge is carried by a single proton. The smallest unit of negative charge is carried by the electron. The proton and electron have opposite charge polarity, but equal charge quantity. The terms positive and negative are chosen arbitrarily.

Electric lines of flux are considered to originate at positive-charge poles and terminate at negative-charge poles. An electric current, likewise, is theoretically considered to flow from the positive pole of a circuit to the negative pole, even though current often consists of electrons that move away from the negative-charge terminal. *See also* CHARGE, NEGATIVE CHARGE.

POSITIVE LOGIC

The way in which logical signals are defined is a matter of convention. Normally, the logic 1 is the more positive of the voltage levels in a binary circuit, and the logic 0 is the more negative. Thus, the logic 1 is high and the logic 0 is low. This is called positive logic.

In some digital circuits, the logic 1 is the more negative (low) and the logic 0 is the more positive (high) of the voltage levels. This is known as negative logic (*see* NEGATIVE LOGIC).

From a practical standpoint, it makes no difference whether a circuit uses positive logic or negative logic, as long as the same form is consistently used throughout the circuit. *See also* LOGIC.

POSITRON

A positron is an atomic particle that has mass identical to the electron, or 9.11×10^{-31} kilogram, but carries a unit positive charge rather than a unit negative charge. If a positron happens to collide with an electron, the two particles annihilate each other and produce energy in the form of a photon. Because the electron (which is matter) and the positron disappear when they collide, the positron has been called a particle of antimatter.

Positrons are generated in numerous atomic reactions. The hydrogen-fusion process (*see* FUSION) causes a positron to be formed, as two protons combine to form a nucleus of deuterium.

Some scientists believe that positrons may orbit negatively charged antimatter nuclei, forming whole atoms of antimatter. It is possible that stars and planets may exist, made up entirely of antimatter; there may even exist antimatter universes! In such an environment, the charge carriers in a wire would be positrons. The entire convention of positive and negative charge would be inside-out, but the effects would be exactly the same as in a matter-based environment. *See also* CHARGE.

POT

See POTENTIOMETER.

POT CORE

A special form of ferromagnetic coil-winding form, designed to provide large inductances for relatively few turns of wire, is known as a pot core. The pot core gets its name from the fact that it is shaped something like a pot. The core consists of two identical sections that are fastened together by means of a nut and bolt through the center (see illustration). The coil-winding space is toroidal in shape, and the coil is wound in circular turns.

Pot cores can be found with various different permeability factors, and they are made in a wide variety of physical sizes. The magnetic field of the coil is contained entirely within the core, as is the case with toroidal coils (*see* TOROID). This allows the pot-core coil to be placed adjacent to other components with practically no interaction. The inductance of a pot-core coil can be as much as 1 henry or more, with a very reasonable number of coil turns. The pot-core can handle quite large amounts of current, in proportion to its physical size.

Pot-core coils exhibit low loss and high Q factor. They are used mostly at audio frequencies, in passive filters and as transformers.

POTENTIAL

See VOLTAGE.

POTENTIAL DIFFERENCE

Two points in a circuit are said to have a difference of potential when the electric charge at one point is not the

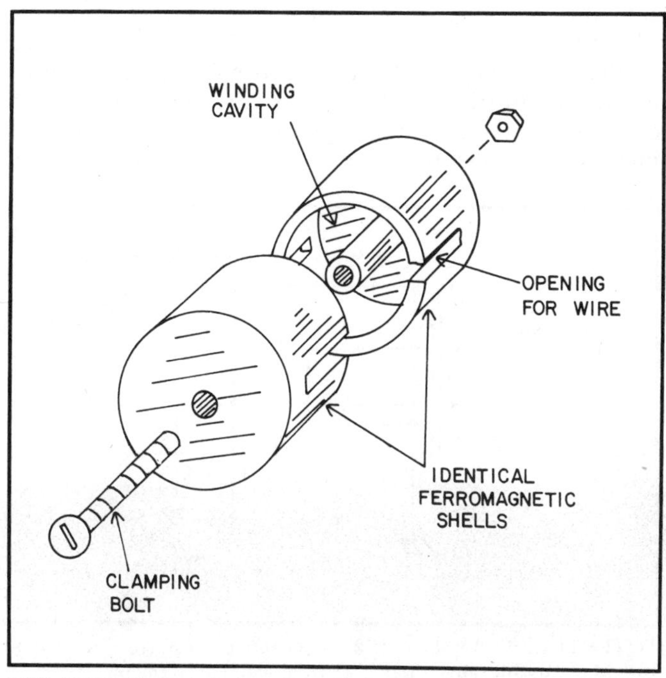

POT CORE: Exploded view of a typical pot core.

same as the electric charge at the other point. Potential difference is measured in volts (*see* VOLTAGE).

When two points having a potential difference are connected by a conducting or semiconducting medium, current flows. Physicists consider the current to flow from the more positive point to the more negative point. If the charge carriers are electrons, they actually move from the more negative point to the more positive point. If the charge carriers are holes, they move from positive to negative. *See also* CHARGE, ELECTRON, HOLE.

POTENTIAL ENERGY

Potential energy is an expression of the capability of a body to produce useful energy. For example, if a weight is lifted up several feet, it gains potential energy; when the weight is dropped, the energy output is realized.

Potential energy can be harnessed for specific purposes. For example, in a hydroelectric power plant, the potential energy of the water at the top of the dam is converted into electricity. Potential energy cannot always be harnessed conveniently, and the conversion from potential energy into useful energy is never 100-percent efficient. *See also* ENERGY.

POTENTIAL GRADIENT

See VOLTAGE GRADIENT.

POTENTIAL TRANSFORMER

A small transformer, having multiple taps, is sometimes used for the purpose of extending the range of an alternating-current voltmeter. Such a transformer is called a potential transformer.

An example of a voltmeter incorporating a potential transformer is shown in the diagram. The actual range of the voltmeter is 0 to 100 volts. The transformer turns ratio can be set for primary-to-secondary values of 1:100, 1:10, 1:1, and 10:1. This provides full-scale metering ranges of 1 volt, 10 volts, 100 volts, and 1 kilovolt, respectively.

The potential transformer is, of course, not the only means of adjusting the full-scale range of an alternating-current voltmeter. Shunt or series resistors can be used to increase the full-scale range. The range can be decreased by reducing the value of the series resistance. Alternatively, a direct-current amplifier, following the rectifier and filter, can be used to decrease the full-scale range. *See also* VOLTMETER.

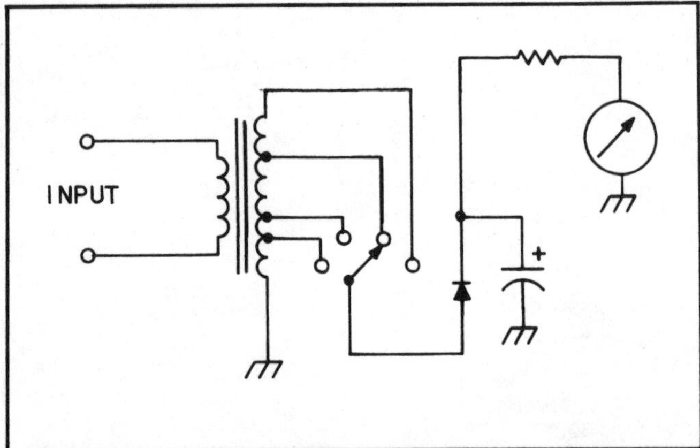

POTENTIAL TRANSFORMER: A potential transformer is used to extend the useful range of an alternating-current voltmeter.

POTENTIOMETER

A potentiometer is a noninductive, tapped resistor. The position of the tap can be varied over a continuous range. This allows the value of the resistance to be adjusted from zero to the maximum. The schematic symbol for the potentiometer is shown at A in the illustration.

Some potentiometers exhibit resistance that varies in exact proportion to the physical position of the tap. This is called linear taper (*see* LINEAR TAPER). Other potentiometers show a resistance value that varies according to the logarithm of the actual tap position; this is called logarithmic taper (*see* LOGARITHMIC TAPER). The choice of tapering depends on the nature of the circuit in which the potentiometer is used.

Potentiometers are used in many different applications. Most volume-control circuits incorporate potentiometers.

POTENTIOMETER: At A, schematic symbol for a potentiometer. At B, a miniature circuit-board potentiometer. Its diameter is only about ⅜ inch.

The devices are also used in such diverse situations as meter calibration, sensitivity adjustment, panel-light brightness control, oscillator frequency, filter bandwidth, and keyer-speed control.

In modern, miniaturized electronic equipment, some potenitometers are designed to be installed directly on a printed-circuit board. Such potentiometers, such as the one shown at B, are extremely small; this unit is located inside a radio transceiver.

Some potentiometers are equipped with a rotating shaft for attaching a knob. This kind of device is used when it is necessary to make adjustments via front-panel controls. Some potentiometers are equipped with internal switches; the familiar "on/off/volume" control makes use of this type of device. Two or more potentiometers may be ganged on a single shaft for multiple-control adjustment.

POUND

The pound is the standard English unit of weight. In the gravitational field of the earth, a mass of 0.4536 kilogram weighs 1 pound. The abbreviation of pound is lb.

A mass that weighs 1 pound on the earth will have different weights in gravitational fields of different intensity. For example, on Mars, a mass of 0.4536 kilogram would weigh only about 6 ounces (6/16 pound). On Jupiter, the same mass would weigh approximately 2.5 pounds. *See also* KILOGRAM, MASS, WEIGHT.

POWDERED-IRON CORE

Powdered-iron cores are used in coils and transformers for the purpose of increasing the value of inductance for a given number of turns. This makes such components much more compact than would otherwise be possible. Powdered-iron cores are used in audio-frequency and radio-frequency applications.

Powdered-iron cores are found in three basic configurations: the solenoid, the toroid, and the pot core (*see* POT CORE, SOLENOID, TOROID). The solenoidal core allows some of the magnetic field to exist outside the windings of the coil. The toroid and pot cores confine the field to the inside of the component.

Powdered-iron cores increase the inductance of a coil because they have high permeability (*see* PERMEABILITY). This causes the magnetic-field intensity within the material to increase, just as it would if the number of coil turns were increased. Powdered-iron cores can be found with various values of permeability for different applications. A material similar to powdered iron with extremely high permeability is known as ferrite (*see* FERRITE CORE).

POWER

Power is the rate at which energy is expended or dissipated. Power is expressed in joules per second, more often called watts.

In a dc circuit, the power is the product of the voltage and the current. A source of E volts, delivering I amperes to a circuit, produces P watts, as follows:

$$P = EI$$

From Ohm's law, we can find the power in terms of the current and the resistance, R, in ohms:

$$P = I^2R$$

Similarly, in terms of the voltage and the resistance:

$$P = E^2/R$$

In an ac circuit, the determination of power is more complicated than in the case of direct current. Assuming there is no reactance in the circuit, the power is determined in the same manner as above, using root-mean-square values for the current and the voltage (*see* ROOT MEAN SQUARE).

If reactance is present in an alternating-current circuit, some of the power appears across the reactance, and some appears across the resistance. The power that appears across the reactance is not actually dissipated in real form. For this reason, it has been called imaginary or reactive power. The power appearing across the resistance is actually dissipated in real form, and it is thus called true power. The combination of real and reactive power is known as the apparent power (*see* APPARENT POWER, REACTIVE POWER).

The proportion of the real power to the apparent power in an alternating-current circuit depends on the proportion of the resistance to the total or net impedance. The smaller this ratio, the less of the power is real power. In the case of a pure reactance, none of the power is real. *See also* POWER FACTOR.

POWER AMPLIFIER

A power amplifier is an amplifier that delivers a certain amount of alternating-current power to a load. Power amplifiers are used in audio-frequency and radio-frequency applications.

The illustration at A is a schematic diagram of a typical audio-frequency power amplifier using bipolar transistors. It is a push-pull circuit. (This is intended only as an example of a power amplifier; there are many different possible circuits.) An amplifier such as this might be used in a radio receiver, tape recorder, or record player. The input requirements are modest; the circuit needs almost no drive power to produce an output of several watts. A transformer should be used at the output of an audio-frequency power amplifier to ensure that the transfer of power is optimized. Audio-frequency power amplifiers are usually operated in Class A, or in a push-pull Class-B configuration, since distortion must be kept to a minimum.

Audio-frequency power is generally measured in terms of the product of the root-mean-square (RMS) current and voltage delivered to a nonreactive load (*see* ROOT MEAN SQUARE). Audio-frequency power amplifiers are rated according to the RMS power they can deliver without producing more than a certain amount of distortion, such as 3 percent or 10 percent. Some audio-frequency power amplifiers can produce several kilowatts of output. Such amplifiers are used, for example, by popular-music bands.

At B in the illustration, a radio-frequency power amplifier is shown. This particular circuit uses a vacuum tube, such as a 3-500Z, in a grounded-grid configuration. (As with the audio-frequency circuit above, this is only an example of a power-amplifier circuit; there are many other types.) The input impedance of this circuit is low, and it provides a good match for most transmitter output

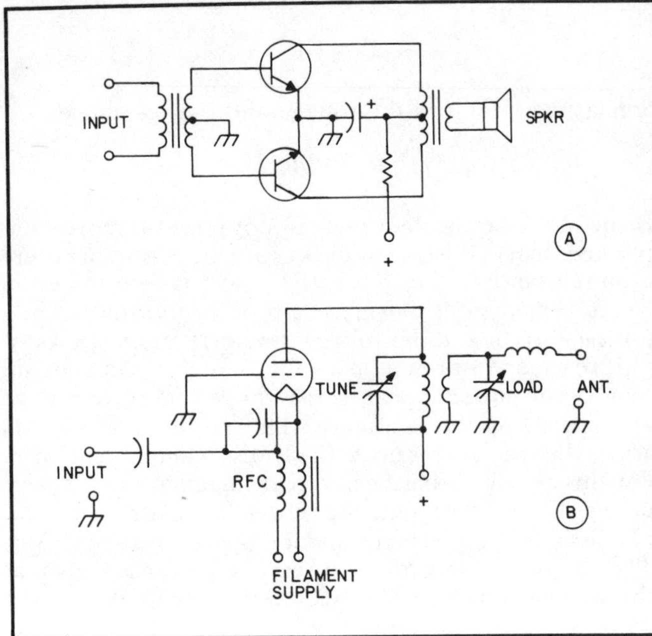

POWER AMPLIFIER: At A, a push-pull audio-frequency circuit using bipolar transistors; at B, a grounded-grid radio-frequency circuit using a vacuum tube.

circuits. The output of the power amplifier incorporates a tuned circuit and an impedance-matching network for optimum power transfer to the antenna. Radio-frequency power amplifiers are usually operated in Class AB, Class B, or Class C, depending on the intended application. *See also* CLASS-A AMPLIFIER, CLASS-AB AMPLIFIER, CLASS-B AMPLIFIER, CLASS-C AMPLFIER, POWER.

POWER DENSITY

Power density is one means of expressing the strength of an electromagnetic field. Power density is expressed in watts per square meter or watts per square centimeter, as determined in a plane oriented parallel to both the electric and magnetic lines of flux.

In the vicinity of a source of electromagnetic radiation, the power density decreases according to the law of inverse squares. *See also* FIELD STRENGTH, INVERSE-SQUARE LAW.

POWER FACTOR

In an alternating-current circuit containing reactance and resistance, not all of the apparent power is true power. Some of the power occurs in reactive form (*see* APPARENT POWER, POWER, REACTIVE POWER, TRUE POWER). The ratio of the true power to the apparent power is known as the power factor. Its value may range from 0 to 1.

Power factor may be determined according to the phase angle in a circuit. The phase angle is the difference, in degrees, between the current and the voltage in an alternating-current circuit. If the phase angle is represented by ϕ, then the power factor, PF, is:

$$PF = \cos \phi$$

The power factor, as a percentage, is:

$$PF = 100 \cos \phi$$

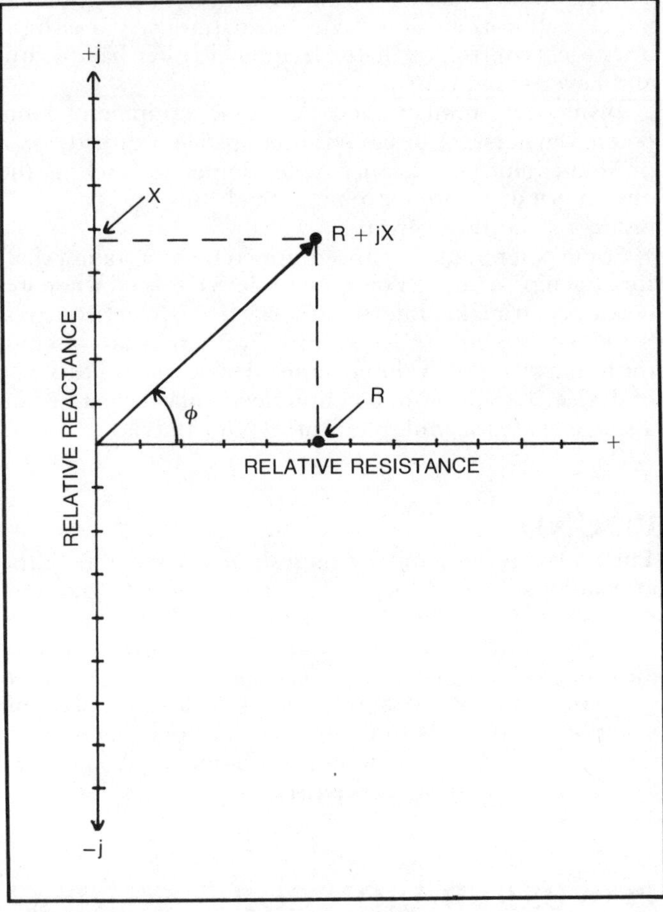

POWER FACTOR: Impedance-plane diagram showing determination of phase angle as a function of the resistive component, R, and the reactive component, jX, of a complex impedance. Once the phase angle is known, the power factor can be determined.

The power factor in an ac circuit can also be determined by plotting the impedance on the complex plane, as shown in the illustration. If the resistance is R ohms and the reactance is X ohms, the phase angle is:

$$\phi = \arctan (X/R)$$

and thus the power factor is:

$$PF = \cos (\arctan (X/R))$$

As a percentage,

$$PF = 100 \cos (\arctan (X/R))$$

See also IMPEDANCE, PHASE ANGLE.

POWER GAIN

Power gain is an increase in signal power between one point and another. Power gain is used as a specification for power amplifiers.

If the input to a power amplifier, in watts, is given by P_{IN} and the output, in watts, is given by P_{OUT}, then the power gain in decibels is:

$$G = 10 \log_{10} (P_{OUT}/P_{IN})$$

See also DECIBEL, GAIN.

POWER INVERTER

See INVERTER.

POWER LINE

A utility line, carrying electricity for public use, is called a power line or power-transmission line. The small line, usually carrying 234 volts, that runs from a house to a nearby utility pole, is a power line. So are the high-tension wires that carry several hundred thousand volts.

The term power line is technically imprecise, since power does not actually travel from place to place. It is the electrons in the conductors, and the electromagnetic field, that travel. Power is dissipated, or used, at the terminating points of the power line (*see* POWER). Some power is dissipated in the line because of losses in the conductors and insulating materials. This dissipation is, of course, undesirable, since power lost in the line represents power that cannot be used at the terminating points. *See also* AC POWER TRANSMISSION, HIGH-TENSION POWER LINE, TRANSMISSION LINE.

POWER-LINE NOISE

Power lines, in addition to carrying the 60-Hz alternating current that they are intended to transmit, also carry other currents. These currents have a broadband nature. They result in an effect called power-line noise.

The currents usually occur because of electric arcing at some point in the circuit. The arcing may originate in appliances connected to the terminating points; it may take place in faulty or obstructed transformers; it can even occur in high-tension lines as a corona discharge into humid air. The broadband currents cause an electromagnetic field to be radiated from the power line, because they flow in the same direction along all of the line conductors.

Power-line noise sounds like a buzz or hiss when picked up by a radio receiver. Some types of power-line noise can be greatly attenuated by means of a noise blanker. For other kinds of noise, a limiter can be used to improve the reception. Phasing can sometimes be used in an antenna system to reduce the level of received power-line noise. *See also* NOISE BLANKER, NOISE LIMITER, NOISE-REDUCING ANTENNA.

POWER LOSS

Power loss occurs in the form of dissipation, usually resulting in the generation of heat. Power loss occurs to a certain extent in all electrical conductors and dielectric materials. Power loss can also take place in the cores of transformers and inductors.

The power loss in a given circuit is usually specified in decibels. If the input power is represented by P_{IN} and the output power by P_{OUT}, then the loss L, in decibels, is:

$$L = 10 \log_{10} (P_{IN}/P_{OUT})$$

A power loss is equivalent to a negative power gain. That is, if the loss (according to the above formula) is x dB, then the gain is −x dB. *See* DECIBEL, LOSS.

In a power-transmission line, loss occurs because of the finite resistance of the conductors, and also to some extent because of losses in the insulating and/or dielectric materials between the conductors. Power loss in a transmission line is sometimes given in watts, as the difference between the available power at the line input and the actual power dissipated in the load. *See also* AC POWER TRANSMISSION, POWER LINE, TRANSMISSION LINE.

POWER MEASUREMENT

The power used by a device, or radiated by an antenna, can be measured in various different ways. The most direct method is to use a wattmeter. A watt-hour meter can also be used; the device is allowed to run for a certain time, and the meter reading is then divided by that period of time (*see* WATTMETER, WATT-HOUR METER).

Direct-current power can be measured by means of a voltmeter and ammeter. The voltmeter is connected in parallel with the power-dissipating device, and the ammeter is connected in series (see illustration). The power in watts is the product of the voltage across the device, in volts, and the current through the device, in amperes.

In certain situations, power can be measured by determining the amount of heat that is liberated. This method works only for devices that dissipate all of their power as heat, such as resistors. The total heat energy can be determined, over a period of time, and the dissipated power then obtained by dividing the total heat energy by the time period.

POWER PLANT

A power plant is an installation that generates electric energy for use by the general public. There are several different types of power plants; all convert some nonelectric form of energy into electricty by means of large generators (*see* GENERATOR).

Some power plants burn fossil fuels, such as coal, oil, or natural gas, to obtain the energy necessary to turn the generators. The energy liberated by the burning of the fuel is used to heat water. This heating causes steam to form, and the steam pressure is used to drive turbines connected to the generators. The basic scheme for converting heat energy to electric energy is illustrated at A.

Nuclear power plants also use heat energy to drive turbines. However, in the nuclear plant, the heat is obtained from atomic reactors, rather than by the burning of fossil fuels (*see* NUCLEAR ENERGY).

Hydroelectric power plants use the force of falling water to operate the turbines (B). This process is somewhat simpler than the heat-and-steam method, but the plant must be located near a river that can be dammed to obtain the necessary mechanical energy (*see* HYDROELECTRIC POWER PLANT).

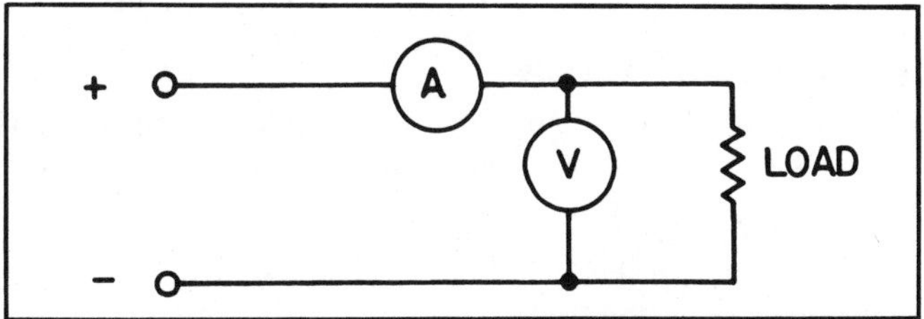

POWER MEASUREMENT: A simple method of power measurement. The ammeter (A) reading is multiplied by the voltmeter (V) reading.

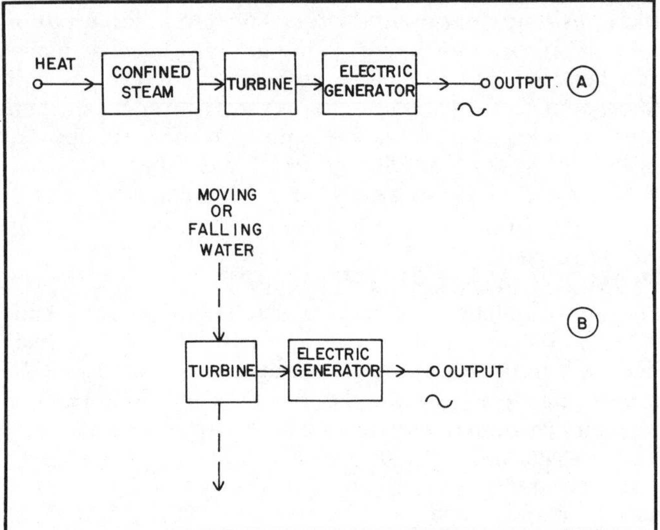

POWER PLANT: At A, block diagram showing the operation of a heat-driven power plant. At B, block diagram showing the operation of a hydroelectric power plant.

Because of the energy crisis of recent years, more attention is being paid to the use of power-generating methods that do not require fossil fuels. Also, more and more people are generating their own power from alternative sources such as the sun and the wind, rather than relying on commercial power plants. *See also* ENERGY, ENERGY CRISIS, SOLAR POWER, WIND POWER.

POWER SENSITIVITY

Certain amplifier circuits, notably Class-A circuits using vacuum tubes or field-effect transistors, consume essentially no power while producing measurable output power (*see* CLASS-A AMPLIFIER). Such amplifiers theoretically draw no current, and need only an alternating voltage at the grid or gate in order to function. The perfomance of such an amplifier is sometimes expressed by a parameter called power sensitivity.

If the root-mean-square (RMS) input-signal voltage is E, and the RMS output power is P, then the power sensitivity S is given by the ratio:

$$S = P/E$$

Power sensitivity is expressed in watts per volt.

The power sensitivity of a Class-A amplifier does not vary with the driving voltage, as long as the amplifier is correctly biased and is not overdriven. If the amplifier is driven into the nonlinear portion of the characteristic curve, the power sensitivity decreases.

POWER SERIES

A power series is an infinite summation of expressions associated with an infinite sequence of numbers. Given the sequence:

$$a_0, a_1, a_2, \ldots$$

the power series is:

$$S = a_0 + a_1 x + a_2 x^2 + \ldots$$

where x is a variable that may attain any real-number value.

A power series may or may not have a finite sum. For certain values of the a-factors, the power series will have a finite sum regardless of the value of x. For other values of the a-factors, the series will have a finite sum for certain values of x, but the sum will not be finite for other values of x. For still other a-values, the power series will be finite only if $x = 0$. When the series is finite, it is said to converge. When it is not finite, the series is said to diverge.

It is often the case that a power series will converge only if x is within a certain range. In such a series, the range of x over which the series converges is called the interval of convergence.

Power series are occasionally encountered in the design of electronic circuits, particularly bandpass filters.

POWER SUPPLY

A power supply is a device that produces electricity for use by electronic equipment, or that converts the electricity from the utility mains to a form suitable for use by electronic equipment. Power supplies generally consist either of batteries or of transformer/rectifier/filter circuits. A generator or set of photovoltaic cells may also be used as a power supply (*see* BATTERY, GENERATOR, SOLAR CELL).

A direct-current power supply is rated according to its voltage output, its current-delivering capacity, and its ability to maintain a constant voltage under varying load conditions. All of these parameters must be considered when choosing a power supply for use with a given piece of electronic apparatus. The voltage must be correct; the supply must be able to deliver the necessary current; the voltage must remain within a certain range as the load impedance or resistance changes. There is no need, however, to use a supply of much greater precision than required. It would be inefficient, for example, to use a 20-ampere power supply with a circuit that draws only 10 milliamperes.

Power-supply voltage can be as small as 1 volt or less, or as large as hundreds of thousands of volts. The current-delivering capacity may be just a few milliamperes, or it might be hundreds of amperes. The ability of a power supply to maintain a nearly constant voltage is called the regulation; some power supplies have excellent regulation, while in other situations the regulation need not be precise.

Some electronic devices have built-in power supplies. Most radio receivers, tape recorders, hi-fi amplifiers, and other consumer apparatus have built-in power supplies tailored to the requirements of the circuit. Some equipment, however, requires an external power supply. A wide variety of commercially manufactured power supplies is available for different specialized uses.

A power supply designed to produce direct current, with alternating-current input, must incorporate a transformer, a rectifier, a filter, and perhaps a regulator circuit. The transformer converts the utility voltage (usually 117 or 234 volts) to the proper voltage for the equipment in use. The rectifier converts the alternating current to pulsating direct current. The filter converts the pulsating direct current to more or less pure direct current. *See also* FILTER, POWER TRANSFORMER, RECTIFIER CIRCUITS, REGULATED POWER SUPPLY, TRANSFORMER.

POWER TRANSFORMER

A power transformer is a device designed especially for changing the voltage of alternating-current electricity. Power transformers can be found in all power supplies designed to be used with household current. Such transformers may be very small, supplying a relatively low voltage, or they may be large, supplying hundreds or even thousands of volts. Massive power transformers are used in the distribution of alternating-current power for utility purposes. Such transformers often must deal with many thousands of volts, and large amounts of current (*see* TRANSFORMER).

The basic construction of a typical power transformer is shown in the illustration. Two coils of wire are wound around a ferromagnetic core. The coil receiving the input current is called the primary, and the coil from which the output is taken is called the secondary. The core of a power transformer is usually made from laminated iron (*see* LAMINATED CORE, TRANSFORMER PRIMARY, TRANSFORMER SECONDARY).

The alternating current in the primary winding generates a fluctuating magnetic field in the core. The magnetic field causes alternating currents to flow in the secondary winding. The voltage across the secondary winding may be larger or smaller than the voltage applied to the primary, depending on the primary-to-secondary turns ratio.

Assume that a given power transformer is 100-percent efficient and has P turns in the primary and S turns in the secondary. Also, suppose that x volts are applied to the primary winding. Then the voltage y, appearing across the secondary winding, is:

$$y = (S/P)x$$

If the same transformer is connected to a circuit that draws x amperes from the secondary winding, then the current y flowing in the primary can be determined according to the above formula.

Of course, no transformer is completely efficient. This results in slightly lower secondary-winding voltages than the above formula would indicate; it also necessitates that the primary-winding current be greater than the above formula dictates. *See also* TRANSFORMER EFFICIENCY.

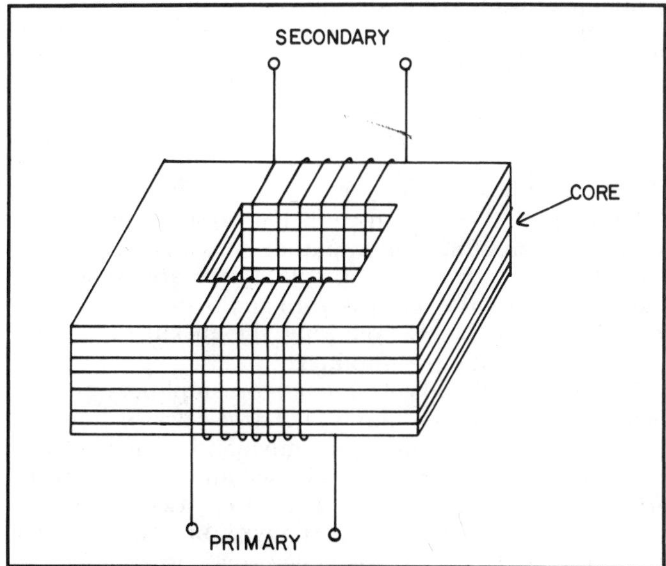

POWER TRANSFORMER: Pictorial diagram of a power transformer.

POWER TRANSISTOR

A power transistor is a semiconductor transistor designed for power-amplifier applications at audio and radio frequencies. Power transistors may also be used for switching purposes at high current levels. Devices with dissipation ratings higher than approximately 1 watt are usually considered power transistors. Bipolar power transistors have available in either the PNP or the NPN form. Certain field-effect power transistors are also available.

The main consideration in the use of a power transistor is the effective removal of heat from the device, especially from the base-collector junction. This is generally accomplished by means of a heat sink (*see* HEAT SINK).

Power transistors are available for operation at frequencies from direct current up to the microwave range. The obtainable power output decreases as the frequency increases.

The photograph shows a typical power transistor installed in a linear amplifier for use in the 440-MHz to 450-MHz range (70 cm). The top of the transistor is visible in the photograph; the case is made of ceramic material so that it can withstand high operating temperatures. The device is bolted firmly to a heat sink under the circuit board. (The heat sink is not visible.) The transistor is about ½ inch in diameter.

Power transistors are usually operated in Class-B or Class-C configurations, since this minimizes the dissipation at the base-collector junction. At audio frequencies, power transistors are usually operated in push-pull. *See also* CLASS-B AMPLIFIER, CLASS-C AMPLIFIER, PUSH-PULL AMPLIFIER, TRANSISTOR.

POWER TRANSISTOR: Closeup view of a power transistor. This device produces about 70 watts of RF output at 440 to 450 MHz.

POWER TRANSMISSION

See AC POWER TRANSMISSION, POWER LINE, TRANSMISSION LINE.

POWER TUBE

A power tube is a vacuum tube designed for use in power amplifiers. A power tube may be a triode device, a tetrode, or a pentode.

Power tubes have large plates, since the plate is where most of the dissipation takes place. The plate of a power tube may be finned, thus maximizing the heat-radiating surface area. A power tube having a plate-dissipation rating of more than a few watts is usually cooled by means of a fan or heat sink (*see* AIR COOLING, COOLING, HEAT SINK, PLATE DISSIPATION).

Power tubes can be used at frequencies from direct current well into the microwave range. At the low, medium, and high frequencies, single tubes can deliver several thousand watts of useful radio-frequency output power. The biggest power tubes are as large as a man, are housed in room-sized enclosures, and are cooled by elaborate air or fluid circulation devices.

Power tubes typically require rather high plate voltages, on the order of several kilovolts in some cases. For this reason, extreme precautions must be taken when servicing equipment containing power tubes. Lethal voltage is present. The power supply should be switched off before any work is done; it should be made certain that the filter capacitors have completely discharged.

Power tubes are usually operated in Class AB, Class B, or Class C. The push-pull configuration is sometimes used. *See also* CLASS-AB AMPLIFIER, CLASS-B AMPLIFIER, CLASS-C AMPLIFIER, PUSH-PULL AMPLIFIER.

POYNTING, JOHN HENRY

A theoretical physist, John Henry Poynting (1852-1914) is best known for his work in the theory of electromagnetic fields and propagation. Poynting demonstrated that fluctuating electric and magnetic fields result in a vector that propagates in a direction perpendicular to the plane containing the lines of flux. *See* POYNTING VECTOR.

POYNTING VECTOR

In an electromagnetic field, the electric lines of flux are perpendicular to the magentic lines of flux at every point in space (*see* ELECTROMAGNETIC FIELD). The electric field and the magnetic field have magnitude as well as direction, and thus they can be represented as vector quantities. Generally, the electric-field vector at a given point in space is called E, and the magnetic-field vector is called H.

The cross product of the E and H vectors (*see* CROSS PRODUCT) results in a vector perpendicular to both, and having length equal to the product of the lengths of E and H.

The vector E × H is called the Poynting vector for an electromagnetic field.

The Poynting vector is significant because its direction indicates the direction of propagation of the field (see

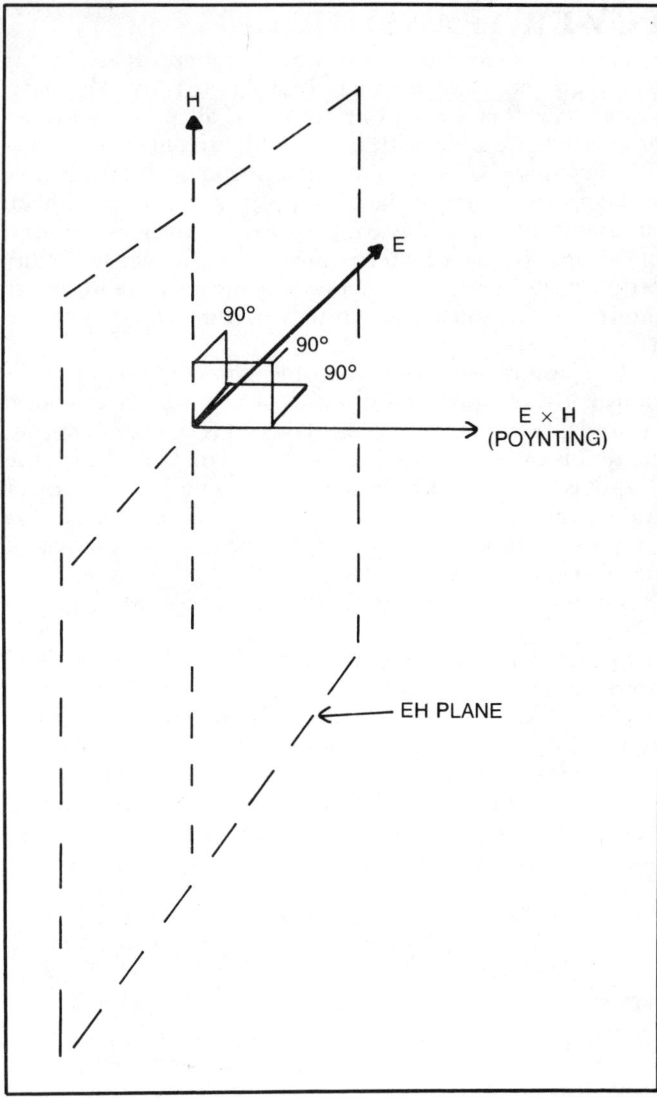

POYNTING VECTOR: The Poynting vector in an electromagnetic field is the cross product of the electric-field (E) and magnetic-field (H) vectors.

illustration), and its length indicates the intensity, or field strength, in terms of power per unit area in the surface containing the electric and magnetic vectors. *See also* ELECTRIC FIELD, ELECTRIC FLUX, FIELD STRENGTH, MAGNETIC FIELD, MAGNETIC FLUX.

PREAMPLIFIER

A preamplifier is a high-gain, low-noise amplifier, intended for boosting the amplitude of a weak signal. In radio-frequency receivers, preamplifiers are sometimes used to improve the sensitivity. In audio applications, the amplifier immediately following the pickup or microphone is called the preamplifier.

Some modern radio-frequency preamplifiers employ field-effect transistors. This type of device draws almost no power, and has a high input impedance that is ideally suited to weak-signal work. The gallium-arsenide field-effect transistor, or GaAsFET, is often used (*see* FIELD-EFFECT TRANSISTOR, GALLIUM-ARSENIDE FIELD-EFFECT TRANSISTOR). For audio-frequency work, either bipolar or field-effect transistors can provide good results. Field-

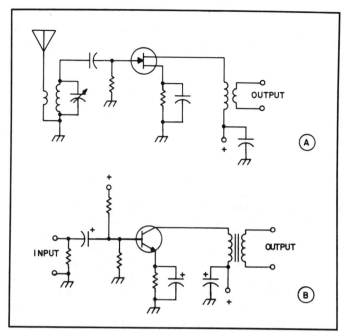

PREAMPLIFIER: Two examples of preamplifier circuits. At A, a radio-frequency preamplifier using a field-effect transistor. At B, an audio-frequency preamplifier using a bipolar transistor.

effect transistors are preferable if noise reduction is a major consideration.

The illustration at A shows a typical circuit suitable for use at radio frequencies. The amplifier operates in Class A. Input tuning is used to reduce the noise pickup and provide some selectivity against signals on unwanted channels. Impedance-matching networks are employed both at the input and at the output, to optimize the transfer of signal through the circuit and to the receiver. This preamplifier will produce about 10 dB to 15 dB gain, depending on the frequency.

A preamplifier suitable for use at audio frequencies is shown at B. This circuit has a broadband response, since it is intended for high-fidelity work. The gain is essentially constant, at about 15 dB to 20 dB, up to frequencies well beyond the range of human hearing.

In the design of any preamplifier, regardless of the intended frequencies of operation, it is important that linearity be as good as possible. Nonlinearity in a preamplifier results in intermodulation distortion at radio frequencies and harmonic distortion at audio frequencies. *See also* CLASS-A AMPLIFIER, INTERMODULATION, LINEARITY, NONLINEARITY.

PRECIPITATION STATIC

Precipitation static is a form of radio interference caused by electrically charged water droplets or ice crystals as they strike objects. The resulting discharge produces wideband noise that sounds similar to the noise generated by electric motors, fluorescent lights, or other appliances.

Precipitation static is often observed in aircraft flying through clouds containing rain, snow, or sleet. But occasionally, precipitation static occurs in base of mobile radio installations. This is especially likely to happen when it is snowing; then the noise is called snow static. Dust storms can also cause precipitation static. Precipitation static can be quite severe at times, making radio reception difficult, especially at the very low, low, and medium frequencies.

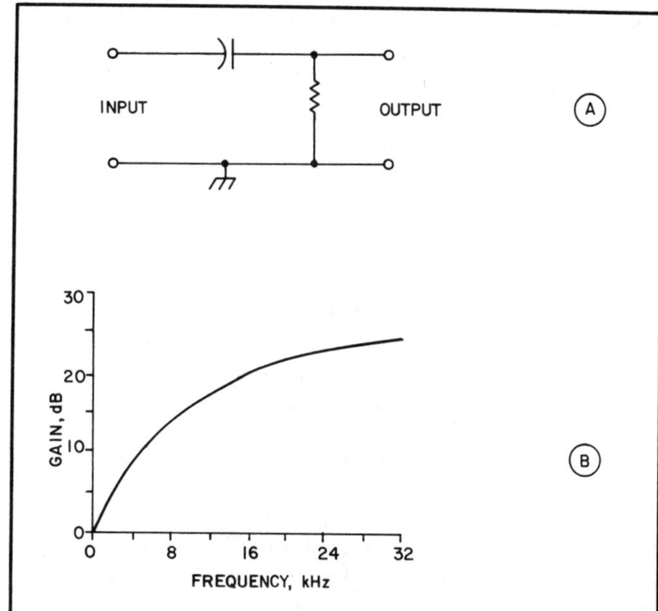

PREEMPHASIS: At A, a simple network for introducing preemphasis into an audio-frequency circuit. At B, a graph of the gain, in decibels, versus the modulating frequency, in kilohertz, for a typical frequency-modulated transmitter incorporating preemphasis.

A noise blanker or limiter is effective in reducing interference caused by precipitation static. A means of facilitating discharge, such as an inductor between the antenna and ground, may be helpful. Improvement may also be obtained by blunting any sharp points in antenna elements. *See also* NOISE BLANKER, NOISE LIMITER.

PRECISION

See ACCURACY, TOLERANCE.

PRECISION POTENTIOMETER

A precision potentiometer is a variable resistor having very accurate calibration and ease of resettability. Precision potentiometers are usually constructed using wirewound elements, in a manner similar to a rheostat (*see* RHEOSTAT, WIREWOUND RESISTOR). Because of this, such devices are not suitable for use in situations where they must carry radio-frequency currents. The windings exhibit significant inductive reactance, as well as resistance, at radio frequencies.

Some precision potentiometers have multi-turn shafts. This makes it possible to set them almost exactly to any desired value within the resistance range. *See also* POTENTIOMETER.

PREEMPHASIS

Preemphasis is the deliberate accentuation of the higher audio frequencies in a frequency-modulated transmitter. Preemphasis is obtained in a transmitter by inserting a highpass filter in the audio stages. Such a highpass filter is shown in the illustration at A.

In a phase-modulated transmitter, the preemphasis occurs automatically, because of the nature of phase modulation as compared with frequency modulation (*see* FRE-

QUENCY MODULATION, PHASE MODULATION). The graph at B shows the attenuation-versus-frequency characteristic of a typical preemphasis network.

Preemphasis in a frequency-modulated transmitter, in conjunction with deemphasis at the receiver, improves the signal-to-noise ratio at the upper end of the audio range. *See also* DEEMPHASIS.

PREFIX MULTIPLIERS

Physical units are often subdivided or multiplied for purposes of mathematical convenience. Special prefixes, known as prefix multipliers, indicate the constant factor by which the unit is changed.

Generally, prefix multipliers are given in increments of three, such as 10^3, 10^6, and so on for large units, and 10^{-3}, 10^{-6}, and so on for small quantities. However, intermediate values are sometimes given. The table shows the prefix multipliers used for factors ranging form 10^{-18} to 10^{18}.

As exponents become more and more negative, the unit size diminishes. As the exponents become more and more positive, the unit size increases. Each exponential increment represents one order of magnitude, or a factor of 10. *See also* ORDER OF MAGNITUDE, SCIENTIFIC NOTATION.

PRESELECTOR

A tuned circuit in the front end of a radio receiver is called a preselector. Most receivers incorporate preselectors. The preselector provides a bandpass response that improves the signal-to-noise ratio, and also reduces the likelihood of receiver overloading by a strong signal far removed from the operating frequency. The preselector also provides a high degree of image rejection in a superheterodyne circuit. Most preselectors have a 3-dB bandwidth that is a few percent of the received frequency.

A preselector may be tuned by means of tracking with the tuning dial. This eliminates the need for continual readjustment of the preselector control, thereby making the receiver more convenient to operate. The tracking type of preselector, however, requires careful design and alignment (*see* TRACKING).

Some receivers incorporate preselectors that must be adjusted independently. Although this is less convenient from an operating standpoint, the control need be reset only when a large frequency change is made. The independent preselector simplifies receiver design. *See also* FRONT END.

THE MOST COMMONLY USED PREFIX MULTIPLIERS.

Prefix	Abbreviation	Multiple
atto-	a	10^{-18}
femto-	f	10^{-15}
pico-	p	10^{-12}
nano-	n	10^{-9}
micro-	μ or u	10^{-6}
milli-	m	10^{-3}
centi-	c	10^{-2}
deci-	d	10^{-1}
deca-	D or da	10
hecto-	h	10^2
kilo-	k	10^3
mega-	M	10^6
giga-	G	10^9
tera-	T	10^{12}
peta-	P	10^{15}
exa-	E	10^{18}

PRESENCE

In high-fidelity systems, presence is the extent to which the reproduced sound seems true-to-life. Voices or music have good presence if they are to sound real.

Excellent presence requires the use of adequate speakers or headphones and good amplifier design. The speakers or headphones must have a flat frequency response over the entire human hearing range; the same must be true of the amplifiers. The linearity and dynamic range of the system must also be excellent.

To a certain extent, a speaker/headphone frequency-response deficiency can be offset by adjustments in the amplifier circuit. An equalizer (also known as a graphic equalizer) can often improve the presence of a marginal system. An equalizer is especially useful in any system, since people do not always exactly agree on what constitutes good presence. *See also* DYNAMIC RANGE, EQUALIZER, FLAT RESPONSE, HIGH FIDELITY, LINEARITY.

PRIMARY

See TRANSFORMER PRIMARY.

PRIMARY COLORS

The primary colors are the three colors that, when combined with equal intensity, result in white. These colors are red, blue, and green for radiant light. By mi-ing the three primary colors in various brightness combinations, all possible hues and saturation values can be obtained (*see* CHROMA, HUE, SATURATION).

For reflecting surfaces, colors are technically called pigments. The three primary pigments, when combined in equal concentration, result in no reflected light (black). The primary pigments are red, yellow, and blue.

In color television, primary-color combinations result in the various shades and hues that you see. Three different sets of phosphor dots on the picture-tube screen are activated to form all colors. One set of dots is red, one is blue, and one is green. The dots are interwoven in such a way that, from a distance, they seem to blend together. *See also* COLOR PICTURE SIGNAL, COLOR TELEVISION.

PRIMARY EMISSION

When electrons emanate directly from an object, such as an electron gun or the cathode of a vacuum tube, the emission is said to be primary. Primary emission results from the application of a negative charge to an electrode. The greater the applied charge, assuming all other parameters remain constant, the greater is the primary emission.

When an electrode is struck by high-speed particles, electrons may be knocked from it. This occurs, for example, in the plate of a vacuum tube or the dynode of a photomultiplier. This kind of emission does not arise directly from a negative charge, and is called secondary emission. *See also* DYNODE, PHOTOMULTIPLIER, SECONDARY EMISSION.

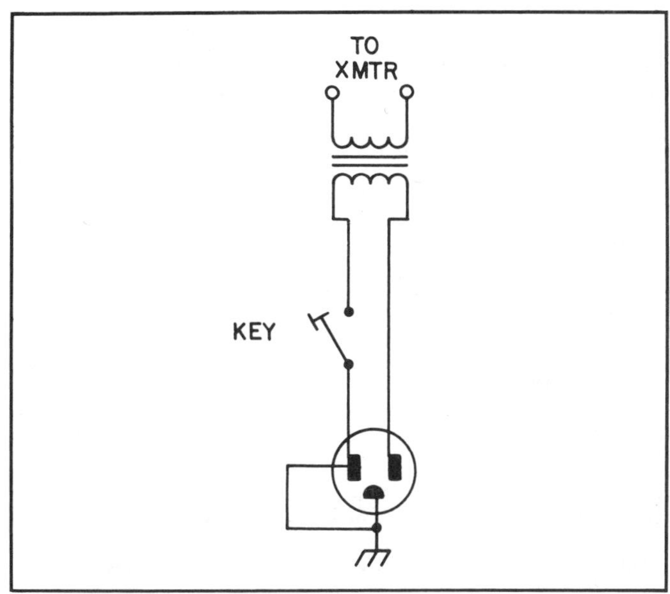

PRIMARY KEYING: Primary keying is accomplished by interrupting the alternating-current line to the power supply of a transmitter.

PRIMARY KEYING

A code transmitter may be keyed by interrupting the power supply. If the power is broken in the primary circuit of the supply transformer, the keying method is known as primary keying. Primary keying may be employed in a vacuum-tube circuit or in a solid-state circuit. The drawing illustrates primary keying in a code transmitter.

Primary keying is seldom used, especially in high-power transmitters. The voltage across the keying terminals in primary keying is usually quite high, either 117 or 234 volts. This presents a shock hazard. The power-supply filter capacitors, especially in a transmitter that draws considerable current, have a time constant that is too slow to follow keying at any reasonable speed. In a vacuum-tube circuit, primary keying cannot be used in the transformer supplying the tube filaments, since the filaments cannot heat up and cool down fast enough to follow any reasonable keying speed.

Primary keying is most practical in low-power transistor circuits. *See also* CODE TRANSMITTER, KEYING.

PRINCIPLE OF UNCERTAINTY

See UNCERTAINTY PRINCIPLE.

PRINCIPLE OF INVERSE SQUARES

See INVERSE-SQUARE LAW.

PRINTED CIRCUIT

A printed circuit is a wiring arrangement that is fabricated by means of foil runs on a circuit board. Printed circuits can be mass-produced inexpensively and efficiently. Printed circuits allow extreme miniaturization and high reliability. Most electronic devices today are built using printed-circuit technology, although high-power circuits still use point-to-point wiring methods. The photograph shows a typical printed circuit.

Printed circuits are fabricated by first drawing an etching pattern. This pattern is then photographed and reproduced on clear plastic. The plastic is placed over a copper-coated glass-epoxy or phenolic board, and the assembly undergoes a photochemical process (*see* ETCHING). This may be done on one side of the board only, resulting

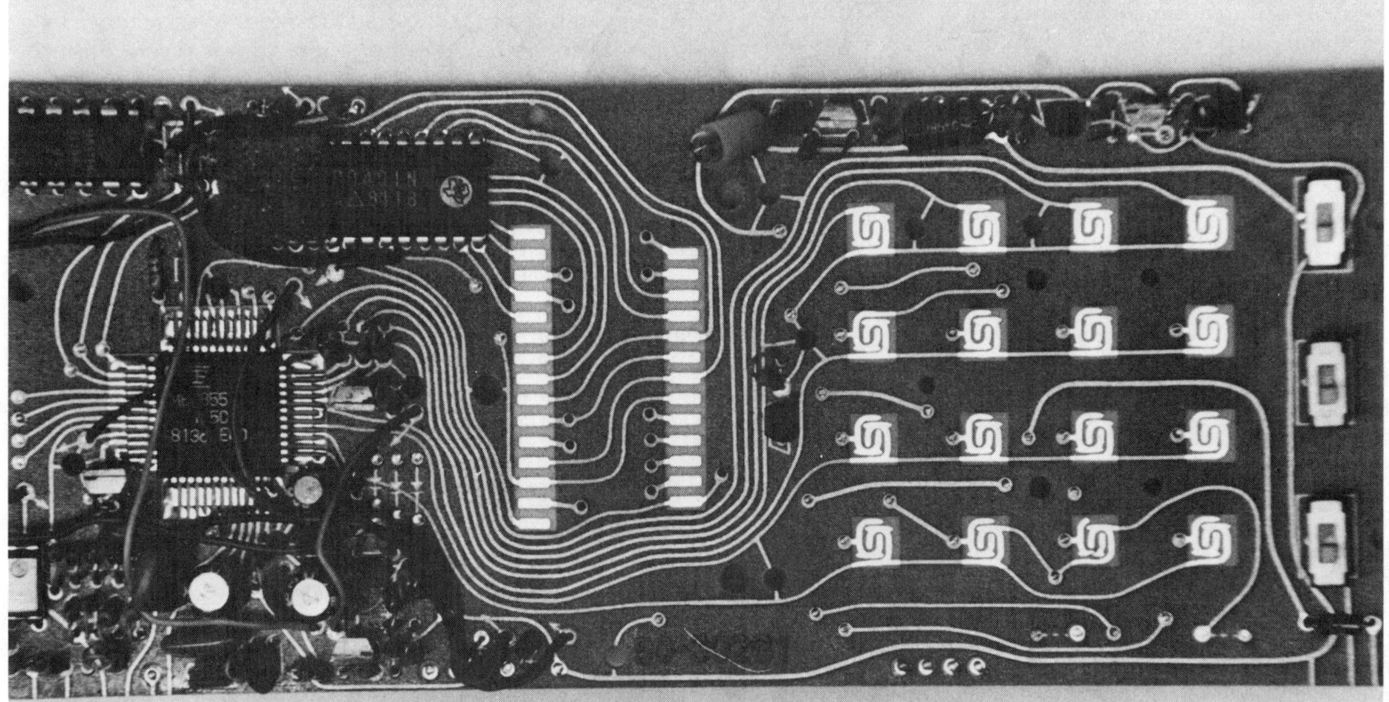

PRINTED CIRCUIT: A modern printed-circuit board.

in what is called a single-sided printed circuit. If the circuit is too complicated to be drawn on a single plane surface without crossing foil runs, the etching is done on both sides of the board. Then, the board is called a double-sided printed circuit.

The use of printed circuits has vastly enhanced the ease with which electronic equipment can be serviced. Printed circuits allow modular construction, so that an entire board can be replaced in the field and repaired in a fully equipped laboratory. *See also* CARD, CIRCUIT BOARD, MODULAR CONSTRUCTION.

PRINTER

A printer is any device that produces a permanent copy of alphanumeric data. Printers are widely used in conjunction with computers, terminal units, word processors, and many other electronic systems.

Most modern printers are of either the daisy-wheel type or the dot-matrix type (*see* DAISY-WHEEL PRINTER, DOT-MATRIX PRINTER). The dot-matrix printer can produce hard copy at an extremely rapid rate; typical speeds are 60 characters per second or more. The daisy-wheel printer is somewhat slower, but produces letter-quality copy.

Older style printers include the ball type, such as is used in many electric typewriters, and the key-strike type, used in most mechanical typewriters. These types of printers are slower than the daisy-wheel and dot-matrix printers.

PRIVATE BRANCH EXCHANGE

A private branch exchange, abbreviated PBX, is a telephone switching network installed in a private business or public office building. Such systems are commonly used in large and medium-sized corporations, and in various government offices. Some small companies also use PBX systems. A private branch exchange may have just a few extensions, or it may have hundreds or even thousands.

The main switchboard of the PBX is connected to the outside telephone line, and has a single number or set of numbers. An operator receives all incoming calls. A caller then requests a certain extension, which has its own number, usually consisting of two, three, or four digits. The operator then dials that extension number within the PBX system, and the appropriate extension telephone is rung. *See also* TELEPHONE.

PRIVATE LINE®

Private Line® is a trademark of Motorola, Inc. The term is commonly used to describe a radiotelephone tone-squelch system. Private-Line, or PL, systems are sometimes used in repeaters to limit the access to certain persons. A PL system may also be used to keep a receiver from responding to unnecessary or unwanted signals. For the receiver squelch to open in the presence of a signal, that signal must be modulated with a tone having a certain frequency. Tone-squelch systems are most often used with frequency modulation. Private-line communications is used by radio amateurs in certain parts of America as well as other countries.

In the United States, most PL systems utilize one of 32 standard subaudible-tone frequencies (see Table 1). In some systems, nonstandard tones are used that do not appear in the table. Each tone is designated by a pair of letters.

PRIVATE LINE®: SUBAUDIBLE-TONE FREQUENCIES.

Designation	Freq., Hz.	Designation	Freq., Hz.
XZ	67.0	2B	118.8
XA	71.9	3Z	123.0
WA	74.4	3A	127.3
XB	77.0	3B	131.8
SP	79.7	4Z	136.5
YZ	82.5	4A	141.3
YA	85.4	4B	146.2
YB	88.5	5Z	151.4
ZZ	91.5	5A	156.7
ZA	94.8	5B	162.2
ZB	97.4	6Z	167.9
1Z	100.0	6A	173.8
1A	103.5	6B	179.9
1B	107.2	7Z	186.2
2Z	110.9	7A	192.8
2A	114.8	MI	203.5

Table 1

PRIVATE LINE: TONE-SQUELCH BURST FREQUENCIES.

Freq., Hz.	Freq., Hz.
1600	2150
1650	2200
1700	2250
1750	2300
1800	2350
1850	2400
1900	2450
1950	2500
2000	2550
2100	

Table 2

In a frequency-modulation PL system, the carrier is usually continuously modulated with the deviation of approximately about 0.5 to 1.5 kHz by a sine-wave tone. The tone cannot be heard on the signal as it is received, since the frequency is below the response cutoff of the voice-amplifier chain.

In European countries, and also in some U.S. communications systems, tone bursts of higher frequency are often used for tone-squelch purposes. The most common burst-tone frequencies appear in Table 2. The burst may consist of just one tone, or it may consist of several tones in sequence, each having a different frequency.

Burst tones must have a certain frequency, duration, and deviation. Tone-burst systems are used by some police departments and business communications systems in the United States; in recent years, tone-burst squelch has gained popularity among radio amateurs as well.

PROBABILITY

Probability is an expression of the mathematical likelihood that a given event will occur within a certain amount of time, or in a certain number of trials or tests. For example, we may speak of the probability that a coin will come up "heads" at least once in three tosses; we may wish to know the probability that our radio transceiver will fail during a ten-day safari in Africa.

The determination of probability is a part of the mathematical field of statistics. Probability is of obvious importance in electronics. *See also* STATISTICAL ANALYSIS.

PROBABLE ERROR

In any measurement, a certain amount of error invariably occurs. The extent of the error depends on the accuracy of the test equipment, and on the precision of the calibration standard. If a measurement is repeated several times, the error will generally be different each time. The probable error is defined as the median error that is expected to occur.

Suppose that a carrier wave has a frequency of exactly 10.000000 MHz, and we attempt to measure its frequency using a shortwave receiver that is not especially well-calibrated. Let us say that the probable error is plus or minus 5 kHz. This means that if we make many measurements of the frequency, we can expect that our determination will be inaccurate by more than 5 kHz just as often as it will fall within 5 kHz. Half of the time, we will obtain an indication between 9.999995 and 10.000005 MHz; half of the time we will get a reading below 9.999995 MHz or above 10.000005 MHz. *See also* ACCURACY, ERROR.

PROBE

A probe is a device intended for picking off a signal from a specific point in a circuit for test purposes. The probe normally does not disturb the operation of the circuit.

Probes may be used to measure direct-current voltages. Probes are used with ohmmeters to determine the resistance between two points in a circuit. Probes may also be used with oscilloscopes to determine the voltage, waveform characteristics, frequency, and phase of an alternating-current signal. Oscilloscope probes are also extensively employed in the testing and troubleshooting of digital equipment.

Most probes have extremely high impedance. An insulated handle prevents shock to personnel and reduces the effects of hand capacitance. A typical low-voltage oscilloscope test probe is illustrated. Probe cables should be as short as possible, and should be shielded to prevent the pickup of hum or radio-frequency interference.

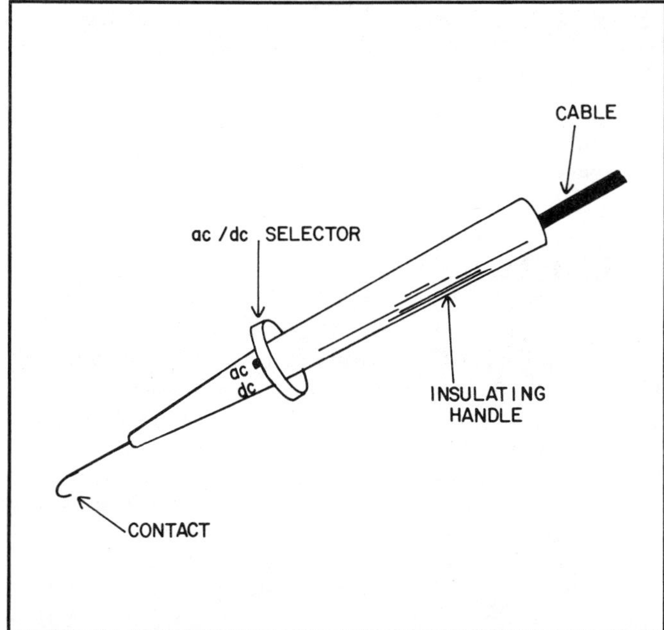

PROBE: An example of a test probe, in this case for use with an oscilloscope.

PROBLEM-ORIENTED LANGUAGE

A higher-order computer language, designed so that the commands have relevance to the intended application, is called a problem-oriented language. Most higher-order languages are problem-oriented to some extent (*see* HIGHER-ORDER LANGUAGE).

An excellent example of a problem-oriented language is COBOL, which is used extensively in business. The computer statements and inquiries, as well as the operator commands, consist of words familiar to business people. The scientific language FORTRAN, and the mathematical/scientific language BASIC, are problem-oriented to a lesser extent. Most word-processor systems use languages that are highly problem-oriented. *See also* BASIC, COBOL, FORTRAN, WORD PROCESSING.

PROD

See PROBE.

PRODUCT DETECTOR

Single-sideband (SSB), continuous-wave (code or CW), and frequency-shift-keyed (FSK) signals require a special form of detector for reception. Such a detector must supply an unmodulated local signal, and must provide for some means of mixing this signal with the incoming signal. This type of detector is called a product detector. The product detector works according to the same principles as the heterodyne detector. Generally, the term heterodyne detector is used for direct-conversion reception, and the term product detector is used with superheterodyne reception, although the terms may sometimes be used interchangeably (*see* DIRECT-CONVERSION RECEIVER, HETERODYNE DETECTOR, SUPERHETERODYNE RECEIVER).

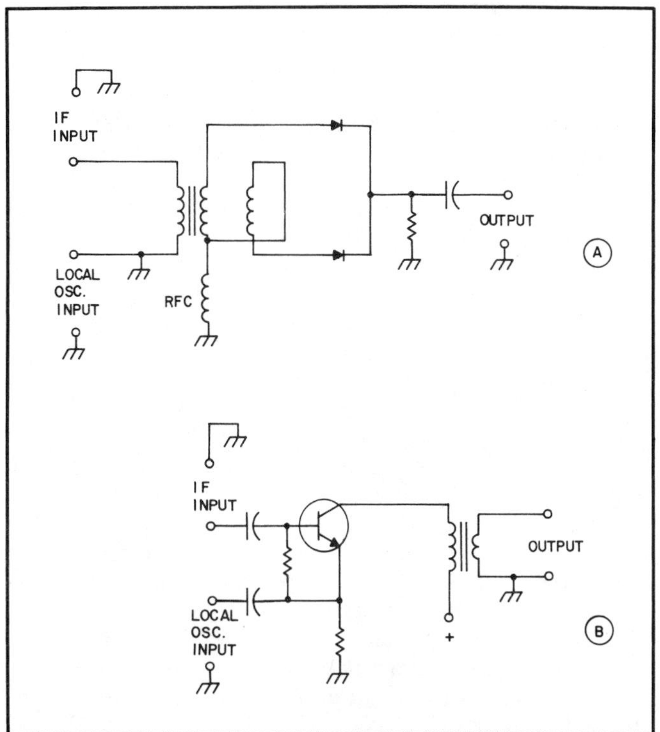

PRODUCT DETECTOR: At A, a passive product detector. At B, an active product detector. The active circuit produces some gain.

Product detectors may be either passive or active. The illustration shows schematic diagrams of a passive product detector (A) and an active product detector (B). These circuits are only examples of product detectors; many different configurations are possible.

In single-sideband reception, the local oscillator must be set to precisely the same frequency as the suppressed carrier of the incoming intermediate-frequency signal. This requires that the local oscillator be very stable, for otherwise, severe distortion will result. In code and FSK reception, the local oscillator is set to a slight different frequency than the incoming intermediate-frequency signal. The difference is usually between 300 Hz and 3 kHz, creating an audible beat note or notes.

Product detectors can be used to receive ordinary amplitude-modulated signals, but an annoying beat note or flutter usually occurs as the local-oscillator signal competes with the unsuppressed carrier. For amplitude modulation, the envelope detector gives much better results than the product detector (*see* ENVELOPE DETECTOR).

A product detector can be used to receive narrow-band frequency-modulated signals, but the same problems are encountered in this case as with amplitude modulation. For frequency modulation, the discriminator, phase-locked loop, or ratio detector are the most preferable circuits (*see* DISCRIMINATOR, PHASE-LOCKED LOOP, RATIO DETECTOR).

PROGRAM

A program is a set of data bits, comprising a unit of information and/or instructions to perform a specific task. In computer practice, a program tells the device what to do. One program, called the assembler, is used to derive the machine language of the computer from the assembly language (*see* ASSEMBLER, AND ASSEMBLY LANGUAGE, HIGHER-ORDER LANGUAGE, MACHINE LANGUAGE). Another program, the compiler, converts higher-order language to machine language.

For a given higher-order language, the assembler and compiler do not change until a different higher-order language is used. The higher-order program can be changed at will by the operator (*see* COMPUTER PROGRAMMING).

A radio or television broadcast, intended for reception by the public and conveying a message, is called a program. Examples of such a program include the news/weather/sports broadcast, a movie broadcast, and a single episode in a continuing series.

PROGRAMMABLE CALCULATOR

Electronic calculators facilitate arithmetic operations with ease and accuracy. However, even with a calculator, certain mathematical procedures can be tedious. An example of such a procedure is the summation of all the positive integers up to 500. Another example is the approximation of an integral by the rectangle method.

A programmable calculator can perform certain functions over and over, saving the operator from having to do the iterations. A programmable calculator is more sophisticated than an ordinary "manual" calculator, but it is not as complex as a computer.

A few programmable calculators use a standard computer card. Holes are punched manually in the card to create the program, which may have dozens of steps and perform a relatively complicated operation. More commonly, electronic memory is used for program storage. Other programmable calculators are small enough to be put into a shirt pocket. *See also* ELECTRONIC CALCULATOR.

PROGRAMMABLE READ-ONLY MEMORY

A programmable read-only memory (PROM) is a form of read-only memory (ROM) used in computers. The information in the PROM is stored permanently. Data is not lost if power is removed from the unit.

In its original form, the PROM integrated circuit is filled with logic blanks. These may be either zeroes or ones; it makes no difference. A device called a programmer allows any desired information to be written into the PROM, provided the data does not exceed the storage capacity of the integrated circuit.

Some PROM devices can be erased and reprogrammed with a different set of data. Such a device is called an EPROM, for erasable PROM. *See also* READ-ONLY MEMORY.

PROGRAMMABLE UNIJUNCTION TRANSISTOR

A programmable unijunction transistor (PUT) is a form of unijunction transistor with a variable triggering voltage. The triggering voltage is set by means of an external voltage divider. Such a divider may consist of resistors, or it may consist of independent sources.

The PUT is similar to the silicon-controlled rectifier. The PUT, silicon-controlled rectifier, and conventional unijunction transistor are all used in similar applications. *See also* SILICON-CONTROLLED RECTIFIER, UNIJUNCTION TRANSISTOR.

PROGRAMMING

See COMPUTER PROGRAMMING, PROGRAM.

PROM

See PROGRAMMABLE READ-ONLY MEMORY.

PROPAGATION

Propagation is the transfer of energy through a medium or through space. Certain disturbances, such as sound waves or electric currents, can propagate only through a material substance. Electromagnetic fields can propagate through empty space (*see* ELECTRIC FIELD, ELECTROMAGNETIC FIELD, MAGNETIC FIELD).

Electromagnetic-wave propagation occurs in straight lines through a perfect vacuum in the absence of intervening forces or effects. In and around the atmosphere of the earth, however, electromagnetic waves are often propagated in bent paths. The science of radio-wave propagation is sophisticated. The following several articles describe the most important aspects of wave propagation at radio frequencies.

PROPAGATION CHARACTERISTICS

Radio waves are affected in a variety of ways as they travel through the atmosphere of the earth. The effects vary with the wavelength. The troposphere, or lowest part of the atmosphere, causes electromagnetic waves to be bent or scattered at some frequencies. The ionized layers (*see* IONOSPHERE) affect radio waves at some frequencies.

A general description of propagation characteristics, for various electromagnetic frequencies, follows.

Very Low Frequencies (Below 30 kHz). At very low frequencies, propagation takes place mainly via waveguide effect between the earth and the ionosphere. Surface-wave propagation also occurs for considerable distances. With high-power transmitters and large antennas, communication can be realized over distances of several thousand miles at frequencies between 10 kHz and 30 kHz. The earth-ionosphere waveguide has a lower cutoff frequency slightly below 10 kHz. For this reason, signals much below 10 kHz suffer severe attenuation and do not propagate well.

Antennas for very-low-frequency transmitting must be vertically polarized. Otherwise, surface-wave propagation will not take place, and the proximity of the ground tends to short-circuit the electromagnetic field.

Propagation at very low frequencies is remarkably stable; there is very little fading. Solar flares occasionally disrupt communication in this frequency range, by making the ionosphere absorptive and by raising the cutoff frequency of the earth-ionosphere waveguide.

Because the surface-wave mode is very efficient at frequencies below 30 kHz, it is possible that the very-low-frequency band might be used for over-the-horizon communication on planets that lack ionized layers concentrated enough to return signals at higher frequencies. The moon and Mars are examples of such planets. *See* SOLAR FLARE, SURFACE WAVE, WAVEGUIDE PROPAGATION.

Low Frequencies (30 kHz to 300 kHz). In the low-frequency range, propagation takes place in the surface-wave mode, and also as a result of ionospheric effects.

Toward the lower end of the band, wave propagation is similar to that in the very-low-frequency range. As the frequency is raised, surface-wave attenuation increases. While a surface-wave range of more than 3,000 miles is common at 30 kHz, it is unusual for the range to be greater than a few hundred miles at 300 kHz. Surface-wave propagation requires that the electromagnetic field be vertically polarized.

Ionospheric propagation at low frequencies usually occurs via the E layer. This increases the useful range during the nighttime hours, especially toward the upper end of the band. Intercontinental communication is possible with high-power transmitters.

Solar flares can disrupt communication at low frequencies. Following such a flare, the D layer becomes highly absorptive, preventing ionospheric communication. *See* D LAYER, E LAYER, SOLAR FLARE, SURFACE WAVE.

Medium Frequencies (300 kHz to 3 MHz). Propagation at medium frequencies occurs by means of the surface wave, and by E-layer and F-layer ionospheric modes.

Near the lower end of the band, surface-wave communications paths can be up to several hundred miles. As the frequency is raised, the surface-wave attenuation increases. At 3 MHz, the range of the surface wave is limited to about 150 or 200 miles.

Ionospheric propagation at medium frequencies is almost never observed during the daylight hours, since the D layer prevents electromagnetic waves from reaching the higher E and F layers. During the night, ionospheric propagation takes place mostly via the E layer in the lower portion of the band, and primarily via the F layer in the upper part of the band. The communications range increases as the frequency is raised. At 3 MHz, worldwide communication is possible over darkness paths.

As at other frequencies, medium-frequency propagation is severely affected by solar flares. The 11-year sunspot cycle and the season of the year also affect propagation at medium frequencies. Propagation is usually better in the winter than in the summer. *See* D LAYER, E LAYER, F LAYER, SOLAR FLARE, SUNSPOT CYCLE, SURFACE WAVE.

High Frequencies (3 MHz to 30 MHz). Propagation at the high frequencies exhibits widely variable characteristics. Effects are much different in the lower part of this band than in the upper part. Let us consider the lower portion as the range 3 MHz to 10 MHz, and the upper portion as the range 10 MHz to 30 MHz.

Some surface-wave propagation occurs in the lower part of the high-frequency band. At 3 MHz, the maximum range is 150 to 200 miles; at 10 MHz it decreases to about the radio-horizon distance, perhaps 15 miles. Above 10 MHz, surface-wave propagation is essentially nonexistent.

Ionospheric communication occurs mainly via the F layer. In the lower part of the band, there is very little daytime ionospheric propagation because of D-layer absorption; at night, worldwide communication can be had. In the upper part of the band, this situation is reversed; ionospheric communication can occur on a worldwide scale during the daylight hours, but the maximum usable frequency often drops below 10 MHz at night. Of course, the transition is gradual, and the range in which it occurs is variable.

Communications in the lower part of the high-frequency band are generally better during the winter months than during the summer months. In the upper part of the band, this situation is reversed.

Some E-layer propagation is occasionally observed in the upper part of the high-frequency band. This is usually of the sporadic-E type. This may occur even when the F-layer maximum usable frequency is below the communication frequency.

Near 30 MHz, there is often no ionospheric communication. This is especially true when the sunspot cycle is at or near a minimum. The extreme upper portion of the high-frequency band then behaves similarly to the very high frequencies.

Solar flares cause dramatic changes in conditions in the high-frequency band. Sometimes, ionospheric communications are almost totally wiped out within a matter of minutes by the effects of a solar flare. *See* D LAYER, E LAYER, F LAYER, MAXIMUM USABLE FREQUENCY, SOLAR FLARE, SPORADIC-E PROPAGATION, SUNSPOT CYCLE, SURFACE WAVE.

Very High Frequencies (30 MHz to 300 MHz). Propagation in the very-high-frequency band occurs along the line-of-sight path, or via tropospheric modes. Ionospheric F-layer propagation is rarely observed, although it can occur at times of sunspot maxima at frequencies as high as about 70 MHz. Sporadic-E propagation is fairly common, and can occur up to approximately 200 MHz.

Tropospheric propagation can occur in three different ways: bending, ducting, and scattering. Communications range varies, but can often be had at distances of several hundred miles.

Meteor-scatter and auroral propagation can sometimes be observed in the very-high-frequency band. The range for meteor scatter is typically a few hundred miles; auroral propagation can sometimes produce communications over distances of up to about 1,500 miles. Code transmission must usually be employed for auroral communications; the moving auroral curtains introduce phase modulation that makes voices practically unintelligible.

Repeaters are used extensively at very high frequencies to extend the range of mobile communications equipment. A repeater, located in a high place, can provide coverage over an area of thousands of square miles.

At the very high frequencies, satellites are often used to provide worldwide communication on a reliable basis. The satellites are actually orbiting repeaters. The moon, a passive satellite, is sometimes used, mostly by amateur radio operators. *See* AURORAL PROPAGATION, E LAYER, F LAYER, METEOR SCATTER, MOONBOUNCE, REPEATER, SATELLITE COMMUNICATIONS, TROP OSPHERIC PROPAGATION, TROPOSPHERIC-SCATTER PROPAGATION.

Ultra High Frequencies (300 MHz to 3 GHz). Propagation in the ultra-high-frequency band occurs almost exclusively via the line-of-sight mode and via satellites and repeaters. No ionospheric effects are ever observed. Auroral, meteor-scatter, and tropospheric propagation are occasionally observed in the lower portion of this band. Ducting can result in propagation over distances of several hundred miles.

The main feature of the ultra-high-frequency band is its sheer immensity—2,700 megahertz of spectrum space. Another advantage of this band is its relative immunity to the effects of solar flares. Since ultra-high-frequency energy has nothing to do with the ionosphere, disruptions in the ionosphere have no effect. However, an intense solar flare can interfere to some extent with ultra-high-frequency circuits, because of electromagnetic effects resulting from fluctuations in the magnetic field of the earth. *See* AURORAL PROPAGATION, LINE-OF-SIGHT COMMUNICATION, METEOR SCATTER, MOONBOUNCE, REPEATER, SATELLITE COMMUNICATIONS, TROPOSPHERIC PROPAGATION, TROPOSPHERIC-SCATTER PROPAGATION.

Microwaves (Above 3 GHz). Microwave energy travels essentially in straight lines through the atmosphere. Microwaves are affected very little by such phenomena as temperature inversions and scattering. Microwaves are unaffected by the ionosphere.

The primary mode of propagation in the microwave range is the line-of-sight mode. This facilitates repeater and satellite communications.

The microwave band is much more vast than even the ultra-high-frequency band. The upper limit of the microwave range, in theory, is the infrared spectrum. Current technology has produced radio transmitters capable of operation at frequencies of more than 300 GHz.

At some microwave frequencies, atmospheric attenuation becomes a consideration. Rain, fog, and other weather effects cause changes in the path attenuation as the wavelength becomes comparable with the diameter of water droplets. *See* LINE-OF-SIGHT COMMUNICATION, REPEATER, SATELLITE COMMUNICATIONS.

Infrared, Visible Light, Ultraviolet, X Rays, and Gamma Rays. Propagation at infrared and shorter wavelengths occurs in straight lines. The only factor that varies is the atmospheric attenuation.

Water in the atmosphere causes severe attenuation of infrared radiation between the wavelenghts of approximately 4,500 and 8,000 nanometers (nm). Carbon dioxide interferes with the transmission of infrared radiation at wavelengths ranging from about 14,000 nm and 16,000 nm. Rain, snow, fog, and dust interfere with the transmission of infrared radiation.

Light is transmitted fairly well through the atmosphere at all visible wavelengths. But rain, snow, fog, and dust interfere considerably with the transmission of visible light through the air.

Ultraviolet light at the longer wavelengths can penetrate the air with comparative ease. At shorter wavelengths, attenuation increases. Rain, snow, fog, and dust interfere.

X rays and gamma rays do not propagate well for long distances through the air. This is primarily because the mass of the air, over propagation paths of great distance, is sufficient to block these types of radiation. *See* GAMMA RAY, INFRARED, OPTICAL COMMUNICATION, ULTRAVIOLET, X RAY.

PROPAGATION DELAY

See PROPAGATION TIME.

PROPAGATION FORECASTING

Ionospheric propagation conditions vary considerably from day to day, month to month, and year to year at frequencies below 60 MHz. To some extent, these changes are predictable.

The sun is constantly monitored during the daylight hours for fluctuations in the intensity of radiation at various electromagnetic frequencies, especially 2800 MHz. The 2800-MHz solar flux, as it is called, provides a good indicator of propagation conditions for the ensuing several hours. A sudden increase in the solar flux means that a flare is occurring. Within a short time, propagation conditions usually beome disturbed following a flare. Flares sometimes wipe out F-layer communications completely, but the E layer may become sufficiently ionized to allow propagation over moderate distances. At higher latitudes, auroral propagation can usually be observed after a solar flare (*see* AURORAL PROPAGATION, SOLAR FLARE, SOLAR FLUX, SPORADIC-E PROPAGATION).

Regular variations in propagation conditions take place. At frequencies below about 10 MHz, propagation is generally better at night than during the day. Above about 10 MHz, this situation is reversed. The winter months are generally better for propagation conditions below 10 MHz, while the summer months bring better conditions above 10 MHz.

The 11-year sunspot cycle produces regular changes in the maximum usable frequency (MUF). The MUF is highest at the time of the sunspot maximum. Occasionally the MUF may reach 60 MHz or perhaps even 70 MHz. When the sunspot level is at or near a minimum, the MUF may fall below 10 MHz (*see* MAXIMUM USABLE FREQUENCY, SUNSPOT CYCLE).

Tropospheric-propagation conditions can be forecast to some extent simply by observing weather maps. At very high frequencies, long-distance tropospheric bending or ducting often occurs along a frontal line (*see* TROPOSPHERIC PROPAGATION, TROPOSPHERIC-SCATTER PROPAGATION).

The likelihood of meteor-scatter propagation is, logically, increased when a meteor shower takes place. Astronomical publications often provide information about meteor showers. Several different meteor showers take place at the same time every year (*see* METEOR SCATTER).

Propagation-forecast bulletins are regularly transmitted by the National Bureau of Standards time-and-frequency stations, WWV and WWVH. Long-term propagation forecasts are also issued each month in amateur-radio magazines and other publications for the electronics hobbyist and professional. *See also* PROPAGATION CHARACTERISTICS, WWV/WWVH.

PROPAGATION MODES

Electromagnetic communications takes place as a result of several different forms of wave propagation.

In free space, propagation occurs in straight lines, in directions perpendicular to the electric and magnetic flux vectors (*see* ELECTROMAGNETIC FIELD). On and near the surface of the earth, however, electromagnetic radiation sometimes follows curved or irregular paths. The effects vary depending on the wavelength.

The most common propagation modes, and the frequency bands at which they are most likely to be observed, are listed in the table. For more detail concerning these propagation modes, consult the following articles: AURORAL PROPAGATION, E LAYER, F LAYER, IONOSPHERE, LINE-OF-SIGHT COMMUNICATION, METEOR SCATTER, MOONBOUNCE, OPTICAL COMMUNICATION, PROPAGATION CHARACTERISTICS, REPEATER, SATELLITE COMMUNICATIONS, SPORADIC-E PROPAGATION, SURFACE WAVE, TROPOSPHERIC PROPAGATION, TROPOSPHERIC-SCATTER PROPAGATION, WAVEGUIDE PROPAGATION.

PROPAGATION MODES: THE MOST COMMON PROPAGATION MODES, AND THE BANDS OF FREQUENCIES AT WHICH THEY ARE MOST LIKELY TO BE OBSERVED. ABBREVIATIONS ARE AS FOLLOWS: vlf = BELOW 30 kHz; lf = 30 kHz TO 300 kHz; mf = 300 kHz TO 3 MHz; hf = 3 MHz TO 30 MHz; vhf = 30 MHz TO 30 MHz; uhf = 300 MHz TO 3 GHz; MICROWAVE = ABOVE 3 GHz.

Mode	Band(s)
Auroral	hf (upper) vhf (all)
E-layer	vlf, lf, hf, vhf
F-layer	lf, mf, hf
Line-of-sight	All bands
Meteor Scatter	vhf
Moonbounce	vhf, uhf
Repeater	vhf, uhf
Satellite	vhf, uhf, Microwave
Sporadic-E	vhf
Surface-wave	vlf, lf, mf, hf
Tropospheric	vhf, uhf
Waveguide	vlf

PROPAGATION SPEED

The speed with which a disturbance propagates depends on many factors. Electromagnetic waves in free space travel at 186,282 miles per second (299,792 kilometers per second). In media other than free space, this speed may be reduced considerably.

In air, electromagnetic waves travel at essentially the same speed as in free space. However, in a transmission line, the speed with which an electromagnetic field propagates is much slower. *See also* VELOCITY FACTOR.

PROPAGATION TIME

The propagation time, also known as the propagation delay, is the time required for a wave disturbance to travel from one point to another. The propagation time depends on the speed with which the disturbance travels, and on the distance between the two points of interest.

In general, if the speed of propagation is given by c and the distance by d, then the propagation time is:

$$t = d/c$$

For electromagnetic waves traveling via ionospheric modes, the propagation time is somewhat longer than we would calculate by setting c = 186,282 miles per second or 299,792 kilometers per second (the speed of light in free space). The reason for this is that the waves require some time to be bent in the ionospheric layers. Similar effects occur in tropospheric propagation. *See also* PROPAGATION SPEED.

PROTECTIVE BIAS

In some vacuum-tube circuits, the grid bias is provided by the incoming signal. If the driving signal to such a circuit is lost, the negative grid bias is also lost. This results in an increase in the plate current. In some instances the plate current can rise to the point where the tube is damaged or destroyed. To prevent destructive plate current in the absence of drive, protective bias can be used.

The diagram illustrates a common method of protective biasing. A negative voltage is applied to the grid circuit.

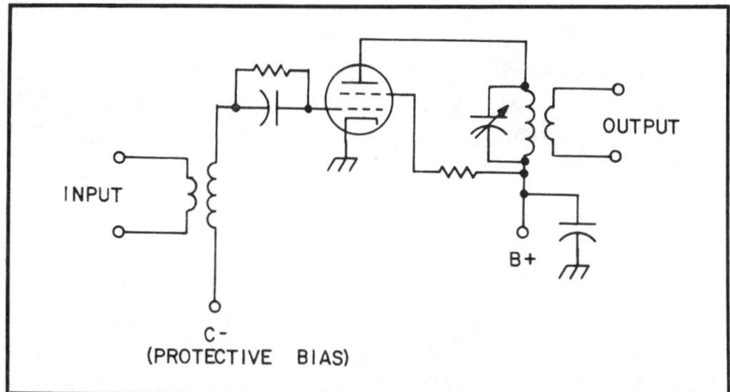

PROTECTIVE BIAS: An applied negative grid voltage prevents damage to the plate.

This keeps the grid sufficiently negative, even in the absence of a driving signal, to keep the plate current at a reasonable level.

Self bias can also serve as protective bias in a vacuum-tube circuit. Self biasing consists of a resistor and capacitor in series with the cathode of a tube. *See also* SELF BIAS.

PROTECTIVE GROUNDING

In an antenna system, a substantial voltage may develop between the active elements and the ground. This often happens in the proximity of a thundershower. It can also result from electromagnetic interaction with nearby power lines. This voltage is often large enough to present a shock hazard.

To keep an antenna system at dc or 60-Hz ground potential, protective grounding is used. The most common method of protective grounding is the installation of radio-frequency chokes between the antenna feed line and ground.

In an unbalanced system, one choke is connected as shown at A in the illustration. In a balanced system, two chokes are needed (B). The chokes should have about 10 times the impedance of the antenna system at the lowest operating frequency. For example, in a 50-ohm unbalanced system, the choke should present 500 ohms of inductive reactance at the lowest frequency on which operation is contemplated. The inductor windings must be made of wire that is large enough to handle several amperes of current.

When an antenna is not in use, it is wise to disconnect it entirely from radio equipment, while the protective-grounding circuit is left intact. This will prevent damage to radio equipment in the event of a direct lightning strike. Alternatively, an antenna grounding switch may be used (*see* LIGHTNING PROTECTION).

When several pieces of equipment are used together, dangerous potential differences can build up between or among the chassis of the individual units unless adequate grounding precautions are taken. In any electronic installation, all equipment chassis should be grounded by means of a common bus. *See also* GROUND BUS.

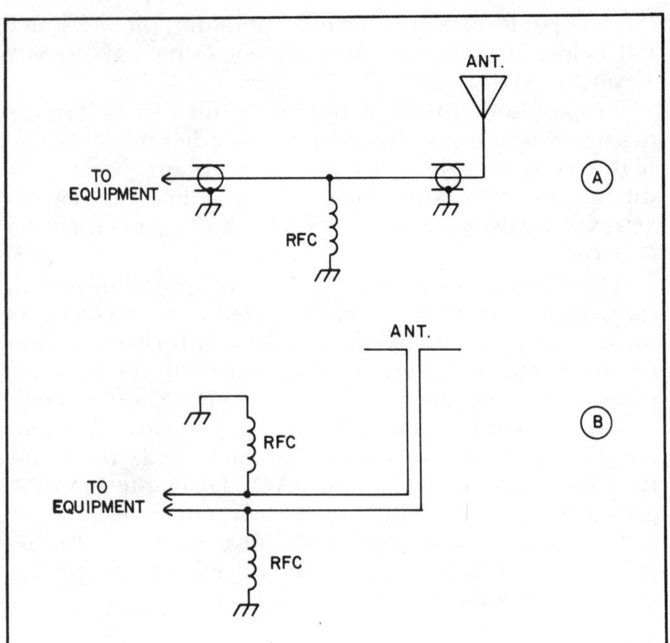

PROTECTIVE GROUNDING: At A, protective grounding for an unbalanced antenna system. At B, protecive grounding for a balanced antenna system.

PROTECTOR

A protector is a device intended to prevent damage to electronic equipment, or to eliminate a shock hazard to personnel who use electronic equipment.

A few examples of protectors for electronic equipment are circuit breakers, current-limiting devices, fuses, lightning arrestors, and overvoltage-protection circuits. Protective devices for personnel include such devices as door-interlock switches, grounding systems, and parallel (bleeder) resistors in high-voltage power supplies.

PROTECTOR TUBE

A protector tube is a glow-discharge device that is used to prevent overvoltage in direct-current power supplies. The protector tube is a cold-cathode device. Protector tubes have largely been replaced by solid-state regulation devices in modern low-voltage power supplies (*see* OVERVOLTAGE PROTECTION). However, the tubes can still be found in some high-voltage apparatus.

Protector tubes are connected across the supply output terminals, with the cathode negative and the anode positive. As long as the voltage remains below the threshold of the protector tube, nothing happens. But if the supply voltage rises past the threshold, the tube ionizes or "fires." This effectively places a short circuit across the supply output, preventing excessive voltage from reaching equipment connected to the supply.

PROTON

The proton is one of three particles that comprise all of the matter in the universe. The other two are the electron and the neutron (*see* ELECTRON, NEUTRON).

A proton at rest has a mass of about 1.67×10^{-27}. The proton carries the same charge quantity as the electron, but of the opposite polarity—positive rather than negative.

Protons are extremely dense. A teaspoonful of protons would weigh tons. Protons and neutrons contribute most of the mass in all substances.

The elements are classified according to the number of protons in their nuclei. This number is called the atomic number (*see* ATOMIC NUMBER). A given element always has the same number of protons in each of its nuclei, although the number of neutrons or electrons in the atom may vary (*see* ION, ISOTOPE).

Protons can be used for the purpose of atom smashing. The positive charge of the proton facilitates its being accelerated, by means of electric fields, to high speeds. *See also* PARTICLE ACCELERATOR.

PROTOTYPE

In the design and manufacture of electronic apparatus, a test unit is generally built before production begins. The test unit is perfected by experimentation, and also to some extent by trial and error. Two or more test units, known as prototypes, may be built, with each succeeding prototype designed to more closely resemble the actual production unit.

The construction of prototypes allows equipment to be debugged before mass manufacture.

PROXIMITY EFFECT

A conductor or component carrying a high-frequency alternating current can induce a current of the same frequency in a nearby conductor or component. The intensity of the current in a component can be affected by nearby objects. These effects are called proximity effects.

Proximity effect can occur as a result of electric coupling between or among objects. Proximity effect can also result from magnetic coupling or from electromagnetic-field propagation. Hand capacitance is an example of an electric proximity effect (*see* HAND CAPACITANCE). All transformers operate because of magnetic proximity effects (*see* MUTUAL INDUCTANCE, TRANSFORMER).

Proximity effect can be manifested in other ways, and used for specific purposes. For example, certain intrusion detectors make use of ultrasonic-wave disturbances caused by nearby moving objects. Another form of intrusion detector senses the infrared radiation given off by a person's body.

PSEUDORANDOM NUMBERS

An algorithm can be used by a calculator or computer for the purpose of generating random numbers. Such numbers are not truly random, however, since they are generated according to an organized formula. Numbers produced in this way are called pseudorandom numbers.

Some electronic calculators have built-in pseudorandom-number generators. The operator of the calculator simply presses one or two buttons on the keyboard, and the display becomes filled with pseudorandom numbers.

Some formulas provide numbers that are almost truly random; other formulas provide numbers that are less nearly random. An example of a good algorithm for generating pseudorandom numbers is the extraction of an irrational square root.

Truly random numbers must be obtained from a random-number table. Tables of this sort are published by various companies. *See also* RANDOM NUMBERS.

PSEUDOSTEREOPHONICS

A monaural, or single-channel, audio system can be made to sound somewhat like a stereophonic system by splitting the sound into two channels and introducing a delay into one of the channels. This is known as pseudostereophonics. When one channel has a delay of a few milliseconds with respect to the other channel, the sound seems to come from a direction that varies with the audio frequency.

Pseudostereophonics can be obtained in several different ways. Perhaps the simplest method is to place one speaker several feet farther from the listener than the other (see A in illustration). The sound volume at the more distant speaker is then adjusted to be greater than the volume at the nearer speaker, by just the right amount so that the listener hears sound of equal intensity from both speakers. Another method of obtaining pseudostereophonics is to play two recordings of a selection simultaneously. The two recordings will not be perfectly synchronized. If the error is very small, however, pseudostereophonic effect will be observed. A third method of obtaining pseudostereophonics is to incorporate a delay circuit in one of the channels, as at B.

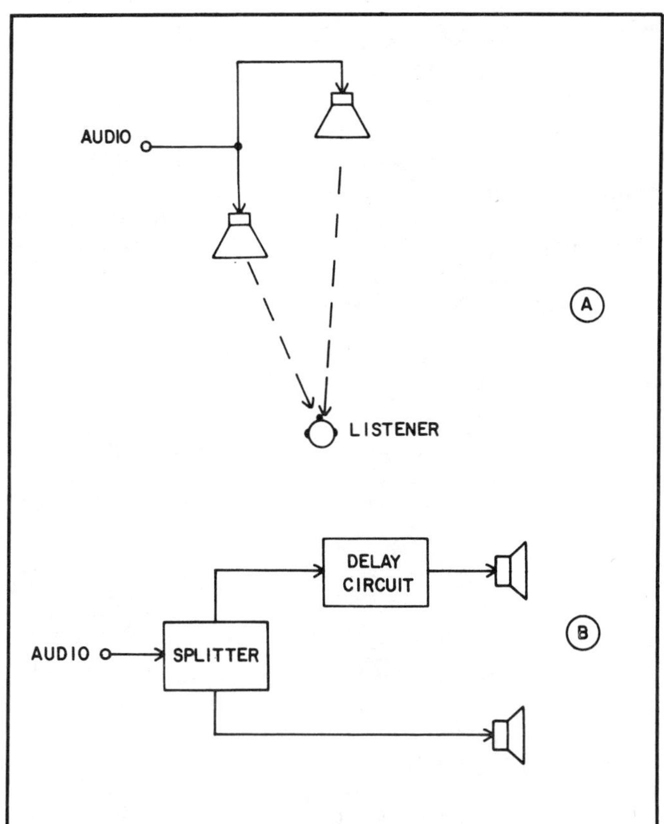

PSEUDOSTEREOPHONICS: Two means of obtaining pseudostereophonic sound. At A, the speaker-placement method; at B, the delay-circuit method.

Pseudostereophonics can be used in a true stereophonic system to provide interesting sound effects. For example, a four-channel reproduction can be obtained by splitting each channel in any of the ways described above. *See also* QUADROPHONICS, STEREOPHONICS.

PSYCHOACOUSTICS

Psychoacoustics is a science devoted to the study of the ways people perceive various sounds. It is a relatively new, but potentially far-reaching, science.

It is well-known that sounds can cause mood changes; a good example of this is the effect of a favorite song being played over the radio. Certain sounds can cause irritability; the noise from a jackhammer is a good example. Ultrasound produces severe headaches in many people. Some sounds can help people fall asleep; wideband noise generators are actually sold as sleeping aids. These are just a few of the many ways in which sounds influence our minds. Psychoacoustic effects are receiving increasing attention. *See also* ACOUSTICS, MUSIC, PINK NOISE, WHITE NOISE.

P-TYPE SEMICONDUCTOR

A P-type semiconductor is a substance that conducts mostly by the transfer of electron deficiencies among atoms. The electron deficiencies are called holes and carry a positive charge. The P-type semiconductor gets its name from the fact that it conducts mainly via positive carriers.

P-type material is manufactured by adding certain impurities to the semiconductor base, which may consist originally of pure germanium, silicon, or other semiconductor element. The impurities are called acceptors, because they are electron-deficient.

The carrier mobility in a P-type material is generally less than that of N-type material.

P-type material is used in conjunction with N-type material in most semiconductor devices, including diodes, transistors, field-effect transistors, and integrated circuits. *See also* DOPING, ELECTRON, HOLE, N-TYPE SEMICONDUCTOR, P-N JUNCTION.

PULLING

When a capacitor or inductor is placed in parallel with a piezoelectric crystal, the natural frequency of the crystal is lowered. This effect is called pulling.

Pulling is routinely used in crystal oscillators and filters for the purpose of adjusting the crystal frequency to exactly the desired value. The frequency of the average crystal can be pulled down by less than 0.1 percent of its natural frequency, or approximately 1 kHz/MHz.

When two variable-frequency oscillators are connected to a common circuit, such as a mixer, one of the oscillators may drift into tune with the other. This effect is sometimes called pulling. It is most likely to occur when tuned circuits are too tightly coupled. *See also* CRYSTAL, CRYSTAL-LATTICE FILTER, CRYSTAL OSCILLATOR, VARIABLE CRYSTAL OSCILLATOR, VARIABLE-FREQUENCY OSCILLATOR.

PULSATING DIRECT CURRENT

In a direct-current power supply having the filter disconnected, the output voltage pulsates. If the alternating-current frequency is 60 Hz at the supply input, then the pulsating direct current has a frequency of either 60 Hz or 120 Hz. The half-wave rectifier circuit produces pulsating direct current with a frequency of 60 Hz; the bridge and full-wave circuits produce 120-Hz pulsating direct current. The pulse waveshape, for a sinusoidal, alternating-current input, is shown at A in the illustration for a half-wave rectifier and at B for the bridge and full-wave circuits.

In general, 120-Hz pulsating direct current is easier to filter than 60-Hz pulsating direct current. This is simply because the filter capacitors and inductors must hold their charge for only half as long when the frequency is 120 Hz, as compared to when the ripple frequency is 60 Hz. *See also* BRIDGE RECTIFIER, FULL-WAVE RECTIFIER, HALF-WAVE RECTIFIER.

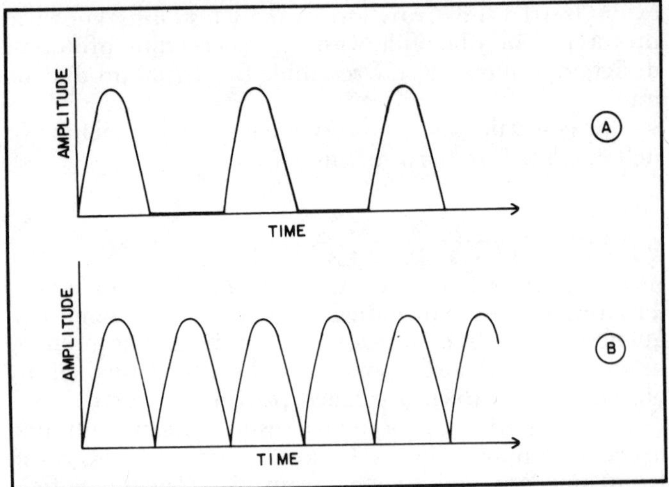

PULSATING DIRECT CURRENT: At A, pulsating direct-current output from a half-wave rectifier. At B, pulsating direct-current output from a full-wave rectifier.

PULSATION

Any fluctuation in a parameter, especially amplitude, is called pulsation. The frequency of a regular pulsation is measured in pulses per second (hertz). The pulsation frequency may vary. Some pulsations occur with such irregularity that no frequency can be defined.

Pulsations occur naturally in many phenomena. Certain stars, for example, pulsate with a period ranging from a few hours to days or even weeks. The magnetic fields of some planets, including the earth, pulsate in intensity.

A single pulsation, or pulse, in a series can have any magnitude-versus-time function. The pulses in a series may all be alike; they may differ slightly; they may differ markedly. *See also* PULSE.

PULSE

A pulse is a burst of current, voltage, power, or electromagnetic energy. A pulse may have an extremely short duration (as little as a fraction of a nanosecond), or it may have a long duration (thousands or millions of centuries).

The amplitude of a pulse is expressed either in terms of the maximum instantaneous (peak) value, or in terms of the average value (*see* AVERAGE VALUE, PEAK VALUE). The average and peak amplitudes are sometimes, but not always, the same.

The duration of a pulse can be expressed either in actual terms or in effective terms. The actual duration of a pulse is the length of time between its beginning and ending. The effective duration may be shorter than the actual duration (*see* AREA REDISTRIBUTION). Pulse duration is also called pulse width.

The interval, or length of time between pulses, has meaning if the pulses are recurrent. The pulse interval is the length of time from the end of one pulse to the beginning of another. The pulse interval may be zero in some cases. An example is the output of a full-wave rectifier.

In electronics, a pulse generally has a well-defined waveshape, such as rectangular, sawtooth, sinusoidal, or square. These pulse shapes, shown in the illustration, are the most common. However, a pulse may have a highly irregular shape, and the number of possible configurations is infinite.

A pulse of current or voltage normally maintains the same polarity from beginning to end, unless droop or overshoot occur. *See also* DROOP, OVERSHOOT, PULSATION.

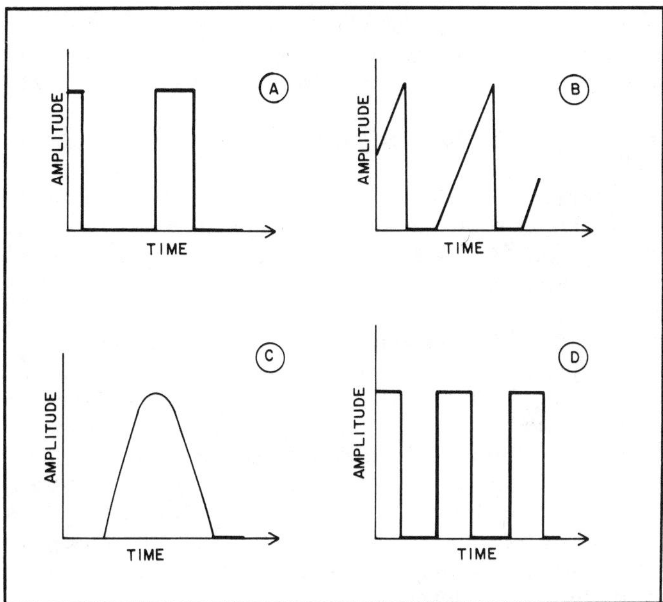

PULSE: Four types of pulses. At A, rectangular, with unequal high and low time; at B, sawtooth; at C, sinusoidal; at D, square, with equal high and low time.

PULSE AMPLIFIER

A specialized form of wideband amplifier, characterized by the ability to respond to extremely rapid changes in amplitude, is called a pulse amplifier. The pulse amplifier must be capable of handling pulses of short duration with a minimum of distortion. The output pulse should have the same waveshape and the same duration as the input pulse.

A simple pulse-amplifier circuit is shown in the illustration. Pulse amplifiers are used in such applications as digital circuits, pulse-modulated transmitters, and radar transmitters. *See also* PULSE.

PULSE AMPLITUDE

See PULSE.

PULSE-AMPLITUDE MODULATION

See PULSE MODULATION.

PULSE-CODE MODULATION

See PULSE MODULATION.

PULSE COUNTER

See COUNTER.

PULSE DROOP

See DROOP.

PULSE DURATION

See PULSE.

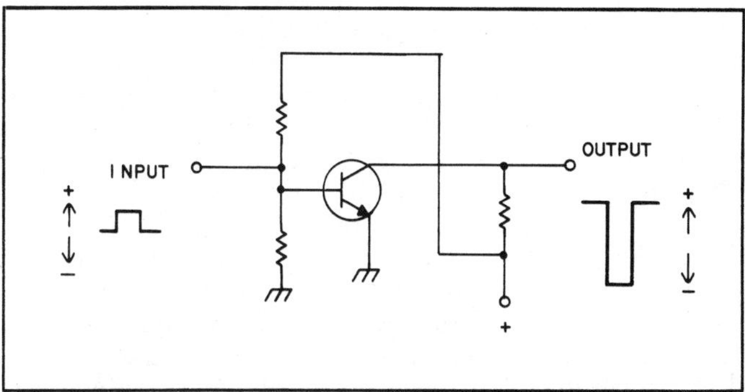

PULSE AMPLIFIER: This is a simple wideband amplifier.

PULSE-DURATION MODULATION

See PULSE MODULATION.

PULSE FREQUENCY

See PULSATION, PULSE.

PULSE-FREQUENCY MODULATION

See PULSE MODULATION.

PULSE GENERATOR

A pulse generator is a circuit specifically designed for producing electronic pulses. Commercially manufactured pulse generators are available for a variety of laboratory test purposes.

A typical pulse generator can produce rectangular, sawtooth, sinusoidal, and square pulses. The pulse amplitude, duration, and frequency are independently adjustable. Some sophisticated devices can reproduce pulses of any desired waveshape. This is done via computer. The operator draws the desired pulse function, using a light pen, on a video-monitor screen. The computer then produces a pulse or pulse train having that waveshape. *See also* PULSE.

PULSE INTERVAL

See PULSE.

PULSE-INTERVAL MODULATION

See PULSE MODULATION.

PULSE-LENGTH MODULATION

See PULSE MODULATION.

PULSE MODULATION

Pulse modulation involves the transmission of intelligence

by means of varying the characteristics of a series of electromagnetic pulses. Pulse modulation can be accomplished by varying the amplitude, the duration, the frequency, or the position of the pulses. Pulse modulation can also be obtained by means of coding. A brief description of each of these pulse-modulation methods follows:

Pulse-Amplitude Modulation. A complex waveform can be transmitted by varying the amplitude of a series of pulses. Generally, the greater the modulating-signal amplitude, the greater the pulse amplitude at the output of the transmitter. However, this can be reversed. The drawing illustrates pulse-amplitude modulation at A.

Pulse-Duration Modulation. The effective energy contained in a pulse depends not only on its amplitude, but also on its duration. In pulse-duration modulation, the width of a given pulse depends on the instantaneous amplitude of the modulating waveform. Usually, the greater the amplitude of the modulating waveform, the longer the transmitted pulse, but this can be reversed. Pulse-duration modulation may also be called pulse-length or pulse-width modulation. The principle is illustrated at B.

Pulse-Frequency Modulation. The number of pulses per second can be varied in accordance with the modulating-waveform amplitude. The duration and intensity of the individual pulses remains constant in pulse-frequency modulation. However, the effective signal power is increased when the pulse frequency is increased.

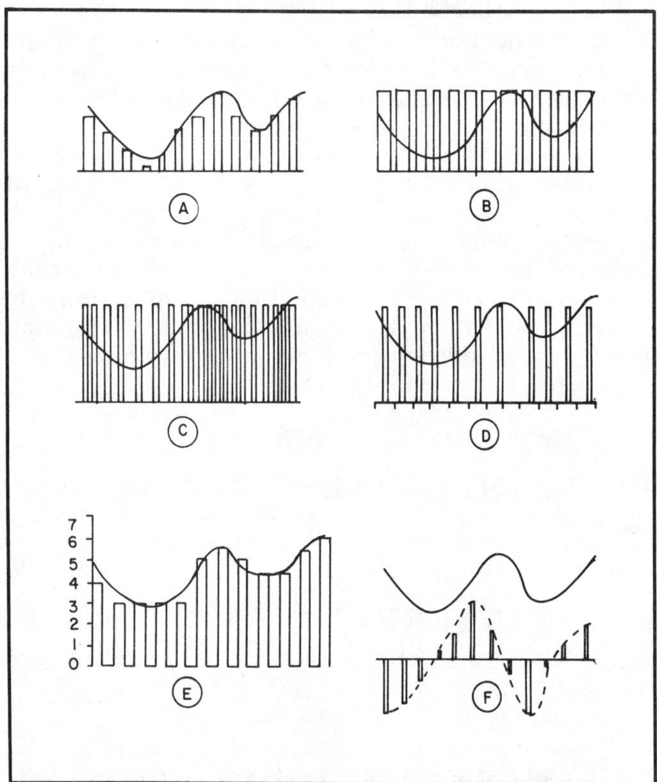

PULSE MODULATION: The vertical scale represents amplitude, and the horizontal scale represents time. The solid line shows the modulating waveform. At A, pulse-amplitude modulation. At B, pulse-duration modulation. At C, pulse-frequency modulation. At D, pulse-position modulation. At E, pulse-code modulation. At F, delta modulation, showing the derivative of the modulating waveform (dotted line).

Usually, the pulse frequency increases as the modulating-signal amplitude increases; but this can be reversed. Pulse-frequency modulation is illustrated at C.

Pulse-Position Modulation. Pulse modulation can be obtained even without varying the frequency, amplitude, or duration of the pulses. The actual timing of the pulses can be varied, as shown at D. This is known as pulse-position modulation. It may also be called pulse-interval or pulse-time modulation.

Pulse-Code Modulation. All of the foregoing methods of pulse modulation are analog methods. That is, the amplitude, duration, frequency, or position of the pulses can be varied in a continuous manner. A digital method of pulse modulation is called pulse-code modulation. In pulse-code modulation, some parameter (usually amplitude) of the pulses may achieve only certain values. The drawing at E illustrates the concept of pulse-code modulation, in which the pulse amplitude may attain any of eight discrete levels.

A special form of pulse-code modulation makes use of the derivative of the modulating-signal waveform. This is known as delta modulation. The pulse amplitude (or other parameter) may attain only certain values, according to the derivative of the modulating-waveform amplitude function, as at F.

Pulse modulation is used for a variety of communications purposes. It is especially well-suited to use with communications systems incorporating time-devision multiplexing. *See also* TIME-DIVISION MULTIPLEX.

PULSE OVERSHOOT

See OVERSHOOT.

PULSE-POSITION MODULATION

See MODULATION.

PULSE RISE AND DECAY

An electric or electromagnetic pulse has a specific, measurable rise time and a measurable decay time.

The rise time of a pulse is defined as the time required for the pulse to reach its maximum, beginning at zero amplitude.The decay time is the length of time needed for the pulse to reach zero amplitude from its maximum value. These definitions, illustrated in the drawing at A, apply to pulses having identifiable geoemtric shapes.

An irregular pulse may have rise or decay times that are difficult to define. One method of defining the rise and decay times of an irregular pulse is to consider the time required for the amplitude to reach the first local peak, beginning at zero amplitude, and the time required for the amplitude to fall to zero following the last local peak (B). *See also* PULSE.

PULSE TRANSFORMER

A pulse transformer is a special kind of transformer, designed to accommodate rapid rise and decay times with a minimum of distortion. A normal transformer may introduce distortion into a square pulse by slowing down the

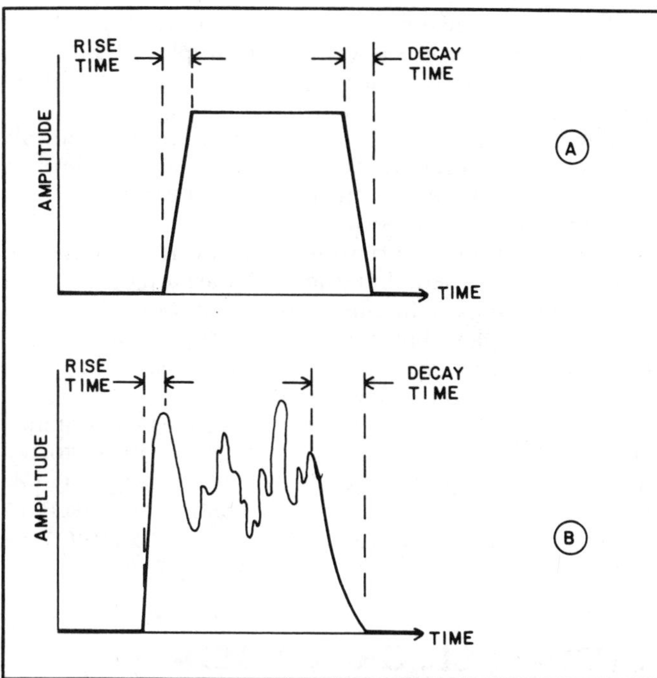

PULSE RISE AND DECAY: Rise and decay time. At A, a well-defined, regular pulse; at B, an irregular pulse.

rise and decay times (*see* PULSE RISE AND DECAY).

Pulse transformers generally have windings with less inductance than the windings of a transformer for sine waves. The core material of a pulse transformer usually has lower permeability than the core material in a typical transformer. *See also* TRANSFORMER.

PULSE WIDTH

See PULSE.

PULSE-WIDTH MODULATION

See PULSE MODULATION.

PUSHDOWN STACK

A first-in/last-out memory is called a pushdown stack. It is a read-write memory; information can be both stored and retrieved. The pushdown stack differs from the first-in/first-out memory (*see* FIRST-IN/FIRST-OUT). In the pushdown stack, the data bits that are inserted first must be retrieved last; the bits inserted last are recalled first.

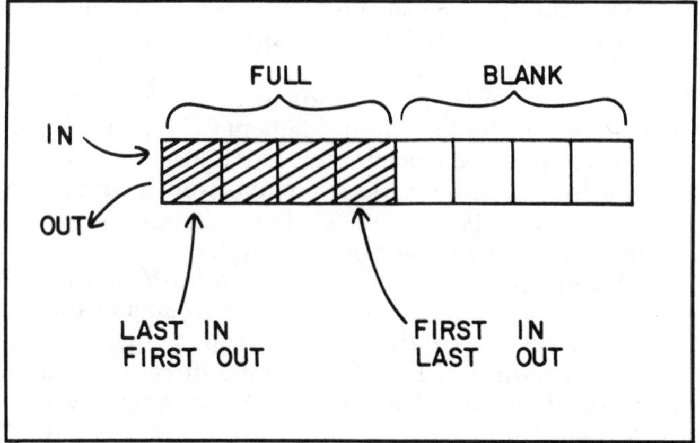

PUSHDOWN STACK: Principle of the pushdown stack.

The illustration shows the principle of the pushdown stack. The term describes the operation quite well; data bits are stored and retrieved as if they are "stacked" in a confined column.

As the data is read from the pushdown stack, the remaining bits all move one space closer to the input.

PUSH-PULL AMPLIFIER

A push-pull amplifier is a specialized, low-distortion amplifier that is often used at audio frequencies and is sometimes used at radio frequencies. The drawing illustrates push-pull amplifiers incorporating tubes (A), bipolar transistors (B), and field-effect transistors (C). These are audio-frequency circuits. The radio-frequency circuits differ only in that they may incorporate a tuned input and output.

Any push-pull amplifier requires a pair of active devices. The cathodes, emitters, or sources are grounded through small resistors. (The resistors help to prevent a phenomenon called current hogging, in which one device tends to do most of the work in the circuit. (*See* CURRENT HOGGING.) The grids, bases, or gates receive the input signal in phase opposition; they are connected to opposite ends of a transformer secondary. The center tap at the transformer helps to balance the system, and also provides a point at which bias may be applied. The plates, collectors, or drains are connected to opposite ends of the primary winding of the output transformer. As with the input, the center tap helps to balance the system, and also provides the point at which voltage is applied.

The input signal may be applied at the cathodes, emit-

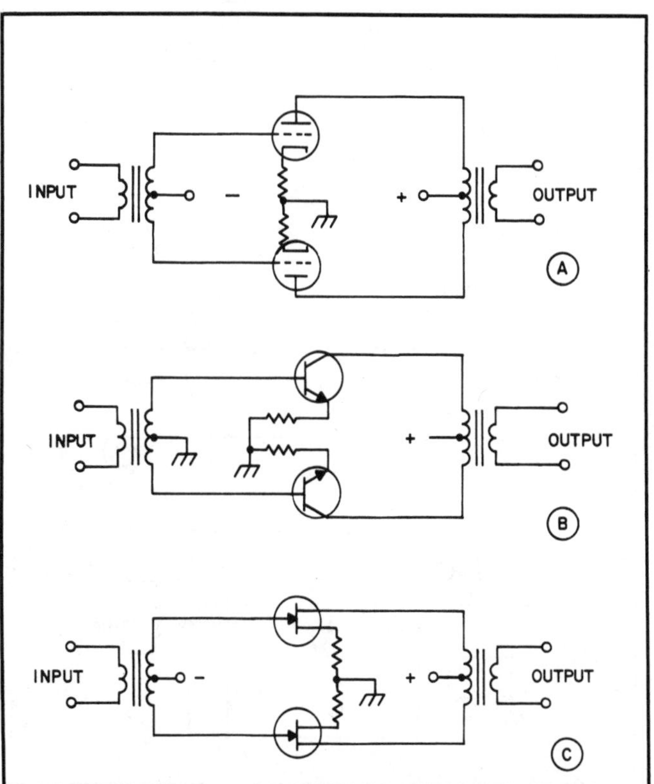

PUSH-PULL AMPLIFIER AND PUSH-PULL CONFIGURATION: At A, a pair of triode tubes is used. At B, bipolar transistors are used. At C, field-effect transistors are used. These are Class-B amplifiers, suitable for audio-frequency use.

ters, or sources of the active devices, leaving the grids, bases, or gates at ground potential. This configuration also provides push-pull operation (*see* PUSH-PULL, GROUNDED-GRID/BASE/GATE AMPLIFIER).

Push-pull amplifiers are operated either in Class AB or in Class B, and are always used as power amplifiers (*see* CLASS-AB AMPLIFIER, CLASS-B AMPLIFIER, POWER AMPLIFIER). The main advantage of a push-pull amplifier is the fact that it effectively cancels all of the even harmonics in the output circuit. This reduces the distortion in audio-frequency applications, and enhances the even-harmonic attenuation in a radio-frequency circuit.

A push-pull radio-frequency amplifier is sometimes used to multiply the frequency of a signal by a factor of 3. Such a circuit is called a tripler. The odd harmonics are not attenuated in the push-pull configuration; in fact, if the output circuit is tuned to an odd-harmonic frequency, the push-pull circuit will favor the harmonic over the fundamental frequency (*see* TRIPLER).

Push-pull circuits can be employed in oscillators, modulators, and passive devices as well as in power amplifiers. *See also* PUSH-PULL CONFIGURATION.

PUSH-PULL CONFIGURATION

Any circuit in which both the input and the output are divided between two identical devices, each operating in phase opposition with respect to the other, is called a push-pull circuit. The push-pull configuration requires two transformers, one at the input and one at the output. The secondary of the input transformer, and the primary of the output transformer, are usually center-tapped. The previous illustration shows three examples of push-pull configuration; in this case the circuits are audio-frequency power amplifiers.

The term push-pull is well-chosen, since it graphically illustrates the operation of the circuit. While one side of the circuit carries current in one direction, the other side carries current in the opposite direction.

In the push-pull configuration, the even harmonics are effectively cancelled, but the odd harmonics are reinforced. This makes the push-pull circuit useful as a radio-frequency tripler. The push-pull configuration is widely used in audio-frequency power amplifiers, and to some extent in radio-frequency amplifiers.

In push-pull, it is important that balance be maintained between the two halves of the circuit. Otherwise, distortion will occur in the output, and current hogging may take place. *See also* CURRENT HOGGING, PUSH-PULL AMPLIFIER, PUSH-PULL, GROUNDED-GRID/BASE/GATE AMPLIFIER, TRIPLER.

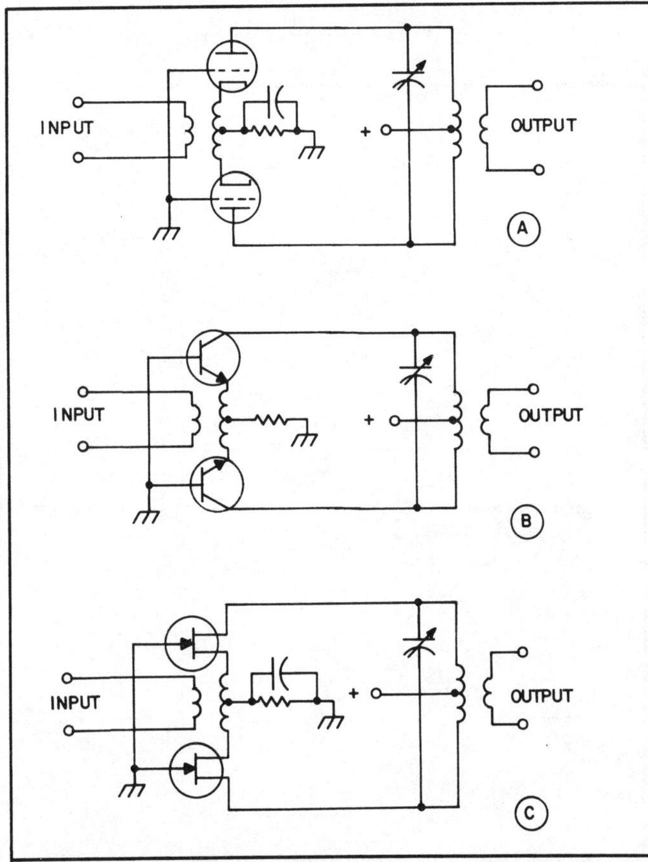

PUSH-PULL, GROUNDED-GRID/BASE/GATE AMPLIFIER: At A, a push-pull, grounded-grid amplifier. At B, a push-pull, grounded-base amplifier. At C, a push-pull, grounded-gate amplifier. These are radio-frequency circuits, and are biased for Class-B operation.

PUSH-PULL, GROUNDED-GRID/BASE/GATE AMPLIFIER

In a push-pull power amplifier, the input is usually applied at the grids, bases, or gates of the active devices, as shown in the previous illustration (*see* PUSH-PULL AMPLIFIER). However, the input may be applied at the cathodes, emitters, or sources. This kind of push-pull circuit is shown here using tubes (A), bipolar transistors (B), and field-effect transistors (C). These are radio-frequency power amplifiers having tuned output circuits.

The amplifiers shown are somewhat more stable than the normal push-pull type. The circuits shown here do not normally require neutralization. The main disadvantage of the grounded-grid, grounded-base, or grounded-gate configuration is that they require more driving power than the grounded-cathode, grounded-emitter, or grounded-source amplifiers. *See also* POWER AMPLIFIER.

PUSH-PUSH CONFIGURATION

A push-push circuit is a form of balanced circuit, incorporating two identical passive or active halves. The push-push configuration is similar to push-pull (*see* PUSH-PULL CONFIGURATION), but there is one important distinction. In a push-push circuit, the inputs of the devices are connected in phase opposition, just as they are in push-pull, but the outputs are connected in parallel. The illustration shows examples of push-push radio-frequency amplifiers using tubes (A), bipolar transistors (B), and field-effect transistors (C).

The push-push circuit tends to cancel all of the odd harmonics (including the fundamental frequency), while reinforcing the even harmonics. This is just the opposite of the push-pull configuration. The push-push circuit is not often used at audio frequencies, but it can be employed as a radio-frequency doubler or quadrupler.

Perhaps the simplest form of push-push circuit is the full-wave rectifier. This is, in effect, a frequency doubler. With 60-Hz alternating-current input, the full-wave rectifier produces a 120-Hz pulsating direct current output. An identical circuit can be used as a passive radio-frequency doubler. *See also* DOUBLER, FULL-WAVE RECTIFIER, QUADRUPLER.

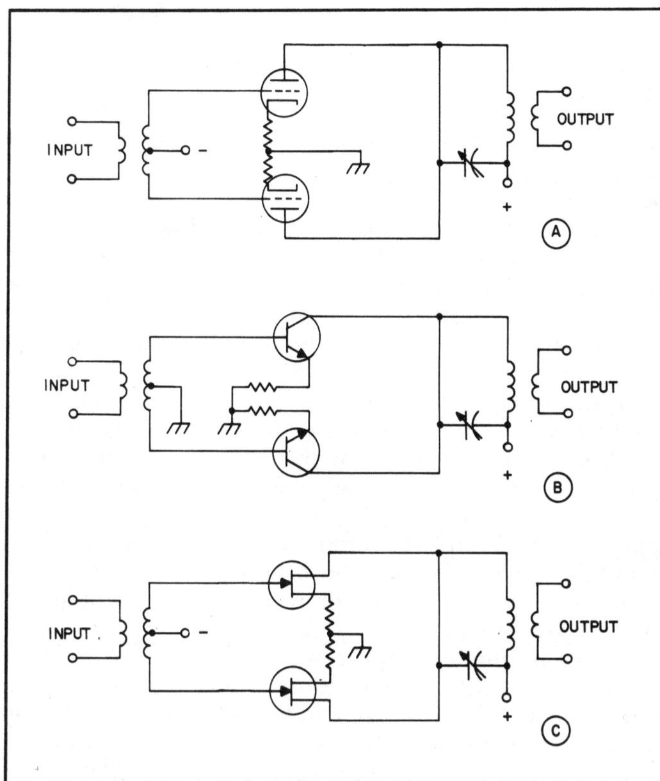

PUSH-PUSH CONFIGURATION: At A, a pair of triode tubes is employed. At B, a bipolar-transistor circuit. At C, a field-effect-transistor circuit. These are Class-B circuits.

PYRAMID HORN ANTENNA

A pyramid horn is an antenna that is used for transmitting and receiving microwaves. The pyramid horn gets its name from the fact that it is shaped like a pyramid; the faces are flat, and they converge to a single point, called the apex. The flare angle may vary, but it is usually about 30 to 45 degrees with respect to the center axis.

The pyramid horn is fed directly at the apex by a waveguide. *See also* HORN ANTENNA.

PYTHAGOREAN THEOREM

The Pythagorean Theorem is a simple rule of plane geometry discovered by the mathematician Pythagoras in ancient times. In a right triangle, the square of the length of the hypotenuse (the side opposite the right angle) is always equal to the sum of the squares of the lengths of the other two sides. This principle applies regardless of the shape of the triangle. The drawing illustrates this principle.

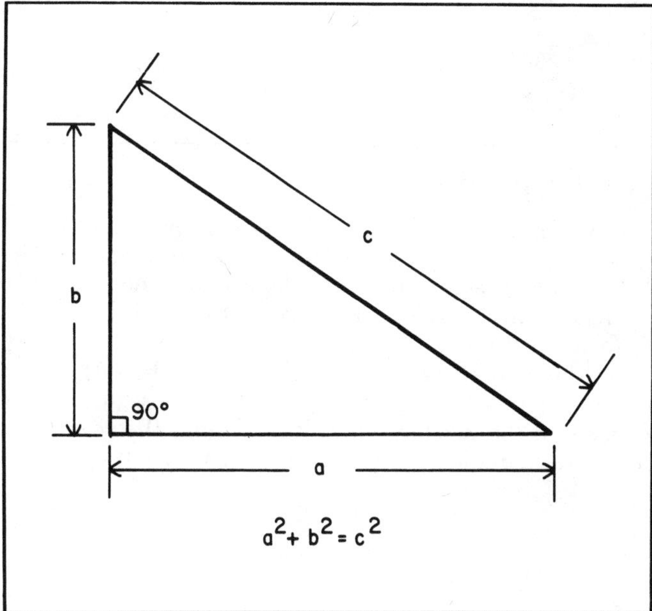

PYTHAGOREAN THEOREM: The theorem of Pythagoras relates the lengths of the three sides of a right triangle.

In analytic geometry, the distance between any two points is given by a formula derived from the Pythagorean theorem. Suppose we are given a Cartesian n-space containing two points:

$$A = (x_{11}, x_{12}, x_{13}, \ldots, x_{1n})$$

$$B = (x_{21}, x_{22}, x_{23}, \ldots, x_{2n})$$

Then the distance d between the points A and B, in the n-dimensional space, is given by:

$$d = \sqrt{(x_{11}-x_{21})^2+(x_{12}-x_{22})^2 + \ldots + (x_{1n}-x_{2n})^2}$$

See also CARTESIAN COORDINATES.

Q FACTOR

For a capacitor, inductor, or tuned circuit, the Q factor (also known simply as the Q) is a figure of merit. The higher the Q factor, the lower the loss and the more efficient the component or tuned circuit. In tuned circuits, the Q factor is directly related to the selectivity.

In the case of a capacitor or inductor, the Q factor is given in terms of the reactance and the resistance. If X represents the reactance and R represents the resistance, in a series circuit, both given in ohms, then

$$Q = X/R$$

for an inductor, and

$$Q = -X/R$$

for a capacitor (capacitive reactance is considered negative).

Since both inductive and capacitive reactance vary with the frequency, the Q factor of an inductor or capacitor depends on the frequency (*see* CAPACITIVE REACTANCE, INDUCTIVE REACTANCE).

In a series-resonant inductance-capacitance (LC) circuit, the Q factor is given by the same formula as above, where X represents the absolute value of the reactance of either the coil or the capacitor (the absolute values are equal at resonance) and R represents the resistive impedance of the circuit at resonance.

In a parallel-resonant LC circuit, the Q factor is given by:

$$Q = R/X$$

where R represents the resistive impedance of the circuit at resonance and X represents the absolute value of the reactance of either the inductor or the capacitor (again, the absolute values are identical at resonance).

The Q factor of a tuned circuit can be determined in another way. Suppose a series or parallel LC circuit has a resonant frequency f Hz. Further suppose that the 3-dB attenuation points are separated by g Hz (that is, the 3-dB bandwidth is g Hz). Then the Q factor of the tuned circuit is given by:

$$Q = f/g$$

This method of determining the Q is convenient from an experimental standpoint. The narrower the 3-dB bandwidth, the higher the factor at a given resonant frequency. For a constant bandwidth, the Q factor increases as the resonant frequency is increased.

The Q factor of a trap circuit can also be determined according to the 3-dB bandwidth. The only difference is that the 3-dB points are taken with respect to the frequency of maximum attenuation, instead of with respect to the frequency of least attenuation. *See also* BANDWIDTH, RESONANCE, TRAP.

Q MULTIPLIER

In early superheterodyne receivers, a circuit called a Q multiplier was occasionally used to enhance the selectivity (that is, to reduce the bandwidth). Such circuits can still be found in some superheterodyne receivers, although crystal-lattice filters, ceramic filters, and mechanical filters are more common today.

The multiplier consists of a tuned intermediate-frequency amplifier in which some of the output signal is fed back to the input through the tuned circuits. This forces the signal to effectively pass through the resonant networks several times. The illustration is a schematic diagram of a simple Q multiplier. The circuit is said to multiply the Q, because the effective resistance of the tuned circuits is made lower with the incorporation of positive feedback.

A potentiometer, connected in the feedback path, is used to control the regeneration. There is a limit to how large the effective Q factor can become before oscillation occurs in the circuit. In a well-designed Q multiplier, the selectivity can be made very sharp before oscillation begins. *See also* Q FACTOR, SELECTIVITY.

Q SIGNAL

In radiotelegraphy, certain statements, phrases, or words are made very often. It becomes tedious to send complete sentences, phrases, and words in Morse code, especially if they are repeated. To streamline code operation, therefore, a set of abbreviations called Q signals has been devised.

Each Q signal consists of the letter Q, followed by two more letters. A Q signal followed by a question mark indicates a query; if no question mark follows the signal, or if data follows, it indicates a statement. For example, "QRM?" means "Are you experiencing interference?" while QRM" means "I am experiencing interference." As another example, "QTH?" means "What is your location?" and "QTH ROCHESTER, MN" means "My location is Rochester, Minnesota."

The table is a comprehensive list of Q signals. While Q

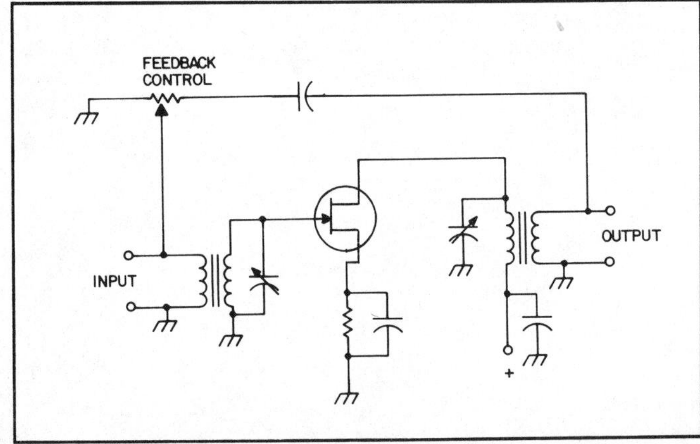

Q MULTIPLIER: A simple Q-multiplier circuit for enhancing the selectivity of a superheterodyne receiver. The circuit is placed in the intermediate-frequency chain.

Q SIGNAL: Q Signals, with Query and Statement Informaton.

Signal	Meaning
QRA	What is the name of your station? The name of my station is ---.
QRB	Approximately how far from my station are you? I am about --- miles or kilometers from you.
QRD	From where to where are you going? I am going from --- to ---.
QRG	What is my exact frequency, or that of ---? Your frequency, or that of ---, is ---.
QRH	Does my frequency vary? Your frequency varies.
QRI	How is the tone of my signal? The tone of your signal is: 1 (good), 2 (fair or variable), 3 (poor)
QRK	How readable are my signals? Your signals are: 1 (unreadable), 2 (somewhat readable), 3 (readable with difficulty), 4 (readable with almost no difficulty), 5 (perfectly readable).
ORL	Are you busy? I am busy.
QRM	Are you experiencing interference? I am experiencing interference.
QRN	Are you bothered by static? I am bothered by static.
QRO	Shall I increase transmitter power? Increase transmitter power.
QRP	Shall I decrease transmitter power? Decrease transmitter power.
QRQ	Shall I send faster? Send faster, or speed up to --- words per minute.
QRS	Shall I send more slowly? Send more slowly, or at --- words per minute.
QRT	Shall I stop transmitting? Stop transmitting
QRU	Have you any information for me? I have no information for you.
QRV	Are you ready? I am ready.
QRW	Shall I tell --- that you are calling him/her? Tell --- that I am calling him/her.
QRX	When will you call me again? Call me again at ---.
QRY	What is my turn? Your turn is number --- in order.
QRZ	Who is calling me? You are being called by ---.
QSA	How strong are my signals? Your signals are: 1 (almost imperceptible), 2 (weak), 3 (fairly strong), 4 (strong), 5 (very strong).
QSB	Are my signals varying in strength? Your signals are varying in strength.
QSD	Is my keying bad? Your keying is bad.
QSG	Shall I send more than one message? Send more than one (or ---), messages.
QSJ	What is your charge per word? My charge per word is ---.
QSK	Can you hear me between your signals? Or, do you have full break-in capability? I can hear you between my signals. Or, I have full break-in capability.
QSL	Can you acknowledge receipt of my message? I acknowledge receipt of your message.
QSM	Shall I repeat my message? Repeat your message.
QSN	Did you hear me on --- MHz? I heard you on --- MHz.
QSO	Can you communicate with ---? I can communicate with ---.
QSP	Will you send a message to ---? I will send a message to ---.
QSQ	Is there a doctor there? Or, is --- there? A doctor is here. Or, --- is here.
QSU	On what frequency shall I reply? Reply on --- MHz.
QSV	Shall I send a series of Vs? Send a series of Vs.
QSW	On which frequency will you transmit? I will transmit on --- MHz.
QSX	Will you listen for ---? I will listen for ---.
QSY	Shall I change the frequency of my transmitter and/or receiver? Change the frequency of your transmitter and/or receiver.
QSZ	Shall I send each word or group of words more than once? Send each word or group of words --- times.
QTA	Shall I cancel message number ---? Cancel message number ---.
QTB	Does your word count agree with mine? My word count does not agree with yours.
QTC	How many messages do you have to send? I have --- messages to send.
QTE	What is my bearing relative to you? Your bearing relative to me is --- degrees.
QTH	What is your location? My location is ---.
QTJ	What is your speed? My speed is --- miles or kilometers per hour.
QTL	In which direction are you headed? I am headed toward --- or at bearing --- degrees.

(Continued)

QTN	When did you leave ---? I left --- at ---.	QUA	Do you have information concerning ---? I have information concerning ---.
QTO	Are you airborne? I am airborne.	QUD	Have you received my urgency signal, or that of ---? I have received your urgency signal, or that of ---.
QTP	Do you intend to land? I intend to land.	QUF	Have you received my distress signal, or that of ---? I have received your distress signal, or that of ---.
QTR	What is the correct time? The correct time is --- UTC.		
QTX	Will you stand by for me? I will stand by for you until ---.		

signals were originally designed for radiotelegraph use, many radioteletype and radiotelephone operators also use them. The Q signals can streamline radioteletype operation in the same way that they make code operation more convenient. There is some question as to how beneficial (if not detrimental) the Q signals are in voice communication.

QUAD ANTENNA

A quad antenna is a form of parasitic array. It operates according to the same principles as the Yagi antenna, except that full-wavelength loops are used instead of half-wavelength straight elements.

A full-wavelength loop has approximately 2 dB gain, in terms of effective radiated power, compared to a half-wavelength dipole. This implies that a quad antenna should have 2 dB gain over a Yagi with the same number of elements. Experiments have shown that this is true.

A two-element quad antenna may consist of a driven element and a director, or a driven element and a reflector. A three-element quad has one driven element, one director, and one reflector. The director has a circumference of approximately 0.97 electrical wavelength; the driven element measures exactly 1 wavelength around; the reflector measures about 1.03 wavelength in circumference (*see* DIRECTOR, DRIVEN ELEMENT, REFLECTOR). The lengths of the director and reflector depend, to some extent, on the element spacing; therefore, these values should be considered approximate.

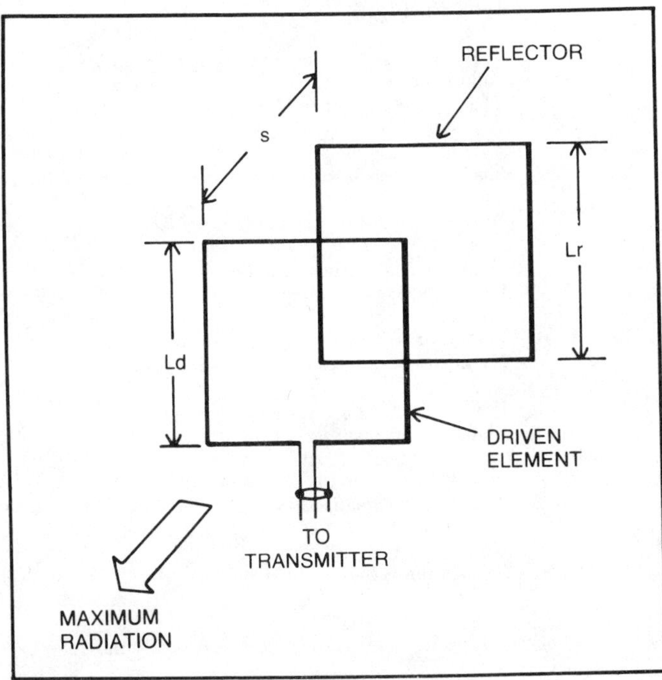

QUAD ANTENNA: A simple two-element quad antenna. The dimensions Ld, Lr, and s are discussed in the text. Effective power gain is approximately 7 dBd.

Additional director elements can be added to form quad antennas having any desired numbers of elements. Provided optimum spacing is used, the gain increases as the number of elements increases. Each succeeding director should be slightly shorter than its predecessor. Long quad antennas are practical at very-high and ultra-high frequencies, but they tend to be mechanically unwieldy at high frequencies.

A complete discussion of all the design parameters of quad-antenna construction is beyond the scope of this book. However, a simple two-element quad antenna can be constructed according to the dimensions shown in the illustration. Such an antenna will provide approximately 7 dBd forward gain. The elements are square. The length of each side of the driven element, in feet, is given by:

$$Ld = 251/f$$

where f is the frequency in megahertz. The length of each side of the reflector is given by:

$$Lr = 258/f$$

The element spacing, in feet, is given by:

$$s = 200/f$$

Geometrically, the quad antenna shown has an almost perfect cube shape. For this reason, two-element quad antennas are often called cubical quads. *See also* PARASITIC ARRAY, YAGI ANTENNA.

QUADRATIC FORMULA

The quadratic formula is a formula for solving quadratic equations. A quadratic equation is an equation of the form:

$$ax^2 + bx + c = 0$$

where a, b, and c are constants.

Every quadratic equation has two solutions. The solutions may be real numbers or complex numbers. In some instances, both solutions are the same; then they are said to be coincident or redundant.

The solutions s and t of the above general quadratic equation are given by:

$$s = (-b + \sqrt{b^2 - 4ac}) / 2a$$

$$t = (-b - \sqrt{b^2 - 4ac}) / 2a$$

The value under the radical sign, $b^2 - 4ac$, is known as the discriminant. If the discriminant is positive, the equation has two different real solutions. If the discriminant is zero, the equation has coincident real solutions. If the discriminant is negative, the equation has two different complex solutions. *See also* COMPLEX NUMBER, REAL NUMBER.

QUADRATURE DETECTOR

A quadrature detector is a special form of detector for frequency-modulated or phase-modulated signals. The quadrature detector, also called a gated-beam detector, converts the instantaneous phase variations of the frequency-modulated signal into amplitude fluctuations.

The quadrature detector makes use of a special vacuum tube known as a gated-beam tube. The gated-beam tube has three grids. The incoming signal is applied to one grid, which is called the signal or limiter grid. An unmodulated carrier, having a frequency identical to that of the incoming signal but differing by 90 degrees of phase, is applied to another grid, which is called the quadrature grid. An intermediate grid is used to accelerate the electrons through the tube from the cathode toward the plate.

When the incoming signal contains no modulation, the waveforms at the signal and quadrature grids are in phase quadrature (they differ by 90 degrees). But when modulation occurs in the incoming signal, the instantaneous frequency or phase changes at the signal grid. Thus, the two signals get more or less in phase according to the modulation of the incoming signal. As this happens, the current through the tube rises and falls, producing amplitude variation in the plate circuit. *See also* FREQUENCY MODULATION, GATED-BEAM TUBE, PHASE MODULATION.

QUADRIFILAR WINDING

A set of four coil windings, usually oriented in the same sense and having the same number of turns on a common core, is called a quadrifilar winding. Quadrifilar windings are sometimes used in the manufacture of broadband balun transformers.

Quadrifilar windings on a toroidal core can be connected to obtain many different impedance-transfer ratios. *See also* BALUN.

QUADROPHONICS

Quadrophonics is another name for four-channel high-fidelity recording and reproduction. The spelling of the term varies considerably; it may be denoted as quadraphonics or quadriphonics. It is often simply called quad.

An ordinary stereophonic system has two channels, commonly called the left channel and the right channel. In a quadrophonic system, there are four channels, usually designated left front, right front, left rear, and right rear.

A true quadrophonic system has four entirely independent channels. This provides true 360-degree reproduction of sound for a listener. Most quadrophonic systems in use today, however, have only two actual recording tracks. The two channels are combined according to some predetermined scheme, obtaining four different sound tracks in the reproduction. An example of such a scheme might be to combine the left and right channels in phase quadrature (90 degrees out of phase), in one instance with the right channel leading the left, and in the other case with the left channel leading the right. Such a phasing circuit can be incorporated into any ordinary stereophonic playback system, enhancing the sound effects. *See also* HIGH FIDELITY, STEREOPHONICS.

QUADRUPLER

A quadrupler is a circuit that multiplies a radio-freuqency signal by a factor of 4. The quadrupler generally consists of a push-push amplifier with tuned input and output circuits. The input circuit is tuned to the frequency of the incoming signal, while the output circuit is tuned to the fourth harmonic of the incoming-signal frequency (*see* PUSH-PUSH CONFIGURATION).

A power-supply circuit that multiplies the incoming voltage by 4 is sometimes called a quadrupler. Such a power supply is used when the secondary of the power transformer will not provide adequate voltage, and when the regulation need not be especially precise. *See also* VOLTAGE MULTIPLIER.

QUALITATIVE TESTING

A qualitative test or expression involves the determination of equipment performance in general, without considering exact numerical values.

An example of a qualitative test of a radio receiver might consist of such exercises as: (1) Switch it on. Does it receive signals? (2) Does it appear to be as sensitive as it should be, or as sensitive as a second receiver used for comparison purposes? (3) Does it seem to be calibrated correctly? (4) Does the selectivity appear adequate?

The above are just a few examples of qualitative tests. More precise testing, in which numerical measurements are made, is called quantitative testing. *See also* QUANTITATIVE TESTING, TEST LABORATORY.

QUALITY CONTROL

Quality control is a part of the manufacturing process of any goods. It is also known as quality assurance. Quality control is necessary in order to ensure that electronic (or other) equipment performs according to the claimed specifications.

The most common methods of quality control are: the complete testing of each unit at various stages during, and after, manufacture; the complete testing of each unit only after manufacture is completed; the complete testing of a certain proportion of units at various stages during, and after, manufacture; and the complete testing of a certain proportion of units only after manufacture is completed.

The first or second method is preferable, but when it is not possible to test every unit, a less rigorous quality-control procedure may be used. The most comprehensive quality-control procedures are used with complicated devices, equipment that must be calibrated to a high degree of accuracy, or equipment in which human lives or safety is involved.

A person who devises and performs quality-control tests is called a quality-control engineer. All manufacturing companies have one or more such engineers. In some companies, the lead quality-control engineer actually has the power to shut down a production line if manufactured units do not meet the specifications.

QUANTITATIVE TESTING

Quantitative testing involves the evaluation of equipment performance in concrete, numerical terms. This is done with standard test equipment such as voltmeters, oscilloscopes, wattmeters, and spectrum analyzers.

Examples of quantitative tests for a radio receiver are: (1) How sensitive is it, in terms of the number of microvolts required to produce a specific signal-to-noise ratio at the speaker terminals? (2) What is the actual frequency coverage? (3) How accurate, in terms of percentage discrepancy, is the frequency calibration? (4) How much current does the unit require, in amperes?

Quantitative testing is invariably subject to some error. The accuracy of the measuring instruments is not perfect; the engineer or technician using the apparatus may not interpolate analog readings just right. The maximum measurement error, as a percentage of the total quantity, must be known if quantitative data are to have any meaning. *See also* ACCURACY, ANALOG METERING, ERROR, INTERPOLATION, QUALITATIVE TESTING, TEST LABORATORY.

QUANTIZATION

Many natural phenomena occur in the form of discrete packets or bits, and not as a continuous, smooth flow. This is known as quantization—the occurrence of discrete, although perhaps tiny, steps or intervals in nature.

The most familiar example of quantization is the particle nature of electromagnetic energy. All electromagnetic radiation consists of discrete packets of energy, called photons. For radiation of any fixed wavelength, the total energy is always an integral multiple (although a huge multiple!) of the energy contained in a single photon. There can be no "in between." (*See* PARTICLE THEORY OF LIGHT, PHOTON.)

Another example of quantization is the flow of electricity. Electric current is an expression of the number of unit charge carriers (generally either electrons or holes) passing a given point in a certain interval of time. Suppose that 1 coulomb of electrons passes a point in one second. Then that represents 1 ampere of electric current. Although 1 coulomb is a very large number, it is still an integer, and represents a discrete, finite number of charge carriers. Electric charge can exist only in integral multiples of a unit charge (*see* CHARGE, COULOMB, ELECTRON, HOLE).

Matter is obviously quantized, since it is made up of discrete particles. We do not know yet, for certain, what the smallest particle actually is, but the discrete nature of molecules, atoms, electrons, neutrons, and protons is undeniable (*see* ATOM, MOLECULE).

Quantization may exist in unknown forms. For instance, it may be that space and time are actually quantized. And, although it may seem like a philosophical or semantical argument, we can say that all physical variables are quantized, in effect, simply because our measurement of them is always based on some discrete unit, and our measurements are always subject to some error.

QUANTUM MECHANICS

The study of phenomena related to the quantum theory, both in the theoretical and practical fields, is known as quantum mechanics. The quantum theory of matter and energy holds that the emission or absorption of energy by matter always occurs in discrete packets, called photons (*see* PARTICLE THEORY OF LIGHT, PHOTON).

When an atom absorbs a photon of a certain wavelength, an electron moves to a higher orbit in the atom. This requires a photon of just the right energy content. Such a photon imparts energy to an electron, allowing it to gain speed and move farther from the nucleus of its parent atom.

Absorption of energy by any given atom can occur at one or more discrete wavelengths. As radiant energy passes through a material, we can determine the identity of the material because an absorption-line "signature" results, unique to that substance (see A in illustration). Astronomers use spectroscopes and radio telescopes to determine the makeup of interstellar gas and dust clouds (*see* ELECTRON ORBIT, RADIO ASTRONOMY, SPECTROSCOPE).

When an electron in an atom moves to a lower energy level, a photon, having a precise wavelength, is given off. This happens when substances are heated to extremely high temperatures. Under such conditions, the electrons gain energy because of heating, and move into higher orbits. The electrons tend to be unstable in the higher orbits, however, and therefore the electrons fall back into lower orbits, emitting photons as they do so. The con-

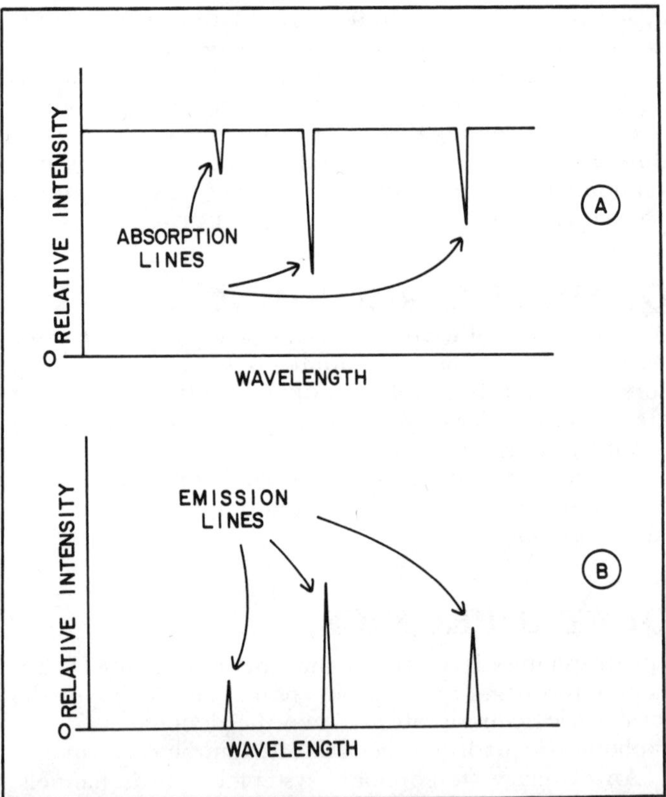

QUANTUM MECHANICS: Quantum effects can be observed in the absorption lines (A) and the emission lines (B) of various substances. In this hypothetical case, the substances at A and B are the same, resulting in absorption and emission lines at coincident wavelengths.

tinued presence of heat energy causes electrons to move alternately into higher and lower orbits. The result is continuous electromagnetic radiation at discrete wavelengths, as at B.

The possible number of electron-orbit changes, and hence the number of emitted wavelengths, is finite for any element. Each element has its own unique emission "signature." Astronomers use radio telescopes and spectroscopes to observe the emission wavelengths of stars. This makes it possible to determine the material composition of distant stars, and even of whole galaxies.

In its theoretical form, quantum mechanics is a highly sophisticated and elegant science. A full discussion of the subject is far beyond the scope of this book. Serious students of quantum mechanics are referred to college textbooks and courses.

QUANTUM THEORY OF LIGHT

See PARTICLE THEORY OF LIGHT.

QUARK

Electrons, protons, and neutrons are evidently not the smallest particles of matter. Each electron, proton, or neutron is believed to be composed of still smaller particles, called quarks. The term quark arises from the book Finnegan's Wake, by James Joyce. (The word was probably chosen at random, for physicists are hard-pressed nowadays to find names for the new particles that are constantly being discovered!)

It is believed that various combinations of quarks make up all the matter in the universe.

Are quarks the smallest particles, or are they themselves composed of discrete parts? This is not yet known, but there are scientists who feel that no smallest particle exists—that the succession of tinier and tinier constituents goes on without end.

QUARTER-WAVE ANTENNA

See MONOPOLE ANTENNA.

QUARTER WAVELENGTH

A quarter wavelength is the distance that corresponds to 90 degrees of phase as an electromagnetic disturbance is propagated (see illustration). In free space, it is related to the frequency by a simple equation:

$$\lambda/4 = 246/f$$

where $\lambda/4$ represents a quarter wavelength in feet, and f represents the frequency in megahertz. If $\lambda/4$ is expressed in meters, then the formula is:

$$\lambda/4 = 75/f$$

for the frequency f given in megahertz.

In media other than free space, electromagnetic waves propagate at speeds less than their speed in free space. The wavelength of a disturbance having a given frequency depends on the speed of propagation. In general, if v is the velocity factor in a given medium (*see* VELOCITY FACTOR),

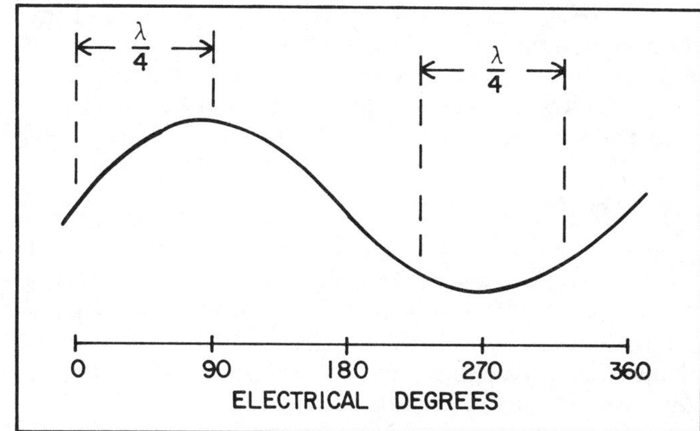

QUARTER WAVELENGTH: A quarter-wavelength corresponds to 90 electrical degrees.

then:

$$\lambda/4 = 246v/f$$

in feet, and

$$\lambda/4 = 75v/f$$

in meters.

The quarter wavelength is important from an electrical standpoint. Conductors and transmission lines have special properties when they have length $\lambda/4$. *See also* CYCLE, MONOPOLE ANTENNA, QUARTER-WAVE TRANSMISSION LINE, WAVELENGTH.

QUARTER-WAVE MATCHING SECTION

See QUARTER-WAVE TRANSMISSION LINE.

QUARTER-WAVE TRANSMISSION LINE

A quarter-wave transmission line is any section of transmission line measuring an electrical quarter wavelength. If the velocity factor of a certain type of line is given by v (*see* VELOCITY FACTOR), then a quarter-wave transmission line measures:

$$L = 246v/f$$

in feet, where f is the frequency in megahertz. The length in meters is:

$$L = 75v/f$$

A quarter-wave section of transmission line can be used for the purpose of impedance transformation. Suppose a quarter-wave line has a characteristic impedance of Z_o ohms. Also imagine that a purely resistive impedance of Z_1 ohms is connected to one end of this section. Then, at the other end, the impedance Z_2 will also be a pure resistance, with a value of:

$$Z_2 = Z_o^2/Z_1$$

Quarter-wave sections are used in a variety of antenna

systems, especially phased arrays, for the purpose of impedance matching. Quarter-wave sections can also act as series-resonant or parallel-resonant tuned circuits. *See also* CHARACTERISTIC IMPEDANCE, COAXIAL TANK CIRCUIT, QUARTER WAVELENGTH.

QUARTZ

Quartz is a naturally occurring, clear mineral, with characteristic hexagonal cleavages. Quartz has a higher index of refraction and a greater hardness than glass. Quartz has a characteristic sparkle, not unlike that of diamond. Quartz can be grown artificially. Chemically, it is composed of silicon dioxide.

Quartz is used in electronics because of its piezoelectric properties. When a piece of quartz is subjected to electric currents, it will produce vibrations. The vibration frequency is determined by the size and shape of the crystal. The frequency can be from audio up to several megahertz.

Probably the most important use of quartz in electronics is in the manufacture of oscillator crystals. Quartz crystals can be made to oscillate at frequencies of several megahertz, and the frequency is extremely stable. *See* CRYSTAL.

QUARTZ CRYSTAL

See CRYSTAL.

QUASAR

In 1960, the astronomers J. L. Greenstein and A. Sandage discovered that a strong celestial source of radio emission was visible as a faint, rather ordinary-looking blue star in visible-light photographs. The object, because of its stellar appearance and strong radio-frequency emission, was called a quasi-stellar radio source—quasar for short. Since the discovery of a quasar by these two astronomers, many more have been found, and they still present one of the greatest mysteries in modern cosmology.

A quasar may emit as much energy as a whole galaxy measuring 100,000 light years in diameter. But the quasar itself, according to deductions made by astrophysicists, is generally less than 1 light year across. Therefore, quasars must be incredibly intense sources of energy. The mechanism by which quasars operate is not yet fully known. *See also* RADIO ASTRONOMY, RADIO TELESCOPE.

QUASI-RANDOM NUMBERS

See PSEUDORANDOM NUMBERS.

QUASI-STELLAR RADIO SOURCE

See QUASAR.

QUASI-STEREOPHONICS

See PSEUDOSTEREOPHONICS.

QUEUING THEORY

In any complicated mass-production arrangement, it is necessary that all steps proceed in the proper order, and with the correct synchronization. Otherwise, problems will occur in the manufacturing process. The science of determining the correct order of processes is called queuing theory. The word "queue" means "a waiting line."

Queuing problems in modern manufacturing plants can be extremely sophisticated, and are generally handled with the aid of a computer. Since many electronic products are mass-produced nowadays, queuing theory is important to production engineers and quality-control engineers in the electronic industry. Optimum queuing is reflected in improved efficiency and reduced cost of manufacture.

QUICK-BREAK FUSE

A quick-break fuse, also known as a fast-blow fuse, is designed to break a circuit almost instantaneously if excessive current is drawn. Quick-break fuses are used in circuits containing components that are easily destroyed, in a short time, by excessive voltages or currents. Such components include many solid-state devices, such as bipolar transistors, diodes, and field-effect transistors.

A typical quick-break fuse consists of a piece of wire enclosed in a cartridge, as shown in the photograph. The wire is of such a gauge that it will immediately melt with current greater than a precise magnitude. Quick-break fuses are available for many different applications; some blow with current flow less than 1 mA, while others will tolerate 10 to 20 A or more.

In some circuits, it is desirable to have a fuse that will break the circuit rather slowly. *See also* FUSE, SLOW-BLOW FUSE.

QUICK-BROWN-FOX MESSAGE

See FOX MESSAGE.

QUICK CHARGE

A rechargeable battery, such as a lead-acid type or a nickel-cadmium type, may be recharged at various current levels. In general, within reason, a rechargeable battery having x ampere hours capacity may be charged at I amperes for t hours, as long as $x = tI$. When the time t is relatively short, the charging process is called quick charging.

In most cases, the shorter the charging period,, the less readily a battery will hold the charge. Therefore, it is desirable, whenever possible, to use a slow-charging process (trickle charge). But this is not always convenient. A quick charge can provide short-term operation of a battery in an emergency situation. *See also* TRICKLE CHARGE.

QUICK-BREAK FUSE: This unit is a cartridge type fuse, commonly used with many different electronic devices.

QUIESCENT-CARRIER TRANSMISSION

See CONTROLLED-CARRIER TRANSMISSION.

QUIETING SENSITIVITY

See NOISE QUIETING.

R

RABBIT EARS

A small indoor television receiving antenna, consisting of two whips attached to a plastic base, is called a set of rabbit ears. This arises from the geometric configuration of the pair of whips (see illustration). Each whip is adjustable in length from less than 1 foot to several feet; both of the whips can be tilted at any desired angle. A ribbon feed line, usually having a characteristic impedance of 300 ohms, is attached to the system. It is evident from the illustration that the rabbit-ear antenna is similar to a dipole antenna (*see* DIPOLE ANTENNA).

Most users of rabbit ears adjust them by random manipulation of the antennas. This is actually not a bad way to do it, as long as both whips are always extended to the same length, so that balance is maintained. The lengths of the whips do not have to be set for electrical resonance. Several different lengths should be tried. At higher-frequency television channels, the longer whip lengths may provide some gain. The angle of tilt may be any value for either whip, as long as the angle between the two whips is at least 30 degrees. The base should be rotated for best reception. (It is extremely important that the tips of the antennas do not present a hazard to the eyes of people near the rabbit ears.)

Rabbit ears are entirely satisfactory for reception of television signals from local stations, and adequate reception can sometimes be realized for distances of more than 100 miles. Some rabbit-ear antennas are equipped with built-in preamplifiers, extending the range. For fringe-area reception, a high-gain, rotatable outdoor antenna, mounted as high as possible, will give the best results in a televison-receiving installation. *See also* TELEVISION RECEPTION.

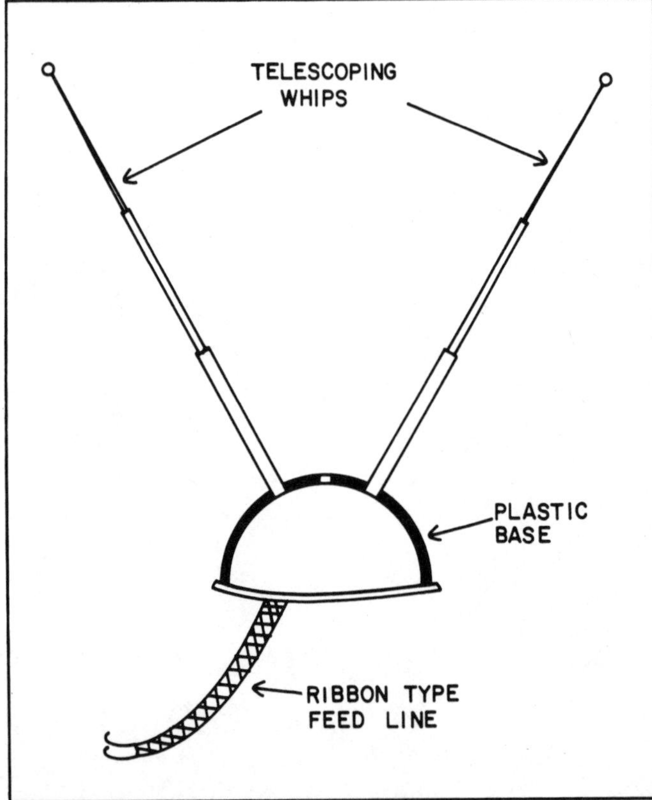

RABBIT EARS: The rabbit-ear antenna resembles a dipole and is used for television reception.

RACK AND PINION

A special form of gear drive, intended to convert rotary motion to linear motion or vice-versa, is called a rack and pinion. Rack-and-pinion drives are used in some tuning-dial mechanisms.

The operation of a rack-and-pinion drive is shown in the illustration. The control shaft is attached to the circular gear. The indicator needle is attached to the straight gear. As the shaft is rotated, the needle thus moves back and forth. Several rotations of the control shaft are necessary to move the indicator needle from one end of the scale to the other.

Rack-and-pinion dial mechanisms provide high resolution and accurate resettability. There are several other commonly used methods of constructing dial-readout systems. *See also* DIAL SYSTEM, SLIDE-RULE TUNING.

RADAR

Almost since the discovery of radio, it has been known that electromagnetic waves are reflected from many kinds of objects, especially metallic objects. Just before and during World War II this property of radio waves was put to use for the purpose of locating aircraft. The term radar is a contraction of the full technical description, Radio Detection And Ranging. Since the war, it has been found that radar is useful in a great variety of applications, such as measurement of automobile speed (by the police), weather forecasting (rain reflects radar signals), and even the mapping of the moon and the planet Venus (*see* RADAR TELESCOPE). Radar is extensively used in aviation, both commercial and military.

A radar system consists of a transmitter, a highly directional antenna, a receiver, and an indicator or display. The transmitter produces intense pulses of microwave elec-

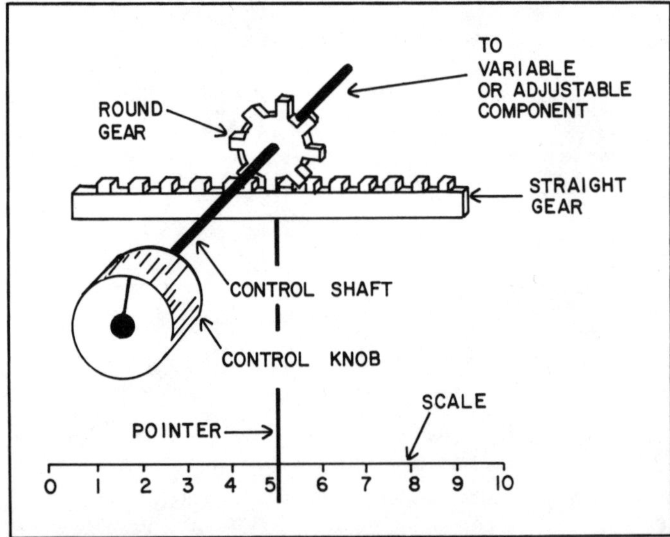

RACK AND PINION: A rack-and-pinion tuning device. An additional gear system may be incorporated in the adjustable component, attached to the control shaft opposite the knob.

tromagnetic energy at short intervals. The pulses are propagated outward in a narrow beam from the antenna, where they strike objects at various distances. The reflected signals, or echoes, are picked up by the antenna shortly after the pulse is transmitted. The farther away the reflecting object, or target, the longer the time before the echo is received. The transmitting antenna is rotated so that all azimuth bearings can be observed (*see* RADAR ANTENNA).

A typical circular radar display consists of a cathode-ray tube, and the face appears similar to A in the accompanying diagram. (The light and dark areas are reversed in this diagram; on an actual radar screen, targets are bright and the surrounding areas are darker.)The observing station is at the center of the display. Compass directions are indicated in degrees clockwise from true north, and are marked around the perimeter of the screen. The distance, or range, is indicated by the radial displacement of the echo; the farther away the target, the farther from the display center the echo or blip will be. The radar display is, therefore, a set of polar coordinates (*see* POLAR COORDINATES).

As the radar antenna rotates, and echoes are received from various directions, the electron beam in the cathode-ray tube is swept around and around. The position of the sweep coincides exactly with the position of the antenna. The period of rotation may vary, but it is typically from 1 to 10 seconds. The transmitted pulse frequency must be high enough so that targets will not be missed as the antenna rotates.

When an echo is recieved, it appears as a spot on the screen, depending on the shape and size of the target. The cathode-ray-tube phopshor has persistence, so that a blip will remain visible for about one antenna rotation. This makes it easy to see the echoes on the screen, and it also facilitates comparison of the relative locations of targets in a group (*see* PERSISTANCE TRACE).

The maximum range of a radar system depends on several factors: the height of the antenna above the ground, the nature of the terrain in the area, the trans-

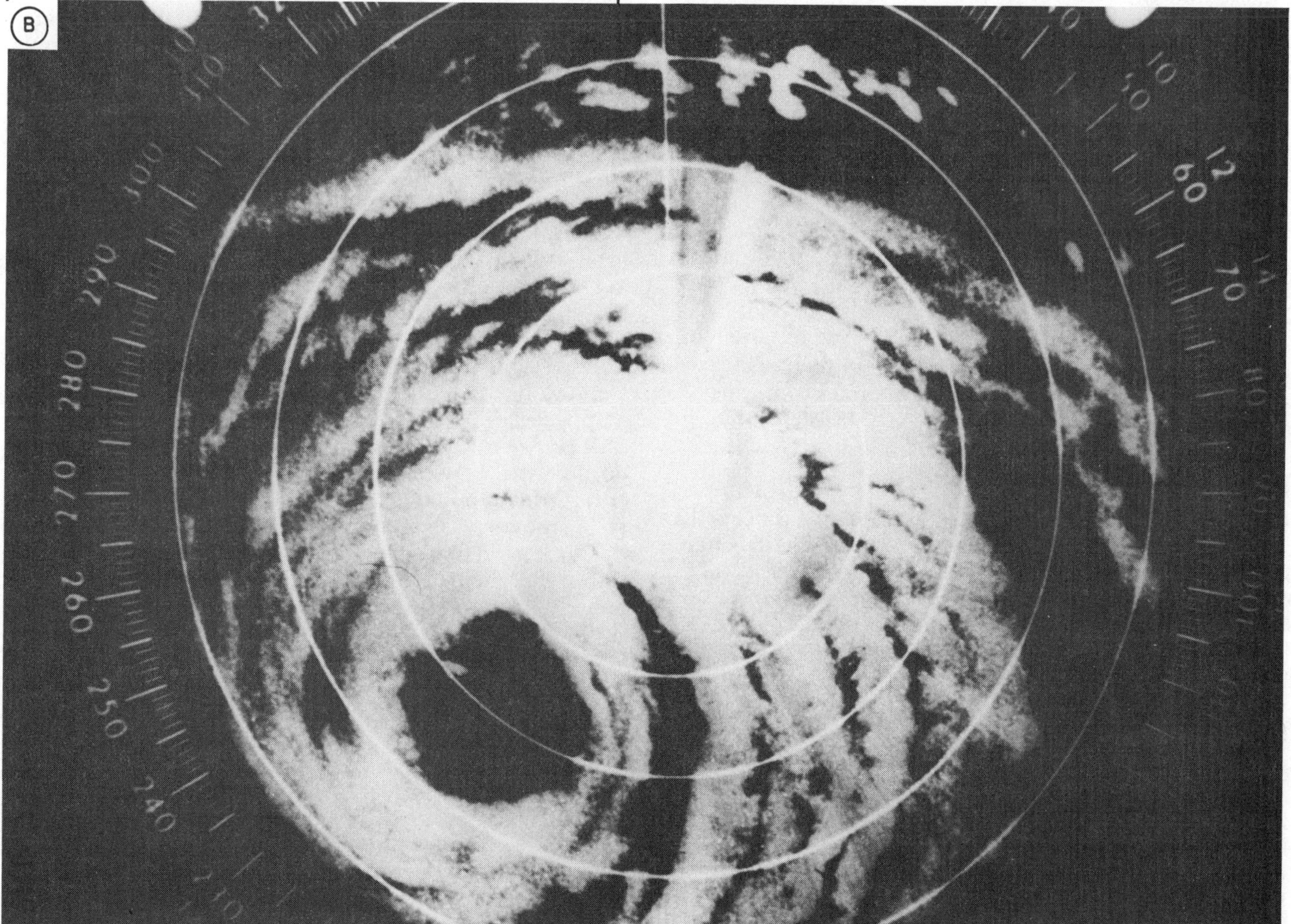

RADAR: At A, a typical radar display, showing an echo with azimuth bearing θ and range r. At B, a radar view of Hurricane Betsy as it passed near Miami, Florida, in 1965. (Courtesy of NOAA/National Hurricane Center.)

RADAR ANTENNA: A radar antenna may be enclosed in a dome-shaped or spherical shell for protection against the elements. This radar antenna, along with other meteorological and communications equipment, is located at the National Hurricane Center, Coral Gables, Florida.

mitter output power and antenna gain, the receiver sensitivity, and the weather conditions in the vicinity. Airborne long-range radar can detect echoes from several hundred miles away under ideal conditions. A low-power radar system, with the antenna at a low height above the ground, may receive echoes from only about 50 miles.

The fact that rain reflects radar echoes is a nuisance to aviation personnel, but it is invaluable for weather forecasting and observation. Radar has made it possible to detect and track severe thunderstorms and hurricanes. The double-vortex type thunderstorm, which is likely to produce tornadoes, causes a hook-shaped echo on radar. The eye of the hurricane is easy to distinguish on radar (see B).

A special form of radar, called Doppler radar, is used to measure the speed of an approching or retreating target. As its name implies, Doppler radar operates by means of Doppler effect. *See also* DOPPLER EFFECT, DOPPLER RADAR.

RADAR ANTENNA

In a radar system, a highly directional antenna is needed. This is because a narrow transmitted beam results in better resolution (ability to discern between targets near each other) than does a wide beam. Radar antennas must also exhibit high gain. The most common type of radar antenna is a dish antenna.

Radar antennas can be of many different sizes, from a foot or two in diameter to several meters across. The larger antennas provide higher resolution and greater range, as a general rule, than smaller antennas provide.

Many radar antennas must be constantly rotated in the azimuth plane. This requires a special junction in the transmission line feeding the antenna. Radar antennas are usually fed with waveguides and operate at microwave frequencies.

When a radar antenna is mounted on an aircraft, or in an environment where high winds or icing are likely to occur, an enclosure is used to protect the antenna. Such an

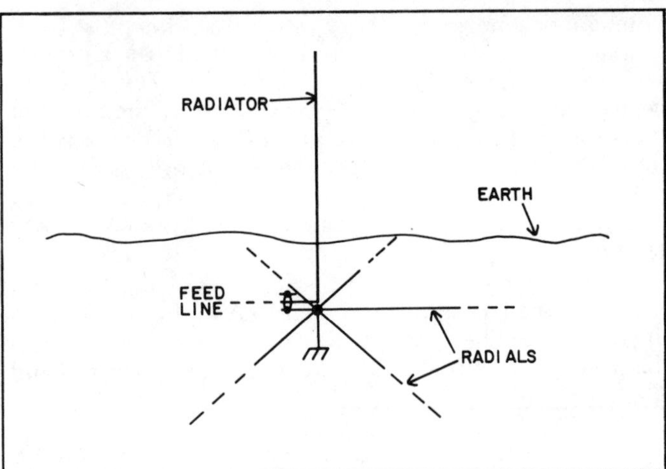

RADIAL: Ground radials, installed at the base of a vertical antenna, can improve efficiency.

enclosure is dome-shaped or spherical, and is transparent to radio-frequency electromagnetic waves. See the photograph of the radar-antenna enclosure at the National Hurricane Center in Coral Gables, Florida. This antenna is atop a relatively tall building, so that the coverage area is fairly large. *See also* DISH ANTENNA, RADAR.

RADAR TELESCOPE

Radar can be used for the purpose of locating, or analyzing, objects in outer space. The use of radar for this purpose is known as radar astronomy, and a high-powered, high-resolution radar set is needed. A radar telescope is a radar transceiver designed for the extreme long-range requirements of radar astronomy.

The first extraterrestrial radar echoes were received from the moon. The resolution obtainable in the microwave spectrum has proven to be equal to that obtainable with optical telescopes. Since the moon is comparatively nearby and has no clouds to obscure it from optical viewing, radar-telescope viewing is not really necessary. A different situation exists for the planet Venus, which is constantly hidden from visible-light view because of its thick layer of clouds. Radar signals can pendtrate the clouds of Venus, and the surface of this planet has been roughly mapped via radar echoes.

Radar telescopes have also been used to precisely determine the rotational periods of both Venus and Mercury. Originally, it was thought that Mercury kept one side facing the sun at all times, and that its rotational period was the same as its year—that is, 88 of our days. The radar has shown that the true rotational period of Mercury is a little more than 59 earth days. Venus was found to have a retrograde rotation; it is the only planet that rotates from east to west.

Radar telescopes are useful for other things besides mapping planetary surfaces and checking rotational periods. Satellites can be precisely located, and their orbits determined with a high degree of accuracy. If an asteroid or comet ever comes close to the earth, radar telescopes will probably be used to plot its path. *See also* RADAR.

RADIAL

A radial is a conductor that is used to enhance the ground system of an unbalanced, vertical monopole antenna. Radials may be constructed from wire or metal tubing. They

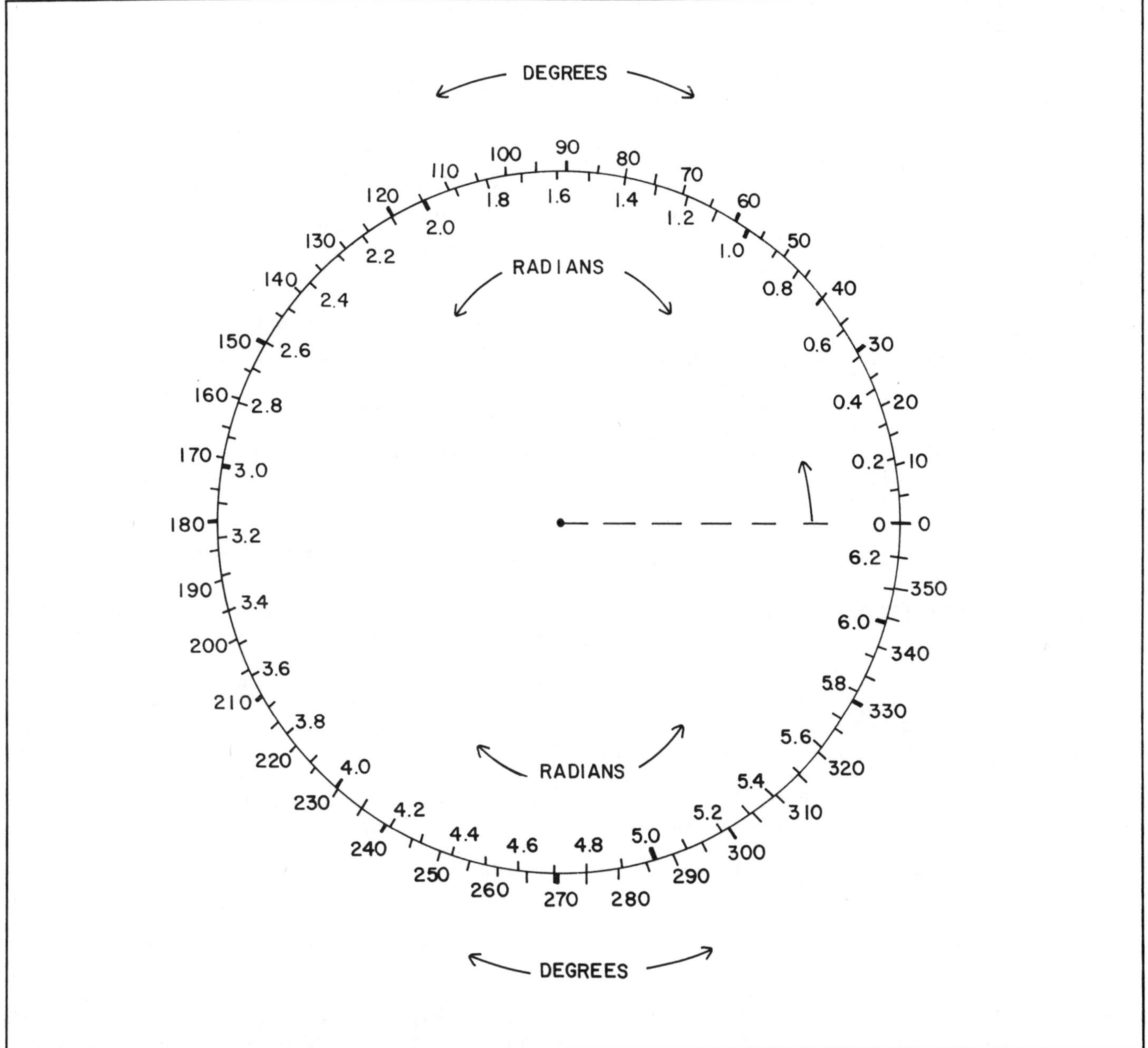

RADIAN: Nomograph showing approximate relationship of radians and degrees of angular measure.

generally measure ¼ wavelength or more.

When a vertical monopole antenna is mounted at ground level, the earth itself may be quite lossy, and will not serve as a very good radio-frequency ground. The ground conductivity is improved by the installation of radials, as shown in the illustration. The radials are run outward from the base of the antenna, and are connected to the shield part of the coaxial feed line. The improved ground conductivity results in better efficiency for any type of ground-mounted monopole, including the conical and Marconi types (*see* CONICAL MONOPOLE ANTENNA, MARCONI ANTENNA, MONOPOLE ANTENNA).

The radials may be installed just above the ground, or buried just under the surface. The greater the number of radials for a given length, the better the antenna will work. Also, the longer the radials for a given number, the better. Some broadcast stations have 360 radials, one for every azimuth degree, each measuring ½ wavelength at the broadcast frequency.

If a vertical monopole is placed well above the ground level, forming a ground-plane antenna, there need only be three or four radials of exactly ¼ wavelength (*see* GROUND-PLANE ANTENNA). The radials in the ground-plane antenna sometimes run down at a slant, rather than horizontally outward; such radials are called drooping radials (*see* DROOPING RADIAL).

Radials may be used to obtain a good radio-frequency ground for a set of communications equipment in any situation. A large number of radials measuring at least ¼ wavelength, attached to a station ground rod, may improve transmitting efficiency and reduce received noise.

RADIAN

A radian is the angle measure subtended around the perimeter of a circle, over a length equal to the radius of the circle. Since the circumference of a circle is 2π times the radius, there are 2π radians in a full circle of 360 degrees.

The radian is used as an angle of measure by mathematicians and physicists. This is because, in many situations, the radian arises naturally, while the degree was fabricated by man and is not based on pure mathematics. To 10 significant digits, a radian is equivalent to 57.29577951 degrees.

For most practical purposes, degrees can be obtained from radians by multiplying by 57.3; an angle measure in radians can be obtained from degrees by multiplying by 0.0175. The accompanying nomograph can be used for quick reference. *See also* DEGREE.

RADIANCE

Radiance is an expression of radiation intensity passing through a surface, or emitted from a surface. At visible-light wavelengths, radiance is identical to luminance, and is measured in candela per square centimeter or candela per square meter (*see* CANDELA, LUMINANCE). At wavelengths other than the visible, radiance is generally specified in terms of watts per square centimeter or watts per square meter.

RADIAN FREQUENCY

See ANGULAR FREQUENCY.

RADIANT ENERGY

Radiant energy is any form of energy capable of propagating through a vacuum. This includes all electromagnetic waves, from the longest to the shortest imaginable. Radiant energy can be manifested as radio waves, infrared, visible light, ultraviolet, x rays, or gamma rays (*see* ELECTROMAGNETIC SPECTRUM, GAMMA RAY, INFRARED, LIGHT, RADIO WAVE, ULTRAVIOLET, X RAY). Radiant energy can also occur in the form of high-speed subatomic particles (*see* ALPHA PARTICLE, BETA PARTICLE, COSMIC RADIATION). Gravitational energy may also be considered radiant.

Radiant energy is emitted by a tremendous variety of sources. The sun, the stars, and galaxies, quasars, collapsing stars, some planets, light bulbs, and radio-transmitting antennas are just a few examples. Radiant energy travels through empty space at a uniform speed of 186,282 miles per second (299,792 kilometers per second). This speed is the same as measured from any point of view in the universe.

Radiant energy may be categorized as either primary or secondary. Primary radiant energy is emitted by a source independent of the action of another source. An example is the light and shortwave (near) infrared from the sun. Secondary radiant energy is emitted by a source as a result of irradiation from another source. An example is a black tile floor, which, after having been heated during the day by the light and near-infrared rays from the sun, gives off longwave (far) infrared at night.

Energy can be transported in other ways besides radiation. These modes are called conduction and convection. Conduction and convection require the presence of a material medium, such as air, water, or metal, in order to transfer energy from one place to another. *See also* CONDUCTION COOLING, CONVECTION COOLING.

RADIATION

Radiation is a mode of energy transport. Energy radiation can occur through some materials, and always through a vacuum. Energy that is transported via radiation is called radiant energy (*see* RADIANT ENERGY). The other two modes of energy transport are conduction and convection (*see* CONDUCTION COOLING, CONVECTION COOLING).

The term radiation is used especially to describe the extremely short wavelengths of ultraviolet, X rays, and gamma rays. Such radiation can be dangerous to living things. High-speed subatomic particles, such as protons, neutrons, helium nuclei, or electrons, produce effects similar to X rays or gamma rays, and are also called radiation. *See also* COSMIC RADIATION, ELECTROMAGNETIC SPECTRUM, GAMMA RAY, RADIOACTIVITY, ULTRAVIOLET, X RAY.

RADIATION COUNTER

See COUNTER TUBE, GEIGER COUNTER.

RADIATION LOSS

In a radio-frequency feed line, there is always some loss; no feed line is 100-percent efficient. The conductor resistance and the dielectric loss contribute to the dissipation of power in the form of heat. Some energy may also be radiated from the line, and this energy, since it cannot reach the antenna, is considered loss.

In an unbalanced feed line such as a coaxial cable, radiation loss can occur because of inadequate shield continuity (*see* COAXIAL CABLE). Radiation loss can also occur because of currents induced on the outer shield by the field surrounding the antenna. Radiation loss in a coaxial feed line can be minimized by using cable with the highest possible degree of shielding continuity, and also by ensuring that the ground system is adequate at radio frequencies. A balun, placed at the feed point where a coaxial cable joins a balanced antenna system, may also be helpful.

In a balanced feed line, such as open wire or twin-lead, radiation loss results from unbalance between the currents in the line conductors. Ideally, the currents should be equal in magnitude and opposite in phase everywhere along a balanced line. If this is not the case, radiation loss will occur. Imbalance in a parallel-conductor line an result from a physically or electrically asymmetrical antenna, lack of symmetry in the antenna/feed configuration, or poor feed-line installation. *See also* BALANCED TRANSMISSION LINE, OPEN-WIRE LINE, TWIN-LEAD.

RADIATION RESISTANCE

When radio-frequency energy is fed into an electrical conductor, some of the power is radiated. If a nonreactive resistor were substituted for the antenna, in combination with a capacitive or inductive reactance equivalent to the reactance of the antenna, the transmitter would behave in precisely the same manner as it would when connected to the actual antenna. The resistor would dissipate the same amount of power as the antenna would radiate. For any antenna, there exists a resistance, in ohms, for which this can be done. The value of such a theoretical resistor is called the radiation resistance of the antenna.

Radiation resistance depends on several factors. The main consideration is the length of the antenna, as measured in free-space wavelengths. The presence of objects near the antenna, such as trees, buildings, and utility wires, can also affect the radiation resistance.

Suppose we place an infinitely thin, perfectly straight, lossless vertical monopole antenna over perfectly conducting ground. Further suppose that there are no objects in the vicinity to affect the radiation resistance. Then the radiation resistance, as a function of the vertical-antenna height in wavelengths, is as shown at A in the illustration.

For a quarter-wavelength vertical, the radiation resistance is approximately 37 ohms. As the conductor length decreases, the radiation resistance also decreases, becoming zero when the conductor vanishes. As the conductor becomes longer than a quarter wavelength, the radiation resistance increases, and becomes larger without limit as the height approaches a half wavelength. These are theoretical values. In practice, the radiation resistance is somewhat lower than the figures given in A. However, the graph represents a good approximation for most practical purposes.

Suppose that we place a conductor in free space and feed it at the exact center. Again, we assume that the conductor is perfectly straight, infinitely thin, lossless, and far from objects that might affect it. Then the value of the radiation resistance, as a function of the antenna span in wavelengths, is given by the graph at B.

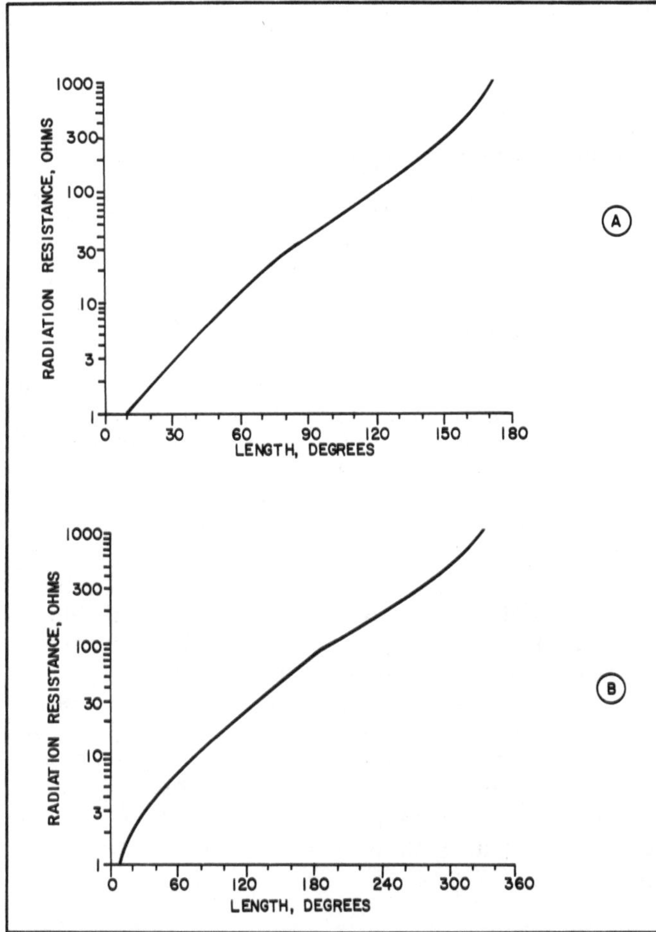

RADIATION RESISTANCE: Radiation resistance of a vertical monopole over perfectly conducting ground (A), and a center-fed radiator in free space (B), as a function of the antenna length in degrees of phase.

When the conductor measures a half wavelength, the radiation resistance is approximately 73 ohms. As the conductor becomes shorter, the radiation resistance gets smaller, approaching zero. As the conductor is lengthened, the radiation resistance increases without bound as the full-wavelength value is approached. As with the vertical system, the values shown are theoretical, but they represent a fair approximation for most center-fed vertical or horizontal antenna systems.

In practice, it is desirable to have a radiation resistance that is very large. This is because the efficiency of an antenna depends on the ratio of the radiation resistance to the total system resistance. If the radiation resistance is R and the loss resistance is S, then the total system resistance, Z, is given by:

$$Z = R + S$$

and the efficiency, in percent, is:

$$\text{Eff (\%)} = 100\ (R/Z)$$

The loss resistance in an average antenna system is a few ohms, but may be as high as 30 to 50 ohms or more. If the radiation resistance is very low, therefore, an antenna tends to be rather inefficient. *See also* ANTENNA EFFICIENCY.

RADIOACTIVITY

Radioactivity is the emission of electromagnetic energy of extremely short wavelength. In particular, radioactivity refers to X-ray and gamma-ray emission, and also to high-speed atomic particles (*see* ALPHA PARTICLE, BETA PARTICLE, GAMMA RAY, X RAY). Intense or excessive radioactivity is hazardous to life because its quanta contain large amounts of energy. High-energy photons, or other particles, can alter the nuclei of atoms upon impact. Radioactivity can cause genetic damage in plants, insects, and animals.

There is a certain amount of radioactivity present in the natural environment. An average person, living to age 70, can expect to receive 10 to 15 roentgens from natural sources. It is difficult to ascertain the extent of radiation contributed by manmade sources over the span of a lifetime. Some scientists believe that manmade radioactivity has increased in recent decades to the point where it presents a health hazard. Others believe that this increase in radioactivity has not been of sufficient magnitude to cause harm.

Radioactivity can be measured in terms of relative intensity, or it can be measured as a total exposure (dose) over a given period of time.The unit of radioactive dosage is called the roentgen. *See also* DOSIMETRY, ROENTGEN.

RADIO AMATEUR

A radio amateur is a person who holds a license to operate an amateur radio station. Radio amateurs are sometimes called "hams." A radio amateur may communicate only with other licensed amateurs, and within designated frequency bands, and may use a limited level of transmitter power. *See* AMATEUR RADIO.

RADIO AMATEUR SATELLITE CORPORATION

The Radio Amateur Satellite Corporation (AMSAT) is an institution established for the purpose of placing amateur transponders aboard active communications satellites. AMSAT publishes a magazine, *AMSAT Satellite Journal,* for the radio-amateur satellite-communications enthusiast.

The first amateur satellite was called OSCAR 1. The acronym OSCAR stands for Orbiting Satellite Carrying Amateur Radio. The satellite did nothing more than transmit "HI" over and over in Morse code. Since OSCAR 1, several more complex amateur-radio transponders have been placed on satellites.

The Radio Amateur Satellite Corporation transmits bulletins at scheduled times on the amateur bands. These bulletins contain orbital information for currently operational amateur satellites. Additional information is given in *AMSAT Satellite Journal* Magazine, and in other amateur-radio magazines. *See also* ACTIVE COMMUNICATIONS SATELLITE, SATELLITE COMMUNICATIONS.

RADIO ASTRONOMY

In the 1930s Karl Jansky, an employee of the Bell Laboratories, discovered that the center of the Milky Way produces radio-frequency energy at a wavelength of 15 meters. Jansky was the first scientist to discover that radio waves come from the cosmos. Many others have followed him; a few are named here, but it is regrettably impossible to name all of the radio astronomers who have added so much to our understanding of the universe.

A few years after Jansky's discovery, a radio amateur, Grote Reber, built a 31-foot dish antenna for the purpose of receiving radio noise from space. Reber also found radio waves coming from the Milky Way, at various wavelengths down to about 2 meters. With his modest apparatus and a lot of hard work, Reber was able to make a rough radio map of our galaxy. By 1940, astronomers began to get interested, and the science of radio astronomy was born.

Modern radio astronomers use large dish antennas; some are several hundred feet across. A phased antenna system, called the interferometer, allows radio astronomers to view the sky with a high degree of resolution. Systems such as the Mills Cross in Australia are also used (*see* INTERFEROMETER, MILLS CROSS). Most of the observations are carried out at ultra-high and microwave frequencies, where the wavelengths are short enough to allow good resolution and the electromagnetic fields are not greatly affected by the atomsphere of the earth.

Radio astronomy has provided a wealth of knowledge that could not have been discovered by optical means. Quasars, pulsars, and radio galaxies, for example, would not have been found by optical telescopes.

An offshoot of radio astronomy has developed in recent years. High-powered transmitters, in conjunction with high-gain antennas, have made it possible to receive echoes from the nearer planets. The distances of these objects can be precisely measured, and their surfaces mapped, by a technique called radar astronomy (*see* RADAR TELESCOPE).

Perhaps the most significant contribution that radio astronomy has made, however, was inadvertent. In 1965, two Bell Laboratories employees, Arno Penzias and Robert Wilson, found that a weak radio-noise signal seemed to be coming from all directions in space. They had discovered the remnants of radiation from the very beginning of the universe 20,000,000,000 years ago. At that time, it is believed, all the matter in the universe exploded in an incredible cataclysm called the "big bang." *See also* COSMOLOGY AND COSMOGONY, RADIO TELESCOPE.

RADIO FREQUENCY

An electromagnetic disturbance is called a radio-frequency (RF) wave if its wavelength falls within the range of 30 km to 1 mm. This is a frequency range of 10 kHz to 3000 GHz.

The radio-frequency spectrum is split into eight bands, each representing one order of magnitude in terms of frequency and wavelength. These bands are called the very-low, low, medium, high, very-high, ultra-high, super-high, and extremely-high frequencies. They are abbreviated, respectively, as vlf, lf, mf, hf, vhf, uhf, shf, and ehf. Super-high-frequency and extremely-high-frequency RF waves are sometimes called microwaves (*see* BAND).

Radio-frequency waves propagate in different ways, depending on the wavelength. Some waves are affected by the ionosphere, troposphere, or other environmental factors (*see* PROPAGATION CHARACTERISTICS).

Radio frequencies represent a sizable part of the electromagnetic spectrum. As the wavelength becomes shorter than 1 mm, we encounter first the infrared, then visible light, ultraviolet, X rays, and gamma rays. *See also* ELECTROMAGNETIC SPECTRUM.

RADIO-FREQUENCY INTERFERENCE

See ELECTROMAGNETIC INTERFERENCE.

RADIO-FREQUENCY PROBE

A radio-frequency probe is a special pickoff device, intended for sampling radio-frequency signals. A radio-frequency probe resembles an ordinary alternating-current or direct-current probe (*see* PROBE), except that precautions are taken to prevent signal leakage.

A radio-frequency probe must have a shielded cable, and the cable length must be as short as possible. The probe imepdance is high, generally about 1 megohm. If the probe is designed to be held with the hand, the handle must be adequately insulated and shielded to prevent hand capacitance from affecting the impedance.

Radio-frequency probes are sometimes inserted into the tuned coil of a transmitter final amplifier, for such purposes as signal monitoring or the operation of radio-frequency-actuated devices. This type of probe consists of a short, stiff wire or a small coil, which allows some energy to be picked up without affecting the tank-circuit impedance.

RADIO-FREQUENCY TRANSFORMER

At radio frequencies, transformers are often used for the

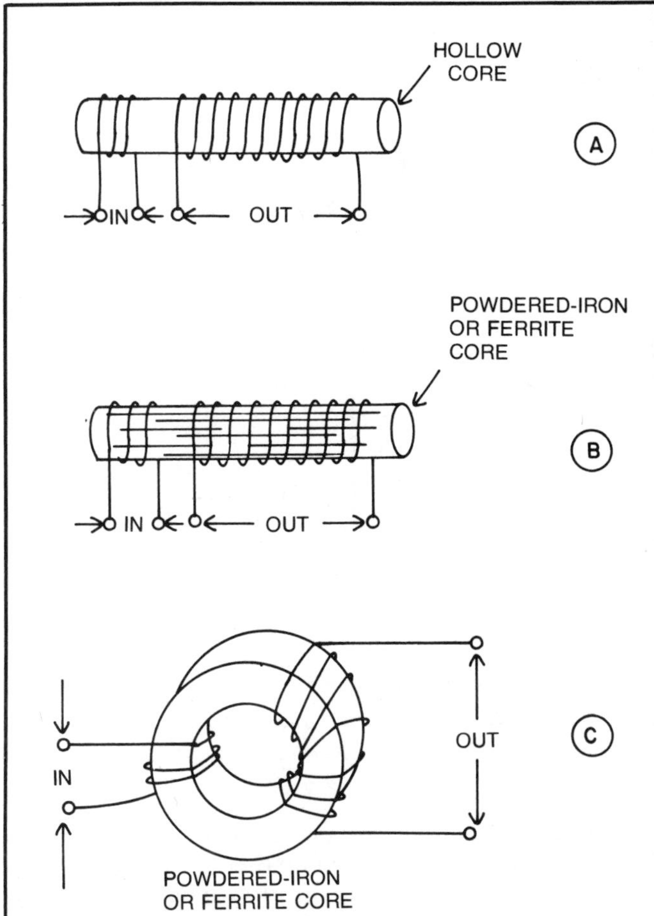

RADIO-FREQUENCY TRANSFORMER: Three types of radio-frequency transformer. At A, an air-core solenoidal transformer; at B, a ferromagnetic-core solenoidal transformer; at C, a toroidal transformer. These are step-up transformers.

purpose of impedance matching. Radio-frequency transformers are also used for coupling between or among amplifiers, mixers, and oscillators. A special form of radio-frequency transformer, known as a balun, is sometimes used in antenna systems to match impedances and optimize the balance (*see* BALUN).

A radio-frequency transformer may consist of solenoidal windings with an air core, solenoidal windings with a powdered-iron or ferrite core, or toroidal windings with a powdered-iron or ferrite core. These three configurations are illustrated.

The air-core transformer presents the least loss, and is therefore the most efficient; however, air-core transformers are rather bulky and are easily affected by nearby metallic objects. Solenoidal transformers with powdered-iron or ferrite cores are less bulky and almost as efficient as the air-core type. The solenoidal core may be moved in and out of the coil to vary the inductances of the windings (*see* PERMEABILITY TUNING). Toroidal transformers offer the advantage of being immune to the presence of nearby metallic objects, since all of the magnetic flux is contained within the powdered-iron or ferrite core. This facilitates miniaturization (*see* TOROID).

When no reactance is present, the impedance-transfer ratio of a radio-frequency transformer is equal to the square of the turns ratio. Thus, a 2:1 turns ratio results in an impedance-transfer ratio of 4:1; a turns ratio of 3:1 produces a 9:1 impedance-transfer ratio. *See also* IMPEDANCE TRANSFORMER.

RADIO GALAXY

Galaxies vary greatly in terms of the ratio of visual brightness to radio-frequency emission intensity. A galaxy that is faint in the visible wavelengths, but very bright (intense) at radio frequencies, is called a radio galaxy. Radio galaxies were not known to exist until the invention of the radio telescope (*see* RADIO ASTRONOMY, RADIO TELESCOPE).

The astronomers Rudolph Minkowski and Walter Baade were the first to notice that an extremely intense source of radio emission, called Cygnus A (the brightest radio source in the constellation Cygnus), can be seen in the optical telescopes. Cygnus A appears as two discrete radio sources, and it is believed that perhaps it is a pair of galaxies in collision. Cygnus A may also be a galaxy in the process of breaking apart.

Many radio galaxies have been discovered, but the reason for their unusual radio-frequency intensity is still not completely known. The radio galaxies closely resemble the quasi-stellar radio sources, or quasars. *See also* QUASAR.

RADIO INTERFERENCE

See INTERFERENCE.

RADIOISOTOPE

An element always has the same number of protons in its nucleus. This number is called the atomic number. The number of neutrons, however, may vary, resulting in changes in the atomic weight. The various different possible nuclei are called isotopes. Normally, one isotope of a given element is by far the most stable, and is therefore the most common (*see* ATOMIC NUMBER, ATOMIC WEIGHT, ISOTOPE). Some isotopes are unstable, and tend to decay, producing radioactivity. Such isotopes exist for most elements. They are known as radioisotopes.

Carbon, with atomic number 6, normally has six neutrons in its nucleus, and thus the atomic weight is 12. However, some atoms of carbon have eight neutrons, resulting in an atomic weight of 14. This isotope, called carbon 14, is a radioisotope. Some carbon 14 exists in nature. Carbon 14 is a relatively stable radioisotope; it requires a long time to decay. Archeologists and other earth scientists can tell how old something is by the amount of carbon 14 still remaining, as determined with a radiation counter.

Heavy elements generally have more radioisotopes than lighter elements. This is simply because there are a larger number of different isotope possibilities as the atomic number increases. Some radioisotopes, such as carbon 14, occur naturally. Others are made by man. *See also* ELEMENT.

RADIOLOCATION

Radiolocation is a process by which the position of a vehicle, aircraft, or ocean-going vessel is determined.

A radiolocation system can operate in different ways. The simplest method is the directional method. Two or three fixed receiving stations, some distance from each other, are used. Direction-finding equipment is used at

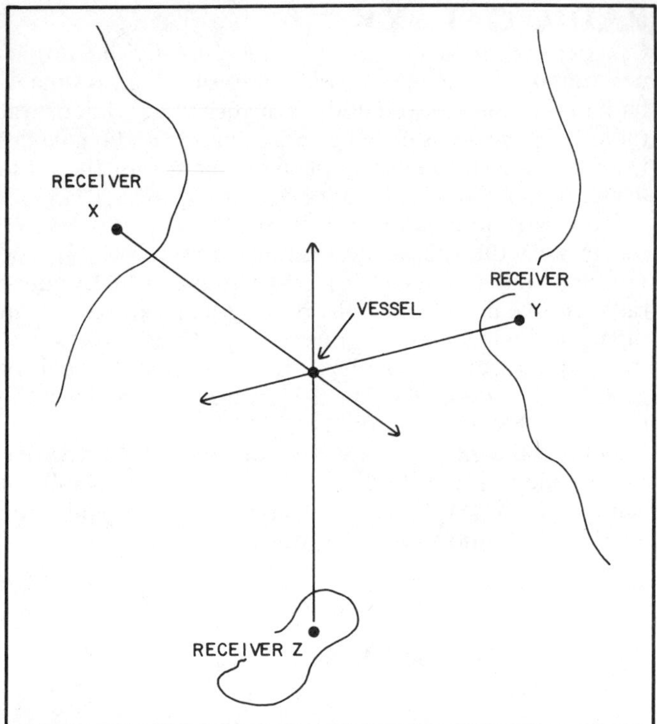

RADIOLOCATION: A simple method of radiolocation, using three land-based direction-finding systems.

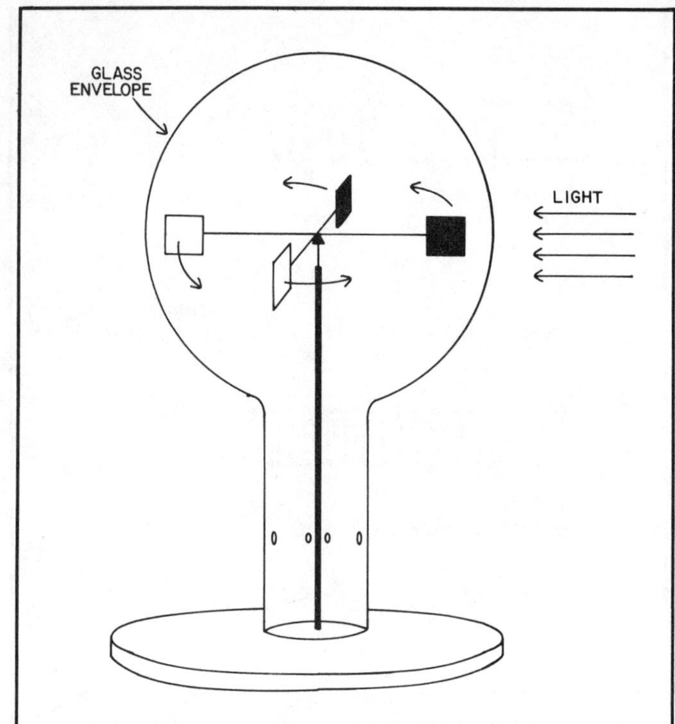

RADIOMETRY: A radiometer measures the ability of light to generate heat energy.

each of these stations, in conjunction with an omnidirectional transmitter aboard the vessel, to establish the bearings of the vessel with respect to each station. The vessel location corresponds to the intersection point of great circles drawn outward from the receiver points in the appropriate directions, as shown in the illustration (*see* DIRECTION FINDER, GREAT CIRCLE).

The above method of radiolocation requires that the vessel be equipped with a transmitter, and that the captain of the vessel wants to be located. If the captain does not want to be located, or if the radio equipment aboard the vessel is not working, other means of radiolocation are necessary. Radar can be used to locate such vessels (*see* RADAR). Sometimes enemy craft can be located by visual or infrared apparatus. In recent years, satellites have been developed that can locate enemy ships and missiles.

A special form of over-the-horizon radiolocation device, operating in the high-frequency radio band, has been recently developed by both the United States and the Soviet Union to detect and locate launchings of intercontinental ballistic missiles. Because of its sound in radio receivers, this device has been called a woodpecker. *See also* RADIONAVIGATION, WOODPECKER.

RADIOLOGY

Radiation, especially X rays, is sometimes used in medicine for the purpose of diagnosing illness. The X-ray photograph is invaluable in locating tumors, for example. But radiation can also be used to help cure some illnesses, particularly cancer. Radiology is the branch of medicine in which radiation is used to ameliorate certain illnesses. A physician who specializes in radiology is called a radiologist.

Radiation therapy is generally used only as a last resort. This is because radiation produces side effects, most notably the alternation of the chromosomes in cell nuclei. However, in some cases radiation therapy will prolong the life of a patient by arresting the growth of a malignancy.

RADIOMETRY

A radiometer is a device that is used to measure the ability of optical radiant energy to produce heating of objects. There are several different types of radiometer. The science of measuring the heat resulting from optical energy is known as radiometry.

The bolometer and thermocouple are two devices that can be used to measure the heating caused by radiant energy (*see* BOLOMETER, THERMOCOUPLE). These devices, in conjunction with peripheral measuring equipment, facilitate quantitative measurement of the heat power in microwatts, milliwatts, or watts.

A radiometer used for qualitative measurements is familiar to most high-school physics students. It is the vane device shown in the illustration. Three or four flat vanes are mounted on a rotating bearing. Each vane is painted black on one side, say the clockwise-facing side, and white on the other side. When light strikes the black side of a vane, heat is produced. This agitates the air molecules adjacent to the vane. When light strikes the white side of a vane, no heat is produced, and the air molecules next to the vane are not affected. Since there is more molecular movement on the black side of each vane, there is greater pressure there, and the vane moves in the direction of the white faces. The whole assembly rotates with a speed that depends on the intensity of the light. Quantitative measurements can be obtained by counting the number of revolutions of the assembly per unit time.

A temperature-sensitive current or voltage source, connected to a precision ammeter or voltmeter, can be used to measure the ability of light to produce heating. Some capacitors exhibit large positive or negative temperature coefficients; a capacitor of this type, in conjunction with a tuned-circuit oscillator and an accurate frequency meter, can be used as a radiometer.

Another science, devoted to the measurement of the intensity of visible light as perceived by the human eye, is known as photometry. *See also* PHOTOMETRY.

RADIO, MOBILE

See MOBILE EQUIPMENT.

RADIONAVIGATION

Radionavigation refers to the use of radio apparatus, by personnel aboard moving vessels, for the purpose of determining position and plotting courses.

There are many means of radionavigation. The intersecting-line method is probably the simplest. Two or three land-based transmitters are needed. Their locations must be accurately known. A direction-finding device on the vessel is used to determine the bearings of each of the transmitters (*see* DIRECTION FINDER). After the bearings have been determined, great circles are drawn on a map outward from each transmitter, so that the lines all intersect at a common point. That point represents the position of the vessel.

An alternative intersecting-line method requires a transmitter aboard the vessel. The position is determined by land-based personnel with the aid of direction finders. The land stations, in communication with each other and with the captain of the vessel, can inform the captain of his position (*see* RADIOLOCATION).

Modern radionavigation systems use low-frequency, land-based transmitters in conjunction with computers aboard the vessel. The signals or pulses from the land-based transmitters arrive at the vessel in varying phase as the vessel follows its course. If the ship or plane strays from its course, the computer indicates the error. These systems are known as hyperbolic radionavigation systems (*see* LORAC, LORAN).

Aircraft radionavigation can be performed by radar, or with direction-finding apparatus at the very-high or ultra-high frequencies. Air-traffic controllers constantly monitor the skies via radar. An airline pilot is automatically told if the plane is off course. *See also* AIR-TRAFFIC CONTROL, CHAIN RADAR, RADAR.

RADIOPAQUE SUBSTANCE

A material is called radiopaque if it offers considerable opposition to the passage of X rays. Radiopaqueness is a relative phenomenon; some materials are more radiopaque than others. All matter blocks X rays to some extent, but nothing will completely stop X rays.

The primary factor that determines the radiopaqueness of a material is its density. The more dense a substance, in general, the more it attenuates X rays. This is simply because, in a dense material, there are many protons and neutrons to get in the way of the high-energy X-ray photons; in a substance with low density, there are fewer nuclear particles to intercept the photons. The best example of a radiopaque material is lead. Other metals, too, are relatively radiopaque.

An X-ray photograph of the human body, or a part of the body, reveals that some tissues are more radiopaque than others. Bones, for example, appear white in an X-ray photograph, since the bones are more radiopaque than other body tissues. Some tumors can be identified because they are more radiopaque than the surrounding tissue.

A substance that offers little or no opposition to X rays is called radiotransparent. Soft flesh is comparatively radiotransparent. So are most plastics, rubber, wood, and other materials with low density. *See also* X RAY.

RADIO, PORTABLE

See PORTABLE EQUIPMENT.

RADIOSONDE

In weather forecasting, it is essential to know the conditions not only at ground level, but at various altitudes in the atmosphere. It is fairly expensive to send airplanes up into the sky to make measurements of upper-atmospheric conditions, so the radiosonde is often used. A radiosonde consists of a helium-filled balloon with various instruments suspended from its base. A radio transmitter sends the data to a ground station.

A typical radiosonde is shown in the illustration. The ballon diameter is chosen on the basis of how high the meteorologist wants it to go. A small radiosonde needs a ballon with a diameter of just a few feet. A large, high-altitude radiosonde may use a balloon with a length of more than 100 feet. The largest radiosonde devices can reach the lower stratosphere.

A typical radiosonde measures the temperature, relative humidity, and barometric pressure. Instruments may be included for measuring such variables as the wind velocity (both vertical and horizontal), the relative amounts of various gases in the air, or the intensity of solar and cosmic radiation. Television cameras may be mounted on a radiosonde for viewing the earth below. Astronomers sometimes use radiosonde devices to observe the heavens at wavelengths to which the lower atmosphere is opaque.

RADIO STATION

See RADIO EQUIPMENT.

RADIOTELEGRAPH

The term radiotelegraph is used to describe any form of radio emission in which Morse-code signals are sent. This may include A1, A2, F1, and F2 emission (*see* EMISSION CLASS).

A transmitter intended for sending messages by Morse code is sometimes called a radiotelegraph (*see* CODE TRANSMITTER).

RADIOTELEGRAPH CODE

See INTERNATIONAL MORSE CODE.

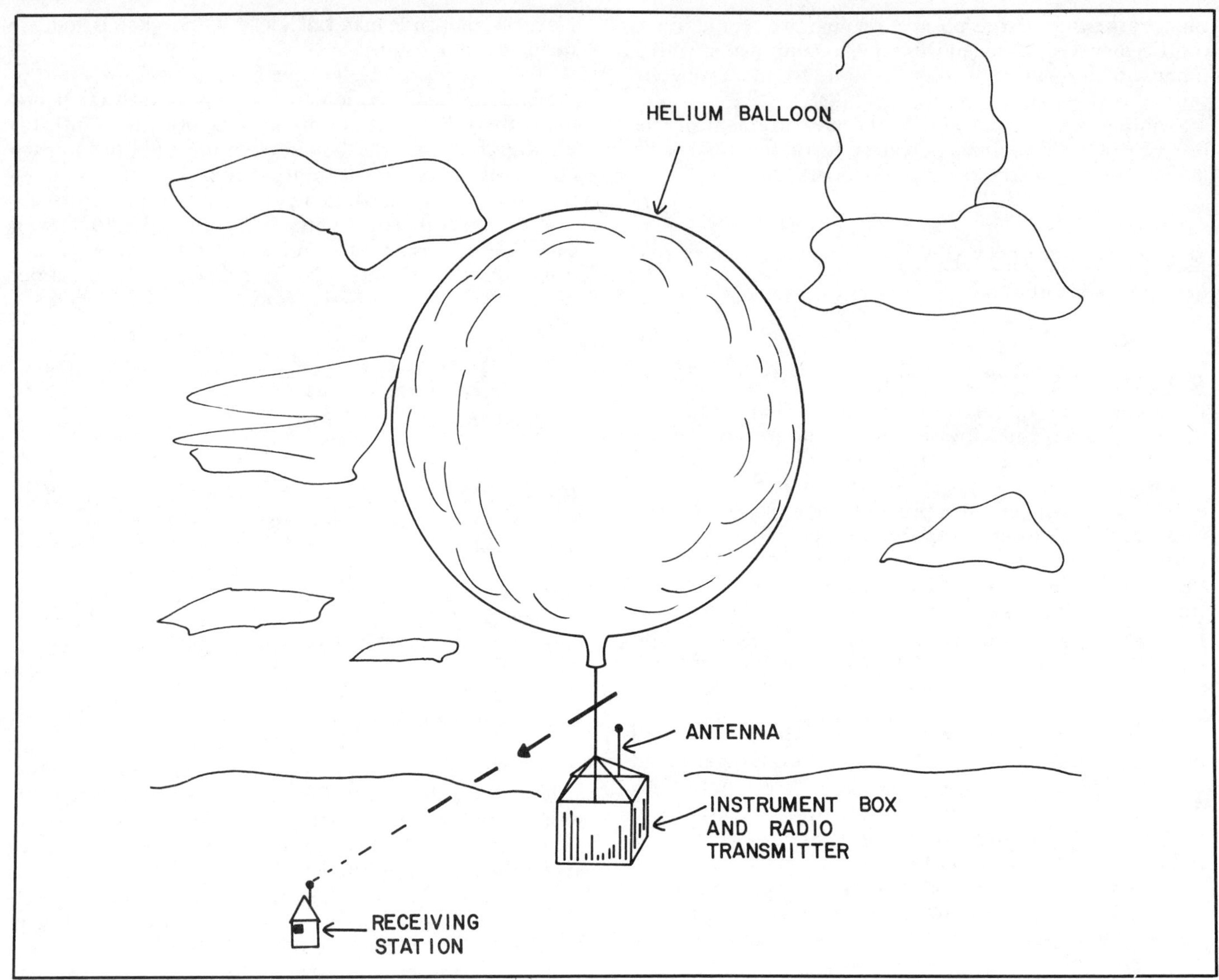

RADIOSONDE: A radiosonde is used to investigate weather conditions at various levels in the atmosphere.

RADIOTELEGRAPH TRANSMITTER

See CODE TRANSMITTER.

RADIOTELEPHONE

Radiotelephone refers to any form of two-way communication by voice. The earliest form of radiotelephone was amplitude modulation, or emission type A3. Today, radiotelephone may use not only A3 emission, but also any of types A3A, A3J, A9, F3, F9, P3, or P9 (*see* AMPLITUDE MODULATION, EMISSION CLASS, FREQUENCY MODULATION, PULSE MODULATION, SINGLE SIDEBAND).

A transmitter or transceiver, designed for the purpose of sending or communicating by voice, is sometimes called a radiotelephone. *See also* MOBILE TELEPHONE, PORTABLE TELEPHONE, VOICE TRANSMITTER.

RADIO TELESCOPE

A radio telescope is a sensitive, highly directional radio receiver, intended for intercepting and analyzing radio-frequency noise from the cosmos. The first radio telescopes were built by Karl Jansky and Grote Reber in the 1930s. Modern radio telescopes use the most advanced antennas, receiver preamplifiers, and signal processing techniques. Radio astronomy is generally carried out at wavelengths shorter than about 10 m, but especially at ultra-high and microwave frequencies.

A block diagram of a radio telescope is illustrated. The installation consists of an antenna, a feed line, a preamplifier, the main receiver, and a signal processor and recorder.

A radio-telescope antenna may consist of a large dish or an extensive phased array. The dish antenna offers the advantage of being fully steerable (usually), so that it can be pointed in any direction without changing the resolution or sensitivity. The phased array, which may consist of a set of yagis, dipoles, or dish antennas, provides better resolution because it can be made very large. However, the steerability is not as flexible; phased arrays can usually be steered only along the meridian (north and south). The rotation of the earth must then be used to obtain full coverage of the sky.

Two or more large antennas, spaced a great distance apart, can be used to obtain extreme resolution with a

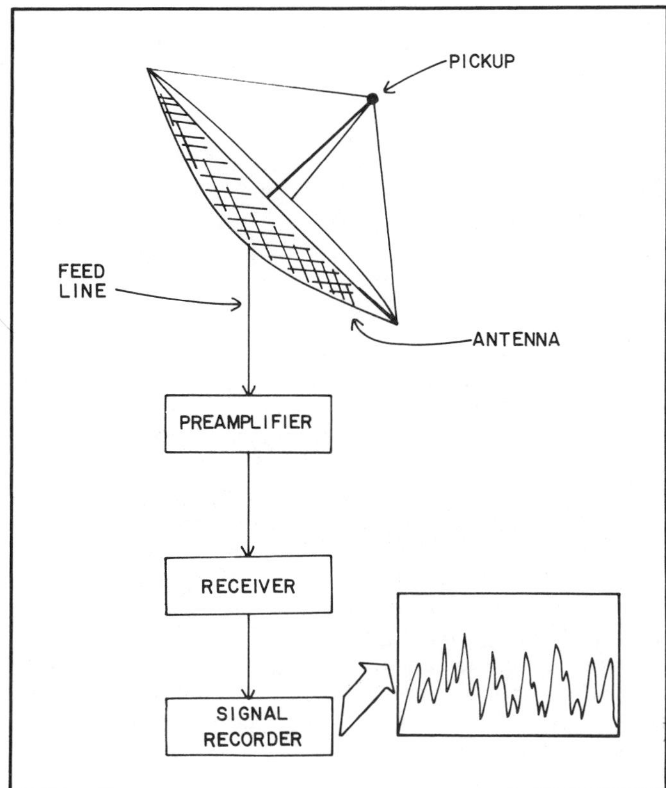

RADIO TELESCOPE: Simplified block diagram of a radio telescope using a dish antenna.

radio telescope. This technique is called interferometry (*see* INTERFEROMETER). The interferometer allows the radio astronomer to probe into the details of celestial radio sources. In some cases, what was originally thought to be a single source has been found to contain multiple components.

The antenna feed line is usually a coaxial cable at very-high and ultra-high frequencies. A waveguide is used for microwave reception. The lowest possible loss, and the highest possible shielding continuity, must be maintained. This minimizes interference from earth-based sources, and ensures that the antenna directivity and sensitivity are optimum.

The radio-telescope receiver differs somewhat from an ordinary communications reciever. Sensitivity is of the utmost importance. A special preamplifier is needed (*see* PREAMPLIFIER). The components may be cooled down to just a few degrees Kelvin to reduce the noise in the circuit (*see* CRYOGENICS, SUPERCONDUCTIVITY).

The main receiver of a radio telescope must have a relatively large response bandwidth, in contrast to a communications receiver, which usually exhibits a very narrow bandwidth. The wavelength must be adjustable over a wide and continuous range.

In a radio telescope, a speaker may be used for effect, but is of little practical value in most cases, since cosmic radio emissions sound like little more than hiss or static. A pen recorder and/or oscilloscope, and perhaps a spectrum analyzer, are used to evaluate the characteristics of the cosmic noise (*see* OSCILLOSCOPE, PANORAMIC RECEIVER, PEN RECORDER, SPECTRUM ANALYZER).

Most radio telescopes are located well away from sources of manmade noise. A city is a poor choice as a location to build a radio telescope. Some radio-telescope observatories, such as the one at Green Bank, West Virginia, are protected against manmade noise by law. The Federal Communications Commission has established a radio quiet zone near this installation. This allows the radio astronomers at Green Bank to carry on their research under the best manageable conditions.

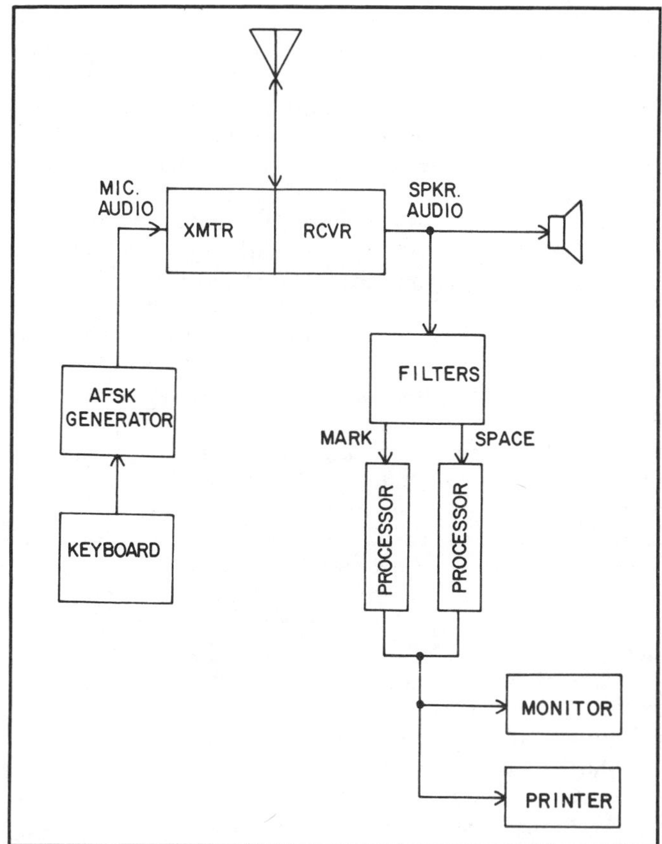

RADIOTELETYPE: Block diagram of a simple radioteletype station.

Radio telescopes can "see" things that cannot be detected with optical telescopes. For example, the hydrogen line at the 21-cm wavelength can be easily observed with a radio telescope, although it is invisible with optical apparatus. It is thought that the 21-cm line might serve as a marker for possible interstellar communication—messages from other worlds.

The radio telescope, with its high-gain, high-resolution antennas, can be used to transmit signals as well as receive them. This has allowed astronomers to precisely ascertain the distances to, the motions of, and the surface characteristics of the moon, Venus, and Mercury. When a radio telescope is used in this way, it is called a radar telescope (*see* RADAR TELESCOPE).

In 1974, Dr. Frank Drake used the massive radio telescope at Arecibo, Puerto Rico, to transmit the first message to the cosmos. This transmission has already passed some of the nearer stars. A binary code was used, telling of our world and our society. Perhaps it has been received by now. Our radio telescopes might someday receive a reply. *See also* RADIO ASTRONOMY.

RADIOTELETYPE

Radioteletype is a mode of communications in which printed messages are sent and received via radio waves. For this to be possible, a special device is needed to properly modulate a transmitter so that the printed characters are conveyed. A demodulator is needed to convert the audio output of the receiver into the electrical impulses that drive the teleprinter or cathode-ray-tube display. Both the modulator and demodulator are usually combined into a single circuit called a terminal unit. The terminal unit is a form of modem (*see* MODEM, TERMINAL UNIT).

The most common method of transmitting radioteletype signals is frequency-shift keying. Two different carrier frequencies are used, called the mark (key-down) and the space (key-up) conditions. The ASCII or Baudot codes are used with frequency-shift keying (*see* ASCII, BAUDOT, FREQUENCY-SHIFT KEYING). Morse code is sometimes used to transmit and receive radioteletype signals.

The illustration is a simple block diagram of a radioteletype station using a single-sideband transceiver. The received signals are fed into the terminal-unit demodulator from the speaker terminals of the transceiver. The modulator generates a pair of audio tones, which are fed into the microphone terminals of the transceiver. This arrangement can provide communication via ASCII or Baudot at speeds up to approximately 300 baud.

Commercially manufactured radioteletype equipment, including terminal units, cathode-ray-tube displays, and printers, is available. A radioteletype station is, however, not hard to build from scratch, or "junk-box," components. Many amateur radioteletype enthusiasts enjoy building their own terminal units and using discarded machines.

RADIOTRANSPARENT SUBSTANCE

See RADIOPAQUE SUBSTANCE.

RADIO WAVE

A radio wave is an electromagnetic field that exhibits a wavelength in the range of 30 km to 1 mm (*see* RADIO FREQUENCY). Radio waves propagate through free space at 186,282 miles per second or 299,792 kilometers per second. This is the same as the speed of light, and all other electromagnetic radiation, in free space.

Radio waves are generated when electrons, or other charged particles, are induced to oscillate back and forth at a rate between 10 kHz and 300 GHz. For communications purposes, electrons are made to accelerate in a circuit called an oscillator (*see* OSCILLATOR). The magnitude of this acceleration is increased by power amplifiers (*see* POWER AMPLIFIER), so that the electromagnetic field can travel long distances and remain strong enough to cause movement of the electrons in a far-off antenna.

Radio waves are modulated to convey information. Modulation can be accomplished in a variety of ways. The most common are amplitude modulation, in which the intensity of the radio wave is varied, and frequency modulation, in which the frequency is varied (*see* EMISSION CLASS, MODULATION). Radio waves can be used to transmit voices, radioteletype signals, still pictures, moving pictures, and other data.

Radio waves are affected by the atmosphere of the earth. At some frequencies, radio waves are trapped or bent by the ionosphere. At other frequencies, radio waves are bent or reflected by air masses. At still other frequencies, radio waves are essentially unaffected by the atmosphere. *See also* ELECTROMAGNETIC FIELD, PROPAGATION CHARACTERISTICS.

RAM

See RANDOM-ACCESS MEMORY.

RAMP WAVE

A ramp wave, or ramp function, is a form of sawtooth wave in which the amplitude rises steadily and drops abruptly (*see* SAWTOOTH WAVE). The rise in amplitude is linear with respect to time, so that the increasing part of the wave looks straight on an oscilloscope display (see illustration). The ramp wave is usually of only one polarity, with the minimum amplitude being zero at the beginning of the linear rise.

The ramp wave is ideally suited for applications that require scanning. For example, a ramp-wave generator is employed in an oscilloscope, spectrum analyzer, television receiver, or camera tube to make the electron beam move across the screen at a constant speed. A ramp wave is also used to operate a device called a sweep generator (*see* SWEEP GENERATOR).

RANDOM-ACCESS MEMORY

A form of memory that allows inputting and removal of data in any order, regardless of the particular data addresses, is called random-access memory (RAM). An integrated circuit that contains this type of memory circuitry is sometimes called a RAM.

Random-access memory is extensively used in computers and other data-processing systems. The RAM offers much greater ease and convenience of retrieval than other forms of memory.

RANDOM NOISE

Certain forms of electromagnetic noise, especially man-made noise, exhibit a definite pattern of amplitude versus time. Other types of noise show no apparent relationship between amplitude and time; such noise is called random noise. An example of non-random noise is the impulse type. Example of random noise are sferics, thermal noise, and shot noise.

Random noise is more difficult to suppress than noise

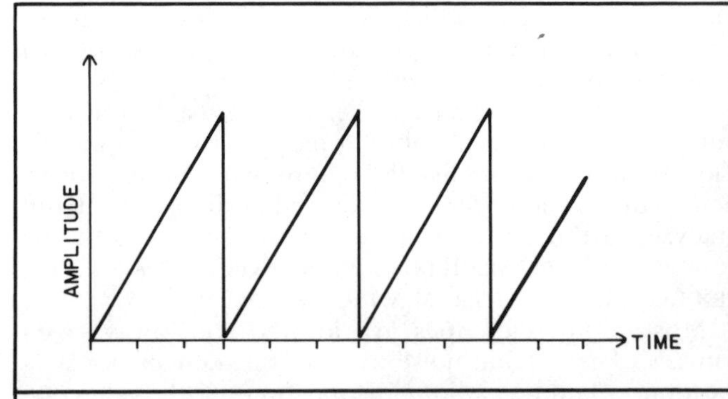

RAMP WAVE: The ramp wave is characterized by a linear rise and an abrupt decay.

having an identifiable waveform. If noise shows discernible amplitude-versus-time patterns, we can design a circuit that "knows" the noise behavior, and that can act accordingly to get rid of it. No circuit exists that "knows" the patterns of random noise: There are no patterns!

A limiting circuit, which at least allows the desired signal to compete with the noise, is a good defense against strong random noise. The use of narrowband emission, in conjunction with a narrow receiver bandpass, is also helpful in dealing with strong random noise. The use of frequency modulation may give better results than amplitude modulation if the receiver is equipped with an effective limiter or ratio detector. Directional or noise-cancelling antenna systems can sometimes improve communications in the presence of random noise. *See also* NOISE, NOISE LIMITER.

RANDOM NUMBERS

A set of digits is called random if the order exhibits absolutely no pattern. Truly random numbers are difficult, perhaps even impossible, to generate. It is essentially impossible for a person to choose sequences of digits 0 through 9 at random. Algorithms can be used to generate numbers that are "almost random," but the existence of the algorithm suggests that such numbers are not truly random.

Irrational numbers are believed to contain digits in random sequence. An example of an irrational number is the natural-logarithm base, e. This number, and all irrational numbers, are nonterminating (infinitely long) and nonrepeating (the digits exhibit no apparent pattern).

Random numbers are used for statistical calculations. A large sample of random digits 0 through 9 has certain properties. For example, as larger and larger samples are taken, the proportion of times each digit occurs, as a fraction of the total number of digits, approaches 0.1. The probability that any given digit will occur in a particular place is 0.1; the probability that any given digit will occur twice in sequence is 0.1×0.1 or 0.01; the probability that any given digit will occur n times in sequence is 0.1^n. There is, however, absolutely no pattern in a sequence of random numbers, so it is impossible to predict which digit will occur in a given place in the sequence.

Statisticians use random-number tables when they need to have truly random digits. (Whether these digits are truly random, or merely pseudorandom, is an open question.) Pseudorandom digits can be obtained by such exercises as extracting the square root of 2, or by using a computer algorithm designed for pseudorandom-number generation. *See also* PSEUDORANDOM NUMBERS.

RANGE

The range of a mathematical function is the set of values that the dependent variable attains. In the Cartesian coordinate system, the range of a function is usually a subset of the values on the vertical axis (*see* CARTESIAN COORDINATES).

The range of a function may be the entire set of real numbers, or it may consist of a subset of the real numbers. Sometimes the range of a function is an interval. Some functions have ranges consisting of discrete numbers. The range of a function may even be the empty set.

As an example, consider the function $f(x) = x^2$, for the domain consisting of all real numbers. Then the range of f is the set of nonnegative real numbers. The range of the function $g(x) = 1$ is simply the set containing the number 1.

The range of a function often depends on the extent of the domain of values that we allow the independent variable to attain. For example, the range of $f(x) = x^2$ is reduced to the set of all real numbers greater than 100, if we insist that the domain contain only real numbers greater than 10. *See also* DEPENDENT VARIABLE, DOMAIN, FUNCTION, INDEPENDENT VARIABLE.

In a radio broadcasting or communications system, the distance over which messages can be sent and received is sometimes called the range. The communications range depends on the transmitter power, the type and location of the antennas, the receiver sensitivity, and various propagation factors. *See also* PROPAGATION CHARACTERISTICS.

In radar, the distance from the station to a target is called the range. Most radar systems display the range radially from the center of a circular screen. *See also* RADAR.

The term range is used in a variety of other situations in electronics. We may speak, for example, of the useful range of a meter, the range over which the value of a component varies or fluctuates, or the range of power-supply voltage that a circuit will tolerate without malfunctioning. A site for measuring the characteristics of an antenna, such as the gain, front-to-back ratio, or front-to-side ratio, is called an antenna test range. *See also* TEST RANGE, TOLERANCE.

RANKINE TEMPERATURE SCALE

An absolute temperature scale, called the Rankine scale, is sometimes (but not often) used to express temperature relative to absolute zero. The coldest possible temperature on the scale is 0 degrees; above this, temperature is specified in increments that are the same size as the degrees on the Fahrenheit temperature scale.

Since absolute zero is −459.69 degrees Fahrenheit, the freezing point of pure water is 491.69 degrees Rankine. Water boils at 671.69 degrees Rankine.

Temperatures on the Rankine scale can be converted to degrees Kelvin by multiplying by 5/9. To convert degrees Rankine to degrees Celsius, first multiply by 5/9, and then substract 273. *See also* CELSIUS TEMPERATURE SCALE, FAHRENHEIT TEMPERATURE SCALE, KELVIN TEMPERATURE SCALE.

RASTER

In a television receiver tuned to a channel on which there is no signal, the rectangle of light, composed of horizontally scanned lines, is called the raster.

In the United States, the television-broadcast raster is standardized at 525 lines per frame. The aspect ratio, or ratio of width to height, is 4 to 3. In some other countries, the number of lines per frame is 625 rather than 525; however, the aspect ratio is almost universally 4 to 3.

In recent years, because of improved television broadcasting and receiving techniques, some thought has been given to increasing the number of lines per frame to get better resolution. *See also* TELEVISION.

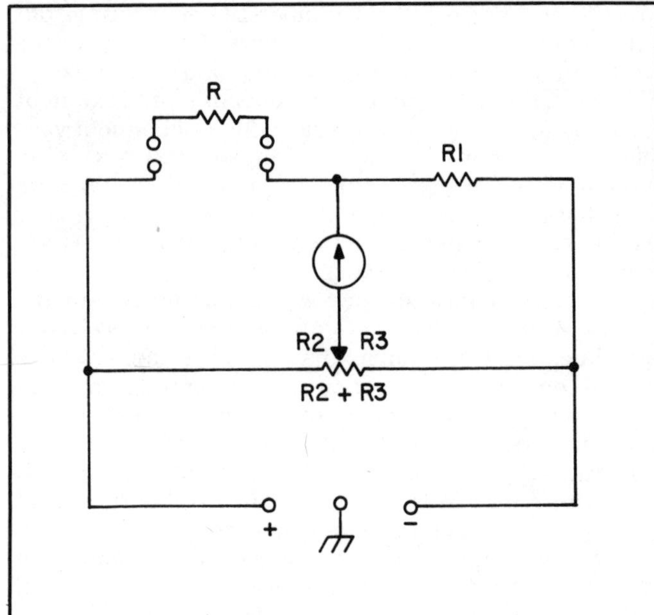

RATIO-ARM BRIDGE: A ratio-arm bridge used for measuring extreme values of resistance.

RATIO-ARM BRIDGE

A radio-arm bridge is a device for measuring resistance. The bridge gets its name from the fact that it compares the value of a standard resistance with that of the unknown resistance by electrically determining their ratio.

A schematic diagram of a simple ratio bridge is shown. A potentiometer is adjusted until a null reading is obtained on the galvanometer. If the standard resistance is designated by R1, and the partial values of the potentiometer are given by R2 and R3, then the unknown is:

$$R = R1R2/R3$$

since the ratio R:R1 is equal to the ratio R2:R3.

RATIO DETECTOR

The ratio detector is a special form of detector for frequency-modulated or phase-modulated signals. The ratio detector, unlike other types of detectors for these two modes, is insensitive to changes in amplitude. This eliminates the need for a preceding limiter stage. The ratio detector was developed by RCA. The ratio detector is used in most modern high-fidelity and television receivers. The ratio detector is not as commonly used in communications equipment.

A typical passive ratio detector is shown schematically in the illustration. A transformer, with tuned primary and secondary windings, splits the signal into two components. A change in amplitude will result in equal and opposite fluctuations in both halves of the circuit; thus they cancel out. But if the frequency changes, an instantaneous phase shift is introduced, unbalancing the circuit and producing output.

The ratio detector is not as sensitive as the discriminator, but additional amplification can be incorporated to make up for this slight deficiency, which amounts to approximately 3 dB. *See also* DISCRIMINATOR, FREQUENCY MODULATION, PHASE-LOCKED LOOP.

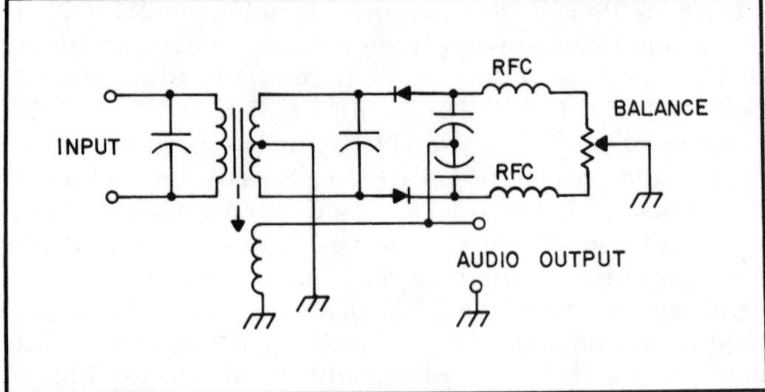

RATIO DETECTOR: A passive ratio detector for frequency or phase modulation. This circuit is insensitive to amplitude changes.

RATIONAL NUMBER

Any number that can be expressed as a ratio between two integers is called a rational number. If x is a rational number, then x can be written in the form:

$$x = m/n$$

where m and n are integers. The set R of all rational numbers is an infinite set. The size of R is the same as the size of the set of integers.

Examples of rational numbers include 4, −7, 3⅜, and −5.232323 . . . Note that the last of these is nonterminating in decimal form. This is called a repeating decimal. All nonterminating-decimal rational numbers are of the repeating type.

Certain fundamental observations can be made concerning the rational numbers. A sum, difference, product, or quotient of two rational numbers is always rational. However, some functions produce irrational or complex numbers when performed on rational numbers. Examples of such functions are square roots, logarithms, and trigonometric functions.

Some numbers are nonterminating in decimal form, and show no apparent pattern of repeating digits. These are the irrational numbers. The rational and irrational numbers, together, form the set of real numbers. *See also* IRRATIONAL NUMBER, REAL NUMBER.

RAWINSONDE

A radiosonde, tracked by radar to accurately determine its horizontal and vertical velocity components, is called a rawinsonde.

The rawinsonde is used to determine wind speed and direction at various levels in the atmosphere. This helps meteorologists locate the jet stream. The position of the jet stream greatly affects weather in various parts of the world. *See also* RADIOSONDE.

RAY

In mathematics, a ray, also known as a half line, is a set of points that propagates straight away from a single end point. In the Cartesian set of coordinates (*see* CARTESIAN COORDINATES), linear equations can be restricted to obtain rays.

In physics and electronics, a ray is a beam of radiant energy. The amount of energy contained in a ray, as well as the diameter of the ray, is not defined. One way to imagine a ray is to consider that it is the path that a single photon follows (*see* PHOTON). Rays always travel in straight lines through free space in the absence of electric, mangetic, electromagnetic, or gravitational disturbances.

A barrage of high-energy electrons, protons, neutrons, atomic nuclei, or other particles is called a ray. For example, the electron stream in a television picture tube may be called a ray. A beam of alpha particles may be called alpha rays.

High-energy electromagnetic waves are often called rays. For example, infrared, visible, ultraviolet, X rays, and gamma rays are generally thought of as rays, while radio signals (having a wavelength greater than 1 mm) are thought of as waves. The distinction is purely intuitive. *See also* ELECTROMAGNETIC FIELD, ELECTROMAGNETIC SPECTRUM.

RAYLEIGH DISTRIBUTION

A two-dimensional form of normal distribution (*see* NORMAL DISTRIBUTION) is known as a Rayleigh distribution. In the Rayleigh distribution, the probability density is determined according to the distance of a point (x,y) from the origin of the Cartesian plane (*see* CARTESIAN COORDINATES). In polar coordinates, the probability density is a function of the radius value r (*see* POLAR COORDINATES).

Let ϕ be the Rayleigh probability-density function, and let s be the standard deviation. Then the function, with the radius r as the independent variable, is:

$$\phi\ (r) = (r/s^2)e^{-(r^2/2s^2)}$$

with r resistricted to nonnegative values. In Cartesian coordinates, we may substitute the values x and y for r:

$$r = \sqrt{x^2 + y^2}$$

This is the standard Pythagorean distance formula (*see* PYTHAGOREAN THEOREM).

The Rayleigh distribution arises in conjunction with certain natural phenomena. It is useful for predicting the behavior of a scattered beam of particles, for example.

RC CIRCUIT

See RESISTANCE-CAPACITANCE CIRCUIT.

RC CONSTANT

See RESISTANCE-CAPACITANCE TIME CONSTANT.

RC COUPLING

See RESISTANCE-CAPACITANCE COUPLING.

REACTANCE

Reactance is the opposition, independent of resistance, that a circuit or component offers to alternating currents Reactance is exhibited by inductors, capacitors, antennas, and transmission lines. Waveguides and cavities also contain reactance. The unit of reactance is the ohm, and the symbol for reactance is X.

Reactances do not actually dissipate power. Instead, reactances affect the phase relationship between the current and the voltage in an alternating-current circuit. In a circuit having inductive reactance, the current lags behind the voltage by an amount ranging from 0 to 90 degrees. In a capacitive reactance, the current leads the voltage by an amount ranging from 0 to 90 degrees (*see* ANGLE OF LAG, ANGLE OF LEAD). The exact phase angle depends on the amount of resistance, as well as reactance, in a circuit or component (*see* PHASE ANGLE).

Mathematically, reactance may be either positive or negative. An inductive reactance is considered positive, and a capacitive reactance is considered negative. This is a matter of convention (*see* CAPACITIVE REACTANCE, INDUCTIVE REACTANCE). In engineering mathematics, reactance takes imaginary values. The ohmic quantities are multiplied by the j factor (*see* J OPERATOR). This results in a model that accurately reflects the way in which reactances behave.

When resistance and reactance occur together, the result is a complex impedance. The resistance forms the real part of a complex impedance, and the reactance forms the imaginary part. A complex impedance may be represented in the form:

$$Z = R + jX \text{ or } Z = R - jX$$

where R and X represent resistance and reactance and $\sqrt{-1}$. *See also* COMPLEX NUMBER, IMPEDANCE.

REACTANCE GROUNDING

When a circuit point is grounded through a capacitor or inductor, that point is said to be reactively grounded. Reactive grounding is used for various purposes in radio-frequency circuits.

A capacitor, connected between a circuit point and ground, allows some signals to pass while others are shorted to ground. Such a capacitor is called a bypass (*see* BYPASS CAPACITOR).

An inductor can be connected between a circuit point and chassis ground, when it is desired to keep the point at a stable dc voltage while allowing radio-frequency signals to exist there. Transistors and tubes, especially in power amplifiers, are often biased in this manner. An antenna may be kept at dc ground, while still accepting radio-frequency energy, by means of reactance grounding (*see* PROTECTIVE GROUNDING).

REACTANCE MODULATOR

A reactance modulator is a circuit that produces a variable reactance from an audio-frequency input signal. The variable reactance is coupled to the crystal or tuned circuit of an oscillator, resulting in frequency modulation or phase modulation.

A reactance modulator can be built using a bipolar transistor, a field-effect transistor or a vacuum tube. However, the most common reactance-modulator circuit today uses a device called a varactor diode. The varactor exhibits a variable capacitance, depending on the amount of reverse bias applied to it (*see* VARACTOR DIODE).

The illustration is a schematic diagram of a crystal oscillator incorporating a varactor diode to obtain frequency modulation. The capacitance of the diode pulls the crystal frequency by an amount that varies with the instantaneous audio-frequency amplitude. The deviation produced by

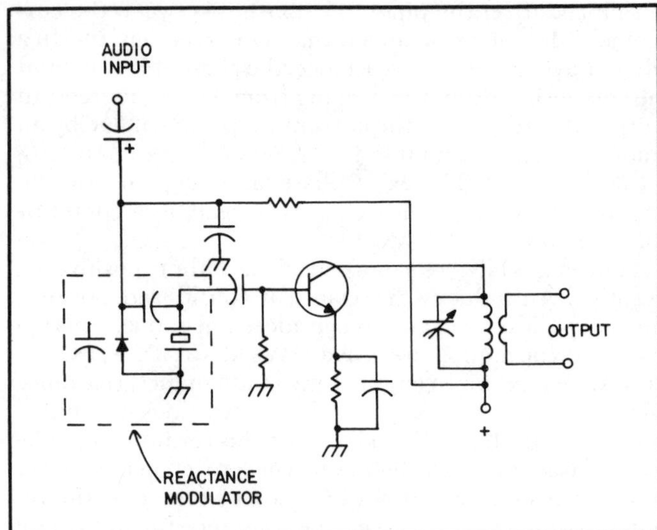

REACTANCE MODULATOR: A reactance-modulated oscillator for generating frequency-modulated signals.

this method is fairly small, but frequency multipliers can be used to increase it. *See also* FREQUENCY MODULATION, PHASE MODULATION.

REACTIVE CIRCUIT

An alternating-current circuit is called reactive if it contains inductance and/or capacitance, in addition to dc resistance.

In a reactive circuit, the resistance dissipates power, but the reactance does not. Examples of reactive circuits include antennas, transmission lines, waveguides, and tuned circuits. At radio frequencies, most circuits contain some reactance as well as resistance, and thus they must be considered reactive circuits.

A reactive circuit may become nonreactive at one or more frequencies. This occurs when there are equal amounts of inductive and capacitive reactance. Such a condition is known as resonance. *See also* CAPACITIVE REACTANCE, IMPEDANCE, INDUCTIVE REACTANCE, REACTIVE POWER, RESONANCE.

REACTIVE POWER

In an ac circuit, power can be dissipated in a resistance, but not in a reactance. Nevertheless, voltages and currents do occur in reactances. The product of the voltage and the current in a reactance is called reactive power. Reactive power is the difference between the apparent power and true power in a complex impedance (*see* APPARENT POWER, TRUE POWER). Reactive power is sometimes spoken of as reactive volt-amperes, or imaginary power.

The proportion of apparent power that is reactive depends on the phase angle in the circuit. The phase angle, in turn, depends on the ratio of reactance to resistance. The larger the phase angle, the more of the apparent power is reactive. In a pure resistance, all of the apparent power is true power. In a pure capacitance or inductance, none of the apparent power is true power (*see* POWER FACTOR).

Reactive power, if not recognized, can cause confusion. A radio-frequency wattmeter, installed in an antenna system containing reactance, will exhibit an exaggerated reading. On a directional wattmeter (reflectometer), the reactive power appears as reflected power and the apparent power appears as forward power. The true power—that which can actually be radiated by the antenna—is the difference between the forward and reflected readings. *See also* REFLECTOMETER.

READ-MAINLY MEMORY

A form of memory, similar in practice to the read-only memory, is known as a read-mainly memory (RMM). The contents of RMM are nonvolatile unless deliberately changed by a special process. The process is much too slow, however, to allow the RMM to be used as a read-write memory.

The read-mainly memory is essentially the same as the erasable programmable read-only memory (EPROM). *See also* PROGRAMMABLE READ-ONLY MEMORY, READ-ONLY MEMORY, READ-WRITE MEMORY.

READ-ONLY MEMORY

A read-only memory (ROM) is a common form of digital memory used in computers, calculators, and data-processing devices. An integrated circuit containing such memory is itself called a ROM.

The ROM is nonvolatile, retaining its data even when power is removed. Some types of ROM can be programmed as desired, but never changed after that; this is called programmable read-only memory (PROM). Some types of ROM can be erased and reprogrammed at will; this is known as erasable PROM (EPROM). *See also* PROGRAMMABLE READ-ONLY MEMORY.

READOUT

See DIAL SYSTEM.

READ-WRITE MEMORY

Any form of memory in which it is possible to both store and retrieve information is called a read-write memory. The contents of a read-write memory can be changed quickly. This makes read-write memory different from the read-mainly or read-only memory (*see* READ-MAINLY MEMORY, READ-ONLY MEMORY).

Read-write memory may be either nonvolatile or volatile; that is, it may or may not be retained if power is removed. In a calculator or computer, a read-write memory used for temporary storage of numbers in the intermediate stages of a computation is called scratch-pad memory (*see* SCRATCH-PAD MEMORY).

There are various forms of read-write memory, used for different purposes. The first-in/first-out (FIFO) memory allows removal of data in the same order it is written. The pushdown stack, or last-in/first-out memory, facilitates removal of data in opposite order from the way it is written. A random-access memory allows retrieval of information in any order. *See also* FIRST-IN/FIRST-OUT, PUSHDOWN STACK, RANDOM-ACCESS MEMORY.

REAL NUMBER

A real number is any number that is either rational or

irrational. This includes integers, such as 3 or −7; it includes numbers that can be written as a fraction, such as 4/5; it includes nonterminating, nonrepeating decimals, such as the natural-logarithm base e or the value of $\sqrt{2}$. Imaginary and complex numbers, however, are not real numbers (*see* COMPLEX NUMBER).

There are an infinite number of real numbers. Basically, a real number is an "ordinary" number, the size of which we can readily imagine. Numbers can be represented geometrically for example, by an infinitely long line known as the number line (*see* NUMBER LINE). On such a line, once the unit size has been decided, every point corresponds to exactly one real number. Furthermore, every real number can be paired up with exactly one point on the line. Mathematicians say that the set of real numbers, and the set of points on a line, have a one-to-one correspondence or homomorphism.

The set of real numbers is infinitely greater in size than the set of rational numbers. This may seem paradoxical, since there are infinitely many rational numbers! Nevertheless, set theoreticians are able to prove that this is in fact the case. *See also* NUMBER SYSTEMS, RATIONAL NUMBER.

REAL POWER

See TRUE POWER.

REAL-TIME OPERATION

In any broadcast, communications, or data-processing system, operation done "live" is called real-time operation. The term real-time applies especially to computers. Real-time data exchange allows a computer and the operator to converse. This is not true of pre-recorded modes.

Real-time operation is a great convenience when it is necessary to store and verify data within a short time. This is the case, for example, when making airline reservations, checking a credit card, or making a bank transaction. But real-time operation is not always necessary. It is a waste of expensive computer time to write a long program at an active terminal. Long programs are best written off-line, tested in real time, and then debugged off-line.

In a large computer system, real-time operation can be obtained from many terminals simultaneously. The most common method of achieving this is called time sharing. The computer pays attention to each terminal for a small increment of time, constantly rotating among the terminals at a high rate of speed. In a powerful system, this provides great convenience for a large number of operators. *See also* TIME SHARING.

REBER, GROTE

During the 1930s, Grote Reber, a radio amateur, constructed a radio telescope in his back yard. This device consisted of a sensitive receiver and a dish 31 feet across. Reber's radio telescope was the first such device used to map the Milky Way galaxy at radio wavelenghts.

Reber's radio telescope was crude compared to the huge modern devices. Reber's work interested astronomers so much that they decided to pursue the investigation further. This gave birth to the science of radio astronomy.*See also* RADIO ASTRONOMY, RADIO TELESCOPE.

RECEIVER

Any circuit that intercepts a signal, processes the signal, and converts it to a form useful to a person, is a receiver. The signal may be in any form, such as electric currents in a wire, radio waves, modulated light, or ultrasound. A receiver converts signals into audio information, video information, or both.

All closed-circuit, or wire-operated, receivers are made up of at least three basic components: a signal filter, an amplifier, and a transducer (see A in illustration). The signal filter eliminates all incoming signals except the desired one. The amplifier boosts the strength of the signal so that it can be detected when applied to the transducer. The transducer converts the signal into a perceivable form, such as sound, a still picture, or a moving picture. The telephone is a common example of a closed-circuit receiver. Some closed-circuit systems use radio-frequency energy; receivers in such systems are described below.

All radio-frequency, visible-light, or other non-wire receivers require five or six fundamental components: an antenna or receptor (except in a closed-circuit wire system), a front end/wideband filter, an amplification chain/narrowband filter, a detector, an audio and/or video amplifier, and a transducer (B). Electromagnetic waves strike the antenna or other receptor, setting up alternating currents in the conductors. (In a closed-circuit system, signals arrive in the form of currents, so no antenna or receptor is needed.) The front end/wideband filter provides amplification and some selectivity. The amplification chain brings the weak signal up to a level suitable for operating the detector, which extracts the modulation information from the radio-frequency energy. The audio or video amplifier gives the detected signal sufficient amplitude to drive the transducer. The transducer converts the detected signal to a form suitable for listening, viewing, driving a set of recording instruments, or some combination of these things. In a sophisticated receiver, there may be additional stages such as converters and mixers.

The most common type of radio-frequency receiver in use today is the superheterodyne circuit. Direct-conversion receivers are occasionally used. *See also* DIRECT-CONVERSION RECEIVER, SUPERHETERODYNE RECEIVER.

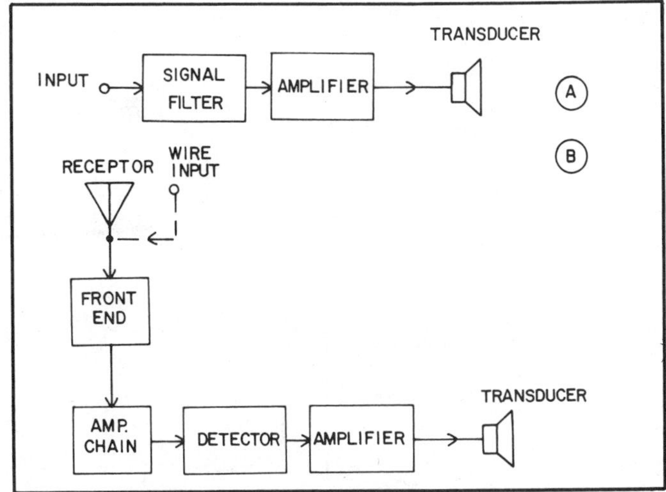

RECEIVER: At A, block diagram of a wire-circuit receiver. At B, block diagram of a radio-frequency receiver. These are basic circuits; many receivers have additional stages.

RECEIVER INCREMENTAL TUNING

In a radio transceiver, it is often desirable to set the receiver to a slightly different frequency from that of the transmitter. This is especially true in code (CW) operation, but it may also be necessary at times in other modes. A receiver-incremental-tuning (RIT) control accomplishes this. The RIT moves the receiver frequency without changing the transmitter frequency. Receiver incremental tuning is also known as receiver offset tuning.

The most common method of obtaining RIT is to change the frequency of the variable-frequency oscillator (VFO). Since most transceivers use a common VFO for both the transmitter and the receiver, it is necessary to disconnect the RIT while transmitting. This is done either by relays or by electronic switching methods.

Generally, an RIT control allows a receiver to be tuned approximately 5 kHz above or below the transmitter frequency. If a larger split is needed, separate VFO units can be used, one for the receiver and one for the transmitter. *See also* TRANSCEIVER.

RECEIVER SELECTIVITY

See SELECTIVITY.

RECEIVER SENSITIVITY

See SENSITIVITY.

RECEIVING ANTENNA

Any transmitting antenna will function well for receiving purposes within the same range of frequencies. However, receiving antennas need not be as sophisticated or large as transmitting antennas in order to function well, especially at very low, low, medium, and high frequencies. Because of the nature of electromagnetic waves, tiny antennas can function very well for reception at these frequencies, even though they are virtually useless for transmission. At very-high, ultra-high and microwave frequencies, large dish antennas may be necessary.

The simplest type of receiving antenna is a random wire, connected to the receiver input terminals either directly or through a tuning network. The antenna may consist of a simple end-fed wire, or a horizontal conductor can be fed by a lead-in wire (see illustration). Such an antenna is usually made as long as possible. It will work well at all radio frequencies into the very-high or even ultra-high range. Such an antenna can provide marginal performance for transmitting.

Tuned receiving antennas provide a high degree of selectivity and some noise suppression. The most common tuned receiving antennas are the ferrite rod and the loop (*see* FERRITE-ROD ANTENNA, LOOP ANTENNA). These antennas can be used at frequencies from about 30 MHz down to 10 kHz with excellent results. Despite their physically tiny size at the longer wavelengths, the ferrite rod and the loop perform very well for receiving at the very low, low, and medium frequencies. Marginal performance can be had up to about 30 MHz. Above that frequency, ferrite becomes rather lossy, and open loop antennas are not significantly smaller than a half-wave dipole or full-wave loop.

At very high frequencies and above, the same antenna is generally used for receiving as for transmitting, since physical size is not a major problem. When the same antenna is used for receiving and transmitting, the directional characteristics work to advantage in both modes.

Receiving antennas will often work fairly well if located indoors. However, it is always best, if possible, to put them outdoors, especially if the receiver is inside a concrete-and-steel structure. *See also* INDOOR ANTENNA, OUTDOOR ANTENNA.

RECEIVING TUBE

A small vacuum tube, capable of dissipating only a few watts of power and intended for audio-frequency or radio-frequency use, is called a receiving tube. Receiving tubes have largely been replaced by solid-state devices in recent years. However, some of the older receivers still use tubes.

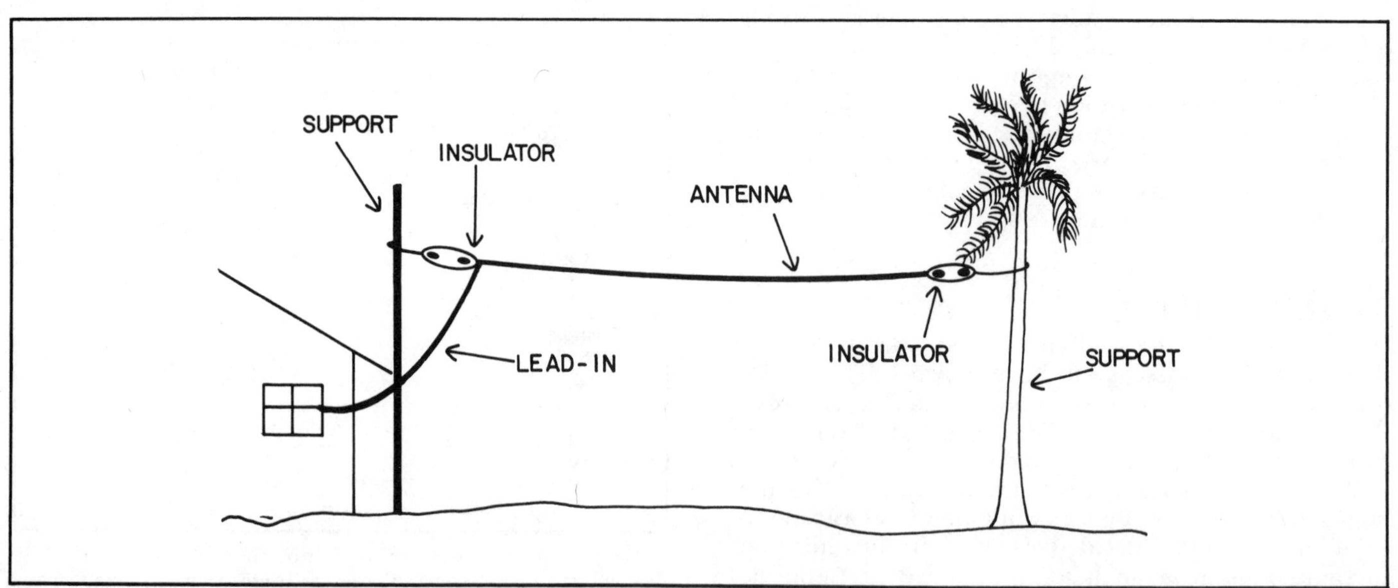

RECEIVING ANTENNA: A random-length wire can be used for receiving at practically any radio frequency.

Most receiving tubes are pentodes. Many are dual tubes; that is, two tubes in a single envelope. The plate voltage requirements range from a few volts up to 400 volts, with an average of perhaps 150 volts. Filament current requirements are modest. Receiving tubes may be found in older tape recorders, phonographs, and the low-power stages of public-address systems and transmitters. *See also* TUBE.

RECHARGEABLE CELL

Some electrochemical cells can accept a charge and deliver the current at a later time. Such cells are called rechargeable, since they can be used over and over. Rechargeable cells offer an obvious advantage over one-time cells. Although rechargeable cells are more expensive than the one-time variety, and a charging unit is needed, the additional expense is more than offset by the reusability.

There are many different kinds of rechargeable cells, but the most common are the lead-acid type and the nickel-cadmium type. Lead-acid cells are used in automobile batteries. Nickel-cadmium cells are employed in a wide variety of electronic devices, such as radio receivers and low-power transmitters, calculators, and memory-backup systems. The lead-acid cell is a high-current cell; the nickel-cadmium cell is designed for low-current use. *See also* CHARGING, LEAD-ACID BATTERY, NICKEL-CADMIUM BATTERY, QUICK CHARGE, TRICKLE CHARGE.

RECIPROCAL

In mathematics, the reciprocal of a number x is that number y such that $xy = 1$. Thus, $y = 1/x$. A unique reciprocal exists for every integer, rational number, or real number except 0. Every complex number also has a unique reciprocal.

We may make certain observations concerning reciprocals:

- The reciprocal of an integer is always a rational number between, and including, -1 and 1.
- The reciprocal of a rational number is always another rational number.
- The reciprocal of an irrational number is always irrational.
- The reciprocal of a real number is always real; if the number is between -1 and 1, the reciprocal is outside that range, and vice-versa.
- The reciprocal of a pure imaginary number is always another pure imaginary number.
- The reciprocal of a nonreal complex number is always nonreal and complex.
- We must be extremely cautious in algebraic calculations, to avoid expressing the reciprocal of a variable that might attain the value 0.

See COMPLEX NUMBER, RATIONAL NUMBER, REAL NUMBER.

In electronics, the term reciprocal is sometimes used with reference to bilateralism or interchangeability. *See, for example,* RECIPROCITY THEOREM.

RECIPROCITY THEOREM

The Reciprocity Theorem is a rule that describes the behavior of currents and voltages in certain multi-branch circuits.

Suppose we have a passive, linear, bilateral circuit with two or more separate, current-carrying branches, such as that shown in the illustration. Suppose a power supply, delivering Ex volts, is placed across branch X of this circuit, resulting in a current Ix in branch X, and a current Iy in branch Y. We assume that the power supply has zero internal resistance. Then the voltage Ey across branch Y is found from the current Iy and the resistance Ry according to Ohm's Law:

$$Ey = IyRy$$

Now imagine that we disconnect the power supply having voltage Ex from branch X of the circuit shown, and connect a different supply, with voltage Ey and zero internal resistance, across branch Y. Suppose the same current as before, Iy, flows through resistance Ry.

According to the Reciprocity Theorem, under these circumstances, the current through the resistance Rx will be the same as before, or Ix. Therefore, the voltage across branch X will be:

$$Ex = IxRx$$

which is precisely the same voltage Ex of the original supply.

The Reciprocity Theorem is a direct result of Kirchhoff's laws and Ohm's law. This theorem can be used to advantage in determining currents and voltages in complicated networks. *See also* KIRCHHOFF'S LAWS, OHM'S LAW.

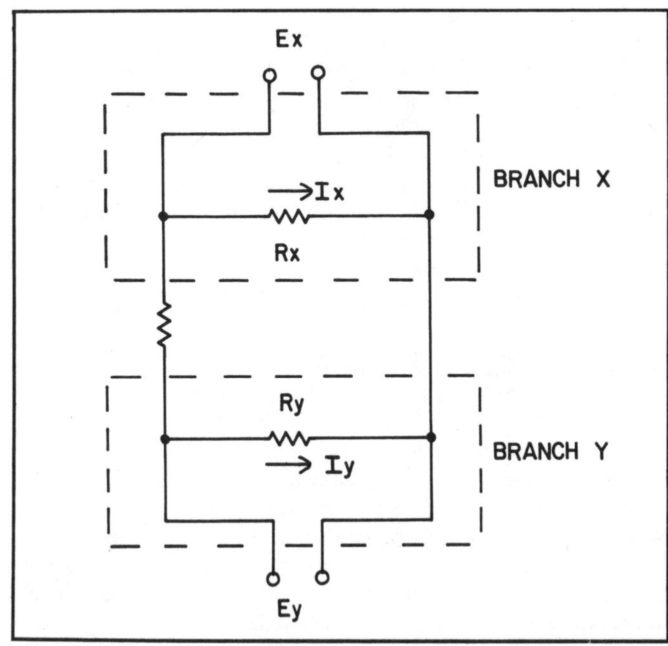

RECIPROCITY THEOREM: Principle of the Reciprocity Theorem. See text for discussion.

RECOMBINATION

When a charge is applied to a semiconductor material, excess holes and/or electrons are present. When the charge is removed, equilibrium is restored, assuming there is no other disturbance present that might produce excess holes and/or electrons. The process of restoring charge-carrier equilibrium is called recombination. The excess holes are filled in by electrons.

Recombination takes a certain amount of time. It occurs exponentially, as do most decay processes. Recombination time is also called the lifetime. For high-speed switching and in ultra-high-frequency applications, it is necessary to have a very short recombination time in a semiconductor device. In other applications, this may not be especially important.

Recombination time varies for different semiconductor materials; it can range from a few microseconds to less than 1 nanosecond, depending on the impurity substances that have been added to the semiconductor. *See also* CARRIER MOBILITY, DOPING, ELECTRON, HOLE, N-TYPE SEMICONDUCTOR, P-N JUNCTION, P-TYPE SEMICONDUCTOR.

RECORDER

Please see listings of devices according to their application or type.

RECORDING DISK

A recording disk is a medium for storing and retrieving information. The data is impressed on the disk in a series of "rings or tracks" that are circular and concentric. This configuration allows a large amount of data to be written in such a manner that no two bits are separated by a distance greater than the diameter of the disk. This makes the disk ideal for high-speed systems.

There are two basic types of recording disks in common use today. The magnetic disk is used in many computers and word-processing systems, and in some video recorders. The plastic disk, in which grooves are etched and where the path is spiral, is used mostly for high-fidelity recording, although some office dictating machines use disks of this type. Data can be erased and rewritten on a magnetic disk. On the plastic disk, data cannot be rewritten.

Some information storage and retrieval systems use tapes instead of disks. However, the disk is replacing the tape in many applications. *See also* DISKETTE, DISK RECORDING, FLOPPY DISK, MAGNETIC RECORDING, MAGNETIC TAPE, RECORDING TAPE.

RECORDING HEAD

A recording head is an electromagnetic transducer that converts alternating electric currents to fluctuating magnetic fields and vice-versa. A recording head is used for the purpose of writing data onto, or retrieving information from, magnetic tapes or disks. Every tape recorder or disk recorder contains one or more such transducers. Usually, there are two: one for recording and one for playback. The illustration shows the typical recording-head configuration in a tape recorder.

The recording head is placed adjacent to a tape or disk. The tape or disk is drawn past the head at a uniform, precise speed. If the head is to be used for recording, an electric current is supplied to it. Fluctuating magnetic fields, which occur in synchronization with the alternating currents, cause polarization of the fine magnetic particles in the tape or disk. If the head is to be used for playback or information retrieval, the polarized particles cause electric currents to flow in the recording head. *See also* MAGNETIC RECORDING.

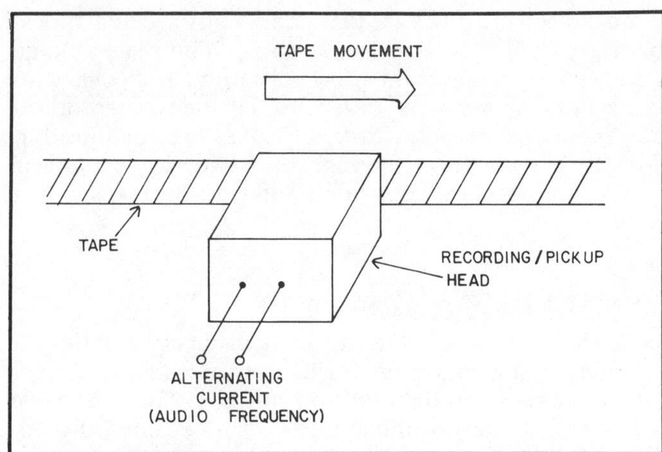

RECORDING HEAD: A recording head produces a fluctuating magnetic field in the record mode, transferring it to the tape. In the playback mode, the head converts the magnetic impulses from the tape to electric currents.

RECORDING TAPE

Recording tape is a linear medium on which information can be stored. The most common and familiar form of recording tape is the magnetic variety. Magnetic tape is used by almost everyone for some purpose (*see* MAGNETIC TAPE).

There are other, less familiar forms of recording tape. In some computer terminals and teletype systems, paper tape is used to record printed data. A machine punches holes in the tape in rows perpendicular to the edges of the tape. Each row of holes represents one character (*see* PAPER TAPE).

Some pen recorders use a long, thin strip of paper, marked in rectangular coordinates, for storing the information. This is a form of recording tape. In some older Morse-code machines optical paper tape is used. This tape is marked with a long, black, crooked line. During silent periods, the black line is on one side of the strip, and it blocks the light beam in an optical coupler. During key-down periods it is on the other side, allowing the beam to pass through the paper or reflect from it, actuating a tone generator.

In recent years disks have begun replacing tapes for recording purposes. This is because disks allow faster access to information. On a long roll of recording tape, two files of data may be separated by thousands of feet. To access two such data files in succession, it is necessary to move the entire tape through the recorder. But on a disk, the maximum possible separation is the diameter of the disk—never more than a few inches. *See also* RECORDING DISK, TAPE RECORDER.

RECTANGULAR COORDINATES

See CARTESIAN COORDINATES.

RECTANGULAR RESPONSE

A bandpass or band-rejection filter is usually designed to have what is called a rectangular response. In theory, the rectangular bandpass response is characterized by zero attenuation between the cutoff frequencies, and infinite attenuation outside the bandpass; a rectangular band-rejection response is characterized by infinite attenuation between the cutoff points and zero attenuation at all other frequencies. These theoretical responses are shown in the illustration.

In a radio receiver, a rectangular bandpass response provides the best possible signal-to-noise ratio. Modulated signals have well-defined bandwidth. The optimum receiver response is therefore rectangular, having a bandwidth equal to that of the incoming signal.

The ideal rectangular response cannot be realized in practice, but engineers have devised filters that have sharp cutoff, high attenuation in the rejection range, and uniform, low attenuation within the pass range. The response of such a filter is said to be rectangular, even though there is some deviation from the ideal.

A single resonant circuit provides a peaked response, while a group of tuned circuits or resonant devices, all having a slightly different natural frequency, can be connected together to produce a rectangular response. Piezoelectric ceramics, and crystals are widely used at radio frequencies to obtain a rectangular response. Mechanical filters are also quite common. *See also* BANDPASS FILTER, BANDPASS RESPONSE, BAND-REJECTION FILTER, BAND-REJECTION RESPONSE, CERAMIC FILTER, CRYSTAL-LATTICE FILTER, MECHANICAL FILTER.

RECTANGULAR WAVEGUIDE

See WAVEGUIDE.

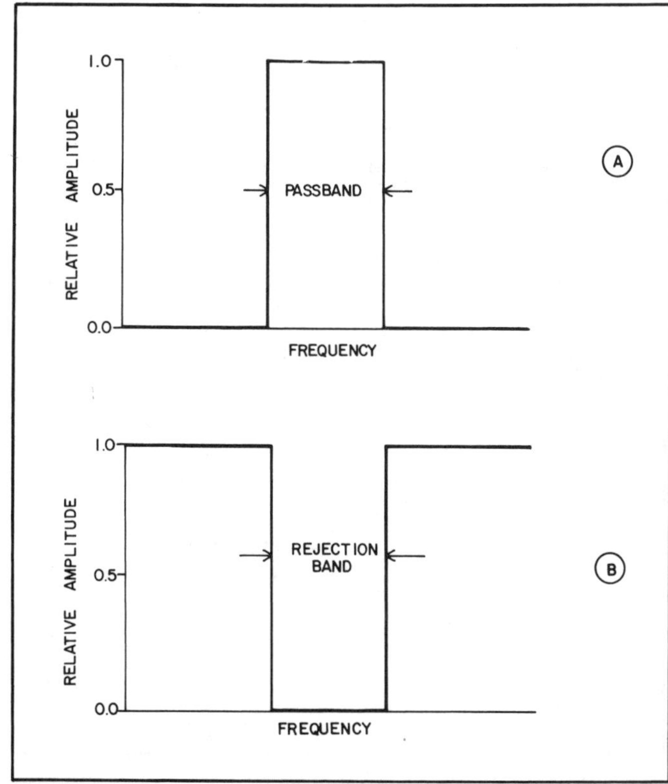

RECTANGULAR RESPONSE: At A, the ideal rectangular bandpass response. At B, the ideal rectangular band-rejection response.

RECTIFICATION

The conversion of alternating current to direct current is called rectification. Rectification can be accomplished in a variety of ways, but the most common methods are the use of diodes or commutators. Diodes are generally used for electronic-circuit rectification; commutators are used in motors and generators when direct current is needed (*see* COMMUTATOR, DIODE, DIODE ACTION).

The diode conducts current in only one direction. Therefore, when a diode is placed in a circuit where alternating current flows, the result is pulsating direct current. Rectification is used in all direct-current power supplies designed for operation with ordinary household current. *See also* RECTIFIER, RECTIFIER CIRCUITS, RECTIFIER DIODE, RECTIFIER TUBE.

At radio frequencies, a process similar to rectification is used to demodulate certain types of signals. This process is called envelope detection. *See* ENVELOPE DETECTOR.

RECTIFIER

A rectifier is a device that allows current to flow in only one direction. Rectifiers may be mechanical or electronic. In some motors and generators, a mechanical device called a commutator is used as a rectifier (*see* COMMUTATOR). In electronic circuits, semiconductor devices or vacuum tubes are used (*see* RECTIFIER DIODE, RECTIFIER TUBE).

A complete circuit that converts alternating current to pulsating direct current is called a rectifier circuit. There are several different types of rectifier circuits. Electronic rectifier circuits generally consist of a transformer and one or more diodes. *See also* RECTIFIER CIRCUITS.

RECTIFIER CIRCUITS

Rectifier circuits are used for the purpose of converting alternating current to pulsating direct current. Rectifier circuits are found in all direct-current power supplies designed to operate from the utility mains. Rectifier circuits are also used in some radio-frequency-actuated devices.

The simplest type of rectifier circuit is the half-wave circuit. A single diode is placed in series with the alternating-current line. The diode allows the current to flow in only one direction (*see* HALF-WAVE RECTIFIER). A half-wave circuit results in a 60-Hz pulsation frequency with 60-Hz alternating-current input. Half-wave circuits are used in applications where some ripple can be tolerated, and in which voltage regulation need not be especially good.

The bridge and full-wave rectifier circuits are probably the most popular (*see* BRIDGE RECTIFIER, FULL-WAVE RECTIFIER). Both circuits require transformers in order to function properly. With 60-Hz alternating current, the output of a full-wave rectifier has a pulsation frequency of 120 Hz. The bridge and full-wave circuits are the most common in modern electronic equipment. The output is fairly easy to filter, and good regulation is easily obtained.

In three-phase systems, various different configurations are used to accomplish rectification. Special transformers are required for three-phase rectifier circuits. The three-phase system offers the advantage of minimal ripple and excellent regulation (*see* POLYPHASE RECTIFIER).

The output of a rectifier circuit must normally be filtered before it is suitable for use by electronic devices. Filtering can be accomplished by means of chokes in series and capacitors in parallel with the rectifier output. *See also* FILTER, FILTER CAPACITOR, POWER SUPPLY, RECTIFIER DIODE, RECTIFIER TUBE.

RECTIFIER DIODE

A rectifier diode is a semiconductor diode designed for use in rectifier circuits. Rectifier diodes are generally fabricated from silicon. They are characterized by a fairly large P-N-junction surface area, resulting in high capacitance under reverse-bias conditions. Rectifier diodes have replaced tubes in most modern power-supply circuits.

Silicon rectifier diodes are available with various peak-inverse-voltage (PIV) ratings, ranging from a few volts to several thousand volts. In high-voltage supplies, two or more rectifier diodes can be connected in series to increase the PIV rating of the combination. (A large-value equalizing resistor is connected in shunt with each diode to distribute the voltage among them. An 0.01-μF capacitor should also be connected in parallel with each diode to protect against transients.) Some rectifier diodes can carry several amperes of current in the forward direction, and a few are available for use at 300 to 500 amperes. High-current rectifier diodes require heat sinks. Two or more diodes can be connected in parallel to increase the forward-current-handling capacity. (A small-value resistor is connected in series with each diode to prevent current hogging.)

The primary advantage of rectifier diodes over rectifier tubes is compactness. Rectifier diodes also require no warm-up and no filament power. *See also* DIODE PEAK INVERSE VOLTAGE, P-N JUNCTION, POWER SUPPLY, RECTIFIER CIRCUITS, RECTIFIER TUBE.

RECTIFIER TUBE

A rectifier tube is a two-electrode vacuum tube intended for use in high-voltage, direct-current power supplies. Rectifier tubes are not generally used in modern circuits, since semiconductor diodes can be used up to several thousand volts or hundreds of amperes. However, rectifier tubes may sometimes still be seen in high-voltage power supplies, especially for high-power transmitters.

There are two basic kinds of rectifier tubes. The high-vacuum rectifier consists of a cathode, a plate, and a filament. Electrons are driven from the cathode by the heat of the filament. Electron flow occurs easily from the cathode to the plate, but not from the plate to the cathode. The high-vacuum rectifier can be thought of as an ordinary tube with no grids (*see* TUBE). The mercury-vapor rectifier has an interior that is not a vacuum, but instead contains gaseous mercury. Because of the characteristics of mercury vapor, this type of rectifier tube conducts well in only one direction. The mercury-vapor rectifier also has a constant forward voltage drop.

Tubes are more "forgiving" than semiconductor devices in the event of a temporary current or voltage overload. This is the main advantage of tubes; a semiconductor diode can be instantly destroyed by spikes or surges. Tubes are, however, more physically bulky than semiconductor diodes, and require a source of power for the filament. *See also* DIODE, POWER SUPPLY, RECTIFIER CIRCUITS.

REDUNDANCY

Redundancy refers to the use of extra devices or methods for the purpose of guarding against possible malfunction of a system.

An example of component redundancy is the use of two diodes in parallel in a power supply, when only one is actually needed for circuit operation. If one diode burns out, the other diode can take over. Component redundancy is used in computer systems so that operation can continue even if part of the system fails (*see* GRACEFUL DEGRADATION).

In data storage, several copies of the information are usually made to guard against damage or loss. For example, two or three backup copies may be made for each disk in a word-processor file. This is called redundancy.

In mathematics, the term has a different meaning. Redundancy in an equation refers to the use of unnecessary parameters or variables. In a computer program, redundancy results in less-than-optimum efficiency, and therefore unnecessary consumption of computer time.

In data transmission, redundancy can be used as a means of checking the accuracy of the information at the receiving end. This is also known as parity. *See also* PARITY.

REED RELAY

A reed relay is a high-speed relay used for such purposes as code-transmitter keying and rapid switching. Reed relays are sometimes used in logic circuits and low-frequency oscillators.

A typical single-pole, double-throw reed relay is shown in the illustration. The relay assembly is enclosed in a glass envelope, which may be evacuated or filled with an inert gas to prevent corrosion of the contacts. The glass envelope is surrounded by a coil of wire, to which a voltage is applied to actuate the relay. If multiple-pole operation is desired, several reed envelopes can be surrounded by a single coil.

The metal reed is pulled from one contact to the other by the magnetic field created by the current in the coil. The small gap and the flexibility of the reed result in the short switching time. Some reed relays can be actuated at a rate of more than 1 kHz (1000 openings and closings per second). Reed relays have excellent longevity and reliability.

Reed relays may be either dry or mercury-wetted. Mercury wetting eliminates contact bounce that sometimes occurs in dry-reed relays. *See also* DRY-REED SWITCH, RELAY.

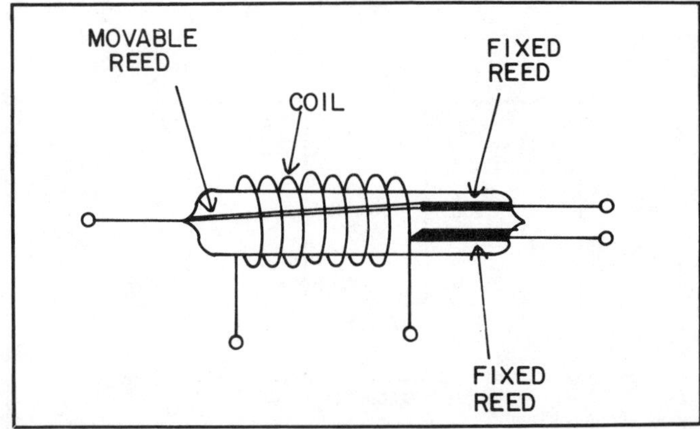

REED RELAY: Pictorial diagram showing the operation of a reed relay. The magnetic field, produced by the coil, causes the movable reed to change position.

REED SWITCH

See DRY-REED SWITCH.

REEL-TO-REEL TAPE RECORDER

In recent years, cassettes have become increasingly popular for use with tape recorders (*see* CASSETTE). However, the first tape recorders all used a feed configuration known as reel-to-reel. While reel-to-reel tape recorders are somewhat less common for personal use today than they were a few years ago, the reel-to-reel arrangement is still used by most professionals because it allows extremely long recording and playing time.

The reel-to-reel tape-recorder feed system resembles that of a movie projector (see illustration). The tape is wound on a plastic reel, usually measuring either 5 inches or 7 inches in diameter. This reel is called the supply reel. The supply reel is inserted at the left, and rotates counterclockwise as side 1 is recorded or played. A second reel, called the takeup reel, is located at the right, and also rotates counterclockwise as it receives the tape. When the takeup reel is full, it is exchanged with the original supply reel for recording or playing on side 2.

A 7-inch tape reel can accommodate about 3000 feet of 0.5-mil magnetic tape. Typical reel-to-reel tape recorders operate at either 3.75 inches per second (IPS) or 7.5 IPS. Professional machines run at 15 or 30 IPS. A few reel-to-reel machines have capability to run at 1.875 IPS. This allows very long continuous playing time (per tape side): 1 hour and 20 minutes at 7.5 IPS, 2 hours and 40 minutes at 3.75 IPS, or 5 hours and 20 minutes at 1.875 IPS. *See also* TAPE RECORDER.

REFERENCE ANTENNA

For antenna power gain to have meaning, a reference antenna is needed. Antenna power gain is generally measured in decibels relative to a reference dipole. Sometimes the reference antenna is an isotropic radiator (*see* ANTENNA POWER GAIN, dBd, dBi, DIPOLE ANTENNA, ISOTROPIC ANTENNA).

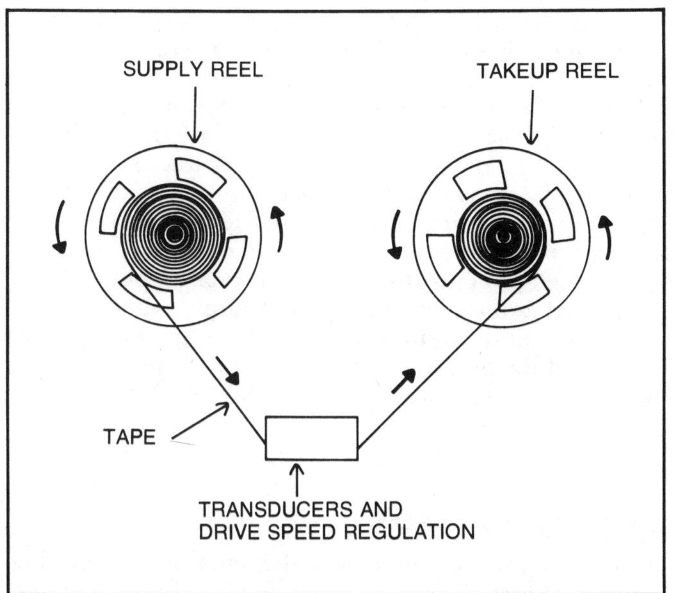

REEL-TO-REEL TAPE RECORDER: A reel-to-reel tape recorder operates in a manner similar to a movie projector.

For the measurement of transmitted-signal gain, the reference antenna and the test antenna are set up side by side. An accurately calibrated field-strength meter is positioned at a distance of at least several wavelengths from both antennas. The reference and test antennas are oriented so that the field-strength meter is located in the favored direction (major lobe) of both antennas. A signal of P watts is applied to the reference antenna, and the field-strength meter is set to show a relative indication of 0 dB. Then the same signal of P watts is applied to the test antenna, and the gain is read from the field-strength meter. *See* TEST RANGE.

To measure received-signal gain, a similar arrangement is used. The reference and test antennas are placed side by side with their major lobes oriented toward a signal-generator/antenna several wavelengths distant. A field-strength meter or receiver, equipped with an accurate S meter, is connected to the reference antenna, and the receiver gain is adjusted until a relative indication of 0 dB is obtained. Then the receiver is connected to the test antenna. A perfect impedance match must be maintained for both antennas. The gain of the test antenna can then be read from the receiver S meter. *See* S METER.

REFERENCE FREQUENCY

In the operation and calibration of radio receivers, a reference frequency is sometimes used. A standard signal generator, such as a crystal oscillator or broadcast station whose frequency is precisely known, can be used for this purpose.

Let us consider an example of the use of a reference frequency in the operation of a shortwave receiver. Suppose we are listening to signals in the band 7.0 MHz to 7.5 MHz. The accuracy of the dial calibration may not be known. However, we do know that the time-and-frequency station, CHU, is operating at 7.335 MHz. We tune the receiver back and forth in the vicinity of this frequency as shown on the dial, and finally we hear the intermittent tones of CHU. The receiver dial shows 7.328 MHz; the indication is 7 kHz (0.007 MHz) too low. We adjust the dial-calibration control until the dial shows 7.335 MHz. We may then be confident that the calibration is reasonably accurate over the range 7.0 MHz to 7.5 MHz.

A reference-frequency source is considered primary if it is received directly from a standard-frequency broadcast station such as CHU or WWV/WWVH. If some other source, such as a commercial broadcast station or crystal calibrator is used, the reference frequency is considered secondary. *See also* CHU, FREQUENCY CALIBRATOR, FREQUENCY MEASUREMENT, WWV/WWVH.

REFERENCE LEVEL

When signal strength is measured in relative terms, a reference level must be specified. A reference level may be established for current, field strength, flux density, power, voltage, or any other quantity that varies. The reference level is assigned a gain factor of 1, corresponding to 0 dB.

Certain standard reference levels are used. A common reference level for current is the milliampere; for power, the milliwatt; for voltage, the millivolt. Other standard values, however, may be specified. If the parameter exceeds the reference level, the gain factor is greater than 1 and the gain figure in decibels is positive. If the parameter is smaller than the reference level, the gain factor is less than 1 and the gain figure in decibels is negative. *See also* DECIBEL, GAIN.

REFLECTANCE

When an electromagnetic field strikes a surface or boundary, some of it may be reflected. The ratio of the reflected energy or field intensity to the incident intensity is known as reflectance. If the incident field intensity (known as incident or forward power) is P1 and the reflected field intensity (called the reflected power) is P2, then the reflectance factor R is:

$$R = P2/P1$$

In percent, reflectance is given by:

$$R \text{ (percent)} = 100P2/P1$$

As an example, consider a mirror. When visible light strikes a mirror, almost all of the energy is reflected. Therefore, the reflectance factor is almost 1, or 100 percent. A pane of glass reflects some light, but most of it is transmitted. Thus, the reflectance of glass is just a few percent. The reflectance of a surface may vary depending on the angle at which the light strikes.

In a power-transmission line and load system, some of the electromagnetic field, traveling from the generator to the load, may be reflected upon reaching the load. This results in less than optimum current in the load. The ratio of actual current to optimum current is called reflectance or, more commonly, the reflection factor. *See also* REFLECTED POWER, REFLECTION COEFFICIENT, REFLECTION FACTOR.

REFLECTED IMPEDANCE

In an impedance transformer having a certain load connected to the secondary winding, the impedance across the primary is called reflected impedance (*see* IMPEDANCE TRANSFORMER). If the secondary, or load, impedance does not contain reactance, then the reflected impedance will contain no reactance. This is generally the situation with impedance transformers. However, if the load does contain reactance, then the reflected impedance will also contain reactance (*see* CAPACITIVE REACTANCE, IMPEDANCE, INDUCTIVE REACTANCE).

In a transmission line operating at radio frequencies, the reflected impedance is the impedance at the transmitter end of the line. In the ideal case, where the antenna exhibits a pure resistance equal to the characteristic impedance of the line, the reflected impedance is the same as the load impedance for any line length. If the load contains reactance, or if the resistance of the load differs from the characteristic impedance of the line, the reflected impedance will vary depending on the line length.

Certain transmission lines have special properties. Neglecting line loss, a transmission line measuring a multiple of a half wavelength will always result in a reflected impedance identical to the load impedance; a line measuring an odd multiple of a quarter wavelength will invert the resistance and reactance. *See also* CHARACTERISTIC IMPEDANCE, HALF-WAVE TRANSMISSION LINE, MATCHED LINE, MISMATCHED LINE, QUARTER-WAVE TRANSMISSION LINE.

REFLECTED POWER

Ideally, the impedance of the load in a power-transmission system should be a pure resistance, and should have the same ohmic value as the characteristic impedance of the line. This is not always the case. When the load impedance differs from the characteristic impedance of the line, the line is said to be mismatched.

The electromagnetic field, traveling along a mismatched line from the generator to the load, is not all absorbed by the load. Some of the electromagnetic field is reflected at the feed point, and travels back toward the generator. Since the electromagnetic field causes power to be dissipated in a resistance, the forward-moving field is sometimes called forward power and the backward-moving field is called reflected power.

The amount, or proportion, of incident power that is reflected depends on the severity of the mismatch between the line and the load. The greater the mismatch, the more of the incident power is reflected at the feed point. In the extreme, when the load is a pure reactance, or if it is a short circuit or open circuit, all of the incident power is reflected. The degree of mismatch is expressed by a figure known as the standing-wave ratio (*see* STANDING-WAVE RATIO).

If the standing-wave ratio is represented by S, the forward power by P1, and the reflected power by P2, then:

$$\begin{aligned} P2/P1 &= ((S-1)/(S+1))^2 \\ &= (S^2-2S+1)/(S^2+2S+1) \end{aligned}$$

The proportion of reflected power to forward power is equal to the square of the reflection coefficient (*see* REFLECTION COEFFICIENT).

Reflected power can be directly read in a radio-frequency transmission line with a calibrated reflectometer (*see* REFLECTOMETER). If the reflectometer is not calibrated in forward and reflected watts, the reflected power can be determined from the standing-wave ratio observed on the meter.

When reflected power reaches the transmitter end of a radio-frequency transmission line, all of the electromagnetic field is reflected again back toward the load, assuming the transmitter is adjusted for optimum output. If the transmitter output circuit is not adjusted to compensate for the reactance at the input end of the line, some of the reflected power is absorbed by the final amplifier. This is observed as a reduction in power output, although the final-amplifier dc input is just as great, or greater, than it would be if the output circuit were adjusted properly. *See also* MATCHED LINE, MISMATCHED LINE, REFLECTION FACTOR.

REFLECTED WAVE

In line-of-sight communication, the signal at the receiving antenna consists of two components. One component, the direct wave, follows the path through the air between the two antennas (*see* DIRECT WAVE). The other component

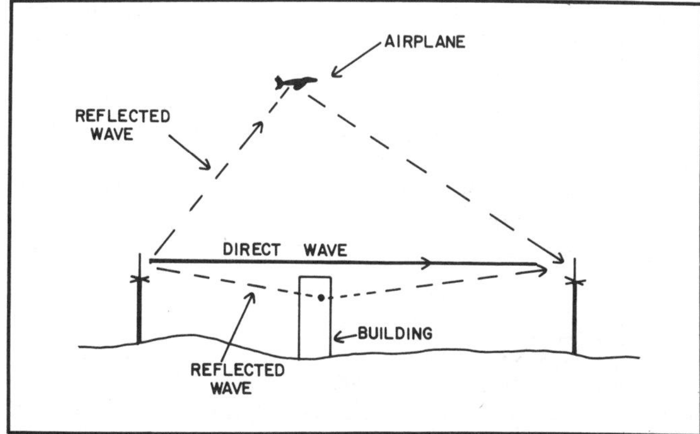

REFLECTED WAVE: The reflected wave combines with the direct wave at the receiving antenna in short-range communications.

consists of one or more waves that are reflected from the ground and/or other objects such as buildings (see illustration). They are called the reflected wave or waves.

The direct wave and the reflected wave(s) add together in varying phase. The reflected-wave signal is usually weaker than the direct wave, but not always. In certain situations, the direct-wave and reflected-wave signals are equal in amplitude and opposite in phase at the receiving antenna. This renders communication difficult or impossible. The remedy is to move either the transmitting antenna or the receiving antenna a fraction of a wavelength, so that the coincidence is eliminated. This kind of phase problem is observed primarily at frequencies above about 10 MHz and is especially acute at microwave frequencies.

The effects of reflected waves can be noticed in a mobile frequency-modulation receiver such as a car stereo. As the received station becomes weak, it begins to flutter. This is the result of alternate phase reinforcement and phase cancellation between the direct-wave and reflected-wave components. The faster the car moves, the more rapid the flutter. If the car comes to a stop in a "dead zone" where total phase cancellation occurs, the signal will disappear until the car is moved again. *See also* LINE-OF-SIGHT COMMUNICATION.

REFLECTION COEFFICIENT

In a power transmission line, not all of the incident power is absorbed unless a perfect match exists (*see* MATCHED LINE, MISMATCHED LINE). When there is a mismatch, not all of the available or "forward" current is absorbed by the load; some current is "reflected." The same thing happens with the voltage. The ratio of reflected current to forward current, or reflected voltage to forward voltage, is called the reflection coefficient. The reflection coefficient is represented by the lowercase Greek rho (ρ).

Let Z_o be the characteristic impedance of a transmission line and let R be the ohmic value of a purely resistive load. Then the reflection coefficient is:

$$\rho = (R-Z_o)/(R+Z_o)$$

if R is larger than Z_o, and

$$\rho = (Z_o-R)/R+Z_o)$$

if R is smaller than Z_o. The above formulas are accurate only for loads that contain no reactance.

The reflection coefficient can also be derived from the standing-wave ratio, S:

$$\rho = (S-1)/(S+1)$$

This second formula is accurate whether or not the load contains reactance. *See also* REFLECTED POWER, STANDING-WAVE RATIO.

REFLECTION FACTOR

In a matched transmission line, all of the incident current is absorbed by the load. This is not the case in a mismatched line (*see* MATCHED LINE, MISMATCHED LINE). The reflection factor is an expression of the ratio of the current actually absorbed by a load, compared with the current that would be absorbed under perfectly matched conditions. The reflection factor may also be called reflectance.

Let R be the ohmic value of a purely resistive load, and let Z_o be the characteristic impedance of the transmission line. Then the reflection factor, F, is given by:

$$F = \sqrt{4RZ_o} / (R+Z_o)$$

This formula holds only if the load impedance is a pure resistance. *See also* REFLECTED POWER, REFLECTION COEFFICIENT.

REFLECTION LAW

If a particle or wave disturbance encounters a flat (plane) reflective surface, the angle of incidence is equal to the angle of reflection, measured with respect to the normal to the surface. If the surface is not flat, then the angles of incidence and reflection are identical as measured with respect to the normal to a plane tangent to the surface at the point of reflection (see illustration). This fact is sometimes called the reflection law or the law of reflection.

The reflection law holds for radio waves impinging against the ground, light striking a mirror, sound waves bouncing off a wall or baffle, and all other situations in which reflection takes place. *See also* ANGLE OF INCIDENCE, ANGLE OF REFLECTION.

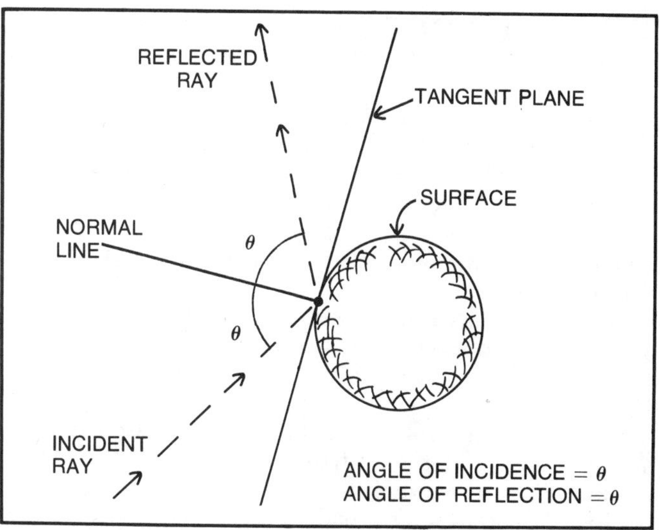

REFLECTION LAW: According to the reflection law, the angle of incidence is equal to the angle of reflection. Here, both angles are indicated by θ.

REFLECTOMETER

A reflectometer is a device that is installed in a transmission line for the purpose of measuring the standing-wave ratio (*see* STANDING-WAVE RATIO). Some reflectometers are calibrated in forward watts and reflected watts. A reflectometer may also be called a directional coupler, directional power meter, or monimatch.

Many different kinds of reflectometers are available from commercial manufacturers for use in coaxial transmission lines. One such device is shown at A. A simple uncalibrated reflectometer, for use at frequencies below 300 MHz, can be constructed from parts that are easy to find. A schematic diagram of a simple reflectometer is shown at B.

An uncalibrated reflectometer is first adjusted so that the meter reads full scale with the switch in the FORWARD position. Then the switch is moved to the REVERSE or REFLECTED position to read the standing-wave ratio. A calibrated reflectometer shows the forward power and the reflected power directly in watts; the standing-wave ratio, S, may be calculated from these readings according to the formula:

$$S = (1+ \sqrt{P2/P1})/(1- \sqrt{P2/P1})$$

where P1 is the forward power and P2 is the reflected power, both specified in the same units.

When the load is a pure resistance with an ohmic value equal to the characteristic impedance of the line, the reflectometer will indicate zero with the switch in the REFLECTED position. If there is a mismatch, the reflected-power reading will not be zero. If the mismatch is very severe, the reflected-power reading will be almost as great as the forward-power reading. When a calibrated reflectometer is used, the true power is determined by subtracting the reflected-power reading from the forward-power reading. *See also* APPARENT POWER, MATCHED LINE, MISMATCHED LINE, REFLECTED POWER, TRUE POWER.

REFLECTOMETER: At A, a typical reflectometer, intended for amateur and CB use. (Courtesy of Gold Line Connector.) At B, simple schematic diagram of an uncalibrated reflectometer.

REFLECTOR

A reflector is a form of parasitic element used in a directional antenna array. The reflector results in signal gain in some directions ,at the expense of gain in other directions. Parasitic arrays increase the efficiency of communications by maximizing the effective radiated power, and also by reducing interference from unwanted directions. *See* ANTENNA POWER GAIN, PARASITIC ARRAY.

An example of the operation of a reflector is found in most Yagi antennas. A half-wave dipole has a gain of 0 dBd (*see* dBd) in free space. A reflector, slightly longer than the dipole and positioned about a quarter wavelength from it, produces gain (see illustration).

The behavior of parasitic arrays in general, and reflectors in particular, is very complex. The gain depends on the spacing between the driven element and the reflector, and also on the length of the reflector. The forward gain and the front-to-back ratio are optimized under slightly different conditions. Many excellent references are available that discuss the design of parasitic antennas in detail. *See also* DIRECTOR, DRIVEN ELEMENT.

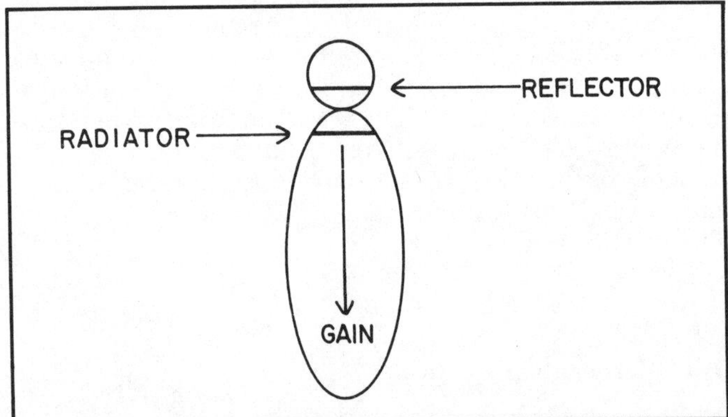

REFLECTOR: Radiation pattern of an antenna incorporating a reflector. The maximum gain is on the side of the driven element (radiator) opposite the reflector.

REFRACTION

When electromagnetic energy passes from one medium to another, the direction of propagation may change at the boundary between the two media. This is known as refraction.

Refraction never occurs when a ray of electromagnetic energy strikes a boundary at a right angle. If the angle of incidence is not 0 degrees, refraction will occur if the two media have different indices of refraction (*see* INDEX OF REFRACTION). When a ray passes from a medium having a given index of refraction to a medium with a lower index of refraction, the angle of incidence, measured with respect to the normal to the plane of the boundary, is smaller than the angle at which the ray leaves the boundary. If the angle of incidence is large enough, total reflection will take place at the boundary. When a ray passes from a medium with a given index of refraction to a medium having a higher index of refraction, the angle of incidence, measured with respect to the normal to the plane of the boundary, is larger than the angle at which the ray leaves the boundary. *See* ANGLE OF INCIDENCE, ANGLE OF REFRACTION, TOTAL INTERNAL REFLECTION.

Radio waves are refracted when they pass from one air mass into another air mass with a different density. Total internal reflection sometimes occurs if the angle of incidence is very large. These effects are called tropospheric bending and ducting, respectively. *See also* TROPOSPHERIC PROPAGATION.

REFRACTIVE INDEX

See INDEX OF REFRACTION.

REGENERATION

Regeneration is another name for positive feedback. Regeneration is deliberately used in oscillators, certain amplifiers, and some detectors. *See also* FEEDBACK, OSCILLATOR, REGENERATIVE DETECTOR.

REGENERATIVE DETECTOR

A regenerative detector is a form of demodulator that can be used for receiving code signals, frequency-shift keying, and single sideband. The regenerative detector can also increase the gain for the reception of amplitude-modulated signals, provided an envelope detector is placed after the circuit. Regenerative detectors are not often used today, although they may still be found in simple direct-conversion receivers.

The regenerative detector consists of an amplifier circuit, with some of the output applied in phase to the input. This increases the gain of the amplifier. If the positive feedback is made large enough, the circuit will oscillate. Incoming signals then mix with the resulting carrier, producing beat notes that can be heard in the speaker or headset. This makes code, frequency-shift-keyed, or single-sideband signals discernible. The schematic diagram shows a simple regenerative detector.

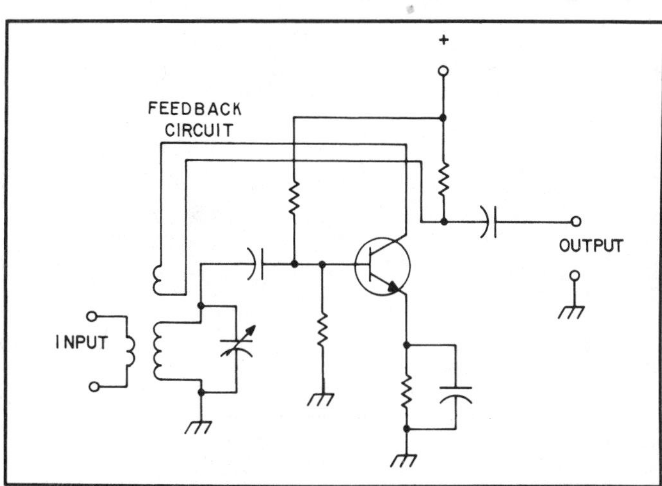

REGENERATIVE DETECTOR: A simple regenerative detector, incorporating inductive feedback.

Regenerative detectors were extensively used in early radio receivers. In many circuits, the regenerative detector was incorporated into the front end of the receiver. This resulted in the transmission of a carrier from the receiver antenna; this carrier could travel several miles and cause interference to other receivers on or near the same frequency. Contemporary circuits never use this arrangement; the regenerative detector is always preceded by at least one tuned circuit to avoid transmission of the carrier. *See also* DIRECT-CONVERSION RECEIVER.

REGENERATIVE RECEIVER

See REGENERATIVE DETECTOR, SUPERREGENERATIVE RECEIVER.

REGENERATIVE REPEATER

A regenerative repeater is a circuit that reshapes pulses. In teletype or code transmission, the pulses often become distorted in transit. This is particularly true for signals that have been propagated via the ionosphere. The regenerative repeater restores the pulses to their original shape and duration.

A regenerative repeater "knows" the appropriate length, spacing, and shape for a pulse. For example, the duration of a Morse-code dot at a given speed might be 0.1 second. At this same speed, the space between code elements in a character is also 0.1 second; the length of a dash is 0.3 second; the space between characters is 0.3 second; the space between words is 0.7 second. If a code signal having some distortion, but the appropriate speed, is applied to the input of the regenerative repeater, perfect code will appear at the output.

Regenerative repeaters can reduce the error rate in digital communications, especially when a machine is used to receive the signals. The higher the data speed, the greater is the chance for distortion in the signal. Therefore, regenerative repeaters become more useful as the data speed is increased.

REGIONAL CHANNEL

A regional-channel broadcast station is a station that serves a large community. Regional-channel stations have higher priority than local-channel stations, but less priority than clear-channel stations.

In the standard AM broadcast band at 535 to 1605 kHz, there are two regional-channel classifications. The high-power regional-channel station operates at a power level of between 1 kW and 5 kW. The low-power regional-channel broadcast station runs between 500 W and 5 kW during the day, and between 500 W and 1 kW during the night.

Many regional-channel stations must use phased antenna systems to modify the coverage area. This prevents interference to other stations operating on or near the same frequency. *See also* CLEAR CHANNEL, LOCAL CHANNEL.

REGISTER

A short-term memory-storage circuit is sometimes called a

register. A register may store any kind of information, and usually has a small capacity. In a computer, a number of registers store data for the central processing unit. A register configuration can be first-in/first-out (FIFO), pushdown stack, or random access (*see* FIRST-IN/FIRST-OUT, PUSHDOWN STACK, RANDOM-ACCESS MEMORY).

The reading of an instrument, in specified units, is sometimes called the register. A meter reading may have to be multiplied by a certain value in order to obtain the correct register (*see* REGISTER CONSTANT).

REGISTER CONSTANT

In some meters and controls, the indication must be multiplied by, or added to, a constant in order to obtain the actual value. This constant is called the register constant.

The illustration shows the scale of a typical ohmmeter. Note that half scale is near the point marked 5. If the register constant is 1, then a half-scale reading indicates approximately 5 ohms. As the value of the resistance increases, it gets more and more difficult to read the meter with a register constant of 1. Therefore, a switch is incorporated into the ohmmeter, allowing us to select various register constants. This particular unit has register constants of 1, 10, 100, 1,000, 10,000, 100,000, and 1,000,000. When the meter needle is at the point marked 5 on the scale, the actual resistance is 5 ohms with the switch in the R×1 position, 50 ohms in the R×10 positon, 500 ohms in the R×100 position, and so on.

In certain meters, indicators, and dial systems, it is sometimes necessary to add a constant to the reading in order to obtain the actual value. For example, a radio transceiver, operating in the range 140.000 MHz to 149.999 MHz, may have a frequency readout that shows values from 0.000 to 9.999. The actual frequency is obtained by adding 140. The value 140 is, therefore, a register constant.

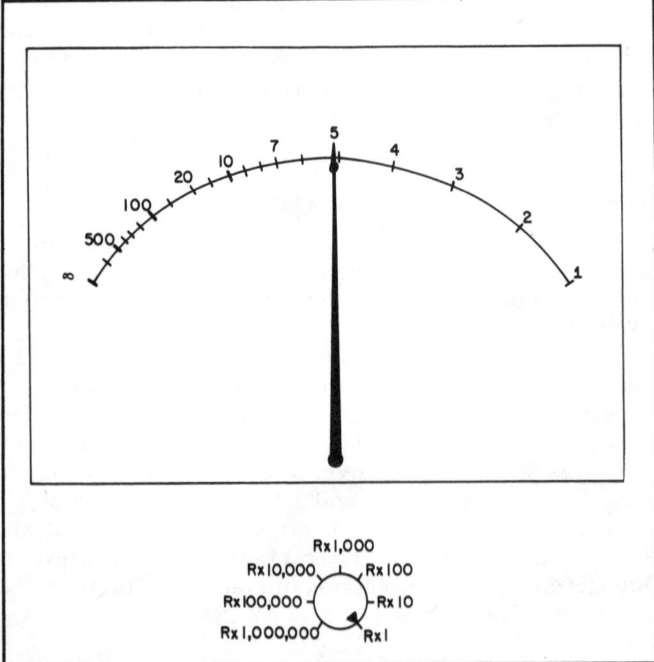

REGISTER CONSTANT: The register constant is the factor by which a meter reading must be multiplied in order to get the actual value of a measured parameter. In this example, the register constant is indicated by the switch position.

REGULATED POWER SUPPLY

A direct-current power supply is said to be regulated if it has a circuit designed for maintaining a precise output voltage under varying load conditions. Many types of electronic equipment will function properly only if the supply voltage is within a narrow range. An especially common operating range is 13.8 V with a tolerance of plus or minus 10 percent.

There are several different methods for obtaining voltage regulation. Zener diodes are sometimes used in low-voltage supplies. Special tubes are more often used in high-voltage power supplies.

In some regulated supplies, failure of the regulating circuit will result in a dramatic rise in the output voltage. This can damage equipment connected to the supply. A well-designed power supply should have some means of overvoltage protection. *See also* OVERVOLTAGE PROTECTION, POWER SUPPLY, VOLTAGE REGULATION.

REGULATING TRANSFORMER

There are two types of regulating transformers, both designed to maintain a constant output voltage under varying load conditions.

Some transformers use resonance and core saturation to maintain a constant output voltage. This type of transformer actually varies in efficiency as the load changes. As the load resistance decreases, the transformer operates more efficiently, compensating for the drop in voltage that would otherwise occur.

A small transformer may be connected in series with the secondary winding of the main transformer in a power supply, for the purpose of adjusting the voltage. This type of arrangment is shown in the schematic. The small transformer is called a regulating transformer. When the current from the regulating transformer is in phase with the current from the main transformer, the output voltage is increased. When the phase of the current from the regulating transformer is opposite to that of the main transformer, the output voltage is decreased. By applying a small voltage of the appropriate phase to the primary of the regulating transformer, the output voltage can be held constant under fluctuating load conditions. *See also* VOLTAGE REGULATION.

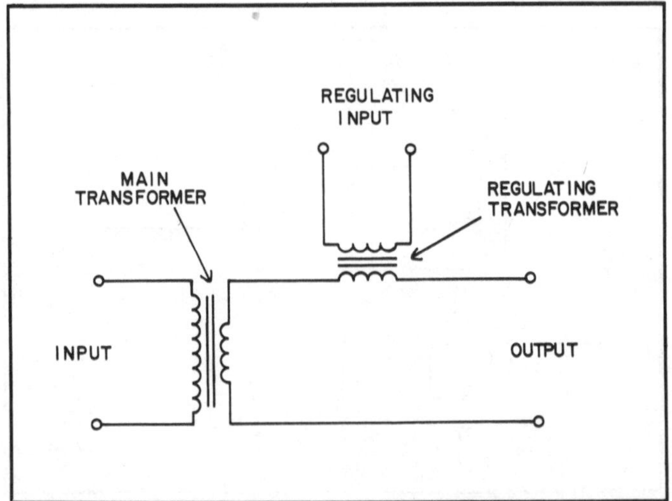

REGULATING TRANSFORMER: A series-connected regulating transformer.

REGULATION

Regulation refers to the maintenance of a constant condition in a circuit. Any parameter, such as current, voltage, resistance, or power, may be regulated.

The extent to which a parameter varies can be expressed mathematically as a percentage. This figure is called the regulation. Let the optimum, or rated, value of a parameter be represented by P, and suppose that the actual value of the parameter fluctuates between the minimum value P1 and the maximum value P2, such that $P1 < P < P2$. Then the regulation R, in percent, is given by:

$$R = 100(P2 - P1)/P$$

Regulation is sometimes expressed as a plus-or-minus figure. In the above example, we might specify the regulation in this manner if the fluctuation is the same above and below the ideal; that is, if:

$$P2 - P = P - P1$$

The plus-or-minus regulation. in percent, is:

$$R\pm = \pm 100(P2 - P)/P$$

$$= \pm\ 100(P - P1)/P$$

Note that the plus-or-minus regulation, as determined by the second method, is exactly half the regulation as determined by the first method. *See also* CURRENT REGULATION, VOLTAGE REGULATION.

REGULATOR

A regulator is a circuit or device that maintains a parameter at a constant value. Many different devices can serve as regulators; a few examples are diodes, transformers, transistors, tubes, and varactors.

Some regulators operate on a "brute-force" principle, by clipping or limiting the output of a power supply. An example of a "brute-force" regulator is the zener diode (*see* ZENER DIODE). Other regulators consist of a sensing circuit that determines whether the value of the parameter is too high or too low, and a control circuit that compensates for any error. An example of this kind of regulator is the phase-locked-loop circuit. *See also* CURRENT REGULATION, VOLTAGE REGULATION.

REINARTZ CRYSTAL OSCILLATOR

A Reinartz crystal oscillator is a special form of oscillator, characterized by high efficiency and little or no output at frequencies other than the fundamental. The Reinartz configuration can be used in oscillators with field-effect transistors, bipolar transistors, or vacuum tubes.

The illustration is a schematic diagram of a Reinartz oscillator that employs a dual-gate MOSFET. A tank circuit is inserted in the source line. This resonant circuit is tuned to approximately half the crystal frequency. The result is enhanced positive feedback, which allows the circuit to oscillate at a lower level of crystal current than would otherwise be possible. *See also* CRYSTAL OSCILLATOR.

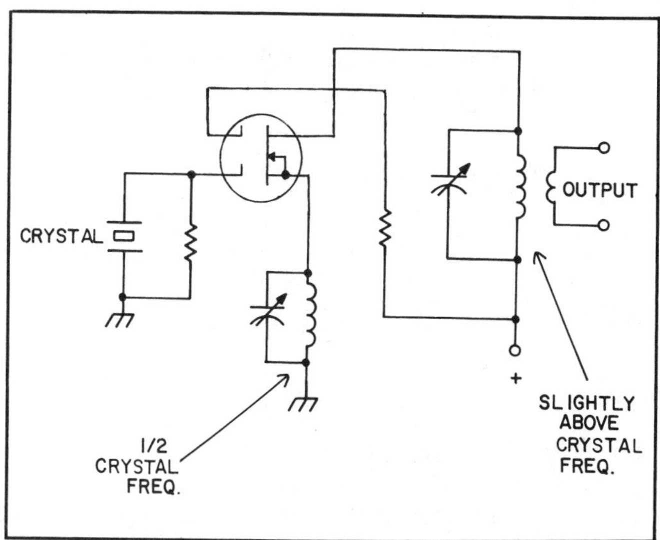

REINARTZ CRYSTAL OSCILLATOR: A Reinartz crystal oscillator using a dual-gate MOSFET.

REINITIALIZATION

Under certain circumstances, a microcomputer may operate improperly because of misprogramming or stray voltages. When this happens, the microcomputer becomes useless until it is reinitialized.

Reinitialization consists of setting all of the microcomputer lines to the low or zero condition. Most microcomputers are automatically reinitialized every time power is removed and reapplied. However, not all microcomputers have this feature; a specific procedure must be followed to reinitialize such devices.

The deliberate removal and reapplication of power, for the purpose of reinitialization, is called a "cold boot." The reinitialization of a microcomputer without total removal of power is called a "warm boot." *See also* INITIALIZATION, MICROCOMPUTER.

RELATION

A relation is a mathematical statement of comparison between or among variables. All mathematical functions are relations. There are, however, some relations that are not functions.

A relation can usually be expressed in different ways in equation form. For example, the unit circle, centered at the origin, can be expressed in any of the following ways:

$$x^2 + y^2 = 1$$

$$x = \pm \sqrt{1 - y^2}$$

$$y = \pm \sqrt{1 - x^2}$$

For any x such that $-1 < x < 1$, there are two correspondent values of y in the relation; for any y such that $-1 < y < 1$, there are two correspondent values of x. Therefore, this relation is not a function. No distinction can be made between the dependent variable and the independent variable.

All relations between two variables can be graphed in Cartesian or polar coordinates. Relations may also be expressed in tabular form, or as a nomograph. *See also* FUNCTION.

RELATIVE ANTENNA GAIN

See ANTENNA POWER GAIN, dBd, dBi.

RELATIVE GAIN AND LOSS

See DECIBEL, GAIN, LOSS.

RELATIVE HUMIDITY

See HUMIDITY.

RELATIVE MEASUREMENT

The magnitude of a quantity may be measured by comparing it with a known standard. This is called a relative measurement. Relative measurements are often made for current, power, radiant energy, sound, and voltage.

A relative measurement can be expressed as a numerical ratio between the unknown and the standard. But the most common method of relative measurement is the decibel system. *See* dBa, dBd, dBi, DECIBEL.

RELATIVE-POWER METER

An uncalibrated voltmeter, installed at the output of a transmitter or in a transmission line, can be used for relative measurement of power. Such a device is called a relative-power meter.

A simple relative-power meter is shown in the schematic diagram. The sensitivity of a relative-power meter usually varies with frequency. The presence of standing waves on a transmission line will also affect the sensitivity of a meter such as the one shown in the illustration. The power at a given frequency is proportional to the square of the meter reading. Thus, if the power is doubled, the meter reading increases by a factor of 1.414; if the power is quadrupled, the meter reading increases by a factor of 2, and so on.

Relative-power meters are extensively used in low-power and medium-power transmitters, especially those intended for Amateur Radio or CB operation. A relative-power meter makes transmitter adjustment easy, and also serves to indicate whether or not the unit is functioning. Many transceivers use the receiver S meter as a relative-power indicator in the transmit mode.

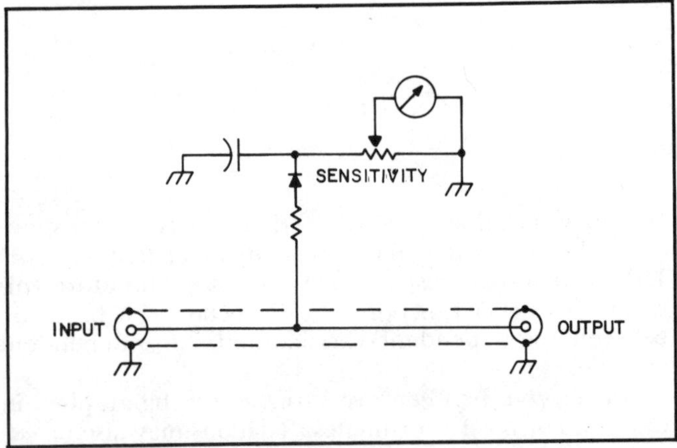

RELATIVE-POWER METER: A simple relative-power meter can be built from components that are easily found in an electronics store.

RELATIVISTIC SPEED

A speed is considered relativistic if it is a sizable fraction of the speed of light. The speed of all electromagnetic radiation is approximately 186,282 miles per second, or 299,792 kilometers per second, in free space. In relativistic calculations, the speed of light is represented by the small letter c.

Relativistic speeds cause a change in the way time progresses. Relativistic speeds also cause objects to become more massive, and physically shorter in the direction of the motion. These effects all occur to the same degree and their magnitudes can be calculated if the speed is known.

Let v represent the speed of an object, and let c represent the speed of light in the same units. Then the relativistic factor is given by:

$$R = \sqrt{1 - v^2/c^2}$$

for time distortion and spatial distortion, and:

$$R' = 1/\sqrt{1 - v^2/c^2}$$

for mass distortion.

The factor R has a value very close to 1 for all speeds at which man has ever traveled. Even the Apollo moon spacecraft did not attain speed sufficient to cause much relativistic effect. But atomic particles have been accelerated to speeds of relativistic values, causing an increase in their mass over the at-rest mass. Experiments have also verified the fact that time progresses more slowly from a moving point of view. *See also* RELATIVITY THEORY.

RELATIVITY THEORY

As the twentieth century began, the science of physics was undergoing radical changes. Many discoveries were made about the nature and behavior of light and other electromagnetic radiation. The most famous of the scientists, Albert Einstein, formulated his special theory of relativity during the first decade of the new century. Several years later, Einstein completed his general theory of relativity.

The principal axiom of the special theory holds that light (and all other forms of electromagnetic radiation) always appears to have the same speed, regardless of the motion of an observer. This speed, represented in the equations by the small letter c, is approximately 186,282 miles per second, or 299,792 kilometers per second. This axiom has been verified by modern scientists, using precision equipment not available to Einstein at the time of his work.

The principal axiom of special relativity leads to the conclusion that time, space, and mass do not have constant proportions. Instead, these parameters vary, depending on the motion of an observer. The relativistic effect does not become important until very high speeds are attained (*see* RELATIVISTIC SPEED).

The special theory of relativity proves that all non-accelerating reference frames can be made equivalent according to a simple mathematical transformation. The general theory goes further, accounting for accelerating viewpoints as well. According to the general theory, space must be curved in four dimensions from an accelerating point of view. The same holds true in gravitational fields. The distortion of space by gravitational fields has been demonstrated by experiments.

The most devastating result of the theory of relativity has, no doubt, been the discovery of atomic fission and fusion. Einstein showed that matter and energy are different manifestations of the same thing, and that one can be transformed into the other. Matter and energy are related by the factor.

$$E/m = c^2$$

where E is the energy is joules, m is the mass in kilograms, and c is the speed of light in meters per second. The above equation is more often seen in the form:

$$E = mc^2$$

and is sometimes called the Einstein equation.

A complete and thorough discussion of the special and general theories of relativity cannot be given here. However, many excellent books are available on the subject. A discussion of relativity theory in layman's terms is given in the books *Understanding Einstein's theories of Relativity: Man's New Perspective on the Cosmos* (TAB book No. 1505) and *E = mc²: Picture Book of Relativity* (TAB book No. 1580). A more technical evaluation of Einstein's theories is given in *Einstein's Theory of Relativity* (Dover Publications, Inc., 1965). *See also* NUCLEAR ENERGY, NUCLEAR FISSION, NUCLEAR FUSION.

RELAXATION OSCILLATION

Relaxation oscillation is a form of oscillation resulting from the gradual buildup of charge in a capacitor, followed by a sudden discharge through a resistance. Various simple circuits can be designed for relaxation oscillation. Such oscillators are often used for audio-frequency applications in which the wave shape is not important. The neon-lamp oscillator is the most common relaxation circuit (*see* NEON-LAMP OSCILLATOR).

At radio frequencies, and in audio applications requiring a pure sine wave, relaxation oscillators are not used. Crystal-controlled or tuned oscillators are necessary if precision or high-frequency signals are needed. *See also* OSCILLATION, OSCILLATOR.

RELAXATION OSCILLATOR

See NEON-GAMP OSCILLATOR, RELAXATION OSCILLATION.

RELAY

A relay is a remote-control switch, actuated by a direct or alternating current. Relays allow control of a signal from a distant point, without having to route the signal path to the control point. Relays are extensively used in the switching of radio-frequency signals. Relays are also useful when it is necessary to switch a very large current or voltage; a smaller (and safer) control voltage can be used to actuate the relay. Relays may have one or two throw positions, and one or more poles.

A relay consists of an iron-core solenoidal coil to which the control voltage is applied, an armature that is attracted to the solenoid by the magnetic field set up when current flows through the coil, and a set of contacts that carry the signal current. A single-pole, double-throw relay is shown in the illustration at A. A small relay, used for switching radio-frequency signals, is shown in the photograph (B).

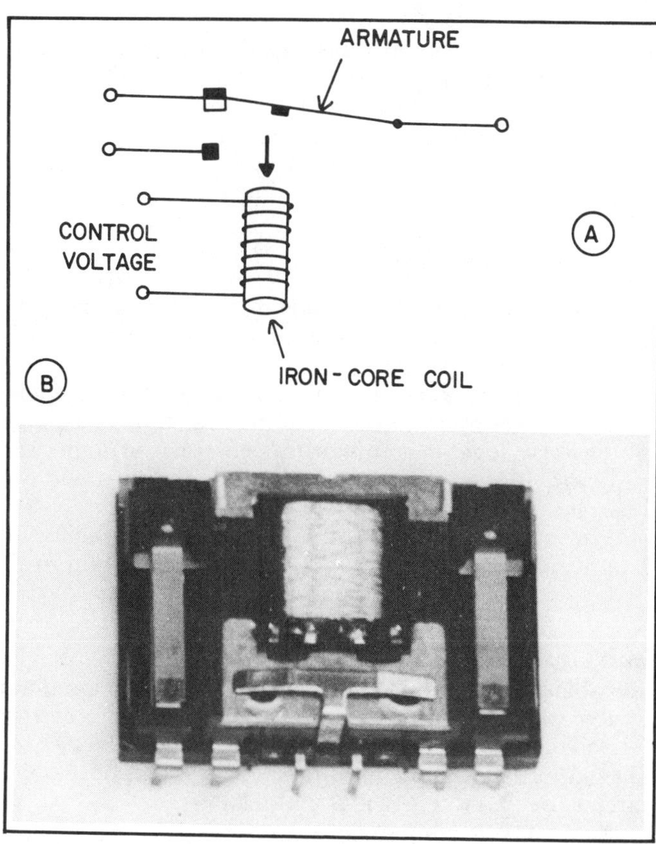

RELAY: At A, pictorial diagram showing the operation of a conventional relay. At B, photograph of a miniature plug-in relay used in a radio transceiver.

Relays vary greatly in specifications. Some relays are designed for high-speed or frequent switching; the reed relay is made especially for such applications (*see* REED RELAY). Other relays are intended for occasional switching of large amounts of current relays are rated according to the number of actuations that can be accomplished per minute or per second. Some relays operate with a small control current, in some cases less than 50 mA; others require several amperes of control current. Certain relays will operate from either alternating or direct control currents, while others must have direct current. Relay contacts are rated according to maximum allowable signal current and voltage.

In modern devices, electronic switching has largely replaced the relay for control purposes. Diodes, silicon-controlled rectifiers, and transistors can be used in place of relays in many applications. Tubes are also sometimes used for switching. Since semiconductor devices and vacuum tubes have no moving parts, they offer better reliability and higher switching speed than relays. *See also* SILICON-CONTROLLED RECTIFIER, SWITCHING DIODE, SWITCHING-TRANSISTOR, SWITCHING TUBE.

In a communications system, an intermediate link between stations is sometimes called a relay. There may be several relays in a single system. In the early days of radio, communications range was limited to only a few miles. In order to send a message over a great distance, relays were used. The orginating station sent the message to the first intermediate station; the first intermediate station sent the message to the second intermediate station; this process was continued until the message reached the receiving station. Today, relays are still sometimes used at the very-high, ultra-high, and microwave frequencies. The relaying is usually done automatically by means of a repeater or satellite. *See also* ACTIVE COMMUNICATIONS SATELLITE, REPEATER, SATELLITE COMMUNICATIONS.

RELAY LOGIC

A digital logic circuit can be built using relays rather than the usual semiconductor devices. Such a logic system is called relay logic, abbreviated RL.

Relay logic is somewhat slower than solid-state logic, because the relay armatures require some time to change state. Relay logic is a fully saturated form of logic; the resistance in the open state is practically infinite, and the resistance in the closed state is practically zero.

The main advantages of relay logic are its ability to handle rather large currents, and the high saturation level.

RELIABILITY

Reliability is a measure of the probability that a circuit or system will operate according to specifications for a certain period of time. Reliability is obviously related, inversely, to the failure rate and the hazard rate of the components in a circuit (*see* FAILURE RATE, HAZARD RATE).

Reliability is sometimes expressed as a percentage, based on the proportion of units that remain in proper operation after a given length of time. Suppose, for example, that 1,000,000 units are placed in operation on January 1, 1990. If 900,000 units are operating properly on January 1, 1991, then the reliability is 90 percent per year. It is assumed that the units are operated under electrical and environmental conditions that are within the allowable specifications.

Reliability is largely a function of the design of electronic apparatus. Even if the components are of the best possible quality, equipment failure is likely if the design is poor. Reliability can be optimized by a good quality-assurance procedure (*see* QUALITY CONTROL).

The reliability of electronic equipment can be tested in the laboratory, provided a large number of units is used. Reliability testing is part of the overall quality-control procedure in a good engineering scheme.

RELUCTANCE

Magnetic resistance is known as reluctance. Reluctance is generally measured in ampere turns per weber. The greater the number of ampere turns required to produce a magnetic field of a given intensity, the greater the reluctance (*see* AMPERE TURN, WEBER).

Magnetic reluctance is a property of materials that oppose the generation of a magnetic field. The unit of reluctance is the rel. Reluctance is usually abbreviated by the capital letter R.

When two or more reluctances are combined in series, the total reluctance is found by adding the individual reluctances. If R1, R2, R3, . . . , Rn represent the individual reluctances, and R is the overall total reluctance, then:

$$R = R1 + R2 + R3 + \ldots + Rn$$

in series. In parallel, the same reluctances total:

$$R = 1/(1/R1 + 1/R2 + 1/R3 + \ldots + 1/Rn)$$

Reluctances add in the same way as resistances in series and in parallel.

Suppose that F is the magnetomotive force in a magnetic circuit, ϕ is the magnetic flux, and R is the reluctance. Then the three parameters are rdlated according to the equation:

$$F = \phi R$$

This formula is known as Ohm's law for magnetic circuits. *See also* MAGNETIC FLUX, MAGNETOMOTIVE FORCE, OHM'S LAW.

RELUCTIVITY

Reluctivity is a measure of the reluctance of a material per unit volume. Reluctivity is usually expressed in rels per cubic centimeter. Reluctivity is the reciprocal of magnetic permeability. *See also* PERMEABILITY, RELUCTANCE.

REMANENCE

Certain materials remain magnetized after the removal of an applied magnetic field which has resulted in saturation, or maximum magnetization. All ferromagnetic substances have remanence. Diamagnetic and paramagnetic substances do not have remanence (*see* DIAMAGNETIC MATERIAL, FERROMAGNETIC MATERIAL, PARAMAGNETIC MATERIAL).

The remanence of a ferromagnetic material is related to the degree of magnetic hysteresis. When a substance remains magnetized, it is more difficult for the magnetic field to be reversed than it would be if the substance did not remain magnetized. The greater the remanance of the material in an inductor core, for example the lower is the maximum frequency at which the coil can be efficiently operated.

The residual magnetism in a material is sometimes called the remanence of that material. Residual magnetism is expressed as a percentage. If the flux density is given by B1 at saturation and by B2 after the removal of the magnetic field, then the residual magnetism Br, in percent, is:

$$Br = 100(B2/B1)$$

Residual magnetism is dependent on the type of material, and on the shape and size of the sample.

REMOTE CONTROL

Electronic equipment can be operated from a control panel located at a distance. This is called remote control. Remote control is useful in a variety of situations.

Remote control can be accomplished either by wire or by radio. For example, a telephone answering system may receive messages and record them; it is possible to call the device and instruct it to play back the messages over the telephone. This is remote control by wire. A synchronous satellite is boosted into orbit by means of radio signals, which actuate the proper engines; this is remote control by radio.

In the design of a system for remote control by radio, a transmitter is needed at the control site, and a receiver is

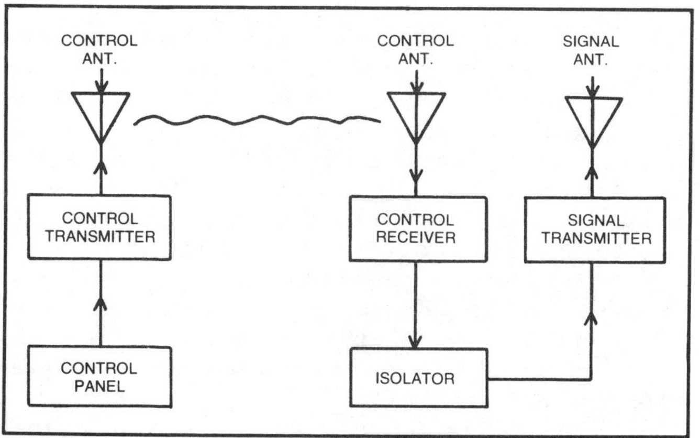

REMOTE CONTROL: Block diagram of a system for remote control by radio.

required at the site of the equipment to be controlled. If the equipment itself contains a transmitter, it is important that the transmitter not interfere with the reception of the control signals. The control signal must have a frequency that is at least 0.3 percent away from the fundamental and all harmonics of the transmitter signal. An isolator is used for this purpose. The illustration is a block diagram of a simple system for remote control of a transmitter by radio. *See also* LOCAL CONTROL.

REPEATER

A repeater is a device that intercepts and retransmits a signal for the purpose of providing wide-area communications. Repeaters are generally used at the very high, ultra-high, and microwave frequencies. Repeaters are especially useful for mobile operation. The effective range of a mobile station is greatly enhanced by a repeater. In the 144-MHz amateur band, for example, direct simplex communication between mobile stations of moderate power ranges from 10 to 30 miles; with a repeater, this range may exceed 100 miles (see illustration).

A repeater antenna should be located at the greatest possible height above average terrain. A vertical collinear array is generally used to maximize the receiving and transmitting gain. The receiver sensitivity and the transmitter power output should be such that the repeater coverage is about the same for both reception and transmission.

A repeater consists of an antenna, a receiver, a transmitter, and an isolator. The transmitter and receiver are operated at slightly different frequencies. The separation is approximately 0.3 percent to 1 percent of the transmitter frequency. The separation of the receiver and transmitter frequencies allows the isolator to work at maximum efficiency, preventing undesirable feedback and desensitization.

Repeaters are sometimes placed aboard satellites. All active communications satellites use repeaters. A satellite, placed in a synchronous orbit, can provide coverage over approximately 30 percent of the globe. *See also* ACTIVE COMMUNICATIONS SATELLITE, MICROWAVE REPEATER, SATELLITE COMMUNICATIONS.

REPEATING DECIMAL

Certain rational numbers, when expressed in decimal form, repeat endlessly. The repetition may consist of any number of digits. The general form of a repeating decimal is:

$$m_0 \,.\, m_1 m_2 m_3 \ldots m_n m_1 m_2 m_3 \ldots m_n \ldots$$

where m_0 is the digit or digits to the left of the decimal point, and $m_1 m_2 m_3 \ldots m_n$ is the repeating sequence of digits.

Any repeating decimal expression can be converted to fractional form. In the above general case, the fractional component, following the digit or digits m_0, is:

$$m_1 m_2 m_3 \ldots m_n \,/\, 999 \ldots 9$$

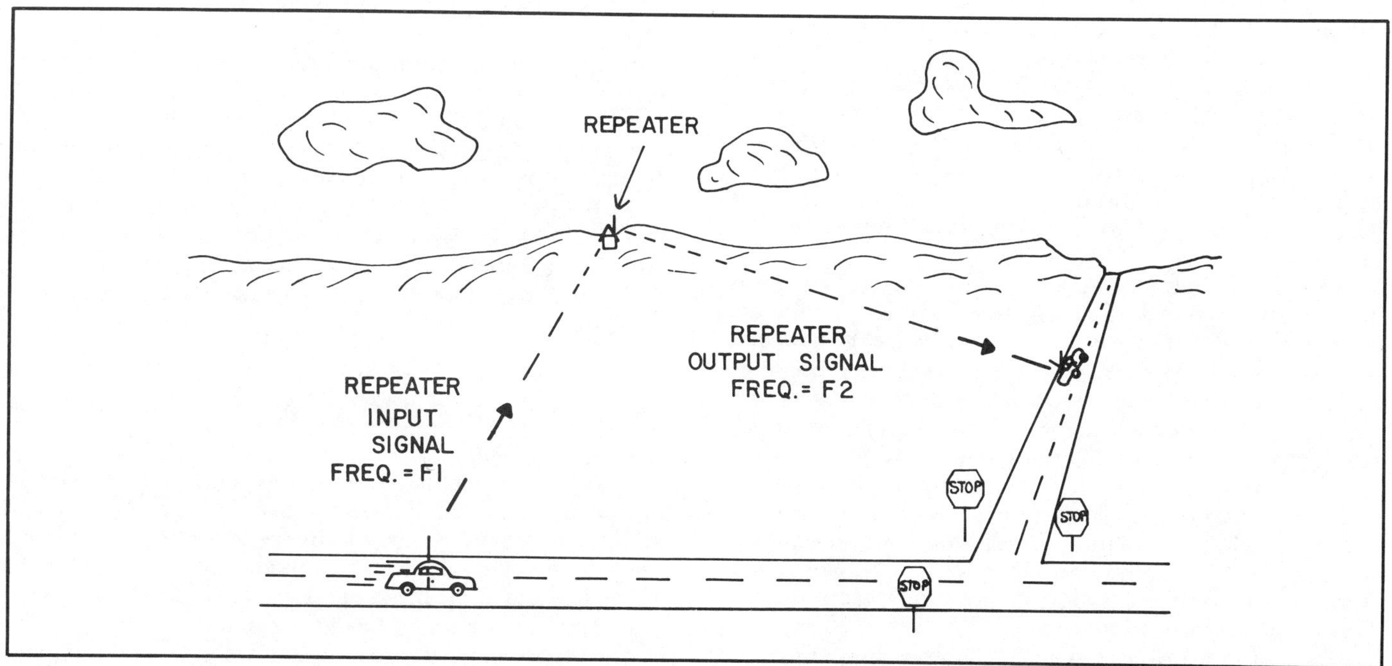

REPEATER: A repeater, located in a high place, intercepts and retransmits the signals from a moderate-power or low-power station. This can enhance the range.

The denominator contains exactly n 9s. *See also* RATIONAL NUMBER.

RESET

When a device is initialized, or placed in some prescribed state, the action is called reset, or reset action. In a microcomputer, reset action is called reinitialization. In a calculator, the CLEAR command constitutes resetting. A count-down timer is reset by adjusting it to show the appropriate number of seconds, minutes, or hours. A count-up timer is reset by placing it at 0. *See also* INITIALIZATION, REINITIALIZATION.

RESIDUAL MAGNETISM

See REMANENCE.

RESISTANCE

No electrical conductor is perfect. The transfer of charge carriers is always finite per unit time. The opposition that a substance offers to the flow of electric current can be precisely defined as the number of amperes that flow for each volt of applied potential (*see* OHM'S LAW). This parameter is known as electrical resistance.

Resistance can be specified either for direct currents or alternating currents. In ac circuits, the average resistance may differ from the instantaneous or dynamic resistance (*see* DYNAMIC RESISTANCE).

The standard unit of resistance is the ohm, often abbreviated by the uppercase Greek letter omega (Ω). In equations, the symbol for resistance is R. The range of possible resistance values, in ohms, is represented by the set of positive real numbers. In theory, it is possible to have resistances that are zero or infinite as well. Resistance is the mathematical reciprocal of conductance. The higher the conductance, the lower the resistance, and vice-versa (*see* CONDUCTANCE).

The resistance of wire is usually measured in ohms per unit length. Silver has the lowest resistance of any wire. Copper and aluminum also have very low resistance per unit length. The dc resistance of a wire depends on the diameter. The table shows the direct-current resistance of copper wire at 20 degrees Celsius, for various sizes in the American Wire Gauge (*see* AMERICAN WIRE GAUGE).

Resistance is often introduced into a circuit deliberately, to limit the current and/or to provide various levels of voltage. This is done with components called resistors (*see* RESISTOR).

An inductor has extremely low resistance for dc but it may exhibit a considerable amount of resistance for ac. This effective resistance is known as inductive reactance; it increases as the alternating-current frequency becomes higher. A capacitor has practically infinite resistance for dc, but it will exhibit a finite effective resistance for ac of a high enough frequency. This effective resistance is called capacitive reactance, and it decreases as the ac frequency gets higher. An ac circuit may contain both resistance and reactance. The combination of resistance and reactance is represented by a complex number, and is known as impedance (*see* CAPACITIVE REACTANCE, IMPEDANCE, INDUCTIVE REACTANCE, REACTANCE).

The resistance of a material is, to a certain extent, affected by temperature. For most substances, the resistance increases with increasing temperature. A few materials exhibit very little change in resistance with temperature fluctuations; some substances become better conductors as the temperature increases. The behavior of a resistance under varying temperature conditions is known as the temperature coefficient (*see* TEMPERATURE COEFFICIENT). Certain materials notably the highly conductive metals, attain extremely low resistance values at temperatures near absolute zero (*see* SUPERCONDUCTIVITY).

A radio transmitting antenna offers a certain inherent opposition to the radiation of electromagnetic energy. This property is called radiation resistance, and is a function of the physical size of the antenna in wavelengths (*see* RADIATION RESISTANCE).

Resistance can be defined for phonomena other than electric currents. For example, a mechanical device may show a certain amount of resistance to changes in position; a pipe resists the flow of water to a greater or lesser degree. The opposition that a substance offers to the flow of heat is called thermal resistance; its reciprocal is thermal conductivity (*see* THERMAL CONDUCTIVITY).

RESISTANCE BRIDGE

See RESISTANCE MEASUREMENT.

RESISTANCE-CAPACITANCE CIRCUIT

A circuit that contains only resistors and capacitors, or only resistance and capacitive reactance, is called a resistance-capacitance (RC) circuit. Such a circuit may be extremely simple, consisting of just one resistor and one capacitor, or it might be a complicated network of components.

An RC circuit can be used for the purpose of highpass or lowpass filtering. An example of an RC highpass filter is illustrated at A in the illustration. A lowpass network is shown at B. The RC filters have a less well-defined cutoff characteristic than inductance-capacitance (LC) circuits (*see* HIGHPASS FILTER, LOWPASS FILTER). The RC type highpass or lowpass filter is generally used at audio frequencies, while the LC type is more often used at radio frequencies.

An RC filter is commonly used in a code transmitter to obtain shaping of the keying envelope (*see* KEY CLICK, SHAPING). Simple RC circuits provide filtering in low-current power supplies. More complicated RC circuits can be used to measure unknown capacitances.

Every RC circuit displays a certain charging and discharging characteristic. This property depnds on the circuit configuration, and also on the values of the resistor(s) and capacitor(s). *See also* RESISTANCE-CAPACITANCE TIME CONSTANT.

RESISTANCE-CAPACITANCE COUPLING

Resistance-capacitance coupling is a form of capacitive coupling that is sometimes used in audio-frequency and radio-frequency circuits. Biasing resistors are used to provide the proper voltages and currents at the output of the first stage and at the input of the second stage. The signal is transferred via a series blocking capacitor.

Resistance-capacitance coupling is simple and inexpensive. However, if a high degree of interstage isolation is required, or if selectivity is needed, transformer coupling

RESISTANCE: Resistance of Solid Copper Wire for American Wire Gauge (AWG) 1 through 40, in Ohms per Kilometer.

AWG	Ohms/km	AWG	Ohms/km
1	0.42	21	43
2	0.52	22	54
3	0.66	23	68
4	0.83	24	86
5	1.0	25	110
6	1.3	26	140
7	1.7	27	170
8	2.1	28	220
9	2.7	29	270
10	3.3	30	350
11	4.2	31	440
12	5.3	32	550
13	6.7	33	690
14	8.4	34	870
15	11	35	1100
16	13	36	1400
17	17	37	1700
18	21	38	2200
19	27	39	2800
20	34	40	3500

is preferable. *See also* CAPACITIVE COUPLING, TRANSFORMER COUPLING.

RESISTANCE-CAPACITANCE TIME CONSTANT

The length of time required for a resistance-capacitance (RC) circuit to charge and discharge depends on the values of the resistance and capacitance. The larger the resistance or capacitance, the longer it takes for a circuit to charge and discharge.

Charging and discharging are actually exponential processes. The full-charge condition is theoretically never reached from a zero-voltage start; the zero-charge condition is theoretically never reached from a partial-charge start. For practical purposes, however, we may speak of zero charge and full charge.

The period of time required for an RC circuit to reach 63 percent of the full-charge condition, starting at zero charge is called the RC time constant. The RC time constant can also be defined as the length of time required for an RC circuit to discharge to 37 percent of the full-charge condition, starting at full charge.

The time constant t, in seconds, of an RC circuit having an effective resistance of R ohms and an effective capacitance of C farads is:

$$t = RC$$

The RC time constant is of importance in the design of some power-supply filtering circuits, code-transmitter shaping circuits, and other RC filtering devices. *See also* RESISTANCE-CAPACITANCE CIRCUIT.

RESISTANCE-INDUCTANCE CIRCUIT

A circuit that contains only resistors and inductors, or only resistance and inductive reactance, is called a resistance-

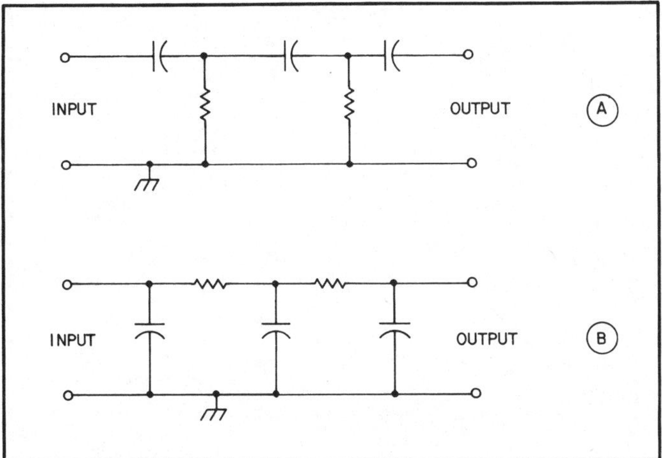

RESISTANCE-CAPACITANCE CIRCUIT: At A, a resistance-capacitance highpass filter. At B, a resistance-capacitance lowpass filter.

inductance (RL) circuit. Such a circuit may be extremely simple, consisting of just one resistor and one inductor, or it might be a complicated network of components.

An RL circuit can be used for the purpose of highpass or lowpass filtering. An example of an RL highpass filter is illustrated at A in the illustration. A lowpass network is shown at B. The RL filters have a less well-defined cutoff characteristic than inductance-capacitance (LC) circuits (*see* HIGHPASS FILTER, LOWPASS FILTER). The RL type highpass or lowpass filter is generally used at audio frequencies, while the LC type is more often used at radio frequencies. Resistance-inductance circuits can be used to measure unknown values of inductance; there are several different types of RL circuit designed for this purpose.

Every RL circuit displays a certain charging and discharging characteristic. This depends on the circuit configuration, and also on the values of the resistor(s) and capacitor(s). *See also* RESISTANCE-INDUCTANCE TIME CONSTANT.

RESISTANCE-INDUCTANCE TIME CONSTANT

The length of time required for a resistance-inductance (RL) circuit to charge and discharge depends on the values of the resistances and inductances. The larger the resistance and inductance, the longer it takes for a circuit to charge and discharge.

Charging and discharging are actually exponential processes. The full-charge condition is theoretically never reached for a zero-voltage start; the zero-charge condition is theoretically never reached from a partial-charge start. For practical purposes, however, we may speak of zero charge and full charge.

The period of time required for an RL circuit to reach 63 percent of the full-charge condition, starting at zero charge, is called the RL time constant. The RL time constant can also be defined as the length of time required for an RL circuit to discharge to 37 percent of the full-charge condition, starting at full charge.

Any resistance-inductance circuit can, ideally, be reduced to a simple equivalent combination of one resistor and one inductor. The simpler equivalent circuit has the same charging and discharging times as the more complicated circuit. The time constant t, in seconds, of an RL

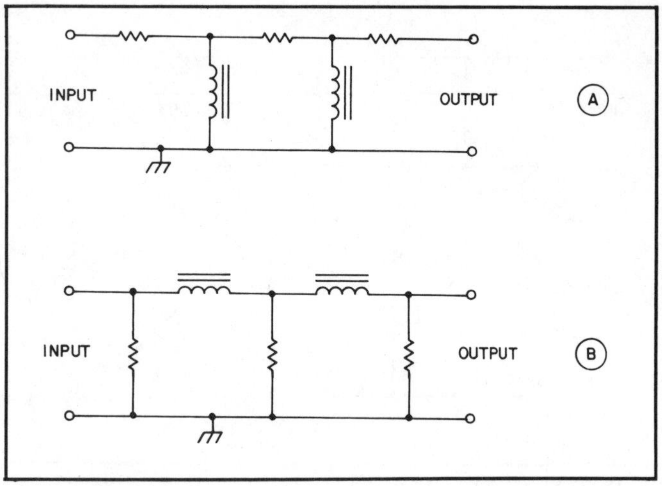

RESISTANCE'INDUCTANCE CIRCUIT: At A, a resistance-inductance highpass filter. At B, a resistance-inductance lowpass filter.

circuit having an effective resistance of R ohms and effective inductance of L henrys, is:

$$t = RL$$

See also RESISTANCE-INDUCTANCE CIRCUIT.

RESISTANCE LOSS

In any electrical or electronic circuit, some power is lost in the wiring because of resistance in the conductors. In an antenna system, some power is lost in the earth because of ground resistance. The amount of power lost because of resistance is called the resistance loss. It may also be called I^2R loss, or ohmic loss.

Resistance loss affects the efficiency of a circuit. The efficiency is 100 percent only if the resistance loss is zero. As the resistance loss increases in proportion to the load resistance, the efficiency goes down. If the loss resistance is R and the load resistance is S, the efficiency E in percent is:

$$E = 100S/(R + S)$$

The resistance loss in any circuit is directly proportional to the resistance external to the load. The loss is also proportional to the square of the current that flows. Mathematically, if R is the resistance in ohms and I is the current in amperes, the resistance power loss P, in watts, is:

$$P = I^2R$$

Resistance loss, unless deliberately introduced for a specific purpose, should be kept to a minimum. This is especially true in ac power transmission, and in radio-frequency transmitting antennas.

In a power-transmission system, resistance loss is minimized by using large-diameter conductors, and by ensuring that splices have excellent electrical conductivity. The loss is also minimized by using the highest possible voltage, thereby reducing the current for delivery of a given amount of power.

In a transmitting antenna system, the resistance loss is reduced by using large-diameter wire or metal tubing for the radiating conductors, by using a heavy-duty tuning network and feed system, by providing excellent electrical connections where splices are necessary, and by ensuring that the ground conductivity is as high as possible. The resistance loss in a transmitting antenna can also be reduced by deliberately maximizing the radiation (load) resistance. *See also* AC POWER TRANSMISSION, ANTENNA EFFICIENCY, HIGH-TENSION POWER LINE, RADIAL, RADIATION RESISTANCE.

RESISTANCE MEASUREMENT

The value of a resistance can be measured in various ways. All methods of resistance measurement involve the application of voltage across the resistive element, and the direct or indirect measurement of the current that flows as a result of the applied voltage. Two often-used schemes are discussed below.

The most common method of resistance measurement is the use of a device called an ohmmeter. The voltage source is usually a cell or battery, which provides between 1.5 and 12 volts in most cases. The ohmmeter is used for measurement of moderate values of resistance. For determination of extremely high values of resistance, a special form of ohmmeter, known as a megger, is used. The megger has a high-voltage generator instead of the low-voltage cell or battery (*see* MEGGER, OHMMETER).

An alternative means of resistance measurement is the comparison, or balance, method. The balance method gives more accurate results than the ohmmeter when the resistance is extremely low or high. The resistance-balance circuit is known as a ratio-arm bridge. The unknown resistance is compared with a known, variable, resistance, resulting in a balanced condition when a specific ratio is achieved (*see* RATIO-ARM BRIDGE).

RESISTANCE POWER LOSS

See RESISTANCE LOSS.

RESISTANCE STANDARD

See STANDARD RESISTOR.

RESISTANCE TEMPERATURE COEFFICIENT

See TEMPERATURE COEFFICIENT.

RESISTANCE THERMOMETER

The resistance of most substances changes with fluctuations in temperature. This makes it possible to measure the temperature with a device similar to an ohmmeter. A length of wire, whose resistance-versus-temperature function is accurately known, is connected in series with a low-voltage power source, a current-limiting resistor, and a microammeter (see illustration). This apparatus is called a resistance thermometer.

As the temperature rises, the resistance of the wire increases, and the current reading falls. Conversely, when the temperature falls, the resistance of the wire drops, and thus more current is indicated. It is important that the temperature fluctuations not affect the current-limiting resistor, however; the use of a standard resistor, having

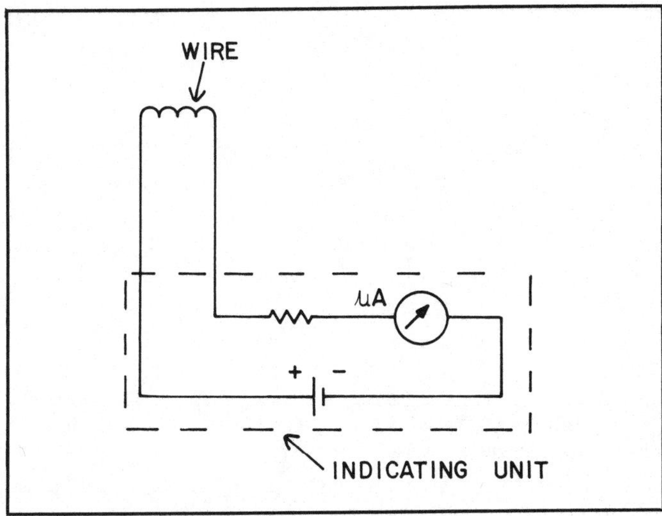

RESISTANCE THERMOMETER: A simple resistance thermometer. The temperature affects the resistance of the wire element.

zero temperature coefficient, is imperative (*see* TEMPERATURE COEFFICIENT).

A resistance thermometer, such as the one in the diagram, can be used to measure very high or low temperatures. The device is not extremely accurate for determination of moderate temperatures. However, greater resolution can be obtained by substituting a thermistor in place of the resistance wire. *See also* THERMISTOR.

RESISTIVE-CAPACITIVE CIRCUIT

See RESISTANCE-CAPACITANCE CIRCUIT.

RESISTIVE-CAPACITIVE COUPLING

See CAPACITIVE COUPLING, RESISTANCE-CAPACITANCE COUPLING.

RESISTIVE-CAPACITIVE TIME CONSTANT

See RESISTANCE-CAPACITANCE TIME CONSTANT.

RESISTIVE-INDUCTIVE CIRCUIT

See RESISTANCE-INDUCTANCE CIRCUIT.

RESISTIVE-INDUCTIVE TIME CONSTANT

See RESISTANCE-INDUCTANCE TIME CONSTANT.

RESISTIVE LOAD

A load is said to be resistive, or purely resistive, when it contains no reactance for alternating current at a specific frequency. A load may be resistive for either of two quite different reasons: It may contain no reactive components, or it may consist of equal but opposite reactances.

A simple, noninductive resistor presents a load having essentially zero reactance. This is true at all frequencies up to the point at which the lead lengths become a substantial fraction of a wavelength. Noninductive resistors are used for the purpose of testing radio transmitters off the air; such a load is called a dummy load or dummy antenna (*see* DUMMY ANTENNA).

A tuned circuit, consisting of a capacitor and an inductor, is resistive at one specific frequency, known as the resonant frequency. The same is true of an antenna system. This type of resistive load is clearly different from the noninductive resistor. When the capacitive and inductive reactances cancel each other, a tuned circuit or antenna is said to be resonant. The remaining resistance is the result of losses in a tank circuit, or the combination of radiation resistance and resistance loss in an antenna.

A resistive load is desirable in any power-transfer circuit. If the load contains reactance, some of the incident power is reflected upon reaching the load. Ideally, the load resistance should be identical to the characteristic impedance of the feed line. *See also* CAPACITIVE REACTANCE, IMPEDANCE, INDUCTIVE REACTANCE, RADIATION RESISTANCE, REFLECTED POWER, RESISTANCE LOSS, RESONANCE.

RESISTIVE LOSS

See RESISTANCE LOSS.

RESISTIVITY

The resistance of a flat surface can be expressed per unit area, and the resistance of a solid can be expressed per unit volume. These parameters are called resistivity. The resistivity of a surface is specified in ohms; the resistivity of a solid is usually given in ohm-centimeters or ohm-inches.

Surface resistivity and volume resistivity depend on the type of material, and also on the temperature. Metals such as aluminum, copper, and silver have very low resistivity. Various other elements, compounds, and mixtures are designed to have certain amounts of resistivity for different applications. The dielectric substances have high resistivity.

The term resistivity is occasionally used loosely as a synonym for resistance, although this is technically imprecise. The resistance of a conductor per unit length may be called resistivity. *See also* RESISTANCE.

RESISTOR

A resistor is an electronic component that is deliberately designed to have a specific amount of resistance (*see* RESISTANCE). Resistors are available in many different forms.

The most often-used type of resistor in electronic devices is the carbon variety (*see* CARBON RESISTOR). Values range from less than 1 ohm to millions of ohms; power-dissipation ratings are usually ¼ watt or ½ watt. The photograph shows a ¼-watt resistor and a ½-watt resistor, next to a dime for size comparison. Some carbon resistors have power-dissipation ratings of more than ½ watt or less than ¼ watt.

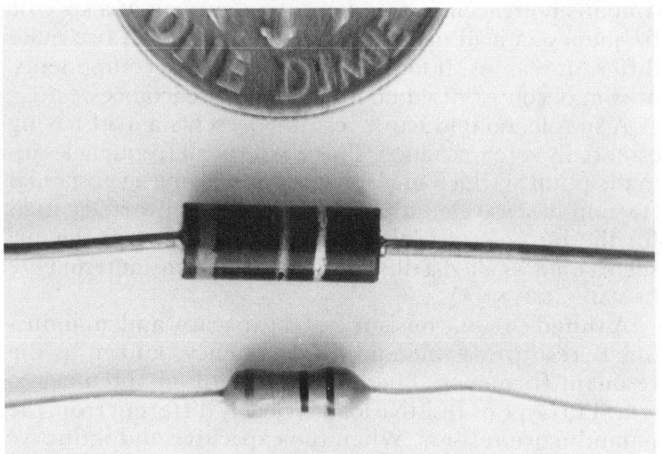

RESISTOR: Small resistors, such as the ones shown here, are extensively used in electronic equipment.

When a fairly large amount of current must be handled, resulting in a power dissipation of several watts, wirewound resistors are used. Such resistors consist of a coil of nichrome or other high-resistance wire wound on a heat-resistant core. Wirewound resistors have inductance as well as resistance (*see* WIREWOUND RESISTOR).

Variable resistors are extensively used in electronic apparatus. The most common type of variable resistor is the potentiometer. For high-current applications, a wirewound type of potentiometer, known as a rheostat, is used. *See also* POTENTIOMETER, RHEOSTAT.

RESISTOR-CAPACITOR-TRANSISTOR LOGIC

Resistor-capacitor-transistor logic, abbreviated RCTL, is a variation of resistor-transistor logic (*see* RESISTOR-TRANSISTOR LOGIC). In RCTL, bypass capacitors are connected in parallel with the base resistors (see illustration). This results in higher operating speed than is possible with conventional resistor-transistor logic.

The RCTL scheme was one of the earliest forms of high-speed logic. The power consumption is low to moderate. This type of logic circuit exhibits sensitivity to noise, as does conventional resistor-transistor logic. *See also* DIODE-TRANSISTOR LOGIC, DIRECT-COUPLED TRANSISTOR LOGIC, EMITTER-COUPLED LOGIC, HIGH-THRESHOLD LOGIC, INTEGRATED INJECTION LOGIC, METAL-OXIDE-SEMICONDUCTOR LOGIC FAMILIES, RESISTOR-TRANSISTOR LOGIC, TRANSISTOR-TRANSISTOR LOGIC, TRIPLE-DIFFUSED EMITTER-FOLLOWER LOGIC.

RESISTOR COLOR CODE

On most carbon-composition resistors, it is not possible to print the value in ohms, because the component is so small and because heat can cause printed characters to fade. A standard color code has therefore been adopted for use on such resistors. The color coding gives the value of the resistor in ohms, and the tolerance above and below the indicated value. On some resistors, the expected failure rate is also shown.

Most resistors have three or four color bands. An example is shown in the illustration. The bands are read by placing the resistor so that the bands are to the left of center, as illustrated in the figure; the bands are read from left to right.

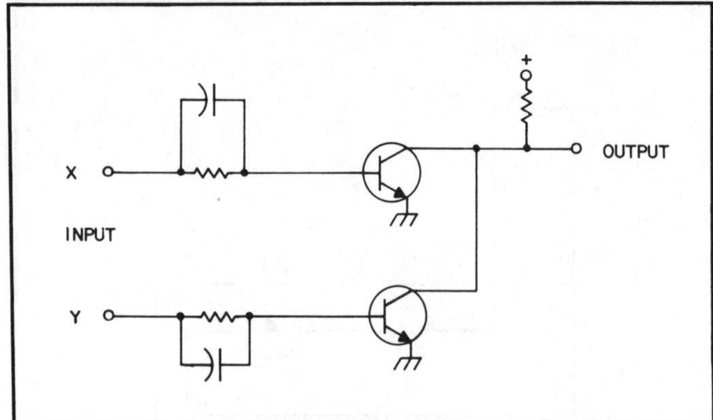

RESISTOR-CAPACITOR-TRANSISTOR LOGIC: Schematic diagram of a resistor-capacitor-transistor logic gate.

The first (left-most) band specifies a digital value ranging from 0 through 9. The second band indicates a digital value ranging from 0 through 9. The color of the band indicates the digit, according to Table A. The third band designates the power of 10 by which the two-digit number (according to the first two bands) is to be multiplied. These powers of 10 are given in Table B.

There may or may not be a fourth band. If there is no fourth band, then the resistor value may be as much as 20 percent greater than or less than the indicated value. A silver band indicates that the tolerance is plus or minus 10 percent. A gold band indicates that the tolerance is plus or minus 5 percent.

A fifth band may exist; this indicates the failure rate. The failure rate is expressed as a percentage of units expected to malfunction within 1,000 hours of operation at the maximum rated power dissipation. Table C indicates the colors for failure rate.

As an example, consider that a resistor has the following sequence of color bands, from left to right: yellow, violet, red, silver, red. The first two bands indicate the digits 47. The third band designates that the number 47 is

RESISTOR COLOR CODE: AT A, DIGITAL COLOR BANDS. AT B, MULTIPLIER BANDS. AT C, FAILURE-RATE BANDS.

A

Color	Digit
Black	0
Brown	1
Red	2
Orange	3
Yellow	4
Green	5
Blue	6
Violet	7
Gray	8
White	9

B

Color	Multiplier
Black	1
Brown	10
Red	10^2
Orange	10^3
Yellow	10^4
Green	10^5
Blue	10^6

C

Color	Failure Rate %/1000 hours
Brown	1
Red	0.1
Orange	0.01
Yellow	0.001

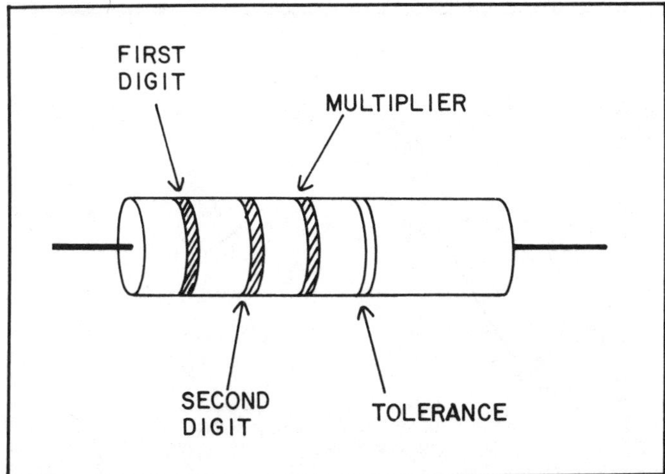

RESISTOR COLOR CODE: Resistor color-code bands.

to be multiplied by 100; thus the rated resistance is 4700 ohms. The fourth band designates a tolerance of plus or minus 10 percent (470 ohms above or below the rated value). The fifth band indicates an expected failure rate of 0.1 percent per 1,000 hours; in other words, one out of 1,000 resistors can be expected to fail within this length of time.

Some precision resistors have 5 color bands in which the first three bands are significant figures and the fourth is the multiplier. Then the fifth is the tolerance. Example: A resistor colored green yellow red red silver is 54200 ohms ±5 percent. *See also* RESISTOR.

RESISTOR-TRANSISTOR LOGIC

Resistor-transistor logic, abbreviated RTL, is a form of bipolar digital logic circuit that uses, as its name implies, resistors and transistors. Resistor-transistor logic was among the first bipolar forms of logic to be extensively used in the manufacture of digital integrated circuits. The operating speed is moderate, but can be increased by placing capacitors in shunt across the resistors (*see* RESISTOR-CAPACITOR-TRANSISTOR LOGIC). A schematic diagram of a RTL gate is shown.

The power-dissipation rate of RTL is fairly high. The design is quite simple and economical. However, RTL is susceptible to noise impulses. *See also* DIODE-TRANSISTOR LOGIC, DIRECT-COUPLED TRANSISTOR LOGIC, EMITTER-COUPLED LOGIC, HIGH-THRESHOLD LOGIC, INTEGRATED INJECTION LOGIC, METAL-OXIDE-SEMICONDUCTOR LOGIC FAMILIES, TRANSISTOR-TRANSISTOR LOGIC, TRIPLE-DIFFUSED EMITTER-FOLLOWER LOGIC.

RESOLUTION

Resolution is an expression of the ability of a radar, radio telescope, or optical telescope to distinguish between objects or targets that are close together. Resolution can be expressed in angular form, in radial (distance) form, or in absolute form.

Angular resolution is defined as the minimum anrle, with respect to the measuring station, that two objects or targets may subtend while still appearing separate. The two objects or targets may be at much different distances from the measuring station, but almost lined up. Angular

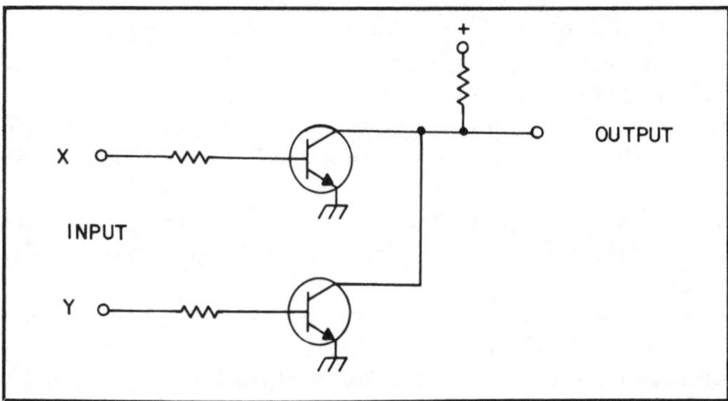

RESISTOR-TRANSISTOR LOGIC: Schematic diagram of a resistor-transistor logic gate.

resolution is the only expression of resolution in optical or radio telescopes. Angular resolution may be given in degrees, minutes of arc (abbreviated by an apostrophe: 1' = 1/60 degree), or in seconds of arc (abbreviated by a quotation mark: 1" = 1/3600 degree).

Radial resolution is defined as the minimum difference in range for which a radar can distinguish between two targets. Radial resolution is usually specified in feet, meters, miles, or kilometers. Radial resolution is of importance in radar, but not in astronomy.

The absolute resolution is defined as the minimum actual separation, in feet, meters, miles, or kilometers, for which a radar can distinguish between two targets. *See also* RADAR, RADAR TELESCOPE, RADIO ASTRONOMY, RADIO TELESCOPE, RESOLVING POWER.

RESOLVING POWER

Resolving power is the angular resolution obtainable with an optical telescope, radio telescope, or camera. Angular resolution is expressed in degrees, minutes of arc, or seconds of arc (*see* RESOLUTION). The fineness, or detail, in a visual image or radio image is related to the resolving power. The better the resolving power, the more detail there will be in the image.

The resolving power of a radio telescope is proportional to the diameter of the aperture in wavelengths (*see* RADIO ASTRONOMY, RADIO TELESCOPE). The resolving power in an optical telescope is proportional to the diameter of the objective lens or the reflecting mirror; the precision of the lens or mirror is also a factor. In television, the resolving power is largely dependent on the number of lines scanned in the frame (*see* TELEVISION).

RESONANCE

Resonance is a condition in which the frequency of an applied signal coincides with a natural response frequency of a circuit or object. In radio-frequency applications, resonance is a circuit condition in which there are equal amounts of capacitive reactance and inductive reactance. There must be nonzero reactances in a circuit for resonance to be possible; a purely resistive circuit cannot have resonance. Resonance occurs at a specific, discrete frequency or frequencies.

In a parallel-tuned or series-tuned inductance-capacitance (LC) circuit, the reactances balance at just one frequency. A parallel-tuned circuit exhibits maximum impedance at the resonant frequency; neglecting conductor losses, the resonant impedance of such a circuit is infinite, and decreases as the frequency departs from resonance. In a series-tuned circuit, neglecting conductor losses, the impedance at resonance is theoretically zero (*see* CAPACITIVE REACTANCE, IMPEDANCE INDUCTIVE REACTANCE). The impedance of a series-tuned circuit rises as the frequency departs from resonance (*see* RESONANCE CURVE).

In an antenna radiator, resonance occurs at an infinite number of frequencies; the lowest of these is the fundamental frequency, and the integral-multiple frequencies are the harmonics (*see* ANTENNA RESONANT FREQUENCY, FUNDAMENTAL FREQUENCY, HARMONIC). The impedance of an antenna at resonance consists of radiation resistance and loss resistance. This is a finite value that depends on many factors (*see* RADIATION RESISTANCE).

Radio-frequency resonance can occur within a metal enclosure known as a cavity. Cavities are used as tuned circuits at ultra-high and microwave frequencies. A cavity exhibits resonance at an infinite number of frequencies, just as an antenna does (*see* CAVITY RESONATOR). A length of transmission line exhibits resonance on the frequency at which it measures ¼ electrical wavelength, and also on all integral multiples of this frequency (*see* HALF-WAVE TRANSMISSION LINE, QUARTER-WAVE TRANSMISSION LINE).

Cavities and rigid objects have acoustic resonant frequencies (*see* REVERBERATION). A tuning fork is an example of an acoustically resonant device. Most musical instruments operate via acoustic resonance; a piano has resonant wires, for example, while horns and woodwinds have resonant air cavities. Mechanical resonance is a form of low-frequency acoustic resonance that occurs in rigid objects. Mechanical resonance is important to civil and mechanical engineers.

In radio-frequency applications, resonance is always the result of equal and opposite reactances. If the inductance and capacitance are known, the fundamental resonant frequency of any circuit can be calculated. *See also* PARALLEL RESONANCE, RESONANT CIRCUIT, RESONANT FREQUENCY, SERIES RESONANCE.

RESONANCE CURVE

A resonance curve is any graphic representation of the resonant properties of a circuit or object. Resonant curves are almost always plotted on the Cartesian plane, with the frequency as the independent variable. The dependent variable may be any characteristic that displays a peak or dip at resonance. In radio-frequency circuits, such parameters include the attenuation, current, absolute-value impedance, and voltage. Two examples of resonant curves are shown in the illustration.

At A, the absolute-value impedance of a lossless parallel-resonant inductance-capacitance (LC) circuit is plotted against the frequency. The absolute-value impedance of a lossless parallel-resonant circuit is given by:

$$Z = X_L X_C / (X_L + X_C)$$

where X_L and X_C represent the inductive and capacitive reactances, respectively. The curve at A is based on an inductance of 1 μH and a capacitance of 1 μF, resulting in a resonant frequency of about 159 kHz. The absolute-value impedance is theoretically infinite at resonance. (In a practical circuit, losses in the components and wiring result in a finite, but very large, absolute-value impedance at resonance.)

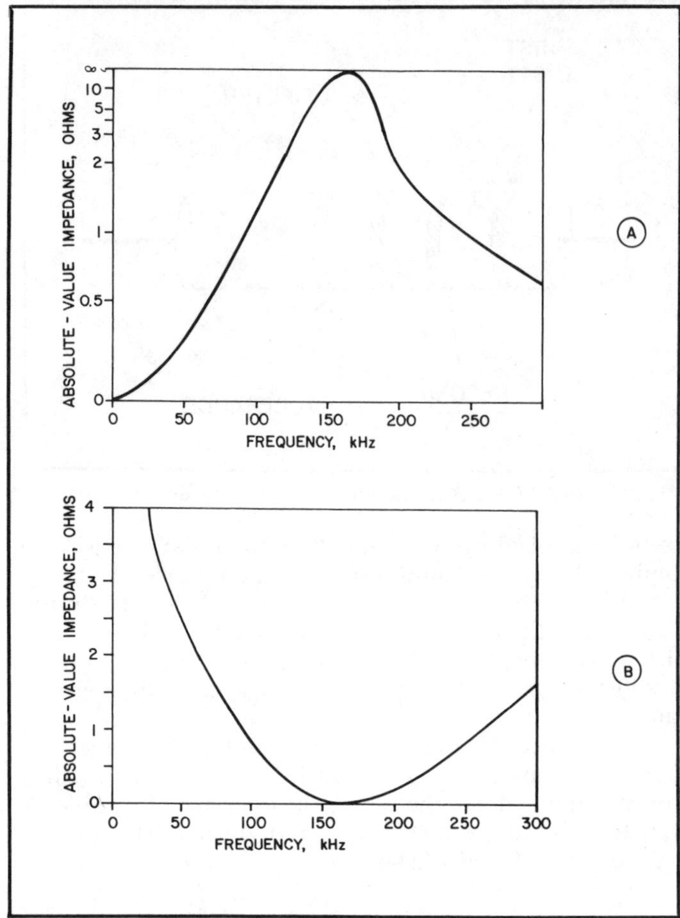

RESONANCE CURVE: Examples of resonance curves. At A, absolute-value impedance versus frequency for a parallel-resonant circuit. At B, absolute-value impedance versus frequency for a series-resonant circuit.

At B, the absolute-value impedance is plotted as a function of frequency for a lossless series-resonant circuit, having an inductance of 1 μH and a apacitance of 1 μF. This yields the same resonant frequency as the parallel configuration, but the absolute-value impedance shows a minimum rather than a maximum. In a lossless series-resonant circuit, the absolute-value impedance is

$$Z = X_L + X_C$$

At the resonant frequency, the absolute-value is theoretically zero. (In a practical circuit, losses in the components and wiring result in a finite, but very small, absolute-value impedance.) *See also* CAPACITIVE REACTANCE, IMPEDANCE, INDUCTIVE REACTANCE, PARALLEL RESONANCE, RESONANCE, RESONANT CIRCUIT, RESONANT FREQUENCY, SERIES RESONANCE.

RESONANT CAVITY

See CAVITY RESONATOR.

RESONANT CIRCUIT

A circuit is considered resonant if it contains finite nonzero reactances which cancel each other. A parallel or series inductance-capacitance (LC) circuit is an example of a resonant circuit at a specific frequency. Antennas, lengths of transmission line, and cavities are resonant circuits at a multiplicity of frequencies. Piezoelectric crystals have resonant properties similar to those of parallel LC circuits. Some ceramics and metal objects act as resonant circuits under certain conditions.

All resonant circuits exhibit variable attenuation, depending on frequency. Resonant circuits are extensively used in audio-frequency and radio-frequency design, for such purposes as obtaining selectivity, impedance matching, and notching. *Related articles include* ANTENNA RESONANT FREQUENCY, CAPACITIVE REACTANCE, CAVITY RESONATOR, HALF-WAVE TRANSMISSION LINE, IMPEDANCE, INDUCTIVE REACTANCE, NOTCH FILTER, PARALLEL RESONANCE, Q FACTOR, QUARTER-WAVE TRANSMISSION LINE, RESONANCE, RESONANCE CURVE, RESONANT FREQUENCY, SELECTIVITY, SERIES RESONANCE, TANK CIRCUIT, TRAP, TUNED CIRCUIT.

RESONANT FREQUENCY

The resonant frequency of a tuned inductance-capacitance circuit depends on the values of the components. The larger the product of the inductance and capacitance, the lower the resonant frequency will be. If a series-tuned or parallel-tuned circuit has an inductance of L henrys and a capacitance of C farads, then the resonant frequency f, in hertz, is given by the formula:

$$f = 1/(2\pi\sqrt{LC})$$

This formula also holds for values of L in microhenrys, C in microfarads, and f in megahertz.

The resonant frequency of a piezoelectric crystal depends on the thickness of the crystal, the manner in which it is cut, and, in some cases, the temperature. The presence of a coil or capacitor in series or parallel with a crystal als affects the resonant frequency to some extent (*see* CRYSTAL).

The resonant frequencies of an antenna radiator, assuming that no loading reactances are present, can be determined from the length of the radiator. The fundamental frequency f, in magahertz, for a free-space radiator can be determined from the length s in feet according to the formula:

$$f = 468/s$$

For s in meters,

$$f = 143/s$$

The radiator is resonant at all positive integral multiples of the fundamental frequency f. For a radiator operating against a ground plane, the fundamental resonant frequency f, in megahertz, is given according to the length s, in feet, by:

$$f = 234/s$$

For s in meters,

$$f = 71/s$$

The radiator is resonant at all positive integral multiples of the fundamental frequency f.

The resonant frequencies of any antenna radiator are affected by the presence of reactive elements. An inductor in series with an antenna radiator causes the resonant frequency to be lower than the above formulas would indicate. A capacitor in series will result in a higher resonant frequency than the above formulas would yield (*see* CAPACITIVE LOADING, INDUCTIVE LOADING).

A cavity exhibits resonance at frequencies that depend on its length. The same is true for half-wave and quarter-wave section of transmission line, when the velocity factor is taken into account. *See also* CAVITY RESONATOR, HALF-WAVE TRANSMISSION LINE, QUARTER-WAVE TRANSMISSION LINE, RESONANCE, RESONANT CIRCUIT.

RESONANT RESPONSE

See RESONANCE CURVE.

RESPONSE TIME

The length of time between the occurrence of an event, and the response of an instrument or circuit to that event, is called response time. Response time is of importance in switching circuits and measuring devices such as meters.

If the response time of a switching circuit is not fast enough, the switch will fail to actuate properly in accordance with an applied signal. Similarly, if the response time of a measuring instrument is too slow, the instrument will not give accurate readings.

If the response time of a switching device is much more rapid than necessary, undesirable effects may sometimes be observed. An example of this is an excessively fast time constant in the shaping circuit of a code transmitter; this can result in key clicks (*see* KEY CLICK, SHAPING). In a measuring instrument, a fast response time may be undesirable if average values must be determined. *See also* DAMPING, OVERDAMPING.

RESULTANT VECTOR

When two or more vectors are added, the resultant vector is the sum of the original vectors. *See* VECTOR, VECTOR ADDITION.

RESUSCITATION

See CARDIOPULMONARY RESUSCITATION, MOUTH-TO-MOUTH RESUSCITATION, MOUTH-TO-NOSE RESUSCITATION.

RETARDING-FIELD OSCILLATOR

See BARKHAUSEN-KURZ OSCILLATOR.

RETENTIVITY

See REMANENCE.

RETRACE

In a cathode-ray tube, the electron beam normally scans from left to right at a predetermined, and sometimes adjustable, rate of speed. At the end of the line, when the beam reaches the right-hand side of the screen, the beam moves rapidly back to the left-hand side to begin the next line or trace. This rapid right-to-left movement is called the retrace or return trace.

The forward trace creates the image that is displayed on the screen. The retrace is usually blanked out, so that it will not interfere with the viewing of the image. This is called retrace blanking. The ratio of the retrace time to the trace time is called the retrace ratio. *See also* CATHODE-RAY TUBE, OSCILLOSCOPE, PICTURE TUBE, TELEVISION.

RETURN CIRCUIT

For a complete, closed circuit to exist, the current must have some way to get back to the generator from the load. The path that the current follows to complete a circuit is called the return circuit or return path.

In some electrical systems, the return current flows through the ground. *See also* GROUND RETURN.

REVERBERATION

Acoustic resonance is usually called reverberation. In an enclosed chamber with sound-reflecting walls, resonance occurs at certain wavelengths because of phase reinforcement (*see* RESONANCE). the result is an increase in the sound intensity at a discrete fundamental frequency, and at all harmonics thereof. Most speaker enclosures exhibit a certain amount of reverberation; a well-engine red speaker can take advantage of this effect to enhance the frequency response.

A repeated sound echo, generated either by actual sound reflection or by electronic means, is called reverberation. Reverberation effects are sometimes used by musical groups. A device that produces reverberation is called a reverberation chamber or reverberation circuit. The time required for the reverberations to die out is called the reverberation time. The number of echoes per second is called the reverberati n frequency or reverberation rate.

REVERSE BIAS

When a semiconductor P-N juncion is biased in the nonconducting direction, the junction is said to be reverse-biased. Under conditions of reverse bias, the P-type semiconductor is negative with respect to the N-type semiconductor. This creates an ion-depletion region at the junction (*see* P-N JUNCTION).

If the reverse bias at a P-N junction is made large enough, the junction begins to conduct. This is called avalanche breakdown (*see* AVALANCHE BREAKDOWN). Some diodes are designed to take advantage of this effect; these devices are called Zener diodes. The minimum reverse voltage at which avalanche breakdown takes place is called the avalanche voltage or peak inverse voltage.

A vacuum tube is reverse-biased when the anode (plate) is negative with respect to the cathode. The resistance of a tube under reverse-bias conditions is extremely high. However, if the voltage becomes too great, conduction will occur. The minimum voltage at which revese-bias conduction takes place is called the reverse-breakdown voltage or peak inverse voltage.

Reverse bias may occur during half of an applied alternating-current cycle; this is the case in detection and rectification. Reverse bias may be applied deliberately by an external voltage source. This is done with varactor diodes and Zener diodes. *See also* DIODE, DIODE ACTION, FORWARD BIAS, PEAK INVERSE VOLTAGE, VARACTOR DIODE, ZENER DIODE.

REVERSE POWER

See REFLECTED POWER.

REVERSE VOLTAGE

See REVERSE BIAS.

RF

The acronym RF stands for radio frequency. In practice, the term RF is used quite loosely; it may serve as a descriptive indicator or all by itself. For example, we may speak of a radio-frequency current as an RF current, or simply as RF. The same is true for power and voltage. *See also* RADIO FREQUENCY, RF CURRENT, RF POWER, RF VOLTAGE.

RF AMMETER

An RF ammeter is a device intended for the measurement of radio-frequency (RF) current. An RF ammeter is sometimes used in an antenna system for the indirect determination of the RF power output from a transmitter.

When a transmission line is perfectly matched (*see* MATCHED LINE), and the antenna impedance is a pure resistance of Z ohms, the RF power P, in watts, can be determined from the current I according to the formula:

$$P = I^2Z$$

This formula holds only with a perfect match. *See also* RF CURRENT, RF POWER.

RF AMPLIFIER

There are various different types of amplifiers for radio-frequency energy. They may be broadly classified as signal amplifiers and power amplifiers.

Signal amplifiers are intended for receiving. A signal amplifier draws essentially no power from the source. The input impedance is extremely high. Signal amplifiers are used in the front ends of all radio-frequency receivers (*see* FRONT END, PREAMPLIFIER).

Radio-frequency power amplifiers are used in the driver and final stages of transmitters. Power amplifiers are usually of the Class-AB, Class-B, or Class-C type (*see* CLASS-AB AMPLIFIER, CLASS-B AMPLIFIER, CLASS-C AMPLIFIER, POWER AMPLIFIER).

Radio-frequency amplifiers may have narrow bandwidth, or they may be broadbanded. Narrow-band RF amplifiers offer superior unwanted-signal rejection in receivers, and provide attenuation of harmonics and spurious signals in transmitters. Broadbanded amplifiers are simpler to operate because no tuning is required, but undesired signals can be amplified along with the primary signal. Highpass and/or lowpass filters are often used in conjunction with broadbanded RF amplifiers to ameliorate this problem. *See also* HIGHPASS FILTER, LOWPASS FILTER.

RF CHOKE

A radio-frequency (RF) choke is an inductor used for the purpose of blocking RF signals while allowing lower-frequency and dc signals to pass. Such chokes are often used in electronic circuits when it is necessary to apply an audio-frequency or direct-current bias to a component without allowing RF to enter or leave.

Radio-frequency chokes typically have inductance values in the range of about 100 μH. The exact value depends on the impedance of the circuit, and on the frequency of the signals to be choked. If the impedance of a circuit is Z ohms at the frequency to be choked, an RF choke is usually selected to have an impedance of approximately 10Z ohms.

Radio-frequency chokes are commercially manufactured in a variety of configurations. Most have powdered-iron or ferrite cores and solenoidal windings. Some have air cores or toroidal windings. *See also* CHOKE, INDUCTIVE REACTANCE.

RF CLIPPING

The intelligibility of an amplitude-modulated or single-sideband speech signal can be increased by either of two basic methods: audio compression or radio-frequency-envelope compression. The latter of these two methods is usually called RF clipping, or RF speech processing.

In RF speech clipping, the signal envelope is increased in average amplitude by a combination of an amplifier, clipper, and filter. The amplifier boosts the signal voltage so that the minima are quite strong; the clipper cuts off the maxima, resulting in a signal with much greater average amplitude than the original. The filter, having a bandpass response with a bandwidth corresponding to the minimum needed for transfer of the modulation information, eliminates the splatter that would otherwise be caused by the clipping. The illustration is a block diagram of a simple RF clipping circuit that can be installed in the intermediate-frequency chain of a single-sideband transmitter.

When RF clipping is used, the readability of a signal can be improved when conditions are marginal, as compared with the readability when no clipping is employed. However, if the clipping is excessive, or if the filtering is inadequate, the signal may be distorted to such an extent that the intelligibility is made worse, not better. *See also* SPEECH CLIPPING, SPEECH COMPRESSION.

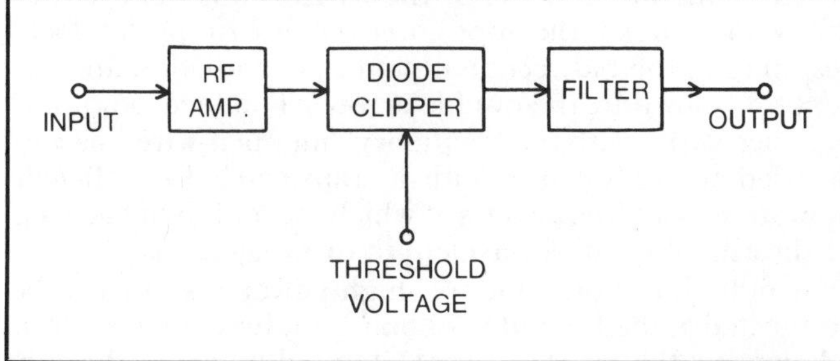

RF CLIPPING: Block diagram of an rf clipping circuit.

RF CURRENT

An alternating current having a frequency of at least 10 kHz is known as a radio-frequency (RF) current. The alternating component of a direct current that pulsates in magnitude at a rate of at least 10 kHz may also be called an RF current.

The intensity of an RF current is measured in root-mean-square amperes (*see* ROOT MEAN SQUARE). In a circuit having a nonreactive impedance of Z ohms and an RF power level of P watts, the rf current I is given by:

$$I = \sqrt{P/Z}$$

Radio-frequency current is measured by means of an RF ammeter. *See also* RF AMMETER.

RF FIELD STRENGTH

See FIELD STRENGTH.

RF FIELD-STRENGTH METER

See FIELD-STRENGTH METER.

RFI

See ELECTROMAGNETIC INTERFERENCE.

RF POWER

At radio frequencies, power can be manifested in three ways: It can be radiated, it can be dissipated as heat and/or light, and it can appear in reactive form.

Power can be radiated or dissipated only across a resistance; this is called true power (*see* TRUE POWER). A current and voltage can exist across a reactance, but no radiation or dissipation can take place in a reactance. This form of power is called reactive power (*see* REACTIVE POWER). The sum of the true power and the reactive power is known as the apparent power (*see* APPARENT POWER).

Power at radio frequencies is usually measured by means of a wattmeter in a transmission line. The wattmeter indicates the apparent power. If reactive power is present in a transmission line and antenna system, it causes an exaggerated reading on the wattmeter. *See also* REFLECTED POWER, REFLECTOMETER, WATTMETER.

RF POWER AMPLFIER

See POWER AMPLIFIER.

RF SPEECH PROCESSING

See RF CLIPPING, SPEECH PROCESSING.

RF VOLTAGE

An RF voltage is an alternating potential difference between two points, having a frequency of 10 kHz or more. If a voltage fluctuates at a rate of 10 kHz or more, but maintains the same polarity at all times, the alternating component constitutes an RF voltage.

The intensity of RF voltage is measured in root-mean-square volts (*see* ROOT MEAN SQUARE). In a circuit having a nonreactive impedance of Z ohms and an RF power level of P watts, the RF voltage E is given by:

$$E = \sqrt{PZ}$$

A radio-frequency voltage is measured by means of an RF voltmeter. An oscilloscope can be used to measure the peak or peak-to-peak RF voltage; the root-mean-square value is obtained by multiplying the peak value by 0.707 or the peak-to-peak value by 0.354. *See also* OSCILLOSCOPE, RF VOLTMETER.

RF VOLTMETER

At radio frequencies, the measurement of voltage is somewhat more difficult than at the 60-Hz ac line frequency. Radio-frequency voltage can be measured directly by means of an oscilloscope, but a meter can only work indirectly.

Most radio-frequency voltmeters operate by rectifying and filtering the signal, obtaining a direct-current potential that is proportional to the radio-frequency voltage. An uncalibrated reflectometer uses an RF voltmeter of this type (*see* REFLECTOMETER).

An RF voltmeter is usually frequency-sensitive. This is because the rectifier diode exhibits a capacitance that depends on the frequency of the applied voltage. When using an RF voltmeter, therefore, it is necessary to be sure that the frequency is within prescribed limits.

An RF voltmeter can be used indirectly to determine RF power. If the load contains no reactance and has an impedance of Z ohms, then the RF power P, in watts, that is dissipated and/or radiated by the load is given by:

$$P = E^2/Z$$

where E is the voltmeter reading in root-mean-square volts. *See also* RF VOLTAGE.

RF WATTMETER

See REFLECTOMETER, WATTMETER.

RHEOSTAT

A rheostat is a form of variable resistor. A rheostat consists of a solenoidal or toroidal winding of resistance wire such as nichrome, two fixed end contacts, and a sliding or rotary contact. A solenoidal rheostat is shown in the illustration at A; a rotary or toroidal type is shown at B.

Rheostats have inductance as well as resistance, and they are therefore not suitable for radio-frequency use. The rheostat is not continuously adjustable; the resistance is determined by a whole number of wire turns, presnting a finite number of discrete values.

Rheostats can be made to dissipate large amounts of power, and this makes them useful for voltage dropping in high-current circuits. *See also* POTENTIOMETER.

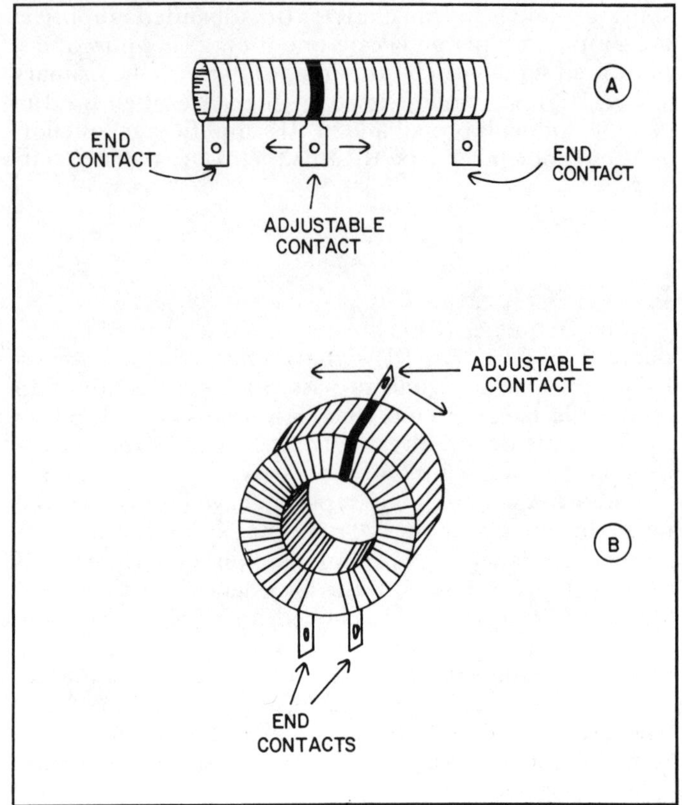

RHEOSTAT: At A, a solenoidal rheostat. At B, a toroidal rheostat. Both types exhibit inductance, and are therefore suitable only for use at low frequencies.

RHOMBIC ANTENNA

A rhombic antenna is a form of longwire antenna that exhibits gain in one or two fixed directions. Rhombic antennas are used mostly at the high frequencies (3 to 30 MHz), and are constructed of wire. The rhombic antenna gets its name from the fact that it is shaped like a rhombus (see illustration).

The amount of gain of a rhombic antenna depends on the physical size in wavelengths, and also on the corner angles. The larger the rhombic antenna in terms of the wavelength, the more elongated the rhombus must be in order to realize optimum gain. A rhombic antenna having legs measuring ½ wavelength can produce approximately 2 dBd of power gain; this figure increases to 5 dBd for legs of 1 wavelength, 10 dBd for legs of 3 wavelengths, and 12 dBd for legs of 5 wavelengths, provided the corner angles are optimized. (A detailed discussion of rhombic-antenna design is beyond the scope of this book, but many references are available.)

The rhombic antenna shown can be fed at either corner at which the angle between the wires is less than 90 degrees. The wires at the opposite corner are simply left free, resulting in the bidirectional pattern of radiation and reception (shown by the double arrow). The feed-point impedance varies with the frequency, but open-wire line can be used in conjunction with a transmatch for efficient operation at all frequencies at which the length of each leg of the rhombus in ½ wavelength or greater.

A 600-ohm, noninductive, high-power resistor can be connected at the far end to obtain a unidirectional pattern (shown by the single arrow). The addition of the terminating resistor results in a nearly constant feed-point

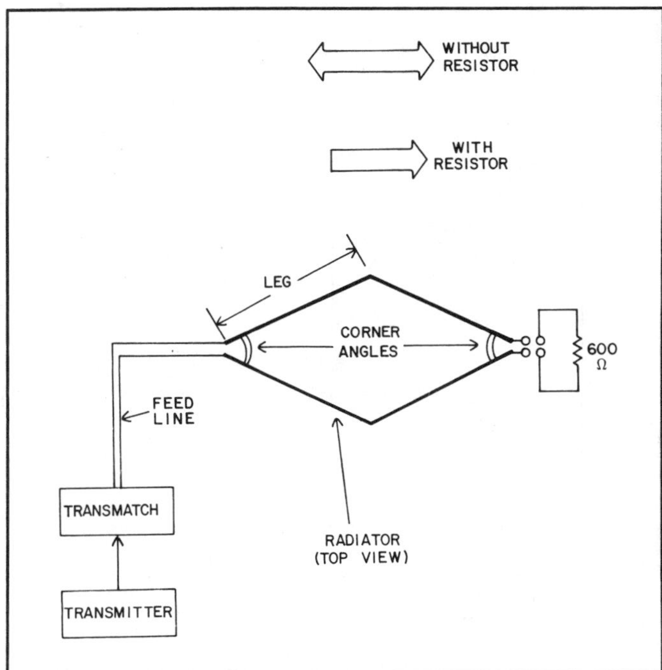

RHOMBIC ANTENNA: The double arrow shows the directinal maxima if a terminating resistor is not used. The single arrow shows the pattern with a terminating resistor.

impedance of 600 ohms, at all frequencies at which the leg length is ½ wavelength or more.

The main advantages of the rhombic antenna are its power gain, and the fact that it can produce this gain over a wide range of frequencies. The principal disadvantages are that it cannot be rotated, and that it requires a large amount of real estate. In recent years, the quad and yagi antennas have become much more common than the rhombic. *See also* QUAD ANTENNA, YAGI ANTENNA.

RIBBON CABLE

A ribbon cable is a flat, flexible, multiconductor cable. Ribbon cables are used in situations where the interconnected components must be frequently moved. Ribbon cables are found in computers, electronic printers and typewriters, and other devices with movable or often-replaced parts.

Ribbon cables are generally made using stranded, specially treated wire. This allows the cable to be flexed thousands of times without conductor breakage.

RIBBON MICROPHONE

See VELOCITY MICROPHONE.

RIGHT ASCENSION

Right ascension is one of two coordinates used by astronomers to define the position of an object in the sky.

Right-ascension coordinates are formed by great circles, all of which pass through the celestial poles. The zero-degree right-ascension circle passes through the Vernal Equinox on the celestial equator. (The sun crosses the Vernal Equinox on approximately March 21 of each year.)

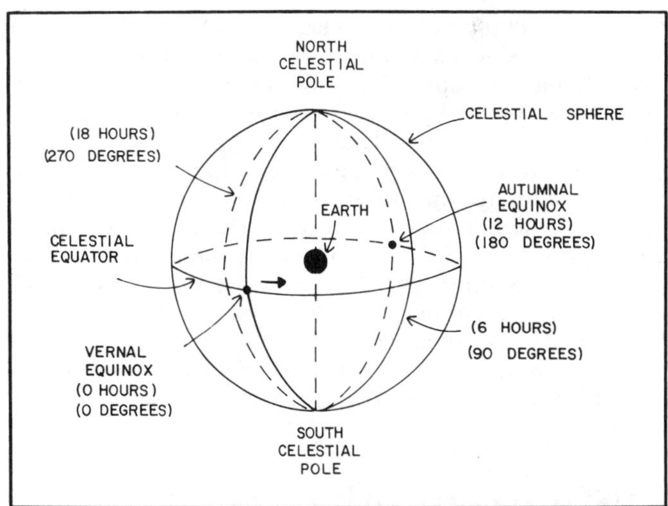

RIGHT ASCENSION: Pictorial illustration of right-ascension coordinates. The right-ascension coordinate is equivalent to celestial longitude, measured eastward from the Vernal Equinox.

Right ascension is measured eastward from this point, either in degrees (from 0 to 360) or in hours, minutes, and seconds (from 00:00:00 to 24:00:00). The hour circles are separated by 15 degrees of arc; the minute circles, by 15 minutes of arc; the second circles, by 15 seconds of arc.

The illustration shows the scheme of right-ascension coordinates on the celestial sphere. The position of an object in the sky is uniquely defined by the right ascension and the declination. *See also* DECLINATION.

RIGHT-HAND RULE

See FLEMING'S RULES.

RING COUNTER

A ring counter is a form of shift register in which data bits move in circular or loop fashion. If a data bit reaches the end of the ring counter, the next pulse will cause the data bit to revert to the beginning of the loop. This circular data movement can occur in either a left-to-right (clockwise) or right-to-left (counterclockwise) direction.

If a ring counter has n data positions, the information pattern is repeated every n pulses, as each data bit completes the circle exactly once. *See also* SHIFT REGISTER.

RINGING

When an alternating-current signal is applied to a tuned circuit at the resonant frequency, circulating currents are set up in the inductance and capacitance. These circulating currents continue for a short time after the signal is removed. This effect is called ringing.

The amount of time for which ringing will continue depends on the Q factor of the tuned circuit; the higher the Q, the longer it takes for the ringing to decay (*see* Q FACTOR). Ringing can be either desirable or undesirable.

In a Class-B or Class-C radio-frequency power amplifier, the tuned output circuit stores energy during the conduction period and releases it during the nonconduction period of the transistor or tube. This is a form of ringing called the flywheel effect (*see* CLASS-B AMPLIFIER,

CLASS-C AMPLIFIER, FLYWHEEL EFFECT). It results in a nearly pure sine-wave output signal from the amplifier, even though the amplifier does not conduct for the full cycle. The flywheel effect also enhances the operation of most types of tuned oscillators.

In an audio filter having extremely narrow bandwidth, ringing limits the maximum data speed that can be received. If the bandpass is too small, the maximum speed will be too slow, and reception will be impaired. Less frequently, the same problem may be encountered in intermediate-frequency bandpass filters. Ringing in a bandpass filter can be reduced by designing the circuit so that the response is rectangular. *See also* BANDPASS FILTER, BANDPASS RESPONSE, RECTANGULAR RESPONSE.

RINGING SIGNAL

In a telephone system, alternating-current or direct-current (nonvoice) signals are sent along the line between two stations for two purposes: to inform the receiving operator of an incoming call, and to inform the sending operator that the distant telephone is ringing. These signals are called ringing signals.

The current that actuates the bell, tone oscillator, buzzer, or other device at the receiving station is called ringdown. The signal is transmitted at a rate of about 20 Hz, and is interrupted at regular intervals. Ringdown consists of a fairly high level of current, since it must actuate a loud transducer. The return signal is called ringback; it is an audio-frequency signal having the same amplitude as the voice signals that are sent and received in a telephone conversation. The ringback signal is usually sent at a frequency of 500 Hz, modulated at 20 Hz to create a rippling sound. *See also* TELEPHONE.

RING MODULATOR

A ring modulator is a circuit that is used for mixing or modulating of radio-frequency signals. The ring modulator consists of four semiconductor diodes connected in a loop, allowing current to circulate around and around. The signal inputs and outputs are taken from the apex points of the loop.

The ring modulator is a passive circuit. The main advantage of the circuit is its simplicity. The ring modulator does not produce gain, but this is not a major problem in most applications, since the ring modulator can be followed by a simple amplifier circuit.

The ring modulator may be used as a balanced modulator for generating single-sideband signals; it can be employed as a mixer in superheterodyne circuits; it can be used as a product detector. *See also* BALANCED MODULATOR, DOUBLE BALANCED MIXER, MIXER, MODULATOR, PRODUCT DETECTOR.

RING WINDING

See TOROID.

RIPPLE

Ripple is the presence of an alternating-current component in a direct-current signal. Usually, the term refers to the residual 60-Hz or 120-Hz ac component in the output of a dc power supply. The ripple in the output of a power supply can result in modulation of a radio-frequency transmitter; power-supply ripple can also cause apparent modulation of received signals. This modulation is, itself, sometimes called ripple.

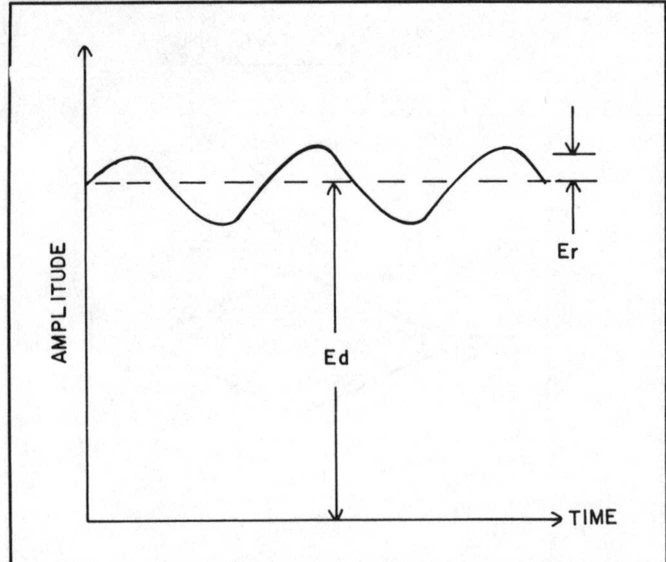

RIPPLE: Ripple voltage. The ripple can be expressed as the root-mean-square ripple, Er, divided by the average direct-current voltage, Ed.

The ripple in the output of a power supply can be expressed quantitatively in two ways. The actual ripple voltage is determined by eliminating the direct-current component, and then determining the root-mean-square value of the remaining alternating current (*see* ROOT MEAN SQUARE). We might call this voltage Er, and the steady direct-current voltage Ed. We can then express the ripple as a percentage, Rp:

$$Rp = 100\ Er/Ed$$

These expressions of ripple are shown in the illustration.

Ripple can be reduced by the use of large-value capacitors and chokes in the filter of a power supply. The amount of filtering depends on whether the ripple frequency is 60 Hz or 120 Hz; less filtering is needed at 120 Hz. The amount of filtering also depends on the load resistance. The lower the load resistance for a given power supply, the more filtering is required. *See also* FILTER, FILTER CAPACITOR, POWER SUPPLY.

RISE

The increase in amplitude of a pulse or waveform, from zero to full strength, is called the rise. In acoustics, and in automatic-gain/level-control systems, rise is called attack (*see* ATTACK, AUTOMATIC GAIN CONTROL, AUTOMATIC LEVEL CONTROL). The rise of a pulse or waveform may sometimes appear instantaneous, such as in a square wave; but it is never actually so. A certain finite amount of time is required for the rise.

The rise in output of some devices, such as high-wattage incandescent bulbs, can be watched and noticed. The rise in amplitude from other devices, such as neon lamps or light-emitting diodes, is too rapid to be seen. But the rise, no matter how rapid, is never instantaneous.

The illustration shows the rise of the waveform in a code

transmitter, following the change from the key-up condition to the key-down condition. The rise curve is logarithmic. The opposite of rise—the drop in amplitude from full strength to zero—is called decay. *See also* DECAY, DECAY TIME, RISE TIME.

RISE TIME

The rise time of a pulse or waveform is the time required for the amplitude to rise from 10 percent to 90 percent of the final value. The time interval begins at the instant the amplitude begins to rise (see previous illustration), and ends when the determined percentage has been attained.

The rise of a pulse or waveform often proceeds in a logarithmic manner, as does the decay (*see* LOGARITHMIC DECREMENT). In theory, the amplitude never reaches the final value; it is always rising a little bit, until decay begins. In practice, a point is reached at which the amplitude can be considered maximum. This point may be chosen for the determination of the rise time interval. *See also* DECAY, DECAY TIME, RISE, TIME CONSTANT.

RL CIRCUIT

See RESISTANCE-INDUCTANCE CIRCUIT.

RL TIME CONSTANT

See RESISTANCE-INDUCTANCE TIME CONSTANT.

RMM

See READ-MAINLY MEMORY.

RMS

See ROOT MEAN SQUARE.

ROBOTICS

Robotics is the relatively new science involving the design, construction, use, and maintenance of robots for various purposes. The idea of the robot or automation is not new; automatic or semi-automatic machinery has been around for years. But today, machines can be built to perform much more complex tasks than ever before; these modern robots are run by computers.

A robot may be constructed in "quasi-human" form, with a head, arms, and perhaps legs. When most people think of a robot, they think of something that looks like a mechancial human. But most robots do not look much like people. The design of a robot is tailored to the application, not to the human body (unless that configuration is ideal for the purpose!).

Robots are being used more and more in industry, to relieve people of mundane chores. For example, in automobile manufacturing, much of the assembly-line work is now done by robots. The Japanese have made extensive use of robots in various assembly-line functions, and may be considered pioneers of the use of robots in manufacturing.

Robotics is still a young branch of science and engineering. Much research is being conducted to learn how to apply robotics to new fields. It is impossible to give a complete description of robotics here. However, several informative volumes are currently available that discuss the various aspects of robotics in detail.

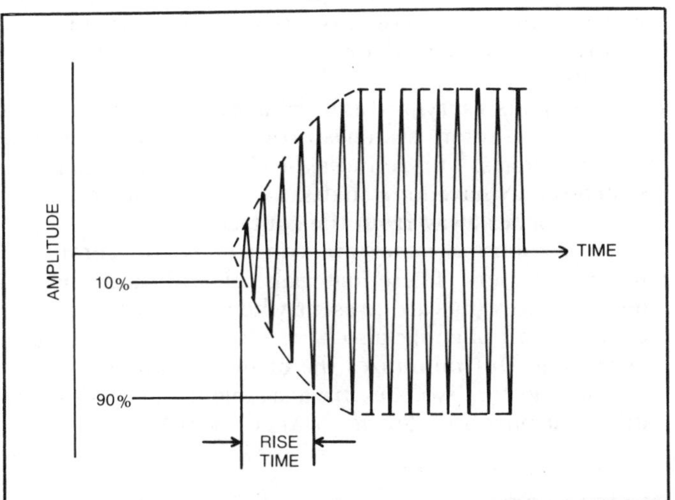

RISE TIME: The rise time of a signal is the time required for transition from 10 percent to 90 percent of the final amplitude.

ROENTGEN

The roentgen is a unit of radioactive flux, generally used to express the total exposure to X rays and gamma rays. An exposure of 1 roentgen results in ionization equivalent to 93 ergs per gram, or 2.58×10^{-4} calorie per kilogram.

A person in an average environment can expect to receive between 10 and 50 roentgens of radiation in a lifetime. An exposure of more than 100 roentgens, received within a few hours or days, can cause radiation sickness. Higher levels can result in violent sickness and death. *See also* GAMMA RAY, X RAY.

ROLL

In a television picture, roll is the result of a lack of vertical synchronization. Roll is sometimes called flip-flop or rolling.

Rolling can generally be eliminated by changing the vertical-hold adjustment in a televison receiver. Adjusting the antenna may help by improving reception in general. *See also* TELEVISION.

ROLLOFF

When a circuit exhibits a gradual increase in attenuation as the input frequency changes, the characteristic is known as rolloff. Rolloff may be the result of an increase in attenuation as the frequency increases; it may be the result of an increase in attenuation as the frequency decreases.

Rolloff is usually mentioned in reference to audio-frequency characteristics. If the attenuation increases as the audio frequency rises, a bass or lowpass response is indicated. If the attenuation increases as the frequency drops, a treble or highpass response is indicated (*see* BASS RESPONSE, HIGHPASS RESPONSE, LOWPASS RESPONSE, TREBLE RESPONSE).

Rolloff is usually expressed quantitatively in decibels per octave. If the amplitude of a signal drops 6 dB per octave (as measured in volts), it means that the amplitude is E volts at 2f, for a given frequency f, if the amplitude is 2E volts at frequency f. *See also* DECIBEL, OCTAVE.

ROM

See READ-ONLY MEMORY.

ROOF MOUNT

A roof mount is a bracket that is used for attaching an antenna mast to a peaked roof. Roof mounts are available in most electronics and hardware stores. Roof mounts are generally used for mounting of television-antenna masts, and masts for smaller amateur and CB antennas.

A typical roof mount is shown in the illustration. This roof mount accepts a mast having a diameter of about 1 inch to 3 inches. Guy wires are required to keep the mast vertical. *See also* GUYING.

ROOT MEAN SQUARE

The current, power, or voltage in an alternating-current signal may be determined in various ways. The most common method of expressing the effective value of an alternating-current waveform is the root-mean-square (RMS) method. The root-mean-square current, power, or voltage is an expression of the effective value of a signal.

The root-mean-square current, power, or voltage is determined via the following procedure. First, the amplitude is squared, so that the negative and positive halves of a waveform are made identical. Then the value is averaged over time. Finally, the square root of the average square value is determined. Mathematically, the RMS value is an expression of the direct-current effective magnitude of an alternating-current or pulsating-direct-current waveform.

For a sine wave, the RMS value is 0.707 times the peak value, or 0.354 times the peak-to-peak value. For a square wave, the RMS value is the same as the peak value, or half the peak-to-peak value. For other waveforms, the ratio varies. *See also* PEAK-TO-PEAK VALUE, PEAK VALUE.

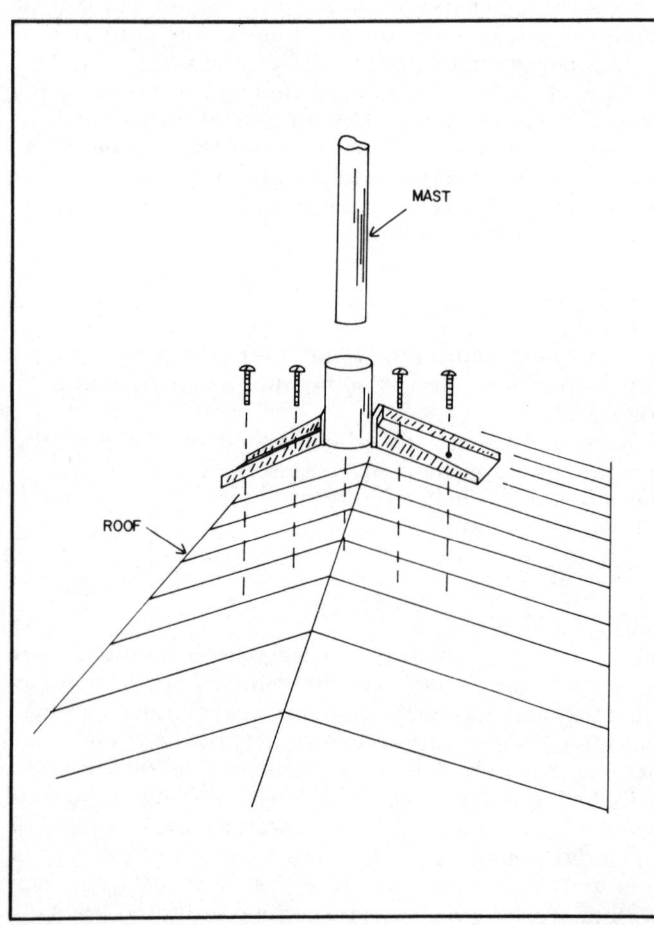

ROOF MOUNT: A roof mount suitable for use with a television-antenna mast.

ROTARY CONVERTER

See DYNAMOTOR.

ROTARY DIALER

In older telephone systems, a rotary dialer is used instead of the more modern tone dialer. The rotary dialer interrupts the telephone circuit a specific number of times for the dialing of a number. If the dialed digit, n, is between 1 and 9 inclusive, the rotary dialer interrupts the system n times. If n = 0, the dialer interrupts the circuit 10 times.

The rotary dialer is a spring-loaded, circular wheel with ten finger holes for the ten digits. To dial a given digit, the finger is placed in the appropriate hole and the dialer is turned clockwise until the finger reaches the stop. The dialer then turns counterclockwise by itself, interrupting the circuit at a rate of about 5 Hz. *See also* TELEPHONE, TOUCHTONE®.

ROTARY SPARK GAP

See SPARK TRANSMITTER.

ROTARY SWITCH

A rotary switch is a switch that is thrown or actuated by turning. Rotary switches are generally used in multiple-throw applications. A rotary switch may have as many as 10 or even 20 throw positions; some rotary switches have several poles as well (*see* MULTIPLE-POLE, MULTIPLE-THROW).

In circuits carrying high voltages, especially at radio frequencies, rotary switches provide wide separation between contacts. This reduces the chance of arcing. Rotary switches may have long shafts, so that the control knob is several inches away from the contacts.

Most rotary switches are constructed on ceramic or phenolic wafers. These materials can withstand the high temperatures that are present when the current is high. Ceramic and phenolic also have excellent dielectric properties, and this minimizes the chance of arcing between contacts. Rotary switches are available in various sizes for different purposes. *See also* WAFER SWITCH.

ROTATOR

See ROTOR.

ROTOR

An antenna rotator is often called a rotor. This device is a slow-operating motor, capable of withstanding and providing a large amount of torque.

Antenna rotors are commercially manufactured in many different sizes and configurations for various purposes. Most rotors turn at the rate of one rotation per minute, or about 6 degrees per second. Small rotors are used with directional television receiving antennas; larger rotors are used in amateur and commercial radio installations in conjunction with log-periodic, quad, or yagi antennas.

A rotor is sometimes installed inside a tower (A in the illustration). This arrangement prevents excessive lateral strain on the mechanism; the only forces occur vertically downward (because of the weight of the antenna) and in a clockwise or counterclockwise direction (torque). Smaller antenna systems may allow the less rugged installation method shown at B. There is some lateral strain on the rotor mechanism when this installation scheme is used.

When choosing a rotor for a particular antenna system, it is important that the device be large enough and rugged enough. However, the use of an excessively heavy rotor is a waste of money. Small television-antenna rotors, available in most hardware stores, can be used with small amateur and CB antennas. For larger antennas at frequencies below the 27-MHz (11-meter) band, more rugged rotors can be found in amateur-radio stores or ordered via catalogs.

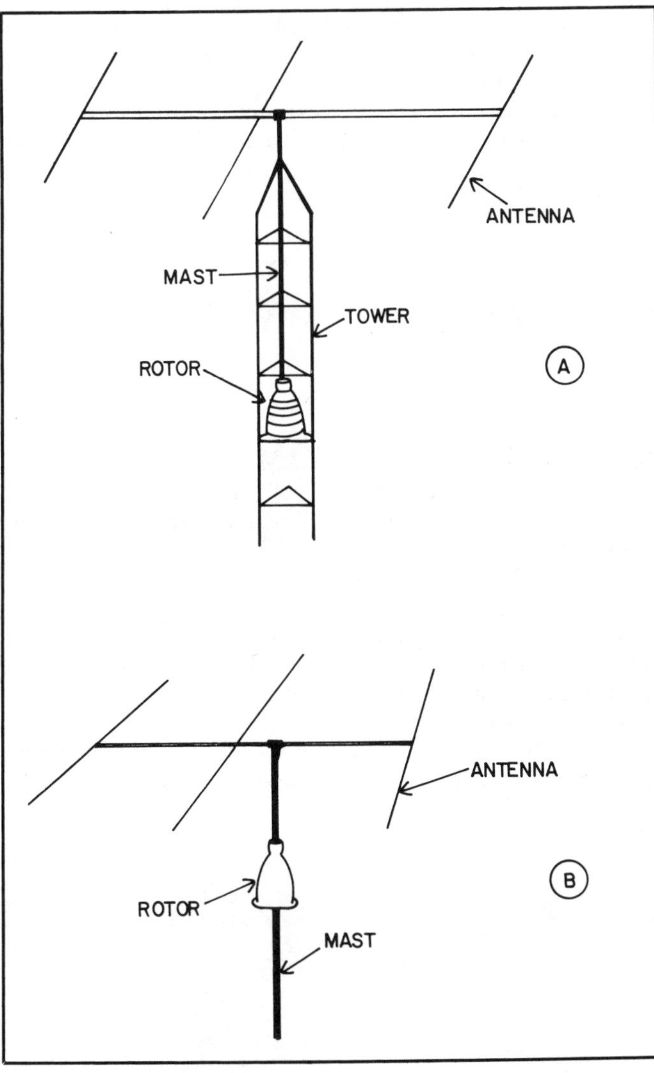

ROTOR: Mounting of an antenna rotor in a tower (A) and on a mast (B).

Most rotors are designed for the purpose of turning an antenna in the azimuth plane. However, some rotors are made especially for moving an antenna in the elevation plane. In a satellite-communications system, one of each type of rotor is employed in an azimuth-elevation (az-el) configuration (*see* AZ-EL). This arrangement can be used with groups of yagi antennas, helical arrays, or dish antennas.

The term rotor may refer to various other components or devices. Certain variable inductors have one fixed and one rotatable coil; the movable coil is called the rotor. The moving, or turning, part of a generator or motor is called the rotor (*see* GENERATOR, MOTOR). The movable plates in a variable capacitor are called the rotor or rotor plates (*see* AIR-VARIABLE CAPACITOR).

ROUNDING OF NUMBERS

It is sometimes desirable to reduce the number of digits contained in a number. This is done by either of two processes. The most common method is called rounding or rounding off. Less often, numbers are truncated.

To round off a number to m digits from n digits ($m<n$), the following procedure is used. First, digits are stricken from the right-hand side of the numeral, one at a time, until the expression contains $m+1$ digits. If the extreme right-hand digit in the remaining expression is 5, 6, 7, 8, or 9, then the digit second from the right is increased by 1. If the extreme right-hand digit is 0, 1, 2, 3, or 4, then the digit second from the right remains the same. Finally, the digit farthest to the right is stricken, leaving m digits in the rounded expression.

As an example, suppose we wish to round off the number 5.7812 to two significant figures. We strike the 1 and the 2, obtaining 5.78. Since the extreme right-hand digit is 8, we increase the digit second from the right by 1, getting 5.88. Finally, we strike the extreme right-hand digit, obtaining 5.8.

Rounding of numbers is done automatically by most small calculators and computers. Some calculators and computers allow the operator to select rounding or truncation. Truncation consists of simply striking all of the unwanted right-hand digits. In the above example, 5.7812 becomes 5.7 when truncated to two significant figures. *See also* SCIENTIFIC NOTATION, SIGNIFICANT FIGURES.

ROUTINE

See ALGORITHM, PROGRAM.

R-S FLIP-FLOP

See FLIP-FLOP.

R-S-T FLIP-FLOP

See FLIP-FLOP.

RST SYSTEM: THE RST (READABILITY-STRENGTH-TONE) SYSTEM.

Readability	
1—Unreadable signals	
2—Barely readable signals	
3—Readable, but with difficulty	
4—Almost perfectly readable	
5—Readable with no difficulty	
Strength	
1—Faint signals	
2—Very weak signals	6—Good signal strength
3—Weak signals	7—Moderately strong signals
4—Fair signals	8—Strong signals
5—Fairly good signals	9—Very strong signals
Tone (Code Only)	
1—AC Note, 60-Hz	
2—Rough AC, 60-Hz or 120-Hz	6—Definite trace of ripple
	7—Some ripple, but not much
3—Rectified but unfiltered note	8—Barely detectable ripple
4—Rough, some filtering	
5—Markedly rippled note	9—No detectable ripple

RST SYSTEM

The RST system is a scheme used by radiotelegraph operators for telling each other about the quality of reception. The acronym RST stands for Readability-Strength-Tone. A variation of the system, expressing only the readability and the signal strength, is used by radiotelephone operators.

The readability is given by a number from 1 to 5. The strength is indicated as a number from 1 to 9. (The higher numbers indicate improving reception or signal strength.) The tone quality figure is an expression of the amount of ripple on a code signal (*see* RIPPLE). The designator ranges from 1 to 9. With modern code transmitters, the tone designator should always be 9; circuit design has improved greatly since the RST system was first implemented. If a signal has a tone-quality figure less than 9, the transmitter is not operating according to the state of the art.

A radiotelegraph operator gives the RST as a three-digit number, according to the definitions shown in the table. Radiotelephone operators use only the readability and strength figures. In a radiotelegraph conversation, the operator at the other end might send to you, "RST 389," which means, "Your signals are readable with difficulty; they are strong; the tone is perfectly pure." The radiotelephone operator would tell you, "You are 3 by 8." (The marginal readability, in spite of the signal strength, implies interference in this situation.)

Most communications receivers are equipped with signal-strength meters, calibrated in units called S units. This type of meter, called an S meter, is usually marked from 1 to 9 and in decibels above the S9 level. The S meter can be used as a guide to the signal-strength figure to be given in an RST signal report. However, many operators prefer to use their own judgment for both readability and signal strength. *See also* S METER, S UNIT.

RUBY LASER

See LASER.

RUMBLE

Rumble is an effect that sometimes occurs in a poorly installed high-fidelity system. Rumble often is the result of vibration of a phonograph turntable, caused by low-frequency sound from the speakers.

Rumbling begins when a loud, low-frequency note causes the turntable to vibrate. This movement, in turn, produces low-freqeuncy modulation of the audio output; this modulation itself leads to more rumble. In a severe case, rumble can get so severe that the stylus jumps across the disk. This can damage the stylus or the disk.

Rumble can be eliminated by the use of shock-absorbing materials set underneath the turntable apparatus. Shock-absorbing materials can also be placed under the speakers. The speakers should be located at a reasonable distance from the turntable. A turntable should never be placed directly on top of a speaker cabinet. Rumble can also be caused by poor bearings on the turntable, or by improper shock mounts for the driving mechanism. *See also* HIGH FIDELITY, TURNTABLE.

RUNAWAY

See THERMAL RUNAWAY.

RUTHERFORD ATOM

In 1912, Ernest Rutherford, a physicist, developed a model of the atom that has led to modern atomic theory and particle physics. Rutherford suggested that charged particles orbit other charged particles having opposite electric charge.

Rutherford realized that opposite electric charges cause a force of attraction. This force would, he knew, keep the orbiting particles from flying away from the nucleus, or central particle.

Rutherford's model of the atom was revised somewhat by Niels Bohr a year later. *See also* BOHR ATOM.

SAFETY FACTOR

Many electrical and electronic components are rated according to the maximum current, voltage, or power they can handle. For example, a resistor may be rated at ½ watt; a vacuum tube may be rated at 100 watts of continuous plate dissipation; a diode might be rated at 500 peak inverse volts. These ratings incorporate a factor called the safety factor.

The safety factor allows for the possibility that a component or device might be operated at, or near, the maximum limit of its rated capability. We might insist that a ½-watt resistor actually dissipate ½ watt; we might design an amplifier so that its tubes are run at their maximum level of rated plate dissipation; we might design a rectifier circuit using diodes that are operated at their maximum peak-inverse-voltage rating. The ratings are purposely under-quoted by the manufacturer for this reason; the actual limits may be from 10 to 20 percent higher than the rated limits. The above-mentioned resistor, for instance, may really be able to handle 0.6 watt before it burns out. The manufacturer includes a safety factor in the ratings. This improves the reliability of the component, because it provides a buffer zone that allows for slight variation in actual ratings from one component to another.

When we design a circuit, we should never subject components to the maximum rated current, voltage, or power. We should incorporate our own safety factor of 10 to 20 percent, in addition to that of the manufacturer. This will ensure that the components will remain operational for the longest possible time, thus greatly minimizing service problems with the equipment.

Some ratings obviously do not incorporate safety factors. An example is a 100-watt incandescent bulb or an 0.1-μF capacitor. Component values of this type may be in error by a certain percentage; the maximum possible error is called the tolerance. If the component value in a particular application is critical, we may want to use a high-precision component. We would then install a component with a tolerance somewhat smaller (more precise) than we think necessary. The difference between the necessary component tolerance and the tolerance of the component we use would then constitute a tolerance safety factor. *See also* TOLERANCE.

SAFETY GROUNDING

When several different pieces of electrical or electronic apparatus are used together, a potential difference can develop between their respective chassis. This can present a shock hazard when moderate or high voltages are used. Experienced electricians and engineers will tell you: "Never touch two grounds at the same time." You may already know what they mean!

Dangerous potential differences among various constituents of a system are eliminated by bonding all chassis together electrically, and connecting them to a common ground. This is called safety grounding. The ground-bus method is the most often-used means of accomplishing safety grounding. Safety grounding should always be used when dangerous voltages might occur. Ground loops, which can cause problems in radio-frequency equipment, are avoided by the bus system. *See also* GROUND BUS.

SAG

When a wire is strung between two points, some sag is inevitable. In longwire antenna systems, sag is important, because it can, if severe, modify the radiation pattern from the system (*see* LONGWIRE ANTENNA, RHOMBIC ANTENNA, V BEAM).

A sagging wire describes a curve called a catenary, which is very similar to the parabola (*see* PARABOLA). For two supports at the same height, wire sag can be calculated from the following formula:

$$s = 0.612 \sqrt{mn - m^2}$$

where s is the sag at the center of the span, m is the distance between the two supports, and n is the length of the cable (all measurements are in feet). The sag is defined as the distance at the center of a span between the actual wire position and the position if the wire were perfectly straight (see illustration). The above formula is based on the assumption that the sagging wire assumes the geometric form of a parabola.

The problem of sag is increased by the use of elastic wire, and also by icing or severe winds. *See also* ICE LOADING, WIND LOADING.

SAINT ELMO'S FIRE

During a thunderstorm, the ends of long metal objects such as antenna elements, flag poles, and airplane wings may acquire a strange glow. This glow results from the accumulation of electric charge on the object with respect to the surrounding environment. It is called Saint Elmo's Fire.

Saint Elmo's Fire was observed on the masts of sailing ships, even in ancient times. Since we have known about the nature of electricity for only the past two centuries, the phenomenon caused bewilderment and even fear in people who saw it. There is some good reason to be afraid of Saint Elmo's Fire: Lightning is likely to strike an object when it is aglow with an excess of electric charge. *See also* LIGHTNING.

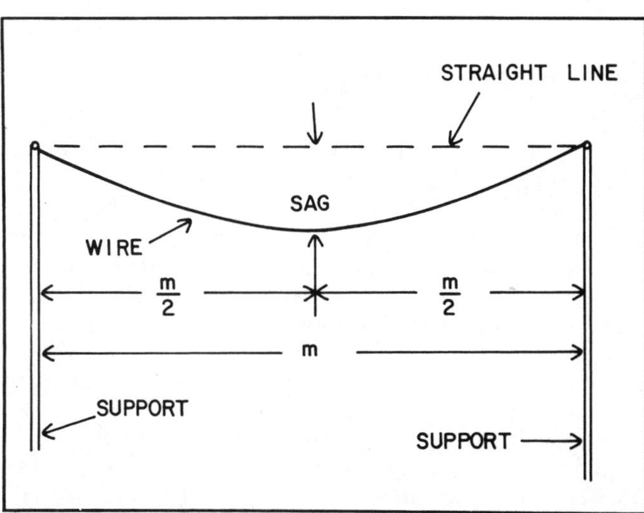

SAG: The sag is the extent to which a wire droops in the middle of a span.

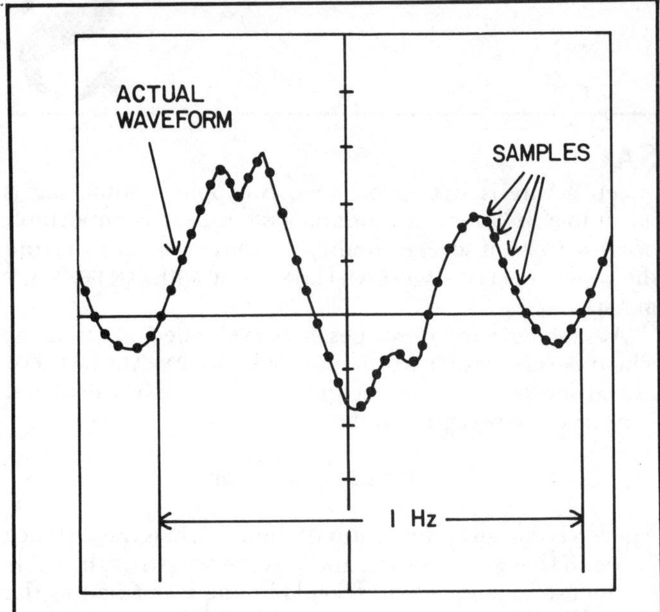

SAMPLING OSCILLOSCOPE: Sampling of a complex waveform. A certain minimum number of samples is required in order to define the waveform without ambiguity.

SAMPLING OSCILLOSCOPE

An oscilloscope can be used to obtain a prolonged view of a regular, repeating waveform simply by adjusting the sweep rate to correspond with the period of the wave. Most oscilloscopes can perform this function automatically, by the use of a triggering circuit (*see* TRIGGERING). If the waveform is very irregular, however, this cannot be done. A different method must be used for the analysis of irregular, repeating waveforms or waveforms of extreme frequency. A sampling oscilloscope provides one means for evaluating such waveforms.

The sampling oscilloscope measures the instantaneous voltage of a waveform at various moments. The samples are taken at the rate of one per cycle, until the entire waveform has been sampled at intervals. The signals are amplified and stored. Then they are displayed simultaneously on the cathode-ray screen (see illustration). The display appears segmented or digitized, since a finite number of samples are used. The result is an approximation of the waveform. The accuracy of this approximation depends on the number of samples taken in a given amount of time, and also on the complexity of the waveform itself. *See also* OSCILLOSCOPE.

SAMPLING THEOREM

When a signal is evaluated by sampling, it is necessary to take at least a certain number of samples per cycle. Otherwise, ambiguity will result, and we cannot be certain that the waveform actually is what we think it is. For example, in the previous illustration, it is apparent that the sampling oscilloscope must obtain at least a certain number of samples per cycle in order to show accurately what the waveform looks like. The minimum sampling rate for accurate representation of a waveform is defined by a rule known as the sampling theorem. The sampling theorem was developed by the engineer Nyquist.

Suppose that the highest-frequency component of a signal has a frequency of f Hz. According to the sampling theorem, we need at least two samples per cycle for this component. That is, the sampling rate must be at least 2f. The maximum allowable interval between samples is therefore 1/2f seconds.

In practice, it is a good procedure to obtain more than the minimum number of samples as defined by this theorem. The greater the number of samples, the more accurate the representation will be. It is rarely necessary, however, to obtain more than 10 to 15 samples for each cycle of the highest-frequency component. *See also* SAMPLING OSCILLOSCOPE.

SATELLITE COMMUNICATIONS

The earth is literally surrounded by artificial satellites today. Many of these satellites carry repeaters, and are used for communications. In recent years, satellites have been placed in synchronous orbits (*see* SYNCHRONOUS ORBIT), providing continuous communications capability among almost all possible points on the globe. We take this for granted now, but such communications was impossible as recently as the 1960s.

The oldest communications satellite, in terms of the time of "launching," is our moon. The moon is a fairly good reflector of radio waves and has been used for communications, mostly by radio amateurs. The moon is a rather inconvenient communications satellite, since it is above the horizon only half of the time, and because it moves across the sky (*see* MOONBOUNCE).

In the 1960s, a series of passive satellites was launched into orbit around the earth. These devices were like large metal balloons, which reflected radio waves sent up to them. You may remember them as the Echo satellites. The Echo satellites were placed in low orbits, since we did not have the technology at that time to put a satellite in a synchronous orbit. The area of coverage for each Echo satellite was limited by the low orbit, and access time was brief.

The active communications satellite was developed after the Echo satellites. An active communications satellite is an orbiting repeater having broadband characteristics (*see* ACTIVE COMMUNICATIONS SATELLITE, REPEATER). The signal from the ground station is intercepted, converted to another frequency, and retransmitted at a moderate power level. This provides much better signal strength at the receiving end of the circuit, as compared with a passive satellite. The first active satellites were placed in low orbits, and thus their users were bothered by the same shortcomings as they had been with the previous Echo satellites. Finally, active communications satellites were placed in synchronous orbits, making it possible to use them with fixed antennas, a moderate level of transmitter power, and at any time of the day or night.

Today, there are so many synchronous satellites orbiting our planet that it is getting difficult to find room for more. Everyone has some need for a synchronous satellite; they are used for television and radio broadcasting, communications, weather forecasting, and military operations. We can now make telephone calls via satellite.

SATURABLE REACTOR

A saturable reactor is an inductor with a ferromagnetic core having special properties. The core of the saturable reactor achieves magnetic saturation at a fairly low level of coil current. This makes it possible to change the effective permeability by passing dc through the windings.

The saturable reactor exhibits its maximum inductance, and therefore the maximum inductive reactance, when there is no direct current passing through the coil. As the current increases, the inductive reactance decreases, reaching a minimum when saturation occurs in the core material.

The saturable reactor is used in a circuit called a magnetic amplifier, which is a form of voltage amplifier. *See also* MAGNETIC AMPLIFIER.

SATURATED LOGIC

All digital logic circuits operate in two defined states, generally called high and low. A variety of different schemes has been developed to accomplish digital-logic switching. Most of the methods use groups of bipolar transistors or metal-oxide-semiconductor devices. Sometimes other components, such as resistors and capacitors, are included.

Although there are only two possible logic states, many semiconductor devices are not always "all the way on" or "all the way off." Some forms of logic do operate this way—the devices are either at cutoff or at saturation—and this form of circuit is known as saturated logic (*see* CUTOFF, SATURATION).

Saturated logic offers definite advantages. It is relatively immune to the effects of noise and external analog signals. The error rate is therefore quite low. Saturated logic circuits are often simpler and less expensive than other forms. However, saturated forms of logic generally consume more power than their nonsaturated counterparts, and the switching speed is somewhat slower. *For examples of saturated logic forms, see* DIODE-TRANSISTOR LOGIC, DIRECT-COUPLED TRANSISTOR LOGIC, HIGH-THRESHOLD LOGIC, RELAY LOGIC, RESISTOR-CAPACITOR-TRANSISTOR LOGIC, RESISTOR-TRANSISTOR LOGIC, TRANSISTOR-TRANSISTOR LOGIC.

SATURATION

In a switching or amplifying device, the fully conducting state is called saturation. The term is especially used in bipolar-transistor and field-effect-transistor circuitry. Saturation may also be mentioned in reference to the operation of a a vacuum tube.

An NPN bipolar transistor becomes saturated when the base voltage is sufficiently positive relative to the emitter; a PNP transistor becomes saturated when the base voltage is sufficiently negative relative to the emitter. An N-channel junction field-effect transistor becomes saturated when the gate voltage is sufficiently positive relative to the source; a P-channel junction field-effect transistor becomes saturated at high negative gate voltages relative to the source. A vacuum tube conducts better and better as the grid voltage gets less and less negative; the actual saturation voltage may be either positive or negative, depending on the tube characteristics.

The voltage at which saturation occurs is called the saturation point or saturation voltage. The current that flows under these conditions is called the saturation current (*see* SATURATION CURRENT, SATURATION VOLTAGE). In the characteristic curve of an amplifying or switching device, saturation is indicated by a leveling off of the collector, drain, or plate current as the base, gate, or grid voltage changes. The saturation voltage and current are affected by the voltage at the collector, drain, or plate of the amplifying or switching device. The illustration at A is an example of saturation in an NPN bipolar transistor. *See also* TRANSISTOR, NPN TRANSISTOR, PNP TRANSISTOR, TUBE.

In a field-effect transistor, the condition of pinchoff is sometimes called saturation (*see* FIELD-EFFECT TRANSISTOR, PINCHOFF).

In a ferromagnetic substance, the flux density generally increases as the magnetizing force increases. However, there is a limit to the flux density that can be obtained in a ferromagnetic material. When this limit is reached, further increases in the magnetizing force do not produce an increase in the flux density (B). This condition is called saturation. *See also* FERROMAGNETIC MATERIAL.

In the expression of color, saturation is the relative purity of a hue. The least saturation occurs when the brightness is the same at all wavelengths; this gives various shades of gray. As the saturation increases, a given color becomes more and more vivid. Various levels of saturation for the color green are illustrated at C. *See also* CHROMA, HUE.

SATURATION CURRENT

When an amplifying or switching device becomes saturated, a certain current flows through its collector, drain, or plate circuit. This current represents the maximum possible current that can flow in the circuit for a fixed collector, drain, or plate voltage.

Generally, the saturation current of a transistor or tube increases as the collector, drain, or plate voltage is made larger. This is simply because large voltage causes greater current flow than small voltages. *See also* SATURATION.

SATURATION CURVE

A saturation curve is a graphical representation of saturation in an amplifying or switching device, or in a ferromagnetic material. The previous illustrations, A and B, show examples of such saturation curves.

The condition of saturation may be affected by parameters such as the collector, drain, or plate voltage in an amplifying or switching transistor or tube. In a ferromagnetic substance, the temperature may have an effect. Sometimes several different saturation curves are graphed on a single set of coordinates, one curve each for specific values of the parameter. This kind of graph is called a family of saturation curves. *See also* SATURATION.

SATURATION LOGIC

See SATURATED LOGIC.

SATURATION POINT

See SATURATION, SATURATION CURRENT, SATURATION CURVE, SATURATION VOLTAGE.

SATURATION VOLTAGE

In an amplifying or switching transistor or tube, the current in the collector, drain, or plate circuit changes with changing voltage at the control electrode. In an NPN

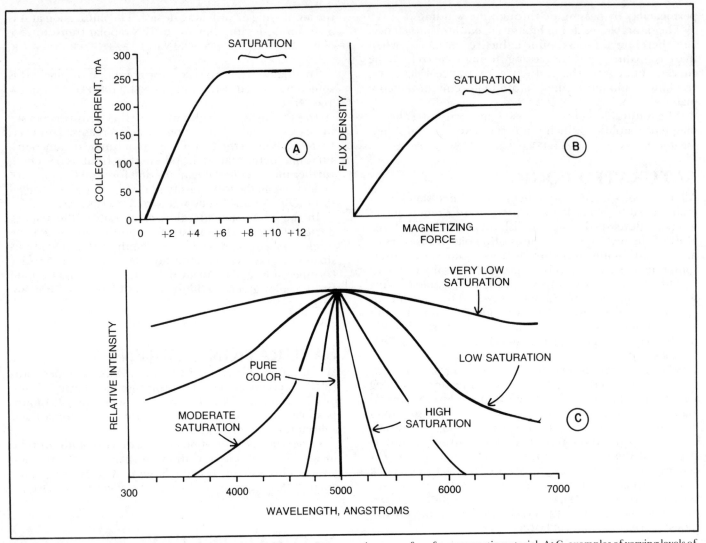

SATURATION: At A, a saturation curve for a bipolar transistor. At B, a saturation curve for a ferromagnetic material. At C, examples of varying levels of saturation for visible green light.

transistor or N-channel field-effect transistor, the current rises with increasing positive voltage at the base or gate. In a PNP transistor or P-channel field-effect transistor, the current rises with increasing negative voltage at the base or gate. In a vacuum tube, the current rises as the voltage at the grid becomes less negative or more positive.

As the voltage is changed so that the collector, drain, or plate current rises, a point is eventually reached at which no further increase occurs. The voltage at the base, gate, or grid, measured with respect to the potential at the emitter, source, or cathode, is called the saturation voltage. The saturation voltage depends on the type of device, and also on the collector-emitter, drain-source, or plate-cathode voltage. The saturation voltage is sometimes called the saturation bias or saturation point. *See also* SATURATION, SATURATION CURRENT.

SAWTOOTH WAVE

A waveform that gradually rises and abruptly falls, or vice-versa, is called a sawtooth wave. When displayed on an oscilloscope, a sawtooth wave appears to have perfectly straight, or linear, rise and decay traces. The illustration shows three examples of sawtooth waves. Sawtooth waves may reverse polarity, such as the example at A, or they may maintain one polarity, as shown at B and C.

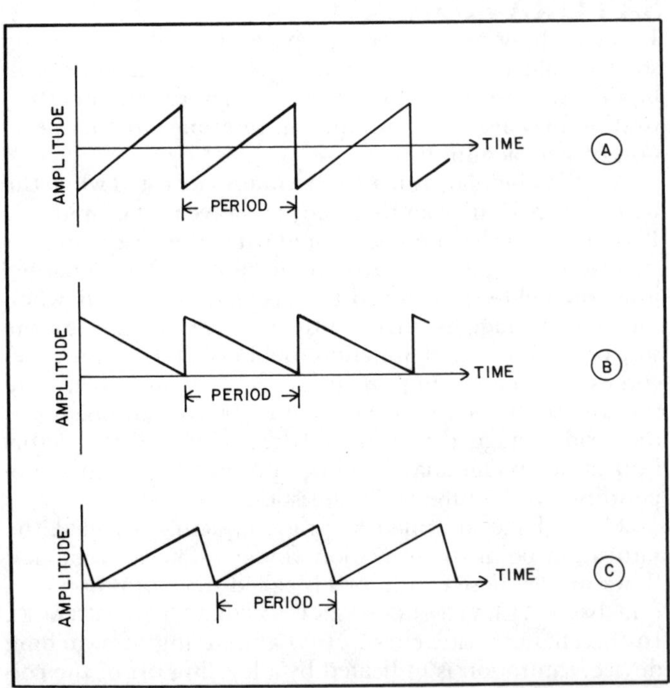

SAWTOOTH WAVE: At A, slow-rise, rapid-decay; at B, rapid-rise, slow-decay; at C, slow-rise, moderate-decay. Note that the amplitude-versus-time function is always linear.

The sawtooth wave is used for scanning in cathode-ray-tube devices. The gradual-rise, rapid-fall wave, called a ramp wave, facilitates controlled scanning from left to right and an almost instantaneous return trace from right to left (*see* RAMP WAVE, RETRACE). Sawtooth waves are used for various test purposes. Modern audio signal generators can produce sawtooth waves as well as square waves and sine waves.

Some sawtooth waves have nonzero rise and fall times. An example of such a wave is shown at C. If the rise and fall times are identical, the wave is called a triangular wave.

The period of a sawtooth wave is the length of time from any point on the waveform to the same point on the next pulse in the train. This period is divided into 360 degrees of phase, with the zero point usually corresponding to the moment of instantaneous change, as at A and B, or the end of the more rapid change (C). Alternatively, the zero-degree point can be set at the time of zero polarity and increasing amplitude. The frequency of a sawtooth wave, in hertz, is the reciprocal of the period in seconds.

SCA

See SUBSIDIARY COMMUNICATIONS AUTHORIZATION.

SCALAR

An ordinary, nondirected number is sometimes called a scalar quantity, or simply a scalar. Examples of scalar quantities are speed, current, and voltage, in which polarity or direction are not expressed. If direction is implied, the quantity is called a vector. Examples of vector quantities are velocity, left-to-right current, and positive voltage. *See also* DIRECTED, NUMBER, VECTOR.

SCALAR PRODUCT

See DOT PRODUCT.

SCAN FUNCTIONS

A scan function is a mode of scanning in a cathode-ray-tube system. Different scan functions are used for different purposes. The waveform used to actuate scanning is often referred to as the scan function. The scan waveform in an oscilloscope or television system, for example, is a ramp wave (*see* RAMP WAVE).

In radar, there are several different scan functions that can be used. The most common is the circular scan. In television, a left-to-right, top-to-bottom scan function is used. In most oscilloscopes, a left-to-right, single-line scan function is used. *See also* OSCILLOSCOPE, RADAR, RASTER, TELEVISION.

In an automatic scanning receiver, the scanning apparatus may work in various different ways. These modes are called scan functions or scan modes. The scanner may search for an occupied channel among empty channels, stopping when a signal is encountered and remaining on that channel until the signal disappears. This is called the "busy" scan mode. The scanner may stop at an occupied channel and remain there only for a predetermined length of time, such as 5 seconds; this is called free scanning. The scanner may search for an empty channel among busy ones; this is called the "vacant" scan mode. *See also* AUTOMATIC SCANNING RECEIVER.

SCANNER

See AUTOMATIC SCANNING RECEIVER.

SCANNING

Scanning is a term that describes two quite different electronic processes.

In a television or facsimile system, the picture is scanned both in the camera tube and in the receiver. The picture is generally scanned from left to right and top to bottom, in the same way you read the lines of this page. Scanning is controlled by an actuating circuit having a ramp waveform (*see* RAMP WAVE). In an oscilloscope, the electron beam in the cathode-ray tube scans from left to right along a single line. The scanning rate can be controlled, and is measured in terms of the time per graticule division (about 1 cm). Scanning in an oscilloscope is sometimes called sweeping. As in television and facsimile, a ramp wave is used to produce the scanning effect in an oscilloscope. *See also* CATHODE-RAY TUBE, FACSIMILE, OSCILLOSCOPE, RASTER, TELEVISION.

In computers or microcomputer-controlled devices, scanning refers to the sampling of data in a continuous manner. If there are n data channels, having addresses 1, 2, 3, . . . , n, then scanning may proceed from channel 1 upward to channel n, or from channel n downward to channel 1. The process of scanning may occur just once across the range of channels, or it may be repeated over and over. A good example of this kind of scanning is illustrated by the operation of an automatic scanning receiver, which constantly checks communications channels for signals. The scanning rate is designated in channels per second. *See also* AUTOMATIC SCANNING RECEIVER.

SCATTERING

When any type of disturbance, such as sound waves, electromagnetic waves, or particle radiation, passes through a material substance, scattering may occur. The effects of scattering may be unnoticeable, or they might be very evident. The atmosphere, for example, scatters blue light to some extent. This is why the sky looks blue to us. Water has the same effect; this is why a swimming pool has such a vivid blue appearance.

Radio waves at very-high and ultra-high frequencies are sometimes scattered by the molecules of the atmosphere. This is called tropospheric-scatter propagation. The ionosphere scatters radio waves at low, medium, high, and sometimes very-high frequencies. *See also* PROPAGATION CHARACTERISTICS, TROPOSPHERIC-SCATTER PROPAGATION.

SCATTER PROPAGATION

See BACKSCATTER, TROPOSPHERIC-SCATTER PROPAGATION.

SCHEMATIC DIAGRAM

A schematic diagram is a technical illustration of the interconnection of components in a circuit. Most schematic diagrams include component values, and perhaps tolerances. Standard symbols are used. A schematic diagram does not indicate the physical arrangement of the components on the chassis or circuit board; it shows only how the components are interconnected.

Schematic diagrams are designed so that they are easy to read. Although a schematic diagram may look complicated to the beginner in electronics, this kind of representation is far simpler than a pictorial diagram. In the case of a complex device such as a superheterodyne radio receiver or transceiver, a pictorial diagram would be utterly impractical, although a schematic diagram of such a device can usually be placed on one page.

SCHEMATIC SYMBOLS

In schematic diagrams, standard symbols are used. These symbols are universal throughout the world. The most commonly used schematic symbols are shown in the front matter of this book.

In some cases, symbols may be found that differ from those shown. However, the differences are usually inconsequential, and the components are usually easy to recognize.

SCHERING BRIDGE

The Schering bridge is a circuit for the measurement of capacitance. The Schering bridge operates by comparing an unknown capacitance with a standard, known capacitance. Balance is indicated either by a meter, as shown in the illustration, or by a set of headphones. The Schering bridge is employed for the determination of relatively large capacitances.

If the unknown capacitance, in farads, is C, and the other component values are as labeled (capacitances in farads, resistances in ohms), then:

$$C = C2R1/R2$$

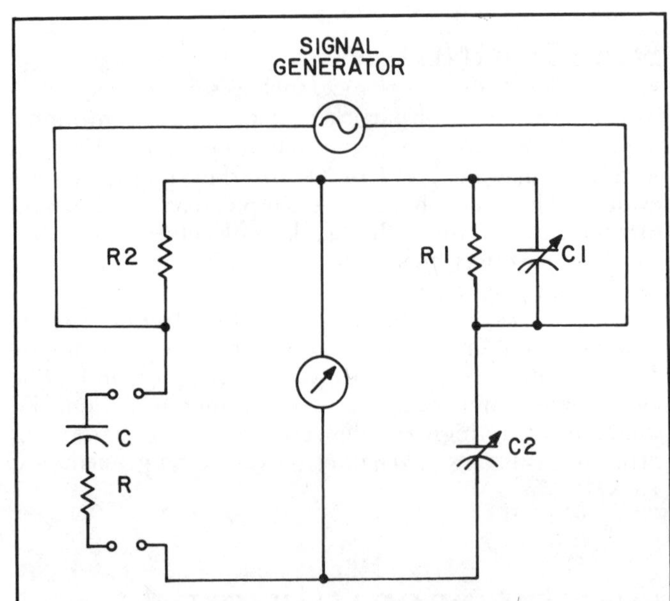

SCHERING BRIDGE: A Schering bridge for determination of capacitance.

and the Q factor of the unknown capacitor, assuming an effective series resistance of R ohms and a frequency at balance of f hertz, is:

$$Q = 1/(2\pi fCR)$$

SCHMITT TRIGGER

A Schmitt trigger is a form of multivibrator circuit that is actuated by means of an input signal having a constant frequency. The Schmitt trigger produces rectangular waves, regardless of the input waveform. The schematic illustrates a typical Schmitt trigger circuit.

The circuit remains off until a specified rise threshold voltage is crossed; then it is actuated, and the output voltage abruptly rises. When the input voltage falls back below the fall triggering level, the output voltage drops to zero almost instantly. The rise and fall thresholds are generally different.

The Schmitt trigger is used in applications where square waves with a constant amplitude are needed. A Schmitt trigger may also be used to convert sine waves to square waves. The Schmitt trigger is a form of bistable multivibrator, or flip-flop. *See also* FLIP-FLOP.

SCHOTTKY BARRIER DIODE

A Schottky barrier diode, also known as a Schottky diode or hot-electron diode, is a solid-state diode in which the junction is formed by metal-semiconductor contact. The Schottky barrier diode is usually fabricated from lightly doped N-type material and a metal such as aluminum. The metal is evaporated or sputtered onto the semiconductor. The hot-carrier diode is a form of Schottky barrier diode (*see* HOT-CARRIER DIODE, POINT-CONTACT JUNCTION).

Schottky barrier diodes are characterized by extremely rapid switching capability. The reverse-bias capacitance is very low. Schottky barrier diodes are useful at very-high and ultra-high frequencies because of their high-speed properties. They can be used as mixers, harmonic generators, detectors, and in other applications requiring diodes. The Schottky-diode design is sometimes used in the manufacture of transistors; this type of semiconductor device is known as a Schottky transistor.

SCHOTTKY LOGIC

Schottky logic is any form of logic that incorporates Schottky diodes or transistors. Schottky transistors are used especially in the manufacture of integrated-injection-logic devices.

Schottky logic is characterized by high switching speed. *See also* INTEGRATED INJECTION LOGIC.

SCIENTIFIC NOTATION

Very large or small numbers are cumbersome to write in the conventional fashion. For example, if we write out the number 7 googol, we have to write a 7 followed by 100 zeroes. Scientific notation provides a shortcut for expressing extreme numerical values.

A number in scientific notation consists of a decimal number with a value of at least 1 but less than 10, followed by a power of 10. The decimal number tells us the first few digits of the actual value as we would write it in longhand, or conventional, form. The power of 10 tells us the factor by which the decimal number is to be multiplied. The exponent is always an integer; it may be positive or negative.

The decimal part of a scientific-notation expression may have any number of significant figures, depending on the accuracy we need. For example, we might write 2,345,678 as 2.3×10^6; this is an approximation accurate to two significant figures.

Many small calculators, and almost all computers, have the capability to work in scientific notation. This makes it possible for them to handle much larger numbers than would otherwise be possible.

When two numbers are multiplied or divided in scientific notation, the decimal numbers are first multiplied or divided by each other. Then the powers of 10 are added (for multiplication) or subtracted (for division). Finally, the product or quotient should be reduced to standard form. That is, the decimal part of the expression should be at least 1 but less than 10. For example,

$$3 \times 10^2 \times 7 \times 10^3 = 21 \times 10^5 = 2.1 \times 10^6$$

When working with scientific notation, it is important to be aware of the number of significant figures in the expressions. Values are often expressed approximately, not precisely. *See also* SIGNIFICANT FIGURES.

SCINTILLATION

The intensity of an electromagnetic field often fluctuates at the receiving point. This fluctuation in strength is called fading at the lower frequencies. At very high frequencies and above, it is sometimes called scintillation, especially if it is rapid.

Scintillation occurs primarily as a result of atomospheric effects. A familiar example of scintillation is the twinkling of a star as its light is refracted in the atmosphere. A similar effect can be noticed when small-diameter celestial objects are observed with the radio telescope; this scintillation is caused by interstellar matter. Radio astronomers have determined the angular sizes of very distant quasars by observing their scintillation.

When a radar set is used, there may be a sudden and dramatic shift in the position of a target, or the momentary appearance of false targets. This effect is known as scintillation. It can occur because of atmospheric refraction or reflection, in the vicinity of severe weather, or during a geomagnetic storm. Meteor showers often produce spectacular scintillation effects on radar.

When a high-speed atomic particle strikes a phosphorescent substance such as sodium iodide, there is a brief flash of visible light. This effect is called scintillation. A special type of radiation counter, known as a scintillation counter, uses phosphor crystals and light-detection apparatus to measure the intensity of radioactivity. *See also* SCINTILLATION COUNTER.

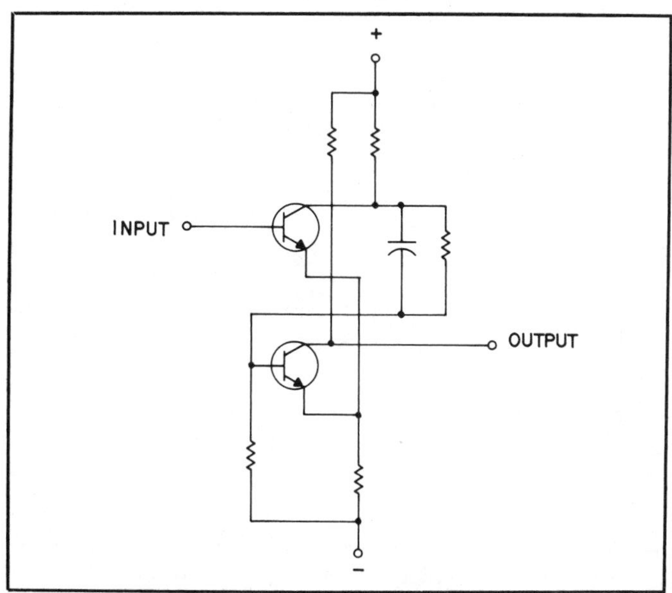

SCHMITT TRIGGER: A Schmitt trigger is a form of multivibrator, actuated by an external signal.

SCINTILLATION COUNTER

A scintillation counter is a device that is used for the purpose of measuring the intensity of atomic (nuclear) radiation. This type of counter is sensitive to all forms of nuclear radiation, including alpha, beta, gamma, and X rays.

A scintillation counter consists of a phosphorescent material such as a sodium-iodide crystal, and a photomultiplier tube (see illustration). When a photon or subatomic particle strikes the phosphor, a momentary flash of visible light is produced. This flash is converted into a weak electrical impulse by the photocathode. The photomultiplier tube amplifies the impulse. The impulse is then transmitted to a counting device or to a speaker, oscilloscope, or other indicator.

Scintillation counters are capable of detecting more than 1,000,000,000 particles per second, compared with only about 2,000 particles per second for a Geiger counter. *See also* ALPHA PARTICLE, BETA PARTICLE, GAMMA RAY, GEIGER COUNTER, PHOTON, X-RAY.

SCR

See SILICON-CONTROLLED RECTIFIER.

SCRAMBLER

A scrambler is a device that is sometimes used to encode signals, making it difficult for unauthorized persons to intercept them. Scramblers can be used for digital or analog signals of any kind, and at any speed.

A common type of scrambler, used with voice signals, inverts the frequency characteristics of the voice. The av-

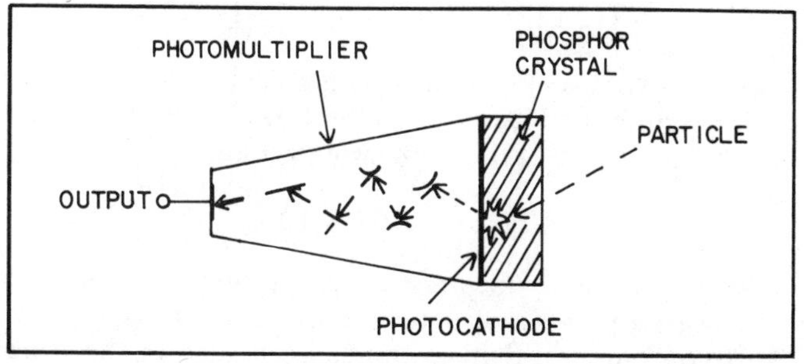

SCINTILLATION COUNTER: A scintillation counter detects the light flashes that occur when high-energy radiation strikes a phosphorescent material.

erage human voice can be transmitted within a passband about 2.7 kHz wide, having frequency components between 300 Hz and 3 kHz. The scrambler turns the voice frequencies "upside down." A given impulse, having a frequency f, in hertz, is converted to a frequency g, in hertz, such that:

$$g = 3300 - f$$

When a human voice signal is applied to the input of such a circuit, the resulting output signal is utterly unreadable. (You are familiar with this effect if you have ever tried to listen to a lower-sideband signal with a receiver set for upper sideband, or vice-versa.) However, by passing the unreadable "monkey chatter" through a second, identical scrambler circuit, the voice is restored to its original form.

The foregoing is just one method of data scrambling. It is actually an oversimplification, and is described for illustrative purposes only. Many different scrambling methods are used by various commercial and government communications services. Most scramblers are more complex than the type described above, and may incorporate such devices as digital-to-analog and analog-to-digital converters, multiple-frequency operation, and other sophisticated schemes, making it essentially impossible for unauthorized persons to decode the signals.

SCRATCH-PAD MEMORY

A low-capacity, random-access memory, used in computers and some calculators for temporary data storage, is called a scratch-pad memory. Scratch-pad memory is used in complex arithmetic computations involving more than one operation.

Suppose, for example, that you want to calculate the value of the following expression:

$$x = (3.443 - 6.474)\ (4.001 + 8.211)$$

If you have a calculator with a scratch-pad memory, you would first compute:

$$y = 3.443 - 6.474 = -\ 3.031$$

and store this number in the memory. You would then compute:

$$z = 4.001 + 8.211 = 12.212$$

Then you would recall the number y from the memory and multiply z by y, obtaining x:

$$zy = (12.212)\ (-3.031) = -37.014572 = x$$

If you are working in scientific notation and are concerned about significant figures (*see* SCIENTIFIC NOTATION, SIGNIFICANT FIGURES), then you would obtain:

$$x = -3.701 \times 10^1$$

If your calculator does not have a scratch-pad memory, you need a paper and pencil (a real scratch pad)!

In computers and programmable calculators, there may be several different scratch-pad memories which are used automatically, as necessary, in more complex computations. *See also* MEMORY, RANDOM-ACCESS MEMORY.

SCREEN GRID

Some vacuum tubes have two or more grids. In most such tubes, the second grid from the cathode is called the screen grid. The screen grid is biased positively with respect to the cathode of a tube. Usually, the screen grid is connected to the B+ (positive power-supply voltage) through a large-value resistor.

Triode-tube amplifier circuits have a tendency to break into oscillation. This is especially true of power amplifiers. A tetrode or pentode tube, having a screen grid, is less subject to oscillation, because the screen grid provides some isolation between the control grid and the plate. The screen grid also serves to accelerate electrons toward the plate, resulting in higher gain than is possible with a triode. The main disadvantage of the screen grid is that it generates wideband electrical noise as the electrons bombard it. In receivers, the additional gain of the tetrode is partially offset by this extra noise.

The cathode, control grid, and screen grid of a multigrid tube form a triode, and can be used as such for certain purposes. This arrangement is found in the electron-coupled oscillator, for example. The screen grid of a multigrid tube can also be used as a signal grid in mixer and modulator circuits. *See also* CONTROL GRID, ELECTRON-COUPLED OSCILLATOR GRID, PENTODE TUBE, TETRODE TUBE, TRIODE TUBE, TUBE.

SECANT

The secant function is a trigonometric function equal to the reciprocal of the cosine function (*see* COSINE). In a right triangle, as shown in the drawing at A, the secant of an angle θ between zero and 90 degrees is equal to the length of the hypotenuse divided by the length of the side adja-

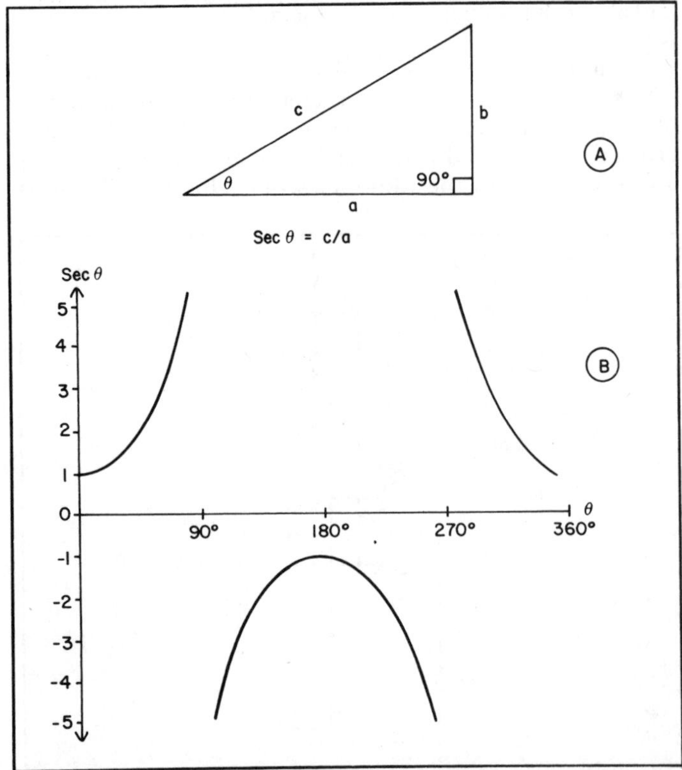

SECANT: The secant function is the ratio of the length of the hypotenuse of a right triangle to the length of the adjacent side, as shown at A. At B, an approximate graph of the secant function for angles between 0 and 360 degrees.

cent to the angle. The larger the angle, the larger the secant becomes for values between zero and 90 degrees. The values of the secant for angles between zero and 360 degrees are shown at B. The secant is undefined for $\theta = 90°$ and $\theta = 270°$.

In mathematical calculations, the secant function is abbreviated sec, and is given by the formula:

$$\sec \theta = 1/\cos \theta$$

where cos represents the cosine function. *See also* TRIGONOMETRIC FUNCTION.

SECOND

The second is the standard international unit of time. A second corresponds to 1/86,400, or 1.1574×10^{-5}, mean solar day. Using modern atomic clocks, a more precise definition has been formulated: The second is represented by 1,420,405,752 oscillations of the hydrogen maser. Seconds are transmitted over time-and-frequency standard stations such as CHU and WWV/WWVH (*see* CHU, WWV/WWVH).

In angular measure, the second is 1/3600, or 2.778×10^{-4}, angular degree. This unit is also known as the arc second or the second of arc. It is abbreviated by a quotation mark or a double apostrophe ("). The arc second is used by astronomers as an indicator of the resolving power of an optical or radio telescope. The angular size of a celestial object may be expressed in seconds of arc.

In the expression of right ascension, one second corresponds to 1/3600 of an hour circle, or 15 angular seconds of arc. *See also* RIGHT ASCENSION.

SECONDARY

See TRANSFORMER SECONDARY.

SECONDARY BATTERY

See STORAGE BATTERY.

SECONDARY COLOR

A secondary color is any color that is not a primary color. Any color can be obtained by combining primary colors in certain proportions, and with certain amounts of saturation. Most of the colors you see are secondary. *See also* HUE, PRIMARY COLORS, SATURATION.

SECONDARY ELECTRON

See SECONDARY EMISSION.

SECONDARY EMISSION

When an electrode is bombarded by high-speed electrons, other electrons are knocked from the atoms of the electrode. Such electron emission is known as secondary emission.

The photomultiplier tube operates via secondary emission. A single electron may knock several electrons free from an electrode under certain conditions. This makes it possible to magnify the intensity of an electron beam by a large factor (*see* DYNODE, PHOTOMULTIPLIER).

Secondary emission may be undesirable. In a tetrode vacuum tube, the electrons may be accelerated to such speed that secondary emission occurs from the plate of the tube. This reduces the efficiency of the tube, because the secondary electrons represent lost plate current. A negatively charged electrode, placed between the plate and the screen grid, reduces this effect. Such a grid is called a suppressor, since it suppresses the secondary emission by driving the electrons back toward the plate. *See also* PENTODE TUBE, SUPPRESSOR GRID, TETRODE TUBE, TUBE.

SECONDARY FAILURE

When an electronic component malfunctions, it may cause other parts of a circuit or system to malfunction also. The failure of the other components is known as secondary failure.

Secondary failure often represents a much more serious problem than the original malfunction. Suppose that a regulated power supply, having no provision or overvoltage protection, malfunctions and produces 25 V at the output instead of the intended 13.8 V. The single component that causes this failure—the pass transistor—might cost $1.00 to replace. But the damage resulting to a radio transceiver, connected to the supply when the overvoltage condition occurs, may result in a far greater repair cost.

In well-designed electronic equipment, the possibility of secondary failure is addressed and minimized. In the above situation, for example, an overvoltage-protection circuit would prevent the secondary damage to the transceiver.

In the troubleshooting and repair of electronic apparatus, a technician must always be aware of the possibility of secondary failure. If a component has burned out because of secondary failure, merely replacing that component will not solve the problem, and it will probably recur. *See also* TROUBLESHOOTING.

SECONDARY FREQUENCY STANDARD

See REFERENCE FREQUENCY, SECONDARY STANDARD.

SECONDARY RADIATION

Electromagnetic fields cause electrons to move back and forth in an electrical conductor. This is how a receiving antenna works; alternating current is generated by the passing waves. The alternating current is not, however, absorbed completely by the receiver. The current causes its own electromagnetic field to be transmitted from the receiving antenna; approximately half of the energy in a receiving antenna is reradiated into space. This radiation is called secondary radiation.

Secondary radiation takes place from all electrical conductors. Utility wires, fences, and even metal clothes lines all radiate electromagnetic fields as a result of excitation by external fields.

The earth is constantly bombarded b y high-speed subatomic particles from space. These particles constitute radiation, and are known as cosmic particles or cosmic rays (*see* COSMIC RADIATION). We observe the effects of cosmic radiation at the surface of the earth, but the particles we see are not the original particles from space. The atmosphere absorbs the original, or primary, particles. But when a primary cosmic particle strikes an atom of air, another particle—a proton, electron, neutron, alpha particle, or photon—is knocked out of the nucleus, or one of the electron shells, of that atom. This particle is a secondary particle. The resulting radiation that we observe is called secondary radiation. *See also* ALPHA PARTICLE, BETA PARTICLE, ELECTRON, NEUTRON, PHOTON, PROTON.

SECONDARY STANDARD

When we make a measurement of any parameter, we must have a standard. This standard must, in turn, be calibrated or set against a universal standard. Universal standards are rarely used directly. The standards we use are called secondary, since they are not the original standards.

As an example, suppose we wish to find out what time it is. We look at our digital quartz watch, which tells us it is 5:00:00 P.M. (17 hours, 0 minutes, 0 seconds) local time. The watch, although perhaps a very accurate timepiece, probably does not exactly agree with the absolute time standard of WWV/WWVH. There is almost certainly an error of at least a few tenths of a second. The watch is a secondary time standard. The broadcasts of WWV/WWVH are a primary time standard.

As another example, consider the measurement of frequency. We might maintain a crystal oscillator under the most ideal possible conditions of temperature and humidity, and use it to calibrate our frequency-measuring apparatus. But the oscillator nevertheless represents a secondary standard. The primary standard is the transmission from WWV/WWVH (*see* REFERENCE FREQUENCY).

Primary or absolute standards are determined by agreement among the societies of the world. All parameters and standards are based on this agreement, which is known as the Standard International (SI) system of units. In the United States, absolute units are maintained by the National Bureau of Standards. *See also* NATIONAL BUREAU OF STANDARDS, STANDARD INTERNATIONAL SYSTEM OF UNITS.

SECONDARY TIME STANDARD

See SECONDARY STANDARD.

SECONDARY WINDING

See TRANSFORMER SECONDARY.

SECOND BREAKDOWN

Second breakdown is a phenomenon that occurs in some bipolar power transistors under certain adverse conditions. Second breakdown can occur as a result of flaws in manufacture. It can also take place if the device is subjected to excessive voltage or current.

Second breakdown causes a dramatic fall in the output impedance of a transistor. The emitter-base voltage, and any input signal, no longer affects the current through the device, and the output therefore drops to practically zero. The collector becomes effectively short-circuited to the emitter. A transistor may be permanently damaged by the occurrence of second breakdown. The probability of this happening can be reduced by ensuring that transistors are operated well within their ratings. *See also* POWER TRANSISTOR, TRANSISTOR.

SECOND LAW OF THERMODYNAMICS

The Second Law of Thermodynamics, also known as Carnot's Principle, states that the entropy of a system cannot be reduced to zero in a finite number of operations. In practical situations, this implies that the entropy of a system cannot be reduced to zero.

In a reversible energy system, the amount of available thermal energy is always greater than the mechanical energy that can be derived from it. The maximum efficiency of a thermal-to-mechanical energy converter is equal to

$$\mathrm{Eff} = 1 - T2/T1$$

where Eff is the efficiency, T2 is the temperature (in degrees Kelvin) of the cold reservoir, and T1 is the temperature (in degrees Kelvin) of the hot reservoir.

The greater the temperature difference, the greater the maximum possible efficiency. Theoretically, in the extreme case, if the cold reservoir were absolute zero, the efficiency could be as great as 100 percent. But temperatures of absolute zero do not exist in the real universe. *See also* ENERGY, FIRST LAW OF THERMODYNAMICS, THIRD LAW OF THERMODYNAMICS.

SECTIONALIZED ANTENNA

A sectionalized antenna is a form of vertical collinear array designed for the purpose of producing omnidirectional gain (*see* COLLINEAR ANTENNA). Two or more straight elements, each measuring ½ electrical wavelength, are fed in

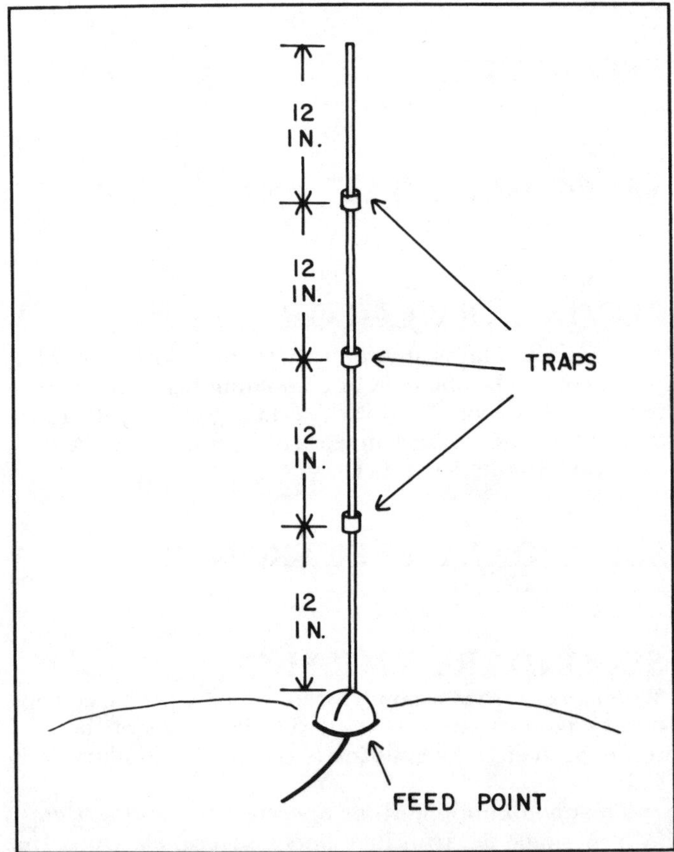

SECTIONALIZED ANTENNA: A sectionalized antenna is a form of collinear array that provides gain in all horizontal directions. Here, a four-element sectionalized antenna is shown for use at ¾ meters.

phase. Sectionalized antennas are used mostly at very high and ultra-high frequencies.

A sectionalized mobile antenna is shown in the illustration. There are four sections, each 12 inches long. This is ½ electrical wavelength in the ¾-meter amateur band. The total length of this particular antenna is just 4 feet, which is entirely practical for mobile operation.

The power gain of a sectionalized antenna, measured relative to a half-wave dipole, increases as the number of elements increases: Each time the number of elements is doubled, the gain becomes larger by 3 dB. In theory, the antenna shown has 6 dBd gain.

In practice, the gain of a sectionalized antenna is slightly less than the theoretical figure; there is a limit to the number of elements that can be connected to form an efficient sectionalized antenna. When a large number of collinear elements are used, the tuning becomes extremely critical, and the losses in the interconnecting traps become considerable. Sectionalized antennas rarely use more than six or eight elements for these reasons.

SEEBECK COEFFICIENT

The Seebeck coefficient is an expression of the amount of potential difference developed by a thermocouple. The Seebeck coefficient is expressed in volts per degree Celsius.

If the temperature difference between the two conductors of a thermocouple is T degrees Celsius, and the voltage between the conductors is E volts, then the Seebeck coefficient S is expressed as a limit:

$$S = \lim_{t \to o} E/T$$

The Seebeck coefficient may be either negative or positive, depending on which conductor is named first. *See also* SEEBECK EFFECT, THERMOCOUPLE.

SEEBECK EFFECT

When two dissimilar metals are joined, a potential difference may develop between them. This is how a thermocouple works (*see* THERMOCOUPLE). The production of such a voltage occurs when the temperature of the junction differs from that of the rest of the metal, or when the two metals have different temperatures. This phenomenon is called Seebeck effect. The resulting voltage is known as the Seebeck EMF, Seebeck potential, or Seebeck voltage.

In general, the greater the temperature gradient in a thermocouple junction, the greater the voltage. Different combinations of metals result in different voltages for a given temperature gradient. The voltage arising from a given temperature differential is expressed as the Seebeck coefficient. *See also* SEEBECK COEFFICIENT.

SELCAL

See SELECTIVE CALLING.

SELECTANCE

Selectance is an expression of the selectivity of a receiver or a resonant circuit (*see* SELECTIVITY).

In a receiver, selectance is given as a sensitivity ratio between two channels, one desired and the other undesired. Suppose the receiver is tuned to a channel at f1 MHz, and the sensitivity is found to be E1 microvolts for 10-dB signal-to-noise ratio. If a signal is applied at some frequency f2 MHz, different from f1, the sensitivity will be less; the voltage E2, needed to produce a 10-dB signal-to-noise ratio, will be larger than E1. The selectance S can be expressed in different ways. The general form is:

$$S = (E2\text{-}E1)/(f2\text{-}f1)$$

if f2>f1, and

$$S = (E2\text{-}E1)/(f1\text{-}f2)$$

if f2<f1. These expression are given in microvolts per megahertz. Of course, other units of voltage and frequency might be used.

A specific expression of selectance is obtained by determining the frequency f2 at which E2 = 2E1. Then:

$$S = (2E1-E1)/(f2-f1) = E1/(f2-f1)$$

if f2>f1, and

$$S = (2E1\text{-}E1)/(f1\text{-}f2) = E1/(f1\text{-}f2)$$

if f2<f1.

In the determination of selectance, there will be two frequencies f2; one larger than f1 and the other smaller than f1. The selectance for f2>f1 may differ from that in the case f2<f1.

Selectance can be expressed for resonant circuits. If the voltage is E1 at the resonant frequency f1, and E2 at some nonresonant frequency f2, then selectance is specified in the same ways as outlined above. There are, again, two selectance figures: one for f2>f1 and one for f2<f1. These two figures may differ for a fixed ratio E2/E1.

The selectance of a radio receiver determines the degree of adjacent-channel rejection. The greater the selectance, the less the possibility of adjacent-channel interference. The selectance of a resonant circuit depends on the Q factor. The higher the Q, the greater the selectance. *See also* ADJACENT-CHANNEL INTERFERENCE, Q FACTOR.

SELECTIVE CALLING

Selective calling is an automatic method of sending messages to specific points, without causing interference to other stations.

The most common example of selective calling is the telephone. When you call a particular number, only the phone or phones at that number will ring (*see* TELEPHONE). Selective calling can be done in a similar fashion in wireless communications. In radio, selective calling is known as SELCAL. It is used in radioteletype systems and in radiotelephones.

Suppose you have ten different messages intended for ten different stations. Without selective calling, you would have to send all ten messages to all ten stations, resulting in a great waste of time. With selective calling, however, this waste does not occur.

Selective calling for radioteletype involves the use of specific opening (start-up) and closing (shut-down) codes for each station. The opening and closing codes may be the same, such as the call letters of the station. When the SELCAL unit "hears" the appropriate sequence of characters, the printer is actuated. When the closing sequence is received, the printer stops. If the opening and closing codes are identical, then the printer is alternately turned on and off each time the SELCAL unit "hears" the code. It is generally best to use different opening and closing codes in a SELCAL system. Otherwise, the accidental transmission of one extra code sequence will throw off the system indefinitely, causing it to be up when it should be down, and down when it should be up.

Radioteletype terminal units, incorporating SELCAL as well as many other features, can be purchased, already assembled, for amateur or commercial use.

SELCAL is also used in selective signalling on radiotelephone circuits, particularly between ground stations and aircraft. Most commercial air-carriers (airlines) have frequencies that are used for conducting business communications (having nothing to do with the safe operation of the aircraft). Rather than have all planes monitor all communications all the time (which could be distracting), selective signalling is used to alert only the particular aircraft involved. This type of signalling uses tone-coded squelch systems such that the proper sequence of tones must be received by the equipment on the aircraft in order for the receiver's audio circuits to become activated. The equipment can then sound a tone, flash a light, or both, to alert the crew that a message requires their attention. *See also* PRIVATE LINE, RADIOTELETYPE, TERMINAL UNIT.

SELECTIVE FADING

When a signal is propagated via the ionosphere, fading is commonly experienced at the receiving station (*see* FADING, PROPAGATION CHARACTERISTICS). This fading usually does not occur at the same time at all frequencies. For a given communications circuit, fading will not normally be observed at 5 MHz and 7 MHz at the same time, for example. This is because fading is a phase-related phenomenon; frequency and phase changes are interdependent.

Normally, the effects of frequency are not noticeable for channels separated by just a few hundred hertz. However, under certain conditions, fading may take place several seconds apart at two frequencies that are very close together. If this occurs, a signal can become distorted. For example, a single-sideband signal may get "bassy" and then "tinny" as the fade moves across the occupied channel.

The narrower the bandwidth of a signal, the less likely it is to be affected by selective fading. Code (continuous-wave type A1 emission) signals are essentially unaffected by selective fading, since the bandwidth is very small. But radioteletype (F1 emission) may be affected; single-sideband transmissions are often distorted; amplitude-modulated signals can be severely affected; and wideband frequency-modulated signals can be mutilated beyond recognition. The effects of selective fading can be reduced by using the minimum bandwidth needed to carry on communications. Selective fading can also be alleviated, to some extent, by the use of diversity receiving systems. *See also* DIVERSITY RECEPTION.

SELECTIVITY

Selectivity is the ability of a radio receiver to distinguish between a signal at the desired frequency and signals at other frequencies. Selectivity is an extremely important criterion for any receiver.

In the front end of a receiver, selectivity is important because it reduces the chances of overloading or image reception (*see* FRONT END, PRESELECTOR). In the intermediate-frequency stages of a superheterodyne receiver, selectivity is important because it reduces the probability of adjacent-channel interference (*see* ADJACENT-CHANNEL INTERFERENCE). Selectivity is obtained by the use of bandpass filters (*see* BANDPASS FILTER, BANDSPACE RESPONSE).

The selectivity of the intermediate-frequency chain in a superheterodyne receiver is often expressed mathematically. The bandwidths are compared for two voltage-attenuation values, usually 6 dB and 60 dB. For example, after we purchase a new receiver, we may read in the specifications:

Selectivity: ± 8 kHz or more at −6 dB
± 16 kHz or less at −60 dB

This gives us an idea of the shape of the bandpass response.

A single receiver may have several different bandpass filters, each with a different degree of selectivity. The ideal amount of selectivity depends on the mode of communication that is contemplated. The above figures are typical of a very-high-frequency receiver for frequency-modulation communications. For amplitude-modulated signals, the plus-or-minus bandwidth at the 6-dB points is usually 3 to 5 kHz. For single-sideband, it may be roughly 1 to 1.5 kHz, representing an overall 6-dB bandwidth of 2 to 3 kHz. For code reception, the overall bandwidth figure can be as small as 30 to 50 Hz at lower speeds, and perhaps 100 Hz at higher speeds. Examples of receiver selectivity curves for various emission modes are shown in the illustration. (These examples are representative, and are not intended as absolute standards.)

The ratio of the 60-dB selectivity to the 6-dB selectivity is called the shape factor. It is of interest, because it denotes the degree to which the response is rectangular. A rectan-

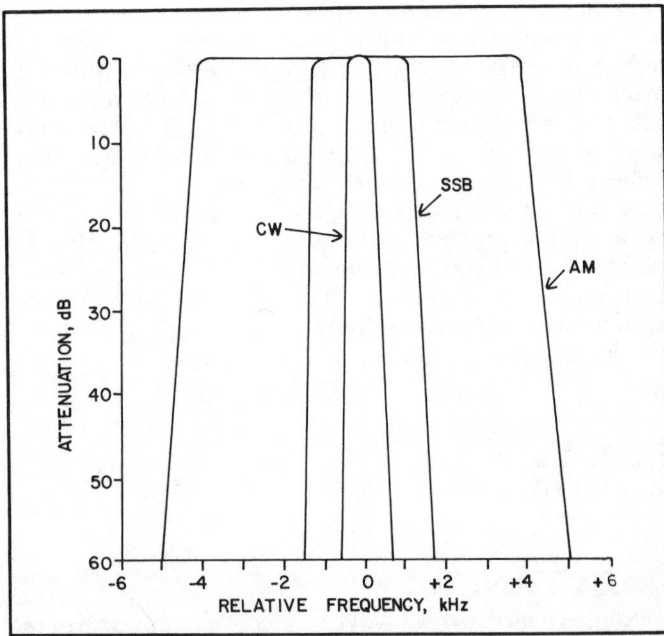

SELECTIVITY: Examples of selectivity curves for various modes of reception.

gular response is the most desirable response in most receiving applications. The smaller the shape factor, the more rectangular the response (*see* RECTANGULAR RESPONSE, SHAPE FACTOR). Filters such as the ceramic, crystal-lattice, and mechanical types can provide a good rectangular response in the intermediate-frequency chain of a receiver (*see* CERAMIC FILTER, CRYSTAL-LATTICE FILTER, MECHANICAL FILTER).

Selectivity is usually provided in the radio-frequency and intermediate-frequency stages of a receiver, but additional selectivity is sometimes added in the audio-frequency amplifier chain. This is especially useful for reception of code and radioteletype signals. Many different commercially manufactured audio filters are available for enhancing the selectivity of a receiver.

SELENIUM

Selenium is an element with atomic number 34 and atomic weight 79. Selenium is a semiconductor material that exhibits variable resistance, depending on the amount of ambient light.

Selenium is used in the manufacture of photocells, and also in certain rectifier diodes. *See also* PHOTOCELL, SELENIUM RECTIFIER.

SELENIUM RECTIFIER

A selenium rectifier is a diode device made by joining selenium and some other metal, usually aluminum. Electrons flow easily from the metal to the selenium, but the resistance in the opposite direction is high (see illustration at A). Thus, the metal forms the cathode, and the selenium forms the anode of the rectifier. A selenium rectifier can be recognized by its heat sink, which makes it look similar to a stack of cards (B).

Selenium rectifiers are useful only at low frequencies, such as the 60-Hz line frequency. Although they are light in weight, they tend to be physically large because of the heat sink that is needed. The forward drop is about 1 V in normal operation. Selenium rectifiers are especially resistant to transient spikes on the power line. This is their main advantage over other types of rectifier diodes. *See also* RECTIFIER DIODE.

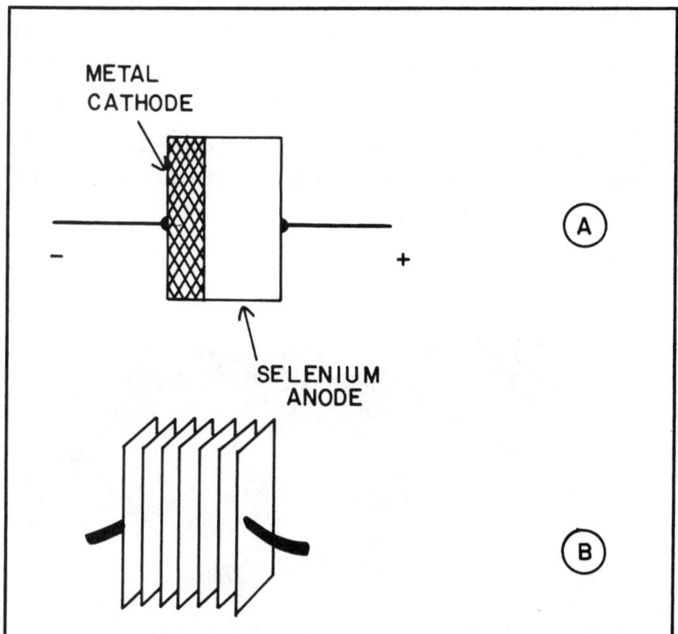

SELENIUM RECTIFIER: At A, pictorial diagram of the construction of a selenium rectifier. At B, illustration of the appearance of a typical selenium rectifier.

SELF BIAS

Self bias is a means of providing effective negative grid or gate bias in a vacuum tube or field-effect transistor. Although negative grid bias can be obtained from a special power supply, it can also be provided by elevating the cathode above direct-current ground potential. A noninductive resistor and a capacitor are used for this purpose (see illustration).

There is a limit to the amount of effect negative gate bias that can be obtained by the method shown. The bias voltage is the product of the current (which is essentially constant) and the ohmic value of the resistor. The greater the bias voltage that is needed, the larger the resistor must be, and thus the more power it must dissipate. Most noninductive resistors have rather low dissipation ratings.

Self bias is often used in low-power tube amplifiers in Class A, Class AB, or Class B. Self bias cannot normally be used to obtain sufficient bias for Class-C operation. Self bias can be used in audio-frequency or radio-frequency circuits.

Biasing can be obtained in circuits using bipolar transistors, by means of the same technique described here. In solid-state circuits, the method is also called automatic bias. *See also* AUTOMATIC BIAS.

SELF OSCILLATION

An amplifier circuit will sometimes oscillate if positive feedback occurs. This type of oscillation degrades the performance of the amplifier. In audio-frequency circuits, self oscillation causes a fluttering or popping sound known as motorboating. In radio-frequency circuits, it can result in the emission of signals on undesired frequencies, and is called parasitic oscillation. *See also* FEEDBACK, MOTORBOATING, PARASITIC OSCILLATION.

SELF RESONANCE

Most electronic components exhibit a certain amount of capacitive reactance and inductive reactance. This may be, of course, because a component is deliberately designed to show reactance. But the physical dimensions of a component, especially the lead lengths, result in some reactance

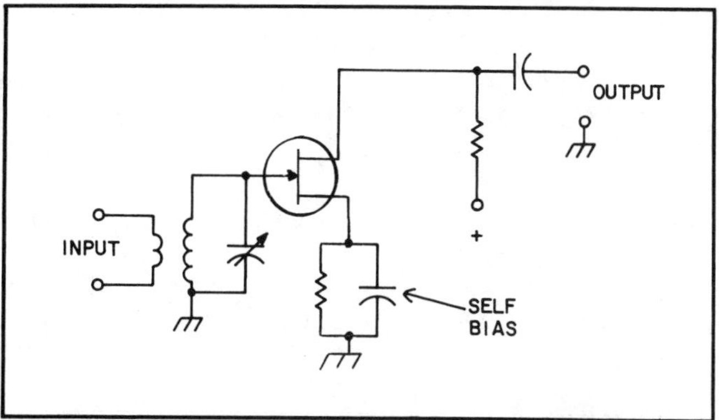

SELF BIAS: Self bias in a field-effect transistor amplifier.

of both types, even if the component is not a capacitor or an inductor.

Self resonance usually occurs at ultra-high or microwave frequencies. In some larger capacitors and inductors, self resonance may exist at the high or very high frequencies. Self resonance may be fairly pronounced, or almost indistinguishable. It is of primary concern in the design of circuits for operation above approximately 300 MHz.

Inductors, because of their inherent inter-winding capacitance, are the most likely type of component to have significant self-resonant effects. Some inductors are deliberately manufactured to be self-resonant. The resonance can take the form of a bandpass, band-rejection, highpass, or lowpass response. *See also* BANDPASS RESPONSE, BAND-REJECTION RESPONSE, HIGHPASS RESPONSE, LOWPASS RESPONSE, RESONANCE.

SELSYN

A selsyn is an indicating device used to show the direction in which an object is pointing. A selsyn is a synchronous motor system, with the transmitting unit at the movable and the receiving unit where the operator can easily see it. A common application of the selsyn is as a direction indicator for a rotatable antenna.

One advantage of a selsyn over other directional indicators in an antenna system is that the selsyn will respond to unexpected changes in the antenna direction, such as might occur if the rotor slips. Assuming the antenna boom-to-mast joint does not slip, the selsyn will always show the antenna direction.

Another advantage of the selsyn is that the indicator needle normally rotates the same number of degrees as the antenna itself. Thus, a selsyn for azimuth bearings will turn 360 degrees; a selsyn for elevation bearings will turn 90 degrees. This gives the station operator a good mental picture of the antenna position. For azimuth indication, a great-circle map of the world, centered at the geographical location of the station, can be placed in the indicator (see illustration). The station operator can then observe, with reasonable accuracy, the part of the world toward which the antenna is beaming. *See also* AUTOSYN, GREAT CIRCLE.

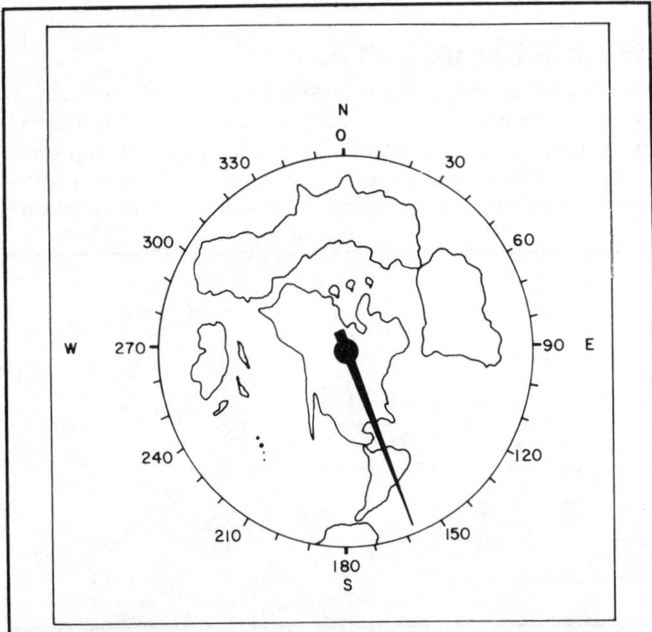

SELSYN: A selsyn directional indicator with a great-circle map of the world on the dial face.

SEMIAUTOMATIC KEY

A semiautomatic key, also popularly known as a bug, is a mechanical device used for sending Morse code at high speeds. The semiautomatic key has been largely replaced, in recent years, by various forms of electronic keyers. Some radio operators, however, still use semiautomatic keys.

With a straight key, most operators find it difficult to send more than about 25 words per minute. This is because, at that speed, the dot frequency is 10 Hz, and it is not easy to pump a straight key up and down that fast!

The semiautomatic key uses a weight, lever, and spring combination to produce mechanical oscillation that forms the dots automatically. The speed is adjusted by moving the weight along the lever. The dashes must be made manually. The illustration shows a typical semiautomatic key.

To form a string of dots, the lever, or paddle, is pressed toward the right and held in position. To make dashes, the lever is intermittently pressed toward the left. This is how a "right-handed" bug works. For "left-handed" operation, these directions are reversed. Semiautomatic keys are usually made for "right-handed" operation, but a few are available for "left-handed" operation.

The dashes must be made manually, but since their frequency is just half that of the dots, this is not especially hard to do at speeds as high as 30 to 40 words per minute. Some operators can send as fast as 50 words per minute using a semiautomatic key. Code sent with a bug has a tempo that varies from one operator to another. The bug is a mechanical device, and different operators make the dashes in various, subtly different ways. *See also* AMERICAN MORSE CODE, INTERNATIONAL MORSE CODE, KEY, KEYER.

SEMI BREAK-IN OPERATION

Semi break-in operation is a form of switching scheme used in many communications installations. Semi break-in can be used for code, radioteletype, or voice operation. In voice systems, it is called voice-operated transmission or VOX (*see* VOX).

In semiautomatic code or radioteletype break-in, the transmitter is actuated the moment the operator presses

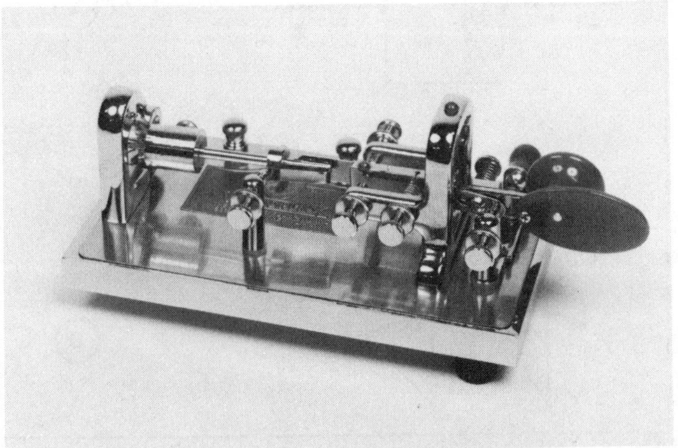

SEMIAUTOMATIC KEY: A semiautomatic key, the original "Presentation" model. (Courtesy of The Vibroplex Company, Inc.)

the key. A relay or electronic switch performs the changeover from the receive mode to the transmit mode. When a pause occurs in the transmission, the transmitter stays on for a predetermined length of time. This delay is adjustable from about 0.1 second to 3 or 5 seconds. Most code operators do not like to have the receiver click on between characters or words of a code transmission, and they set the delay according to the speed at which they most often send the code. In radioteletype operation, the delay is not of much concern, since the carrier is on all the time.

Semi break-in operation does not allow the code operator to hear between dots and dashes, as full break-in does. However, semi break-in is easier and less expensive to implement. Some operators prefer semi break-in over full break-in. *See also* BREAK-IN OPERATION.

SEMICONDUCTOR

A semiconductor is an element or compound that exhibits a moderate or large resistance. The resistance of a semiconductor is not as low as that of an electrical conductor, but it is much lower than that of an insulator or dielectric. The resistance of a semiconductor depends on the particular impurities, or dopants, added to it (*see* DOPING). Semiconductors conduct electricity via two forms of charge carrier: the electron and the hole (*see* ELECTRON, HOLE).

Common semiconductor elements include germanium, selenium, and silicon. Other semiconductor elements are arsenic, antimony, boron, carbon, sulfur, and telurium. Many compounds are classified as semiconductors; the most often-used of these include gallium arsenide, indium antimonide, and various metal oxides.

Semiconductors have revolutionized electronics. A semiconductor device may perform the function of a vacuum tube having hundreds of times its volume. Integrated circuits, manufactured on wafers of semiconductor material, have increased the miniaturization even more.

For more specific information on various semiconductor devices and applications, please refer to articles according to subject.

SEMICONDUCTOR JUNCTION

See P-N JUNCTION, POINT-CONTACT JUNCTION.

SEMICONDUCTOR RECTIFIER

See P-N JUNCTION, RECTIFIER DIODE, SELENIUM RECTIFIER.

SEMILOGARITHMIC GRAPH

See LOGARITHMIC SCALE.

SENSE AMPLIFIER

Many types of memory circuits retain logic states at very low voltage levels. A logic circuit requires a difference of several volts—normally 4 or 5 V—between the low and high conditions. A sense amplifier is a circuit that magnifies the relatively small voltages of the memory, so that they differ by the required amount for use by the logic circuit.

A sense amplifier is a straightforward, low-level, direct-current amplifier. It is advantageous for the sense amplifier to draw as little current as possible from the memory. A sense amplifier must therefore have a high input impedance. Sense amplifiers are usually low-noise circuits; this reduces the chances of interference from stray signal sources. *See also* DC AMPLIFIER.

SENSITIVITY

Sensitivity is a general term that applies to many different electronic devices. In general, sensitivity is an expression of the change in input that is necessary to cause a certain change in the output of a device. Usually, sensitivity is measured in terms of voltage.

In an amplifier or a radio receiver, sensitivity is a measure of the ability of the device to distinguish between a signal and noise at the output. Sensitivity may also be given for microphones and other forms of transducers, and for such electromechanical devices as relays and meters.

In an amplifier, sensitivity is expressed in terms of the input-signal voltage that is needed to produce a certain signal-to-noise ratio, in decibels, at the output. The sensitivity is indirectly related to the gain of the amplifier circuit or amplifier chain. But the noise figure is just as important (*see* GAIN, NOISE FIGURE, SIGNAL-TO-NOISE RATIO). Sensitivity is limited absolutely by the level of the noise that is generated in the amplifier, especially the first amplifier in a chain.

In a microphone or other transducer, sensitivity is expressed as the acoustic pressure, light intensity, or other flux required to produce a certain amount of signal at the output.
Alternatively, the sensitivity may be expressed as the minimum actuating flux that results in satisfactory circuit operation. The sensitivity of a relay is the smallest coil current that will reliably close the armature contacts.

In a radio receiver, sensitivity is expressed as the number of microvolts at the antenna terminals that is required to produce a certain signal-to-noise ratio or level of noise quieting at the speaker. In amplitude-modulation, continuous-wave, frequency-shift-keying, and single-sideband applications, the sensitivity is usually given as the number of microvolts needed to produce a 10-dB signal-to-noise ratio. In frequency-modulation equipment, the sensitivity is specified either as the number of microvolts needed to cause 20-dB noise quieting, or as the number of microvolts that results in a 12-dB signal-to-noise-and-distortion ratio (*see* NOISE QUIETING, SINAD).

The sensitivity of a piece of equipment is usually given in the table of specifications. Sensitivity is an important criterion for any amplifier or receiver, but it is not the only one. The audio distortion, dynamic range, image rejection, intermodulation distortion, and selectivity are also important figures of merit for receivers and amplifiers. *See also* DISTORTION, DYNAMIC RANGE, FRONT END, IMAGE REJECTION, INTERMODULATION, SELECTIVITY.

SENSOR

See TRANSDUCER.

SEPARATION

See CHANNEL SEPARATION.

SEPARATOR

A separator is a layer of material, in a battery or capacitor, intended to prevent the negative and positive poles of the cells, or the opposite plates of the capacitor, from coming into direct physical contact.

In a battery, the separator is soaked with an electrolyte substance, so current conduction does occur through it. But a potential difference, ranging from 1 to 2 V, is present across the separator material (*see* BATTERY, STORAGE BATTERY). If the separator should fail and the poles come into contact, the action of one cell will be lost, and the battery voltage will drop by 1 to 2 V.

In a capacitor, the separator is usually called the dielectric. It is a nearly perfect insulator. If the separator fails, and the plates come into contact, the capacitor is short-circuited, and no longer will hold a charge (*see* CAPACITOR, DIELECTRIC).

In an air-dielectric feed line, separators are used to keep the conductors apart. In this application, the separators are usually called spacers. *See also* SPACER.

SEQUENCE

A sequence is an ordered set of numbers, objects, or pieces of data. The specific order in which data bits, characters, or words appear is known as the sequence. An algorithm may also be called a sequence (*see* ALGORITHM).

In mathematics, a sequence of numbers is an ordered finite set:

$$S = \{x1, x2, x3, \ldots, xn\}$$

or an ordered infinite set:

$$S = \{x1, x2, x3, \ldots\}$$

such that for every term xk, the following term x(k+1) is related to xk by a specific function f. The sum of a sequence is known as a series. *See also* FUNCTION, SERIES.

SEQUENTIAL ACCESS MEMORY

A sequential-access memory is a form of memory in which data channels can be recalled only in a certain predetermined order. Any form of memory may have sequential access. The most common forms of sequential-access memory are the first-in/first-out (FIFO) memory and the pushdown stack (*see* FIRST-IN/FIRST-OUT, PUSHDOWN STACK).

If the addresses in a memory are numbered, such as m1, m2, m3, . . ., mn, then the most common sequence for recall is upward from m1 to mn. A downward sequence, from mn to m1, is somewhat less common. Still less common are odd sequences in which the addresses do not ascend or descend in numerical order.

The data from a sequential-access memory may sometimes be recalled in two or more specific sequences. If the data can be recalled in any sequence possible, then the memory is called a random-access memory. *See also* RANDOM-ACCESS MEMORY.

SERIAL DATA TRANSFER

Information may be transmitted simultaneously along two or more lines, or it may be sent bit by bit along a single line. The first method is known as parallel data transfer (*see* PARALLEL DATA TRANSFER). The latter method is called serial data transfer. The term serial is used because the data is sent in numerical order, according to some predetermined sequence.

In computer practice, serial data transfer requires that a word be split up into all of its constituent bits, then sent over the line, and reassembled at the other end of the line in the same order as originally.

Serial data transfer requires only one transmission line, while parallel data transfer requires several lines. However, serial data transfer is somewhat slower than parallel transfer. *See also* PARALLEL DATA TRANSFER.

SERIES

A series is the sum of a sequence of numbers (*see* SEQUENCE). A series may consist of a finite number of numbers or an infinite number of numbers. In electronics, series are sometimes encountered in conjunction with waveform analysis (*see* FOURIER SERIES, POWER SERIES, TAYLOR SERIES).

A finite series always has a finite value. An infinite series may have a finite value, or it may not, depending on the nature of the sequence function. If an infinite series has a finite sum, then it is said to converge. If the sum is not finite, the series is said to diverge.

Series are generally denoted in either of two ways: the sequential method and the summation method. Let S be the value of a series of n individual numbers denoted by xk, where x is a real number and k is a positive integer ranging from 1 to n. Then we write:

$$S = x1 + x2 + x3 + \ldots + xn$$

Let S now be the value of an infinite series of numbers denoted by xk, where k may be any positive integer. Then we write the infinite series as:

$$S = x1 + x2 + x3 + \ldots$$

This is the sequential method. The summation method makes use of the upper-case Greek letter sigma (Σ), to write the above finite series as:

$$S = \sum_{k=1}^{n} xk$$

and the above infinite series as:

$$S = \sum_{k=1}^{\infty} xk$$

Certain infinite series have characteristic equivalent values. When such a series is encountered, the equivalent value can be substituted for it. Some of the more frequently encountered convergent infinite series, and their equivalent values, are shown in the table.

The theory of infinite series is quite involved, and a complete discussion cannot be given here. College-level calculus textbooks are an excellent source of further information concerning infinite series.

SERIES: Examples of Frequently Encountered Series.

$$1 + 1/2 + 1/4 + 1/8 + 1/16 + \ldots = 2$$

$$1 - 1/2 + 1/4 - 1/4 + 1/5 + \ldots = \ln 2 = 0.693147$$

$$1 + 1/2^2 + 1/3^2 + 1/4^2 + 1/5^2 + \ldots = \pi^2/6 = 1.64493$$

$$1 - 1/2^2 + 1/3^2 - 1/4^2 + 1/5^2 + \ldots = \pi^2/12 = 0.822467$$

$$1 - x + x^2 - x^3 + x^4 - x^5 + \ldots = 1/(1 + x) \text{ for } -1 < x < 1$$

$$1 + x + x^2/2! + x^3/3! + x^4/4! + x^5/5! + \ldots = e^x$$

$$x - x^2/2 + x^3/3 - x^4/4 + x^5/5 + \ldots = \ln (1 + x) \text{ for } -1 < x < 1$$

$$2\left(\frac{x-1}{x+1}\right) + \frac{2}{3}\left(\frac{x-1}{x+1}\right)^3 + \frac{2}{5}\left(\frac{x-1}{x+1}\right)^5 + \frac{2}{7}\left(\frac{x-1}{x+1}\right)^7 + \frac{2}{9}\left(\frac{x-1}{x+1}\right)^9 + \ldots = \ln x$$

$$x - x^3/3! + x^5/5! - x^7/7! + x^9/9! + \ldots = \sin x$$

$$1 - x^2/2! + x^4/4! - x^6/6! + x^8/8! + \ldots = \cos x$$

$$x + x^3/3! + x^5/5! + x^7/7! + x^9/9! + \ldots = \sinh x$$

$$1 + x^2/2! + x^4/4! + x^6/6! + x^8/8! + \ldots = \cosh x$$

SERIES CONNECTION

Components are said to be connected in series when the current follows a single path through them all. The diagram illustrates series connections of resistors (at A), inductors (at B), and capacitors (at C).

When components are connected in series, the current through each is the same. The voltage across each component depends on its resistance, reactance, or absolute-value impedance. (The absolute-value impedance is the distance from the origin to the impedance point in the complex plane.) The voltage across a given component is equal to the product of the current and the absolute-value impedance, according to Ohm's law (*see* IMPEDANCE, OHM'S LAW).

Series connections can be used for the purpose of increasing voltage-handling capability. A series connection of five rectifier diodes, each having a peak-inverse-voltage (PIV) rating of 100 volts, will yield a PIV rating of 500 volts.

Series connection may result in unequal distribution of voltages across components that should, ideally, have identical impedances. This phenomenon is called voltage hogging. In the case of the rectifier diodes mentioned above, voltage hogging could result in the destruction of one diode, and a possible chain reaction leading to the destruction of them all. To prevent this, equalizing resistors are placed across each diode. Similar precautions must be taken in any series circuit (*see* VOLTAGE HOGGING).

In a household circuit, the fuses or circuit breakers are wired in series with each branch. Therefore, the total current drawn by all of the appliances in the branch is forced through the fuse or breaker. If too many appliances are used in a single branch, the fuse or breaker will blow (*see* CIRCUIT BREAKER, FUSE). The appliances or loads themselves are connected in parallel. *See also* PARALLEL CONNECTION.

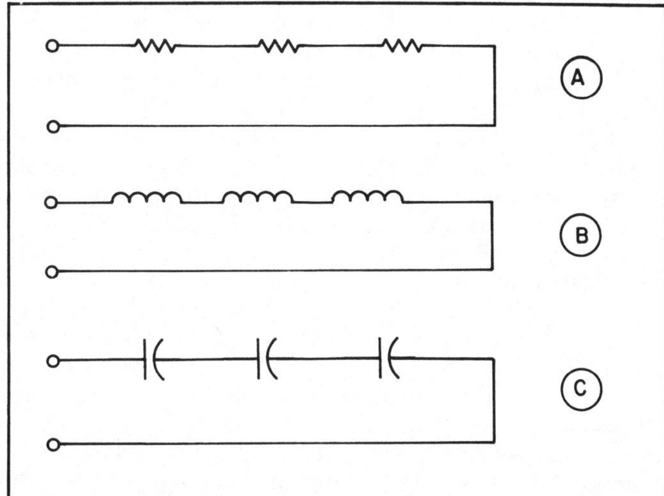

SERIES CONNECTION: Examples of series connections. At A, resistors; at B, inductors; at C, capacitors.

SERIES MODULATION

Amplitude modulation can be obtained in a radio-frequency amplifier by the insertion of a variable-resistance component in the power-supply lead. This method of modulation is known as series modulation. Series modulation can be used with bipolar-transistor circuits, field-effect-transistor circuits, or vacuum-tube circuits.

The drawing illustrates two methods of series modulation in a bipolar-transistor amplifier circuit. At A, the modulating component is in series with the emitter of the amplifying transistor. At B, the modulating component is in series with the collector lead. Either method will provide satisfactory results.

Series modulation can be used in the final amplifier circuit of a transmitter to obtain amplitude modulation. However, this requires that the modulating component handle a large amount of current or voltage. It also requires a high level of audio power. These disadvantages are offset by the fact that the amplifier itself can be operated in Class C, yielding higher radio-frequency efficiency than would be possible with a Class-AB or Class-B configuration. Also, no attention must be paid to the linearity of the radio-frequency amplifier circuits when series modulation is used at the final amplifier. Series modulation

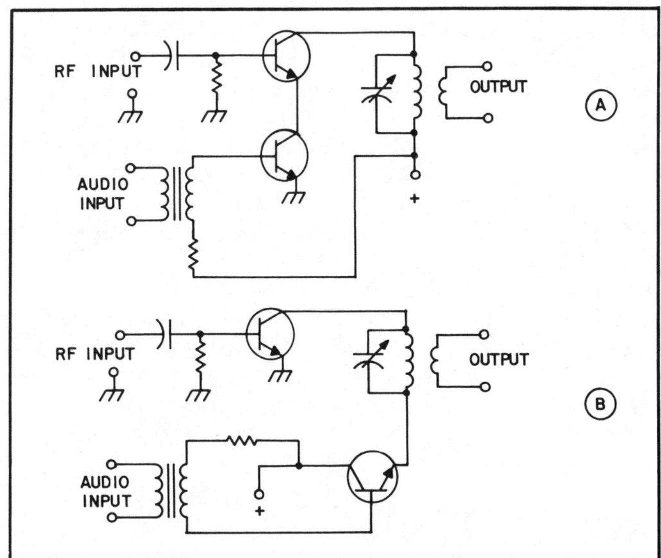

SERIES MODULATION: At A, series modulation in the emitter lead of a bipolar-transistor amplifier. At B, series modulation in the collector circuit.

SERIES-PARALLEL CONNECTION: Examples of series-parallel connections. Assuming a value of R ohms and a power-dissipation rating of P watts for each resistor, either configuration (A or B) will yield a net resistance of R ohms with a power-dissipation rating of 9P watts.

shares these advantages (and disadvantages) with plate modulation in vacuum-tube power amplifiers. *See also* PLATE MODULATION.

SERIES-PARALLEL CONNECTION

A series-parallel connection is a method of increasing the power-handling capacity of a component. Series-parallel connections are often used with resistors to obtain a higher wattage capability. Series-parallel connections may also be used with other components, such as inductors, capacitors, and diodes.

Series-parallel connections may be made in two basic configurations. One method is as follows: A certain number (say n) of components may be connected in series, and n of these combinations assembled; the resulting n combinations are then connected in parallel. This method is the most common, and is shown at A in the illustration for $n = 3$. Alternatively, n components may be connected in parallel, and n of these combinations made up; then these n combinations are connected in series. This scheme is shown at B for $n = 3$. In both of these arrangements, if the value of each resistor is R ohms, then the value of the composite is R ohms. If the power-dissipation rating of each resistor is P watts, then the composite will be capable of dissipating 9P watts. In general, the power-dissipation rating is increased by a factor of n^2.

Series-parallel configurations can be assembled in other ways, where the number of parallel elements differs from the number of series elements. This causes a change in the composite value, depending on the particular arrangement used. *See also* PARALLEL CONNECTION, SERIES CONNECTION.

SERIES REGULATOR

In a power supply, voltage regulation can be obtained by means of a component (usually a transistor) connected in series with the output. Such a component is called a series

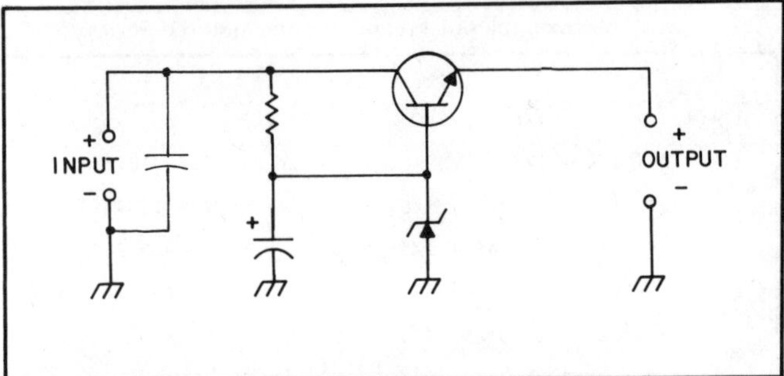

SERIES REGULATOR: A transistor, in conjunction with a zener diode, can be used as a series regulator.

regulator; this method of regulation is known as series regulation.

The schematic diagram shows a transistor series-regulation circuit. The transistor is called a pass transistor. The Zener diode determines the output voltage; normally the input voltage is about 1.5 to 2 times the output voltage for best regulation. A standard 13.8-V power supply, for example, uses a 14-V Zener diode, and an input voltage of 20 to 25 V.

Integrated-circuit series regulators, containing the transistor, Zener diode, and associated components shown, are available for various regulated voltage levels.

With any regulation circuit, it is a good idea to have overvoltage protection. In the example shown, the output voltage would rise dramatically if the pass transistor were to become shorted. *See also* OVERVOLTAGE PROTECTION, SHUNT REGULATOR.

SERIES RESONANCE

When an inductor and capacitor are connected in series, the combination will exhibit resonance at a certain frequency (*see* RESONANCE, RESONANT FREQUENCY). This condition occurs when the inductive and capacitive reactances are equal and opposite, thereby cancelling each other and leaving a pure resistance.

The resistive impedance that appears across a series-resonant circuit is very low. In theory, assuming that the inductor and capacitor are lossless components, the impedance at resonance is zero (see illustration). But in practice, since there is always some loss in the components, the impedance is greater than zero, but may be less than 1 ohm. The exact value depends on the Q factor; the higher the Q, the lower the impedance at resonance. The Q factor is a function of the losses in the components (*see* IMPEDANCE, Q FACTOR).

Below the resonant frequency, a series-resonant circuit shows capacitive reactance. Above the resonant frequency, it shows inductive reactance. Resistance is also present under nonresonant conditions, because of the losses in the components; therefore, the impedance is complex at frequencies other than the resonant frequency.

Series-resonant circuits are sometimes used as traps (*see* TRAP). They are also used in some oscillators and amplifiers for tuning purposes.

A section of transmission line can be used as a series-resonant circuit. A quarter-wave section, open at the far end, behaves like a series-resonant circuit. A half-wave section, short-circuited at the far end, also exhibits this property. Such resonant circuits can be, and sometimes are, used in place of coil-capacitor (LC) combinations,

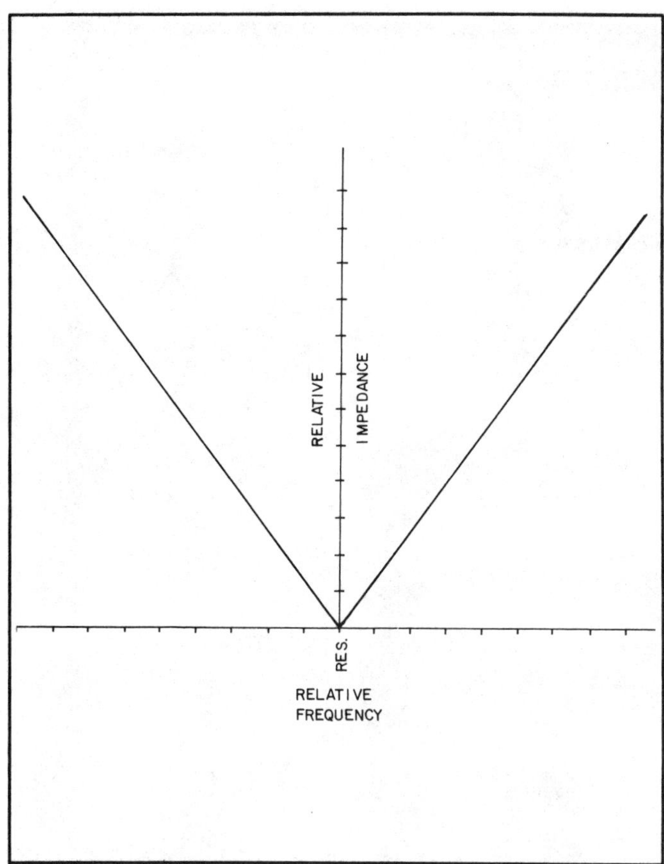

SERIES RESONANCE: A series-resonant circuit displays a pure resistance of a very low value.

especially at the very-high and ultra-high frequencies, where the physical lengths of such sections are reasonable.

SERVICE EQUIPMENT

In the servicing of electronic apparatus, a well-equipped service bench is essential. A service bench for audio or communications equipment must have certain test instruments.

A multimeter, or volt-ohm-milliammeter, is a necessity for checking of various currents, resistances, and voltages (*see* VOLT-OHM-MILLIAMMETER). An oscilloscope is useful for evaluating the waveform distortion, peak-to-peak voltages, and gain (*see* OSCILLOSCOPE).

In the servicing of radio-frequency receivers, transceivers, or transmitters, a complete signal generator and monitor is valuable. Such a device contains equipment necessary for checking the frequency response of a receiver, and for testing various aspects of transmitter performance. The monitor/generator contains an accurate frequency meter, an oscilloscope for observing modulation, and a signal generator that can be modulated. Some service monitors have built-in dummy loads and wattmeters (*see* SERVICE MONITOR/GENERATOR).

Other instruments that are valuable for the troubleshooting and servicing of electronic apparatus include a frequency counter, a dummy antenna, a wattmeter, an audio-distortion analyzer, a SINAD meter, and, of course, a soldering iron. *See also* DESOLDERING TECHNIQUE, DISTORTION ANALYZER, DUMMY ANTENNA, FREQUENCY COUNTER, SINAD, SINAD METER, SOLDERING IRON, SOLDERING TECHNIQUE, TEST LABORATORY, TEST INSTRUMENT.

SERVICE MONITOR/ GENERATOR

A service monitor/generator is a self-contained unit that is used in the troubleshooting, alignment, and repair of electronic equipment, particularly radio-frequency devices such as amplifiers, receivers, and transmitters.

An example of a service monitor/generator is shown in the illustration. This device can operate from the very low frequencies up to the microwave range. The frequency is set by means of the set of 10-position rotary switches (A), for the testing of transmitters. A meter (B) indicates the amount by which the signal frequency departs from the channel center. An oscilloscope (C) allows the technician to observe the modulation characteristics of the signal. A switch provides high sensitivity or low sensitivity, as desired (D). Another switch selects wideband or narrowband response (E).

For the testing of receivers, a built-in signal generator can be adjusted to any frequency by means of the rotary switches (A). The level of the output is set by a control knob with a dial calibrated in microvolts (F). The signal can be amplitude-modulated or frequency-modulated at various audio frequencies (G) and levels (H).

A service monitor is used in conjunction with other test equipment when necessary. *See also* SERVICE EQUIPMENT.

SERVOMECHANISM

A servomechanism is a form of feedback-control device. Servomechanisms are used in the remote control or automatic control of mechanical apparatus such as motors, steering mechanisms, radio controls, and valves. A servomechanism is sometimes called a servo. A complete electromechanical system, incorporating one or more servomechanisms, is called a servo system (*see* SERVO SYSTEM).

An example of the use of a servomechanism is shown in the diagram. This device is an automatically tuned antenna. A rotary inductor is adjusted by means of an electric motor, which is in turn moved by the output signal from a sensing circuit. The sensing circuit and motor form the servomechanism; the complete antenna-tuning unit is a servo system.

When the transmitter is actuated and radio-frequency energy is applied to the antenna at a frequency f, the antenna is initially tuned to some higher or lower frequency g. Therefore, there is reactance as well as resistance present at the feed point. If $f > g$, the reactance is inductive, and the current lags the voltage. If $f < g$, the reactance is capacitive, and the current leads the voltage. The sensing circuit detects either condition, and causes the motor to turn the coil in the required direction to bring the current and the voltage into phase. Then the rotary inductor remains fixed in that position, until a change is made in the transmitter frequency. The response time depends on the difference between the initial frequencies f and g.

Servomechanisms can be operated in conjunction with computers. For example, a computer program may tell a servomechanism to move according to a certain function. Several servomechanisms, when interconnected and controlled by a computer, can do quite complicated things, such as cook a meal automatically. This type of servo system is called a robot. The technology of robot design and construction is known as robotics. *See also* ROBOTICS.

SERVICE MONITOR/GENERATOR: A typical service monitor-generator. Arrows show controls as follows: A, frequency adjust; B, carrier centering; C, modulation monitor; D, sensitivity adjust; E, passband adjust; F, generator output adjust; G, generator modulation frequency; H, generator modulation level.

SERVO SYSTEM

A complete electromechanical set of one or more servomechanisms, including all of the associated circuits and hardware, and intended for a specific task or purpose, is called a servo system. Servo systems may be used for a variety of purposes, such as temperature regulation, frequency control, voltage regulation, or transmitter tuning. The following illustration shows an automatic antenna tuner, which is a form of servo system.

In recent years, servo systems have become more and more complex and sophisticated. Small computers can be used to control a servo system consisting of several different servomechanisms. For example, an unmanned plane can be programmed to take off, fly a specific mission, return, and land. (Such an airplane is called a drone.) Servo systems can be programmed to do assembly-line work and other mundane things. Such servo systems are called robots. *See also* ROBOTICS, SERVOMECHANISM.

SETTLING TIME

When a digital meter is used to measure current, frequency, power, resistance, or voltage, a certain amount of time must elapse before the meter comes to rest and shows the value. The length of time between the connection of the meter leads and the stabilization of the reading is called the settling time.

Settling time varies, depending on the type of instrument and on the number of significant figures that it shows. In general, the more accurate the meter, the longer the settling time. Typical digital test meters have a settling time of about 1 to 2 seconds.

In a servo system, equilibrium is reached after a certain length of time. When a parameter changes, the servomechanism does not compensate for it instantaneously (although it may do so very fast). The length of time required for the system to reestablish equilibrium is called settling time. *See also* DIGITAL METERING, SERVOMECHANISM, SERVO SYSTEM.

SEVEN-SEGMENT DISPLAY

When seven independently controlled, bar-shaped light-emitting diodes, liquid crystals, or other two-state visual

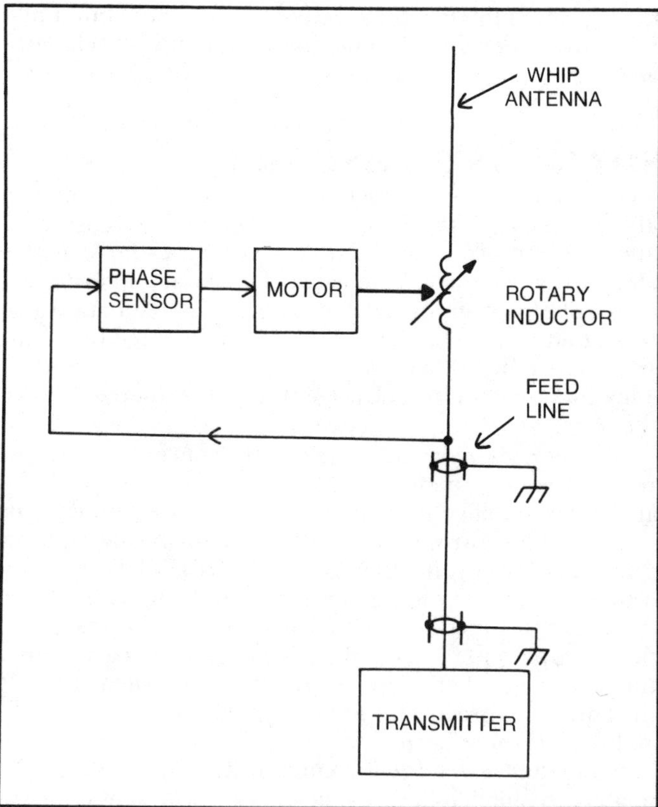

SERVO SYSTEM: An example of a servomechanism and servo system. This device is an automatic antenna tuner.

devices are arranged in the pattern shown in the illustration, the combination is known as a seven-segment display. The seven-segment display is universally used in digital devices of all kinds, since most English letters, and all numerals, can be represented by some combination of states.

Although the seven-segment display is the most commonly used configuration in digital clocks, calculators, meters, and other indicators, there are other methods of obtaining an alphanumeric display. The dot-matrix method, for example, is often used, especially in languages having characters that cannot be represented on a conventional seven-segment display. *See also* DOT-MATRIX PRINTER.

SEVEN-UNIT TELEPRINTER CODE

The seven-unit code is identical to the Baudot teleprinter code (*see* BAUDOT), except that each character is preceded by a start signal and followed by a stop signal. Thus, instead of five bits or units per character, there are seven. The start unit is a space, and the stop unit is a mark (*see* MARK/SPACE).

The advantage of the seven-unit teleprinter code is that it provides synchronization between the transmitter and receiver. If some data bits are missed, a single character may be lost, but the error will not continue for several characters. All modern Baudot systems use the seven-unit code.

SFERICS

Electromagnetic noise is generated in the atmosphere of

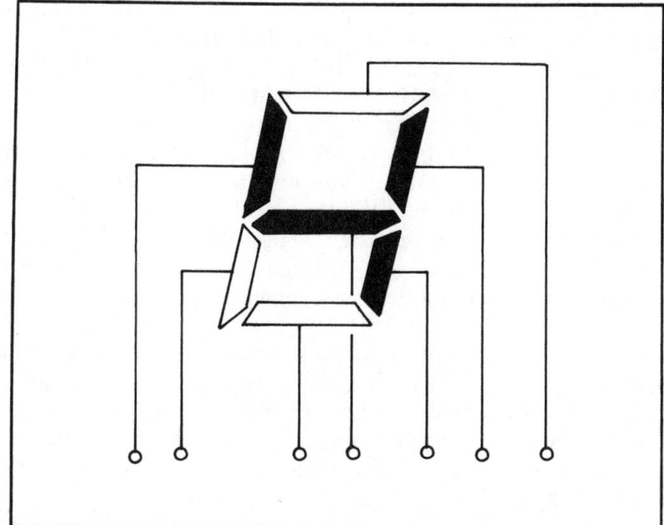

SEVEN-SEGMENT DISPLAY: A seven-segment display can show most English characters and all numerals.

our planet, mostly by lightning discharges in thundershowers. This noise is called sferics. In a radio receiver, sferics produce a faint background hiss or roar, punctuated by bursts of sound we call static.

A potential difference of about 300,000 volts exists between the surface of the earth and the ionosphere. The earth and ionosphere therefore act like a huge capacitor, with the troposphere and stratosphere serving as the dielectric (*see* ATMOSPHERE). Sometimes this dielectric develops "holes," or pockets of imperfection, where discharge takes place. Such "holes" are the thundershowers. There are normally about 700 to 800 such areas at any given time, concentrated mostly in the tropics. Sand storms, dust storms, and volcanic eruptions also produce some lightning, contributing to the overall sferics level.

An individual lightning stroke produces a burst of electromagnetic energy from the very low frequencies through the microwave spectrum—and even in the infrared, visible, and ultraviolet ranges (*see* LIGHTNING).

The current that flows across the atmospheric capacitor averages 1500 amperes. Since the potential difference across this capacitor is 300,000 volts, the sferics generator on our planet is a 450-megawatt, wideband radio transmitter! At very low and low frequencies, the sferics propagate around the world, as the electromagnetic field is trapped between the "plates" of the capacitor. As the frequency increases, sferics become a more local phenomenon until, at very high frequencies and above, sferics travel only a few miles. This is why the general level of background noise decreases as the frequency gets higher, making low-noise circuit design of such importance in the very-high-frequency, ultra-high-frequency, and microwave spectra.

Sferics are not confined to the earth. Some noise is generated by the immensely powerful storms in the atmosphere of the giant planet Jupiter. Astronomers have heard this noise with radio telescopes. Sferics probably also occur on Saturn, and perhaps on Uranus, Neptune, Venus, and even Mars. In the cases of Venus and Mars, dust storms and volcanic eruptions would be the cause of sferics.

Not all of the background hiss that we hear in our radio receivers is the result of sferics; the sun generates a great deal of electromagnetic noise, as does the center of our galaxy, a powerful object in the constellation Cygnus, and many other celestial sources. Noise of extraterrestrial origin is not, technically, categorized as sferics.

You can hear the sferics from a distant thundershower on the standard AM broadcast band. If you have a general-coverage communications receiver, you can listen at progressively higher frequencies as the storm or storm system approaches. When you hear the static bursts at 30 MHz, the storm is probably less than 100 miles away.

All storm systems produce sferics, although the amount of noise varies. Using directional antenna systems and special receivers, some meteorologists have been able to locate and track large storm systems. A receiver designed especially for listening to atmospheric noise is called a sferics receiver. In normal radio communications, however, sferics are just a nuisance, since they cause interference. *See also* INTERFERENCE, INTERFERENCE FILTER, INTERFERENCE REDUCTION, NOISE, NOISE LIMITER, SFERICS RECEIVER.

SFERICS RECEIVER

A wideband radio receiver, designed especially for intercepting atmospheric noise (sferics), is called a sferics receiver. A series receiver may also be called a lightning receiver.

A typical sferics receiver can intercept signals from the very low frequencies to the very high frequencies. A directional antenna system is used. As a storm system approaches, the sferics can be heard at higher and higher frequencies. This makes it possible to gauge the distance from the receiver to the storm center. The compass bearing of the system is determined by means of the directional antenna.

Sferics receivers have been used with some success in tracking hurricanes, squall lines, and other severe weather phenomena in which lightning occurs. However, the occurrence of lightning in storms varies, and often the greatest amount of lightning is observed at considerable distances from the storm center. Modern weather satellites provide much more precise and detailed information, and have largely replaced the sferics receiver for the purpose of storm location and tracking. *See also* SFERICS, WEATHER SATELLITE.

SHADOW EFFECT

Electromagnetic fields sometimes pass around obstructions with very little or no attenuation; under other circumstances the obstruction causes a large amount of attenuation. The extent to which this shadow effect occurs depends on the size, or diameter, of the obstruction relative to the wavelength of the electromagnetic field.

Generally, the shadow effect is produced by objects having a diameter of at least several wavelengths. When the diameter of the obstruction is small compared with the wavelength of the electromagnetic field, diffraction occurs and the field is essentially unaffected (*see* DIFFRACTION). As the obstruction becomes larger, or the wavelength becomes shorter, less and less diffraction is observed. The shape of the object has some effect, as well: A rounded or smooth object casts a deeper shadow, for a given ratio of wavelength to obstruction diameter, than does an object having sharp corners.

Shadow effect is most likely to affect radio communications at very high frequencies and above, where buildings, hills, and other obstructions measure many wavelengths across.

SHANNON'S THEOREM

The amount of noise present in a communications channel affects the ease with which information can be transmitted and received accurately. The speed of transmission also affects the accuracy. These two factors are interdependent. The probability of obtaining error-free data transfer, based on the channel noise and the speed of transmission, is specified by a theorem called Shannon's Theorem. This theorem is also known as the noisy-channel coding theorem.

Shannon's Theorem is based on the concept of entropy. In information theory, entropy is an expression of the amount of intelligence contained in each communications symbol. (The entropy function is mathematically sophisticated, and is beyond the scope of this book. A text on information theory is a good source of details concerning the entropy function. Shannon's Theorem states that if the entropy is no greater than the channel capacity, then the probability of error in data transmission is negligible. If the entropy is greater than the channel capacity, the probability of error is large.

No matter how large the entropy, the error rate can be made very small, as long as the data speed is slow enough. *See also* INFORMATION THEORY.

SHAPE FACTOR

The attenuation-versus-frequency response of a bandpass filter, especially in a communications receiver, is often evaluated according to an expression known as the shape factor. The shape factor determines the extent to which the response is rectangular.

The shape factor is generally given as the ratio of the

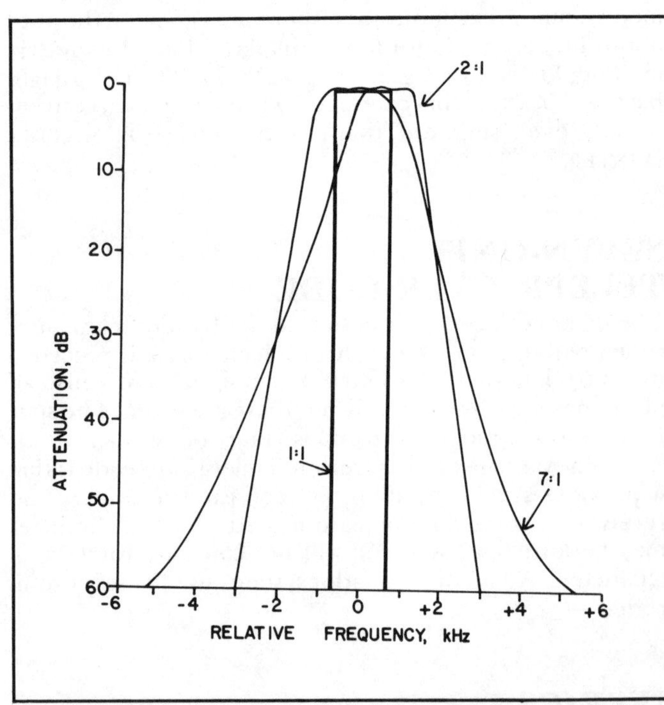

SHAPE FACTOR: Examples of selectivity shape factors. The 1:1 shape factor represents a theoretically perfect rectangular response.

bandwidth of a filter at the −60-dB points to the bandwidth at the −6-dB points. The signal levels are measured in terms of voltage attenuation in the intermediate-frequency chain.

Various shape factors are illustrated graphically in the illustration. In general, the smaller the shape factor, the more rectangular the response. (A shape factor of 1:1 indicates a perfectly rectangular response—a theoretical ideal, but not obtainable in practice.)

A rectangular response is desirable in a bandpass filter, since a signal generally has a well-defined band over which the information is carried. It is essential that all of the signal components get through the filter, but it is undesirable for frequencies outside the signal band to be passed. *See also* BANDPASS RESPONSE, BANDWIDTH, RECTANGULAR RESPONSE, SELECTIVITY.

SHAPING

In continuous-wave (code) transmission, the rise and decay times must be fast enough to convey the necessary information at the required speed. However, if the rise or decay times are too rapid, key clicks will occur (*see* KEY CLICK). The envelope of a code signal is called the shaping of the signal. The process of tailoring, or adjusting, the envelope is also called shaping.

Shaping is usually accomplished by means of a simple resistor-capacitor combination in the keying circuit of a code transmitter. The resistor is connected in series with the keying line, and the capacitor in parallel, as shown at A in the illustration. This type of circuit is suitable for use in blocked-gate or blocked-base circuits. In the case of source keyed or emitter-keyed circuits, a large-value choke may be substituted for the resistor, as at B.

In general, the larger the value of the resistor or choke, the slower the rise time for a given capacitance. The larger the value of the capacitor, the slower both the rise and decay times will be. The values are selected so that the shaping is optimum for the range of code speeds to be transmitted. Usually, the rise time is on the order of a few milliseconds, and the decay time is approximately twice the rise time.

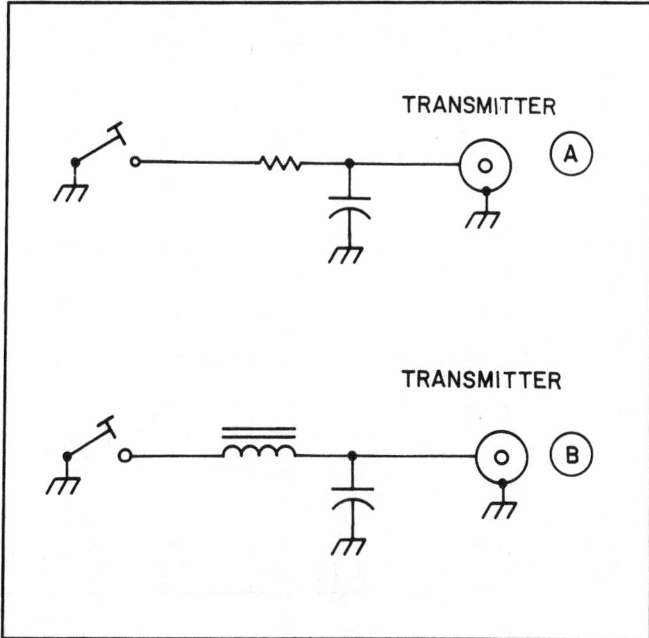

SHAPING: At A, a resistance-capacitance (RC) circuit; at B, an inductance-capacitance (LC) circuit.

SHELF LIFE

A battery will, of course, last longer if it is not used than if it is used in a circuit or device. Even if a battery is kept in storage, it will not retain its charge or charge-holding capacity forever. Eventually, a cell or battery will become unusable even if it is stored under the best possible conditions. The length of time during which a cell or battery is usable is called the shelf life. Ordinary dry cells, and most other types of cells and batteries, have shelf lives of 2 or 3 years.

The shelf life can be specified for any type of component. Some devices, such as resistors, diodes, and transistors, will last for decades (or even centuries!) in storage under ideal conditions. Other components, such as electrolytic capacitors, oil-filled capacitors, potentiometers, switches, and some tubes, have more meaningful shelf lives.

SHELL

See ELECTRON ORBIT.

SHIELD CONTINUITY

See SHIELDING EFFECTIVENESS.

SHIELDING

The intentional blocking of an electric, electromagnetic, or magnetic field is known as shielding. The deliberate blocking of high-energy radiation, such as X rays, gamma rays, or subatomic particles, is also called shielding. Different types of fields or rays are shielded in different ways.

Electric shielding is also called electrostatic shielding or Faraday shielding. A grounded metal screen or plate will block the lines of flux of an electric field. Electromagnetic shielding requires a completely enclosed cage or box made of a conducting material (*see* ELECTROMAGNETIC SHIELDING, ELECTROSTATIC SHIELDING). Magnetic shielding necessitates the use of a ferromagnetic metal enclosure, such as iron or steel, which will block the lines of flux of a nonfluctuating magnetic field. *See also* ELECTRIC FIELD, ELECTROMAGNETIC FIELD, MAGNETIC FIELD.

To obtain shielding against X rays, gamma rays, or high-speed subatomic particles, a certain amount of matter must be placed in the way of the radiation. Lead is generally used for radiation shielding; concrete, earth, water, or other materials can also be used. Beta particles can pass through only a paper-thin layer of lead. Alpha particles and X rays are blocked by about 1 to 3 mm of lead. Gamma rays, if they have sufficient energy, can penetrate through a layer of lead 1 cm or more thick. If materials lighter than lead are used for radiation shielding, the required thicknesses are greater. The most energetic radiation can pass unimpeded through practically anything; such high-energy gamma rays are, fortunately, encountered only in outer space. *See also* ALPHA PARTICLE, BETA PARTICLE, GAMMA RAY, X RAY.

SHIELDING EFFECTIVENESS

No shielding barrier stops 100 percent of the field or radiation that it is designed to block, although in many cases the figure is very close to 100 percent. The percentage of energy that is blocked is called the shielding effectiveness. In electronics, this term is used mostly with electromagnetic shielding, and is of particular importance in coaxial cables.

The effectiveness of an electromagnetic shield is a function of the continuity, or the physical completeness, of the barrier. The shield continuity is defined as the percentage of the total surface area of the shield enclosure that is actually covered by metal. If the total surface area is A, and the shield surface area is B, then the continuity C is given in percent by the equation:

$$C = 100B/A$$

A solid metal enclosure, with absolutely no holes or gaps and with an excellent ground, provides 100-percent shielding continuity, and practically 100-percent shielding effectiveness if a good ground is used. A screen or wire cage offers less shielding continuity, but the effectiveness will be near 100 percent as long as the holes or gaps are very small compared to the wavelength of the electromagnetic field.

In a coaxial cable, the shield continuity is of increasing importance as the frequency becomes higher. Some types of coaxial cable, designed for use at audio frequencies only, have very poor shield continuity. Some radio-frequency cables have marginal shield continuity; such cables are satisfactory for most high-frequency applications, but are not suitable for use at very high frequencies and above. The best shield continuity is afforded by hard line, which has a solid metal outer conductor. Double-shielded coaxial cable is also good for use at very-high and ultra-high frequencies (*see* COAXIAL CABLE, HARD LINE).

Even if the shield continuity of a coaxial cable is 100 percent, the shielding effectiveness may be poor if an installation is faulty. Radio-frequency currents can be induced on the outer conductor of any coaxial cable. These so-called antenna currents can be minimized, and the shielding effectiveness thereby optimized, by taking these precautions: (1) Ensure that the ground system is adequate; (2) Use a balun at the feed point if the antenna is balanced; (3) Avoid using resonant lengths of coaxial cable in the feed line; (4) Run the cable away from the antenna at right angles to the radiating element for a distance of at least ¼ wavelength. *See also* ELECTROMAGNETIC SHIELDING.

SHIFT

Shift is the difference between the mark and space signals in a radioteletype transmission. When F1 type emission (frequency-shift keying) is employed, the mark and space frequencies may both be on the order of hundreds of megahertz, but the difference is usually less than 900 Hz. The most common values are 170 Hz, 425 Hz, and 850 Hz (*see* FREQUENCY-SHIFT KEYING). When F2 type emission (audio frequency-shift keying) is used, the audio frequencies are normally 2125 Hz and 2295 Hz, with a shift of 170 Hz. Occasionally, tones of 1275 Hz and 1445 Hz are used. Sometimes, audio tone pairs of 2125 Hz and 2975 Hz, or 1275 and 2125 Hz, are used, giving 850-Hz shift. In two-way modem systems, standard frequencies are 1270 Hz (mark) and 1070 Hz (space) in one direction, and 2225 Hz (mark) and 2025 Hz (space) in the other. Other shift values are less common.

In a memory or other circuit containing data in specific order, a shift is the displacement of each data bit one place to the right or one place to the left. In the case of a shift to the right, the far right-hand bit moves to the extreme left; in the case of a shift to the left, the far left-hand bit moves to the extreme right. This movement of data is also known as a cyclic shift. *See* CYCLIC SHIFT, SHIFT REGISTER.

SHIFT REGISTER

A shift register is a form of digital memory circuit that is extensively used in calculators, computers, and data-processing systems. Information is fed into the shift register at one "end," and emerges from the other "end." These "ends" are usually called the left and the right.

A clock provides timing pulses for the shift register. Each time a clock pulse is received, every data bit moves one place to the right or left. As data moves to the right, the right-hand bits leave the shift register one by one (see illustration). The opposite happens with the left-hand movement of data. Some shift registers can operate in only one direction, and are therefore called unidirectional. Other shift registers can operate in either direction, and are called bidirectional.

There are two basic categories of unidirectional or bidirectional shift register. These are known as the dynamic type and the static type. The dynamic shift register uses temporary storage methods. Data is lost if the clock operates at an insufficient rate of speed. In the static shift register, flip-flops are used to store the information (*see* FLIP-FLOP). The clock rate may be as slow as desired, because the data remains as long as power is supplied to the flip-flops. Shift registers are available in integrated-circuit form.

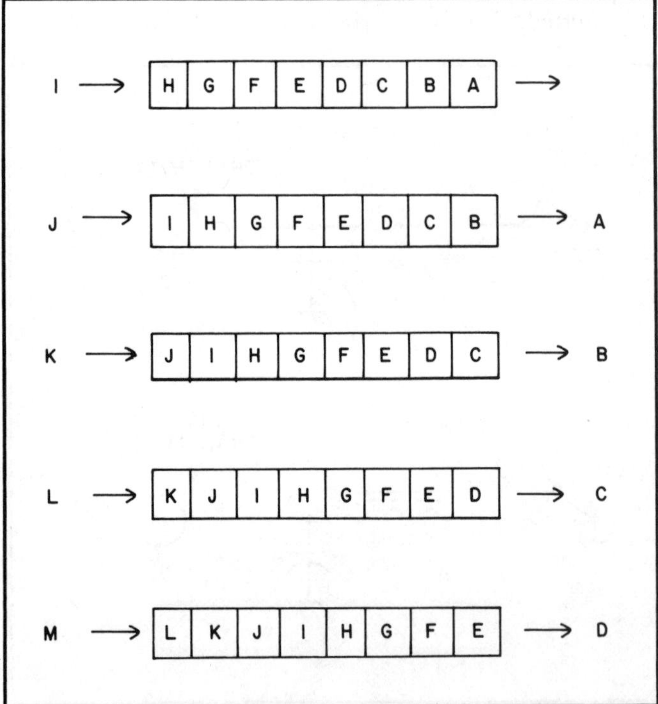

SHIFT REGISTER: Operation of a shift register. The letters represent consecutive data bits. In this case, the shift is from left to right.

SHOCK
See ELECTRIC SHOCK.

SHOCK HAZARD
A shock hazard exists whenever there is a sufficient potential difference between two exposed objects to cause harmful or lethal currents in a person who touches both objects at the same time. Even the 117-V potential in household appliances can, under some circumstances, be lethal. Sometimes a hazard can exist at far lower voltages.

A current of just a few milliamperes causes pain. If a current of approximately 100 to 300 mA flows through the heart, there is a high probability that fibrillation will occur (*see* HEART FIBRILLATION). Fibrillation is possible even with currents as low as 30 mA in some cases. This can cause death. Large amounts of current can result in severe burns.

Generally, a shock hazard is considered to exist if there is sufficient voltage to cause a current of 5 mA to flow through a resistance of 500 ohms. The resistance of the human body may be much less than 500 ohms, especially if water is present. Sweaty hands, for example, greatly lower the effective resistance of the body. *See also* ELECTRIC SHOCK.

SHOCKLEY DIODE
A Shockley diode is a form of switching diode. It has four semiconductor layers, instead of the usual two. *See* FOUR-LAYER SEMICONDUCTOR.

SHORE EFFECT
Radio waves at the very low, low, and medium frequencies travel more rapidly over salt water than over land, because the surface-wave conductivity of sea water is better than that of land (*see* SURFACE WAVE). This difference results in a change of direction in the path of a radio wave crossing a shoreline.

The extent of the bending depends on the angle with which the radio waves cross the shoreline. No refraction occurs when the waves cross at a right angle. As the crossing angle becomes smaller, the refraction increases. Radio waves traveling parallel to the shore tend to curve inland. These effects are shown at A in the illustration.

At the very-high and ultra-high frequencies, another type of shore effect occurs. During the daylight hours, the air over the land is heated and becomes less dense than the air over the water. As a result, radio waves are refracted upon crossing the shoreline. The effect is most pronounced at large angles of incidence (B). At night, the temperature gradient is reversed, and the sense of the refraction is therefore inverted (C). The shore effect is increased as the difference in temperature gets larger. *See also* REFRACTION, TROPOSPHERIC PROPAGATION.

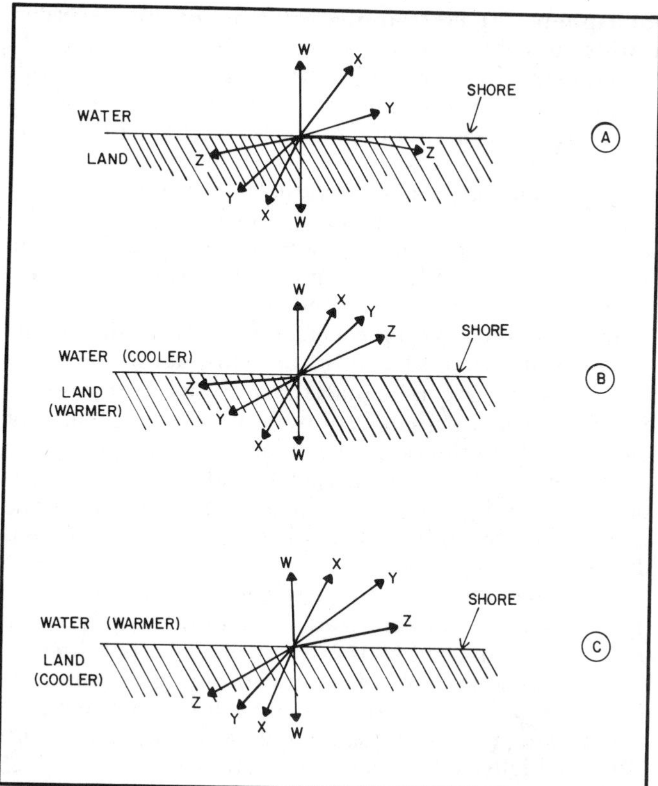

SHORE EFFECT: At A, radio waves at very-low to medium frequencies are bent because of differences in surface conductivity. Very high frequency and ultra-high-frequency waves are refracted because of temperature differences during the day (B) and at night (C).

SHORT CIRCUIT
A short circuit occurs when current is partially or wholly diverted from its normal or proper path. It is called a short circuit because the path of the current is often shorter than it otherwise would be. A short circuit is sometimes called a short out, or simply a short.

In a 117-V household utility system, or in a power supply, a short circuit causes the tripping of a circuit breaker or the blowing of a fuse. This is known as short-circuit protection (*see* CIRCUIT BREAKER, FUSE, SHORT-CIRCUIT PROTECTION). In an audio-frequency or radio-frequency system, a short circuit causes a reduction in the level of the signal at the output. A short circuit is sometimes deliberately inserted into a circuit for a specific purpose.

When a single component is short circuited either directly or by another component, the connection is called a shunt. *See also* SHUNT.

SHORT-CIRCUIT PROTECTION
When a short circuit occurs in the utility lines or in the output of a power supply, overheating will occur unless there is some means of protection for the components. This overheating can damage wiring, transformers, and series-connected resistors and chokes. If the current flow is large enough, fire can result.

The most common methods of short-circuit protection are the circuit breaker and the fuse. These devices open the circuit when the current exceeds a certain level (*see* CIRCUIT BREAKER, FUSE).

Current limiting, also called foldback, is provided in some power supplies to protect the components against a short circuit. The current-limiting circuit inserts an effective resistance in series with the output of the supply (*see* CURRENT LIMITING).

SHORT-PATH PROPAGATION
At the very low, low, medium, and high frequencies, the

ionosphere affects radio waves and often results in worldwide communications. This is sometimes also observed in the lower part of the very-high-frequency band. In long-distance ionospheric communication, the waves travel over, or near, great-circle paths (*see* GREAT CIRCLE).

There are two possible great-circle paths between any two points on the globe. Both paths lie along the same great circle, but one path is generally much shorter than the other. The shorter path is known, appropriately, as the short path. It is this path that usually results in the best signal propagation between the two points.

The term short-path propagation generally applies only to signals that are returned by the ionosphere. The direct wave, ground wave, or signals that travel via tropospheric modes, are short-path in an obvious sense, but they are not short-path ionospheric propagation. For example, if you communicate with a friend via CB, and your two stations are only 1 mile apart, the propagation occurs over the short path, obviously, but it is not usually referred to as short-path propagation.

Sometimes radio waves will travel the long way around the world to get from one station to another. This is called long-path propagation. It is most likely to be observed between stations that are separated by a great distance. Sometimes a signal is propagated via both the short path and the long path at the same time. This takes place most often at high frequencies, especially between 5 and 25 MHz, and is the result of interaction of the signals with the F layer of the ionosphere. *See also* F LAYER, IONOSPHERE, LONG-PATH PROPAGATION, PROPAGATION CHARACTERISTICS.

SHORT WAVES

The high-frequency band, from 3 to 30 MHz (100 m to 10 m), is sometimes called the shortwave band. The wavelengths are not really very short; compared to microwaves, for example, the short waves are extremely long.

The term shortwave originated in the early days of radio, when practically all communication and broadcasting was done at frequencies below 1.5 MHz (200 m). It was thought that the higher frequencies were useless. The range "200 meters and down"—that is, 1.5 MHz and above—was given to the radio amateurs. In fact, radio amateurs were restricted to the frequencies above 1.5 MHz. Within a few years, the radio amateurs discovered that these frequencies were anything but useless. The shortwave-radio era was thus born, and the short waves are still extensively used today.

A high-frequency, general-coverage, communications receiver is sometimes still called a shortwave receiver. Most shortwave receivers cover the range from 1.5 MHz through 30 MHz. Some also operate in the standard broadcast band at 535 kHz to 1.605 MHz. A few shortwave receivers can operate below 535 kHz, and into the so-called longwave band.

SHORTWAVE LISTENING

Anyone can build or obtain a shortwave or general-coverage receiver, install a modest random-wire antenna, and listen to signals from all around the world. This interesting hobby is called shortwave listening. Millions of people in the United States enjoy this hobby.

There are many commercially manufactured shortwave receivers on the market today, some for very modest prices. A wire antenna costs practically nothing. Most electronics stores carry one or more models of shortwave receiver, along with antenna equipment, for a complete installation. Most electronics stores and book stores carry periodicals and books for the beginner as well as the seasoned shortwave listener.

A shortwave listener need not obtain a license to receive signals. At the time of writing, no formal licensing procedure exists for Class-D Citizen's-Band (CB) radio transmissions, which are made on the shortwave bands in a narrow segment at about 27 MHz (*see* CITIZEN'S BAND). Some low-power transmissions are also authorized at some other frequencies. In general, however, a license is required if shortwave transmission is contemplated. Shortwave listeners often get interested enough in communications to obtain amateur radio licenses (*see* AMATEUR RADIO).

SHOT EFFECT

In a transistor or tube, or in any current-carrying medium, the individual charge carriers cause noise impulses as they move from atom to atom. The individual impulses are extremely faint, but the large number of moving carriers results in a hiss that can be heard in any amplifier or radio receiver. This effect is called shot effect. The noise is known as shot-effect noise or shot noise.

Shot-effect noise limits the ultimate sensitivity that can be obtained in a radio receiver. This is because a certain amount of noise is always produced in the front end, or first amplifying stage (*see* FRONT END), and this noise is amplified by succeeding stages along with the desired signals. The amount of shot-effect noise that a device produces is roughly proportional to the current that it carries. In recent years, low-current solid-state devices such as the gallium-arsenide field-effect transistor (GaAsFET) have been developed for optimizing the sensitivity of a receiver front end.

Shot-effect noise is not the only form of noise that is generated in an electronic circuit. All atoms and subatomic particles are constantly in motion. The rate of motion is proportional to the temperature of the conducting media. The higher the temperature, the more the particles move, and the more electrical noise they make. This noise is called thermal noise. It, too, limits the sensitivity that can be obtained in the front end of a radio receiver. *See also* NOISE FIGURE, SENSITIVITY, SIGNAL-TO-NOISE RATIO, THERMAL NOISE.

SHOT-EFFECT NOISE

See SHOT EFFECT.

SHOT NOISE

Shot noise is a term that is used to describe three different forms of noise.

Whenever current flows, the movement of the electrons or holes results in electrical noise. This is called shot effect, and the noise is known as shot-effect noise or shot noise (*see* SHOT EFFECT). All semiconductor devices produce this kind of noise; the amount of noise depends on many factors, primarily the current and the area of the P-N junction(s) through which the current flows.

In a vacuum tube, noise is produced when the electrons bombard the plate. A smaller amount of noise also results from electrons striking the screen grid in a pentode or tetrode tube. This noise is called shot noise.

Whenever an electric spark occurs, an electromagnetic field is produced. Many different types of appliances, as well as internal-combustion engines employing spark plugs, produce sparks. The noise is usually called impulse noise, but it may sometimes be referred to as shot noise. *See also* IMPULSE NOISE.

SHUNT

The term shunt refers to a parallel connection of one component across another component or group of components. When one component is placed across another for a specific purpose, the inserted component is sometimes called a shunt.

A shunt can be any type of component. Examples of shunt components include bypass capacitors, bleeder resistors, diode limiters, and various types of voltage regulators. Shunting resistors or coils are often used in ammeters to increase the indicated current range. *See also* PARALLEL CONNECTION, SHUNT RESISTOR.

SHUNT CAPACITOR

See BYPASS CAPACITOR.

SHUNT FEED

Shunt feed is a method of supplying radio-frequency energy to a grounded mast or tower. The principle of shunt feed is the same as that of the gamma match (*see* GAMMA MATCH), except that the system is unbalanced rather than balanced.

The drawing illustrates the shunt-feed configuration. The mast or tower should be well-grounded at the base; a set of radial wires will enhance performance (*see* RADIAL). The shunt element may be a wire, held away from the mast or tower by means of standoff insulators. Aluminum or copper tubing can also be used. The spacing distance should be approximately 1 to 3 feet. The tapped coil at the base can be a solenoidal air-core inductor or a toroidal inductor, but it must be capable of carrying the required current without suffering damage.

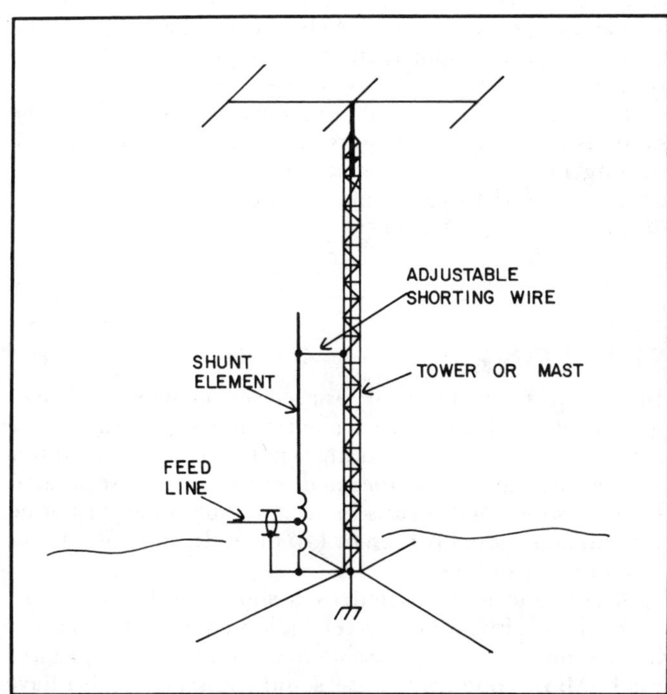

SHUNT FEED: Shunt feed can be used to make an excellent vertical antenna out of a grounded mast or tower.

For purposes of shunt feeding, a mast or tower should measure at least ¼ electrical wavelength in height. Since a sizable antenna or set of antennas is often mounted at the top of a mast or tower, the physical height may be considerably less than ¼ wavelength. Guy wires result in greatly increased electrical height.

Adjustment of the shunt-feed system is done by trial and error. The coil is first disconnected entirely from the system, and the feed point is directly connected to the transmission line. The shunt point is moved up and down the mast or tower by means of a shorting wire as shown, until the standing-wave ratio is minimum, indicating zero reactance. The coil is then inserted, and the tap is adjusted for a perfect match. As an alternative to the use of the matching inductor, the spacing between the tower and the shunt wire or tubing can be varied until the standing-wave ratio is 1:1.

Shunt feed is especially useful in the lower part of the high-frequency band, and in the medium-frequency band. It is used extensively by radio amateurs in the 160-meter and 80-meter (1.8-MHz and 3.5-MHz) bands. Many amateur radio operators have obtained amazing results in DX (distance) work using shunt-fed towers on these bands.

SHUNT REGULATOR

A shunt regulator is a voltage-regulation device, connected in parallel with the output of a power supply. Various types of transistors and tubes can be used as shunt regulators. In low-current power supplies, zener diodes are sometimes used for shunt regulation. *See also* OVERVOLTAGE PROTECTION, SERIES REGULATOR, VOLTAGE REGULATION, ZENER DIODE.

SHUNT RESISTOR

The current-indicating range of an ammeter can be increased by placing small-value resistors across the coil of the ammeter. Such resistors can be noninductive, or they can consist of coils of resistance wire such as nichrome.

A general example of the use of a shunt resistor is shown in the illustration. Let the resistance of the meter coil be Rc (in ohms), and the full-scale range of the ammeter be A (in amperes). Suppose that a shunt resistor, having an ohmic value Rs, is placed across the meter coil as shown. Then the new full-scale range, A^1, will be:

$$A^1 = A(Rc+Rs)/Rs$$

When shunt resistors are used, care must be exercised to ensure that they can handle the current without overheating. In the above case, the power rating P, in watts, for the shunt resistor should be at least:

$$P = (A^1-A)^2Rs$$

See also AMMETER.

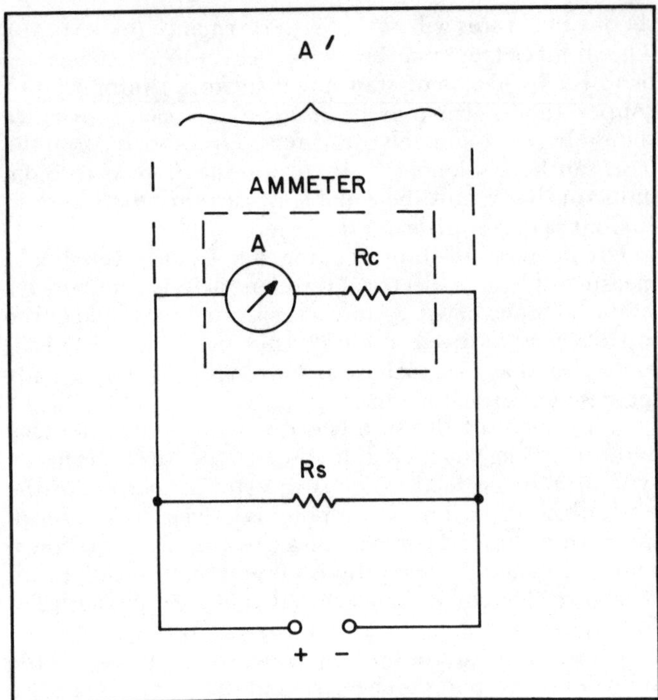

SHUNT RESISTOR: A shunt resistor is connected in parallel with an ammeter to increase the current-indication range.

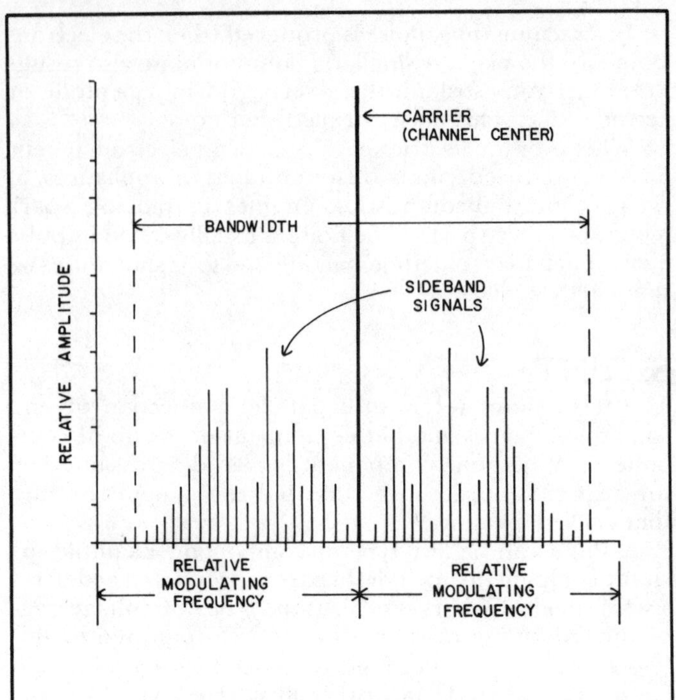

SIDEBAND: Sidebands result from mixing of the modulating signal with the carrier wave in a transmitter.

SHUNT TUNING

See PARALLEL RESONANCE.

SIBILANT

A sibilant is a high-frequency component of speech. Sibilants result from the pronunciation of certain consonants. These consonant sounds are the soft C, the soft G, and the sounds of the letters J, S, X, and Z. Sibilants contain frequency components up to 10 kHz or higher.

In high-fidelity audio applications, it is important that all of the sibilants, as well as the other components of a voice, be passed with little or no attenuation. In communications, however, it has been found that a cutoff frequency of approximately 3 kHz will allow a voice to be understood well, including the sibilant sounds, even though some of the frequencies are not passed. *See also* VOICE FREQUENCY CHARACTERISTICS.

SIDEBAND

When a carrier is modulated in any manner, for conveying intelligence of any sort and at any speed, sidebands are produced. Sideband signals occur immediately above and below the carrier frequency.

Sideband signals are the result of mixing between the carrier and the modulating signal. The greater the data speed, the higher the frequency components of the modulation signal, and the farther from the carrier the sidebands will appear (see illustration).

The data speed is not the only factor that affects the overall band of frequencies occupied by a carrier and its sidebands. Other influences on bandwidth include the type of modulation, the percentage or index of the modulation, and the efficiency of the data-transmission method. *See also* AMPLITUDE MODULATION, BANDWIDTH, EMISSION CLASS, FREQUENCY MODULATION, MODULATION, MODULATION METHODS, PHASE MODULATION, SINGLE SIDEBAND.

SIDESTACKING

Sidestacking is a method of combining two or more parasitic arrays or phased arrays to obtain greater power gain than would be possible with a single anntenna.

The sidestacking technique involves the connection of antennas in the horizontal plane, with the elements (and thus the polarization) oriented vertically (see illustration). The antennas are fed in phase by means of a harness. The technique is used mostly at the very-high and ultra-high frequencies.

The gain that can be obtained with sidestacking is the same as the gain that results from vertical stacking. Assuming identical antennas are used, the power gain increases by 3 dB every time the number of antennas is doubled. *See also* STACKING.

SIDETONE

In a code transmitter or transceiver in which receiver muting is used, the actual transmitted signal cannot be heard. Even if receiver muting is not used, the transmitter and receiver are often operated on frequencies separated by an amount that results in an inaudible tone, or a tone with an objectionably high or low pitch. In such situations, a sidetone is employed.

A sidetone is generated by a simple audio oscillator, such as a multivibrator or relaxation circuit. Most keyers have built-in sidetone oscillators, some with adjustable pitch. Many code transmitters and transceivers also have built-in sidetone oscillators, which are keyed along with the transmitter. *See also* SIDETONE OSCILLATOR.

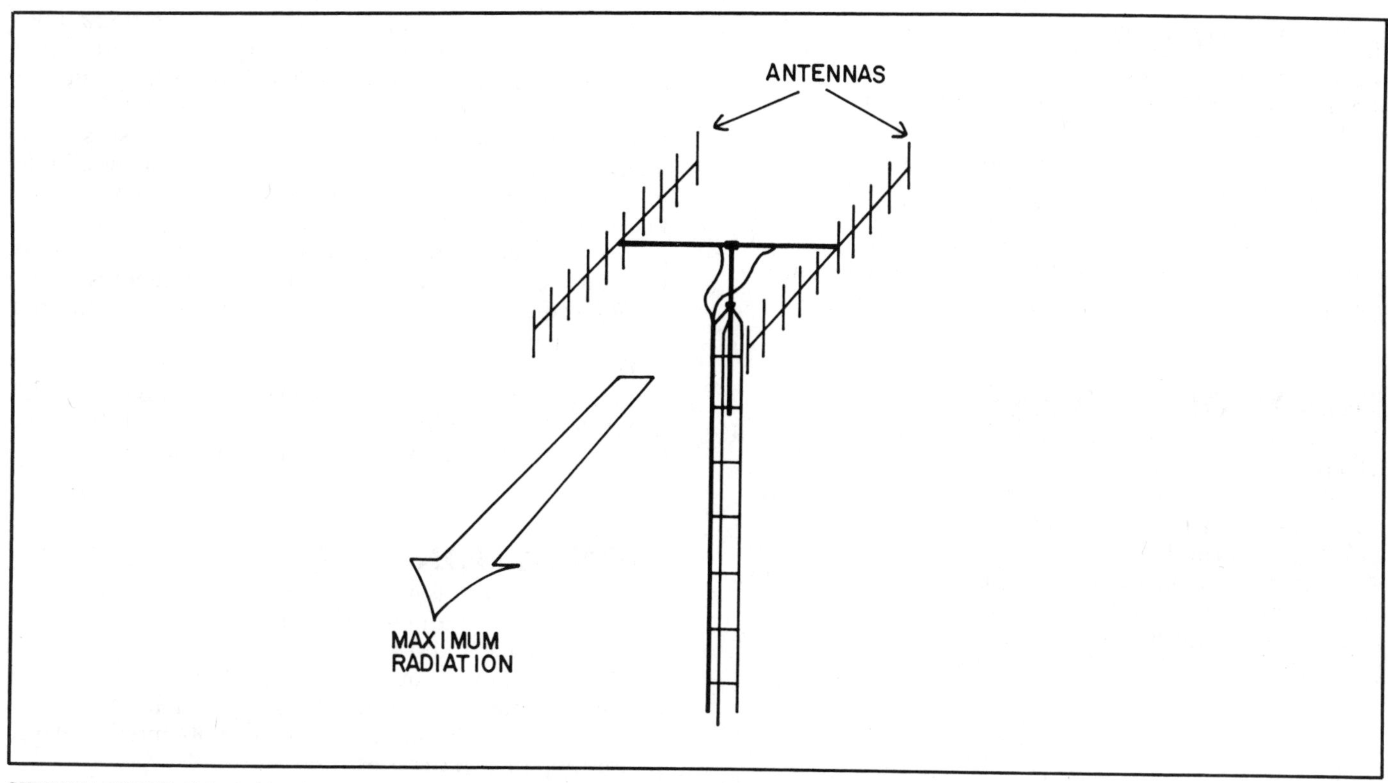

SIDESTACKING: Sidestacking is sometimes used to increase the power gain of an antenna system.

SIDETONE OSCILLATOR

A sidetone oscillator is a simple audio-oscillator circuit used with keyers and code transmitters for the purpose of generating a monitor tone or sidetone (*see* SIDETONE). Many different types of circuit will function well as a sidetone oscillator.

A sidetone oscillator should have certain characteristics. First, it is imperative that the oscillator function every time the key is closed; that is, the oscillator must start instantly 100 percent of the time. Second, the frequency (pitch) of the tone must not change as the circuit is keyed. Third, the tone should have a pleasing sound that is not tiring to the ear. A sine-wave tone is undesirable because it causes fatigue; an excessively raw or "buzzing" sidetone is equally objectionable. A typical sidetone oscillator is shown in the illustration.

If a transmitter/keyer combination has no sidetone oscillator, a radio-frequency-actuated device can be built. This type of oscillator uses a circuit similar to that shown, in conjunction with a radio-frequency-actuated switch.

SIEMENS

The siemens is the unit of electrical conductance. It is also sometimes called the mho. Given a resistance of R ohms, the conductance S in siemens is simply 1/R. *See also* MHO.

SIEMENS, ERNST WERNER VON

A 19th-century physicist and experimenter, Ernst Werner von Siemens (1816-1892) is known for his work with electric generators. The unit of conductance is named after him (*see* MHO, SIEMENS).Siemens developed a standard for determining electrical resistances.

SIGNAL AMPLITUDE

See AMPLITUDE.

SIGNAL CONDITIONING

Signal conditioning is the process of modifying, or changing, a signal for a certain purpose. Signal conditioning is also called signal processing. It is used extensively in audio-frequency and radio-frequency equipment.

An example of signal conditioning is the detection and filtering of an alternating-current wave for indirect voltage measurement using a direct-current meter. Another example is the use of an equalizer in a high-fidelity system (*see* EQUALIZER).

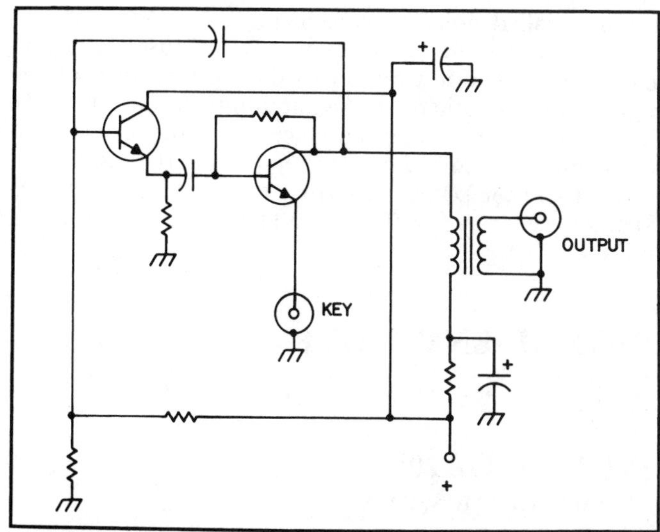

SIDETONE OSCILLATOR: A sidetone oscillator suitable for use with a code transmitter or keyer. A radio-frequency-actuated switch can be used in place of the key if desired.

SIGNAL DIODE

A semiconductor diode, used for low-power applications at audio or radio frequencies, is sometimes called a signal diode. Signal diodes are used for envelope detection, peak clipping or limiting, mixing, and many other purposes.

Signal diodes are characterized by relatively small P-N-junction surface area. Signal diodes have small to moderate junction capacitance. The current-handling capacity and peak-inverse-voltage (PIV) rating are usually low. *See also* DIODE, POINT-CONTACT JUNCTION, P-N JUNCTION.

SIGNAL DISTANCE

Signal distance is a quantitative expression of the extent to which two binary words differ.

All binary words are composed of sequences of the numerals 0 and 1. For example, we may have two binary words, 100100 and 001111. These are 6-bit binary words. To determine signal distance, we simply compare the two binary words and count the number of digits that do not match. This is facilitated by writing the two words, one above the other:

100100

001111

We see, in this example, that four of the six digits do not match. The signal distance is therefore equal to 4. The smallest possible signal distance is 0, representing identical words; the greatest possible signal distance is the number of bits in the word (in this case, 6).

Sometimes the signal distance is given as the proportion, rather than the actual number, of digits that fail to match. In the above example, four of the six digits are different; we would then say the signal distance is 4/6, or 0.667. Using this method, the smallest possible signal distance is 0, representing identical words; the greatest possible signal distance is 1. *See also* BIT, WORD.

SIGNAL ENHANCEMENT

Any method of improving the quality of intelligibility of a signal is called signal enhancement.

In radio-frequency receiving apparatus, signal enhancement generally refers to the optimization of the signal-to-noise ratio or the signal-to-interference ratio. In high-fidelity audio, signal enhancement is the reduction of noise and the tailoring of the sound to suit the taste of the listener. *See also* EQUALIZER, HIGH FIDELITY, SELECTIVITY, SIGNAL-TO-INTERFERENCE RATIO, SIGNAL-TO-NOISE RATIO.

SIGNAL ENVELOPE

See ENVELOPE.

SIGNAL GAIN

See AMPLIFICATION, GAIN.

SIGNAL GENERATOR

A signal generator is an oscillator, often equipped with a modulator, that is used for the purpose of testing audio-frequency or radio-frequency equipment. Most signal generators are intended for either audio-frequency or radio-frequency applications, but not both.

In its simplest form, a signal generator consists of an oscillator that produces a sine wave of a certain amplitude in microvolts or millivolts, and a certain frequency in hertz, kilohertz, or megahertz. Some audio-frequency signal generators can produce several different types of waveforms. The more sophisticated signal generators for radio-frequency testing have amplitude modulators and/or frequency modulators.

For the testing, adjustment, and servicing of radio-frequency transmitters and receivers, a combined signal generator and monitor is often used. *See also* FUNCTION GENERATOR, SERVICE MONITOR/GENERATOR.

SIGNAL MONITOR

A signal monitor is a test instrument that is used for analyzing radio-frequency signals.

An ordinary radio receiver can be used as a signal monitor in many situations. The technician simply listens to the characteristics of the incoming signal. For a more detailed view of the signal than can be obtained by ear, a panoramic receiver can be used.

An oscilloscope is often used for the purpose of signal monitoring. In the test laboratory, a service monitor/generator is used. A spectrum analyzer is a form of signal monitor. *See* OSCILLOSCOPE, PANORAMIC RECEIVER, SERVICE MONITOR/GENERATOR SPECTRUM ANALYZER.

SIGNAL-PLUS-NOISE-TO-NOISE RATIO

In a receiver, the sensitivity is sometimes specified in terms of the ratio of the audio signal-plus-noise strength to the noise strength for a given input. This is called the signal-plus-noise-to-noise ratio. It is abbreviated (S+N)/N or (S+N):N. (The parentheses are mathematically important, but they are often omitted, giving the imprecise expressions S+N/N or S+N:N.)

The (S+N)/N ratio is always specified in decibels. A certain signal level is required to cause, say, a (S+N)/N ratio of 10 dB.

If the root-mean-square (rms) signal level in microvolts is Es, and the rms noise level in microvolts is En, then the S+N/N ratio, in decibels, is

$$(S+N)/N = 20 \log_{10} ((Es + En)/En)$$

The signal-to-noise-plus-noise (S+N)/N ratio is always greater than the signal-to-noise ratio. However, the latter is more frequently used. *See* SIGNAL-TO-NOISE RATIO.

SIGNAL PROCESSING

See SIGNAL CONDITIONING, SIGNAL ENHANCEMENT.

SIGNAL STRENGTH

See AMPLITUDE.

SIGNAL SYNTHESIZER

See FUNCTION GENERATOR, SERVICE MONITOR/GENERATOR, SIGNAL GENERATOR.

SIGNAL-TO-IMAGE RATIO

In a superheterodyne receiver, images are always present. The image signals should, ideally, be rejected to the extent that they do not interfere with desired signals. However, this is not always the case.

The signal-to-image (S/I) ratio is usually given in decibels at the output of a receiver. *See also* IMAGE REJECTION, SUPERHETERODYNE RECEIVER.

SIGNAL-TO-INTERFERENCE RATIO

In radio reception, the signal-to-interference ratio is sometimes mentioned. This ratio is measured in decibels, and is given as S, where:

$$S = 20 \log_{10} (a/b)$$

where a is the strength of the desired signal in microvolts at the antenna terminals, and b is the sum of the strengths of all of the interfering signals in microvolts at the antenna terminals.

The signal-to-interference ratio is important as a specification of sensitivity and selectivity in a radio receiver. *See also* SELECTIVITY, SENSITIVITY, SIGNAL-TO-NOISE-PLUS-NOISE RATIO, SIGNAL-TO-NOISE RATIO.

SIGNAL-TO-NOISE RATIO

The sensitivity of a communications receiver is often specified in terms of the audio signal-to-noise ratio that results from an input signal of a certain number of microvolts. This ratio is abbreviated S/N or S:N.

If the root-mean-square (rms) signal strength at the antenna terminals of a receiver is Es, given in microvolts, and the rms noise level is En, also in microvolts, then the ratio S/N, in decibels, is:

$$S/N = 20 \log_{10} (Es/En)$$

Usually, the sensitivity is specified as the signal strength in microvolts that is necessary to cause a S/N ratio of 10 dB, or 3.16:1.

Modern communications receivers generally require about 0.5 μ V, or less, to produce a S/N ratio of 10 dB at the high frequencies in the continuous-wave or single-sideband modes. For amplitude modulation, the rating is usually 1 μ V or less. The S/N sensitivity of a communications receiver is usually mentioned in the table of specifications. In the case of frequency-modulation (FM) receivers, the noise-quieting figure or the SINAD figure are standard. *See also* NOISE QUIETING, SINAD.

SIGNAL VOLTAGE

The amplitude of a signal is often specified in terms of its voltage. There are three methods of expressing the voltage of an ac signal: the peak, peak-to-peak, and root-mean-square values. *See* PEAK-TO-PEAK VALUE, PEAK VALUE, ROOT MEAN SQUARE.

SIGNIFICANT FIGURES

For the mathematician, all numbers have values that are precisely as written. In physics and related fields of material science, however, we must always approximate. The degree of precision to which we specify a quantity is indicated by the number of significant figures in the expression.

In their calculations, physicists and engineers use a scheme of expressing quantities called scientific notation (*see* SCIENTIFIC NOTATION). A decimal number is followed by a power of 10. The number of digits in the decimal expression is the number of significant figures.

When mathematical operations are performed between or among numbers, the resultant cannot have more significant digits than the numbers themselves. If the original values have different numbers of significant figures, then the resultant must be rounded off to have the smallest number of significant figures consistent with the original values.

As an example of the use of significant figures, suppose we wish to find the product of 1.5535×10^2 and 7.4×10^3. Then we multiply

$$1.5535 \times 7.4 = 11.4959$$

using a calculator. The powers of 10 multiply as follows:

$$10^2 \times 10^3 = 10^5$$

The complete product is thus:

$$\begin{aligned} 1.5535 \times 10^2 \times 7.4 \times 10^3 &= 11.4959 \times 10^5 \\ &= 1.14959 \times 10^6 \end{aligned}$$

We are not justified in claiming six significant figures for this product. The original decimal expressions, 1.5535 and 7.4, have only five and two significant figures, respectively. We must round off the answer to two significant figures, obtaining 1.1×10^6. *See also* ROUNDING OF NUMBERS.

SILICON

Silicon is an element with atomic number 14 and atomic weight 28. In its pure state, silicon appears as a light-weight metal similar to aluminum. Silicon is a semiconductor substance; it conducts electric currents better than a dielectric, but not as well as the excellent conductors such as silver and copper.

Silicon is found in great quantities in the crust of the earth. In its natural state, silicon is almost always combined with oxygen or other elements. Pure silicon is extracted from the various compounds for use in the manufacture of semiconductor diodes, transistors, integrated circuits, and other devices.

Silicon has replaced germanium in the manufacture of many types of semiconductor components, because silicon can withstand higher temperatures than germanium. However, germanium is still used in some devices. *See also* GERMANIUM, SEMICONDUCTOR.

SILICON-CONTROLLED RECTIFIER

The silicon-controlled rectifier (abbreviated SCR) is a four-layer semiconductor device that is used primarily for power control. The SCR is the solid-state counterpart of the thyratron tube (*see* THYRATRON). The SCR is a commonly used form of thyristor. The terms SCR and thyristor are sometimes used interchangeably (*see* THYRISTOR).

The SCR has three electrodes, called the cathode, anode, and gate. When a forward bias is applied between the cathode and the anode, no current flows until a pulse is applied to the gate. Then the SCR continues to conduct until the bias between the cathode and anode is reversed or reduced below a certain threshold value.

Silicon-controlled rectifiers are available with many different voltage and current ratings for various applications. The devices are widely used in power-control circuits for ac or dc. An example of an alternating-current power-control circuit is shown at A in the illustration. A direct-current controlled rectifier is shown at B.

SILICON DIODE

A silicon diode is any semiconductor diode consisting partly or wholly of silicon. Most semiconductor diodes today are fabricated from silicon, either in junction form or in point-contact form. *See also* P-N JUNCTION, POINT-CONTACT JUNCTION, SILICON.

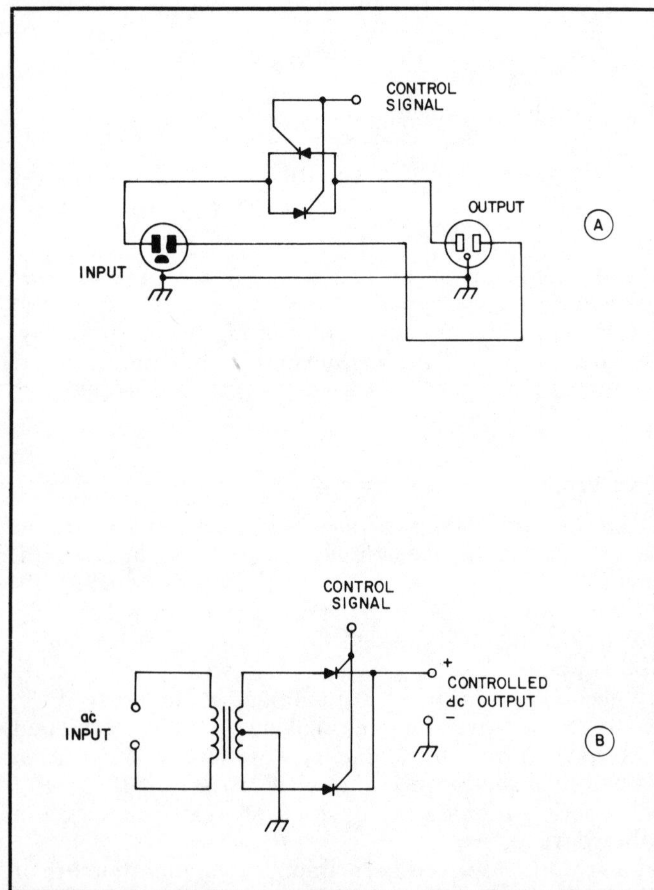

SILICON-CONTROLLED RECTIFIER: At A, an example of the use of silicon-controlled rectifiers in an alternating-current power-control circuit. At B, silicon-controlled rectifiers are used in an adjustable full-wave direct-current power supply.

SILICON DIOXIDE

Silicon dioxide is a common compound consisting of silicon and oxygen. In the earth's crust, much of the silicon is found in this form. The chemical formula is SiO_2.

In the manufacture of silicon semiconductor devices, particularly certain metal-oxide-semiconductor (MOS) integrated circuits, the outer surface of the silicon wafer is oxidized. This prevents corrosion that would otherwise eventually ruin the device. It also facilitates the doping (etching) of the various components in an integrated circuit.

Silicon dioxide allows fabrication of integrated circuits having extremely high density and low current consumption. Individual components measuring about 2 microns across can be fabricated on a silicon-dioxide film by means of a photographic process. Some MOS devices draw only a few nanoamperes of current. Silicon dioxide may be used in conjunction with other metal oxides or silicon nitride. *See also* METAL-OXIDE SEMICONDUCTOR, METAL-OXIDE-SEMICONDUCTOR LOGIC FAMILIES, SILICON NITRIDE.

SILICONE

Silicone is a polymerized material consisting of silicon and oxygen atoms. It is an excellent electrical insulating material, but is a good conductor of heat. Silicone can withstand very high temperatures.

Silicone is commercially available in various forms. One common product comes in a small tube (like toothpaste), is white in color, and has the consistency of petroleum jelly.

Silicone is commonly used as a heat-transfer agent for semiconductor power transistors and diodes. The silicone is applied between the heat-conducting metal base of the device and a metal heat sink. This ensures efficient transfer of heat away from the semiconductor device. Small particles of dust or dirt would otherwise interfere with the heat bond.

SILICON MONOXIDE

Silicon monoxide is a compound consisting of silicon and oxygen. Its chemical formula is SiO. In the earth's crust, some of the silicon is found in this form.

Silicon monoxide is an excellent insulating material. It is useful as a capacitor dielectric, especially in the manufacture of integrated circuits.

SILICON NITRIDE

In the manufacture of integrated circuits, silicon nitride is sometimes used in conjunction with silicon dioxide to facilitate etching of components. The main advantage of a silicon-nitride/silicon-dioxide combination is that it prevents migration of impurities. The current drain is also reduced, compared with devices using pure silicon dioxide. *See also* SILICON DIOXIDE.

SILICON RECTIFIER

A silicon rectifier is a form of rectifier diode in which P-type silicon is diffused into N-type silicon, resulting in a P-N junction. *See* P-N JUNCTION, RECTIFIER DIODE.

SILICON SOLAR CELL

See PHOTOVOLTAIC CELL, SOLAR CELL.

SILICON STEEL

A small amount of silicon can be added to steel, forming an alloy called silicon steel. Silicon steel normally contains about 4 percent silicon and 96 percent steel.

Silicon steel is used as a core material for some inductors, chokes, and transformers at ac frequencies. The core sections are laminated to minimize eddy-current loss (*see* EDDY-CURRENT LOSS, LAMINATED CORE). Silicon-steel cores exhibit very good efficiency at frequencies up to several kilohertz. At much higher frequencies, silicon steel becomes lossy because of hysteresis. *See also* TRANSFORMER.

SILICON TRANSISTOR

A silicon transistor is a bipolar or field-effect transistor consisting partly or wholly of silicon. Most transistors today use silicon as the semiconductor material. Some other substances, such as indium antimonide and gallium arsenide, are less commonly used. *See also* BIPOLAR TRANSISTOR, FIELD-EFFECT TRANSISTOR, METAL-OXIDE-SEMICONDUCTOR FIELD-EFFECT TRANSISTOR, SILICON.

SILICON UNILATERAL SWITCH

A silicon unilateral switch (SUS) is a four-layer semiconductor device, identical to the silicon-controlled rectifier (SCR) except that a zener diode is placed in the gate lead. The zener diode reduces the sensitivity of the switch, by blocking trigger pulses having voltages smaller than a certain predetermined level.

The SUS is less likely than an ordinary SCR to be triggered accidentally by stray noise pulses. *See also* SILICON-CONTROLLED RECTIFIER, ZENER DIODE.

SILVER

Silver is an element with atomic number 47 and atomic weight 108. In its pure form, silver is a shiny, almost white metal. Silver is considered one of the precious metals.

Silver is an excellent conductor of electricity, and is highly resistant to corrosion. Silver has a relatively high melting point. For these reasons, component leads, as well as switch and relay contacts, are often plated with silver. Silver is a constituent metal in a high-quality hard solder called silver solder. Silver wire is sometimes used in the manufacture of high-Q coils. Silver is also used in the manufacture of low-loss capacitors. *See also* SILVER-MICA CAPACITOR, SILVER SOLDER.

SILVER-MICA CAPACITOR

A silver-mica capacitor is a form of mica capacitor (*see* MICA CAPACITOR) in which silver foil or plating is used on the mica sheet. Silver-mica capacitors have basically the same characteristics as other mica capacitors. The loss is somewhat lower when silver is used, as compared with the loss when other metals are employed; this difference arises from the fact that silver is an excellent electric conductor.

SILVER-OXIDE CELL

A silver-oxide cell is a form of electrochemical dry cell. A silver-oxide cell can be fabricated with tiny physical dimensions. Silver-oxide cells are employed in cameras, electronic watches and clocks, miniature calculators, and in other low-voltage, low-current devices.

Silver-oxide cells are not rechargeable. They produce approximately 1.5 V. Silver-oxide cells have excellent milliampere-hour capacity for their size. They are available in various different physical sizes; a typical camera cell of this type is shown in the photograph. (Note the dime for size comparison.) *See also* CELL.

SILVER SOLDER

Silver solder is a hard solder that is sometimes used in the manufacture of electronic devices. Silver solder consists of copper, zinc, and silver.

Silver solder melts at a higher temperature than ordinary solder, which is composed of tin and lead. This makes silver solder an advantage in circuits where large amounts of current must be handled. Silver solder maintains a better electrical bond than tin-lead solder, primarily because the metals used are better conductors. *See also* SOLDER.

SILVER-ZINC CELL

A silver-zinc cell is a form of electrochemical dry cell. The positive electrode is silver, and the negative electrode is zinc. Various different electrolyte substances may be used.

SILVER-OXIDE CELL: Silver-oxide cells can be made exceptionally small, for use in various miniaturized devices.

A silver-zinc cell can be made very small, similar in size to a silver-oxide cell (see previous photograph). Silver-zinc cells produce about 1.5 V, and are not rechargeable. Silver-zinc cells have excellent milliampere-hour capacity for their size. They are used in low-voltage, low-current electronic devices such as calculators, cameras, and watches. *See also* CELL.

SIMILARITY

Similarity is a mathematical expression that means two objects have identical shape. This does not necessarily mean that they are the same size. Similarity is of importance in plane and space geometry, and in many engineering fields.

Certain objects are always similar. Examples are the square, the cube, the circle, and the sphere. Other objects, such as triangles, may or may not be similar.

Two plane polygons are similar if the measures of corresponding angles are identical (see illustration). In general, any two objects in any number of dimensions are similar if one can be "shrunk" or "enlarged" to become identical, or congruent, to the other.

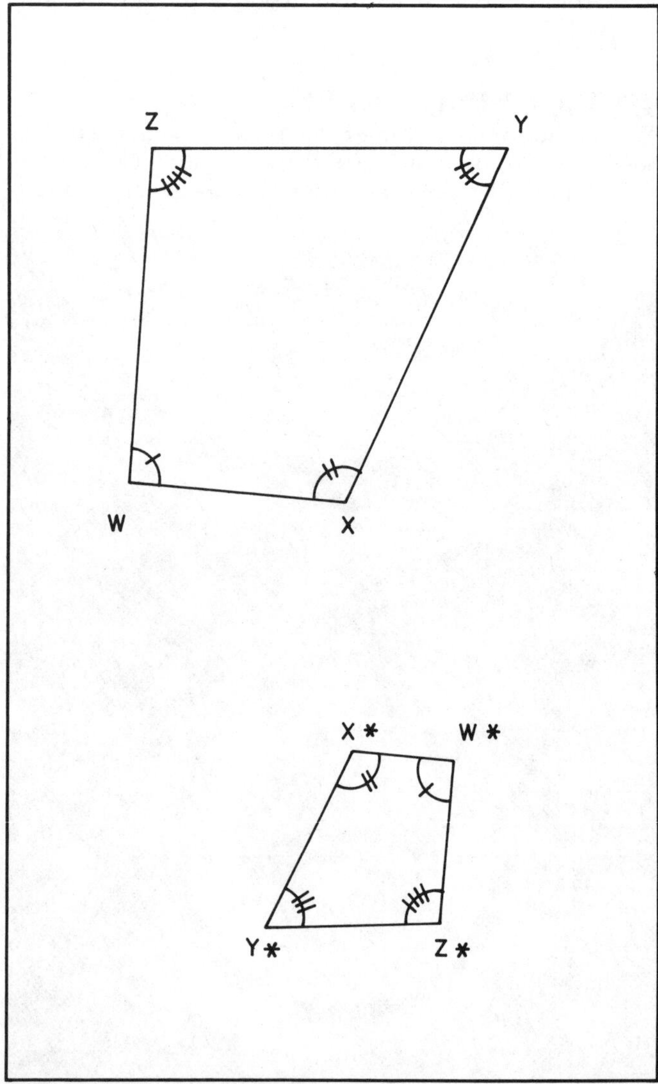

SIMILARITY: Two plane polygons are similar if, and only if, they have corresponding angles of identical measure. Here, the two polygons are upside-down with respect to each other, but the corresponding angles have equal measure.

SIMPLE TONE

A simple tone is an audible tone having essentially no harmonic content. It is a sine-wave tone. *See* SINE WAVE.

SIMPLEX

In a two-way communications system, both transmitters and receivers are often operated on a single frequency. This is known as simplex or simplex operation. The two stations communicate directly with each other; no repeater or other intermediary is used.

At very-high and ultra-high frequencies, simplex operation may not provide enough communications range. This is especially true if one or both stations are mobile. To increase the effective range, repeaters are used. Some repeaters are placed on satellites, greatly increasing the communications range (*see* ACTIVE COMMUNICATIONS SATELLITE, REPEATER, SATELLITE COMMUNICATIONS).

In simplex operation, it is possible for only one station to transmit at a time. This is because neither station can receive signals at the same time, and on the same frequency, as they are transmitting. If it is necessary to send and receive data simultaneously, two different frequencies must be used. This is called duplex operation (*see* DUPLEX OPERATION).

It is not normally possible for one station to interrupt the other station in simplex operation. However, it can be accomplished if continuous-wave (type A1) or single-sideband, suppressed-carrier (type A3J) emissions are used. This is called break-in operation. It requires the use of a special switch and muting system. The receiver is activated during brief pauses in a transmission (*see* BREAK-IN OPERATION).

SIMULATION

A computer can often be programmed to simulate certain conditions or situations, such as driving a car or flying an airplane. This is of obvious benefit in the training of personnel to do complicated tasks: Simulation is much safer than the real thing!

Simulation is used, in some form, in almost every modern computer game. The circumstances simulated do not have to be the sorts of things we would expect in the real world. This use of simulation can make fantasy come to life. *See also* ELECTRONIC GAME.

SIMULCASTING

Simulcasting is the transmission of a single program via two or more different channels or modes at the same time.

A play-by-play broadcast of a football game, for example, may be transmitted over an amplitude-modulated (AM) broadcast station at 830 kHz and a frequency-modulated (FM) station at 90.5 MHz.

In television broadcasting, an enhanced (stereo) sound track is sometimes broadcast on the standard FM band. This form of simulcasting is often done for such events as orchestra concerts. The viewer can turn to the appropriate television channel, turn the volume control of the television receiver down, and tune in to the appropriate station on an FM stereo receiver to hear the sound.

SINAD

The term SINAD is an abbreviation of the words SIgnal to Noise And Distortion. This expression is frequently used to define the sensitivity of a frequency-modulation receiver at the very-high and ultra-high frequencies. The SINAD figure takes into account not only the quieting sensitivity of a receiver, but its ability to reproduce a weak signal with a minimum of distortion.

Usually, the SINAD sensitivity of a receiver is given as the signal strength in microvolts at the antenna terminals that results in a signal-to-noise-and-distortion ratio of 12 dB at the speaker terminals. The signal is modulated with a 1-kHz sine-wave tone, at a deviation of plus-or-minus 3 kHz. This is the standard test modulation for a frequency-modulated system (*see* STANDARD TEST MODULATION).

The SINAD sensitivity of a receiver is measured by using a calibrated signal generator and a distortion analyzer or SINAD meter. *See also* NOISE QUIETING, SINAD METER.

SINAD METER

A distortion analyzer, designed especially for the purpose of measuring the signal-to-noise-and-distortion (SINAD) sensitivity of a frequency-modulation receiver, is called a SINAD meter or SINAD distortion analyzer.

The SINAD meter contains a notch filter centered at an audio frequency of 1 kHz. For determination of the SINAD sensitivity of a receiver, the procedure is as follows:

1. The meter is connected to the speaker terminals of the receiver.
2. The receiver squelch is opened, and the volume set near the middle of the control range.
3. The meter-level control is set for a meter reading of 0 dB. (Many SINAD meters have an automatic level control that sets the meter at 0 dB regardless of the volume-control setting of the receiver.)
4. A signal generator is connected to the receiver antenna terminals, and the generator is set to the same frequency or channel as the receiver.
5. The signal is modulated at an audio frequency of 1 kHz, with a deviation of plus-or-minus 3 kHz.
6. The signal level is adjusted until the SINAD meter indicates 12 dB. The SINAD sensitivity, in microvolts, is then read from the calibrated scale of the signal-generator amplitude control.

Most SINAD meters can be used for measuring the quieting sensitivity, as well as the SINAD sensitivity, of a frequency-modulation receiver. *See also* DISTORTION ANALYZER, NOISE QUIETING, SINAD.

SINE

The sine function is a trigonometric function. In a right triangle, the sine is equal to the length of the far or opposite side, divided by the length of the hypotenuse, as shown in the illustration at A. In the unit circle $x^2 + y^2 = 1$, plotted on the Cartesian (x,y) plane, the sine of the angle θ, measured counterclockwise from the x axis, is equal to y. This is shown at B. The sine function is periodic, and begins with a value of 0 at $\theta = 0$. The shape of the sine function is identical to that of the cosine function (*see* COSINE), except that the sine function is displaced to the right by 90 degrees.

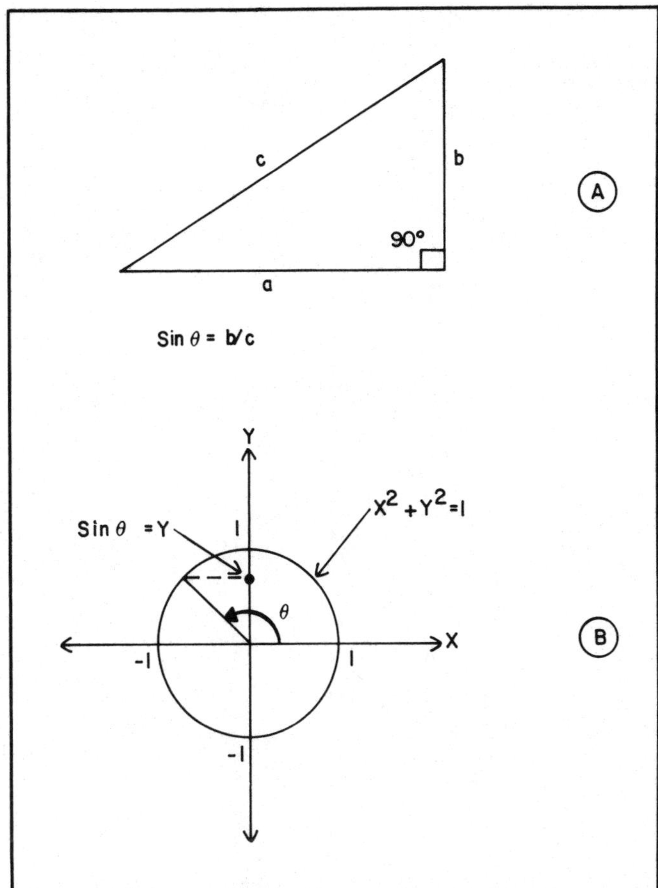

SINE: The sine function is the ratio of the length of the far (opposite) side of a right triangle to the length of the hypotenuse, as shown at A. At B, the unit circle model for the sine function is shown; sin θ = y (see text).

The sine function represents the waveform of a pure, harmonic-free, alternating-current disturbance. In mathematical calculations, the sine function is abbreviated sin. Values of sin θ for various angles θ are given in the table. *See also* SINE WAVE, TRIGONOMETRIC FUNCTION.

SINE LAW

See LAW OF SINES.

SINE WAVE

A sine wave, also called a sinusoidal waveform, is an alternating-current disturbance that has only one frequency. The harmonic content, and the bandwidth, are theoretically zero (*see* BANDWIDTH, HARMONIC). The sine wave gets its name from the fact that the amplitude-versus-time function is identical to the trigonometric sine function. The cosine function also is a perfect representation of the shape of a sine wave (*see* COSINE).

When displayed on an oscilloscope, the sine wave has a characteristic shape as shown in the illustration. Each cycle is represented by one complete alternation as shown. The cycle of a sine wave is divided into 360 electrical degrees.

The waveform produced by an oscillator is never a perfect sine wave. There is always some harmonic content.

SINE: VALUES OF SIN θ FOR VALUES OF θ BETWEEN 0° AND 90°. FOR 90° < θ ≤ 10°, CALCULATE 180° − θ AND READ FROM THIS TABLE. FOR 180° < θ ≤ 270°, CALCULATE θ − 180°, READ FROM THE TABLE, AND MULTIPLY BY −1. FOR 270° < θ ≤ 360°, CALCULATE 360° − θ, READ FROM THE TABLE, AND MULTIPLY BY −1.

θ, degrees	**sin θ**	**θ, degrees**	**sin θ**	**θ, degrees**	**sin θ**
0	0.000				
1	0.017	31	0.515	61	0.875
2	0.035	32	0.530	62	0.883
3	0.052	33	0.545	63	0.891
4	0.070	34	0.559	64	0.899
5	0.087	35	0.574	65	0.906
6	0.105	36	0.588	66	0.914
7	0.122	37	0.602	67	0.921
8	0.139	38	0.616	68	0.927
9	0.156	39	0.629	69	0.934
10	0.174	40	0.643	70	0.940
11	0.191	41	0.656	71	0.946
12	0.208	42	0.669	72	0.951
13	0.225	43	0.682	73	0.956
14	0.242	44	0.695	74	0.961
15	0.259	45	0.707	75	0.966
16	0.276	46	0.719	76	0.970
17	0.292	47	0.731	77	0.974
18	0.309	48	0.743	78	0.978
19	0.326	49	0.755	79	0.982
20	0.342	50	0.766	80	0.985
21	0.358	51	0.777	81	0.988
22	0.375	52	0.788	82	0.990
23	0.391	53	0.799	83	0.993
24	0.407	54	0.809	84	0.995
25	0.423	55	0.819	85	0.996
26	0.438	56	0.829	86	0.998
27	0.454	57	0.839	87	0.999
28	0.469	58	0.848	88	0.999
29	0.485	59	0.857	89	0.999
30	0.500	60	0.866	90	1.000

However, the amount of harmonic energy is often so small that, when the waveform is displayed on an oscilloscope, it appears as a perfect sine wave. *See also* SINE.

SINGLE BALANCED MIXER

A single balanced mixer is a mixer circuit that is easily built for a minimum amount of expense. The circuit operates in a manner similar to a balanced modulator (*see* BALANCED MODULATOR, MIXER, MODULATOR). The input and output ports are not completely isolated in the single balanced mixer; some of either input signal leaks through to the output. If isolation is needed, the double balanced mixer is preferable (*see* DOUBLE BALANCED MIXER).

A typical single-balanced-mixer circuit is shown in the illustration. The diodes are generally of the hot-carrier type. This circuit will work at frequencies up to several gigahertz. The circuit shown is a passive circuit, and therefore some loss will occur. However, the loss can be overcome by means of an amplifier following the mixer. *See also* HOT-CARRIER DIODE.

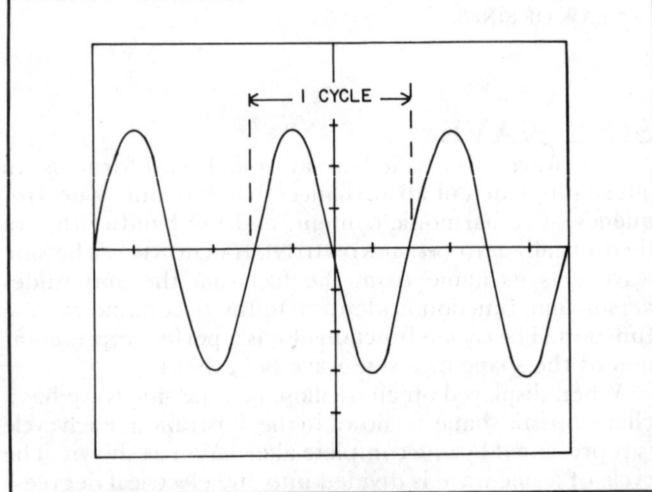

SINE WAVE: Characteristic shape of a sine wave, as viewed on an oscilloscope.

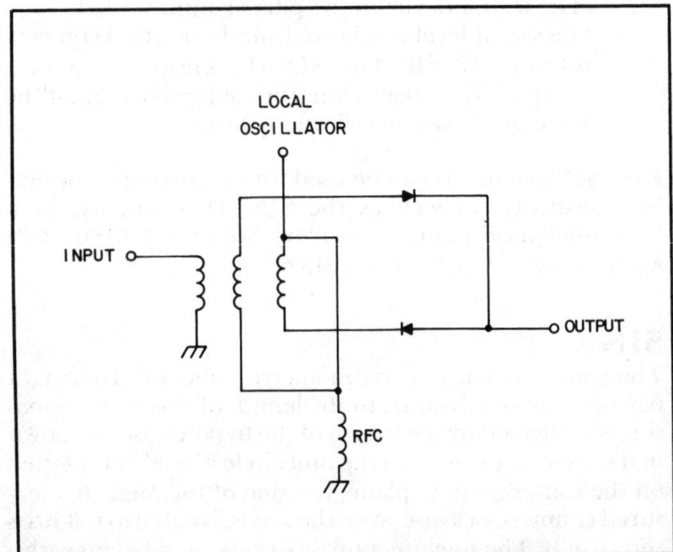

SINGLE BALANCED MIXER: An example of a single-balanced mixer using hot-carrier diodes.

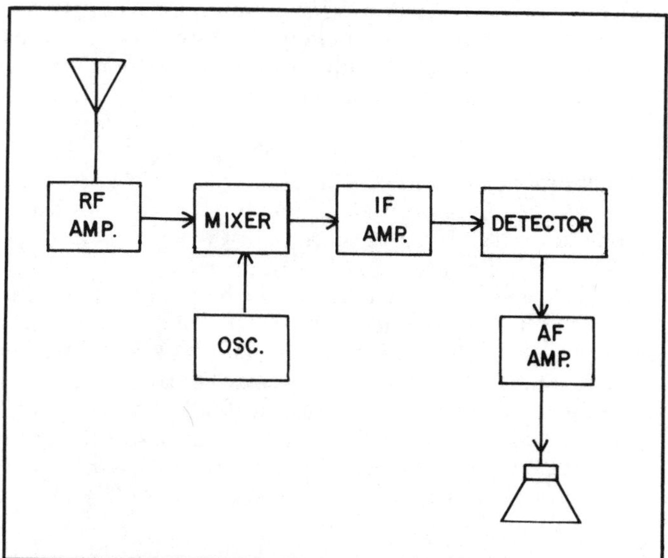

SINGLE CONVERSION RECEIVER: Block diagram of a single-conversion receiver.

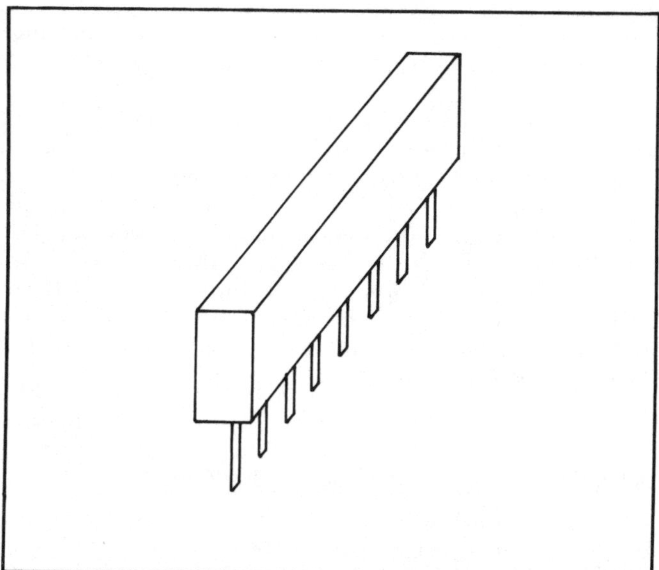

SINGLE IN-LINE PACKAGE: A single in-line package integrated circuit has all the pins along one edge.

SINGLE-CONVERSION RECEIVER

A single-conversion receiver is a form of superheterodyne receiver that has one intermediate frequency. The incoming signal is heterodyned to a fixed frequency. Selective circuits are employed to provide discrimination against unwanted signals. The output from the mixer is amplified and fed directly to the detector. A block diagram of a single-conversion receiver is illustrated.

The main advantage of the single-conversion receiver is its simplicity. However, the intermediate frequency must be rather high—on the order of several megahertz for a typical high-frequency communications receiver—and this limits the degree of selectivity that can be obtained. In recent years, excellent ceramic and crystal-lattice filters have been designed, improving the performance of single-conversion receivers. *See also* DOUBLE-CONVERSION RECEIVER, INTERMEDIATE FREQUENCY, MIXER, SUPERHETERODYNE RECEIVER.

SINGLE-ENDED CIRCUIT

See UNBALANCED SYSTEM.

SINGLE IN-LINE PACKAGE

The single in-line package (SIP) is a form of housing for integrated circuits. A flat, rectangular box containing the chip is fitted with lugs along one side, as shown in the illustration. There may be just a few pins, or as many as 12 or even 15 pins.

The single in-line package is very easy to install in, and to remove from, a circuit board. *See also* DUAL IN-LINE PACKAGE, INTEGRATED CIRCUIT.

SINGLE-PHASE ALTERNATING CURRENT

Single-phase alternating current is found in most 117-V household utility systems. Single-phase current consists of only one 60-Hz alternating sine wave, having a root-mean-square (RMS) amplitude of about 117 V.

SINGLE SIDEBAND

Single sideband is a form of amplitude modulation. An ordinary amplitude-modulated signal consists of a carrier and two sidebands, one above the carrier frequency and one below the carrier frequency (*see* AMPLITUDE MODULATION). A single-sideband signal results from the removal of the carrier and one of the sidebands.

Single-sideband, suppressed-carrier emission, also called A3J emission, provides greater communications efficiency than ordinary amplitude modulation, or type A3 emission. This is because ⅔ (67 percent) of the power in an amplitude-modulated signal is taken up by the carrier wave, which conveys no intelligence. In a single-sideband signal, all of the power is concentrated into one sideband, and this yields an improvement of about 8 dB over A3 emission. A single-sideband signal has a bandwidth of approximately half that required for amplitude modulation. Therefore, it is possible to get twice as many A3J signals as A3 signals into a given amount of spectrum space.

In order to obtain A3J emission, a balanced modulator must be used, followed by a filter or phasing circuit. The balanced modulator produces a double-sideband signal with a suppressed carrier (*see* BALANCED MODULATOR, DOUBLE SIDEBAND). The filter or phasing network then removes either the lower sideband or the upper sideband. If the lower sideband is removed, the resulting A3J signal is called an upper-sideband (USB) signal. If the upper sideband is removed, a lower-sideband (LSB) signal results (*see* LOWER SIDEBAND, UPPER SIDEBAND).

Most voice communication at the low, medium, and high frequencies is carried out via single sideband. Generally, the lower sideband is used at frequencies below about 10 MHz; the upper sideband is preferred at frequencies higher than 10 MHz. This is simply a matter of convention; either sideband will provide equally good communication at a particular frequency.

Reception of single-sideband signals requires a product detector. The product detector contains a local oscillator and a mixer. The local oscillator supplies the "missing" carrier, so that the signals are intelligible (*see* PRODUCT DETECTOR).

The tuning of a single-sideband receiver is rather critical. Stability is therefore extremely important, both for the transmitter and the receiver. The receiver local-oscillator frequency must be within a few hertz of the suppressed-carrier frequency of the intermediate-frequency signal. If the receiver is too far off frequency, the result is an unnatural, sometimes humorous-sounding, and perhaps even unintelligible, garbled noise. If the receiver is set for the wrong sideband, a completely unreadable signal will be heard (*see* MONKEY CHATTER).

SINGLE-SIGNAL RECEPTION

In a receiver having a product detector, and intended for the reception of A1 (continuous-wave), F1 (frequency-shift-keyed), or A3J (single-sideband) signals, it is an advantage to have single-signal reception. Most modern superheterodyne receivers have this feature. Direct-conversion receivers and regenerative receivers generally do not.

In a receiver with single-signal reception, the signal can be heard only on one side of zero beat. In the case of A1 or F1 type emission, a beat note or pair of notes is audible on either the upper or lower side of zero beat, but not on both sides. As the receiver is tuned through zero beat, the pitch of the sound drops until the note or notes disappear. The notes do not recur as tuning is continued. If the receiver does not have single-signal reception, the beat notes will reappear on the other side of zero beat, gradually rising in pitch. A similar effect is observed with A3J type emission.

The advantage of single-signal reception over so-called double-signal reception is that it cuts the interference among signals in half. With single-signal reception, the selectivity can be much sharper, while still maintaining intelligibility, as compared with double-signal reception. *See also* PRODUCT DETECTOR, SUPERHETERODYNE RECEIVER.

SINGLE THROW

A relay or switch is called single-throw if it simply connects and disconnects a circuit or group of circuits. The single-throw device cannot be used for switching a common line between two or more other lines.

Single-throw switches are commonly used for power switching. A single-pole switch or relay can be used as a circuit interrupter in any system. A single-throw device may have any number of poles, facilitating the opening and closing of several circuits at once. A single-pole, single-throw switch is often called an SPST switch; a double-pole, single-throw switch is called a DPST switch. *See also* MULTIPLE THROW.

SINGLE-WIRE LINE

A radio antenna can be fed with a single wire for either receiving or transmitting purposes. The single wire forms an unbalanced feed line. Single-wire lines are sometimes used at medium and high frequencies, especially for reception.

A single-wire feed line exhibits a free-space characteristic impedance of approximately 600 ohms to 800 ohms. The return circuit is provided by the earth; therefore, an excellent ground system is needed if single-wire line is used.

A single-wire feed line inevitably picks up or radiates a considerable amount of electromagnetic energy. This makes the single-wire line unsuitable for use with directional antennas. If the feed line must be kept from intercepting or radiating signals, a balanced or shielded line must be used. The most common such lines are open wire, twin-lead, four-wire line, and coaxial cable (*see* BALANCED TRANMISSION LINE, COAXIAL CABLE, FOUR-WIRE TRANSMISSION LINE, OPEN-WIRE LINE, TWIN-LEAD).

Single-wire line can be used to feed a horizontal or vertical radiator at either end, in the center, or off center. The true Windom antenna uses a single-wire line in an off-center configuration. *See also* WINDOM ANTENNA.

SINK

See LOAD.

SINUSOIDAL WAVEFORM

See SINE WAVE.

SKEW

In video transmission, skew is distortion that results from failure of the receiver to be exactly synchronized with the transmitter. Skew results from a constant discrepancy in

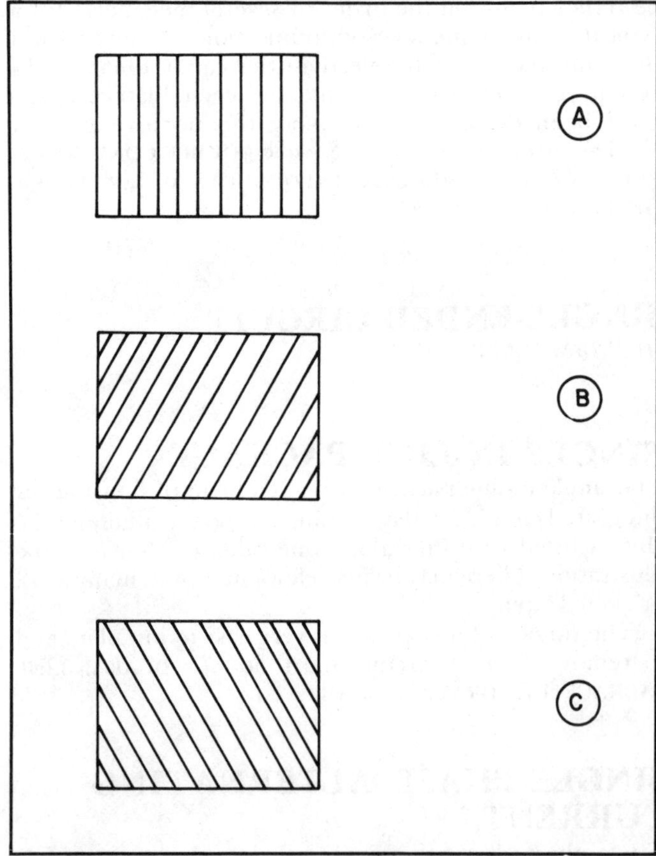

SKEW: Examples of skew in a video signal. At A, zero skew; at B, positive skew; at C, negative skew.

the horizontal scanning rates of the receiver and transmitter. Skew may also occur in an improperly operating video recording system.

Skew can be detected in a facsimile or television system by transmitting a set of vertical, straight lines, such as the pattern shown at A in the illustration. If the receiver and transmitter are in precise synchronization, then the received picture will be identical to this pattern. However, if skew is present, the lines will appear slanted. If the receiver scans faster than the transmitter, the lines are tilted clockwise, as at B. If the receiver scan rate is slower than that of the transmitter, the lines appear tilted counterclockwise, as at C. The severity of the skew is given in degrees. The perfectly vertical lines at A representing a skew of 0 degrees; the effect at B is called positive skew; the effect at C is called negative skew.

If the synchronization discrepancy becomes nonconstant, or if it is especially severe, a picture becomes completely unrecognizable. This effect is familiar to all television viewers who have inadvertently misadjusted the horizontal-hold control. *See also* HORIZONTAL SYNCHRONIZATION.

SKIN EFFECT

In a solid wire conductor, direct current flows uniformly along the length of the wire. The number of electrons passing through a given cross section of the wire does not depend on whether the cross section is near the surface of the wire or near the center. The same holds for alternating currents at relatively low frequencies. The conductivity of the wire is proportional to the cross-sectional area, which is in turn proportional to the square of the diameter.

At radio frequencies, the conduction in a solid wire becomes nonuniform. Most of the current tends to flow near the outer surface of the wire. This is called skin effect. It increases the effective ohmic resistance of a wire at radio frequencies. The higher the frequency becomes, the more pronounced the skin effect will be. At high and very high frequencies, the conductivity of a wire is more nearly proportional to the diameter than to the cross-sectional area.

In the design of radio-frequency transmitting antennas, tubing is generally used rather than wire, if such construction is mechanically feasible. A large-diameter tubing conducts at the high and very-high frequencies almost as well as a solid metal rod of the same diameter. The use of tubing thus provides very low ohmic loss in an antenna system, at reasonable cost.

SKIP

Long-distance ionospheric propagation is sometimes called skip, although this is a technically inaccurate use of the term.

Skip is actually the tendency for signals to pass over a certain geographical region. At high frequencies, skip is sometimes observed. This effect is shown in the illustration. A transmitting station X is heard by station Z, located thousands of miles distant, but not by station Y. The ionization of the F layer is insufficient to bend the signals back to earth at the sharper angle necessary to allow reception by station Y. *See also* PROPAGATION CHARACTERISTICS, SKIP ZONE.

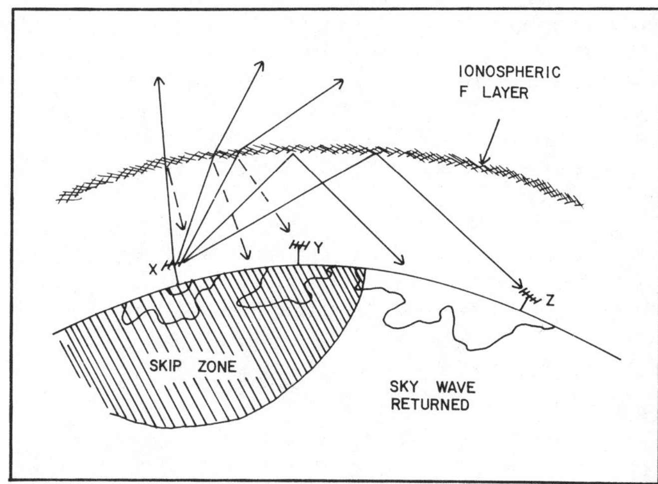

SKIP ZONE: The skip zone is the geographic region, outside the ground-wave radius, in which F-layer signals are not returned to the earth. Here, transmitter X is heard at receiver Z, but not at receiver Y.

SKIP ZONE

When skip occurs in an ionospheric F-layer communication circuit, the signals from a transmitting station cannot be received by other stations located within a certain geographical area. This "dead" area is called the skip zone.

Under most conditions, the skip zone begins at a distance of about 10 to 15 miles from the transmitting station, or the limit of the range provided by the direct wave, the reflected waves, and the surface wave (*see* DIRECT WAVE, REFLECTED WAVE, SURFACE WAVE). The outer limit of the skip zone varies considerably, depending on the operating frequency, the time of day, the season of the year, the level of sunspot activity, and the direction in which transmission is attempted. A pictorial example is shown in the previous illustration.

At the very low, low, and medium frequencies, a skip zone is never observed. In the high-frequency spectrum, however, a skip zone is often present. In the upper part of the high-frequency band, a skip zone is almost always observed.

At times, certain conditions arise that allows signals to be heard from points within the skip zone. Densely ionized areas may form in the E layer, causing propagation over shorter paths than would normally be expected (*see* E LAYER, SPORADIC-E PROPAGATION). Auroral propagation and backscatter sometimes allow communication between stations that would otherwise be isolated by the skip zone (*see* AURORAL PROPAGATION, BACKSCATTER). Above about 20 MHz, tropospheric effects can partially "fill in" the skip zone at times (*see* TROPOSPHERIC PROPAGATION).

If the frequency of operation is increased, the skip zone widens. The outer limit of the skip zone may be several thousand miles away. The same widening effect takes place at a constant frequency as darkness falls. At frequencies above a certain maximum, the outer limit of the SKIP zone disappears entirely, and no F-layer propagation is observed. *See also* F LAYER, IONOSPHERE, PROPAGATION CHARACTERISTICS, SKIP.

SKIRT SELECTIVITY

Skirt selectivity is an expression of the cutoff sharpness of a receiver bandpass filter, and also of the maximum or ultimate attenuation. Good skirt selectivity is an important feature of any communications receiver.

Skirt selectivity can be expressed qualitatively in terms of relative "steepness." Skirt selectivity can be specified quantitatively according to the shape factor and ultimate attenuation of the response (*see* SHAPE FACTOR, ULTIMATE ATTENUATION).

In general, steep skirts represent a nearly rectangular bandpass response. This is desirable in most receiving situations. Steep skirts provide excellent adjacent-channel rejection, and optimize the signal-to-noise ratio. *See also* ADJACENT-CHANNEL INTERFERENCE, RECTANGULAR RESPONSE, SELECTIVITY.

SKY WAVE

An electromagnetic signal is called a sky wave if it has been returned to the earth by the ionosphere. Sky-wave propagation can be caused by either the E layer or the F layer of the ionosphere. Sky-wave propagation can also result from reflection off the auroral curtains, or from scattering by the ionized trails left by meteors (*see* AURORAL PROPAGATION, E LAYER, F LAYER, METEOR SCATTER, SPORADIC-E PROPAGATION).

Sky waves are observed at various times of day at different frequencies. It is this form of propagation that is responsible for most of the long-distance communication that takes place on the so-called shortwave bands. Sky-wave propagation occurs almost every night on the standard amplitude-modulation broadcast band. Sky waves are affected by the season of the year and the level of sunspot activity, as well as the time of day and the operating frequency. *See also* PROPAGATION CHARACTERISTICS.

SLAVE

A slave is a component that is operated, or controlled, by another component. The controlling component is called the master (*see* MASTER).

In a servomechanism or other mechanical device, the slave follows the master according to a certain set of instructions. In a selsyn, for example, the master is the synchronous motor at the rotor, and the slave is the indicator motor at the station console. *See also* SERVOMECHANISM, SERVO SYSTEM.

SLEEVE ANTENNA

See COAXIAL ANTENNA.

SLEW RATE

In a closed-loop operational-amplifier circuit, there is a limit to how rapidly the output voltage can change as the input voltage changes. This maximum limit, under linear operating conditions, is called the slew rate. It is measured in volts per second, volts per millisecond, or other units of voltage per unit time.

The slew rate determines the maximum frequency at which an operational amplifier will function in a linear manner. If the frequency is increased so that the instantaneous rate of change of the input signal exceeds the slew rate, distortion will occur, and the operational amplifier will not perform according to the theoretical specifications. *See also* OPERATIONAL AMPLIFIER.

SLIDE POTENTIOMETER

A slide potentiometer is a form of variable resistor, designed for low-current applications in direct-current and audio-frequency circuits. Most potentiometers are of the rotary type, and are adjusted by turning a knob, shaft, or metal plate. The slide potentiometer is set by means of a lever that moves back and forth, or up and down.

Slide potentiometers give visual reinforcement of a control position in certain situations. A common example is the graphic equalizer. *See also* EQUALIZER, POTENTIOMETER.

SLIDE RULE

A slide rule is a mechanical calculating device that can be used for determining the values of products, quotients, logarithms, trigonometric functions, and other functions. The slide rule has been largely replaced in recent years by the electronic pocket calculator, which provides greater accuracy and versatility.

Slide rules range in complexity from very simple devices, available in drug stores and stationery stores, to highly specialized units designed for various engineering and mathematical fields. A rather sophisticated slide rule is

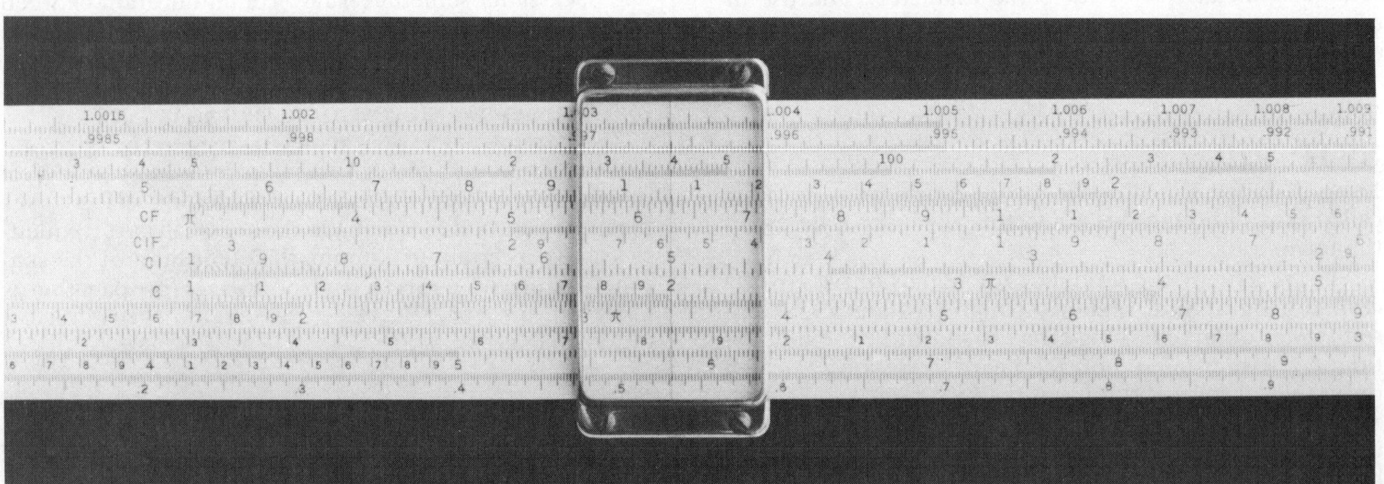

SLIDE RULE: A typical engineering slide rule.

shown in the photograph. This particular slide rule has numerous calibrated scales. It can be used for several different mathematical operations, including multiplication and division, logarithmic functions, exponential functions, square roots and cube roots, and trigonometric functions. Slide rules cannot be used for addition and subtraction (except indirectly via logarithms).

The accuracy of a typical slide rule is three or four significant figures. This compares with 10 significant figures for the average pocket calculator. A few engineers keep a slide rule as a backup for the pocket calculator. But nowadays, calculators are just about as cheap as slide rules, so if a calculator goes bad, it's easy to buy another!

SLIDE-RULE TUNING

Slide-rule tuning is a method of indicating the frequency of a radio receiver or transmitter. It is a form of analog display. Slide-rule tuning is often seen in broadcast receivers, such as the AM/FM stereo tuner shown in the illustration. Some general-coverage (shortwave) receivers also use this method of frequency indication.

Slide-rule tuning is usually accomplished by means of a string-driven or wire-driven arrangement. A rack-and-pinion system may also be used.

The advantage of slide-rule tuning is the ease of readability and approximate resettability. This makes slide-rule tuning desirable for applications such as the AM/FM stereo tuner, in which operating simplicity is very important.

SLIDE SWITCH

A slide switch is a small, one-pole or two-pole switch that is often used in communications equipment. Slide switches are employed mostly in situations not requiring frequent changing of the switch position.

The slide switch has two positions, and may be either a single-throw or a double-throw device. The switch is thrown by pressing a lever up, down, or sideways (see illustration). Some slide switches are equipped with a plate that can be screwed down over the lever to prevent accidental changing of the switch position.

SLIP

An electric motor will run at a certain speed, expressed in revolutions per minute, with no load connected to it. This is called the synchronous speed or the free-running speed. When a load is connected to a motor, the speed decreases. The difference between the synchronous speed and the running speed, expressed as a percentage, is called slip.

If the free-running speed of a motor is F revolutions per minute and the running speed under load is R revolutions per minute, then the slip, S, is given in percent by the equation

$$S = 100(F-R)/F$$

The slip may also be expressed simply in revolutions per minute, as F-R.

The slip in a motor depends on the size of the motor, the application for which it is intended, and the amount of load that is connected to its shaft. The greater the mechanical load resistance, the larger the slip. *See also* MOTOR.

SLOPE

In a Cartesian coordinate plane, the slant of a straight line can be defined in terms of a numerical value called the slope of the line. A straight line in the Cartesian system has an equation of the form:

$$y = mx + b$$

where m and b are real-number constants, x is the independent variable, and y, also called f(x), is the dependent variable. In this equation form, the number m is the slope of the line.

Linear equations do not always appear in the above form, which is called the slope-intercept form. For the slope to be determined it is necessary to manipulate the constants and variables of a linear equation to obtain the slope-intercept form. Alternatively, if one point (x_1, y_1) on the line is known, the slope m can be obtained by putting the equation into following form:

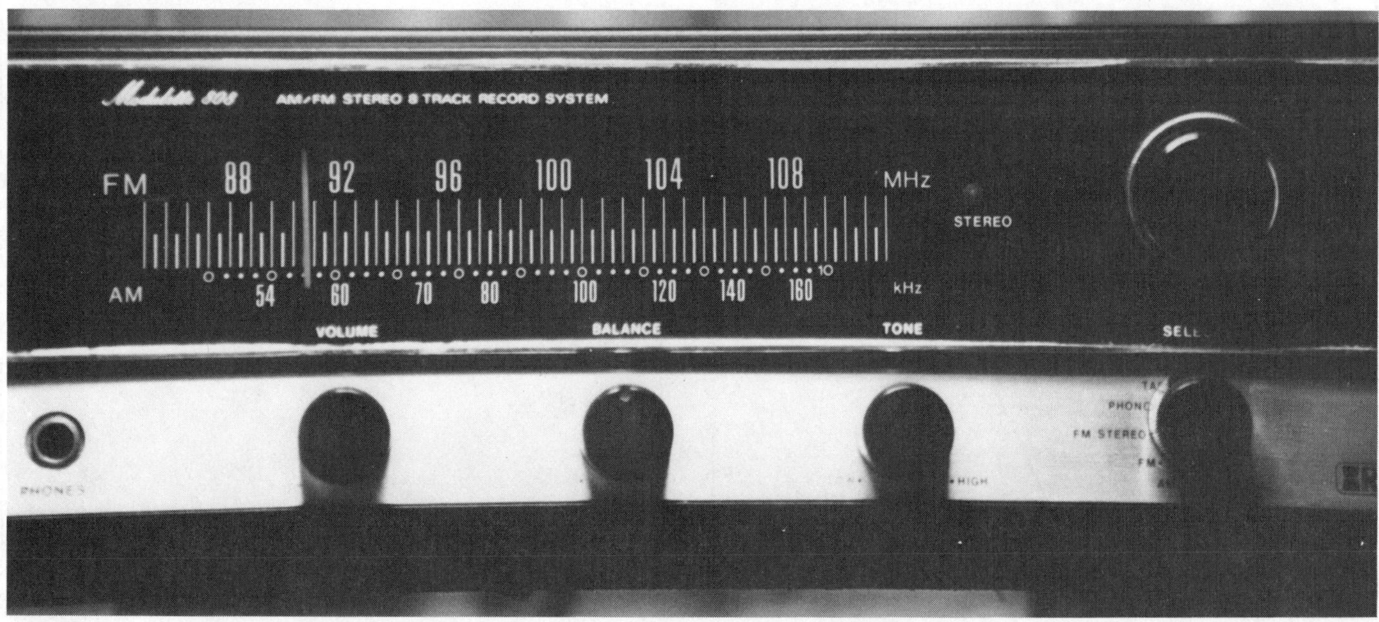

SLIDE-RULE TUNING: Slide-rule tuning is popular as a method of frequency display in AM/FM receivers.

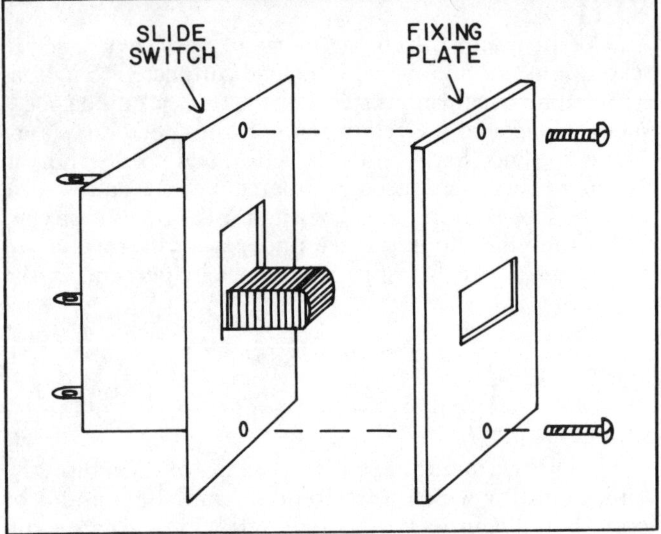

SLIDE SWITCH: A typical slide switch, with a cover plate for preventing unwanted, accidental switching.

$$y - y_1 = m(x - x_1)$$

This is called the point-slope form of a linear equation in two variables.

The slope of a straight line is the ratio of the increase in the dependent-variable value, for a given span, to the increase in the independent-variable value for the same span. If x and y are the independent and dependent variables, respectively, and two points on the line are given as (x_1,y_1) and (x_2,y_2), then the slope is:

$$m = (y_2 - y_1)/(x_2 - x_1)$$

If a line has a slope $m = 0$, then the line is parallel to the independent-variable axis. This axis is usually represented horizontally; thus the line with $m = 0$ is a horizontal line in most cases.

If a line has a positive slope, then it slants upward toward the right in the typical graphic representation. The larger the slope, the steeper the slant of the line. The slope m is equal to the tangent of the angle that the line subtends counterclockwise from the positive independent-variable axis.

If a line has a negative slope, then it slants downward toward the right in the typical graphic representation. The more negative the value of the slope, the steeper the slant of the line. The slope m is, again, equal to the tangent of the angle that the line subtends, measured counterclockwise with respect to the positive independent-variable axis. Lines having various different slope values are shown in the illustration.

If a line happens to run parallel to the dependent-variable axis, its slope is undefined, since $m = \tan (90°)$ which has no real-number value. *See also* CARTESIAN

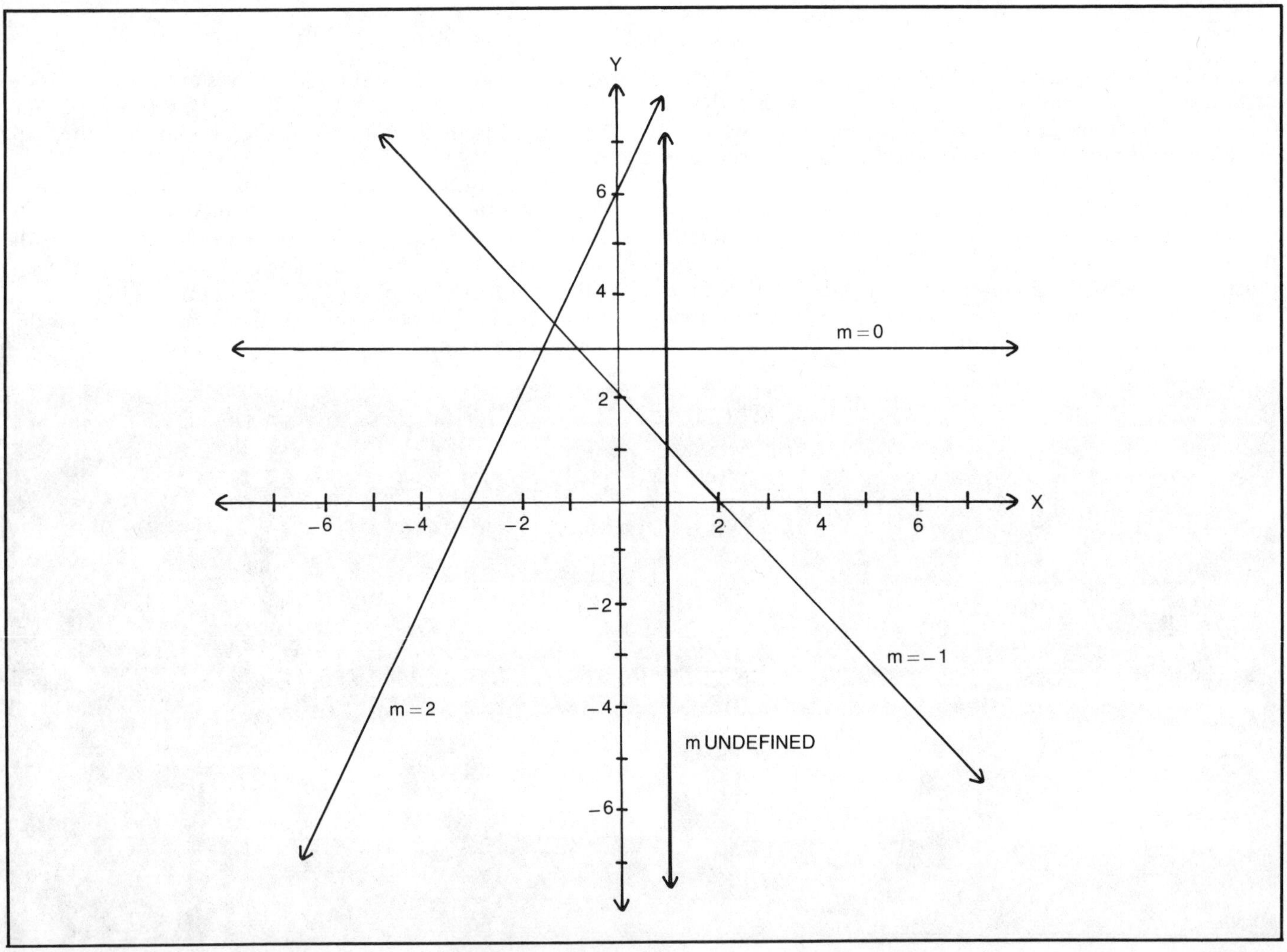

SLOPE: Lines with various slopes (denoted by m) in the Cartesian plane.

COORDINATES, DEPENDENT VARIABLE, INDEPENDENT VARIABLE, LINEAR FUNCTION.

SLOPE DETECTION

A narrow-band frequency-modulated (FM) or phase-modulated (PM) signal can be demodulated using a receiver designed for amplitude modulation (AM). This is done by means of a technique called slope detection.

A standard FM and PM communications signal has a deviation of plus-or-minus 5 kHz. If an AM receiver, with its typical passband of plus-or-minus 3 to 5 kHz, is tuned slightly above or below the carrier frequency of the FM or PM signal, the signal will move in and out of the receiver passband with each cycle of modulation. The receiver should be tuned approximately 5 kHz above or below the FM or PM carrier for optimum results (see illustration).

Slope detection provides excellent sensitivity for reception of FM or PM signals. However, some distortion is likely to occur, because the skirt selectivity curves of most AM receivers do not have the gradual, uniform rolloff that would be necessary for distortion-free slope detection. Also, the signal will have an exaggerated treble (tinny) sound unless a deemphasis circuit is employed in the audio stages of the receiver—a feature that AM equipment does not generally have. *See also* DEEMPHASIS, DETECTION, FREQUENCY MODULATION, PHASE MODULATION, PREEMPHASIS.

SLOT COUPLING

Slot coupling is a method of coupling between a waveguide and a coaxial transmission line. Slot coupling is used at radio frequencies in the ultra-high and microwave range.

Slot coupling is accomplished by cutting identical rectangular openings in the waveguide and the coaxial-cable outer conductor. The rectangular openings are placed in precise alignment. The waveguide and the coaxial cable may be either parallel or perpendicular with respect to each other. The parallel method is shown at A in the illustration; the perpendicular method is shown at B.

Slot coupling can be used to transfer radio-frequency energy from a coaxial cable into a waveguide, or to transfer energy from a waveguide into a coaxial cable. Thus, slot coupling is useful for transmitting and receiving with a single physical installation. Slot coupling can also be used

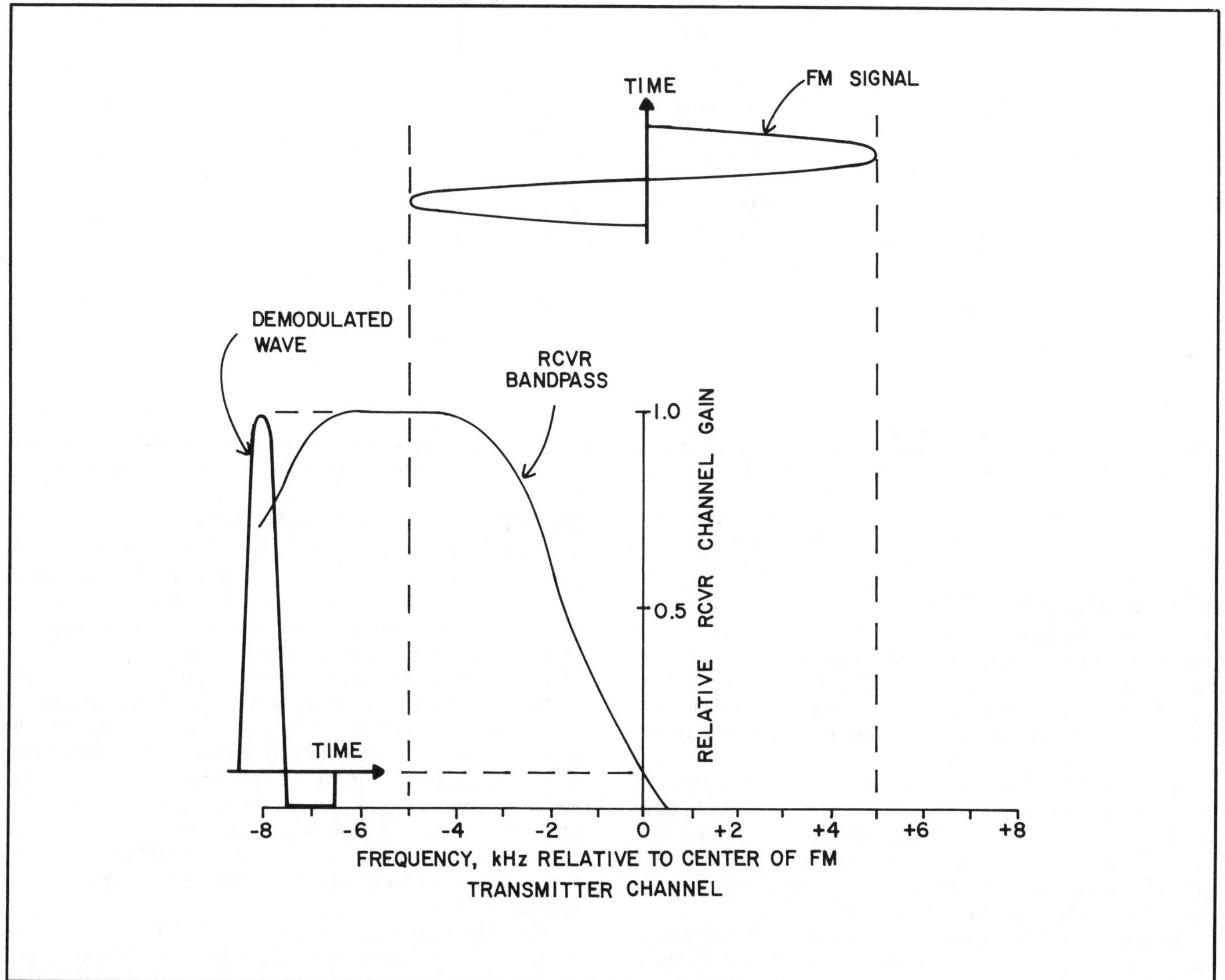

SLOPE DETECTION: Slope detection allows reception of an FM signal with an AM receiver.

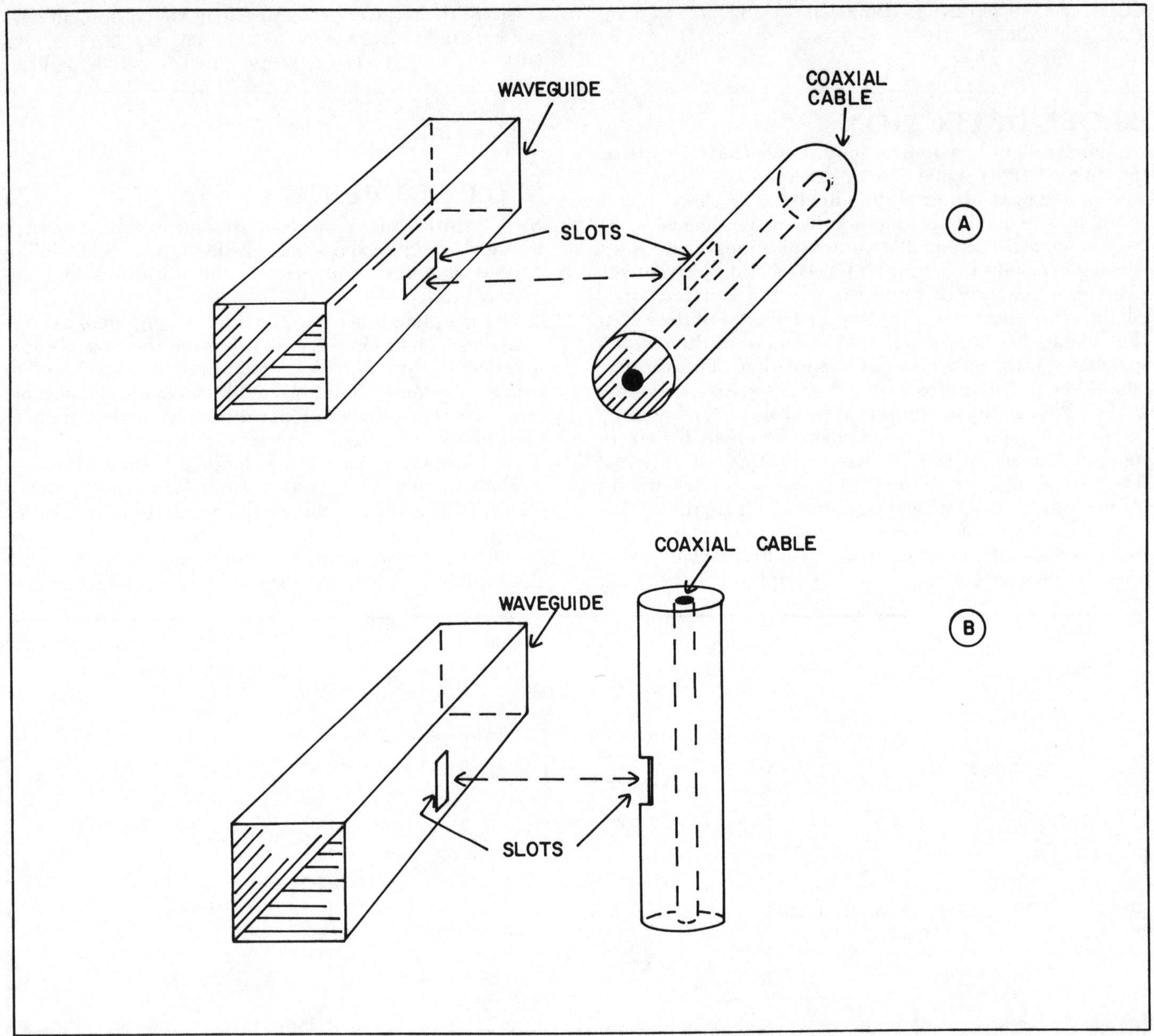

SLOT COUPLING: Two methods of slot coupling. At A, the cable and the waveguide are parallel at the coupling point; at B, they are perpendicular.

to provide signal transfer between two sections of waveguide. *See also* COAXIAL CABLE, WAVEGUIDE.

SLOW-BLOW FUSE

It is often desirable to have a fuse that does not blow instantly, but instead allows some time before breaking a circuit. This is the case, for example, in equipment that draws a heavy initial current surge when the power is first applied. A slow-blow fuse is designed for this type of application.

The slow-blow fuse has a wire filament that softens if too much current passes through it. The wire is held taut by a spring. If the current exceeds the fuse rating for several seconds, the wire gets soft enough for the spring to pull it apart, breaking the circuit. The photograph shows a typical cartridge-type slow-blow fuse.

In low-current solid-state devices, slow-blow fuses may not react fast enough to protect the components in case of a malfunction. For this sort of situation, fast-blowing or quick-break fuses are used. *See also* QUICK-BREAK FUSE.

SLOW-SCAN TELEVISION

Slow-scan television (SSTV) is a method of transmitting pictures within a narrow band of frequencies. Slow-scan television is used at high and very-high frequencies, where spectrum space is limited. Slow-scan television has recently become very popular among radio amateurs.

The ordinary fast-scan television signal must be modulated very rapidly to convey the fast-changing picture that allows us to see motion. This rapid modulation results in a signal that is several megahertz wide. Slow-scan modulation reduces the occupied spectrum space at the expense of conveying motion in the picture. Some detail is also sacrificed. The slow-scan signal is 3 kHz wide. An SSTV system transmits and receives one still picture every 8 seconds, compared with 30 frames per second for fast-scan television.

The slow-scan transmitting system consists of a fast-scan camera, a fast-scan-to-slow-scan converter, a modulator that produces audio-frequency tones ranging up to 3 kHz, and a single-sideband transmitter. The output of the

SLOW-BLOW FUSE: A slow-blow cartridge fuse. This photograph is enlarged several times.

modulator is fed directly to the microphone input of the transmitter (see A in the illustration).

A slow-scan receiving system consists of an ordinary single-sideband radio receiver, a demodulator that converts the audio tones into the impulses needed by the monitor, a slow-scan-to-fast-scan converter, and a television set or composite-video monitor, as at B. Equipment for both transmitting and receiving SSTV is available from several commercial manufacturers.

Slow-scan pictures can be recorded on an ordinary tape recorder, since the modulating-signal components are all within the audio range. A small cassette tape recorder can be used to take signals directly off the air; they can then be retransmitted by simply connecting the tape-recorder output to the transmitter microphone terminals.

Slow-scan signals have a characteristic sound. Once you have heard the fluctuating tones of such a signal, you will be able to recognize SSTV instantly every time you come across it. Most amateur SSTV operation is found at or near 3.84-3.85 MHz and 14.23 MHz.

For the proper reception of slow-scan signals, receiver tuning is critical, just as it is with single sideband. It is especially important that the synchronization pulses have the proper audio frequency (1.2 kHz) in the receiver output. If the receiver is tuned just a little bit above or below the proper frequency, the picture will "tear" and will not be recognizable. It is, however, not very difficult to get the receiver on the right frequency; the monitor screen is watched until a sensible picture materializes. Most radio amateurs using SSTV also use single sideband at the same frequency, and talk to each other in between picture exchanges. Thus, simply tuning in the voice signal will usually result in proper receiver adjustment for SSTV as well.

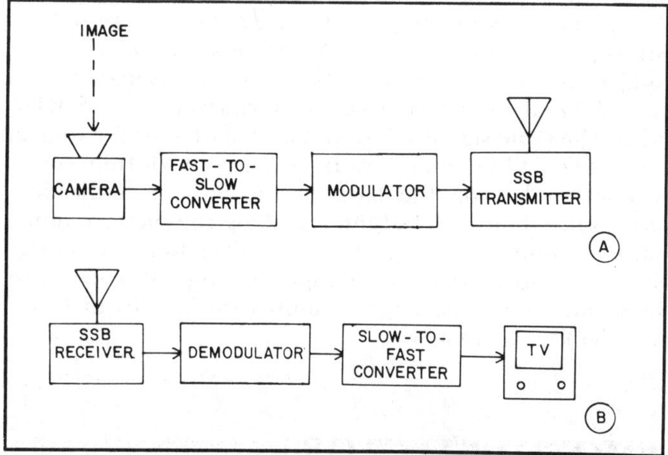

SLOW-SCAN TELEVISION: At A, block diagram of a slow-scan television transmitting setup. At B, block diagram of a slow-scan receiving station.

The technology for sending and receiving SSTV signals has improved since the first use of this mode. Microcomputers can be used to process a received picture, getting better detail and more levels of shading. Some SSTV signals have even been sent and received in color. At the high frequencies, SSTV allows worldwide video communication. *See also* SINGLE SIDEBAND, TELEVISION.

SLUG TUNING

See PERMEABILITY TUNING.

SMALL SIGNAL

A small signal is an alternating-current signal which, when applied to the input of an amplifier or other device, will not cause nonlinearity. A circuit designed to operate with small signals is called a small-signal or low-level circuit.

All radio receivers contain small-signal circuits at the front end, intermediate-frequency chain (in superheterodyne receivers), and audio stages. The detector and mixer stages are nonlinear, and are therefore not small-signal circuits. Radio transmitters usually have small-signal circuits at all stages except for the mixers, modulator, driver, and final amplifier.

Small-signal circuits do not require significant input-signal power for their operation. Small-signal amplifiers must be operated in Class A, and generally have a high input impedance. Small-signal probes, likewise, must have a high input impedance to minimize the current drawn. *See also* CLASS-A AMPLIFIER, LARGE SIGNAL.

SMITH CHART

The Smith chart is a special form of coordinate system that is used for plotting complex impedances (*see* IMPEDANCE). Smith charts are especially useful for determining the resistance and reactance at the input end of a transmission line when the resistance and reactance at the antenna feed point are known. The Smith chart is named after the engineer P.H. Smith, who first devised and used it.

The resistance coordinates on the Smith chart are eccentric circles, mutually tangent at the bottom of the circular graph. The reactance coordinates are partial circles having variable diameter and centering. Resistance and reactance values can be assigned to the circles in any desired magnitude, depending on the characteristic impedance of the transmission line. The illustration is an example of a Smith chart intended for analysis of feed systems in which the characteristic impedance of the line is 50 ohms ($Z_o = 50$).

Complex impedances appear as points on the Smith chart. Pure resistances (impedances having the form R + jo, where R is a nonnegative real number) lie along the resistance line; the top of the line represents a short circuit and the bottom represents an open circuit. Pure reactances (having the form 0 + jX, where X is any real number except 0) lie on the perimeter of the circle, with inductance on the right and capacitance on the left. Impedances of the form R + jX, containing finite, nonzero resistances and reactances, correspond to points within the circle. Several different complex impedance points are illustrated.

The Smith chart can be used to determine the standing-wave ratio (SWR) on a transmission line, if the characteristic impedance of the line and the complex antenna impedance are known (*see* STANDING-WAVE RATIO). For determination of SWR, a set of concentric circles is added to the Smith chart. These circles are called SWR circles.

The center point of the chart corresponds to an SWR of 1:1. Higher values of SWR are represented by progressively larger circles. The radii of the SWR circles on a given Smith chart can be determined according to points on the resistance line. In the example, the 2:1 SWR circle passes through the 100-ohm point on the resistance line; the 4:1 SWR circle passes through the 200-ohm point on the resistance line, and so on. In general, for any SWR value x:1, the x:1 SWR circle passes through the point on the resistance circle corresponding to 50x ohms.

The Smith chart illustrates why an SWR of 1:1 can be obtained only if the impedance of the load is a pure resistance equal to the characteristic impedance of the feed line. The SWR cannot be 1:1 if reactance exists in the load. A given SWR greater than 1:1 can occur in infinitely many ways, corresponding to the infinite number of points on a circle. An SWR of "infinity" exists when the load is a short circuit, an open circuit, a pure capacitive reactance, or a pure inductive reactance.

For a thorough discussion of the use of the Smith chart in feed-line applications, a book on antenna theory or communications engineering should be consulted. *See also* CAPACITIVE REACTANCE, INDUCTIVE REACTANCE.

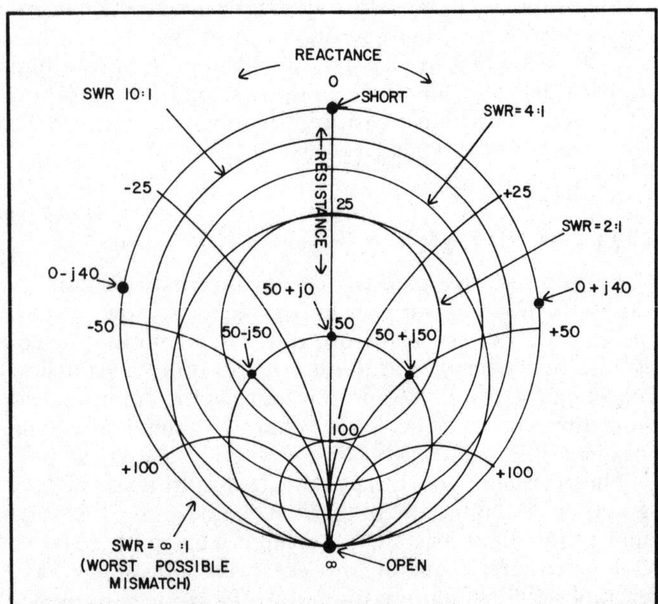

SMITH CHART: The Smith chart for denoting complex impedances. Various impedance points and SWR circles are shown.

S METER

An S meter is a device in a radio receiver that indicates the relative or absolute amplitude of an incoming signal. Many types of receivers are equipped with such meters. There are many different styles of S meter, but they can be categorized as either uncalibrated or calibrated.

Uncalibrated S meters may consist either of an analog meter or a digital meter (*see* ANALOG METERING, DIGITAL METERING). The meter indicates relative signal strength. The signal for driving the meter can be obtained from the i-f stages of the receiver. The most common method of obtaining this signal is by monitoring the automatic-gain-control (AGC) voltage. The stronger the signal, the greater the AGC voltage, and the higher the meter reading.

Uncalibrated S meters are found in many FM stereo tuners, and also in most FM communications equipment. Some general-coverage communications receivers employ uncalibrated S meters.

Calibrated S meters are generally of the analog type. This facilitates marking the scale in definite increments. The standard unit of signal strength is the S unit (*see* S UNIT), which is a number ranging from 0 to 9 or from 1 to 9. A meter indication of S9 is defined as resulting from a certain number of microvolts at the antenna terminals. Most manufacturers agree on the figure of 50 μV for a meter reading of S9, as determined at a frequency in the center of the coverage range of the receiver. Some manufacturers use a different standard level to represent S9.

Since many receivers exhibit greater gain at the lower frequencies than at the higher frequencies, a reading of S9 often results from a weaker signal at the low frequencies (perhaps 40 μV) and a stronger signal at the higher frequencies (perhaps 60 μV). Each S unit below S9 represents a signal-strength change of 3 dB or 6 dB, depending on the manufacturer. This value is independent of the frequency to which the receiver is tuned.

Calibrated S meters are usually marked off in decibels above the S9 level. The meter readings above S9 are thus designated S9 + 10 dB, S9 + 20 dB, and so on; most meters can register up to S9 + 30 dB. These scales are accurate for typical communications. If signal-strength levels must be known with great accuracy, a laboratory test instrument should be used for measuremnt.

A receiver S meter provides a mental picture of how strong an incoming signal is. All signal-strength values are subjective, however, because band conditions may vary. A 50 μV signal might be masked by sferics (static) at 1.8 MHz, while the same signal stands out as that of a local station at 28 MHz. While many radio operators define signal strength solely on the basis of the S-meter indication, others use their own judgment, taking the meter reading into account as a subjective quantity. Relative signal-strength information is exchanged among radio operators by means of an S (strength) number ranging from 1 to 9. *See also* RST SYSTEM.

SMOKE DETECTOR

A smoke detector is an electronic device that produces a loud buzz or tone when exposed to smoke. This helps to protect people against fires in homes and other buildings.

Smoke detectors can operate in several different ways. Some are sensitive to the dielectric properties of the air, which change when smoke is present. Others use optical devices, sensitive to the light transmittivity of the air. Smoke detectors should not be confused with heat detectors, which are actuated by the infrared radiation or the rise in temperature that accompanies a fire.

Smoke detectors are generally powered by dry cells. This is because the utility power often fails before the smoke gets thick enough to set off the device. The cell or cells in a smoke detector should be checked regularly; most detectors have a built-in test button to facilitate this.

SMOOTHING

Smoothing is the elimination of rapid fluctuations in the strength of a current or voltage. A good example of smoothing is the removal of the ripple in the output of a power supply. Smoothing may also be called filtering. Smoothing is usually accomplished by means of a capacitor (*see* FILTERING CAPACITOR, RIPPLE).

In an envelope detector, a small capacitor is used to eliminate the radio-frequency fluctuations of the carrier wave, leaving only the audio-frequency signals. This process is called smoothing (*see* DIODE DETECTOR).

Smoothing is used in automatic-gain-control (AGC) systems of all kinds. A capacitor and resistor are used to smooth out the audio-frequency components of a received or transmitted signal, while still providing a fast enough time constant to allow effective compensation for changes in the signal intensity (*see* AUTOMATIC GAIN CONTROL, RESISTANCE-CAPACITANCE TIME CONSTANT).

SNAP DIODE

See STEP-RECOVERY DIODE.

SNELL'S LAW

Snell's law is a rule of physics that applies to light waves passing through two media, one of which is air.

Suppose that a light beam strikes a surface at an angle measuring θ degrees relative to the perpendicular (see illustration). Suppose also that the angle at which the light beam leaves the surface, measured with respect to the normal to the surface, is given by ϕ. Then Snell's law states that:

$$\sin \theta / \sin \phi = c$$

where c is the index of refraction of the non-air substance. *See also* INDEX OF REFRACTION.

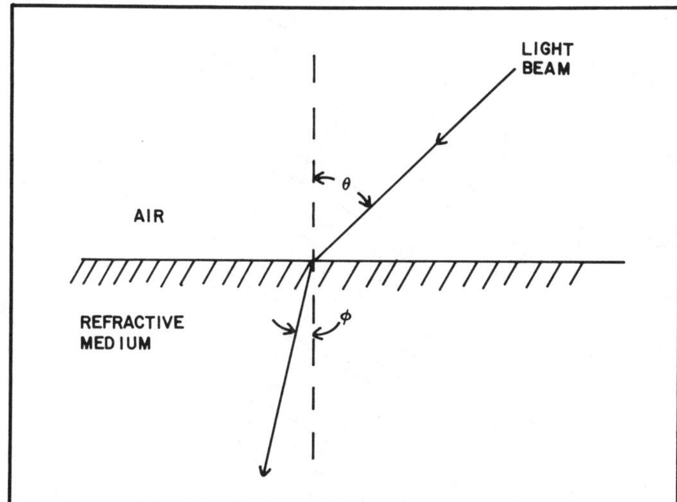

SNELL'S LAW: Snell's law for the refraction of light is a relation between the angles of incidence and refraction, and the refractive index.

SNOOPERSCOPE

A snooperscope is an infrared-sensitive television device. In total darkness, certain objects (such as the bodies of warm-blooded animals, including humans) emit infrared radiation. This makes it possible to literally see in the dark with the appropriate equipment. Infrared radiation passes through many barriers that are opaque to visible light. A snooperscope can be used to "see through walls."

The snooperscope obtains its name from the fact that it can be, and often is, used for the purpose of spying. An infrared camera/recorder can, however, be used for other purposes. Infrared photographs are valuable in determining the places from which heat is lost from a house or business structure. This makes the snooperscope useful for optimizing energy efficiency. Infrared photographs are taken from orbiting satellites to aid in weather forecasting; various cloud systems exhibit different infrared-reflecting properties. *See also* INFRARED, INFRARED DEVICE.

SNOW STATIC

See PRECIPITATION STATIC.

SOCKET

A socket is a form of jack into which a device with many prongs is designed to fit. Sockets are used in some electronic circuits for easy installation and replacement of integrated circuits (see illustration). Sometimes a jack, accommodating just two or three conductors, is called a socket. *See also* JACK.

SODAR

The thermal characteristics of the atmosphere can often be evaluated by analyzing the manner in which the air carries

SOCKET: A multi-prong socket used for an integrated circuit (in this case a microcomputer).

acoustic waves. Scientists use a technique called sound detection and ranging (abbreviated SODAR) to accomplish this.

The SODAR system uses an audio signal generator, a transducer for sending out the sound in the range 20 Hz to 20 kHz, a receiving transducer, a set of amplifiers, and an oscilloscope. The sound waves are reflected when they encounter the boundary between a cold air mass and a warm air mass.

SODAR is useful for determining the depth of the warm-air layer near the surface of the earth. Meteorologists use it to determine the potential instability of the atmosphere, and thus the probability of stormy weather. SODAR is also useful in aviation for locating regions of potential clear-air turbulence.

SODIUM

Sodium is an element with atomic number 11 and atomic weight 23. Sodium appears in its natural form as a corroded metal. This is because sodium combines with oxygen very readily. If a section of sodium is scraped free of its oxide coating, it will corrode so fast that the process can be observed.

Sodium is most familiar as one of the elements in common table salt (sodium chloride). Sodium and chlorine combine very easily; sodium chloride is abundant in sea water.

In vapor form sodium produces a bright yellow-orange light when energized. *See* SODIUM-VAPOR LAMP.

SODIUM-VAPOR LAMP

A sodium-vapor lamp is a device that emits visible light by means of energized, vaporized sodium atoms. It is an electroluminescent lamp. It produces visible light at several discrete wavelengths. The main emission lines occur in the yellow-orange visible spectrum. The sodium-vapor lamp is much more energy efficient than the common incandescent lamp.

Sodium-vapor lamps are often used for lighting streets and highways. *See also* MERCURY-VAPOR LAMP.

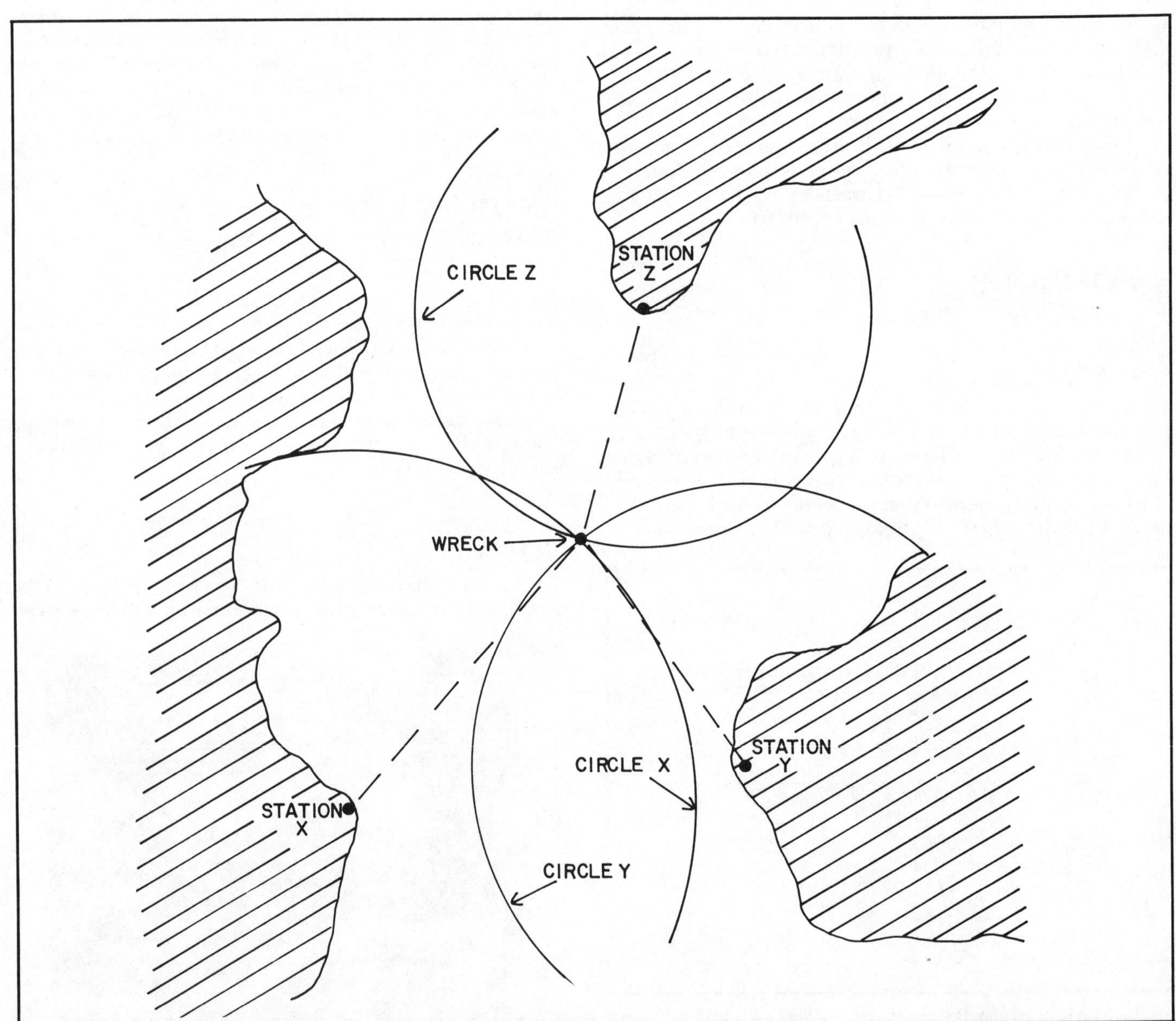

SOFAR: SOFAR is a sonic method of locating a ship or plane that has been wrecked at sea.

SOFAR

When a ship or airplane is wrecked at sea, the location of the survivors can be determined by a technique called sound fixing and ranging (abbreviated SOFAR).

Acoustic waves propagate very well under water. The survivors set off depth charges which produce acoustic waves that travel for hundreds or even thousands of miles. The survivors maintain radio contact with two or three land-based or ship-based stations, equipped with devices for receiving the sound impulses from the depth charges. The time required for the acoustic waves to reach each station indicates the distance to the explosions. The personnel at the stations draw circles showing the distance to the depth charges. The point where the circles intersect (see illustration) is the location of the wreck.

SOFTWARE

In a computer system, the programs are called software. Software may exist in written form, as magnetic impulses on tapes or disks, or as electrical or magnetic bits in a computer memory. Software also includes the instructions that tell personnel how to operate the computer.

There are several types of computer-programming languages, each with its own special purpose. The most primitive form of software language is called machine language, since it consists of the actual binary information used by the electronic components of the computer.

Software may be programmed temporarily into a memory, or it may be programmed permanently by various means. When software is not alterable (that is, it is programmed permanantly), it is called firmware. *See also* ASSEMBLER AND ASSEMBLY LANGUAGE, COMPUTER, COMPUTER PROGRAMMING, FIRMWARE, HARDWARE, HIGHER-ORDER LANGUAGE, MACHINE LANGUAGE, MICROCOMPUTER, MINICOMPUTER.

SOLAR CELL

A solar cell is a photovoltaic cell (*see* PHOTOVOLTAIC CELL), designed for supplying electricity directly from the light of the sun. Solar cells are generally made of silicon, and resemble semiconductor diodes. The P-N junction, which has a very large surface area, produces direct current when sunlight strikes it.

In direct sunlight at noon on a midsummer day, a typical solar cell will produce about 0.1 watt for each square inch of surface area. The power output is decreased if clouds obscure the sun, or if the sun is very close to the horizon. However, solar cells will produce some power even on a cloudy day, or just after sunrise or before sunset.

Most solar-power systems have some means of storing the energy obtained during the hours of maximum sunlight, for use at night or when the sunlight is very weak. Energy storage can be done by means of storage batteries; many active communications satellites and other space vehicles use this method of obtaining power from the sun. A solar-energy system can also be interconnected with commercial power sources. Power is sold to the utility company in times of excess, and bought back during times of shortage.

Solar cells can be used in conjunction with ammeters for light-intensity determination. Many light meters operate on this principle. Solar cells can also be used to receive modulated-light signals (*see* MODULATED LIGHT, OPTICAL COMMUNICATION).

Most solar cells are rather inefficient. They convert less than 20 percent of the incident solar power into electrical power. New types of solar cells are being developed, however, that may provide much better efficiency and lower production cost. Solar cells are available in large batteries called solar panels, which can be connected in series-parallel to obtain large amounts of electrical power. *See also* SOLAR ENERGY, SOLAR POWER.

SOLAR ENERGY

The sun radiates a tremendous amount of energy throughout the electromagnetic spectrum. Most of the energy occurs at visible wavelengths, but significant energy occurs in the infrared and ultraviolet ranges. Relatively little energy is radiated by the sun at radio, X-ray, and gamma-ray wavelengths.

Because of the large amount of energy that we receive from the sun, much thought has been given in recent years to harnessing this radiant energy. Solar energy can be converted directly into electricity by means of solar cells (*see* PHOTOVOLTAIC CELL, SOLAR CELL). A large set of solar cells can provide enough electricity for the needs of an average residence.

Solar radiation can be used, directly or indirectly, for heating. Whenever the sun shines in your window, it heats your room to a certain extent (whether you like it or not). This is direct solar heating. Large black panels absorb the visible and short-wave infrared radiation from the sun, causing the panels to become hot; water can be heated by passing it through pipes in the panels. This is indirect solar heating. The current from photovoltaic cells can be passed through resistance wires to obtain heat indirectly.

The equatorial latitudes receive far more energy from the sun than the polar latitudes. This is true for two reasons. First, the sun shines more directly at lower latitudes than at higher latitudes. Second, places at higher latitudes generally have more cloud cover than places at lower latitudes. Solar energy is therefore more easily harnessed in a location such as Florida, as compared with a location such as Montana. Nevertheless, solar energy can be used in any part of the world to reduce consumption of conventional fuels. *See also* SOLAR POWER.

SOLAR FLARE

A solar flare is a violent storm on the surface of the sun. Solar flares can be seen with astronomical telescopes equipped with projecting devices to protect the eyes of observers. A solar flare appears as a bright spot on the solar disk, thousands of miles across and thousands of miles high. Solar flares also cause an increase in the level of radio noise that comes from the sun (*see* SOLAR FLUX).

Solar flares emit large quantities of high-speed atomic particles. These particles travel through space and arrive at the earth a few hours after the occurrence of the flare. Since the particles are charged, they are attracted toward the geomagnetic poles. Sometimes a goemagnetic disturbance results (*see* GEOMAGNETIC FIELD, GEOMAGNETIC STORM). Then we see the aurora at night, and experience a sudden, dramatic deterioration of ionospheric radio-propagation conditions. At some frequencies, communications can be completely cut off within a matter of a few seconds. Even wire communications circuits are sometimes affected.

Solar flares can occur at any time, but they seem to take place most often near the peak of the 11-year sunspot cycle. Scientists do not know exactly what causes solar flares, but they are evidently associated with sunspots, which are another type of solar storm. *See also* PROPAGATION CHARACTERISTICS, SUNSPOT, SUNSPOT CYCLE.

SOLAR FLUX

The amount of radio noise emitted by the sun is called the solar radio-noise flux, or simply the solar flux. The solar flux varies with frequency. However, at any frequency, the level of solar flux increases abruptly when a solar flare accurs (*see* SOLAR FLARE). This makes the solar flux useful for propagation forecasting: A sudden increase in the solar flux indicates that ionospheric propagation conditions will deteriorate within a few hours.

The solar flux is most often monitored at a wavelength of 10.7 cm, or a frequency of 2800 MHz. At this frequency, the troposphere and ionosphere have no effect on radio waves, making observation easy.

The 2800-MHz solar flux is correlated with the 11-year sunspot cycle. On the average, the solar flux is higher near the peak of the sunspot cycle, and lower near a sunspot minimum. *See also* SUNSPOT, SUNSPOT CYCLE.

SOLAR POWER

Electrical power can be obtained directly from sunlight by means of solar cells. Most modern solar cells can produce about 0.1 watt for each square inch of surface area exposed to bright sunlight. This is about 15 watts per square foot, or 150 watts per square meter, of photovoltaic-cell surface area. Large solar panels can be assembled, having hundreds of square feet of surface area, and producing thousands of watts of power in direct sunlight. This makes the sun a potential source of power for residences and businesses. Even motor vehicles can be powered by sunlight (see photographs).

There are basically two types of solar-power system: the stand-alone system and the interconnected system. A stand-alone system uses batteries to store solar energy during the hours of daylight; the energy is released by the batteries at night. An interconnected system operates in conjunction with the utility company; energy is sold to the utility during times of daylight and minimum usage, and is bought back at night or during times of heavy usage.

Solar cells produce direct current, and batteries store direct current. But the typical household appliance will not operate from direct current; 117-V alternating current, having a frequency close to 60 Hz, is required. All solar-power systems, therefore, must incorporate power inverters (*see* INVERTER). The inverter produces 117-V alternating current from the direct current of the solar cells and/or batteries. A typical stand-alone solar-power system is shown at A in the illustration; an interconnected system is shown at B.

Large solar panels are expensive; a typical stand-alone or interconnected solar-power system costs several thousand dollars. However, if conventional fossil fuels continue to increase in price and decline in supply, solar power will become an attractive alternative for more and more people. *See also* PHOTOVOLTAIC CELL, SOLAR CELL, SOLAR ENERGY.

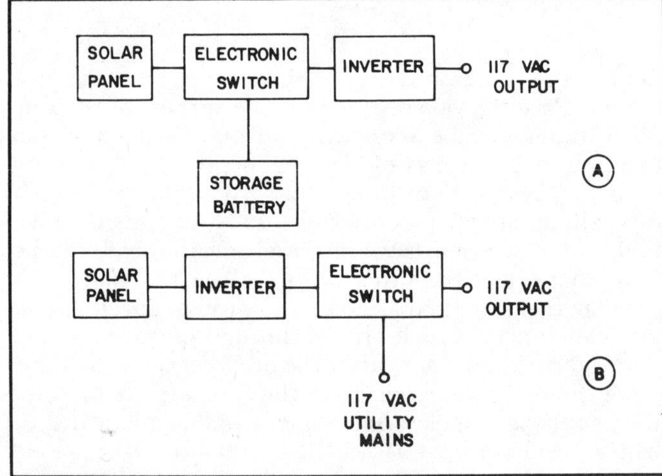

SOLAR POWER: Two types of solar-power systems. At A, a stand-alone system; at B, a system that interconnects with the utility lines.

SOLAR SYSTEM

Our sun has nine known planets orbiting within the influence of its gravitational field. The sun, the planets, and the moons of the planets comprise the solar system.

The four planets closest to the sun, and the most distant planet from the sun, are similar in composition to the earth and moon. They are known as the terrestrial planets. The terrestrial planets are Mercury, Venus, Earth, Mars, and Pluto. The other planets—Jupiter, Saturn, Uranus, and

A

B

SOLAR POWER: At A, a solar-powered "three-wheeler." At B, a solar-powered electric car.

SOLAR SYSTEM:
CHARACTERISTICS OF THE SUN AND PLANTES.

	Distance from sun, AU*	Mass, kg	Diameter, km	Year, Earth Days
Sun	0.00	2.0×10^{30}	1,400,000	-
Mercury	0.39	3.2×10^{23}	4,800	88
Venus	0.72	4.9×10^{24}	12,000	220
Earth	1.0	6.0×10^{24}	13,000	365
Mars	1.5	6.4×10^{23}	6,800	690
Jupiter	5.2	1.9×10^{27}	140,000	4,300
Saturn	9.5	5.7×10^{26}	120,000	11,000
Uranus	19	8.7×10^{25}	52,000	31,000
Neptune	30	1.0×10^{26}	48,000	60,000
Pluto	39	5.4×01^{24}	14,000?	91,00

*1 AU = 93,000,000 miles = 150,000,000 km.

Neptune—are much different from the terrestrial planets. They are known as the gas giants.

Various characteristics of the planets are illustrated in the table. It is evident that the solar system is extremely vast, compared to distances with which we are familiar. Astronomers generally use the astronomical unit, equivalent to the distance of the earth from the sun, to express distances in the solar system.

In the future, people may travel among the planets. Such interplanetary travel will result in certain communications problems because of the distances involved. For example, a radio signal requires about 4 minutes to get to Mars when that planet is at opposition (opposite the sun in our sky). When the planet is at conjunction (lined up with the sun), the propagation time increases to 20 minutes. *See also* SPACE COMMUNICATIONS, SPACE TRAVEL.

SOLDER

Solder is a metal alloy that is used for making electrical connections between conductors. There are several types of solder, intended for use with various metals and in various applications.

The most common variety of solder consists of tin and lead, with a rosin core. Some types of solder have an acid core. In electronic devices, rosin-core solder should be used; acid-core solder will result in rapid corrosion. Acid-core solder is more commonly used for bonding sheet metal.

Solder Type	Melting Point °F/°C	Principal Uses
Tin-lead 50:50, Rosin-core	430/220	Electronic Circuits
Tin-lead 60:40, Rosin-core	370/190	Electronic Circuits, Low-heat
Tin-lead 63:37, Rosin-core	360/180	Electronic Circuis, Low-heat
Tin-lead 50:50, Acid-core	430/220	Non-electronic metal bonding
Silver	600/320	High-current, High-heat

The ratio of tin to lead in a rosin-core solder determines the temperature at which the solder will melt. In general, the higher the ratio of tin to lead, the lower the melting temperature. For general soldering purposes, the 50:50 solder can be used. For heat-sensitive components, 60:40 solder is better because it melts at lower temperature. An ordinary soldering gun or iron can be used for applying and removing all types of tin-lead solder. *See* DESOLDERING TECHNIQUE, SOLDERING GUN, SOLDERING IRON, SOLDERING TECHNIQUE.

Tin-lead solders are suitable for use with most metals except aluminum. For soldering to aluminum, a special form of solder, called aluminum solder, is available. It melts at a much higher temperature than tin-lead solder, and requires the use of a blowtorch for application.

In high-current applications, a special form of solder, known as silver solder, is used, since it can withstand higher temperatures than tin-lead solder. A blowtorch is usually necessary for applying silver solder. Silver solder must be applied in a well-ventilated area because it produces hazardous fumes.

The table lists the most common types of solder, and their principal characteristics and applications.

SOLDERING GUN

A soldering gun is a quick-heating soldering instrument. It is called a gun because of its shape, which resembles that of a handgun (see photograph). A trigger-operated switch is pressed with the finger, allowing the element to heat up within a few seconds.

Soldering guns are convenient in the assembly and repair of some kinds of electronic equipment. Soldering guns are available in various wattage ratings for different electronic applications. The unit shown has two power levels: 100 watts (for electronic connections and components) and 140 watts (for such purposes as wire splicing).

Some electronic engineers and technicians prefer soldering irons to soldering guns. Soldering irons are generally easier to use with extremely miniaturized equipment, such as handheld and compact radio transceivers. Soldering guns are most often used in the assembly and repair of large, point-to-point wired apparatus such as power amplifiers. *See also* SOLDERING IRON, SOLDERING TECHNIQUE.

SOLDERING IRON

A soldering iron is a device that is used by engineers and technicians in the assembly and maintenance of electronic apparatus.

A typical soldering iron consists of a heating element and a handle. The soldering iron requires from about 1 minute to 5 minutes to warm up after it is turned on; the larger the iron, the longer the warm-up time. Soldering irons are available in a wide range of different wattages. The smallest irons are rated at only a few watts, and are used in highly miniaturized equipment; the largest irons draw hundreds of watts, and are used for such purposes as wire splicing and sheet-metal bonding.

Some electronic engineers and technicians prefer soldering guns to soldering irons. Soldering guns are often more convenient in situations where moderately high heat is needed, and where point-to-point wiring is used. The soldering gun heats up and cools down very quickly. *See also* SOLDERING GUN, SOLDERING TECHNIQUE.

SOLDERING GUN: This unit can be used for point-to-point wiring and splicing.

SOLDERING TECHNIQUE

Most electrical and electronic circuits have soldered connections. In the assembly of such equipment, or in component replacement, the proper soldering technique assures optimum performance with a minimal chance for connection failure.

Choosing Solder. In electronic-circuit work, a tin-lead solder of 50:50 or 60:40 ratio, having a rosin core, should be used. The 50:50 type is suitable for most wiring, but if heat-sensitive components are to be wired on a printed-circuit board, the 60:40 type will reduce the chance of heat damage (*see* SOLDER), since it melts at a lower temperature. Acid-core solder is not suitable for wiring electronic components, but can be used to bond sheet metal that will not readily adhere to rosin-core solder.

The Soldering Instrument. The proper soldering instrument for a given soldering application is, to some extent, a matter of choice. Soldering irons of low wattage (10 to 40 W) are generally best for printed-circuit wiring or highly miniaturized apparatus. For point-to-point electronic wiring, or for splicing of small wires, a soldering gun or iron can be used; the power rating should be between about 50 and 150 W. For heavy-gauge wire splicing or sheet-metal bonding, a high-wattage iron (200 W or more) is best.

In all cases, it is essential that the soldering instrument provide sufficient heat to prevent "cold" solder joints. However, too much heat can result in component damage (*see* SOLDERING GUN, SOLDERING IRON).

Soldering on Printed-Circuit Boards. Most printed-circuit soldering is done from the non-component (foil) side of the printed-circuit board. The component lead is inserted through the appropriate hole, and the soldering iron is placed so that it heats both the foil and the component lead (A in illustration). If the component is heat-sensitive, a needle-nosed pliers should be used to grip the lead on the component (nonfoil) side of the board while heat is applied. The solder is allowed to flow onto the foil and the component lead after the joint becomes hot enough to melt the solder; heating normally takes only 1 to 3 seconds. The solder should completely cover the foil dot or square in which the component lead is centered. Excessive solder should not be used. After the joint has cooled, the component lead should be snipped off, flush with the solder, using a diagonal cutter.

If the circuit board is double-sided (foil on both sides, rather than on just one side), it will usually have plated-through holes, so that the soldering procedure described above will be adequate. However, if the holes are not plated-through, solder must be applied, as described above, to the foil and component lead on either side of the circuit board.

Some printed-circuit components are mounted on the foil side of the board. This is the case, for example, with flat-pack integrated circuits. In such cases, the component lead and the circuit board foil are first coated, or "tinned,"

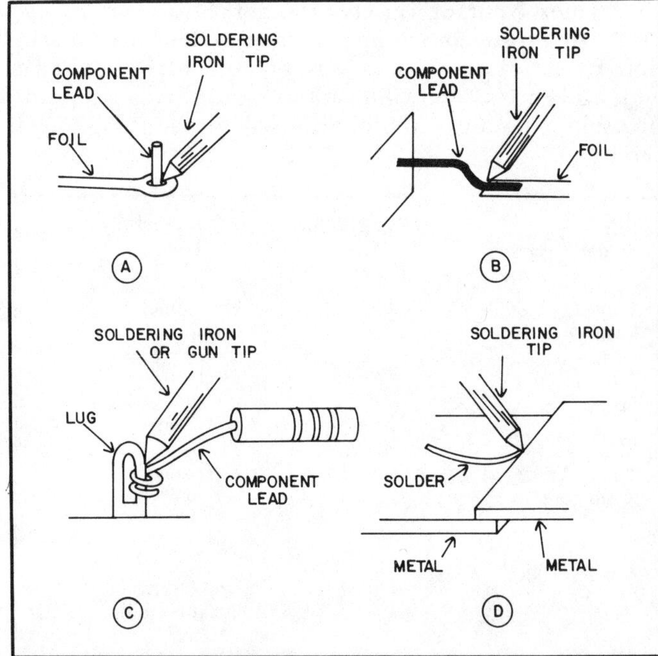

SOLDERING TECHNIQUE: Soldering techniques for printed-circuit wiring (A and B), point-to-point wiring (C), and sheet-metal bonding (D).

with a thin layer of solder. The component lead is then placed flat against the foil, and the iron is placed in contact with the component lead as at B. The heat melts the solder by conduction.

Soldering a Point-to-Point Connection. Most point-to-point wiring is accomplished by means of tie strips, where one or more wires terminate (*see* TIE STRIP).

In tie-strip wiring, the lug should first be coated with a thin layer of solder. The actual soldering of the connection should not be carried out until all of the wires have been attached to the lug. Wires are wrapped two or three times around the lug using a needle-nosed pliers. Excess wire is cut off using a diagonal cutter. When all wires have been attached to the lug, the soldering instrument is held against each wire coil, one at a time, and the connection is allowed to heat up until the solder flows freely in between the wire turns each time, adhering to both the wire and the lug as at C. Sufficient solder should be used so that the connection is completely coated. However, solder should not be allowed to ball up or drip from the connection.

If a heat-sensitive component is wired to a tie strip, a needle-nosed pliers should be used to conduct heat away from the component lead as long as heat is applied to the connection.

Splicing Wires. In temporary or permanent wire splicing, the wires should be twisted together and soldered according to the procedure described under WIRE SPLICING.

Bonding Sheet Metal. When soldering sheet metal, rosin-core solder should be used if possible, but sometimes the rosin does not allow a good enough mechanical bond. Then, acid-core solder can be used. Special solder is available for use with aluminum, which does not readily adhere to most other types of solder.

The edges to be bonded should be sanded with a fine emery paper, and cleaned with a non-corrosive, grease-free solvent such as rubbing alcohol. A high-wattage iron or blowtorch should be used to heat the metal while both sheets are "tinned" with a thin layer of solder. Then the sheets should be secured in place. Sufficient heat should be applied so that the solder will flow freely. Some additional solder should be applied on each side of the bond, working gradually along the length of the bond from one side to the other (see D in illustration). The bond will require some time to cool, and it should be kept free from stress until it has cooled completely. Water or other fluids should not be used in an attempt to hasten the cooling process.

"Cold" Solder Joints. If sufficient heat has not been applied to a solder connection, a "cold" joint may result. A properly soldered connection will have a shiny, clean appearance. A "cold" joint looks dull or rough. Many equipment failures occur simply because of "cold" solder joints, which exhibit high resistance.

If a "cold" joint is found, the solder should be removed as much as possible, using a wire braid (solder wick). Then the connection should be resoldered.

Removing Solder. *See* DESOLDERING TECHNIQUE, SOLDER WICK.

SOLDER WICK

Solder wick is a specially treated wire braid that is used to draw solder away from a connection for desoldering purposes. Solder wick appears somewhat similar to braided grounding wire, except that solder wick is smaller in size.

Solder wick is commercially available in easy-to-use rolls, and in various different gauges for different purposes. An example is the product shown in the photograph. *See also* DESOLDERING TECHNIQUE.

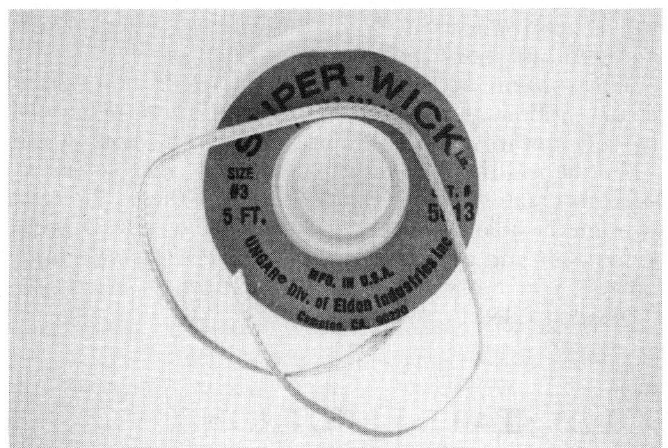

SOLDER WICK: Solder wick typically comes on a small, plastic spool like the one shown here.

SOLENOID

A solenoid is a single-layer or multi-layer, helically wound coil or wire, usually having a movable ferromagnetic core. A helically wound choke or inductor is generally called a solenoidal inductor, while the term solenoid is more often used to describe the electromagnet in a relay, ringer, or other electromechanical device.

An example of the use of a solenoid is shown in the illustration. This type of device is used in many businesses as a telephone ringer. The coil consists of several layers of wire around an iron core with a cylindrical opening. An

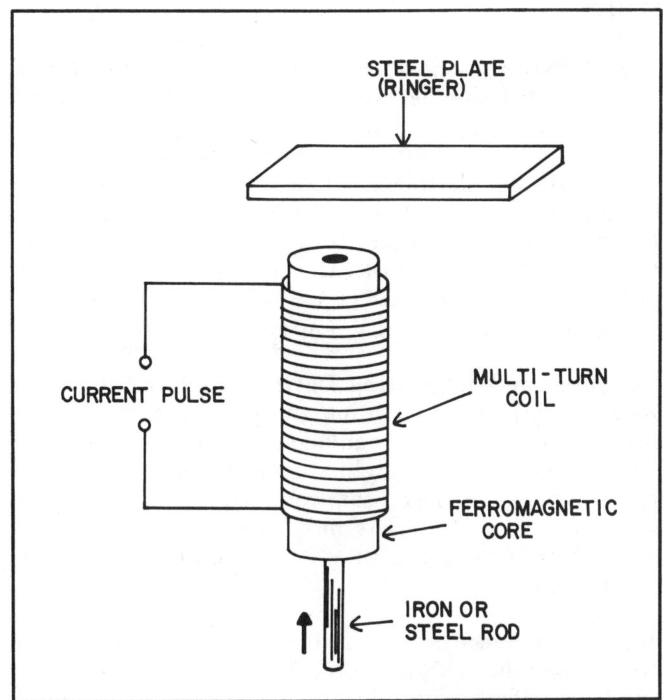

SOLENOID: Construction of a solenoid type telephone ringer.

iron or steel rod rests just below the coil core. A steel plate is mounted just above the coil.

The iron core becomes highly magnetized when a pulse of current flows through the coil. The magnetic field pulls the rod upward, accelerating it through the hole in the core. The rod then strikes the metal plate with sufficient force to create a loud "ding!" The rod then falls back through the hole following the current pulse. This process occurs over and over at intervals of several seconds until somebody answers the telephone. *See also* COIL, ELECTROMAGNET, INDUCTOR, RELAY.

SOLID-STATE ELECTRONICS

During the 1960s, the transistor came into widespread use as an active device in amplifiers, oscillators, and mixers. The semiconductor diode replaced the diode tube for rectification and envelope detection. Since that time, many different kinds of semiconductor devices have been developed. All of these devices are called solid-state components. The science involved with the development and application of these components, and the design of circuits using them, is known as solid-state electronics.

Solid-state components get their name from the fact that the flow of charge carriers is entirely confined to solid substances. A vacuum tube is not a solid state device, since electrons flow through free space. A transistor, though, operates via charge carriers that flow inside semiconductor material. There are two types of charge carriers in a solid-state device: the negatively charged electron and the positively charged hole (*see* ELECTRON, HOLE).

The Diode. The solid-state diode consists of a junction between an N-type semiconductor and a P-type semiconductor, or a semiconductor and a wire. In either type of junction, current flows readily in one direction, provided the voltage is large enough (about 0.3 to 0.6 V). In the other direction, almost no current flows (*see* N-TYPE SEMICONDUCTOR, P-N JUNCTION, P-TYPE SEMICONDUCTOR).

Many types of solid-state diodes are available for various different purposes. *See also* DETECTION, DIODE, DIODE ACTION, DIODE CAPACITANCE, DIODE CLIPPING, DIODE DETECTOR, DIODE FEEDBACK RECTIFIER, DIODE FIELD-STRENGTH METER, DIODE MATRIX, DIODE MIXER, DIODE OSCILLATOR, DIODE-TRANSISTOR LOGIC, DIODE TYPES, DOUBLE BALANCED MIXER, GUNN DIODE, LIGHT-EMITTING DIODE, MIXER, PHOTOVOLTAIC CELL, PIN DIODE, POINT-CONTACT JUNCTION, RECTIFICATION, RECTIFIER DIODE, SINGLE-BALANCED MIXER, VARACTOR DIODE, ZENER DIODE.

Switching and Regulating Devices. Bipolar transistors, zener diodes, and certain integrated circuits can be used for current and voltage regulation (*see* SERIES REGULATOR, SHUNT REGULATOR). Light dimmers and motor-speed controls use four-layer semiconductor devices such as silicon-controlled rectifiers and thyristors (*see* FOUR-LAYER SEMICONDUCTOR, SILICON-CONTROLLED RECTIFIER, THYRISTOR).

Bipolar Transistors. Bipolar transistors are found in two basic types: the NPN and PNP. Bipolar transistors can be used as ocillators, amplifiers, detectors, mixers, modulators, and switching and logic devices. They behave in a manner similar to vacuum tubes, except that they require much less power for operation, are physically smaller, and amplify current rather than voltage. *See also* BIPOLAR TRANSISTOR, NPN TRANSISTOR, PNP TRANSISTOR.

Field-Effect Transistors. Field-effect transistors are made of either N-type and P-type material, or of certain metal oxides that behave in a similar manner. New, improved types of field-effect transistors are constantly being developed. They are generally used in oscillators and low-noise amplifiers. *See also* FIELD-EFFECT TRANSISTOR, GALLIUM ARSENIDE FIELD-EFFECT TRANSISTOR, METAL-OXIDE-SEMICONDUCTOR FIELD-EFFECT TRANSISTOR, VERTICAL METAL-OXIDE-SEMICONDUCTOR FIELD-EFFECT TRANSISTOR.

Integrated Circuits. The variety of currently available integrated circuits is tremendous. An integrated-circuit catalog should be consulted to obtain details concerning specific integrated circuits. Some of the functions of integrated circuits are amplification, counting, logic operations, oscillation, and switching. *See* INTEGRATED CIRCUIT.

Other Solid-State Devices. Semiconductor devices are not the only types of solid-state components. Resistors, capacitors, inductors, and other common components are also solid-state. An entire circuit is considered solid-state only if all of its components are solid-state components. (An exception is sometimes made for devices containing a picture tube or camera tube.)

Advantages of Solid-state Technology. Solid-state semiconductor components generally require far less power for their operation than did the older vacuum tubes. Solid-state components have no heaters or filaments, and are also physically smaller than vacuum tubes. A single integrated circuit, smaller in size than a dime, can house the equivalent of a whole building full of vacuum-tube circuitry. Solid-state equipment is vastly more reliable than vacuum-tube equipment. Solid-state electronics has given us things that we now take for granted—such as battery-powered television sets, handheld calculators, and home computers—that were unheard of just a few decades ago.

Further Information. Solid-state devices are used in so many different applications, and are found in so many different forms, that the science of solid-state technology can only be touched upon here. For detailed information, please consult the appropriate article according to the device, circuit, or subject. A textbook on solid-state technology is recommended for a complete and thorough discussion of this science.

SONAR

SONAR is a device that is used for echo ranging, or for depth finding in bodies of water. A SONAR system uses sound waves, reflected from the bottom, to determine the depth. Sonar is an abbreviation for Sound Navigation And Ranging.

A SONAR system consists of a sound-pulse transmitter, a transmitting transducer (underwater speaker), a receiving transducer (underwater microphone), a receiver, and a

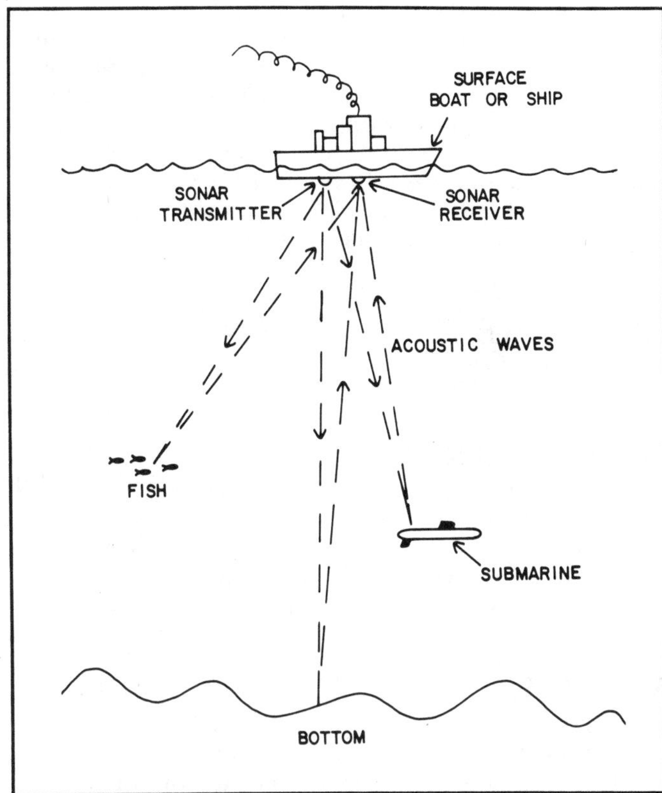

SONAR: SONAR can be used to determine the depth of a body of water, or the presence of objects in the water such as fish or a submarine.

delay timer. Acoustic waves, sent out by the transmitter, are picked up by the receiver after being reflected from the bottom or from some object in the water (see illustration). The depth of the water, or the distance to the object, can be determined from the time the sound pulse takes to return to the receiver.

A SONAR system must be calibrated according to the amount of salt in the water. Sometimes layers of water having different salinity or temperature result in false readings on a SONAR system. Experienced SONAR operators can differentiate between these false readings and the true indication.

Some SONAR devices are equipped with pen recorders that make a continuous plot of the contour of the bottom. Other devices simply provide a readout of the depth in feet, meters, or fathoms (a fathom is 6 feet). SONAR has been used to map the ocean floor, as well as the bottoms of lakes and rivers. SONAR can also be employed to locate submarines. Fishermen have used SONAR to locate large schools of fish.

SONAR devices operate on the premise that the speed of sound in fresh water is 4600 feet, or 770 fathoms, or 1400 meters, per second. In salt water, the speed of sound is approximately 4900 feet, or 820 fathoms, or 1500 meters, per second.

A device similar to SONAR is used aboard some aircraft to measure altitude, to locate other aircraft, and to detect clear-air turbulence. *See also* SONIC ALTIMETER.

SONE

The sone is a unit of perceived loudness of sound. A sound of 1 sone is defined as a pure sine-wave tone at a frequency of 1 kHz, at a level 40 dB above the threshold of hearing. Sounds louder or softer than 1 sone are determined subjectively by the listener. Since people generally perceive loudness on a logarithmic scale, a sound of 2 sones for a 1-kHz tone is approximately 43 dB above the threshold of hearing, a sound of 0.5 sones is 37 dB above the threshold of hearing, and so on.

For most people, the optimum hearing sensitivity is at a frequency of approximately 1 kHz. Therefore, at higher or lower frequencies, more audio power is required to produce a sound level of 1 sone. A hearing test is needed to obtain a precise determination of the audio power required to produce a perceived sound of 1 sone over the complete hearing range. Every individual is unique in this respect. *See also* DECIBEL, LOUDNESS.

SONIC ALTIMETER

A sonic altimeter is a device used for measuring the altitude of an aircraft. The sonic altimeter works according to the same principle as SONAR.

The functioning of a sonic altimeter is affected by many factors. The situation is more complicated than that for SONAR. The air speed of the aircraft, the barometric pressure, and the wind directions at various altitudes all affect the reading obtained with a sonic altimeter. The boundary, or front, between two air masses that differ in temperature can produce false readings. Other aircraft can reflect the sound sent out by the device, resulting in interference.

A sonic altimeter can be used for purposes other than the measurement of altitude. Since sound is reflected by other aircraft, the sonic altimeter can be employed to determine the approximate range (distance to) such aircraft. The speed of the aircraft can be found by measuring the Doppler shift of the reflected sound. A sonic altimeter can also be used to locate areas of potential clear-air turbulence.

Most sonic altimeters operate according to the premise that the speed of sound is 1,100 feet, or 335 meters, per second. *See also* SONAR.

SONOPTOGRAPHY

See ULTRASONIC HOLOGRAPHY.

SORTING

Data is sometimes organized in categories, or groups, based on some particular characteristic. This process is called sorting.

A familiar example of the sorting process is the arrangement of terms according to the first letter (A through Z). This process results in 26 different categories, assuming there is at least one term beginning with each letter.

Computers can be instructed to sort data according to any desired characteristic. When a computer is programmed to sort, it is called a sorter. Sorters are extensively used in data processing by businesses, scientists, and engineers. Sorting devices are also used by the U.S. Post Office to speed mail service. These sorters operate according to zip code.

SOS

See DISTRESS SIGNAL.

SOUND: Speed of sound through varios substances.

Medium	Speed of sound, Feet per second	Speed of sound, Meters per second
Fresh water	4600	1400
Salt water	4900	1500
Petroleum	4200	1300
Turpentine	4600	1400
Aluminum	17000	5200
Copper	12000	3600
Silver	12000	3600
Gold	6600	2000
Lead	4100	1290
Tin	8900	2700
Wood	3300-16000	1000-5000
Glass	9800-20000	3000-6000

SOUND

Sound is an acoustic disturbance that can be heard by the average person. The frequency of sound waves ranges from about 16 or 20 Hz to 20 kHz; this corresponds to a wavelength range of 55 feet to ⅝ inch, or 17 meters to 1.7 centimeters. Acoustic disturbances at frequencies below 20 Hz are called infrasound. Disturbances at frequencies above 20 kHz are known as ultrasound (*see* INFRASOUND, ULTRASOUND).

Sound travels through the air by compression of the molecules. The molecules themselves move back and forth parallel to the direction of sound propagation. Thus, sound waves in air are longitudinal disturbances (*see* LONGITUDINAL WAVE). Sound waves expand outward from a source in spherical wavefronts.

Sound can consist of a single wave disturbance at a single frequency. However, sound is more often made up of disturbances at many frequencies at the same time. The waveforms of most sounds are extremely complex.

Sound propagates differently in liquids and solids, such as water or metal, than in air. Water cannot be compressed, so sound waves propagate by means of lateral motion of the molecules, in a manner similar to wind waves on the surface of a lake. In water, sound is a transverse wave. Sound is also a transverse wave in most solid substances (*see* TRANSVERSE WAVE).

The speed of sound in air depends on the temperature and also on the pressure. In dry air at room temperature at a pressure of 1 atmosphere (30 inches of mercury), sound travels at about 1,100 feet, or 335 meters, per second. In other mixtures of gases, even at the same pressure, the speed of sound is different than it is in air, which is a mixture of about 78 percent nitrogen, 21 percent oxygen, and 1 percent other gases.

In liquids and solids, the speed of sound is generally greater than in air. The table gives the speed of sound in various common liquids and solids.

The wavelength λ of a sound wave is a function of the frequency f and the velocity v. The general relation is:

$$\lambda = v/f$$

The frequency f is always specified in hertz. If λ is given in feet, then v must be given in feet per second; if λ is to be specified in meters, then v must be given in meters per second.

The intensity of sound is usually specified in decibels above the threshold of hearing. Sound loudness may also be expressed in units called sones, or in terms of the pressure it exerts (*see* DECIBEL, LOUDNESS, SONE, SOUND PRESSURE LEVEL, THRESHOLD OF HEARING).

Some sounds are pleasing or soothing to a listener, while others are irritating or even infuriating. For example, most people like the sound of waves on a beach, which is a form of "pink noise." Most people also like certain kinds of music. But the sounds from such things as jackhammers, bulldozers, and buzz saws, heard over an extended period, can cause distcration and irritation. *See also* DISSONANCE, HARMONY, MUSIC, PINK NOISE, WHITE NOISE.

SOUND ARTICULATION

See ARTICULATION.

SOUND BAR

When an amplitude-modulation or single-sideband signal interferes with a television signal, sound bars may occur.

Sound bars appear as horizontal bands of light and dark across the picture when a television signal is received. *See also* TELEVISION INTERFERENCE.

SOURCE

The term source is used to refer to the emitting electrode of a field-effect transistor. The output of a field-effect-transistor amplifier is sometimes taken from the source. The input signal may be applied to the source. The source of a field-effect transistor corresponds to the cathode of a tube or the emitter of a transistor (*see* FIELD-EFFECT TRANSISTOR).

The originating generator of a signal is called the source of the signal. We may, for example, speak of a radio-frequency source, a sound source, or a light source.

SOURCE COUPLING

When the output of a field-effect-transistor amplifier or oscillator is taken from the source circuit, or when the input to a field-effect-transistor amplifier is applied in series with the source, the amplifier or oscillator is said to employ source coupling. Source coupling may be capacitive, or it may make use of transformers. The illustration shows an example of source coupling in a two-stage amplifier. Capacitive coupling is employed in this case. The output of the first stage is taken from the source.

Source coupling generally results in a low input or output impedance, depending on whether the coupling is in the input or output of the amplifier stage. Source coupling is often used with grounded-gate amplifiers. If the following stage needs a low driving impedance, source coupling can be used; this is called a source follower. *See also* SOURCE FOLLOWER.

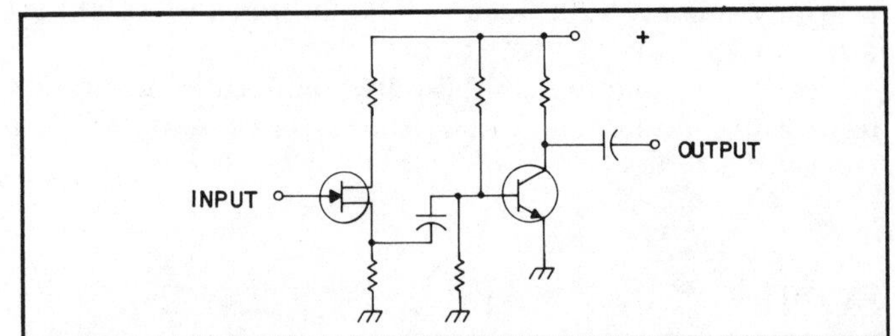

SOURCE COUPLING: An example of source coupling in a two-stage amplifier, in which the first stage uses a field-effect transistor.

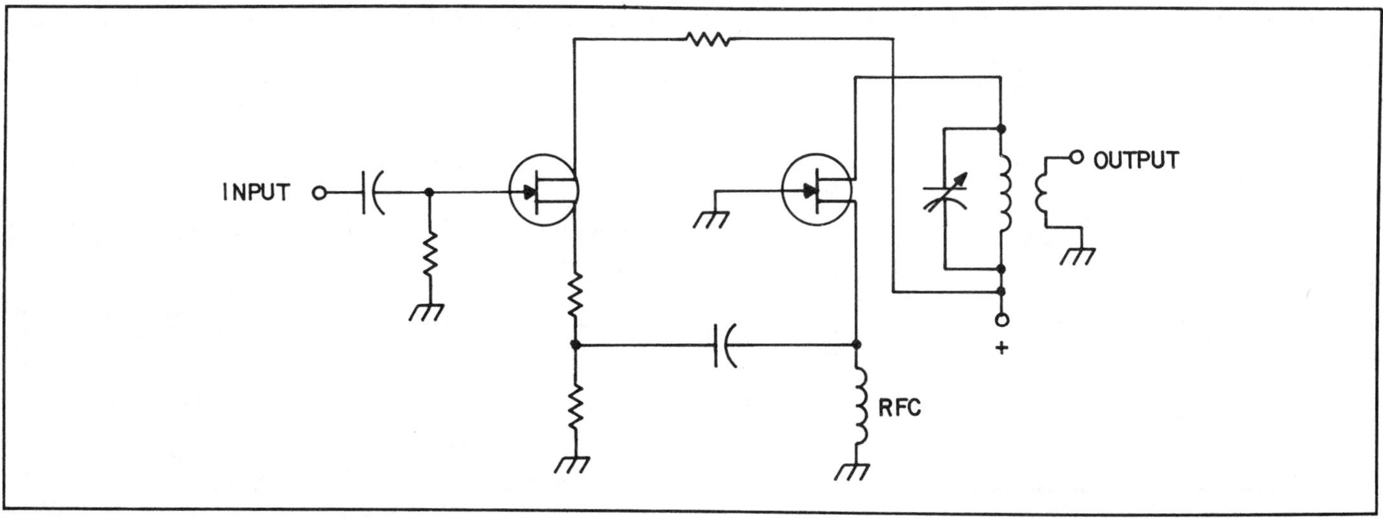

SOURCE FOLLOWER: A source-follower circuit provides an impedance match for driving a low-impedance amplifier or load with the output from a field-effect-transistor amplifier.

SOURCE FOLLOWER

A source follower is an amplifier circuit in which the output is taken from the source circuit of a field-effect transistor. The output impedance of the source follower is low. The voltage gain is always less than 1; in other words, the output signal voltage is smaller than the input signal voltage. The source-follower ciruit is generally used for impedance matching. The amplifier components are often less expensive, and offer greater bandwidth, than transformers.

The diagram illustrates a two-stage field-effect-transistor amplifier circuit in which the first stage is a source follower. The output of the source follower is in phase with the input. The impedance of the output circuit depends on the particular characteristics of the field-effect transistor used, and also on the value of the source resistor. Sometimes, two series-connected source resistors are used. In this case, the output is taken from between the two resistors. A transformer output may also be employed. Source-follower circuits are useful because they offer wideband impedance matching at low cost. *See also* SOURCE COUPLING.

SOURCE KEYING

In a field-effect-transistor oscillator or amplifier, source keying is accomplished by placing the key in series with the source. This is done for the purpose of obtaining code transmission. Source keying is analogous to emitter keying in a bipolar-transistor circuit (*see* EMITTER KEYING).

In a source-keyed oscillator, the key-up condition cuts off the power and stops the oscillation. In a source-keyed amplifier, only a negligible amount of signal leaks through when the key is up.

The drawing illustrates source keying in a radio-frequency amplifier circuit. Source keying is the most common method of keying a field-effect-transistor oscillator or amplifier. Source keying is always done either in the low-level amplifier stages of a transmitter, or in the local-oscillator circuit. Source keying is sometimes done at more than one stage simultaneously.

A shaping circuit, shown in the schematic as a series resistor and a parallel capacitor, slows down the source voltage drop when the key is closed. This allows a controlled signal rise time. When the key is released, the capacitor discharges through the resistor, slowing down the decay time of the signal. The rsistor and capacitor values are chosen for the desired rise and decay times in the range of keying speeds to be used. *See also* KEY CLICK, SHAPING.

SOURCE MODULATION

Source modulation is a method of obtaining amplitude

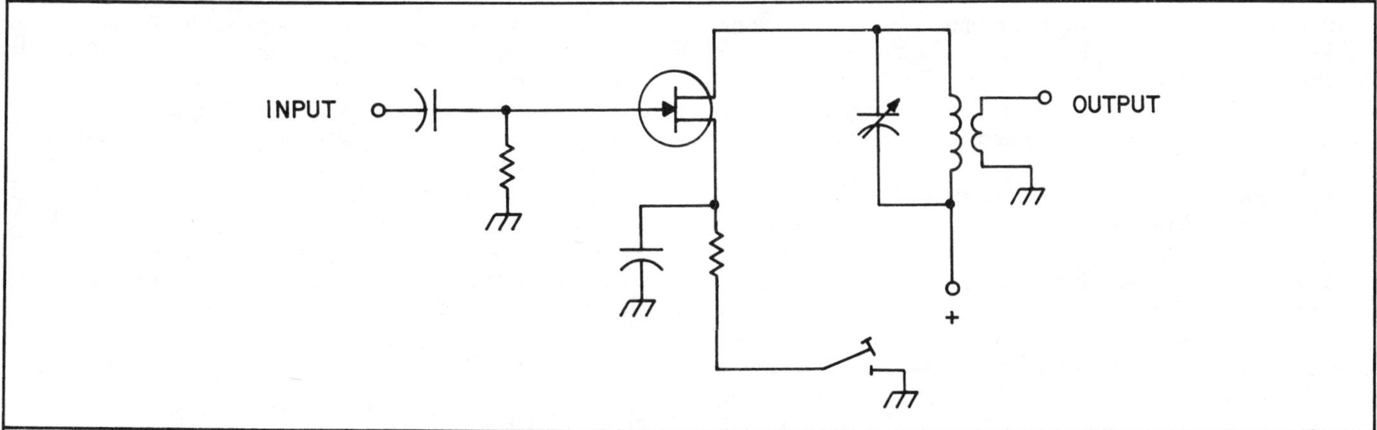

SOURCE KEYING: An example of source keying in a field-effect- transistor amplifier.

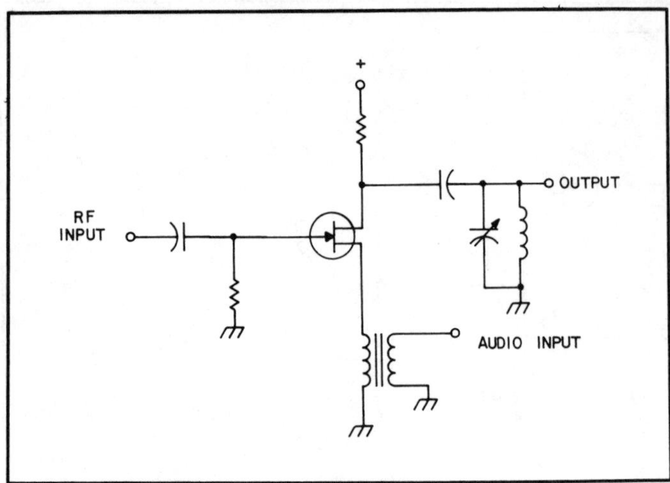

SOURCE MODULATION: An example of source modulation in a field-effect-transistor circuit.

modulation in a radio-frequency amplifier. The carrier is applied to the gate of the device in most cases. Sometimes the carrier is applied to the source. The audio signal is applied in series with the source; this can be done across a resistor, or by means of an audio-frequency transformer (see illustration). Source modulation is the counterpart of cathode modulation in a vacuum-tube circuit, or emitter modulation in a bipolar-transistor amplifier (*see* CATHODE MODULATION, EMITTER MODULATION).

As the audio-frequency signal at the source swings negative, the instantaneous radio-frequency output voltage riss if an N-channel field-effect transistor is used. If a P-channel device is used, the instantaneous signal output voltage ncreases when the audio signal swings positive. Under conditions of 100-percent modulation, the instantaneous amplitude just drops to zero at the negative radio-frequency signal peaks.

Source modulation requires very little audio power for 100-percent amplitude modulation, provided that the modulation is done at a low-level stage and not in the final amplifier circuit of the transmitter. *See also* AMPLITUDE MODULATION, MODULATION.

SOURCE RESISTANCE

The source resistance of a field-effect transistor is the effective resistance of the source in a given circuit. The source resistance depends on the bias voltages at the gate and the drain. It also depends on the input-signal level and on the characteristics of the particular field-effect transistor used. The source resistance can be controlled, to a certain extent, by inserting a resistor in series with the source lead.

The external resistor in the source circuit of a field-effect-transistor oscillator or amplifier is usually called the source resistor, although its value is sometimes called the source resistance. A source resistor can be used for impedance-matching purposes, for biasing, for current limiting, or for stabilization. The source resistor often has a capacitor connected across it to bypass radio-frequency energy to ground. *See also* FIELD-EFFECT TRANSISTOR, SOURCE FOLLOWER, SOURCE STABILIZATION, SOURCE VOLTAGE.

SOURCE STABILIZATION

Source stabilization is a method of providing bias control in a power-FET circuit using two or more FETs in parallel.

A resistor, having a value that depends on the circuit application and input impedance, is connected in series with the source of each FET. This reduces the effects of minor differences in the characteristics of the FETs. Bypass capacitors may be connected across each resistor if it is necessary to keep the sources at signal ground.

If the channel current increases because of a temperature rise in an FET, and there is no resistor in series with the source lead, the change in current will reduce the gain of the amplifier, and thereby reduce its efficiency. If a source-stabilization resistor is connected in series with the current path, however, any increase in the current will cause an increase in the voltage drop across the resistor. This will change the bias in such a way as to stabilize the current through the channel.

Stabilization is more important in bipolar-transistor circuits than in FET circuits. In fact, stabilization is used almost universally in bipolar-transistor amplifiers and oscillators, which in the case of the FET, stabilization is usually necessary only in power amplifiers. *See also* COMMON CATHODE/EMITTER/SOURCE, EMITTER STABILIZATION.

SOURCE VOLTAGE

In a field-effect-transistor circuit, the source voltage is the direct-current potential difference between the source and ground. If the source is connected directly to chassis ground, or is grounded through an inductor, then the source voltage is zero. But if a series resistor is used for stabilization, as is the case in some common-source power amplifiers, the source voltage is positive in an N-channel device and negative in a P-channel device.

If I is the channel current under no-signal conditions and R is the value of the source resistor (assuming there is a source resistor), then the source voltage V is obtained by Ohm's law as:

$$V = IR$$

A capacitor, placed across the source resistor, keeps the source voltage constant under conditions of variable input signal. If no such capacitor is used, the instantaneous source voltage will vary along with the input signal because of changes in the instantaneous current through the channel. *See* SOURCE STABILIZATION.

In common-gate and common-drain circuits, it is usually necessary to provide a voltage at the source. In a common-gate amplifier, the source voltage is normally positive in the N-channel case and negative in the P-channel case. In a common-drain configuration, the source voltage must be negative if an N-channel device is used, and positive if a P-channel device is employed. In these situations, the source voltage is provided directly by the power supply. *See also* COMMON CATHODE/EMITTER/SOURCE, COMMON GRID/BASE/GATE, COMMON PLATE/COLLECTOR/DRAIN, FIELD-EFFECT TRANSISTOR.

SO-239

See UHF CONNECTOR.

SPACE

See MARK/SPACE.

SPACE ATTENUATION

See FREE-SPACE LOSS.

SPACE CHARGE

In a vacuum tube, the cathode initially emits more electrons than the other electrodes drain off. This results in a negatively charged cloud of electrons in the space among the electrodes. The total negative charge of these electrons is called the space charge.

The space charge in a vacuum tube depends on the voltages at the cathode, grids, and plate. In general, the greater the potential difference between the cathode and the plate, the greater the space charge will be. The grid voltages have a lesser effect on the space charge. *See also* ELECTRON, TUBE.

In a solid-state device, the depletion region near the P-N junction is often called the space-charge region (*see* P-N JUNCTION).

When electrically charged particles or objects are scattered in space, there exists a certain electric charge per unit volume. This charge concentration is sometimes called space charge. It is expressed in coulombs per cubic meter. *See also* CHARGE, COULOMB.

SPACE COMMUNICATIONS

Until quite recently, all long-distance radio communications took place via ionospheric or tropospheric propagation. If conditions were unfavorable (and they often were), a communications circuit became unreliable, and perhaps even unusable (*see* PROPAGATION CHARACTERISTICS). We no longer have this problem, because of the advent of space communications.

Space communications is, in general, the transmission and reception of signals over paths that lie partially or entirely above the atmosphere of our planet. Space communications can be categorized in three ways: earth-to-earth, earth-to-space, and space-to-space.

Earth-to-Earth Communications. Earth-to-space communications has actually been possible since the early days of radio. Radio amateurs bounced signals off the moon when it was discovered that radio waves at very high frequencies would travel through the ionosphere unaffected. Some radio amateurs still converse this way today (*see* MOONBOUNCE).

Since the first satellite was launched into space in the late 1950s, satellite communications has become increasingly widespread. The earliest artificial communications satellites were passive devices, launched in the 1960s and known as the Echo satellites. Since then, active repeaters have been placed aboard satellites. Today, satellites provide constant communications capability among almost all points on the globe (*see* ACTIVE COMMUNICATIONS SATELLITE, SATELLITE COMMUNICATIONS).

Earth-to-Space and Space-to-Space Communications. As we make our first tentative ventures to the other planets—and perhaps someday even to other stars or galaxies—earth-to-space and space-to-space communications will become more and more important. There will be certain difficulties to overcome.

The first, and most obvious, difficulty arises from the sheer distances involved. The planet Pluto is about 4,000,000,000 miles, or 6,000,000,000 kilometers, from the earth. The nearest star is over 6,000 times further away than Pluto! High-power transmitters and extremely directional antenna systems will be necessary to carry out communications over such great distances.

When the Apollo astronauts visited the moon, their conversations with earth-based personnel were complicated by the fact that the earth and the moon are separated by about 1.34 light seconds. This caused a delay after every transmission before a reply was received. This 2.68-second round-trip propagation delay might have been bothersome, but it did not seriously affect communications. But when astronauts go farther into space, two-way conversations will become difficult, and ultimately impossible, because the speed of radio signals is only 186,282 miles (299,792 kilometers) per second. This problem has already beeen experienced, in a sense, by technicians controlling the distant space probes. Pluto is about 6 light hours away; the nearest star is over 4 light years from us.

If space vessels achieve very high speeds, Doppler shifts will cause changes not only in the frequency of transmission, but also in the rate at which the modulation is received. At speeds approaching the speed of light, relativistic effects will also become significant (*see* DOPPLER EFFECT, RELATIVISTIC SPEED). The frequency and data-transmission rate may be altered so greatly that special equipment is needed to receive and demodulate the signals.

The Future of Space Communications. Some scientists believe that civilizations might exist on planets in orbit around other stars, and that some of these civilizations might attempt to transmit radio signals into space. Attempts have already been made to receive such signals. Dr. Frank Drake has even made a transmission of his own (*see* RADAR TELESCOPE, RADIO ASTRONOMY, RADIO TELESCOPE).

Satellite-communications technology can be expected to improve continually in the coming years. Someday it will be possible for two people on opposite sides of the world to converse via satellite, using handheld radios, at any time of the day or night.

There may be as-yet undiscovered modes of communication that will prove useful for space travelers. Some of these modes might eliminate the time-lag, Doppler, and relativistic problems associated with long-distance communications. *See also* SPACE TRAVEL.

SPACE-DIVERSITY RECEPTION

See DIVERSITY RECEPTION.

SPACER

A spacer is an insulator that is used for keeping the conductors of a transmission line separated. Spacers are used in air-dielectric coaxial and open-wire lines to maintain the proper conductor separation. In a coaxial line, spacers are

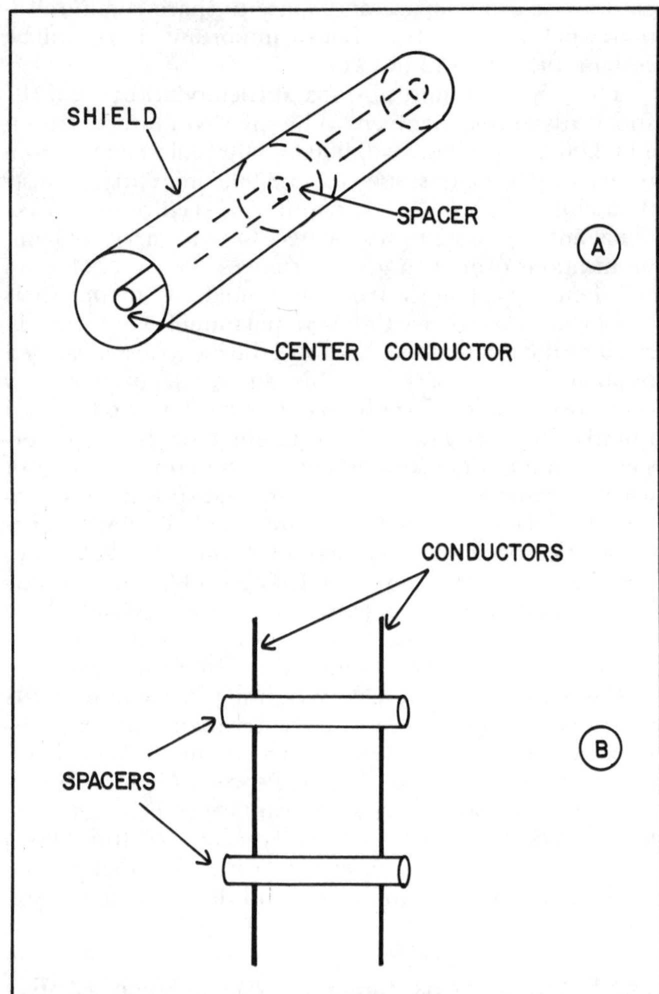

SPACER: At A, a spacer in an air-dielectric coaxial cable. At B, spacers in an open-wire transmission line.

shaped like disks or flattened spheres (see A in illustration). In an open-wire line, the spacers are rod-shaped as at B.

Spacers are positioned at regular intervals along a feed line. The spacers have a small effect on the dielectric constant of the medium between the conductors; this tends to lower the characteristic impedance of the line by a few percent. The more spacers that are used, the greater this effect becomes.

Spacers should be made of a low-loss, waterproof dielectric, such as plastic or glass. When making an open-wire transmission line, the smallest possible number of spacers should be used. *See also* COAXIAL CABLE, OPEN-WIRE LINE.

SPACE TRAVEL

For centuries, people have dreamed of traveling to the moon, to other planets, and to other star systems. Space travel has just recently become reality. Men have been to the moon and back several times, and have orbited the earth literally thousands of times.

The distances to the planets are far greater than the distance to the moon. It takes about 3 days to get to the moon using conventional rocket-powered space vehicles; a trip to Mars would require several months, and a journey to any of the outer planets would take years. Communications over such vast distances would become complicated because of time lag. The technology does exist, however, for interplanetary travel. Unmanned probes have been sent as far as Uranus.

The stars are immensely farther away than the planets. The nearest star, Proxima Centauri, is more than 4 light years from our sun. If the earth's orbit around the sun were reduced in scale to the size of a nickel, Proxima Centauri would be about 2 miles away. This has prompted some scientists to say that we will never reach the stars.

Space travel among the stars would require speeds so great that relativistic time dilation would occur. The technology for attaining these speeds is not yet available to us. Various scientists have devised space-vehicle designs that might make relativistic speeds possible.

An acceleration of 1 gravity—just 32 feet per second per second—would result in near-light speeds within only a few months. Space travelers would age slowly when traveling at these speeds, and a journey around the entire known universe could be completed within a single lifetime. The earth, however, would age billions of years during that time. The aspiring star traveler will have to contend with psychological, as well as physical, problems.

The prospects, challenges, and physical details of space travel and communications are addressed in many books on astronomy and cosmology. *See also* SPACE COMMUNICATIONS.

SPADE LUG

A spade lug is a simple connector that is sometimes used in conjunction with binding posts. The spade lug eliminates the problem of wire slippage when binding posts are used. The spade lug also improves the quality of the connection, because the metal-to-metal surface area is increased.

The drawing shows a spade lug, and how it fits onto a binding post. Spade lugs can be purchased at most electronic stores in various sizes. The spade lug is simply soldered or crimped onto the end of the wire. *See also* BINDING POST.

SPAGHETTI

In the construction of electronic equipment, it is sometimes necessary to insulate component leads to prevent a possible short circuit with other wiring. This is often the case in point-to-point wiring when a component must be mounted between two tie strips located a considerable distance apart. Plastic tubing, called spaghetti, is used to insulate component leads.

Spaghetti is available in many different sizes for different gauges of wire. In circuit wiring, the smallest size of spaghetti should be used that will fit easily over the component lead. The spaghetti is simply cut to the necessary length with a scissors.

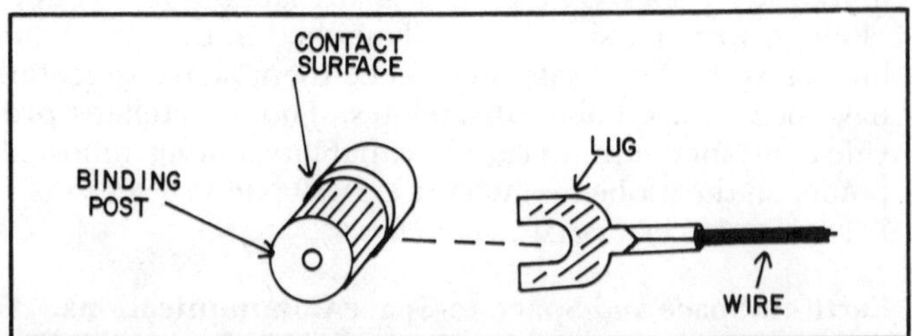

SPADE LUG: A spade lug provides an easy method of attaching a wire to a binding post.

SPARK

When the voltage gets very large between two points that are separated by an air dielectric, a momentary discharge, or arc, will occur through the air. This discharge is called a spark. If a spark continues for more than a brief instant, or through some medium other than the air, it is usually called an arc (*see* ARC). The exact voltage that will cause a spark depends on the shapes of the charged objects, the distance between them, and amount of impurities (such as water vapor, dust, and pollutants) in the air.

Sparks may contain very little energy, or they may contain a tremendous amount. The energy in a spark is expressed in joules, or watt seconds. Sometimes a spark occurs repeatedly within a short interval of time; this is called a multiple spark.

A spark is a flow of electrons. The electrons are accelerated as they pass from one charged object to another; this acceleration sets up a wideband electromagnetic field. The electromagnetic energy set up by a spark generally decreases as the frequency increases. The energy-versus-frequency curve of a particular spark is called the spark spectrum or spark spectral distribution. The electromagnetic field causes static in nearby radio receivers. When a thunderstorm is close by, the static you hear on the radio is caused by this acceleration of charged particles. This static is known as sferics (*see* SFERICS).

Early radio transmitters employed sparks to generate their signals. The electromagnetic energy in repeating sparks was concentrated in a narrow band of frequencies by means of resonant circuits (*see* SPARK TRANSMITTER).

If a spark carries a great amount of electric charge, flammable materials in the path of the spark may be ignited. Some materials, such as gunpowder or gasoline, will explode if a spark passes through them. Internal combustion engines use gasoline explosions, ignited by sparks, to generate mechanical energy. Lightning causes many forest fires each year, in addition to numerous fires in homes and businesses (*see* LIGHTNING). If lightning strikes a tank containing a flammable liquid, the result can be disastrous.

Sparks can be generated simply by walking over a carpet with hard-soled shoes. The charge on your body gets great enough to generate a spark when you come within 1 or 2 millimeters of a grounded metallic object. Physicists use laboratory-generated sparks, generated in a manner similar to the carpet-shuffling technique, to study the nature of lightning. The devices that create these sparks, which can be several feet long, are called Van de Graaff generators. *See also* VAN DE GRAAFF GENERATOR.

SPARK TRANSMITTER

Around the turn of the century, physicists noticed that an electric spark, occurring on one side of a room, induced sparks in isolated circuits elsewhere in the room. This is how electromagnetic propagation was discovered (*see* ELECTROMAGNETIC FIELD, SPARK). The earliest radio transmitters generated their energy by means of electric sparks. Such a transmitter was called a spark transmitter.

The typical spark transmitter consisted of a charging coil, similar to the spark coil in an internal combustion engine. The coil was connected in parallel with a capacitor and a rotary device to allow repeated discharging of the spark. The rotary wheel was driven by an electric motor. A telegraph key allowed the transmitter to be turned on and off at intervals for code transmission.

Although a spark transmitter looks and sounds rather silly today, the rotary spark gap was considered a precision electronic device in its time. It was capable of generating more than 1 kW of electromagnetic power, concentrated in a narrow band of frequencies by the resonance effects of the coil and capacitor. The coil-and-capacitor tank circuit was connected to an antenna, resonant at the same frequency, usually between about 0.5 MHz and 1.5 MHz (600 m and 200 m).

Using spark transmitters and crystal-set receivers, radio amateurs began communicating over distances of several miles. Radio amateurs set up communications links across the country. They called this organization the American Radio Relay League (*see* AMATEUR RADIO, AMERICAN RADIO RELAY LEAGUE).

Spark transmitters are illegal today. They produce signals that are rough and broad by modern standards. The signals from a spark transmitter would cause terrible interference to other stations in the vicinity. The signal from such a transmitter would sound like the interference from an electric razor.

By the 1920s, there were so many radio amateurs and so many spark transmitters that the situation had reached a state of pandemonium. Radio amateurs were restricted to frequencies above 1.5 MHz (200 m). Spark transmitters did not function well as these high frequencies. The development of the vacuum tube and the continuous-wave oscillator, however, made it possible to transmit at shortwave frequencies. This led to the discovery of ionospheric effects and worldwide communications. *See also* SHORT WAVES.

SPEAKER

A speaker, also known as a loudspeaker, is a form of electroacoustic transducer. The speaker converts alternating electric currents into sound waves having the same frequency characteristics and the same waveforms, within the range of human hearing (approximately 16 Hz to 20 kHz).

There are several types of speakers intended for different applications. The woofer is designed to operate at the lower audio frequencies. The midrange speaker is designed to be used at frequencies near the middle of the human hearing range. The tweeter is intended for reproduction of sound at the upper end of the range; some tweeters will operate at ultrasonic frequencies (*see* MIDRANGE, TWEETER, WOOFER). Many high-fidelity speaker systems consist of one of each of these three transducer types. Communications speakers usually consist of a simple midrange speaker.

The construction of a typical midrange speaker, suitable for communications purposes and low-fidelity music reception, is shown at A in the illustration. Front-view and back-view photographs of a typical 3-inch speaker are shown at B. A coil of wire, called the voice coil, is held in the field of a permanent magnet. When alternating currents flow in the coil, a fluctuating magnetic field is produced around the coil. This field interacts with the field from the permanent magnet to produce back-and-forth forces on the coil. The coil is mounted so that it can move as these forces occur. The coil is physically attached to a cone-shaped diaphragm, which vibrates along with the coil. The moving diaphragm produces sound waves in the air.

There are other speaker configurations besides that shown. Electrostatic speakers operate via electric fields, rather than magnetic fields (*see* ELECTROSTATIC SPEAKER).

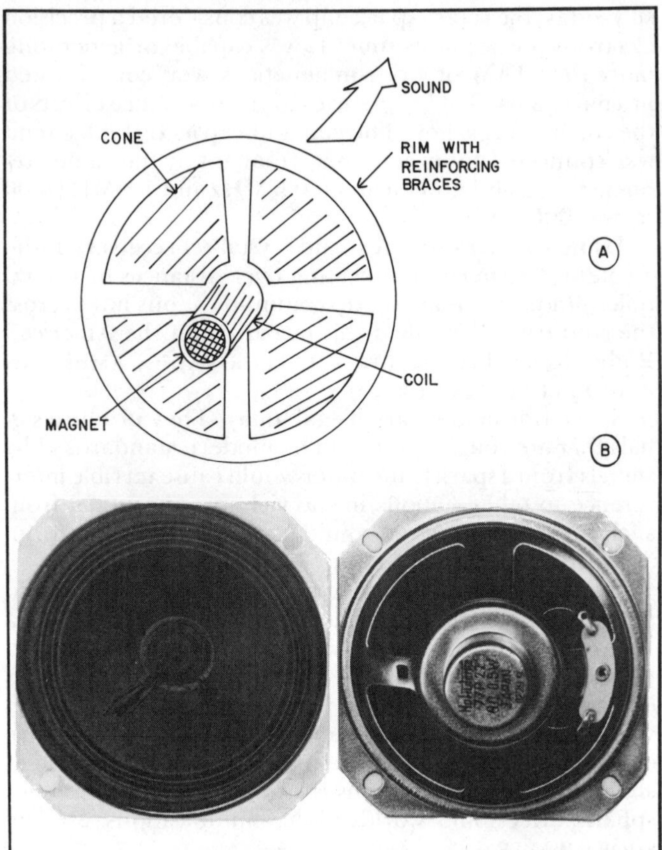

SPEAKER: At A, construction of a typical speaker, as viewed from behind. At B, front and rear views of a common type of communications speaker.

Some speakers use electromagnets, rather than permanent magnets, to produce the stationary magnetic field. In public-address systems and some high-fidelity midrange and high-frequency applications, horn-shaped speakers are used. Various other forms of transducers are used for SONAR and other acoustic devices. *See also* TRANSDUCER.

SPECIFICATIONS

The specifications of an electronic device are the operating characteristics, expressed in tabular form. Specifications give information of importance to operators of the equipment.

In a radio receiver, important specifications include the sensitivity, selectivity, tuning range, emission modes that can be received, and frequency stability. In a transmitter, the specifications might include such factors as the frequency range, emission types, frequency stability, and radio-frequency power output. (The table is a list of specifications for a commercially manufactured FM transceiver.) For test instruments, the degree of accuracy is the most important specification. For power supplies, the voltage output, regulation, and current-delivering capability are important specifications. These examples are, however, only representative of the many different specifications that can be expressed for a particular piece of electronic equipment.

SPECIFIC GRAVITY

Specific gravity is an expression of the relative density of a material.

If a given volume of material has a mass of x grams, while an equal volume of water has a mass of y grams, then the specific gravity, S, of the non-water material is given by:

$$S = x/y$$

provided the two materials are at the same temperature.

The table gives the specific gravities of some substances at room temperature.

SPECIFIC HEAT

The specific heat of a substance is an expression of the

SPECIFICATIONS: SPECIFICATIONS FOR A TYPICAL (BUT HYPOTHETICAL) VHF FM TRANSCEIVER.

GENERAL	
Frequency Range	144-148 MHz
Display Type	LED
Frequency Control	Microcomputer PLL VCO
Emission Type	F3
Memory Channels	8
Temperature Tolerance Range	−20 to +60 degrees C
Power-Supply Requirements	12 to 15 V dc, 5 A
Semiconductors	17 IC, 20 FET, 29 Tr, 59 Di
Dimensions	HWD 2.5 × 6 × 9 in (64 × 152 × 229 mm)
Weight	3 lbs (1.4 kg)
RECEIVER SECTION	
Intermediate Frequencies	17.0 MHz, 455 kHz
Sensitivity	Better than 0.35 μV for 20-dB noise quieting
Selectivity	Plus/minus 5 kHz at −6 dB Plus/minus 15 kHz at −60 dB
Autio Output	2 watts or more
Speaker Requirements	Impedance 4-8 ohms
TRANSMITTER SFCTION	
RF Output	15 watts
Frequency Deviation	Plus/minus 5 kHz
Spurious Radiation	Less than −60 dB
Antenna Requirements	50 ohms resistive, SWR < 2:1
Microphone Rquirements	Impedance 300-600 ohms

SPECIFIC GRAVIY AND SPECIFIC HEAT: ALPHABETICAL LIST OF SOME COMMON SUBSTANCES, WITH SPECIFIC GRAVITIES AND SPECIFIC HEATS.

Substance	Specific Gravity	Specific Heat
Aluminum	2.7	0.22
Antimony	6.6	0.050
Arsenic	5.7	0.082
Calcium	1.5	0.14
Carbon	.2	0.17
Cesium	1.9	0.052
Copper	9.0	0.092
Gallium	5.9	0.079
Germanium	5.4	0.073
Gold	19	0.031
Helium	0.17	1.3
Hydrogen	0.084	3.4
Iodine	4.9	0.052
Indium	22	0.058
Iron	7.9	0.032
Lead	11	0.030
Magnesium	1.7	0.25
Manganese	7.5	0.11
Mercury	14	0.034
Nickel	8.9	0.11
Nitrogen	1.2	0.24
Oxygen	1.3	0.22
Selenium	4.8	0.078
Silicon	2.4	0.18
Silver	10	0.057
Tin	7.3	0.054
Tungsten	19	0.033
Water	1.0	1.0
Zinc	7.0	0.10

capacity of that material to be raised in temperature by the application of heat energy. Specific heat is expressed in calories per degree Celsius, for 1 gram of the material in question.

Water has a specific heat of 1; this means that the application of 1 calorie of heat energy will raise 1 gram of water by 1 degree Celsius. Most materials have a specific heat of less than 1. The preceding table gives the specific heats of some substances.

SPECIFIC RESISTANCE

The resistance of an electrical conductor is sometimes expressed in terms of a factor called the specific resistance. Specific resistance is given in ohms per foot per circular mil.

The specific resistance of a conductor depends on the material and the temperature. The specific resistance of a material is generally expressed for direct current. *See also* RESISTANCE.

SPECTRAL DENSITY

A radio-frequency signal or other electromagnetic emission often contains more than one wavelength component. For example, an amplitude-modulated signal contains not only the carrier frequency, but sideband frequencies as well. Spectral density is an expression of the concentration of energy in terms of frequency.

In general, for a given signal power, the spectral density increases as the bandwidth of the signal gets smaller. The maximum possible spectral density exists when a signal is a steady carrier (emission type A0 or F0). The faster the rate of data transmission, the lower the spectral density. Also, the spectral density decreases as the efficiency of data transmission decreases.

Spectral density is given in terms of the average number of watts per kilohertz of spectrum space at the fundamental frequency of the signal. Spectral density may be expressed in relative terms, as the percentage of the total signal power per kilohertz of spectrum space at the fundamental frequency.

SPECTRAL ENERGY DISTRIBUTION

The electromagnetic output of an energy source can be expressed graphically, in terms of the energy-versus-frequency function. This function is called the spectral energy distribution.

Most natural phenomena exhibit a characteristic spectral energy distribution, with a peak at a certain wavelength. This is called blackbody radiation (*see* BLACK BODY). Some manmade devices, such as incandescent light bulbs, have a similar spectral energy distribution. Radio transmitters have a spectral energy distribution that is different from natural electromagnetic-energy sources; the energy is concentrated within a very small band of frequencies.

Spectral energy distribution is sometimes given in terms of power rather than energy. Then, it is called the spectral power distribution. The difference is purely semantical. *See also* ELECTROMAGNETIC SPECTRUM.

SPECTRAL RESPONSE

The spectral response of a sensing device is the function of its relative sensitivity versus electromagnetic or acoustic frequency or wavelengths. Spectral response is often expressed for the human eye (for visible-light wavelengths) or for the human ear (for sound frequencies).

The human eye is most sensitive in the yellow and green parts of the visible spectrum, near the center of the visible-light range. The human ear is most sensitive at a frequency near 1 kHz. Other animals have eyes and ears with different spectral responses. For example, flies see light at shorter wavelengths than humans see, and dogs hear sound at a higher average frequency than humans hear. *See also* LIGHT, SOUND.

Most radio receivers exhibit variable sensitivity throughout the range of coverage. The function of sensitivity versus frequency for a radio receiver is sometimes called the spectral response of the receiver. Spectral response may also be given for transducers. *See also* SENSITIVITY, TRANSDUCER.

SPECTROSCOPE

A spectroscope is a device that displays the spectral energy distribution of a visible-light source.

The spectroscope splits visible light into its constituent wavelengths by means of a slit and a prism or diffraction grating, as shown in the illustration. The resulting spectrum can be photographed or viewed directly.

The spectroscope is used by astronomers to evaluate the spectral energy distribution of light from distant stars and planets. Certain absorption lines and emission lines are characteristic of various elements and compounds. The patterns formed by these lines allow astronomers to de-

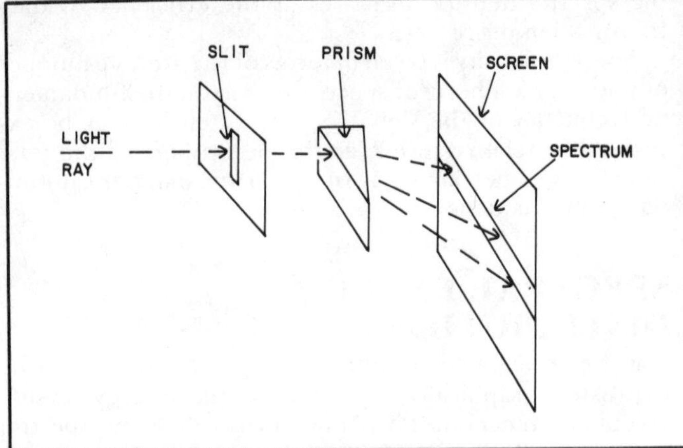

SPECTROSCOPE: Construction of a spectroscope.

termine the material composition of planetary atmospheres, stellar surfaces, and diffuse nebulae in space.

Doppler shifts in the wavelengths of spectral absorption lines make it possible to determine whether an object is moving toward or away from us, and if it is moving, to determine its radial speed. Edwin Hubble and Milton Humason were the first astronomers to recognize that consistent red shifts (increases in the wavelength) of the spectral lines of distant galaxies meant that the universe is generally expanding in size. *See also* DOPPLER EFFECT.

SPECTRUM

When visible light is passed through a prism or diffraction grating, the light is bent at varying angles depending on wavelength. This creates a rainbow-like pattern called a spectrum. The spectral colors are generally recognized as red, orange, yellow, green, blue, indigo, and violet, with intervening shades (*see* VISIBLE SPECTRUM).

Any range or band of frequencies can be called a spectrum. We may speak, for example, of the very-high-frequency spectrum; we would then be talking about the range of frequencies from 30 MHz to 300 MHz. The whole range of electromagnetic wavelengths, from the longest to the shortest imaginable, is called the electromagnetic spectrum. *See also* ELECTROMAGNETIC SPECTRUM.

SPECTRUM ANALYSIS

Spectrum analysis is the process of the evaluation of the spectral energy distribution from a source of electromagnetic waves (*see* SPECTRAL ENERGY DISTRIBUTION). Astronomers analyze the spectra of planetary atmospheres, stars, galaxies, and other celestial objects to determine their material composition (*see* SPECTROSCOPE). Electronic engineers evaluate the spectral energy distribution of radio transmitters by means of a device known as a spectrum analyzer (*see* SPECTRUM ANALYZER).

SPECTRUM ANALYZER

A spectrum analyzer is an electronic test instrument that graphically displays the spectral energy distribution of radio-frequency signal generators (*see* SPECTRAL ENERGY DISTRIBUTION). Engineers use spectrum analyzers extensively in the design, alignment, and troubleshooting of radio transmitters and receivers.

A typical spectrum analyzer consists of an oscilloscope and a circuit that provides the spectral display for the cathode-ray tube. The photograph is of a typical spectrum analyzer, suitable for work with equipment at frequencies up to 1 GHz. The frequency is displayed horizontally from left to right. Signal amplitude is displayed vertically. In the spectral display shown, there are four signals on the screen, in addition to the local-oscillator signal at the extreme left. (The second signal is barely visible above the noise floor.)

Some spectrum analyzers are designed for the audio frequency range. These include narrowband analyzers of fixed bandwidth, such as 5 Hz, and devices having a bandwidth that is a constant percentage of the frequency of the signal being evaluated. Examples of the latter are octave, half-octave, and 1/3-octave analyzers. Real-time analyzers (RTAs) commonly have visual displays of response in 1/3-octave bands covering the audio range.

To operate the spectrum analyzer, a technician first chooses, by means of a selector switch, the band pass of frequencies to be displayed. This band may cover the whole electromagnetic spectrum from direct current to 1 GHz or more, or it may cover any desired part of this range. The gain (vertical-scale sensitivity) of the circuit is adjusted as desired. The resolution and sweep rate must also be adjusted for the application intended.

The spectrum analyzer is useful for determining the spurious-signal and harmonic content of the output of a radio transmitter. In the United States, radio equipment must meet certain government-imposed standards regarding spectral purity. The spectrum analyzer gives an immediate indication of whether or not a transmitter is functioning properly. The spectrum analyzer can also be used to observe the bandwidth of a modulated signal. Such improper operating conditions as splatter or overmodulation can be easily seen as excessive bandwidth.

Spectrum analyzers are often used in conjunction with sweep generators, for evaluating the characteristics of bandpass, band-rejection, highpass, or lowpass filters. We may, for example, need to adjust a helical filter to obtain the desired bandpass at the front end of a radio receiver. We may wish to determine whether a lowpass filter is providing the right cutoff frequency and attenuation characteristics. The sweep generator produces a radio-frequency signal that varies in frequency, exactly in synchronization with the display of the spectrum analyzer (*see* SWEEP-FREQUENCY FILTER ANALYZER, SWEEP GENERATOR).

Some radio receivers are equipped with narrow-band spectrum analyzers for monitoring an entire communications band at once. Such a receiver is called a panoramic receiver. *See also* PANORAMIC RECEIVER.

SPECTRUM MONITOR

A spectrum monitor is a narrow-band spectrum analyzer that can be connected in the intermediate-frequency chain of a receiver to observe signals at or near the operating frequency. The spectrum monitor converts any ordinary superheterodyne receiver into a panoramic receiver.

The spectrum-monitor display is contered at the intermediate frequency of the receiver. Signals appear as pips to the left or right of the center of a cathode-ray-tube screen. The bandwidth is adjustable. This makes it possible to observe an entire communications band, such as 7.000—7.300 MHz, at once; or the operator can choose to "zero in" on one signal and observe its modulation charac-

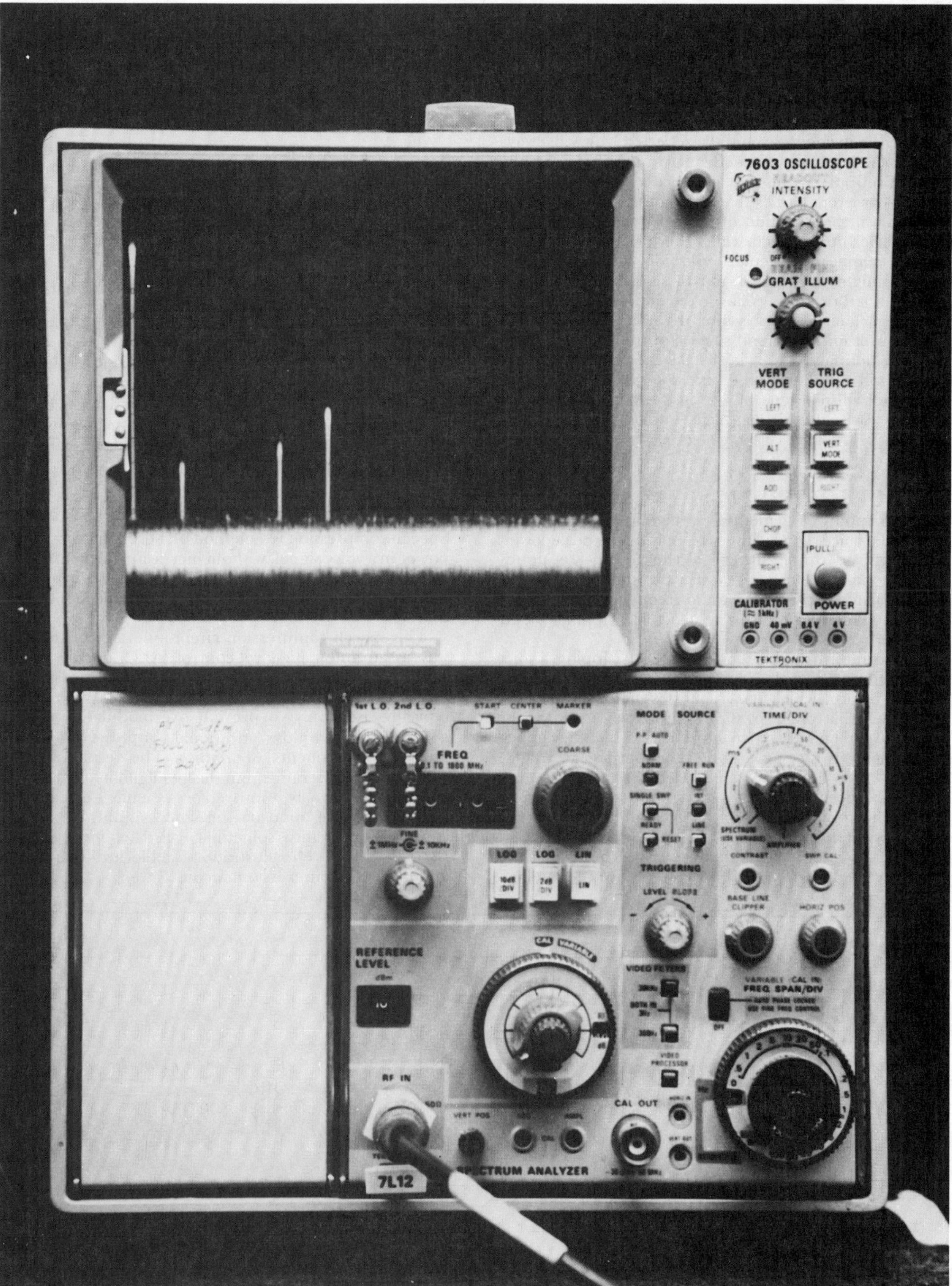

SPECTRUM ANALYZER: This instrument graphically displays amplitude versus frequency in the electromagnetic spectrum.

teristics. *See also* PANORAMIC RECEIVER, SPECTRUM ANALYZER.

SPECULAR REFLECTION

When electromagnetic waves are reflected from a smooth surface, the incident ray and the reflected ray will always lie in a single plane, perpendicular to the surface at the point of reflection. This is called specular reflection. If the surface has irregularities that are a substantial part of the wavelength, the direction of the reflected ray will fluctuate; this is non-specular reflection.

An example of specular reflection can be observed when sunlight shines on a mirror-smooth lake surface. The reflected rays always stay in the same orientation; the angle of incidence is always equal to the angle of reflection with respect to the general surface of the lake.

If a wind makes the lake surface choppy, specular reflection will not occur; the reflected light rays will be scattered and will not generally lie in the same place as the incident waves. *See also* REFLECTION LAW.

SPEECH CLIPPING

Speech clipping is a method of increasing the average power of a voice signal without increasing the peak power. This can be done either in the audio-frequency microphone-amplifier stages of a transmitter, or in the intermediate-frequency or radio-frequency (RF) stages. The latter method is usually called RF clipping (*see* RF CLIPPING).

To accomplish audio-frequency speech clipping, a voice signal is first amplified. Then the signal is passed through a limiting device, which has a cutoff voltage substantially below the peak voltage of the amplified voice signal. The output of the limiter is then amplified, so that the clipped peak voltage is the same as the unclipped peak voltage ahead of the limiter. A simplified circuit for audio-frequency speech clipping is shown in the illustration.

Speech clipping, if done properly, can improve the intelligibility of an amplitude-modulated or single-sideband signal. If done improperly, however, speech clipping can cause severe distortion, and actually reduce the intelligibility as well as cause objectionable splatter (*see* SPLATTER). Audio-frequency speech clipping can be accomplished only to a certain extent before distortion occurs. If the clipping threshold is made too low, or the gain of the first amplifier is made too high, the resulting distortion will partially or totally offset the effects of the average-power increase.

Whenever speech clipping is used in a transmitter, a bandpass filter, having steep skirts and high ultimate attenuation, is an absolute necessity. A spectrum analyzer or panoramic receiver should always be used to check the output of a transmitter in which a speech clipper is employed.

Since speech clipping increases the average power in a transmitter, the final amplifier is forced to run at a higher duty cycle than would be the case without speech clipping. This may place excessive strain on the transistors or tubes in that circuit. Before any type of speech clipper is used in a transmitter, therefore, it is a good idea to check the ratings of the final amplifier to avoid possible damage.

Speech clipping is not the only method that can be used to increase the ratio of average power to peak power in a voice transmitter; speech compression is also used for this purpose. *See also* SPEECH COMPRESSION.

SPEECH COMPRESSION

Speech compression is a method of increasing the average power in a voice signal, without increasing the peak power. Speech compression is used in many amplitude-modulated and single-sideband communications transmitters.

The speech compression circuit operates in the same way as an automatic-level-control (ALC) circuit (*see* AUTOMATIC LEVEL CONTROL). But speech compression carries the process farther than ordinary ALC. While ALC is typically used only to prevent overmodulation, speech compression employs additional amplification of the low-level components of a voice. The result is greatly reduced dynamic range, but the intelligibility of the signal is often considerably improved because more effective use is made of the modulating-voice signal. A speech-compression circuit is sometimes called an amplified-ALC (AALC) circuit. The illustration is a block diagram of an audio speech-compression circuit.

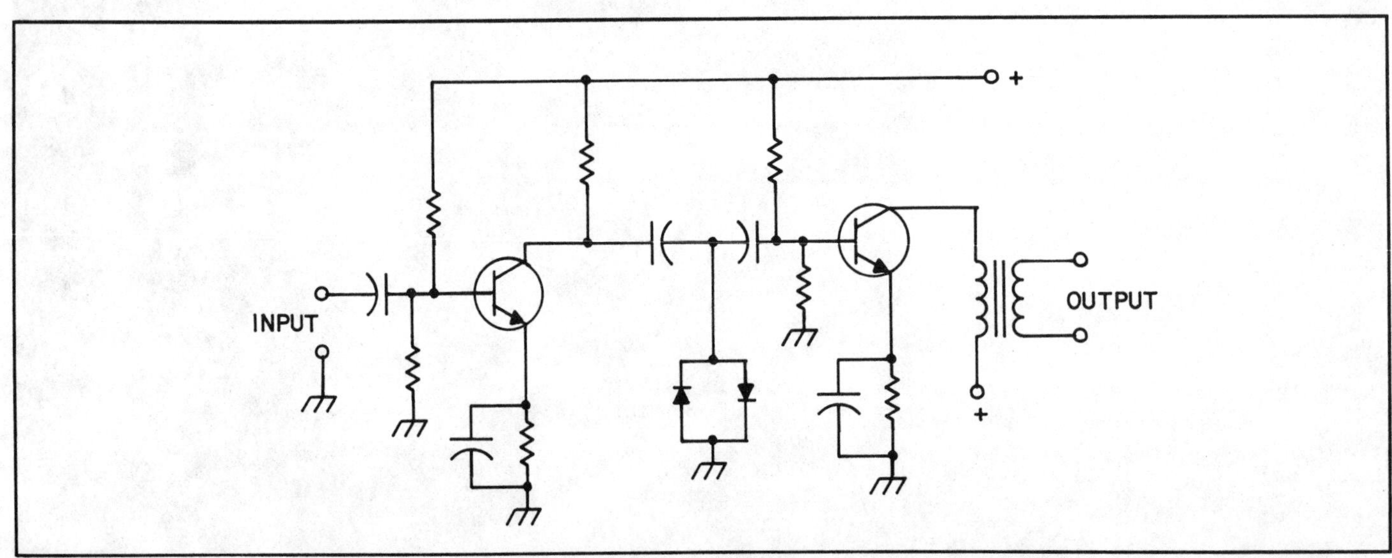

SPEECH CLIPPING: A two-stage speech-clipping circuit. The actual clipping is perormed by the pair of diodes connected in reverse parallel.

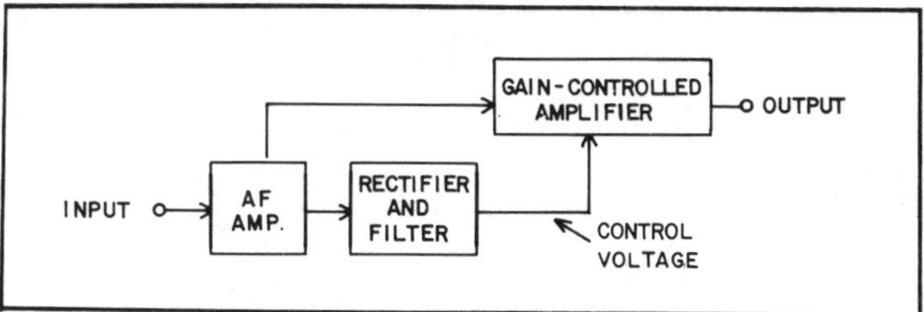

SPEECH COMPRESSION: Block diagram of an audio speech-compression circuit.

Speech compression is usually done in the audio (microphone-amplifier) circuits of a transmitter. But it can be done in the intermediate-frequency or radio-frequency (RF) stages. Accordingly, we may speak of either audio speech compression or RF speech compression. Radio-frequency speech compression is often called envelope compression.

Speech compression, like speech clipping, can cause problems if it is not done correctly. There is a certain maximum increase in the ratio of average power to peak power that can be realized without objectionable envelope distortion. Too much speech compression will actually degrade the intelligibility of a voice signal.

When speech compression is used in a transmitter, the transmitter output should be monitored with a spectrum analyzer or panoramic receiver to ascertain that splatter is not taking place (*see* SPLATTER). Also, precautions should be taken to ensure that the average-power increase will not put too much strain on the final amplifier. *See also* SPEECH CLIPPING.

SPEECH INVERSION

See SCRAMBLER.

SPEECH PROCESSING

See RF CLIPPING, SPEECH CLIPPING, SPEECH COMPRESSION.

SPEECH RECOGNITION

With some computer devices, it is possible to give commands by simply talking into a microphone. In recent years, circuits have been designed to recognize certain speech sounds, and process them into digital information that a computer can recognize and work with.

Speech-recognition techology is still rather immature. Its uses are limited to basic commands. However, the state of the art is rapidly becoming more sophisticated. In the near future, we can expect that computers will have speech-recognition vocabularies exceeding that of the average person. Such capability will make computers immensely more versatile and useful.

Speech recognition has already been used in conjunction with speech synthesis to converse, in simple terms, with computers. Someday it will be possible to carry on intelligent, stimulating, and perhaps even emotional conversations with computers! *See also* SPEECH SYNTHESIS.

SPEECH SYNTHESIS

Speech synthesis is a process by which electrical impulses, such as digital signals, are converted into sounds that people can understand as words and sentences. Speech synthesis is, like speech recognition, still in its infancy.

Most speech synthesizers have a limited vocabulary, and the voice sounds "electronic." However, the technology has greatly improved in the past few years, and synthesized voices are beginning to sound almost like real people.

Speech synthesizers are used in educational devices for children. Speech synthesizers, in conjunction with speech-recognition devices, will someday make it possible to converse verbally with a computer at a high level of efficiency. This will make computers usable by more and more people, for increasingly varied purposes. *See also* SPEECH RECOGNITION.

SPEED OF LIGHT

See LIGHT.

SPEED OF SOUND

See SOUND.

SPEED OF TRANSMISSION

The speed at which digital data is transmitted can be expressed in a variety of different ways. The most common methods are the baud rate and the word-per-minute (WPM) rate.

A baud consists of one pulse or element of a digital signal. The number of bauds sent in 1 second is called the baud rate. A data word is usually considered to consist of six characters, including spaces if present. Thus the speed in words per minute is equal to the number of characters and spaces in 1/6 minute (10 seconds). Both the baud rate and the WPM rate are determined by averaging over a period of time.

The ratio of the baud rate to the WPM rate is not exactly the same for all types of signals. Some codes are a little more efficient, and require fewer bauds power word, than other codes. In ASCII code, the baud rate is the same as the WPM rate. In BAUDOT code, the baud rate is approximately 75 percent of the WPM rate. In the International Morse code, the baud rate is approximately 83 percent of the WPM rate. *See also* ASCII, BAUDOT, BAUD RATE, INTERNATIONAL MORSE CODE, WORDS PER MINUTE.

SPHERE GAP

A sphere gap is a special form of spark gap. Two electrodes, each having a sphere-shaped tip, are placed near each other. The exact spacing can be adjusted by means of thumb screws. Sphere gaps are used to facilitate the discharge of excessive voltages. Such spark gaps can be used, for example, as lightning arrestors.

The sphere gap will withstand a higher voltage before arcing occurs, compared with a sharp-tipped pair of electrodes. Thus, a sphere gap can be made physically smaller than other types of spark gaps. The spherical tips can also withstand more arcing before damage occurs to the electrodes.

SPHERICAL ANTENNA

A spherical antenna is a form of dish antenna in which the

reflecting surface is a portion of a geometric sphere in three dimensions. A spherical antenna appears the same as a paraboloid dish antenna (*see* PARABOLOID ANTENNA). The difference is not readily apparent.

A spherical antenna has a focal point that lies approximately midway between the surface and the center. At the focal point, rays arriving parallel to the antenna axis converge almost, but not quite, perfectly. An electromagnetic field originating at the focal point produces rays that are nearly parallel to the axis of the antenna.

Spherical reflectors can be constructed from wire mesh, or they may be made of solid sheet metal. Spherical antennas are used primarily at ultra-high and microwave frequencies. Wire-mesh construction is satisfactory at the longer wavelengths, but at short microwave frequencies, solid sheet metal is preferable. *See also* DISH ANTENNA.

SPHERICAL COORDINATES

The position of an object can be uniquely specified in three dimensions by means of a spherical coordinate system. This coordinate system is a three-dimensional extension of the polar-coordinate plane (*see* POLAR COORDINATES).

The spherical coordinate scheme requires a reference point, a geometric plane passing through that point, a reference ray that lies in the plane and originates at the point, and a reference ray that lies perpendicular to the plane and originates at the reference point. The position of a point in space is determined according to its distance r from the central point, and also according to two angles θ and ϕ, measured with respect to the two rays as shown in the illustration at A.

The distance r may be any nonnegative real number. The angle ϕ is measured counterclockwise from the ray in the reference plane. Accordingly, ϕ may range between 0 degrees and 360 degrees. The angle θ is measured between the perpendicular reference ray and a line connecting the origin point with the point in question. The value of θ may range from 0 to 180 degrees. The constraints, expressed mathematically, are:

$$r \geq 0$$

$$0 \leq \phi < 360$$

$$0 \leq \theta \leq 180$$

Using this scheme, each point in space corresponds to exactly one set of values (r, ϕ,θ), and each set of values (r, ϕ,θ) corresponds to exactly one point in space.

It is possible to convert three-dimensional Cartesian coordinates (x, y, z) to spherical coordinates (r, ϕ,θ), according to the illustration at B, by means of the following set of formulas:

$$r = \sqrt{x^2 + y^2 + z^2}$$

$$\phi = \arctan (y/x)$$

$$\theta = \arccos (z / \sqrt{x^2 + y^2 + z^2})$$

The inverse of this conversion is given by the formulas:

$$x = r \sin \theta \cos \phi$$

$$y = r \sin \theta \sin \phi$$

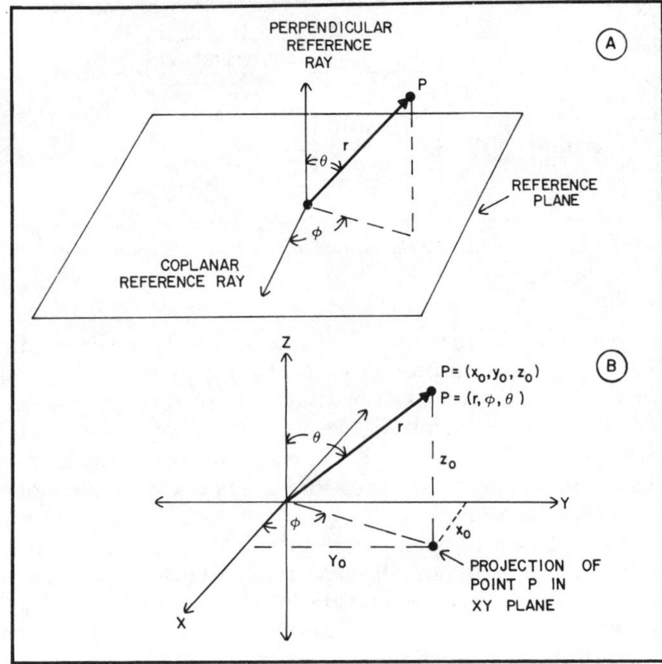

SPHERICAL COORDINATES: At A, the spherical-coordinate scheme. At B, coordinate orientation for transformation between spherical and Cartesian coordinates.

$$z = r \cos \theta$$

Spherical coordinates are used in various forms by astronomers. This coordinate scheme is also employed in satellite communications and tracking. *See also* CARTESIAN COORDINATES, DECLINATION, LATITUDE, LONGITUDE, RIGHT ASCENSION.

SPIDERWEB ANTENNA

A spiderweb antenna is a broadband form of antenna, consisting of several dipoles connected in parallel at a common feed point. The dipoles have varying lengths, and all lie in a single horizontal plane. The dipoles run radially outward from the feed point, as at A in the illustration.

The bandwidth of a spiderweb antenna depends on the lengths of the shortest and longest dipoles. If the longest dipole measures x feet in length, and the shortest dipole measures y feet, then the lower and upper limit frequencies f and g, in megahertz, are approximately:

$$f = 468/x$$

$$g = 468/y$$

If x and y are given in meters, then:

$$f = 143/x$$

$$g = 143/y$$

Spiderweb antennas can be built to operate at any frequency. In the shortwave bands, such antennas are sometimes constructed from wire, using multiple supports in a radial pattern around a central support. The ends of the wires can also be run downward at a slant, terminating at ground stakes, and using insulators to provide the proper

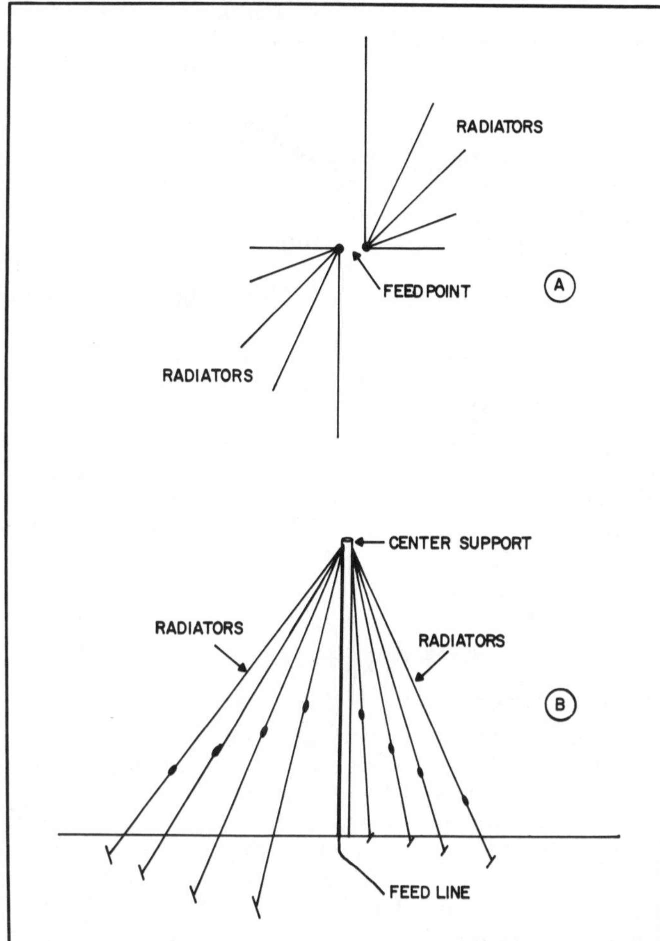

SPIDERWEB ANTENNA: At A, the principle of the broadband spiderweb antenna. At B, construction of a spiderweb antenna for use at high frequencies.

resonant lengths, as at B. At very high frequencies, spiderweb antennas are made from aluminum tubing for broadband television reception. *See also* DIPOLE ANTENNA.

SPLATTER

Splatter is a term that is used to describe the effects of a voice signal having too much bandwidth. Splatter can occur with amplitude-modulation, frequency-modulation, phase-modulation, or single-sideband emission types.

When the amplitude, frequency, or phase of a signal is varied, sidebands are produced (*see* SIDEBAND). The more rapid the instantaneous change in amplitude, frequency, or phase becomes, the farther the sidebands occur from the carrier frequency. If modulation is excessive, peak clipping will occur in amplitude modulation or single sideband, and overdeviation will occur in frequency modulation and phase modulation. This causes sidebands to be generated at frequencies too far removed from the carrier.

In a receiver, splatter sounds like a crackling noise at frequencies up to several hundred kilohertz from the carrier frequency of the transmitter. Splatter can cause serious interference to numerous communications circuits.

In an amplitude-modulated or single-sideband transmitter, splatter can be eliminated by:

1. Avoiding overmodulation
2. Ensuring that the audio-amplifier circuits are operating with a minimum of distortion
3. Ensuring proper bias and drive for transmitter amplifiers, especially the final amplifier
4. Ensuring that any external linear amplifiers are in fact operating in a linear manner
5. Proper operation of speech-processing circuits and automatic level control

In a frequency-modulated or phase-modulated transmitter, splatter can be avoided by:

1. Keeping the deviation within rated system limits
2. Ensuring that the audio-amplifier stages are operating with minimum distortion

A form of splatter is sometimes observed in continuous-wave (code) transmissions. If the rise and/or decay times are too rapid, clicks can be heard at frequencies far removed from the carrier frequency. These sidebands are called key clicks. *See also* KEY CLICK, SHAPING.

SPLICING

See TAPE SPLICING, WIRE SPLICING.

SPLITTER

A splitter is an impedance-matching device that allows more then one receiver to be used with a single antenna system. Splitters are commonly employed in television antenna systems and in cable-television installations where multiple receivers are desired.

Twin-lead television antenna systems are generally of 300 ohms impedance. If two receivers, having input impedances of 300 ohms, are connected in parallel to a single 300-ohm line, a 2:1 impedance mismatch occurs. This will result in degraded reception at both receivers. A two-way splitter will eliminate this mismatch and improve the reception, especially if signals are weak.

Cable television systems have a typical characteristic impedance of 75 ohms. (A 1:4 impedance step-up transformer is used at television receivers to provide a match to the 300-ohm television input impedance.) Cable installations often have numerous receivers hooked up to a single incoming feed line. Each receiver has its own branch cable of 75 ohms impedance. A splitter provides a 1:1 match to all of the receivers, and to the incoming signal at the main line. Without a splitter, the mismatch at the junction might be 10:1, 20:1, or even greater, resulting in very poor reception at each receiver.

Two-way splitters are generally available in electronics stores for use with either 300-ohm or 75-ohm systems. A four-way splitter can be built up by combining three two-way splitters as shown in the illustration. A three-way splitter can be obtained by terminating one of the four-way splitter outputs with either a 300-ohm or 75-ohm noninductive resistor, as appropriate. For splitting a signal more than four ways, a professional should be consulted.

Splitters are sometimes used in audio circuits. This is especially true in public-address systems using many speakers, or in multiple-listener systems having several sets

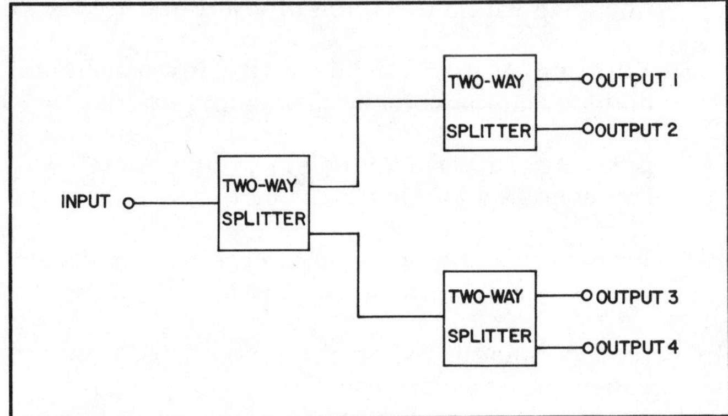

SPLITTER: Three two-way splitters can be combined, as shown here, to form a four-way splitter.

of headphones connected to the output of a single audio amplifier. *See also* CHARACTERISTIC IMPEDANCE, IMPEDANCE MATCHING.

SPONGE LEAD

Sponge lead is a black or gray, lead-saturated sponge material that is used for storing metal-oxide-semiconductor (MOS) components. Sponge lead is conductive, and this reduces the chances for static charges to build up in a MOS component during storage.

Many integrated circuits and transistors use MOS technology. These components are extremely susceptible to destruction by small static charges. The leads of MOS components should always be inserted into sponge lead when such devices must be stored for any period of time.

Sponge lead is available at most electronics stores. *See also* METAL-OXIDE SEMICONDUCTOR.

SPONTANEOUS EMISSION AND ABSORPTION

When an electron in an atom falls from a higher-energy shell to a lower-energy shell, a photon is produced. The wavelength of this photon depends on the extent of the energy transition through which the electron passes. This emission is called spontaneous emission.

If a photon, having just the right amount of energy, is absorbed by an electron, that electron will jump to a higher-energy shell in the atom. This is known as spontaneous absorption (*see* ELECTRON ORBIT, PHOTON).

All energized matter, especially gases, produces emissions at specific wavelengths. Cooler gases exhibit absorption at certain electromagnetic wavelengths. These wavelengths show up as bright lines or dark lines in the spectrum of the gas. These lines are, appropriately, called emission lines or absorption lines.

Astronomers use spectroscopes to observe the emission and absorption lines from distant planets, stars, galaxies, and diffuse nebulae. At radio wavelengths, the radio telescope can be used to observe spontaneous emission and absorption, such as the 21-cm hydrogen emission line. This enables scientists to ascertain the material composition of distant celestial objects. *See also* RADIO ASTRONOMY, RADIO TELESCOPE, SPECTROSCOPE.

SPORADIC-E PROPAGATION

At certain radio frequencies, the ionospheric E layer occasionally returns signals to earth. This kind of propagation tends to be intermittent, and conditions can change rapidly. For this reason, it is known as sporadic-E propagation.

Sporadic-E propagation is most likely to occur at frequencies between approximately 20 MHz and 150 MHz. Rarely, it is observed at frequencies as high as 200 MHz. The propagation range is on the order of several hundred miles, but occasionally communication is observed over distances of 1,000 to 1,500 miles.

The standard frequency-modulation broadcast band is sometimes affected by sporadic-E propagation. The same is true of the lower television channels, especially channel 2. Sporadic-E propagation often occurs on the amateur bands at 21 MHz through 148 MHz.

Sporadic-E propagation is sometimes thought to exist when, in fact, tropospheric propagation is taking place. The confusion can also occur in the opposite sense. *See also* E LAYER, IONOSPHERE, PROPAGATION CHARACTERISTICS, TROPOSPHERIC PROPAGATION.

SPREAD-SPECTRUM TECHNIQUES

Until recent years, communications engineers have striven to minimize the bandwidth of transmitted signals. The motivation for this is based on simple arithmetic: The narrower the signal bandwidth, the more signals can be fit into a given frequency band. Also, the narrower the bandwidth, the less noise is received in proportion to the signal. However, these axioms are a little oversimplified. Other factors must also be considered. Such considerations have led to the idea of spread-spectrum communications techniques, in which the bandwidth is made deliberately very large rather than very small.

Spread-spectrum transmission and reception are achieved by means of frequency modulation of a transmitter and receiver in exact synchronization. The variety of possible deviation values and modulating waveforms is theoretically infinite. A signal with any type of emission—such as single sideband, frequency-shift keying, or any other mode—can be frequency modulated, according to a specific scheme or function. The only requirement is that the receiver frequency follow along, exactly, with the changes in the transmitter frequency.

Different communications circuits can use different spread-spectrum functions, so that no two will ever match. As more and more signals are placed into a given band, the apparent noise level rises, but the concentrated interference common to fixed-channel systems will not take place.

The illustration is a block diagram of a simple spread-spectrum transmitting and receiving setup. The local oscillators should, ideally, have the same nominal frequency. Both local oscillators are frequency-modulated by a function generator and reactance modulator. The function generators are synchronized so that the transmitter and the receiver are always on the same frequency. The deviation may be any value, although there is a practical limit to how large it can be without exceeding the transmitter-tuning or receiver-front-end passbands.

The primary advantage of spread-spectrum communications is that strong, concentrated interference from another station is practically impossible. The main disadvantage is that selective fading, especially at low, medium, and high frequencies, is likely to be more severe than

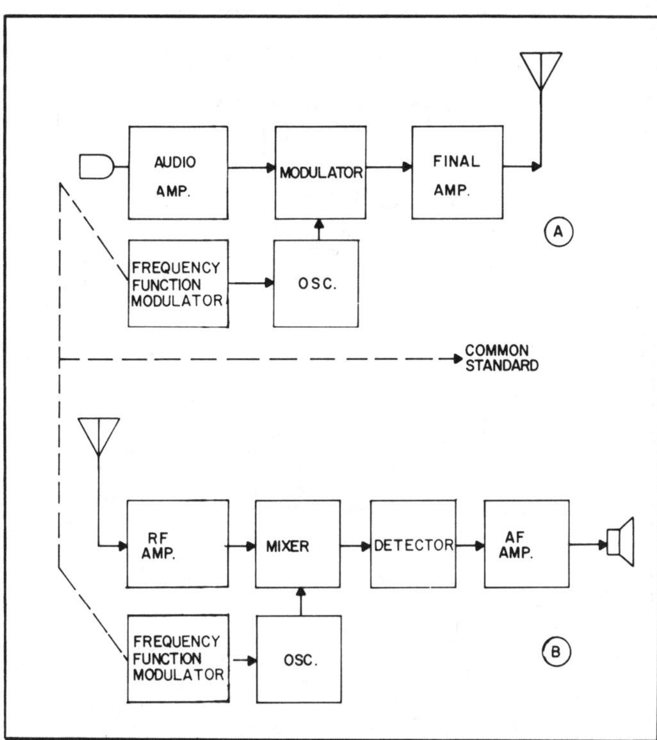

SPREAD-SPECTRUM TECHNIQUES: Spread-spectrum techniques. At A, block diagram of a spread-spectrum transmitter; at B, block diagram of a spread-spectrum receiver.

would be the case in single-frequency communications.

A complete discussion of various spread-spectrum design techniques can be found in a textbook on modern communications engineering.

SPURIOUS EMISSIONS

Spurious emissions, also known as spurious radiation or spurious signals, are emissions from a radio transmitter that occur at frequencies other than the desired frequency.

Harmonic emissions are a particular example of spurious radiation. Harmonics are signals that occur at integral multiples of the fundamental frequency of a radio transmitter (*see* HARMONIC, HARMONIC SUPPRESSION).

Spurious emissions sometimes occur as a result of parasitic oscillation (*see* PARASITIC OSCILLATION, PARASITIC SUPPRESSOR). Spurious emissions can also result from inadequate selectivity in the output stages of a transmitter.

In the United States, the maximum allowable level of spurious emissions is dictated by the Federal Communications Commission. This protects the various radio services from interference that could be caused by improperly operating radio transmitters.

A transmitter can be checked for spurious emissions by means of a device called a spectrum analyzer. *See also* SPECTRUM ANALYZER.

SPURIOUS RESPONSES

A radio receiver sometimes picks up signals that exist at frequencies other than the frequency to which the receiver is tuned. Such false signals are known as spurious responses.

Spurious responses in a receiver can take place as a result of mixing between or between two or more external signals. Such mixing may occur in the front end of a receiver, especially if one signal is strong enough to cause nonlinear operation of this stage. Spurious signals can sometimes result from mixing in nonlinear junctions external to the receiver (*see* FRONT END, INTERMODULATION).

Image signals in a superheterodyne receiver are an example of spurious responses. These signals can cause severe interference in an improperly designed receiver (*see* IMAGE FREQUENCY, IMAGE REJECTION).

A receiver can be checked for image responses by using a signal generator and for intermodulation distortion by using two signal generators.

SQUARE-LAW DETECTOR

A square-law detector is a form of envelope detector (*see* ENVELOPE DETECTOR). The square-law detector gets its name from the fact that the root-mean-square output-signal current is proportional to the square of the root-mean-square input-signal voltage.

The square-law detector operates not by rectification, but because of the nonlinear response of the circuit. This type of detector is sometimes called a weak-signal detector, since there is no threshold level above which the signal strength must be in order to accomplish demodulation.

SQUARE-LAW METER

Some analog meters respond to an applied quantity according to the square of that quantity. This is the case, for example, in wattmeters that actually measure the voltage across a particular resistance. Such a meter is called a square-law meter.

The square-law meter has a characteristic nonlinear scale. The divisions become progressively closer together toward the right-hand end of the scale, and are widely spaced near the left-hand end (see illustration). *See also* ANALOG METERING.

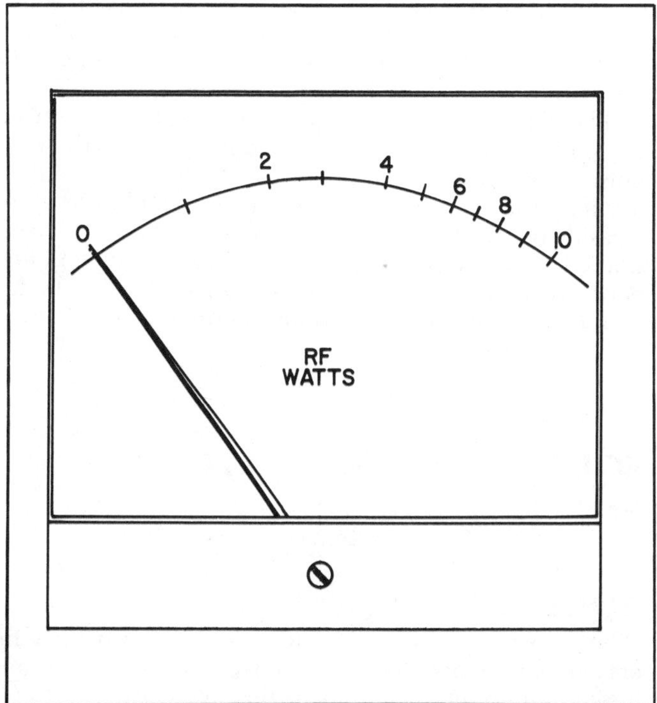

SQUARE-LAW METER: The scale of a square-law meter is nonlinear.

SQUARE-LAW RESPONSE

A circuit or device is said to have a square-law response when the output is proportional to the square of the input, or when the deflection of a meter is proportional to the square root of the input signal magnitude. A square-law curve is parabolic (*see* PARABOLA).

In general, if x represents the input quantity for a square-law circuit and y represents the output magnitude, then:

$$y = kx^2 + c$$

where k and c are constants. In the case of a square-law meter, if x represents the input-signal magnitude and y represents the meter-needle deflection, then:

$$x = ky^2 + c$$

where k and c, again, are constants. In both cases, the value of k depends on the units specified for input and output or scale deflection, and the value of c depends on the bias of the circuit or the starting point of the meter scale. *See also* SQUARE-LAW DETECTOR, SQUARE-LAW METER.

SQUARE WAVE

A square wave is a special form of alternating-current or pulsating direct-current waveform. The amplitude transitions, both rise and decay, take place instantaneously. When displayed on an oscilloscope, a square wave appears as two parallel, dotted lines, sometimes with faint vertical traces connecting the ends of the line segments (see illustration at A).

Square waves may have equal positive and negative peaks, or the peaks might be unequal. Unequal peaks result from the combination of an alternating square-wave current and a direct current. If the magnitude of the direct-current component is exactly equal to the peak magnitude of the square-wave current, a train of pulses results as shown at B. If the direct-current amplitude exceeds the peak amplitude of the square wave, both peaks of the waveform have the same polarity, as at C.

The period of a square wave is the length of time from any point on the waveform to the same point on the next pulse in the train. This period is usually considered to begin at the instant during which the amplitude is increasing positively (B); sometimes some other point is considered to mark the beginning of the period (C). The period is divided into 360 degrees of phase. The frequency of a square wave, in hertz, is equal to the reciprocal of the period in seconds.

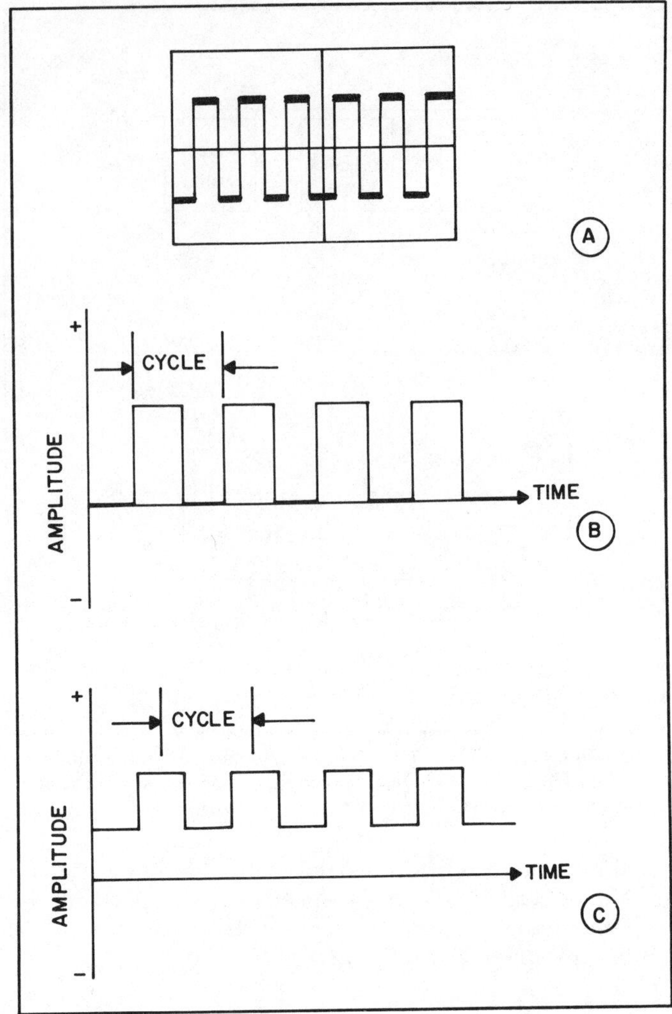

SQUARE WAVE: At A, a square wave as it appears on an oscilloscope screen. At B, a square wave with a direct-current component equal to the peak value of the alternating voltage. At C, a square wave with a direct-current component greater than the peak alternating voltage.

SQUEGGING OSCILLATOR

See BLOCKING OSCILLATOR.

SQUELCH

When it is necessary to listen to a radio receiver for long periods, and signals are present infrequently, the constant hiss or roar becomes fatiguing to the ears. A squelch circuit is used to silence a receiver when no signal is present, while allowing reception of signals when they do appear.

Squelch circuits are most often used in channelized units, such as Citizen's-Band and very-high-frequency communications transceivers. Squelch circuits are less common in continuous-tuning radio receivers. Most frequency-modulation receivers use squelching systems.

A typical squelching circuit is shown in the diagram. This squelch circuit operates in the audio-frequency stages of a frequency-modulation receiver. The squelch is actuated by the incoming audio signal. When no signal is present, the rectified hiss produces a negative direct-current voltage, cutting off the field-effect transistor and keeping the noise from reaching the output. When a signal appears, the hiss level is greatly reduced, and the field-effect transistor conducts, allowing the audio to reach the output. The cut-off squelch circuit is said to be closed; the conducting squelch is said to be open.

In most receivers, the squelch is normally closed when no signal is present. A potentiometer facilitates adjustment of the squelch, so that it can be opened if desired, allowing the receiver hiss to reach the speaker. This control also provides for adjustment of the squelch sensitivity. Most squelch controls are open when the knob is turned fully counterclockwise. As the knob is rotated clockwise, the receiver hiss abruptly disappears; this point is called the squelch threshold. At this point, even the weakest signals

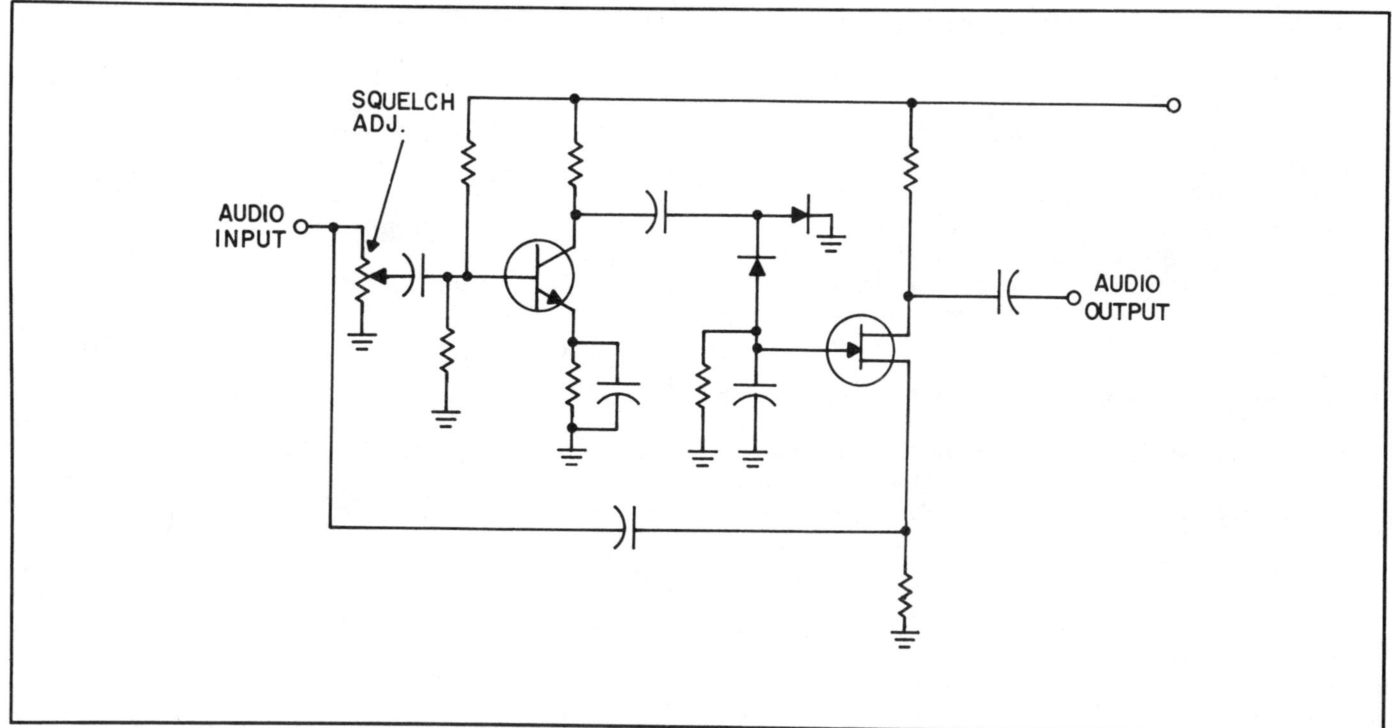

SQUELCH: An audio-operated squelch circuit for use with frequency-modulation receivers.

will open the squelch. As the squelch knob is turned farther clockwise, stronger and stronger signals are required to open the squelch.

In some receivers, the squelch will not open unless the signal has certain characteristics. This is called selective squelching. It is used in some repeaters and receivers to prevent undesired signals from being heard. The most common methods of selective squelching use subaudible-tone generators or tone-burst generators. Selective squelching is also known as private-line (PL) operation. *See also* PRIVATE LINE, SQUELCH SENSITIVITY.

SQUELCH SENSITIVITY

The squelch sensitivity of a squelched receiver is the signal level, in microvolts at the antenna terminals, that is required to keep the squelch open (*see* SQUELCH). The squelch sensitivity of a receiver depends on several factors.

Virtually all squelch controls allow for some adjustment of the squelch sensitivity by means of a control knob. The squelch sensitivity is greatest (the least signal is required to open it) at the threshold setting of the control.

An unmodulated carrier will usually open a squelch at a lower level than will a modulated carrier. If a frequency-modulated signal has excessive deviation, the squelch may cut off on modulation peaks. This effect is known as squelch blocking.

The squelch sensitivity of a frequency-modulation receiver is related to its noise-quieting sensitivity. The better the quieting sensitivity, the less signal is required to actuate the squelch at the threshold.

Squelch sensitivity is measured using a calibrated signal generator. The signal is modulated at a frequency of 1 kHz, with a deviation of plus-or-minus 3 kHz. The generator output is increased until the squelch remains continuously open at the threshold setting. *See also* NOISE QUIETING.

STABILITY

Stability is an expression of how well a component, circuit, or system maintains constant operating conditions over a period of time. We may speak, for example, of the frequency stability of a radio receiver, the stability of the output power of a radio transmitter, or the voltage-output stability of a power supply. The stability of an electronic component is usually called tolerance, and the stability of a power supply is known as the regulation (*see* REGULATION, TOLERANCE).

Stability can be expressed mathematically in absolute terms or as a percentage. Suppose the intended, or desired, value of a parameter is k units. Suppose that the actual value fluctuates between values of x units and y units, where $x < k < y$. Then the absolute stability is represented by the larger of two values S1 and S2, such that:

$$S1 = k - x$$

$$S2 = y - k$$

Under the same circumstances, the percentage stability is represented by the larger of two values:

$$SP1 = 100(k - x)/k$$

$$SP2 = 100(y - k)/k$$

Alternatively, the stability percentage may be given by:

$$SP = 100(y - x)/k$$

This denotes the maximum fluctuation in the parameter.

Stability is obviously a desirable trait in any component, circuit, or system. Receiver frequency stability should be on the order of a few hundred hertz over a period of hours following initial warmup. The same holds for radio transmitters. The output voltage of a power supply should not change significantly over a period of time, as long as the supply is operated within rated limits.

In some circuits, such as radio-frequency power amplifiers, stability is used to describe normal operating conditions. If the gain constantly changes, or if an amplifier breaks into oscillation, the circuit is said to be unstable. Good circuit design and proper operation are obviously important for maintaining normal function.

The stability of a device is affected by the environment in which it is operated. Temperature changes often cause a change in the characteristics of a component, circuit, or system. Mechanical vibration, the presence of external electromagnetic fields, and other factors can also affect stability. Circuit design, and the quality of the components used, are important. In recent years, electronic circuit-stability has greatly improved because of advances in the state of the art.

STABISTOR

A stabistor is a form of semiconductor diode that has a specific forward-conduction (breakover) voltage. All ordinary diodes have this property; a germanium diode breaks over at about 0.3 V and a silicon diode at 0.6 V. The stabistor characteristically has a much higher breakover voltage.

If a forward voltage is applied to a stabistor, the device will not conduct until the breakover voltage is reached. At forward voltages greater than the breakover voltage, the stabistor exhibits a very low resistance. A stabistor behaves, in the forward-biased mode, very much like a zener diode behaves in the reverse-biased mode. The stabistor is thus used in many of the same applications as the zener diode. *See also* ZENER DIODE.

STACK

A stack is a computer memory that is used for the purpose of short-term data storage. In a stack type memory, information must be entered and retrieved in a specific sequence, such as first-in/first-out or first-in/last-out. Random access is not possible.

A magnetic-core memory is often called a stack. *See also* CORE MEMORY, FIRST-IN/FIRST-OUT, MEMORY, PUSHDOWN STACK, SEQUENTIAL ACCESS MEMORY, SHIFT REGISTER.

STACKED ANTENNAS

See STACKING.

STACKING

Stacking is a method of combining two or more identical antennas for the purpose of obtaining directional characteristics, especially enhanced effective-power gain. Stacking can be done with any type of antenna, but it is most often done with dipoles, halos, or Yagis (*see* DIPOLE ANTENNA, HALO ANTENNA, YAGI ANTENNA).

The usual stacking configuration consists of two, three, or four horizontally polarized antennas placed vertically above each other and spaced apart by at least ½ wavelength (see illustration). The optimum spacing is between ¾ and 2 wavelengths. The individual antennas, if they are directional, each point in the same direction. The antennas are fed in phase. If two antennas are used, the power gain is about 3 dB over that of a single antenna. If three antennas are used, 4.8 dB gain can be realized; if four antennas are used, 6 dB gain is possible compared with the gain of a single antenna.

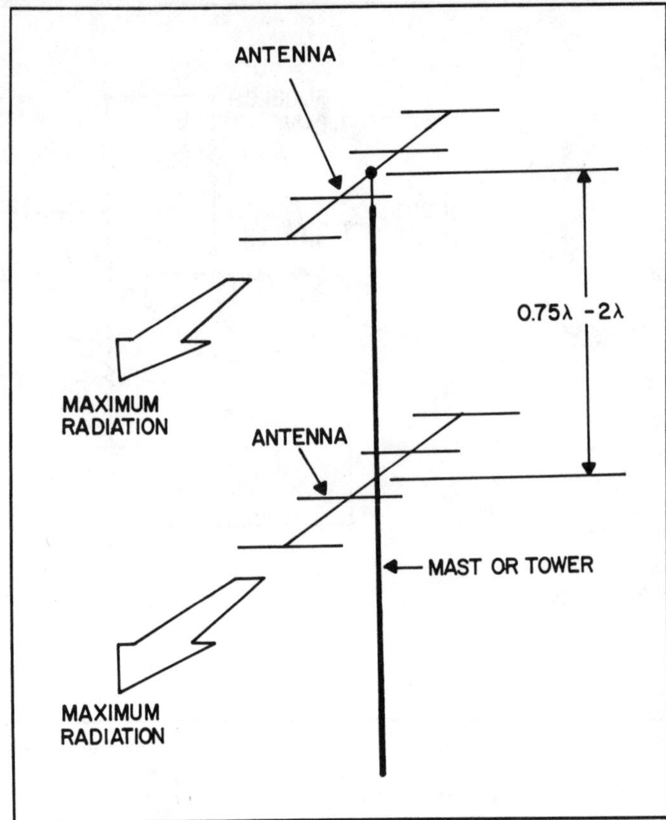

STACKING: An example of antenna stacking.

The above gain figures are theoretical. In practice, the gain will probably be slightly less. To optimize the gain, it is essential that the phasing harness be properly designed to provide the antennas with signals that are exactly in phase. The impedances must also be matched by means of quarter-wave feeder sections.

There are other stacking configurations besides that shown. Vertical dipoles are sometimes stacked in collinear fashion to obtain gain in the horizontal plane. Other types of antennas can be stacked either vertically or horizontally in various ways. *See also* COLLINEAR ANTENNA, SIDESTACKING.

STAGGER TUNING

Stagger tuning is a method of aligning a bandpass filter or the intermediate-frequency chain of a receiver or transmitter. In a stagger-tuned filter, there are several tuned circuits, each set for a slightly different frequency. In the stagger-tuned amplifier chain, each stage is tuned to a higher or lower frequency than the stage immediately before or after it.

In a bandpass filter, stagger tuning results in a more nearly rectangular response than would be obtained if all of the resonant circuits or devices were tuned to the same frequency. Stagger tuning broadens the bandpass response and provides steep skirts (*see* BANDPASS RESPONSE, RECTANGULAR RESPONSE, SKIRT SELECTIVITY).

In a multi-stage, radio-frequency amplifier, stagger tuning reduces the possibility of interstage oscillation. If all of the tuned circuits were set for exactly the same resonant frequency, the positive feedback among the various amplifiers might be enough to cause instability (*see* FEEDBACK).

STAND-ALONE SOLAR-POWER SYSTEM

See SOLAR POWER.

STANDARD ANTENNA

A standard antenna is an antenna used for comparison purposes in the determination of effective power gain. Such an antenna may be either a dipole or an isotropic radiator. *See* ANTENNA POWER GAIN, DIPOLE ANTENNA, ISOTROPIC ANTENNA.

An unbalanced, end-fed radiator having an overall length of exactly 13 feet (4 meters) is sometimes called a standard antenna.

STANDARD BROADCAST BAND

See BROADCAST BAND.

STANDARD CELL

A standard cell is an electrochemical cell that is used to provide a constant, known voltage. There are various different types of standard cell used for electronic measurements. The most common is the Weston cell. *See* WESTON STANDARD CELL.

STANDARD CAPACITOR

A standard capacitor is a capacitor that is manufactured to have a known, stable value. Standard capacitors are used in precision test equipment, particularly frequency meters and impedance bridges, to ensure the highest possible degree of accuracy. The tolerance of a standard capacitor is very low. The temperature coefficient is practically zero. *See also* ACCURACY, TEMPERATURE COEFFICIENT, TOLERANCE.

STANDARD DEVIATION

Standard deviation is an expression of the extent to which a set of data values varies above and below the mean value. Standard deviation is often specified in statistical analysis. The standard deviation is usually abbreviated by the Greek lower-case letter sigma (σ), although other symbols, such as the capital S or the small s, are also frequently used.

The standard deviation of a distribution having a finite number of values is defined as the root mean square of the deviations of all the individual values. For a large number of samples, standard deviation must be calculated by a computer. In the case of a continuous distribution, the standard deviation is determined by integration.

Let the data samples in a finite collection be represented by $x_1, x_2, x_3, \ldots, x_n$. Let μ be the mean value. Let p_k be the probability that a chosen sample will have the value x_k, where k is some integer between 1 and n inclusive. Then the standard deviation, s, is given by the formula:

$$s = \sqrt{\sum_{k=1}^{n} \; || \; p_k(x_k - \mu)^2}$$

If the distribution is a continuous function of x, where x may attain any of an infinite number of values, the standard deviation is determined differently. Let p(x) be the probability density function. Then the standard deviation is

$$S = \sqrt{\int_{-\infty}^{\infty} (x-\mu)^2 p(x)\, dx}$$

The larger the standard deviation in a probability function, the more scattered (that is, the less concentrated) the values will be. If the standard deviation is zero, then all of the values are the same. *See also* NORMAL DISTRIBUTION, STATISTICAL ANALYSIS.

STANDARD FREQUENCY

See REFERENCE FREQUENCY.

STANDARD INDUCTOR

A standard inductor is an inductor that is manufactured to have a precise, known, and essentially unchanging value. Standard inductors are used in precision test equipment, especially frequency-measuring apparatus or impedance bridges. The tolerance is very low, and the temperature coefficient is essentially zero. *See also* ACCURACY, TEMPERATURE COEFFICIENT, TOLERANCE.

STANDARD INTERNATIONAL SYSTEM OF UNITS

In 1960, the scientists of the world convened to agree on a universal scheme of measurement units. The system they adopted is known as the Standard International (SI) System of Units. This system is summarized in the table.

The meter-kilogram-second (MKS) measurement scheme is considered a subsystem of the SI System. Some scientists, especially in Europe, use a nonstandard scheme in which the basic units are the centimeter, the gram, and the second. This is called the CGS system. *See also* AMPERE, CANDELA, KELVIN TEMPERATURE SCALE, KILOGRAM, METER, MOLE, SECOND.

STANDARD INTERNATIONAL SYSTEM OF UNITS.

Quantity	Unit	Most Common Symbol
Displacement	Meter	m
Mass	Kilogram	kg
Time	Second	s
Current	Ampere	A
Temperature	Degree Kelvin	K
Luminous Intensity	Candela	cd
Quantity of Atoms	Mole	M

STANDARD RESISTOR

A standard resistor is a noninductive, fixed-value resistor having a precise value that does not change significantly with time. Standard resistors are commonly used in precision electronic circuits, especially resistance-measuring bridges. The standard resistor has a low tolerance and a temperature coefficient of practically zero. *See also* ACCURACY, TEMPERATURE COEFFICIENT, TOLERANCE.

STANDARD TEST MODULATION

In the evaluation of the distortion characteristics of radio receivers, a modulated signal must be used. Engineers and technicians use a signal generator that produces modulation at a certain frequency and extent. This is called standard test modulation.

For amplitude-modulation receivers, a test signal is 100-percent modulated by a pure sine-wave audio tone having a frequency of 1 kHz. For frequency-modulation and phase-modulation communications receivers, the carrier is modulated at a deviation of plus or minus 3 kHz by a pure sine-wave tone of 1 kHz. For single-sideband equipment, two sine-wave tones of equal amplitude, both within the passband (between 300 Hz and 3 kHz), and spaced 1 kHz apart, are used. (A common combination is 1 kHz and 2 kHz.) The audio output of the receiver is then analyzed with a distortion analyzer and/or oscilloscope. *See also* DISTORTION ANALYZER.

STANDARD TIME THROUGHOUT THE WORLD

The common worldwide time standard is based on the solar time at the Greenwich Meridian. This meridian is assigned 0 degrees of longitude and passes through the city of Greenwich, England. This common time is known as Coordinated Universal Time, and is abbreviated UTC (*see* COORDINATED UNIVERSAL TIME). The various countries of the world have their own local standard times, which differ from UTC.

The continental United States is divided into four time zones, known as the Eastern, Central, Mountain, and Pacific zones. During the part of the year in which standard time is used, these zones are behind UTC by 5, 6, 7, and 8 hours, respectively. During the part of the year in which daylight-savings time is used, these zones run behind UTC by 4, 5, 6, and 7 hours, respectively. (Some states and counties do not use daylight time, and thus their time lags behind UTC by the same amount all year.)

In general, locations in the western hemisphere of the world run from 0 to 12 hours behind UTC, while places in the eastern hemisphere run from 0 to 12 hours ahead of UTC. An approximate idea of the local standard time at any given location can be determined from its longitude (*see* LONGITUDE). Consider west longitude values to be negative, ranging from 0 to −180 degrees. Consider east longitude values to be positive, from 0 to +180 degrees.

For the western hemisphere, divide the west longitude in degrees by 15; you will get a number between 0 and −12. Round this number off to the nearest whole integer, and add the result to the time in UTC. (Use 24-hour time, in the manner of the military: 0000 for 12:00 midnight, 0100 for 1:00 A.M., and so on up to 2400 for 12:00 midnight the following night.) If the final value is less than 0000, add 2400 and subtract one day.

For the eastern hemisphere, divide the east longitude in degrees by 15; you will get a number between 0 and +12. Round this number off to the nearest whole integer, and add the result to the time in UTC. If the final value is greater than 2400, subtract 2400 and add one day. This will give the approximate local time in the part of the world of interest.

The above scheme is not completely reliable, because time zones are not strictly based on longitude. Some peoples have chosen to use the same standard time everywhere in their country, even though the land spans much more than 15 degrees of longitude. Other peoples run a fractional number of hours ahead of UTC or behind UTC. (Tonga, for example, runs 12 hours and 20 minutes ahead of UTC.) However, in astronomical terms, the above method is accurate.

STANDING WAVE

A standing wave is an electromagnetic wave that occurs on a transmission line or antenna radiator when the terminating impedance differs from the characteristic impedance.

Standing waves are generally present in resonant antenna radiators. This is because the terminating impedance at the end of a typical radiator, which has a finite physical length, is a practically infinite, pure resistance. This means that essentially no current flows there. This sets up standing-wave patterns as shown in the illustration.

The pattern of standing waves on a half-wave resonant radiator, fed at the center, is shown at A. The pattern of standing waves on a full-wave resonant radiator, fed at the

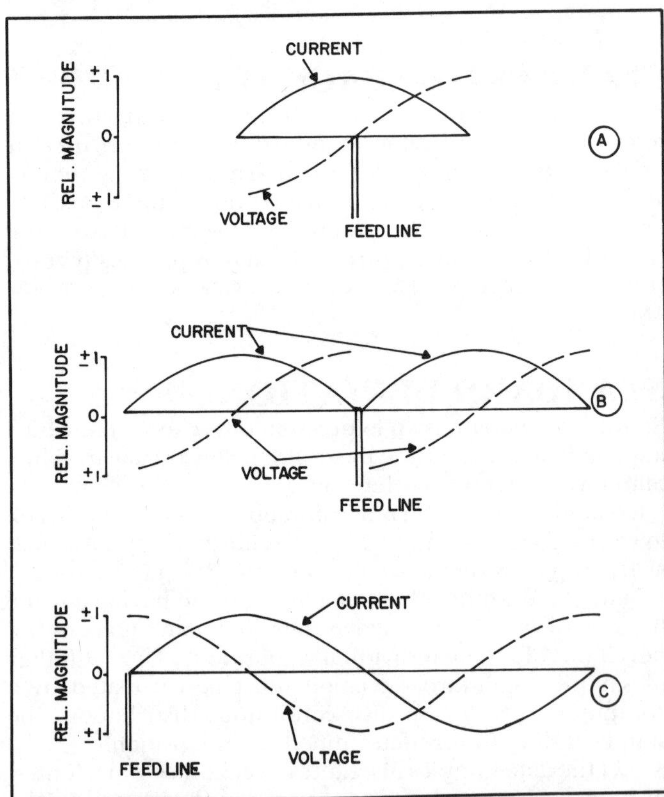

STANDING WAVE: At A, pattern of standing waves on a half-wave dipole. At B, standing waves on a center-fed, full-wave radiator. At C, standing waves on an end-fed, full-wave radiator.

center, is shown at B. The pattern of standing waves on a full-wave resonant radiator, fed at one end, is shown at C. Notice the difference between the pattern for current and the pattern for voltage.

Some antenna radiators do not exhibit standing waves; instead, the current and voltage are uniform all along the length of the radiating element. This is the case in antennas that have a terminating resistor with a value equal to the characteristic impedance of the radiator (about 600 ohms). Some longwire and rhombic antennas employ such resistors to obtain a unidirectional pattern (*see* LONGWIRE ANTENNA, RHOMBIC ANTENNA).

In a transmission line that is terminated in a pure resistance having a value equal to the characteristic impedance of the line, no standing waves occur. This is a desirable condition, because standing waves on a transmission line contribute to power loss. *See also* CHARACTERISTIC IMPEDANCE, STANDING-WAVE RATIO.

STANDING-WAVE RATIO

On a transmission line terminated in an impedance that differs from the characteristic impedance (*see* CHARACTERISTIC IMPEDANCE), a nonuniform distribution of current and voltage occurs along the length of the line. The greater the mismatch, the more nonuniform this distribution becomes.

The ratio of the maximum voltage to the minimum voltage, or the maximum current to the minimum current, is called the standing-wave ratio on the transmission line. The standing-wave ratio is 1:1 if the current and voltage are in the same proportions everywhere along the line. Standing-wave ratio is usually abbreviated SWR.

The SWR can be 1:1 only when a transmission line is terminated in a pure resistance having the same ohmic value as the characteristic impedance of the line. If the load contains reactance, or if the resistive component of the load impedance is not the same as the characteristic impedance of the line, the SWR cannot be 1:1. Then, standing waves will exist along the length of the line (*see* STANDING WAVE). In theory, there is no limit to how large the SWR can be. In the extreme cases of a short circuit, open circuit, or pure reactance at the load end of the line, the SWR is theoretically infinite. In practice, line losses and loading effects prevent the SWR from beoming infinite, but it may be as great as 100:1 or more.

The SWR is important as an indicator of the performance of an antenna system, because a high SWR indicates a severe mismatch between the antenna and the transmission line. This can have an adverse effect on the performance of a transmitter or receiver connected to the system. An extremely large SWR may result in significant signal loss in the transmission line. If a high-power transmitter is used, the large currents and voltages caused by a high SWR can actually damage the line. *See also* IMPEDANCE, REFLECTED POWER, REFLECTOMETER, SMITH CHART, STANDING-WAVE-RATIO LOSS.

STANDING-WAVE-RATIO LOSS

No transmission line is lossless. Even the most carefully designed line has some loss because of the ohmic resistance of the conductors and the imperfections of the dielectric material. In any transmission line, the loss is smallest when

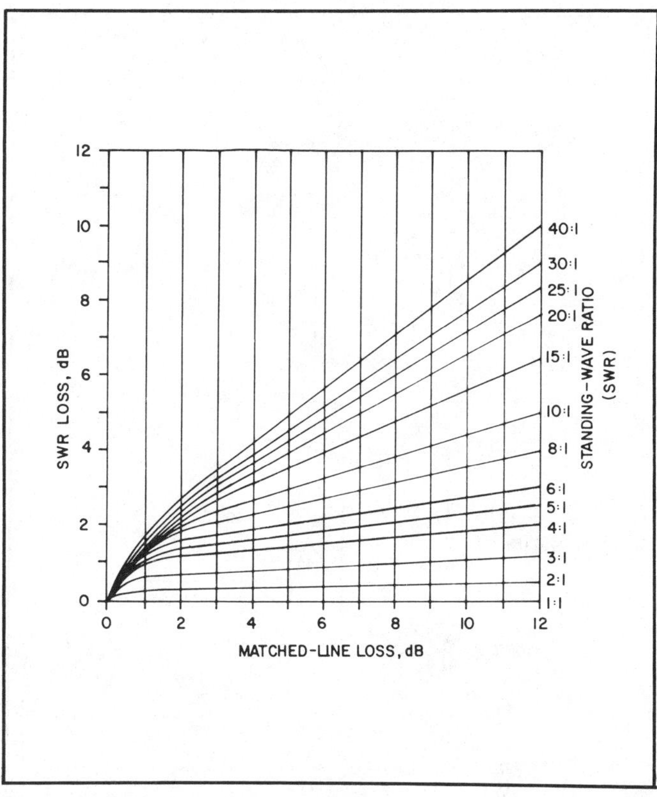

STANDING-WAVE-RATIO LOSS: Standing-wave-ratio loss can be determined with this chart. Find the matched-line loss on the horizontal axis, and the standing-wave ratio (as determined at the antenna feed point) among the family of curves. The corresponding point on the vertical axis represents the loss caused by standing waves.

the line is terminated in a pure resistance equal to the characteristic impedance of the line: that is, when the standing-wave ratio (SWR) is 1:1.

If the SWR is not 1:1, the line loss increases. This additional loss is called standing-wave-ratio loss or SWR loss. The illustration shows the SWR loss that occurs for various values of matched-line loss and SWR. The SWR loss is generally not serious until the SWR becomes greater than 2:1. *See also* STANDING-WAVE RATIO.

STANDING-WAVE-RATIO METER

See REFLECTOMETER.

STANDOFF INSULATOR

A standoff insulator is an insulator that is used to hold an electrical conductor or feed line away from a surface. Standoff insulators are used with open-wire or twin-lead type feed lines.

A typical standoff insulator, suitable for use with television twin-lead, is shown at A in the illustration. The feed line is run through a slot in the insulating material. This type of standoff insulator is commonly available, in various lengths, at most hardware and electronics shops.

A standoff insulator for use with open-wire line must usually be homemade. One method of construction is illustrated at B. A board is painted or coated with wax to protect it against rotting. Two cone-shaped insulators are attached to one end of the board. The entire board is

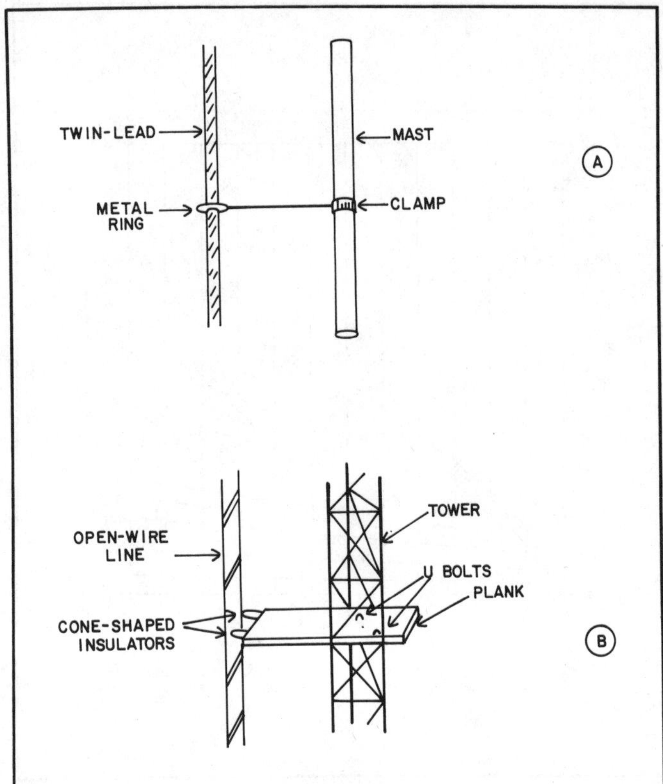

STANDOFF INSULATOR: At A, a standoff insulator suitable for use with television twin-lead. At B, an example of the construction of a standoff insulator for use with open-wire line.

clamped to the tower with U bolts. The feed-line wires are attached to the screws at the ends of the cone-shaped insulators.

Standoff insulators keep the electromagnetic field in an open-wire or twin-lead line from being affected by objects near which the line passes. This keeps the line loss at a minimum, and keeps the characteristic impedance constant all along the length of the line. Standoff insulators are not generally used with coaxial cables. The electromagnetic field is entirely contained within the shield of such a cable, and standoff insulators are therefore not necessary. *See also* OPEN-WIRE LINE, TWIN-LEAD.

STATIC

See SFERICS, STATIC CHARACTERISTIC, STATIC ELECTRICITY.

STATIC CHARACTERISTIC

The steady-state behavior of a component, circuit, or system is known as the static characteristic. The static characteristic of a component is determined using direct currents that do not fluctuate rapidly with time. For example, we may determine the characteristic curve of a transistor or tube by making measurements as we apply various direct currents or voltages. The static characteristic of a circuit or system is an expression of its behavior with no signal input or with a direct-current signal input.

When a component carries an alternating current or a fluctuating direct current, the characteristics may differ from the static condition. In fact, this is usually the case. A diode does not function the same way at 1 GHz as it does with the application of direct current. Even such simple components as resistors, capacitors, or wire conductors behave differently for alternating or fluctuating currents than for direct currents. The behavior of a component under rapidly changing conditions is called the dynamic characteristic. *See also* DYNAMIC CHARACTERISTIC.

STATIC ELECTRICTY

The existence of a potential difference between two objects without the flow of current, or the existence of an excess or deficiency of electrons on an object, is called static electricity. This expression arises from the fact that no charge transfer takes place.

A static electric charge generates an electric field, the intensity of which depends on the potential difference and distance between two objects, or the quantity of charge and its concentration on a single object (*see* CHARGE, ELECTRIC FIELD).

If a static electric potential between two objects becomes large enough, a discharge will occur. The greater the separation distance between the objects, the larger the potential difference must be in order to cause the discharge. *See also* LIGHTNING, SPARK.

STATIC MEMORY

A static memory is an electronic memory that retains its information as long as power is supplied. The data does not change position unless a command is given to that effect. Many integrated-circuit memory devices are of the static type. Some static memory devices draw almost no current, and thus memory can be stored for months or even years by means of the power from a small battery. *See also* MEMORY, MEMORY BACKUP BATTERY.

STATIC SHIFT REGISTER

See SHIFT REGISTER.

STATION

A collection of radio or wire equipment, intended for receiving, broadcasting, or two-way communication, is called a station. There are many different kinds of stations: telephone, radio broadcast, television broadcast, teleprinter, facsimile, short-wave-listening, and amateur-radio, to name just a few.

A receiving station must be equipped with a power supply, a wire or radio receiver, and a wire input or antenna system. A broadcasting station always has a power supply, a transmitter, and a wire output or antenna system. A two-way communications station has a power supply, a receiver, a transmitter, and a wire input/output or antenna system. There may be accessory devices such as terminal units, speech processors, television cameras, and other peripheral equipment.

For information on particular types of station equipment or accessories, please refer to the appropriate individual articles.

STATISTICAL ANALYSIS

Statistical analysis, also known as statistics, is a branch of mathematics that deals with the large-scale behavior of things. In electricity and electronics, statistical analysis is used for quality control, the evaluation of failure and hazard rates, analysis of communications circuits, and many other situations. Statistical analysis is based on the probability that certain things will happen or not happen (*see* PROBABILITY).

A function that indicates the probability of a given event, in terms of some other parameter such as time, frequency, or voltage, is called a probability-density function. A probability-density function may consist of a number (finite or infinite) of values, or of a continuous range of values. The first type of function is called discrete; the second type is called indiscrete or continuous.

Discrete Probability Functions. In the discrete case, the parameter is represented by variables $x_1, x_2, x_3, \ldots, x_n$, where n is some positive integer, or by an infinite number of variables $x_1, x_2, x_3, \ldots$ The probability-density function is expressed as $p(x_k)$, for all integers k ranging from 1 to n in the finite case, or from 1 upward in the infinite case. The sum of all $p(x_k)$ is equal to 1, which is the greatest possible probability.

Let us consider an example of a discrete probability function. Suppose we flip a coin over and over. How likely is it that we will get "tails" every time? The answer obviously depends on the number of times we flip the coin. Let x_1 represent a single flip of the coin; then $p(x_1) = ½$. (The chance of getting "tails" is 1 out of 2.) Let x_2 represent two consecutive coin flips. Then $p(x_2) = ¼$. (The probability that we will get "tails" twice in a row is 1 out of 4.) In general, let x_k represent k consecutive coin flips. Then $p(x_k) = (½)^k$. The sum of all the values $p(x_k)$, for all positive integers k, is equal to 1:

$$\sum_k p(x_k) = 1$$

This probability function is illustrated graphically in the illustration at A.

Indiscrete or Continuous Functions. In the indiscrete case, the value of the parameter x may be any real number, perhaps with one or two limiting constraints. The probability-density function is expressed simply as p(x).

As an example of an indiscrete probability-density function, consider an ordinary incandescent light bulb. If we apply the proper voltage to the bulb, illuminating it to its normal brilliance, how long will the bulb last? We let the parameter x represent time, say in hours. Then p(x), for a given value of x, is the probability that the bulb will be aglow after x hours. When $x = 0$ (the bulb is first turned on), p(x) is very close to 1 (most bulbs do not burn out the instant you apply a voltage). As x increases, p(x) decreases toward zero. We conduct this experiment with a large number of bulbs, and eventually we come up with a meaningful, continuous probability-density function p(x) such as that illustrated at B.

For any continuous probability-density function p(x), there exists a cumulative function P(x) such that:

$$P(x) = \int_{-\infty}^{x} p(s)\, ds$$

The area under the density function is always equal to 1:

$$\int_{-\infty}^{\infty} P(x)\, dx = 1$$

One kind of continuous function, called the normal distribution, is frequently observed in the behavior of things. This curve is bell-shaped (*see* NORMAL DISTRIBUTION).

For Further Information. A complete course on statistical analysis is far beyond the scope of this book. The examples here are intended only to illustrate the general nature of discrete and indiscrete statistical functions. The applications of statistical analysis are widely varied. Many excellent college-level texts are available on statistical analysis and its applications in physics and electronics. A concise and rigorous discussion of statistical analysis, and its applications in electronics, is provided in *Reference Data for Radio Engineers* (Howard W. Sams & Co., Inc., 1975), in the chapters on information theory, probability and statistics, and reliability and life testing.

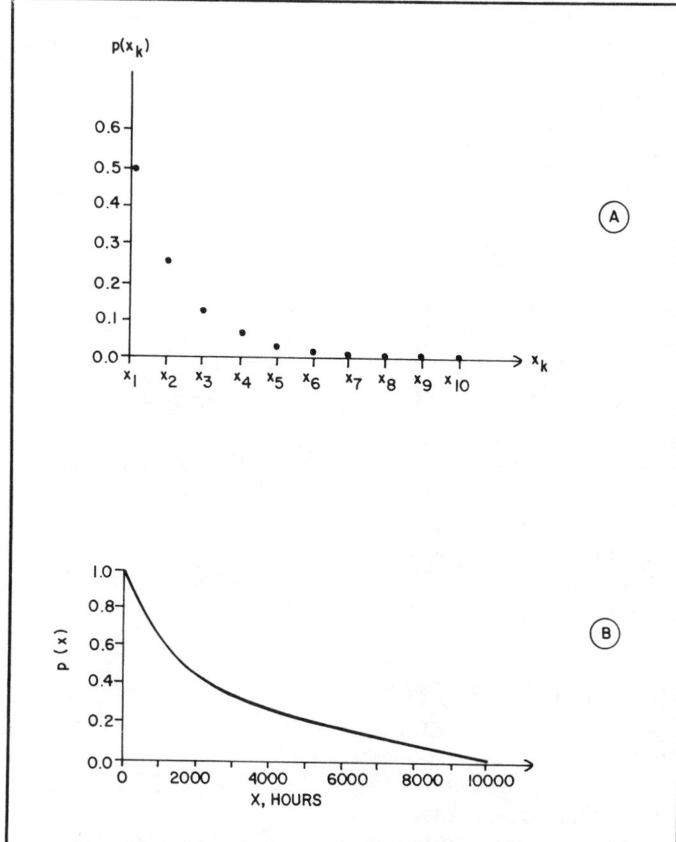

STATISTICAL ANALYSIS: At A, a discrete probability function. At B, a continuous probability function.

STATISTICAL COMMUNICATIONS THEORY

See INFORMATION THEORY.

STATISTICS

See STATISTICAL ANALYSIS.

STATOR

In a variable capacitor, the plate or set of plates that remains fixed is called the stator. The stator is usually connected to a terminal that is insulated from the frame of the capacitor. In a motor, the stationary coil is called the stator. In a rotary switch, the nonmoving set of contacts is sometimes called the stator. The moving set of plates, coils, or contacts is called the rotor. *See also* AIR-VARIABLE CAPACITOR, MOTOR, ROTARY SWITCH, ROTOR.

STAT UNITS

Electrostatic units are usually given in the Standard International (SI) system (*see* STANDARD INTERNATIONAL SYSTEM OF UNITS). However, the centimeter-gram-second (CGS) system is occasionally used as the basis for determining electrostatic quantities. The CGS units are called stat units.

The statampere is equal to about 0.336 ampere, the statvolt is approximately 300 volts, the statcoulomb is about 0.336 coulomb, and the statohm is about 8.99×10^{-12} ohm. A CGS electromagnetic system of units also exists; these units are called ab units. Neither ab units nor stat units are generally used in electronics. *See also* AB UNITS.

STATUTE MILE

The statute mile is a common English unit of distance that is used on land. A statute mile is equivalent to 5,280 feet, or 1.609 kilometers.

At sea, a unit called the nautical mile is used. The nautical mile is a little longer than the statute mile. To obtain a distance in statute miles when it is specified in nautical miles, multiply by 1.151. Conversely, multiply by 0.8688. *See also* NAUTICAL MILE.

STEATITE

Steatite is a ceramic material made from magnesium silicate. Steatite has excellent dielectric properties, and is extensively used in insulators for radio-frequency equipment. The dielectric constant is approximately 5.8 throughout the radio-frequency spectrum. Steatite appears somewhat like porcelain. *See also* DIELECTRIC.

STEEL

Steel is a metallic alloy made from iron and carbon. Steel is noted for its high physical resiliency and strength. It is a fairly good conductor of heat and electricity.

Steel is sometimes used as a core material for copper wire. The steel core provides tensile strength for the wire, and reduces the tendency for stretching in long spans (*see* COPPER-CLAD WIRE).

Steel is commonly used in the construction of antenna supports. When exposed to the environment, steel will rust unless it is protected in some way. The most common method of protecting steel is to coat it with zinc. This process is called galvanizing. Steel may also be coated with chromium, or protected by paint.

Steel is a ferromagnetic material. A piece of steel is attracted by a magnet. Steel can be permanently magnetized or used as the core for an electromagnet. *See also* FERROMAGNETIC MATERIAL.

STEEL WIRE

Steel is extensively used in the manufacture of wire requiring high tensile (breaking) strength. In antenna systems, copper-clad steel wire is often used (*see* COPPER-CLAD WIRE). Steel is also commonly employed in the manufacture of guy wire (*see* GUYING).

Steel wire is generally coated with zinc (galvanized) to prevent it from rusting and becoming weak. Steel wire is available in solid or stranded form, in the common American Wire Gauge (AWG) sizes. Steel wire has a maximum breaking strength of approximately twice that of annealed copper wire, or 1.5 times that of hard-drawn copper wire. *See also* AMERICAN WIRE GAUGE.

STEFAN-BOLTZMANN LAW

The Stefan-Boltzmann law is a function that relates the temperature of a black body, or perfectly radiating object, with the amount of power it radiates per square meter of surface area. If W represents the radiant emittance in watts per square meter, and T represents the absolute temperature in degrees Kelvin, then:

$$W = \sigma T^4$$

where σ is a constant called the Stefan-Boltzmann constant. The value of this constant is approximately 5.67×10^{-8} in the above formula. *See also* BLACK BODY.

STENODE CIRCUIT

In a superheterodyne receiver, a resonant crystal circuit can be used to obtain a high degree of selecitivty. This can be done in various ways. The stenode circuit is a special form of piezoelectric resonant circuit.

The stenode actually operates by cancelling out all signals at frequencies other than the desired frequency. The crystal is connected between the output of one intermediate-frequency amplifier and the input of the same stage or a previous stage, in such a way as to produce negative feedback. The negative feedback causes the gain of the stages to be greatly reduced, except at the crystal frequency, where the crystal acts as a trap to block the feedback.

The advantage of a stenode circuit is that it is very stable. Oscillation is unlikely. The stenode circuit is simple to construct and align.

STEP-DOWN TRANSFORMER

A step-down transformer is a transformer in which the output voltage is smaller than the input voltage. The primary-to-secondary turns ratio is the same as the input-to-output voltage ratio. The input-to-output impedance ratio is equal to the square of the input-to-output voltage ratio. *See also* IMPEDANCE TRANSFORMER, TRANSFORMER, VOLTAGE TRANSFORMER.

STEP FUNCTION

A step function is a form of discontinuous mathematical function (*see* DISCONTINUOUS FUNCTION). There exist many discontinuities at regular intervals. The step function is sometimes called the U.S. Post Office function, since postal rates are graduated according to a step function (see illustration).

The value of a step function may generally increase or

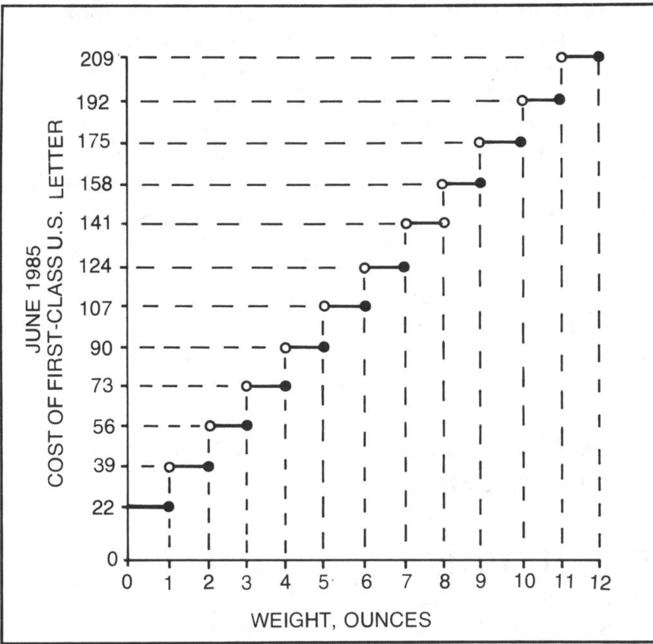

STEP FUNCTION: An example of a step function, in this case the U.S. Post Office function for first-class mail in 1985.

generally decrease as the value of the independent variable gets larger. *See also* FUNCTION.

STEPPED LEADER

See LEADER, LIGHTNING.

STEP-RECOVERY DIODE

A step-recovery diode, also called a charge-storage or snap diode, is a special form of semiconductor device that is used chiefly as a harmonic generator. When a signal is passed through a step-recovery diode, literally dozens of harmonics occur at the output. A step-recovery diode, connected in the output circuit of a very-high-frequency transmitter, can be used in conjunction with a resonator to obtain signals at frequenices up to several gigahertz. Step-recovery diodes are also used to square off the rise and decay characteristics of digital pulses.

A step-recovery diode is manufactured in a manner similar to the PIN diode (*see* PIN DIODE). Charge storage is accomplished by means of a very thin layer of intrinsic semiconductor material between the P-type and N-type wafers. The charge carriers accumulate very close to the junction.

When the step-recovery diode is forward-biased, current flows in the same manner as it would in any forward-biased P-N junction. However, when reverse bias is applied, the conduction does not immediately stop. The step-recovery diode stores a large number of charge carriers while it is forward-biased, and it takes a short time for these charge carriers to be drained off. Current flows in the reverse direction until the charge carriers have been removed from the P-N junction, then the current falls off with extreme rapidity. The transition time between maximum current and zero current may be less than 10^{-10} second. It is this rapid change in the current that produces the harmonic energy. *See also* P-N JUNCTION.

STEP-UP TRANSFORMER

A step-up transformer is a transformer in which the output voltage is larger than the input voltage. The input-to-output voltage ratio is equal to the primary-to-secondary turns ratio. The input-to-output impedance ratio is equal to the square of the input-to-output voltage ratio. *See also* IMPEDANCE TRANSFORMER, TRANSFORMER, VOLTAGE TRANSFORMER.

STEP-VOLTAGE REGULATOR

A step-voltage regulator is a special form of alternating-current regulating transformer that allows manual adjustment of the output voltage to compensate for changes in the load (*see* REGULATING TRANSFORMER).

The step-voltage regulator is connected in series with the secondary winding of the main transformer, just as is the case with an ordinary series regulating transformer. The phase may either add to, or buck against, the output of the main transformer. Regulation is accomplished without interrupting the power to the load, and without creating transient surges.

STERADIAN

The steradian, also called a unit solid angle, is a three-dimensional angle subtended by the vertex of a cone having a certain shape.

Suppose we have a sphere of any size, and a circle on the surface of the sphere that encloses $1/4\pi$ (7.96 percent) of the surface area of the sphere. This area is equal to the square of the radius of the sphere. If we construct a cone that passes through this circle and has its vertex at the center of the sphere, then the angle at the vertex of the cone has a solid measure of one steradian (see illustration).

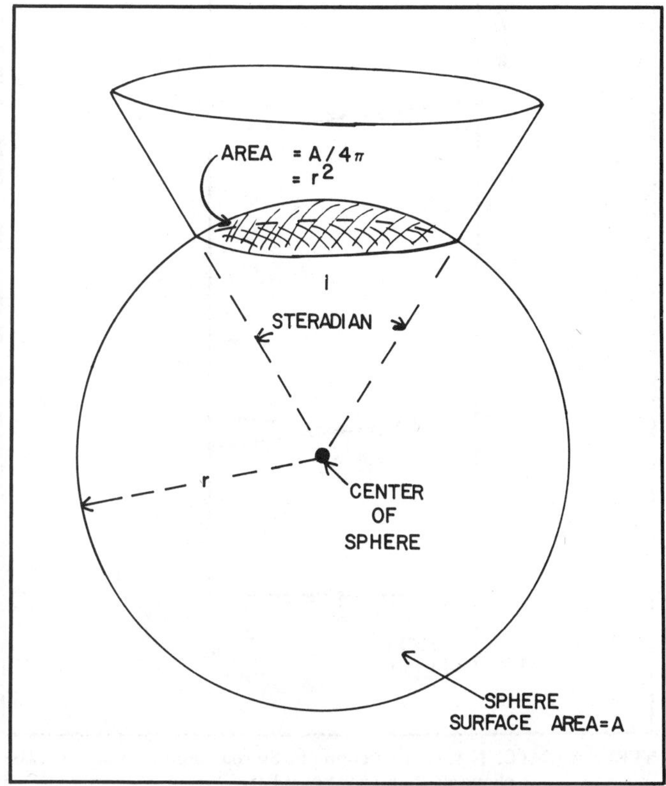

STERADIAN: The steradian is a unit of solid angular measure.

The steradian is often used in determination of luminous intensity. *See also* CANDELA, LUMEN, LUMINOUS FLUX, LUMINOUS INTENSITY.

STERBA ANTENNA

A Sterba antenna, also called the Sterba-curtain array, is a directional array that exhibits gain with respect to a dipole antenna. The Sterba-curtain antenna consists of several elements fed by transposed sections of transmission line (see illustration). The feed point is normally at the end of the array.

The two end elements of the Sterba curtain array are ¼ wavelength long, providing an impedance transformation for the input and a termination for the last section. When fed as shown, the input is at a current maximum; points A are current minima.

The Sterba antenna has a bidirectional response pattern, with the maxima perpendicular to the plane of the elements; therefore, it is a broadside array. The radiation (or response) is in and out of the page as you look at the illustration (*see* BROADSIDE ARRAY). A screen could be placed ¼ wavelength behind the array to provide a unidirectional response.

Sterba-curtain arrays can be used at any frequency up through vhf. Then were once popular for point-to-point communications systems on the hf bands, where their directional characteristics provided an advantage. The array can be mounted with the elements vertical or horizontal.

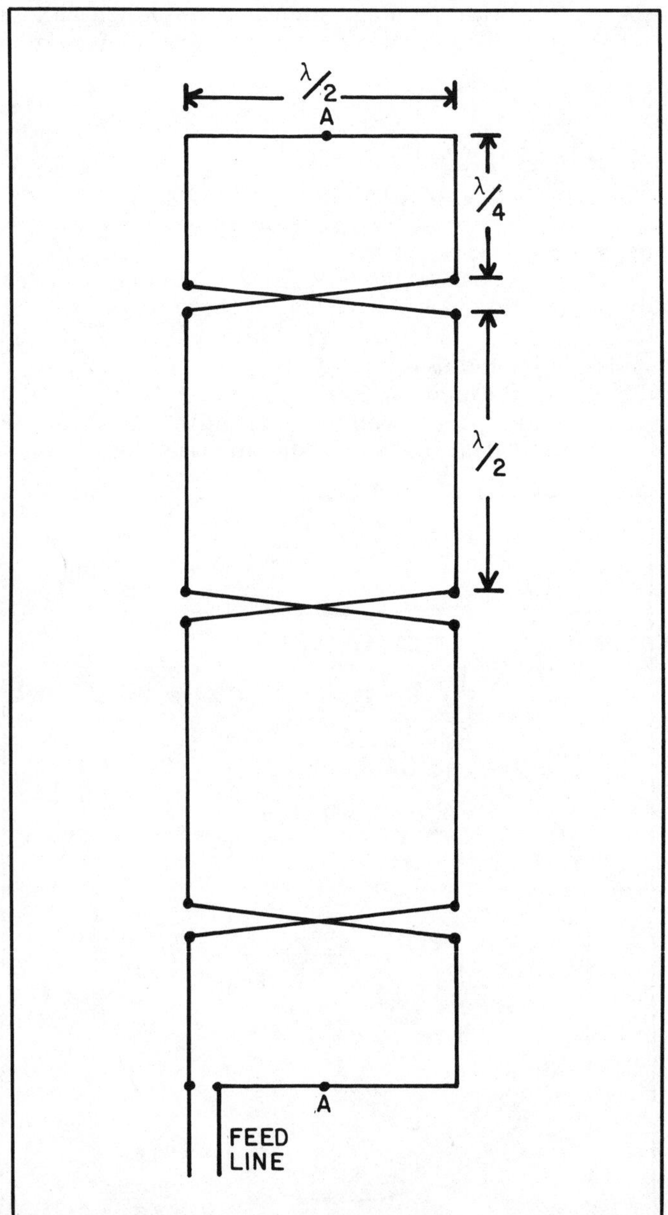

STERBA ANTENNA: Construction of a Sterba antenna. There may be more or fewer elements than are shown here.

STEREO

See STEREOPHONICS.

STEREO BROADCASTING

See FM STEREO, STEREO MULTIPLEX.

STEREO DISK RECORDING

It is possible to impress two independent sound channels onto a single groove in a disk, thereby obtaining stereo sound. This is done by a device called a stereo disk recorder.

The stereo disk recorder cuts a groove by vibrating in the plane perpendicular to the groove. The left-hand-channel information is impressed in the groove by vibrations that occur at a 45-degree angle with respect to the surface of the disk. The right-hand-channel information is impressed in the groove by vibrations at a right angle to those of the left-hand channel (see illustration).

Once the disk recording process is complete, a mold is made from the disk. Mass production consists of pressing this metal mold into disks of vinyl plastic.

When a stereo disk is played back, the stylus follows in

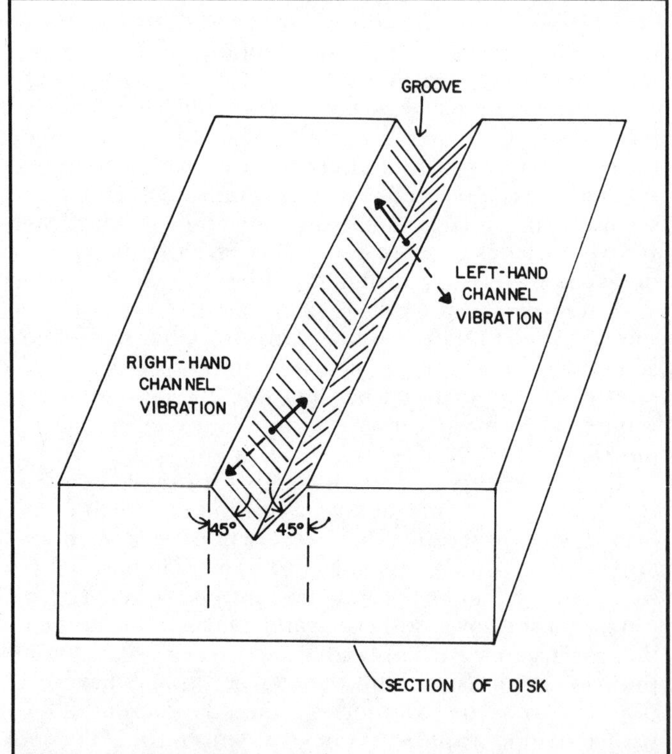

STEREO DISK RECORDING: A stereo groove contains two independent channels of information, impressed on faces oriented at right angles to each other.

the groove, vibrating in the same way as did the original cutting stylus. Since the left-hand and right-hand channel vibrations occur at right angles to each other, their instantaneous amplitudes are entirely independent. *See also* STEREOPHONICS.

STEREO MULTIPLEX

Stereo multiplex is a method of broadcasting two independent channels of sound over a single radio-frequency carrier, thereby obtaining stereo sound in a specially designed receiver.

To accomplish stereo multiplex, the left-hand and right-hand audio signals modulate low-frequency carriers (subcarriers). The two subcarriers have different frequencies. The main carrier is modulated by both subcarrier signals at the same time. Linearity in the circuits is extremely important to maintain separation between channels. A block diagram of a stereo-multiplex transmitter is shown at A.

The stereo-multiplex receiver uses a demodulator to obtain the two subcarrier signals from the main carrier. A pair of selective circuits separates the subcarrier channels. A demodulator is then used in each channel to obtain the original left-hand and right-hand audio signals (B). *See also* FREQUENCY-DIVISION MULTIPLEX.

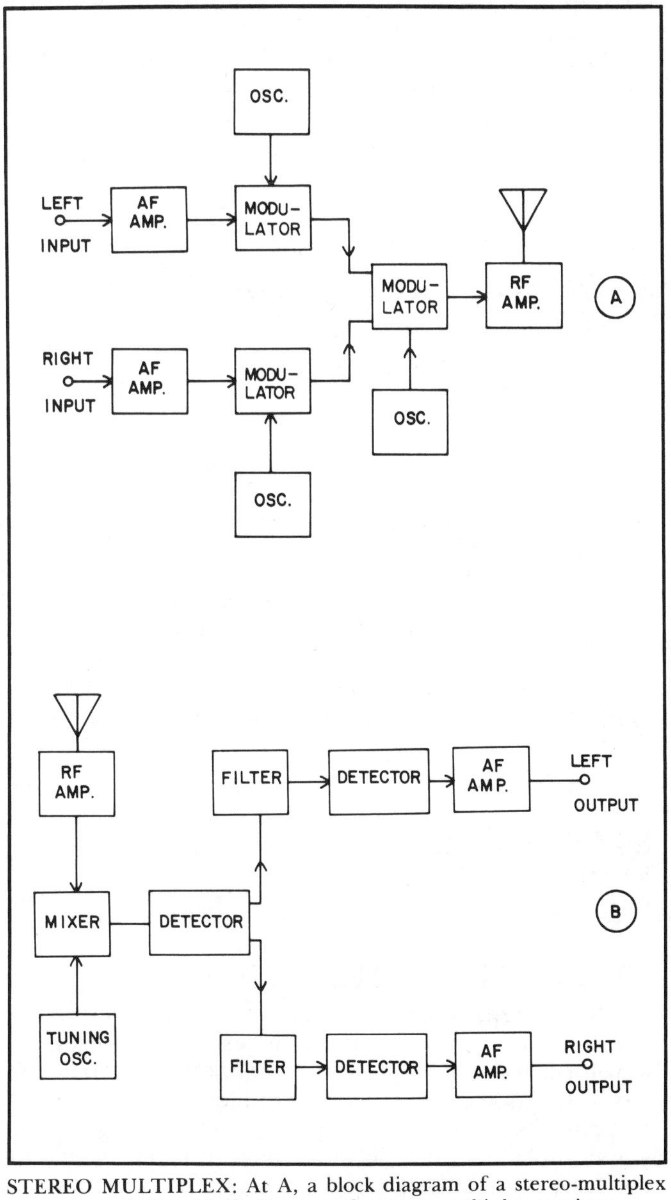

STEREO MULTIPLEX: At A, a block diagram of a stereo-multiplex transmitter. At B, a block diagram of a stereo-multiplex receiver.

STEREOPHONICS

Stereophonics, also known as stereo, is the science of two-channel sound reproduction, especially of high-fidelity music. Stereo music sounds more realistic than monaural music because the listener gets a sense of space orientation.

A stereo recording has two channels, generally called the left-hand and right-hand channels. There may also be a center channel, obtained by combining the left and right channels. The two channels are independent. The two channels are recorded with separate microphones and played back through separate speakers or earphones. This requires two sets of audio recording and reproduction systems.

Stereo is almost universally used in the recording of commercial music (*see* STEREO DISK RECORDING, STEREO TAPE RECORDING). Stereo can be used for creating strange and interesting sound effects, such as the noise of a jet airplane taking off or a train passing close by. Some stereo systems employ four channels for even more enhanced effects. A tremendous variety of stereo systems and accessories are manufactured commercially. *See also* HIGH FIDELITY, QUADROPHONICS.

STEREO TAPE RECORDING

Most stereophonic recording is done, at least initially, on magnetic tape. The left-hand and right-hand channels are recorded on parallel paths or tracks simultaneously.

A stereo tape recorder has two separate and independent sets of microphones, audio amplifiers, and recording heads. Some stereo tape systems have room for two or four simultaneous stereophonic recordings. In some stereo tape recorders, the input (microphone) gain can be controlled separately for each channel. All stereo tape recorders have high-fidelity audio circuits for reproduction of music.

In stereo recording, it is important that the sound be synchronized for each channel. Otherwise, an echoing or false-motion effect takes place. The heads must be precisely aligned, both while recording and while playing back.

The tape speed for good stereophonic recording must be fast enough to allow excellent reproduction of high-frequency sounds. The most commonly used speeds for commercial recording are 7.5 and 15 inches per second. Some home systems use a speed of 3.75 inches per second. It is extremely important that the tape speed be constant. The slightest variation will result in noticeable and objectionable flutter.

Stereo tape must be thick enough so that stretching does not occur while recording, playing back, or rewinding the tape. Even a tiny bit of tape stretching will seriously degrade the quality of sound reproduction. The shorter-playing tapes are generally thicker, for a given cassette or reel size, than longer-playing tapes. *See also* EIGHT-TRACK RECORDING, FOUR-TRACK RECORDING, MAGNETIC TAPE, STEREOPHONICS, TAPE RECORDER, TWO-TRACK RECORDING.

STIMULATED EMISSION

When electrons move from a higher energy level to a lower energy level, photons are emitted at discrete wavelengths. The energy contained in the photon, and consequently the wavelength of the photon, depends on the difference between the electron energy levels before and after the transition. The farther an electron falls, the more energy it gives up, and the shorter will be the wavelength of the emitted photon.

Stimulated emission is the basis for the operation of the mercury-vapor, neon, and sodium-vapor lamps. Rarefied gases are raised to high temperatures, exciting the electrons. Each electron is driven away from the atomic nucleus, and wanders around freely among other nuclei. Occasionally a nucleus captures the electron, causing stimulated emission. Then the electron is again driven away by the applied energy.

Diffuse gases in space, excited by the energy from stars in the vicinity, exhibit stimulated emission. Each element or compound has its own unique set of emitted wavelengths. These wavelengths can be determined with a radio telescope or spectroscope. *See also* ELECTRON ORBIT, PHOTON, PLANCK'S CONSTANT, SPECTROSCOPE.

STOCHASTIC PROCESS

Any deliberately randomized process is called stochastic. Stochastic processes are applied in diverse ways in all scientific fields, including electronics. Such processes are needed in quality control, and in evaluating the behavior of molecules, atoms, and subatomic particles.

Consider the problem of determining the number of faulty diodes in a large lot. Suppose that 1,000,000 diodes have just been made. We surely cannot test them all, so we choose 1,000 of them at random—by stochastic process—and find that three of them do not function. Thus, the hazard rate, or proportion diodes that are bad, is 0.3 percent. We should expect, then, that there will be about 3,000 bad diodes in the lot.

The implementation of a stochastic process requires the use of random numbers. Truly random numbers are not easily obtained. Usually, pseudorandom numbers, generated by a computer from a special algorithm, are used. *See also* PSEUDORANDOM NUMBERS, RANDOM NUMBERS.

STOPBAND

A selective filter causes attenuation of energy at some frequencies. The band of frequencies at which a filter has high attenuation is called the stopband. Generally, the stopband is that band for which the filter blocks at least half of the applied energy. In other words, the stopband is that part of the spectrum for which the power attenuation is 3 dB or more. Sometimes a 6-dB attenuation figure is specified instead.

We may speak of the stopband of any kind of filter. But this term is most often used with band-rejection filters. *See also* BANDPASS FILTER, BANDPASS RESPONSE, BAND-REJECTION FILTER, BAND-REJECTION RESPONSE, HIGHPASS FILTER, HIGHPASS RESPONSE, LOWPASS FILTER, LOWPASS RESPONSE.

STORAGE

Storage refers to the retention of data or information, either analog or digital, and by any means. The electrical or magnetic storage of digital information is usually called memory.

The deliberate accumulation of an electric charge is called storage. Electric charge is stored in such things as batteries and capacitors. *See also* MEMORY, STORAGE BATTERY, STORAGE CELL, STORAGE OSCILLOSCOPE.

STORAGE BATTERY

A storage battery, also known as a secondary battery, is a battery capable of accepting and releasing a large quantity of electric charge. The battery contains two or more cells that each have a positive plate and a negative plate immersed or embedded in an electrolyte material. A common automobile battery is an example of a storage battery.

The charge in a storage battery is usually expressed in terms of ampere hours. An ampere hour is equivalent to 3,600 coulombs. A typical automotive storage battery can hold approximately 50 to 100 ampere hours of electric charge.

Storage batteries can be charged in a short time at a high level of current, or over a longer period of time using correspondingly less current. *See also* QUICK CHARGE, TRICKLE CHARGE.

STORAGE CATHODE-RAY TUBE

See STORAGE OSCILLOSCOPE.

STORAGE CELL

A storage cell is a single electrochemical cell that is capable of accepting and releasing a quantity of electric charge. Storage cells supply approximately 1 V to 1.5 V. The amount of charge that can be stored depends on the physical size of the cell, and on the materials that make up the cell.

Two or more storage cells, connected in series, form a storage battery. *See also* CELL, STORAGE BATTERY.

STORAGE LIFE

See SHELF LIFE.

STORAGE OSCILLOSCOPE

In the design, alignment, and servicing of electronic equipment, it is sometimes necessary to observe a waveform for a long period of time, or to compare two waveforms that occur at different times. A storage oscilloscope is designed with this in mind. The storage oscilloscope can be operated in the same way as a conventional oscilloscope if desired.

The heart of the storage oscilloscope is a special

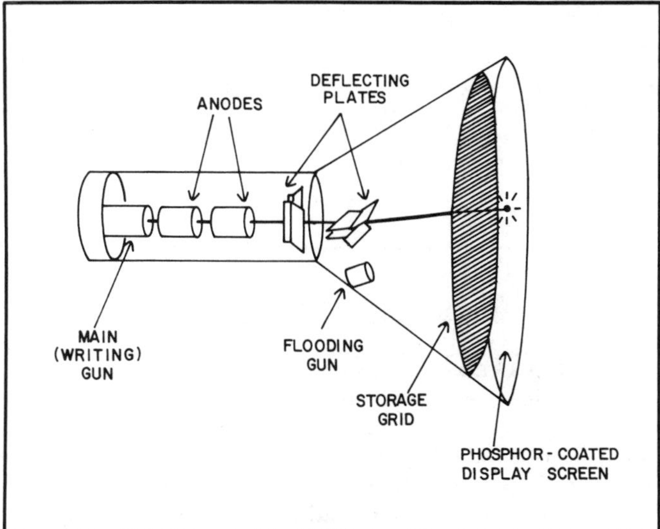

STORAGE OSCILLOSCOPE: Construction of a storage cathode-ray tube for use in a storage oscilloscope.

cathode-ray tube that can hold an electric charge in a pattern corresponding to the waveform displayed. The illustration is a simplified cutaway view of a cathode-ray tube. There are two electron guns, called the writing or main gun and the flooding gun.

A fine grid of electrodes, just inside the phosphor screen, is saturated with negative charge prior to storing a waveform. Then the oscilloscope is adjusted until the waveform appears as desired. The saturating charge is removed from the grid, and it acquires localized positive charges wherever it is struck by the electron beam from the main gun. This positive charge decays very slowly unless it is deliberately removed.

When a stored waveform is to be observed, the grid of electrodes is irradiated by electrons from the flooding gun. The grid prevents these electrons from reaching the phosphor, except in the areas of localized positive charge. Thus the original waveform is reproduced on the phosphor screen. The waveform can be viewed for as long as the operator wants, up to several minutes. The stored waveform is erased by saturating the grid with negative charge. *See also* OSCILLOSCOPE.

STORAGE TIME

When the voltage at the base of a bipolar transistor is beyond the saturation level, an excess current flows in the emitter-base circuit. This current is greater than the amount required to keep current flowing in the collector circuit. Because of this excess current, an electrical charge accumulates in the base of the transistor.

If the bias is suddenly removed from the base of the transistor, the accumulated charge must dissipate before the transistor will come out of saturation. This requires a small amount of time, known as the storage time, abbreviated t_s.

Storage time depends on several factors, including the size of the base electrode, the type and concentration of impurity in the semiconductor material, and the extent to which the transistor is driven beyond saturation. The storage time limits the speed at which switching can be accomplished by means of saturated logic. *See also* SATURATION, SATURATED LOGIC, SATURATION VOLTAGE.

STORE

A memory file is sometimes called a store. A store may contain any amount of information, usually specified in bytes, kilobytes, or megabytes.

The act of putting information into memory is called a STORE operation. *See also* BYTE, FILE, MEMORY, STORING.

STORING

Storing is the process of putting information into a memory circuit (*see* MEMORY). Storing is usually accomplished by a simple sequence of operations, or even a single operation, such as pushing a button.

For example, in a small electronic calculator, we may wish to store a number in the memory. We simply press a button, perhaps labeled M IN (memory input), and the displayed number is stored in the memory. If there is more than one memory, we must press two buttons, such as M IN and the memory address (*see* ADDRESS).

In computers, the storing process usually involves several operations. We must tell the computer which file we want to store, and where we want to store it.

STORM LOADING

See ICE LOADING, WIND LOADING.

STRAIGHT RADIATOR

An antenna radiator is straight if it lies along or nearly along a single geometric line. Straight radiators are used in most dipole antennas, phased arrays, and yagi antennas (*see* DIPOLE ANTENNA, PHASED ARRAY, YAGI ANTENNA).

Some antennas do not use straight radiators. At low, medium, and high frequencies, there may not be enough physical space for a straight antenna. Then a bent antenna is used (*see* BENT ANTENNA). A special form of bent antenna is called a loop (*see* LOOP ANTENNA).

STRAIGHT-LINE EQUATION

See LINEAR EQUATION.

STRAIGHT-LINE FUNCTION

See LINEAR FUNCTION.

STRAIN

When an object is subjected to physical forces, that object may change in shape. Strain is an expression of the extent to which an object changes shape under stress (*see* STRESS).

Strain is an important consideration for wires run in long spans. Some metals change shape quite easily since they are highly elastic. An example of such a substance is annealed, or soft-drawn, copper wire. A long antenna of solid, soft-drawn copper wire will stretch. The stretching is clearly undesirable if an antenna is to remain resonant at the same wavelength.

Some materials are relatively strain-resistant. Hard-drawn or stranded copper wire stretches much less than solid, soft-drawn copper wire. Steel wire stretches less than copper. This is part of the reason why copper-clad steel wire is preferred in the construction of wire antennas (*see* COPPER-CLAD WIRE, WIRE ANTENNA).

Strain is of importance in other applications besides the construction of wire antennas. A piezoelectric device operates by producing electrical impulses as a result of mechanical strain, but the crystal will break if it is put under too much strain. Electrostriction and magnetostriction are examples of strain that occurs in various substances in the presence of strong electric or magnetic fields (*see* ELECTROSTRICTION, MAGNETOSTRICTION).

Strain can be measured by means of a device called a strain gauge. *See also* STRAIN GAUGE, STRESS, WIRE.

STRAIN GAUGE

A strain gauge is a device capable of quantitatively measuring the extent to which a body changes shape under stress. Strain can be measured in various different ways (*see* STRAIN).

One common method for determining strain is to attach one or more resistive wires to the object under test. The wires stretch along with the body as it changes shape. The stretching of wire causes a change in the dc resistance. This resistance can be measured, and the amount of strain thereby ascertained.

The strain on a long span of wire can be determined in much the same way. If the wire stretches, becoming thinner and longer, the resistance will increase. If the characteristics of the wire metal are known, the amount of stretching can be calculated from the change in the direct-current resistance of the complete span.

The strain on some objects can be measured by means of light interference patterns. A change in the physical shape of an object affects the way it reflects visible light. The amount of stress can be directly measured in some cases, and the strain determined indirectly from the force on the object under test.

STRAIN INSULATOR

Insulators are often used in guying systems to eliminate resonances that can affect the radiation pattern of an antenna. Such insulators are subjected to tremendous stress in the guy wires of a large tower. This stress can pull a conventional insulator apart. This can cause a catastrophe!

A strain insulator is designed so that the stress compresses the material instead of pulling at it. Most substances can endure far more stress in the form of compression rather than tension. Strain insulators are thus prefered in guying systems.

A typical strain insulator is shown. The two wire loops are oriented at right angles with respect to each other. Stress causes the insulator to be squeezed. If the stress becomes too great, and the insulator breaks, the wire span will not fall, since the wire loops interlock.

Strain insulators have a fairly large amount of leakage capacitance, and are not used in active wire-antenna elements unless conventional insulators are unable to withstand the stress. *See also* GUYING, INSULATOR.

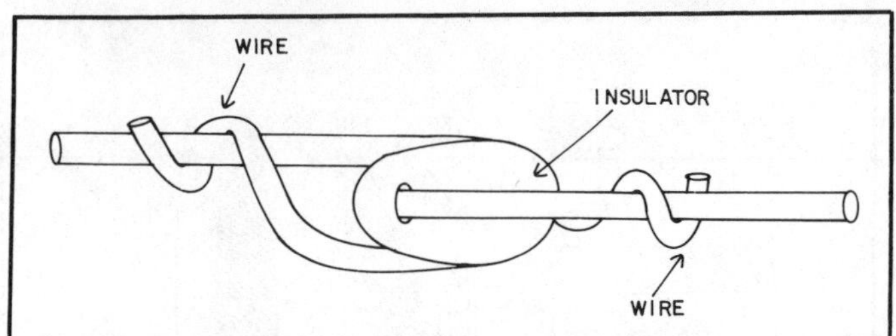

STRAIN INSULATOR: A strain insulator is compressed by the tension of the attached wires.

STRANDED WIRE

Several thin wires can be wound, or spun, together to form a single, larger conductor. Wire is often made in this way. Such wire is called stranded. Most stranded wires have seven individual conductors; some have 19 conductors.

Stranded wire has some advantages over solid wire in electrical and electronic work. Stranded wire is less likely to break when subjected to flexing. Stranded wire adheres more readily to solder. When used in antennas, stranded wire stretches less than solid wire. Stranded wire is less susceptible to radio-frequency skin effect than solid wire, because the total surface area is larger than that of a solid conductor of the same diameter (*see* SKIN EFFECT).

Stranded wire also suffers from certain shortcomings compared with solid wire. Uninsulated stranded wire corrodes more quickly than uninsulated solid wire. Stranded wire generally cannot be used in the winding of coils; finely stranded copper wire does not hold its position when bent. The cross-sectional area of stranded wire is less than that of solid wire having the same outside diameter; therefore, stranded wire has slightly higher direct-current resistance than solid wire.

The size of stranded wire can be specified in either of two ways. The gauge of the entire conductor can be specified, such as AWG No. 12. Alternatively, the number of strands and the gauge of each strand can be indicated. A wire made from seven strands of AWG No. 20 wire, for example, might be called 7 × AWG No. 20. Wire having very fine strands is sometimes called Litzendracht (Litz) wire. *See also* AMERICAN WIRE GAUGE, BIRMINGHAM WIRE GAUGE, BRITISH STANDARD WIRE GAUGE, LITZ WIRE.

STRAPPING

Some printed circuits can be used for more than one purpose, depending on the way in which external wires are connected. For example, an amplifier and filter can be wired to operate as an oscillator, or an oscillator can be wired to operate at any of several fixed frequencies. The manner in which the wires are connected among designated points, for obtaining the desired operation, is called strapping.

In a magnetron tube, oscillation sometimes occurs at unwanted frequencies. To prevent this, metal strips are connected among the resonator segments. This is known as strapping. *See also* MAGNETRON.

STRESS

Stress is any static force exerted against a body. Stress is normally measured in pounds or kilograms per unit of

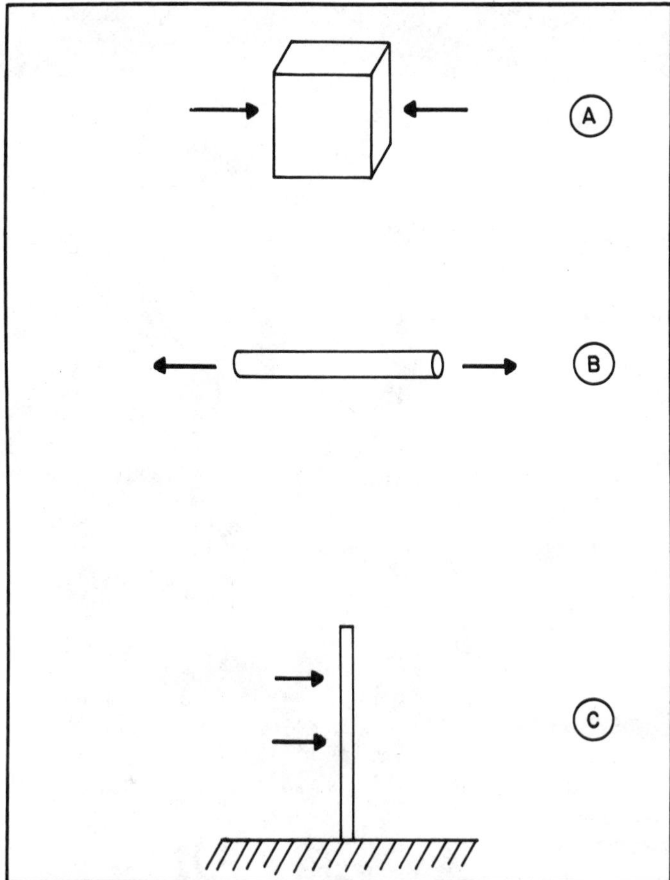

STRESS: Stress may take the form of compression (A), tension (B), or lateral force (C).

cross-sectional area. Stress can occur in three different ways: constriction (pushing in), tension (pulling apart), or laterally (sideways). These three kinds of stress are shown schematically in the illustration.

Stress causes distortion in the shape of an object. A dielectric material changes shape because of electrostriction. Wire stretches when it is subjected to tension. A metal-tubing antenna bends because of the lateral stress caused by wind. The change in the shape of a body, induced by stress, is called strain. *See also* STRAIN.

STRING ELECTROMETER

A string electrometer is a sensitive device for measuring electrostatic charge with practically zero current drain. The string electrometer operates in a manner much different from that of a conventional electrometer. The principle more closely resembles that of the electroscope or electrostatic instrument (*see* ELECTROMETER, ELECTROSCOPE, ELECTROSTATIC INSTRUMENT).

A fine, flexible, conducting wire is tightly strung midway between two parallel metal plates. The wire is electrically connected to one of the plates, and is mechanically connected to an indicator needle.

When an electrostatic charge is applied to the plates, so that one plate is negative and the other is positive, the wire is repelled away from the plate to which it is connected, and attracted toward the other plate. This causes the needle to move upscale. The amount of deflection is proportional to the electrostatic charge applied to the plates. *See also* ELECTROSTATIC DEFLECTION, ELECTROSTATIC FORCE.

STRIPLINE

Stripline, also (somewhat imprecisely) called microstrip, is a method of constructing radio-frequency tuned circuits at short wavelengths. A section of transmission line, measuring a quarter wavelength or a half wavelength, exhibits resonant properties. If the wavelength is very short, a quarter or half wavelength may be smaller than the length or width of a typical printed-circuit board. Then, it is practical to construct tuned circuits from parallel foil runs.

An example of stripline construction is shown in the photograph. This circuit is a power amplifier that delivers about 40 watts of radio-frequency output at 440 MHz. At this frequency, the wavelength is about 75 centimeters in free space. The electrical quarter wavelength, considering the velocity factor of the circuit-board dielectric, is only 15 centimeters, or a little less than 6 inches.

Stripline construction results in a high Q factor and excellent efficiency. *See also* HALF-WAVE TRANSMISSION LINE, Q FACTOR, QUARTER-WAVE TRANSMISSION LINE.

STRIP TRANSMISSION LINE

A specialized form of parallel-conductor radio-frequency transmission line, made of metal strips rather than wires, is called a strip transmission line. The strip transmission line is a balanced line (*see* BALANCED LINE).

The illustration shows the strip-transmission-line configuration. The spacing between the strips, as measured between the two inside surfaces, is d, and the width of each strip is w. (It is assumed that the two strips are identical in thickness and width.) The characteristic impedance of an air-dielectric strip transmission line is given by:

$$Z_o = 377d/w$$

Strip transmission lines are generally used at very-high and ultra-high frequencies. *See also* OPEN-WIRE LINE.

STROBOSCOPE

A stroboscope is a device that is used to measure the rate of rotation or revolution of a motor, or of an object attached to the shaft of a motor.

A stroboscope emits bright, extremely brief flashes of light at regular intervals. The flash rate can be adjusted. The stroboscope flash rate is adjusted until the turning object appears to stand still. This represents one flash per rotation or revolution. The angular speed can then be directly read from a calibrated scale.

STUB

A stub is a section of transmission line, usually either a half-wave or quarter-wave section, connected in parallel or in series with the feed system of an antenna. Stubs can be used as impedance-matching transformers, as notch filters, or as bandpass filters.

For impedance matching between a transmission line and an antenna having purely resistive impedances of different values, a quarter-wave stub can be inserted in series with the feed system. The section is placed between the antenna and the feed point (see A). The characteristic impedance of the matching stub is equal to the geometric

STRIPLINE: An example of stripline construction in a UHF radiofrequency amplifier.

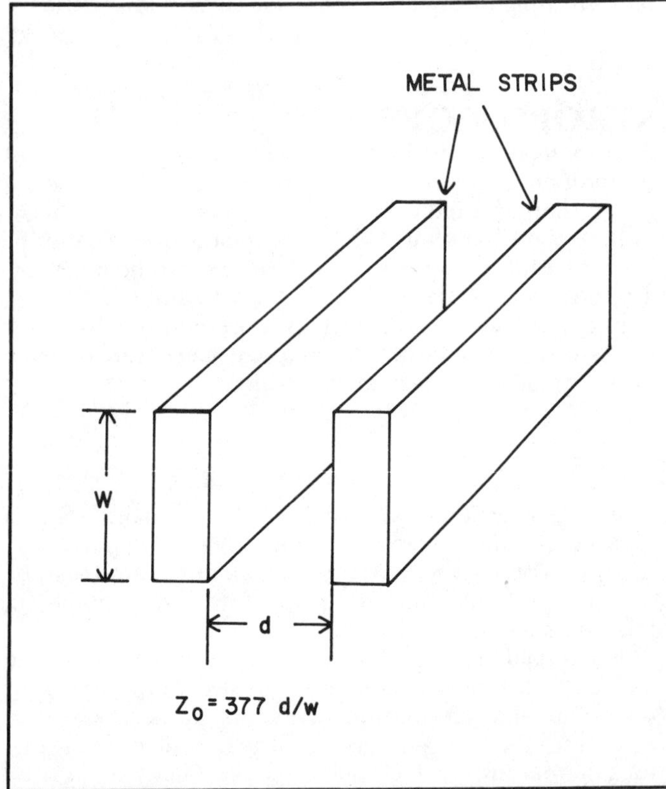

STRIP TRANSMISSION LINE: A variation of parallel wire line.

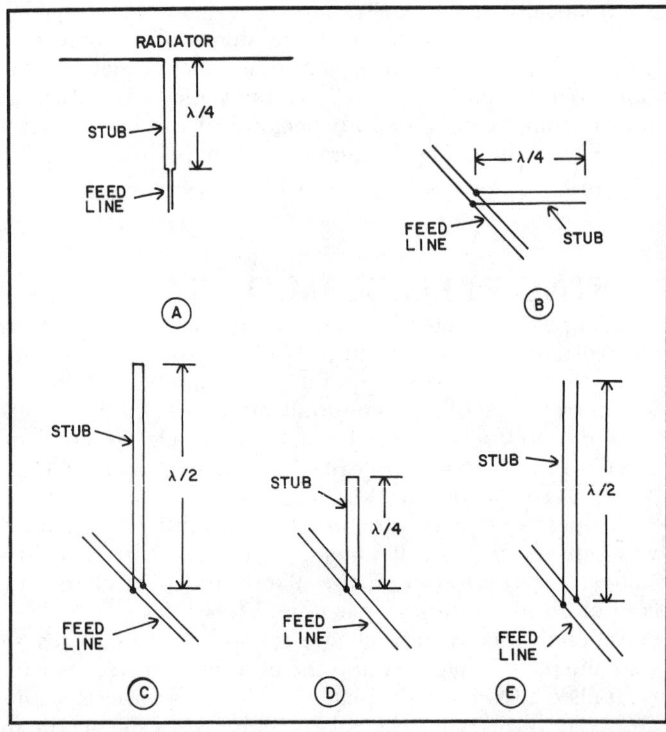

STUB: Examples of stubs. At A, a quarter-wave matching section; at B, a quarter-wave, series-resonant stub; at C, a half-wave, series-resonant stub; at D, a quarter-wave, parallel-resonant stub; at E, a half-wave, parallel-resonant stub.

mean of the two impedances to be matched (*see* GEOMETRIC MEAN).

A trap circuit can be constructed from a quarter-wavelength stub open at the far end (B) or from a half-wavelength stub shorted at the far end (C). This type of circuit is placed in parallel with a transmission line to eliminate signals at an unwanted frequency.

A narrowband resonant stub can be connected in parallel with a feed line to provide additional selectivity. A short-circuited, quarter-wave stub (D) or an open half-wave stub (E) will facilitate this. *See also* HALF-WAVE TRANSMISSION LINE, QUARTER-WAVE TRANSMISSION LINE.

STUDIO-TRANSMITTER LINK (STL)

Most radio and television broadcasting stations have transmitters that are located some distance away from the studio. This is a matter of convenience: The transmitter must be close to the antenna, so that the antenna feed-line length can be minimized. Most broadcasting antennas are located in rural areas while studios are usually in urban areas. A studio-transmitter link gets the modulating signal from one place to the other.

Studio-transmitter audio links are sometimes made via wire transmission. Most modern TV systems, however, use microwaves. A microwave transmitter, located at the studio, is modulated by the station program. The resulting signal is beamed by means of line-of-sight propagation to the transmitter location. The program is retrieved from the microwave signal by a demodulator. Then the program is used to modulate the signal of the main transmitter. *See also* MICROWAVE.

STYLUS

A stylus is a device that is used in disk recording and reproduction, as an electromechanical transducer.

In disk recording, the stylus is supplied with electrical impulses at audio frequencies. The stylus vibrates as the disk rotates underneath it. As a result, the stylus cuts a "wiggly" groove in the disk. The "wiggles" correspond to the audio-frequency vibrations. In monaural disk recording, the stylus moves back and forth, parallel to the surface of the disk and perpendicular to the direction of movement, as at A in the illustration. In stereo disk recording, the stylus moves along two independent axes, as shown at B (*see* STEREO DISK RECORDING).

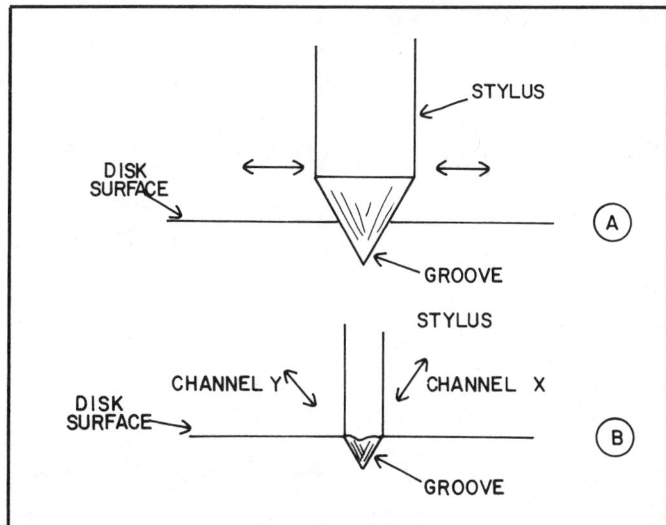

STYLUS: At A, a monaural stylus. At B, a stereophonic stylus.

When a disk is played back for reproduction, the stylus vibrates as it moves along the groove in the disk. These movements are translated into electrical impulses, usually by means of a dynamic or piezoelectric transducer (*see* CERAMIC PICKUP, CRYSTAL TRANSDUCER, DYNAMIC PICKUP).

A stylus, also known as a needle, is fabricated from a hard substance such as quartz, diamond, or carborundum. This maximizes its life, since such materials are much harder than the vinyl plastic from which disks are made.

SUBAUDIBLE TONE

A subaudible tone is a tone that is used to modulate a communications transmitter for the purpose of obtaining private-line or tone-squelch operation. Subaudible tones all have frequencies below 300 Hz, which is the approximate lower audio cutoff for communications. *See* PRIVATE LINE, TONE SQUELCH.

SUBBAND

A given frequency allocation may be subdivided for various types of radio transmission. A subband is literally a band within a band.

Consider, for example, the amateur band at 144-148 MHz. The lowest 100 kHz of this band—that is, 144.000-144.100 MHz—is allocated for nonvoice emission types only. This 100-kHz subband is an example of a legislated subband.

In the amateur band at 1.8-2.0 MHz, there were, at one time, eight different subbands in the United States. Each subband was 25 kHz wide. Various parts of the country had their own set of power-output limitations for each subband.

Sometimes a subband is agreed upon among the operators who use the band, but it does not carry the force of law. In the non-voice subbands of the high-frequency amateur bands, for example, the users of A1 and F1 emissions stay out of each other's way by operating within different frequency ranges. *See also* BAND.

SUBCARRIER

A radio-frequency carrier is sometimes modulated by another radio-frequency signal. Such a signal has a frequency much lower than that of the main carrier. The lower-frequency carrier wave is called a subcarrier.

Subcarriers are used in some multiplex transmissions (*see* FREQUENCY-DIVISION MULTIPLEX, MULTIPLEX, STEREO MULTIPLEX). A main carrier may be modulated by more than one subcarrier. Subcarriers generally have frequencies between 20 kHz and 75 kHz. This allows the subcarrier to be modulated at audio frequencies, and prevents it from interfering with the main program (since listeners cannot hear modulation above 20 kHz).

In a receiver, the subcarriers are retrieved from the main carrier by a detector. A tuned circuit filters out all subcarrier signals except the desired one. A second detector retrieves the information that modulates the selected subcarrier.

In FM broadcasting, subcarriers are used or leased for various purposes. Many of the stations that are heard on stereo receiver are modulated by inaudible subcarrier signals, as well as by the voices and music. *See also* SUBSIDIARY COMMUNICATIONS AUTHORIZATION.

SUBHARMONIC

A subharmonic is a signal with a frequency that is an integral fraction of the frequency of a specified signal. Any signal has infinitely many subharmonics. If the main-signal frequency is f, then the subharmonic frequencies are f/2, f/3, f/4, and so on. Subharmonic signals occur at wavelengths that are integral multiples of the main-signal wavelength.

While most alternating-current signals inherently contain some energy at harmonic frequencies (*see* HARMONIC), signals do not naturally contain energy at subharmonic frequencies. Subharmonics must be generated by means of a frequency divider (*see* FREQUENCY DIVIDER).

The term subharmonic is sometimes used, imprecisely, in reference to certain signals in the output of a frequency multiplier. Some radio transmitters, for example, use doublers, triplers, or quadruplers (*see* FREQUENCY MULTIPLIER). Suppose that a tripler is used, so that the input frequency is f and the output frequency is 3f. A small amount of energy will appear in the tripler output at frequencies of f and 2f. These signals are occasionally referred to as subharmonics. The signal at frequency 2f is not, however, a true subharmonic of the signal at frequency 3f.

SUBLIMINAL COMMUNICATIONS

We are affected not only by what we consciously see and hear, but also by things we see and hear subconsciously. We may not be aware of the ticking of a clock, for example, but we might get a headache as a result of the constant background noise. Things that affect us subconsciously are called subliminal.

It is possible to electronically generate, or transmit, a signal that a person does not consciously see or hear, because it is of short duration, low amplitude, or is otherwise imperceptible. Experiments have shown that such subliminal transmissions can influence people.

In one subliminal-communications experiment, a message, such as "BUY XYZ CANDY," was flashed momentarily on the screen at a movie theater. The duration was too short to be consciously noticed by the people watching the picture. The number of people buying the product at the concession stand was continuously monitored. Following the message, a sharp increase was noticed in the number of people buying "XYZ candy."

Modern electronic devices provide an almost unlimited variety of ways to use (and abuse) subliminal communication.

SUBMARINE CABLE

A submarine cable is a cable that is used to transmit signals, by wire, for long distances under water. Submarine cables have largely been replaced by satellite links today, but a few such cables still carry signals under the oceans of the world.

The illustration is a simplified diagram of a submarine-

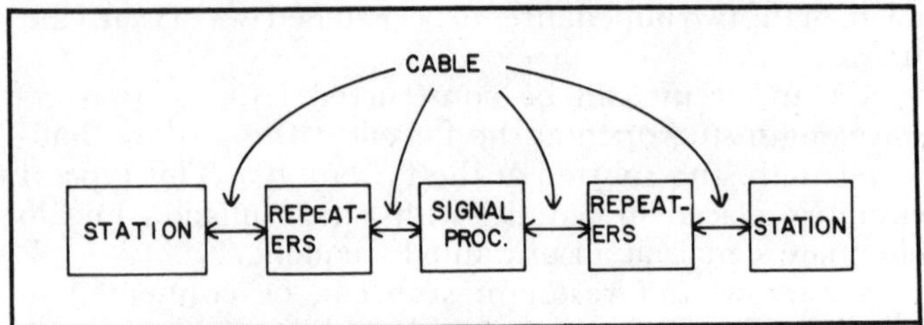

SUBMARINE CABLE: A submarine-cable system includes repeaters and signal processors for optimum signal-to-noise ratio.

cable communications system. The signals are transmitted in both directions via a single cable, using different frequencies. One cable may carry several hundred single-sideband signals, facilitating many simultaneous two-way conversations. The signal frequencies range from just above the human hearing range (20 kHz) to approximately 5 MHz.

Special amplifiers, called two-way repeaters, are placed at intervals of several miles along the length of the cable. These repeaters boost the signal level in either direction. In general, more repeaters are required in a high-frequency cable than in a low-frequency cable. In addition to the repeaters, shipboard signal processors are used. *See also* CARRIER-CURRENT COMMUNICATION.

SUBMINIATURIZATION

See MINIATURIZATION.

SUBROUTINE

A subroutine is a computer program within a program, and it is intended to perform a specific function. A subroutine is often used when it is necessary to perform a certain calculation numerous times.

An example of a subroutine is an algorithm that solves a quadratic equation. Certain data is inputted to the subroutine, in this case the coefficients of the equation, and the subroutine calculates the roots of the equation. Another example of a subroutine is a program that calculates the energy in joules produced by a certain process. Such a program, in BASIC, might be:

```
 10  REM  THIS PROGRAM COMPUTES ENERGY
     IN JOULES
 20  REM  ALSO IN ELECTRON VOLTS
 30  PRINT "ENTER THE MASS IN KILOGRAMS"
 40  INPUT K
 45  GOSUB 160
 50  GOSUB 180
 60  PRINT "THE ENERGY EQUIVALENT OF"; K;
     "KILOGRAMS"
 70  PRINT "IS"; E; "JOULES"
 80  PRINT "WHICH IS THE SAME AS"; V; "ELEC-
     TRON VOLTS"
 90  PRINT
100  PRINT "TO CONTINUE TYPE 1, TO STOP
     TYPE O"
110  INPUT C
120  IF C X1 THEN 140
130  STOP
140  PRINT
150  GO TO 30
160  LET E=K*8.9874E16
170  RETURN
```

```
180 LET V=E/1.6022E-19
190 RETURN
200 END
```

Once the energy has been calculated the computer will leave the subroutine program and go back to the main program.

A computer program might have no subroutines, or just a few, or perhaps hundreds of them. Some subroutines might even have subroutines within them. When subroutines are contained within other subroutines, the programs are said to be nested.

Generally, a subroutine can be called from any point within a computer program, as needed.

SUBSIDIARY COMMUNICATIONS AUTHORIZATION

In the frequency-modulation (FM) broadcast band, a carrier wave may be modulated not only by the main-program signal, but by modulated carriers between 20 kHz and 75 kHz. The subcarriers are used for such purposes as the transmission of background music, telemetry, and station-to-station information transmission.

A broadcast station must obtain a Subsidiary Communications Authorization (SCA) from the Federal Communications Commission in order to use subcarrier signals in the United States. The legal details regarding SCA application and use are found in Section 73 of the Rules and Regulations of the Federal Communications Commission. *See also* SUBCARRIER.

SUBSOUND

See INFRASOUND.

SUBSTITUTION METHOD FOR SOLVING EQUATIONS

A pair of two-variable equations can often be solved easily using a method called substitution. Substitution is a process in which one of the equations is solved for one of the variables in terms of the other; then the solution is substituted into the other equation. Consider the pair of linear equations:

$$x + 5y = 3$$

$$-2x - 3y = 0$$

There are four possible ways to solve this pair of equations by substitution. Suppose we manipulate the first equation to get:

$$x = 3 - 5y$$

The we can substitute $3 - 5y$ for x in the second equation:

$$-2(3 - 5y) - 3y = 0$$

$$-6 + 10y - 3y = 0$$

$$-6 + 7y = 0$$

$$7y = 6$$

$$y = 6/7$$

We can now solve for x by substituting 6/7 for y in either equation. We will get the same result in either case. Substituting in the first equation gives:

$$x + 5(6/7) = 3$$

$$x + 30/7 = 3$$

$$x = 3 - 30/7 = -9/7$$

Substituting in the second equation, we get:

$$-2x - 3(6/7) = 0$$

$$-2x - 18/7 = 0$$

$$-2x = 18/7$$

$$x = (18/7)/(-2) = -9/7$$

The solution to this set of equations is therefore:

$$x = -9/7$$

$$y = 6/7$$

You might want to try the other three substitution options as an exercise.

Substitution can sometimes (but not always) be used for nonlinear equations as well as for linear equations. Substitution can also be used for any set of n linear equations in n variables, no matter how large n may be—although you can correctly imagine that, if n is very large, the substitution process becomes lengthy. For a set of 100 equations in 100 variables, for example, we might want to obtain the solution by using a computer!

SUBSTITUTION OF COMPONENTS

See INTERCHANGEABILITY.

SUDDEN IONOSPHERIC DISTURBANCE

When a solar flare occurs, ionospheric propagation is almost always affected (*see* SOLAR FLARE). A short time after an intense solar flare, medium-frequency and high-frequency propagation may be almost totally wiped out. This is called a sudden ionospheric disturbance.

A solar flare causes a dramatic, abrupt increase in the number of charged subatomic particles emitted by the sun. These particles travel much slower than light, so they arrive at the earth a few hours after we see the solar flare. Astronomers can predict sudden ionospheric disturbances by watching the sun.

When the charged particles reach the earth, they cause an increase in the ionization density of the D layer (*see* D LAYER). This results in absorption of radio-frequency energy at medium and high frequencies. Long-distance propagation is seriously degraded by dense ionization in the D layer. At night, the ionization may be so dense that it is visible as an auroral display (*see* AURORA).

The magnetic field of the earth is also affected by the arrival of many charged particles from the sun. The magnetic field deflects large numbers of the particles toward the poles. Since the particles are accelerated, they produce a fluctuating magnetic field of their own. This magnetic field affects the overall geomagnetic field, producing a so-called geomagnetic storm (*see* GEOMAGNETIC FIELD, GEOMAGNETIC STORM).

Many of our communications circuits today are via satellite. This has helped to ameliorate the problem of sudden ionospheric disturbances, since the ultra-high-frequency and microwave spectra are not affected by the ionosphere. *See also* IONOSPHERE, PROPAGATION CHARACTERISTICS.

S UNIT

The S unit is a unit of signal strength. Generally, 1 S unit represents a change in signal strength amounting to 6 dB of voltage, or a ratio of 2:1. A root-mean-square signal level of 50 μV is considered to represent 9 S units. A few engineers use other values to represent S9, or other voltage ratios to represent 1 S unit.

The S-unit scale ranges from 1 to 9. A signal strength of S8 means that the signal voltage at the antenna terminals is 50/2, or 25, μV; a signal strength of S7 means that the voltage is 25/2 , or 12.5, μV, and so on. The signal strength is sometimes expressed in fractions of an S unit, such as S5.5. The illustration is a graph showing the actual RMS signal voltages for S units ranging from 1 to 9, according to the most common definition of the scale.

Many radio operators prefer to use their own intuition regarding signal strength. The strength (S) expression in the RST system is a subjective, rather than quantitative, variable. *See also* RST SYSTEM, S METER.

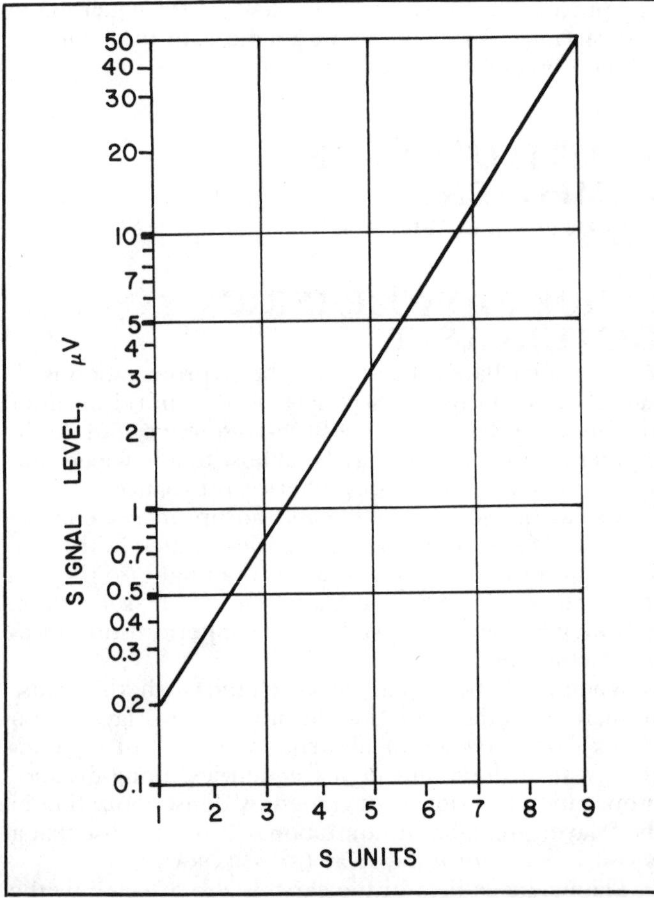

S UNIT: Graph showing relation between S units and signal strength in microvolts at the antenna terminals of a radio receiver.

SUNSPOT

A sunspot is a massive magnetic storm that occurs on the sun. Sunspots are visible through a telescope equipped with a special projection system. Sometimes sunspots are large enough to be seen with a simple pinhole projection device. (The sun should never be viewed directly, even through dark film negatives, since the ultraviolet light can damage the eyes.) Sunspots appear as dark regions on the sun's surface.

A typical sunspot is several times the diameter of the earth. The darkest part of the spot is at, or near, the center; this is called the umbra. The periphery of a sunspot is less dark, and is known as the penumbra.

Sunspots are more numerous in certain years than in other years. The number of sunspots affects the density of the earth's ionosphere, and therefore we observe changes in propagation conditions over the years. The maximum usable frequency tends to be much higher when the sunspots are numerous, as compared to when they are sparse. *See also* IONOSPHERE, MAXIMUM USABLE FREQUENCY, PROPAGATION CHARACTERISTICS, SUNSPOT CYCLE.

SUNSPOT CYCLE

In the past century, scientists have noticed that the number of sunspots is not constant, but changes from year to year. The fluctuation is fairly regular, and very dramatic. This fluctuation of sunspot numbers is called the sunspot cycle. It has a period of approximately 11 years (see illustration). The rise in the number of sunspots is generally more rapid than the decline, and the maxima and minima vary from cycle to cycle.

The sunspot cycle affects propagation conditions, especially at the high frequencies and in the very high frequency spectrum up to about 70 MHz for F-layer propagation and 150-200 MHz for sporadic-E propagation (*see* E LAYER, F LAYER, SPORADIC-E PROPAGATION). When there are few sunspots, the maximum usable frequency is low,

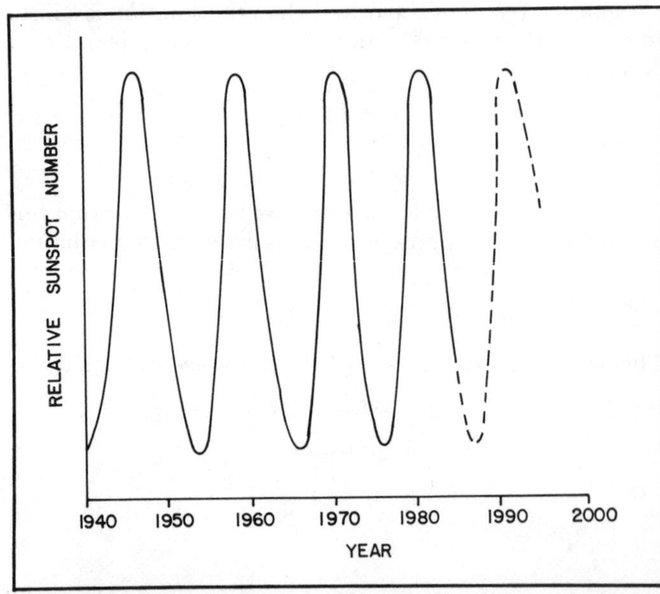

SUNSPOT CYCLE: Relative graph of sunspot numbers versus year.

because the ionization of the upper atmosphere is not dense. At or near a sunspot maximum, the maximum usable frequency is high, since the upper atmosphere is densely ionized. The changes in ionization result from the fact that the number of sunspots is correlated with the solar flux, and the solar flux directly affects the degree to which the upper atmosphere is ionized. *See also* MAXIMUM USABLE FREQUENCY, PROPAGATION CHARACTERISTICS, SOLAR FLUX, SUNSPOT.

SUPERCONDUCTIVITY

The resistance of an electrical conductor depends on the temperature. Most materials exhibit increasing resistance with increasing temperature.

At extremely low temperatures, near absolute zero, some conducting materials have extremely low resistance. This is known as superconductivity. When a closed loop of wire is cooled with liquid helium to a temperature of a few degrees Kelvin, an electric current will flow around and around the loop almost indefinitely. This is unheard-of at ordinary temperatures. Magnetic fields will, however, interfere with this effect.

Superconductivity results in low levels of thermal and shot-effect noise in an electronic circuit. The ohmic losses are greatly reduced, and the Q factor is thereby increased. This makes it possible to build receiver preamplifiers that are very sensitive. Superconductivity may someday be used to make tiny antennas efficient for transmitting. Theoretically, it should be possible to make a paper clip an effective transmitting radiator for high-frequency use! *See also* CRYOGENICS, Q FACTOR, SHOT EFFECT, THERMAL NOISE.

SUPERHETERODYNE RECEIVER

A superheterodyne receiver is a receiver that uses one or more local oscillators and mixers to obtain a constant-frequency signal. A fixed-frequency signal is more easily processed than a signal that changes in frequency.

In the superheterodyne circuit, the incoming signal is first passed through a tunable, sensitive amplifier called the front end (*see* FRONT END). The output of the front end is mixed with the signal from a tunable, unmodulated local oscillator (*see* LOCAL OSCILLATOR, MIXER). Either the sum or the difference between the frequencies of the two signals is then amplified. The sum or difference signal is called the first intermediate-frequency (I-F) signal (*see* INTERMEDIATE FREQUENCY, INTERMEDIATE-FREQUENCY AMPLIFIER). The signal is filtered to obtain a high degree of selectivity.

If the first I-F signal is detected, we have a single-conversion receiver (*see* SINGLE-CONVERSION RECEIVER). Some receivers use a second mixer and second local oscillator, converting the first I-F signal to a much lower frequency called the second I-F. This type of receiver is known as a double-conversion circuit (*see* DOUBLE-CONVERSION RECEIVER).

The superheterodyne receiver offers several advantages over direct-conversion receivers. Probably the most important advantage is the excellent selectivity that can be obtained, along with single-signal reception (*see* SELECTIVITY, SINGLE-SIGNAL RECEPTION). Sensitivity is enhanced because the fixed I-F amplifiers are easy to keep in optimum tune. A superheterodyne circuit can be used to receive signals having any emission type.

The main disadvantage of the superheterodyne circuit is that it sometimes receives signals that are not on the desired frequency. The signals may be external to the receiver, in which case they are called images, or they may come from the local oscillator or oscillators. If the local-oscillator frequencies are judiciously chosen, false signals are generally not a problem. *See also* DIRECT-CONVERSION RECEIVER, IMAGE FREQUENCY.

SUPERHIGH FREQUENCY

The superhigh-frequency (SHF) range of the radio spectrum is the range from 3 GHz to 30 GHz. The wavelengths corresponding to these limit frequencies are 10 cm and 1 cm, and thus the superhigh-frequency waves are sometimes called centimetric waves. Most of the superhigh frequencies are allocated by the International Telecommunication Union, headquartered in Geneva, Switzerland.

Superhigh-frequency waves are not affected by the ionosphere of the earth. Signals in this range pass through the ionosphere as if it were not there. Centimetric waves behave very much like waves at higher frequencies. Superhigh-frequency waves can be focused by medium-sized parabolic or spherical reflectors. Because of their tendency to propagate in straight lines unaffected by the atmosphere, centimetric waves are extensively used in satellite communications. Centimetric waves are also used for such purposes as studio-to-transmitter links. Because the frequency of a centimetric signal is so high, wideband modulation is practical. *See also* ELECTROMAGNETIC SPECTRUM, FREQUENCY ALLOCATIONS.

SUPERPOSITION THEOREM

The Superposition Theorem is a rule that describes the behavior of currents and voltages in linear multi-branch circuits.

Suppose we have a linear circuit with three or more separate, current-carrying branches. An example is shown in the illustration. Suppose that a power supply, delivering Ex volts, is placed across branch X of this circuit, while the terminals at Ey are shorted. This results in a certain current, call it Izx, in branch Z. Now suppose we remove the voltage source from terminals Ex and short them, and connect a voltage Ey to terminals Ey. This will also cause a current, call it Izy, to flow in branch Z. (The currents Izx and Izy will not necessarily be the same; in fact, they probably will not be.)

The superposition theorem tells us what current will flow in branch Z of the circuit when the supplies Ex and Ey are connected at the same time. It is simply the sum of the currents Izx and Izy. That is, if the current in branch Z is Iz with both Ex and Ey connected, then

$$Iz = Izx + Izy$$

This theorem can be used to advantage in determining currents and voltages in complicated circuits. *See also* KIRCHOFF'S LAWS.

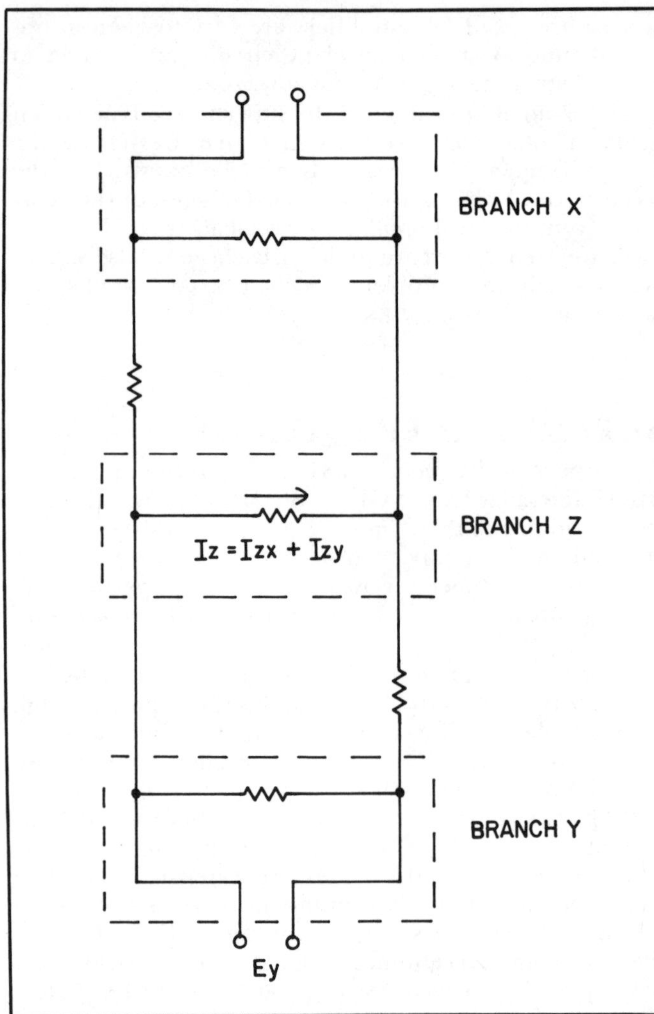

SUPERPOSITION THEOREM: Schematic diagram showing principle of the Superposition Theorem.

SUPERREGENERATIVE RECEIVER

A special form of regenerative detector is used in a circuit called a superregenerative receiver. The superregenerative circuit provides greater gain, and better sensitivity, than does an ordinary regenerative circuit. A superregenerative receiver is a form of direct-conversion circuit (*see* DIRECT-CONVERSION RECEIVER).

In the superregenerative detector, the positive feedback is periodically increased and decreased. This makes it possible to increase the feedback without resulting in unwanted oscillation of the detector. The changes in feedback cause fluctuations in the circuit gain, and this modulates incoming signals. However, if the feedback is varied at a rate of more than 20 kHz, the modulation cannot be heard, and it does not affect the sound of an incoming signal.

An example of a superregenerative detector is shown in the illustration. The amount of feedback is controlled by a varactor diode. The diode capacitance fluctuates with the applied ultrasonic signal. The changes in capacitance cause the amount of feedback to vary. The detector is adjusted by manipulating the series variable capacitor and the amplitude of the applied ultrasonic signal. *See also* REGENERATIVE DETECTOR.

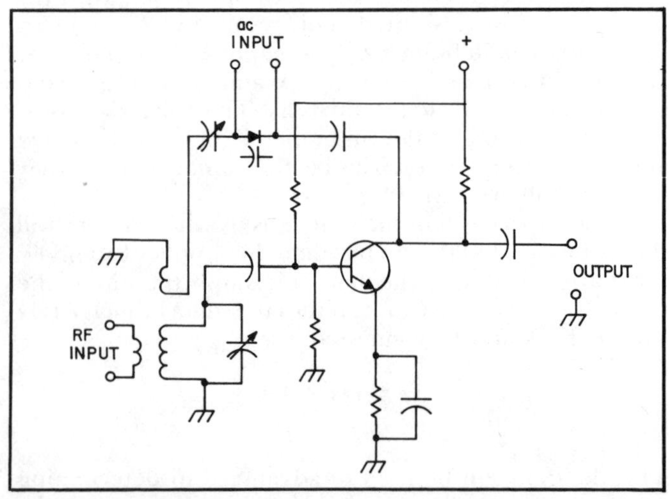

SUPERREGENERATIVE RECEIVER: A superregenerative detector uses an external oscillator to maximize the gain.

SUPPRESSED CARRIER

In single-sideband transmission, the carrier wave is usually reduced greatly in amplitude. This is also sometimes done in double-sideband transmission. A signal is said to have a suppressed carrier in such instances.

The advantage of suppressed-carrier transmission is that all of the signal power goes into the sidebands, resulting in optimal efficiency. Suppressed-carrier signals are generated by means of a balanced modulator. *See also* BALANCED MODULATOR, SINGLE SIDEBAND.

SUPPRESSOR GRID

Some vacuum tubes have three grids. Such a tube is called a pentode (*see* PENTODE). In the pentode tube, the outermost grid, nearest the plate, is called the suppressor grid. The suppressor grid consists of a few fine wires, widely spaced.

The suppressor grid reduces the effects of secondary emission as electrons bombard the positively charged plate (*see* SECONDARY EMISSION). The suppressor grid is generally at the same potential as the cathode of the tube, so it is negative with respect to the plate. Since the wires of the suppressor are widely spaced, high-speed electrons from the cathode can get to the plate easily. But the secondary electrons move much more slowly. The negative charge on the suppressor is sufficient to repel most of these electrons back toward the plate. This increases the efficiency of the tube, since more of the electrons contribute to the output.

In some pentode tubes, the suppressor grid is internally connected to the cathode. In other tubes, the suppressor grid is connected to a pin on the base, but is not internally shorted to the cathode, so that the tube can be used as a triode or tetrode if desired. *See also* TETRODE, TRIODE, TUBE.

SURFACE WAVE

At some radio frequencies, the electromagnetic field is propagated over the horizon because of the effects of the ground. This is observed at very low, low, and medium frequencies, and to a limited extent at high frequencies. Surface-wave propagation does not take place at very high frequencies or above. This is because the ground becomes progressively more lossy as the frequency increases.

Surface-wave propagation occurs when the electromagnetic field is vertically polarized: that is, when the electric lines of flux are perpendicular to the surface of the earth. This requires a vertical radiator (*see* VERTICAL POLARIZATION). If the electromagnetic field is horizontally polarized, the electric lines of flux are short circuited in the ground, and surface-wave propagation does not take place.

When you listen to a broadcast station that is more than a few miles away during the daytime, the signal is reaching you because of surface-wave propagation. The same is true of daytime signals at low frequencies. At very low frequencies, surface-wave propagation occurs in conjunction with waveguide propagation (*see* WAVEGUIDE PROPAGATION).

The moon has no ionosphere. The same is true of many other planets. Over-the-horizon communication, without satellites, will nevertheless be possible on such planets if the ground conductivity is fairly good: Surface-wave propagation might be used at low and very low frequencies for planet-wide communication. *See also* PROPAGATION CHARACTERISTICS.

SURFACE-WAVE TRANSMISSION LINE

Surface-wave propagation can be deliberately made to take place along wire conductors at very-high, ultra-high, and microwave frequencies. We never see surface-wave propagation along the ground at these wavelengths because the ground is not an excellent conductor (*see* SURFACE-WAVE PROPAGATION), but the situation is different when a good conductor is used. A single-wire line, designed to transfer surface-wave signals along its length, is called a surface-wave transmission line.

A surface-wave line consists of a wire conductor surrounded by a thick layer of low-loss dielectric material (see illustration). The signals are coupled into and out of the line by horns called launchers. The surface-wave line does not radiate significantly because the electromagnetic field travels in contact with the conductor. The line is unbalanced since there is only one conductor. This type of line is also called G-line, or GOUBAU line, after its inventor, Dr. George Goubau.

The surface-wave line is more efficient than coaxial cable or a parallel-wire line with two or four conductors. However, the wavelength must be short of the electromagnetic field will tend to be radiated rather than propagated along the line. Surface-wave transmission lines are not effective at the very low, low, medium, and high frequencies. Shielded or balanced feed lines are used at those wavelengths.

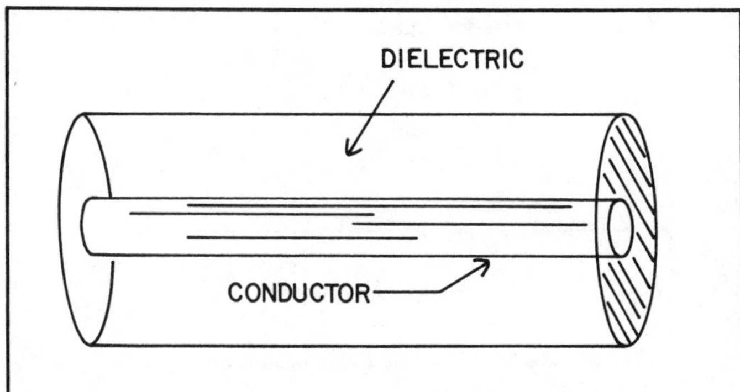

SURFACE-WAVE TRANSMISSION LINE: A surface-wave transmission line is a single-wire line surrounded by a dielectric material.

SURGE

When power is first applied to a circuit or device that draws considerable current, a surge often occurs. For a moment, excessive current flows. This surge can result in damage to the circuit or appliance.

You have probably experienced the effects of a surge following a general power failure. People tend to leave appliances switched on during a power failure. When power is restored, thousands or millions of devices are connected to the utility mains, and a massive start-up current is demanded. The result is a momentary initial surge. If you are lucky, nothing happens, or perhaps fuses or circuit breakers blow or trip off.

In the event of a power failure, it is a good idea to switch off or unplug circuits and appliances that might be damaged by a surge—and that includes just about everything. If you have a breaker box, you should switch off all of the breakers, including the main breakers. Then, even if nobody else unplugs their appliances, yours will not be damaged. After power returns, switch on the main breakers first, and then the auxiliary breakers, one at a time.

Surges occur on a small scale in electronic circuits, especially those that draw significant current and operate from utility mains. *See also* SURGE SUPPRESSOR.

SURGE IMPEDANCE

See CHARACTERISTIC IMPEDANCE.

SURGE PROTECTION

All electronic equipment should have some means of protection against damage from surges. Surge protection can be accomplished by means of fuses, circuit breakers, or suppressors. *See also* CIRCUIT BREAKER, FUSE, SURGE, SURGE SUPPRESSOR.

SURGE SUPPRESSOR

A surge suppressor is a device that is inserted in series or in

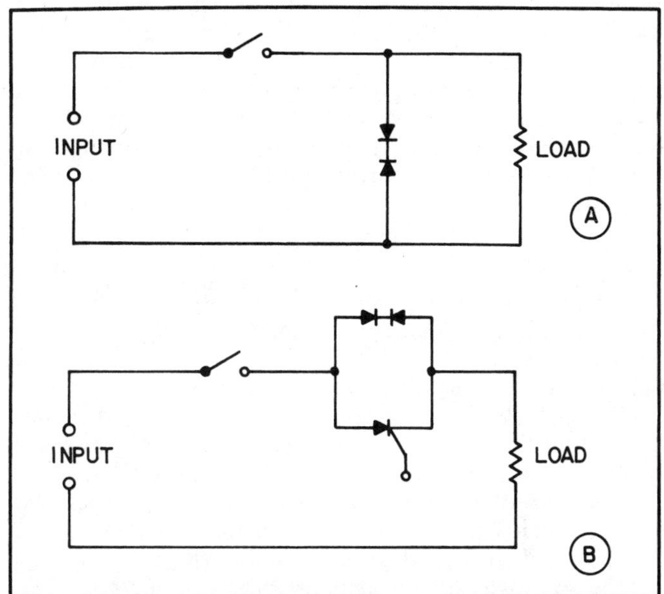

SURGE SUPPRESSOR: Two types of surge suppressors. The device at A uses two selenium rectifiers. The circuit at B uses two selenium rectifiers and a silicon-controlled rectifier.

parallel with the power line to a circuit, protecting the circuit from excessive voltages or currents (*see* SURGE).

There are various kinds of surge suppressors. Perhaps the simplest is a Zener diode, with a voltage rating equal to the voltage for which the equipment is designed. Current-regulating and voltage-regulating circuits can be used for surge protection.

A common method of surge suppression is the use of selenium rectifiers connected back-to-back. Selenium rectifiers are sometimes used in conjunction with silicon-controlled rectifiers to obtain surge suppression. These two configurations are shown in the illustration. *See also* CURRENT REGULATION, SELENIUM RECTIFIER, SILICON-CONTROLLED RECTIFIER, VOLTAGE REGULATION, ZENER DIODE.

SUSCEPTANCE

Susceptance is the alternating-current equivalent of conductance, and the reciprocal of reactance. Susceptance is a mathematically imaginary quantity.

Given an inductive reactance $X_L = jX$ the susceptance Y_L is given by:

$$Y_L = 1/X_L = 1/(jX) = -j\ (1X)$$

where X is a positive real number.

Given a capacitive reactance $X_C = -jX$, the susceptance Y_C is given by:

$$Y_C = 1/X_C = 1/(-jX) = j(1/X)$$

where X is, again, a positive real number. *See also* REACTANCE.

SUSCEPTIVENESS

A telephone circuit, especially a long-distance circuit, often picks up significant hum from power lines. Telephone wires and power lines are frequently run side-by-side along the same path for a considerable distance. Telephone lines can also be affected by sferics (thunderstorm static) and geomagnetic storms. Susceptiveness is an expression of the vulnerability of a telephone circuit to interference from external sources.

In general, the susceptiveness can be reduced, and the system performance thereby optimized, by ensuring that the telephone lines are properly balanced. *See also* TELEPHONE.

SUSPENSION GALVANOMETER

A suspension galvanometer is a device that is sometimes used to measure weak currents. The device operates on the principle of interaction between magnetic fields.

A coil of wire is connected to two terminals. The coil is hung in the field of a permanent magnet. When a voltage is applied to the terminals, causing a current to flow in the coil, the coil turns as a result of interaction between the magnetic fields of the coil and the permanent magnet. The deflection of the coil is proportional to the current.

The sensitivity of the suspension galvanometer can be increased by attaching a mirror to the coil, and reflecting a beam to light off the mirror. The beam of light forms an optical lever (*see* OPTICAL LEVER).

SWEEP

Sweep is a continuous, usually linear change of the value of a variable from one defined limit to another. The movement of an electron beam across the face of a television picture tube is an example of sweep. An oscillator or radio transmitter can be made to change in frequency; this is another example of sweep.

If the limits of parameter values are defined as A and B, sweep is usually characterized by a controlled change from limit A to limit B, and an almost instantaneous return from limit B to limit A. Sweep is sometimes called scanning. *See also* SCANNING.

SWEEP-FREQUENCY FILTER OSCILLATOR

A sweep-frequency filter oscillator is a specialized form of sweep generator (*see* SWEEP GENERATOR). The sweep-frequency filter oscillator produces a signal that varies rapidly in frequency between two defined and adjustable limits. The oscillator is used in conjunction with a spectrum analyzer for testing and alignment of selective filters (*see* SPECTRUM ANALYZER).

The interconnection scheme for a sweep-frequency filter oscillator, spectrum analyzer, and test filter is shown at A in the illustration. The oscillator frequency sweeps between two limits, f1 and f2, where f1 < f2. The frequency f1 is well below the cutoff or passband; the frequency f2 is well above the cutoff or passband. The sweep rate is very rapid, so that the spectrum-analyzer display appears continuous.

An example is shown at B of a spectrum-analyzer display that might be obtained in this manner for a 3-kHz single-sideband filter in the intermediate-frequency chain of a superheterodyne receiver. This test arrangement gives a graphic illustration of the performance of the filter.

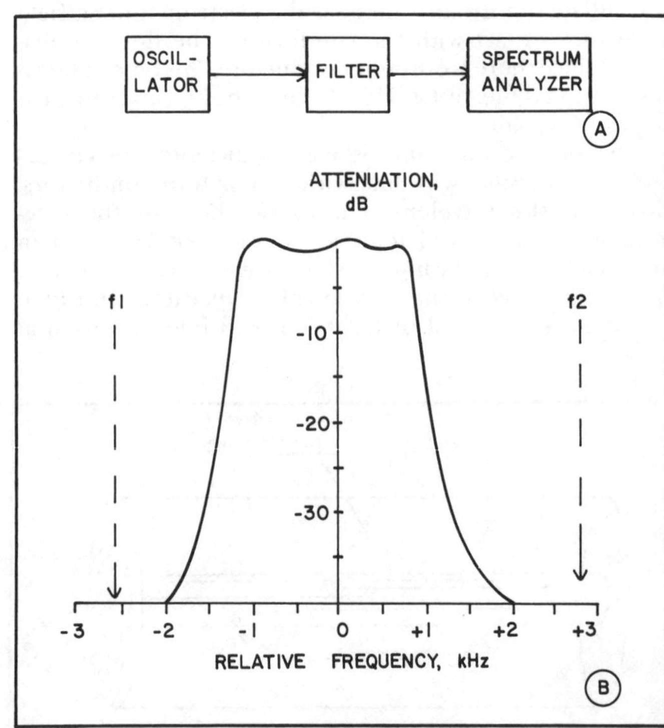

SWEEP-FREQUENCY FILTER OSCILLATOR: At A, connection of a sweep-frequency filter analyzer. At B, typical display on a spectrum analyzer connected to a sweep-frequency filter analyzer.

The filter can, if necessary, be adjusted for optimum rectangular response while the oscillator and spectrum analyzer are connected.

The sweep-frequency filter oscillator can be used with all types of filters for checking the response. *See also* BANDPASS RESPONSE, BAND-REJECTION RESPONSE, HIGHPASS RESPONSE, LOWPASS RESPONSE, RECTANGULAR RESPONSE.

SWEEP GENERATOR

A sweep generator is a device that generates the fluctuating voltages or currents that operate the deflection apparatus in a cathode-ray tube. The changing voltages or currents cause the electron beam to move from left to right, causing a trace to appear on the screen. The voltage or current generally varies with time as shown in the graph at A. *See also* CATHODE-RAY TUBE.

A wideband oscillator, with a frequency output that fluctuates continously between two limits in a sweeping manner, is sometimes called a sweep generator. Sweep generators are used in conjunction with spectrum analyzers to evaluate the frequency-response characteristics of receivers, tuned circuits, and selective filters of various types. The frequency varies with time in a manner similar to that illustrated by the function at A. *See also* SWEEP, SPECTRUM ANALYZER, SWEEP-FREQUENCY FILTER OSCILLATOR.

SWINGING CHOKE

A swinging choke is a special form of power-supply filter choke, designed to exhibit an inductance that varies inversely with the flow of current. When the current level is low, the swinging choke has maximum inductance, but the core saturates easily as the level of current rises.

In the average swinging choke, the inductance is halved when the current is tripled; the inductance is cut to ¼ its original value when the current increases by a factor of 9. Some swinging chokes have more or less variability. The low-current inductance of a swinging choke depends on the application, but is generally in the range of 1 to 15 H.

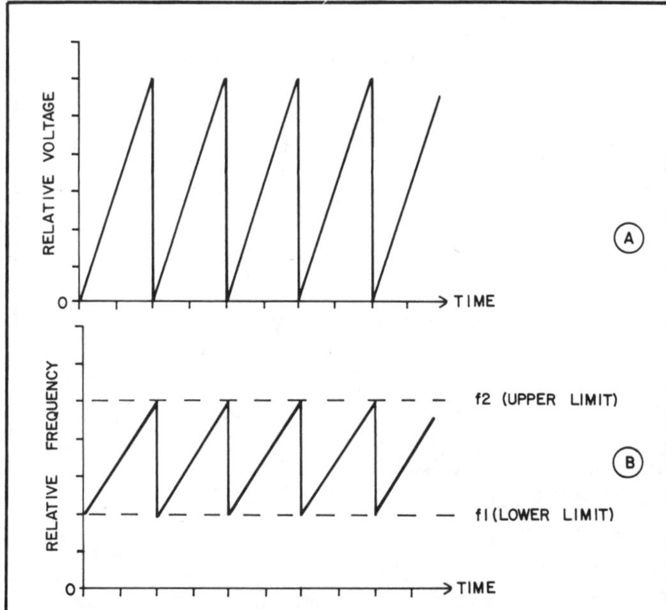

SWEEP GENERATOR: Sweep-generator waveforms. At A, voltage-versus-time waveform for a sweep generator used with a cathode-ray tube. At B, frequency-versus-time function for a sweep-frequency oscillator.

SWITCH

A switch is a mechanical or electronic device that is used to deliberately interrupt, or alter the path of, the current through a circuit. When we think of a switch, we usually think of an ordinary household wall switch, or the power switch on a piece of electronic apparatus. These are mechanical switches. There are many types of mechanical switches. *See* MULTIPLE POLE, MULTIPLE THROW, ROTARY SWITCH, SINGLE THROW TOGGLE SWITCH, WAFER SWITCH.

A relay is a form of switch that is operated by remote control. Various semiconductor devices are used as switches. Vacuum tubes can also be used for switching, although they are rarely employed nowadays. *See* DIAC, RELAY, SILICON-CONTROLLED RECTIFIER, SWITCHING, SWITCHING DIODE, SWITCHING TRANSISTOR, THYRATRON, TRIAC.

SWITCHING

Switching is the action of making and breaking , or diverting the path of the current through, an electrical or electronic circuit. Switching can be done manually or automatically. Switching may be accomplished by a command issued directly at the location of the current to be controlled, or it may be done by remote control. The speed of switching may be relatively slow or extremely rapid. The type of switching to be used depends on the application. Any switching arrangement uses one of the following four schemes.

Single-pole, single-throw (SPST) switching makes and breaks a single current path (see illustration at A). Switching diodes, transistors, and other semiconductor devices are SPST switches.

Multiple-pole, single-throw switching opens and closes two or more current paths simultaneously, (as at B). Many mechanical switches and relays operate in this mode. Two or more semiconductor devices can be operated in tandem

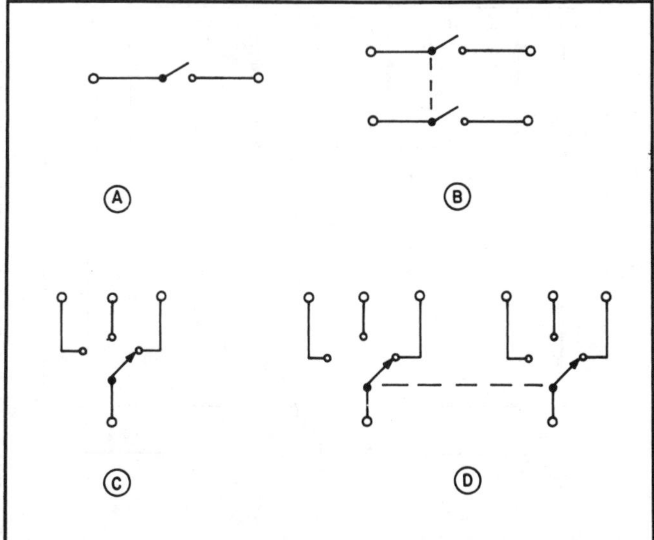

SWITCHING: A: single-pole, single-throw. B: multiple-pole, single-throw. C: single-pole, multiple-throw. D: multiple-pole, multiple-throw.

to obtain multiple-pole, single-throw operation.

Single-pole, multiple-throw switching diverts a current through two or more circuits (C). Many mechanical switches work this way, and relays often have two throw positions.

Multiple-pole, multiple-throw switching is accomplished by operating two or more single-pole, multiple-throw switches together (D). This type of switching is done mechanically. Some rotary switches have several poles and several throw positions. *See also* MULTIPLE POLE, MULTIPLE THROW, SINGLE THROW, SWITCH.

SWITCHING AMPLIFIER

See DC AMPLIFIER.

SWITCHING DIODE

Semiconductor diodes are often used for switching purposes, especially at high speeds or at radio frequencies. Diodes conduct when they are supplied with a forward bias of a few volts; they do not conduct when no bias, or a reverse bias, is applied (*see* DIODE, P-N JUNCTION).

For direct-current or low-frequency ac switching, most germanium or silicon diodes are satisfactory. For switching of radio-frequency signals, diodes must exhibit the smallest possible amount of capacitance. The PIN diode is often used for radio-frequency switching (*see* PIN DIODE). An example of diode radio-frequency switching is shown in the illustration. The radio-frequency signal can be directed along any of four paths by applying a direct-current forward bias to the appropriate diode. The bias is applied by remote control.

Diodes are used in some digital-logic devices for switching purposes. An example of this is found in diode-transistor logic, or DTL (*see* DIODE-TRANSISTOR LOGIC).

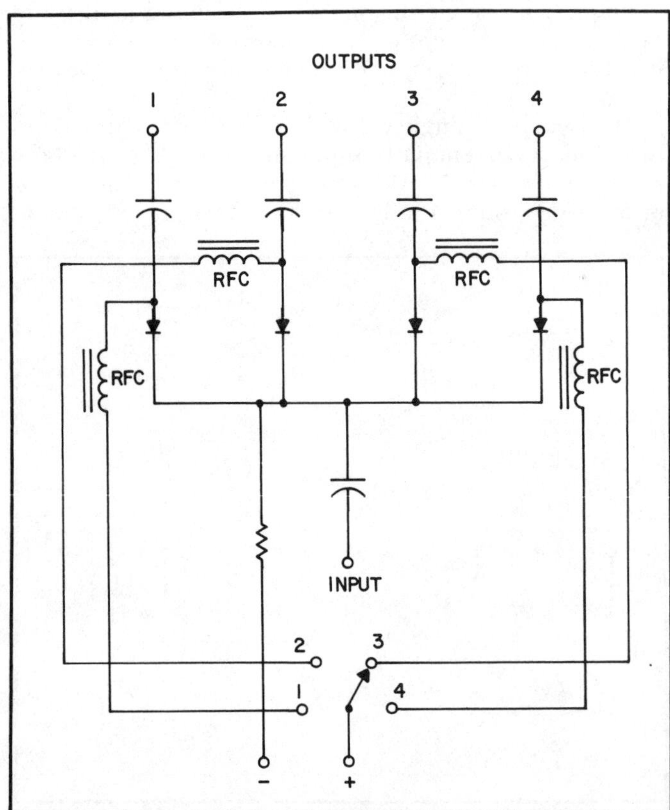

SWITCHING DIODE: A switching-diode circuit for radio-frequency signals.

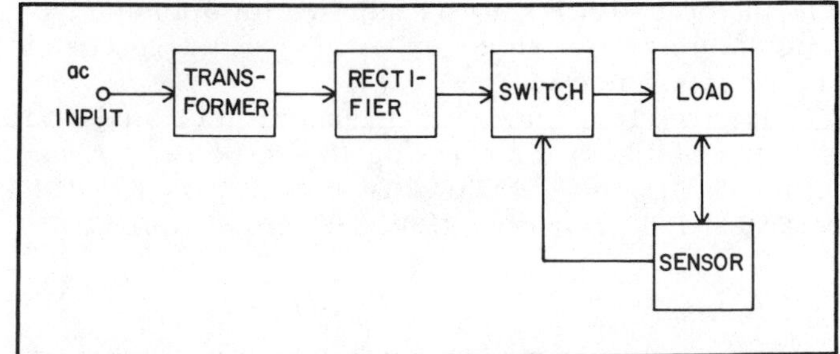

SWITCHING POWER SUPPLY: Block diagram of a switching power supply.

SWITCHING POWER SUPPLY

In a power supply, voltage or current regulation can be achieved in various ways. An unusual, but effective, method of regulation is the switching scheme.

In the switching power supply, the output is not continuous, but instead it is switched on and off at a high rate of speed. The output duty cycle varies, depending on the load resistance. The supply is either fully on or fully off at any given instant. The output is smoothed by means of a conventional power-supply filter.

The illustration is a block diagram of a switching power supply. The load resistance is determined by means of a sensing device. If the load resistance decreases, the sensing device tells the switch to increase the duty cycle. If the load resistance increases, the duty cycle decreases. *See also* DUTY CYCLE, POWER SUPPLY, REGULATION.

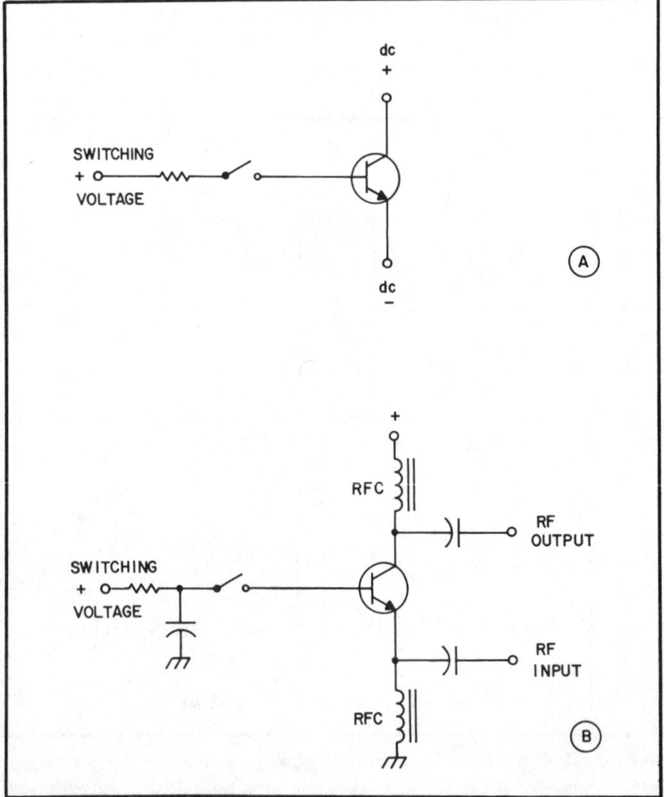

SWITCHING TRANSISTOR: At A, a direct-current switching-transistor circuit. At B, a radio-frequency switching-transistor circuit.

SWITCHING TRANSISTOR

Transistors are often used for switching purposes. A transistor conducts when the emitter-base junction is provided with sufficient forward bias to cause saturation. If no bias, or a reverse bias, exists at the junction, the transistor will not conduct.

Transistors can be used for switching direct-current or alternating-current signals. Most transistors will operate well as switches for direct current and low-frequency alternating current. For switching radio-frequency signals, transistors should have the smallest possible capacitance.

The schematic illustrates direct-current and radio-frequency transistor switching schemes. In these examples, NPN bipolar transistors are used. (If PNP transistors were used, the polarities would be reversed.) A positive voltage, applied at the base of a transistor, results in conduction.

Switching transistors are extensively used in digital integrated circuits. Hundreds, or even thousands, of individual switching transistors can be fabricated on a single chip. *See* DIODE-TRANSISTOR LOGIC, DIRECT-COUPLED TRANSISTOR LOGIC, EMITTER-COUPLED LOGIC, HIGH-THRESHOLD LOGIC, INTEGRATED INJECTION LOGIC, METAL-OXIDE-SEMICONDUCTOR LOGIC FAMILIES, RESISTOR-CAPACITOR-TRANSISTOR LOGIC RESISTOR-TRANSISTOR LOGIC, TRANSISTOR-TRANSISTOR LOGIC, TRIPLE-DIFFUSED EMITTER-FOLLOWER LOGIC.

SWITCHING TUBE

A diode, triode, tetrode, or pentode tube can be used as a switching device. Tubes are seldom used as switches nowadays, since semiconductor devices are available that can handle high voltages. But tubes were common a few decades ago. The earliest computers used tubes as switches to perform digital operations.

A tube normally conducts under forward-bias conditions—that is, when the plate is positive with respect to the cathode. A tube will not conduct when the cathode is positive relative to the plate. This diode action makes it possible to use a rectifier tube as a switch, in the same way as a semiconductor switching diode is used (*see* SWITCHING DIODE).

A triode, tetrode, or pentode tube conducts in the forward direction when the control-grid voltage is zero or slightly negative with respect to the cathode. If the negative control-grid voltage is high, cutoff occurs and the tube fails to conduct. This makes it possible to use a tube as a switch. The diagram shows how triode tubes can be used to direct current to any of three different circuits. *See also* TUBE.

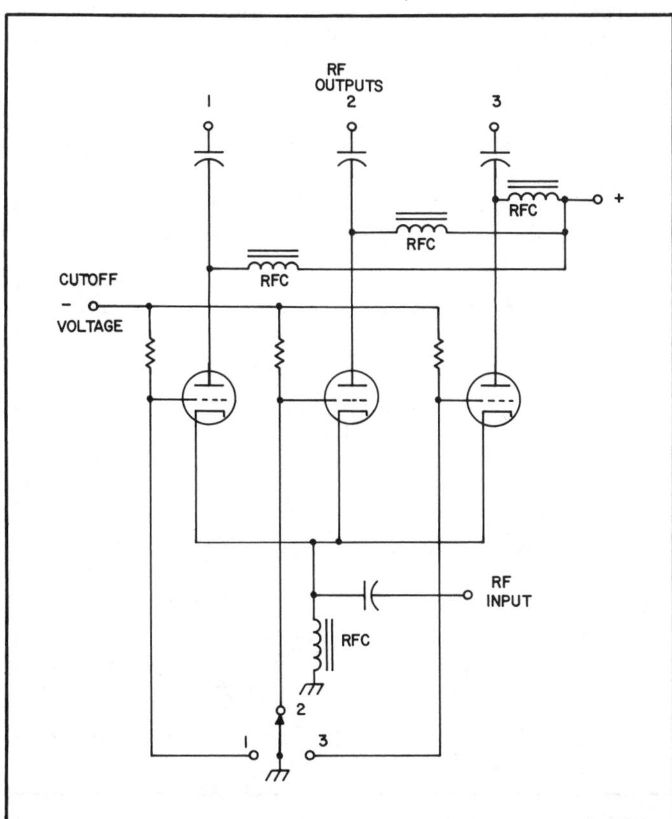

SWITCHING TUBE: A switching-tube circuit for radio-frequency signals.

SWL

See SHORTWAVE LISTENING.

SWR

See STANDING-WAVE RATIO.

SWR LOSS

See STANDING-WAVE-RATIO LOSS.

SWR METER

See REFLECTOMETER.

SYMBOLIC LOGIC

See MATHEMATICAL LOGIC.

SYMMETRICAL RESPONSE

A bandpass filter has a symmetrical response if the attenuation-versus frequency curve is balanced. Mathematically, a symmetrical response can be defined in terms of the center frequency.

Suppose that f is the center frequency of a bandpass filter. Let f1 be some frequency below the center frequency, and let f2 be some frequency above the center of the passband, such that $f2 - f = f - f1$. If the response is symmetrical, then the filter attenuation at frequency f1 will be the same as the filter attenuation at f2, no matter what values we choose for f1 and f2 within the above constraints.

A symmetrical filter response is illustrated at A. A non-symmetrical response is shown at B. *See also* BANDPASS FILTER, BANDPASS RESPONSE.

SYNC

See SYNCHRONIZATION.

SYNCHRO

A synchro is a special type of motor, used for remote control of mechanical devices. A synchro consists of a generator and a receiver motor. As the shaft of the generator is turned, the shaft of the receiver motor follows along exactly.

In antenna rotors, a form of synchro may be used to indicate the rotor position. The generator is connected mechanically to the antenna mast, and the receiver is located in the station and connected to a directional pointer. This type of synchro is called a selsyn (*see* SELSYN).

Some synchro devices provide a digital indication of the angular position of the generator shaft. Some antenna rotor systems use this arrangement. Rather than a pointer, the display box has a digital readout of the azimuth bearing. Some synchro devices are programmable. The operator inputs a number into the synchro generator, and the receiver changes position accordingly.

SYNCHRONIZATION

When two signals or processes are exactly aligned, they are said to be in synchronization. Two identical waveforms, for example, are synchronized if they are in phase. Two clocks are synchronized if they agree more or less precisely.

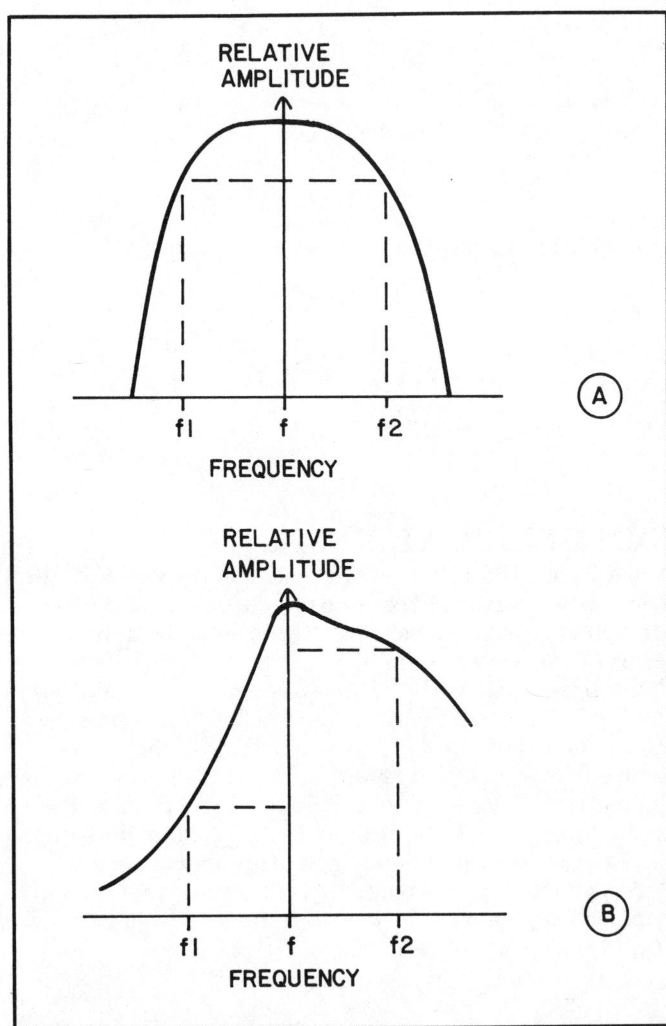

SYMMETRICAL RESPONSE: At A, a symmetrical response. At B, a nonsymmetrical response.

In a television transmitting and receiving system, the electron beam in the picture tube must move in synchronization with the beam in the camera tube. Otherwise, the picture will appear to be split, rolling, or tearing (*see* HORIZONTAL SYNCHRONIZATION, VERTICAL SYNCHRONIZATION). Pulses are sent by the transmitting station to keep the receiver in synchronization, so that we see a clear picture.

In an oscilloscope, the sweep rate can be synchronized with the frequency of the applied signal to obtain a motion-free display. This is called triggering (*see* TRIGGERING).

In some communications systems, the transmitter and receiver are synchronized against an external, common time standard. This is sometimes called synchronous or coherent communications (*see* SYNCHRONIZED COMMUNICATIONS).

SYNCHRONIZED COMMUNICATIONS

One of the most important problems in communications is the maximization of the number of signals that can be accommodated within a given band of frequencies. This has traditionally been done by attempting to minimize the bandwidth of a signal. There is a limit, however, to how small the bandwidth can be if the information is to be effectively received (*see* BANDWIDTH).

Digital signals, such as Morse code, occupy less bandwidth than analog signals, such as voice. Perfectly timed Morse code consists of regularly spaced bits, each bit having the duration of one dot. The length of a dash is three bits; the space between dots and/or dashes in a single character is one bit; the space between characters in a word is three bits; the space between words and sentences is seven bits. This makes it possible to identify every single bit by number, even in a long message (see A in illustration). Therefore, the receiver and transmitter can be syn-

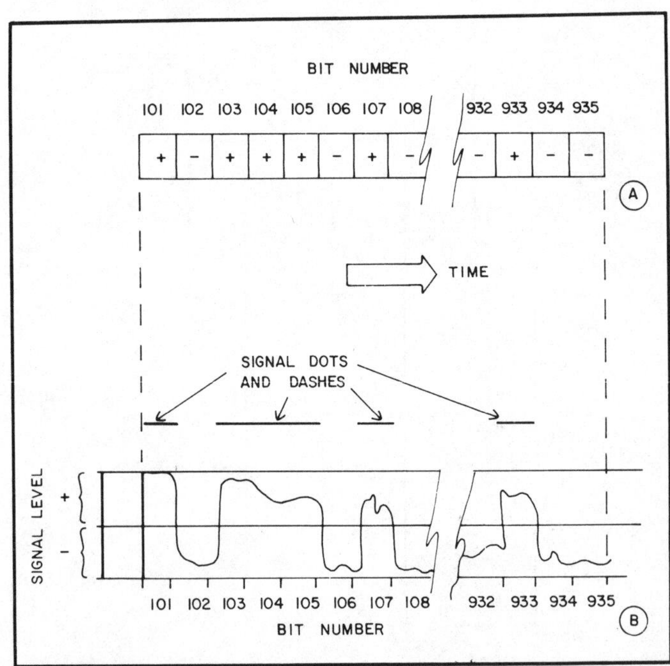

SYNCHRONIZED COMMUNICATIONS: The high/low patterns at A result in the signal at B. Integration circuits, synchronized between the receiver and transmitter, optimize accuracy.

chronized, so that the receiver "knows" which bit of the message is being sent at a given moment.

In synchronized or coherent Morse code, the receiver and transmitter are synchronized, so that the receiver hears and evaluates each bit individually. This makes it possible to use a receiving filter having extremely narrow bandwidth. The synchronization requires the use of an external, common frequency or time standard. The broadcasts of WWV/WWVH are used for this purpose (*see* WWV/WWVH). Frequency dividers are used to obtain the necessary synchronizing frequencies. A tone is generated in the receiver output for a particular bit, if, and only if, the average signal voltage exceeds a certain value over the duration of that bit, as at B. False signals, such as might be caused by filter ringing, sferics, or other noise, are generally ignored since they do not result in sufficient average voltage.

Synchronization can be used with other codes besides Morse. For example, a system might be built for use with Baudot or ASCII teletype transmitters and receivers. Experiments with synchronized communications have shown that the improvement in signal-to-noise ratio, compared with nonsynchronized systems, is nearly 20 dB at low speeds (on the order of 15-20 words per minute). The reduced bandwidth of coherent communications allows proportionately more signals to be placed in any given band of frequencies.

SYNCHRONIZED VIBRATOR POWER SUPPLY

A mechanical vibrator can be used to generate high-voltage ac from low-voltage dc (*see* VIBRATOR POWER SUPPLY). Generally, if a direct-current output is wanted, the resulting alternating current is rectified by means of semiconductor diodes; however, the rectification can be done by the vibrator. Then the device is called a synchronized or synchronous vibrator.

The synchronized vibrator has two sets of contacts. One set performs the chopping (interrupting) function. The other set of contacts reverses the polarity of the alternating current in the transformer secondary, inverting either the negative or positive pulses to obtain pulsating direct current. This pulsating direct current is then filtered in the conventional manner.

SYNCHRONOUS COMMUNICATIONS SATELLITE

The world is literally ringed with active communications satellites placed in synchronous, or geostationary, orbits *see* SYNCHRONOUS ORBIT). The satellites orbit the earth about 22,500 miles above the equator. They revolve in the same direction, and with the same period, as the rotation of the planet. Therefore, a synchronous satellite stays above the same place all the time. The bearings (azimuth and elevation) of a synchronous satellite are always the same from any point on the surface of the earth.

The advantage of a synchronous satellite is obvious. A highly directional, high-gain antenna can be adjusted to point at the satellite, assuming the satellite is above the horizon as seen from a particular place. The antenna azimuth or elevation never have to be changed.

Synchronous satellites are used not only for communications, but for weather observation. Three geostationary satellites, placed at 120-degree intervals of longitude, can see the whole world except for the immediate polar regions. *See also* ACTIVE COMMUNICATIONS SATELLITE, SATELLITE COMMUNICATIONS, WEATHER SATELLITE.

SYNCHRONOUS CONVERTER

Alternating currents can be converted to pulsating dc by means of a mechanical device called a synchronous converter. The synchronous converter consists of a commutator and a synchronous motor (*see* COMMUTATOR, SYNCHRONOUS MOTOR).

For mechanical rectification of 60-Hz alternating current, the synchronous motor turns at a rate of 60 revolutions per second or 3,600 revolutions per minute. The commutator reverses the polarity of the output twice for each cycle. This results in inversion of either the negative or positive parts of the waveform in the output.

SYNCHRONOUS DATA

Synchronous data is any data, usually digital, that is transmitted according to a precise time function. Baudot and ASCII are examples of codes that can be sent as synchronous data. Each bit has the same predetermined duration. Mechanically sent Morse code can also be synchronous.

In some synchronous-data communications circuits, the receiver is kept in step with the transmitter by means of pulses sent at the beginning and/or end of each character, or at regular intervals. In radioteletype, some codes use such pulses to indicate the start and finish of each character. Television transmitters send out synchronizing pulses to keep the receiver scanning properly.

A special form of synchronous-data communication uses an independent reference standard to lock the receiver and transmitter precisely. This results in great improvement in the signal-to-noise ratio. *See also* SYNCHRONIZATION, SYNCHRONIZED COMMUNICATIONS.

SYNCHRONOUS DETECTOR

A synchronous detector is a circuit that operates in a manner similar to a product detector (*see* PRODUCT DETECTOR). A local-oscillator signal is mixed with the incoming signal to recover the sidebands.

The synchronous detector uses a phase-locking system that keeps the local-oscillator output in exact synchronization with the carrier wave of the incoming signal. This ensures that the output of the detector will be identical with the original modulating signals.

Synchronous detectors are used in color television receivers to recover the chrominance sidebands with optimum precision. If the local-oscillator signal were not locked with the carrier or subcarrier, distortion would result from even the smallest frequency difference.

SYNCHRONOUS MOTOR

A synchronous motor is an ac motor,
varies with the frequency of the applied current. The motor speed is not affected by the voltage. Most synchro-

nous motors are designed to operate at a frequency of 60 Hz. Many run at 60 revolutions per second, or 3,600 revolutions per minute. But some synchronous motors run at other speeds. In general, the speed S of a synchronous motor, in revolutions per second (RPS), is given by the formula S = 120/P, where P is the number of poles in the motor.

Synchronous motors exhibit no slip under load, as compared to the free-running speed (*see* SLIP). If the mechanical load resistance becomes excessive, the motor will simply fail to operate.

Synchronous motors are used in conjunction with commutators to rectify alternating currents for certain purposes (*see* SYNCHRONOUS CONVERTER). Synchronous motors are also used in electric clocks and other devices requiring constant motor speed. *See also* MOTOR.

SYNCHRONOUS MULTIPLEX

Synchronous multiplex is a form of time-division multiplex, in which two or more signals are transmitted over a single circuit at the same time. The signals are split into discrete intervals having equal duration, and interwoven at a higher speed. Synchronous multiplexing is a form of serial data transfer (*see* SERIAL DATA TRANSFER, TIME-DIVISION MULTIPLEX).

Suppose there are n different signals, designated s1, s2, s3, . . ., sn, that are to be combined in series by means of synchronous multiplexing. Let each signal be divided into intervals having duration t. When the signals are recombined, they are sent one interval at a time, in sequence s1, s2, s3, . . ., sn, s1, s2, ... In order to maintain synchronization of the composite signal with each of the original signals, the information in each interval must be speeded up by a factor of n, resulting in interval durations of t/n. This situation is illustrated for n = 3.

SYNCHRONOUS ORBIT

The period of revolution of a satellite around the earth depends on the altitude of the satellite. Objects in low orbits go around the earth fast; the Mercury and Gemini spacecraft, orbiting at altitudes of a few hundred miles, had periods of only about 90 minutes. Objects in high orbits revolve slowly; the moon, at a distance of 250,000 miles, takes almost a month to complete one orbit around the earth. An object at an altitude of 22,500 miles takes 24 hours to complete one orbit. This type of orbit is called synchronous, since the period is the same as the length of the earth day.

If an object is placed in a synchronous orbit above the equator, so that the object revolves in the same direction the earth rotates, that object will remain above a fixed point on the surface. Many satellites have been placed in such orbits. *See also* SYNCHRONOUS COMMUNICATIONS SATELLITE.

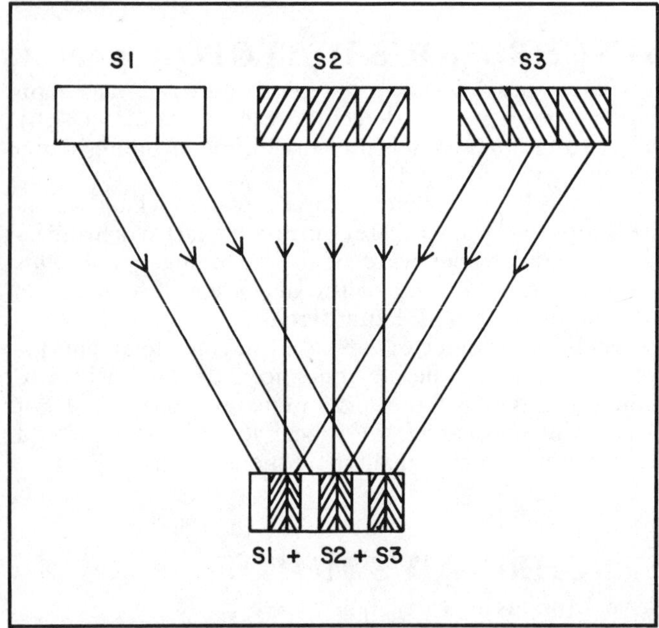

SYNCHRONOUS MULTIPLEX: An example of synchronous multiplex.

SYNCHROSCOPE

A synchroscope is an oscilloscope that shows a waveform or train of pulses in a motionless display, by means of triggering. *See* OSCILLOSCOPE, TRIGGERING.

SYNCHROTRON

The synchrotron is a form of atomic-particle accelerator similar to the cyclotron. The synchrotron can accelerate charged particles, such as electrons, protons, and alpha particles, to extremely high speeds. This gives the particles tremendous energy. The synchrotron is used for converting the nuclei of atoms to different forms (atom smashing).

In the synchrotron, particles move in a circle of constant radius. The particles are moved along by a modulated magnetic field after being injected by a smaller, linear accelerator tangent to the synchrotron itself (see illustration). The synchrotron has a radius of several hundred feet. *See also* CYCLOTRON, PARTICLE ACCELERATOR.

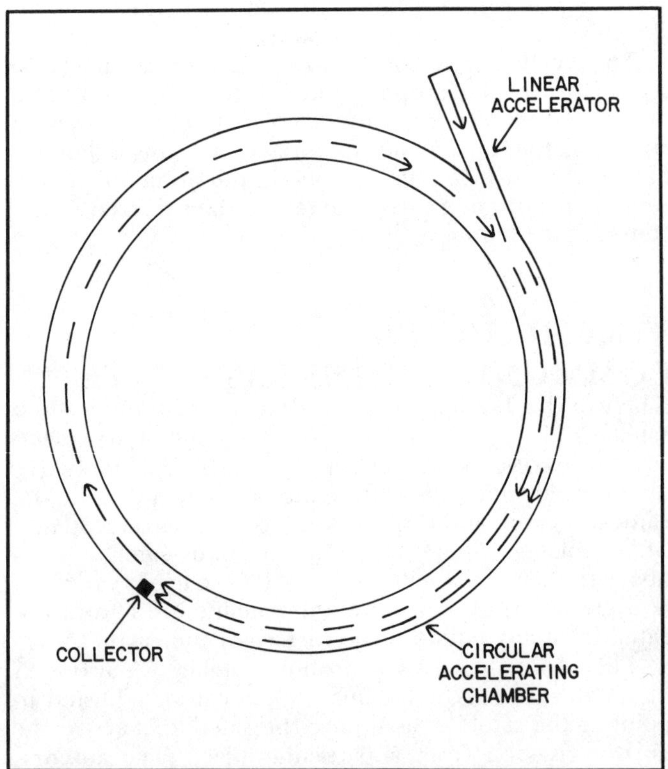

SYNCHROTRON: Diagram of a synchrotron.

SYNTHESIZER

See FREQUENCY SYNTHESIZER, SPEECH SYNTHESIS.

SYSTEMS ENGINEERING

There are many different electrical and electronic engineering specialties. Some engineers design radio-frequency circuits, and some design digital circuits. Software engineers devise the programs that make the hardware work in a computer. The systems engineer is more of a generalist. Systems engineers figure out the best way to interconnect and arrange the individual pieces of equipment to perform a certain job.

The field of systems engineering has many subspecialties. The knowledge required to design a telephone system is different from that needed to put together a broadcasting station. It takes a still different sort of expertise to lay out an optimal computer system. We therefore speak of telephone systems engineers, broadcast systems engineers, computer systems engineers, and so on.

T

TABULATOR

A tabulator, also called a tab, is a function found on most typewriters and word processors. The tabulator moves the carriage or cursor to the right by a certain number of spaces.

Some tabulators can be preset and cleared as the operator wishes. Other tabulators are not adjustable or programmable.

The tabulator function is useful when it is necessary to prepare lists, tables, and charts.

TACAN

See TACTICAL AIR NAVIGATION.

TACHOMETER

A tachometer is an electromechanical or electronic device that is used for measuring the angular speed of a shaft. The speed may be indicated in revolutions per second (RPS) or revolutions per minute (RPM).

A typical tachometer consists of an electric generator, connected via a belt drive or a gear system to the motor shaft, in conjunction with a rectifier/filter and direct-current voltmeter. The faster the motor shaft rotates, the more voltage appears at the output of the generator. The voltmeter is calibrated in rps or rpm. The drawing illustrates this principle. *See also* GENERATOR.

TACTICAL AIR NAVIGATION

Tactical Air Navigation (abbreviated TACAN) is a form of radionavigation used by commercial and military aircraft. The TACAN system operates at ultra-high or microwave frequencies, where the atmosphere has a minimal effect on electromagnetic propagation.

The TACAN system provides constant information on the distance and bearing to one or more ground stations with respect to an aircraft. The distance (range) and bearing (azimuth) can be directly read out on a display in the aircraft. Altitude is not indicated.

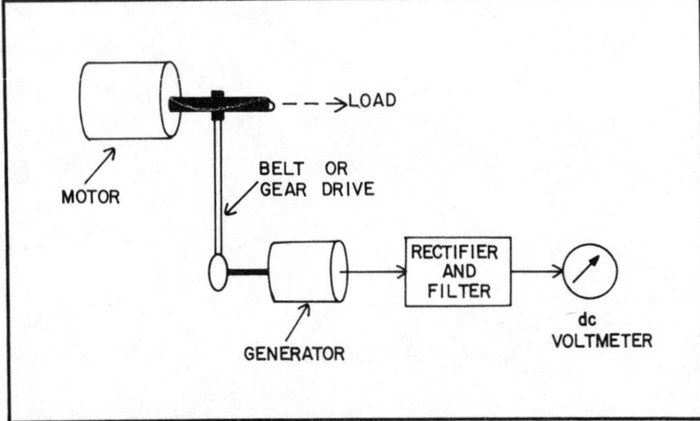

TACHOMETER: A tachometer measures the angular speed of a motor shaft.

TAIL

A pulse always has a finite decay time, even though the amplitude may decay almost instantly. When a pulse is observed on an oscilloscope, the decay time can be measured by using a fast sweep rate. The decay part of a pulse is called a tail if it lasts for a significant length of time (*see* DECAY, DECAY TIME).

TANDEM CONNECTION

Two or more devices are said to be in tandem if they are electrically or mechanically connected together so that they operate one after the other. Active devices such as transistors or tubes, operated in series, are another example. *See also* SERIES CONNECTION.

TANGENT

The tangent function is a trigonometric function. In a right triangle, the tangent is equal to the length of the opposite side divided by the length of the adjacent side (see illustration). In the unit circle $x^2 + y^2 = 1$, plotted on the Cartesian (x,y) plane, the tangent of the angle θ, measured counterclockwise from the positive x axis, is equal to y/x. This is illustrated at B. For values of θ that are an odd

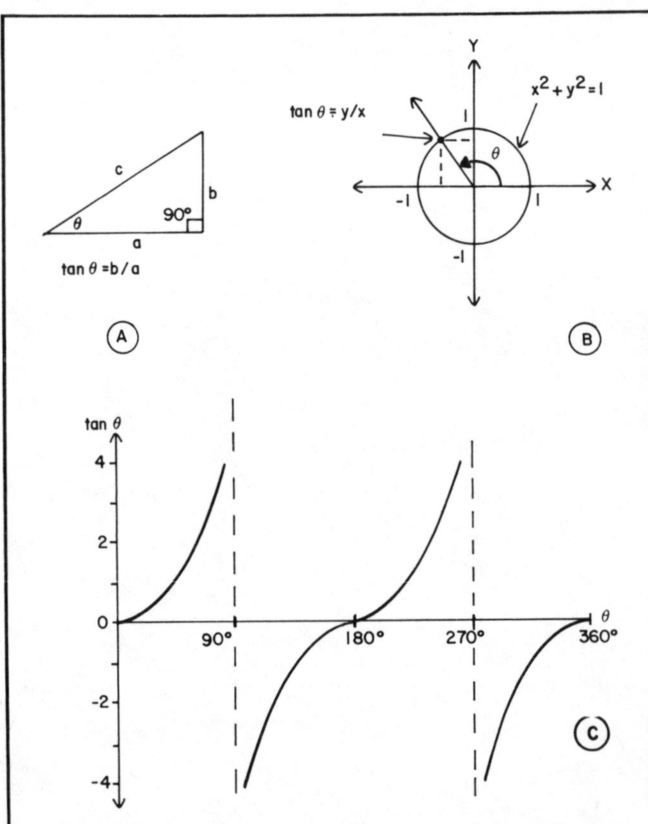

TANGENT: The tangent function is the ratio of the length of the opposite side of a right triangle to the length of the adjacent side, as shown at A. At B, the unit-circle model for the tangent function. At C, an approximate graph of the tangent function for angles between 0 and 360 degrees.

multiple of 90 degrees, the tangent is not defined, since for those angles, $x = 0$.

The tangent function is periodic and discontinuous. The discontinuities appear at odd multiples of 90 degrees. The tangent function ranges through the entire set of real numbers shown at C.

In mathematical calculations, the tangent function is abbreviated tan. Mathematically, the tangent function is always equal to the value of the sine divided by the value of the cosine:

$$\tan \theta = (\sin \theta) / (\cos \theta)$$

Values of tan θ for various angles θ are given in the table. *See also* TRIGONOMETRIC FUNCTION.

TANGENTIAL GALVANOMETER

A sensitive galvanometer can be made from a coil of wire and a magnetic compass. The compass is oriented so that the needle lines up with the 0-degree azimuth marker (north), and the coil is placed around the compass as shown in the illustration.

When a current is passed through the coil, a magnetic field is set up by the current. This field is oriented east and west, at a right angle to the geomagnetic field. This causes the compass needle to be deflected easterly or westerly, depending on the polarity of the voltage applied to the coil terminals. The meter can be used only with direct current.

The coil field and the geomagnetic field add together vectorially to determine the exact position at which the compass needle will come to rest. The stronger the current in the coil, the farther the needle will be deflected. The tangent of the angle of deflection is proportional to the current in the coil. For this reason, the compass galvanometer is called a tangential galvanometer.

A tangential galvanometer can be calibrated against a reference ammeter. A large-value resistor, placed in series with the coil, allows the device to be used as a calibrated direct-current voltmeter. *See also* AMMETER.

TANGENTIAL SENSITIVITY

The sensitivity of a receiver is usually expressed in terms of the signal voltage that results in a certain signal-to-noise ratio or amount of noise quieting in the audio output (*see* NOISE QUIETING, SENSITIVITY, SIGNAL-TO-NOISE RATIO). The most frequently used figures are a 10-dB signal-to-noise ratio or 20 dB of noise quieting. These specifications are not always a precise indication of how well a receiver will respond to weak signals. A more accurate indication of this is given by the tangential-sensitivity figure. Tangential sensitivity is sometimes called threshold sensitivity or weak-signal sensitivity.

Tangential sensitivity is defined as the signal level, in microvolts at the antenna terminals, that results in a barely discernible signal at the output under conditions of

TANGENT: VALUES OF TAN θ FOR VALUES OF θ BETWEEN 0° AND 90°. FOR $90° < 0 \leq 180°$, CALCULATE $180° - \theta$, READ FROM THIS TABLE, AND MULTIPLY BY -1. FOR $180° \leq 0 < 270°$, CALCULATE $\theta - 180°$ AND READ FROM THIS TABLE. FOR $270° < \theta \leq 360°$, CALCULATE $360 - \theta$, READ FROM THIS TABLE, AND MULTIPLY BY -1.

θ, degrees	tan θ	θ, degrees	tan θ	θ, degrees	tan θ
0	0.000	31	0.601	61	1.80
1	0.017	32	0.625	62	1.88
2	0.349	33	0.649	63	1.96
3	0.052	34	0.675	64	2.05
4	0.070	35	0.700	65	2.15
5	0.087	36	0.727	66	2.25
6	0.105	37	0.754	67	2.36
7	0.123	38	0.781	68	2.48
8	0.141	39	0.810	69	2.61
9	0.158	40	0.839	70	2.75
10	0.176	41	0.869	71	2.90
11	0.194	42	0.900	72	3.08
12	0.213	43	0.933	73	3.27
13	0.231	44	0.966	74	3.49
14	0.249	45	1.00	75	3.73
15	0.268	46	1.04	76	4.01
16	0.287	47	1.07	77	4.33
17	0.306	48	1.11	78	4.70
18	0.325	49	1.15	79	5.14
19	0.344	50	1.19	80	5.67
20	0.364	51	1.23	81	6.31
21	0.384	52	1.28	82	7.12
22	0.404	53	1.33	83	8.14
23	0.424	54	1.38	84	9.51
24	0.445	55	1.43	85	11.4
25	0.466	56	1.48	86	14.3
26	0.488	57	1.54	87	19.1
27	0.510	58	1.60	88	28.6
28	0.532	59	1.66	89	57.3
29	0.554	60	1.73	90	–
30	0.577				

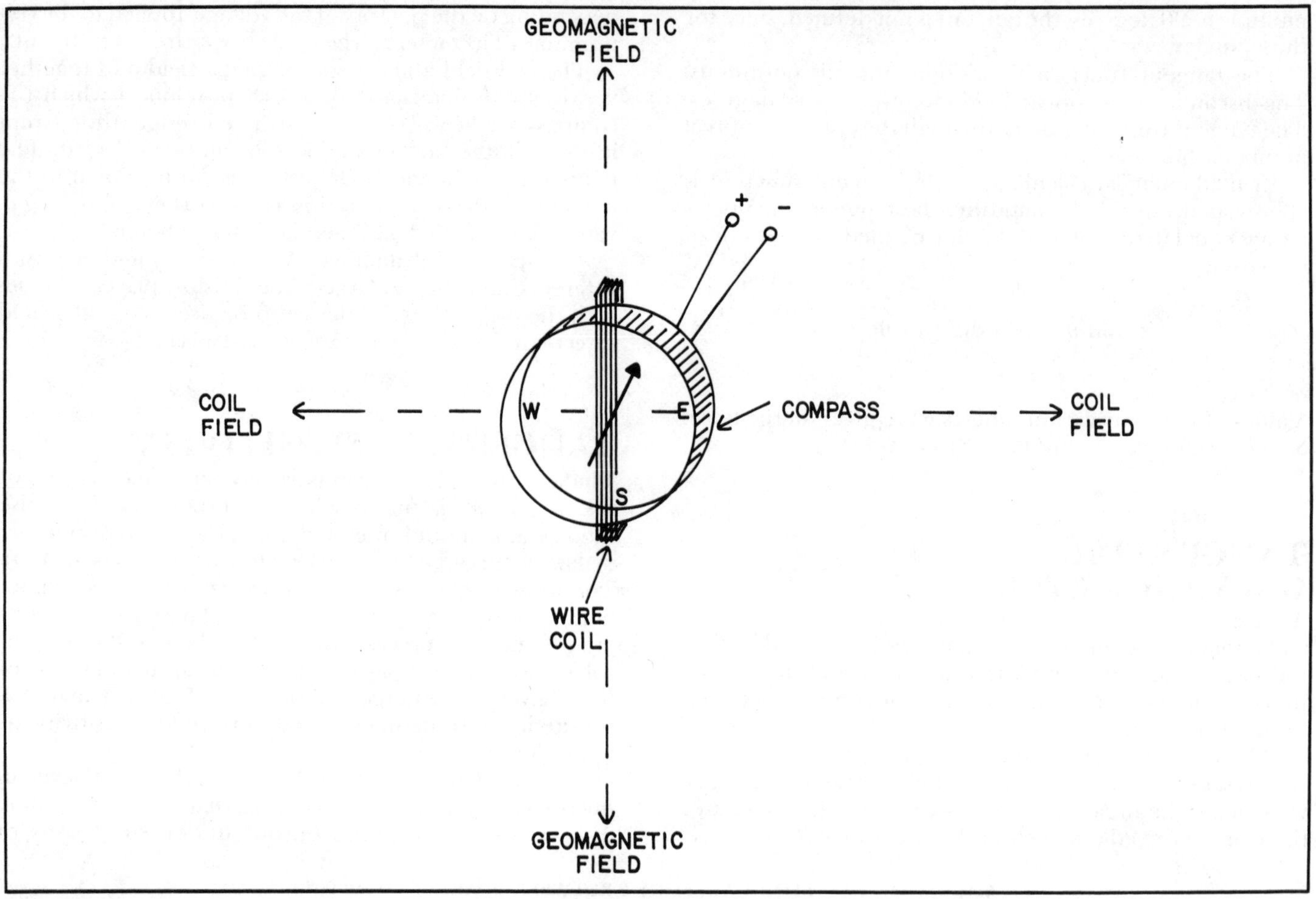

TANGENTIAL GALVANOMETER: A magnetic compass can be used as a galvanometer.

minimum external noise. Occasionally, the tangential sensitivity is defined as the signal level that produces a 3-dB signal-to-noise ratio, or 3 dB of noise quieting, at the output. These voltages are considerably smaller than the voltages required to produce a 10-dB signal-to-noise ratio or 20 dB of noise quieting. Two different receivers, both having the same sensitivity in terms of the 10-dB signal-to-noise or 20-dB noise-quieting figures, may differ in tangential sensitivity.

TANGENT LINE

A tangent line is a straight line that passes through a point on a curve, lies in the plane of the curve in the vicinity of the point, but does not cross the curve in the vicinity of that point.

In a function f(x) = y, defined in the Cartesian coordinate plane, the tangent line has special properties. Suppose we specify a point (x_o, y_o) on the curve. A tangent line at that point can be uniquely defined as having a slope m, equal to the derivative of the function at the point (x_o, y_o):

$$m = f'(x_o)$$

Since the line passes through a known point and has a known slope, we can define the line in point-slope form as

$$y - y_o = (x - x_o) f'(x_o)$$

This is shown in the illustration. *See also* DERIVATIVE, DIFFERENTIATION, FUNCTION.

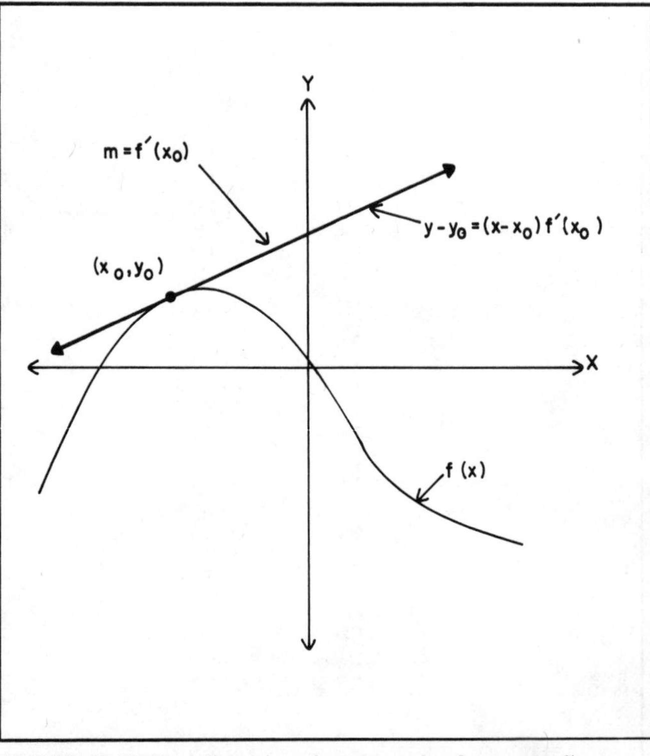

TANGENT LINE: A Cartesian-plane example of a tangent line.

TANK CIRCUIT

A tank circuit is an electrical circuit that stores energy by passing it alternately between two reactances. A tank circuit operates at one frequency, and perhaps at integral multiples of that frequency. All parallel-resonant circuits are tank circuits. A resonant antenna, cavity, or length of transmission line can also act as a tank circuit if it is fed at a voltage loop. At resonance, the impedance of a tank circuit is theoretically infinite. In practice there are some component losses, but the impedance is still extremely high.

Many radio-frequency power amplifiers employ parallel-resonant tuned circuits in the output. This circuit is often called the tank circuit or simply the tank. Either the inductor or capacitor, or both, are adjustable so that the tank circuit can be made resonant at the desired frequency. In the tuned power amplifier, tank-circuit resonance is indicated by a dip in the collector, drain, or plate current. *See also* PARALLEL RESONANCE, RESONANCE, TUNED CIRCUIT.

TANTALUM CAPACITOR

A tantalum capacitor is a polarized device similar to an electrolytic capacitor. Tantalum capacitors characteristically have large values and small size, and this is their main advantage. Tantalum capacitors are extensively used in miniaturized equipment.

A tantalum capacitor is constructed as shown in the illustration. A tantalum electrode is placed in an electrolyte solution. This causes a thin layer of tantalum oxide to form on the surface of the electrode. Tantalum oxide is an insulating, or dielectric material, so a capacitor is formed by the electrolyte and the tantalum. Tantalum oxide has a very high dielectric constant—about three times that of the aluminum oxide found in conventional electrolytic capacitors. This makes it possible to obtain more capacitance per unit volume.

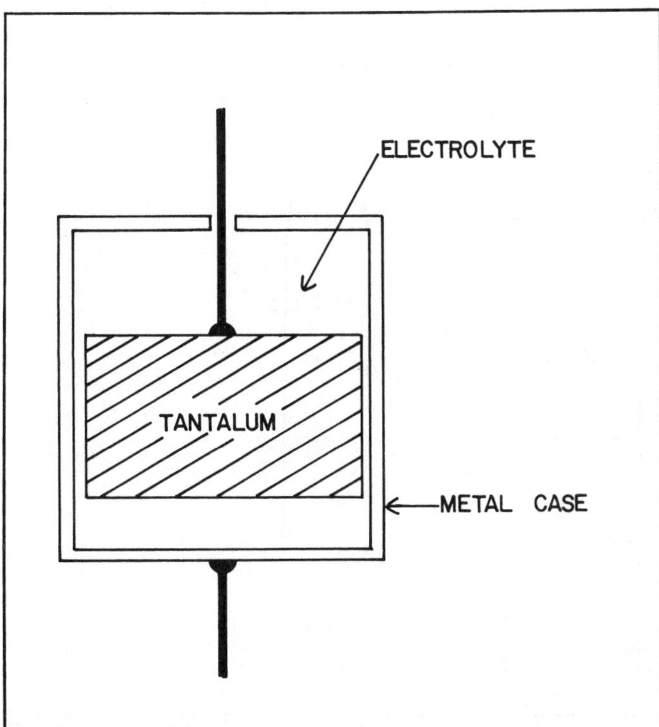

TANTALUM CAPACITOR: Cutaway view of a tantalum capacitor.

Tantalum capacitors are used at low voltages, and primarily at audio frequencies. When tantalum capacitors are used, it is necessary to observe the proper polarity, just as with conventional electrolytics. *See also* CAPACITOR, ELECTROLYTIC CAPACITOR.

T ANTENNA

A T antenna is a wire antenna with a vertical radiator and a capacitance hat consisting of a horizontal wire. The vertical radiator is connected to the center of the horizontal element as shown in the illustrations. The T antenna radiates, and responds to, vertically polarized electromagnetic fields.

The main advantage of the T antenna is that it is shorter, for a given frequency of operation, than a straight quarter-wave vertical antenna. The resonant frequency of the T antenna depends on the height h of the vertical radiator and the radius r of the horizontal radiator. An approximate formula for determining the resonant frequency f, in megahertz, is:

$$f = 234/(h + 1.5r)$$

for h and r in feet, and:

$$f = 71/(h + 1.5r)$$

for h and r in meters.

For optimum performance, a system of radials should be used with the T antenna. The feed-point impedance varies from about 20 ohms to 35 ohms, depending on the ratio h/r. The larger the ratio h/r, the higher the feed-point impedance will be. *See also* VERTICAL ANTENNA, VERTICAL POLARIZATION.

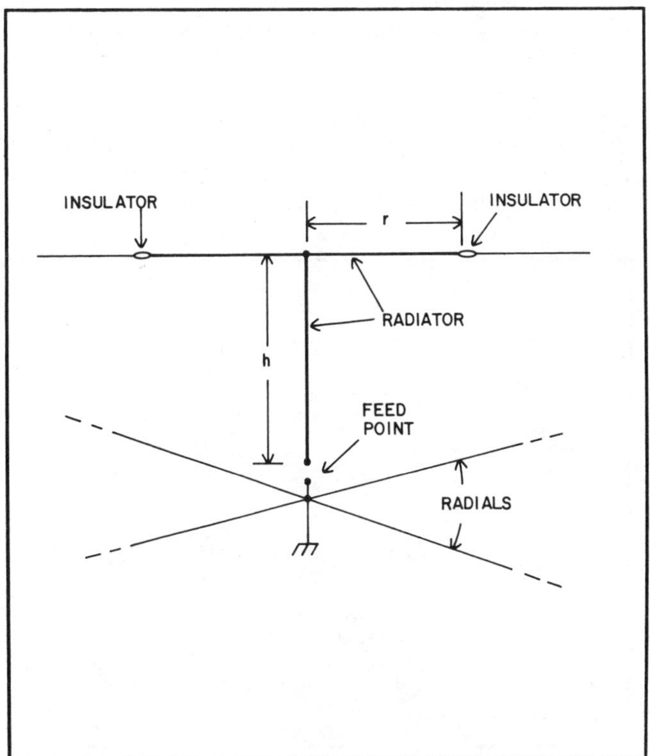

T ANTENNA: Schematic diagram of a T antenna.

TAP

A potentiometer, rheostat, inductor, transformer primary, or transformer secondary may have one or more leads connected to an intermediate point in the element or winding, as well as at either end. Such a connection is called a tap. An element or winding can have several taps.

In a potentiometer, a tap is used to provide adjustment of the resistance of the device. The same is true of a rheostat (*see* POTENIOMETER, RHEOSTAT). The tap position is adjustable.

In an inductor or transformer, a tap can be used to obtain various values of inductance or impedance. Some inductors have variable taps, must most coil taps are fixed. A center tap can be used to obtain two signals or opposite phase and equal amplitude. Center-tapped transformer secondaries are used for balanced-signal output at audio and radio frequencies. In a power transformer, a center-tapped secondary facilitates full-wave rectification. *See also* BALANCED OUTPUT, FULL-WAVE RECTIFIER.

TAPE

See MAGNETIC TAPE.

TAPE RECORDER

A tape recorder is a device that is designed for impressing audio or video signals onto, and recovering them from, magnetic tape. Many types of tape recorders are available for various applications, such as voice reproduction, music reproduction, picture reproduction, and the storage of data for computers and word processors.

Tape recorders can be classified as either cassette type or reel-to-reel type (*see* CASSETTE, REEL-TO-REEL TAPE RECORDER). Cassettes are made in several sizes, ranging from the tiny containers used in dictating machines to the large ones used for recording video signals. Reel-to-reel tape is also available in various sizes.

All tape recorders operate according to the same principle: They act as transducers between electrical impulses and fluctuating magnetic fields. In the RECORD mode, a tape recorder produces an alternating magnetic field that causes polarization of fine particles in the tape. In the PLAYBACK mode, the tape is pulled at constant speed through a device that converts the fields around the particles into electrical impulses. The illustration is a block diagram of a tape recorder.

In recent years, a new form of magnetic recording, called disk recording, has been developed. Disk recording has several advantages over tape recording. Disks do not stretch or jam, as tape occasionally does. Data is more quickly retrieved or modified using disks. A magnetic disk takes up less space, for a given amount of recorded material, than a magnetic tape. Disks are now used almost universally in computer systems. The disk is becoming increasingly common in video recording. Most audio systems still use tape recorders. *See also* DISKETTE, FLOPPY DISK, MAGNETIC RECORDING, MAGNETIC TAPE, VIDEO DISK RECORDING, VIDEO TAPE RECORDING.

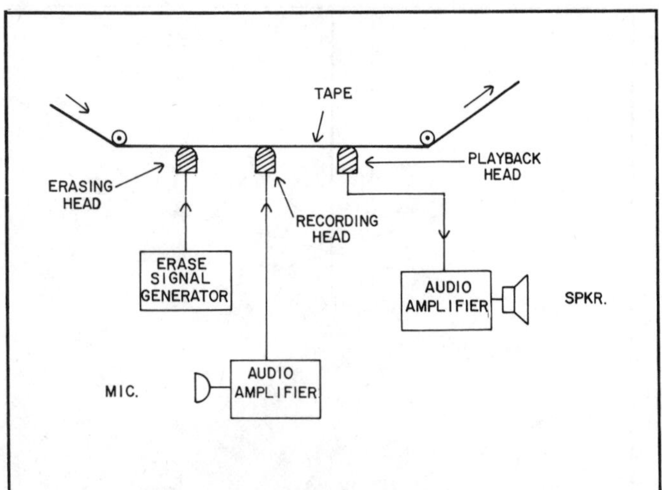

TAPE RECORDER: Block diagram of a tape recorder.

TAPERED FEED LINE

In a radio-frequency transmission line, the characteristic impedance can be made to vary with length. This is accomplished by means of a gradual change in the spacing between the conductors. Such a transmission line is called a tapered line.

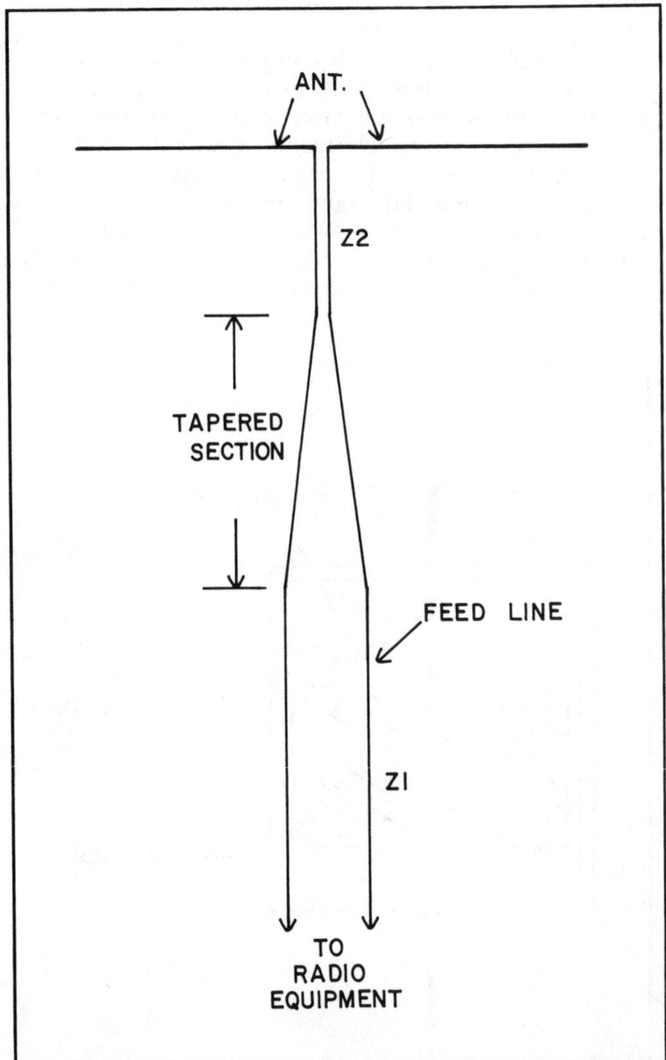

TAPERED FEED LINE: Example of the use of a tapered feed line for impedance-matching purposes.

A tapered line can be used for impedance-matching purposes. The principle is similar to that of the delta match (*see* DELTA MATCH). Suppose that an impedance Z1 is to be matched to an impedance Z2. A tapered line is connected between the two points. The characteristic impedance of the tapered line is equal to Z1 at one end and Z2 at the other end (see illustration).

The main advantage of a tapered matching section over a stub type matching section is that no impedance "bumps" are created. Therefore, there are no standing waves on the line. This results in lower loss, and consequently the efficiency is better. Tapered parallel-wire sections can be tailor made for a particular antenna system. *See also* OPEN-WIRE LINE, STUB.

TAPERED POTENTIOMETER

See AUDIO TAPER, LINEAR TAPER, LOGARITHMIC TAPER, POTENTIOMETER.

TAPE SPLICING

In making up audio or video programs with magnetic tape, it is often necessary to edit something in or out after the recording is complete. This requires tape cutting and splicing.

When cutting magnetic tape, the ends are not cut off straight, but at an angle of approximately 45 degrees. This reduces the thumping noise that sometimes occurs when a tape splice passes the playback head. To make a splice, the two pieces of tape are brought together so that the angled cuts line up, and the magnetic (dull) sides face in the same direction (see illustration). The pieces of tape should not overlap, nor should there be any gap between them. The angles of the cuts should be such that the two pieces of tape come together perfectly. A short length of thin adhesive tape, having the same width as the magnetic tape, is applied to the nonmagnetic (shiny) side of the tape to keep the two sections together. *See also* MAGNETIC TAPE.

TAUT-BAND SUSPENSION

The taut-band suspension is a fairly common type of analog meter movement. The principle is similar to the D'Arsonval movement (*see* D'ARSONVAL), except that a different method is used to obtain suspension and turning tension.

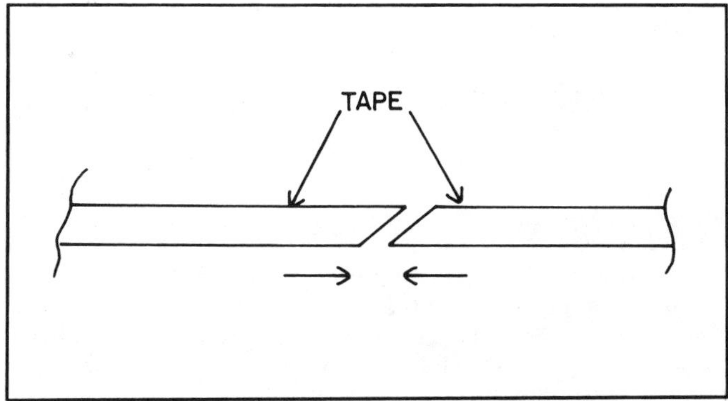

TAPE SPLICING: Magnetic tape is spliced by cutting at an angle. This reduces the effect of the splice on the playback quality.

In a taut-band meter, a coil is rigidly attached to an indicating needle. The coil is placed within the field of a permanent magnet. The coil end of the indicating needle is suspended by two metal strips (see illustration).

When a current passes through the coil, a magnetic field appears around the coil. This causes torque between the coil and the magnet. The needle therefore moves upscale until the torque is balanced by the resistance of the metal strips.

The main advantage of the taut-band suspension is that it is simpler and generally less expensive than the D'Arsonval type suspension. Also, there are no pivot bearings to become loose or sticky.

TAYLOR SERIES

A Taylor series is a summation of expressions associated with a sequence of numbers or variables. Each term in the Taylor series involves a derivative one order greater than that of the previous term.

Suppose we have a function f, a constant value a, and a variable x. Let the Taylor series be represented by T(x). Then:

$$T(x) = f(a) + f'(a)\,(x-a) + f''\,(a)\,(x-a)^2/2! + \ldots + f^{(n)}\,(a)\,(x-a)^n/n! + R,$$

where n is an integer representing the number of terms in the series, and R is a remainder that depends on the integer n.

The larger the number of terms in a Taylor series, the smaller the remainder, and the closer T(x) becomes to its actual value. As n becomes larger without limit, R approaches 0.

Taylor series occasionally arise in electronic engineering, especially in waveform and filter analysis. *See also* DERIVATIVE, DIFFERENTIATION, SERIES.

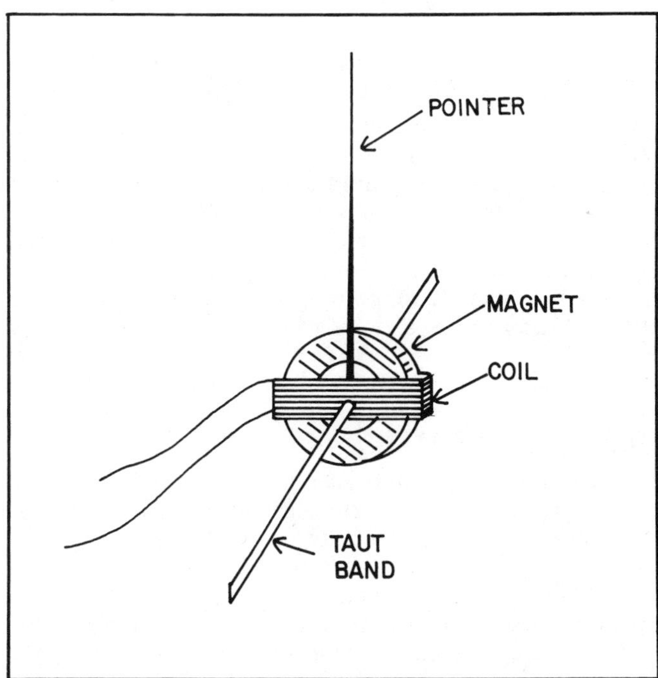

TAUT-BAND SUSPENSION: The taut-band meter suspension is similar to the D'Arsonval movement.

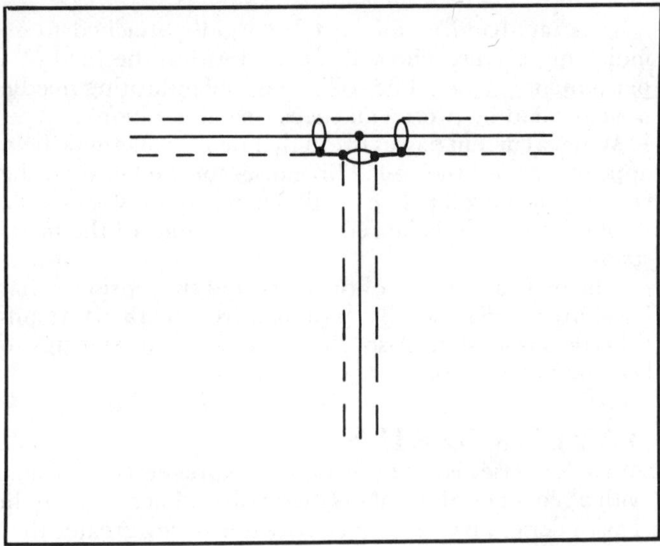

TEE CONNECTION: Schematic diagram of a coaxial tee connection.

TCHEBYSHEV FILTER

See CHEBYSHEV FILTER.

TEE CONNECTION

When a conductor or cable feeds two other conductors or cables, the junction is sometimes called a tee connection. A tee is used for such purposes as splitting (when impedance matching is not critical), phasing, and the connection of stubs to a transmission line. The connection gets its name from the fact that is looks like a capital letter T in schematic diagrams (see illustration).

Electrically, a tee is identical to a wye (Y). The wye expression is more often used in polyphase alternating-current systems. *See also* WYE CONNECTION.

TELECOMMUNICATION

Any form of electromagnetic transmission or communication, not involving the use of wires or cables between stations, is known as telecommunication. Telecommunication can be carried out at any wavelength from the audio frequencies to the ultraviolet and above. Telecommunication is often called wireless communication.

For information concerning facets of telecommunication, please refer to articles according to subject.

TELEGRAPH

A telegraph is a system for sending Morse code signals via wire. The telegraph was the earliest method of electrical communication over long distances. It has been in use since the 1800s. A few telegraph systems still exist.

The simplest form of telegraph consists of a power supply, a key, a relay or other indicating device, and a long cable or two-wire line (see illustration). The range of the telegraph is limited by the loss in the line. The earliest telegraph systems used direct current. There was no such thing as an amplifier. If a system did not have enough

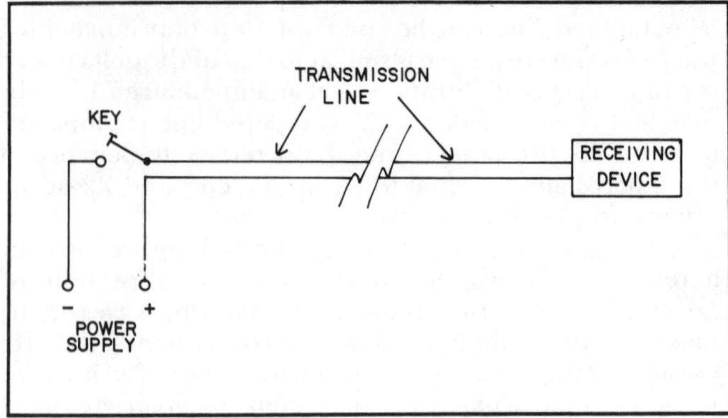

TELEGRAPH: A simple direct-current telegraph circuit.

range, it was necessary to relay the message.

A modern telegraph system uses amplifiers along the line, greatly increasing the range. Although the telegraph is still in limited use today, it has largely been replaced by telephone and telex systems. *See also* TELEPHONE, TELEX.

TELEGRAPH CODES

See AMERICAN MORSE CODE, INTERNATIONAL MORSE CODE.

TELEGRAPH KEY

See KEY, KEYER, SEMIAUTOMATIC KEY.

TELEGRAPHY

Morse-code transmission is often called telegraphy. In telecommunication, telegraphy is regarded as the simplest mode of transmission, since a code transmitter is easy to construct. Many radio operators regard telegraphy as the most efficient and accurate means of communication in high-frequency, ionospheric circuits. The human ear and brain make an excellent information processor! Telegraphy signals can still be heard in the high-frequency band today.

Proficiency in telegraphy is required for anyone who intends to obtain an amateur radio license in the United States. Telegraphy is also taught in the military. Telegraphy can be sent with a simple straight key, or with any of a variety of other keying devices. *See also* AMATEUR RADIO, AMERICAN MORSE CODE, CODE TRANSMITTER, INTERNATIONAL MORSE CODE, KEY, KEYER, SEMIAUTOMATIC KEY.

TELEMETERING

See TELEMETRY,

TELEMETRY

Telemetry is the transmission of quantitative information from one point to another by electromagnetic (radio) means. Telemetry is extensively used by weather balloons, satellites, and other environmental monitoring apparatus. Telemetry is used in space flights, both manned and unmanned, to keep track of all aspects of the equipment and the physical condition of astronauts.

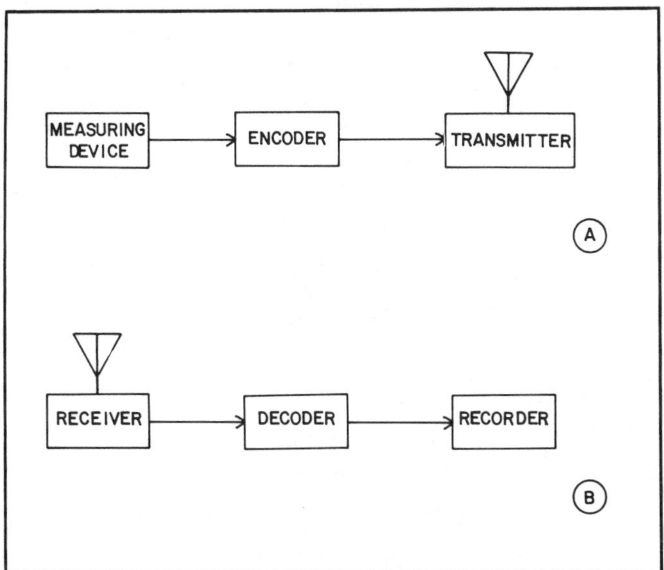

TELEMETRY: At A, a block diagram of a telemetry transmitter. At B, a block diagram of a telemetry receiver.

A telemetry transmitter consists of a measuring instrument, an encoder that translates the instrument readings into electrical impulses, and a modulated radio transmitter with an antenna (see illustration at A). A telemetry receiver consists of a radio receiver with an antenna, a demodulator, and a recorder (B). A computer may also be used to process the data received.

TELEPHONE

Any voice communications system constitutes a telephone, but the term is usually applied to a wire-and-radio system having many users, all of whom can contact any other.

The earliest telephone consisted of a power supply, a variable-resistance transducer (microphone) for converting sound into electrical impulses, an earphone, and a long length of wire. The direct-current power supply was modulated, generating audio-frequency pulsations. The range was limited by the loss in the wire. Switching was done manually by operators at a central office.

Modern telephones use single-sideband-modulated carriers, digital multiplexing, and other sophisticated techniques for transmission. A large system is usually operated by a computer located at a master control center. The computer performs all switching functions. It also has a recorder-announcer that informs subscribers of disconnected or changed numbers, system overload, and other problems. Operators are still used to handle special problems.

Operating Features of Telephone Systems. One of the most important considerations in designing and building a telephone system is the problem of switching. The complexity of the switching network depends on the number of users or subscribers. A telephone system normally has the following two characteristics:

1. Accessibility: Any user can call any other user most of the time.

2. No redundancy: Whenever a subscriber X calls a subscriber Y, only Y will receive the call.

Some telephone systems have various other features, such as:

- Conference calling: Three or more subscribers can communicate on a single line.
- Call forwarding: A subscriber X may arrange to have his calls directed to any other subscriber Y.
- Call waiting: If subscriber X calls subscriber Y and the line is busy, subscriber Y is informed that someone is trying to call. Subscriber Y may then communicate with X at any time.
- Radio links: Subscribers may have mobile or portable telephone sets, linked by wireless radio into the system. An example of this is the cellular radio network.
- Local switching networks: A subscriber may have a single line split up into local lines (extensions). This arrangement is often used by businesses.

Interconnection. A telephone system consists of one or more main lines, called trunks, one or more central-office switches, numerous branch lines, and the telephone sets themselves. There may also be operators, radio links, satellite links, and local switching systems. There may also be a provision for connecting with other systems. This is illustrated in the block diagram.

When a subscriber X wishes to contact a subscriber Y, X dials a number consisting of several digits. The number of digits depends on the complexity of the switch. (In the United States telepohone numbers have ten digits, consisting of a three-digit area code and a seven-digit subscriber number.) If the subscriber Y has a switching system, an operator answers the call and directs it to the appropriate extension.

Most telephone systems cannot operate properly if every subscriber attempts to use it at the same time. The system can handle a limited number of calls, based on the assumption that only a certain percentage of users will try to make calls at a given time. In areas of rapid population growth, or during holidays such as Christmas, a master control center may become overloaded. Callers are then informed that all lines are busy.

Long-distance telephone circuits require the use of low-noise amplifiers and signal processors. The farther a signal travels along a wire, the worse the signal-to-noise ratio becomes. Thermal noise, shot-effect noise, sferics, and geomagnetic-field fluctuations all contribute to the signal-to-noise problem. Modern technology has largely overcome this problem, but long-distance circuits occasionally suffer when conditions ar adverse.

For Further Information. A detailed discussion of telephony is behond the scope of this book. For more information, consult a textbook on telephony. *Related articles in this book include* AUTOPATCH, CARRIER-CURRENT COMMUNICATION, CENTRAL OFFICE SWITCHING SYSTEM, HANDSET, MODEM, MULTIPLEX, RINGING SIGNAL, ROTARY

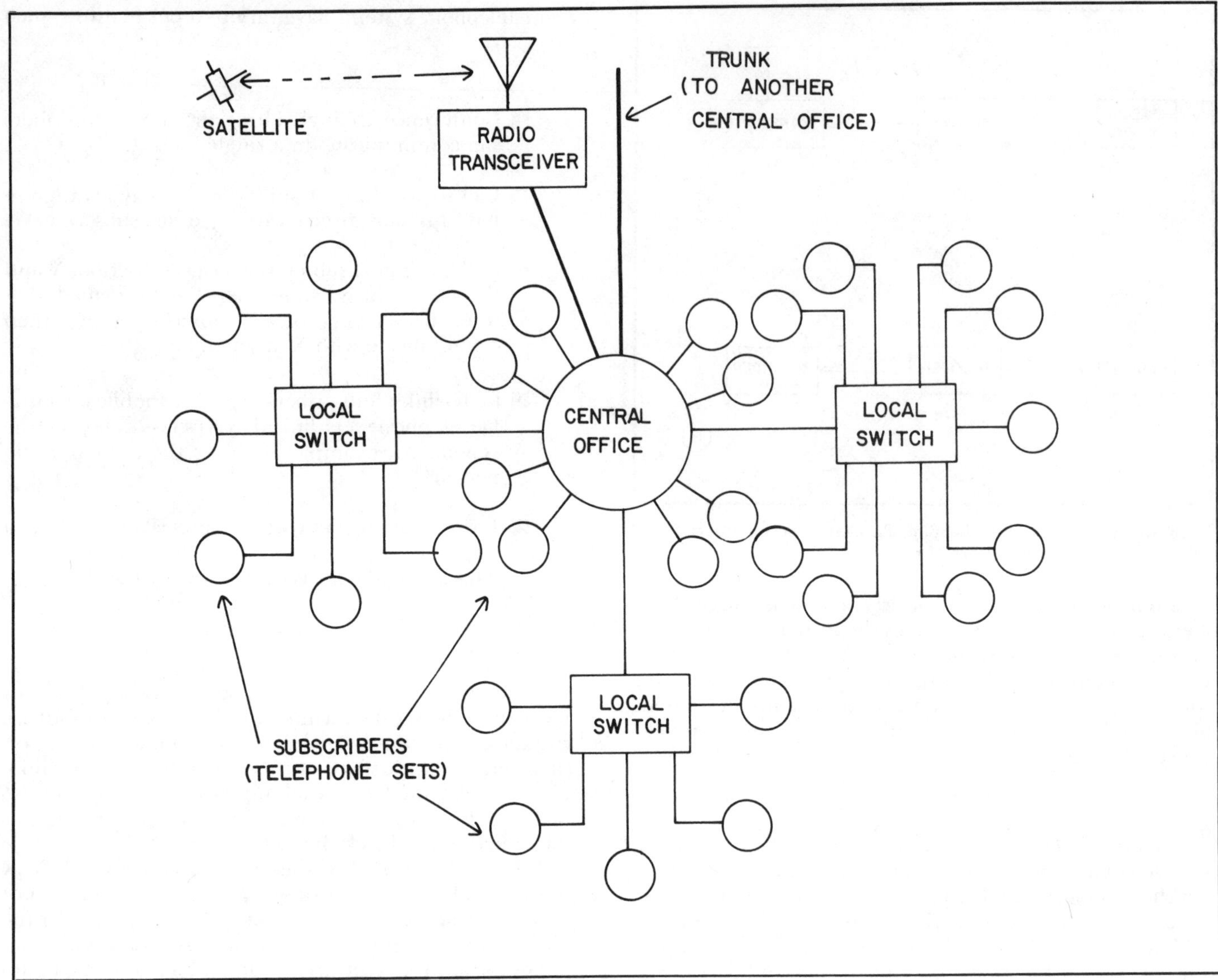

TELEPHONE: Block diagram of a telephone switching system. There may be other features not shown here.

DIALER, SATELLITE COMMUNICATIONS, SUBMARINE CABLE, TELEGRAPH, TELETYPE, TELEX, TOUCHTONE®, TRUNK LINE.

TELEPHONE DIALER

See ROTARY DIALER, TOUCHTONE®.

TELEPHONE MODEM

See MODEM.

TELEPHONY

See TELEPHONE.

TELETYPE®

Teletype is a tradename of the Teletype Corporation. A teletype system may use wire transmission only, a combination of wire and radio circuits, or radio circuits exclusively. The transmission of printed material via radio-only circuits is called radioteletype and sometimes abbreviated RTTY. (*See* RADIOTELETYPE.)

Teletype signals are digital, consisting of two levels or tones corresponding to on (high) and off (low) conditions. In wire systems, these conditions are represented by direct currents. In radio systems, the high and low conditions are represented by different carrier frequencies. The on state is called mark, and the off state is called space (*see* FREQUENCY-SHIFT KEYING, MARK/SPACE).

Teletype systems normally use either the Baudot code or the ASCII code. The speed may vary, although the most often used speeds are 45.45 baud for Baudot and 110 baud for ASCII (*see* ASCII, BAUDOT, BAUD RATE, SPEED OF TRANSMISSION).

The heart of a teletype installation is the teleprinter, which resembles a typewriter. Messages are sent by typing on the keyboard. The message is printed out simultaneously in the transmitting station and at all receiving stations. For reception, the attending operator need do nothing at all, except make sure that the paper piles up neatly if the message is long!

Many teleprinters have a means of preloading and storing a message prior to transmitting it. In modern sys-

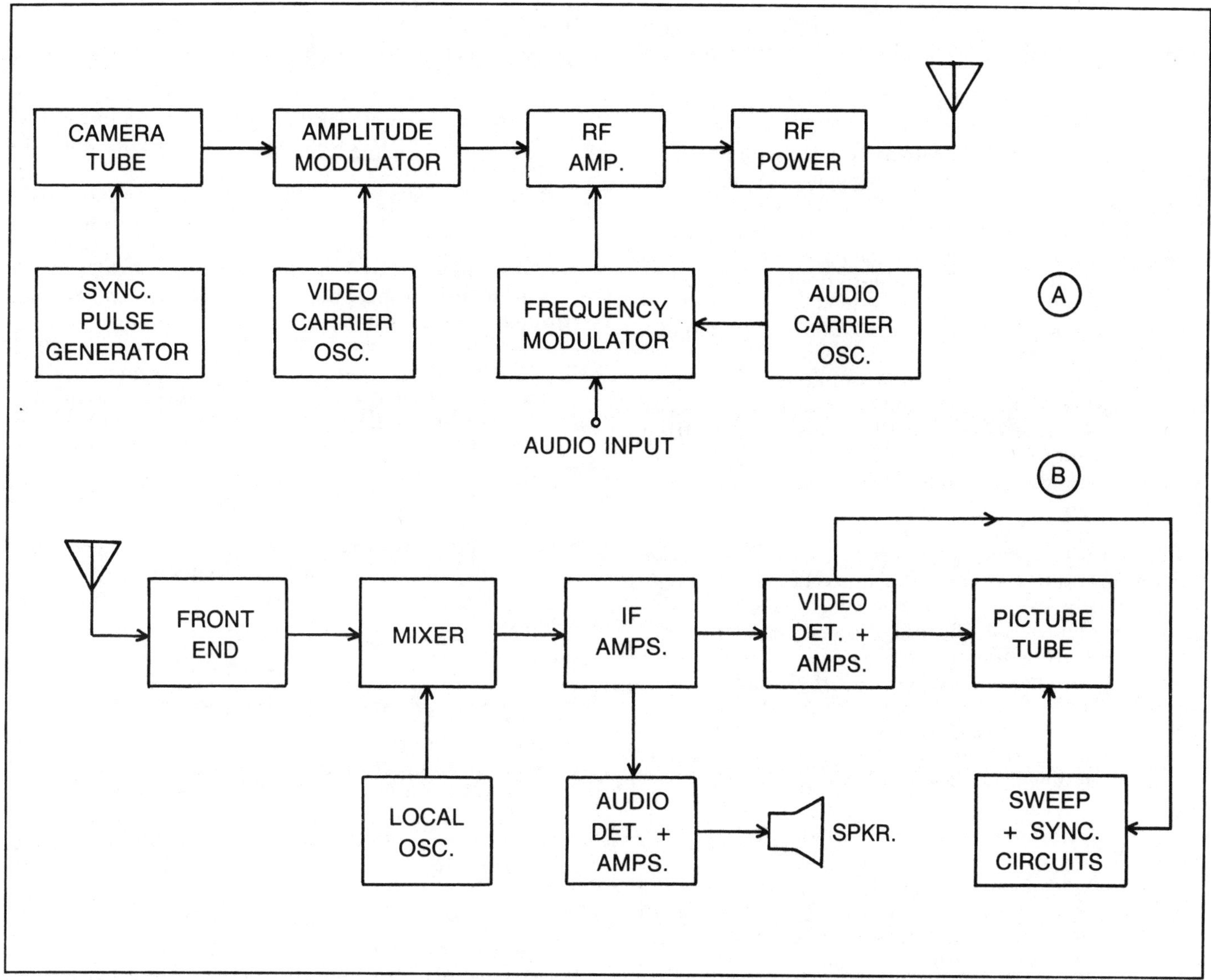

TELEVISION: At A, a simplified block diagram of a television transmitting system. At B, a simplified block diagram of a television receiver.

tems, this is done with electronic memory. In older teleprinters, messages are encoded on paper tape by a device called a reperforator. The reperforator punches holes in the tape according to a five-level or eight-level code (*see* PAPER TAPE). When the paper tape is run through the sending apparatus, the message goes out at the normal system speed.

Home computers can be used as teleprinters. There are various software packages available for this purpose. A teleprinter, conversely, can be interfaced with the telephone lines via a modem, and thereby interconnected with a computer (*see* MODEM).

Some teletype systems operate over telephone lines, making possible the quick sending of messages at moderate cost. Two-way communications can also be accomplished. This is called a telex system. It is extensively used by businesses throughout the world. *See also* TELEX.

TELEVISION

Television is the transfer of moving visual images from one place to another. Television has existed for only a few decades, but it has greatly changed our lives since broadcasting stations first began to use it around 1950. Television is used not only for broadcasting, but for two-way communications. The video signals of television are normally sent and received along with audio signals.

Television systems can be categorized as either fast-scan or slow-scan. In broadcasting, fast-scan television is always used. Slow-scan television is used for communications when the available band space is limited (*see* SLOW-SCAN TELEVISION).

The Television Picture and Signal. In order to get a realistic impression of motion, it is necessary to transmit at least 20 still pictures per second, and the detail must be adequate. A fast-scan television system provides 30 images, or frames, each second. There are 525 or 625 lines in each frame, running horizontally across the picture, which is 1.33 times as wide as it is high. Each line contains shades of brightness in a black-and-white system, and shades of brightness and color in a color system. The image is sent as an amplitude-modulated signal, and the sound is sent as a frequency-modulated signal.

Because of the large amount of information sent, the fast-scan television channel is wide. A standard television-broadcast channel in the North American system takes up

6 MHz of spectrum space. All television broadcasting is done at very high and ultra-high frequencies for this reason.

The Transmitter and Receiver. A television transmitter consists of a camera tube, an oscillator, an amplitude modulator, and a series of amplifiers for the video signal. The audio system consists of an input device (such as a microphone), an oscillator, a frequency modulator, and a feed system that couples the radio-frequency output into the video amplifier chain. There is also, of course, an antenna or cable output. A simplified block diagram of a television transmitter is illustrated at A.

A television receiver is shown in simplified block form at B. The receiver contains an antenna, or an input having an impedance of either 75 ohms or 300 ohms, a tunable front end, an oscillator and mixer, a set of intermediate-frequency amplifiers, a video demodulator, an audio demodulator and amplifier chain, a picture tube with associated peripheral circuitry, and a speaker.

In order for a television picture to appear normal, the transmitter and the receiver must be exactly synchronized. The studio equipment generates pulses at the end of each line and at the end of each complete frame. These pulses are sent along with the video signal. In the receiver, the demodulator recovers the synchronizing pulses and sends them to the picture tube. The electron beam in the picture tube thus moves in exact synchronization with the scanning beam in the camera tube. If the synchronization is upset, the picture appears to "roll" or "tear."

New Trends in Television. In recent years, more and more television transmission has been done via cable. In most major metropolitan areas, cable television is available to anyone who wants it. An increasing number of television stations also broadcast through geostationary satellites. People can select from dozens, or even hundreds, of different channels.

Fast-scan television is used by radio amateurs for communications at ultra-high and microwave frequencies.

Because of the use of cable and satellite systems, television picture quality is much better than ever before. Some thought has been given to increasing the number of lines per frame in the television picture to get a sharper image.

Television can be used along with a computer. Experiments are being conducted in some areas, allowing viewers to interact with a program on television—for such things as shopping and opinion polls.

For further information about fast-scan television, consult the following articles: ASPECT RATIO, CABLE TELEVISION, CAMERA TUBE, COLOR PICTURE SIGNAL, COLOR TELEVISION, COMPOSITE VIDEO SIGNAL, HORIZONTAL SYNCHRONIZATION, ICONOSCOPE, IMAGE ORTHICON, PICTURE SIGNAL, PICTURE TUBE, RASTER, TELEVISION BROADCAST BAND, TELEVISION INTERFERENCE, TELEVISION RECEPTION, VERTICAL SYNCHRONIZATION, VIDICON.

TELEVISION ANTENNA

See LOG-PERIODIC ANTENNA, RABBIT EARS, TELEVISION RECEPTION.

TELEVISION BROADCAST BAND

In the United States, fast-scan television broadcasts are made on 68 different channels in the vhf and uhf ranges. Each channel is 6 MHz wide, including both the video and audio information.

There is no channel 1. Channels 2 through 13 are sometimes called the very-high-frequency (vhf) television channels (see Table 1). Channels 14 through 69 are called the ultra-high-frequency (uhf) channels (Table 2).

Ionospheric propagation occasionally affects channels 2 through 6. Long-distance propagation can take place because of dense ionization in the E layer (*see* SPORADIC-E PROPAGATION) on these channels. Tropospheric propagation is occasionally observed on all of the vhf channels (*see* TROPOSPHERIC PROPAGATION). The uhf channels are unaffected by the ionosphere, although tropospheric effects can occur to a small extent.

Cable television signals are transmitted on a variety of wavelengths. Since the cable prevents signals from getting in or out, the entire electromagnetic spectrum can be used. Satellite television systems also use frequencies other than those listed in the tables. *See also* CABLE TELEVISION, TELEVISION.

TELEVISION INTERFERENCE

Radio transmitters and electrical equipment occasionally cause interference to television receivers. This is called television interference (TVI). Television interference can take place for any of several different reasons. The fault may be in the radio transmitter, the television receiver, both, or neither.

Visual and Audible Effects. An unmodulated carrier, beating against a television picture carrier, causes a diagonal pattern of lines on the screen. This is called cross hatching (see photograph A). Cross hatching is usually accompanied by lightening or darkening of the picture. The stronger the interfering signal, the more washed-out the picture appears. If the interfering signal is very strong, the picture is totally blacked out.

An amplitude-modulated carrier produces cross hatching and horizontal areas of dark and light on a television screen, as at B. The horizontal lines are called modulation bars. A single-sideband signal produces modulation bars but no cross hatching, as at C. A frequency-modulated or phase-modulated signal does not produce modulation bars, but the cross hatching appears to move as the frequency or phase of the interfering signal changes.

A radio signal may interfere with the sound as well as the picture in television reception. An unmodulated carrier may not be noticed in the sound channel at all, unless it is strong, in which case the volume will fluctuate. An amplitude-modulated signal sounds distorted, but can sometimes be understood. A single-sideband signal sound muffled and unintelligible. A frequency-modulated or phase-modulated signal can be understood.

Causes and Cures. Radio transmitters operating at subharmonics of a television channel may cause interference

TELEVISION BROADCAST BAND: STANDARD VHF TELEVISION BROADCAST CHANNELS IN THE UNITED STATES.

Channel Designator	Frequency, MHz	Channel Designator	Frequency, MHz
2	54.0-60.0	8	180-186
3	60.0-66.0	9	186-192
4	66.0-72.0	10	192-198
5	76.0-82.0	11	198-204
6	82.0-88.0	12	204-210
7	174-180	13	210-216

Table 1

TELEVISION BROADCAST BAND: STANDARD UHF TELEVISION BROADCAST CHANNELS IN THE UNITED STATES.

Channel Designator	Freq., MHz	Channel Designator	Freq., MHz
14	470-476	42	638-644
15	476-482	43	644-650
16	482-488	44	650-656
17	488-494	45	656-662
18	494-500	46	662-668
19	500-506	47	668-674
20	506-512	48	674-680
21	512-518	49	680-686
22	518-524	50	686-692
23	524-530	51	692-698
24	530-536	52	698-704
25	536-542	53	704-710
26	542-548	54	710-716
27	548-554	55	716-722
28	554-560	56	722-728
29	560-566	57	728-734
30	566-572	58	734-740
31	572-578	59	740-746
32	578-584	60	746-752
33	584-590	61	752-758
34	590-596	62	758-764
35	596-602	63	764-770
36	602-608	64	770-776
37	608-614	65	776-782
38	614-620	66	782-788
39	620-626	67	788-794
40	626-632	68	794-800
41	632-638	69	800-806

Table 2

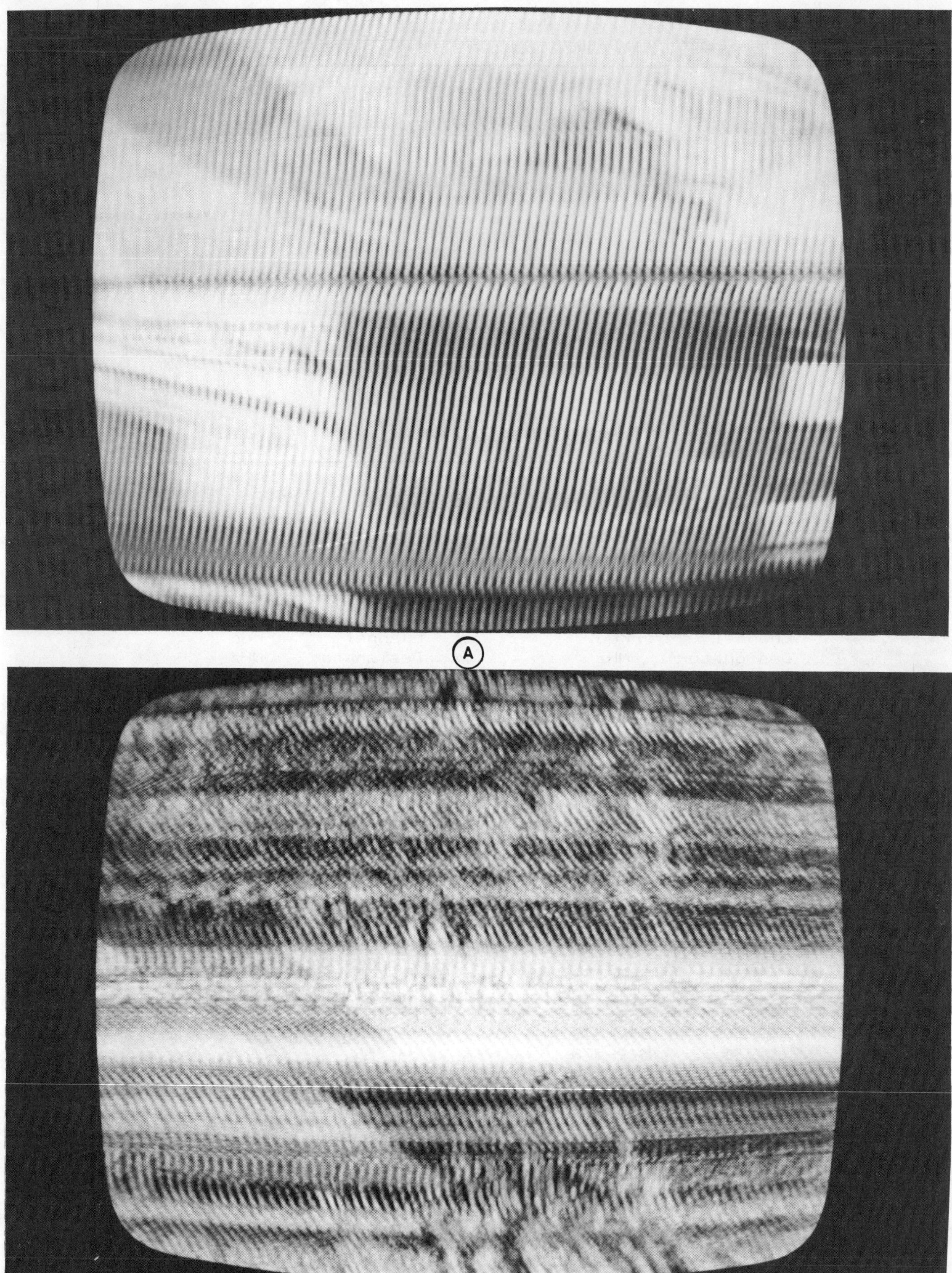

TELEVISION INTERFERENCE: At A, cross hatching; at B, modulation bars accompanied by cross hatching.

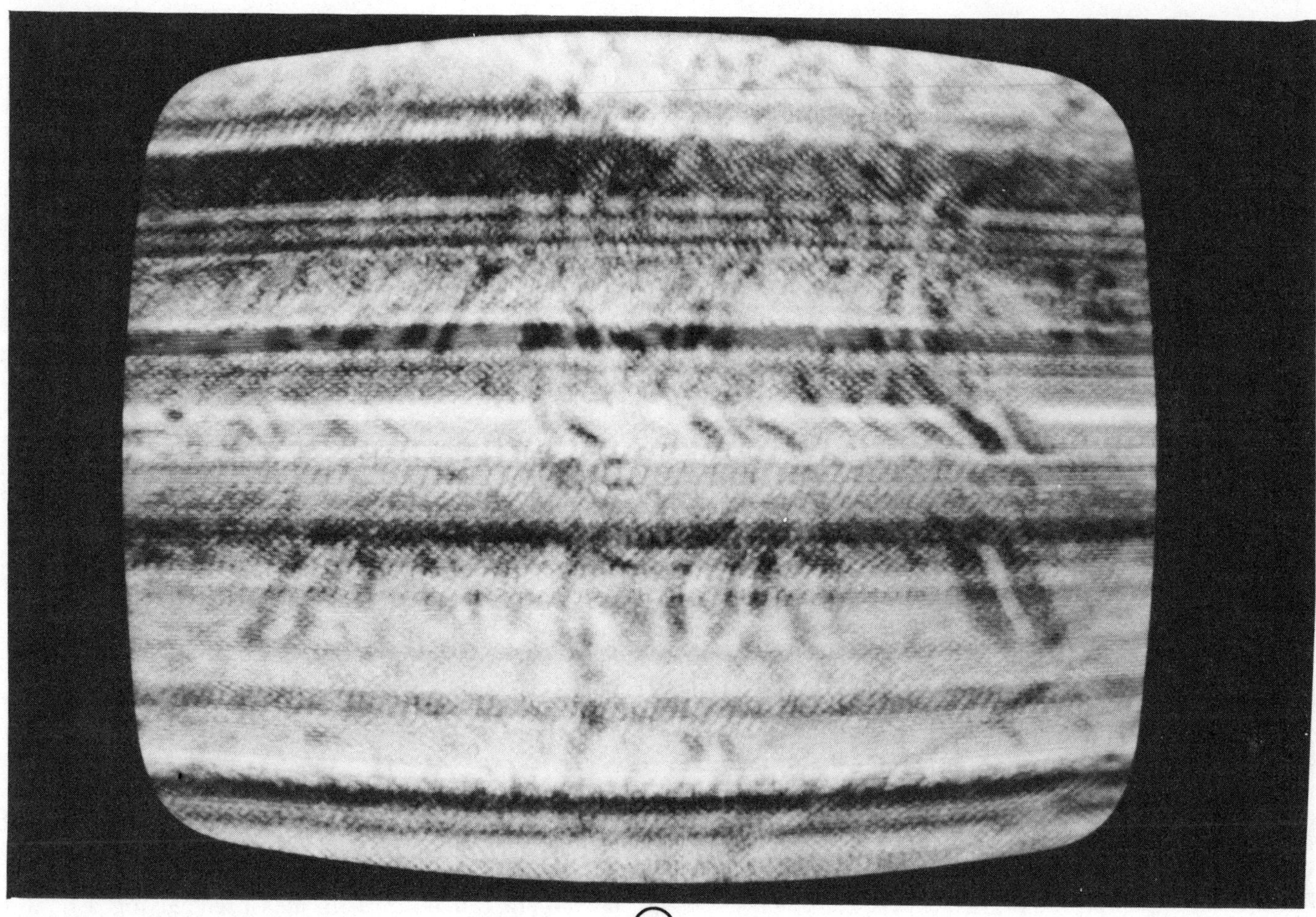

TELEVISION INTERFERENCE: At C, modulation bars without cross hatching.

because of harmonic radiation from the transmitter. The solution is the reduction of harmonic energy at the transmitter output. A lowpass filter is generally used for this purpose with high-frequency transmitters (*see* LOWPASS FILTER). The filter is placed in the antenna feed line at the transmitter output.

Television receivers are sometimes overloaded by a strong signal at any frequency. There may be nothing wrong with the radio transmitter causing the interference. If the frequency of the transmitter is known, a trap circuit can be used at the receiver to eliminate the interference. If the frequency is not known precisely, but is lower than the television frequency, a highpass filter can be used at the television-receiver antenna terminals (*see* HIGHPASS FILTER). Overloading is rarely caused by signals at frequencies higher than that of the television receiver, but if this does happen, a lowpass filter, having the appropriate cutoff, should be used.

Harmonic radiation by a transmitter, and overloading of a receiver, can happen at the same time. Then, both of the above measures will be necessary to eliminate TVI.

The feed line of a television antenna can act as a high-frequency receiving antenna, resulting in TVI from a properly operating transmitter. The cure for this type of TVI is to wind the television feed line into a small coil of about 10 to 20 turns at some point near the television set. This will choke off the unwanted currents on a twin-lead or coaxial line. Grounding the chassis of the television will also help.

Harmonics can be generated in objects not connected to a radio transmitter, resulting in TVI. Any poor electrical connection can be nonlinear at radio frequencies. Such a junction may be in the television antenna, a wire fence, utility wiring, or even metal roofing. A strong radio signal produces currents in nearby electrical conductors. When these currents flow through a nonlinear junction, harmonics are generated. Even if a lowpass filter is used at the transmitter and a highpass filter is used at the receiver, TVI may occur. The cure for this type of TVI is to find the nonlinearity and correct it. That is not always easy, and sometimes it borders on the impossible! Fortunately, this cause of TVI is uncommon. *See also* HARMONIC, TELEVISION.

TELEVISION RECEPTION

Television reception is affected by many things, including the type of set used, the location, the conditions of the ionosphere and troposphere, and the presence of interfering signals.

Antennas. For many people television reception has been simplified in recent years by the proliferation of cable systems. But in rural areas and some cities, cable television does not yet exist. Then it is necessary to use a television receiving antenna.

Many television sets are equipped with built-in whip antennas. These antennas are adequate for local reception, but they do not work well in fringe areas. Even when receiving local signals, ghosting and fading are common because of signal reflection from nearby objects. A set of rabbit ears will improve the performance of a television receiver. Some rabbit ears have preamplifiers built in (*see* RABBIT EARS). Ghosting and fading sometimes occur even when rabbit ears are used.

The best type of antenna for television reception is an outdoor antenna. Many different types of outdoor antennas are made commercially. Some use 75-ohm coaxial cable, and some use 300-ohm twin-lead line. Outdoor television antennas are directional, and sometimes require rotors. An outdoor antenna should be placed as high as possible, and equipped with some means of lightning protection (*see* LIGHTNING PROTECTION). Some municipalities have zoning restrictions that may affect those who wish to erect television antennas outdoors.

For reception of satellite television, special antennas are required. These dish antennas measure several feet across. Various satellite-television receiving systems are available from commercial sources.

Propagation Anomalies. The very-high-frequency (vhf) television channels 2 through 13 are subject to tropospheric effects. At times, signals may be received from hundreds of miles away because of bending or ducting. On channels 2 through 6, sporadic-E propagation occasionally results in reception of stations more than 1,000 miles distant. At or near the time of a sunspot maximum, F-layer propagation has been known to take place on frequencies as high as that of channel 2. On a few occasions, stations have been received from other continents on this channel (*see* E LAYER, F LAYER, PROPAGATION CHARACTERISTICS, SPORADIC-E PROPAGATION, TROPOSPHERIC PROPAGATION).

The ultra-high-frequency (uhf) channels are somewhat affected by the troposphere, but there are no ionospheric effects. Propagation is rarely observed over distances of more than approximately 100 miles.

Interference and Other Problems. Various devices can cause interference to television reception. Anything that makes a spark, such as an electric motor or gasoline-powered engine, will produce interference to nearby television receivers. Radio transmitters can also cause television interference (*see* TELEVISION INTERFERENCE).

A weak television signal is accompanied by flecks on the screen known as snow. There is also a hissing sound in the background of the audio signal.

Electromagnetic reflection causes ghosts, or false images. Reflection also results in severe fading. These problems can usually be eliminated by the use of an outdoor antenna. Cable systems are less affected than conventional television systems by interference, weak-signal effects, ghosting, and fading. *See also* TELEVISION.

TELEVISION SIGNAL

See COLOR PICTURE SIGNAL, PICTURE SIGNAL, RASTER.

TELEVISION TRANSMISSION

See COLOR PICTURE SIGNAL, PICTURE SIGNAL, SLOW-SCAN TELEVISION, TELEVISION.

TELEX

Teletype® signals can be sent over the telephone by means of a frequency-shift-keyed audio tone. This is known as teletype exchange or telex. A telex system consists of a conventional teleprinter and a modem (*see* FREQUENCY-SHIFT KEYING, MODEM) connected into a telephone line leased specifically for the purpose of telex operation.

Most telex systems operate at 100 words per minute. Older systems may use a speed of 60 words per minute. Either speed facilitates the sending and receiving of short messages in just a few seconds.

A modern telex machine has a electronic memory in which text can be stored before it is sent. The operator loads the text into the memory, dials the desired telex number, and sends the message at full system speed by pressing a single memory-recall button. If necessary, text can be sent manually. The machine at the receiving end is switched on automatically, no matter whether it is day or night, and the message is printed.

Telex systems are used by businesses throughout the world. An importer in the United States, for example, can send a message instantly to a manufacturer overseas, and a reply can be expected within a few hours. *See also* TELEPHONE.

TEMPERATURE

The atoms of all substances are in constant motion. The rate at which the atoms and molecules in a material move is proportional to the amount of energy present. This movement can be directly measured, giving a quantity called molecular temperature. All substances radiate energy to a certain extent; the wavelength of this radiant energy can be measured, yielding a quantity known as spectral temperature.

Molecular Temperature. There is a limit to how cold it can get: absolute zero is the complete lack of all movement among the atoms or molecules of a material. No place in the known universe is this cold, although in intergalactic space the temperature is very nearly absolute zero. At the other extreme, there is no theoretical limit to how hot it can get. The highest known temperatures occur at the centers of large stars, where the atoms are in such violent motion that elemental changes take place.

Molecular temperature can be measured directly using a thermometer. The molecular temperature-measurement method is familiar to everyone.

Temperature is expressed as a number, based on a scale at which zero represents some specific phenomenon or condition. In the United States, the Fahrenheit scale is most often used by civilians. In other countries of the world, and among scientists, the Celsius scale is used. The Fahrenheit and Celsius scales are based on specific properties of water, and are therefore most often used for expressing molecular temperature.

Spectral Temperature. In deep space, there are practically no molecules, and molecular temperature therefore becomes meaningless. Moreover, it is impossible to measure the temperature of a distant celestial object using a thermometer. Astronomers use a radio telescope or spectrocope to determine the wavelength distribution of the radiation from a planet, star, or gas cloud in outer space. From this, the temperature can be determined according to Wiens Displacement Law.

Astronomers have determined that the surfaces of the sun and distant stars have temperatures of thousands of degrees Fahrenheit. These are spectral temperatures, since no one has ever put a thermometer in such places!

The Kelvin scale is most often used for expressing spectral temperature. This scale is based on absolute zero, which is the complete lack of radiant energy. Occasionally, spectral temperature is expressed according to another absolute scale called the Rankine scale. *See also* BLACK BODY, CELSIUS TEMPERATURE SCALE, ENERGY, FAHRENHEIT TEMPERATURE SCALE, KELVIN TEMPERATURE SCALE, RADIO TELESCOPE, RANKINE TEMPERATURE SCALE, SPECTROSCOPE, THERMOMETER, WIEN'S DISPLACEMENT LAW.

TEMPERATURE COEFFICIENT

Many electronic components are affected by fluctuations in temperature. Resistors and capacitors, especially, tend to change value when the temperature varies over a wide range. The tendency of a component to change in value with temperature variations is known as temperature coefficient.

If the value of a component decreases as the temperature rises, that component is said to have a negative temperature coefficent. If the value increases as the temperature rises, a component has a positive temperature coefficient. A few components exhibit relatively constant value regardless of the temperature; these devices are said to have a zero temperature coefficient. The temperature coefficient is usually expressed in percent per degree Celsius. For piezoelectric crystals, the temperature coefficient is often expressed in hertz or kilohertz per degree Celsius.

As an example, consider a capacitor that has a value of 100 pF at room temperature (20° C). If the component has a positive temperature coefficient of 0.1 percent per degree Celsius (0.1%/° C), then the value will rise by 0.1 pF as the temperature rises by 1° C. At 21° C, the value will be 100.1 pF; at 30° C, it will be 101 pF; at 120° C, it will be 110 pF.

Temperature coefficients are an important consideration in radio-frequency circuit design. This is especially true of oscillators. Sometimes, a component having a certain temperature coefficient is deliberately placed in a circuit to offset the effects of the temperature coefficients of other components. This is known as temperature compensation. *See also* TEMPERATURE COMPENSATION.

TEMPERATURE COMPENSATION

Changes in temperature can cause instability in electronic circuits, especially oscillators. While crystal oscillators are generally more stable than variable-frequency oscillators, either type of circuit will exhibit some drift as the ambient temperature rises or falls. To compensate for such changes, special components can be added to an oscillator-circuit.

Suppose that a crystal has a negative temperature coefficient of 10 Hz/° C. This means that, for a rise in temperature of 1° C, the frequency will fall by 10 Hz; for a rise of 10° C, the frequency will drop by 100 Hz. Conversely, as the temperature falls, the frequency increases.

To compensate for this, a small capacitor, having a known negative temperature coefficient, can be installed in parallel with the crystal. As the temperature rises, the value of the capacitor decreases, effectively raising the frequency of the crystal-capacitor combination. As the temperature falls, the value of the capacitor gets larger, pulling the frequency down. If the capacitor is carefully selected, the temperature-coefficient effects of the capacitor will just offset those of the crystal, and the frequency will not change as the temperature fluctuates.

In any oscillator circuit, it is desirable to use components with temperature coefficients as close to zero as possible. This makes temperature compensation simpler than when components having large positive or negative temperature coefficients are used. It also makes sense to keep the temperature as constant as possible. *See also* TEMPERATURE COEFFICIENT.

TEMPERATURE DERATING

Some electronic equipment, especially power amplifiers, generate significant heat. If this heat becomes excessive, components may be damaged.

The temperature is more likely to exceed the maximum limits if the weather is hot, as compared to when it is cold. A component can dissipate more power when the ambient temperature is low, and less power when it is hot. For this reason, some electronic devices must be operated at a reduced power level when the temperature is high. The hotter it gets, the more the power must be reduced. This deliberate reduction of power is called temperature derating.

Temperature derating is specified according to a graph of power level versus temperature. A radio transmitter, for example, might normally be run at 100 watts output. However, if the temperature exceeds a certain limit, the power input is reduced according to a derating curve such as that

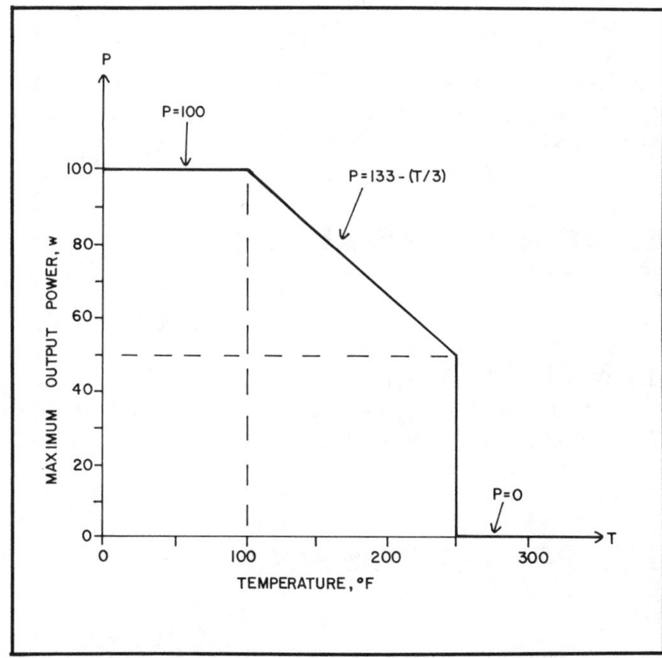

TEMPERATURE DERATING: An example of a temperature-derating curve.

shown in the illustration. Note that, above 250 degrees Fahrenheit, the curve drops to zero, indicating that the equipment should not be operated at all when the temperature gets that high.

Extreme temperatures do not usually occur in nature. However, inside a cabinet containing many heat-generating components, or inside an enclosed automobile on a hot, sunny day, the temperature may reach nearly 200 degrees Fahrenheit. This can have a significant effect on electronic equipment.

TEMPERATURE MEASUREMENT

See CELSIUS TEMPERATURE SCALE, FAHRENHEIT TEMPERATURE SCALE, KELVIN TEMPERATURE SCALE, RANKINE TEMPERATURE SCALE, TEMPERATURE, THERMOMETER, WIEN'S DISPLACEMENT LAW.

TEMPERATURE PROTECTION

See THERMAL PROTECTION.

TEMPERATURE REGULATION

Many types of electronic equipment, especially complicated apparatus, is temperature-sensitive. Such equipment requires some form of temperature regulation, so that it can function in an unchanging environment.

The most common form of temperature regulation consists of heating and air-conditioning systems. In buildings where computers are located, the temperature should be kept as constant as possible. Atmospheric humidity may also be regulated. These measures serve to minimize down time and service costs, as well as to optimize accuracy and reliability.

Temperature regulation is used to keep reference oscillators operating at constant frequency. Such oscillators are crystal-controlled, and the crystals are housed in temperature-regulated chambers called ovens (*See* CRYSTAL OVEN).

All temperature-regulation devices use a heating or cooling system, or both, as well as a thermostat. *See also* THERMOSTAT.

TEMPERATURE RUNAWAY

See THERMAL RUNAWAY.

TEMPERATURE SCALE

See CELSIUS TEMPERATURE SCALE, FAHRENHEIT TEMPERATURE SCALE, KELVIN TEMPERATURE SCALE, RANKINE TEMPERATURE SCALE.

TEMPERATURE SHOCK

See THERMAL SHOCK.

TEN CODE

Some radiotelephone operators use a set of abbreviations called the ten code. These signals consist of the spoken numeral 10, followed by a one-digit or two-digit numeral that indicates the meaning. The ten code is used by police and fire officials in their radio communications, and also by Citizen's Band (CB) operators.

Any ten signal may be sent as a statement or asked as a question. For example, "My 10-20 is Phoenix, Arizona," means that an operator is located at Phoenix, Arizona. The question "What is your 10-20?" means "What is your location?"

The most common ten signals are listed in Tables 1 and 2. Amateur radio operators use a similar set of code symbols known as the Q code. *See also* Q SIGNAL.

TENSILE STRENGTH

The amount of tension that a wire can withstand is known as the tensile strength of the wire. Tensile strength is sometimes called the breaking load. Tensile strength is usually indicated in pounds per square inch, or in kilograms per square centimeter, of cross-sectional area. Tensile strength may also be expressed in terms of the breaking load in pounds or kilograms. The larger the size of the wire for a given material, the greater the tensile strength.

Steel has the greatest tensile strength of common wire materials. The breaking load varies from about 75,000 to 150,000 pounds per square inch of cross-sectional area. For this reason, steel wire is preferred for use in tower-guying systems. Copper and aluminum wire have somewhat less tensile strength; typically it ranges from 30,000 to 60,000 pounds per square inch for copper and 20,000 to 40,000 pounds per square inch for aluminum.

The average breaking load for various gauges of steel, copper, and aluminum wire is indicated in the accompanying table. *See also* WIRE.

TENSION

See STRESS.

TERA

Tera is a prefix multiplier that means 1,000,000,000,000 (10^{12}). For example, 1 terahertz is 10^{12} Hz or 1,000,000 MHz. (The wavelength of a signal at this frequency is just 0.3 mm.) The abbreviation for tera is the capital letter T. *See also* PREFIX MULTIPLIERS.

TERMAN-WOODYARD AMPLIFIER

See DOHERTY AMPLIFIER.

TERMINAL

A terminal is a point at which two or more wires are connected together, or where voltage or power is applied or taken from a circuit. We may speak, for example, of the antenna terminals of a receiver or transmitter, or the speaker terminals of a stereo high-fidelity amplifier. Input and output terminals are generally equipped with binding posts or connectors.

TEN CODE: THE 10-CODE SIGNALS USED BY CITIZEN'S-BAND RADIO OPERATORS, WITH QUERY AND STATEMENT/INFORMATION.

Signal	Meaning
10-1	Are you having trouble receiving my signals?
	I am receiving you poorly.
10-2	Are my signals good?
	Your signals are good.
10-3	Shall I stop transmitting?
	Stop transmitting.
10-4	Have you received my message? OK?
	Affirmative?
	I have received your message; OK; affirmative.
10-5	Shall I relay a message to ---?
	Relay a message to ---.
10-6	Are you busy?
	I am busy; stand by until ---.
10-7	Is your station out of service?
	My station is out of service.
10-8	Is your station in service?
	My station is in service.
10-9	Shall I repeat my message?
	Repeat your message; reception is poor.
10-10	Are you done transmitting?
	I am done transmitting.
10-11	Am I talking too fast?
	You are talking too fast.
10-12	Do you have visitors?
	I have visitors.
10-13	How are your weather and road conditions?
	My weather and road conditions are ---.
10-14	What time is it?
	The local time is ---.
10-15	Shall I pick up --- at ---?
	Pick up --- at ---.
10-16	Have you picked up ---?
	I have picked up ---.
10-17	Do you have urgent business?
	I have urgent business.
10-18	Have you anything for me?
	I have --- for you.
10-19	Have you nothing for me?
	I have no information for you.
10-20	Where are you located?
	I am located at ---.
10-21	Shall I call you on the telephone?
	Call me on the telephone.
10-22	Shall I report in person to ---?
	Report in person to ---.
10-23	Shall I stand by?
	Stand by until ---.
10-24	Are you finished with your last assignment?
	I am finished with my last assignment.
10-25	Are you in contact with ---?
	I am in contact with ---.
10-26	Shall I disregard the information you just sent?
	Disregard the information I just sent.
10-27	Shall I move to channel ---?
	Move to channel ---.
10-30	Is this action legal?
	This action is illegal.
10-33	Do you have an emergency message?
	I have an emergency message.
10-34	Do you have trouble?
	I have trouble.
10-35	Do you have any confidential information?
	I have confidential information.
10-36	Is there an accident?
	There is an accident at ---.
10-37	Is a tow truck needed?
	A tow truck is needed at ---.
10-38	Is an ambulance needed?
	An ambulance is needed at ---.
10-39	Is there a convoy at ---?
	There is a convoy at ---.
10-41	Shall we change channels?
	Change channels.
10-60	Please give me your message number.
	My message number is ---.
10-63	Is this net directed?
	This net is directed.
10-64	Are you clear?
	I am clear.
10-65	Do you have a net message for ---?
	I have a net message for ---.
10-66	Do you wish to cancel your previous messages numbered --- through ---?
	I wish to cancel my previous messages numbered --- through ---.
10-67	Shall I clear for a message?
	Clear for a message.
10-68	Shall I repeat my messages numbered --- through ---?
	Repeat your messages numbered --- through ---.
10-70	Have you a message?
	I have a message.
10-71	Shall I send messages by number?
	Send messages by number.

10-79	Shall I inform --- regarding a fire at ---?	10-92	Are my signals distorted?
	Inform --- regarding a fire at ---.		Your signals are distorted.
10-84	What is your telephone number?	10-94	Shall I send you a test transmission?
	My telephone number is ---.		Send me a test transmission.
10-91	Are my signals weak?	10-95	Shall I key my microphone without speaking?
	Your signals are weak.		Key your microphone without speaking.

Table 1

TEN CODE: THE 10-CODE SIGNALS USED BY LAW-ENFORCEMENT AGENCIES WITH QUERY AND STATEMENT/INFORMATION.

Signal	Meaning	Signal	Meaning
10-1	Are you having trouble receiving my signals?	10-18	Shall I hurry to finish this assignment?
	I am receiving your poorly.		Hurry to finish this assignment.
10-2	Are my signals good?	10-19	Shall I return to ---?
	Your signals are good.		Return to ---.
10-3	Shall I stop transmitting?	10-20	What is your location?
	Stop transmitting.		My location is ---.
10-4	Have you received my message?	10-21	Shall I call --- by telephone?
	I have received your message.		Call --- by telephone.
10-5	Shall I relay a message to ---?	10-22	Shall I ignore the previous information?
	Relay a message to ---.		Ignore the previous information.
10-6	Are you busy?	10-23	Has --- arrived at ---?
	I am busy; stand by until ---.		--- has arrived at ---.
10-7	Is your station out of service?	10-24	Have you finished yoru assignment?
	My station is out of service.		I have finished my assignment.
10-8	Is your station in service?	10-25	Shall I report in person to ---?
	My station is in service.		Report in person to ---.
10-9	Shall I repeat my message?	10-26	Are you detaining a subject?
	Repeat your message.		I am detaining a subject.
10-10	Is there a fight in progress at your location?	10-27	Do you have information about driver's license number ---?
	There is a fight in progress at my location.		Here is information concerning driver's license number ---.
10-11	Do you have a case involving a dog?		
	I have a case involving a dog.		
10-12	Shall I stand by until ---?	10-28	Do you have information concerning vehicle registration number ---?
	Stand by until ---.		Here is information concerning vehicle registration number ---.
10-13	How are your weather and road conditions?		
	My weather and road conditions are ---.		
10-14	Have you had a report of a prowler?	10-29	Shall I check records to see if --- is a wanted person?
	I have had a report of a prowler.		
10-15	Is there a civil disturbance at your location?	10-30	Is --- using a radio illegally?
	There is a civil disturbance at my location.		--- is using a radio illegally.
10-16	Is there domestic trouble at your location?	10-31	Is there a crime occurring at your location, or at ---?
	There is domestic trouble at my location.		There is a crime occurring at my location, or at ---.
10-17	Shall I meet the complainant?		
	Meet the complianant.		

Signal	Meaning
10-32	Is there a man with a gun at your location, or at ---?
	There is a man with a gun at my location, or at ---.
10-33	Do you have an emergency?
	I have anemergency, here or at ---.
10-34	Is there a riot at your location, or at ---?
	There is a riot at my location, or at ---.
10-35	Do you have an alert concerning a major crime?
	I have an alert concerning a major crime.
10-36	What is the correct time?
	The correct time is --- (local) or --- (UTC).
10-37	Shall I investigate a suspicious vehicle?
	Investigate a suspicious vehicle.
10-38	Are you stopping a suspicious vehicle?
	I am stopping a suspicious vehicle of type ---.
10-39	Is your situation urgent?
	This situation is urgent; use lights and or siren.
10-40	Shall I not use my light or siren?
	Do not use your light or siren.
10-41	Are you just starting duty?
	I am just starting duty.
10-42	Are you finishing with duty?
	I am finishing with duty.
10-43	Do you need, or are you sending, information about ---?
10-44	Do you want to leave patrol?
	I want to leave patrol to go to ---.
10-45	Is there a dead animal at ---?
	There is a dead animal at ---.
10-46	Shall I assist a motorist at ---? or Are you assisting a motorist at ---?
	Assist a motorist at ---; or I am assisting a motorist at ---.
10-47	Are road repairs needed immediately at ---?
	Road repairs are needed immediately at ---.
10-48	Does a traffic standard at --- need to be fixed?
	A traffic standard at --- needs to be fixed.
10-49	Is a traffic light out at ---?
	A traffic light is out at ---.
10-50	Is there an accident at ---?
	There is an accident at ---.
10-51	Is a tow truck needed at ---?
	A tow truck is needed at ---.
10-52	Is an ambulance needed at ---?
	An ambulance is needed at ---.
10-53	Is the road blocked at ---?
	There is a roadblock at ---.
10-54	Are there animals on the road at ---?
	There are animals on the road at ---.
10-55	Is there a drunk driver at ---?
	There is a drunk driver at ---.
10-56	Is there a drunk pedestrian at ---?
	There is a drunk pedestrain at ---.
10-57	Has there been a hit-and-run accident at ---?
	There has beeen a hit-and-run accident at ---.
10-58	Shall I direct traffic at ---?
	Direct the traffic at ---.
10-59	Is there a convoy at ---? or Does --- need an escort?
	There is a convoy at ---; or --- needs an escort.
10-60	Is there a squad at ---?
	There is a squad at ---.
10-61	Are there personnel in the vicinity of ---?
	There are personnel in the vicinity of ---.
10-62	Shall I reply to the message of ---?
	Reply to the message of ---.
10-63	Shall I make a written record of ---?
	Make a written record of ---.
10-64	Is this message to be delivered locally?
	This message is to be delivered locally.
10-65	Do you have a net message assignment?
	I have a net message assignment.
10-66	Do you want to cancel message number ---?
	I want to cancel message number ---.
10-67	Shall I clear for a net message?
	Clear for a net message.
10-68	Shall I disseminate information concerning ---?
	Disseminate information concerning ---.
10-69	Have you received my messages numbered --- through ---?
	I have received your messages numbered --- through ---.
10-70	Is there a fire at ---?
	There is a fire at ---.
10-71	Shall I advise of details concerning the fire at ---?
	Advise of details of the fire at ---.
10-72	Shall I report on the progress of the fire at ---?
	Report on the progress of the fire at ---.
10-73	Is there a report of smoke at ---?
	There is a report of smoke at ---.
10-74	(No query)
	Negative.
10-75	Are you in contact with ---?
	I am in contact with ---.

Signal	Meaning	Signal	Meaning
10-76	Are you going to ---?	10-90	Is there a bank alarm at ---?
	I am going to ---.		There is a bank alarm at ---.
10-77	When do you estimate arrival at ---?	10-91	Am I, or is ---, using a radio without cause?
	I estimate arrival at --- at --- local time.		You, or ---, is/are using a radio without cause.
10-78	Do you need help?	10-93	Is there a blockade at ---?
	I need help at ---.		There is a blockade at ---.
10-79	Shall I notify a coroner of ---?	10-94	Is there an illegal drag race at ---?
	Notify a coroner of ---.		There is an illegal drag race at ---.
10-82	Shall I reserve a motel or hotel room at ---?	10-96	Is there a person acting mentally ill at ---?
	Reserve a motel or hotel room at ---.		There is a person acting mentally ill at ---.
10-85	Will you, or ---, be late?	10-98	Has someone escaped from jail at ---?
	I, or ---, will be late.		Someone has escaped from jail at ---; or, There is a jail escapee at ---.
10-87	Shall I pick up checks for distribution?	10-99	Is --- wanted or stolen?
	Pick up checks for distribution; or I am picking up checks for distribution.		--- is wanted or stolen; or, There is a wanted person or stolen article at ---.
10-88	What is the telephone number of ---?		
	The telephone number of --- is ---.		

Table 2

TENSILE STRENGTH: TENSILE STRENGTH FOR VARIOUS AMERICAN WIRE GAUGE (AWG) SIZES OF STEEL, COPPER, AND ALUMINUM WIRE. ALL FIGURES ARE GIVEN IN POUNDS, AND ARE APPROXIMATE.

AWG No.	Steel Wire	Copper Wire	Aluminum Wire
1	5,000-10,000	2,000-4,000	1,300-2,600
2	3,900-7,800	1,600-3,200	1,000-2,000
3	3,100-6,200	1,200-2,400	830-1,700
4	2,500-4,900	980-2,000	650-1,300
5	2,000-3,900	780-1,600	520-1,000
6	1,600-3,200	630-1,300	420-840
7	1,200-2,500	490-980	330-650
8	980-2,000	390-780	260-520
9	770-1,500	310-620	210-410
10	610-1,200	240-490	160-330
11	490-970	190-390	130-260
12	380-770	150-310	100-200
13	305-610	120-240	80-160
14	240-480	97-190	65-130
15	190-380	77-150	51-100
16	150-300	61-120	41-81
17	120-240	48-97	32-64
18	96-190	38-77	26-51
19	76-150	30-61	20-41
20	60-120	24-48	16-32

In a large computer, the operating console or consoles are located separately from the actual computer circuitry. Such a console, which consists of a keyboard and a teleprinter or cathode-ray-tube display, is called a terminal. Some computer terminals can be connected to a telephone and used with a distant computer.

In a Teletype or radioteletype station, the keyboard, modem, and printer or cathode-ray-tube display are called the terminal. The modem alone is sometimes called a terminal unit. *See also* MODEM, TERMINAL UNIT, VIDEO DISPLAY TERMINAL.

TERMINAL RESISTANCE HEATING

When current flows through a terminal, heat is sometimes generated because of terminal resistance. This heating is normally very minor, but if excessive current flows, or if the terminal contacts are dirty, a large amount of power may be dissipated.

Consider a terminal that carries 10 A of current. If the resistance of the junction is just 0.1 ohm, the power dissipated will be 10 watts. This much power can raise a small terminal to a dangerously high temperature. Terminal resistance heating can cause damage to electrical and electronic equipment, and can also present a fire hazard. (The preceding example is typical of high-current household appliances. You have probably noticed terminal resistance heating when pulling out the plug of some heavy appliance.)

To minimize the chances and hazards of terminal resistance heating, all plugs, connectors, and terminals should be kept clean. Terminals should not be forced to carry more current than they can handle. Flammable materials should be kept well away from terminals that carry large currents.

TERMINAL STRIP

See TIE STRIP.

TERMINAL UNIT

In a Teletype system, a modem is used to encode signals for transmission and to decode signals for reception (*See* MODEM). Such a device is called a terminal unit.

Teletype systems today normally use either the ASCII or Baudot codes at various speeds (*See* ASCII, BAUDOT, BAUD RATE). Sometimes, the International Morse code is used. A terminal unit consists of a receiving section and a transmitting section. A terminal unit may have a built-in cathode-ray-tube monitor and an attachable keyboard, forming a complete communications terminal (see illustration).

The receiving demodulator separates the mark and space tones from the output of the radio receiver (*See* MARK/SPACE). This is normally done with a pair of highly selective audio filters. The signals are then processed into digital pulses. The pulses are used to operate a high-speed switch that controls the printer. Alternatively, the pulses are fed to a cathode-ray-tube display.

The transmitting section of a terminal unit is quite simple. The digital pulses from the keyboard can be fed to a switching transistor that keys a reactance in and out of the oscillator of the radio transmitter, thereby changing the

TERMINAL UNIT: A terminal unit, designed for amateur and commercial use.

frequency. This is know as frequency-shift keying (FSK). If the output is in Morse code, the switching transistor is connected directly to the key jack of the transmitter. Some terminal units have audio oscillators that produce mark/space tones that can be fed into the microphone jack of a transmitter. This is called audio frequency-shift keying (AFSK). *See also* FREQUENCY-SHIFT KEYING, RADIO-TELETYPE, TELETYPE.

TERMINATION

The point at which a transmission line is connected to a load is called the termination. The load itself may also be called the termination.

A termination may consist of any type of load, such as an antenna, a dummy antenna, a telephone set, or an electrical appliance. Sometimes a transmission line is terminated in a short or open circuit. In a radio-transmitting antenna system, the termination is usually called the feed point.

Ideally, all of the available power at a termination is radiated, absorbed, or dissipated by the load. In a radio-transmitting antenna, a cable-television system, a telephone line, an audio-frequency amplifier/speaker system, and many other types of circuits, this requires that the load impedance be a purely resistive, with an ohmic value equal to the characteristic impedance of the transmission line. *See also* CHARACTERISTIC IMPEDANCE, IMPEDANCE, IMPEDANCE MATCHING.

TERTIARY COIL

Some audio-frequency or radio-frequency transformers have a small coil, coupled to the main windings, that can be used for feedback purposes (see illustration). Such a coil is called a tertiary coil. In a radio-frequency transformer, a tertiary coil is often called a tickler coil.

A tertiary winding can be used to facilitate either positive feedback or negative feedback. The Armstrong oscillator employs a tertiary coil to obtain the positive feedback needed to maintain oscillation. Some regenerative detectors use tertiary windings to obtain positive feedback. A negative-feedback tertiary winding is sometimes used to keep audio-frequency or radio-frequency power amplifiers from oscillating. *See also* FEEDBACK.

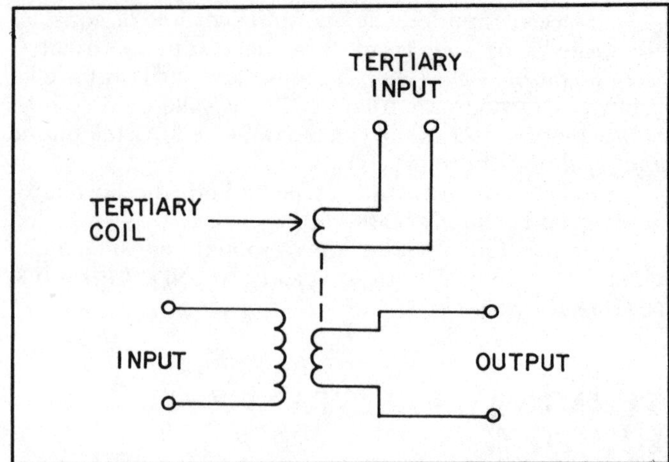

TERTIARY COIL: A tertiary coil is used for feedback purposes.

TESLA

The tesla is the standard international (SI) unit of magnetic flux, equivalent to 1 weber per square meter. Other units are also used quite often to express magnetic flux density. The most common of these is the gauss. A flux density of 1 tesla is equivalent to 10,000 gauss. *See also* GAUSS, MAGNETIC FLUX, WEBER.

TESLA, NIKOLA

An inventor and engineer, Nikola Tesla (1856-1943) is best known for his work on improving methods of power transmission. Tesla was the first scientist to propose wireless transmission of electrical power. While his goal has not yet been realized, it is possible that someday it will, probably by means of high-energy beams of electromagnetic or radiant energy.

Tesla performed research on numerous other data devices, including motors and generators and telephone systems.

TESLA COIL

A Tesla coil is a special form of transformer and spark gap that is used to generate high voltages at high frequencies. A typical Tesla coil is an air-core device with a small primary winding and a large secondary winding. It operates at radio frequencies.

A Tesla coil produces a spark up to several feet in length. Although the voltage is high, the current flow is moderate to low.

TEST INSTRUMENT

In the design, alignment, and troubleshooting of electrical and electronic equipment, test instruments are used extensively. Test instruments vary in complexity from the simple ammeter or voltmeter to highly sophisticated devices such as monitor/generators and even computers.

The most commonly used test instrument is a volt-ohm-milliammeter. This meter facilitates measurement of moderate to large voltages, currents, and resistances in typical electronic apparatus. Volt-ohm-milliammeters are relatively inexpensive, and are usually powered by batteries, so they can be used in the field as well as in the laboratory.

Other common test instruments, found in most electronic test laboratories, are indicated in the table. For detailed information about these devices, please refer to articles according to the name of the instrument as given in the table.

TEST LABORATORY

All firms involved in the design and manufacture of electronic equipment have a test laboratory where apparatus is tested, debugged, and repaired. The types of instruments used in a test laboratory depends on the nature of the equipment to be tested, but most test labs are equipped with the instruments shown in the preceding table.

The photograph shows an engineer in the test labora-

TEST INSTRUMENT: COMMONLY USED TEST INSTRUMENTS AND FUNCTIONS OR QUANTITIES MEASURED.

Instrument	Function or Quantities Measured
Distortion analyzer	Audio-signal distortion
Field-strength meter	Electromagnetic-field intensity
Frequency counter	Frequency
Frequency meter	Frequency
Function generator	Circuit response to various waveforms
Oscilloscope	Frequency, voltage, wave shape
Signal generator	Receiver and filter testing
Signal monitor	Radio-frequency circuit testing
SINAD meter	Receiver sensitivity and distortion
Spectrum analyzer	Transmitter and filter testing
Transceiver	Receiver, transmitter, and amplifier testing
Transistor tester	Checking transistor performance
Tube tester	Checking tube performance
Vacuum-tube voltmeter	Voltage
Volt-ohm-milliammeter	Current, voltage, resistance
Wattmeter	Power

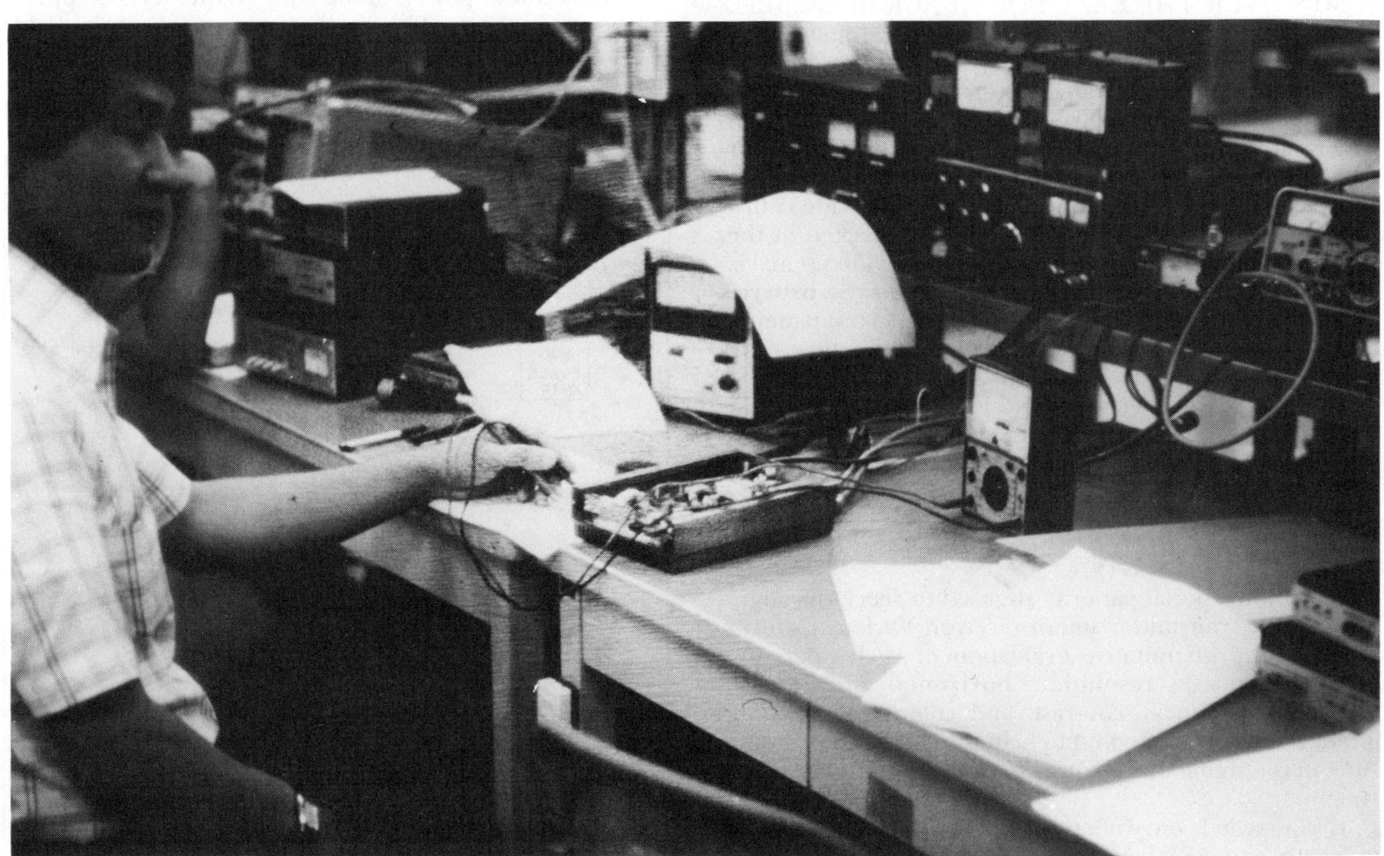

TEST LABORATORY: A typical test-laboratory bench.

tory of a typical research-and-development firm. This engineer is designing a power amplifier for very-high frequency use. The laboratory layout is convenient and functional. All of the large test instruments are placed on shelves elevated above the workbench. (A signal monitor/generator, a transceiver, and wattmeters can be seen at the top right.) This leaves room on the table for tools, such as pliers, screwdrivers, and soldering equipment, and for smaller test instruments such as the volt-ohm-milliammeter at right center.

In any test laboratory, it is important that the work area be uncluttered. This keeps things from getting misplaced, provides plenty of room for data sheets and other papers, and reduces the chance of damage to equipment by accidentally dropped tools or components. *See also* TEST INSTRUMENT.

TEST MESSAGE

A test message or test signal is sent by a radio or television transmitter operator to give receiving operators time to adjust their equipment for optimum reception. There are various different test messages in common use.

In Morse-code communications, the test message "VVV" is used. The sending operator keeps repeating the message until the receiving operator tells him to stop. Sometimes, a continuous, unmodulated carrier is sent for a period of a few seconds.

In radioteletype, the test signal RYRYRY . . . is used in Baudot mode, and U*U*U*. . . is used in ASCII mode. These signals are used because they cause rapid transitions between the mark and space states. Another test message that is used in radioteletype is THE QUICK BROWN FOX JUMPS OVER THE LAZY DOG'S BACK 0123456789. This sentence contains all of the letters of the alphabet, and also a figures and letters shift.

In voice communications, the transmitting operator usually says something like "TESTING 1, 2, 3, 4." In single-sideband transmission, some operators whistle or blow into the microphone for test purposes.

In television communication, the audio and video must be tested separately. The audio signal is checked in the same manner as any voice transmitter. The video signal is checked by sending a specific pattern called a test pattern. There are several different kinds of television test patterns used for checking various aspects of a picture (*see* TEST PATTERN).

In any mode of telecommunications, test signals should not be sent unnecessarily, since non-essential transmissions increase the chances of interference among stations.

TEST PATTERN

In television, special patterns are used to check the alignment of the transmitter and/or receiver. Various patterns facilitate the quantitative evaluation of such operating characteristics as resolution, horizontal and vertical linearity, brightness, contrast, and color reproduction. Test patterns are also used to help the receiving operator tune in the signal in a two-way television communications system.

If you switch on your television set very early in the morning, you may see stations transmitting video test patterns. Each station has its own distinctive pattern. The audio channel may have a steady tone, music, or other test signals. The test pattern is used by the station technicians to facilitate regular checking of the transmitter alignment. Test patterns can be picked up by a television camera, or they may be electronically generated.

Test patterns are used by technicians who repair television receivers. The most common of these patterns is the bar pattern. *See also* BAR GENERATOR.

TEST PROBE

See PROBE.

TEST RANGE

A test range is a large area, usually located outdoors, in which transmitting and receiving antennas are tested for such properties as power gain and front-to-back ratio.

Testing of Transmitting Antennas. Illustration A shows a typical test range for evaluating the performance of a transmitting antenna. A field-strength meter is placed at a distance of several wavelengths from the antenna under test. A reference antenna, such as a dipole or isotropic radiator, is placed near the antenna under test. The distance between the test antenna and the field-strength meter is identical to the distance between the reference antenna and the field-strength meter. The test and reference antennas are resonant at the same frequency.

For testing of antenna power gain, the reference antenna is supplied with a certain amount of radio-frequency power at its resonant frequency. The field-strength-meter sensitivity is adjusted until the scale reads 0 dB. Then the same amount of power, at the same frequency, is supplied to the test antenna. The reading on the scale of the field-strength meter, in decibels, indicates the power gain of the test antenna in dBd (if a dipole is used as the reference) or dBi (if an isotropic antenna is used as the reference).

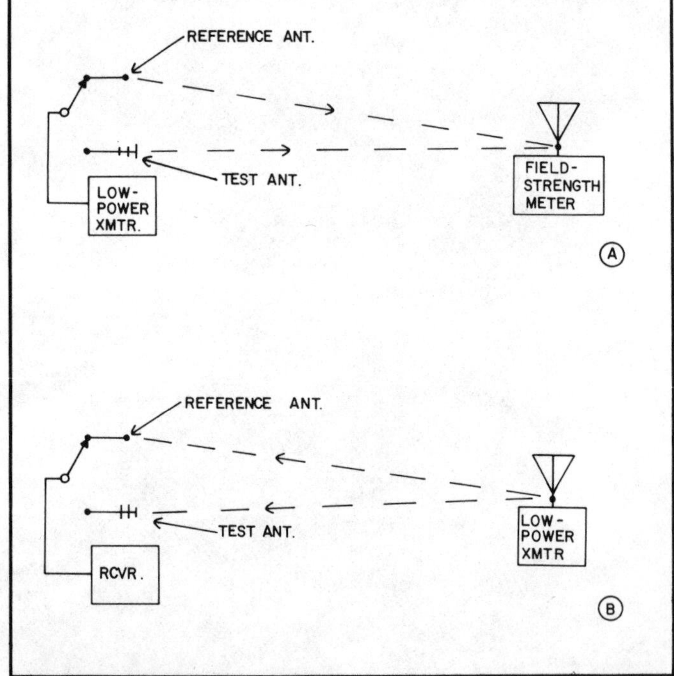

TEST RANGE: Ranges for testing transmitting antennas (A) and receiving antennas (B).

For the testing of front-to-back ratio, if applicable, the test antenna is turned so that it points straight away from the field-strength meter. The scale of the field-strength meter is set to read 0 dB when a given amount of power is applied to the test antenna. Then the test antenna is rotated so that it points directly at the field-strength meter. The reading on the meter scale, in decibels, indicates the front-to-back ratio. The antenna can be continuously turned, and meter readings taken at intervals of a few degrees, to get a complete directional pattern.

Testing of Receiving Antennas. Illustration B shows a test-range layout for the evaluation of receiving-antenna performance. A low-power transmitter is located several wavelengths from the test and reference antennas. The test and reference antennas are both resonant at the same frequency, and are both located at the same distance from the transmitter.

For the testing of effective receiving gain, the test antenna is pointed directly at the transmitter, and the transmitter delivers a constant amount of power to its antenna. The receiver is switched between the reference antenna and the test antenna. The receiver has a calibrated S meter that allows precise determination of relative signal levels in decibels. The effective receiving gain of the test antenna is the increase, in decibels, that occurs in the signal when the receiver is switched from the reference antenna to the test antenna. (If the signal is weaker with the test antenna, the gain is negative.)

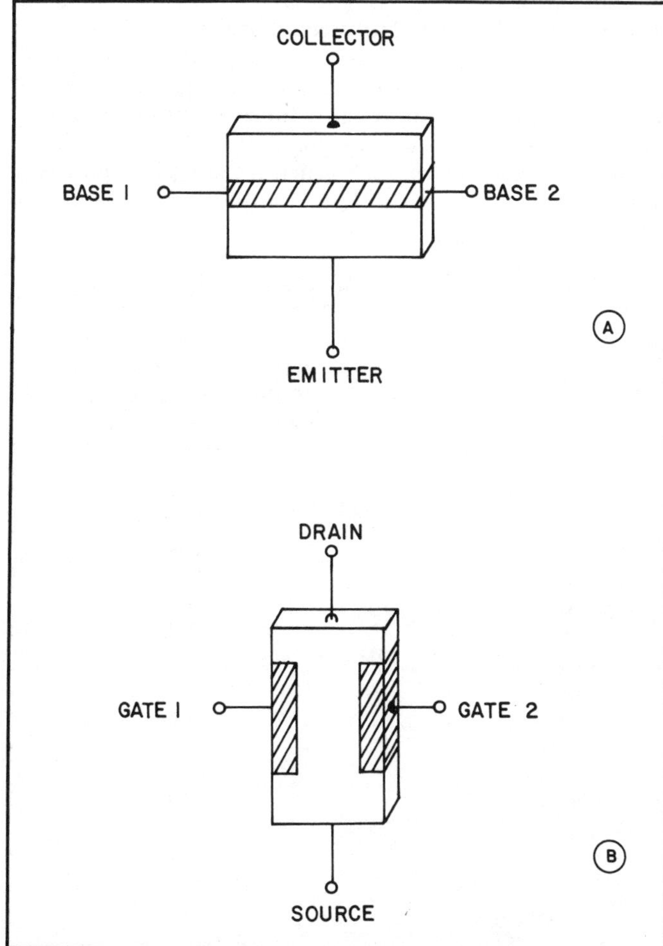

TETRODE TRANSISTOR: Simplified renditions of the construction of tetrode bipolar (A) and field-effect (B) transistors.

For measurement of front-to-back ratio, the receiver is connected to the test antenna, and signal levels are compared with the antenna pointed directly toward and directly away from the transmitter. A plot of the directional response can be obtained by checking the received signal level at numerous compass points.

For Further Information. A complete discussion of all aspects of antenna testing cannot be given here. For detailed information, a textbook on antenna theory or communications theory should be consulted. *Related articles in this book include:* ANTENNA EFFICIENCY, ANTENNA PATTERN, ANTENNA POWER GAIN, dBd, dBi, DIPOLE ANTENNA, FIELD-STRENGTH METER, FRONT-TO-BACK RATIO, ISOTROPIC ANTENNA, REFERENCE ANTENNA, S METER.

TETRODE TRANSISTOR

Some bipolar transistors have two leads connected to the base, for the purpose of reducing the lead inductance at the base (see illustration). This scheme can be used for either the NPN or PNP type of bipolar transistor. Some field-effect transistors have two gates, each with its own lead connected, as at B. Transistors with two base or gate leads, since they have four leads in all, are called tetrode transistors.

Tetrode bipolar transistors are used at very high frequencies where lead inductance and loss, especially at the input, must be kept to a minimum. Tetrode field-effect transistors, also known as dual-gate field-effect transistors, are often used as mixers and modulators. *See also* FIELD-EFFECT TRANSISTOR, TRANSISTOR.

TETRODE TUBE

A tetrode tube is a vacuum tube with four elements. In addition to the cathode and the plate or anode, there are two grids. The two grids serve different purposes.

The control grid operates as the input-signal electrode in most tetrode-tube circuits (*see* CONTROL GRID). The sec-

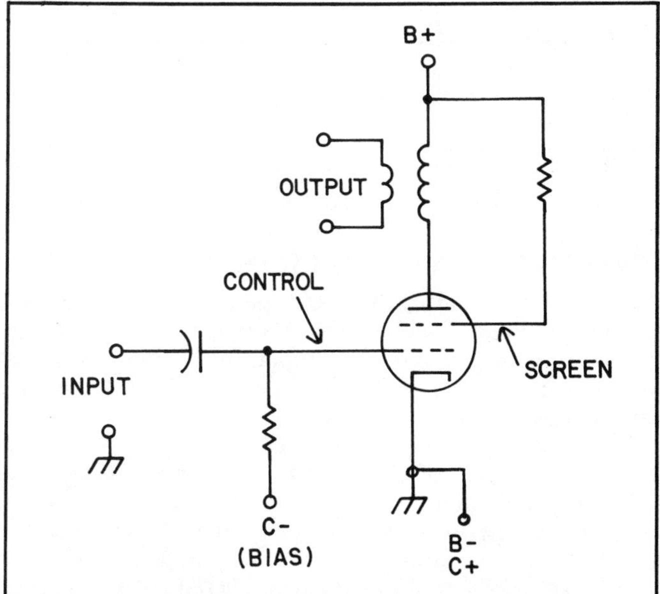

TETRODE TUBE: Tetrode tube showing general biasing arrangement.

ond grid, known as the screen grid, can serve any of three functions. In an amplifier, the screen causes electrons, en route from the cathode to the plate, to be accelerated, resulting in enhanced gain. The screen can be used as an auxiliary plate. The screen grid may also be used as a second input grid in mixers and modulators (*see* SCREEN GRID).

The control grid in the tetrode tube is usually biased negatively with respect to the cathode. The screen grid is biased at a voltage more positive than the cathode, but less positive than the plate, in a tetrode amplifier circuit. The same is true of the electron-coupled oscillator, in which the screen grid serves as the plate of the oscillator circuit (*see* ELECTRON-COUPLED OSCILLATOR). In a tetrode mixer or modulator, the screen grid is supplied with a bias that is about the same as that of the control grid.

The illustration shows the schematic symbol for a tetrode tube, showing a typical biasing arrangement for an amplifier or electron-coupled oscillator. The two grids are concentric, with the control grid on the inside, nearer the cathode.

The electron acceleration in a tetrode tube causes some secondary emission from the plate in power-amplifier circuits. A third grid is often added to repel the secondary electrons back to the plate. This grid, which is negatively charged, is called a suppressor. *See also* PENTODE, SUPPRESSOR GRID, TUBE.

TE VALUE

An insulator or semiconductor generally shows a variable resistance. The resistance depends on the temperature.

The TE value of a substance is the temperature at which a cube of that substance, measuring 1 cm × 1 cm × 1 cm, has a resistance of 1 megohm.

TEXT EDITING

See WORD PROCESSING.

T FLIP-FLOP

See FLIP-FLOP.

THD

See TOTAL HARMONIC DISTORTION.

THERMAL AGITATION

In any substance, all of the atomic particles are in constant motion. The speed with which the atomic particles move is proportional to the absolute temperature. This movement of molecules, atoms, and subatomic particles is known as thermal agitation.

Thermal agitation creates electrical noise, since the charged particles—especially the electrons—are accelerated. This noise limits the ultimate sensitivity of electronic amplifiers. Thermal noise can be reduced by cooling the conductors and components of an amplifier to nearly absolute zero. *See also* CRYOGENICS, NOISE, SENSITIVITY, SIGNAL-TO-NOISE RATIO, SUPERCONDUCTIVITY, THERMAL NOISE.

THERMAL BREAKDOWN

In an electric conductor or semiconductor, the flow of current always generates some heat. This is because no conductor is completely lossless, and the resistance causes dissipation of power in the form of heat. If the flow of current is excessive, the dissipated power may cause the temperature to rise so high that the substance deteriorates. This is called thermal breakdown. It can occur in any device that carries current. *See also* THERMAL RUNAWAY.

An electric field in a dielectric substance, or a magnetic field in a ferromagnetic substance, causes a rise in the temperature of that substance. This is because no dielectric or ferromagnetic material is completely without loss. If the field strength becomes great enough, the dielectric or ferromagnetic material may get so hot that it is physically damaged. This is another form of thermal breakdown. *See also* DIELECTRIC, DIELECTRIC LOSS, EDDY-CURRENT LOSS, FERROMAGNETIC MATERIAL, HYSTERESIS LOSS.

THERMAL COEFFICIENT

See TEMPERATURE COEFFICIENT.

THERMAL CONDUCTIVITY

All substances conduct heat to some extent, but some conduct heat much better than others. The extent to which a substance can transfer heat efficiently from one place to another can be expressed quantitatively. This expression is known as the thermal conductivity of the material.

In general, substances are good thermal conductors if they are good electrical conductors, and vice-versa. Electrical insulators tend to be poor thermal conductors. But there are exceptions to this rule. For example, silicon, while a poorer electrical conductor than platinum, is a better thermal conductor. The best thermal insulator is a

THERMAL CONDUCTIVITY: THERMAL CONDUCTIVITY OF SOME COMMON ELEMENTS.

Element	Thermal conductivity, Milliwatts per meter per degree Celsius
Aluminum	22
Carbon	2.4
Chrominum	6.9
Copper	39
Gold	30
Iron	7.9
Lead	3.5
Magnesium	16
Mercury	0.85
Nickel	8.9
Platinum	6.9
Silicon	8.4
Silver	41
Thorium	4.1
Tin	6.4
Tungsten	20
Zinc	11

perfect vacuum; since there are no molecules, no conduction can take place. The best thermal conductor of the known elements is silver.

Thermal conductivity is expressed in watts, milliwatts, or microwatts per degree Celsius. The thermal conductivity figures for some common materials at room temperature are given in the table.

THERMAL ENERGY

Thermal energy, also called heat energy, is the kinetic energy of the moving atoms and molecules in matter. Thermal energy is proportional to the molecular temperature of a substance (*see* TEMPERATURE). The higher the temperature, the more rapidly the particles move, and the greater is the thermal energy.

Thermal energy can be converted into other forms of energy by manmade devices. This makes thermal energy important, because of the shortages of conventional fuels. The heat from the earth's interior can be tapped; infrared and visible radiation from the sun can be used to generate thermal energy. *See also* ENERGY, GEOTHERMAL ENERGY, SOLAR ENERGY, THERMODYNAMICS.

THERMAL INSTABILITY

See THERMAL STABILITY.

THERMAL METER

See HOT-WIRE METER.

THERMAL NOISE

In all materials, the electrons are in constant motion. Since the electrons move in curved paths, they are always accelerating, even if they stay in the same orbital shell of a single atom (*see* ELECTRON, ELECTRON ORBIT). This acceleration of charged particles produces electromagnetic fields over a wide spectrum of wavelengths. To some extent, the random movement of the positively charged atomic nuclei has the same effect. In electronic circuits, this charged-particle acceleration causes a form of noise known as thermal noise.

The level of thermal noise in any material is proportional to the absolute temperature. The higher the temperature gets, the more rapidly the charged particles are accelerated. The lower the temperature becomes, the slower the particles move. As the temperature approaches absolute zero—the coldest possible condition—the particle speed and acceleration approach zero, and so does the level of thermal noise.

Thermal noise imposes a limit on the sensitivity that can be obtained with radio-frequency receivers. This noise can be minimized by placing a preamplifier circuit in a bath of liquid gas, such as helium or nitrogen. Helium has the lowest boiling point of any element: just a few degrees Kelvin. Such cold temperatures cause the atoms and electrons to move very slowly, thus greatly reducing the thermal noise compared with the level at room temperature. This technique is used in the front ends of radio-telescope receivers using FETs.

Thermal noise is not the only form of noise that limits the sensitivity of receiving circuits. Some noise is caused by shot effect, and some noise comes from the environment outside the circuitry itself. *See also* CRYOGENICS, NOISE, SFERICS, SHOT EFFECT, SUPERCONDUCTIVITY, THERMAL AGITATION.

THERMAL PROTECTION

Thermal protection is a method of preventing damage to circuit components because of overheating that can result from an excessive long-term demand for current or power, or from the failure of an equipment-cooling system.

A thermal-protection device operates in a manner similar to a circuit breaker. However, rather than being actuated by the excessive current itself, a thermal-protection device is triggered by high temperature.

The illustration is a simplified diagram of a thermal-protection circuit that might be used in a power amplifier. A small thermostat is attached to, or placed in the vicinity of, the components to be protected. The thermostat is connected to a relay or switching transistor in the power-supply line of the protected components. If the temperature reaches the critical level, the thermostat changes state, removing power from the circuit. *See also* THERMAL BEAKDOWN, THERMAL RUNAWAY, THERMOSTAT.

THERMAL RUNAWAY

Some semiconductor devices, especially power transistors and high-current diodes, may be destroyed by a phenomenon called thermal runaway.

Thermal runaway can take place only in a component that exhibits increased resistance with a rise in temperature. The problem begins as the current heats up the device, raising the resistance slightly. This increase in resistance may or may not produce a significant decrease in the flow of current, depending on the external circuitry. If the current does not decrease, the power dissipated in the

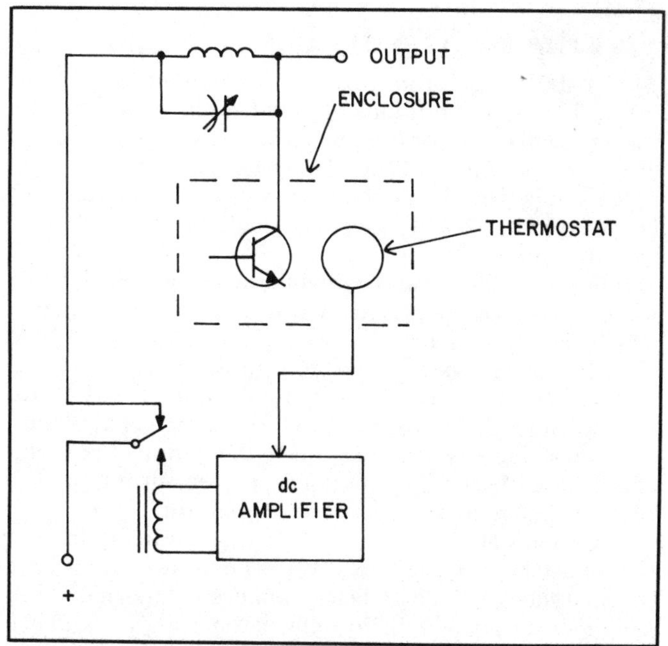

THERMAL PROTECTION: A simple thermal-protection circuit.

component will increase in direct proportion to the resistance. This increased dissipation heats the device even faster, and a vicious circle ensues. If allowed to go on unchecked, the end result will be thermal breakdown.

Thermal runaway can be prevented by placing a small-value, high-power resistor in series with the diode or transistor. This reduces the current that flows through the component as the temperature rises. A thermal-protection circuit can also be used; it will shut off the power if thermal runaway starts to occur. *See also* THERMAL BREAKDOWN, THERMAL PROTECTION.

THERMAL SHOCK

Thermal shock is a rapid, marked increase or decrease in the temperature of the environment in which an electronic circuit or component is operated. Thermal shock often causes dramatic changes in the characteristics of a device. This is usually the result of temperature-coefficient effects, although some electronic components completely fail if the temperature gets too cold or too hot.

Thermal shock is sometimes inflicted deliberately on a piece of equipment as part of a testing or troubleshooting procedure. Suppose that a transceiver is received at a service shop for repair, with the customer complaining that the radio will not operate at temperatures below freezing. The unit is subjected to thermal shock by being placed in a freezer for a few hours and then tested. If failure occurs, the radio is allowed to warm up again, or is deliberately warmed using a heater. The defective component can then be found using a cooling spray.

Some electronic equipment can withstand very little thermal shock without failure. A large computer, for example, requires fairly constant temperature conditions. Other circuits, such as mobile or portable devices, are designed to withstand severe thermal shock since they must function both indoors and outdoors. *See also* THERMAL STABILITY.

THERMAL STABILITY

Thermal stability is the capability of a circuit to function properly under conditions of variable temperature. The thermal stability of component values is expressed in terms of the temperature coefficient (*see* TEMPERATURE COEFFICIENT). The thermal stability of a circuit depends, to a large extent, on the temperature coefficients of the individual components.

Thermal stability can be quantitatively described in two ways. The simpler way is to state the temperature range in which the equipment will always operate normally. The specifications for a small computer might state an ambient-temperature range of 10 to 50 degrees Celsius, for example. A more comprehensive way of expressing thermal stability is to indicate how the various operating characteristics (such as frequency, power output, or sensitivity) change as a function of temperature.

Thermal stability is especially important in equipment that must be operated in diverse environments, where thermal shock is likely to be encountered. Thermal stability is not as important for equipment that is always used in a controlled environment. *See also* TEMPERATURE COMPENSATION, THERMAL SHOCK.

THERMAL TIME CONSTANT

When the ambient temperature around an object changes suddenly, the object gradually gets hotter or cooler until the object and the environment are at the same temperature. The rapidity with which this occurs depends on the substance, and also on the mass of the object. The heating or cooling process is exponential, but the relative speed of change can be expressed in terms of the thermal time constant.

Suppose the ambient temperature around an object suddenly changes from T1 to T2 (in the Kelvin scale). The thermal time constant is the time t required for the object to reach a temperature of

$$Tc = T1 + 0.63\ (T2 - T1)$$

if $T2 > T1$, that is, heating; or

$$Tc = T1 - 0.63\ (T1 - T2)$$

if $T2 < T1$, that is, cooling.

The thermal time constant is sometimes expressed for electronic components such as resistors and transistors. If the current suddenly changes, the temperature of the component will change exponentially, starting at T1 and approaching a final value of T2. The thermal time constant for an electronic component is determined as above.

THERMIONIC EMISSION

When an electrical conductor is heated to a high temperature in a vacuum, the electrons move with such speed that they are easily stripped from the material. This is called thermionic emission. It is the principle of operation of a vacuum-tube cathode.

Thermionic emission from a surface is expressed in watts per square meter or watts per square centimeter. The thermionic emission from a cathode depends on the temperature, the voltage supplied to the cathode, and the type of material from which the cathode is made. Oxide-coated cathodes generally have the greatest emission at a given absolute temperature and voltage. Specially treated nickel and tungsten also have high thermionic emissivity. *See* CATHODE, TUBE.

THERMISTOR

Most materials exhibit variable electrical conductivity with changes in temperature. A resistor, designed especially to change value with temperature, is called a thermistor. The term thermistor is a contraction of THERMally sensitive resISTOR.

Thermistors are made from semiconductor materials. The most common substances used are oxides of metals. The resistance may increase as the temperature rises, or it may decrease as the temperature rises. The resistance is a precise function of the temperature. Usually, the resistive temperature coefficient is large and negative (*see* TEMPERATURE COEFFICIENT). The construction of a typical thermistor is shown in the illustration.

Thermistors are used in various circuits for detecting temperature. The resistance-versus-temperature characteristic makes the thermistor ideal for use in thermostats and thermal-protection circuits. Thermistors are operated

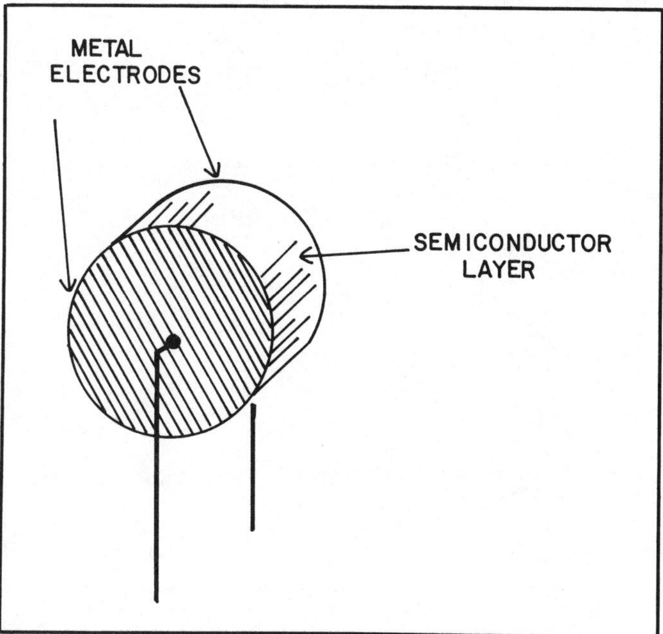

THERMISTOR: Construction of a thermistor. Other shapes may be used, but the principle is the same.

at low current levels, so that the resistance is affected only by the ambient temperature, and not by heating caused by the applied current itself. *See also* THERMAL PROTECTION, THERMOSTAT.

THERMOCOUPLE

A thermocouple is a device that generates a voltage by the heating of a junction between two metal electrodes placed in physical contact. The electrodes consist of dissimilar metals. Commonly used combinations include constantan/copper, constantan/iron, and carbon/silicon carbide. The illustration shows the construction of a typical thermocouple.

A thermocouple normally produces only a few millivolts of potential at a temperature of 100 degrees Celsius. Some thermocouples can be used at temperatures of more than 1,000 degrees Celsius, and generate almost 0.5 volt. In general, the voltage produced by a given thermocouple is proportional to the temperature. The voltage is produced because of Seebeck effect (*see* SEEBECK COEFFICIENT, SEEBECK EFFECT).

Thermocouples can be used to measure high temperatures. A thermocouple thermometer consists of an even number of thermocouples and a remote indicating device, usually a voltmeter or ammeter. Thermocouples can survive much higher temperatures than ordinary thermometers.

THERMODYNAMICS

Thermodynamics is a branch of physics, concerned with the interaction between thermal energy and mechanical energy. In particular, thermodynamics involves the behavior of matter under conditions of variable temperature.

Scientists recognize three laws of thermodynamics. They are:

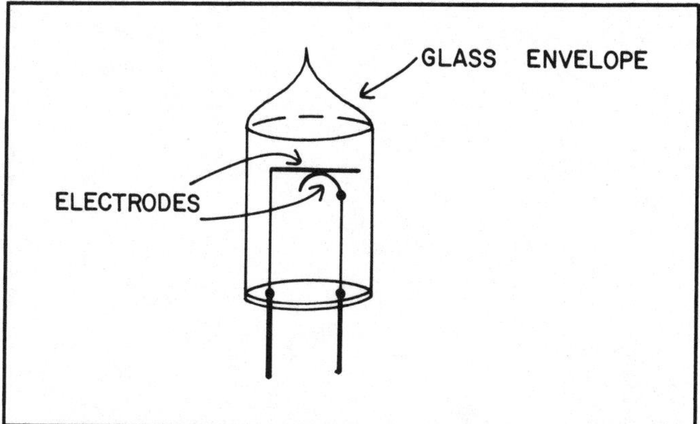

THERMOCOUPLE: Construction of a thermocouple.

1. Thermal energy can be converted to mechanical energy, and vice-versa;
2. In a reversible system, not all of the available thermal energy is converted to mechanical energy;
3. As the temperature of an isothermal process approaches absolute zero, the entropy approaches zero. At absolute zero, the entropy is zero.

Thermodynamics is of importance in electricity and electronics. Thermal energy is extensively used for generating electricity. Steam turbines convert the thermal energy into mechanical energy, and this energy is used to drive electric generators. The thermal energy may come from atomic reactors, from inside the earth, or from the burning of fossil fuels. It may also come from solar radiation. *See also* ENERGY, FIRST LAW OF THERMODYNAMICS, SECOND LAW OF THERMODYNAMICS, THERMOELECTRIC CONVERSION, THIRD LAW OF THERMODYNAMICS.

THERMOELECTRIC CONVERSION

Thermal energy can be converted into electricity. This process is known as thermoelectric conversion.

Many power-generating plants make use of thermoelectric conversion to bring us our electricity. The heat from atomic reactions, the burning of fossil fuels such as oil or coal, the interior of the earth, or the radiation from the sun can be harnessed to drive a steam turbine. The turbine provides the mechanical energy needed to turn the electric generators (*see* POWER PLANT). This is an example of indirect thermoelectric conversion, since the thermal energy is converted first to mechanical energy, then to electrical energy.

Direct thermoelectric conversion can be achieved using certain semiconductor materials. A junction between dissimilar metals sometimes acquires a potential difference at high temperatures. *See also* SEEBECK COEFFICIENT, SEEBECK EFFECT, THERMOCOUPLE.

THERMOELECTRIC EFFECT

See SEEBECK COEFFICIENT, SEEBECK EFFECT, THERMOCOUPLE.

THERMOMETER

A thermometer is a device that is used to measure molecu-

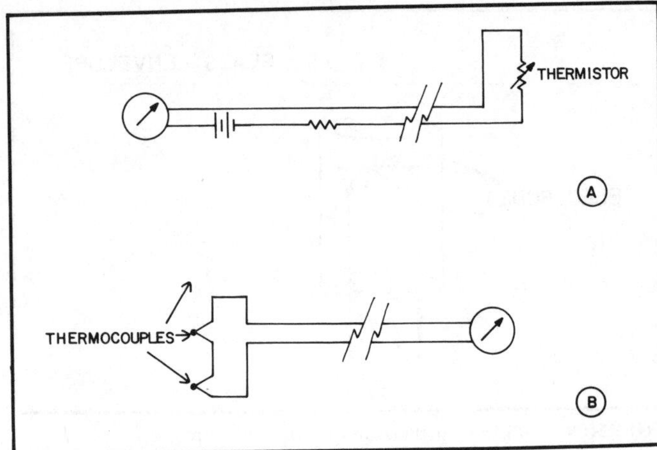

THERMOMETER: Electronic thermometers. At A, a thermistor is used; at B, thermocouples form the sensing device.

lar temperature (*see* TEMPERATURE). There are three basic types of thermometers: fluidic, metal-strip, and electronic.

The fluidic thermometer consists of a glass tube with a reservoir (bulb) of mercury or alcohol at one end. As the temperature rises, the fluid expands and rises upward inside the tube. As the temperature falls, the fluid contracts and falls back down the column toward the bulb.

The metal-strip thermometer consists of a spirally wound, thin length of metal, to which a pointer is attached. Changes in temperature cause the strip to expand and contract, and this turns the pointer.

Electronic thermometers can be made using various different components. The thermistor and thermocouple are the most commonly used.

A thermistor, a battery or other source of direct current, a resistor, and a microammeter can be connected in series to form a thermometer (see illustration at A). As the temperature changes, the resistance of the thermistor fluctuates, causing the meter reading to vary.

A thermocouple thermometer consists of a voltmeter or galvanometer attached to an even number of series-connected as at B. As the temperature rises, the potential difference increases and the meter needle moves upscale. Direct-current amplifiers can be used with either the thermistor thermometer or the thermocouple thermometer to enhance the sensitivity and resolution. *See also* THERMISTOR, THERMOCOUPLE.

THERMOPILE

Two or more thermocouples can be connected in series to obtain higher voltages than a single thermocouple delivers. Such a series connection of thermocouples is called a thermopile.

Thermopiles are commonly used in thermocouple type thermometers. *See also* THERMOCOUPLE, THERMOMETER.

THERMOPLASTIC CAPACITOR

See POLYSTYRENE CAPACITOR.

THERMOPLASTIC RECORDING

Thermoplastic recording is a method of impressing data onto plastic tape. Thermoplastic recording is similar to magnetic recording, except that physical distortion is used instead of magnetization.

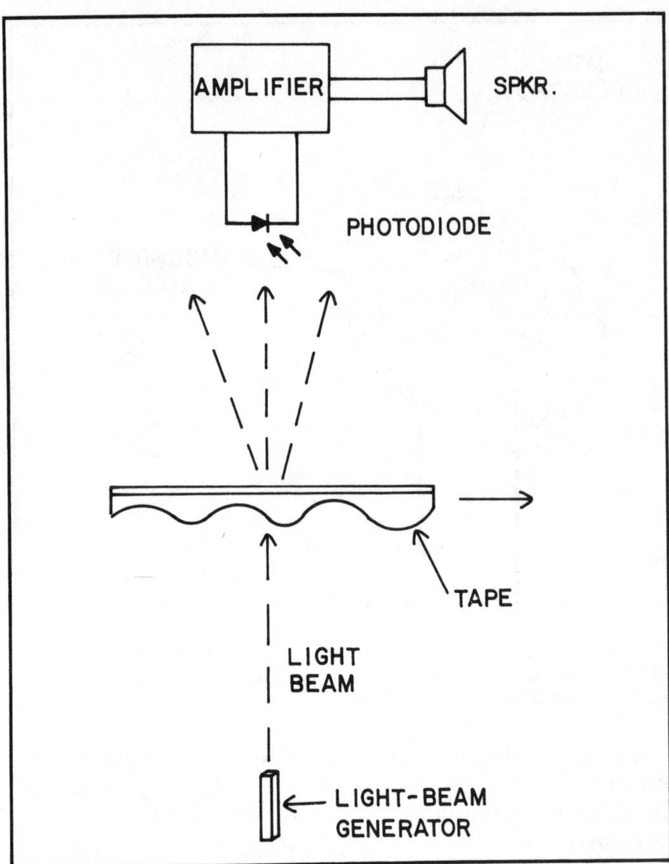

THERMOPLASTIC RECORDING: A thermoplastic tape can be played back by shining a light beam through the material, and using a detector to sense the modulation caused by refraction.

A thermoplastic recording tape consists of a clear film that has a clear coating with a relatively low melting temperature. In the recording process, the coating is first softened by heat. Then it is run at a constant speed through a device that produces a modulated electric field. The electric field fluctuates according to the waveform to be recorded. The plastic, which is a dielectric material, is subjected to electrostriction (*see* ELECTROSTRICTION) and is thus deformed or warped to a greater or lesser extent, depending on the instantaneous electric-field strength.

A thermoplastic tape can be played back in various ways. The most common method is shown in the illustration. A beam of light is passed through the tape as the tape moves by at a constant rate of speed. The distortion in the thermoplastic coating causes the beam to be refracted at an angle that varies according to the waveform impressed on the tape. A photosensitive device, placed in the path of the light beam, receives a modulated ray that produces a current identical to the original signal.

The main advantage of thermoplastic recording, compared with magnetic recording, is that a thermoplastic tape will not be erased by a magnetic field.

THERMOSTAT

A thermostat is a temperature-sensing device that is used to actuate some piece of apparatus for regulation or pro-

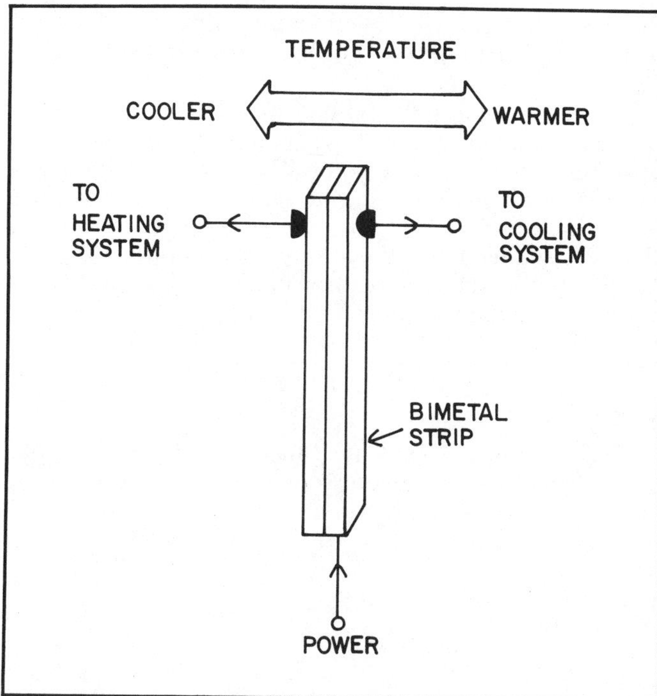

THERMOSTAT: A bimetal-strip thermostat.

tection. The most common application of a thermostat is for temperature control in buildings. Almost every home heating/cooling system has a thermostat. A thermostat can be constructed in various ways. All thermostats act as switches, opening and closing a circuit according to the temperature.

A bimetal-strip thermostat opens and closes because of the change in curvature of a strip made of two dissimilar metals. In a heating system, the thermostat switches the circuit on when the temperature falls below a certain point; in a cooling system, the system is switched on when the temperature rises above a certain point. Thermistors and thermocouples can be used to sense temperature in electronic thermostats.

A typical bimetal-strip thermostat operates as shown in the illustration. There are two modes: heating and cooling. In the heating mode, the furnace is switched on when the temperature falls below a preset value. In the cooling mode, an air conditioner is switched on when the temperature rises above a preset value. There is a certain amount of hysteresis, or sluggishness, in the thermostat, so that it will not cycle on and off rapidly and continuously.

A thermostat can be used for thermal-protection purposes. If a device gets too hot, the thermostat actuates a switch that removes power. *See also* TEMPERATURE REGULATION, THERMAL PROTECTION.

THEVENIN'S THEOREM

In any linear circuit having more than one source and/or more than one load, simplification is possible. This simplification can be carried out theoretically according to a rule known as Thevenin's theorem.

Suppose there is a circuit, no matter how complex, consisting of power sources and linear loads. Then the circuit will operate in the same manner as a hypothetical circuit consisting of one source and one load of the appropriate values. This hypothetical circuit is called the Thevenin equivalent circuit.

Thevenin's theorem can be stated in another way. Suppose an impedance Zx is connected into a linear system in which the open-circuit voltage is E and the existing impedance is Z. Then the current Ix in the circuit will be:

$$Ix = E/(Z+Zx)$$

Thevenin's theorem is used in network analysis for determination of current and voltage. *See also* KIRCHHOFF'S LAWS.

THICK-FILM AND THIN-FILM INTEGRATED CIRCUITS

An integrated circuit can be fabricated in various ways. Film processes are commonly used to obtain the passive components in an integrated circuit. There are two types of film integrated circuits, known as thick-film (TF) and thin-film (tf).

Thick Films. A thick-film circuit is manufactured by depositing specially treated inks and pastes onto a ceramic wafer (substrate) by means of a photographic technique called silk screening. Several layers are generally applied. The layers are thicker than 0.001 mm. Each layer contains an intricate pattern of inks. The inks have different properties; some are excellent conductors, others are more or less resistive, and still others are dielectrics. The deposited materials form the passive circuit elements—wiring, capacitors, and resistors—in the thick-film circuit.

Active devices, such as transistors, are found in many thick-film integrated circuits. The active components are fabricated from microminiature semiconductor chips, and are welded or soldered to the passive components. Large-valued capacitors are usually added in discrete form, rather than thick-film form. A thick-film circuit containing active or discrete devices is called a hybrid circuit.

Thin Films. Thin-film (tf) integrated circuits are manufactured by means of depositing layers of conducting, semiconducting, and insulating materials onto a ceramic wafer (substrate). The layers are less than 0.001 mm thick, and are formed by an evaporation process.

The materials used in the thin-film process are metals and metal oxides, rather than inks or pastes. Different combinations of resistive or semiconducting metals are used to obtain the wiring, resistors, and capacitors. As in the thick-film circuit, active components may be attached by means of soldering or welding. Large-valued capacitors are added as descrete microminiature components. A thin-film integrated circuit containing active or discrete components is called a hybrid integrated circuit.

Additive and Subtractive Processes. The layers in a thick-film or thin-film integrated circuit are usually applied one after the other to the insulating substrate. This method is known as the additive process, since materials are deposited in a defined sequence, and only in the places where components are to be located. The interconnections, and the lower electrodes of capacitors, are deposited first. Then the capacitor dielectric is applied, followed by the resitors, the top electrodes of the capacitors, and finally the discrete and active components.

In the subtractive process, layers of electronic material are first applied, and the components are then etched from the layers. The top layer is etched first, and then progressively lower layers. The etching is accomplished by photographic means, similar to the process used to make a printed-circuit board (*see* ETCHING). Finally, the discrete and active components are attached.

For Further Information. A textbook on semiconductor technology is recommended for a detailed discussion of thick-film and thin-film processes. *See also* INTEGRATED CIRCUIT.

THIRD LAW OF THERMODYNAMICS

The Third Law of Thermodynamics is a theorem concerning the behavior of substances at low temperatures. The principle involves entropy, an expression of the distribution of thermal energy among substances or objects.

In any isothermal process involving a solid or liquid, the entropy approaches zero as the temperature approaches absolute zero. Also, the entropy is zero at absolute zero. This means simply that the temperature distribution becomes more and more uniform as the temperature gets lower and lower.

In practice, absolute zero is never actually reached anywhere in the universe. If this temperature, about −459 degrees Fahrenheit or −273 degrees Celsius, were attained, all molecular and atomic motion would cease. *See also* FIRST LAW OF THERMODYNAMICS, SECOND LAW OF THERMODYNAMICS.

THOMSON BRIDGE

See KELVIN DOUBLE BRIDGE.

THOMSON EFFECT

When a current flows in a conductor having an uneven temperature distribution, heat is either produced or absorbed. This is called the Thomson effect. The extent to which heat is produced or absorbed depends on the material and also on the direction of current flow. If heat is liberated when current flows in a given direction, heat will be absorbed when the current flows in the opposite direction, and vice-versa.

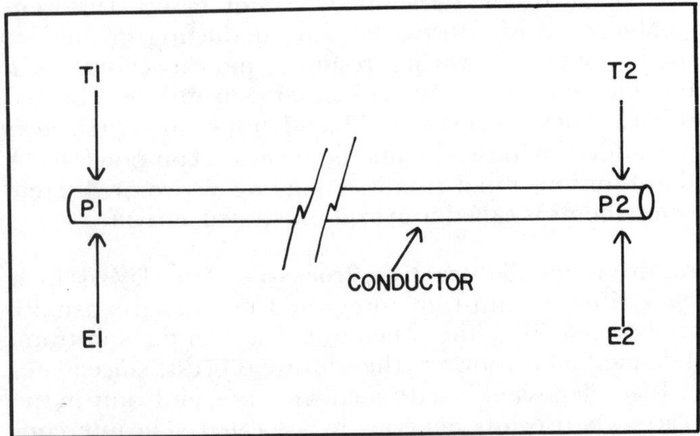

THOMSON EFFECT: A voltage E2-E1 between two points P1 and P2, differing in temperature by T2-T1, causes heating or cooling.

The tendency for the Thomson effect to occur is expressed as a quantity called the Thomson coefficient. Let E1 be the voltage at some point P1 along a metal conductor, and let E2 be the voltage at some other point P2. Let T1 be the absolute temperature (in degrees Kelvin) at point P1, and let T2 be the absolute temperature at point P2. This is illustrated in the drawing. The Thomson coefficient is given approximately by:

$$C = (E2 - E1)(T2 - T1)$$

Considering the voltage E as a function of the temperature T, the Thomson coefficient is expressed precisely by:

$$C = dE/dT$$

in volts per degree Kelvin.

THORIATED CATHODE/ FILAMENT

See THORIUM.

THORIUM

Thorium is an element with atomic number 90. The most common isotope has atomic weight 232. Thorium is a heavy metallic substance in its pure form.

Thorium compounds, especially thorium oxides, are added to the cathodes of some vacuum tubes. This is done to improve the thermionic emission. Under conditions of high temperature, thorium and its compounds are excellent emitters of electrons. A filament or cathode is called thoriated when a small amount of this element is present. *See also* THERMIONIC EMISSION.

THREE-PHASE ALTERNATING CURRENT

The alternating current that appears at 117-V utility outlets is single-phase current. We normally think of alternating current as a simple sine-wave current. However, the 234-V utility mains generally carry three such waves simultaneously. The waves are identical except that they are separated by 120 degrees of phase (⅓ cycle). The current is supplied through three wires, with each wire carrying a single sine wave differing in phase by ⅓ cycle with respect to the currents in the other two wires (see illustration). Some three-phase circuits incorporate a fourth wire, which is kept at neutral (ground) potential.

Three-phase alternating current has certain advantages over single-phase current. When a three-phase current is rectified, the filtering process is much easier than when single-phase current is used, although the rectifier circuit is somewhat more complicated (*see* POLYPHASE RECTIFIER). Three-phase systems provide superior efficiency for the operation of heavy appliances, especially electric motors.

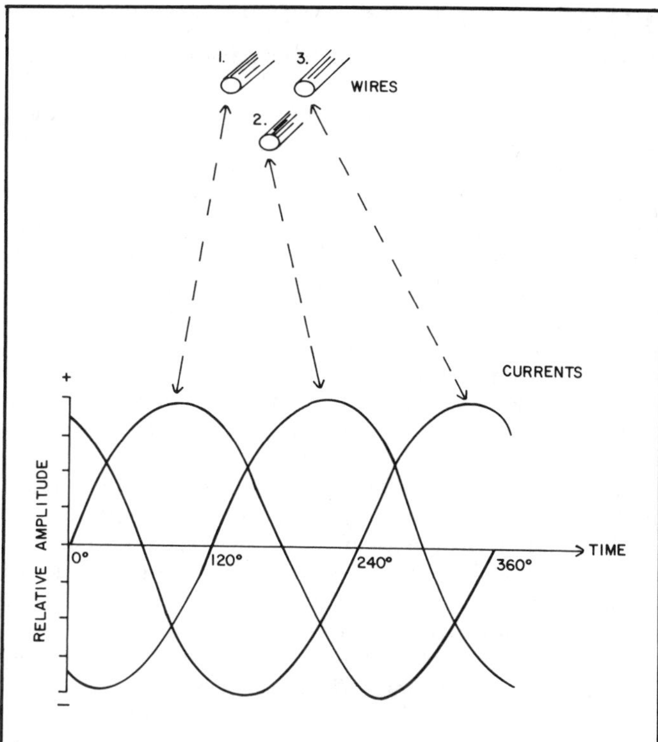

THREE-PHASE ALTERNATING CURRENT: The waves differ by 120 degrees of phase.

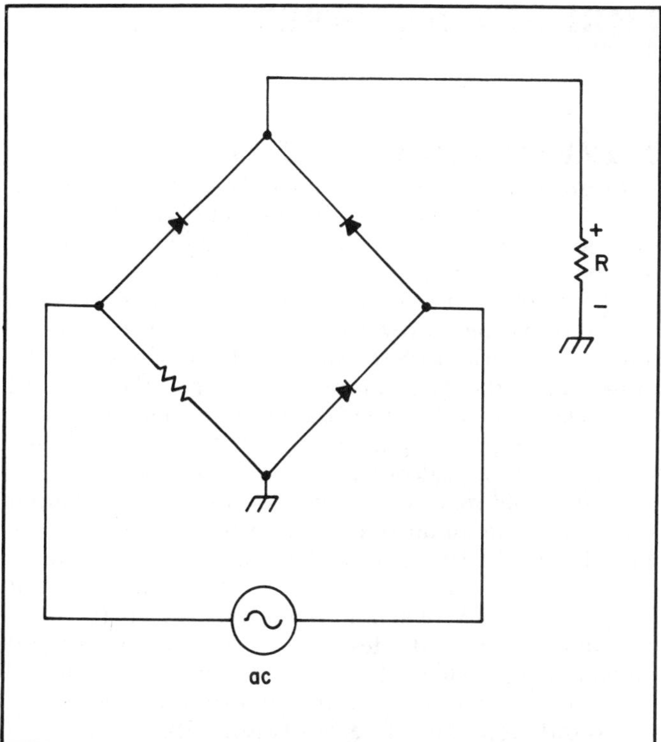

THREE-QUARTER BRIDGE RECTIFIER: A three-quarter bridge circuit may be used in unbalanced power supplies.

THREE-PHASE RECTIFIER

A three-phase rectifier is a special form of polyphase rectifier, used with three-phase alternating current. *See* POLYPHASE RECTIFIER.

THREE-QUARTER BRIDGE RECTIFIER

A three-quarter bridge rectifier is a form of alternating-current rectifier circuit identical to a bridge rectifier, except that one of the diodes is replaced by a resistor. The three-quarter bridge has three diodes while a conventional bridge rectifier has four. The three-quarter configuration performs in essentially the same way as a normal bridge rectifier in power supplies with unbalanced outputs. The illustration is a schematic diagram of a three-quarter bridge. *See also* BRIDGE RECTIFIER.

THREE-WIRE SYSTEM

A three-wire system is a scheme for transmitting electric power, especially in alternating-current form. Three-wire systems are sometimes used in utility wiring. In recent years, the three-wire system has become increasingly popular because it affords better shock-hazard protection than the older two-wire system (*see* TWO-WIRE SYSTEM).

A three-wire system consists of one grounded (neutral) wire and two live wires. The live wires sometimes carry equal and opposite currents; in most three-wire systems, one of the live wires is kept at neutral potential, the same as the grounded wire, while the other carries 117 V of ac potential.

Three-wire electrical systems facilitate grounding of the chassis of radio equipment and appliances. This keeps lethal voltages from appearing at places where personnel might be electrocuted.

THRESHOLD

As the amplitude of a signal increases from zero, it is not detectable until it reaches a certain level. The minimum level at which a signal of any kind can be detected, either by the human senses or using electronic instrumentation, is known as the threshold level. We may speak of the threshold of hearing, the threshold of visibility, the threshold sensitivity of a meter or receiver, or the threshold sensitivity of some other device or component (*see* TANGENTIAL SENSITIVITY, THRESHOLD OF HEARING, THRESHOLD OF VISIBILITY).

For certain adjustable controls, the point at which a defined transition takes place is called the threshold setting. A squelch control, for example, will cut off the internally generated receiver noise when it is set past a certain point. A regenerative detector will begin to oscillate if the feedback is increased past a certain threshold, called the threshold of oscillation.

Diodes, transistors, and tubes require a certain minimum forward voltage or current in order to conduct. This voltage or current is called the breakover or threshold voltage or current. For a germanium semiconductor diode, the threshold signal level is near 0.3 V. For a silicon diode it is near 0.6 V. In the case of a vacuum tube, it varies from less than 1 V to several volts. The threshold property of a diode, transistor, or tube can be used for peak clipping (limiting) or for elimination of signals weaker than a certain amplitude. *See also* SPEECH CLIPPING, THRESHOLD DETECTOR.

THRESHOLD CURRENT

See THRESHOLD.

THRESHOLD DETECTOR

A threshold detector is a circuit that allows strong signals to pass, but blocks weaker signals. The transition, or threshold, is sharply defined. A squelch circuit is a form of threshold detector (*see* SQUELCH).

The simplest form of threshold detector consists of two diodes, connected in reverse parallel with respect to each other and in series with the signal path. Signals will not be passed until the peak amplitude exceeds the forward breakover voltage of either diode. If germanium diodes are used, the peak voltage must be 0.3 V or greater; for silicon diodes, a peak signal of 0.6 V is necessary.

A threshold detector can be used to obtain enhanced selectivity in an audio filter, as shown in the illustration. The first tuned circuit provides a sharp resonant audio response. The threshold detector cuts off the skirts at levels below 0.3 V or 0.6 V (depending on whether germanium or silicon diodes are employed). The second tuned circuit, having values L and C identical to those of the first, reduces the waveform distortion caused by the threshold detector. The transistor circuit provides amplification if needed.

THRESHOLD OF HEARING

The lowest level of sound that a person can hear is called the threshold of hearing. This level varies from one person to another; it also depends on the frequency and wave shape of the sound.

An absolute threshold of hearing has been defined, based on auditory tests of a large number of people. This level has been determined as 10^{-16} watt per square centimeter at a frequency of 1 kHz. A sound of this intensity is considered to have a loudness of 0 dB. Sound intensity is measured in decibels with respect to this level. *See also* DECIBEL, LOUDNESS, SOUND.

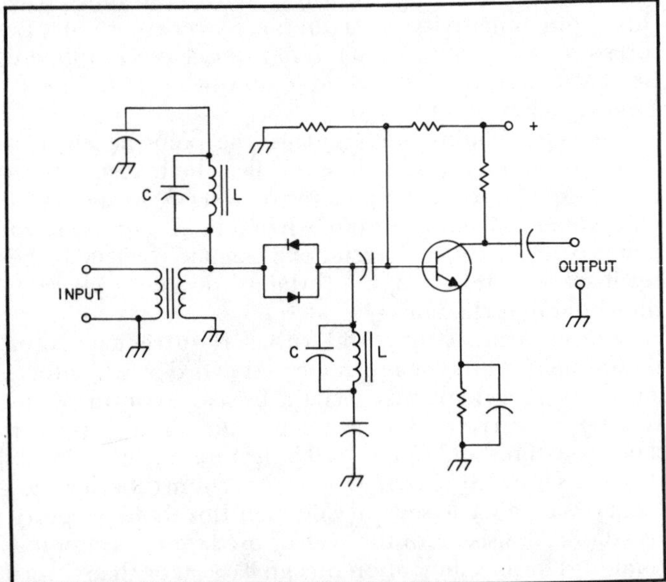

THRESHOLD DETECTOR: A threshold detector in an audio-filter circuit.

THRESHOLD OF VISIBILITY

The faintest intensity of light that a person can see is called the threshold of visibility or the threshold of sight. The threshold of visibility varies from one person to another, and it also varies with the wavelength of the light and the apparent diameter of the light source.

The retina of the eye contains two kinds of light-sensing cells, called cones and rods. The cones allow us to distinguish between colors if the light is sufficiently intense. The threshold of cone visibility is lowest (that is, the cones are the most sensitive) at a wavelength corresponding to green light. The rods have a lower threshold of visibility than the cones, but they cannot perceive color. (This is why colors appear washed-out or invisible in near darkness.) The rods are most sensitive at about the same wavelength as the cones.

The threshold of visibility, in terms of watts per square meter, depends on the apparent size of the light source observed. For a point source in the green part of the spectrum, the average person can detect an intensity of about 10^{-12} watt per square meter, corresponding to the visible radiation from a star of the 6th visual magnitude. The larger the apparent diameter of a light source, the less sensitive the eye becomes. *See also* PHOTOMETRY.

THRESHOLD SENSITIVITY

See TANGENTIAL SENSITIVITY.

THRESHOLD VOLTAGE

See THRESHOLD, THRESHOLD DETECTOR.

THROAT MICROPHONE

In high-noise environments, an ordinary microphone picks up too much background sound to be useful in voice communications. In some instances, a highly directional microphone can be used with adequate results. However, in extreme situations, a throat microphone is needed.

A throat microphone operates by conduction. It is worn around the neck of the user. The diaphragm is placed next to the operator's voice box. Vibration of the throat provides the audio input to the microphone. Since sound propagates much more efficiently by conduction than it does through the atmosphere, the background noise is greatly attenuated relative to the voice. *See also* MICROPHONE.

THROUGHPUT

The speed at which a computer can process a given amount of data is called the throughput. Throughput is generally measured in problems per minute.

Suppose a computer is given n problems to solve, and it takes t minutes to process them and provide the answers. The throughput T is then:

$$T = n/t$$

in problems per minute.

The throughput varies depending on the type of computer and the complexity of the problems. Therefore, throughput is a subjective expression of data-processing speed.

THUNDERSTORM

See LIGHTNING, SFERICS.

THYRATRON

A thyratron is a form of electron tube that is used for switching purposes. It operates in a manner similar to the silicon-controlled rectifier. The envelope of the thyratron is filled with a gas such as argon, neon, or xenon for low-voltage applications. Mercury vapor is generally used for high-voltage applications.

Initially, the thyratron does not conduct. When a positive pulse is applied to the grid, the device conducts because the gas becomes ionized. The tube continues to conduct until the plate voltage is removed or made negative. The voltage required to trigger the thyratron depends on the plate voltage.

The main advantage of the thyratron over semiconductor devices is its higher tolerance for momentary excessive current or voltage. *See also* SILICON-CONTROLLED RECTIFIER, THYRISTOR.

THYRECTOR

A thyrector is a semiconductor device, consisting essentially of two diodes connected in reverse series. The thyrector is used for protection of equipment against transients (*see* TRANSIENT).

A thyrector is connected across the alternating-current power terminals of a piece of electronic apparatus (see illustration). The thyrector does not conduct until the voltage reaches a certain value. If the voltage across the thyrector exceeds the critical value, the device conducts. The conducting voltage does not depend on the polarity.

THYRISTOR

A thyristor is a four-layer semiconductor device, similar to an ordinary rectifier except that it has a control electrode, called the gate, in addition to the anode and cathode.

The thyristor initially does not conduct. When a pulse is supplied to the gate electrode, the device conducts in one direction but not the other, in a manner identical to a diode. Conduction continues until the voltage between the anode and cathode is removed.

The thyristor behaves very much like a thyratron tube. The thyristor, however, operates much more rapidly than the thyratron. The forward-voltage drop of a typical thyristor is approximately 0.9 V.

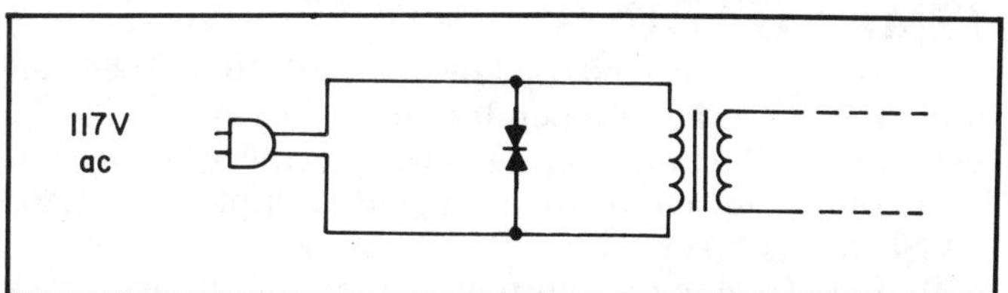

THYRECTOR: A thyrector can be used for transient protection.

Thyristors are used primarily for switching purposes. The most common type of thyristor is the silicon-controlled rectifier. *See also* SILICON-CONTROLLED RECTIFIER, THYRATRON.

TICKLER COIL

See TERTIARY COIL.

TIE STRIP

A tie strip, also called a terminal strip, is a set of contacts used for connecting wires in a point-to-point-wired circuit (*see* POINT-TO-POINT WIRING). A tie strip consists of two or more lugs, attached to a strip of phenolic material. Some of the lugs may be connected to a mounting terminal, and thus grounded (see illustration). The lugs are designed to adhere easily to solder.

Tie strips are available in a wide variety of sizes for different applications.

TIMBRE

Timbre is an expression that is used in music. Different musical instruments, all playing the same note, produce different sounds. A clarinet sounds different, for example, from a trombone at middle C.

The timbre of a note depends on the waveform. Every musical instrument produces its own unique waveform; this waveform may or may not change with pitch. Electronic synthesizers such as the Moog use various oscillators, working together, to create tones having various timbre characteristics.

TIME BASE

See CLOCK, TIME STANDARD.

TIME CONSTANT

Certain circuits change state exponentially as a function of time. Combinations of inductances, capacitances, and/or resistance exhibit this property. Materials heat up and cool down exponentially. The rate at which a change of state occurs, in a given situation, is known as the time constant.

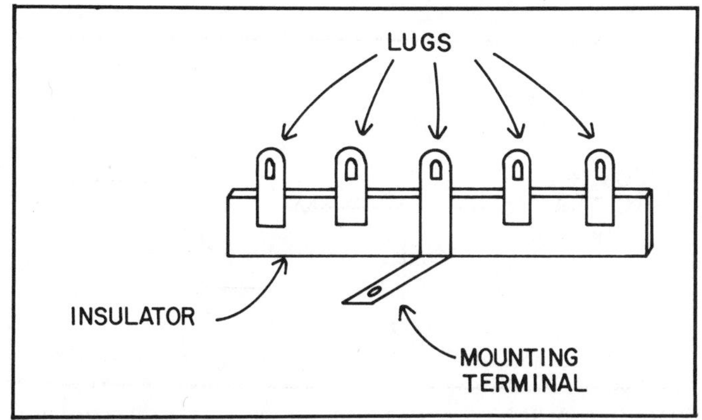

TIE STRIP: A typical tie strip for use in point-to-point wiring.

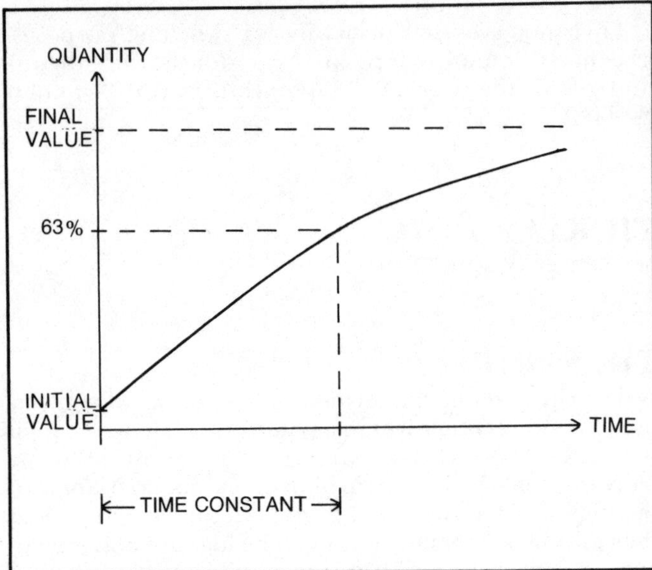

TIME CONSTANT: Graphical illustration of time constant.

The time constant is generally defined as the time, in seconds, required for a change of state to reach 63 percent of completion.

This is illustrated graphically in the drawing. *See also* RESISTANCE-CAPACITANCE TIME CONSTANT, RESISTANCE-INDUCTANCE TIME CONSTANT, THERMAL TIME CONSTANT.

TIME-DELAY RELAY

A time-delay relay is a special form of relay in which the armature closes or opens after the current appears in, or is removed from, the coil. The delay is a constant, known, definite quantity.

All relays have some inherent time delay. In the case of a reed relay, the delay is very small, sometimes less than 0.001 second. Large relays may require several hundredths of a second to close. This is normal. A time-delay relay, however, may require up to several seconds to close following application of current, and/or several seconds to open after the removal of current.

The delay in opening or closing of a relay can be increased by placing a large capacitance across the coil. A resistor in series with the relay coil, in conjunction with a capacitor across the coil and the resistor, will result in even more delay (see illustration). Any ordinary relay can be modified in this manner so that it will operate as a time-delay relay. *See also* RELAY.

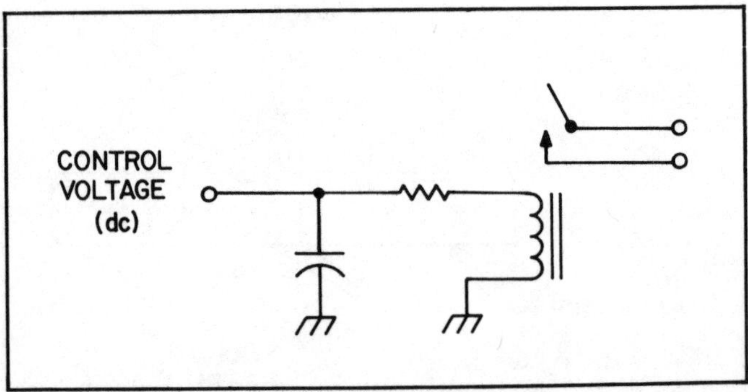

TIME-DELAY RELAY: A resistor and capacitor can be connected in a dc relay circuit to obtain time delay.

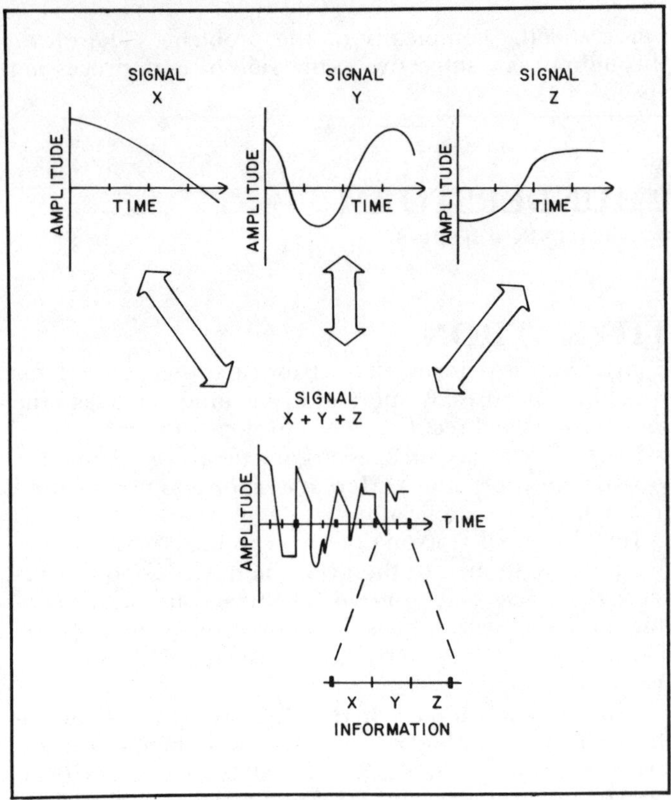

TIME-DIVISION MULTIPLEX: An example of the combination of three signals (X, Y, and Z) in time-division multiplex.

TIME-DIVISION MULTIPLEX

Time-division multiplex is a method of combining analog signals for serial transfer (*see* SERIAL DATA TRANSFER). The signals are sampled at intervals, and interwoven for transmission. At the receiving end, the process is reversed; the signals are separated again (see illustration).

Time-division multiplex is used for transmitting two or more channels of information over a single carrier. The speed of the multiplexed signal is increased, with respect to the individual channel speed, by a factor equal to the number of signals combined. For example, if 10 signals are to be multiplexed by time-division process, the data speed of each signal must be multiplied by 10 to maintain synchronization.

Time-division multiplexing results in an increase in the bandwidth of a signal. The increase takes place by a factor equal to the number of signals combined. This is because of the increased data speed.

Time-division multiplex can be used with digital signals as well as analog signals. Digital time-division multiplex is usually called synchronous multiplex, since the data from each channel is split bit by bit, in accordance with the speed of transmission. *See also* MULTIPLEX, SYNCHRONOUS MULTIPLEX.

TIME SHARING

Large computers can be used by many different operators at once, each at an independent terminal, by means of a technique called time sharing. The principle of time sharing is similar to that of time-division multiplex (*see* TIME-DIVISION MULTIPLEX).

In time sharing, the computer communicates with each terminal for a moment, then moves to the next terminal, and so on in a repeating sequence. If there is just one terminal, the computer spends all of its time communicating with that terminal; if there are n terminals, the computer works with each terminal for 1/nth of the time.

The sequencing process is very rapid, so it seems to each operator that the computer is communicating only with that terminal. The computer appears to operate more slowly, however, when many terminals are connected, as compared to when only a few terminals are connected.

Time-sharing systems are often operated via telephone. A subscriber can work with a computer hundreds of miles away, simply by calling a certain number and using a telephone modem in conjunction with a terminal. *See also* MODEM, TERMINAL.

TIME SIGNALS

See CHU, WWV/WWVH.

TIME STANDARD

In order for clocks to be synchronized, they must be set according to a universal standard. For everyday purposes, most of us simply set our clocks according to the information broadcast by local radio stations. (The news is often preceded by a tone that marks the hour to within a few seconds.) For greater accuracy, the time signals from stations such as CHU or WWV/WWVH are used. *See also* CHU, WWV/WWVH.

TIME SWITCH

A time switch is a device that opens and/or closes a circuit at predetermined times. Time switches are used in homes and businesses for controlling various appliances, especially electric lights.

TIME SWITCH: A commercially manufactured time switch.

The illustration shows a typical adjustable time switch, suitable for use with 117-V appliances. The device consists of a clock connected to a switch. The circuit-closing (on) and opening (off) times are adjustable within a 24-hour period. The on/off cycle is repeated daily. Time switches of this type are available in most hardware and electronics stores.

More sophisticated time switches may have a cycling period of a week, a month, or even a year. Some communications terminals are equipped with electronic time switches, allowing transmission or reception of messages at any time. Home computers can be programmed to switch several appliances on or off at different times.

TIME ZONE

People prefer to have the sun rise and set at similar hours no matter what the longitude. For this reason, time zones have been devised.

The earth rotates at an angular rate of 15 degrees per hour. The time zones of the world are thus about 15 degrees wide, so there is one zone for each hour of the day. The situation, in theory, is quite simple. But in practice it is complicated by the preferences of individual counties, states, and countries.

In the continental United States, there are four times zones, called the Eastern, Central, Mountain, and Pacific. Eastern Standard Time (EST) is 5 hours behind Coordinated Universal Time (UTC). Eastern Daylight Time (EDT) is 4 hours behind UTC. Central, Mountain, and Pacific time are 1, 2, and 3 hours behind Eastern time, respectively. *See also* COORDINATED UNIVERSAL TIME, STANDARD TIME THROUGHOUT THE WORLD.

TIN

Tin is an element with atomic number 50 and atomic weight 119. In its pure form, tin is a soft, malleable, silver-colored metal.

Tin is used in the manufacture of solder. The solder used in electronic circuits generally consists of 50 or 60 percent tin, and 40 to 50 percent lead. *See also* SOLDER..

TIN-LEAD SOLDER

See SOLDER.

TINNING

The leads of discrete electronic components are often covered with a thin layer of tin or solder. This practice, known as tinning, is done to prevent corrosion and to make the leads adhere easily to solder.

The tip of a soldering gun or iron should always be kept coated with a thin layer of solder. This is called tinning. Wires and sheet metal are generally tinned prior to splicing. *See also* SOLDERING TECHNIQUE, WIRE SPLICING.

T MATCH

A T match is a device for coupling a balanced transmission line to a balanced antenna element such as a half-wave,

T-MATCH: A T match is used for impedance matching between a radiator and a balanced line.

center-fed radiator. The radiating element consists of a single conductor. A rod or wire is run parallel to the radiator, and the feed line is connected to the center of the rod or wire. The ends of the rod or wire are connected to the radiator at equal distances from the ends of the radiator. The drawing illustrates a T match.

The impedance-matching ratio of the T match depends on physical dimensions such as the spacing between the radiator and the rod or wire, the length of the matching element, and the relative diameters of the radiator and the matching element. The T match allows the low feed-point impedance of a yagi antenna driven element to be matched to a 75-ohm balanced line. If a balun is used at the feed point, the antenna can be fed with coaxial line. Sometimes a half-T, or gamma match, is used with a coaxial line (*see* GAMMA MATCH).

There are several variations of the T match that can be used when large impedance-step-up ratios are needed. If the rod or wire of a T match has the same length and diameter as the radiating element, the result is a folded dipole antenna. This configuration produces a step-up impedance ratio of 4:1, relative to a single element alone. Three conductors can be connected in parallel to obtain a step-up ratio of 9:1, or four conductors can be used to obtain a factor of 16:1. *See also* FOLDED-DIPOLE ANTENNA, IMPEDANCE MATCHING, YAGI ANTENNA.

T NETWORK

A T network is an unbalanced circuit, used in the construction of filters, attenuators, and impedance-matching devices. Two series components are connected on either side of a single parallel component.

The diagram at A illustrates a T-network circuit consisting of non-inductive resistors. This circuit can be used as an attenuator. B shows a T-network circuit that can be used for impedance matching. The T network gets its name from its shape in the schematic diagrams.

In a T-network impedance-matching circuit, the capacitor and/or inductors may be adjustable to vary the impedance-transfer ratio. A fairly wide range of load impedances can be matched to a fixed source impedance. The T-network circuit at B might also be used as a lowpass filter. *See also* IMPEDANCE MATCHING.

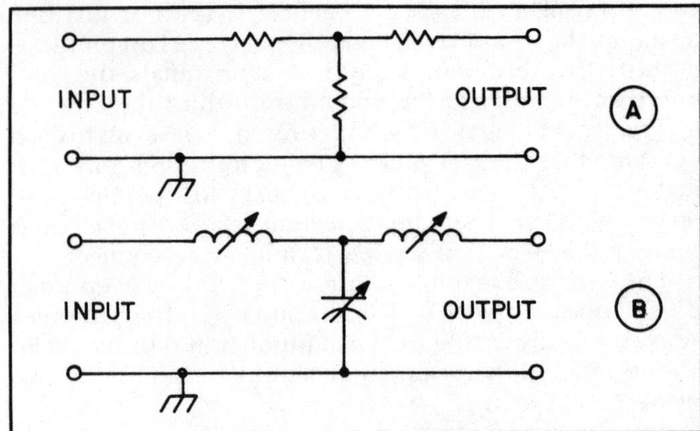

T NETWORK: At A, a resistive network, suitable for use as an attenuator. At B, an inductance-capacitance network for use as a filter or impedance-matching circuit.

TOGGLE SWITCH

A toggle switch is a lever type panel switch (see illustration) that is often found in electronic equipment. Toggle switches usually have one or two poles, and one or two throw positions. Toggle switches are available in many sizes, ranging from subminiature to very large. The voltage and current ratings are low to moderate.

Toggle switches are used for switching something on and off, or for selecting between two circuits or states.

TOLERANCE

The actual value of a fixed component is never exactly the same as the rated value; there is always some error. When components are mass produced, some end up having values less than the rated value, and others end up with larger values. The manufacturer will guarantee, however, that a component from a given lot will have a value that is within a certain range. This range is expressed as a figure called tolerance.

For a component with a rated value r, the tolerance T is expressed as a plus-or-minus percentage that is the larger of the following two figures:

$$T1 = \pm 100(r - s)/r$$

$$T2 = \pm 100(t - r)/r$$

where s represents the lowest expected value and t represents the highest expected value. (While it is true that $s < r$

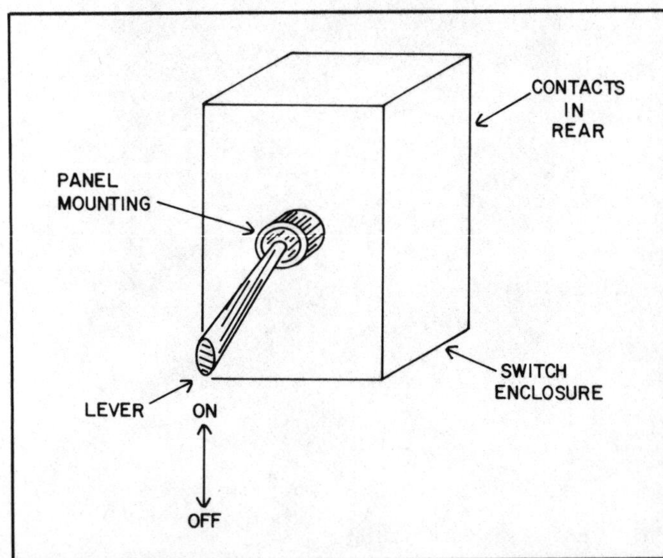

TOGGLE SWITCH: Drawing of a toggle switch.

< t, it may not be true that r–s = t–r. The tolerance T represents the greatest possible departure of the actual value from the rated value.)

If we have a component with a rated value r and a tolerance of plus-or-minus T percent, we can expect that the component will have a value larger than:

$$s = r - rT/100$$

and smaller than:

$$t = r + rT/100$$

Tolerance is an important consideration in the design and manufacture of any electronic device. Some equipment requires components with low tolerance, that is, high precision. But the use of low-tolerance components where they are not needed is a waste of money, because in general, the price of a component goes up as the tolerance goes down.

In electronic measurement, the counterpart of tolerance is known as accuracy. *See also* ACCURACY.

TONE BURST

See PRIVATE LINE®, TONE SQUELCH.

TONE CONTROL

The amplitude-versus-frequency characteristics of a high-fidelity sound system are adjusted by means of a tone control. Tone controls vary greatly in complexity.

In its simplest form, a tone control consists of a single knob or slide device. The counterclockwise or lower position boosts the bass and attenuates the treble; the clockwise or upper position boosts the treble and attenuates the bass (*see* BASS, TREBLE). This is accomplished by means of a capacitor and potentiometer, as shown in the schematic diagram at A.

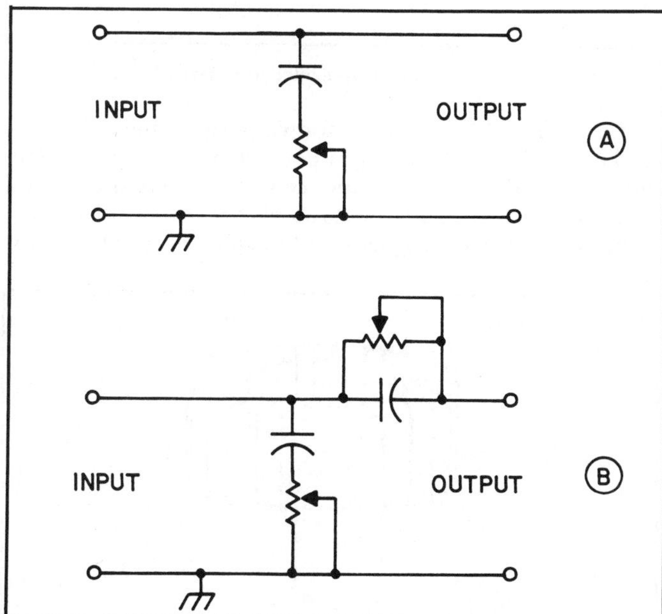

TONE CONTROL: Two examples of tone control. At A, a single control; at B, a dual control.

A somewhat more sophisticated tone control has two potentiometers and two capacitors, one combination in series and the other in parallel. One potentiometer controls the treble and the other controls the bass, as at B.

The most sophisticated tone controls have several filters that allow the listener to tailor the circuit response. This type of system is called an equalizer or graphic equalizer (*see* EQUALIZER).

TONE DIALER

See TOUCHTONE®.

TONE ENCODING

See PRIVATE LINE®, TONE SQUELCH, TOUCHTONE®.

TONE SQUELCH

Some receivers are equipped with a special squelching circuit that stays closed unless an incoming signal is modulated with a tone of a certain frequency and/or duration. This type of squelch is called a tone squelch.

Tone squelching is used mostly in frequency-modulated communications systems operating at very high and ultra-high frequencies. The tone may be continuous and low-pitched, so it cannot be heard by the receiving operator. This is known as subaudible-tone squelching. Alternatively, a brief tone, having a midrange audio frequency, can be used at the beginning of the transmission. This is called tone-burst squelching.

Tone squelching is sometimes used in repeater-input receivers. This prevents unauthorized use of a repeater, since only the properly modulated signals can actuate it. This use of tone squelching is called Private-Line operation. *See also* PRIVATE LINE®.

TOP-LOADED VERTICAL ANTENNA

Top loading is a method of feeding a vertical antenna at the top, rather than the bottom, of the radiating element. This scheme is used when a transmitter or receiver must be located well above the ground and nearby support structures. Top loading can also be used as an alternative, in any situation, to other feed methods. Top loading can be accomplished in various ways. Two examples follow.

A dweller in a high-rise building may install a top-loaded, more-or-less vertical wire antenna as shown at A. The grounding system can consist of the cold-water plumbing in the building. The antenna is tuned to resonance at the desired frequency by means of a transmatch.

Another method of top loading, in this case for a half-wave vertical element, is shown at B. The antenna is suspended from a support such as a tree branch. This feed system is similar to that used in the zeppelin antenna (*see* ZEPPELIN ANTENNA).

The resonant frequency of a vertical antenna can be lowered by a technique called top loading. A capacitance hat, placed at the top of the radiating element, allows the height of the antenna to be reduced for operation at a

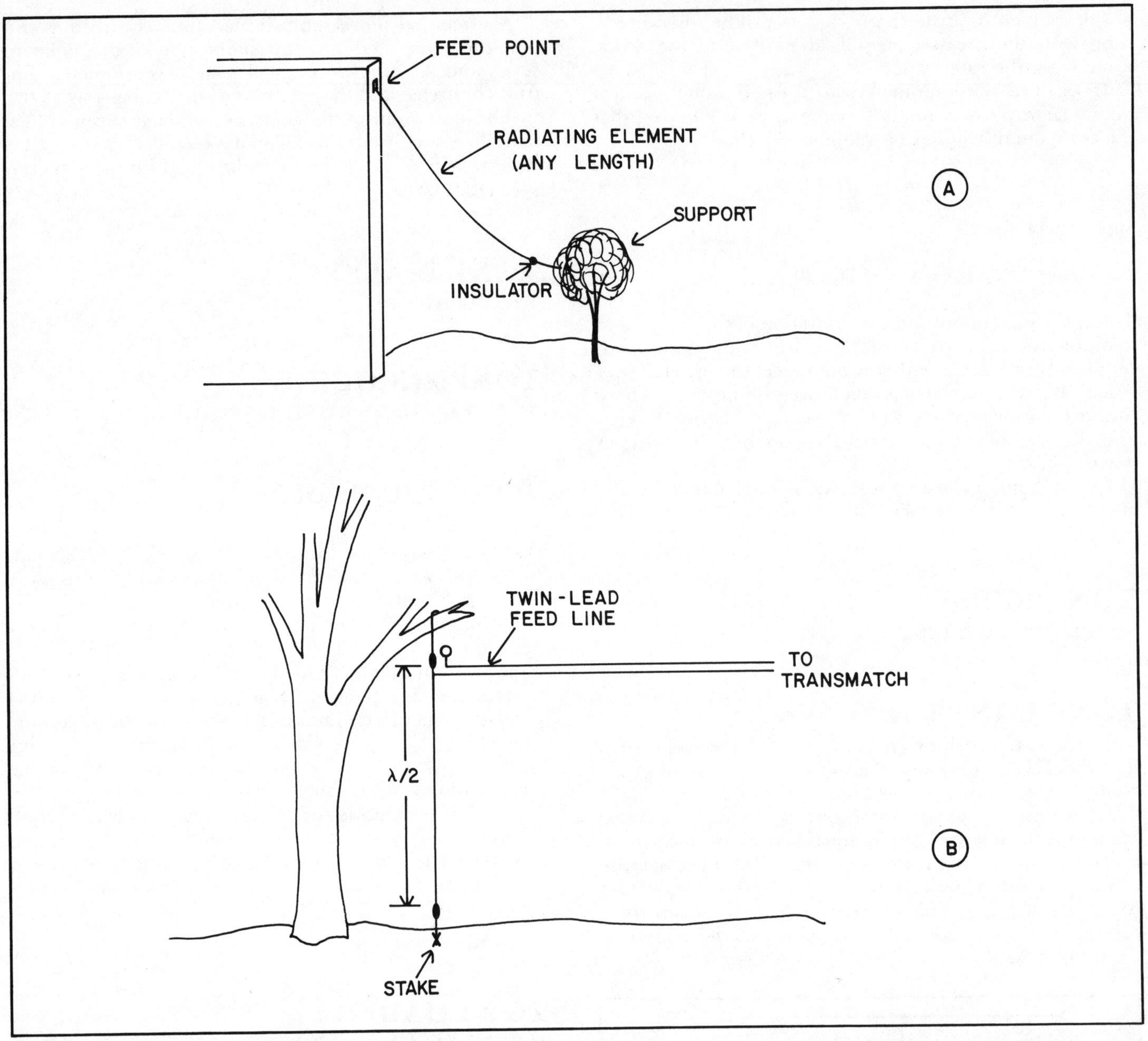

TOP-LOADED VERTICAL ANTENNA: At A, a random wire is fed at the top for use as an antenna. At B, a top-loaded (top-fed) half-wave antenna.

given frequency. This is sometimes called top loading, but is more often called capacitive loading (*see* CAPACITIVE LOADING).

TOROID

A toroid is a coil wound on a doughnut-shaped ferromagnetic core (see illustration). Toroid coils and cores have become increasingly popular in recent years, replacing the traditional solenoidal coils and cores.

The main advantage of the toroidal coil is that all of the magnetic flux is contained within the core material. Therefore, a toroid is not affected by surrounding components or objects. Also, nearby components are unaffected by the coil field. Another advantage of the toroid is that it has a smaller physical size, for a given amount of inductance, than a solenoidal coil. Still another advantage is that fewer coil turns are required to obtain a given inductance, compared with a solenoidal coil. This enhances the Q factor.

Toroidal coils can be used for winding simple inductors, chokes, and transformers. They are used at audio frequencies, and at radio frequencies up to several hundred megahertz. Ferromagnetic toroidal cores are available in a tremendous variety of sizes and permeability ratings, for

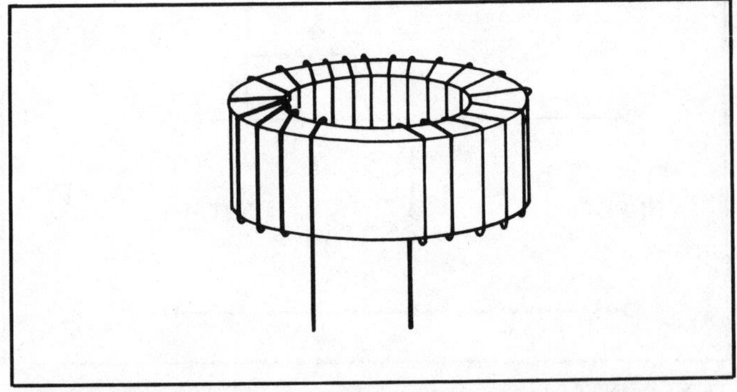

TOROID: A toroid coil is wound on a doughnut-shaped form.

use at various frequencies and power levels. *See also* INDUCTOR.

TORQUE

Torque is an expression of twisting or angular force. The turning force produced by a motor is a form of torque. Antenna towers and masts may be subjected to torque because of wind effects.

Torque occurs around a defined axis, in either a clockwise or a counterclockwise direction, depending on the point of view.

TOTAL HARMONIC DISTORTION

The total root-mean-square (rms) harmonic voltage in a signal, as a percentage of the voltage at the fundamental frequency, is called total harmonic distortion (THD). Total harmonic distortion may vary from zero percent (no harmonic energy) to theoretically infinite (no fundamental-frequency energy).

The THD figure is often used as an expression of the performance of an audio-frequency amplifier. The amplifier is provided with a pure sine-wave input, and the harmonic content is measured using a distortion analyzer or SINAD voltmeter (*see* DISTORTION ANALYZER).

Ideally, an audio-amplifier circuit should have as little THD as possible at the rated audio-output level. Most amplifiers are rated according to the amount of root-mean-square power they can deliver to a speaker of a certain impedance (usually 8 ohms) with less than a specified THD (usually 10 percent).

TOTAL INTERNAL REFLECTION

When a ray of light passes from one medium to another, and the index of refraction of the first substance is different from that of the second, refraction occurs (*see* INDEX OF REFRACTION, REFRACTION). Under certain circumstances, however, reflection takes place instead. This phenomenon, known as total internal reflection, occurs only when the following two conditions exist:

- The first medium must have a higher index of refraction than the second
- The angle of incidence must measure more than a certain limiting value

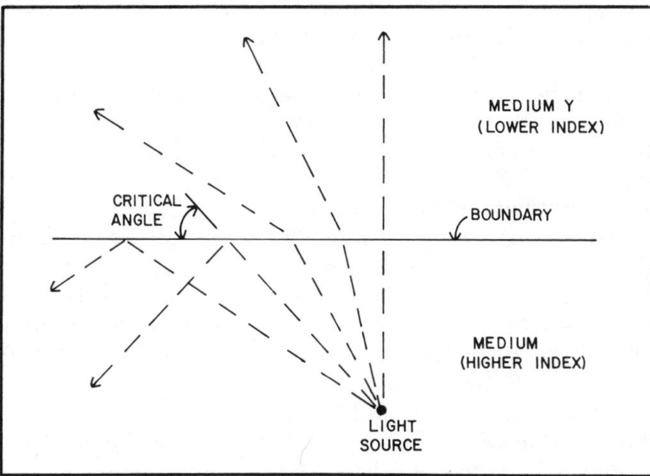

TOTAL INTERNAL REFLECTION: Total internal reflection occurs at the boundary between two substances having different indices of refraction, provided the angle of incidence is large enough.

This principle is illustrated in the accompanying drawing.

As the incidence angle gets larger, the angle at which total internal reflection begins depends on the ratio of the indices of refraction of the two substances. If the indices are nearly the same, reflection will occur only if the incidence angle is very large. If the first medium has a much greater index of refraction than the second, total internal reflection will be observed for fairly small incidence angles.

Optical fibers operate on the principle of total internal reflection. Since glass or plastic has a higher index of refraction than air, a light beam will stay inside the fiber because the light will be reflected from the inside surfaces (*see* FIBER, FIBER OPTICS).

Under some circumstances, total internal reflection affects radio signals. The boundary between two air masses, one much more dense than the other, can produce reflection of electromagnetic fields. This is called the duct effect (*see* DUCT EFFECT). It is observed primarily at very high frequencies and above. Total internal reflection can also affect sound waves in a similar way.

TOTAL REGULATION

The output current or voltage of a power supply is affected by several factors. The load resistance and the input voltage are the most significant, but other variables, such as temperature or mechanical vibration, can also have an effect. Sometimes these variables partially cancel each other, but in some cases their effects may be additive. Total regulation is an expression of the worst case, in which all of the contributing factors work together to increase or decrease the output of the power supply.

Total regulation is determined by first determining how each variable affects the output of the supply. Then the variables are adjusted so that they all tend to reduce the supply current or voltage. Finally, the variables are all adjusted so that they tend to maximize the output current or voltage. The total regulation is calculated in terms of the nominal, minimum, and maximum currents or voltages. *See also* CURRENT REGULATION, REGULATION, VOLTAGE REGULATION.

TOUCHTONE®

In recent years, an increasing number of telephone sets have been using pushbutton dialing systems instead of the rotary dialer. The buttons actuate tone pairs that cause automatic dialing of the numbers. This is called Touchtone® dialing (the term is a trademark of American Telephone and Telegraph Company).

The original Touchtone dial, still found on most telephone sets, has 12 buttons corresponding to digits 0 through 9, the star symbol (*) and the pound symbol (#). Some dialers have four additional keys, designated A, B, C, and D. The tone-pair frequencies for the 16 designators are listed in the table.

TOUCHTONE®: TOUCHTONE® AUDIO FREQUENCIES.

LOW GROUP, HZ	HIGH GROUP, HZ 1209	1336	1477	1633
697	1	2	3	A
770	4	5	6	B
852	7	8	9	C
941	*	0	#	D

Although the original push-button keypads were called Touchtone pads, other manufacturers have produced similar arrangements known by different names. All use the common combinations of frequencies listed in the table in order to access the telephone systems. These systems offer what is called DTMF (Dual-Tone, Multiple Frequency) access. Proper operation requires that both tones be present. The DTMF system can be used to remotely control distant objects or electronic equipment.

Tone-dialing operation has several advantages over the older rotary dial. Tone dialing is much faster, and it can be done automatically with extreme speed. The tones can also be transmitted into a telephone set from an external source, which is not possible with a rotary dial. Many radio repeaters, for example, are interconnected with the telephone lines, so that mobile radio operators, equipped with Tone keypads in their transceivers, can gain access to the lines. *See also* AUTOPATCH, ROTARY DIALER.

TOWER

A reinforced structure, intended for support of power lines or antenna systems, is called a tower. Towers vary in height and size from the small assemblies used for home television reception to massive structures that support broadcasting antennas or high-tension power lines.

Types of Towers. Towers can be categorized as either self-supporting or guyed. The self-supporting towers require more rigid construction than the guyed types. One common method of tower construction uses three vertical members having a triangular cross-section. Reinforcing rods are welded to the three vertical members at intervals. An example is shown at A in the illustration. A less common scheme is the use of a single, heavy vertical mast, similar to a flagpole. A guyed tower may have one or more sets of supporting wires (*see* GUYING).

Self-supporting towers are sometimes equipped with tilt-over devices. This allows an antenna to be installed with comparative ease at the top of the tower, as at B. Other self-supporting towers may have telescoping sections for height adjustment, as at C; this is known as the crank-up configuration.

Passive and Active Towers. A tower can be used either as a support for one or more antennas, or as an antenna itself. In the former case, the tower is called passive; in the latter case, it is called active. At longer wavelengths, active towers are common, but at the shorter wavelengths, towers are used only to put the antennas at the height necessary for reliable communications.

Sometimes a passive tower, supporting antennas of a short wavelength, is fed with radio-frequency energy at a longer wavelength. This is a popular practice among radio amateurs. A tower, supporting antennas for 14, 21, 28, 144, and 220 MHz, for example, can be used as an active radiator at 1.8, 3.5, and 7 MHz. Active towers produce vertically polarized fields. Shunt feed is a common method of feeding a tower directly (*see* SHUNT FEED).

Safety. Any massive antenna structure is potentially dangerous. When installing or maintaining a tower, personal safety is the most important consideration.

The choice of a tower depends on the environment to be expected, the size of the antennas to be supported, and the height desired. Any tower should be periodically checked for signs of deterioration that might weaken the structure and eventually result in collapse.

Tall structures are more frequently hit by lightning than shorter ones, and for this reason, lightning protection is imperative with any tower system (*see* LIGHTNING, LIGHTNING PROTECTION). An excellent electrical ground should be provided at the base of a tower to prevent dangerous induced voltages.

Tower climbing should be done by professionals only. Backyard towers can present a special hazard to children, who may attempt to climb the structure. Some means should be devised to prevent this from happening.

A professional structural engineer should be consulted before installing any tower. All applicable safety codes, as

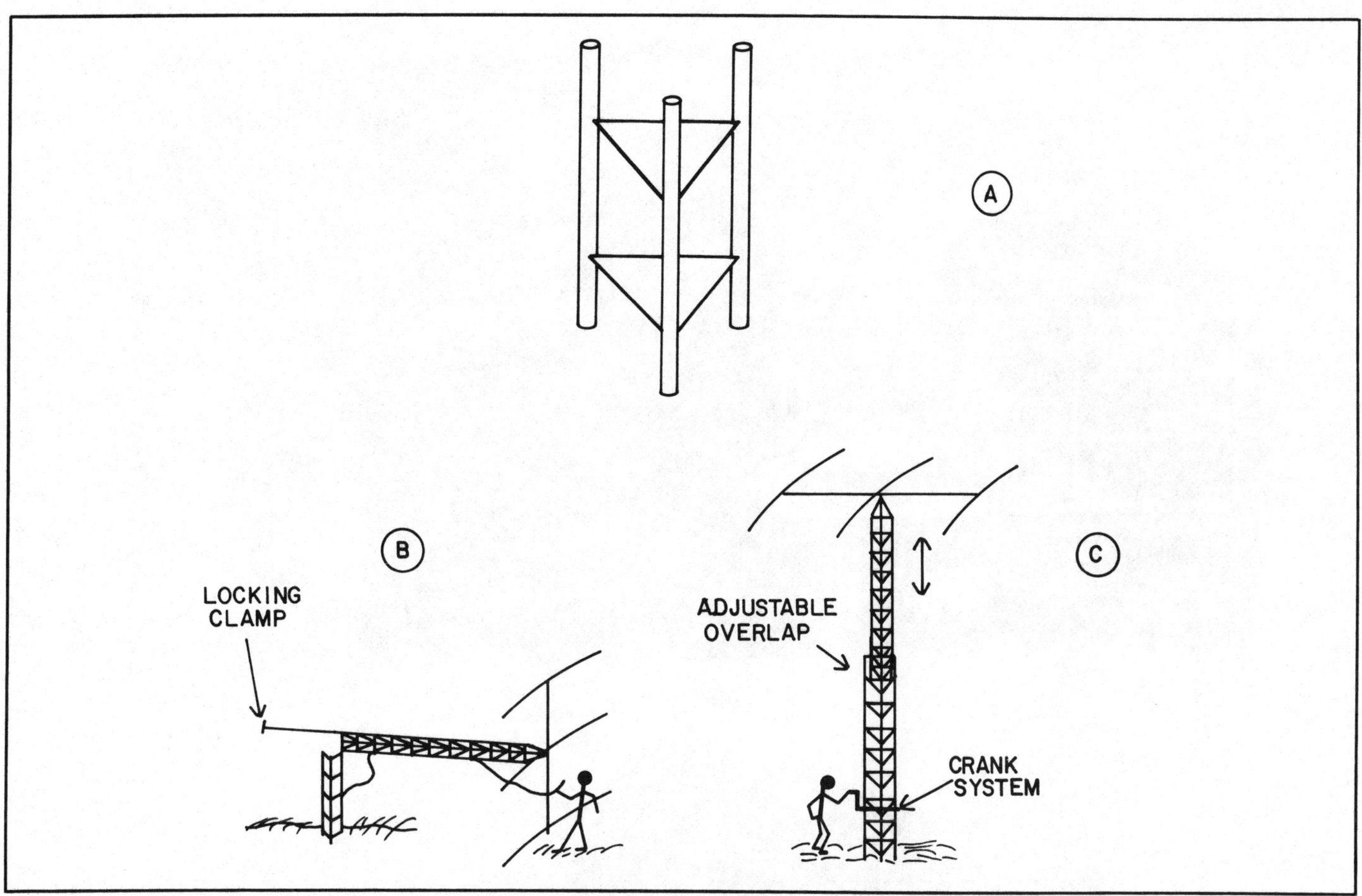

TOWER: At A, typical tower construction. At B, a tilt-over tower. At C, a crank-up tower.

well as zoning laws, must be strictly followed.

TRANSCEIVER

A transceiver is a combination of a transmitter and a receiver, having a common frequency control and usually enclosed in a single package. Transceivers are extensively used in two-way radio communication at all frequencies, and in all modes.

The main advantage of a transceiver over a separate transmitter and receiver is economical: Many of the components can be used in both the transmit and receive modes. Another advantage is that most two-way operation is carried out at a single frequency, and transceivers are more easily tuned than separate units.

The main disadvantage of a transceiver is that communication must sometimes be carried out on two frequencies that differ greatly. Also, duplex operation is not possible with most transceivers. Some transceivers have provisions for separate transmit-receive operation, however, overcoming these difficulties.

Transceivers are extensively used by amateur and Citizen's-band radio operators. Illustration A shows a typical amateur transceiver for use in the 10-meter frequency-modulation (FM) band. Illustration B is a photograph of a Citizen's Band radio transceiver. Both units are compact, and can be used for fixed or mobile operation. The variety of commercially manufactured transceivers is enormous, and these units are shown only as examples.

Illustration C shows a block diagram of a transceiver. The principal components are a variable-frequency oscillator or channel synthesizer, a transmitter, a receiver, and an antenna-switching device. The simplest transceivers employ direct-conversion techniques; this scheme is sometimes used for high-frequency Morse-code (CW) communication. More sophisticated transceivers use superheterodyne design. *See also* RECEIVER, TRANSMITTER.

TRANSCONDUCTANCE

Transconductance is a parameter that is sometimes used as an expression of the performance of a vacuum tube or transistor. The symbol for transconductance is g_m, and the unit is the mho or siemens (*see* CONDUCTANCE, MHO, SIEMENS).

Transconductance is defined as the ratio of the change in plate, collector or drain current to the change in grid, base or gate voltage over a defined, arbitrarily small interval on the plate-current-versus-grid-voltage collector-current-versus-base-voltage, or drain-current-versus-gate-voltage characteristic curve. If ΔI represents a change in plate, collector or drain current caused by a change in the grid, base or gate voltage ΔE, then the transconductance is approximately:

$$g_m = \Delta I/\Delta E$$

The precise value at a given point on the characteristic curve is given by:

$$g_m = dI/dE$$

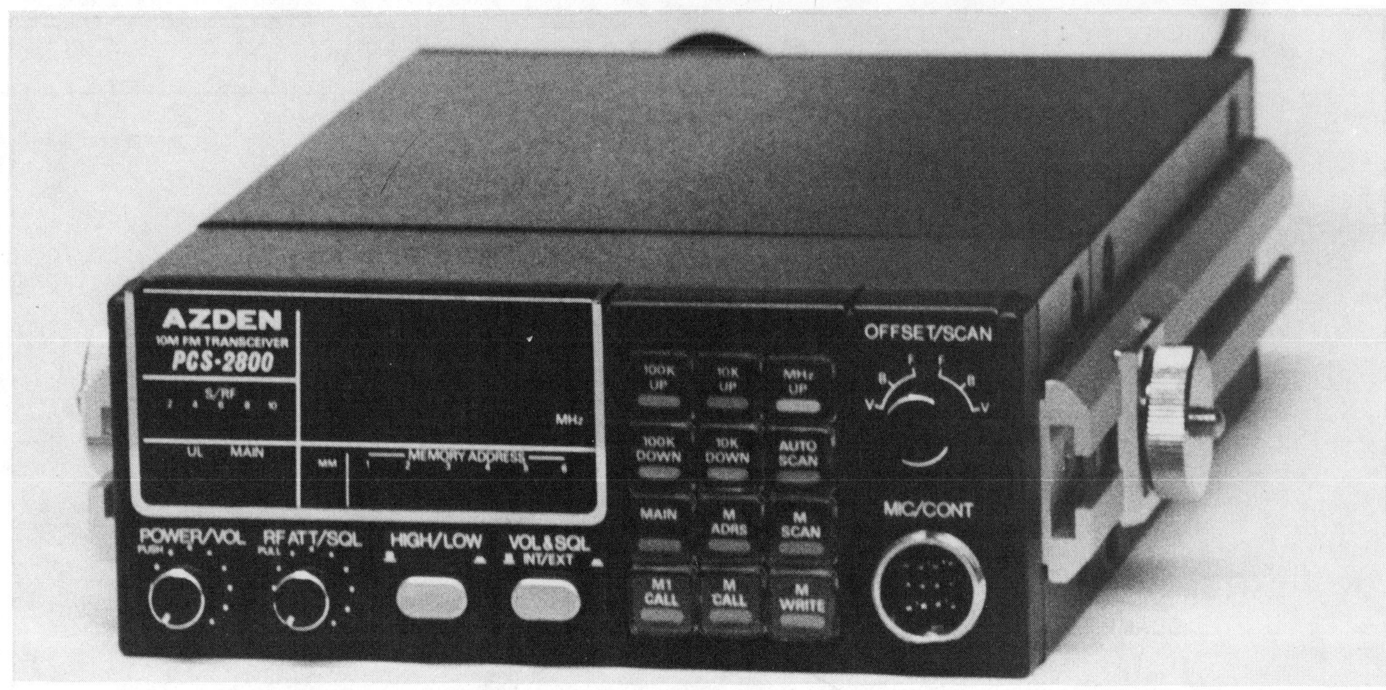

AZDEN
10M FM TRANSCEIVER
PCS-2800
S/RF
UL
MAIN
MM
MEMORY ADDRESS
MHz
100K UP
10K UP
MHz UP
100K DOWN
10K DOWN
AUTO SCAN
MAIN
M ADRS
M SCAN
M1 CALL
M CALL
M WRITE
OFFSET/SCAN
MIC/CONT
POWER/VOL
RF ATT/SQL
HIGH/LOW
VOL&SQL
INT/EXT

A

AWI
VOLUME
SQUELCH
MOD
40
CH9
REALISTIC
TRC-427
CH9
ANL
PA
OUT
OUT
MON
CB
RF GAIN
TONE

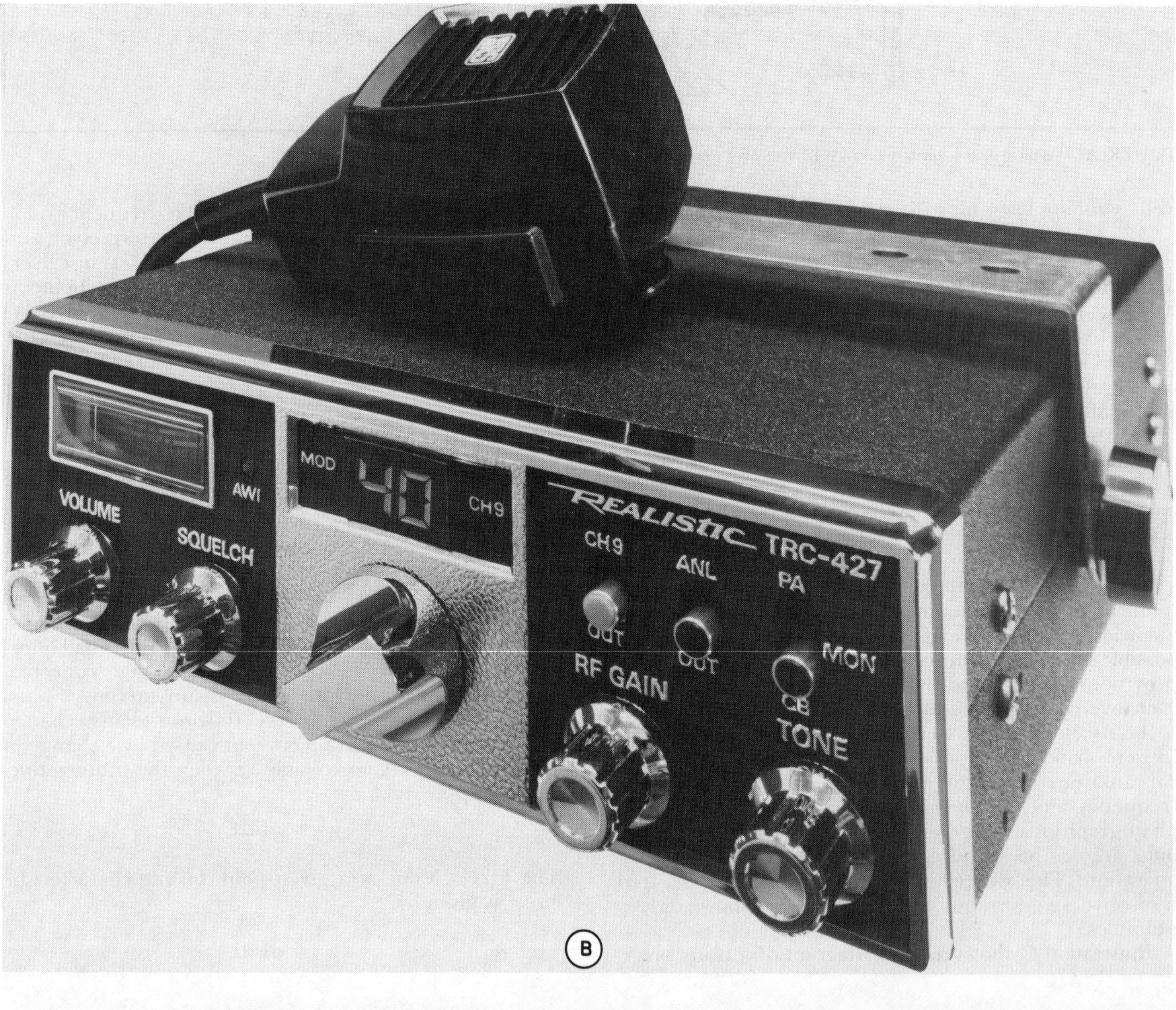

B

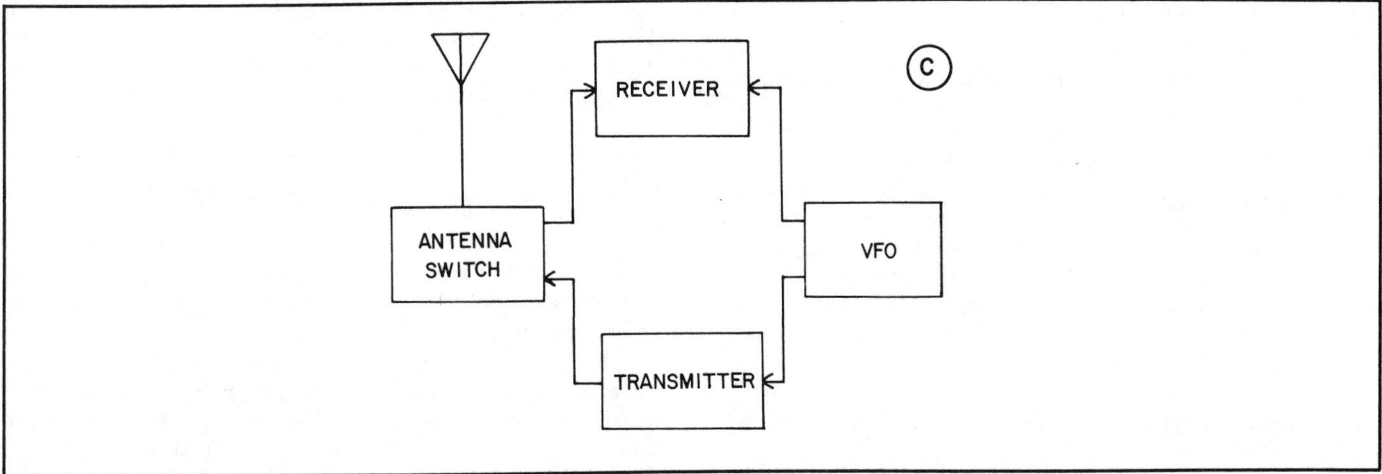

TRANSCEIVER: At A, a transceiver for use in the 28-MHz amateur band. (Courtesy of International Electronic Systems, Inc.) At B, a typical Citizen's Band transceiver. (Courtesy of Radio Shack, a division of Tandy Corporation.) At C, block diagram of a basic transceiver.

and corresponds to the slope of the tangent line to the curve at that point (see illustration). *See also* CHARACTERISTIC CURVE.

TRANSDUCER

A transducer is a device that converts one form of energy or disturbance into another. In electronics, transducers convert alternating or direct electric current into sound, light, heat, radio waves, or other forms. Transducers also convert sound, light, heat, radio waves, or other energy forms into alternating or direct electric current.

Common examples of electrical and electronic transducers include buzzers, speakers, microphones, piezoelectric crystals, light-emitting and infrared-emitting diodes, photocells, radio antennas, and many other devices. For specific information concerning a particular kind of electrical or electronic transducer, please refer to the article according to the name of the device.

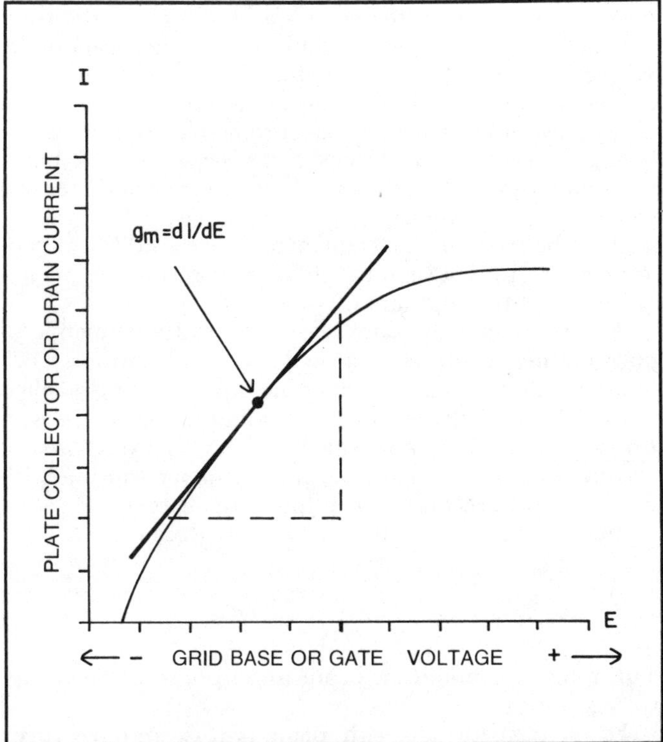

TRANSCONDUCTANCE: Transconductance is the mathematical slope of the characteristic curve of a tube or field-effect transistor, evaluated at a given point.

TRANSFER FUNCTION

See FREQUENCY RESPONSE.

TRANSFER IMPEDANCE

In a direct-current or alternating-current circuit, a voltage applied at one point may result in a current at another point. The ratio E/I of the applied voltage to the resultant current is known as the transfer impedance.

Consider the example of a transformer with a step-down ratio of 5:1, an input voltage E of 120 volts, and a load impedance Z of 12 ohms, as shown in the illustration. The output voltage, Eo, is:

$$Eo = 120/5 = 24 \text{ V}$$

and the load current I is therefore:

$$I = Eo/Z = 24/12 = 2 \text{ A}$$

therefore, the transfer impedance, Zt, is given by:

$$Zt = E/I = 120/2 = 60 \text{ ohms}$$

This example assumes that the efficiency of the transformer is 100 percent. *See also* IMPEDANCE.

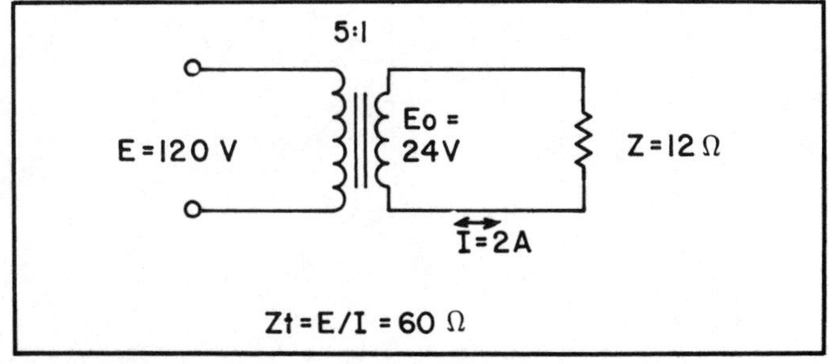

TRANSFER IMPEDANCE: An example of transfer impedance. See text for details.

TRANSFORMATION FUNCTIONS

The relationship between two sets of coordinates is expressed by a set of mathematical functions known as transformation functions (*see* FUNCTION). A set of these functions can be defined for any two coordinate systems, provided the number of dimensions is the same in both systems.

The general case is defined as follows: Suppose we have two sets of coordinates in n dimensions, defined as

$$S1 = \{x_1, x_2, x_3, \ldots, x_n\}$$

$$S2 = \{y_1, y_2, y_3, \ldots, y_n\}$$

where x and y are real or complex numbers. Then for any mathematical functions

$$x_n = f(x_1, x_2, x_3, \ldots, x_n) \text{ and}$$

$$y_n = g(y_1, y_2, y_3, \ldots, y_n)$$

there exists a set of transformation functions $\{T_1, T_2, T_3, \ldots T_n\}$ such that:

$$y_1 = T_1\ (x_1)$$

$$y_2 = T_2\ (x_2)$$

$$y_3 = T_3\ (x_3)$$

$$\vdots$$

$$y_n = T_n\ (x_n)$$

We will consider a simple example. Suppose we have two Cartesian planes defined by $S_1 = (x_1, x_2)$ and $S_2 = (y_1, y_2)$, as shown in the illustration. (*See* CARTESIAN COORDINATES). The coordinate-system origins are displaced, but the units in both systems are linear and equal in size. The transformation functions are:

$$y_1 = T_1\ (x_1) = x_1 + 2$$

$$y_2 = T_2\ (x_2) = x_2 + 3$$

This means that any ordered pair (x_1, y_1) in system S1 will be represented by the ordered pair $(y_1, y_2) = (x_1 + 2, x_2 + 3)$ in the system S2.

The most common two-dimensional (plane) coordinate systems are the Cartesian system and the polar system. The most common three-dimensional (space) coordinate schemes are the Cartesian and spherical systems. The relationship between a Cartesian plane system (x,y) and a polar plane system (r,θ) is given by the equations specified in the article POLAR COORDINATES. The relationship between a Cartesian space system (x,y,z) and a spherical coordinate system(r,θ,ϕ) is given as specified in SPHERICAL COORDINATES.

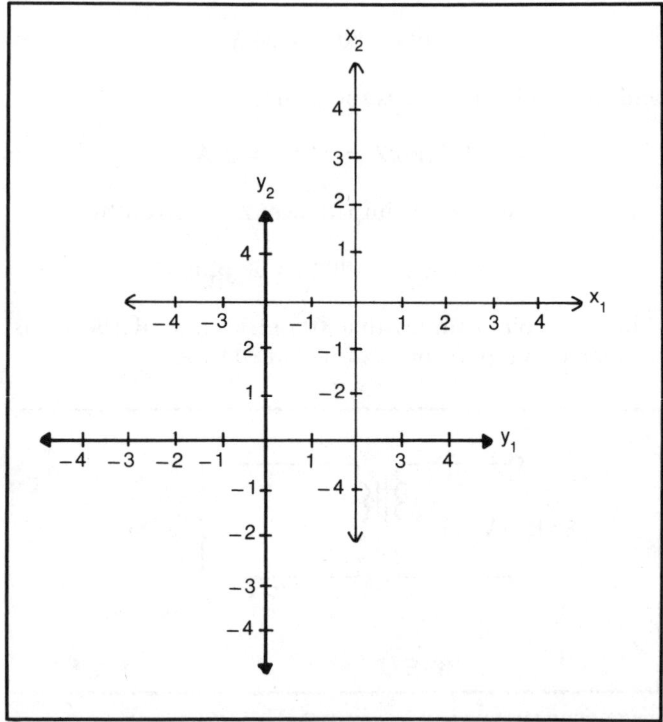

TRANSFORMATION FUNCTIONS: Two Cartesian coordinate systems, related by a linear transformation function (see text).

TRANSFORMER

A transformer is a device that is used to change the voltage of an alternating current, or the impedance of an alternating-current circuit path. Transformers operate by means of inductive coupling. The windings of a transformer may have a ferromagnetic core or a dielectric core.

Transformers, such as the one shown in illustration A, are extensively used in electronic power supplies and audio circuits. The unit shown is an audio-frequency transformer, intended for matching the output impedance of a transistorized amplifier to the 8-ohm impedance of a typical speaker. This transformer is about the size of a walnut. Larger transformers are used in high-current power supplies. Still larger ones provide the 117-V utility power to which we are accustomed in our homes and businesses (see B). The largest transformers handle many thousands of volts and thousands of amperes, and are used at power-transmission stations.

At radio frequencies, transformers are used for impedance matching between amplifying stages, or between a power amplifier and an antenna.

A transformer operates according to the principle of magnetic induction, as illustrated at C. Transformers work only with alternating currents. The input winding is called the primary, and the output winding is called the secondary (*see* TRANSFORMER PRIMARY, TRANSFORMER SECONDARY). If a transformer has a primary winding with n1 turns and a secondary winding with n2 turns, then the primary-to-secondary turns ratio, t, is defined as:

$$T = n1/n2$$

This ratio determines the input-to-output impedance and voltage ratios.

For a transformer with primary-to-secondary turns ratio T and an efficiency of 100 percent (that is, no power loss), the output voltage Eo is related to the input voltage Ei by the equation:

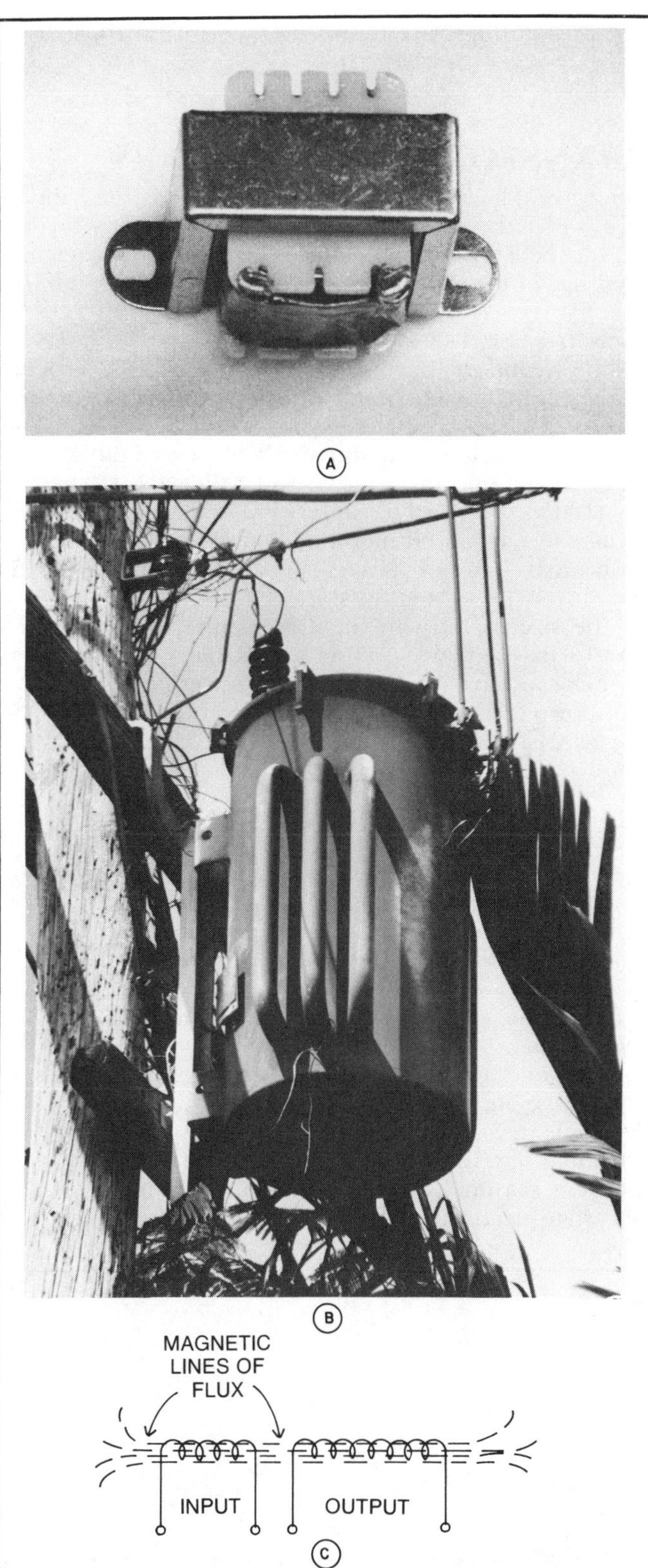

TRANSFORMER: At A, a small audio transformer, often found in radio equipment. At B, a large power transformer. At C, diagram showing the principle of operation of a transformer.

$$Eo = TEi$$

The output, or load, impedance Zo in such a transformer is related to the input impedance according to the equation

$$Zo = T^2 Zi$$

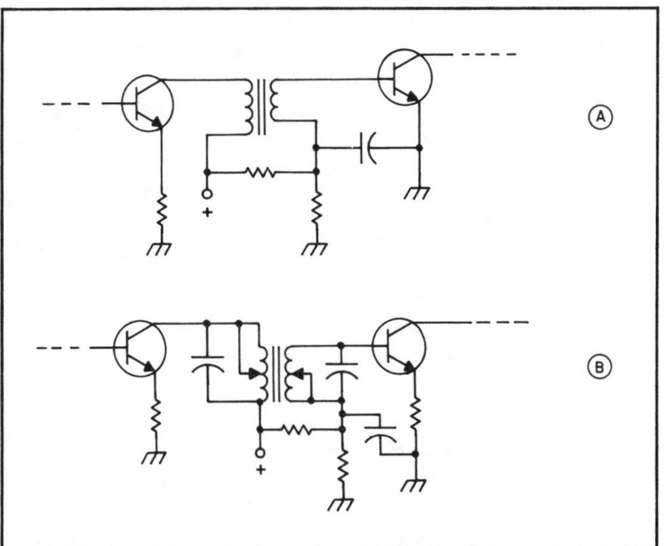

TRANSFORMER COUPLING: At A, transformer coupling in an audio-frequency circuit. At B, transformer coupling in the intermediate-frequency circuit of a radio receiver.

See also IMPEDANCE TRANSFORMER, TRANSFORMER EFFICIENCY.

TRANSFORMER COUPLING

In an audio-frequency or radio-frequency system, the signal can be transferred from stage to stage in various ways. Transformer coupling was once a commmonly used method.

The illustration at A illustrates transformer coupling between two audio-frequency amplifier stages. The transformer is chosen so that the output impedance of the first stage is matched to the input impedance of the second stage. B shows transformer coupling in the intermediate-frequency chain of a radio receiver. In this arrangement, the windings are tapped for optimum impedance matching. Capacitors are connected across the transformer windings to provide resonance at the intermediate frequency.

An important advantage of transformer coupling is that it minimizes the capacitance between stages. This is especially desirable in radio-frequency circuits, because stray capacitance reduces the selectivity. In transmitter power amplifiers, the output is almost always coupled to the antenna by means of a tuned transformer. This minimizes unwanted harmonic radiation. An electrostatic shield provides additional attenuation of harmonics by practically eliminating stray capacitance between the primary and secondary windings (*see* ELECTROSTATIC SHIELDING).

The main disadvantage of transformer coupling is cost: It is more expensive than simpler coupling methods. Transformer coupling is not often seen in modern circuitry. *See also* CAPACITIVE COUPLING.

TRANSFORMER EFFICIENCY

In any transformer, some power is lost in the coil windings. If the transformer has a ferromagnetic core, some power is lost in the core. Conductor losses occur because of ohmic resistance and sometimes skin effect (*see* SKIN EFFECT).

Core losses occur because of eddy currents and hysteresis (*see* EDDY-CURRENT LOSS, HYSTERESIS LOSS). The efficiency of a transformer is therefore always less than 100 percent.

Let Ep and Ip represent the primary-winding voltage and current in a hypothetical transformer, and let Es and Is be the secondary-winding voltage and current. In a perfect transformer, the product EpIp would be equal to the product EsIs. However, in a real transformer, EpIp is always greater than EsIs. The efficiency, in percent, of a transformer is:

$$Eff = 100EsIs/(EpIp)$$

The power P that is lost in the transformer windings and core is equal to the difference between EpIp and EsIs:

$$P = EpIp - EsIs$$

The efficiency of a transformer varies depending on the load connected to the secondary winding. If the current drain is excessive, the efficiency is reduced. Transformers are generally rated according to the maximum amount of power they can deliver without serious degradation in efficiency.

In audio and radio circuits, the frequency affects the efficiency of a transformer. In transformers with air cores, the efficiency gradually decreases as the frequency becomes higher. This occurs because of skin effect in the windings. In a ferromagnetic-core transformer, the efficiency gradually decreases as the frequency increases, until a certain critical point is reached. As the frequency rises beyond this point, the efficiency drops rapidly because of hysteresis loss in the core material.

Transformer efficiency is maximized by the use of the proper type of transformer in a given application. *See also* TRANSFORMER.

TRANSFORMER LOSS

See EDDY-CURRENT LOSS, HYSTERESIS LOSS, SKIN EFFECT, TRANSFORMER EFFICIENCY.

TRANSFORMER PRIMARY

The primary winding of a transformer is the winding to which a voltage or signal is applied from an external source. In a step-down transformer, the primary winding has more turns than the secondary. In a step-up transformer, the primary has fewer turns than the secondary (*see* STEP-DOWN TRANSFORMER, STEP-UP TRANSFORMER, TRANSFORMER SECONDARY).

Some transformers have center-tapped primary windings. This type of transformer is used in the output of push-pull circuits (*see* PUSH-PULL CONFIGURATION) and in various amplifiers and oscillators. A center-tapped primary may be used for obtaining a balanced input for an audio-frequency or radio-frequency amplifier.

The size of the wire used for a primary winding depends on the amount of current it must carry. For a given amount of power, a step-up transformer must use larger primary-winding wire than a step-down transformer requires. *See also* TRANSFORMER.

TRANSFORMER SECONDARY

The secondary winding of a transformer is the winding from which the output is taken. In a step-up transformer, the secondary winding has more turns than the primary winding; in a step-down transformer, the secondary has fewer turns (*see* STEP-DOWN TRANSFORMER, STEP-UP TRANSFORMER, TRANSFORMER PRIMARY).

The secondary winding of a transformer may have one or more taps for obtaining different voltages or impedances. A center-tapped secondary winding is used in full-wave rectifier circuits, and in oscillators or amplifiers having balanced outputs (*see* BALANCED OUTPUT, FULL-WAVE RECTIFIER). A tapped secondary is used in the input transformer of a balanced modulator, push-pull amplifier, or push-push circuit (*see* BALANCED MODULATOR, PUSH-PULL CONFIGURATION, PUSH-PUSH CONFIGURATION).

The size of the wire in the secondary winding of a transformer depends on the turns ratio. A step-up transformer can have smaller secondary-winding wire than a step-down transformer for a given amount of power. *See also* TRANSFORMER.

TRANSIENT

A transient is a sudden, very brief surge of high voltage on a utility power line. Normally, the peak voltage at a 117-V root-mean-square (RMS) household utility outlet is about 165 V. However, lightning or arcing can result in momentary surges of several hundred volts (see illustration), because of the electromagnetic pulse generated by such phenomena.

Transients, or "spikes," do not normally affect ordinary appliances, but they can cause damage to some types of electronic devices that are not properly protected. Transients may also blow fuses and trip circuit breakers for no apparent reason. Transients are quite common, and all solid-state electronic equipment should have some sort of

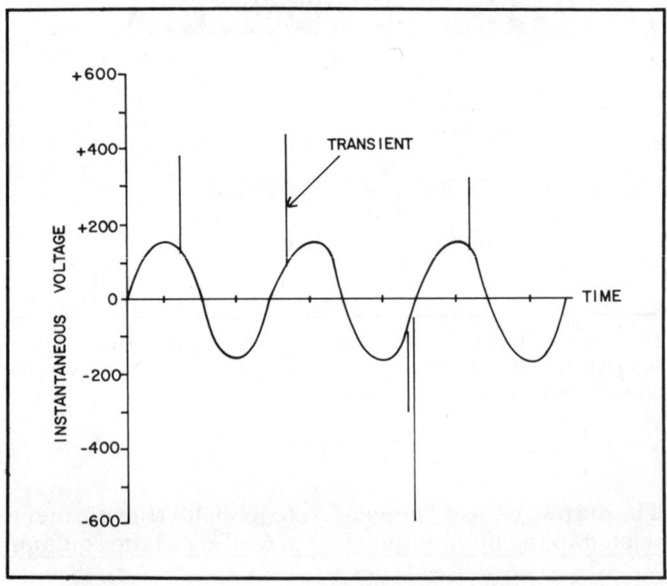

TRANSIENT: Transients may reach instantaneous peak values of several hundred volts on a household power line.

protection circuit to reduce the chances of damage.

A transient is not the same as a surge. A surge results from a sudden demand for current. *See also* ELECTROMAGNETIC PULSE, SURGE, TRANSIENT PROTECTION.

TRANSIENT PROTECTION

Certain types of electronic apparatus, especially solid-state equipment, can be damaged by transients (*see* TRANSIENT). Some form of protection should be provided for such devices.

The simplest method of transient protection is the connection of a moderate-sized capacitor (approximately 0.01 uF) in parallel with the ac power line. A more effective method is the installation of moderate-sized chokes (approximately 10 mH) in series with the lines, in addition to the capacitor. The most effective transient protection is afforded by a thyrector (*See* THYRECTOR).

No form of transient protection is 100-percent reliable. A nearby lightning strike can result in a transient of such severity that the protection device, as well as the equipment connected to the power line, may be destroyed. Fortunately, transients of such magnitude do not occur very often.

TRANSISTOR

A transistor is a semiconductor device consisting of three or four sections, or layers, of N-type and P-type material (*see* N-TYPE SEMICONDUCTOR, P-TYPE SEMICONDUCTOR). There are many types of transistors, manufactured for various applications. Transistors are used as oscillators, amplifiers, and switches at frequencies ranging from direct-current to ultra-high.

Transistors were invented by scientists at the Bell Laboratories in the late 1940s. Their discovery has revolutionized electronics. Ultimately, it led to miniaturization, making possible the computers and handheld calculators that we have today.

Types of Transistors. There are two main types of transistors: the bipolar device and the field-effect device. Bipolar transistors are classified as NPN or PNP (*see* NPN TRANSISTOR, PNP TRANSISTOR). Field-effect transistors may be either N-channel or P-channel (*see* FIELD-EFFECT TRANSISTOR).

Bipolar transistors are made from germanium or silicon; field-effect transistors may be fabricated from germanium, silicon, or metal oxides (*see* METAL-OXIDE-SEMICONDUCTOR FIELD-EFFECT TRANSISTOR). Methods of construction vary considerably, depending on the application for which the device is intended.

The photograph illustrates two small transistors that are commonly used in miniaturized equipment. (A dime is placed next to the transistors for size comparison.) The device with the metal case and four leads is a dual-gate, metal-oxide-semiconductor field-effect transistor (MOSFET). It can be used as a radio-frequency amplifier or mixer in receiving circuits. The device with the black case is an NPN bipolar transistor, suitable for digital switching or amplification at low to medium frequencies. These are just two examples among many different kinds of field-effect or bipolar transistors.

Some transistors are massive, and are capable of handling hundreds or even thousands of watts at audio and radio frequencies. These transistors are of the bipolar type, usually NPN, and are called power transistors (*see* POWER TRANSISTOR). Other devices are far smaller than the ones shown, consisting of microminiature silicon chips that are soldered, welded, or etched onto an integrated circuit (*see* INTEGRATED CIRCUIT, THICK-FILM AND THIN-FILM INTEGRATED CIRCUITS).

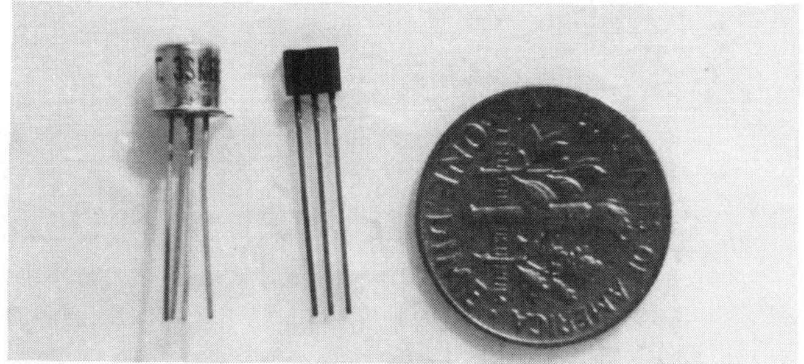

TRANSISTOR: Transistors are often small, such as the ones shown here (note the dime for size comparison). This has contributed to miniaturization in electronic circuits.

Biasing and Operation. Bipolar transistors consist essentially of two diodes, connected in reverse series, and having a common center element. This results in variable resistance through the device. An electrode is attached to each of the three semiconductor wafers. The electrode at the center is called the base, and the outer two electrodes are called the emitter and collector (*see* BASE, COLLECTOR, EMITTER). A source of direct-current voltage, generally between 3 and 30 V, is applied to the emitter and collector. In the NPN transistor, the collector is positive with respect to the emitter; in the PNP device, the emitter is positive with respect to the collector. The base is supplied with a voltage somewhere in between the emitter and collector potentials in most applications. Sometimes the base is made negative with respect to the emitter in the NPN transistor, or positive in the PNP device. The base-emitter voltage depends on the class of amplification desired (*see* CLASS-A AMPLIFIER, CLASS-AB AMPLIFIER, CLASS-B AMPLIFIER, CLASS-C AMPLIFIER).

The signal is applied either in series with the emitter, or between the emitter and the base. The resulting fluctuations in the emitter-base current cause large changes in the collector current. This is how a transistor amplifies: It produces large signal changes as a result of small ones. The base draws some current, and the input impedance of a bipolar transistor is therefore usually low to moderate.

The field-effect transistor is constructed somewhat differently. A single layer of semiconductor material, called the channel, is surrounded by sections of the opposite type semiconductor. The outer sections are called the gates. An electrode is attached to each end of the channel; one of these is the drain and the other is the source. The gates may be tied together, in which case they are attached to a single electrode; or they may be separated and connected to different electrodes (*see* CHANNEL, DRAIN, GATE, SOURCE). If the channel is fabricated from N-type material, the device is called an N-channel field-effect transistor. If the channel is made from P-type material, the device is a P-channel field-effect transistor.

In the N-channel device, the drain is biased positively with respect to the source; in the P-channel field-effect transistor, the polarity is reversed. The gate is biased at

TRANSISTOR BATTERY: A 9-V transistor battery, next to a dime for size comparsion.

some intermediate potential, or may be negative in the N-channel device or positive in the P-channel device. The gate bias depends on the class of amplification for which the field-effect transistor is to be used.

The signal is applied either in series with the source, or between the gate and the source. Small changes in the input signal cause large fluctuations in the current through the channel. (The principle of operation is similar to that of a water faucet.) The gate draws almost no current. Therefore, the input impedance is high. In metal-oxide-semiconductor devices, the input impedance is extremely large. This makes the field-effect transistor especially useful as a weak-signal amplifier.

Characteristics. Important characteristics of bipolar transistors include the alpha, the alpha-cutoff frequency, the beta, and the beta-cutoff frequency. In the field-effect transistor, the principal characteristic is the transconductance. All of these parameters are related to the gain. *See* ALPHA, ALPHA-CUTOFF FREQUENCY, BETA, BETA-CUTOFF FREQUENCY, TRANSCONDUCTANCE.

For More Information. Various types of bipolar and field-effect transistors and their construction, operation, and uses are discussed in more detail in other articles of this book. *In addition to the subjects referenced above, related articles in this book include* CHARACTERISTIC CURVE, COMMON CATHODE/EMITTER/SOURCE, COMMON GRID/BASE/GATE, COMMON PLATE/COLLECTOR/DRAIN, COMPLEMENTARY METAL-OXIDE SEMICONDUCTOR, CUTOFF, CUTOFF VOLTAGE, DOPING, EPITAXIAL TRANSISTOR, GALLIUM-ARSENIDE FIELD-EFFECT TRANSISTOR, IMPURITY, OPERATIONAL AMPLIFIER, PINCHOFF, PINCHOFF VOLTAGE, P-N JUNCTION, SEMICONDUCTOR, SEMICONDUCTOR DEVICE, SOLID-STATE ELECTRONICS, SWITCHING TRANSISTOR, VERTICAL METAL-OXIDE-SEMICONDUCTOR FIELD-EFFECT TRANSISTOR.

TRANSISTOR BATTERY

A small battery, supplying 9 V and used in various low-current transistorized circuits, is called a transistor battery. The transistor battery has a rectangular case, measuring about 1.8 × 1.0 × 0.6 in (46 × 26 = 15 mm). The contacts are brought out at one end, as shown in the photograph. (A dime is placed next to the battery for size comparison.)

Transistor batteries are fabricated from six miniature dry cells, generally either alkaline type or zinc type. *See also* ALKALINE CELL, BATTERY, CELL, ZINC CELL.

TRANSISTOR TESTER

A transistor tester is a device that quantitatively determines whether or not a transistor is functioning properly, and if so, gives a quantitative indication of the operating characteristics.

Sophisticated transistor testers are commercially available for checking the performance of bipolar and field-effect transistors. Such devices measure the alpha and beta of a bipolar device and the transconductance of a field-effect device (*see* ALPHA, BETA, TRANSCONDUCTANCE, TRANSISTOR).

If it is suspected that a bipolar transistor has been destroyed, an ohmmeter can be used for a quick check. All ohmmeters produce some voltage at the test terminals. The ohmmeter should be switched to the lowest resistance (R×1) scale, and the polarity of the test leads determined by means of an external voltmeter. Designate P+ as the lead at which a positive voltage appears, and P– the lead at which negative voltage appears. (Note: In a volt-ohm-milliammeter, the ohmmeter-mode polarity may not correspond to the red/black lead colors used for measurement of voltage and current!) To test an NPN transistor:

- Connect P+ to the collector and P– to the emitter. The resistance should appear infinite.
- Connect P+ to the emitter and P– to the collector

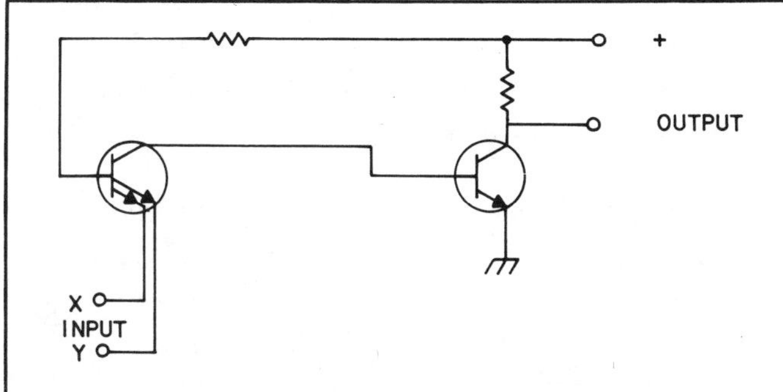

TRANSISTOR-TRANSISTOR LOGIC.

The resistance should again appear infinite.

- Connect P+ to the emitter and P– to the base. The resistance should appear infinite.
- Connect P+ to the base and P– to the emitter. The resistance should appear finite and measurable.
- Connect P+ to the collector and P– to the base. The resistance should appear infinite.
- Connect P– to the collector and P+ to the base. The resistance should appear finite and measurable.

If any of the above conditions is not met, the transistor is suspect.

To check a PNP transistor, the procedure is just the same as above, except that every "P+" should be changed to "P–," and every "P–" should be changed to "P+." Again, if any of the above conditions is not met, the transistor is suspect.

TRANSISTOR-TRANSISTOR LOGIC

Transistor-transistor logic, abbreviated TTL, is a bipolar logic design in which transistors act on direct-current pulses. Transistor-transistor logic is similar to diode-transistor logic, except that a multiple-emitter transistor replaces several diodes.

Transistor-transistor logic is characterized by high speed and good noise immunity. For these reasons, TTL is widely used in digital applications.

The illustration is a schematic diagram of a simple TTL gate. Many TTL logic gates can be fabricated into a single integrated-circuit package. Transistor-transistor logic can be used for either negative or positive logic circuits. *See also* DIODE-TRANSISTOR LOGIC, DIRECT-COUPLED TRANSISTOR LOGIC, EMITTER-COUPLED LOGIC, HIGH-THRESHOLD LOGIC, INTEGRATED INJECTION LOGIC, METAL-OXIDE-SEMICONDUCTOR LOGIC FAMILIES, NEGATIVE LOGIC, POSITIVE LOGIC, RESISTOR-CAPACITOR-TRANSISTOR LOGIC, RESISTOR-TRANSISTOR LOGIC, TRIPLE-DIFFUSED EMITTER-FOLLOWER LOGIC.

TRANSITION ZONE

The electromagnetic field surrounding a transmitting antenna can be divided into three distinct zones. They are known as the far field, the near field, and the transition zone. The far field exists at a great distance from the antenna, and is sometimes called the Fraunhofer region. The near field occurs very close to the transmitting antenna; it is sometimes called the Fresnel zone. The region between the far field and the near field is known as the transition zone.

The far field, near field, and transition zone are not precisely defined. They exist in qualitative form. In general, the transition zone occurs at distances where the electric and magnetic lines are somewhat curved. The lines of flux are decidedly curved in the near field, but they are practically straight in the far field. In communications practice, the far field is the only important consideration; the near field and transition zone fall within a few wavelengths of the radiating element. *See also* FAR FIELD, NEAR FIELD.

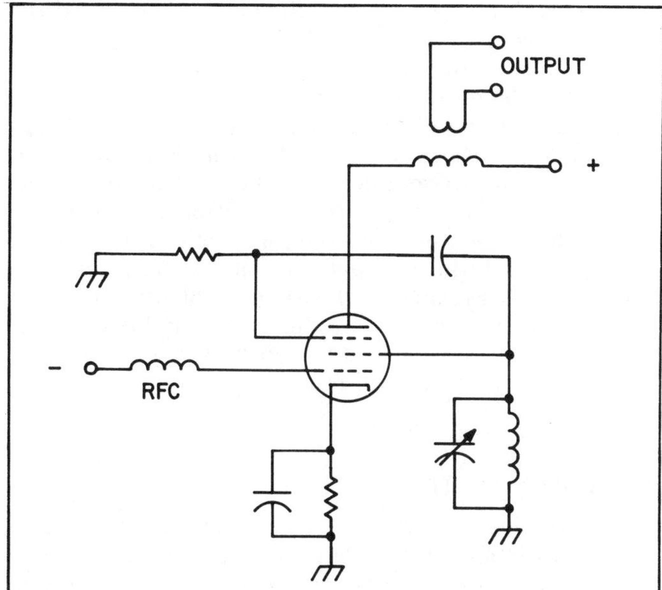

TRANSITRON OSCILLATOR: A transitron oscillator uses a pentode vacuum tube to obtain isolation in the output circuit.

TRANSITRON OSCILLATOR

A transitron oscillator is a vacuum-tube circuit that operates on a principle similar to an electron-coupled oscillator. A pentode vacuum tube is used. The control grid is kept at a fixed bias, and the output is taken from the plate. The tuned circuit is connected between the screen and suppressor grids as shown in the illustration. As currents circulate in the tank circuit, the plate current varies.

The transitron oscillator is less affected than most other types of oscillators by changes in the load impedance. *See also* ELECTRON-COUPLED OSCILLATOR.

TRANSIT TIME

Transit time is the time required for a charge carrier to get from one place to another. Transit time is of significance in the operation of vacuum tubes and semiconductor devices.

In a tube, the electrons travel from the cathode to the plate through a vacuum or near vacuum. Although the electrons move very fast, they require a small amount of transit time to cross the gap. The transit time depends on the type and size of the tube, the plate voltage, and the current drawn.

In a semiconductor diode or bipolar transistor, electrons and holes move and diffuse through the wafers. The transit time is the effective time required for a charge carrier to get to the anode from the cathode in a diode, or from the emitter-base junction to the collector-base junction in a bipolar transistor. In a field-effect transistor, the transit time is the time required for the electrons or holes to pass through the channel. Transit time in a semiconductor device is related to the carrier mobility, which depends on the type of semiconductor material and the extent of doping. *See also* CARRIER MOBILITY, FIELD-EFFECT TRANSISTOR, TRANSISTOR, TUBE.

TRANSLATION

The conversion of data from one language into another is called translation. A translation is a definable, constant mathematical function between two sets of data. Although the two sets may differ greatly in format and speed, the translation preserves all of the information without altering the content or meaning.

Encoding and decoding are forms of translation; most often, however, the term is used to denote the changing of languages in, or by, a computer. The assembler, for example, is responsible for translating assembly language into machine language. A modem translates between printed symbols and audio-frequency tones. An analog-to-digital or digital-to-analog converter is a form of translator.

Computers are extensively used for translating data to achieve compatibility between or among systems. This allows one computer to "talk to" another, even if they use different languages. The computer that performs the data conversion is called the translator or the interface computer. The operation of the interface computer is governed by a special program called the translator program. *See also* ANALOG-TO-DIGITAL CONVERTER, ASSEMBLER AND ASSEMBLY LANGUAGE, DATA CONVERSION, DATA PROCESSING, DECODING, DIGITAL-TO-ANALOG CONVERTER, ENCODING, HIGHER-ORDER LANGUAGE, INTERFACE, MACHINE LANGUAGE, MODEM.

TRANSMATCH

The output impedance of a radio transmitter, or the input impedance of a receiver, should ideally be matched to the antenna-system impedance. This requires that the antenna system present a nonreactive load of a certain value, usually 50 or 75 ohms. Very few antenna systems meet this requirement, but almost any load impedance can be matched to a transmitter or receiver by means of a device called a transmatch.

A transmatch consists of one or more inductors and capacitors arranged so that they cancel any existing reactance in the antenna system, and convert the remaining resistance to the appropriate value. Transmatches vary in complexity from simple L or pi networks to sophisticated adjustable circuits.

The illustration shows four transmatch circuits. The arrangements at A and B are for use with unbalanced loads; the circuits at C and D are for use with balanced loads. Most transmatch circuits incorporate reflectometers in the input circuit to facilitate adjustment.

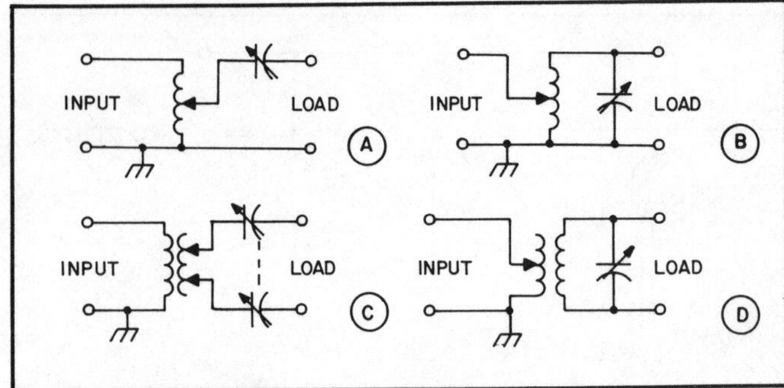

TRANSMATCH: Examples of transmatch circuits. At A, a circuit for unbalanced loads having low impedance; at B, a circuit for unbalanced loads of high impedance; at C, a circuit for balanced loads of low impedance; at D, a circuit for balanced loads of high impedance.

When a transmatch is used with a radio transmitter and receiver, adjustments should be made using an impedance bridge if possible. This eliminates the need for transmitting signals over the airwaves. If an impedance bridge is not available, transmatch adjustments should be performed according to the following procedure:

- Set the receiver and transmitter to the same frequency.
- Adjust the transmatch until the signals and/or noise are maximized in the receiver.
- Preset the transmitter tuning and loading controls to the approximate positions for the frequency in use.
- Switch the transmitter on, and apply a low-power signal to the transmatch.
- Adjust the transmatch for minimum standing-wave ratio (SWR) as indicated on a reflectometer placed between the transmitter and the transmatch.
- Retune the transmitter for normal operating output power.
- Identify your station.
- Record the positions of the transmatch controls for future reference.

Transmatches are extremely useful for portable station operation. A random-wire antenna can be strung in any fashion and tuned to resonance by using a transmatch. Generally, such antennas must measure at least ¼ electrical wavelength if good results are to be obtained, although some transmatches can provide an impedance match with much shorter wires.

A well-designed and properly operated transmatch acts as a bandpass filter. This provides additional front-end selectivity for a receiver, improving the image response and reducing the chances of front-end overload (*See* FRONT END, IMAGE REJECTION). The transmatch also attenuates spurious outputs from a transmitter, including the harmonics (*see* HARMONIC SUPPRESSION, SPURIOUS EMISSIONS).

TRANSMISSION LINE

A power generator is seldom located in the same place as the load or loads. Some method must be employed to provide power to distant loads. A transmission line is used for this purpose. A transmission line makes power available at distant points by means of electromagnetic-field propagation (*see* ELECTROMAGNETIC FIELD).

Transmission lines generally consist of one or two conductors in a balanced or unbalanced configuration. The most common type of balanced transmission line has two conductors running parallel to each other. The most common type of unbalanced line consists of a wire surrounded by a cylindrical shield (*see* BALANCED TRANSMISSION LINE, UNBALANCED TRANSMISSION LINE). In either the balanced or unbalanced configuration, the electric and magnetic components of the propagated field are always perpendicular to the line conductors. The result is that the electromagnetic field follows the transmission line.

In an ac power system, a transmission line is balanced. In a radio-frequency system, the transmission line is either balanced or unbalanced, depending on the load. A radio-frequency transmission line is often called a feed line, especially if the load is an antenna (*see* FEED LINE).

In order for a transmission line to provide the greatest possible available power to the load, certain conditions must be met: First, the load must not have reactance, but only resistance. Second, the characteristic impedance of the line must have the same ohmic value as the resistive impedance of the load. If these conditions are not met, the load will not convert all of the electromagnetic field into power. Some of the field will be reflected at the load, and sent back toward the generator. *See also* AC POWER TRANSMISSION, IMPEDANCE MATCHING, REFLECTED POWER.

TRANSMISSION SPEED

See SPEED OF TRANSMISSION.

TRANSMIT-RECEIVE SWITCH

When a transmitter and a receiver are used with a common antenna, some method must be devised to switch the antenna between the two units. It is especially important that the receiver be disconnected while the transmitter is operating. A transmit-receive (T-R) switch accomplishes this.

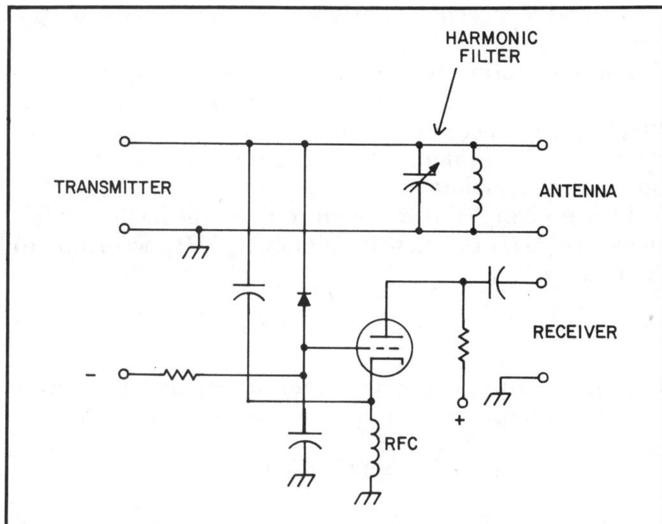

TRANSMIT-RECEIVE SWITCH: An example of a tube type T-R switch.

The simplest form of transmit-receive switch is a relay. When the transmitter is keyed, the relay connects the antenna to the transmitter. When the transmitter is unkeyed, the relay connects the antenna to the receiver. This type of arrangement is used in many transceivers.

A more sophisticated type of T-R switch consists of a radio-frequency-actuated electronic circuit. The transmitter is always connected to the antenna, even while receiving. But when the transmitter is keyed, the receiver is disconnected. The schematic diagram shows this kind of T-R switch.

The T-R switch shown allows full break-in operation if A1 emission (continuous-wave Morse code) is used. *See also* BREAK-IN OPERATION.

TRANSMITTANCE

Transmittance is an expression of the extent to which a substance or circuit passes energy. Transmittance is defined as a ratio between 0 and 1, or between 0 percent and 100 percent. In radio-frequency circuits, transmittance is sometimes called transmittivity.

If e1 represents the incident energy and e2 represents the energy transmitted through a substance or circuit, the transmittance T is given in percent by:

$$T = 100e2/e1$$

A transmittance value of 0 indicates that a material is perfectly opaque, or that all of the available power is absorbed by a circuit. A transmittance figure of 100 percent indicates that a substance is perfectly transparent, or that none of the power is absorbed by a circuit.

TRANSMITTER

A transmitter is a device that produces a signal for broadcasting or communications purposes. The signal may consist of an electric current, radio waves, light, ultrasound, or any other form of energy. A transmitter converts audio and/or video information into a signal that can be sent to a distant receiver (*see* RECEIVER).

A basic transmitter consists of an oscillator, a transducer, a modulator, and a signal amplifier (see illustration). The amplifier output is connected to a wire transmission line or an antenna system. The oscillator provides the carrier wave. The transducer converts audio and/or video information into electrical impulses. The modulator impresses the output of the transducer onto the carrier wave. The amplifier boosts the signal level to provide sufficient power for transmission over the required distance.

Many transmitters incorporate more sophisticated designs than the basic scheme shown. For example, mixers are often used to obtain multiband operation; several different modulators may be used for subcarrier operation. The design of a transmitter depends to some extent, therefore, on the application. *See also* AMPLITUDE MODULATION, CARRIER-CURRENT COMMUNICATION, CODE TRANSMITTER, FREQUENCY MODULATION, MODULATED LIGHT, MODULATOR, OPTICAL COMMUNICATION, OSCILLATOR, PHASE MODULATION, POWER AMPLIFIER, PULSE MODULATION, SINGLE SIDEBAND.

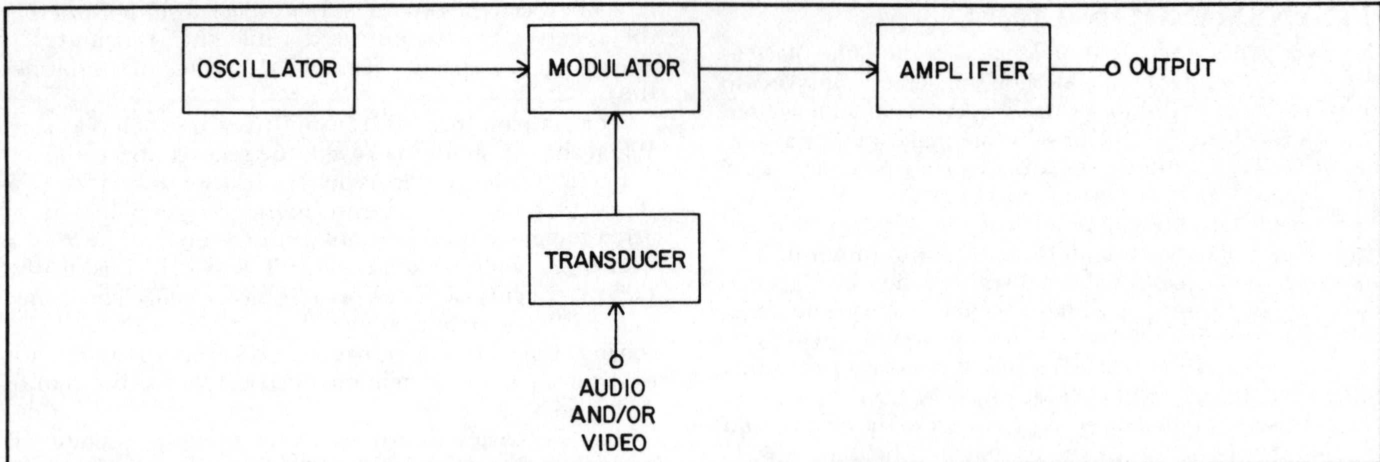

TRANSMITTER: Block diagram of a radio transmitter. This is a basic circuit; many transmitters are more complex.

TRANSMITTING TUBE

A vacuum tube, designed for operation in a radio-frequency power amplifier, is sometimes called a transmitting tube. Transmitting tubes are characterized by their ability to dissipate large amounts of power.

Transmitting tubes are physically larger than receiving tubes, simply because the high power level dictates that the plate and cathode be large. Some transmitting tubes have finned plates to maximize the amount of power they can dissipate. Others have graphite plates for the same reason.

Although tubes have been largely replaced by transistors in recent years, tubes are still used in some transmitters. This is especially true of high-power circuits operating at microwave frequencies. Tubes are also fairly tolerant of momentary excessive power dissipation. *See also* TUBE.

TRANSMITTIVITY

See TRANSMITTANCE.

TRANSPONDER

A transponder is a device that sends a radio signal whenever it receives a certain command from a distant station. The command signal is called an interrogation. The signal sent out by the transponder is called the response.

Transponders are used for radionavigation, especially in aviation. A transponder aboard an aircraft allows continuous monitoring of the aircraft position by ground-based personnel. A ground-based interrogator, similar to a radar set, transmits pulses at a rate of 400 Hz. Transponders aboard aircraft reply at the same rate. The interrogator determines the position of each aircraft just as a radar does. The transponder reply pulses contain information about the altitude and identity of each aircraft. *See also* AIR-TRAFFIC CONTROL, RADAR.

A repeater aboard an orbiting satellite is sometimes called a transponder. Such a device receives signals in one band of frequencies, converts them to another band of frequencies, and retransmits them. *See* ACTIVE COMMUNICATIONS SATELLITE, REPEATER, SATELLITE COMMUNICATIONS.

TRANSVERSE WAVE

A transverse wave is a disturbance in which the displacement is perpendicular to the direction of travel. All electromagnetic waves are transverse waves, since the electric and magnetic lines of flux are perpendicular to the direction of propagation. In some substances, sound propagates as transverse waves.

The waves on a lake or pond illustrate the principle of transverse disturbances. When a pebble is dropped into a mirror-smooth pond, waves form and are propagated outward. The effects of the disturbance travel at a defined and constant speed. The waves have measurable length and amplitude. The actual water molecules, however, move up and down, at right angles to the wave motion.

Some waves are propagated by movement of particles in directions parallel to the wave motion. This kind of disturbance is called a longitudinal wave. *See also* LONGITUDINAL WAVE.

TRANSVERTER

A transverter is a device that allows operation of a transceiver on a frequency much different from the design frequency. A transverter consists of a transmitting converter and a receiving converter in a single package (*see* CONVERTER).

The illustration is a block diagram of a hypothetical transverter. The transceiver operates in the band 28.0 to 30.0 MHz. The actual operating frequency band is 144.0 to 146.0 MHz. A common local oscillator at 116.0 MHz provides the conversion.

The receiving converter heterodynes the incoming frequency fr_1 MHz to a new frequency fr_2 MHz according to the relation:

$$fr_2 = fr_1 - 116.0$$

The transmitting converter heterodynes the transceiver output frequency ft_1 to a new frequency ft_2 according to:

$$ft_2 = 116.0 + ft_1$$

Transverters are commercially manufactured for use with most kinds of transceivers. *See also* MIXER, RECEIVER, TRANSCEIVER, TRANSMITTER.

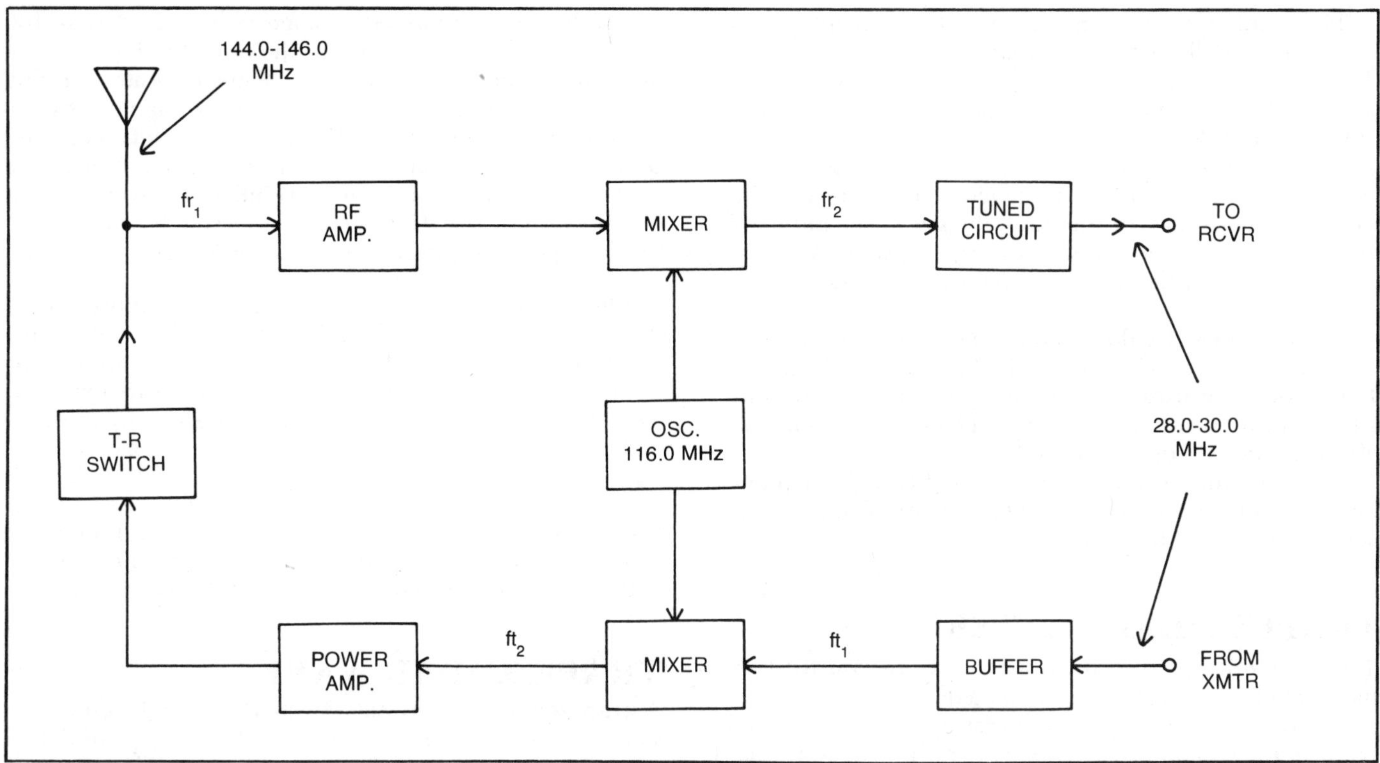

TRANSVERTER: Block diagram of a transverter for operation of a 28.0-30.0-MHz transceiver in the band 144.0-146-0 MHz. A common local oscillator is used.

TRAP

A trap is a form of band-rejection filter, designed for blocking energy at one frequency while allowing energy to pass at all other frequencies. A trap consists of a parallel-resonant circuit in series with the signal path (see illustration A) or a series-resonant circuit in parallel with the signal path, as at B. The configuration at A is the more common.

Traps can be constructed from discrete inductors and capacitors, as shown, or they can be fabricated from sections of transmission line. In the latter case, the trap is usually called a stub (*see* STUB).

Traps are used extensively in the design and construction of multiband antenna systems. *See also* TRAP ANTENNA.

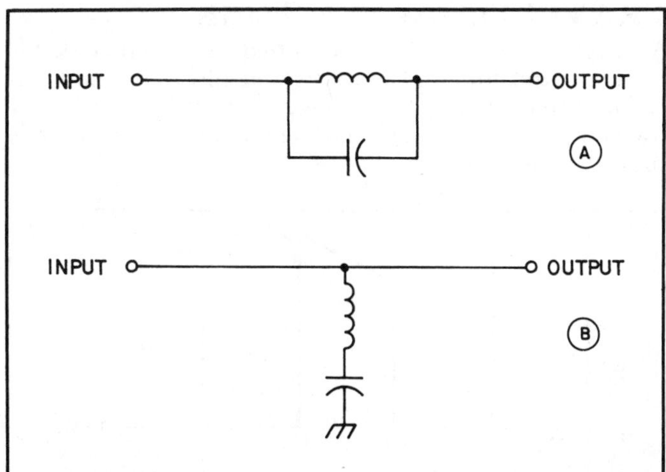

TRAP: Trap circuits. At A, a parallel-resonant circuit connected in series with the signal path; at B, a series-resonant circuit connected in parallel with the signal path.

TRAP ANTENNA

A trap antenna is a form of multiband antenna in which traps are used to obtain resonance on two or more different frequencies (*see* TRAP). The traps are placed at certain points along the radiating element and parasitic elements (if any), as shown in the illustration.

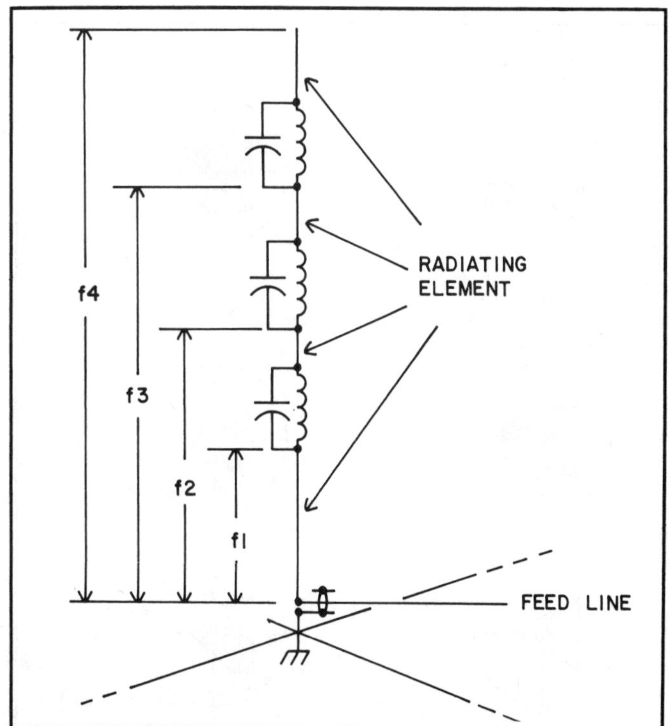

TRAP ANTENNA: A three-trap antenna. Various portions of the radiating element, as shown, provide resonance at four frequencies (f1, f2, f3, and f4).

The number of frequencies at which a trap antenna is resonant is usually one greater than the number of traps in the radiating element. For example, an antenna with three traps will be resonant on four frequencies. This is the case in the example shown.

The design of a trap antenna is rather complicated, because several interacting parameters are involved: the spacing and positions of the traps, the size of the inductor in each trap, and the number of traps. The greater the number of traps, the more complicated the antenna design.

Antenna traps are designed so that they have the least possible amount of loss, and the highest possible Q factor. This involves the use of large wire for the coil of a trap, and a low-loss, high-voltage capacitor. The capacitor may be placed inside or outside the coil.

Trap antennas, while commonly used at high frequencies, are not the only means of obtaining multiband operation. *See also* MULTIBAND ANTENNA.

TRAPEZOIDAL PATTERN

The percentage of modulation of an amplitude-modulated signal can be determined using an oscilloscope, connected in such a way that a trapezoid-shaped pattern appears on the screen. The radio-frequency signal is applied to the vertical deflection plates of the oscilloscope, and the modulating audio signal is applied to the horizontal deflection plates (see illustration A).

The modulation percentage is determined according to how much the shape of the trapezoid differs from a perfect rectangle. If there is no modulation, the pattern appears as a perfect rectangle. As the modulation percentage increases, the pattern becomes more and more distorted until, with 100-percent modulation, it appears as a triangle.

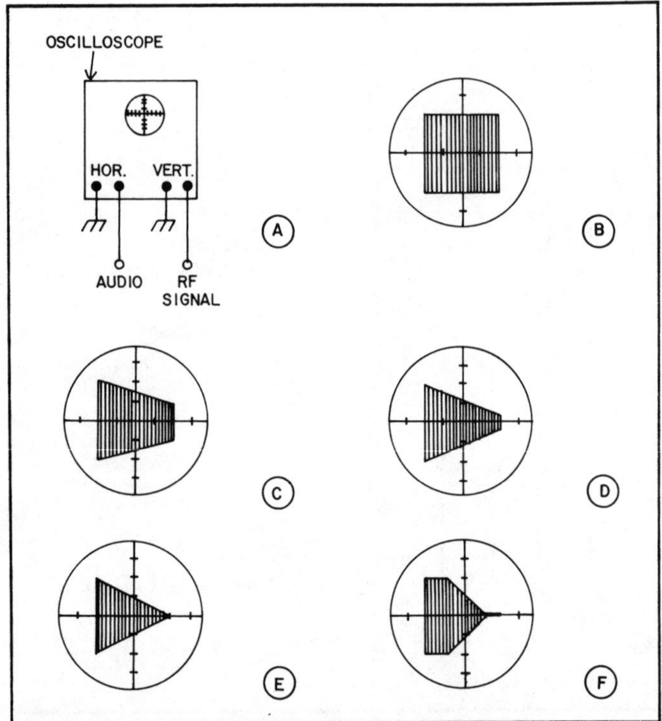

TRAPEZOIDAL PATTERN: At A, oscilloscope connection for obtaining trapezoidal patterns. At B through F, trapezoidal patterns for various percentages of amplitude modulation (see text).

As the modulation percentage is increased past 100 percent, the triangle shrinks in the horizontal dimension, and a line appears at the right. The line indicates clipping of the negative peaks of the modulated signal. Positive peak clipping shows up as flattening of the left-hand end of the triangle. Examples of trapezoidal patterns are shown at B, C, D, E, and F for modulation percentages of 0, 33, 67, 100, and 150 percent respectively. At F, both negative peak clipping and positive peak clipping are present.

The modulation percentage can be determined mathematically from the lengths of the vertical sides of the trapezoid. If the long side measures L graduations on the oscilloscope screen and the short side measures S divisions, then the modulation percentage, m, is given by the formula:

$$m = 100(L-S)/(L+S)$$

within the limits of 0 to 100 percent. The formula does not hold for modulation percentages over 100. *See also* AMPLITUDE MODULATION, MODULATION PERCENTAGE.

TRAPEZOIDAL WAVE

A trapezoidal wave results from the combination of a square wave and a sawtooth wave, having identical frequencies but perhaps different amplitudes. The trapezoidal wave gets its name from the fact that its shape somewhat resembles a trapezoid.

The period of a trapezoidal wave is the length of time from any point on the waveform to the same point on the next pulse in the train. This period is divided into 360 degrees of phase, with the zero point usually corresponding to the moment at which the amplitude is zero and rapidly changing (see illustration). The frequency of the trapezoidal wave, in hertz, is the reciprocal of the period in seconds.

In a cathode-ray tube, the voltage applied to the deflecting coils always has a trapezoidal waveshape. This provides the necessary sawtooth-wave variation in the currents through the coils, and thus in the magnetic fields, ensuring a linear forward sweep and a fast return sweep. *See also* CATHODE-RAY TUBE, SAWTOOTH WAVE.

TRAVELING-WAVE TUBE

A traveling-wave tube is a specialized vacuum tube that is used in oscillators and amplifiers at ultra-high and microwave frequencies. There are various different types of traveling-wave tubes. The most common configuration is shown in the illustration.

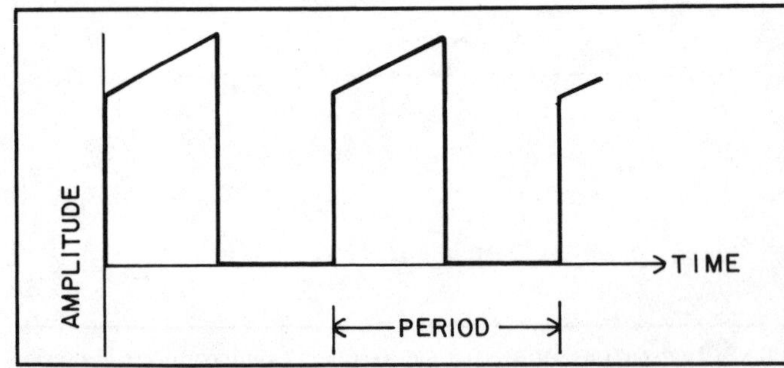

TRAPEZOIDAL WAVE: Combination of square and sawtooth waveforms.

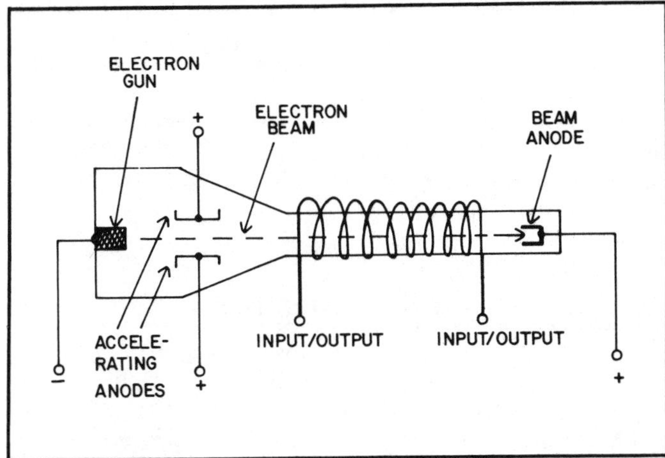

TRAVELING-WAVE TUBE: Diagram of a traveling-wave tube.

The electron gun produces a high-intensity beam of electrons that travels in a straight line to the anode. A helical conductor is wound around the tube. The distributed inductance and capacitance of this winding result in a very low velocity factor (*see* VELOCITY FACTOR), typically from 10 to 20 percent, approximately matching the speed of the electrons in the beam inside the tube. A signal is applied at one end of the helix and is taken from the other end.

When the helical winding is energized, the electron beam inside the tube is phase-modulated. Some of the electrons travel in synchronization with the wave in the helix, because of the low velocity factor of the winding. This produces waves in the electron beam. The waves may travel in the same direction as the electrons (forward-wave mode) or in the opposite direction (backward-wave mode). In either case, energy from the electrons is transferred to the signal in the winding, producing gain when the beam voltage is within a certain range. Gain figures of 15 to 20 dB are common, and some traveling-wave tubes can produce more than 50 dB of gain.

A traveling-wave tube can be used to produce energy at ultra-high and microwave frequencies by coupling some of the output back into the input. This type of oscillator is called a backward-wave oscillator, because the feedback is applied opposite to the direction of movement of the electrons inside the tube. The backward-wave oscillator produces about 20 to 100 mW of radio-frequency power at frequencies up to several gigahertz. The backward-wave arrangement can also be used for amplification, but the most common traveling-wave amplifier configuration is the parametric amplifier, which uses the forward-wave mode. *See also* PARAMETRIC AMPLIFIER.

TREBLE

The term treble refers to high-frequency sound energy. On a piano or musical staff, any note at or above middle C (261.6 Hz) is considered to be in the treble clef. Any note below middle C is in the bass clef (*see* BASS).

A tone control in a high-fidelity sound system can have separate bass and treble adjustments, or it may consist of a single knob or slide type potentiometer. The amount of treble, especially at frequencies above 1000 Hz, affects the "crispness" of musical sound. Too little treble results in a muffled quality, and too much treble makes music sound "tinny." *See also* AUDIO RESPONSE, TONE CONTROL, TREBLE RESPONSE.

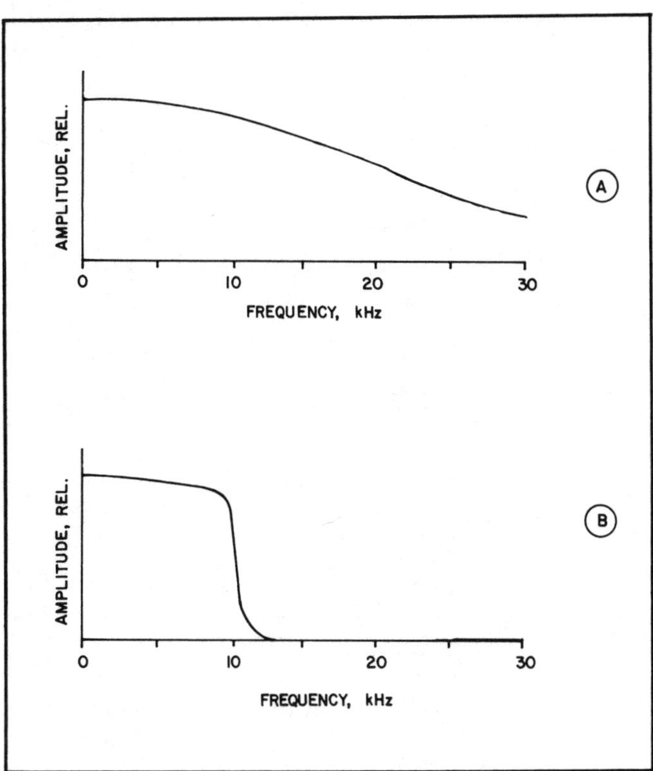

TREBLE RESPONSE: At A, a good treble response for music reproduction. At B, a poor response.

TREBLE RESPONSE

Treble response is the ability of a sound amplification or reproduction system to respond to treble audio-frequency energy. The treble response can be defined for microphones, speakers, radio receivers, record players, and recording tape—anything that involves the transmission of audio energy. In most high-fidelity sound systems, the treble response is adjustable by means of a tone control or a treble gain control (*see* TONE CONTROL).

A good treble response involves more than the simple ability of a sound system to reproduce high-frequency audio. The sound will not be accurately reproduced unless the audio response is optimized (*see* AUDIO RESPONSE). Many systems have a gradual roll-off starting at 3-5 kHz (see illustration A). Some high-fidelity amplifiers have a flat response at frequencies as high as 40 kHz or more. If the gain drops off anywhere within the human hearing range, the quality of the reproduction will be degraded (B).

A device called an equalizer (*see* EQUALIZER) facilitates compensation for differences in sound-system response. The bass and treble response can be tailored for the best possible sound reproduction with such a device.

TRIAC

A triac is a semiconductor device that is used for controlling ac voltage. The triac is constructed in a manner similar to a PNP transistor, except that no distinction is made between the emitter and collector. A triac is identical to a diac, except that a control electrode is connected to one of the anodes via an N-type section of semiconductor. Illustration A shows the construction of the triac, and B illustrates the schematic symbol.

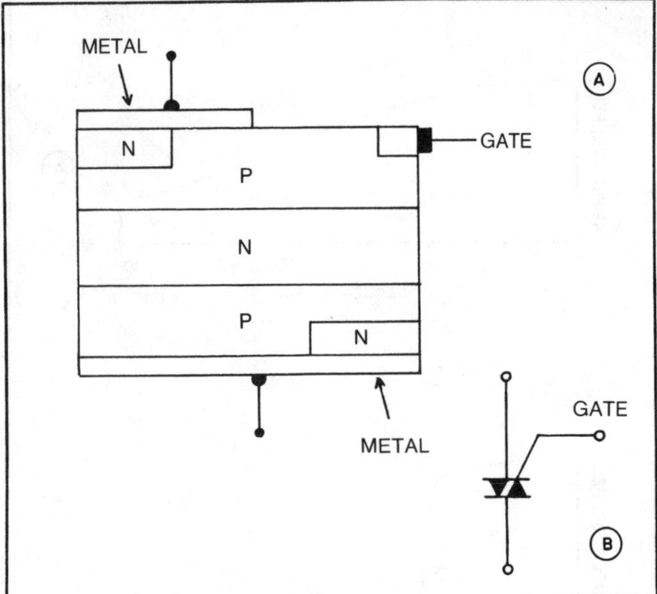

TRIAC: At A, the construction of a triac. At B, schematic symbol.

The triac is used as a light dimmer and motor-speed control in common household circuits. *See also* DIAC, MOTOR-SPEED CONTROL.

TRIANGULAR WAVE

A triangular wave is a form of sawtooth wave in which the rise and decay times are identical. The triangle wave is therefore symmetrical with respect to the amplitude peaks. *See* SAWTOOTH WAVE.

TRIAXIAL® SPEAKER

Speakers characteristically have limited frequency response. A speaker that responds well to bass sound will not work well at higher audio frequencies, for example. There are three basic types of speakers used in high-fidelity sound reproduction: the midrange speaker, the tweeter for extremely high audio frequencies, and the woofer for bass sounds (*see* MIDRANGE, TWEETER, WOOFER). High-fidelity speaker enclosures often contain one of each of these three types of speaker. The speakers are connected in parallel with a three-way crossover network. Such a set of speakers, and its enclosure is called a Triaxial speaker. (The term is a trademark of Jensen Sound Laboratories, a division of Pemcor, Inc.) It may also be called a three-way speaker.

The main advantage of a Triaxial speaker, or any similar combination of woofers, midrange speakers, and/or tweeters, is that the audio response range is optimized. A good two-way or three-way speaker will reproduce signals from well below to well above the human hearing range. A good set of speakers is necessary if excellent hi-fi sound is to be realized, no matter what kind of amplifier is used. *See also* AUDIO RESPONSE, CROSSOVER NETWORK, HIGH FIDELITY, SPEAKER.

TRIBOELECTRIC EFFECT

When certain substances are rubbed together, an electrostatic charge develops on each object. One object becomes positive because electrons get transferred to the other object, which becomes negative. This is called the triboelectric effect.

The most common example of the triboelectric effect is the transfer of charge from a carpeted floor to a person shuffling across the floor. *See also* STATIC ELECTRICITY, TRIBOELECTRIC SERIES.

TRIBOELECTRIC SERIES

A triboelectric series is a list of substances in order of their tendency to become positively or negatively charged as a result of triboelectric effect (*see* TRIBOELECTRIC EFFECT).

The table is an example of a triboelectric series. An object consisting of a given substance becomes negatively charged when rubbed against an object made from something higher on the list. The farther apart two substances appear on the list, the greater the tendency for transfer of charge.

Consider the following two examples, based on the table. When silk is rubbed against glass, the glass attains a positive charge and the silk attains a negative charge. This example is commonly used in elementary-school science classes to illustrate static electricity. The friction of rubber-soled shoes against a wool carpet results in a negative charge on the person wearing the shoes; this is a familiar annoyance in many regions in the wintertime. *See also* STATIC ELECTRICITY.

TRICKLE CHARGE

A rechargeable battery may be charged at a low current level or at a high current level. In general, a rechargeable cell or battery having a capacity of x ampere hours can be fully charged, starting at zero charge, with I amperes of current for t hours, such that $x = tI$. When I is very small, t must be large. Low-current, long-term charging is known as trickle charging.

Trickle charging is the best method for maintaining the energy in rechargeable cells and batteries. The nickel-cadmium memory backup batteries in some microcomputer-controlled devices are trickle-charged whenever power is on (*see* MEMORY BACKUP BATTERY). The nickel-cadmium battery packs in some handheld transceivers and small calculators are trickle-charged with plug-in transformer/rectifier devices.

When a battery is initially in a state of nearly complete discharge, there is not always time for trickle charging. Then quick charging is used. *See also* CHARGING, QUICK CHARGE.

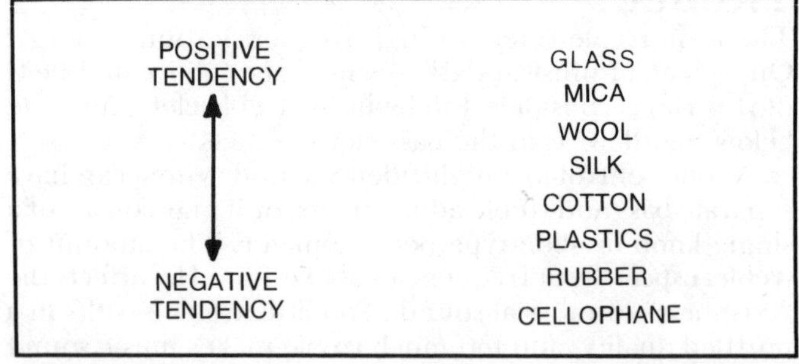

TRIBOELECTRIC SERIES: Substances near the top of the list tend to acquire positive charge; those near the bottom, negative charge.

TRIFILAR WINDING

A set of three coil windings, usually oriented in the same sense and having the same number of turns on a common core, is called a trifilar winding. Trifilar windings are sometimes used in transformers at audio or radio frequencies.

Trifilar windings on a toroidal core can be connected to obtain various different impedance-transfer ratios. This is useful in radio-frequency impedance-matching applications. *See also* TRANSFORMER.

TRIGGER

A trigger is a digital circuit that changes state or initiates conduction when a pulse is received. Trigger circuits may consist of relays, tubes, diodes, transistors, resistors, capacitors, and/or other components. *See also* FLIP-FLOP.

TRIGGERING

Triggering is a means of synchronizing the sweep rate of an oscilloscope with the frequency of the applied signal. Most oscilloscopes have triggering capability. Triggering keeps the display steady on the screen so that the waveform can be easily analyzed. Some oscilloscopes can be triggered according to a signal different from the incoming signal, or according to the 60-Hz power-line voltage.

When a triggered oscilloscope is used, the ratio of the sweep frequency to the signal frequency can be varied by adjusting the sweep rate. In this way, a single cycle can be viewed for "closeup" analysis, or many cycles can be viewed for determination of peak voltage, frequency, and general wave shape. *See also* OSCILLOSCOPE.

TRIGONOMETRIC FUNCTION

A trigonometric function, also called a circular function, is a mathematical function that relates angular measure to the set of real numbers. There are six trigonometric functions. The three most common ones are the cosine, sine, and tangent. Less well known are the cosecant, cotangent, and secant (*see* COSECANT, COSINE, COTANGENT, SECANT, SINE, TANGENT). The trigonometric functions are abbreviated in equations and formulas, as in the table.

TRIGONOMETRIC FUNCTION: ABBREVIATIONS FOR COMMON TRIGONOMETRIC FUNCTIONS.

cosine: cos	hyperbolic cosine: cosh
sine: sin	hyperbolic sine: sinh
tangent: tan	hyperbolic tangent: tanh
cosecant: csc	hyperbolic cosecant: csch
cotangent: cot	hyperbolic cotangent: coth
secant: sec	hyperbolic secant: sech

If we wish to specify the sine of a certain angle θ, then, we say sin θ.

Although trigonometric values can be expressed for negative angles and angles measuring 360 degrees or more, the domains of the trigonometric functions are usually restricted to angles θ, in degrees, such that $0 \leqslant \theta < 360$, or in radians such that $0 \leqslant \theta < 2\pi$. When the domain of a trigonometric function is deliberately restricted in this way, the first letter of the abbreviation is sometimes capitalized. (This semantical distinction is important to mathematicians, but not as important to engineers.)

All of the trigonometric functions can be expressed in either of two ways: the unit-circle model and the triangle model.

The Unit-Circle Model. Mathematicians usually express the trigonometric functions in terms of values on a circle in the Cartesian plane. The circle has a radius of 1 unit and is centered at the origin. (see illustration A). The equation of the circle is:

$$x^2 + y^2 = 1$$

If θ represents an angle, measured counterclockwise from the positive x axis, then the trigonometric-function values of this angle can be determined from the values of x and y

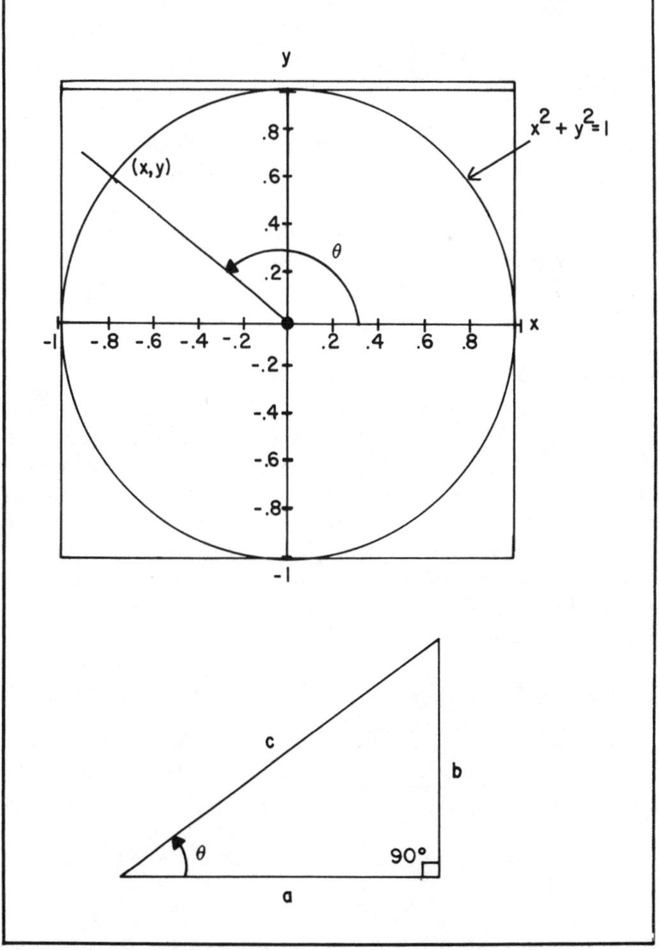

TRIGONOMETRIC FUNCTION: The trigonometric functions are defined according to a unit circle (A). For angular measures between 0 and 90 degrees, the trigonometric functions can be defined as ratios of the sides of a right triangle (B).

at the point (x,y) on the circle that corresponds to the angle θ, as shown. The functions are:

$$\begin{aligned} \cos\theta &= x \\ \sin\theta &= y \\ \tan\theta &= y/x \\ \csc\theta &= 1/y \\ \cot\theta &= x/y \\ \sec\theta &= 1/x \end{aligned}$$

A special category of trigonometric functions exists, based on the unit hyperbola having the equation:

$$x^2 - y^2 = 1$$

These are known as hyperbolic trigonometric functions (*see* HYPERBOLIC TRIGONOMETERIC FUNCTIONS).

The Triangle Model. A right triangle (B) can be used to express trigonometric-function values. The angle θ corresponds to one of the angles not measuring 90 degrees. If the sides of the triangle have lengths a, b, and c, as shown in the figure, and the angle θ is opposite the side of length b, then:

$$\begin{aligned} \cos\theta &= a/c \\ \sin\theta &= b/c \\ \tan\theta &= b/a \\ \csc\theta &= c/b \\ \cot\theta &= a/b \\ \sec\theta &= c/a \end{aligned}$$

These formulas are valid no matter how large or small the right triangle may be, in terms of the magnitudes of the numbers a, b, and c.

The triangle model is normally used only for angles θ measuring between 0 and 90 degrees.

Inverse Trigonometric Functions. When the domain of a trigonometric function is properly restricted, an inverse function exists that maps a real number into a unique angular value. This type of function is called an inverse trigonometric function. Each of the six trigonometric functions has an inverse. The inverse of a given function can be denoted by the prefix "arc–" or by an exponent −1. The inverse trigonometric functions are defined only when the angles are restricted to certain values, some of which are negative:

If $\cos\theta = z$, then $\arccos z = \theta$ for $0 \leqslant \theta \leqslant 180$

If $\sin\theta = z$, then $\arcsin z = \theta$ for $-90 \leqslant \theta \leqslant 90$

If $\tan\theta = z$, then $\arctan z = \theta$ for $-90 \leqslant \theta < 90$

If $\csc\theta = z$, then $\operatorname{arccsc} z = \theta$ for $-90 \leqslant \theta < 0$ and $0 < \theta \leqslant 90$

If $\cot\theta = z$, then $\operatorname{arccot} z = \theta$ for $0 < \theta < 180$

If $\sec\theta = z$, then $\operatorname{arcsec} z = \theta$ for $0 \leqslant \theta < 90$ and $90 < \theta \leqslant 180$

Significance and Uses of Trigonometric Functions. Trigonometric functions, especially the sine and cosine functions, are manifested in many different natural phenomena. These functions are representations of circular motion, which occurs frequently in the physical universe.

The most common example of a trigonometric function in electricity and electronics is the classical alternating-current wave. A wave disturbance, containing energy at only one frequency, can be perfectly represented by a sine or cosine function. The wave cycle is divided into 360 degrees of phase, beginning at the point where the amplitude is zero and increasing in a positive direction (*see* PHASE ANGLE, SINE WAVE). Trigonometric functions arise in certain antenna and feed-line calculations, frequency and phase modulation, and many other instances. Trigonometric functions are important in polar and spherical coodinate systems. *See also* COORDINATE SYSTEM, FUNCTION, INVERSE FUNCTION, POLAR COORDINATES, SPHERICAL COORDINATES, TRIGONOMETRIC IDENTITIES.

TRIGONOMETRIC IDENTITIES

In calculations involving trigonometric functions, it is often necessary to simplify an expression or to substitute one expression for another. Trigonometric identities are used for this purpose. A trigonometric identity is simply an equation that holds for all possible angles, or for angles within a specified range.

Table 1 lists common trigonometric identities for circular functions. Table 2 lists some identities that hold for inverse of circular functions. *See also* TRIGONOMETRIC FUNCTION.

TRIGONOMETRY

Trigonometry is a branch of mathematics that is concerned with angles and straight lines in a plane or in space. *See* COSECANT, COSINE, COTANGENT, SECANT, SINE, TANGENT, TRIGONOMETRIC FUNCTION, TRIGONOMETRIC IDENTITIES.

TRIMMER CAPACITOR

A trimmer capacitor is a small, variable capacitor used for the purpose of adjusting the characteristics of a tuned circuit. A trimmer capacitor can, for example, be placed in parallel with the fixed or variable capacitor in the tank circuit of a radio-frequency amplifier or oscillator. Trimmer capacitors are also used in filters and in other networks containing capacitance.

Trimmer capacitors are often used in the same manner as padder capacitors (*see* PADDER CAPACITOR). A trimmer capacitor contains two sets of plates separated by thin layers of dielectric material, usually mica. The plates are pressed together more or less tightly by means of an adjustable screw (see illustration). The whole assembly is mounted on a porcelain base.

The capacitance depends on how the screw is set. When the screw is turned clockwise, pressing the plates more tightly together, the capacitance increases. When the screw is turned counterclockwise, the plates move apart and the capacitance decreases.

TRIGONOMETRIC IDENTITIES: SOME COMMON TRIGONOMETRIC IDENTITIES FOR CIRCULAR FUNCTIONS.

$\csc\theta = 1/\sin\theta$
$\sec\theta = 1/\cos\theta$
$\tan\theta = \sin\theta/\cos\theta$
$\cot\theta = \cos\theta/\sin\theta$
$\sin^2\theta + \cos^2\theta = 1$
$\sec^2\theta - \tan^2\theta = 1$
$\csc^2\theta - \cot^2\theta = 1$
$\sin(\theta+\phi) = \sin\theta\cos\phi + \cos\theta\sin\phi$
$\sin(\theta-\phi) = \sin\theta\cos\phi - \cos\theta\sin\phi$
$\cos(\theta+\phi) = \cos\theta\cos\phi - \sin\theta\sin\phi$
$\cos(\theta-\phi) = \cos\theta\cos\phi + \sin\theta\sin\phi$
$\tan(\theta+\phi) = (\tan\theta + \tan\phi)/(1 - \tan\theta\tan\phi)$
$\tan(\theta-\phi) = (\tan\theta - \tan\phi)/(1 + \tan\theta\tan\phi)$
$\sin -\theta = -\sin\theta$
$\cos -\theta = \cos\theta$
$\tan -\theta = -\tan\theta$
$\csc -\theta = -\csc\theta$
$\sec -\theta = \sec\theta$
$\cot -\theta = -\cot\theta$
$\sin 2\theta = 2\sin\theta\cos\theta$
$\cos 2\theta = \cos^2\theta - \sin^2\theta$
$\tan 2\theta = 2\tan\theta/(1 - \tan^2\theta)$
$\sin\theta + \sin\phi = 2\sin(\theta/2 + \phi/2)\cos(\theta/2 - \phi/2)$
$\sin\theta - \sin\phi = 2\cos(\theta/2 + \phi/2)\sin(\theta/2 - \phi/2)$
$\cos\theta + \cos\phi = 2\cos(\theta/2 + \phi/2)\cos(\theta/2 - \phi/2)$
$\cos\theta - \cos\phi = 2\sin(\theta/2 + \phi/2)\sin(\theta/2 - \phi/2)$

Table 1

Trimmer capacitors are available in various sizes, ranging in maximum value from about 3 pF to 150 pF. Typical trimmer capacitors can handle around 100 V, although some can tolerate higher voltages. The frequency and loss characteristics are similar to those of silver-mica capacitors. *See also* CAPACITOR, SILVER-MICA CAPACITOR.

TRIODE TUBE

A triode tube is a vacuum tube having three elements.

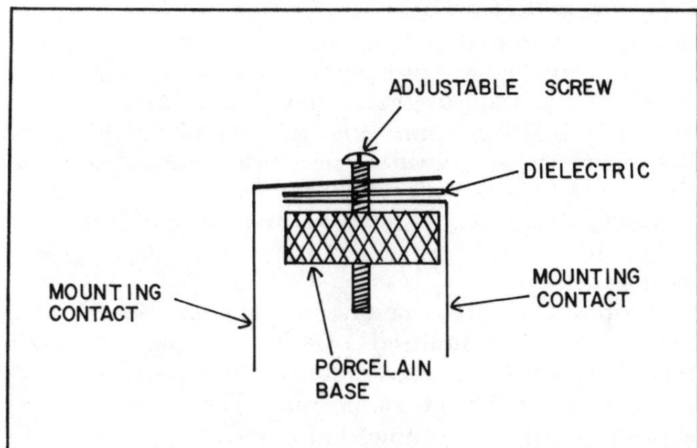

TRIMMER CAPACITOR: Construction of a trimmer capacitor.

TRIONOMETRIC IDENTITIES: SOME COMMON TRIGONOMETRIC IDENTITIES FOR INVERSES OF CIRCULAR FUNCTIONS.

$\arcsin z + \arccos z = 90°$
$\arctan z + \mathrm{arccot}\, z = 90°$
$\mathrm{arcsec}\, z + \mathrm{arccsc}\, z = 90°$
$\arcsin z = \mathrm{arccsc}\,(1/z)$
$\arccos z = \mathrm{arcsec}\,(1/z)$
$\arctan z = \mathrm{arccot}\,(1/z)$
$\arcsin(-z) = -\arcsin z$
$\arccos(-z) = 180° - \arccos z$
$\arctan(-z) = -\arctan z$
$\mathrm{arccot}(-z) = 180° - \mathrm{arccot}\, z$
$\mathrm{arccsc}(-z) = -\mathrm{arccsc}\, z$
$\mathrm{arcsec}(-z) = 180° - \mathrm{arcsec}\, z$

Table 2

Early in the twentieth century, engineer and inventor Lee deForest discovered that the conductance of a diode tube could be controlled by putting an electrode, called the grid, between the cathode and the plate. A signal applied to the grid caused large changes in the plate current, resulting in amplification; if some of the output was coupled back to

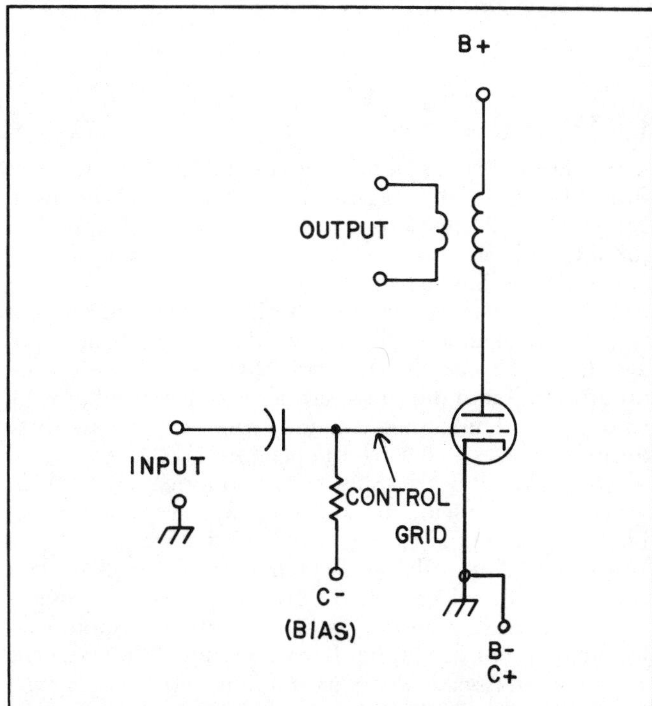

TRIODE TUBE: Triode tube, showing typical biasing arrangement.

the input and shifted 180 degrees in phase, oscillation occurred.

The control grid in a triode is usually biased at a voltage somewhat negative with respect to the cathode. The bias can be supplied by a resistance-capacitance network (*see* SELF BIAS), or by means of a separate power supply called the C supply. The illustration is the schematic symbol for a triode tube, showing a typical biasing arrangement.

As the instantaneous grid voltage becomes more negative, the instantaneous plate current decreases. As the instantaneous grid voltage becomes less negative, the instantaneous plate current rises. The change in the plate current, passing through a suitable load, results in an output-signal voltage that is many times larger than the input-signal voltage.

Some tubes have extra grids added to enhance the gain and improve the stability. A tube with two grids is called a tetrode, since it has four elements altogether. A tube with three grids is called a pentode. *See also* CONTROL GRID, PENTODE, SCREEN GRID, SUPPRESSOR GRID, TETRODE, TUBE.

TRIPLE-DIFFUSED EMITTER-FOLLOWER LOGIC

A bipolar logic family, similar to emitter-coupled logic but using a simple triple-diffusion process in manufacture, is known as triple-diffused emitter-follower logic (3DEFL). A 3DEFL gate operates below saturation, allowing fast operating speed.

The main advantage of 3DEFL is that it allows many switching gates to be impressed on a chip. Manufacturing is simple and relatively inexpensive. *See also* DIODE-TRANSISTOR LOGIC, DIRECT-COUPLED TRANSISTOR LOGIC, EMITTER-COUPLED LOGIC, HIGH-THRESHOLD LOGIC, INTEGRATED INJECTION LOGIC, RESISTOR-CAPACITOR-TRANSISTOR LOGIC, RESISTOR-TRANSISTOR LOGIC, TRANSISTOR-TRANSISTOR LOGIC

TRIPLER

A tripler is a radio-frequency circuit that produces an output signal at a frequency three times that of the input signal. There are various different methods to obtain frequency multiplication by 3; two common methods are shown in the illustration.

The circuit at A uses a push-pull configuration. This causes reinforcement of the odd harmonics of the input signal, and suppresses the even harmonics. The output circuit is tuned to the third harmonic of the fundamental input signal. This circuit, if tuned properly, provides some amplification as well as frequency multiplication.

The circuit at B makes use of the nonlinear characteristics of semiconductor diodes to obtain harmonic energy. The output tank is tuned to the third harmonic of the fundamental input signal. This circuit has a small amount of insertion loss since only passive components are used.

Triplers are commonly used with radio transmitters to obtain operation at a higher frequency. Two triplers can be connected in cascade to obtain multiplication by 9. *See also* FREQUENCY MULTIPLIER, HARMONIC, PUSH-PULL CONFIGURATION.

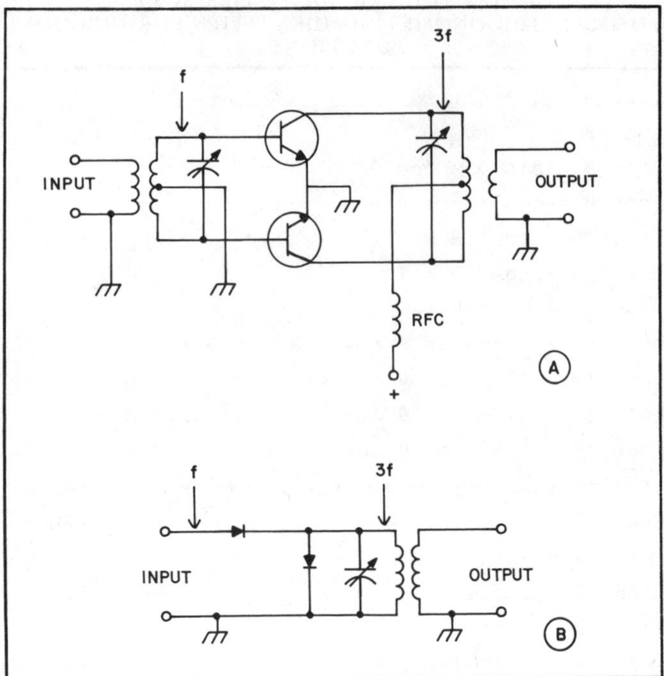

TRIPLER: At A, a push-pull tripler. At B, a tripler that makes use of the nonlinear characteristics of diodes.

TRI-TET OSCILLATOR

A crystal-controlled electron-coupled oscillator is sometimes called a tri-tet oscillator (*see* ELECTRON-COUPLED OSCILLATOR). The tri-tet oscillator uses a tetrode or pentode tube, with the crystal connected between the control grid and the cathode. The screen grid acts as the anode for the oscillator. The output is taken from the plate.

The tri-tet oscillator produces energy at harmonic frequencies of the crystal. If an untuned output circuit is used, signals will be generated at harmonics of large order. For example, if a 100-kHz crystal is used, harmonic energy will appear at 200 kHz, 300 kHz, and so on up to several megahertz.

The tri-tet oscillator was extensively used for providing calibration markers in tube type radio receivers. Nowadays, of course, calibrator circuits are designed using semiconductor components.

TROPOSPHERIC PROPAGATION

The lower part of the earth's atmosphere has an effect on electromagnetic-field propagation at certain frequencies. At wavelengths shorter than about 15 m (or a frequency of 20 MHz), refraction and reflection take place within and between air masses of different density. The air also produces some scattering of electromagnetic energy at wavelengths shorter than 3 m (or a frequency of 100 MHz). All of these effects are generally known as tropospheric propagation.

Tropospheric propagation can result in communication over distances of hundreds of miles. The most common type of tropospheric propagation occurs when radio waves are refracted in the lower atmosphere. This takes place to a certain extent all the time, but is most dramatic in the vicinity of a weather front, where warm, relatively light air lies above cool, denser air. The cooler air has a higher

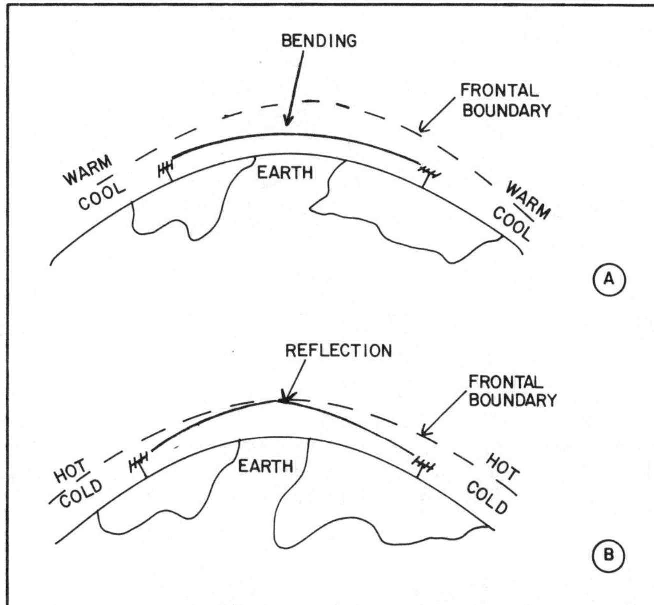

TROPOSPHERIC PROPAGATION: At A, tropospheric bending. At B, tropospheric reflection.

index of refraction than the warm air, causing the electromagnetic fields to be bent downward at a considerable distance from the transmitter (see illustration A). This phenomenon is called tropospheric bending.

If the boundary between a cold air mass and a warm air mass is extremely well-defined, and an electromagnetic field strikes the boundary at a near-grazing angle of incidence from within the cold air mass, total internal reflection occurs. This is known as tropospheric reflection, as at B. If a cold air mass is sandwiched in between warm air masses, the energy may be propagated for long distances because of repeated total internal reflection. This is called the tropospheric ducting or duct effect (*see* DUCT EFFECT, TOTAL INTERNAL REFLECTION).

Tropospheric propagation is often responsible for anomalies in reception of television and frequency-modulation broadcast signals. A television station hundreds of miles away may suddenly appear on a channel that is normally dead. Unfamiliar stations may be received on a hi-fi radio tuner. Sometimes two or more distant stations come in on a single channel, causing interference. Sporadic-E propagation has similar effects, and sometimes it is difficult to tell whether tropospheric or sporadic-E propagation is responsible for long-distance reception or communication (*see* SPORADIC-E PROPAGATION). Usually, sporadic-E events result in propagation over longer distances than tropospheric effects, but this is not always the case.

At very high, ultra-high, and microwave frequencies, the atmosphere scatters electromagnetic energy in much the same way as it scatters visible light. This can result in propagation over the horizon. It is called tropospheric scatter or troposcatter. *See also* PROPAGATION CHARACTERISTICS, TROPOSPHERIC-SCATTER PROPAGATION.

TROPOSPHERIC-SCATTER PROPAGATION

At frequencies above about 150 MHz, the atmosphere has a scattering effect on electromagnetic fields. The scatter-

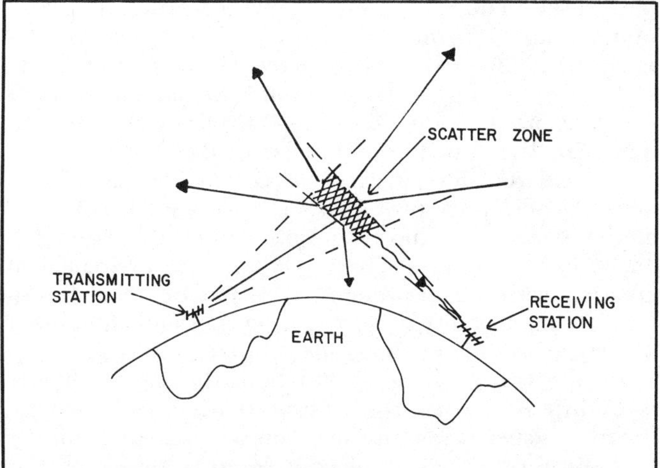

TROPOSPHERIC-SCATTER PROPAGATION: Random scattering allows over-the-horizon communication.

ing allows over-the-horizon communication at very-high, ultra-high, and microwave frequencies. This mode of propagation is called tropospheric scatter, or troposcatter for short.

Dust and clouds in the air increase the scattering effect, but some troposcatter occurs regardless of the weather. Troposcatter takes place mostly at low altitudes, but some effects occur at altitudes up to about 10 miles. Troposcatter propagation can provide reliable communication over distances of several hundred miles when the appropriate equipment is used.

Communication via troposcatter requires the use of high-gain antennas. Fortunately, the size of a high-gain antenna is manageable at ultra-high and microwave frequencies. The transmitting and receiving antennas are aimed at a common parcel of air, ideally located midway between the two stations and at as low an altitude as possible (see illustration). The maximum obtainable range depends not only on the gain of the antennas used for transmitting and receiving, but on their height above the ground: the higher the antennas, the greater the range. The terrain also affects the range; flat terrain is best, while mountains seriously impede troposcatter propagation.

In order to realize communication via troposcatter over long distances, sensitive receivers and high-power transmitters must be used, because the path loss is high.

Tropospheric scatter, while occurring for different reasons than bending, reflection, or ducting, is often observed along with other modes of tropospheric propagation. While communicating via troposcatter, a sudden improvement in conditions may actually be caused by bending, reflection, or duct effect. *See also* DUCT EFFECT, TROPOSPHERIC PROPAGATION, PROPAGATION CHARACTERISTICS.

TROUBLESHOOTING

Troubleshooting is the process of finding the faulty component or components in a malfunctioning piece of equipment. In electronics, troubleshooting saves time and expense in the repair of equipment. Faulty components can be located without "shotgun-style" removal and replacement when the proper troubleshooting procedures are followed.

The Basic Concept. Most complicated electronic devices

have service manuals. A service manual contains instructions on how to troubleshoot for specific problems. The most common, and most effective, troubleshooting instructions are given in the form of flowcharts for various potential malfunctions (*see* FLOWCHART). Troubleshooting data can also be written in tabular form.

Some digital devices, especially computers, can be used to find their own problems. The two most common methods are the diagnostic program and the service read-only-memory (ROM). These methods are used in conjunction with a comprehensive service manual. Troubleshooting of faulty computer programs and equipment prototypes is called debugging (*see* DEBUGGING).

In any troubleshooting situation, the most effective procedure is to work from output to input, from general systems to specific circuits and components, or from the most obvious to the least obvious potential problems. For example, if a radio transmitter does not work, the troubleshooting procedure would be as follows, assuming a transmitter design as shown in the accompanying block diagram.

- Check to be sure equipment is operated within the quoted specifications
- Check power supply
- Check output at antenna terminals (point X1)
- Check output of driver stage (point X2)
- Check output of modulator stage (point X3)
- Check modulating-signal output (point X4)
- Check oscillator output (point X5)

When the faulty stage is found, the bad component(s) can be located using standard test equipment (*see* TEST INSTRUMENT, TEST LABORATORY).

Intermittent Problems. Electronic equipment does not always malfunction catastrophically; sometimes the problem occurs intermittently. Intermittent problems are well known to the experienced technician. In order to find the source of a problem, the equipment must be evaluated while the malfunction can be observed. The process is sometimes time-consuming, tedious, and frustrating.

Intermittents usually occur either because of a faulty connection, or as a result of temperature sensitivity. Some intermittents, however, occur for no apparent reason. A bad connection can often be located visually, or by wiggling the leads in a suspected area. Thermal intermittents require manual temperature control. Small heaters, along with "freeze mist," can be used to subject individual components to thermal shock until the source of a thermal intermittent is found. In some stubborn cases, a technician must use intuition or perhaps even guesswork to solve an intermittent problem!

T-R SWITCH

See TRANSMIT-RECEIVE SWITCH.

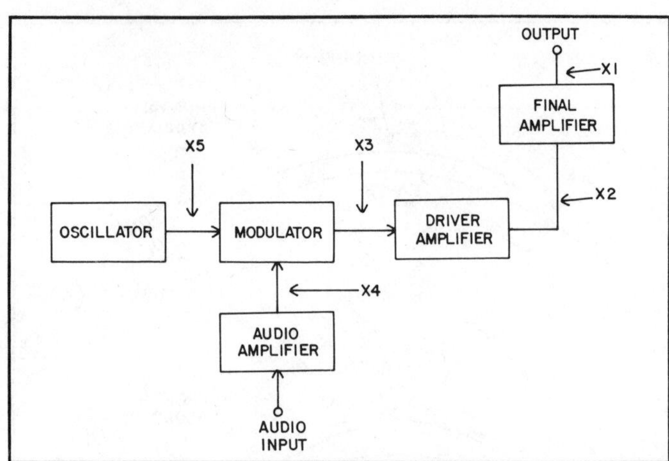

TROUBLESHOOTING: An example of troubleshooting of a typical voice transmitter. The points are checked in the order X1, X2, X3, X4, X5.

TRUE POWER

In a direct-current circuit, or in an alternating-current circuit containing resistance but not reactance, the determination of power is relatively simple. Given a root-mean-square (RMS) voltage E across a load of resistance R, and an RMS current I through the load, the dissipated or radiated power P_t is given by the formulas:

$$P_t = EI = E^2/R = I^2R$$

In this case, the value P_t is called the true power because it represents power that is actually manifested as dissipation or radiation. True power is often called real power, since it is represented by a real number.

In an ac circuit containing reactance, the true power cannot be found by the above formulas. If we try to determine the power according to the above formulas, we get an inflated (artifically large) value called the apparent power. The greater the reactance in proportion to the resistance, the greater the difference between the true power and the apparent power. In the extreme, if a load is a pure capacitive or inductive reactance, the true power is zero. A pure reactance will not radiate or dissipate power, but can only store power. The difference between the apparent power and the true power is called reactive or imaginary power. It is represented in the equations by an imaginary number.

If the apparent power is represented by P_a, the reactive power by P_r, and the true power by P_t, then:

$$P_t^2 + P_r^2 = P_a^2$$

See also APPARENT POWER, POWER, POWER FACTOR, REACTIVE POWER.

TRUE POWER GAIN

The true power gain of an amplifier is an expression of the extent to which the amplifier can increase the realizable power of a signal. True power gain is generally given in decibels.

Suppose the true power input to an amplifier is Pti watts. This means that, if the amplifier input circuit is replaced by a dissipating or radiating load of the same impedance, the load will consume Pti watts. Let the true power output of the amplifier, in watts, be denoted Pto. Then the true power gain, in decibels, is:

$$G = 10 \log_{10} (Pto/Pti)$$

The true power gain of an amplifier is always maximized when the input and output impedances are perfectly matched. This means there must be no reactance, and the resistance must be of the proper value. Any impedance mismatch results in a reduction in the true power output, and therefore in the true power gain. *See also* DECIBEL, GAIN, TRUE POWER.

TRUNCATION OF NUMBERS

See ROUNDING OF NUMBERS.

TRUNK LINE

A trunk line is a transmission line that interconnects two central offices in a telephone system. Subscribers communicate via their respective central offices, and through a trunk line for long distances (*see* CENTRAL OFFICE SWITCHING SYSTEM).

A trunk line carries many conversations at once. This makes it necessary to multiplex signals. The most common method is the use of frequency-division multiplex, using a carrier-current system (*see* CARRIER-CURRENT COMMUNICATION, FREQUENCY-DIVISION MULTIPLEX). In some systems, time-division multiplex may be used (*see* TIME-DIVISION MULTIPLEX).

A trunk line does not necessarily have to be a wire or cable. With the proliferation of microwave technology and communications satellites in recent years, it has become feasible to use telecommunication methods in telephone trunk systems. *See also* TELEPHONE.

TRUTH TABLE

A truth table is an expression of a logical, or Boolean, function. Truth tables provide a useful means of showing logical equivalences, and for analyzing the meanings of complicated logical expressions.

The table is an example of a truth table. *See also* BOOLEAN ALGEBRA, LOGICAL EQUIVALENCE.

TRUTH TABLE: AN EXAMPLE OF A TRUTH TABLE, DENOTING A LOGICAL FUNCTION (X+Y)′ + XZ.

X	Y	Z	X+Y	(X+Y)'	XZ	(X+Y)' + XZ
0	0	0	0	1	0	1
0	0	1	0	1	0	1
0	1	0	1	0	0	0
0	1	1	1	0	0	0
1	0	0	1	0	0	0
1	0	1	1	0	1	1
1	1	0	1	0	0	0
1	1	1	1	0	1	1

TTL

See TRANSISTOR-TRANSISTOR LOGIC.

TUBE

A tube is an electronic component that is used for amplification, rectification, and as the active device in some forms of oscillators. Tubes are also used as voltage regulators and display indicators.

There are two basic types of tubes: the vacuum tube and the gas-filled tube.

Vacuum Tubes. A vacuum tube consists of two or more metal electrodes, enclosed in a glass envelope that has been evacuated to allow electrons to flow freely. The electrode at the center is charged negatively and emits electrons. It is called the cathode. A cylindrical electrode surrounds the cathode, and is charged positively; it is called the plate. There may be one or more intermediate electrodes, charged variously to control the flow of electrons from the cathode to the plate; these electrodes are called grids (*see* CATHODE, CONTROL GRID, PLATE, SCREEN GRID, SUPPRESSOR GRID).

Vacuum tubes can be used for rectification, since electrons readily flow from the cathode to the plate but not in the other direction (*see* DIODE TUBE). However, semiconductor diodes are much more commonly used today for rectification.

When an alternating-current signal is applied to the control grid of a vacuum tube, large plate-current fluctuations occur (*see* TRIODE TUBE). This makes amplification possible. Amplification is enhanced by the addition of screen and suppressor grids (*see* TETRODE TUBE, PENTODE TUBE).

Vacuum tubes can be used for weak-signal amplification in receivers, or for power amplification in transmitters at all radio frequencies. The illustration is a photograph of a typical transmitting pentode. This tube is about 2.5 inches (6.4 cm) high, and has a continuous plate-dissipation rating of approximately 25 watts.

There are other types of vacuum tubes that are used for diverse purposes such as television and oscilloscope displays, oscillators and amplifiers at very short wavelengths, and the generation of X rays. *See also* CATHODE-RAY TUBE, ICONOSCOPE, IMAGE ORTHICON, KLYSTRON, MAGNETRON, TRAVELING-WAVE TUBE, VIDICON, X-RAY TUBE.

Gas-filled Tubes. Gas-filled tubes generally have no grids, but may have several cathodes or anodes. Gas-filled tubes

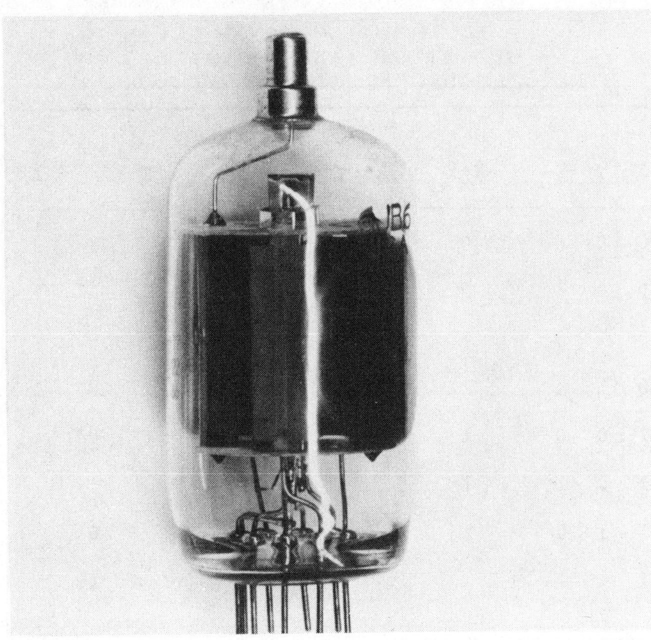

TUBE: A typical vacuum tube.

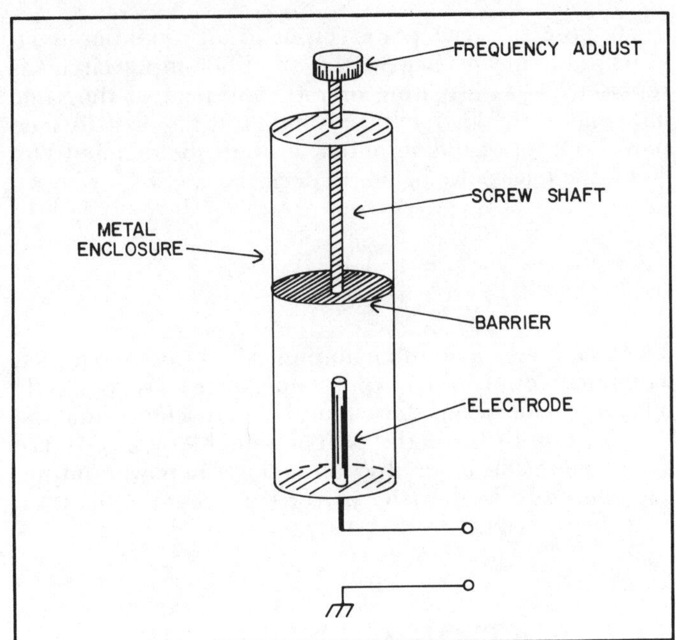

TUNABLE CAVITY RESONATOR: A form of resonant circuit.

are used for voltage regulation and the generation of visible light. All gas-filled tubes operate on the principle of ionization of elements in vapor form.

A gas-filled tube normally exhibits a high resistance at voltages below a certain level. When the anode or anodes are sufficiently positive with respect to the cathode or cathodes, conduction occurs because the rarefied gas ionizes. The electrons then flow easily among the atoms, from the cathode to the anode but not vice-versa.

A common form of gas-filled tube, the mercury-vapor rectifier, is still used in some high-voltage power supplies today. Gas-filled tubes are also used for regulation in high-voltage supplies. Semiconductor devices have taken the place of rectifier and regulator tubes, however, in power supplies designed for low and moderate voltages.

Various elements emit colored light when they are heated and ionized. A common example of this is the neon tube, which produces pink-orange light. Mercury-vapor and sodium-vapor lamps, as well as common fluorescent lamps, are examples of gas-filled tubes. *See also* FLUORESCENCE, FLUORESCENT TUBE, MERCURY-VAPOR LAMP, MERCURY-VAPOR RECTIFIER, NEON LAMP, NIXIE® TUBE, SODIUM-VAPOR LAMP, XENON FLASHTUBE.

TUBULAR CAPACITOR

A capacitor can be made by rolling layers of metal foil between layers of flexible dielectric material. This results in a cylindrical or tube-shaped package, and for this reason, such a capacitor is called a tubular capacitor.

Tubular capacitors are less common today than they were a few years ago. Paper is the most common dielectric material used in tubular capacitors. *See* PAPER CAPACITOR.

TUNABLE CAVITY RESONATOR

Metal enclosures are often used as tuned circuits at frequencies above about 200 MHz. Such enclosures exhibit excellent selectivity characteristics. Such devices are known as cavity resonators.

A cavity resonator can be either fixed or tunable. In a tunable resonator, the physical length is mechanically adjustable. The illustration shows one method of adjusting the length of a cavity resonator. The wavelength at resonance is directly proportional to the physcial length of the cavity. Therefore, to increase the frequency, the cavity is shortened, and to lower the frequency, the cavity is lengthened.

Tunable cavity resonators can be used in power amplifiers at very high, ultra-high, and microwave frequencies. Tunable resonators are used in some types of frequency meters. An adjustable cavity may also be used to provide selectivity at the front end of a receiver. *See also* CAVITY FREQUENCY METER, CAVITY RESONATOR.

TUNED CIRCUIT

Any circuit that displays resonance at one or more frequencies is called a tuned circuit. A tuned circuit may be either fixed or adjustable.

Physical Forms. Tuned circuits are found in many different physical forms, such as inductor-capacitor combinations, sections of transmission lines, or metal enclosures. The most familiar tuned circuit consists of a combination of descrete inductors and capacitors. A resonant section of transmission line is called a stub (*see* STUB). A resonant metal enclosure is called a cavity resonator (*see* CAVITY RESONATOR). An antenna is a form of tuned circuit.

Electrical Forms. All tuned circuits operate according to the same principle: the interaction of capacitive reactance and inductive reactance at different frequencies (*see* IMPEDANCE, REACTANCE, RESONANCE).

There are four electrical types of tuned circuit. Some act

as bandpass or band-rejection filters, and are classified as either series-resonant or parallel-resonant. Other tuned circuits may exhibit highpass or lowpass properties (*see* BANDPASS FILTER, BAND-REJECTION FILTER, HIGHPASS FILTER, LOWPASS FILTER).

A series-resonant tuned circuit acts as a short circuit at the resonant frequency or frequencies. A parallel-resonant tuned circuit acts as an open circuit at resonance. At non-resonant frequencies, a series- or parallel-resonant tuned circuit behaves as a pure reactance, the value of which depends on the frequency. A parallel-resonant tuned circuit is sometimes called a tank circuit (*see* PARALLEL RESONANCE, SERIES RESONANCE, TANK CIRCUIT).

TUNED FEEDERS

An antenna feed line is usually operated with a standing-wave ratio that is as low as possible. This involves operating the antenna at resonance, and matching the characteristic impedance of the feeder to the radiation resistance of the antenna (*see* CHARACTERISTIC IMPEDANCE, RADIATION RESISTANCE, RESONANCE, STANDING-WAVE RATIO). This type of feed system is called untuned, because the length of the feed line does not affect the antenna-system impedance at the transmitter/receiver.

An antenna system having a low-loss feed line can be operated with a high standing-wave ratio. This is sometimes done with an open-wire line (*see* OPEN-WIRE LINE). The electrical feed-line length is important, and therefore the line is said to be tuned. This type of feed system was popular in the early days of radio, when open-wire line was almost universally used, and is still employed today by some radio amateurs.

A tuned feeder is "pruned" to such a length that the input impedance is a pure resistance. This necessitates that the system have the following properties:

- Each side of the whole system must measure an integral multiple of ¼ electrical wavelength.
- The lengths of the two sides of the system must either be identical, or differ by an integral multiple of ½ electrical wavelength.

Ideally, both halves of the system should measure the same.

The illustration shows four examples of antennas with tuned feeders. The system at A is symmetrical, with each side measuring 2 wavelengths. The system at B incorporates off-center feed, with one side measuring 1¾ wavelength and the other side measuring 2¼ wavelength. The system at C uses off-center feed, with sides measuring 1½ and 2 wavelengths. The antenna system at D uses end feed, in which the feed line measures 2 wavelengths and the radiating element measures 1 wavelength.

The impedance at the input of a tuned feed system can range from a few ohms to several hundred ohms in practice. It is, therefore, usually necessary to use an impedance-matching transformer between a transmitter and a tuned feeder system (*see* IMPEDANCE TRANSFORMER).

If a transmatch, capable of tuning out reactance, is used at the input of a tuned-feeder system, the feed line length can be varied at will. This simplifies the installation procedure by eliminating the need for "pruning" the feed line. The two halves of the antenna must, however, still differ in length by some integral multiple of ½ wavelength (*see* TRANSMATCH). A center-fed radiator, fed with open-wire line and tuned with a transmatch, can be operated over a wide range of frequencies.

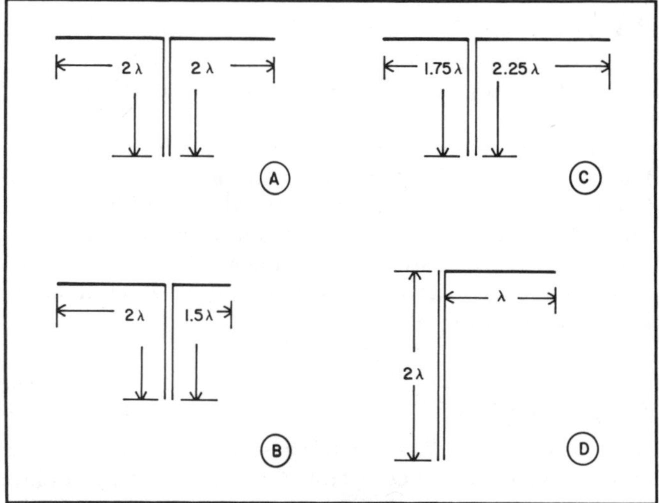

TUNED FEEDERS: Antennas with tuned feeders. At A, a symmetrical antenna system; at B and C, examples of off-center tuned feed; at D, an example of a system using tuned end feed.

TUNED-INPUT/TUNED-OUTPUT OSCILLATOR

A radio-frequency amplifier, having tuned input and output circuits resonant at the same frequency, often breaks into oscillation. This can be avoided by neutralization (*see* NEUTRALIZATION). However, it can also be used to advantage if oscillation is wanted.

The illustration is a schematic diagram of a tuned-input/tuned-output circuit that incorporates positive feedback to obtain oscillation. The feedback is provided by a capacitor between the collector and the base. The frequency is adjusted by means of a dual-gang variable capacitor, so that the input and output circuits are always resonant at the same frequency.

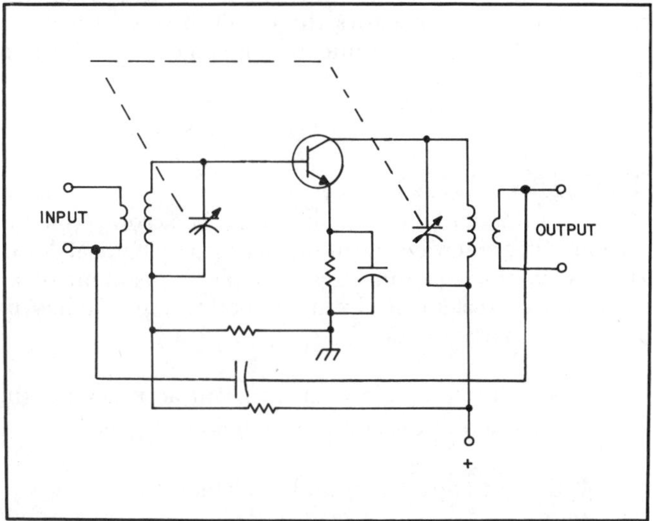

TUNED-INPUT/TUNED-OUTPUT OSCILLATOR: A tuned-input/tuned-output transistorized oscillator.

TUNGSTEN

Tungsten is an element with atomic number 74 and atomic weight 184. In its pure form, tungsten is a rather heavy, shiny metal.

Tungsten is characterized by a high melting point and fair conductivity. Most incandescent bulbs have tungsten filaments. Thoriated tungsten is used in the filaments of tubes (*see* FILAMENT). Some wirewound resistors are fabricated from tungsten (*see* WIREWOUND RESISTOR). Tungsten is also used for switch and relay contacts in high-current applications.

TUNING

The process of adjusting the resonant frequency of an inductance-capacitance circuit, a stub, or a cavity resonator for the purpose of obtaining certain operating characteristics, is called tuning. In an oscillator, tuning is used to control the frequency. An amplifier is tuned for maximum efficiency, power output, or gain. An antenna system is tuned for resonance at a particular frequency or set of frequencies.

Tuning of an inductance-capacitance circuit may be accomplished by the use of variable inductors or capacitors. Inductors can be tapped in various places and different portions switched in; some inductors have continuously variable taps (*see* TAP). Inductors with ferromagnetic cores can be permeability-tuned (*see* PERMEABILITY TUNING). If the capacitance in a tuned circuit is less than 0.001 μF (1000 pF), air-variable capacitors can be used to obtain continuous tuning (*see* AIR-VARIABLE CAPACITOR).

Tuning of a stub involves the use of a movable shorting bar. The longer the stub, the lower the resonant frequency or frequencies (*see* STUB). Tuning of a cavity resonator is accomplished by a movable barrier inside the enclosure (*see* CAVITY RESONATOR, TUNABLE CAVITY RESONATOR). An antenna is tuned either by cutting the radiator to a certain length, or by adding reactances in series. Antenna tuning is sometimes called loading (*see* CAPACITIVE LOADING, INDUCTIVE LOADING).

To facilitate adjustment of a tuned circuit, some mechanical means must be devised that is convenient for operators of the equipment. In tuned circuits having variable inductors or capacitors, this usually involves the use of a calibrated scale, in conjunction with a gear or belt drive. *See also* TUNING DIAL.

TUNING DIAL

The adjustment of a tuned circuit can be accomplished in various different ways. A tuning dial is a mechanical device that facilitates convenient and simple adjustment of an inductance-capacitance circuit. A good tuning dial has the following characteristics:

- A calibrated scale with sufficient accuracy for the intended application, or a digital display
- Ease of adjustment and readability
- Ease of resettability
- Minimal backlash
- Immunity to accidental misadjustment because of vibration

A tuning dial may operate continuously, or in defined increments. The display may be analog or digital. Analog tuning mechanisms are usually mechanical, and operate via a gear drive or belt drive. Even if the tuning control is an analog device, a digital display may be used. Some tuning dials are entirely electronic, using keyboard entry and a digital display. *See also* ANALOG CONTROL, DIAL SYSTEM, DIGITAL CONTROL, RACK AND PINION, SLIDE-RULE TUNING, TUNING, VERNIER.

TUNING FORK

A tuning fork is a metal object, shaped something like a fork, and cut to a precise physical length. When a tuning fork is struck so that the tines vibrate, a pure sine-wave sound is generated at a frequency that depends on the size of the fork. A typical tuning fork is shown in the illustration. A tuning fork serves as an audio-frequency reference standard.

Tuning forks maintain their frequency characteristics almost indefinitely, and under a wide variety of temperature conditions. Musicians sometimes use tuning forks to align their instruments. Some clocks use tuning forks as a frequency standard from which the seconds, minutes, and hours are mechanically or electronically derived. *See also* REFERENCE FREQUENCY.

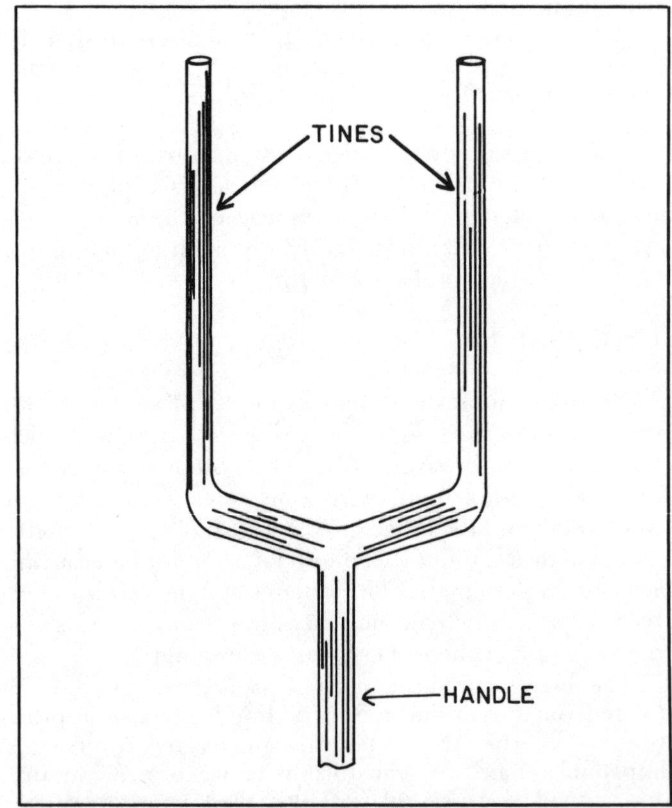

TUNING FORK: The resonant frequency of a tuning fork depends on the length of the tines and the substance from which they are made.

TUNNEL DIODE

A tunnel diode is a special form of semiconductor device that can be used as an oscillator and amplifier at ultra-high and microwave frequencies.

A tunnel diode does not rectify, as does an ordinary semiconductor P-N junction. This is because the semiconductor materials contain large amounts of impurities. The current-versus-voltage curve has a negative-resistance region that makes oscillation and amplification possible when a suitable bias is applied.

In recent years, Gunn diodes have largely replaced tunnel diodes as oscillators and amplifiers at ultra-high and microwave frequencies. Gunn diodes are more efficient. *See also* DIODE OSCILLATOR, GUNN DIODE.

TURNS RATIO

In a transformer, the turns ratio is defined as the number of turns in the primary winding divided by the number of turns in the secondary winding. In a step-up transformer, the turns ratio is between 0 and 1; in a step-down transformer, the turns ratio is greater than 1.

The primary-to-secondary turns ratio of a transformer is usually denoted by two integers, separated by a colon to indicate the ratio. The ratio is reduced to lowest terms. A transformer with 500 turns in the primary and 200 turns in the secondary, for example, has a turns ratio of 5:2. *See also* STEP-DOWN TRANSFORMER, STEP-UP TRANSFORMER, TRANSFORMER, TRANSFORMER PRIMARY, TRANSFORMER SECONDARY.

TURNSTILE ANTENNA

A turnstile antenna is a pair of horizontal half-wave dipoles, oriented at right angles to each other and fed 90 degrees out of phase. This results in radiation in all possible directions. The polarization is linear in the horizontal plane, and elliptical in directions above the horizon. The polarization is circular toward the zenith (*see* CIRCULAR POLARIZATION, ELLIPTICAL POLARIZATION). The illustration is a pictorial diagram of a turnstile antenna.

Turnstile antennas are commonly used in satellite communications when high gain is not needed. They are also used at VHF television transmitting stations. A turnstile antenna presents a reasonable match to 50-ohm coaxial cable. *See also* DIPOLE ANTENNA.

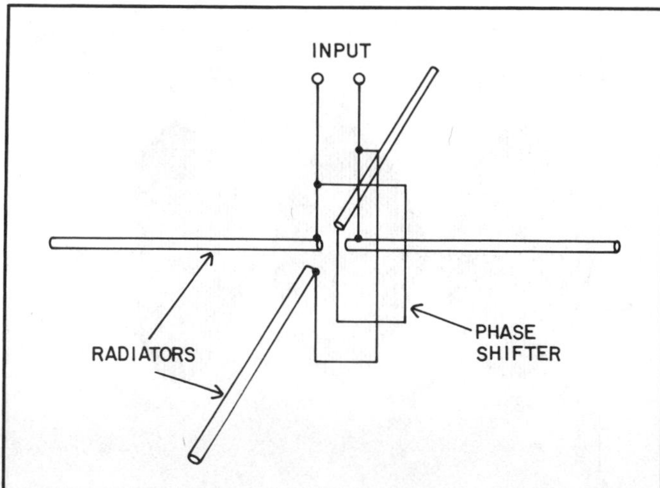

TURNSTILE ANTENNA: A turnstile antenna consists of two horizontal dipoles fed 90 degrees out of phase.

TURNTABLE

A turntable is a mechanical device that is used for reproduction of disk recording. A turntable consists of a motor, a rotating, round table, a spindle (perhaps equipped with a disk changer), a speed control, and a tone arm with a stylus (see illustration).

The standard turntable speed for recordings in the United States is 33⅓ revolutions per minute (RPM). This is 0.555 revolutions per second (RPS). Most turntables can also operate at 45 RPM or 0.750 RPS; a few can run at 78 RPM or 1.30 RPS. The standard disk size for 33⅓-RPM recording is 12 inches in diameter, and the playing time per side is 15 to 20 minutes. The standard size for a 45-RPM disk is 7 inches in diameter and the playing time per side is 3 to 5 minutes. The 78-RPM disk is the same size as the 33⅓-RPM disk, and the playing time is approximately 5 minutes per side. (The 78-RPM speed is not commonly used in disk recording today, although many older disks were recorded at that speed.)

In high-fidelity reproduction, it is important that the turntable maintain a constant speed. This requires a precision motor and drive system. Most turntables are equipped with shock-absorbing devices that minimize the effects of mechanical vibration.

Many turntables incorporate high-fidelity amplifiers, so that they can be connected directly to speakers for disk playing. More sophisticated high-fidelity systems, however, use separate amplifiers. Turntables are commercially manufactured in an almost unlimited variety of forms, to suit the casual listener as well as the serious hi-fi enthusiast. *See also* HIGH FIDELITY, STEREO DISK RECORDING, STYLUS.

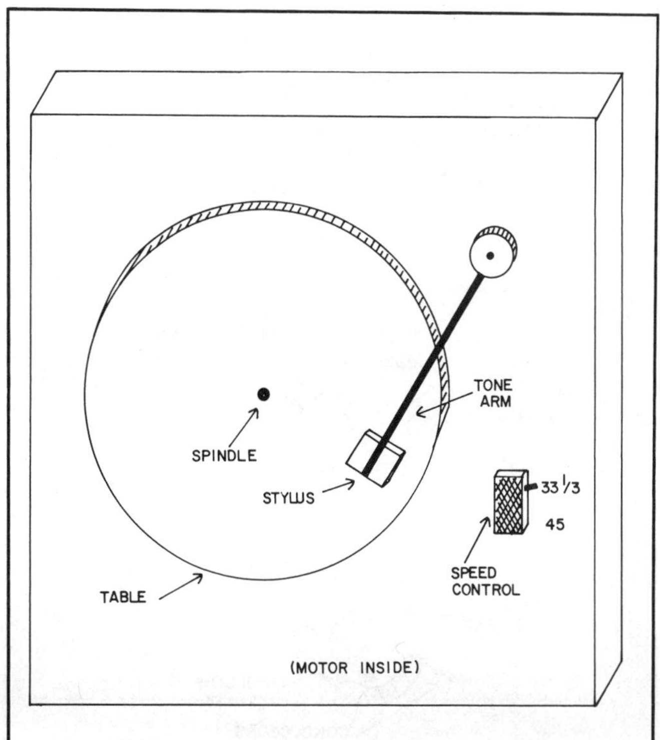

TURNTABLE: Drawing of a typical turntable showing the rotating table with spindle, the tone arm, the head with stylus, and the speed control.

TV

See TELEVISION.

TVI

See TELEVISION INTERFERENCE.

TWEETER

A tweeter is a speaker that is designed to radiate at high audio frequencies. A typical tweeter has a nearly flat response from about 3 kHz to well above the human hearing range. Tweeters are used in high-fidelity systems, in conjunction with midrange speakers and woofers, to obtain a flat audio response (*see* MIDRANGE, SPEAKER, WOOFER).

The most common types of tweeters are either dynamic or electrostatic units. Some tweeters can be used as ultrasonic transducers.

TWIN-LEAD

Twin-lead is a prefabricated, parallel-wire, balanced transmission line, molded in polyethylene and extensively used for television reception. Most twin-lead lines have a characteristic impedance of 300 ohms. A few have a characteristic impedance of 75 ohms. Twin-lead is sometimes called ribbon line.

Twin-lead is available in a wide variety of different forms. The illustration at A is a cross-sectional illustration of the most common form of twin-lead. The conductors are stranded copper, of size AWG (American Wire Gauge) No. 18 or No. 20. Illustration B shows a form of twin-lead with a foamed-polyethylene dielectric for reduced loss.

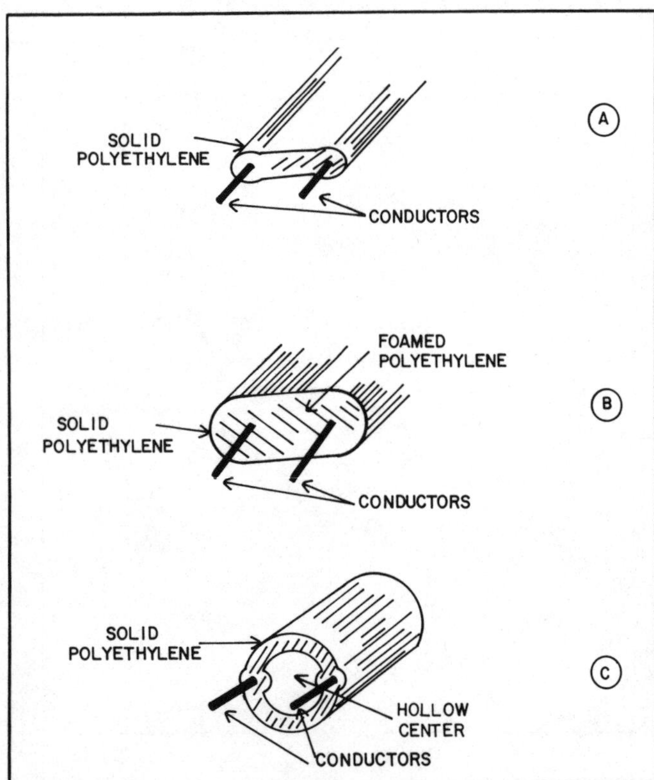

TWIN-LEAD: Examples of prefabricated twin-lead line. At A, the most common type; at B, foamed-dielectric type; at C, tubular type.

Illustration C is a cross-sectional view of a tubular twin-lead, with even lower loss.

Twin-lead can be used in low-power transmitting applications in place of open-wire line. However, the loss is somewhat greater in twin-lead than in a conventional open-wire line. This is attributable to the higher dielectric loss of polyethylene as compared with air. *See also* BALANCED TRANSMISSION LINE, OPEN-WIRE LINE.

TWO-TONE TEST

The distortion in a single-sideband transmitter can be evaluated using a procedure called a two-tone test. The intermodulation distortion of a receiver can be evaluated by means of a similar procedure, also called a two-tone test.

Testing a Single-sideband Transmitter. The single-sideband (SSB) transmitter two-tone test consists of the application of a pair of audio tones, both within the audio passband, to the microphone terminals. Any two audio frequencies between about 300 Hz and 3 kHz may be used; the most common are 1 kHz and 2 kHz. The output of the transmitter is analyzed with an oscilloscope. The ideal pattern is shown in the illustration. Any deviation from this pattern indicates distortion, which may result in unnecessarily large signal bandwidth. *See also* SINGLE SIDEBAND.

Testing a Receiver. For testing the intermodulation distortion of a receiver, two unmodulated signals are applied at the antenna terminals. These signals are usually close together in frequency; the spacing may vary between a few kilohertz and about 100 kHz.

If the two applied frequencies are f1 and f2, then the intermodulation products will be most noticeable at frequencies f3 and f4, such that:

$$f3 = 2f1 - f2$$

$$f4 = 2f2 - f1$$

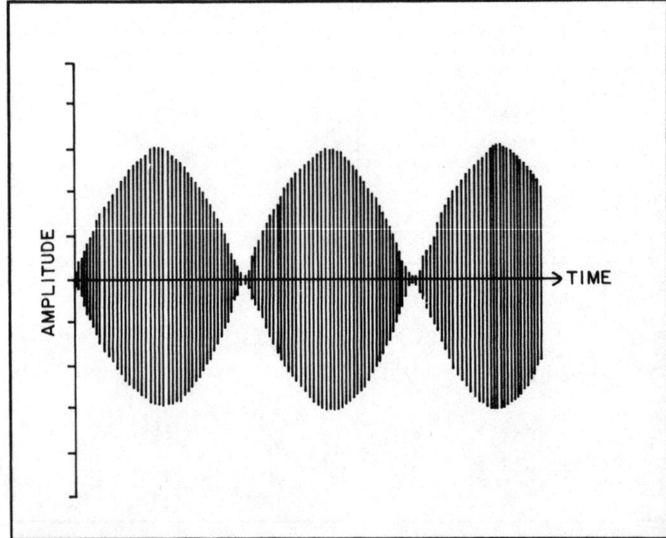

TWO-TONE TEST: Proper oscilloscope display for a two-tone test of a single-sideband transmitter.

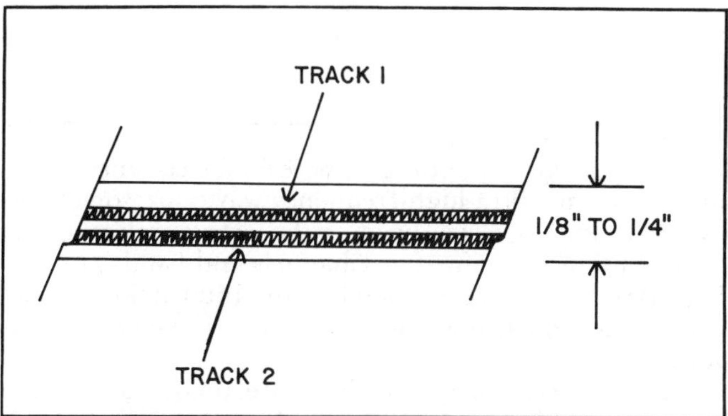

TWO-TRACK RECORDING: The two magnetic paths may be used together for stereo recording, or separately (usually in opposite directions) for monaural recording.

The lower the levels of signals at frequencies f3 and f4, the better the intermodulation-distortion performance of the receiver. *See also* INTERMODULATION.

TWO-TRACK RECORDING

Two-track recording is a common method of recording stereo audio signals on magnetic tape. Some reel-to-reel recorders use two-track recording. Smaller cassette recorders also sometimes use this method. In stereo recording one track is used for the left-hand channel and the other is used for the right-hand channel. The illustration shows the principle of two-track recording.

Excellent stereo channel separation is obtainable with modern two-track tapes. If desired, the recording time can be doubled by using each track separately in monaural form. Two-track magnetic tapes can be stored almost indefinitely, provided the temperature and humidity are kept within reasonable limits. *See also* MAGNETIC RECORDING, MAGNETIC TAPE.

TWO-WAY RADIO

A two-way radio is a transceiver designed to be used for telecommunication with other similar transceivers. In particular, the expression is used to describe a voice transceiver that operates via amplitude modulation, frequency modulation, or single sideband.

Amateur-radio transmitter/receiver combinations and Citizen's Band radios are examples of two-way units (*see* AMATEUR RADIO, CITIZEN'S BAND, TRANSCEIVER). Two-way radios are used extensively in businesses such as bus and taxi lines, aviation, and police and fire departments. Two-way radio communication takes place at practically all frequencies, from the very low to the ultra high.

Some two-way radio units operate on specific channels or sets of channels. Others can be used at any frequency within an assigned band. Fixed-channel radios are crystal-controlled, while continuous-coverage units incorporate variable-frequency oscillators. Commercially manufactured two-way radios are available in many different sizes and price ranges for each radio service.

In the United States, two-way radio communication requires, with a few exceptions, that the operator have a license for the specific service to be used. There are limitations on the type of emission that can be used, the maximum power input or output, and in some cases, on the length of transmissions. Two-way radio is regulated by the Federal Communications Commission in the United States. The worldwide regulatory body is the International Telecommunication Union (*see* FEDERAL COMMUNICATIONS COMMISSION, INTERNATIONAL TELECOMMUNICATION UNION).

TWO-WIRE FEED LINE

See BALANCED TRANSMISSION LINE, OPEN-WIRE LINE, TWIN-LEAD.

TWO-WIRE SYSTEM

A two-wire system is a scheme for transmitting electric power, either in direct-current form or in alternating-current form at any frequency. Two-wire systems were originally used in house wiring. Most modern house wiring uses the three-wire arrangement (*see* THREE-WIRE SYSTEM).

A two-wire electrical system may be either unbalanced or balanced. It usually has one grounded wire and one live wire that carries the voltage, commonly 117 V at 60 Hz. Some two-wire systems operate with both conductors floating (that is, above ground potential), but carrying 117 V at 60 Hz relative to each other.

A two-wire system for transmitting radio-frequency energy is called a parallel-wire or balanced line. Parallel-wire lines include the open-wire and the twin-lead types of feed line. *See also* BALANCED TRANSMISSION LINE, OPEN-WIRE LINE, TWIN-LEAD.

UHF
See ULTRA-HIGH FREQUENCY.

UHF CONNECTOR
The uhf connector is a form of connector that is widely used with coaxial cables. The term uhf, which actually means ultra-high-frequency, is somewhat imprecise, because this type of connector is normally used at low, medium, and high frequencies. At very-high and ultra-high frequencies, the N-type connector is preferred (*see* N-TYPE CONNECTOR).

A uhf connector has a central pin or receptacle and an outer screw-on shell, separated by hard plastic dielectric. The dimensions are designed to present a characteristic impedance of 50 ohms, so that there will be no impedance "bump" when 50-ohm coaxial cable is used. This scheme works very well for most applications up to 150 or 200 MHz. A uhf connector can handle from a few hundred watts to several kilowatts of radio-frequency power, depending on the particular type of connector and the standing-wave ratio on the feed line.

Illustration A shows a male uhf connector on the end of a length of RG-58/U coaxial cable. The male connector is sometimes called a PL-259, which is the catalog number of a specific male uhf connector manufactured by Amphenol. (The connector at A is not a true PL-259, but instead is a solderless male uhf connector that can be interchanged with a PL-259 for use with small-diameter cable.) Illustration B shows a chassis-mounted female uhf connector. It is sometimes called an SO-239, which is the catalog number of a specific chassis-mounted female uhf connector manufactured by Amphenol. (Not all female uhf connectors are true SO-239 types.) There are various other sorts of uhf connectors, such as male-to-male, female-to-female, right-angle, and tee configurations.

All uhf connectors are suitable for use indoors. When a uhf connector is used outdoors for splicing cable or for connecting a cable to an antenna, some means must be devised to keep water from entering the cable through the connector. Electrical tape can be used to wrap a connection. Various special tapes, for sealing splices in radio-frequency cable, are available from commercial sources.

UHF TELEVISION
See TELEVISION BROADCAST BAND.

UL
See UNDERWRITERS' LABORATORIES, INC.

ULSI
See ULTRA-LARGE-SCALE INTEGRATION.

ULTRA-HIGH FREQUENCY
The ultra-high-frequency (uhf) range of the radio spectrum is the band extending from 300 MHz to 3 GHz. The wavelengths corresponding to these limit frequencies are 1 m and 10 cm. Ultra-high-frequency waves are sometimes called decimetric waves because the wavelength is on the order of tenths of a meter. Channels and bands at ultra-high frequencies are allocated by the International Telecommunication Union, headquartered in Geneva, Switzerland.

At uhf, electromagnetic fields are unaffected by the ionosphere of the earth. Signals in this range pass through the ionized layers without being bent or reflected in any way. The uhf waves behave very much like waves at higher frequencies. In the upper portion of the uhf band, waves can be focused by moderate-sized dish antennas for high gain and directivity. Because of their tendency to propagate through the ionosphere, uhf waves are extensively used in satellite communications. Because the frequency of a decimetric signal is so high, wideband modulation is practical.

Signals in the lower portion of the uhf band are sometimes bent or reflected within or between air masses of different temperatures. A certain amount of scattering also takes place in the atmosphere. *See also* DUCT EFFECT, ELECTROMAGNETIC SPECTRUM, FREQUENCY ALLOCATIONS, TROPHOSPHERIC PROPAGATION, TROPOSPHERIC-SCATTER PROPAGATION.

ULTRA-LARGE-SCALE INTEGRATION
Miniaturization of digital electronic circuits has led some engineers to speculate that we will someday have the capability to put well in excess of 1,000 gates on an individual chip. A chip having from 1,000 to 10,000 logic gates would be referred to as an ultra-large-scale integration (ULSI) chip.

Many modern integrated circuits contain from 100 to 1,000 gates; this level of miniaturization is called very-large-scale integration. *See* VERY-LARGE-SCALE INTEGRATION.

ULTRASONIC DEVICE
Any apparatus that detects or generates ultrasound, for any purpose, is an ultrasonic device. Ultrasonic devices include transducers, holograph-making equipment, motion detectors, and scanners.

Ultrasonic devices are used for such diverse purposes as bonding, brazing, cleaning, medical diagnosis, drilling, finding defects in materials, making holograms, plating, soldering, and welding. Some of the characteristics and applications of ultrasound are described in the articles immediately following.

ULTRASONIC HOLOGRAPHY
Ultrasonic waves can be used to obtain three-dimensional representations of objects, in much the same way as visible light is used to make holograms. This technique is known as ultrasonic holography. It is sometimes also called sonoptography.

Because ultrasound can penetrate materials that are

(A)

(B)

UHF CONNECTOR: Coaxial uhf connectors. At A, male; at B, female. These views are magnified to about 3 times life size.

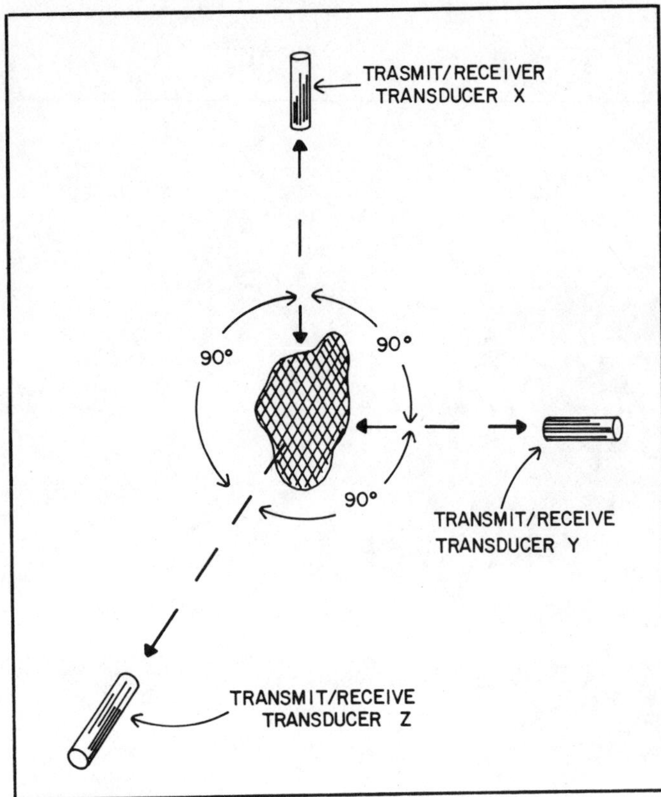

ULTRASONIC HOLOGRAPHY: Simplified diagram showing the operation of an ultrasonic-holography system. The signals from the three transducers are combined by a computer-graphics system to provide an image.

visually opaque, ultrasonic holography is capable of providing three-dimensional details of the interior of an object such as a concrete wall. This makes it possible to locate flaws in a structure such as a bridge or building. Ultrasonic holography can also be used to create three-dimensional images of body organs (*see* ULTRASONIC SCANNER).

The ultrasonic hologram is generated by means of transducers that transmit and receive ultrasonic waves in various phase. The received sounds are recorded on magnetic tape or disks, and analyzed by a computer. The hologram is displayed as graphics on a monitor screen. By varying the phases of the ultrasound, views may be obtained from any angle the operator desires. The illustration is a simplified block diagram of an ultrasonic holography system. *See also* ULTRASONIC TRANSDUCER, ULTRASOUND.

ULTRASONIC MOTION DETECTOR

Ultrasonic transducers can be used to detect motion in an area in which there are normally no moving objects. The most common application of an ultrasonic motion detector is in burglar-alarm systems. Ultrasonic methods are extremely effective for this purpose.

The illustration shows the principle of operation of an ultrasonic motion detector. There are two transducers: one emits an ultrasonic wave and the other picks up reflections from various objects in the vicinity. The reflected waves arrive in constant phase, provided that no reflecting object moves. But if something moves, the received signal is shifted in phase. A phase comparator detects the shift, and a triggering pulse is sent to the alarm.

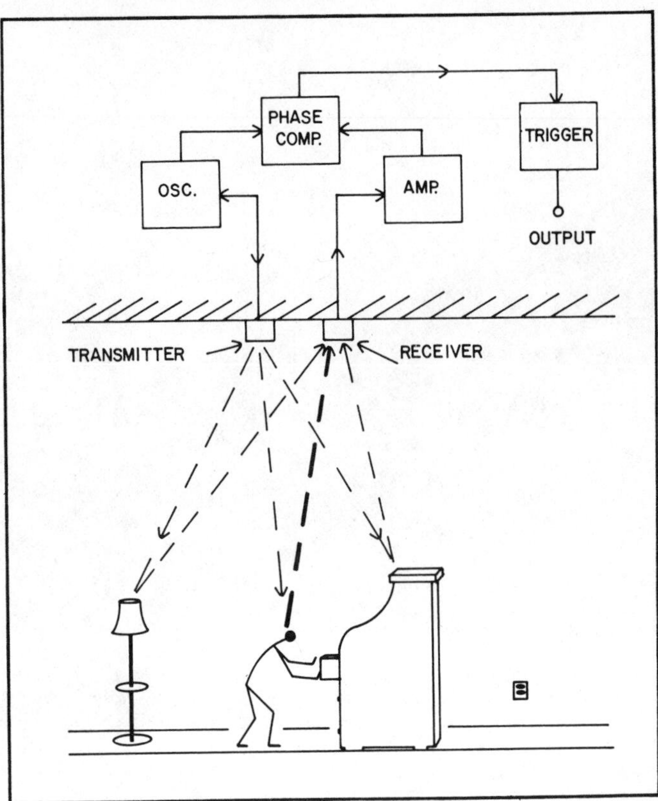

ULTRASONIC MOTION DETECTOR: Operation of an ultrasonic motion detector. The heavy dotted line shows the variable-phase echo from the moving subject.

Ultrasonic motion detectors have certain advantages and disadvantages compared with other types of motion detectors. The main advantage is that they are extremely sensitive and fast-acting. The main problem is that they sometimes respond to normal environmental vibration, such as might be caused by a passing train, plane, or truck. Some types of motion detectors use infrared sensors, electric eyes, or other devices to avoid the problem of unintentional actuation. However, all alarm systems have some drawbacks. *See also* ALARM SYSTEM, INFRARED DEVICE, ULTRASONIC TRANSDUCER, ULTRASOUND.

ULTRASONIC SCANNER

An ultrasonic scanner is a device that makes use of high-frequency acoustic waves to observe the interiors of objects that are opaque to visible light. In recent years, the use of such scanners has become increasingly widespread in medical practice, because of concern over radiation hazards from the traditional X-ray machine. Many medical scientists believe that ultrasound presents less potential than radiation for damage to human tissue.

Ultrasonic scanners operate on the principle that different substances reflect and transmit ultrasonic waves in varying degrees. An ultrasonic scanner can provide a two-dimensional image on a cathode-ray-tube monitor. A special technique called ultrasonic holography makes it possible to construct a three-dimensional image. *See also* ULTRASONIC HOLOGRAPHY, ULTRASONIC TRANSDUCER, ULTRASOUND.

ULTRASONIC TRANSDUCER

An ultrasonic transducer is a device that converts electrical impulses into ultrasound, or ultrasound into electrical impulses. Generally, the electrical waves have the same frequency as the acoustic waves that are produced or picked up by the transducer. An ultrasonic transducer thus resembles a speaker or microphone, depending on whether it is used to generate ultrasound or receive ultrasound.

Most ultrasonic transducers operate via one of three principles: dynamic interaction, magnetostriction, or piezoelectric effect (*see* CRYSTAL TRANSDUCER, DYNAMIC LOUDSPEAKER, DYNAMIC MICROPHONE, ELECTROSTRICTION, MAGNETOSTRICTION). A common hi-fi tweeter can operate as an ultrasonic transducer in some applications (*see* TWEETER).

Ultrasonic transducers are used in practically all apparatus that makes use of ultrasound. *See also* ULTRASONIC DEVICE, ULTRASONIC HOLOGRAPHY, ULTRASONIC MOTION DETECTOR, ULTRASONIC SCANNER, ULTRASOUND.

ULTRASOUND

Ultrasound is an acoustic disturbance that occurs at frequencies too high to be heard by humans. Ultrasound ranges in frequency from near 20 kHz to several hundred kilohertz, or even 1 MHz. (There is no well-defined upper limit.) In air, ultrasonic waves measure less than ⅝ in., or 1.7 cm, in length. Ultrasound propagates through materials in the same way, and at the same speed, as audible sound (*see* SOUND).

Ultrasound is produced along with audible sound by many natural phenomena. Ultrasound can be heard by certain animals, notably dogs, who can detect frequencies up to 30 or 40 kHz. Ultrasound can be generated and detected by means of ordinary oscillators, amplifiers, and special transducers for various purposes. *See also* ULTRASONIC DEVICE, ULTRASONIC HOLOGRAPHY, ULTRASONIC MOTION DETECTOR, ULTRASONIC SCANNER, ULTRASONIC TRANSDUCER.

ULTRAVIOLET

Ultraviolet is a term used to describe electromagnetic radiation at wavelengths somewhat shorter than visible light. Ultraviolet waves measure between approximately 390 nanometers (nm) and 4 nm. The ultraviolet spectrum is sometimes divided into two categories, known as long-wave ultraviolet, abbreviated UV (390 nm to 50 nm) and short-wave or extreme ultraviolet, abbreviated XUV (50 nm to 4 nm). Ultraviolet light propagates through a vacuum in the same way, and at the same speed, as visible light. The photons of ultraviolet contain more energy than those of visible light, but less energy than photons of X rays and gamma rays (*see* LIGHT, PHOTON).

Although we cannot see ultraviolet radiation, it can have a pronounced effect on us. Intense ultraviolet can be damaging to the eyes; for this reason, you should never look directly at a source of ultraviolet radiation. A familiar effect of ultraviolet is the tan or burn that it produces on our skin. (Many doctors and scientists believe that long-term overexposure to ultraviolet contributes to skin cancer in humans.) Ultraviolet also aids in the synthesis of vitamin D in our skin. Ultraviolet can be seen by certain insects and other creatures, and is important in their environment. Ultraviolet from the sun is responsible for the ionization of the upper atmosphere. This makes long-distance radio communication possible at some frequencies (*see* D LAYER, E LAYER, F LAYER, IONOSPHERE, PROPAGATION CHARACTERISTICS).

Excessive exposure to ultraviolet will kill most forms of life. The rarefied ozone in the earth's atmosphere blocks out much of the ultraviolet radiation from the sun. This minimizes damaging effects to life at the surface (*see* OXYGEN AND OZONE).

Ultraviolet can be generated by various types of gas-filled tubes. Specialized arc lamps and mercury-vapor lamps are the most common devices for generating ultraviolet (*see* ARC LAMP, MERCURY-VAPOR LAMP). Ultraviolet is generated or absorbed when the electrons of certain atoms change orbital shells. Any electric arc produces energy at ultraviolet wavelengths (*see* ENTLADUNGSSTRAHLEN).

Ultraviolet exposes most types of camera film; specially designed pinhole type cameras are used by astronomers to photograph the sun and other celestial objects in ultraviolet. Some camera tubes are also sensitive to ultraviolet light.

Ultraviolet sources can be modulated, and thus used for communications purposes. Aviators sometimes use ultraviolet devices to measure altitude. *See also* ULTRAVIOLET DEVICE.

ULTRAVIOLET DEVICE

Any device that produces, or is sensitive to, radiation between 390 nanometers (nm) and 4 nm can be considered an ultraviolet device.

Most lamps produce some ultraviolet light. Certain lamps, especially arc lamps and mercury-vapor lamps, generate large amounts of energy at ultraviolet wavelengths (*see* ARC LAMP, MERCURY-VAPOR LAMP). Certain types of light-emitting diodes and lasers produce ultraviolet (*see* LASER, LIGHT-EMITTING DIODE).

Ultraviolet can cause changes in the resistance through some materials by means of photoelectric effect. Some devices can convert ultraviolet directly into electricity (*see* PHOTOCELL, PHOTODIODE, PHOTOFET, PHOTORESISTOR, PHOTOTRANSISTOR, PHOTOTUBE, PHOTOVOLTAIC CELL). Specialized cameras and camera tubes allow ultraviolet images to be observed or recorded.

Aviators sometimes use a device called an ultraviolet altimeter to ascertain their height above the terrain. An ultraviolet altimeter is essentially immune to jamming. Ultraviolet devices can be used for communications purposes. *See also* ULTRAVIOLET.

UMBRELLA ANTENNA

An umbrella antenna is a wire antenna that is used mostly at low and very low frequencies. An umbrella antenna consists of several wire conductors and a metal mast. The mast and the wires all radiate, and serve to support each other. The conductors extend from the top of the mast to points at ground level (see illustration). The wires and the mast are inductively loaded so that they are ¼ electrical wavelength at the operating frequency. The feed point is at the base of the vertical mast. The umbrella antenna produces a vertically polarized signal.

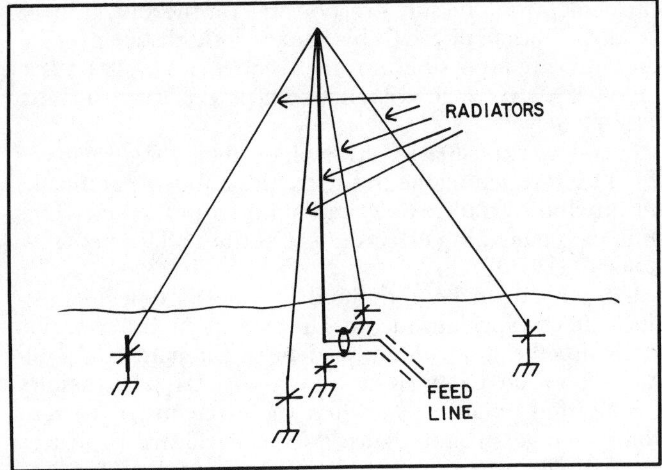

UMBRELLA ANTENNA: Simplified diagram of an umbrella antenna.

The advantage of the umbrella antenna is that it increases the radiation resistance as compared with a single radiator. At very low frequencies, even a fairly large tower has a radiation resistance so small that efficiency is severely impaired by ground loss (*see* ANTENNA EFFICIENCY, RADIATION RESISTANCE). When conductors are connected in parallel in a configuration such as that shown, the radiation resistance increases in proportion to the square of the number of conductors. If the radiation resistance of a single conductor is x ohms, the radiation resistance of n conductors in parallel is n^2x ohms.

Consider the example of an antenna for transmission at a frequency of 12 kHz. A single radiator, 500 feet high, would have a radiation resistance of approximately 0.01 ohm at this frequency. If an umbrella antenna with 10 radiators (including a 500-foot support mast) is used, the radiation resistance will increase to $0.01 \times 10^2 = 1$ ohm; if 70 conductors are used, the radiation resistance will theoretically be $0.01 \times 70^2 = 49$ ohms (a good match for 50-ohm cable). In practice, the radiation resistance does not increase quite this much, because of ground-current effects. However, the umbrella is the antenna of choice at frequencies below 300 kHz.

UNBALANCED CIRCUIT

See UNBALANCED SYSTEM.

UNBALANCED LINE

Any single-conductor electrical line, having a defined potential (either alternating-current or direct-current) with respect to ground, is called an unbalanced line. A shield may or may not be present.

The advantage of an unbalanced line is that it is simple. The main disadvantage of unbalanced line is that it must be shielded if leakage is to be prevented. The loss of a shielded, balanced line is usually greater, per unit length, than the loss in a balanced line. *See also* BALANCED LINE, TRANSMISSION LINE, UNBALANCED TRANSMISSION LINE.

UNBALANCED LOAD

An unbalanced load is a load with one side or terminal at ground potential and the other side above ground potential. An unbalanced load is required at the termination of an unbalanced transmission line, such as coaxial cable, to ensure that unwanted currents do not flow on the ground side of the transmission line (*see* UNBALANCED TRANSMISSION LINE).

A good example of an unbalanced load is an end-fed or asymmetrical antenna. Ground-plane and vertical antennas are the most common types of unbalanced antennas (*see* GROUND-PLANE ANTENNA, VERTICAL ANTENNA).

UNBALANCED OUTPUT

An unbalanced-output circuit is designed to be used with an unbalanced load and an unbalanced line. There is only one output terminal in an unbalanced circuit. The output is usually shielded.

A balanced output can be used as an unbalanced output by simply grounding one of the terminals. However, if an unbalanced output is to be used with a balanced load or line, a special transformer, called a balun, is required. *See also* BALANCED OUTPUT, BALUN, UNBALANCED LINE, UNBALANCED LOAD.

UNBALANCED SYSTEM

An unbalanced system is any circuit in which one side is at ground potential and the other side is at some ac or dc voltage different from ground potential.

Unbalanced systems are common in electronics. Many types of audio-frequency and radio-frequency oscillators, amplifiers, and transmission lines are unbalanced. Unbalanced systems are somewhat more stable than balanced systems, since a condition of near-perfect balance is difficult to maintain. *See also* BALANCED CIRCUIT, BALANCED LINE, BALANCED LOAD, BALANCED OUTPUT, BALANCED TRANSMISSION LINE.

UNBALANCED TRANSMISSION LINE

An unbalanced transmission line is a form of unbalanced line used in radio-frequency antenna transmitting and receiving applications. Such a line is usually a coaxial cable (*see* COAXIAL CABLE). In some instances, an unbalanced

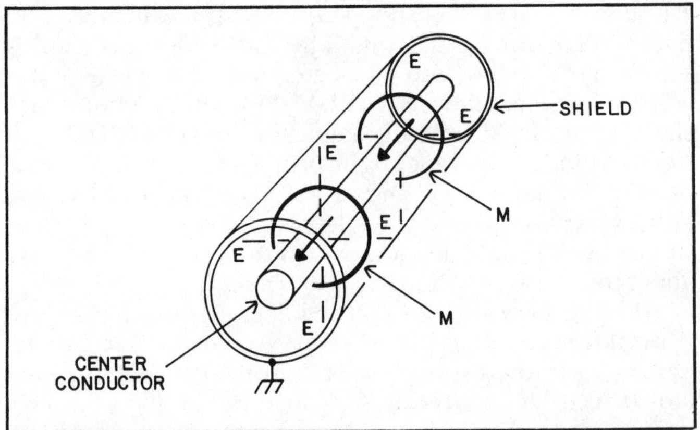

UNBALANCED TRANSMISSION LINE: The electric field (E) is represented by the dotted lines. The magnetic field (M) is represented by the circles. The current in the center conductor is shown by the arrows. The E and M fields are perpendicular within the shield, so the electromagnetic (EM) field travels parallel with the line conductors. No field exists outside the shield.

transmission line consists of a single wire, or a parallel-wire line in which one conductor is at ground potential. The illustration shows the principle of operation of a coaxial unbalanced transmission line.

The current in the center conductor flows back and forth at a certain frequency (heavy arrow). This current sets up an electric, or E, field, shown by the dotted lines, and a magnetic, or M, field, shown by the solid circles. The E and M fields are mutually orthogonal. They are prevented by the shield from leaking outside the line.

Since the E and M fields are perpendicular within the coaxial line, an electromagnetic field exists there. This field propagates in a direction perpendicular to both the E and M lines of flux. This direction is along the length of the transmission line. The speed of propagation depends on the dielectric material between the center conductor and the shield. If the dielectric is air, the speed of propagation is about 95 percent of the speed of light. If the dielectric is polyethylene, the speed varies between 66 and 80 percent of the speed of light (*see* VELOCITY FACTOR).

With an unbalanced transmission line such as that shown, it is important that the shielding be as complete as possible. This keeps electromagnetic fields from being radiated or picked up by the center conductor. In the case of an unshielded, unbalanced transmission line, radiation is inevitable, and is accepted as a characteristic of the line. Radiation from, or reception of signals by, a transmission line degrades the directional characteristics of the antenna with which the line is used. This may or may not be important, depending on the type of antenna.

Unbalanced transmission lines usually have more loss than balanced transmission lines. This is because the electromagnetic field in an unbalanced transmission line must pass through more dielectric material than it does in a balanced transmission line. Shielded, unbalanced lines are easier to install, however, than are balanced transmission lines, and this is their main advantage.

UNCERTAINTY PRINCIPLE

The uncertainty principle is a theory fact pertaining to the behavior of electrons in an atom. The exact position of an electron, at any given instant, cannot be determined because of difficulties in determining the momentum of the electron.

The accuracy of measurement of physical quantities is limited by certain mathematical realities. This can be proven, and the resulting theorem is sometimes called the uncertainty principle. The basic idea is that all real quantities, such as voltage, current, or mass, are irrational, and any irrational number, since it does not have a finite decimal expression, is impossible to measure exactly. *See also* IRRATIONAL NUMBER, RATIONAL NUMBER.

UNDERGROUND CABLE

In recent years, concern for the environment has led to the increased use of underground cables as an alternative to earlier above-ground cabling for telephone and television purposes. Some utility wires are also put underground.

An underground wire or cable is usually placed in a conduit to protect it from damage and corrosion. The conduit may be made of metal or plastic. Several wires and cables may be run through a single conduit. Underground wires and cables are sometimes run adjacent to sewer lines in cities and suburbs. This allows fairly easy access.

The major advantage of an underground cable is that it does not present an eyesore. However, underground wires and cables are considerably more difficult to maintain, although less likely to be damaged, than above-ground wires and cables.

UNDERWRITERS' LABORATORIES, INC.

In the United States, certain types of electrical equipment and appliances are tested for safety by an independent organization before distribution and sale. The organization is called Underwriters' Laboratories, Inc, abbreviated UL. The types of equipment checked by UL include wires, motors, fuses, circuit breakers, outlets, and similar devices.

The UL organization is primarily concerned with the safety of users of electrical equipment. Fire and shock hazards must be minimized. An appliance or device carries a UL approval tag or sticker if it meets UL standards. Some attention is also given by UL to the proper and consistent use of definitions. *See also* NATIONAL ELECTRIC CODE.

UNIDIRECTIONAL ANTENNA

Any antenna that exhibits maximum radiation or sensitivity in one direction, with less radiation or sensitivity in all other directions, is called a unidirectional antenna. Many types of antennas have this property (*see* UNIDIRECTIONAL PATTERN).

At high frequencies, the most common unidirectional antennas are the log-periodic, terminated longwire and rhombic, quad, and Yagi configurations. Various types of phased arrays operate as unidirectional antennas. These antennas will also work well at very-high and ultra-high frequencies (*see* LOG-PERIODIC ANTENNA, LONGWIRE ANTENNA, PHASED ARRAY, QUAD ANTENNA, RHOMBIC ANTENNA, YAGI ANTENNA).

As the wavelength gets shorter, other antenna arrangements become reasonable in size, and can provide excellent unidirectional operation. These include the billboard, Chireix, helical, horn, parabolic, spherical, and Sterba antennas. Some of these antennas are bidirectional unless a reflector is added (*see* BIDIRECTIONAL PATTERN, BILLBOARD ANTENNA, CHIREIX ANTENNA, HELICAL ANTENNA, HORN ANTENNA, PARABOLOID ANTENNA, SPHERICAL ANTENNA, STERBA ANTENNA, STERBA CURTAIN).

UNIDIRECTIONAL PATTERN

Any transducer that performs well in one direction, but poorly in all other directions, is said to exhibit a unidirectional pattern. Such a transducer can be an antenna, microphone, speaker, or other device designed for converting one form of energy to another. Illustration A shows a unidirectional pattern in the azimuth (horizontal) plane.

Devices with unidirectional radiation or response are quite common. A unidirectional pattern is usually associated with gain in the favored direction. For example, a Yagi antenna has gain compared with a dipole antenna.

A unidirectional pattern is generally symmetrical with respect to at least one plane passing through the line of the

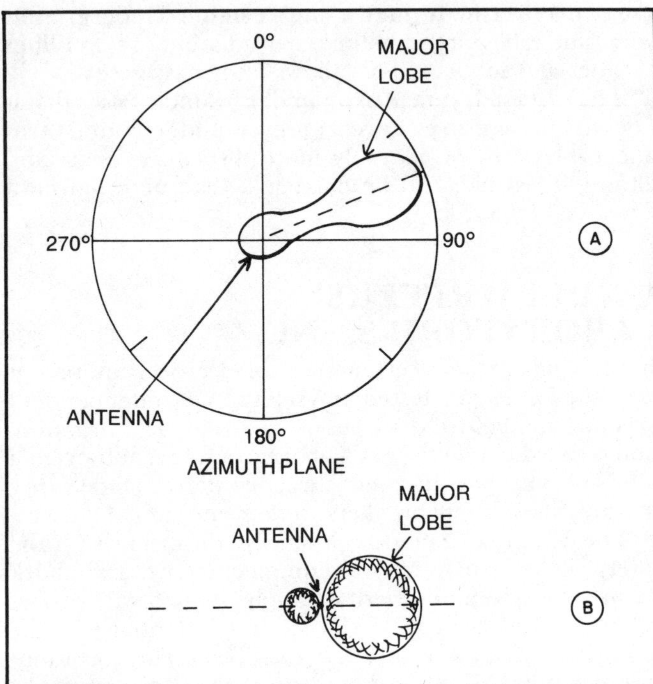

UNIDIRECTIONAL PATTERN: At A, a unidirectional pattern in the azimuth plane, showing a peak at 70 degrees. At B, a unidirectional pattern in three dimensions, symmetrical with respect to the major axis (dotted line).

favored direction. At A, that plane is horizontal. The pattern may be symmetrical in all planes that lie in the axis of the favored direction, as at B.

For antennas at low, medium, and high frequencies, only the azimuth plane is generally used to define unidirectionality. Both the azimuth and elevation become increasingly significant as the wavelength gets shorter. *See also* BIDIRECTIONAL PATTERN, UNIDIRECTIONAL ANTENNA.

UNIJUNCTION TRANSISTOR

A unijunction transistor (UJT) is a semiconductor device similar to a silicon-controlled rectifier. The UJT is sometimes called a double-base diode. The UJT is a three-terminal device.

The physical construction of a UJT is illustrated at A. Two metal contacts, called the bases, are attached to either end of a silicon rod. A third contact, called the emitter, is attached to the side of the rod. When a voltage is applied at the base terminals, the emitter-substrate junction becomes reverse-biased. The current through the UJT can be controlled by varying the voltage at the emitter.

The schematic symbol for the UJT (B) resembles the symbol for an N-channel field-effect transistor. The two devices are used in much different situations. Typical applications of the UJT include relaxation oscillators, multivibrators, and voltage-control devices.

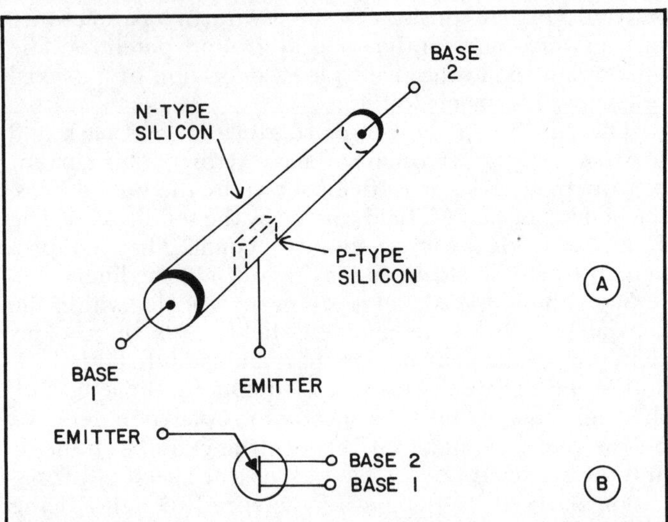

UNIJUNCTION TRANSISTOR: At A, construction of a unijunction transistor. At B, schematic symbol.

UNIPOLAR TRANSISTOR

See FIELD-EFFECT TRANSISTOR.

UNITS

See STANDARD INTERNATIONAL SYSTEM OF UNITS.

UNITY

Unity is a term that refers to a situation in which two quantities are identical or equal. The term unity is sometimes used interchangeably with the number 1. We may speak, for example, of unity gain; this expression means that the gain is 1. A perfect (1:1) standing-wave ratio is often called a unity standing-wave ratio. When two inductors are coupled to the greatest possible extent, we have a condition of unity coupling. These are just a few examples of the use of the term.

UNIVERSAL COUPLER

See TRANSMATCH.

UNIVERSAL TIME

See COORDINATED UNIVERSAL TIME.

UNIVIBRATOR

See MONOSTABLE MULTIVIBRATOR.

UNMODULATED CARRIER

See CARRIER, CONTINUOUS WAVE.

UNSATURATED LOGIC

See NONSATURATED LOGIC.

UP CONVERSION

Up conversion refers to the heterodyning of an input signal with the output of a local oscillator, resulting in an intermediate frequency that is higher than the frequency of the incoming signal. Technically, any situation in which this occurs is up conversion. But usually, the term is used in reference to converters designed for reception of low-frequency and very-low frequency signals, using receivers designed for higher frequencies.

Up conversion is accomplished using heterodyne techniques common to all mixers and converters. *See also* DOWN CONVERSION, MIXER.

UPLINK

The term uplink is used to refer to the bank on which an active communications satellite receives its signals from earth-based stations. The uplink frequency is much different from the downlink frequency, on which the satellite transmits signals back to the earth. The different uplink and downlink frequencies allow the satellite to receive and retransmit signals at the same time, so it can function as a repeater.

At a ground-based satellite-communications station, the uplink antenna can be relatively nondirectional if high transmitter power is used. But for more distant satellites, and especially for satellites in geostationary orbits, directional antennas are preferable. *See also* ACTIVE COMMUNICATIONS SATELLITE, DOWNLINK, REPEATER.

UPPER SIDEBAND

An amplitude-modulated signal carries the information in the form of sidebands, or energy at frequencies just above and below the carrier frequency (*see* AMPLITUDE MODULATION, SIDEBAND). The upper sideband is the group of frequencies immediately above the carrier frequency. Upper sideband is often abbreviated USB.

The USB frequencies result from additive mixing between the carrier signal and the modulating signal. For a typical voice amplitude-modulated signal, the upper sideband occupies approximately 3 kHz of spectrum space (see illustration). For reproduction of television video information, the sideband may be several megahertz wide.

The upper sideband contains all of the modulating-signal intelligence. For this reason, the rest of the signal (carrier and lower sideband) can be eliminated. This results in a single-sideband (SSB) signal on the upper sideband. *See also* LOWER SIDEBAND, SINGLE SIDEBAND.

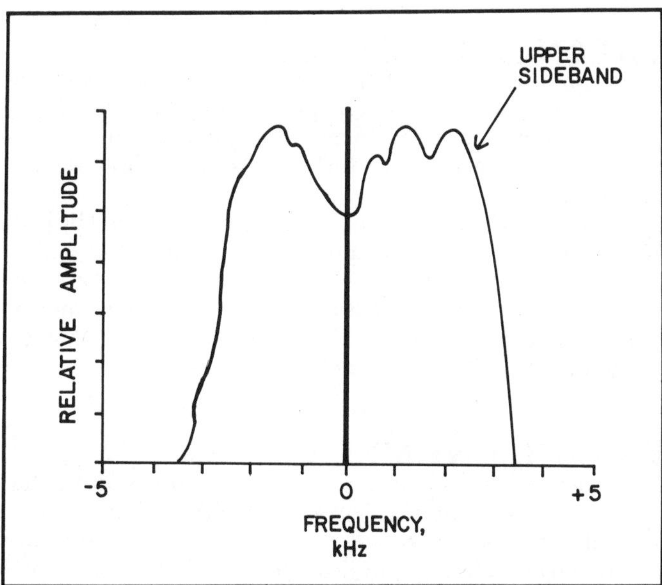

UPPER SIDEBAND: An amplitude-modulated signal as seen on a spectrum-analyzer display, showing the upper sideband.

UPWARD LEADER

An upward leader is the initial flow of electrons that occurs in a ground-to-cloud lightning stroke. *See* LEADER, LIGHTNING.

UPWARD MODULATION

If the average power of an amplitude-modulated transmitter increases when the operator speaks into the microphone, the modulation is said to be upward. Ordinary amplitude modulation is of the upward type. The average amplitude is minimum when there is no modulation.

Upward modulation is not the only means by which a carrier wave can be amplitude-modulated. Information can also be impressed on a carrier by decreasing the amplitude. *See also* AMPLITUDE MODULATION, DOWNWARD MODULATION.

URANIUM

Uranium is an element with atomic number 92. In its natural form, uranium has an atomic weight of 238. Various other isotopes can be made in the laboratory; of these, the most common has an atomic weight of 235. Uranium has the highest atomic number of any naturally occurring element. Pure uranium is a heavy, metallic substance.

All of the known isotopes of uranium are radioactive. Uranium 238 decays very slowly; it has a half life of many centuries. Other isotopes are less stable and decay more quickly. Uranium 235 can be used to make atomic weapons. *See also* RADIOACTIVITY.

USASCII

See ASCII.

USA STANDARD CODE

See ASCII.

USB

See UPPER SIDEBAND.

UTILITY LINE

See POWER LINE.

UTILIZATION FACTOR

The extent to which an electrical or electronic system is loaded, compared with its maximum capability, is called utilization factor. The utilization factor is generally expressed as a percentage.

Suppose, for example, that a generator can deliver a maximum continuous power of 5 kW, and appliances are connected that draw 3 kW. Then the utilization factor is 3/5, or 60 percent. As another example, consider a microcomputer memory that has a maximum capacity of 16 kilobytes. If 12 kilobytes are filled with data, then the utilization factor is 12/16, or 75 percent.

VACUUM TUBE

See TUBE.

VACUUM-TUBE VOLTMETER

A vacuum-tube voltmeter (VTVM) is a device that allows measurement of voltages in electronic circuits without drawing significant current. The input impedance is theoretically infinite, although in practice it is finite but very large. The vacuum-tube voltmeter gets its name from the fact that it uses a tube in a Class-A configuration to obtain the high input impedance.

The illustration is a simplified schematic diagram of a vacuum-tube voltmeter. The ranges are selected by varying the gain of the tube, or by switching various resistances in parallel with the microammeter.

Some vacuum-tube voltmeters have provision for measuring current and resistance. The tube is not required for these modes; the operation is identical to that of the volt-ohm-milliameter (*see* VOLT-OHM-MILLIAMMETER). A VTVM with the capability to measure current and resistance is sometimes called a vacuum-tube volt-ohm-milliammeter (VTVOM).

In recent years, field-effect transistors have replaced vacuum tubes for measurement of low and medium voltage when it is essential that the current drain be small. The VTVM is still sometimes used, however, for measuring high voltages. *See also* FET VOLTMETER.

VACUUM-TUBE VOLT-OHM-MILLIAMMETER

See VACUUM-TUBE VOLTMETER.

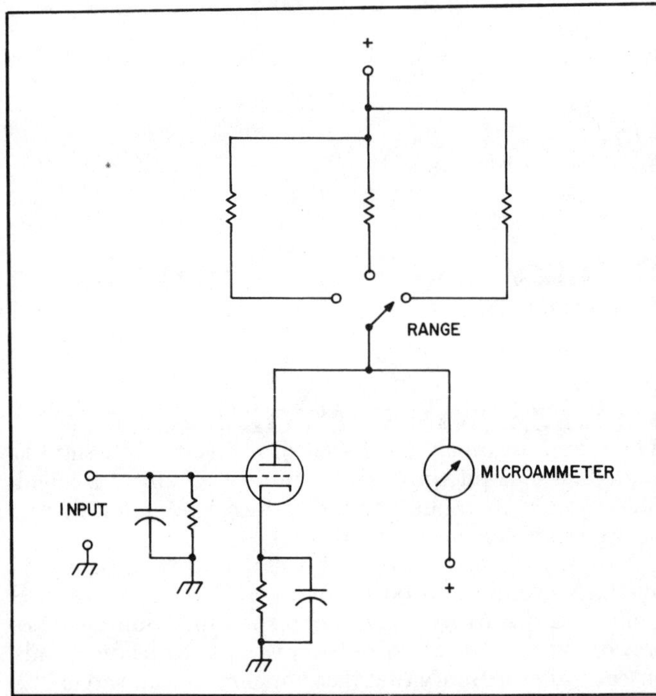

VACUUM-TUBE VOLTMETER: Schematic diagram of a vacuum-tube voltmeter.

VACUUM-VARIABLE CAPACITOR

A vacuum-variable capacitor is a variable capacitor that is enclosed in an evacuated chamber. The construction of the vacuum variable is similar to that of an air variable capacitor.

Vacuum-variable capacitors have extremely low loss and high voltage ratings, because a vacuum cannot become ionized. This makes the vacuum variable ideal for high-power radio-frequency transmitting applications. *See also* AIR-VARIABLE CAPACITOR, VARIABLE CAPACITOR.

VALENCE BAND

The electrons in an atom orbit the nucleus at defined distances, following average paths that lie in discrete spheres called shells. The more energy an electron possesses, the farther its orbit will be from the nucleus. If an electron gets enough energy, it will escape the nucleus altogether and move to another atom. This is how conduction occurs, and it is also responsible for reactions between or among various elements.

An electron is most likely to escape from a nucleus when the electron is in the outermost shell, or conduction band, of an atom. Electrons are stripped easily from nuclei in good conductors such as copper and silver, and less readily in semiconductors such as germanium and silicon. In insulators, the nuclei hold onto the electrons very tightly. The willingness with which a nucleus will give up its outermost electrons depends on how "full" the outer shell is. The shell just below the outer shell of an atom is called the valence shell. The range of energy states of electrons in this shell is known as the valence band. *See also* ELECTRON ORBIT, VALENCE ELECTRON, VALENCE NUMBER.

VALENCE BOND

See COVALENT BOND.

VALENCE ELECTRON

In an atom, any electron that orbits in the partially filled outermost shell, without escaping from the nucleus, is called a valence electron. In an atom having a "full" outer shell (*see* ELECTRON ORBIT), there are, by definition, no valence electrons.

Some atoms have valence shells that are occupied by just one or two electrons, and thus the electrons are given up easily. Other atoms have nearly full valence shells; such elements tend to attract electrons. The various elements have properties that depend, to some extent, on the state of the outer shell. *See also* VALENCE NUMBER.

VALENCE NUMBER

The valence number of an atom is an expression of the conditions in the outer (valence) shell of that element in its natural state (*see* VALENCE ELECTRON). The valence numbers of various elements indicate how their atoms conduct an electric current, and how they behave in chemical reactions and mixtures. In some cases, relative valence numbers are important for elements combined as mixtures.

Definition. An element with a filled valence shell has a valence number of 0. If the outer shell contains either an excess or a shortage of electrons, the valence number is the number of electrons that must be added or taken away to result in an outer shell that is completely filled.

Valence numbers are sometimes expressed as positive or negative quantities, indicating an excess or deficiency of electrons compared to the most stable possible state. These numbers are called oxidation numbers. If the outer shell is less than half full, containing n electrons, the oxidation number is positive, and is defined as +n. If the outer shell is more than half full, containing n electrons but having a capacity to hold k electrons, then the oxidation number is negative, and is defined as −(k−n). Substances with positive oxidation numbers tend to give up electrons; substances with negative oxidation numbers tend to accept electrons from outside the atom.

Sometimes both the oxidation values +n and −(k−n) are given, especially if the shell is approximately half filled. Such an element has two effective valence numbers. A few elements behave in such a way that they can be considered to have several different oxidation numbers.

Conduction and Chemical Reactions. The oxidation number of an element determines how readily it will conduct an electric current. Metallic substances have low positive oxidation numbers, indicating that the outer shell has just a few electrons. Semiconductors have outer shells that are approximately half full. Poor conductors have valence shells that are almost completely filled, and thus their oxidation numbers are small and negative. Those elements that have totally filled outer shells are the worst conductors, but the best dielectric substances.

The oxidation number of an element also indicates how readily it will react with certain other elements. When the nonzero oxidation numbers of two or more atoms add up to 0, the elements will react very easily when brought together. For example, two atoms of hydrogen (each with oxidation number +1) will combine readily with one atom of oxygen (oxidation number −2), since the three values add up to 0.

When an element has two or more oxidation numbers, it may react with other atoms according to any of the values. An example of such a substance is hydrogen, which can be considered to have an outer shell that is either half full or half empty, having oxidation value −1 in some cases and +1 in other cases. Hydrogen thus reacts with many other elements; it can either accept or give up an electron.

Relative Valence in Semiconductors. In a semiconductor material, impurity substances are added to modify the conducting characteristics. The valence of the impurity, relative to that of the original semiconductor, determines the major and minor charge carriers.

If an impurity has more electrons than the semiconductor, it is called a donor. If the impurity has fewer electrons than the semiconductor, it is called an acceptor. The addition of a donor impurity causes conduction mostly via electrons, and the resulting material is called N-type. The addition of an acceptor impurity causes conduction mostly in the form of electron-deficient atoms called holes, and the material is called P-type. *See also* ELECTRON, HOLE, IMPURITY, N-TYPE SEMICONDUCTOR, P-TYPE SEMICONDUCTOR.

VALVE

See TUBE.

VAN ALLEN RADIATION BELTS

The earth is surrounded by a magnetic field, known as the geomagnetic field (*see* GEOMAGNETIC FIELD). This field causes charged particles to be accelerated and deflected toward the poles. The particles are mostly protons and electrons, although there are some alpha particles and a few heavier nuclei. The charged particles come primarily from the sun, and to a lesser extent from other stars (*see* ALPHA PARTICLE, COSMIC RADIATION, ELECTRON, PROTON).

The deflection of charged particles around the earth results in zones of radiation in outer space. High-energy protons and alpha particles circulate at an altitude of approximately 2,000 miles. High-energy electrons are accelerated at an altitude of about 10,000 miles. These particles, especially the protons and alpha particles, produce gamma rays if they strike heavy materials such as rock or metal. The regions in which the particles move are known as the Van Allen radiation belts, after one of the scientists who first theorized their existence. The gamma radiation may present a hazard to space travelers if they spend a long time at altitudes of 2,000 or 10,000 miles.

At one time, it was believed that the Van Allen radiation might seriously endanger, or even kill, anyone who ventured more than a few hundred miles above the surface of the earth. Some thought that the belts might render space travel impossible. However, the radiation level was determined by the Apollo astronauts, in the late 1960s and early 1970s, to be lower than was originally believed.

VAN DE GRAAFF GENERATOR

A Van de Graaff generator is a device that generates large

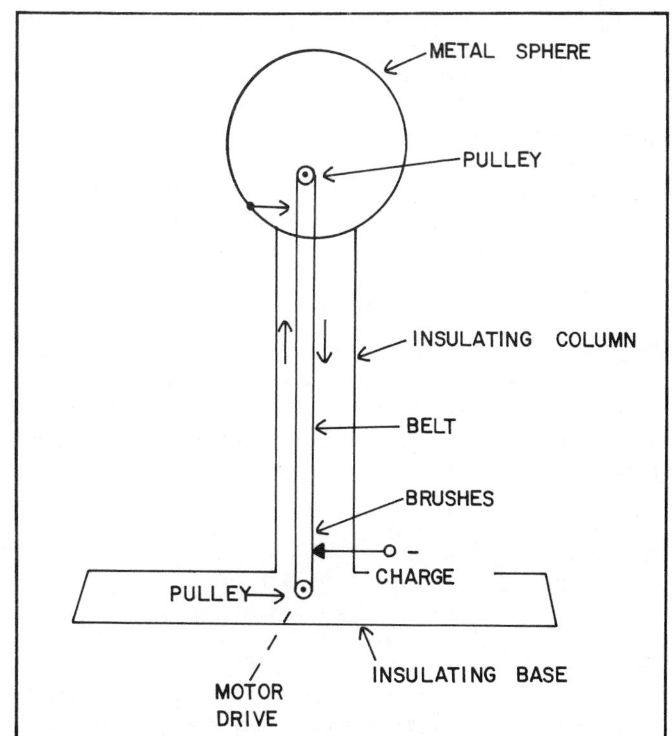

VAN DE GRAAF GENERATOR: A Van de Graaff generator operates by means of triboelectric effect. The charge is stored on a metal sphere.

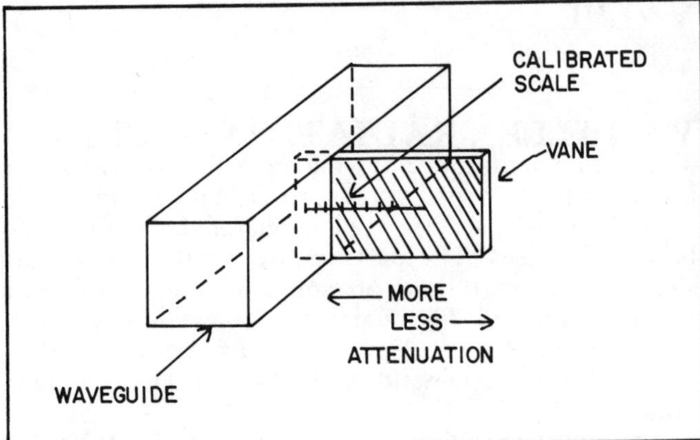

VANE ATTENUATOR: A vane attenuator is used to control the electromagnetic-field propagation through a waveguide.

electrostatic charges. The Van de Graaff generator operates by means of triboelectric effect, in much the same way as your shoes pick up a charge when you shuffle along a carpet (*see* TRIBOELECTRIC EFFECT).

A typical Van de Graaff generator is shown in the illustration. The friction of the belt against the brushes results in the accumulation of a large quantity of charge. The charge is transferred to the metal sphere. Van de Graaff generators vary in size from small tabletop devices to units that occupy a whole room. The smaller physics and electricity experiments, generate 100 to 300 kV. The larger devices, used by scientists for making artificial lightning, testing insulators, and providing the voltage needed by certain types of particle accelerators, can produce several megavolts.

VANE ATTENUATOR

A piece of resistive material, placed in a waveguide to absorb some of the electromagnetic field, is called a vane attenuator. The device can be moved back and forth to provide varying degrees of attenuation, from zero to maximum (see illustrations). The resistivity of the vane is chosen so that it does not upset the characteristic impedance of the waveguide.

Vane attenuators can provide exact resettability by means of a calibrated scale marked on the sliding element. Vane attenuators are extensively used at ultra-high and microwave frequencies in various applications. *See also* ATTENUATOR, WAVEGUIDE.

VAR

The unit of reactive power is called the reactive volt-ampere, abbreviated VAR. Reactive power does not represent power actually dissipated, although it has an effect on power measurement. *See* REACTIVE POWER.

VARACTOR DIODE

All diodes exhibit a certain amount of capacitance when they are reverse-biased. This capacitance limits the frequency at which a diode can be used as a rectifier or detector. But the reverse-bias capacitance properties of a semiconductor diode can be used to advantage in certain situations. A varactor diode is deliberately designed to provide an electronically variable capacitance. A varactor diode is sometimes called a variable-capacitance diode, or varicap, for this reason.

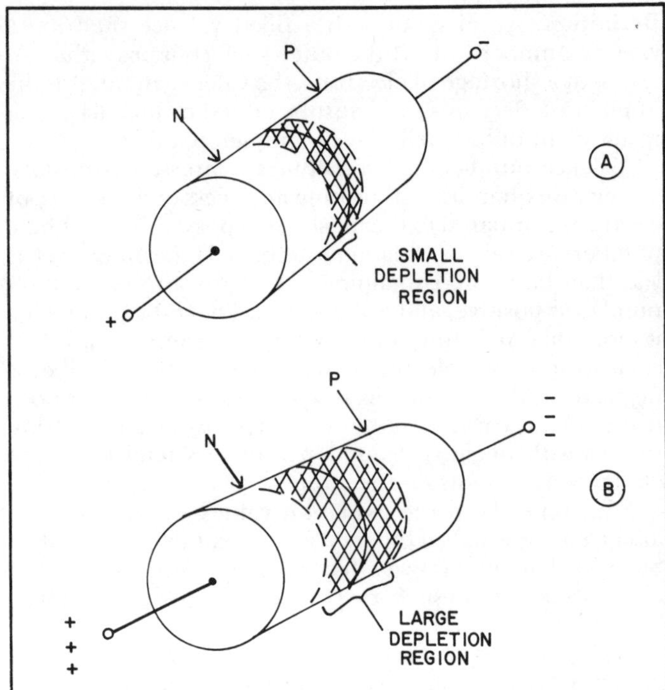

VARACTIOR DIODE: At A, a small reverse-bias voltage produces a narrow deletion region with a relatively large capacitance between conducting layers. At B, a large reverse-bias voltage results in a wide depletion region with a smaller capacitance between conducting layers.

The illustration shows the principle of a varactor diode. At A, the reverse-bias voltage is very small, and the depletion region is therefore narrow, resulting in a fairly large capacitance. At B, the reverse-bias voltage is greater, and the depletion region is thus larger. The result is a smaller value of capacitance. Varactor diodes exhibit widely varying reverse-bias-versus-capacitance characteristics, depending on the method of manufacture. Typical minimum capacitance values range from less than 1 pF to 20 or 30 pF; maximum values may be as much as 200 pF.

Varactor diodes can be used to provide frequency modulation of an oscillator (*see* REACTANCE MODULATOR). Varactor diodes can also be used to tune radio-frequency amplifiers at practically all wavelengths. Varactor diodes operate efficiently as frequency multipliers. *See also* DIODE CAPACITANCE, P-N JUNCTION, REVERSE BIAS.

VARIABLE

In a mathematical equation, a variable is a numerical quantity the value of which is not specified. A variable may also be called an unknown. Variables are used, along with constants (known or specific quantities), to express mathematical equations and functions.

Variables may be denoted by any symbol or letter in an equation or function. Letters from the second half of the alphabet, however, are preferred. The letters n through q are most often used to represent variables that can attain only integer values. The letters r through z are more often used to represent rational or real values. Greek letters are also sometimes used to represent variables. *See also* CONSTANT, EQUATION, FUNCTION.

VARIABLE CAPACITOR

A variable capacitor is any device that exhibits a capacitance that can be adjusted over a continuous range. Variable capacitors are extensively used for tuning oscillators and amplifiers, and for adjusting transmatches and other impedance-matching networks.

Variable capacitors can be found in a wide variety of voltage and capacitance ratings for various applications. Most variable capacitors can be adjusted within a range of values having a maximum-to-minimum ratio of between 10:1 and 40:1. Very few variable capacitors have maximum values of more than a few hundred picofarads.

A special form of variable capacitor, known as a varactor diode, allows electronic adjustment of capacitance. The varactor is unique because it facilitates rapid variation of capacitance—in some cases at ultra-high or microwave frequencies. *See also* AIR-VARIABLE CAPACITOR, PADDER CAPACITOR, TRIMMER CAPACITOR, VACUUM-VARIABLE CAPACITOR, VARACTOR DIODE.

VARIABLE CRYSTAL OSCILLATOR

The oscillating frequency of a piezoelectric crystal is normally fixed. However, it can be varied to a certain extent by the addition of a small capacitor or inductor. The components are connected in parallel or in series with the crystal. This is called pulling (*see* PULLING). A variable crystal oscillator (VXO) is a crystal oscillator in which the frequency can be adjusted over a small range by means of added reactances.

Variable crystal oscillators are sometimes used in radio transmitters or transceivers to obtain operation over a small part of a band. Any crystal oscillator can be made into a VXO by adding the proper parallel or series reactances. Two examples of VXO circuits are shown in the illustration.

The main advantage of the VXO over an ordinary variable-frequency oscillator is excellent frequency stability. The main disadvantage of the VXO is its limited frequency coverage. Usually, the frequency of the VXO can be varied up or down by a maximum of approximately 0.1 percent of the operating frequency without crystal failure or loss of stability. *See also* CRYSTAL OSCILLATOR, OSCILLATOR, PIERCE OSCILLATOR, VARIABLE-FREQUENCY OSCILLATOR.

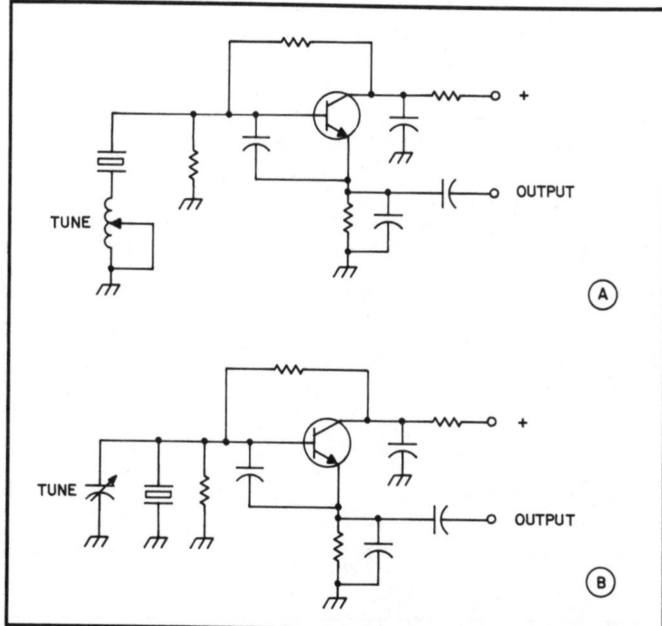

VARIABLE CRYSTAL OSCILLATOR: Two examples of crystal oscillators in which the frequency is adjustable. At A, a series inductor is used; at B, a parallel capacitor is employed.

VARIABLE FREQUENCY

A variable-frequency device is an oscillator, receiver, or transmitter designed to operate on more than one frequency. Variable-frequency devices are usually tuned by means of continuously adjustable inductance-capacitance circuits, or with synthesizers. Variable-frequency operation may be conducted on discrete channels, or over a continuous range of frequencies.

The main advantage of variable-frequency communication is that interference from other stations can be avoided by choosing an unoccupied channel or frequency. The primary disadvantage of variable-frequency operation is that a scheduled contact is not feasible unless the operators have previously agreed on the frequency. *See also* FIXED FREQUENCY, VARIABLE CRYSTAL OSCILLATOR, VARIABLE-FREQUENCY OSCILLATOR.

VARIABLE-FREQUENCY OSCILLATOR

Any oscillator in which the frequency can be adjusted, either in discrete channels or over a continuous range, is called a variable-frequency oscillator (VFO). Many different types of oscillators can be used for variable-frequency operation.

Some VFO circuits use inductance-capacitance tuning to obtain oscillation over a band of frequencies. The Colpitts and Hartley configurations are probably the most common. The tuned-input/tuned-output circuit is also sometimes used. The frequency of a crystal-controlled oscillator can be varied to a certain extent (*see* COLPITTS OSCILLATOR, HARTLEY OSCILLATOR, TUNED-INPUT/TUNED-OUTPUT OSCILLATOR, VARIABLE CRYSTAL OSCILLATOR).

In recent years, another type of VFO has evolved: the frequency synthesizer. This device contains a reference oscillator, usually crystal-controlled. Various multiplier and divider circuits, in conjunction with a phase-locked loop, allow operation at hundreds, thousands, or millions of discrete frequencies over a wide band. *See also* FREQUENCY SYNTHESIZER, PHASE-LOCKED LOOP.

VARIABLE INDUCTOR

A variable inductor is a device having an inductance that can be adjusted over a continuous range. The inductance of a coil can be varied in several ways. Basically, any variable inductor can be considered to fall into one of three categories: adjustable-turns, adjustable-core, or adjustable-phase.

The number of turns in a coil can be varied by means of a roller device. Roller inductors are sometimes used in transmatch circuits (*see* INDUCTOR TUNING, TRANSMATCH). The permeability of the core can be controlled by moving the slug in and out of the coil (*see* PERMEABILITY TUNING). Two inductors can be connected in series and the relative phase controlled by mechanical means, resulting in variable net inductance (*see* VARIOMETER).

VARIABLE-MU TUBE

See EXTENDED-CUTOFF TUBE.

VARIABLE RESISTOR

See POTENTIOMETER, RHEOSTAT.

VARIAC

A Variac is a variable transformer that is sometimes used to control the line of voltage in a common household utility circuit. Any variable transformer may be loosely termed a Variac. However, a true Variac is a special type of adjustable autotransformer consisting of a toroidal winding and a rotary contact (see illustration). The name Variac was originally used by the General Radio Company for their adjustable transformer.

A Variac provides continuous adjustment of the root-mean-square output voltage from 0 to 134 V in a 117-V line. The output voltage is essentially independent of the current drawn. A Variac can be used at the input to a high-voltage power supply to obtain adjustable direct-current output. This scheme is sometimes used in tube type radio-frequency power amplifiers.

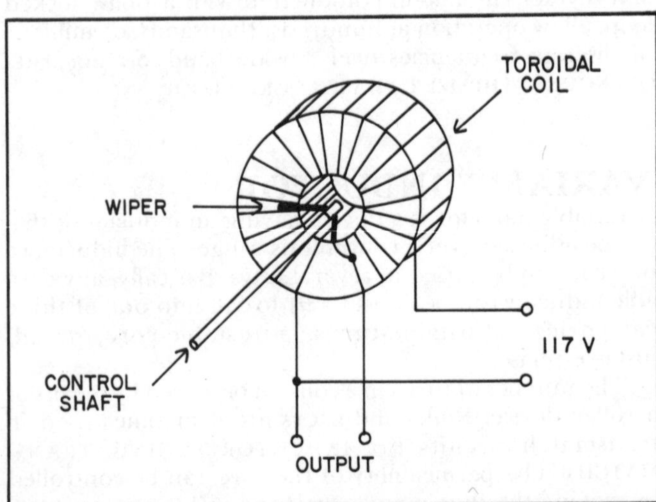

VARIAC: Principle of the Variac. The voltage is controlled by turning the shaft, moving the wiper along the wire turns.

VARICAP

See VARACTOR DIODE.

VARIOMETER

A variometer is a type of adjustable inductor that was popular at one time in transmatches and in the output circuits of radio transmitters. Today, the variometer is not often seen.

The principle of the variometer is shown in the illustration. There are two windings, one inside the other. The inner winding is mounted on a bearing, and can be turned as shown. The outer winding is not movable. The mutual inductance is maximum when the axes of the two coils coincide, and is minimum when the axes are at right angles. Continuous adjustment is possible by turning the inner coil through 90 degrees. This facilitates variable coupling between two circuits, such as a radio-frequency power amplifier and an antenna system.

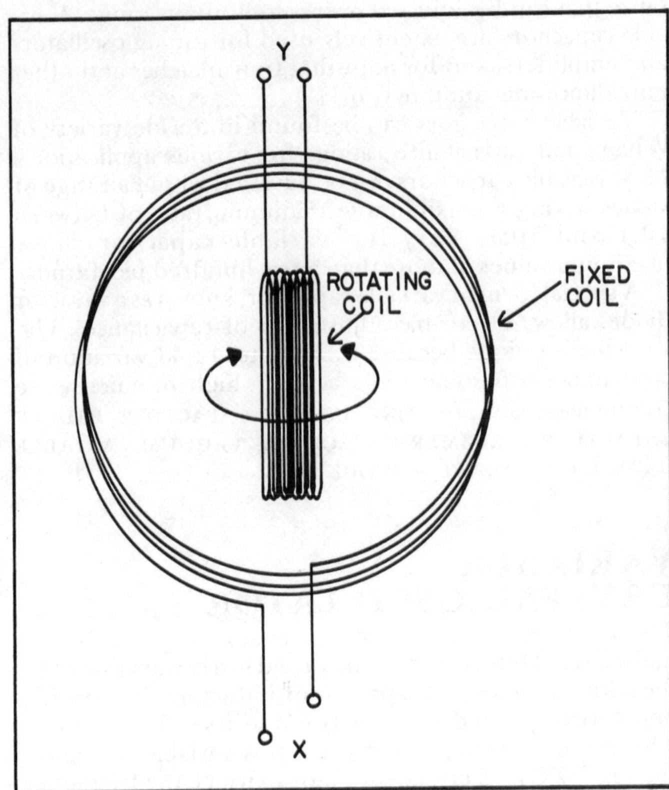

VARIOMETER: Operation of the variometer. The inner coil can be rotated to control the mutual inductance between X and Y.

The two coils of a variometer can be connected in series to obtain adjustable inductance. The inductance is maximum when the currents in the two coils are in phase, and is minimum when they are out of phase. The phase is controlled by turning the inner coil through 180 degrees. *See also* VARIABLE INDUCTOR.

VARINDOR

See SATURABLE REACTOR.

VARISTOR

See VOLTAGE-DEPENDENT RESISTOR.

VAR METER

The reactive power in a radio-frequency circuit can be

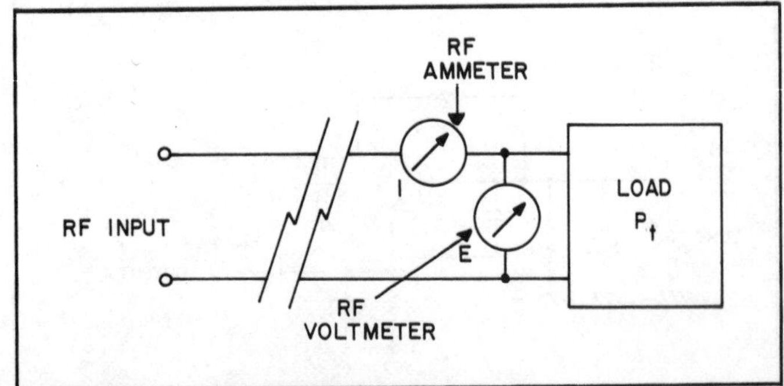

VAR METER: An ammeter and voltmeter can be used to measure VAR, provided the true power, Pt, is known. The ammeter and voltmeter must be located at the same point along the line between the input and the load—preferably at the load.

determined by a device called a VAR meter. A simple VAR meter consists of a radio-frequency ammeter in series and a radio-frequency voltmeter in parallel with a circuit (see illustration). If the true power P_t is known, the reactive power P_r, in VAR, is given by:

$$P_r^2 = E^2I^2 - P_t^2$$

where E and I are the radio-frequency voltage and current as indicated by the meters, in volts and amperes respectively.

A reflectometer can be used as a VAR meter if the scales are calibrated in watts. The reactive power in VAR is simply the wattmeter reading on the reflected-power scale, or with the function switch set for reflected power. *See also* POWER, REACTIVE POWER, REFLECTED POWER, REFLECTOMETER, TRUE POWER, VAR.

V BEAM

A V beam is a form of longwire antenna that exhibits gain in one or two fixed directions. V-beam antennas are used mostly at the high frequencies (3 MHz to 30 MHz), although they are sometimes used at medium frequencies (300 kHz to 3 MHz). The V beam is essentially half of a rhombic antenna (*see* RHOMBIC ANTENNA). The illustration is a diagram of the configuration.

The gain and directional characteristics of a V beam depend on the physical lengths of the wires. The longer the wires in terms of the wavelength, the greater the gain and the sharper the directional pattern. The gain and directional characteristics of the V beam are similar to those of the rhombic.

The V beam operates as two single longwire antennas in parallel. As the wires are made longer, the lobe of maximum radiation becomes more nearly in line with the wires (*see* LONGWIRE ANTENNA). The apex angle is chosen so that the major lobes from the wires coincide; thus, long V-beam antennas have small apex angles and short V beams have larger apex angles.

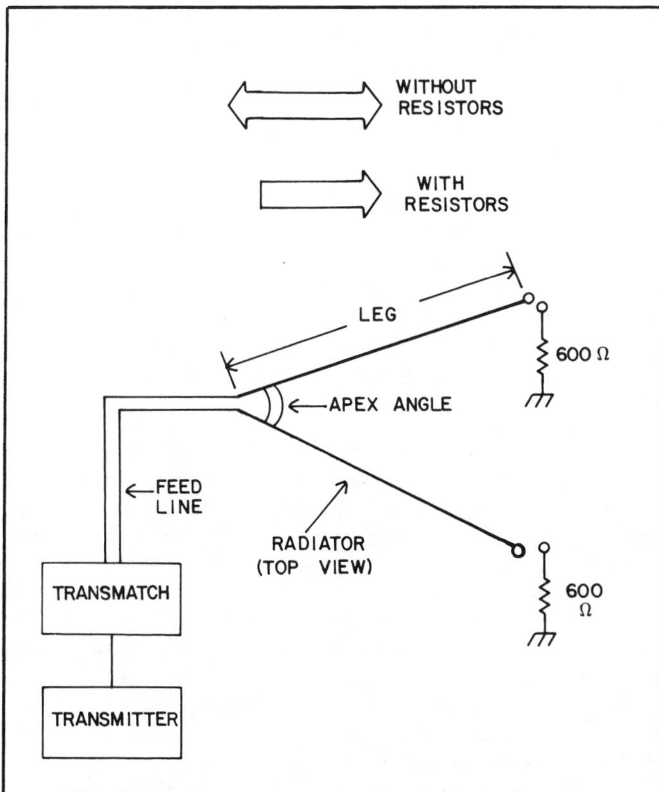

V BEAM: The double arrow shows the dirctinal maxima if terminating resistors are not used. The single arrow shows the directional pattern when terminating resistors are used.

The V beam shown is fed at the apex with open-wire line. A bidirectional pattern results, as shown by the double arrow. The feed-point impedance varies with the frequency, but this is of no concern if low-loss line is used in conjunction with a transmatch. Efficient operation can be expected at all frequencies at which the lengths of the legs are at least ½ wavelength. Optimum performance will be at the frequency for which the angle of radiation, with respect to either wire, is exactly half the apex angle.

To obtain unidirectional operation with a V beam, a 600-ohm, noninductive, high-power resistor can be connected between the far end of each wire and ground. This eliminates half of the pattern, as shown by the single arrow. The addition of the terminating resistors results in a nearly constant feed-point impedance of 300 to 450 ohms at all frequencies for which the wires measure ½ wavelength or longer.

The main advantage of the V beam is its power gain and its simplicity. The main disadvantages are that it cannot be rotated, and that it requires a large amount of space. In recent years, the quad and Yagi antennas have become more popular than the V beam for high-frequency operation. *See also* QUAD ANTENNA, YAGI ANTENNA.

VDR

See VIDEO DISK RECORDING.

VECTOR

A vector is a quantity that has magnitude and direction. Examples of vectors include the force of gravity, the movement of electrons in a wire, and the propagation of an electromagnetic field through space. If direction is not specified, a quantity is called a scalar (*see* SCALAR). A vector for which just two directions can exist is sometimes called a directed number (*see* DIRECTED NUMBER).

A vector is usually represented in equations by a capital, such as X. Vectors may also be represented in equations by capital letters in boldface or italics, or both.

Vectors are geometrically represented by arrows whose length indicate magnitude and whose orientation indicate direction. Some examples of vectors are shown in the illustration. These are velocity vectors, corresponding to various speeds in miles per hour (MPH) and azimuth bearings in degrees (°) as follows:

X = 30 MPH at azimuth 45°
Y = 60 MPH at azimuth 180°
Z = 20 MPH at azimuth 280°

The examples shown are vectors in polar coordinates. Vectors can also be expressed in rectangular coordinates. Complex numbers are sometimes denoted as ordered pairs in a Cartesian plane (*see* CARTESIAN COORDINATES, COMPLEX NUMBER, POLAR COORDINATES).

Sometimes it is necessary to consider only the length of a vector. The magnitude alone is denoted by placing vertical lines on either side of the designator, such as |X|, |Y|, and

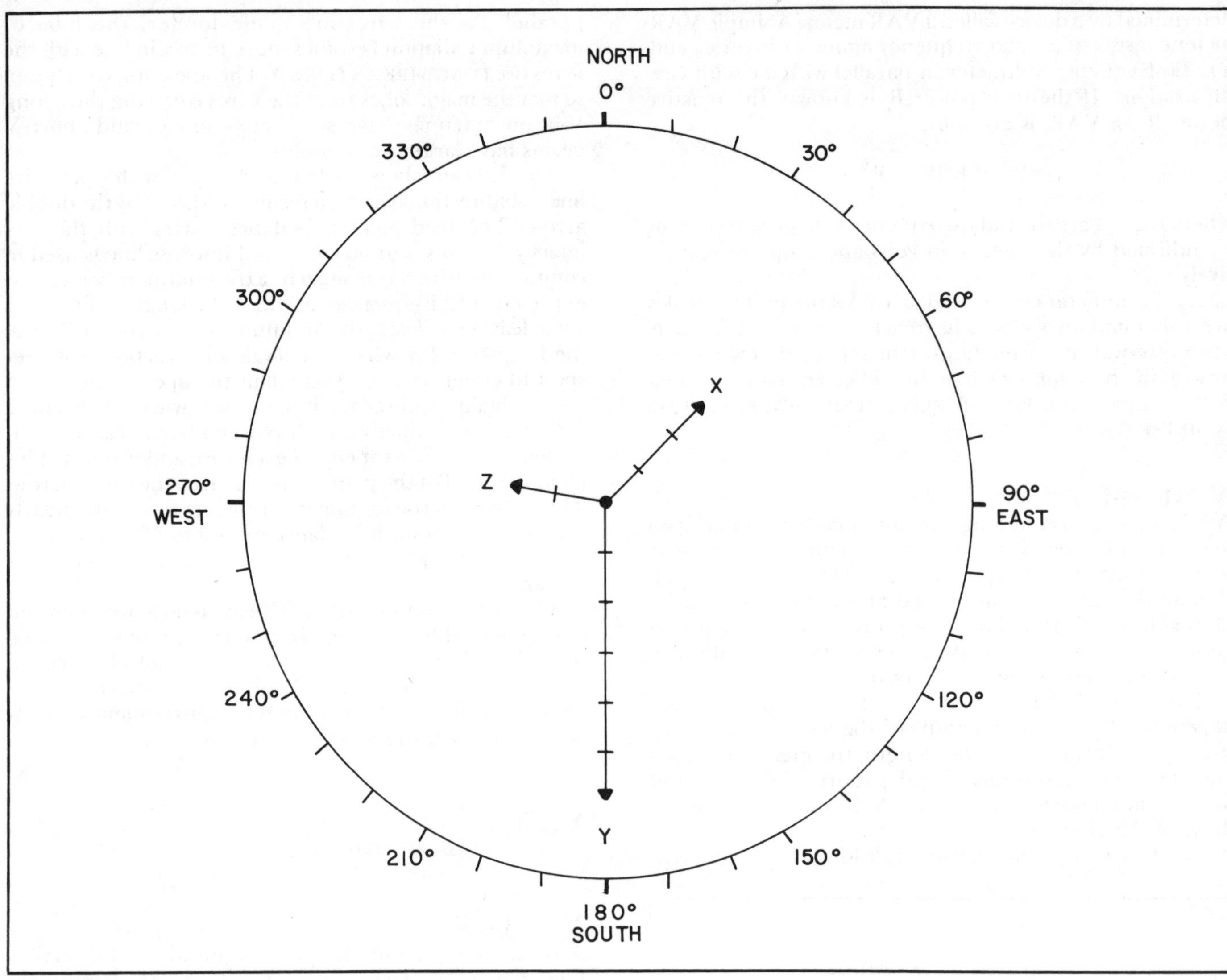

VECTOR: Examples of velocity vectors in polarcoordinates.

|Z|. Alternatively, the length of a vector can be denoted by removing the arrow above the capital letter, or by eliminating the boldface or italic notation.

Vectors are important in applied mathematics. In electronics, vectors simplify the definitions and concepts of such abstract things as impedance and the behavior of fields. Vector quantities can be added, subtracted, and multiplied, although the procedures differ from the familiar arithmetic operations used with scalars. *See also* CROSS PRODUCT, DOT PRODUCT, VECTOR ADDITION, VECTOR DIAGRAM.

VECTOR ADDITION

It is often necessary to add two or more vectors together in order to determine certain net effects. For example, we might have two magnetic fields acting in different directions and having different intensities; their net effect would be found by means of a vector sum.

When vectors are expressed in Cartesian coordinates, sums are quite simple to determine. For two vectors X and Y in n dimensions, represented by ordered n-tuples:

$$X = (x_1, x_2, x_3, \ldots, x_n)$$
$$Y = (y_1, y_2, y_3, \ldots, y_n)$$

the sum is simply

$$X+Y = (x_1 + y_1, x_2 + y_2, x_3 + y_3, \ldots, x_n + y_n)$$

For example, if X = (3,5) and Y = (−2,−4), then

$$X+Y = (3-2, 5-4) = (1,1)$$

In Cartesian coordinates, vectors are added or subtracted by adding or subtracting their corresponding components. But in polar coordinates, a different method must be used for adding vectors. The geometric method, which works in any coordinate system, consists of the construction of a parallelogram (*see* VECTOR DIAGRAM).

Vector addition is commutative and associative. That is, for any two vectors X and Y, X+Y = Y+X; and for any three vectors X, Y, and Z, (X+Y)+Z = X+(Y+Z).

The commutative and associative properties do not always hold for vector subtraction unless the subtraction is defined in terms of adding a negative (inverse) vector. The vector X−Y can be obtained by reversing the direction of Y and adding the resultant to X. For a given vector:

$$Y = (y_1, y_2, y_3, \ldots, y_n)$$

the inverse vector −Y is the result of multiplying each component by −1:

$$-Y = (-y_1, -y_2, -y_3, \ldots, -y_n)$$

The difference X−Y is equal to the sum X+(−Y). This definition of vector subtraction allows us to use the commutative and associative properties in expressions containing both vector addition and subtraction. *See also* VECTOR.

VECTOR DIAGRAM

Vector quantities are geometrically represented by means of a vector diagram. In the vector diagram, the magnitude is denoted as the length of a line segment, and the direction is indicated by the orientation of the line segment and an arrowhead.

Vector diagrams can be easily drawn in two dimensions to find the sum or difference of two vectors. An example is shown in the accompanying illustration. Two vectors, X and Y, are added by completing a parallelogram as shown. The difference, X−Y, is determined by constructing the vector −Y, having the same length as Y but opposite direction, and then finding the sum X+(−Y) by the construction of a parallelogram.

Vector diagrams can be drawn in three or more dimensions by means of a computer. A three-dimension vector diagram can be visualized using computer graphics. Diagrams in four or more dimensions can be visualized only by looking at cross sections. *See also* VECTOR, VECTOR ADDITION.

VECTOR MULTIPLICATION

See CROSS PRODUCT, DOT PRODUCT.

VECTOR SUBTRACTION

See VECTOR ADDITION.

VELOCITY FACTOR

The speed of electromagnetic progagation through a vacuum is approximately 186,282 miles (299,792 kilometers) per second. In material substances, however, the speed is reduced somewhat. The velocity factor, v, is expressed as a percentage by:

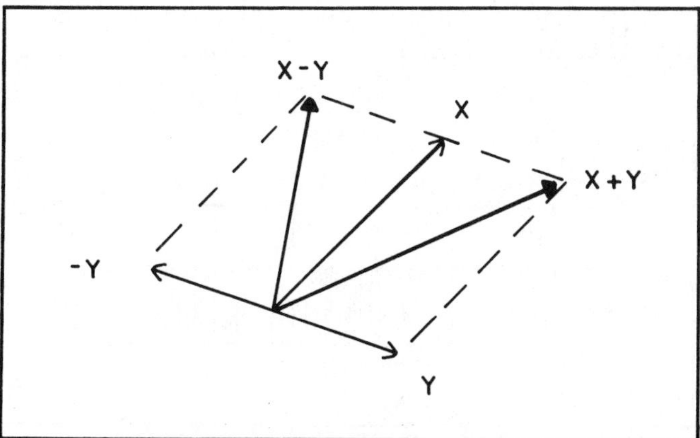

VECTOR DIAGRAM: Addition and subtraction of vectors using the parallelogram method.

$$v = 100c'/c$$

where c′ represents the speed of propagation in a particular substance, and c is the speed of light in a perfect vacuum. (Both c′ and c must be specified in the same units.)

The velocity factor affects the wavelength of an electromagnetic field at a fixed frequency. For a disturbance having a frequency f in megahertz, the wavelength λ in a substance with velocity factor v is given in feet by:

$$\lambda = 984v/f$$

and in meters by:

$$\lambda = 300v/f$$

Velocity factor is an important consideration in the construction of antennas. An electromagnetic field travels along a single wire in air at a speed of about 0.95c. An electrical wavelength is therefore only 95 percent as long as in free space. As the conductor diameter is made larger, the velocity factor decreases a bit, perhaps to 92 or 93 percent for large-diameter metal tubing at very-high frequencies. The velocity factor must be taken into account when cutting an antenna to a resonant length for a particular frequency. The radiating and parasitic elements must be cut to length vm, if m represents the length determined in terms of a free-space wavelength.

Velocity factor is also important when it is necessary to cut a transmission line to a certain length. To make a half-wavelength stub, for example, from a transmission line with velocity factor v, we must multiply the free-space half wavelength by v to obtain the proper length for the line. We must do the same thing if tuned feeders are used with an antenna (*see* STUB, TUNED FEEDERS).

VELOCITY FACTOR: VELOCITY FACTORS OF VARIOUS TYPES OF TRANSMISSION LINES.

Line Type or Manufacturer's No.	Velocity Factor, Percent
Coaxial cable, RG-58/U, solid dielectric	66
Coaxial cable, RG-59/U, solid dielectric	66
Coaxial cable, RG-8/U, solid dielectric	66
Coaxial cable, RG--58/U, foam dielectric	75 - 85
Coaxial cable, RG-59/U, foam dielectric	75 - 85
Coaxial cable, RG-8/U, foam dielectric	75 - 85
Twin-lead, 75-ohm, solid dielectric	70 - 75
Twin-lead, 300-ohm, solid dielectric	80 - 85
Twin-lead, 300-ohm, foam dielectric	85 - 90
Open-wire with plastic spacers, 300-ohm	90 - 95
Open-wire with plastic spacers, 450-ohm	90 - 95
Open-wire, homemade, 600-ohm	95

In an open-wire transmission line having few spacers, the velocity factor is near 95 percent. But when dielectric material is present, for example in a twin-lead line, the velocity factor is reduced. In a coaxial line, the velocity factor is still smaller. The velocity factor of a transmission line depends on:

- The type of dielectric material used in construction
- The amount of the electromagnetic field that is contained within the dielectric material
- The presence of objects near a parallel-wire line
- The extent of contamination of the dielectric because of aging in a coaxial cable

The most common dielectric materials used in fabrication of feed lines are solid polyethylene, with a velocity factor of 66 percent, and foamed polyethylene, with a velocity factor of 75 to 85 percent.

The table lists the velocity factors of some common types of transmission lines, assuming ideal operating conditions. These values are approximate. For precise velocity-factor ratings in critical transmission-line applications, the manufacturer should be consulted.

VELOCITY MICROPHONE

A velocity microphone is a transducer that converts sound waves into electric currents by means of the interaction of a moving conductor with a magnetic field. The principle of operation of the velocity microphone is similar to that of the dynamic microphone (*see* DYNAMIC MICROPHONE), but the construction is different. A velocity microphone is sometimes called a ribbon microphone.

The illustration shows the construction of a velocity microphone. A thin ribbon, made from a nonmagnetic metal such as aluminum or copper, is suspended in a strong magnetic field. The magnetic lines of flux are parallel with the plane of the ribbon. When an acoustic disturbance is present, the ribbon moves back and forth, cutting across the magnetic lines in the ribbon. The current has the same waveshape as the sound that moves the ribbon. The output terminals are connected to the ends of the ribbon.

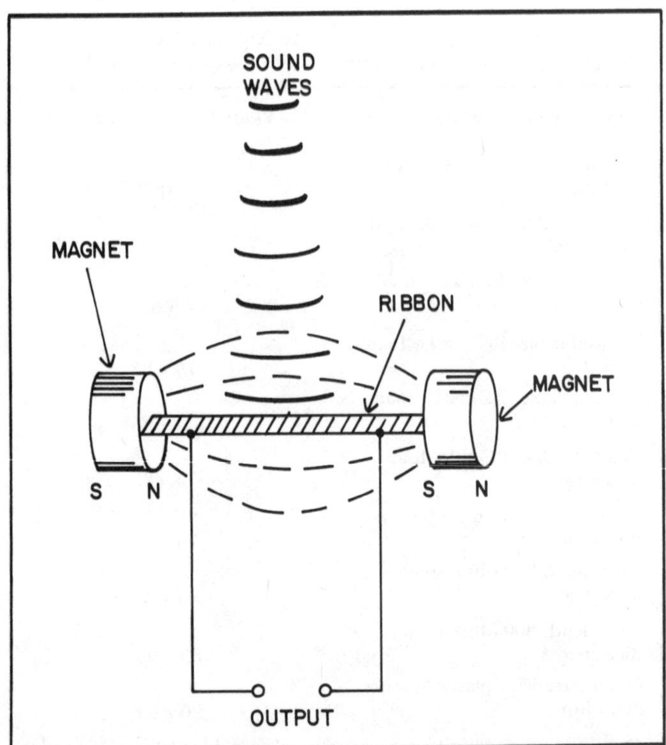

VELOCITY MICROPHONE: Principle of operation of a velocity microphone. The impinging sound waves vibrate the ribbon in the magnetic field, producing alternating current at the output.

VELOCITY-MODULATED OSCILLATOR

See TRAVELING-WAVE TUBE, VELOCITY MODULATION

VELOCITY MODULATION

A beam of electrons can be modulated by varying the speed at which the electrons move. This is the principle of operation of certain types of vacuum tubes. When the electrons are accelerated, their energy is increased, and when they are decelerated, their energy is reduced. Although the number of electrons per unit volume fluctuates, the actual current remains constant or nearly constant.

When an electron beam is velocity-modulated, a pattern of waves is produced. The density of the beam is inversely proportional to the acceleration of the electrons, and 90 degrees out of phase with their instantaneous velocity (see illustration). The pattern of waves may stand still along the length of the beam, move toward the anode, or move toward the cathode. If the pattern of waves moves toward the anode, we say that the beam is forward-modulated; if it

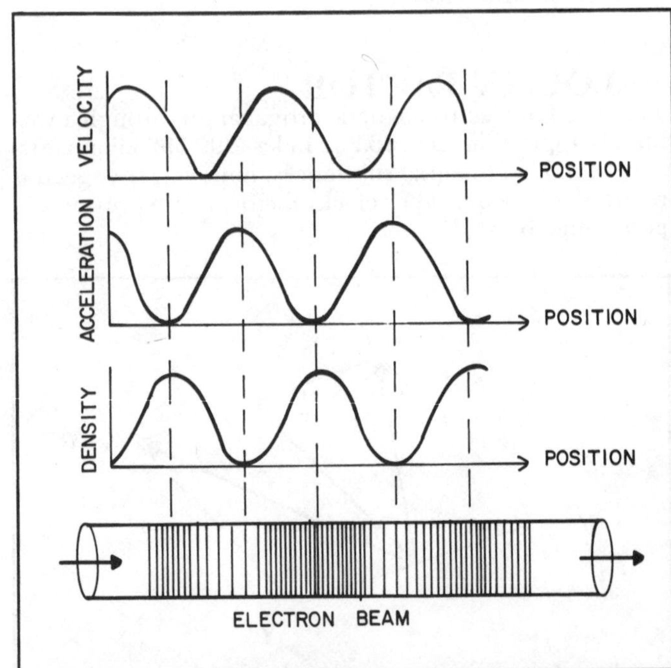

VELOCITY MODULATION: Relationship between electron density, acceleration, and velocity for a velocity-modulated beam.

moves toward the cathode, we say the beam is backward-modulated.

Velocity modulation is used in traveling-wave tubes to obtain amplification and oscillation at ultra-high and microwave frequencies. *See also* KLYSTRON, PARAMETRIC AMPLIFIER, TRAVELING-WAVE TUBE.

VELOCITY OF LIGHT

See LIGHT

VELOCITY OF SOUND

See SOUND.

VENN DIAGRAM

A Venn diagram is a simple, graphic method for illustrating the relationships amoung sets. Venn diagrams can be used to find the unions and intersections of various sets and their complements.

The union of two sets is the set of elements in one set or the other, or both. There is a clear resemblance between set union and the Boolean OR operation. The intersection of two sets is the set of elements in both sets; this resembles the logic AND. The complement of a set is the set of all elements outside tht set; there is a resemblance to the logic NOT.

The illustration demonstrates a Venn diagram showing three sets X, Y, and Z, with elements as follows:

X = {x1, x2, x3, x4, x5}

Y = {y1, y2, y3, y4, y5, y6}

Z = {x4, x5, y1, y2, z1, z2}

There are also three elements, w1, w2, and w3, that do not belong to any of the sets X, Y, or Z. The set of all elements in the diagram is called the universal set.

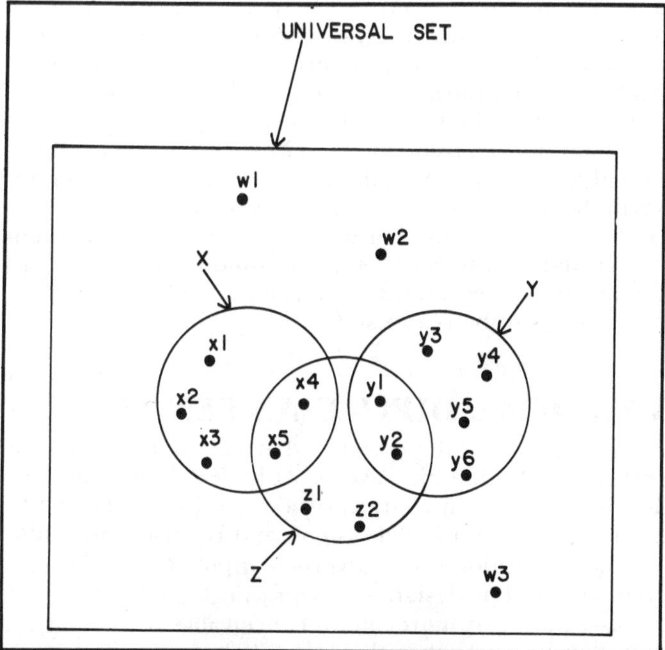

VENN DIAGRAM: An example of a Venn diagram, showing the relationship among elements of various sets.

From the Venn diagram, it is easy to find the various union and intersection sets. Complements can also be observed. If we let union be represented by addition, intersection by multiplication, and complementation by an apostrophe ('), then we can clearly see that:

X' = {w1, w2, w3, y1, y2, y3, y4, y5, y6, z1, z2}

Y' = {w1, w2, w3, x1, x2, x3, x4, x5, z1, z2}

Z' = {w1, w2, w3, x1, x2, x3, y3, y4, y5, y6}

XY = ϕ (the empty set)

XZ = {x4, x5}

YZ = {y1, y2}

X+Y = {x1, x2, x3, x4, x5, y1, y2, y3, y4, y5, y6}

X+Z = {x1, x2, x3, x4, x5, y1, y2, z1, z2}

Y+Z = {x4, x5, y1, y2, y3, y4, y5, y6, z1, z2}

There are many other examples that could be listed.

Venn diagrams are sometimes applied to Boolean algebra, because of the similarities between set union and logic OR, set intersection and logic AND, and set complementation and logic NOT. *See also* AND GATE, BOOLEAN ALGEBRA, INVERTER, OR GATE.

VENTILATION

See AIR COOLING

VENTRICULAR FIBRILLATION

See HEART FIBRILLATION.

VERIFIER

After information has been recorded on magnetic tape, disks, paper tape, cards, or any other storage medium, it is desirable to ascertain that the data is free from errors before running it through a computer. A verifier is a device that checks the recording to ensure that the data is properly stored. This saves expensive computer time if errors are found.

A verifier operates by comparing the recorded data, bit by bit, with the original data. The verifier identifies any discrepancies according to location in the list or program.

VERNIER

Vernier is a special form of gear-drive and scale that is used for precision adjustment of variable capacitors, inductors, potentiometers, and other controls.

The vernier drive is a simple gear mechanism. The control knob must be rotated several times for the shaft to turn through its complete range. The gear ratio determines the ratio of the knob and shaft rotation rates.

The vernier scale is designed to aid an operator in

interpolating readings between divisions. An auxiliary scale, located above the main scale, is marked in divisions that are 90 percent of the size of the main-scale divisions (see illustration). The auxiliary-scale lines are numbered 0 through 9, representing tenths of a division on the main scale.

When the dial is set to a point between two divisions on the main scale, one mark on the auxiliary scale will line up with some mark on the main scale. The operator looks for this coincidence, and notes the number of the auxiliary-scale mark at which the alignment occurs. This is the number of tenths of the way that the control is set between the two main-scale divisions. In the illustration, for example, the control is set to 75.3 units.

VERTICAL ANTENNA

A vertical antenna is any antenna in which the radiating element is perpendicular to the average terrain. There are many types of vertical antennas, and they are used at all frequencies from the very low to the ultra-high.

Quarter-Wave Verticals. The simplest form of vertical antenna is a quarter-wave radiator mounted at ground level. The radiator is fed with a coaxial cable. The center conductor is connected to the base of the radiator, and the shield is connected to a ground system. The feed-point impedance of a ground-mounted vertical is a pure resistance, and is equal to 37 ohms plus the ground-loss resistance. The ground-loss resistance affects the efficiency of athis type of antenna system. A set of grounded radials helps to minimize the loss (*see* ANTENNA FREQUENCY, ANTENNA GROUND SYSTEM, EARTH CONDUCTIVITY, RADIAL).

The base of a vertical antenna can be elevated at least a quarter wavelength above the ground, and a system of radials used to reduce the loss. This is called a ground-plane antenna (*see* GROUND-PLANE ANTENNA). The impedance can be varied by adjusting the angle of the radials with respect to the horizontal (*see* DROOPING RADIAL).

At some frequencies, the height of a quarter-wave vertical is unmanageable unless inductive loading is used to reduce the physical length of the radiator. This technique is used mostly at very low, low, and medium frequencies, and to some extent at high frequencies (*see* INDUCTIVE LOADING).

A vertical antenna can be made resonant on several frequencies by the use of multiple loading coils, or by inserting traps at specific points along the radiator (*see* TRAP ANTENNA).

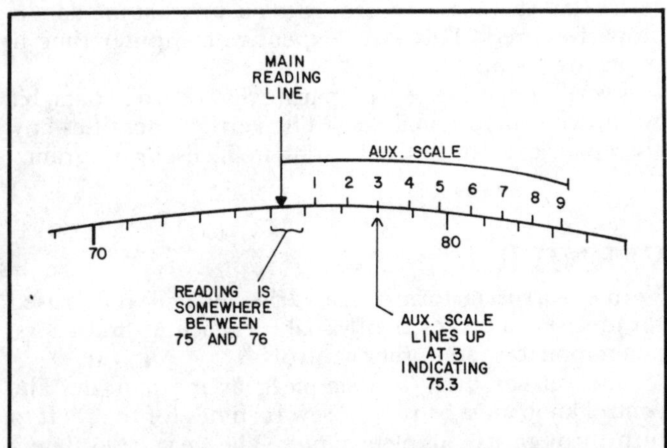

VERNIER: The Vernier scale allows precise interpolation of dial readings to the tenth of a division.

Half-Wave Verticals. A half-wave radiator can be fed at the base by using an impedance transformer. This provides more efficient operation for a ground-mounted antenna, as compared with a quarter-wavelength radiator, because the radiation resistance is much higher. A quarter-wave matching transformer can be attached to a half-wave vertical to form a J antenna (*see* J ANTENNA). A half-wave radiator can be fed at the center with coaxial cable to form a vertical dipole (*see* VERTICAL DIPOLE ANTENNA). If the feed line is run through one element of a vertical dipole, the configuration is called a coaxial antenna (*see* COAXIAL ANTENNA). A half-wave element can even be fed at the top with open-wire line (*see* TOP LOADING).

Multiple Verticals. Two or more vertical radiators can be combined in various ways to obtain gain and/or directivity. Parasitic elements can be placed near vertical radiators to achieve the same results (*see* PARASITIC ARRAY, PARASITIC ELEMENT, PHASED ARRAY). Two or more vertical antennas can be stacked in collinear fashion to obtain an omnidirectional gain pattern (*see* COLLINEAR ANTENNA). A Yagi can be oriented so that all of its elements are vertical (*see* YAGI ANTENNA).

Advantages and Disadvangates of Vertical Antennas. Vertical antennas radiate, and respond to, electromagnetic fields having vertical polarization (*see* VERTICAL POLARIZATION). This is an asset at very low, low, and medium frequencies, because it facilitates efficient surface-wave propagation (*see* SURFACE WAVE). Vertical antennas provide good low-angle radiation at high frequencies, an attribute for long-distance ionospheric communication. At medium and high frequencies, vertical antennas are often more convenient to install than horizontal antennas because of limited available space. Self-supporting vertical antennas require essentially no real estate. At very high and ultra high frequencies, some communications systems use vertical polarization; a vertical antenna is obviously preferable to a horizontal antenna in such cases.

One of the chief disadvantages of a vertical antenna, especially at high frequencies, is its susceptibility to reception of man-made noise; most artificial noise sources emit vertically polarized electromagnetic fields. Ground-loss problems present another disadvantage of vertical antennas at high frequencies. A ground-mounted vertical antenna usually requires an extensive radial system if the efficiency is to be reasonable.

VERTICAL DIPOLE ANTENNA

A half-wave vertical radiator, fed at the center, is called a vertical dipole antenna. A vertical dipole exhibits excellent low-angle radiation characteristics, good efficiency without the need for a radial system, and relative simplicity.

The illustration shows a vertical dipole for use at high frequencies. The design is very practical at frequencies of about 10 MHz or more, and is conceivable at frequencies considerably lower than that, provided adequate guying is used. The radiating element is constructed from aluminum tubing. The feed line should be run away from

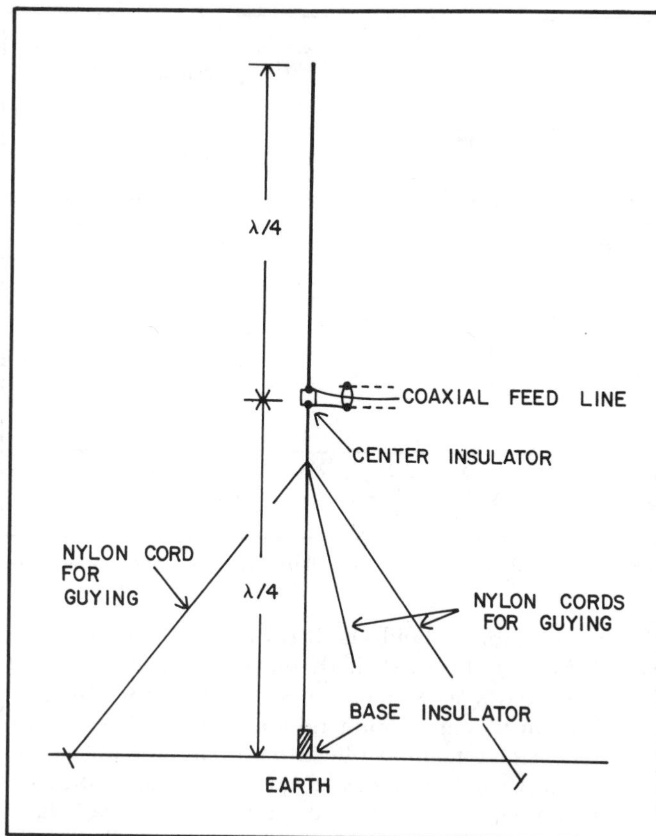

VERTICAL DIPOLE ANTENNA: This is a balanced antenna.

the radiator at a right angle for a distance of at least a quarter wavelength. The feed-point impedance is approximately 73 ohms at the resonant frequency of the antenna.

A vertical dipole can be fed with coaxial line through one of the halves of the radiating element. This type of arrangement is called a coaxial antenna. *See also* COAXIAL ANTENNA, DIPOLE ANTENNA.

VERTICAL LINEARITY

In a television video system, vertical linearity refers to the faithful reproduction of the image along the vertical axis. Ideally, all of the scanning lines in the raster are equally spaced. If this is not the case, the image will appear distorted.

Most television receivers have a vertical-linearity control, usually placed on the chassis inside the cabinet to prevent accidental tampering. *See also* RASTER.

VERTICAL METAL-OXIDE SEMICONDUCTOR FIELD-EFFECT TRANSISTOR

A relatively new form of metal-oxide semiconductor field-effect transistor (MOSFET) is the vertical MOSFET or VMOSFET. The VMOSFET can be used in power-amplifier circuits at audio and radio frequencies.

The VMOSFET operates in the enhancement mode (*see* ENHANCEMENT MODE). The source is located at the top of a four-layer semiconductor. The drain is connected to a substrate located at the bottom. The current therefore moves in a vertical direction, relative to the construction of the device.

The VMOSFET can be used with high efficiency at frequencies well above 100 MHz. Several watts of radio-frequency output can be obtained at considerably higher frequencies. The efficiency of VMOSFET devices is excellent; in high-frequency power amplifiers, values of 70 to 75 percent are realizable with Class-B and Class-C arrangements. The VMOSFET is more stable than most other semiconductor devices at very high and ultra high frequencies. *See also* FIELD-EFFECT TRANSISTOR, METAL-OXIDE SEMICONDUCTOR FIELD-EFFECT TRANSISTOR.

VERTICAL POLARIZATION

Vertical polarization is a condition in which the electric lines of flux of an electromagnetic wave are vertical, or perpendicular to the surface of the earth. In communications, vertical polarization has certain advantages and disadvantages, depending on the application and the wavelength.

At low and very-low frequencies, vertical polarization is ideal, because surface-wave propagation, the major mode of propagation at these wavelengths, requires a vertically polarized field. Surface-wave propagation is effective in the standard amplitude-modulation (AM) broadcast band as well; most AM broadcast antennas are vertical.

Vertical polarization is often used at high frequencies, mainly because vertical antennas can be erected in a very small physical space (*see* VERTICAL ANTENNA). At very high and ultra high frequencies, vertical polarization is used mainly for mobile communications, and also in repeater communications.

The main disadvantage of vertical polarization is that most man-made noise tends to be vertically polarized. Thus, a vertical antenna picks up more of this interference than would a horizontal antenna in most cases. At very high and ultra high frequencies, vertical polarization results in more "flutter" in mobile communications, as compared with horizontal polarization. *See also* CIRCULAR POLARIZATION, HORIZONTAL POLARIZATION, POLARIZATION.

VERTICAL STACKING

See STACKING.

VERTICAL SYNCHRONIZATION

In television communications, the picture signals must be synchronized at the transmitter and receiver. The electron beam in the television picture tube scans from left to right and top to bottom, just as you read the pages of a book. At any given instant of time, in a properly operating television system, the electron beam in the receiver picture tube is in exactly the same relative position as the scanning beam in the camera tube. This requires synchronization of the vertical and the horizontal positions of the beams.

When the vertical synchronization of a television system is lost, the picture appears split along a horizontal line. The picture may appear to "roll" upward or downward continually, much like the way a movie picture "rolls" when the film slips in the projector. Even a small error in synchronization results in a severely disturbed picture at the receiver.

The transmitted television picture signal contains vertical-synchronization pulses at the end of every frame (complete picture). This tells the receiver to move the electron beam from the end of one frame to the beginning of the next.

Both vertical and horizontal synchronization are necessary for a picture signal to be transmitted and received without distortion. *See also* HORIZONTAL SYNCHRONIZATION, PICTURE SIGNAL, TELEVISION.

VERY HIGH FREQUENCY

The very-high-frequency (vhf) range of the radio spectrum is the band extending from 30 MHz to 300 MHz. The wavelengths corresponding to these limit frequencies are 10 m and 1 m. Very-high-frequency waves are sometimes called metric waves because the wavelength is on the order of several meters. Channels and bands at very high frequencies are allocated by the International Telecommunication Union, headquartered in Geneva, Switzerland.

At vhf, electromagnetic fields are somewhat affected by the ionosphere and the troposphere. Propagation can occur via the E and F layers of the ionosphere (*see* E LAYER, F LAYER, PROPAGATION CHARACTERISTICS) in the lower part of the vhf spectrum. Tropospheric bending, ducting, reflection, and scattering take place throughout the vhf band (*see* DUCT EFFECT, TROPOSPHERIC PROPAGATION, TROPOSPHERIC-SCATTER PROPAGATION). Exotic modes of signal propagation, such as auroral, meteor-scatter, and moonbounce, are also feasible at vhf (*see* AURORAL PROPAGATION, METEOR SCATTER, MOONBOUNCE).

The vhf band is very popular for mobile communications and for repeater operation. Some satellite communications is also done at vhf. Wideband modulation is used to some extent ; the most common example is television broadcasting. *See also* ELECTROMAGNETIC SPECTRUM, FREQUENCY ALLOCATIONS.

VERY-LARGE-SCALE INTEGRATION

A common kind of integrated-circuit chip incorporates from 100 to 1,000 individual logic gates. Such a chip is called a very-large-scale-integration (VLSI) chip. VLSI technology is widely used in the manufacture of microcomputers and peripheral circuits.

VERY LOW FREQUENCY

The range of electromagnetic frequencies extending from 10 kHz to 30 kHz is known as the very-low-frequency (vlf) band. The wavelengths corresponding to these frequencies are 30 km and 10 km. Some engineers consider the lower limit of the vlf band to be 3 kHz, or a wavelength of 100 km.

The ionosphere returns all vlf signals to the earth, even if they are sent straight upward. If a vlf signal arrives from space, we will not hear it at the surface because the ionosphere will completely block it.

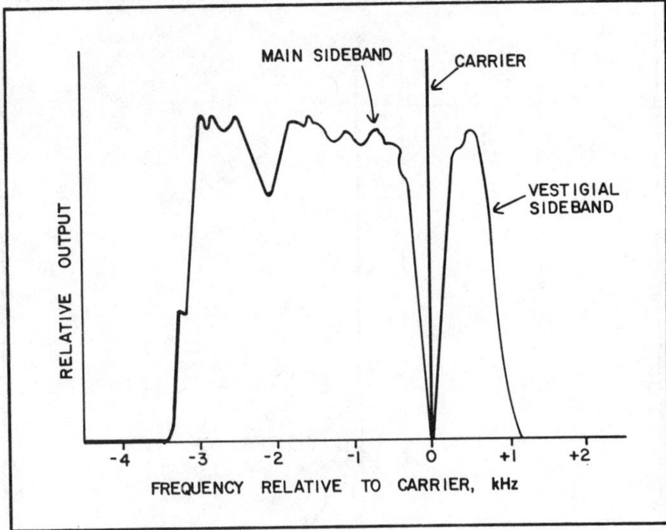

VESTIGIAL SIDEBAND: Spectral illustration of a vestigial-sideband signal.

Electromagnetic fields at vlf propagate very well along the surface of the earth if they are vertically polarized. Very-low-frequency signals may someday be used by space travelers on planets having no ionosphere. Even on the moon, which has no ionosphere to return radio waves back to ground, vlf might be useful for over-the-horizon communication via the surface wave. Another mode of signal travel, known as waveguide propagation, takes place in the vlf band on our planet. *See* PROPAGATION CHARACTERISTICS, SURFACE WAVE, WAVEGUIDE PROPAGATION.

VESTIGIAL SIDEBAND

Vestigial-sideband transmission is a form of amplitude modulation (AM) in which one of the sidebands has been largely eliminated. The carrier wave and the other sideband are unaffected. Vestigial-sideband transmission differs from single sideband in that the carrier is not suppressed (*see* SIDEBAND, SINGLE SIDEBAND).

In television broadcasting, vestigial-sideband transmission is used to optimize the efficiency with which the channel is utilized. This mode may also be used for ordinary communications. A spectral illustration of a typical vestigial-sideband signal is shown in the illustration. *See also* AMPLITUDE MODULATION.

VFO

See VARIABLE-FREQUENCY OSCILLATOR.

VHF

See VERY HIGH FREQUENCY.

VIBRATOR

A vibrator is a device that interrupts a direct current, producing square-wave pulses. A vibrator operates on the same principle as a solid-state chopper. The terms chopper and vibrator are often used interchangeably. Vibrators are employed in certain direct-current power supplies. *See also* CHOPPER, VIBRATOR POWER SUPPLY.

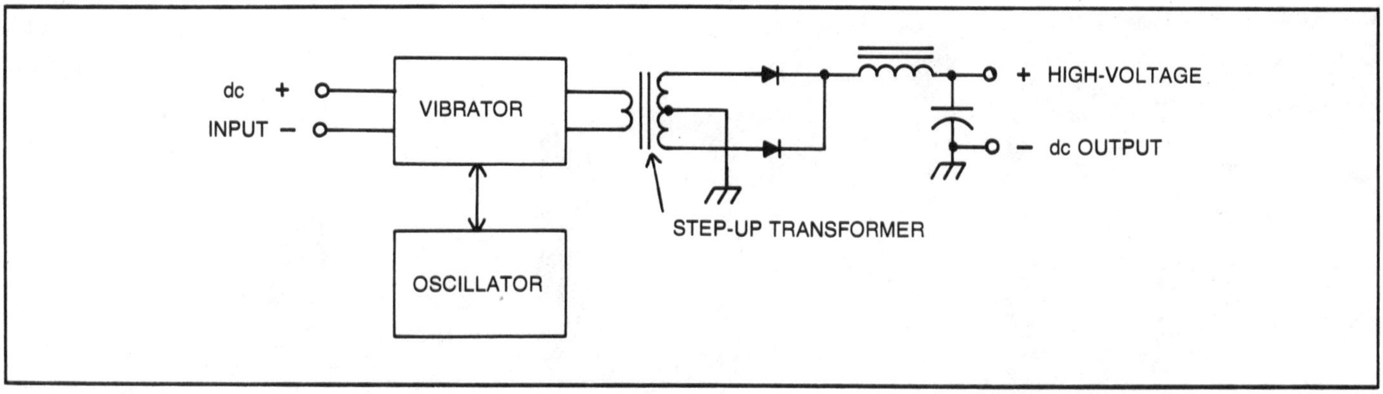

VIBRATOR POWER SUPPLY: Block diagram of a vibrator power supply.

VIBRATOR POWER SUPPLY

A vibrator power supply is a circuit that produces a high direct-current voltage from a lower direct-current voltage. Such devices are often used in the operation of mobile or portable equipment containing vacuum tubes. Some vibrator power supplies can provide 117 V of alternating current, allowing certain types of household appliances to be used with low-voltage, direct-current sources of power.

A block diagram of a vibrator power supply is shown in the illustration. The principle of operation is the same as that of the chopper supply (*see* CHOPPER POWER SUPPLY). The vibrator consists of a relay connected in a relaxation-oscillator configuration (*see* VIBRATOR). The relay contacts chatter, interrupting the current from the source. The resulting direct-current pulses are converted into true alternating current, as well as boosted in voltage, by a transformer. If the output voltage is 117 V root-mean-square (RMS) with a frequency near 60 Hz, the vibrator supply will provide satisfactory power for simple appliances. (Devices that require a good sine wave, or a frequency of precisely 60 Hz, cannot generally be used with vibrator supplies.)

The supply shown incorporates a rectifier and filter after the transformer. Thus, the supply operates as a dc-to-dc converter (*see* DC-TO-DC CONVERTER). The voltage is suitable for use with medium-power tube type amplifiers, which are still used in some radio transmitters today.

VIDEO DISK RECORDING

In recent years, magnetic recording technology has advanced to the point where video signals can be reproduced with quality and resolution comparable to that of fast-scan television. Magnetic disks, capable of storing large amounts of information, are used to record and reproduce video signals in a device called a video disk recorder (VDR). Video disk recorders make it possible to view a given television program at any time.

The video disk recorder operates in a manner similar to a computer disk drive. The television picture is demodulated, resulting in a composite video signal that is impressed on the magnetic disk (*see* COMPOSITE VIDEO SIGNAL). The program is played back in the same manner as any magnetic recording.

Video programs are sometimes recorded on a special form of magnetic tape. The advantage of the disk recorder is that any program, or part of a program, can be reproduced without the need for fast-forward or rewinding operations. This saves time. *See also* DISK DRIVE, FLOPPY DISK, MAGNETIC RECORDING, VIDEO TAPE RECORDING.

VIDEO DISPLAY TERMINAL

A video display terminal is a device that facilitates transmission and reception of digital and graphic information. A typical video terminal contains a cathode-ray tube, which shows the data to the operator as it is sent and received, and a keyboard for sending or storing the information. Video display terminals are used in conjunction with modems or terminal units for various communications purposes (*see* MODEM, TERMINAL UNIT). A printer can be used for obtaining hard copies of displayed data.

The photograph shows a typical video display terminal, used in conjunction with a home computer (in this case, the computer is built into the keyboard cabinet).

VIDEO GAME

The development of high-resolution video graphics, along with more and more sophisticated small computers, has led to the recreational use of computers for game playing. Computerized video games are extremely popular among young people.

Some video games are entirely self-contained. This is typical of the "arcade" type game. Other video games are available in the form of devices that can be attached to a home television set. Still others are provided as software packages for home computers.

While video games are fun and interesting, they can also be educational. Certain video games are used to help people learn to drive automobiles, fly airplanes, or operate complex machinery, without endangering lives. This type of video game is called a simulator.

The technology of video games is improving rapidly. Some commercial manufacturers have begun to develop games in which animated characters can be manipulated by the operator, interacting with other animated characters in an almost infinite variety of situations. What the future will bring is limited only by the imagination of the users and manufacturers.

VIDEO SIGNAL

See COMPOSITE VIDEO SIGNAL, PICTURE SIGNAL.

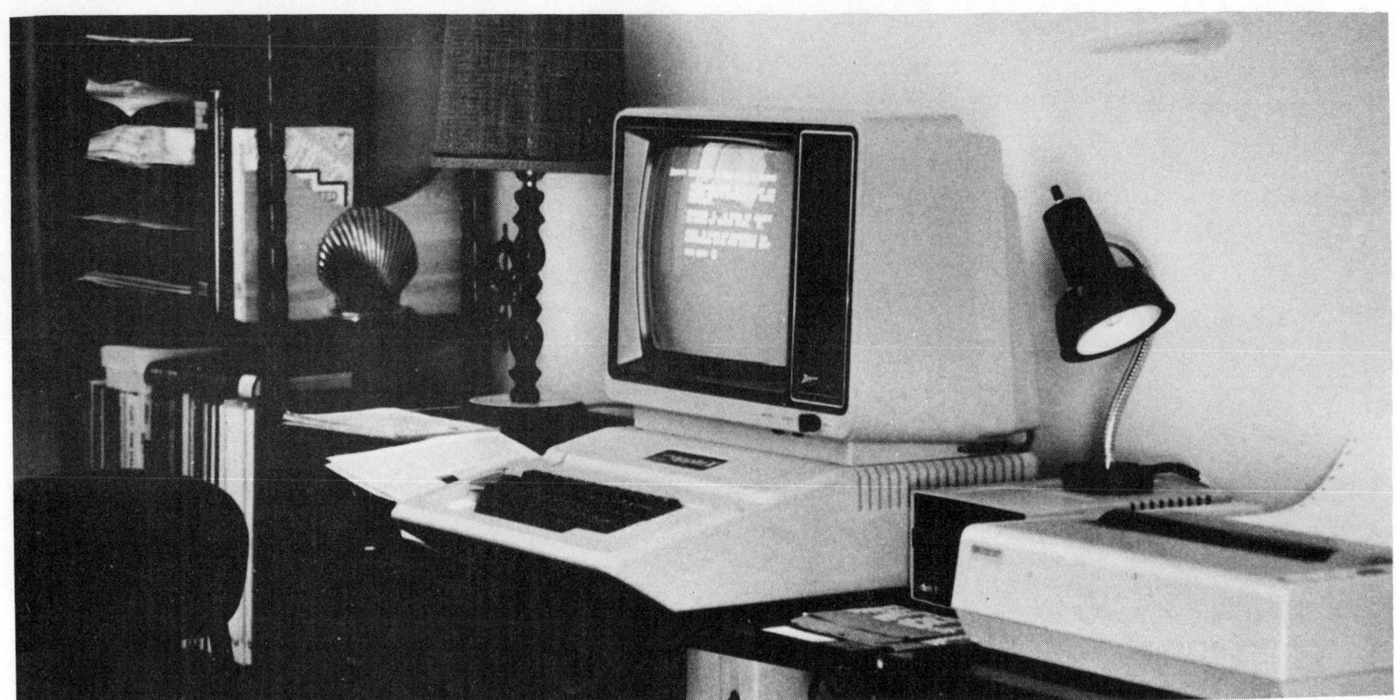

VIDEO DISPLAY TERMINAL: A video display terminal consists of a keyboard and a cathode-ray-tube monitor.

VIDEO TAPE RECORDING

The technology for recording and reproducing information by magnetic means has improved greatly in recent years. This has made it possible to store video information on magnetic tapes and disks, rather than on film.

A video tape recorder (VTR) operates in a manner similar to an audio tape recorder. The main difference is the higher tape speed, which is necessary to allow storage and reproduction of the high-frequency components of a video signal. The VTR uses somewhat more sophisticated techniques, and a different tape track configuration, than an audio tape recorder.

The principle of video-tape recording is shown in the illustration. The sound track runs along one edge of a tape 2 in (51 mm) wide. The opposite edge contains a control track. The video information is recorded at an angle with respect to the direction of the tape, thus greatly increasing the effective speed of the tape. The actual forward speed is 15 inches per second (IPS). Some VTR units run at 7.5 IPS.

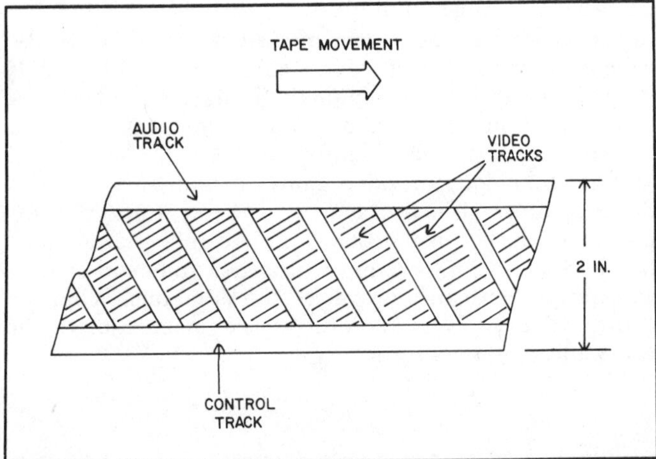

VIDEO TAPE RECORDING: A video tape contains separate tracks for audio, video, and control information.

The composite video signal from the camera or television receiver is used frequency modulate a carrier for recording on the tape. In the playback mode, a demodulator is used to recover the composite signal. This optimizes the signal-to-noise ratio in the recording and playback processes.

In the past few years, magnetic disks have begun to take the place of magnetic tape in some video recording applications. The VTR is still widely used, however, especially in television broadcasting. *See also* COMPOSITE VIDEO SIGNAL, MAGNETIC RECORDING, TAPE RECORDER, VIDEO DISK RECORDING.

VIDICON

A vidicon is a small, relatively simple televison camera tube. The vidicon is preferred over other types of camera tubes in portable and mobile applications because it is less bulky and heavy. The vidicon is extensively used in closed-circuit television surveillance systems in business and industry.

The operation of a vidicon is illustrated by the diagram. The image is focused, by means of a lens, onto a layer of photoconductive material. The electron gun and grids

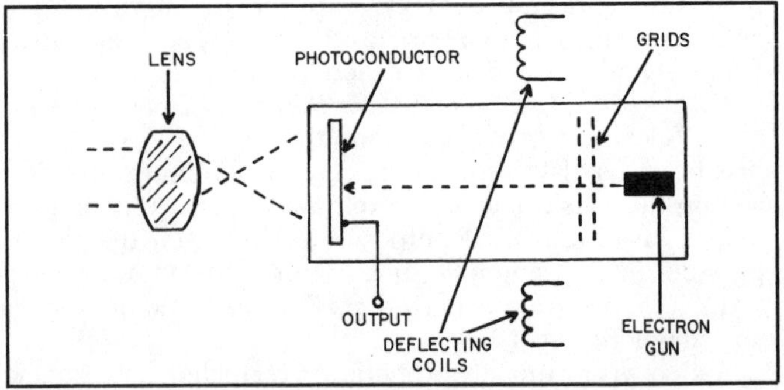

VIDICON: Simplified diagram showing the operation of a vidicon camera tube.

generate an electron beam that scans the photoconductor via the action of external deflecting coils. This beam causes the photoconductive material to become electrically charged. The rate of discharge is roughly proportional to the intensity of the light in a particular part of the photoconductor. The result is different conductivity for different portions of the visual image.

The vidicon produces very little noise because it operates at a low level of current. Thus, the sensitivity is very high. Because of capacitive effects in the photoconductor, the vidicon has some lag. The image orthicon is preferable if a rapid image response is needed. *See also* CAMERA TUBE, ICONOSCOPE, IMAGE ORTHICON.

VIRTUAL HEIGHT

Virtual height is an expression of the altitude of the ionosphere in terms of radio-frequency energy sent directly upward. The virtual height Hv of an ionized layer is always greater than the actual height H, because electromagnetic energy requires some time to be turned around by interaction with the ionized molecules. The D, E, and F layers have different virtual heights as well as different actual altitudes (*see* D LAYER, E LAYER, F LAYER).

The virtual height of the ionosphere is determined from the delay in the return of a pulse sent straight upward (see illustration). If the pulse returns after t milliseconds, then the virtual height is given in miles by:

$$Hv = 186t$$

and in kilometers by:

$$Hv = 300t$$

The virtual height of an ionized layer depends to some extent on the frequency. As the frequency rises, the virtual height increases very slightly up to a certain point. As the frequency increases further, the virtual height rapidly grows greater. At a certain frequency called the critical frequency, the ionosphere no longer returns signals sent directly upward. *See also* CRITICAL FREQUENCY, IONOSPHERE.

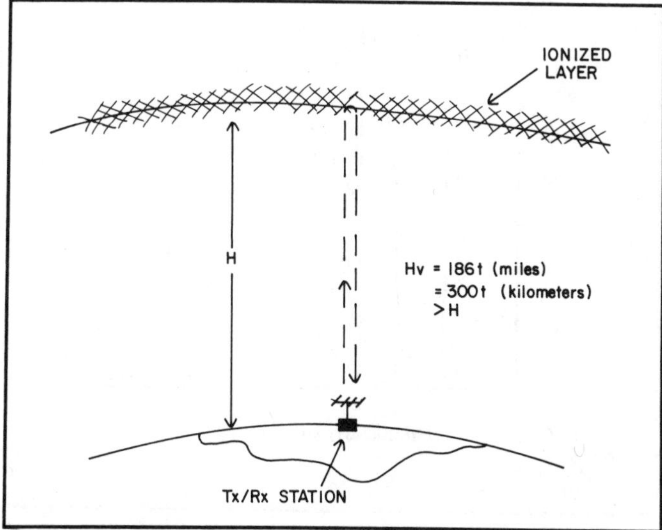

VIRTUAL HEIGHT: Virtual height is determined by bouncing a radio wave off of the ionosphere.

VISIBLE LIGHT

See LIGHT, VISIBLE SPECTRUM.

VISIBLE SPECTRUM

The visible spectrum covers the range of electromagnetic wavelengths from 750 nanometers (nm) to 390 nm. This corresponds to frequencies of 400 THz and 770 THz. These wavelength and frequency values are approximate, since some people can visually detect energy at wavelengths slightly outside this range. The visible spectrum is bordered by the infrared and ultraviolet (*see* INFRARED, ULTRAVIOLET).

The wavelength of visible light is a function of the energy contained in each photon. The photons contain the least energy at the longest wavelengths of lowest frequencies, and the most energy at the shortest wavelengths or highest frequencies (*see* PHOTON, PLANCK'S CONSTANT).

The wavelength of light is correlated with the colors we see. For monochromatic light, the longest wavelengths appear deep red. As the wavelength is shortened, the colors progress from red to red-orange, orange, yellow-orange, yellow, yellow-green, green, blue-green, blue, indigo (blue-violet), and violet. There are infinitely many colors possible in the visible spectrum. The eyes of the average person are sensitive in the green part of the spectrum; the sensitivity decreases toward the red and violet.

Various different beams of monochromatic light can be combined to obtain colors that do not appear in the spectrum. By combining red, green, and blue light in the proper proportions, any hue can be obtained. *See also* LIGHT, PRIMARY COLORS.

VLSI

See VERY-LARAGE-SCALE INTEGRATION.

VMOSFET

See VERTICAL METAL-OXIDE SEMICONDUCTOR FIELD-EFFECT TRANSISTOR.

VOCODER

A vocoder is a device that greatly reduces the occupied bandwidth of a voice signal. The unprocessed voice signal needs 2 to 3 kHz of spectrum space for intelligible transmission. Using a vocoder, a voice can be compressed into a much smaller space, with some sacrifice in inflection. The vocoder was originally invented by W. H. Dudley of the Bell Laboratories.

There are different types of vocoders, and a detailed discussion of the technology involved is beyond the scope of this book. The basic principle of the vocoder is the replacement of certain voice signals with electronically synthesized impulses. A typical vocoder reduces the signal bandwidth to approximately 1 kHz with essentially no degradation of intelligibility. Greater compression ratios are possible with some sacrifice in signal quality. *See also* VODER, VOICE FREQUENCY CHARACTERISTICS.

VODER

A voder is a speech synthesizer that was originally designed by H. W. Dudley of the Bell Laboratories. The intended purpose was the transmission of recognizable voice signals over a channel having a narrow bandwidth. The voder is similar to the vocoder, except that the voder produces entirely synthesized signals.

A voder generates the various voice sounds by means of a keyboard, in somewhat the same way a Moog synthesizer operates. In theory, the voder could reduce the occupied bandwidth of a voice signal by a factor of several hundred. This might result in a bandwidth of as little as a few hertz. But in practice, the quality of a synthesized voice, compressed to such a tiny part of the spectrum, is degraded. Virtually all inflection, and therefore the emotional content of the voice, is lost in the process. Therefore, the realizable compression factor is smaller than that predicted by theory.

The voder experiments by Dudley eventually led to the development of the vocoder, which partially synthesizes speech and provides a significant reduction in occupied bandwidth. *See also* VOCODER, VOICE FREQUENCY CHARACTERISTICS.

VOICE FREQUENCY CHARACTERISTICS

We easily recognize, and mentally decode, the sounds of speech, but the electrical characteristics of a voice wave are complicated. A voice signal transmits information not only as characters and words, but via emotional content as a result of subtle inflections.

A voice signal contains energy in three distinct bands of frequencies. These bands are called formants. The lowest frequency band is called the first formant, and occupies the range from a few hertz to near 800 Hz. The second formant ranges in frequency from approximately 1.5 to 2.0 kHz. The third formant ranges from near 2.2 to 3.0 kHz and above. The exact frequencies differ slightly, but not greatly, from person to person. The illustration demonstrates the basic nature of the formants of speech.

It has been discovered that speech information can be adequately conveyed by restricting the audio response to two ranges, approximately 300 to 600 Hz and 1.5 to 3.0 kHz. A unique method of doing this was first put forth by R. W. Harris and J. C. Gorski in the December, 1977 issue of *QST* magazine, published by the American Radio Relay League, Inc. (*see* NARROW-BAND VOICE MODULATION).

There is another way to look at speech. There are five basic types of sounds made by humans. These include the vowels, the semivowels, the plosives, the fricatives, and the nasal consonants. The lowest audio frequencies are generated by the utterance of vowels and semivowels: A, E, I, L, O, R, U, V, W, and Y. Medium frequencies are generated by nasal sounds: M, N, and NG. The highest frequencies are generated by the utterance of fricatives and plosives: B, C, D, F, G, H, J, K, P, PL, Q, S, SH, T, TH, X, and Z. (There are a few irregularities because of language anomalies.)

The analysis of speech characteristics is important in communications, especially for the purpose of minimizing the occupied bandwidth of voice signals. Devices such as the vocoder and the voder have been used to synthesize some voice sounds, resulting in improved communications efficiency. When speech synthesis is used, however, some degradation occurs in voice inflection. Thus, while the bandwidth may be reduced, the subtle meanings, conveyed by emotion, are sacrificed or even lost. This may or may not be important, depending on the intended communications applications. *See also* SPEECH SYNTHESIS, VOCODER, VODER.

VOICE-OPERATED TRANSMISSION

See VOX.

VOICE PRINT

A voice print is a display of the spectral characteristics of a human voice. The voice print is usually given as a function of frequency and amplitude versus time, such as is shown in the illustration. Modern computer graphics can produce three-dimensional voice print images, showing the relationship between frequency, amplitude, and time.

Voice prints exhibit remarkable similarities from person to person. This had made it possible to reduce the bandwidth of a voice signal by electronic means (*see* NARROW-BAND VOICE MODULATION, VOCODER, VODER). However, every person has certain voice-print characteristics that no one else has. This fact has led to the use of voice prints for identification purposes, in much the same way as finger prints are used. *See also* VOICE FREQUENCY CHARACTERISTICS.

VOICE RECOGNITION

See SPEECH RECOGNITION.

VOICE SYNTHESIS

See SPEECH SYNTHESIS, VOCODER, VODER.

VOICE TRANSMITTER

A voice transmitter is any radio transmitter intended for the purpose of conveying information by means of the human voice. The most common methods of voice transmission are amplitude modulation, frequency modulation,

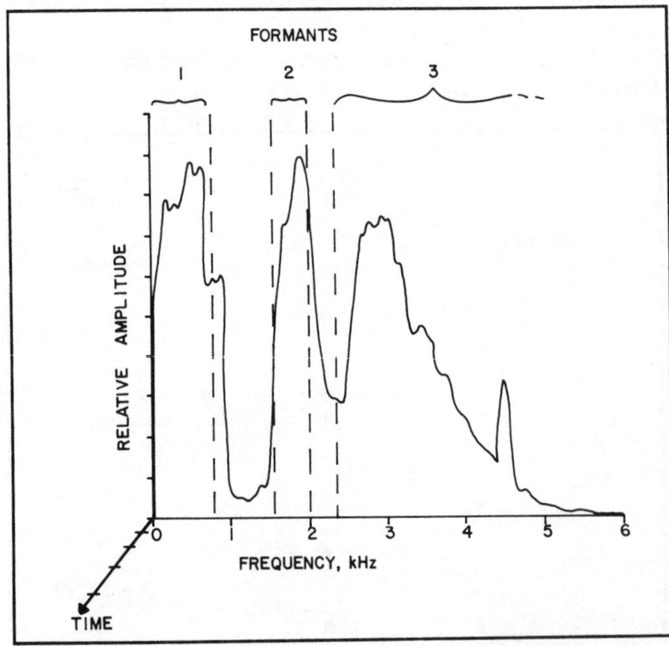

VOICE PRINT: A typical voice pattern, showing the relationship between amplitude and frequency. If time is plotted, the pattern becomes a three-dimensional voice print.

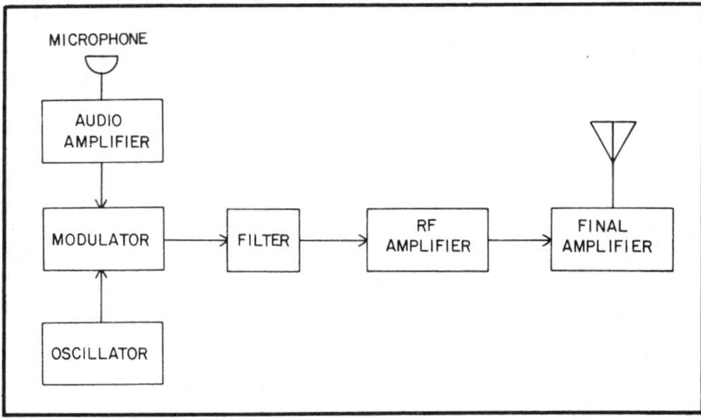

VOICE TRANSMITTER: Block diagram of a voice transmitter.

and single sideband. Less often used are phase modulation, polarization modulation, pulse modulation, and other methods.

A voice transmitter consists of an oscillator, a modulator, a filter (if needed), and a chain of amplifiers (see illustration). In variable-frequency voice transmitters, mixing circuits are often used for multiband operation. *See also* AMPLITUDE MODULATION, FREQUENCY MODULATION, PHASE MODULATION, POLARIZATION MODULATION, PULSE MODULATION, SINGLE SIDEBAND.

VOLT

The volt is the unit of electric potential. A potential difference of one volt across a resistance of one ohm will result in a current of one ampere, or one coulomb of electrons per second (*see* AMPERE, COULOMB, OHM, OHM'S LAW).

Various units smaller than the volt are often used to measure electric potential. This is especially true for weak radio signals, which may have root-mean-square magnitudes of less than a millionth of a volt (*see* ROOT-MEAN-SQUARE). A millivolt (mV) is one thousandth of a volt. A microvolt (μV) is one millionth of a volt. A nanovolt (nV) is a billionth of a volt. It is not likely that you will ever encounter a voltage smaller than 1 nV and be able to detect it.

Some voltages commonly encountered in electricity and electronics are as follows: electrochemical cell, 1.0 to 1.8 V; automobile battery, 12 to 14 V; household utility outlet, 117 or 234 V (alternating-current, root mean square). *See also* ALTERNATING CURRENT, DIRECT CURRENT.

VOLTA, ALESSANDRO

In the late 18th and early 19th centuries, Alessandro Volta (1745-1827) performed various electrochemical experiments. He is best known for the battery he made using various layers of copper and zinc in a solution of brine.

A similar battery can be made using pennies and nickels stacked alternately one on top of the other, with brine-soaked cloths in between. The internal resistance is rather high, and the battery therefore cannot deliver much current. The nickel forms the negative electrode, and the penny forms the positive electrode.

VOLTA EFFECT

Certain dissimilar metals develop a potential difference when they are brought into contact. This phenomenon is known as the Volta effect. The Volta effect is responsible for the operation of electrochemical cells. The Volta effect also causes corrosion in some metallic junctions over a period of time. *See also* CELL, CORROSION.

VOLTAGE

Voltage is the existence of a potential (charge) difference between two objects or points in a circuit. Normally, a potential difference is manifested as an excess of electrons at one point, and/or a deficiency of electrons at another point. Sometimes, other charge carriers, such as holes or protons, can be responsible for a voltage between two objects or points in a circuit (*see* ELECTRON, HOLE, PROTON).

Electric voltage is measured in units called volts. A potential difference of one volt causes a current of one ampere to flow through a resistance of one ohm (*see* AMPERE, COULOMB, OHM, OHM'S LAW). Voltage may be either alternating or direct. In the case of alternating voltage, the value can be expressed in any of three ways: peak, peak-to-peak, or root-mean-square (*see* PEAK-TO-PEAK VALUE, PEAK VALUE, PEAK VOLTAGE, ROOT MEAN SQUARE). Voltage is symbolized by the letter E or V in most equation involving electrical quantities.

If a given point P has more electrons than another point Q, the point P is said to be negative with respect to Q, and Q is said to be positive with respect to P. This is relative voltage. Relative voltage is not the same as absolute voltage, which is determined with respect to a point at ground (neutral) potential.

When two points having different voltage are connected together, physicists consider the current to flow away from the point having the more positive voltage, although the movement of electrons is actually toward the positive pole. *See also* CURRENT.

VOLTAGE AMPLIFICATION

Voltage amplification is the increase in the magnitude of a voltage between the input and the output of a circuit. It is also sometimes called voltage gain.

Some circuits are designed specifically for the purpose of amplifying a direct or alternating voltage. Other circuits are intended for amplification of current, while still others are designed to amplify power.

In theory, a voltage amplifier can operate with zero driving power and an infinite input impedance. In practice, the input impedance can be made extremely high, and the driving power is therefore very small, but not zero. A Class-A amplifier, using a field-effect transistor, is the most common circuit used for voltage amplification (*see* CLASS-A AMPLIFIER). Bipolars can also be used for this purpose.

Voltage amplification is measured in decibels. Mathematically, if E_{in} is the input voltage and E_{out} is the output voltage, then;

$$\text{Voltage gain (dB)} = 20 \log_{10} (E_{out}/E_{in})$$

See also DECIBEL, GAIN.

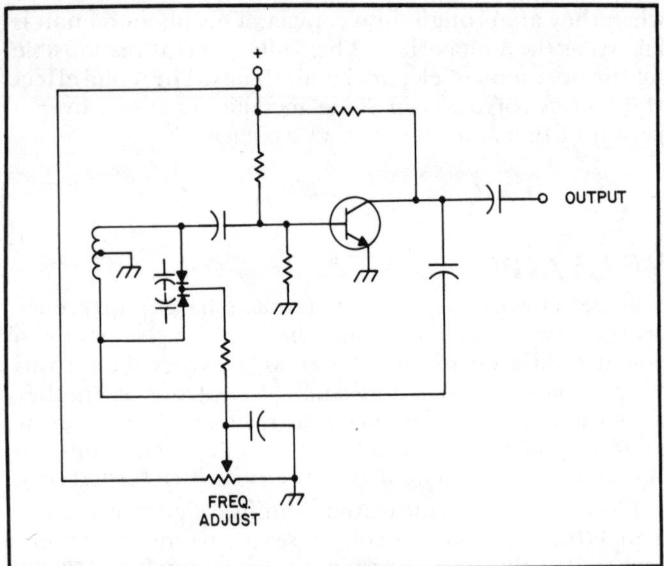

VOLTAGE-CONTROLLED OSCILLATOR: Example of a voltage-controlled oscillator. This circuit uses the Hartley configuration.

VOLTAGE-CONTROLLED OSCILLATOR

Varactor diodes can be used to adjust the frequency of an oscillator circuit. This is commonly done to provide frequency modulation (*see* FREQUENCY MODULATION, REACTANCE MODULATOR, VARACTOR DIODE). A direct-current voltage, supplied to the varactor diode or pair of diodes, facilitates variable-frequency operation in a circuit called a voltage-controlled oscillator (VCO).

In a VCO, one or two varactor diodes are connected in parallel with, or in place of, the tank capacitor of the oscillator. A pair of diodes can be connected in reverse series, replacing the tuning capacitor, as shown in the illustration. Frequency adjustment is accomplished by means of a potentiometer. The tuning range depends on the ratio of inductance to capacitance in the tank circuit. The maximum-to-minimum oscillator frequency ratio can be made as high as 2:1 without difficulty; considerably greater ratios are possible.

VOLTAGE-DEPENDENT RESISTOR

Some resistors exhibit variable ohmic value, depending on the voltage across them. Such a resistor is said to be voltage-dependent. The current through a voltage-dependent resistor is a nonlinear function of the voltage.

Voltage-dependent resistors are used in certain current-regulating circuits. *See also* CURRENT REGULATION.

VOLTAGE DIVIDER

A voltage divider is a network of passive resistors, inductors, or capacitors, that is used for the purpose of obtaining different voltages for various purposes.

A resistive voltage divider is commonly employed for providing the direct-current bias in an amplifier or oscillator circuit (see illustration A). The desired bias is obtained by selecting resistors having the proper ratio of

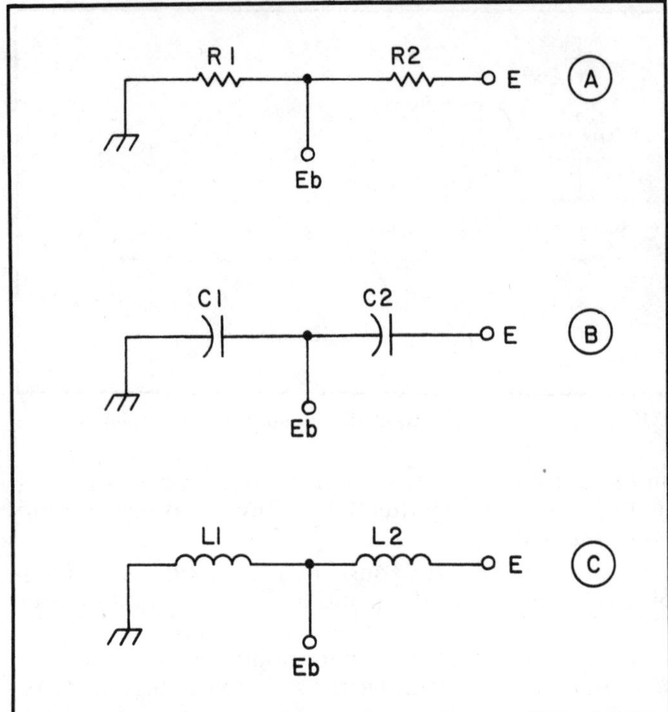

VOLTAGE DIVIDER: At A, a resistive circuit; at B and C, reactive circuits.

values. If the supply voltage is E and the resistors have values R1 and R2 as shown, then the bias voltage is:

$$Eb = E \times R1/(R1+R2)$$

Although there exists an infinite number of values R1 and R2 that will provide a given bias voltage Eb, the actual values are chosen on the basis of the circuit impedance.

A capacitive voltage divider is shown at B, and an inductive divider is shown at C. These types of dividers are used for ac voltages. The ratio of reactances determines the voltage at the tap point, in exactly the same manner as with resistances (see above equation). An example of a capactive voltage divider is found in the Colpitts oscillator, and an example of inductive voltage division is found in the Hartley oscillator (*see* COLPITTS OSCILLATOR, HARTLEY OSCILLATOR.

VOLTAGE DOUBLER

For obtaining high voltages needed in vacuum-tube circuits, a voltage-doubler power supply is often used. The voltage double makes it possible to use a transformer having a lower step-up ratio than would be needed if an ordinary full-wave supply were used. Voltage doublers are sometimes used in radio-frequency-actuated circuits to obtain the control voltage. Voltage-doubler circuits are not generally used when excellent regulation is needed, or when the current drain is high.

The illustration shows two types of voltage-doubler supplies. The circuit at A is called a conventional doubler; the circuit at B is known as a cascade doubler. In both circuits, the dc output voltage is twice the peak alternating-current input voltage, or 2.8 times the root-mean-square input voltage. The conventional doubler provides superior regulation and less output ripple, but the cascade circuit can be used without a transformer. Two

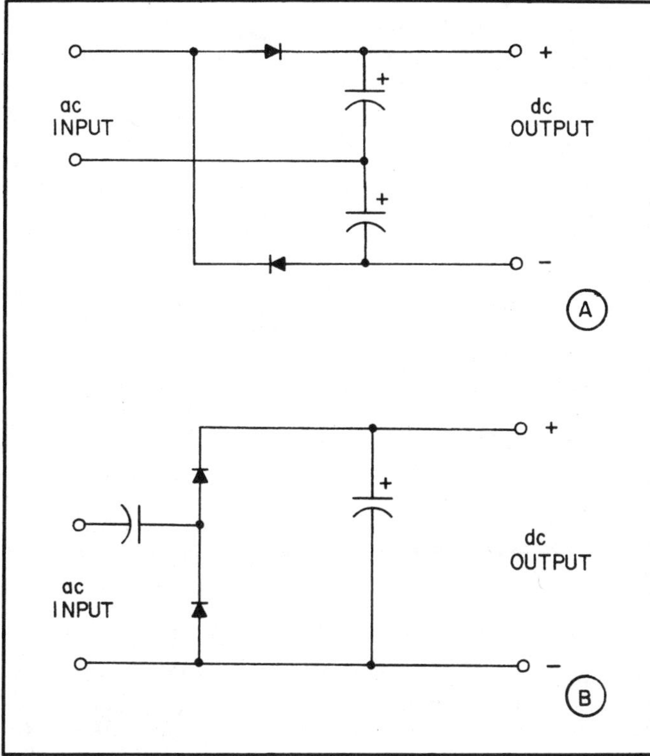

VOLTAGE DOUBLER: At A, a conventional voltage-doubler supply. At B, a cascade type voltage doubler.

or more cascade circuits can be connected in series to obtain voltage multiplication by any power of 2. *See also* VOLTAGE MULTIPLIER.

VOLTAGE DROP

Voltage drop is the potential difference that appears across a current-carrying impedance.

For direct currents and resistances, the voltage drop E across a resistance of R ohms that carries a current of I amperes is given by Ohm's law as:

$$E = IR$$

For alternating currents and reactances or complex impedances, the root-mean-square voltage drop E is determined from the root-mean-square current I and the absolute-value impedance R+jx as

$$E = I\sqrt{R^2+X^2}$$

See also CURRENT, IMPEDANCE, OHM'S LAW, REACTANCE, RESISTANCE, VOLTAGE.

VOLTAGE FEED

Voltage feed is a method of connecting a radio-frequency feed line to an antenna at a point on the radiator where the voltage is maximum. Such a point is called a voltage loop or a current node (*see* CURRENT NODE, VOLTAGE LOOP). In a half-wavelength radiator, the voltage maxima occur at the ends. Therefore, voltage feed for a half-wavelength an-

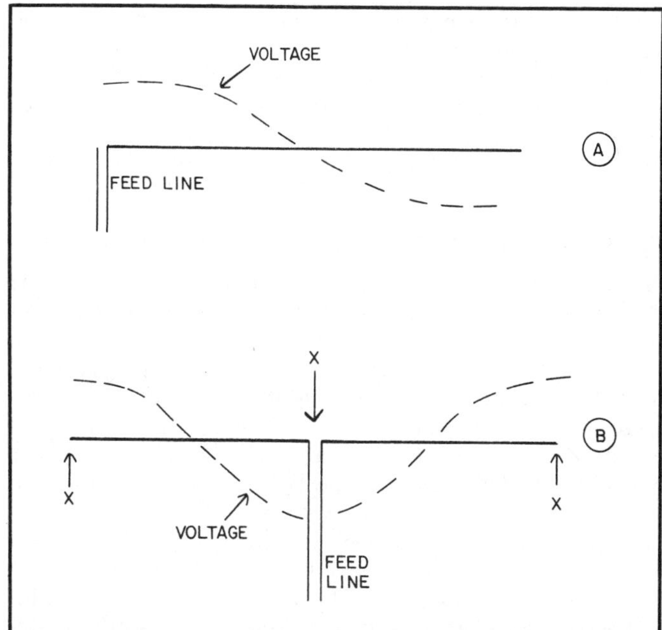

VOLTAGE FEED: At A, end feed always constitutes voltage feed. At B, a 1-wavelength radiator may be fed at any of the points marked X.

tenna can only be accomplished by connecting the line to either end of the radiating element (see illustration A). In an antenna longer than ½ wavelength, voltage maxima exist at multiples of ½ wavelength from either end. There may be several different places on a resonant antenna that are suitable for voltage feed, as at B.

The feed-point impedance of a voltage-fed antenna is high. It is a pure resistance of 200 ohms or more, depending on the amount of end-effect capacitance and the diameter of the radiating element. In wire antennas, the feed-point impedance is difficult to predict with certainty; it may range from several hundred ohms to 2000 ohms or more. Because of the uncertainty, low-loss open-wire line is generally used in voltage-fed antennas. Voltage feed results in good electrical balance for a two-wire line, provided the current and voltage distribution are reasonably symmetrical in the antenna. *See also* CURRENT FEED.

VOLTAGE GAIN

See VOLTAGE AMPLIFICATION.

VOLTAGE GRADIENT

Whenever a potential difference exists between two points, such as at either end of a long wire, resistor, or semiconductor wafer, there are points at which the voltage is at some intermediate value. The rate of voltage change per unit length is called the voltage gradient.

The voltage gradient in a given circuit or conductor depends on the potential difference between the two end points, and on the distance between these points. In general, if the end points have voltages E1 and E2, and the length of the conductor or component is d meters, then the voltage gradient, in volts per meter, is:

$$G = (E2-E1)/d$$

Voltage gradient may also be specified in volts per foot, centimeter, millimeter, or micron.

The voltage gradient is sometimes specified for semiconductor materials. This is especially true for field-effect transistors, in which a voltage gradient exists along the channel (*see* FIELD-EFFECT TRANSISTOR).

VOLTAGE HOGGING

Voltage hogging is a phenomenon that sometimes occurs in resistive components connected in series. Normally, if several components having identical resistances are connected in series, the voltage is divided equally among them. But if the components have slightly different resistances and a positive temperature coefficient, the component with the highest initial resistance will "hog" most of the voltage. This can cause improper operation of a circuit and, in some cases, component failure.

Voltage hogging sometimes occurs with semiconductor devices, especially diodes. In a high-voltage supply, several semiconductor diodes may be connected in series to increase the peak-inverse-voltage rating of the combination. Ideally, if n diodes, each with a peak-inverse-voltage rating of E volts, are connected in series, the combination will tolerate En volts. However, voltage hogging can substantially reduce the peak-inverse-voltage rating of the combination.

Voltage hogging is prevented by placing large-value resistors across components connected in series. All of the resistors have the same value, and they act as a voltage divider to maintain a constant voltage across each component. *See also* VOLTAGE DIVIDER.

VOLTAGE LIMITING

Voltage limiting is a method of preventing a direct-current or alternating-current voltage from exceeding a certain value. In the case of alternating current, voltage limiting is usually called clipping (*see* CLIPPER).

Voltage limiting is sometimes used as a means of regulation in a direct-current power supply. A Zener diode, connected across the source of voltage, serves this purpose. *See also* VOLTAGE REGULATION, ZENER DIODE.

VOLTAGE LOOP

In an antenna radiating element, the voltage in the conductor depends on the location. At any free end, the voltage is maximum. At a distance of ½ wavelength from a free end, or any multiple of ½ wavelength from a free end, the voltage is also at a maximum. The points of maximum voltage are called voltage loops.

A ½-wavelength radiator has two voltage loops, one at each end. A full-wavelength radiator has three voltage loops: one at each end and one at the center. In general, the number of voltage loops in an antenna radiator is 1 plus the number of half wavelengths. The illustration shows the voltage distribution in an antenna radiator of 3/2 wavelength, showing the locations of the voltage loops.

Voltage loops will occur along the length of a feed line not terminated in an impedance identical to its characteristic impedance. These loops occur at multiples of ½ wavelength from the resonant antenna feed point when the antenna impedance is greater than the characteristic impedance of the line. The loops exist at odd multiples of ¼ wavelength from the feed point when the resonant antenna impedance is less than the characteristic impedance of the line. Ideally, the voltage on a transmission line should be the same everywhere, and equal to the product of the characteristic impedance and the line current. *See also* VOLTAGE NODE, STANDING WAVE.

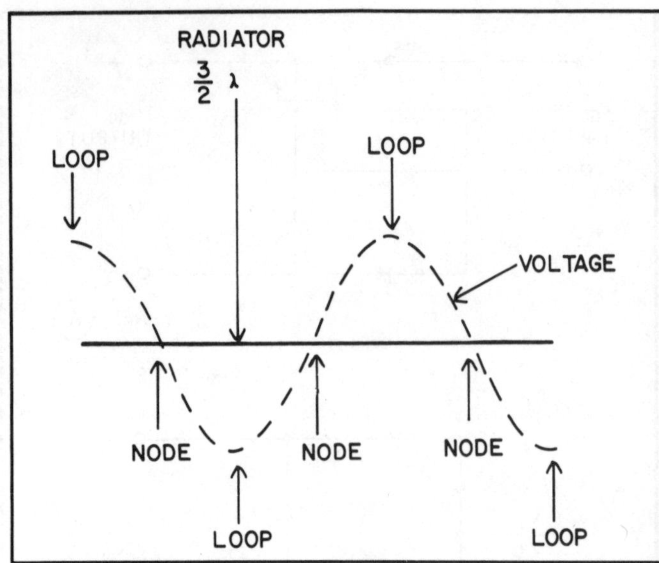

VOLTAGE LOOP: Locations of voltage loops and nodes on a radiator of 3/2 wavelength.

VOLTAGE MULTIPLIER

A voltage multiplier is a form of power supply designed to deliver a very high voltage using a transformer with a secondary that supplies only a moderate voltage. The most common type of voltage multiplier is the voltage doubler (*see* VOLTAGE DOUBLER). However, some supplies have multiplication factors of 3, 4, or more.

The illustration demonstrates the general principle of operation for a voltage multiplier. Capacitors and diodes are connected in a lattice configuration. For a multiplication factor of n, the circuit requires 2n capacitors and 2n diodes. The direct-current output voltage is n times the peak input voltage, or about 1.4n times the root-mean-square input voltage.

In practice, there is a limit to the extent to which a voltage can be multiplied using the circuit shown. The larger the multiplication factor n, the worse the voltage regulation becomes. For this reason, high-order voltage multipliers are not often used except in situations where the current drain is extremely small.

VOLTAGE NODE

A voltage node is a voltage minimum in an antenna radiator or transmission line. The voltage on an antenna depends, to some extent, on the location of the radiator. In general, voltage nodes occur at odd multiples of ¼ wavelength from the nearest free end of a radiating element. The preceding illustration shows the locations of voltage nodes along a 3/2-wavelength radiator. The number of voltage nodes is equal to the number of half wavelengths.

Voltage nodes may occur along a transmission line that is not terminated in an impedance identical to its charac-

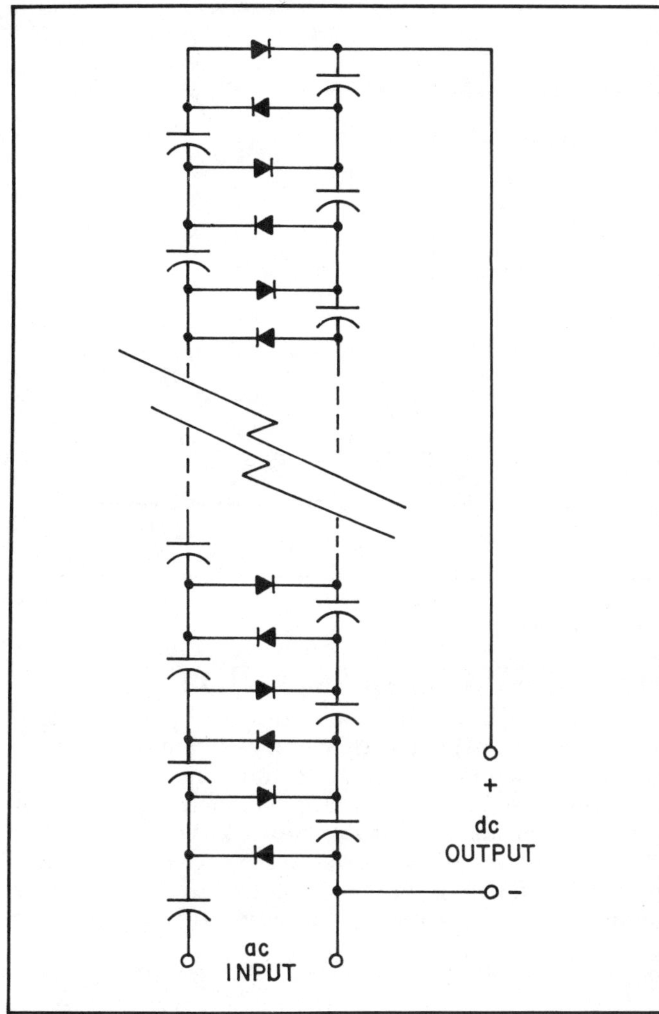

VOLTAGE MULTIPLIER: General configuration of a voltage-multiplier power supply.

teristic impedance. These nodes occur at multiples of ½ wavelength from the resonant antenna feed point when the antenna impedance is smaller than the feed-line characteristic impedance. Voltage nodes exist at odd multiplies of ¼ wavelength from the resonant antenna feed point when the antenna impedance is larger than the characteristic impedance of the line.

Voltage nodes are always spaced at intervals of ¼ wavelength from voltage loops. Ideally, the voltage on a transmission line is the same everywhere, being equal to the product of the characteristic impedance and the line current. *See also* VOLTAGE LOOP, STANDING WAVE.

VOLTAGE REGULATION

Voltage regulation is the process of maintaining the voltage at a constant value across a load. This is done by means of a regulated, or constant-voltage, power supply. In some applications, good voltage regulation is very important.

Voltage regulation is expressed as a percentage (*see* REGULATION). In most power supplies, the output voltage falls slightly as the load resistance decreases to a certain minimum tolerable load resistance. As the load resistance continues to drop past the critical point, the power-supply output voltage decreases rapidly. The regulation percentage indicates the extent to which the voltage drops while the load resistance remains within the tolerable range.

In low-voltage power supplies, good voltage regulation can be realized in various different ways. The simplest method is the shunt connection of a Zener diode across the output terminals. The Zener voltage is about half the output voltage of the supply without the diode (*see* ZENER DIODE). A more sophisticated means of obtaining good regulation is the use of a series transistor that inserts more or less resistance into the circuit as the load resistance changes.

In high-voltage power supplies, a vacuum tube is most often used for regulation. Various gas-filled tubes are available for use as voltage regulators in such supplies. Voltage regulation can be obtained manually in any power supply by means of a small transformer in series with the alternating-current supply input. *See also* REGULATING TRANSISTOR, SERIES REGULATOR, SHUNT REGULATOR, STEP-VOLTAGE REGULATOR.

VOLTAGE TRANSFORMER

A voltage transformer is a two-winding device, intended especially for stepping an alternating-current voltage up or down. Voltage transformers are used in almost all power supplies designed for operation from the 117-V utility mains. Voltage transformers are also extensively employed in power transmission.

In general, the input-to-output voltage ratio of a voltage transformer is identical to the primary-to-secondary turns ratio. *See also* STEP-DOWN TRANSFORMER, STEP-UP TRANSFORMER, TRANSFORMER.

VOLT-AMPERE

In any electrical or electronic circuit, the volt-ampere is an expression of the apparent power. If the voltage between two points is E and the current in amperes is I, then the apparent power in volt-amperes is:

$$P_a = EI$$

In direct-current circuits, the volt-ampere is identical to the watt, since the apparent power is the same as the true power. However, in alternating-current circuits, the volt-ampere may not be a true indication of the dissipated power. *See also* APPARENT POWER, POWER, REACTIVE POWER, TRUE POWER, VAR, WATT.

VOLT-AMPERE-HOUR METER

A volt-ampere-hour meter is a device that determines the product of the voltage, current, and time in an electrical circuit. The common utility meter operates as a volt-ampere-hour meter. Such a device integrates the current in amperes over a period of time, obtaining the reading in watt hours or kilowatt hours, under the assumption that the voltage is constant (234 V in most cases). *See* WATT-HOUR METER.

VOLT BOX

A volt box is a precision voltage-divider circuit, used in the

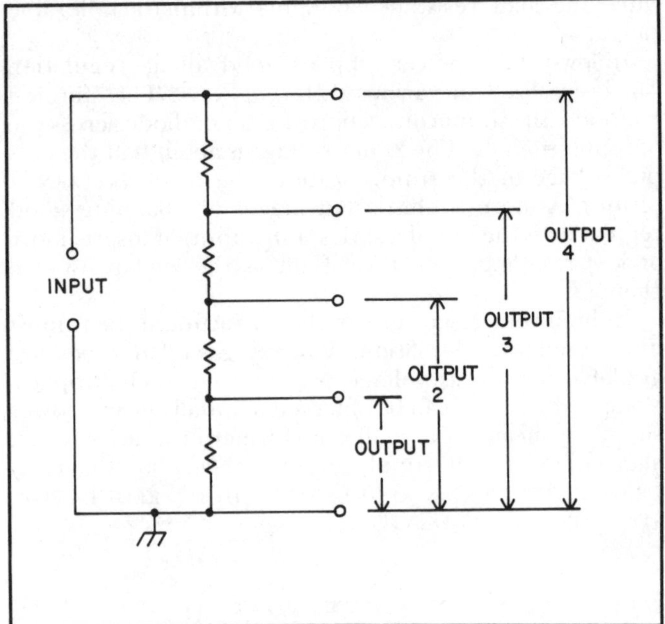

VOLT BOX: This circuit provides four different voltages.

calibration of meters and other test instruments. A typical volt box is constructed using series-connected, fixed resistors, as shown in the diagram. There is one pair of input terminals and several output terminals. The various voltages are obtained from the appropriate output terminals, provided the load resistance is significantly higher than that of the series combination in the box. This is always the case for voltmeters and oscilloscopes. *See also* VOLTAGE DIVIDER.

VOLTMETER

A voltmeter is a device that is used for measuring the voltage between two points. There are many types of voltmeters, used for various purposes in equipment design, testing, operation, and troubleshooting.

The simplest type of voltmeter consists of a microammeter in series with a large-value resistor. The size of the resistor determines the voltage required to produce a full-scale indication on the meter. Most voltmeters contain several resistors that can be switched in to obtain the desired meter range.

The illustration is a schematic diagram of a typical direct-current voltmeter. The meter is a 100-uA device. Full-scale ranges of 1V, 10V, 100V, and 1kV are obtained using resistors of 10K, 100K, 1M, and 10M respectively (K = 1,000 ohms; M = 1,000,000 ohms). The device can be used to measure ac voltage by switching in a diode and capacitor to rectify and filter the incoming voltage. (The meter scale must be calibrated differently for alternating current.)

In some applications, it is necessary that the internal resistance of a voltmeter be extremely high. The internal resistance of a voltmeter is expressed as a figure called ohms per volt (*see* OHMS PER VOLT). The meter illustrated has an internal resistance of 10K ohms per volt—a low enough value to cause problems in high-impedance circuits. More sophisticated voltmeters are available for use in such cases. *See also* FET VOLTMETER, VACUUM-TUBE VOLTMETER.

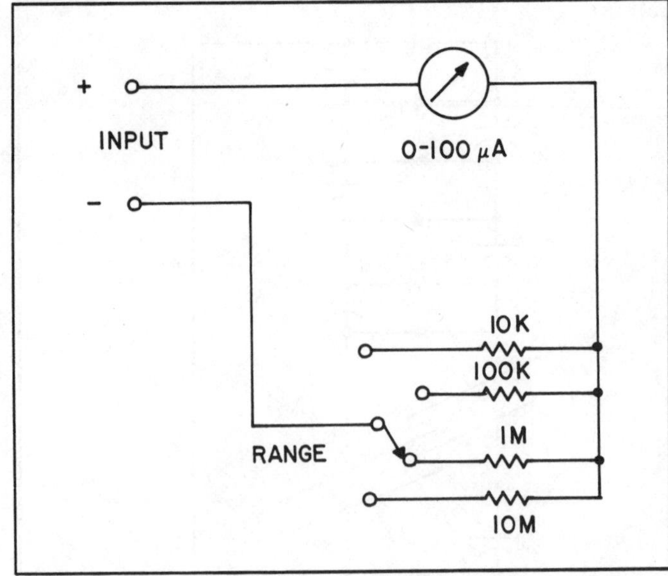

VOLTMETER: A typical voltmeter, with four range settings: 0-1 V, 0-10 V, 0-100 V, and 0-1 kV.

VOLT-OHM-MILLIAMMETER

A common type of test meter consists of a combination voltmeter, ohmmeter, and milliammeter, and is known as a volt-ohm-milliammeter (VOM) or multimeter.

A multimeter contains a microammeter, a source of direct current, and a switchable network of resistors. In the voltmeter mode, series resistances are selected to determine the meter range (*see* VOLTMETER). In the ohmmeter mode, the meter is connected in series with the test leads

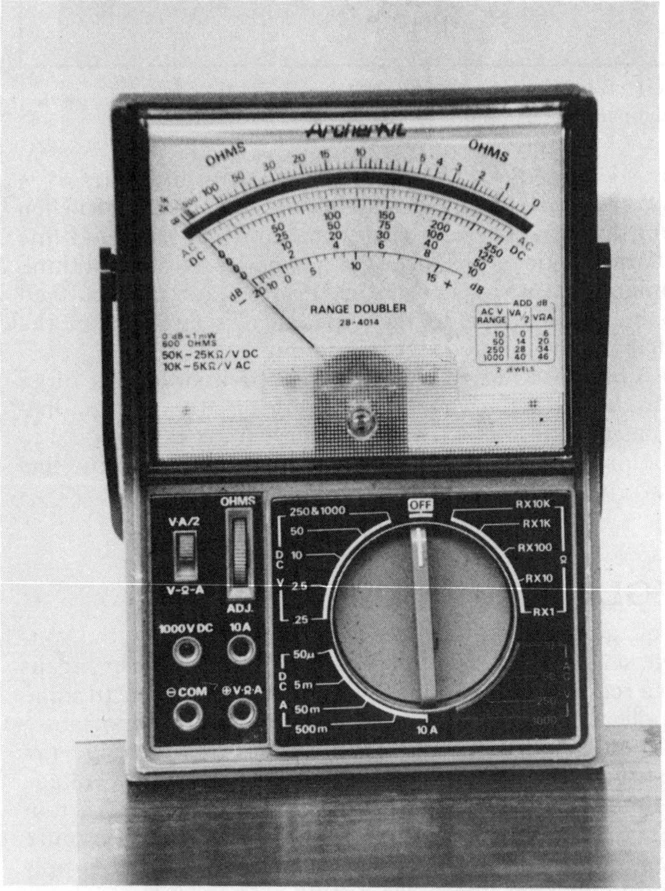

VOLT-OHM-MILLIAMMETER: Commonly used in the lab.

and the power source (*see* OHMMETER). In the milliammeter mode, shunt resistances are placed across the meter to obtain the desired full-scale range (*see* AMMETER, SHUNT RESISTOR).

The photograph shows a typical VOM. This device can measure voltages up to 1 kV (either dc or ac), direct currents to 10 A, and resistances to several megohms. A range-doubler switch allows for optimal meter reading.

While the instrument shown is adequate for most testing and service purposes, the ohms-per-volt rating may not be high enough in some cases. Then, a special voltmeter must be used. *See also* FET VOLTMETER, VACUUM-TUBE VOLTMETER.

VOLUME

Volume is an expression of the or loudness of the sound produced by an electronic device. In particular, volume refers to the level of the sound coming from the speakers of a radio, tape recorder, or record player. This is a function of the audio-frequency current and voltage supplied to the speakers.

Volume can be expressed in decibels relative to the threshold of hearing. Volume is also specified in terms of electrical decibels relative to a power level of 2.51 mW in a load of 600 ohms resistive impedance. These units are called volume units. (*see* DECIBEL, LOUDNESS, SOUND, VOLUME UNIT, VOLUME-UNIT METER).

VOLUME COMPRESSION

Volume compression is a technique that is sometimes used for improving the signal-to-noise ratio in a communications, high-fidelity, or public-address system. The basic principle is to enhance low-level sounds at some or all frequencies in the transmitting or recording process. There are several volume-compression schemes, used for various purposes. *See* AUTOMATIC GAIN CONTROL, AUTOMATIC LEVEL CONTROL, COMPRESSION, COMPRESSION CIRCUIT, DOLBY, SPEECH COMPRESSION).

VOLUME CONTROL

A volume control is a component, usually a potentiometer or set of potentiometers, that is used for adjusting the volume in an audio circuit.

There are two basic types of volume control: input adjustment and output adjustment. These two methods of volume control are shown in the illustration. Either method will provide satisfactory results. In high-power audio circuits, input adjustment is more commonly used.

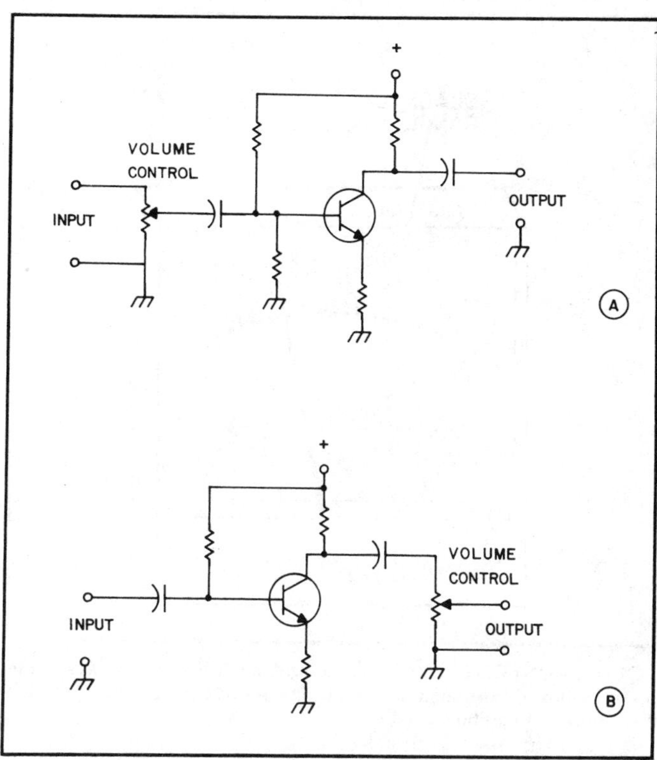

VOLUME CONTROL: Examples of volume control. At A, the adjustment is accomplished at the input of an audio amplifier; at B, the adjustment is done at the output of an amplifier.

VOLUME INDICATOR

See VOLUME-UNIT METER.

VOLUME UNIT

In high-fidelity applications, the amplitude of a music signal is generally expressed in terms of volume units (VU). The VU is a relative indicator of the root-mean-square power output from an audio amplifier.

A level of 0 VU corresponds to +4 dBm, or 2.51 mW, across a purely resistive load of 600 ohms impedance. This is a root-mean-square voltage of 1.23 V and a root-mean-square current of 2.05 mA. In general, a level of x VU corresponds to a signal x dB louder than the reference level. Thus, for a power level of P mW, the VU level is:

$$x = 10 \log_{10} (P/2.51)$$

Volume units are measured by means of a special meter, called a volume-unit meter, or VU meter, at the output of an audio amplifier. *See also* DECIBEL, VOLUME, VOLUME-UNIT METER.

VOLUME-UNIT METER

A volume-unit (VU) meter is a device that is used for measuring the root-mean-square volume level in an audio amplifier. The VU meter is calibrated in decibels relative to +4 dBm (*see* VOLUME UNIT).

Most VU meters are fast-acting devices, with just enough damping to allow easy reading. In a sophisticated stereo high-fidelity amplifier, each channel has a VU meter. The scale is marked off in black and red numerals, with a black and red reference line, (see illustration). The amplifier gain should normally be set so that the meter needle never enters the red range. If the needle moves into the red range, it indicates that distortion is likely to occur on audio peaks.

VOM

See VOLT-OHM-MILLIAMMETER.

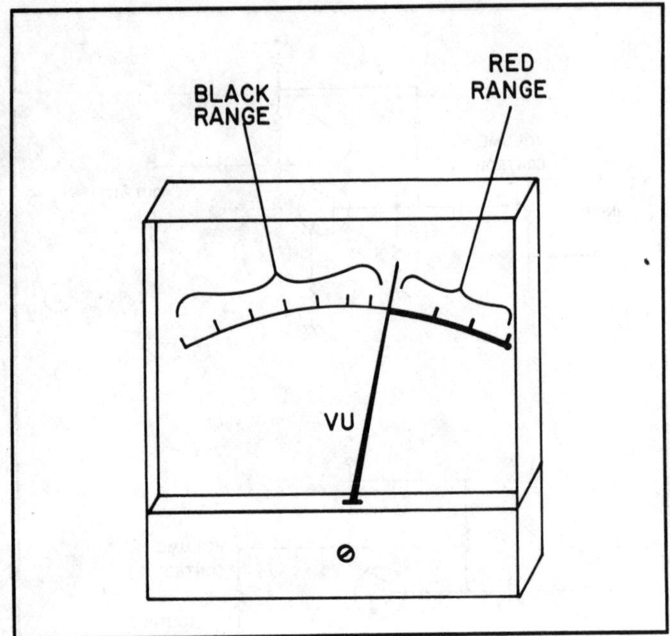

VOLUME-UNIT METER: A volume-unit meter has a black range, representing low distortion, and a red range, representing the risk of high distortion in an audio amplifier.

VOX

VOX is a means of actuating a circuit, such as a radio transmitter, using the electrical voice impulses from a microphone or audio amplifier. Such a voice-actuated system may use a relay or an electronic switch.

VOX circuits are commonly used in communications transceivers. This is especially true of single-sideband units. VOX eliminates the need for pressing a lever or switch to change from receive mode to transmit mode. This feature is especially useful in mobile operation, or when both hands are needed for such things as taking notes. The VOX circuit can be disabled and push-to-talk switching used instead if the operator desires.

In VOX operation, the transmitter is actuated within a few milliseconds after the operator speaks into the microphone. The transmitter remains actuated for a short time after the operator stops speaking; this prevents unwanted "tripping out" during short pauses.

The schematic diagram shows a VOX circuit. The sensitivity and delay are adjustable. An anti-VOX circuit prevents the received signals from actuating the transmitter if a speaker is used for listening.

VOX is a form of semi break-in operation (*see* SEMI BREAK-IN OPERATION). If full break-in operation is wanted

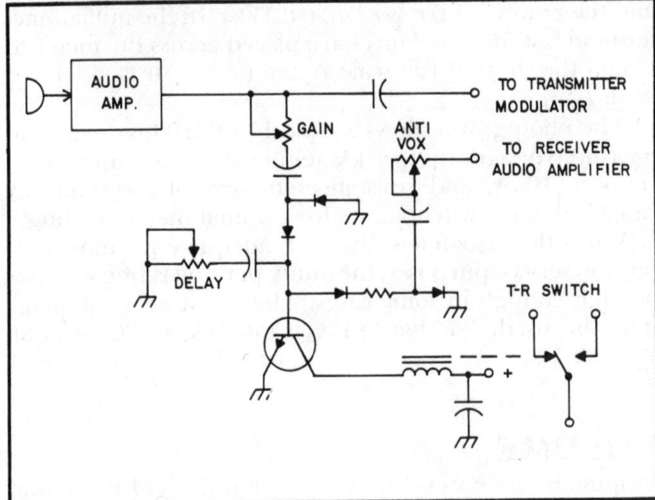

VOX: An example of a VOX circuit for voice-actuated transmission.

with single-sideband communications, a more sophisticated circuit is needed. *See also* BREAK-IN OPERATION.

VSWR

See STANDING-WAVE RATIO.

VTR

See VIDEO TAPE RECORDING.

VTVM

See VACUUM-TUBE VOLTMETER.

VTVOM

See VACUUM-TUBE VOLTMETER, VOLT-OHM-MILLIAMMETER.

VU METER

See VOLUME-UNIT METER.

VXO

See VARIABLE CRYSTAL OSCILLATOR.

WAFER

See CHIP, INTEGRATED CIRCUIT.

WAFER SWITCH

A wafer switch is a type of rotary switch, having one or more poles and one or more throw positions for each pole. The shaft passes through the center of one or more parallel, roughly circular wafers of plastic porcelain, or other dielectric material. The terminals are attached around each wafer (see illustration).

As the shaft is turned, a rotating terminal makes contact with each of the fixed terminals, one after the other in a circle. If the center shaft breaks contact with a terminal before making contact with the next, the switch is called nonshorting. If the shaft maintains contact with a terminal until after contact has been established with the next, the switch is called shorting. Either type has certain advantages and disadvantages, depending on the application.

Wafer switches are commercially manufactured in various sizes and configurations. Larger wafer switches can be used at radio frequencies. Wafer switches are commonly used for such purposes as bandswitching in receivers, transmitters, and tuning networks. Wafer switches are also used in some types of test equipment.

WALKIE-TALKIE

A handheld radio transceiver is sometimes called a walkie-talkie. This term is used especially with the lower-power transceivers that are popular among children. *See also* PORTABLE EQUIPMENT.

WATS

See WIDE-AREA TELEPHONE SERVICE.

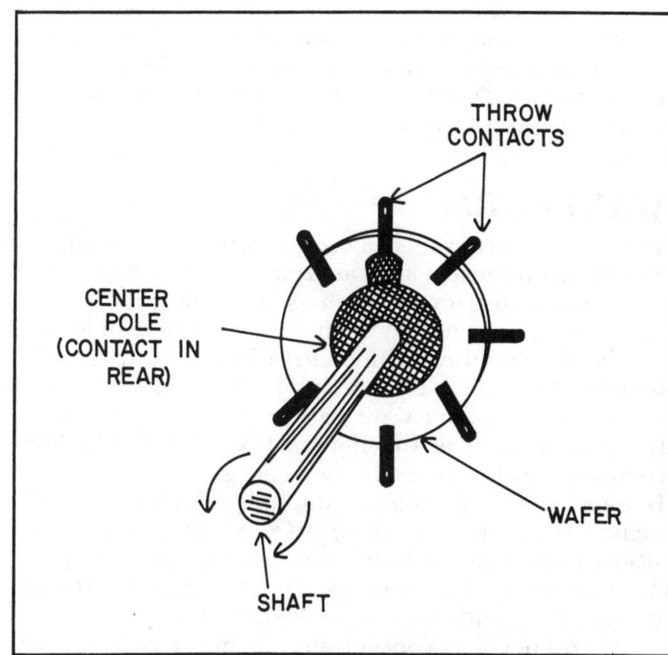

WAFER SWITCH: Construction of a wafer switch.

WATT

The watt is the unit of true, or real, power. The abbreviation is W. In a dc circuit, or in an ac circuit containing no reactance, the power P, in watts, can be found from any of the formulas:

$$P = EI = I^2R = E^2/R$$

where E is the rms voltage in volts, I is the rms current in amperes, and R is the resistance in ohms. In an alternating-current circuit containing reactance as well as resistance, the true power in watts must be determined in a different way (*see* POWER, TRUE POWER).

A power level of 1 W represents the expenditure of 1 joule of energy per second. A flashlight or small lantern bulb dissipates approximately 1 W of power. The audio output from a typical small portable radio is about 1 W when the volume is turned to maximum. An average fluorescent light bulb consumes 15 to 50 W; an incandescent bulb, as much as 500 W; a heavy appliance, as much as 2,000 W.

Fractions of the watt are sometimes used to express low power. The milliwatt (mW) is a thousandth (0.001) watt; the microwatt (uW) is a millionth (0.000001) watt. For larger power levels, the kilowatt (kW) or megawatt (MW) are used: 1 kW = 1,000 W and 1 MW = 1,000,000 W.

The radio-frequency output of a transmitter can be expressed in watts. This is the amount of power that would be dissipated in a pure resistor having a value equal to the matched output impedance of the transmitter—usually 50 ohms. *See also* APPARENT POWER, JOULE, REACTIVE POWER, TURE POWER, VAR, VOLT-AMPERE, WATTMETER.

WATT HOUR

The watt hour is a unit of energy equivalent to 3,600 joules (watt seconds). The watt hour is sometimes used to express the energy consumed by an electrical device over a period of time. The watt hour may also be used to express the storage capacity of a battery. Energy in watt hours is the product of the average power drawn, in watts, and the time, in hours. The abbreviation for watt hour is Wh.

Suppose that a 60-W bulb burns for one hour. Then the energy consumed by the bulb is 60 watt hours. In two hours the same bulb consumes 120 watt hours; in three hours it uses up 180 watt hours, and so on. A 12-V battery having 50 ampere hours of storage capacity can deliver 12x50, or 600, watt hours of energy. This might constitute 1 watt for 600 hours, 2 watts for 300 hours, or any of a number of other situations.

In utility applications, the kilowatt hour is more often specified than the watt hour. This is equivalent to 1,000 watt hours. The abbreviation for kilowatt hour is kWh. *See also* ENERGY, JOULE, KILOWATT HOUR, WATT-HOUR METER.

WATT-HOUR METER

A watt-hour meter is an instrument that is used to measure consumed energy. Your electric meter is a form of watt-hour meter, although it actually registers kilowatt hours

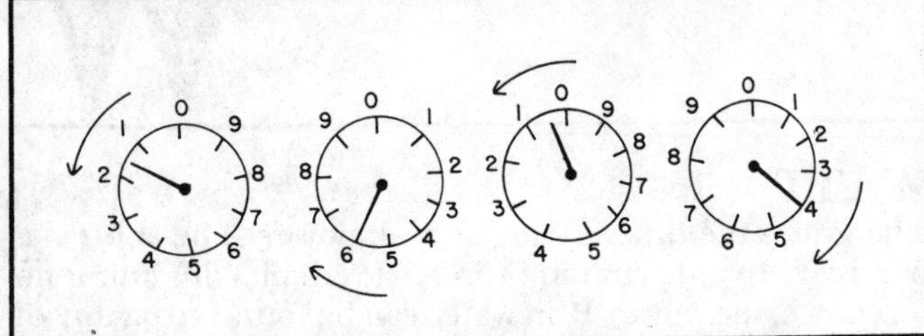

WATT-HOUR METER: A typical electric meter reads kilowatt hours. This illustration shows the scales of a meter that indicates 1604 kWh.

(*see* KILOWATT HOUR, WATT HOUR). The abbreviaton for watt hour is Wh, and the abbreviation for kilowatt hour is kWh.

Most watt-hour meters actually register the integral of the root-mean-square current in a circuit over a period of time. Thus, such a meter is suitable for operation only at a single, rated voltage, such as 234 V. A motor runs at a speed that is directly proportional to the current drawn at any given time. The motor is connected to a set of gears, in turn attached to pointers or drum indicators that numerically indicate the energy used up to a given time.

A typical electric meter appears like the accompanying drawing. The pointers turn in opposite directions as you read the digits. An electric meter is somewhat tricky to read; the unit illustrated indicates 1604 kWh. A few electric meters have rotating drums with the numerals painted directly on their surfaces, allowing direct and simple reading.

WATTMETER

A wattmeter is an instrument that measures power. There are various types of wattmeters; some measure true power, while others measure only the apparent power (*see* APPARENT POWER, POWER, REACTIVE POWER, TRUE POWER).

The simplest wattmeter consists of an ammeter in series with a circuit, and a voltmeter in parallel with the component or set of components for which the power is to be determined. The power (in watts) is given by the product of the voltage (in volts) and the current (in amperes). This method is satisfactory for direct-current circuits and for alternating-current circuits in which no reactance is present. However, if reactance exists, the product of current and voltage is artificially large, and does not represent true power.

In a common household circuit, there is usually no reactance. Therefore, a root-mean-square ammeter can be used to measure the power consumed by a device at a particular voltage. The ammeter is simply connected in series with the appliance. The consumed power, P, is given (in watts) in terms of the current reading, I (in amperes) and the line voltage, E (in volts) as:

$$P = EI$$

In radio-frequency circuits, direcional wattmeters are generally used to measure true power. Such wattmeters, such as the device shown in the illustration, can measure forward or reflected power. The true power is the difference between the forward and reflected readings. The greater the amount of reactance in the antenna system, the larger the reflected reading will be, in proportion to the forward reading. *See also* REFLECTED POWER, REFLECTOMETER.

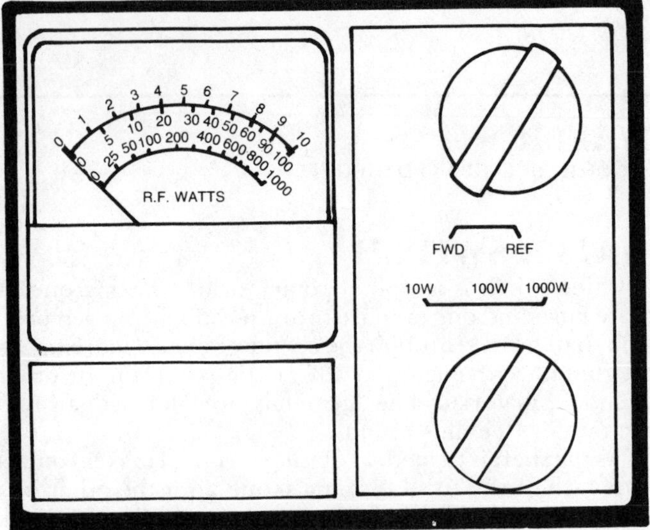

WATTMETER: A typical radio-frequency wattmeter, with forward-power and reflected-power modes (courtesy of Gold Line Connector).

WAVE

A wave is any form of disturbance that exhibits a periodic (repeating) pattern. Examples of waves include acoustic disturbances, pulsating or alternating currents or voltages, and electromagnetic fields.

All waves have a definable period T (usually given in seconds), a specific frequency f, a propagation speed c (usually expressed in units per second), and a wavelength λ. These quantities are related according to the formulas:

$$c = f\lambda = \lambda/T$$

Some disturbances exhibit changes in amplitude that have no identifiable period. These disturbances are not true waves, although they may produce noticeable effects such as acoustic or electromagnetic noise over a wide band of frequencies.

A wave does not, in itself, necessarily appear "wave-like." The wave nature of some disturbances, such as ripples on a pond or swells in the ocean, is obvious. But in most cases, some means of artificial display is required to see the wave nature of a disturbance. *See also* ALTERNATING CURRENT, ELECTROMAGNETIC FIELD, LONGITUDINAL WAVE, SOUND, TRANSVERSE WAVE, WAVEFORM.

WAVEFORM

Waveform is an expression used to describe the "shape" of a wave disturbance, either as seen directly or as observed on a display instrument. such as an oscilloscope. A wave disturbance can have any of an infinite number of forms.

The simplest waveform occurs when a disturbance has only one frequency. The resulting waveform is sinusoidal, and is known as a sine wave because its shape is identical to the graph of the sine function (*see* SINE, SINE WAVE). More complex waveforms result when energy is concentrated at frequencies that are integral multiples of a lowest, fundamental frequency (*see* FUNDAMENTAL FREQUENCY, HARMONIC). Examples include the sawtooth, square, trapezoidal, and triangular waves (*see* SAWTOOTH WAVE, SQUARE WAVE, TRAPEZOIDAL WAVE, TRIANGULAR WAVE).

When energy is concentrated at many different frequencies, waveforms become exceedingly complicated.

Various musical instruments produce waveforms ranging from a near-perfect sine wave to a complex jumble.

All periodic waves, no matter how complicated they appear, have one common characteristic: they repeat themselves at defined intervals. Thus, it is possible to predict the instantaneous amplitude of the waveform for any specificed moment in the future. However, some waveforms do not have this property. Natural electromagnetic noise, for example, has a waveform that is random. Although such disturbances are not true waves, they nevertheless can be observed on display instruments such as the oscilloscope. *See also* WAVE.

WAVEGUIDE

A waveguide is a feed line that is often used at ultra-high and microwave frequencies. A waveguide consists of a hollow metal pipe, usually having a rectangular or circular cross section (see illustration). The electromagnetic field travels down the pipe, provided that the wavelength is short enough. A waveguide provides excellent shielding and low loss.

Cross-Sectional Size Requirements. In order to efficiently propagate an electromagnetic field, a rectangular waveguide must have sides measuring at least 0.5 wavelength, and preferably more than 0.7 wavelength. A circular waveguide should be at least 0.6 wavelength in diameter, and preferably 0.7 wavelength or more.

Consider the example of a waveguide operating at a frequency of 3 GHz. The wavelength is 3.9 in (10 cm). Thus, a rectangular waveguide must be at least 2.0 in (5.1 cm) wide, and preferably 2.7 in (7.0 cm). A circular waveguide should be at least 2.3 in (6.0 cm) in diameter, but preferably 2.7 in (7.0 cm). The frequency at which the length of the side, or the diameter, is ½ wavelength, is known as the cutoff frequency. The waveguide exhibits a highpass response (*see* HIGHPASS RESPONSE).

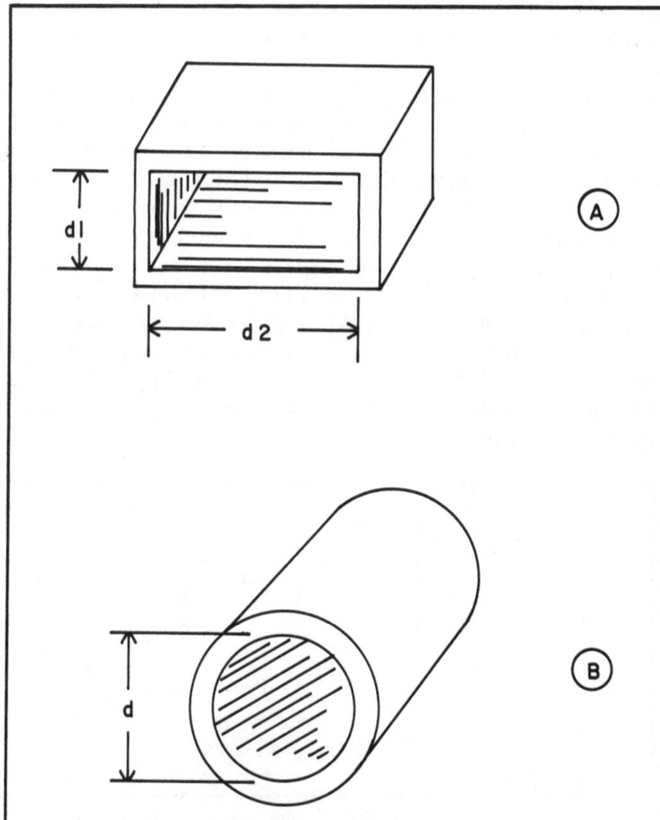

WAVEGUIDE: Waveguide cross sections. At A, a rectangular waveguide; d1 and d2 represent the minor and major internal dimensions, respectively. At B, a circular waveguide; d represents the inside diameter.

Propagation. An electromagnetic field can travel down a waveguide in various ways. If all of the electric lines of flux are perpendicular to the axis of the waveguide, the propagation mode is called transverse-electric (TE). If all of the magnetic lines of flux are perpendicular to the axis, the mode is called transverse-magnetic (TM). The electromagnetic field can be coupled into the waveguide via either the electric or the magnetic components.

In a waveguide, electromagnetic fields tend to circulate in eddies. Depending on the frequency of the applied energy, there will be one or more eddies. In general, as the frequency becomes higher, the number of eddies in the cross section increases.

Operation. When a waveguide is used, it is important that the impedance of the antenna be purely resistive, and that the resistance be matched to the characteristic impedance of the waveguide (*see* CHARACTERISTIC IMPEDANCE). Otherwise, there will be standing waves in the line, and the loss will be increased compared with the perfectly matched condition (*see* STANDING WAVE, STANDING-WAVE RATIO, STANDING-WAVE-RATIO LOSS). The waveguide behaves exactly like other types of feed lines in this respect.

The characteristic impedance of a waveguide varies with frequency. In this sense, it differs from coaxial or parallel-wire lines. In most cases, matching transformers are needed to achieve a low standing-wave ratio on a waveguide, because few antennas present impedances that are favorable to a waveguide if direct coupling is employed. A quarter-wave section of waveguide, coaxial cable, or parallel wires or rods, can be used for this purpose (*see* QUARTER-WAVE TRANSMISSION LINE).

Because of the highpass characteristics of waveguides, they can be used as filters and attenuators. They are very effective for this purpose because of their high efficiency.

It is important that the interior of a waveguide be kept clean and free from condensation. Even a small obstruction can seriously degrade the performance of a waveguide.

For Further Information. The theory of waveguides cannot be thoroughly treated here. For more details, a volume on communications technology or microwave theory and practice is recommended.

WAVEGUIDE PROPAGATION

Waveguide propagation is an effect that is observed at long wavelengths. The ionosphere acts as a total reflector at very low and low frequencies. If the wavelength is great enough, the earth and ionosphere act like the surfaces of a waveguide, and propagation occurs in a manner similar to that in a waveguide transmission line at ultra-high and microwave frequencies (*see* WAVEGUIDE).

In the waveguide mode, the transmitting antenna, which must be vertical, couples energy into the space between the earth and ionized upper atmosphere. The receiving antenna picks up the energy in a manner similar to a probe in a waveguide feed line. The earth-ionosphere waveguide has a cutoff (lower limit) of about 10 kHz. Below this frequency, propagation is very poor because of severe attenuation.

WAVELENGTH

All wave disturbances have a certain physical length in space. This length depends on two factors: the frequency of the disturbance, and the speed at which it is propagated (*see* WAVE). The wavelength is inversely proportional to the frequency, and directly proportional to the speed of propagation. Wavelength is denoted in equations by the lowercase Greek letter lambda (λ).

In a periodic disturbance, the wavelength is defined as the distance between identical points of two adjacent waves. In electromagnetic fields, this distance can be millions of meters or a tiny fraction of a millimeter, or anything in between. The wavelength determines many aspects of the behavior of a wave disturbance.

For electromagnetic waves in free space, the wavelength, in feet, is given in terms of the frequency by the formula:

$$\lambda = 9.84 \times 10^8/f$$

where f is in hertz. The wavelength in meters is:

$$\lambda = 3.00 \times 10^8/f$$

For sound waves in air at sea level, wavelength is given in feet by:

$$\lambda = 1.10 \times 10^3/f$$

where f is in hertz; the wavelength in meters is:

$$\lambda = 3.35 \times 10^2/f$$

Wavelength is an important consideration in the design of antenna systems for radio frequencies. Wavelength is also important to the designers of optical apparatus and various acoustic devices. *See also* ELECTROMAGNETIC FIELD, ELECTROMAGNETIC SPECTRUM, FREQUENCY, SOUND, VELOCITY FACTOR.

WAVEMETER

A wavemeter is a device that makes use of the properties of resonant circuits to determine the frequency or wavelength of a radio signal. The most common type of wavemeter is called the absorption wavemeter (*see* ABSORPTION WAVEMETER). However, there are other methods of using resonant circuits to measure frequency. *See also* CAVITY FREQUENCY METER, COAXIAL WAVEMETER, LECHER WIRES, WAVE TRAP.

WAVE POLARIZATION

See POLARIZATION.

WAVESHAPE

See WAVEFORM.

WAVE THEORY OF LIGHT

See ELECTROMAGNETIC THEORY OF LIGHT.

WAVE-TRAP FREQUENCY METER

A trap can be used to measure the frequency of a radio signal by means of a circuit called a wave trap or wave-trap frequency meter. The device consists of a tunable trap (parallel-resonant circuit) in series with the signal path, and an indicating device such as a meter (see illustration).

The tunable trap is adjusted until a dip is observed in the meter reading. This condition occurs when the trap is set for resonance at the frequency of the applied signal. The frequency is then read from a calibrated scale.

WAVE TRAP

See TRAP, WAVE-TRAP FREQUENCY METER.

WEAK-SIGNAL SENSITIVITY

See TANGENTIAL SENSITIVITY.

WEATHER SATELLITE

In April of 1960, the first weather satellite was launched. It was called TIROS, and it sent the first photographs of the earth from space. Since the launch of TIROS, many other satellites have been placed into orbit for the purpose of aiding us in forecasting the weather.

Weather satellites orbit the earth at various altitudes. Some satellites travel around the earth in polar orbits, while others remain over the same place all the time.

Two Weather Satellites. The photographs show artist's conceptions of two satellites used for weather observation.

The satellite at A is the Improved Tiros Operational Satellite (ITOS). It was first launched on January 23, 1970. Ultimately, the National Oceanic and Atmospheric Administration will place seven ITOS satellites into polar orbits, allowing meteorologists to observe weather throughout the world. The ITOS satellites weigh approximately 600 pounds and measure 3 × 3 × 4 feet. The three rectangular fin-like structures are solar panels that produce up to 400 watts of power from sunlight. The orbit is synchronized with the sun, so that the panels receive light constantly.

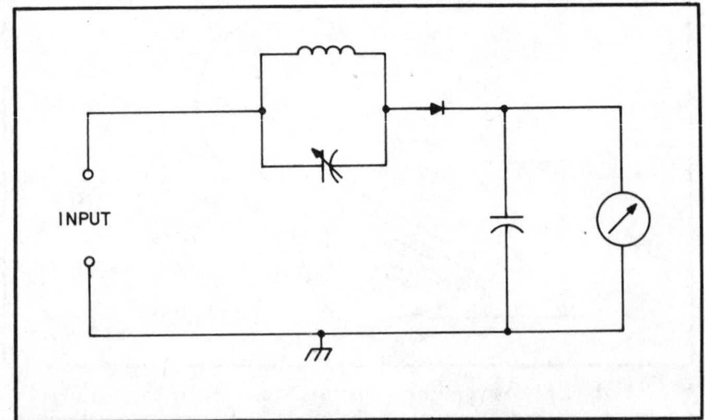

WAVE-TRAP FREQUENCY METER: The variable capacitor facilitates adjustment of the resonant frequency of the trap.

WEATHER SATELLITE: At A, an artist's conception of ITOS.

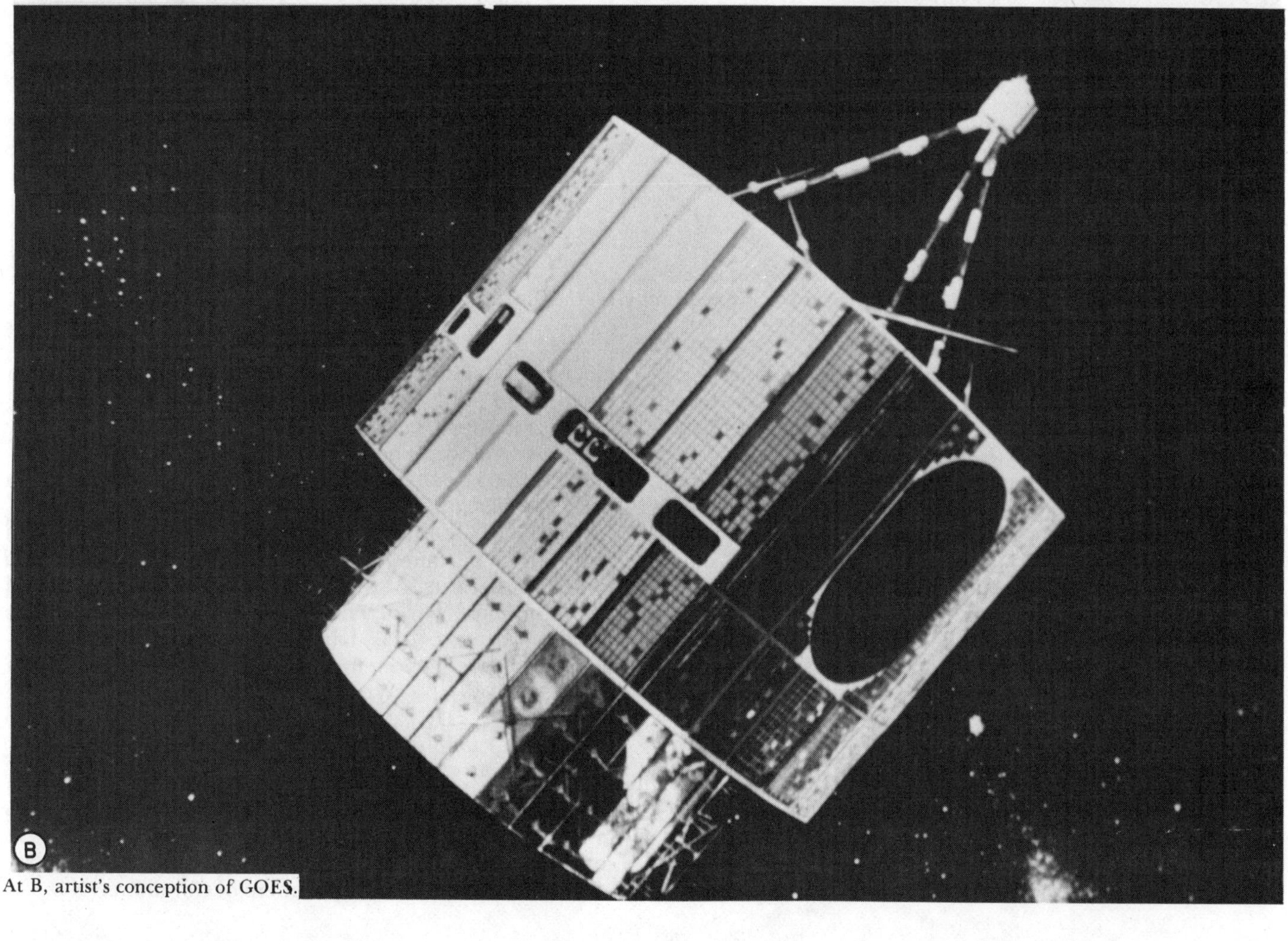

At B, artist's conception of GOES.

At C, a photograph of Hurricane David, taken on August 31, 1979 as the storm churned the waters southeast of the island of Hispaniola. (Courtesy of NOAA, National Environmental Satellite Service.)

The satellite at B is the Geostationary Operational Environmental Satellite (GOES). It gets its name from the fact that its orbit is geostationary, or synchronous with respect to the rotation of the earth (*see* SYNCHRONOUS ORBIT). The first GOES satellite was put into orbit on October 16, 1975, over the equator at 55 degrees west longitude. The GOES satellites weigh nearly 660 pounds and measure approximately 6 feet in diameter and 7 feet high, not including projections. Solar cells cover the exterior of the cylindrical spacecraft. In addition to monitoring weather on the earth, the GOES satellites contain sensors for monitoring the geomagnetic field and solar emissions at various wavelengths.

Gathering Information and Transmitting the Images. Weather satellites usually contain cameras for both the visible spectrum and the infrared (*see* INFRARED, LIGHT). They resemble ordinary television camera tubes. The images are transmitted to earth-based stations by means of facsimile. The resolution and magnification can be adjusted. The quality is excellent in all cases, (C).

Geostationary satellites facilitate observation of only part of the globe. The pictures from polar-orbiting satellites can be assembled into "mosaics" of the entire earth over a period of a half day.

The GOES satellites not only take pictures of the earth and monitor certain space conditions; they also collect data from thousands of observation platforms located throughout the world. These stations monitor such things as tides, seismic effects, barometric pressure, wind direction, wind speed, and temperature. All of the information is relayed to the World Weather Building in the Washington, DC area, and thus made available to meteorologists in cities around the country.

For Further Information. It is not possible, in this book, to give a detailed account of all aspects of weather satellites. A complete and informative booklet on the subject can be obtained from the U.S. Department of Commerce/NOAA,

Satellite Data Services Division, World Weather Building, Room 100, Washington, DC 20233.

WEBER

The weber is a unit of magnetic flux, representing one line of flux in the meter-kilogram-second (MKS) or standard international (SI) system of units. The weber is equivalent to a volt second.

Magnetic flux is sometimes expressed in units called maxwells. The maxwell is equal to 10^{-8} weber. *See also* MAGNETIC FIELD, MAXWELL.

WEBER-FECHNER LAW

Our senses do not perceive things in direct proportion to their actual intensity. Instead, we sense effects according to the logarithm of their actual intensity. This rule is called the Weber-Fechner Law. The Weber-Fechner Law is of special significance for audible and visible effects. *See also* DECIBEL.

WEIGHT

Weight is a measure of the force exerted on an object in a gravitational field. In any constant gravitational field, weight is directly proportional to mass. For a constant mass, the weight is directly proportional to the gravitational acceleration.

Consider the example of a person who weighs 100 pounds on the earth, where the gravitational acceleration is about 32 feet per second per second, or 9.8 meters per second per second. This amount of gravitational acceleration is called 1 gravity (abbreviated 1 g). On Mars, that same person would weigh only 38 pounds, because the acceleration of gravity is just 0.38 g. On Jupiter, the person would weigh 250 pounds, since the gravitational acceleration is 2.5 g. When a space rocket takes off, the acceleration may be as great as 330 feet, or 100 meters, per second per second. This is 10 g, and a 100-pound astronaut riding in such a vessel will weigh half a ton! *See also* MASS.

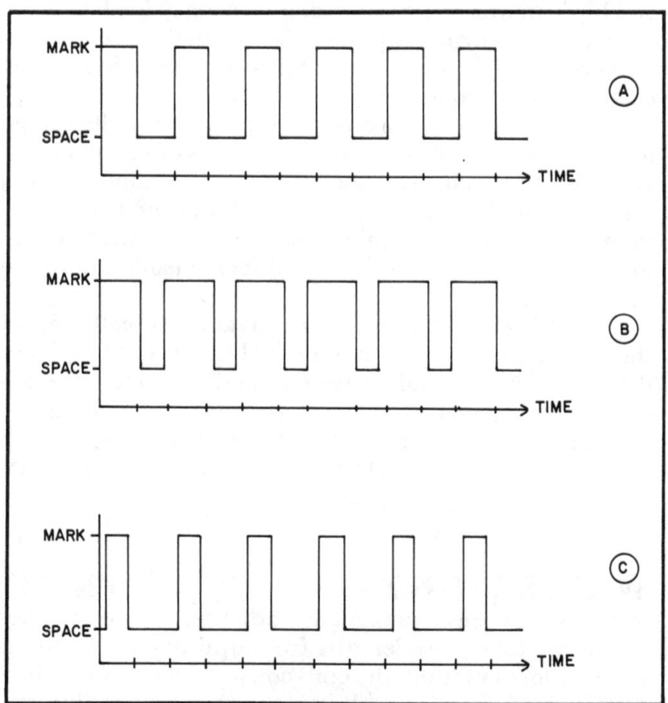

WEIGHT: Examples of strings of dots in Morse code, showing normal weight (A), heavy weight (B), and light weight (C).

In digital transmission, the duration ratio of mark to space is sometimes called weight. This expression is especially used in Morse-code transmission. A string of dots has a weight ratio of 1:1 if the duration of a dot is identical to the duration of the spaces between the dots (see illustration A). If the dots last longer than the spaces, the weight ratio is greater than 1:1, and is said to be heavy, as at B. If the dots are briefer than spaces, the ratio is less than 1:1, and is called light. (C). Normally, a weight ratio of 1:1 is used, although some telegraphers prefer a slightly heavier weight at slow speeds and a slightly lighter weight at very high speeds.

Some manufacturers define weight according to a different formula than that described above. The dot-to-dash ratio is sometimes specified. Normally it is 1:3. Small ratios such as 1:4 represent lighter weight, and larger values such as 1:1.2 represent heavier weight. *See also* MARKSPACE, MORSE CODE, WEIGHT CONTROL.

WEIGHT CONTROL

Many keying devices, especially for Morse-code transmission, have weight controls. A weight control is used to compensate for the varying tastes of receiving operators, and for the effects of shaping circuits in code transmitters (*see* SHAPING).

A typical weight control allows adjustment of the dot-to-space ratio between limits of about 1:2 and 2:1. Some weight controls have larger ranges. In certain digital terminals, the weight is adjustable in discrete steps; in many electronic keyers, the weight can be varied continuously. *See* MORSE CODE, WEIGHT.

WEIGHT RATIO

See WEIGHT CONTROL.

WESTON STANDARD CELL

A Weston standard cell is an electrochemical cell that provides a constant 1.0183 V. The electrolyte material is cadmium sulfate liquid. The positive electrode consists of mercury, and the negative electrode is fabricated from a combination of mercury and cadmium.

The Weston standard cell is generally housed in a pair of glass enclosures connected by a tube. The cell must be kept upright for proper operation. This type of cell is most often used as a voltage standard. The illustration is a cross-sectional diagram of a Weston standard cell. *See also* CELL.

WHEATSTONE BRIDGE

A Wheatstone bridge is a device for measuring unknown resistances. The Wheatstone bridge operates by comparing the ratio of the unknown resistance to three other known resistances.

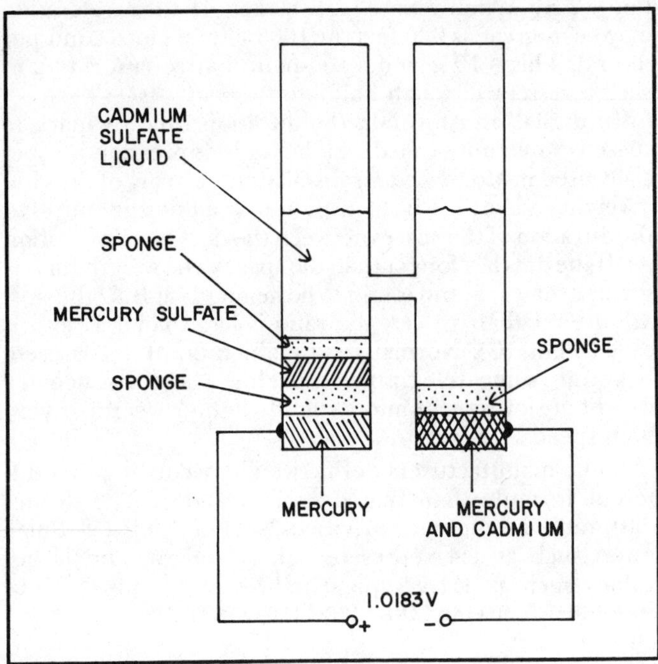

WESTON STANDARD CELL: Construction of a Weston standard cell.

The illustration is a schematic diagram of a Wheatstone bridge. The indicator may be a galvanometer if a direct-current supply is used. If all the resistances are nonreactive, an audio-frequency generator can be used in conjunction with a headphone. The unknown resistance is R. Two fixed resistors, R1 and R2, and an adjustable resistor, R3, are used for comparison.

The value of R3 is adjusted until the galvanometer reads zero, or until a null is heard in the headset. This indicates that the ratio R:R3 is equal to the ratio R1:R2. The unknown resistance R is given by the formula:

$$R = R3R1/R2$$

The Wheatstone bridge is similar to the ratio-arm bridge. *See also* RATIO-ARM BRIDGE, RESISTANCE MEASUREMENT.

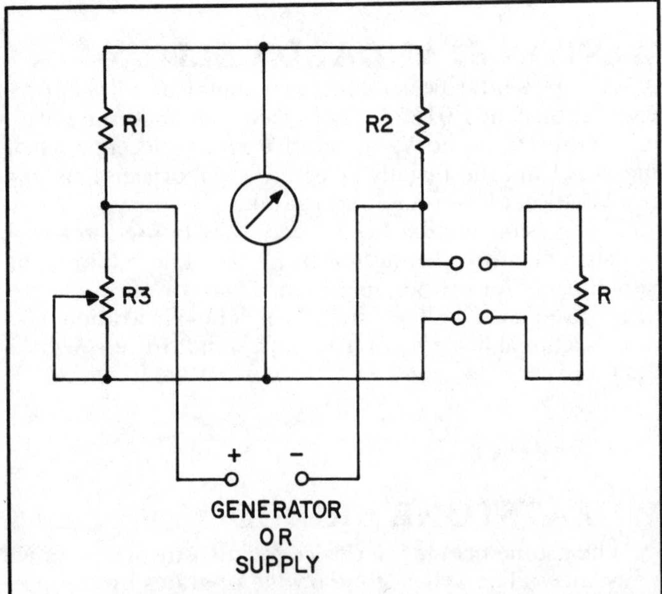

WHEATSTONE BRIDGE: The Wheatstone bridge is used for measuring small values of resistance.

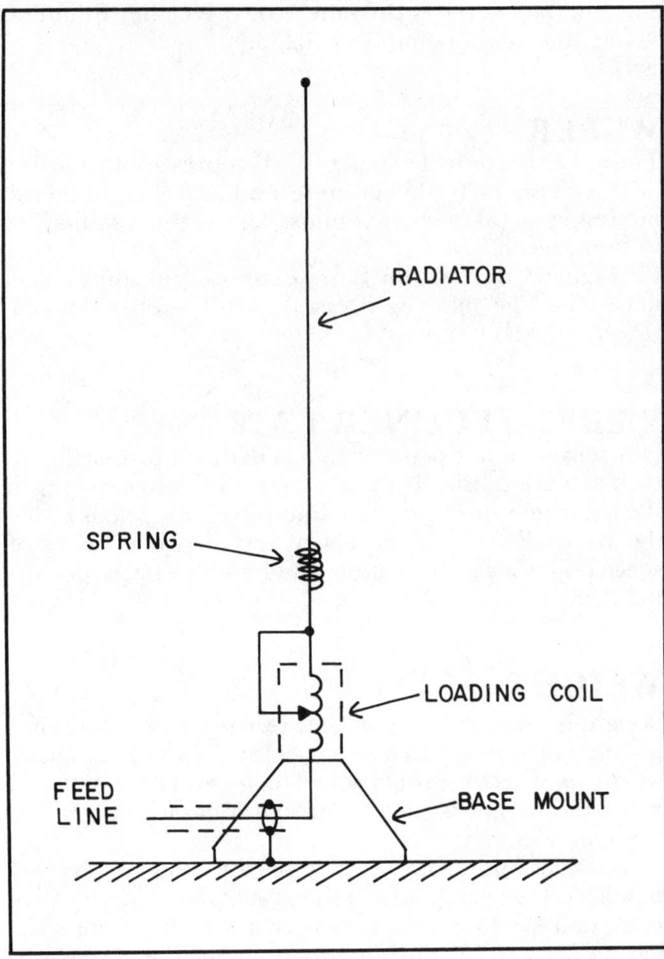

WHIP ANTENNA: Construction of a whip antenna, with inductive base loading.

WHIP ANTENNA

A whip antenna is a short radiator, usually loaded at the base and measuring ¼ physical wavelength or less. Whip antennas are often used in mobile communications, especially at high frequencies.

The construction of a typical whip antenna is illustrated in the drawing. A tapered rod of stainless steel forms the radiator. A tapped coil at the base facilitates adjustment of the resonant frequency. A spring allows for wind resistance. A magnetic mounting can be used for small whips. For antennas longer than about 3 feet, a ball mounting is generally used.

The efficiency of a whip antenna depends on the size of the vehicle on which it is mounted. The larger the vehicle, the better the grounding system, and the better the antenna performance. In general, efficiency improves as the frequency is increased, and worsens as the frequency is decreased. *See also* ANTENNA EFFICIENCY, VERTICAL ANTENNA.

WHITE LIGHT

When visible electromagnetic radiation is concentrated uniformly at all wavelengths from approximately 390 to 750 nanometers (nm), the emission appears gray or white. An equal combination of the primary colors (red, blue, and green) also results in white light. *See also* LIGHT, PRIMARY COLORS.

WHITE NOISE

Broadband noise is sometimes called white noise. It gets its name from the fact that white light contains energy at all visible wavelengths, and thus is wideband electromagnetic energy; wideband noise is therefore, in a sense, "white."

White noise at audio frequencies consists of a more or less uniform distribution of energy at wavelengths from 20 Hz to 20 kHz. If energy is wideband in nature but concentrated toward one end or the other, the noise is called pink or violet. White noise sounds like a rough hiss and roar, similar to the sound of ocean breakers on a beach. White noise is pleasing to the ear, and is even used in some cases to induce sleep.

At radio frequencies, the term white noise is quite general. If a given band of frequencies contains noise of essentially equal amplitude at all points, the noise may be called white. *See also* NOISE.

WHITE TRANSMISSION

White transmission is a form of amplitude-modulated facsimile signal (*see* FACSIMILE) in which the greatest copy density, or darkest shade, corresponds to the minimum amplitude of the signal. White transmission is the opposite of black transmission, in which the darkest shade corresponds to the maximum signal amplitude (*see* BLACK TRANSMISSION). White transmission may be considered, in a sense, right-side-up.

In a frequency-modulated facsimile system, white transmission means that the darkest copy corresponds to the highest transmitted frequency.

WIDE-AREA TELEPHONE SERVICE

Wide-area telephone service, abbreviated WATS, is a special form of long-distance telephone operation. Most long-distance calls are charged on a timed basis, such as 35 cents per minute. A WATS line allows a subscriber to obtain long-distance service for a flat monthly charge. In some cases, surcharges are applied for use of the line for more than a specified amount of time in a month.

A WATS line may be either incoming or outgoing. The most familiar example of an incoming WATS line is an "800" line, so named because the area code is 800. Outgoing WATS lines are used by many businesses that engage in extensive long-distance telephone correspondence. *See also* TELEPHONE.

WIDEBAND MODULATION

See BROADBAND MODULATION, SPREAD-SPECTRUM TECHNIQUES.

WIEN'S DISPLACEMENT LAW

The electromagnetic radiation from a black body is distributed over a wide band of frequencies. However, there is always a wavelength at which the intensity is greater than at any other wavelength. This wavelength is related to the absolute temperature of the black body according to a formula called Wien's Displacement Law.

If λ represents the peak wavelength in microns (millionths of a meter) and T represents the temperature in degrees Kelvin, then:

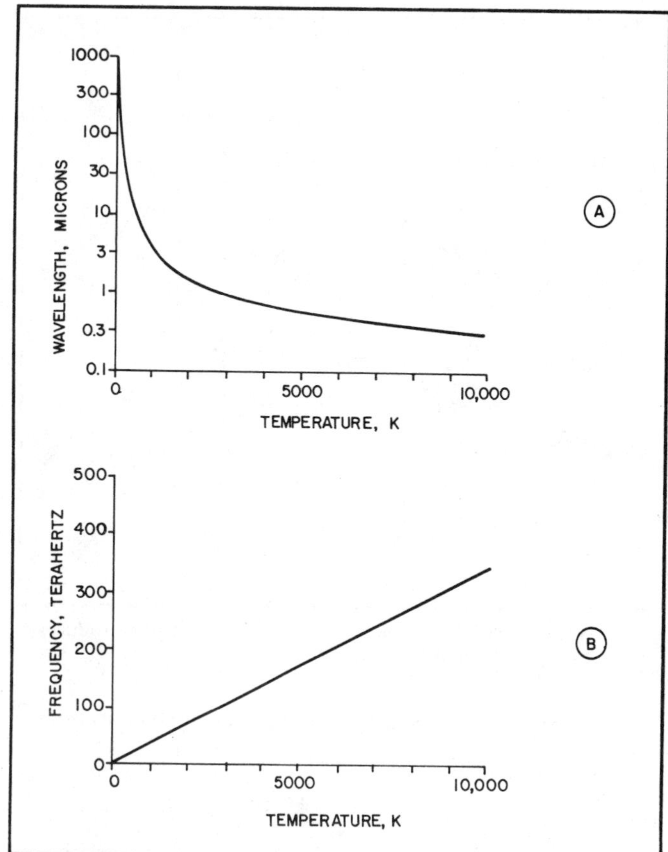

WIEN'S DISPLACEMENT LAW: At A, peak wavelength versus absolute temperature for blackbody radiation. At B, peak frequency versus absolute temperature.

$$\lambda = 2898/T$$

Another way of representing Wien's Displacement Law is:

$$f = T/28.98$$

where f is the frequency in terahertz (1 THz = 10^{12} Hz).

As a black body gets hotter, the peak wavelength gets shorter, and the peak frequency becomes higher. As a black body gets cooler, the wavelength becomes greater and the frequency decreases. The relationship between temperature and peak wavelength is illustrated in the graph at A; the relationship between temperature and peak frequency is shown at B. *See also* BLACK BODY, KELVIN TEMPERATURE SCALE.

WIEN-BRIDGE OSCILLATOR

A Wien bridge can be connected in the feedback circuit of an oscillator, resulting in a nearly pure sine wave at audio frequencies. This type of oscillator is a two-stage circuit, and is known as a Wien-bridge oscillator (see illustration).

In order for the Wien-bridge oscillator to produce a sine wave, the components must have values such that the Wien bridge is balanced. This requires that the values, as shown in the diagram, satisfy the equation:

$$C = C1(R2/R1 - R3/R)$$

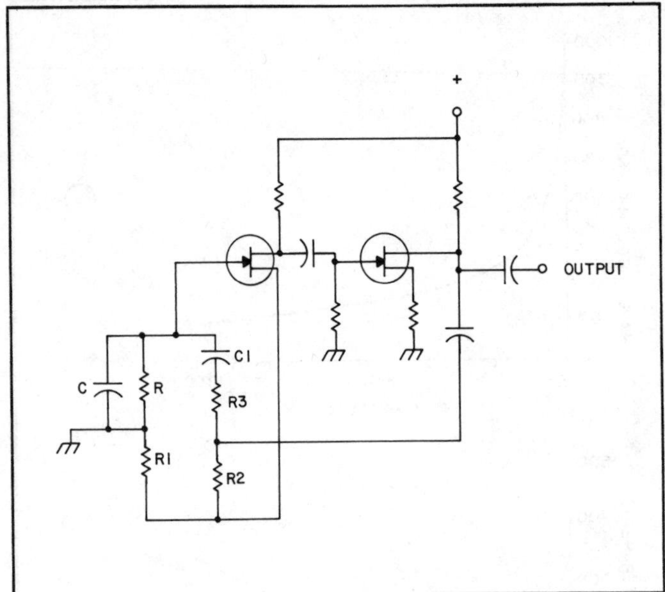

WIEN-BRIDGE OSCILLATOR: A Wien-bridge oscillator produces a sine-wave output at audio frequencies.

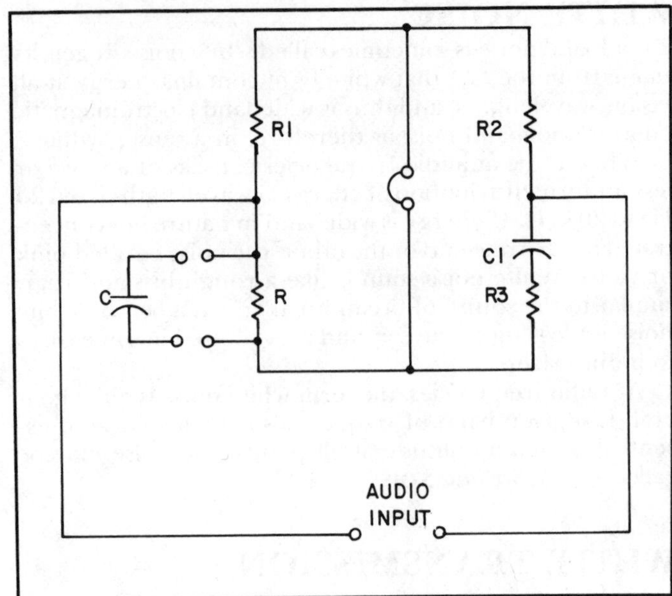

WIEN BRIDGE: A Wien bridge can be used for measurement of capacitance.

Under these conditions, the resonant frequency f, in hertz, is:

$$f = 1/(2\pi C1R3)$$

The primary advantage of the Wien-bridge circuit is its relatively pure output. The circuit is very stable. *See also* WIEN BRIDGE.

WIEN BRIDGE

A Wien bridge is a resistance-capacitance circuit that can be used for measurement of unknown values of capacitance. The Wien bridge can also be used as a resonant circuit.

The configuration of a Wien bridge is shown in the diagram. The unknown capacitance C is connected across a known resistance R. Other known resistances and capacitances, R1, R2, R3, and C1, are connected as shown. If all resistances are given in ohms and all capacitances are given in farads, then the value of the unknown capacitance C, with the circuit in a balanced condition, is:

$$C = C1(R2/R1 - R3/R)$$

The resonant frequency f, in hertz, is:

$$f = 1/(2\pi \sqrt{C1R3})$$

The Wien bridge requires an audio-frequency generator, and an indicating device such as a meter or headset. *See also* WIEN-BRIDGE OSCILLATOR.

WILSON, ROBERT

See PENZIAS, ARNO

WIND LOADING

Wind loading is an expression of the maximum wind speed, in miles per hour or meters per second, that an antenna can withstand without damage. In general, the larger antennas exhibit more wind resistance than smaller ones. Therefore, larger antennas require more rugged physical construction than small antennas for a given wind-loading figure.

Wind loading is affected by the lengths and diameters of conductors used in an antenna. In some cases, wind loading depends on the direction of the wind with respect to the antenna. If an antenna becomes laden with ice or wet snow, the wind loading is greatly reduced (*see* ICE LOADING).

Wind-loading figures are sometimes given for towers. The wind speed that a tower can withstand is affected by the sizes and types of antennas it supports. The ability of a tower to resist high winds can be enhanced by guying (*see* GUYING).

WINDOM ANTENNA

A Windom antenna is a multiband wire antenna that uses a single-wire feed line. The antenna gets its name from the radio amateur who devised it. The Windom is a half-wave horizontal antenna, fed slightly off center (see illustration A).

For a given fundamental frequency f in megahertz, the length d of a windom, in feet, is:

$$d = 468/f$$

If d is expressed in meters, then:

$$d = 143/f$$

The feed line is attached at a point 36 percent of the way from one end of the radiator to the other end. For a given frequency f in megahertz, the distance r, in feet, from the near end of the radiator to the line is:

$$r = 168/f$$

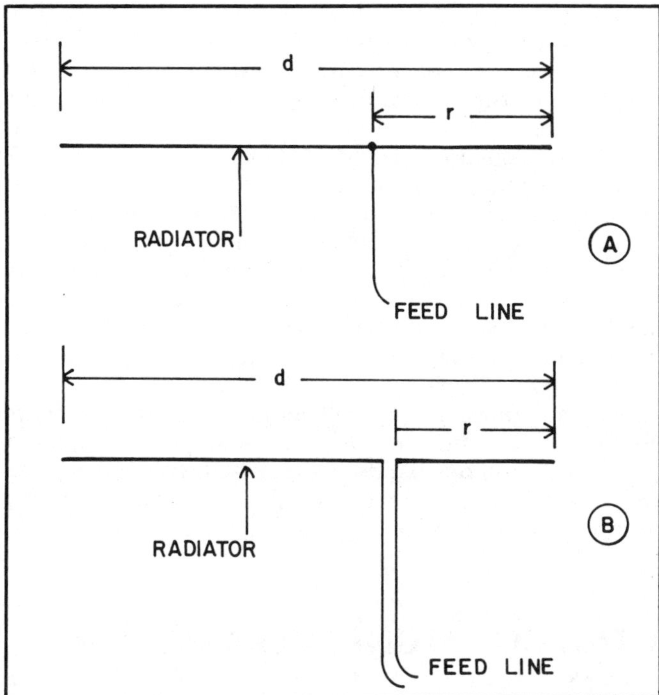

WINDOM ANTENNA: At A, a true Windom antenna; at B, a variation of the Windom. Dimensions d and r are discussed in the text.

If r is expressed in meters,

$$r = 51/f$$

The Windom, thus constructed, will operate satisfactorily at all of the even harmonics, as well as the fundamental frequency. An antenna designed for a frequency f can be used at 2f, 4f, 6f, and so on. A transmatch can be used to match the antenna to a transmitter output circuit. Some radiation occurs from the feed line, since it is unbalanced and not shielded.

A parallel-wire line can be used with off-center feed in an antenna that is sometimes called a Windom. (Actually, this antenna is not a true Windom; it could be called a quasi-Windom.) The design is illustrated at B. The radiator measures ½ electrical wavelength at the fundamental frequency. The length in feet or meters is found by the same formula as given earlier. The feed line is attached ⅓ of the way from one end of the radiator to the other. For a fundamental frequency f, the distance r, in feet is:

$$r = 156/f$$

and in meters:

$$r = 48/f$$

This configuration results in a feed-point impedance of about 300 ohms at frequencies f, 2f, 3f, 4f, 5f, and so on. Some radiation occurs from the line because the system is not perfectly balanced. A 4:1 balun or a transmatch can be used at the transmitter end of the feed line to obtain a more or less purely resistive impedance of approximately 75 ohms.

WIND POWER

The wind can be used to generate electrical or mechanical

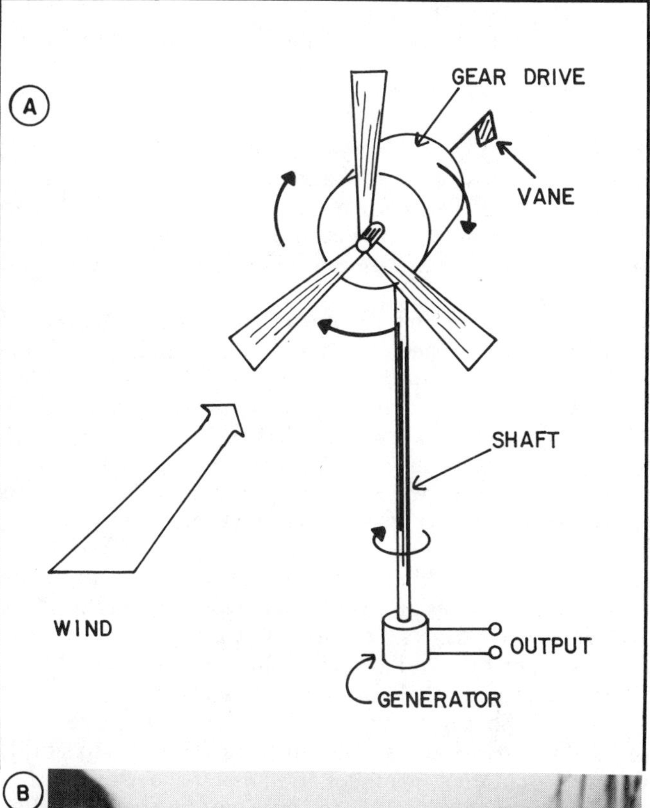

WIND POWER: At A, pictorial diagram of a simple wind-power generator. At B, a commericially manufactured wind-power generator, mounted on the roof of a residential dwelling.

power. Farmers have been harnessing the wind for centuries; in recent years, some attention has been given to the possibility of using wind as an alternative source of energy, as fossil fuels have become more and more expensive.

A simple wind-power "plant" can be constructed by attaching an ordinary electric generator to the shaft of a windmill. As the wind turns the blades of the mill, the shaft turns the coils or magnets in the generator (*see* GENERATOR). The resulting ac power is proportional to the angular speed of the shaft. The frequency, too, is a direct function of the angular speed of the shaft. The principle is illustrated in the drawing (A). Commercially manufactured windmill generators are available (B).

The main advantage of wind power is that a generating system is simple and easy to maintain. Rectifiers and choppers can be used to convert the variable-frequency current to 117 V at 60 Hz. The main problem with a wind-power generator is that the wind does not always blow! Nevertheless, a wind-power generator can provide considerable energy savings if it is interconnected with the utility mains so that power is available when the weather is calm.

WIRE

Wire is the most common and universal way of getting electrical energy from one place to another. Wires are also used for physical reinforcement in a variety of situations. Some types of wire are used to make lamps, resistors, and heating elements.

The most common type of wire is a cylindrical length of metal such as iron, steel, copper, or aluminum. Some wires consist of just one conductor, while others may have several. A single-conductor is called solid, and a multiconductor wire is called stranded (*see* STRANDED WIRE). Some wires have square or rectangular cross sections. Two or more wires can be combined to form a cable, cord, or transmission line (*see* CABLE, CORD, TRANMISSION LINE).

Wires are sometimes bare, and sometimes coated with a layer of enamel, rubber, plastic, or other insulating material. Insulated wires are used when a short circuit must be avoided (*see* INSULATED CONDUCTOR). Uninsulated wires are less expensive, and are preferable in situations where a short circuit is unlikely or of no consequence.

For wire having a circular cross section, the size is expressed in various ways. The simplest method is to simply state the diameter in inches, millimeters, or some other units. The cross-sectional area may also be specified. The most common way of expressing wire size is by means of standard gauges (*see* AMERICAN WIRE GAUGE, BIRMINGHAM WIRE GAUGE, BRITISH STANDARD WIRE GAUGE).

For electrical purposes, the current that a wire can handle is a function of the material used and the diameter of the conductor. In general, larger diameters allow for more current. Copper is the most efficient conductor for reasonable cost, and has the highest carrying capacity of the common types of wire (*see* CARRYING CAPACITY, COPPER). The resistance of a wire, per unit length, is also a function of diameter (*see* RESISTANCE). The tensile strength of a wire is also related to material and size, and to the method of manufacture (*see* TENSILE STRENGTH).

For futher information regarding various wire types and applications, see the following articles: ALUMINUM-CLAD WIRE, BELL WIRE, COPPER-CLAD WIRE, ENAMELED WIRE, HOOKUP WIRE, GUYING, NICHROME, SAG, STEEL WIRE, STRAIN, TUNGSTEN, WIRE ANTENNA, WIRE SPLICING, WIREWOUND RESISTOR.

WIRE ANTENNA

At very low, low, medium, and high frequencies, antennas are often constructed of wire. Any type of antenna can, theoretically, be made from wire. The dipole, longwire, rhombic, V beam, and Windom are the most common examples of antennas that are usually constructed using wire (*see* DIPOLE ANTENNA, LONGWIRE ANTENNA, RHOMBIC ANTENNA, V BEAM, WINDOM ANTENNA).

When erecting a wire antenna that must be a certain length, it is best to use wire that will not stretch significantly. Hard-drawn copper or copper-clad steel wire are best for such applications. These types of wire also have excellent tensile strength, and are preferable for use in long spans.

If the antenna length is not critical, or if the span is relatively short, soft-drawn or annealed copper wire can be used. It is not necessary to use insulated wire unless short circuits with other objects must be avoided. The ends of the antenna should be electrically isolated, however, by means of suitable insulators (*see* INSULATOR).

Any convenient supports can be used to anchor the ends of a wire antenna: trees, towers, and the wides of buildings are all excellent for this purpose. Utility poles should not be used; this is illegal in some municipalities, and can present an electrocution hazard. A wire antenna should never be run over or under a utility wire; if the antenna or the wire falls, the antenna may become live with hundreds or thousands of volts.

Wire antennas are especially popular among shortwave listeners and radio amateurs in the medium-frequency and high-frequency bands. At very high frequencies and above, antennas are small enough to make metal tubing practical for construction.

WIRE COMMUNICATION

See CARRIER-CURRENT COMMUNICATION, SUBMARINE CABLE, TELEGRAPH, TELEPHONE.

WIRE GAUGE

See AMERICAN WIRE GAUGE, BIRMINGHAM WIRE GAUGE, BRITISH STANDARD WIRE GAUGE.

WIRELESS

In the early days of radio, telecommunication was known as wireless. A code transmitter was called a wireless telegraph, for example. Nowadays, the term is not often used. *See* TELECOMMUNICATION.

WIRELESS MICROPHONE

A small radio transmitter can be installed in a microphone enclosure, and a receiver attached to the audio input of the device with which the microphone is used. This eliminates the necessity for using a cord for short distances. This type of device is called a wireless microphone. Wireless microphones are sometimes used in newscasting and entertainment, when personnel must move around a studio or stage.

Most wireless microphones operate at very-high or ultra-high frequencies. This is primarily because an efficient short antenna is practical at these wavelengths. Frequency modulation is usually employed for optimum immunity to noise.

In the United States and some other countries, the use of wireless microphones is regulated by government agencies. The power level of the transmitter, and the frequency at which it operates, are such that there is little chance of interference to other services.

WIRELESS TELEGRAPH

See CODE TRANSMITTER, TELECOMMUNICATION.

WIRELESS TELEPHONE

In the early days of radio, a voice transmitter or transceiver was called a wireless telephone (*see* TRANSCEIVER, VOICE TRANSMITTER). Today, the term is used to describe a special form of voice transceiver.

Radio equipment can be interconnected with the telephone lines in a variety of ways. The simplest method is called an autopatch system (*see* AUTOPATCH). A more sophisticated system, recently developed, is sometimes called cellular radio.

A cellular system operates through a network of repeaters to allow a subscriber to call any telephone number from any place at any time, using mobile or portable equipment. The cellular radio field is in its infancy.

Wireless telephone systems are used by businesses, especially traveling sales people. Wireless telephones are also valuable to anyone who has a vehicle break down or get stuck in an isolated area.

Wireless telephones should not be confused with cordless telephones, which are designed for use within a range of a few hundred feet of a main unit. The two terms are sometimes used interchangeably, but a true wireless telephone has unlimited range, and it is not associated with a base unit other than the repeaters through which it functions. *See also* CELLULAR TELEPHONE SYSTEM, CORDLESS TELEPHONE, REPEATER, TELEPHONE.

WIRE SAG

See SAG.

WIRE SIZES

See AMERICAN WIRE GAUGE, BIRMINGHAM WIRE GAUGE, BRITISH STANDARD WIRE GAUGE.

WIRE SPLICING

In wiring of electrical circuits and antennas, it is often necessary to splice two lengths of wire. There are various splicing methods; the two most common are described here.

Twist Splice. The simplest way to splice two wires is the method illustrated at A, B, and C. The wires are brought parallel with each other and twisted together. This method can be used with solid wire or stranded wire. About six to eight twists should be made. If the wires are of unequal diameter, the smaller wire is twisted around the larger wire. Electrical tape can be put over the connection if insulation is important. This type of splice has very poor mechanical strength unless solder is used. Even if a twist splice is soldered, the connection cannot withstand much strain.

The twist splice is used in most household electrical systems with AWG No. 12 or No. 14 solid copper wire. Special caps are placed over the splices for insulation purposes.

Western Union Splice. When it is desired that a splice have high breaking strength, the method shown at D, E, and F is preferable. This type of splice is called a Western Union splice. The wires are brought together end-to-end, over-

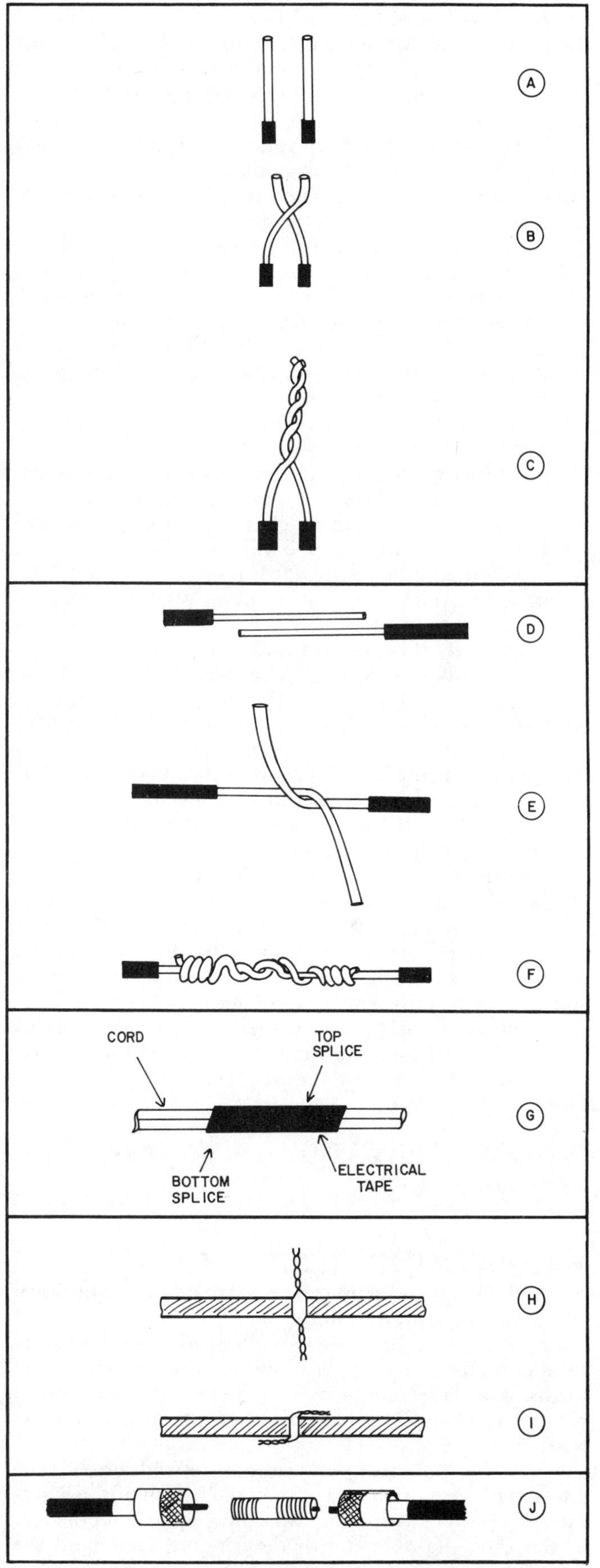

WIRE SPLICING: Wire-splicing techniques. At A, B, and C, twist splice. At D, E, and F, Western Union splice. At G, a method of splicing two-wire cord. At H and I, twist splicing method for twin-lead transmission line. At J, preferred method for joining sections of coaxial cable.

lapping a couple of inches. They are hooked around as shown at E, and twisted several times as shown at F. For guy-wire splicing, each end should be twisted around 10 to 12 times. Any protruding ends are removed using a diagonal cutter. The splice is then soldered if a good electrical bond is needed, and a layer of electrical tape or other insulation is applied. For large-diameter wires, the ends can be tinned with solder before the splice is made to facilitate the best possible electrical bond.

This splicing method can be used with solid or stranded wire of any gauge, but it is rather difficult to splice wires having diameters that differ by more than 4 in the AWG system (*see* AMERICAN WIRE GAUGE). The Western Union splice is also somewhat awkward when one wire is solid and the other is stranded. For maximum strength, both wires should be the same size and the same type.

Cord and Cable Splicing. When splicing cords or lengths of open-wire transmission line, the Western Union splice is preferable because the wires remain in line. Insulation is very important when splicing electrical cords; both splices should be thoroughly wrapped with electrical tape, and the combination wrapped afterwards (G). For additional insulation, the splices may be made at slightly different points along the cord.

Two lengths of twin-lead transmission line can be effectively and conveniently twist-spliced, as illustrated at H and I. After the twists have been soldered, the splices are trimmed to about ¼ inch and folded back. The whole area is then carefully wrapped with electrical tape.

Multi-conductor cables can be spliced wire by wire, using Western Union splices in each instance. All of the splices must be individually wrapped with insulating material to prevent short circuiting.

Coaxial cables, and other cables in which a constant characteristic impedance must be maintained, are generally spliced by using special connectors. The N-type or UHF connectors are the most common for coaxial cable. A male connector is soldered to each of the two ends to be spliced; a female-to-female adaptor is used between them. This is illustrated in J. The whole joint must be thoroughly wrapped with insulating tape. *See also* COAXIAL CABLE, N-TYPE CONNECTOR, UHF CONNECTOR.

WIRE STRETCH

See STRAIN.

WIREWOUND RESISTOR

A length of resistive wire can be wrapped on a cylindrical core to obtain a specific ohmic value. Such a device is called a wirewound resistor. Wirewound resistors are commonly used in high-current applications, because of their ability to dissipate large amounts of power: some wirewound resistors can handle several hundred watts on a continuous basis.

The value of a wirewound resistor depends on the type and size of wire used, and also on the length of the winding. The power-dissipating capability depends on the size of the wire, the diameter of the winding form, and the material from which the winding form is made. Ceramic forms are common.

Wirewound resistors are not generally suitable for use at radio frequencies because they exhibit inductive reactance, as well as resistance, for high-frequency alternating currents. *See also* RESISTOR.

WIRE WRAPPING

Wire wrapping is a method of circuit wiring in which solder is not used. Components are mounted on lugs that are inserted into a perforated board. The lugs protrude underneath the board. Small-gauge, solid wire is used for interconnection among the lugs. A connection is made by neatly wrapping the wire several times around the lug (see illustration). A special tool is used for wrapping the wire. The lugs are long enough so that three or more connections can be made on top of each other.

Wire wrapping is extremely convenient, and makes it possible to assemble complicated circuits in a small amount of physical space. Wire wrapping is extensively used in conjunction with computer systems and large switching networks in controlled environments. Wire wrapping is also ideal for construction of experimental circuits, since the wires are easy to remove.

WIRING DIAGRAM

Wiring diagrams are designed to show point-to-point connections or printed circuit patterns from one point to another on system or circuit boards, including cabling connections between boards. They are completely separate from schematics and have no electrical symbols indicated.

Schematic diagrams do not locate parts, but are the electrical signal flow and operating voltage charts giving all component values, waveforms, test conditions, and I/O points as important references. Whenever engineers and technicans troubleshoot defective electronic equipments, they always refer to schematics first for service information

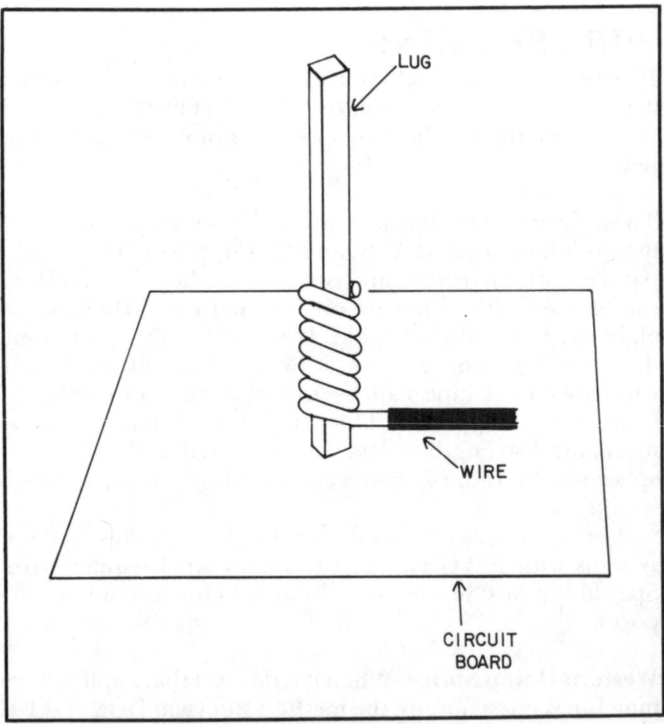

WIRE WRAPPING: Technique of wire wrapping. The corners on the lug provide a good electrical contact with the wire.

and electrical measurements. Afterwards, PC board or wiring diagrams may be consulted for specific cabling or physical parts locations on the equipment's plug-in boards or chassis. *See also* SCHEMATIC DIAGRAM.

WIRING METHODS

See POINT-TO-POINT WIRING, PRINTED CIRCUIT, WIRE SPLICING, WIRE WRAPPING.

WOLFRAM

See TUNGSTEN.

WOODPECKER

In the 1970s, a previously nonexistent signal began to be heard on the high-frequency radio bands. Because of its constant pulse rate and the way the pulses sound, it has become known as the woodpecker. It has also been called the buzz saw.

At times, the woodpecker signal is very strong and causes interference to radio communications over a wide band of wavelengths. A well-designed noise blanker can provide some reduction in the interference. Switching off the automatic gain control (AGC) and using an audio clipper can also help to alleviate the interference.

The woodpecker is believed to have originated in the Soviet Union as an over-the-horizon radar system (*see* RADAR). Some engineers have theorized that the woodpecker might be used to modify the ionosphere for the purpose of gaining some control over the propagation conditions at shortwave frequencies. The signal consists of sharp, regular pulses at a rate near 10 Hz. The bandwidth is usually several hundred kilohertz. The transmitter power is on the order of several megawatts. The center frequency changes, apparently following the maximum usable frequency (MUF) to take advantage of optimum long-distance ionospheric propagation (*see* MAXIMUM USABLE FREQUENCY).

WOOFER

A woofer is a speaker that is designed to respond to low audio frequencies, and also to subaudible frequencies. A woofer basically resembles an ordinary speaker (*see* SPEAKER), except that it is physically larger. The large size is necessary because of the length of sound waves at low frequencies.

Woofers are used in high-fidelity speaker systems to enhance the reproduction of bass sound (*see* BASS RESPONSE). Many high-fidelity speakers contain a woofer, a midrange speaker, and a tweeter in a single enclosure. *See also* MIDRANGE, TRIAXIAL SPEAKER, TWEETER.

WORD

In digital and computer practice, a group of bits or characters, having a specified length or number of units, is called a word. A word conveys a certain amount of information, and occupies a certain amount of space in memory. Words can be used as data, instructions, designators, or numerical values. *See also* BIT, BYTE, CHARACTER.

In teletype and telegraph communications, a word consists of five characters plus one space. To obtain the total word count in a message, the number of characters, symbols, and spaces is counted. (In teletype, symbols and punctuation marks count as single characters, but in Morse code, they count as two characters each.) The result is divided by 6. Speed of transmission may then be expressed in terms of the number of words sent per minute. *See also* WORDS PER MINUTE.

WORD PROCESSING

Computers and microcomputers can be used as an aid in writing and editing letters and manuscripts. This is known as word processing. Some computers are designed specifically for this purpose. These computers are called word processors. They are often completely self-contained units, including a video terminal, keyboard, disk drive, and printer.

Some Common Word-Processing Capabilities. Different writing styles require different word-processing functions. For example, letters require a different set of functions than magazine articles require. Tables and charts require still another set of functions. Any of several specialized software packages can be used with a single word processor for various kinds of writing. A few of the most common word-processing functions are listed below.

- Left-margin adjustment and justification: all rows are aligned along the left side of the page with a margin width the operator can select.
- Right-margin adjustment and justification: If desired, all rows can be aligned along the right side of the page. The margin width is adjustable.
- Insertion and deletion of characters, words, lines, and paragraphs: Text can be added or removed and the continuity of the manuscript will be readjusted.
- Exchange of characters, words, lines, and paragraphs: Text can be altered as necessary.
- Row and column alignment: Used for making tables and charts.
- Word wrap around: Automatically moves a word to next line in the text if the word runs past the right-hand margin limit.
- Double spacing: Inserts an extra space between each line of text.

Data Storage and Printing. Almost all word processors allow storage of information on magnetic disks. The floppy disk is by far the most commonly used (*see* FLOPPY DISK). Information is easily erased or rewritten onto the disks.

Every word processor uses a printer. There are various types of printers available for use with word processors. The dot-matrix printer operates at high speed, and is useful when a lot of text must be printed in a short time. Ball type and daisy-wheel printers give crisper copy, but run at a somewhat slower speed than dot-matrix printers (*see* DAISY-WHEEL PRINTER, DOT-MATRIX PRINTER).

An important feature of a word processor is the memory capacity for each file of text. All computers have a capacity for each file of text. All computers have a certain maximum memory limit. In general, each page of double-spaced text, containing 220 words, has 1400 to 1500 bytes. The number of such pages that can be stored in a file is thus about 70 percent of the number of kilobytes memory. Larger computers, of course, contain more of memory space than small computers. Small files are adequate for letters and magazine articles. For maximum flexibility in book writing, larger files are desirable.

As our society becomes progressively more information-oriented, the need for word processors in everyday life will increase. Many moderately priced word processors and software packages are now available. *See also* KEYBOARD, KEYPUNCH, VIDEO DISPLAY TERMINAL.

WORDS PER MINUTE

The speed of transmission of a digital code is sometimes measured in words per minute. A word generally consists of five characters plus one space (*see* WORD). The abbreviation for words per minute is WPM.

For teleprinter codes, speeds range from 60 WPM to hundreds or even thousands of words per minute. The original Baudot speed was 60 WPM, and this speed is still in widespread use (*see* BAUDOT). The standard ASCII speed is 110 WPM, and the highest commonly used ASCII speed is 19,200 WPM (*see* ASCII). Sometimes the speed is given in bauds rather than words per minute. For Baudot transmissions, the speeds in words per minute is about 33 percent greater than the baud rate. For ASCII transmissions, the speed in words per minute is the same as the baud rate (*see* BAUD RATE).

The speed of an International Morse Code transmission, in words per minute, is about 1.2 times the baud rate, where one baud is the length of a single dot or the space between dots and dashes in a character. If a string of dots is sent, the speed in words per minute is equal to 2.4 times the number of dots in one second (*see* INTERNATIONAL MORSE CODE).

WORK

Work is an expression of mechanical energy. The most common units are the foot pound and the kilogram meter. In either case, work is expressed as a product of force (pounds or kilograms) and distance (feet or meters). For example, when a weight of 1 pound is lifted through a distance of 1 foot, the amount of work done is 1 foot pound. This represents 1.356 joules. If 1 foot pound is expended every second over a period of time, 1.356 watts of power are generated. *See also* ENERGY, ENERGY UNITS.

WORK FUNCTION

For electrons to be stripped from atoms, extracted from a substance, or moved from one place to another, a certain amount of energy must be expended. A work function is the amount of energy, in electron volts, required to move one electron from some specified place to some other specified place. Usually, this means the energy needed to move the electron from the valence band to the conduction band.

The work function is sometimes given for electrons in vacuum tubes and semiconductor materials.

WOW

Wow is a slow warbling, or periodic variation in pitch, that sometimes occurs in audio recording or reproduction. Wow may also be called flutter, although the term flutter is more often used to describe a rapid change in the pitch of reproduced sound (*see* FLUTTER).

Wow is most often caused by placing a phonograph record off center on a turntable. Ordinarily, if the proper adaptors are used, this will not occur. Sometimes, however, a record is actually manufactured poorly, so that the grooves are not centered. Then wow will be noticed even when the required adaptor or spindle is used. The greater the centering error, the more pronounced the wow will be.

Wow is not often a problem with tape recorders, since most recorders have reasonably constant motor speed. Flutter is, however, sometimes observed when attempting to record music on a tape recorder that runs at a slow speed. *See also* DISK RECORDING.

WRITING GUN

A writing gun is an electron gun that is used in a storage cathode-ray tube. The writing gun is basically an ordinary electron gun, which emits a narrow beam of electrons that strikes the phosphor surface at a certain point, resulting in a bright spot (*see* CATHODE-RAY TUBE, ELECTRON-BEAM GENERATOR).

In a storage oscilloscope, the writing-gun beam has constant intensity, and scans from left to right at a constant speed. The vertical position is varied by the incoming signal (*see* OSCILLOSCOPE, STORAGE OSCILLOSCOPE). The intensity of the writing-gun beam may be modulated, and the beam moved through a standard raster, for storage of television pictures. The display is observed using another electron gun, known as the flood gun (*see* FLOOD GUN).

WWV/WWVH

The National Bureau of Standards maintains two shortwave time-and-frequency broadcasting stations. Station WWV is located at Fort Collins, Colorado, and WWVH is located on the island of Kauai, Hawaii.

Both stations transmit signals on frequencies of 2.5, 5, 10, 15, and 20 MHz. Station WWV also broadcasts on 25 MHz. The frequencies are accurate to a tiny fraction of 1 Hz. Standard amplitude modulation is used.

Time broadcasts are sent continuously, 24 hours a day, with brief exceptions. Voice announcements indicate the hour and minute. Audible clicks indicate seconds. Standard audio-frequency tones are transmitted at 440, 500, and 600 Hz according to a prescribed format.

The time and frequency broadcasts of WWV and WWVH are generally accurate to within 1 part in 10^{12}, and never vary up or down by more than 1 part in 10^{11}.

In addition to time-and-frequency information, WWV and WWVH transmit data about weather and radio-propagation conditions. Propagation conditions are expressed according to a two-character code consisting of a phonetic designator and a number (see table). Geophysical

(A)	Indicator	Geomagnetic Conditions
	W (Whiskey)	Disturbed
	U (Uniform)	Unsettled
	N (November)	Normal or Stable

(B)	Indicator	Propagation Quality
	1	Completely Useless
	2	Very Poor
	3	Poor
	4	Poor to Fair
	5	Fair
	6	Fair to Good
	7	Good
	8	Very Good
	9	Outstanding

WWV/WWVH: Phonetic indicators of geomagnetic conditions (at A) and numerical indicators of expected propagation quality (at B).

information, if of great significance, is also sent.

In addition to WWV and WWVH, a low-frequency station, WWVB, is operated at 60 kHz from Fort Collins, Colorado. Time information is sent using a carrier-shift code. The signals from WWVB can be used directly with a phase-locked loop to obtain an extremely accurate frequency standard.

WYE CONNECTION

In a three-phase alternating-current system, a wye connection is a method of interconnecting the windings of a transformer to a common point. The configuration gets its name from the fact that, in a schematic diagram, it appears like a capital letter Y.

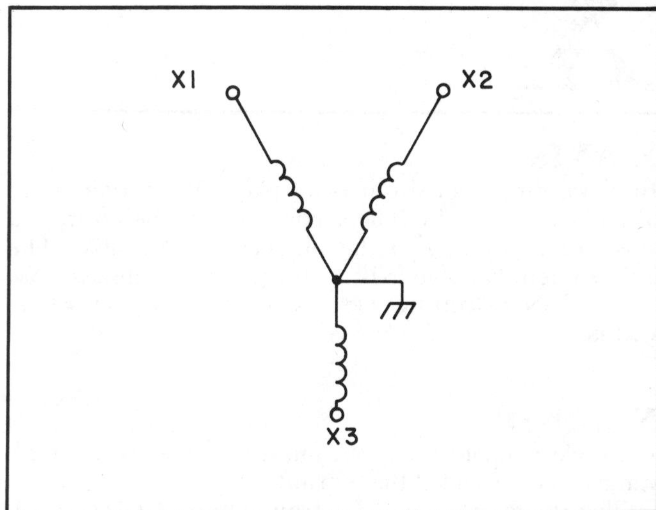

WYE CONNECTION: Wye transformer winding for three-phase alternating current. The currents at X1, X2, and X3 differ in phase by 1/3 cycle.

The diagram illustrates a wye transformer connection. The phases at points X1, X2, and X3 are 120 electrical degrees (⅓ cycle) out of phase with respect to each other (*see* THREE-PHASE ALTERNATING CURRENT). The center point is neutral.

Wye configurations are often used in polyphase rectifiers (*see* POLYPHASE RECTIFIER).

WYE MATCH

See DELTA MATCH.

X AXIS

In a two-dimensional Cartesian plane, the variables are often labeled x and y. The x axis is usually the horizontal axis, corresponding to the independent variable. The independent-variable axis is also called the abscissa. *See* CARTESIAN COORDINATES, INDEPENDENT VARIABLE, Y AXIS.

X BAND

In the electromagnetic spectrum, either of two wavelength ranges may be called the X band.

The range of frequencies from 5.2 to 11 GHz, corresponding to wavelengths of 5.8 and 2.7 cm, is known as the X band. This band is used for radar, line-of-sight communications, and satellite communications. The X band falls in the superhigh range, which is considered to be in the microwave part of the electromagnetc spectrum (*see* SUPERHIGH FREQUENCY).

At much shorter wavelengths, ranging from 4 nm (nanometers) down to about 0.01 nm, electromagnetic fields have extreme energy and penetrating properties. These are the X rays (*see* X RAY). This band is occasionally called the X band.

XENON

Xenon is an element with atomic number 54 and atomic weight 131. Xenon is a gas at room temperature. Because its outer electron shell is completely filled, xenon is an inert, or noble, gas.

Xenon is used in certain types of flash tubes, and in some voltage-regulator tubes. *See also* XENON FLASHTUBE.

XENON FLASHTUBE

A xenon flashtube is a device that generates an extremely brilliant, white, incoherent light. Two electrodes are enclosed in a glass envelope that has been evacuated and then filled with xenon gas (*see* XENON). Xenon flashtubes are used to energize certain types of lasers (*see* LASER).

The light from a xenon tube originates from two sources: the hot electrodes, and the ionized gas. Although the xenon tube emits energy at all visible wavelengths, the greatest amount of energy occurs at about 500 nanometers (nm) and 800 nm. These wavelengths are approximately in the blue and red ranges. The two wavelengths combine to produce a whitish light.

Xenon can be used as a continuous light source, in the same way as neon gas (*see* NEON LAMP).

X-RAY TUBE

When high-speed electrons strike a massive barrier such as lead or tungsten, X rays are generated (*see* X RAY). This effect is put to use in a device called an X-ray tube.

The illustration is a simplified drawing of an X-ray tube. A glass envelope is evacuated, and contains a heated cathode and an anode. The anode is usually made from tungsten so that it will resist destruction by heating as energetic electrons bombard it. The anode surface is turned at an angle. The gap between the cathode and the anode is a few inches.

When a positive voltage is supplied to the anode, electrons are emitted by the hot cathode, just as in an ordinary vacuum tube. A high-voltage power supply results in acceleration of the electrons to extreme speed. The electrons strike the anode at an angle of approximately 45 degrees. The X-rays are thrown off in various directions, but mostly from the side of the tube, at approximately 90 degrees with respect to the electron beam.

X-ray tubes are used mostly in medical diagnosis, for such purposes as locating tumors, broken bones, and cavities in the teeth. In recent years, ultrasonic devices have been used in place of X-ray devices in some medical practice (*see* ULTRASONIC HOLOGRAPHY, ULTRASONIC SCANNER).

X-RAY

The term X-ray is used to describe electromagnetic radiation at wavelengths shorter than ultraviolet, but longer than gamma rays. The precise cutoff between ultraviolet and X-rays is not universally agreed upon, but it can be considered to be in the neighborhood of 4 nanometers (nm). The lower wavelength limit is around 0.01 nm. X-rays propagate through a vacuum in the same way, and at the same speed, as other forms of electromagnetic radiation, including visible light. The photons of X-rays contain considerable energy, however, which gives X-rays their penetrating power.

We cannot see X-rays, but they cause some phosphor materials to glow. X-rays also expose camera film. This, in conjunction with the ability of X-rays to penetrate soft tissues of the body, has led to the use of X-rays in medicine (*see* FLUOROSCOPE, X-RAY TUBE).

Overexposure to X-rays can be dangerous, and if too much radiation is received in a short time, illness can occur (*see* DOSIMETRY, RADIOACTIVITY). Excessive exposure to X-rays will kill most forms of life. Our atmosphere blocks almost all of the X radiation from the sun and other stars.

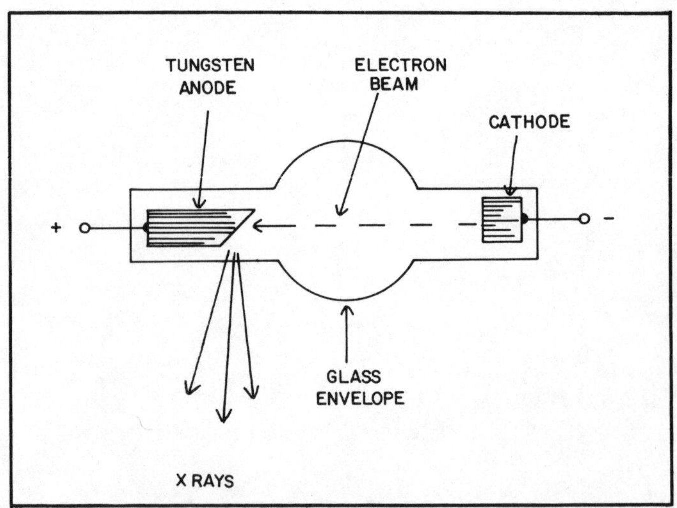

X-RAY TUBE: Cross-sectional view of an X-ray tube.

Y

YAGI ANTENNA

A yagi antenna is a form of parasitic array. It operates by means of electromagnetic coupling between a driven element and one or more separate, parallel conductors called parasitic elements (*see* PARASITIC ARRAY, PARASITIC ELEMENT). The yagi antenna gets its name from the Japanese engineer who discovered the properties of parasitic elements and put them to use. Yagi antennas are used at high, very high, and ultra-high frequencies for receiving and transmitting radio signals.

In the yagi, the power is applied to a radiator called the driven element. The parasitic elements act to produce power gain and directional characteristics (*see* ANTENNA POWER GAIN, FRONT-TO-BACK RATIO). The driven element is usually an electrical half wavelength. The parasitic elements are somewhat shorter or longer. In general, a reflector is slightly longer than the driven element, and a director is somewhat shorter (*see* DIRECTOR, DRIVEN ELEMENT, REFLECTOR).

A yagi antenna may have one or two reflectors, and one or more directors. There is usually only one drive element, although two or more can be phased to provide some gain. A common configuration at high frequencies is one driven element, one director, and one reflector. The director measures approximately 0.48 electrical wavelength, and the reflector near 0.52 wavelength. Since the exact lengths depend on the element spacing, these figures must be considered approximate.

Additional director elements can be added to obtain enhanced forward gain and front-to-back ratio. Provided optimum element spacing is used, the gain increases as the number of elements increases. Each succeeding director should be slightly shorter than its predecessor. Long yagi antennas are practical at very-high and ultra-high frequencies, but they are difficult to construct at high frequencies because of their large size.

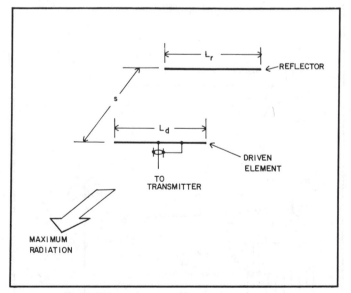

YAGI ANTENNA: A simple two-element yagi antenna. The dimensions L_d, L_r, and s are discussed in the text. Effective power gain is approximately 5 dBd.

A complete discussion of all the design parameters of yagi antennas cannot be given here. However, a simple two-element yagi antenna can be built according to the dimensions shown in the illustration. Such an antenna will provide about 5 dB of forward gain. The elements are straight. The length L_d of the driven element, in feet, is given in terms of the frequency f, in megahertz, by the equation:

$$L_d = 460/f$$

The length of the reflector, in feet, is given by:

$$L_r = 480/f$$

(These formulas are based on the assumption that the element diameter is very small compared with the element length. If heavy tubing is used, the elements will be slightly shorter.) The element spacing s, in feet, is:

$$s = 200/f$$

Some adjustment of L_d, L_r, and s are required to obtain optimum forward gain.

Resonant loops can be used in a configuration similar to that of the yagi antenna. This type of antenna is called a quad. *See also* QUAD ANTENNA.

Y AXIS

In a two-dimensional Cartesian plane, the variables are often labeled x and y. The y axis is usually the vertical axis, corresponding to the dependent variable. The dependent-variable axis is also called the ordinate. *See* CARTESIAN COORDINATES, DEPENDENT VARIABLE, X AXIS.

Y CONNECTION

See WYE CONNECTION.

Y MATCH

See DELTA MATCH.

YOKE

In a cathode-ray tube, the electron beam may be deflected by means of charged plates, or by means of coils. A complete set of magnetic deflection coils is called the yoke of the tube (*see* CATHODE-RAY TUBE, PICTURE TUBE).

In an electric motor or generator, a ring-shaped piece of ferromagnetic material is used to hold the pole pieces together, and also to provide magnetic coupling between the poles. This assembly is known as a yoke (*see* GENERATOR, MOTOR).

Z

ZENER DIODE

All semiconductor diodes conduct poorly in the reverse-bias condition as long as the voltage remains below a certain critical value called the avalanche voltage (*see* AVALANCHE VOLTAGE). If the reverse bias exceeds the avalanche voltage, the resistance of the diode decreases greatly. The avalanche voltage of a particular diode depends on the way in which it is constructed.

A zener diode is manufactured specifically to have a constant, predictable avalanche voltage. This makes it possible to use reverse-biased diodes as voltage regulators, voltage dividers, reference-voltage sources, threshold detectors, transient-protection devices, and in other situations where a controlled avalanche response is needed.

Zener diodes are rated according to their avalanche voltage, and also according to the amount of power they can dissipate on a continuous basis. Zener diodes are available at ratings from less than 3 V to more than 150 V, and at power-dissipation ratings of up to 50-60 W. The high-voltage, high-power zener diodes can be used in place of regulator tubes in most types of medium-voltage power supplies.

A typical zener-diode characteristic curve is illustrated. In the forward-biased condition, the zener diode behaves very much like an ordinary silicon rectifier. In the reverse-biased condition, the impedance is practically infinite up to the critical, or zener, voltage (in this case 14 V). Then the impedance drops to practically zero. The reverge-bias voltage drop across this particular zener diode is a constant 14 V. This diode might be connected in parallel with the output of a power supply to obtain voltage regulation.

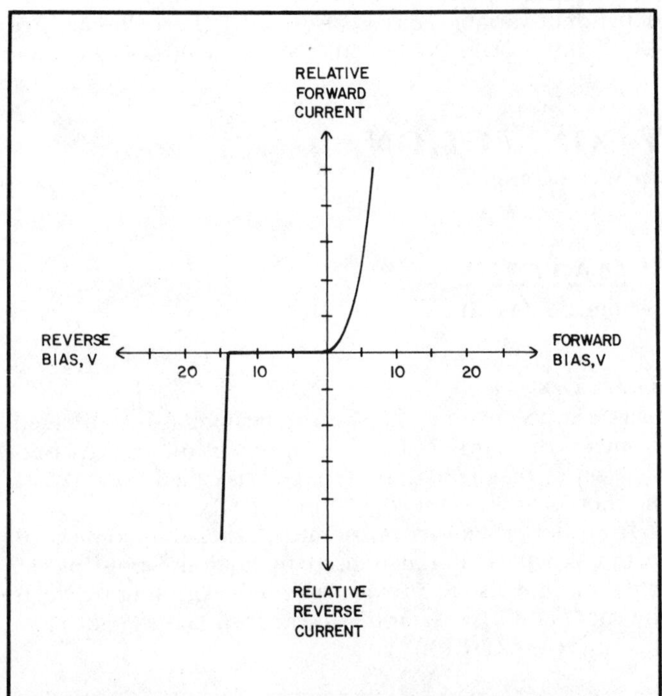

ZENER DIODE: Characteristic curve of a 14-volt zener diode.

ZEPPELIN ANTENNA

A zeppelin antenna is a half-wavelength radiator, fed at one end with a quarter-wavelength section of open-wire line. The antenna gets its name from the fact that it was originally used for radio communications aboard zeppelins. The antenna, also called a zepp, was dangled in flight (see illustration A).

The zeppelin antenna may be thought of as a current-fed, full-wavelength radiator, part of which has been folded up to form the transmission line. The impedance at the feed point is extremely high, and at the transmitter end of the line, the impedance is practically zero. Any desired resistive impedance can be obtained by short-circuiting the end of the line opposite the feed point, and connecting the transmitter/receiver at some intermediate point, as at B. This configuration is sometimes used to feed a vertical half-wavelength radiator (*see* J ANTENNA). A zeppelin antenna will operate well at all odd harmonics of the design frequency.

If a transmatch is available, a half-wavelength radiator can be end fed with an open-wire line of any length (C). This arrangement is popular among many radio amateurs. The antenna will operate well at all harmonics of the design frequency. Thus, if the antenna radiator is cut for 3.5 MHz, the antenna will also work (as a longwire) at 7, 14, 21, and 28 MHz.

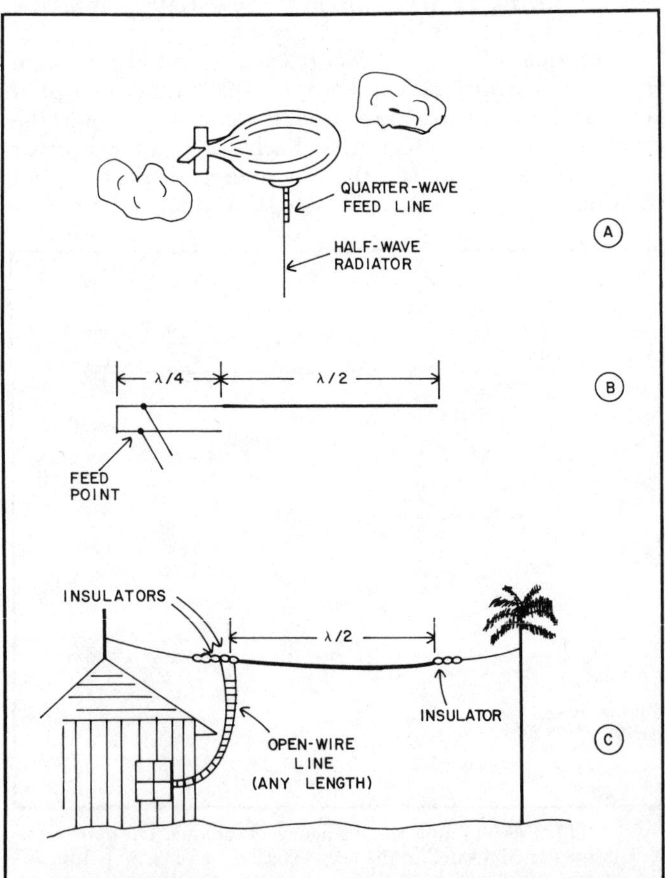

ZEPPELIN ANTENNA: At A, the zeppelin antenna was originally dangled from air vehicles. At B, a method of obtaining a desired pure resistance at the feed point. At C, a variation of the zeppelin antenna used by communications hobbyists.

The primary advantage of the zeppelin antenna, especially the configuration at C, is ease of installation. The main difficulty with the zepp is that the feed line radiates to some extent because the system is not perfectly balanced. Feed-line radiation can be kept to a minimum by carefully cutting the radiator to ½ wavelength at the design frequency, and by using the antenna only at this frequency or one of its harmonics. For a wire radiator, properly insulated at the ends and placed well away from obstructions, the length L, in feet, at the design frequency f, is approximately:

$$L = 468/f$$

The length is meters is:

$$L = 143/f$$

Some "pruning" will probably be necessary to minimize radiation from the line. A field-strength meter can be used to observe the line radiation while adjusting the length of the antenna for optimum performance at the design frequency.

The radiation pattern from a zepp antenna is the same as that from a dipole at the design frequency, provided the radiation from the line is not excessive. At harmonic frequencies, the radiation pattern is identical with that of an end-fed longwire of the same length. *See also* DIPOLE ANTENNA, LONGWIRE ANTENNA.

ZERO BEAT

When two signals are mixed, beat notes, or heterodynes, are generated at the sum and difference frequencies (*see* BEAT, HETERODYNE). As the frequencies of two signals are brought closer and closer together, the difference frequency decreases. When the two signals have the same frequency, the difference signal disappears altogether. This condition is called zero beat.

In a receiver having a product detector, a carrier produces an audible beat note (*see* PRODUCT DETECTOR). As the receiver is tuned, the frequency of the audio note varies. When the receiver is tuned closer to the carrier frequency, the beat note gets low-pitched and then disappears, indicating zero beat between the carrier and the local oscillator.

In communications practice, two stations often zero beat each other. This means that they both transmit on exactly the same frequency, reducing the amount of spectrum space used.

ZERO BIAS

When the control grid of a vacuum tube is at the same potential as the cathode, the tube is said to be operating at zero bias. In a field-effect transistor, zero bias is the condition in which the gate and the source are provided with the same voltage. Similarly, in a bipolar transistor, zero bias means that the base is at the same potential as the emitter.

In vacuum tubes and depletion-mode field-effect transistors, zero bias usually results in fairly large plate or drain current. In enhancement-mode field-effect transistors and in bipolar transistors, little or no drain or collector current normally flows with zero bias. There are a few exceptions, however.

Some devices are designed to operate at zero bias in certain applications. Some Class-AB and Class-B vacuum-tube power amplifiers operate in this way. Bipolar transistors at zero bias are often used as Class-C amplifiers. *See also* BIAS, CHARACTERISTIC CURVE, CLASS/AB AMPLIFIER, CLASS/B AMPLIFIER, CLASS-C AMPLIFIER, CUTOFF VOLTAGE, DEPLETION MODE, ENHANCEMENT MODE, FIELD-EFFECT TRANSISTOR, GRID BIAS, NPN TRANSISTOR, PINCHOFF VOLTAGE, PNP TRANSISTOR, TUBE.

ZERO-CENTER METER

Some meters are capable of indicating both positive and negative values, with the zero reading at the center. Such a meter is called a zero-center meter. Zero-center meters are used to measure negative and positive voltage or current, frequency centering, plus-or-minus errors, and other parameters that may fluctuate to either side of a zero or balanced condition. Zero-center meters are universally employed to indicate balance in bridge circuits.

The illustration shows an example of a zero-center meter for determining the carrier frequency of a radio transmitter. If the carrier is on the correct frequency, the meter needle will be at center scale, indicating zero error. If the transmitter frequency is too high, the needle position is to the right of center; if the transmitter frequency is low, the needle position is left of center. The meter shows that the transmitter frequency is approximately 2.5 kHz below the center of the channel.

The most widely used type of zero-center meter is the galvanometer. *See also* GALVANOMETER.

ZERO LEVEL

See REFERENCE LEVEL.

ZINC

Zinc is an element with atomic number 30 and atomic weight 65. In its pure form, zinc appears as a grayish metal of low luster. Zinc is a fairly good conductor of electric current.

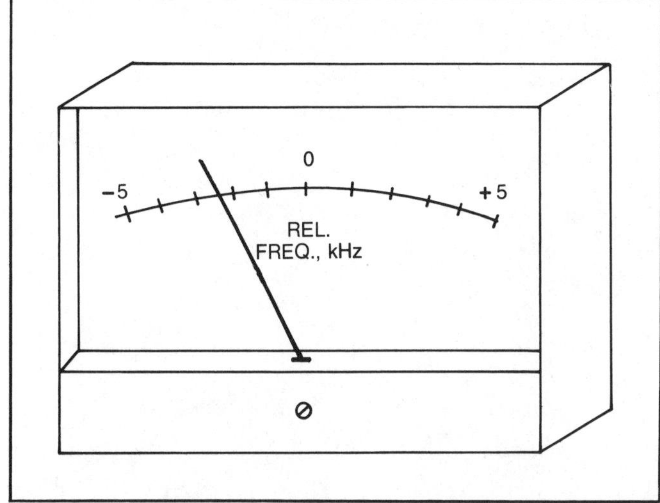

ZERO-CENTER METER: An example of a zero-center meter, indicating a transmitter operating 2.5 kHz below the channel center.

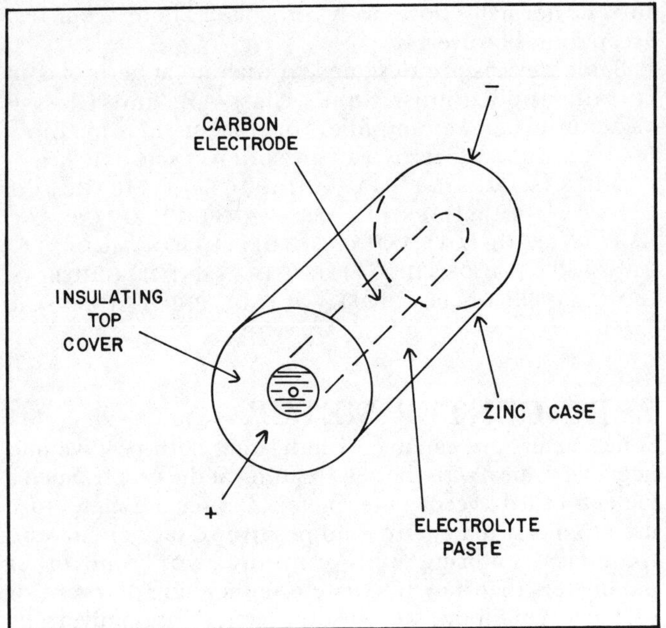

ZINC CELL: Construction of a zinc cell. Regardless of physical size, the cell provides about 1.5 V.

Zinc is extensively used to coat iron and steel structures. This reduces the tendency for such materials to rust, since zinc corrodes less easily then many other metals. Zinc is also used in the manufacture of various types of electrochemical cells (*see* ZINC CELL).

ZINC CELL

Many different types of electrochemical cells use zinc as the negative electrode. Such cells are sometimes called zinc cells. The most well-known type of zinc cell is the zinc-carbon dry cell (the common flashlight cell).

The zinc-carbon cell consists of a zinc case, usually cylindrical in shape, filled with an electrolyte paste. The zinc case forms the negative electrode of the cell. A carbon rod, immersed in the paste, forms the positive electrode (see illustration). This type of cell provides approximately 1.5 V of direct current. Several such cells are often connected in series in a common case, forming a battery. *See also* BATTERY, CELL, DRY CELL.

ZONE OF SILENCE

See SKIP ZONE.

INDEX

This index contains article titles, cross references, and illustration and table titles. It also contains many terms that are not article titles. **Boldface** page numbers indicate that a term is represented by an article on that page. *Italicized* page numbers indicate that the term is represented by an illustration and/or a table on that page. For terms that are not represented as articles, page numbers are listed for occurrences of that term that are of special significance.

INDEX

A

B

C

D

E

F

G

H

I

J

K

L

M

N

O

P

S

T

U

V

W

X

Y

Z